Developments in Economic Geology, 19

empirical metallogeny

Depositional Environments, Lithologic Associations and Metallic Ores

Vol. 1: Phanerozoic Environments, Associations and Deposits

PART A

Further titles in this series

1. *I.L. ELLIOTT and W.K. FLETCHER (Editors)*
GEOCHEMICAL EXPLORATION 1974

2. *P.M.D. BRADSHAW (Editor)*
CONCEPTUAL MODELS IN EXPLORATION GEOCHEMISTRY
The Canadian Cordillera and Canadian Shield

3. *G.J.S. GOVETT and M.H. GOVETT (Editors)*
WORLD MINERAL SUPPLIES
Assessment and Perspective

4. *R.T. SHUEY*
SEMICONDUCTING ORE MINERALS

5. *J.S. SUMNER*
PRINCIPLES OF INDUCED POLARIZATION FOR GEOPHYSICAL EXPLORATION

6. *R.A. RICH, H.D. HOLLAND and U. PETERSEN*
HYDROTHERMAL URANIUM DEPOSITS

7. *J.G. MORSE (Editor)*
NUCLEAR METHODS IN MINERAL EXPLORATION AND PRODUCTION

8. *M. KUŽVART and M. BÖHMER*
PROSPECTING AND EXPLORATION FOR MINERAL DEPOSITS

9. *C.R.M. BUTT and I.G.P. WILDING (Editors)*
GEOCHEMICAL EXPLORATION 1976

10. *G.B. FETTWEIS*
WORLD COAL RESOURCES
Methods of Assessment and Results

11. *R.G. TAYLOR*
GEOLOGY OF TIN DEPOSITS

12. *H.K. GUPTA*
GEOTHERMAL RESOURCES
An Energy Alternative

13. *C.R.M. BUTT and R.E. SMITH (Editors)*
CONCEPTUAL MODELS IN EXPLORATION GEOCHEMISTRY, 4
Australia

14. *G.BÁRDOSSY*
KARSTIC BAUXITES

15. *A.W. ROSE and H. GUNDLACH (Editors)*
GEOCHEMICAL EXPLORATION 1980

16. *R.W. BOYLE*
GEOCHEMICAL PROSPECTING FOR THORIUM AND URANIUM DEPOSITS

17. *G.R. PARSLOW (Editor)*
GEOCHEMICAL EXPLORATION 1982

18. *M. KUŽVART*
INDUSTRIAL MINERALS AND ROCKS

DEVELOPMENTS IN ECONOMIC GEOLOGY, 19

empirical metallogeny

Depositional Environments, Lithologic Associations and Metallic Ores

Vol. 1: Phanerozoic Environments, Associations and Deposits

PART A

Peter Laznicka

Department of Earth Sciences, University of Manitoba, Winnipeg, Man. R3T 2N2, Canada

ELSEVIER, Amsterdam – Oxford – New York – Tokyo 1985

ELSEVIER SCIENCE PUBLISHERS B.V.
Sara Burgerhartstraat 25
P.O. Box 211, 1000 AE Amsterdam, The Netherlands

Distributors for the United States and Canada:

ELSEVIER SCIENCE PUBLISHING COMPANY INC.
52, Vanderbilt Avenue
New York, N.Y. 10017, U.S.A.

ISBN 0-444-42530-6 (Vol. 19A)
ISBN 0-444-42553-5 (Vol. 19B)
ISBN 0-444-42554-3 (Set)
ISBN 0-444-41250-6 (Series)

Printed in The Netherlands

DEDICATION

Dedicated to the prospectors who use science and logic to find a mine.

Посвящено геологам-разведчикам, которые используют науку и логику к откритии рудных месторождений.

Dédié aux prospecteurs qui utilisent la science et la logique pour découvrir les gisements metalliques.

Dedicado a los exploradores que utilizan la ciencia y la lógica para el descubrimiento de yacimientos metálicos.

Den Prospektoren gewidmet, die Wissenschaft und Logik nützen, um Erzlagerstätten zu finden.

Dedicato a tutti i geologi che usano la scienza e la logica per scoprire i giacimenti metallici.

Esta obra é dedicada aos prospectores que usam a ciência e a lógica para encontrar uma mina.

Věnováno geologům-prospektorům kteří užívají vědu a logiku k objevu rudních ložisek.

Bagi para pemburu cebakan yang menggunakan ilmu dan penalaran.

مهدى الى المنقبين الذين يستخدمون العلم والمنطق لاكتشاف رواسب معدنية .

献给用科学和逻辑找矿的勘探者

...and to my family,
co-workers in many ways.

CONTENTS

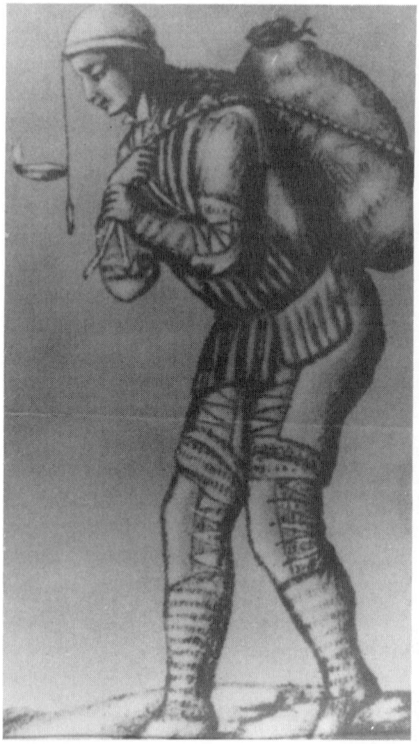

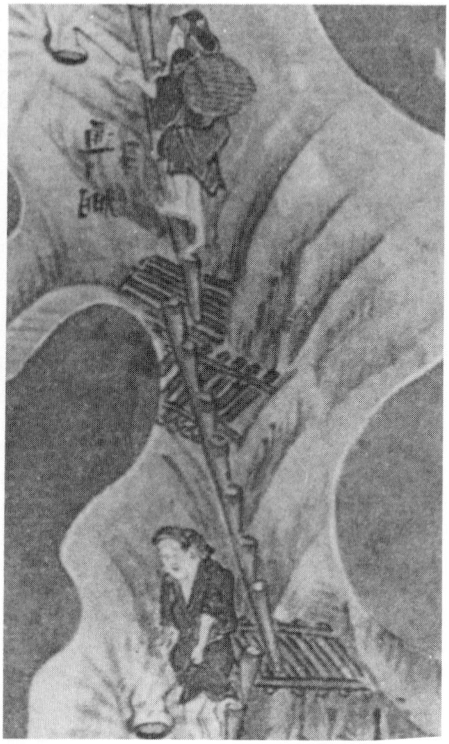

Plate 0.
The international tradition of metal mining.

TOP LEFT: Indian porter at Cerro de Pasco, Peru.

TOP RIGHT: Gold miners in the Sado Island mines, Japan.

BOTTOM: Silver miners in Annaberg, Saxony.

PREFACE AND ACKNOWLEDGEMENTS

"It is not as embarassing to know little about a subject as it is to not realize that such a subject exists at all". (Quoted from a letter of a former student).

This book contains a large body of information on Phanerozoic metallic ore deposits. A second, companion volume will cover metallic ores of Precambrian age and unusual associations of all ages, and will contain search tables covering both volumes. In these volumes all materials containing ferrous, base, rare and precious metals of economic value are considered metallic ores; they include accumulations of Ag, Al, As, Au, Be, Bi, Cd, Co, Cr, Cu, Fe, Ga, Ge, Hg, Mn, Mo, Nb, Ni, Pb, Pt metals, Re, REE, Sb, Se, Sn, Ta, Te, Th, Ti, Tl, U, V, W, Zn and Zr. A nonconventional organization of ore occurrences of the world, by environments in which they presently form or lithologic associations in which they occur, is used, an organization so far tried within a limited scope only or in a framework too rigid to be acceptable here. Considerable effort has been made to achieve a balanced coverage, both in the selection and treatment of subjects, and on the regional basis. The work is based on the literature, consulted in the original in over thirty languages, and on field work, and it treats locality examples in some one hundred and fifty countries and regions, many of them checked in the field.

As an academic, I have a deep admiration for the hardy prospectors of the past as well as the dynamic and imaginative exploration geologists of the present, and this book has been primarilly prepared to serve their practical needs. Its organization parallels the approach usually employed in regional mineral-potential evaluation and exploration. The book answers the call for empirical data frequently expressed by the geological community (for example, as reported by J. Eidel at the 1980 Geological Society of America Meeting). My ideal was to prepare an all-inclusive book and, although "everything" in metalliferous geology is an unattainable dream, I believe this work goes farther than the rest. Given the global coverage, the text unfortunately can give only an overview and cannot consider details of individual environments, associations, and deposits. The theory of rock and ore petrology and related subjects had to be reduced to a minimum, but this topic is extensively covered in textbooks. My dream will come true if this book becomes a convenient "first aid" to exploration practitioners in the bush and in the office, in which at least some information might be found on almost every imaginable mineralization style and its setting. In writing this book,

I attempted to reverse, temporarily at least, the changing nature of geologists "from natural scientists with a broad outlook and knowledge of foreign languages to specialized technicians conversing in one language and competent in a single field that seems to become smaller with time" (A.V. Carozzi, 1975, p.6). As, however, Gill (1981) stated in the introduction to his book, there are serious drawbacks: "The single-authored approach is intimidating in scale and can result in loss of thoroughness or authority on individual topics. The alternatives lack scope or integration or both".

The present compendium is my mid-life project for which I have been collecting literature and field data since my student days in the 1950's. I was fortunate to have examined, however briefly, over 2,000 mineralized localities in 85 countries, thus acquiring a rather broad, first-hand feeling about most mineralized associations and environments, other than the submarine ones, which are outside of my means. Much of the factual material, in the form of rock and ore samples, has been assembled to become part of a "rock library" (LITHOTHEQUE; Laznicka, 1974, 1975; Fig.1). This information system, now consisting of over 1,600 sets and 30,000 specimens, was actively used in all stages of preparation of the present book, often acting as a final arbiter on interpretation of controversial lithologic associations or mineralization styles.

In contrast to some of my more fortunate peers who are narrowly researching in style supported by grants or expense accounts, my family and I bore the brunt of the research expenditures. I operated on a super-economic basis, biking, hitch-hiking, busing to localities; sleeping in the bush, on airport floors or in rented cars; and dining out on canned food. As a fringe benefit, I learned about the world as it really is. The results came slowly and there were some setbacks, like being apprehended by police as a suspected spy in one southeastern European country and having all my notes, maps, photographs and over 2,000 samples confiscated (I have never had these returned).

The greatest assistance in easing up, temporarily, the financial burden of the field travel, came from several mining companies (Amax Inc., Falconbridge Ltd., B.P.Minerals Ltd., Selection Trust Ltd., Western Mining Corporation Pty. and W.M.C. Mineracão, Ltda., for whom I did short-term seasonal work, consultations, or who purchased duplicate sets of LITHOTHEQUE. The consulting work itself involved study of large areas in Canada, Australia, the U.S.A., Mexico, Bolivia and Brazil, and I invested the earnings into further fieldwork of my own. I am indebted to former party chiefs, managers and co-workers: Dave Silversides, Aldo Gambardella, Chris Bates, the late Paul Andrews, Jim Martin, Antonio de Castro; Drs Tom Pedersen, Ray Lett, Bill Hancock, Geoff Loftus-Hills, Douglas Haynes, Owen Bavinton, and others for their support. Particular thanks are due to Don Mustard, Jim Lalor, and Roy Woodall for encouragement and inspiration.

In my institution, past general assistance came from Professors H.D.B. Wilson and W.C. Brisbin during their tenure as Heads of the Department of Earth Sciences, University of Manitoba, between 1967 and 1977. The warm collegial encouragement received afterwards from Professor J.T. Teller in the difficult period of solitary book writing, helped considerably to rise my spirits. Partial contribution by the University of Manitoba towards the cost of research done during my study leave in 1980/81 is gratefully acknowledged.

Several hundred mining company executives kindly gave permission and support to visit their operations, and a similar number of mining and exploration geologists, university personnel, and officers of geological surveys and U.N. missions spent time with me and contributed their knowledge and hospitality. To all these colleagues and friends I have a debt of gratitude. Special thanks are due to Dr. R. Crossley, Peter Freeman, the Gallon family, Dr. and Mrs. Magnus Garson, Dr. G. Kautsky, Dr. J. Kouřimský, I. Kusnir, Dr. Jan Kutina, Prof. Takeshi Nakamura, M. Motys, Prof. K. Paděra, N. C. Pathirana, Dr. O. Pluskal, Prof. Z. Pouba, Dr. M. Prosová, H. Pudjowalujo, Prof. R. Reddi, the Sibiya family, Prof. R. L. Stanton, Peter Stuart-Smith, Dr. Teh Guan Hoe, Dr. and Mrs. Phil Seccombe, Dr. V. Šípek, Dr. J. Švenek, Dr. K. Tuček, J. Vandenberghe, Dr. B. N. Wahju, and many additional persons too numerous to name. Professors Chen Guoda, A.M. Al-Shanti, C. Povoledo; Dr. O.A. Bavinton; Mr. H. Pudjowalujo; and Mmes. G. Lacroix and L. Baydack, kindly supplied or checked translations of my dedication.

To produce a camera-ready copy, a requirement of our times, I purchased word processing equipment and did all the typing, layout, and drafting. I hope the reader finds these acceptable. The photographs included are mine unless stated otherwise. My geological wife Sarka deserves thanks for taking over my share of dishwashing and other household chores, for looking after Dennis and Paul almost singlehandedly, and for helping in the field and during the compilation stage. My late father Dr. Jiří Láznička helped considerably to procure the hard-to-get Russian literature.

Dr. K.H. Wolf of Sydney, Australia, the "helpful editor" from the time of my involvement with the "Handbook of Strata-Bound and Stratiform Ore Deposits", and a respected friend, volunteered graciously to undertake the formidable task of reading the entire work. To him belong my sincere thanks. Elsevier Publishers, and in particular Drs. H. Frank, Director, are thanked for their remarkable tolerance regarding the increase in the size of this book, unforseen in the original contract agreement. Intended initially as a "DC 9"-size modern substitute of the Lindgren (1933) book, a "DC 3" of economic geologic literature in terms of versatility, the book turned into a "Boeing 747". Whereas, however, in the Lindgren's days the geological literature of the world amounted to some 50,000 items, the count is now some 1.5 to 2 million. This provides some justification of the book size.

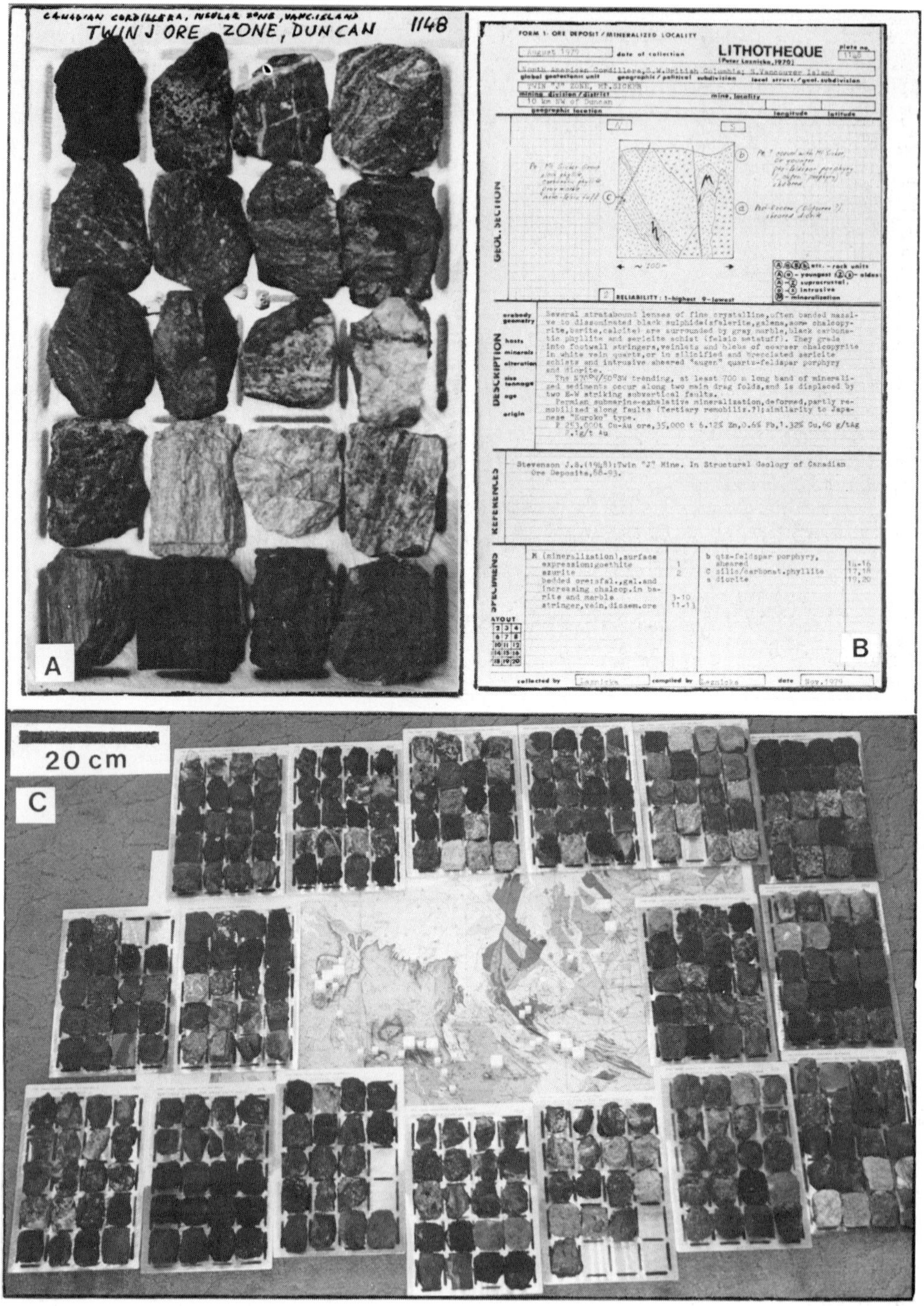
CANADIAN CORDILLERA, INSULAR ZONE, VANC.ISLAND
TWIN J ORE ZONE, DUNCAN
1148
A
FORM 1- ORE DEPOSIT / MINERALIZED LOCALITY
LITHOTHEQUE
(Peter Laznicka, 1970)
plate no.
date of collection
North American Cordillera, S.W.British Columbia; S.Vancouver Island
global geotectonic unit
geographic / political subdivision
local struct. / geol. subdivision
TWIN "J" ZONE, MT.SICKER
mining division / district
mine, locality
10 km NW of Duncan
geographic location
longitude
latitude
GEOL. SECTION
RELIABILITY: 1-highest 9-lowest
etc. - rock units
youngest
oldest
supracrustal,
intrusive
mineralization
DESCRIPTION
orebody geometry
hosts
minerals
alteration
size tonnage
age
origin
Several stratabound lenses of fine crystalline, often banded massive to disseminated black sulphide(sfalerite, galena, some chalcopyrite, barite, calcite) are surrounded by gray marble, black carbonatic phyllite and sericite schist (felsic metatuff). They grade into footwall stringers, veinlets and blebs of coarser chalcopyrite in white vein quartz, or in silicified and brecciated sericite schists and intrusive sheared "augen" quartz-feldspar porphyry and diorite.
The N70°W/50°SW trending, at least 700 m long band of mineralized sediments occur along two main drag folds, and is displaced by two E-W striking subvertical faults.
Permian submarine-exhalative mineralization, deformed, partly remobilized along faults (Tertiary remobiliz.?); similarity to Japanese "Kuroko" type.
REFERENCES
Stevenson J.S.(1948):Twin "J" Mine. In Structural Geology of Canadian Ore Deposits, 88-93.
SPECIMENS
M (mineralization), surface expression: goethite
azurite
bedded ore: sfal., gal. and increasing chalcop. in barite and marble
stringer, vein, dissem. ore
1
2
3-10
11-13
b qtz-feldspar porphyry, sheared
C silic/carbonat. phyllite
a diorite
14-16
17,18
19,20
LAYOUT
collected by
Laznicka
compiled by
Laznicka
date
Nov. 1979
B
20 cm
C

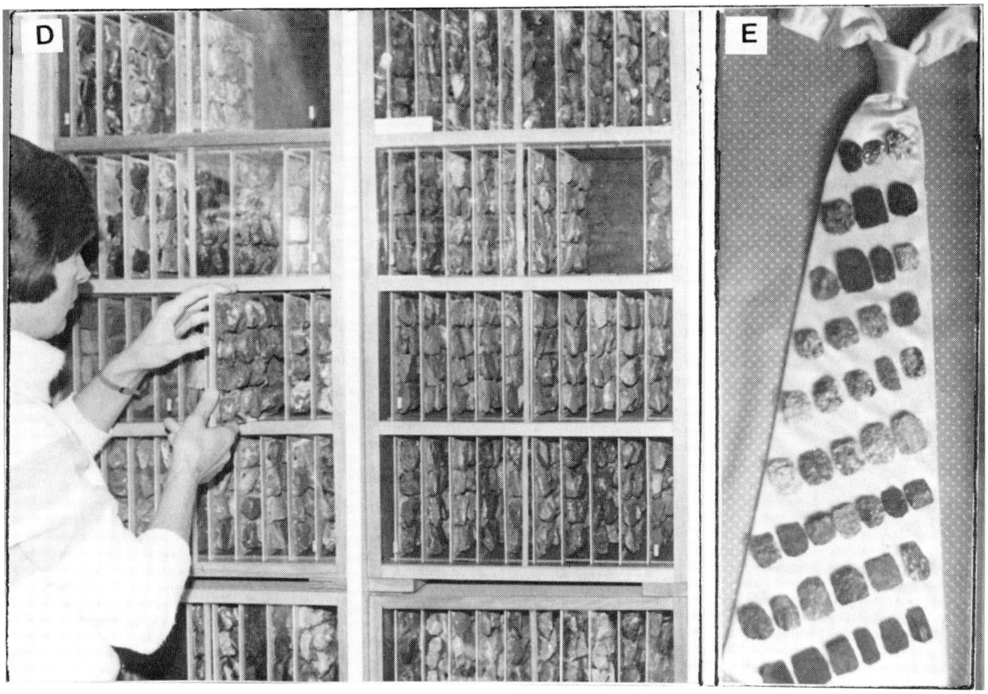

Fig. 1. LITHOTHEQUE ("rock library"), a portable documentation system covering metallic deposits of the world and their lithologic associations. At the end of 1984, the system consisted of 1,600 sets (at 20 samples/set, about 32,000 specimens) and it was extensively applied to supplement the literature data used in compilation of this book, to strenghten its realism.
A,B: An average locality set, with samples cemented on an aluminium plate 18x28 cm.
C: Ore-hosting lithologic associations and selected deposits in part of the Pine Creek "Geosyncline", Australia.
D: Space-economical, "user-friendly", and instantly accessible LITHOTHEQUE storage.
E: LITHOTHEQUE is a serious bussiness, but there are exceptions such as this LITHOTIE (TM), representing lithostratigraphy of the Canadian Shield. A retirement gift to H.D.(Bruce) Wilson.

I am most grateful to the publishers and authors who gave permission to reprint their material, and to the countries and institutions (the U.S.S.R. prior to 1973 and the U.S. Government agencies) who permit the public to use excerpts from their publications without written authorization. In addition to reprints, many previously published figures have been modified and redrafted. Their authors, always carefully referenced, deserve the credit due to them.

In an extensive and largely "home-made" book like this, errors are almost inevitable ("if errors can occur, they will!"; an adaptation of the Murphy's Law). I beg my readers to treat such errors with a broadminded tolerance and I will appreciate everyone who will let me know.

ABBREVIATIONS, EXPLANATIONS

GRADE AND TONNAGE DATA

Unless stated otherwise, metal tonnages in ore deposits are uniformly expressed as metric tons (tonnes) of contained metal/average grade, as follows: 2 Mt Cu/1%. The quantity of ore can be calculated, using the equation:

$$\text{ore (tons)} = \frac{\text{metal (tons)} \times 100}{\text{grade (\%)}}$$

Grade of precious metals is expressed in grams per ton (gt)= parts per million.

All numeric units of tonnage and grade (long and short tons, lbs, units WO_3, flasks of mercury, ozs, dwt's, etc.) have been converted to metric equivalents for easy comparability. Grades/tonnages expressed in compounds (Al_2O_3, MoS_2, SnO_2, UO_3) have been recalculated to pure metals.

Abbreviations: t=tons, Tt=thousand tons, Mt=million tons, Bt=billion tons.

The tonnage figures listed correspond ideally to the "geological tonnage", that is, the tonnage of ore present before the mining has started. Reserves of new deposits and total production figures of mined-out deposits approach this figure, but the rest of available data is incomplete. Incomplete or missing figures have been sometimes extrapolated or estimated and if so, this has been clearly stated. It should be realized that grade/tonnage data are based on the mixture of geologic and economic factors and that the economic factors change substantially with time. The basic purpose of the tonnage figures in this book is to give a more accurate idea about the economic importance of a deposit than by using expressions like "large", "important", "minor", etc., but the figures need not be final, complete, and may change with time.

EXPRESSION OF ORDERS OF MAGNITUDE

Many quantitative data for which accurate figures are not available, are expressed in the form of orders of magnitude. The letter "X" substitutes for "several". For example, X deposits (=several deposits); X00 t (=several hundred tons); X0 Tt Cu (several tens of thousand tons of copper). The "X" abbreviation is most widely used in tables where the space is at premium.

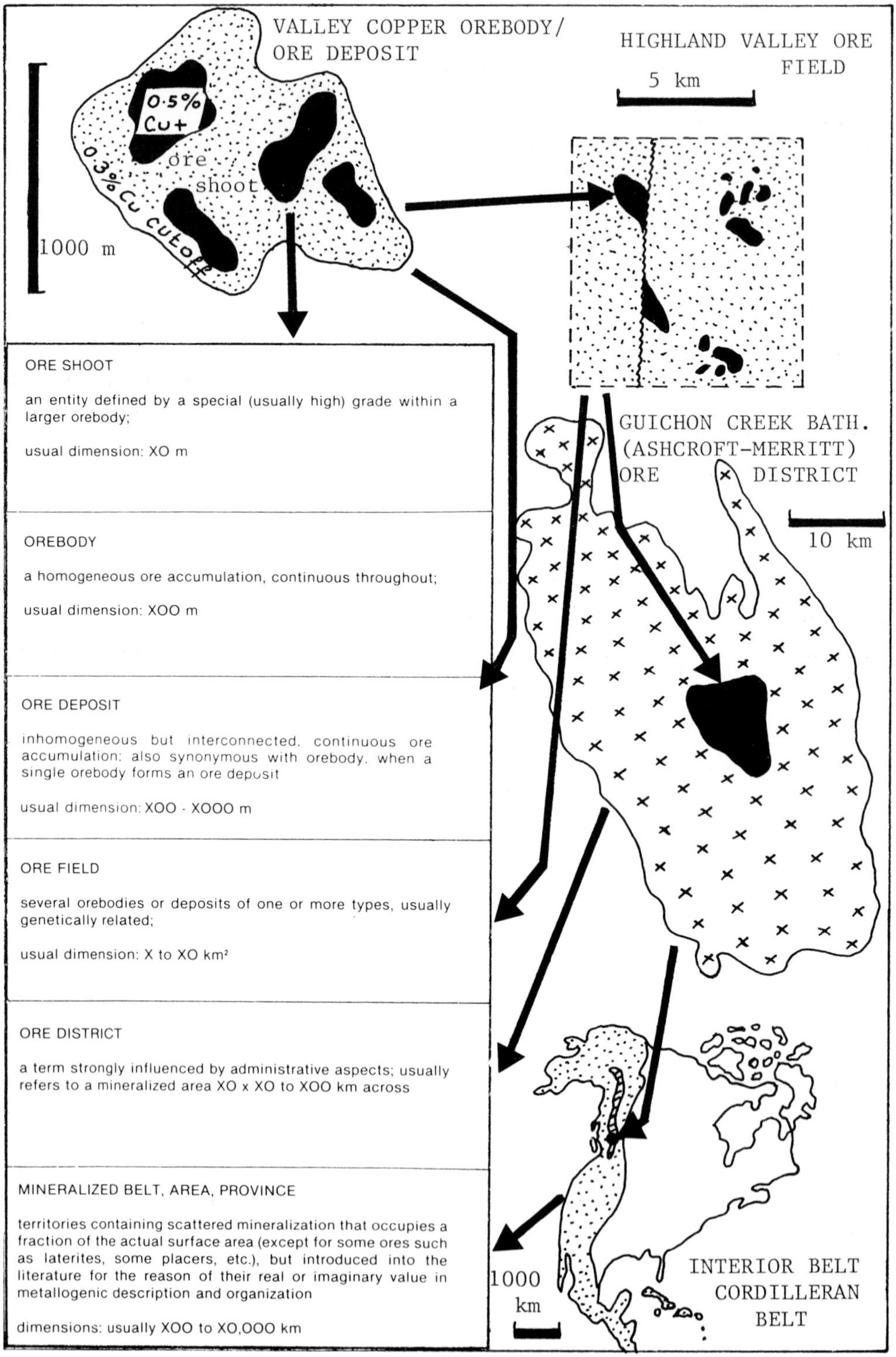

ORE SHOOT

an entity defined by a special (usually high) grade within a larger orebody;

usual dimension: XO m

OREBODY

a homogeneous ore accumulation, continuous throughout;

usual dimension: XOO m

ORE DEPOSIT

inhomogeneous but interconnected, continuous ore accumulation; also synonymous with orebody, when a single orebody forms an ore deposit

usual dimension: XOO - XOOO m

ORE FIELD

several orebodies or deposits of one or more types, usually genetically related;

usual dimension: X to XO km²

ORE DISTRICT

a term strongly influenced by administrative aspects; usually refers to a mineralized area XO x XO to XOO km across

MINERALIZED BELT, AREA, PROVINCE

territories containing scattered mineralization that occupies a fraction of the actual surface area (except for some ores such as laterites, some placers, etc.), but introduced into the literature for the reason of their real or imaginary value in metallogenic description and organization

dimensions: usually XOO to XO,OOO km

HIERARCHICAL TERMINOLOGY OF MINERALIZED UNITS

Terminology shown in Fig.2 is used, with some exceptions. This may occasionally differ from the established usage. For example, "District" is applied in North America to administrative units of variable size (e.g. the Robinson District, Nevada, is here listed under the Ely ore field).

REFERENCING

Reference "from LITHOTHEQUE" (in captions to figures) indicates that the locality has been visited and sampled by this author, and the sample record is a part of the LITHOTHEQUE system (Fig. 1) and was consulted during this compilation. Reference to the original source (if any) is given, but the figure may have been slightly to considerably altered as a result of field observation or "modernization". The original author cannot be responsible for such a modified figure.

STANDARD HEADING OF MINERALIZATION STYLE TABLES (e.g. Tables 5-4, 5-7)

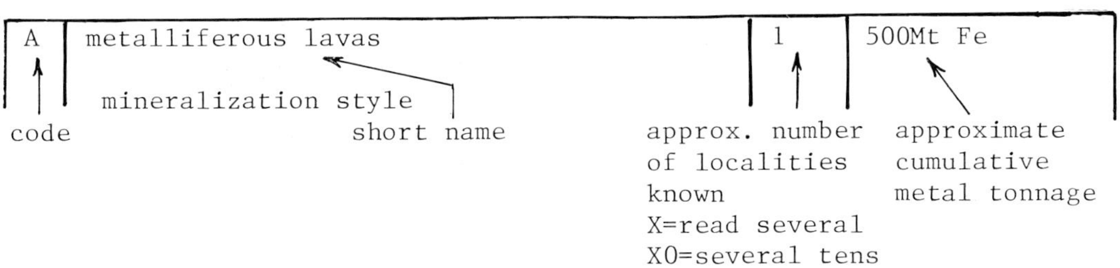

Fig. 2. Hierarchy and terminology of mineralized bodies and areas, illustrated by examples from the porphyry copper province of British Columbia. From Laznicka (1984).

GEOLOGICAL AGE ABBREVIATIONS

Ages (in tables and figures) are expressed as chronostratigraphic units (letter codes) or in absolute age figures, as follows:

Era				Code	Name	
				Q	Quaternary	
CZ	T	T_3		Pl	Pliocene	
				Mi	Miocene	
		T_2		Ol	Oligocene	
				Eo	Eocene	
		T_1		Pc	Paleocene	
MZ				Cr	Cretaceous	
				J	Jurassic	
				Tr	Triassic	
PZ	PZ_3			Pe	Permian	
			Cb_2	Ps	Pennsylvanian	Carboniferous
			Cb_1	Ms	Mississippian	Carboniferous
		PZ_2		D	Devonian	
				S	Silurian	
	PZ_1			Or	Ordovician	
				Cm	Cambrian	

Era			Age
PC			0.6 b.y.
	Pt	Pt_4	
			1.0 b.y.
		Pt_3	
			1.4 b.y.
		Pt_2	
			1.7 to 1.8 b.y.
		Pt_1	
			2.4 to 2.55 b.y.
		Ar	
			about 3.0 b.y.
		KAr	

Indices 1,2,3 (e.g. Cm_1, Pt_3, Cr_{1-2}) are for lower, middle and upper, respectively (e.g. Cm_2 = middle Cambrian).

"Go, have a look!"
Roy Woodall (speaking
about the art of exploration)

CHAPTER 1

Introduction

1.1. EMPIRICISM, SCIENCE AND METALLIFEROUS GEOLOGY

Roy Woodall (Director of Explorations, Western Mining Corporation, Australia) was the Distinguished Lecturer of the Society of Economic Geologists in 1984. He gave a series of lectures on succesful methods of mine finding and management. Western Mining Corporation, among other achievements, discovered and developed Kambalda, the first major nickel sulphide discovery in Archaean rocks and the first significant nickel discovery in Australia. The Corporation's exploration team subsequently found the giant Olympic Dam deposit, a blind U, Cu, Au, REE, etc. orebody beneath 300 metres of younger, unmineralized sediment, an orebody destined to be one of the world's most valuable deposits. Mr. Woodall attributed much of this exploration success to the utilization of empirical data on global ore distribution, extended by scientific reasoning and prediction. His priorities contrast sharply with the spirit presently prevalent at some university geoscience departments. There, in a strange twist of Orwellian newspeak, fragmentation of science, narrowness of approach, irrelevant ("hobby") research, disinterest in the accumulated knowledge and reading about geology instead of doing it, are presented to the public as major achievements. A wholesale commitment to empiricism and a broad, realistic synthesis, approaches suicide.

Empiricism is the "belief that knowledge is derived from experiment or experience" (Webster's Dictionary). In exploration geology, empirical attributes are usually considered to be those obtained by observation in the field, as opposed to attributes based on a theory, hypothesis or a model which involves speculation.

Science is defined as "knowledge, especially of facts and principles, gained by systematic study" (Webster), so empiricism is an ingredient of science, not a "primitive" approach contrasted with an "educated" one, as some might think. Metallogeny initiated by De Launay was originally empirical, without this being explicitly emphasized. His monumental work Traité de Métallogenie (De Launay, 1913) was essentially a global summary of distribution of ore deposits based on observation and literature, and it is still usable today.

Increasingly, however, the metallogeny has become theoretical and hypothetical, enhanced but also distorted, by temporary models. In the geosynclinal model of metallogeny of the 1950s, ore deposits in "eugeosynclines" were related to a hypothetical linear zone of vertical tectonism that subsided, was filled by mafic volcanics and sediments, reversed, deformed, intruded by plutons, uplifted, eroded and levelled. Plate tectonics shattered the earlier paradigm and in its early stage of development (1968-1975) attributed virtually all ore deposits in volcanic sedimentary and plutonic orogenic belts to subduction processes taking place above Benioff zones, at convergent continental margins. Proponents of the static model of the origin of comparable belts, not all of them traditional "geosynclinalists" (e.g. Krebs, 1975; major segment of the Soviet school; van Bemmelen, etc.) were generally excluded from the exuberant gatherings of devotees of the "pop" version of plate tectonics, about which most of us had only a very superficial understanding.

Bandwagonism was rampant and the Archean "greenstone belts" of the Canadian Shield were "old island arcs" produced by subduction. With the plate tectonics 16 years old, we are now more sober. It is quite obvious that there are both andesite-dominated orogenic belts which the available evidence indicate are in some way associated with subduction (e.g. the late Triassic suite of the Canadian Cordillera), as well as basalt only and bimodal belts, in some way associated with vertical tectonism (the Harz Mts., the Urals, etc., as advocated by Krebs and the Soviets). In the latest version of the continuous saga of the Canadian "greenstone belts" interpretation, they are now "equivalents of the present rifts". What are they going to be in the year 2000 ?

The above modelling on a global scale has its parallel in more local situations, involving genesis and ore affiliation. The metavolcanics and metasediments- hosted massive sulphide deposits as in the Abitibi Belt, Canadian Shield were, in the 1950s, considered as hydrothermal replacements affiliated to granite plutons. In the 1960s and 1970s they were "volcanogenic" and "submarine exhalative", unrelated to granites. Now, the granites are back, in the form of sills or stocks in the footwall, credited with driving the convecting hydrothermal fluids that precipitated the ores after the peak of volcanism. The genetic interpretation went almost full circle (Fig. 1-1) in about 30 years.

The hypothetical models of metallogeny cannot be entirely rejected. We want to know how ores formed and when this cannot be observed, interpretation based on indirect data has to be made. When even this fails, we settle for a hypothesis. Most ore-forming hypotheses and models have had some credibility and usually a minor to major success in the area or a setting, for which they were developed. The rate of ore discovery jumped significantly, following the introduction of several initially hypothetical models (Fig. 1-2). The main harm was usually caused by an uncritical promotion of a local interpretation to assume universal validity, or by a careless application of a correct model to the wrong situation. Trümpy (1982) made another point regarding inherent dangers of succesful models, which includes the readiness of scientists to apply models to cases in which facts are poorly understood.

Hypothetical metallogenic models appear dated and most of those over 10-15 years old are now consulted for historical purposes only. Objective empirical local or regional geological and deposit descriptions, on the other hand, lose their value much more slowly and

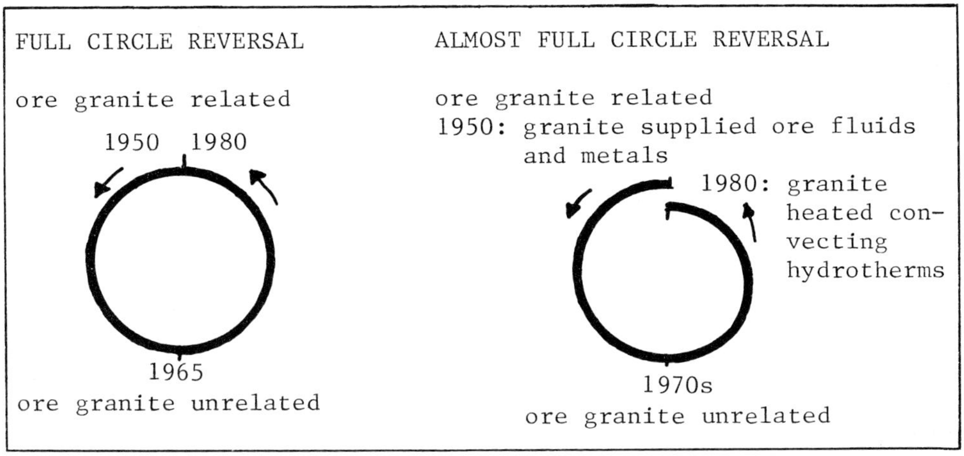

Fig. 1-1. "Full circle" and "almost full circle" reversals in geological interpretation. In full circle reversals, interpretations return exactly to their earlier state (e.g. hydrothermal plutonic body later reinterpreted as sedimentary, now back in the earlier version). Almost full circle interpretations return to the major genetic attribute, but their actual function has been reinterpreted.

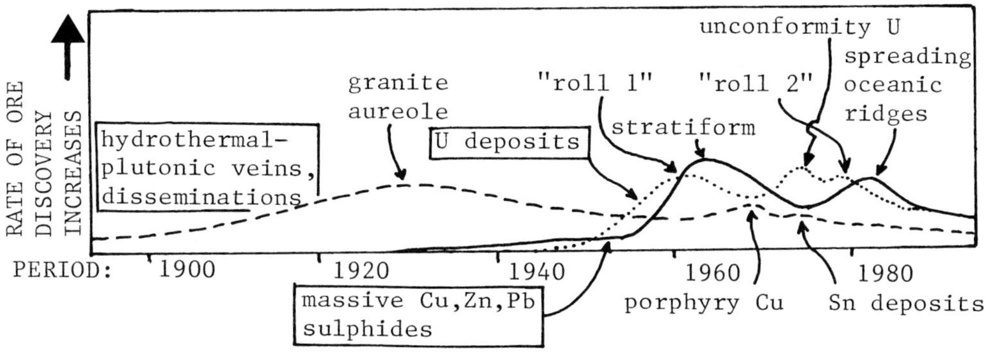

Fig. 1-2. Influence of succesful geological models on the rate of ore discovery, 1900-1984.

some (those actually describing the geology of presently abandoned underground mines; terrains flooded by dams or overrun by cities; etc.) are permanent. Many masterpieces of geological description and objective documentation, remarkably free of cheap contemporary hypothesizing and jargon, have been produced since the turn of the century as part of the U.S. Geological Survey Professional Papers series (e.g. Emmons et al., 1927, memoir on Leadville; Butler and

Burbank, 1929, memoir on the Michigan native copper region; Gordon et al., 1958, memoir on the Arkansas bauxite region; compare review in Laznicka, 1985a). Empirical vintage text- and reference- books such as Beyschlag et al. (1916) or Lindgren (1933) or others, are still useful sources of information on many mineralization styles and deposits of the past economic interest, rarely described now. Theoretical or hypothetical books from the same period are, on the other hand, rarely consulted.

Compilation of this book took a long time and a major effort, so I would like it to serve to the public longer than what appears to be a common tenure of a contemporary reference book (one year honeymoon followed by few years of productive life and an early retirement at the age of 8-10 years). For this reason, I selected the longer lasting empirical style although where factual data were unavailable, I took the liberty of venturing moderately into the realms of hypotheses. The second reason for an empirical approach is the fact that empirical units are directly and objectively recognizable, while the hypothetical categories have first to be interpreted (starting with the empirical associations anyway !); this involves a double approach as well as a considerable loss of accuracy. A prospector advised to search for porphyry coppers above former Benioff zones has first to outline the facies belts believed to correspond to a former magmatic front, and then proceed with a search for orebodies within this zone. Why not forget about the Benioff zone, and start by outlining the favourable lithologic association right away ?

1.2. SOME PROBLEMS OF PRESENTATION, ORGANIZATION AND STYLE IN METALLOGENIC WRITING

There are about 1.5 million citations of geological literature accumulated by now and this number grows at a rate of 150,000 new citations per year (Laznicka, 1985a). About 30% of this is in English, and over 30% in Russian. At least 50% of the literature involves, directly or indirectly, ore deposits or their host areas. Wolf (1976) presented forceful reasons for the need for translating, reviewing and summarizing this data as a basic means of coping with this flood.

In the organization of data, two "end member" approaches are available: (1) detailed (intensive) study and interpretation of a single locality (or a small population of localities), the results of which are then incorporated into an "ore type" model believed to be applicable to other localities not yet identified; (2) rapid and partial (extensive) study of a large number (thousands or tens of thousands) localities, that are subsequently mentally sorted and incorporated into a system of usually very transitional "mineralization styles". Joralemon (1952) reviewed very succintly the advantages and pitfalls of the second approach, which he called "Comparative Economic Geology", and concluded that both approaches (methods) have their place in exploration.

The intensive models (1) are put together much faster and appear more dynamic. The bulk of the recent Western writing on mineral deposits is close to this end-member. It is also very much alive in exploration (one year it is a search for the "Carlin-type" disregarding the rest of deposits, the next year it is the "Olympic Dam-type"). Text and reference books prepared in this spirit, in places go into a substantial depth, but they are limited in the extent of coverage and are full of gaps. This follows graphically from Fig. 1-3.

The extensive method (2) is more popular with the "mature" generation of geologists, both academic and practical, who have seen enough real geology in their life to question the elegance of many striking models. They are aware of the imperceptible category transitions (almost a rule in geology) and they remember the rise and fall of many formerly celebrated models. Reference books written under the influence of the extensive approach have a considerable range but a restricted depth of coverage. It is, however, easier to pursue supplementary reading vertically (into a greater depth) than laterally (into adjacent subjects), provided that the base book contains abundant references. Good, extensive books cover the field evenly and there is minimum of gaps. The classical compendia on economic geology (Lindgren, 1933; Schneiderhöhn, 1941 and 1955, and others) are like this, but post—1970 examples are rare. One of them is the exhaustive treatment of gold by Boyle (1979). The present book has been designed along similar lines.

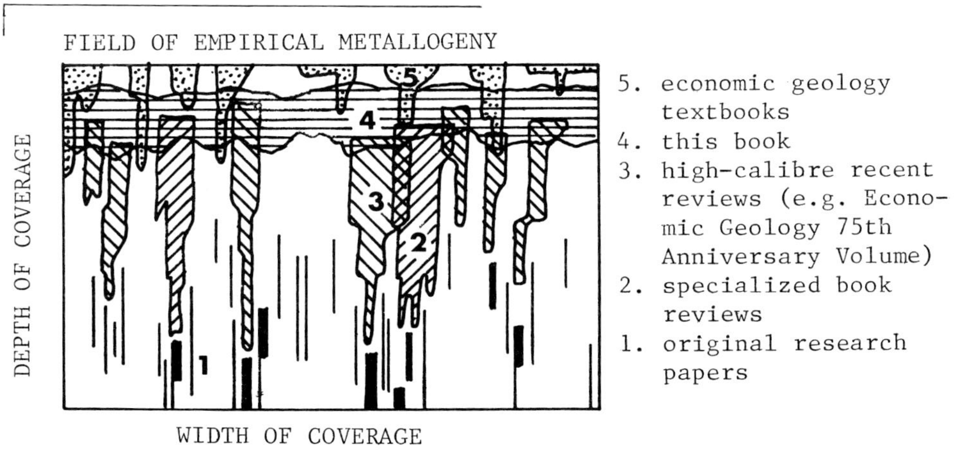

Fig. 1-3. Subjective diagrammatic representation of the depth versus width of coverage of the subject of global distribution of ore deposits, by five categories of literature (including this book).

The exponential growth of science (including metalliferous geology) that includes narrowing specialization, is increasingly countered by anthologies prepared under group authorship. Two contemporary volumes on carbonatites published in 1966 (Heinrich, 1966; Tuttle and Gittins, eds., 1966), offer an unusual comparison of a single-author treatise and a multi-author anthology on the same subject. Both works have their advantages and disadvantages but my heart is with the former which is more homogenous, balanced and complete.

As can be seen, the present book reviews deposits and occurrences many of which are "unknown", many of which seem to "belong to the past", and many of which "will never make a mine" (points listed as negatives in several recent reviews). In defence it should be said that the purpose of describing "unknown" deposits is to make them known. As far as the making of a mine goes, one never knows. Before Olympic Dam, hematitized breccias were not considered possibly to herald a significant mineralization. The "past" deposits (e.g. bog irons; small limonite bodies precipitated in soil profiles; etc.) are indeed unlikely to become economically important in the future but they are still of interest scientifically. Furthermore, in mineral exploration and evaluation it is often necessary to decide which showing is of possible importance and which is not, and the "past" mineralizations have to be recognized, identified and included in this context. Back in my days as a junior consultant, a client summarized my efforts as follows: "well, your recommendations have not resulted in a new discovery, but at least they kept us from the wrong leads the others followed". The "not following the wrong leads" was worth about $ 1 million. Poor targets do not look as such in the field, unless one knows something about them.

"Mineral exploration programs should be based on observation and experience, rather than on genetic theories if only because the latter have a nasty habit of proving to be half-truths at best"
Paul Gilmour (1977, p.295)

CHAPTER 2

Depositional Environments and Lithologic Associations

2.1. GENERAL

Depositional sedimentary environment is "a geomorphic unit in which deposition takes place" (Reineck and Singh, 1975). Other rock and ore groups (magmatic, metamorphic) also form in characteristic environments, although these are better called "environments of generation", "formational environments" and similar. Those located at some depth are hypothetical because they have never been observed directly. Most directly recognizable depositional environments described in the literature are "active" (modern), or recently active. Inactive (past) environments cannot, as a rule, be observed, but they are represented by characteristic lithologic (rock) associations. The terms "environments" (or "paleoenvironments") and "lithologic associations" are often used interchangeably in the literature, but this is erroneous.

An ancient lithologic association ("formation" in the Soviet literature) can presently be observed in outcrop and subcrop and directly, objectively studied, whereas paleoenvironments have to be interpreted. In this book, metallogenies of modern environments and ancient lithologic associations are treated in separate chapters, although references are always made as to their linkage. The term "facies" (in particular "lithofacies"), indicates "a body of rock with specified characteristics", which include colour, bedding, composition, texture, fossils, and sedimentary structures (Reading, ed., 1978). Its meaning is close to that of a lithologic association, but facies is slightly broader. The art and science of recognition of ancient sedimentary environments is now a well-established discipline (e.g. Rigby and Hamblin, eds., 1972; Reineck and Singh, 1975; Reading, ed., 1978), whereas magmatic and metamorphic environments are usually treated in the framework of corresponding petrology and petrogenesis.

Metallogeny or mineralization as related to specific depositional environments have recently been treated fairly frequently (e.g. in the Handbook of Strata-Bound and Stratiform Ore Deposits edited by Wolf; volumes 1-10, 1976 and 1981), but no work is available that would review all (or at least most) of them.

Lithologic associations ("formations") have been standardized and systematically broken down into the minutest units in the U.S.S.R. (Kuznetsov, 1964). This is helpful and manageable in the case of the "clear line" associations, but difficult and subjective when hybrids are involved. Several impressive cartographic works based on this concept have been produced, such as the Atlas of Lithologic-Paleogeographic Maps of the U.S.S.R. (Vinogradov, ed., 1968), and the Map of Magmatic Associations of the U.S.S.R. (Kharkevich, ed., 1968). Taken together, these two map sets provide the most objective and complete representation of the regional distribution of lithologic associations published anywhere, and presumably are widely used in exploration and in mineral potential evaluation in the U.S.S.R. These maps also provide objective data for paleotectonic and paleogeographic interpretation, including those based on the plate tectonic model (e.g. Laznicka, 1976a).

This phase of regional interpretation has been virtually bypassed in western countries (with the exception of sedimentogenic facies maps, particularly those used in petroleum exploration), and plate tectonic compilations have been made directly from incompletely pre-interpreted, unorganized data fragments.

The Soviets also created a massive organization of "ore associations" (rudnye formatsii). A "rudnaya formatsiya" is a "group of deposits with corresponding and compositionally persistent mineral associations, connected with repetitive geological conditions represented at all deposits"(Konstantinov and Sirotinskaya, 1974). Selected examples: quartz - cassiterite, hydrothermal ("formation"); copper - molybdenum, hydrothermal (=porphyry Cu-Mo here); tin-rhyolitic, subvolcanic; manganese - metamorphogenic; etc. This is fairly similar to the European concept of "paragenesis". Summaries of "ore formations" (associations) together with their basic characteristics, have been provided in the works of Strona (1978) and Chaikovskii (1978).

Recognition and manipulation of "ore formations" for metallogenetic purposes is usually termed "formational analysis" in the U.S.S.R. (e.g. Kuznetsov, 1972). To my liking, the Soviet approach to the organization of lithologic and ore associations is too rigid and in some respects considerably subjective (to outright incorrect), when applied to localities or terrains with which the authors were not personally familiar.

In the English literature, Stanton (1972) was a pioneer in treating ore deposits in the framework of lithologic associations, based on the thesis that "ores are but rare rocks". His anorthosite-iron-titanium oxide association; quartz monzonite - granodiorite - copper - molybdenum sulphide association; and similar associations, are logical units recognizable in the field.

In this book, lithologic associations have been used exhaustively as the basic framework for the treatment of field ore occurrences. They are much broader and less rigid than the Soviet units, and most are transitional or overlapping. For that reason, they are not classes, but merely practically convenient (utilitarian) organizational units. The subject arrangement in this book therefore should not be confused with a classification.

2.2. ORES AND HOST ASSOCIATIONS: PARALLEL TO NON-PARALLEL DEPOSITIONAL HISTORIES, MULTISTAGE ORES AND INTERACTION METALLOGENY

Cumulus grains of chromite in an ultramafic layer or hematite ooliths in a sedimentary ironstone, have a "parallel history" with their host rocks and lithologic associations. Most stratiform deposits are genetically parallel but some disseminated, non-stratiform ores (such as magnetite or ilmenite in gabbro dikes) may also be parallel. Parallel mineralizations can be syngenetic (e.g. heavy mineral placers in sands; cumulus chromite); syngenetic-diagenetic (oolithic hematite); diagenetic, internally segregated (e.g. some sphalerite nodules); epigenetic, autometasomatic (cassiterite in apogranite); or take other forms. Genetic parallelism facilitates prediction, and all voluminously restricted parallel ores are associated with the more abundant characteristic "common rocks".

Massive sulphide deposits formed on the seafloor by precipitation from hydrothermal fluids introduced from below, have a semi-parallel history shared with their host association of marine volcanic and sedimentary rocks. There is, as a rule, much greater variation among the composition of the host rocks and ores, and the only aspect the host rocks have in common with the ores is that both are seafloor deposits. Hydrothermal fissure veins hosted by shales, sandstones, andesite flows, etc., are genetically non-parallel. There is no genetic link between the host and the ore, and ore predictability on the basis of composition of the host rock only is impossible. Any rocks can host hydrothermal epigenetic veins.

It can be concluded that lithologic host associations are a suitable framework for the description and prediction of parallel and semi-parallel mineralizations, but are unsuitable for the non-parallel ores. Although the non-parallel ores are genetically unrelated to their immediate host rock units, they must be related to an ore forming event which, in turn, is usually genetically parallel with a higher-rank process, that frequently produced a lithologic association of its own but in more or less distant location. Epigenetic deposits in "granite" aureoles are directly to indirectly genetically parallel with intrusions present at greater depths.

Modern, rational interpretation of ore deposits (reviewed in Laznicka, 1985e) indicates, that a few formed instantly during a single stage (such as meteoritic irons, provided they are found in economically interesting quantities). Most ores are the result of a stage-by-stage origin, marked by a gradual increase in the metal concentration and accumulation. Such stages can either be parallel with the formational history of the wallrock (e.g. metals in late diagenetic nodules are the product of diagenetic fractionation of the original sediments; chromite in ultramafics is the product of differentiation of the parent magma), or non-parallel (e.g. tropical weathering superimposed on ultramafics, to generate a 1.4% Ni "Ni laterite" ore out of the 0.2% Ni "common ultramafic"). The non-parallel mineralizations require the assistance of an outside agent or influence and are a result of interaction between such an agent or influence and a rock or of an earlier ore.

Interaction ore deposits can be treated (classified, organized, predicted) from two or more directions (Fig. 2-1). When, for example, an ultramafic association is selected as a base line of reasoning, then one may attempt to predict the possible mineralizations that form as a result of interaction with a variety of other rock associations, processes or depositional environments. Ultramafics interacting with a "granite" aureole marked by widespread hydrothermal activity may be subject to alteration dolomitization, and the dolomite, in turn, replaced by Pb-Zn sulphides (e.g. Dundas, Tasmania; Janjevo, Yugoslavia) or pyrrhotite, cassiterite (Razorback mine, Tasmania).

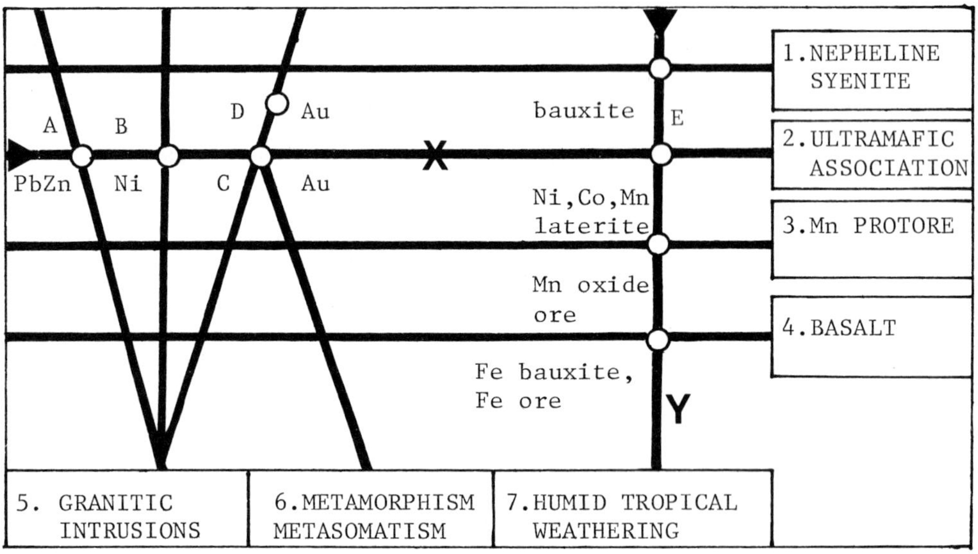

Fig. 2-1. Diagrammatic representation of ore formation as a consequence of interaction among several lithologic associations, environments or processes. Focus on line X (ultramafic association): Interaction with granite-heated convecting fluids (point A) produces alteration dolomite, which in turn is replaced by galena and sphalerite. At point B, hydrothermal sulphurization produces minor Ni sulphides. Further to the right, hydrotherms causing remobilization of trace gold in the ultramafics may produce Au deposits within (C) or outside (D) the ultramafics. At point (E), ultramafics interacting with humid tropical weathering result in Ni,Co,Mn "laterite".

The humid tropical weathering, as a leading factor, is represented by another interaction line (Y), and a variety of residual deposits form at intersections with lines of four example lithologic associations.

Hydrothermal remobilization of some ultramafics is believed to be one mechanism of generation of gold deposits, particularly those containing abundant fuchsite (Cr-sericite). Such deposits could be located within the ultramafics, or outside them.

When the tropical humid weathering (reviewed here as an environment) is selected as the base line of reasoning, it becomes possible to predict the possible locations of economic orebodies at the points of interaction of weathering (strenghtened by other favourable attributes within the humid tropical environment, such as a suitable geomorphology) and particular lithologic associations.

The above examples of predictive metallogeny are rather trivial; the rest, however, are more complicated, less apparent (sometimes not aparent at all), and increasingly hypothetical (Fig. 2-2). It is a challenge for the future to unravel the complex interactions as one of the components of the art and science of ore prediction. In this book, the problem of metallogenic interactions (and ores of dual or multiple affiliation) has been countered by extensive cross-referencing and brief descriptions biased towards the line of reasoning being covered in the particular chapter. These, however, are little short of an outright repetition and description of the same locality and mineralization style, made in a different context. Nickel "laterites", for example, are treated in two major (Chapter 7 on ophiolites; Chapter 23 on weathering), and several minor, positions here.

I anticipate an occasionally strong reaction from some of the readers against some transgressions of the conventional order committed here, such as the partial treatment of the Bolivian plutonic and subvolcanic tin province in Chapter 19, under the heading "shallow marine detrital sediments" (the regional host association). The indexes, however, have been designed to ease this burden.

2.3. FRAMEWORK OF ORGANIZATION

Thirty of the chapters that follow (Fig. 2-3) treat metallic mineralizations in the framework of recent depositional, formational, weathering, etc. environments (Chapter 3: marine; 4: oceanic; 5: continental margins; 23: continental weathering; 24: continental deposition) and ancient lithologic associations (7: ophiolites; 8: melanges; 9: zoned mafic-ultramafic complexes; 10: unimodal mafic; 11: unimodal felsic; 12: bimodal; 13: andesite; 14: sequentially differentiated; all volcanic-sedimentary marine sequences; 15: pelagic sediments; 16: deeper marine sediments; 17: deeper marine slate and carbonate; 19: shallow marine detritals; 20: shallow marine carbonate; 21: marine evaporite; 22: petroleum and bitumen; 25: ancient continental sediments; 26: ancient calc-alkaline volcanics; 28: granitic rocks; 29: high-grade metamorphics; 31: plateau basalts; 32: diabase dikes, layered intrusions; 33: alkaline association).

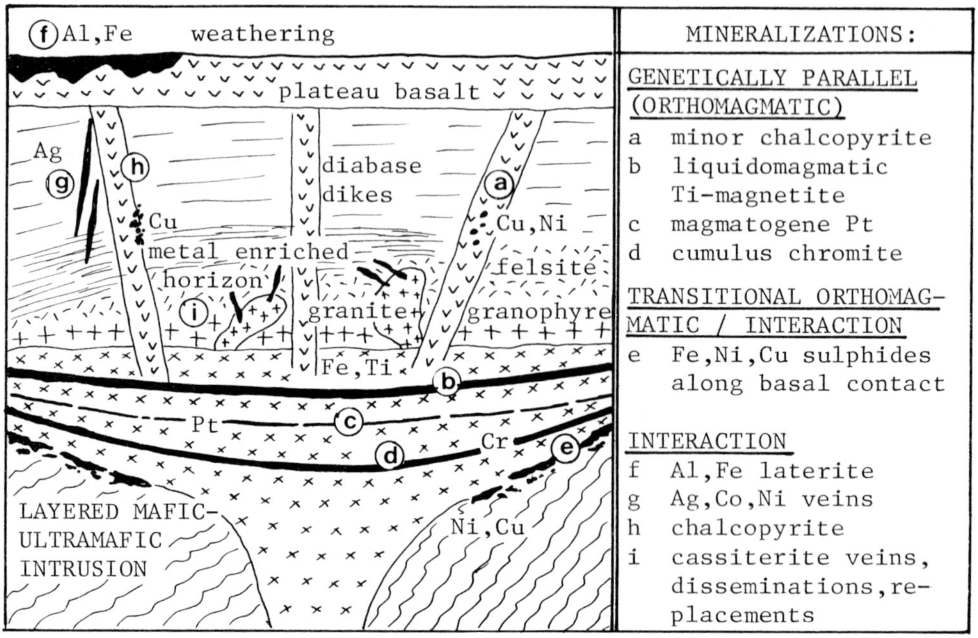

Fig. 2-2. Hypothetical cartoon indicating genetically parallel ("straight line", orthomagmatic); semi-parallel, transitional; and non-parallel, interaction mineralizations affiliated with a multi-level mafic/ultramafic complex emplaced into the continental crust. The orthomagmatic ores formed largely by internal differentiation within the closed magmatic system. Laterites (f) formed by weathering interaction with basalt; Ag-veins (g) by precipitation from hydrotherms in a convecting cell, driven by heat supplied by the dike. Sn (i) is associated with granites probably formed by a high-level anatexis, in the intrusive roof. Ni-Cu sulphides (e) may have formed by separation of immiscible liquid, as well as by a contact sulphurization.

Chapters 3 (world ocean); 5 (continental margins); 6 (orogenic belts); (18) platforms; (27) intracrustal environments; and (30) continental fragmentation, are introductory and directory chapters. Their function is to introduce features common to a group of depositional environments or lithologic associations, and/or relate them to large geotectonic units.

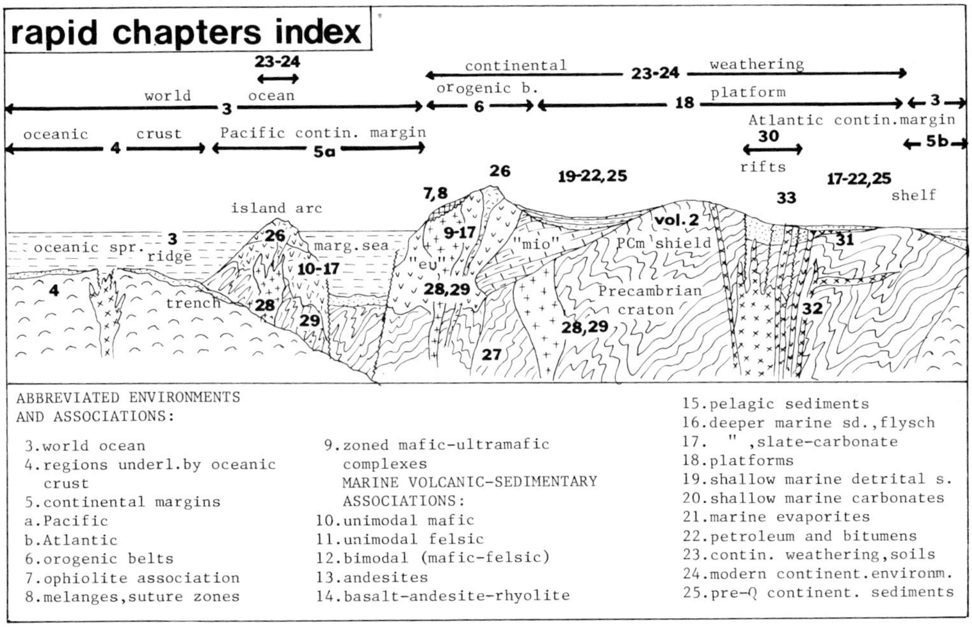

Fig. 2-3. Rapid graphic index, showing organization of chapters in this book. Continuation of the list of chapters: (26) Pre-Q calc-alkaline volcanics and sub-volcanics. (27) Intracrustal environments. (28) Granitic intrusions. (29) High-grade metamorphics. (30) Continental fragmentation, rifting, activation. (31) Plateau basalts, bimodal continental volcanics. (32) Diabase dikes, layered intrusions. (33) Alkaline magmatic association. Numbers refer to book chapters.

Plate 1. Some marine organisms accumulate substantial quantities of trace metals in their bodies. Organic farming and harvesting is one of the possibilities of future metal recovery. The photograph shows a starfish colony resting on the granitic seafloor off N.E. Tasmania.

"Indeed one has the feeling that the whole field of trace metal marine geochemistry would have been a completely dull one over the past fifty years if it weren't for analytical errors"
K. K. Turekian (1977)

CHAPTER 3

The World Ocean

3.1. INTRODUCTION

The world ocean (or "the sea") is an all-embracing term for the interconnected salt-water oceans and seas of the world that cover 71% of the Earth's surface. This vast domain is considered the last frontier of the metal-producing industry and metalliferous geology, and considerable progress in understanding was achieved in the past 30 years. The economic geology of the World Ocean as a whole is treated in the pioneering book of Mero (1965), and good geological, oceanographic and other summaries from different viewpoints are also available (e.g. Lisitzin, 1971; Kennett, 1982; Hill, ed., 1963).

A two-fold division of the world ocean into the regions underlain by the oceanic crust, and regions underlain by transitional or continental crust (continental margins, "geosynclines") has been used by most geologists. The two above megadomains have been further subdivided as indicated in Table 3-1, and each subdivision is treated here in a separate chapter.

The seawater, suspended matter and biota are common to all the geological domains of the world ocean and are, increasingly, looked upon as a future resource of numerous metals.

3.2. SEAWATER AS A METAL SOURCE

Seawater contains all the naturally occurring elements in trace quantities (Table 3-2) and is already being utilized for the extraction of halite (and other salts); magnesium; and bromine (Mero, 1965). Mero also gave a history of attempts (so far unsuccesful) to recover base, rare and precious metal from seawater and demonstrated that such extraction would be quite feasible if done in conjunction with the production of fresh water from the ocean. Brown and Dey (1975) stated that in Britain, extraction of uranium from seawater might become feasible when U reaches the value of $44/kg. The near-uniformity of seawater as a metal source in comparison with common rocks and ores is remarkable.

It appears that the regional variation in the trace metal content in seawater in most cases does not exceed one order of magnitude, except in the immediate vicinity of an active supply of anomalous quantities of trace metals such as river outlets and sea-floor

Table 3-1. Geological subdivision of the world ocean

<table>
<tr><td colspan="5">(3) World Ocean
(the seawater)</td></tr>
<tr><td colspan="3">(4) Oceans: regions underlain by the oceanic crust</td><td colspan="2">(5) Continental margins and inland seas</td></tr>
<tr><td>(4.2)
Oceanic spreading ridges</td><td>(4.3)
Deep ocean floor</td><td>(4.4)
Oceanic islands</td><td>(5.1)
Pacific-type continental margins</td><td>(5.2)
Atlantic-type continental margins</td></tr>
<tr><td colspan="3">NOTE: numbers refer to book chapters</td><td colspan="2">transitional and difficult to classify environments</td></tr>
</table>

Table 3-2. Dissolved trace metals in seawater

	CRUICKSHANK (1974) ppm	MERO (1965) ppm	RESIDENCE TIME, YEARS (VINOGRADOV, 1967)
U	0.0015-0.003	0.003	1.2×10^5
Th	0.000001-0.0005	0.00005	2.0×10^4
Cr	0.00005-0.002	0.00005	n.a.
Co	0.0001-0.0007	0.0005	1.96×10^4
Fe	0.001-0.15	0.01	5.8×10^2
Mn	0.0007-0.01	0.002	7.8×10^3
Mo	0.0005-0.016	0.01	4.7×10^5
Ni	0.0001-0.0002	0.002	1.6×10^4
V	0.0002-0.002	0.002	1.2×10^6
Al	0.1-1.9	0.01	7.8×10^3
Sb	0.0002-0.0005	0.0005	n.a.
As	0.003-0.035	0.003	3.9×10^4
Bi	0.00002-0.0002	0.00002	n.a.
Au	0.000004-0.4	0.000004	n.a.
Pb	0.0003-0.005	0.00003	1.2×10^3
Hg	0.00003-0.00027	0.00003	n.a.
Ag	0.00015-0.0003	0.00015-0.0003	5.9×10^4
Sn	0.003-0.0003	0.003-0.0003	n.a.
Ti	0.001-0.009	0.001-0.009	n.a.
Zn	0.001-0.021	0.001-0.021	2.0×10^4

springs. The latter, however, can provide temporary "high-grade" seawater areas, as discussed by Cruickshank (1974). Depth variation in the dissolved trace metal content seems to be more widespread. Rona et al. (1959) reported a seventeen-fold increase in the trace Mn content in the Pacific waters (Chinook Station), ranging from 0.65 mg/l Mn at the surface to 14.2 mg/l Mn in the depth of 5,440m.

Chemical elements supplied to the seawater from a variety of sources rapidly undergo partition into those that prefer a long-term residence in seawater (e.g. Br, J, Na, Cl, Mg, etc.), and those that are rapidly sequestered by sedimentary particles and added to the seafloor sediments (most base, rare and precious metals). Mo, U, V, In, and few other metals tend to have relatively long seawater residence times, and therefore relatively high contents in the seawater; for example, the continental crust Cu:Mo clarke ratio of about 25 has a corresponding seawater ratio of only 0.3.

3.3. METALS IN MARINE ORGANISMS

Macro- and micro- organisms are known to accumulate during their lifetime many chemical elements from dilute solutions, including the seawater. Enormous concentration factors (up to several million) are frequently achieved (Mero, 1965; Bowen, 1966). Following the death of the metal-accumulating organisms, portion (or the entire content) of the metal is usually returned to the seawater, and added to the seafloor sediments. In most cases (but not exclusively) the returned portion of metals is the one bound in the soft organic tissues and in body fluids (e.g. blood); the seafloor-accumulated metals, on the other hand, usually resided in mineral skeletons. Considered within the geological cycle, the organic metal accumulation is transient and only the portion cumulatively added to the seafloor sediments may ultimately achieve economic importance.

At any time during their lifetime, however, certain trace element-concentrating organisms could be harvested for element extraction. In the past, small quantities of potash and iodine have been obtained from seaweeds along the British coasts (Mero, 1965). There is a theoretical possibility that farming of metal - concentrating marine organisms may be attempted in the future, as one of the means of metal extraction from seawater.

Plate 2. Heteromictic breccia, Udot Island in the Truk Lagoon, Micronesia, a lithologic anomaly in the oceanic association.

"Just think: enough copper to last us for 1,100 years, enough nickel for 23,000 years, enough cobalt for 260,000 years! It's all there under the ocean and it belongs to nobody or to everybody".
Cooperation Canada (CIDA), March-April 1974

CHAPTER 4

Oceans: The Regions Underlain by Oceanic Crust

4.1. GENERAL

Regions directly underlain by oceanic crust (lithosphere) account for about 60% of the Earth's surface (306 sq.kms; Drake and Burk, 1974). Over 99% of it is under water, while presently subaerially exposed samples of "in situ" (=not tectonically displaced) oceanic crust in the form of oceanic islands, are areally insignificant. The bulk of the oceanic lithosphere forms the extensive floor of the major oceans (Pacific, Atlantic, Indian and Arctic), but small basins floored by the oceanic crust occur within the continental (Caspian Sea) and transitional (continental margin) domains (e.g. the Tyrrhenian Sea, Black Sea, some marginal seas behind island arcs) do exist and are difficult to classify.

The oceans considered as a depositional mega environment, and the oceanic crust as a lithologic mega-association, are unique among the remaining environments and associations of the Earth's crust in that the bulk of the oceanic rocks is still within the broad environment of its origin (that is, oceanic rocks are within the ocean). The ancient oceanic rocks displaced after their generation so that they lost completely connection with their formative environment (e.g. the ophiolite association), are volumetrically and areally rare. The ratio of the present oceanic crust to ophiolites outcropping in the continents is about 99.9 : 0.1. The plate tectonic model (e.g. Dewey and Bird, 1970) offers an elegant explanation: the bulk of the oceanic crust created in the geological past was subducted and magmatically recycled in subcrustal depths. Only occasionally tectonic slices obducted onto the continental margin and possibly some taphrogenically (=by crustal extension) generated small oceanic basins evolutionary "frozen" before they managed to develop into full oceans, escaped subduction and survived. The geology and mineralization in oceans is usually discussed in the framework of the three fundamental subdivisions, listed in Table 3-1.

4.2. OCEANIC SPREADING RIDGES AND FRACTURE ZONES

4.2.1. Geology

Oceanic spreading ridges are intraoceanic boundaries of accreting lithospheric plates and a pillar of the plate tectonic theory. The recognition of their extent and continuity is usually attributed to the pioneering studies of Ewing and Heezen (1956), Heezen (1960), and others. At the present level of knowledge the global oceanic ridge system extends for a distance of 60,000 km and the ridges commence as a 2,000 km or so wide topographic rise, ascending from about 5 km deep ocean floor to a level about 2,000 m below sea level. Exceptionally, an oceanic ridge crops out subaerially as in Iceland. The axial portion of the ridge is disturbed by elongated second-order ridges and valleys parallel to the axis, and sometimes by an axial rift valley. The presence or absence of the rift valley usually depends on the spreading rate. Prominent valleys are characteristic for the slow spreading ridges (Basalt Volc.Study Project, 1981). The suboceanic rift valley as summarized in the above citation is typically between 30 - 35 km wide and 1-2 km deep, with step-like inner walls. In the deepest part of the valley is 2-10 km wide relatively flat region, in which the volcanic activity is most intensive. Most of the volcanics are lenticular flows of pillowed or autobrecciated basalts, mid-ocean ridge tholeiites (plagioclase + clinopyroxene, some olivine, magnetite, ilmenite).

Pigeonite andesite, rhyodacite, tholeiitic picrite, ultramafics, gabbros, anorthosites, hornblende diorite, monzonite and granite have been occasionally reported (e.g. Hékinian, 1982; Wilkinson, 1981). At most sites basalt flows are interlayered with, or completely masked by, deep sea sediments so that ocean drilling rather than a simple visual study has to be applied (Hall and Robinson, 1979).

The subsurface nature of spreading ridges is imperfectly known from deep ocean drilling, from seismic refraction studies, and from presumed analogy with ancient ophiolite complexes, which are exposed subaerially. The following vertical sequence of rock units based on the data obtained in the FAMOUS area (Mid - Atlantic Ridge near 36° N), Iceland, and Afar (Ethiopia), has been presented in the Basaltic Volcanism Study Project (1981, p.806):

Layer 1: unconsolidated oceanic floor sediments;

Layer 2: upper oceanic crustal layer of about 2 km thick, mostly pillowed basaltic subaqueous lava, as well as fractured and rubbly basalt changing downward into intrusive or extrusive basaltic rocks (or a sheeted dike complex);

Layer 3: lower oceanic crustal layer of about 4.5 km thick gabbro, grading into pyroxene rich cumulate gabbroids and serpentinized peridotite.

Upper Mantle, usually in depths of 5 to 10 km below the ocean floor.

Recent studies of the spreading ridges and continent based ophiolites have demonstrated the widespread metamorphism and metasomatism of the oceanic crustal units, caused primarily by the convective percolation of heated seawater through highly permeable rocks in the submarine environment. The water penetration reach to depths of at least 3 to 5 km, causing igneous cooling and pervasive hydration (Andrews and Fyfe, 1976; Silantyev and Chernysheva, 1981; Hékinian, 1982)

The intensity of alterations corresponds to the metamorphic grades, ranging from zeolite and prehnite-pumpellyite facies in the Layer 1 and top of the Layer 2, through greenschist facies in the bottom of the Layer 2 and top of Layer 3, to amphibolite facies affecting the bulk of the layer 3. From there, the intensity again decreases downward. Highly permeable zones of episodic vigorous brine discharge along subvertical faults and fracture systems are of particular economic importance and so are subhorizontal units of fractured and scoriaceous lava flowtops.

Oceanic fracture zones have recently received a comprehensive review in the book of Hékinian (1982, his Chapter 7). They are defined as structural lineations offsetting accreting plate boundary regions, and evolutionary closely associated with the history of oceanic ridges. Along the fracture zones, tectonic activity, such as faulting, prevails over volcanism. The largest zones (e.g. the Romaneche zone in the equatorial Atlantic, 8,000 km long) are associated with a deep "transform valley".

Compared with the spreading ridges, oceanic fractures are characteristically enriched in ultramafics such as serpentinized harzburgite and lherzolite, while the mafics are represented by fresh and metamorphosed basalts and gabbroic rocks. Dolerite sills and dikes are particularly abundant and so are breccias and mylonites.

4.2.2. Metal geochemistry and metallogeny

Oceanic basalts have a supra—clarke enrichment of the "basaltophile" trace metals (Cr, Ni, Co, Cu, Mn), quite comparable with the continental tholeiites. Keays and Scott (1976) reviewed the trace gold content of the Mid-Atlantic Ridge basalts and concluded that although the basalts are not substantially enriched in gold and other precious metals (0.21 - 0.76 ppb Au), the metal contents are up to seven times higher in the glassy rims of pillow basalts, than in their interiors. The frequent occurrences of serpentinized tholeiites and ultramafics on spreading ridges and along the oceanic fracture zones, have high trace Cr (0.36 - 0.64 % Cr) and Ni (0.15 - 0.35 % Ni) contents (Hékinian, 1982). Some basalts are enriched in TiO_2 (up to 5.5%), but the greatest TiO_2 enrichment is in ferrogabbros, as for example in those dredged from the Carlsberg Ridge (up to 8% TiO_2; Hékinian, 1982).

In contrast to the continental tholeiite, however, oceanic basalts and to a considerable degree the subsurficial members of the oceanic suite, suffer from pervasive interaction with hydrothermal fluids (in time of emplacement) and subsequently with seawater of ordinary temperature (during seafloor residence). This interaction is a powerful agent of leaching, transportation, and redeposition of the trace (as well as major) elements in the ocean floor and subfloor environment. As pointed out by Andrews and Fyfe (1976), water in the sub-seafloor hydrothermal systems may be supercritical due to the hydrostatic pressure of the overlying seawater. At a depth of 3,000 meters, fluids could discharge at temperatures of 400° C, boiling will not occur, and dissolved gasses such as CO_2 are prevented from rapid escape. As a consequence, high-temperature fluids from the sub-ocean floor depth could reach the floor practically intact and the dissolved constituents will have the tendency to precipitate at or near the sea-floor interface.

Bonatti et al. (1972) genetically subdivided the metalliferous seafloor deposits into hydrothermal (=precipitated directly from depth-derived hydrothermal fluids upon entering the seafloor); and hydrogenous (=slowly precipitated from metals dissolved in seawater). Obviously, both categories are only end - members of a continuous series. Halmyrolitic mineralizations contain metals leached by seawater of ordinary temperature from basaltic pyroclastics, and subsequently precipitated (Meylan et al., 1981).

The study of metallogenesis at oceanic ridges is no more than 15 years old, yet impressive results have been obtained (compare the reviews and summaries in Bonatti, 1975; Rona, 1978; Meylan et al., 1981; Hékinian, 1982). Several tens of occurrences of metal-enriched sediments, metalliferous crusts, scattered ore minerals and recently massive sulphide bodies, have been recorded and described. Selected examples are in Table 4-1. So far, however, only one major deposit has been discovered. GEOTIMES (January 1982, p.22) reported a find of a 1,000 x 218 x 43 m lens of massive sulphides on the Galapagos Ridge, made by a National Oceanic and Atmospheric Administration (U.S.A.) partly using the submersible Alvin. The orebody may contain in excess of 25 Mt ore with 10% Cu, 0.1% Mo, 0.1% V and 300 ppm Ag. GEOTIMES reported that "during the four week investigation the team spent almost 48 hours at the bottom of the rift", so if the estimated tonnage and grade is basically correct, this must have been the quickest case of ore discovery, exploration, and reserve calculation ever recorded in the history of economic geology.

Expectations run high and in February 1984 the U.S. Department of the Interior and the Minerals Management Service proposed a Draft Environmental Impact Statement to asses the possible consequences of mining of oceanic massive sulphide deposits in the Gorda Ridge area (offshore Oregon and Northern California). It is assumed that at least three 30 Mt orebodies might be found over the 180,000 square kilometers of ocean floor. The variety and setting of metallic mineralizations in the spreading ridge environment is summarized in Fig. 4-1 and Table 4-2.

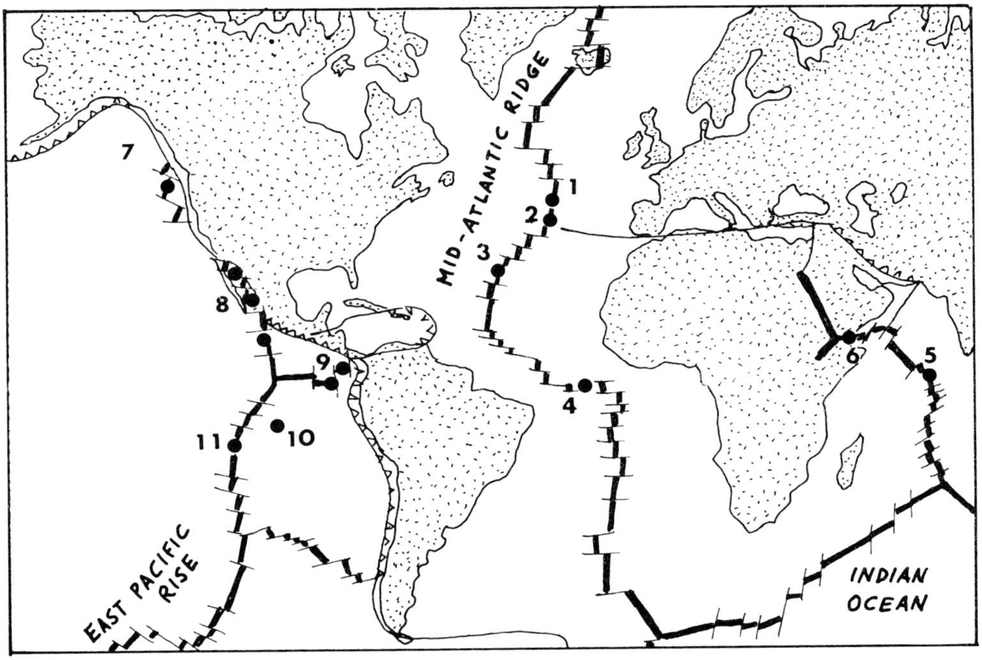

Fig. 4-1. Distribution of metalliferous sediments and ore occurrences at Recent oceanic spreading ridges and transform faults (see Table 4-1 for names and descriptions of the localities shown).

By far the most frequently reported are the mineralization styles A, B, D, and E, represented by goethite, todorokite, birnessite and manganite crusts, amorphous and gel-like masses, clays, bands, pisoliths and concretions. In the styles B and F the Fe-Mn oxides are associated with nontronitic clays. The "metalliferous sediments" recognized by Bostrom and Peterson (1966), Field et al. (1976) and others at the East Pacific Rise and the adjacent Nazca Plate, correspond to the styles B and C. They are extremely fine-grained sediments of orange-yellow to chocolate-brown colour, strongly resembling pulverized gossan and often diluted by calcareous oozes. The highest trace-metal contents recorded are 104 ppm As, 240 ppm Co, 0.124% Cu, 145 ppm Pb, 310 ppm Mo, 820 ppm Ni, 5.1 ppm Ag, and 760 ppm Zn.

The presently inactive Fe-Mn mineralized hydrothermal mounds hosted by pelagic sediments (style F) have been described in detail by Honnorez et al. (1981) from the vicinity (south of) the Galapagos Ridge axis. The mounds as much as 25 m in height have developed in a layer of nannofossil oozes (dominantly carbonate and siliceous foraminifer and diatom oozes), resting on basaltic floor.

Table 4-1. Metalliferous sediments and ore occurrences at Recent oceanic spreading ridges and transform faults

NO.	LOCALITY	STYLE	DESCRIPTION	REFERENCES
MID-ATLANTIC RIDGE				
1	Transform fault A near 37° N	B,F	Seafloor hydrotherms produce black,stratified concretionary Fe-Mn oxides	Hékinian (1982)
2	FAMOUS area (near 36°N)	B,D	Mn oxide encrustations and nodules form in the vicinity of small basaltic volcanoes and palagonite tuffs	Hékinian (1982)
3	TAG area , 26-30° N	D,E	Up to 5 cm thick birnessite and todorokite crusts cover volcanic breccia in 4.2 to 4.5 km depth	Hékinian (1982)
4	Romaneche fracture zone near intersection with the Mid Atlantic Ridge	H	Blocks of metabasalts cut by veinlets and disseminations of chalcopyrite and atacamite dredged from depths of 2.7 - 3.1 km	Bonatti et al. (1976) Rona (1978)
INDIAN OCEAN				
5	Carlsberg Ridge near 9° N	D,H	Cu minerals form veinlets and disseminations in altered mafic magmatites; Mn-Fe oxide crusts	Hékinian (1982)
6	Gulf of Aden	B,F	Fe-Mn oxide concretions, ferruginous clays (up to 34-42% Mn) form as hydrothermal sediments in a small median valley	Cann et al. (1977)
PACIFIC OCEAN				
7	Juan de Fuca Ridge at 45-49° N	G	Hydrothermal sphalerite, wurtzite, minor pyrite, galena,chalcopyrite,anhydrite,gypsum form mounds and encrustations in axial valley on the Ridge, surrounded by benthonic organisms	Koski et al. (1982)

8	RITA, East Pacific Rise 20° 54' N	G	7,000 x 100 m zone of active hydrothermal vents in depth of 2,610 - 2,650 m. "Black smokers" composed of massive sphalerite, pyrite and chalcopyrite mixed with opal, anhydrite; friable, rest on pillowed basalts. Minor Fe oxide gossans. Up to 29% Zn, 6% Cu, 0.07-0.18% Pb	Francheteau et al. (1979) Hékinian (1982) Brevart et al. (1981)
9	Galapagos Spreading Centre near 86°09' W	B,F	Active hydrothermal area; small elongated mounds surrounded by abundant organisms. Mn-Fe crusts, concretions; patches of nontronite	Hékinian (1982)
	same area, near 96° W	F	Hydrothermal mounds intermixed with pelagic sediments; Fe-Mn oxide encrustations	Hékinian (1982)
	"Mounds" hydroth. field 20-30 km s. of Galapagos Spreading Centre	F,G	1,000 x 218 m, 43 m thick seafloor sulphide ore lens in 2.4 km depth, containing over 25 Mt ore with 10% Fe; 10% Cu; 0.1% Mo; 0.1% Pb; 0.1% V 300 ppm Ag; encrustations of Mn oxides	Geotimes, January 1982
10	Bauer Basin adjacent to East Pacific Rise 10° S, 100° W	B,C	Metalliferous sediments in a zone of convergence of transform faults and anomalous heat flow; smectite, goethite, todorokite; 9-18% Fe; 2-7% Mn; 0.07-0.12% Cu; 0.04-0.17% Ni	Rona (1978)
11	East Pacific Rise at 10-15° S	B,C,D	Metalliferous sediments with goethite, nontronite, Mn-hydroxides; 0.6-8.8% Mn; 0.05-0.18% Cu 0.025-0.12% Ni	Rona (1978)

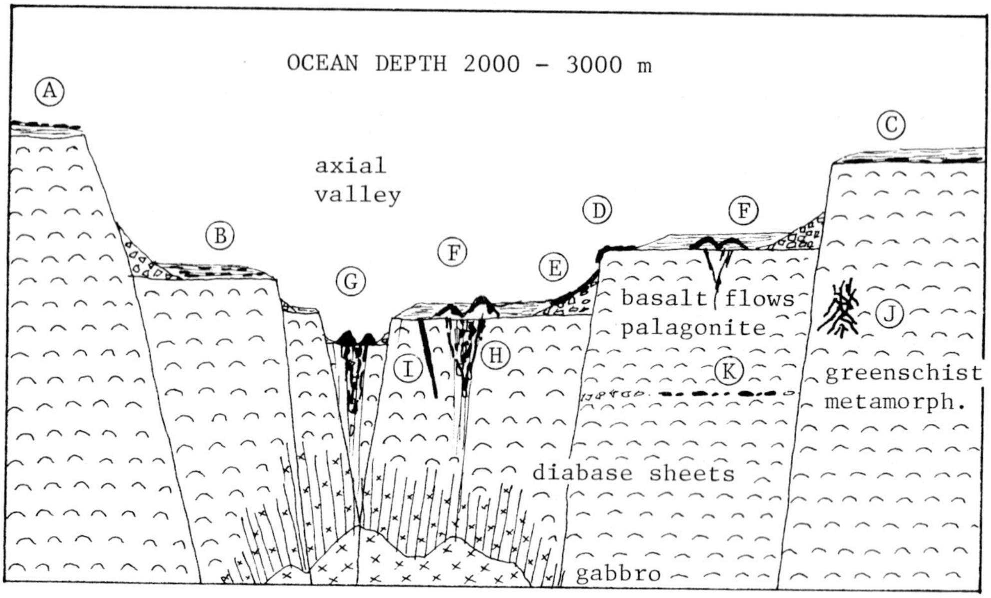

Fig. 4-2. Principal mineralization styles at oceanic spreading ridges (diagrammatic, not to scale); see Table 4-2 for the letter code explanations.

The hot-springs precipitates (hydrothermal sediments) are represented by green nontronitic clays and by Mn oxide crusts. Such crusts occur on the top of hydrothermal mounds, and are up to 1.4 m thick. The mounds occur often as chains, oriented sub-parallel to the spreading axes and are from 5 to 20 m higher than the general level of the surrounding seafloor.

The seafloor hydrothermal massive sulphide deposits (style G) are by far the most interesting metal accumulations (e.g. at the RITA site), and also potentially the most economically valuable ones. The RITA occurrences have been described by Francheteau et al. (1979), and reviewed by Hékinian (1982, p.298 - 302). Active hydrothermal vents, some emitting black mineral suspension (the "black smokers"), are aligned along a 7 km long, 100 m wide zone, and at a depth of 2,610 to 2,650 m on the axial zone of the ridge.

Hékinian (1982) distinguished three categories of vents:
(1) Gently sloped hills or mounds (10-50 cm in diameter, up to 1 m high) of dark-brown colour, composed of sphalerite, wurtzite and pyrite. The black precipitate from the "smoke" is scattered as a powdery Zn sulphide material at the base of the vents and on the surrounding pillow lavas.

Table 4-2. Principal mineralization styles at oceanic spreading ridges
(see also Fig. 4-2)

(A)	Hydrogenous Fe,Mn oxide crusts or nodules (precipitated from seawater)
(B)	Low - grade Fe-Mn oxide layers, concretions, etc.,in hydrothermal and pelagic clays, precipitated from low - temperature, distal springs ("metalliferous sediments")
(C)	Metal (Fe,Mn,Cu) enrichment in pelagic clays immediatedly above the basalt floor ("metalliferous sediments")
(D)	Fe,Mn oxide encrustations on bedrock and sediment surface
(E)	Fe,Mn oxide encrustations or cements of talus
(F)	Mineralized (Fe,Mn oxides or Fe,Cu,Zn sulphides) hydrothermal mounds in sediments
(G)	Hydrothermal Fe,Cu,Zn sulphide, silica, anhydrite chimneys, mounds, masses, lenses, surrounding seafloor hydrothermal vents (e.g. Galapagos Ridge; 2.5 Mt Cu, 7.5 Tt Ag)
(H)	Fe,Cu sulphides, native Cu, etc. veinlets, disseminations, in altered basaltic vent breccia, fractured basalt, etc.
(I)	Cu,Zn, rare Pb sulphide - mineralized fissure, fracture veins
(J)	Stockworks, veinlets of Cu minerals in basalts, along transform offsets
(K)	Remobilization Cu sulphide, gold, etc. veins in hydrothermally altered and/or sheared structures in the deep levels of spreading ridges (hypothetical)

(2) Tall, columnar cylindrical chimneys, rising up to about 7 - 10 m above the surrounding basaltic floor, and composed of anhydrite, chalcopyrite, bornite, cubanite and digenite.
(3) Encrustations of amorphous silica and Fe oxide stains on the surface of basalt.

Powdery and friable accumulations of goethite found about 2 km from the site of sulphide generation have been interpreted as a product of hydrothermal oxidation of sulphides, followed by hydrodynamic reworking.

The mineralization styles at RITA have many general points of similarity with massive sulphide occurrences in ancient ophiolites, particularly in the Troodos Complex in the Cyprus; this comparison has been made several times in the literature (e.g. Andrews and Fyfe, 1976).

The epigenetic copper mineralization in altered sub - seafloor basalts (styles H and J) is exemplified by the material dredged from a site along the Romaneche offset zone in the equatorial Atlantic (Bonatti et al., 1976). There, chalcopyrite, pyrite, pyrrhotite and secondary atacamite, delafossite and other minerals form disseminations and fillings of thin, criss - crossing chlorite and zeolite veinlets. The occurrence of atacamite $Cu_4Cl_2(OH)_6$ in the above environment is unusual; some textbooks of mineralogy state that atacamite is restricted to deserts.

4.3. DEEP OCEAN FLOOR

4.3.1. Geology

The ocean floor is situated between the continental margins and the mid-ocean ridges, and is usually subdivided into the following physiographic units (Kennett, 1982): (1) abyssal floor (a: abyssal plains; b: abyssal hills); (2) oceanic rises; (3) seamounts and seamount groups.

The abyssal plains are most extensive and are flat subaqueous territories characterized by great thickness of sediments covering the original topographic irregularities. The abyssal hills are small (1 - 50 km in plane) sharply defined elevations, rising no more than 1,000 above the abyssal plain floor. They are most widespread in the Pacific.

Seamounts are active or inactive volcanoes that rise above the ocean floor to elevations of more than 1,000 m. They are solitary or in groups, often overlapping to form volcanic ridges and seamount chains (e.g. the Hawaiian Chain). Guyots or tablemounts are subsided volcanic islands with erosion-truncated tops.

Aseismic oceanic ridges are considered to be portions of the stable oceanic region (Hekinian, 1982). They are widespread (about 25% of the world's ocean floor area), morphologically similar to active ridges (except for the generally smooth crests), and usually attached to continental margins. Their geological age ranges from about 50 to 110 m.y. Their composition is basalt and/or oceanic andesite, and the best known localities are the Iceland - Faeroe Ridge, Cocos Ridge, Walwis Ridge, and others. They may be comparable in origin with oceanic intra-plate islands.

The deep ocean floor is underlain by the oceanic crust. The dominant hard rock is the oceanic basalt of the Layer 2 generated at spreading ridges, and to a lesser degree basalts and other volcanics of the intra-plate volcanoes. Subaqueous outcrops of basalts on the deep-ocean floor are, however, very rare.

The unconsolidated and semi - consolidated oceanic sediments have a variety of derivations, being purely "oceanic" (=generated at or above the seafloor underlain by the oceanic crust by biologic, volcanic or authigenic processes); transitional (derived from active continental margin, e.g. from the island arc volcanoes); continental (terrigenous), derived from the continental crust - floored domain, and extraterrestrial (cosmic dust and meteorites).

Most sediments in the ocean are mixtures of components derived from more than one provenance and a sharp separation by provenance is impossible. This is particularly so in the case of terrigenous sediments, which cover substantial portion of the deep ocean floor in oceans bordered by the "Atlantic - type" continental margins (e.g. the Atlantic Ocean; Kennett, 1982; Sheppard, 1973).

The Pacific Ocean, bordered by trench / island arc systems, contains a substantially lesser proportion of terrigenous sediments because the trenches represent an efficient barrier to their dispersion. The thickness of unconsolidated sediments in the oceanic interiors ranges from 0 to several hundred metres, and it increases to several thousand metres adjacent to the continents (Heezen and Hollister, 1971).

Although the surficial portions of most of the sediments and volcanics on the ocean floor are "Recent" (Quaternary), some are "Ancient": that is, Tertiary and older. Because there is little difference in the degree of consolidation of the recent and ancient oceanic deposits, recognition is difficult and tentative. The most conspicuous are relict sediments formed under substantially different conditions than those prevalent at the site of their present preservation: for example, shallow - water sediments now preserved in substantial oceanic depths due to subsidence.

The oceanic sediments have a voluminous literature (Berger, 1974; Heezen and Hollister, 1971; Hsü and Jenkyns, eds., 1974; Arrhenius, 1963; Lisitzin, 1971; and others; Fig. 4-3). Berger (1974) suggested a detailed classification.

"Red clay" (pelagic clay, oxypelite, deep - sea clay) is very widespread (38% of oceanic abyssal sediments) and most characteristic. It slowly accumulates in oceanic regions of low organic fertility, at a distance from land. It is a chocolate-brown, extremely fine-grained sediment composed of clay minerals (mainly montmorillonite, illite, chlorite, kaolinite); authigenic zeolites (phillipsite, clinoptilolite, heulandite); Fe-Mn oxides and hydroxides. In addition, there is usually a variable proportion of lithogenetic minerals (feldspars, pyroxene, quartz, magnetite, etc.), reworked from oceanic volcanics or transported (blown) from the continent. Recent research (Berger, 1974) indicates significant, even dominant (greater than 70%; Moore and Cruickshank,1973) terrigenous input in the formation of the red clay, in addition to the intrabasinal mafic volcanics and authigenic sources.

Organic calcareous oozes (Globigerina, Diatom and Pteropod) form a milky white to brown sediment composed of microfossil skeletons. The oozes are widely distributed (about 48% of oceanic pelagic sediments) in mid - ocean areas at depths less than 4,000 m. In greater depths, the carbonate is generally unstable and undergoes dissolution.

Organic siliceous ooze (Radiolarian and Diatom; about 15% of oceanic sediments) is light, greenish to greenish - yellow, composed of over 30% of organic skeletons and less than 30% of carbonate. The grain size ranges from fine silt to fine sand. The siliceous oozes form three latitudinal belts in the World Ocean (particularly in the Pacific), that correlates with zones of high organic productivity.

Mid - plate oceanic volcanics.

In addition to subaerially exposed volcanic islands (treated in the following section) and subaqueous volcanic bases to coral islands, volcanic rocks are present in seamounts. The results of the Glomar Challenger sampling summarized by Hékinian (1982) indicate a wide

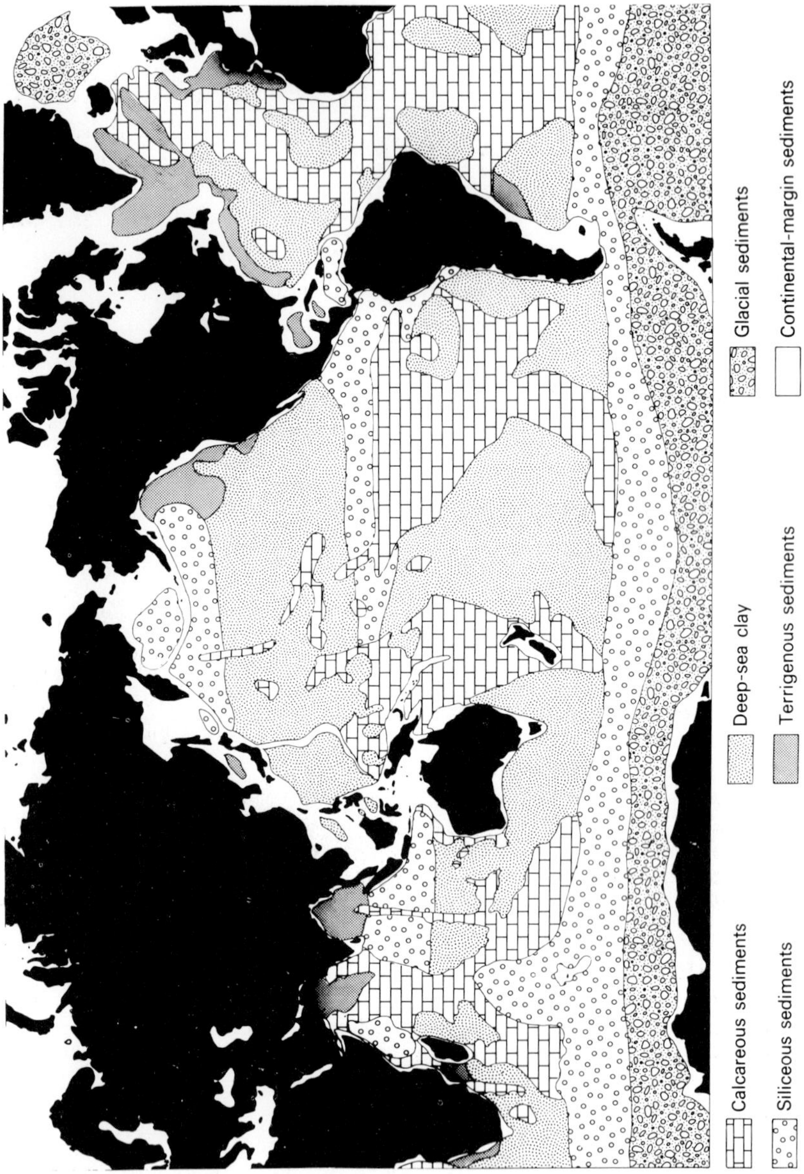

Fig. 4-3. Distribution of principal types of pelagic sediments on the ocean floor. After Davies and Gorsline (1976), courtesy of the Academic Press.

range of compositions of the mid-plate basalts from the relatively primitive tholeiites, to highly fractionated alkali basalts. Mugearites and trachytes have been recorded locally.

Volcanic sediments.

Volcanic ejecta and eroded volcanic detritus of a variety of sizes are found in the vicinity of intra-oceanic volcanoes, as well as downwind from the continental margin volcanoes. Locally they are a major component of the seafloor sediments.

Hemipelagic and terrigenous muds are substantially coarser than the "red clay" (over 50% of the material is coarser than 30 microns), and they cover continental slopes and rises. Their colour is usually green to dark gray, and the sediments are locally rich in organic matter. Contourite is a hemipelagic sediment transported and deposited by ocean contour currents. Terrigenous turbidites composed of mixed sand, silt and clay, cover the oceanic floor in, around, and in the extension of submarine canyons.

Glacial marine sediments are heterogeneous and range from poorly sorted reworked gravely moraines in the high latitudes (dominantly along Antarctica) to laminated silts and muds. These have a variable proportion of ice-rafted dropstones. The dropstones reach beyond the limit of distribution of the glaciogenic matrix material, and are locally abundant in the environment in which diatom and radiolarian oozes form.

4.3.2. Metal geochemistry and metallogeny

The basal portion of the Layer 1 (=sedimentary layer) and top of the Layer 2 (basaltic layer) may contain preserved metallic enrichment or mineralization established in the process of lithosphere generation at the spreading ridge. This is a relict mineralization the most consistent variety of which are the so called metalliferous or iron-rich basal sediments (Kennett, 1982; Rona, 1978). These sediments often contain more than 20% Fe in goethite and nontronite on a carbonate - free basis, and they are carried with the moving lithospheric plate away from the ridge. Under special conditions, "metalliferous sediments" can precipitate from metal - bearing fluids generated at spreading ridges, but transported for a considerable distance by bottom currents. The Fe and Mn - rich sediments of the Bauer Basin and the Central Basin of the Nazca Plate are interpreted to have been derived from the East Pacific Rise, 1,000 km away (Heath and Dymond, 1977; Edmond et al., 1979).

In the course of the subaqueous soujourn, ocean-floor sediments and volcanics are subject to subaqueous weathering and diagenesis (halmyrolysis). Hékinian (1982) devoted a chapter to the review of this phenomena and pointed out that the potash exchange between seawater and the rock (as well as the volcanics undergoing devitrification and hydratation) is one of the most impressive parameters of alteration. Leaching of trace metals (Cu, Ni, Au, Mn, Co) out of basalts and their possible transfer into the sediments and seawater has been recorded, but accurate data are still insufficient

The non-organic pelagic clays and the non-organic component of the organogenic oceanic muds are geochemically highly fertile, and have anomalous contents of several metals. The average Al_2O_3 content of 15% locally increases to 25% Al_2O_3 (Mero, 1965), which makes the oceanic high-alumina clay occurrences comparable with the continental high-alumina clays, considered as a future source of aluminium.

The average / highest trace metal contents in the Pacific red clays listed by Goldberg and Arrhenius (1958), are: Mn 0.7 / 3%; Ni 200 ppm / 830 ppm; Cu 190 ppm / 0.2%; Co 100 ppm / 310 ppm; V 240 ppm / 430 ppm; Pb 70 ppm / 170 ppm; Mo (highest value only) 370 ppm. The red clays with 0.2% Cu are close to the cut-off grade of some low grade / large tonnage copper accumulations on the continent considered as resources for the future, and may themselves become a future metallic resource particularly if other components (Al, Mn, Co, Ni) are recovered simultaneously (Mero, 1965). Whereas most of the base metals occur finely dispersed in the sediment, considerable proportion of the manganese is concentrated in discrete particles of Mn oxides. These average about 0.1 mm in diameter and could be mechanically separated with relative ease. These particles constitute a link between the geochemically anomalous pelagic muds and the ocean floor ferromanganese nodules. The total resource of Mn in oceanic red clays has been estimated by Mero (1965) to be about 125 trillion tons (average grade 1.25% Mn).

FERROMANGANESE NODULES ON THE OCEAN FLOOR

The ferromanganese nodules are the best studied and economically most important metalliferous substance, widely distributed on the oceanic floor. Discovered more than a century ago by the Challenger expedition, the nodules have by now been studied from all possible angles and have an extensive literature (Murray and Renard, 1891; Mero, 1965, 1977; Crerar and Barnes, 1974; Glasby and Read, 1976; Emery and Skinner, 1977; Bischoff and Piper, eds., 1979; and others). The nodules are typically zoned, spherical objects with the average size of about 3 cm, and they are an oversize equivalent of the small Mn - oxide particles dispersed in pelagic clays. They are of hydrogenous and authigenic origin and their growth rate is usually quoted as 1 mm in 10,000 years. In addition to the concentric nodules, Fe-Mn oxide slabs, rock cappings, impregnations of porous materials, replacements of coral, and organic debris occur. The bulk of the nodules studied rests loosely on the seafloor surface, most of them in the red clay environment. Nodule occurrences associated with the pelagic oozes or with hemipelagic sediments are also widespread (Fig. 4-4).

The unconsolidated oceanic sediments under the seabed / water interface contain sharply decreased proportion of Fe-Mn nodules, typically 1/4 of the amount of surficial nodules in the first meter of the sediment, and 1.3% in the depth of 10m. Anomalous density of nodules in the subsurface, however, may appear at unconformities (Glasby and Read, 1976). The nodule colour varies from bluish - black (those high in Mn) to tan (those high in Fe). Most nodules are porous and soft. The component minerals of nodules and crusts are manganite, birnessite, buserite, todorokite, ramsdellite, psilomelane and goethite. Many of the species are unstable and change upon dehydratation. The nodules are rarely pure and pores and fractures are filled by clay. Many nodules also have a non-metalliferous core.

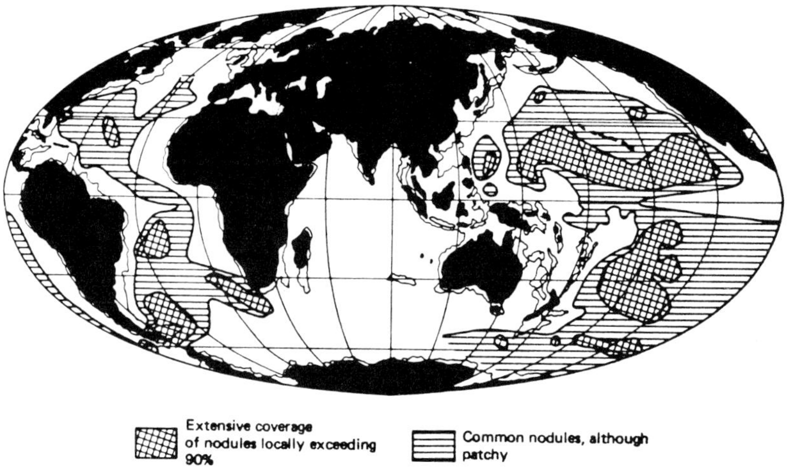

Fig. 4-4. Distribution of manganese nodules in the Pacific and Atlantic oceans. After Cronan (1977), courtesy of the author.

The manganese content in the nodules ranges from 1.32% to 53% (average about 18%); the iron content ranges from 5 to 42% (average about 17%). In addition to the major metals, the nodules are significantly enriched in minor elements, such as Ni, Co, Cu, Pb, Mo, Tl, Sn, and others (Table 4-3). Out of these elements, Ni, Co, and Cu are the most economically valuable, and the recent feasibility studies of nodule mining focussed on the above metals, with the manganese considered as a by-product. The compositional variation of ferromanganese nodules is local, bathymetric and regional (Fig. 4-5). Mero (1965) suggested four compositional classes of the nodules and his classification is now widespread in the literature:

A-type nodules are rich in iron and are widespread in the South Pacific. Average composition: 28.3% Fe; 21.7% Mn; 0.35% Co; 0.46% Ni; 0.32% Cu; 0.21% Pb.

B-type nodules are manganese-rich and depleted in Cu, Ni and Co. They are interpreted as rapidly precipitated, so that there was insufficient time to extract and incorporate the trace metals from the seawater. Mo content, however, could be relatively high. These nodules are widespread along the continental borderland of the United States and Mexico, particularly in the Gulf of California. The average composition is: 49.8% Mn; 2.3% Fe; 0.055% Co; 0.26% Ni; 0.14% Cu; 0.047% Pb.

C-type nodules are copper and nickel-rich and widespread in the central part of the Pacific. The average composition is 33.3% Mn; 17.7% Fe; 0.39% Co; 1.52% Ni; 1.13% Cu; 0.18% Pb.

Table 4-3.
Minor elements in manganese nodules

METAL	AVERAGE,%	MAXIMUM,%
Ti	0.67-0.8	1.7
V	0.054-0.07	0.11
Co	0.31-0.35	2.3
Ni	0.42-0.99	2.0
Cu	0.2-0.53	1.6
Zn	0.05	0.08
Mo	0.035-0.052	0.15
Pb	0.09-0.1	0.36
Sn	267 ppm	320 ppm
Tl	170 ppm	300 ppm
Th	50 ppm	143 ppm

Data from Mero (1965)

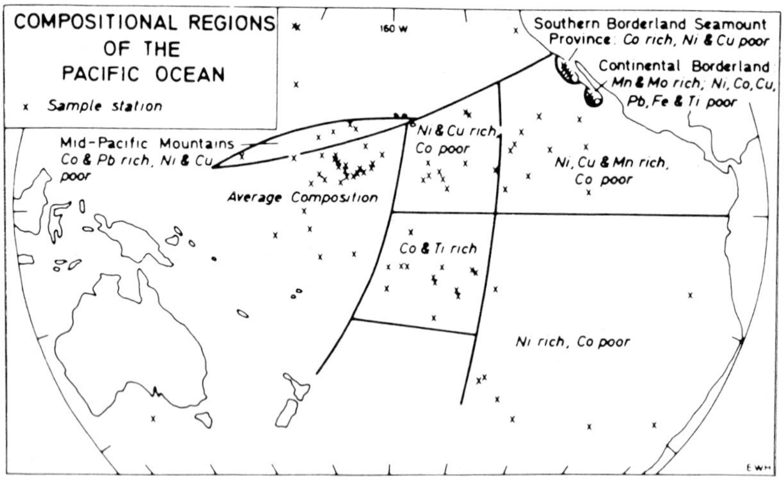

Fig. 4-5. Compositional regions of trace elements in the Pacific manganese nodules. After Cronan (1977), from Kennett (1982); courtesy of the author.

D-type nodules are cobalt - rich and their favourite occurrence are topographic highs in the central part of the Pacific. The highest cobalt values have been obtained from nodule fields topping localized seamounts and guyots. The average composition is: 28.5% Mn; 22.6% Fe; 0.66% Ni; 1.2% Co; 0.21% Cu; 0.3% Pb.

Economic considerations involving the ferromanganese nodules have to take into account the content of the valuable trace elements (the highest grade of nodules so far discovered is in the equatorial Pacific: 6°N and 170°W; on a dry weight basis the nodules averaged 2.3% Cu; 1.9% Ni; 0.2% Co; and 35% Mn; Mero, 1965, 1977), as well as the size of the nodules and their distribution density. The range of concentration of the nodules on the seafloor varies from 0.05 g/cm^2 to 3.8 g/cm^2 in the Pacific (Mero,1965); the concentration is substantially lower in the Atlantic and the Indian oceans. At present, two "high-grade" nodule areas are known in the Pacific: the North Pacific area south of Hawaii (between the Clarion and Clipperton fracture zones) of about 6 million sq.km, and the South Pacific area (6.8 million sq.km). Thiry et al.(1977) recently calculated the metal resources available in the North Pacific nodule zone, based on a very detailed sampling by a French group. They assumed a total tonnage of wet nodules to be between 5 and 10.6 Bt. These contain between 42 and 90 Mt Ni and 35-76 Mt Cu.

THE REMAINING OCEAN FLOOR CONCRETIONS AND NODULES

Barite concretions have been reported from the bathyal seafloor off Sri Lanka, California and Indonesia (Cruickshank, 1974). They contain between 50 and 85% $BaSO_4$, but trace metal contents have not been reported. The radiolarian ooze in the equatorial Pacific contains up to 10% barite (Arrhenius and Bonatti, 1965).

Concretions of authigenic phosphorite have been reported in areas of upwelling, at depths varying from 60 to 3,000 m (Mero, 1965); that is, dominantly from the continental margin domain rather than from the oceanic floor. Bezrukov et al. (1969) described widespread phosphatized abyssal oceanic sediments as well as basalts west of Hawaii, and over the Jan Mayen Ridge in the northern Atlantic. Trace metals are rarely reported, but by analogy with the ancient phosphates mined on continents, increased trace uranium can be expected. Karns (1976) described nodular phosphorite deposits at Chatham Rise, a broad submarine platform southeast of New Zealand. The nodules are associated with Miocene Globigerina ooze, and the potential recoverable reserves are approximately 200 Mt of 21.3% P_2O_5 material. Additional data on oceanic phosphates are available from the compilation of Baturin (1982).

4.4. OCEANIC ISLANDS

4.4.1. Introduction

Islands located within the vast territory underlain by the oceanic crust are usually subdivided into (A) volcanic islands generated at spreading ridges (Recent and earlier), and (B) islands generated by mid plate volcanism. To these have to be added (C) microcontinents of small island dimensions, and (D) "sedimentary islands" composed entirely or largely of biogenic sediments such as coral carbonates. All members of the D category in the oceanic realm have also developed on volcanic (A,B) or rarely microcontinental (C) pedestals. Many oceanic islands including the largest one, Iceland, are composite and consist of several categories listed above. Additional to the Recent islands and their rocks are relict rock associations formed in the past, that could have been produced in a geotectonic / environmental setting distinct from the one in which they occur now (e.g. the Tertiary subaerial flood basalts that build-up the greater part of Iceland). Insufficient discrimination among the recent and relict rocks has been responsible for several errors in metallogenetic interpretation. The relict lithologic associations of oceanic islands are not treated in this chapter, unless their interaction with the genuinely oceanic lithogenetic and metallogenetic processes resulted in "new", interesting rocks and ores.

4.4.2. (A) Islands generated at spreading ridges

ICELAND

Iceland lies astride the Mid-Atlantic spreading ridge and has been the site of volcanic activity since at least the Miocene (Ward, 1971; Palmason and Saemundsson, 1974). Eastern and western Iceland is covered by as much as 10 km thick sequence of subaerial flood basalts, interbedded with occasional layers of lignite and sandstone. They are interpreted as a product of none or slow crustal spreading between 10 and 20 m.y. ago, and are lithologically and metallogenically indistinguishable from plateau basalts of the Brito - Arctic Province (compare Chapter 31).

The volcanic rocks produced during the relatively recent fast spreading episode (about 3 m.y. ago to present) occur in the Neovolcanic Zone. This is about 50 - 100 km wide zone of recent tectonic activity doubled in the south, that divides the island in half. The oldest rock here is the "Old Gray Basalt". It is similar to the plateau basalt, but contains intercalations of glaciogenic rocks and fragmental volcanics formed as a result of subglacial eruptions (pillow lavas covered by palagonite breccias). Basalt or diabase dikes are common. The recent basaltic eruptions associated with spreading are controlled by fissures and rifts.

Large central volcanoes with calderas (Katla, Hekla, Askja) are located in the Neovolcanic Zone, but some may occur more than 50 km from its centre, controlled by fracture zones. Thorarinsson (1967) pointed out that about 10% of the Icelandic eruptions were acid or intermediate, composed of rhyolite, granophyre, dacite, andesite and other rocks. Rhyolites are associated with most recent geothermal areas in Iceland. The geothermal activity, thermal brines and hydrothermal alterations have been treated by Kononov and Polyak (1976) and Tomasson and Kristmannsdottir (1972). Surprisingly, no significant trace-metal anomalies or metallic occurrences have been reported so far from the Neovolcanic Zone of Iceland, except for minor mercury showings in the Great Geyser sinters.

THE AZORES

The Azores Islands (37°-39° N) are interpreted as situated on the Mid-Atlantic Ridge, broken by several transform faults (Machado et al.,1972). Submarine and subaerial central volcanoes are composed of basalt lavas and pyroclastics with subordinate trachyte. No metallic occurrences are known.

4.4.3. (B) Islands generated by mid-plate volcanism

Volcanic islands that rise from the oceanic floor outside of the spreading ridges are interpreted as a consequence of partial melting at a localized "hot spot" within the mantle (Basalt.Volc.Study Proj., 1981). Most are simple or complex shield volcanoes dominated by tholeiites (oversaturated quartz tholeiites, slightly undersaturated olivine tholeiites, rare picrites) that are chemically distinct from tholeiites of the oceanic spreading ridges. They are particularly notable for their relatively high proportion of the large ion lithophile elements such as K, Rb and U. Some islands are exclusively tholeiitic (e.g. the Galapagos, most of the Samoan islands). Others are composed of a dominantly tholeiitic early association, followed by late-stage members of the alkali and alkaline series (alkali olivine basalt, trachybasalt, basanite, melilite basalt, andesite, nephelinite, trachyte, etc.). In Hawaii, considered a type locality, the alkaline rocks form a thin cap over the eroded tholeiitic base (e.g. Macdonald, 1949). In a few cases, for example at Tristan da Cunha, the undersaturated rocks comprise the entire exposed part of the oceanic volcano (Borley, 1974).

A special category of present oceanic islands has probably originated in a continental rift setting on or in the vicinity of the continental crust (e.g. Fernando Po; the most easterly islands of the Canary Islands group; Cape Verde Islands; etc.), and were rafted as passive travellers atop a lithospheric plate. The oceanic islands in the Atlantic off the Brazilian coast (Fernando de Noronha, Trinidade) are composed of alkaline-ultramafic association comparable with and in the extension of a similar province on the Brazilian mainland (de Almeida, 1955).

The dominant form of the volcanics are lava flows with the usual cooling — generated zoning, often grading into autobreccias. Pyroclastics are subordinate. Intrusive rocks are known, but are very rare. They are most common as (1) inclusions in lavas (e.g. the abundant dunite protoenclaves of Oahu, Hawaii); (2) components of vent breccias (gabbro, monzonite, quartz monzonite; Truk Island in the Caroline Islands, the Pacific; Stark and Hay, 1963); (3) dikes and sills of gabbro, mugearite, syenite, etc.; (4) plugs and stocks in cores of eroded volcanoes or in calderas (e.g. essexites in Gomera, Canary Islands); and others.

The intensity of weathering and rock freshness depends on geological age and location in a particular climatic belt. Young, subarctic and arid islands are usually fresh, while the older islands located in the humid tropical belt are lateritically weathered. This is impressively demonstrated in the Hawaiian chain in which the geologically oldest major island, Kauai (4.5-5.6 m.y.) is deeply lateritized in contrast to most of the outcrop on the youngest island, Hawaii (0.8 m.y. to Recent), which is virtually fresh.

Volcanics - derived continental and nearshore sediments of predictable types (talus rubble, alluvial sand to mud, beach sand, submarine fan sand to clay) are moderately common. The tropical islands have usually a coral reef fringe.

METAL GEOCHEMISTRY AND MINERALIZATION

Oceanic island basalts have a slightly supra-clarke contents of major metals (Al, Fe, Ti) compared with other basalts. The hypersthene - bearing basalts of Mauna Loa, Hawaii, contain from 2.8 to 9.27% ilmenite (i.e., 1.5 to about 4.89% TiO_2; Macdonald, 1949). Lateritic weathering is capable of upgrading these contents into a potentially mineable lateritic ore, and reserves of 125 Mt of low-grade ferruginous and titaniferous lateritic bauxite have formed over basalts in the two geologically older islands of Kauai and Maui (Patterson, 1971). The largest bauxite deposits on Kauai are Pleistocene and Recent soils and saprolites, overlying Pliocene basalts. Gibbsite occurs as concretions, veinlets and vesicle filling and about 100 Mt of the materials with grades of 25.9% Al_2O_3; 39.4% Fe_2O_3; and 6.7% TiO_2, is available. Small amounts of ilmenite and magnetite occur in heavy mineral streaks in streams and on beaches. Comparable lateritic profiles containing gibbsite nodules formed on trachyte bedrock, and lateritic nodular goethite formed on nepheline basalt bedrock, at several localities in the Truk Lagoon (Stark and Hay, 1963).

Copper and nickel are the only trace elements of other than petrological interest, present in oceanic basalts. Patterson (1971, Table 10) gave 0.15% Cu (average 0.029% Cu) and 0.15% Ni (average 0.084%) as the maximum trace contents of these metals in the "hard" basalts of Kauai. Macdonald (1949) found 0.12% NiO in olivine gabbro porphyry dikes at Kilauea, but substantially less in the Hawaiian lavas. Sulphides (pyrrhotite, chalcopyrite, pentlandite) that crystallized from immiscible sulphide melt, were found in the Hawaiian basalts (Skinner and Peck, 1969; Desborough et al.,1968), but in quantities not exceeding 0.05% by volume. The olivine-rich inclusions (protoenclaves) in the Hawaiian basalts have high trace Cr and Ni contents.

Fumaroles and hot springs in Hawaii and few other active volcanic islands (e.g. the Volcan Azufre, Galapagos; Colony and Nordlie, 1973) deposit abundant sulphur sublimate in the near-surface regions, episodically remelted to form sulphur lava flows. Murata (1966) reported anomalously high trace selenium contents associated with the Kilauea Iki fumaroles (3% Se in the fumarolic sulphur; 1.1% Se in the altered pumice; 5.18% Se in sulphur-impregnated lava). Hot springs and products of subvolcanic hydrothermal activity have been rarely described from the oceanic islands environment. Numerous but small deposits of psilomelane in nodules, veins and reworked sinters have been known from the island of St. Helena in the tropical Atlantic (Mitchell-Thomé, 1970). They are hosted by basaltic tuffs and are probably related to hot springs activity.

Propylitic hydrothermal alteration (chlorite, albite, epidote, minor quartz) is superimposed on a variety of rock fragments in a volcanic breccia (former vent ?) exposed in the Udot Island (Truk Lagoon). Rare specks of native copper have been found in amygdules filled by prehnite, pumpellyite, chlorite and epidote. Minor chalcopyrite is present in epidote or specularite veinlets in the vicinity of pink albite syenite schlieren (Laznicka, 1981 visit).

The rare volcanic sediments exposed in small coves, beaches and in subaqueous fans are usually rich in detrital titanomagnetite. Kort (1970) reported the presence of up to 15% magnetite and ilmenite in several kilometers long, several meters thick, bodies of sand in the Hawaiian offshore.

4.4.4. (C) Microcontinent islands

The "granite" islands (e.g. Mahé) of the Seychelles Archipelago in the Indian Ocean are considered a fragment of the Mozambique belt, displaced by continental drift (Baker, 1967). At Mahé, the dominant rock type is a 515 m.y. old coarse red hornblende-microperthite-quartz-oligoclase granite, locally intruded by diabase dikes. The granite islands are fringed by a coral reef. No metallic accumulations are known, except for the occurrences of gibbsite nodules in the weathering profiles, and possibly some bauxitic clays.

4.4.5. (D) "Sedimentary" oceanic islands

Sedimentary oceanic islands are widespread in the tropical belt and almost always composed of a coral reef framework attached to a volcanic (seamount, ridge, etc.) or rarely microcontinental pedestal in depth. The coral material either rims a lagoon (atolls), or forms a continuous platform. Deep drilling on several Pacific islands, such as Midway (Ladd et al.,1970), disclosed a considerable thickness of biogenic carbonates grading downwards into tuffaceous carbonates, volcanogenic sediments and ultimately alkaline and tholeiitic pedestal basalts. The more than 400 m thick pile of sediments at Midway, and an over 1,400 m thick pile at Eniwetok Atoll (Marshall Islands, Schlanger et al.,1963) contain frequent leached intervals, solution unconformities and dolomitized intervals. The basal carbonates at Eniwetok are of Eocene age. Lignitic intervals have been found in the Midway subsurface.

The surface areas of the oceanic coral islands are devoid of metallic accumulations, if one disregards the pockets of the residual red soils some of which contain gibbsite. The discovery of sizeable deposits of young bauxite at the Rennell Island (see Chapter 5) formed by conversion of a blown-in volcanic ash, makes it equally possible that comparable deposits could occur on some oceanic islands as well. The only non-metallic commodity presently profitably mined on the oceanic islands is calcium (rarely aluminium) phosphate, formed by metasomatic alteration of limestone and rarely silicate bedrock by descending phosphatic meteoric waters. The phosphatic content was leached from accumulated guano (bird droppings). The best known deposits mined are on the tiny (5 x 4.5 km) island of Nauru. They consist of a subvitreous collophane with some dahllite and crandallite, filling depressions in a "Karrenfeld-type" terrain of dolomitized limestone (Fig. 4-6). The original reserves were 90 Mt of 38-40% P_2O_5 (White and Warin, 1964). Some of the Pacific phosphate deposits have an increased trace content of uranium; 50-70 ppm U on Nauru, 80 ppm on the Christmas Island, and 60 ppm on the Makatea Island (Cathcart, 1978). This could possibly be recovered as a by-product. No metallic occurrences have been reported from the subsurface of the coral islands, or from their pedestals.

Fig. 4-6. "Karrenfeld-type" topography of the mined-out phosphate field, Nauru Island, western Pacific.

"The zones of earthquakes dipping deep into the mantle from the trenches mark the descent of plates of oceanic lithosphere subducted beneath island arcs and active continental margins"
Warren Hamilton (1979,p.17)

"Subduction exists only in the minds of its creators"
Warren Carey (1976, p.16)

CHAPTER 5

Continental Margins

The term continental margin has a variety of meanings. Drake and Burk (1974) considered continental margins to be boundaries between deep oceanic basins and the subaqueous extension of the continental masses. In the very comprehensive volume edited by Burk and Drake, eds. (1974), only the subaqueous regions are considered. Other writers also included the dry-land extension of the margin and considered the belts in which the present-day seismic and magmatic activity is taking place; for example, the "Andean-type" margins. In the present book, a compromise convenience division is applied. The basic approach used in Burk and Drake, eds. (1974) is followed and the additional divisions are reviewed only briefly here. Reference is made to those sections where more detailed coverage is provided.

In contemporary literature, it has became customary to treat the continental margins in terms of two characteristic "types": Pacific and Atlantic (compare Figures 5-2, 5-14 and 5-14 later in the text). Heezen (1974) wrote: "Continental margins can be characterized as of "Atlantic" or "Pacific" type, essentially depending on whether they have experienced a relatively long period of stability, in which case they are called Atlantic type, or whether they have suffered active tectonism during latter geological times, in which case they are assigned to the Pacific type". Although transitional environments do exist (one margin "type" often has components of the other "type", and margin "types" that do not exactly correspond to either of the two types are also known), the Pacific versus Atlantic margins division is convenient and established, and consequently is used in this book.

5.1. PACIFIC-TYPE (CONSUMING, ACTIVE, SUBDUCTIVE) CONTINENTAL MARGINS

5.1.1. General

Consuming (subduction-influenced) continental margins are defined geophysically (by Benioff seismic zones); physiographically or bathymetrically; and geologically. They start at, and are separated from, the oceanic lithosphere by oceanic trenches. Two principal varieties of continental margin are caused by different configuration of the margin elements, particularly the magmatic arc. In (1) the island arc margins, the magmatic arc is bounded on both sides by the sea; in (2) "Andean-type" margins, the magmatic belt is situated wholly on dry land.

The main physiographic and geotectonic subunits of the island arc margins (1) are: (a) oceanic trench; (b) arc-trench gap; (c) volcanic or magmatic arc with its basement; (d) marginal sea; (e) the edge of a contiguous continent. The main subunits of the "Andean-type" margin include: (a) oceanic trench; (b) arc-trench gap, that usually includes (i) the subaqueous portion and (ii) the subaerial portion; (c) volcanic or magmatic arc with its basement; (d) terrestrial clastic wedge; (e) the continent, generally unaffected by the processes that took place along the margin. The principal varieties of continental margins and their subdivisions appear in Fig. 5-2, later in this chapter.

5.1.2. Oceanic trenches

Trenches mark the sites of downturns of oceanic lithospheric plates beneath other plates, oceanic or continental. A typical trench, such as the Java Trench off Indonesia studied by Hamilton (1979, p.19), is about 5,000 km long and 70-100 km wide. The trench is 6-7 km deep off Java, 4.5 km deep off northern Sumatra and less than 3 km deep off Burma. The north-westward decreasing depth is proportional to the increasing thickness of terrigenous sedimentary fill in an extension of the Bay of Bengal abyssal fan. The Indian Ocean lithospheric plate subducts beneath the Asian continental margin at a very gentle dip, and the Benioff zone is traceable for about 800 km downdip in the Java sector. The mafic oceanic "bedrock" (Layer 2) is covered only by a thin veneer of pelagic sediments in the vicinity of Java. The same sediments off Sumatra are covered by turbidites flowing longitudinally along the trench, and they, in turn, are topped by terrigenous and continental sediments and pyroclastics derived from the volcanic arc. As the oceanic lithosphere and the topping sediments move conveyor belt– style onto the trench and the subduction zone, a considerable portion of the sediments is scraped off against the outer wall of the trench and added to the basinward-thickening sedimentary wedge.

The modern trench sediments have recently been reviewed by Scholl (1974) and von Huene (1974). They pointed out that trenches adjacent to larger continents, particularly those recently glaciated, are filled by a thick sequence of terrigenous and hemi-terrigenous turbidites to the point that they may be completely filled and lose their bathymetric expression (e.g. the Washington-Oregon Trench). Trenches adjacent to island arcs (such as the Kurile - Kamchatka Trench and most trenches of the western Pacific), on the other hand, frequently lack terrigenous turbidites and their floor is underlain by thin pelagic or hemipelagic sediments. Trenches containing abundant terrigenous fill show a distinct coarsening upward (from pelagic clay or organic ooze to conglomeratic turbidites) and a rapid upward increase of the terrigenous / pelagic-derivation ratio.

Mud, silt and sand are the most common sediments. Mafic and ultramafic members common in the oceanic lithosphere (hydrated basalts, serpentinized and mylonitized peridotites, troctolites, gabbros, rodingite-altered troctolitic gabbros) and even shoshonitic potassic rocks have been dredged from several Pacific (e.g. Marianas, Tonga) and Atlantic trenches (e.g.Puerto Rican Trench; Hékinian,1982).

No metallic occurrences have been reported so far from the recent trenches, and few can be anticipated (except for the ores generated within the oceanic domain and transported into the present trench such as the Fe-Mn nodules, metalliferous muds or ophiolite - hosted minerals). Sections on the trench and outer arc metallogeny in the literature (e.g. in Mitchell and Garson, 1981) consider the ancient rock associations now preserved in the dry land and interpreted as having formed in the trench setting. These are treated here in Chapter 8.

5.1.3. Arc-trench gaps

An arc-trench gap (Dickinson, 1973; alternatively called deep-sea terrace, outer high, trench slope break, etc.), contains a generally low-density accumulation of sediments located between the outer (i.e. continentward) edge of an oceanic trench, and an inner (i.e. oceanward) edge of a volcanic island arc. Its width ranges from about 50 to about 200 km and a variety of physiographic or bathymetric variants has been summarized by Dickinson (1974a; p.235). Two end-member variants are (1) a terrace or a shelf-slope complex, having a persistent though variable dip from an island arc towards the trench; and (2) an arc-trench gap subdivided into an outer-arc ridge (sometimes taking the form of subaerially exposed islands such as the Mentawai Islands off Sumatra), and outer-arc basin. The latter is typical in regions with an abundant supply of terrigenous detritus. The rock body under the arc-trench gap surface usually takes the form of a wedge.

The dominant component rocks are lithic sand(stone) to mud(stone), oozes, etc., which have two contrasting directions of supply: (a) oceanward supply of terrigenous components (incl. pyroclastics from island arc volcanoes); and (b) continentward supply of oceanic (pelagic, hemipelagic) components. The supply (a) is the result of a variety of processes contemporary with the wedge growth, turbidite dumping being most prominent; the supply (b) is tectonic, and involves sediments and magmatic rocks deposited earlier in the oceanic domain, carried on top of a moving lithospheric plate, scraped off at the edge of a trench (thus freed from subduction), and added to the arc-trench gap in the form of a tectonic mélange.

The mechanism of mélange generation in Indonesia has recently been interpreted by Hamilton (1979). The thin wedge of deformed predominantly oceanic sediments beneath the landward slope of the Java Trench is imbricated at a dip substantially steeper than the gentle inclination of the subducting plate. As the deformation progresses further in the continental direction, increasingly greater amounts of the terrigenous sediment are incorporated into the mélange so that a sharp separation of rocks of oceanic and continental provenances becomes almost impossible. Away from the zone of intensive shearing, the mélange grades into imbricated slices. Hamilton (1979, p.30) explained the common occurrence of slabs of metamorphics generated in depth (such as glaucophane schist ; depth of formation about 25 km) with sheared but unmetamorphosed sediments, as an extreme product of uplift and mixing by imbrication and trenchward spreading of the mélange wedge.

In the literature, tectonic mélanges are commonly confused with olistostromes: chaotically mixed rocks of paraconglomeratic appearance, interpreted as having formed by submarine slumping. Although olistostromes are also members of the arc-trench gap association, Hamilton (1979) pointed out that mélanges are ubiquitous and olistostromes rather rare. No metallic mineralization has been reported from the contemporary arc-trench gap environment, with the possible exception of some of the magnetite beach placers located at the interface of this environment with the magmatic arc, treated in the next paragraphs. The depositional processes, low heat flow, and rock types produced are unfavourable as metal accumulators, so that this environment seems to be of little economic interest. The ancient equivalents of arc-trench gaps and their mineralizations are treated in Chapter 16.

5.1.4. Volcanic island arcs

INTRODUCTION

Volcanic island arcs (see Fig. 5-3 later in the text) are the most conspicuous and most extensively studied, yet also the most complex subenvironments of the Pacific-type continental margins. A portion of the arcs is subaerially exposed but the rest is subaqueous, and island arc deposits merge with those of the arc-trench gaps, and marginal seas, on either side. "Active" (meaning seismically and magmatically active) and "inactive" arcs are sometimes distinguished (e.g. Mitchell and Bell, 1973), but although the "inactive" arcs need not have a contemporary volcanism they still influence the sedimentogenesis along and under the arc/sea interface. The "active" arcs, in turn, are not necessarily active as a whole. Large islands that are members of island arc chains (such as the Japanese islands, Java, etc.) have only a portion of their territory recently tectono-magmatically active, which conforms to the actual plate tectonic and metallogenetic models. The rest of the islands are a geological relict, not even necessarily formed in an older island arc setting. Most of the larger islands were once parts of continents and there is little qualitative difference between the recent magmatic activity taking place in the interiors of such islands, and magmatic activity along the Andean-type continental margins. The large islands usually have sizeable areas underlain by sialic crust and consequently much of the recent volcanism is ensialic and evolutionarily rather mature. This contrasts with the ensimatic character of the small islands scattered in the Pacific and elsewhere.

Island arcs with a prolonged history of development are often incredibly complex and ill-suited to summary generalizations of the arc geology and metallogenesis. Lesser-order environments or lithological associations (local rifts, ophiolites) are included in higher-order environments (that is, island arcs), but are not always present everywhere. The geological and metallogenic complexity of island arcs has been admirably outlined in the pioneering paper by Mitchell and Bell (1973; also in Mitchell and Garson, 1976; 1981), but the above publications treat jointly the recent processes and supposedly corresponding ancient mineralizations, which in this book are treated separately. Table 5-1 lists selected examples of mineralized volcanic island arcs.

MAGMATISM

Island arcs with active volcanism closely parallel the traces of oceanic trenches (being located about 300 km farther continentward), and their surface is some 80-300 km above the geophysically determined Benioff zone. This association provided material for the early petrogenetic models (e.g. Hatherton and Dickinson, 1969), that explained the generation of andesitic magmas in terms of melting of the descending lithosphere along Benioff zones at depth from which the magmas rose to the ground surface without undergoing much contamination during their ascent. It has been demonstrated empirically that the increasing depth to the Benioff zone often correlates with the increase of K_2O content across the magmatic arc and although the geochemical polarity of magmatic arc rocks has been firmly established (compare Condie, 1982, for review), the assumption that magmas suffer no contamination during their ascent is now widely

Table 5-1. Late Cainozoic volcanic island arcs - selected examples

NAME OF ARC, LENGTH	RECENT VOLCANISM, SEDIMENTATION	ANCIENT ARC, BASEMENT	QUATERNARY MINERALIZATION	REFERENCES
Aleutian arc E-W, 2,500 km	80 volcanoes, basalt, andesite comparable activity since upper Cretaceous	Cr_3 basement, continental in the east, oceanic in the west	fumarolic encrustations, magnet. sands	Marsh (1982)
Lesser Antilles N-S, 700 km	basalt, andesite, minor dacite lavas, pyroclastics	Pc to Ol oceanic or quasi-continental crust	fumarolic encrustations, magnet. sands	Rea (1982)
Aeolian Islands, Aegean Sea Islands The Mediterranean	high Al basalt, andesite, dacite, rhyolite; some shoshonites, high K suite	Mi_3-Pl tholeiite, continental fragments ?	Fe, Mn sedim. in caldera; Fe sulphides from hot springs	Keller (1982)
Kurile Islands NE-SW 1,200 km	basalt, andesite	oceanic crust	Hg, Zn, Mo, pyr. hot springs	Aver'yanov et al. (1968)
Japanese Islands SW-NE, 1,800 km	200 volc. centres, entire compos. range (basalt to rhyolite); andesite dominant	contin. crust 1.98 b.y. Paleoz., Mesoz. ophiolite volc.-sedim., granites	S, pyr., Fe hydr. related to hot springs magnet. sands	Aramaki and Ui (1982)
Marianas, Volcano, Izu arc, N-S, 1,000 km	olivine - augite basalt, pyroxene andesite; island arc tholeiite series	Eoc.-Recent oceanic crust	Fe-Mn ox. on seafloor from hot springs	Meijer (1982)
Philippine Islands composite arc NNW-SSE, 1,500 km	jumble of arcs / subduction complexes; basalt, andesite most common	late PZ - Recent transitional volc.- sedim. foldbelts	magnetite sands	Hamilton (1979)

Sumatra - Java arc WNW, 2,500 km	andesite,dacite,rhyodacite in Sumatra; pyrox. andesite, basalt, very minor dacite in Java	PZ-MZ contin. metasediments in Sumatra, Ol-Q arc volcanis, sediments in Java	Fe;Au-Ag hot springs precip.,magnet. sands	Hamilton (1979)
Banda Arc, E-W to N-S, NW-SE; 3,500 km	inner volc.ridge, andesite,basalt, dacite; outer mélange ridge	south: continental crust north: oceanic crust undergoing collision	magnetite sands	Hamilton (1979)
New Guinea - New Britain arc, E-W 1,000 km	basalt, andes.on islands, "Andean" volc.on New Guinea minor peralkal.,alkaline	oceanic cr.under islands continental under New Guinea	magnetite sands,Fe-Mn springs seds.	Johnson (1982)
Solomons Islands NW-SE, 1,800 km	basalt 60%, andesite 30%, dacite 10%, lavas, pyroclastics in places reef fringe	Cr_3-Pc ophiolitic oceanic basem.,Ol-Q island arc volc.-sedim.assembl.	gray bauxite on limestone magnet.sand	Coleman and Hackman (1974)
New Hebrides, NNW, 1,400 km	mostly basalts, minor andesite subaerial, submarine; reef fringe	Mi_{1-2} ophiolite and pelagic seds, Mi_3-Q isl. arc assemblage	Mn oxides at tuff / limestone cont.	Warden and Mitchell (1974)
New Zealand (North Island) SW-NE 300 km	basalt to rhyolite, full range; andesite most common lavas,pyroclastics,pumice	Mesoz. load metam.graywacke	Au,Sb,W,Zn, hot springs magn.sands	White (1981) Browne (1969)
Tonga - Kermadec arc, N-S, 2,000 km	olivine basalt, andesite, rare dacite,rhyolite; reefs	lower Tertiary ophiolite association	magnet.sands	Karig (1970)

disputed. Hamilton (1979, p.307) argued that the composition of magmas reaching the surface above Benioff zones is controlled by the composition of the crust through which the magma rose. This is supported by the striking contrast in magmatic composition along the continuous Sumatra – Java-Banda arc. In Sumatra which has an old continental crust, the modern volcanics are intermediate and sialic; on Flores, formed on an oceanic crust, the modern volcanics are mostly mafic. The increasing potassium content across the magmatic arc could thus be proportional to the increasing thickness of the lithosphere through which the magma had to pass and which probably supplied the large, light elements. This subsequently governs the evolutionary advancement of island arcs. The most primitive arcs are ensimatic, developed on an earlier oceanic crust. Their basement is therefore oceanic (ophiolitic) and ophiolites or their dismembered components are often infaulted with the products of the younger superimposed arc volcanism and magmatism (e.g. Tonga, Solomon Islands, New Hebrides, the Marianas, etc.). The evolutionarily advanced arcs are underlain by a continental crust often of considerable antiquity (e.g. 1.88 b.y. in Honshu; Aramaki and Ui, 1982).

The island arc magmatism is dominated by strongly porphyritic rocks, that can be further subdivided into three gradational subseries (Jakeš and White, 1972; Condie, 1982): (1) arc tholeiite; (2) low-K and high-K calc-alkaline rocks, sensu stricto; and (3) shoshonite (Fig.5-1). The arrangement of these subseries across an arc often complies with the principle of compositional polarity (arc tholeiites are dominant near volcanic fronts facing the ocean; shoshonites appear further inland), but the regularity is frequently lost due to the interference of numerous migrating magmatic phases, each of which generates its own polarity pattern even within a single volcano.

The most common volcanic rock type in island arcs is andesite (calc-alkaline as well as tholeiitic and shoshonitic andesite), followed by basalt, dacite, rhyodacite and rhyolite. Olivine-rich volcanics approaching ultramafics appear locally, e.g. in the Solomon Islands and New Hebrides. Stanton and Ramsay (1980) described high-magnesium lavas spatially associated with Pliocene – Pleistocene basaltic andesite emissions from the New Georgia Group of the Solomons that had a high K_2O and P_2O_5 contents, which they interpreted as subalkaline and hybrid. Alkaline mafics and ultramafics on the Malaita Island (Solomon Islands; Arthurs, 1979) include alnöite breccias, ankaratrites and melanite ouachitites. Unusual felsic and aluminous varieties of island arc volcanics include cordierite-bearing dacites and rhyolites from Hitu in the Banda Arc (Hutchison and Jezek, 1978), probably a result of continental crust contamination in a microcontinental setting.

The subaerial island arc volcanism has a high degree of explosivity. The Rittman explosive indices (i.e. quantity of fragmental material / total erupted material x 100) vary from 41 in the Mediterranean to 99 in Indonesia and Central America (Condie, 1982). As a consequence, pyroclastics and a variety of reworked volcaniclastics and volcanic-sedimentary mixed rocks dominate the

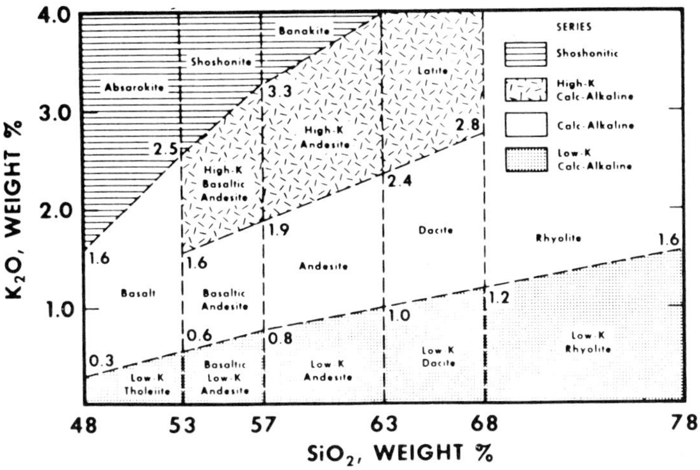

Fig. 5-1. Terminology of island arc volcanic rocks. From Basaltic Volcanism Study Project (1981), courtesy of the Pergamon Press.

island arc lithology and lavas are relatively rare. The arc volcanism is both subaqueous and subaerial, and the ratio of both decreases with time and with the degree of evolution. The typical forms of subaqueous volcanics are basaltic pillow lavas or autobrecciated submarine andesitic flows grading to pillow breccias and hyaloclastites. These are frequently topped by submarine lahar and mudflow deposits generated by rapid erosion of unconsolidated pyroclastics on flanks of subaerial volcanoes.

Subaerial volcanism produces steep-sided stratovolcanoes composed of fragmental rocks with a minor proportion of lavas. Calderas are common. Laharic breccias, siliceous ash-flow tuffs and ash-fall tuffs in various stages of sedimentogenic reworking are laterally extensive, and blanket the territory between the volcanoes.

The deep-seated equivalents of the surficial volcanics range from gabbros through tonalites and granodiorites to quartz monzonites, but they are rarely in sight in recent arcs. The surficially outcropping intrusive rocks are the relics of earlier phases of magmatism. The youngest dated presently outcropping intrusive rocks in island arcs are several million years old (1.5 m.y. old youngest phases of the Koloula Complex, Guadalcanal, Solomon Islands; Arthurs, 1979; 3.3 m.y. old cordierite granite, Latimor, Banda Arc; Hutchison and Jezek, 1978). The Recent volcanics in island arcs are unmetamorphosed, except for the hydrothermally altered patches in the vicinity of hot springs or in the porous steam reservoirs. Metamorphism of course appears in the deeply buried or dynamometamorphosed sections, but these are almost always pre-Quarternary.

SEDIMENTOGENESIS

The high degree of explosivity of the island arc volcanoes generating an immense quantity of loose rock fragments, coupled with the proximity of marine sedimentary basins, made island arcs a prolific sedimentary depositional environment. In the subaerial portion of island arcs, the freshly deposited unconsolidated ash, tephra and blocks, are readily reworked by water: initially on flanks of volcanic cones as poorly sorted lahars, subsequently as alluvial sands, gravels and/or beach sediments. The large, heavy and massive blocks from volcanic ejecta resist rapid transportation, so they often accumulate as lag gravels at volcanic flanks (Fisher and Schmincke, 1984).

Instant reworking of the volcanic explosive material produces sediments of pyroclastic derivation (e.g. Pettijohn, 1975), in which the clasts are, at least initially, fresh and unsorted so that they approximate to the composition of their volcanic source. Lava flows, surge deposits, tuffs, etc. subjected to weathering and surficial fragmentation after a period of residence in the crust, however, respond to the exogenic agents as any other rock and yield an epiclastic sediment. In such a sediment the unstable constituents (volcanic glass, olivine, pyroxene, hornblende, plagioclase) are absent and the relatively resistant components only (quartz and various forms of silica; a variety of contact hornfelses, e.g. those rich in epidote; magnetite) are being transported and redeposited. In most island arcs, the epiclastic and pyroclastic sediments overlap and additional epiclastic detritus is being derived from rocks of the older basement.

Rapid erosion of stratovolcanoes rising directly from the sea or within a short distance from the shore, leads to the deposition of subaqueous mass-flow deposits. These form extensive aprons of quartz-poor boulder breccias, conglomerates and sandstones. Much of the "agglomerate" described from many of the Pacific islands probably consists of mass flow deposits (Mitchell and Reading, 1978). These poorly sorted, proximal deposits rapidly change facies into a volcanic gravel (conglomerate), turbidite sand(stone), mud(stone) and clay(stone). In lithified assemblages of island arcs, "graywackes" (dark gray to green, matrix-rich litharenites), are the dominant sediments. Graywackes form from the orthomatrix-rich, poorly sorted proximal sediments, and virtually identical rocks can form from well-sorted pyroclastically derived sands (e.g. of beaches) in which the matrix is an epimatrix formed by diagenetic breakdown of the unstable clasts (Pettijohn, 1975; Galloway, 1974).

A significant proportion of island arc volcanic sediments formed wholly as a by-product of subaqueous eruptions, without even reaching the surface. Mixed volcanic-sedimentary rocks (such as peperites) abundant along the interface of autobrecciated, granulated lavas and local sediments (e.g. mud, sand, biogenic detritus, etc.) are particularly characteristic. Hydrothermal sediments (or mixed detrital and hydrothermal sediments) formed at sites of subaqueous emissions of hot springs.

In the tropical belt, limestone reefs and bioclastic deposits formed widely around coastal fringes of volcanic arc islands. They frequently occur in the form of several generations of raised terraces, or they completely top a subsided volcanic island. A wide variety of rock types resulted from mixing of the organic carbonate, volcanic detritus, lava, etc., under both subaerial and submarine conditions.

LESSER-ORDER TROUGHS ("RIFTS", GRABENS, ETC.) WITHIN ISLAND ARCS

Some island arcs are made locally more complex by the presence of extensional (taphrogenic) structures of lesser order than the arc itself. Such structures can appear in a variety of settings: in the axial zone of double arcs (e.g. the Lau-Havre Trough, Tonga-Kermadec Arc); along the arc margin; etc., and can be either subaqueous (New Hebrides) or subaerial (Taupo zone, New Zealand). Usually, they are zones of exceptionally high heat flow and they are metallogenically significant.

The subaqueous troughs are insufficiently understood. The troughs situated at the rear of the New Hebrides Arc (Dubois et al., 1978) occupy two 300 km long collapse zones about 35 to 55 km from the recent volcanic axis. Internally, they are a system of steep-walled horsts and grabens up to 45 km wide and 3 km deep. In the grabens, flat-lying to deformed unconsolidated sediments rest on a volcanic basement. A limited number of dredge hauls recovered black mud, fine volcaniclastics, tuffs with foraminifera and Fe-Mn oxide crusts. The Lau-Havre Trough (interarc basin) in the Tonga - Kermadec arc system (Karig, 1970) has a steep relief of up to 1,500 m, high heat flow and an oceanic crust floor. The meagre floor sediment includes calcareous ooze and volcanic mud.

The 300 km long, 16 to 24 km wide, north-east trending Taupo volcanic zone in the North Island, New Zealand, occurs in the extension of the Lau-Havre Trough and is exposed subaerially. It is a young tensional graben, that includes five volcanic centres having a full range of calc-alkaline rocks (basalt, andesite, dacite, rhyolite), but rhyolite is dominant. The central graben is flanked by flat-lying sheets of ignimbrite, and a Mesozoic graywacke and argillite is in the basement (Cole, 1979). The dominant graben fill consists of continental near - horizontal air-fall, ash-flow and water-laid tuffs and breccias, interbedded with rhyolite and dacite flows. Portions of the fill act as an excellent reservoir for steam and hydrothermal solutions (Ewers and Keays, 1977).

5.1.5. Recent "Andean-type" continental margins

Andean (or Cordilleran) -type magmatic arcs are located above Benioff zones at continental edges. In the most frequently quoted example areas: the Cascade Mountains of North America (McBirney and White, 1982) and the Andes of South America (Thorpe et al., 1982), the presently active magmatic belt consists of a narrow range of stratovolcanoes resting on an older orogenic belt basement.

The Andean volcanic chain can be divided into three segments (Colombia + Ecuador; S.Peru—N.Chile—N.W. Bolivia; S. Chile and Argentina), separated by gaps. In the northern and central zones the Recent conical stratovolcanoes and compound volcanoes lacking the radial symmetry, tower above a high plateau (3-4,000 m) and reach altitudes of 5,000 to 6,800 m. About 20 large dacite—rhyolite ignimbrite sheets or low shields which erupted contemporaneously with andesites, are known in the central Andes. The volcanics have been emplaced into a 40-70 km—thick continental crust about 140 km above an eastward dipping Benioff zone (Thorpe et al.,1982), and the magmatic activity has occurred episodically at or near the present location since about the Mesozoic age. Compared with the evolutionarily immature island arc volcanism in which there is, at least to some extent, a predictable connection between the recent volcanics and the basement into which and over which they have been emplaced, no similar connection exists in the Andean belts and the recent volcanics are completely unrelated to the older basement.

The Andean belt volcanoes are morphologically and compositionally comparable with volcanoes of the ensialic island arcs, and the main difference is in the type of associated sediments and the degree of dissection and weathering. The volcanoes of the central Andes are in an arid region far from the sea, with much of their ejecta locally accumulated under subaerial conditions. The volcanic contribution to the sedimentogenesis in the Pacific ocean has been minimal. Table 5-2 lists selected example regions.

5.1.6. Intracontinental calc-alkaline volcanic belts and centres

The parallel nature of trenches, Benioff zones and calc-alkaline volcanic belts suggesting causality does not hold in some instances, in which calc-alkaline volcanic or magmatic belts occur far from continental margins (e.g. partly the Yellowstone-Absaroka Plateau; San Juan Mountains, western United States), or where trenches and Benioff zones are not apparent. Voluminous literature has been produced discussing the geotectonic affiliation of such occurrences which, although often controlled by grabens ("rifts"), lack the dominant alkaline rock association of the East-African-like rifts. The volcanic and magmatic style and lithology are comparable with the continental margin magmatic belts, or transitional into or overlapping with, the continental alkali basalt provinces (see Chapter 31).

Table 5-2. Late Cainozoic continental margin (Andean) magmatic belts, examples

NAME OF BELT	GEOLOGY	QUATERNARY MINERALIZATION	REFERENCES
Cascade Mts., U.S.A and Canada	large composite stratovolcanoes, isolated, built on Eo-Mi volcanics and Paleozoic - Mesozoic metamorphics; explosive volcanicity mostly	minor detrital alluvial magnetite	McBirney and White (1982)
California, Northern Mexico	Scattered small magmatic centres, associated with hydrothermal area, very limited Quaternary volcanism; strike slip and extensional faulting	numerous hot springs fields, metall. brines and sinters; Hg, Sb, Au, Ag, Mn, Cu, Zn, Pb deposited	White (1981)
Central Mexican Volcanic Belt, E-W	10 major central volcanoes, thousands of simple cones, lava plateaus; andesite dominant, lesser basalt, dacite, rhyolite	minor alluvial magnetite	Robin (1982)
Central American volcanic chain	several strings of composite stratovolcanoes; basaltic andesite dominant, lesser high Al basalt, dacite, rhyolite	metalliferous fumarole encrustation, magnetite sands	Carr et al. (1982)
The Andes, South America	3 zones of stratovolcanoes separated by gaps; basaltic andesite, lesser high Al basalt, andesite, dacite; lavas, widespread ignimbrites; playa lakes	Mn, W precip. from hot springs; fumarolic wood tin, Mn	Thorpe et al. (1982)
Turkey, Caucasus, Iran	scattered volcanic centres, calc-alkaline, some shoshonites, minor alk. basalts; andesite, dacite most common	cinnabar showings in hot springs precipitate	Innocenti et al. (1982) Brinkmann (1976)

5.1.7. Metal geochemistry and metallogeny

There is little purpose in applying separate treatments to the trace metal geochemistry and metallogeny of island arcs, the Andean, and the intercontinental magmatic belts because of their different geotectonic position. Corresponding magmas under corresponding emplacement and interaction conditions produce corresponding mineralizations. Consequently, the local environment of emplacement is of greater importance to an economic geologist than the geotectonic setting (which, in many cases, is controversial anyway), so that the former is used as a framework in the review that follows.

The modern petrologic literature provides abundant data on the composition of the "orogenic" volcanics in terms of major and petrogenetically significant minor elements (such as Rb, Sr, light and heavy rare earths, U, Th, Nb, Ni, Cr, etc.). Data on the transition elements (Cu, Zn, Pb, Co, etc.) that may actually form ore concentrations associated with the magmatites are, however, much rarer. Ewart (1982) recently compiled the available data on the composition of the "orogenic" andesites and basalts of the world, from which the following range of contents of the principal ore-forming elements results: 13-20% Al_2O_3; 0.77-2.21% TiO_2; 0.11-0.24% MnO; 60-116 ppm Zn; 30-175 ppm Cu; 20-332 ppm Ni; 18-71 ppm Co; 9-869 ppm Cr; 100-370 ppm V; 2.3-34 ppm Pb.

Table 5-3 shows the range of Cu, Zn and Pb contents in mafic orogenic volcanics by the major petrochemical series. Although one can discern a general trend of copper decrease and lead increase with magmatic evolution (that correlates with the empirical experience that Cu deposits tend to be more common in the "primitive" magmatic associations and Pb deposits in the more evolved ones), there are numerous exceptions and inconsistencies. Only the arc tholeiites have consistently high trace Cu content and a low trace Pb content. It has been demonstrated repeatedly that the Pb content increase follows the

Table 5-3. The range of trace contents of Cu,Zn,Pb in the major petrochemical groups of mafic Quaternary orogenic volcanics (in ppm). Based mostly on data in Ewart (1982)

PETROCHEMICAL GROUP	Cu	Zn	Pb
island arc tholeiite	125-181	60-80	1-2
low K calc-alkaline basalt	40-99	78	4-5.3
high K calc-alkaline basalt	47-122	69-96	4-15.5
low K andesite	49-138	84-94	2.3-4
high K andesite	36-54	73-104	17-27
shoshonitic basalt	30-130	73-106	6.2-12.1
shoshonitic andesite	27-81	71	19.4

increase in K during magmatic evolution and crystallization, whereas Cu together with Cr, Ni, Co and V become erratically depleted with increasing differentiation index.

During magmatic differentiation, major Ni and minor Co, Zn are partitioned into olivine; major Cu and minor Zn into pyroxene; Pb into feldspars; and major V, Cr, Co, Zn and minor Ni, Cu into the Fe-Ti oxides. In a closed system metallogenesis, this has a significant bearing on the possibility of depositional accumulation of the various metals. Many ore-forming events coincide with episodes of accumulation, destruction, modification, etc., of the trace-metal carrier minerals. Several examples are discussed later, as well as in Chapters 6, 26 and 28.

Figures 5-2 , 5-3 and Table 5-4 show and list the principal mineralization styles in both the ensimatic and ensialic varieties of magmatic arcs. A more detailed description follows.

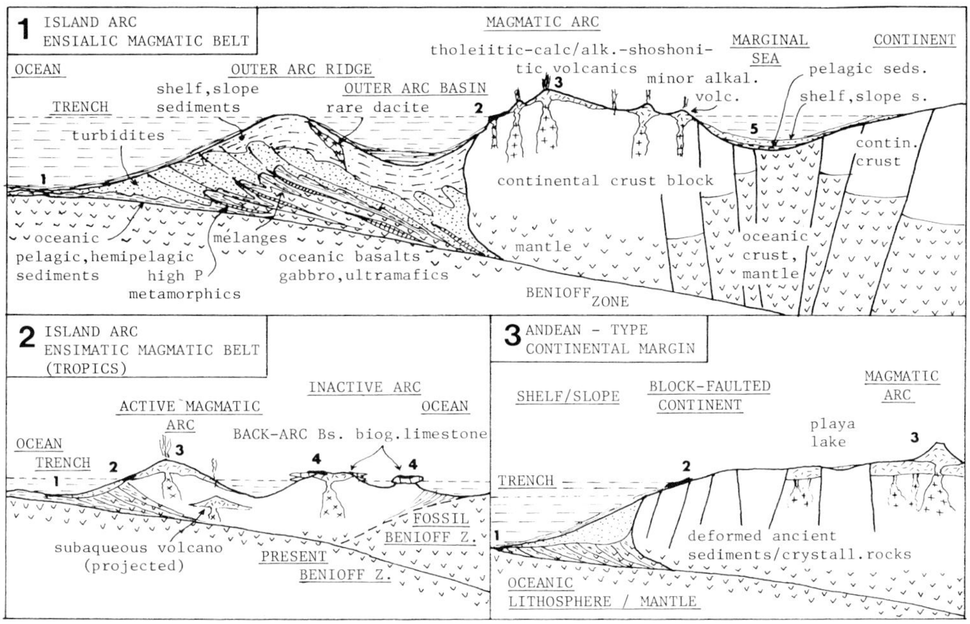

Fig. 5-2. Active (Pacific-type) continental margins, showing the Recent mineralizations forming at the surface or in shallow subsurface. Numbers mark the prominent mineralization groups: (1) ores generated in the oceanic domain and tectonically brought into the trench (e.g. Mn nodules, ores formed at spreading ridges; see Chapter 2); (1a) as above, but incorporated into a mélange and tectonized, metamorphosed with it; (2) beach sand magnetite; (3) ores associated with active volcanism and hydrothermal activity; (4) weathering-generated deposits: e.g. bauxite, Mn ores; (5) back arcs, minor Fe-Mn submarine hydrothermal precipitates.

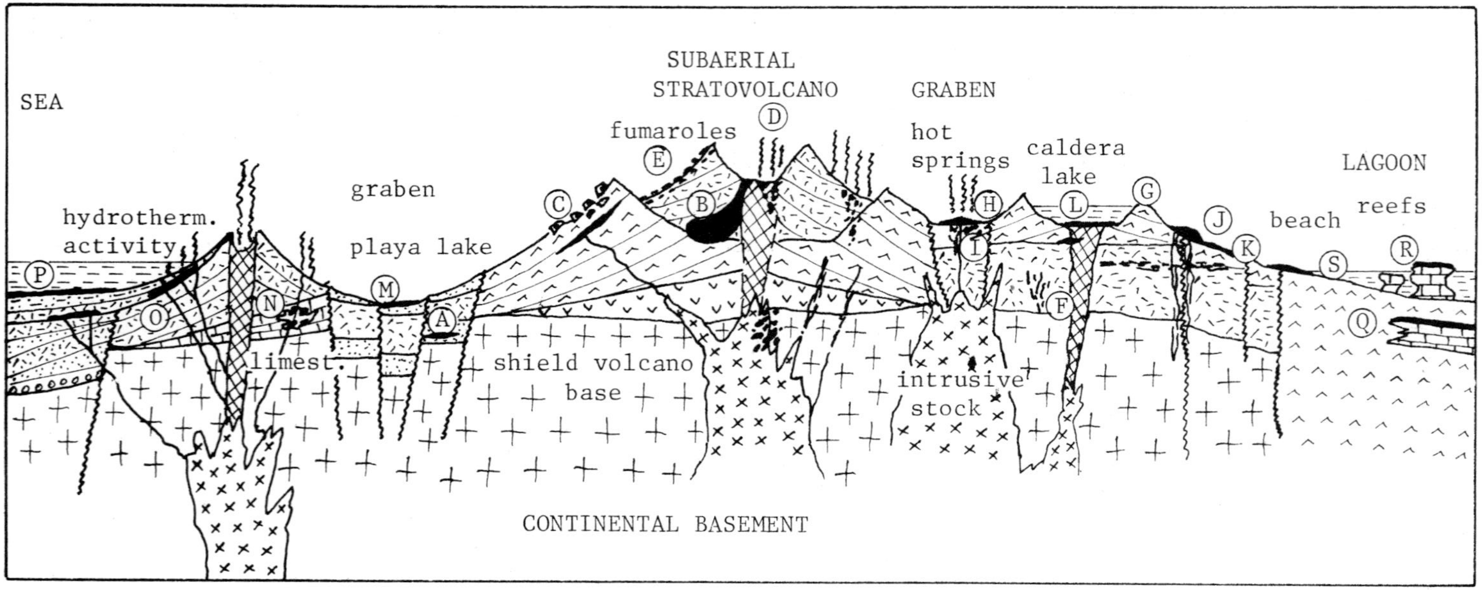

Fig. 5-3. Principal Quaternary mineralization styles in a calc-alkaline, ensialic continental margin magmatic (volcanic) arc (island arcs). See Table 5-4 for explanation of the letter symbols.

Table 5-4. Principal Quaternary mineralization styles in an ensialic continental margin magmatic (volcanic) arc (e.g. island arc)

(A)	phenocrysts - rich layers in ash falls and tuffs (e.g. magnetite)	X	---
(B)	metalliferous lavas (magnetite,El Laco)	1	500 Mt Fe
(C)	mineralized ejecta	X	---
(D)	fumarole encrustations	XO	---
(E)	soluble metals in volcanic ash	XO	---
(F)	metalliferous precipitates in fumarole subsurface	XO	---
	METALLIFEROUS HOT SPRINGS (G-P)		
(G)	anomalous metal concentrations dissolved in water (e.g. Salton Sea,California)	X	5 Mt Zn 900 Tt Pb 60 Tt Cu 10 Tt Ag
(H)	metalliferous precipitates in boiling water pools and hot mud	X	Au,Sb,As,W
(I)	minerals precipitated in the shallow subsurface of hot springs systems	X	70 Mt Fe 7.5 Tt Hg
(J)	metalliferous surficial springs precipitates (sinters,travertine)	XO	100 t Hg
(K)	alluvial metalliferous precipitates from hot springs	XO	5 Mt Fe
(L)	subaqueous hydrothermal deposits in crater lakes	1	py,Hg
(M)	lacustrine springs precipitates	XO	500 Tt Mn 2 Tt W
(N)	hot springs filling or interacting with karst	X	60 Tt Hg
(O)	submarine-hydrothermal replacements in unconsolidated rocks, and seafloor crusts, sinters	X X	--- ---
(P)	submarine hydrothermal-sedimentary metalliferous occurrences (gradational to normal sedim.)		
(Q)	weathering-mobilized ores (e.g.Forari,Mn)	X	1.1 Mt Mn
(R)	residual bauxites on volcanics and limestones	XO	22 Mt Al
(S)	Ti-magnetite, beach and alluvial sands	XO	800 Mt Fe 70 Mt Ti

METALS IN PHENOCRYSTS

In the absence of cumulus concentrations of ore minerals in calc-alkaline volcanics (ultramafic cumulus blocks containing minor chromite, Ti-magnetite, and other minerals, however, occur locally as ejecta and inclusions in some recent basaltic andesite volcanoes, e.g. Sufrière on the St.Vincent Island; Rea, 1982), phenocrysts become the most conspicuous sites of metal accumulation in volcanic rocks. The phenocrysts are, however, only in exceptional cases capable of becoming concentrated during active volcanism to such an extent that they develop into an "ore occurrence". The most feasible mechanism of phenocryst separation exists in ash-fall tuffs. There, frequently observed thin bands of Fe-Ti oxide phenocrysts have probably accumulated as a result of gravity sorting during the fall.

Hydrodynamic reworking of fresh or weathered volcanics postdating the active volcanic events, on the other hand, is an efficient mechanism for accumulation of the resistate, heavy mineral phenocrysts (such as Ti-magnetite or ilmenite) in alluvial and beach placers.

METALLIFEROUS LAVAS

Recent native sulphur lavas have been observed, described and mined at several places in the world, most prominently in Japan (e.g. at the Shiretoko Peninsula, Hokkaido; Saito, 1960). Pleistocene magnetite lava flows have been discovered more recently, associated with a secondary crater of an andesite-rhyodacite El Laco volcano, northern Chile (Frutos and Oyarzún, 1975; Fig. 5-4). The metalliferous lava flows are about 20 m thick, and consist of magnetite with minor hematite and apatite, containing some actinolite and scapolite gangue minerals. The adjacent silicate rock lavas are lightly epidote and chlorite-altered. Although the ore emplacement as a lava flow is not disputed, the melt generation and the iron source remain controversial. There are about 1 billion tons of a 50% iron ore.

MINERALIZED EJECTA

Explosive volcanism produces large amounts of ejecta, consisting of (1) members of the volcanic suite and its deep-seated equivalents; (2) genetically unrelated rocks from the basement. Both ejecta types may occasionally contain ore minerals and a variety of them (galena, sphalerite, molybdenite, chalcopyrite, magnetite, Mo-sodalite, etc.) have been described from the classical locality Monte Somma (part of Vesuvius, Italy; Zambonini, 1910). There, the minerals occur in sedimentary carbonate blocks, contact metamorphosed under high temperature, low-pressure conditions. Mineralized ejecta in themselves can hardly be expected to produce economic ore accumulations, but they can indicate the presence of an economic mineralization at depth.

FUMAROLES, SOLUBLE METALS IN ASH, ENCRUSTATIONS

Fumaroles are vents and conduits that emit or conduit volcanic gases and vapours. They occur in active volcanic areas in the vicinity of vents or along faults. Most geologists apply the term in a broader sense to include the gases and vapours as well. Solfataras are fumaroles conspicuously high in sulphur. Fumaroles have a broad temperature range (900 to 100° C) and with decreasing temperature and increasing fluid phase, grade into hot springs. Active fumaroles often excavate conduits (pipes, tubes, fissures) in freshly deposited

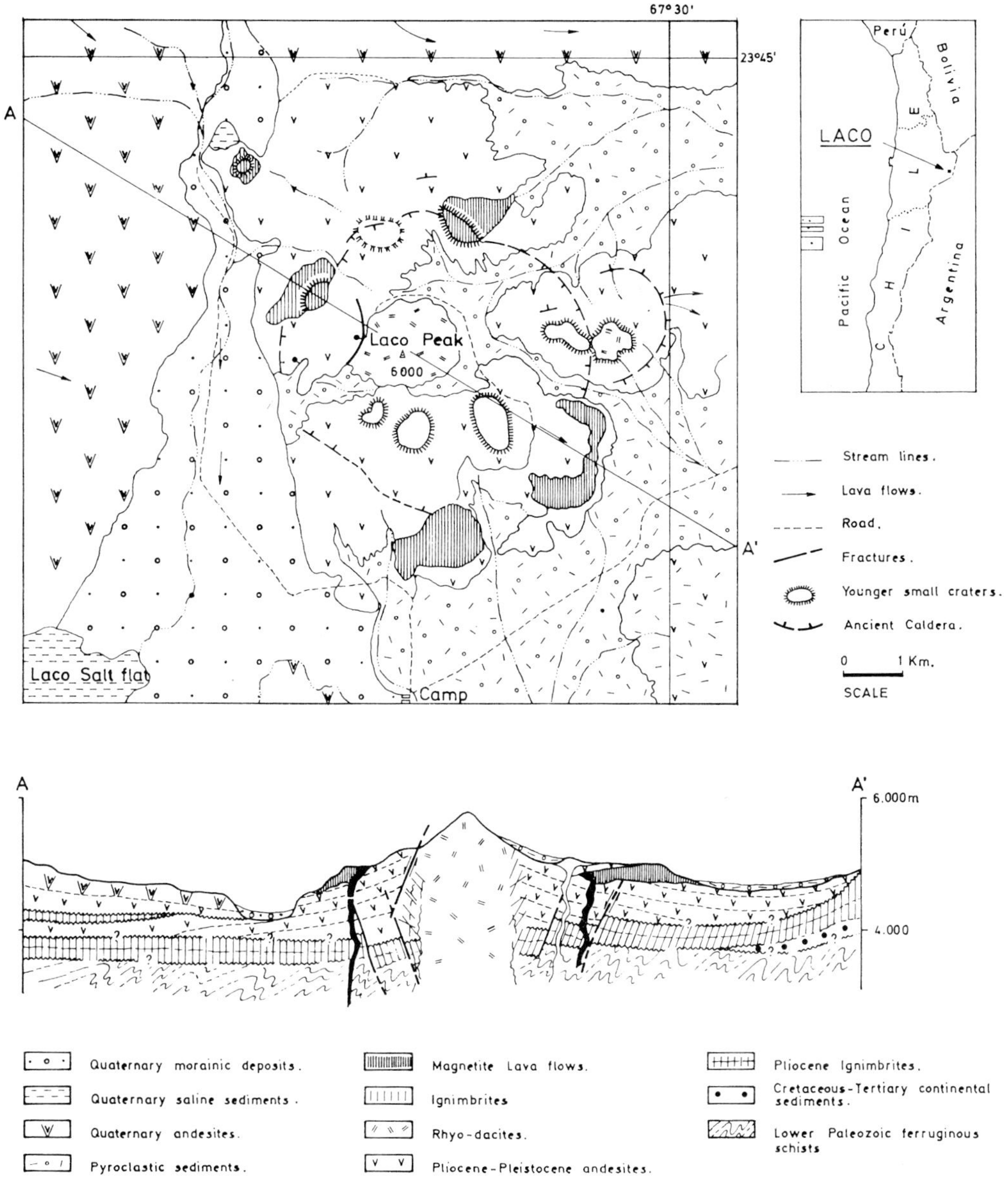

Fig. 5-4. Geological map and cross-section of El Laco, northern Chile, deposit of magnetite lavas. From Frutos and Oyarzún (1975), courtesy of Economic Geology and Professor Frutos.

pyroclastics, and this serves as one criterion for distinguishing volcanic ash flows from mudflows.

Fumarole gases have been sampled and the geochemistry of gas condensates summarized in White and Waring (1963) and Stoiber and Rose (1970). The fumarole gases that reach the surface rapidly dissipate into the atmosphere and hydrosphere, and calculations made during the Mt.Katmai, Alaska, activity in the 1910's (e.g. Allen and Zies, 1923) provided data on the immense quantities of volatiles released after volcanic eruptions: 1.3 Mt HCl; 0.2 Mt HF; and 0.3 Mt H_2S, during the year 1919 alone.

Fumaroles also carry trace metals most of which precipitate at the surface or in the subsurface of the volcano in the following forms: (1) conspicuous encrustations of salts on altered lavas or pyroclastics on the surface and in the shallow subsurface (within several centimetres or tens of centimetres), in the immediate vicinity and on walls of the fumarole orifice; (2) invisible (but analytically detectable) soluble metals on a fresh volcanic ash; and (3) metalliferous precipitates, both open space fillings and replacements, in the shallow subsurface of the recent volcanic terrains.

The encrustations (1) have been summarized recently by Stoiber and Rose (1974). They have been recorded from several localities in Central America (Izalco, Santa Ana, Cerro Negro, San Miguel); Alaska (Valley of 10,000 Smokes near Mt.Katmai); Japan (e.g. Showa-Shinzan, Hokkaido); Vesuvius, the Andes, New Zealand, Kuriles and Kamchatka, and elsewhere. The encrustations are usually temperature zoned and in addition to native sulphur, gypsum, Na, K, Ca, Mg, etc. sulphates and chlorides, they contain metallic minerals such as hematite, magnetite, native selenium, realgar, orpiment, cottunite, chalcocite, covellite, vanadinite, galena, sphalerite, Bi minerals, etc. At least 50 mineral species have been recorded from the Vesuvius alone (Zambonini, 1910).

Fe, Mn, Cu, V, Mo, Zn, Pb, As, Se, Tl, Ag, and Bi are distinctive metals associated with fumaroles related to the calc-alkaline volcanism. A selection of remarkably rich metalliferous encrustations include those from the White Island, New Zealand (1% Pb, 0.4% Cu, 0.3% As, 0.3% Zn, 0.5% Sn); Shirane Volcano, Japan (1.64% Ni); Shiveluch Volcano, Kamchatka (up to 12% V_2O_5); Klyuchevskii and Bezym'yannyi volcanoes, Kamchatka (0.3-1% V); White and Waring (1963), Popov (1979). The life-time of most encrustations is only few days or weeks, although there are exceptions. Byers (1959) described a 2.5 cm thick buried layer of basaltic ash resulting from the 1926-1927 eruption on the Bogoslov Island, the Aleutians, impregnated and encrusted by gypsum, paraatacamite and caledonite.

The soluble material on ash overlaps with (1). The considerable quantity of metals so accumulated has been calculated by Zies (1929), Taylor and Stoiber (1973) and others. 8×10^{-4} cubic kilometres of ash resulting from two 1970 eruptions of the San Miguel volcano (El Salvador) alone, contained 14 t Mn, 5 t Cu, 0.7 t Zn in soluble form. The residence time of soluble metals in the porous volcanics is very short, particularly in the humid tropics, so that hardly any "in situ" metal accumulations could be expected to survive. Leaching and transportation of these metals by surficial and ground waters, however, can conceivably result in redeposition if suitable sedimentogenic traps are present, for example along the continental / marine interface; hence (2) is a potentially important metal source to be considered in metallogenic interpretations.

Metalliferous precipitates formed in shallow subsurface and attributed to fumarolic activity are known from several localities hosted by Quaternary to Pliocene volcanics dissected by erosion. Many such occurrences are transitional to epithermal orebodies. The occurrences of cassiterite having the form of "wood tin", or fine crystalline cassiterite – hematite coatings on the surface or in fractures of dacites and rhyolites showing virtually no hydrothermal alteration, are probably of fumarolic origin. The known Quaternary occurrences include the Cerros Pircas, Pululus, San Pedro, etc., in the Puno region, northern Argentina (Ahlfeld, 1949). The economic importance has been small so far and some cassiterite has been recovered from placers adjacent to the bedrock occurrences. The vein and impregnation magnetite and hematite associated with the El Laco magnetite lavas, are probably also of fumarolic origin, and so are the short, discontinuous veinlets of stibnite in dacitic lavas found at several places in the Sud Lípez and Jujuy provinces of Bolivia and Argentina.

SUBAERIAL HOT SPRINGS AND GEYSERS

Hot springs are manifestations of hydrothermal activity associated with any area of anomalous heat flow and groundwater circulation, so that they are by no means confined to areas of active continental volcanism (hot springs are also discussed in Chapters 24 and 26). Even those springs that correlate spatially with the presence of recent volcanism usually occur far away from the positive volcanic edifices and are most common in calderas (as in Yellowstone); in fault-bounded structural depressions filled by permeable rocks (as in the Taupo zone, New Zealand), and in intensively faulted areas (as in western California). The hot springs activity usually postdates the cataclysmatic phase of volcanism and is related to the declining (waning) stage of volcanism. The hydrothermal activity persists long after the extinction of the active volcanoes, and particularly abundant hot springs are found in regions of late Tertiary volcanism (e.g. in the Basin and Range Province, western United States).

There is extensive literature dealing with many aspects of the hot springs activity, and the following works present convenient reviews from the volcanological (Williams and McBirney, 1979); geochemical (Ellis, 1977), and metallogenetic (Weissberg et al., 1979; White, 1981) points of view. Simple compositional classifications of hot springs usually distinguish (1) carbonate rich waters that dominate regions underlain by limestone and dolomite (e.g. Mammoth Hot Springs, Yellowstone); (2) acid sulphate waters, gradational into solfataras and fumaroles and rich in volatiles (B, CO_2, H_2S, NH_3, HCl, etc.), as well as metals (Hg, Pb, Bi, Sn, As, Ag, Au, Sb); (3) alkaline chloride waters, lower in volatiles but rich in dissolved Na, K, Ca, Mg, Cl, SO_4, HCO_3, CO_2. The category (2) is most closely associated with active calc-alkaline volcanism.

Several of the hot springs observed on the surface and in the shallow subsurface appear to be manifestations of the uppermost levels of the deeper-seated "subvolcanic" hydrothermal systems, believed to be actively depositing metallic ores at some depth. Some such springs are "spent" solutions that have already deposited their metal load to form epithermal deposits in the 300 to 2,500 m depth interval, and bear little resemblance to the "pregnant" solutions they were once while in depth. Other surficially discharged hot waters such as those in the Taupo zone, New Zealand (Weissberg, 1969; Ewers and Keays, 1977) still have high contents of metals, and are actively depositing Sb, As, Hg, Au, Ag, Tl. They appear to be in equilibrium with the hydrothermal fluids currently generating the epithermal showings of Zn, Pb and Cu in the Broadlands field, at a depth of 300 m or more. The sphalerite, galena, chalcopyrite, etc. mineralization at Broadlands was discovered by drilling (Browne, 1969) and is a unique example of an epithermal ore in the process of formation. The extensive though low-grade pyrite and minor Zn, Pb, Ag, Sb, etc. mineralization in silicified Pleistocene felsic volcanic breccia layers exposed near San Fernando, Ecuador (Goossens, 1972) appears to be similar although no longer actively forming.

Because epithermal mineralization formed at some depth, the epithermal deposits now exposed on the surface and mined (compare Chapter 26) required uplift and dissection to be exhumed. This required time so the bulk of known and mined epithermal deposits is of Tertiary and older age. Quaternary examples are exceptional. At only a few places in the world, particularly in the High Andes of northern Argentina (Nevados de Pastos Grandes; Sillitoe, 1975), Pleistocene stratovolcanoes have been dissected by glacial erosion shortly after their formation. The dissection exposed an unusually wide depth span of contemporary mineralizations that include surficially formed ores (e.g. manganiferous thermal springs aprons and sulphur accumulations), as well as epithermal Pb-Ag veins precipitated some 400-500 m beneath the original surface.

The recent hot springs contributed significantly to the understanding of hydrothermal ore genesis, but not all are the tops of metal-forming epithermal systems in depth. Many are just very shallow, convectively circulating meteoric waters. Despite their scientific interest, hot spring systems are unimportant as repositories of economically recoverable metals, at least at present (Table 5-5).

MINERALIZATION STYLES IN THE HOT SPRINGS ASSOCIATION

(1) Hydrothermal fluids (spring waters) as sources of recoverable metals

The common hot spring waters contain a variety of dissolved elements in low concentrations (see Table 1 in White, 1981), and small quantities of Na, K, Mg, Li, B, and other salts are recovered by many local industries using a simple evaporation process. Some springs show a significant increase in the usually low trace metal content, such as in Fe and Mn (160 and 43 ppm, respectively; Matsao, Taiwan); As and Sb (5.7 and 0.2 ppm; Broadlands, New Zealand); and Pb and Zn (0.36 and 3.2 ppm; Mendeleev Volcano). Hg, Tl, Mo, Bi, W, Ag, Au, Se, etc. traces have also been reported in many springs. Selective metal recovery is a future possibility, particularly if this is done as a by-product of energy harvesting in geothermal power plants. There does not seem to be any immediate interest, though.

Such interest, however, does exist in case of the highly metalliferous deep hydrothermal brines considered to be the parent fluids of epithermal deposits, when such brines are tapped by deep drilling. The Salton Sea brines, California, are so far the best known and a unique example. They contain 0.23% Fe, 0.14% Mn, 12 ppm As, 8 ppm Cu; 102 ppm Pb, 540 ppm Zn, 1.4 ppm Ag, 1.5 ppm Tl. The hot brine reservoir has been estimated by White (1981) to hold 11.6 cubic kilometres of brine containing 5 Mt Zn, 900 Tt Pb, 120 Tt As, 60 Tt Cu, 20 Tt Cd, 10 Tt Ag, and 1.4 Tt Tl.

(2) Metalliferous precipitates in boiling water pools and boiling mud

The scenery of a hydrothermal area located on the flanks of recently active volcanoes (e.g. Soufrière, St.Lucia) or in volcanic-tectonic depressions (e.g. Rotorua, Wairakei, etc., New Zealand), is characteristic. Bright white or beige ridges of hydrothermally argillized and silicified volcanics with yellow sulphur afflorescences alternate with gray boiling mud pools. Hot creeks move downslope and as they cool off, abundant green algal growths cover their floor. Clear, often boiling hot water pools dammed by sinter occur in depressions. Rusty limonitic infiltrations form by oxidation on sites where the hydrothermal activity has ceased.

The hot gray muds contain dispersed pyrite and troilite and locally have a high trace content of metals. Those at Steamboat Springs, Nevada, contain up to 15 ppm Au, 150 ppm Ag, 0.01% Hg, and 3.9% Sb (White, 1981). The cinnabar–mineralized "boiling kettles" in the Mendeleev Volcano hydrothermal area (Kurile Islands; Lebedev, 1970) probably correspond to the mud pools.

The clear water-filled pools deposit siliceous sinter. Darker (orange, brown, gray, black) coatings and streaks parallel with lamination often have high metal contents. A sinter collected from beneath 15 cm of water in the Frying Pan Lake, Waimangu, New Zealand

Table 5-5. Near surface Quaternary mineralizations related to hot springs in recently active volcanic areas

LOCALITY	STYLE*	HOSTS, BASEMENT ROCKS	MINERALIZATION, ALTERATION	REFERENCES
Steamboat Springs, Nevada	1,4,3	Q silic. sinters, andesite, T_1 granodiorite	cinnab. in silic. sinter, metastibnite Au, Ag, Sb, Hg high trace contents in siliceous muds	White (1968, 1981)
Coso Springs, California	4,5,3	Q silic. sinter, alluvium, resting on granitic batholith	cinnabar dissem. in silic. sinter, 1.8 Mt of low grade ore, about 8 t Hg produced	Ross and Yates (1943)
Sulphur Bank, California	3	Q andesite, landslide talus, over MZ ophiolite, sediments	cinnabar, sulphur, pyrite dissem., veinlets in argillized, zeolitized hosts, peneconc. mineraliz. zone; 7,000 t Hg	White and Roberson (1962)
Wilbur Springs, California	3	MZ ophiolite, sedim. melange, Q opalized rocks	cinnabar, metacinnabar, marcasite, opal impregnations, stockworks, veinlets in opalized and silica-carbonate altered hosts	White (1981)
Uncia, Bolivia	4	PZ slate, sandstone, capped by Q travertine	W bearing psilomelane-opal lenses in travertine; 1,000 t ore with 0.5% WO_3	Ahlfeld and Schneider - Scherbina (1964)
Ochaguí, Salta, N.W. Argentina	4	Quaternary sediments, travertine, Paleozoic basement rocks	psilomelane, pyrolusite impregnations in travertine and veins in the basement	Haude and Weber (1975)
Huachipato, etc., Altiplano, Chile and Bolivia	4,7	Q felsic tuff, ignimbrite, lake sediments	0.3-4 m thick tabular bodies of cryptomelane and hollandite in silicified volcaniclastics and sediments	Cruzat (1970)
Ngawha Springs, N. New Zealand	4,3,7	MZ sediments	cinnabar coatings, impregnations in hot spring throat, sediments, wood fragm.	White (1981)

Taupo zone, N.New Zealand; several localities (Broadlands, Waimangu)	1-4	T_3-Q rhyolite, ignimbrite, lake sedim. resting on Mesozoic graywacke	metastibnite, cinnabar as flocculant in hot spring pools; high Au in altered sinters; epithermal Pb, Zn, Cu in depth	Browne (1969) Weissberg (1969)
Uzon Caldera, Kamchatka, USSR	4,3	Plioc.- Pleist. bimodal basalt rhyol. asc., ignimbrite, tuff, lavas, gravels	0.3-0.5 m thick altered, silicified and mineral veined or impregnated horizon in unconsol. rocks contains realgar, stibnite, cinnabar, Au, Ag	Popov (1979) Kuznetsov (1974)
Burevestnik Volc., Iturup Isl., Kuriles, USSR	3	Quaternary island arc andesites	sulphur and molybdenite are disseminated and in veinlets in opalized andesites	Aver'yanov et al. (1968)
Mendeleev Volcano Kunashir Isl., Kuriles, USSR	6,3	Q basalt, dacite, andesite, rhyolite, diatomite	Fe sulphide precipitate on the floor of crater lake is veined by cinnabar, stibnite, realgar, etc.; opal, alunite cinnabar, etc. in veinlets and replacements in pyroclastics	Smirnov (1968) Popov (1979)
Gumma Mine, W.C. Honshu, Japan	5	Q terrace gravel, alluvium, sinters	goethite, jarosite form 2.2 km long layer overlying a stream terrace in a valley	Saito (1960)
Tatun volc. field (Matsao), Taiwan	3,1	Pleist.+ Recent andesites	galena and As, Sb, Ag minerals precipitate from hydrothermal brines in pipes; epithermal Fe, Cu sulphides in altered andesites at depth	White (1981)
Monte Amiata, Italy (partly)	8	Pleist. trachyte, Mesozoic carbonates	cinnabar, pyrite impregnations and replacements in residual clays, carbonates, rarely volcanics	Dessau (1977)

* see text for explanation of the mineralization style codes

(Weissberg, 1969), contained under 1.5% Sb, 400 ppm W and around 1 ppm Au. A one-centimetre-thick metalliferous layer in a siliceous sinter in the Champagne Pool, Waiotapu, New Zealand, (Fig. 5-5) contained 80 ppm Au, 175 ppm Ag, 2% As, and high Sb content. The Ohaki Pool, one of the hot spring-fed pools in the Broadlands geothermal field, New Zealand, produced an orange-red flocculent with over 10% Sb, 85 ppm Au, 500 ppm Ag, 0.2% Hg and 630 ppm Tl (Fig. 5-6). After deposition, it has been encrusted by an opaline sinter.

(3) Mineralizations formed by hydrothermal alteration and precipitation in the shallow subsurface

This is gradational into the epithermal mineralization treated in Chapter 26. The "true" hot spring precipitates form within a few tens of metres under the surface and are usually exposed in erosion-dissected zones of existing hot springs areas. The common host rocks are young volcanics (porous tuffs as well as fractured lavas); a variety of basement rocks; or fractured or leached earlier spring deposits.

In Lomitas near Laurani, Bolivia (Ahlfeld and Schneider-Scherbina, 1964), spring-deposited mimetesite in chalcedony gangue forms an irregular stockwork or a breccia cement in Tertiary dacite pyroclastics and lavas.

The Sulphur Bank, California (White and Roberson, 1962; Fig. 5-7), is considered to be the world's most productive Quaternary hot spring precipitated mineralization. There, cinnabar, sulphur, pyrite, marcasite, zeolites, montmorillonite and carbonates form disseminations, impregnations and masses in argillized and silicified Quaternary andesite flow, landslide rubble and talus unconformably resting on Mesozoic sediments and ophiolites of the Franciscan Supergroup. The ore zone is peneconcordant with the surface and the water table, and mineralization persists to a depth of 75 m. Sulphur Bank is just one deposit in the Geysers-Clear Lake geothermal area, that contains several steam and hot-water reservoirs in fractured Mesozoic graywacke, shale, chert, blueschist, serpentinite and metabasalt. It is notable that slabs of serpentinite constitute a cap rock to the system. The thermal area is believed to be heated and controlled by a silicic magma body, the top of which is within 7 km of the surface (McLaughlin, 1981; Fig.5-8).

Two moderate tonnage, complexly zoned sulphur-Fe sulphide-limonite deposits in Japan (Matsuo Mine, Honshu; Abuto Mine, Hokkaido; Takeuchi et al., 1966; Fig. 5-9), are bedding replacements in andesites formed in the zone of mixing of ascending hot springs and descending groundwater in the shallow subsurface. The deepest mineralization facies (up to 250 m under the surface) has sulphur and it changes upwards into pyrite and marcasite. Limonite formed under oxidizing conditions at a depth of about 50 m. The mineralization is almost symmetrically enveloped by alteration zones ranging from the most proximal kaolinization, montmorillonitization and saponitization to the most distal chloritization.

Fig. 5-5. Auriferous sinter in the Champagne Pool, Waiotapu, New Zealand.

Fig. 5-6. Dried-up Ohaki Pool, Broadlands geothermal field, New Zealand. Site of Sb, Au, Ag, Hg and Tl-rich flocculant in geysirite.

Fig. 5-7. Sulphur Bank mercury mine, California.

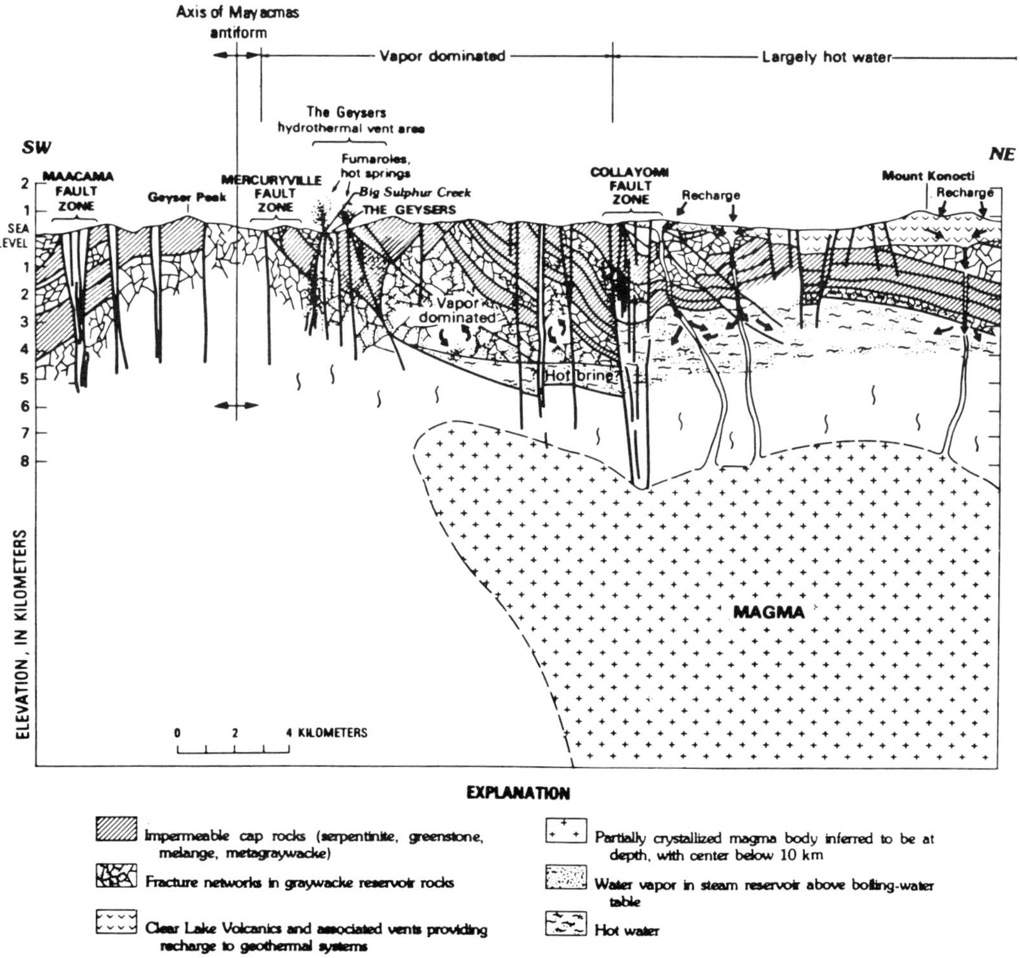

Fig. 5-8. The Geysers geothermal system, California: a structural model. From McLaughlin (1981).

Intensive alteration (argillization, silicification, pyritization) is commonplace in the vicinity of most acid and sulphurous hot springs, but rarely carries base-metal mineralization. The fault-controlled alteration zone on Savo Volcano, Solomon Islands (Taylor, 1976) is exceptional since it contains disseminated chalcopyrite (up to 1.33% Cu).

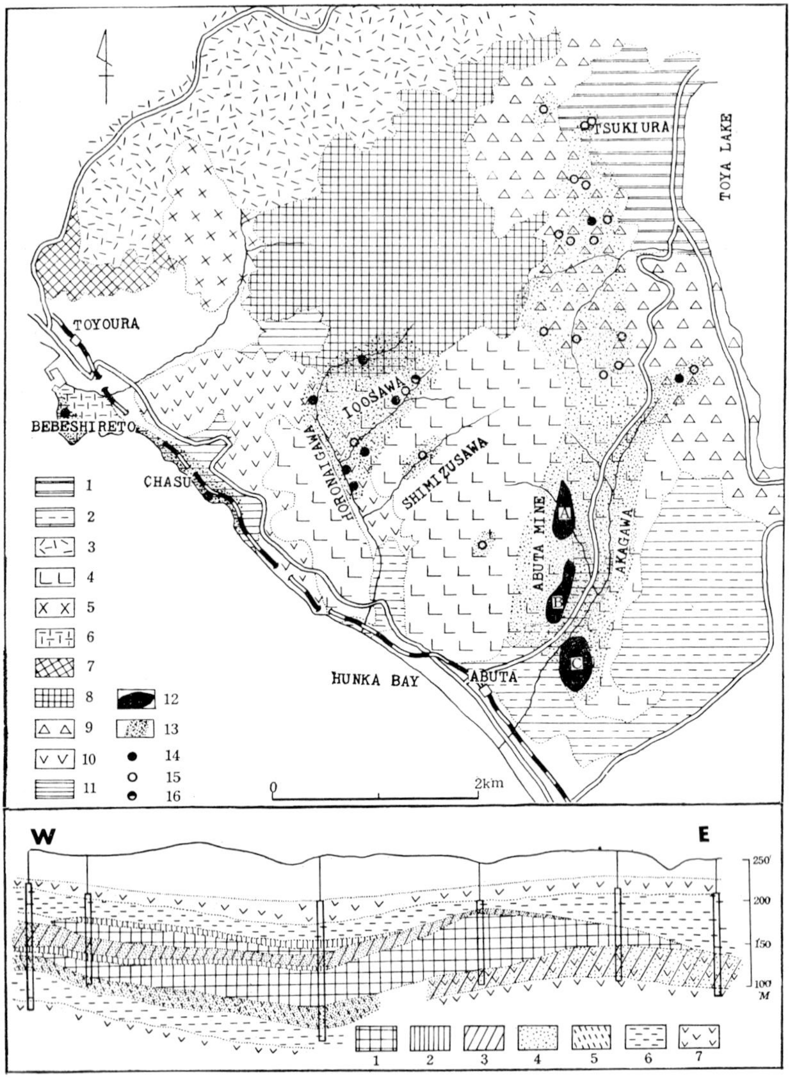

Fig. 5-9. Map and section of the Abuta mine area, Hokkaido, example of an economic zoned sulphur, pyrite, limonite hot springs replacement deposit in Quaternary andesites. EXPLANATIONS, MAP: (1-10) Pliocene to Pleistocene andesitic lavas and pyroclastics, designated by local formational names; (11) Miocene felsic volcanics; (12) Abuta deposit: (A) the second iron sulphide deposit; (B) the first iron sulphide deposit; (C) sulphur deposit; (13) altered zone; (14) iron sulphide ore zone; (15) limonite zone; (16) sulphur zone. SECTION: (1) Fe sulphide; (2) limonite; (3) silicified rock; (4) alunite alteration; (5) kaolinite alteration; (6) montmorillonite alteration; (7) Pliocene - Pleistocene hypersthene andesite. From Takeuchi et al. (1966), courtesy of Professor Takeuchi and Tohoku University, Sendai.

(4) Mineralization in surficial spring precipitates

The mineral matter dissolved in hot springs discharged on the surface soon precipitates to form mounds, terraces, platforms and similar bodies. Calcitic travertine is the most common precipitate, followed by opaline silica (geysirite) and Fe-Mn oxides. Ore minerals may be present in small quantities and when they are, they may form invisible, low-grade dispersions (e.g. Hg minerals in the Coso Hot Springs, California, opaline sinter); local disseminations (cinnabar and metastibnite, Steamboat Springs, Nevada); and coatings, impregnations, veinlets, replacements and breccia cement (e.g. sulphur and metastibnite, Steamboat Springs).

In the Uzon Caldera, Kamchatka (Kuznetsov, 1974; Popov, 1979), widespread As-Sb-Hg sulphide mineralization is in geysirite spring aprons; in altered proluvial sediments; and in Quaternary volcanics. The "ore horizon", a string of coalescing lenses, reaches from the surface to a depth of 7.5 m and is vertically zoned (from the surface downwards: siliceous sinter - orpiment and sulphur - realgar, stibnite, cinnabar - pyrite, marcasite).

Iron hydroxides are ubiquitous near orifices of many hot springs. Their economic importance as iron ores is limited, but the ferruginous gels may contain adsorbed base or precious metals. Taylor (1977) quoted 68 ppm Ag (a moderate Ag ore grade in itself!) in a gel from the Konjuku hot springs, Solomon Islands.

Manganese oxides are fairly common hot springs precipitates. Psilomelane sinter with 0.5% WO_3 accompanied by opal and barite forms lenses in a travertine, capping a hot spring near Uncia, Bolivia (Ahlfeld and Schneider-Scherbina, 1964). About 1,000 t of ore are present. A similar but larger deposit is found in Ochagui, northern Argentina (Haude and Weber, 1975). There, psilomelane and pyrolusite impregnates travertine and surrounding sediments, and forms veins in the basement. Numerous occurrences of Mn oxides in travertine mounds (e.g. Saline Valley, California; Rose Creek, Nevada; Abraham Hot Springs, Utah; Beef Hollow, Utah) have been recorded by Hewett et al. (1963).

(5) Alluvial precipitates from spring waters

The more mobile dissolved components in hot springs are carried for some distance from the site of discharge and precipitate in stream beds and terrace gravels and sands. The small Gumma deposit, west-central Honshu (Saito, 1960) is a layer of porous goethite and jarosite, overlying terrace gravel in a river valley in the vicinity of hot springs. Small deposits of limonite, jarosite and iron phosphates form a small valley fill near Ciater, Java. Mn oxide deposits of similar origin have been described at Lava Hot Springs, Idaho, by Hewett et al. (1963). A combination of (5) and (3) controlled a portion of the recent mineralization in the Coso Hot Springs, California (Ross and Yates, 1943). There, cinnabar is in alluvium that has been kaolinized and silicified by hot springs and "solfataric action". The occurrence of U- and Cu-impregnated peat interbedded with recent alluvial sediments near La Piadosa (east-south-east of Copiapó, Chile; Ruíz, 1965) is attributed to hot springs activity.

(6) Lacustrine hot springs precipitates

Hot spring waters discharged subaqueously on lake floors, or discharged subaerially but added to lakes, result in a variety of mineralizations. The less-mobile metals precipitate near the site of spring discharge in the form of a subaqueous lake floor ore lens, comparable with the submarine-hydrothermal orebodies. The lens can be underlain by altered and mineralized conduits in the bedrock. A Quaternary, subaqueous - hydrothermal deposit of massive pyrite and marcasite formed in the crater lake in the Mendeleev Volcano, Kurile Islands (Lebedev, 1970). There, a fine crystalline syngenetic iron sulphide lens is veined and partly replaced by a younger stage metasomatic pyrite and minor opal, sphalerite, cinnabar, metacinnabar, galena, stibnite, realgar and orpiment.

Less exotic examples of mineralization include the Fe and Mn oxide aprons, partly grading into lacustrine ores, reported in several places in the Basin and Range Province, Nevada (e.g. the Golconda deposit; see Chapter 25) and in the South American Altiplano. The Huachipato, Coronel Alcerreca and other deposits located in Chile in the Bolivia and Peru border region (Cruzat, 1970) are Mn oxide mantos in Pleistocene lake sediments that coincide with the region of former hot springs activity. There are over 1 million tons of 24% Mn ore, and the orebodies have been protected from erosion by capping of rhyolite ignimbrite and andesite, dacite and basalt lavas.

(7) Hot springs-limestone karst interaction

Continental volcanics and hot-spring systems interacting with bedrock limestones result in a "hydrothermal karst" that sometimes contains a metallic mineralization. The Monte Amiata Hg-Sb district in Tuscany, Italy, associated with a 0.43 m.y. old volcanic centre, may serve as an example. There, metamorphic basement is overlain by a deformed Mesozoic and Cainozoic carbonate, shale, sandstone sequence, subdivided by numerous unconformities and topped and intruded by Pleistocene rhyolite, quartz latite and trachyte. Numerous hot springs and CO_2 emissions exist in the area. In the largest mine near Abbadia San Salvatore, cinnabar with pyrite and marcasite occurs in a black residual clay and on fractures of karsted carbonates, under the impervious cover of an ignimbrite and a residual clay. Only a small quantity of cinnabar is hosted by the volcanics. The carbonate karsting is considered pre-volcanic, and many investigators consider the mercury mineralization to be hot springs generated, superimposed over the karst.

SUBMARINE-HYDROTHERMAL SYSTEMS IN REGIONS OF CONTINENTAL MARGIN VOLCANISM

Several occurrences of active, metal depositing hot springs located on the seafloor or in the intertidal zone, have recently been recorded. In contrast to the subaqueous hydrothermal systems in rifts (Chapter 30) and on the oceanic ridges (Chapter 4), no impressive metal accumulations have so far been discovered in the island arc setting.

Honnorez et al. (1973) described the contemporary formation of an subaqueous "exhalative" sulphide deposit in the nearshore area of the Vulcano Island (Aeolian Islands, S. Italy; Fig. 5-10). There, pyrite and marcasite precipitated from percolating fumaroles and hot springs in the porous permeable intermediate and felsic pyroclastics and in volcaniclastic sands that cover the submerged flank of a stratovolcano. Ferguson et al. (1974) described a bacteria- assisted precipitation of hydrotroilite, pyrite and marcasite in an intertidal thermal pool near Talasea, New Britain. The resulting mineralization was both conformable and crosscutting with respect to the unconsolidated host muds (Fig. 5-11). Butuzova (1969) studied, and described in detail, the contemporary precipitation of iron hydroxides and sulphides taking place on the seafloor in the submerged Thíra

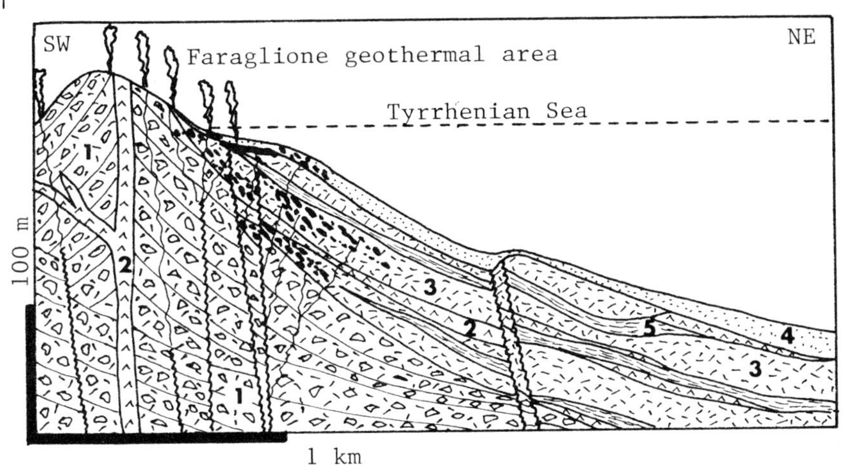

Fig. 5-10. Schematic section of the geothermal area, Vulcano Island, Italy, showing the Recent submarine pyrite- marcasite replacements in unconsolidated volcanics (in black); (1) Holocene trachyte pyroclastics; (2) trachyte lavas; (3) Recent porous tuffs; (4) volcanic sand; (5) volcanic siltstone, mudstone. Modified after Honnorez et al. (1973).

Fig. 5-11. Hydrothermal alteration in intertidal muds along a zone of hot springs. Talasea, New Britain, Niugini.

(Santorini) caldera in the Aegean Sea (Greece; Fig. 5-12). There, bright orange limonitic clays underlain by black and greenish iron sulphide-containing clays, form a stratiform horizon, 60-70 cm thick adjacent to seafloor springs and fumaroles. Siderite, vivianite, and iron silicates have been reported by Puchelt et al. (1973) from the same locality, but no enrichment in base metals has been found.

The hot springs subaqueously discharged on the floor of the breached and flooded Rabaul caldera (New Britain, Niugini; Ferguson and Lambert, 1972) are Fe- Mn- and Zn- enriched and mix with the indigenous bottom sediments to form reddish-brown silts. No sulphides have been detected. Ferruginous and manganiferous hot springs precipitates have been further described in the vicinity of the Simbo Island, Solomon Islands (Taylor, 1977).

WEATHERING-GENERATED ORES

Island arcs and the Andean-type margins carry their normal complement of weathering-generated ores (e.g. Ni laterites in the Indonesian islands; bauxite on Palau Island, etc.), that formed on pre-Quaternary rocks and are unrelated to the Recent development of the magmatic arcs. Such ores are described under different headings

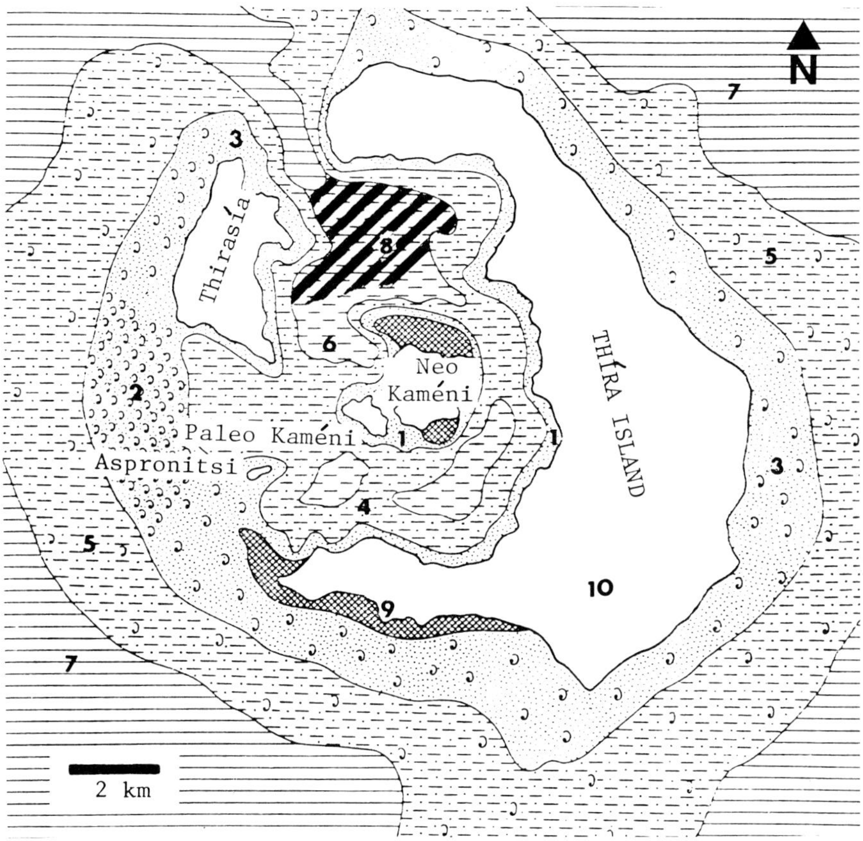

Fig. 5-12. Thíra (Santorin) caldera, Aegean Sea, Greece, a map showing Recent seafloor sediments and Fe-Mn precipitates from hot springs. After Butuzova (1969). (1) volcanic-terrigenous sands; skeletal sands with (2) 50-70% $CaCO_3$; (3) 10-30% $CaCO_3$; (4) coarse silt; (5) skeletal coarse silts (10-30% $CaCO_3$); (6) fine silt; (7) limey clay; (8) ferruginous and manganiferous clay; (9) subaqueous rocky outcrops; (10) subaerial islands.

in Chapters 10-14 and 26. Here, attention is paid only to the young weathering crusts formed on Quaternary rocks, and these are only exceptionally of economic importance.

Poor quality, immature bauxitic tropical weathering profiles containing nodular or clayous gibbsite are locally developed over Quaternary basalts and andesites in some tropical islands (e.g. Solomon Islands, New Britain). Arthurs (1979) recorded 12 small to medium size deposits of gibbsitic bauxite, recently formed on uplifted

Pleistocene reef limestones fringing or capping the volcanic islands in the Solomons. The bauxite probably formed by reconstitution of volcanic ash layers trapped in lagoons, floored by limestone. About 85 Mt of ore with 47-50% Al_2O_3 is available, of which about 30 Mt are on the Vaghena Island and 25 Mt on the Rennell Island. Politically, Rennell is a part of the Solomons, but it is an oceanic island situated outside the island arc. The bauxite there represents an unusual occurrence. It is a white to gray, often pyritic, sticky clay resting on the biogenic limestone floor in a brackish lake. The occurrence has been described by Taylor and Hughes (1975).

Tropical weathering-assisted leaching obviously played an important rôle in the formation of the Mn oxide deposits located at the interface of the now emergent, leached submarine andesite pyroclastics, and reef limestones, at Forari on the Efate Island (Vanuatu, formerly New Hebrides; Warden, 1970). The Forari deposit consists of todorokite and psilomelane impregnations, nodules and massive lenses accompanied by clay and peneconcordant with layering of the pyroclastics and with the limestone contact. The tonnage is about 1 Mt of ore with 42% Mn, and there is a similar deposit on the nearby Erromango Island (480 Tt Mn/30%).

SEDIMENTARY ORES

In regions with andesitic, explosive volcanism, magnetite phenocrysts present in ash and in semi-consolidated tuffs are easily concentrated by hydrodynamic sorting. They accumulate in streams and beaches in the form of placers (Fig. 5-13). Table 5-6 lists the major example localities of magnetite placers. Some of the deposits listed are probably continuous offshore (e.g. the New Zealand ironsands; Carter, 1980) but little information is available, so far.

Ilmenite is not a common accessory mineral in the explosive island arc volcanics, and instead is present in basalts, gabbros, alkaline rocks, metamorphics, etc. These are massive rocks that require weathering decomposition first, after which the resistate ilmenite present in regoliths can be reworked and possibly accumulated hydrodynamically in placers. Although minor ilmenite placers do occur in the island arc setting, they are not typical.

The marine sediments flanking island arcs rarely contain significant syngenetic-diagenetic metal enrichment, except in the vicinity of hot springs systems.

5.1.8. Marginal seas and back-arc basins

Modern island arcs are separated from the margin of a contiguous continent by a marginal sea, or from another island arc or an oceanic plate by a back-arc basin. Both the above terms are commonly used interchangeably. Western Pacific is the classical area of distribution of marginal and back-arc basins (Karig, 1971; Watts et al., 1977), but equivalent basins also occur in the Caribbean, in the

Fig. 5-13. Magnetite beach placer near Denpasar, Bali, Indonesia.

South Atlantic and in the Mediterranean Sea. Both basin categories are narrow, moderately deep, high heat-flow regions partly floored by oceanic lithosphere. They have been interpreted as (1) fragments of larger oceans trapped behind an island arc; (2) "oceanized" (basified) and foundered segments of the continental crust; or (3) basins developed by back-arc rifting, spreading and oceanic crust generation (a miniature equivalent of spreading along mid-ocean ridges). Mitchell and Garson (1981; p. 244-248) gave a more detailed genetic review.

The oceanic lithosphere usually floors the axial portions of the marginal or back-arc basins, but in most instances it is covered by young sediments. In the Sea of Japan, these are about 2 km thick (Melankholina and Kovylin, 1977). In the few instances where oceanic-type rocks have been recovered (e.g. from the East Scotia Sea; Saunders and Tarney et al., 1979; Lau Basin; Mariana Basin) they were found to be geochemically close to, or corresponding to, oceanic ridge basalts, and to be equally geochemically heterogeneous. Inasmuch as the present marginal seas are increasingly considered to be the depositional environment of many ancient ophiolites, they can be expected to carry the ophiolitic mineralization as well.

Table 5-6. Example localities of detrital magnetite reworked from Quaternary andesitic ash and tuff

LOCALITY	TYPE	TONNAGE	REFERENCES
Kurile Islands (Rucharskoe dep., Iturup Isl.)	B	7.2 Mt Fe/14.5% 630 Tt Ti/1.2%	Rundkvist ed. (1978)
S.Hokkaido, N.Honshu, S.E.Kyushu, Jappan	B,A	41 Mt Fe/56% (conc.) 2.88 Mt Ti/7.2%	Mero (1965)
Philippine Isl. (30 localities)	B	38 Mt Fe/50-60% (concentrate)	Wolff (1978)
Cilatap and Progo, S.Java Island, Indonesia	B	87 Mt Fe/12-57% 7 Mt Ti	Schmidt (1976)
Papuan Delta sands, S.E.New Guinea Island	B,A	32 Mt Fe/8%	Berkman (1975)
Solomon Islands (21 localities)	B	60 Tt Fe/60% (conc.)	Arthurs (1979)
W.Auckland and Taranaki Provinces, New Zealand	B	570 Mt Fe/12-20% 54 Mt Ti/2-5%)	Williams (1965)

ABBREVIATIONS, TYPE: B = beach placer
A = alluvial placer

The oceanic basalts are probably topped by a thin veneer of pelagic sediments in some marginal or back-arc basins, but the bulk of the sedimentary fill is of terrigenous or continental - pyroclastic derivation and supplied from both the continent and the island arc. Lithic sands, often turbiditic, form thick wedges along both sea margins. Mud, clay and biogenic carbonate cover the floor in the central part of the basin. Blocks and rafts of continental crust, submarine and island volcanoes, seafloor troughs and grabens, have been recorded from many marginal or back-arc basins.

Despite the general favourability of the marginal or back-arc basins for containing "oceanic-style " ores, few occurrences have been reported so far. Mitchell and Garson (1981) mentioned the Fe-Mn oxides dredged from the Mariana back-arc trough. Kobayashi and Nomura (1972) described the widespread occurrence of authigenic pyrite and magnetite formed under reducing conditions in the shallow subsurface in the organic-rich sediments in the Sea of Japan. Their piston samples came from water depths of 3,390 to 3,575 m, but the metallic minerals concentrations were just "ordinary".

Zelenov (in Schott, 1976) described the recent precipitation of iron hydroxides in the Okhotsk Sea. The iron compounds were supplied by a small stream draining the hot springs discharging on the Paramushir Island (Kuriles Islands). Over 10 Tt of suspended and dissolved iron are added per year.

5.1.9. Inland seas of the Black Sea-type

The Black Sea (Degens and Ross, eds., 1974) is an oval basin with a maximum water depth of 2,206 m, surrounded by two branches of the Alpine Orogenic Belt and by segments of stable platform. In terms of the geotectonic setting and crustal structure, the Black Sea has some points of similarity with marginal seas (it is positioned at the margin of the East European Platform behind the Anatolian magmatic and seismic belt) and it has an oceanic ? basalt floor topped by 8-12 km of sediment. It has also many points of dissimilarity, the chief one being its isolation and a lack of tectonic activity.

From the sedimentogenic point of view, the Black Sea is a prototype anoxic basin, having the present interface between anaerobic and aerobic water at a depth of 200 m. Bathymetrically, the Sea is usually subdivided into four major provinces (shelf, slope, basin apron or rise, abyssal plain). The bulk of the modern sediments are terrigenous sands, silts and muds, brought in by rivers from the surrounding landmass. The euxinic sediments in the Black Sea have been forming only for the past 9,000 years and are best developed on the abyssal plain where they take the form of black, microlaminated organic rich black clays topping the earlier lighter muds formed in a better aerated environment. The seawater at a depth of 1,500 to 2,000 m contains 5-6 ml H_2S/1.

The geochemistry of the Black Sea sediments has been treated in numerous, mostly Russian publications (e.g. Strakhov, 1962), and the sediments seem to have an ordinary, "continental" trace element distribution. The recent beach and coastal plain mineralizations are infrequent and predictable (e.g. the magnetite beach sands near Gagra, Georgia and elsewhere). Substantial quantities of Mn, Fe and V have been precipitated in the shelf zone of the Oligocene and Pliocene Black Sea precursor.

Geochemically, the most interesting recent sediments are the euxinic muds known to have a high U, P and possibly also V, Mo, Zn and Pb metal content. They contain widespread authigenic pyrite. Ferromanganese nodules forming in small amounts on the sediment / seawater interface have been described by Sevastyanov and Volkov (1966).

The southern (Trans-Caucasian) basin of the Caspian Sea believed to be underlain by oceanic crust topped by unconsolidated sediments, is similar to the Black Sea.

5.1.10. Submerged "quasi-continental" terrains

An extensive area in the south-western Pacific located west of the Kermadec-Tonga Trench and east of the Australian continental shelf, is composed of depressed "quasi-continental" plateaus and rises. These rises such as the Campbell Plateau and Chatham Rise (Katz, 1974) are composed of a continental crust of reduced thickness and are probably products of crustal stretching and continental breakup (possibly comparable with the Basin and Range Province) and oceanization. Similar terrains are known in the Mediterranean.

The geology is poorly understood; generally, blocks of peneplained sialic basement subdivided by ridges and troughs with oceanic rocks, are covered by Mesozoic and Cainozoic sediments. The youngest, unconsolidated seafloor sediments are pelagic and hemipelagic clays and oozes. Cainozoic volcanics are represented by oceanic and alkaline basalts and andesites. No ore occurrences, other than thinly scattered Fe-Mn nodules, have been reported so far.

5.2. ATLANTIC-TYPE CONTINENTAL MARGINS

5.2.1. Introduction

Atlantic-type continental margins are those that have experienced a relatively long period of stability (Heezen, 1974). The "typical" margins have passed through two fundamental evolutionary stages (Heezen, 1974; Sheridan, 1981): (1) Rift stage, that generated two opposing continental edges separated by a tectonic graben floored by thinned continental, and later oceanic, crust; and (2) seafloor spreading stage that involved the creation of the oceanic crust, drifting away of the continental edges, and subsidence of the truncated continental margin.

The first stage produced rugged variations in the basement structure presently buried in the subsurface of the passive margins and known from deep drilling and geophysical data, as well as a partly preserved "rift" association of rocks and possibly ores. The second evolutionary stage generated a thick wedge of terrigenous, biogenic (in warm climates) and minor chemogenic sediments thinning oceanward and prograding over the oceanic lithosphere.

The above model is based on the example of the Atlantic margin of North America, where the rift stage was active during the Triassic period and the seafloor spreading took place between the Jurassic and the present time. The model is applicable to many other continental margins throughout the world (e.g. the Atlantic margins of South America, western Europe, Africa; the Indian Ocean margins of Africa,

India, Australia, etc.). Association of sediments structurally, lithologically and metallogenically identical to products of the "seafloor spreading stage" of the eastern American margin, however, could have developed along any elevation gradient caused by faulting, uplift, downwarp, collapse, etc., formed along a continental edge or within a continent; for example along transform fault margins (e.g. Schlee, 1980). Lesser-order passive continental margins, moreover, are a part of a higher-order active (Pacific-type) continental margin, such as on the continental side of marginal seas.

In this book, the emphasis is on local environments and demonstrable lithologic associations and although the rifting and spreading model is accepted as a convenient framework, it is not implied that all the present passive continental margins have formed in the same way. Because of the considerable heterogeneity of the subenvironments, a summary section on trace metal geochemistry and metallogenesis has not been included and the subject is discussed in the context of each subenvironment. Figure 5-14 and Table 5-7 list and show the principal mineralization styles.

5.2.2. Rift-stage generated rock association under passive continental margins

The buried continental / oceanic boundary under the Atlantic-type margins can be best imagined as one half of the present Red Sea profile: a series of step-like, oceanward thinning fault blocks of continental lithosphere ultimately terminating against the continuous oceanic crust (compare Chapter 30). In the actual situation, however, numerous variations of the basement structure have been detected, the most notable of which involves wide tracts of mixed (transitional) crust in which pervasive vertical oceanic mafic dikes invade continental rocks, or where horizontal oceanic sills are densely interlayered with the rift sediments (e.g. Sheridan, 1974, 1981; particularly his Fig. 15).

The rift tectonism produced a rugged basement topography with numerous elongated fault basins filled by continental clastics (typically in the red beds facies), followed upwards by evaporites and marine sediments deposited in a gradually deepening sea. The elongated basins in the subsurface of the North American Atlantic shelf contain up to 5 kilometres of upper Triassic to lower Jurassic red beds associated with basalt and diabase, and masked by thick salt deposits (Schlee, 1980; Fig. 5-15). Comparable rocks are subaerially exposed, for example, in the Newark Graben. A similar situation exists along the West African coast (Schlee, 1980); off Brazil (Asmus and Porto, 1980); off Western Europe (Naylor and Mounteney, 1975) and elsewhere.

The evaporites topping the red beds and volcanics have usually flowed plastically after loading to form diapirs. Discontinuous carbonate banks and reef complexes also formed during the early phase of continental separation and subsidence, and many later acted as

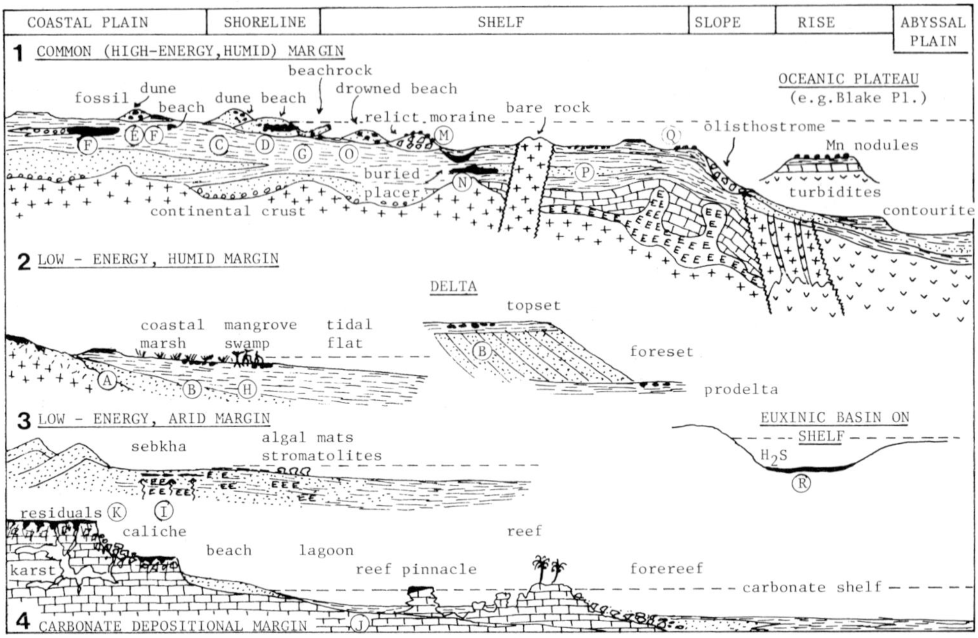

Fig. 5-14. Atlantic-type continental margins, Recent environments and principal Quaternary mineralization styles (see Table 5-7 for explanation of the letter codes). NOTE: distances and thicknesses are grossly telescoped and distorted.

barriers to the terrigenous sediment preventing its dispersion into the deep sea. The rift-stage generated lithologic association is metallogenically by far the most interesting and variegated divisions of the Atlantic-type margins and its subaerially outcropping equivalents are discussed in Chapters 25 and 30. No metallic mineralizations have been reported from its buried equivalents.

5.2.3. Sediments generated in the coastal plain, shelf, slope environments of the passive continental margins

GENERAL

Passive continental margins are the classical regions of sedimentogenic and stratigraphic studies on which a voluminous (and often highly specialized) literature has accumulated (Reading, ed., 1978; Reineck and Singh, 1975). Among the numerous frameworks of

Table 5-7. Atlantic-type continental margins: principal Quaternary mineralization styles

HUMID COASTAL PLAINS AND DELTAS			
(A)	bauxite and high - alumina clays	X	X Mt Al
(B)	bog Fe,Mn ores in swamps and marshes	?	---
DUNES, BEACHES AND BARRIER ISLANDS			
(C)	heavy mineral concentrations in active dunes	X0	see below
(D)	heavy mineral concentrations in active beaches	X00	see below
(E)	heavy mineral concentrations in fossil dunes	X0	see below
(F)	heavy mineral concentrations in fossil beaches	X00	see below

TONNAGES FOR (C-F)						
	ILMENITE	RUTILE	ZIRCON	MONAZITE	GOLD	CASSITERITE
(C)	minor quantities, included in (D)					
(D)	108.5 Mt	33 Mt	4.52 Mt	3.55 Mt	30 t	X0 Tt Sn
(E)	3 Mt	1 Mt	500 Tt	20 Tt	?	?
(F)	112.5 Mt	11.6 Mt	28.3 Mt	770 Tt	180 t	X0 Tt Sn
TOT.	225 Mt	45.6 Mt	33.32 Mt	4.34 Mt	210 t	X00 Tt Sn

MISCELLANEOUS ENVIRONMENTS			
(G)	heavy minerals accumulated in beachrocks	?	?
(H)	metals dispersed or adsorbed on clays in coastal marshes and swamps	X 1 1	100 Tt Ni ? 300 t Cu X0 t Au
(I)	arid coastal salt flats (sebkhas)	?	Fe,Mn,Cu
(J)	bottom sediments of backreef lagoons (possible)	?	no data
(K)	residual metalliferous enrichment over sub-aerially exposed carbonates (e.g.bauxite)	-	---
(L)	guano / limestone interaction (low U phosphate)	X	X00 t U
(M)	detrital gold in relict sediments (sands,gravels) on shelf floors	X0	10 t Au ?
(N)	cassiterite in buried relict elluvial / alluvial placers	X0	3 Mt Sn ?
(O)	heavy minerals in drowned beaches on the shelf	X	53 Mt ilm. 4.5 Mt zr. 1.5 Mt rt.
(P)	Fe-Mn oxide nodules on shelf floor, Mn-rich bottom sediments (muds)	X0	?
(Q)	nodules and encrustations of phosphorite,barite chamosite on shelf floors (low U in phoshorite)	X0	150 Tt U
(R)	U-V-Mo enriched euxinic muds	X	---

description, the most durable seems to be the simple bathymetric progression which has the following zones (environments) between a continent and an abyssal ocean floor: coastal plains, deltas and shorelines, continental shelf, continental slope, continental rise (Fig. 5-14).

Coastal plains are regions of low land composed of fluvial and marine deposits presently above the sea level which, in the context of geological history, have been the site of repeated marine transgressions and regressions triggered by subsidence and uplift or drop and rise in the sea level of the order of several metres. Their width varies from several kilometres to several hundreds of kilometres. The plains are interrupted by the lower reaches of rivers and are gradational into deltas. Coastal plains may be missing along the mountainous and rugged coasts.

Shorelines have a variety of appearances ranging from rocky shores through high-energy beaches to tidal flat marshes and lagoons. The continental shelf, a gently-sloping shallow portion of the seafloor with a smooth relief, varies from 30 to over 300 km in width and changes into a continental slope along a conspicuous gradient break called "slope break" (Heezen, 1974). The depth of the shelf break often varies with latitude, from under 100 m to 600 m. The continental slope is substantially steeper and interrupted by submarine canyons. At the foot of the continental slope the continental rise starts and this is often the last mass of sediments of continental derivation resting directly on the oceanic crust of the abyssal plain. The continental rise sedimentation is dominated by both turbidity and contour currents. Numerous local variations frequently disturb the regularity of the profile, as presented above.

Many other factors in addition to the bathymetric zonation, influence the nature of the resulting rock association. Climate is of paramount importance. It not only controls the presence or absence of carbonates or evaporites, but also influences the type of detritus and the chemical components in solution supplied from continent to the world ocean. The rock forming components derived from basaltic terrains undergoing erosion in the glaciated Arctic will essentially be a fresh basaltic detritus in a wide range of clast sizes. Basalts exposed in the humid tropics, on the other hand, will contribute reworked lateritic clays only, and an irregular supply of chalcedonic silica from the lower portions of dissected tropical regoliths. Basinal restriction, upwelling currents, volcanic supply (often from volcanoes thousands of kilometres away, via ash clouds drifting in the atmosphere or pumice floating on the surface of the sea), are additional factors influencing sedimentogenesis along the continental margins.

In comparison to the high degree of variability of the rock-forming processes active in environments that include magmatism or elevated temperatures (at least above 150° C), the sedimentogenesis at passive continental margins largely devoid of the more "exotic" processes tends to produce petrographically and geochemically monotonous suites. The bulk of sedimentary rocks are simple in terms of mineralogical

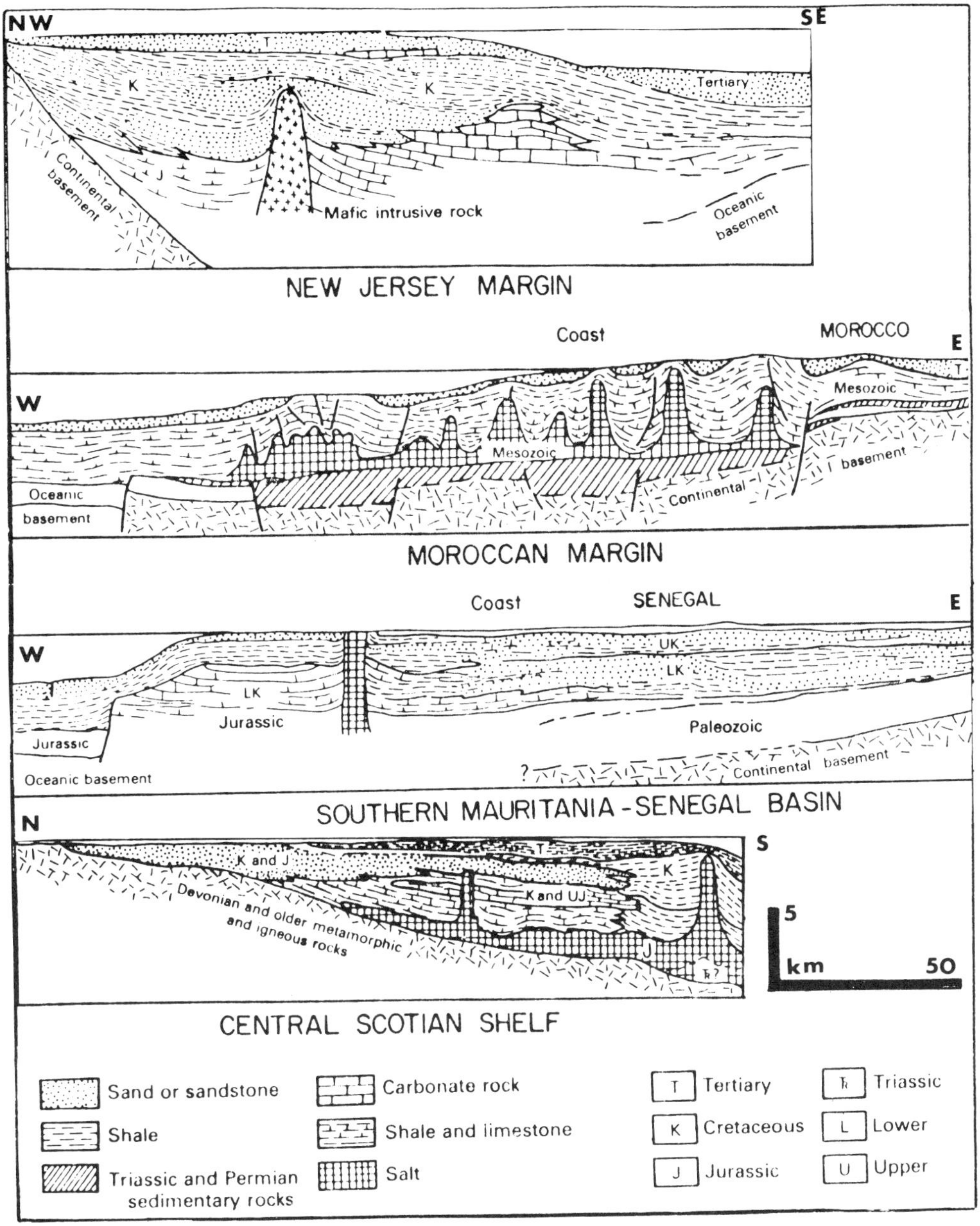

Fig. 5-15. Schematic sections of examples of the present Atlantic-type continental margins. A compilation from Schlee (1980).

and chemical composition; their complexity is textural. A series of sedimentary rocks of increasing grain size: claystone - shale - siltstone - sandstone - conglomerate, may have an almost identical chemical composition and provenance. Because metallic mineralization typically correlates with compositional contrasts and reversals in the "common rocks" suites, metallic deposits do not form frequently during the stages of the early sedimentogenesis (syngenesis). Those infrequent syngenetic ores that are known are invariably associated with an anomalous metal source in the earlier (source) rocks (e.g. cassiterite beach placers); with an endogenic metal source (the subaqueous – hydrothermal deposits) or, alternatively, represent "technological ores" having a low to extremely low factor of metal concentration.

The latter case is demonstrated by the rutile, ilmenite and zircon-producing beach and dune sands, Stradbrook Island (Queensland, Australia). The cut-off grade of this "ore" is 1% of the heavy minerals which represents about 0.5% TiO_2. This is lower than the TiO_2 content of most igneous rocks, but the detrital minerals present in the unconsolidated dune sand located in an almost uninhabited area, can be profitably recovered using the advanced technology.

The recent literature treating the ores in sediments is often overly optimistic and emphasizes certain processes, associations, structures, etc., observable in the recent environments, as possibly responsible for the generation of ores present in the ancient lithologic associations. One example, of many, is the widespread belief in the role of biogenic processes as agents of formation of the stratiform base metal sulphide deposits, based on the known facts that bacteria, algae, microorganisms, etc. generate sulphur compounds including H_2S as well as small amounts of pyrite. Trudinger (1981) recently concluded that there were no known examples of purely biogenic base-metal sulphide deposits primarilly because of the lack of anomalous, concentrated sources of the constituent metals.

In the text that follows, contemporary (Quaternary) depositional environments recognized and often meticulously documented by sedimentologists, will be reviewed briefly. Many, if not most, are completely devoid of metallic occurrences (e.g. the present carbonate reef environments), but their ancient equivalents are often repetitively associated with certain mineralization styles (e.g. the ancient "reef facies" frequently hosting the Zn-Pb deposits of the "Mississippi Valley-type"). Such mineralizations have mostly been interpreted recently as diagenetic or post-lithification epigenetic. Repeated association of certain ancient ores with certain ancient rock types (facies),even if interpreted as post-depositional or epigenetic, however, can also serve as a reminder that the Recent sediment equivalents might be in the process of ore (or protore) generation and could have been overlooked. In such a case, the past could serve as a key to the present, a reversed principle of uniformitarianism.

HUMID COASTAL PLAINS

Coastal plains are generally composed of extensive fluvial deposits intercalated with marine sediments representing sea transgressions. The "most normal", broad and thick coastal plains and their sequences represented by the Gulf Plain in the Southern United States, are located in the lowermost reaches of long, voluminous streams (Mississippi in the above example), so that most of the sediments deposited are fine grained and brought in as a suspended load. Sand accumulates only in the bottom of the river channels (Reineck and Singh, 1975). Swamps are commonly present and peat occurs frequently in floodbasins. The detritus derived from large basins and distant source rocks is evolutionarily mature, thoroughly mixed and compositionally "neutral".

Narrow coastal plains adjacent to mountains and fed or crossed by numerous short streams (e.g. the discontinuous Pacific coastal plain of British Columbia) contain granulometrically (gravel to clay) and compositionally a much broader range of sediments. The sediments often reflect faithfully the source area of their detritus.

From the economic point of view, coastal plains are more important as an environment of formation and hosting of ancient mineral deposits (e.g. the bauxite of the Guyanas; coal in paralic basins) than as a present ore-accumulating environment. Only minor occurrences of low-quality bauxite formed during the Quaternary period on the Suriname coastal plain (Aleva, 1979). The high-alumina (30-34% Al_2O_3) plastic clays composed of halloysite, montmorillonite, illite and minor gibbsite that seem to be rather common worldwide in the poorly drained tropical coastal plains, could become bauxite substitutes in the future.

Many present coastal plains contain "fossil beaches"; that is, raised former beach tracts. These are becoming increasingly more important as a low-grade source of several heavy minerals, and are discussed later.

DELTAS

Deltas are transitional into coastal plains, fluvial environments and clastic shorelines. The simpliest generalizations emphasize the coarsening upward delta profile, starting with the fine-grained, gently dipping prodelta and bottomset, overlain by a moderately dipping sand and gravel foreset, which, in turn, is topped by a topset of essentially flat-lying sands and gravels (Fig. 5-14). Modern research (Elliott, 1978a; Morgan and Shaver, eds., 1970) established the important variation in the structure and lithology of deltas, and four or six end-member models have been introduced.

Briefly, high-constructive deltas exemplified by the Mississippi River delta, are subject to minimum interference from the sea, into which they advance. Delta plains dissected by numerous distributary channels pass directly into the delta front. The reported indications of metallic mineralizations are mostly chemical or biochemical precipitates initiated during sedimentation and later diagenetically fixed; e.g. Fe hydroxides and pelosiderite nodules. They are genetically equivalent to the "bog ores" treated in Chapter 24.

The high – destructive deltas are influenced by waves, that redistribute most of the sediment supplied to the delta front. The beach shoreline is thus prominent and the most common mineralizations are the heavy mineral beach sands (e. g. the beaches adjacent to the Nile delta; Nakhla, 1958; see also the next section).

A significant portion of the ancient coal deposits associated with the deltaic sequences is believed to be related to delta abandonement and overgrowing by vegetation.

BEACHES AND BARRIER ISLANDS

Beaches are long and narrow sand (and occasionally gravel) accumulations parallel to the shoreline and attached to the land. Most beaches have an associated dune ridge or a dune field. Barrier islands are equivalent to beaches, but are separated from the land by a shallow lagoon and are frequently dissected by tidal inlets (Komar, 1976). The "most normal" beaches (of the kind treated in the sedimentogenic literature) are composed of epiclastic mineral and rock clasts, derived by weathering and fragmentation of the continental (and rarely oceanic) crust. In the tropical belt, terrigenous beaches usually contain a significant proportion of carbonate skeletal particles and grade to purely carbonate beaches. These have been reviewed recently by Inden and Moore (1983). Beaches located in areas of active explosive volcanism consist of reworked pyroclastic material and have been discussed earlier.

Terrigenous beaches and barrier islands require, in order to form, (1) a steady supply of sand to the shoreline either directly by stream input or by longshore drift; (2) a hydrodynamic setting of variable wave energy and moderate-to-low tidal range; and (3) a moderately stable coastal plain with a low gradient (Elliott, 1978b). Interplay of these factors results in a typical beach profile, shown in Fig. 5-16. There are numerous variations of this profile, both permanent and temporary. The temporary variations result from destruction of the original full profile by cyclonic storms, and it usually takes several months to rebuild.

Beaches and bars are composed of the most weathering and abrasion resistant particles locally available, and beach sediments are among the best sorted and rounded sediments known. Quartz and siliceous fragments are by far the most common epiclastic beach constituents, followed by feldspars, quartz - feldspar intergrowths ("aplites"), fine crystalline volcanics, and a variety of contact metamorphic hornfelses. These common "light" clasts (specific weight under 2.8) contain a heavy mineral fraction the normal proportion of which is usually of the order of 0.03 to 0.05% by weight. Heightened wave

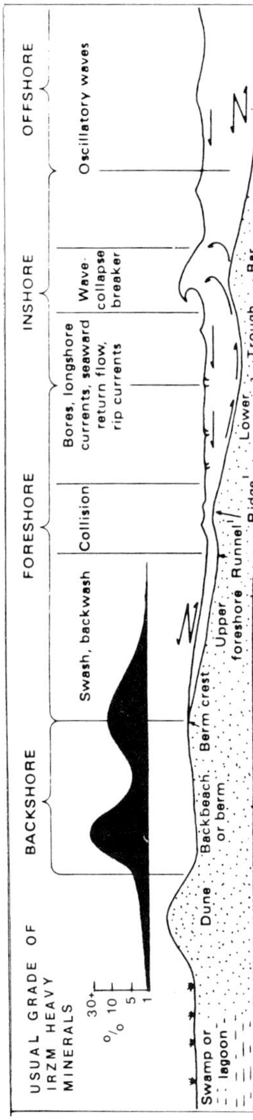

[1] Ridge-and-runnel system may be replaced by low-tide terrace or by planar slope. At low tide, swash uprush and backwash occur on seaward edge of foreshore.

[2] Step is present at lower edge of foreshore on low-energy sand beaches and on most pebbly beaches.

Facies	Characteristics
Swamp or lagoon	Burrowed lagoonal silt and clay; rich in organic material; disrupted laminations; small scale low angle trough and planar cross-lamination; ripple marks; oyster and coquinoid lime deposits
Dune	Very fine sand[1]; trough cross-laminations; vertical root traces; laminated to structureless
Backbeach	Fine sand; low-angle landward, seaward, or horizontal, dipping wedge planar cross-lamination; channels common; nonfossiliferous
Berm crest	Fine sand; convex laminae; low- to moderate- to high-angle planar cross-lamination; nonfossiliferous
Upper foreshore	Fine sand; low-angle seaward-dipping wedge planar cross lamination; rare micro-cross-lamination; burrows present but not abundant
Ridge and runnel	Fine sand; locally coarser and bimodal on ridge; high-angle cross-lamination; trough cross lamination; burrows uncommon
Lower foreshore	Fine sand; low-angle wedge-planar cross lamination dipping seaward; burrows common. Coarse sand at lower edge of foreshore
Trough	Fine sand, high-angle trough and planar cross-lamination; shell fragments
Bar	Locally coarser sand; high-angle cross-lamination in some examples. Sand in trough may be coarser than on bar
Offshore	Very fine sand, thin bedded; undulatory wave ripples; plants and animals in growth position; structureless where bioturbated

[1] Grain size may vary in all facies because it is, in part, dependent on material supplied.

Fig. 5-16. Common subdivisions (sub-environments) in a beach profile. After Ingle, from Houston and Murphy (1977); heavy minerals grade curve added.

activity generated by squalls or storms at sea is capable of partial separation and sorting of the light and heavy fraction, so that local accumulations of heavy minerals within a beach may result. Similar, although substantially less efficient, separation takes place as a result of wind action (winnowing) in dunes.

Waves generated by low intensity winds tend to produce enrichment in the heavy minerals on the foreshore. There, the minerals accumulate in thin laminae on the relatively steep slope of the foreshore. These accumulations are rarely permanent and are repeatedly reworked. The preserved foreshore heavy mineral accumulations (for example, along regressive coasts), tend to be low-grade (2-5% of heavies).

Waves of high and exceptional intensity carry the heavy minerals to and over the berm crest. Such crests are frequently draped by accumulations of heavy minerals having an anticlinal cross-section, and several sets of truncated and redeposited mineral sands are usually present. The most concentrated bodies of heavy minerals (up to 70% of the heavy fraction) are formed by particularly large waves on the backshore at the dune foot. There, they appear as seams up to 50 cm thick (e.g. the conspicuous black seams composed of opaque minerals), or flat lenses. The grade rapidly drops as the orebody continues into the foredune.

On the active Australian beaches and dunes (Gardner, 1955), the heavy mineral bodies located in the upper beach are some 5-20 m wide and may extend along the entire length of a beach (Fig. 5-17). In a cross-section, the orebodies are wedge-shaped, tapering towards the foreshore and dipping seaward at a low angle.

Fig. 5-17. Heavy mineral (mostly ilmenite) black beach sand near Gladstone, Queensland.

The "fossil beaches" are partly denudation-flattened, but still usually well recognizable, remnants of beach tracts along ancient shorelines situated on the present coastal plains. They have the form of a low ridge and up to nine subparallel ancient shorelines have been recognized along the Atlantic Seaboard of the United States, reaching up to an altitude of 33 m above the present sea level (Pirkle et al., 1974). In north-eastern Florida, the heavy mineral orebodies in the inland ridges formed at the height of marine transgression and occur parallel with the main fossil ridges.

In eastern Australia, the "high dunes" (Fig. 5-18), located some 6-8 km inland from the present position of the shoreline are prominent and contain low-grade heavy mineral deposits. Fossil dunes along the belt of raised beaches in south-eastern Madagascar (de la Roche et al., 1956) are vegetated, weathering-modified, and contain a large volume of sand with up to 25-30% of heavy minerals, 1.2-1.5% of which is monazite. Table 5-8 gives examples of dimensions, grade and tonnage of selected heavy mineral orebodies.

The sand of the fossil beaches and dunes could be loose, semiconsolidated or almost consolidated into a hardpan by infiltrated cement and matrix. In Florida and Australia, the fossil beach orebodies are often covered by peat and the peat material cements the uppermost layers of the mineralized sand. The "coffee rock" of the Australian miners is a hard, rather massive goethite-cemented sand up

Fig. 5-18. A "high dune" along the N.W. coast of Stradbrook Island, Queensland, is dredged for its low-grade content of ilmenite and rutile.

Table 6-2. Dimensions, composition and tonnage of selected ilmenite, zircon, rutile and monazite heavy minerals placer orebodies

LOCALITY	STYLE	OREBODY SHAPE	DIMENSIONS (m)	GRADE, TONNAGE	MINERALS %	REFERENCES
Starke Orebody, Trail Ridge, Florida	FB	lens	4800x2000x11	200 Mt 4%	45% ilmenite, leucoxene 20% staurolite 15% zircon	Overstreet (1967)
Jacksonville South northern Florida	AB AD	lens	600x22x8	35 Mt 0.75%	55% ilmenite, leucoxene 20% zircon 6% rutile 2% monazite	Overstreet (1967)
South Capel deposit, Geographe Bay Western Australia	FB FD	several lenses	14500x600x5 (average)	80 Mt 5-30% 10.88 Mt heavies	84% ilmenite, leucoxene 9% zircon 1% rutile 0.5% monazite	Welch et al. (1975)
Rosetta sand spit Nile Delta, Egypt	AB	single lens	2500x430x?	X00 Mt X-90%	55% ilmenite 18% magnetite 8% zircon 1.5% rutile 0.5% monazite	Nakhla (1958)

STYLE ABBREVIATIONS: AB=active beach; AD=active dune
FB=fossil beach; FD=fossil dune

to 1 m thick, marking successive strand lines at Eneabba (Lissiman and Oxenford, 1975), and having usually a high heavy metal content. A few Cainozoic heavy mineral sands are partly or completely buried under recent alluvial or windblown sand, or laterite (e.g. Eneabba, Australia).

VARIETIES AND ECONOMIC IMPORTANCE OF BEACH AND DUNE HEAVY MINERAL DEPOSITS

Magnetite beach placers of pyroclastic derivation, chromite placers of ultramafic derivation, cassiterite placers of granite derivation, and others, have already been discussed and tabulated in Chapters 4, 5, 7, 28 and 29. All the above placers formed in close proximity to their rather distinct bedrock sources, some of which had been ore deposits themselves; that is, occurrences of an already concentrated ore substance. Most of the placers treated in this section display a much looser source relationship; in most cases a variety of source rocks contributed to the detritus and in some cases the source cannot be pinpointed. The hydrodynamic processes along beaches are thus the principal factor responsible for the mineralization.

Ilmenite, rutile, zircon and monazite (IRZM) placers

These are the most widespread type of beach placers and, at present, the principal source of titania, Zr, Th, REE, etc., for the industry (Fig. 5-19). The first three minerals are widespread although in terms of volume, usually insignificant accessory minerals of a variety of crystalline rocks, from which they are released by deep weathering and stream reworking. Monazite is more restricted. Although accessory monazite is a rather common mineral in granites, pegmatites, gneisses, etc., its content increases sharply in the rocks metamorphosed in the granulite facies, granulites and charnockites. Overstreet (1967) reviewed the worldwide distribution of monazite; its correlation with high-grade metamorphic terrains is overwhelming.

Where strong longshore currents are active, the heavy minerals are transported down the coast and deposited hundreds of kilometres from their original point of entry into the sea. The Florida placers generated from the Appalachian detritus are up to about 500 km away from their bedrock source, and much of the Nile delta detritus was brought from a distance of several thousands kilometres.

The heavy minerals usually form rounded to subrounded grains within the size range 1/16 mm to about 1 mm. Ilmenite is usually the dominant mineral and a portion of it is usually altered to leucoxene (by iron removal). Zircon, rutile, monazite, kyanite, sillimanite, staurolite, tourmaline, monazite, etc., are additional accessories recovered in the heavy mineral concentrate. The semi-stable heavy minerals, such as garnet, epidote and magnetite, are usually solution removed.

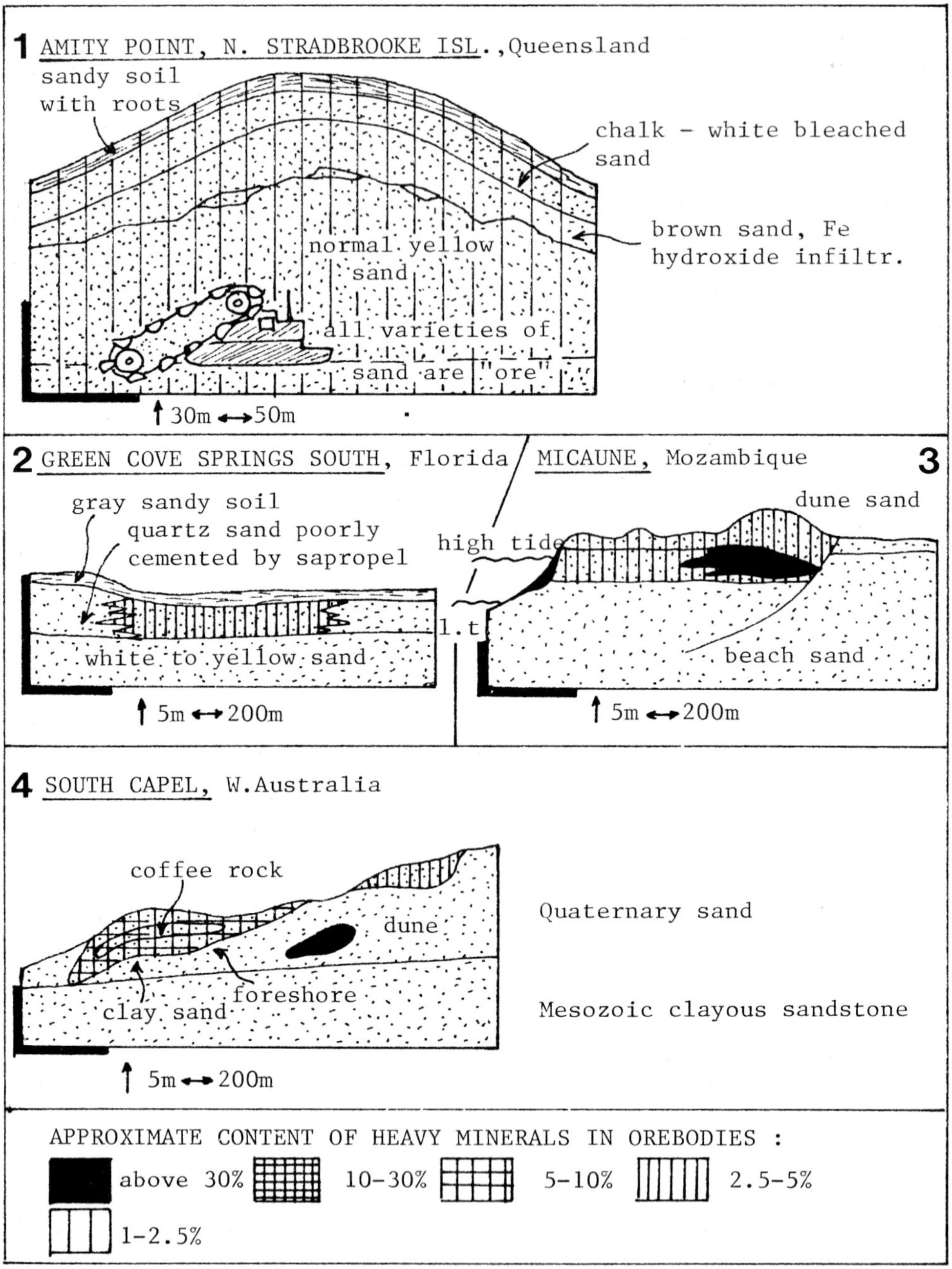

Fig. 5-19. Examples of ilmenite, rutile, zircon, monazite beach placers. (1,2) from LITHOTHEQUE, based on visits; (3) after Jaritz et al. (1977); (4) after Welch et al. (1975).

In the early days of small-scale mining, small layers and patches of high-grade, visibly conspicuous sands (black or brown sands) were being recovered. Such sands contained from about 40 to 90% of the "heavies", in the case of the black sands mostly ilmenite. At present, however, the bulk of IRZM concentrate produced in industrialized countries, comes from low-grade (3-5% of heavies) to very low-grade (0.5-2% heavies) deposits in which the "ore" is visually unrecognizable from the "waste". This is because of the extremely cheap recovery using the modern technology. Table 5-9 lists the principal beach placers of the ilmenite, rutile, zircon and monazite suite of the world.

Multistage placer origin

It has been known for some time, and strenghtened by recent research, that many modern beach placers were formed partly or entirely by the reworking of fossil placers, both unconsolidated and consolidated. The reworking usually resulted in heavy mineral enrichment in the youngest generation of placers, and loosening of consolidated placers that made the industrial recovery cheaper. Several localities of heavy mineral placers of multistage origin are characterized briefly in Table 5-10. This feature is important for exploration, because the fossil placers may provide additional heavy mineral resources once the recent placers are exhausted.

Leonardos (1974) described an interesting case of multistage origin of the monazite-bearing placers found along the Atlantic coast of Brazil. The original source of the monazite is believed to be the Archean migmatite-charnockite terrain. From there, a portion of the monazite was reworked into late Precambrian quartzites, now also metamorphosed in the granulite facies. Both the Precambrian sources then supplied monazite into the Recent Atlantic beach placers either directly, or via a brief residence in the Tertiary coastal plain sandstones.

Gold beach placers

Compared with the alluvial gold placers (Chapter 24), marine beach gold placers have been, so far, economically insignificant, but their importance may increase in the future as the alluvial placers are rapidly becoming exhausted. Because detrital gold, coarse enough to accumulate by hydrodynamic processes and recoverable by the present technology, does not travel far from its bedrock source, the placers have to be source-proximal. This requirement, however, has to be modified in the glaciated terrains in which gold could be carried for a considerable distance in dispersed form in the moraine, without undergoing sorting and further comminution (compare Chapter 24). The hydrodynamic sorting, gold winnowing and concentration then takes place at the location at which the moraine came to rest, along streams and beaches.

Table 6-3. Cainozoic unconsolidated ilmenite, rutile, zircon and monazite-producing beach and dune sands: selected localities

LOCALITY	AGE	STYLE	PROBABLE SOURCE	MINERALOGY, TONNAGE, GRADE	REFERENCES
S.E.Georgia - N.E. Florida recent beach placers (centre: Jacksonville)	Q	AB AD	Pt_3-PZ metamorphics in the Appalachians (schist, gneiss, amphibolite, gabbro)	4.5 Mt ilmenite and leucoxene 1.0 Mt zircon 110 Tt monazite 0.5-5% heavies in quartz sand	Pirkle et al. (1974) Overstreet (1967)
N.E.Florida fossil beach placers (Trail Ridge, Boulougne, Green Cove Springs	Q_1 Mi_3-Pl	FB	as above	35 Mt ilmenite and leucoxene 6 Mt zircon 200 Tt monazite 1.5-5% heavies in quartz sand	as above
Atlantic beaches of Brazil and Uruguay	Q	AB AD	Precambrian high grade metam.of Brazilian Shield,quartzite, Tert.sandstone	2 Mt ilmenite, 500 Tt zircon, 100 Tt monazite (minimum) in quartz sands	Leonardos (1974) Overstreet (1967)
Senegal and Gambia Atlantic coast, Africa	Q	AB AD	Precambrian metamorphics	7.7 Mt ilmenite 800 Tt zircon	de Kun (1965)
Sherbo deposit, Sierra Leone,Africa	Q	AB	Precambrian metamorphics,laterites	30 Mt rutile	de Kun (1965)
Nile Delta, Abu Quir to Port Said, Egypt	Q	AB	crystall.rocks in Soudan,Ethiopia,Egypt	possible resources 540 Mt ilmenite, 16.2 Mt zircon 5.4 Mt monazite	Higazy and Naguib (1958)
Namaqualand dune sands (W.S.W. of Garies), S.Africa	Q	AD	Precambrian high grade metamorphics (granulites)	300 Tt monazite	von Backström (1976)

South African coast (Zululand, Natal, Transkei)	Q	AB FB FD	Precambrian metamorphics, Karoo basalts	68 Mt ilmenite 5 Mt zircon 2.8 Mt rutile	Hammerbeck, ed. (1976)
Mozambique coast (Micaune, Pebane)	Q	AB AD FB	high grade metamorphics, Precambrian	22 Mt ilmenite 2.4 Mt zircon, 600 Tt rutile 300 Tt monazite	Jaritz et al., (1977)
S.E. Madagascar coast (Mananara River to Fort Dauphin)	Q	FB FD AB	Precambrian granulite charnockite terrain	2 Mt ilmenite 100 Tt monazite 100 Tt zircon	Chantraine and Radelli (1970)
Kerala coast (centre Quilon), S. India	Q	AB	Precambrian granulite, charnockite terrain	300 Mt ilmenite 10 Mt zircon, 2.9 Mt monazite	Overstreet (1967)
Sri Lanka beaches (Pulmoddai, Induruwa)	Q	AB	Precambrian high grade metamorphics	2.5 Mt ilmenite 600 Tt rutile 20 Tt monazite	Overstreet (1967)
Pacific coast of Queensland, N.S.W., E. Australia	T_3-Q	AB FB FD	Tr quartz sandstone, PZ + MZ metamorphics, T volcanics	18 Mt ilmenite, 9 Mt rutile 8 Mt zircon 60 Tt monazite	McKellar (1975)
W. Australia coast north of Perth (Eneabba, Cataby, Gingin)	T_3-Q	FB FD	MZ detrital sedim. derived from Precambr. high-grade metam.	18 Mt ilmenite and leucoxene 9.1 Mt zircon 2.6 Mt rutile 260 Tt monazite	Lissiman and Oxenford (1975)
W. Australia coast south of Perth (Geographe Bay, Capel)	Q	FB	Archean metamorphics, Mesozoic sediments	25 Mt ilmenite 2.2 Mt zircon 160 Tt monazite	Welch et al. (1975)

ABBREVIATIONS OF STYLE: AB= active beach AD= active dune
FB= fossil beach FD= fossil dune

Table 6-4. Selected examples of multistage placer origin

LOCALITY	MINERALS	PRESUMED ORIGINAL MINERAL SOURCE	INTERMEDIATE SOURCE (PALEOPLACER)	ECONOMIC IMPORTANCE OF INTERMEDIATE SOURCE	REFERENCES
Atlantic Coast of Georgia and Florida	ilm, lcx, zircon, rtl, mnz	metamorphics in the Piedmont and Blue Ridge provinces, Appalachians	Cr_3 and Eo sandstone	minor	Overstreet (1967) Pirkle et al. (1974)
Pacific coast of New South Wales Australia	as above	crystalline rocks of the Tasman orogenic belt (PZ)	Tr Hawkesbury quartz sandstone	none at present	McKellar (1975)
Western Australia (Eneabba, Geographe Bay)	as above	Precambrian high grade metamorphics	Mesozoic platformic sandstones	minor	Lissiman and Oxenford (1975)
Kerala beaches, southern India	as above	Precambrian charnockite - granulite terrain	Tertiary coastal plain sandstones	minor	Overstreet (1967)
Atlantic coast of Brazil	as above (mainly monazite)	Archean granulite-charnockite; 2 b.y. metaquartzite	Tertiary quartz sandstones of the coastal plain	minor to moderate	Leonardos (1974)

The coast of Alaska has the largest number of marine gold placers (Popof Island, Nome, Yakataga), followed by the Sea of Okhotsk coast (N.E. USSR; Denisov, 1979); Oregon and California coasts (Moore and Silver, 1968); Queen Charlotte Islands, Canada; Nova Scotia; Bermagui, New South Wales (Noakes and Jones, 1975); and Westland and Nelson Provinces, New Zealand (Williams, 1965).

At most localities, the detrital gold occurs in the present beach, in the same position as the IRZM heavy minerals; that is, in the upper foreshore and backshore. Gravel, rather than sand, however, is the major host sediment. Locally, drowned (submarine) and raised (subaerial) or buried beaches carry the gold. At Nome, Alaska, twelve such auriferous beaches have developed over a horizontal section of about 17 km across the coastal plain and the adjacent shelf, having a vertical relief of about 80 m (Moffit, 1913; Nelson and Hopkins, 1972). The description that follows is based partly on the summary given by Boyle (1979; p. 371-373).

The bedrock at Nome is composed of deformed Paleozoic schists, graphitic quartzites and slates, phyllites, greenstones and marbles intruded by Mesozoic granitic rocks. Numerous quartz, carbonate, sulphide veins and small lenses that carry gold are regionally widespread, but few have been of much importance economically. The bulk of the 170 t of gold produced in the Nome area came from alluvial and beach placers, the beach placers having the lion's share. The 12 auriferous beaches near Nome represent succesful periods of late Tertiary and Quaternary coastal uplift (Fig. 5-20). In the modern beach, fine—grained gold associated with magnetite, garnet, limonite, hematite, scheelite and pyrite, was concentrated on clay or mica-rich false bottoms. The paystreaks were discontinuous lenses of stratified quartz, granite and other pebbles, 83 to 100 m wide and 5 to 50 cm thick.

Five raised beaches lie inland from the modern beach, some under the cover of glacial drift and organic-rich muck. Some gravels are frozen (permafrost). The paystreaks were located more frequently on false bottoms than on the bedrock, and were comparable with the modern beach placers except for the frequent presence of arsenopyrite and a coarser gold.

The three offshore beaches developed during standstills between the numerous Quaternary transgressions and regressions, and are auriferous, but their gold content has not been disclosed. In addition to the beaches representing distinct and linear strand lines, gold in the Nome area is widespread in thin relict gravel blankets, described below.

Cassiterite beach placers

The bulk of the world's placer cassiterite is in elluvial, colluvial and proximal alluvial deposits (Chapter 24). Marine beach placers are rare and account for no more than 1% of the world's placer cassiterite production. Most are the result of reworking of earlier continental placers or stanniferous regoliths that involved very

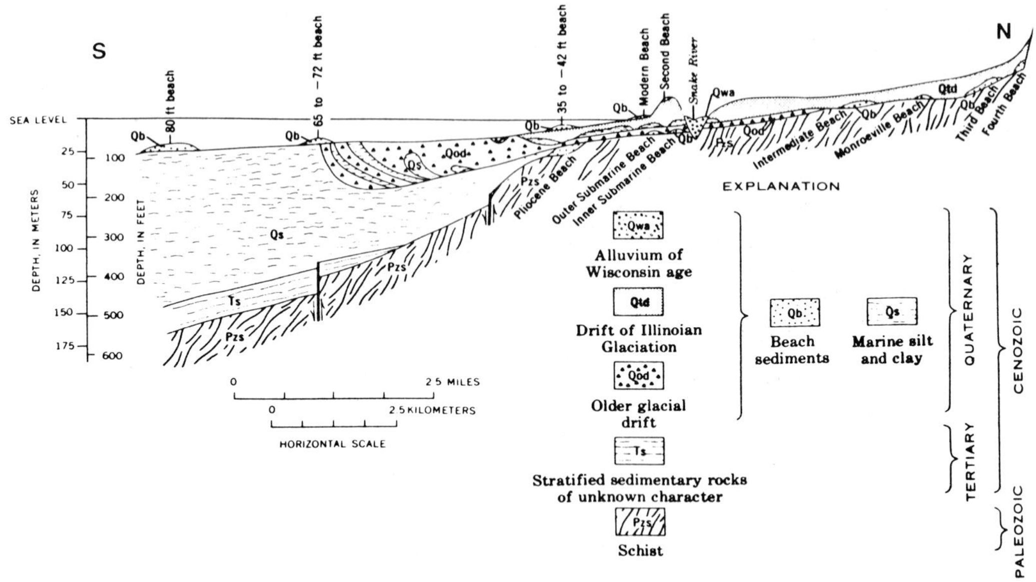

Fig. 5-20. Geological section across the coastal plain area at Nome, western Alaska, showing the modern beach and the former beaches (containing auriferous gravel) in the offshore and on land. From Nelson and Hopkins (1972).

little cassiterite transportation. Several beach placers occur along the west coast of Malaya, for example near Dindings in Perak (Batchelor, 1979). There, the placers formed along relatively stationary strand lines and are quite rich, but thin.

Hematite granule and sand beaches

In New Caledonia, thick ferricrete capping (cuirasse) tops lateritically weathered ultramafics in a high plateau and is being slowly dissected and eroded. Blocks of the cemented ferricrete and even-sized hematite granules (microconcretions, buckshot) from the underlying zone of "red laterite" move by gravity and by periodic streams into valleys and ultimately onto the seashore. At several places along the eastern New Caledonia coast, beaches composed of well sorted, well rounded hematite granules have formed. A narrow beach south-east of Monéo (Figures 5-21 and 5-22) is composed of a body of pure hematite granule sand about 6.5 km long, 10-30 m wide and 0-3 m

GEOLOGICAL MAP , NORTH - WEST OF MONÉO , N.C.

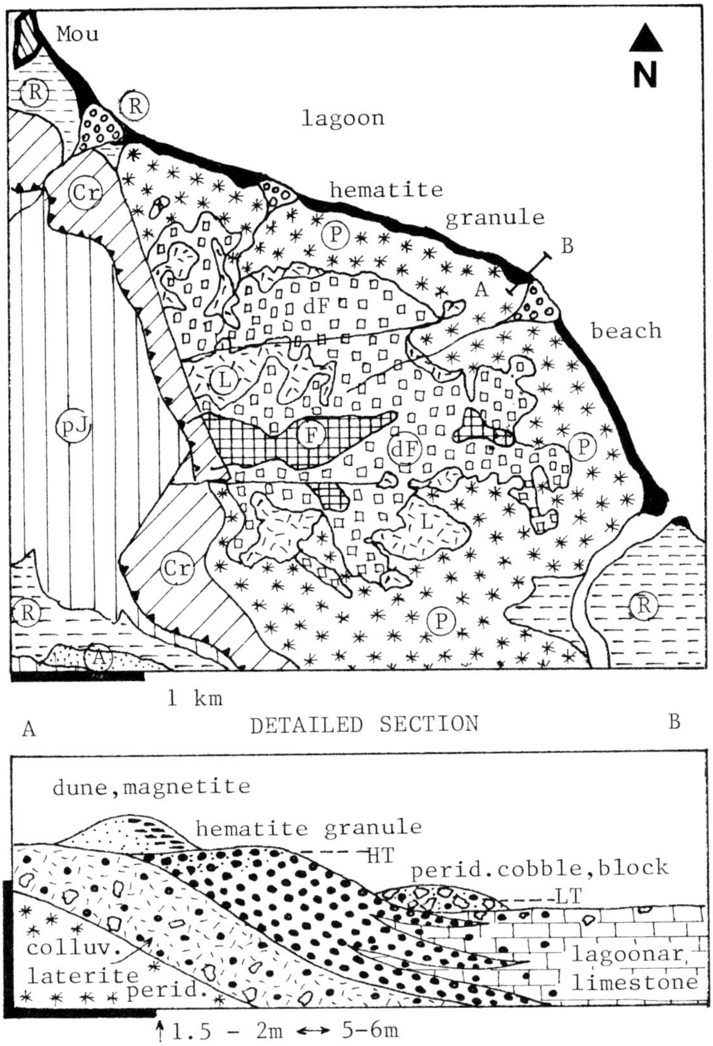

Fig. 5-21. Hematite granule beach near Monéo, eastern New Caledonia. The geological map is slightly modified from Lozes et al. (1977), the cross-section is from LITHOTHEQUE, based on 1981 fieldwork. LEGEND: (A) alluvial deposits; (Cr) upper Cretaceous sandstone, shale; (dF) dissected cuirasse (F) ferricrust (cuirasse), intact; (L) yellow laterite; (P) Oligocene-emplaced peridotite; (pJ) Jurassic and earlier volcanic - sedimentary association; (R) Recent littoral and estuarine sediments.

thick, that represents several million tons of about 50% Fe ore. Loose hematite accumulation is best developed in the upper foreshore and lower backshore and it contains lenses of carbonate - cemented, hematite beachrock. The upper backshore and the foot of a low, vegetated dune ridge, is enriched in fine magnetite and chromite sand. These minerals are resistates released from the decomposed ultramafics.

The low-energy beach faces a lagoon, and the hematite sand progrades across a cemented bioclastic carbonate beachrock pavement in the lower foreshore. Fans of leached and silicified peridotite cobbles and ferricrete blocks mark the point of entry of small periodic streams onto the lagoon.

Beachrocks

In the tropics, portions of the foreshore (intertidal zone) often undergo penecontemporaneous cementation to produce massive slabs of beachrock (e.g. Kaye, 1959). The cementation is selective and favours gravelly portions of beaches, particularly near the mouths of youthful streams entering the shoreline and bringing in pebbles and cobbles. As a consequence many beachrocks are polymictic conglomerates with aragonitic, calcitic or rarely Fe-hydroxide cement.

During storms the slabs of beachrock are undercut by wave action, they collapse, and are repeatedly reworked. Beachrock formation does not appear to be associated with any particular metallic mineralization, but it can conceivably cause an early cementation of loose detrital ore fragments thus protecting them from dispersal. Beachrocks along the northern Jamaica coast often contain streaks of detrital magnetite-ilmenite sand; beachrocks along the Pacific coast of New Caledonia contain blocks of ferricrete and hematite granules; beachrocks east of Santiago de Cuba contain abundant magnetite blocks brought in from skarn deposits developed in the Gran Piedra and Sierra Maestra massifs. No economic ore deposits associated with beachrocks are known.

HUMID, CLASTIC TIDAL FLATS, MARSHES AND LAGOONS

Tidal flats are extensive low-energy environments, dominated by fine-grained sediments often having a high proportion of organic materials (Reineck and Singh, 1975; Ginsburg, ed., 1975; Elliott, 1978b). Lagoons are shallow depressions behind barrier islands that have a restricted communication with the open sea, but in other

---

Fig. 5-22. Derivation of hematite granule beach sand near Monéo, New Caledonia. (A) dissected cuirasse of lateritized peridotite; (B) clay and granule colluvium; (C,D) hematite beach sand.

respects are comparable with and transitional into tidal flats. Coastal marshes are located in the supratidal zone of tidal flats, so they are influenced to a considerable degree by the local climate; usually they are densely vegetated.

Salt marshes have a fairly slow sedimentation rate. The sediments are exceptionally organic rich and bound by plants. During spring tides and storms a marsh is flooded by the seawater, which is soon diluted by rainwater and fresh groundwater. The fresh- and salt-water interaction is a process considered favourable, under certain conditions, for metal precipitation. Certain ancient copper deposits are believed to have formed in such a way, and brackish coastal marshes are one of the environments in which coal is believed to have formed.

Evidence for metal accumulation in modern humid salt marshes is, however, still rare. Swanson et al. (1972) studied the geochemistry of tidal marsh sediments in north-western Florida and along the Gulf of Mexico coast and found neutral, or only slightly increased, content of trace metals accumulated with the organic carbon. Although the trace element content is bound to increase with the degree of compaction and diagenesis after burial, it was still far from anomalous. It appears that an anomalous source or supply of the trace metals is needed to cause a marked metal accumulation in the salt marsh environment.

The Tantramar Swamp near Sackville, New Brunswick, described by Fraser (1961), is an example of a Recent environment located in the temperate climatic zone and transitional between the freshwater peat bogs (Chapter 24) and coastal marshes, anomalously enriched in copper. Up to 10% (dry weight) of megascopically invisible copper is adsorbed in peat to form an orebody about 1 m thick, containing about 300 t Cu. The copper is brought in by groundwaters containing from 0.005 to 1 ppm Cu, derived from the Pennsylvanian red-beds known to contain small quantities of copper. Boyle (1979) reported up to 2.3 ppm Au and 5 ppm Ag in the peat.

Lelong et al. (1976) gave an example of an anomalous Fe and Ni enrichment in the tropical coastal marshes and mangrove swamps along the coast of New Caledonia at sites where streams and seepages draining the interior "Ni-laterite" region enter the shore. Nickel-containing goethite brought in suspension into the swamp reacted with the dissolved silica to generate neoformed nickeliferous nontronite. The nontronite deposits contain locally 2-3% Ni, and up to 5% exists in horizons particularly rich in organic matter.

An unusually low-grade but extensive accumulation of finely dispersed (colloidal) gold is known to exist in the mangrove swamp muds between Tenasserim and Mergui, lower Burma (Boyle, 1979, p. 85). The swamp sediments located in the South-East Asian Tin Belt average 0.285 ppm Au and there is presumably a significant gold resource, but the gold is not recoverable by the techniques presently available.

ARID, LOW-ENERGY SHORELINES WITH EVAPORITES

Low-energy coastal environments in climates where evaporation greatly exceeds precipitation (Reineck and Singh, 1975; Till, 1978), are different in many respects when compared with the humid environment treated in the previous section: (1) surficial fresh water is almost non-existent and groundwater limited, so higher vegetation is sparse or even missing. Consequently, the humid salt-water marsh is substituted by sebkha, a barren salt flat located in the supratidal zone. (2) Blue-green algae are the most widespread organic form, and they actively assist in the sediment binding. In addition to that, they also generate a special diagenetic microenvironment during the early diagenesis. (3) Restriction of sedimentary basins and the high rate of evaporation results in hypersalinity, and evaporite minerals crystallize both on the surface from ephemeral brine pools, and in the subsurface from highly concentrated saline pore waters drawn towards the surface. (4) The lack of surficial water means a low clastic input, although this is often offset by an abundant aeolian sediment, blown into the sea from the adjacent desert.

Along the Persian Gulf coast of south-eastern Arabia which is the most thoroughly studied arid seacoast used in many sedimentogenic models (e.g. Purser, ed., 1973), the supratidal zone adjacent to a sand or rocky desert is represented by sebkha composed of a red-brown quartz sand partially or fully cemented by anhydrite, topped by gypsum and an algal mat-rich sand, and sometimes covered by an ephemeral halite crust. Halite, dolomite, celestite, huntite, and magnesite are other cementing minerals, that are occasionally present. In the upper intertidal zone, sedimentogenesis is dominated by the abundance of algal mats or Recent stromatolites, that generate a thinly laminated sediment composed of alternating algal tissue and trapped sand. This sediment is being destroyed locally by burrowing and by diagenetic precipitation of gypsum crystals. Aragonitic mud, black fetid mud, faecal pellets and bioclastic sediments of lagoons grade into the subtidal facies of gray muddy sands, having a high proportion of biogenic and possibly chemically precipitated carbonate.

Understanding the present arid shorelines is essential for the interpretation of the ancient sedimentary sequences that are members of, or are facially transitional into, the red-beds and/or evaporite association. Although the thick, relatively pure saline deposits ("saline giants") are of little interest to the geologist looking for metallic ores, the alternating detrital / evaporite and/or carbonate sections and particularly some rock members located along facies contrasts (e.g. the Kupferschiefer black marl, the basal member of a marine sequence transgressive over continental red—beds), are of considerable significance. They contain distinct and economically important Cu, Zn, Pb and less significant Ag, Au, Mo, V, U, Pt, etc. accumulations described in Chapters 19 to 21. Several models to explain the ancient mineralization by means of the recent equivalents have been offered in the literature, and the Renfro (1974) sebkha model has been most widely applied.

Despite the important mineralization in what is believed to be the ancient equivalent of the present arid coasts, no significant ore accumulation and only a few minor geochemical anomalies have been found in the modern environments. There may be two reasons: (a) the Kupferschiefer and similar ores formed diagenetically at some depth, so that one cannot compare them with the present depositional equivalent exposed on the surface or in the shallow subsurface; (b) the majority of the recent sedimentogenic models are based on conditions that exist along the Persian Gulf coast, which is an area underlain by monotonous and metallogenically sterile Mesozoic and Cainozoic platformic sediments, with sub-clarke contents of trace metals. There probably was insufficient metal to be supplied to an environment otherwise very favourable for the entrapment and precipitation of the ore minerals.

The point (b) seems to be supported by the fact that the known localities of at least small metal enrichments in the recent arid coastal environments are located outside south-eastern Arabia, in environments sedimentologically less impressive, but adjacent to metallogenically more productive "backyards". Ferguson and Burne (1981) studied the interaction of continental groundwaters with the recent nearshore sediments along the Spencer Gulf of South Australia. Although the Gulf has a climate substantially less arid than south-eastern Arabia, it is flanked by the Flinders Ranges, a mobile belt containing numerous small base metal occurrences. In Fischerman Bay, a groundwater discharges on a tidal flat, and precipitates a lens of a manganiferous iron oxide (46.5% Fe_2O_3; 3-10% Mn), that has high trace Co (0.16-0.2%) and Ni (0.023-0.05%) contents. At Wood Point, ephemeral iron sulphides precipitate in the supratidal zone (a sebkha equivalent) from groundwater in the summer, when bacterial reduction of the gypsiferous sediments is active. In the winter, the sulphide is destroyed by the oxidizing flood waters.

SHALLOW-WATER CARBONATE ENVIRONMENTS

Carbonate depositing environments have been extensively reviewed in the literature (Bathurst, 1975; Scholle et al., 1983), and their study has became a highly specialized and frequently also (by virtue of location in the tropical seas) a romantic and pleasant activity. By far the most spectacular recent carbonate-depositing environment is the reef. Its uniqueness has been summarized by James (1983) as follows: "unlike other sedimentary deposits, a reef is not entirely the product of physical sedimentation but is almost entirely the physical expression of a community of organisms, growing in one place for an extended period of time". These organisms (corals, algae, benthos) leave an impressive record after their death in the form of carbonate buildups. The numerous cavities that remain within the organic carbonate framework are gradually filled by debris, so that the resulting lithosome is essentially a carbonate megabreccia.

The central tract of a barrier reef is flanked on the open seaside by the fore reef facies, a limestone talus, and the back reef facies on the land-facing side. The back reef facies contains low-energy sediments deposited in a sheltered lagoon of predominantly carbonate mud, sand and small patch reefs.

In an extensive carbonate - depositing system, such as the Bahamas (Newell and Rigby, 1957) a variety of subenvironments exists, including low to high-energy beaches where a sorted carbonate sand and gravel accumulates; sand dunes composed of carbonate aeolianite (McKee and Ward, 1983); carbonate deltas, tidal flats and marshes and others. Carbonate cliffs and terraces are exposed along the rising carbonate coasts, undergoing subaerial diagenesis, dissolution and degradation. Esteban and Klappa (1983) distinguished two facies that develop as a product of carbonate subaerial exposure: caliche (the rubbly carbonate weathering crust) and the subterranean karst.

The main disappointment facing a metalliferous geologist, however, is the fact that the "most normal" modern shallow water carbonate environments contain virtually no ore minerals, and give little geochemical indication of possible metal-forming processes in action. Yet the fossil carbonates are important hosts to metallic ores. This is because the pure carbonates acted almost exclusively as reservoir rather than source rocks during the diagenetic and post-lithification processes. This theme is discussed in more detail elsewhere (Chapter 20). If metals do accumulate in the modern carbonate environments, they do so in other than the "most normal" settings and the following sites or processes should be watched because of their highest metal-accumulating potential:

(1) Organic-rich carbonate muds in the back reef lagoons. There is a theoretical possibility that the high trace metal content in the soft tissues and skeletons of some marine organisms e.g. oysters (discussed in Chapter 3) may locally become preserved and ultimately converted diagenetically into sulphides in a suitably reducing, sulphur-rich environment. Alternatively, trace metals can be supplied to the lagoon from the continent.

(2) The seawater / freshwater reaction, evaporation, precipitation, etc. at the arid continental and marine interface as in the seepage refluxion dolomitization model of Adams and Rhodes (1960) and similar models. Bush (1970) demonstrated the generation of chloride-rich brines in the sebkha environment, compositionally similar to the brines found in fluid inclusions in the "Mississippi Valley Zn-Pb type". Unfortunately, the metals to form actual ores were missing in the region (Abu Dhabi) Bush studied.

(3) Diagenetic conversion of aragonite and high-Mg calcite (the unstable $CaCO_3$ polymorphs) into the low-Mg calcite (the stable polymorph). This is widespread in the Quaternary carbonates, particularly those subaerially exposed. The conversion is known to release Mg and Sr and possibly other trace elements such as Pb and Zn present in the originally skeletal or micritic carbonate, and set them on a path of migration.

(4) Residual enrichment of a metalliferous insoluble residuum on the surfaces of carbonate dissolution. The red- or brown-coloured residual soil composed of alumina hydrosilicates and gibbsite is commonplace in the tropically and subtropically weathered carbonate terrains, including the Quaternary ones (e.g. on the Florida Keys). Further upgrading of the soil may produce bauxite but demonstrably carbonate-derived Quaternary bauxite has not yet been found.
(5) Areas of interacting volcanism, hot springs activity, and anomalous metal sources (e.g. preexisting metalliferous deposits undergoing oxidation and leaching) with carbonate deposition.

SHELVES AND SHALLOW EPICONTINENTAL SEAS

Heckel (1972) differentiated between pericontinental marine environments that occupy shelf seas around the continents, and the broad epicontinental (or epeiric) seas located on the continent. The latter are shallower, with a gentler bottom slope than the former, and are relatively rare now, although they produced a great thickness of sediments in the geologic record.

The "most normal" shelf or epicontinental sea (later referred to only as "shelf" for brevity) sediments start to appear beyond the belt of coastal sand in water at a depth of some 10 to 20 m (Reineck and Singh, 1975). In the sand to shelf mud transition zone the most common sediment is clayey silt to silty sand. With increasing water depth, this grades into a monotonous shelf mud that contains locally layers of coarse silt or sand accumulated during storms. Adjacent to carbonate depositing shorelines, the sand and mud are of carbonate composition (Ginsburg and James, 1974). It is important to realize that the shelf sediments older than about 7,000 years had been laid down under conditions of a rapid global transgression following the latest glacial, and are "atypical". About 70% of the present continental shelves are covered with relict sediments (Emery, 1968); that is, sediments formed under conditions different from those that are characteristic of the present environment.

The relict sediments: mostly coastal, alluvial or glacial deposits, have undergone various degrees of reworking in their present shelf environment. Many were affected surprisingly little, due to the virtual absence of bottom currents in many shelf environments, so that beach ridges, moraine lobes, old river channels, etc. can frequently be recognized and mapped on the shelf seafloor.

Emery (1968) also distinguished the following types of shelf sediments and their constituents: (a) detrital or terrigenous, supplied by water, wind and ice; (b) biogenic, composed mostly of carbonate skeletal remains; (c) authigenic, chemically precipitated on the seafloor or in the uppermost layers of the unconsolidated bottom muds; (d) residual (more correctly subaqueous-residual), formed by the in situ weathering on the seafloor. To this should be added the relict or palimpsest sediments and volcanogenic sediments.

METALLIC MINERALIZATIONS

Although the present shelves are among the geologically most monotonous regions of the continental crust and ore minerals occur rarely, a moderately long list of mineral occurrences has accumulated now largely because of the sheer size of the shelf environment. Two major classes of shelf mineralizations can be recognized: (1) detrital; (2) chemical and biochemical precipitates.

(1) Detrital ore accumulations

Hydrodynamically concentrated detrital ore minerals on shelves are hosted only by sediments coarser than silt: that is, sands and gravels. Although occasional sand bodies are members of the "normal" shelf facies, at least 90% of the detrital mineralizations are hosted by the relict sediments. This is important for prediction and search for ores, which is usually based on paleogeographic reconstructions, and on tracing of extensions of the continent-based ore-bearing features (e.g. alluvial channels, beaches, moraines) onto the shelf domain, rather than on the expected granulometric and compositional zonality of the "ordinary" shelf sediments. Ainemer and Konshin (1982) recently reviewed the shelf-based placers and discussed the applicable metal exploration techniques.

Detrital mineralization in relict sediments on the shelf that is physically continuous with the subaerially exposed mineralization is the easiest one to predict, and many examples have been known for a long time. The submarine extension of the eluvial and alluvial tin placers in south-eastern Asia is the prime example. Earney (1980) stated that in 1977 about 6% of the world's total tin production was obtained by offshore dredging. One of the largest operations was 8 km off Bangka Island, Indonesia, in a depth of 10 to 18 m of water. The future potential of the S.E. Asian offshore tin is substantial. Several extensive deposits have been outlined off the West coast of Malaya (Batchelor, 1979)and the Kuala Langat placer contains at least 300,000 t Sn in reserves. Over 50% of the Indonesian tin resource potential is in the offshore areas.

Batchelor (1979) pointed out the absolute need for a thorough understanding of the stratigraphy and development history of terrestrial tin placers, before one attempts to search for the adjacent offshore tin accumulations. The richest Malayan and Indonesian continental placers are members of the late Miocene to early Pliocene granite regolith and the "Old Alluvium". The superimposed landforms (e.g. meandering stream valleys filled by the sediments of the "Young Alluvium") caused an almost total destruction of the earlier cassiterite accumulations in Malaya, and impoverishment (although not a complete destruction) of the placers located in offshore locations near the Indonesian islands. As a consequence, the offshore exploration based on tracking the submarine meander valleys resulted in failure (except off Bangka), and some offshore placers were found "where they were not supposed to occur according to the hypothesis". The latter are located close to the granite basement, near the base of the regolith and sedimentary succession, buried by 10-50 m of younger sediments. The placers do not have a conspicuous physiographic expression on the seafloor, so that the exploration has to depend on seismic interpretation and drilling (Fig. 5-23). The S.E. Asian placers are discussed further in Chapters 24 and 25.

In addition to the S.E. Asia, offshore cassiterite placers are known off Cornwall (England) coast (e.g. in St. Ives Bay, Earney, 1980; there, the attempts to dredge were unsuccesful mainly because of the unfavourable physical conditions such as the rough sea, rather than for geological reasons); off the Shirokastan Peninsula, Siberia; off the Primor'ye coast, U.S.S.R.; in the Sea of Japan (Sivoplyas, 1979); off the north-eastern Tasmania coast including offshore locations near the Furneaux Islands; and off the northern Queensland coast (Noakes and Jones, 1975). The Tasmanian placers in the offshore continuation of the Ringarooma River are believed to contain some 23 million cubic metres of sand, grading about 148 g Sn/m^3, to a maximum water depth of 40 m. The thickness of the tin-bearing sediments rarely exceeds 13 m.

Offshore gold placers have been explored or exploited in the seaward extension of the auriferous beaches already discussed earlier. To these could be added the offshore occurrences near Ovens, Nova Scotia (gold in gravels of buried stream channels; 30.1 million cubic metres of gravel with 0.03 to 3 ppm Au; Boyle, 1979), as well as the occurrences on the Soviet Pacific coast, for example off western Kamchatka (Ainemer and Konshin, 1982).

The extensive gold mineralization in the surficially exposed as well as drowned beaches in the Nome area, Alaska, has already been mentioned. The vicinity of Nome, however, is but a small portion of an extensive region of offshore occurrences of Quaternary relict sediments located in the northern Bering Sea, adjacent to eastern Alaska and the Soviet Chukotka Peninsula (Nelson and Hopkins, 1972). Relict patches of glacial till reach almost to the centre of the Chirikov Sea basin, and several kilometres beyond the present shoreline of the Seward Peninsula (Figures 5-24, 5-25). There, post-glacial transgressions and regressions have reworked the exposed glacial drift, and left relict gravel as a thin lag layer overlying the glacial deposits. This gravel veneer is richly auriferous along parts of the Seward Peninsula coast and relict gold flakes 1 mm in diameter or larger constitute the bulk of the gold present. Small gold particles (0.255 mm or less), by contrast, have been dispersed by waves and the bottom currents and appear in small quantities in the fine-grained bottom sediments of the northern Bering Sea.

"Light" heavy minerals (ilmenite, rutile, zircon, monazite, etc.) probably form extensive accumulations in drowned offshore beaches off the Queensland and New South Wales Pacific coasts, for example off Tweed Heads. The ore grades that have been detected so far, however, were quite low (0.8% of the heavies along a fossil strand in about 30-36 m of water; the "ore" material, moreover, was topped by 3-5 m of barren sand; Noakes and Jones, 1975). Low heavy-mineral contents

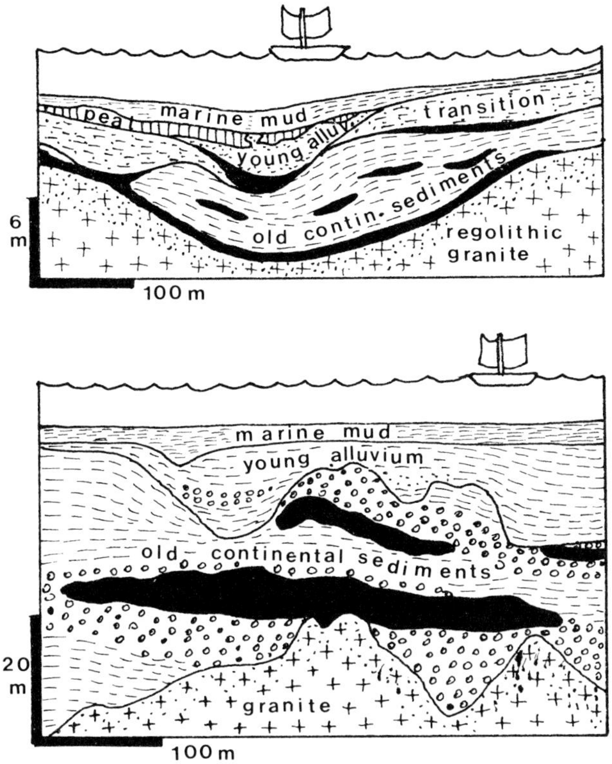

Fig. 5-23. Two varieties of cassiterite placers (in black) on the shelf off Bangka Island, Indonesia. TOP: reworked placer in alluvium of a young meandering stream; the former valley has a physiographic expression; BOTTOM: buried placer in an old alluvium, colluvium and regolith; no sign of its presence on the seafloor. Modified after Batchelor (1979).

have been also detected in "blanket" sands, probably representing the "ordinary" (rather than relict) shelf sediments, or in completely reworked, subaqueous beach ridges. Similar types of offshore sand bodies bearing detrital ilmenite and zircon, have recently been drilled in the German Bight (part of the North Sea off West Germany; Ludwig et al., 1979). There, 32-38 Tt of zircon and 111-128 Tt of ilmenite occur in bodies of well-graded, fine sand and the concentrations range from 2.6% to 6.6% of heavy minerals. Additional offshore accumulations of the "light" heavy minerals are known to be present on the Mozambique Shelf off the Zambezi estuary (51.25 Mt ilmenite; 4.05 Mt zircon; 900 Tt rutile; Schott et al., 1980); on

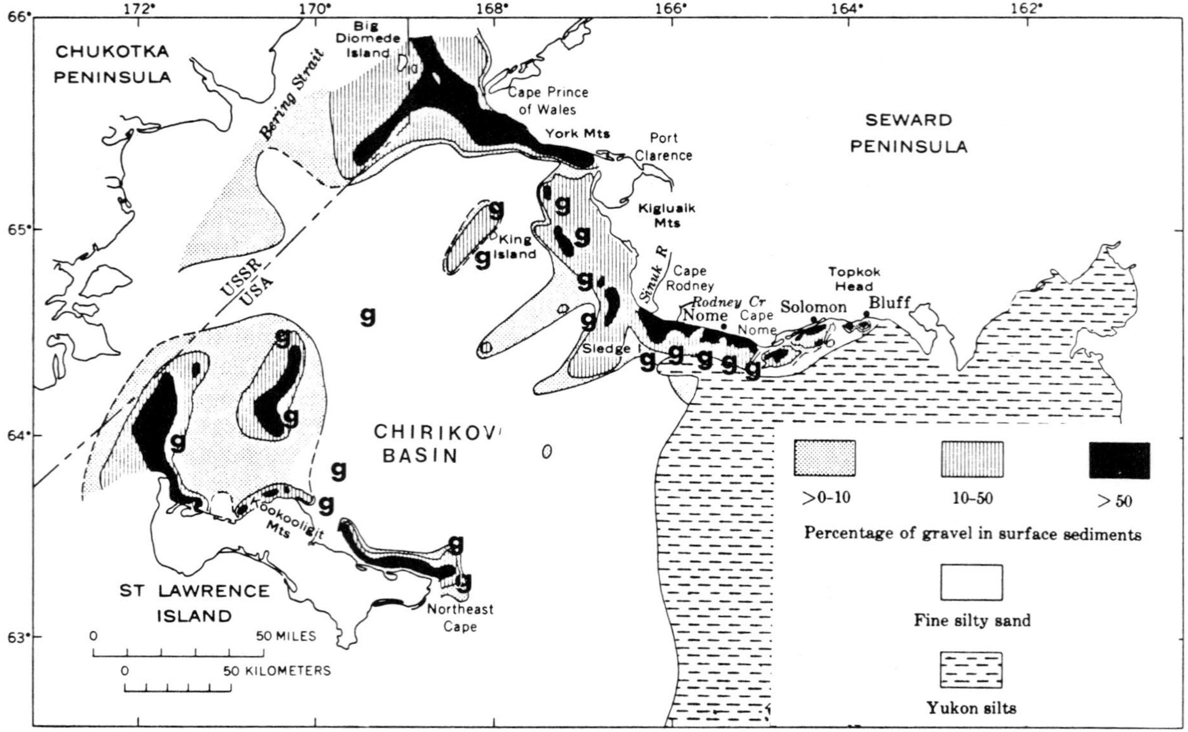

Fig. 5-24. Distribution of gold-bearing relict sediments on the northern Bering Sea shelf off Alaska and eastern Siberia. (g) = occurrences of detrital gold. From Nelson and Hopkins (1972), slightly modified.

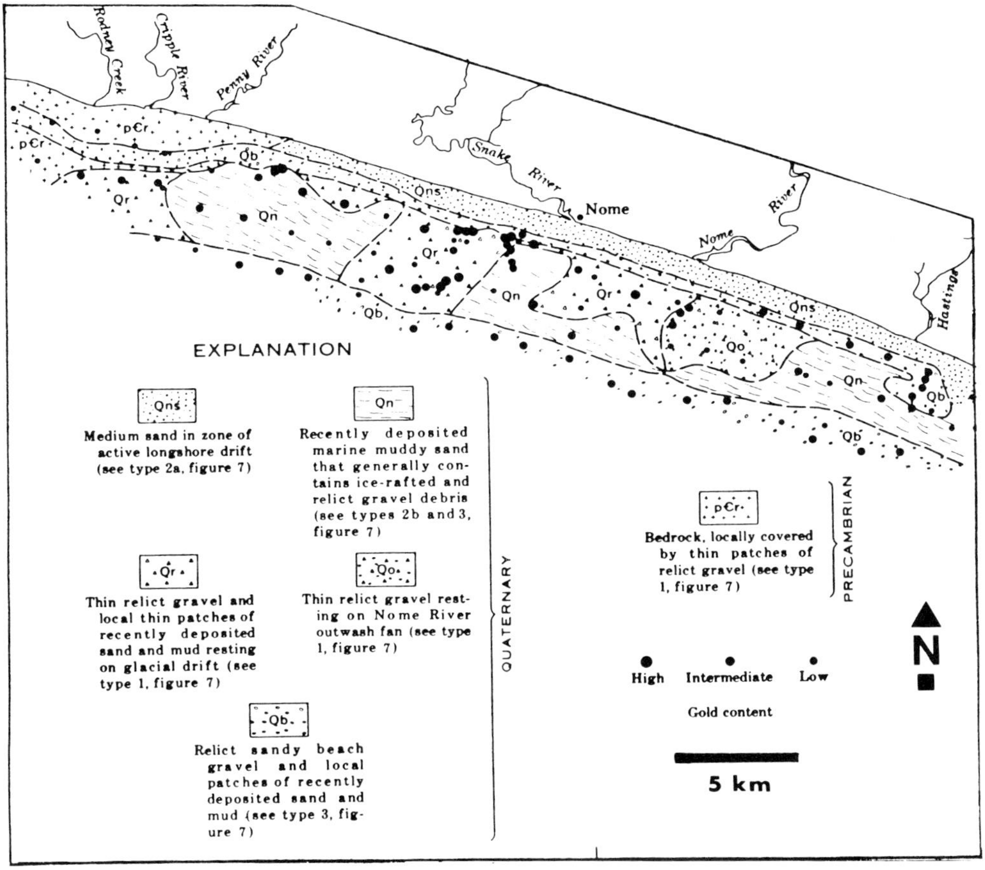

Fig. 5-25. Map of gold distribution in the nearshore relict gravels, Nome area, W. Alaska. From Nelson and Hopkins (1972).

the Atlantic shelf off the Pará State, north-eastern Brazil (in a probable Amazon relict channel; up to 28% of heavy minerals; Milliman et al.,1972), and on the Atlantic shelf off Richards Bay, Namibia (Hammerbeck, ed., 1976).

(2) Chemical and biochemical ore precipitates

Metalliferous segregations (concretions, nodules, crusts) and anomalous contents of certain trace metals, have been described from several regions of the shelf and epicontinental seas. The metalliferous substances are believed mostly to have precipitated from the seawater (hence they are hydrogenous), in areas of limited to no detrital deposition. Euxinic environments and areas of upwelling of nutrients-rich seawater from greater depth are particularly favourable sites. Four varieties of chemical or biochemical ores are described below.

(i) Manganese and ferromanganese nodules and crusts

(Ferro) manganese nodules having an appearance and mineralogical composition similar to the oceanic nodules (Chapter 4), have been reported from the Baltic sea (Manheim, 1961; Winterhalter, 1966); Barents, White and Chukchi Seas (Strakhov et al., 1968); from the Loch Fyne estuary of the Atlantic, Scotland (Calvert and Price, 1970); from the British Columbia coast and elsewhere. Manheim (1965) and Calvert and Price (1977) reviewed the subject.

Compared with the oceanic nodules, the shelf or epicontinental nodules have a patchy distribution, are thinly scattered (in the Gulf of Bothnia off Sweden, max. 10-20% of the seafloor was covered by the nodules), are smaller, and have a substantially lower trace Ni, Co and Cu contents. In addition to the nodules, dark gray marine muds and clays are locally enriched in Mn (up to 2.84% Mn in the east-central Baltic Sea; Strakhov et al., 1968).

(ii) Phosphorite

Widespread occurrence of phosphorite nodules and laminated slabs was first recorded by Emery (1960) from the shelf off California. There, the phosphorite densely covers about 15,000 km^2 and Emery estimated that about 1 billion tons of the phosphate was available. The phosphorite occurs in association with glauconite and rests on the topographically higher sites of the shelf, where the detrital sedimentation is restricted. The nodules, slabs, small grains and ooliths are hard and dense and have a smooth, glazed, brown surface. In deep water, the surface is usually covered by a thin film of Mn oxides. The main constituent mineral is a microcrystalline collophanite (fluorapatite), and it is considered truly authigenic, being currently precipitated from the seawater.

Similar phosphorite was later found to be abundant along and outside the Atlantic continental shelf off the United States, at depths of between 200 and 500 m, and also off Somalia, Australia, India, South Africa, etc. The phosphorite occurrences off the South African coast have been reviewed recently by Fuller (1979). There, most of the nodules occur in Neogene relict sediments, or they were reworked from Tertiary into Quaternary sediments in a manner reminescent of the subaerial Florida Land Pebble phosphorites (Chapters 23, 25).

From the point of view of a metalliferous geologist, phosphorites are interesting for their trace uranium content, that can be recovered as a by-product during processing. Few data are available on the U content of the present shelf phosphorites, but it is probably comparable with the content in the Miocene and Pliocene deposits of Florida, where it is 0.015% U. If so, the phosphorites off California might contain some 150,000 t U.

(iii) Chamosite

Chamosite is a locally important constituent of Paleozoic and Mesozoic iron ores. Several scattered occurrences of chamosite are known to exist and to form actively by authigenic precipitation on the Atlantic tropical shelves off the Niger, Orinoco and Amazon deltas (Porrenga, 1965). In the Baia de Aratu, off the Bahia coast of Brazil (Bittencourt and Boas,1977), authigenic chamosite forms pellets and internal moulds of Halimeda, gastropods and foraminiferas in clay, deposited in a restricted, 5-30 m deep bay. Chamosite is early diagenetic.

The Loch Etive occurrence off Scotland where chamosite is present as faecal pellets in the shelf sediments is exceptional, because of its location outside the tropical belt. Although scientifically interesting, the Recent chamosite occurrences have no practical importance.

(iv) Uraniferous and iron sulphide-rich euxinic muds

In several deep Norwegian fjords (e.g. Indre Topdalsfjord), in the Gotland Deep of the Baltic Sea, and off Walvis Bay, South-West African coast (Swanson, 1961; Schott et al., 1980), an euxinic environment exists in which the bottom waters contain up to 200 ccm/liter of H_2S. As a consequence, pyrite and uranium compounds precipitate in the black or greenish-gray, organic-rich bottom muds. The uranium content ranges between 5 and 93 ppm, which is below the lowest grade of presently mined "straight" uranium ores (about 300 ppm). The muds, however, are an interesting Recent equivalent of the ancient uraniferous shales such as the Chattanooga or the Kolm Shale (Chapter 19). Locally, a high gold content (up to 0.7 ppm Au) has also been detected. In the shelf muds off Walvis Bay (von Backstrom, 1974), Mo (aver. 62 ppm; range 20-120 ppm) and V (aver. 112 ppm; range 60-180 ppm), accompany the uranium.

OCEANIC PLATEAUS

The Blake Plateau located in the Atlantic Ocean and separated from the continental shelf off Florida by a deepwater channel, is a broad, flat marginal plateau. It is bordered in the east (against the abyssal Atlantic) by a steep escarpment (Sheridan, 1974). The Plateau is covered by Recent pelagic carbonates (Globigerina and Pteropod oozes), and underlain by Pliocene and Miocene carbonates and turbidites, topping a thick oceanic crust basement. Abundant Fe-Mn nodules and crusts occur in water depths between 750 and 1,050 m (Emery, 1965).

CONTINENTAL SLOPE AND RISE

In the literature describing depositional environments, sediments of continental slopes and rises are usually treated under the heading "deep clastic seas" (Heezen and Hollister, 1971; Reineck and Singh, 1975; Rupke, 1978). The most conspicuous are detrital sediments of the large turbidity currents that flow downslope towards the continental rise and the abyssal plain. On the continental rise, the turbidites are, to a considerable degree, substituted by contourites: sediments transported and deposited by strong ocean bottom currents. Tectonic mega-environments influence, to some degree, the pattern and composition of the slope and rise sediments, and an "Atlantic model" (numerous, recurrent and parallel turbidity flows debouching into the abyssal plain and covering a substantial portion of it in the form of a sheet of an almost constant thickness), and a "Pacific model" (turbidite-filled long individual channels and fans) have sometimes been recognized in the literature (Rupke, 1978).

The "Pacific-type" turbidite fans often occur in a setting transitional between the passive and active continental margins. The Bengal Deep-Sea Fan (Curray and Moore, 1974) is the world's largest deep-sea fan, extending for approximately 3,000 km with a width of 1,000 km. Its sediments were brought in by the Ganges and Brahmaputra Rivers from the rising Himalayas, reached a thickness of up to 16 km beneath the continental shelf of Bangladesh, and were ultimately added to the developing Indonesian island arc.

Despite the considerable volume of the Recent turbidites and contourites and their importance as a model of formation of the flysch facies (Chapter 16), the bulk of these sediments are lithologically monotonous and metallogenically sterile. The sedimentary piles consist of alternating, graded, usually quartz-rich lithic sand, silt and mud, containing minor conglomerate interbeds. Geochemically, such sediments approach the average composition of the continental crust, and no anomalous metal contents have been reported. Authigenic mineral accumulation is atypical, because it is overwhelmed by the tremendous supply of detritus. Rarely, it could take place in the fore-fan facies muds (equivalent to prodelta muds), or in sites shielded from the detrital influx (e.g. on dividing ridges). Abundant authigenic (?) goethite has been reported (Milliman et al., 1972) from the Atlantic continental slope near the Amazon underwater canyon, off Brazil. The interesting variety of debris flows (rock fall, sliding, slumping, debris flows; Rupke, 1978, p. 374-379) described on the continental slopes, disturbs the lithological monotony by adding exotic blocks of rocks derived from the extra-formational, older "basement" sources (e.g. from submarine canyon walls). These are most common in olisthostromes and in the wildflysch association. The olistoliths (i.e. clasts in olisthostromes) are often of gigantic dimensions and their lithologic composition is variable and often unpredictable. The blocks, as well as the "bedrock" outcrops in the seafloor, constitute a discrete "mini-association", having their own metallic mineral association.

"Neither volcanism nor plutonism can be understood until we understand the formation of mountain chains".
R. A. Daly (1925)

"Compressional orogenesis is essentially an English-language obsession"
Warren Carey (1976)

CHAPTER 6

Orogenic Belts
(introduction to Chapters 7-17 and 27-29)

6.1. GENERAL

6.1.1. Definition and variations

An orogenic belt has been characterized (Spencer, ed., 1974) as "a well defined area (measurable in tens of thousands of square kilometres) which has shown, or is showing, marked mobility typically characterized by deformation structures (e.g. folds, foliations, lineations and reverse, thrust and transcurrent faults). The belt may also have been affected by regional metamorphism and magmatism, and either has undergone or may be undergoing uplift and sub-aerial erosion".

The Pacific and some Atlantic continental margins (Chapter 5) as well as some rifts (Chapter 30), are either (1) simultaneously bona fide orogenic belts where the present lithogenetic activity is a continuation of comparable earlier activity in which the earlier rocks have already been deformed or where deformation is taking place contemporaneously (e.g. all the ensialic continental margin magmatic belts and associated environments); or (2) embryonal (precursor) orogenic belts in which orogenic (compressional) deformation is, by comparison with the older equivalents, expected in the future (e.g. the "primitive", ensimatic island arcs); or (3) potential orogenic belts which would be deformed should the present configuration of the lithospheric plates change (Atlantic-type continental margins and rifts). There is, consequently, little qualitative difference between the supracrustal sequences of many present continental margins and ancient orogenic belts, except in timing and in the level of exposure and consequently emphasis. In Chapters 3 to 5 we have been considering Quaternary developments only, whereas the present chapter deals with the pre-Quaternary rocks. In the same chapters, surficial to shallow subsurface lithogenesis and metallogeny were emphasized; here, the main subjects are rocks and ores which, although exposed on the present erosional surface, formed in the subsurface (e.g. the intrusive rocks), or were modified at some depth (lithified sediments, metamorphics).

It could be said that the Mesozoic-Cainozoic orogenic belts have now reached their "prime time". They are the most typical belts that fit best the numerous proposed models. This is because they have already completed one (or more) "geotectonic cycles" and are suffering, at present,from the process of degradation that inevitably follows (erosion, denudation and ultimately levelling). The degradation has worked long enough to expose locally some of the deeper levels of the belt (e.g. intrusive batholiths, metamorphic cores), but it has still left a good deal of the higher-level products such as the remnants of volcanic edifices, preserved and available for reconstruction of the original supracrustal depositional environment.

With increasing geological age, the ratio of exposed high level / low level products in orogenic belts decreases. In the Paleozoic belts, for example, well-preserved volcanic edifices (e.g. calderas) comparable with those in the Cainozoic belts, are rare, although several remarkably well preserved paleovolcanic terrains have been described and interpreted (e.g. in the Tokrau belt, Kazakhstan). The ratio of infracrustal (formed within the crust) to supracrustal (formed subaerially or subaqueously) rocks is close to 1:1; moreover, substantial proportion of supracrustals is, at least lightly, metamorphosed. Paleozoic orogenic belts do not, as a rule, form physiographically continuous uninterrupted mountain ranges parallel with the original structural grain and facies belts arrangement, unless they have been recently rejuvenated by a vertical uplift (e.g. the Appalachians, the Urals). Even so, such uplift exposed only a portion of the former belt.

In Precambrian orogenic belts the denudation was statistically the deepest (although exceptions occur), so that the root regions of the former orogenic mountain belts are now preserved in outcrop. Supracrustals account for substantially less than 50% by area. The latter form discontinuous relicts of originally linear belts (e.g. greenstone belts), surrounded and subdivided by a dominant granite-gneiss mass. Continent-wide connections have been severed, and isometric outcrop blocks, shields, are the fundamental unit of organization and description. Although there is little qualitative difference among the Phanerozoic and Precambrian orogenic belts, both have been traditionally treated separately in the literature and this tradition has been retained here.

Orogenic belts have been the subject of numerous geotectonic classifications and interpretations. In the global geosynclinal orogenic model used before 1968, orogenic belts had usually been subdivided into "geosynclinal" and "non-geosynclinal" ones. The former had passed through most stages of the geotectonic cycle, starting with the marine basin sedimentation and culminating with deformation, metamorphism and intrusive emplacement. The latter lack the marine basin sedimentation and tholeiitic or calc-alkaline volcanism, immediately preceding the deformation (e.g. the Colorado-New Mexico Rocky Mts.), and were usually treated under the heading "tectono-magmatically activated belts" in the U.S.S.R. (compare review in Laznicka, 1981b).

In the recent plate tectonic framework, orogenic belts have been placed, from the genetic point of view, into the following major categories (e.g. Dewey and Bird, 1970): (1) island arc belts; (2) Cordilleran-type belts; (3) continental / island arc collision belts; (4) continent/continent collision belts. To this can be added arc / arc collision belts, intracratonic belts, and the many lesser order styles of belts such as the obducted ophiolite slices.

Intracratonic belts are of particular interest, because they are widely represented in the Paleozoic orogens (e.g. in the Variscan belt of central Europe: Harz Mts., Barrandian Basin, Erzgebirge, Sudeten Mts., Rio Tinto belt, Ural Mts., etc.) as well as in the Precambrian "greenstone belts". In contrast to the "Cordilleran-type" (subductive) orogenic belts, the intracratonic belts completely lack andesites and the volcanism is bimodal. The belts may (and are expected to) have a bilateral symmetry, but this is not always obvious because of the lack of exposure or dismemberment due to subsequent continental fragmentation. Krebs (1975; Fig. 6-1) attributed similar belts to an "orogenic diapirism", triggered by a global vertical tectonism and a rise of astenoliths.

It is not the purpose of this book to engage in the subject of plate tectonic setting, development and classification of orogenic belts; there is an abundant recent literature doing so (e.g. Mitchell and Garson, 1981; Hutchison, 1983; Sawkins, 1984). Instead, we are going to focus on the characteristic lithologic associations of orogenic belts which may or need not have a mutually comparable geotectonic setting and origin, but which can be recognized in the field by a practising geologist regardless of the interpretational model in force.

6.1.2. Classical facies megadomains

The subdivision of orogenic belts into the "eugeosynclinal" (eugeoclinal) and "miogeosynclinal" (miogeoclinal) domains in the now defunct geosynclinal model (e.g. Auboin, 1965) had, in fact, a lithofacies basis. This was obscured by the rather unconvincing genetic terminology, difficult to demonstrate in the present environments.

Eugeoclinal domains had their early supracrustal suite dominated by marine volcanics and sediments. In the miogeoclinal domain early volcanics were supposed to be missing. The dual megafacies division was most applicable in a very broad context, and it was frequently applied in the preparation of tectonic and metallogenic maps. It had, however, proven insufficient for the purposes of detailed metallogenic analysis of territories on a scale of 1: 200,000 or less.

Eugeoclinal and miogeoclinal megafacies can seldom be separated by a sharp line in an actual completed orogenic belt. Most actual "eugeoclinal" continental margin belts consist of a mosaic of facies and "terrains", many of which developed outside of their present location in a "stretched" complex environment as in the present-day

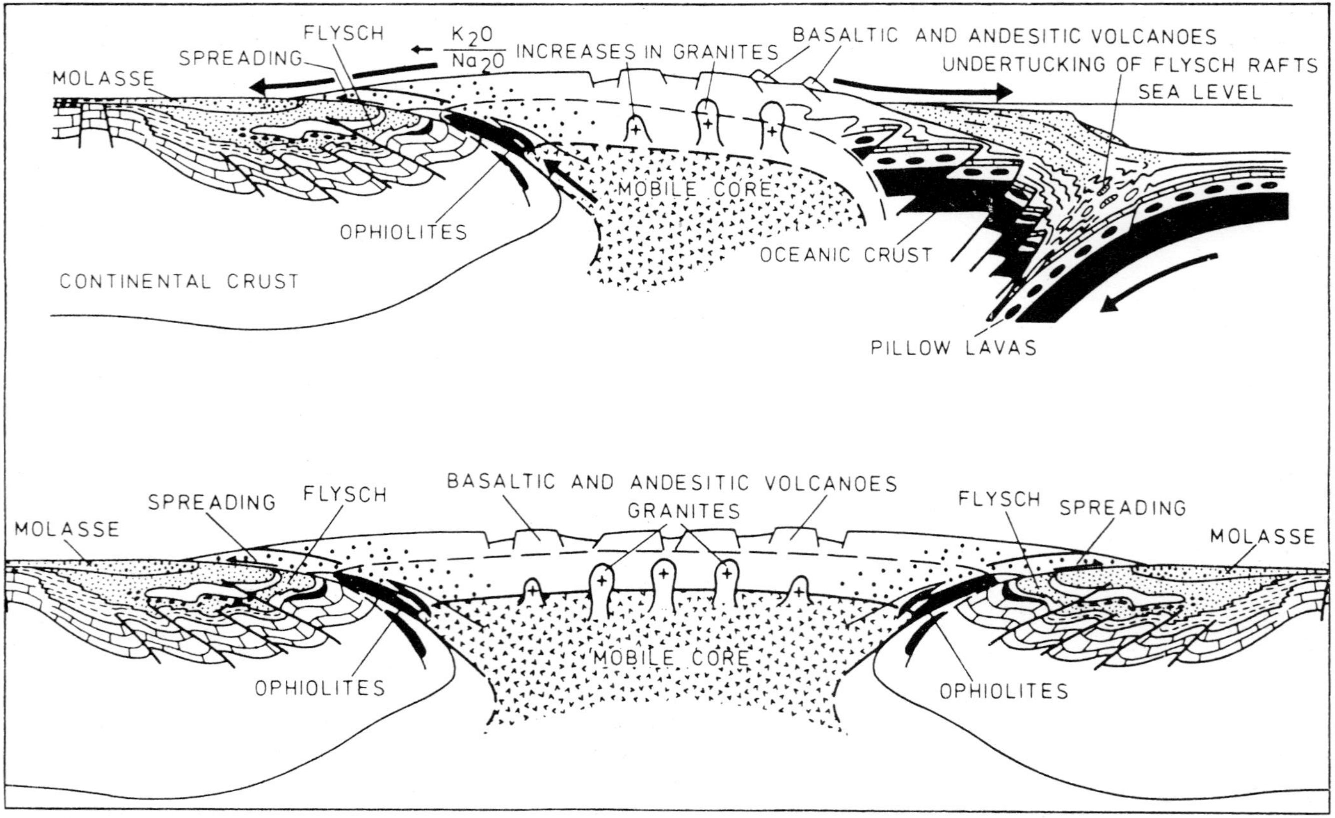

Fig. 6-1. Comparison of the Dewey and Bird (1970) subductive model of formation of andesite-dominated orogenic belts (a) with the orogenic-diapiric bilaterally symmetrical model, common in intracratonic setting. From Krebs (1975).

Courtesy of the American Association of Petroleum Geologists.

Indonesian region (Hamilton, 1979). Such terrains, presently largely exposed in islands, were later added to a continuous continental margin in a series of collisions. In Indonesia, the presence of most supracrustal rock associations can be demonstrated, ranging from oceanic crust through transitional crust (with mafic to felsic volcanic composition) to entirely non-volcanic deposits of temporarily stable (micro)continental margins. Yet the entire system is likely to be added (welded) in the future to the Asian continent, the present margin of which is already marked by a substantial lithologic heterogeneity due to its past development including island additions. The Pacific margin of North America composed of a variety of terrains that include oceanic, volcanic arc, flysch wedge, stable platform, etc. facies (e.g. Churkin, 1973) bears many similarities to the anticipated post-collision situation in south-eastern Asia. Small scale units formed in microenvironments, identical with those in which the "miogeoclinal" suite was produced (e.g. biogenic carbonates) have been recorded in most "eugeoclinal" domains.

As a consequence, the classical facies zoning (polarity) of continental margin orogenic belts generalized by Kay (1951), Auboin (1965) and others is, in fact, mainly a statistically supported general trend. Along the Pacific coast of the Americas, the volcanics-rich (eugeoclinal) facies are dominant in the west (close to the present ocean), the non-volcanic (miogeoclinal) facies prevails in the east, adjacent to the craton. The facies belts that usually produce the strongest impression in a superficial observer, are usually of a different age. In the Cordilleran and Andean belts, for example, the strongest "eugeoclinal" volcanics in the west are Mesozoic, whereas the striking carbonate, sandstone, shale, etc. dominated terrains in the east are Paleozoic. Regularity of many orogenic belts is, furthermore, disturbed by a variety of superimposed features, that include: (1) "exotic terrains" (drifters), emplaced along strike-slip faults; (2) alternating structural "lows" (basins) and "highs" (ridges, uplifts). This resulted in a side-by-side arrangement of units earlier generated at profoundly different crustal levels, that disturb the facies continuity; (3) development of thrusts and nappes that disturb (sometimes even reverse as in the Alps) the depositional facies assembly; (4) establishment of extensional structures (grabens, troughs, "rifts") responsible for local deeper-water depositional environments and lithofacies within a predominantly shallow water environment, as well as for the local appearance of mafic volcanics.

Taphrogenesis can cause emplacement of mafic volcanics in every domain of an orogenic belt. Since certain metal accumulations, such as copper, tend to be affiliated with the mafic volcanism, basalts, gabbros and associated or derived sediments along tensional structures deep within the "miogeoclinal" megafacies may be parallel with localized "copper belts" situated on a wider background of 'lead belts" (Laznicka and Wilson, 1972). Overgeneralization of miogeoclines as "lead belts" may result in oversights, criticized by Routhier (1980).

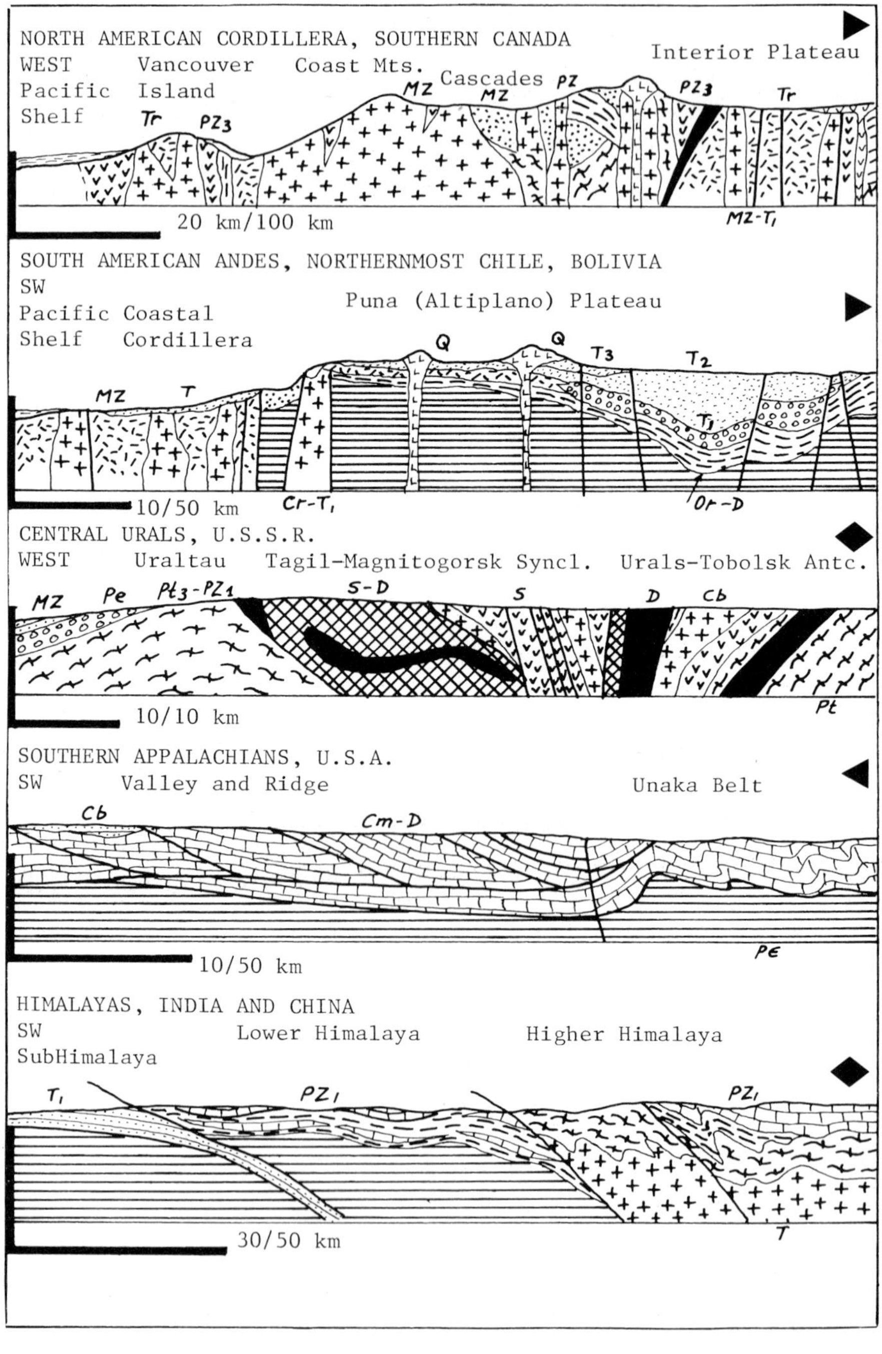

Fig. 6-2. See legend on page 124.

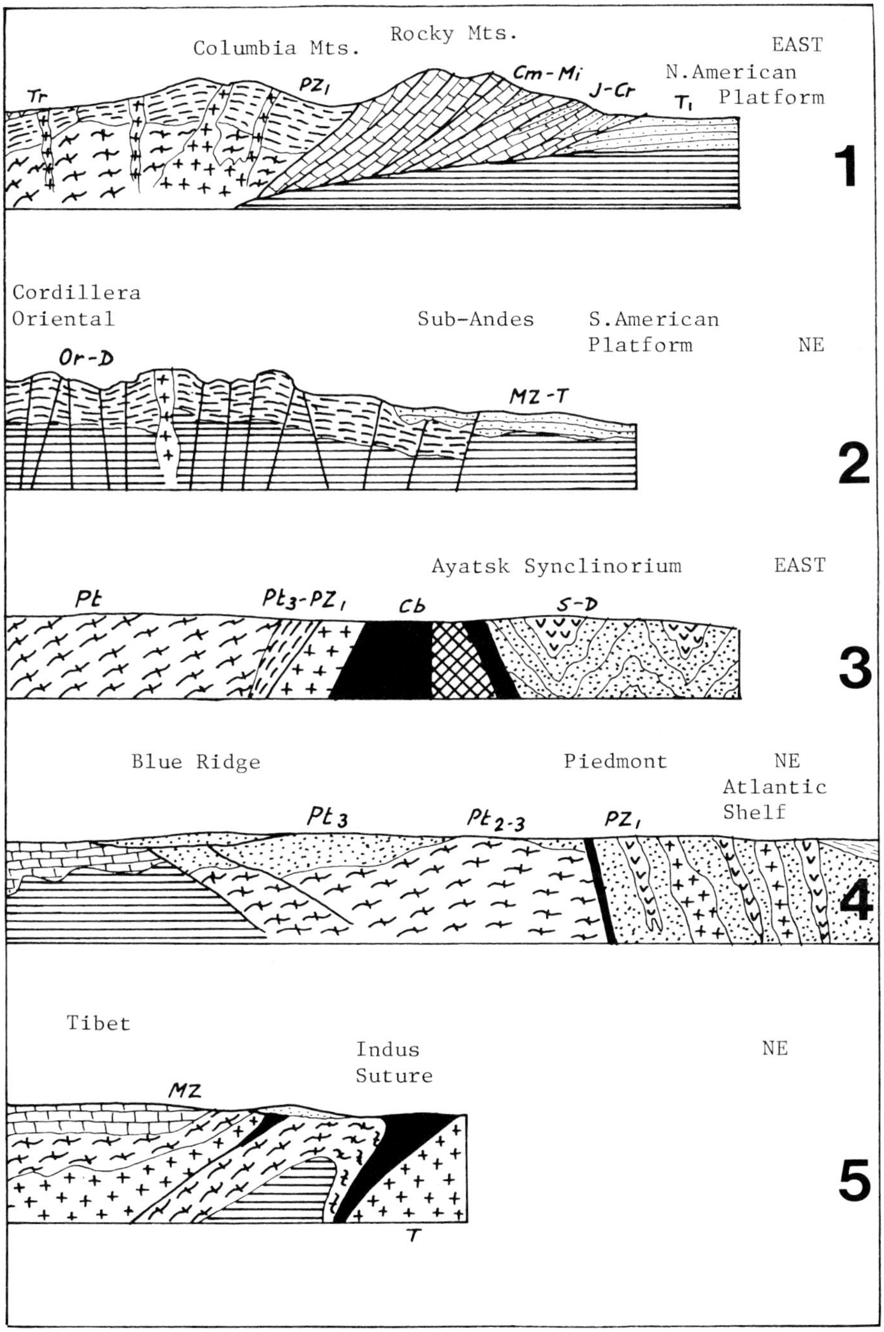

Columbia Mts.
Rocky Mts.
EAST
N.American Platform
Tr
PZ1
Cm-Mi
J-Cr
T1
1
Cordillera Oriental
Sub-Andes
S.American Platform
NE
Or-D
MZ-T
2
Ayatsk Synclinorium
EAST
Pt
Pt3-PZ1
Cb
S-D
3
Blue Ridge
Piedmont
NE
Atlantic Shelf
Pt3
Pt2-3
PZ1
4
Tibet
Indus Suture
NE
MZ
T
5

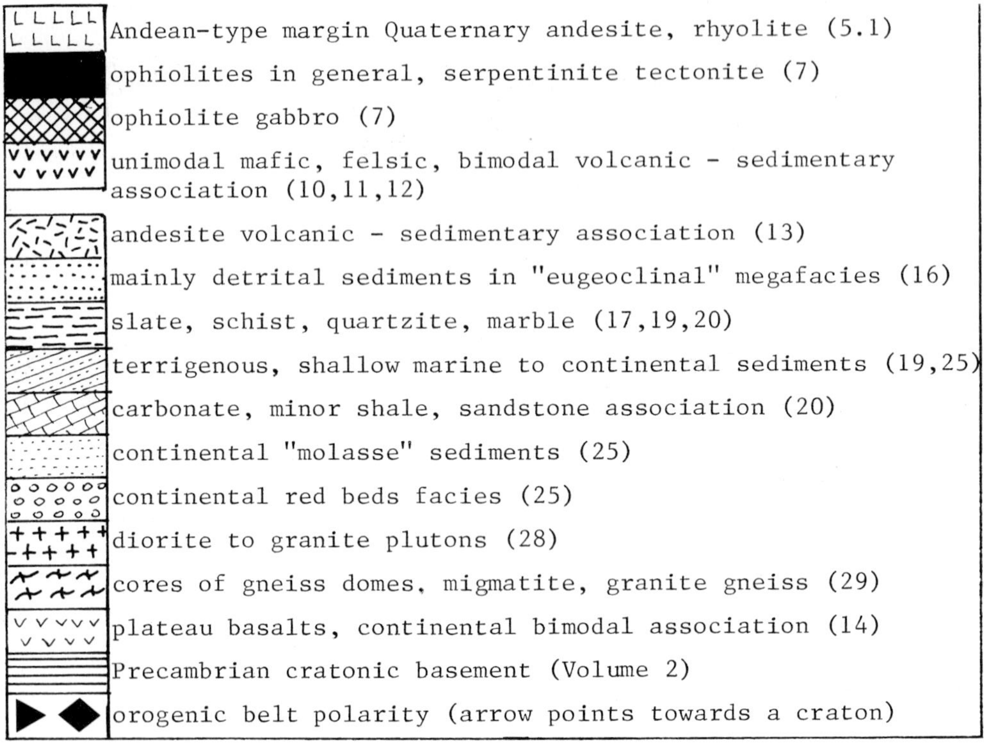

Fig. 6-2. Generalized example sections of orogenic belts. 1,2: Belts presently and in the Mesozoic-Cainozoic situated at convergent continental margins, marked by the presence of andesites. 3,4: Intracratonic belts, in which the "geosynclinal" volcanism is represented by bimodal associations. 5: Belts resulting from recent continent—continent collision. Numbers in brackets (in legend) indicate chapters in which an association is reviewed.

Generalized and modified after sections in Douglas, ed. (1970); Rutland (1974); Perfilieff and Kheraskoff (1964); Gansser (1974) and Bryant and Reed (1970).

Fig. 6-2 gives a cross-section of several orogenic belts that demonstrate the high degree of overall heterogeneity resulting from the lengthy history of development and incorporation of units formed under different geotectonic conditions. Despite this, the majority of belts display a very general gradient of change in a section going from the "internal" (eugeoclinal) zones to the "external" (miogeoclinal) zones and the former craton.

6.1.3. Orders of magnitude of lithologic associations

One of the frequent causes of misunderstanding in geological interpretation and writing is mixing and mutual comparison of units of vastly different scale and complexity (compare Laznicka, 1981a, b; 1985c). Recently, Krebs (1979) distinguished three orders of magnitude of Devonian depositional basins in the Central European Variscan Basin (first order: over 100 km; second order: over 10 km; third order: 100m-10 km), and his approach has subsequently been widely used in North America.

Because most ore deposits form and are controlled by very local anomalous conditions, it would be advisable to treat them in terms of the smallest-order "basins". This, unfortunately, is not always possible because of space limitations in the text, and also because it is impossible to isolate many smallest-scale paleoenvironments during both the literature research and fieldwork. One can only achieve a certain degree of perfection. Although the utmost care has been taken not to mix scales, some mixing has been unavoidable. Many such cases certainly exist in the sections dealing with the volcanic-sedimentary associations, broken down into those containing: (a) mafic; (b) felsic; (c) bimodal; (d) sequence-differentiated, volcanics. Many bimodal associations may have been exposed, preserved and described in the literature in such a way, that only the mafic or the felsic suite has been reported. This then led to a situation in which an area X, practically identical with the area Y, but fragmentarily described (or described as a lesser-scale unit), is reviewed in a different section. The only solution to alleviate this problem lies in overlapping and frequent cross-referencing.

6.1.4. The problem of lithologically identical (or nearly identical) associations located in different geotectonic settings

Black shales or slates, turbidites, carbonates, basalts, etc., can occur in different geotectonic settings: for example in platforms, in the eugeoclinal or miogeoclinal domains, and in the various plate tectonic environments. Sometimes, the lithologically corresponding associations also have identical geochemistry and affiliated deposits. Sometimes, the latter are different. Although the observable lithologic association forms the basis of the organization in this book, geotectonic setting may become a valuable supplementary attribute provided it can be determined with a reasonable degree of credibility in the field, or identified from a description.

Several "flysch" categories treated in this book (island arc flysch, mixed flysch, alpine flysch) actually contain compositionally different litharenites. These can be broken down into three major categories based on the quartz content (Crook, 1972): quartz-poor graywackes with under 15% quartz; intermediate-quartz graywackes with 15-65% quartz; and quartz-rich graywackes with over 65% quartz.

These three classes are, in chemical composition, approximately equivalent to andesite, granodiorite and tectite, respectively, and it is likely that these compositional differences also influence variations in metallogeny. The black shale or slate association included here in several Chapters (7, 10-17, 19, 25) is more difficult to subdivide on compositional grounds.

Some lithologically equivalent associations formed in different environments or under different post-depositional conditions, contain a distinct rock type or a "special facies" which, although minor, is unique and helpful in establishing the identity of an association. It has been termed "guide rock" and sometimes used in this book (e.g. the distinct "Schallstein" volcaniclastics diagnostic of the mafic volcanic sedimentary association).

6.1.5. Metallogenic aspect (e.g. Fe aspect)

Certain lithologic associations (e.g. black phyllite; basalt-black phyllite) carry, locally, a distinct assemblage of ore metals that differentiates them from compositionally indistinguishable lithologic associations elsewhere. The cause of the metallogenic variation is usually unclear or at best only hypothetical. In most cases it is due to the geological style and structure outside the limits of the unit considered (e.g. black slate association floored by an oceanic mafic assemblage will likely have an Au, Cu, Sb or Fe, Mn, aspect. The same rocks floored by continental crust (e.g. granites) may display Sn, Pb, U, etc. enrichment).

6.1.6. Sequential evolution, geotectonic cycle, and multistage orogenic belts

Geotectonic cycle: "a fundamental abstraction in earth sciences" (Coney, 1970), is a deterministic sequence of events in space and time incorporated within the geosynclinal model of mountain building. Such a sequence begins with a thick, usually linear, marine accumulation of sediments and volcanics (early or geosynclinal phase); passes through a period of intensive deformation, metamorphism and plutonism (orogenic, tectogenic or batholitic phase); and ends with a differential uplift or collapse, subaerial volcanism and incorporation into a continental block, craton (late, final phase); e.g. Beloussov (1962). Geotectonic cycle provided a convenient framework to metallogeny most thoroughly organized by the Soviet school (e.g. Bilibin, 1955) and highly acclaimed, until recently, by many exploration geologists (e.g. Gilmour, 1976).

Coney (1970), Hamilton (1979) and other investigators convincingly argued that the geotectonic cycle concept is obsolete, and instead they suggested a multiple, uncoordinated origin for mountain systems. Although the rigid organization of the classical geotectonic cycle may be outdated, the subject of evolutionary changes within various segments of orogenic belts and interaction among deposition, tectonism and magmatism is alive and applied, as the former cycle, in a framework of description and organization of ore deposits (e.g. Hutchinson, 1973; Mitchell and Bell, 1973; Mitchell and Garson, 1981).

All orogenic belts are multistage. Even the most primitive, embryonic ones, the ensimatic island arcs formed on the oceanic lithosphere, are the product of two successive stages: (1) formation of the oceanic lithosphere; (2) formation of the arc basalts, each represented by a characteristic lithologic association. As the younger (2) association forms, it either causes destruction or modification of (1), or, alternatively (1) survives almost intact and is buried under (2). In most cases, both alternatives coexist, not necessarily directly vertically under each other.

Continuous development of the completed two-stage arc will result in an increased facies diversification assisted by intracrustal processes (partial melting of the earlier formed rocks), and supracrustal sedimentogenesis. It has been repeatedly demonstrated (e.g. by Crook, 1972) that a detrital sediment (e.g. "graywacke") tends to be about one step more silicic and alkaline than the igneous precursor from which it was derived. For example, detrital recycling of an arc basalt will produce a "graywacke" of andesitic composition, and detrital recycling of an andesite or diorite complex containing a minor quantity of quartz-bearing rocks (e.g. quartz diorite), will produce "graywacke" of granodioritic composition. The lost (solution-removed) elements Na, Mg and Ca are probably added to seawater.

The sedimentogenic differentiation goes hand in hand with the magmatic differentiation and follows the same trend: increase in Si, Al, K, etc.; a "continentization". A fragment of an earlier generated evolutionarily mature continental crust can become involved in a superimposed magmatic event in an island arc (e.g. Japan) or Andean-type (the American Cordillera) setting, and undergo a supplementary "continentization". This is usually manifested by widespread rhyolitic and rhyodacitic volcanism and quartz monzonite intrusive magmatism.

A reverse process, "oceanization", has, from time to time, been invoked as a mechanism of mafic (ultramafic) crust generation, on sites formerly occupied by the continental crust (e.g. Beloussov, 1962). Although Beloussov specifically explained oceanization as a metasomatic basification, there is certainly a wide range of process variations, including mafic dike or sill injection into dilations in the attenuated continental crust (e.g. Dickinson, 1974b). Katz (1974) proposed additional mechanisms giving similar results.

Although a uniform, regularly timed and almost global geotectonic cycle is no longer considered credible, the common evolutionary sequence of active continental margin associations not very different from the end-product treated in the geosynclinal model, is a reality. Its final result is a material transformation of oceanic crust into the continental crust (Peyve et al., 1977). Fig. 6-3 shows a

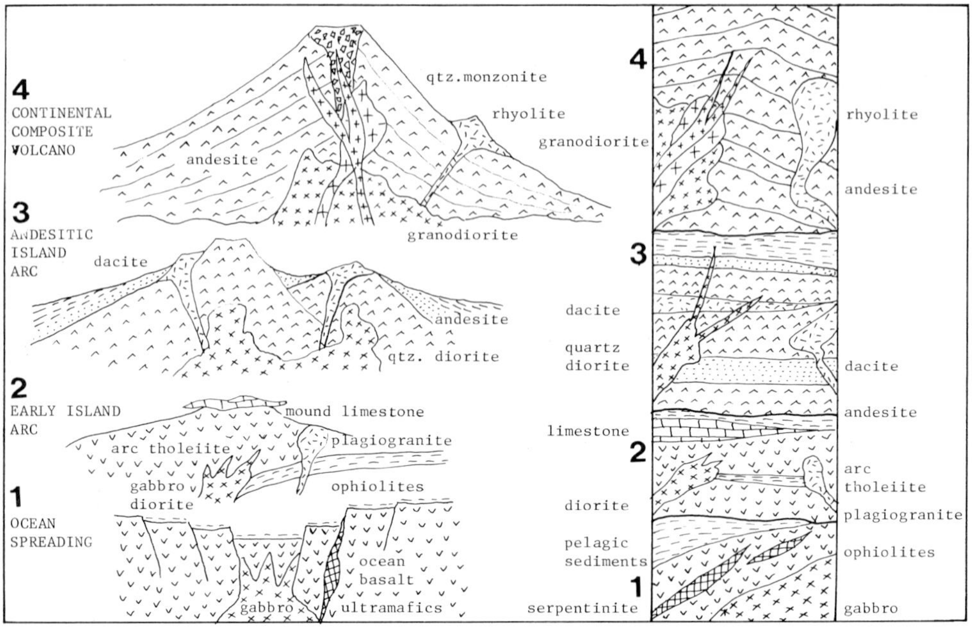

Fig. 6-3. Idealized evolutionary sequence of an orogenic belt. LEFT: at the time of deposition; RIGHT: resulting geological column (assuming vertical stacking without destruction of the earlier association).

hypothetical evolution of a continental margin, in which the progressively younger and more evolved associations accumulated by vertical superposition. A more realistic situation, however, includes lateral shifts so that, for example, the association (3) is not buried under the association (4), but is displaced laterally. Vertical stacking, lateral shifts, facies variation, cyclic repetition, faulting, continental drift, collisions, etc. are responsible for a large number of variations of local sections in orogenic belts.

Fig. 6-4 shows a representative composite section in the East Shasta District, California (Albers and Robertson, 1961). The section is a result of a prolific lithogenesis in the period from Devonian to late Triassic, followed by a near-stagnation between Jurassic and Recent. From the metallogenic point of view it is notable that in this area two independent volcanic sedimentary units (middle Devonian and middle-late Triassic) carry Fe-Cu-Zn(Pb) massive sulphide deposits believed to be contemporary with the supracrustal lithogenesis. The only other mineralized supracrustal unit in the District, the Permian limestone, hosts skarn deposits of magnetite that are the result of interaction of the limestone with Jurassic-Cretaceous granitic rocks.

6.1.7. Precambrian units within Phanerozoic orogenic belts

Although this chapter emphasizes Phanerozoic rocks deformed and metamorphosed by Phanerozoic orogenies, Precambrian basement always underlies the "miogeoclinal" portion of all the Paleozoic, and some Mesozoic-Cainozoic orogens. Initially buried Precambrian metamorphics frequently appear in block uplifts within orogenic belts; in cores of mantled gneiss domes; or in median massifs. The latter are large, rigid blocks floored by Precambrian metamorphics, surrounded on all sides by belts of less intensively metamorphosed supracrustals (e.g. the Massif Central of France, Bohemian Massif, Menderes Massif, Rhodopen Massif; Dimitrov, 1977; Abdullaev and Borisov, 1963).

Geologically, median massifs are quite comparable with shields ("mini-shields"), or with segments of stable platforms covered by a variable thickness of undeformed to little deformed Phanerozoic supracrustals (e.g. the Colorado Plateau). They differ from shields by the greater incidence of marginal effects and more intensive "activation". The activation mostly includes high-level brittle deformation (normal faulting, brecciation, mylonitization, horst and graben formation), as well as emplacement of "granitic" intrusions, continental bimodal suite or alkaline rocks. A special category of median massifs takes the form of large depressions filled by flat-lying basinal sediments, resting on thinned continental or locally even oceanic crust. The Pannonian Basin of Hungary is the principal example.

6.2. SUPRACRUSTAL LITHOLOGIC ASSOCIATIONS OF OROGENIC BELTS

6.2.1. "Geosynclinal" volcanics

The nature, petrochemistry and evolution of continental margin volcanics has already been discussed briefly in Chapter 5. The volcanics generated in ancient island arcs and, to a lesser degree, along the subaerial Andean-type magmatic belts, can now be found incorporated into orogenic belts. Paleoenvironmental interpretation of such volcanics, although commonplace in the literature, has to cope with the effects of collisions, strike-slip movements, faulting, folding, metamorphism and igneous intrusion. Some of the modifications are capable of altering radically the petrochemistry of the original rocks. Despite this, some categories of volcanics appear to have a high degree of correspondence with their modern counterparts (e.g. the unimodal mafic volcanic - sedimentary association, Chapter 10, closely corresponds to volcanics of the primitive or first stage ensimatic volcanic arcs formed on the oceanic crust), while others are more controversial.

The bimodal mafic-felsic association (Chapter 12) is substantially more widespread among the ancient volcanics of orogenic belts than it is in the modern environments. It is in the forefront of genetic

Thickness in feet	Lithologic description	Economic value	Number on diagrammatic section	Diagrammatic section
0-100 Q	Includes alluvial material in stream valleys, rock and soil mantle on hill slopes, talus, local terrace deposits, and landslide debris.	None	16	
0-200 Q Pl	Small bodies of basalt that locally cap hilltops and ridges in southeastern part of the map area.	None	15	
0-360 Pl	Tuff breccia, consisting of poorly lithified aggregate of blocks, lapilli, and volcanic ash.	None	14	
JCr	Includes fine-grained mafic quartz diorite, medium-grained augite quartz diorite, fine-grained diorite, albite diabase, and minor granodiorite.	None	13	
JCr	Leucocratic granodiorite, albite granite and quartz diorite, with minor aplite.	None	12	
0-250 Tr 3	Thick- to thin-bedded light-gray limestone containing abundant fossils.	Potential source of lime products.	11	
2000-4400 Tr 2	Predominantly medium-gray to black shale and siltstone, with abundant lenses of metadacite tuff and quartz keratophyre tuff. Also includes lenses of limestone and lava flows.	Host rock for sulfide deposits.	10	
100-2500 Tr	Chiefly quartz keratophyre lava flows and pyroclastic rocks but includes metadacite, keratophyre, and shaly tuff. Also includes quartz keratophyre dikes in the Dekkas andesite.	Host rock for sulfide deposits.	9	
1000-3500 Pe	Chiefly keratophyre and spilite lava flows and pyroclastic rocks, with lenses of siliceous mudstone and quartz keratophyre in upper part.	None	8	
0-2000 Pe	Chiefly tuffaceous mudstone and tuff interlayered with subordinate tuff breccia. Fossils abundant in some tuffaceous mudstone beds.	None	7	
500-2500 Pe	Thin- to very thick-bedded light-gray limestone, with local layers of chert and chert nodules. Many beds contain abundant fossils.	Host for contact metasomatic magnetite deposits. Potential source of lime products.	6	
3000-5000 Ms	Lower part consists dominantly of pyroclastic rock with subordinate mudstone and keratophyre flows; middle part chiefly siliceous mudstone, with subordinate tuff and minor limestone; upper part is keratophyre, lava and volcanic breccia.	None	5	
0-3000 Ms	Mainly shale and mudstone, with numerous lenses of grit and chert conglomerate; locally a few lenses of tuff and sandstone.	None	4	
0-400 D2	Chiefly dark gray siliceous mudstone, tuff, and brownish-gray tuffaceous mudstone containing fossils; small lenses of light gray limestone in upper part.	Potential source of lime products.	3	
1000± D2	Quartz keratophyre lava flows and pyroclastic rocks.	Host for massive sulfide deposits in West Shasta copper-zinc district.	2	
2000± D?	Keratophyre and spilite lava flows and breccias; includes minor quartz keratophyre tuff.	None	1	

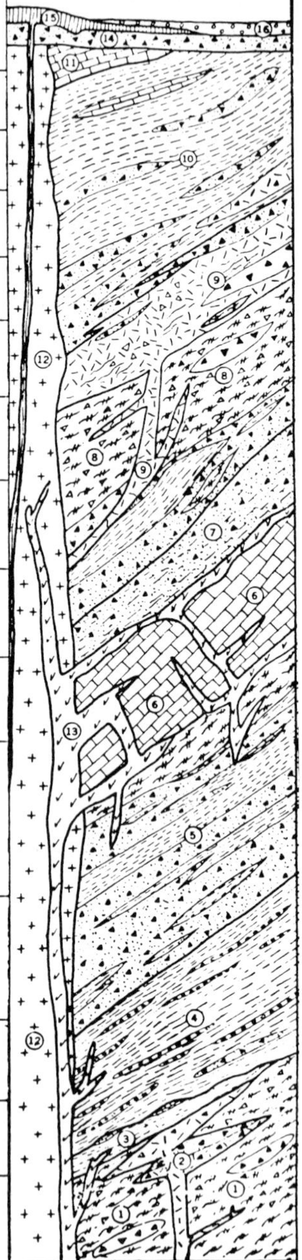

controversy. The most widespread generalization (e.g. Genshaft et al., 1975) states that sequentially - differentiated magmatic associations along continental margins originate by magma fractionation in regions of crustal compression, whereas unimodal (basalt) and bimodal (basalt-rhyolite) associations require crustal extension to form. This triggered a search for extentional regimes operational during the development of an orogenic belt, and a variety of terms for corresponding paleoenvironments has been introduced (for example, the "volcanic rift" of Scheibner and Markham, 1976).

The following varieties of crustal extension and bimodal volcanism have been recorded from the recent environments, and applied to the interpretation of the ancient associations: (a) initial rifting during ocean opening as in the Red Sea region; (b) crustal extension contemporary with island arc development, as in the inter-arc or back-arc basins; (c) crustal extension in the advanced phase of development of magmatic arcs, where taphrogenic grabens often form within a relatively thick and heterogenous crust; and (d) rifting and graben formation in the final stages of development and adjustment of orogenic belts, accompanied by extrusion of plateau basalts, rhyolites or trachytes (e.g. in the Rio Grande Rift). Few ancient taphrogenic events and paleoenvironments, however, can be accurately interpreted. Alternating compression and extension (Genshaft et al., 1975) or mixing and hybridization of magmas high in the crust (Eichelberger, 1975) smooth the differences. Fyfe and McBirney (1975) demonstrated that even the physiographically high-standing, sequentially-differentiated volcanic chains at continental margins such as the Cascades, are in fact situated in a tectonically subsiding area (graben or semi-graben), probably qualitatively not substantially different from a "volcanic rift".

The problem of bimodal assemblages partly overlaps with another petrogenetic problem of long standing, involving the spilite keratophyre association. Spilite (Na-plagioclase, usually albite; augite or more frequently its altered equivalents actinolite or chlorite; lesser epidote, calcite, hematite); keratophyre (albite, lesser quartz, epidote, chlorite, hematite, etc.); and associated rocks such as albite diabase, weilburgite, etc., have an extensive literature (e.g. Hughes, 1973; Amstutz, ed., 1974; Carmichael et al., 1974). They are also archetypes of the "geosynclinal" volcanism.

Fig. 6-4. Piling of a variety of lithologic associations apparent in a stratigraphic column, as in the East Shasta Cu-Zn field, California. There, two or three volcanic-sedimentary units (D, Unit 2 and Tr, units 9, 10) host stratabound massive sulphide deposits. From Albers and Robertson (1961).

Almost all recorded occurrences of spilites-keratophyres have been emplaced subaqueously, or into a water-saturated unconsolidated rock environment. Four principal petrogenetic hypotheses explain spilites-keratophyres as: (a) primary magmatic differentiates; (b) products of syn-emplacement metasomatism, particularly as a result of the hot lava / seawater interaction; (c) products of localized, post-lithification hydrothermal alteration; (d) products of superimposed metamorphism. Hypothesis (b) is presently most commonly applied to "typical" spilites.

Spilites-keratophyres are most common in the bimodal volcanic associations, and rare in suites containing andesite and dacite. Moreover, the majority of spilite - keratophyre occurrences (some 80-90%) have been described from prehnite-pumpellyite to greenschist metamorphosed terrains. It is interesting to note that products of greenschist metamorphism of basalts and andesites (actinolite or chlorite, albite, lesser quartz) as well as dacite and rhyolite (sericite, lesser quartz, albite, chlorite) are compositionally almost equivalent to spilites and keratophyres. In zones of intensive dynamometamorphism (penetrative deformation, shearing) where complete obliteration of relict igneous textures took place, andesite and dacite practically disappear as recognizable species due to the metamorphic convergence. Even worse, sediments are often indistinguishable from volcanics. Alternating zones of a darker, chlorite - rich greenschist and a light sericite schist common in "greenstone" terrains regardless of age, visually mimic the bimodal volcanic assemblage and tend to be routinely interpreted as such. Often it may be demonstrated that there is a corelation between the rock colour and composition and the intensity of deformation. Massive zones composed of a "greenstone" with well preserved pillows or hyaloclastite breccia change gradually or sharply into strongly schistose sericite (muscovite) or quartz-sericite schists. The latter are probably sheared, altered greenstones rather than felsic volcanic members.

In this book, no major reinterpretations of literature statements have been made, and the "visual bimodality" was accepted whatever its cause as an objectively recognizable field attribute. An effort, however, has been made to separate, when possible, the strongly schistose (dynamometamorphosed) terrains from the weakly schistose to unmetamorphosed ones, because of the metallogenic implications. The major drawback here is that minor localized schistose zones occur in most massive metavolcanic terrains and tend to be preferentially mineralized.

6.2.2. Volcanic-sedimentary associations

Volcanoes, particularly the subaqueous ones or those surrounded by a marine depositional basin, shed detritus which is then spread by gravity and currents. Volcanic terrains thus frequently change facies into sedimentary terrains but such a change rarely has sharp boundaries. In the literature, it is customary to differentiate between the proximal and distal facies of volcanic - sedimentary associations. The proximal facies forms in the immediate vicinity of the site of discharge of volcanic material (vent, fissure), whereas distal facies forms on the opposite end of the system under consideration, farthest from the source. This concept is most useful in the description of explosive pyroclastics and pyroclastic-sedimentary suites, because with the increasing distality the pyroclastics to sediments ratio and the particle size decrease. Franklin et al. (1981) pointed out that the terms listed above should not be confused with proximal and distal volcanogenic deposits. Proximal deposits occur over or immediately adjacent to sites of discharge (vents) of hydrothermal solutions, whereas distal deposits lack these vents. The ore distality is independent of the facies distality of the host units.

In most orogenic belts, "eugeoclinal" sequences have a dual provenance of detritus supplied by (1) approximately contemporary volcanism; (2) weathering and erosion of older exposed outcrops (usually a continent, supplying a terrigenous detritus). To this should be added intrabasinally generated (e.g. biogenic carbonates) and recycled or reworked sediments. Most "geosynclinal" sequences are lithologically monotonous (e.g. sandstone, shale, limestone), and difficult to subdivide into objectively recognizable smaller units. In the literature, the "strongest impression" has often been used to name facies or associations (sample of common names: flysch, turbidite, detrital wedge, shelf, shale-sandstone, etc. sequence). A single rock or a set of rocks could be a member of any of the above categories. The present inconsistent practice is hard to improve substantially, unless one resorts to extremely lenghty composite unit names that list all the rock types involved; a practical impossibility here.

6.2.3. Sedimentary associations

In the "eugeoclinal" megafacies of an orogenic belt, the bulk of sediments were derived from volcanics and subsequently from unroofed intrusive rocks of the magmatic arc, or from rapidly rising geanticlines. The latter can have a very distant and, in ancient environments, obscure provenance. Often turbiditic litharenite (graywacke), polymictic conglomerate, siltstone and shale are the most common (Chapter 16). Carbonate in the "eugeoclinal" domain occur either as "pelagic limestones" (Chapter 15) associated with or gradational into the ophiolite suite (Chapter 7), or as shallow-water buildup ("reef") limestones capping volcanic piles. Limestones are most common, dolomites rare. Shallow marine to continental detrital sediments in the volcanic- sedimentary belts are most common in "successor" or "molasse" basins superimposed on an earlier folded complex.

In the "miogeoclinal" domain adjacent to a continent (craton), volcanics are nonexistent or exceptional and generally thick. An extensive sedimentary sequence rests on a cratonic basement. These sediments have continental (terrigenous) and intrabasinal (biogenic carbonates, evaporites) provenances, and range from subaerial (Chapter 26) through shallow marine (Chapters 19, 20) to fairly deepwater ones (Chapter 17). The latter formed in a variety of structural troughs within a broader shelf; at continental slopes; or in marginal seas. In orogenic belts, a very regular facies progression of the "miogeoclinal" sediments can be observed, although in some cases (e.g. in the Alps) this is disrupted by nappes and thrust faults.

6.2.4. Variations due to metamorphism

GENERAL

It is common practice among many field geologists to designate metamorphic rocks by their pre-metamorphic names (e.g. "graywacke" and "shale" instead of phyllite or schist; "basalt" instead of greenschist or amphibolite), as long as they believe interpretation can be made safely. Petrographers, on the other hand, use the scientifically more correct metamorphic terms. Unfortunately, there is no unity and, as a consequence, terrains of comparable lithology and style are often described using different terminological emphasis and this often makes correlations and comparisons difficult. This problem has recently been amplified rather than alleviated, because in the 1970s vintage plate tectonic literature, there was rarely a place left for metamorphism (except the high-pressure one).

In this book, for the sake of consistency, all chapters based on lithologic associations include rocks ranging from unmetamorphosed to moderately metamorphosed. Further subdivision by intensity or style of metamorphism (e.g. "massive" versus "schistose" associations) was made within chapters, where warranted. High-grade metamorphic rocks where metamorphism is the most fundamental aspect and the pre-metamorphic rock composition often obscured, are treated in Chapter 29.

METAMORPHIC CONVERGENCE

Some rocks (quartz arenite, limestone, etc.) are merely recrystallized by the effects of metamorphism and pre-metamorphic equivalents are relatively easy to interpret. Other rocks (lithic arenites, shales, volcanics, etc.) substantially change compositionally as well as texturally, during metamorphism. Convergence is frequent, and a variety of original rocks (e.g. rhyolite, porphyry, arkose, granite, felsic tuff) can produce a single metamorphic rock type (e.g. quartz-sericite schist). Genetic convergence of albite, albite-chlorite and similarly composed rocks has already been discussed earlier, in connection with the spilite keratophyre problem. Metabasalts (amphibolites) have frequently been recorded in the literature as "andesites", due to Ca and Mg depletion during metamorphism. Meta-"rhyolites" often formed by feldspathization and Mg, Ca, Fe removal from intermediate or mafic volcanics. "Greenstones" and "greenschists" are sometimes metasomatically chloritized ("basified") felsic volcanics. Sericite schists with quartz augen have been interpreted either as porphyritic (meta)rhyolites or as metavolcanics or metasediments with quartz porphyroblasts. Many "gabbros" or "diorites" are hornblende metasomatites rather than primary magmatic rocks, and so are many "syenites" (particularly albitic syenites).

STATIC AND DYNAMIC METAMORPHISM AND METAMORPHICS

Static metamorphism is least destructive with regard to preservation of the depositional features of rocks and ores, and generally, in the past, was considered isochemical. Although several examples to the contrary have been described (e.g. by Hower et al., 1976, from the burial metamorphosed Tertiary sediments in the Gulf of Mexico), the mobility of elements, including ore metals, in statically metamorphosed sequences is limited and rarely results in metamorphogenic ore deposition.

In contrast, varieties of metamorphism associated with penetrative deformation, shearing and movement (dynamometamorphism) are demonstrably accompanied by a significant chemical transfer, that could result in the generation of metallic ores. It is quite obvious that terrains considered comparable in pre-metamorphic composition, but modified by a different style of metamorphism, tend to have different intensity and style of mineralization.

In California, the virtually unmetamorphosed Mesozoic Great Valley Sequence (Davis, 1966) is composed of bedded lithic sandstones, mudstones and minor basalts resting on ophiolites. Almost no metallic mineralization is known, except from the ophiolites (e.g. podiform chromite) and from fractured ophiolite contacts where hot springs deposited cinnabar orebodies. In Klamath Mountains, similar original rocks have been penetratively deformed, low-grade metamorphosed and intruded by granitic massifs. The rocks are dominantly lustrous slates and greenstones. The proportion of sandstone appears reduced, but this can often be an optical illusion. In a sequence of "slates" (or schists), the original graywackes composed of unstable (volcanic) clasts loose their depositional granularity by clast conversion into metamorphic silicates. Consequently, slate or schist formed in place of alternating shale and graywacke appear homogeneous and uniform after dynamometamorphism, at least to the casual observer. The slates or schists of the Klamath Mountains host widespread gold mineralization. The gold lodes appear syn- to post- metamorphic, and many are pre-granite.

A similar relationship exists among quartz-rich unmetamorphosed flysch sequences (e.g. the Alpine flysch), one of the most ore-deficient associations known, and the "slate flysch" in places richly gold mineralized (e.g. the Victoria goldfields, Australia; Meguma gold region, Nova Scotia; etc.).

Hopwood (1976) in his provocative paper on "quartz-eye porphyroids", recognized two fundamental lithologic assemblages hosting massive sulphide deposits, different in style and degree of metamorphism: (1) single-phase folded and unmetamorphosed to low-grade metamorphosed "flysch" with persistent stratigraphy and units continuity (e.g. the middle Proterozoic Aldridge Formation, hosting the Sullivan Pb-Zn orebody, British Columbia); (2) repeatedly deformed, greenschist-metamorphosed volcanic-sedimentary associations, characterized by steep schistosity and an almost vertical mineral-streaking style of lineation which is often parallel to steep fold axes. The stratigraphy is complex and lenticular, with many discontinuous units and the massive orebodies are conformable with the structure and could be metamorphogenic or substantially remobilized. "Cherts" in the latter case are probably the result of post-metamorphic silicification, rather than syn-depositional "exhalites".

6.3. INTRACRUSTAL ROCKS AND SUPRACRUSTAL / INTRACRUSTAL INTERACTIONS

Intrusive bodies of magmatic rocks are depth-generated lithologic associations treated later in this book. Except for the relatively rare but distinct, mantle-derived differentiated layered intrusions of the "Bushveld-type" (Volume 2), diabase dikes and sills (Chapter 32), and to a lesser degree alkaline intrusions (Chapter 33), the bulk of intrusions in orogenic belts are varieties of "granitic rocks", generated within the crust (Barker, 1981) (Chapter 28). This has important lithogenetic and metallogenetic implications. While it is possible to treat mafic and some alkaline intrusions once they reached the crust, as a closed system in which the variety of rock and ore types is essentially the product of internal differentiation, such approach would be insensitive for the treatment of "granites". Because the latter formed from the crust and occur within the crust, the character of the crust has a substantial bearing on the nature of the "granite" and has to be considered in interpretations.

Some "granite" massifs formed metasomatically "in situ" by thermal transformation of preexisting rocks "without, in the main, passing through a liquid state" (Weaver, 1958, for the Utuado Pluton, Puerto Rico). In cases such as the above one, the generative (source) site of the "granite" coincides with its site of emplacement and the mutual interaction is of a relatively uncomplicated type. The majority of "granites", however, crystallized from a magmatic melt generated at deeper crustal levels so that the sites of granite generation and emplacement are different and intrusion / roof interaction significant.

The classical endogenous metallogeny of the 1930s to 1960s (e.g. Bateman, 1951) paid serious attention only to the internal differentiation processes supposedly taking place within the "granitic" magma chamber considered as a closed system, and to the "export" of the various mobile fluids including ore-bearing solutions into the above-intrusion roof, to produce ores. Little attention was paid to the roof itself, except from the structural control viewpoint. By now, considerable advances have been made in demonstrating the importance of interaction between the intrusion and its surroundings, one of the fundamental prerequisites of hydrothermal ore genesis (compare Pavillon, 1964; Franklin et al., 1981; Laznicka, 1985c,e).

The manifold function of the roof rocks is (1) to receive and accomodate the granite-supplied metals; (2) to act as a "screen" that traps and retains ore metals, but allows the spent mineralizing fluids to pass through and escape; alternatively, it deflects and returns the circulating fluids as in convecting systems; (3) to influence physicochemically the fluids so that they precipitate their metallic load; and (4) to supply the ore metals and / or the constituents of the mineralizing fluids. Dodge and Bateman (1977) studied the spatial relationship of mineral deposits to the Sierra Nevada Batholith (California) and concluded that the source of most metals in ore deposits there (W,Mo,Au,Cu) was in country rocks adjacent to the deposits. Barbier (1970) even argued that granites tend to inherit the geochemical background of the site into which they were emplaced, and produce superimposed compatible mineralization (particularly of uranium).

Although at present the "granite" metallogeny is passing through a period of accelerated evolution and many concepts are in flux, the wealth of observations and literature data makes it clear that certain metals and certain mineralization styles are repeatedly hosted by distinct lithologic associations in a "granite" roof or aureole. This is an important empirical lead in exploration that deserves to be harnessed and tabulated, and a major effort has been made to do so in this book. To accomplish this, however, an emphasis split was necessary and attention to "granites" had to be paid twice: (1) in chapters dealing with the dominantly supracrustal rocks in orogenic belts (7-17,19,20,26), the emphasis is on the supracrustal association and "granites" are treated as one of the agents of interaction. (2) in Chapters 28 and 29 that treat the calc-alkaline intrusive systems, the emphasis is reversed. Sharp separation of both emphases is impossible and as usually, there is an overlap in coverage to prevent a loss of the border cases.

In considering and interpreting the "granite" and supracrustals interaction and metallogeny, it is of considerable importance to know whether the supracrustals (in this case, volcanics) are broadly contemporary and comagmatic with the intrusive, or whether the intrusive was emplaced independently later. In the first case, the volcanics show a gradational system continuity and change gradually with the increasing depth into subvolcanic, hypabyssal and plutonic equivalents, derived from the same magma. Comagmatic suites need not always be strictly contemporaneous at the level of exposure under consideration. The deeper-seated intrusive stocks or plutons are almost always slightly younger than the volcanics, into which they may have risen by the process of magmatic stoping (e.g. Myers, 1975). "Granites" independently emplaced at high crustal levels into sediments or substantially older volcanics, on the other hand, display a relatively sharp and compositionally contrasting contacts.

Recently, several igneous-looking rocks closely associated with metamorphosed, mostly subaqueous volcanic piles, have been interpreted as metasomatites. Examples include the quartz and feldspar rich "trondhjemites" in the Norwegian Caledonides (Waltham, 1968) and similar rocks formed by "porphyrization" (=metasomatic porphyroblast growth) of volcanics in the southern Urals (Vakhromeyev, 1969). Both rocks are affiliated with massive sulphide deposits.

6.4. GENERAL PROBLEMS OF MINERALIZATION IN OROGENIC BELTS

6.4.1. Hosting of ores

"Hosting" (e.g. volcanics-hosted massive sulphide) records an empirical observation. There is no genetic interpretation. A volcanic-hosted deposit may have been created as a result of volcanic activity (volcanogenic deposit) and be contemporary with the volcanism (synvolcanic, volcanic-syngenetic). Alternatively, it may have been generated from fluids produced or driven by an intrusive body in depth (plutonic, postmagmatic, intrusive-hydrothermal, etc.), in which case the ore emplacement would postdate the deposition and solidification of the host volcanics (epigenetic deposit). The genetic interpretation is rarely known with certainty and it changes with time, whereas the hosting term is straightforward and permanent. In many cases, ore is hosted by a certain rock, and another rock believed to have possibly had a genetic influence occurs nearby. In this case, it may be advantageous to note such association in a descriptive term, for example: sediment-hosted, volcanic-associated mineralization.

Franklin et al. (1981) in their review of classifications of massive sulphides, recommended distinction between (1) volcanic-hosted; (2) volcanic-associated (but sediment-hosted); and (3) sediment-hosted (without associated volcanics) deposits. All such classes are, however, transitional and any sediments deposited within a radius of several hundred or even thousand kilometres from an explosive volcanic centre (particularly downwind), may have received an airborne volcanogenic contribution. This, however, is difficult to demonstrate in fine grained, clastic sediments in which the volcanic component underwent diagenetic and metamorphic reconstitution and converges with the detrital or authigenic sediment components.

6.4.2. Metal zoning across orogenic belts

Buddington and Chapin (1929) outlined several mineralized belts in the Cordilleran orogen in S.E. Alaska and adjacent British Columbia, within each of which "deposits valuable for a certain metal or a group of metals may predominate and the type of mineralization may differ for each belt. The tendency toward localization is, however, only a tendency, not a rule, for mineral deposits that differ in character from the prevalent ones are found in each belt, and mineralization is by no means confined to these localized belts". This has been a remarkably restrained and scientifically sensible statement. In subsequent years, a zoned distribution of metallic deposits usually having a form of a belt and elongated parallel with the axis of an orogenic belt, has been widely recognized and described (compare Laznicka, 1985c, for review). Progression of metal- or mineralization-style zones is transversal to the trend of a belt, and is either assymetric or symmetric.

Gabelman (1976) recognized the following zoned sequences of ore metal culminations, in direction from the paleo-ocean to the continent:

European Hercynian Belt:	magnetite-Sn-W-As-Au-Cu
Appalachian Belt:	magnetite-Au-W-Mo-As-Au-Cu
the Andes:	magnetite-high temp.Cu,Zn,Pb-Cu sulphides - Pb,Cu,As,Sb sulphosalts; low temp. Au - Cu in sandstone - U,V in sandstone.

There certainly is not a standard uniform zoning pattern of orogenic belts which is not surprising, given the variety of origins, assembly and depth of erosion of such belts. Most metal belts, moreover, are laterally discontinuous (Fig. 6-5) with many subtle variations, so that establishment of a continent-wide zoning pattern is highly speculative. Fig. 6-6 plots semiquantitatively the zones of intensity of mineralization and of the prominently accumulated metals, based on the five example orogenic belts shown earlier on Fig. 6-2. The unequal intensity of mineralization in belts of various types and within the belts as related to facies and the depth of erosion is immediately apparent.

Belts dominated by collisional effects (the Himalayas, partly the Alps), are, as a rule, lightly mineralized or barren. Belts related to subduction or intracratonic fragmentation are most intensively mineralized in moderately eroded segments rich in "geosynclinal" volcanics, black metasediments and epi- to mesozonal granites. Deeply eroded batholiths, migmatitic domes, flysch wedges and platformic flanks are weakly mineralized to barren.

Culminations of depositional concentration of metals in ore deposits within orogenic belts are controlled by facies and megafacies zones. Some metals are diagnostic for certain zones (e.g.Cr,Ni,Co for ophiolites). Other metals indicate a broader set of conditions (Sn: anatexis and extreme "refinement" of continental lithosphere). Copper and lead, two metals that have clarkes of the same order of magnitude

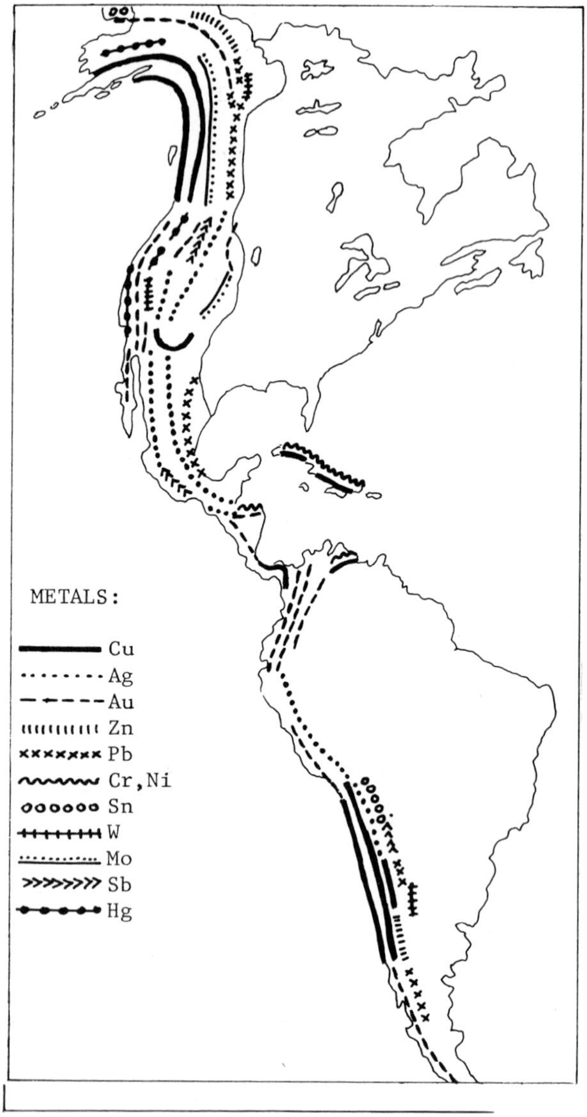

Figure 6-5. Trends (axes) of maximum geochemical accumulation of metals in ore deposits located in the Cordilleran and Andean orogenic belts From Laznicka (1970).

(35 and 16 ppm, respectively) and which accumulate by similar processes (hydrothermal and sedimentogenic) tend to, nevertheless, be distinctly separated in deposits of composite orogenic belts. The bulk of the Cu deposits is in the "internal" zones of orogens, the bulk of Pb deposits is in the "external" zones (Fig. 6-7). Copper may reappear in the "external" zones, almost exclusively affiliated with taphrogenesis that in turn introduced mafic magmas into the crust. The "external" Cu zones in orogenic belts and in their fringe are, however, considerably discontinuous.

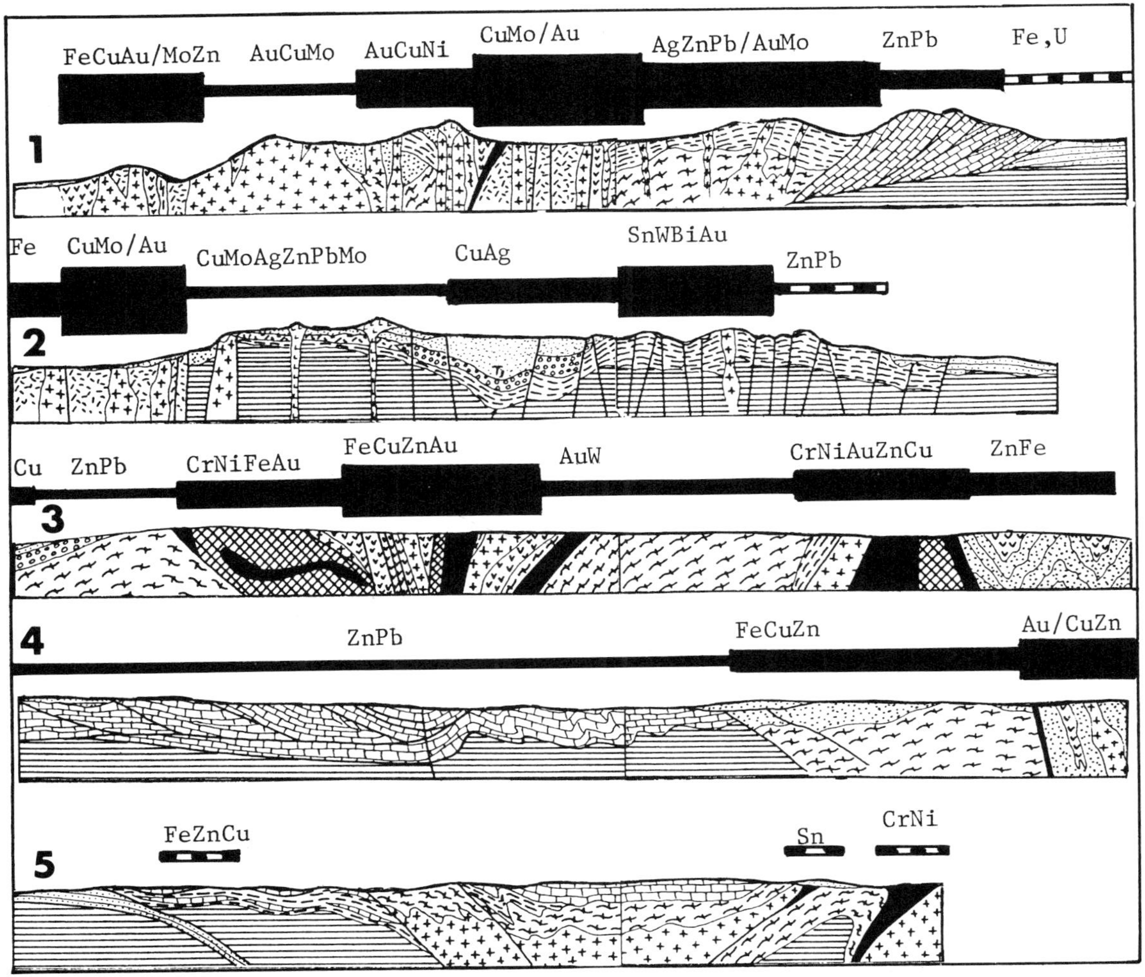

Figure 6-6. Zoning in mineralization intensity and major ore metals across the five orogenic belts shown on Fig. 6-2:
(1) North American Cordillera in southern British Columbia;
(2) the Andes in N. Chile and Bolivia;
(3) central Urals;
(4) southern Appalachians;
(5) Himalayas.

See references in Fig. 6-2.

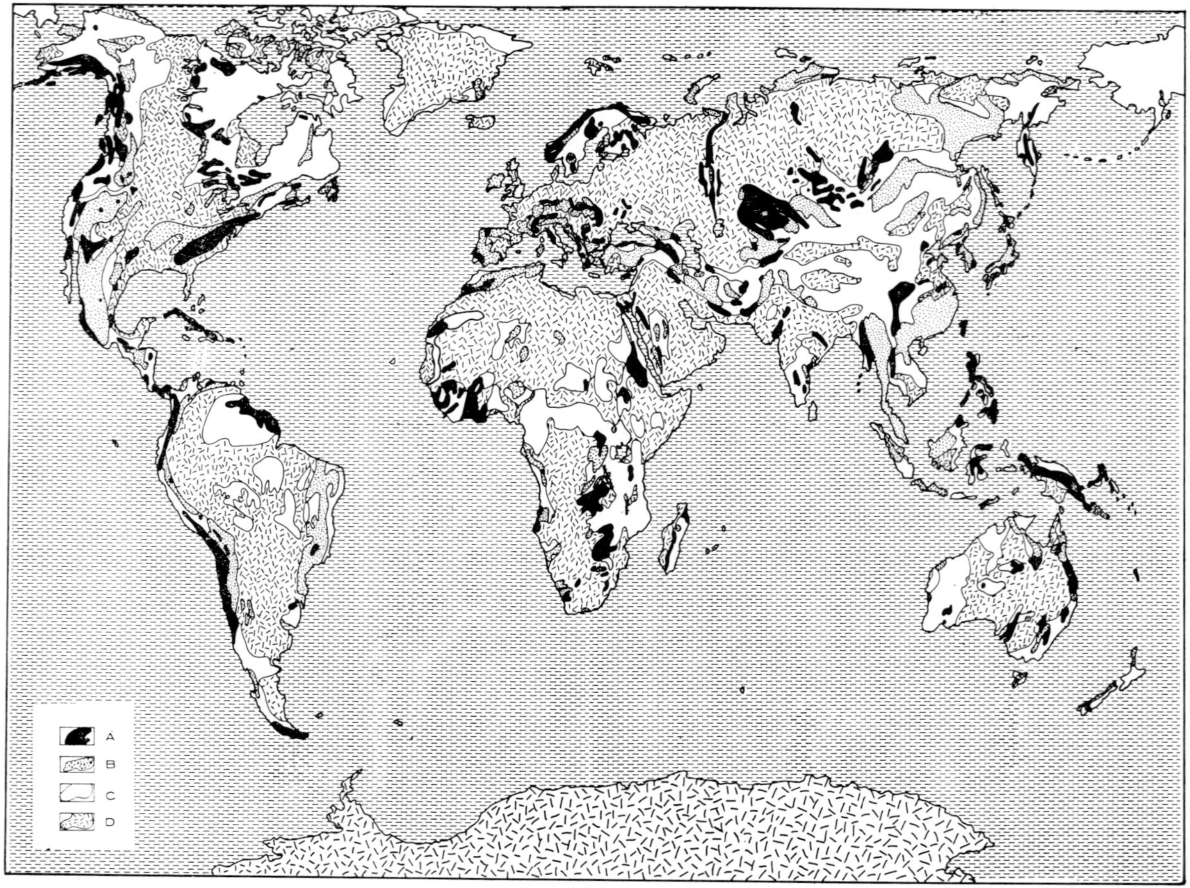

Fig. 6-7. Map showing that the bulk of copper deposits in orogenic belts occurs in the "internal" (perioceanic), the bulk of lead deposits in the "external" (pericratonic) zones.

A: Cu greater than Pb, B: Pb greater than Cu; C: basement areas with little or no Cu or Pb; D: post-mineralization cover. From Wilson and Laznicka (1972).

Laznicka and Wilson (1972) discussed the metallogenic importance of a Cu:Pb line, a statistically derived boundary between mineralized belts dominated by either metal. Although the causes of its presence are numerous, it appears that the line (in fact a plane) marks the proximity of the edge of an ancient continental crustal block in the subsurface, incorporated into the orogenic belt. Pb suddenly appears above such a block while Cu, most abundant in the transitional crust megafacies located oceanward from the continental edge, tends to disappear in the continentward direction more gradually.

The position of the remaining metals in the zoning pattern across orogenic belts is less predictable and strongly influenced by local conditions. Fig.6-8 shows, semiquantitatively, the preferential accumulation of base, rare and precious metals in a symmetrical, facially divergent orogenic belt (as in the Alps; Fig. 6-9).

6.4.3. Stratiform, stratabound and concordant deposits compared with the discordant ones

Although ore deposits interpreted as "metalliferous sediments" (e.g. placers; oolithic Fe ores; the Kupferschiefer) have been known and correctly interpreted for a long time (e.g. Lindgren, 1933), the importance of stratigraphic control to some of the more controversial deposits (e.g. massive sulphides; Mississippi Valley "type") has been recognized only relatively recently (e.g. Stanton, 1955).

In its early phase, the reinterpretation essentially took form of protest movement against the established monolithic interpretation of virtually all hydrothermal deposits as precipitates from "granite"-supplied postmagmatic fluids. The stratiform or stratabound concept rejuvenated exploration, contributed to a rash of new discoveries, and generated an extensive literature (e.g. Wolf, ed., 1976, 1981; Klemm and Schneider, eds., 1977). In some instances, however, the concept was unduly stretched, due to its heightened popularity, to the point of becoming restrictive. An exploration geologist should be aware of it; several examples of the excess stretching are treated in Laznicka (1985b).

A universally accepted definition of the stratiform or stratabound relationship is missing. The majority of geologists consider stratiform deposits to have the form of a concordant sheet or a lens, formed contemporaneously with bedded sediments, volcanics or layered intrusions. The European "Kupferschiefer", a sharply outlined, less than one metre thick bed of a metalliferous marl distributed in subcrop over an area of several tens of thousand kilometres, is a well known example. The Merensky Reef, a magmatic layer of similar dimensions and traceable along a strike for over 250 km, is another.

Stratabound mineralization is merely stratigraphically confined and need not be concordant or contemporary with its hosts. The limiting thickness of the reference strata (a unit to which the orebodies are bound) has never been agreed upon; for example, about 10% of ores designated as "stratabound" in the literature are confined to intervals over 500 m thick.

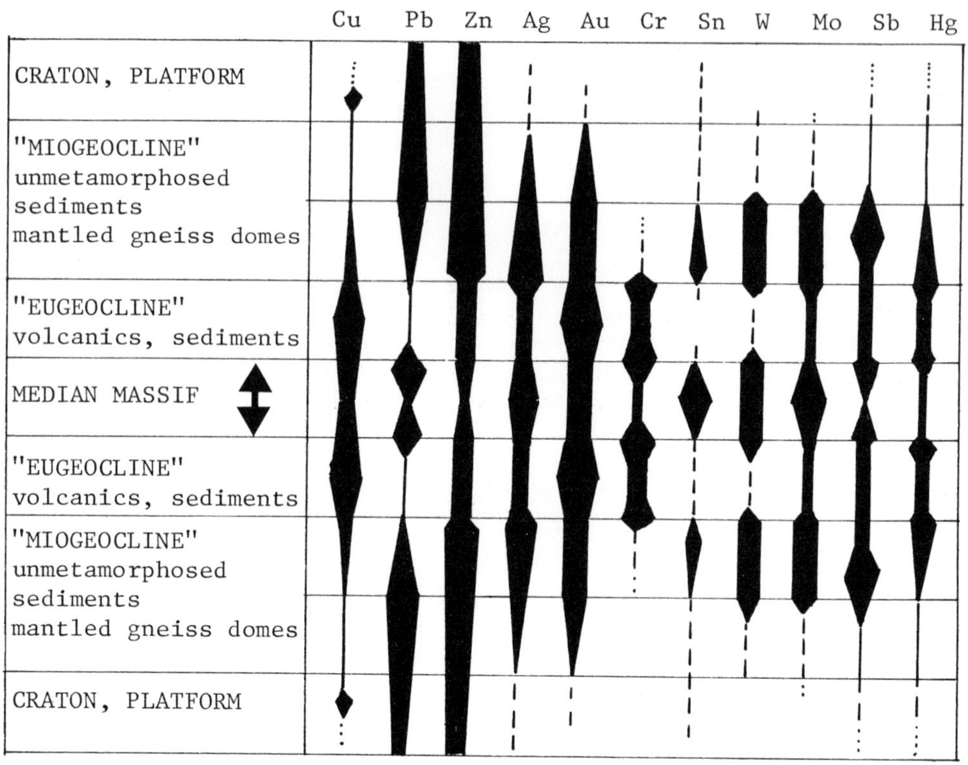

Fig. 6-8. Preferential accumulation of base, rare and precious metal deposits in a symmetrical, facially divergent orogenic belt. Modified from Laznicka (1970).

In metamorphic terrains, pre- and syn—metamorphic orebodies are almost always concordant with the tectonic fabric (schistosity, foliation), but need not originally have been stratiform; in fact, many such orebodies probably did not even exist before the metamorphism. Unless there is a good evidence that metamorphosed deposits were originally stratiform ores in supracrustal rocks, they should not be so named.

Orebodies that do not satisfy the criteria for stratiformity, are non-stratiform, or simply without designation. Most are discordant, crosscutting or transgressive when they are hosted by a bedded or layered host. In a massive, non-layered rock (e.g. a "granite"), their position is neutral.

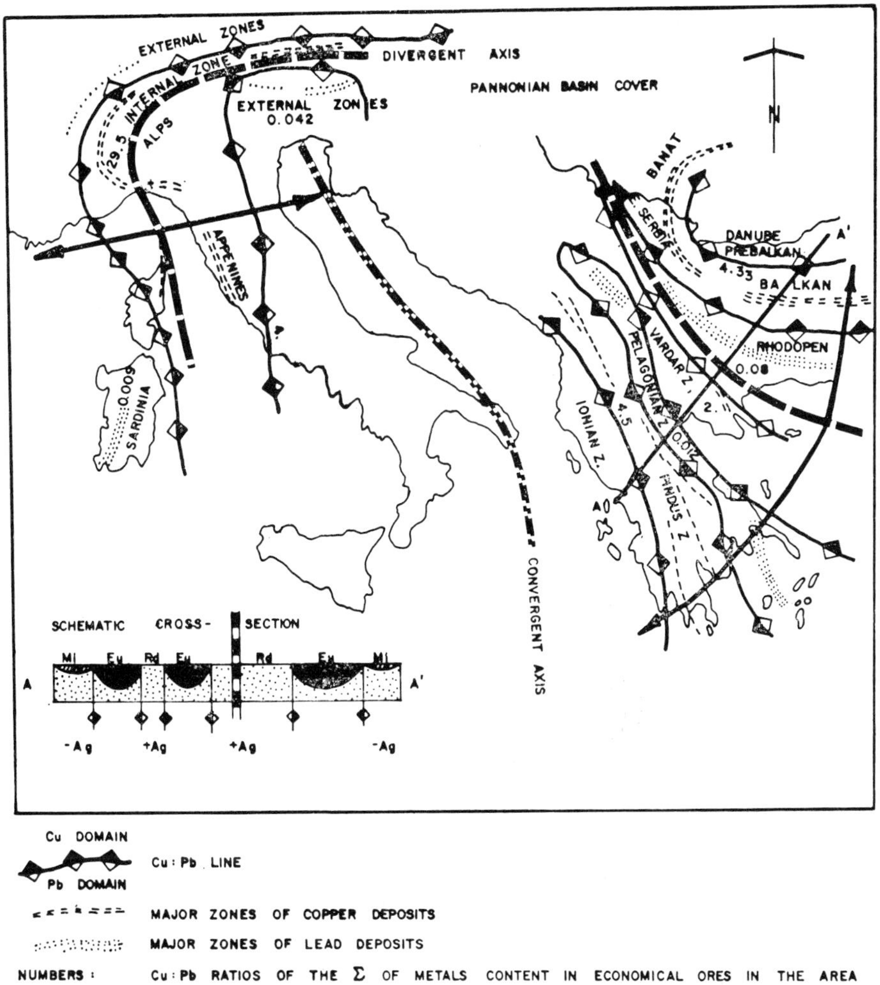

Fig. 6-9. The concept of divergent (d) and convergent (c) facies and metallogenic zones (facies zones after Auboin, 1965), in the Italo-Dinaric region of the Alps-Mediterranean orogen of Europe. Cu dominates the internal, Pb the external zones. From Laznicka and Wilson (1972).

Many field associations of ore deposits consist of concordant and discordant members. Laznicka (1985b) coined the term doubles (contemporary discordant and concordant orebodies forming a continuous, composite set; e.g. a concordant massive sulphide lens underlain by a discordant feeder pipe), and twins (two close, contemporary and genetically related, but discordant orebodies); Fig. 6-10. Several known examples of discordant orebodies cut across an older concordant mineralization, or occur in its vicinity. They may have formed by remobilization (Fig. 6-11). Remobilization has been defined (Laznicka, 1985b) as "any natural process of transformation of a preexisting orebody or a geochemically enriched segment into a new orebody, without a substantial gain or loss of the metal concentration". Mobilization, on the other hand, is only a geological process that extracts metals previously dispersed in a solid rock, channels them and increases their local concentration to such a degree that a significant geochemical accumulation or an orebody may form.

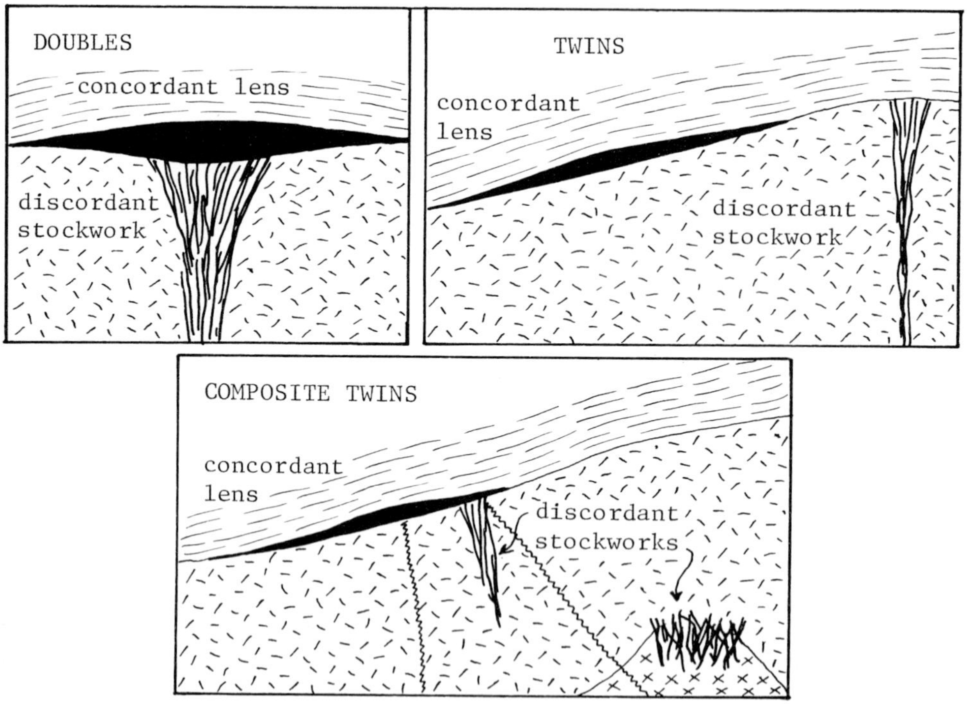

Fig. 6-10. Mineralization twins and doubles (see the text for explanation).

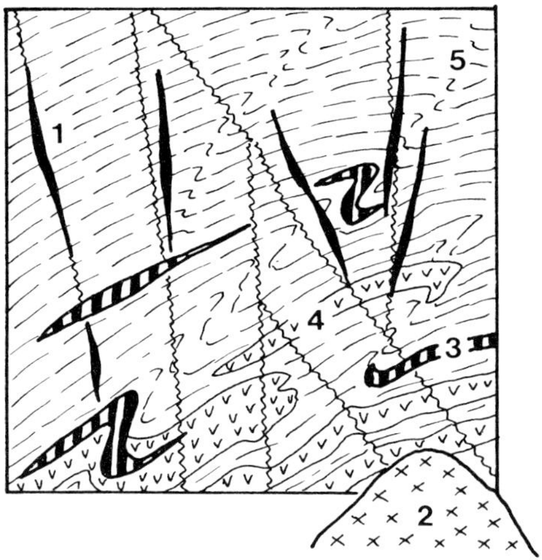

POST-TECTONIC INTRUSIVE ASSOCIATION

1. Pb-Zn fissure veins
2. granodiorite

EARLY "GEOSYNCLINAL" VOLCANIC-SEDIMENTARY ASSOCIATION

3. stratiform Pb-Zn orebodies
4. volcanics (greenstone)
5. gray or black phyllite

Fig. 6-11. Diagrammatic model of epigenetic, hydrothermal remobilization veins hosted by a marine volcanic sedimentary association that also contains earlier stratiform orebodies.

Remobilization orebodies, mostly veins, are usually of limited economic importance, although significant exceptions are known. Suspected remobilization veins, mineralized faults, breccias, etc., however, are indicators of a possible presence of an earlier stratabound ores in the region, so these are important in exploration. Ore veins located in the vicinity of stratiform sulphide occurrences could either be (1) feeder veins; (2) remobilization veins; or (3) genetically unrelated veins. Laznicka (1985b) tabulated the criteria that can be applied to distinguish the first two categories (Table 6-1).

6.4.4. Metallogenic cycle, relic, transformed and neoformed mineralizations

The metallogenic cycle is an evolutionary sequence of ore-forming processes considered in time and space, resulting in ore formation or a significant geochemical enrichment. The classical metallogenic cycle of the 1950s and 1960s (Bilibin, 1955) was based on the geosynclinal — orogenic model. A variety of usually partial metallogenic cycles has been suggested, based on the plate tectonic model (e.g. Mitchell and Garson, 1981).

Table 6-1. Criteria to interpret mostly discordant veins (veinlets, stockworks, etc.), spatially associated with mostly concordant orebodies. From Laznicka (1985b). V=veins; C=concordant bodies.

FEEDER VEINS	REMOBILIZATION AND SECRETION VEINS
TIMING	
V are contemporary with the concordant orebody, determined by: -lack of crosscutting relationships between both styles of orebodies; -when hosted by metamorphics, C and V have the same metamorphic grade; -evidence indicative of additive movement of metals from V to C, e.g. overlaps, penetrations. Vector of movement is indicated by textures and zoning;	V are younger than C, superimposed; determined by: -veins crosscut the concordant ore; -in metamorphics, V and C may be of the same grade, or V have lower grade than C (or are completely unmetamorphosed); -evidence indicative of subtractive movement of metals from C to V, e.g. depletion haloes in C along contacts with V;
STRUCTURE, TEXTURE	
If C is of subaqueous-hydrothermal origin, V's are developed in the C footwall only; hydrothermal feeder veins tend to be more persistent, sometimes display a vertical zoning with a P/T gradient increasing with depth; V's are not stratigraphically confined, usually cross several units	V's are in the hangingwall, in C, in the footwall; V's are impersistent, form and die out within a narrow vertical interval; mineralized V tend to be confined to the stratigraphic or lithologic interval that hosts C;
MINERALOGY, GEOCHEMISTRY	
V tend to be mineralogically more complex than C; hydrothermal V's tend to have equivalent or higher temperature mineralogical assemblage than C; the more mobile components (under given P/T conditions) are enriched in C, the less mobile in V; intensive hydrothermal wallrock alteration is usual along V's, also in the footwall of C; when V's contain metallic minerals, mineralization is vertically persistent; gangue minerals in V and C may be unrelated.	remobilization V's are either mineralogically simplier or more complex than C (accumulating various trace metals from C); V's tend to have equivalent or lower temperature mineralogical assemblage than C; the more mobile components are enriched in V's; weak (discoloration, bleaching) to absent wallrock alteration; metallic minerals in V's are usually confined only to the interval where they actually cross C; cements correspond to C.

The criticism levelled against the geotectonic cycle, is also applicable to the metallogenic cycle. It is a deterministic scheme with numerous exceptions, yet it demonstrates convincingly that as local geological conditions change and evolve, the styles of mineralization change too. As this happens, different metals may form concentration maxima sequentially and, under very favourable conditions, generate regional zoning patterns. An evolutionary pattern of oceanic crust (mantle)-continental crust transformation, is approximately as follows: Cr-Ni,Pt,Co-Fe,Cu,Au-Zn,Pb,Mo-W,Sn-Be,Ta,U. Alternatively, a single ore metal (or a group of metals) may persistently accumulate locally in a sequence of evolutionarily more advanced mineralization styles (Fig. 6-12). The French school (Lab.Geol.Appliq.,1973; Routhier, 1980) attributed this (l'heredite) to a repeated tapping and reconstitution of a primordial subcrustal inhomogeneity, possibly in the upper mantle.

Repetitions such as the one illustrated above are very common, and they may serve as a helpful exploration aid if sensibly interpreted. In this book, reference is made to three categories of ore deposits, considered from the point of view of a certain mineralization period or an ore forming process: (1) relic ores; those generated earlier and preserved in the new association; (2) transformed ores; those formed earlier but modified into compatibility with the new conditions; (3) neoformed ores; those newly generated.

6.4.5. Structural control to mineralizations

In the classical period of economic geology (1930s to 1950s), structural control to orebodies was overemphasized and treated in a deterministic way. This was a by-product of the pervasive adherence to the endogenic postmagmatic hydrothermal model, where ore fluids were automatically considered to emanate from a "granitic" pluton undergoing differentiation, or from the upper mantle. This interpretation is still valid for many classes of ore deposits (e.g. truly postmagmatic veins), where the ore-forming fluids require avenues and channels to move, and a space to deposit their load. In such a case, a search for faults is one of the more reliable exploration techniques. It should be realized, however, that major faults, active or ancient, do not necessarily have a better chance of being mineralized than insignificant branch faults or local fractures. Generally it is the other way around. The considerable variety of mineralized and ore-controlling structures is extensively treated in textbook of economic geology, and cannot be reviewed here.

Recently, the concept of ore-generating structures has been applied to ore interpretation. Such structures do not simply guide the ore fluids or provide a space for ore deposition, but also confine and control the agents that mobilize and upgrade the dispersed metals from the surrounding rocks, and redeposit them in the form of an economic mineralization presumably within the structure and its surrounding. Axial fold zones in penetratively deformed terrains and shears are the best known ore-generating and hosting structures.

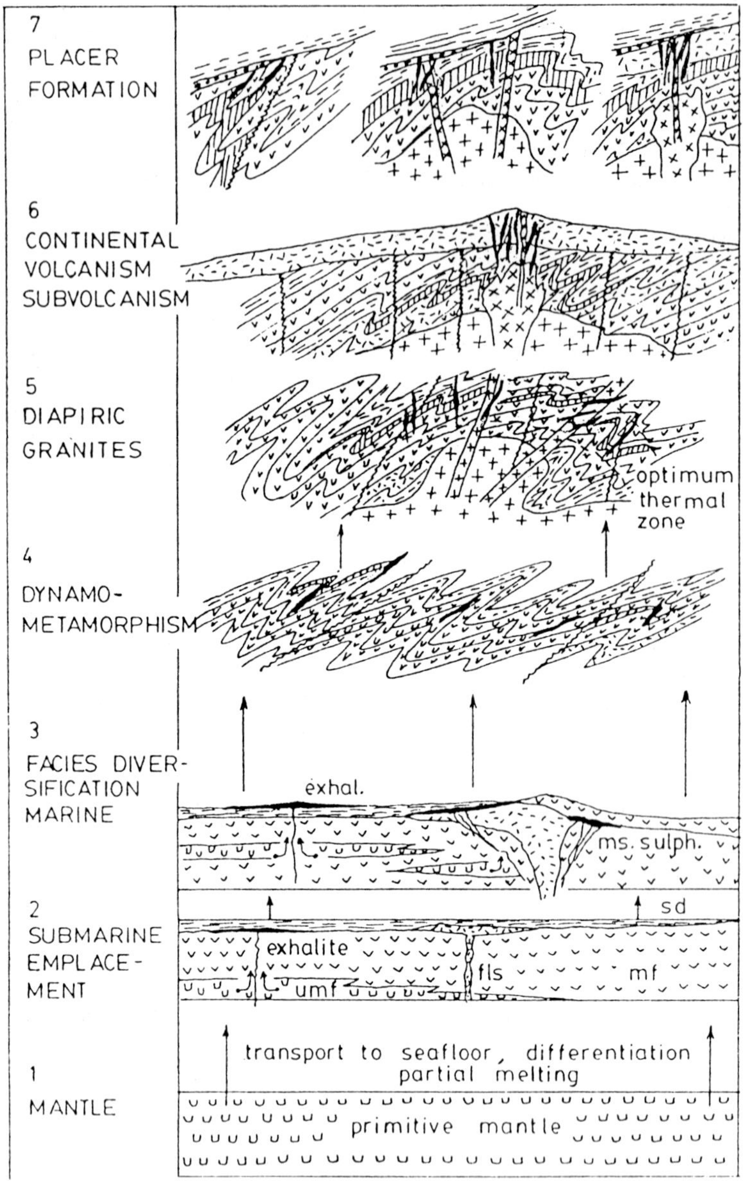

Fig. 6-12. Metallogenic cycle of a single metal (gold), illustrated by a hypothetical scenario of a possible sequential derivation of gold from initial mantle materials in the course of an orogenic belt development, as in the "eugeosynclinal" greenstone belts. From Laznicka (1985e).

SCENARIO

(1) "Primitive" (undepleted) mantle is considered gold enriched, relative to the present crust. This is based on indirect evidence of trace gold content in supposedly derived rocks (e.g. the komatiites) and on the modern interpretation of the geological history of the large gold belts. The gold content in the "primitive" mantle is probably of the order of 0.0X ppm Au, and could be rather evenly distributed.
(2) Submarine emplacement of "primitive" partial melts from the mantle and initial lithologic diversification. Orthomagmatic gold may accumulate slightly preferentially in certain rocks (ultramafics and basalts; in these komatiitic, Mg-rich suite is considered to be higher in gold than the tholeiitic suite. Boyle (1979) gave trace gold means as 0.0114 ppm Au in ultramafics, 0.0174 ppm in basalts, 0.0230 in mafic intrusives). Highest gold contents approaching ore or protore have been detected in zones of localized alteration, flushing and redeposition, e.g. in the "exhalites".
(3) Further submarine to partly subaerial facies diversification. Sequence differentiated volcanic — intrusive association (basalt, andesite, dacite, rhyolite) and derived sediments form. Gold is locally and selectively withdrawn from earlier rocks, enriched and reprecipitated in "exhalites" that include ankeritites, iron formations, cherts and Fe (Fe,Cu,Zn) massive sulphides. Gold content of "exhalites" is of the order of 0.X to X ppm, Au-enriched bodies tend to be conformable to the depositional pattern.
(4) Dynamometamorphism. Metamorphic fluids leach trace "rock" gold as well as gold present in earlier ores and protores, and reprecipitate it in structurally and lithologically suitable suites, in dilations and in replaceable host rocks. Gold-mineralized shear zones, saddle reefs, etc., result. These are usually conformable with the local tectonic fabric. The gold grade reaches X to X0 ppm Au.
(5) Emplacement of granites and open fracturing in their roof partly "opened" the system and increased metal leaching, migration, redeposition from earlier rocks and ores, by hydrothermal fluids. Redeposition tends to be most efficient in an "optimum thermal zone". Discordant gold-bearing fissure veins and replacements (e.g. Au-skarns) tend to have higher grades than the shear lodes (X0 ppm Au), but are usually shorter and less predictable.
(6) Continental volcanism. This high-level, low-pressure system is open to such an extent that tracing the gold source in depth is rarely possible. The gold grade may reach its recorded maxima over short sections (X00-X000 ppm bonanzas), but is impersistent and unpredictable.
(7) Placer formation. Some placers are later buried and lithified to form paleoplacers.

6.5. TRACE METALS AND ORE EVOLUTION

Chapter 5 contains a brief outline of trace metal variations and the processes leading to metal accumulation along active continental margins, the precursors to orogenic belts. Most of the data presented are equally applicable to the orogenic belt situation (e.g. the petrochemical groups of volcanics and their trace metals, Table 5-3), and appear to have been little modified by evolution. Partly because of the much greater availability of data on the pre-Quaternary volcanics, however, the range of the trace metal contents shown in Table 6-2 is much greater. Table 6-3 provides data on the trace metal contents of pre-Quaternary "geosynclinal" sediments.

There was little evolutionary change in comparable lithologic associations of different ages (e.g. Ordovician and Miocene basaltic andesites are geochemically comparable). Substantial change was, however, due to (1) hydrothermal alteration, and (2) in-depth modification of the rocks.

(1) The effects of visible alteration of volcanics (as well as the minor selective leaching that left little, if any, visible indication) on the movement, release and subsequent ore accumulation in the calc-alkaline volcanic sequences of island arcs, have recently been studied by Stanton and Ramsay (1980). These authors proposed that the calc-alkaline and alkali-enriched island arc volcanic lineages may be produced from an appropriate partial melt by fractional crystallization, upon which is superimposed an increment of loss in an aqueous volatile phase. The released metals, suitably channeled and trapped, may result in the generation of subvolcanic-hydrothermal and subaqueous hydrothermal-sedimentary ore deposits. Alternatively and in addition to this, the loss probably causes the distinct trace metal enrichment in surrounding sediments, and a wide range of variation in the trace metals range in the parent volcanics.

Sodic metasomatism capable of converting a variety of volcanic rocks into a bimodal spilite - keratophyre suite, has already been discussed. The metallogenic importance of this process, particularly when it takes place under submarine conditions, has been recognized relatively early (by the German school, trying to explain the origin of the Lahn-Dill "type" iron ores; Schneiderhöhn, 1941; Amstutz, 1959). Recently, Stephens (1980) reviewed spilitization as a process forming some base metal massive sulphides, and demonstrated that spilitization of about 12 km^3 of Silurian volcanics would release the amount of Cu, Fe and Zn present in the Stekenjokk (Sweden) deposit.

(2) As for the metamorphism, the current petrochemical literature pays little attention to its effects on the magmatic petrochemistry of rocks. Amphibolite facies metavolcanics are plotted on AFM diagrams beside non-metamorphosed equivalents and are believed to "fit". Much depends on the sample selection. Load metamorphosed, homogeneous, massive lavas tend to be least affected by metamorphism, while volcanic fragmentals, densely fractured lavas and particularly sheared and penetratively deformed rocks, show substantial variations in their

trace metal contents (e.g. Schau, 1973). The chemical changes associated with Barrowian metamorphism tend to make metavolcanics more "felsic"; for example, the most widespread "greenstone" of the Precambrian Canadian Shield formerly generally interpreted as original andesite, is now considered to had been basalt. The same is probably true in the Phanerozoic belts.

Migration of trace metals under metamorphic conditions has been demonstrated with increasing frequency. This is probably reflected in the extremely wide range of trace metals to exist in the unevenly metamorposed paleovolcanic terrains. Schau (1973) listed the trace Cu range in the Triassic Nicola Group (British Columbia) andesites to be between 1 and 725 ppm. In regions with generally high trace metal contents (as the 200-300 ppm Cu basaltic andesites of Chile; Ruiz, 1965), rapid variation in trace rock geochemistry appears to correlate with an increased frequency of economic orebodies.

In one of the sections that follow, attention is paid to the repeated (although often loose) regional coexistence of Co,Hg,Sb,W as well as Au,Ag deposits with "geosynclinal" metavolcanics and metasediments and minor occurrences of mafic or ultramafic (frequently highly magnesial) lavas and intrusions. Although the genetic explanation offered (Holl and Maucher, 1976) usually follows the "exhalative" model, other alternatives are possible. One of them is based on the concept of mafics as the source of ore metals, released and redistributed by the metamorphogenic migration. In this context the finding of Stanton and Ramsay (1980) that the Solomon Islands magnesian lavas suffered no volatile loss may be significant; without such loss, the mafics (and probably the adjacent sediments) may have preserved their metallogenic fertility until a later metal release, migration or concentration event comtemporary with an orogeny, metamorphism or superimposed igneous intrusion. Although Co, Hg, Au, Ag and possibly Sb are known usually to reach their peaks of geochemical accumulation in basalts, the source and position of tungsten remains mysterious.

6.6. MINERALIZATION STYLES

6.6.1. Phenocrysts, cumulus crystals and porphyroblasts

PHENOCRYSTS

As in Quaternary volcanics (Chapter 5), magnetite is the most widespread metalliferous mineral of the volcanic phenocrysts present in amounts between 0.1 and 2% in most mafic to intermediate flows and tuffs. No economically important "primary" phenocryst accumulations are known.

As in the Recent volcanic terrains, the phenocryst magnetite weathering - released from volcanics accumulates in alluvial and beach placers. The magnetite is thus epiclastic, and is associated in placers with epidote, silicified volcanic clasts, chert, etc. rather

Table 6-2. Trace metal contents in major igneous associations in Phanerozoic orogenic belts

	Mn	Cu	Zn	Pb	Cr	Ni	Co	V
tholeiitic arc basalts	930-1600 m1300	25 - 207	70 - 150	1.4 - 6.4	69 - 540 m190	21 - 240	2.4 - 82 m50	120 - 340
continental olivine basalts	m1200	50 - 200	70 - 150	0.4 - 10	100-180	25 - 120	m40	30 - 150
island arc andesites	m1200	20 - 90 m37	m70	0.3 - 11	m55	15 - 28 m18	m28	m175
continental andesites	1000-1300	20 - 90	m70	2 - 16	m55	3 - 15	13 - 20	m175
dacite,rhyolite	400 - 600	5 - 30	30 - 70	4 - 50	m4	2 - 6		
ultramafics	10500	20 - 180	50 - 80	m0.3	2400-4000	1450	90 - 130	40 - 150
gabbros	1400	50 - 200	80 - 140	3 - 5	160-400	30 - 220	30 - 80	120-400
diorites	620-14000	30 - 70	m70	9 - 11	m25	15 - 30	10 - 20	50 - 150
granodiorites	m540	m18	10 - 70	7 - 70	m55	m15	m7	20 - 300
granites	380-465	m11	20 - 70	10 - 60	1 - 4	m4.5	m1	3 - 310

ABBREVIATION: m=mean all values are in ppm

Table 6-2 (continued)

	Sn	W	Mo	Zr	Nb	Ta	U	Au	Ag
tholeiitic arc basalts	0.05-1		0.3-1	10-250			0.1-0.6	0.2-19 m5.5	
continental olivine basalt		0.7-1.2					0.4 - 1	0.4-10	
island arc andesites	0.5 - 2	0.7-1.2		40-120			0.8	1 - 40 m4.5	30-190
continental andesites	1 - 2	0.1 - 1		80-260			0.8	1 - 17	
dacite,	3-30	2.4		72			0.3-0.7-	1 - 3.5	25-90
rhyolite							15		
ultramafics	0.1-1	0.1-0.8	0.1-1.4	45-30	0.1-4	0.1	0.003-0.05	0.6-30	30-80
gabbros	0.1-1	0.5-0.8	0.1-1.5		13	0.9	0.8	2.5-50	68-140
diorites	0.01-2	1.4-2	0.1-2.5	26-190	1-8	1.5	0.5-2	1-18	20-85
granodiorites	2-30	m1.5	1.3-2.2	123-350	11-27	1.4-2.4	1-2	1.2-4.2	20-90
granites	2-30	m1.5	1.3-2.2	28-232	1-27	2-16	2-6	0.5-2.3	

ABBREVIATIONS: m=mean all values are in ppm; Au and Ag in ppb

Table 6-3. Trace metal contents in Phanerozoic sedimentary associations of orogenic belts

	Mn	Cu	Zn	Pb	Cr	V	Sn	Mo	Be	Sb	Ag	Au	Hg
black shale	u 2%	u500	u5000	u700		u5000		u300		u500	u20ppm	3.5-30	u100 ppm
shale or mudstone	200-800	20-70	90-200	18-28	60-200	100-300	2.5-11	0.4	1-9	3		1-8	500
lithic arenites	500-800	9-60	70-116	5-50	40-200	10-70			1-2.5	1		2-50	
arkose	380	5-40								7-135			
carbonates	1-16	1.5-10	15-30	0.4-10	1-16	3-80			0.2-2	0.3		1-40	

ABBREVIATIONS: u=up to; all values are in ppm, for Au,Ag,Hg in ppb than with reworked, fresh pyroclastic detritus. Not all placer magnetite (and ilmenite) with provenance in orogenic belts came from volcanics; granitic rocks and particularly gabbros were also important contributors. This type of placer is substantially less productive than placers produced from Quaternary explosive volcanics.

CUMULUS

Cumulus crystals of chromite, magnetite and ilmenite occur in accessory quantities in tectonite serpentinites and dunites, located in obducted sheets and along sutures in orogenic belts. Cumulus (segregation) as well as liquidomagmatic magnetite and ilmenite are frequent in gabbro and diabase dikes and in small mafic ultramafic differentiated intrusions (Chapter 9) in orogenic belts.

PORPHYROBLASTS

Rutile and monazite are the chief metallic resistate minerals that form porphyroblasts in metamorphics in orogenic belts, from which they could be reworked into placers. They reach the peak of development in the high-grade metamorphic belts, discussed in Chapter 29.

6.6.2. Early postvolcanic or diagenetic, dispersed metallic minerals in lavas

Mafic (basalt and basaltic andesite) lavas frequently contain minute quantities of disseminated and scattered native copper. Copper is most common in the scoriaceous and amygdaloidal flowtops, but it also occurs in pillows and at fractures of massive volcanics. Although in subaerial lavas the native copper formed largely by mobilization after burial (e.g. in the Keweenawan basalts; Volume 2), copper in submarine basalts as in the Philippines, Solomon Islands, etc., probably formed in the seafloor or subfloor region by hydrothermal alteration. Several occurrences interpreted in this way have been reported from the Recent ocean floor basalts (Nagle et al., 1973).

Although several tens of occurrences of native copper in submarine mafic lavas have been reported so far, no economic deposits have been outlined mainly because of the low grade (in the region of 0.1-0.2% Cu), irregular distribution and unpredictability. During exploration, scattered native copper occurrences are generally considered to be a nuisance, because they produce "false" anomalies. On the other hand, native copper-enriched lavas may provide excellent "source beds" to superimposed mineralizations such as hydrothermal veins or porphyry coppers. At the Cebu Island, the Philippines (Bryner, 1969; Wolff, 1978), widespread native copper occurrences regionally coexist with porphyry copper deposits.

Accumulations of Mn oxides within or above mafic flows are usually accompanied by jasper and chert. They are most prominent in the ophiolite association (Chapter 7).

6.6.3. Massive, stockwork to disseminated ores in submarine volcanics and/or sediments

Massive orebodies should contain at least 60% of fine to coarse crystalline metallic minerals (sulphides or Fe oxides), the rest being a gangue typically corresponding, compositionally, to the adjacent wallrock. Although some hydrothermal fissure veins may be partly or wholly filled by massive sulphides or iron oxides, the majority of massive ores occur as sheets, lenses - to - irregular bodies in "geosynclinal" marine volcanics and sediments at, or close to, the interpreted former seafloor.

The bedded hematite, magnetite or siderite bodies spatially associated with spilites and/or keratophyres (Lahn-Dill "type"), were the first ores genetically interpreted as "submarine - exhalative", precipitated at the seafloor from metalliferous fluids released during subaqueous volcanic extrusions (Schneiderhöhn, 1941). Massive sulphide deposits previously attributed to hydrothermal replacement of solid rocks in "granite" aureoles, were similarly interpreted in several pioneering papers in the late 1950s (Stanton, 1955; Oftedahl, 1958b) and "en masse" in the literature of the 1960s to 1980s (see reviews in Hutchinson, 1973; Large, 1977; Franklin et al., 1981). About one half of the examples lie in Precambrian terrains, mostly in "greenstone belts" of the Canadian Shield (treated in Volume 2).

The fundamental correctness of the "exhalative" (more correctly subaqueous-hydrothermal) model originally based on ancient and mostly deformed or metamorphosed example deposits, has been confirmed later by the growing list of localities of modern massive sulphide equivalents, some of them in the process of generation on the floor of advanced intercontinental rifts (the Red Sea); at mid-ocean ridges; and in island arcs.

The standard model of the 1970s applicable to submarine volcanogenic massive sulphide systems (Fig. 16-13) is a hypothetical reconstruction of a situation as it presumably was shortly after ore deposition but before folding and metamorphism. Fig. 16-14 is a restored section across the Devonian Urup massive sulphide zone (Caucasus, U.S.S.R.; Skripchenko, 1972), believed closely corresponding to the model. The following subdivisions are recognizable in the model:

(1) Ceiling (topping, hangingwall) to the orebody. This consists of a variety of volcanics or sediments generally free of alteration, that cover the orebody, with which they usually have a sharp, transgressive contact. The most common cover rocks are red, green, gray or black "shales", siliceous "shales", mafic volcanic flows, felsic to mafic pyroclastics, etc. Sometimes, the sediments are slightly ferruginous, manganiferous, or they enclose thin stratiform horizons of Fe and Mn-sediments.

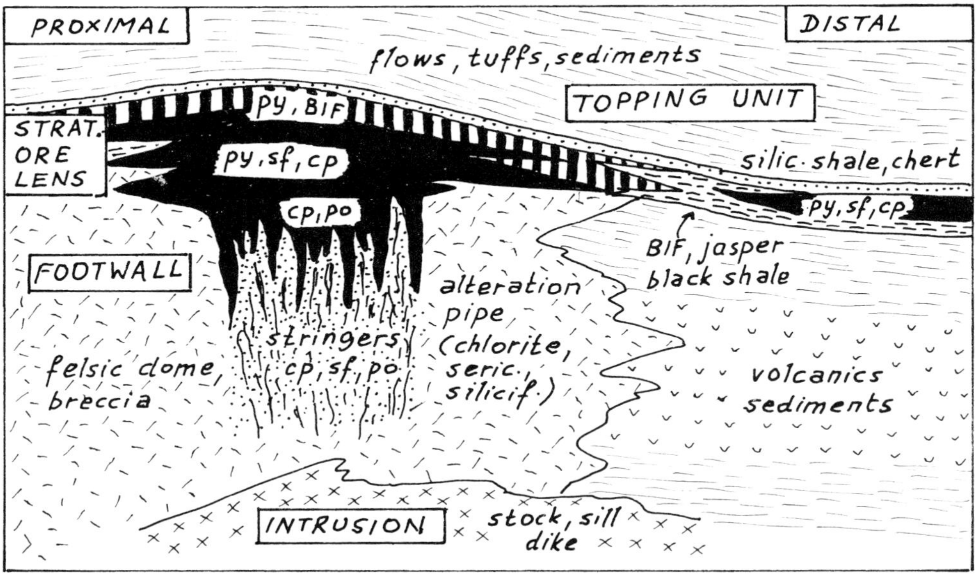

Fig. 6-13. Volcanogenic and volcanics-hosted massive sulphide deposits, conventional model (before deformation and metamorphism). See text for description.

(2) Peneconcordant (stratiform) massive sulphide layer or a lens. This lens appears to be a facies of the sediments or volcanics located in its lateral extension that could be either compositionally neutral (corresponding to the hangingwall rocks, often "chert"), or alternatively there is an elaborate, zoned facies transition of the base metal-rich sulphide core into pyrite, magnetite, siderite or hematite. The massive sulphide lens is either structureless, fine to coarse crystalline, banded, or brecciated. Sometimes it is topped by clastically reworked sulphide conglomerate or arenite or by a laterally persistent pyritic or Fe oxide concordant band. The upper contact of a massive sulphide lens is either conformable to disconformable, the bottom (footwall) contact is in most cases an intraformational unconformity.

(3) Floor (footwall) to many (but not all) massive sulphide deposits contains the former conduit(s) (feeder) by which the metal-bearing hydrothermal fluids ascended and presumably spread on the seafloor. Such conduits can be either a single, sharply outlined fracture (fault) now having a form of a vein, or more commonly a fracture stockwork, pipe or a porous zone (typically a breccia). The conduit and the rocks adjacent to it (typically sericitized, chloritized or silicified) and the footwall mineralization, when

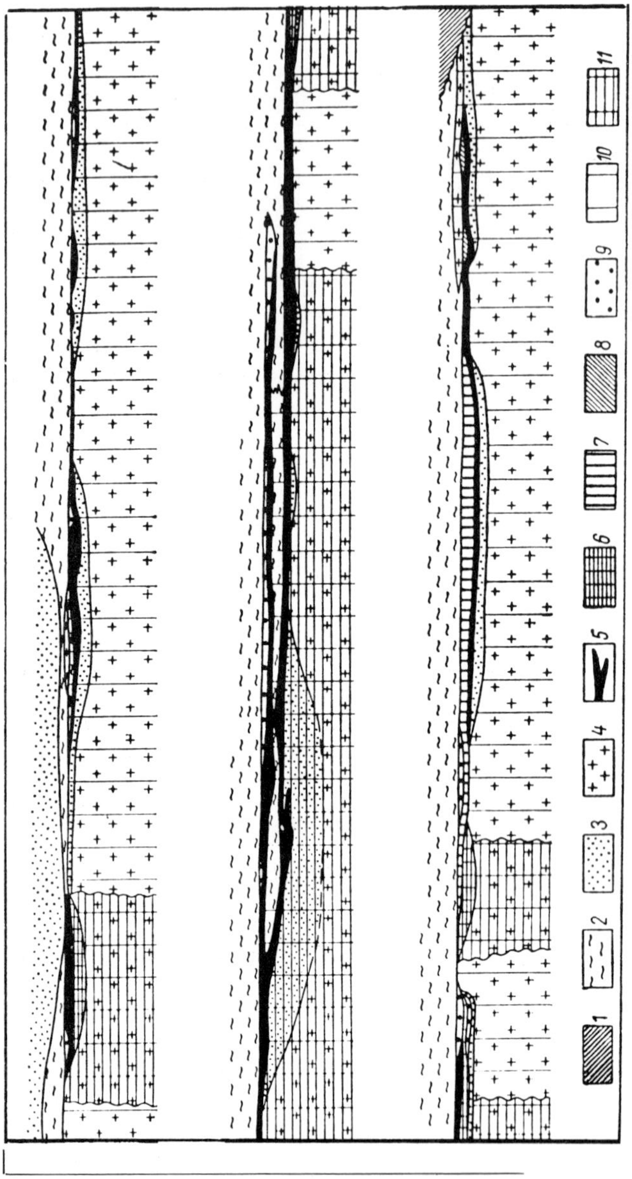

Fig. 6-14. Restored section across the Urup (Caucasus) sulphide orebody, showing the facies variation of Devonian rocks in the ore zone. LEGEND: (1) J_1 conglomerates and sandstones; (2) Devonian, red and green siliceous slates; (3) volcanic arenites, breccias and conglomerates; (4) quartz keratophyres; (5) Cu and Cu-Zn massive sulphide; (6) massive pyrite; (7) lightly altered rocks; (8) pyritic Zn ores; (9) silica-sulphide ores; (10) lightly altered rocks; (11) intensively altered rocks. After N.S. Skripchenko and A.S. Tambiev, from Skripchenko (1972).

present, take the form of stockwork veinlets, stringers and ore disseminations. Discrete veins are rare. A great variety of footwall rocks has been recorded, but by far most common are domal or stock-shaped felsic (rhyolite, quartz keratophyre, quartz porphyry) breccias or brecciated rocks.

(4) The last component of the model, a subvolcanic to hypabyssal sill, dike, stock, etc. broadly contemporary with the mineralization, has been added in the 1980s (e.g. Campbell et al., 1982).

The intrusion is believed to have been a "heater" driving the hydrothermal convective system.

Massive sulphide systems are usually characteristically geochemically and mineralogically zoned. In Cu,Zn,(Pb) systems, Cu (in chalcopyrite, usually with pyrrhotite) dominates the footwall stockwork and the bottom part of the massive sulphide lens, while Zn(Pb) prevails in the central and upper portions of the lens.

The usual massive sulphide composition depends considerably on the lithologic association in which the ore resides, and its geotectonic setting. The ophiolite and mafic associations (Chapters 7, 10) carry almost exclusively "barren" (pyrite or pyrrhotite only), or Fe-Cu sulphides. These may have increased trace contents of Co,Ni and Au, sometimes recovered as by-products. Bimodal associations (Chapter 12) typically produce Fe-Zn-Cu sulphides. Sequentially differentiated basalt-rhyolite association (Chapter 14) hosts (Fe)-Zn-Pb-Cu sulphides.

Genetic interpretation of massive sulphide systems (particularly their metal source and the nature of mineralized fluids) changes with time, but recently the majority opinion has favoured metal leaching from the footwall volcanics or sediments by convecting hydrotherms, followed by sulphide precipitation at the seafloor, an important geochemical interface. Lesser quantities of the ore fluids could have been brought "from depth". A detailed genetic review is in Large (1977) and Franklin et al. (1981).

The simple conventional subaqueous hydrothermal massive sulphide deposit model assumes mineralization during a single, even prolonged, ore-forming period. Essentially no remobilization, replacement, veining, etc. of older-stage minerals by younger-stage phases is considered. This is unrealistic. Fig. 6-15 from Cheprasov et al. (1972) shows compositional and ore-style zoning of the Ridder Sokol'noe Zn,Pb,Ag volcanics hosted massive sulphide deposit (Rudnyi Altai, U.S.S.R.), interpreted as resulting from a multi-stage mineralization (Fig. 6-16). The majority of the "lens-and-stockwork" orebodies described earlier also appear to be multistage, although this is rarely emphasized.

SEDIMENTS-HOSTED MASSIVE SULPHIDES

These occur entirely within sediments, and no substantial accumulations of volcanics (except for the tuffaceous component of sediments) are in sight. The overall configuration of the deposits (peneconcordant or stratiform lens and stockwork feeder) may be identical with the volcanics-hosted deposits, and there can be an intrusive stock or pluton nearby. The most common host association is the "flysch" (Chapter 16), or the deeper marine, typically black argillite, slate, phyllite, etc. (Chapter 17). Compositionally, the sediment—hosted massive sulphides may parallel the volcanics—hosted systems (Fe,Cu sulphides are in the lithologic association of ophiolites; Fe,Cu,Zn sulphides are in the bimodal

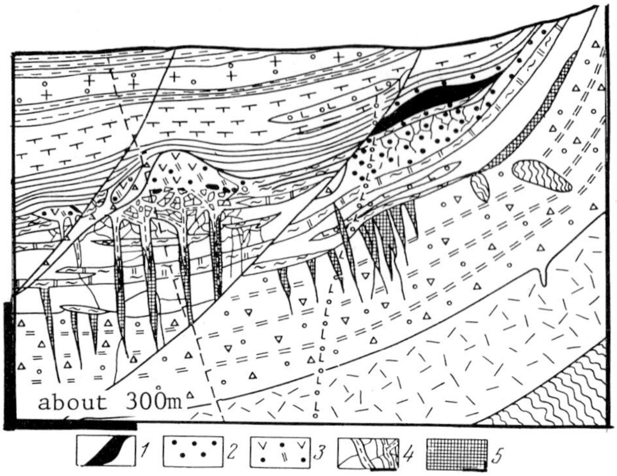

ORE STYLES:
1. banded,massive ZnPb
2. disseminated ZnPb
3. barite ZnPb
4. crustification thin banded ankerite quartz veins with dissem.and stockwork ZnPb
5. vein, ZnCu

Fig. 6-15. Vertical zoning in the Devonian Ridder-Sokol'noe volcanics-hosted Zn,Pb,Ag deposit (Rudnyi Altai, U.S.S.R.) resulting from several mineralization phases (as shown on Fig. 16-15). From Cherpasov et al. (1972).

volcanic sedimentary association), although the "black sediments"-hosted ores (e.g. Rammelsberg, Meggen) are commonly rich in Pb, poor in Cu, and accompanied by large concordant bodies of barite.

Proximal and distal massive sulphides.
The deposits described above are proximal to the site of discharge (feeder) of mineralizing solutions (orebodies are located directly above the feeder). The distal deposits lack identifiable feeders, and are usually more laterally persistent (Fig. 6-17).

Iron ores of the Lahn-Dill "type".
These are bedded, usually truly stratiform, tabular orebodies, lacking obvious feeders so that morphologically they correspond to the distal massive sulphide style.

DEFORMED AND METAMORPHOSED MASSIVE AND DISSEMINATED OREBODIES

It is a characteristic feature that the great majority of massive sulphide deposits (including the Precambrian ones), are modified by polyphase deformation and at least a greenschist metamorphism (the Kuroko deposits of Japan are the most notable undeformed,

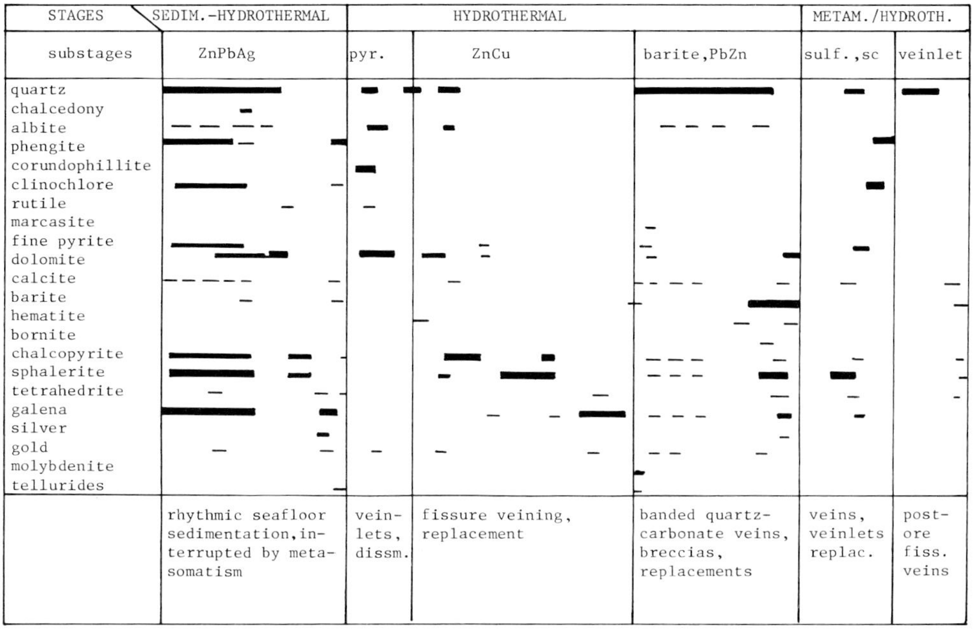

Fig. 6-16. Sequence of mineralization (paragenesis) at the Ridder-Sokol'noe Zn,Pb,Ag massive sulphide deposit. After Cherpasov et al. (1972).

unmetamorphosed exception). A substantial proportion of massive sulphide bodies ocur in shear zones. Almost all the Precambrian deposits are vertical to steeply dipping, hosted by "keels" of supracrustal metamorphics enveloped by granite-gneisses or by granite batholiths and a proportion of Phanerozoic deposits approaches such an image (Fig. 6-18). Detailed studies of deformed and metamorphosed massive sulphide deposits made (e.g. Addy and Ypma, 1977 at Ducktown; compare Chapter 29) often indicate the polygenetic nature of the orebodies, in which pre-deformation "seed deposits" were enlarged during the metamorphism, by sulphides mobilized from the surrounding rocks.

Other authors (e.g. Hopwood, 1976) concluded that the deposits they studied (most of them Precambrian; compare Volume 2) were discordant to the original bedding, or emplaced epigenetically during the late stages of deformation. Massive sulphide banding, long considered to be the best piece of evidence for syndepositional origin, was demonstrated to form equally well by metamorphic foliation of initially structureless ore. Seafloor deposited orebodies; deposits formed by replacement of unconsolidated sediments or volcanics under

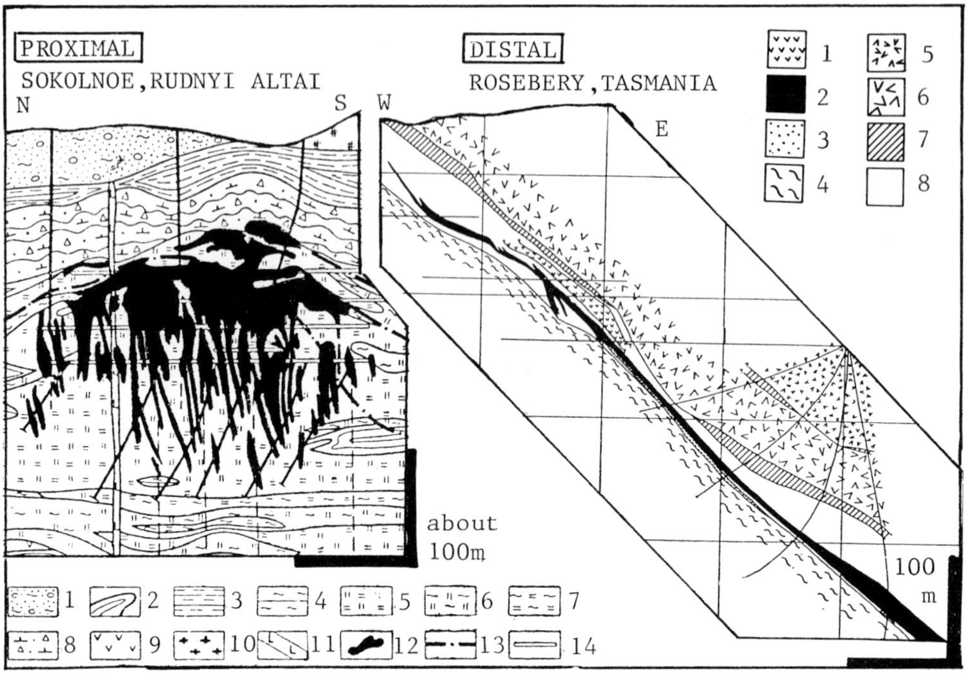

Fig. 6-17. Proximal and distal volcanics-hosted massive sulphide deposits.

PROXIMAL, Sokol'noe Zn,Pb,Ag deposit (Rudnyi Altai, U.S.S.R.; from Bublichenko et al., 1972). LEGEND: (1) Cainozoic cover rocks. Devonian, (2) siltstones; (3) limy siltstones; (4) sericitized mudstones; (5) silicified rocks (microquartzites); (6) sericite, quartz, chlorite alterites; (7) sericitic microquartzites; (8) volcanic conglomerates; (9) amygdaloidal andesites; (10) quartz albitophyres; (11) basalts; (12) polymetallic ore; (13) faults; (14) mine workings.

DISTAL, Rosebery Zn,Pb,Ag deposit, Tasmania (from Brathwaite, 1974). LEGEND: Cambrian; (1) vitric and vitric-crystal tuff; (2) crystal and lithic tuff; (3) black slate; (4) shale; (5) crystal tuff; (6) Zn-Pb ore; (7) Cu ore; (8) footwall schist.

the seafloor; deposits generated metamorphogenically; deposits formed by hydrothermal plutonic- replacement of reactive rocks, e.g. carbonates, are genetically convergent; that is, comparable looking orebodies may have formed by a variety of processes. The possibility of genetic alternatives should be kept in mind, because it can influence the exploration philosophy.

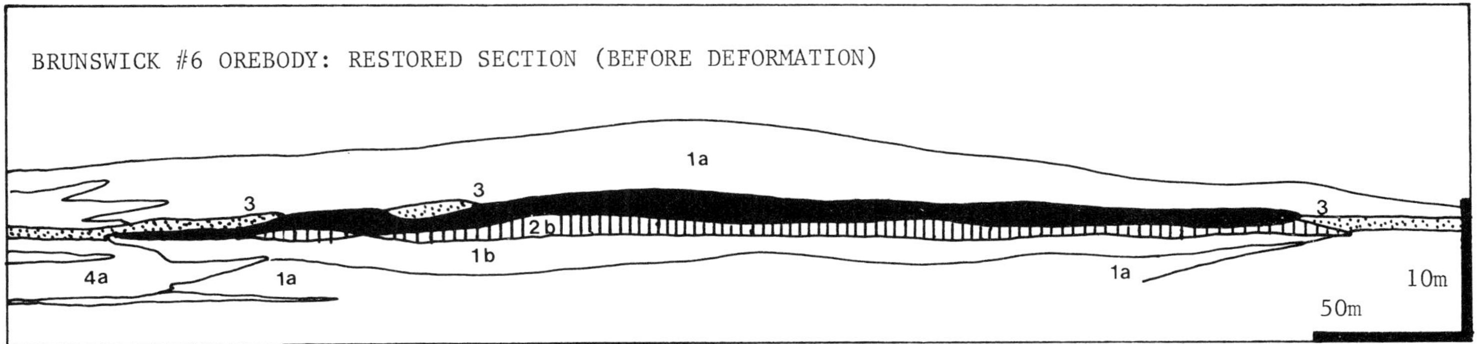

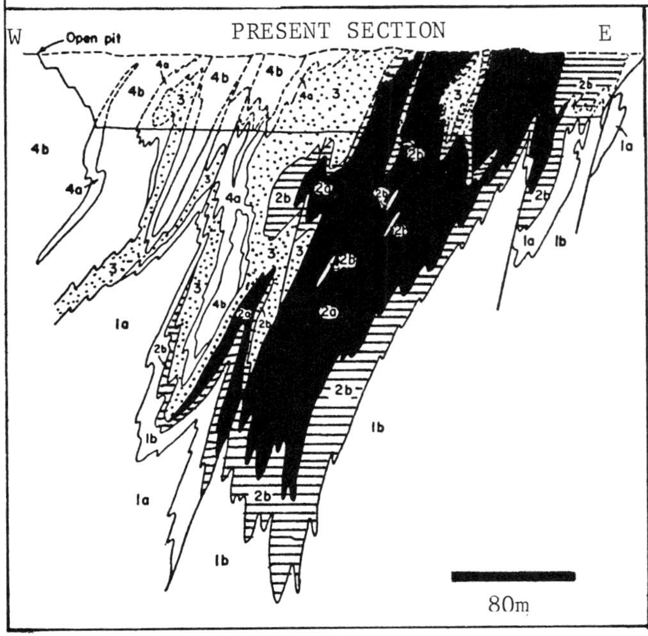

Fig. 6-18. Brunswick No. 6 orebody, Bathurst-Newcastle district, Canada (Ordovician), deformed and greenschist-metamorphosed massive Zn,Pb,Cu sulphide deposit. A: actual section; B: restored section (gaps interpolated). After Pertold (1970; unpublished), reprinted with permission. LEGEND (all rocks are Or, Tetagouche Group): (1a) quartz-augen schist; (1b) quartz, chlorite, sericite and chlorite, sericite schist; (1c) quartz, feldspar augen schist; (2a+b) massive to disseminated pyrite, sphalerite, galena, chalcopyrite, pyrrhotite; (3) iron formation; (4a+b) felsic metavolcanics; (5) meta-basalt.

6.6.4. Syngenetic-diagenetic ore accumulations in (meta)sediments

The bulk of "geosynclinal" sediments are remarkably free of ore mineralization. Directly mineable, bedded ores of ferrous metals (Fe,Mn) are most common in the nearshore regions, where they are controlled by sedimentary facies. Mn-enriched sediments also accumulated in the deep-water regions of slow sedimentation (in pelagic sediments), but they rarely make up industrial ores without a superimposed residual enrichment.

"Black" (carbon-rich) sedimentary facies usually developed in fault-bound troughs or near small volcanic centres, carry the bulk of the early, sediment-hosted Zn,Pb,Ag and lesser Sb,Hg,W,Au,U deposits having the form of bedded or banded sulphide masses; ore nodules; or dispersed ("invisible") metallic substance. The metals are usually believed to have been introduced by hydrothermal fluids from the subsurface, or coprecipitated with organic substance or with phosphates (e.g. U,V). More common than economic deposits are anomalously metal-enriched sediments that require a superimposed concentrating process to form mineable orebodies.

6.6.5. Metamorphic modification of ores and metamorphogenic ores

Deformation and metamorphism are usually thought of as agents of destruction of existing ores. While this is true in many respects, it is not absolutely true. Metamorphism and in particular dynamic metamorphism that results in penetrative deformation of rocks, are important agents of expulsion of water in sediments which, in turn, can extract, transport and redeposit metals. Axial zones of cleavage and schistosity in greenschist-metamorphosed sediments and volcanics are favourable sites of increased occurrence of ores (particularly Au) in otherwise monotonous suites. The effects of a high-grade metamorphism on earlier orebodies and the process of metamorphogenic ore formation, are discussed in Chapter 29.

6.6.6. "Granite" / supracrustals interaction and resulting ores

Increasingly, hydrothermal deposits are interpreted as products of convection of fluids in thermal aureoles of intrusive bodies (compare Chapter 28), and the source of metals is sought in the host association rather than in the intrusion. If this is the case, the trace metal enrichment of the host suite or just the ease of a selective release of the ore metals from it, predetermines the composition of the ores. Orebodies thus formed are mostly exocontact veins, stockworks, disseminations and replacements, including skarns. Typically, they are most common in the "fertile" supracrustal associations known also to contain other, pre-"granite" mineralization styles, and the concentrated elements are the same, or similar, in both categories of orebodies. Examples include Cu veins, skarns and disseminations in high-Cu basalt and andesite terrains; Zn-Pb-Ag or U veins in "black slate" association; etc.

6.6.7. Weathering-generated ores

Tropical weathering acts as a low-efficiency process, capable of upgrading metal-anomalous source rocks or "protores" (not yet economic ores) into economically recoverable residual metalliferous ore. In orogenic belts, the most important residual mineralizations are Ni-laterites" over ultramafics, "limonitic" iron ores over ironstones and pyritic sediments or volcanics, and ferruginous bauxites over basalts. In the "miogeoclinal" megadomain, the "Mediterranean-type" bauxites located at carbonate unconformities, are most important.

6.6.8. Deposits formed by reworking of older ores into younger sediments

Gold placers are by far the most important example in orogenic belts, and the bulk of placers derived from Phanerozoic bedrocks is adjacent to gold-veined supracrustals (slates, greenstones). Detrital (cobbly, gravelly, sandy) iron ores hematite, siderite or magnetite, are frequent in basal sediments of transgressive successor sequences. These are sometimes accompanied by precipitated ferruginous cements.

Plate 3. Dismembered ophiolite terrain exposed in the rocky cliffs near Port Macquarie, N.S.W., Australia. Two members of the Steinmann's Trinity are shown: a ribbon chert (C) and a blueschist-metamorphosed basalt (B).

"Die ophiolitischen Massengesteine sind in ihrem Auftreten an die Tiefseezone gebunden".
G. Steinmann in Ber. Naturf. Ges. Freiburg, 1905.

"I have been accused of setting up deposits whose chief characteristic is variability".
T. P. Thayer (1964) referring to podiform chromite deposits.

CHAPTER 7

Ophiolite Association

7.1. GENERAL

Steinmann, working in the Alpine-Mediterranean region, recognized a repetitive association of serpentinite, metabasalt and chert ("Steinmann's Trinity"), which he named ophiolite suite (Steinmann, 1926). Other widely used terms "Alpine (or Alpine-type) serpentinite" or "Alpine ultramafic complex" refer to what is possibly the most conspicuous member of the ophiolite suite, the tectonized peridotite. In most cases, other members of the Trinity are found nearby as well. It was, however, only a few years ago when the "Alpine-type peridotite - serpentinite association" (Wyllie, ed., 1967) was clearly distinguished and separated from the many other forms of ultramafic associations known.

In the late 1960s the idea that ophiolites are fragments of the former ocean floor probably generated along spreading ridges was first expressed (Gass, 1968) and rapidly gained wide acceptance (Dewey and Bird, 1971; Coleman and Irwin, 1974; Coleman, 1977). The 1972 Penrose Conference on Ophiolites (GEOTIMES, 1972, p.25) established the following definition of ophiolite:

"Ophiolite refers to a distinctive assemblage of mafic to ultramafic rocks. It should not be used as a rock name or as a lithologic unit in mapping. In a completely developed ophiolite the rock types occur in the following sequence, starting from the bottom and working up:
(1) Ultramafic complex, usually with a metamorphic tectonic fabric;
(2) Gabbroic complex, with cumulus textures and commonly containing cumulus ultramafics, usually less deformed than (1);
(3) Mafic sheeted dike complex;
(4) Mafic volcanic complex, commonly pillowed.
(5) Associated rocks including overlying ribbon chert, shale, limestone; podiform chromite; sodic magmatites".

If the oceanic origin of most ophiolites is accepted, two different sets of problems will appear: (1) the original formation of ophiolite in the oceanic crust; and (2) the emplacement of ophiolite onto the present site within the continental crust. This two-fold approach to ophiolite interpretation was not recognized until the early 1970's and this accounts for the considerable degree of confusion in the older literature. Many uncertainties still persist. In keeping with the spirit of this book, the interpretational aspects have been reduced to the bare minimum and the reader is refered to the excellent reviews mentioned earlier, and also to Ruzhentsev, (1976); Panayiotou, ed., (1980); UNESCO (1981).

7.2. ORIGIN AND EMPLACEMENT OF OPHIOLITES

There is a surprising correspondence between the rocks dredged, observed and drilled at the presently active oceanic spreading ridges, and rocks known from the geologically older ophiolite terrains exposed within continents. Consequently, few geologists now doubt the oceanic origin of at least the "typical" ophiolites. In addition to major oceanic ridges, however, similar associations could have been generated by sea-floor spreading in substantially harder to interpret environments such as in narrow rift seas (like the Red Sea; e.g. the interpretation of the Turkish ophiolites by Brinkmann, 1976); inland and marginal seas (such as the Black and Caspian Seas); back-arc and inter-arc basins and marginal seas (Dewey, 1974); and similar. The availability of the oceanic data significantly fertilized the interpretation of ophiolites, but the reverse flow of ideas and information was equally significant. The understanding of the ophiolite stratigraphy as seen in land outcrops helped to explain the lithologic composition of the geophysically determined oceanic crustal layers.

It is now believed that, at the end of a spreading episode in the ocean, a more or less stratified complex may result. This probably requires special conditions, such as a differentiation of a 2-3 km deep and some 25 km long magma body, collected at shallow depth below the seafloor (Gass, 1977). It has been suggested that discrete magma bodies exist at present beneath the fast-spreading ridges (e.g. the East Pacific Rise), but are absent under the slower-spreading ridges (e.g. the Mid-Atlantic Ridge). It is doubtful that a satisfactory rock stratification will result in the second case.

Glennie et al.(1974) used the term deuteric metamorphism for all the low-temperature reactions that occurred at depths of less than 5 km under high water-vapour pressure and 200-500° temperature still within the oceanic crust, within the spreading ridge, before the end of igneous activity. This includes serpentinization, rodingite generation, pilite reactions, prehnitization and albitization of plagioclase, and chloritization of various Fe-Mg minerals.

Following its generation, the oceanic crust is carried away from the spreading ridge during which it accumulates an increasing thickness of oceanic (as well as hemipelagic and terrigenous) sediments, and undergoes "aging" (volcanic and sedimentary diagenesis; initial load metamorphism; interaction with mid-plate volcanics). When a portion of such an aged oceanic lithosphere reaches the trench site, its lithologic association is greatly expanded and the sedimentary Layer 1 thickened. The plate tectonic model elegantly explains the cause of disappearance of much of the oceanic crust generated in the past, by lithospheric subduction. Preservation of this crust requires a malfunction of the subduction mechanism, or a temporary to permanent cessation.

Obduction (Coleman, 1971) is the converse of subduction and it is usually interpreted as the thrusting of an oceanic slab over or onto an island arc or a continental margin. Several variations of this process and alternatives, however, have been proposed (e.g. Gass, 1977, p.74). The most realistic alternative process proposed is a vertical uplift of a piece of the oceanic floor, that eliminates much of the usually severe lateral deformation associated with obduction. The Troodos Complex of Cyprus, and Macquarie Island in the southern Pacific, are the two most frequently quoted examples. Some ophiolites may have resulted from uplift or reversal of a partially subducted slab; from a reversed island arc collision; or from a continent-continent collision. In the latter case, mantle lherzolite slices have been thrust onto the collision suture, representing continental root zones (Coleman and Irwin, 1974). Saunders et al. (1979) described a late Jurassic autochthonous ophiolite in southern Chile, which they interpreted as having been formed by rifting in a narrow marginal basin floored by oceanic crust.

Dominant features of most ophiolite complexes are their allochthonous nature, intensive deformation, tectonic dismemberement, and a variable degree of metamorphism. Few ophiolites are complete. This, coupled with partial erosion, and in the tropical belt by intensive lateritic weathering, often makes it difficult to distinguish true ophiolites from similarly modified ultramafic - mafic associations of different origin. This is particularly difficult in the Precambrian terrains.

7.3. TECTONIC SETTING AND DISTRIBUTION OF OPHIOLITES

Ophiolites are exposed along belts of intensive deformation (geosutures), believed to mark the edges of former continental margins and sites of oceanic and continental crust interaction (Coleman and Irwin, 1974; Coleman, 1977). In some complexes, (1) tectonic facies of the intensively dynamometamorphosed suture zone, and (2) facies of a thrust sheet or nappe with relatively little internal deformation, can be distinguished. The suture zones are often marked by the occurrence of blueschist metamorphic belts and by mélanges, and are treated in greater detail in Chapter 8. The ophiolite sheets or nappes (2) often have an impressive continuity and their allochthonous nature is particularly well apparent in the geologically young (Mesozoic-Cainozoic) terrains. The extensive Semail Ophiolite in Oman (Alleman and Peters, 1972; Glennie et al., 1974) is about 10 km thick, 400 km wide, and transported 100-200 km westward over another allochthonous unit of deep water sediments. Interpretation of the old

(Paleozoic, Proterozoic) ophiolites tends to be more difficult, because repeated tectonism and metamorphism commonly obscure the ophiolite stratigraphy and allochthony. Along certain long-lasting continental margins, a number of subparallel ophiolite belts exist. Along the Pacific coast of North America, five to six discontinuous ophiolite belts, generally subparallel with the Cordilleran structural grain, range from Ordovician to late Jurassic.

7.4. METAL GEOCHEMISTRY AND METALLOGENY

It is generally assumed that the ophiolite suite originated in the mantle peridotite, so mantle peridotite geochemistry is the starting point of considerations involving the fate of metals in the course of ophiolite evolution. Compared with the crustal clarke, the mantle peridotite (approximated by chondritic data; Vinogradov in Rösler and Lange, 1972) is significantly enriched in Au (0.17 ppm); Co (800 ppm); Cr (0.25%); Hg (3 ppm); Pt (2 ppm); and Ni (1.3%). It is moderately enriched in Cu (100 ppm); Fe (25%); Mn (0.2%), and depleted in the lithophile metals Sn, Pb, Th, U, Ta, Nb, REE, Be, etc.

The first partition of the "chondritic" metal abundances was probably accomplished during partial melting of the mantle peridotite, believed to have taken place at some depth under oceanic spreading ridges. There, separation of a refractory residue (now the tectonized basal metaperidotite) and a partial basaltic (?) melt (now represented by the mafic - ultramafic cumulates lying stratigraphically above the tectonite) took place. Data are still incomplete, but it appears that the residue became slightly enriched in Cr (in the form of chromite) and possibly Hg, and strongly depleted in Au, Cu and S. Ni, Co, Fe, Pt metals and Mn behaved ambiguously, but in general they tend to be enriched in the partial melt. There are significant local variations. Page et al. (1982) in their study of platinoids in the New Caledonian chromitites hosted by tectonized peridotite, registered a strong depletion in Pt and Pd and mild depletion in Ir and Ru, with respect to an average chondrite composition. Oshin and Crocket (1982) in their study of trace noble metals in the Thetford, Quebec, ophiolites, reported a significant contrast in the noble metal content between the residual tectonite dunites and cumulate dunite, respectively: Au, 0.27 versus 1.3 ppm; Ir, 3.6 versus 2.4 ppb; Pt, 2.1 versus 28 ppb; Pd, 0.29 versus 21 ppb. The initial partition predisposed the character of either unit as an active or passive (in the sense of Laznicka, 1985e) metal source: the economic accumulations of chromite in ophiolites are more frequent and larger in the tectonite peridotite and substantially less frequent in the cumulates, while Au, Cu, Pt, Fe dominate the cumulate metallogeny. Ni seems to be significantly enriched in any ultramafic (aver. 0.1-0.2% Ni; max. 0.43% Ni in Galax, Virginia; Worthington, 1964), but it is almost exclusively a "silicate" nickel substituting for Mg^{2+} in olivine in the tectonites, but silicate as well as sulphide Ni in the cumulates. The strong partition of sulphur into the partial melt gave the melt a significantly increased capability to generate metallic sulphides.

Differentiation of the magmatic melt in separate magma chambers produced a metal enrichment comparable with the one in the Bushveld Complex (such as chromite in basal ultramafics; Pt-enrichment; Ti-magnetite; e.g. in the Bay of Islands ophiolite, Newfoundland). Unfortunately, it rarely resulted in economically mineable accumulations. The bulk of the differentiates produced have a gabbro or diabase composition, and high trace copper. Intermediate to felsic rocks of the sodic series (trondhjemite, keratophyre, albitite, etc.) as well as spilite are products of the latest magmatic phase of ophiolitic magmatism and considered as either orthomagmatic or metasomatic by a variety of investigators.

Retrogressive alteration or mobilization superimposed on lithified (or almost lithified) ophiolite at and under the oceanic spreading ridges and again during or after their tectonic emplacement on the continent, caused a far-reaching major and trace element movement. The exact timing and site of the alteration or mobilization is usually difficult to pinpoint, but increasingly it is attributed to the intra-oceanic locations. Hydrothermal hydration (serpentinization) and/or sulphurization of peridotites set in motion some of the trace elements previously "frozen" in the ultramafics (Ni, Co, Cu, Fe, Mn, Au, platinoids). Veins, stockworks, disseminations of Ni sulphides (millerite, violarite), Ni-Co arsenides (gersdorffite, rammelsbergite, smaltite), platinoids, gold, etc., formed within and outside of the ophiolite. Trace metals (e.g. Mn, Co, Ni, Hg) could also have been supplied to the concurrently forming seafloor metalliferous sediments.

Similar alteration or mobilization processes superimposed on gabbros and basalts accumulated copper (and minor zinc) in the highly pyritic "Cyprus-type" seafloor massive sulphide deposits; low-pyrite stringer, stockwork and vein copper mineralizations; vein and impregnation gold deposits; and Mn oxide / silicate / carbonate deposits. A variety of Cu, Ni, Co, Hg, Mn, Au, Ag formed within, and close to, ophiolite complexes by numerous processes of ensialic interactions, following the ophiolite emplacement onto a continent. Most such processes still remain obscure although it is increasingly apparent that there is a spatial correlation between most Hg deposits, many Au and Cu deposits, and some Zn-Pb and Sb deposits, and ophiolites.

Continental tropical weathering and sedimentary recycling are the latest in the chain of metal-accumulating interactions superimposed on ophiolites. Lateritization of ultramafics caused a low concentration factor (4 to 10) upgrading of the high initial trace Ni content to form important "Ni-laterites". Co, Mn and Fe co-precipitated in different depth zones of lateritic profiles. Chromite, a resistate mineral, is sometimes recovered as a by-product during lateritic nickel mining and small quantities of chromite accumulated in placers.

Ophiolite complexes account for some 60-70% of the world nickel resources in presently economic ores (laterites), and the ordinarily high Ni trace metal content in ultramafics is an important resource for the future. The podiform chromite deposits in ophiolites represent only about 3% of the world Cr resources, but at present account for over 50% of the world's chromite production because the huge southern African magmatic - stratiform chromite deposits have barely been tapped. Similar relationship applies to platinum. About 25% of the world's mercury deposits are hosted by the ophiolite ultramafics, but these are in most cases members of mélanges. Despite the scientific publicity given recently to the "Cyprus-type" Cu (Zn) massive sulphide deposits, less than 1% of the world's copper resources are in ophiolites. Less than 0.2% of the world's manganese is ophiolite-associated, and there is not enough data to quantify the ophiolite and gold relationship.

The principal mineralization styles in the ophiolite association are assembled in Table 7-1, and depicted diagrammatically in Fig. 7-1.

7.5. "ALPINE-TYPE" ULTRAMAFIC TECTONITES

Alpine ultramafic tectonites occur either independently as lenses, slivers, sheets or irregular masses several metres to kilometres thick along prominent faults, or under the cumulate ophiolite piles. They are compositionally monotonous and dominated by harzburgite (olivine-bronzite peridotite). Dunite forms schlieren and lenses in harzburgite and there are sometimes dikes of pyroxenite. Most Alpine-type ultramafics are serpentinized and range from the massive varieties in which relict outlines of bronzite crystals can easily be recognized, to slivery sheared varieties. Small (1 mm) crystals of chromite and magnetite are widespread, and apparent on the light, weathering-bleached outcrop surfaces.

7.5.1. Podiform chromite

The single most important metallic mineralization styles in the ultramafic tectonites are the "podiform" chromite deposits (Thayer, 1964), so-named to distinguish them from the stratiform chromitites as in the Bushveld. In the most popular image, these are high-grade massive chromitites; aggregates of chromite nodules and single crystals in silicate matrix; bunches composed of interlocking grains; etc., and they form irregular, pencil-shaped or tabular masses in dunite or, rarely, in harzburgite. In the latter case, the chromite has a narrow dunite aureole anyway. Thayer (1968) gave an excellent review of the ore petrography.

Many chromite deposits in ophiolite ultramafics, however, appear to conform to the planar fabrics of their hosts and some have the form of layers or bands sometimes of remarkable persistency. In the voluminous literature on podiform chromites, numerous theories of their origin have been proposed of which two have been applied most frequently (compare Panayiotou, ed., 1980): (1) chromite had crystallized in the mantle and accumulated in the refractory mantle residuum; it was later dismembered and tectonically emplaced into its present site (e.g. Hawkins, 1980); (2) chromite developed as magmatic cumulate in an ophiolitic magma chamber and generally remained undisturbed during the tectonic emplacement onto the continental crust (e.g. Juteau and Whitechurch, 1980).

It would appear that the chromites (1) are the "unconformable" ones hosted by the ultramafic tectonites, and the chromites (2) the "stratiform" ones hosted by the cumulate dunite layers. While both divisions of ophiolite complexes demonstrably carry chromites, Cassard et al. (1981) pointed out, using the example of the chromite occurrences in New Caledonia, that the chromites formed originally in the mantle as discordant pods, were subsequently plastically deformed and tectonically reoriented, so that the most penetratively deformed chromites now conform most faithfully to the banding of their hosts, the banding being metamorphic. With few exceptions, the literature data regarding the position of particular chromite deposits in either of the two ophiolite divisions mentioned above are conflicting, so distinction has not been made here. Several thousand of podiform chromite occurrences are known worldwide, most of them small. Table 7-2 lists the more important ones, and Fig. 7-2 shows four typical examples of podiform chromite deposits. The deposits tend to occur in clusters, but there is no correlation between the size of an intrusive mass and chromite content. Thayer (1964) noted that several major chromite deposits occur close to contacts with ophiolite gabbros, but this rule has not been universally confirmed.

7.5.2. "Podiform" (or similar) magnetite

Conceivably, processes equivalent to those generating the podiform chromites may also be applicable to other, rarer and less distinct mineralization styles. The "podiform" magnetite is one of them. At the example locality of Lipovac (Yugoslavia; Janković, 1982), elongated lenses of disseminated to massive magnetite and minor chromite and spinel are irregularly distributed in dunite schlieren, enclosed in a Jurassic-lower Cretaceous ophiolite. The grade ranges from 25-45% Fe; 2.7-3.4% Cr and there is up to 1% Ni and Co.

7.5.3. Nickel sulphides and platinoids

Bryner (1969) described an occurrence of blebs of intergrown pyrrhotite, troilite, pentlandite and violarite in the Acoje chromite mine, Philippines, in what is now believed to be an ultramafic tectonite. The mineralization, erratically distributed for at least 5 km, was mined locally on a small scale and the quantity of approximately 1 Mt ore mined had an average grade of 1.5% Ni and almost 9 ppm platinum metals.

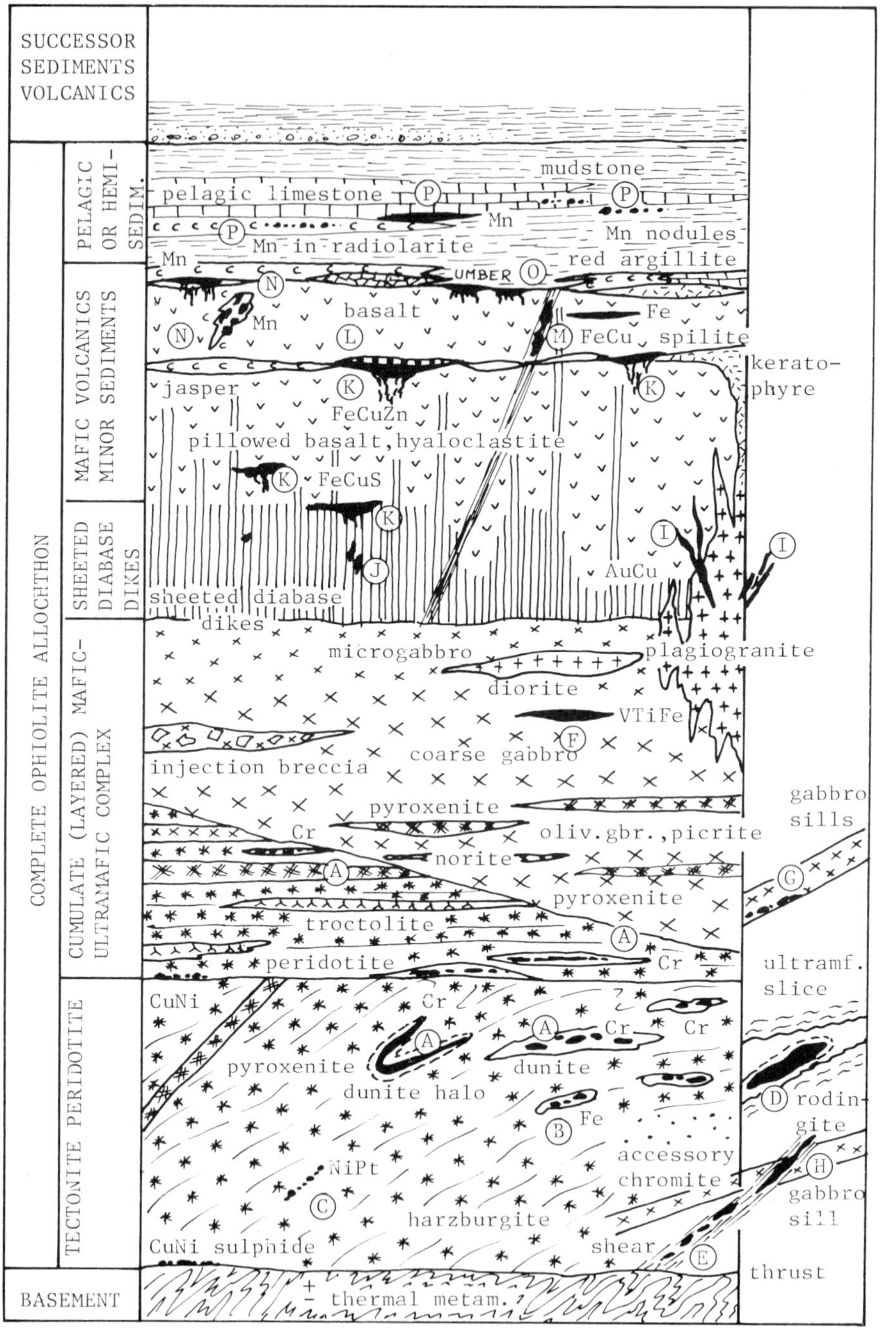

Fig. 7-1. Lithologic sub-associations and mineralization styles in the ophiolite association (see Table 7-1 for the explanation of letter codes).

Table 7-1. Principal mineralization styles in the ophiolite association

TECTONITE PERIDOTITE			
(A)	podiform chromite in dunite and harzburgite	X000	90 Mt Cr
(B)	podiform magnetite in ultramafics	X	---
(C)	blebs of pyrrhotite,pentlandite,violarite,etc. in ultramafics, accessory Pt metals	X	X00 t Ni Pt
(D)	mobilized ? magnetite affiliated with sheared and serpentinized ultramafics	X0	20 Mt Fe
(E)	Ni,Cu sulphides, access.Pt metals, in sheared and rodingite - altered ultramafics	X00	X0 Tt Ni X0 Tt Cu
LAYERED CUMULATE GABBRO, ULTRAMAFICS			
(F)	vanadium - bearing Ti-magnetite cumulate in gabbros	X0	---
(G)	massive to disseminated pyrrhotite,pentlandite, chalcopyrite, in gabbro sills or along contacts	X0	X0 Tt Ni X0 Tt Cu
(H)	magnetite,pyrrhotite,pentlandite in sheared gabbros intruding peridotites	X	X00 Tt Fe X Tt Ni
(I)	hydrothermal Au or Cu deposits associated with plagiogranites	X00	X0 t Au X0 Tt Cu
SHEETED DIKES			
(J)	pyrite, chalcopyrite, veins and veinlets	X	---
VOLCANICS AND INTERBEDDED SEDIMENTS			
(K)	stratabound Fe,Cu (Zn) massive sulphide deposits with footwall stockworks (Cyprus "type")	X00	1.5 Mt Cu 150 Tt Zn
(L)	Fe-hydroxide ochers	X	---
(M)	Fe,Cu sulphide stringers, masses, disseminations in mafic metavolcanics along shears	X00	500 Tt Cu
(N)	Mn oxides, minor silicates, carbonates in jasper or chert within or on top of basalts	X00	500 Tt Mn
(O)	Fe-Mn enriched shales (umbers) on base of marine sediments topping basalts	X	---
(P)	stratiform Mn oxides, carbonates, in chert (jasper) or silic.shale, in pelagic sediments on top of an ophiolite sequence	X0	X0 Tt Mn
SUCCESSOR SEDIMENTS			
(R)	weathering enriched and reworked ores derived from ophiolites (treated in subsequent chapters)	-	---

Table 7-2. Selected localities of podiform chromite in ophiolitic ultramafics

LOCALITY	AGE	GEOLOGY	TONNAGE, GRADE	REFERENCES
Piedmont Upland Cr Belt, Maryland	PZ_1	irregular masses, pods, veins, stringers, sacks of chromite in tectonized dunite in an "Alpine - type" ultramafic massif	98.5 Tt Cr/33%	Pearre and Heyl (1960)
Camagüey district, N.C. Cuba	J-Cr	lenticular, nodular, irregular to layered chromite in dunite and peridotite at base of a troctolite horizon	350 Tt Cr+/35%	Thayer (1942)
Kempirsai Massif, S.Urals (Kazakh-stan), USSR	380-404 m.y.	160 chromite deposits in a 920 km^2 laccolith-shaped massif of ophiolite; schlieren, pene-concordant chromite bodies in dunite	30 Mt Cr/34%	Pavlov and Grigor'yeva (1974)
Klyuchevsk-Revda Massif, S.Urals, USSR	D_{2-3}	200 small chromite lenses, schlieren in harzburgite and dunite; some are banded	200 Tt Cr/10.3%	as above
Duboštica deposit, Konjuh-Krivaja Ms., N.W.Yugoslavia	J	layers and lenses of chromite concordant with banding in lherzolite (cumulus ?) in a complete ophiolite complex	200 Tt Cr/30%	Sunarić and Olujić (1968)
Ljuboten (Raduša) Ms., Macedonia, S.E. Yugoslavia	J-Cr	400 chromite occurrences in dunite; most are irregular or pipe-like, one is banded (Nada) 80 km^2 peridotite-gabbro ophiolite	685 Tt Cr/27.4%	Janković (1982)
Djakovica Dist., S.E.Yugoslavia	J-Cr	lenses and schlieren of massive chromite in dunite and harzburgite	110 Tt Cr/27.4%	as above
Mirdita-Korab zone ophiolites, Albania	J	several hundred chromite occurrences in 3000 km^2 ophiolite belt; most oreb. are irregular; largest (Bulqiza) dep. have per-sistent layers in dunite	1.3 Mt Cr/26%	Pumo et al. (1982)
Democos-Tsangli, Greece	J	pods, irreg. masses, cylindrical orebodies of chromite in dunite	276 Tt Cr/28%	Marinos (1982)

Vourinos Complex, N.Greece	J_{2-3}	chromite schlieren and massive bodies in dunite bands in peridotite; over 30 deposits; differentiated ophiolite massif 30x15 km	411 Tt Cr/14%	Marinos (1982)
Orhaneli Massif, N.W.Turkey	MZ	0.5-8 m thick chromite layers, irregular pods, disseminations // with layering in dunite and harzburgite in an ophiolite thrust sheet	1.3 Mt Cr	Tankut (1980)
Guleman-Soridağ district, S.E.Turkey	MZ	tabular, pencil-shaped, irregular chromite bodies in dunite	11.2 Mt Cr/34%	Thayer (1964)
Troodos Complex, Cyprus	Cr	coarse podiform, massive to disseminated chromite in dunite; most deposits are at base of cumulus sequence, in ultrabasic tectonite	136 Tt Cr/27%	Hadjistavrinou and Constantinou (1982)
Semail Nappe, Oman	Cr	numerous elongated pods, stringers, schlieren, in dunite on base of layered ultramaf.unit	X Mt Cr	Peters and Kramers (1974)
Zambales Massiv, Luzon, Philippines	Cr-Eo	many podiform, lenticular, pencil-shaped chromite bodies in a 200 m thick dunite horizon above 2 km thick harzburgite unit	7 Mt Cr	Hawkins (1980)
Tiébaghi Massif, N.W. New Caledonia	Eo_3	chromite pods, schlieren, bands mostly // with tectonic layering in dunite within harzburgite-lherzolite transitional zone	972 Tt Cr/36%	Cassard et al. (1981)
Massif du Sud, S. New Caledonia	Eo_2	chromite pods, bands, disseminations, most // with tectonic foliation in tectonite dunite and harzburgite	122 Tt Cr/31%	as above

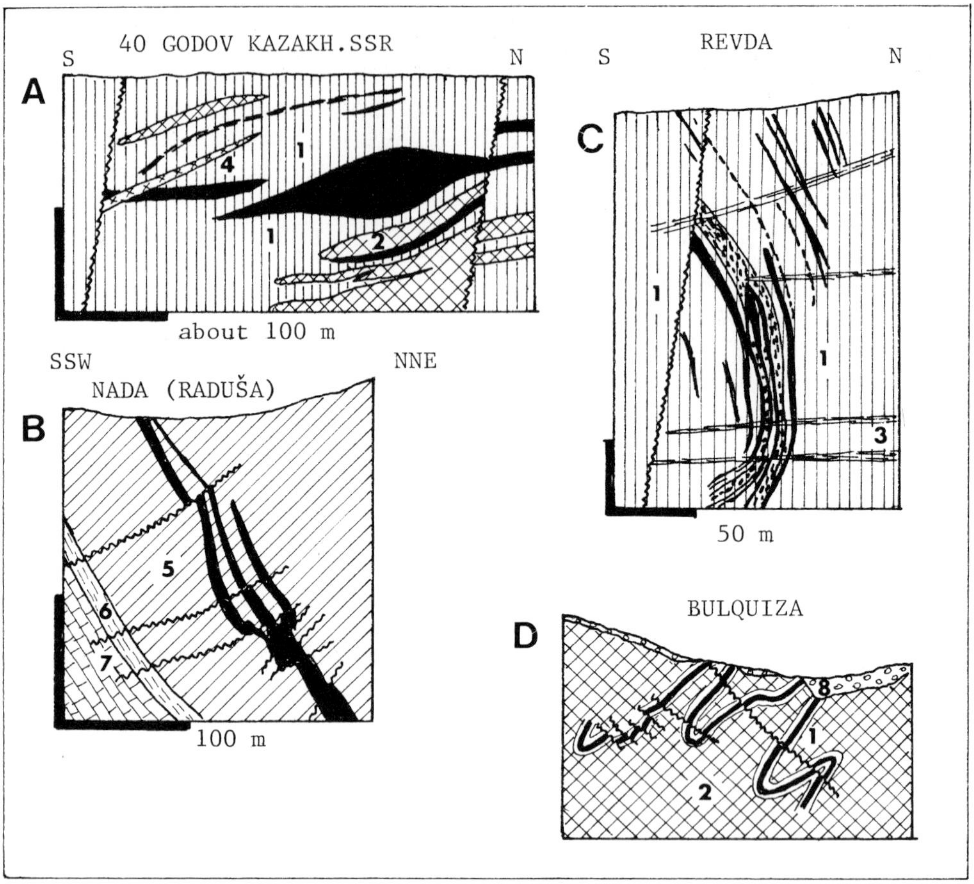

Fig. 7-2. Examples of podiform chromite deposits. 40 Godov Kazakhskoi SSR and Revda in the Urals, U.S.S.R., are late Paleozoic; Nada, Raduša Massif (Yugoslavia) and Bulquiza (Albania), are both Jurassic. LEGEND: (1) dunite; (2) peridotite (harzburgite); (3) talc carbonate rock; (4) gabbro; (5) serpentinite; (6) schist; (7) limestone; (8) Cainozoic cover sediments. Modified after Pavlov and Grigor'eva (1974), Grafenauer (1977), and Pumo et al., (1982).

7.5.4. Mineralizations resulting from internal mobilization within ultramafic tectonites, following deformation and hydration

The early stage of hydration (serpentinization) and metasomatism (e.g. generation of rodingite with hydrogarnets, vesuvianite, diopside, prehnite) believed to have taken place in the sub-oceanic domain, has produced a variegated suite of mineralizations involving Fe, Ni, Cu, Au, Co, platinoids and other elements. The known occurrences have, so far, been of little economic value.

The small but productive Fe deposit, Cogne, in the Aosta Valley (Italian Alps; Compagnoni et al., 1981; Fig.7-3C) is a magnetite body believed to have been formed by a metasomatic mobilization of iron from lherzolite. The massive magnetite forms a lens 600 m long and 60-70 m thick, enclosed in a serpentinite tectonite slice. The ore lens has a narrow rodingite halo and the reserves are about 20 Mt of medium-grade ore.

The massive magnetite sheet containing up to 4.25% Cr_2O_3, 0.63% Ni and 0.09% Co and topping sheared serpentinite near Pukhpur (Nagaland, north-eastern India), could have a similar origin (Singh and Srivastava, 1980).

Blebs of nickel sulphides, rare arsenides and Ni-alloys (heazlewoodite, millerite, niccolite) with some copper minerals (cubanite, chalcopyrite, bornite) are relatively frequent in sheared, serpentinized and rodingite-altered ultramafic tectonites (e.g. at Coolac, N.S.W., Ashley, 1973; Heazlewood Complex, Tasmania; Fig. 7-3D; New Caledonia, Guillon and Lawrence, 1973; Kellaki, Cyprus, Antun et al., 1966; Balangero, Italian Piedmont, Zuffardi, 1977). In the Khabarnyi ultramafic massif, southern Urals (Yakovlev et al., 1970; Fig.7-4), serpentinized dunite is sheared and fractured and some of the tectonic zones are marked by dikes of wehrlite and pyroxenite. Ore lenses and schlieren consisting of disseminated pyrrhotite, pentlandite, chalcopyrite, molybdenite and chromite are concentrated in a 2 x 1 km long, north-east trending zone.

Sometimes Pt-metals are associated; elsewhere copper sulphides are the only metallic minerals present (e.g. Mt. Ramazzo and Rossola, Liguria, Italy; Ferrario and Garuti, 1980). In the small Tasmanian bedrock occurrences of platinoids (Elliston, 1953), osmiridium transecting serpentinites was found as irregularly scattered grains along shears and foliated zones.

7.6. DIFFERENTIATED GABBRO-ULTRAMAFIC CONSTRUCTIONAL PILES

In the simple ophiolite complexes, the basal tectonite ultramafics are overlain by cumulate gabbro which, in turn, passes into hypabyssal gabbroic rocks that include gabbro porphyrite and diabase. These grade further up into swarms of vertically arranged diabase dikes (Glennie et al., 1974). Elsewhere, the cumulate assemblage is more complex. It is rhythmically layered and it includes ultramafic

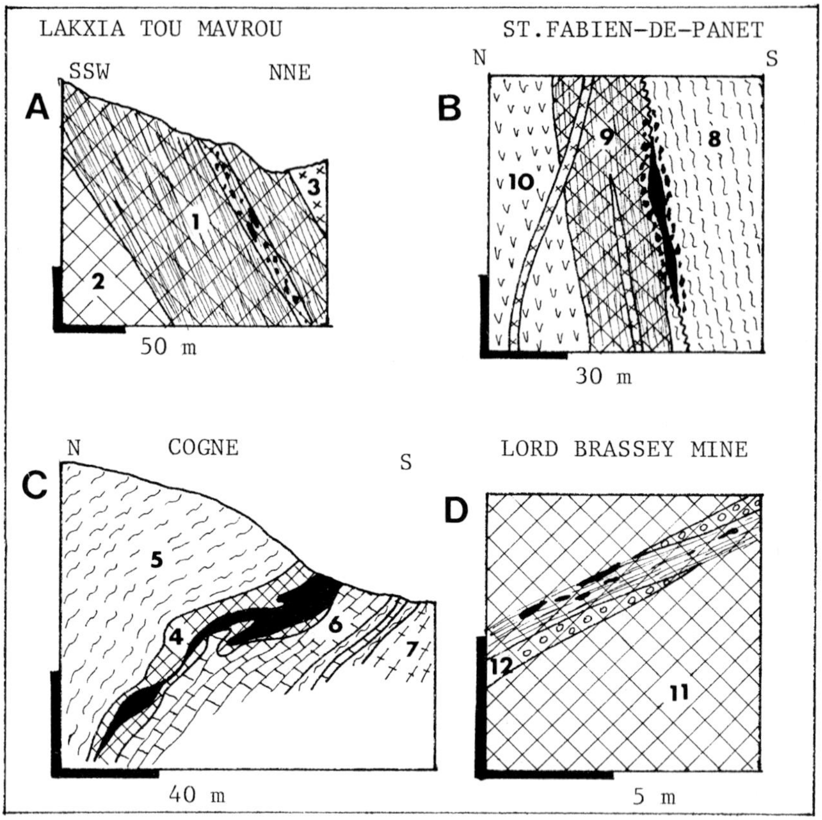

Fig. 7-3. Examples of hydrothermal-mobilization ores associated with sheared ophiolite ultramafics.
(A) Lakxia tou Mavrou, Cyprus; (1) lenses, disseminations, masses of pyrrhotite, chalcopyrite, cubanite, valeriite, etc. in sheared Cretaceous serpentinite; (2) massive harzburgite; (3) gabbro. After Panayiotou (1980).
(B) St. Fabien-de-Panet, Québec. Massive and disseminated pyrite, chalcopyrite, sphalerite, millerite, gersdorffite, along shears in silicified, middle Ordovician slate (8), and carbonatized Cambro Ordovician serpentinite (9). (10) Cambrian mafic metavolcanics. From LITHOTHEQUE, data after Béland (1957).
(C) Cogne, Aosta Valley, Italian Alps. (4) Massive magnetite lens in a sheared and rodingite-altered J-Cr serpentinite; (5) Mesozoic calcareous phyllites; (6) Triassic limestone, dolomite; (7) albitic gneiss. From LITHOTHEQUE, after Compagnoni et al. (1981).
(D) Lord Brassey prospect, Heazlewood River Complex, Tasmania. (11) Blebs, disseminations, stringers of heazlewoodite in sheared peridotite, near (12) rodingite-alteration zones. From LITHOTHEQUE.

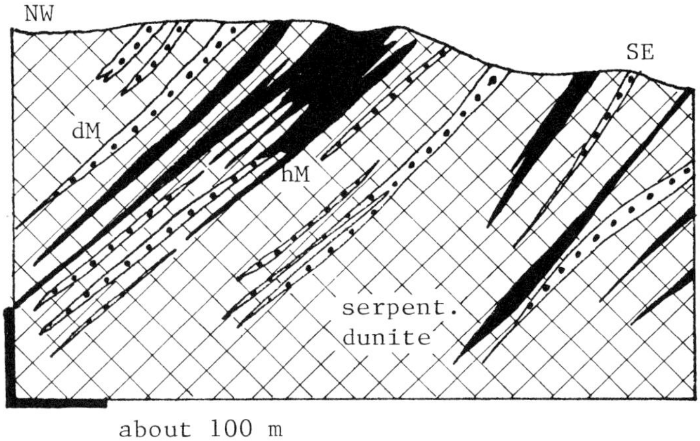

Fig. 7-4. Goryunskoe showing, Khabarnyi massif, S. Urals. (dM) low, under 3%, and (hM) intermediate, 3-10%, density disseminated pyrrhotite, pentlandite, chalcopyrite mineralization in serpentinized dunite. After Yakovlev et al. (1970).

members (such as cumulate harzburgite, dunite or lherzolite with layers of websterite, wehrlite, pyroxenites: both clino- and ortho-pyroxenites, and troctolite) and mafic members (dominant eucritic clinopyroxene gabbro, minor olivine gabbro, picrite, norite, hornblendite and diorite). Rare anorthosite has been reported from California (Bailey et al., 1970). Glennie et al. (1974) described a variety of magmatic breccias and mafic migmatites from the Semail Nappe in Oman, in which fragments of ultramafics are enclosed in a leucocratic gabbro or in which one type of layer is the product of injection of a younger melt.

Trondhjemite (plagiogranite) bodies are frequent in the top portion of the cumulus zone or in the sheeted dike complex, but are largely post-cumulus. They form irregular small massifs, schlieren and dikes, and sometimes grade into keratophyres (Brown et al., 1979). Because of their high water content, they often cause hydrothermal alteration of the adjacent older members. Trondhjemites are chemically distinct from the "continental" granitoids and are considered by most to be true magmatic differentiates, although seawater introduction played a role in their evolution (Coleman and Donato, 1979). Independent gabbro sills or dikes commonly intruding a variety of rocks in the vicinity of ophiolite complexes, are probably offshoots from the cumulate zone.

As expected, the cumulates carry a variety of magmatogene Cr, Fe, Ti, V, Pt and Ni mineralizations, but with the exception of chromites no significant ore deposits have so far been reported. Cumulate chromite in crystalline lenses or in bends generally parallel with the igneous layering, occurs in dunite layers near the base or at the basal contact. They are rare near the top of the ultramafic unit as, for example, in north-western Newfoundland (Church, 1972). The cumulate-hosted chromites are parallel with the true igneous layering and tend to have higher Fe/Mg ratios.

Magnetite and ilmenite occur frequently in cumulate gabbros in small quantities. Hamilton (1979) reported olivine - ilmenite cumulate from ophiolites near Soroako (Sulawesi, Indonesia). In the Jurassic ophiolites in the Drocea Mountains, Rumania (Ianovici and Borcoş, 1982), vanadium-bearing titanomagnetite cumulate forms low-grade (16% Fe) lenses and sheets in rhythmically layered gabbros. The gabbros are usually enveloped by metabasalts. Magnetite crystals contain exsolved ilmenite and lamellae and blebs of pyrite. Rutile and sphene are present in small proportions.

Pyrrhotite and/or pyrite with Cu and Ni sulphides are moderately common in gabbroic rocks. At Pukë, Albania (in the Mirdita ophiolite zone; Ilavský, ed., 1979), massive pyrite - pyrrhotite - chalcopyrite lenses occur in the gabbro and along its contacts with serpentinite. In the Cuni deposit near Zeehan, Tasmania (Blissett, 1962), lenses of disseminated to massive pyrrhotite, pentlandite, chalcopyrite and millerite are aligned along a contact of a Cambrian gabbro sill, probably related to the adjacent ophiolite. 1,600 t Ni and 800 t Cu were produced. At the Adau River nickel prospect (Papuan ultramafic belt; Spratt, 1975), pentlandite, bravoite and Ni-pyrite are disseminated in a lens of massive magnetite, aligned along a 320°-trending shear zone in a Cretaceous gabbro. This, in turn, intrudes ultramafics. Obviously, the mineralization is post-magmatic. Hassan and Al-Sulaimi (1979) reported an epigenetic copper mineralization hosted by a layered gabbro in the Semail Nappe, Oman. There, secondary copper minerals are widespread in the form of disseminations and criss-crossing veinlets in a 1 km wide, N 70°E trending kaolinized zone.

The trondhjemites, although geochemically consistent with having originated from high-level differentiation of ophiolitic magmas (Saunders et al.,1979, contradicted by Brown et al., 1979, who assumed metasomatic origin of trondhjemites involving seawater hydration), are, nevertheless, late-stage intrusions. They often postdate intraoceanic shearing and cleavage development. Trondhjemites are typically associated with hydrothermal mineralization located in both endo- and exocontact. In the classical metallogenic model of Bilibin (1955), the plagiogranite-associated ores were treated as a prominent example of the early stage mineralization in geosynclines. The important group of Uralian gold deposits (Berezovsk, Kachkar, Pyshminsk; the first two credited with having produced over 100 t gold) involves hydrothermal vein systems related to Silurian-lower Devonian plagiogranites. Some veins are hosted by serpentinites (Pronin, 1962). In Berezovsk, more than 70,000 thin quartz, pyrite, ankerite, tourmaline, albite, fuchsite, gold veins cut the plagiogranite dikes and ophiolites. The most widespread alteration in the ultramafics is "listvenitization", a talc, carbonate, fuchsite assemblage. Similar gold deposits are known in the Kuznetsk Alatau, U.S.S.R.; in several small goldfields (e.g. Cawarral, Canoona) in the Marlborough serpentinite belt, Queensland; and elsewhere.

7.7. SHEETED DIABASE DIKES

In many ophiolite complexes (Troodos, Semail, Vourinos, N.W.Newfoundland), the tops of the cumulate gabbros grade upwards into a unit composed entirely of diabase dikes intrusive into one another, and inclined at a high angle to the layering of gabbros. The dikes, in turn, feed the overlying pillow basalts and in the excellent cliff outcrops near Betts Cove, Newfoundland, the dike feeder and basalt flow relationship can be seen in spectacular detail. Metallic mineralization in the diabase is sparse. Zuffardi (1977) reported veinlets and disseminations of quartz, pyrite, bornite and chalcopyrite from Bargone and Monte Rossola in the ophiolite complex of Eastern Liguria. Upadhyay and Strong (1973) demonstrated that most of the small subaqueous-hydrothermal pyrite, chalcopyrite, sphalerite deposits associated with the Newfoundland ophiolites occur immediately at the sheeted dikes - pillow lavas contact. Their remobilized equivalents extend into the tops of the sheeted dikes unit.

A single diabase dike (probably not a member of the sheeted dike unit) intrudes an ophiolitic peridotite and amphibolite in Čavka (Bosnia; Janković, 1982) and it carried disseminated to massive pyrrhotite, pentlandite and chalcopyrite along the contact.

7.8. TRANSITION BETWEEN META-OPHIOLITES AND BUSHVELD-STYLE COMPLEXES

The interesting Ivrea-Verbano mafic - ultramafic complex in the Italian Alps has been listed in the literature as an example of an "Alpine-type" peridotite or ophiolite. Recently, the complex has been re-interpreted (Ferrario et al., 1982) as a deep-seated layered intrusion emplaced at 8 to 9 kb and 1,000 to 1,200°C C into the surrounding crustal sequence causing its granulite metamorphism. The presence of slices of tectonite peridotite along the zone that locally forms the base of the layered intrusion, indicates a possible relationship with ophiolites. According to the recent interpretations, the tectonic zone is a segment of the upper mantle / deep crust transition, and a deep root of an orogen. The Ivrea-Verbano Complex is composed of cyclic units of interlayered peridotite, pyroxenite, gabbro and anorthosite in its lower part, overlain by a large body of homogeneous gabbro-norite. This grades upward into diorites. Ferrario et al. (1982) recorded 14 small occurrences and deposits of Ni-Cu sulphides with traces of platinum metals. They occur in two main settings: (1) as low-grade disseminated to massive sulphides in ultramafic layers of the cyclic units (garnet pyroxenite, orthopyroxenite, etc.) in the vicinity of metasedimentary septa, and (2) high-grade disseminated sulphides in the pyroxenitic dikes.

7.9. VOLCANICS AND INTERBEDDED SEDIMENTS ON TOP OF OPHIOLITE COMPLEXES

Maxwell (1970) fittingly described the top of the Appenine (Italy) ophiolite complexes as follows: "The (cumulate) gabbro gives way abruptly upward to coarse diabase, which in turn grades upward to finer grained diabase, then to microlithic (basalt) lava with breccia and pillow structure characteristic of an extrusive environment. Bedded, usually red, radiolarian cherts overlie the extrusive lavas, though locally the (basalt) lavas and cherts may be interbedded. On Elba, I found angular fragments of red chert in the cores of pillows. The base of the chert sequence is commonly a chippy chloritic mudstone similar to the filling between pillows; it frequently contains fragments of variolites, derived from adjacent pillows, and appears to have been produced by the reaction of hot lava with sea water. In many areas, the radiolarian cherts pass upward through a thin transition zone of gray, calcareous, radiolarian chert to pale gray or cream-coloured finely grained limestone with abundant calpionellids of Tithonian to earliest Cretaceous age".

This is representative of many ophiolites of the world, but there are numerous variations. In the Troodos Complex of Cyprus, there are at least seven mafic volcanic units with interbedded sediments, separated by unconformities, on top of the sheeted dike complex (Adamides, 1980). Although pillowed basalts (tholeiite, olivine basalt, rare picritic basalt), their hyaloclastites and autoclastic breccias are the most common form of volcanics, spilites (albitic basalts) and keratophyres (albite, minor quartz) are frequently associated and/or substituted for basalts. Despite the voluminous literature (compare Amstutz, ed., 1974, and the summary in Brown et al.,1979), the origin of the albitic rocks remains as enigmatic as ever, although seawater metasomatism of oceanic basalts under hydrothermal conditions to form spilite (e.g. Spooner and Fyfe, 1973) is now the preferred hypothesis. Keratophyres appear to be the volcanic equivalents of plagiogranites. Intensive deformation, alteration and metamorphism further increases the lithologic complexity. Fe-Cu-(Zn) sulphide, Fe oxide, and Mn oxide (silicate, carbonate), are the main mineralizations associated with the submarine tops of ophiolite sequences.

7.9.1. Iron-copper-(zinc) sulphide deposits

Massive to disseminated iron, copper and occasionally zinc sulphide deposits hosted by mafic volcanics, located on top of ophiolite sequences, have recently been treated extensively in the literature under the heading of "Cyprus-type" (Sawkins, 1976; Franklin et al., 1981, p.510-515, present a good review). In the type area located in the Mesozoic Troodos Complex (Searle, 1972; Constantinou and Govett, 1973; Constantinou, 1980), about 20 orebodies contain some 900 Tt Cu, 16 Mt Fe and 4.2 Tt Zn (Fig. 7-5A).

The mineralization is distinctly vertically zoned (Fig. 7-6). The main sulphide mass is located at the original seawater basaltic pillow lava interface and consists of hard angular blades of fine-grained, metacolloform and banded pyrite with chalcopyrite and rare sphalerite, in a matrix of soft, friable pyrite with a dark sooty discoloration. This grades downward through a transition zone of increasing silica content into a stockwork zone. The latter contains stringers, disseminations and coalescing veins of coarser-grained pyrite, often combined with quartz and jasper and surrounded by strongly silicified, chloritized and argillized metabasalt. At some localities, the massive sulphide bodies are topped by the "Ochre Group" (a goethite-rich sediment that commonly contains sulphide bands and fragments). The ochre does not extend beyond the boundaries of the sulphide and is interpreted (Constantinou and Govett, 1973) as resulting from submarine leaching and secondary enrichment of the primary pyritic deposits. Similar ocher topping a "Cyprus-type" massive sulphide was described by Taylor (1977) from the Hanesawa deposit, Solomon Islands.

Table 7-3 lists selected Fe-Cu-(Zn) sulphide occurrences associated with the ophiolitic mafic volcanics. Certain localities in the Semail Nappe (e.g. the Lasail deposit; Alabaster et al., 1980); in Liguria (e.g. Libiola, Reppio; Ferrario and Garuti, 1980); in the Lake Izabal ophiolite, Guatemala (Oxec orebody; Petersen and Zantop,1980) and in the Cambro-Ordovician ophiolites of Newfoundland (Betts Cove, Tilt Cove; Upadhyay and Strong, 1973; Fig. 7-5D) are, in many respects, comparable with the Cyprus model in both the general setting and ore appearance. Other localities are significantly different, mostly due to intensive tectonization.

In the Whalesback-Little Deer deposit, Newfoundland (Kanehira and Bachinski, 1968; Fig.7-5E), pyrrhotite - chalcopyrite stringers and small sulphide masses occur in chlorite-sericite altered and silicified metabasalts along a narrow shear zone. The impregnations of secondary Fe, Mn and Cu minerals in sheared ophiolite metavolcanics in the Al Fujairah area, Oman (Hassan and Al-Sulaimi, 1979), may have had a comparable pre-oxidation appearance.

The ores hosted by "chaotic breccias" including ophiolite members, red mudstones, serpentinites, and similar rocks (such as Ergani Maden, Turkey and the Balabac Island, the Philippines; Fig. 7-5C) are more properly consided to be in mélanges than in type ophiolites and they are treated in Chapter 8.

7.9.2. Manganese deposits

Manganese oxides, carbonates and silicates are frequently present in the vicinity of the basalt, or topping sediments contact. In the Nicoya ophiolite complex in Costa Rica (Kuÿpers and Denyer, 1979),

Table 7-3.

Fe-Cu-(Zn) sulphide deposits in mafic volcanic ophiolite setting ("Cyprus"- or "Kieslager"-"type")

LOCALITY	AGE	MINERALOGY, GEOLOGY	TONNAGE, GRADE	REFFRENCES
York Harbour, west. Newfoundland	Or_1	py,sf,cp; irregular masses in metabasalt, underlain by stringer zone	17 Tt Zn/4.7% 7 Tt Cu/2%	Franklin et al. (1981)
Tilt and Betts Cove N.W. Newfoundland	Or_1	py,cp; semiconf.,irreg.masses,sheets,stringers in chloritiz.shears; in pillow basalts above sheeted dikes.	217 Tt Cu/3-10%	Franklin et al. (1981)
Whalesback, Little Bay M., Newfoundland	Or_1	cp (py); stringers along shears in m-basalt and breccia	50 Tt Cu/1-2.5%	Kanehira and Bachinski (1968)
Oxec m., L. Izabal, area, Guatemala	Cr	cp,py; dismembered lens, mass./dissem.sulph. in chloritiz.,silicif.pillow basalts	20 Tt Cu/3%	Petersen and Zantop (1980)
Nicoya Complex, Costa Rica	Cr	py,cp; 3 small lenses in pillow basalts	n.a.	Kuÿpers and Denyer (1979)
Saint Véran, W. Alps, France	J_2-Cr_2	born,cp,covell,mag,py; several lenses along 3 km long zone; dissem.,stringers,masses in chlor.,glaucophane,silic.sch.,metabasalt	1.7 Tt+	Blondel and Raguin (1935)
E. Liguria (e.g. Libiola), Italy	J	py,cp,sf,mag; mass.lenses near tops of spilitized basalts	n.a.	Zuffardi (1977)
Troodos Complex, Cyprus	Cr_3	py,cp; rare sf; peneconc.massive lenses,underlain by stringer stockworks,in pillow basalts Over 10 stratiform orebodies.	900 Tt Cu/2.5% 16 Mt Fe/40% 4.7 Tt Zn	Constantinou and Govett, (1973)
Lasail dep., Semail Nappe, Oman	Cr_3	py,cp; massive stratif.lens between two pillow lava units	about 300 Tt Cu/2%	Alabaster et al. (1980)
Sohar area, Semail Nappe, Oman	Cr_3	second.Cu miner.; veinlets, infiltr.,in m-basalt, m-chert, gabbro. 8 km long zone, 3 dep.	315 Tt Cu/2.1%	Hassan and Al-Sulaimi (1979)
Balabac Island, Philippines	Cr_3-Eo_1	pb,cp,sf; mass.sulph.,irreg.stratabd.masses on top of pillow basalts,in melange	20 Tt Cu/4.65%	John (1963)

Barlo m., W.Luzon Philippines	Cr_3-Pc	py,cp,sf; massive to disseminated,several stratabound lenses	43 Tt Cu/1.6% 5.4 Tt Zn	Bryner (1969)
Shimokawa m., Hokkaido, Japan	J	ph,py,cp,sf; en echelon mass.sulph.lenses, grading to stockworks; in chlor.schist	182 Tt Cu/2.28 107 Tt Zn/1.34	Mitsubishi Comp.materials
Woolomin Beds localities, N.S.W., Australia	S-D	fine grained py,cp,sf; massive sulphide lenses in jasper above pillowed m-basalt;	8 Tt Zn/4.65% 5 Tt Cu/2.7%	Scheibner and Markham (1976)
Hanesawa Isl., Solomon Islands	Cr ?	py,cp,chalcoc.; mass.to dissem. in altered pillow basalts topped by ochre	1 Tt Cu+/5.3%	Taylor (1977)

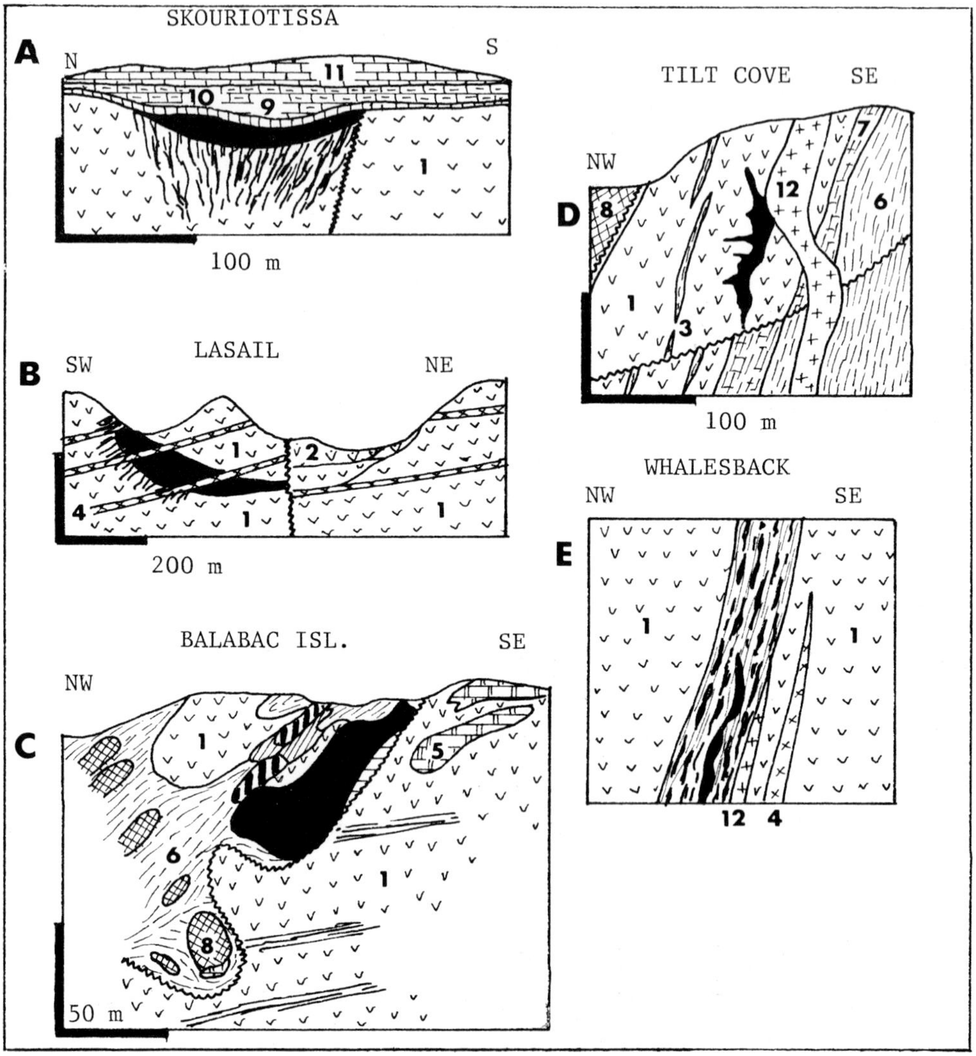

Fig. 7-5. Ophiolite-hosted pyritic Cu deposits (Cyprus "type"); Skouriotissa, Cyprus (Cretaceous); Lasail, Oman (Cretaceous); Balabac Isl., Philippines (Cretaceous to Eocene); Tilt Cove and Whalesback, Newfoundland (Ordovician). LEGEND: (1) pillowed metabasalt; (2) porphyritic basalt; (3) basalt breccia and interflow sediment; (4) diabase, gabbro dikes; (5) chert; (6) mudstone, siltstone, graywacke; (7) red argillite; (8) serpentinite; (9) umber; (10) marl; (11) limestone; (12) quartz - feldspar porphyry; BLACK, massive to stringer sulphide; VERTICALLY RULED, disseminated sulphide. Modified after Constantinou and Govett (1973); Alabaster et al. (1980); John (1963); Maritime Mining, oral commun. (1973); Kanehira and Bachinski (1968); (D) and (E) are from LITHOTHEQUE.

Genesis	Mode of formation	Mode of deposition	Ore type	Geol. column	Ore zones	Mineralogical description	Type of mineralization	Minerals other than pyrite	Average Ore grade %S	Average Ore grade %Cu
					Post-mineralization lava					
Magmatic	Oxidation / Sedimentary	Exhalative-sedimentary	Sulphides and iron mudstone		Zone 1	Iron-rich and siliceous mudstones; oxidation of sulphides; lenses of friable black ore. Graded bedding, with grains of pyrite, and slumping boulders of colloform sulphide	Type 1	Sediment: goethite, silica, montmorillonite. Sulphides: covellite, sphalerite, chalcopyrite, marcasite	45–50	0.5–1.5
	Fumarolic		Massive sulphide		Zone 2	Fragmentary ore consisting of angular and hard blocks of yellow ore in matrix of black sandy ore	Type 2	Marcasite, sphalerite, po(v.r.), bn(v.r.), galena (v.r.)*, chalcopyrite	40–45	1.0–4.0
					Zone 2A	Fragmentary sulphide, blocks of pyrite in siliceous matrix with decreasing amount of black, friable sulphide	Type 3	Marcasite (r), chalcopyrite, sphalerite (r), po (v.r.) quartz, jasper, chalcedony	30–40	0.5–1.5
		Hydrothermal	Stockwork, cavity, fracture-filling, disseminations		Original lava surface	More coarsely crystalline splendent pyrite as cavity-filling with much quartz and jasper	Type 4	Chalcopyrite, sphalerite, quartz, illite, jasper, chlorite	15–30	0.4–1.2
					Zone 3	Discrete fracture fillings with pyrite and quartz and jasper. Disseminations of pyrite in vesicles and cracks	Type 5	Chalcopyrite (r) sphalerite, (v.r.), rutile (r.), illite, quartz, chlorite, jasper	5–15	0.2–0.5

* From Agrokipia B only   r, rare   vr, very rare

Fig. 7-6. The characteristics and vertical zoning in the Cyprus massive sulphide deposits. From Searle (1972), courtesy of the Institute of Mining and Metallurgy, London.

several varieties of Mn deposits composed of braunite, psilomelane and pyrolusite, occur: (1) The deposits in basalt are in fact hosted by jasper and jasper-basalt breccia, and these siliceous rocks form cross-cutting pipes, or lenses of interlayered manganiferous jasper controlled by a specific stratigraphic horizon. The Mn minerals are most common as cross-cutting veinlets in the upper part of the stratabound jasper lenses and contain locally replaced radiolarian tests. Footwalls of the mineralized lenses are usually brecciated and intensively propylitized. (2) The deposits described as in radiolarian jasper are stratiform. They are situated 15 m or more above the basalt, and consist of alternating millimetre-thick laminae of Mn-oxides and jasper. (3) Former submarine Mn oxide crusts and nodules occur along the strike and slightly above the Mn-radiolarite lenses.

Almost identical Mn occurrences are abundant in jasper and argillite above the basalt in the Franciscan Assemblage of California (Crerar et al.,1982); in the Silurian - Devonian Woolomin Beds, New South Wales (e.g. in the Walcha area; Fitzpatrick, 1975); in the Port Macquarie area, (Figures 7-7 and 7-8) and in the Mesozoic ophiolites in Liguria, Italy (Burckhardt and Falini, 1956). In New South Wales, however, rhodochrosite and rhodonite rather than braunite appear to be the primary minerals. Although genetically interesting,

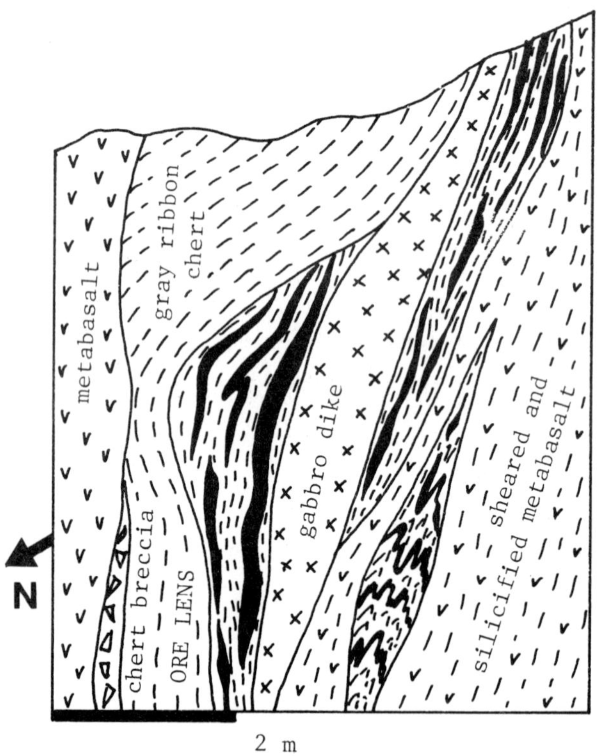

Fig. 7-7. Mn ore lens in the Permian dismembered ophiolite complex, Port Macquarie, N.S.W.; an outcrop plan. The ore is composed of bands of pink rhodonite (black), enclosed in gray chert, minor jasper and siliceous hematite. From LITHOTHEQUE, based on a 1981 field sketch.

Fig. 7-8. Bedded Fe and Mn ores associated with chert in the dismembered ohiolite complex near Port Macquarie, N.S.W., Australia. (A,B): light-green ribbon chert; (C) banded rhodonite lens; (D) banded magnetite lens.

ophiolite associated Mn deposits are of low economic importance (Table 7-4).

The Cyprus umbers (Robertson, 1975) are Fe, Mn, and several trace metals-enriched mudstones located at the base of a marine sedimentary sequence topping the youngest basalt cycle of the Troodos ophiolite complex. They are interpreted as ancient subaqueous - hydrothermal "metalliferous sediments" and they have the same stratigraphic position as the Mn-bearing jaspers.

In the Florida group of orebodies, Solomon Islands (particularly at Hanesawa Island; Taylor, 1977), Mn-oxide lenses occur stratigraphically above the "Cyprus-type" sulphide bodies and have a high Au (3.2 ppm) and Ag (35 ppm) content.

7.10. MINERALIZATIONS DUE TO INTERACTION OF OPHIOLITES WITHIN THE CONTINENTAL CRUST

Mineralizations resulting from simple and predictable interaction of various crustal processes, such as weathering, sedimentary reworking, etc., have been known and reported for a long time. Others, more complex, are in the process of being recognized. Still other associations and interactions are obscure and uncertain and can be unravelled only with a great deal of imagination. Continuing systemization, however, is of considerable importance for the purpose of exploration prediction.

7.10.1. Ophiolites and Barrovian (high temperature / low pressure) metamorphism

Coleman (1977, p.123) summarized briefly the effects of the Barrovian metamorphism on ophiolites. While amphibolite—grade metamorphosed peridotite remains indistinguishable from a peridotite metamorphosed under high pressure / low temperature conditions (or a mantle tectonite), gabbros, diabases and basalts are converted to amphibolites. The influence of metamorphism on the ophiolite - hosted mineralizations has rarely been discussed in the literature. Presumably, few new (metamorphogenic) mineralizations form (except, perhaps, the rutile-rich pockets as in the Dinning deposit, Maryland; Pearre and Heyl, 1960), but existing ores are modified. Chromite, magnetite and most sulphides are usually modified only texturally, but metamorphic silicates may form in the gangue. Dietrich (1972) described metamorphically remobilized Fe-Cu-Ni sulphide ore from Graubunden (Switzerland) ophiolites, conspicuous by the presence of an abundant ilvaite gangue. The Fe and Mn oxide and carbonate deposits associated with ophiolites responded in by far most spectacular manner to the regional metamorphism. Mn silicates (rhodonite, spessartite, Mn-pyroxenes, Mn-amphiboles, Mn-epidote, etc.) are abundant in the meta-ophiolites located in the Penninic zone of the French and Italian Alps (e.g. Saint Marcel). Sturt et al. (1980)

Table 7-4. Mn oxide, silicate, carbonate lenses in ophiolite basalt, jasper and chert

LOCALITY	AGE	MINERALOGY, GEOLOGY	TONNAGE, GRADE	REFERENCES
Franciscan Assembl. California	J_3-Cr_1	Mn oxides; 1 x 15 m lenses in jasper, separated by radiolarites, above basalts	53 Tt Mn/40%	Crerar et al. (1982)
Nicoya Peninsula Costa Rica	Cr	Mn oxides; stockworks in chert interbedd. with basalt; stratif. lenses in jasper above basalt	17 Tt Mn	Kuÿpers and Denyer (1979)
Gambatesa, etc., Liguria, Italy	J	qtz, braunite, rhodochr., rhodonite, pyrolus. compact, banded, brecciated in jasper; stratabound.	131 Tt Mn/19%	Burckhardt & Falini (1956)
Drocei Mts (e.g. Pirnesti), Rumania	J_3-Cr_1	psilm., pyrolus., rhodochr., hematite; stratif. lenses in jasper or hemat. slate, above ophiolitic basalt	X0 Tt Mn	Petrulian (1973)
Lasbela State, S. Pakistan	J-Cr	psilom., pyrolus.; stratabd. lenses or beds conform. with basalt, silic. slate, serpentin.	210 Tt Mn/42%	Master (1956)
Netea, etc., W. coast of N. Caledonia	Pc	Mn oxides; in jasper, red silic. argillite; up to 70 m long stratabound lenses.	26 Tt Mn/47%	Paris (1981)
Walcha, etc., Woolomin beds, N.S.W. Australia	S-D	Mn oxides, rhodonite; stratabound lenses in jasper above basalt, often dismembered, cont. metamorphosed by granitic intrusions.	8 Tt Mn/35%	Fitzpatrick (1975)
Hanesavo Isl., etc., Florida Group, Solomon Islands	Cr_3-	Fe-Mn oxides; stratigraphically above Cu-Zn massive sulphide dep., on top of metabasalts	10 Tt Mn/47% 32 kg Au/3.2ppm	Taylor (1977) Arthurs (1979)

interpreted the Mn-garnet porphyroblasts and magnetite-rich bands in the Karmøy (Norway) ophiolite schists as metamorphosed equivalents of the Cyprus umbers.

In the small, enigmatic Ransko ophiolite mass in western Czechoslovakia, a block of an intensively metamorphosed massive sphalerite, barite, pyrite, pyrrhotite, chalcopyrite mineralization of economic importance associated with cordierite-anthophyllite-gahnite assemblage, is surrounded by layered ultramafic plutonites. The genetic interpretations so far offered a range from metamorphosed intra-ophiolitic epigenetic replacements, to a metamorphosed raft brought in from outside and genetically unrelated to the ophiolite (Watkinson et al., 1978).

7.10.2. Interaction ophiolites / calc-alkaline magmatism (plutonic level)

The effects of the thermal metamorphism or metasomatism on ophiolites due to a "granitic" intrusion are often indistinguishable from the regional metamorphic effects and in most cases both effects merge. The most fundamental function of an igneous intrusion is that it provides a localized heat source, capable of driving hydrothermal fluids. Such fluids can carry metals released in the process of differentiation of the intrusion, or alternatively the metals can be leached from the rocks located in the "granite" exocontact. If ophiolites are in the exocontact, they can become both a passive host (recipient of ore metals derived from outside), and a metal supplier through leaching.

Hydrothermal alteration of ophiolite ultramafics readily produces a dolomitic, ankeritic or sideritic carbonate that can be replaced in the subsequent mineralization pulse by base or rare metal minerals. If this is the case, the altered ultramafic rock behaves as a "normal" (metasedimentary) carbonate and when mineralized, there is no particular difference in the composition of the altered ultramafic-hosted ores and other mineralization styles in the area, hosted by more common rocks. Several examples are known from Pb-Zn, Sn and other mineralized regions. In the Koporić mine, Yugoslavia (Janković, 1982), low-grade stockworks and metasomatic impregnations of galena and sphalerite are distributed in a quartz-carbonate alteration zone in serpentinite, along its contact with Cretaceous flysch sediments, and in a "granite" aureole. In the Magnet mine (Waratah, Tasmania; Cox, 1975a), narrow veins and aggregates of galena, sphalerite and manganosiderite are hosted by ferrodolomite, an alteration product of a Cambrian ophiolitic spilite and albitic diabase, in the aureole of a Devonian granite. Comparable occurrences are known in the Dundas field near Zeehan, Tasmania and these are notable for the presence of the rare Cr minerals, stichtite and crocoite, found in accessory quantities in gossans of galena-sphalerite veins and replacements (Fig. 7-9A).

In the Razorback mine, Dundas field, Tasmania (Blissett, 1962; Fig. 7-9B) disseminations, replacement veinlets and masses of pyrrhotite, pyrite, lesser chalcopyrite, arsenopyrite and cassiterite, are hosted by silicified talcose dolomite. The dolomite, in turn, is a former Cambrian serpentinite hydrothermally altered along a N.N.W.-trending fault zone. The Fe-As-Sn mineralization grades zonally into sphalerite-galena replacements, and there is a prominent Sn-enriched gossan. About 2,500 t Sn are present in an ore with a grade close to 0.3% Sn, probably coeval with a Devonian granite cupola in depth.

A variety of minor Au, Cu, Ni, Co hydrothermal veins, impregnations or replacements in regions of coexisting older ophiolite and younger "granite" suggest empirically that the ophiolite was, at least in part, the metal source. In the Ciungani deposit, Drocea Mts., western Rumania (Ianovici and Borcoş, 1982), magnetite, pyrrhotite, pentlandite, bravoite and pyrite form impregnations and veinlets in Jurassic gabbro, situated close to a contact with Cretaceous granitoids. Elsewhere in the Drocea ophiolite complex (Cazanesti, Roşia Nouă), pyrite and chalcopyrite in quartz, calcite, epidote gangue form east-west and northeast – southwest trending veins and stockworks in sheared Jurassic metabasalts, several hundred metres

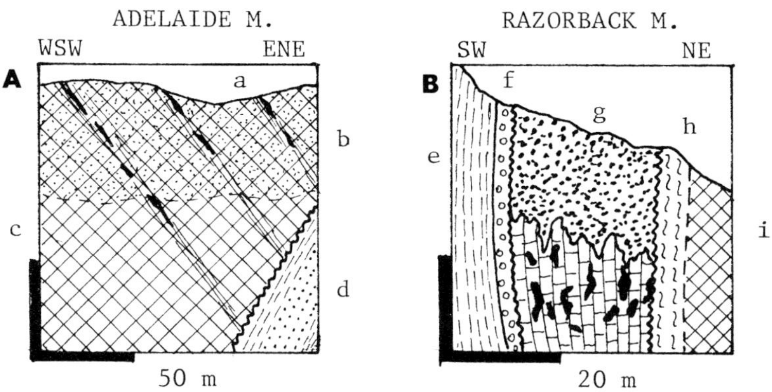

Fig. 7-9. Two serpentinite-hosted but Devonian granite – coeval ore deposits, Dundas field, N.W. Tasmania. Adelaide Pb-Zn and crocoite mine: (a) sulphide mineralized talc - carbonate altered shears; (b) gossan and regolith; (c) Cambrian serpentinite; (d) Eocambrian quartzite and slate; Razorback Sn-pyrrhotite and Pb-Zn mine: (e,f) Cambrian black slate and conglomerate; (g) Sn-bearing limonitic gossan (top), grading downward into Fe,Pb,Zn sulphide and cassiterite replacements in talcose dolomite; (h) talc, chlorite, carbonate schist grading to (i) serpentinite. From LITHOTHEQUE, based on data in Blissett (1962).

from contact with a Cretaceous granodiorite. In the Komori mine, Japan (Imai, 1966), chalcopyrite - cubanite veins occur in serpentinite near granitic intrusions. In the Dobšiná siderite-ankerite vein and replacement deposit, Slovakia (Ilavský, 1976), gersdorffite appears in some veins generated from hydrotherms that had probably passed through a nearby body of serpentinite. Over 600 t of cobalt have been produced.

The Rolette mine (St. Fabien-de-Panet, Québec; Béland, 1957; Fig. 7-3B) contains two zones of massive and disseminated pyrite, chalcopyrite, sphalerite, millerite and accessory gersdorffite in silicified Ordovician slates along a contact with carbonatized serpentinite. Granodiorite dikes are moderately common in the field. It has been suggested that the mineralization might be a subaqueous hydrothermal deposit, later remobilized; the nickel remobilized from the serpentinite was added later. The reserve was 25 Tt Cu, 5 Tt Zn and 1 Tt Ni.

Hydrothermal gold occurrences are frequent in regions where ophiolites and granitoids interact (e.g. in the Klamath Mts., Oregon; the Urals; Bralorne district, British Columbia; etc.), but a direct evidence pinpointing the gold source has rarely been presented. In the Zod ore field in Armenia (Lesser Caucasus, U.S.S.R.; Borodaevskaya and Rozhkov, 1974), upper Cretaceous ophiolite is locally sheared and intruded by quartz-feldspar porphyry dikes. The gold mineralization is localized by north-south shears, and consists of quartz-sulphide veinlets and disseminated pyrite, arsenopyrite, sphalerite, chalcopyrite, tetrahedrite, stibnite, Bi, Ni, Ag tellurides and gold in a quartz, Fe-Mg carbonate, talc alteration envelope. Elsewhere, known "granite" intrusions are missing, as in the Ligurian ophiolites west of Genoa, Italy (Pipino, 1980). There, ultramafics carry unevenly distributed gold reaching local concentrations of 0.1 to 2 ppm. Most of the gold actually mined on a small scale, however, is in stockworks of quartz, ankerite, chalcopyrite and sphalerite filling a crush breccia, as well as in alluvial placers derived from the stockworks.

7.10.3. Interaction ophiolites / calc-alkaline magmatism (subvolcanic level)

There is little qualitative difference between this and the previous association, as the mobilizing hydrotherms are generated by the deeper—seated intrusive equivalents of surficial volcanism, but the latter is highly visible and conspicuous. In the Apuseni Mts. of western Rumania and at several places in the internal zones of the Yugoslav Dinarides, Mesozoic ophiolites coincide spatially with Miocene continental andesite, dacite, rhyolite volcanic and subvolcanic suites. Several polymetallic and Au-Ag vein deposits (e.g. Baiţa and Musariu, Rumania; Ianovici and Borcoş, 1982; Žuta Prlina, Janjevo, Badovać and other Zn-Pb localities in Yugoslavia; the antimony deposit Rajičeva Gora, Yugoslavia; Janković, 1982) are ophiolite hosted, but their mineralogy is the same as in the neighbouring localities hosted by volcanics, continental sediments or the basement metamorphics. This suggests a lack of metal contribution from the ophiolites.

Elsewhere, the "common" mineral association of the hydrothermal deposits (such as sphalerite, galena, barite, etc.) contains a variety of additional Ni and Cu minerals (cubanite, valeriite, breithauptite, maucherite, millerite, in the Pb-Zn-Bi-Ag deposit Rudnik; cubanite and valeriite in the Pb-Zn-Au-Ag deposit Kišnica; 0.2-0.4% Ni in the orpiment deposit Lojane, hosted by serpentinite; all the deposits are in Yugoslavia; Janković, 1982) which indicates derivation from ophiolites. Mulina (1973) interpreted the genesis of the fracture controlled garnierite and Ni-saponite deposit in peridotitic saprolite in Goleš (Yugoslavia) in terms of lateral secretion and redeposition of nickel by hydrothermal solutions derived from Tertiary dacitic volcanism.

Although direct evidence is lacking, it is interesting to note that the world's major occurrences of gold and silver tellurides (Brad, Săcărîmb and others in the Apuseni Mts., Rumania; Cripple Creek, Colorado; Jílové - Bohuliby, Czechoslovakia; Kalgoorlie, W.Australia) are in hydrothermal veins or stockworks associated with intermediate to felsic magmatism, emplaced in a basement unit containing ophiolites or mafic and ultramafic bodies.

The interaction of ophiolites with hot springs resulting in the distinct silica-carbonate alteration and cinnabar mineralization as in California and elsewhere, is treated in Chapter 8 because the ultramafics are almost always members of a mélange. Except for mercury, it has been noted (McLaughlin, 1981) that the interaction of Quaternary volcanics and intrusives with Mesozoic ophiolitic cap rocks in the Clear Lake-The Geysers area, results in a significant contamination of the young magmatic rocks by MgO and Cr.

7.10.4. Ophiolites and tropical weathering

Tropical weathering superimposed on members of the ophiolite suite is capable, under favourable conditions, of selectively removing some of the major elements (Mg, Si, Ca, alkalies) and so, correspondingly, enriching the weathering profile in the residual elements (Al, Fe, Ti). The general conditions and variations of tropical weathering are treated in Chapter 23; here only the metallogeny of weathering crusts superimposed on ophiolites will be reviewed briefly. Tropically weathered ophiolite gabbros, diabases and basalts are metallogenically equivalent to any gabbros and basalts (e.g. Chapters 10, 12, 31) and mineralization such as ferruginous bauxite could be expected. However, no significant bauxite deposits have, so far, been reported from tropically weathered ophiolitic basics.

Tropically weathered ultramafics have the most interesting profiles metallogenically, containing industrial accumulations of Fe, Mn, Ni and Co. The nickel accumulations ("Ni-laterites") are most unique and important, and contain the greatest proportion of the world's nickel resources. Any ultramafics can produce identical metalliferous laterite, but ophiolitic ultramafics contain the greatest share.

METALLIFEROUS WEATHERING PROFILES OVER ULTRABASICS

The term "metalliferous laterite" is inaccurate, because it refers not only to the metalliferous tropical soil (the actual laterite), but to the decomposed and leached bedrock as well (the saprolite). It is, nevertheless, widely used. Metalliferous "laterites" have an extensive literature, much of which is devoted to the nickel "laterites". Golightly (1981) prepared a recent summary.

The metalliferous weathering profiles are located on ultramafic massifs with dissected (often bold) relief, capping plateaus and exposed in section on hill and plateau slopes, in stream valleys, and in road cuts. A typical image of a metalliferous "laterite" is an extensive blanket ("areal type" in the Soviet terminology; Fig.7-10A), but its floor is rarely even. It frequently has a karst topography (Fig. 7-11) and the tropical weathering effects extend to a considerable depth along fault zones. In many geologically old weathering profiles much of the areal laterite is erosion—removed and only the deep-reaching portions remain ("linear type" of the Soviet school; Fig. 7-10B). Although most of the known "laterites" have never been covered by younger rocks, a few were once buried and have subsequently been exhumed (e.g. most of the Urals deposits; the Ni "laterites" of Greece and Albania; Riddle, Oregon; etc.).

The "most normal" tropical weathering profile over ultramafics is mineralogically, texturally and geochemically zoned (Fig. 7-12), and the following zones are recognizable from the surface down (e.g. Golightly, 1981; Meillon, 1978):

(1) An iron-rich, reddish-brown, indurated duricrust, that breaks into distinct blocks along plateau edges, generating abundant rubble. The iron concentration is highest here, and the material constitutes a medium-grade iron ore.

(2) The ferricrust grades downward into a red concretionary layer and further down into a "red laterite": a brownish-red, ferralitic clay. Patches, blocks and discontinuous horizons of opaline or chalcedonic silica accumulate near the base.

(3) The "red laterite" is transitional downward into a brownish-yellow and ultimately yellow laterite (soil), which in turn is rather inconspicuously transitional into a yellow saprolite (residual, decomposed parent rock). Original bedrock structures gradually become discernible, and the colour changes to greenish-gray with depth.

(4) The underlying boulder saprolite contains spheroids of weathered and leached peridotite surrounding fresh peridotite cores. The solid hard bedrock below is usually karsted into a system of pinnacles and ridges, subdivided by depressions filled by greenish clay. Fragments of opaline or chalcedonic silica are common, and masses or veinlets of

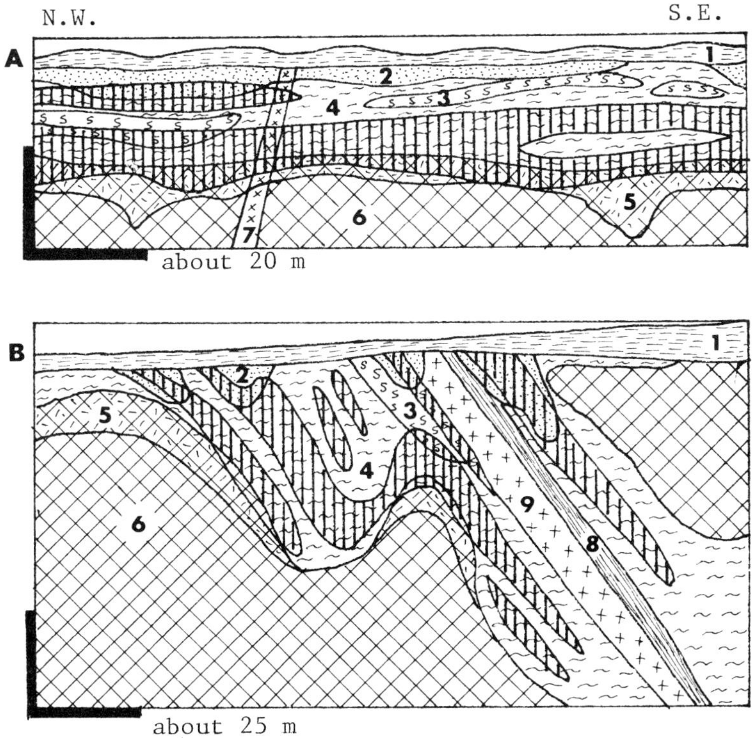

Fig. 7-10. Example of an "areal" (top; Kempirsai), and "linear" (bottom; Lipovskoe deposit) Ni-hydrosilicate deposits, the Urals, U.S.S.R. LEGEND: (1) Tertiary clays; (2) ocherous laterite; (3) silicified laterite; (4) nontronitic laterite or saprolite; (5) bleached, bouldery serpentinite; (6) fresh Paleozoic serpentinite; (7) gabbro dike; (8) talc - carbonate schist along shear; (9) granodiorite dike. Vertically ruled: economic Ni orebodies. After Pustovalov, ed., (1965).

green "garnierite" (mixture of Ni-hydrosilicates, mostly smectite) are common. This zone represents the most valuable nickeliferous horizon.

(5) Fresh peridotite, usually fractured. The rock along fractures is bleached and silica, chrysoprase (silica and "garnierite" mixture) and "garnierite" infiltration veins are common.

In lateritic profiles, nickel is bound in hydrous magnesium silicates (Ni-montmorillonite or pimelite; Ni-chlorite or schuchardite; Ni-serpentinite; "garnierite"), or in goethite. In the "A-type" nickeliferous "laterite" of Zeissinck (1969) the Ni

Fig. 7-11. Karst-like morphology of tropically weathered and bleached ophiolitic serpentinite, filled by laterite. South of Mayagüez, W. Puerto Rico.

values are in the silicates. In the "B-type" "laterite" located higher in the profile in the zone of "yellow laterite", goethite is the principal nickel carrier. The "A-type" tends to have a higher grade (1.5-7% Ni), but of lesser tonnage and limited predictability. The "B-type" has lower Ni grade (1.5% Ni down to a cutoff-grade of typically 1% Ni), but is more extensive, regular and cheaper to mine. In most deposits both types coexist, but one tends to be dominant. "A" prevails in the "fossil", uplifted and eroded deposits situated in terrains of youthful topography (New Caledonia, Riddle); "B" is found in the less dissected, piedmont and low-ridge terrains as in Cuba, Greenvale, etc. Local variations and departures from the usual, "Ni-laterite" model characterized above, include:

(a) Profiles rich in silcrete (silica duricrust). Silcrete occurs just under a thin crust of ferricrete, and may entirely replace the "red" and "yellow" laterite. Several localities described in Brazil (Trescases et al., 1981) are over non-ophiolite ultramafics, but ophiolite-associated localities are also known.

(b) "Contact karst" Ni "laterites" developed at several localities in the Gornyi Ufalei district, central Urals, U.S.S.R. (Glazkovsky et al., 1974; Fig. 7-13). There, late Paleozoic serpentinite was tectonically emplaced into Silurian marbles. "Linear" Ni "laterite" profiles formed along the shear and fault zones, and abundant karst depressions were filled with mechanically reworked and chemically

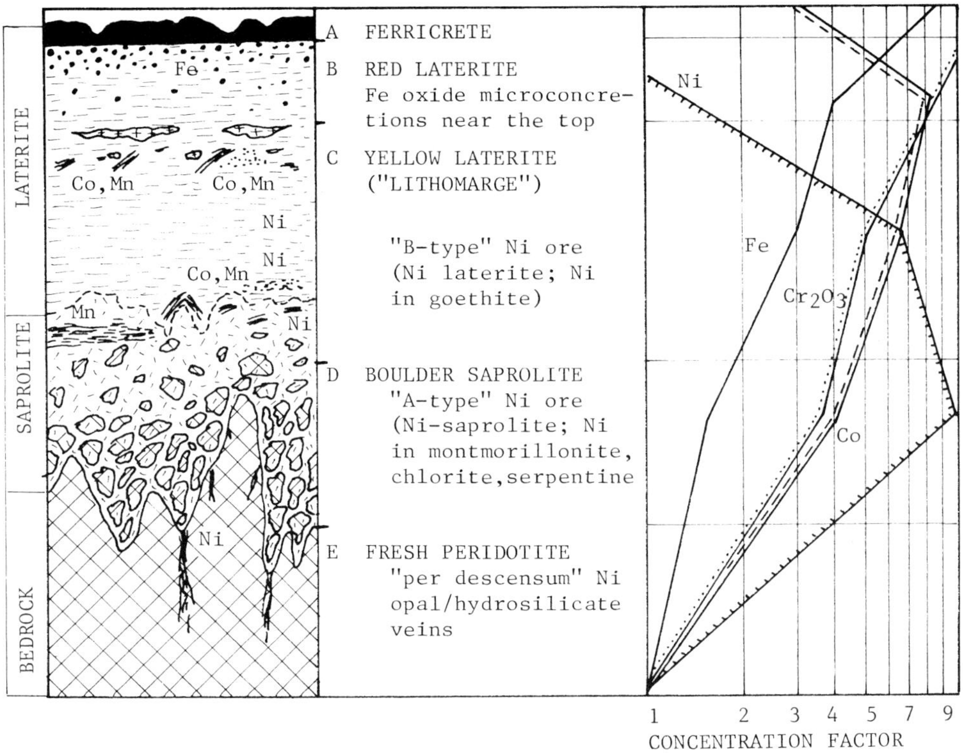

Fig. 7-12. A "typical" zoning in a metalliferous tropical weathering profile "laterite" over ophiolite ultramafics. The concentration factor is based on 6% Fe, 0.4% Cr_2O_3, 0.2% Ni and 0.02% Co in the fresh peridotite. The depth shown ranges between about 10 and 70 m in nature.

infiltrated Ni-minerals ("garnierite", "revdinskite"), topped by clays. In the Cheremshan deposit, the karst mineralization extends to a depth of 423 m, and contains 2.62% Ni.

(c) Deformed and low-grade metamorphosed lateritic weathering profiles over ultramafics. These are represented by the Cle Elum deposit 100 km east of Seattle, Washington. It is a 5 m thick folded lens of fine-grained meta-laterite containing magnetite, spinel, diaspore and chromite. The orebody overlies peridotite and is topped by Eocene sediments. The composition is transitional between "Ni-laterite" and bauxite, and there is about 2.56 Mt Fe; 150 Tt Ni; 102 Tt Cr and 800 Tt Al (Lamey and Hotz, 1952).

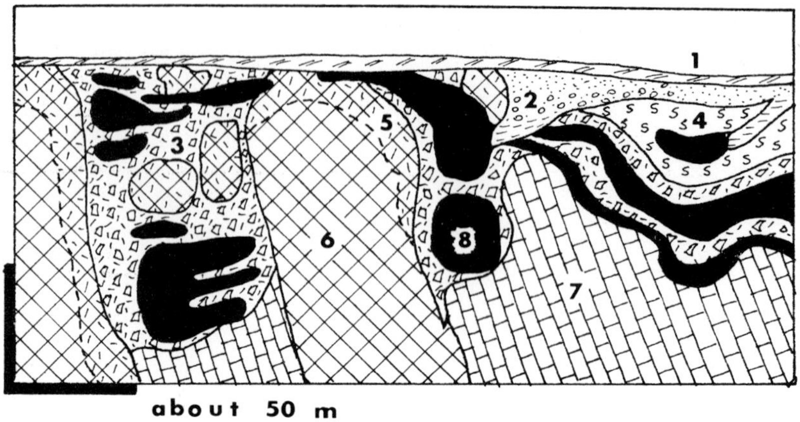

Fig. 7-13. "Contact karst" Ni silicate deposit, Gornyi Ufalei district, central Urals, U.S.S.R. (1) Quaternary cover; (2) Mesozoic Cainozoic sand, clay, conglomerate; (3) reworked laterite and saprolite filling karst depressions; (4) silicite; (5) saprolitic late Paleozoic serpentinite; (6) fresh serpentinite; (7) Silurian marble; (8) economic orebodies. After Pustovalov, ed., (1965).

Table 7-5 gives a selection of the more important "Ni-laterite" areas of the world. Although the distribution of Fe, Cr, Mn and Co in typical tropical weathering profiles over ultramafics (Fig. 7-12) shows peaks of substantial metal concentration, these metals are rarely recovered as a by-product of nickel mining, but this could occur in the future.

Anomalously accumulated cobalt usually resides in asbolite, a bluish-black mixture of Mn oxides with adsorbed Co, containing up to 8% Co. In New Caledonia, asbolite is present either as nodules (2-8% Co), powdery disseminations or fracture coatings in the red and yellow laterites (0.3-0.4% Co). Both varieties are very irregularly distributed. Only about 100 Tt of asbolite with an average grade of 5% Co have been produced there up to 1927, and Co production from the 0.15-0.25% Co laterites near Noumea soon ceased because of technological problems (Paris, 1981). New Caledonia, however, has substantial reserves of cobalt in the nickeliferous weathering crust (2.5 Mt of cobalt content in materials with an average Co grade of 0.075% Co), possibly recoverable in the future.

Chromium in lateritic profiles is bound in chromite, an original accesory mineral in peridotite, that behaves as a resistate mineral. Consequently, its content in laterites increases in proportion to the volume reduction due to the removal of Mg and Si. Most laterites contain around 1 or 2% Cr_2O_3. Friedrich (1982) discussed the widespread chromite–bearing lateritic soils developed over the Philippine ophiolites as a potential future reserve. The chromite grains range in size from 0.03 to 5 mm, and the weathering profile contains from 6.2 to 14.8% Cr_2O_3 and 42-66% Fe_2O_3. Similar laterites were explored in the California–Oregon border region (e.g. Gasquet area); Medellín, Colombia; Larona, central Sulawesi, and south-eastern Kalimantan, Indonesia; Surigao, Mindanao Island, Philippines; Macedonia, Albania, and elsewhere.

PRECIOUS METAL ACCUMULATIONS IN WEATHERING PROFILES OVER ULTRAMAFICS.

Ottemann and Augustithis (1967) described Pt-Fe, Os-Ir, and minor gold nuggets from the lateritic cover over dunite and pyroxenite massif of Yubdo, Ethiopia (the geotectonic affiliation of the massif is uncertain). The authors argued that the nuggets formed by in situ precipitation ("agglutination") in the weathering profile, derived from the dispersed trace content of platinoids in dunite.

7.10.5. Sedimentogenic reworking of ophiolites in the continental crust

Surficial agents of sedimentogenesis particularly running water, are capable of reworking, transporting, sorting, and redepositing the various constituents of fresh or weathered ophiolites. In most cases, the sedimentogenic interaction results in a dispersal and loss of identity of the source rock but in some instances, valuable metallic components accumulate. The considerable variety of products depends on (1) the form of the material at source (concentrated primary ore accumulations in the ophiolites, such as podiform chromite; dispersed resistate minerals, such as accessory chromite or magnetite; metalliferous particles of granular consistency such as the Fe-oxide concretions or clay-like consistency such as Al-hydroxides, formed in the weathering profile); (2) the mode of transportation (in traction, suspension, solution); (3) the proximity to the source; (4) the depositional environment. Obviously, the products accumulated closest to the source are easiest and safest to interpret.

RECENT PLACERS

Alluvial chromite is abundant in streams draining ophiolite ultramafics, particularly those hosting the podiform chromite deposits. This chromite has been widely mined in the past, but production figures are rarely available because the alluvial chromite is usually included in district figures that make no distinction between the hard-rock and placer chromite. Alluvial chromite derived from chromite pods tends to be coarser and easier to recover than the accumulated accessory chromite from ultramafics. Much of the latter

Table 7-5. Ni, Fe, Co and Cr "lateritic" deposits in weathering crusts over ophiolitic ultramafics

LOCALITY	AGE OF WTH.	AGE OF BRK.	STYLE	TONNAGES, GRADES	REFERENCES
NORTH AMERICA					
Riddle, Oregon	Eo	J	A//B (saprolite) siliceous stockwork	330 Tt Ni/1.5%	Chace et al., (1969)
Gasquet area, California/Oregon	T	J?	B/A ; 7 m thick laterite	384 Tt Ni/0.8% 960 Tt Cr/2%	Hotz (1964) E/MJ June 1981
Mayarí, E.Cuba	Ol	Cr	B//A , blanket and ferricrete layer	240 Mt Fe/49% 100 Tt Co/0.1% 1.3 Mt Ni/1.3%	Boldt (1967)
Nicaro, E.Cuba	Ol	Cr	B/A , laterite blanket	1.4 Mt Ni/1.4% 100 Tt Cr/1% 70 Tt Co/0.07%	Boldt (1967)
Moa Bay, E.Cuba	Ol	Cr	B/A, extensive later. blanket	7 Mt Ni/1.4% 500 Tt Co/0.1%; Cr, Fe	Boldt (1967)
Bonao, Dominican Republic	T,Q	MZ	A+B	1.05 Mt Ni/1.5%	E/MJ Nov.1977
Mayagüez, Puerto Rico	T	Cr	B/A	910 Tt Ni/1.3% 70 Tt Co/0.1%	E/MJ Nov.1977
Lago Izabal, Guatemala	T,Q	Cr	B/A 30m	1.5 Mt Ni/1.5%	Cornwall (1973)
SOUTH AMERICA					
Loma Hierro, Venezuela	T	MZ	B/A 30m thick	885 Tt Ni/1.58%	Putzer (1976)
Cerro Matoso, N.Colombia	Q	MZ	B/A	1.4 Mt Ni/2%	Putzer (1976)
Medellín to Belmira, Colombia	T,Q	MZ	B; compact ferricrete	190 Mt Fe/47.5% 5.4 Mt Cr/1.35% 1.4 Mt Ni/0.35%	Putzer (1976)

EUROPE					
Żabkowice, Silesia, Poland	T	Cm-S	A//B	100 Tt Ni/1.4 ?	Bohdanowicz (1952)
Rezh,Gor.Ufalei, C.Urals,USSR	Tr_3 J_1	D_3	A//B linear oreb.at faults, contact karst	360 Tt Ni+/1.5%	Glazkovsky et al. (1974)
Khalilovo (Orsk), S.Urals, USSR	J_1	Cb	A/B ; also reworked laterite	600 Tt Ni+/1.5 85 Mt Fe/30% 2.83 Mt Cr/1%	Sokolov and Grigoryev (1974)
Kempirsai Massif, S.Urals, USSR	Tr_3 J_1	Cb	B/A	377 Tt Ni/1.3% 20 Tt Co/0.05%	Glazkovsky et al. (1974)
Goleš Mt.,Priština Yugoslavia	Cr	J	B,A	102 Tt Ni/1.36%	Min.Yearbook 1967
Ržanovo,s. Macedonia, Yugoslavia	Cr?	J	B,reworked laterite	1.2 Mt Ni/1.09% 2.2 Mt Cr/2% 36.5 Mt Fe/32%	Berthold (1980)
Mirdita zone, E. Albania	Cr T	$J_{2,3}$	A,B,reworked ochers	X Mt Ni/0.6-2.6%	Pumo et al. (1982)
N.Euboea Isl., Greece	Cr_3	J	B,A,reworked ochers	2.4 Mt Ni/1.3%	Min.Mag.July 1973
ASIA					
Soroako,Sulawesi, Indonesia	Q	Cr_3T	B/A; 20 m thick laterite	4.37 Mt Ni/1.75% 340 Mt Fe/49% 350 Tt Co/0.12%	Schmidt (1976)
Pomalea, S.E.Sulawesi, Indonesia	T,Q	Cr_3T	B/A; over mélange	1.95 Mt Ni/1.35%	Hamilton (1979)
Molukkas and Waigeo,Gag Isl.,Indonesia	T,Q	Cr,T	B/A	min. 3.6 Mt Ni/1.4-1.6% 70 Mt Fe/36% 2oo Tt Co/0.12	Schmidt (1976)
Rio Tuba,C.Palawan Philippines	T,Q	Cr,T	B/A	6.3 Mt Ni/1.8%	Wolff (1978)
Zambales,etc.,Luzon,Philippines	T,Q	Cr,T	B/A	3.45 Mt Ni/1.5%	Wolff (1978)

Table 7-5 (continued); Ni, Fe, Co and Cr "lateritic" deposits in weathering crusts over ophiolitic ultramafics

LOCALITY	AGE OF		STYLE	TONNAGES, GRADES	REFERENCES
	WTH.	BRK.			
Surigao del Norte, N.E. Mindanao, Philippines	T	Eo	B; prominent ferricrete	207 Mt Fe/47.8% 5 Mt Ni/0.8-1.1 12.26 Mt Cr/2.85	Santos-Ygiño (1952)
Davao Oriental pr.	T,Q	T	B/A	4.8 Mt Ni/1.2%	Wolff (1978)
AUSTRALIA, OCEANIA					
Cykloop Mts., W. Irian, Indonesia	T,Q	Cr,T	A/B	660 Tt Ni/1.1% 90 Tt Co/0.15%	Boldt (1967)
New Caledonia	Ol-Mi	Eo	A//B; prominent ferricrete	50 Mt Ni/1.2-9% 2.5 Mt Co/0.15% 15 Bt Fe	Paris (1981)
S. Isabel, S. Jorge Isl., Solomon Isl.	T,Q	Cr_3-	A,B	571 Tt Ni/1.3-1.46% 13.5 Mt Fe/30% 31 Tt Co/0.06%	Arthurs (1979)
Greenvale, Queensland, Australia	T	D	B; Ni in ocher	628 Tt Ni/1.57 48 Tt Co/0.12%	Fletcher and Couper (1975)
Marlborough-Yaamba Qld., Australia	T	PZ	A/B	371 Tt Ni/1.28%	INAL Staff (1975)

EXPLANATIONS:

In the column AGE OF: WTH = weathering; BRK = bedrock

In the column STYLE: A and B are the laterite types of Zeissinck (1969)

A = Ni is in hydrosilicates in the basal saprolite

B = Ni is in goetite in the yellow laterite

/ = greater than; A/B = A is greater than B

is below the threshold of the present recovery (approximately under 0.1 mm). The chromite placers near Co Dinh, northern Vietnam (Fontaine and Workman, 1978) are an example of the accessory chromite accumulations. They are adjacent to a serpentinite massif, the chromite is extremely fine-grained, and the grade is low: 2-5% chromite. Between thirty and forty thousand tons of concentrate are produced annually. Marine chromite placers of similar derivation have been reported from the Solomon Islands (Arthurs, 1979).

The chromite placers in the Camarones Sur Province, Philippines (Mayon, Lagonoy, etc.; Wolff, 1978) are adjacent to known podiform chromite deposits and are one of the few examples where its alluvial reserves (3 Mt chromite) exceed the bedrock chromite reserves (220 Tt chromite). Marine placers of chromite ("black sands") in both recent and raised beaches are known from the vicinity of Bandon on the Oregon Pacific coast (Griggs, 1945; about 3.5 Mt of sand with more than 3% Cr_2O_3 are available); from Ulcinj in south-western Yugoslavia (Janković, 1982); from New Caledonia; and from elsewhere.

Placers of gold derived from ophiolite-hosted primary orebodies are probably moderately common, but the bedrock gold source is rarely known with certainty. The substantial placer gold production from the Urals partly belongs here. The small gold placers near Nundle and Hanging Rock, New South Wales (Fitzpatrick, 1975; about 6.2 t Au produced) have probably been derived from the disseminated and vein primary gold mineralization in altered serpentinites, diabase dikes, mafic and felsic volcanics and plagiogranites of the Great Serpentine Belt. The same applies to the gold placers of Liguria, mentioned earlier. Portion of the Klondike, Yukon, alluvial gold (over 300 t recovered) could have been derived from altered and quartz, ankerite, talc, fuchsite-veined ultramafics, such as those cropping out in the bed of the Hunker Creek.

Platinum metal placers derived from ophiolites are known in Tasmania (near Adamsfield and in the Heazlewood Complex; less than 1 t of Os-Ir produced; Elliston, 1953), as well as possibly in the Chocó region of Columbia.

PALEOPLACERS

Accessory detrital chromite is common in many basal sandstones and conglomerates, indicating partial ultramafic provenance (e.g. in the Ordovician Pioneer Beds, Mt.Lyell, Tasmania), but mineable chromite paleoplacers are rare. The most extensively studied example is at Anderson's Creek near Beaconsfield, Tasmania (Summons et al., 1981). There, lateritically weathered and chromite-enriched erosion-bevelled Cambrian serpentinite is topped by patches of Eocene quartz sands, containing lenses of chromite-rich basal sandstone. The orebodies are 1-6 m thick and the indicated reserves are about 75 Tt chromite concentrate (Figures 7-14 and 7-15).

Platinum-metal paleoplacers are represented by small occurrences near Adamsfield, Tasmania (Elliston, 1953). There, Ordovician basal

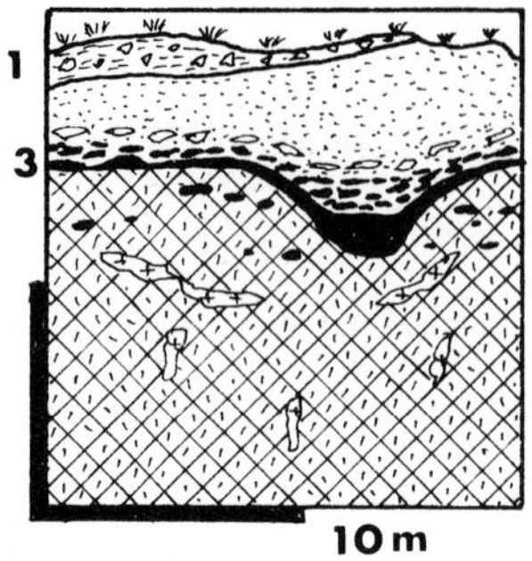

1 Holocene soil
2 Neogene quartz sand
3 chromite-rich sand, septa of silicified sand, quartz-chromite breccia
4 regolithic Cambrian serpentinite

Fig. 7-14. Anderson's Creek, Beaconsfield, Tasmania; chromite paleoplacer. Pit wall, from LITHOTHEQUE, based on a 1981 visit.

sandstones contain small quantities of detrital osmiridium reworked from bedrock showings in an elongated ultramafic body considered to be a dike.

SEDIMENTARY Fe, Ni, Cr, AND Co DEPOSITS GENERATED BY REDEPOSITION OF METALLIFEROUS LATERITES OVER ULTRAMAFICS

This is a quite distinct mineralization style that represents some 260 Mt Fe; 8 Mt Cr; 10.6 Mt Ni; and 100 Tt Co worldwide in produced and reserve ore (Table 7-6, Figure 7-16). The ore has an appearance of pisolithic or earthy "limonite" composed predominantly of goethite and clay mineral impurities. Cr is present in dispersed detrital chromite, Ni and Co in goethite and locally in "garnierite", particularly in the basal parts of orebodies. Locally, diagenetic sulphides (millerite in Trstenik and Aidyrlinskoe) or carbonates (siderite in Akkermanovskoe) formed. At Blewett, Washington (Lamey, 1950), limonitic ore forms minor lenses within a transported serpentinite conglomerate with lateritic matrix. Similar mineralization forms lenses grading to persistent beds (that may be traced for up to 20 km in eastern Albania), unconformably topping regolithic ultramafics or karsted carbonates. The orebodies in carbonates are located in the vicinity of (or down the paleoslope from) ultramafics. Some deposits in Greece and Albania have limonitic lenses resting at an unconformity between two carbonate units, a setting structurally identical to the "Mediterranean-type" bauxites (compare Chapter 20). The above ores are interpreted as deposits of low-energy sheltered lagoons, restricted bays and continental lakes. Recent deposition of comparable sediments is currently taking place at several localities along the sheltered western coast of New Caledonia, e.g. in the Nepoui lagoon. Granular Ni, Cr, Co-enriched hematite sand

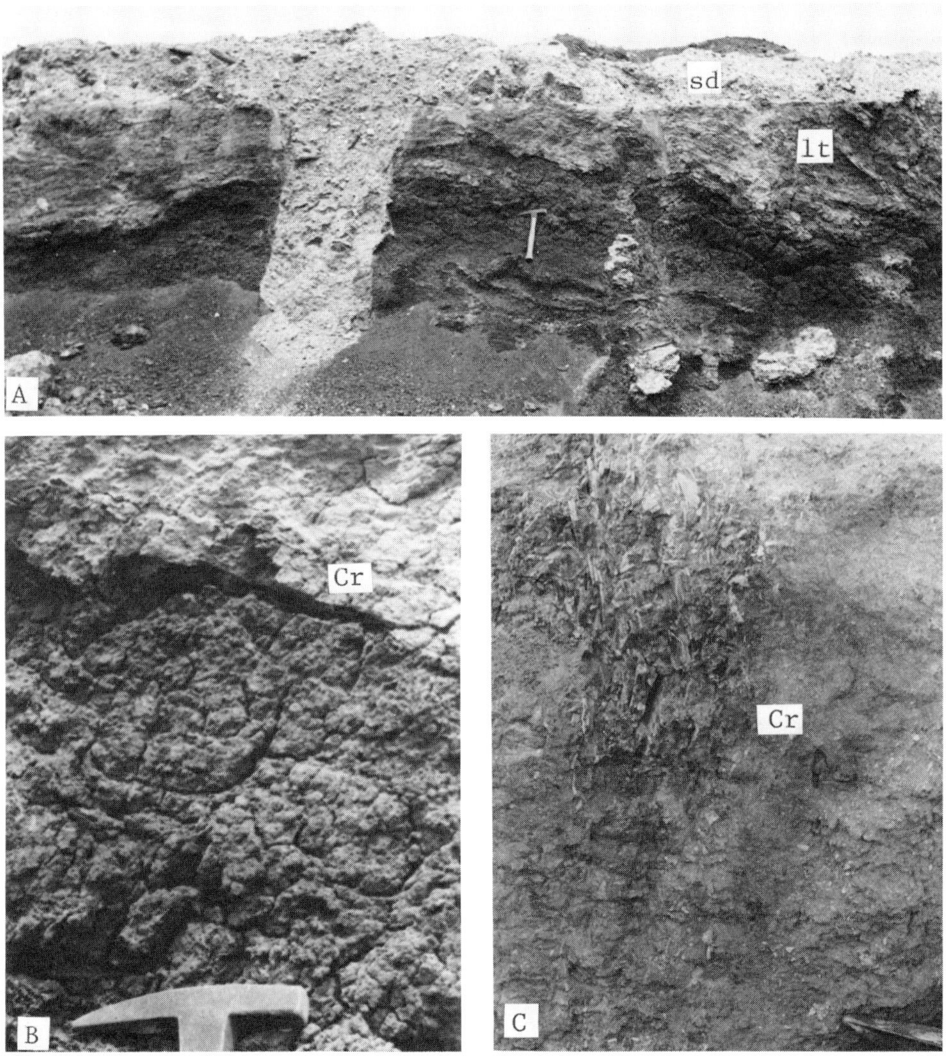

Fig. 7-15. Anderson's Creek chromite paleoplacer, Tasmania (A), showing bevelled ultramafic saprolite and laterite (lt) topped by friable Tertiary quartz sandstone (sd) with detrital chromite near base. (B,C) detail of the detrital chromite zone (Cr).

Table 7-6. Sedimentary Fe, Ni, Cr, Co deposits generated by redeposition of metalliferous laterites over ultramafics

LOCALITY	AGE 1	AGE 2	GEOLOGY	TONNAGE, GRADE	REFERENCES
Blewett, E.of Seattle, Washington, USA	J?	Eo	ocher lenses enclosed in peridotite conglomerate with ferruginous laterite matrix	66 Tt Ni/0.47%	Lamey (1950)
Trsteník, Kosovo, S.E. Yugoslavia	PZ	Cr_3	pelitic, concret. goethitic ironstone; minor millerite	1.2 Mt Fe/40% 105 Tt Cr/3.5% 30 Tt Ni/1%	Halbach (1972)
Librazhdi - Pogradeci d., Albania	J-Cr	Cr_3	stratiform ironstone layer at unconf. between two limest., up to 20 m thick	? 1 Mt Ni 3.6 Mt Cr 50 Mt Fe	Pumo (1973)
Larymna (Hagios Ioannis), Greece	J-Cr	Cr	pisolithic ironstone layer resting on karsted unconformity between J_{1-2} and Cr_3 limestones	1.54 Mt Ni/1.4%	Min.Mag.1973
Centr. Euboea Isl., Greece	J-Cr	Cr	conform. ironstone horizon on base of Cr sediments unconformably overlying J-Cr limestone, adjac. to J ophiolite	7.2 Mt Ni/1.2%	Min.Mag.1973
Akkermanovskoe, Serovskoe, Khalilovo, etc., S.Urals	PZ_3	$Cr-T_1$	pisolithic to conglomeratic ironstone overl. lateritized PZ ultramafic and at unconformities within sediments	77 Mt Fe/27% 4.05 Mt Cr/4% 820 Tt Ni/0.3%	Sokolov and Grigor'yev (1974)

EXPLANATIONS: AGE 1 = age of the lateritized source ultramafics
AGE 2 = age of ironstone sedimentary deposition

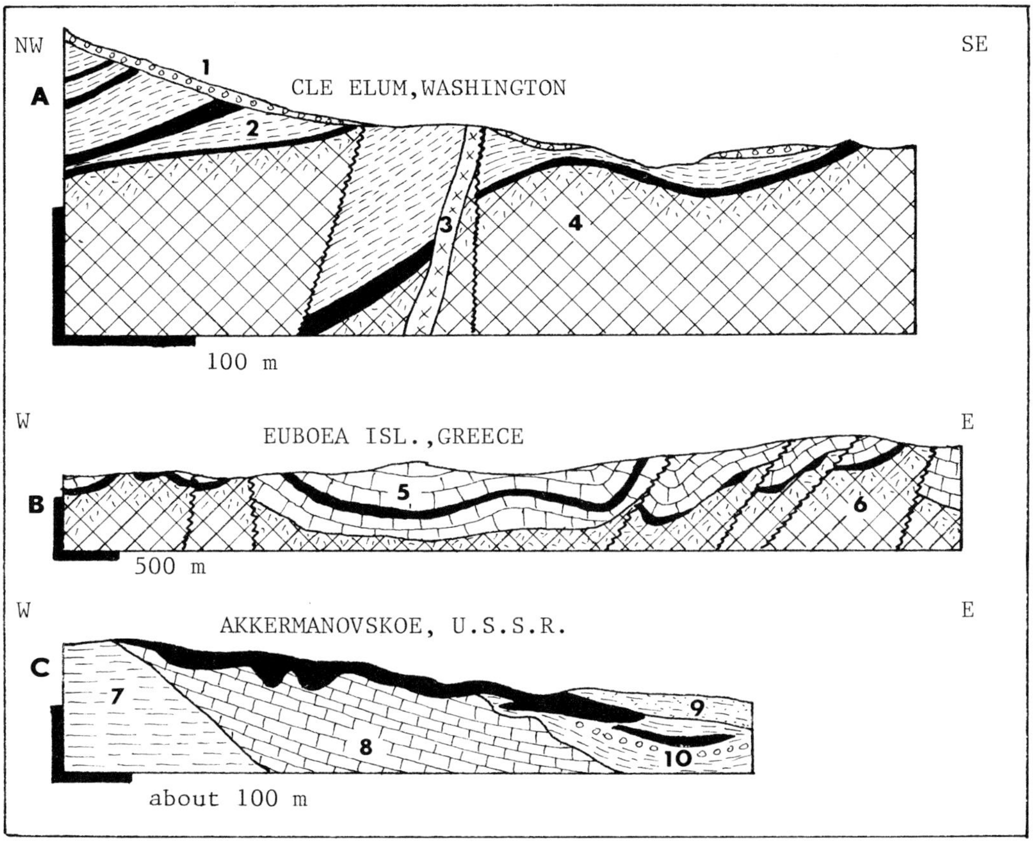

Fig. 7-16. Unconformity-controlled, Ni,Cr-bearing limonitic ore (in black), derived by reworking of lateritic profiles over ultramafics. From the source-proximal to the distal deposits: Cle Elum; (1) Quaternary talus; (2) Eocene sandstone, siltstone, conglomerate; (3) diabase dike; (4) Jurassic peridotite. Euboea Island: (5) Jurassic and Cretaceous limestone; (6) Jurassic ophiolites and pelagic sediments. Akkermanovskoe: (7) siliceous shales; (8) Carboniferous limestone; (9) Cainozoic clays; (10) Jurassic sand, clay. After Lamey and Hotz (1952), Albandakis (1981) and Pustovalov, ed. (1965).

currently accumulates at beaches in the slightly higher energy environment along the eastern (Pacific) New Caledonia coast, for example near Monéo. The material was derived from the ferruginous "cuirasse" undergoing dissection.

Mn-rich sediments generated by redeposition of manganese laterite horizons have rarely been recorded. The small Shaitantass deposit in Kazakhstan (Pustovalov et al., 1965) has stratiform lenses and persistent horizons of cobaltiferous Mn oxides (asbolite) mixed with goethite and clay minerals. The ore tops regolithic serpentinite and grades into Tertiary lacustrine clays. The reported metal contents are 0.15-0.18% Co and 0.4-0.6% Ni.

Nickeliferous lignites ("kerzinites")

Oligocene lignites containing up to 8% Ni in the form of authigenic sulphides (mostly millerite) overlie "Ni-laterite" and "contact karst" nickel deposits at several localities in the Verkhnyi Ufalei district, central Urals (Glazkovsky et al., 1974; Fig. 7-13). Most of them are nothing more than a metallogenic curiosity, although in the Cheremshan deposit they probably account for several percent of the overall nickel reserves. Pipino (1980) reported several comparable occurrences from Liguria (Italy).

7.11. EXAMPLES OF MINERALIZED OPHIOLITE PROVINCES

Two mineralized ophiolite provinces are summarized briefly below: the Troodos Complex, Cyprus, as an example of a little disturbed ophiolite complex dominated by basalts; and New Caledonia, as an example of tropically weathered ophiolite nappes, dominated by tectonite peridotites. Essential data and references on additional ophiolites are listed in Table 7-7 and the locations are shown in Fig. 7-17.

7.11.1. Troodos Ophiolite Complex, Cyprus

The Troodos Complex (Constantinou, 1980; Constantinou and Govett, 1973; Searle, 1972) covers an area of about 2000 km^2 in the west-central part of Cyprus (Fig. 7-18).The bulk of it was emplaced during the upper Cretaceous (85 m.y.) under a carapace or basement of an earlier Tethyan oceanic crust (Searle and Panayiotou, 1980), and was subsequently block-faulted. The original upward succession of ultrabasic tectonites, mafic plutonic rocks, sheeted diabase, mafic pillow lavas and marine sediments, is now concentrically arranged from the centrally exposed ultramafics and gabbros to sheeted diabase and several groups of basalts. The total stratigraphic thickness is over 10,000 m (Fig. 7-19).

The tectonite core of partly serpentinized harzburgite and minor lherzolite contains an economic asbestos deposit, and small occurrences of Fe,Ni,Cu,Co sulphides along the faults. The tectonite

Table 7-7. Examples of mineralized Phanerozoic ophiolites

LOCALITY	AGE	GEOLOGICAL ASPECTS	MINERALIZATIONS	REFERENCES
Eastman-Thetford Belt, S.E. Quebec	Cm-Or_1	very complete, but UT most conspicuous; mélange	domin. chrysotile, minor pCr; small mCus	Laurent (1975)
Bay of Islands, W. Newfoundland	Cm-Or_1	very complete, well exposed as thick, horiz. allochth. sheet	minor Cr in CG dunite; small mCu, Zns	Smith (1958)
Tilt / Betts Cove N. Newfoundland	Cm-Or_1	CG, SD, BS excell. exposures in shore outcrops, subhorizontal	small partly remobiliz. mCus	Upadhyay and Strong (1973)
British Columbia, Yukon	Cb-J	several parallel N.N.W. belts; UT elongated // to faults are most conspicuous	small pCr; interaction Hg (melanges, faults); Au (plutonic)	Douglas, ed., (1970)
Pinchi Fault zone centr. Br. Columbia	Pe? Tr	suture marked by UT, minor blueschist	minor pCr, mCus; important Hg (Pinchi L.)	Paterson (1977)
Klamath Mts., Oregon and California	Or	complete, UT most conspicuous; interaction with plutonites	minor pCr, lNi, Fe, Cr; Hg interact. Au lode/placers	Albers (1966)
California Coast Ranges (Franciscan Supergroup)	J_3-Cr_1	complete but dismembered, UT most conspicuous; mélanges	800 small sMn occur., pCr, one mCuS; important Hg at UT / springs interaction	Bailey et al. (1970)
Eastern Guatemala ophiolite belt	Cr	dismembered but almost complete	small mCuS, pCr; important lNi (Lake Izabal)	Petersen and Zantop (1980)
Nicoya Peninsula, Costa Rica	Cr	complete ophiolite, BS and chert most conspicuous	minor mCuS; numerous small sMn occurrences	Kuypers and Denyer (1979)
Cuba ophiolite belt	J-Cr_3	1000 km long belt (discont.), UT most important and conspic.	minor mCus, sMn; important pCr; very imp. lNi, Fe, Co	Thayer (1942) Gealey (1980)
Venezuelan Coast Ranges	Cr	discontinuous, probably complete UT most conspicuous	minor mCus, important lNi, Fe	Putzer (1976) Gealey (1980)
Karmøy Ophiolite, Haugesund, Norway	Pt_4 Or_2	complete, metam. in upper greens. facies	small mCus; metam. sMn pyritic phyllites	Sturt et al. (1980)
Ural Orogenic Belt USSR	Or-Cb_1	several discont. // N.-S. belts over a dist. of 3000 km; UT most conspicuous	small mCus, sMn; important pCr; moderate buried lNi, Fe; karst lNi; redeposited FeNi	Pronin (1962)

ABBREVIATIONS: UT=ultramafic tectonite; CG=cumulate complex (mostly gabbro) SD=sheeted diabase dikes; BS=pillow basalts; pCr=podiform chromite; mCus=massive Cu sulphides; sMn=siliceous Mn deposits; lNi,Fe,Co="lateritic" Ni,Fe,Co ores over ultramafics

Table 7-7 (continued); Examples of mineralized Phanerozoic ophiolites

LOCALITY	AGE	GEOLOGICAL ASPECTS	MINERALIZATIONS	REFERENCES
Appenine Mts.ophiolites, Italy	J–Pc	complete ophiolite; widespread mélanges, tect.interaction	small mCus,cp veining in gabbros; small sMn,pCr,Au	Sestini (1974)
Western Alps and Corsica	J_2–Cr_1	complete ophiolite; BS,UT most visible; mélanges, blueschists	small mCus; sMn; pCr; mag.in serp.(Cogne); NiCu sulph.in CG,in UT // shears	Bernoulli et al. (1974)
Drocea z.,Apuseni Mts.,W.Rumania	J_3–Cr_1	W-E belt,complete ophiol.,topped by Cr flysch; volcanic and plutonic interaction	small cumul.magnetite in CG; py,cp stockworks	Ianovici and Borcoş (1982)
Inner Dinarides oph.belt,Yugoslavia	J_{2-3}	complete ophiolite; UT most conspicuous	important pCr; minor mCus, sMn; Ti magnet.in CG; small lNi,Fe; reworked FeNiCo	Janković (1982)
Mirdita-Korab zone eastern Albania	J_{2-3}	complete ophiolite, widespread mélange	important pCr, lNiFeCo, reworked FeNiCo; minor mCus	Pumo et al. (1982)
Vardar & Pelagonian zones, Greece	J_2–Cr_1	complete ophiolite; classical differ,massifs (Vourinos)	important pCr, lNiFeCo, reworked NiFeCo; min.mCus	Marinos (1982)
Troodos Complex, Cyprus	Cr	complete ophiolite,classical well pres.almost unmetam.ophiol.	important mCus; minor pCr, NiCus in CG, along faults	Constantinou (1980)
Turkey ophiolite belts	Tr–Cr	numerous discont.belt,many complete; dismemberement, mélanges; volc.,plut.interaction	very important pCr; mass.Cu in melange (Ergani Maden) minor mCus, sMn	Brinkmann (1976)
Lake Sevan belt, Armenia, USSR	J–Cr	complete ophiolite,mélange, blueschist, volc.,plut. interact.	moder.pCr; tect./plut.interaction Au veins (Zod)	Knipper (1980)
Zagros Thrust Belt Iran	J–Cr_3	complete ophiolite, mélanges	important pCr	Falcon (1974)
Afghanistan, W.& S. Pakistan ophiolites	Cr_3–Eo_1	probably complete,dismembered	important pCr (Zhob Valley) minor sMn	Master (1956) Auden (1974)
Semail Nappe, Oman, Arabian Pen.	Cr_3	complete ophiolite and mélange almost flat thrust sheet	important mCus and Cu stockworks; pCr, minor sMn	Glennie et al. (1974)
Indus Suture Zone, Himalayas	J,Cr Mi-Pl	2400 km long discont. belt,locally complete	showings of mCus, sMn	Gansser (1974)

Naga Hills, N.E. India	Cr_3-Eo	dismembered incompl.ophiolite in a N.N.E. suture; mélanges	Ni,Cr,Co magnet.at ultramf.-sedim.contact; minor mCus	Singh and Srivastava (1980)
Burma-Andaman-Nicobar I.-Sumatra b.	Cr_3	2000 km long discont.belt,in places complete	small occur. of pCr, mCus, lNiFe	Banerji et al. (1981)
Nui Na Son massif northern Vietnam	pre-Tr	400 km long N.W. belt, mostly Alpine ultramafics	alluvial chromite workings	Fontaine and Workman (1978)
North-eastern Borneo (Sabah)	Cr_3-	complete but dismembered ophiolite, UT most conspicuous	minor lNiFe	Hamilton (1979)
S.E. and N.E.Kalimantan, Indonesia	Cr	discontinuous ophiolites; mélange, glaucophane schists	minor Pt placers; lNiFeCo	van Bemmelen (1949)
W. and S.E.arm of Sulawesi,Indonesia	Tr-Cr_3	numerous ophiolite masses,some complete; UT most conspicuous	important lNiFeCo (Soroako) minor sMn; oliv.-ilm.cumul.	Hamilton (1979)
Timor Island, Indonesia	Pe_3-Eo	dominated by mélanges and pelagic sedim.,UT,BS	Mn nodules in pelag.sedim. minor Cu in basalts	Hamilton (1979)
W.Irian and adjac. Islands,Indonesia	Cr_3 Mi_1	2 ophiolite belts, locally complete; UT most conspicuous	important lNiFeCo	Hamilton (1979)
Papuan Ultramafic Belt, Niugini	J_3-Eo_2	up to 15 thick complete ophiol. thrust sheet	minor pCr; sMn; mCus; lNi,Fe,Co	Davies (1971)
Solomon Islands (Malaita,S.Isabel)	Cr_3-T_1	discontinuous ophiolites,UT, CG,BS,pelagic sediments	minor pCr; sMn; lNiFe; mCus; chromite placers	Taylor (1977)
New Caledonia	MZ-Ol	extensive thrust sheet domin.by UT peridotites; minor CG,BS	very important lNiCoFe important pCr; minor sMn	Paris (1981)
Great Serpentine Belt,NSW,Australia	S-D Pe	dismembered,discont.ophiolites, some complete; UT most apparent	small occurr. of mCus; sMn lNiFeCo; Au with plagiogr.	Benson (1913-1918).
Coolac Ultram.belt NSW,Australia	D_{1-2}	steeply dipping sheet,most UT minor CG,BS; plut.interaction	small pCr; minor Cu,Au assoc.with plagiogr.interact.	Ashley (1969)
Heazlewood and Adamsfield,Tasmania, Australia	Cm_2	tectonized dunite, overl.by GC complex; dismembered, BS and sedim. adjacent; plut.interact.	minor Ni,Os-Ir occur.,reworked to placers	Brown et al. (1980)
Hitaka,Kamuikotan b.,Hokkaido,Japan	J-T_1	discontin.200 km N.-S. belt,in places complete; mélanges	minor pCr; Cr and Pt placers mCus, NiCu sulphides	Saito (1960)
Zambales R.,etc., ophiol.Philippines	Cr-Mi_3	several ophiolite complexes, many complete	important pCr, lNiFeCo minor mCus	Bryner (1969)

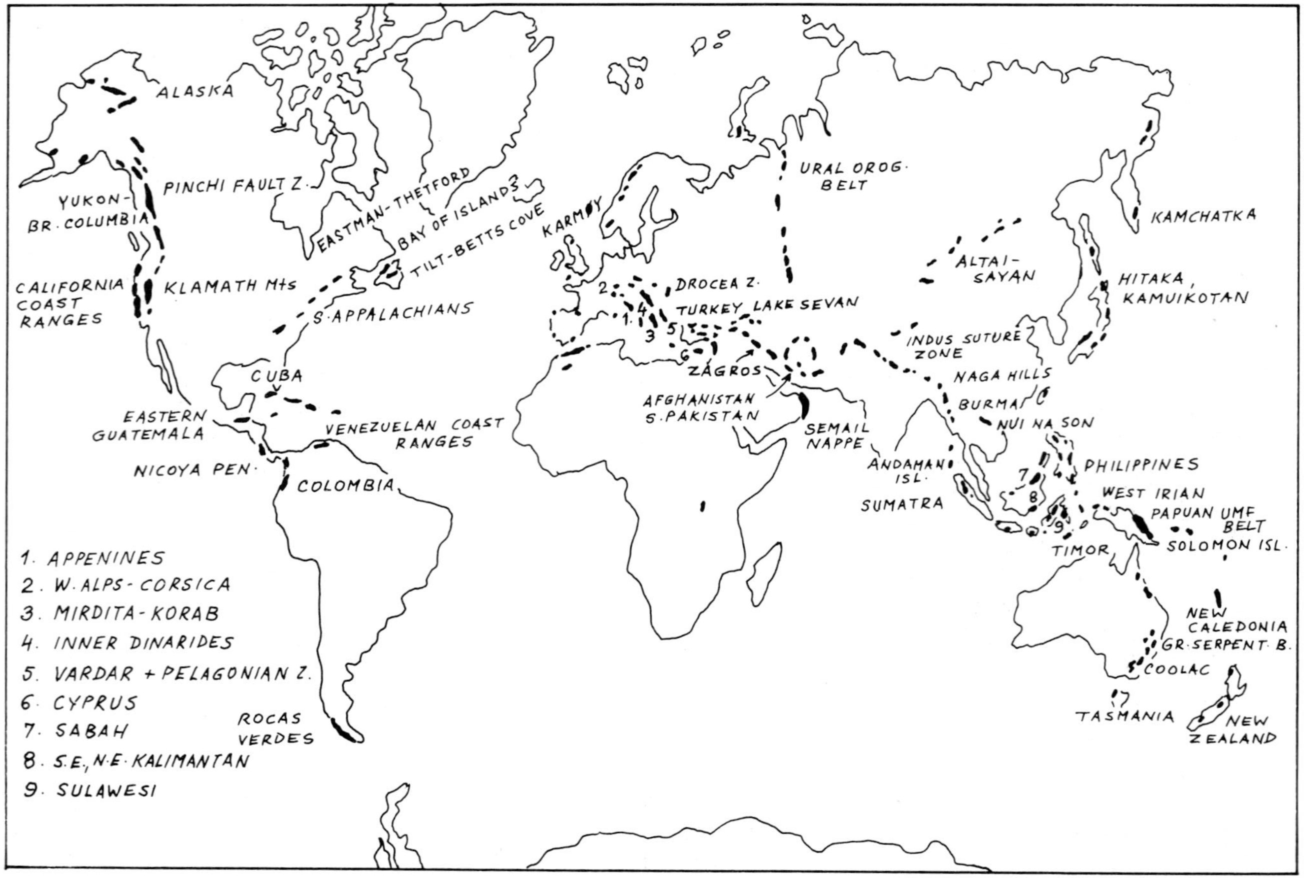

Fig. 7-17. Global distribution of Phanerozoic ophiolites.

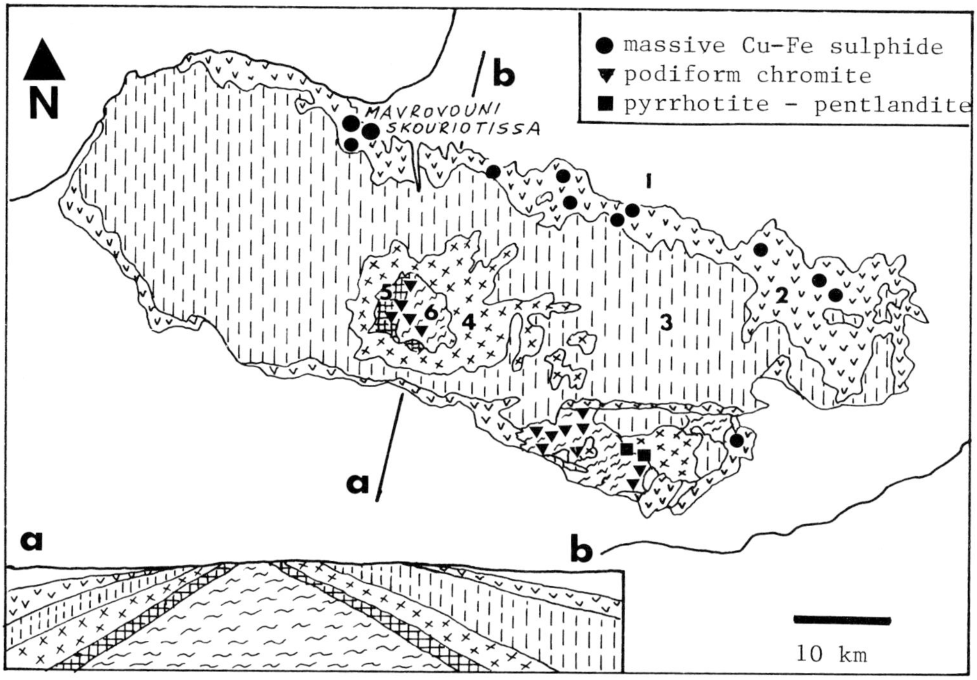

Fig. 7-18. Troodos Complex, Cyprus (Cretaceous), a simplified geological map and section. After Gass (1980), mineralized localities added. LEGEND: (1) post-ophiolite cover sediments; ophiolites: (2) pillow basalts; (3) sheeted diabase dikes; (4) cumulate gabbro; (5) cumulate peridotite; (6) peridotite tectonite.

is separated from the dominantly gabbroic cumulate plutonic complex by a rim containing dunite, wehrlite and pyroxenite. The affiliation of these rocks is controversial. Podiform chromite deposits are situated in dunite in the vicinity of the harzburgite contact and to a lesser degree stratigraphically above the contact.

The plutonic zone includes olivine, pyroxene and uralite gabbro cumulates and plagiogranites occur marginally along the gabbro-sheeted dike complex. The gabbros, cut by microdiorite and plagiogranite dikes, contain several sulphide showings (Ayios Ioannis and Agros). Pyrrhotite with subordinate chalcopyrite, sphalerite, pyrite and rare bravoite, valeriite and mackinawite, form fracture fillings and quartz-sulphide veins in chloritized gabbros. Small pyrite-chalcopyrite vein occurrences along breccia and shear zones also occur along the dike contacts in the sheeted diabase dike complex, surrounding the plutonic rocks.

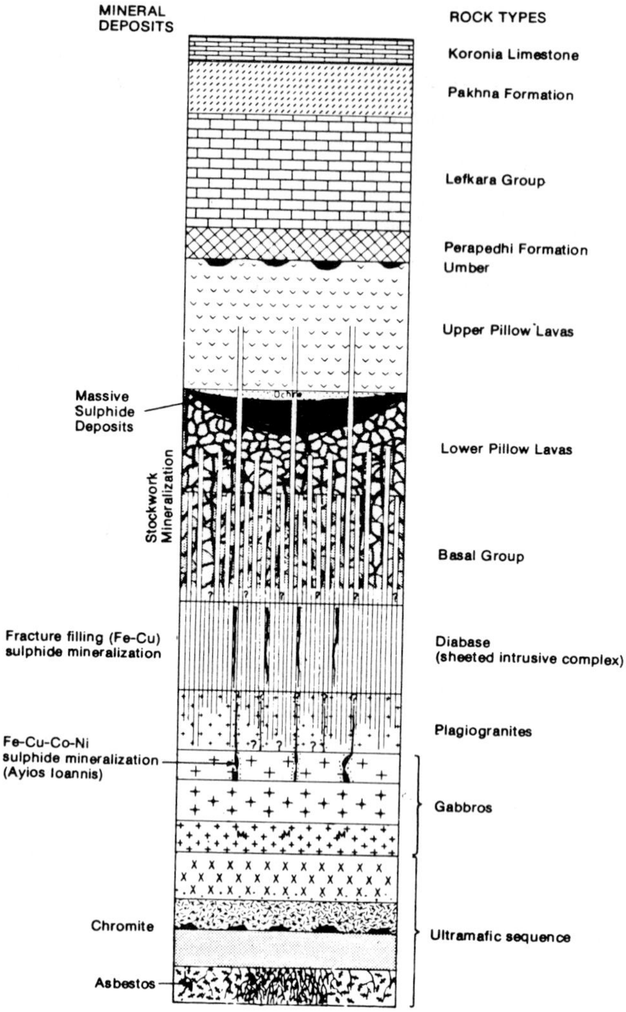

Fig. 7-19. Stratigraphic section of the Troodos Complex, showing the setting of the major metallic mineralizations. From Constantinou (1980), courtesy of the Cyprus Geological Survey Department.

Basalt topping the intrusive core includes three major emplacement phases. The Basal Group changes upward from indurated metabasalts into pillowed, greenschist-metamorphosed basalts. They, in turn, are unconformably overlain by oversaturated, zeolite facies metamorphosed Lower Pillow Lavas. The stratigraphically highest Upper Pillow Lavas are undersaturated olivine-bearing basalts, grading into picrites. They contain sporadic layers of tuff and mudstone.

Massive and stockwork Fe-Cu-(Zn) sulphide deposits of the "Cyprus-type" are the most important and most thoroughly studied metallic commodity in the Troodos Complex. Most of the orebodies (except Skouriotissa, which is in the Upper Pillow Lavas) are located along the former altered lava / seawater interface on top of the Lower Pillow Lavas, buried under the Upper Pillow Lavas and intersected by diabase feeders to the topping lavas. Most of the massive pyrite, chalcopyrite, sphalerite orebodies are topped by goethitic ochre, interpreted as being a product of subaqueous sulphide oxidation and indicating intravolcanic diastems. There are at least 20 orebodies that vary in size from 50 Tt to 20 Mt of ore, and their copper content ranges from about 0.2% to 4.5% Cu. The massive, more or less stratabound, sulphide lenses are underlain by a stockwork zone up to 700 m deep, mineralized by pyrite, chalcopyrite and quartz, sulphide veins and breccia cement.

Umber, a ferruginous and manganiferous marine shale, forms pockets on top of the Upper Pillow Lavas. It is often associated with red shale, argillite and radiolarian chert. It is overlain by upper Cretaceous to Eocene chalk, marl and chert. The Cyprus sulphide deposits have moderate gossans, resulting from late Tertiary and Quaternary subaerial weathering.

7.11.2. New Caledonia

The island of New Caledonia is located in the "quasi-cratonic" area of the Pacific ocean, 1200 km north-east of Australia. Two recent reviews of the New Caledonian geology in English were prepared by Avias and Gonord (1975) and by Guillon (1974). An excellent French memoir was written by Paris (1981).

The 400 x 50 km island (Fig. 7-20) is composed of Permian to lower Tertiary continental margin sediments, volcanics and metamorphics. These are topped locally by Eocene tholeiitic basalts on which, in turn, rest numerous allochthonous peridotite massifs tectonically emplaced from the north-east. They are probably erosional remnants (klippen) of a former, more extensive sheet, believed to have been obducted during the late Eocene orogeny. The ultramafics cover nearly 30% (6,000 km^2) of the island's area. The basalts distributed along the western coast contain numerous manganese oxide bodies associated with jasper lenses on top of pillowed tholeiites; for example, the Raymond mine (Arnould and Routhier, 1956). Minor occurrences of Fe-Cu sulphides of the "Cyprus type" are also known (e.g. Edison mine).

In the recent regional literature, a distinction is usually made between serpentinites and peridotites. The former occur as relatively small slices along the strongly deformed east coast, and as a basal layer (sole) in the numerous west coast ultramafic massifs and in the Grand Massif du Sud. The peridotite massifs are composed of thinly layered harzburgite with minor dunite and lherzolite considered to be mantle tectonites. They are locally overlain by volumetrically very subordinate, cumulate series, composed of a basal ultramafic cumulate

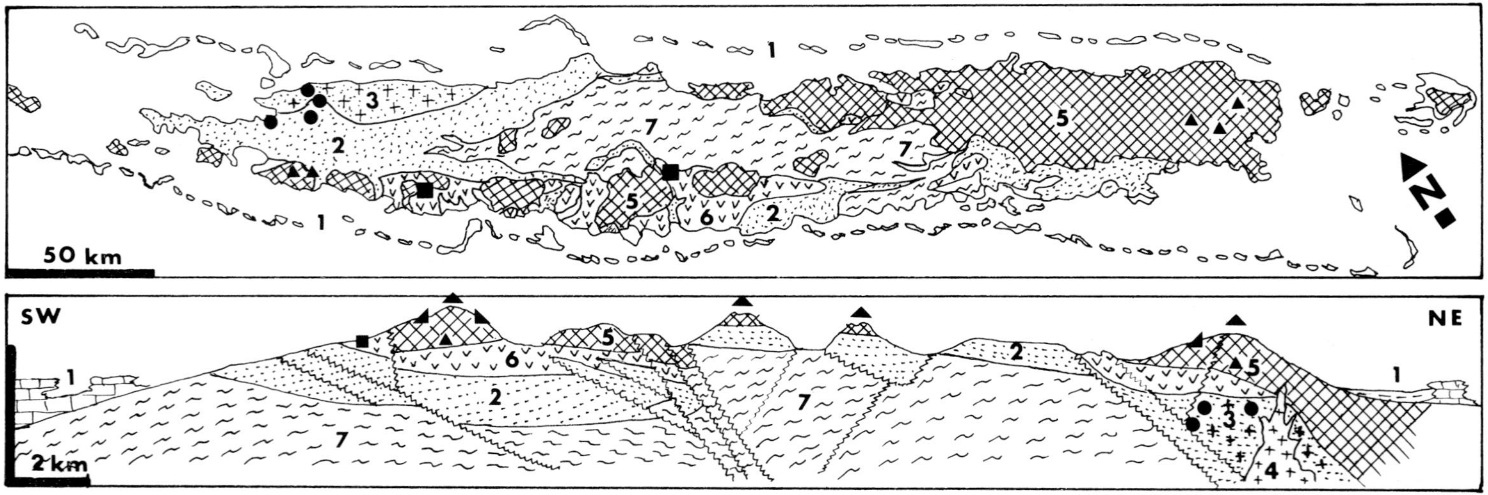

▲ Fe in ferricrete crusts; ◢ Ni,Co laterite and saprolite (not shown on map; they occur in all the ultramafic massifs (5); ▲ podiform chromite; ● volcanogenic and hydrothermal Cu,Zn,Pb sulphides; ■ bedded Mn in jasper and chert

Fig. 7-20. New Caledonia, simplified geological map (from LITHOTHEQUE, based on Paris, 1981) and a diagrammatic section located approximately in the central part of the island (considerably modified after Guillon, 1974). LEGEND: (1) Cainozoic reefs and other sediments; (2) upper Cretaceous to Eocene detrital sediments; (3) metamorphosed equivalents of (2), now schists; (4) Tertiary granite to quartz diorite; (5) allochthonous ophiolite ultramafics, emplaced in Oligocene; (6) Cretaceous ? basalts; (7) Permian and pre-Permian basement complex, Permian to Jurassic sediments, volcanics.

superseded by gabbro, minor quartz diorite and rare stilpnomelane bearing rocks. The main layered ultramafic mass in the Grand Massif du Sud is cut by several large dunitic pipes arranged perpendicularly to layering. In the highest part of the dikes, dunite changes progressively to noritic gabbros. In the same massif, injections of granitic rocks are common.

No particular mineralization (except for the regionally distributed laterite) is associated with the serpentinites. The harzburgites and dunites host podiform chromite deposits, several tens of which are known on the island. The largest cluster (2.7 Mt chromite produced) is in the Tiebaghi Massif in the north. The chromite deposits have the form of sheets, lenses, rods, pods, etc., which are discordant, subconcordant to concordant with respect to layering in the ultramafic tectonites. Several small showings of Cu sulphides and secondary oxides associated with magnetite have been reported from fault zones in the ultramafics.

Economically the most important Ni, Cu and Fe mineralization of New Caledonia is associated with tropical weathering crusts over the ultramafics and the island contains the world's greatest nickel resource estimated to be at least 30 Mt Ni recoverable under the present economic conditions (Paris, 1981; Figure 7-21). The weathering probably started in the Oligocene, and the most intensive early lateritization formed the oldest Cycle 1 weathering crusts ("carapace") protected by a ferruginous duricrust (cuirasse) up to 30 m thick. The cuirasse, composed of concretionary to rubbly hematite, contains some 15 Bt of 30-65% Fe ore. The "Ni-laterite" deposits are controlled by geomorphology and occur only on the edges of the dissected Cycle 1 peneplaination surface at relatively high altitudes.

The Ni hydrosilicates ("garnierite", pimelite) are richest in the greenish-brown saprolite, filling a karst-like surface on a silicified peridotite. The ore zone may extend locally to a thickness of up to 100 m, but it is extremely uneven and there are frequent deep-reaching, fissure controlled "per-descensum" type chalcedony-"garnierite" veins. The ore produced in the past had a high average grade of 3% Ni and 0.075% Co. The overlying "yellow laterite" is of substantially lower grade (1.3-1.6% Ni), but it contains numerous patches of asbolite mined locally as an ore of Mn and Co.

At present, a variety of mineralizations form by redeposition (degradation) of the Cycle 1 "laterites", topographically below in the valley alluviums (reworked alluvial clays with about 1% Ni, some detrital chromite, magnesite concretions), at moderate - energy beaches (hematite and chromite beach sands), and in low-energy lagoons at the mouths of rivers (ferruginous clays with around 1% Ni). Trescases and Guillon (1977, their Table II) reported up to 0.8 to 1% NiO in nontronite - rich alluvial clays filling some valleys in the Grand Massif du Sud, and up to 5-6% NiO in peat lenses locally present in the same association.

Minor bauxitic weathering profiles have developed over the mafic members of the ophiolite suite in the same area.

Fig. 7-21. New Caledonia, supergene Ni (Fe,Co) mineralization environments and associations near Népoui. (1) K=Kopéto peridotite massif, NL=Népoui lagoon; (2) Ni-"laterite" mine; (3) top of a weathering profile, ferricrete grading to the "red laterite"; (4) Ni ore zone in "boulder saprolite"; (5) infiltration opal-"garnierite" vein and breccia cement.

"The evidence so far passed in review illustrates a most important principle in geology: that of recognizing evolutionary series of rocks, each "arrested" member of which can be seen to be genetically related to its neighbours, although the end members of the series might be so different that no relationship between them would be suspected if they were seen alone, without the connecting links".
Arthur Holmes (1965, p. 180)

CHAPTER 8

Mélanges, Suture Zones, Blueschist Metamorphic Belts and Serpentinite Filled Faults

8.1. GENERAL

Mélanges (French for "mixtures") are chaotic, unsorted mixtures of a great variety of rock fragments ranging from milimetre - sized grains to rafts and slices several kilometres long. The fragments are set in a matrix of foliated, scaly and slickensided clay, shale, volcanic ash (e.g. on Timor) or serpentinite (Fig. 8-1). Although most mélanges are polymictic and have a considerable variety of exotic blocks, some are homomictic and composed entirely of serpentinite, claystone or a shale. Mélanges have been reported under different names in the literature and have puzzled geologists for over a century. Their present popularity dates back to the advent of plate tectonics. Hsü (1968) and later many other writers (see review in Hamilton, 1979, p.11) studied, described and interpreted the best known Franciscan Mélange of the California Coast Ranges: "a tectonic chaos formed primarily of abyssal clastic sediments and oceanic crustal and mantle rocks, expressed in a belt as wide as 75 km" (Fig. 8-2). The mélange contains materials of the Jurassic and Cretaceous ages dominated by volcanogenic litharenite ("graywacke"), siltstone and argillite. Present locally are fragments of ophiolites, abyssal pillow basalt or spilite, pelagic red or white radiolarian chert, limestone, serpentine and peridotite. In various places slivers of visually conspicuous high-pressure metamorphic rocks occut such as the grayish-blue glaucophane schist and/or red and green spotted eclogite. The inconspicuous high-pressure metamorphics include prehnite-pumpellyite, lawsonite - jadeite and aragonite bearing rocks, mostly originally litharenites. The components of the Franciscan Mélange are Jurassic and Cretaceous, believed to have been successively generated in the (1) mid-ocean ridge setting (ophiolites); (2) abyssal oceanic plain (the pelagic sediments); (3) trench environment (the arenites). The age of tectonic emplacement is considered to be late Cretaceous or early Tertiary. The Franciscan Mélange forms an eastward thickening wedge above the oceanic crust and mantle dipping eastward, in contrast

a
b
c

to the internal structures of the mélange that dip steeply to moderately.

Although the Franciscan and comparable mélanges are believed to be associated with subduction and are internal members of the plate tectonic model, opinions vary as to their exact origin. Hsü (1968) considered the Franciscan Mélange as a product of successive westward migration of the plate juncture (trench). Hamilton (1979); see also Chapter 5, interpreted mélanges as a part of the subduction complex, forming a wedge between the subducting oceanic plate and continental slope, and grading upward into an imbricated flysch of the outer arc ridge. This view has been strenghtened by the interpretation of many currently forming mélanges in the Indonesian region, as convincingly apparent on a series of seismic reflection profiles. Fox (1983) recently developed a model in which "mélange" was interpreted as material that was broken and sheared "as it was plowed aside and either coasted or rammed inland at a dynamic (Humboldt-type) triple junction, migrating along the edge of the continental plate" (Fig. 8-3).

Windley (1977) distinguished two types of mélange, tectonic and sedimentary (or olistostrome). Both form on the outer side of the arc-trench gap and both are related to underthrusting regime at convergent plate margins. Olistostromes are sedimentary units notable for numerous exotic and indigenous blocks usually resting in a pelitic or psammitic matrix. They formed by subaqueous gravity sliding. The present description is applicable largely to tectonic mélanges, but olistostromes may also be included. Many are indistinguishable from tectonic mélanges, in particular when tectonized in the post-depositional stage.

One of the major puzzles concerning the mélange composition, the juxtaposition of blueschist metamorphics formed at depths of the order of 25 km and sheared but unmetamorphosed sediments, has been explained by Hamilton (1979, p.30) as an extreme product of uplift and mixing by imbrication. The "normal" (Barrovian) metamorphics, such as greenschist, amphibolite, gneiss, even retrogressed granulites (as in Timor; Hamilton, 1979), are interpreted as exotic relicts metamorphosed in their original environment within the oceanic or continental crust.

Mélanges in the regional sense, such as the Franciscan, usually contain many individually mappable lithosomes (rock types) and overlap with the high-pressure metamorphics so the terms mélanges and

---

⇦

Fig. 8-1. Ophiolite and blueschist mélange (Permian age of emplacement) exposed in rocky cliffs near Port Macquarie, N.S.W., Australia. Blocks of massive rocks (metabasalt, blueschist, graywacke, gabbro, chert) rest in scaly serpentinite or shale matrix.

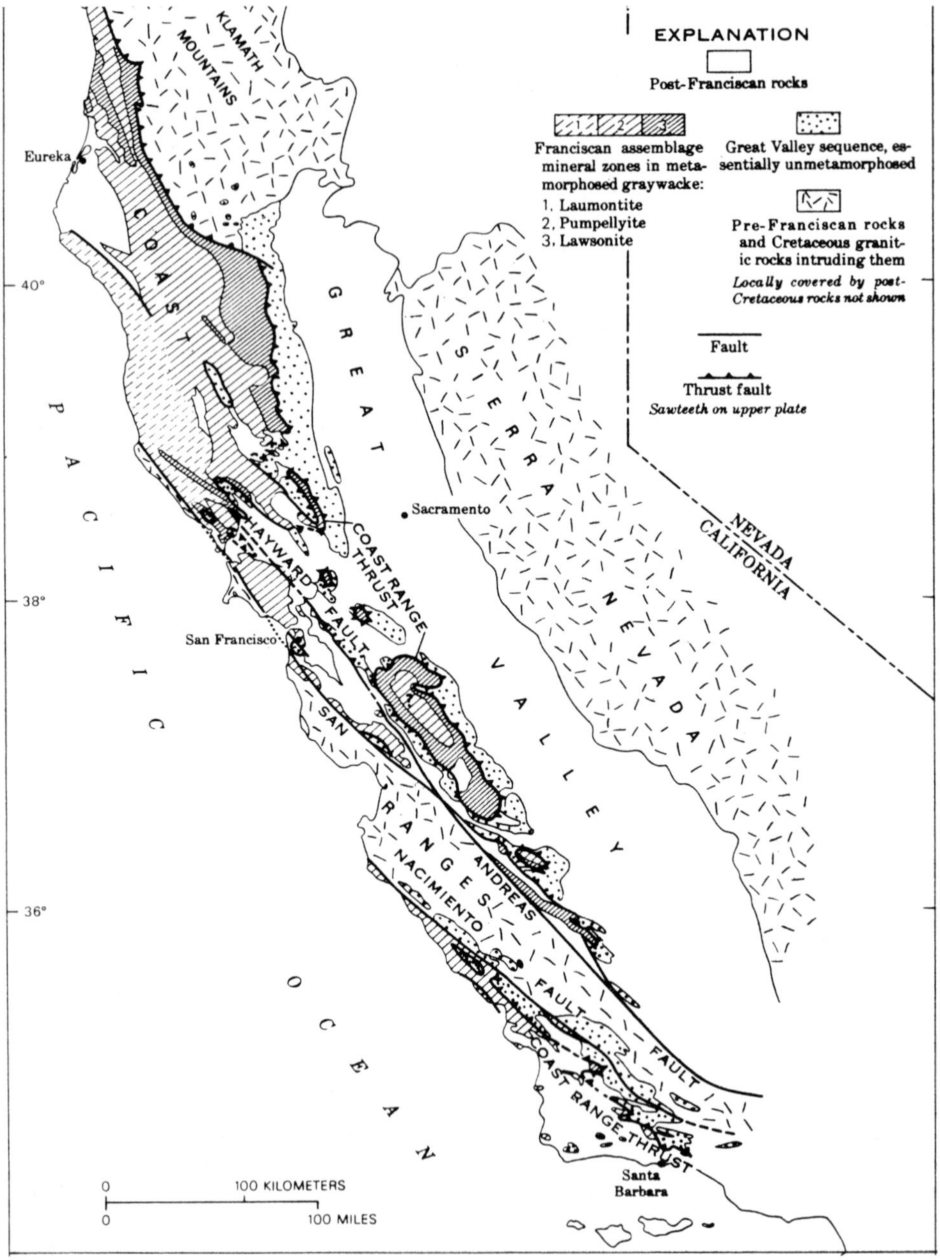

Fig. 8-2. Map of the California Coast Ranges and adjacent areas, showing the distribution of mélanges, allochthonous units and high-pressure metamorphic zones (from Bailey et al., 1970). Hsü (1968) classified the entire Jurassic-Cretaceous Franciscan assemblage as "mélange", a Mesozoic subduction complex.

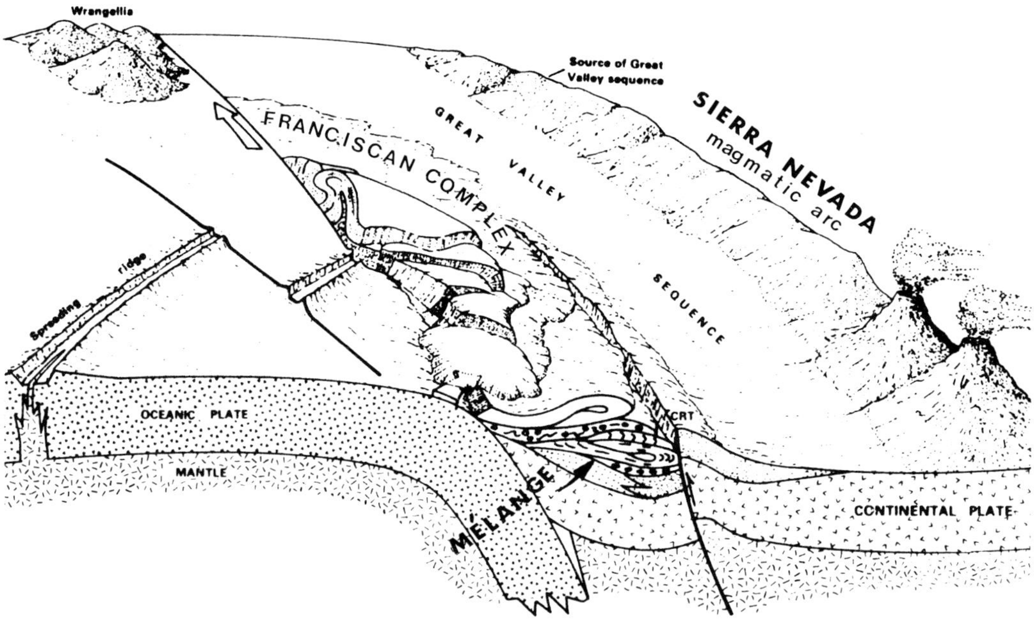

Fig. 8-3. The Franciscan mélange complex, California, in the lower Cretaceous as interpreted by Fox (1983). The mélange resulted from breaking, shearing and inland ramming of a repeatedly deformed ophiolitic and trench sequence along a triple junction, migrating along the edge of the continental plate. From Fox (1983)

blueschist (or high-pressure metamorphosed) belts are often used interchangeably. The regional units should more properly be termed mélange complexes or belts to distinguish them from the "type" mélange as seen in an outcrop, which is a giant breccia consisting of lenses of resistant or coherent rocks resting in a highly sheared, scaly clay, phyllite or soft breccia matrix. In some areas (e.g. along the Pinchi Fault Zone, British Columbia; Paterson, 1977), high-pressure metamorphics are distributed along fault zones containing elongated blocks of contrasting petrographic composition and metamorphic grade and style, but "typical" mélange breccias have not been reported.

Recently, several generally low-pressure, high-temperature metamorphic units have been interpreted as being subduction mélanges, buried and metamorphosed within the continental crust after their formation. The Sierra Foothills Mélange of Duffield and Sharp (1975) in California, is one of them (Fig. 8-4).

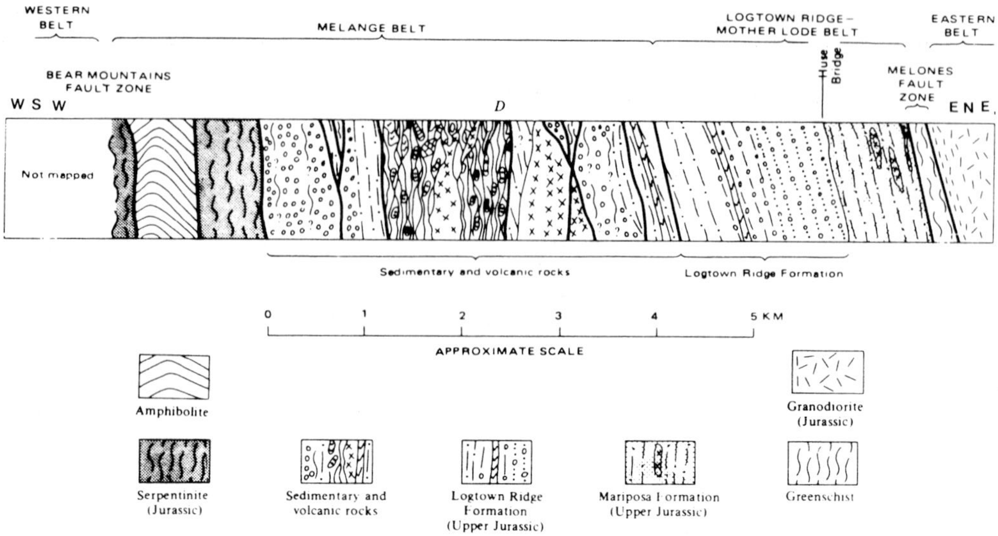

Fig. 8-4. Sierra Nevada Foothills Mélange, a belt, 4 km wide, of pervasively sheared tectonic breccia of Jurassic or older age, considered originally horizontal, subsequently tilted to the present subvertical position. From Duffield and Sharp (1975).

It is conceivable that lithologic associations comparable with the subduction mélange can form (or earlier mélanges can be modified) by other mechanisms as well, such as by (1) tectonic fragmentation and transport during thrusting and nappe generation; (2) continental collision; and (3) tectonization along transcurrent faults, or by combination of subduction with any of the above processes. Such associations often occur in suture zones, for example in the Taurus Suture Zone of south-eastern Turkey (Hall, 1976); in the Zagros Suture of southern Iran (Haynes and McQuillan, 1974); in the Himalayan Suture Zone (Gansser, 1974); and elsewhere. The genetic interpretation of suture zones is in flux and it is not the purpose of this book to review it in detail. The chaotic, mixed lithology, however, is there, available for mapping and visual inspection by a field geologist and the metallogeny in subduction mélanges and tectonic sutures is believed to be comparable in many respects.

The most modest (but most widespread) tectonic structures indicating communication with the oceanic crust or the upper mantle are faults filled by the "Alpine type" (i. e. tectonite) serpentinites. The ultramafic filling is usually discontinuous, having the form of slivers and lenses. In the Melones Fault Zone, Sierra Nevada, California, that partly coincides with the Mother Lode

Gold Belt (Clark, 1964), ultramafic bodies form thin lenses and pods concordant with the fault and with the regional structural grain. The steeply dipping ultramafic is intensively slickensided and altered, grading to talc, talc-antigorite or talc-carbonate-fuchsite schist.

8.2. METALLOGENY AND MINERALIZATION

The metallic mineralization hosted by mélanges displays considerable variety and is of moderate importance; the most important accumulated metals are mercury, nickel, copper, arsenic, antimony and gold. Many mélange mineralizations are identical to those in ophiolites (Chapter 7) and the proper pigeonhole is a matter of scale and subjective preference. The mineralizations treated here can be genetically placed into the following categories (Fig. 8-5; Table 8-1): (1) relict mineralizations formed outside the present mélange and unchanged during the process of incorporation (except for a dismemberment); (2) relict mineralizations, modified after incorporation into mélange; (3) mineralizations newly formed within a mélange largely by "internal" processes; (4) mineralizations formed within a mélange as a result of interaction with outside agents. Table 8-2 presents a selection of the mineralized mélanges of the world.

Category (1) involves ore occurrences formed at, under or near oceanic spreading ridges and hosted by ophiolites. Examples include (a) podiform and disseminated chromite in dunites (Sierra Nevada Foothills, California Front Ranges); (b) "Cyprus-type" massive pyrite-chalcopyrite deposits hosted by metabasalt, spilite or diabase (e.g. Balabac Island, the Philippines; see Fig. 7-5C); (c) subaqueous hydrothermal lenses of Mn oxide, silicate or carbonate, hosted by chert or jasper lenses in or at a contact with metabasalts (e.g. the Ladd-Buckeye Mines, California Coast Ranges; Davis, 1966). Magnetite and Mn silicate lenses, have been described by Barron et al. (1976) from the ophiolite mélange near Port Macquarie, New South Wales (compare Fig. 7-7).

Oceanic-style ferromanganese oxide nodules hosted by upper Cretaceous red radiolite clays, now a member of a Miocene mélange in Timor, Indonesia, have been reported by Audley-Charles (1972). The nodules range in size from several millimetres to 5 cm, are spherical, ellipsoidal or irregular, and contain around 12.5% Mn; 21% Fe; up to 1.3% Ti and increased trace contents of Ni, Cu, Co and Ba. Comparable occurrences have been noted by Hamilton (1979) on the Samar Island, Philippines.

Category (2) includes several small occurrences of the "Cyprus-type" massive and stringer pyrite–chalcopyrite sulphide deposits hosted by blueschist—metamorphosed metabasalts or incorporated into a mélange, dismembered and slightly remobilized. The Island Mountain mine in the Franciscan Assemblage, California,

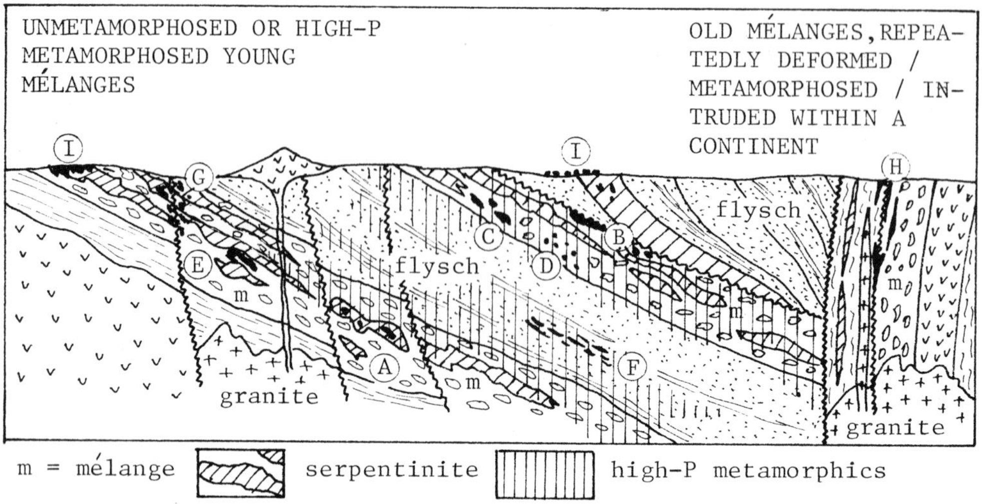

Fig. 8-5. Principal mineralization styles in mélanges (see Table 8-1 for explanation of the letter codes).

is the best-known example (Davis, 1966; 4,475 t Cu, 4 t Ag, 2.1 t Au). Several small deposits in the Diahot district, New Caledonia (particularly the La Ballade mine; Paris, 1981) are in a zone of blueschist metamorphism, but hosted by what was originally a late Cretaceous black phyllite and keratophyre. Pyrite, pyrrhotite, chalcopyrite and minor sphalerite form massive sulphide lenses, sulphide stockworks, and veins. The small bornite, chalcopyrite, magnetite, siderite, etc. orebodies in the Saint Véran zone in the French Alps (Blondel and Raguin, 1935) are in chloritic quartzite and glaucophane-riebeckite schist at contact with dismembered mafic and ultramafic members of the ophiolite association. The important Besshi massive sulphide deposit, Japan (Chapter 10) is partly hosted by piedmontite-actinolite-glaucophane schist (Banno et al., 1970).

Category (3). Several copper deposits including the important Ergani Maden field in Turkey, are located in mélanges or suture zones, but depart considerably from the "Cyprus type". The final act of mineralization appears to be post-mélange, but it is difficult to determine whether the ore is a pre-existing remobilized one, or a newly generated one. Ergani Maden is located in the Taurus Mountains of south-eastern Turkey (Griffitts et al., 1972), and its total production plus remaining reserves are about 680 Tt Cu/ 1.34-6.5%, 25 Tt Co, 170 t Ag, and 10 t Au. Several massive, stringer and disseminated pyrite, chalcopyrite and subordinate magnetite and pyrrhotite bodies are hosted by Eocene red and black marly mudstone

Table 8-1. Principal mineralization styles in mélanges and high-pressure metamorphic belts

RELICT MINERALIZATION, UNMODIFIED AND MODIFIED AFTER EMPLACEMENT			
(A)	Podiform chromite	X0	200 Tt Cr
(B)	"Cyprus-type" massive Cu sulphides in mafic metavolcanics	X0	50 Tt Cu
(C)	Mn oxides and silicates in jasper, chert	X00	200 Tt Mn
(D)	Oceanic-style Mn nodules	X	
MINERALIZATIONS FORMED WITHIN A MÉLANGE			
(E)	Massive Fe,Cu (Zn,Co) deposits	X0	900Tt Cu 30Tt Co 15t Au, 200t Ag
(F)	Rutile in high-grade schist, eclogite	X	1.8Mt Ti
INTERACTION MINERALIZATIONS WITHIN A MÉLANGE			
(G)	Serpentinite hosted or associated cinnabar dep.	X0	90Tt Hg
	stibnite dep.	X	20Tt Sb
	As sulphides	X0	30Tt As
	Au deposits	X	30t Au
(H)	Hydrothermal veins in Barrovian metamorphosed mélanges, in proximity to granitic intrusions;	X0	Au,Sb
(I)	Metalliferous weathering crusts over mélanges	X0	Ni,Co,Fe
(J)	Sedimentary reworked metalliferous components of mélanges (e.g. detrital chromite)	X0	Cr,Fe,Ti

and altered serpentinite breccia, located near intrusive gabbro bodies within a chloritized east-west trending belt, 20 km long. The mineralization is believed to be middle Oligocene.

The copper deposits at Montecatini, Italy (chalcocite, bornite and native copper located in "clay pipes"; veins and lenses of pyrite and chalcopyrite in a Mesozoic basalt mélange; Zuffardi, 1977); Bone, Timor (oxidized and secondary enriched small high-grade massive pyrite chalcopyrite orebodies in a serpentinite mélange; 2.4 Tt Cu/4.7%; Schmidt,1976); Rehova, Albania (Fig. 8-6); Bingara, New South Wales (fracture veining of Cu carbonates in serpentinite and siltstone in an altered late Paleozoic mélange; Loudon and Bowman, 1967); Vigonzano in the Appenines (Zuffardi, 1977) and others are, in many respects, similar to Ergani Maden and poorly understood.

In the Küre deposit, northern Turkey (Gümüs, 1964; 120 Tt Cu/4%), massive and stockwork pyrite and chalcopyrite occur along the interface of Jurassic basal pillowed metabasalt and a black slate, intruded by Eocene diorite. There is no uniform genetic

Table 8-2. Selected examples of subduction mélanges associated with metallic mineralization

LOCALITY	AGE OF FORMATION	AGE OF EMPLACEMENT	GEOLOGY	MINERALIZATION	REFERENCES
Franciscan Mélange California Coast Ranges	J-Cr_1	Cr_3-T_1	lenticular fragments of abyssal grayw.,argillite, jasper; rafts of mafics, serpentinites; zoned high grade metamorphics	relict podiform Cr; Mn in radiolarite; small Cu "Cyprus type"; only Hg at mélange/hot springs interaction important	Davis (1966)
Sierra Nevada Foothills Mélange, California	J	J_3 ?	folded, metam.heterogenous megabreccia; slate, graywacke, greenstone, serpentinite	small podiform Cr; Au lodes along faults,shears, granite coeval	Duffield and Sharp (1975)
Pinchi Fault Zone, British Columbia	Pe,Tr	Tr_3	elongate blocks of contrasting ohiolite lithology, limest.,grayw.;local blueschists	minor podiform Cr; Hg impregnations at Pinchi L. mine important; interaction Sb,Au	Paterson (1977)
St.Daniel Formation Eastern Townships, Quebec, Canada	Cm	Or_1	polygenic breccia, blocks of ophiolites in slate matrix; local ultramafics	small podiform chromite	Laurent (1975)
Liguria,north - western Italy	Cr-Ol	Ol	ophiolite mélange,scaly shale matrix,serpent.block flysch,pelagic sediments, sheared Hercynian granite	Cu "Cyprus type" and similar moder.important; minor Hg,Au	Sestini (1974)
Turkey, numerous mélanges (e.g. Bitlis Massif)	S_3-D_1 Cr,Eo	Cr_3 Mi	ophiolitic association, abundant ultramafics,minor blueschists	relict/remobiliz.Cu,Co at Ergani Maden important; podiform Cr, minor Mn, interaction Hg important	Brinkmann (1976) Hall (1976)

Zagros Suture zone southern Iran	J,Cr	Cr_3	ophiolitic and pelagic association, ultramafic slices	podiform chromite in dunite tectonite	Haynes and McQuillan (1974)
Arakan-Chin-Naga belt, Burma; Andaman and Nicobar Islands	Pc Mi	Eo Mi	jumble of ophiolite and pelagic associations	Ni laterite over ultramafic blocks; minor Au placers	Hamilton (1979)
Palawan and Balabac Islands, Philippines	Cr, T_1	T_1	mélange of ophiolite and Paleogene flysch clastics	Cu "Cyprus type" moderately important; Hg,Cr,Mn	Hamilton (1979)
Samar Island, Philippines	Cr_3	Cr_3	ophiolite, abyssal mudstone, pelagic assoc., greenschist, garnet schist	Mn nodules in pelagic mudstone	Hamilton (1979)
Western Sarawak, Borneo Island Malaysia	J-Cr_1	$Cr_{2,3}$	slivers of shale, graywacke, ophiolites in clayey groundmass; intruded by Cr_3 quartz monzonite	Mn oxides in chert Sb veins - interaction miner. related to quartz monzonite	Hamilton (1979)
Meratus Mountains, s. Kalimantan, Indonesia	Cr_1	Cr_2	polymictic mélange, ophiolite and pelagic members; blueschists, qtz.monzonite	minor podiform Cr; Fe,Cr, Ni laterite over ultramafics; minor Pt placers	Hamilton (1979)
Sulawesi Island, Indonesia; several mélanges	Cr_3-Mi	Mi	ophiolite, pelagic assoc., widespread ultramafics; greensch. to amphibolite, blueschists	large Ni (Fe,Cr,Co) laterites over ultramafics	Hamilton (1979)
Timor Island, Indonesia	Pe-Q	Q	polymictic mélange and imbricate terrains, incl. pelagic sed., ophiolites, high-grade contin. metam. blocks; undergoing collision	Mn nodules in pelagic clay; minor oxidized Cu in metabasalts	Audley-Charles (1972) Hamilton (1979)

Table 8-2 (continued); Selected examples of subduction mélanges associated with metallic mineralization

LOCALITY	AGE OF		GEOLOGY	MINERALIZATION	REFERENCES
	FORMATION	EMPLACEMENT			
Northern Molukkas (Halmahera, Obi, Waigeo), Indonesia	Cr-Mi	Mi	mélange, pelagic sediments and ophiolite assoc., widespread ultramafics	large Ni laterite deposits	Hamilton (1979)
Medial Mts., Cykloop Mts, western New Guinea island	Cr_3-Mi	Mi_1	mélange with ophiolite, greenschist, serpentinite rafts	Ni laterite over serpentinites	Hamilton (1979)
South - Western Papuan Peninsula, New Guinea Island	Cr_3-Eo	Eo_3 or Mi	deepwater argillite, siltstone, shale, chert intermingled with gabbro, diabase, basalt, ultramafics, pelagic sediments	small massive pyrite-Cu occurrences; showings of Mn oxides in chert	Hamilton (1979)
New Caledonia Island, the Pacific	Pe-Eo	Ol	zoned blueschist belt, high-grade metasedim., metavolcanics	relict and remobilized massive sulphides Fe, Cu, Zn	Paris (1981)
Woolomin Beds (partly), Great Serpentinite Belt N.S.W., Australia	Or-D_1 S-Cb	Pe	discontin. mélange segments ophiolite, chert, graywacke, ultramafic slivers; grade to non-disrupted units	podiform Cr, Mn in chert Ni, Co laterite; small Cu, Zn "Cyprus-type" showings; Au placers	Leitch (1980)

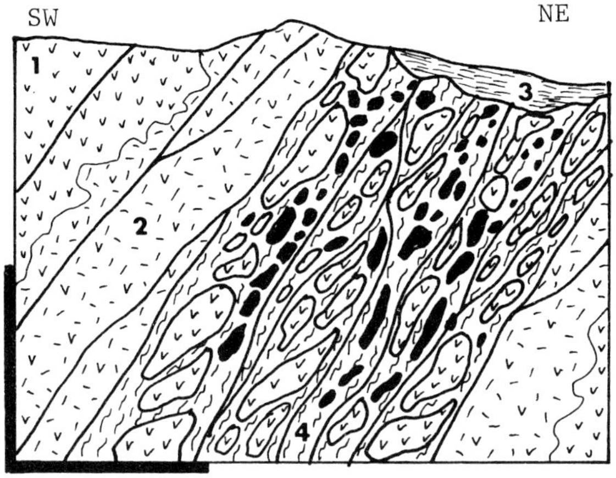

Fig. 8-6. Massive Fe-Cu sulphide-mineralized Jurassic ophiolite mélange or a fault zone, Rehova, Albania. After Čechovič (1937).

interpretation available, but a "Cyprus-type" volcanogenic deposit incorporated into a mélange and later modified by an igneous intrusion, seems to be most plausible.

The Barlo Cu-Zn mine in the Philippines, has usually been considered to be of "Cyprus" or "Kuroko" massive sulphide type, but a more detailed description given by Bryner (1969) suggests a mélange setting. He reported the existence of massive low-grade lenses of pyrite with small, randomly distributed zones of chalcopyrite and sphalerite-rich ore associated with a "boulder gouge": a mass of 1 cm to 2 m ophiolitic boulder fragments irregularly distributed in clayey matrix in a maze of faults.

RUTILE IN HIGH-PRESSURE METAMORPHICS

Metamorphogenic rutile forms at the expense of ilmenite, Ti-magnetite, and pyroxene in the mafic members of ophiolites that underwent a high grade high-pressure metamorphism during which eclogites, garnet - glaucophane schists, and amphibolites formed. Rutile usually forms millimetre - sized porphyroblasts evenly disseminated in the metamorphics, or large crystals present in irregular secretions and coarse-grained schlieren in the host rock. The rutile content is usually of the order of 1-2%, but locally may reach 5-10% or more. The dispersed rutile may accumulate physically in placers, and the accumulated "bedrock" rutile may form economic orebodies locally.

Such orebodies forming lenses 150-200 m thick and up to 1 km long are currently mined near Shubino in the southern Urals, U.S.S.R. (Chesnokov, 1960). Blake and Morgan (1976) reviewed the additional occurrences known from the Franciscan Mélange of California (e.g. on Catalina Island), from coastal Venezuela, New Caledonia, and the Alps. Zuffardi (1977) reported a newly discovered rutile occurrence near Val di Vara in western Liguria, Italy.

8.3.MÉLANGE/THERMAL SPRINGS INTERACTION

By far the most interesting metallic ores hosted by mélanges are formed by a high-level crustal interaction of mélanges with Na-bicarbonate, Na-chloride and CO_2-rich thermal springs. Several tens of young (Pliocene to Quaternary) cinnabar occurrences have been recorded from the Franciscan Assemblage in the Coast Ranges of California, and the most important localities are listed in Table 8-3.

The majority of the California mercury deposits are hosted by serpentinite (which is a member of the mélange), or occur in the proximity of a serpentinite, usually in fractured flysch litharenites and mudstones. Henderson (1969) reviewed the "serpentinite type" mercury deposits and the associated alteration. His "type" locality, Mt. Jackson mine, contains cinnabar and minor metacinnabar, marcasite, millerite, bravoite, ankerite and other minerals disseminated in a "silica-carbonate" gangue, located in a footwall of a fault breccia, or in gangue-filled faults. The silica-carbonate rock is a hydrothermally silicified and carbonatized serpentinite and the silica is either low quartz, or chalcedony sometimes pigmented white by magnesite.

The largest Californian mercury producer, the New Almaden field near San José (38,090 t Hg, aver. grade 2.21% Hg), has been described meticulously by Bailey and Everhart (1964; Fig. 8-7). There, a fine-grained to microcrystalline cinnabar replaces silica-carbonate, or less commonly forms open space fillings in a variety of fractured rocks. Extensive replacement stockworks formed along dense systems of feeding fractures and the richest orebodies contained up to 50% HgS. The orebodies mined were relatively large and rich. The largest was about 67 m wide and 5 m thick, and extended down the dip for about 500 m. The ore grade in stopes ranged from 0.5 to 20% Hg. New Almaden is believed to be the alteration-mineralization product of Pliocene thermal springs percolating through the plumbing system controlled largely by fracturing in the serpentinite-rich Franciscan mélange, and confined by the impervious zones of clay fault gauge. The ore extended from the surface to a depth of about 867 m. An unusual placer deposit containing nuggets of high-grade cinnabar in a muddy gravel (Fig. 8-7), yielded about 20t Hg.

New Idria, the second-largest California mercury producer (about 19 Tt Hg; Linn, 1968), is a north-west trending zone containing separate

Table 8-3. Cainozoic cinnabar deposits generated by interaction of thermal springs with fragmented and fractured serpentinites and ophiolite assemblages in mélanges

LOCALITY	HOST ROCKS	MINERALIZATION	REFERENCES
Pinchi Lake Mine, central British Columbia, Canada	Intermingled Penns.chert,marble, ophiolites,serpentinite and Tr_3-J_1 graywacke,siltsone,limestone. Blueschist metam.in spots	cinnabar,fracture coating and replac.in marble,graywacke,silica-carbonate-altered serpentinite. Neogene ?,4.2 Tt Hg/0.75%	Paterson (1977) Armstrong (1949)
New Almaden,Coast Ranges,California	J-Cr Franciscan mélange: serpentinite, graywacke, shale	cinnabar as replacement and fracture filling veinlets in silica-carbonate altered serpentinite;38,090 t Hg/2.21%	Bailey and and Everhart (1964)
Mayacmas district, Coast R.,California	as above	cinnabar in quartz veinlets in silicified and ordin.silica-carbonate altered rocks along fault 3 m.y.-Rec.miner.,13 Tt Hg	Dickson and Tunell (1968)
Wilbur Springs, Coast Ranges, California	as above; widespread "detrital serpentinite" (mélange ?) active hot springs in the area	cinnabar with opal,quartz,magnesite,calcite,hydrocarbons form fine grained replac.and fracture fillings; Pl-Rec.,2,700 t Hg	Moiseyev (1968)
New Idria,Coast R., California	Cr_3 shale,litharenite,at contact with serpentinite	cinnabar in fracture zones and stockworks,in silica-carbonate altered rocks under reverse fault. Plioc.,19 Tt Hg	Linn (1968)
Avalja,Šuplja Stijena,S.of Belgrade Yugoslavia	J_{2-3} fractured serpentinite, ophiolites, sediments	dissem.and stockwork cinnabar in silica-carbonate-altered rocks Miocene miner.? 229 t Hg/0.2-1%	Janković (1982)
Kaleçik deposit,Karaburun P.,Turkey	S_3-D_1 ophiolite,limest.,graywacke in mélange,Neogene volcanics	cinnabar,pyrite impregn.,disseminations in silicif.sediments	Süzen (1977)

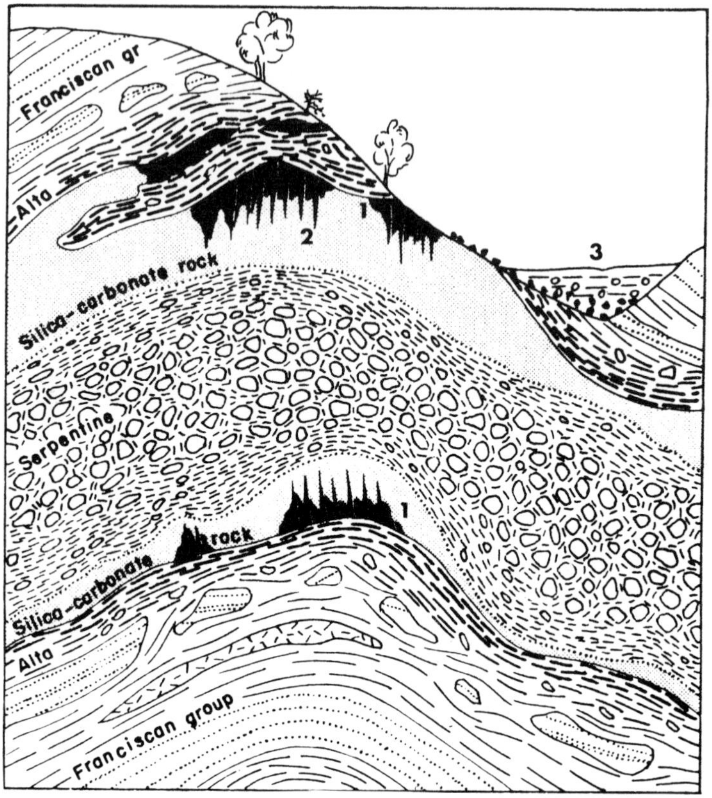

Fig. 8-7. New Almaden mine, California; schematic representation of the cinnabar mineralization styles: (1) cinnabar replacing silica-carbonate along fractures; (2) cinnabar fills open fractures; and (3) reworked cinnabar in placer. From Bailey and Everhart (1964).

cinnabar-mineralized silica-carbonate lenses along the contact of Franciscan serpentinite and an upper Cretaceous shale, beneath a steep reverse fault.

Geologically young cinnabar deposits hosted by a serpentinite-rich mélange, are known in several other places in the world (Table 8-3). The producing mines include Pinchi Lake, British Columbia (Fig. 8-8); Kaleçik, western Turkey (Sözen, 1977) and Šuplja Stena, Yugoslavia (Janković, 1982). Henderson (1969) argued that the main role of serpentinite in the formation of mercury deposits was the influence that its distinct chemistry exerted on the percolating hydrothermal fluids, causing them to precipitate the mercury they carried. Convincing data are not available. However, the global correlation of

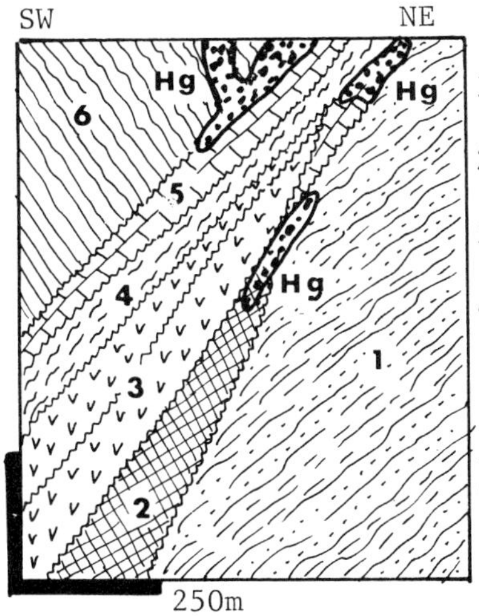

1 Tr_3 arkosic graywacke, laminated siltstone, conglomerate, limestone
2 altered serpentinite grading to talc-carbonate schist
3 high-pressure metavolcanics (e.g. pumpellyite-aragonite greenstone)
4 blueschist-metamorphosed sediments: metachert, metaargillite, lawsonite-glaucophane schist, aragonite marble
5 meta-limestone
6 Ps-Pe ribbon chert with argillite parting, interbedded with partly dolomitized and silicified limestone

Fig. 8-8. Pinchi Lake cinnabar deposit, British Columbia; diagrammatic section. From LITHOTHEQUE, data from Armstrong (1949) and Paterson (1977).

mercury and serpentinites can be extended to several deposits where the ultrabasics are not in sight but can be anticipated in depth (or in the vicinity), suggesting a possible source relationship.

In addition to mercury–antimony, arsenic and gold may also accumulate in virtually the same mélange setting. Albers (1981) reported the recent discovery of a ? Pleistocene gold deposit containing at least 6 Mt ore with 4.76 ppm Au (equivalent to 28.6 t Au), near the Manhattan Hg mine, Clear Lake region, California. There, the gold is in opalite or in silica–carbonate rock, related to the hot springs interaction with the Franciscan ultramafic mélange and also with the Pliocene continental volcanics.

Low-temperature arsenic deposits consist mostly of realgar and orpiment. In the Matra deposit of north-eastern Corsica (Feraud et al., 1977; 10-20 Tt As), quartz, chalcedony, dolomite, realgar and minor orpiment, bravoite and other minerals form banded lenses up to 4 m thick, hosted by a quartz-carbonate-altered and argillized Mesozoic serpentinite, chlorite schist and glaucophane schist. Portions of the orebodies are rich in stibnite. Orebodies, predominantly of stibnite, hosted by a silica–carbonate altered serpentinite and mélange, are represented at the Stuart Lake mine located near the Pinchi Lake Hg deposit, British Columbia (Armstrong, 1949).

8.4. MÉLANGE/GRANITIC INTRUSIONS INTERACTION

Several mélanges now exposed within the continental crust, metamorphosed and intruded by granitic rocks, contain hydrothermal veins, stockwork and replacement orebodies of Au, Hg, Sb, As and other metals. Quartz,carbonate, arsenopyrite, gold veins are locally hosted by mélanges in the Mother Lode Belt of California, particularly along the Melones fault. Similar, but very small gold occurrences are found in the Los Burros district, California Coast Ranges; Kilkivan, Queensland; Bingara, N.S.W.; and elsewhere. A portion of the stibnite orebodies in the Bau district (Sarawak, Malaysia) and probably in the Nakase Mine, Japan, is also mélange-hosted in the vicinity of a "granite".

8.5. MINERALIZATIONS GENERATED BY WEATHERING AND BY SEDIMENTOGENIC REWORKING OF MÉLANGES

The ultramafic members of mélanges altered by tropical weathering can predictably produce "lateritic" Fe, Ni, Cr, and Co deposits already described and tabulated in Chapter 7. Of the localities listed there, the nickel "laterites" of Soroako (Sulawesi), the Molukkas and the Cykloop Mts., as well as the Fe-Cr-Ni "laterites" of the Meratus Mts. (all the localities are in Indonesia), have all developed over ultramafics-dominated mélanges.

8.6. MINERALIZATIONS IN AND ADJACENT TO SERPENTINITE-FILLED FAULTS

Serpentinite-filled faults in a variety of geotectonic settings are metallogenically comparable with the mélanges, described above. Cinnabar occurrences close to the "serpentinite Hg-type" are of particular interest, and several tens of occurrences are known worldwide. The four example localities described by Kuznetsov (1974) in the U.S.S.R. (Chagan-Uzun and Aktash in the Gornyi Altai; Chazadyr in Tuva; Chonkoi in South Ferghana) are all located along prominent faults that separate tectonically thick sections of Paleozoic supracrustal rocks. Slivers of serpentinite filling the faults are talc-carbonate (listvenite) or talc-carbonate-silica altered, and replacements or fissure filling impregnations, veinlets and stockworks of cinnabar, pyrite and minor As, Sb, Cu, Pb, Zn sulphides occur in the listvenite bordering the faults.

Several mercury deposits located in intensively faulted sedimentary associations (e.g. Idrija, Yugoslavia; Chapter 17), have many points of similarity with the above localities, except that they lack serpentinites at the present level of exposure; they are probably transitional.

"One of the disadvantages of our geological terminologies is their inability to incorporate intermediate cases or situations between two or more end-members, thus not emphasizing the transitional or gradational characteristics of all natural phenomena and their products". Karl H. Wolf in Handbook of Strata-Bound etc. Deposits, 1981.

CHAPTER 9

Zoned Mafic/Ultramafic Complexes in Phanerozoic Orogenic Belts (Alaska or Ural "Type")

9.1. GENERAL

Small zoned mafic/ultramafic intrusions of Mesozoic-Cainozoic age have been recognized in South-East Alaska and singled out as distinctive types of ultramafics in the 1960's (Taylor and Noble, 1960; Taylor, 1967). Later it was realized that similar, Paleozoic, intrusions have been known in the Urals, U.S.S.R., for some time and the term "Alaska and/or Ural type" of intrusion is now widely used in the literature. Gabbro and/or norite bodies without ultramafics that are contemporary with the ultramafic intrusions and located in the same areas are also included in this chapter, although in the petrological literature they are often treated separately. Table 9-1 lists the better known mineralized complexes worldwide.

The mafic-ultramafic intrusive complexes range from simple, small ultramafic plugs (less than 2 km in diameter; e.g. Kane Peak, Annette Island, Red Bluff Bay; all in S.E. Alaska) through small composite circular to ellipsoidal gabbro-peridotite intrusions (e.g. Union Bay, S.E. Alaska) to 10-15 km large composite massifs. Although they are comparable with the alkaline-ultramafic intrusions described in Chapter 33 as far as their shape and distribution of rock types is concerned, the Alaska-Ural complexes are tholeiitic, and are consistently members of the continental margin association. The gabbroic intrusions tend to be larger. In the western Cordillera, the intrusions appear to be comagmatic with the thick, tholeiitic, partly subaerial basalt to basaltic andesite flows, described in Chapter 10.

In the classical area of South-Eastern Alaska (Fig. 9-1), about 39 ultramafic bodies and about twice as many gabbro bodies occur in a broad N.W.-S.E. trending belt about 500 km long, extending down the Alaska Panhandle. Most of the small ultrabasic bodies are composed largely of hornblende pyroxenite, but eight larger complexes contain peridotite and they are consistently zoned. Almost all are in contact with, or surrounded by, the regionally more widespread gabbro and the gabbro to ultramafics age relationship remains controversial in most cases (compare McTaggart, 1971, for discussion). Most authors, however, consider the gabbros to be slightly older than the ultramafics.

Table 9-1. Small zoned mafic - ultramafic complexes in Phanerozoic orogenic belts; selected examples

LOCALITY	AGE	GEOLOGY	MINERALIZATION	REFERENCES
Yakobi Island, S.E.Alaska	Tr-Cr	composite stock of norite enveloped by gabbro,qtz.diorite	trough-shaped body of dissem. ptl,cp,po above norite base 66 Tt Ni/0.33% 44 Tt Cu/0.21%	Kennedy and Walton (1946)
Klukwan,n.of Haines, S.E.Alaska	Cr_2	5x1.6 km stock of magnetite hornbl.pyroxenite intruded gabbro-diorite	15-20% Ti-magnetite uniform dissem. in pyroxenite, local segreg.of massive magnetite 70 Mt Fe/14% 375 t Pt/0.75 ppm	Taylor and Noble (1969)
Union Bay, S.E.Alaska	Cr-T_1	dunite,peridotite,oliv.pyroxenite,hornbl.pyrox.,gabbro zoned intrusion	large mass of magnetite-hornbl. pyroxenite, uniform texture; aver. 18.03% Fe, 1.36% TiO_2	as above
Duke Island, S.E.Alaska	Cr-T_1	two separate intrusions; dunite or perid.core enveloped by pyroxenite,intr.to gabbro	concordant layers of segregated magnetite in pyroxenite, 10-60% Fe	as above
Salt Chuck Intrusion Kasaan Penn.,S.E. Alaska	Cr-T_1	small pyroxenite body with minor gabbro and diorite emplac. into S metamorphics	dissem.and veinl.of bornite, cp,nat.Cu,palladium in pyroxenite; 2.4 Tt Cu 280 kg Pd	Wolff and Heiner (1971)
Brady Glacier, S.E.Alaska	T_1	layered gabbro intrusion emplaced into MZ metamorphics has an olivine,pyroxene,chromite cumulate near base	po,ptl,cp,dissem.to massive in troctolite cumulate and along contacts; 500 Tt Ni/0.5% 300 Tt Cu/0.3%, 18 t Pt met.	Himmelberg and Loney (1981)
Aiken L.-McConnell Cr.area, British Columbia	Tr	about 10 small,zoned dunite, wehrlite,pyroxenite,hornblendite,gabbro intrusions	ilmenite segreg.pods in gabbro (Axelgold); magnetite dissem. in pyroxenite	Irvine (1974)

Hope, 130 km E. of Vancouver, British Columbia	Cr_2	3.2x1.6 km irregular zoned peridotite, pyroxenite, hornblendite body envel. by diorite and norite, in metamorphics	po, ptl, cp, disseminations to massive pipes in peridotite, sulphide masses on contact of pyroxenite and peridotite	Clark (1969) McTaggart (1971)
Tulameen complex, British Columbia	Tr_3	zoned small intrusion; dunite core envel. by olivine and hornbl. pyroxenite, gabbro	Pt in dunite, highest value 0.225 ppm assoc. with chromite; small Au and Pt placers	McTaggart (1971)
Kochkanar Complex, Isovsk area, the Urals, USSR	S_3	110 km^2 laccolithic pluton, dunite, perid., oliv. pyrox., oliv. gabbro, gabbro, norite, intruded into amphibolites	dissem. Ti-magnetite in olivine magnet. pyroxenite, 6 Bt Fe/16.64%; Pt, Pd dissem. in pyroxenite, Fe ore, placers; minimum 500 t Pt metals	Reshit'ko (1967)
Gusevogorsk, adjac. to Kochkanar, Urals, USSR	S_3	like Kochkanar	coarse to fine Ti-magnetite segreg., lenses, sheets; small quant. of nat. Au in magnetitite unevenly distributed Pt	Razin (1974)
Nizhnyi Tagil intrusion, the Urals USSR	S_3-D_1	dunite, pyroxenite, gabbro, empl. into Or-S metavolcanics	pockets, lenses, veinlets of Pt alloys in dunite, accumulated in placers; min. 259 t Pt	Razin (1974) Rundkvist, ed., (1978)
Volkovskoe deposit, part of Tagil intr. the Urals, USSR	S_3-D_1	n.n.w. elong. gabbro, gabbro-diorite, diorite, quartz dior. layered intrusion	3 km long ore zone, 200 lens-like bodies of dissem. born., cp overlapping with layered Ti-magnetite, in pyrox. gabbro about 600 Tt Cu/0.85%, 300 t Pd	as above Samonov and Pozharisky (1974)
Chocó Province, Colombia	MZ-CZ	small serpentinized ultra-mafic - mafic bodies	poor bedrock outcrop; 75 t Pt plus produced from placers	Putzer (1976)

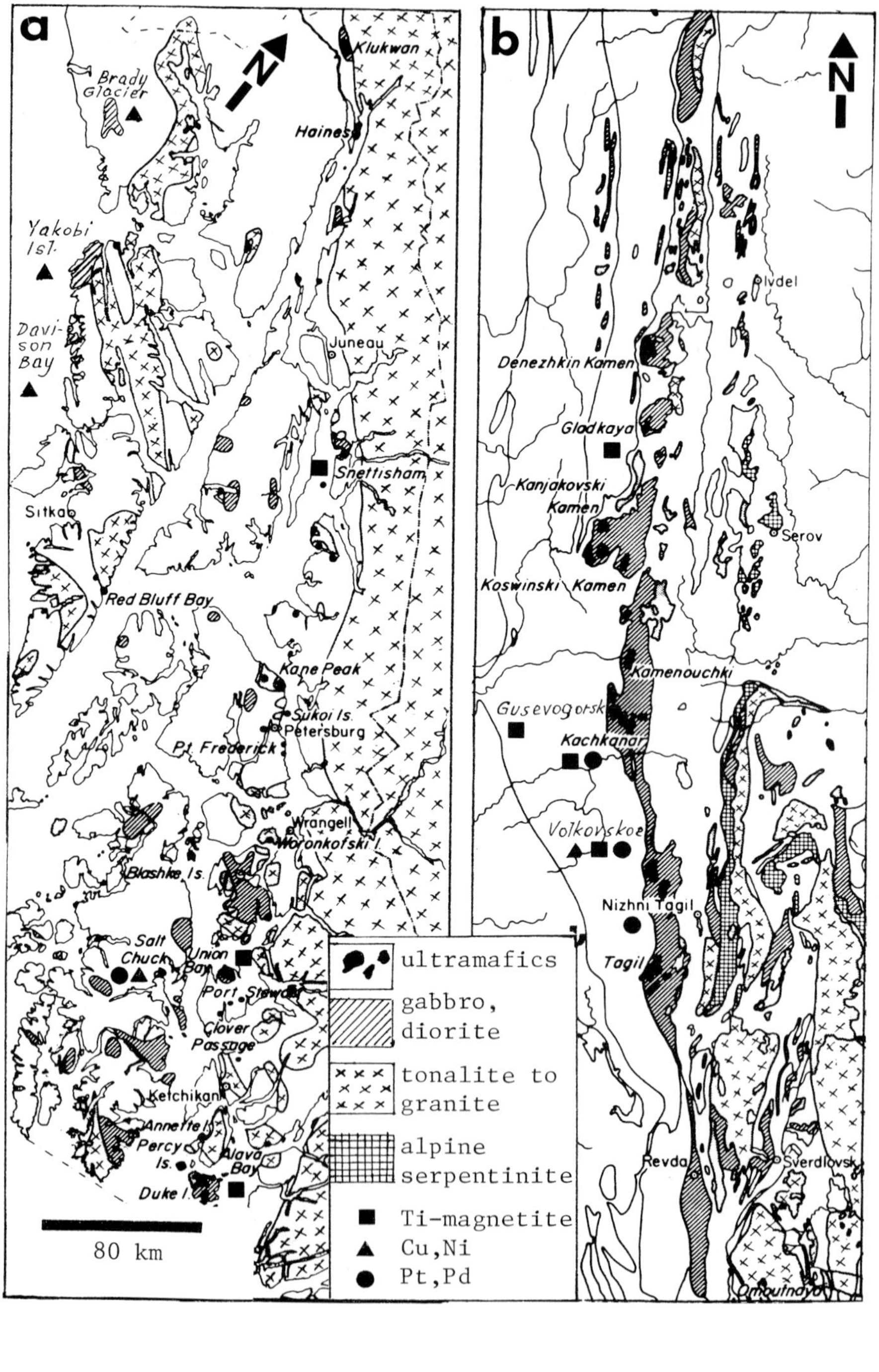
a
N
Brady Glacier
Klukwan
Haines
Yakobi Isl.
Davison Bay
Juneau
Snettisham
Sitka
Red Bluff Bay
Kane Peak
Sukoi Is.
Petersburg
Pt. Frederick
Wrangell
Woronkofski I.
Blashke Is.
Salt Chuck
Union Bay
Port Stewart
Clover Passage
Ketchikan
Annette I.
Percy Is.
Alava Bay
Duke I.
80 km
ultramafics
gabbro, diorite
tonalite to granite
alpine serpentinite
Ti-magnetite
Cu,Ni
Pt,Pd
b
N
Ivdel
Denezhkin Kamen
Gladkaya
Kanjakovski Kamen
Serov
Koswinski Kamen
Kamenouchki
Gusevogorsk
Kachkanar
Volkovskoe
Nizhni Tagil
Tagil
Revda
Sverdlovsk
Omoutnaya

The most common zoning pattern (Taylor, 1967) has a small dunite or peridotite core, surrounded by hornblendite or hornblende pyroxenite, that sometimes contain units of olivine pyroxenite. Gabbro or norite occur marginally and are sometimes missing. In such cases, the ultramafics are emplaced into the basement metamorphics, causing little contact alteration. The Duke Island Complex described by Irvine (1967; Fig. 9-2B), is composed of two ultramafic bodies separated by altered gabbro. The ultramafics comprise peridotite and dunite showing cumulate layering, graded bedding, local unconformities, slump structures and fragmental layers that are often remarkably persistent. The contact with the surrounding gabbro is represented by hornblende clinopyroxenite and hornblendite, interpreted as metasomatites by McTaggart (1971).

The Brady Glacier intrusion described recently by Himmelberg and Loney (1981; Fig. 9-3A), had been an example of one of the S.E. Alaskan gabbroic intrusions. It nevertheless contains basal olivine-chromite; plagioclase-augite-olivine; and olivine-bronzite-chromite, ultramafic cumulates, discontinuously interlayered. Several intrusions located on Baranof, Chicagof and Yakobi Islands are dominated by norite (Kennedy and Walton, 1946).

In the north-central Ural, a north-south trending belt of gabbro about 600 km long, encloses about 20 ultramafic plutons, most of them zoned (Fig. 9-1). The intrusions are of the late Silurian age (440-455 m.y.) and occur along the western margin of the Magnitogorsk Trough, a thick Paleozoic folded mafic volcanic-sedimentary facies belt. Compositionally, they correspond to the dunite, peridotite, gabbro association (Pronin, 1962). The Kachkanar gabbro-peridotite complex (Reshit'ko, 1967), is an aggregation of three lens-shaped massifs surrounded by amphibolites and chlorite schists. The cores of the massifs are occupied by dunite, that grades outward into peridotite, olivine pyroxenite, diallag pyroxenite, olivine gabbro, gabbro-norite, diorite, syenite, and amphibolite. Reshit'ko (1967) argued that the massifs are relicts of a much larger intrusive body probably of phacolithic shape, that survived with little modification the dynamometamorphism that had converted their environs into foliated metamorphics

9.2. METALLIC MINERALIZATIONS

The zoned mafic/ultramafic intrusions contain essentially a "Bushveld-style " mineralization (compare Volume 2), produced by differentiation of mafic magma. Five categories of ores may be

---

◁

Fig. 9-1. Distribution of the zoned mafic-ultramafic intrusions in the "eugeosynclinal" orogenic belts of south-eastern Alaska (left) and the central Urals (right). From Taylor and Noble (1969); mineralization symbols added. Courtesy of Economic Geology.

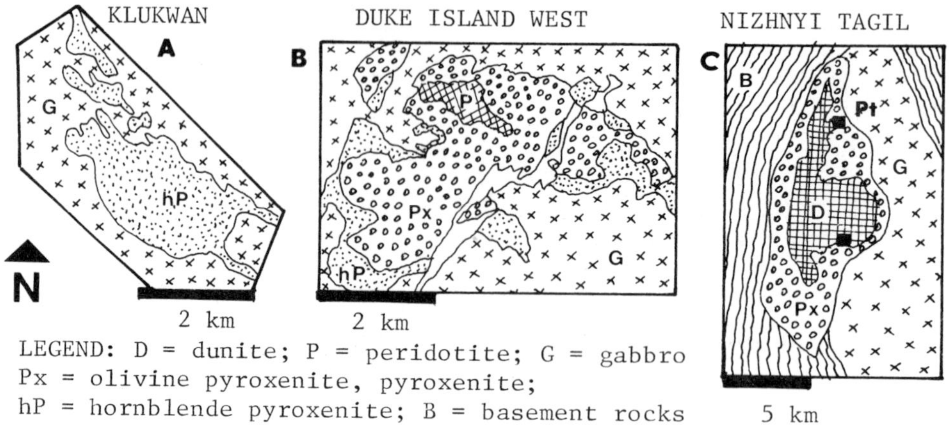

Fig. 9-2. Zoned mafic-ultramafic intrusions. After Taylor, Stebbins and Irvine, in Taylor and Noble (1969); Magak'yan (1968).

distinguished: (1) chromite (enriched in dunite and peridotite); (2) platinum metals (enriched in either dunite as in the Tagil Massif, or in pyroxenite, as in Kachkanar); (3) vanadiferous titanomagnetite, accumulated in pyroxenite; (4) Ni-Cu sulphides which display substantially less regularity in distribution than the above mentioned cumulus concentrations, and are found mostly in contact or hybrid settings; (5) copper sulphides that often contain significant palladium. The latter have been recorded in several localities.

The estimated economic (and potentially economic) metal content in the known deposits, worldwide, is of the order of 2.5 Bt Fe; 200 Mt Ti; 600 Tt Ni; 600 Tt Cu; 25 Mt V; and 2,200 t of platinum metals.

(1) Chromite is a widespread accessory in all dunites and peridotites, but nowhere it is sufficiently concentrated to constitute an ore. Himmelberg and Loney (1981) found 3-5% chromite in the olivine-chromite cumulate in the Brady Glacier deposit.

(2) Platinum metals have been produced in significant quantities (estimated 1,500 t Pt metals) from the Uralian placers, derived in turn from the zoned ultramafic complexes. Rundkvist, ed., (1978) distinguished two subtypes of bedrock platinum deposits in the Urals: Nizhnyi Tagil, and Kachkanar.

In the Tagil subtype (Fig. 9-2C), platinoids are bound in polyxene (Pt, platinoids and Fe alloy) and Os-Ir alloys, that form minute, irregular disseminations in dunite. These disseminations are most common within chromite-rich schlieren and in steeply dipping pipes.

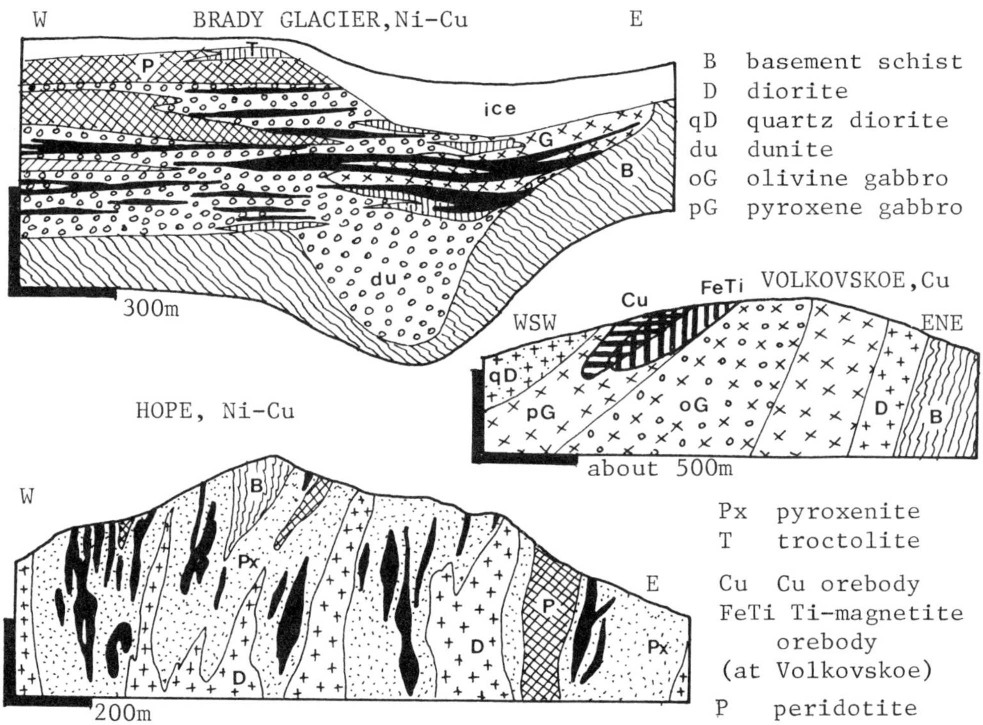

Fig. 9-3. Sulphide ores in differentiated ultramafic-mafic complexes of the Alaska-Ural "type". Modified and generalized sections after Himmelberg and Loney (1981); Timokhov in Samonov and Pozharisky (1974); and Clark (1969); from LITHOTHEQUE.

The chromite and platinoids are interpreted as epigenetic, postdating crystallization of the host dunite, and are probably genetically equivalent to the platiniferous pipes in the Bushveld (Volume 2). The chromite-rich zones are most common in the faulted sections, and are sometimes associated with dunitic pegmatites.

In the Kachkanar subtype, Pt-metals have been recorded only in those massifs in which pyroxenites are the dominant ultrabasic, but only one intrusion in the Urals contained enough Pt-metals to form a rich placer. Polyxene and Pt-Pd alloys are irregularly disseminated in olivine pyroxenites and wehrlites, whereas Pt-Pd sulphides are in diallagites and magnetite-rich pyroxenites. The latter do not accumulate in placers, but can be recovered as a by-product of recovery of the very low-grade titanomagnetite.

The platinum placers of Chocó, Columbia; Tulameen Complex, British Columbia; and Yubdo, Ethiopia, are believed to have derived from "Alaska-type" complexes, but the primary occurrences are not well known.

(3) Vanadiferous titano-magnetite with subordinate quantities of ilmenite, occurs as uniform disseminations in pyroxenite layers, forming extensive sheets. The majority of such layers contain around 20-30% of magnetite, which gives the rock grades between 15-25% Fe and 1-2% Ti, too low to be of much economic interest at present. Rundkvist, ed., (1978) quoted the ore reserves in Kachkanar to be 12.2 Bt of 16.64% Fe ore (2.03 Bt Fe). Within the low-grade layers, however, local high-grade segregations are present that contain from 60 to 80% magnetite.

Taylor and Noble (1969) discussed, in detail, the petrology of the Alaskan magnetite pyroxenites that also contain large tonnages of a low-grade titanomagnetite ore.

(4) Pyrrhotite, pentlandite and chalcopyrite in the Brady Glacier layered intrusion (Himmelberg and Loney, 1981), are irregularly disseminated in a variety of mafic and ultramafic types. Small bodies of massive to "matrix" sulphides tend to occur along the boundaries between mafic and ultramafic layers (Fig. 9-3A). The textures indicate that the mineralization formed as an immiscible sulphide melt. The large (20.7 Mt) but low-grade (0.33% Ni;0.21% Cu; Kennedy and Walton, 1946) pyrrhotite, pentlandite, chalcopyrite disseminations on Yakobi Island occur as disseminated segregations in the basal portion of a basin-shaped norite intrusion.

The Hope, British Columbia Ni-Cu deposit (Clark, 1969; Fig. 9-3C) was the only locality actually mined. It has 28 subvertical pipe-like disseminated, matrix to massive pyrrhotite - pentlandite orebodies of replacement origin, located near intrusive or fault contacts of pyroxenite and peridotite.

(5) Several occurrences of copper sulphides are associated with the Uralian zoned mafic-ultramafic massifs (Tagil, Kytlym, Verkhneuralsk), but only the Volkovo (Volkovskoe) deposit in the northernmost part of the Tagil Massif is of economic importance. The Volkovo intrusion (Rundkvist, ed., 1978) is a N.N.W. - elongated olivine gabbro, gabbro, gabbro-diorite, diorite, and quartz diorite zoned intrusion, emplaced into mafic metavolcanics (Fig. 9-3B). The copper is in disseminated bornite and chalcopyrite, accumulated in the form of elongated lenses overlapping with disseminated Ti-magnetite-apatite orebodies. About 200 individual ore-grade lenses have been recorded in pyroxene gabbros, along the western margin of the intrusion, along a strike length of 3 km. Palladium and gold are recovered as by-products. The small Salt Chuck deposit in the Kasaan Peninsula, S.E. Alaska, is very similar to Volkovo. At the Kolpakovsk showing in the Urals (Pronin, 1962), native copper disseminations are in gabbro of a differentiated complex.

"Every deposit brings with it the evidence that it is no independent and separate mass, but that it is very closely related to the geological formations enclosing it".
F. Beyschlag, P. Krusch, and J.H.L. Vogt (1916).

CHAPTER 10

Unimodal Mafic Volcanic-Sedimentary Association

10.1. INTRODUCTION

"Geosynclinal" (that is marine, continental margin-based) volcanic associations in which basalts (and spilites, diabases, gabbros, etc.) are the only or predominant magmatic rocks, formed in four basic environments: (1) within the oceanic lithosphere; (2) within an island arc/marginal sea system; (3) along the extremities of the Atlantic-type continental margin, and in extensional troughs within the margin and (4) in the "quasi-continental" segments of the oceans.

In the ancient rock associations, however, the depositional environment cannot always be determined with certainty and the geochemical evidence is often inconclusive. This accounts for the extreme transitionality and overlap of the mafic-dominated lithologic associations, shown graphically on Fig. 10-1. Of the four major environments, (1) and (2) are dominant and are most thoroughly treated in the literature. Because basalts and sediments form the two topmost units of an ophiolite complex corresponding to the oceanic layers 1 and 2, it is conceivable that these layers alone may crop out independently over large areas, without the remaining members of the ophiolite association (in particular the most diagnostic ones, the ultramafic tectonites), or containing them as minor accessories only. Such occurrences are treated more comprehensively in Section 10.2a.

The ensimatic island arc association is also dominated by widespread basalts ("arc tholeiites") that differ from the ophiolitic basalts by their frequent gradations into basaltic andesites, alkaline basalts and calc-alkaline associations. Although widespread coeval ultramafics are absent in units composed of the arc basalts, some may be represented as members of the Alaska-Ural type of differentiated ultramafic-mafic complexes (Chapter 9), that are now considered as feeders to the arc tholeiites. When tectonized or dismembered, they are easily confused with ophiolites. The arc association is mostly autochthonous in contrast to ophiolites, that are mostly allochthonous. Its occurrences and mineralizations will be reviewed in Section 10.2b.

In orogenic belts, the island arc associations are usually younger than the ophiolites and in young, still ensimatic island arc systems (e.g. the Solomon Islands) the arc association formed on top of ophiolites. Elsewhere, for example in the Rocas Verdes Complex on the Chilean side of the Magalhães "Geosyncline" studied by

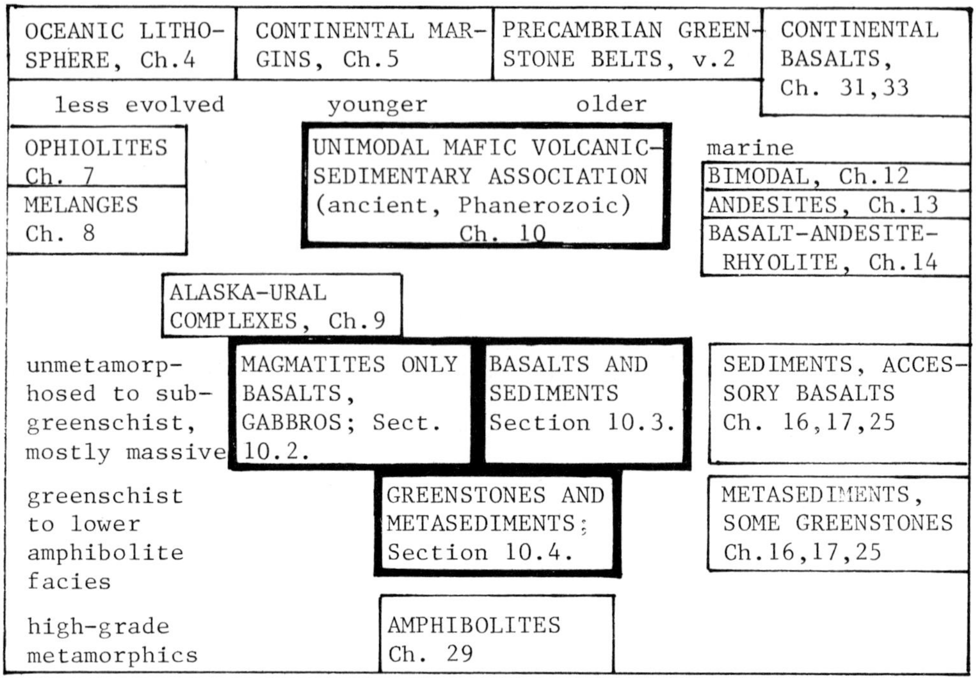

Fig. 10-1. Organization and transitionality of basalts treated in Chapter 10.

Tarney et al. (1976), it appears that ophiolites formed in marginal or interarc basins, are marked by a rapid facies change as well as the presence of hemipelagic and terrigenous sediments, and transitions into contemporary island arc and Andean-type continental margin rock suites (basalts, andesites, felsic members), formed on flanks. In keeping with the emphasis on field recognition of the lithologic associations that forms the backbone of this book, the descriptive subdivisions are based on the obvious and recognizable ratio of the mafic magmatites and sediments (magmatites only or dominant: Section 10.2; magmatites and sediments about equally important: Section 10.3; sediments dominant: treated in Chapters 16, 17 and 25). Similarly, the intensity of metamorphism or deformation that profoundly alter the rock appearance and also influence to a considerable degree the mineralization, has been used as a basis for subject subdivision (Sections 10.2. and 10.3. deal with dominantly massive, little metamorphosed rocks; Section 10.4. deals with schistose, metamorphosed rocks). Again it has to be emphasized that these categories overlap generously, that there are no sharp boundaries, that much depends on the scale (local areas versus extensive belts), and that inclusion or exclusion of certain mineralizations is often highly subjective.

10.2a. MASSIVE BASALTS (GABBROS), ABSENT OR MINOR SEDIMENTS, "OCEANIC" AFFILIATION

10.2a.1. Introduction

The simplest and most predictable local sections are represented by massive flows of pillowed tholeiitic basalts, breccias and hyaloclasties, usually yielding "oceanic" petrochemical indicators. Interflow sediments are minor, and mostly represented by red chert (jasper), green chert, or siliceous shale (slate). Numerous comagmatic diabase dikes (probably feeders) and small stocks of gabbro intervene. Basalt flow units frequently display the predictable zoning from massive fine-grained basal rocks through amygdule-rich intermediate level rocks to scoriaceous flowtops. The flowtops are either green (reduced) or red (oxidized, often hematite rich). The porous portions of basalts are usually altered (metamorphosed) to the prehnite-pumpellyite facies, or spilitized. The spilites tend to be fine-grained (aphanitic), pale green, and abundantly veined by white crystalline calcite, prehnite and chert. Table 10-1 lists example localities of this association, but significant mineralizations are uncommon.

10.2a.2. Ore minerals in basalts

Nagle et al. (1973) described an occurrence of native copper and minor cuprite in the form of small blebs in hematite-rich flowtops in the La Desirade Island, in the Caribbean. Native copper in pillow basalts is also widespread in the lower Cretaceous region of the Philippines, e.g. in the central Cebu Island (Bryner, 1969). Fracture and fault-controlled Cu sulphide occurrences are found at several places in the Eocene basalts and gabbros of the south-western Vancouver Island (e.g. Sooke, River Jordan; Carson, 1969).

Chalcopyrite, pyrrhotite and pyrite occur in stockworks of stringers forming zones 1-30 m wide along fractures and shears. Minor amphibolization of the pyroxene, feldspathization and scapolitization is associated with this. Interflow and flow-top siliceous shales and sometimes carbonates associated with the basalts, contain Mn oxide lenses comparable with those already described in Chapter 7 (Ophiolites). The frequently mentioned Mn belt in the Olympic Peninsula, Washington, that extends for about 240 km, contains numerous but small lens- and bed-like stratiform occurrences of braunite, hausmannite, rhodochrosite, etc. in spilitic flows and red pelagic limestones grading to limey shales (Park, 1946). The largest Crescent deposit produced only 25 Tt of Mn.

Table 10-1. Massive basalt, gabbro, minor sediments association of "oceanic" affiliation: mineralized example areas or localities

LOCALITY	AGE	METAM. GRADE	GEOLOGY	MINERALIZATION	REFERENCES
Crescent Fm. and Metchosin Volcanics, Vancouver Island, Olympic Peninsula	Eo	zeol. ph-pp	tholeiitic basalt flows diabase, gabbro, minor jasper, felsic tuff	dissem. and veinlet cp, po, py in sheared and fractured basalt, gabbro; Mn oxides in and above flowtops	Northcote and Muller (1972)
Shickshock and Maquereau Groups, Gaspé P., Québec	Cm	sub-gsch cont.	basalt, spilite, diabase, gabbro grading to slate graywacke; D "granite"	nat. Cu and Cu sulph. widespread dissem. in mafics Cu veins, skarns, in granite aureole	Dugas et al. (1969)
La Desirade Island Lesser Antilles	Cr	ph-pp	basalt and spilite amygdaloidal, pillow fl.	dissem. nat. Cu in hematitic basalt flowtops	Nagle et al. (1973)
Karamalytash Suite, Ural Mts, USSR	D	sub-gsch	basalt, spilite fissure eruptions, jasper	braunite lenses in jasper, second. oxides; 1.06 Mt Mn	Roy (1976)
Caucasus Front Ranges belt (partly), USSR	D_{1-2}	sub-gsch	basalt, spilite flows, chert, siliceous slate	Khudes - mass. and stockwork py, cp, sf deposit; Mn oxides	Skripchenko (1972)
Mbirao Group, Solomon Islands	Cr?	sub-gsch (loc.)	oceanic basalts, diabase, gabbro; intruded by uralite gabbro stocks	quartz, py, cp veins and dissem. in altered gabbro, basalt; quartz-gold veins in gabbro, placers	Arthurs (1979)

10.2a.2. Massive Fe, Cu, (Zn) sulphides

Massive sulphide deposits hosted entirely by basalts or spilites correspond to the "Cyprus-type", already described in Chapter 7. Typically they consist of dominant massive pyrite and subordinate chalcopyrite and sphalerite lenses usually peneconcordant with the volcanic layering, underlain by a footwall zone of brecciation and quartz-sericite or chlorite alteration. The majority of deposits, however, are hosted by schistose and metamorphosed volcanics, treated in a later section. The Khudes deposit in the Caucasus is, judging from the literature description (Skripchenko, 1972; Fig. 10-2) one of the few examples known, where the hosts are sub-greenschist metamorphosed and relatively massive.

Khudes is located in the Devonian volcanic-sedimentary belt of the Caucasus Front Ranges which, over its 450 km length, is mostly a bimodal (mafic-felsic) association. The ore zone is represented by a series of thin massive pyrite, chalcopyrite and sphalerite lenses located along an interflow horizon surrounded by spilites. The sulphide in some orebodies changes facies into pyritic siliceous slates or into cherts and quartz-hematite banded iron formation. The sulphide orebodies are underlain by a quartz-sericite altered feeder pipe, that carry a minor stockwork sulphide mineralization.

10.2a.3. Mineralized massive gabbros

Gabbros very probably comagmatic with the extrusive basalts occur in the volcanic infrastructure, or intrude into the volcanics. The gabbroic bodies have a lens-like, layered or stock--like form. The gabbros occasionally carry small quantities of Ni-Cu sulphides, variously interpreted as a cumulus or an intercumulus phase, the latter most probably a product of hybridization or sulphurization near contacts with pyritic metasediments. A gabbro body (dike ?) mineralized by sub-economic quantities of Ni-, Cu- and Pt-bearing pyrrhotite, pyrite and magnetite, occurs in the Skorovas bimodal massive sulphide field in Norway. In the Iskut River area of north-western British Columbia (Jeffery, 1966), several possibly economic but small Ni-Cu ore occurrences are associated with coarse-grained gabbros interacting with the probably broadly contemporary ? Permian bedded cherts, with tuffaceous cherts, and with thinly-banded shaley argillites and tuffs. Pyrrhotite, pentlandite and chalcopyrite with lesser pyrite and magnetite, occur as coarse blebs distributed fairly evenly throughout the gabbro. Almost massive sulphide sections occur in the chilled gabbro endocontact near the chert, and to a lesser degree in the exocontact cherts and hornfelsed argillites. The orebodies have a pipe-like form and there are about 26 Tt Ni/0.8% and about 20 Tt Cu/0.62%.

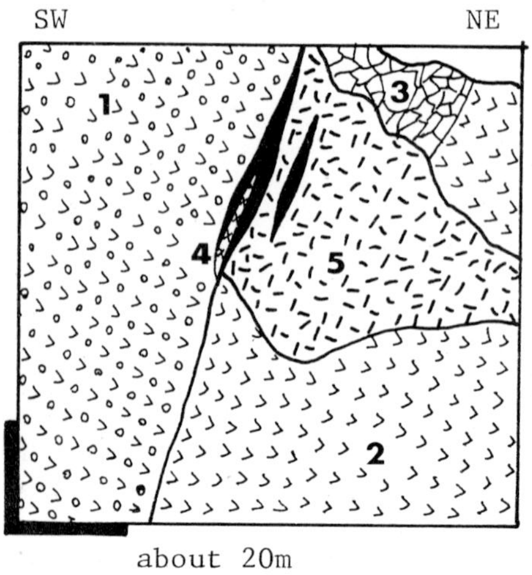

DEVONIAN ROCKS

1 spilitic variolite
2 spilite
3 mafic breccia, hyaloclastite, pyroclastics
4 diorite porphyry
5 quartz-sericite-chlorite footwall alteration pipe
BLACK: massive pyrite, pyrrhotite, chalcopyrite, sphalerite orebodies

Fig. 10-2. Khudes ore field, northern Caucasus, U.S.S.R.: example of a volcanogenic massive sulphide deposit hosted entirely by mafic arc volcanics. After Skripchenko (1972).

Small, scattered occurrences of copper minerals are most widespread in gabbros. Native copper has been reported at several Uralian localities (e.g. the Kolpak Intrusion), but chalcopyrite and bornite are more common. They come as disseminations, stringers, schlieren and veinlets, most abundant in the hybrid and metasomatized segments of gabbros; for example, in zones of pyroxene amphibolization, biotitization or silicification (e.g. in Guadalcanal; Arthurs, 1979; near Sooke, Vancouver Island; in the Gaspé Peninsula; etc.).

The Katahdin Iron Works ore zone in the Piscatequis County, Maine (Miller, 1945) contains massive to disseminated bodies of cobaltiferous and slightly cupriferous pyrrhotite, concordant with a raft of hornfelsed metasediments enclosed in the gabbro. The limonite gossan was mined on a small scale in the past, and it is estimated that there is potentially a recoverable resource of about 60 Tt Co in the sulphide zone.

Epigenetic Fe, Cu and Co mineralization in, and in the vicinity of, gabbro bodies is widespread in the Paleozoic mafic "eugeosynclinal" belts of the Urals, such as in the Magnitogorsk Synclinorium (Ivanov et al., 1972). Some occurrences are at contacts with the "Alpine" serpentinites (e.g. the Ivanovskoe showing), while the others are veins, disseminations or nests in fractured or faulted gabbros, or along the gabbro or intermediate intrusions contacts. The ore is usually dominated by pyrrhotite, and there are lesser quantities of pyrite, pyrrhotite, sphalerite, chromite, pentlandite, and/or cobaltite. Talc and chlorite are the most common alteration minerals. The economic importance is small.

10.2a.4. Interaction of basalts or gabbros with granitic intrusions

Thermal aureoles of younger "granitic" intrusions emplaced into basalts and gabbros of the present association are marked by frequent occurrences of essentially the same mineralization styles described earlier, and almost always distributed along altered fault or fracture zones. Copper showings are common both in and close to, the Cambrian mafics intruded by Devonian "granites" in the Gaspé Peninsula, Québec (Dugas et al., 1969); in the Cretaceous mafic belts in the Philippines (Bryner, 1969) and elsewhere. Gold in quartz, pyrite, and arsenopyrite veins in gabbros also occur in similar settings, for example in the Sutakiki field in central Guadalcanal (Arthurs, 1979).

Geologically, the most intriguing situations, potentially important because of their predictive value in exploration, arise when the older basalt or gabbro, and younger "granite" interaction, takes place at a depth. Under these circumstances metals may be released and migrate to the higher crustal levels to accumulate as veins, stockworks or replacements chiefly along faults. This may have taken place in the Gaspé Peninsula, where Cambrian basalts and gabbros are buried under a thick pile of slate, graywacke and minor limestone and intruded by Devonian diorites and granodiorites. The Madelaine Cu sulphide deposit (Girard, 1969) may have formed by accumulation of so released copper. It is a chimney-like, 60° south-plunging mass of hornfelsed slate, criss-crossed by chalcopyrite and bornite veinlets localized along fissure planes. Some porphyry copper deposits may have similar source relationships.

10.2b. MASSIVE THOLEIITIC BASALT FLOWS, ARC AFFILIATION

10.2b.1. Geology

Thick, massive and very homogeneous sequences of Mesozoic and Cainozoic tholeiitic subaerial basalt to basaltic andesite flows changing into submarine basalts have developed in several areas along the Pacific continental margin of the Americas (e.g. in Alaska, Queen Charlotte and Vancouver Islands, Oregon, Baja California, Peru and Chile). Probable Paleozoic equivalents are known on Novaya Zemlya Island, in the Urals, in Kazakhstan and in Kuznetsk Alatau in the U.S.S.R.; in the Komodor Islands, on Guam in the Pacific and elsewhere (Table 10-2). The subaerial basalt equivalents usually form dense volcanic piles of multiple, uniformly thick flows, with amygdaloidal, scoriaceous and often oxidized (brownish-red) flowtops. Interflow sediments rarely exceed 10% and are represented by arkosic sandstone, shale, siliceous shale or chert. The recorded occurrences are fairly close to the "green" plateau basalts, described in

Table 10-2. Massive tholeiitic basalt flows, minor gabbros, sediments, near continental / marine facies boundary; mineralized example localities (units)

LOCALITY	AGE	GEOLOGY	MINERALIZATION	REFERENCES
Nikolai Greenstone Wrangell Mts., S.E.Alaska	Tr_{2-3}	Subaerial to submar.tholeiit. basalt flow compl.3,500 m thick Amygd.sections,prehn.-pump. metam.,topped by limestone	numerous nat.Cu,cc,born.,cp, dissem.,veinlets; possible source rock to Kennecott Cu, porph. coppers	Jones et al. (1977)
Karmutsen Group, Queen Charlotte Isl., Canada	Tr_{2-3}	over 4,000 m thick,massive amygd.,pillow bas.,breccia, minor limest.,topped by black argill.,limest."Granite" intr.	minor native Cu in basalts; magnetite and Cu skarns: Tasu, 16 Mt Fe, 80 Tt Cu	as above Sutherland Brown (1968)
Karmutsen Group, Vancouver Island Brit.Columbia	Tr_{2-3}	as above, 7000 m thick unit minor interflow carbonate;shale,litharenite; pillow breccia intr.by J-T "granites"	nat.Cu,cp,born,cc,common in terrestr.flowtops of basalts volborthite in interflow sedim. Fe,Cu skarns	Surdam (1968) Northcote and Muller (1972)
Comondú Volcanics, Punta Concepción Baja California,Mx.	Mi_2	massive tholei.to alkal.cont. basalt flows,1,200 m thick, interstrat.contin.sedim.,basalt porphyry intrusions	fracture and stockw.veinlets of Mn oxides	Noble (1950)
La Negra Fm.betw. Antofagasta and Tocopilla, Chile	J	thick subaer.to submar.flows of basalts to bas.andesites, minor interflow sandst. Intr. by J-T diorite,granodiorite	Cu sulphide mantos in flowtops (Buena Esperanza), min. 300 Tt Cu	Ruíz (1965)
Tokrau Synclinorium Centr. Kazakhstan USSR	Cb-Pe	gently dipping flows of basalts grading to trachybasalt	Annenskoe dep.,3 horiz. 2.5-3.5 m thick of dissem.nat.Cu in basalts	Kolesnikov (1975)
Kuznetsk Alatau, Siberia, USSR	Pt_3-Cm_1	spilite-basalt complex,assoc. gabbro, intruded by Cm_{2-3} gr.	native Cu in basalt (Taimetskoe),FeCu skarn,Au,Cu veins	Kuznetsov (1967)

Chapter 31. The submarine volcanics are mostly pillow lavas, pillow breccias, aquagene tuffs, etc., and are often spilitic.

The Triassic Karmutsen Group basalts of Vancouver Island (Monger et al., 1972) resemble oceanic basalts chemically, but have pronounced alkaline tendencies. They are evolutionarily superseded by arc andesites, so they most probably formed within an island arc setting. The Miocene Comondú volcanics of Baja California, on the other hand, appear to be related to the initial rifting phase of the Gulf of California. The basalts are occasionally intruded by dikes and sills of basalt porphyry of equivalent composition, believed to be magma feeders. On Vancouver Island and in the Wrangell Mountains of Alaska, the outpouring of basalts was followed by an interval of quiescence, during which limestones formed on top of the basalts.

10.2b.2. Mineralization in basalt units

Fig. 10-3 shows the principal mineralization styles. The basalts invariably have high trace copper contents (150-300 ppm), that were further redistributed and locally accumulated during the load metamorphism. The highest trace-copper contents have been reported in thinly layered amygdaloidal flows in the Buttle Lake area on Vancouver Island (Surdam, 1968). Native copper associated with abundant hematite and minor chalcocite, bornite, calcite, quartz, epidote, prehnite, and pumpellyite is widespread in amygdules, in fractures and along the flowtops, but with the exception of the deposits between Tocopilla and Antofagasta, Chile, in subeconomic quantities. In the Nizina area, Alaska, nuggets of native copper derived from the amygdaloids were abundant locally in glaciofluvial sediments.

In the Buena Esperanza Mine south of Tocopilla, Chile, stratabound copper mineralization is confined to the altered tops of Jurassic basalts to basaltic andesites, and 28 mineable parallel horizons of average thickness 7 m are distributed within a stratigraphic interval of 270 m. Dominant chalcocite and bornite and minor specularite and chalcopyrite occur as disseminations, veinlets and blebs in prehnite-pumpellyite, chlorite, epidote and specularite gangue (Ruíz, 1965). The mineralization is probably a result of post-volcanic redistribution of the initially high trace-copper content during the load metamorphism, and particularly during the thermal metamorphism due to the emplacement of diorite. The tonnage is of the order of 300 Tt Cu/3%. A string of similar deposits (Mantos de la Luna; Mantos del Pacífico) occurs farther south.

Minor interflow sediments in the Karmutsen Basalt, Vancouver Island, occur in the upper half of the thick lava sequence as well as along the base of the basalts, and are represented by black, laminated pyritic siliceous shales with minor bioclastic limestone (Carlisle and Susuki, 1974). Rose (1973) recorded several occurrences of yellowish-green volborthite, formed by oxidation of the V- and Cu-enriched seams of black tuffaceous argillite near Menzies Bay, Vancouver Island. The best assays gave 1.8% V, but the mineralization

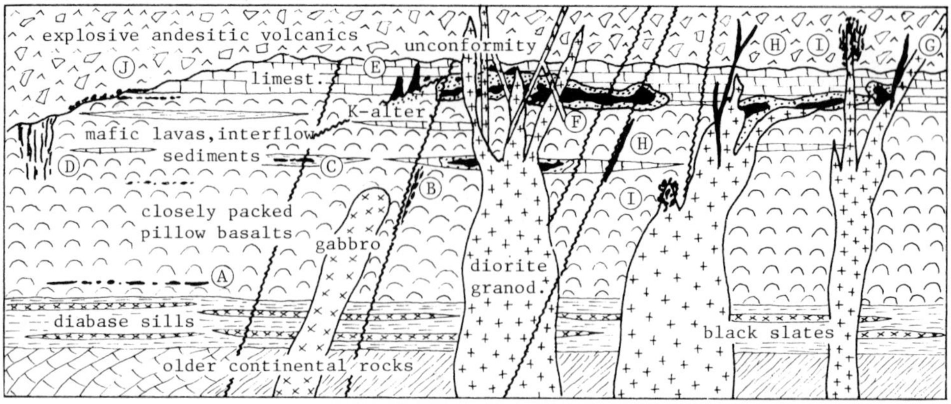

Fig. 10-3. Massive tholeiitic basalt flows near continental/marine facies boundary, principal mineralization styles. (A) native Cu, chalcocite, etc., concordant ore zones along basalt flowtops; (B) native Cu, sulphides, along faults and shears; (C) V, Cu, etc. minerals in interflow sediments; (D) infiltrated ? Mn oxides in basalts under unconformity; (E) chalcocite, etc., replacements in carbonates above basalts (Kennecott); (F) magnetite-chalcopyrite skarns; (G) hydrothermal gold veins; (H) hydrothermal chalcopyrite veins; (I) porphyry copper deposits; (J) native copper nuggets in placers.

was erratic.

Widespread fracture veinlets to stockworks of crystalline pyrolusite, ramsdellite, hollandite, aragonite and calcite, occur in the lightly chloritized Miocene Comondú basalts in the Punta Concepción area, Baja California (Noble, 1950). The volcanics are subaerial tholeiites. The density of the Mn oxide veinlets is proportional to the density of fracturing and their origin is uncertain, although precipitation from descending solutions under unconformity is most likely.

10.2b.3. Interaction of basalt & limestone with deformation & "granitic" intrusions

The Triassic basalts of the Pacific coast of North America were faulted, deformed and intruded by several generations of plutonic rocks in the period between the Triassic and Oligocene. Many faults and shears within the volcanics contain traces of copper (mostly chalcopyrite), irregularly scattered in quartz or carbonate gangue or silicified and carbonatized host rocks. No important ore deposits, however, have been discovered.

Substantially more important interaction-type mineralization formed in the over-intrusion zone of diorite, quartz diorite and granodiorite bodies emplaced into basalts topped by Triassic limestones. The immediate basalt/limestone contact and basal portions of the limestones in the vicinity of the intrusive bodies, are most intensively mineralized by magnetite-chalcopyrite skarns, and a few magnetite-free chalcopyrite skarns. Unique and rich chalcocite "bonanza" bodies formed in the same setting at Kennecott, Alaska.

The skarns have the usual garnet, diopside – hedenbergite, actinolite, calcite, and magnetite composition. Copper is almost exclusively contained in chalcopyrite that in turn is disseminated in magnetite or in the silicate skarn, or accumulates in the late-stage quartz, chlorite, epidote, etc. veins. Some of the skarns are located in the marble several metres to tens of metres stratigraphically above the upper basalt contact (JIB deposit, Queen Charlotte Islands), but most occur along the marble/basalt contact, replacing both rocks (e.g. Texada Island, Cowichan, Tasu; Fig. 10-4), and rarely even the pre-ore phase of the plutonic rocks (Tasu). The intermediate mineralizing intrusions are accompanied by a swarm of pre- and post-ore dikes of diabase, andesite porphyry, diorite porphyry, quartz-feldspar porphyry, etc. Most of the intrusions are Jurassic and some may be comagmatic with a subaerial explosive andesite, dacite, rhyolite association.

The largest single deposit in the Tasu Sound, Queen Charlotte Islands (Sutherland Brown, 1968; 16 Mt Fe/37%, 80 Tt Cu/0.66%), has orebodies that take the form of irregular sheets or mantos that thicken locally, and also vein-like protrusions along faults and porphyry dikes. At Benson Lake, Vancouver Island (Wilkins, 1971; 370 Tt Fe/29%, 21 Tt Cu/1.8%), chalcopyrite and minor bornite are found in a garnet-magnetite skarn near the contact of diorite and andesite sills with a marble unit located near the upper contact of the basalt sequence. Minor basaltic flows and sills are interlayered with the marble itself. The total production and reserves in several tens of skarn deposits in the Insular Belt of British Columbia, has been about 25 Mt Fe, 120 Tt Cu and small amounts of Au and Ag.

Gold-bearing magnetite and copper skarns and quartz, arsenopyrite and pyrite veins (e.g. Zeballos) are found in Jurassic explosive volcanics and sediments intruded by granitoids stratigraphically above the Triassic basalt basement, but close to the granite contact. So are several porphyry copper deposits (e.g. Orange Hill, Alaska; Port Hardy, Vancouver Island).

The Kennecott mines in Alaska (Bateman and McLaughlin, 1920), have produced about 540 Tt Cu and 340 t Ag from ores grading up to 70% Cu. The lodes composed of chalcocite and subordinate covellite, enargite and malachite are fault-controlled replacement masses, veins, disseminations and stockworks. They are all hosted by brecciated Triassic crystalline dolomite overlying the copper-rich Nikolai basalt. The origin of the Kennecott deposits is unclear, and probably

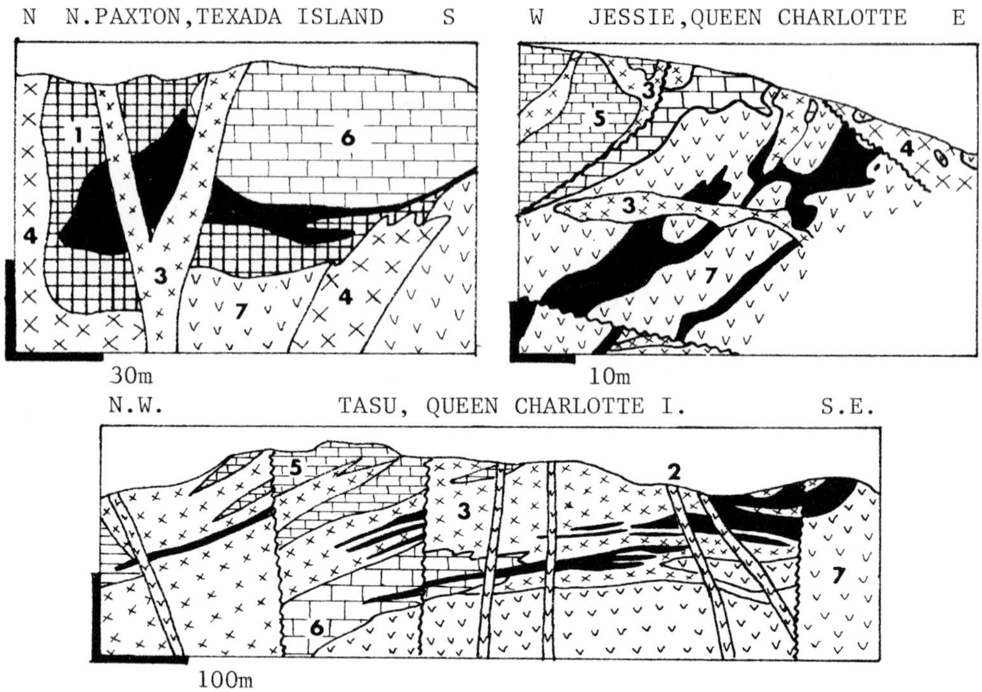

Fig. 10-4. Examples of Fe-Cu skarn deposits near the top of Triassic metabasalts, British Columbia. BLACK: orebodies; (1) skarn; (2) andesite, basalt dikes; (3) diorite porphyry; (4) quartz diorite, diorite; (5) flaggy argillite and argillaceous limestone; (6) massive limestone; (7) greenstone (metabasalt). Triassic supracrustals, Jurassic intrusions. After Sutherland Brown and Merrett (1965) and Sutherland Brown (1968).

involved copper transfer from the footwall basalts by fluids driven by the heat of Tertiary granodiorite porphyry intrusions, their sulphurization and precipitation in a suitable trap. Light potash alteration is developed along the basalt/carbonate contact.

In the Tocopilla region of Chile, the Jurassic copper-bearing basalts are intruded by diabase dikes and several diorite massifs, considered to be of late Jurassic age (Ruíz, 1965). The diorites and their thermal aureole contain numerous hydrothermal veins mineralized by chalcopyrite and minor Cu, U and Mo minerals.

10.3. MASSIVE SUBMARINE BASALTS, SHALE, GRAYWACKE, CHERT, LIMESTONE ASSOCIATION

10.3.1. Fe aspect: Lahn-Dill "type" iron ores associated with varicoloured mafic tuffs ("Schallstein")

"Geosynclinal" submarine basalts and spilites formed in subsiding troughs during the "initial phase of geosynclinal sedimentation" (e.g. in the classical area of the Variscan = Late Paleozoic, orogenic belt of Europe) show a rapid facies transition into marine, non-volcanic sediments (Fig. 10-5). This facies transition is an important control of the occurrence of characteristic bedded iron ores, designated as Lahn-Dill "type" (Schneiderhöhn, 1941; Quade, 1976). These ores, interpreted as of "submarine-exhalative" origin, are associated with mafic volcanics, pyroclastics and sediments. Felsic volcanics that are often members of the host association, are only rarely found directly accompanying the ore.

Petrographically, the mafic volcanics are tholeiitic basalts or sodic spilites; the latter are particularly common. They take the form of submarine pillowed lava flows, vesicular and massive flows, and flows gradational into aquagene breccias, hyaloclastites and peperites. Varicoloured (purple to green), strongly calcareous, coarse to fine fragmentals such as tuffs or redeposited tuffs alternating with coarse basalt agglutinates and basalt flows and black argillites, are a characteristic lithofacies known by the German term "Schallstein". This facies grades laterally and vertically into black or gray volcanic arenites, shales and marls (in the Czech Barrandian Basin); into greenish or red varicoloured shale and sandstone (in the Dill district, Germany); or into initially thin-bedded, gray or red "pelagic" limestones, often topped by massive, light-coloured "reef" limestones. Siliceous shales and cherts may but need not be present.

The most proximal iron ores occur (1) within the volcanic pile itself, usually at the contact of flows and reworked volcaniclastics (e.g. Langenaubach; Krebs, 1966; Fig. 10-6), or (2) within the banded volcaniclastics (e.g. Čilina ore horizon, western Bohemia; Svoboda, ed., 1966). These ores are usually composed of massive hematite, and are often pyritic. In the Dědova Hora mine near Hořovice, Czechoslovakia, barite-ankerite fracture veinlets in the ore and in the adjacent basalts, carried abundant cinnabar. The bedded ores located at (3) tuff/black shale or siltstone contact (e.g. Krušná Hora, Ejpovice; western Bohemia; Fig. 10-6), or (4) within black to gray shale several tens to hundreds of metres from the volcanics (e.g. most ores in the lower Ordovician Šárka Beds, Barrandian Basin, Bohemia; ores of the Nízký Jeseník Mts., north-central Czechoslovakia; Svoboda, ed., 1966), are most widespread and economically most important. Orebodies within and at contact with carbonates (5), are least common.

The iron orebodies usually form stratiform lenses to sheets several meters thick, and range from massive through fragmental (reworked) to granular and oolithic ores. Although hematite is the most common

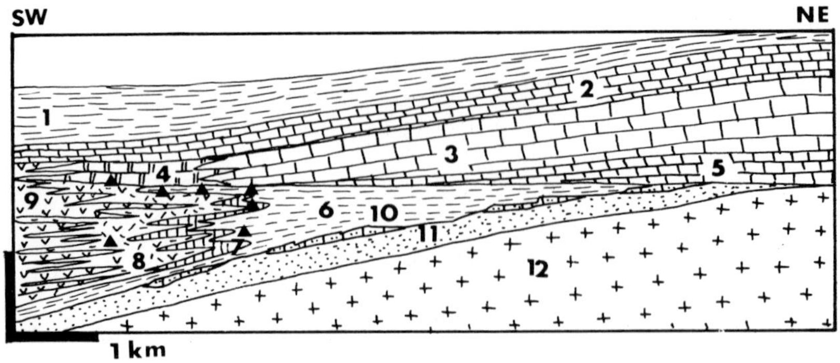

Fig. 10-5. Mafic volcanic to sedimentary facies change progression common in the Variscan orogenic belt of Europe, showing the setting of iron ores. Based on the Devonian facies of northern Moravia, Czechoslovakia. Modified after Chlupáč and Svoboda, in Svoboda, ed. (1966).

Upper Devonian: (1) siliceous shale with radiolarites; (2) chert, cherty limestone; (3) light, massive coral limestone; (4) thin-bedded gray limestone; (5) Stringocephalus limestone. Middle Devonian: (6) calcareous shale; (7) thin-bedded limestone; (8) basalt tuff; (9) basalt or spilite; (10) sandy limestone; (11) basal conglomerate, sandstone. (12) Lower Paleozoic crystalline basement. BLACK TRIANGLES: common setting of iron deposits.

mineral, magnetite, chamosite and/or thuringite (Lahn-Dill region, Nízký Jeseník); magnetite, quartz; magnetite, thuringite, siderite; siderite and chamosite, etc., varieties are frequently present. The grade of shipping ores ranges from about 20 to 40% Fe, and individual orebodies are of small to medium size (several hundred thousand tons to 10 Mt of ore). Fortuna, the largest deposit in the Lahn district, contained 10 Mt ore with 35% Fe. The aggregate district tonnages of these ores range from several tens to about 100 Mt ore.

In the 500 km long, north-east trending Uda River-Shantar Islands mineralized zone of eastern Siberia (Shkolnik, 1972), several hundred deposits and occurrences of siliceous iron and manganese ores are associated with Eocambrian and/or lower Cambrian submarine basalts and cherts, closely comparable with the Lahn-Dill "type". The economic potential is estimated to be several billion tons of Fe and several million tons of Mn. Table 10-3 lists the example areas of this mineralization style.

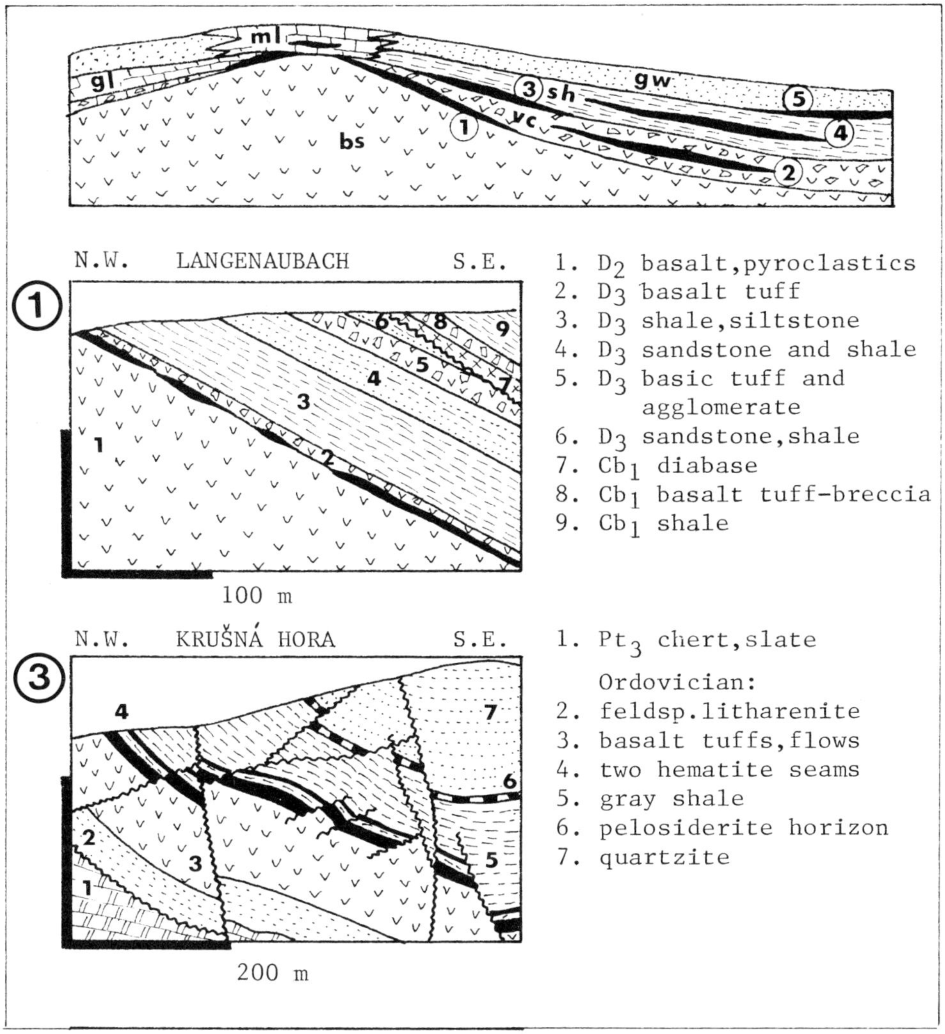

Fig. 10-6. Iron ores associated with basalt-sediment facies transition. TOP: general scheme; (bs) basalt; (vc) volcaniclastics; (sh) shale; (gl) gray, thin bedded limestone; (ml) massive limestone; (gw) graywacke. Numbers in circles indicate mineralization styles, two of which are shown below. Langenaubach mine, Dill district, W.Germany; from LITHOTHEQUE, after Krebs (1966); Krušná Hora, Czechoslovakia; after Petránek (1974).

Table 10-3. Lahn - Dill "type" iron and associated Mn ores affiliated with varicoloured mafic tuffs ("Schallstein"); example localities (regions)

LOCALITY	AGE	STYLE	GEOLOGY	MINERALIZATION	REFERENCES
Lahn-Dill district, West Germany	D-Cb	n-m mass.	basalt (spilite) flows grade to tuffs, chert, shale, limest.; minor keratophyre	banded silic. hemat. to siderite stratiform orebodies 43 Mt Fe/35%	Quade (1976)
Elbingerode Complex Harz Mts., Germany	D_3-Cb_1	sub-grsch mass.	amygdal. basalt, tuff, shale limest., qtz. keratophyre tuff	silic. hemat., siderite lenses on top of spilite pillow lava or tuff; 5 Mt Fe/40%	Zitzmann, ed., (1976)
Komárov, Čilina ore horiz., Barrandian Bas., Czechoslovakia	Or_1	n-m mass.	basalt flows, tuffs, grading to shale	mass. to oolit. hematite, minor siderite, at volc.-sedim. contact; 3 ore hor. 8-25 m thick; 28 Mt Fe/27%; cinnabar	Ilavský et al. (1976)
Nízký Jeseník Mts., c. Czechoslovakia	D_{2-3}	zeol. grsch mass	basalts, tuff, grading to shale, limestone; minor keratophyre	several ore layers at differ. strat. positions; hemat., thuringite, magn., siderite 15 Mt Fe/30-36%	Svoboda, ed., (1966) Ilavský et al. (1976)
Yeravna, 150 km n.w. of Chita, Siberia, USSR	Cm_1	grsch mass	spilite, keratoph., tuff, slate, limestone	massive, banded qtz., magn., hemat., sider., pyr. stratif. lenses, access. Pb-Zn; 80 Mt Fe	Smirnov (1968)
Uda R.-Shantar Isl. zone, E. Siberia, USSR	Pt_3-Cm_1	grsch mass scst	meta-basalt, spilite, tuff, Schallstein, jasper, slate, graywacke	sheet to lens-like stratif. banded magn., hemat., quartz, thuring., chamos., Mn minerals XO Bt 40-46% Fe ore, 35 Mt Mn	Shkolnik (1972)
Hitaka belt (partly) n.e. Hokkaido, Jap.	J_3	grsch	m-basalt, Schallstein, chart, jasper, carbonate	bedded quartz-hemat., rhodochr., piedmont. body	Suzuki and Ohmachi (1956)

10.3.2. Chert, shale, graywacke, carbonate / basalt, gabbro, minor ultramafics association: Mn and minor base metals aspect

Lithologic associations dominated by widespread ribbon chert interbedded with cherty argillite, argillite, minor volcanic sandstone and mafic volcanics, and intersected or bound by faults (sutures) filled by serpentine tectonites, gabbros, blueschist metamorphics, etc., have been interpreted as "oceanic" and this topic has already been partly treated in Chapters 7 and 8. The late Paleozoic Cache Creek-Bridge River Assemblage in British Columbia (Monger, 1977) has been selected as an example area, because of the generally low degree of metamorphism (sub-greenschist facies; Fig. 10-7).

This assemblage forms a linear, north-west trending belt up to 1,400 km long and 75 km wide, in which three principal facies have developed. They are (1) chert-argillite, (2) mafic volcanics, and (3) carbonate. Here, the emphasis is on facies (1). In this, the chert is usually gray or black, locally (in the proximity of volcanic accumulations) red or varicoloured. It is a ribbon chert, marked by numerous chert layers a few centimetres thick separated by argillite laminae. The chert is slightly manganiferous, and black Mn oxide coatings or dendrites are conspicuous on the surface of exposed outcrops. The thicker interbedded argillite sections are dark gray to black, slightly pyritic, usually encrusted by secondary gypsum and jarosite on exposed surfaces. Dark gray limestone forms pods in both the argillite and chert.

Basaltic pillow lavas, breccias and hyaloclastites usually have gradational contacts with the sediments through a wide range of tuffs (both varicoloured "Schallstein" and grayish-green to black tuffs), volcaniclastics (breccias, dark green "graywacke", argillite), and volcanichemical sediments. The mafic volcanics are often interbedded with, or contain lenses of limestones, and all the fragmental volcanic rocks tend to be strongly calcitic. Although the Cache Creek Assemblage lacks significant metallic occurrences, the chert-associated "bedded siliceous Mn ores" widespread in comparable sequences elsewhere, could be expected.

The Carboniferous to Permian Chichibu foldbelt of Japan is more intensively metamorphosed and closely equivalent to Cache Creek (greenschist facies); phyllite and metachert are the most common rocks. In several regions of Honshu (e.g. in the Ashio area: Hamayokogawa, Ippo, Nakanoyama mines; in the Kitakami Mts.: Nodatamagawa mine), numerous conformable bodies of Mn oxides, carbonates and silicates accompany lenses of metachert. Rhodochrosite, Mn-ankerite and rhodonite are the most common minerals, followed by braunite and hausmannite. The purely manganese lenses and beds are transitional into Fe-Mn orebodies (e.g. Kunimiyama mine, Shikoku). The ore is composed of small particles of manganiferous hematite and magnetite finely dispersed in chert and in meta-pyroclastics. The ore grade is 5-10% Mn and 25-30% Fe (Saito, ed., 1960). Bedded Mn ores in chert have a global distribution (e.g. the Ordovician and Pennsylvanian units of Nevada; the Bugulygyr

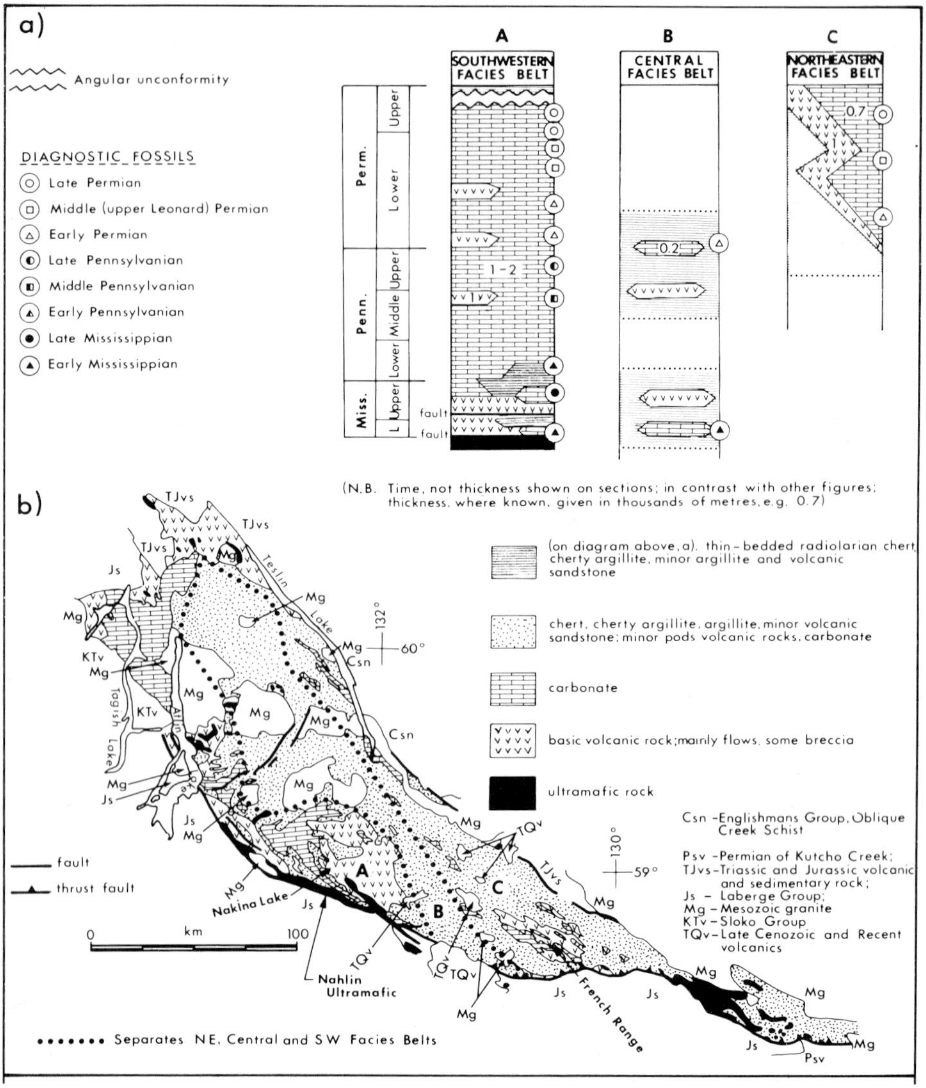

Fig. 10-7. The late Paleozoic Cache Creek-Bridge River assemblage of British Columbia and Yukon: example of chert, argillite, minor carbonate, basalt, etc. "oceanic" association. From Monger (1977), courtesy Canad. J. Earth Sciences.

Formation of the Urals, and the Campbellton Sequence of Newfoundland; Table 10-4).

TECTONIC AND INTRUSIVE INTERACTION

Cherts, argillites and carbonates adjacent to serpentine and/or blueschist assemblages along major sutures, frequently host cinnabar impregnations, fracture coatings and replacements (e.g. along the Pinchi Fault, British Columbia; Chapter 8). Hydrothermal occurrences (veins, veinlets and derived placers) occur in the same setting (e.g. near Manson Creek, British Columbia) particularly when sodic granitoids (trondhjemites) are also present (e.g. the Bridge River field; Chapter 28). The important Carlin, Nevada, deposit of "invisible" replacement gold (Chapter 28) is located in the tectonic footwall (under an overthrust) of the Ordovician chert, black slate, basalt-composed Valmy Formation, and this unit may have been the source of gold. Antimony deposits often occur in similar setting.

Thermal metamorphism of the "syngenetic" ores in cherts (Mn oxides and carbonates) usually produces a silicate assemblage of Mn minerals (rhodonite, dannemorite, spessartite, tephroite, etc.) and generates magnetite from an earlier hematite. The contact metamorphic Mn/Fe assemblage is common in the Japanese Chichibu belt (e.g. Kaso, Taguchi, Nodatamagawa, etc. mines). Hamachi (1962) reported occurrences of uranium in the Japanese thermally metamorphosed Mn cherts, in either dispersed, or uraninite form. Pyrrhotite, Ni-Co arsenides and high trace Mo and V contents are associated and are suggestive of an early syndepositional enrichment in the host sediments.

Several important hydrothermal-plutonic deposits in the Chichibu belt (Akenobe, Ashio, Chichibu, etc., fields) are hosted by the chert, phyllite, metabasalt assemblage, and metal contribution of the supracrustals to the evolving superimposed, "granite"-related hydrothermal system is a possibility. This seems to be strongest in Ashio (Nakamura, 1970), where the "Kajika" orebodies are peneconcordant zones of siliceous massive chalcopyrite, pyrrhotite, sphalerite, chlorite, etc. hosted by a sedimentary chert bed close to the contact with the copper-mineralized Cretaceous felsic stock. The "Kajika" is suggestive of a hydrothermally remobilized syndepositional ore. At Mountain City, Nevada (Coats and Stephens, 1968), bedded black chert and slate of the Ordovician Valmy Formation intruded by a Cretaceous quartz monzonite, host a largely conformable lens of a pyrite - chalcopyrite massive ore that could be an earlier syndepositional mineralization. The last example to be mentioned is the Mount Cleveland Sn-Cu mine in north-eastern Tasmania (Palmer, 1976). There, massive to disseminated pyrrhotite with lesser quantities of chalcopyrite and cassiterite, occur as a series of ? replacement lenses along a chert horizon. The chert is a member of a Cambrian argillite, chert, spilite sequence, located in the proximity of ophiolites. The lithophile element mineralization

Table 10-4. (Meta)chert, argillite, graywacke, lesser basalt, gabbro, ultramafics, carbonate association; mineralized example areas

LOCALITY	AGE	GEOLOGY	MINERALIZATION	REFERENCES
Cache Creek Group, Brit.Columbia	Pe-Cb	4,500 m thick sequence of chert,black slate,limest., fine basalt,diabase,serpent.	cinnab.,Au,Sb interaction along sutures,and near granitic intrusions	Monger (1977)
Valmy Formation, Nevada, USA	Or	ribbon chert, black slate to phyllite, greenstone	small bedded Mn dep.in chert interaction Cu,Au	Coats and Stephens (1968)
Campbellton Sequence, Newfoundland	Cm-Or	mafic pillow lava,breccia, agglom.,overl. by silic. argillite, chert	stratif.bands of rhodonite and rhodochrosite alternate with black silic.argillite	Kay (1975)
Bugulygyr Suite, e.slope of Ural Mts USSR	D_2	150-200 m thick unit of radiolarian chert,jasper,silic. shale, mafic tuffs	numerous stratiform lenses of Mn oxide,carbonate,silicate	Pronin (1962)
Chichibu belt (partly),Honshu,Shikoku, Japan	Cb-Pe	slate to schist,chert,scattered greenstone, Schallstein, limestone	many stratif.bedded Mn oxide, carbonate,silicate dep., some contact metamorphosed	Saito, ed. (1960); Hamachi (1962)
Hall's Formation, n.w. Tasmania, Australia	Cm_1	slate,chert,graywacke,metaspilite,tuff adjac.to ultramaf.mass; intr.D granite	pyrrhot.,cp,cassiterite mass. sulphide lenses in granite exocontact; could be partly remobiliz. Fe,Cu sulphide, Sn added	Palmer (1976)

(Sn, and W, Mo, Bi, present in veins and stockworks in a buried granite body) are without doubt epigenetic, granite introduced. The Fe-Cu sulphides, however, could be remobilized "syngenetic" ores.

10.4. GREENSTONE, PHYLLITE (OR SCHIST), META-ARENITE, CHERT AND CARBONATE ASSOCIATION

10.4.1. Fe/Mn aspect

Progressive metamorphism of the mafic volcanic - sedimentary association treated in Section 10.3.1. results in a rock suite composed of greenstones and greenschists, phyllites, marbles, and metaquartzites. In the Devonian Rakovec Series in the Slovak Carpathians (Mahel' and Buday, eds., 1968) located in the eastern continuation of the Lahn-Dill and Nízký Jeseník iron ore regions discussed above, the volcanics-dominated unit consists mostly of violet or green mafic tuffitic phyllites (metamorphosed equivalents of the "Schallstein"), grading into a massive greenstone on one side, and into green or gray phyllites on the other. There are subordinate lenses of meta-arenite, quartzite (meta-chert) and meta-carbonate. Minor intrusive equivalents are represented by gabbro and gabbrodiorite. An increase in the quantity of the hematite pigment in the violet metatuffs may result in the development of thin (max. 2 cm), conformable lenses of red hematite. Alternatively, hematite or magnetite quartzites may have formed by metamorphism of the siliceous banded chert-iron ores of the Lahn-Dill "type" (e.g. Švedlár, Hnilec). Near Dobšiná, concordant to peneconcordant deposit of crystalline siderite grading into, or interchangeable with, ankerite, have been interpreted recently as "stratiform" (that is, coeval with the submarine volcanism; Ilavský, 1976).

In the lower Cambrian greenschist (including Schallstein), phyllite, black phyllite and carbonate association of the Malyi Khingan Mts., eastern Siberia (Yegorov and Timofeeva, 1972), three large conformable deposits of banded quartz, magnetite and hematite are hosted by metachert near mafic contact. Lenses of braunite, hausmannite and rhodochrosite 1-10 m thick and several hundred metres long, are often associated (Varentsov and Rakhmanov, 1974). Banded hematite-magnetite ores hosted by jaspers at or near the contact with mafic submarine metavolcanics have also been recorded from other places in the U.S.S.R., such as Kazakhstan (Novokhatskii, 1972). The Cambrian iron ores at Burultas, hosted by metamorphosed mafic volcarenites which are interbedded with greenstone, black meta—chert and marble, contain intergrained or interlaminated magnetite-actinolite and quartz. The Ordovician-Silurian sequence at Karashat intruded by Carboniferous granodiorites, contains about hundred 0.5-15 m thick, concordant manganiferous magnetite - hematite lenses (23-57% Fe; 1.4-10.24% Mn), associated with quartzites and red meta-jaspers, interbedded with metabasalts.

The large Hajigak iron deposit in central Afghanistan (1.24 Bt Fe/62%; Bouladon and de Lapparent, 1975) is composed of hematite, magnetite, quartz, chlorite and albite units 5-10 m thick interbedded with Silurian-Devonian greenschists (metaspilites ?).

The economically important stratiform Mn deposits of eastern Rumania (Bistriţa Mts.; e.g. Iacobeni, Dealul Rusului, Brosteni; Ianovici and Borcoş, 1982) are lenses of rhythmically banded Mn carbonate and in places silicates (rhodonite, spessartite) associated with barite and hosted by black meta-quartzites (former cherts) and marbles at and near contacts with greenschists. The age of the Mn ores and their host sequence is now considered to be lower Cambrian.

IRON ORES IN A SUCCESSOR BASIN ASSEMBLAGE, OVERLYING A GREENSTONE-PHYLLITE BASEMENT.

In the Spiš-Gemer Mts. of Slovakia (Mahel' and Buday, eds., 1968), Devonian greenstone-phyllite basement is unconformably overlain by a slightly less metamorphosed marine Carboniferous assemblage, formed after a short hiatus in a "successor basin". There, the Bindt-Rudňany facies starts with a thick polymictic conglomerate, containing clasts derived from the basement, as well as from a granitic terrain. The conglomerate contains lenses of litharenite and shale, and passes upward into graphite, sericite and chlorite phyllites and slates, interlayered with greenstone. In the Dobšiná field, the Carboniferous basal conglomerate contains locally abundant siderite clasts, derived from the mineralized Devonian basement. Stratigraphically higher, concordant lenses of ankerite and siderite are hosted by the basal Carboniferous black phyllites, and it is debatable whether these are stratiform deposits, or metasomatized Ca or Mg carbonates.

The most valuable metallic mineralization in the Carboniferous Bindt-Rudňany facies are, however, fissure-controlled siderite-barite veins in the Rudňany field (Ilavský, 1976). Two major east-west trending, steeply dipping veins range in thickness from several metres to over 30 m, and the largest Droždiak Vein has been mined for a length of 4 km. The vein filling is a medium-to-coarse crystalline siderite intergrown with, or substituted by, barite. Flakes of specular hematite form veinlets and bunches and pyrite, chalcopyrite, Hg-tetrahedrite form irregularly scattered grains, blebs and veinlets in siderite. The production and reserves are of the order of 100 Mt of ore, containing some 31 Mt Fe, 2.5 Mt Mn, 100 Tt Cu, 30 Tt Sb and 1 Tt Hg. The veins continue into the Devonian greenstone and phyllite basement. Widespread siderite, ankerite, and magnesite mineralization is a most characteristic regional metallogenic feature of the Spiš-Gemer region. It is not stratigraphically confined and occur in other units as well.

10.4.2. Massive Fe, Cu, Zn sulphides aspect

The east-north-east trending Sambagawa Metamorphic belt in the northern Shikoku Island, Japan (Banno et al., 1970; Fig. 10-8), is composed of approximately equal proportions of mafic volcanic and sedimentary metamorphics, containing widespread Fe-Cu sulphide mineralization. The rocks are penetratively deformed and metamorphosed within the prehnite - pumpellyite to epidote-amphibolite facies range. Blueschist metamorphics occur locally along sutures. The dominant rock types are gray to brown pelitic and psammitic schist, graphitic schist, chlorite or actinolite greenschist. These enclose narrow bands of quartz schists to quartzites (metacherts) and there are irregular scattered masses of serpentinite and gabbro. The southern border zone of the Sambagawa belt adjacent to a prominent fault has a thick belt of gabbro-amphibolite and metabasalt with numerous thin interbeds of metacarbonates.

The massive and disseminated Fe-Cu (Zn) sulphide deposits sometimes associated with magnetite quartzite (metamorphosed banded cherty iron formation) are usually concordant with schistosity, and most are believed to be stratiform. The immediate hosts to ore are either metasediments (e.g. Ikadatsu mine), metavolcanics (Yokei mine), or both (Besshi mine). The Besshi mine contains the largest deposit, and has been a type locality to one massive sulphide "type". The assignment of various localities to this type in the literature is, however, controversial and of low accuracy.

The Besshi sulphide orebody (Banno et al., 1970; Fig. 10-9), strikes N 60° W, and its dip changes from 45° to 70° N. Its extent is about 2,500 m, the thickness varies from 0.3 to 7 m. The bulk of the orebody is hosted by a black to gray quartz, muscovite, chlorite, epidote, etc. schist with albite porphyroblasts, and locally also by a quartzose and basic actinolite, glaucophane, chlorite, epidote, albite schist. In the central, dynamometamorphically most disturbed portion of the orebody, glaucophane, piemontite-quartz, and calcitic schist are developed. The deposit is conformable with the planar structure of its wallrocks, and its plunge coincides with the high--angle lineation. The ore is composed of pyrite, chalcopyrite and sphalerite, with lesser quantities of pyrrhotite, bornite, magnetite and hematite, and ranges from massive to disseminated. Although discordant tongues of copper minerals often branch out of the concordant layer, they are both in the footwall and hangingwall and appear to be the product of tectonomagmatic remobilization, rather than a footwall feeder.

The Shirataki and Sazare deposits in Shikoku (Takeda, 1970) and Iimori on Honshu (Kanehira and Tatsumi, 1970), are hosted largely by spotted greenschists (=mafic metatuffs). The Okuki mine (Imai, 1960; Fig. 10-9), on the other hand, consists of a series of ore lenses aligned along a mobile horizon of a green schist, sandwiched between a massive greenstone and metagabbro. The richest orebodies were formed by the plastic flow of ore into the axial zones of folds during metamorphism. In addition to the sulphide mineralization,

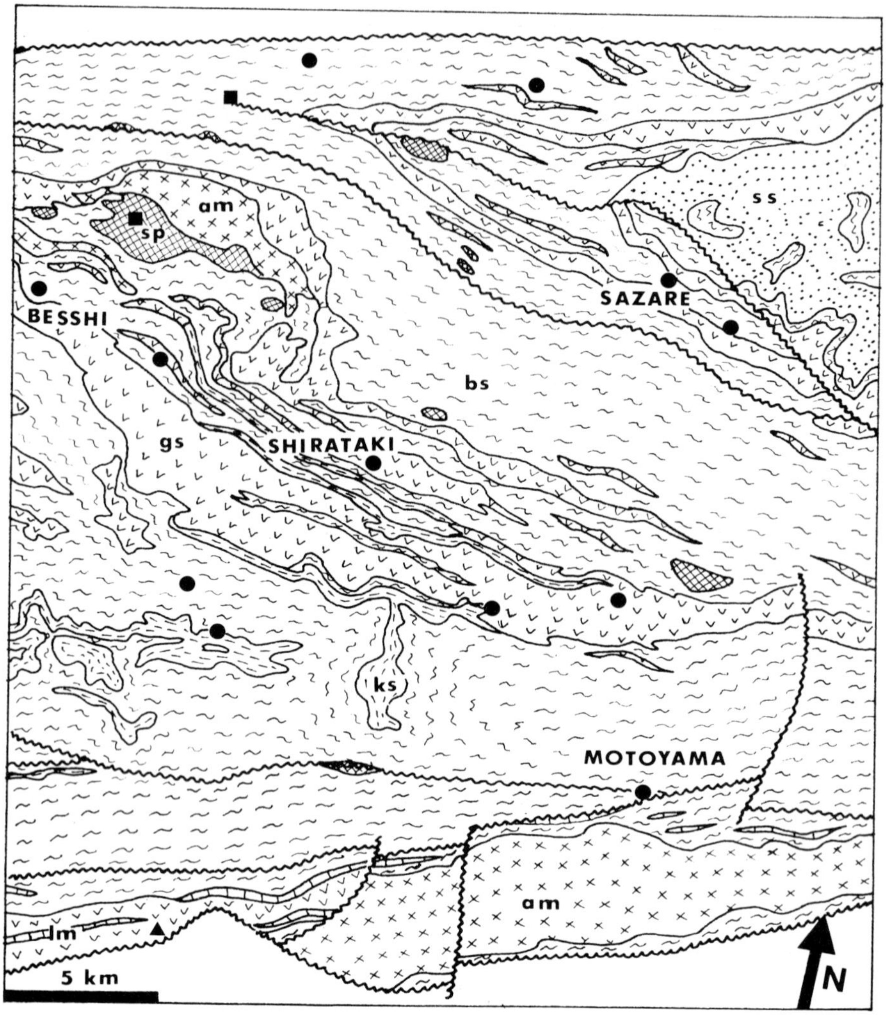

Fig. 10-8. Portion of the Carboniferous - Permian Sambagawa Belt, Shikoku Island, Japan. LEGEND: (am) amphibolite; (sp) serpentinite; (lm) limestone; (ks) quartz schist; (gs) greenschist; (bs) black schist; (ss) "sandstone" schist. Symbols of ore deposits and occurrences: dot = Cu; square = chromite in serpentinites; triangle = Mn in cherts. From LITHOTHEQUE, after Geol. Survey of Japan, geol. map 1:200,000, sheet Kochi,1958.

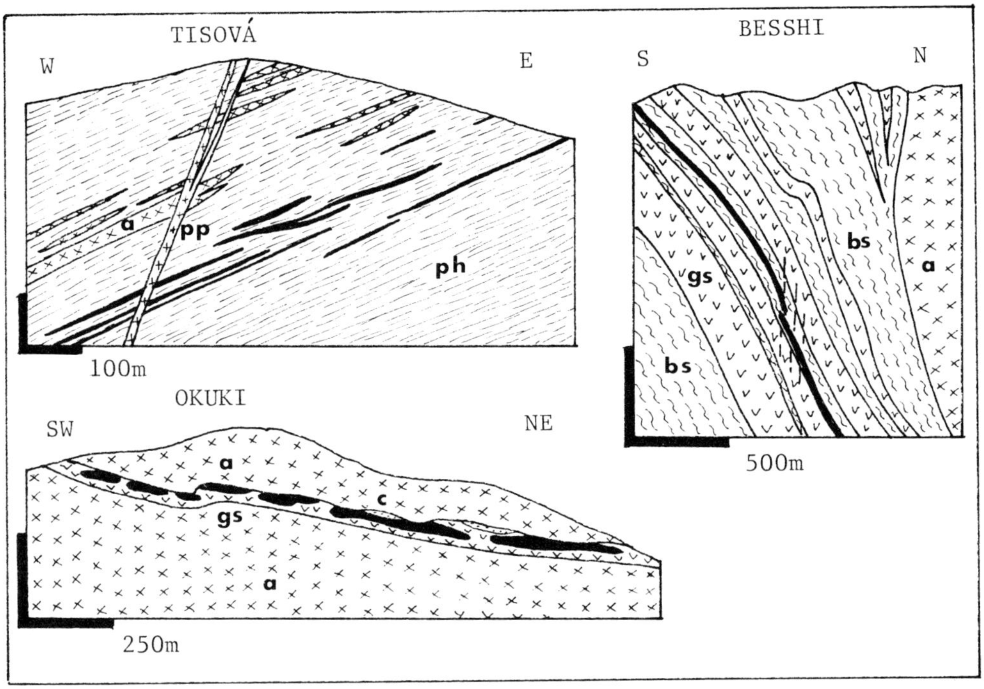

Fig. 10-9. Examples of massive Fe-Cu deposits in a greenstone and schist (phyllite) association. (a) amphibolite or massive greenstone; (gs) greenschist; (ph) sericite-chlorite phyllite; (bs) black schist; (c) metachert; (pp) granodiorite porphyry. From LITHOTHEQUE, after Chrt et al.,(1972); Banno et al. (1970), and after Imai (1960).

the Sambagawa belt is enriched in Mn. The garnet porphyroblasts in schists have up to 10% MnO, and there are scattered, small stratiform lenses of Mn silicates and carbonates enclosed in the siliceous schist.

Local mafic metavolcanic piles in the Chichibu belt on Honshu, Japan, frequently contain small pyrite - chalcopyrite massive orebodies of the "Kieslager" type. Several such small orebodies occur in the vicinity of the Akenobe hydrothermal vein field.

Additional terrains and localities listed in Table 10-5 (Girilambone and Copper Mine Range Beds, N.S.W., Australia; Acton Belt, Quebec; the Cambro-Ordovician schists in the Kraslice-Klingenthal region in the Erzgebirge; Fig. 10-9), are closely comparable with the Sambagawa belt in both the geology and mineralization. In Girilambone, Suppel (1974) reported the presence of pyrite, pyrrhotite, chalcopyrite-rich stringer or veinlet zone in chlorite-altered metamorphics, and a significant remobilization controlled by the axial cleavage. In other regions, such as the Caledonides of Scandinavia, the Grauwackenzone of Austria and elsewhere, most of the local mineralized associations also contain felsic metavolcanics; so that these areas are treated in the following sections although individual localities might qualify for inclusion here.

This applies in particular to the Lökken deposit in central Norway, which is one of the largest massive sulphides in the Caledonides (Rutter et al., 1967; Fig. 10-10). There, a 4 km long zone of several separate, fine-grained massive pyrite, minor sphalerite, chalcopyrite orebodies is parallel with the fold axes of the enclosing penetratively deformed lower Ordovician greenstones and chlorite schists. Lökken is a frequently quoted example of massive sulphides (e.g. in Franklin et al., 1981), because of the presence of black cherty schists ("vassskis") or reddish jaspers, interbanded with pyritic and magnetitic bands, in the meta-basalts stratigraphically above the main massive sulphide lens. The footwall stockwork is indistinctly developed, and substituted by small lenses of albitic quartzite, usually interpreted as quartz keratophyre.

With the increasing metasedimentary / metavolcanic ratio, this association grades into the "schist belts" many of which have a "flyschoid" character, and are treated in Chapter 16. They, too, contain massive sulphides,but they are much rarer than in the presently discussed sequence. Even so, the majority of the massive sulphide orebodies in the "schists" occur within or at contacts, or in the vicinity of the "accessory" metavolcanic bodies anyway.

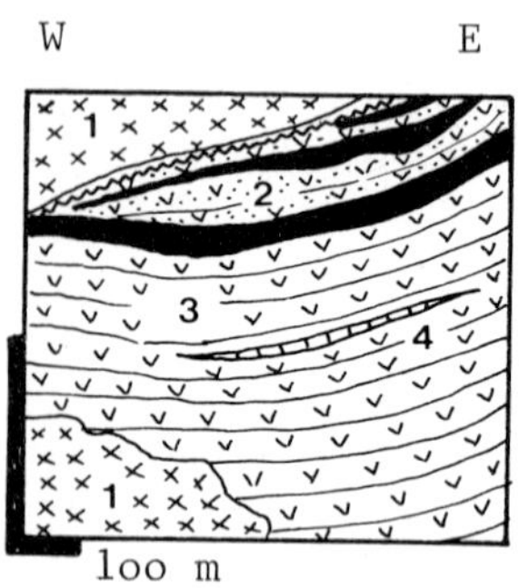

1 Cm_3-Or_1 coarse metagabbro
2 greenstone to chloritic schist (meta-basalt); minor schist interbeds
3 sodic greenstone (meta-spilite)
4 jasper and jasper conglomerate

All the metavolcanics and metasediments are members of the Cm_3-Or_1 Støren Group

Fig. 10-10. Lökken, central Norway; mafic metavolcanics-hosted massive pyrite, minor chalcopyrite, deposit. From LITHOTHEQUE, based on data in Rutter et al. (1967).

10.4.3. Greenstone and phyllite (schist)-hosted gold belts

Elongated, penetratively deformed belts of steeply dipping gray to black slate or phyllite interbedded with meta-graywacke, greenstone, greenschist, minor quartzite, marble and serpentinite, strongly resemble in style a portion of the Precambrian greenstone belts. Characteristically they occur on flanks of major linear deep-seated multiphase batholiths such as the North American Coastal and Sierra Nevada Batholiths and are intruded by minor satellite plutons. Numerous strike faults and sutures, some with serpentinite slices, tectonically subdivide these belts and usually also form boundaries with the adjacent facies. The recent geotectonic interpretations usually place similar belts into the fore—arc (accretionary prism and mélange), or back-arc (marginal sea) domains. Slices of the oceanic lithosphere may be included.

Many "gold belts" (elongated terrains with numerous occurrences of a "bedrock" (=primary) as well as placer gold mineralization parallel with the regional structural grain), are members of this association (Fig. 10-11; Table 10-6). The Mother Lode Belt in Sierra Nevada, California, is an easily accessible and thoroughly studied example. The Belt is a N.N.W. trending, eastward dipping slice of metamorphics, traceable for at least 190 km along the strike, between Mariposa and Georgetown (Fig. 10-12). It is situated between a former mélange on the western (outer) side (Duffield and Sharp, 1975) and a magmatic arc (Sierra Nevada) on the inner side. The gold mineralization is most widespread in the late Jurassic Mariposa Formation: a dark gray, lustrous slate interbedded with sheared graywacke and greenstone, adjacent to the Melones Fault. Most rocks are strongly cleaved approximately parallel with axial planes. There are small plutons of hornblende gabbro and diorite parallel with the structural grain (Clark, 1964). The gold deposits have been described in detail in the classical memoir by Knopf (1929), who distinguished the following mineralization styles: (1) quartz veins, (2) orebodies in mineralized country rocks, (a) mineralized greenstones, (b) mineralized schists, the latter comprising of (b1) hard slates and schists laced with gold-bearing veinlets, (b2) soft schists replaced by ankerite.

(1) The quartz veins are usually of the shear lode variety, that is subparallel with the local structural grain. Despite this, the dip of most veins is less steep than the cleavage, within the range of 50-70° (Albers, 1981). The vein filling is a coarse milky-white quartz, which forms lenticular masses as much as 17 m thick and 2,200 m long. At the end of such lenses the orebody often frays out into a stringer zone, consisting of slate penetrated by a large number of parallel quartz stringers. Wide zones of quartz stringers usually also envelope the solid quartz ore lens. The vein quartz hosted by slates is usually ribboned, with widespread screens of slate. Large quartz masses in greenstone hosts are, on the contrary, homogeneous and clean. Ribboning, a grayish tint and the presence of sulphides distinguishes the gold-bearing quartz from the locally abundant "secretion" milky quartz variety. The vein quartz is associated with beige ankerite, sericite and albite and occasionally the conspicuous

Table 10-5. Massive and disseminated Fe,Cu,Zn sulphide deposits in greenschist-schist (or phyllite) association; example localities or areas

LOCALITY	AGE	GEOLOGY	MINERALIZATION	REFERENCES
Prince William Sound, S.Alaska	Pc	weakly metam.flysch of graywacke, slate,interbedded with pillowed metabasalt (greenstone)	100 occur. of mass.py,po,minor cp,sf, in peneconc.lenses in metasedim., rarely volcanics; 97 Tt Cu, 3.86 t Au	Moffit (1954)
Acton Belt and Huntingdon mine, E.Townships,Quebec	Cm_1-Or_1	greenschist,metagraywacke, marble, peridotite	mass.to dissem.stratabound lenses of po,py,cp,magn. 20 Tt Cu/2%	Béland et al., (1962)
Lökken,Central Norway	Or_1	greenst.,chlor. schist, minor keratoph.lenses; intruded, overthrusted by gabbro	4 km long zone of mass.sulph. lenses conform. with fold axes; py,lesser cp,sf; pyritic and Mn-ox.stratif.horiz. in hangingwall; 400 Tt Cu/2% 360 Tt Zn,320 t Ag, 4 t Au	Rutter et al., (1967)
Joma, Grong distr. Central Norway	Or_1	greenst.,metachert,phyllite, keratoph. metatuff, serpent.	med.cryst., mass.to dissem. peneconc. py,cp,sf lenses 289 Tt Zn/1.7% 216 Tt Cu/1.27	Oftedahl (1958)
Kraslice-Klingenthal,Erzgebirge Mt. Czechoslovakia-East Germany	Cm-Or	phyllite interb.with greenstone,minor quartz, graphite phyllite	several concordant massive and dissem. po,py,cp lenses 3 north-south horizons	Chrt et al. (1972)
Altyn Têpe,Dobrogea eastern Rumania	Pt_4	greenschist, metatuff, chert phyllite	fine grained, massive to disseminated py,cp peneconcord. bodies alternate with bedded magnetite iron formation. Min. 4 Mt Fe, 100 Tt Cu	Superceanu (1967)

Sambagawa Belt, N.Shikoku Isl., S.W. Honshu, Japan	Cb-Pe	greenschist, schist, graphitic schist, silic. schist, minor meta-chert, gabbro, serpentinite; local blueschists	numerous scattered peneconcordant massive sulphide bodies; Mn oxide, silicate, carbonate lenses	Banno (1970) Takeda (1970) Imai (1960) Kanehira and Tatsumi (1970)
Besshi mine		black schist, lesser greenschist, silic. schist	py, lesser cp, sf; 1.14 Mt Cu/3.8%	
Makimine mine		black phyllite	py, lesser cp, sf, po; access. Ni-Co sulphides, thermally metam. by "granite"	
Shirataki mine		greenschist, pelitic schist	po, lesser cp, sf, bornite 66.5 Tt Cu/1.33%	
Okuki mine		greenschist	py, lesser cp; 24 Tt Cu/4.5%, 1.4 t Au/3 ppm	
Iimori mine		greenschist (metaspilite)	py, lesser cp, sf, bornite 32.5 Tt Cu/1.3%	
Girilambone and Copper Mine Range Beds, N.S.W., Australia		quartz micasch., phyllite, greenschist, minor quartzite, gabbros, serpentinite	peneconcordant massive sulphide lenses in Girilambone (76 Tt Cu/2.5%), Tottenham and Hermidale	Suppel (1974)

GOLD BELTS - GENERAL SCHEME

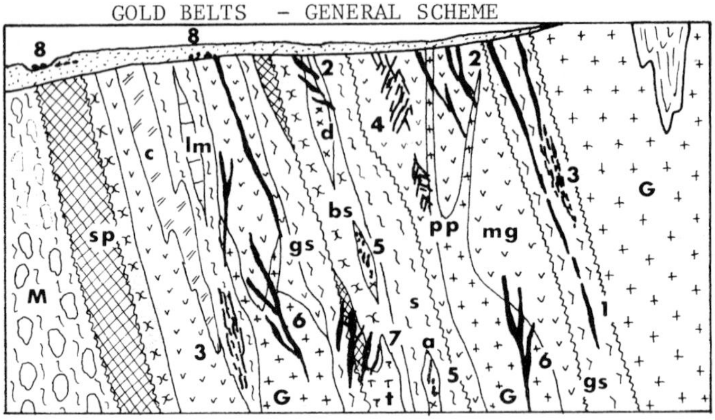

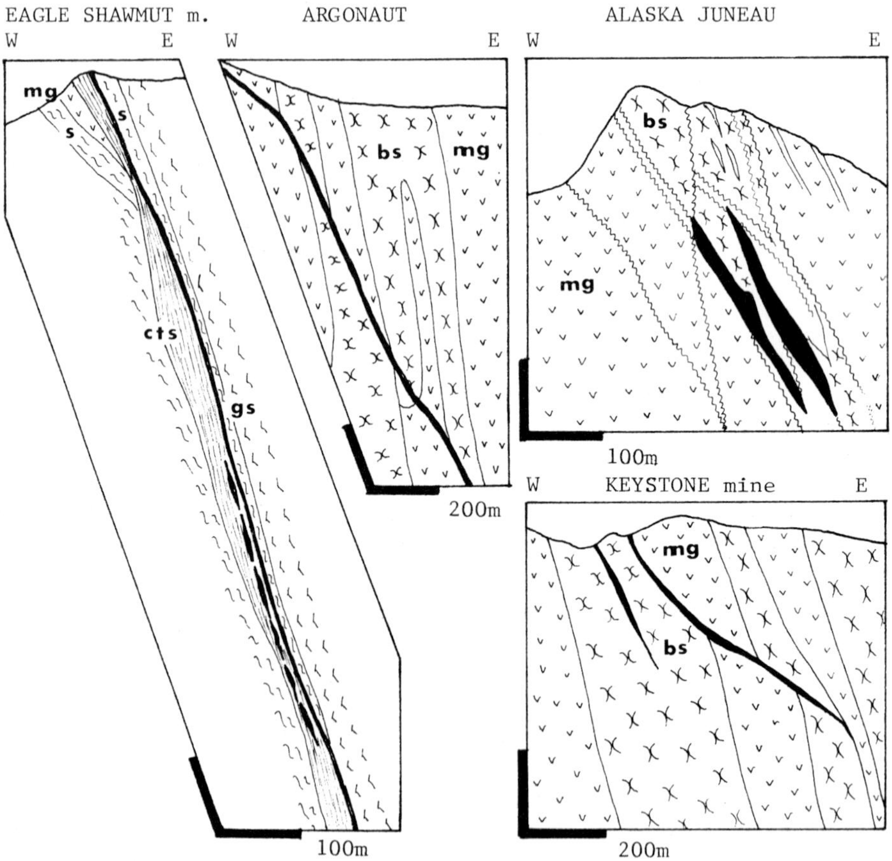

green fuchsite. Pyrite is the most common sulphide, followed by arsenopyrite and rare chalcopyrite, galena, sphalerite. Gold occurs mostly as free gold, although petzite has occasionally been found. The average ore grade of the Mother Lode deposits was about 9 ppm Au, fineness 839-899. The wallrock alteration is relatively light: bleaching in greenstones, and ankerite impregnation or veining in the slate. In the latter case, however, the slate is not discoloured.

(2a) The mineralized greenstone ("gray ore"; e.g. Fremont, Keystone, Bunker Hill mines) consists of bleached ankerite, sericite, albite and quartz impregnated or veined metabasalt that also contains 3-4% of pyrite and arsenopyrite. The grade averaged 7 ppm Au, and this ore style was usually adjacent to large quartz vein systems.

(2b) The "mineralized schists" (e.g. Melones, Carson Hill mines) are schistose (sheared) metavolcanics, now having the characteristics of amphibolites and chlorite schists. They usually carry a low-grade (about 3 ppm Au) mineralization in a pyrite-sericite altered zone, ramified by quartz and ankerite veinlets. The gold is contained in the pyrite, which has the form of veinlets, thin stringers and scattered cubes. In the "soft schist" the metavolcanics were largely replaced by ankerite.

Albitites form dikes and small intrusive-looking masses of a white rock resembling aplite, composed almost exclusively of albite, with minor riebeckite. These rocks are, in places, slightly auriferous, but attempts to mine them as a gold ore were largely unsuccesful (e.g. the Willieta mine near Jacksonville).

Knopf (1929) pointed out that the vein quartz was derived from the

---

◁■

Fig. 10-11. Mineralization styles in the "Gold Belts" hosted by elongated, steeply dipping slate (phyllite), greenstone, serpentinite, etc. Phanerozoic association. MINERALIZATION STYLES: (1) lode parallel with schistosity or shears; (2) crosscutting veins; (3) schist zones veined by parallel stockworks of veinlets; (4) altered and disseminations, veinlets-mineralized massive greenstones; (5) disseminations in albitites, albite diorites, quartz-eye porphyries; (6) ore veins in granitic rocks; (7) veins in a complex setting in the vicinity of major faults; (8) elluvial and alluvial placers. LITHOLOGIC ABBREVIATIONS: (a) albitite; (bs) black slate, phyllite; (c) metachert; (cts) chlorite-talc schist; (d) diorite, gabbro; (G) "granitic" batholiths; (gs) greenschist; (lm) limestone; (M) mélange; (mg) massive greenstone; (pp) porphyry; (s) slate, phyllite; (Sp) serpentinite; (t) trondhjemite.

Eagle Shawmut, Argonaut and Keystone mines in the Mother Lode Belt (Jurassic and late Paleozoic supracrustal rocks) are from LITHOTHEQUE, after Knopf (1929). Alaska Juneau mine (Triassic supracrustals) is after Wernecke (1932).

Table 10-6. "Gold Belts" hosted by steeply dipping penetratively deformed phyllite (schist, slate) - greenstone, minor serpentinite, association

LOCALITY	AGE OF		GEOLOGY	MINERALIZATION	REFERENCES
	HOSTS	ORE			
Juneau gold belt S.E.Alaska	Tr	J-Cr	green and black slate to phyllite, greenst., minor limest., intr. by gabbro and dior. sills along edge of Coastal Batholith	quartz, calcite, py, po, minor ars, cp, ga, in all rock types; shear veins grading to stringer lodes; fissure veins; placers. 200 t Au	Spencer (1906) Berg and Cobb (1967)
Klondike Goldfield Yukon; "Nasina Series" host assoc.	PZ_1?	Cr-T_1	phyllite to schist, greenstone, graph. phyllite, gray quartzite, serpent.	numerous small qtr-carbonate veins // shears, Au free and in pyrite; placers	Green (1972)
Bridge R. (Bralorne) distr., S.W. British Columbia	Pe,Tr	Cr_3-T_1	chert, silic. slate, greenst. serpent., intruded by trondhjemite // faults	30 persistent qtz, py, ars, Au, minor tellurides, fault filling veins in all rocks, rel. to intrus. 132 t Au/16 ppm 25 t Ag/3 ppm	Joubin (1948)
Klamath Mountains, N. California	Or-J	J_3	slate, phyllite, chert, marble, greenst., intr. by gabbro, dior., trondhjem.	variety of qtz, carb., py, ars. veins in all rocks, along shears, faults, coeval with plutons. 185 t Au, most from placers	Albers (1966)
Mother Lode belt, Sierra Nevada Foothills, California	PZ, J_3	J-Cr	gray to black slate to phyllite, metagraywacke, greenst., serpent. // parallel strike faults dior.-granod. intrusions	variety of quartz, anker., py, ars, gold veins along shears in black phyll., greenstone; fissure veins in plutons; placers; 803 t Au, 150 t Ag	Knopf (1929) Clark (1964) Duffield and Sharp (1975)

Piedmont gold belt, S.Appalachians,USA	Pt_3-PZ_1	PZ_1	greenschists,chloritic phyllites,silic.schist deeply weathered,intrud. by gabbro,dior.,granod.	Au-bearing massive py,lesser cp,sf,lenses; gold-qtz veins along shears. placers 8o t Au	Tull and Stow (1982)
Jílové-Bohuliby, W.Czechoslovakia	Pt_3	D-Cb	greenst.,chlor.sch.,slate qtz.keratoph.,plagiogran. intr.by dior.- granod. pluton	qtz,py,ars,minor scheel.,Au, tellurides shear and fract. veins to stockworks in supracrustals; min. 12 t Au	Sattran et al. (1966)
Berezovsk and Pyshma near Sverdlovsk, the Urals,USSR	Cm-D_3	Cb	greenst.,schist,serpent. intr.by qtz porph.dikes, gabbro,plagiogranite, granodiorite	qtz,seric.,tourm,py,Au,telluride veins in altered ultramaf.and intr.dikes min. 110 t Au	Borodaevskaya and Rozhkov (1974)
Kochkar goldfield, the Urals, USSR	D_2-Cb_1	Cb_1	greenst.,slate,carbonate, ultramafics, intr.by plagiogran.and granite dikes	large number of fiss.fill. and replac. qtz,ars,py,cp, tetr,Au,tellurides veins in blastite along fissures in intrus.dikes. Min.63 t Au	as above
Westland and Nelson Prov.goldf., South Island,New Zealand	Pt, Cm-S	PZ_3 MZ	phyllites and schists interb.with greenst.,intruded by granitic plutons	qtz,carbonate,py,ars,gold veins along shears;most gold from placers. 210 t Au	Williams (1965)
Otago Schist Belt, South Island New Zealand	Pe-J	J	chlor.and biot.schist (metagrayw.,slate),greenstone,intr.by granitic plutons	qtz,carb,py,ars,minor scheel. shear and fissure lodes, placers. 240 t Au	Williams (1965)

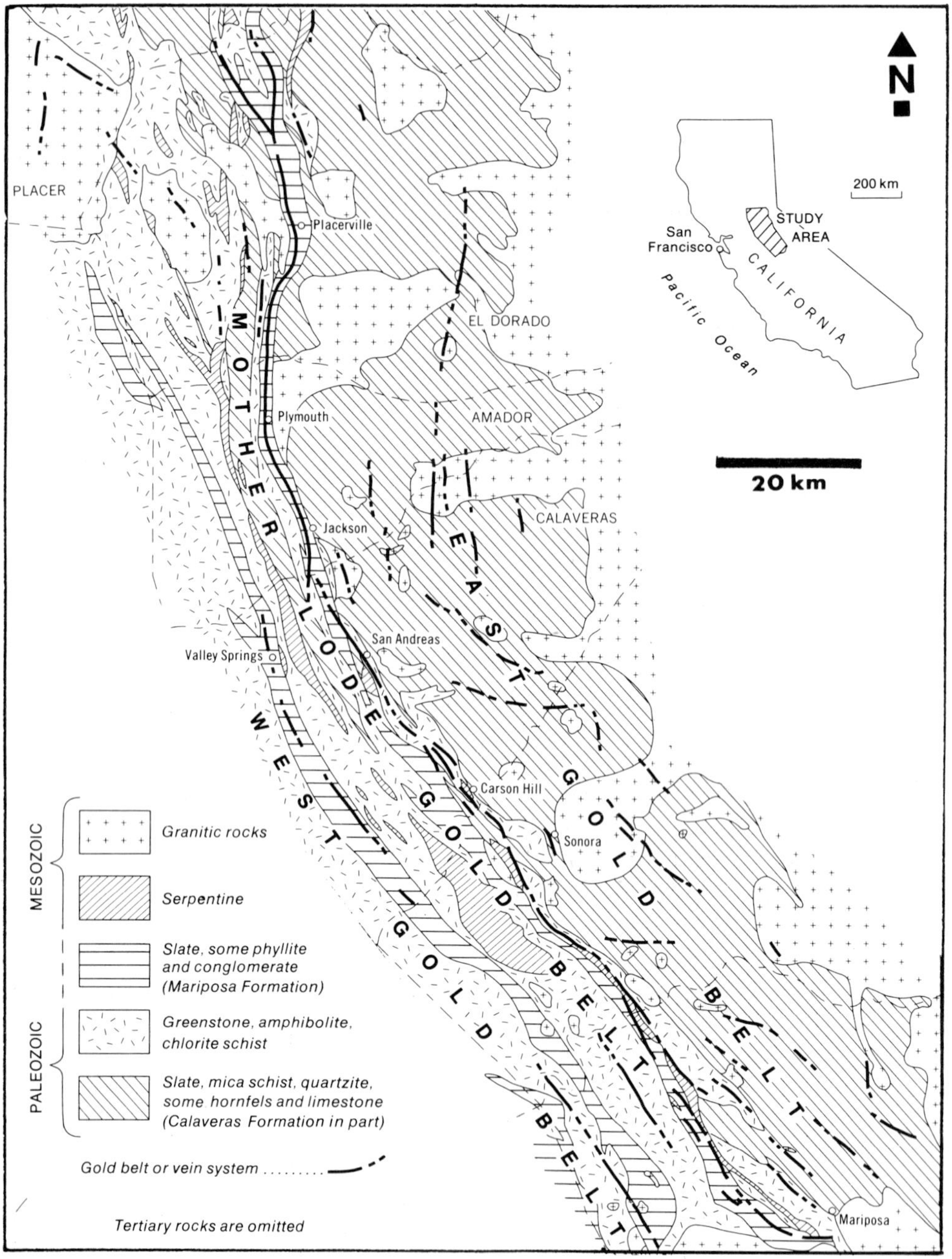

Fig. 10-12. Mother Lode and adjacent gold lode systems, Sierra Nevada, California. After Clark (1970) from Boyle (1979); courtesy of the Geological Survey of Canada.

wallrocks during ankeritization, and attributed the gold to a deepd-seated magmatic source. Derivation of all the components (including gold) from the metallogenically fertile broad host assemblage during deformation and "granite" development in depth, appears most likely.

The Klamath Mountains gold province (northern California; Albers, 1966) is comparable with the Mother Lode belt in style and age. The "primary" gold mineralization occurs as (1) steeply dipping lodes in black slate; (2) steeply dipping fissure veins in greenstone under slate; (3) gently dipping vein-like deposits along faulted contacts between slate and greenstone; (4) steeply or gently dipping veins in the "birdseye" or quartz porphyry dikes and (5) steeply dipping veins in schists and miscellaneous rock types. All the auriferous bodies have quartz and usually also ankerite gangue, pyrite and arsenopyrite, and are fracture-controlled, postdating the emplacement of gabbro, tonalite and diorite batholiths. The intimate association of many gold lodes with the birdseye porphyry (=diorite or dacite porphyry containing subrounded feldspar and/or quartz phenocrysts) has been accepted as evidence of a hydrothermal granite-related origin.

The Juneau Gold belt in South- Eastern Alaska is about 160 km long and 6 km wide, a N.N.W.-trending band of Triassic metamorphics, located along the south-western margin of the Coast Range Batholith. The major rock types are slate, black slate (correctly phyllite), greenstone and greenschist, and there are minor lenses of marble and quartzite. In the immediate vicinity of the Juneau cluster of mines, massive diorite sills and sheets peneconcordant with the schistosity intrude the supracrustal metamorphics. The gold deposits are hosted by almost all the rocks listed and, as in the Sierra Nevada, there is a considerable variety of mineralization styles. About 200 t of gold was produced from the belt, of which over 80% came from the Juneau city area. Spencer (1906) recognized discrete schistosity- concordant ("shear") veins grading to "stringer lodes" (stockworks of veinlets parallel with schistosity in a narrow shear zone); rare cross-cutting veins, impregnated rock masses, mixed deposits, and placers.

The largest producer, Alaska Juneau mine (Spencer, 1906), is an example of a black phyllite ("slate") hosted low-grade "bulk" mineralization. The biotited-ankerite altered and/or albitized slate alternates with greenstone and is intruded by several diorite stocks. The ore zone is 530 m long and 250 m wide, and consists of numerous veinlets of quartz distributed throughout the zone. Most of the veinlets and veins are parallel with the cleavage strike of about N 70° W, but cross the cleavage dip. Major quartz and minor calcite, sericite, pyrrhotite, pyrite, sphalerite, galena, chalcopyrite assemblage carries 14 to 35 ppm gold, but the actual "ore" including the finely intergrown slate bulk mined in a quarry, had an overall grade of about 1.23 ppm Au. Altogether, 88.5 Mt ore containing 67 t Au and 24 t Ag was produced. In the adjacent Perseverance mine, galena was abundant in the veinlets and 3.3 Tt Pb was recovered.

The mineralized altered rock masses in the Juneau belt are well represented in the Douglas Island mines, where greenstone contains large amounts of auriferous pyrite in the chlorite-sericite altered and carbonatized rock. The amount of sulphides is in the region of 2-4%, and 28.8 Mt of ore with a grade of about 3 ppm Au was produced. This corresponds to 32.5 t Au and 3.3 t Ag. The altered greenstone appears to be independent of the fractured-filling veins that also occur in the area.

Diorite dike-hosted orebodies in the Juneau belt are infrequent and developed in some of the Douglas Island mines. There, the diorite and albitized diorite is impregnated by auriferous pyrite, or cut by quartz, calcite and pyrite veinlets.

MINERALIZATION DUE TO INTERACTION OF SUPRACRUSTALS AND "GRANITE"

First of all, it should be pointed out that the Mother Lode-style gold deposits reviewed above are, without doubt, themselves a product of a post-depositional tectono-metamorphic interaction to which the generation and emplacement of intrusive bodies in depth is a complementary process. But with few exceptions the orebodies treated earlier are not hosted by, or located at, immediate contacts with the granitoids. This, however, can be expected and indeed in most of the areas considered the "slate"-hosted gold belts are interrupted or terminated by intrusive plutons, and many such plutons contain hydrothermal gold deposits.

The rich gold veins of Grass Valley-Nevada City (California) are in a granodiorite body interpreted as a batholithic cupola, located in the northern termination of the Mother Lode belt (Albers, 1981). The Berners Bay field in the northern termination of the Juneau belt, has gold bearing quartz-pyrite lodes hosted by a gabbrodiorite stock intrusive into the schistose greenstones (Spencer, 1906).

Several gold fields or deposits are hosted rather indiscriminately by a complex rock association involving altered "slate" and greenstone as well as serpentinite, gabbro, diorite and widespread albite, chlorite, silica and carbonate metasomatites. These are usually developed along important faults and probably correspond, in part, to subduction mélanges, modified within the period of their residence in the continental crust by tectonism and plutonism. Example localities include the Bridge River goldfield in British Columbia; the Berezovsk, Pyshminsk, Kochkar and other fields in central Urals and others.

In the Bridge River (Bralorne-Pioneer; Barr, 1980) camp, quartz veins averaging 1 m in width fill very persistent faults and tension fractures in highly deformed Permian metacherts, greenstones and serpentinites intruded by Mesozoic or Cainozoic " soda granites" (trondhjemites) along a major lineament. The ore is a milky white or a ribboned gray quartz, containing 2-3% of auriferous pyrite and arsenopyrite. In the Berezovsk and Pyshminsk fields of the central Urals (Borodaevskaya and Rozhkov, 1974), swarms of subparallel or ladder quartz, sericite, tourmaline, pyrite, galena, chalcopyrite, etc. gold-bearing veins are hosted by a set of numerous quartz porphyry and microgranite dikes emplaced into a schist, greenstone and serpentinite basement. The region has produced at least 200 t Au and is noted for the intensive sericite, pyrite and sometimes feldspar alteration ("berezitization") in the granitoids, and talc, magnesite, quartz, fuchsite alteration ("listvenitization") in the ultramafics.

WEATHERING AND SEDIMENTARY REDEPOSITION

Gold placers are widespread in most of the areas listed in Table 10-6 (except the Bridge River camp, where they were probably removed by glaciation). With very few exceptions, alluvial placers such as those described in Chapter 24, are represented.

10.4.4. Sb (Hg) aspect

In the border region of Austria, Hungary and Czechoslovakia, lower Paleozoic (? Silurian) or ? Cretaceous greenstones and black phyllites host important antimony deposits in a discontinuous belt. In the Malé Karpaty Mts. north-east of Bratislava, Slovakia, the Harmónia Series (Mahel' and Buday, eds., 1968) contains a thick sequence of amphibolites and gabbro-amphibolites near the base. These grade upward into actinolite schists alternating with black graphitic phyllites. Despite the position of these rocks in the roof of a granite-migmatite dome, a special lithofacies of multiple alternation of thin pelitic and psammitic layers in the graphitic rocks succesfully resisted the granitization, so that occasional limestone lenses in the phyllites often contain preserved organic remains. The mineralization consists of stratabound thin lenses and bands of massive to disseminated pyrite dominant within, or immediately along, the greenstone contact. The graphitic phyllites stratigraphically above the greenstones near Pezinok contain stratabound lenses of finely disseminated stibnite. Porphyroblasts of actinolite in the phyllites indicate a volcanogenic component. The ore lenses contain between 1.7 and 2.5% Sb; 0.3-0.8 ppm Au; 5-7 ppm Ag; 0.15-0.3% As; and traces of Zn and Pb. In the nearby Pernek deposit, locally rich but minor quartz, pyrrhotite, stibnite, gudmundite, etc. veins probably formed by remobilization of the stratabound stibnite in the roof of a late Paleozoic granite (Ilavský 1976, Ilavský, ed., 1979).

The stibnite mineralization, probably stratigraphically equivalent near Stadt Schlaining in eastern Austria (Maucher and Höll, 1968; 40 Tt Sb), contains conformable bodies of stibnite disseminations in graphitic phyllite, black quartzite (meta-chert) and chlorite schist, as well as superimposed veins, stringers and stockworks. These bodies may be up to 6 m thick. Tectonite serpentinites are present locally, and are associated with small cinnabar occurrences. Numerous though small showings of stibnite and minor scheelite, arsenopyrite, cinnabar, and other sulphides are hosted by ? Silurian graphitic phyllites and schists interlayered with greenstones in the Kreuzeckgruppe, in the Rabant district and in the Lessnig area of the Austrian Alps (Höll and Maucher, 1976). The same authors and also Höll (1977) report numerous small cinnabar occurrences in the greenstone-schist associations in the Alps.

10.4.5. Magnesite and scheelite aspect

Black to gray phyllites interbedded with greenstone (metabasalt) submarine flows, pyroclastics and reworked pyroclastics in the Paleozoic sequences of the Alps and Carpathians, contain frequent lenses of carbonates. Limestones and dolomites are dominant, but ankerite, siderite and sparry magnesite are widespread. In the Carboniferous region near Jelšava, Slovakia, crystalline magnesite replaces a bedded limestone lens, enclosed in black to dark-gray graphitic slates grading to conglomerates (Mahel' and Buday, eds., 1968).

In the Lanersbach magnesite deposit in Tuxenthal, Austria (Höll and Maucher, 1968; Fig. 10-13), a 3 m thick horizon of black carbonaceous phyllite located on the base of a magnesite-dolomite body and in close proximity to a metabasalt, contains bands and stretched nodules of scheelite. Both the magnesite and scheelite are considered to be stratiform, diagenetically generated, subsequently metamorphosed in greenschist facies, and locally remobilized during the Alpine Orogeny (in Cretaceous). The alternative genetic interpretation of the Alpine magnesites maintains that they are products of a syn-metamorphic Mg metasomatism of limestone and dolomite.

The scheelite occurrence Kleinarltal, Austrian Alps (Höll and Maucher, 1976), has some features comparable with Lanersbach, except that the scheelite is hosted by a Silurian ferroan dolomite interlayered with black phyllite, rather than by a magnesite.

10.4.6. Tungsten (scheelite) aspect

The discovery of the Felbertal scheelite deposit (near Mittersil, Austria; Höll, 1977; 2.25 Mt ore with 0.75% WO_3, the first moderate tungsten mineralization known lacking "granites" in its proximity), focussed attention on its host association of supracrustal metamorphics. The host unit ("Habachseries") is of lower Paleozoic age, was metamorphosed in the Oligocene to greenschist or lower amphibolite facies and is a part of the mantling complex enveloping the crystalline core of the Hohe Tauern gneiss dome. In it, predominantly fine-detrital metasediments (phyllites, schists) with a considerable proportion of black phyllites, are interbedded with subordinate marble and with up to 3,000 m of mafic metavolcanics (chlorite - actinolite amphibolite = prasinite, amphibolite, hornblendite). Rare but characteristic rock members are Cr-rich

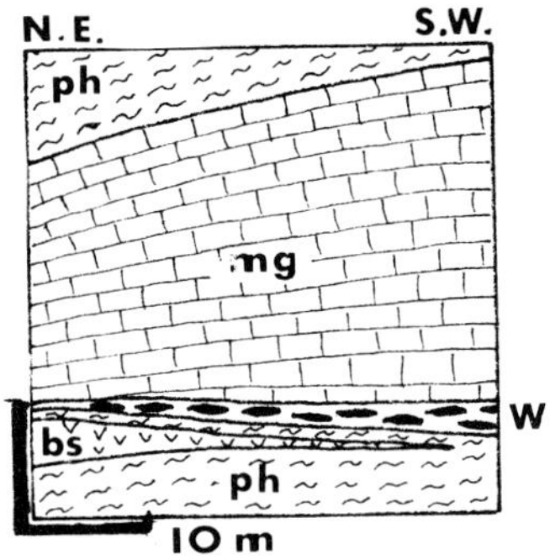

LANERSBACH, TUXENTHAL, AUSTRIA
S_3-D_1 Tuxer Phyllitseries

ph phyllite
mg magnesite-replaced dolomite
bs metabasalt (greenschist)
W lens of black phyllite containing bands and stretched nodules of scheelite

Fig. 10-13. Magnesite and scheelite deposit Lanersbach. From LITHOTHEQUE, after Höll and Maucher (1968).

hornblendites (metapicrites ?), talc - fuchsite schists and serpentinites, reminescent of the komatiitic association. Lenses of black or white metaquartzite, ankerite, marble, etc., may partly represent seafloor hydrothermal sediments or altered seafloor volcaniclastics.

The scheelite mineralization is stratigraphically confined to a 300 m wide section of the metavolcanic unit, in which a variety of the rocks mentioned above repeatedly alternate. Scheelite occurs as fine disseminations (or is substituted by an anomalous W content) in most rock varieties, but mineable orebodies are confined to two horizons of fine crystalline metaquartzites interbedded with hornblendite, about 2 to 25 m thick. There, disseminated scheelite congregates into bands and stringers, and a lesser proportion of it forms scattered porphyroblasts and remobilization veinlets along fractures and joints. The pure scheelite orebodies (in the Eastern Field) are visually extremely inconspicuous. In the Western Field, accessory wolframite, molybdenite, pyrrhotite, pentlandite, sphalerite and numerous other base and rare metal minerals serve as visual indicators of the presence of scheelite and powellite.

Höll (1977) interpreted the mineralization as seafloor-hydrothermal, and emphasized the abundance of syn-depositional and diagenetic fabrics. Numerous additional minor scheelite occurrences in the Eastern Alps are reported in the same paper.

Plate 4. Mount Lyell near Queenstown, Tasmania, an important copper ore field hosted by strongly cleaved and altered Cambrian Mount Read felsic metavolcanics.

"Massive sulphides are the cadillacs of ores".
R. W. Hutchinson

CHAPTER 11

Unimodal Felsic Volcanic-Sedimentary Association

11.1. GENERAL

Lawrence (1977) published a paper entitled "The syngenetic-epigenetic transition, an Australian example", in which he elaborated on the origin of the Mount Morgan pyritic Cu-Au deposit. The same heading would be equally appropriate to this entire chapter, because in only a few other lithologic associations is the interpretation and classification so transitional, uncertain and controversial as it is here. The fundamental problem in the interpretation of the association of contemporary felsic volcanics and the dominantly marine "geosynclinal" sediments, is the scarcity of modern examples. Compared with the widespread submarine basalt and andesite eruptions, wholly subaqueous felsic eruptions are rare, although some have been documented in the Pacific-type continental margins, and in the oceanic spreading ridges.

The first environment is exemplified by the Tuluman Volcano off New Guinea (Reynolds et al., 1980). There, the felsic submarine volcanism started with the production of floating pumice, followed by the development of largely subaerial vent composed of felsic lavas, breccias and tuffs. Recent keratophyres are also known to have formed in the Kurile Islands and in Kamchatka (Yazeva, 1977). The second environment is exemplified by Iceland, where both subaerial and subglacial Quaternary felsic volcanoes occur (e.g. Furnes et al., 1980).

In the ancient marine sequences, the felsic volcanics have two major forms: (1) a relatively distal facies of concordant layers or lenses of pyroclastics, grading into reworked volcaniclastics and ultimately nonvolcanic seafloor sediments. It is believed that most of the pyroclastic material was subaerially erupted and reached the seafloor as an ash-fall or (a very proximal) ash-flow material (e.g. in the submarine ignimbrites and pyroturbidites), as well as an on-land initiated mudflow, stream sediment or a beach sediment. (2) A very proximal near-vent facies of usually brecciated convex bodies (domes) and short flows, usually intensively hydrothermally altered and associated with hydrothermal sediments.

The environmental interpretation of felsic volcanics is greatly complicated by deformation and metamorphism. A greenschist facies dynamometamorphism is capable of causing a far-reaching convergence of the originally submarine (generally "gray"-reduced), and originally subaerial (generally "red"-oxidized) felsic volcanics and associated sediments, so that both might appear as uniformly coloured (greenish, beige) quartz-sericite schists. This is very convincingly demonstrated by the Verrucano facies of the southern European Permian continental felsic volcanics, originally members of the red beds volcanic-sedimentary association, that have lost their "red" appearance in the strongly tectonized and metamorphosed segments in the Alps and Carpathians. Such occurrences are grossly dissimilar when compared with the broadly contemporary, yet undisturbed, pink and red rhyolites that are members of the central and northern European Rothliegende facies continental red-beds (e.g. von Bubnoff, 1956).

It is likely that many felsic volcanic terrains, traditionally treated as "geosynclinal", are partly or wholly tectonized continental volcanics. One such terrain, the Cambrian Mt.Read Volcanics of Tasmania, has recently been interpreted as a product of the largely subaerial "Andean-type" volcanism (Corbett, 1981). If so, the volcanics are metallogenically comparable with, and visually indistinguishable from, the many other similarly mineralized felsic occurrences of the world generally considered as "marine", so that there was no reason for their separate treatment in this book. The major genetic controversy here involves the massive sulphide deposits commonly hosted by the felsic volcanics. If the sulphides are indeed "exhalative" (that is hydrothermal, precipitated on the seafloor), a depth of some 1,000 m of water was necessary to prevent boiling of the metal-bearing fluids; a point difficult to reconcile with the subaerial or nearshore volcanic emplacement. The greenschist-metamorphosed Devonian felsic volcanics, associated with the important iron ore deposits in the Altai Mts., U.S.S.R. (Kalugin, 1972), appear to represent a link between the continental, subaerial ignimbrites and nearshore marine sediments. They are reviewed in Chapter 25.

TRANSITIONS AND OVERLAPS

The "geosynclinal" felsic association is transitional into the subaerial felsic volcanics and their intrusive equivalents, treated in Chapter 26, and into the volcanic red-beds association (Chapter 25). Felsic volcanics similar to, or identical with, those treated here also form an integral part of the bimodal (Chapter 12); andesitic (Chapter 13); and sequentially differentiated (Chapter 14) "geosynclinal" assemblages, and their Precambrian counterparts (Volume 2), and a sharp separation is impossible. Inclusion of many of the example localities in either of the above-mentioned chapters is highly subjective and essentially influenced by the scale (focussing) of treatment. The localities considered in this chapter are predominantly related to felsic volcanic and subvolcanic rocks, but intermediate or mafic volcanics could still be present in "accessory" quantities, or could be widespread in the broader rock unit.

11.2. FELSIC META-VOLCANICS, BLACK PHYLLITE (SCHIST), CARBONATE ASSOCIATION: Fe (Mn) ASPECT

This association is prominently developed in the Ordovician to Carboniferous strata in the eastern Alps (in Austria) and western Carpathians (in Slovakia). The Cambrian to Silurian Gelnica Series in the Spiš-Gemer Mts., eastern Slovakia (Mahel' and Buday, eds., 1968; Ilavský, 1974; Fig. 11-1) represents a good example area. There, epimetamorphosed metasediments (black, gray and green phyllites; black meta-cherts and meta-carbonates) are interstratified with felsic metavolcanics ("porphyroids"; quartz-sericite schists, quartz-augen sericite schists) in an east-west trending arch, 80 km long. The sediments and volcanics are intimately interstratified, and subvolcanic to plutonic equivalents appear to be rare. The mineralization in the Gelnica Series (as is also the case of other units in the Alps and Carpathians) is remarkable for the compositional variety of carbonates (limestone, dolomite, ankerite, siderite, magnesite) that usually have the form of lenses as well as discordant veins. In the past this variety had been explained as a product of hydrothermal metasomatism of the original sedimentary limestones by granite-generated hydrothermal fluids, but the recent opinion in both the Carpathians (Ilavský, 1974, 1976) and Alps (Höll and Maucher, 1976) gives a greater prominence to the "primary" (i.e. syn-depositional) carbonate diversification and origin. The "epigenetic" features are then explained as products of the Alpine remobilization. The latter interpretation has recently been applied to two major operating iron ore (siderite) deposits Nižná Slaná, Slovakia (Ilavský, 1974; 10.5 Mt Fe/32%, 640 Tt Mn/2%); and Erzberg, Austria (Höll and Maucher, 1976; Holzer and Stumpfl, 1980; 214 Mt Fe/39%, 13.75 Mt Mn/2.5%).

At both localities, lenses of a well-bedded and often thinly laminated, finely crystalline siderite interstratified with thin black phyllite partings, change laterally into ankerite and/or dolomitic limestone. Superimposed are younger metasomatic layers, schlieren and lenses of a coarse crystalline siderite, that tend to have a variable but generally low content of scattered sulphides (pyrite, chalcopyrite, tetrahedrite, arsenopyrite) and abundant specularite. Compositionally identical coarse siderite, specularite and sulphide fissure veins form separate orebodies in both the Spiš-Gemer Mts. (e.g. Rudňany), and in the Alps.

The typical ore hosting unit in Nižná Slaná contains all the five principal rock types listed above, that have the form of multiple thin alternations within a stratigraphic thickness of about 50m (Fig. 11-1). At Erzberg (Fig. 11-1), a relatively homogeneous unit of "porphyroids" forms a footwall to a thick lens of mineralized carbonate. Ilavský (1974) reported numerous horizons of stratabound arsenopyrite hosted by the black phyllites in a hangingwall of the Nižná Slaná siderite.

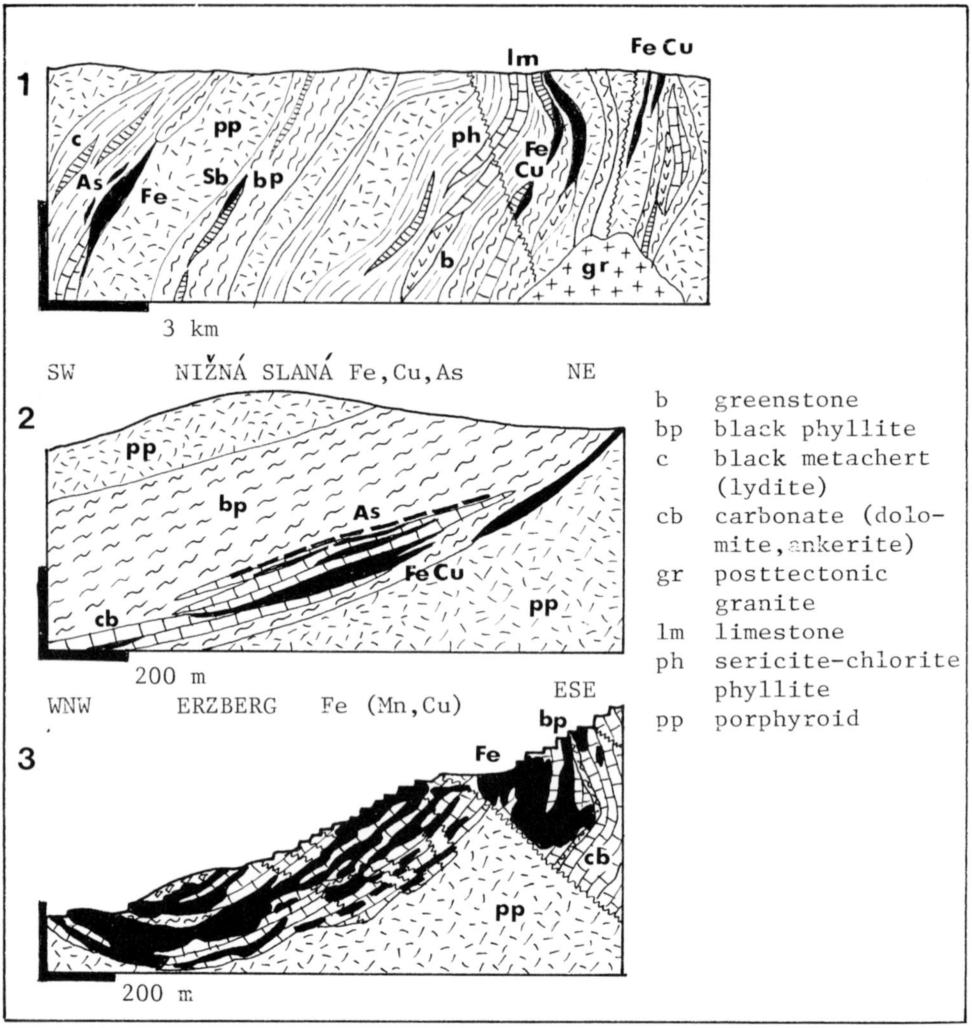

Fig. 11-1. Felsic metavolcanics, black phyllite, carbonate association; Fe (Cu,Mn) aspect. (1) General scheme based on the Cambro-Silurian Gelnica Series, Slovak Carpathians. Fe,Cu: stratabound and discordant vein deposits of siderite (ankerite), lesser chalcopyrite; As: stratabound arsenopyrite; Sb: stratabound stibnite.

(2) Nižná Slaná, Slovakia, siderite-ankerite deposit, in Ordovician to Silurian rocks; simplified after Ilavský (1974, 1976).

(3) Erzberg, Austria, Ordovician to Devonian siderite-ankerite deposit. Simplified after Holzer and Stumpfl (1980).

11.3. Mn ASPECT

Several manganese deposits hosted by the "porphyry-chert formation" of the Soviet geologists are members of this category, and the Durnovsk(oe) deposit in the Salair Range (Siberia; Varentsov and Rakhmanov, 1974; Fig. 11-2) is a representative example. There, sheared and schistose lower-middle Cambrian quartz porphyry body contains interstratified lenses of quartzites (meta-jaspers), some of which are mineralized by conformable layers of massive braunite. The braunite lenses are, on the average, 4 m thick, and traceable downdip for more than 150 m. The porphyry is enveloped by marbles interbedded with felsic meta-volcanics and with thin lenses of a low-grade Mn carbonate ore. The reserves are 120 Tt Mn contained in 20% Mn ores.

11.4. MASSIVE Fe, Zn, Pb, Cu SULPHIDE ASPECT

Massive sulphide orebodies hosted by felsic meta-volcanics and associated sediments typically occur in linear facies belts characterized by rather persistent horizons of alternating gray to black phyllites (schists) and light, originally well-bedded and equigranular quartz-sericite schists (probable original felsic meta-pyroclastics and volcaniclastics). Meta-arenites (graywackes) may be present, and carbonate lenses as well as "cherts" occur in

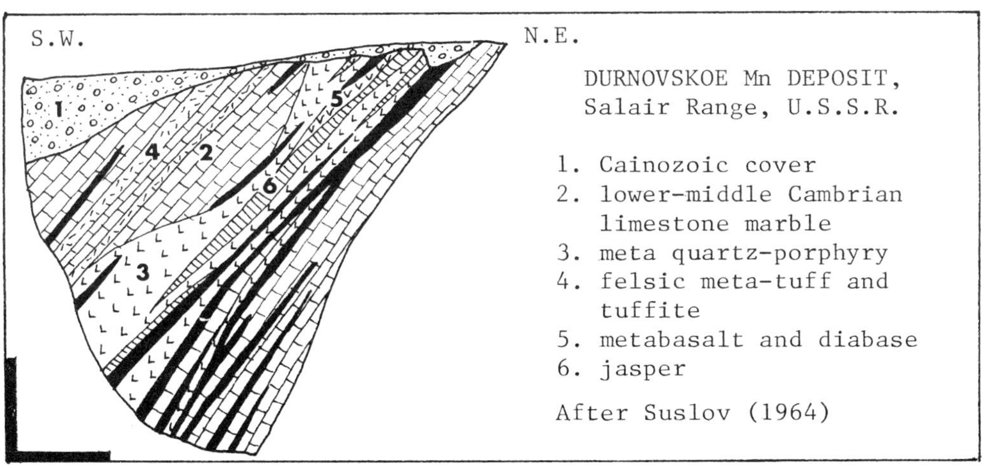

Fig. 11-2. Durnovskoe deposit, an example of felsic meta-volcanics and carbonates hosted Mn deposit.

minor quantities. This association may contain pyritic horizons grading into thin concordant lenses of massive or disseminated pyrite with or without base-metal sulphides, lacking a conspicuous alteration modification in the footwall ("distal type" of Large, 1977). These ores are usually of limited importance.

Significant sulphide accumulations, on the other hand, tend to be associated with localized discontinuous and usually brecciated felsic eruptive centres, found from time to time in the laterally persistent association described above. Such centres are semiconcordant to discordant in respect to the original bedding, although this is not always obvious in the deformed, schistose terrains because of the metamorphic "conformization". The deformation and metamorphism probably also caused partial or complete convergence of the originally potassic (rhyolitic) or sodic (keratophyric) affiliation. The sulphide orebodies, when present, conform more or less to the general "mineralized felsic subaqueous volcanic centre" model (Chapter 6), and can usually be subdivided into the peneconcordant ore lens located within, on top or at a lateral contact of the felsic (meta)volcanic centre, or at a short distance above or laterally (downslope) from it. There, the ore is usually hosted by rocks designated in the literature as "chert", "fine felsic tuff", "metacarbonate", or by a black slate or schist. The fine host rocks could be both "exhalites" (i.e. volcanogenic hydrothermal seafloor precipitates), or "alterites" (i.e. hydrothermally altered solid, or in the time of deposition unconsolidated, rocks). Both varieties are transitional and difficult to distinguish. Typical stockworks of sulphide stringers, veinlets or disseminations in the altered footwall rocks are usually developed. In mineralized regions that contain Fe-Zn-Pb-(Cu) massive sulphide systems, a variety of coeval copper-only deposits or occurrences may be present; these will be discussed in the following section.

Table 11-1 lists selected example localities. Although all more or less conform to the general model and all occur in more or less the same lithologic association as reviewed above, each locality has its own personality and has some features that make it different from the rest. The deposits Avoca and others, are localized in units where metasedimentary schists are abundant and continuous. Rosebery, on the other hand, is hosted by a small metasedimentary lens, enclosed in dominant metavolcanics.

The Rosebery deposit (N.W. Tasmania; Adams et al., 1976), consists of several ore lenses extending along a north-south strike for about 1,650 m, and continuing 800 m down the 45° easterly dip. The orebody consists of a relatively high grade, fine grained massive sulphide (sphalerite, pyrite, galena, chalcopyrite; minor As, Sb, Bi, Ag, Au minerals), zoned in the model pattern. A semi-separate barite, sphalerite and galena lode occurs stratigraphically above the main orebody. The orebodies are hosted by tuffaceous metasediments (now sericite-quartz schist) containing frequent silicified zones, and are topped by a cleaved black slate. Massive pyrite or pyrite-chalcopyrite rich ore marks the base of mineralization, and is in sharp contact with the chlorite-altered footwall pyroclastics interpreted as felsic ash flows.

The Rudnyi Altai metallogenic belt of the eastern Kazakhstan (Bublichenko et al., 1972; Cherpasov et al., 1972), contains the largest accumulation of massive sulphide deposits in the U.S.S.R. The ores are considered largely contemporary with, and hosted by, middle and upper Devonian marine sediments (meta-shale, siltstone, arenite, minor limestone), dominantly felsic pyroclastics, and numerous subvolcanic intrusions of rhyolite and keratophyre porphyries. The largest Leninogorsk (Ridder-Sokol'noe) ore field (Fig. 11-3 and 11-4) is in a N.N.W.-trending fault-bounded graben filled by calcareous meta-siltstones, layered rhyolite lavas, felsic tuffs, breccias and volcaniclastics, resting on a basement of Ordovician metamorphics. The supracrustal suite is gently dipping to almost flat-lying, which greatly facilitates the environmental interpretation. The faulting is intensive, however, and the suite is intruded by numerous small bodies of quartz-feldspar porphyries and albitophyres. The subvolcanic intrusions were emplaced in several phases, although the bulk appears to be contemporary and comagmatic with the volcanics, with which the intrusive bodies often merge imperceptibly. The most prevalent form of the intrusions is a small dome, stock, stub or lens, and most are intensively altered (mostly silicified).

The massive, disseminated and stockwork Zn-Pb-Ag (Cu) mineralization is intimately associated with the felsic intrusions and occurs in their roof region, or within thick felsic lenses that do not seem to be continuous to any depth. Massive sphalerite, pyrite, galena and minor chalcopyrite, tetrahedrite, arsenopyrite, etc., bodies form peneconcordant lenses on the top and flanks of the subvolcanic domes, laterally intertonguing with silicified felsic pyroclastics and topped by unaltered meta-pelites or felsic volcaniclastics. Downward, they grade into typical stockworks rich in pyrite (with high Au content) and chalcopyrite, enclosed in quartz-sericite altered rocks. Heavy silicification often producing masses of almost pure "secondary quartzites" is a characteristic feature of the Rudnyi Altai mineralization.

11.5. Cu ASPECT

Copper-only occurrences and sometimes major deposits are often located in the same units and regions that host the complex polymetallic massive sulphides described above. Many such occurrences are comparable in their appearance with the stockwork or disseminated pyrite-chalcopyrite ores present in the alteration footwall regions under the polymetallic sulphide lenses, others are different. With the exception of copper mineralization associated with minor occurrences of basalts or andesites (these are comparable with the deposits treated in Chapter 10), the bulk of copper orebodies appears to be affiliated with felsics, having probably formed at deeper levels of magmatic systems. There appears to be a complete gradation between such deposits and the "porphyry coppers" (Chapter 28).

Table 11-1. Massive Fe,Zn,Pb,Cu sulphides in unimodal felsic volcanics, slate (schist), black slate, minor carbonate association; selected example localities

LOCALITY	AGE	GEOLOGY,MINERALIZATION	REFERENCES
Ambler distr., Brooks Range, N. Alaska,USA	D_3	numerous peneconcordant mass.sulph.lenses up to 20 m thick in qtz.-muscovite and graphitic schist; py,sf,cp,gal. Arctic deposit only: 1.92 Mt Zn/5.5% 1.4 Mt Cu/4% 350 Tt Pb/1%; 1,505 t Ag/43 ppm	A.Findlay,oral comm.(1980)
Campo Morado, Guerrero,Mexico	Cr_1	4 lentic.peneconc.massive py,sf,cp,gal,lenses underlain by qtz,seric.,chlor.,altered zone,in felsic laharic breccia; slate,grayw.,rhyol.porph.association. 270 Tt Zn/3.12; 180 Tt Cu, 80 Tt Pb; 125 t Ag; 4 t Au	Lorinczi and Miranda (1978)
Avoca,eastern Ireland	Or_3	bedded massive py,lesser cp,topped by sf,cp,minor gal. and Fe dolomite-siderite beds. Dissem.and veinlet py,cp in chlorite altered footwall; assoc.with metarhyol.and breccia, n.e.-s.w. string of vents in dolom.metatuff, black phyllite,dolomite. 605 Tt Cu,bulk in stringer z.	Platt (1977)
Parys Mountain, Anglesey, w.c. Great Britain	Or_3-S_1	lentic. peneconc.bodies of fine massive py,cp,sf,gal, along contact of silicif. meta rhyol.lavas and tuffs (fw) and black to gray slates (hw);gradational to qtz,py,cp stringer zone. 130 Tt Cu/4%	Wheatley (1971)
Tulgheş Series E.Carpathians Rumania	Cm_1	4 major fields: Baia Borsa, Fundu Moldovei, Lesu Urşului, Burloaia.Massive peneconc.py,lesser cp,sf, lenses in chlor. and sericite altered felsic metavolcanics and graphite schists,black quartzites; dissem.py,cp in altered footwall. Min. 1 Mt Zn, 500 Tt Cu	Ianovici and Borçoş (1982)
Khandiza dep., Gissar Range,Uzbekistan,USSR	Cb_{1-2}	peneconcordant massive py,lesser gal,sf,cp sulphide lens in metarhyolite and dacite tuffs,brecc.,underlain by subvolc.felsic vent facies and py,cp stockw.of veinlets	Smirnov and Gorzhevsky (1974)

Rudnyi Altai miner. zone, E.Kazakhstan, USSR	D_{2-3}	variety of peneconc. and discord. massive,vein,dissem. and stockw.dep. related to felsic volc.and subvolc. centres, modified by D_3-Cb deformation and calc-alkal. plutonic emplacement	Bublichenko et al.(1972) Smirnov and Gorzhevsky (1974)
Leninogorsk field (Ridder-Sokolnoe)		massive peneconc.sf,gal,py,minor cp,tetr,barite bodies underl.by qtz,cp,Au rich stockw.in silicified felsic subvolcanic body. Minim. 2.4 Mt Zn, 1.43 Mt Pb, 4.8 Tt Ag	
Zyr'yanovsk field		series of parallel massive sf,py,gal,minor cp lenses along cleavage in silicif.and sericitiz.rhyol. to dacite metavolc. and metasiltstone intr.by qtz keratophyre dome Minim. 327 Tt Zn, 65 Tt Pb, 13 Tt Cu	
Orlovskoe dep.		banded massive py,lesser cp,sf,gal,tetr. body topped by a barite,gal,sf lens; cp rich stockwork in footwall; hosted by meta qtz-keratophyre, intruded by subvolcanic albite porphyry	
Woodlawn, Lachlan belt,N.S.W., Australia	S_3	zoned, banded mass. py,lesser sf,gal,cp body in qtz-feldspar porph., felsic agglom.,tuff,black slate. Small barite,sf,gal lenses in hw, cp rich stockwork in altered fw. 840 Tt Zn/14%; 300 Tt Pb/5.5% 160 Tt Cu/1.5% 528 t Ag	Gilligan et al. (1979)
Captain's Flat, Lachlan belt,N.S.W. Australia	S_3	fine grained, banded py,sf,lesser gal,cp,ars,tetr. are in fine felsic crystal metatuff, assoc.with black slate, metagrayw.,dior.and microdior.intrusions. Silicif.zone in fw. 450 Tt Zn/10%, 270 Tt Pb/6% 30 Tt Cu	as above
Rosebery deposit, N.W.Tasmania Australia	Cm	zoned,fine grained sf,gal,peneconc.mass.lens grades to py,cp basal zone, and barite, Mn carb.rich top zone. Hosted by a meta-siltst.lens encl.in meta-rhyol.tuff; 2.9 Mt Zn/18% 880 Tt Pb/5.5%; 128 Tt Cu; 3 Tt Ag; 49 t Au	Adams et al. (1976)

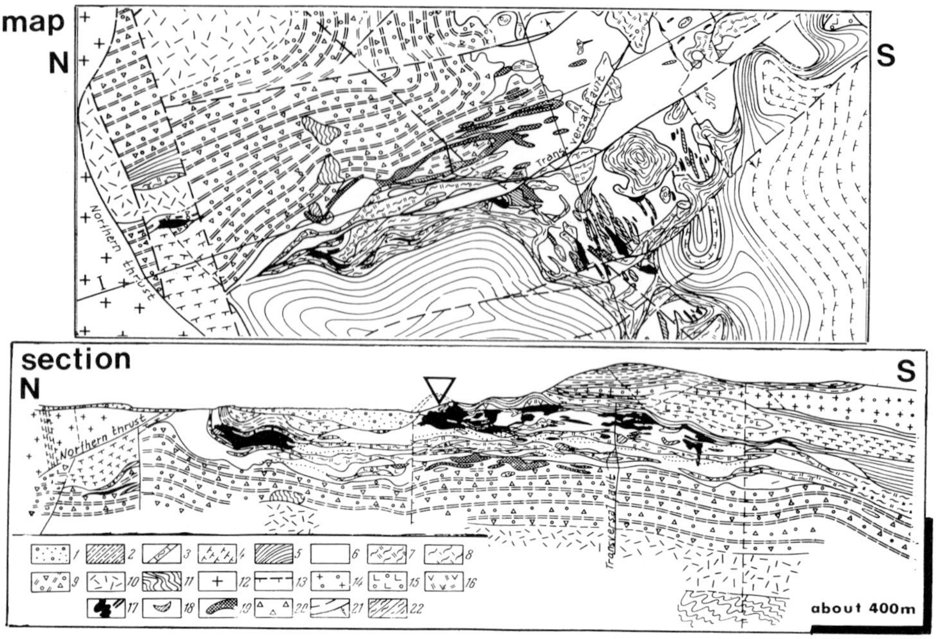

Fig. 11-3. Ridder-Sokol'noe deposit, Leninogorsk, Rudnyi Altai, map of the 13th horizon (top) and a N.-S. section (bottom). LEGEND: (1) Quaternary cover; upper Devonian rocks: (2) limy siltstones; (3) bedded felsite porphyry lavas; (4) felsic tuffs, volcaniclastics; (5) screen of siltstones; (6) alteration microquartzites and sericite-microquartzites; (7) sericite, chlorite, quartz alterites; (8) sericite-carbonate alterites; (9) volcanic conglomerate (lahars ?); (10) quartz porphyry lavas and ignimbrites; (11) Ordovician metamorphics; Devonian to Carboniferous: (12) granodiorite; (13) diabase; (14) quartz keratophyre; (15) amygdaloidal andesite-basalt porphyries; (16) quartz-barite ; (17) Zn-Pb ores; (18) gold-rich ores; (19) Zn-Cu ore; (20) eruptive breccias; (21) faults; (22) dark gray organic rich pelitic sediments. From Cherpasov et al. (1972). Triangle in the section: approximate location of the detailed section in Fig. 11-4.

Central Sequence of the Mt.Read Volcanics (Tasmania) and the Mount Lyell copper field

The Cambrian Mt. Read Volcanics (Corbett, 1981), form a 90 km long, 9-12 km wide north-south trending belt. Within the belt felsophyric lavas and ash flows of subaerial derivation with a predominantly rhyolitic to rhyodacitic and minor andesitic composition are common. Sediments, however, are rare and probably formed in a marine embayment. Although rare, the sedimentary units host all the

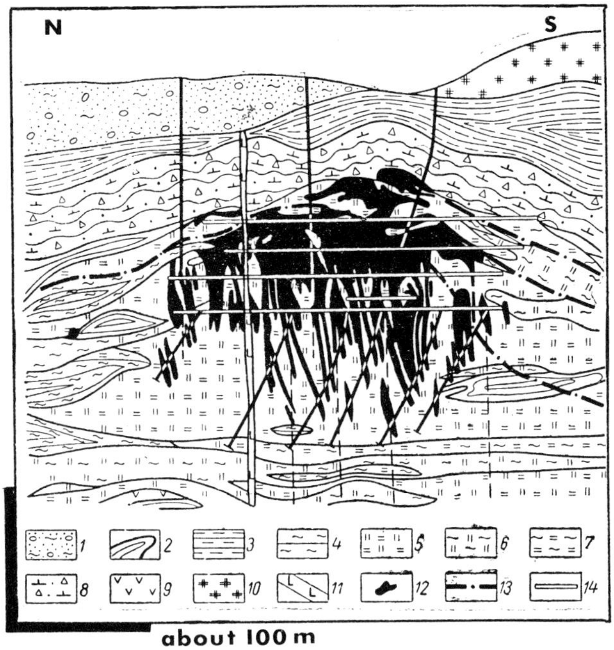

Fig. 11-4. Sokol'noe Zn-Pb massive sulphide deposit, Leninogorsk Rudnyi Altai, U.S.S.R.; second south-western orebody. LEGEND: (1) Quaternary cover; middle-upper Devonian: (2) siltstone, argillite; (3) limey argillite; (4) sericitized slate; (5) alteration microquartzite; (6) sericite, chlorite, quartz alterites; (7) quartz-sericite alterite; (8) mixed volcanic breccia to conglomerate; (9) amygdaloidal andesite porphyry; (10) quartz albite porphyry; (11) diabase; (12) Zn-Pb ore; (13) faults; (14) mine workings. From Bublichenko et al. (1972).

polymetallic stratabound sulphide deposits (Rosebery, Hercules, Que River), reviewed earlier. The volcanics formed over and within the continental crust of the Precambrian Tyennan Block in the east, and they change facies westward into a flysch wedge containing some submarine basalt flows. The felsic volcanics range from massive, well-preserved and lightly metamorphosed varieties to intensively sheared and cleaved quartz—sericite or quartz-chlorite schists.

In the Mt. Lyell ore field near Queenstown (Reid, 1975; 1.5 Mt Cu/1.15-1.45%, 972 t Ag/8 ppm, 48 t Au; Fig. 11-5, 11-6), fifteen individual deposits are found in 10 km^3 of hydrothermally altered rhyolite porphyry flows, banded tuffs, tuffaceous siltstone and chert, intensively cleaved by the superimposed Devonian deformation

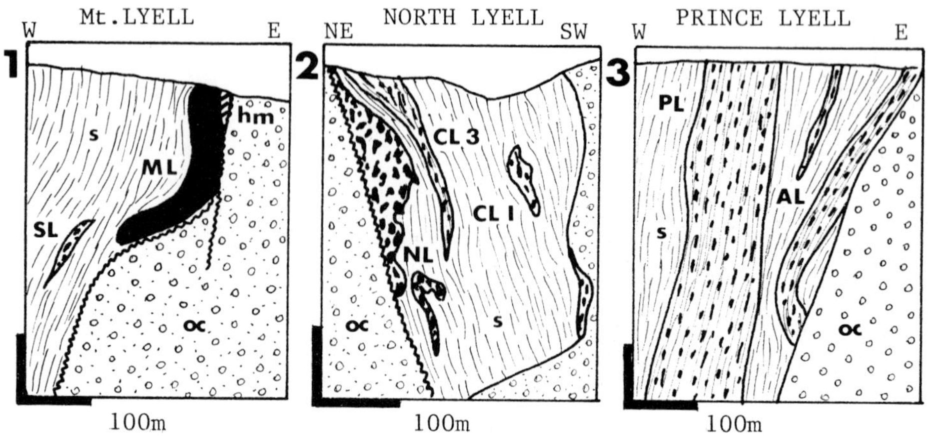

Fig. 11-5. Three principal varieties of Mt. Lyell, Tasmania, ores: (1) massive pyrite ore; (2) pods and irregular masses of bornite-chalcopyrite ore in chert, chert breccia, quartz - sericite schist (North Lyell only); (3) disseminated pyrite, chalcopyrite ore in schistose metavolcanics. LEGEND: (s) Cambrian schistose felsic metavolcanics; (oc) Ordovician molasse conglomerate; (hm) hematite-barite lens. OREBODIES: SL=South Lyell; ML=Mount Lyell; NL=North Lyell; CL=Crown Lyell 1, 3; PL=Prince Lyell; AL="A" Lens. From LITHOTHEQUE, after Reid (1975) and Walshe and Solomon (1981).

(Walshe and Solomon, 1981). The majority of orebodies and the largest of these (Prince Lyell) are lenses of disseminated to stringer-like pyrite and chalcopyrite in quartz-sericite or chlorite schist. The ore lenses are broadly concordant with the Cambrian stratigraphy and subparallel with the Devonian cleavage, but far from stratiform. The disseminated deposits are interpreted as having formed from hydrothermal solutions percolating laterally along permeable horizons in a felsic volcanic centre, and several small massive sulphide deposits formed at the top of the sequence. The medium-sized massive sulphide deposits (The Blow, North Lyell) had dominant pyrite, minor chalcopyrite, and siliceous barite, bornite, chalcopyrite composition, respectively. The small but rich Zn,Pb,Cu peneconcordant bodies in the Comstock mine group provided a link with the Rosebery deposit.

Other regions

Conditions strongly reminescent of those in the Mt. Lyell field exist in the little known Madneuli– Poladauri-Tsitelsopel mineralized area in Soviet Georgia (Smirnov, 1969). There, highly altered (silicified, up to the degree of "secondary quartzites")

Fig. 11-6. Mount Lyell, Tasmania, copper field. (A) general view from the N.W., showing the Owen Conglomerate cover sequence (top left, light colour); (B) North Lyell siliceous bornite-chalcopyrite orebody; (C) strongly cleaved quartz-sericite and quartz, sericite, chlorite host felsic meta-volcanics.

upper Cretaceous quartz keratophyres and tuffs contain irregular massive pyrite-chalcopyrite, as well as barite, galena, sphalerite stratabound bodies in what appears to be a top of a volcanic pile. In the old Uspenski ore field of central Kazakhstan (Borukaev and Shcherba, eds., 1967) about 300 Tt Cu was produced from massive and veinlet, disseminated quartz, barite, pyrite, chalcopyrite orebodies, hosted by strongly foliated Devonian quartz-sericite schists, black phyllites and black argillaceous marbles within a shear zone.

In the Cambrian Carolina Slate Belt in the Appalachians of the United States (Bell, 1982), small massive sulphide deposits associated with silicified and aluminous silicate-altered felsic volcanics have been mined mainly for gold in the past; some also contain tin. In the Brewer mine, fine-grained rhyolite tuff and breccia have been converted to quartz-sericite schist. In places the schist is completely silicified, and contains irregular replacement bodies of gold-bearing pyrite-topaz ore. 194,000 t of ore containing 15% topaz and 1.4 ppm Au were outlined.

11.6. U-Th ASPECT

At the Rexspar uranium-fluorite deposit near Birch Island, British Columbia (Morton et al., 1978), a sequence of brecciated, mylonitized and schistose porphyritic trachyte flows probably dating from the Permian is interlayered with pyroclastics, tuffs, pyritic schists and phyllites. The radioactive and rare-earths mineralization is confined mainly to dark coloured tuffaceous or argillaceous pyritic lenses, and is both concordant and discordant with the schistosity. Finely dispersed pitchblende, U-thorite, bastnaesite, etc., occur mostly within phlogopite, sericite, pyrite and fluorite. The mineralization is multistage, probably initiated during a syn-volcanic hydrothermal episode, and completed after cataclasis, when fluorite, celestite, pyrite veins formed.

11.7. POST-DEPOSITIONAL MODIFICATION OF MASSIVE SULPHIDES, AND INTERACTION MINERALIZATION

Many volcanics-hosted massive sulphide deposits have been substantially modified by the post-depositional deformation, metamorphism and igneous activity. The ore at the Hercules mine near Rosebery, Tasmania, now consists of a series of ore pods and sulphide lenses up to 15 m thick, parallel with the 70° east dipping cleavage. This orientation which is at an angle to the original bedding, is interpreted as being due to deformation realignment (Adams et al., 1976).

The major Zyr'yanovsk field in the Rudnyi Altai, U.S.S.R., contains a series of sub-parallel lens and sheet-like sulphide orebodies, parallel with the zone of axial cleavage in the intensively silicified Devonian felsic volcanics. At other localities in the same area, it is common for zones of slaty cleavage and shearing to become the main controlling features, considerably weakening or even obliterating the stratigraphic control. Hydrothermal remobilization of the early ores along faults in the aureole of post-ore intrusions is also commonplace.

The small discordant deposit Mt. Farrell, Tasmania, compositionally similar to Rosebery and hosted by the same unit, is considered to be a product of remobilization of an earlier stratabound sulphide body, or mobilization of metals from stratigraphically controlled trace-metal enriched rocks. This, and similar examples of remobilization, are discussed in Laznicka (1985b).

Quartz-siderite fissure veins with a variable content of Cu, Zn or Pb sulphides, tetrahedrite, specularite, and other minerals, are widespread in all the massive sulphide mineralized regions listed in Table 11-1, as well as in other regions of coexisting felsic volcanics and marine (meta)-sediments, as in the Rheinische Schiefergebirge of Germany (Fig. 11-7).

Mineralized felsic supracrustal volcanics gradational with increasing depth into subvolcanic and plutonic equivalents and/or intruded by a younger intrusive phase, show the most complex interaction, transitions and variations in the mineralization style. The Mount Morgan Cu-Au deposit in Queensland is unique, because its ores are believed to have been generated at a variety of crustal levels, ranging from the seafloor to the plutonic zone (Lawrence, 1977). The Mount Morgan orebody (Frets and Balde, 1975; 600 Tt Cu/0.75%, 374 t Au/3.8 ppm; Fig. 11-8) is a brecciated mass (pipe) of pyrite, that contains variable amounts of chalcopyrite and minor

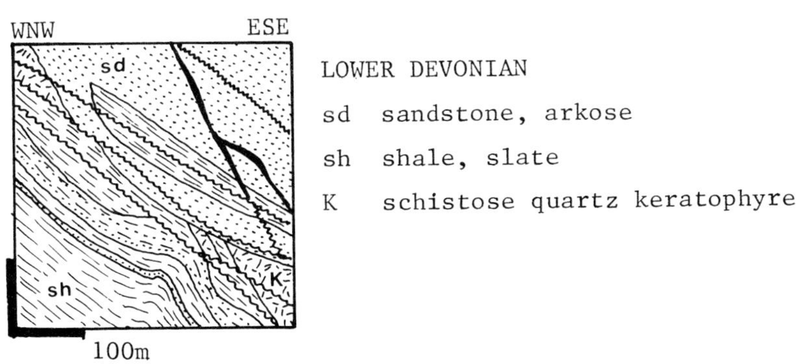

Fig. 11-7. Silberg near Müsen, Siegerland, Germany; an example of a small quartz, siderite, galena, sphalerite fissure vein deposit common in faulted sandstone, keratophyre or bimodal terrains. From LITHOTHEQUE, after Quiring (1936).

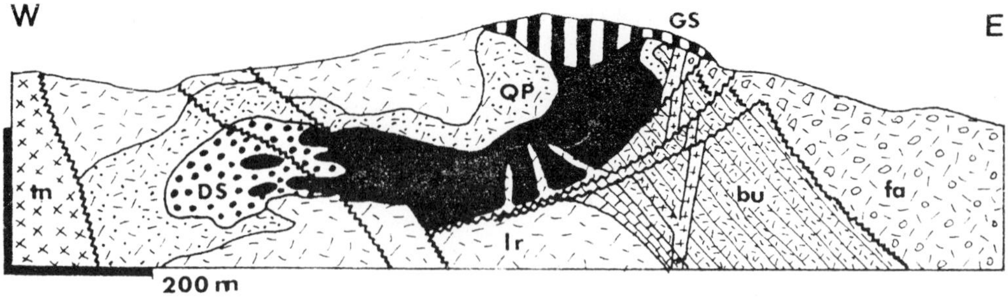

Fig. 11-8. Mount Morgan, Queensland: transitional volcanogenic massive sulphide/plutonic disseminated Cu-Au deposit. LEGEND: Middle-Upper Devonian volcanics and sediments: (bu) banded rhyolite, chert, jasper, limestone; (lr) quartz latite and quartz-feldspar porphyry; (fa) felsic agglomerate; (tn) upper Devonian tonalite. OREBODIES: (GS) goethitic gossan; (QP) silicification and pyritization envelope; (DS) disseminated and stockwork pyrite, chalcopyrite; BLACK, massive pyrite-chalcopyrite ore. From LITHOTHEQUE, after Frets and Balde (1975) and Lawrence (1977).

magnetite, pyrrhotite and sphalerite. Chalcopyrite and pyrite also contain small inclusions of gold and Au-tellurides. Chalcopyrite either forms narrow films interstitial to polygonal pyrite grains, or is scattered and disseminated in quartz veins and stockworks transecting the pyrite mass.

The massive sulphide body at Mt. Morgan grades into a siliceous rock containing disseminated pyrite, chalcopyrite and gold, and both were topped by a thick goethite gossan extremely rich in residual gold in places. The orebody is hosted by several phases of Devonian felsic magmatic rocks ranging from thinly-bedded rhyolites alternating with chert, jasper and minor magnetite-hematite bands (interpreted as seafloor volcanic flows and sediments), to massive quartz-feldspar porphyries and a quartz latite intrusion. All the earlier rocks have been subsequently intruded by a tonalite, quartz diorite and gabbro complex.

Lawrence (1977) found remnants of a distinctive stratified pyrite and a pyritic "shale" ore and concluded that the Mt. Morgan mineralization was polyphase: (1) initiated by a submarine hydrothermal sedimentation of pyrite on the seafloor; (2) followed by formation of a massive pyrite pipe by the collapse of bedded pyrite or otherwise; (3) fracturing of the pyrite mass and filling of the fractures by a hydrothermal assemblage near contact with the existing quartz-latite plug and (4) partial alteration/mineralization of the latite itself, resulting in a porphyry copper-like mineralization. Broadly comparable mineralizations appear to be relatively frequent, yet unrecognized due to insufficient study. Some are treated in Chapter 13 below.

"Gold deposits in the ancient rocks occur in true beds, and the precious metal was deposited with the sedimentary strata. It is not now, however, confined to these beds, but is met with both accumulated in their joints, and in regular quartz veins with sulphurets, traversing alike the auriferous beds and their accompanying strata, showing that it has changed its position since it was first deposited".
S. F. Emmons in Rept. on the Geology of Midland Counties of North Carolina, 1856.

CHAPTER 12

Bimodal (Mafic-Felsic) Volcanic-Sedimentary Association

12.1. GENERAL

Bimodal magmatic associations in orogenic belts result from juxtaposition, interlayering and mixing of members of the mafic and felsic associations (described above) on such a scale that both members cannot be treated separately. Intimately interbedded mafic and felsic volcanics are often underlain by a thick foundation dominated by either member. The Silurian spilite-keratophyre association in the south-western Ural Mts. is underlain by the mafic member (Zavaritsky, 1946); the Devonian spilite-keratophyre suite of Sauerland (Germany), is underlain by the felsic member (Quade, 1976). Both the contrasting volcanic members could be either "normal" (tholeiitic to calc-alkaline basalt; potassic calc—alkaline rhyolite to rhyodacite), or sodic (spilite, keratophyre). The sodic varieties are particularly characteristic of the "geosynclinal" setting, mostly because interaction of the submarine eruptions with seawater appears to be the most logical mechanism for sodic metasomatism. Alternative mechanisms for the origin of spilites-keratophyres including the orthomagmatic one, have been proposed (e.g. Amstutz, 1968). The subvolcanic and plutonic equivalents of the mafic volcanics are diabase and albitic diabase to gabbro; the felsic volcanics equivalents are quartz-feldspar porphyry (K-feldspar), quartz-albite or albite porphyry (albitophyre), albite syenite and quartz-albite syenite.

The submarine volcanics are typically associated with deep-water ("pelagic") sediments: dark gray to black shale, siliceous shale, chert, "graywacke", pelagic limestone, etc., with which they are often intermixed (intergraded). The following is a "typical" mixed series, but not by any means the only one: pillow basalt- pillow breccia-hyaloclastite or aquagene breccia with calcitic or hematitic cement-"Schallstein", varicoloured volcaniclastics — tuffaceous limestone-limestone. Shallow-water facies or even subaerial members, however, are also widespread. As it has been already pointed out in previous chapters, the field appearance and, to a considerable degree the mineralization in volcanic-sedimentary associations are strongly influenced by deformation and metamorphism. Massive, gently open folded and in places almost flat-lying spilites—keratophyres in the south-western Urals were petrographically described in detail by Zavaritsky (1946). There, spilites form pastel green pillow lava flows rarely grading into flow breccias and heteromictic interflow sediments. Variolites are widespread. Keratophyres are almost equally divided into flows and breccias ("agglomerates"). They are light-gray, fine-grained equigranular rocks lacking quartz phenocrysts. In quartz keratophyres, inconspicuous quartz is found in the matrix. Felsitic keratophyres, yellowish--gray aphanitic rocks with conchoidal fractues, are characteristic.

In strongly schistose terrains, mafic volcanics have been converted into greenstones or greenschists (chloritic or actinolitic), felsic volcanics into beige to yellow sericite or quartz-sericite phyllites or into muscovite schists. The majority of the bimodal volcanic terrains are schistose and metamorphosed, and they are widespread in the Paleozoic, substantially rarer in the Mesozoic, and very rare in the Cainozoic. Along shears, massive rocks can change into schistose ones over a distance of a few centimetres. Hydrothermal alteration can substantially modify the original rock appearance and make identification difficult. Sericitization and silicification causes colour lightening and many rock bodies mapped as "felsic volcanics" are altered metasediments or even mafic metavolcanics. Mg-Fe metasomatism results in chloritization and is capable of converting a light felsic metavolcanic into a dark green chloritic rock, often visually identical to a metabasalt.

Fig. 12-1 and Table 12-1 show and list the principal mineralization styles in the bimodal volcanic-sedimentary association.

12.2. Fe (Mn) ASPECT

In textbooks and reference books (e.g. Schneiderhöhn, 1955), the Lahn-Dill "type" of volcanogenic bedded iron ore is often placed in the spilite-keratophyre, or keratophyre host association. The bulk of such ores including those in the type area in Germany (with the exception of the rare and economically unimportant orebodies forming lenses within keratophyres as, locally, in the Harz Mts.) are hosted by, or show close relationships to, the mafic volcanics (basalts, spilites, "Schallstein") and associated sediments. Similar ores also occur in regions where the felsic members are completely absent. Consequently, the felsic volcanics do not seem to be indispensable in the formation of the Lahn-Dill "type", so that this has been treated above in Chapter 10 describing the mafic association.

Felsic (meta) volcanics, usually keratophyres, together with basalts, however, appear to have been part of the process that generated certain siderite ores. The Poiana Ruscǎ Mountains of western Rumania contain a series of siderite bodies hosted by upper Devonian greenschists (basic metavolcanics) and metasediments (black sericite-chlorite phyllite, carbonate), and intercalations of quartz-keratophyre metatuffs (Kräutner, 1977). In the two largest deposits Teliuc and Ghelar, lenses of massive or banded siderite interpreted as submarine-hydrothermal and locally associated with quartz, magnetite, stilpnomelane, thuringite and pyrite, grade into ankerite and sideroplesite. The lenses are 100-200 m thick. The iron grade is about 35% Fe, and the largest orebodies each contained about 20 Mt ore (Fig. 12-2).

LITHOLOGIC AND ORE FACIES TRANSITION

In many regions of the world (e.g. in the Triassic in the Vareš region, Yugoslavia; near Kremikovtsi, Bulgaria), association of "geosynclinal" sediments similar to those present in the Lahn-Dill, Poiana Ruscă, and other regions is well represented, and concordant to peneconcordant bodies of siderite, ankerite and/or hematite have developed (Chapter 17). Submarine volcanics of the bimodal mafic or felsic associations are, however, present sporadically if at all, but their presence is important because of the possible relationship between the volcanic lithogenesis and ore formation.

Stratabound Zn-Pb and Cu mineralization as well as barite are frequently present in regions that contain "exhalative" iron ores, closely associated with the volcanics (e.g. Horní Město, Horní Benešov), or outside the iron-ore and volcanic regions, hosted entirely by sediments (e.g. Rammelsberg, Meggen). "Exhalative" iron ores, therefore, have some value indicating the possible presence of the base metal sulphides. The discovery site of the important Bathurst-Newcastle district, New Brunswick, was in a small ironstone pit.

TECTONOMAGMATIC INTERACTION

In Ruschiţa, located in the Banat metalliferous region of south-western Rumania, magnetite and Pb-Zn-Cu sulphide-bearing skarns and carbonate replacements are located at the contacts of late Cretaceous granodiorites and Devonian marbles interbedded with amphibolites (e.g. Ilavský, ed., 1979). Most of the orebodies are lenses peneconcordant with the depositional fabric of the supracrustal rocks, and they are probably thermally metamorphosed or metasomatized Devonian volcanic-sedimentary ores situated in the extension of the Poiana Ruscă iron ore belt (Kräutner, 1977).

Anzas iron ore field (western Sayan, U.S.S.R.; Sokolov and Grigor'ev, 1974; 167 Mt of 38.2% Fe) comprises ten east-west trending lens-like bodies of magnetite up to 1.7 km long and 70 m wide, hosted by albite-actinolite and scapolite-altered gabbro and metaspilite. The mineralization is believed to have formed in the Ordovician or Silurian periods, by intrusive interaction with

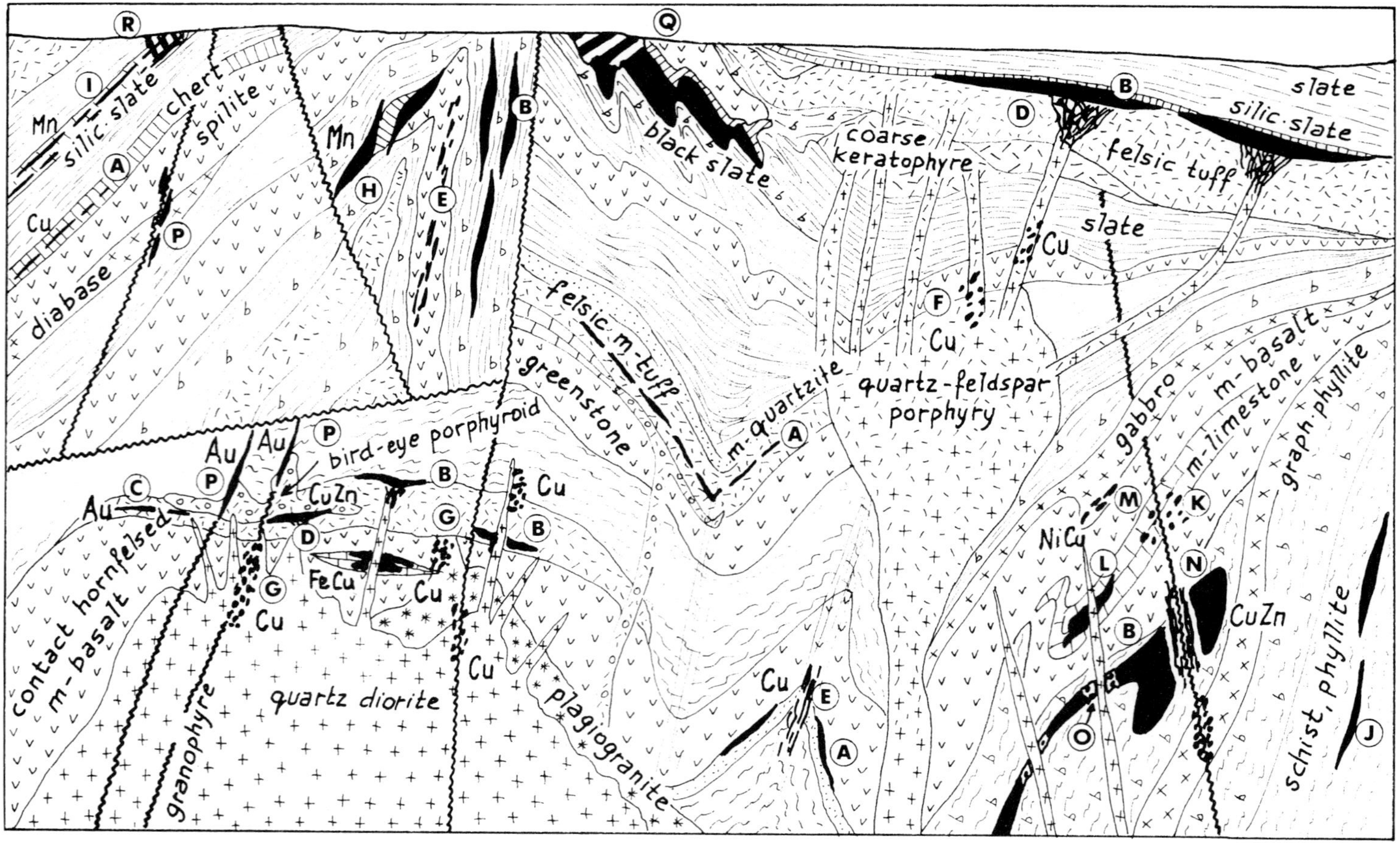

Fig. 12-1. Bimodal volcanic-sedimentary association, principal mineralization styles (see Table 12-1 for an explanation of the letter codes).

Table 12-1. Bimodal volcanic - sedimentary association: principal mineralization styles

(A)	concordant, disseminated pyrite, Cu (Zn,Pb) deposits in volcanics and sediments; X0 localities; concordant to peneconcordant siderite or hematite, magnetite deposits; X0 localities; 350 Mt Fe
(B)	conformable volcanogenic massive sulphide lenses; X00 localities; 40 Mt Zn, 24 Mt Cu, 7 Mt Pb, 40 Tt Ag, 1.6 Tt Au
(C)	concordant to peneconcordant gold - pyrite deposits in silicified metavolcanics, metasediments; X0 localities, 150 t Au
(D)	pyrite or pyrrhotite, chalcopyrite stockworks in altered footwall of massive sulphide deposits; X00 localities, 13 Mt Cu
(E)	pyrite - chalcopyrite stringers, stockworks, veins, in chloritized metavolc.,along shears, cleavage; X0 localities, 3 Mt Cu
(F)	stockwork and disseminated pyrite, chalcopyrite in altered subvolcanic felsic bodies ("porphyry coppers"); X localities; 800 Tt Cu
(G)	disseminated and stockwork pyrite, chalcopyrite in sheared, altered sodic plutonic (or metasomatic) plagiogranites; X0 localities, 2 Mt Cu, 40 t Au
(H)	Mn silicate, oxide, carbonate lenses in volcanics affiliated jasper; X0 localities, 2 Mt Mn
(I)	bedded Mn oxides, carbonates in shales, cherts, siliceous limestones; X0 localities; 80 Mt Mn
(J)	concordant, rhythmically banded Fe,Zn,Pb,Cu sulphides in sediments; X localities, 1 Mt Zn, 300 Tt Pb
(K)	Cu,Pb,Zn, sulphide replacements in silicified carbonates; X0 localities, 3 Mt Zn
(L)	magnetite - chalcopyrite skarns; X0 localities, 1.5 Mt Cu
(M)	pyrrhotite, pentlandite, chalcopyrite disseminated and massive in gabbros; X localities, uneconomic
(N)	remobilization Fe,Cu,Zn,Pb sulphide veins within massive sulphide bodies; minor and local
(O)	thermally metamorphosed volcanogenic massive sulphides at intrusive contacts; minor and local
(P)	hydrothermal gold - quartz veins and replacements in plutonic aureoles
(Q)	gossans over pyritic deposits, containing residual auriferous goethite; X0 localities, 100 t Au
(R)	gossans over "primary" concordant Mn lenses (H,I); enriched Mn oxides

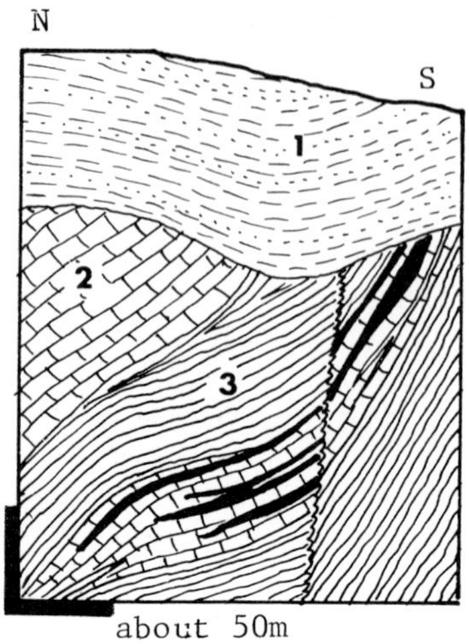

1 Miocene sedimentary cover

2 Upper Devonian dolomite and limestone

3 Graphitic sericite - chlorite schist, greenschist

BLACK: crystalline siderite

Fig. 12-2. Teliuc, western Rumania. A siderite deposit in Devonian bimodal volcanic-sedimentary association. After Kräutner (1977).

the lower Cambrian "geosynclinal" association of a quartz-albite schist (meta-keratophyre ?), amphibolite, quartzite and marble, including possibly volcanogenic iron orebodies.

12.3. Mn ASPECT

Stratabound volcanic-sedimentary Mn ores hosted by "chert", have been recorded in a wide variety of associations, including the bimodal one. In the lower-middle Devonian spilite-keratophyre assemblage near Magnitogorsk, southern Urals (Bikkulovsk Mn Horizon; Varentsov and Rakhmanov, 1974; Popov, 1979), Mn silicates and oxides form lenses 300-500 m long, 1-2 m thick in jasper and in siliceous slates interbedded with a greenschist-metamorphosed basalt, spilite, mafic and felsic tuffs. In the 300 km long north-south trending mineralized belt, over 200 Mn occurrences are known, of which 23 have some economic importance. The deposits rarely exceed several hundred thousand tons of ore, although a few exceed 5 Mt. The grade is between 15 and 45% Mn, and the overall recoverable Mn content is of the order of 600 Tt Mn.

In the Rhiw Mn field, northern Wales (Woodland, 1956), faulted and deformed stratiform lenses of Mn-chlorite, rhodochrosite, chamosite, tephroite and jacobsite are hosted by Ordovician meta-shale and keratophyre tuff, associated with metabasalt lavas. 65,000 t Mn were produced from a 27% Mn ore. In the Rio Tinto pyrite belt, Spain and Portugal (a member of the Carboniferous bimodal association), small Mn oxide deposits form stratiform lenses located on top of felsic metatuffs, several hundred metres stratigraphically under the pyrite.

12.4. TRANSITIONAL Fe,Mn-Zn,Pb,BARITE,Cu, ASPECT

ATASU DISTRICT, KAZAKHSTAN

A remarkable series of transitions is represented in the Atasu ore district of central Kazakhstan, (1) between the earlier bimodal basalt-rhyolite (or trachyte) association, and a later more evolved magmatic sequence also involving andesite and diorite members; (2) between volcanism and chemical, biochemical and detrital sedimentation; (3) between continental and "geosynclinal" deposition; (4) between Fe,Mn and Zn,Pb, barite suites; and (5) between the "primary" seafloor ore rhythmites and superimposed ore metasomatites. Atasu (Borukayev and Shcherba, eds., 1965; Shcherba, 1968; Strakhov et al., 1968; Fig. 12-3) is part of the Devonian to Carboniferous Zhailymin Synclinorium, which in turn is located in the Uspenski Tectonic Zone: a 400 x 80 km E.N.E.-trending structure that contains

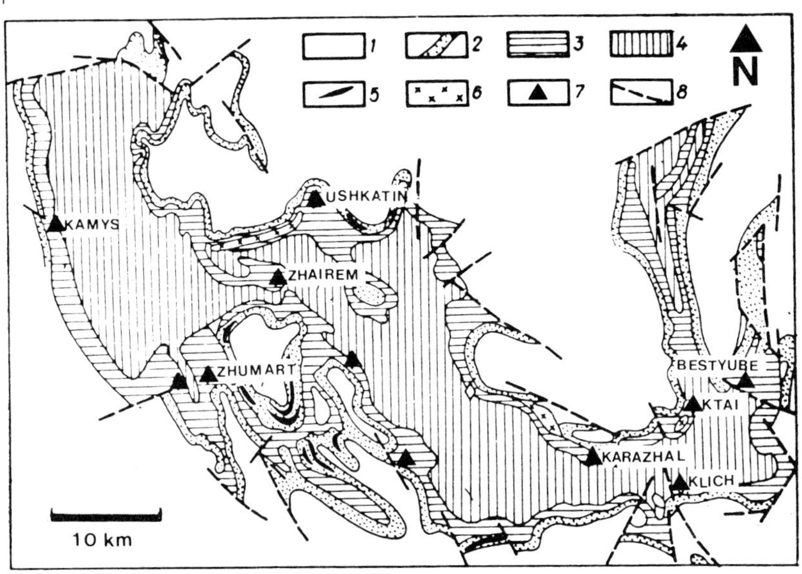

Fig. 12-3. Atasu Fe,Mn and Zn,Pb,barite district, Zhailimin Basin, Kazakhstan, U.S.S.R. (1) Pre-late Ordovician deposits; (2) lower Devonian red sandstones and conglomerates; (3) upper Devonian carbonate and siliceous sediments; (4) Carboniferous carbonate and argillaceous sediments; (5) basalts; (6) small gabbro-diabase intrusions; (7) Fe-Mn (and Zn,Pb) deposits; (8) faults. From Novokhatskii (1972), scale added.

over 400 ore deposits and showings (including 4 large and 4 mediumd-size deposits). Total reserves originally present in the district are estimated to be of the order of 80 Mt Mn, 300 Mt Fe, 3 Mt Zn, 2 Mt Pb, and over 30 Mt barite

The Atasu district proper is in a block floored by a great thickness of the continental crust (50-53 km), topped by the lower Devonian spilite-keratophyre association composed of lavas, tuffs, agglomerates, etc., interstratified with conglomerate, sandstone and siltstone. These are interpreted as continental—marine deposits (molasse facies; Shcherba, 1968). The proportion of terrigenous rocks increases upwards. In the upper Devonian, a 500 to 1,500 m thickness of fine detrital, chemical and biochemical sediments with a significant volcanogenic component formed. These sediments were deposited in a partially stagnant, shallow sea. They are carbon-rich, rhythmically laminated, and include a variety of argillaceous, siliceous, tuffaceous and black limestones, dolomitized limestone, red and black chert and black argillite. The continuity of the sedimentary unit is interrupted by infrequent volcanic centres and dikes, close to which concordant felsic lava and ash flows and resedimented volcaniclastics intertongue with the sediments. The volcanics are mostly basalts and trachytes, but diorites are common among the dikes and stocks.

The metallic (and barite) mineralization is stratigraphically controlled, and its bulk is confined to the 300-500 m thick upper Devonian black argillite, chert and carbonate unit. Two principal mineralization styles can be distinguished (see Fig. 12-4): (1) the earlier stratiform ores, synchronous with the sedimentation and (2) the latter peneconcordant to discordant orebodies synchronous with tectonism and the intrusive activity, superimposed on the former.

Style (1) has the form of rhythmically banded layers or lenses of sedimentary Fe-Mn and Pb-Zn-barite ores, gradually passing into the enclosing rocks without any obvious hydrothermal alteration. Several depositional cycles have been recorded, and each cycle usually starts with an iron ore (siliceous hematite, magnetite), followed by manganese beds (hausmannite, braunite, minor and local tephroite, rhodonite, Mn carbonates, psilomelane, pyrolusite), and zinc-rich beds (sphalerite, pyrite, melnikovite, galena, barite). Mixed Fe-Mn bands and lateral facies transitions of iron to manganese ores are common, but there is no intermixing of Fe, Mn and Zn, Pb, barite ores. The siliceous iron ores in the Atasu district contain up to 27 ppm Ge, and are the most important Soviet germanium source (Rundkvist, ed., 1978).

Style (2) occurs as peneconcordant lenses to irregular masses and veins that destroy the fine rhythmic lamination of their precursors and are controlled by fractures and by axial zones of antiforms and synforms. The ores are massive to banded, brecciated, streaky, veined, medium-to-coarse crystalline and their wallrocks are irregular and complicated by numerous branches and veinlets. The wallrocks are hydrothermally altered; low-temperature K-feldspar alteration (adularization), silicification, chloritization and propylitization is widespread in the silicate rocks. The carbonates are silicified,

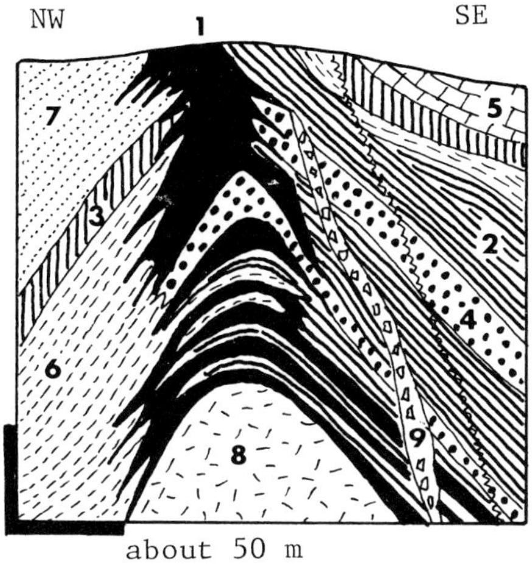

all rocks are upper Devonian

1 metasomatic barite, galena, sphalerite, chalcopyrite
2 thin bedded, stratabound Zn, Pb, barite ore
3 bedded Fe-Mn ore
4 quartz - pyrite metasomatite
5 siliceous black limestone
6 siliceous black shale
7 red sandstone
8 trachyte porphyry
9 breccia pipe

Fig. 12-4. Contrast between the early stratiform Zn-Pb and Fe-Mn ore bands (2,3) synchronous with sedimentation, and superimposed metasomatites (1,4). Zhairem deposit, Atasu district, Kazakhstan. LEGEND (all rocks are upper Devonian): (1) metasomatic barite, galena, sphalerite, minor chalcopyrite; (2) Pb-Zn, barite, "syngenetic" thin-bedded, stratabound ores. (3) Fe-Mn bedded ores; (4) quartz-pyrite metasomatites; (5) siliceous and carbonaceous limestone; (6) black carbonaceous and siliceous shale; (7) lower Devonian, red sandstone; (8) "trachyte porphyry"; (9) breccia pipes. After Borukayev and Scherba, eds., (1967).

baritized, dolomitized and pyritized. The main ore minerals are sphalerite, galena, pyrite and minor chalcopyrite; the gangue minerals are barite, quartz and calcite.

The individual deposits of the Atasu district (Table 12-2) contain variable proportions of the mineralization styles and metals. Karazhal is the largest Mn and Fe mine (Mn horizons are up to 6 km long and 20 m thick; the reserves of the West Karazhal mine were 173 Mt Fe, 13.2 Mt Mn/30%, and at least 100 Tt Zn); Dzhumart has the best Mn ore almost free of iron; Zhairem (Smirnov and Gorzhevsky, 1974) is the largest Zn, Pb, barite deposit, with important Mn and Fe by-products.

SOMOVA, ŠANKOVCE, OZERNOE

In the Tulcea-Somova mineralized zone of Dobrogea, N.E. Rumania (Ianovici et al., 1961), a large lens of pure barite and a smaller lens of barite impregnated by galena, sphalerite and chalcopyrite

Table 12-2. Principal ore deposits in the Atasu district, Uspenski zone, Kazakhstan, U.S.S.R. Based on data in Borukaev and Scherba (1967) and in Sokolov and Grigor'yev (1974)

Karazhal, 260 km S.W. of Karaganda	6.5 km long, 30-40 m thick, magnetite, hematite, braunite, hausmannite stratiform layer in black siliceous limestone with jasper bands. 173 Mt Fe/46%; 13.2 Mt Mn/27.2%
Zhairem, 56 km N.W. of Karazhal	4 bedded stratiform horizons of fine disseminated pyrite, sphalerite, lesser arsenopyrite, galena (total thickness 83 m), in siliceous black carbonate. Several Fe-Mn oxide layers (like in Karazhal); 3 superimposed irregular to lens - like metasomatic barite, sphalerite, galena orebodies. Min. estim. 2 Mt Zn, 400 Tt Pb, 10 Mt Fe
Ushkatyn, N.E. of Zhairem	one stratiform Fe-Mn oxide bed 15-40 m thick; two composite pyrite, sphalerite, galena horizons in black siliceous carbonate, beneath the Fe-Mn zone. Superimposed metasomatic veinlets, disseminations, masses of Zn,Pb sulphides. Prominent cerussite-rich gossan. 6 Mt Fe, 2 Mt Mn, estim. 200 Tt Pb+Zn.
Ktai, E.N.E. of Karazhal	magnetite, hematite, pyrite, siderite, braunite stratiform layer in black siliceous limestone; 2.95 Mt Mn/20%
Bestyube, N.E. of Ktai	several separate stratiform horizons of magnetite, hematite, pyrite, siderite, and barite, sphalerite, galena in black argillite. Superimposed metasomatic barite, galena, sphalerite bodies. Gossan along fissures reaches to 330 m depth; estim. 1.6 Mt Zn, 1.2 Mt Pb
Kairakty, 150 km S. of Karaganda	4 metasomatic barite, dolomite, calcite, galena, sphalerite, minot tetrahedrite, chalcopyrite bodies, up to 20 m thick, 260 m long, in tectonized black slates. High Ag,Au,Bi content; min.estim. 500 Tt Pb, 200 Tt Zn

in the footwall, are hosted by Triassic limestones. These limestones rest on, or are interbedded or interstratified with broadly coeval quartz porphyries, basaltic pillow lavas and breccias. The quartz porphyry in the ore footwall is brecciated and interpreted as a laccolithic body, grading into granite porphyry.

The small iron and copper deposit Šankovce in south-eastern Slovakia (Ilavský, 1976), contains evaporites in addition to the Atasu-style elements. There, peneconcordant lenses of hematite are hosted by lower Triassic varicoloured shales. Acid to intermediate synsedimentary breccias and tuffs, marls, limestones and mafic lava flows are found in the vicinity. A Permian copper-bearing gypsum and anhydrite layer occurs several meters stratigraphically below the iron ores.

The interesting Ozernoe deposit in Buryatia (eastern Siberia, U.S.S.R.; Skripchenko, 1972; Smirnov and Gorzhevsky, 1974; Fig. 12-5), contains up to twelve 5-30 m thick pyritic Zn-Pb and siderite stratiform layers and lenses, conformably interbedded with lower Cambrian felsic tuffs, eruptive breccias and quartz porphyries, black tuffitic slates, limestones, cherts and greenstones. The ore layers are rhythmically banded, composed of fine-grained pyrite and lesser sphalerite and galena, and they often change facies into siderite, ankerite, dolomite or limestone. Superimposed (or feeder stockwork ?) copper-rich mineralization occurs as veins and veinlets.

ATASU TRANSITIONS

Although close counterparts of the Atasu "type" are rare in the rest of the world, numerous links occur and some may constitute important ore leads. In some of the Alpine Zn-Pb fields (particularly Bleiberg-Kreuth; Chapter 20); in the Irish Zn-Pb belt (Chapter 19); and elsewhere, the ore-hosting sedimentary facies strongly resemble those in Atasu, but there are either no coeval volcanics or these are rare. The magmatogene models of mineralization have often been prematurely discarded in favour of the pan-sedimentogenic models, but the former should still be kept in mind, perhaps in a modernized form.

The Magnet Cove deposit in Nova Scotia (Boyle and Jambor, 1966; 4.5 Mt barite, about 20 Tt Pb, Zn, Cu, some Mn, Fe) is an interesting deposit that shares with Atasu the combination of the economic commodities produced. It is a raking pipe-like body of barite containing a massive sulphide orebody in the footwall. Substantial Mn and Fe most clearly visible in the oxidation zone, have been derived from siderite. The deposit occurs along a contact between Mississippian detrital and topping carbonate sediments. No coeval magmatic rocks, however, have been reported in the vicinity, but Triassic mafic intrusions related to rifting are known.

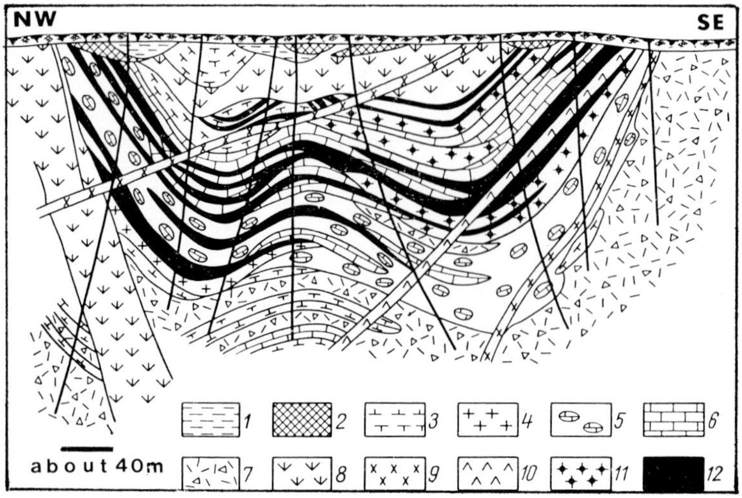

Fig. 12-5. Ozernoe (Buryatia, eastern U.S.S.R.), a bedded massive Fe,Zn,Pb,Cu sulphide and siderite deposit in a Cambrian bimodal volcanic-sedimentary association. (1) weathering crust; (2) limonite; (3) felsic tuffs; (4) black slates with tuffs, volcaniclastics, limestones; (5) limestone breccia with tuff material; (6) limestone; (7) quartz porphyry eruptive breccias; (8) greenstones, (9) syenite porphyries; (10) leucocratic diabase porphyrites; (11) siderite ore; (12) pyritic Pb-Zn ores. After Fil'ko, Babkin and Tarasova, from Skripchenko (1972); scale and orientation added.

12.5. Au (Ag) ASPECT

The Pueblo Viejo mine near Bonao, Dominican Republic, is an interesting example of a gold mineralized spilite-keratophyre – black slate association, for which a modern description is available (Kesler et al., 1981; Fig. 12-6). There, the ore–bearing lower Cretaceous Los Ranchos Unit consists of a black slate topping a lens of spilite (greenstone) and carbonaceous sandstone, resting on a thick unit of felsic "quartz-eye" conglomerate or agglomerate. The black slate contains stratiform pyrite layers and lenses, and frequent although short pyrite-sphalerite gash veins, but lacks gold values. The gold is in a funnel-shaped alteration (quartz, alunite, pyrophyllite, kaolinite, calcite) and mineralization zone, and occurs in a network of numerous thin quartz, pyrite, sphalerite, barite and pyrophyllite veins and veinlets. The gold takes the form of extremely fine electrum, closely associated with sphalerite and tellurides. The hypogene ore has a grade of 3 ppm Au, 23 ppm Ag, 0.8% Zn, 0.2% Cu,

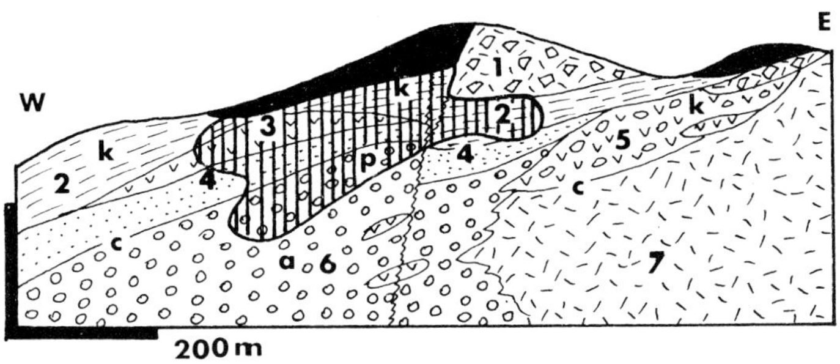

Fig. 12-6. Pueblo Viejo Au-Ag deposit, Dominican Republic. LEGEND (all rocks are lower Cretaceous Los Ranchos Fm.): (1) felsic agglomerate; (2) black slate; (3) spilite; (4) sandstone; (5) spilite conglomerate; (6) quartz-eye conglomerate; (7) felsic volcaniclastics. Alteration zones: (k) kaolinite; (c) calcite; (p) pyrophyllite. Black: oxidized Au-Ag ore; vertically ruled: hypogene ore. From LITHOTHEQUE, after Kesler et al. (1981) and N. Russell, S. Kesler, guided tour (1980).

and is not mined at present. All the contemporary (1980) operations are in the oxidation zone, in which 27 Mt of ore with 4.23 ppm Au and 21.6 ppm Ag, are available. In the oxidation zone, the metals are in the form of a submicroscopic spongy gold, electrum, and silver chlorides, iodides and bromides. Pueblo Viejo is an example of a virtually "invisible" gold mineralization, recognizable only by assay and suggested by the highly silicified nature of the host assemblage.

Gold mineralization in strongly schistose (or sheared) bimodal volcanics and sediments is probably relatively common, but the deformation and metamorphism obscure the original rock composition to such a degree that the original lithology is difficult to interpret. In the Southern Piedmont region in the Appalachians, U.S.A. (Fig. 12-7), many gold deposits are located in several N.E.-S.W.-striking "gold belts", parallel with the schistosity of the lower Paleozoic host metamorphics. The deposits are hosted by a quartz-sericite schist usually interpreted as rhyolite or keratophyre tuffs, or by a greenschist (metabasalt). Pardee and Park (1948) distinguished two principal styles of the gold lodes: (1) tabular, sharply outlined veins and (2) irregular to lens-like impregnation mineralized zones, subparallel with the regional schistosity. The mineralized zones are ill-defined bands, 15 to 70 m wide and up to 330 m long, of silicified and weakly pyritic strata, that together constitute a low—grade gold ore. Worthington and Kiff (1970) concluded that in North Carolina, the zones are almost entirely within the basal felsic volcanic unit, conformable with the original bedding and, because no "granites" are

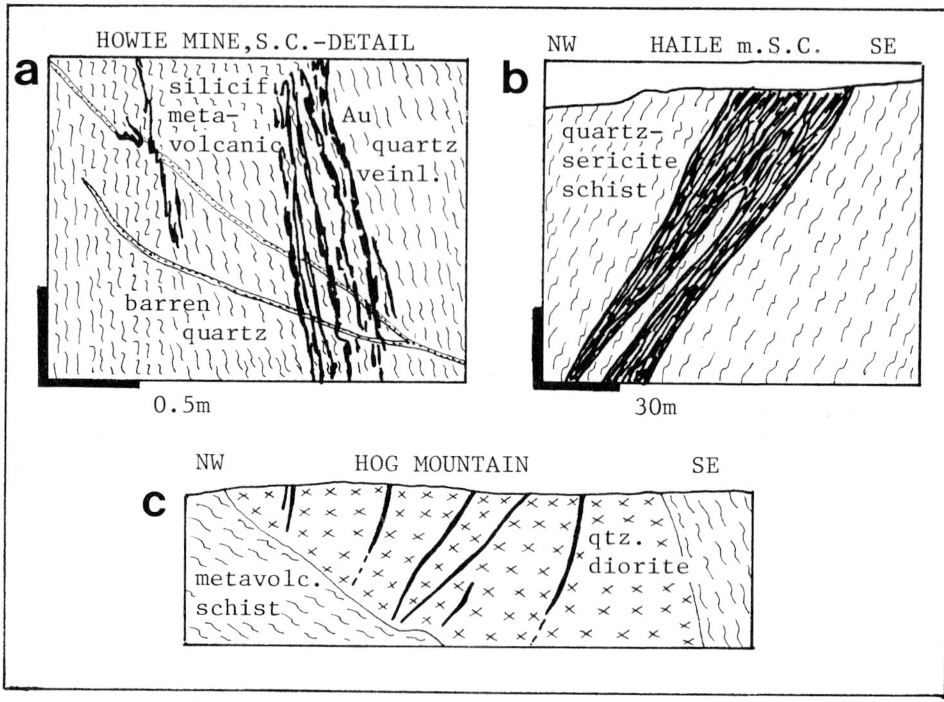

Fig. 12-7. Principal styles of gold mineralization in southern Piedmont, U.S.A. (a,b) silicified and quartz-sulphide veined zones in schistose felsic metavolcanics; (c) gold-quartz fissure veins in diorite intruding metavolcanic schists. After Pardee and Park (1948).

in sight, are likely to be volcanogenic in origin. Elsewhere, for example in central Virginia and in Georgia and Alabama, gabbros, quartz diorites and granodiorites intervene with the ultramafics and some are themselves mineralized by gold fissure veins (e.g. Hog Mountain, Alabama; Fig. 12-7C).

One of the largest deposits of the "mineralized zone" style is the Haile mine near Kershaw, South Carolina (Fig. 12-7B; Pardee and Park, 1948), which has produced 4.5 t Au. There, the host rock is a light-gray silicified quartz-sericite schist (probably a felsic tuff). The gold-bearing lodes are zones in which the schist has been more or less impregnated by a very fine ("cherty") quartz and pyrite. In one place, the zone is gradational into a pyrite-sphalerite massive sulphide. The Howie mine near Waxhaw (production about 1.5 t Au; Fig. 12-7A) is similar.

The Piedmont gold deposits and their lithologic association are closely comparable with some gold districts hosted by the felsic metavolcanics in the Precambrian greenstone belts (Volume 2). Comparable Phanerozoic gold mineralization is widespread in the Ural Mts. and Kuznetsk Alatau, U.S.S.R., in the Klamath Mountains, California, in the Tasman "Geosyncline", Australia and elsewhere.

12.6. MASSIVE Fe, Cu, Zn (Pb) SULPHIDE ASPECT

Massive sulphide deposits rich in pyrite and producing zinc, copper and sometimes lead as the most valuable commodities (pyrite concentrate alone is difficult to sell these days), are prominently developed in the bimodal volcanic-sedimentary association. The introductory paragraphs devoted to the massive sulphides in the felsic association (Chapter 11) are applicable here with only a minor change, because most deposits in the bimodal association are intimately connected with the evolutionary history of the felsic members and the mafic members just appear to "stand by". There are, of course, exceptions and generally the mafic volcanic piles in the deeper sub-ore region may have been the source of metals extracted by hydrothermal convection.

The massive sulphide occurrences, as a group, are more or less compatible with the general volcanogenic massive sulphide model reviewed in the introductory section. In detail, however, there is a large number of local variations and no two localities are exactly alike. For this reason, reference to the mineralization "types" proposed in the literature is used sparingly and example localities are reviewed briefly within a regional framework. Table 12-3 contains a brief list of the more important or interesting massive sulphide deposits or regions hosted by the bimodal association, and Fig. 12-8 shows their global distribution.

NORTH AMERICAN CORDILLERA

Scattered occurrences of the bimodal volcanic-sedimentary association of late Paleozoic age hosting massive sulphide deposits, can be traced from Alaska to northern California. There a similar Mesozoic suite takes over. In the Pelly Mountains of the Yukon, Mississippian bimodal metavolcanics contain several massive sulphide occurrences. The "MM" deposit 55 km south of Ross River (Morin, 1976), consists of three extensively deformed separate ore lenses hosted by a pyritic metaquartzite. The footwall is a dark green chlorite-garnet amphibolite mineralized by pyrrhotite and minor chalcopyrite. A 60 cm thick zone of sphalerited-rich massive sulphide is underlain by several layers of a quartz-rich schist containing a variable amounts of disseminated sulphides, and topped by several barited-pyrite units. A sericite, chlorite, quartz, calcite schist occurs in the hangingwall.

Table 12-3. Massive Fe,Cu,Zn (Pb) sulphide deposits hosted by Phanerozoic bimodal volcanic - sedimentary association; selected examples

LOCALITY	AGE	DEFORMATION METAMORPH.	FOOTWALL	OREBODY	HANGINGWALL	TONNAGE	REFERENCES
MM deposit, Pelly Mts, Yukon	Ms	PD,S,F/A	garnet amphibolite (MV)	MS sf/py/po topped by py,ba,sf,ga	sc-chl-qtz schist (FV)		Morin (1976)
Kutcho Cr., Cassiar Mt, British Columbia, Canada	Ms-Pe	PD,AxF/1A	sc-chl schist ("grit")	mass.to ds. sever.beds py/cp,sf/ bornite	m-pelite, argillite	586 Tt Zn/ 3.53% 357 Tt Cu/ 2.15%	Panteleyev and Pearson (1977)
Buttle L., C.Vancouver Island, Brit.Columbia	Ps	S,F/Gr	felsic m-volc., tuffs	MS to STW qtz,ba,sf, py/cp,gal, tetr,born qtz-sc sch.	black phyllite, schist	412 Tt Zn/ 7.5% 82 Tt Cu 66 Tt Pb 622 t Ag	Thompson and Panteleyev (1976)
Twin J (Mt.Sicker) Vancouver Island Brit.Columbia	Ps	AxF,C,F/Gr	sc sch, qtz augen porph.	MS sf,gal/ cp,ba in marble, black phyl. STW cp	phyllite	34 Tt Zn/ 12% 5.3 Tt Cu 2.9 Tt Pb	Stevenson (1948)
West Shasta distr., California	D_2	Fd,F/Gr	felsic brecc.volc.	MS py/sf,cp in m-tuff STW qtz,chl py,cp	augen porph. m-dacite	230 Tt Zn 200 Tt Cu 300 t Ag 68 t Au	Kinkel et al. (1956)
East Shasta distr. California	Pe-Tr	Fd,F/Gr	m-qtz keratoph.	MS py,sf, gal,cp; in qtz kerat.	m-shale	38 Tt Zn 30 Tt Cu 3 Tt Pb	Albers (1981)

Sierra Nevada Foot-hills belt, California	J_3	Fd,F/Gr	m-rhyolite	MS py,sf, gal,cp in sheared greenst., black sch.	massive m-basalt	90 Tt Cu 18 Tt Zn 5 Tt Pb 300 t Ag 2 t Au	Albers (1981) Reid (1907)
Point Leamington, Newfoundland	Or	Fd,F/Gr	coarse rhy-olite m-pyrocl.	MS py//sf, cp; siliceous	silic. m-argillite	280 Tt Zn/ 2% 70 Tt Cu/ 0.5%	Noranda Mines Ltd.(1974)
Gull Pond, west. Newfoundland	Or_1	AxF,C/ Gr-Am contact metam.	qtz-sc sch. (m-rh.tuff) topping rh. breccia	MS py,po// cp in cord. anthophyll. schist	pillowed m-basalt	30 Tt Cu/ 1%	Upadhyay and Smitheringale (1972)
Rambler group,S.S.W. of Baie Verte, N.W. Newfoundland	Cm-Or	PD/Gr schist, massive	qtz-sc sch mass.rhyol.	MS py,sf, cp; MS,DS py,cp,po	m-tuff mafic m-lava	80 Tt Cu 10 Tt Zn 34 t Ag 4 t Au	Tuach and Kennedy (1978)
Stoke belt,Sher-brooke area,S.E. Quebec	Or_1	PD,AxF/Gr	felsic,ma-fic m-volc.	MS py/sf, cp; in si-lic.m-tuff, BIF	felsic m-volc., m-sedim.	225 Tt Zn 185 Tt Cu 26 Tt Pb	Sauve et al. (1972)
Sterling m.,Fram-boise,Nova Scotia Canada	Pt_3 Cm	PD,S/Gr	felsic m-pyrocl.	MS py/sf, ga,cp in qtz,carb, talc gang.	felsic m-pyrocl.	66 Tt Zn/ 5.5% 15.6 Tt Pb 8.4 Tt Cu	Watson (1956)
Ely,Elizabeth m., Vermont, USA	Or	PD/1A	amphibolite biot.sch.	MS po/cp/ sf,sili-ceous	schist	55 Tt Cu	Howard (1959)

Table 12-3 (continued). Massive Fe,Cu,Zn (Pb) sulphide deposits hosted by Phanerozoic bimodal volcanic-sedimentary association; selected examples

LOCALITY	AGE	DEFORMATION METAMORPH.	FOOTWALL	OREBODY	HANGINGWALL	TONNAGE	REFERENCES
Bathurst-Newcastle district, New Brunswick, Canada	Or_2	PD,AxF/Gr				13.3 Mt Zn 5.5 Mt Pb 650 Tt Cu 11 Tt Ag	
Brunswick No.6,No. 12,mines	"	"	qtz,sc,chl schist	MS py,sf/ gal/cp,po topped BIF	felsic m-volc., m-tuffs	9.4 Mt Zn 3.75 Mt Pb 235 Tt Cu	Potter et al. (1972)
Caribou mine	"	"	qtz-seric. schist	MS py//sf/ cp,gal	phyllite	1.4 Mt Zn/5% 560 Tt Pb/2% 795 t Ag	Davis (1972)
Heath Steele m.	"	"	chl.metased. m-porphyry	MS sf,py, po/gal,cp	silic. fragment. volcanics	1.8 Mt Zn/ 7.2% 750 Tt Pb 275 Tt Cu 2,343 t Ag	Lusk (1969)
Murray Brook m.	"	"	chloritic schist	MS py/sf, gal,cp gossan	felsic m-volc., m-sedim.	448 Tt Zn 200 Tt Pb 101 Tt Cu	Potter et al. (1972)
Central Virginia gold-pyrite belt USA	Cm_1-Or_1	PD/Gr,A	phyllite amphibolite silic.fels.v	MS py//sf cp gossans	qtz-eye porphyry phyllite	4 t Au	Pavlides et al. (1982)
Skorovass m.,Grong distr., C.Norway	Or_{1-2}	PD/Gr	m-rhyodac. and breccia gabbro intr.	MS py//sf, cp; STW py, cp	iron fm., m-basalt	100 Tt Cu/ 1% 150 Tt Zn	Reinsbakken (1980)
Tverrfjelett m., C. Norway	Or_1	PD/A	amphib.to chlor.sch.	MS py//po, cp,sf; 6 ore lenses	garnet micaschist	271 Tt Zn 200 Tt Cu 240 t Ag	M.Motys,oral comm.(1975)

Stekenjokk, Caledonides, western Sweden	S	PD/Gr	qtz m-keratoph.	MS py/po, cp,sf,gal fels. m-tuffite	black phyllite	458 Tt Zn 220 Tt Cu 45 Tt Pb 800 t Ag	Zachrisson (1971)
Brévenne group (Sain Bel,Chessy) C.France	D_3-Cb_1	PD,AxF/ Gr	m-spilite chlor.sch.	MS py//sf, cp	seric.sch. black sch.	60 Tt Zn/0.3% 40 Tt Cu/0.2%	Bardin (1971)
Pierrefitte, N. Pyrenees,France	Or-S	Fd/Gr	m-spilite subvolc. m-qtz-kert.	MS py/sf, gal; STW felsic m-pyrocl.	seric. phyllite	160 Tt Zn/7% 48 Tt Pb/2.1% 109 t Ag	Pouit et al. (1979)
Rio Tinto belt, S. Spain and Portugal	D_3-Cb_1	Fd,AxC PP to lgr		850 t Au 1 Mt Mn	2 Mt Pb 4 Mt As 17.5 Tt Ag	415 Mt Fe 20 Mt Cu 8 Mt Zn	
Rio Tinto field	"	"	fels.pyrcl. subvolc. fels.stocks	MS py//cp, sf; in fels.volc.	fine volc. slate	10.2 Mt Cu 178 t Au 8 Tt Ag	Strauss et al (1977)
Cerro Colorado	"	"	"	STW ore in fw pyrocl. qtz,chl,py, cp	mass.sulph. gossan goethite	420 Tt Cu 50 t Au 940 t Ag	as above
Tharsis,Spain	"	"	black + tuff.slate	MS py//cp, sf; STW py, cp,chl	m-spilite	2.275 Mt Cu	as above
La Zarza,Spain	"	"	felsic volcanicl., black slate	MS py//qtz, cp,sf,gal STW	coarse pyrocl., jasper	1.5 Mt Zn 700 Tt Cu 600 Tt Pb	Strauss et al. (1981)

Table 12-3 (continued). Massive Fe,Cu,Zn (Pb) sulphide deposits hosted by Phanerozoic bimodal volcanic-sedimentary association; selected examples

LOCALITY	AGE	DEFORM., METAMORPH.	FOOTWALL	OREBODY	HANGINGWALL	TONNAGE	REFERENCES
Aljustrel, Port.	"	"	qtz.keratf. pyroclast.	4 MS lenses py//cp,sf in silic. slate	shale, jasper,BIF	1.86 Mt Cu	Strauss et al. (1977)
Aznalcollar, Spain	"	"	felsic volcanics	MS very fine; py//sf, gal,cp	slates	1.6 Mt Zn/3.3 775 Tt Pb 400 Tt Cu 5,360 t Ag	as above
Lousal, Port.	"	"	black slate interb. fels.tuff	MS py//cp, sf; in bl. slate Mn horiz.	spilite	100 Tt Cu 140 Tt Zn 80 Tt Pb	as above
Zlaté Hory, Jeseník Mts, Czechoslovakia	D_{1-2}	AxF/Gr	greensch. graph.phyl.	DS to MS py/sf,gal, in quartzt.	quartzite phyllite porphyroid	1.1 Mt Zn 700 Tt Cu 250 Tt Pb 160 t Au	Gruzczyk and Pouba (1968)
Horní Benešov, Jeseník Mts., Czechoslovakia	D_1-Cb_1	Fd/PP to 1Gr	seric. slate	MS to DS py/sf,gal barite	limestone	300 Tt Zn 100 Tt Pb	as above
Smolník, eastern Czechoslovakia	S	PD/Gr	chlorit. phyllite	MS,DS py// cp,tetr,sf	chloritic phyllite	45 Tt Cu 12 Tt Zn	Ilavský (1976)
Levikha, S.of Nizh. Tagil, the Urals, USSR	S-D_1	PD?Gr	silic.seric. chlor.sch.	MS,DS py// cp,sf; STW	qtz-seric. schist	305 Tt Zn 198 Tt Cu	Yefremov (1934)
Degtyarka, S.W. of Sverdlovsk, the Urals, USSR	S_3-D_1	Fd,PD/Gr	m-keratoph.	MS lens py//sf,cp chl-sc.sch.	chlor.-seric.schist	1.5 Mt Cu	Magak'yan (1968)

Uchaly, southern Urals, USSR	D_1	Fd/Gr	keratoph. qtz-seric. schist	MS py/cp sf,born, tetr.	m-basalt	300 Tt Cu	Borodaevskaya (1962)
Madenköy,eastern Turkey	Cr_3	Fd,Ft	dacite pyroclast.	MS sf,cp,py STW py in silic.,sc.	basalt spilite pyroclast.	1.3 Mt Zn 840 Tt Cu 33 Tt Pb	Franklin et al. (1981)
Urup, Caucasus, USSR	D_{1-3}	Fd/1Gr	propylitiz. qtz.keratf.	thin MS py//sf,cp silic.slate	jasper m-tuff	260 Tt Zn 200 Tt Cu	Skripchenko (1972)
Berezovsk-Belous-sovka,Rud.Altai, USSR	D	Fd/1A	m-qtz. keratoph., amphibolite	thin MS py/po,sf,cp gal; STW	iron fm. felsic m-tuff	840 Tt Zn 280 Tt Cu 70 Tt Pb	Ivankin et al. (1971)
Salair field, C.Siberia,USSR	Cm_{1-2}	PD/Gr	qtz m-keratf. subvolc. stock	MS ba,py// sf,gal; STW py,cp,chl.	m-basalt marble black slate	900 Tt Zn 100 Tt Cu	Lapukhov (1975)
Kyzyl Tashtyg, N.E.Tuva, eastern USSR	Cm_1	PD,S/Gr	m-spilite m-basalt m-pyroclast.	MS py/sf,cp, gal	black slate silicif. slate, jasper		Smirnov ed. (1968)
Hitachi mine,E.C. Honshu, Japan	Cb_1	PD,S,AxF/ A	greenschist	MS py/po, cp,sf/ga	greensch., biot.sch.	435 Tt Cu/1.4 321 Tt Zn	Nippon Mining (1966)
Yanahara m., S. Honshu, Japan	Pe	Fd/1Gr contact m.	qtz.porph. intr.diabase diorite	MS py//po, magn,cp; in pyroclast.	slate, diabase altered pyroclast.	70 Tt Cu 96 t Ag 13 Mt Fe	Oshima (1964)

Table 12-3 (continued). Massive Fe,Cu,Zn (Pb) sulphide deposits hosted by Phanerozoic bimodal volcanic-sedimentary association; selected examples

LOCALITY	AGE	DEFORM., METAMORPH.	FOOTWALL	OREBODY	HANGINGWALL	TONNAGE	REFERENCES
Hixbar mine, Rapu Rapu Isl., Philippines	Pe	Fd/Gr,A	chlorite schist	MS py//cp, sf;in sc, paragonite schist	felsic m-volc.	11.4 Tt Cu 10 t Ag	Bryner (1969)
Bagacay mine, Philippines	Cr_3-Pc		m-keratoph.	MS py//cp, sf	jasper, m-tuff	134 Tt Cu	as above
Peak Downs mine, Clermont, Queensland, Australia	PZ_1	PG/Gr	qtz-chlor. schist	DS+stringer cp,py in qtz-chl.sch. magn.BIF	quartz-sericite schist gossan	17 Tt Cu	Cox (1975b)
Bleïda, central Anti-Atlas Mts, Morocco	Pt_3	Fd/Gr,1A	jasper, m-kerat. tuff	MS,DS born, cp,py in black,green phyllite	m-arkose	181 Tt Cu	Leblanc and Billaud (1978)
Poli, northern Cameroun	Cm	Fd/1A	felsic m-tuffs	DS cp,py in sc-chl. m-quartzite	felsic m-tuffs	28 Tt Cu	Le Fur (1971)

ABBREVIATIONS

DEFORMATION,METAMORPHISM: Fd=folded; AxF, AxC=axial foliation, cleavage PD=penetrative deformation; S=shearing; F=faulting; M=mylonitization A=amphibolite; Gr=greenschist; PP-prehnite - pumpellyite; Z=zeolite metamorphic facies; 1=lower; UM=unmetamorphosed

SUBSEQUENT CATEGORIES: MV=mafic volcanics; FV=felsic volcanics m=meta; MS=massive sulphide DS=disseminated sulphide; STW=stockwork fw=footwall; hw=hangingwall

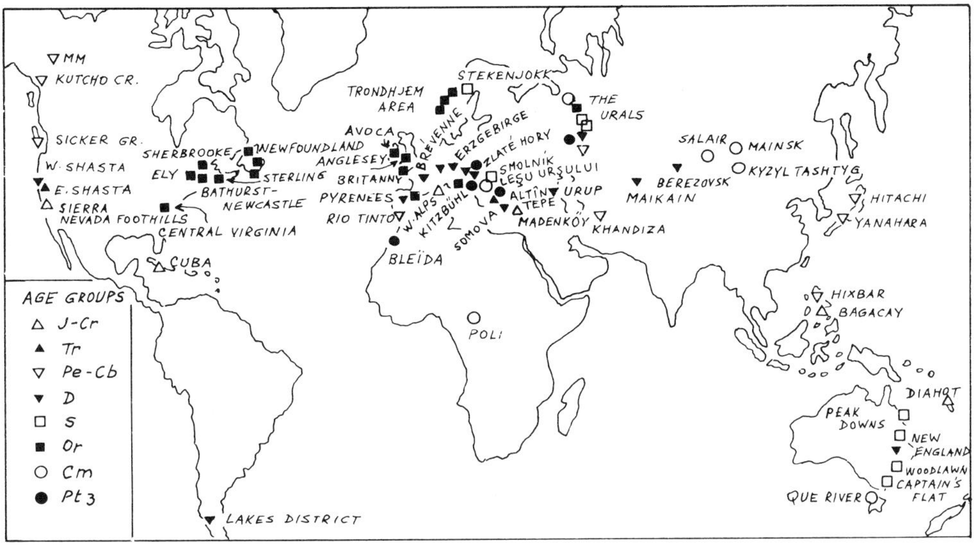

Fig. 12-8. Worldwide distribution of the major Phanerozoic massive sulphide deposits hosted by the bimodal volcanic-sedimentary association.

The Kutcho Creek deposit in the Cassiar Mountains, north-entral British Columbia (Panteleyev and Pearson, 1977; Fig. 12-9), is a concordant lens about 2 km long consisting of massive to disseminated pyrite, sphalerite, chalcopyrite and locally abundant bornite, hosted by a fine dolomitic quartz-sericite schist which in turn is enclosed in a unit of a quartz-eye sericite and chlorite schist. The polymetallic lens is at the end of a 15 km long concordant pyritic horizon in the schist with a pronounced axial foliation. The immediate hosts to the ore are interpreted as felsic volcaniclastics ("grits"), and they are underlain by a sodic bimodal volcanic unit now represented by a chlorite and sericite schist. The mineralization has a style comparable with the Besshi ore zone in Japan.

The Pennsylvanian Sicker Group of Vancouver Island (Monger, 1977), consists of relatively deep-water deposited mafic and felsic metavolcanics (now slates to schists), that include lithic and crystal tuff, volcanic breccia, amygdaloidal basalt, and an epiclastic volcanic sandstone. Complex interfingering is displayed. This is topped by a turbiditic meta-graywacke and argillite, containing local limestone lenses. Two groups of massive sulphide deposits occur at Buttle Lake (Western Mines) and at Mt. Sicker (Twin J). The Buttle Lake deposits (Thompson and Panteleyev, 1976) are massive siliceous barite, sphalerite, pyrite, chalcopyrite, bornite and tetrahedrite lenses, deformed in a complex manner, dismembered and localized

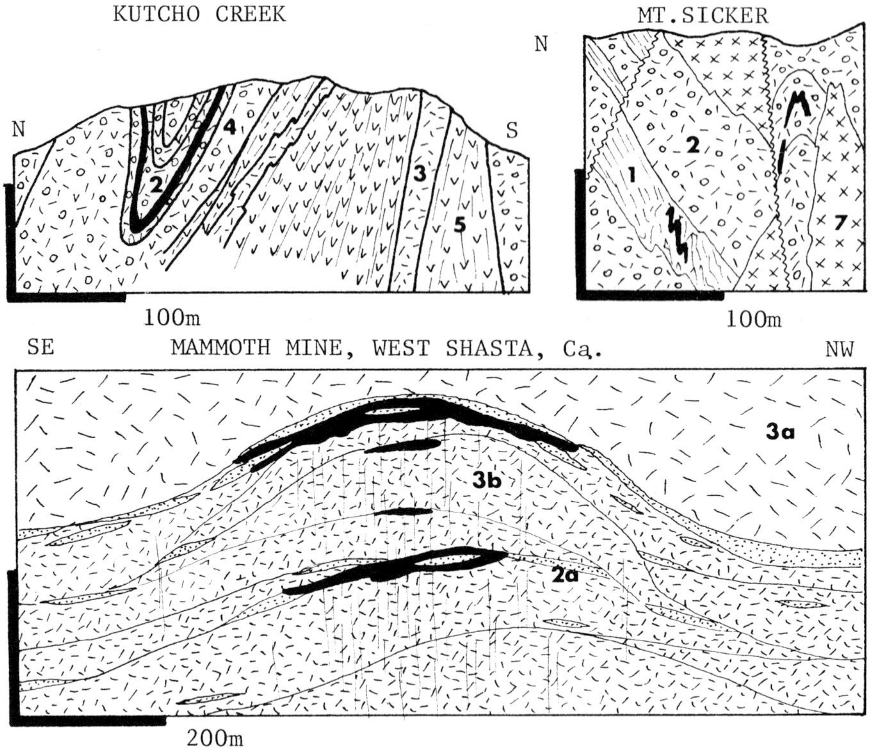

Fig. 12-9. Selection of massive Fe,Zn,Cu sulphide deposits in the bimodal association, North American Cordillera. Kutcho Creek, N. British Columbia, is Permian; Mt. Sicker, Vancouver Island, B.C., is Permian; West Shasta district, California, is Devonian. LEGEND: (1) black phyllite, marble and felsic metatuff; (2) quartz-eye sericite schist; (3) massive rhyolite in general; (3a) coarse phenocrystic rhyolite; (3b) medium phenocrystic rhyolite; (4) quartz-eye chlorite-sericite schist; (5) chlorite schist; (6) limestone, marble; (7) Tertiary sheared diorite. After Panteleyev and Pearson (1977); Stevenson (1948), from LITHOTHEQUE; and Kinkel et al. (1956).

in a shear, and underlain by a copper-enriched zone. The smaller Twin J deposit (Fig. 12-9) has several lenses of fine massive to disseminated black Zn-Pb minor Cu sulphide ore in a barite and calcite gangue, hosted by a gray marble, black phyllite and a sericite schist. This ore grades into footwall stringers, veinlets and blebs of coarser chalcopyrite associated with a white vein quartz, or included in a sheared augen quartz-feldspar porphyry and diorite. The meta-sedimentary band hosting the ore is strongly deformed in the axial zone of folds, and displaced by faults. If this is a volcanogenic mineralization at all, it was considerably modified by tectonism and intrusive magmatism.

In the Shasta district of northern California, two age groups of massive sulphides occur. In the West Shasta district 15 km north-west of Redding (Kinkel et al., 1956; Fig. 12-9), nine orebodies are found in the middle Devonian Balaklala Rhyolite Formation: a metamorphosed fine-grained to porphyritic dacite grading into trondhjemite. The immediate host rock to the massive and stockwork Fe,Zn,Cu sulphide orebodies is a "quartz-eye rhyolite" believed to be located on the flanks of a former dacite dome. The felsic unit is underlain by a greenstone. The much less productive East Shasta District is hosted by the silicic facies of the Permian and Triassic bimodal volcanic-sedimentary sequence (Albers, 1981).

The northwest-trending Sierra Nevada Foothills Copper Belt contains about ten small massive sulphide deposits and several times as many occurrences, located in a chlorite schist (sheared greenstone; e.g. Copperopolis; Albers, 1981), or in a meta-quartz keratophyre. The Copperopolis pyrite--chalcopyrite deposit takes the form of a narrow sheet to lens about 2 km long, surrounded by massive mafic metavolcanics, and intruded by a granodiorite. It is similar to the mafics-hosted deposits in the Sambagawa belt of Japan. The meta-keratophyre hosted deposits in the area are zinc-rich.

THE APPALACHIAN BELT

Numerous massive sulphide deposits hosted by bimodal Paleozoic metavolcanic suite occur along a 3,000 km long stretch between Newfoundland and Alabama. The deposits located in the Canadian Maritime Provinces, eastern Quebec and New England, a region affected by the Quaternary glaciation, are fresh, well-exposed and base metal dominated. The numerous deposits and ore occurrences in the southern Appalachians particularly in the Piedmont Belt, are deeply weathered together with their host rocks and still imperfectly understood. Most were mined for gold in the past.

In the Ordovician region of Newfoundland, the low-grade Point Leamington orebody (Noranda Mines, Ltd., 1974; Fig. 12-10), is a conformable siliceous lens of fine pyrite with subordinate sphalerite and chalcopyrite, topping a coarse rhyolite meta-pyroclastic unit that in turn overlies pillow basalts. The Gull Pond mine (Upadhyay and Smitheringale, 1972; Fig. 12-10) is a conformable lens of massive to disseminated pyrite, pyrrhotite and chalcopyrite located at the contact of the footwall pilow meta-basalt, and the hangingwall meta-pyroclastics. The felsic unit becomes coarser moving upwards so that the sequence may be overturned. The orebody is surrounded by a tabular zone of cordierite, andalusite, chlorite schist that in turn is enveloped by a cordierite-anthophyllite rock. This envelope is believed to be the composite result of (1) synvolcanic alteration and mineralization, (2) regional metamorphism that produced tight folds

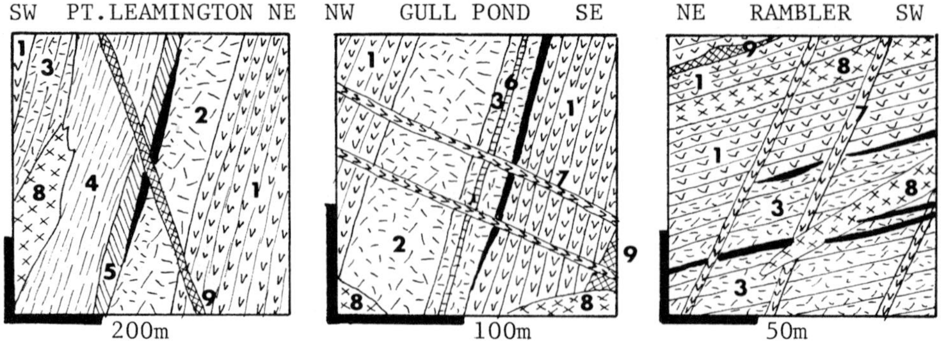

Fig. 12-10. Newfoundland bimodal volcanogenic massive sulphide deposits (Ordovician). LEGEND: (1) pillowed metabasalt; (2) metarhyolite pyroclastics (breccia), coarse, fining upward; (3) felsic metatuff (quartz sericite schist); (4) meta-argillite; (5) siliceous argillite; (6) chert, jasper; (7) diabase; (8) diorite; (9) ultramafics. From LITHOTHEQUE, based on data in Noranda Ltd. (1974), Upadhyay and Smitheringale (1972), and Tuach (1974).

with a strong axial plane schistosity and (3) contact metamorphism by a Silurian-Devonian granodiorite-diorite intrusion.

The Rambler group of orebodies (Tuach and Kennedy, 1978; Fig. 12-10) is in tightly folded, overturned and greenschist-metamorphosed felsic volcanic pile underlain by Ordovician ophiolite basalts, and intruded by a Devonian granodiorite. The individual orebodies include a peneconcordant pyrite, sphalerite, chalcopyrite lens on top of the felsic sequence (Rambler), as well as a massive, disseminated and stringer pyrite, pyrrhotite, chalcopyrite ore in quartz-chlorite schist (East, Ming mines).

In eastern Quebec and adjacent Vermont and Maine (Sauvé et al., 1972), the Cambrian and lower Ordovician volcanic-sedimentary sequence is in four parallel belts and include a belt of ophiolite, melange, and a former island arc suite. Bimodal association is developed in the Stoke Range between Sherbrooke and Stratford, and it hosts five small size Fe-Zn-Cu massive sulphide deposits, and numerous additional showings. Most of the deposits are ordinary massive sulphide lenses hosted by a siliceous rock ("metachert", "iron formation") between felsic and mafic metavolcanics or metasediments, and all orebodies are parallel with schistosity that in most cases is also parallel with the original lithology. The Weedon mine is remarkable for disruption of its orebody by an intrusion of a post-tectonic Devonian granite (Fig. 12-11). The disrupted mineralized zone under the granite is marked by frequent changes in the orientation of beds, a jumble of variously

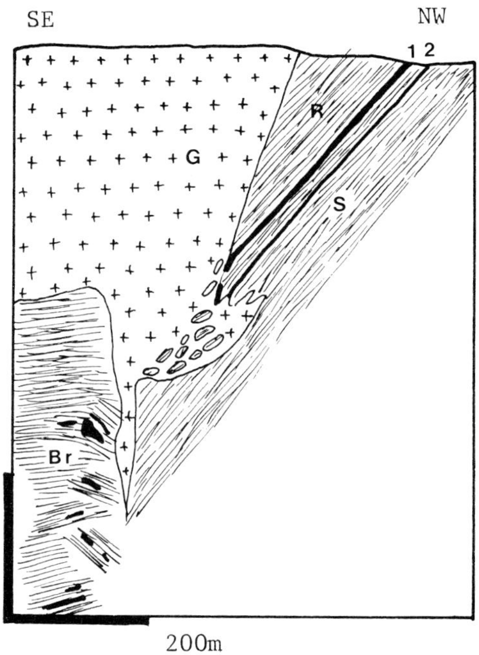

Cambrian - Lower Ordovician

R meta-rhyolite, quartz-sericite schist, siliceous biotite hornfels

S hornblende hornfels to greenschist

1 two parallel massive
2 Fe,Cu,Zn sulphide lenses

Devonian

G post-tectonic, biotite granite

Br disturbed and brecciated zone

Fig. 12-11. Weedon massive sulphide deposit, Eastern Townships, Quebec. A section showing, diagrammatically, the interruption, disruption and brecciation of a persistent ore horizon by granitic intrusion. After Sauvé et al. (1972).

oriented blocks, and diking. Cordierite and anthophyllite are important minerals formed in places where the thermal granite aureole was superimposed on the synvolcanic alteration envelope.

The Bathurst-Newcastle ore district in New Brunswick was discovered in the late 1950's and it contains over 25 Zn, Pb, Cu and Ag massive to disseminated sulphide deposits hosted by the Ordovician Tetagouche Group. There is an extensive recent literature describing the district (e.g. Davis, 1972; McAllister and Lamarche, 1972; Potter et al., 1972; Harley, 1979), as well as summary reviews (e.g. Franklin et al., 1981). The Tetagouche Group in the district (Fig. 12-12) consists of four supracrustal units: (1) metasedimentary slate, argillite, arenite now converted to phyllites and schists; (2) mafic metavolcanics, dominantly meta-spilite and chloritic phyllite; (3) quartz-feldspar augen schist and associated meta-tuffs; this includes an iron formation and a pyritic schist and (4) felsic metavolcanics, originally mostly rhyolite and dacite, now quartz-sericite and biotite-chlorite schists. To this should be added a variety of syn- and post- depositional intrusive rocks.

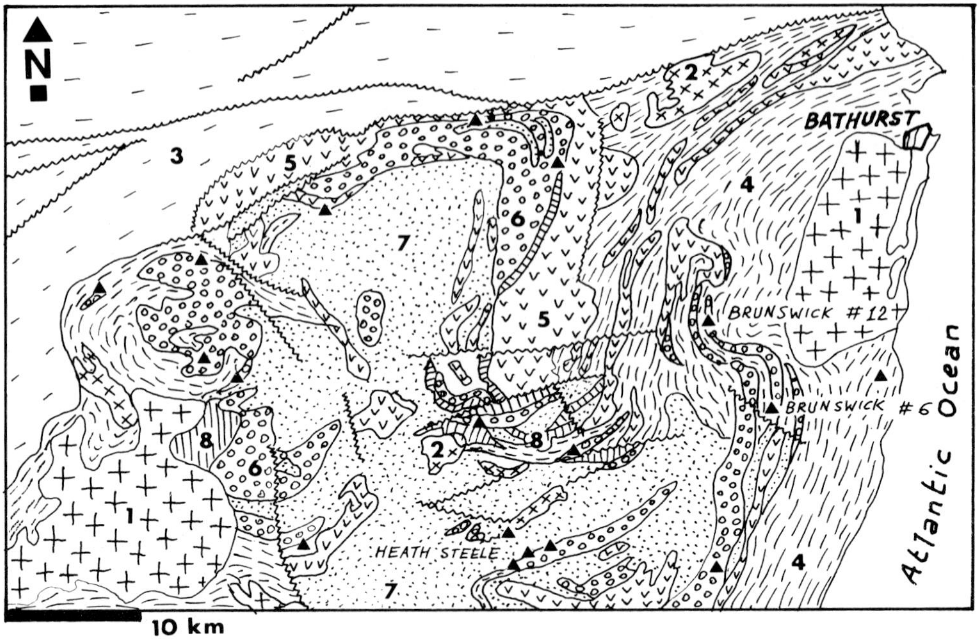

Fig. 12-12. Bathurst-Newcastle Zn,Pb,Cu massive sulphide district, hosted by the Ordovician Tetagouche Group. LEGEND: (1) Devonian biotite granite, granodiorite; (2) Devonian diabase, gabbro; (3) Ordovician to Devonian sediments and minor volcanics; Ordovician Tetagouche Group: (4) gray and varicoloured shale, graywacke, quartzite, phyllite; (5) mafic metavolcanics, minor graphitic schist, banded iron formation; (6) augen quartz-feldspar porphyroid; (7) felsic metavolcanics; (8) siliceous metasediments. Triangles: massive sulphide locations. Simplified after a map in Potter et al. (1972).

The rocks have been penetratively deformed and greenschist metamorphosed and this makes the stratigraphic interpretation controversial. Harley (1979) assumed an evolutionary history starting with an initial continent-derived sedimentation (unit 1), grading into early felsic (units 3,4) and later dominantly mafic (unit 2) volcanic episodes, and terminating with immature clastic sedimentation.

The massive sulphide deposits occur in a broadly circular, domal volcanic-sedimentary complex, interpreted by Harley (1979) as a resurgent caldera. Although the sulphide deposits occur at several stratigraphic levels, most are hosted by the metasediments situated near the outer or upper contacts of the felsic volcanic pile. The majority of the pyrite, sphalerite, galena, chalcopyrite and pyrrhotite massive sulphides are interpreted as having originally been stratiform, though modified by several deformation events to such a degree that the early history cannot be directly interpreted. The largest fine-grained massive sulphide bodies are banded and usually located at or near a contact between the augen schist and some other rock, commonly chloritic metasediments. Several smaller orebodies form discontinuous lenses in brecciated felsic lavas. Small vein Zn-Pb deposits (Keymet, Nigadoo) occur in Silurian rocks on the fringe of the massive sulphide region.

Numerous but small and poorly exposed base-metal gold-rich deposits occur in several parallel runs in the greenschist to amphibolite facies — metamorphosed Cambrian to Ordovician Chopawamsic Formation in central Virginia (Pavlides et al., 1982).

THE CARIBBEAN

The lower Cretaceous metavolcanic and metasedimentary belt of Cuba extending to Hispaniola has, at least locally, bimodal characteristics. Numerous small gossans and oxidized Cu occurrences, partly indicating the presence of volcanogenic pyritic massive sulphides, are known on both islands. The Buena Vista mine in western Cuba and the Antonio mine in central Cuba, representing a total of some 20 Tt of copper (Bogdanov et al., 1965), are the better known examples.

THE CALEDONIAN OROGENIC BELT OF EUROPE

Within the Caledonian belt of north-western Europe, bimodal associations hosting massive sulphide deposits are known in western Britain, eastern Ireland and in several regions in Norway and Sweden. Avoca and Anglesey are very closely associated with a felsic suite, so they have been treated in Chapter 11. Portions of the Scandinavian Caledonides represent a classical area of study and interpretation of metamorphosed massive sulphide deposits covereed by an extensive literature, including several recent reviews in English (e.g. Vokes, 1976; Franklin et al., 1981). In the mineralized area of central Norway (Trøndelag) and adjacent western Sweden, several hundred recorded ore occurrences and small-to-medium sized massive sulphide deposits occur in mostly schistose metavolcanics and metasediments at several stratigraphic levels, ranging from Proterozoic to Silurian. The total production and reserves are of the order of 1.5 t Cu and 2.2 Mt Zn. The most important host unit is the lower Ordovician Støren Group, followed by the Silurian Lasterfjall Group. Most of the host units are allochthonous, members of the easterly-transported complex of nappes.

The Støren Group in Trondhjem region (Nilsen, 1978) is metamorphosed within the range of lower greenschist to upper amphibolite facies, and it consists of an interbedded sequence of greenstone (metabasalt) pillow lavas, pillow breccias, tuffites, felsic metatuffs and graywacke, mudstone, volcanogenic metasediments. This is penetratively deformed in a complex manner. Siliceous metasediments (quartzite, metachert), siliceous iron formations, minor limestone and lenses and pods of ultramafic tectonites, also occur. The felsic metavolcanics, as elsewhere, occur either as rather persistent volcaniclastic bands, or localized "bulges". Sodic intrusions coeval with volcanism (trondhjemites, gabbros) are widespread. Waltham (1968) argued that many are metasomatically recrystallized keratophyre lavas. The Lasterfjall Group in Stekenjokk area (Zachrisson, 1971) is similar, except for its greater proportion of graphitic and calcareous phyllites and impure carbonates.

The numerous ore deposits are impossible to summarize in the short review here. Some have been treated in other chapters, such as Lökken, Killingdal and Joma hosted entirely by mafic metavolcanics (Chapter 10), or Bleikvassli, hosted by high-grade metamorphics with signs of initial granitization (Chapter 29). Waltham (1968) distinguished six principal "types" of stratabound mineralization, some including iron oxides in addition to sulphides: (1) banded magnetite-quartz ore with a low content of accessory chalcopyrite and sphalerite, located along mafic/felsic metavolcanic contacts, e.g. Fosdalen; (2) banded magnetite, pyrite, quartz ore, with accessory sphalerite and chalcopyrite, Tverrfjellet; (3) banded, concordant coarse-grained massive pyrite with minor magnetite, chalcopyrite and sphalerite, found along acid pyroclastics in the footwall, and "iron formation" and mafic metavolcanic in the hangingwall, contacts; Folldal, Tverrfjellet, Sulitjelma, partly Røros; (4) fine-grained pyrite with interstitial silicate gangue and sphalerite, chalcopyrite forming conformable lenses in horizons of graphite-pyrrhotite schist or a pyritic schist, in dominantly mafic association; Lökken, Joma, Gjersvik; (5) pyrrhotite-chalcopyrite ore, a variation of the coarse pyritic orebodies, confined to their margins, footwalls, or cementing the fragmented massive pyrite; Alvdal, Joma, Folldal, etc. and (6) Independent pyrrhotite-chalcopyrite orebodies hosted by greenstones, close to felsic intrusions (Røros, partly).

All the Caledonian deposits now popularly treated as stratabound are penetratively deformed (typically they are in axial, fold and cleavage zones and are often rod or ruler-shaped), metamorphosed and dismembered. Their original appearance has rarely been convincingly interpreted, and the stereotype "submarine-exhalative" origin is questioned by some geologists (e.g. Rui, 1973).

FRANCE

Mineralized metamorphosed Devonian-Lower Carboniferous spilites and keratophyres are present in the Brévenne group (eastern border of the Massif Central; Bardin, 1971). The large but almost exclusively pyritic Sain Bel orebody is strongly reminescent of Rio Tinto, Spain and there is a small but copper-rich Chessy deposit in the same region.

Pierrefitte (Pouit et al., 1979), and several additional small occurrences on the French side of the Pyrenees, are massive to disseminated stratabound lenses located along contacts of meta-spilites and meta-sediments, intruded by subvolcanic bodies of quartz meta-keratophyres. Grunerite, stilpnomelane, chlorite, ankerite and barite "iron formation" mark the ore horizon.

RIO TINTO (SPAIN) - LOUSAL (PORTUGAL) PYRITE BELT

This approximately east-west trending, 230 x 40 km, Devonian to lower Carboniferous volcanic--sedimentary belt, straddles the Spanish/Portuguese border and has a reputation as the world's largest pyrite accumulation (about 1 Bt pyrite). It is also a classical region treated in most textbooks on economic geology and it was recently studied in considerable detail (Strauss and Madel, 1974; Strauss et al., 1977). The belt is a portion of a late Paleozoic "geosyncline", a marine succession deposited on an older crystalline basement. The basal " slate-quartzite group" is conformably overlain by the "volcanic-sedimentary complex" of variable thickness (up to 800 m). This complex hosts all the sulphide and Mn ore deposits. This sequence is topped by a lower Carboniferous flysch, a monotonous unit of alternating slate and graywacke.

The volcanic-sedimentary complex is marked by rapid facies change. It is dominated by gray, black and varicoloured slate, siliceous slate, chert, jasper, and minor limestone. These are interlayered with fine-grained spilites, most of which are pillow lavas, and with quartz keratophyre pyroclastics and rare autobrecciated lavas. Felsic eruptive centres consist of coarse, massive pyroclastics controlled by east-west to S.E.-N.W. faults.

The dominant styles of mineralization are thick to thin, folded lenses of fine-grained massive pyrite with subordinate quantities of chalcopyrite, sphalerite and galena. About sixty ore lenses have been mined. The ore is massive or banded, usually homogeneous, and the orebodies are confined to the vicinity of the felsic extrusive centres where they top the acid pyroclastics, or are interbedded with black slates. The massive sulphide lenses may be topped by siliceous shales to jaspers, enriched in manganese. About three hundred small lenses composed of rhodochrosite, rhodonite, pyrite and sometimes spessartite hosted by radiolarian chert or ankerite horizons have been mined.

The massive sulphide lenses at some localities are underlain by disseminated sulphides, located in chlorite-altered footwall pyroclastics. At Louzal, the ore contains pyrite, pyrrhotite, chalcopyrite, sphalerite and galena in quartz gangue (Strauss and Madel, 1974), that is, the same minerals present in the massive sulphide lenses. The footwall stockwork ores are not particularly common except in the Rio Tinto (town) group of deposits. There, the Nuevo Planes orebody is a vertical mushroom-shaped feeder channel composed predominantly of pyrite. The Cerro Colorado and Alfredo deposits near Rio Tinto are stockworks rich in chalcopyrite, superimposed on an earlier pyritic stockwork, and hosted by chloritized acid volcanics. The stockworks are interpreted either as feeders to the massive sulphides, or they represent a subsequently emplaced hydrothermal mineralization of the "stockwork" or "porphyry copper" style.

The Rio Tinto deposits are deeply weathered, having a thick goethitic gossan (Mn oxide gossans are found over the manganiferous lenses). The gossans are rich in residual gold and silver and have been mined since Roman times. Even now, the remainder of the gossan topping the Cerro Colorado stockwork contains some 50 t Au and 940 t Ag.

BOHEMIAN MASSIF

Scattered small massive sulphide occurrences associated with bimodal volcanics and sediments, occur in the Erzgebirge Mountains (e.g. Zlatý Kopec; there, magnetite, pyrrhotite, sphalerite and chalcopyrite ore lens is contact metamorphosed to "skarn" by granitic intrusions and veined by cassiterite). Economically important orebodies, however, have recently been outlined in the Jeseník Mts. of north-central Czechoslovakia, associated with a Devonian metavolcanic and metasedimentary suite. In the Zlaté Hory field (Gruzczyk and Pouba, 1968), several lenses of massive to disseminated pyrite, sphalerite, chalcopyrite, galena and pyrrhotite are hosted by sericite quartzite and greenschist, interbedded with quartz-sericite phyllites, porphyroids and black phyllites. The best orebodies lie in the crests of folds.

The smaller Horní Benešov deposit contains a 1,500 m long zone of peneconcordant lenses of massive to disseminated pyrite, sphalerite and galena in meta-keratophyre (sericite schist), marble and calcareous slate. The sulphide ore is gradational into a barite orebody, and several lenses of siliceous stratiform hematite of the Lahn -Dill "type" are found at contacts with the associated spilitic tuffs ("Schallstein"). Horní Benešov thus represents an interesting transition. The sulphide ores are partly remobilized into veins. The smaller Horní Město deposit is similar. Several Carboniferous Pb-Zn vein deposits in the Nízký Jeseník Mts. (e.g. Nová Ves u Rýmařova) may represent remobilized Devonian volcanogenic ores.

THE ALPS AND CARPATHIANS

Several clusters of pyritic massive sulphide deposits are associated with metamorphosed Paleozoic and minor Mesozoic bimodal volcanic-sedimentary terrains, distributed in the internal zones of the Alpine and Carpathian mobile belts in Austria, Czechoslovakia, and Rumania.

In the "Grauwackenzone" of Austria (Höll and Maucher, 1976; Schulz, 1972), small deposits in the Kitzbühel, Mittersil and Oblarn areas, form fine-grained bedded pyrite, pyrrhotite, chalcopyrite and sphalerite masses hosted by sericite, quartz-sericite and chlorite schist. Basic metavolcanics are usually in the hangingwall.

In eastern Slovakia, the Cambrian to Silurian Gelnica Series, already discussed in Chapter 11, also contains local areas with substantial development of mafic metavolcanics, resulting in bimodal centres. The Smolník pyritic copper ore field is the largest (Ilavský, 1976). There, Silurian chloritic schists (metavolcanics) interbedded with a graphitic phyllite, chert and porphyroid, contain six conformable massive sulphide lenses, partly cut by remobilized epigenetic veins.

In the Carpathians of Rumania, an economically significant massive sulphide province is hosted by the upper portion of the lower Cambrian Tulgheş Series, dominated by felsic rhyodacitic volcanics and pyroclastics interbedded with chlorite-sericite phyllites and minor chlorite schists (Superceanu, 1967; Ianovici and Borcoş, 1982). The orebodies in the three most important fields (i.e. Lesu Urşului, Fundu Moldovei and Bălan), are massive lenses of pyrite and subordinate sphalerite, chalcopyrite, grading into disseminations in the footwall chloritic schist.

The late Proterozoic Altîn-Tepe field in Dobrogea (N.E. Rumania; Ianovici and Borcoş, 1982), has a stratabound horizon of an alternately banded magnetite and pyrite, minor chalcopyrite, quartzite, enveloped by quartz, sericite, chlorite schists. Local massive pyrite lenses contain some sphalerite, galena and barite.

THE URAL OROGENIC BELT, U.S.S.R.

The Ural belt contains the world's most extensive Phanerozoic volcanogenic massive sulphide province 2,000 km long and containing about 100 producing deposits and thousands of showings (Smirnov, ed., 1968; Fig. 12-13). The massive sulphides formed in Cambrian, Ordovician, Silurian, Devonian and lower Carboniferous periods and are hosted by a 7,500 m thick sequence. The bulk of the deposits are in the north-south trending Tagil-Magnitogorsk synclinorium. The age of the Uralian deposits generally decreases from north to south and from west to east, and so does the intensity of metamorphism. The Cambro-Ordovician deposits in the Polar Urals are metamorphosed in the amphibolite facies, the Silurian and Carboniferous deposits in the southern and south-eastern Urals suffered only a lower greenschist metamorphism.

The Uralian massive sulphides can be subdivided into those associated with the bimodal ("contrast differentiated") suite and with the basalt, andesite, rhyolite ("sequence-differentiated") suite. The former is dominant and neither suite is sharply separated geographically. Deposits affiliated with either suite, as well as transitional deposits, may be present in the same mining field. Silurian and Devonian deposits are the most common and are the most important economically. The Silurian massive sulphide province in the

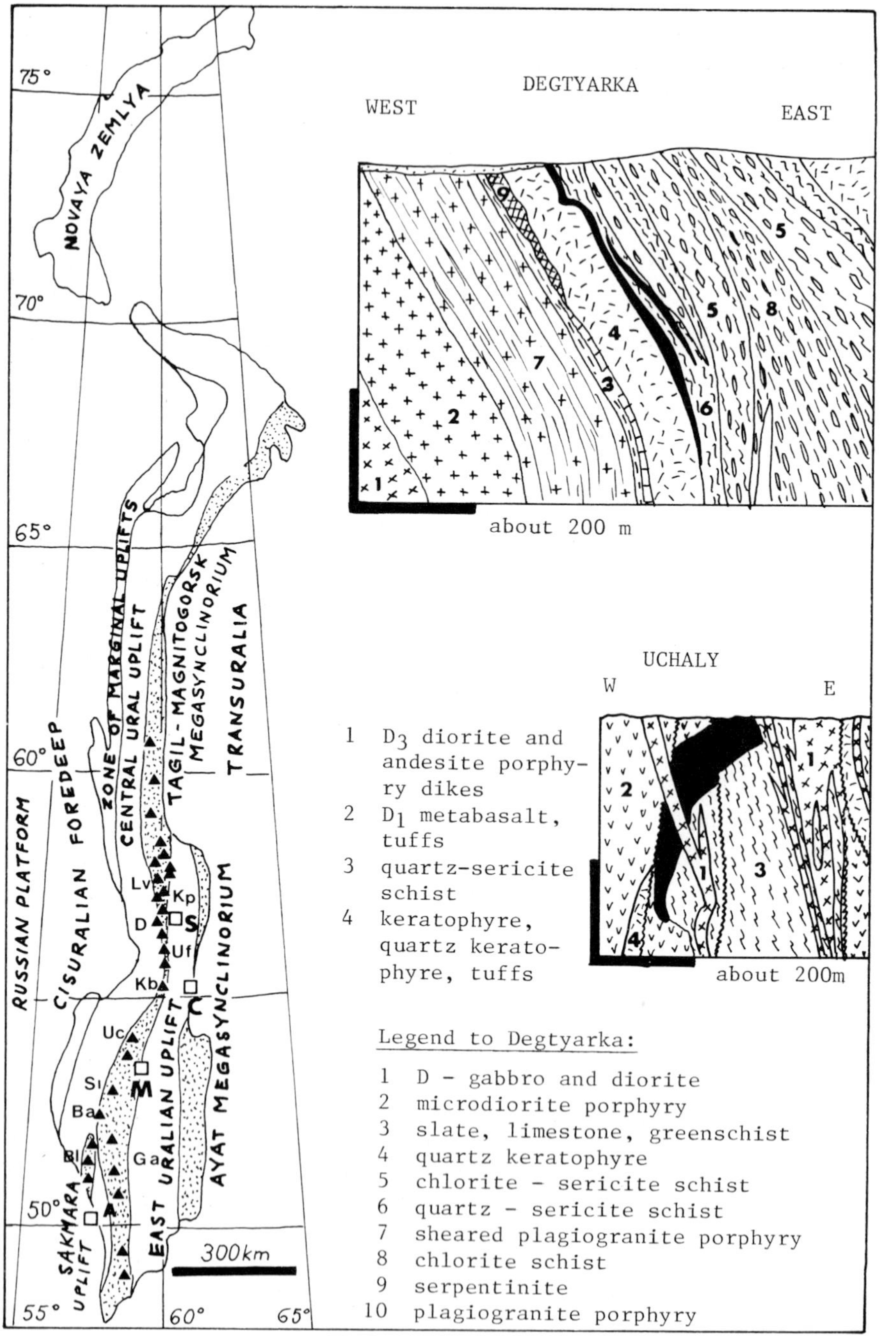

75°
70°
65°
60°
50°
55°
60°
65°
NOVAYA ZEMLYA
RUSSIAN PLATFORM
CISURALIAN FOREDEEP
ZONE OF MARGINAL UPLIFTS
CENTRAL URAL UPLIFT
TAGIL-MAGNITOGORSK MEGASYNCLINORIUM
TRANSURALIA
EAST URALIAN UPLIFT
AYAT MEGASYNCLINORIUM
SAKMARA UPLIFT
Lv
Kp
D
S
Uf
Kb
C
Uc
Si
M
Ba
Bl
Ga
300km
DEGTYARKA
WEST
EAST
about 200 m
UCHALY
W
E
about 200m
1 D3 diorite and andesite porphyry dikes
2 D1 metabasalt, tuffs
3 quartz-sericite schist
4 keratophyre, quartz keratophyre, tuffs
Legend to Degtyarka:
1 D - gabbro and diorite
2 microdiorite porphyry
3 slate, limestone, greenschist
4 quartz keratophyre
5 chlorite - sericite schist
6 quartz - sericite schist
7 sheared plagiogranite porphyry
8 chlorite schist
9 serpentinite
10 plagiogranite porphyry

central Urals (around Sverdlovsk) contains a marginal part composed of the products of spilite-keratophyre volcanism, grading into the andesite-dominated central portion of the "geosynclinal" trough. The spilite-keratophyre volcanism was cyclic (2-3 cycles), and each cycle started with the outpouring of submarine spilite lavas and closed with the emplacement of keratophyre lavas and their localized subvolcanic equivalents. Massive sulphides formed towards the end of each cycle, and are often located at intravolcanic unconformities.

A typical sequence of the middle Devonian volcanics in the southern Urals (south of Magnitogorsk) starts with basalts with subordinate spilites. This is followed by dominant spilites interbedded with jasper and siliceous slate, and terminated with keratophyres and their tuffs. The massive sulphide deposits are located near the base of the keratophyre suite and are covered by essentially unmineralized upper Devonian cherts, tuffs and mafic lavas.

As in the rest of the world, the Uralian massive sulphide deposits tend to congregate in clusters, usually surrounding a felsic volcanic centre. A typical centre posesses the characteristics of a "brachyanticlinal uplift" (Ivanov et al., 1962). Within such an uplift, orebodies are located on flanks or on the top. The anatomy of the Uralian massive sulphides is identical to those in the rest of the world: thick stratabound massive pyrite and minor sphalerite, chalcopyrite and exceptionally galena lenses, underlain by sulphide disseminations or stringers in the quartz, sericite, chlorite altered footwall felsic pyroclastics. Almost all the deposits are situated in strong zones of dynamometamorphism. Tonnage and grade figures on the Soviet base and precious metal deposits have not been released since the outbreak of the World War II, so that the data given in Table 12-3 are extrapolated figures from the 1930's or pure estimates based on geological descriptions.

The Degtyarka ore field with its original content of at least 120 Mt of massive sulphide ore, is probably the largest one. It is located 40 km S.W. of Sverdlovsk in the central Urals and is hosted by the

---

Fig. 12-13. The Paleozoic Ural orogenic belt, U.S.S.R., and its massive sulphide province. The present outline of the belt is due to a Cainozoic uplift, so that only some facies zones are represented and the eastern culmination of the orogen is buried under Mesozoic and Cainozoic sediments. Within the present bedrock outcrop, two (three in the south) parallel "eugeosynclinal" belts (stippled) only host volcanogenic massive sulphides, shown by triangles.

Locality abbreviations: Lv=Levikha; Kp=Karpushikha; D=Degtyarka; Uf=Ufalei; Kb=Karabash; Uc=Uchaly; Si=Sibay; Ba=Baymak; Ga=Gai. Large centres: S=Sverdlovsk; C=Chelyabinsk; M=Magnitogorsk; A=Aktyubinsk. The two representative examples of the Uralian massive Fe,Cu,Zn sulphide deposits: Degtyarka and Uchaly, are modified after Ivanov in Magak'yan (1968), and Borodaevskaya (1962).

upper Silurian-lower Devonian bimodal volcanic-sedimentary sequence. The orebody (Magak'yan, 1968) is a flat, 4,800 m long lens of fine-grained pyrite and subordinate chalcopyrite and sphalerite. There is some accessory bornite, tetrahedrite, galena and arsenopyrite, located along the contact of quartz-albite greenstone (i.e. chlorite-altered felsic pyroclastic ?), and a quartz-sericite schist. A substantial quantity of the copper produced (about 1.5 Mt Cu) came from the zone of secondary enrichment, rich in chalcocite.

NORTHERN CAUCASUS

The massive sulphide deposits in the Devonian volcanics of the northern Caucasus (Urup, Beskes and Khudes) have been well described in the literature (Samonov and Pozharisky, 1974; Smirnov, 1968; Skripchenko, 1972). In the largest Urup field (Fig. 12-14), the ore-bearing sequence has a phyllite on the base, topped by basalts, quartz keratophyres, mafic tuffs and quartz keratophyre tuffs. Massive pyrite and lesser chalcopyrite, sphalerite lenses are located in all the volcanic horizons, but are most widespread along the contacts of quartz keratophyres (footwall) and mafic tuffs (hangingwall). The gently dipping (10-40°) Urup main ore zone is a thin, persistent lens conformable with the hangingwall tuffs and peneconcordant with the footwall rocks; the footwall is probably an intravolcanic unconformity. The footwall volcanics are highly altered (propylitized), but the most intensive chlorite alteration containing pyrite-chalcopyrite stringer stockworks has the form of several discreet alteration pipes. The main Urup ore layer consists of bands of fine grained, massive ore, interlayered with a siliceous slate. The orebodies are compositionally zoned, from footwall to hangingwall: (1) pyrite band; (2) pyrite, minor sphalerite, chalcopyrite banded ore; (3) banded chalcopyrite-sphalerite; (4) banded and brecciated siliceous chalcopyrite ore and (5) thinly banded siliceous pyrite, minor sphalerite, chalcopyrite ore. These bands are topped by a layer of mechanically redeposited ore conglomerate, and finally by red siliceous hematitic slates. The Khudes deposit is hosted entirely by metabasalt, and has been reviewed in Chapter 10.

EASTERN TURKEY

The large Madenköy deposit hosted by upper Cretaceous dacite pyroclastics is interesting as one of the youngest Cu-Zn deposits associated with bimodal volcanics, but transitional into the sequentially differentiated basalt, andesite, rhyolite association. It was reviewed by Franklin et al. (1981). The mineralization consists of an upper sphalerite, chalcopyrite, pyrite, barite zone underlain by a pyrite, chalcopyrite ore. Adjacent to the massive sulphide is a massive gypsum zone, and under the ore lens is a copper-rich pyritic stockwork zone in predominantly sericite—siderite altered and silicified rocks. There is a considerable similarity to the Japanese "Kuroko"-style ores.

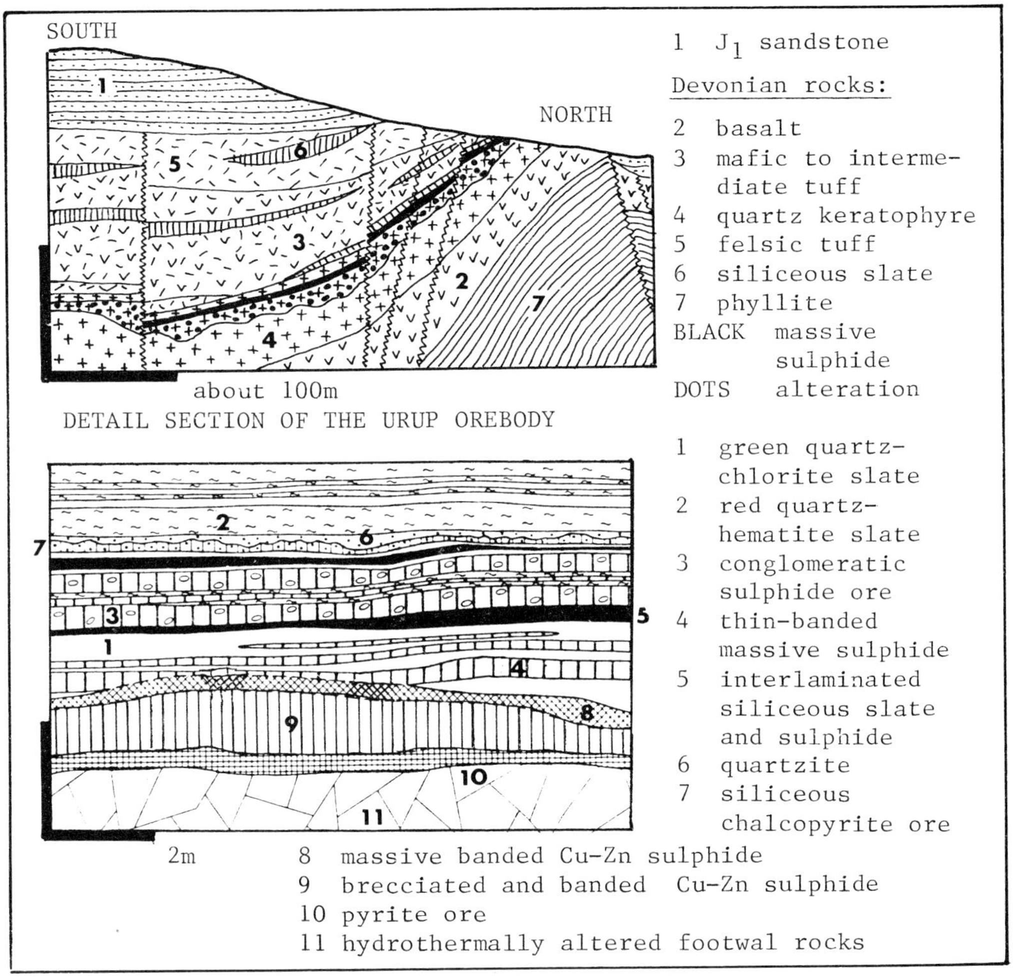

Fig. 12-14. Urup massive sulphide deposit, hosted by a Devonian bimodal association. After Skripchenko (1972).

SOVIET CENTRAL ASIA AND SIBERIA

Scattered occurrences of volcanogenic massive sulphides are known to occur in Soviet Central Asia, but the literature coverage available here is patchy and incomplete.

The Maikain deposit in northern Kazakhstan (Nekrasoff, 1935) is a massive pyrite-chalcopyrite lens resting on top of a sheared and altered Devonian quartz keratophyre. It started as a gold mine producing residual gold from the limonitic gossan, and in style this deposit has often been compared with Rio Tinto.

Several Devonian ore fields in the Rudnyi Altai mineralized belt (described in Chapter 11 because of the predominant felsic association of most orebodies) are members of the bimodal suite. In the Berezovsk-Belousovka field (Ivankin et al., 1961), more than ten long, persistent stratabound lenses of massive to disseminated pyrite and lesser pyrrhotite, chalcopyrite, sphalerite and galena occur along a 8 km long zone, and are spatially associated with a quartz, ankerite, hematite iron formation and a quartz-sericite schist. Several chlorite-altered stockworks with a stringer mineralization are located in the footwall felsic pyroclastics. A granitic intrusion within the ore field converted a portion of the ore horizon into a "skarn" and generated numerous hydrothermally remobilized veins mined mostly for their gold content.

The Karchiga deposit south of Zyr'yanovsk (Ivankin et al., 1961) occurs at the contact of a Devonian amphibolite (metabasalt) in the footwall, and a quartz, feldspar, muscovite schist (felsic metatuff ?) in the hangingwall. A post-ore granitic intrusion produced an anthophyllite-cordierite alteration envelope and converted a portion of the volcanogenic pyrite into pyrrhotite.

The important Salair ore field in Siberia (Lapukhov, 1975; min. 900 Tt Zn, 200 Tt Cu; Fig. 12-15) consists of several irregular lenses of massive pyrite, barite, lesser sphalerite, chalcopyrite and galena, located in a strongly silicified apical portion of a Cambrian quartz keratophyre dome (or a stock), or along its contact with keratophyre metatuffs, marbles, black siliceous slates, and a porphyritic metabasalt. The volcanogenic ore underwent a substantial regeneration during the Devonian deformation and granite emplacement.

The Kyzyl Tashtyg field in Tuva near the Mongolian border (Smirnov, ed., 1968) is a banded massive sulphide lens in a black and light coloured quartz, sericite, chlorite schist associated with a Cambrian hydrothermally altered felsic dome, surrounded by mafic metavolcanics.

JAPAN

In the Hitachi mine situated in the Abukuma Plateau, eastern Honshu (Nippon Mining, 1966), twelve groups of massive to disseminated pyrite, lesser pyrrhotite, chalcopyrite and sphalerite lenses occur parallel to schistosity of Carboniferous amphibolites and greenschists. Biotite schists (metasediments) and quartz-sericite schists (felsic metavolcanics ?) are less important hosts.

The Yanahara mine in southern Honshu (Oshima, 1964; Fig. 12-16), is the largest pyritic deposit of Japan. There, several peneconcordant massive pyrite lenses are confined to a horizon of Permian felsic metatuffs, topped by a thick, flysch-like Permian slate and minor arenite sequence. Chalcopyrite and sphalerite are present in accessory quantities through most of the deposit (grades are 0.1-0.2% Cu) and only locally form richer sections. The genetic interpretation has been rendered difficult by the presence of large

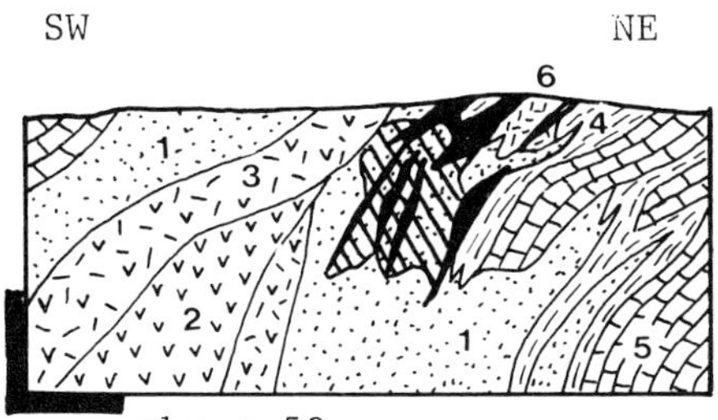

Fig. 12-15. Salair ore field, Siberia, Kvartsitovaya Sopka deposit. Massive and stockwork pyrite, barite, Pb,Zn,Cu sulphide mineralization is in a quartz keratophyre centre in a Cambrian bimodal volcanic-sedimentary association. Modified after Ivankin, ed. (1973).

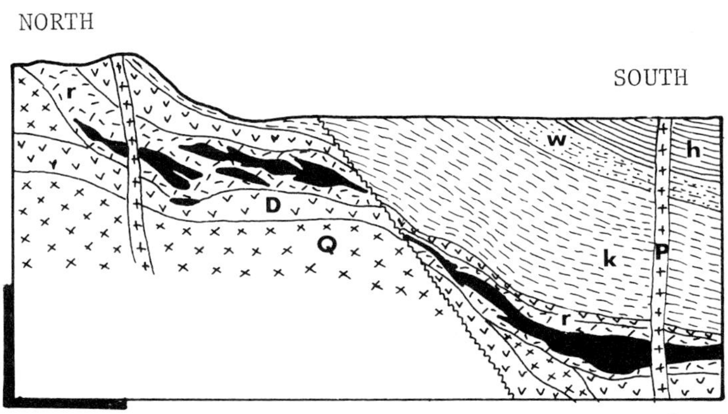

Fig. 12-16. Yanahara massive pyrite deposit, S.W. Honshu, Japan. From LITHOTHEQUE, slightly modified after Oshima (1964).

masses of greenstone interpreted as an intrusive diabase; diorite (epidiorite) and diorite, felsite and quartz-feldspar porphyry cross-cutting dikes. It is possible that the "diabase" is a retextured basalt.

PHILIPPINES

Bimodal volcanic-sedimentary association is widely developed in the Philippines (Bryner, 1969) and it contains several small massive sulphide deposits. The Basement Complex of Permian or Carboniferous age (a correlate of the Sambagawa and Chichibu belts of Japan) is dominated by amphibolite to greenschist and a quartz-felspar micaschist to phyllite, with local bodies of felsic metavolcanics, marble and serpentinite. The small Hixbar Cu-Zn massive sulphide deposit is hosted by quartz-sericite and paragonite schist (altered felsic metavolcanics), enveloped by a chlorite schist.

The upper Cretaceous to Paleogene "Spilite-Graywacke" sequence (Bryner, 1969) is up to 5,000 m thick and in a typical section consists of low-grade metamorphosed pillowed spilite and mafic lava flows, intercalated with volcanic arenites, slates and cherts. Towards the top, the proportion of felsic volcanics, volcaniclastics and limestones increases. The massive sulphide occurrences are represented by the Bagacay copper deposit containing a pyritic chalcopyrite-sphalerite orebody, located on top of an altered keratophyre and overlain by bedded jasper, tuff, and younger volcanics.

AUSTRALIA

The largest Australian Phanerozoic massive sulphide deposits (Rosebery, Woodlawn, Captain's Flat, Benambra) are intimately associated with dominant felsic volcanics and subvolcanic rocks, so that they have been reviewed in Chapter 11. The small Peak Downs copper deposit near Clermont, Queensland (Cox, 1975) appears to be hosted by a bimodal suite of lower Paleozoic metamorphics (quartz-chlorite and quartz-sericite schist), containing intercalations of metaquartzite and a banded iron formation. The iron formation is a concordant and rhythmically banded lens up to 2.5 m thick, composed of alternating quartz-pyrite and quartz-magnetite laminae. Chalcopyrite is disseminated in small quantities in the iron formation, as well as in the lenticular bodies of a quartz-chlorite schist. The 17 Tt Cu produced in the past came entirely from the oxidized and secondary enriched ore.

AFRICA

Only two regions composed of late Proterozoic to early Paleozoic bimodal volcanic-sedimentary associations containing massive sulphide deposits of any importance, have been described in Africa.

In the central Anti-Atlas of Morocco, the greenschist to amphibolite facies-metamorphosed late Proterozoic Bleïda Formation (Leblanc and Billaud, 1978) is composed of mafic lavas intruded by metagabbros and topped by massive spilites and hyaloclastites. In the upper part of the sequence there exist banded keratophyres and quartz keratophyre tuffs rich in magnetite, followed by a jasper, limestone marble and a 10-80 m thick sulphide-mineralized zone. There, interbanded gray-green phyllites (metamorphosed felsic tuffs) alternating with black phyllites, contain six stratabound lenses of massive, banded or disseminated bornite, chalcopyrite and pyrite. Low-grade copper mineralization is widespread in the area in several other volcanic units.

In northern Cameroun, the Cambrian Poli Group (Le Fur, 1971) contains several stratabound horizons of disseminated chalcopyrite, pyrite, bornite and magnetite hosted by sericite-chlorite metaquartzites sandwiched between felsic metatuffs, and interpreted as "exhalites" (the Kodeala-Koti group). There appears to be a marked similarity to the Zlaté Hory deposit in Czechoslovakia. At the Kakat showing, a 50-70 m thick horizon of felsic metatuff resting on amphibolites (meta-spilites), contains disseminated chalcopyrite, pyrite and bornite.

12.7. TRANSITIONAL AND INTERACTION MINERALIZATIONS

12.7.1. Facies transitions of supracrustal host rocks

Volcanic rock associations in the immediate vicinity of a volcanic centre change into volcanic-sedimentary and ultimately, purely sedimentary associations with the increasing distance. Although many shales, slates and phyllites may have a volcanic provenance, this can rarely be demonstrated in pelitic rocks only, unless coarser interbeds are present. The purely sedimentary associations treated in Chapter 17 generally have a higher probability of hosting massive sulphide deposits when they are a facies equivalent of volcanics; the Devonian slates and partly impure carbonates hosting the Rammelsberg and Meggen deposits in Germany, are facies equivalents of the "geosynclinal" bimodal suite. In addition to the pyritic deposits, the sedimentary facies frequently contains bedded barite, bedded Mn ores, bedded pyrite (pyrrhotite) and "metalliferous shales" with an increased trace content of Fe, Mn, Cu, Zn, Pb, Ag and Au.

12.7.2. Metamorphic transitions

All the massive sulphide occurrences treated here are hosted by metamorphosed rocks, but it is assumed that despite the metamorphic overprint the original depositional characteristics of orebodies are still recognizable. When this becomes difficult or impossible due to the high grade of metamorphism or granitization, the deposits are treated in Chapter 29.

12.7.3. Depth transition of volcanic to subvolcanic and plutonic equivalents

Basalt (spilite) changes with depth into diabase (albite diabase) and gabbro (rare albite gabbro). Rhyolite (keratophyre, quartz keratophyre) changes into a quartz-feldspar porphyry or albitic porphyry, plagiogranite and syenite. The subvolcanic mafic rocks rarely contain metallic mineralization of any importance, although small disseminations, veins and veinlets are fairly common. In the Nundle goldfield, N.S.W., Australia (Weber, 1979) and elsewhere, small quartz, gold, minor stibnite and scheelite veins and mineralized shears occur within, or are related to, Devonian sodic diabase dikes.

Subvolcanic felsic rocks are substantially more important economically. Many take the form of quartz-augen porphyries, rocks commonly associated with massive sulphides. Equivalent rocks can probably form by metamorphic blastesis. Generation of some massive sulphide bodies by replacement under subvolcanic conditions is a possibility and an alternative to the seafloor generation. The formation of stockwork and disseminated pyrite-chalcopyrite mineralization hosted by felsic subsurface magmatic rocks, however, has been established beyond doubt. Such copper deposits are known not only in the immediate footwall of some massive sulphides (e.g. at Cerro Colorado, Rio Tinto), but may also occur separately in districts known to contain massive sulphides, as well as outside them. The Mount Lyell field, already reviewed in Chapter 11, has several close equivalents affiliated with the bimodal suite.

In some polyphase, deeply eroded volcanic-sedimentary and intrusive complexes that contain a wide range of volcanogenic (surficial), subvolcanic and plutonic mineralizations of the same metals (e.g. Fe, Cu, Zn, Au, Ag as in the massive sulphide fields), coeval (contemporary) deposits formed at different levels within the crust, and deposits formed sequentially may be practically identical and usually difficult (or even impossible) to distinguish. This can be rather convincingly demonstrated using the example of the Blyava mineralized centre in the south-eastern Urals. The Blyava centre adjacent to Mednogorsk, near Orsk (Fig. 12-17) is an ellipsoidal polyphase mineralized brachysynclinal structure, considered to be an eroded central volcano (Yakovlev et al., 1965; Skripchenko, 1972; Kropachev et al., 1973). The structure is elongated in the north-south direction and has dimensions of about 21 x 14 km.

There, the earliest phase of Silurian basalts (probably arc tholeiites) formed over and within an earlier (lower Silurian) chert-basalt (ophiolite ?) and Ordovician tuff, sandstone and shale association. Gabbro, plagiogranite and minor ultramafics probably represent its deep-seated equivalents. Central explosive vents probably at least partly subaerial, formed during several phases between upper Silurian and lower Devonian, and they are represented mostly by proximal rhyolite (minor andesite) pyroclastics and volcaniclastics grading downward into subvolcanic and plutonic equivalents.

Blyava is particularly notable for the overlapping felsic and mafic subvolcanic and plutonic phases; the former represented by rhyolite and quartz-feldspar porphyries and breccias, the latter by diabase porphyries, gabbros and diorites. Within the Blyava structure, several tens of Fe, Cu and Zn sulphide deposits and occurrences are known, which include three significant massive sulphide orebodies (two at Blyava and Komsomolskoe), one group of stockwork to vein ("porphyry Cu-style") occurrences (Tsessovka), and numerous Cu showings of the same style plus minor Zn,Pb,Ag and gold-bearing veins.

The massive sulphides (Fig. 12-18a) typically top and replace brecciated and hydrothermally (quartz-sericite and chlorite) altered lower Devonian rhyolite porphyries under the confining screen of Silurian porphyritic basalts. If the stratigraphy is correct, this precludes the "exhalational" seafloor origin of the sulphide, and the orebodies are true subvolcanic-hydrothermal metasomatites.

The vein, disseminated and stockwork mineralization (Fig. 12-18b), is represented by pyrite, chalcopyrite and minor sphalerite disseminated in quartz, epidote; quartz, chlorite, albite; chlorite; pumpellyite, zeolite, calcite, etc., veinlets, as well as in pervasively altered (propylitized) zones in the lower Devonian gabbro and diorite complex. This complex is believed to be a member of a younger intrusive phase than the one hosting the massive sulphides. The veinlet to disseminated ores are controlled by faults (steep, cross-cutting mineralization zones) and by intrusive contacts (gently dipping lens or cap-like bodies). The material is low grade (0.02-2% Cu) and the grades are irregularly distributed.

No tonnage figures have been published for the Blyava field since the late 1930's, when the metal reserves were 450 Tt Cu, 385 Tt Zn and 45 t Au, all of them in the "old" Blyava massive sulphide deposit. The post-war discoveries must at least have tripled these figures, so that the field may be credited with at least 1.3 Mt Cu.

Additional stockwork and "porphyry" mineralizations similar to Tsessovka and present in the plutonic and subvolcanic phases located in massive sulphide mineralized bimodal terrains, include the Voznessenskoe deposit in the Urals (Pronin, 1962); partly Boschekul' in northern Kazakhstan (Kolesnikov, 1975); Bukhtarma and other localities in the Rudnyi Altai and others.

Gold vein deposits spatially associated with massive Fe,Cu,Zn sulphides, are affiliated with the same type of deeper-seated magmatic bodies as the stockwork Cu occurrences. The gold deposits in the

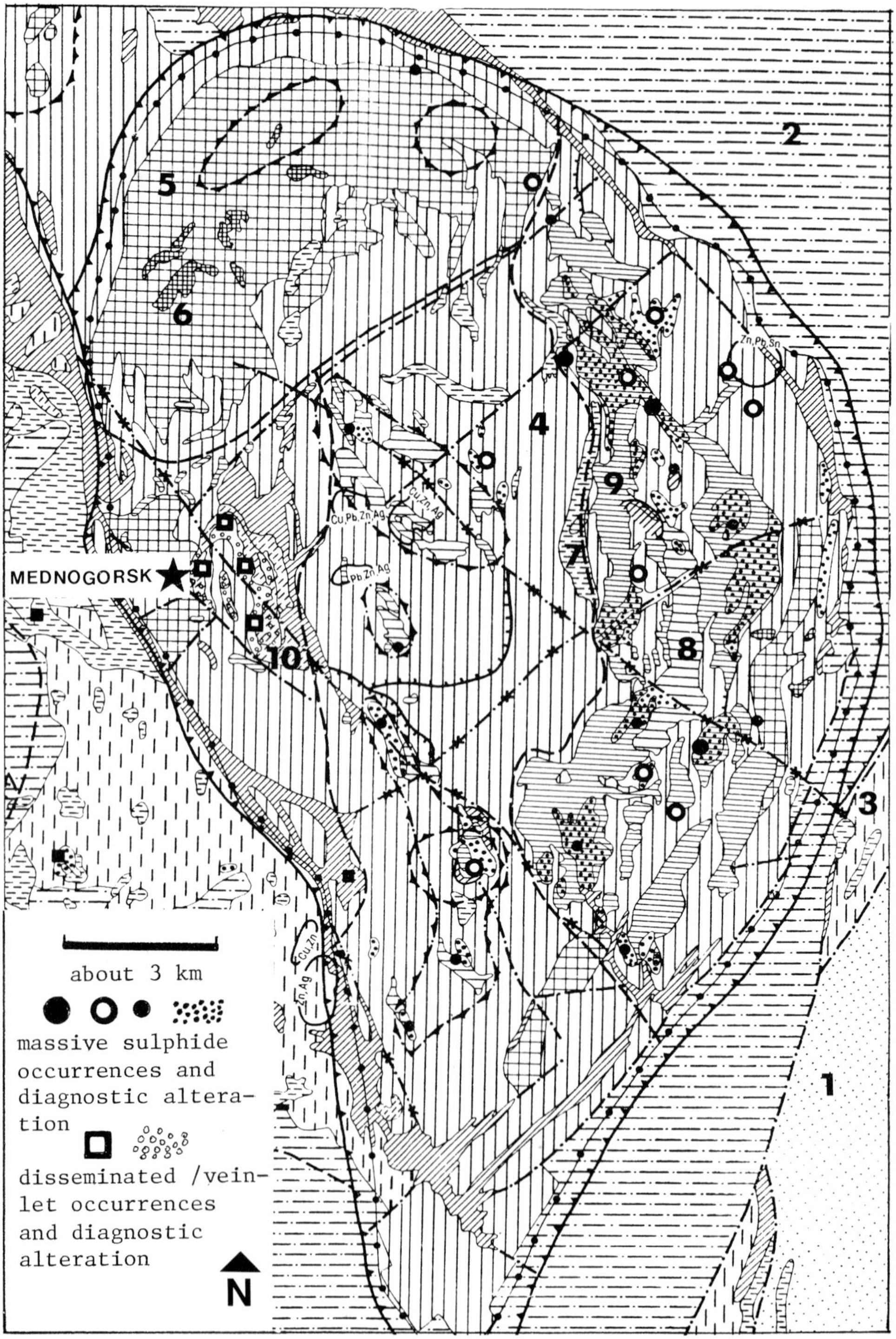
2
5
6
4
9
7
8
10
3
1
MEDNOGORSK
Zn,Pb,Sn
Cu,Pb,Zn,Ag
Cu,Zn,Ag
Pb,Zn,Ag
Cu,Zn
Zn,Ag
about 3 km
massive sulphide occurrences and diagnostic alteration
disseminated /veinlet occurrences and diagnostic alteration
N

French Gulch-Deadwood district, northern California (Albers, 1961; 26 t Au) are in the vicinity of, and hosted by, the same units as the West Shasta massive sulphide bodies (reviewed earlier). There, quartz, calcite, pyrite, arsenopyrite, minor galena, sphalerite and gold veins are gently or steeply dipping, are controlled by faults or thrusts and are more numerous in the vicinity of the "Birdseye porphyry" of Devonian age: a unit closely related to the massive sulphide. The veins, however, cut the porphyry so they must be younger.

Many gold deposits in the Appalachian Piedmont (Pardee and Park, 1948) and in the Urals, are similar; Pronin (1962) emphasized their association with differentiates of the "gabbroic magma", namely with plagiogranites and their porphyries.

12.7.4. Mineralizations resulting from interaction of plutonic intrusions with massive sulphide deposits and their host rocks

Hypabyssal intrusions that have extended into the formerly higher crustal levels occupied by earlier massive sulphide deposits may have modified the earlier ores. The modifications may have involved a physical disruption (e.g. at the Weedon mine, reviewed earlier; Fig. 12-11); contact metamorphism or metasomatism (e.g. conversion of pyrite into pyrrhotite or magnetite as at intersections of quartz porphyry and quartz diorite dikes with massive pyrite as in the Yanahara mine, Japan); or they may have involved a local remobilization. Remobilization veins which probably derived directly from massive sulphide bodies are relatively rare and considered alone they are only economically important in exceptional instances. Examples include quartz, siderite, chalcopyrite,, terahedrite, galena, sphalerite fissure veins in the Smolník, Slovakia, pyritic massive sulphide deposit; Ilavský (1976); similar veins near Kitzbühl, Austria; partly probably the important chalcopyrite-siderite vein

---

Fig. 12-17. Blyava mineralized centre, southern Urals, showing the distribution of subvolcanic Cu-Zn massive sulphide, and plutonic veinlet/disseminated ("porphyry") copper mineralization in Devonian magmatic complexes. LEGEND: (1) Ordovician sandstone, slate; (2) upper Ordovician-lower Silurian tuffitic shale; (3) Silurian chert, basalt; (4) Silurian basalt complex; (5) upper Silurian extrusive rhyolite, minor andesite, vent facies; (6) upper Silurian subvolcanic intrusions; (7) Silurian-Devonian subvolcanic diabase and gabbro-diabase; (8) lower Devonian effusive and vent facies rhyolite; (9) lower Devonian late felsic subvolcanic bodies (ore host); (10) upper Devonian gabbro, diorite complex; associated Cu disseminations, veinlets. Slightly modified after Starostin and Yakovlev in Kropachev et al. (1973).

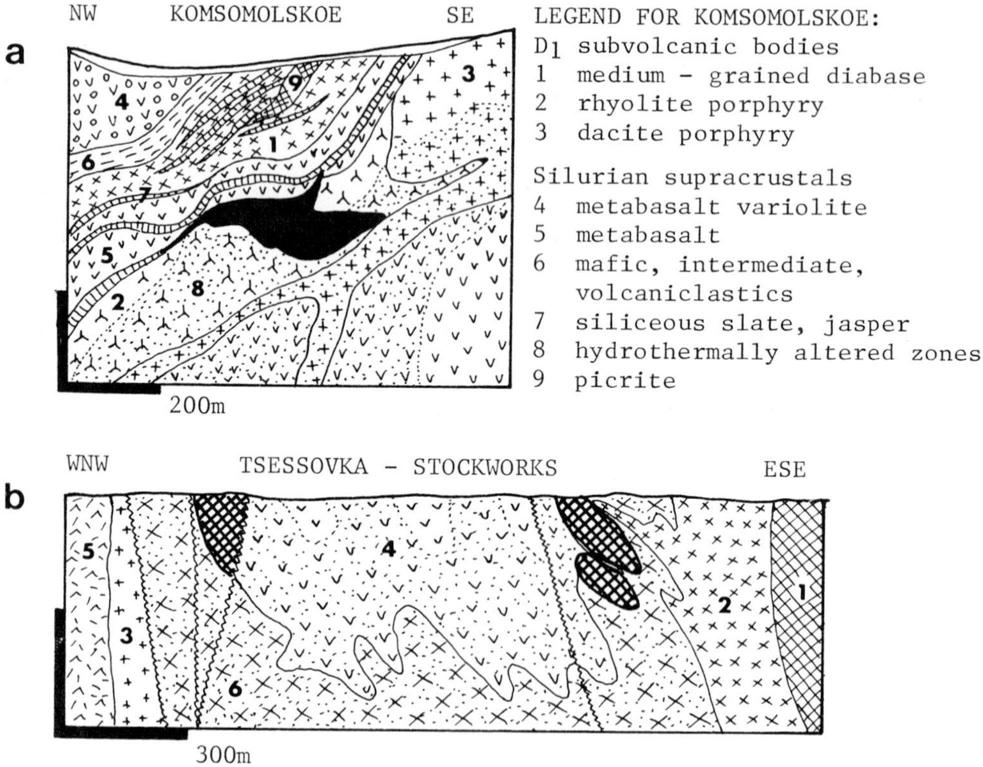

Fig. 12-18. Two main mineralization styles, Blyava ore centre, S. Urals. (a) Komsomolskoe, massive Fe,Cu,Zn sulphide body at contact of subvolcanic rhyolite porphyry and metabasalts; (b) Tsessovka, disseminated and stockwork pyrite, chalcopyrite in fractured and propylitized diorite. LEGEND FOR TSESSOVKA: Silurian (1) ultramafics; (2) gabbro; (3) subvolcanic andesite-dacite porphyry; (4) diabase, variolite, basalt porphyrite; lower Devonian intrusions: (6) diorite. Black mesh: zones of Cu disseminations. After V.I. Starostin in Yakovlev et al. (1965).

deposit Mitterberg, Austria (170 Tt Cu); parts of Lesu Urşului, Tulgheş, and other localities in Rumania (compare Laznicka, 1985b).

Of greater economic importance appear to be younger mineralizations that seem to be closely comparable geochemically with the regionally developed association of volcanogenic massive sulphide deposits, which could be the product of areal mobilization or remobilization. The quartz, siderite, Ag-tetrahedrite stockworks hosted by a Devonian dolomite in Schwaz, Austria (Lukas, 1971; 170 Tt Cu, 2.5 Tt Ag, min. 50 Tt Sb) and similar ores near Brixlegg are probably a product of Alpine regeneration of a Paleozoic supracrustal geochemical anomaly or a stratabound mineralization. In any case, the presence of a young epigenetic mineralization in bimodal volcanic-sedimentary terrains may indicate the presence of older, stratigraphically controlled deposits and this possibility should be carefully investigated.

Skarn deposits spatially associated with rocks of the bimodal suite are relatively rare, compared with skarns in the other associations reviewed in this book (Chapters 10, 13,14). Most of the quoted examples are from the mafic metavolcanic belts of the Urals, where magnetite and copper skarns occur in nine ore belts (Baklaev, 1972). There, andradite-hedenbergite or andradite-epidote skarns are hosted by carbonate horizons, by tuffaceous carbonates, or by pyroclastics rich in carbonate matrix. They often form sheet-like stratabound bodies. This, and the frequently observed fragmental nature of the skarn, led some investigators to propose that these ores were of synvolcanic origin. Baklaev (1972) argued for the validity of the contact metasomatic interpretation, in which the skarns formed in the roof zones of diorite, quartz diorite and plagiogranite intrusions. It thus appears that most of the skarn-generating intrusions have been members of the evolved, sequence-differentiated magmatic suite, so that the Uralian skarns are not substantially different from the rest worldwide.

12.7.5. Residual and sedimentation-reworked deposits.

These are of the "common" predictable nature treated in Chapters 23 and 24. Weathering of massive sulphide deposits may yield gossans composed of powdery or cellular goethite invariably having high residual gold and silver contents. Because of the finely dispersed nature of the gold, placers rarely form. Gold placers, however, result from vasting of the gold-bearing veins that in turn often accompany massive sulphide deposits.

Plate 5. Outcrop of the Sustut orebody, British Columbia. Disseminated copper sulphides in Triassic low-grade metamorphosed andesitic volcaniclastics.

"Andesitic magmatism seems intimately related to the formation of many ore deposits. Andesites, whether as magmas or as previously crystallized volcanic rocks, seem the source of these metals".
J. B. Gill (1981)

CHAPTER 13

Andesite-Dominated, Marine to Continental Volcanic-Sedimentary Association

13.1. GENERAL

Andesites are much in evidence in the present volcanic island arcs and along the Andean-type continental margins (Chapter 5). They occur frequently as dominant constituents of numerous superimposed cycles, in the basement of the modern island arcs, as well as in the ancient orogenic belts that are now welded to continents. In some terrains, andesite is the only volcanic rock. Elsewhere, andesite coexists with basalt and/or dacite. Naturally, there is no sharp boundary between the basalt-dominated (Chapter 10) and andesite-dominated (this Chapter) terrains; the former often occur in the basement of the latter. A sharp boundary is also lacking between the andesite-dominated terrains and occurrences of the basalt, andesite, rhyolite suite (Chapter 14), as well as between "marine" and "continental" andesites.

In the primitive segments of the Earth underlain by oceanic crust, subaqueous "oceanic andesite" (Maaløe and Petersen, 1981) is the first andesite to form by partial melting of the subducted oceanic crust. These andesites are relatively pristine, because they could not have assimilated material from the continental crust (except the one dissolved in seawater). With continuing arc evolution, both magmatic and sedimentogenic diversification and island growth took place, which resulted in volcanism being increasingly subaerial. The subaerial volcanics and their intrusive equivalents in island arcs are indeed little different compared with volcanics at the margins of large continents (Chapter 26), except that isotopic and trace element evidence usually indicates a much greater degree of continental crustal input to the latter.

Despite the numerous transitions, the andesite-dominated orogenic terrains constitute a fairly distinct, repetitive entity with, to a certain degree, predictable metallogeny. In some of the well-developed, well-described Mesozoic-Cainozoic arc andesite dominated regions of the world, such as the Aleutians (e.g. Drewes et al., 1961), the Fiji region (Green and Cullen, 1973; Colley and Greenbaum, 1980), Puerto Rico (Glover, 1971), the Nicola Group of southern British Columbia (Schau, 1970), and others (Table 13-1), the following evolutionary pattern and a corresponding local stratigraphy have emerged:

Table 13-1. Phanerozoic andesite-dominated, marine to continental volcanic - sedimentary association; selected mineralized examples

LOCALITY	GEOLOGY	MINERALIZATION	REFERENCES
Miocene-Pliocene rocks of the Aleutians	Mi,Pl 17 km thick pile of andesite flows, pyroclastics, incl. pillowed subaqueous flows; graywacke, argillite intruded by Mi-Pl granodiorite	Mi-Pl: small Zn,Cu,Pb hydrothermal veins in thermal aureoles of intrusions	Drewes et al. (1961)
Nicola and Takla Groups, C. British Columbia ; Tr_3-J_1	7 km thick sequence; andesite to basalt massive flows to breccias, pyroclastics, lahar conglomerates. Grade to graywacke, argillite. Widespread augite andesite. Intruded by comagmatic diorite, qtz.diorite, syenite and MZ-CZ granodiorite, qtz.monzon.	widespread Cu showings: massive sulphides (700 Tt Cu), disseminations (350 Tt) veins. Minor Zn,Pb,Au veins. Numerous porph.Cu and skarns (about 13 Mt Cu, some Mo,Au)	Schau (1970) Church (1974) Thompson and Panteleyev (1976)
Robles Fm. (Cr_1) Puerto Rico	calc-alkaline to subalkaline andesite pillow lavas, tuffs, limestone overlain by andesite pyroclastics and volcaniclastics, mudstone, limestone. Minor comagmatic diorite, microdiorite intrusions, later diorite-qtz.monzonite.	Mineralogical occurrences of chalcopyrite along shears, veinlets, disseminations in intrusive aureoles	Glover (1971)
Paleogene complex of eastern Cuba (El Cobre Form.)	4 km thick suite of dominant andesite flows and pyroclastics, lesser basalt, dacite; localized volcanic centres; interbedded marine limestone and argillite; zeolite metam., intruded by diorite to granodiorite.	disseminations and stockworks of Cu sulphides near vents (El Cobre; 150 Tt Cu); widespread stratiform Mn oxides at tuff/limestone contacts; magnetite skarns in intrusive aureoles	Bogdanov et al. (1965) Simons and Straczek (1958)

Andesite belts of the Andean orogenic belt, Chile; J_1-T_1 ; (e.g. Veta Negra, Las Chilcas, Lo Prado, Farrelones, Arqueros, etc. Formations)	several cycles of thick submarine andesite lavas, pyroclastics, volcaniclastics, shale, limestone, changing facies to continental volcanics and sediments; massive, low metam. grade. Intruded by comagmatic gabbros, diorites, younger granodior., qtz. monzonite	Cu sulphides widespread in volcanics and pyroclastics as mantos in subaer. lavas; impregnations; mantos in sediments. Very important interaction porphyry coppers. About 150 Mt Cu is in deposits directly to indirectly related. Also Fe, Mn.	Ruíz (1965) Ruíz et al. (1971)
Philippines, Eo-Mi associations	basaltic andesite lavas, volc. breccias, tuffs, volcanic sediments, marine limestones. Intruded by widespread diorites, gabbros, quartz diorites	native Cu and Cu sulphides occur. in lavas and pyroclastics; Cu, Fe, Au deposits at andesite/diorite interactions; disseminated Cu deposits, skarns, veins in intrusive aureoles	Bryner (1969) Wolfe (1973)
Suta and Umasani Volcanics (Ol-Mi) Solomon Islands	basaltic andesite and graywacke pile resting on oceanic basement; some reef limestone tops. Intruded by small comagm. gabbro and diorite bodies, large later diorite to granodiorite stocks and massifs	small Cu sulphide and gold showings in lavas and pyroclastics; vein and disseminated Cu in intrusive aureoles (e.g. Koloula)	Arthurs (1979)
Fiji Platform, Eo_3-Pl sequence	tholeiitic, spilitic to andesitic pillow lavas, agglomerates, breccias with graywacke, mudstone and limestone, are topped by Mi basalt, andesite flows and agglom., pyroclastics, volcaniclastics. Gabbro, tonalite, minor granodior. intrusions, shoshonites	volcanogenic Cu-Zn mass. sulphides (Undu P.); Fe-Cu skarns, disseminated Cu, Cu, Au, veins, in intrusive aureoles; bedded Mn, magnetite beach placers	Colley and Greenbaum (1980)

(1) The basement is represented by oceanic lithosphere (ophiolite), or a "quasi-continental" crust (that is, a compositional hybrid of remnants of continental lithosphere and oceanic elements) or a bimodal volcanic-sedimentary suite.

(2) The andesitic cycle starts with a submarine basalt or basaltic andesite pillow lavas, volcanic breccias, hyaloclastites, pépérites, tuffs, and interbedded deepwater sediments (volcanic mudstones, radiolarian cherts, siliceous shales, dark gray limestones). Snyder and Fraser (1963) described from the Unalaska Island, the Aleutians, a variety of rock bodies transitional between intrusive sills and subaqueous lava flows, developed by andesitic magma intruding semi-consolidated muds. The Unalaska andesites are thoroughly albitized, and so is a considerable proportion of the early oceanic andesites elsewhere ("andesitic spilites"). Members of the high-magnesian lava suite, possibly comparable with the Archean komatiites, are known in several Mesozoic and Paleozoic andesitic belts, including the still existing island arcs, such as the Solomon Islands (New Georgia Group) or the New Hebrides (Aoba Island).

Stanton and Ramsay (1980) characterized the Melanesian high-Mg volcanics as primarilly lavas erupted in shallow water, and spatially and temporally associated with basaltic andesites. Petrographically, alkali olivine basalts and picritic basalts occur. They are relatively high in K_2O and P_2O_5 and should not be confused with ophiolites. In contrast to the Precambrian komatiites that contain numerous economic Ni and Au deposits, no appreciable accumulation of Ni has been reported in the Phanerozoic high-Mg lavas, although the maximum of 0.2% Ni reported by Stanton and Ramsay (1980) from New Georgia demonstrates at least a possible Ni potential.

Intrusive rocks, probably comagmatic with the early andesites, include pyroxene diorite dikes and small stocks. Some intrusives contain explosion breccias, indicating possible eruption sites. Small syenite intrusions related to potash-rich andesite lavas are occasionally associated. Tese have an alkaline tendency and are unusually high in trace copper (200 ppm plus; e.g. in the Robles Fm., Puerto Rico; in the Nicola Group, British Columbia).

(3) Completion of Stage 2 was frequently followed by an "orogeny" causing deformation, slight metamorphism and intrusive emplacement of chiefly dioritic rocks. The superimposed rock association usually rests unconformably on the Stage 2 rocks, sometimes initiated with a thick basal volcanic conglomerate. Stage 3 has a much greater diversity of rocks (basalt flows, andesite lavas, pyroclastics, lahars, coarse to fine volcaniclastics, carbonates) that had their source in at least partly subaerial volcanoes. The proportion of volcanics more felsic than andesites (particularly dacites) increased.

In some orogenic belts, there is a marked evolutionary change of the andesitic composition and appearance from feldspar-phyric andesites (white tablets of plagioclase on the dark green fine porphyritic background) to augite andesites (black, sharply developed augite phenocrysts on a gray or green background). This can best be seen in thick sequences of andesitic lavas, as in the Triassic

Nicola Group (there, the "feldspathic andesite cycle" is topped by the "augite cycle"; Schau, 1970), or in the Andean belt of Chile (the Jurassic-Cretaceous Formacia Porfirítica has mainly feldspar-phyric andesite, the Eocene Farellones Formation has augite andesite; Ruíz, 1965; Camus, 1975).

The Stage 3 rocks are extensively intruded by differentiated calc-alkaline massifs having a gabbro to alaskite range, but dominated by quartz diorite to granodiorite. It has been repeatedly demonstrated in geologically young island arcs that generation and emplacement of calc-alkaline intrusive complexes commences when the arc becomes well established and its crust thickens to 25-40 km. This requires a minimum of several tens of millions of years since the commencement of volcanic activity.

The proportion and absolute volume of andesites in the preserved geological record decreases sharply with geological time. This may be due partly to (1) the overall evolution of the lithosphere (initiation of subduction); (2) erosion. Since andesites have a tendency to be associated with positive areas more than the volcanics discussed in Chapters 10 and 12, many have been eroded so that they are now represented by their plutonic equivalents only (mostly diorites); and (3) metamorphic conversion of andesites into greenstones, chlorite schists and amphibolites, often indistinguishable from the same metamorphics generated from basalts.

13.2. ECONOMIC IMPORTANCE AND METALLOGENY

The andesitic terrains listed in Table 13-1 as well as the unlisted ones, contain about 500 Mt Fe, 12 Mt Mn, 6.5 Mt Cu, 900 Tt Zn, 400 Tt Pb, 800 t Ag and 50 t Au in currently economic deposits other than those associated directly and intimately with intrusive activity. If the later deposits, dominated by porphyry coppers, are included, the copper content will increase to several tens of million tons, and Mo, Ag and Au would also increase proportionally. Overall, andesitic belts are clearly copper provinces and copper accumulations overshadow the rest. This correlates with the large variety of copper mineralization styles, and also with the high geochemical copper backgrounds in many rock types, particularly in basaltic andesite lavas.

The metallogeny in andesitic terraines (Fig. 13-1; Table 13-2) is rather difficult to organize logically and present in a coherent form. The main reason is the overlapping effect of numerous ore-concentrating factors and an abundance of multistage mineralizations. In many cases it is impossible to tell when the ore was emplaced in its present form, and what was the major factor or agent of emplacement. Colley and Greenbaum (1980) in their very complete review of metallogeny of the Fiji Platform, chose to treat the subject in terms of pre-, syn-, and post- orogenic mineralizations. Orogeny, however, is highly relative and in Fiji, massive sulphide deposits, for example, formed in both the pre- and

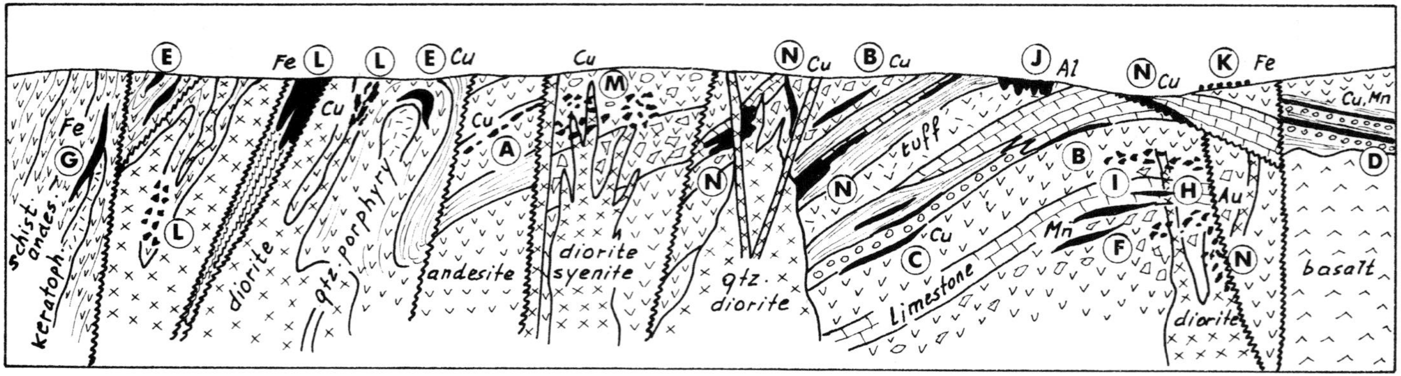

Fig. 13-1. Principal mineralization styles in the andesite-dominated volcanic and sedimentary association (see Table 13-2 for letter code explanations).

Table 13-2. Principal mineralization styles in the andesite dominated volcanic - sedimentary association

(A)	native Cu in basaltic andesite lavas	X0	X00 Tt Cu
(B)	Cu sulphide disseminated mantos in hydrotherm. altered andesites	X0	2 Mt Cu
(C)	Cu sulphide mantos in the contact region of andesite flows, tuffs, breccias and sediments	X0	---
(D)	Cu,Mn(Zn,Co) sulphide and oxide mantos in argillized andesitic tuff above basalt (Boleo)	1	700 Tt Cu
(E)	Fe,Cu volcanogenic massive sulphides	100	2 Mt Cu
(F)	stratiform Mn oxide beds at or near contacts of andesitic tuff and limestone	X0	10 Mt Mn
(G)	banded siliceous iron formations in schistose andesites	X	100 Mt Fe
(H)	disseminated gold in tuffs in the vicinity of explosive andesite centres	X	10 t Au
(I)	Au-bearing mantos in bedded metasomatites in and.	X	5 t Au
(J)	weathering-generated residual bauxite on andesite	X	10 Mt Al
(K)	magnetite alluvial and beach placers	X0	50 Mt Fe
(L)	ores in the intimate andesite - diorite assoc.	X X0 X	200 Mt Fe 700 Tt Cu 2 t Au
(M)	"volcanic" (diorite,syenite affiliated) porphyry copper deposits	20	6.5 Mt Cu
(N)	interaction deposits: gabbro,diorite,plagiogranite / andesite supracrustal association (veins, skarns, stockworks, etc.)	X00	3.5 Bt Fe 3 Mt Cu 100 t Au

post- orogenic stages of development. The latter, however, were evolutionarily more mature.

Many, probably most, ore deposits in andesitic terrains owe their origin to interaction of the supracrustal rocks enriched or even pre-mineralized in trace elements (particularly copper) followed by intrusive upgrading. Many intrusions are at least broadly comagmatic with the supracrustal volcanics. "Volcanic" and "intrusive" metallogeneses cannot be sharply separated, and mineralizations are treated, with an overlap, both in this chapter and in Chapters 26 and 28.

13.3. Cu ASPECT

13.3.1. Copper occurrences in massive andesitic lavas

Megascopically visible irregularly scattered grains, blebs, veinlets, etc., of bornite and/or chalcocite, chalcopyrite or native copper are extremely common in mineralogical quantities in andesitic lavas that show the evidence of burial or thermal metamorphism. This category is a close equivalent to similar copper occurrences in plateau basalts (Chapter 31) and the "geosynclinal" metabasalts (Chapter 10). The copper deposits south of Tocopilla, Chile, treated in the "Basalts" chapter, are alternatively designated in the literature as andesites.

In the Ingenika Range, northern British Columbia (Chris Bates, guided visit, 1976), a thick sequence of upper Triassic alternating flows of prehnite-pumpellyite to chlorite metamorphosed feldspar-phyric andesites and brown or green tuffs contain a variety of small copper showings (Fig. 13-2). Bornite is the dominant sulphide and it forms small grains in the epidote, lesser albite, quartz, calcite and laumontite patches, amygdules and veinlets along a variety of depositional and deformational dilations.

In the White River area, Yukon (Cairnes, 1915), the dominant mineral in a similarly set mineralization was native copper. Continuous blocks of copper over 3 m long have been found filling fractures. Near Rannes, Queensland, finely disseminated, virtually invisible bornite and chalcopyrite correlate with the diffuse zones of epidotization in Permian massive to amygdaloidal andesites.

In the Mesozoic andesites north-west of Santiago, Chile, similar copper occurrences are particularly common in andesitic vesicular lava flowtops and also in the topping limestones, in the Lo Prado and Veta Negra Formations. Levi (1969) attributed the copper accumulation to migration of copper released from pyroxenes during burial metamorphism.

Although this style of copper mineralization is usually dismissed as having a low potential, small-to—medium-sized occurrences have been found and economically exploited. In Las Cuevas, 100 km west of Mendoza, Argentina (Haude and Weber, 1975), Jurassic to lower Cretaceous flows of porphyritic andesite interbedded with tuff and marine sediments, contain irregular lenses of scattered chalcocite, bornite and copper carbonate 5-15 m thick. The reserves comprised 242 Tt of ore with 1% Cu.

The Lo Aguirre orebody near Pudahuel (near Santiago Airport, Chile; Comp. Min. Pudahuel, oral communication, 1977; 200 Tt Cu; Fig. 13-3), consists of pervasively disseminated and less commonly veinlet pyrite, chalcopyrite, bornite and chalcocite. These form a loosely defined but broadly stratabound manto in slightly metamorphosed (prehnite, pumpellyite, chlorite) and slightly hydrothermally altered (silicified, carbonatized) massive to sheared porphyritic andesites and dacites. Mineralogical zonality has developed locally, and runs from the centre outward: chalcocite, bornite-chalcopyrite – pyrite.

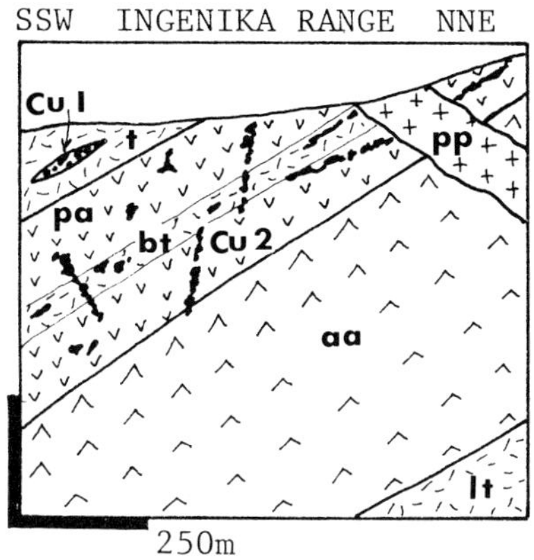

Tr$_3$ t = pyroclastics
pa = bladed plagioclase-phyric basaltic andesite
bt = fine reddish-brown tuff
aa = augite andesite flow
lt = dark green lapilli tuff
J pp = feldspar and diorite porphyry
Cu 1 Cu-sulphide impregnations in tuff
Cu 2 bornite,etc.in epidote, albite,quartz,calcite, laumontite secretions

Fig. 13-2. Diagrammatic representation of Cu sulphide occurrences in (meta) andesites, Ingenika Range, British Columbia. From LITHOTHEQUE.

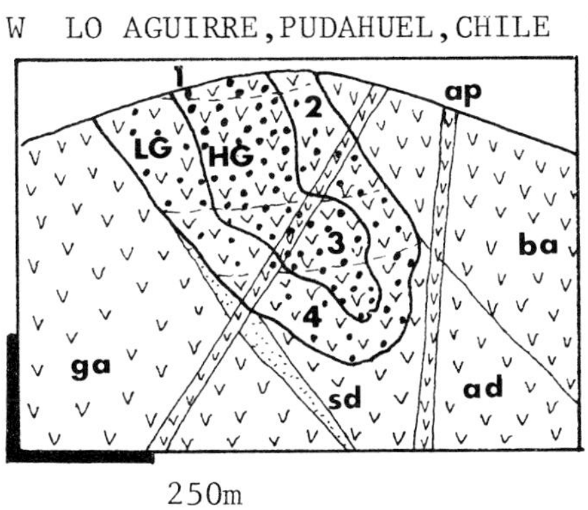

Upper Cretaceous:
ba = brown porphyritic andesite
ad = porphyritic andesite,minor dacite,ignimbrite
sd = interflow volcanic arenite
ga = gray porphyritic andesite
ap = andesite porphyry dikes

OREBODY: HG = higher grade zone
LG = lower grade zone
1 = leached capping; 2 = oxidation zone; 3 = sulphide enriched zone; 4 = hypogene zone.

Fig. 13-3. Lo Aguirre, andesite-hosted Cu manto, Chile. From LITHOTHEQUE, data from Compañía Minera Pudahuel, guided tour (1977).

The volcanics form a roof pendant over the intrusive rocks of the coastal batholith. This orebody is most probably a product of local remobilization of a high local trace copper content, coeval with the intrusive emplacement. More than one half of the ore reserve is in the oxidation zone.

El Soldado and Punta del Cobre (70 Tt Cu/2%) are two other Chilean andesite-hosted Cu mantos. In the El Soldado field 65 km N.E. of Valparaíso (Ruíz, 1965), finely disseminated bornite, chalcopyrite and pyrite form several irregular columns and mantos in a lower Cretaceous andesite and in intrusive subvolcanic bodies of trachyte and porphyritic trachyte, indicating that this is undisputably an epigenetic interaction mineralization. The economic Cu content is 712 Tt Cu/2%. The El Cobre copper deposit near Santiago, Cuba (Bogdanov et al., 1965; about 150 Tt Cu), is very similar in style. It is hosted by Paleocene-Eocene andesites and pyroclastics, and disseminated, veinlet, vein, etc. pyrite and chalcopyrite accumulated in zones of crushing and shearing in the vicinity of andesite porphyry-filled vents. Propylitic alteration is dominant and most of the past production came from the oxidation zone.

In the Abbas-Abad region, Iran (Khadem, 1964), patchy occurrences of chalcocite and lesser bornite, covellite, cuprite and malachite are traceable over a distance of 30 km, hosted by a 15–20 km thick top of an Eocene andesite flow, topped by limestone and detrital sediments. The orebody worked in the Abbas-Abad mine was 250 m long and had a grade of 1-1.5% Cu.

13.3.2. Disseminated Cu sulphides in alternating andesite flows, tuffs and sediments

With increasing proportions of interbedded volcaniclastics and marine (to partly continental) sediments, the earlier association passes into a sequence of massive andesite flows, coarse to fine varicoloured tuffs and breccias, volcanic arenites, and gray to black siltstones, mudstones, pyritic argillites, siliceous shales, cherts and thin-bedded limestones. Concordant to discordant copper sulphide (bornite, chalcocite, chalcopyrite) mineralization occurs in a variety of host rocks; a portion of it appears to be syndepositional-diagenetic, and the rest is epigenetically superimposed. A variety of copper occurrences have been described from the Cretaceous Lo Prado and Las Chilcas Formations in the Cabildo district, central Chile (Carter and Aliste, 1964; Ruíz et al., 1971; Pélissonnier, 1971). In the Guayacán mine, the mineralization consists dominantly of bornite, present as fine disseminated grains in the andesite groundmass and filling vesicles and minor fractures. The andesite is topped by a calcite-cemented volcanic arenite, that contains low-grade disseminations of pyrite and chalcopyrite.

In the Cerro Negro-Diablo field (about 250 Tt Cu/2%; Compañía Minera Cerro Negro, on-site information, 1977; Fig. 13-4), stratabound, fault-segmented thin manto orebodies are hosted by fine-to-coarse continental - to - lagoonal volcaniclastics of andesitic derivation. Chalcocite, bornite, chalcopyrite and pyrite form crudely zoned, fine-grained, even disseminations, locally interrupted by rich sulphide pockets. The host rocks underwent prehnite-pumpellyite metamorphism, and are gradational into rocks containing abundant epidote. In the Pirquitas mine, copper sulphides form concordant bands and disseminations in a pyritic and calcareous black argillite, and in the Los Maquis mine the ore is hosted by thirteen parallel impure limestone beds, subdivided by argillites and volcanic conglomerates. The mineralized beds are 0.5 to 3 m thick, the dominant minerals are pyrite and chalcopyrite, and the presence of garnet indicates contact metasomatism in the roof of an intrusion.

The Los Maquis mine represents a link between slightly metamorphosed andesite and sediments-hosted copper mantos, and the intensively contact metamorphosed mantos located in the roof zone of a dacite porphyry to granodiorite intrusion, as in the Mala field, coastal Peru. There (Ripley and Ohmoto, 1977; Comp. Minera Condestable Staff, on-site information, 1977; Fig. 13-4), a submarine pile of upper Cretaceous porphyritic andesite lavas, pyroclastics and minor interbedded sediments intruded and contact-metamorphosed by granodiorite and dacite, contains a variety of copper mineralization styles. Economically most important are the conformable mantos 2-3 m thick, trending N 20° W/45° W (up to 30 parallel mantos), hosted by volcanic metasiltstones (now chlorite-actinolite schists) and calc-silicate hornfelses. The ore ribbons in the mantos contain pyrite, chalcopyrite, and lesser pyrrhotite, magnetite and accessory sphalerite. The ore is interpreted as submarine-hydrothermal, syn-depositional. Second in importance are the discordant chalcopyrite and pyrite stringers, breccia cement or disseminations in meta-andesites of units III and IV, that could be broadly contemporary with volcanism but deeper seated. Alternatively, they may be the products of contact metasomatism. Discrete ore veins are of limited importance. The ore reserves were of the order of 100 Tt Cu/0.7%.

Numerous additional localities resemble the "disseminated copper mantos" described above. In the Talcuna ore field, central Chile (Ruíz et al., 1971), the ore deposits are tabular bodies and veins, hosted by alternating lenses of finely laminated volcanic sandstone and andesite sedimentary breccia, both calcite-cemented. Bornite, chalcocite, chalcopyrite and minor galena replace the cement, and the ore horizon has dimensions of 3x2 km and is up to 15 m thick. The numerous cross-cutting veins have a calcite and barite gangue and identical ore minerals.

At Sustut, British Columbia (Church, 1974; Fig. 13-4), a zone of peneconcordant ore lenses containing zonally arranged disseminated chalcopyrite, bornite and chalcocite, is up to 80 m thick and 800 m long and hosted by upper Triassic andesitic rocks. The immediate host to the ore is a green, coarse, volcanic conglomerate interpreted as lahar. In addition there are andesite tuffs, and all the rocks are pervasively epidotized. The reserve is estimated to be about 30 Mt ore with 1% Cu.

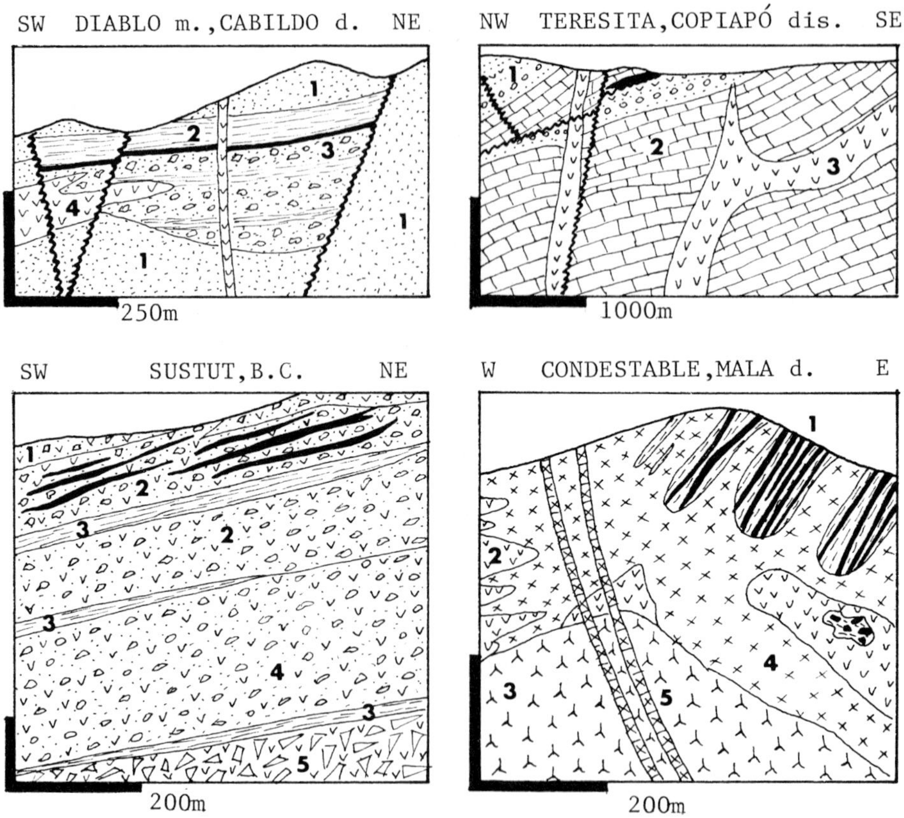

Fig. 13-4. Disseminated Cu sulphides in alternating andesite flows, volcaniclastics, and other sediments.

Diablo mine, Cabildo district, Chile; upper Cretaceous rocks. (1) volcanic arenite to siltstone; (2) gray to black calcareous siltstone to siliceous limestone, in places cherty; (3) alternating andesitic breccia/conglomerate and volcanic sandstone/siltstone. (4) andesite.

Teresita mine, Copiapó district, Chile; lower Cretaceous supracrustals, upper Cretaceous andesitic intrusions. (1) volcanic sandstone, conglomerate, chert, andesitic flows; (2) gray limestone; (3) andesite dikes and sills.

Sustut deposit, northern British Columbia. Upper Triassic rocks (diagrammatic section only). (1) red andesitic tuff, volcaniclastics; (2) green lahars, andesitic tuffs, volcaniclastics; (3) laminated volcanic shale/siltstone; (4) green and red pyroxene andesite tuff, volcanic sediments; (5) augite andesite flow breccia.

In the Teresita mine near Copiapó, Chile (Segerstrom, 1960; 40 Tt Cu/2%; Fig. 13-4), roughly stratabound as well as cross-cutting stockworks of stringers, fracture coatings and veinlets of quartz, calcite, pyrite, chalcopyrite, bornite, chalcocite and native copper, are hosted by fractured and brecciated gray to black early Cretaceous limestone. The mineralized zone is in shattered rocks in the hangingwall of an overthrust, underlain by Cretaceous volcanogenic cherts, volcarenites, and andesite flows and pyroclastics. The sediments are interpreted as volcanic facies of the host limestones. Furthermore, the mineralized area is intruded by Cretaceous andesite dikes and it is located within the contact aureole of a late Cretaceous-early Tertiary granodiorite batholith. This is probably a "granite-coeval" hydrothermal mineralization formed by mobilization of the copper content contained in the andesites.

The Boleo copper (and also Mn, Zn, Co)field in the vicinity of Santa Rosalía (Baja California, Mexico; Wilson, 1955; 760 Tt Cu/4.8%, min. 1 Mt Mn, 150 Tt Zn, 20 Tt Co; Fig. 13-5), is an unique and unusual mineralization style, difficult to classify. Briefly, five major stratabound and very persistent 0.8-5 m thick horizons (mantos) are located within a N.W.-S.E.-elongated area 11 x 3 km in size, on the shore of the Gulf of California. The host unit, lower Pliocene Boleo Formation (on the average 140 m thick), consists of interbedded shallow marine intermediate tuffs and tuff conglomerates usually underlain by a non-marine (? deltaic) basal conglomerate, sandstone and gypsum. This unit rests unconformably on upper Miocene Comondú Volcanics of continental tholeiite flows, and is cyclic.

A full cycle starts with the conglomerate or with a manganiferous basal limestone, and changes upward into a tuff or tuffaceous arenite (now bentonitic). The copper-bearing horizons are typically located along the base of the volcanics, topping the conglomerate. They are relatively impervious, moist, soft, argillized (montmorillonitic) ash-fall beds. The ore minerals: very fine chalcocite with minor chalcopyrite, bornite, covellite, native copper, pyrite and galena can be virtually invisible in the darkd-coloured, unoxidized host rock, despite the very high average grade (4.81% Cu). In the oxidation zone, chrysocolla and malachite infiltrations and coatings are more clearly visible. There is a considerable Mn content in the entire

---

⇦

Fig. 13-4 (legend continued)

Condestable mine, Mala, Peru. Upper Cretaceous supracrustals, "Laramide" intrusive rocks. (1) metamorphosed slate, siltstone, minor tuff, limestone, chert; (2) porphyritic andesite flows; (3) amphibolite; (4) porphyritic dacite to granodiorite; (5) diabase dikes.

All sections are from LITHOTHEQUE, based on data from Compañía Minera Cerro Negro, guided tour (1977); Segerstrom (1960); Church (1974) and Compañía Minera Condestable, oral comm. (1977).

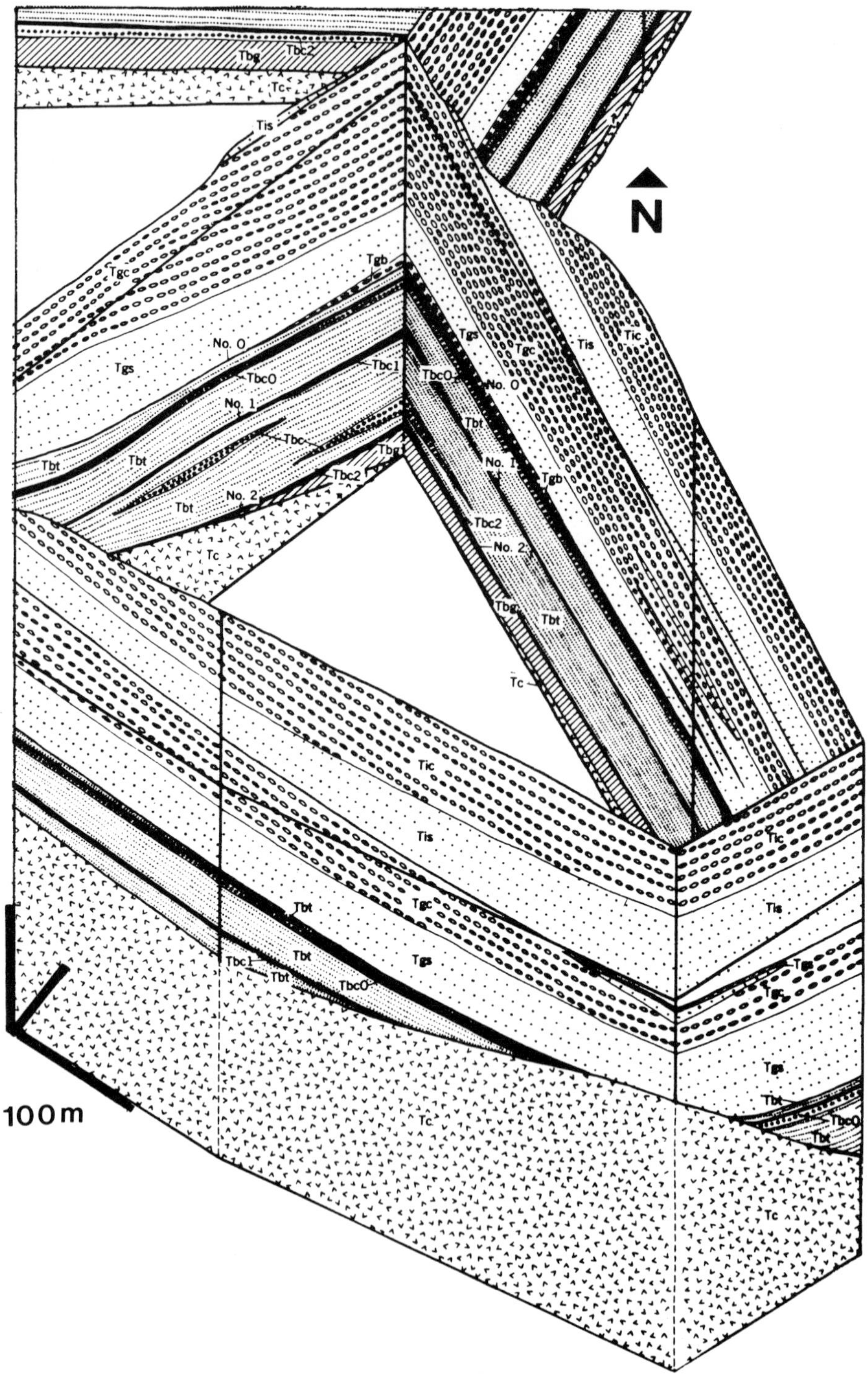

N
Tbg
Tbc2
Tc
Tis
Tgc
Tgb
No. 0
Tgs
Tbc1
TbcO
No. 1
Tbt
Tbc
No. 2
Tic
100m

Boleo Formation, present mostly as oxide impregnations and nodules, but it is not being recovered. Nor are the minor elements associated with copper; an average ore contains 3.5% MnO, 7-12 ppm Ag, 0.8% Zn and 0.12% Co.

The ore genesis is puzzling. The close proximity to the Gulf of California makes it tempting to invoke the "rifting" as somehow involved. This, very probably, controlled the emplacement of the Comondú Volcanics, that have a high trace-copper content. It is, however, unclear, how this (or other) copper got into the very evenly mineralized ash-fall beds overlying the Comondú basalts without signs of a significant interaction.

13.3.3. Massive Fe,Cu (Zn,Pb) sulphides

The massive sulphide deposits in the andesite-dominated associations can be placed into two transitional subdivisions: (1) thin, concordant, persistent sulphide layers, usually hosted by sediments or pyroclastics interbedded with andesites; and (2) peneconcordant to irregular sulphide lenses or masses located on tops of volcanic complexes and floored by hydrothermally altered rocks. The subdivision (1) appears sometimes to be transitional into the disseminated bedded ores treated earlier. In the small, recently described Denali deposit in Alaska (Seraphim, 1975), delicate bands of chalcopyrite and pyrite alternate with bands of black argillaceous limestone and the sulphide constitutes up to 30% of the rock. The Triassic black limestone together with black argillite and andesite tuffs overlie, and change facies into, augite andesite flows. The orebody is much deformed, but practically unmetamorphosed. Seraphim (1975) suggested that the mineralization formed syn-depositionally in shallow water along a shoreline by precipitation from copper-bearing solutions, derived from volcanic flowtops undergoing leaching or from solfataric vents.

Two medium-sized copper deposits, i.e. the Granduc mine in north-western British Columbia and the Cadia deposit in New South Wales, fit only imperfectly into the subdivision (1). Both are associated with magnetite; the former is in schistose, the latter in contact-metamorphosed hosts. Both localities are clearly polygenetic.

◁━

Fig. 13-5. Boleo copper field, Santa Rosalía, Mexico. Fence diagram showing the south-eastern portion of the field. Cover to the ore-bearing unit: Tic, Tis, Tgc, Tgs = middle-upper Pliocene conglomerate and sandstone. Ore bearing lower Pliocene Boleo Formation: Tbt = tuff and tuffaceous sandstone; Tbc = conglomerate bands (numbered); Tbg = gypsum. BLACK (No. 1,2,3) = Cu ore beds. Basement: Miocene Comondú basalts. From Wilson (1955).

The Granduc mine (Dudas and Grove, 1970; 749 Tt Cu/1.73%) is a long zone of conformable massive to disseminated magnetite and lesser pyrite, pyrrhotite, chalcopyrite lenses in a band of a Triassic or Jurassic quartz-biotite schist (metatuff ?), underlain by a siliceous graphitic schist, metaquartzite and greenstone (metamorphosed augite andesite flow). The ore is extensively remobilized locally.

Near Cadia (Welsh, 1975; 317 Tt Cu/0.72%, 224 t Ag, 22t Au, about 20 Mt Fe; Fig. 13-6), an Ordovician unit of andesite flows with interbedded tuff, shale, and limestone, contains a thick body of banded magnetite generally considered stratiform. Superimposed over the magnetite are probably epigenetic pyrite, chalcopyrite, specularite, calcite, chlorite, quartz, epidote replacement patches, disseminations, bands and transgressive veinlets. The ores are in the exocontact of a late Ordovician or Devonian diorite-monzonite stock.

In the subdivision (2) of massive sulphides, submarine andesites and basaltic andesites are the regionally dominant rocks, but significant amounts of felsic rocks (dacites, rhyolites or keratophyres) usually occur as well. Colley and Greenbaum (1980) described several examples of small deposits and showings in the Eocene to Miocene volcanics of Fiji that are only lightly metamorphosed. There, dominant pyrite with subordinate sphalerite and chalcopyrite, are the principal sulphides. Quartz and barite are the most common gangue. The orebodies are either massive, flat-lying "exhalative" pods accompanied by a strong argillization and silicification, or pyrite-silica stringers and replacements in a feeder vent. At Kula Creek, a bedded agglomerate contains fragments of massive chalcopyrite, pyrite and sphalerite as well as dacite, and it is interpreted as a product of penecontemporaneous reworking of a previously consolidated massive sulphide body.

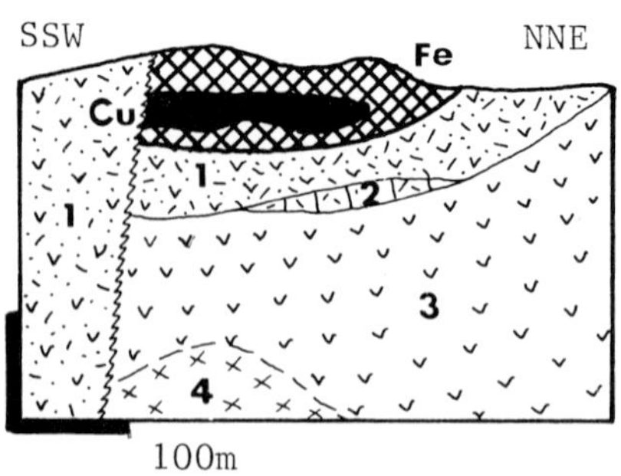

1 Or$_3$ andesite tuff, volc. arenite, calcareous tuff
2 tuffaceous limestone
3 augite andesite porphyry lavas, volc. breccia, agglomerate, tuff
4 Or$_3$? D ? composite diorite, quartz monzonite, granodiorite stock
Cu magnetite-chalcopyrite orebody
Fe magnetite orebody

Fig. 13-6. Cadia, N.S.W.; gold-chalcopyrite mineralization in massive magnetite. From LITHOTHEQUE, modified after Welsh (1975).

The andesite-associated Fijian massive sulphides probably contain no more than 30 Tt Cu and 50 Tt Zn.

In the poorly known Takara mine 80 km west of Tokyo, Japan (Mitsubishi Mining Company, oral communication, 1972), four peneconcordant lenses of massive chalcopyrite, pyrite and gypsum are located in Miocene propylitized green andesite tuffs and volcaniclastics, overlain by Quaternary Mt. Fuji lavas, and intruded by Pliocene ? dolerite, andesite and quartz diorite dikes. 1.5 Mt of 1.13% Cu ore was produced before 1962. Some ore lenses have a high sphalerite content (up to 19.7% Zn) and the total economic tonnage is of the order of 20 Tt Cu and 8 Tt Zn. This deposit bears some similarity with the "Kieslager-type" ores in the bimodal or mafic suite (Chapters 10 and 12), as well as to the "Kuroko-type" ores in the sequentially differentiated suite (Chapter 14).

In the geologically older regions, andesite-associated massive sulphides are more controversial because of the superimposed metamorphism and deformation. In the Granby Bay (Anyox) field, north-western British Columbia (Thompson and Panteleyev, 1976), twelve intensively deformed pipe to sheet-like, broadly concordant massive pyrite, pyrrhotite and sphalerite bodies, are hosted by Jurassic pillowed meta-andesites and meta-siltstones. Most orebodies are at or near the andesited-siltstone contact, and are enveloped by a quartz-biotite schist and floored by a footwall stockwork of disseminated and stringer mineralization. 24 Mt of 1.5% Cu, 7.7 ppm Ag, and 0.14 ppm Au ore was produced.

Granby Bay is exceptional because of the lack of recorded felsic volcanics. More massive sulphides are closely associated with the youngest felsic members of an andesitic eruptive cycle. Mount Chalmers, Queensland (70 Tt Cu/1.6%, 8.03 t Au/1.9-4 ppm; Large and Both, 1980), is hosted by Permian felsic pyroclastics. Two massive peneconcordant complex sulphide bodies consist of a well-banded pyrite, chalcopyrite, sphalerite and minor galena, grading upward into a barite-rich and fragmental sulphide ore. The lenses are underlain by a stringer and stockwork zone, and have formed on top of a rhyodacite pyroclastic dome. The massive lens gradually changes facies into a pyritic dolomite or limestone, and chlorite-pyrite siltstone.

Several massive sulphide deposits rather similar to Mount Chalmers, occur within the Jurassic and Cretaceous volcanics of the Somkhit-Karabakh belt in the Lesser Caucasus, U.S.S.R. In the Atkhala deposit (Smirnov, ed., 1968), a massive pyrite and lesser sphalerite, galena and chalcopyrite lens rests on hydrothermally altered quartz porphyries, and is topped by andesite tuffs and volcanic breccias. A barite-sulphide body is located in the hangingwall of the ore lens. In the Altyn-Tyube and Kyzyl Cheku deposits in central Kazakhstan (Kolesnikov, 1975) stratabound pyrite, chalcopyrite and bornite lenses are found in a gray Devonian limestone along a contact with silicified andesite porphyries.

Numerous small stratabound massive sulphide occurrences (the largest one is near Burraga; 18.6 Tt Cu), are to be found in the late

Ordovician to Silurian belt of meta-andesites and associated sediments south of Bathurst, N.S.W. Stanton (1955) pointed out that most orebodies are hosted by metashales (now biotite schists) in zones of silicification adjacent to felsic or intermediate volcanic centres.

In the Brskovo mine near Mojkovac (Montenegro, Yugoslavia; 487 Tt Zn, 32 Tt Pb, 20 Tt Cu, 19 Tt Sb; Cissarz, 1957),a massive to disseminated pyrite, galena, sphalerite and lesser chalcopyrite and tetrahedrite lens is hosted by a lower Triassic meta-keratophyre (sericite schist) and meta-argillite, within an andesitic belt. The nearby deposit Šuplja Stijena (104 Tt Zn, 40 Tt Pb), contains several high-grade massive Fe, Zn, Pb, Cu sulphide bodies in a sericite-chlorite altered Triassic meta-andesite and marble, intruded by granodiorite.

In the Akbastau-Kusmurun field in Kazakhstan (Mukanov et al., 1966) massive pyrite, chalcopyrite and sphalerite lenses surrounded by a halo of chlorite and sericite alteration, occur in sheared and schistose middle to late Ordovician meta-andesites and pyroclastics.

As in the massive sulphide belts of other volcanic affiliations, the stratabound sulphide masses are often accompanied by pyrite-chalcopyrite or pyrrhotite stockwork or impregnation zones in schistose, cleaved and sheared felsic metavolcanics. This style is clearly exemplified by the Mt. Lyell field reviewed in Chapter 11. Numerous comparable occurrences are known in andesitic belts, and are sometimes listed in the Soviet literature under the heading of "propylitic copper association".

In the east-west trending Spass metallogenetic zone of Kazakhstan (Kolesnikov, 1975), several ore fields (Spass, Koktas) coincide with large middle-upper Devonian basalt-andesite domes, located at intersections of deep—reaching faults. The ore is represented by disseminated and veinlet chalcopyrite and bornite concentrated in irregular stocks, pipes and lenses. The host rocks are sericite, chlorite and epidote-altered and intensively metamorphosed andesitic agglomerates, flows and volcaniclastics. The small Annidale deposit in New Brunswick (stringers of chalcopyrite and quartz-chalcopyrite veinlets in chloritized Ordovician or Silurian greenschists, i.e. andesite meta-pyroclastics) is representative of some one hundred similar small occurrences recorded worldwide.

13.4. Mn ASPECT

After copper, manganese is the metal most commonly and most predictably accumulated in ore deposits hosted by the dominantly marine andesite-sediments association. Bedded, stratiform manganese deposits are most commonly located along, or near, the contacts of andesite pyroclastics and volcaniclastics with marine carbonates (sometimes substituted by evaporites). This setting, closely comparable with the setting of the Quaternary deposit Forari (New Hebrides; see Chapter 5), is repeated in the large Mn provinces of eastern Cuba, central Chile, the Philippines, and many smaller provinces of the world (Table 13-3). In places, this style of Mn mineralization coincides, overlaps or is situated close to, the deposits of disseminated Cu sulphides in andesites and intervening sediments (e.g. in the Santa Rosalía area, Mexico; Coquimbo region, Chile; Pozharevo region, Bulgaria)

In the Mn belt of eastern Cuba located along the northern foothills of the Sierra Maestra Range (Simons and Straczek, 1958) several hundred deposits and occurrences are present in an area of 8,000 square kilometres. Most are hosted by upper Cretaceous to Eocene Cobre Formation, a 4,000 m-thick unit of andesite with minor basalt and andesite tuffs, agglomerates and lavas. Interbedded are subordinate horizons of marine volcaniclastics and biogenic limestones up to 150 m thick. All the productive Mn deposits display a good stratigraphic control, being located closely above or below the base of the distinct Charco Redondo Limestone Member. The variety of the Mn orebodies and their local setting recorded by Woodring and Daviess (1944) from the Guisa-Los Negros area (Fig. 13-7) is representative of the entire province. All the "primary" bedded ores of volcanogenic-diagenetic or hydrothermal (hot springs ?) - sedimentary origin, are now composed of pyrolusite and minor hausmannite and braunite, almost always partly or fully altered to the "secondary" psilomelane, cryptomelane, manganite, and similar species. The ore beds in limestone (e.g. the Ponupo field, Taratana mine) start as impregnations, stringers and nodules, and evolve into massive beds. Jasper or chert are a frequent gangue or an associate. The tuff-hosted Mn minerals initially impregnate then later replace the tuff, and are mixed with argillized, chloritized and zeolitized remnants of the volcanic fragments and matrix. Thick beds of jasper are often found in the footwall of the ore lenses. There is no sharp difference among the tuff- and limestone-hosted orebodies, because the latter frequently formed by selective replacement of a thin tuff lens or a layer enclosed in the limestone, and are particularly common at sites of a tuff/limestone facies change. Young reworked deposits, such as rubble and pellets of Mn oxides in lateritic soils ("granzon ore"), and reprecipitated ore in karst caves in limestone, are of little economic importance.

In central Chile (N.E. and S.E. of La Serena), the Coquimbo Province Mn Belt is hosted by volcanic arenite and occasionally by limestone horizons, interbedded with andesite tuff, breccia and subaqueous vesicular flows (Aguirre and Mehech, 1964). Braunite and pyrolusite are the most common ore minerals, psilomelane and manganite are partly "secondary". Calcite and ? authigenic piemontite are found in the gangue. The mineralization style is closely comparable with that in Cuba, except that the host association is transitional into a gypsum-rich lagoonal and red-beds facies and some Mn ore lenses have a gypsum gangue.

Table 13-3. Bedded Mn ores in andesite pyroclastics - interbedded or topping sediments; selected example localities

LOCALITY	AGE	GEOLOGY	MINERALIZATION	REFERENCES
Santa Rosalía area Baja California, Mexico	Pl	fine andesite tuff resting on basal conglomerate, overlying Mi contin.basalt	cryptomelane and pyrolusite replace argilliz.tuff,impregnate congl.,with or without chert and jasper; tot.resource min.1.15 Mt	Wilson and Veytia (1949)
Northern flanks of Sierra Maestra, Cuba	Cr_1-Eo	4 km thick sequence of andesite and minor basalt, dacite tuffs,minor limest. Block faulted,zeol.facies	over 100 stratif.lenses and layers of mass.pyrolusite,braunite,hausmannite; second.psilomelane,manganite; in tuff and limest. 2.6 Mt Mn/20-40%	Simons and Straczek (1958)
Coquimbo Province Mn belt (N.E. and S.E. of La Serena) Chile	Cr_1	submarine porph.andesite flows interb.with tuff, limest.,calcar. sandst., change facies into red-beds	braunite,pyrolusite lenses and layers in andesitic volcanic arenite,less frequently limestone 7.1 Mt Mn/32-46%	Aguirre and Mehech (1964) Ruíz (1965)
Pozharevo, western Bulgaria	Cr_{2-3}	submarine andesite breccia, tuff,interbedded with marl, limestone	stratabd.lenses of pyrolusite, psilomelane, braunite with opal to chert gangue. 35 Tt Mn/35%, minor Cu occurrences	Iovchev (1961)
Tasikmalaya Mn distr.,central Java Indonesia	Mi_{2-3}	marine andesite pyroclastics overlain by limestone	35 small deposits; pyrolusite, hausmannite, psilomelane lenses and impregn.in tuff; about 100 Tt Mn/28%	Schmidt (1976) own visit (1980)
Western Viti Levu Island, Fiji	Mi_{1-2}	andesite pyroclastics, minor interbedded carbonate, calcareous tuff	mass.to dissem.stratabd.lenses of pyrolusite,psilomelane,minor hausmannite,manganite.Most assoc. with jasper, argillic alteration. 40 Tt Mn	Colley and Greenbaum (1980)

1 bedded ore in limestone
2 bedded ore in tuff, no chert
3 bedded ore in tuff, with chert
4 nonbedded ore in limestone
5 nonbedded ore in tuff
6 rubble and elluvial ore
7 karst cave ore

Fig. 13-7. Mineralization styles of the eastern Cuban manganese deposits in the Guisa-Los Negros area. From Woodring and Daviess (1944).

13.5. Fe ASPECT

Major iron ore deposits except those formed by intrusive or weathering / resedimentation interaction, are rarely hosted by a marine andesite association. Several examples are cited in the Soviet literature. The Kosagaly-Tuyak field in the north-western Karatau Range, central Kazakhstan (Novokhatskii, 1972), appears to be a middle-upper Ordovician counterpart of cherty banded iron formations, hosted by schistose meta-andesites. Hematite and magnetite form numerous 2-40 m thick stratiform lenses interbedded with ferruginous quartzites (metajaspers), enveloped by chlorited-sericite and hornblende, epidote, chlorite phyllites and schists. The ore grade varies from 17-43% Fe and 1.3-3.2% Mn. There are locally superimposed Cu,Pb,Zn sulphides.

The Abakanskoe deposit in the western Sayan (Sokolov and Grigor'yev, 1974) consists of two peneconcordant bodies of magnetite in actinolite, chlorite and calcite gangue. The ore is hosted by a Cambrian limestone and meta-shale enveloped by a meta-keratophyre, a minor member of a dominantly augite andesite sequence. A ? subvolcanic albitite body occurs in the vicinity. The genetic interpretations offered, range from syn-depositional hydrothermal (equivalent to the Lahn-Dill "type") to post-depositional metasomatic.

13.6. Au ASPECT

Small quantities of by-product gold are associated with virtually all the massive sulphide deposits hosted by the andesite association. Substantially less predictable and of much lower grade is the association of gold with the disseminated copper deposits in andesite flows and pyroclastics or sediments. It would be interesting to determine whether gold—only or predominantly gold equivalents of the above mineralization styles do actually exist. Although several occurrences of low--grade disseminated (dispersed) gold in unaltered andesitic volcanics have been mentioned in the literature, none has yet been studied using modern methods and described in detail. The remaining localities that will be reviewed briefly here do not seem to fit any popular model and are probably polygenetic, having required a significant interaction of tectonism and igneous intrusion with the hosting andesitic supracrustals to form.

The Gold Ridge goldfield 27 km S.E. of Honiara, Guadalcanal (Solomon Islands; Purvis, 1975; Arthurs, 1979) is hosted by a Miocene-Pliocene (or even Pleistocene ?) shallow subaqueous to subaerial andesitic explosive volcanic--sedimentary centre, developed over a Cretaceous and Oligocene basement of gabbro and ultramafics. The host unit shows a rapid facies change, dominated by volcanic conglomerates to breccias, grading into a tuff, volcanic arenite and siltstone. The fine sedimentary lithofacies is clearly water-laid, thinly bedded and pyritic and it covers a portion of a former vent.

Gold occurs as free, tiny discrete particles and microscopic inclusions within pyrite or arsenopyrite, over an area of 3 x 4 km^2. The upper portion of the mineralization is hosted by altered tuffs and volcanic breccias, takes a form of a lens with diffuse boundaries, and is at least peneconcordant with bedding of its host rocks. This grades downwards into fracture-controlled gold-bearing zones of silicification and stockworks of quartz veinlets in sericite-chlorite altered rocks. Visible gold appears in some quartz veins. The general tenor is between 0.25 to 2 ppm Au, and the best values (5-10 ppm) occur in the quartz veins. The best recent drilling intersection gave 4.35 ppm Au over 49 m. Placer gold has been recovered from proximal alluvial placers located downstream, and a reserve of 10.7 Mm^3 of 0.109 g/m^3 Au is available. The mineralization is usually interpreted as hydrothermal (hot spring) syn-volcanic, to early (waning stage) post-volcanic, probably formed in a vent region flooded by a crater lake (Peter Dunkley, oral communication, 1981).

A somewhat similar gold occurrence in the Vunda Valley, N.W. Viti Levu Island, Fiji, has been described by Colley and Greenbaum (1980). There, native gold is sporadically disseminated in intensively altered (quartz, sericite, diaspore, argillic, propylitic alteration) Miocene-Pliocene andesite flows and pyroclastics, and accumulates locally in fracture zones also containing visible copper sulphides. Lawrence (1978) interpreted this occurrence as a "porphyry gold" associated with a buried shoshonitic intrusion, but the presence of such an intrusion is only conjectural. The extensive but very low-grade (1.5-2 ppm Au) Mt. Rawdon gold zone (Queensland) hosted by an altered acid volcanic agglomerate in a Devonian andesitic volcanic-sedimentary association, could be a Paleozoic near-equivalent of the two Melanesian occurrences discussed above.

Many "epithermal" epigenetic gold-bearing veins and stockworks described in Chapter 26, are hosted by submarine andesites and associated sediments.

The Junction Reefs goldfield near Mandurama, New South Wales (Wilson, 1965), is usually interpreted as a skarn, but it could also be a thermally metamorphosed disseminated stratabound gold-sulphide deposit comparable with Mala, Peru (described above). There, Ordovician meta-andesites interbedded with tuffs contain a 42 m thick interval of well-bedded and laminated shale, siltstone, volcarenite, tuff and impure limestone, now hornfelsed near contact with a Devonian granodiorite dike swarm. Gold-bearing massive to disseminated pyrite, pyrrhotite, arsenopyrite and chalcopyrite are present in at least twenty 10-30° N.W.—dipping horizons. Here 1.37 t of gold were produced.

13.7. TRANSITIONS AND INTERACTIONS

13.7.1. Facies transition volcanics/sediments

Island arc or continental margin andesites located in the nearshore area predictably change facies in a direction away from a volcanic centre from lava flows, coarse pyroclastics and volcaniclastics into increasingly finer-grained sediments (arenites, siltstones, shales or carbonates). In contrast to basalts capable of forming small local flows in a predominantly sedimentary environment, small andesite occurrences are rare at a distance from the magmatic arc in a particular geological time. As a consequence, distal andesite-derived sediments (arenites, shales, flysch-like arenite and shale alternation) are devoid of mineralization and any appreciable trace metal enrichment.

13.7.2. Recent weathering and redeposition generated ores

Tropical lateritic weathering superimposed on andesitic terrains has little economic significance. Poor quality ferruginous laterite deposits have formed mostly on basaltic andesites in Indonesia, the Philippines, Fiji and elsewhere, but their chances of preservation are slight. The bauxite deposits in Fiji (Colley and Greenbaum, 1980) are usually blankets of lateritic clay containing gibbsite nodules, located in flat-lying areas. The largest Wainunu deposit contains about 8 Mt of washed ore with 45% Al_2O_3 and the total tonnage of bauxites in Fiji does not exceed 35 Mt.

"Iron sands" containing detrital magnetite, derived mostly from weathering-released and stream or surf reworked phenocrysts, are widespread along andesitic tropical coasts for example of Fiji (about 10 Mt of 57-77% Fe concentrate), the Solomon Islands, New Guinea, etc. Gold placers, most of them small, are commonplace in the vicinity of hydrothermal gold occurrences, worldwide.

13.7.3. High-grade metamorphic transitions

Andesites are converted into amphibolites by high-grade metamorphism, and meta-andesitic amphibolites are practically indistinguishable from meta-basalts. The latter, however, are substantially more frequent. As a consequence, andesitic terrains virtually lose their identity in high-grade metamorphic regions.

13.7.4. Volcanic/plutonic transitions and interactions, 1: the "intimate" andesite/diorite association

Andesite and diorite are practically identical in composition and differ only in grain size and texture (andesites are dominantly fine crystalline and porphyritic, diorites are holocrystalline). Coexisting andesite and diorite could be either (1) contemporary comagmatic rocks solidified at different crustal levels; (2) diorite intrusions emplaced into older andesites; or (3) diorite-like metasomatites formed by metasomatic recrystallization (dioritization) of older andesites (dominantly pyroclastics) or other rocks of suitable composition, including sediments. The alternative (3) has recently been postulated to be a significant ore-concentrating process, probably responsible for the formation of certain Fe, Cu and Au deposits. Even if the genetic interpretation reviewed later is incorrect, the mineralized intimate diorited/andesite association is an empirical reality that deserves to be reported.

EL ROMERAL AND SIMILAR MAGNETITE OREBODIES

El Romeral (Bookstrom, 1977; 110 Mt Fe/55%; Fig. 13-8) is the largest and best studied locality in a belt of magnetite deposits hosted by the lower Cretaceous andesitic Arqueros Formation in central Chile. Here, a group of steeply dipping to vertical lenses and pods of a high-purity massive magnetite with accessory actinolite, clinozoisite, apatite, sphene, etc., are hosted by actinolitized and epidotized andesites and schists, resting as a roof pendant on a Cretaceous diorite pluton. Bookstrom (1977) concluded that the ore formed by a metasomatic reconstitution of an earlier banded siliceous hematite during metasomatic dioritization, at temperatures around 550-475° C, and that the ore-forming event was a retrogressive one, postdating the regional metamorphic culmination.

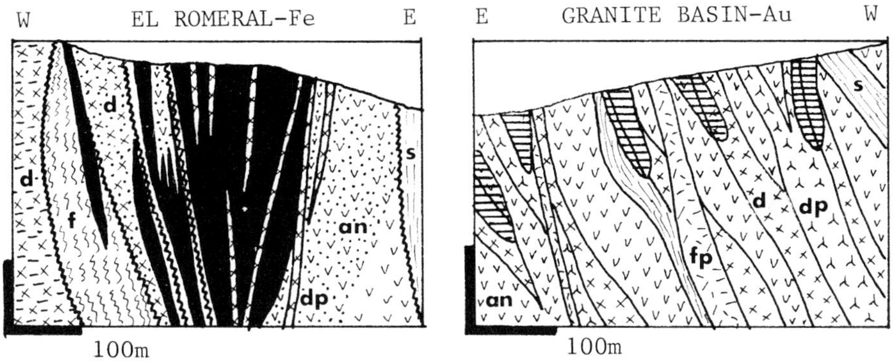

Fig. 13-8. Examples of ores in the "intimate andesite /diorite" association. El Romeral, Chile, massive magnetite. Lower Cretaceous (d) diorite, (an) andesite, (dp) diorite porphyry, aplite dikes, (f) fault breccia, mylonite. Carboniferous (s) phyllite, schist; BLACK = magnetite ore; dots = actinolite alteration; lines = feldspathization.

Granite Basin, British Columbia, gold. Upper Triassic (an) porphyritic andesite; (s) tuff, volcanic arenite, argillite, limestone; (d) feldspar-phyric diorite; (dp) diorite porphyry, hornblende-phyric. Ruled = Au bearing pyritized zones; (fp) feldspar porphyry.

From LITHOTHEQUE, after Bookstrom (1977), Mina El Romeral staff, guided tour, (1977); Roots (1954).

GOLD DEPOSITS

In the Granite Basin showing near Aiken Lake, northern British Columbia (Roots, 1954; Fig. 13-8), an intensively fractured complex of Triassic porphyritic andesites and associated marine sediments is intruded (or replaced) by several generations of porphyritic diorite and feldspar porphyry dikes. In all the surrounding rocks and particularly in the porphyritic diorite, five to six irregular zones of quartz-sericite alteration and pyritization have developed, which in places are gold-bearing. The average assay over a 13 m thickness was 5.6 ppm Au.

COPPER DEPOSITS

Widespread association, overlapping and gradational relationship of diorite and andesite, is a characteristic feature of the geology of the Philippines, particularly its lower Tertiary sequences (Bryner, 1969). Diorites and andesite/diorite mixed rocks are frequently copper-mineralized, and the Marcoper porphyry copper orebody (Loudon, 1976) is an example of a mineralization hosted by the intermediate magmatites. Wolfe (1973) introduced the "Viscayan type" of disseminated pyrite-chalcopyrite deposits hosted by the andesite/diorite set, the peculiarity of which is that the mineralized rocks are fine-grained and equigranular. There are no associated hypabyssal rocks except the diorites, and the orebodies are subhorizontal to gently dipping and undulating. As such, the "Visayan-type" deposits tend to be rapidly removed by erosion. The ore substance is mostly in fracture filling veinlets.

Wolfe (1973) concluded that this style of mineralization formed from hydrothermal solutions at small depths (about 1 km), in upper Miocene or later time, and in structures associated with the Philippine Fault. The following deposits mentioned by Wolff (1978) are probably members of this category: Boneng-Lobo, Luzon (314 Tt Cu/0.34% , 3.4 t Au/0.37 ppm); Kennon, Luzon (109 Tt Cu/0.45%); Tabbac, Luzon (120 Tt Cu/0.4%, 1.17 t Au/0.39 ppm); and Balatoc, Luzon.

13.7.5. Volcanic/plutonic transitions and interactions, 2: the "volcanic" porphyry copper systems

"Volcanic" (more properly subvolcanic) porphyry copper deposits have been defined by Sutherland Brown (1976) as "those formed adjacent to variably shaped porphyry intrusions that have risen high in the crust to intrude a coeval and at least partly consanguinous volcanic pile in an orogenic setting that is commonly marine". The porphyry bodies have variable shapes (tabular, lensoid, cylindrical, irregular bodies or anastomosing dikes and sills), and the intrusions tend to be small in comparison with the relatively large orebodies (Fig. 13-9).

In dominantly mafic andesite terrains, consanguinous intrusions are mafic to intermediate (gabbrodiorite, diorite, quartz diorite) or "alkaline" (monzonite, syenite). The alkalinity in the latter may also be due to alkaline metasomatism akin to spilitization or potash metasomatism. On the basis of their intrusive affiliation, disseminated copper deposits, such as those above, are treated in the North American literature either under the heading "alkaline suite" (Barr et al., 1976), or "diorite model" porphyry coppers (Hollister, 1978). Similar deposits occur in the rest of the world (e.g. in the U.S.S.R.; Vorob'yev et al., 1978), but were not uniformly typified so that their inclusion in this category is rarely quoted in the literature. Not all "diorite"/"alkaline suite" porphyry copper deposits are simultaneously "volcanic". Some are hosted by deep-seated, plutonic diorites, and not all volcanics-hosted porphyry coppers are "volcanic" in the sense of Sutherland Brown (1976); that is, consanguinous with the volcanic pile.

The plutonic and "phallic" porphyry deposits are treated in Chapter 28, although later in this chapter a brief account is devoted to those deposits emplaced into older andesitic associations with which

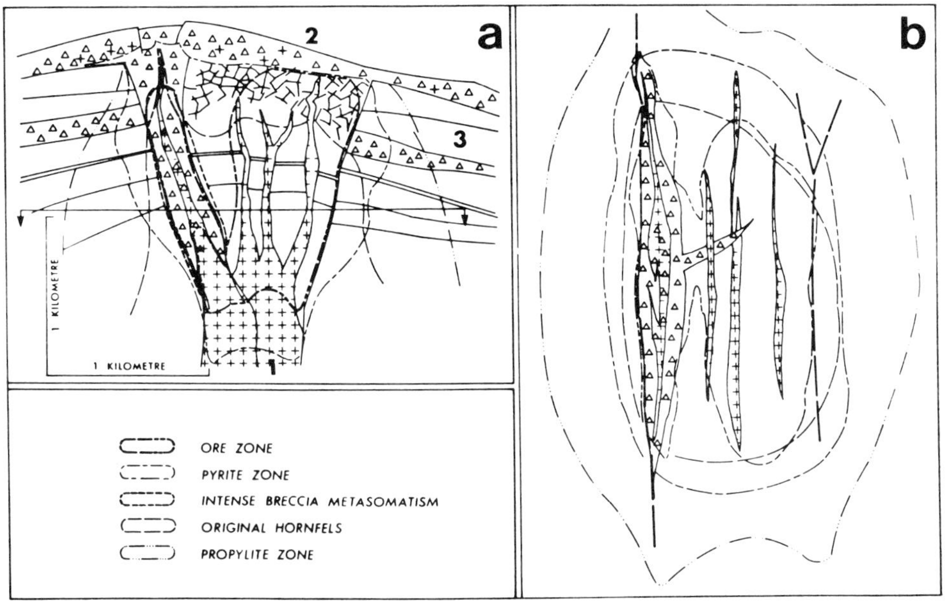

Fig. 13-9. "Volcanic" porphyry copper model, section (a) and plan (b). (1) intrusive body (syenite, diorite); (2) igneous breccias; (3) surrounding rocks (volcanics and sediments). From Sutherland Brown (1976), courtesy of the Canadian Institute of Mining and Metallurgy and the author.

they are not consanguinous. The "phallic" porphyry deposits are those centered on small cylindrical plutons that have risen high in the crust, intruding unrelated host rocks in a late or post-orogenic setting; Sutherland Brown (1976).

The "volcanic" porphyry deposits, consanguinous with dominantly marine mafic andesitic terrains, are treated in this chapter and example localities are listed in Table 13-4. The "volcanic" deposits, related to evolutionarily more advanced systems that include magmatic rocks predominantly derived from the remelting of an older continental crust (granodiorites, quartz monzonites) are discussed in Chapter 14. Figure 13-10 illustrates diagrammatically the environmental systematics of the porphyry coppers.

DIORITE, SYENITE, ETC.-AFFILIATED "VOLCANIC" PORPHYRY COPPER DEPOSITS

These deposits are located in thick sequences of augite- and plagioclase-phyric andesites, basaltic andesites and sometimes tholeiitic basalt flows and pyroclastics (as in the Karmutsen-Nikolai

Table 13-4. "Volcanic" (consanguinous) porphyry copper deposits in marine andesitic terrains, affiliated with intermediate (or sodic, potassic "alkaline") subvolcanic intrusions; selected example localities

LOCALITY	AGE	HOST SUPRACRUSTALS	MINERALIZING INTRUSION	OREBODIES,TONNAGE	REFERENCES
Gnat Lake, N.W. British Columbia	Tr_3-J_1	andesite,dacite	syenite porphyry	dissem.cp in KF and propyl.altered volcanics; 110 Tt Cu/ 0.44%	Barr et al. (1976)
Galore Creek, Stikine R.area, N.W. British Columbia	"	aug.andes. and basaltic andes.lavas breccias,volcaniclastics	potass.alk.complex; phonol.,trachyte, syenite	intens.K-altered li- near N.N.E. brecc. zones in metavolc., dissem.,veinlets of py,cp/born.,in biot, KF,anhydr.,garnet 1.46 Mt Cu/1.06% 870 t Ag, 4.3 t Au	Allen et al. (1976)
Cariboo Bell, S.C. British Columbia	"	augite trachybasalt flows,breccias,volcaniclast.	multiple syenitic laccolith,subvolc., monzonite, syenodiorite	dissem.magn,cp,and veinlets in intrus. brecc.near laccol. top; KF,bio,diops. alter.,122 Tt Cu/ 0.49%	Hodgson et al.(1976)
Afton N.W. of Kamloops, S.C. British Columbia	"	andes.and basalt flows,breccia, tuff,minor argil., limest,picrite bs.	diorite to syenite pluton, latite porphyry,breccia	dissem.cp,born,in shattered volc., irreg.orebody,supergene enrichm. 310 Tt Cu/1%	Carr and Reed (1976)

Copper Mountain - Ingerbelle, Princeton, Br. Columbia	Tr_3-J_1	andes.brecc.flows, tuffs, volcanic siltstone	albitized porph.syenite, monz., diorite; concentric differ. intrus., dike swarm	dissem.py, cp, born, in biot.altered m-volc. and porph. 691 Tt Cu	Fahrni et al. (1976)
El Arco, Baja California, Mexico	Cr	andesite pyroclastics, flows	diorite-syenite intrusion	py, cp, Au dissem. to stockw.in K-alter.+ propyl.z.	Hollister (1978)
Boshchekul', N.E. Kazakhstan, USSR	Cm	basalt.andes.porph. and amygd.flows, tuffs, volcaniclast.	gabbro-monzon., diorite, plagiogranite porphyry	veinl., dissem., cp in endo- and exocon. propyl.alter., secon. enr. 2.16 Mt Cu/ 0.4-0.81%; Au, Ag	Laznicka (1976)
Marian dep., Luzon, Philippines	Mi?	trachyand.and and. brecc.empl.into arenites, siltst.	monzon., trachyand., syenite complex	py, born, magn, cp, Au in KF, biot, propyl. alt. 87 Tt Cu/0.45%	Wolff (1978)
Lumbay General Nakar, Luzon, Philippines	Mi?	basalts, andesites, volc.sediments	diorite, fault controlled intrusion	py, cp, born, magn, in bio-sc altered zone 282 Tt Cu/0.41%	as above
Waivaka Prospect, Fiji	Mi-Pl	andesitic volcan.	diorite porphyry	cp, born, magn, small but high grade veins	Colley and Greenbaum (1980)
Sambeto-Mbalevuto	Mi_3	subvolc.andesite, volcaniclastics	monzonite, diorite	dissem.cp, born, Au in altered volc. and intrusives	as above

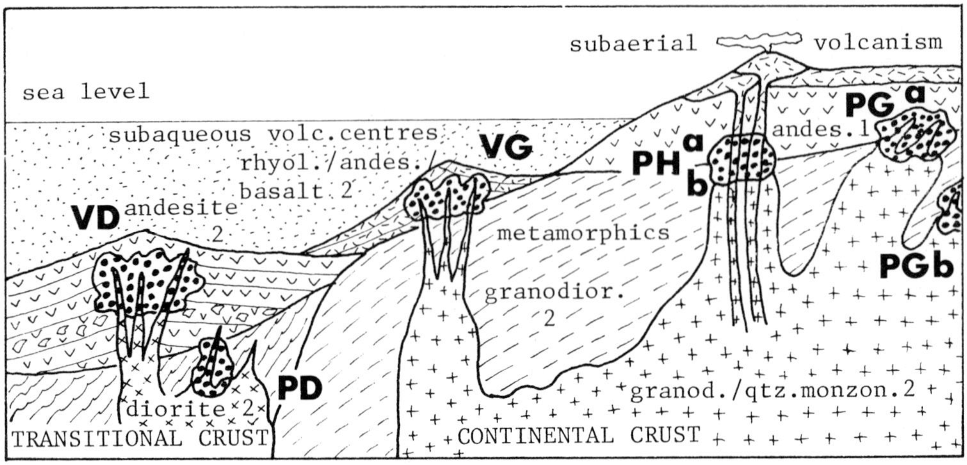

Fig. 13-10. Diagrammatic representation of the porphyry coppers environmental systematics. The "diorite" affiliated porphyries are either "volcanic" (VD), hosted by andesites comagmatic, consanguinous and contemporary with the diorite, or "plutonic" (PD).

"Granitic" (evolved) porphyries can be also "volcanic" (VG), but most are "phallic" (PH) and plutonic (PG). "Phallic" and plutonic (a) deposits have been emplaced into older buried andesites, so andesitic volcanics may, after exhumation, be found on the surface hosting younger, "alien" porphyry deposits. The (b) deposits crop out within the older basement.

assemblage, reviewed in Chapter 10), disturbed by numerous faults of regional magnitude. These volcanics have anomalous trace Cu contents (150-300 ppm), often megascopically indicated by frequent occurrences of native copper grains. At least partly submarine volcanic centres marked by abundant breccia bodies usually of linear outline and distinctly fault-controlled, subvolcanic intrusive bodies occur at frequent intervals along with alteration and mineralization. The breccias often exhibit mixing of fragments, and volcanic fragments are dominant along the margins, with intrusive fragments in the central regions. Stratified volcaniclastic rocks in the vicinity of intrusions often contain widespread porphyry fragments that decrease in size and number as their distance from the intrusive center increases.

The igneous intrusions cause a relatively light contact metamorphism (hornfelsing) but in the mineralized complexes this is usually destroyed by a later hydrothermal chloritization. Hydrothermal alteration zoning and related mineralization are usually confused. They are asymmetrically and irregularly distributed, diffuse, patchy, and of variable intensity. The visual impression of an altered/mineralized "volcanic" porphyry center gained by a first time visitor is that of a jumble of minor pink patches (syenite porphyries, feldspar metasomatites) in a dark-green chloritized background.

In the "volcanic" porphyry deposits affiliated with syenitic or dioritic intrusions, the broadest alteration zone consists of pervasive albite, zoisite and chlorite, lesser garnet, pyroxene, sphene and apatite alteration (Sutherland Brown, 1976). The anhydrous silicates are transitional into skarns, that are associated at many localites with the volcanics or intrusions-hosted porphyry deposits. In the more restricted internal alteration zones, the early alteration is destroyed or overprinted by potassic alteration producing K-feldspar and biotite. Light potassic alteration usually advances along fractures and voids in breccias, but with increasing intensity the volcanic hosts can be wholly converted into pink or spotty massive pseudo-syenites or hydrothermal pegmatites.

Sulphide mineralization (bornite, chalcopyrite, pyrite) with some magnetite, is usually found pervasively disseminated or as veinlets, veins in the matrix of altered breccias and along fractures. It is usually quartz-free, although late carbonates and zeolites (laumontite) are often present. The known "diorite" (syenite) model "volcanic" porphyries are typically of small to medium size and yield copper usually with quite high gold and silver contents. Molybdenite is hardly ever present. If skarns are developed, magnetite is abundant and can be recovered as a by-product.

At Galore Creek, the largest of the British Columbia "volcanic" porphyry coppers (Allen et al., 1976), ten orebodies occur in altered upper Triassic augite andesite and basaltic andesite breccias, flows and tuffs with minor sandstone and siltstone interbeds. These have been intruded by a 198 m.y. old potassic alkaline complex, including pseudoleucitic phonolite, trachyte, a variety of syenite porphyries and extensive breccia zones. The orebodies are hosted by feldspathized and biotitized volcanic rocks and pipe-like breccias. They are tabular to manto-shaped, trending north to north-east. The Central Zone is composed of several parallel en echelon zones of disseminated pyrite, chalcopyrite, minor bornite associated with hydrothermal biotite, K-feldspar, anhydrite and garnet, and is 1,950 m long and up to 518 m wide.

Cariboo Bell (Hodgson et al., 1976; Fig. 13-11) is associated with a multiphase syenitic laccolith emplaced into upper Triassic augite trachybasalt and trachyandesite pillowed lavas, volcanic breccias and volcaniclastics, including abundant lahars. The mineralization of disseminated magnetite and chalcopyrite most commonly occurs in a network of irregular veinlets, pods and drusy cavities in a crackle breccia superimposed on an intrusion breccia as well as on a lapilli tuff and monzonite porphyry. About 25% of the reserves are in supergene minerals (malachite, native copper, chrysocolla, chalcocite). The gangue and alteration minerals are represented by biotite, potash feldspar and diopside, and the central mineralization-alteration zone has a distinct, intense salmon-pink colour. The copper zone is surrounded by a pyrite halo.

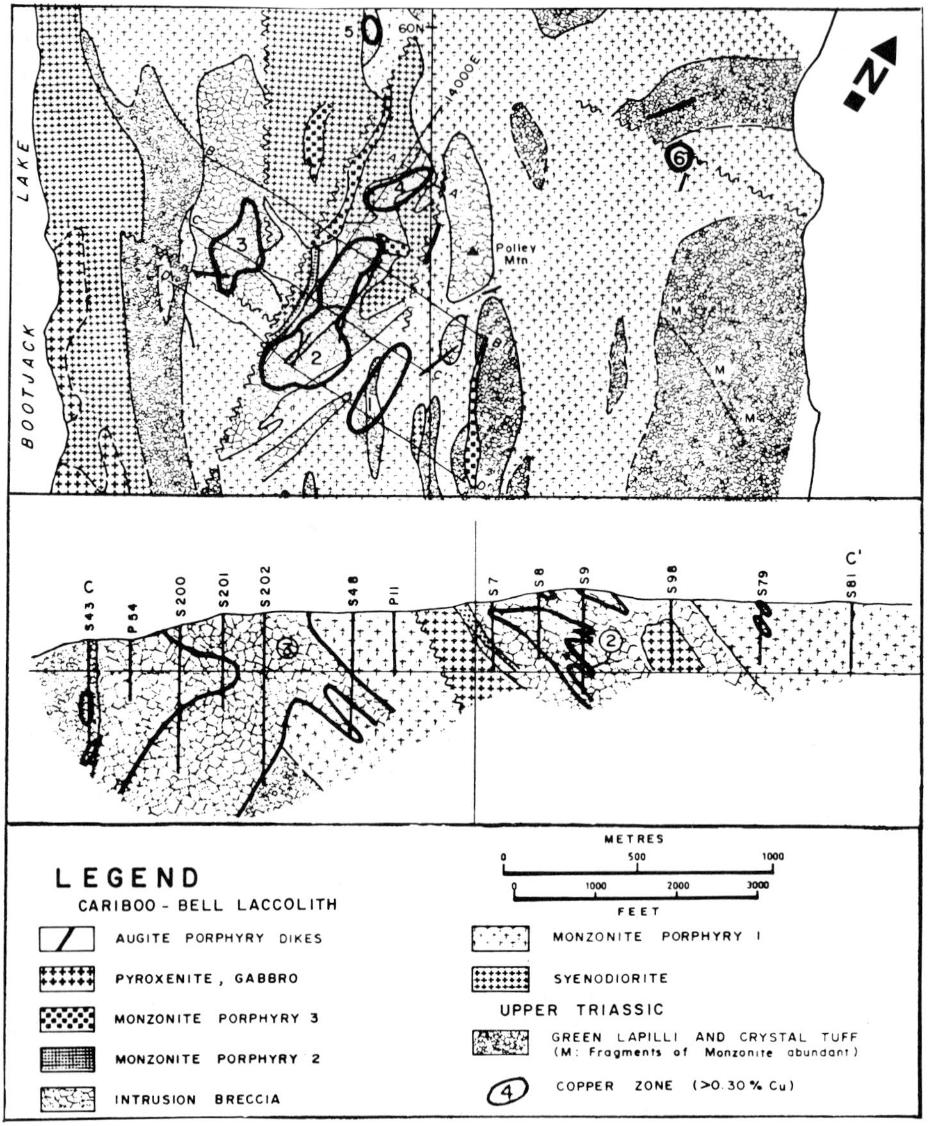

Fig. 13-11. Cariboo Bell, British Columbia. Example of a "volcanic" porphyry copper deposit Map and section are from Hodgson et al. (1976), courtesy of the Canadian Institute of Mining and Metallurgy and the author.

The Copper Mountain-Ingerbelle field near Princeton, British Columbia (Fahrni et al., 1976) contains numerous orebodies in an 11 km long N.W.-S.E. trending, fault-controlled zone that also coincides with the subvolcanic (dikes) fringe of a concentrically zoned diorite, syenite and monzonite intrusion. The mineralization consists of disseminations and stockworks largely of chalcopyrite, pyrite, lesser bornite, in altered andesite tuffs, andesite fragmental rocks, and to a lesser degree in monzonite and diorite in the Ingerbelle deposit. At Copper Mountain, the ore is comprised of disseminations to stockworks of bornite and chalcopyrite in a sheeted, brittle, fractured fine-grained bedded tuff. Spectacular, but small high-grade bornite and chalcopyrite lenses were associated with hydrothermal potash feldspar veins ("pegmatite") in the Copper Mountain stock intrusives. A typical feature of mineralization in this field is an extreme irregularity in the grade distribution and a lack of alteration zoning. Massive, K-feldspar, scapolite, and locally garnet altered andesitic volcanics and volcaniclastics resemble skarns (and have been interpreted by some geologists as such).

Examples of deposits and showings of the "volcanic" or "diorite" porphyry copper style outside British Columbia, include the following localities:

Fiji (Colley and Greenbaum, 1980)

In the Sambeto - Mbale – Vuto area, potassic ("shoshonitic") monzonite and diorite intrude coeval andesites and volcaniclastics containing disseminated, gold-bearing chalcopyrite and bornite. This occurrence is fairly close to the British Columbia example localities. In the Waivaka Prospect, copper sulphide and magnetite associated with diorite porphyry differ from the B.C. model due to the presence of quartz gangue. They may be transitional into an evolutionarily more advanced suite. The porphyry deposits of northern Sulawesi, Indonesia, are of similar transitional nature.

U.S.S.R.

In the southern Urals, the Salavatskoe and Voznessenskoe copper deposits corresponding to the "diorite" model (Vorob'yev et al., 1978), consist of chalcopyrite, bornite, pyrite and magnetite veinlets and disseminations located in the endocontact of porphyritic quartz diorites and gabbrodiorites. These intrusive rocks are comagmatic with upper Silurian-lower Devonian basaltic andesites.

In northern Kazakhstan, the important Boshchekul' deposit has veinlet and disseminated chalcopyrite associated with propylitic alteration in the endo- and exo-contact of a Cambrian gabbro-monzonite complex, emplaced into a basaltic andesite suite. Several deposits in the Kuznetsk Alatau Range (e.g. in Mariin Taiga), occur in Cambrian mafic andesites, associated with subalkaline to alkaline (nepheline syenite) stocks. It is interesting to note that in the belt of British Columbia syenites associated with the "volcanic" porphyry coppers, minor occurrences of feldspathoidal rocks have also been described (e.g. the natrolite, cancrinite, aegirine, melanite syenite

N.E. of Kamloops ; Currie, 1974).

Seven occurrences of the "diorite model" porphyry coppers are tabulated briefly in Hollister (1978).

TRANSITIONS

The "volcanic" or "diorite" porphyry coppers are similar in many respects to the disseminated copper deposits in the "intimate andesite-diorite" association as in the Philippines. These have been reviewed briefly in an earlier paragraph. The andesite and diorite of the latter association, however, do not seem to be contemporary (the diorite is later), and the emplacement or dioritization probably took place at a deeper level. The appearance and increase in quantity of associated dacites and rhyolites, widespread quartz gangue and silicification, widespread sericite alteration, abundant Mo, etc., mark a transition into the more evolved (granodiorite, quartz monzonite-affiliated) porphyry deposits. Many of the Philippine porphyry coppers (e.g. those on the Negros Island) are transitional in this manner.

13.7.6. Volcanic/plutonic transitions and interactions, 3: alien porphyry copper systems hosted largely or in part by dominantly marine andesites

In statistical ratings of porphyry coppers (e.g. De Geoffroy and Wignall, 1972), andesites are listed as the most frequent hosts of the exocontact porphyry copper mineralizations. Many are continental andesites treated in Chapter 26, and several are marine andesites interbedded with sediments. If the "volcanic" porphyry deposits are excluded, there is always a time difference between the age of the host andesite emplacement and age of the ore-producing intrusion. This time gap ranges from several million to several hundred million years and it indicates that the supracrustals and the intracrustals are not coeval. Even so, however, the affiliation of porphyry coppers with andesites is probably not accidental; the igneous intrusion may be just one of the numerous possible events superimposed on the trace copper-rich andesites, causing the broadly distributed copper to concentrate and accumulate locally. This need not have taken place at the present level of exposure, and the intrusive "concentrator rock" need not look overly conspicuous; often it takes a form of minor dikes.

Experienced prospectors have a "feel" for andesite-hosted porphyry copper provinces. There, the dominantly basaltic and pyroxene andesites in flows, pyroclastics (typically ash-flow tuffs) and volcaniclastics are characteristically massive (in contrast to the schistose rocks most characteristic of the massive sulphide provinces), but have a distinct hydrothermal metamorphic overprint represented by abundant conspicuous patches of epidote and/or albite, quartz veining and scattered pyrite. The alteration assemblage is close to propylitization and is fairly well distributed throughout the region. The high trace-copper content in andesites is megascopically revealed by frequent mineralogical occurrences of chalcopyrite grains typically scattered in the various alteration patches, veinlets, or native copper most common in amygdules and in scoriaceous lava tops. The actual porphyry copper mineralization when developed, however, is in zones of potassic or quartz-sericite alteration that are located within or in the immediate vicinity of the concentrator intrusion.

Porphyry coppers hosted by marine andesites are known in almost all the example regions listed in Table 13-4 such as Melanesia, the Philippines, Fiji, British Columbia, Panama and others. They are reviewed and tabulated in Chapter 28.

13.7.7. Volcanic/plutonic transitions and interactions, 4: miscellaneous ore-bearing veins and replacements

Plutonic rocks intrusive into the marine andesite association belong to two end-member categories of derivation: (1) those generated within the relatively "primitive" assemblage of ensimatic island arcs and (2) those generated within the continental crust over which the arc andesites are draped or to which they have been welded during collisions. There is a complete analogy in setting and igneous petrology with the porphyry deposits (Fig. 13-10): association (1) corresponds to the generator rocks of the "diorite" porphyries; assemblage (2) corresponds to the granodiorite-quartz monzonite suite. Several diorite-associated mineralizations are reviewed briefly below. Ores affiliated with the more evolved suite (2) are discussed in Chapter 14.

In a mafic island arc assemblage, the igneous intrusions have derived largely by partial melting of the subducted oceanic crust and the mantle, and they passed during their ascent into the upper level represented by the andesite association through the basement ophiolites and/or metabasalts, bimodal volcanics and marine sediments (including rare limestones). In orogenic belts such as the Urals, S.E. Alaska, Philippines, Fiji, etc., the basement metamorphics tend to be intermixed in a complex manner with the topping andesite sequence due to faulting and folding, and are collectively invaded by intrusive rocks ranging from gabbro to quartz diorite and plagiogranite. The lithologically diverse multiphase terrains also tend to be most intensively mineralized, and they contain the greatest variety of mineralization styles although the range of the accumulated metals is restricted: Fe, Cu and Au deposits are the most common; Co, Ni, Ag, Zn and Hg are present in lesser amounts.

By far the most conspicuous mineralization style are skarns. Although most are hosted by carbonates (both pure carbonates and mixed carbonate/volcanic fragments hybrids), skarns have also developed in andesitic rocks, and endoskarns in diorites are also relatively common. The majority of exoskarns are calcic, Fe-rich and

Mg-poor, composed of the prograde assemblage of andradite, salite, epidote and magnetite, and retrograde amphibole, chlorite and sometimes ilvaite (Einaudi et al., 1981). Magnetite is the main mineral mined, but almost all magnetite skarns have an irregular and variable content of pyrite. Pyrrhotite and chalcopyrite are also common. Magnetite skarns with accessory chalcopyrite grade through deposits where chalcopyrite is an important by-product, into skarns where chalcopyrite is the only product. Such variations are rather unpredictable and occur within most ore fields. The bulk of chalcopyrite in skarns is paragenetically late, superimposed on magnetite and gradational into chalcopyrite and/or chlorite, epidote, actinolite, calcite, etc., veins, veinlets, irregular masses, scattered blebs, disseminations, etc., in both the intrusive exo- and endo-contact. Copper skarns often form common orebodies with porphyry coppers.

In the Tur'ya River district in the Urals (Chernyshev, 1960; Fig. 13-12), the following Fe-Cu mineralization styles have been recorded in the contact aureole of largely dioritic intrusions with Devonian andesites, tuffs, volcaniclastics and marble lenses, underlain by a Silurian bimodal metavolcanic suite:

(1) purely magnetite skarns (Auerbakh deposit);
(2) magnetite skarns with disseminated chalcopyrite and pyrite (e.g. Vasilevskoe);
(3) pyrite-chalcopyrite orebodies in pyroxene-garnet skarns (e.g. Frolovskoe);
(4) pyrrhotited-chalcopyrite bodies in pyroxene-epidote skarns (Bogoslov);
(5) massive and disseminated pyrite, lesser chalcopyrite, sphalerite bodies in quartz-sericite altered metavolcanics (Razovskoe); and
(6) quartz-carbonate and quartz-barite veins with chalcopyrite and minor sphalerite, galena.

Style (5) is probably a relict of an older (Silurian) volcanogenic massive sulphide, unrelated to the diorite/andesite interaction cycle.

The La Africana Cu mine near Pudahuel, Chile (min. 25 Tt Cu/2.4%; N. R. Sarić, oral commun., 1977; Fig. 13-13), is an example of a copper-bearing fissure vein filled by quartz, specularite, lesser chalcopyrite, magnetite, pyrite and bornite, and hosted by a late Cretaceous quartz diorite intrusive into lower Cretaceous andesites. This vein is up to 1.7 km long, 30 m wide, fringed by a narrow alteration rim of epidotization, potash feldspathization and chloritization, and upgraded by secondary enrichment. The Talcuna deposit in Chile is similar. Thousands of veins of comparable style have been reported from andesite/diorite contacts worldwide. Table 13-5 lists selected examples of ore districts or fields in which magnetite and chalcopyrite skarns and other hydrothermal deposits are associated with intermediate intrusions emplaced into mafic andesite associations.

Amongst those metals other than Fe and Cu, hydrothermal gold occurrences are the most widespread and characteristic. In the Lucknow Goldfield near Orange, New South Wales (Harper, 1920;

Table 13-5. Magnetite - chalcopyrite skarns affiliated with intermediate intrusions emplaced into mafic andesite association; selected example localities

LOCALITY	SUPRACRUSTAL HOSTS	INTRUSIONS	MINERALIZATION	REFERENCES
Kasaan Penins.near Ketchikan, S.E. Alaska	S-D m-andes.lavas, volcanicl.,graywacke,limest.lenses	Cr gabbro,diorite qtz diorite,pyroxenite,dior.porph.	numerous small skarn bodies; magnetite,chalcopyrite, bornite ; 6 Tt Cu	Wolff and Heiner (1971)
Tur'ya R.distr., near Karpinsk, the Urals, USSR	D aug.andes.,pyrocl.,volc.sed.,limest.lenses,overl.S_2 spil.-keratoph.	D-Cb gabbro,dior., gabbrodiorite	numerous magn. and magn.-cp.skarns; po,py,cp skarns, veins; min. 600 Tt Cu	Chernyshev (1960)
Magnitogorsk dist. S.Urals, USSR	S_1-D_3 m-basalt, andes.,phyll,sch, limst. over ophiol.	D-Cb gabbro to qtz.diorite, granodiorite	magnetite,minor Cu skarns at diorite - marble contacts min. 200 Mt Fe	Magak'yan (1968)
Turgai Fe province N.W. of Kustanai, W.Kazakhstan,USSR (Kachar,Sarbai, Sokolovsk dep.)	Cb_1 andes.,tuffs, interb.bitum.limest.	Cb gabbro,dior, granodior.;dior. porphyry	several large magnetite bodies,some peneconc.,layer-like, in skarns,pyrox.-albite and anhydr.,scapolite metas. 1.9 Bt Fe in 3 largest dep.	Sokolov and Grigor'ev (1974)
Dashkesan,Azerbaidzhan, USSR	J_{2-3} andes.,basalt, pyrocl.,argill., marl,limestone	Cr_1 gabbro,gb-syenite,qtz.diorite	magnetite skarns,peneconcordant bodies; high sulphide content; superimp.vein Co	as above
Kuznetsk Alatau Range,Siberia, USSR	Cm basalt.andes., minor limest.resting on basalt-gabbro assoc.	Cm_{2-3} gabbro, syenite,diorite	small magnetite skarns; cp,tetradymite,gold skarns (Natal'yevskii), and Au veins superimp.on skarns	Kuznetsov (1967)
Larap, Philippines	Cr-Mi metavolc., sedim.,andes.,marl volc. arenites	dior.,syen., monzon. stocks and dikes	small magn.,lesser py,po,cp bornite skarns	Bryner (1969)

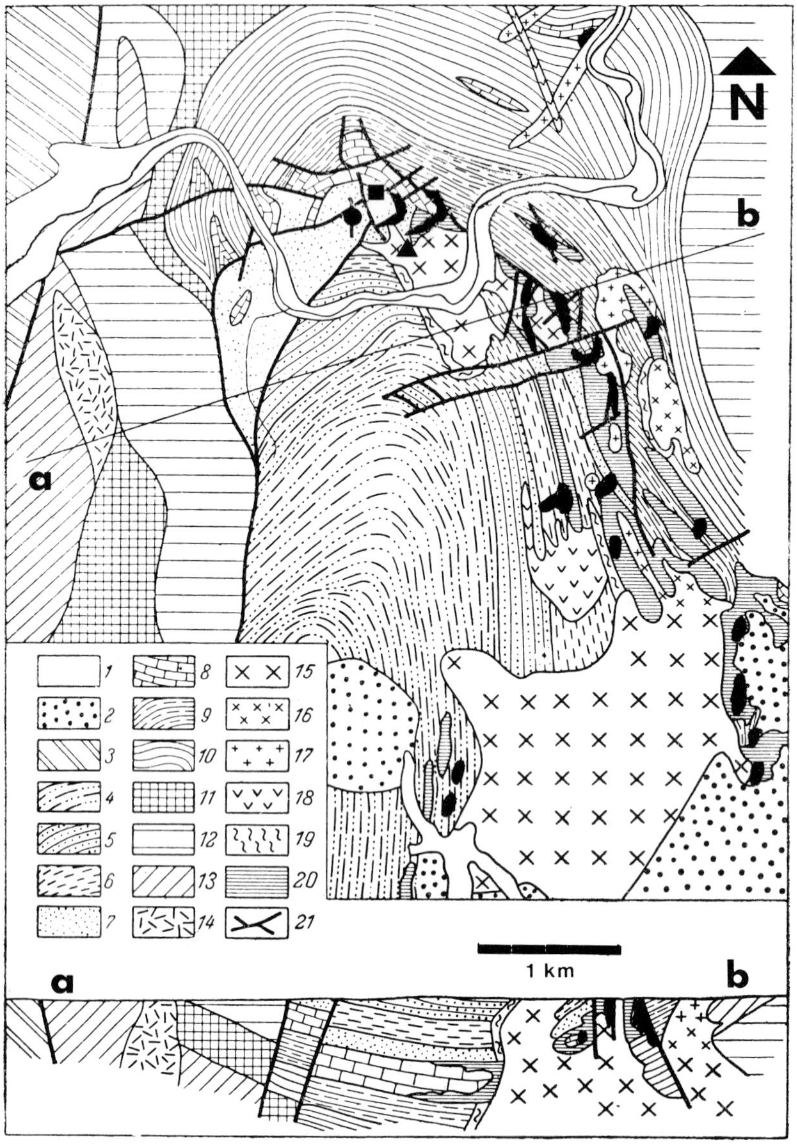

Fig. 13-12. Geology and mineralization styles in the Tur'ya Fe,Cu (Au,Ag,Zn) polygenetic ore field, central Urals, U.S.S.R. LEGEND, GEOLOGY: (1-2) Cainozoic and Mesozoic cover sediments; Devonian rocks: (3) siliceous slates, tuffs, volcaniclastics; (4) augite and uralite andesite and tuffs; (5) tuffaceous sediments and hornblende andesites; (6) hornblende andesites, tuffs, minor limestone; (7) bedded andesite tuff with limestone horizons; (8) massive "reef" limestones; (9) hornblende andesites and tuffs; (10) andesites and tuffs, minor limestone lenses; Silurian rocks: (11) slate, sandstone,

15.4 t Au), series of quartz-carbonate veins in a Devonian meta-andesite occur along a sheared contact with tectonite serpentinite. The veins carry auriferous arsenopyrite, pyrite, pyrrhotite, some stibnite and gold, and are believed to be related to a "syenite" body. Here, the association with a shear and the ultramafics is probably of greater genetic importance than the andesite and many features at Lucknow resemble those in the Mother Lode zone of California.

In the Mariinskaya Taiga district in the Kuznetsk Alatau, Siberia (Kuznetsov, 1967), gold-quartz vein deposits are found in fractured quartz diorites emplaced into a Cambrian basalt and andesite association (Tsentral'noe field), or superimposed on fractured barren and magnetite skarns (Kommunard, Natal'yevski fields). The Kommunard deposit (Borodayevskaya and Rozhkov, 1974) consists of east-west striking, steeply dipping stockworks of quartz, pyrite and calcite veinlets, hosted by loosely outlined irregular "diorite" bodies located within a Cambrian marine andesite association. The mineralization is adjacent to a major strike-slip fault and is in the thermal aureole of a large intrusion. The host diorites could be metasomatized andesitic pyroclastics.

Economic cobalt occurrences in andesitic association include the Dashkesan field in Azerbaidzhan (Causasus, U.S.S.R.; Magak'yan, 1968). This is a polygenetic field with a variation of mineralization styles, hosted by Jurassic andesites, basalts, their volcaniclastics and minor limestones, intruded by gabbro, norite, diorite and in places granodiorite massifs. Several magnetite skarn bodies formed in the carbonates. All the rocks and earlier orebodies in the region are intersected by N.E. and N.W. faults, filled by diabase porphyry dikes. All the cobalt deposits are related to the diabase dikes so, again, the andesite association is only accidental and the cobalt ores show points of similarity to examples described in Chapter 32. Lenticular orebodies 0.5-5 m thick controlled by a 2 km long fault zone, consist of cobaltite, glaucodote, arsenopyrite and base metal sulphides disseminated in chloritized and epidotized diabase. In the Southern Sector of Dashkesan, the same minerals located in veinlets and pockets are hosted by skarns in the hangingwall of a magnetite orebody.

Fig. 13-12 (legend continued)
conglomerate; (12) spilites, porphyritic basalts; (13) keratophyres and tuffs; (14) intrusive keratophyres; Devonian-Carboniferous intrusions: (15) granodiorite, quartz diorite; (16) diorite; (17) gabbro-diorite and gabbro; (18) diabases, lamprophyre. Altered rocks: (19) contact hornfels; (20) skarns and epidosites. (21) faults.

MINERALIZATION STYLES: BLACK = orebodies in scale; pyrite, chalcopyrite or pyrrhotite, chalcopyrite replacements in skarns; triangles = magnetite skarns; squares = volcanogenic massive Fe,Cu,Zn sulphides; rounded symbol = quartz, carbonate, barite, Cu,Zn,Pb sulphide veins. Modified after Chernyshev (1960).

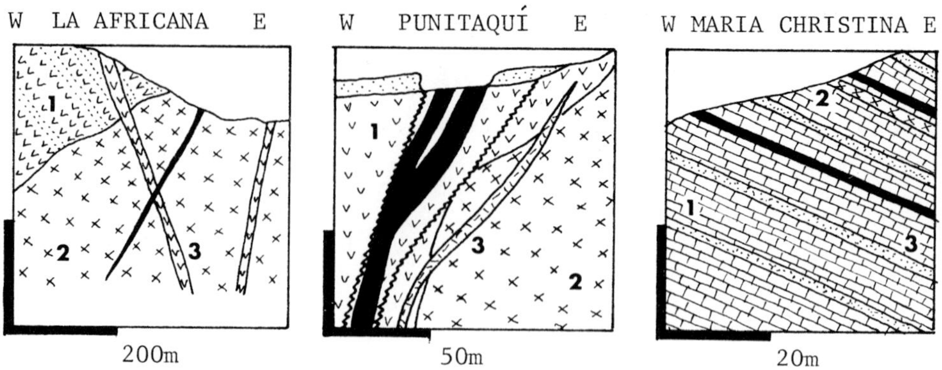

Fig. 13-13. Examples of andesite supracrustals/intrusion interaction, orebodies in Chile. La Africana mine, Pudahuel; (1) lower Cretaceous andesite flows, pyroclastics, volcaniclastics; (2) middle-upper Cretaceous diorite, granodiorite; (3) andesite dikes.

Los Mantos Au-Hg mine, Punitaquí. (1) Jurassic massive to schistose porphyritic andesite; (2) "Laramide" diorite complex; (3) aplite dike.

Maria Christina mine, 60 km s.s.w. of Copiapó. (1) Lower Cretaceous limestone, interbedded with (3) limestone-cemented andesitic volcarenite; (2) "Laramide" diorite sill and dioritized andesite pyroclastics.

From LITHOTHEQUE, after N.R. Sarić, pers. comm. (1977); McAllister et al. (1950); J. Carrozi C., pers. commun.(1977).

Mercury occurs in several enigmatic deposits, partly associated with the mafic andesites. In the Punitaquí field south of Ovalle, Chile (Mc. Allister et al., 1950; Fig. 13-13), a Jurassic massive-to-schistose feldspar-phyric andesite is faulted and intruded by an upper Cretaceous-lower Tertiary diorite-granodiorite complex. At the Los Mantos mine, thick N.N.E.-striking replacement veins follow a fault in both the major rock types, and contain quartz, calcite, specularite, magnetite, pyrite, chalcopyrite, Hg-tetrahedrite and cinnabar. The total production amounted to at least 240 t Hg plus some gold. In the small Azogues and Culebra showing, thin quartz, specularite, tetrahedrite and cinnabar veins carry the mineralization.

In the important Terligkhai mercury field in Tuva, Siberia (Lapin, 1977), cinnabar veins are hosted by lower Devonian basalts and basaltic andesites transected by diabase dikes.

Zinc and lead accumulations are not typical in the andesite, sediments and diorite association, but several deposits do occur in this setting. In the Maria Christina mine 60 km S.S.W. of Copiapó, Chile (minimum 200 Tt Zn+Pb; J. Carrozi, oral commun., 1977; Fig. 13-13), the lower Cretaceous sequence of gray, banded limestone interbedded with limy andesitic volcarenite and intruded by several diorite sills, contains several concordant Pb-Zn mantos. A typical manto is 1-2 m thick and galena, sphalerite, pyrite and lesser barite form disseminated to massive replacements in the limestone. The limestone is a facies equivalent of augite andesites and some or all of the diorites probably formed by contamination of more felsic magmatic melts by the enclosing supracrustals.

Plate 6. Rice paddies and mines scenery in the Hokuroku ore district, N.E. Honshu, Japan. The Kuroko deposits (as in the Shakanai Mine shown) are hosted by altered Tertiary sequentially differentiated volcanics.

"In the course of my nearly twenty years studies of ore deposits... I have yet met with no (sulphide) deposits...whose ores are contemporaneous with the country rock and which form a regular interstratified bed between other rock strata". F. Pošepný in Archiv f. prakt. Geologie, 1880

"Almost everyone accepted the evidence that they (volcanogenic massive sulfide deposits) formed on the sea floor, that the massive bodies are largely syngenetic in origin..." H. Ohmoto and B. J. Skinner in Preface to "The Kuroko and related volcanogenic massive sulfide deposits", 1983.

"Current publication regarding ores now emanates largely from a generation caught in pressures of the breathless modern tempo, which tempts scant reflection upon earlier truth". L. C. Graton in a paper presented to the 20th Intern. Geol. Congress, 1956.

CHAPTER 14

Basalt, Andesite, Rhyolite Sequentially-Differentiated Marine to Continental Volcanic-Sedimentary Association (BAR)

14.1. GENERAL

Volcanics-dominated lithologic associations which have developed along continental margins on an old continental crust, or on an evolutionarily advanced earlier island arc, combine the petrologic and metallogenetic aspects of all the dominantly unimodal (Chapters 10, 11, 13) and bimodal (Chapter 12) associations treated earlier with aspects of continental subaerial volcanic and magmatic rocks (Chapter 26). This is an association with the greatest possible range of facies variations and the greatest variety of accumulated metals and mineralization styles. These variations are the consequence of a tremendous lithologic diversification, the roots of which lie in the three possible principal modes of the parent magma derivation: (1) basalts are probably generated in the subcrustal domain (in the mantle) and ascend into the upper crust along tensional structures; (2) "andesitic" magmas are attributed, in the plate tectonic model, to melting along subduction zones, although the mechanism of the magmatic ascent into the 100-300 km vertically distant upper crust has not yet been satisfactorily explained; and (3) "rhyolitic" (granitic) magmas are now usually explained as products of the remelting of the lower and upper continental crust. It is not the purpose of this chapter to engage in global petrogenetic arguments. Empirical observations indicate that if the three magmatic derivations listed above are correct, they must, in many cases, have operated in accord and/or interacted mutually because basalts, andesites and rhyolites often coexist within a single volcanic centre; a phenomenon originally explained by fractional differentiation of basaltic parent magma.

The recent depositional environments producing sequentially differentiated magmatic rocks in abundance, together with associated sediments, are the evolutionarily mature island arcs, such as Japan or Sumatra, and to a lesser degree the "semi-mature" arcs such as the Philippines, the rest of Indonesia and Melanesia (Chapter 5). The latter, although originally established on the oceanic crust, have reached a considerable crustal thickness and an overall andesitic composition, so that repeated remelting now yields calc-alkaline granodioritic and quartz monzonitic melts locally.

The ancient rock associations now preserved in orogenic belts have a variety of "basements". Most "basements" correspond to the andesite-dominated association and their intrusive equivalents (Chapter 13; e.g. the Hazelton Group, British Columbia; post-Robles sequence, Puerto Rico), but older ophiolite, metabasalt, metarhyolite, bimodal, etc., basements are also common (e.g. in places in Japan and in the Urals). "Cratonic" basements including old crystalline massifs and platformic or marginal cover sediments, are less common in the marine BAR associations, but widespread in the subaerial ones. Examples include the Mesozoic and younger volcanics or sediments developed over the Hida block in north-central Honshu.

The basement variation overprints into the topping sequence in the form of both tectonically emplaced or erosion-exposed older assemblage, but it is also responsible for the magmatic variations established in the uppermost crustal levels. Most importantly, the basement composition appears to be a fundamental metallogenetic factor.

As a class, BAR associations consist of approximately equal proportions of volcanics and sediments most of them marine, but including continental facies. The volcanism is predominantly explosive and the overall average composition is close to granodiorite. The overall deformation and metamorphism is, as a rule, substantially less severe than in the earlier treated associations of continental margins, due to the more rigid basements. The proportion of the BAR association in the global geological record decreases with age (most common in the Mesozoic-Cainozoic age, almost nonexistent in the Archean). This seems to be the result partly of evolution and partly of metamorphogenic convergence. Table 14-1 lists selected examples of mineralized BAR associations.

HAZELTON GROUP, BRITISH COLUMBIA

The lower-middle Jurassic Hazelton Group of British Columbia, described in detail in a recent memoir by Tipper and Richards (1976; Fig. 14-1), has been selected as a fairly typical local example. The Group has now reached the "prime time" of its preservation: the erosional dissection has been sufficient to expose locally the "mesozonal" intrusions but has not been pervasive, so fairly high levels of several former volcanic centres are still preserved. Due to the recent glaciation, the outcrops are virtually unweathered and well

Table 14-1. Phanerozoic basalt, andesite, rhyolite (sequentially differentiated), marine to continental volcanic - sedimentary association; selected mineralized example areas

LOCALITY	GEOLOGY	MINERALIZATION	REFERENCES
Hazelton Group (J_{1-2}) and comparable units, British Columbia	marine shale, siltstone, graywacke; andesitic to rhyolitic tuff and volcaniclastics; local basaltic centres; subaer. explosive and., dac., rhyol. piles. Andesitic basement. Intrud. by granod. to quartz monzonite	rare Fe, Cu, Zn mass. sulph. deposits in sheared m-volcanics (Britannia); abundant interaction miner. in intrusive aureoles: porph. Cu, stockw. Mo, Cu, Au, Ag, Pb, Zn veins	Tipper and Richards (1976)
Bonanza Volcanics, J_{1-2}, Vancouver Island, British Columbia	basalt. andes., andes., lesser basalt, rhyodac., rhyol., explosive volc. with interb. graywacke, mudstone, limest. Diorite to granodiorite plutons, basalt to intermed. basement	ores are almost exclusively of the interaction category, assoc. with subvolc. centres (volc. porph. Cu) or with intrus. aureoles (porph. Cu, skarns, veins Cu, Au)	Northcote and Muller (1972)
Roberts Arm, Buchans, and other Or_3-S units of Newfoundland	mafic flows, andes. and dacite tuffs, felsic volcaniclastics, graywacke, shale; formed on mafic arc-style basement, dior. to granodior. intrus.	massive Pb, Zn, barite sulph. (Buchans) assoc. with volc.-intrus. centres; minor Cu, Au, etc. veins in intrus. aureoles	Strong (1977)
Green Tuff Province Mi-Q, Japan	marine to contin. andes. pyroclastics, volcanicl., minor mudstone, fill many small block-faulted basins on a variety of PZ and MZ crystalline and volc.-sedim. basements. Pervasive propylitization; many volc.-intrus. centres	"Kurokō-type" mass. Zn, Pb, Cu zoned sulphides, numerous "epithermal" Cu, Zn-Pb, Mn, Au-Ag, veins near volc.-intrusive centres	Ishihara (1974) Lambert and Sato (1974) Ohmoto and Skinner, eds. (1983)

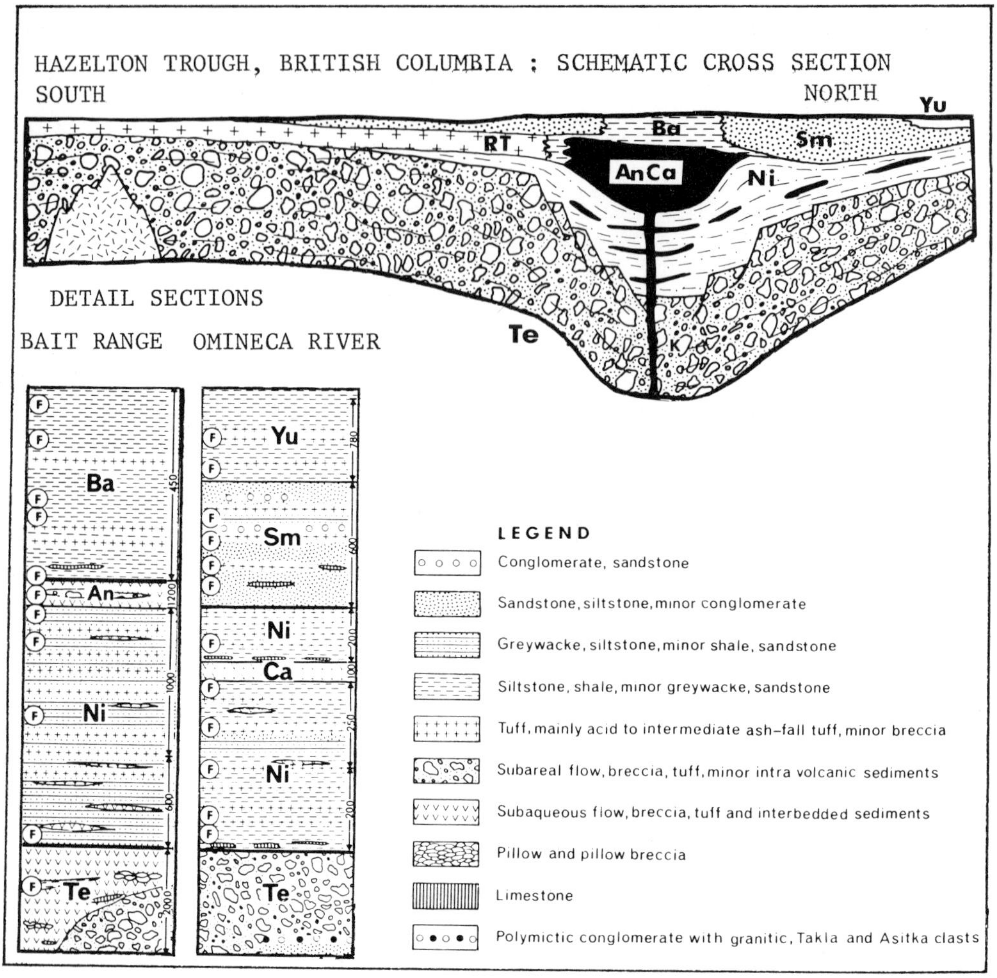

Fig. 14-1. Sinemurian to Bajocian (Jurassic) portion of the Hazelton Group, British Columbia, as an example of basalt, andesite, rhyolite volcanic-sedimentary association. ABBREVIATIONS: Sm = Smithers Fm., Ba = Bait Member; Yu = Yuen Member; Ni = Nilkitkwa Fm.; RT = Red Tuff Member; An = Ankwell Member; Ca = Carruthers Member; Te = Telkwa Fm. From Tipper and Richards (1976), courtesy of the Geological Survey of Canada and the authors.

exposed in the alpine country. The bulk of the Group as currently preserved was deposited in an initially continental but later marine trough, underlain by an earlier (Triassic) volcanic arc; Permo-Carboniferous "oceanic" terrain; and fragments of a consolidated Paleozoic orogenic belt.

The subaqueous facies (e.g. the Nilkitkwa Formation, about 1,000 m thick) includes well-bedded, laminated-to-massive shale, argillite, siltstone, fine-grained graywacke, andesite to rhyolite tuff, fine to medium-grained volcanic breccias, minor limestone and green aquagene tuff. The thinly laminated members weather rusty, which is a consequence of abundant disseminated diagenetic pyrite grains and concretions. Sedimentary structures include graded beds, rip-up clasts, flame structures, syndepositional faulting and flow rolls. Some facies contain immature poorly-sorted volcanogenic sediments, derived as in subaqueous mudflows. Areally restricted rocks include black argillite and shale (up to 150 m thick in the Smithers area) and reefoid limestone. The dominant sedimentary assemblage of the central trough is often interrupted by local, subaerial or subaqueous volcanic piles, a substantial proportion of which are basaltic.

In the Kotsine facies the following are present: basalt and andesite amygdaloidal flows, coarse breccias to finely laminated tuffs, pépérite breccias and tuffs, pillow and broken pillow breccias in various shades of green and gray weathering to brown. Massive flows and beds grade to finely laminated layers. Many breccias are polymictic, composed of fragments ranging in composition from basalt to rhyolite. The clasts are usually set in a fine-grained chloritic matrix. Amygdules and spaces between the rock fragments are often filled with calcite, quartz, albite, prehnite, epidote, and minor pumpellyite. Lenticular limestone bodies up to 20 m thick are common, and often contain volcanic clasts. Interbedded rusty weathering tuffaceous shale, graywacke, siltstone and grit with slump features and flame structures indicate deposition on a paleoslope.

The subordinate subaerial facies (e. g. Talkwa Formation; Tipper and Richards, 1976), consists predominantly of bright red, maroon, purple, gray, green, well-bedded andesite and dacite. Pyroclastics predominate: they include dense, fine-grained tuff, crystal-, lithic-, vitric-tuff, lapilli tuff, accretionary lapilli tuff, lahar and fine-to-coarse-grained breccia. Subordinate acidic volcanics correspond to rhyolite and rhyodacite, and include flow-banded and spherulitic flows, welded tuffs, eutaxitic welded tuff, dense vitric, vitric-crystal, and vitric-crystal-lithic tuff and also unwelded tuff and breccia with an abundance of altered pumice fragments. The thickest accumulations of acidic volcanics coincide with plutonic centres. Elsewhere, rhyolites form blanket-like beds of probable ignimbritic origin. Domes and small plugs of rhyolite and quartz-feldspar porphyry intrusive into the volcanic pile occur frequently.

Lava flows are subordinate to pyroclastics, and include olivine basalt, tholeiite basalt and feldspar-phyric andesite. They are usually found in areas distal from centres of acidic volcanism, probably deposited in topographic lows. Subaerial intravolcanic sediments comprise conglomerate, sandstone, siltstone, mudstone and marl, probably representing stream and lacustrine deposits derived from the associated volcanics. The subaerial volcanics have been extensively altered in the sub-greenschist facies. Zeolites in pyroclastics and sediments have developed regionally and epidote, prehnite and metasomatic K-feldspars are almost always present. The alteration has been interpreted by Tipper and Richards (1976) as being due to a paleo-hot springs system, occurring contemporaneously with volcanism which in turn was probably heated by evolving intrusive masses at some depth.

Intrusions contemporary with volcanism.

In central British Columbia, the Topley Intrusions are believed to be contemporary with the felsic volcanic piles of the Telkwa Formation, described above, into which they are emplaced. They are typical epizonal intrusions, with well-preserved roofs of volcanic ejecta, common porphyry bodies and sharp contacts. The larger intrusive bodies are composite, unfolded, composed of megacrystic potash feldspar, hornblende, quartz monzonite; leucocratic quartz monzonite; granite; tonalite and granodiorite. They are gradational into rhyodacite stocks and dikes. The contact effects are mainly hornfelsing and minor potash metasomatism.

OTHER AREAS

The massive character and sub-greenschist (hydro)thermal metamorphism of the volcanics-dominated (proximal) facies of BAR associations comparable with the Hazelton are characteristic of many other Meso-Cainozoic belts, for example the Cretaceous post-Robles assemblage in Puerto Rico (Glover, 1971), or the "Green Tuff Region" of Japan (Matsuda and Kitamura, 1974).

Paleozoic and older belts tend to be more schistose and usually greenschist-metamorphosed. In the Ural Mts., U.S.S.R., the Silurian-to-Carboniferous BAR association is rather indistinct and merges with the bimodal and andesitic suites (Ivanov et al., 1972). It often occurs as rounded or more commonly elongated volcanic centra marked by several vents, which in turn are flanked by the more distal lithofacies and ultimately merge with the "normal" marine clastics and volcaniclastics. The structurally positive (elevated) character of the BAR centres is often retained and they appear as "brachyanticlines" elongated with the local structural grain (e.g. the Sibai and Gai "brachyanticlines" in southern Ural; Samonov and Pozhariski, 1974). This is very reminescent of the style prevalent in some Archean greenstone belts (Volume 2).

14.2. ECONOMIC IMPORTANCE AND METALLOGENY

The BAR terrains (Table 14-2; Fig. 14-2) contain about 16 Mt Zn, 8.5 Mt Cu, 9 Mt Pb, 30 Tt Ag; 320 t Au in deposits that may currently be mined economically. This is exclusive of the mineralizations related to superimposed intrusions. If the latter deposits are included, the approximate economic content of copper in BAR terrains would increase to several hundreds of millions of tons of Cu, several million tons of Mo, and significantly increased quantities of the metals listed above. Compared with the volcanic-sedimentary associations reviewed so far which, evolutionarily, range from slightly to moderately advanced, the leap in crustal evolution of this association is strikingly apparent in the suite of metals represented in ore deposits. The basaltophile metals (Fe, Cr, Ni, Co) are reduced to non-existent. Copper lost much of its significance in the "pre-orogenic" phase of accumulation (e.g. in the volcanogenic massive sulphide deposits), but has achieved a peak of accumulation in the "orogenic" porphyry copper deposits associated with intrusive massifs. This, in a very general way, is probably a consequence of a massive recycling ("expulsion") of copper from the earlier, evolutionarily more primitive supra- and intracrustals (such as basalts and andesites). Due to burial and subduction, these had reached the crustal levels of magma generation.

Gold remains practically unchanged, but silver accumulation has reached a peak, again probably in a very general way, by endogenic recycling of "black shales" and volcaniclastics in the shallow-to-intermediate crustal levels. Zinc and lead reached one of their accumulation maxima in both the "pre-orogenic" (e.g. volcanogenic massive sulphides) and "orogenic" (hydrothermal veins and replacements) phases. Mo (Re) and W have accumulated significantly in granite-associated "orogenic" deposits which could indicate a selective supracrustal enrichment of these metals in certain facies of the "basement" assemblages: a pre-enrichment still imperfectly known and controversial. The typical lithophile metals (Sn, Be, Ta-Nb, U) have just started to appear.

The BAR terrains can be designated as belts of the first significant appearance and widespread development of lead accumulation, but they are far from being "lead belts"; copper not only interferes, but culminates. Another evolutionary step was needed to eliminate copper and further enhance the lead role; a step that has taken place during the subsequent events of recycling and stabilization of the continental crust. In detail, the occurrence and distribution of metallic deposits in the BAR terrains are highly heterogeneous and influenced by a number of very local factors, so that generalization is difficult.

14.3 MASSIVE (Fe),Zn,Pb (Cu) SULPHIDES ASPECT

Volcanogenic massive sulphide deposits in BAR terrains are less numerous than those in the bimodal association (Chapter 12), but this may be due partly to metamorphic convergence. Apart for being dominated by lead and zinc (against copper and zinc) and being hosted, as a group, by less-deformed or metamorphosed and generally younger rocks, BAR-associated massive sulphide deposits are not qualitatively different from those in the bimodal (and partly felsic and mafic) suites. The general model of a "typical" massive sulphide deposit: the "stratiform" banded lens floored by discordant alteration pipe with stockwork or disseminated mineralization, is identical. The compositional zoning is comparable too: Pb-Zn-rich top grading into

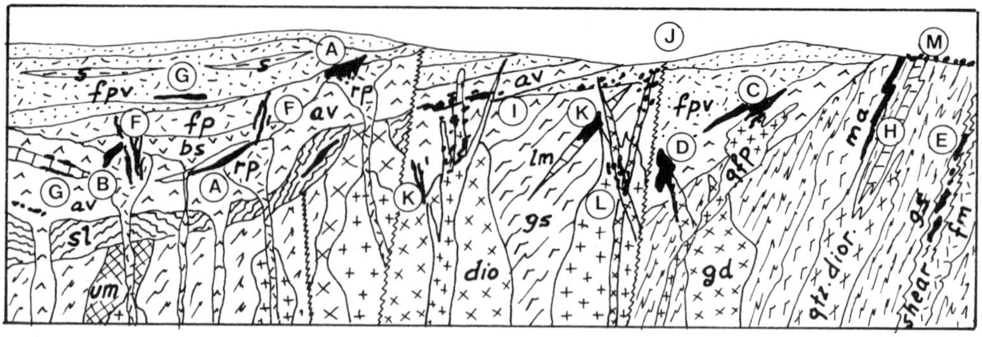

Fig. 14-2. Principal mineralization styles in the basalt, andesite, rhyolite sequentially differentiated volcanic-sedimentary association (see Table 14-2 for explanation of the mineralization letter codes). ABBREVIATIONS: (dio) diorite; (gs) greenschist, greenstone; (fm) felsic metavolcanics; (ma) meta-andesite; (qfp) quartz feldspar porphyry; (gd) granodiorite; (um) ultramafics; (bs) basalt; (rp) rhyolite porphyry; (av) andesite volcanics and pyroclastics; (fpv) felsic volcanics, pyroclastics; (s) sediments; (sl) slate, marble, etc. (lm) limestone.

Cu-rich base. The massive sulphide model has been mainly developed on, and is most relevant to, the Japanese "Kuroko-type" deposits, that are members of the BAR association. For this reason, these deposits and their area of distribution will be treated first.

14.3.1. Cainozoic deposits

GREEN TUFF REGION OF JAPAN AND ITS "KUROKO-TYPE" ORES

The Green Tuff Sequence (Fig. 14-3) comprises a pile up to 5,000 m thick of Miocene epiclastic and volcanic marine sandstone and siltstone with associated submarine to subaerial volcanics. Several cycles of basalt, basaltic andesite, andesite, dacite, and rhyolite (dominant pyroclastics, subordinate lavas, and minor volcanic sediments such as black pyritic and siliceous shales) crop out intermittently in a belt traceable from Sakhalin (U.S.S.R.) to northern Taiwan. The type area of the sequence and also the bulk of the massive sulphide deposits are located in north-east Honshu near Odate and Kosaka (Hokuroku District). There, the Green Tuff rests on Permian meta-graywacke, phyllite to slate (with many black members), and probably subordinate bimodal volcanics of the Chichibu Group locally intruded by Cretaceous granites, in several block-faulted basins (Matsuda and Kitamura, 1974).

Table 14-2. Principal mineralization styles in the basalt - andesite rhyolite sequentially differentiated volcanic - sedimentary association

(A)	young, unmetamorphosed, little deformed, volcanogenic massive Zn, Pb,Cu sulphide deposits (Kuroko-"type"); X0 localities; 5.6 Mt Zn; 3 Mt Cu; 1.8 Mt Pb; 9,700 t Ag; 152 t Au
(B)	auriferous massive pyritic deposits and stockworks, equivalent to (A); X localities; 5 t Au
(C)	Paleozoic and Mesozoic stratabound volcanogenic massive Zn,Pb,Cu sulphide deposits, lightly metamorphosed, deformed; X0 localities; 6 Mt Zn; 3 Mt Cu; 2 Mt Pb; 12 Tt Ag; 160 t Au
(D)	semistratabound to discordant Zn,Pb,Cu sulphide masses closely associated with volcanism and subvolcanism (Bawdwin); 1 locality; 4.5 Mt Pb; 3.6 Mt Zn; 218 Tt Cu; 218 Tt Sb; 130 Tt As; 48 Tt Ni 15 Tt Co; 8 Tt Af
(E)	schistose (sheared) massive and dissem. sulphides Cu,Zn,Pb, etc., believed to be equivalents of (C,D), but significantly modified (remobilized, disrupted); e.g. Britannia Mine; X localities; 900 Tt Cu; 500 Tt Zn; 30 Tt Pb; 180 t Ag; 17 t Au
(F)	discordant hydrothermal metalliferous veins, stockworks, replacements in massive sulphide fields (Pb-Zn,Cu,Au-Ag,Mn,etc.)
(G)	Mn veins, stockworks, replacements, impregnations; X0 localities, X0 Tt Mn
(H)	stratabound hematite, magnetite, siderite in rhythmic felsic pyroclastics interlayered with andesites (Kholzunskoe); X localities, 350 Mt Fe
(I)	transitional massive sulphide / "porphyry" Cu deposits (e.g. Lahanos); X localities; 2.5 Mt Cu; 100 Tt Zn; 4 Tt Ag; 50 t Au
(J)	"volcanic" (consanguinous) porphyry Cu deposits (e.g. Island Copper); X localities; 1.35 Mt Cu; 44 Tt Mo; 146 t Re
(K)	miscellaneous hydrothermal deposits, related to intrusives (skarns, veins); X,00 localities; Fe,Cu,Zn,Pb,Mo,Au,Ag,etc.
(L)	alien porphyry copper deposits hosted by the BAR association; X00 localities; X00 Mt Cu; X Mt Mo; Au, Ag
(M)	magnetite, gold placers; X0 localities

The supracrustal volcanics are intruded by numerous subvolcanic equivalents (feeders and domes) of diabase, andesite, dacite, and rhyolite porphyry, quartz diorite, and are favourably dissected by erosion, so that the base of the metallogenically most fertile middle Miocene volcanic pile near its contact with the basement, is exposed. The name Green Tuff reflects the colour of the regionally pervasive propylitization. This resulted from a combination of load and hydrothermal metamorphism and metasomatism that generated abundant chlorite, prehnite, pumpellyite, zeolites, montmorillonite, etc. mainly in the porous andesitic tuffs.

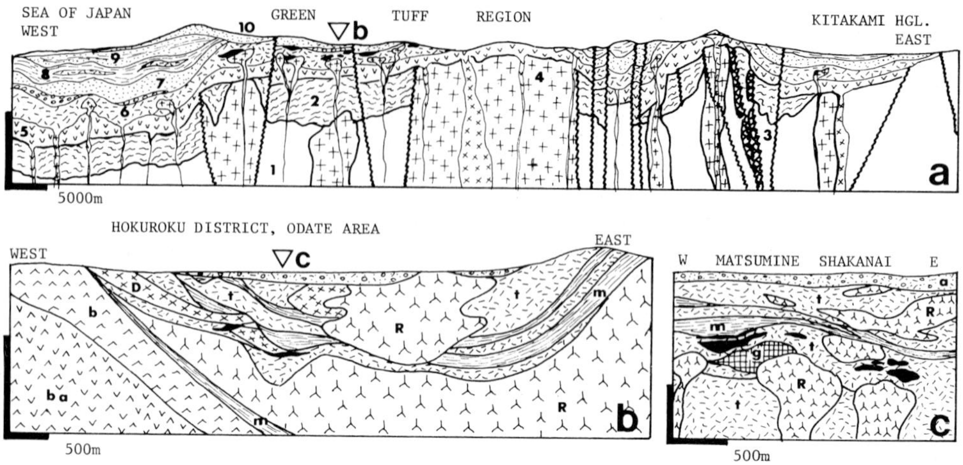

Fig. 14-3. Green Tuff region of northern Honshu, showing setting of the Zn,Pb,Cu "Kuroko" deposits.

a: Diagrammatic east-west section. (1) pre-Silurian crystalline basement; (2) Permian black phyllite, sandstone, bimodal volcanics; (3) ophiolites; (4) Cretaceous granodiorite; Miocene: (5) andesite, basalt flows and tuffs, mudstone; (6) felsic (rhyodacitic) tuff, rhyolite, dacite, andesite flows, tuffs, breccia, rhyolite domes; (7) sandstone, conglomerate; (8) shale, tuff; (9) sandstone, siltstone, shale; (10) Pliocene sediments. BLACK: approximate position of "Kuroko" orebodies (not to scale). b, c: Hokuroku district, Matsumine and Shakanai mines. (R) rhyolite; (ba) propylitized basalt, andesite; (b) basalt; (m) mudstone; (t) felsic tuff; (D) dolerite; (a) alluvium, recent ash; (g) gypsum.

From LITHOTHEQUE, after data in Matsuda and Kitamura (1974), and materials provided by the Dowa and Nippon mining companies.

The Hokuroku district (about 500 km^2, total economic content of 5.1 Mt Zn, 2.04 Mt Cu, 1.53 Mt Pb, 9.7 Tt Ag, 152 t Au) and its major massive sulphide deposits (e.g. Kosaka, Hanaoka), as well as the "Kuroko-type" of massive sulphides in general, are extensively covered in recent literature in both Japanese and English (e.g. Ishihara, ed., 1974; Lambert and Sato, 1974; Urabe and Sato, 1978; Ohmoto and Skinner, eds., 1983). The subject is also prominently treated in numerous general reviews (e.g. Franklin et al., 1981; Mitchell and Garson, 1981; Hutchinson, 1973).

Briefly, the word "kuroko" means "black ore", the galena and sphalerite-rich variety of the "bedded" massive sulphide lens, and not a particular locality. The orebodies are repeatedly associated with white rhyolite domes about 13 m.y. old. These are relatively sporadic rock bodies emplaced as a late-stage product of a typical mafic to felsic volcanic cycle, into the dominant dacite and andesite volcanics with some mudstones and basalt flows. The orebodies are located on flanks of brecciated rhyolite domes or in a tuff breccia. The massive sulphide orebodies tend to be short, thick lenses composed of fine-grained pyrite, sphalerite, galena, chalcopyrite and minor tetrahedrite. They grade downward into a pyrite, chalcopyrite massive ore and pyrite, chalcopyrite, quartz stockwork in an argillized (kaolinite, pyrophyllite, diaspore), sericite-chlorite or sericite, chlorite and carbonate-altered footwall. Some orebodies are associated with abundant gypsum or anhydrite. Others contain almost monomineralic barite zones in the uppermost part of the massive ore. In the immediate hangingwall of a sulphide lens there is frequently a thin, but laterally persistent (up to 1 km) horizon of hematitic or manganiferous chert.

Within a "Mine" (=ore field), orebodies usually occur in clusters. In the Kosaka Mine (Urabe and Sato, 1978), individual major orebodies range from 1.3 to 8 Mt of ore (Table 14-3).

The "Kuroko-type" massive sulphides are not the only mineralization style in the Hokuroku district (and other Japanese and adjacent districts), and cross-cutting vein deposits occur extensively. In Hokuroku the economic significance of the veins is limited, but elsewhere veins are the dominant or indeed the only style of Miocene volcanogenic mineralization (e.g. in S.W. Hokkaido; Toyoha Pb-Zn mine). The vein deposits are probably associated with largely subaerially extruded complexes and are treated in Chapter 26. The presence of the stratabound massive sulphide lenses thus appears to be mainly a function of a relatively deep subaqueous environment. "Kuroko-type" deposits outside the Hokuroku region in Japan, the Kurile Islands and Sakhalin, are listed briefly in Table 14-3.

OTHER CAINOZOIC DEPOSITS

The "Kuroko-type" massive sulphides in north-eastern Vanua Levu Island, Fiji (Undu Peninsula; Colley and Greenbaum, 1980), are similar to those in Japan in their general style, but differ in detail. The only deposit mined, Nukundamu, contains supergene-enriched chalcopyrite, bornite, covellite, enargite, tennantite, sphalerite and pyrite with gypsum and barite, disseminated in argillized and silicified dacitic pumice breccia. Small spheroidal massive pyrite pods containing the same minerals occur locally. The Wainikoro showing is a quartz, pyrite, minor chalcopyrite stockwork only, whereas Mouta is an example of a distal "exhalative" deposit formed at the seafloor. There, the ore is massive, layered and barite rich. It is hosted by argillized dacitic pumice breccia and is topped by ferruginous chalcedony and bedded Mn oxide horizons.

The Cu and Zn sulphide deposits hosted by a Tertiary andesite, dacite and rhyolite (keratophyre) sequence on the Samar Island, the Philippines (Bryner, 1969; Wolff, 1978), have been assigned to the

Table 14-3. Massive Zn,Cu,Pb sulphides ("Kuroko-type") in Phanerozoic basalt, andesite, rhyolite, ensialic volcanic - sedimentary association; selected example localities

LOCALITY	AGE	DEFORMATION METAMORPH.	FOOTWALL	OREBODY	HANGINGW.	TONNAGE	REFERENCES
Tulsequah m.,Taku River,N.W. British Columbia	Tr_3	Fd/Gr	felsic m-volc.qtz-seric.sch.	MS py,sf,gal/cp lens; STW py,cp,qtz	m-tuff	63 Tt Zn/6% 13.6 Tt Cu/1.32% 13.5 Tt Pb	Thompson and Panteleyev (1976)
Britannia m.,N.of Vancouver, British Columbia	J_2-Cr_1	PD,S/Gr contact metam.	dac.m-tf. chl-sc schist	MS,DS,VEIN py/cp,sf,born stringer z. part.remob.	m-andes. tuff, argill.	605 Tt Cu/1.1% 357 Tt Zn/0.65 15.5 Tt Pb 14 t Au 140 t Ag	Payne et al. (1980)
Buchans, Newfoundland,Canada	Or_2-S	Fd,Ft,S/sub Gr some cont.	dac.brc., conglom. chl-sc schist	MS,fine gr. sf,ga,cp/py, tetr,barite STW,string. same miner.	siltst., intermed. tuff	2.68 Mt Zn/14.9% 1.39 Mt Pb/7.7 245 Tt Cu/1.36 1,908 t Ag	Thurlow et al. (1975)
Karabash distr. C.Urals, USSR	D	PD,S/Gr	dacite pyrocl.	MS py/cp,sf STW chl,sc, qtz schist	felsic m-tuff	573 Tt Cu/1.4% 311 Tt Zn	Rakchayev (1962)
Sibay centre, s.Urals, USSR	D_2	Fd,Ft,S/Gr,sub Gr	MS to brc. rhyodac.	MS py/cp,sf STW py,cp silicif.	felsic m-tuff m-basalt	500 Tt Cu 450 Tt Zn Au,Ag	Zhabin et al. (1974)
Kosaka m.,Motoyama deposit,Hokuroku d.,Japan	Mi_2	F/UM	fels.vol. breccia, rhyolite	MS gal,sf; py,cp; gypsum	felsic tuff	360 Tt Zn/4.5% 176 Tt Cu/2.2% 64 Tt Pb/0.8%	Ishihara,ed., (1974) Urabe and Sato (1978)

Kosaka m.,Uchino-tai deposit,Hoku-roku d.,Japan	Mi_2	F/UM	fels.vol. breccia	MS gal,sf; py,cp; STW py,cp,silicif.	felsic tuff, breccia	740 Tt Zn 400 Tt Pb 220 Tt Cu	Urabe and Sato (1978)
Shakanai m.Matsu-mine oreb.,Hoku-roku d., Japan	Mi_2	F/UM	felsic tuff	MS sf,gal; py,cp; underl. by gypsum	mudstone felsic tuff		Ishihara,ed. (1974)
Hanaoka oreb.,Ho-kuroku, Japan	Mi_2	F/UM	brecc.rh. sc,argill alter.	MS gal,sf; py,cp; gypsum STW py,cp, silicif.	felsic tuff	1.8 Mt Zn/3.7% 1.13 Mt Cu/2.2% 420 Tt Pb	as above
Furutobe-Ainai, Hokuroku d., Japan	Mi_2	F/UM	rhyol. expl.brc.	MS sf,gal; py,cp; DS py, cp in silicif. fels.footwall	argillite pumice tf silic.sh.	680 Tt Zn/4% 340 Tt Cu/2% 170 Tt Pb/1%	as above Skripchenko, (1972)
Hanawa group, Hokuroku, Japan	Mi_2	F/UM	fels.tuff rhyolite dome	MS gal,sf; py,cp; STW qtz,py,cp	basalt, bas.tuff	40 Tt Zn 20 Tt Cu 8.3 Tt Pb	Nikko Min.C. (1970)
Kunashir Isl., Kurile Islands USSR	Mi_2	F/UM	silic.-seric. fels.vol.	MS sf,gal; py, cp;gypsum STW py,cp, silicif.	felsic tuff	Zn,Pb,Cu	Smirnov et al. (1968)
Undu Penins., Vanua Levu Isl., Fiji	Mi_3-Pl_1	F/UM	altered dacite breccia	MS,DS cp,sf, py,gypsum and barite	tuff gossan	2.2 Tt Zn 1.95 Tt Cu	Colley and Greenbaum (1980)
Cavallo (El Aoua-na) Massif, Algeria	Mi	F/UM	altered dacite breccia	MS gal,sf,qtz, barite/cp,enrg STW py,cp	silic.shl felsic tuff	9 Tt Cu/1.2%	Glaçon (1971)

"Kuroko-type", but they seem to be completely lacking in lead. The Bagacay mine (134 Tt Cu) consists of a series of gently dipping pyritic lenses distributed along a 2 km long east-west belt, and hosted by argillized and propylitized dacite porphyry. The orebodies are topped by jaspers and tuffs and overlain by a coal-bearing sequence containing reworked copper minerals in places. The Sulat mine east of Bagacay (Wolff, 1978; 240 Tt Cu/1.2%) in middle Miocene dacitic lavas and pyroclastics, contains pyrite-chalcopyrite orebodies capped by basalts and pyroclastics. The massive stratabound orebodies overlain by a red siliceous shale grade downward into breccias with a sulphide matrix and disseminated ore.

In the Cavallo (El Aouana) massif, Algeria (Glaçon, 1971), a middle Miocene andesite, dacite, rhyodacite volcanic pile rests on Tertiary and Mesozoic flysch and on a folded basement. Stratabound massive Pb, Zn, Cu and barite lenses (Qued el Kebir) are found in volcaniclastics, siliceous shales, "chert" and an altered dacite breccia. Independent discordant stockwork pyrite, chalcopyrite, enargite, stibioluzonite and barite mineralization exists in a quartz-sericite altered subvolcanic dacite. Moreover, some disseminated copper showings are found in upper Miocene microdiorites.

14.3.2. Mesozoic deposits

Several massive sulphide deposits in the marine BAR association are known in the Canadian Cordillera. The Tulsequah deposit north-east of Juneau (Thompson and Panteleyev, 1976) lies in Triassic greenschist-metamorphosed felsic volcanics and pyroclastics. Massive, banded pyrite, sphalerite, galena, barite, etc. lens conformable with schistosity, is floored by quartz, pyrite, chalcopyrite stockwork on the footwall side.

The economically insignificant but geologically interesting Seneca massive sulphide occurrence south of Harrison Lake (southern British Columbia; Pearson, 1974) is located in little disturbed and metamorphosed middle Jurassic rocks. A small, irregular mass of fine-grained sphalerite, pyrite, galena and chalcopyrite with some barite rests on a coarse silicified and pyritized rhyolite breccia, considered to be a vent facies. The ore is overlain by intermediate basic flows and pyroclastics containing massive sulphide fragments or a sulphide matrix around volcanic fragments. Laterally, the ore grades into a black chert.

One of the largest western Cordilleran Cu-Zn sulphide deposits, the Britannia mine north of Vancouver (Payne et al., 1980; 55 Mt ore containing 605 Tt Cu, 357 Tt Zn and 14 t Au; Fig. 14-4), has recently been reinterpreted as being a deformed volcanogenic mineralization. There, the ore-bearing zone lies in Mesozoic greenschists and contact hornfelses interpreted as meta-dacites, meta-andesites and their volcaniclastics, and argillites, that form a small roof pendant surrounded by granodiorites of the Coast Range pluton. Within the zone, all the orebodies consisting of massive and stringer pyrite,

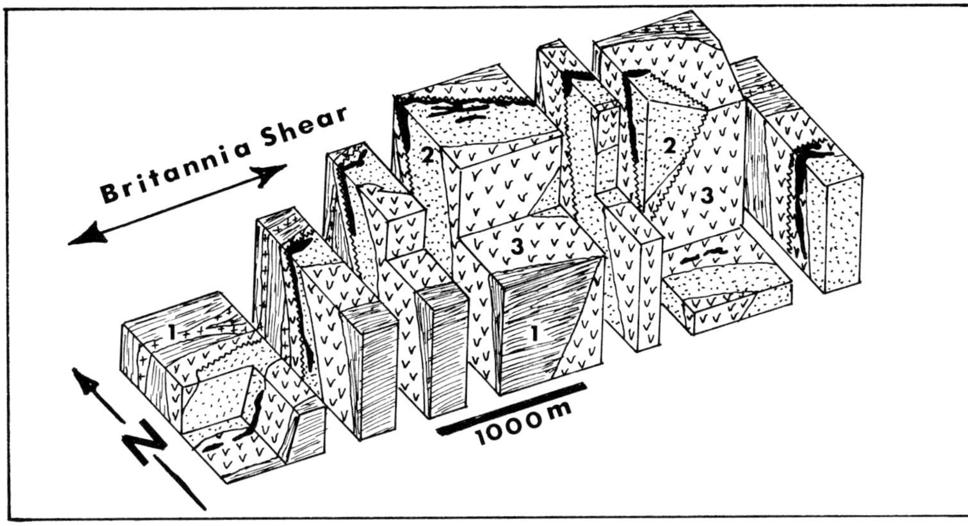

Fig. 14-4. Britannia mine, southern British Columbia. Deformed, shearing-modified, and partly remobilized originally stratabound massive, stringer and vein Fe,Cu,Zn sulphide deposits hosted by a Jurassic-Cretaceous volcanic-sedimentary assemblage along the Britannia Shear, in a roof pendant in the Coastal Batholith. (1) unsheared but thermally hornfelsed argillite, porphyry sills; (2) sheared, silicified, dacitic tuff with subvolcanic dacite intrusions (now quartz, chlorite, sericite schist); (3) sheared andesite tuff (chlorite-sericite greenschist). From LITHOTHEQUE; a block diagram after Irvine (1948) modified by new data of Payne et al. (1980).

chalcopyrite, lesser sphalerite and occasional barite, occur in a north-west trending shear, causing extensive dynamometamorphic modification and partial remobilization of the original ore.

14.3.3. Paleozoic deposits

The Buchans Pb-Zn field in Newfoundland "bears more resemblance to the Kuroko deposits of Japan than any other type of massive sulphide ore" (Thurlow et al., 1975; Fig. 14-5). The mineralization is hosted by a cyclic (4 cycles) assemblage of alternating andesite, dacite pyroclastics and breccias (? lahars), siltstones and granite (epiclastic ?) conglomerates probably of the Ordovician and Silurian age. The metamorphic grade is sub-greenschist. Sheared or schistose zones are present locally and the rocks are thermally metamorphosed near the contact with Devonian granodiorites. Massive, fine-grained sphalerite, galena and chalcopyrite are located here with lesser

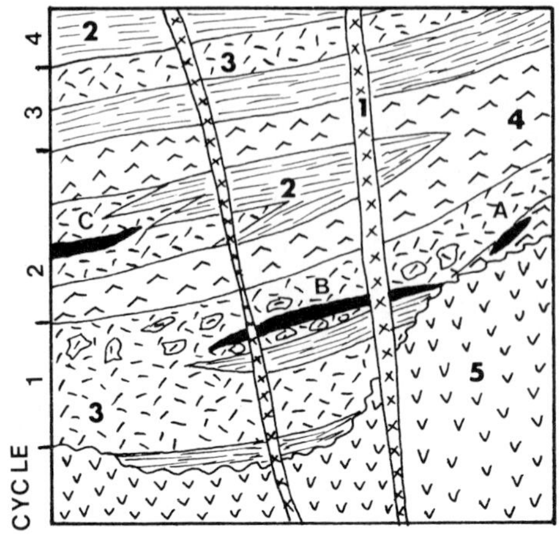

ALL ROCKS: Or_3-S Buchans Group

1 diabase dikes
2 volcanogenic siltstone
3 dacite pyroclastics, volcanogenic, minor terrigenous sediments
4 andesite
5 footwall basalt (andesite)

BLACK: ORE DEPOSITS

A stockwork to massive ore in intermediate footwall andesites (low importance)
B massive ores in I.cycle dacites (important)
C massive ores in II cycle of dacites (important)

DIAGRAMMATIC; NOT TO SCALE

Fig. 14-5. Buchans Zn,Pb,Ag,Cu, barite massive sulphide deposit, Newfoundland, hosted by cyclic sequentially differentiated volcanics. From LITHOTHEQUE, based on Swanson and Brown (1962), Thurlow et al. (1975); a mine tour, 1973.

pyrite and tetrahedrite in barite and sometimes quartz, calcite, sericite, chlorite gangue, and they form conformable lenses, distributed around a former volcanic centre. The orebodies usually top, or are emplaced within, the dacite breccia or granite conglomerate and are overlain by siltstone or tuff. Stockwork and disseminated mineralization that contains the same Pb, Zn, Cu minerals as the massive ore, is found both in the footwall of the massive orebody, and, independently, near the top of the host lithologic units. Moderate to intensive silicification and/or chloritization of the host rocks is associated with this ore and it has been demonstrated that at least a portion of the stockworks predate the emplacement of the massive ore.

Except for Buchans and several fields in the U.S.S.R. which it is impossible for a western geologist to visit, few important massive sulphide deposits hosted by the BAR suite are known, or have been reinterpreted recently. One of the most interesting mineralized regions appears to be in the Patagonian Lakes District along the Chile-Argentina border (e.g. Lago Buenos Aires, Lago Fontana; Angelelli, 1950), where several Pb-Zn massive sulphide and vein deposits are hosted by dacites and porphyritic trachytes.

Several small massive sulphide deposits in the Tasman orogenic belt of Australia are members of the BAR suite. The recently discovered Que River deposit in north-western Tasmania (Webster and Skey, 1979) lies in the Cambrian Mt. Read volcanics that also host the Rosebery and Mt. Lyell fields described in Chapter 11. At Que River, however, the entire differentiation suite is represented from andesite through dacite to rhyolite. There, fine-grained, faintly banded masses of galena and sphalerite grade into brecciated and banded pyrite and are interlayered with sericite schists. These, in turn, are floored by pyrite-chalcopyrite veins and impregnations in a sheared "tuffaceous shale" (chlorite-sericite schist).

The small Hall's Peak Zn-Pb deposit in New England containing a high Sn content, is enveloped by Permian black argillites with albite porphyroblasts, situated above a rhyolite-andesite volcanic pile.

The important but enigmatic Bawdwin Pb,Zn,Ag,Sb field in Burma (4.5 Mt Pb/21.2%, 3.6 Mt Zn/15.9%, 218 Tt Cu, 218 Tt Sb, 130 Tt As, 48 Tt Ni, 15 Tt Co, 8 Tt Ag), has been studied recently by Goossens (1978), and Brinckmann and Hinze (1981; Fig. 14-6), who concluded that this was probably a deformation and remobilization-modified volcanogenic massive sulphide comparable with the "Kuroko", or alternatively a non-stratiform, structurally controlled equivalent of "Kuroko" (that is, a subvolcanic deposit). Despite its lack of other than silicic volcanics and intrusives, the locality is treated here because of its stated resemblance to "Kuroko".

The 4 km long, N.W.-S.E. trending, narrow Bawdwin ore zone contains three massive orebodies of galena, sphalerite, lesser chalcopyrite, pyrite, lollingite, tetrahedrite, bismuthinite, boulangerite, pyrargyrite, gersdorffite, cobaltite, etc. These orebodies are surrounded by disseminated pyrite-chalcopyrite halo in silicified and sericitized rocks. The host rocks are late Cambrian-early Ordovician rhyolite meta-volcaniclastics (agglomerates, volcanic graywackes) containing probably subvolcanic intrusive rhyolite domes. These are enveloped by red and green volcanic sandstone and siltstone. The volcanic-sedimentary suite is underlain by Precambrian high-grade metamorphics (gneiss, quartzite, graphitic schist, marble) intruded by syenite.

In the Ural Mountains, BAR associations occur sporadically in the relics of large Devonian stratovolcanoes, established on the foundation of broadly contemporary or slightly older rocks of the bimodal volcanic-sedimentary association and surrounded by these. Such stratovolcanoes possibly represent evolutionarily the most advanced, partly subaerial products of the bimodal suite, which makes their setting, lacking a thick continental basement, substantially different from the typical "Kuroko" environment. This is almost certainly reflected in the metal ratios of the Uralian ores, which are predominantly copper and zinc-rich with hardly any lead. Most of the identified former stratovolcanoes are situated in a north-south belt parallel with the structural grain and deep faults trend in the Magnitogorsk Megasynclinorium of the southern Urals.

The Sibay "stratovolcano" (12 x 3 km; Ivanov et al., 1962; Fig. 14-7) floored by more than 400 m of spilite and basaltic andesite, tuff and tuff breccia, is overlain by 200 to 800 m of dacitic volcanics and pyroclastics. The sequence is terminated by mafic

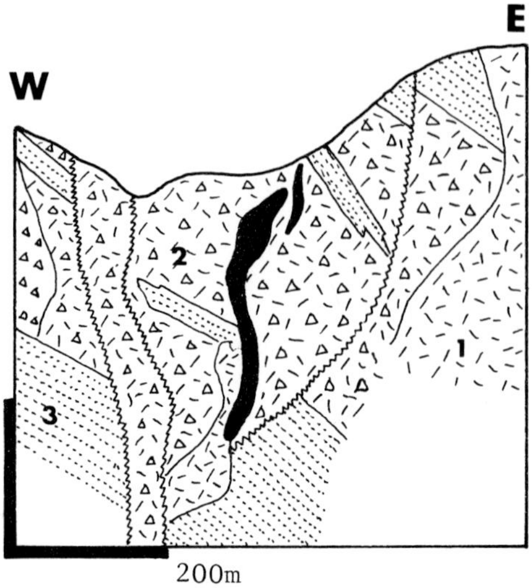

Fig. 14-6. The Bawdwin, Burma, massive Pb,Zn,Ag deposit. After Brinckmann and Hinze (1981).

volcanics. The vent facies is represented by rhyodacite and dacite breccia domes and subvolcanic bodies, grading into rhyodacite lava flows and intermediate ash flows on the flanks. Pervasive silicification and chlorite-sericite and epidote alteration is widespread. The rocks range from massive to schistose and sheared. The Sibay "stratovolcano" contains several sulphide deposits, the most important of which is the Novo-Sibayskoe (Samonov and Pozharisky, 1974). This takes a form of a steep, thick lens to irregular body composed of massive pyrite with minor sphalerite and galena. A recent reinterpretation (Zhabin et al., 1974) has established the composite nature of the body consisting of three vertically stacked and partly coalescing lenses. Each lens thins laterally and changes into a proximal coarse fragmental facies consisting of sulphide clasts (usually massive pyrite fragments with sphalerite cement). This, in turn, changes into thin, concordant pyrite layers gradational into fine tuff. Silicified rhyolite is found in the lowermost footwall, rhyolite tuff is the lateral equivalent, and basaltic agglomerates and tuffs cover the lens set. The tonnage and grade of the Soviet base-metal deposits is a classified information; a prewar reserve figure of 9 Mt ore with 110 Tt Zn/1.16% and 146 Tt Cu/1.63%, was representative of the "Old Sibay" orebody. The "New Sibay" appears to be 3 to 4 times larger, so the field may now represent some 40-60 Mt ore.

Gai, Tanalyk-Baymak, near-Orsk, Dzhusa, and other massive sulphide groups in southern Urals are of the same age and of similar nature

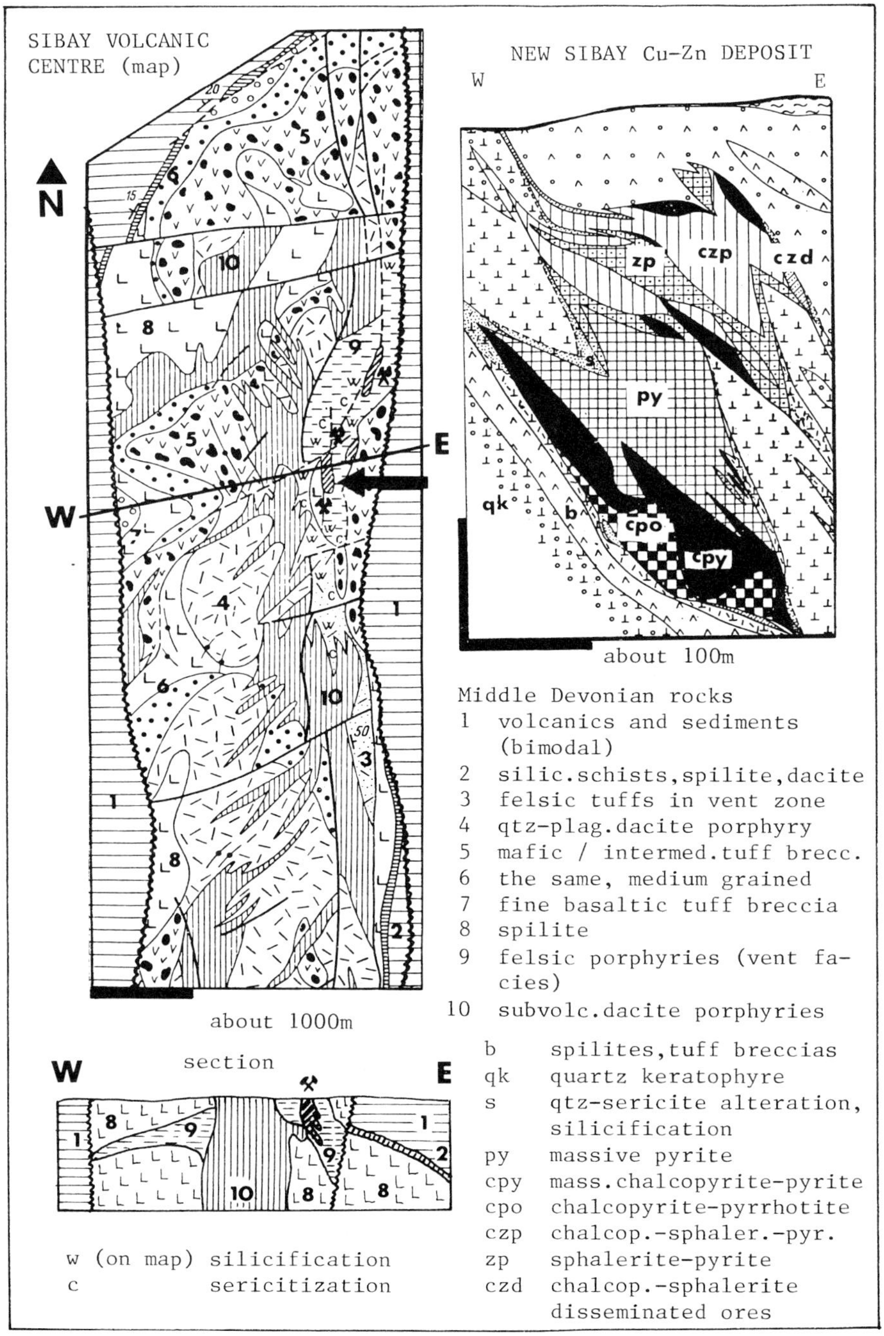

Fig. 14-7. Sibay mineralized volcanic centre, s. Ural Mts., U.S.S.R. After Sharfman (1969), Prokin (1961) and Zhabin et al. (1974).

as are the Karabash and Sultanovo fields in the Chelyabinsk area farther north and east, respectively. Brief data on these deposits have been compiled in Table 14-3.

14.4. GOLD AND SILVER DEPOSITS

Although all "Kuroko-type" and similar massive sulphide deposits often contain significant quantities of gold and silver recovered as a by-product (e.g. 188 to 1530 ppm Ag, mean 312 ppm; 0.7-8.2 ppm Au, mean 3 ppm, in the black ore from Kosaka and Hanaoka; Lambert and Sato, 1974), largely conformable purely Au/Ag deposits are exceptional. Sato (1974) coined the terms "Kuroko-type network deposits" and "Kuroko-type pyrite-gold deposits". The former are cross-cutting hydrothermal stockwork deposits in altered host volcanics resembling the feeder stockwork sulphide lenses, but occurring independently. Both "types" are a nightmare to a classifier, because they are located in the transitional region of subaqueous and subaerial (Chapter 26) BAR associations. The bulk of similar deposits are epithermal or "subvolcanic", formed in surficially-venting volcanics (e.g. the Japanese vein deposits such as Chitose, hosted by subaerial members of the Green Tuff association), but subaqueous counterparts are conceivable and are of some importance in the BAR-like associations in the Precambrian greenstone belts (Volume 2).

The "Kuroko-pyrite gold deposits" (Sato, 1974) have been reported in Hokkaido, Izu Peninsula of central Honshu, and in southermost Kyushu. The orebodies are pipe-like and range from massive pyrite bodies with low gold content (Oage mine) to bodies containing disseminated electrum with only small amounts of pyrite, rare enargite, and luzonite (Akeshi and Kasuga mines, Kyushu). Some deposits carry gold as tellurides (Date and Suzaki mines). The host rocks are usually felsic pyroclastics that are strongly silicified, and affected by advanced argillic (kaolinite, diaspore, sulphur, alunite) alteration.

14.5. Hg ASPECT

Many "Kuroko-type" Pb,Zn,Ag deposits contain high trace mercury contents (e.g. 2.6 ppm in the Buchans ores), and Quaternary hot springs deposits formed recently in the area of distribution of the Green Tuff sequence (e.g. Matsuo, Japan; Mendeleev Volcano, the Kuriles; see Chapter 5). These are particularly high in Hg with contents approaching economic quantities.

14.6. Mn ASPECT

Occurrences of small amounts of manganese oxides particularly in tuffs or in volcanic-associated sediments are as common in the BAR association as in the associations treated earlier, but large deposits have not been reported. In the Eocene dacitic volcanics and Eocene to Miocene limestone of Saipan, Mariana Islands (Cloud et al., 1956), Mn oxides occur as narrow veins, pore-fillings, replacements and stockworks in dacite and tuff, in association with jasperoid. In limestone, powdery Mn oxides fill solution cavities and coat joints. Only about 5,000 t Mn/40% were produced.

In the Green Tuff region of Japan, manganese is enriched in the hangingwall siliceous mudstones topping the "Kuroko-type" deposits, but is not recovered. Numerous epithermal rhodochrosite veins, some with by-product Pb,Zn,Ag minerals (e.g. Jokoku mine, Hokkaido; Saito, ed., 1960), have been an important source of Japanese manganese.

14.7. Fe ASPECT

Although significant quantities of iron are contained in the pyrite of the BAR association-hosted massive sulphide deposits and although some iron occurs in jaspers and "banded iron formations" in facies transitions of sulphide deposits, industrially important iron accumulations are not characteristic for the marine BAR suites. The Kholzunskoe field in the Gornyi Altai, U.S.S.R. (Korzhnev, 1980) is a possible exception. There, a Devonian transgressive, nearshore association of rhythmically interbanded rhyolite, dacite, andesite flows, tuffs and volcaniclastics, is intruded by granosyenite, plagiogranite and rhyolite. A stratabound horizon of ore lenses consisting of hematite, magnetite and minor siderite, disseminated in felsic pyroclastics and rhythmically interlayered with trachyandesite and andesite porphyry lavas, contains 940 Mt of 30% Fe ore. The ore zone has been partly regenerated and upgraded in the intrusive exocontact. This deposit shows many points of similarity with the Proterozoic Kiruna field, Sweden (Volume 2).

14.8. TRANSITIONS AND INTERACTIONS

14.8.1. Lateral facies transitions

The marine BAR volcanic-sedimentary association readily changes facies into the entirely marine sedimentary association (Chapters 17, 19) on one side and into the dominantly volcanic continental facies (Chapter 26) on the other. As in the case of the associations discussed above, detrital magnetite derived from volcanic phenocrysts may accumulate in beach sands, some of which may appear as paleoplacers in the pre-Quaternary sequences. Numerous occurrences of such paleoplacers are known in the ancient sequences, but none is of much economic importance. The proximal and distal marine sediments derived from the BAR suite are likely to inherit its trace-metal signature: that is, the "fertile" metal-accumulating sediments, such as carbonaceous shales, are expected to be mainly Zn-Pb enriched.

14.8.2. Recent weathering and redeposition generated mineralizations

Tropical lateritic weathering superimposed on the BAR suite is not known to contain any valuable metallic occurrences, and few are expected. The low-grade alluvial "iron sands" of Japan (they should not be confused with the magnetite beach sands reworked from the Recent volcanic products; Chapter 5) have a variety of sources, among which are the BAR volcanics.

14.8.3. High-grade metamorphic transitions

Following the introduction of the stratiform massive sulphide models in the 1960's, numerous sulphide deposits conformable with foliation in the high-grade metamorphic terrains have been interpreted as marine-volcanogenic in origin, subsequently metamorphosed (e.g. Bleikvassli, Ballangen, and others in the Norwegian Caledonides; Bayindir and other localities in the Menderes Massif, Turkey). If true, however, the metamorphism and ultra-metamorphism (incl. granitization) produced very significant changes. High-grade metamorphics hosted sulphide deposits are discussed in Chapter 29.

14.8.4. Volcanic/plutonic transitions and interactions, 1: "stratabound" to "porphyry" sulphide deposits transition

A growing number of sulphide deposits (mostly of copper), combine the characteristics of the stratabound deposits confined to a certain host supracrustal unit and magmatogene-hydrothermal deposits clearly coeval with a cross-cutting intrusive body. The stratiform/epigenetic genetic controversy became particularly strong and emotional in the 1960's, but has since weakened. As it becomes increasingly apparent that many "stratiform" massive sulphide systems like the "Kuroko" are composite and consist of a stratabound lens and a discordant feeder, and that many are coeval with the "domal rhyolites" or vent breccias that are essentially intrusive rocks, the difference between many "stratiform-syngenetic" and "crosscutting-epigenetic" mineralizations is becoming more one of degree than of substance (compare also Laznicka, 1985ba). The two most frequently quoted transitional example localities are the Turkish deposits of Murgul and Lahanos.

Murgul (Schneiderhöhn, 1955; 880 Tt Cu, 600 t Ag, 11 t Au) located on the Turkish side of the Caucasus, consists of pyrite and chalcopyrite disseminations in a late Cretaceous felsic volcanic arenite and argillite, above a brecciated dacite body which itself contains a disseminated and stockwork ("porphyry") Cu mineralization. Eocene granodiorite and quartz monzonite are also present, and it is not clear from the literature whether the ore-bearing dacite is coeval with the volcanic sediments or with the granitoids.

Lahanos on the Turkish Black Sea coast is similar (Leitch, 1981). There, an oval stratabound lens of massive-to-disseminated dominant pyrite and subordinate sphalerite, chalcopyrite (128 Tt Cu/1.6%, 53 Tt Zn/2.3%) is hosted by an "ore dacite", a member of an upper Cretaceous to Eocene volcanic sequence composed of spilitic and basalt lavas overlain by dacite, andesite and pyroclastics. The exact nature of the altered host is enigmatic, but it is probably similar to the subvolcanic felsic domes in the Hokuroku district. The dacite contains a disseminated and stockwork pyrite-chalcopyrite mineralization and some ore is also contained in a contaminated quartz-monzonite porphyry.

The Monywa copper deposit in Burma (Goossens, 1978) is in a upper Pliocene ? subaerial volcanic pile, constructed on Miocene and Cretaceous marine pillow basalts and hyaloclastites intruded by diorite. Disseminated and veinlet chalcocite, pyrite and bornite grade to massive sulphide bodies and are capped by 330 m of supergene enriched material. The deepest (hypogene ?) mineralization is in alunitized "biotite porphyry".

The enigmatic Sam Goosly (Equity) deposit near Houston, central British Columbia (Church, 1971; 107 Tt Cu/0.38%, 3,000 t Ag/90 ppm, 27 t Au; Fig. 14-8), has been studied and interpreted several times, with inconclusive results. There, two N.— S.— trending, 45-60° west dipping, tabular zones of coarse-grained, disseminated to locally massive metacolloform pyrite, minor chalcopyrite, pyrrhotite, magnetite, tetrahedrite, etc. lie in Cretaceous marine-to-terrestrial felsic to intermediate lapilli tuff and volcanic breccia, intruded by quartz monzonite and gabbro-monzonite. The host rocks are greenschist and contact metamorphosed. The orebody is surrounded by an aluminous alteration envelope (andalusite, corundum, pyrophyllite, dumortierite, scorzalite, augelite) and the genetic interpretation ranges from "exhalative" to epigenetic replacement (Church, 1971). The subvolcanic syenodiorite bodies (Currie, 1974) suggest a possible link with the "volcanic porphyry deposits" discussed earlier, but in a more felsic association.

The last enigmatic example: the Mantos Blancos copper deposit 45 km N.E. of Antofagasta, Chile (1.183 Mt Cu, 262 t Ag; Anonymous, 1981a), is currently exposed mostly in the oxidation and secondary enriched zone (atacamite, chrysocolla, chalcocite), so that until recently little has been known about the hydrothermal mineralization. There, a late Mesozoic sequence of andesite flows and breccias with porphyritic and augen dacite is intruded by granodiorite and albite, quartz, and epidote-chlorite altered. Hypogene chalcopyrite-pyrite gradational into bornite (partly supergene ?) forms an irregular blanket in the quartz-eye dacite covered by dacite and andesite flows and flow breccias; a setting reminescent of the "Kuroko" footwall unit.

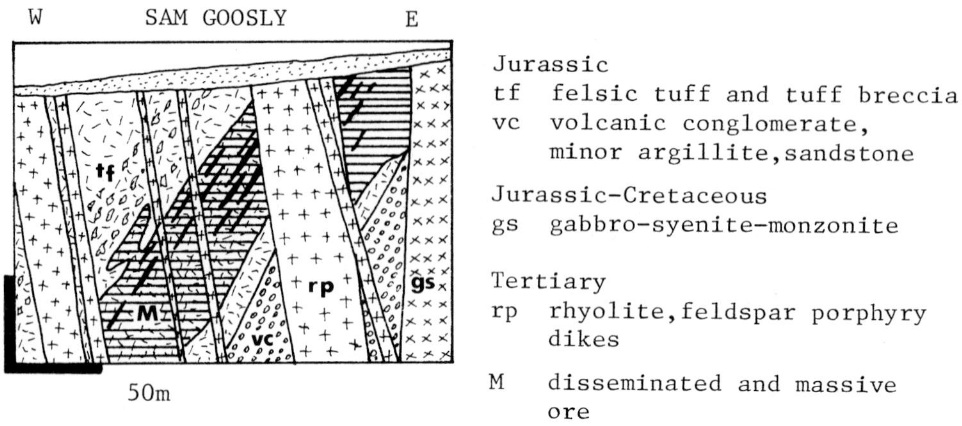

Fig. 14-8. Sam Goosly Cu-Ag deposit, British Columbia. From LITHOTHEQUE, after Church (1971).

14.8.5. Volcanic/plutonic transitions and interactions, 2: "volcanic" porphyry copper systems

Sutherland Brown (1976) listed the Island Copper disseminated Cu-Mo deposit near Port Hardy, Vancouver Island, as the only example of the "volcanic" porphyry copper deposits in British Columbia associated with the calc-alkaline altered felsic subvolcanic porphyry bodies consanguinous with Jurassic BAR volcanics.

At Island Copper (Cargill et al., 1976; 1.34 Mt Cu/0.52%, 43.7 Tt Mo/0.017%, 146 t Re; Fig. 14-9), faulted Jurassic pyroclastics and volcaniclastics are intruded by a comagmatic quartz-feldspar (granodiorite) subvolcanic porphyry dike, bordered by a marginal breccia containing a high proportion of quartz-feldspar porphyry fragments. The dike is intensively altered and at its northwestern end it is capped by tabular bodies of pyrophyllite and dumortierite breccia. The orebody is subdivided by the dike into two halves, and it consists of pyrite, magnetite, chalcopyrite and minor molybdenite present in a stockwork of fracture veinlets, slip surface coatings and minor disseminations. The orebody is associated predominantly with biotite and chlorite alteration and to a lesser degree with a ferroan dolomite veining. Quartz is relatively infrequent.

Island Copper is not the only disseminated Cu-Mo deposit hosted by BAR (but dominantly andesitic) volcanics, formed in response to an emplacement of a comagmatic (but more felsic than the volcanics) subvolcanic intrusive body. Several hundred superficially similar deposits or occurrences have been described worldwide (for example in

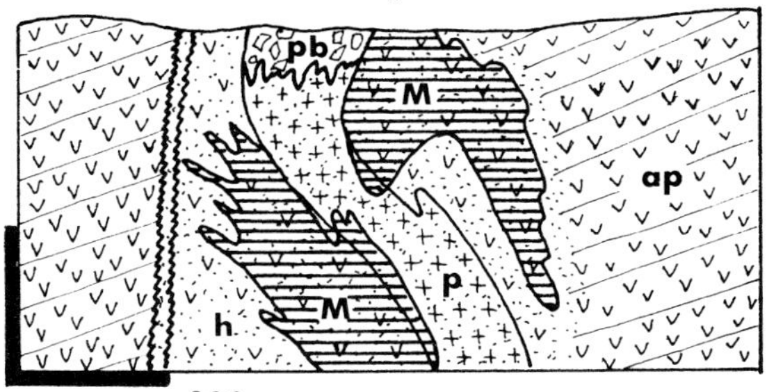

Fig. 14-9. Island Copper, Vancouver Island, B.C., a "volcanic" porphyry Cu-Mo deposit. From LITHOTHEQUE, after Cargill et al. (1976) and Utah Mines Ltd., guided tour (1979).

Arizona, Mexico, the Andes, Fiji and Melanesia, the Urals, Kazakhstan, Puerto Rico, etc). Some of them, however, are hosted by subaerial volcanics (probably former stratovolcanoes) and others are related to intrusive rocks substantially younger than the host volcanics. Although the latter are mentioned briefly in the next paragraph, porphyry coppers are reviewed as a class in Chapter 28. The marine BAR associations are one of the principal host terrains to porphyry coppers, both coeval and alien, and an exploration geologist should be on the lookout.

14.8.6. Volcanic/plutonic transitions and interactions, 3: alien porphyry Cu and Mo deposits, hosted largely or in part by predominantly marine BAR associations

A characteristic and highly productive porphyry copper configuration consists of an older, more mafic (dominantly andesitic) (meta)volcanic unit, intruded by a younger, more felsic and evolved intrusive body (predominantly granodiorite, quartz monzonite). Disseminated and stockwork mineralization is hosted by both the altered volcanics and the parent intrusion. Prominent example deposits include El Teniente (Braden) in Chile (75 Mt Cu), Cerro Colorado in Panama (at least 18 Mt Cu) and other localities. Although the ore-generating intrusions are members of the "evolved" suite, which usually produce the BAR association when it reaches the surface, many intrusions do not have a surficial expression or equivalents. Porphyry (stockwork) molybdenum deposits associated with intrusive bodies emplaced into the rocks of the BAR association, show little correlation with the surrounding supracrustals, although hornfelsed marine (meta) sediments are more common than volcanics.

14.8.7. Volcanic/plutonic transitions and interactions, 4: miscellaneous ore-bearing veins and replacements

Most plutonic rocks intrusive into the BAR association are members of the granodiorite-granite association, believed to have derived from the continental crust. They are the same rocks with which the "alien" porphyry coppers are associated and a considerable variety of hydrothermal replacement and vein deposits occur in their aureole.

Skarns are widespread and by comparison with skarns in the andesite only association (Chapter 13), the former carry a much greater variety of ores. Einaudi et al. (1981) distinguished two depth levels of generation of skarns associated with subduction-related granodiorite and quartz-monzonite plutons. The deeply emplaced (5-15 km) batholiths causing extensive metamorphism but lacking low temperature alterations produced mostly scheelite and minor Cu, Mo, Zn and Pb skarns. Intermediate-to-shallow plutons (1-6 km) often roofed by cogenetic volcanics, produced a much greater variety of Cu, Cu-Fe, Zn-Pb, Au, Mo, etc. skarns. Medium-to-large chalcopyrite and/or magnetite skarns occur in the same setting and often overlap with the porphyry coppers mentioned briefly in the previous paragraph.

Pb-Zn skarns occur relatively rarely. The majority are hosted by the nonvolcanic carbonate association in the "miogeoclinal" realm. Molybdenite-bearing skarns in exocontacts of the BAR suite are represented, for example, by the Red Mountain deposit near Rossland, British Columbia. Gold-bearing arsenopyrite skarns in carbonates interbedded with andesites and intruded by granodiorite, were mined near Headley, in southern British Columbia.

Hydrothermal vein deposits contain a comparable variety of metals. A variety of transgressive hydrothermal veins in the Green Tuff region of Japan (Cu, Pb-Zn, Au-Ag, Mn) have already been mentioned above. Veins of the same metals formed in sligtly deeper (plutonic) settings, for example within the Hazelton Group or in central and northern Chile, where the predominantly andesitic supracrustals are intruded by evolutionarily more advanced granodiorites to quartz monzonites. High-grade ("bonanza") silver vein and replacement deposits in both areas (e.g. the Premier mine near Stewart, B.C.; Chañarcillo south of Copiapó, Chile) occur close to the volcanic-sedimentary and continental-marine facies transitions.

"Numerous so-called descriptive terms are in reality "genetically flavoured". Hence, as soon as we see a descriptive term, we "jump to conclusion" as to genesis".
Karl H. Wolf, in Handbook of Stratiform etc. deposits, 1981.

CHAPTER 15

Pelagic Sediments

Pelagic depositional environments and pelagic sediments are covered by an extensive literature (Berger, 1974; Hsü and Jenkyns, eds., 1974; Jenkyns, 1978) but there is no agreement on their definition and, consequently, the limits of the class. Jenkyns (1978) used the word "pelagic" as essentially synonymous with "of the open sea", but excluding reefs and reef-associated settings. Although the most typical pelagic sediments are those on the abyssal oceanic floor (such as the Globigerina and Pteropod calcareous oozes; Radiolarian, Diatom oozes; pelagic clays), Jenkyns also included continental margin facies and deposits of epeiric seas: that is, an almost complete geotectonic and bathymetric spectrum of sediments.

Recent pelagic environments are much easier to identify and define than the ancient pelagic sediments and the former have already been reviewed to some degree in Chapter 4. In the ancient sequences, pelagic sediments tend to be volumetrically restricted (because of their slow deposition) and many of their identification features (e.g. the microfossil content) are obliterated by recrystallization or deformation. In several cases, the surest way to distinguish pelagic sediments is by their association with a more conspicuous rock suite, such as ophiolites.

Given the emphasis on field recognition of lithologic associations in this book, pelagic sediments represent a marginal category because of the ambiguity of classification and a rather inconspicuous appearance. Most have been treated together with the more abundant associated rocks in Chapters 7, 16 and 17. The brief review presented here gives the basic characteristics and metallic occurrences of the various pelagic sediments using the environmental and facies categories applied by Jenkyns (1978).

(1) Ophiolite association (compare Chapter 7). Pelagic sediments are interbedded with, and overlie, ophiolitic pillow lavas and hyaloclastites. They are usually varicoloured (green, red, black) siliceous shales grading into laminated radiolarian chert and jasper. Gray-to-black micritic limestones may be present. Bedded or nodular manganese oxide deposits associated with the siliceous shales or cherts are the most conspicuous and persistently present metallic ores (e.g. Liguria, Italy; Olympic Peninsula, Washington). Mn carbonates, when developed, accumulate stratigraphically higher above the

volcanics or sediments contact, usually as a component of pelagic shales.

(2) Marine diatomite/chert association, exemplified by the Miocene Monterey Formation of California. The pelletal phosphorites and phosphatic mudstones common near the base of the sequence are slightly uraniferous.

(3) Black graptolithic shales or slates and interbedded limestones and/or cherts, will be covered in Chapter 17.

(4) Condensed carbonate, lesser shale sequences developed on submarine elevations ("Schwellen") in the Devonian-Carboniferous marine basins of central and western Europe, or interdigitating with the Triassic and Jurassic platform carbonates in the Alps and Carpathians. In the Alps, the condensed facies has developed characteristically as red nodular biomicrites containing rich fauna and frequent Fe-Mn oxide encrustations ("hardgrounds"). These may be substituted or gradational into ferromanganese oxide nodules, comparable with the recent oceanic nodules (e.g. Sicily, Montagne Noire), or into Mn-rich stratiform beds or horizons. The latter are usually composed of Mn carbonate and many constitute locally economically mineable deposits.

In the lower-upper Jurassic manganese belt of the Northern Calcareous Alps (e.g. Lechtaler Alpen, south of Oberstdorf, Austria; Germann, 1978), "trough" and "ridge" facies can be distinguished. The trough facies is composed of gray marls and marly pelagic limestone with terrigenous components. The Mn accumulation takes the form of a thinly laminated gray radiolarian shale and marl, containing fine crystalline (micritic) Mn carbonates (Ca-rhodochrosite, oligonite) with minor siderite, chamosite and pyrite. An accessory celadonite is interpreted as an indicator of a volcanic input. A braunite, rhodochrosite, and hematite horizon occurs near Berchtesgaden, Bavaria.

The ridge facies consists of red micritic limestones containing lenses of siliceous radiolarites, associated with or grading into, Mn carbonate or oxide nodules. The Austrian Mn deposits represent about 7 Mt of 10-14% Mn ore and have never been mined on a large scale.

Similar Mn carbonates forming a stratiform horizon in Jurassic marls, radiolarian marls and red limestones in western Hungary (Urkút and Eplény deposits; Drubina-Szabó, 1959; Fig. 15-1), maintained a steady production because the low-grade primary ores have been residually upgraded during the upper Jurassic-middle Cretaceous subaerial exposure. At present, bodies and pockets of concretionary pyrolusite, psilomelane, vernadite and goethite fill karst under the lower Jurassic-middle Cretaceous unconformity, partly topped by upper Cretaceous or Eocene shales, marls and limestones. The economic Mn content is of the order of 3 Mt Mn.

Closely comparable Mn occurrences are known in the Liassic limestones of the Polish Tatra Mts (Chochołów); in Jurassic foraminiferal limestone and limy shale in the Cayetano Formation of

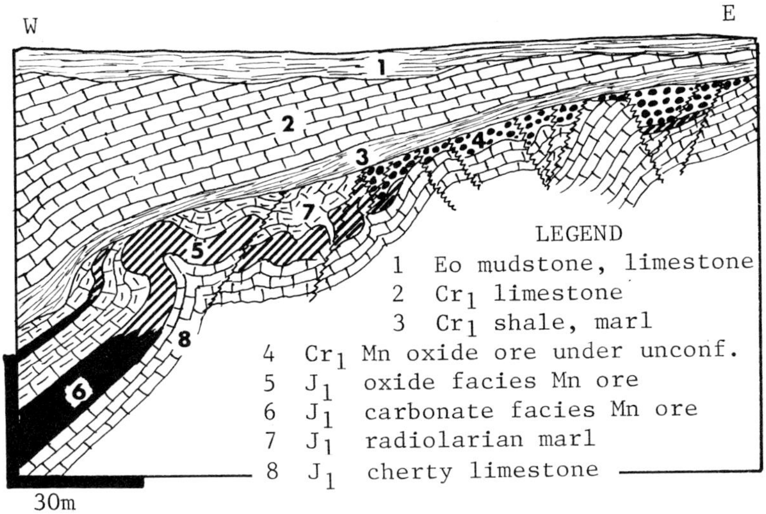

Fig. 15-1. Urkút Mn deposit, Hungary. Stratiform Mn carbonate in Liassic radiolarian marls (6) changes facies into a Mn oxide ore (5), which, in turn, has been upgraded and converted into a lumpy oxide ore (4) under the Cretaceous unconformity. Modified after Barnabás et al. (1966).

western Cuba (Gramales-Las Acostas district; about 25 Tt Mn; Simons and Straczek, 1958); from thin beds of Mn shale in the Cretaceous pelagic limestone on the Santa Isabel, Malaita and San Cristobal Islands (Solomon Islands; Taylor, 1977); from Timor and Sulawesi (Indonesia; Hamilton, 1979); and from elsewhere.

At several places in the canton of Graubünden, eastern Switzerland, Mn and Fe occurrences are hosted by Jurassic radiolarian jaspers or siliceous shale horizons interbedded with pelagic limestones and sometimes associated with spilite flows. The Gonzen deposit near Sargans (de Quervain and Zitzmann, 1976; 1.86 Mt Fe/53%, 120 Tt Mn/60%) is the best-known locality. There, a massive, folded, 1.5-2 m thick bed of hematite with subordinate magnetite and pyrite and local rich hausmannite lenses, is traceable for several kilometres, interbedded with middle Jurassic thin-bedded limestones.

(5) Chalks and glauconitic or phosphatic pelagic sediments of epeiric seas. Although some phosphatic and manganiferous hardgrounds and submarine pavements have been recorded in the northern European Cretaceous chalk (Compare Jenkyns, 1978), nothing of any possible economic significance has been reported. The epeiric sediments are treated in Chapters 19 and 20.

Plate 7. Flysch lithofacies of rhythmically alternating shale and sandstone. Near Medzilaborce, eastern Slovakia.

"Il n'y pas d'autre exemple peut-être dans l'histoire de notre science d'un nom qui depuis son introduction, ait causé plus de confusion que ce malhereux flysch".

"There is no general agreement on the meaning of the term "flysch".

B. Studer, 1848 and Ph. H. Kuenen, 1958 quoted in Hsü (1970).

CHAPTER 16

Deeper-Marine, Sandstone-Shale Association: "Flysch Facies"

16.1. GENERAL

The word "flysch" originated in the Alps, where it was applied to the thick and monotonous association of rhythmically alternating Mesozoic and Cainozoic sandstone and shale. At present, the term is formally used to designate lithofacies with the following typical features (Sestini, 1970): (1) rhythmic alteration of massive sandstones and calcarenites and shales; (2) poorly sorted sandstones, containing variable amounts of clay; (3) sandstones (or calcarenites) having sharply defined bottom surfaces with sole markings and indistinct top surfaces with transitions into a shale; (4) common graded bedding; (5) little lateral variation in thickness and sandstone composition; (6) current sedimentary structures which tend to display a constant preferred orientation; (7) megafossils which are rare, and a fauna which is pelagic or deepwater benthonic; (8) large-scale cross-bedding which again is rare. Lastly, (9), there are lateral transitions into non-flysch associations, but basal unconformities are rarely obvious.

Although flysch is convincingly interpreted as having formed by turbidity currents, flysch and turbidite are not identical terms because turbidites may form episodically in most subaqueous environments including shelves and even lakes and artificial dams. The recent turbidity-current environments and the flysch association have an extensive sedimentologic literature (see Rupke, 1978, and Dzulynski and Walton, 1965, for reviews) and the basic sedimentologic premises will not be repeated here. As a lithofacies, flysch (and other turbidites) can be subdivided into the proximal (near-source) and distal subfacies. The former has massive beds of irregular thickness and is coarse grained (sandstone to conglomerate). The latter is thin-bedded, finely grained and composed of parallel-sided, regularly bedded layers of frequently graded sandstone and mudstone. It is gradational into a predominantly shale facies. Fluxo-turbidites are particularly thick-bedded and coarse-grained beds, that sometimes occur as part of the proximal turbidite. They are most frequently interpreted as products of combined subaqueous slides and turbidity currents.

Interpretation of the geotectonic setting of the flysch sequences received a great impetus after the introduction of plate tectonics, and there is a voluminous literature on the subject. Reading (1972) distinguished and named six recent flysch successions corresponding to various continental margins (Atlantic, Japan Sea, island arc, Andean, Mediterranean, Californian) but it is doubtful that ancient flysch successions can easily be assigned to any of them. The paper, however, demonstrated that flysch is equally widespread along Atlantic and Pacific-type continental margins and that there is, as expected, a degree of correlation between the petrographic composition of the flysch sand(stones) and their setting. The Atlantic flysch, derived from an area completely free of contemporary volcanism, corresponds to a "compositionally mature" phyllarenite (high quartz, metamorphic rock fragments, some feldspar). The andesitic island arc flysch is dominated by compositionally immature volcarenite (intermediate rock fragments) that is ideally free of quartz. This type of turbidite is proximal to the island arc volcanics (e.g. of the Philippines, Melanesia, etc.; Chapter 5) and often loses its identity in geologically old terrains due to a partial diagenetic conversion into mudstone. The quartz-free flysch is the end-member of the Pacific margin turbidites, relatively restricted in occurrence.

The most widespread Pacific margin turbidite, however, does contain quartz, some feldspar and even mica, derived from dacite and rhyolite phenocrysts and from unroofed granodioritic plutons. In geologically complex arcs, such as Japan, the old crystalline basement contributes detritus identical to the Atlantic-margin suite.

In the recent literature dealing with flysch sequences, two example (model) sub-associations and regions are invoked most frequently: (1) volcanic-terrigenous flysch formed along the Pacific-type margin and best studied in the Mesozoic-Cainozoic sequences of North America (e.g. Dickinson, 1974b); and (2) terrigenous quartz-rich flysch, widespread in the Alps and Carpathians ("Alpine flysch"). The sub-association (1) is metallogenically more productive, so that it is the only flysch turbidite discussed in Mitchell and Garson (1981).

16.2. TRACE METAL GEOCHEMISTRY AND MINERALIZATION

The basic characteristic of the flysch metallogenesis is that the "most normal" facies (the undeformed sandstone alternating with shales) has an almost ideal average - to-sub-clarke trace metal content and virtually no currently economic ore deposits. The exceptions include small and low-grade iron and manganese deposits mined mainly for local consumption in the past.

The bulk of the metallic mineralizations listed and shown as hosted by the flysch association (Fig. 16-1; Table 16-1), are either associated with "accessory" rocks (e.g. volcanics) or with superimposed features (cleavage, schistosity, igneous intrusions) that either caused remobilization of the trace metals contained in the flysch rocks or introduced the metals from outside.

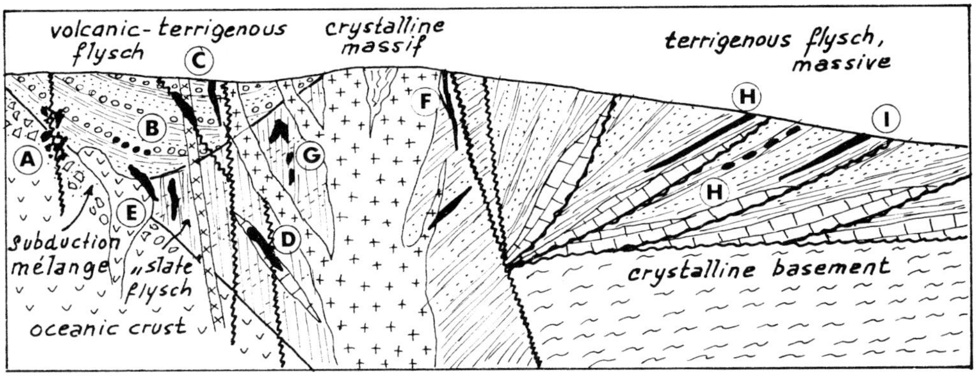

Fig. 16-1. Principal mineralization styles in the flysch association (see Table 16-1 for explanation of the letter codes).

Table 16-1. Principal mineralization styles in the flysch association

MASSIVE, VOLCANIC - TERRIGENOUS FLYSCH	
(A)	Hg,Au,Mn,Cu-Zn,etc. ores hosted by / formed in, mélange, ophiolite, blueschist, etc. along a contact with flysch association, and partly within it
(B)	detrital gold in flysch conglomerate / sandstone
(C)	hydrothermal epigenetic vein,stockwork,impregnation, etc. ores in flysch along faults near diabase dikes (Hg), granitic rocks (Au, Cu, etc.) and similar; X0 localities, 1.5 Tt Hg
DEFORMED, LIGHTLY METAMORPHOSED FLYSCH (SLATE OR SCHIST ASSOC.)	
(D)	siderite or ankerite impregnations in arenites, replacements in carbonate lenses; X0 localities, 600 Mt Fe
(E)	massive Fe,Cu,Zn sulphide deposits in the proximity of mafic or bimodal volcanics, often cleavage controlled; X00 localities, 1.5 Mt Cu; 6 Mt Zn; 2.5 Mt Pb; 100 t Au; 6 Tt Ag
(F)	hydrothermal fissure Zn,Pb,Cu,Au, etc. veins in flysch, in a plutonic thermal aureole; X,000 localities; X Mt Cu,Pb,Zn,Au
(G)	gold lodes, and derived placers, in flyschoid slate belts; X00 localities, 5 Tt Au
TERRIGENOUS FLYSCH, MASSIVE	
(H)	stratiform pelosiderite occurrences; X0.000 localities; X00 Mt Fe (low grade)
(I)	stratiform Mn-Fe carbonate deposits; X0 localities, 1 Mt Mn

Such mineralizations are transitional and together represent an economic potential of some 5,100 t Au, 600 Mt Fe, 6 Mt Zn, 2.5 Mt Pb, 1.5 Mt Cu, 6 Tt Ag; 100 Tt Sb; etc. These figures, however, have little meaning because some localities on which they are based (e.g. the Cobar district) have been placed into alternative genetic categories by different investigators.

16.3. VOLCANIC-TERRIGENOUS FLYSCH

The Mesozoic Great Valley Sequence of California (Maxwell, 1974; Dickinson, 1969) is probably the most widely publicized example of a partly volcanogenic, continental margin flysch association believed to have formed in a forearc basin (arc-trench gap; compare Chapter 5). The sequence tops and is to some extent interdigitated with an ophiolite mélange. Up to 15 km thick, the rock pile is composed of clastic turbidite sandstone flows intercalated with or enclosing a mudstone phase. It was derived from a complex volcanic-plutonic source in the ancestral Sierra Nevada, and its volume closely matches the estimated volume of material eroded from above the present Sierra Nevada core. Because of the manner of its derivation, the sandstone contains the greatest proportion of volcanic fragments near the base, but the proportion of feldspar, mica and quartz increases upward and correlates with the increasing erosional dissection of the plutonic levels in the source area. The bulk of the sequence is composed of massive sediments only lightly load metamorphosed, but it is locally disturbed by imbricate thrusts, strike-slip faults and zones of superimposed high-pressure metamorphism.

In contrast to the adjacent units (the Franciscan assemblage: an ophiolite terrain, Chapter 7; Sierra Nevada and Foothills: volcanic-sedimentary and batholithic belt, Chapters 10 and 28), the Great Valley Sequence is metallogenically remarkably infertile. The bulk of the monotonous flysch sediments are completely devoid even of traces of a metallic mineralization, and the high-pressure metamorphics are no exception. The few metallic occurrences of Au, Mn, Hg, Cu and Cr are in transitional settings, typically associated with the "accessory" rock types syngenetically deposited or tectonically emplaced into the turbidites, or along the contact of the turbidites with the more fertile adjacent units.

The cinnabar orebodies of the New Idria, New Almaden, and other mines (Chapter 8) exist partly in the rocks of the Great Valley Sequence where it overlaps with a serpentinite mélange or where it is tectonically intermixed with ultramafics, which are faulted and hydrothermally altered. The only small Island Mountain Fe,Cu,Zn massive sulphide deposit occurs in a similar setting.

Similar contrast in the density of metalliferous occurrences is apparent in the lower Tertiary continental margin subduction complex in the Olympic Peninsula, Washington (Fig. 16-2), recently mapped and described by Tabor and Cady (1978 a,b). There, the core of the complex is a flysch association composed of a monotonous alternation

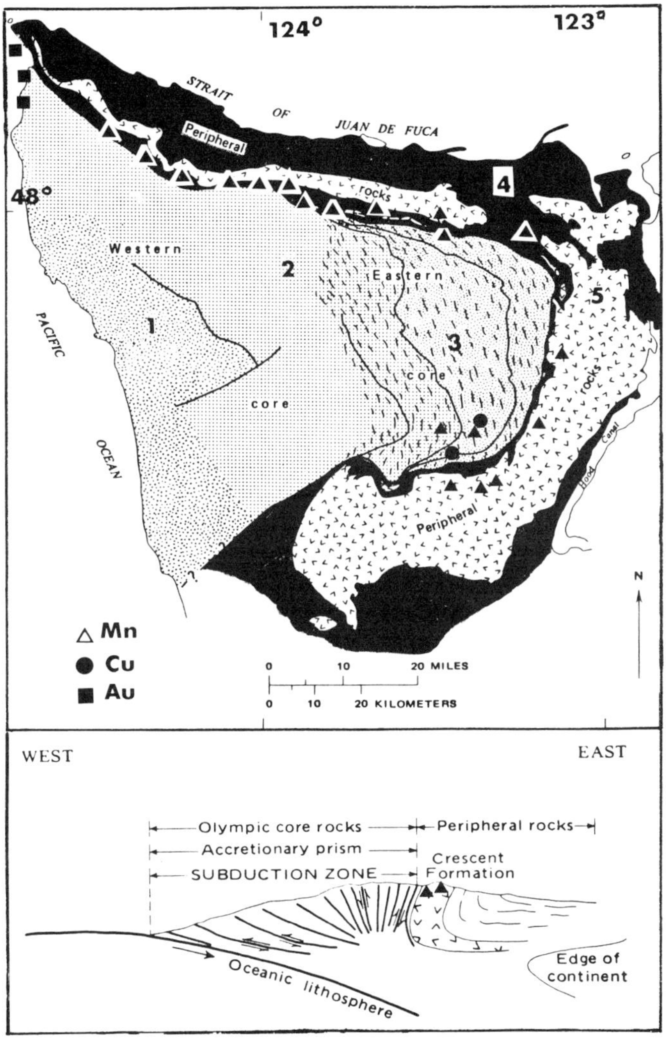

LEGEND

1 Ol_3-Mi_2 litharenite, slate, phyllite

2 Eo_3-Ol, ditto

3 Eo slate, phyllite, litharenite minor basalt, gabbro

4 Eo-Mi_3 shale, sandstone, conglomerate

5 Eo_{1-2} basalt, diabase, gabbro, pelagic red limestone, siliceous slate

Fig. 16-2. Lower Tertiary subduction margin rock association of the Olympic Peninsula, Washington, composed of a monotonous detrital flysch suite in the core, surrounded by a basalt-pelagic sediments peripheral belt. All the rare Cu and Mn occurrences in the flysch core are in, or in the vicinity of, the "accessory" basalts. The more numerous Mn occurrences in the peripheral zone are at or near contacts of basalt with the pelagic sediments. From Tabor and Cady (1978 a,b); metallic occurrences added.

of volcanic-terrigenous litharenite, slate (phyllite), minor conglomerate, and exceptional limestone. Minor (meta)basalt, diabase and gabbro crop out in several belts of scattered occurrences, and major outcrops of basalt and pelagic sediments (red siliceous shale and limestone) are found in the peripheral oceanic assemblage. The rocks are deformed and metamorphosed in zeolite to greenschist facies. The core flysch has no metallic occurrences, despite the frequent presence of unmineralized quartz, carbonate, laumontite, chlorite, etc. fissure and shear veins. The rare mineralized localities shown in the core are all within, or in the vicinity of, an accessory (meta)basalt, and are quartz-chalcopyrite veins, malachite coatings, and stratiform Mn oxide lenses. Widespread Mn oxide mineralization, however, is found in the (meta)basalts and red siliceous pelagic limestones and slates in the peripheral belt.

The general pattern of metallic mineralization (or the lack of it) in the Great Valley sequence is repeated worldwide (Table 16-2). Mitchell and Garson (1981) suggested the possibility that the detrital flysch sediments may have some potential as traps of detrital gold, derived from its "primary" source in the magmatic arc, and they quoted the small Recent alluvial gold placers in the Western Trough of Burma and in the Great Valley Sequence as possible examples. Empirical experience, however, indicates that most of the gold occurrences identified in the flysch setting (e.g. in central and western Alaska) are members of the interaction hydrothermal mineralization, associated with orogenic to post-orogenic igneous intrusions, with zones of cleavage (dynamo-metamorphism) or with both. Syndepositional or diagenetic mineralizations in flysch are often difficult to distinguish from the interaction ores, particularly when exposure is poor or when the metallic commodity (e.g. gold) is present in surficially reworked deposits derived from the flysch bedrock, for example in placers.

16.4. CLEAVED / METAMORPHOSED VOLCANIC-TERRIGENOUS FLYSCH: SLATE AND SCHIST ASSOCIATION

Flysch sediments modified by intensive axial cleavage and schistosity development change into a monotonous slate-to-schist assemblage. To some degree the metamorphic fabrics mask the pre-metamorphic lithology, so that slates (and particularly schists) may correspond to a range of the original petrographic compositions from mudstone to conglomerate. Like the massive flysch, the slate/schist metasedimentary association is frequently transitional into marine volcanic-sedimentary suites. Example terrains include the lower Paleozoic Rosjø Formation in the Scandinavian Caledonides; Neranleigh—Fernvale Group of Queensland; the Post-spilite Group of the Bohemian Massif; and others.

The mineralization in the cleaved/schistose meta-flysch is comparable with the massive flysch in one important point: the bulk of the ore occurrences again occur in or near the "accessory" rocks.

Fig. 16-2. Volcanic - terrigenous flysch association, selected example areas

LOCALITY	AGE	GEOLOGY	MINERALIZATION	REFERENCES
Kobuk Trough, Kuskokwim R., Koyukuk, etc. W.-C. Alaska	Cr_{1-3}	up to 9 km thick sequence of monotonous volc. graywacke/slate, topped by paralic coal assoc.; intr. by diabase dikes, Cr-T granitic massifs	scattered cinnabar dep. near altered diabase bodies (e.g. Red Devil); 1.2 Tt Hg; numerous Au placer, derived from veins in plutonic aureoles	Cady et al. (1955); Berg and Cobb (1967)
Olympic Peninsula, Washington, USA	Eo	graywacke-shale sequence with minor gabbro, diabase, basalt; unmet. to greensch. A forearc complex adjacent to basalt, gabbro "seamount"	minor Mn and Cu occurrence confined to the accessory and fringe occur. of basalt	Tabor and Cady (1978)
Great Valley Sequence, Coast Ranges, California	J_3-Cr_3	thick prism of alternating quartz - rich litharenite and shale; rest on ophiolite and mélange; zoned high-pressure metamorphics	cinnabar dep. along altered faults in ophiolite mélange reach into flysch; Island Mt. mine, mass. Cu sulph., in blueschist metam. flysch and mélange; minor Au placers	Albers (1981)
Kulm Facies, PZ orogene of Central Europe (Harz, Nízký Jeseník Mts.)	Cb_1	thick sequence of graywacke altern. with shale to slate; overl. D bimodal suite, some basalts, keratoph. persist to basal flysch units; intruded by Cb_3 granitic plutons	widespread Zn-Pb veins hosted by flysch in thermal aureole of plutons	v. Bubnoff (1956); Svoboda, ed., (1966)
Koryak Mts., eastern Siberia, USSR	Cr	15-20 km thick turbidite wedge; dominant graywacke, some mafic volc., silic. slate, jasper interbeds. Interpr. as trench fill and arc-tr. g.	Kirpavaam zone: qtz, ars, cp, gold veins in faulted flysch cut by dior. porph. dikes; numerous small Au veins, placers	Bogdanov (1969)

16.4.1. Fe (Mn) aspect

Iron ores are moderately common in both undeformed and deformed (cleaved) volcanic-terrigenous flysch associations. Lenses or horizons of sandstone with ankerite or siderite cement and iron contents ranging from 20 to below 30% are widespread, but of little economic importance, so they are rarely recorded. Economic deposits are almost invariably a product of multistage metallogenesis, usually of uncertain origin. The economically important Ljubija district in the north-western Bosnia, Yugoslavia (Čičić, 1980; min. 500 Mt Fe/38%), contains seventy six siderite or limonite deposits or occurrences hosted by a middle Carboniferous meta-graywacke alternating with slates and locally present carbonates and mafic volcanics with tuff horizons. Most of the siderite bodies are structurally controlled replacements of limestone lenses, but some could be syn-depositional volcanogenic deposits.

16.4.2. Massive Fe,Cu,Zn,Pb sulphide aspect

Several massive sulphide deposits have been denoted in the literature as hosted by the "flysch" but coeval volcanics are usually close by so that we are dealing with a transitional situation between the flysch and one of the volcanic-sedimentary assoiations treated in Chapters 10 to 14. The base-metal ratios in the orebodies are usually identical to those contained or expected to occur in the coeval volcanics: that is, Cu(Zn) if such volcanics are mostly basalts and Zn,Pb(Cu) when such volcanics are members of the BAR suite. The Prince William Sound district, southern Alaska (Moffit, 1954; 200 Tt Cu, 7.5 t Au), is an example frequently quoted. There, several tens of massive sulphide deposits and occurrences are hosted by a thick Paleocene sequence of weakly metamorphosed flysch that contains abundant metabasalt flows in the lower part of the sedimentary pile. Of the four principal orebodies, three occur entirely within black slate, graywacke and argillite and one (the Rua Cove deposit) entirely within basalts. The orebodies comply with the common volcanogenic--stratiform model: conformable massive pyrite, pyrrhotite, chalcopyrite, minor sphalerite lenses are underlain by disseminated sulphides.

Portion of the Cambro-Silurian succession of the Norwegian Caledonides (e.g. the Rosjo Formation in the Røros District; Rui and Bakke, 1975) display the characteristics of a monotonous metamorphosed flysch composed of alternating grayish-brown biotite schists and dark phyllites, with subordinate thin greenstone interlayers. The massive sulphides composed of pyrite and chalcopyrite occur as thin, extensive concordant sheets or irregular, small lenses comparable with the Besshi "type". Much of the Japanese Sambagawa belt is almost identical lithogically.

The Jurassic flysch association exposed along the southern slopes of the Great Caucasus Range, U.S.S.R., contains a 500 km long belt of massive sulphide deposits and occurrences hosted by cleaved and sheared metasediments (Filizchai, Kizil-Dere, Belokany, Laura, etc.; Samonov and Pozharisky, 1974; Smirnov ed., 1968). The orebodies are stratabound massive pyrite, lesser pyrrhotite, sphalerite, galena, chalcopyrite, etc. lenses to sheets, hosted by undisturbed, cleaved and sheared pyritic meta-argillite that appears to be a facies equivalent of a bimodal volcanic-sedimentary suite. A footwall stockwork in lightly chloritized slates is reported in Filizchai, while no conspicuous alteration is associated with the other localities.

The enigmatic, but economically important Cobar Au,Cu,Zn,Pb district, N.S.W., Australia (Brooke, 1975; 4.4 Mt Zn, 2 Mt Pb, 920 Tt Cu, 4.5 Tt Ag, 70 t Au; Fig. 16-3) contains about twelve groups of orebodies hosted by members of a late Silurian to early Devonian turbidite fan system. The fan consists of a transitional sequence of sediments ranging from polymictic conglomerate through graded, bedded feldspathic graywacke to siltstone and slate. The rocks are metamorphosed in the lower chlorite schist grade, strongly cleaved and disturbed by a series of N.N.W. trending shears, parallel with the strike of the depositional trough. The only known igneous rocks in the vicinity of the major deposits are small bodies of acid porphyries in the southernmost end of the mineralized belt. The ore belt is confined to the Cobar Trough and is traceable with minor interruptions for 120 km.

The ore deposits correspond to two major compositional types: (1) siliceous, low-or high-sulphide gold-bearing chalcopyrite or pyrite deposits in silicified meta-siltstone; (2) massive pyrite, sphalerite, galena, chalcopyrite tabular and lens-like bodies, hosted by rhythmically interbedded quartz, chlorite, sericite meta-siltstones and argillites. All the known tabular and lens-like ore occurrences are located within 70-80° east dipping shear zones or in zones of strong axial cleavage, with about 20° discordancy between the bedding and the present position of the orebodies.

The recently discovered Elura deposit at the northern end of the Cobar ore zone (Adams and Schmidt, 1980; 27 Mt ore with 8.3% Zn, 5.6% Pb, 140 ppm Ag; Fig. 16-3) is a discrete, well-outlined vertical pipe with elliptical cross-section (200 m N.-S., 120 m E.-W.), continuing for at least 500 m under the surface. It shows gradational mineralogical zoning. In the core is a mass of pyrite with about 20% pyrrhotite and a variable proportion of sphalerite and galena. Away from the core pyrrhotite dissapears, galena and sphalerite become dominant and minor arsenopyrite, tetrahedrite-tennantite, and chalcopyrite appear. The outer fringe of the orebody contains the same minerals in massive to disseminated form in a siliceous matrix. The host siltstones are increasingly deformed (folding, cleavage, shearing) in the vicinity of the orebody but there is virtually no recognizable alteration outside it except for some quartz veining and light scattering of ankerite and siderite.

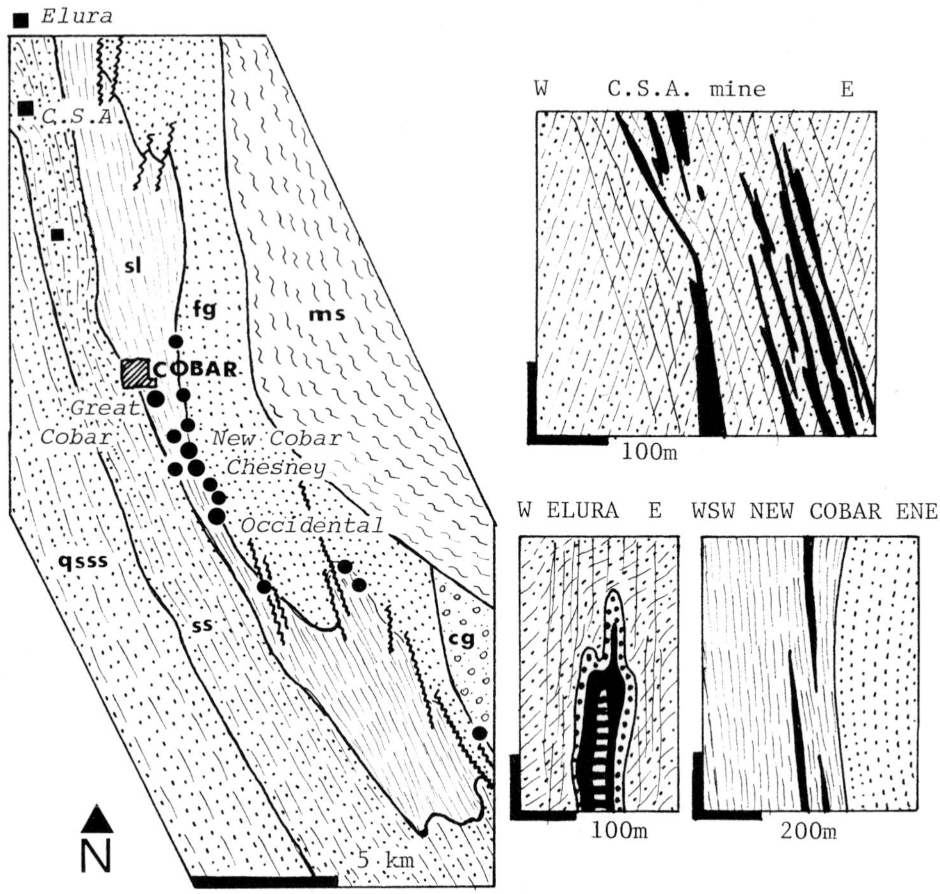

Fig. 16-3. Cobar Cu,Zn,Pb,Au,Ag district, N.S.W., Australia. LEGEND: Ordovician (ms) quartz-micaschist. Upper Silurian-lower Devonian: (cg) polymictic conglomerate; (fg) feldspathic graywacke; (sl) cleaved slate; (ss) slate and siltstone; (qsss) quartzite, sandstone, shale, siltstone. Black dots: dominantly Cu-Au siliceous ore deposits; black squares: Cu,Zn,Pb massive sulphide deposits.

Elura orebody: core - pyrrhotite, pyrite, galena, sphalerite; middle zone - pyrite, galena, sphalerite (main ore zone); outer fringe - disseminated sulphides in siliceous gangue. Map after Brooke (1975), mine sections are from LITHOTHEQUE, modified after Brooke (1975) and Adams and Schmidt (1980).

The origin of the Cobar field is uncertain but recently Sangster (1979) reinterpreted the Cobar deposits as originally stratiform volcanogenic seafloor hydrothermal mineralization transposed into the prominent regional cleavage during the post-ore deformation: a view not generally shared by geologists working in the area.

Several other massive sulphide and barite orebodies in a (meta) sedimentary host association have been interpreted as in "flysch"; for example the Devonian-Mississippian sediments in the Macmillan Pass area, Canada; the Devonian hosts to the Rammelsberg orebody, Germany (Morganti, 1981); the Mississippian Stanley Shale of Arkansas and others. Although it is admitted that the above associations show some of the characteristics of a "distal flysch" (almost as many as the rocks in the Cobar Trough) their inclusion would stretch the flysch application too much. Instead, the above associations and deposits will be treated in Chapter 17.

16.4.3. Hg (Sb) aspect

Volcanic and volcanic-terrigenous flysch is a frequent host to cinnabar and/or stibnite deposits, but interaction with ultramafic or mafic rocks and low-temperature, high-level hydrothermal systems, seems to be the necessary prerequisite. Mercury deposits associated with ultramafic mélange and overlapping into the adjacent flysch in California, have been described already in Chapter 8.

In the Kuskokwim River cinnabar district in south-western Alaska (1,190 t Hg; Sainsbury and MacKevett, 1965), several cinnabar deposits are hosted by thick upper Cretaceous massive graywacke-slate flysch, near or within regionally distributed altered diabase or minette dikes. Cinnabar is often accompanied by stibnite, realgar and orpiment and quartz, dolomite and calcite gangue that fills open fractures in the competent rocks. At the largest Red Devil mine near Sleetmute (Fig. 16-4) the orebodies formed at and near intersections of north-east trending diabase dikes with north-west trending faults, are pencil-shaped and enveloped by the characteristic silica-carbonate alteration (listvenitization). Six mineralization zones arranged en echelon, have been recognized. The ore is most likely to have precipitated from CO_2 - rich springs in shallow subsurface, in the late Miocene-early Pliocene time. A portion of the Alaskan cinnabar came from placer operations.

Two Soviet cinnabar deposits: Polyanskoe in the Chukotka Peninsula, and Sakhala in the north-western Caucasus (Kuznetsov, 1974), are hosted by terrigenous flysch of upper Triassic and Jurassic-Cretaceous age, respectively. In both cases, cinnabar forms impregnations, veinlets, disseminations, joint coatings, etc. along faults in shattered and lightly-altered brittle rocks. The Silurian-Devonian quartzites and shales hosting the world's largest mercury deposit Almaden, Spain, have occasionally been called "flysch". Although a Proterozoic flysch occurs under the ore-bearing Silurian-Devonian sedimentary series and a Carboniferous flysch above,

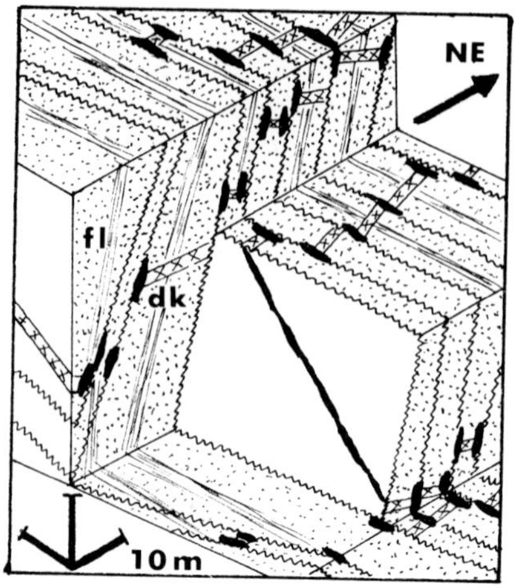

RED DEVIL Hg MINE, SLEETMUTE
ALASKA
isometric block diagram

fl = Cretaceous graywacke / shale
dk = altered diabase dikes
black = cinnabar orebodies

Fig. 16-4. Red Devil Hg mine, Alaska. After Sainsbury and MacKevett (1965).

it does not seem to have a flysch character and is treated in Chapter 19.

16.5. INTERACTION MINERALIZATION IN DEFORMED AND METAMORPHOSED VOLCANIC-TERRIGENOUS FLYSCH, INTRUDED BY GRANITIC ROCKS

Because of its widespread areal occurrence in orogenic belts, the flysch association is often found deformed, metamorphosed and invaded by calc-alkaline granitic intrusions. Here, the style of mineralization is basically controlled by structure and by thermal aureoles, while the facies and lithologic variation within the flysch plays a limited role. Most of the orebodies are hydrothermal veins and stockworks emplaced along faults, shears and zones of axial cleavage. The metallogenic predictability is limited; in other words, one cannot generally tell which metals will accumulate in the hydrothermal orebodies, judging by the lithology of the flysch rocks alone. Statistically, gold-bearing quartz veins appear to be most numerous, followed by Cu, Zn-Pb, Sb and other mineralizations. Locally, a distinct metal distribution pattern appears that is usually attributed to the local ore provincionalism: in regions considered, for example, as Pb-Zn provinces, the hydrothermal veins in flysch are likely to carry Pb-Zn. This is well demonstrated in the Harz Mts district, Germany and to a lesser degree in the Opava region of Czechoslovakia. There, the monotonous lower Carboniferous "Kulm" facies of volcanic-terrigenous turbidite overlies a lithologically more variegated Devonian suite that contains black slate, carbonate, bimodal volcanics and coeval Fe and Zn-Pb(Cu) massive sulphide orebodies. Late Carboniferous granodiorite – to – quartz monzonite plutons emplaced in both the Devonian volcanics-slate level and the lower Carboniferous flysch level, generated numerous vein fields. The Clausthall-Zellerfeld field in the Harz produced Zn,Pb,Ag from several tens of quartz, siderite, dolomite, calcite, sphalerite, galena, etc. veins and the large Grund deposit is still in operation. There, the mineralization in the flysch-hosted hydrothermal veins is completely alien to the enclosing supracrustals, and appears to be more or less an extension of the consanguinous metallogeny of the buried, older sequence so we are probably dealing with a case of regional heritage. Similar heritage seems to be in effect in the "Pb-Zn provinces" in which massive sulphide orebodies are known, for example in the Sierra Morena of Spain. There, (1) the late Proterozoic (Alcudian) flysch contains major Pb-Zn veins (San Quintin, Diogenes mine), lacking a visible association with granites; (2) the lower Carboniferous (Kulm) flysch contains veins in the aureole of a Hercynian granite (El Soldado-Las Morras); and (3) the Hercynian granites themselves host important Pb-Zn veins; e.g. Linares (Ovtracht and Tamain, 1972).

In several regions that have been designated as copper metallogenic provinces, the flysch is preferentially mineralized by copper-bearing hydrothermal vein and disseminated orebodies. In eastern Cuba, the 5.5 km thick low-grade metamorphosed middle-upper Jurassic San Cayetano Formation is composed of alternating slate, phyllite and metagraywacke, so that it is transitional between the "slate" and "flysch" associations. Several small stratiform pyrite and minor sphalerite, galena, chalcopyrite orebodies have been recorded situated in its basal portion (e.g. the Mono deposit in the Matahambre field; 1 Mt ore with 5.2% Zn, 2.4% Pb and 0.17% Cu). However, economically the most important are the rich quartz-chalcopyrite veins, such as Matahambre (Beneš and Hanuš, 1967; min. 600 Tt Cu). The veins are controlled by a/c joints and cross-faults, and are coeval with the regional emplacement of granodiorite plutons during the upper Cretaceous-lower Tertiary periods.

In south-western British Columbia, the Canam (AM) copper property near Hope is located in a sheared and cleaved lower Jurassic-lower Cretaceous flysch argillite intruded by a quartz diorite stock. Metallic mineralization is represented by a pyrrhotite, chalcopyrite-mineralized chlorite-tourmaline altered breccia pipe and by a hydrothermally mineralized shear zone (lenses and bands of quartz, sphalerite, galena, minor jamesonite, arsenopyrite, etc.). A gauge vein traversing the breccia pipe contains lenses of massive arsenopyrite, pyrite, chalcopyrite and minor uraninite, molybdenite and other minerals. Eastwood (1967) reported a widespread halo of scattered pyrite and arsenopyrite crystals in the hornfelsed argillites in the vicinity of the breccia. Similar, distinct occurrences of arsenopyrite are common worldwide. It thus appears probable that arsenic (and possibly some of the "strange" elements: U, Mo, etc.) have been locally derived from the argillite.

A very broad generalization can be offered: a deformed, plutons-intruded flysch association in orogenic belts, will probably host a granite-coeval (1) Zn,Pb,Ag mineralization when the flysch rests on a shale and/or carbonate, bimodal, or BAR suite, probably floored by a continental crust; (2) Cu mineralization when the underlying sequence are island arc basalts or andesites; and (3) Au hydrothermal deposits when the flysch rests on or interdigitates with, ophiolites or "geosynclinal" marine metabasalts. These and similar patterns are discussed further in Chapter 28.

16.6. MASSIVE, UNMETAMORPHOSED TERRIGENOUS FLYSCH

The type area of flysch in the Alps is a thick synorogenic detrital unit of lower Cretaceous to Eocene age, exposed in a series of south-dipping thrust sheets. Autochthonous occurrences, however, also occur. Lithologically, this is typically unmetamorphosed quartz-rich terrigenous flysch, peculiar due to its relatively high calcareous content and predominantly derived from advancing nappe and thrust-sheet fronts. Calcareous shales, sandstones, marls and impure limestones (some corresponding to the pelagic suite of epeiric seas) are present locally. The even more extensive Carpathian flysch, widely distributed in eastern Czechoslovakia, southern Poland and throughout much of Rumania, is contemporary and is comparable lithologically with the Alpine flysch (Fig. 16-5).

As elsewhere, the Alpine and Carpathian flysch is generally a coarsening-upwards sequence with a shale-rich lower section, grading into a sandstone-rich upper section. Usually, however, several sedimentation cycles are represented in addition to the numerous tectonic units often present. Submarine mafic volcanics are occasionally interbedded with the lower shales. In the Bezkydy Mts. in north-central Czechoslovakia (Roth, 1962), pillowed and amygdaloidal flows and sills of teschenites and picrites alternate with lower Cretaceous flysch shales and sandstones. Black siliceous shales and stratiform, low-grade (10-33% Fe) siderite lenses are present locally in the volcanics—rich facies.

Pelosiderite, ankerite and, less frequently Fe-Mn carbonate concretions, sandstone cement, and thin beds, are quite widespread in the Bistriţa and Suceava valleys (Rumanian East Carpathians; Ianovici and Borcoş, 1982). Low-grade (aver. 17-19% Fe) "pelosiderite" layers, 10-15 cm thick, are rhythmically interlayered with black shales, sandstones and marls and interpreted as diagenetic; no volcanism is reported. Near Bicaz Lake the ironstone seams are 60 cm thick. In the Capus deposit near Cluj, Rumania (15.5 Mt/31% Fe), concordant beds of oolithic goethite with ankerite-leptochlorite cement are hosted by Eocene flysch sediments.

Slightly pyritic shales are quite common in flysch. Anomalously high trace uranium contents have been reported in some Paleogene Menilite shales (brown shales with opal or chert concretions) in the Polish Carpathians.

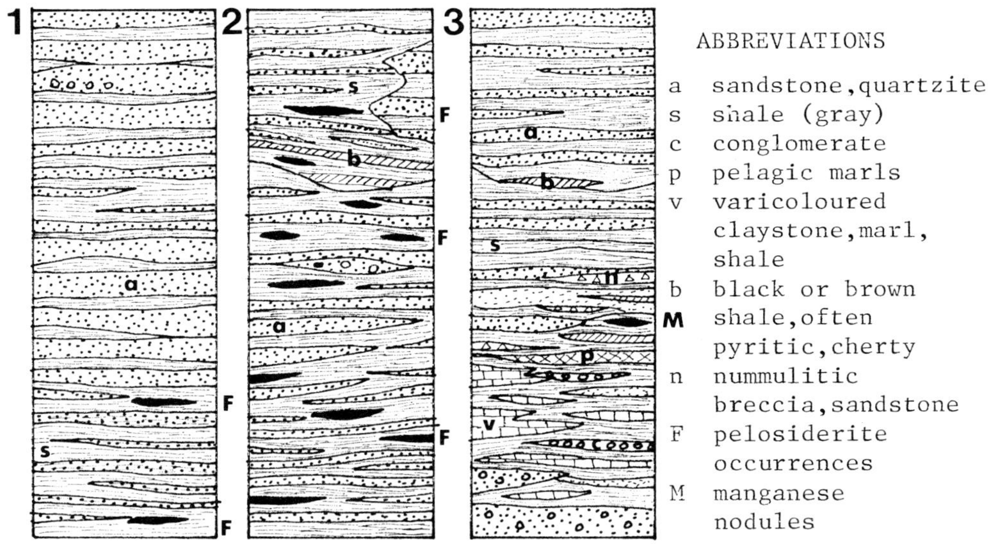

Fig. 16-5. Carpathian flysch, eastern Slovakia. Three schematic stratigraphic sections to illustrate the usual lithology and setting of pelosiderites (F) and Mn nodules (M). Each section represents about 1.2 km of strata. (1) Dukla Nappe, northern part (Cr_3-Eo_2); Dukla Nappe, Smilno facies (Cr_3-Eo_3); (3) Magura nappe, Poprad R.valley (Pc-Eo_3). Modified after Stráník, Hanzlíková and Nemčok in Mahel' and Buday, eds. (1968).

Phosphates and iron-phosphate sediments occur at several localities in the eastern branch of the Canadian Cordillera (Christie, 1979). In the northern Richardson Mountains, Yukon, a middle Cretaceous "flyschoid" sequence contains thin beds of ironstone separated by ferruginous shale. The ironstone is a medium-grained to pebbly wackestone and packstone, containing compact microcrystalline siderite, pyrite and collophanite. The average grade is 33% Fe_2O_3, 5% MnO and up to 33% P_2O_5. Black pelletal and nodular phosphate form the basal 1-2 m of the Jurassic Fernie Group in the Rocky Mountains of British Columbia.

In the Oligocene-Miocene Numidian flysch of north-central Sicily (Alaimo et al., 1970), oxide ferromanganese nodules occur in a gray-brown clay interbedded with shales and sandstones. The host sequence has been interpreted as a detrital wedge deposited at the foot of the African Platform.

Stratiform bedded manganese deposits and occurrences are known in several localities hosted by terrigenous flysch. In the Kyšovce-Švábovce field, eastern Slovakia (Ilavský, 1976;

900 Tt Mn/18%), about 1/2 m thick horizon of finely crystalline Mn carbonate (oligonite) interbedded with Mn oxides and shale, is a component of Eocene autochthonous shale-sandstone flysch. The Mn source has been traced into Permian continental basalts in the basement and the Mn horizon is clearly a shallow marine sediment locally transgressive over coal. A flysch, then, is not always of deep-sea origin. The Mn-bearing suite at Kyšovce is more likely to be a "molasse" than a flysch. This illustrates the transitionality and uncertainty in usage of the term flysch; many of the mineralizations treated in Chapters 7, 17 and 19 have also been described under the heading "flysch" in the literature.

Typically, a flysch is poorly sorted, so that detrital heavy mineral accumulations (paleoplacers) are not to be expected. They could, however, be present in associations transitional into the flysch (e.g. in deltaic sequences), that are often treated with a flysch in summary regional descriptions. Table 16-3 gives brief characteristics of selected terrigenous flysch associations of the world.

16.7. GOLD-BEARING "SLATE BELTS" of "FLYSCHOID" CHARACTER

The extensive lower Paleozoic Meguma Group of Nova Scotia is a 10 km thick sequence of gray argillites to slates alternating with quartz-rich beds (quartzite to endurated sublitharenite). The sequence is, in general, fining upwards and quartzites are dominant in the lower half (Goldenville Formation), slates in the upper half (Halifax Formation). The Meguma sediments have been interpreted recently as abyssal plain turbidites and continental rise contourites, derived from a metasedimentary and plutonic source area (Harris, 1975). The Meguma Group rocks were folded into a set of asymmetric to isoclinal folds in Devonian time and intruded by granodiorite to quartz monzonite plutons. The metamorphic grade varies from sub-greenschist to almandine-amphibolite facies, the latter being confined to the vicinity of the intrusions. Several hundred occurrences and small quartz-gold deposits are scattered within the Meguma Group, thus making it a distinct gold provinces, although the total gold production to-date (about 35 t) is not particularly encouraging. Although all the gold occurrences are hosted by the Meguma meta-sediments and the majority of them occur near the contact of the two Formations, the orebodies are clearly epigenetic veins composed of quartz with variable amounts of pyrrhotite, arsenopyrite with a local scheelite or stibnite and are not stratiform.

Most of the veins are located in the greenschist-metamorphosed sediments, generally along the crests of anticlines, in areas of intensive axial cleavage. Saddle reefs subparallel with the fold surfaces as in the Goldenville field (Fig. 16-6) are the most famous, but schistosity-concordant and cross-cutting veins were more common. Cameron and Hood (1975) demonstrated that most veins lie in the vicinity of lithologic (slate/sandstone) and intrusive contacts;

Table 16-3. Terrigenous flysch association : selected example areas

AREA / UNIT	AGE	GEOLOGY, MINERALIZATION	REFERENCES
The Alps, flysch lithofacies	Cr_3-Ol_1	synorogenic alternation of quartz - rich and carbonate rich arenite and shale, exposed in several south-dipping thrust sheets, itself overriden from the south by the Mesozoic limestone nappes; local wildflysch facies = boulder beds. Remarkably free of mineralization	Trümpy (1960)
Carpathian Flysch (Czechoslovakia, Rumania, Poland, USSR)	Cr_3-Ol_1	marine, synorogenic alteration of quartz-rich litharenite and shale, grades downward into shale-dominant facies (with minor mafic volcanics), and upward into shallow marine to continental coal association; locally carbonate rich, incl. limestone conglomerate. Small, low-grade ironstone (mostly siderite), Mn oxide and carbonate, bedded and nodular, deposits	Mahel' and Buday eds. (1968) Ianovici and Borcoş (1982)
Northern Rocky Mts. Foreland Basins Canada	Cr_3	up to 4 km thick clastic sedimentary wedge; non-marine alternating sandst., shale, congl. in the west, grade into shallow marine shale in the east. Synorogenic deposits derived from the rising Rocky Mountains; locally sideritic or pyritic shale, no economic ore deposits	Douglas, ed. (1970)
N.W. Appalachian Front, E. Canada	Cm_1-Or_1	3-6 km thick wedge of interbedded sandstone, shale, conglomerate; a synorogenic deposit derived from rising basement cordilleras and blocks of the Canadian Shield; minor siderite occurrences	as above
S.W. Thailand-Northern Malaya	D_2-P_1	thick synorogenic clastic wedge of litharenite, feldspathic litharenite, shale, derived from crystalline basement in the west; host to tin granites	Hamilton (1979)

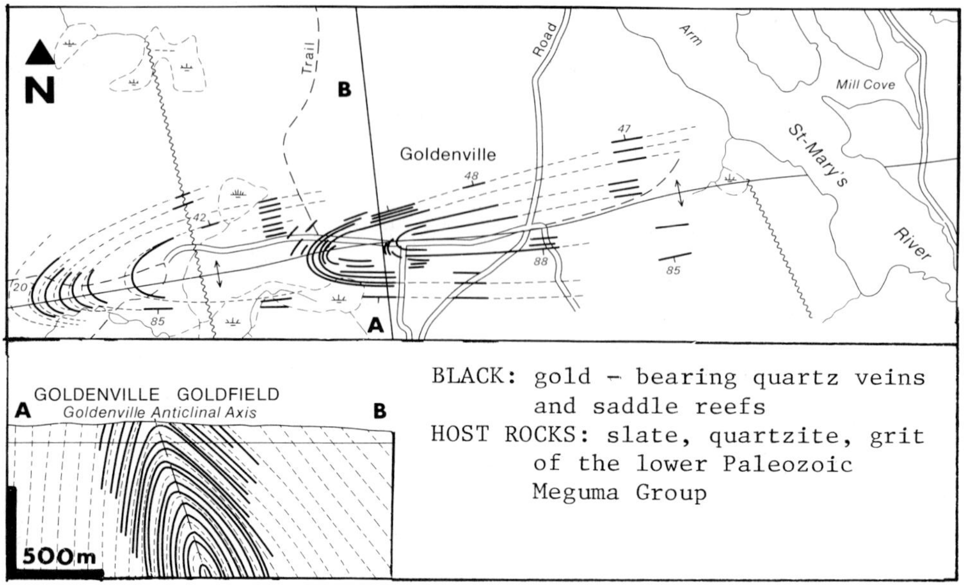

Fig. 16-6. Goldenville Goldfield, Nova Scotia, a system of bedding-conformable gold-quartz "reefs" in slates and quartzites along an anticlinal axis. From Boyle (1979), based on data in Malcolm and Faribault (1929). Courtesy of the Geological Survey of Canada.

that pyrrhotite together with arsenopyrite are the sulphides most frequently associated with gold (in contrast with the locally abundant diagenetic pyrite in the slates) and that the ores are epigenetic (Devonian) although not necessarily a direct product of dilation infilling by hydrothermal fluids emanating from the granites (granite apophyses sometimes cut earlier veins).

Thorpe and Thomas (1976) found generally low trace gold contents in the Meguma sediments (of the order of 0.X ppb Au), although several lenses of pyrite and arsenopyrite-bearing slate in the Mt. Uniacke field ran as high as 0.28 ppm Au. The question of origin of the Meguma gold veins remains open, although a two-stage origin is most commonly proposed. This comprises (1) sedimentogenic or diagenetic enrichment of gold with pyrite and arsenopyrite in organic-rich fine detrital sediments and (2) hydrothermal remobilization of this gold (as well as the more dispersed gold) during deformation / metamorphism / granitic emplacement. The currently economic deposits thus form at sites of maximum coincidence of the controlling factors (that is, in tightly folded, cleaved zones superimposed on Au-As enriched bands near intrusive contacts) but there is a possibility that the anomalously gold-enriched slate bands might become an ore of the future.

The west-central Victoria Goldfields province, Australia (Bowen and Whiting, 1975), is a cleaved and low-grade metamorphosed flysch contemporary with the Meguma Group (Cambrian and lower Ordovician), intruded by Ordovician to Devonian granodiorite and quartz monzonite plutons. The overall structural character and mineralization style is also remarkably similar to that of the Nova Scotia goldfields, except that the Victoria "graywackes" appear to have a lower quartz content, higher trace gold content (5 ppb Au before cleavage development; Glasson and Keays, 1978) and may be partly volcanogenic. The gold production has also been substantially higher: about 2,000 t Au, substantial portion of which came from placers. The structural grain of the Province runs north-south, and strings of goldfields are aligned in this direction along anticlinal axes. Much of the Ordovician bedrock is masked by Tertiary flood basalts.

The gold-bearing veins are mineralogically simple, composed of white to gray massive or ribboned quartz diluted with common relics of country rocks and containing minor pyrite and arsenopyrite, chlorite, ankerite, etc. The gold is fine, or bound in the sulphides. The veins are found as saddle reefs along the crests of tight, cleaved anticlines, as bedding lodes, and as fracture-filling veins. In the Bendigo field (239 t Au produced from lodes only), saddle reefs occur along thirteen adjacent parallel anticlines. As in Nova Scotia, the exact origin of the veins and their relation to the granites is unclear, although the same genetic scenario probably applies. In contrast to the Meguma Group-hosted gold deposits, several ore-grade conformable slate horizons containing finely disseminated gold have been recorded in Victoria (e.g. Amherst, Chiltern). The difficulties associated with finding and recovering the fine, non-panneable gold, obviously discouraged the early miners from a more intensive follow-up.

A slaty flysch lithofacies, granites, and Victoria-style gold-quartz lodes are also represented in the Mathinna Belt, north-eastern Tasmania. The 1,000 km long, 75 km wide Yana-Kolyma gold belt in eastern Siberia (Firsov, 1966) is of the "Victoria type" and comparable or even better in terms of gold production. There, thick upper Triassic and Jurassic flysch unit envelopes the Kolyma median massif along its western side. The deformation and metamorphism are comparable with those described above. Intrusive activity is represented by Triassic diabase and diorite dikes and by late Jurassic-late Cretaceous granodiorite. Several hundred bedrock gold occurrences comprise mineralogically simple quartz-filled fissures, bedding and saddle reefs (the latter in the Rodionovsky field); zones of parallel quartz stringers several kilometres long (Natalkinskiy); brecciated zones cemented by gold quartz in the cleaved metasediments; and similar. Gold-bearing veins and stockworks in diorite dikes (Utinskiy) and granitoids are rare but some apical granitic intrusions are associated with endocontact and proximal exocontact Sn, W and Mo orebodies. Most of the gold, so far, has come from placers and a significant proportion of the placer gold derived from regionally widespread gold-bearing quartz stringers in the metasediments, uneconomic in themselves.

Gold mineralization in slates, sharing many features with the present category and transitional into it, occurs in many other lithologic associations, treated in Chapters 10, 17 and 19. Table 16-4 summarizes briefly the auriferous flyschoid slate belts.

16.8. TERRIGENOUS FLYSCH IN PLUTONIC TIN PROVINCES

Many tinfields of the world recovering primary or secondary (weathering and/or sedimentation reworked) cassiterite, coeval with granitic intrusions, are underlain by terrigenous flysch. In the Herberton field in Queensland (Blake, 1972), the Silurian-Devonian Hodgkinson Formation hosts most of the cassiterite orebodies in the granite exocontact. The Formation consists of alternating thin-bedded sub-litharenite to sub-arkose, siltstone, thick-bedded sandstone, shale, and minor lenses of conglomerate, limestone, chert and basalt. Graded bedding is common and the formation is a proximal turbidite, derived from (and partly overlying) Precambrian high-grade metamorphics and muscovite-albite granite gneiss.

Middle Devonian-lower Permian quartz-rich flysch is the host rock for the tin granites of south-western Thailand and northern Malaya (Hamilton, 1979) and all the gold-bearing flyschoid "slate belts" described in the previous section also contain tin deposits coeval with granites. The Devonian-Carboniferous slates and sandstones (kilas) hosting many tin veins in Cornwall are sometimes interpreted as flysch ("distal flysch"); elsewhere, tin-hosting sediments are alternating slates and quartzites lacking turbidite characteristics (e.g. in the Paleozoic of Bolivia).

The flysch / granite / tin correlation is too systematic to be a mere coincidence, but anomalous tin contents in the flysch sediments have not, so far, been demonstrated. It seems more likely that both the granites and the flysch sediments had the same source in the basement and adjacent metamorphics.

Table 16-4. Gold - bearing "slate belts" of original flysch character; selected examples

LOCALITY	AGE	GEOLOGY	MINERALIZATION	REFERENCES
Nova Scotia (Meguma Group host) Canada	PZ_1	10 km thick sequence of interbedded slate and qtz rich sublitharenite, intruded by Devonian granodiorite. Greensch.to amphibolite m.	structurally controlled qtz, ars,py,lesser scheelite veins, saddle reefs, miner.shears, placers; predating the "granite"	Cameron and Hood (1975)
Yana-Kolyma gold belt, E.Siberia, USSR	Tr_3-J_3	1000 x 75 km N.W.-S.E. trending belt along western edge of Kolyma median massif. Altern. slate,grayw.,intr. by Tr gabbro,dior.,diabase dikes and stocks; J_2-Cr_3 granodiorites	several hundred Au (and some W,Sn,Mo) occurrences; quartz fissure and shear veins, some saddle reefs. Estim. 1,800 t Au, most from placers	Firsov (1966)
Victoria Goldfields, Australia (e.g. Bendigo, Ballarat)	Or_1	12 km thick folded and cleaved sequence of graded graywacke interb.with lustrous pyritic slate; intruded by Or-D granitic rocks	gold-quartz veins with minor pyrite and arsenopyrite, forming saddle reefs, bedding lodes, crosscutting veins; important placers; 2,000 t Au	Bowen and Whiting (1975)
Mathinna slate belt, N.E. Tasmania, Australia	Or_1-D_1	alternating slate - quartz wacke, siltstone, mudstone; cleaved, intr. by Devonian granite	shear, fissure gold,quartz veins with pyrite,arsenopyrite; placers; 8.3 t Au	Williams (1978)

Plate 8. Late Proterozoic slate terrain near Rabštejn nad Střelou, western Bohemia, Czechoslovakia.

"Geologists of the past worked seaward until the exposures were covered by water; then they turned landward again... Marine geologists...developed indirect techniques which allowed them to work succesfully in the deep oceans...but when the study of shallower-water areas became unconfortably complex, they turned seaward again. There has thus always existed a "belt of ignorance"...marking the transition from the exposed continents to the deep oceans".
C. L. Drake and C. A. Burk (1974, p.3).

CHAPTER 17

Deeper Marine Argillite (Slate, Schist) - Lesser Chert, Carbonate, Arenite Association ("Black" Sediment Emphasis)

17.1. GENERAL

This association, an ancient product of deposition in the modern "belt of ignorance" of Drake and Burk (1974), includes a fairly widespread portion of the "geosynclinal" sediments free of nearshore influences. Many are transitional into, or associated with, turbidites. In such a case, most are stratigraphically below the turbidite, although some are above. Many "slate", etc. associations contain some accessory volcanics, and are gradational into the volcanics-dominated sequences described in Chapters 10 to 14. The visitor's impression of these terrains is "slate, slate, slate", occasionally spiced with some of the associated rocks such as bedded chert; thinly-bedded, gray to black limestone; litharenite; intraformational conglomerate or breccia, or a volcanic rock. The bulk of the slates are monotonous, gray, greenish and sometimes purple. Of particular interest to a prospector are the carbon-rich "black" equivalents of all the above rocks. Their colour ranges from gray to deep sooty black. The extent of the black facies ranges from single beds to units several thousand meters thick. The "slate", etc. association has a mixed provenance; usually the greatest proportion of the fine detrital constituents is terrigenous, derived from land. The remainder is volcanogenic (usually "distal"), and pelagic. Much of the slates would be called hemipelagic if found in the modern environments.

The interpretation of depositional environments of the "slate" sequence has changed considerably in the past 20 years. In the earlier geosynclinal model, these sequences were usually placed either in the nonvolcanic "eugeosynclinal" or "miogeosynclinal" troughs. At present, their place of deposition is sought mostly on continental slopes (of both Atlantic and Pacific continental margins) and on deep basin floors. Because of the inherent problems of preservation of sediments deposited in a vast and subduction-prone oceanic basin, the deep basins depositing the "slate" precursors are believed to have been restricted and fault-bound. The present marginal seas (e.g. Churkin, 1974, interpretation of the Paleozoic graptolite shale and chert belts in the American Cordillera); the present Black Sea; rift seas such as the Red Sea; fault-bound tectonic grabens formed within an earlier shelf (Krebs, 1976); subsided shelf (Tempelman-Kluit et al., 1976, for black shales in the Pelly Mts., Yukon) and other models have been proposed. Particularly common is the observation that the "black shale" basins formed on, or within, an earlier stabilized crust.

Although I have been using the words "slate belts" for brevity, a whole range of metamorphic intensities is represented, ranging from argillites to schists. Slates and argillites, however, denote the "prime facies" in which the most extensively studied mineralized belts have developed.

17.2. TRACE METAL GEOCHEMISTRY AND METALLOGENY

The "geosynclinal" argillite, slate,etc. sequences are an important host to metallic orebodies (as well as barite, phosphates, etc.), and to anomalous geochemical accumulations that are not yet ores at present but may become so in the future. Approximately 110 Mt Zn, 60 Mt Pb, 2 Mt Cu, 30 Tt Ag, 1 Bt Fe, 90 Mt Mn, and 1 Bt barite are accumulated in currently economic stratiform and stratabound deposits, hosted by and intimately affiliated with, this lithologic association. An additional 2 Bt fluorine, 15 Mn V, 8 Mt U, and many other trace metals are accumulated in supra-clarke quantities in phosphorites and black "slates", and could possibly be recovered as a by-product.

A further 12 Mt Zn, 8 Mt Pb, 45 Tt Ag, 4 Mt Sb, 500 Tt Hg, minimum 1 Mt W, 100 Tt U, 850 t Au, and some other metals, are found in epigenetic bodies hosted by the (meta) sediments but cogenetic with the superimposed magmatic events. The metal source relation to the host (meta) sediments, however, is likely (Fig. 17-1; Table 17-1).

Because of the virtual absence of physically accumulated ores (with the exception of the intraformationally reworked ores), all mineralizations are chemical or partly biochemical precipitates derived from fluids of variable provenance, temperature and metal concentration. The affinity of many ore deposits for the "black" sediments reflects the favourability of the carbon-rich rocks generating environments for metal precipitation (because of the strongly reducing conditions and the usual availability of sulphur), but the reducing environment in itself does not supply the metals. Some metals concentrated in the black sediments probably precipitated from the seawater (U, V, Mo, Mn; hydrogenous source). Other metals precipitated from more or less heated solutions (brines), discharged on the seafloor. This is best demonstrated in areas of submarine volcanism contemporary with sedimentation. Because the metalliferous hydrotherms are emitted in the waning (or sometimes initial) stages of such volcanism and there is no correlation between the size of the volcanic pile and intensity of the hydrothermal systems, many heated brines reached the seafloor along faults located in sediments-depositing domains showing little or no traces of volcanism.

Orebodies, usually massive sulphides or iron ores proximal (usually directly above) to such a source of hydrotherms but distal in regard to the large volcanic centres, tend to show morphology and facies zoning identical with the volcanics-hosted massive sulphides (Chapters 10-14); that is, a "stratiform" concordant lens is underlain by an altered footwall feeder zone. The "distal" orebodies will lack any adjacent feeder.

Some alternatives to the volcanogenic derivation of seafloor brines have been proposed. Morrow et al. (1978) postulated a model of origin of some of the bedded barite deposits in the "miogeoclinal" shale and carbonate association of the eastern Cordillera, in which barite precipitated at sites of mixing of euxinic seawater with the Ba-enriched meteoric groundwater near the seaward edge of a large coastal aquifer. Krebs (1976; and oral communication, 1975) pointed out the structural position of the Rammelsberg deposit on the mobile flank between a strongly subsiding basin and a stationary-to-rising "structural high" that probably involved an active fault (growth fault), acting as a channelway to the hydrothermal solutions. The solutions, in turn, were contemporary with the metamorphic processes and evolution of a granodiorite pluton within the rise, and were either (1) immediately derived as hydrothermal solutions from the acid magma, or (2) generated by heating of the compaction waters from the metal-bearing marine sediments.

The postulated subfloor source of ore-forming brines controlled by faults that have, by now, largely dissapeared, accounts for the extremely low degree of geological predictability of the occurrence of the argillite, slate, etc. hosted ores. Coincidence of known volcanism, represented by submarine dikes, flows, etc.; tuffaceous interbeds; growth faulting; and euxinic environment, is helpful, when it can be demonstrated. Empirical data suggest that many "black" argillite/slate, etc. regions that failed to produce syn-depositional metallic ores, fared better later and ores related to post-depositional interaction were often formed.

The geochemistry of "black shales" has been treated in many publications (e.g. Davidson and Lakin, 1961; Tourtelot, 1970; Vine and Tourtelot, 1970), but the example localities considered are rarely separated according to the broader association or environment, of which they are members. Table 17-2 gives examples of some "geosynclinal black shale" members of this category, for which anomalous trace metal contents have been reported.

17.3. Fe ASPECT

Hydrothermal-sedimentary ("exhalative") stratiform iron ore bodies typically form along contacts of submarine volcanics (e.g. spilites,

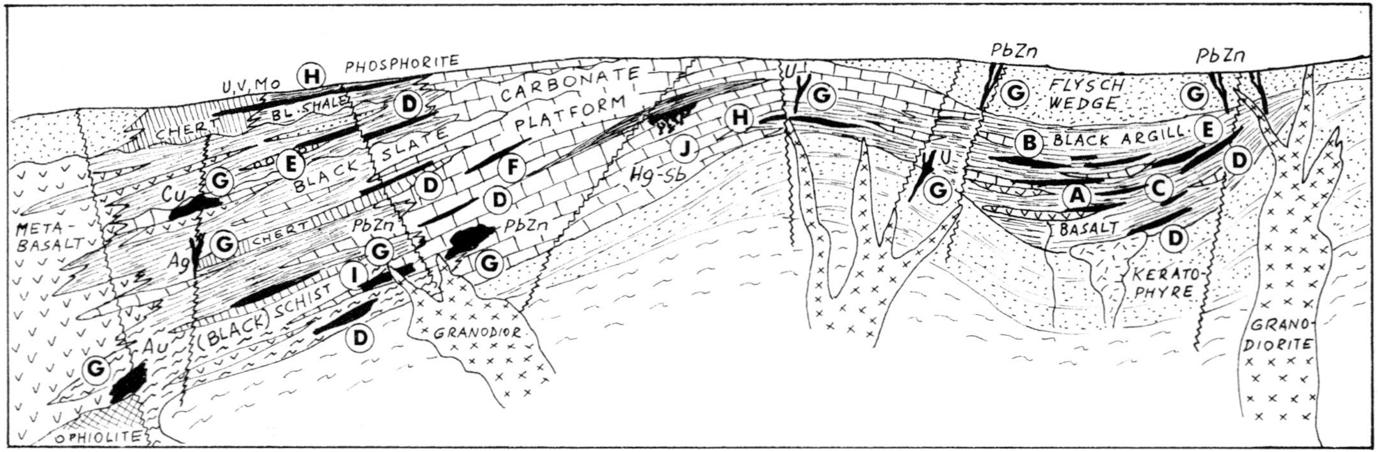

Fig. 17-1. Principal mineralization styles in the deeper marine argillite (slate), lesser chert, carbonate, arenite association (see Table 17-1 for explanation of the letter codes).

Table 17-1. Deeper marine argillite (slate), lesser chert, arenite, carbonate association: principal mineralization styles

(A)	stratabound Fe (Mn) hematite, siderite deposits in argillite, chert, carbonate (e.g. Vareš); X00 localities, 1 Bt Fe
(B)	stratiform pyrite in black slate; X,000 localities, X Bt pyrite
(C)	stratiform Mn carbonates, oxides, silicates in black slate; X0 localities, 90 Mt Mn
(D)	massive Zn,Pb (Cu) sulphide and/or barite concordant lenses, in black slates; X00 localities, 110 Mt Zn; 60 Mt Pb; 2 Mt Cu; 30 Tt Ag
(E)	bedded barite in setting as above; X00 localities, min. 1 Bt barite
(F)	Zn-Pb deposits in transitional black sediments / limestone platform association; X00 localities
(G)	discordant vein and replacement metalliferous bodies in the "black" association, in intrusive aureoles: W,Sb,Hg (cinnabar,stibnite,ferberite,hubnerite,scheelite) association; X0 localities; Pb-Zn sulphides; X00 localities, min. 12 Mt Zn, 8 Mt Pb, 10 Tt Ag silver minerals; X localities, 35 Tt Ag; gold mineralization; X localities, 850 t Au; uraninite bodies; X localities, 100 Tt U
(H)	metalliferous phosphorite and associated black argillite,chert; X0 localities; 2 Bt F (in fluorite); 8 Mt U (low grade resource) 15 Mt V (low grade resource); Mo,Ag,Au, etc.
(I)	scheelite "skarns"; X00 localities
(J)	Hg,Sb, epigenetic replacements, impregnations, in carbonate, black shale, minor mafics / ultramafics faulted and thrusted association; X0 localities, 500 Tt Hg, 4 Mt Sb

keratophyres; Chapters 10-12) and marine sediments that correspond to the presently considerd lithologic association. Orebodies originally formed at some distance from a conspicuous volcanic pile may have their volcanic affiliation obscured at the present level of exposure and this is further complicated by deformation, erosion, etc. The Vareš iron ore district north-west of Sarajevo, Yugoslavia (Janković, 1982; min. 260 Mt Fe, 10 Mt Mn; Fig. 17-2), is an example. There, a series of hematite and siderite stratabound orebodies interpreted as "exhalative" occurs in an allochthonous middle-Triassic, marly pink limestone, dolomite, chert and black slate unit also containing tuffaceous interbeds and rare occurrences of subaqueous andesite flows. This unit has been thrusted over a Liassic limestone and shale unit, and is, in turn, tectonically covered by a lower Triassic sandstone sheet. The iron orebodies form thick, tectonically dismembered lenses showing a distinct facies-zoning (from hangingwall to footwall): (1) red siliceous slate to jasper; (2) siliceous to

Table 17-2. Trace metal contents in selected "geosynclinal" black shales and slates (in ppm)

LOCALITY	AGE	Zn	Pb	Cu	Ni	V	Mo	other metals	REFERENCES
"Vanadiferous Shale" Phosphoria Formation Montana, Idaho	Pe	u10000	u50	u300	u1500	u7000	u500	u30 Ag; u675 Se; u19 Ga	Davidson and Lakin (1961)
Comus Fm., near Golconda, Nevada	Or	u3000	u30	u500	u700	u3000	u10	u30 Ag; u300 Cr	as above
Paleozoic shale near Eureka, Nevada	PZ	u5000			u100	u2000	u50	u2 Ag	as above
Shale Member, Deseret Limestone, Mercur, Utah	Ms	uX000			uX00	uX000	uX0	X0 Ag; X00 Cr	as above
Chainman Shale, Nevada, E. California	Ms	u2000		u200		u1000	u50	u20 Ag; u1000 Cr	as above
Western Facies "shales", Nevada	Or-S	a300				a200		a0.5 Ag; 15 Co; 15 Ga	Vine and Tourtelot (1970)
Ouachita Mts. (Arkansas) "shales"	D-Ps	200				a200		a100 Cr; a50 Ga	as above

ABBREVIATIONS: u = up to; a = average

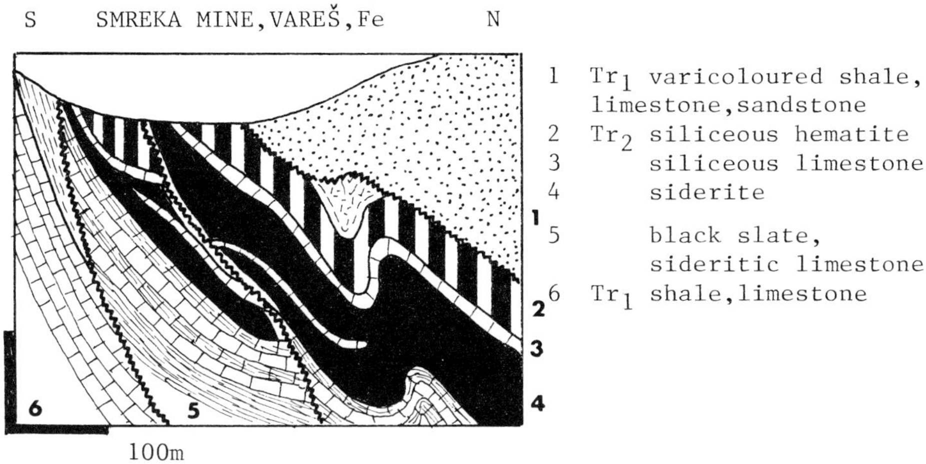

Fig. 17-2. Vareš Fe district, Bosnia, Yugoslavia, an "exhalative" middle Triassic mineralization. Modified after Čičić (1980).

massive, fine-grained specular hematite; (3) reddish-brown calcareous shale to nodular pelagic limestone; (4) fine crystalline, gray manganoan siderite, interbanded with black slate; and (5) limestone with slate interbeds. Facially, the ore is reminescent of both the Lahn-Dill and Teliuc-Ghelar "types" of volcanic-sedimentary deposits, described earlier (Chapters 10, 11).

Cissarz (1957) argued convincingly in favour of the "exhalative" origin of the Vareš siderite, in his period summarily interpreted as a postmagmatic-plutonic replacement of limestones or dolomites. Demonstrably, siderite or ankerite-replaced Ca or Mg carbonate bodies and also siderite veins, however, are widespread in the argillite/slate association as well. Numerous such deposits and occurrences are widespread in the Cambro-Silurian, Devonian and Permo-Carboniferous slate units in the Gemeride zone of Slovakia (e.g. Železník; Ilavský, 1976), transitional into volcanic units and already treated earlier.

Pyrite is widespread in the "black slate" belts, in quantities ranging from the omnipresent scattered single pyrite grains, to tens of kilometres long stratiform horizons of "pyritic shales". In the pyritic shales, pyrite usually forms thin lenses parallel with bedding. These frequently coalesce to form thin laminae or bands of pyrite alternating with the barren argillite/slate. In the formerly mined Chvaletice deposit, Czechoslovakia (described later) late Proterozoic slates with 10-80% pyrite (500 Mt ore with about 30% pyrite), have been traced for 18 km. Similar, but leaner deposits of pyritic argillites/slates of late Proterozoic age were mined in the past for the production of sulphuric acid in the Radnice region (e.g. Hromnice), western Czechoslovakia. In both cases, the sediments are distal members of a mafic (spilite, basalt, gabbro) volcanic-sedimentary association.

17.4. Mn ASPECT

Manganese deposits are often cogenetic and gradational into the iron and/or pyrite deposits described above. The Vareš siderites have a high Mn content (3.5% Mn), and in Chvaletice, pyritic slate and Mn carbonate form separate horizons mined from a single pit.

The Chvaletice Mn deposit (Gruzczyk and Pouba, 1968; Fig. 17-3) is hosted by a unit of zeolite- to greenschist— metamorphosed late Proterozoic graphitic slate, interbedded with sandy slates to slaty meta-graywackes. Volcanic and subvolcanic members corresponding to the bimodal spilite-keratophyre association occur locally. A stratiform, up to 30 m thick body (lens to sheet) of Fe-Mn carbonates is located stratigraphically below the horizon of strongly pyritic slates. The Mn horizon is an inconspicuous light-to-dark gray finely crystalline rock gradational into the graphitic shale composed of oligonite and pyrite, and virtually megascopically unrecognizable except where oxidized at outcrop, or where contact metamorphosed or remobilized along a contact with a granodiorite massif or along fractures. In the thermally metamorphosed sections a hard hornfels formed, composed of pyrrhotite and Mn silicates (rhodonite, tephroite, spessartite, dannemorite, quartz, etc.). Along fractures, mineralogically interesting veinlets of quartz, pyrite, green apatite, kutnahorite, cronstedtite, etc. are widespread. The Mn tonnage contained in the 12-18% Mn ore is probably of the order of 5 to 10 Mt.

The Usa (Usinskoe) Mn deposit in the Kuznetsk Alatau range, Siberia (Varentsov and Rakhmanov, 1974; 20.25 Mt Mn/10-30%), is associated with a zone of lower Cambrian metacarbonates and black slates, underlain by mafic metavolcanics and gradational into pyritic slates, meta-andesites and meta-volcaniclastics. This makes it similar in many respects to the Mn deposits in eastern Cuba, described in Chapter 13. In the ore zone, stratabound lenses of finely bedded gray calcian rhodochrosite, oligonite, Mn stilpnomelane with accessory pyrite and pyrrhotite, occur in the N.N.W.-striking horizons, traceable for 4.6 km. A supergene enriched pyrolusite and psilomelane ore marks the present outcrop.

The extensive but low-grade iron and manganese bedded deposits hosted by Silurian slates in the Aroostook (Maine, U.S.A.) and Woodstock (New Brunswick, Canada) areas, are the largest continental Mn accumulation in North America. In the Aroostook County (Pavlides, 1962; Fig. 17-3), the Fe-Mn deposits lie in a discontinuous belt about 110 km long. The host unit is dominated by a monotonous gray-green and green shale, that contains several lenticular units of felsic and mafic (bimodal) volcanic members (mostly devitrified ash

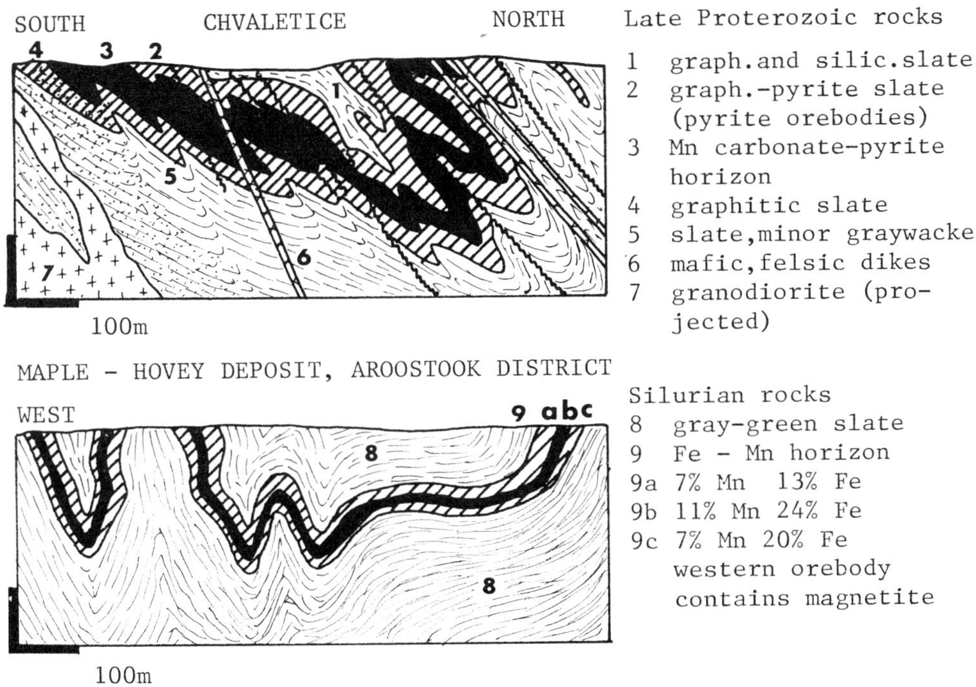

Fig. 17-3. Mn-Fe orebodies in slate association, Chvaletice (Czechoslovakia) and Aroostook (Maine, U.S.A.). Modified after Gruzczyk and Pouba (1968) and Pavlides (1962).

beds). The chemically precipitated orebodies form thin, finely laminated layers, interbanded with the slate and are of oxide and carbonate facies.

In the oxide facies, hematitic shale grades into banded hematitic ironstone composed of crystalline hematite. Braunite and rhodochrosite are present in lesser amounts and spessartite and bementite are still rarer. In the carbonate facies, fine rhodochrositic carbonate is the chief Mn mineral, associated with iron chlorite and stilpnomelane. Magnetite may be but need not be, present in both facies. The ore horizons are folded, faulted and cleaved, and cut by numerous epigenetic veinlets containing a suite of remobilized minerals. The oxide facies deposits are much more widespread; on Maple and Hovey Mountains, they contain 354 Mt of 4% Mn and 17% Fe ore, compared with the 3.3 Mt of the carbonate facies ore. The total known reserves in the district are about 25.2 Mt Mn/9% and 43 Mt Fe/17%.

In the 48 km long S.W.-N.E. trending Woodstock belt in Canada (north of Aroostook) five stratiform units of siliceous hematite, braunite, lesser rhodochrosite, pyrite and piedmontite in Silurian thin-bedded siliceous slate and chert, contain about 19.26 Mt Mn/9% and 27.82 Mt Fe/13% (Gross, 1967).

Bedded manganese deposits in the deeper-water sedimentary belts have a worldwide distribution and are gradational into (1) jasper-hosted siliceous deposits typical of ophiolitic and marine volcanic terrains (Chapters 3, 10-14), on one side, and into Mn-mineralized carbonates, on the other. The transition (1) includes, for example, the braunite and secondary Mn oxide deposits in the Ordovician-Silurian Curtis Island Group near Gladstone, Queensland (Murray, 1975); the Čevljanovići field near Sarajevo, Yugoslavia (about 100 Tt Mn/37%; Janković, 1982); and others.

Some of the upper Devonian rhodonite and bustamite deposits in the chert and jasper belts in the Magnitogorsk Synclinorium in the Urals (Rundkvist, ed., 1978); some of the rhodochrosite, oligonite, rhodonite, dannemorite, spessartite and Mn oxide deposits associated with the lower Cambrian epizonally metamorphosed black graphitic quartzites (metacherts) and schists in the Vatra Dornei district, Rumania (e.g. Iacobeni, Ciocanesti, and other deposits; min. 800 Tt Mn/29% Mn; Wagner, 1977), also belong into this category

Transition (2) is represented by the Takhta-Karacha field south of Samarkand, Uzbekistan (Buroš et al., 1977) hosted by an upper Silurian chert-limestone association. There, several 6-8 m thick stratiform lenses of rhodochrosite, Mn calcite and siderite are gradational into a dark-gray and black limestone with accessory rhodochrosite and manganocalcite. Vernadite and psilomelane are present in the oxidation zone. The Mn carbonate horizon near Molango, Mexico "looking like most somewhat organic limestones that could be easily overlooked" (Dorr et al., 1973), probably belongs in this association and so do the slightly manganoan marbles in the Appalachian belt of Georgia (Cartersville) responsible for the locally developed pockets of residual Mn oxides.

17.5. Zn-Pb(Cu) ASPECT: DOMINANTLY STRATABOUND MASSIVE SULPHIDE AND BARITE DEPOSITS

HARZ AND RHEINISCHE SCHIEFERGEBIRGE, WEST GERMANY

The lower Devonian shelf sequence in the German Rheno-Hercynian facies zone was tectonically subdivided in the middle Devonian age into several subsiding basins (Becken) and submarine structural highs (Schwellen) in which two contrasting facies of marine sediments formed (von Bubnoff, 1956). The subsidence was accompanied by local mafic or bimodal volcanism. In the Goslar Trough (Harz Mts.) a unit of greenish-gray and black argillites with interbedded siltstone, sandstone and several tuff horizons (Wissenbach Schiefer) formed, with a thickness of over 1,500 m. The large, well-known, massive Zn,Pb,Cu sulphide deposit Rammelsberg located on the outskirts of Goslar, formed. The deposit is situated in the transitional zone between the Goslar Trough and Rise, and is perhaps related to the boundary growth fault (Krebs, 1976).

In the Rammelsberg mine (Kraume et al., 1955; Fig. 17-4), two separate massive sulphide lenses and an intervening barite deposit ("Grauerz") are conformable with a north-east – striking, overturned sequence of middle Devonian dark argillites. A second barite lens occurs in the stratigraphic footwall (but apparent hangingwall). The 10-15 m thick massive sulphide lenses (thickened to 50 m in the synformal axial zone) are zoned. Barite-rich Pb-Zn ore near the top grades into a banded and massive pyrite, sphalerite, galena and chalcopyrite ore and this, in turn, is enveloped in massive to disseminated pyrite.

Fracture-filling and disseminated pyrite-chalcopyrite mineralization (Kniest) hosted by silicified and carbonatized slate, occurs in the footwall near the nose of the syncline. Rammelsberg, now practically exhausted, has been a large and rich deposit (22 Mt ore with 18% Zn, 8% Pb, 1.3% Cu, 22-25% barite) mined for several centuries, and frequently used as a type locality for the "geosynclinal" sediment-hosted massive sulphides.

The other West German massive sulphide deposit Meggen (Ehrenberg et al., 1954; Krebs, 1972) is in the Sauerland part of the Rheinische Schiefergebirge. The orebody is a 3.8 km long, S.W.-N.E. trending stratiform sheet 6-8 m thick, composed of a pyrite, sphalerite, barite, minor galena central zone enveloped by a dark massive barite. The sulphide ore is very fine-grained and thinly laminated.

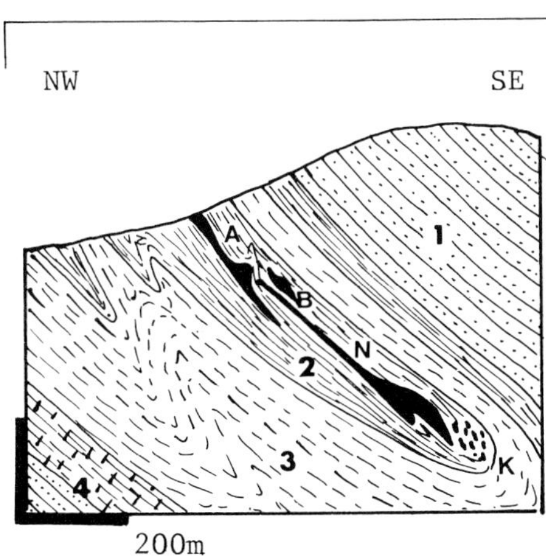

1 D_1 quartzite, arenite
2 D_2 black argillite to slate
3 D_2 argillite with sandstone bands
4 calcareous slate to argillite
A Alte Lager massive sulphide
B baryte orebody
N Neue Lager massive sulphide
K Kniest - stockwork to disseminated sulphides in altered slate

Fig. 17-4. Rammelsberg mine near Goslar, Germany. Massive Zn,Pb,Cu deposit in black slates. From LITHOTHEQUE, after Kraume et al. (1955).

There is no marked footwall alteration except for barite haloes extending for 300-500 m eastward (Krebs, 1976). The host rock is a middle Devonian argillite, deposited on the flanks of a subsiding basin and the ore horizon is overlain by a thin tuffaceous shale and pelagic limestone. Keratophyres known in the area south of Meggen and sometimes, in the literature, genetically associated with the sulphide mineralization are older, lower Devonian, and buried under at least 1,000 m of middle Devonian sediments (Krebs, 1976).

The Devonian metalliferous belt of West Germany extends, with interruptions, westward. Several small stratiform massive sulphide Zn,Pb,Cu deposits have recently been discovered in the Armorican Massif in north-western France (Porte-aux-Moines, Bodennec; Laville-Timsit and Wilhelm, 1979).

THE ALPS-MEDITERRANEAN REGION

Portions of the stratabound volcanogenic massive sulphide deposits of the Alps (e.g. Kitzbühel); Carpathians (e.g. Smolník, Burloaia, etc.) and the Dinarides already discussed in Chapters 10-14, occur in the metasedimentary "shale" facies. The metasediments are the dominant host rocks in the districts described briefly below.

In the Ordovician schists in the Grazer Paleozoikum, Austria (graphitic phyllite, quartz phyllite, limestone, subordinate greenstone; Siegl, 1974), massive sphalerite, pyrite and galena in siderite or barite gangue form thin, persistent concordant ore layers (e.g. Guggenbach deposit), or short lenses. Almost identical and contemporary (Ordovician) ores have been reported by Ilavský (1976) from Švedlár in the eastern Slovakia.

The "exhalative" middle Triassic iron ore deposits in the Vares district of Yugoslavia, have been described earlier. Some of the iron orebodies contain widespread accesory sphalerite, chalcopyrite and galena and one (Droškovać), contains a massive pyrite, sphalerite and galena zone along the iron ore footwall (Janković, 1982). Several strata-bound barite lenses, considered to have been deposited syngenetically with the enclosing Ladinian chert, slate and tuffaceous slate, occur in the same area (Borovica) and were compared with the Rammelsberg "type" by Cissarz (1957). The dark-gray banded barite ore at Borovica with dispersed galena, chalcopyrite and tetrahedrite, closely resembles the "Grauerz" from Rammelsberg.

In the axial zone of the French and partly Spanish Pyrenees, widespread Pb-Zn, barite and minor copper mineralization occurs at several stratigraphic levels, hosted by a great variety of Ordovician to Devonian "geosynclinal" sediments (Pouit et al., 1979). The Ordovician sequence consists essentially of low-grade metamorphosed pelites (quartz phyllites, black siltites), containing local calcareous phyllites ("calceschistes"), black and white limestones, and intraformational conglomerates. Submarine volcanic members of the bimodal suite occur locally. The metallic mineralizations are dominated by zinc with some lead and copper. They are confined to sites with a lithologic change or a rapid facies variation. In the Pierrefitte field Ordovician limestones and quartz phyllites are interbedded with several pillow basalt and rhyolite flow and tuff horizons. A variety of massive-to-disseminated, concordant and discordant sphalerite, galena, siderite and ankerite orebodies occur, that together represent about 2.5 Mt of 6.2% Zn and 1.7% Pb ore.

In the structural dome of Bosost in the French-Spanish border region (ore fields Sentein and Bosost), Ordovician quartz phyllites contain lenticular intercalations of rhyolitic conglomerate to sandstone, limestone and black meta-siltstone (Pouit et al., 1979). Numerous massive-to-disseminated strata-bound sphalerite, lesser pyrite lenses hosted mostly by the black siltstones occur at several levels. Some are substantially modified by dismemberment and thermal metamorphism in the granitic aureole. The Silurian of the Pyrenees is represented by rather monotonous black pyritic schists with rare marble lenses, lacking any significant mineralization. The Devonian has numerous intercalations of metabasalts near the base, topped by minor gray, banded limestones interbedded with gray or black siltstone and several tuff horizons. Several Zn,Pb,Ba occurrences have been recorded, Nerbiou (1.4 km long galena, sphalerite, magnetite stratiform horizon along a contact of black shale with the overlying dolomites) being the most promising.

IRAN

The Kuchke deposit in the Bafq district (Wright, 1964; min. 500 Tt Zn/10%, 300 Tt Pb/6%) is a lens up to 60 cm thick composed of extremely fine-grained (sooty) sphalerite, galena and pyrite mixture, hosted by a black argillite. The ? Devonian host sequence also includes gray shales, dolomite, sandstone, basic lavas, tuffs, and even a slice of serpentinite in the immediate footwall. The lithologic association is unusual and suggests a possibly allochthonous host unit.

SOVIET CENTRAL ASIA

The Koksu-Tekeli mineralized belt in S.E. Kazakhstan hosted by lower-middle Devonian black (meta)sediments is reminescent of the Harz Mts. in its facies distribution: a black fine detrital "trough" facies is found adjacent to a carbonate "rise" facies. As in the Harz region, late Devonian to Carboniferous granites are emplaced into the structural highs (Veits, 1972). Two belts of Zn-Pb (Cu) deposits follow closely the black shale/carbonate facies transition. The largest Tekeli deposit (Fig. 17-5) is hosted by vertical to steeply-dipping black siliceous or calcareous slates to phyllites, gray carbonates and minor arenites. The orebodies are stratiform lenses of thinly-bedded globular pyrite, containing only accessory quantities of sphalerite and galena. The industrial massive

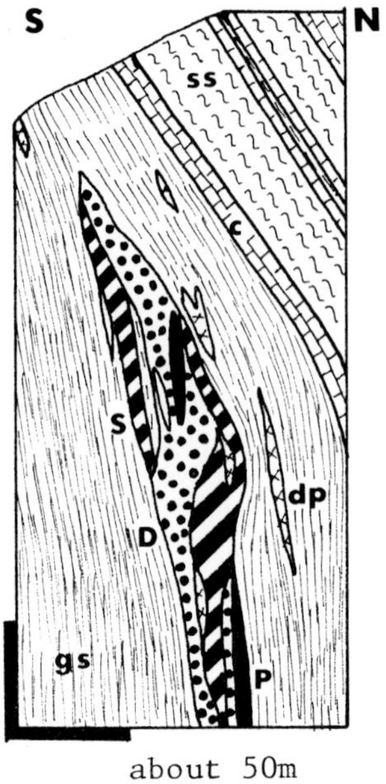

TEKELI Zn-Pb MINE,DZHUNGARIAN ALATAU

All rocks are Cambro-Ordovician (Tekeli Suite)

c carbonates

ss siliceous slate

gs graphitic slate (phyllite)

dp diorite porphyry

OREBODY :

P massive pyrite - polymetallic ore

S galena-sphalerite lenses in black slate

D disseminated ore in silicified slates

Fig. 17-5. Tekeli mine, Kazakhstan; a black slate-hosted massive Zn,Pb sulphide. After Veits (1972).

sphalerite and galena bodies are peneconcordant, controlled by bedding shears and confined to segments of silicified slates or recrystallized dolomites and limestones. They are interpreted as metasomatites superimposed on the stratiform pyrite. Several diorite and quartz-feldspar porphyry bodies present in the mine field are believed to indicate a buried intrusion considered to be the agent of hydrothermal metasomatism. The Tekeli deposit is thus polygenetic.

The Achik-Tash massive pyritic deposit in Kirghizia (Baibulatov, 1964) is hosted by black schists in contact with dolomitic marbles.

NORTH AMERICAN CORDILLERA

The "black sediments" or "graptolitic shale" facies is prominently developed in a series of discontinuous "basins" (that is, sedimentary sequences believed to have been deposited in a common sedimentary basin) of the Paleozoic age traceable, with interruptions, from Alaska to Mexico. Despite the numerous local variations, the "basins" possess (1) a comparable geotectonic setting; they are positioned along the western edge of a shallow-water carbonate platform, on the fringe of the "miogeosyncline", and east of the "eugeosynlinal" facies belt with widespread submarine volcanics; (2) comparable stratigraphy and facies and (3) comparable mineralization styles. Our understanding of the Cordilleran "black sediments facies" has advanced considerably in the past 20 years, largely as a consequence of numerous discoveries of stratabound zinc-lead deposits in the extremely well exposed (though poorly accessible) segment of the belt between Alaska and northern British Columbia. This is virtually free of granite and young volcanic interference. The known metal reserves in this segment alone now stand at between 50 and 180 Mt Zn and 30 and 70 Mt Pb. The Howard's Pass field, discovered in 1972, is now probably the world's largest zinc and possibly also lead accumulation.

Brooks Range, Alaska

The Brooks Range (Tailleur and Brosqe, 1974) is a major east-west trending orogen in northern Alaska, consisting of a core of greenschist-facies metamorphosed late Paleozoic "schist belt" with some metavolcanics in the south, changing facies in the northward direction into deeper water and subsequently shallow-water carbonates. Cretaceous granodiorites were emplaced into the core zone, which is flanked by a system of imbricate thrust sheets of progressively younger sediments tectonically transported in the northward direction over the northern continental margin. The south-western flank of the core zone north of Kobuk contains the extensive but enigmatic 700 km long "copper belt", culminating in the Ruby Creek deposit (Runnels, 1969; 1.2 Mt Cu/1.2%).

There, pyrite, chalcopyrite, bornite, tennantite-tetrahedrite, chalcocite and minor sphalerite are disseminated in middle Devonian dolomitized limestone, enveloped by gray phyllite and marked by numerous occurrences of metabasalt, mafic intrusions and gneissic granite in the vicinity. The mineralization appears to be largely stratigraphically controlled but the ore appears to have been epigenetically emplaced. There is some similarity to the Schwaz-Brixlegg field in Austria. The barium alumosilicate cymrite is an unusual but widespread and characteristic alteration mineral that constitutes an interesting link with the Zn, Pb and barite-dominated metallogeny of the "black detrital" sequence.

Stratabound Zn-Pb deposits associated with Devonian felsic volcanics in the Ambler district have already been mentioned in Chapter 11. Several large stratabound Zn, Pb and barite deposits were discovered recently in the Mississippian sequence developed along the northern flanks of the Brooks Range, in a setting formerly considered unfavourable (Nokleberg and Winkler, 1982). There, the host unit (Kagvik Sequence) is a typical "dark facies" about 250 m thick of a black siliceous shale and radiolarian ribbon chert. It contains thin beds and laminae of light-gray turbidites and tuffaceous material. Locally, intermediate-to-mafic tuff and massive subaqueous mafic flows grading to breccia occur.

The sulphide deposits in the Drenchwater Creek area (Nokleberg and Winkler, 1982) occur along an east-west zone 6 to 45 m wide and traceable for 1,830 m. Disseminated fine sphalerite, galena, pyrite, marcasite, and abundant barite, form lens-like orebodies principally in a dark gray chert, shale, meta-chert, tuffaceous sandstone, and tuff. Their outcrops in the recently glaciated alpine tundra zone were marked by iron oxide staining, but Nokleberg and Winkler have cautioned against a complete reliance on staining because sphalerite and galena occurrences lacking iron sulphides weather to shades of dark-gray to black and could be overlooked.

The Lik and Red Dog deposits located in the same unit 120 km west of Drenchwater Creek (Plahuta et al., 1978), together possibly contain some 18 Mt Zn and 16 Mt Pb. Both deposits consist of several concordant sphalerite, galena, pyrite lenses up to 5 m thick, interdigitated with quartz-barite alteration envelopes up to 50 m thick and lenses of vuggy hydrothermal quartz. Sulphide veins occur in the footwall. The ores are interpreted as subaqueous-hydrothermal ("proximal exhalite"), and are hosted by a little disturbed and lightly metamorphosed rhythmically interbedded argillite, siliceous argillite and bedded chert, with minor lenses of fine-grained felsic tuffaceous sandstone.

Selwyn and Kechika Basins, Yukon and northern British Columbia

Selwyn Basin is an informal, but widely used name given to the late Proterozoic to Mississippian deeper-water sedimentary assemblage (facies) located between the western side of the Mackenzie Mts. carbonate platform, and the Pelly-Cassiar Platform and/or the Tintina-Rocky Mountains Trench system of north-westerly strike-slip faults (Fig. 17-6). It is a zone at least 1,300 km long of completely unweathered rocks, often superbly exposed in the barren mountain ranges. Three stratigraphic sequences are usually recognized: (1) late Proterozoic "Grit Unit"; (2) Cambrian-to-middle-Devonian "black slate", minor limestone association ("Selwyn Basin" as interpreted by Carne and Cathro, 1980); and (3) middle Devonian to Mississippian "Black Clastic Group". To the casual observer, however, there is little difference between (2) and (3) and both assemblages have frequently been confused in the field.

(1) The Grit Unit is a somewhat "flyschoid" unit of dominant greenish-gray argillite to slate, litharenite (phyllarenite) and minor carbonate. Several minor Pb-Zn sulphide occurrences are known, one of the largest being the MacMillan (Quartz Lake) deposit, S.E. Yukon. There, a stratabound lens of reddish-brown sphalerite, galena and pyrite is enclosed in a gray phyllite interbanded with an ankeritic carbonate.

(2) In the central part of the Basin (Summit Lake area), the Grit Unit is overlain by a gray Cambrian slate grading upward into a limestone conglomerate, nodular limestone, dolomite and siltstone. These, in turn, are topped by a black argillite or slate interbanded with a black chert and accessory submarine basalt flows. The latter (Road River Formation) is a condensed sequence about 200-300 m thick, laid down within a time span of about 150 m.y. Along the western edge

of the Basin, the proportion of volcanics as well as the metamorphic grade increases, reaching greenschist to lower amphibolite facies. This sequence contains the most important Zn-Pb deposits in the Selwyn basin (Fig. 17-6; Table 17-3).

The Howard's Pass (Summit Lake) field in the central part of the Basin (Carne and Cathro, 1980; Morganti, 1981) consists of two major deposits hosted by a lower Silurian cyclic sequence of slightly siliceous argillite, carbonaceous argillite, chert and limestone. Very finely grained sphalerite, galena and minor pyrite, form diffuse bands and laminae in the argillite, inconspicuous and virtually unrecognizable in an outcrop when the ore surface is dusted over by white powdery hydrozinckite. Hydrozinckite is practically identical in appearance to the regionally widespread calcite precipitate coatings. Both orebodies are extensive (7,600 x 2,400; "XY" deposit only) folded stratiform sheets, free of barite, copper zones and footwall stockworks. The orebodies are interpreted as "distal exhalites" formed by precipitation from hydrothermal brines in a local seafloor depression (a sub-basin). Other than being slightly more rhythmic than the rest of the sequence, the ore-bearing unit is visually inconspicuous and easy to overlook in casual mapping so that the predictability of these extremely large deposits (a possible reserve of the order of 2 Bt with about 7.5% Zn, 3% Pb and 6 ppm Ag), is very low.

The Faro-Anvil district (Tempelman-Kluit, 1972; Carne and Cathro, 1980; 8.3 Mt Zn, 7.4 Mt Pb, 5.9 Tt Ag; Fig. 17-6), is in the western part of the Selwyn Basin in close proximity to the Tintina strike-slip fault. Seven orebodies have been found to-date along a 25 km long arcuate belt and all are hosted by a lower Cambrian, penetratively-deformed graphitic phyllite to schist unit that shows minor local variations into a sericite-chlorite schist, garnet schist and greenschist, and is intruded by Cretaceous granodiorite. A mid-Ordovician mafic volcanics-rich unit exists several tens of metres stratigraphically above the black phyllite unit.

The orebodies are stratiform sulphide lenses with identical vertical and lateral zonation (from hangingwall to footwall): (1) barite, sphalerite, galena; (2) pyrite, sphalerite, galena, minor pyrrhotite, magnetite, arsenopyrite, massive sulphide; (3) alteration metaquartzite with disseminated pyrite and minor Zn, Pb sulphides and (4) ribbon-banded graphitic quartzite. Chalcopyrite-pyrrhotite stringer zones have developed beneath some deposits, but hydrothermal alteration, except for silicification and the local appearance of andalusite and staurolite, is indistinct, probably having been obliterated during deformation and metamorphism.

(3) The "Black Clastic Group" of middle Devonian to Mississippian age was deposited on the association (2), following a local uplift, extensive block-faulting, and subaqueous erosion. It consists of thick units of interbedded gray to black chert arenite, siltstone and argillite with occasional conglomerate tongues interpreted by some as "flysch", subdivided by units of monotonous black carbonaceous siliceous shale. Metallogenically most conspicuous is the widespread

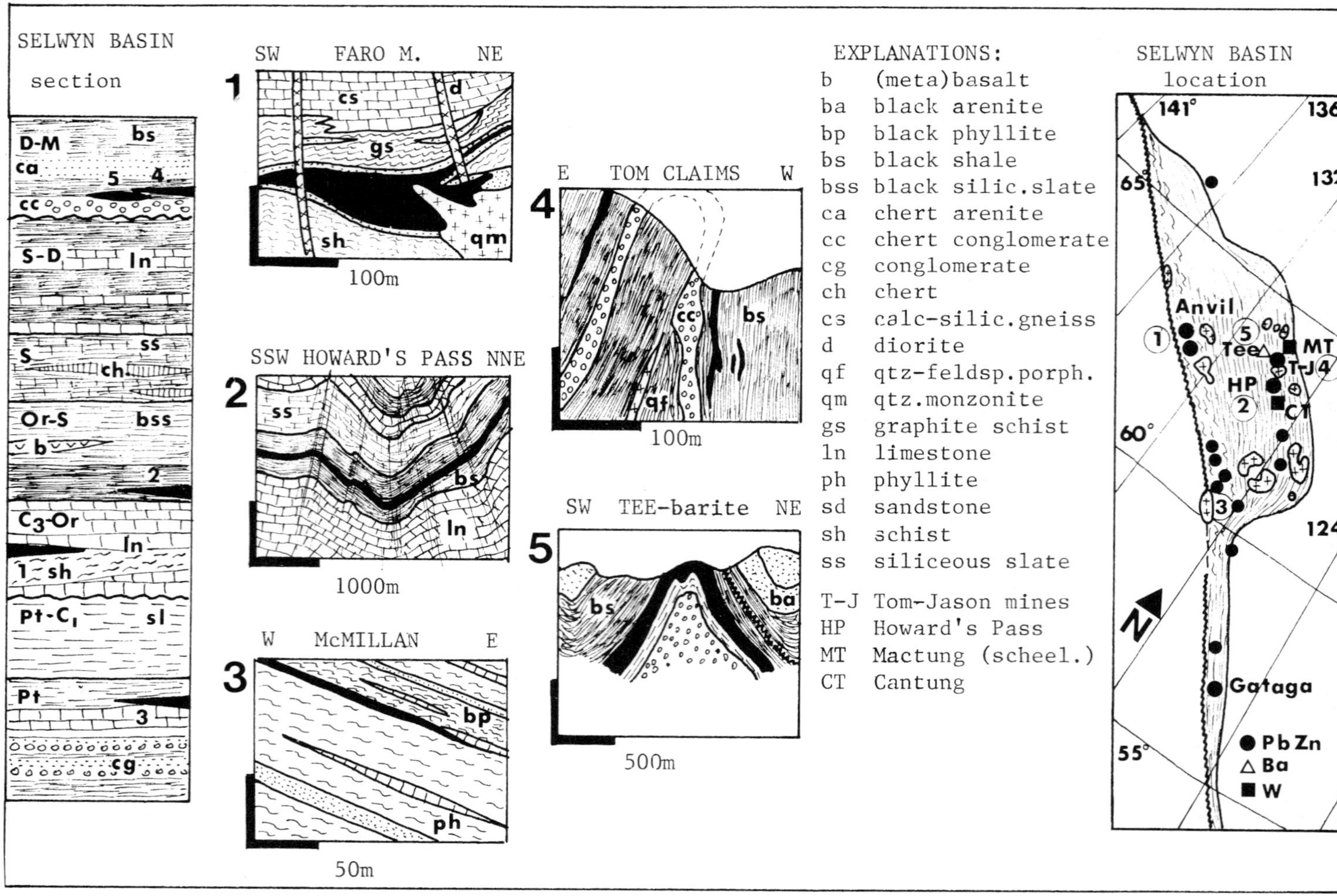
SELWYN BASIN
section
D-M
bs
ca
5
4
cc
S-D
ln
S
ss
ch
Or-S
bss
b
2
C3-Or
ln
1
sh
Pt-C1
sl
Pt
3
cg
1
SW FARO M. NE
cs
d
gs
sh
qm
100m
2
SSW HOWARD'S PASS NNE
ss
bs
ln
1000m
3
W McMILLAN E
bp
ph
50m
4
E TOM CLAIMS W
cc
bs
qf
100m
5
SW TEE-barite NE
bs
ba
500m
EXPLANATIONS:
b (meta)basalt
ba black arenite
bp black phyllite
bs black shale
bss black silic.slate
ca chert arenite
cc chert conglomerate
cg conglomerate
ch chert
cs calc-silic.gneiss
d diorite
qf qtz-feldsp.porph.
qm qtz.monzonite
gs graphite schist
ln limestone
ph phyllite
sd sandstone
sh schist
ss siliceous slate
T-J Tom-Jason mines
HP Howard's Pass
MT Mactung (scheel.)
CT Cantung
SELWYN BASIN
location
141°
136°
132°
124°
65°
60°
55°
Anvil
Tee
MT
T-J
HP
CT
Gataga
1
2
3
4
5
N
Pb Zn
Ba
W

presence of barite, having the form of scattered grains in sediments, spherulitic nodules (Laznicka, 1976) and gray, banded crystalline beds megascopically almost indistinguishable from a carbonate. Although the majority of barite deposits are completely free of sulphides other than pyrite in the enclosing sediments (although they may contain witherite and fluorite) some carry Pb-Zn sulphides.

In the Macmillan Pass field, two medium-sized strata-bound deposits Tom and Jason (Carne and Cathro, 1980; Fig. 17-6) and several showings, appear to have been deposited in a N.N.W.-trending fault-bound trough. The orebodies are lens-like, 3-60 m thick, composed of interlaminated black chert, pyrite, sphalerite, gray-to-white crystalline barite and black argillite.

The most recently discovered Gataga district in northern British Columbia contains two major stratiform deposits of galena, sphalerite and minor pyrite disseminated in fine-grained, thin to medium-bedded gray barite.

SOUTHERN BRITISH COLUMBIA - NORTHERN WASHINGTON

A late Proterozoic and lower Paleozoic slate/phyllite association corresponding to the subunits (1) and (2) in the Selwyn basin, crops out in several regions of southern British Columbia and northern Washington: in the Purcell Mountains, Kootenay Lake area, Salmo-Northport,etc. Several deposits, probably initially stratiform, suffered a high-grade metamorphism in the Shuswap Complex and are treated in Chapter 29. Other localities have been considerably modified tectonomagmatically or newly created by interaction. These are reviewed later in this chapter. Of the few localities with still recognizable stratabound characteristics and a probably synsedimentary origin, one could mention the Mineral King mine (massive-to-disseminated sphalerite, galena, pyrite in barite gangue hosted by dolomitic marble, surrounded by black phyllite and sericite-chlorite phyllite); Duncan mine (disseminated galena, sphalerite, pyrite and pyrrhotite in peneconcordant lenticular zones in a lower Cambrian dolomitized marble and siliceous dolomite, alternating with black phyllite); and the Remac mine south of Salmo (Sangster, 1970).

The bedded barite deposits in Paleozoic quartz siltstones, argillites and limestones near Northport in north-eastern Washington (Mills, 1971) carry only small quantities of lead and zinc.

⇦

Fig. 17-6. Selwyn Basin, northern Cordillera, Canada. Zn-Pb and barite mineralization in the "black argillite/slate" sequence. From LITHOTHEQUE, compiled using data from Tempelman-Kluit (1972); A. Clendenon, property tour, 1974; Noranda Ltd., oral communication, 1975; R.T. McIntosh, oral communication, 1974.

Table 17-3. Argillite, slate, schist hosted concordant to peneconcordant massive Zn,Pb (Cu) deposits; selected example localities

LOCALITY	AGE	GEOLOGY	MINERALIZATION	REFERENCES
Lik and Red Dog deposits, N. Brooks Range, N.W.Alaska	Ms	thrusted black argillite,silic.argillite,minor fine grained tuffaceous sandstone	conformable sf,gal,py,barite lenses in qtz-barite alter. envelope; barite body in hw. Lik only: 1.8 Mt Zn,600 Tt Pb	Plahuta et al. (1978)
Howard's Pass (Summit Lake) ore zone, N.E.Cordillera,Can.	S_1	monotonous silic.gray-to-black argillite,chert,limestone, rhythmic.banded	thin,persistent concord.lenses of very fine grained sf, gal/py; est.up to 150 Mt Zn 60 Mt Pb, 12 Tt Ag resource	Morganti (1981) Carne and Cathro (1980)
Macmillan Pass (Tom,Jason), N.E. Cordillera, Canada	D_3-Ms_1	black argillite to slate, minor interb.chert arenite, intraform.congl.to breccia	rhythm.banded stratabound ba, chert,sf,gal,py, folded lens 1.45 Mt Pb, 1.55 Mt Zn,1050 t Ag	as above
Faro-Anvil distr., Yukon, Canada	Cm_1	graphitic phyllite unit grading to chlor.,seric.,biot. garnet schist; intruded by Cr granodiorite	5 major oreb.,concordant mass. sf,gal,py,po,locally magn,ars, barite rich bodies;locally in dist.fw stockw.py,cp.8.3 Mt Zn 7.4 Mt Pb; 5.9 Tt Ag; 122 Tt Cu; locally contact metam.	as above Tempelman-Kluit (1972)
Cirque dep.,Gataga R.distr.,N.E. British Columbia	D_3-Ms_1	black argillite, black silic. argill.interbedded with chert arenite,intrafm.congl.	concordant fine barite lens contains dissem.gal,sf/py	Carne and Cathro (1980)
Mineral King m.,Toby Creek,S.E. British Columbia	Pt_3	graphitic, calcareous and seric.-chlor.phyllite, marble, greenschist	dissem. sf,gal,py in peneconc. deformed,partly remob.barite; 86 Tt Zn,37 Tt Pb, 52 t Ag	Olson and Fyles (1968)

Duncan m.,Kootenay Lake,S.British Columbia	Cm_1	dolomitiz.and silic.marble, black phyllite, greenschist	dissem.gal,sf,py,po in pene-concordant lenticular ore zone in marble	Fyles (1964)
Mountain City,n.of Elko, Nevada,USA	Or	black and gray phyllite, chert,black marble,green-stone	py,qtz/cp//sf,gal,concord.mass. lenses grading to dissem., stringer zone in chl.alter.fw. 107 Tt Cu,most from enr.zone	Coats and Stephens (1968)
Carboire,near S.Gi-rons,French Pyrenees	Or_3	calc-silicate schist, limy phyllite,dolomite, limestone marble	massive concord.to dissem. sf,ga,po,py/cp lens; deformed 187 Tt Zn/6.25% 25 Tt Pb/0.85%	Bois (1972)
Bosost Dome (Sen-tein,Bosost dep.) Fr.& Sp.Pyrenees	Or	black metasiltst.,marble,qtz. phyllite,greensch.,intr.by granodiorite; thermal metam.	mass. to dissem.sfl,po,gal,py, peneconc.to bedding and veins, replac. Min.150 Tt Zn,50 Tt Pb	Pouit et al. (1979)
Porte-aux-Moines, Armoricain Massif, N.France	D_1	black argillite to siltst. topped by dark sandst.,20-50 m above bimodal volc.pile	conc.lens of mass.py,sf,gal.in chert banded argillite, underl. by py,cp/sf,gal.vein,dissem. bodies. Tect.dismembered.500t Ag 275 Tt Zn, 150 Tt Pb, 77 Tt Cu	Laville-Timsit and Wilhelm (1979)
Meggen,Sauerland, West Germany	D_2	black argillite with pyrite nodules,topped by shale, siltst.with tuff horizons and pelagic limestone	fine grained concord.py/sf//gal. mass.,banded horiz.envel.by ba-rite. 15 Mt barite, 6 Mt Zn/10% 780 Tt Pb/1.3%	Ehrenberg et al.(1954), Krebs (1972)
Rammelsberg mine, Goslar,Harz Mts., West Germany	D_2	monotonous gray and black slate, tuffac.horizons, litharenite	2 concordant mass.py,sf,gal/cp lenses,a hw barite body, and fw py,cp stockw.in silicif.slate 3.96 Mt Zn/18%; 1.76 Mt Pb/8% 286 Tt Cu/1.3%; 5.17 Mt barite	Kraume et al. (1955)

Table 17-3 (continued). Argillite, slate, schist-hosted concordant to peneconcordant massive Zn,Pb (Cu) deposits; selected example localities

LOCALITY	AGE	GEOLOGY	MINERALIZATION	REFERENCES
Arzberg-Haufenreith distr.near Graz, Austria	Or	quartz phyllite,graph.phyllite,black limest.,calcar. phyllite, minor greenstone	thin,concord.massive py,sf,gal and barite sheets and lenses	Siegl (1974)
Agordo,Venetian Alps, N.Italy	Pe?	micasch.,graph.schist,meta qtz.porph.,deformed, amphibolite facies	massive to dissem. py,qtz,cp/gal,sf,Co-Ni arsenides; conc. with schist. 109 Tt Cu/1.7%	Beyschlag et al.(1916)
Borovica, Vareš d., Bosnia,Yugoslavia	Tr_2	black and varicol.slate, chert,tuff.slate,graywacke, dolom., minor spilite	dissem.gal,sf,cp,tetr. in concord.barite lenses. 30 Tt Zn/3.5%; 30 Tt Pb	Janković (1982)
Kuchke, 50 km N.E. of Bafq, Iran	D_1	black argillite,slate,dolom. limest.,assoc.mafic meta-volc., serpentinite	very fine grained,concordant massive sf,gal,py lens. Min. 500 Tt Zn/10%, 300 Tt Pb	Wright (1964)
Tekeli dep.,Koksu-Tekeli z.,S.E. Kazakhstan, USSR	D_{1-2}	black slate to phyllite,gray limest., dolom.,minor arenite; access.gabbro and diabase intrusions	metasomatic peneconc.massive sf,gal/cp bodies superimp.on pyritic phyllites; silicification; min. 1 Mt Zn	Veits (1972)
Achik Tash,Tian Shan,Kirghizia, USSR	Pt_3	graph. schist, garnet micasch. dolomite,marble,amphibolite	massive, strongly deformed py/sf,gal, concordant sheets along black schist - marble contact	Baibulatov (1964)

NEVADA

In Nevada, the Cambrian to Devonian and Devonian-Mississippian black argillite, chert, carbonate sequence is facially very similar to the one in the Selwyn Basin, except for the greater degree of dismemberment due to igneous intrusions and the Basin-and-Range tectonism. Sulphide-free, large stratiform barite deposits are well developed, e.g. in Devonian and Ordovician regions of the Shoshoni Mts, and in the Ordovician region in the Northumberland Canyon (Zimmermann, 1969). In the Mountain Springs mine in the Shoshone Range (8 Mt barite), pods and lenses of gray, fine-grained strata-bound barite are rhythmically interlaminated with a gray chert.

Stratiform massive sulphide bodies are rare; the two best known examples (Big Mike in Pershing County and Mountain City), are essentially copper mines. The Rio Tinto copper deposit near Mountain City (Coats and Stephens, 1968), is hosted by black and gray phyllite of the Ordovician Valmy Formation that also contains interbeds of fine-grained graywacke, black chert, and mafic volcanics. It is unconformably overlain by the Devonian to Mississippian "Black Detrital Sequence". Hypogene ore (massive pyrite, quartz, lesser chalcopyrite, accessory sphalerite, galena) occurs in several conformable lenses, and grades into disseminated and stringer mineralization in a chlorite and argillic-altered footwall. Most of the production to-date, however, came from the rich supergene oxidation and secondary enrichment zones.

OUACHITA OROGEN, ARKANSAS

In the Ouachita Range, the deep marine argillite-chert association is well represented by the Devonian-Mississippian Novaculite (massive gray chert) grading upward into siliceous shale and dark fissile argillite and containing many compact sandstone beds near the base (Stanley Shale; Mississippian-Pennsylvanian; Morris, 1974). Bedded barite is widespread and near Magnet Cove forms a horizon 10 to 33 m thick, composed of barite beds separated by black shale and sandstone. The light-to-dark-gray barite is very fine-grained, resembling limestone or a limy shale and is pyritic in places. No base metal sulphides are associated.

Table 17-3 gives the essential data on the major "slate"-hosted massive sulphide deposits.

Pb-Zn-BARITE DEPOSITS IN TRANSITIONAL "BLACK SLATE" AND CARBONATE ("ALPINE TYPE") SETTINGS

Many Pb-Zn or barite deposits in the "black argillite/slate" sequence are actually hosted by a carbonate, or are at a carbonate/slate contact. Carbonate is a minor, but frequently present, member of this sequence. It is not surprising, then, that a Pb-Zn (barite) ore not very different from those described earlier, is also hosted by sedimentary sequences in which the carbonate is dominant and the argillite/slate subordinate. Most such sequences are members of the "carbonate platform" unit, treated in Chapter 20.

In orogenic belts where the "black detrital" and "carbonate platform" associations exist side by side in long, continuous orogens and are connected by facies transitions, Pb-Zn-barite mineralization occurs in both facies, for example in the Cordilleran belt of North America. Elsewhere, the textbook regularity of facies distribution is disturbed by embryonal development, tectonism, erosion or magmatism, and interrupted belts, local carbonate "terrains" enveloped by "black detritals" and other rocks, occur.

In the Cambrian to Carboniferous sequence in the Montagnes Noires (south-eastern margin of the Massif Central, France; Brevart et al., 1982) numerous Pb-Zn-barite deposits with transitional characteristics occur. In the Salmo-Metaline area, British Columbia and Washington, "Mississippi Valley-type" Zn-Pb ores in dolomitized limestones are surrounded by a "sea of black slates". Sangster (1970) suggested an environmental analogy with the Blake Plateau to justify the presence of the carbonates in their present location and coined the phrase "Remac-type" for Pb-Zn ores hosted by deeper-water euxinic, graphite-bearing carbonates often interbedded with ribbon chert.

The "lesser-order basin" superimposed on a "greater order basin" arrangement initially advocated by Krebs but by now widespread, provides an opportunity to place a "deep-water environment" almost straight into a shallow-water one.

17.6. Zn-Pb (BARITE) ASPECT: INTERACTION OF THE "BLACK SLATE" ASSOCIATION WITH TECTONISM AND MAGMATISM

17.6.1. Introduction

Several instances of demonstrated or suspected influence of plutonism on generation or modification of the stratiform or stratabound Pb-Zn deposits have already been mentioned above. After all, if one accepts the current model of formation of the "exhalative" seafloor deposits from heated brines driven or generated by concealed magmatic bodies, the ore and granites are not as far apart in this class of mineral deposits, as was previously thought. In this section we are going to review several examples where the "black" argillite, slate, schist, etc. association, deformed and intruded by magmatic bodies, hosts demonstrably epigenetic and usually cross-cutting Pb-Zn veins or carbonate replacements. These mineralizations, broadly coeval with the intrusive emplacement and with the cooling period postdating the "granites", or sometimes located along faults even where "granites" are not known, are, nevertheless, compositionally close, or even identical, to the "exhalative" ores one would expect in the "black" metasedimentary hosts. In fact, stratabound orebodies formed by earlier events have been found, with increasing frequency, in mining fields where mining started on cross-cutting, epigenetic hydrothermal vein deposits. Anticipation of similar discoveries elsewhere should become an ingredient of exploration strategies.

Having visited thousands of mineralized occurrences worldwide, I have been impressed by the frequency of epigenetic Zn-Pb veins hosted by the "black slate": a frequency that was far beyond a mere coincidence. Such a correlation is one of the useful empirical experiences that help to a prospector to define "favourable" hosts to ores, although satisfactory genetic explanations are still scarce (remobilization ?; heritage ?; metallogenic provincionalism where deposits of the same metal but of a variety of types and ages reflect a geochemical anomaly in the mantle ?). The subject of genesis has been reviewed in Laznicka (1985e) and will not be repeated here.

17.6.2. Mines (fields) containing coexisting "sediment-coeval" and "granite-coeval" Zn-Pb orebodies, hosted by the "black slate" association

The high plateau of Mexico (Mesa Central) studded by hundreds of small-to-large vein and replacement deposits of Ag, Pb, Zn, Cu (and minor Sb, Mn, Hg, Au, W, etc.) has long been considered a "metallogenetic province" supportive of the postmagmatic-hydrothermal concept of metallogeny completely controlled by the plutonic or subvolcanic magmatism. Following the rapid increase in popularity of the many forms of stratiform mineralizations contemporary with the formation of marine supracrustal rocks, several hitherto enigmatic or unnoticed "manto" orebodies have been interpreted as volcanogenic, or sediment-hosted "exhalites". De Cserna (1976a) presented a whole list of them. The late Jurassic-early Jurassic shale, some graywacke, limestone and minor subaqueous volcanics facies belts located close to the N.W.-S.E. axis of Mexico, contain most of them. Probably syndepositional stratabound mantos in black argillites or limestones coexisting with younger intrusions-coeval and fault-controlled veins or replacements, were recorded in this belt in the following mines or fields: Parral-Santa Barbara, Sombrerete, Fresnillo, Zacatecas, Concepción del Oro, San Martín. Comparable examples of concordant/discordant sets of orebodies have also been recorded in the fields hosted by the "black" basinal sequence (Plomosas) or by Precambrian schists (Taxco).

A very similar situation exists in the well-known Pb,Zn,Ag belt of central Peru (Bellido and de Montreuil, 1972) but despite the empirically obvious repetitive spatial association of the Tertiary hydrothermal deposits with black phyllites of the Devonian Excelsior Formation or with the gray limestone/black argillite of the Triassic-Jurassic Pucará Group, no stratiform deposits were identified from there until the publication of the "Amstutz Volume" in 1982 (Amstutz, ed., 1982). In this volume, Kobe (1982) demonstrated the presence of stratabound and/or volcanogenic pyrrhotite, sphalerite and galena along the contact of the black phyllite and greenschist members of the Excelsior, and Ni, Co, Cu, As, etc. stratabound occurrences hosted by ophiolitic volcanics, in the Yauli Dome (s.e. of Morococha). In the same volume, Lehne and Amstutz (1982) reinterpreted the well-known silver-rich galena, sphalerite, tennantite, enargite, etc. mantos, hosted by the lower Tertiary shallow-marine black shaly carbonates near Colquijirca (and underlain by the Excelsior), as syndepositional.

Colquijirca has been a moderate silver producer (some 2,000 t Ag recovered). The real giant ore deposit in the area, the Cerro de Pasco (8 km from Colquijirca; 100 Mt pyrite, 7 Mt Zn, 3 Mt Pb, 48 Tt Ag, etc.; Einaudi, 1977; Fig. 17-7) is a massive replacement of Pucará limestone in a Miocene vent area underlain by the Excelsior black phyllite. It would indeed be tempting to seek the source of the sulphides in the metallogenically fertile footwall region, but no evidence has, so far, been presented.

17.6.3. Mines (fields) with "granite-coeval" (and post-granite) hydrothermal cross-cutting Zn-Pb orebodies, hosted by the "black slate" association

This style is extremely common, and Table 17-4 lists selected examples. In the important Příbram district 55 km S.W. of Prague, Czechoslovakia (Bernard et al., 1968; Fig. 17-8), around 100 hydrothermal Pb,Zn,Ag and U veins postdate the ? Carboniferous

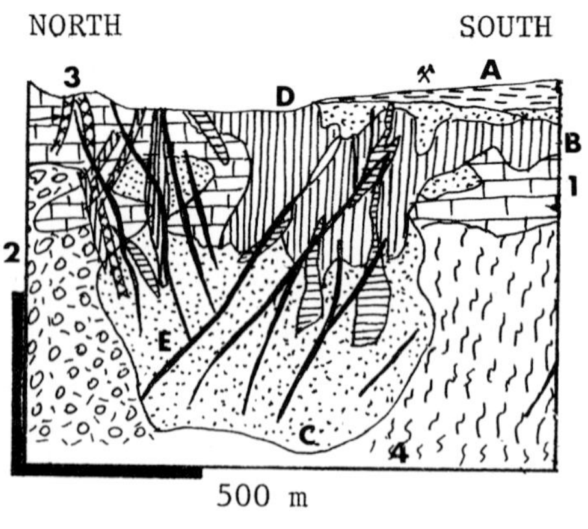

A gossan
B Zn-Pb orebody
C massive pyrite body
D pyrrhotite bodies
E Cu veins and replacements
1 Tr_3-J_1 thick bedded, dark limestone and dolomite
2 Mi Rumiallana Agglomerate (diatreme ?)
3 Mi quartz monzonite porphyry
4 D - Excelsior Gr. black phyllite

Fig. 17-7. Cerro de Pasco, Peru. Massive pyrite, pyrrhotite, Zn,Pb,Cu,Ag, etc. replacement in limestone, and veins, above black phyllite. From LITHOTHEQUE, after Einaudi (1977) and Abele (1977).

Table 17-4. "Granite" - coeval hydrothermal Pb-Zn deposits hosted by the deeper marine black argillite/ slate/ schist association

LOCALITY	"BLACK SLATE" UNIT			PLUTONIC INTRUSIONS			MINERALIZATION, TONNAGE	REFERENCES
	AGE	LITHOLOGY	ORES	AGE	LITHOLOGY	ORES		
Cerro Gordo, E.California	S-Ms	black slate phyllite marble	+++	Cr-T_1	andesite, dacite porphyry	0	qtz,gal,sf,Ag minerals veins,replacements 37 Tt Pb,12 Tt Zn, 1,250 t Ag	Merriam (1963)
Santa Barbara, Parral d.,Chih., Mexico	Cr_1	black argillite limestone	+++	Eo-Ol	rhyol., andes. porph.	0	qtz,gal,sf,py fissure veins; 1.5 Mt Zn; 1.1 Mt Pb; 5.2 Tt Ag/130 ppm	Busch (1980)
Cerro de Pasco C.Peru	D Tr-J	black phyllite;black argillite limestone	+++	Mi	qtz monz. porphyry	+	py mass.replac.in lim. superimp.po,sf,gal, ars,enargite pipes, veins; 100 Mt py; 7 Mt Zn; 3 Mt Pb; 48 Tt Ag; 1.5 Mt As	Einaudi (1977)
Saint Salvy,Montagnes Noires, France	Cm	black slate,limest.	+++	Cb-Pe	granod. qtz.monz.	0	sf,gal,fissure veins 300 Tt Zn	Brevart et al. (1982)
Příbram,W.Czechoslovakia	Pt_3	black slate,minor spilite	++	Cb? Mz?	qtz.dior. granod. diabase	++	qtz,sider,sf,gal, boulanger.,tetr. fissure veins	Bernard et al. (1968)
Stříbro,W.Czechoslovakia	Pt_3	black phyllite	+++	Cb? MZ?	granod.	+	qtz,gal,sf,fissure veins	as above

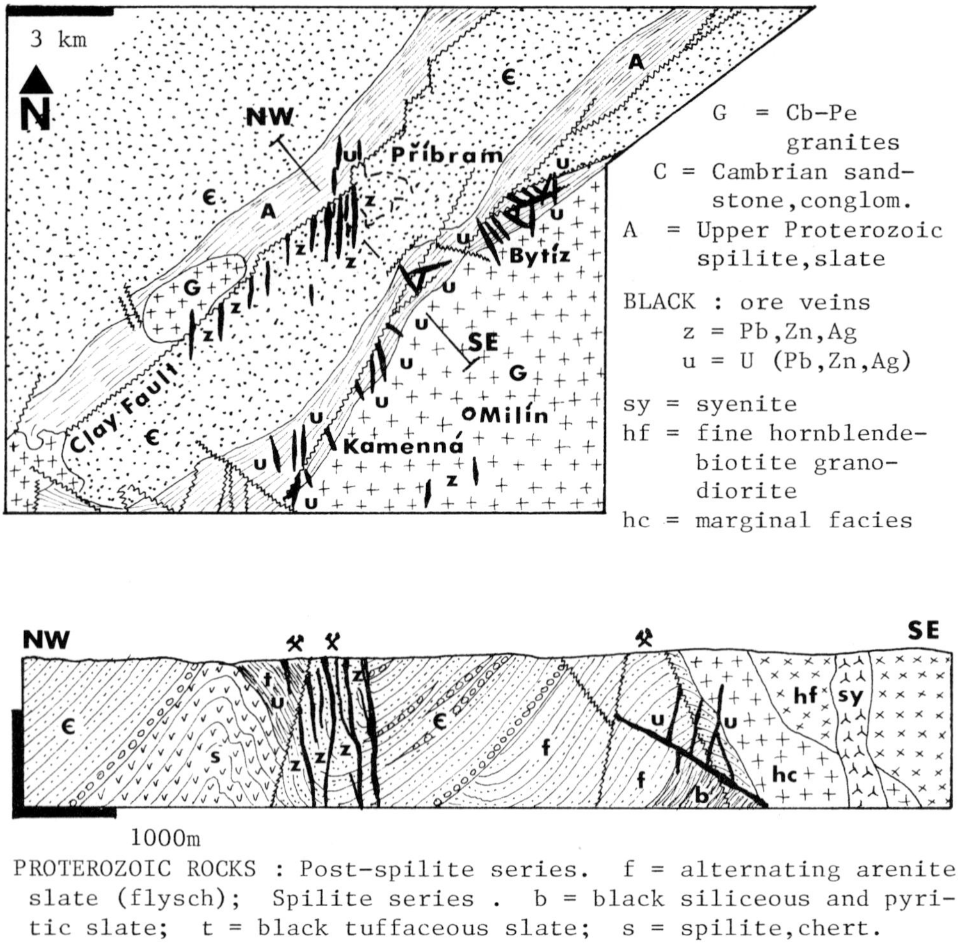

Fig. 17-8. Příbram ore district, western Czechoslovakia: example of Pb,Zn,Ag and U hydrothermal vein mineralization showing a possible source affiliation with late Proterozoic black slates. The ore veins dimensions are exaggerated. From LITHOTHEQUE, based on data in Kutina (1963), Ruzicka (1971), and own visits.

emplacement of the Central Bohemian Pluton and its satellitic intrusions. The vein distribution there is a textbook example of structural control: the important veins fill fractures branching from two N.E.-S.W. trending boundary and axial faults, along which two narrow belts of Cambrian shallow-water sediments were preserved under overthrusts of a late Proterozoic black slate-minor spilite association. Most of the veins in the older, classical Březové Hory field (predominantly quartz, siderite, galena, sphalerite, boulangerite, Ag-sulphosalts) are hosted by the Cambrian sediments or by diabase dikes that intrude them but some persist into the Proterozoic black slates and the entire vein system seems to be "rooted" in these rocks.

The more recently discovered and exploited Bytíz-Kamenná (eastern) vein system is mostly in the Proterozoic slates and also in the granitic rocks. It contains largely dolomite, hematite, calcite, pitchblende, minor silver, Co-Ni arsenides orebodies. Superimposed, a younger generation of siderite, calcite, sphalerite, galena, veins is also present locally. The orebodies are controlled by relatively open zones of mylonitization and brecciation.

In another Czech Pb,Zn,Ag field Stříbro, predominantly quartz, galena, sphalerite veins are hosted by lustrous black late Proterozoic phyllites, interbedded with rare meta-spilitic greenstones. Although the veins are in the supposed aureole of a Variscan granitic intrusion, they have now been isotopically dated as Mesozoic, a product of "activation" (Bernard et al., 1968). Many of the Pb-Zn vein and replacement deposits of the "mesothermal" category treated in textbooks belong partly or fully here (e.g. some localities in the Rheinische Schiefergebirge; Iberian Meseta of Spain; Santa Barbara and other localities in the Mexican Mesa Central; Salmo and Slocan districts, British Columbia; Eureka, Nevada; etc.). Some of the localities listed and some additional ones are transitional into the shallow-water (Chapter 19), and into Precambrian "black slate" associations. All tend to be high in silver and sometimes gradational into "bonanza" silver veins.

17.7. Ag ASPECT

Black argillites/slates appear frequently to be enriched in trace silver. Contents of 20-30 ppm Ag have been reported occasionally (compare Table 17-2) in "slates" not mined for industrial purposes (this represents a geochemical concentration factor of the order of 100-500). These contents are comparable with silver grades in porphyry coppers, many massive sulphide deposits and hydrothermal veins, from which silver is usually recovered as a by-product. If the trace silver determinations are accurate, certain black "slate" horizons may constitute the silver ores of the future. The highest Ag content reported, 210 ppm Ag, occurred in phosphorites hosted by the black slates in the Karatau Range, U.S.S.R. (Kholodov, 1973a). It is conceivable that systematic exploration along high silver "black slate" horizons might, in the future, result in the discovery of economically mineable stratabound silver deposits. Their existence may, under favourable conditions, be revealed megascopically by the presence of short gash veins and fracture veinlets containing silver minerals. Some such veinlets with argentite and stromeyerite have been reported in Soviet Central Asia.

So far, however, industrial argentiferous "black slates" have not been reported but many significant epigenetic silver orebodies are hosted by, or occur in the vicinity of, the "black slates", and their Ag content could be the result of a tectonomagmatic interaction involving the black sediments. All the "interaction Pb-Zn deposits" listed in Table 17-3 are high in silver. "Straight" silver deposits in black slates are best represented in the Guanajuato field in Mexico, which has recently been revitalized (Gross, 1975; 30 Tt Ag, 115 t Au; Fig. 17-9). There, a series of persistent N.W.-trending thick quartz-carbonate, lesser adularia veins with randomly scattered sulphides and a low content of precious metals (10 ppm Ag, 0.1 ppm Au) transect a variety of rocks ranging from upper Tertiary volcanics and sills through lower Tertiary red conglomerate, to Mesozoic basement black slates.

The largest vein zone, the Veta Madre, is approximately 25 km long and up to 90 m wide. The main veins are uneconomic to mine in themselves, but they contain irregularly distributed "bonanza" ore shoots with grades of about 450 ppm Ag and 2.5 ppm Au, mineralized by acanthite, aguilarite, naumannite, electrum, and gold. Gross (1975) demonstrated that the basement shale is anomalously enriched in trace silver and gold (8.3 ppm Ag, 0.15 ppm Au) and sulphur, and that its trace metal ratios are closely comparable with the "Lower Orebodies". He concluded that the mineralization formed during the

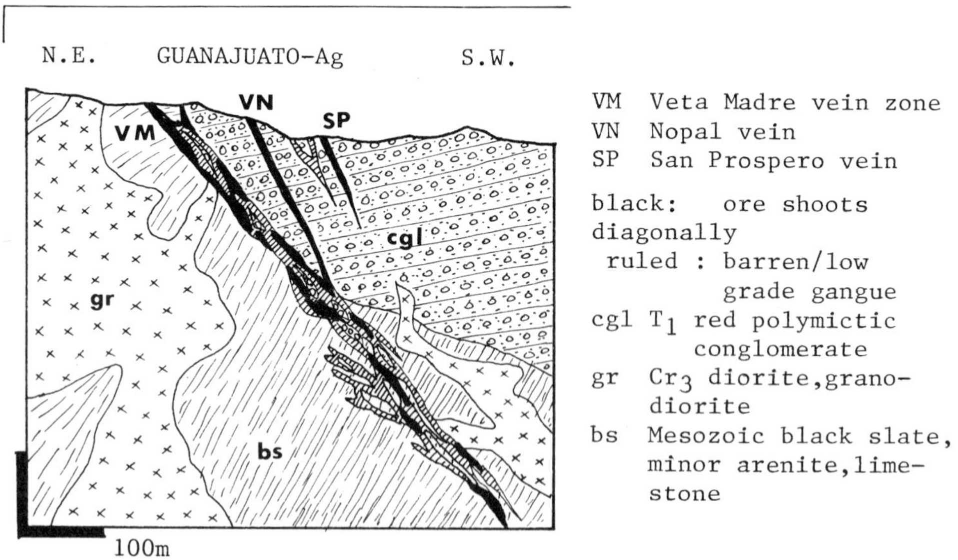

Fig. 17-9. Veta Madre Ag vein system, Guanajuato, Mexico, partly hosted by and probably source related to, a Mesozoic black slate. From LITHOTHEQUE, based on data in Guiza (1956) and Gross (1975).

upper Oligocene, by reprecipitation of the Ag and Au from the black (meta)sediments, leached by alkaline waters heated by the contemporary magmatic activity. Comparable basement slates and also greenstones underlie the Zacatecas and probably also Pachuca silver districts and the silver sources may be comparable.

Many predominantly uranium vein deposits hosted by black slates or schists contain silver, particularly in association with Ni-Co arsenides. Silver may accumulate selectively in some ore shoots. Such mineralization has been recorded in the eastern belt of the Příbram district where the hosts are Proterozoic black slates. In the Jáchymov (Joachimsthal) mining district on the Czech side of the Erzgebirge (Mrňa and Pavlů, 1967) U and Ag, As, Co, Ni veins are hosted by lower Paleozoic micaschists in the roof zone of Permian tin granites. Zones of graphitic phyllites are preferentially mineralized.

17.8. Au ASPECT

Black slate/schist sequences are frequent hosts to epigenetic hydrothermal gold deposits (Table 17-5), many of which are comparable, transitional or identical in style to those described in Chapters 10, 11, 12, 15 and 16. Probably the most common mineralized styles are cleaved, sheared or fractured black slate, siltstone or limestone hosts, healed by simple or multiple quartz or quartz-ankerite veins carrying scattered pyrite, arsenopyrite and native gold. This type of ore is particularly common in sedimentary troughs overlying ophiolites or other mafic-ultramafic assemblages, such as in the upper Triassic black sediments near Manson Creek, British Columbia.

Extensive zones of gold-bearing veinlets often partly stratigraphically or lithologically controlled (confined to "favourable" horizons of black sediments) may develop into "linear stockworks" and constitute an important type of large tonnage, low-grade gold deposit exemplified by the Soviet deposit Muruntau (Zeravshan). This deposit, located in the Kyzyl-Kum desert of central Uzbekistan, has been partly summarized in Boyle (1979). Briefly, late Paleozoic or Silurian carbon-rich slates, phyllites and meta-siltstones are located in the roof of a Permian pluton, and are cut by quartz porphyry and diorite dikes. The meta-sediments themselves are anomalous in gold (in the region of 0.1 ppm Au) particularly in the pyrite-rich lenses, but the economic mineralization takes the form of densely spaced gold-bearing quartz, pyrite, arsenopyrite, minor scheelite veinlets and disseminations in altered metasediments. Quartz, albite, K-feldspar metasomatites are described as the most common. The ore zone is subhorizontal, several kilometres long, and has assay boundaries. The grade and tonnage have not been reported, but could be of the order of 1-3 ppm Au and 600-800 t Au, respectively.

The recently outlined Great Basin (Basin-and-Range) sediments-hosted "invisible" gold province in Nevada and Utah, spans

Table 17-5. Interaction gold mineralization associated with deeper marine "black" argillite, shale, schist association

LOCALITY	SEDIMENTARY UNIT		INTERFERENT			MINERALIZATION		REFERENCES
	AGE	LITHOLOGY	ORE	AGE	TYPE	ORE		
Manson Creek, British Columbia	Tr_3	black phyll. siltst, lim. overlying ophiolites	+++	MZ	tectonism cleavage shearing		qtz, carb, py, ars, gold veinlets, veins; most prod. from placers; 0.5 t Au	own work
Muruntau (Zeravshan), Uzbekistan USSR	Pt_3 or S	black slate siltst, phyl, therm. met.	+++	Pe	qtz dior, granod, dikes, pluton		qtz, py, ars, gold, veinlets, veins, concord. stockw. in K-altered metased. 800 t Au ?	Boyle (1979)
El Oro, Mexico	J? Cr_1	black slate limest. zones, minor andesite	++	Cr?	qtz dior, qtz porph dikes	+	qtz, calc, ars, py, gold, Ag sulphosalts, N.N.W. fract. veins 142 t Au, 710 t Ag	Wisser (1966)
Getchell and Pinson mines, Nevada USA	Cm	black slate limestone chert	++	Cr	granod, diorite faulting	+	fine gold in py, marc, black mylonite, along N.N.W. fault zone 27 t Au	Hotz and Willden (1964)
Goldfield, Nevada USA	Cm	black slate	+	Mi	volc. centre; andes, rhyodac, rhyolite	+++	famatinite-enargite, Bi, gold, Au-Ag tellur. in qtz gangue in veins; alunite etc. alter. 118 t Au	Ashley and Keith (1976)

FREQUENCY OF MINERALIZATION:
- absent
- \+ very low
- ++ moderate
- +++ high

the facies boundary between the "deep water" (in the west) and "shallow water" (in the east) Paleozoic host sediments. The largest deposit (Carlin) is in the "shelf" facies, under a thrust along which the "deep" facies sediments were tectonically emplaced from the west, and will be described in the next Chapter. The mines near Winnemucca, Nevada (Getchell and Pinson), however, are partly hosted by a Cambrian gray-to-black slate, limestone and chert association.

In the Getchell mine (Hotz and Willden, 1964; 14 t Au/5 ppm; Fig. 17-10), orebodies are controlled by a fault zone in black hornfelsed slate containing lenses of thin-bedded limestone, in the vicinity of a granodiorite stock. The orebodies are sheet-like and tabular masses of fractured and mineralized rock, swelling and pinching. Pyrite, marcasite and carbonaceous matrix of the fault gumbo (mylonite) carry most of the invisible, finely dispersed gold and gold also occurs sparsely in realgar, arsenopyrite and coarse vein quartz. Brightly red realgar and yellow orpiment are the most conspicuous minerals and ore indicators. The Pinson mine (min. 12 t Au/3.4 ppm) is about 10 km south of Getchell, along the same fault and in the same host rock. The mineralization is hydrothermal, low temperature and attributed to a middle Cretaceous thermal episode.

In Goldfield, Nevada (Ashley and Keith, 1976; Fig. 17-10), numerous gently dipping "ledges" (zones of highly altered: silicified, alunitized, argillized, pyritized) Miocene dacites and andesites, occur densely along a 1,200 m long, 35° east-dipping zone along the western edge of the Goldfield volcanic centre. The ledges contain ore shoots of cryptocrystalline quartz with specks and lenses of famatinite, enargite, bismuthinite, fine gold, and

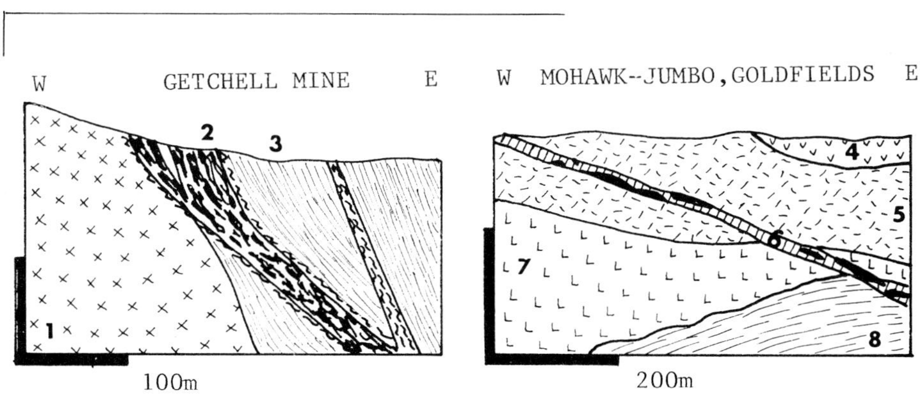

Fig. 17-10. Interaction gold deposits hosted or underlain by a black slate. (1) Cretaceous granodiorite; (2) mineralized fault zone; (3) Cambrian argillite and contact hornfels; (4) lower-middle Miocene subaerial andesite; (5) porphyritic rhyodacite; (6) alteration ledge containing ore shoots; (7) latite; (8) Cambrian slate. From LITHOTHEQUE, after Hotz and Willden (1964) and Ransome (1909).

Au-Ag tellurides. The mineralization reaches into the Cambrian black slate basement, concealed under a thin cover of neovolcanics.

17.9. PHOSPHATE, BLACK ARGILLITE AND CHERT ASSOCIATION: V, Mo, U ASPECT

17.9.1. Introduction

Recent marine phosphorites (e.g. Bentor, ed., 1980; Burnett et al., 1980) form in areas of upwelling along the shelf/slope hinge. This creates a classification problem, because phosphorites are thus both shallow-water (shelf) and deep-water (slope) sediments. Some of the largest deposits of ancient sedimentary phosphorites of the world (the Western Phosphate Field, U.S.A.; the Karatau Range, U.S.S.R.) are prominently associated with black argillites or slates of orogenic belts, so that they are treated here. The phosphorites associated more closely with the platform carbonates (Florida, Morocco) are treated in Chapter 20.

In both the above mentioned fields and elsewhere, the phosphorite is a brown-to-almost—black, bedded, massive and fine-grained or oolithic, granular rock. It is gradational into, or substituted by, nodules of phosphorite and sometimes by hardfloors (phosphate-encrusted or impregnated hardened portions of a seafloor representing a diastem). In all instances, the "primary" phosphorite that could be "diagenetic" (generated by replacement of a carbonate precursor) or "authigenic" (formed by precipitation from seawater or by pelletal accretion), is composed of a single phosphorus-bearing mineral, fluorine-carbonate apatite (Bentor, ed., 1980).

The phosphorite horizons have a characteristic lithologic association: chert (or "porcellanite" in geologically young associations) and organic-rich (brown to black) argillite or carbonate. The chert is usually interpreted as having derived from pelagic diatom or radiolarian oozes in Cretaceous and younger units, and from spongiae in the earlier sequences. Phosphorites are gradational into slightly phosphatic argillites, cherts and limestones and ultimately into entirely non-phosphatic equivalent rocks.

Marine phosphorites contain high fluorine and are known to be a repository of several trace elements. Of these, increased trace uranium is almost constant (50-150 ppm). Other metallic elements are unpredictable, showing extreme concentration ranges (e.g. Ag up to 210 ppm, As up to 80 ppm, Cd up to 200 ppm, Cr up to 3,000 ppm, Cu up to 350 ppm, Ga up to 12 ppm, REE up to 600 ppm, Mo up to 820 ppm, Sc up to 30 ppm, V up to 7,260 ppm, Ni up to 900 ppm, Y up to 1,100 ppm, Se up to 110 ppm, Zn up to 5,900 ppm (Altschuler, 1980, for the Georgina Basin, Australia; McKelvey et al., 1959; U.S. Dept. of Interior Releases, for the western U.S. phosphorites). In terms of the median trace element contents based on 18 regional averages (Altschuler, 1980), only the following metals have average contents in the phosphorites that are greater than 2 clarke values: Ag (1.5 ppm),

As (15 ppm), Mo (6 ppm), Sc (10 ppm), Se (3.7 ppm), U (105 ppm), Cd (18 ppm), V (70 ppm), Zn (190 ppm) and REE (about 450 ppm).

Many exploration geologists recognize U, V and Mo as the most "economic" metals associated with phosphorites which is clearly the consequence of the publicity given to some depositional concentrations of these metals in the Phosphoria Formation (V,U) and in Karatau (V, Mo). It is interesting to note that these metals concentrate in both regions, preferentially in the black argillite associated with phosphorite rather than in the actual phosphorite, and that the same metals could accumulate in completely non-phosphatic argillites. The metal precipitation obviously took place from seawater, either through the organic metabolism or postmortem due to the reducing environment. The local anomalies of metals probably reflect the local environmental variations (e.g. a contemporary volcanism).

At present, several phosphatic associations around the world are mined for their phosphorus content and certain trace metals such as uranium are occasionally recovered (or are potentially recoverable) as by-products. From the economic geologic standpoint such materials are thus "technological ores". Other phosphatic associations are on standby, awaiting improved market conditions or a breakthrough in processing technology but ready to supply metals during national emergencies. It appears that the large Central Asian deposits in the U.S.S.R. are most advanced in terms of development as metal producers.

17.9.2. Permian Phosphoria Formation in the Western Phosphate Field, U.S.A.

The Permian Phosphoria Formation (McKelvey et al., 1959; Cressman and Swanson, 1964; Swanson, 1973) formed on both sides of the shelf-to-slope transition along the stable continental margin of North America, and is a classical textbook example of a facies change progression (Fig. 17-11). The Formation also holds a large phosphate resource estimated to be in the region of 15-20 billion tons of P_2O_5 above the grade of 18% P_2O_5, 1.5-2 Bt fluorine and some 3 to 10 Mt of contained uranium (in Montana, the U:P_2O_5 ratio is about 1:3,000). In addition to this, there is a considerable tonnage of possibly recoverable trace metals for which the information is patchy. In the central Montana division only (about 7,000 km^2) the Phosphoria is developed as an oil shale containing 180 Bbl of crude petroleum, plus up to 0.5% Zn, 0.09% Ni, 0.082% Mo, 0.726% V_2O_5 and 0.011% Se (U.S. Dept. of Interior Release, 1981).

The Permian strata of the Phosphoria Formation are distributed over an area of 346,000 km^2 in western Montana, south-eastern Idaho, north-eastern Utah and south-eastern Wyoming and their thickness rarely exceeds 50 cm. The Formation consists of a variety of rocks, resulting from the mixing of (1) carbonate fluorapatite, (2) calcite and dolomite, (3) clay minerals and detrital quartz, and (4) carbonaceous matter. The mineable phosphate is confined to only two members: the stratigraphically lower Meade Peak Member and the higher

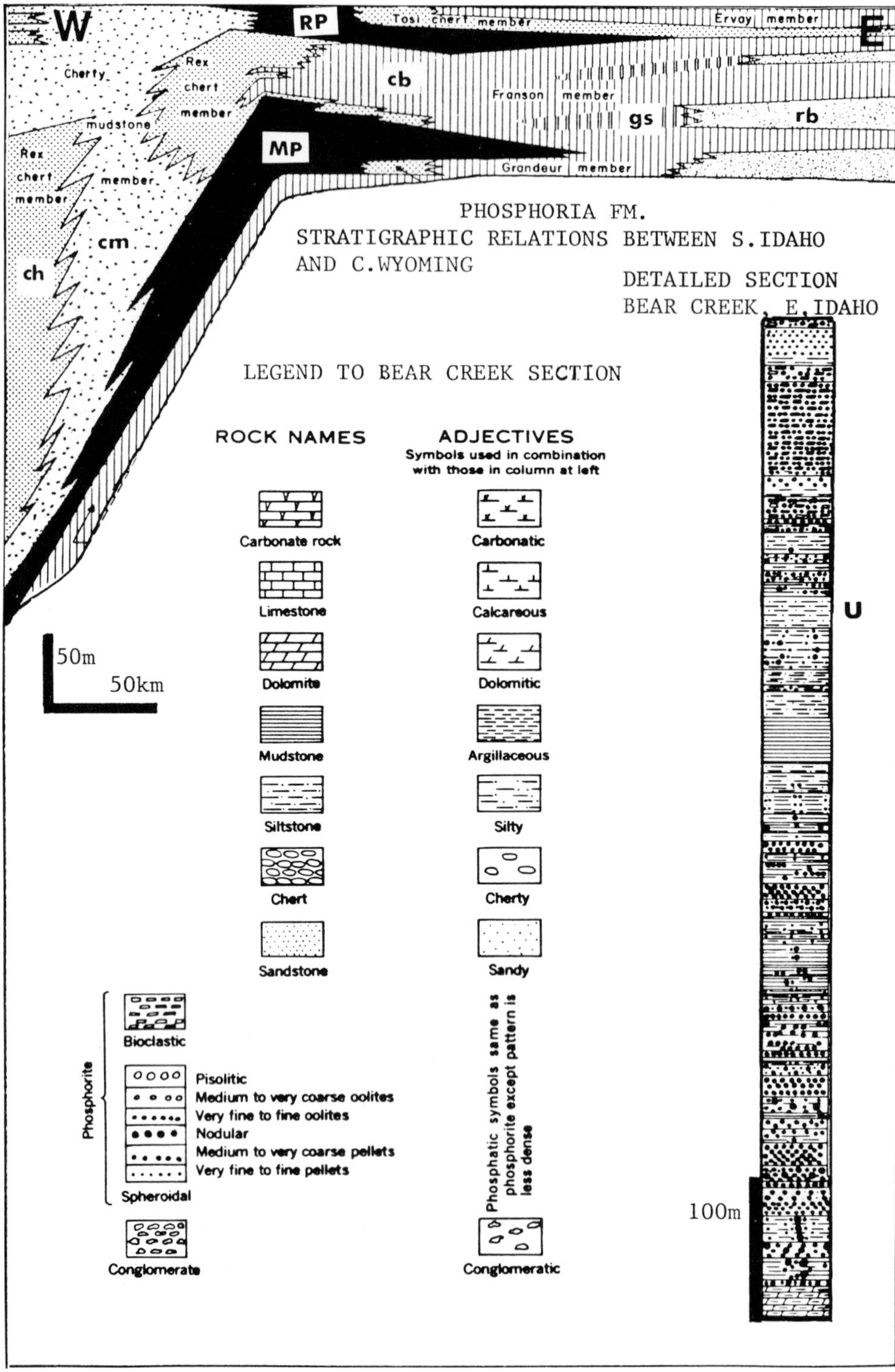
W
E
RP
MP
cb
gs
rb
cm
ch
Tosi chert member
Ervay member
Franson member
Grandeur member
Rex chert member
Cherty mudstone member
Rex chert member
PHOSPHORIA FM.
STRATIGRAPHIC RELATIONS BETWEEN S.IDAHO AND C.WYOMING
DETAILED SECTION
BEAR CREEK, E.IDAHO
U
100m
50m
50km
LEGEND TO BEAR CREEK SECTION
ROCK NAMES
ADJECTIVES
Symbols used in combination with those in column at left
Carbonate rock
Carbonatic
Limestone
Calcareous
Dolomite
Dolomitic
Mudstone
Argillaceous
Siltstone
Silty
Chert
Cherty
Sandstone
Sandy
Phosphorite
Bioclastic
Pisolitic
Medium to very coarse oolites
Very fine to fine oolites
Nodular
Medium to very coarse pellets
Very fine to fine pellets
Spheroidal
Phosphatic symbols same as phosphorite except pattern is less dense
Conglomerate
Conglomeratic

Retort Phosphatic Shale Member. In the Meade Peak Member, dark mudstone to argillite and phosphorite are the chief end-members of the mixed rocks (finely pelletal phosphatic mudstone, structureless phosphorite, bioclastic phosphorite composed of fish scales, pelletal phosphatic dolomite or limestone). The member ranges in thickness from zero to about 15 m. In the Conda area, S.E. Idaho, a large buried deposit of up to 10.6 m thick 20% P phosphorite was outlined recently (Gulbrandsen and Krier, 1980).

The Retort Member has developed largely in Montana and is similar, but its constituents tend to be thin-bedded, more variegated (sandstone, chert and carbonates interbedded) and may contain up to 25% of organic matter. A portion of the member is an oil shale. Near Melrose, Montana, the member is 9 m thick and rich in uranium (average 0.015% U). Vanadium is not particularly abundant in the phosphatic members, except in places in Idaho (a 2.3 m thick phosphate bed located near the footwall of the Meade Peak Member, Conda, contains 0.28% V_2O_5). The phosphate-free metalliferous shale bed within the Phosphoria Formation in Idaho and Wyoming (near Afton and Coal Canyon; McKelvey et al., 1959) investigated during World War II as a source of vanadium, contained as much as 1% V_2O_5, 0.1% Mo_2O_3, 1.3% ZnO, 0.3% NiO_2 and 0.02% U. In Coal Canyon this bed is 1.1 m thick.

17.9.3. Cambrian phosphate province of Soviet Central Asia.

Sedimentary phosphorites are constituents of an extensive, 2,000 km long but discontinuous Cambrian "black facies" belt recorded, so far, in eight segments in the Tian Shan orogenic system of the Soviet Central Asia (Kholodov, 1973; Garkovets et al., 1979; Fig. 17-12). In addition to phosphorite, the unit contains a black siliceous shale, argillite to slate, black chert, dolomitized limestone and dolomite, that is rocks comparable with those in the Phosphoria Formation. The most easily accessible, best explored and most fully described portion of the belt is exposed in the north-western Malyi Karatau Range in southern Kazakhstan. There, the phosphatic horizon which is up to 70 m thick, is a member of the lower Cambrian Chulaktau suite and the beds of the medium-to-high grade phosphorite (17-33% P_2O_5) are comparable in all respects to those in the western United States. The phosphate, to some degree, and especially the low-grade phosphorus and phosphorus-free "black" sediments are metalliferous, and

---

Fig. 17-11. Phosphoria Formation (Permian) in the Western Phosphate Field, U.S.A., stratigraphy and lithology. (cm) cherty mudstone; (ch) chert; (cb) carbonate; (gs) greenish-gray shale; (rb) red beds. Black = phosphorite. RP = Retort Phosphatic Shale Member; MP = Meade Peak Phosphatic Shale; U = uraniferoue shale. From McKelvey et al. (1959).

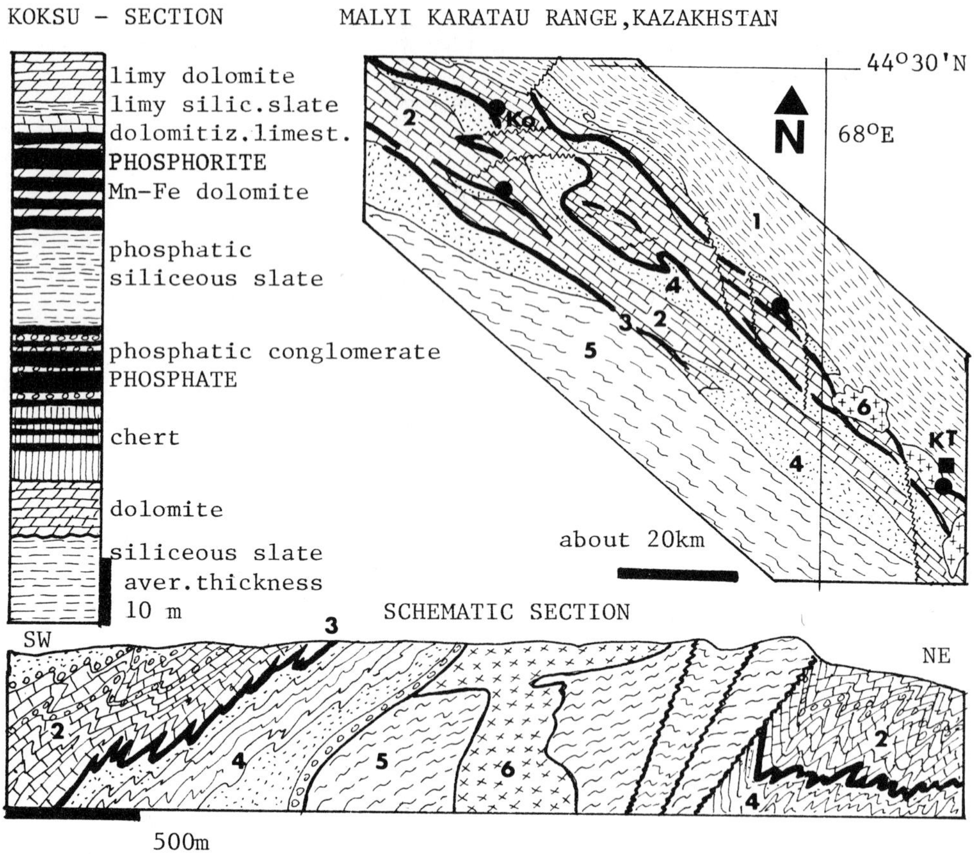

Fig. 17-12. V,Mo,U-bearing phosphorite and black shale, Karatau Range, U.S.S.R.; stratigraphy and lithology. Ko = Koksu (section location); KT = Karatau (town); (1) upper Devonian-Carboniferous terrigenous sediments with subordinate limestone horizons; (2) upper Cambrian-Ordovician dolomitic limestone; (3) middle-upper Cambrian chert, dolomite, phosphorite, black slate; (4) lower Cambrian siliceous slate, sandstone, minor dolomite; (5) upper Proterozoic sericite-chlorite phyllite, quartzite, minor marble; (6) late Proterozoic or lower Paleozoic granitic rocks. Modified after Bespalov (1971) and Sagunov (1971); approximate scales, coordinates added.

(1970) quoted an average of 0.38% V_2O_5 for the entire unit. The highest V values are reported in a horizon of black cherts immediately underlying the phosphatic rocks and up to 30 m thick (1.12% V_2O_5), interbedded with subordinate dolomite. Vanadium is present as V-anthraxolite, V-sericite and roscoelite. The same horizon is enriched in Ba (1.2-2.8% BaO), Cr (0.12-0.19% Cr_2O_3), Mo (no data but estimated at about 500-800 ppm), Se, Re, Cu and U.

In the black chert horizon, uranium is carried by the organic matter and occasional lenticular diagenetic pods (nodules) and is accompanied by pyrite. The uranium content appears to be slightly higher in the chert horizon than in the Phosphoria Formation phosphate (0.01-0.06% U; Adyshev and Kalmurzayev, 1968). Another minor peak of U enrichment lies in the upper black cherty argillite. No grade or tonnage data have been released but using the fragmentary information available with gaps filled by the Phosphoria Formation data considered analogous, the Karatau segment alone may be conservatively estimated to contain some 300 Tt U, 1 Mt Mo and 10 Mt V. The Balasanskandyk deposit is probably the largest operation. The ore constituents are normally dispersed ("invisible") and only occasionally indicated by the presence of minor metallic mineral grains in the epigenetic gash veinlets cutting the metalliferous horizon, or by rare V oxide coatings on weathered outcrops. The sequence is folded, thrust-faulted and intruded by pre-upper Devonian granitic rocks.

17.9.4. Other occurrences of metalliferous phosphates and/or associated black argillites / cherts

Additional Soviet occurrences of sedimentary phosphates, some of them metalliferous, have been reviewed by Gimmel'farb (1958). The Cambrian V and Mo-bearing phosphorite unit in the Baikonur area, Kazakhstan, is developed along the western flank of the Ulutau Precambrian uplift and closely comparable with the Karatau occurrences. Pustovalov et al. (1965) reported the presence of zirconiferous phosphorites from northern Kazakhstan. The main minerals there are microcrystalline varieties of zircon ("gelzircon" and arshinovite). The Zr content is of the order of X0 Tt Zr. In the Kuznetsk Alatau Range of Siberia (Krasil'nikova et al., 1965), Cambrian black siliceous slates interbedded with submarine basalts, volcaniclastics and subordinate limestones, contain lenses and interbeds of phosphorites. These are either siliceous, gradational to cherts or carbonatic, gradational to dolomites and limestones. Some of the limestones are manganiferous. In the Tamalyk phosphate deposit (50 Mt 12-30% P_2O_5 phosphorite) up to 1% V, 0.25% Zn and 1 ppm Ag are present in the phosphorite and associated rocks. On the eastern slopes of the Kuznetsk Alatau, the same phosphatic unit is enriched locally by Mo (up to 0.1% Mo), in addition to V.

In the Gataga River area, north-eastern British Columbia, Ordovician gray and black metalliferous argillites and black limestones contain Ba, Zn, Ag, Mn, and V, and are located in the

Zn, Pb, barite province of Selwyn/Kechika Basins described earlier. The striking orange-red gossans containing sulvanite so impressed Rose (1973), that he dedicated a short poem to them.

17.9.5. Phosphate or black sediments / tectonomagmatic interaction (V, Mo, U aspect)

Mineralizations directly associated with any evidence of tectono-magmatic interaction with sedimentary phosphates are rare and controversial. Bedded phosphates and postmagmatic-hydrothermal Zn-Pb deposits coexist at several localities in Mexico (Concepción del Oro, Zimapán). The phosphates are known to have anomalous Zn, Pb and Ag contents, but no evidence that the sediments acted as a source rocks has been presented.

The Cambrian Mo- and V-enriched phosphorites of the eastern Kuznetsk Alatau correlate with a Devonian Mo (Cu) "porphyry" province, so that a regional heritage is possible but not proven.

Uranium mineralization spatially associated with the coincidence of a black slate or schist basement and tectono-magmatic interaction appears to be relatively common in Phanerozoic sequences (and even more so in the Precambrian; Volume 2) and included in the Ruzicka (1971) classification of uranium deposits. His example locality is an anonymous Soviet deposit hosted by Paleozoic pyritic graptolithic black shales and dolomitic limestones, and containing pitchblende, U-pyrobitumen, and associated clay minerals veins. Somewhat similar occurrences were investigated south-east of Zdice, Czechoslovakia, along faults disrupting Silurian graptolitic shales with cherty interlayers. Short mobilization veins filled by uranoan anthraxolite are also found in the thermally metamorphosed late Proterozoic black slates on the fringe of the Chvaletice pyritic and Mn carbonate zone, reviewed earlier.

17.10. W, Sb, Hg ASPECT

EASTERN ALPS

The association of tungsten, antimony and mercury in ore deposits and fields (in the classical postmagmatic-hydrothermal model, these metals are located at different ends of the temperature spectrum) until recently, had been rarely emphasized in textbooks. There is, however, a rare but distinct group of deposits and metallogenic belts where these metals coexist. Most of the examples are associated with black slates or schists, that usually also contain some accessory volcanics or grade into (meta)volcanics-dominated terrains. In rare cases (e.g. the Yellow Pine field, Idaho) the host association is dominantly plutonic but the "black sedimentary" ancestry can usually be demonstrated. The credit for bringing widespread public attention to the W-Sb-Hg association in slate and schist belts goes to the

Munich University research group (e.g. Höll and Maucher, 1976).

In the study area of the Munich school, the Eastern Alps of Austria, the W-Sb-Hg occurrences are confined to the Paleozoic "geosynclinal" schist belts in the Alpine core. The stratigraphy and lithology of these belts is comparable with those in the Paleozoic terrains of the Variscan Europe and in the Carpathian core and both have already been reviewed to some extent in Chapters 10-12.
The bulk of the W-Sb-Hg occurrences are hosted by Ordovician and Silurian portions of these belts composed of low-to-high grade metamorphosed, deeper-marine metasediments and metavolcanics. The dominant rock types are gray-to-green phyllites and schists, followed by greenstones (mafic metavolcanics), porphyroids (felsic metavolcanics), black phyllites or schists, and carbonate marbles. Although the East Alpine core has occasionally been designated as a "W-Sb-Hg province", these metals usually do not form common occurrences and are locally separated. Höll and Maucher (1976) recognized several parallel metallogenic belts marked by maximum representation of these three and other (Au, As, Cu, Pb-Zn, Mo) metals. They attributed the apparent metallic zoning to different depth zones above a former subduction zone; an interpretation that appears rather unconvincing.

The scheelite deposits in the Alpine schist belts are remarkable in that they are not visibly associated with granitic contacts and skarn host rocks; consequently, they are visually very inconspicuous. The localities in which scheelite is hosted by mafic volcanics or metasediments in their immediate proximity (e.g. Lanersbach and Felbertal) have already been treated in Chapter 10. In the Kleinarltal occurrence, although the scheelite-hosting unit is also underlain by metabasalts, the volcanics are not much in evidence and the host rocks are Silurian dolomites interlaminated with black phyllites. There, graphitic, gray-banded dolomite contains finely disseminated, megascopically unrecognizable scheelite considered of diagenetic origin (Höll and Maucher, 1976). This is accompanied by small quantities of pyrite and pyrrhotite. Remobilization masses several centimetres across of almost pure beige to wax-coloured scheelite occur in the same locality with or without quartz in a coarse-grained, light dolomite.

The stratabound stibnite occurrences of the Alps and Carpathians consist of peneconcordant stibnite, pyrite, lesser arsenopyrite, scheelite lenses and short cross-cutting remobilization veins, confined to units dominated by interbedded mafic metavolcanics and black phyllites (Chapter 10). The actual host rocks to the stibnite orebodies, however, are usually the black metasediments. A 5 m thick horizon in the Rabant field is almost 7 km long and within the Kreuzeckgruppe stibnite zone the deposits are confined to a unit up to 100 m thick and traceable for 30 km (Höll and Maucher, 1976).

Some of the Alpine cinnabar occurrences are, like stibnite, hosted by black schists interbedded with mafic metavolcanics but the most important ones including the Yugoslav deposit, Idrija, are in an interaction situation involving black sedimentary carbonates, accessory mafics and an intensive deformation. Despite the complexity, this is a setting repeated in several important mercury fields of the world, treated in a separate section below.

17.10.2. Ferberite or hübnerite, stibnite, cinnabar mineralization in black slates to schists

A genetically very instructive tungsten province showing a wide range of black schist-hosted ferberite orebodies, is in Rwanda and Uganda. Being of Proterozoic age, however, it is described in Volume 2. The Phanerozoic equivalents are of lesser importance. In the Barun-Shiveye deposit (East Transbaikalia, U.S.S.R.; Denisenko and Rundkvist, 1977), Paleozoic or late Proterozoic black meta-quartzite and graphitic schist contain fracture and breccia-filling bodies of ferberite, lesser stibnite, cinnabar and minor common Fe, Cu, Zn and As sulphides. The ore zone is 1,700 m long, stratabound and the ore lenses average 0.15% WO_3.

Black slate/phyllite-hosted, stibnite-ferberite veins have been further recorded from the Gemer metalliferous region of Slovakia (Spišská Baňa, Chyžné); from Devonian slates of Bolivia (Conde Auque); and elsewhere.

17.10.3. Scheelite skarns

Scheelite skarns evolved from the mineralogical curiosity that they were fifty years ago into a dominant tungsten mineralization type, and more than a thousand occurrences have been recorded worldwide. It is becoming increasingly apparent that the bulk of scheelite skarns exposed in continuous, well-studied orogenic belts, are preferentially located in the vicinity of former hinge lines marking the shelf to slope transition, and near the base of the slope (as in the Macmillan Pass and Flat River districts, north-central Canadian Cordillera; there, the "tungsten belt" overlaps with the "black argillite" to platform carbonate facies transition and with Zn-Pb and barite deposits described earlier). Alternatively, scheelite skarns may occur along the edges of former continental blocks (e.g. in the Sierra Nevada, California). In "confused", multiphase orogens, scheelite skarns typically congregate along boundaries separating "continental highs" from deep-water sediments and mafic volcanics containing "troughs".

In a "typical" scheelite skarn field (compare Einaudi et al., 1981), the following rock types or units mutually interact: (1) carbonates; usually pure limestone, less dolomite marbles. These are the actual hosts to scheelite orebodies. (2) Hornfelsed "argillites", particularly graphitic ones; these rarely host scheelite, but may be mineralized by a quartz-molybdenite or molybdenite-pyrite veins or stockworks. In many scheelite skarn regions, minor metabasalt or picrite flows and tuffs or small diabase or gabbro bodies, are present. (3) Intrusive "granitic" body, usually quartz monzonite or granodiorite.

Scheelite skarns come in a variety of local settings: as xenoliths or screens within the intrusive body, at contacts, or at some distance from the contact ("distal skarns"). Because the coeval intrusions producing "distal skarns" may be hidden, these skarns may appear in a fully sedimentary association of dominant slate (schist; the metasediments are usually thermally hornfelsed) and usually minor carbonate beds and lenses. In this case, commonly the first carbonate bed or lens at the base of the sequence (nearest to the postulated "granite" in the subsurface) becomes mineralized. Thin carbonate beds or lenses could be completely skarn-replaced so that they appear concordant with the enclosing "shales". This led to a wave of interpretations of such skarns as "stratiform" in the 1960s-1970s (e.g. King Island, Tasmania; Sangdong, Korea), but this interpretation has recently lost much credibility.

The tungsten source in scheelite skarns remains enigmatic, but the repetitive association of the fine detrital graphite-rich metasediments (pure carbonate sequences never host scheelite skarns !) is indicative of their relationship to a tungsten source. Such a relationship need not be direct (e.g. a lateral secretion of W) but rather it may take the form of a simultaneous interaction and element exchange among at least three of the major rock types listed as suggested, for example, by Dick and Hodgson (1982) for the Mactung (northern Canada) skarn. Scheelite skarns are also discussed as a genetic group in Chapter 28, devoted to plutonic intrusions.

17.10.4. Hydrothermal stibnite deposits in black argillite to schist

Hydrothermal stibnite fissure or shear veins and carbonate replacements are common in locally interacting associations of black slates or schists, limestones or dolomites, and subvolcanic to plutonic intrusions. The complex interaction of the rocks involved is somewhat reminescent of the scheelite skarns although it appears to have taken place at shallower crustal levels. The repetitive association with the black (meta)sediments suggests a source relationship. Example localities include the Zajača (Krupanj) district, Serbia (Janković, 1982; about 90 Tt Sb). There, Paleozoic black slates and phyllites tectonically interdigitated with Triassic and Jurassic limestones along a series of low-angle thrusts, are intruded by Miocene dacite and granodiorite. Stibnite with pyrite and/or Pb, Zn, As sulphides form replacement bodies in limestones surrounded by alteration jasperoids, or fissure veins.

In the Alšar deposit in southern Macedonia (Janković, 1982; min. 15 Tt Sb, 3 Tt As, X00 t Tl) stibnite, realgar, orpiment and minor arsenopyrite and pyrite form a series of replacement orebodies in silicified Triassic limestones and dolomites, enveloped locally by black schist and intruded by Pliocene trachyte and andesite. The presence of the thallium minerals lorandite and vrbaite, locally in economic quantities, makes Alšar unique as the only high-grade Tl deposit of the world.

In the Wadley antimony field, Sierra de Catorce, Mexico (White and Gonzales, 1946; about 70 Tt Sb) Mesozoic limestone is overlain by a black limey shale and Tertiary andesite porphyry dikes, suggesting the presence of a buried pluton. High-grade stibnite ore forms peneconcordant mantos along the limestone and shale contact and a low-grade ore comprises scattered needles of stibnite interspersed in the black chert (jasperoid ?). A significant tonnage of a low grade oxidized ore is also available.

17.10.5. Carbonates, black sediments, minor mafic or ultramafic bodies/ faults, thrusts interaction. Hg (Sb, As) aspect

Highly mobile metals, most notably mercury but to a lesser degree antimony, arsenic and occasionally base and precious metals, form important accumulations in intensively tectonized terrains. One example are the subduction mélanges and their environs (Chapter 8), a further example is represented by the often chaotically coexisting following rock types: (1) shallow to deep marine (pelagic) limestones and dolomites; (2) shallow to deep marine shales and argillites, frequently organic-rich ("black"); flysch or non-graded sandstones are sometimes present; and (3) accessory, but conspicuously present, ultramafics (mainly serpentinite tectonites) and/or basalts, spilites, diabase dikes, or small gabbro stocks, of oceanic affiliation.

Contemporary with deformation calc-alkaline magmatic rocks are usually absent, the metamorphism is light to nonexistent, and the tectonic style is a combination of imbricate lateral thrusting with deep vertical tectonism and is long-lived. Breccia zones, fault gouges, open faults, crackle breccias, infiltration and fracture mineral veining, products of low-temperature hydrothermal metasomatism, and karsting, are common. The dimension of these terrains and of the mineralization they carry ranges from extensive tectonic belts several thousands of kilometres long, to localized disturbed centres.

The geotectonic interpretation of the above terrains received an important boost with the introduction of the plate tectonic model. It appears most likely that we are dealing with collisional systems, in which a (micro)continental block with a previously stable Atlantic-type continental margin sedimentary facies progression collided with an island arc or with another continent. In this process, the shelf and slope carbonates and black detrital rocks overrode and buried the subduction mélange; a possible analogy of the present process taking place near Timor, Indonesia (Hamilton, 1979; p. 175).

The available evidence is compatible with such an interpretation applied to the South Ferghana Hg-Sb belt in the Soviet Central Asia (e.g. Zonenshain et al., 1976): a likely product of a Permo-Carboniferous collision of a northward-travelling Alai microcontinent with the Tian Shan (Ferghana) continent (or block). The colisions, as depicted above, probably produced the metallogenically fertile rock mixture, but the often-present mercury and antimony deposits probably formed later following terminal adjustments (and often alkaline magmatism), and near-surface hydrothermal activity ("hot springs"). The Hg (Sb) metallogeny of the terrains described above appears to be an extension (overprinting, transition) of the metallogeny of ophiolitic mélanges and affiliated units as in California (Chapter 8), and indeed an almost uninterrupted series of transitions between the (1) ophiolite and (2) limestone, shale end-members association, can be assembled using many of the presently known Hg (Sb) districts of the world. The problem of metal sources is probably identical to the problem discussed in Chapter 8. If the ophiolite ultramafics are the Hg (Sb) source, it is easy to explain the lift of the released mercury from the deeper level of the system where the ophiolites are more plentiful, to the higher portions dominated by excellent traps in fractured and altered carbonates. The silica-carbonate (listvenite) alteration in the Hg-mineralized ophiolites is substituted by silicification in the carbonates, resulting in jasperoid generation. Shales (argillites) present in the system probably have a variety of derivations: from the ophiolite-coeval and ophiolite-derived ones (these may have acted as subsidiary or temporary traps for mercury, and they may carry stratiform cinnabar as probably the Skonca shales in Idrija) to the shelf/slope and post-collisional ones. The latter may have acted as a sulphur source (from dispersed pyrite) or just as an impervious screen to the hydrotherms, topping the open and replaceable footwall carbonates. The latter interpretation is popular with the Soviet school with some justification, because in the South Ferghana belt and elsewhere, many peneconcordant Hg-Sb orebodies are hosted by silicified limestone just below shale.

The above lithologic/tectonic association and Hg-(Sb) mineralization (Table 17-6) is not very well known in the West, because the economically significant occurrences are located mainly in the U.S.S.R. and in Yugoslavia. This association and its mineralization style, however, may further combine with a younger volcanism or plutonism (e.g. Monte Amiata, Italy; Huancavelica, Peru) to produce polygenetic ores.

SOUTH FERGHANA MERCURY-ANTIMONY BELT, KIRGHIZIA, U.S.S.R.

This is an east-west trending belt 500 km long, located along the southern margin of the Ferghana Intermontane Basin, within the Tian Shan orogenic belt in Soviet Central Asia (Fig. 17-13). The belt contains the largest Soviet antimony and mercury deposits and probably substantial reserves of both metals which, in the absence of any officially published data, appear to exceed 200 Tt Hg and 300 Tt Sb. Briefly, the belt (Nikiforov, 1968, 1970) is underlain by middle-upper Paleozoic (Silurian to middle Carboniferous) supracrustal rocks.

Table 17-6. Hg,Sb deposits in regions of interaction of carbonates, black detrital sediments, minor mafics - ultramafics, faults and thrusts; selected locality examples

LOCALITY		HOST AGE	GEOLOGY,MINERALIZATION	REFERENCES
South Ferghana Hg-Sb belt	Khaidarkan	Cb_{1-2}	nests of rich cinnabar,lesser stibnite,py,realgar,orpiment and quartz,calcite in jasperoid breccia,replacing limestone under shale cover; underlain by veinlet,dissem.cinnabar in footwall limest.along faults	Nikiforov et al. (1962)
	Kadamzhay	Cb_1	stibnite,lesser py,marc,realgar,orpiment in calcite,qtz, fluorite gangue,envel.by jasperoid replac.limest.breccia under a thrust; host massive limest.under shale,faulted	as above
	Symap	$D-Cb_1$	cinnabar,lesser stibn.,realg,orpim.,veinlets,dissem., in gray and dolomitiz.limest. and clay fault gauge under overthrust	Kuznetsov (1974)
Aktash,Gornyi Altai, USSR		Cm-Or	stockworks of calcite,cinnabar,py,minor realgar,orpiment, schwazite,in dolomitiz.,fractured and brecc.limest.under overthrust; minor greenst.,serpentinite slivers, subordinate cinnabar in silica - carbonate altered serpentinite	as above
Idrija,N.W. Yugoslavia		Cb-Eo	14 large oreb.in 1,500 x 400 m zone contain epigenetic,disseminated and veinlet cinnabar,py,calc,quartz, and minor low-grade stratiform ? cinnab.dissem.in Tr_2 black and tuffac. shale,argillite; 200 Tt Hg/0.1-3%	Berce (1958)
Si Kon Shan, Hunan, China		D	1-12 thick peneconc.stibnite,lesser py,qtz,fluor.,replac. oreb. in silicif.brecc.limestone (jasperoid), immediately under black argillite - slate cover; controlled by third order anticlines in footwall of a major overthrust; 2 Mt Sb/3.5%	Jovanović and Ramović (1981)

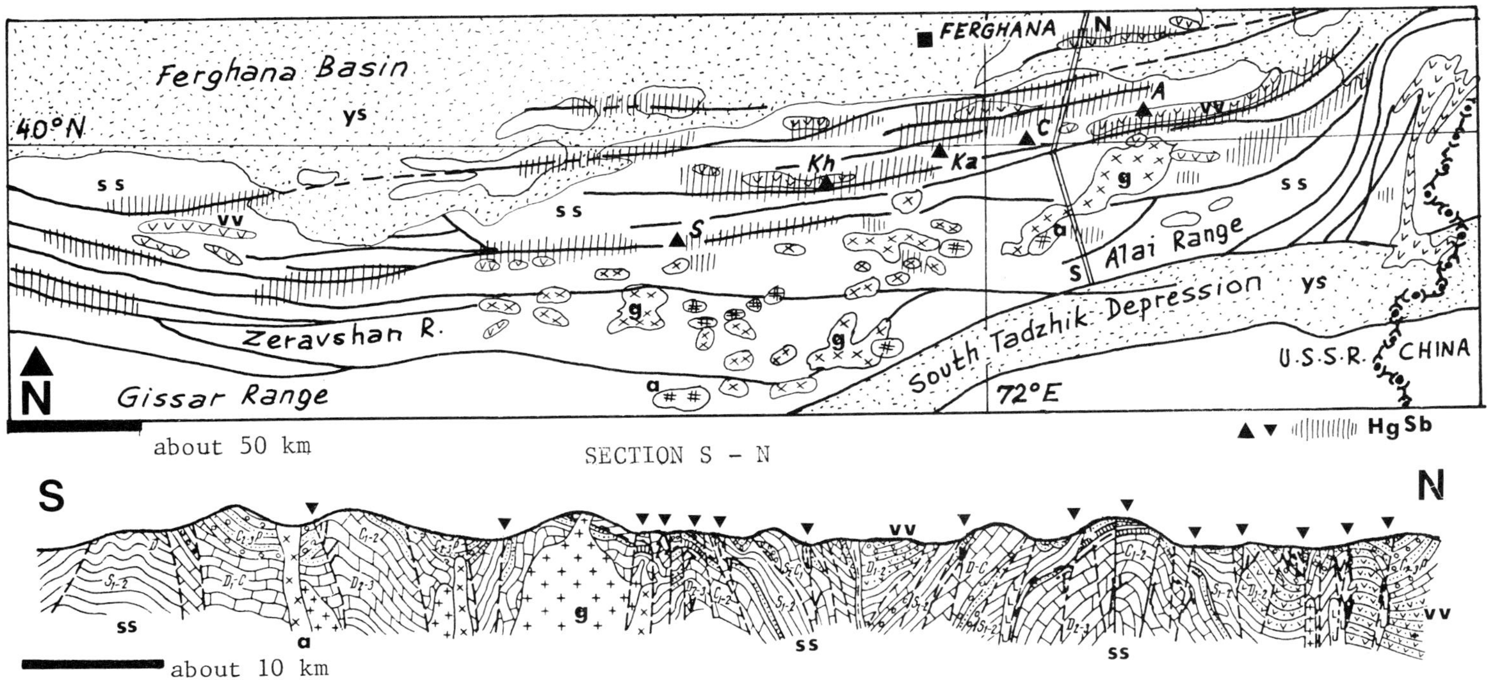

Fig. 17-13. South Ferghana Hg-Sb belt, southern Tian Shan, U.S.S.R. LEGEND: (ys) Tertiary detrital sediments (molasse); (vv) D1 to Cb1 submarine mafic metavolcanics, ophiolite, serpentinite; (ss) S_1-Cb_2 carbonate, slate, litharenite; (g) Cb_2-Pe diorite, granodiorite, granite; (a) Pe alkaline intrusive rocks. MAJOR DEPOSITS: S = Simap; Kh = Khaidarken; Ka = Kadamzhai; C = Chauvay; A = Aktash. Section after Nikiforov (1968), map compiled from data in Nikiforov (1968), Nikiforov et al. (1962), Magak'yan (1968) and others. Scale and orientations added.

The northernmost facies zone contains a thick sequence of Devonian spilites and metabasalts (oceanic assemblage). South of it, Devonian-to middle - Carboniferous dolomites and limestones rest conformably on the Silurian terrigenous sediments of a former stable continental margin. Middle Carboniferous terrigenous rocks (successor basin facies) unconformably top both earlier facies.

The orogenic system is subdivided into a series of narrow east-west slices by numerous parallel faults that originated during three principal phases of tectogenesis: (1) Devonian normal faulting controlling the volcanic trough; (2) Lower Carboniferous collision, resulting in a series of overthrusts and tectonic emplacement of serpentinite tectonite slices; and (3) late Carboniferous and Permian block faulting. The latter event was accompanied by an intrusion of calc-alkaline plutons into the southern ("continental") zone, and minor alkaline intrusions (syenodiorite, monzonite). The Hg and Sb deposits postdate the late Paleozoic block faulting but predate the Alpine tectogenesis. Thus they are probably Mesozoic. In both facies belts the ores are intimately controlled by faults and hosted by both the carbonates and the mafic metavolcanics. The ores, however, do not continue into the Carboniferous terrigenous sediments which are believed to have acted as an impervious cap to the ascending hydrotherms. On a regional scale, the ore distribution has a pattern of "coulisses": long and narrow east-west screens of concentrated mineralization on both sides of faults. Nikiforov (1970) distinguished three parallel mineralization zones. The northernmost zone (e.g. Chonkoi, Sarytash deposits) contains Hg deposits hosted by the mafic volcanics, ultramafics and interbedded sediments. The most common styles are calcite, ankerite, cinnabar veins and stockworks in the sediments, as well as cinnabar impregnations in talc-carbonate altered mafics and ultramafics identical with the "California Hg type". This zone is the least productive.

Economically most important is the Central Zone which contains the bulk of the Hg-Sb deposits. Most of it is confined to the hydrothermally silicified roof of fractured carbonates directly under the Carboniferous shales, in the cores of thrust-modified anticlines (e.g. Khaidarkan, Kadamzhai). The orebodies are peneconcordant with the bedding (stratabound), sub-unconformity and their distribution is strongly reminescent of anticlinal petroleum pools. The ore consists of impregnations, stockworks and aggregates of calcite, dolomite, stibnite and cinnabar. A less productive mineralization style in the Central Zone occurs along faults and crush zones in the limestone monoclines, and consists of simple calcite-cinnabar replacements, stockworks, and veins (e.g. Symap).

The Southern Zone has complex ores resulting from superimposition of Hg and Sb on an earlier W, As, Pb, Zn, and Cu hydrothermal-plutonic suite. The main carrier of antimony and mercury is tetrahedrite and the ores seem to be of limited importance.

The largest ore field, Khaidarkan (Nikiforov et al., 1962) lies in the Central Zone and is an example of the "conformable" type hosted by silicified (jasperoid) limestone breccia in the footwall of the northerly-dipping Ishmetau Thrust. The ore zone is about 15 km long (east-west) and about 1 km wide. The orebodies are controlled by third-order anticlines (Fig. 17-14) and occur along faults in the culminations of lower Carboniferous limestone just under the confining cover of the middle Carboniferous shales. About eighty percent of the rich orebodies (about 2-5% Hg, some Sb) are located within 30-40 m under the shale / limestone contact. The rich ore is composed of nests and aggregates of cinnabar and stibnite in quartz, fluorite and calcite gangue. Realgar and orpiment are common and dominate the small orebodies present in the hangingwall shale. The rich ores are usually floored or enveloped by a low-grade disseminated and veinlet cinnabar and stibnite. The Aktash field is very similar.

Kadamzhay (Nikiforov et al., 1962) represents the largest antimony deposit in the South Ferghana belt (and probably in the entire U.S.S.R.). The style and host association are almost identical to that in Khaidarkan (sub-unconformity orebodies in lower Carboniferous limestones under Devonian shales) but stibnite is the principal mineral associated with rare pyrite, marcasite, realgar and orpiment. Quartz, fluorite, calcite and barite are the main gangue minerals and a silicified limestone breccia (jasperoid) is the principal host of the peneconcordant, 5 km long ore zone. The ore distribution in the breccias is uneven and a subsidiary lower-grade stibnite mineralization occurs as veins and limestone replacements along normal faults.

SI KON SHAN (SHUIKONSHAN) Sb DEPOSIT, CHINA

The deposit Si Kon Shan (200 km west of Changsha, Hunan; Jovanović and Ramović, 1981) with its original reserve of 2.105 Mt Sb (aver. grade 3.5%) is the world's largest antimony accumulation and is virtually identical in style with Kadamzhay. Mineralogically simple stibnite, lesser pyrite, quartz, galena, calcite and fluorite peneconcordant bodies are hosted by jasperoid, replacing fractured Devonian limestone immediately under an impervious cover of Devonian black shales. The orebodies are up to 12 m thick, several metres long, and confined to a N.N.E.-S.S.W. ore zone controlled by Devonian lesser order anticlines in the footwall of a major overthrust. The age of mineralization is lower Cretaceous (120 m.y.).

IDRIJA, NORTH-WESTERN YUGOSLAVIA

The Idrija mercury field (Bercé, 1958; about 200 Tt Hg) is, without doubt, the most frequently mentioned (Janković, 1982; Höll and Maucher, 1976) major representative of the present association. The geology near Idrija is extremely complex with the stratigraphic sequence ranging from Carboniferous to Eocene and involving a variety of black shale, sandstone, shale and limestone. Mafic and bimodal "geosynclinal" volcanics as well as tuffs and radiolarites suggesting

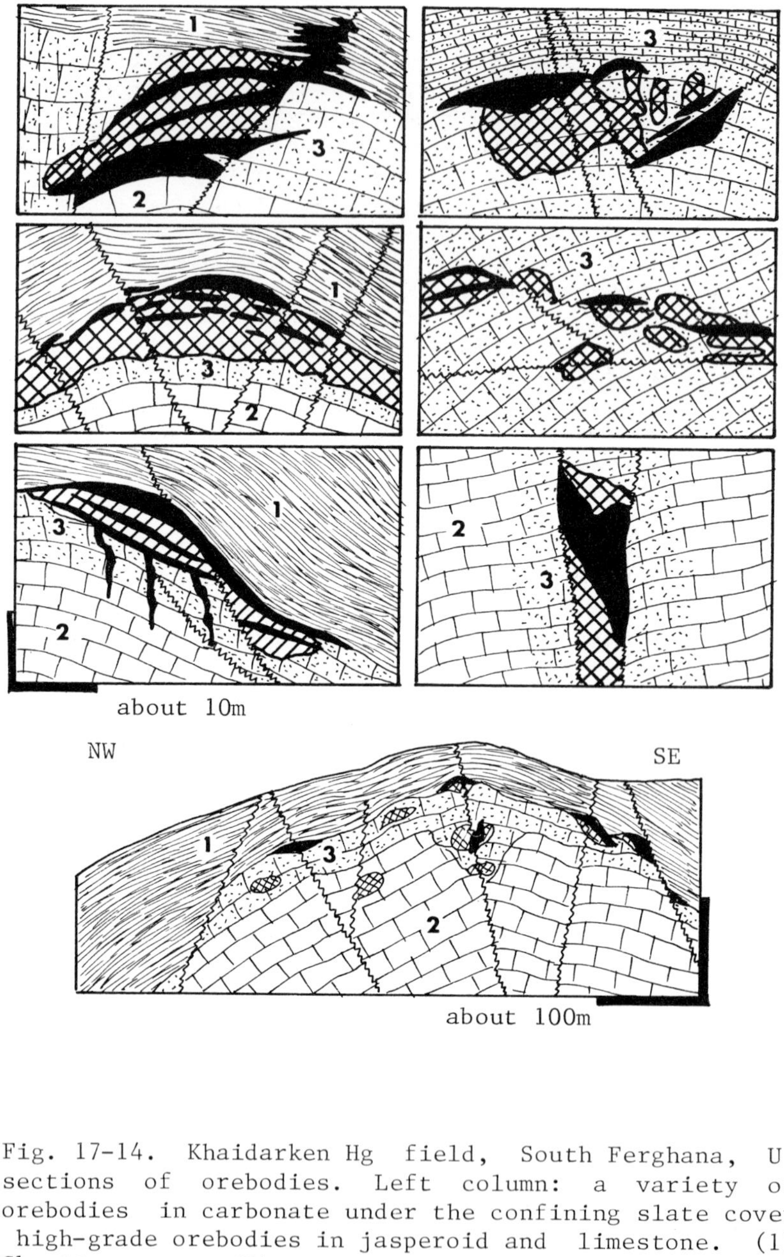

Fig. 17-14. Khaidarken Hg field, South Ferghana, U.S.S.R. Detailed sections of orebodies. Left column: a variety of high-grade Hg orebodies in carbonate under the confining slate cover. Right column: high-grade orebodies in jasperoid and limestone. (1) Cb_2 slate; (2) Cb_2 limestone; (3) silicified limestone; BLACK = massive Hg ore. Net pattern = rich veinlet-disseminated Hg ore. After Nikiforov et al. (1962).

ophiolitic affiliation, occur in the mine field. The deformation is severe and takes a form of a series of epigenetic imbricate overthrust sheets modified by younger block faults. Numerous orebodies of disseminated, veinlet, vein, replacement cinnabar, pyrite, minor mercury and metacinnabar with or without dolomite, calcite, quartz and chalcedony gangue, occur in most of the rock types present (Fig. 17-15). They are most abundant on the footwall-side of the thrusts. The mineralization is the product of several successive stages, and the middle Triassic stage, contemporary with the very restricted deposition of spilites and keratophyres in the region, is generally considered to be the most important. Several tuff, shale and chert horizons contain minor concordant cinnabar mineralization.

The black, calcite-veined Carboniferous argillites have locally developed pyrite lenses containing native mercury. Most of the "stratified" orebodies, however, are hosted by the middle Triassic Sconca black shales and consist of uniform cinnabar impregnations parallel to the shale lamination. This is sometimes cut by transverse cinnabar and calcite-cinnabar veinlets. The observation that the richest cinnabar orebodies formed in brecciated carbonates under impervious shales, so pervasive in the South Ferghana belt, also holds true in Idrija. It is possible that a portion of the "stratiform" cinnabar impregnations are actually epigenetic replacements of pyritic or limey beds.

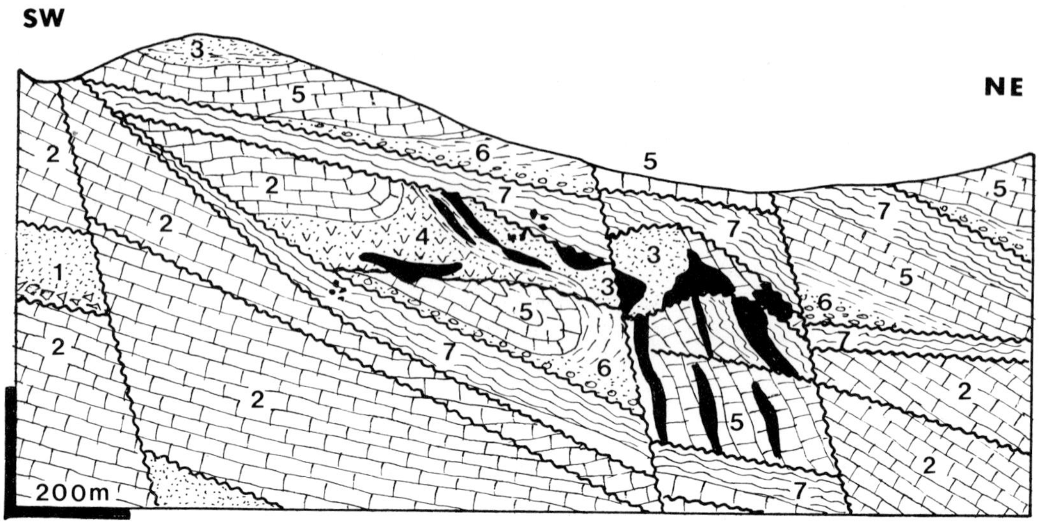

Figure 17-15. Idrija, Slovenia (Yugoslavia) polygenetic mercury deposit, a cross-section. LEGEND: (1) Eo flysch sandstone, marl, limestone breccia; (2) Tr_3-Cr limestone and dolomite; (3) Tr_2 conglomerate, sandstone, shale, black shale; (4) Tr_2 grayish-green volcanic arenite, shale, siliceous radiolarite; (5) Tr_1 limestone, dolomite, shale; (6) Pe-Tr_1 red-beds; red conglomerate, sandstone, arkose, shale, bituminous dolomite; (7) Cb black argillite to slate. BLACK: cinnabar, locally native mercury, orebodies.

From LITHOTHEQUE, modified after Mlakar and Drovenik (1972).

"All features of sedimentogenesis are developed with greatest clarity and completeness under passive tectonic conditions, satisfied in the greatest degree by the tectonic conditions of the platform".
N. M. Strakhov in Principles of Lithogenesis, 1962.

CHAPTER 18

Continental Platforms
(introduction to Chapters 19-27 and 32-36)

18.1. GENERAL

Continental platforms or cratons (not to be confused with oceanic platforms or thalassocratons) are first-order geotectonic units of the continental crust. They have many definitions. According to the A.G.I. Glossary of Geology, platform is "a part of a continent which is covered by flat-lying or gently tilted strata, mainly sedimentary, which are underlain, at varying depths, by basement rocks that were consolidated during early deformations". Craton is "a part of the Earth's crust which has attained stability and which has been little deformed for prolonged periods". Others consider platforms as crustal blocks moving as rigid units, or an entity otherwise called "geoblocks".

As major constituents of continents, the isometric platforms contrast conspicuously in style with the linear mobile belts. The present platforms are either subaerially exposed or covered by seawater as part of shelves and inland seas. The exposed platforms mostly have the physiographic character of plains, relatively smooth sections of the continental surface, occupied largely by gentle rather than steep slopes and exhibiting only small local differences in elevation. This appearance changes in erosion-dissected plains, where excellent outcrops are available in canyon walls or on mesa (table mountain) slopes (Fig. 18-1)

The "most normal" (model) platforms have a consolidated, deeply eroded Precambrian crystalline basement, unconformably covered by flat-lying to gently dipping Phanerozoic sediments. Shields are large domal outcrop areas of the Precambrian basement, completely surrounded (e.g. Arabian Shield, Siberian Platform) or partly surrounded (on three sides; Canadian Shield) by the cover sediments. Other shields are exposed along margins of large platforms, usually in fault contacts with adjacent mobile belts (e.g. Aldan Shield, Siberian Platform). Smaller-size equivalents (or severed continuations) of "typical" platforms are represented by some median massifs (e.g. the Pannonian Basin of Hungary) or by large plateaus (e.g. Colorado Plateau), surrounded by mobile belts.

1
2
3

Fig. 18-1. Variations in physiography of platforms. (1) Flat featureless plain smoothed by glaciation, totally lacking outcrops. Prairies in southern Saskatchewan. (2) Flat featureless arid plain, terminating in sea cliffs. Nullarbor plain, southern Australia. (3) Initial stage of plain dissection by streams; outcrop is in the valley slopes. The Red Deer Badlands, Alberta. (4) Relict mesa in the non-glaciated portion of the American prairies. South Cave Hills, South Dakota. (5) Common outcrop in a "canyonland"-style topography. Near Kayenta, Arizona, Colorado Plateau.

"Young platforms", "quasicratons", "pseudoplatforms", "epiplatforms", "plates", etc. are terms usually given to basement crystalline blocks stabilized in the post-Precambrian period (usually in the Paleozoic), and covered by younger sediments.

Platforms are sufficiently well covered in most text and reference books on general geology, structural geology, stratigraphy and metallogeny. Salop and Scheinmann (1969) reviewed their tectonic history from the Soviet point of view; Conybeare (1979) summarized the lithostratigraphy; Laznicka (1981b) contrasted the metallogeny of platforms with the metallogeny of mobile belts. In the literature, platforms are treated with a different emphasis. To a stratigrapher or a sedimentologist, essentially platforms represent only the sedimentary cover sequences. A structural geologist, on the other hand, is concerned more with the crystalline basement, as well as with the structural modifications of the cover. The latter are numerous, ranging from gently sagged circular to elliptical "basins" (=thickened sequences) and gently elevated atectonic "ridges", through fault-bound horsts and grabens; aborted rifts (aulacogenes) to rifts and intensively activated areas (the latter are treated in Chapter 30).

The student of metallogeny is caught in the middle, and papers or chapters dealing with "platformic metallogeny" are usually uneven in scope. Some include the metallogeny of shields, a subject treated in Volume 2 of this book. Others treat the sedimentary cover metallogeny only. Still others cover "everything", including the Precambrian basement, the cover, and the effects of "activation" ranging from the initial (e.g. propagation of mineralized faults from the basement to the cover), to the far-reaching ones (including layered mafic differentiated complexes and alkaline intrusive provinces). No wonder, then, that the metallogenic appreciation of platforms varies widely, from statements such as "platforms are virtually unmineralized" to accounts extolling, in the platformic context, the virtues of important but lonely giants such as the Bushveld or Witwatersrand to such an extent that it might appear that platforms in general are significantly mineralized domains.

18.2. BRIEF GEOLOGY AND LITHOLOGIC ASSOCIATIONS OF PLATFORMS

Subaerial continental platforms underlie 96.3 Mkm^2 of the Earth's surface (of this, 29.4 Mkm^2 are shields, 66.9 Mkm^2 sedimentary cover), and their underwater regions account for an additional 26 Mkm^2 for a total of 122.3 Mkm^2 (about 24% of the crust). In comparison, "geosynclinal" orogenic belts cover 90.7 Mkm^2 (Ronov and Yaroshevsky, 1969). The average thickness of the platformic sediments is 1,800m, their volume 135 Mkm3 (25% of sediments on continents). Among this, 49.5% are mudrocks; 23.6% sands, sandstones and conglomerates; 22% carbonates; 1% evaporite salts; and 3.9% volcanics (mostly plateau basalts). Compared with mobile belts, platformic sediments are usually unmetamorphosed, undeformed to gently folded, and there is a high proportion of semiconsolidated (porous) and unconsolidated sediments. Many sediments are still filled by the original formational waters. The sediments underwent diagenesis of moderate intensity over a long period (Siever, 1968). The ratio of the first cycle sediments to the repeatedly reworked ones is substantially lower on platforms than the comparable ratio in orogenic belts, and platformic sediments have a higher overall maturity. Compared with the "geosynclinal" suite, platformic sediments are lower in their average content of Al,Fe,Mn,Na and higher in Ca,K,Mg,CO_2 (this, basically, reflects the greater proportion of carbonates). They are also significantly depleted in virtually all the metallic trace elements which are present in sub-clarke quantities.

There are local exceptions, however, due to geochemically favourable environments (e.g. black shales) combined with anomalous "first cycle" supplies (e.g. airborne volcanic ash). In most other respects, platformic sediments do not substantially differ from those in mobile belts and, moreover, both are transitional so that the shallow marine to continental lithologic associations treated in Chapters 19 to 26 are common to both mega-domains.

Many platformic sedimentary associations are cyclic. Khain (1971; Fig. 18-2) pointed out that the following sedimentary cycles are typical of the fully developed platformic sequence: (1) basal continental association, "gray" and often coal-bearing (humid), or "red" (arid); (2) transgressive marine terrigenous association, often

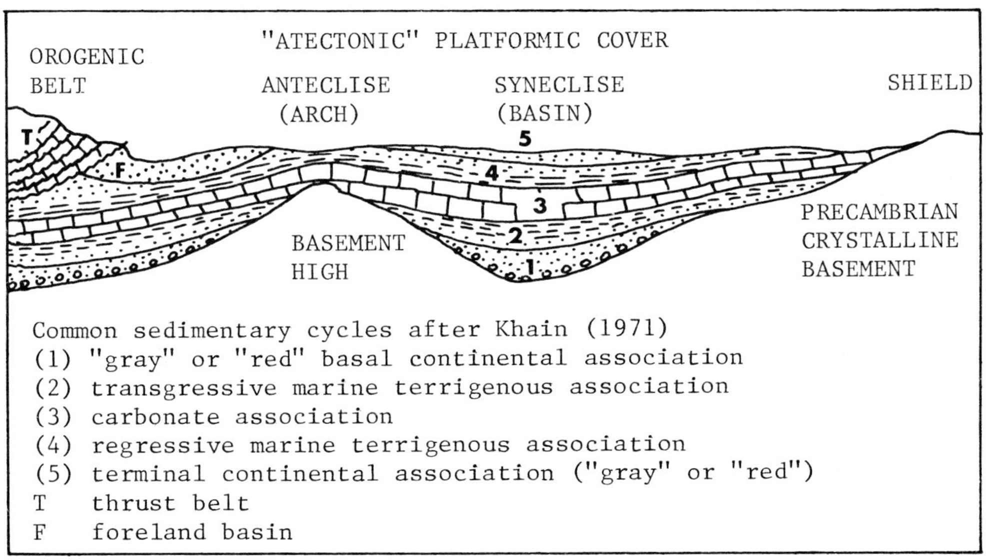

Fig. 18-2. Diagrammatic section of a "most normal" Phanerozoic platformic association. Vertical scale and dips greatly exaggerated.

with glauconite; (3) carbonate association, predominantly limestone in humid climates, dolomite in arid climates; (4) regressive marine terrigenous association; and (5) terminal continental association, "gray" or "red". In the generally "atectonic" (unfaulted, undeformed) portions of platformic cover sequences the thickness of the sedimentary section and the proximity to the surface vary. "Basins" (syneclises; e.g. Michigan, Williston) and "highs", "arches" (anteclises; e.g. Cincinnati Arch), alternate. Foreland basins usually filled by a thick wedge of terrigenous detrital sediments rest on depressed platforms in the foreland of an orogenic belt. A fairly intensive non-orogenic deformation of platformic sequences can be the result of salt diapirism, or alternatively of collapse due to groundwater leaching of salt horizons.

Cratons including sediment-covered platforms are prone to becoming modified by the effects of extensional tectonism (taphrogenesis) sometimes accompanied by magmatism that, unless it stops, ultimately results in a continental fragmentation and drift apart. A wide variety of products of incomplete or aborted fragmentation, however, is preserved in many shields and platforms of the world and treated in Chapter 30. A variety of normal faults, grabens and horsts result, many of which are confined to the basement and sometimes to the lower portion of the sedimentary sequence, are draped by undisturbed younger units. Such buried structures are being detected with increasing frequency mainly geophysically during oil searches. Vast regions formerly considered monotonous (e.g. floor of the North Sea; Ziegler, 1978; portions of the Russian Platform), are now known to consist of a dense mosaic of fault blocks.

Aulacogenes are the most persistent and distinct, deep (8-10 km) and often wedge-like fault-bound grabens. Horsts (uplifts) of comparable dimensions (e.g. the buried Kujave-Pomerania horst in N. Poland; the Boothia Uplift in the Arctic Canada) are designated as para-anticlinoria in the Soviet literature (Khain, 1971). Plateau basalts reach the surface in large quantities along rifts and aulacogenes, but because of their high fluidity they spread over large areas that may be underlain by virtually undisturbed platformic sediments and basement metamorphics (e.g. in Saudi Arabia and Egypt). Plateau basalts are treated in Chapter 31.

18.3. METALLOGENY OF PLATFORMS

Selection of mineralizations usually associated with platforms is given on Fig. 18-2 and Table 18-1. This diorama-style cartoon, however, is unrealistic in giving the impression of a tremendous mineralization density on platforms, which is seldom the case. Large tracts of the "most normal" platforms are virtually devoid of ores. There is, for example, not a single presently mined metallic deposit in the North American Interior Plains between Pine Point and the Upper Mississippi Valley district (an area of about 4 Mkm^2), and only a few metallic occurrences or "resources of the future" (such as uraniferous

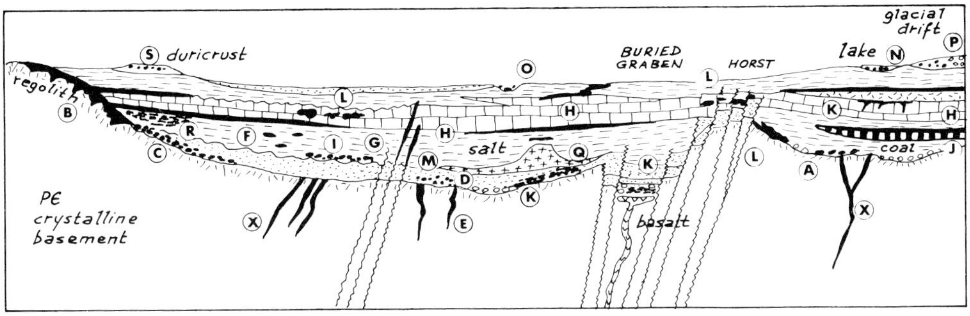

Fig. 18-3. Major mineralization styles of Phanerozoic platforms. See Table 18-1 for explanation of the letter codes.

lignites or scattered Mn nodules).

On the other hand, there are vast regions containing one or more persistent sedimentary ore horizons controlled by facies, paleogeography and an unusually prolific metal supply (e.g. the West Siberian Iron Ore Basin, containing the world's largest calculated iron ore reserves of over 900 Bt of 30% Fe; Mn ores in the South Ukrainian Basin). Other densely mineralized platformic regions, such as the "Mississippi Valley-type" Zn-Pb districts in the central United States (S.E.Missouri, Tri State) or the Kentucky-Illinois vein fluorite, Pb,Zn district, seem to be controlled by the coincidence of several factors one of which is focussed fracturing and faulting ("activation") and/or magmatism in the basement. The aulacogenes are particularly good ore controlling features.

All the mineralizations in platformic sequences (except those in the Precambrian basement described in Volume 2) are treated within the framework of depositional environments and lithologic associations in which they immediately occur (Chapters 19-27 and 32-36). Consequently, there is no need to repeat such descriptions here, beyond the brief summary presented in Table 18-1.

18.4. REGIONAL EXAMPLES

There are about fifteen named platforms with Precambrian basement, shield outcrops and Phanerozoic undeformed sedimentary cover, in the world. In addition to this, there are numerous substantially smaller epiplatforms developed over rigid blocks within orogenic belts, some of which are also shown. Their limits and names are highly subjective. Table 18-2 briefly summarizes and compares metallogenies of four example areas. Fig. 18-4 shows diagrammatically the stratigraphic and lithologic setting of the major ore occurrences in sediments of the North American Platform.

Table 18-1. Metallic mineralizations usually associated with platforms (letter codes correspond to those on Figure 18-3)

(A)	Reworked basement mineralizations on basal unconformities, e.g. paleoplacers (Au, Deadwood-Lead; cassiterite; ilmenite, zircon, monazite, etc.). See also Chapters 19,24.
(B)	Relict metalliferous regolith in the basement. Residual mineralization formed on the subaerially exposed bedrock shortly before burial by the platformic sediments. Examples of exhumed, presently exposed orebodies, include: Ni laterites over Precambrian shield rocks (Ora Banda, W. Australia; Brazil); bauxites on charnockite, S. India; ironstones on banded iron formations; Mn-oxides on metamorphosed protores; etc. Examples of presently buried deposits: enriched Fe ores and bauxite on banded iron formation, Belgorod, U.S.S.R. Regoliths on intraformational beds include some blanket bauxite deposits formed on clays or feldspathic sandstones (e.g. in Guyana). Compare Chapter 23.
(C)	Ores on basement unconformity, such as basement regoliths proximally reworked, now present on base of the sedimentary sequence. This mechanism produces ores resting unconformably on the usually crystalline basement, but conformable with the stratigraphically higher beds. The orebodies are presently exposed (exhumed) or still buried, and the following varieties have been recorded: ore conglomerates (e.g. hematite/magnetite conglomerates adjacent to iron formations; bauxite conglomerates, Arkansas); resistates in sandstone or conglomerate (e.g. ilmenite, leucoxene, rutile, zircon); physically reworked clay/silt consistency ores (e.g. bauxite, ocherous limonite with Ni,Cr,Co, etc.). Compare also Chapters 19, 23, 24.
(D)	Ores on basement unconformity, chemical or biochemical precipitates. Some "U-V sandstone" ores, exotic deposits. Compare Chapters 23,24.
(E)	Ore infiltrations below unconformity (usually in crystalline rocks), formed after sedimentary burial. U infiltrations, Mn oxides. Chapters 23, 24.
(F)	Intraformational stratiform or unconformity - covering ore horizons, formed by reworking of ore from the (regolithic) basement. Examples include ilmenite, zircon, monazite, etc. sands and sandstones; "Mediterranean-type" and other bauxites; limonitic ochers; etc. Chapters 19, 24.
(G)	Intraformational stratiform or unconformity-covering ore horizons, formed by reworking of earlier ore beds, nodules, etc. within the cover sequence (e.g. Cretaceous goethite conglomerate at Salzgitter, formed by reworking of Jurassic diagenetic nodules; Florida "land-pebble" phosphate deposits; etc.). Chapters 19, 20, 24.
(H)	Stratiform metalliferous syngenetic/diagenetic beds or lenses. Examples include cupriferous gray or green horizons in the Permian red beds of Oklahoma; metalliferous beds along facies

	change intervals (e.g. the Kupferschiefer); uraniferous shales (Chattanooga); particulate iron ores; uraniferous phosphorites; and others. Chapters 19, 20, 24.
(I)	Stratiform syngenetic/diagenetic nodular ores (e.g. sphalerite nodules, Bruce Peninsula, Ontario; Mn nodules in tuffaceous Cretaceous shale, Chamberlain, S. Dakota; concretionar chalcocite in the lagoonar Permian deposits of Cisuralia; U phosphate nodules in Miocene carbonate, Florida; etc.). Chapters 19, 20.
(J)	Stratiform syngenetic/diagenetic geochemically contrasting sediments (e.g. coal) containing infiltrated metals. Examples: uraniferous lignites; germaniferous coals.
(K)	Epigenetic, groundwater metalliferous infiltrations. U oxides in sandstones and conglomerates (e.g. Lake Frome, Australia); Cu in sandstone and conglomerate in the red beds association, Cisuralia, U.S.S.R.; U infiltrations in stream calcrete (W. Australia, Niger). Chapters 19, 20, 24.
(L)	Zn-Pb ores (MVT style) of complex origin in carbonates (see a detailed description of the possible varieties in Section 20.7 and Figure 20-11).
(M)	Epigenetic hydrothermal fissure/fault veins. Veins formed after the platformic cover deposition are about equally frequent in the basement as in the cover sediments. In most cases, the veins are interpreted deposited from convecting fluids, heated by minor intrusions that can be concealed in depth. Example: fluorite, sphalerite, galena veins and carbonate replacements, Illinois-Kentucky district, U.S.A. Chapters 19, 20, 24.
(N)	Recent bog, lake and cold spring-precipitated ores (Fe and Mn ochers and nodules; Mn-oxide sinters; U infiltrations in peat bogs). Chapter 24.
(O)	Recent stream placers (e.g. Au, Saskatchewan River, Alberta; Danube, S. Slovakia). Chapter 24.
(P)	Glacial sediments, e.g. Quaternary drift over the Canadian Shield containing minor detrital gold, boulder trains of sulphides,etc. Chapter 24.
(Q)	Metallic mineralization affiliated with salt dome tectonism (e.g. U-oxide infiltrations, Palangana). Chapter 21.
(R)	Metallic mineralization carried by petroleum, gas, bitumens (e.g. trace Hg, U, V). Chapter 22.
(S)	Mineralized tropical duricrusts (e.g. U in calcrete, Namibia; Zr, Nb, Ti in silcrete, Australia; U-V in gypsicrete, Somalia). Chapter 23.

Table 18-2. Comparative Phanerozoic metallogeny of selected platforms (emphasis on sedimentogenic ores; mineralizations in the Precambrian basement are treated in Volume 2, rift / activation metallogeny in Chapter 30, magmatogene ores in Chapters 31-33)

AGE	NORTH AMERICAN PLATFORM	RUSSIAN PLATFORM	AUSTRALIAN PLATFORM	N.W. EUROPEAN EPIPLATFORM
Q	cool springs Mn,Fe (a) bog ores Fe,Mn	bog ores Fe,Mn alluvial ilmenite etc. placers	Au placers, lag residual Fe oxide concretions	bog Fe ores
T_2	reworked U phosphate nodules, Florida (b,G) Mi phosphate nodules in limestone, Florida (C,I)	ilmenite paleoplacers, Centr.Dnieper (C)	Ti-silcrete, Ni,Fe,Mn,V laterite(B) U calcrete (Western Australia ;K) U infiltrations	
T_1	U lignites and infiltrations in sandstone (d,J,K) nodular goethite, N.E.Texas (e); reworked bauxite in shales, Arkansas (f,C)	Ol,diagenetic Mn carbonate and oxide, Nikopol (H) ilmenite, zircon, monazite paleoplacers (C)	residual bauxite	
Cr	Cr-T residual bauxite on nepheline syenite, Arkansas (g,B); Illinois-Kentucky fluorite veins (h,M) particulate ironstone, N.Alberta (i); beach Fe paleoplacers, western U.S. (j); Mn nodules in shale, Chamberlain (k,I); minor Au paleoplacers, Wyoming	ilmenite placer,Korosten concretionary and cementing limonite, Choper (H)	Mn beds and nodules, Groote Eylandt (H)	conglomeratic Fe ores, Peine Ilsede and Salzgitter (G) infiltration U-Zr deposits, N.Bohemia; U coals, Elbe Valley

J Tr	deposition of some MVT Zn-Pb ores (q,L)	reworked Ni,Fe,Cr ochers C. and S.Urals; Kerch Basin particulate ironstones (H)	Dawsonvale particulate ironstones (H)	particulate ironstone (Minette) (H); Silesia-Krakow MVT ores
Pe	Cu shales in red beds assoc., Oklahoma (m,H)	Cu shales, Cu sandstones, Mn oxides in red beds-evaporite assoc.,Cisural, Baltic Republics (H,K)	coal seams, minor Ge	Kupferschiefer, Zn,Pb shales Cu sandstones U,Cu,Ge coals
Cb	Zn,Ge rich coals, Illinois, Ohio, etc.(n); pelosiderite in or close to coal Zn rich marine (paralic) shale, Kansas (o,H) U shale (Chattanooga) (p,H)	Pb-Zn veinlets,concretions in carbonate, sandstone; Valui-Markovo zone (M,I,K) pyrite, alunite in coal association, Moscow Basin reworked bauxite, Tikhvin, N.Onega,Timan (C,F)		U,Ge,Cu rich coals pelosiderites
D	Zn-rich shales, Pine Point area	Yagersk, leucoxene paleoplacer (C); residual buried bauxite and enriched Fe ore, Belgorod,KMA (B)		
S Or	MVT assoc.with Ordovician unconformities; Elmwood, Tenn. (q,L)			
Cm	MVT in carbonates, S.E.Missouri (q,L) Pb,Ni,Co sulphides replac. matrix of sandstone, S.E. Missouri (r); small Au paleoplacers, Black Hills	V and low U bearing oil shale, Estonia (H)	low U phosphorites, Georgina Basin (H)	uraniferous Alum Shale (H)
		NOTE: letters in brackets indicate ore styles as shown in Figures 18-2 (lower case) and 18-1 (upper case)		

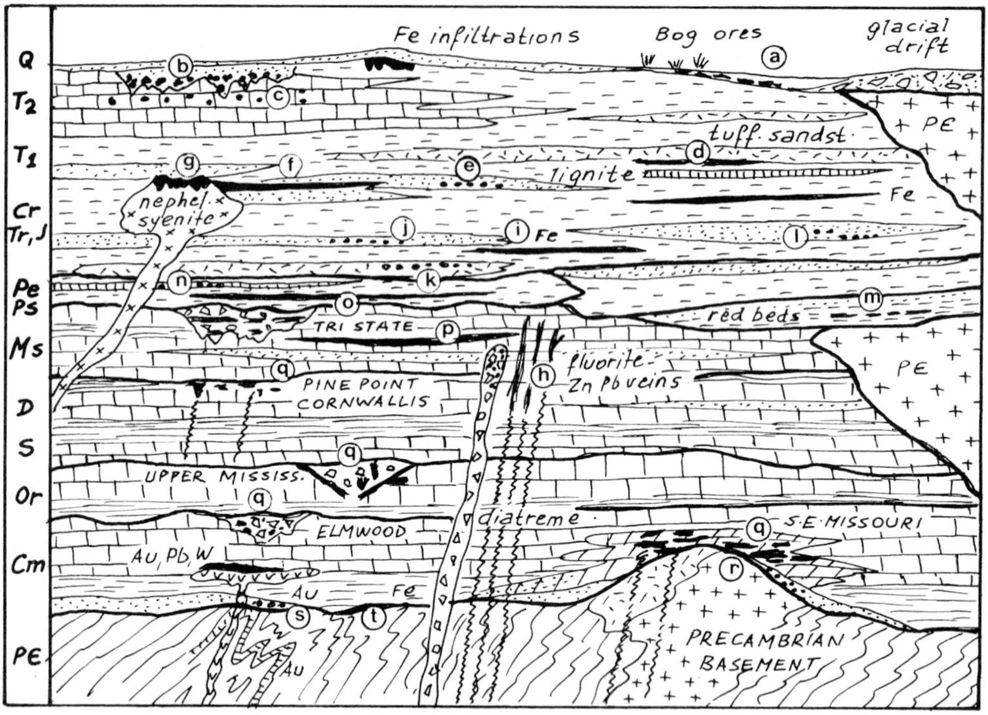

Fig. 18-4. Diagrammatic setting of metallic mineralizations hosted by and associated with the Phanerozoic platformic cover of the North American Platform.

"It is common practice to say of any deposit, not distinctly sedimentary, that it has come from below, and to rest content with this statement, which, even if not susceptible to direct proof, has the merit that in one sense it cannot be disproved".
S. F. Emmons in the Leadville Monograph, 1886.

CHAPTER 19

Shallow-Marine Detrital Sedimentary Association of Orogenic Belts and Platforms

19.1. INTRODUCTION

During the depositional phase, sediments are influenced by climate, topography, bathymetry, and composition of the source area. Consequently, there is no major compositional difference between the sediments currently forming in the same climatic zone in the shallow waters of the "inland seas", such as the Baltic (considered as a model of "basins" in which the widespread "platformic" sediments formed), and the present Atlantic-type continental margins, such as the eastern coast of North America (considered as a model of formation of the "miogeoclinal" sequences of orogenic belts). The obvious qualitative differences in distribution and style of the shallow-water sediments of orogenic belts, as compared with those of platforms, are the consequence of the post-depositional tectonism or a lack of it. The striking visual contrast amongst the linear orogenic terrains and the areal platformic cover areas impressed upon us by the persuasive presentation on tectonic maps, becomes insignificant if one attempts to make a comparison on a lesser scale; say, on an outcrop scale. A shallow-water limestone, examined in the midst of the impressive barrier of imbricate thrust sheets in the Rocky Mountains, is petrographically the same rock that one may encounter in one of the few generally poor exposures scattered over the North American Platform.

Although virtually all the examples of shelf depositional environments and the resulting sequences treated in the sedimentologic literature are (or were) located along the Atlantic-type margins, shelf-type sediments unrecognizable from the Atlantic model also form in the inactive segments of the island arc systems, as members of the Pacific-type margins. A description of several examples can be found in Hamilton (1979). They include the present-day shelf sediments depositing environment of the Java Sea and the Sarawak Basin, the latter containing a shelf sequence reaching back to the Eocene period. Following continental collisions, such sediments are deformed and intermingled with the rocks of magmatic arcs, marginal seas and continents.

In assigning a unit to either "platformic", "geosynclinal", "stable" or "orogenic" suites, there is no sharp boundary; there is plenty of overlap and an overwhelming inconsistency. This has been discussed at some length in Laznicka (1981b). For this reason "platformic" and "geosynclinal" shallow-marine sediments have not been artificially separated at a hierarchically high level, but separation has frequently been made at a lower level of organization whenever it was considered helpful from the point of visual field recognition and ore deposits distribution.

19.2. THE STRUCTURAL / GEOTECTONIC TYPES OF SHALLOW-MARINE "BASINS"

It is common practice to refer to sedimentary sequences accumulated in certain named depositional basins also as "basins". To avoid confusion, the basinal deposits will be designated here as "basinal sequences". A uniform classification of sedimentary basins is nonexistent because most of the classifications proposed and applied so far have been subordinated to a particular geotectonic model (geosynclinal, plate tectonic) or a special purpose. Overindulgence in the classification of sedimentary basins in which ancient sediments were deposited and discussion of global metallogeny in such a framework is unfortunate because it necessarily involves a great deal of imagination and dogma. The same lithologic associations may have been generated in a variety of named basin types and metallic ores correlate much more closely with lithologic associations than with basinal settings although there are some exceptions. The main advantage of having an idea about the basinal type is for prediction of the regional facies change which, in turn, may influence the change in mineralization, and for prediction of the presence, development and interaction of the intracrustal rocks and tectonic interaction.

Several frequently quoted categories of depositional basins are reviewed briefly below. The main emphasis is on lithofacies distribution and transitions.

Continental shelves

This is the least controversial example of a depositional "basin", marginal (pericontinental) to a continent. The most common simple shelves gradually accumulate an oceanward dipping and thickening prism of sediments. More complex shelves are associated with a "dam" (barrier), a positive relief feature that prevents the sediment spillage down the continental slope. Ancient shelf-deposited sequences are characterized by their unidirectional facies change generally from coarser nearshore sediments to the finest shelf-edge sediments. The presence or absence of carbonates is, as at present, climatically controlled.

Epicontinental basins

Epicontinental or epeiric seas lie on the stable continents (cratons) and, although rare today, they produced an extensive sedimentary cover in the past. Such seas are interpreted as having extended for 1,000 to 3,000 km over cratons. Their depth ranged from several metres to 200 m (the average depth was about 30 m) and the bottom slopes were less than 20 cm per 1 km. Many basins were almost circular or elliptical with the sediment thickness gradually decreasing from the centre towards the flanks. Dominant sediments of epeiric basins are micritic carbonates (with occasional organic buildups) or shales with mature sandstones in the basal portion and around the crystalline basement uplifts. Evaporites are common. The epicontinental basinal sequences tend to be extremely persistent laterally. The rocks are flat-lying or at best several degrees-dipping on the flanks of uplifts, and there was normally no syndepositional volcanism.

Successor basins in orogenic belts

Successor basins (e. g. Eisbacher et al., 1974; almost synonymous with the Kay's term "epieugeosynclines") are secondary basins that formed within a composite "eugeosyncline" (i.e. a volcanic-sedimentary depositional domain along a continental margin) following its early stage of development and deformation. As a consequence, they are elongated troughs typically filled with conglomerate, litharenite, or less commonly arkose (when "granitic" batholiths were unroofed) and mudstone detrital association. The successor basin lithofacies are always more mature and less metamorphosed (or unmetamorphosed) than the (meta)volcanic or (meta)sedimentary rocks on which they rest and they often range from a deepwater to a continental facies. Syndepositional basalt, andesite, rhyolite volcanics are common (Chapter 14).

The presence of successor basins usually coincides with a progressive "continentization" (intracrustal spreading of sialic materials; Eisbacher et al., 1974). The mineralization hosted by successor basins has been also treated in Chapters 14, 16, 17.

Examples of successor basin sequences include the Jurassic marine sediments filling the Bowser Basin and the largely continental upper Cretaceous-lower Tertiary sediments filling the Sustut and Sifton Basins in the Canadian Cordillera (Eisbacher et al., 1974); some Paleogene and Miocene units of Borneo (Hamilton, 1979) and others.

Foreland basins

Foreland basins formed (1) in front of a rising cordillera, orogenic belt or advancing thrust belt, and were floored by largely crystalline rocks of a depressed continental craton (e.g. Cretaceous to Tertiary basins in the American Cordillera foreland) or (2) along the flanks of mega-uplifts (geanticlines) following an orogeny (e.g. many Tertiary basins in Indonesia). The basins accumulated a thick sequence of clastic conglomerates, sandstones and shales derived from the cordillera and marked by thick coalescing fans shifting along the mountain front. The sediments are both marine and continental and the proportion of continental sediments increases upward. Important coal deposits often occur in the paludal facies, and minor volcanics or tuff contributions are frequent.

The synorogenic sediments of foreland basins usually take the shape of a clastic wedge and some are overriden by the thrusts that provided the detritus earlier. They largely coincide with the flysch lithofacies reviewed in Chapter 16. The postorogenic, largely continental sediments, are usually termed molasse facies (Chapter 25).

Well-known sedimentary sequences deposited in foreland basins include the upper Cretaceous Alberta Group of the Rocky Mountains (Eisbacher et al., 1974); Eocene and Oligocene sediments of N.E. Sumatra and northern Java (Hamilton, 1979) and others.

19.3. TRACE METAL GEOCHEMISTRY AND MINERALIZATION

19.3.1. Introduction

Despite their apparent simplicity, sedimentary rocks are fairly difficult to treat as a set, using an uncomplicated approach. Traditionally, the first-order classification attributes of detrital sediments have been their fragment size, and for chemical or biochemical sediments their composition. This tends to obscure the features that both classes have in common and also the transitionality; an average consolidated "detrital" sediment (e.g. sandstone) is not usually 100% detrital, because only the clasts and the matrix constitute the detrital fraction, whereas the cements (and often portions of the matrix) are chemically precipitated.

The modern sedimentogenic studies of the past 25 years (e.g. Folk, 1957; Dimroth, 1977) have slightly lessened the detrital / chemical constituents' contrast by demonstrating that many biochemical or purely chemical sediments are also bimodal and detrital (although the detritus is mostly intrabasinally-derived and skeletal). The numerous recent reviews of diagenesis, particularly those with the ore genesis in mind (e.g. Wolf, 1976), demonstrated the lenghty and complex development through which an average lithified sedimentary rock had to pass, and demonstrated too the need to separate carefuly the depositional and diagenetic events or facies in interpretations. In contrast to some of the igneous-generated and hosted metallic concentrations, we tend to consider as "instant" (that is formed by a single, well-defined event at least from the viewpoint of the latest ore emplacement episode) most sedimentogenic and sediment-hosted ore deposits kept evolving over a considerable time, slowly changing their appearance and composition. This makes the interpretation of the "ordinary" (widespread) sediments a lenghty procedure and interpretation of the rare ores often an impossibility.

As far as the ore deposits hosted by the shallow-marine sediments are concerned, they are a very heterogeneous group (Fig. 19-1; Table 19-1). Some of them are the result of (1) sedimentogenic development that paralleled, stage-by-stage, the development of the volumetrically dominant "common" sediments. Heavy mineral paleoplacers, for example, are the result of the same hydrodynamic sorting processes that separated gravel from sand and mud along facies profiles and their study and prediction using the sedimentological approach is both proper and successful (the heavy minerals, however, require a special source rock region). Diagenetically concentrated and fixed dissolved metals constitute another example.

Other mineralizations were superimposed on the completed sediments, and (2) metals precititated from essentially cool solutions circulating along generally predictable hydrogeologic paths, controlled by the measurable rock properties and gradients (porosity, permeability, etc.). Again, "special" metal source rocks or regions were needed but except for this the study and exploration methods used by the hydrogeologists or petroleum geologists are applicable to interpretation of such deposits. Here, the importance of the original rock composition is slightly less than in the case of syndepositional mineralizations (e.g. both sandstone and carbonate could contain "epigenetic" oil as well as lead-zinc or uranium, as long as they are porous). As a consequence, the post-depositional modification of the sediments and hence often their present geotectonic setting, becomes of greater importance.

The remaining category of mineralizations hosted by sediments is due to (3) external, non-sedimentogenic factors. There is a great variety of them and most are related to tectonism, metamorphism and magmatic activity. Heated metalliferous solutions may circulate along the same porosity channels as the unheated fluids mentioned earlier. The crustal heating, moreover, may be due to a variety of causes buried intrusions being one of them, but many hydrothermal solutions were probably heated, nonmagmatically, by the geothermal gradient alone. Tectonism in itself does not generate metallic deposits, but prepares ore sites, creates openings along which the ore fluids move, and triggers or channels magmatism that, in turn, supplies metals or just the heat to the mineralizing system.

It appears logical to expect that the least disturbed, most "orderly" flat-lying sediments of platforms, will contain only the ores formed under the set of conditions (1). The partly disturbed (e.g. faulted, gently-folded) sediments will contain (1) and (2); the most intensively disturbed (folded, faulted, intruded) sediments of orogenic belts, will contain (1), (2) and (3), often (1) as relics or subsequently modified developments. Although the above expectation is, in general, confirmed statistically, many fundamental problems remain that can adversely affect metalliferous exploration. Some of these problems have been continuously blurred by "concept stretching".

The overall economic metal content in deposits hosted by the shallow-marine detrital sediments is extremely difficult to express quantitatively for four major reasons: (1) uncertainty as to how far

Table 19-1. Principal mineralization styles in shallow marine detrital sediments

"ORDINARY" (GRAY) ASSOCIATION	
(A)	heavy mineral paleoplacers: cassiterite (X localities, X0 Tt Sn, low grade) gold (X0 localities, 10 t Au) Pt metals (X localities, 0.1 t Pt) magnetite, hematite (X00 localities, 100 Mt Fe) ilmenite, rutile, zircon, monazite (X00 localities, 10 Mt Ti 1 Mt Zr; 100 Tt REE; 20 Tt Th)
(B)	detrital, proximally redeposited Fe or Mn ores (X localities, 100 Mt Fe)
(C)	detrital (proximally reworked) bauxites in marine sediments
(D)	superfine detrital to chemically precipitated, bedded syngenetic - diagenetic ores:
(D1)	bedded ironstones in poorly indurated sequences; X00 localities, 350 Bt Fe, 50 Mt V (low grade)
(D2)	bedded Mn ores (oxides and carbonates); X0 localities, 2.65 Bt Mn
(D3)	nodular (unconcentrated) Fe,Mn impure carbonate in shales; X00 localities, 50 Mt Mn (low grade)
(D4)	bedded particulate ironstones in well indurated, but unmetamorphosed sequences (Clinton "type"); X0 localities, 15 Bt Fe
(D5)	bedded indurated Mn ores; X localities, 250 Tt Mn
(E)	deformed, metamorphosed bedded ironstones; X0 localities, 600 Mt Fe
(E2)	deformed, metamorphosed bedded Mn ores; X localities, 150 Tt Mn
(F)	shales with a dispersed, anomalous trace metal content; X0 localities, 66 Mt U; 100 Mt V; 50 Mt Mo (hypothetical resources); phosphorites with anomalous metal content recoverable as a by-product; 1 Mt U
(G)	ores generated / modified at a paleosurface, now resting at unconformities (e.g. bauxite)
(H)	bituminous (oil) sands; X localities, 60 Mt V (hypothetical by - product)
(I)	surficially weathered / reworked earlier mineralizations (e.g. unconsolidated placers over paleoplacers)
(J)	metalliferous weathering crusts over detrital sediments (e.g. bauxite over shale)
"GRAY" to "BLACK" MARINE DETRITAL SEDIMENTS RESTING ON (AND TRANSITIONAL INTO) CONTINENTAL RED BEDS AND EVAPORITES	
(K)	"Kupferschiefer" - style conformable metalliferous shales, marls, etc.; X0 localities, 22 Mt Cu; 3 Mt Zn; 15 Mt Pb; 40 Tt Ag
(L)	manganiferous shales and sandstones; X localities, 2.5 Mt Mn
(M)	Pb (Zn) basal marine sandstones (Laisvall); X0 localities, 4 Mt Pb

INTERACTION MINERALIZATIONS	
(N)	diagenetic or epigenetic cinnabar in quartzite enveloped by shale, minor mafics (Almaden); X localities, 550 Tt Hg
(O)	epigenetic orebodies in shallow marine sediments in thermal aureoles of plutons; X0,000 localities, 10 Mt Pb+Zn; 6 Mt Sn; 40 Tt Sb; 200 t Au
(P)	epigenetic orebodies in sediments interacting with volcanic and subvolcanic bodies; X000 localities, 4.5 Mt Sn; Pb,Zn,Sb,Ag,Sb
(Q)	very fine (invisible) precious metal mineralization in faulted shallow marine sediments (Carlin "type"); X localities, 350 t Au; 500 t Ag; 50 t Hg

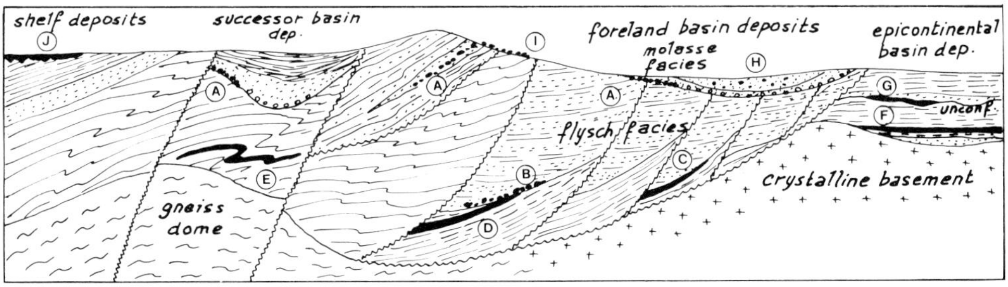

Fig. 19-1. Principal mineralization styles in shallow marine detrital sediments (see Table 19-1 for an explanation of the letter codes).

to stretch the "hosting" concept; (2) uncertainty as to what metal concentration constitutes an "ore"; (3) uncertainty as to whether the literature-quoted figures are reserves or resources; and (4) lack of data. The figures given below and in Table 19-1 should thus be considered only as extremely rough approximations. The "typical" (that is, mostly syngenetic or diagenetic, usually stratiform, not profoundly deformed) and "normal" (generally "gray") shallow marine sequences contain some 365 Bt Fe, 2.7 Bt Mn, 10 Mt Ti, 1 Mt Zr, 300 Tt U, and lesser quantities of Au, Ag, REE, Th, and other metals in deposits both mined in the past and currently economic.

The "gray-beds" transitional into the red-beds and evaporites, and the "lead sandstones", contain an additional 22 Mt Cu, 19 Mt Pb, 3 Mt Zn, 2.5 Mt Mn and 40 Tt Ag. If the "ores of the future" (i.e. supra-clarke accumulations of trace metals probably recoverable by future technology) are added, then the shallow-marine sediments may host some 70 Mt U, 220 Mt V, 100 Tt Mo, and unknown quantities of Sn, Ag, Sb, etc.

If the epigenetic mostly hydrothermal ores probably generated by interaction of the shallow-water sediments with the superimposed "high energy" agents are considered, some 10 Mt Sn, 550 Tt Hg, 10 Mt Pb-Zn, etc., will have to be added. Overall, shallow-marine detrital sediments are important repositories of metallic ores, but this is not surprising given the enormous outcrop area and volume they represent. Fig. 19-1 and Table 19-1 show and list the principal mineralization styles.

DETRITAL DERIVATION OF RESISTATE METALLIC COMPONENTS

Purely physical incorporation of a valuable metallic component into a sediment to make up an ore is the most obvious and least controversial variety of sedimentary metallogenesis. Paleoplacers are the most frequently discussed examples for which excellent actualistic models are available (Chapter 5). Two factors are of greatest importance in the prediction of and exploration for, paleoplacers: (1) former existence of the beach depositional environment, particularly the "upper beach" setting that can be fairly accurately determined by facies analysis, and (2) the source of the valuable metal detritus that tends to be very proximal (up to 5-10 km away) in the case of Au, Sn, Pt, magnetite, etc., but often quite distal (tens to hundreds of kilometres) as in the case of rutile, ilmenite, zircon, monazite, etc. The proximal sources are the most helpful for prediction and exploration. If a bedrock ("primary") gold mineralization is topped by younger detrital sediments, the sediments should be examined for the possible presence of a gold paleoplacer. If, on the other hand, a paleoplacer is found first, a search should be conducted for its possible "primary" source.

Up to now Phanerozoic heavy mineral paleoplacers have been of little economic value but this could change in the future. A significant proportion of the present Soviet production of ilmenite, rutile, zircon, and monazite comes from Mesozoic and Cainozoic paleoplacers resting on the crystalline basement of the Russian and Siberian Platforms. There is promise of future discoveries of low-grade gold paleoplacers.

DETRITAL DERIVATION OF NON-RESISTATE METALLIC COMPOUNDS

"Soft" materials, such as iron ores or bauxites, are rarely considered as detrital at least at the textbook level of thoroughness, but examples from recent environments have been treated in Chapter 5. Virtually any rock or mineral regardless of its resistance can be locally reworked and incorporated into a younger generation of a sediment, very close to the original source. In ancient sediments, such materials usually appear in the lowest layer of a transgressive sequence above an unconformity, having their origin in the sub-unconformity rocks (compare Laznicka, 1985f). Iron ores (such as the iron oxide conglomerates of Salzgitter) and reworked bauxites (e.g. those in the Eocene Berger Formation, Arkansas) are most common.

When the reworked ore mineral is of a very fine particle size (clay), the detrital derivation is often obscured by the early diagenetic modifications. In consolidated sediments, originally fine detrital, colloidal and authigenic derivations are often mutually indistinguishable. These problems closely parallel those involving the non-metalliferous clay minerals.

Recently, Dimroth (1977) pointed out the analogy between biogenic carbonates and iron ores, and in particular coined a class of "detrital chemical sediments" including the cherty and aluminous iron formations hitherto treated as purely chemical precipitates. Like the carbonates, many of the iron ores formed by an intrabasinal redeposition of earlier-precipitated matter. Many bauxites, Mn ores, phosphorites, etc. are of similar origin. Most of the detritally redeposited ores are fairly proximal to their sources and this, again, underscores the exploration usefulness of trying to determine the "primary" (earlier) sources of the ore matter (or vice-versa).

SYNGENETIC AND EARLY DIAGENETIC METAL DERIVATION AND FIXING

Syngenetic processes are those operating during the active sediment deposition and they change into syndiagenesis (Fairbridge, 1967) once the major phase of sedimentation ceases. The coarse detrital minerals of paleoplacers are undisputably syngenetic and do not change much with the diagenesis. The superfine clastic or colloidal ore material (Fe and Mn hydroxides or oxides) is strongly diagenetically modified, so that its original composition and constitution is usually unrecognizable. The ironstones and iron formations of Dimroth (1977) considered as a ferruginous equivalent of shale and mudstone, are probably the best example. The bedded shallow-marine Mn deposits, arranged zonally around regolithic crystalline basement uplifts, considered to be their Mn source (as in the Oligocene South Ukrainian Basin; Strakhov et al., 1968) probably also formed from a superfine "cloud" of suspended manganiferous particles, rather than by precipitation from solution

Deposition of the rare metallic substances, such as Zn, Pb, Cu, U, etc. from a suspended superfine matter, is probably more complicated because rather than being in suspension on their own, the metals are more likely to be bound to suspended carriers: clay minerals, Fe/Mn oxides, organic matter. Degens and Ross (1976) demonstrated such suspension in the case of sphalerite, bound to resin globules in Lake Kivu. Virtually nothing is known about the fate of superfine detrital gold, cassiterite, wolframite and other resistate minerals that certainly exist along the fringe of their equivalent coarser (recoverable by panning) accumulations.

The rarer metal accumulations that settled on the seafloor together with their carriers are usually indistinguishable from the same metals chemically precipitated syndiagenetically from seawater and from sub-floor heated brines. The latter, however, are a special category. Layers of fine volcanic ash which have settled on the seafloor from an ash cloud generated by a volcano hundreds to thousands of kilometres away, provide an instant trace-metal source undepleted by leaching during continental weathering, and thus are geochemically distinct and metallogenically more fertile than the bulk of terrigenous detritus. Diagenetic alteration, however, modifies many ash horizons beyond recognition. In consolidated sediments, the latest event of lithogenesis was diagenesis, that usually modified the syn-depositional character of the constituents and to a lesser degree added or removed some constituents. The diagenetic addition of K_2O during mud lithification is well publicized (e.g. Pettijohn, 1975); the addition of metallic transition elements is more controversial but probable, particularly during compaction, when large volumes of interstitial fluids are set in motion (often at elevated temperatures at depth), leaching, transporting and reprecipitating metals. Diagenetic metal accumulations or orebodies may form in suitable traps. Wolf (1976) reviewed this subject extensively.

Typical representatives of mineralizations containing diagenetically fixed (but syngenetically or diagenetically introduced) metals are metalliferous mudstones, shales, argillites and slates, phosphates, ironstones, Mn sediments, etc. (Vine and Tourtelot, 1970; Swanson, 1961; Table 19-2). As in the previous chapters, carbon-rich "shales" are the most frequent repositories of metals. The almost incredibly high local content of gold (up to 0.3% Au), Pt, Pd and other elements reported by Kucha (1982) from the Polish "Kupferschiefer", were interpreted as initially bound in kerogen and diagenetically concentrated at the border between oxic and anoxic conditions.

The metals accumulated in the fine sediments range from dispersed matter bound in carriers (e.g. in the organic matter such as U in carburan; in clay minerals or hydromica such as V in roscoelite; in Fe/Mn oxides) through scattered grains of their own minerals (mainly sulphides), nodules or concretions, to massive bands. Massive sulphide bodies are, however, uncommon in the shallow water sediments.

SUBAERIAL INTRAFORMATIONAL EMERGENCE

The various stages of sedimentation, deposition and early diagenesis (terminating in lithification) are sometimes considered progressive or prograde. Post-lithification sedimentary rock modification to destruction are, on the other hand, considered retrogressive or retrograde (see Laznicka, 1985b for discussion). The most common and obvious retrograde modification of sediments is taking place at the subaerially exposed surface due to oxidation, hydration, leaching and physical and chemical removal of substances.

Table 19-2. Trace metal contents in selected shallow marine black shales (in ppm)

LOCALITY	AGE	Zn	Pb	Cu	Ni	Co	V	Mo	U	Ag	Au	Th	REFERENCES
Chattanooga Shale, Tennessee, USA	Ms								79			6	Conant and Swanson (1961)
Alum Shale, southern Sweden	Cm						1100-2000	80-300	168-320				Armands (1973)
"Grosse Lette", Kupferschiefer, Mansfeld, Germany	Pe_3	1.85%	0.86%	2.9%		144	914	309		191			Jung and Knitzschke (1976)
Noble Metal Shale, Lubin, Poland*	Pe_3	0.04-0.1%	0.01-0.3%	0.2-0.8%	40-0.4%	60-1.75%	10-0.5%	206-0.48%		10-0.58%	up to 0.3%		Kucha (1982)
Horn River Fm. shale Pine Point, N.W.T., Canada	D	0.32%	82						51				Macqueen et al. (1975)

* additional elements: 10-370 ppm Pt; 10-1000 ppm Pd; 100-2000 ppm Bi; 200-1500 ppm Hg

When the modification took place during a period of a subaerial exposure (emergence) of a marine rock suite, followed by transgression and a renewed marine sedimentation, its products could become preserved either in part or entirely at an unconformity. The metallogenetic role of unconformities has been discussed in detail by Laznicka (1985f). A portion of the residual bauxites, some Fe and Mn oxide ores, barite nodules and uraniferous phosphates are now found occasionally at unconformities directly overlying or in the proximity of, the "primary" sub-unconformity sediment-hosted mineralizations considered later in this chapter. They formed during the period of emergence.

SURFICIAL WEATHERING AND MINERALIZATION

Here, the premises are the same as those in the previous paragraph but the subaerial modification of the sedimentary "bedrock" is taking place on the present surface and its products are exposed. Inactive, fossil weathering profiles are relics that are bound ultimately to be removed sometime in the future. Intensive (mostly tropical) weathering in itself superimposed on ordinary rocks of the shallow-marine detrital association, generated virtually no new economic metal deposits. The only exceptions are residual bauxites formed on "shales" but many (if not most) such "shales" are continental (alluvial, deltaic or lacustrine).

Weathering, however, caused modification of the earlier-formed sediment-hosted deposits which, in some cases, resulted in (1) residual enrichment (e.g. of siderite or leptochlorite iron ores, Mn carbonates and silicates, uraniferous phosphates); (2) cement dissolution and consequently softening (loosening) of the "primary" ore (e.g. several deposits of heavy mineral sands formed over earlier sandstones on the Russian Platform); and (3) dissolution of dispersed ore substances in rocks (e.g. Mn, U, Cu, etc.) followed by a local reprecipitation.

TECTONOMAGMATIC AND METAMORPHIC INTERACTION WITH SEDIMENTS

As in the lithologic associations treated in the preceding chapters, tectonism, metamorphism, and magmatism superimposed over and interacting with the shallow-water sediments are, in some cases, an important factor in the upgrading of existing ores, or in the formation of new ore concentrations. As in other cases, the second possibility still remains largely obscure, but empirical associations strongly suggest that hydrothermal epigenetic ore veins located in sediments anomalously enriched in certain trace metals (e.g. Ag, Au, U, etc.), may also have a source relationship with such sediments. The subject, so far, cannot be generalized and the reader is referred to the discussions that follow on the individual localities.

19.4. MARINE CONGLOMERATE-SANDSTONE ASSOCIATION, CONTAINING RESISTATE PALEOPLACERS

Common resistate detrital minerals such as magnetite, ilmenite, zircon, and monazite, are a usual accessory of well-sorted conglomerates and sandstones, but their contents rarely exceeds 0.2%. Under conditions of a prolific supply of either of the minerals (e.g. adjacent to a highly productive provenance region) or in especially favourable depositional environments (beaches), the heavy resistates may accumulate initially in the form of streaks or pockets of mineral grains parallel with bedding or cross-lamination and grading into heavy mineral beds. The size of the heavy mineral grains ranges from silt size to about 1 mm; coarser grains are exceptional. Zimmerle (1973) briefly summarized the conditions of fossil heavy mineral concentrations of the Phanerozoic ilmenite-zircon suite, and extensive literature exists concerning the Precambrian Au (U) paleoplacers (Volume 2) many of the conclusions on which are applicable to the Phanerozoic placers as well.

19.4.1. Magnetite (hematite) paleoplacers

Titano-magnetite paleoplacers containing a variable proportion of ilmenite, zircon, rutile and monazite, are extensively developed in the shallow marine Cretaceous deposits of Alberta, Montana, Wyoming, Utah, and other territories of the western North American interior (Houston and Murphy, 1977). They are found in regressive "black sandstones", which are the most reliable shoreline marker beds in the region. A typical black sandstone deposit is underlain by a white nearshore marine sandstone and is overlain by a non-marine fine-grained sandstone, carbonaceous shale, and coal. The black sandstone unit consists of alternating laminae of black and white sandstone averaging about 15% magnetite at the base, and a massive black sandstone composed of layers containing up to 90% magnetite at the top (Fig. 19-2). The mineralogical composition of the placers may vary locally; a paleoplacer north of Riverton, Wyoming, contains chiefly ilmenite, anatase and zircon and averages 21.5% TiO_2, 3% ZrO_2 and 0.01% ThO_2.

Houston and Murphy (1977) explained the paleoplacer formation as being a consequence of repeated uplift and subsidence in the source area in turn related to igneous activity and deformation related to batholithic emplacement. It is likely that the periodic explosive volcanism contemporary with the paleoplacer formation that from time-to-time blanketed the area with ash, was also a contributing factor and the source of the bulk of the Ti-magnetite. The sandstones of the Western Interior are chiefly sublitharenites. Although numerous occurrences of black sandstones are known in the area, none has been seriously mined or explored except, perhaps, the one at Dungarvan Creek, Alberta (Mellon, 1961; 6.3 Mt ore with over 27% Fe, 3-12% TiO_2). The largest deposit Choteau, Montana, was estimated to

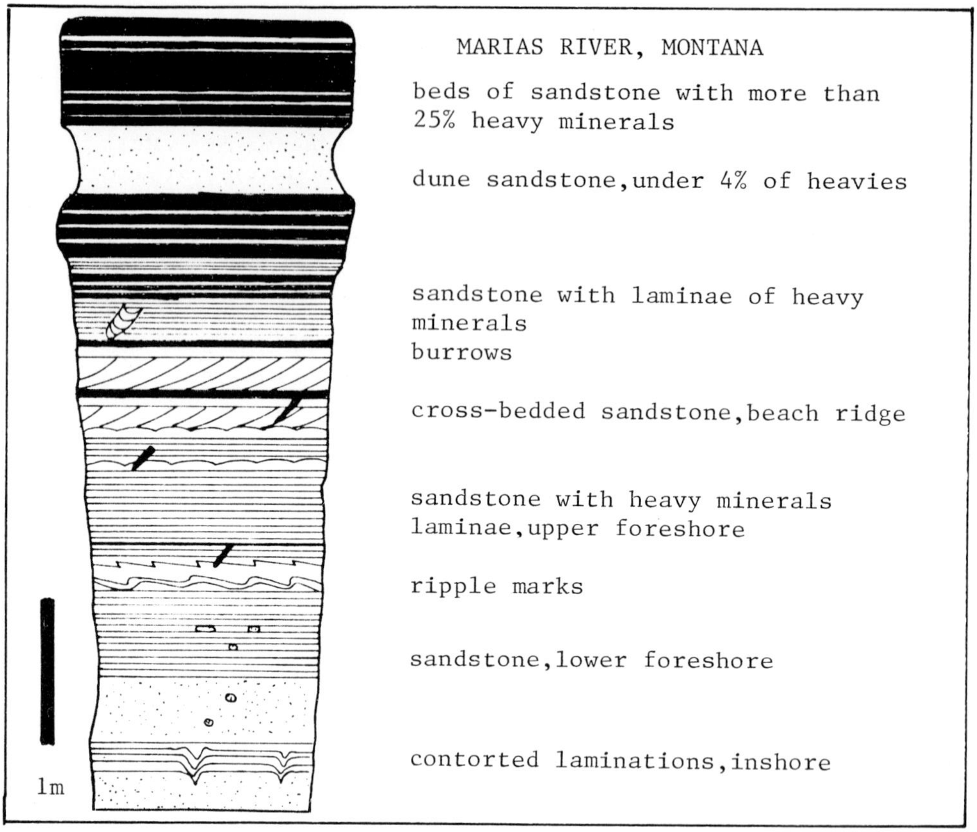

Fig. 19-2. Cretaceous magnetite/ilmenite beach paleoplacer, Montana. From Houston and Murphy (1977).

contain about 10 Mt ore with 44% Fe. The sandstones are not competitive as iron ores but may be of more interest for their Ti, Zr and Th content. The Wyoming localities of the black sandstones are estimated to contain about 20 Mt ore with 5.22% TiO_2, 0.55% ZrO_2 and 0.015% ThO_2.

Elsewhere, similar magnetite paleoplacers have been recorded near Stroud, N.S.W., Australia, where they are concentrated along former beach lines in a Carboniferous unit, dominated by tuffaceous sandstone. The reserve is small, though, a mere 1.5 Mt of ore with 43-49% Fe. The Talladega, Alabama "gray iron ores" (Reed, 1949; 6.8 Mt Fe/34%) also interpreted as a paleoplacer, consist of a series of thin lenticular orebodies present over a distance of 20 km in a lower Cambrian quartzite. Detrital magnetite and specular hematite are the principal minerals. This is one of the few paleoplacers actually mined, and 100 Tt of 50% Fe ore have been produced.

19.4.2. Ilmenite, rutile, zircon, monazite paleoplacers

This association clearly has an epiclastic derivation and usually occurs in close proximity to the crystalline basement. The Ti paleoplacer near Buena Vista, Virginia (Bloomer and de Witt, 1941), occurs as black bands, 8 mm to 15 cm thick, composed of rutile and ilmenite (up to 50% of the rock), interbedded with a reddish-brown arkosic sandstone. The mineralized material forms lenses in the basal portion of the lower Cambrian sedimentary sequence overlying a regolithic Proterozoic migmatite basement and it may be a very proximal, continental deposit. The Snowbird Mountain paleoplacer in Tennessee (Wedow and Hobbs, 1968) lies in the late Proterozoic rocks of the Ocoee Series. It is interpreted as a marine beach sediment, and is higly metamorphosed. Ilmenite, rutile and zircon form beds and laminae in meta-quartzite, up to 3 m thick and containing up to 60% of the heavy minerals. Ilmenite is dominant, but zircon occurs in quantities ranging from 4.5-9% of the rock. The two Appalachian localities reviewed above are not mined.

The Devonian leucoxene paleoplacer near Yagersk, southern Timan (north-eastern corner of the Russian Platform; Smirnov, 1969; Fig. 19-3), however, seems to be an important producing deposit. There, the basal horizon of quartz arenite grading locally to conglomerate and argillite, rests unconformably on regolithic late Proterozoic schists. The orebodies take the form of elongated lenses several metres thick and hundreds of metres long, and the rich ore has a grade of 8-10 % TiO_2.

From the economic point of view, the geologically young (Mesozoic-Cainozoic) paleoplacers of platforms are substantially more

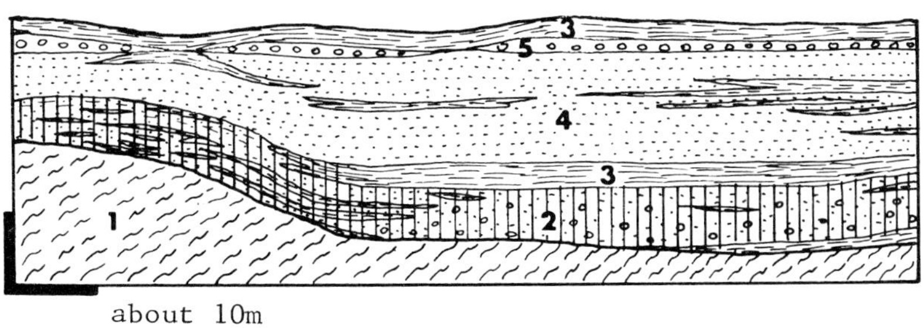

Fig. 19-3. Yagersk deposit, Timan, U.S.S.R. Devonian leucoxene paleoplacer. (1) Pre-Devonian schist; (2) Devonian marine sediments: quartz-leucoxene sandstones (pattern indicates grain size from coarse to fine); the ore horizon; (3) shale; (4) fine to coarse quartz sandstone; (5) polymictic and leucoxene-quartz conglomerate, coarse grained sandstone. After Smirnov (1969).

important economically than the Paleozoic paleoplacers of orogenic belts, because of their continuity, better predictability, and a lesser degree of consolidation that facilitates processing. They are widespread at several places on the Russian and Siberian Platforms, where they represent a significant resource base, although a low grade one (e.g. 2.5-3.5% of heavy minerals in the 6.5 m thick upper Cretaceous Central Placer in the Voronezh Anteclise; Rundkvist, ed., 1978). The greatest concentration of paleoplacers is found along the fringe of the Ukrainian Shield and in outliers of platformic sediments resting on the Shield metamorphics. The best-known paleoplacers are the occurrences of the Cretaceous and Paleogene ages fringing the Korosten anorthosite complex; the Cretaceous and Miocene Central Dnieper ilmenite, rutile, zircon sands (e.g. the Samotkan deposit); and Paleogene paleoplacers in the Glushkevichi-Slovenno area, S.W. Byelorussia (21-100 m thick, aver. 1.86% Ti, 0.68% Zr; Makarov and Bordon, 1970). In the Azov Sea region (Ainemer and Konshin, 1982), buried late Oligocene-early Miocene sands overlying regolithic Precambrian albitized granites, carry detrital columbite.

Paleoplacers containing predominantly the "noble" heavy minerals (monazite, zircon, columbite, etc.) with little or no ilmenite, are visually inconspicuous. The Bald Mountain paleoplacer in the Bighorn Mts., Wyoming (Mc.Kinney and Horst, 1953), lies in quartz-pebble conglomerate at the base of a Cambrian sandstone unit, resting on a Precambrian granite wash. Abundant hematite and limonite pigment in the matrix visually obscures the presence of heavy minerals. About 20 Mt of material with 0.13% of monazite containing 1,760 t Th and 8,000 t REE, have been outlined. The upper Cretaceous paleoplacers in the Tuscaloosa Formation, South Carolina (Overstreet, 1967) have a much lower monazite content between about 100 and 400 ppm.

19.4.3. Phanerozoic gold paleoplacers

Although some 60% of the world's economically accumulated gold is in coarse Proterozoic metasediments interpreted as paleoplacers (Witwatersrand), equivalent Phanerozoic occurrences have, so far, been of scientific interest only. There is no logical reason for it and economic paleoplacers may be found in the future. Boyle (1979, p. 376-381) in his very detailed book on gold deposits discussed the Phanerozoic gold paleoplacers. From the prospector's point of view, they can be divided into two categories: (1) minor paleoplacers formed by reworking (recycling) of gold from a known, "bedrock" deposit and present on top of, or in the immediate proximity of, such a deposit and (2) paleoplacers without a close and distinct source affiliation.

The deposits (1) should be anticipated and searched for wherever a younger sedimentary unit transgresses a gold-mineralized paleosurface. The least controversial example listed by Boyle (1979) is located in the vicinity of Gays River, Nova Scotia, where Mississippian basal conglomerates unconformably overlie lower Paleozoic auriferous slates and graywackes of the Meguma Group (described in Chapter 16). The basal polymictic conglomerate consists of pebbles of shale and quartz in a gritty quartz, slate particles and iron oxides matrix. The gold occurs as bright thin flakes and small nuggets in the matrix of the conglomerate and is associated with a considerable amount of iron oxide. Fifty three kilograms of gold have been produced from a conglomerate having a grade of 5.4 ppm.

The internationally most frequently quoted example of a Phanerozoic gold paleoplacer in the Deadwood-Lead field (Black Hills, South Dakota) is more controversial. There, gold occurs in a Cambrian basal conglomerate and sandstone, transgressive over a Proterozoic schist belt hosting the large Homestake gold deposit (reviewed in Volume 2). The gold-bearing conglomerate, typically 1 m thick, is composed of poorly-sorted quartz, schist and iron formation pebbles, cemented by pyrite. The gold occurs as well-rounded, small nuggets on or near the Precambrian bedrock and also as flakes along fractures and cleavage planes in the schists under unconformity. The controversy over the origin of the post-Proterozoic gold in the area stems from the fact that several additional gold occurrences in the Paleozoic sediments stratigraphically above the basal conglomerate also contain gold, frequently accompanied by Pb-Zn sulphides or by tungsten minerals, and hydrothermally introduced (possibly related to Tertiary rhyolite and phonolite dikes). It appears, nevertheless, that the conglomerate-hosted mineralization is a true paleoplacer and that the variety of small Phanerozoic gold occurrences found near Deadwood is a consequence of a polygenetic and multiphase remobilization of a large (1,000 t Au plus) and extensive bedrock mineralization.

The known paleoplacers of category (2) are all low-grade, and the gold is erratically distributed but when a large volume of rock is available, new methods of handling and recovery may eventually be developed to justify mining. The slightly auriferous Cretaceous and lower Tertiary conglomerates in north-western Wyoming (these are partly continental) were recently investigated by Antweiler and Love (1967) with such a possibility in mind. There, gold occurs in quantities ranging from 47 to 222 ppb Au, being highest in the Eocene Wind River Formation (222 ppb). The gold is associated with quartzite clasts, derived from the now-buried Precambrian uplift north-west of the Teton Range. The rock volume exceeding 400 km^3 may contain some 100,000 t of gold.

19.4.4. Platinum paleoplacers

Small (i.e. several kilograms) amounts of platinum have been recovered from the lower Paleozoic basal conglomerates transgressive over the slightly platiniferous Cambrian ultramafics near Adamsfield, Tasmania.

19.5. MUDROCK, ARENITE, MINOR CARBONATE ASSOCIATION

19.5.1. Introduction

This is the most common and widespread lithologic association of both the "miogeoclinal" megafacies of orogenic belts, and of platforms. Assuming that the completely unconsolidated muds (clays) and sands are Quaternary or at most late Tertiary (Chapter 6), the next least indurated group of sediments includes mudstones (claystones) grading to shales, and sandstones (Section 19.5.2.). Petrographically, the above sediments have been diagenetically lithified but, as a rule, not completely recrystallized. Orthoquartzites are uncommon. These sediments dominate the geologically young (Mesozoic being most typical) sequences of platforms and the Atlantic-type continental margins.

Well indurated sediments of the argillite-orthoquartzite rank (Section 19.5.3.) are best represented by the Mesozoic and Paleozoic rocks of the orogenic belts and by Paleozoic rocks in the deeply buried platformic sequences. They are usually diagenetically (early metamorphically) recrystallized either entirely (argillites) or only in the matrix (arenites).

The most intensively post-depositionally modified category of rocks treated in section 19.5.4., is almost entirely confined to orogenic belts and consists of a variety of metamorphics, of which slates and phyllites interbedded with quartzites are the most common.

19.5.2. Least indurated mudstone (claystone, shale) - sandstone, minor carbonate association

IRON ORE ASPECT

Ironstones of the "sandy clayey and oolithic, shallow inland sea iron formation" (Kimberley, 1978), or "Minette-type", are the most common metallic ores to be found in the present category. In extent and in terms of the iron tonnage they contain these come second after the Proterozoic banded iron formations and like the latter, they formed in a distinct paleogeographic setting, fringing former "highs" (subaerial elevations undergoing weathering and erosion). Although Kimberley (1978) emphasized the epicontinental, inland sea depositional environment, shelf or even continental lake settings have been proposed as alternatives. Most authors, however, agree that the ironstones formed in a very shallow water, but one sufficiently agitated to be suitable for the widespread oolite generation (compare James, 1966).

The overwhelming majority of ironstone deposits are hosted by the flat-lying Cainozoic and Mesozoic platformic sequences, and the lack of severe disturbance makes their paleoenvironmental interpretation relatively reliable. The most important iron minerals are goethite, chamosite and siderite; hematite is less common, and pyrite occurs locally. Mud and sand fractions, carbonate cement or bioclasts, glauconite and fluorapatite, are the most common non-ore components.

West Siberian Iron Ore Basin

This basin (basinal sequence) is located in the interfluve of the Ob, Yenisei and Irtysh Rivers in western Siberia (centre: Tomsk). It has an area of 66,000 km^2 underlain by upper Cretaceous strata, that host up to 6 stratigraphic horizons of sedimentary ironstones (T.K.E., 1964). The overal ore resources are estimated to exceed 900 Bt of a relatively low-grade ore (30% Fe), of which 400 Bt are assured reserves. The iron-bearing suite is 80-500 m thick, and it consists of:

(1) a Basal Unit, 20-50 m thick, composed predominantly of continental friable sandstones with thin intercalations of lignites and siderite-cemented sandstones. Thin lenses of nearshore-marine iron ores occur locally.

(2) A Middle Unit, 10-400 m thick, composed of cyclically alternating tongues of sandy, shaly continental and marine sediments. The marine tongues contain five important horizons of iron ores.

(3) An Upper Unit, composed of greenish-gray, marine, thin-bedded mudstones with sporadic siderite lenses.

Each iron ore horizon, 2-35 m thick, displays a distinct facies and grade distribution pattern (Fig. 19-4), starting with nearshore sandstones containing less than 15% Fe in the cement, and culminating with 30% Fe ironstones in the centre. The ironstones formed during regressions, had an easterly provenance and originated in an extensive (up to 250 km wide) tract of shallow water. The named and outlined ore deposits are local segments of the iron-bearing suite in which a single iron ore horizon has reached an anomalous thickness and grade, or in which several iron ore horizons in their most favourable ore facies development congregate. In the largest Bakchar deposit 180 km N.W. of Tomsk (10.5 Bt Fe/37.4%, 3.64 Mt V/0.13%; Figure 19-5), three exceptionally favourably developed iron ore horizons coalesce and produce an orebody with an aggregate thickness of between 60 and 120 m. The flat-lying, lithologically quite variable marine ore-bearing sequence is of Santonian to Maastrichtian age, and overlies Turonian sandstones and lignites with a slight disconformity. Within the orebody, several cycles separated by short hiati can be recognized. Each cycle displays a distinct fining-upwards progression of sediments from basal terrigenous granule conglomerates to sandy or oolithic ironstones to shales or claystones on the top. The most common ore variety has an earthy dark — brown or greenish-black colour, is medium-grained and composed of goethite, leptochlorite (chamosite-strigovite) and glauconitic ooliths, granules and clastic fragments (about 70%), cemented by leptochlorite. Additional ore varieties include reworked, loose "sand" or conglomeratic ore at the base of the overlying Paleogene sediments; sideritic ore and siderite or limonite-cemented sandstone. Lenses of sediment rich in glauconite, hissingerite, vivianite and phosphorite, occur locally.

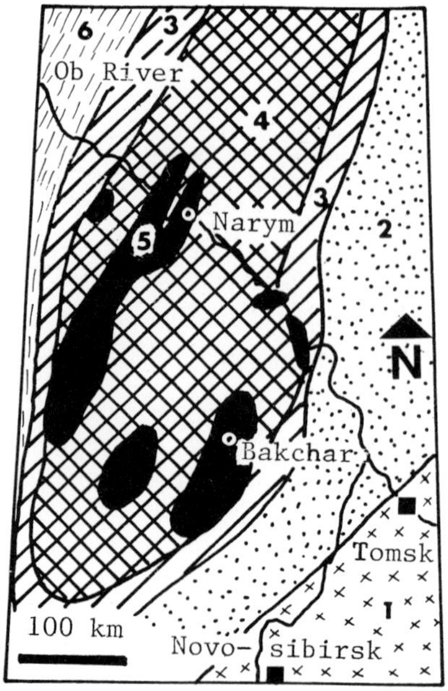

EXPLANATIONS

1 Area of shallow Paleozoic basement
2 continental deposits
3 ferruginous sands and sandstones of the nearshore-marine facies; under 15% Fe
4 sandstones with 15-30% Fe
5 iron ore horizon with 30% Fe plus and bed thickness greater than 2 m
6 offshore sediments (shales)

Fig. 19-4. Facies and grade distribution pattern for the Narym iron ore horizon (Santonian-Cenomanian), West Siberian Iron Ore Basin. After Tomsk. Kompleks. Eksped. (1964).

Western European Jurassic iron ores

Paleogeographic configuration and depositional environments similar to those in the West Siberian Basin, existed in the Jurassic period over large tracts of western Europe, and resulted in the deposition of an ironstone facies that became an important basis for the heavy industry. The ironstone deposits are located in Britain, France, Luxembourg, Belgium, Germany and Switzerland and the political fragmentation is reflected in the nonuniform descriptions and interpretations of the ores in the respective national literatures.

In Britain (Slater and Highley, 1976), Jurassic marine ironstones have been the dominant national iron ore for over a century. Three major ironstone horizons are developed. The most important economically (470 Mt Fe/31%), the Northampton Sand Ironstone, extends for about 130 km. It is an ironstone bed 0-8m thick composed of chamosite ooliths set in a groundmass of siderite, calcite or chamosite, floored by claystone and overlain by calcareous sandstone and sandy limestone. Enriched limonitic ores have developed along the outcrop. The remaining horizons (Frodingham Ironstone, Cleveland Ironstone, etc.; Table 19-3) are comparable but thinner and of lower grade. The total economic (but low grade) iron content in the British Jurassic ironstones has been of the order of 1 Bt Fe.

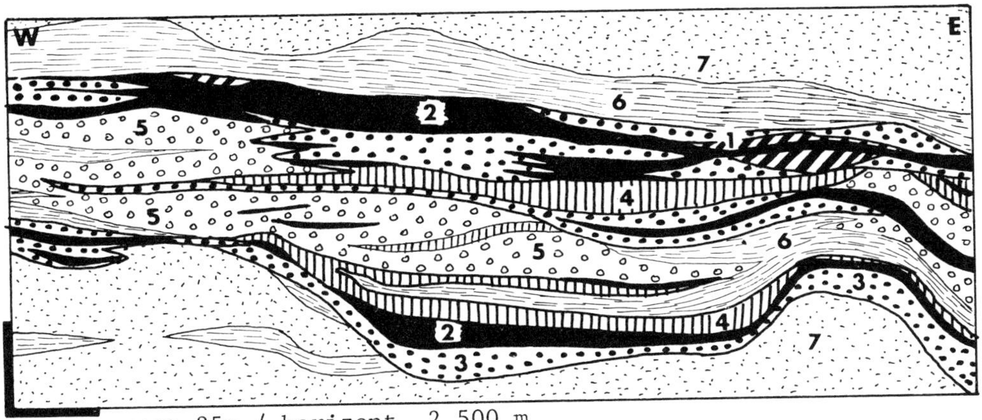

Fig. 19-5. Bakchar Fe ore deposit, West Siberian Basin - a section. Cretaceous (1) siderite; (2) chlorite-siderite ore; (3) ferruginous arenite; (4) leptochlorite mudstone or siltstone (sub-grade); (5) glauconitic sandstone; (6) marine mudstone; (7) continental sandstone, minor lignite. Modified after Tomsk. Kompleks. Eksped. (1964).

The Lorraine (France) - Luxembourg "Minette" ironstones (Horon, 1976) formed during the uppermost Lias in a shallow-water epicontinental basin located in the vicinity of the Vogesen crystalline massif. Some 11.3 Bt of 25-30% Fe ore has accumulated in six to ten 3-4m thick ore beds composed of oolithic goethite and/or chamosite, cemented by siderite. The average thickness of the iron-bearing unit is about 40m, and the beds are almost flat lying.

In southern Germany and Switzerland (Frank et al., 1975), Jurassic ironstones are distributed as a discontinuous arc of deposits fringing the crystalline uplift of the Schwarzwald. The total iron content represents some 1 Bt Fe, but the ores are extremely low-grade (17-23% Fe) and were mined chiefly in the times of emergency. In the Kahlenberg mine 30 km north of Freiburg i.B., a 10-11.5 m thick bed of oolithic red goethite, hematite, minor chamosite, siderite and calcite ironstone (Dogger) rests on a claystone, and is topped by a sandy limestone and clay.

Cretaceous iron ores of northern Germany

The Salzgitter iron ore district is located in the northern German plain north of the Harz uplift (Neumann-Redlin et al., 1976). The Salzgitter deposit proper is geologically most interesting. It consists of lower Cretaceous brown (goethitic) iron ore, produced by the physical hydrodynamic accumulation of oxidized ironstone concretions and particles reworked from Jurassic sediments in submerged valleys in the Neocomian sea and augmented by the supply of

Table 19-3. Oolithic and particulate goethite, leptochlorite, siderite ironstones in poorly indurated, shallow marine mudstone, sandstone, minor carbonate association

LOCALITY	AGE	GEOLOGY, MINERALIZATION	TONNAGE, GRADE	REFERENCES
Clear Hills, Swift Creek, N.W.Alberta Canada	Cr_3	up to 10 m thick horizon of oolithic chamosite, goethite, nontronite, siderite, interbedded with shale	520 Mt Fe/33% 3 Mt V/0.2%	Petruk (1977)
N.E.Texas iron ore region, USA	T_1	lenticular layers of nodular goethite and siderite form a 1.6-10 m thick erosional remnants hosted by glauconitic sandstone	60 Mt Fe/40%	Perkins and Lonsdale (1955)
Frodingham Ironstone, Santon area, England	J_1	lenticular seam of chamosite oolithic ironstone with minor siderite and calcite cement, interb.with claystone and limestone	382 Mt Fe/23%	Slater and Highley (1976)
Cleveland Ironstone, England	J_1	four seams of siderite, chamosite - bearing mudstone separated by shale bands	170 Mt Fe/28%	as above
Northampton Sand Ironstone, England	J_1	130 km long, 2-5 m thick horizon of chamosite, siderite, goethite ironstone, interbedded with claystone, clayous limest., sandst.	470 Mt Fe/31%	as above
Lorraine - Luxembourg Iron Ore Basin	J_1	6-10 3-4 m thick ironstone beds cover an area over 600 km^2; goethite, lesser siderite, chamosite ooliths rest in siderite, clay or carbonate matrix	3.7 Bt Fe/29%	Horon (1976)
South German and Swiss Jurassic ironstone region	J_2	discontinuous beds of low grade sandy and clayey oolithic goethite, hematite, chamosite, are distributed along an arc about 200 km long, fringing the crystalline "highs"	1 Bt Fe/18-30%	Neumann-Redlin et al. (1976)
Gifhorn Iron Ore Basin, N.W. Germany	J_2	up to 55 km long, 1-34 m thick horiz. of oolithic goethite and siderite	500 Mt Fe/32%	as above

Salzgitter distr., N. West Germany	Cr_1	5-30 m thick ore horizon is composed of detrital goethite concretions, reworked from Jurassic sediments; this is augmented by oolithic goethite facies	500 Mt Fe/31.5%	Neumann-Redlin et al. (1976)
Peine-Ilsede distr., N. West Germany	Cr_3	2-20 m thick, steeply dipping beds on flanks of salt domes, composed of detrital, reworked and enriched clay ironstone concretions	40 Mt Fe/28%	as above
Choper (Khoper), Russian Platf., USSR	Cr_3	discontinuous horizon of concretionary and sandstone - cementing goethite	292 Mt Fe/42%	Sokolov and Grigor'yev (1974)
Kerch Fe ore Basin, Black Sea r., USSR	J	single, 25-40 m thick layer of oolithic chamosite, goethite, minor rhodocrosite, vivianite and kerchenite, is underlain by limestone and overlain by siltstone	600 Mt Fe/35% 57 Mt Mn/3% 11.4 Mt V/0.67%	as above
Ayat (Turgai) Fe distr., N.W. Kazakhstan, USSR	Cr_3	coarse to fine oolithic and concretionary chamosite, siderite, ironstone forms layers and lenses in glauc. sandst. and claystone	2.5 Bt Fe/37.1%	as above
West Siberian Iron Ore Basin (around Tomsk), USSR	Cr_3	6 regressive shallow marine Fe ore horizons, up to 50 m thick each, changing facies to sandstone with goethitic cement; oolithic goethite, leptochlorite, siderite or goethite cement; abundant phosphates, vanadiferous	300 Bt Fe/30%+ 3.64 Mt V/0.13% (Bakchar dep.)	TKE (1964)
Dawsonvale, S.W. Queensland, Austr.	J-Tr_3	1.2-3 m thick oolithic hematite ironstone grading to ferrug. siltstone and shale	67.5 Mt Fe/37.5%	McLeod, ed., (1965)
Wādí Shati, near Brakk, S. Libya	Cb_1	oolithic and granular hematite, chamosite, lesser goethite, siderite ironstone lenses grade to hematitic sandstone and siltstone	1.52 Bt Fe/40%	Goudarzi (1970)

chemically precipitated goethite ooliths. The predominantly detrital ore in the south (Salzgitter) changes facies into the predominantly oolithic ore in the north. The Salzgitter ore beds are between 5 and 30 m thick, but in deep syntectonic troughs the overall thickness of the ferruginous facies may exceed 100 m.

Lower Cretaceous clay ironstones, in turn, were physically reworked and redeposited to generate the upper Cretaceous ironstones in the Peine-Ilsede district of northern Germany (see Table 19-3 for quantitative data).

Other iron ore regions

In the Kerch Iron Ore Basin (Black Sea region, U.S.S.R.), Jurassic shallow-marine sediments developed along the southern flank of the Precambrian Ukrainian Shield, contain a single layer 25-40 m thick of oolithic chamosite, ferri-montmorillonite and goethite formed along the terrigenous and carbonate facies boundary. The ore has a high Mn content, present in manganosiderite and rhodochrosite; phosphorus (in vivianite and kerchenite); and arsenic (in realgar). The vanadium content is also unusually high (0.67%). About 10% of the ore formed in depressions on the flanks of mud volcanoes (e.g. Novoselovski ore field; Schnyukov et al., 1974).

In the Ayat iron ore district of north-western Kazakhstan (Sokolov and Grigor'ev, 1974) the ironstones are very proximal to or directly adjacent to their iron source, located in a thick weathering crust topping Carboniferous volcanics. The Fe ore layer covers an area of 2,500 km^2, and is environmentally and facially transitional: the base contains pisolithic goethite changing upward into a ? paludal clay with carbonized plant remains and siderite concretions. The upper part of the layer has an oolithic goethite, chamosite and siderite marine ironstone. The latter then changes into a glauconitic sandstone.

In North America, young marine ironstones are sporadically developed in the Cordilleran foreland. In the Peace River region of north-western Alberta (Clear Hills and Swift Creek deposits; Petruk, 1977) an extensive upper Cretaceous ironstone horizon is composed of chamosite and goethite ooliths in a matrix of quartz, goethite, siderite and minor collophane. The horizon is enveloped by gray shale and is gradational into a pyritic or limonitic sandstone.

Additional examples of ironstone deposits in the rest of the world are listed in Table 19-3.

Mn ASPECT

A shallow-marine sandstone-shale association hosts the most important manganese accumulations outside the oceanic nodules. The style and mineralogical composition of the Mn ores are strongly influenced by sedimentary facies which, in turn in a climatically uniform region, is a function of water depth. The Mn sedimentology and metallogenesis have been extensively treated in Strakhov (1962),

Strakhov et al. (1968), Roy (1976) and others.

Briefly, the Mn facies sequence starts at the continent, where Mn laterite, karst-hosted Mn, and bog Mn ores formed (Chapters 23, 24); these are exclusively oxides. In the transitional continental-marine environment (lagoonal-deltaic) Mn oxides formed, often by reworking of the subaerially generated weathering crusts. The most significant Mn accumulation took place in the nearshore facies of alternating sands (sandstones) and clays (claystones), often marked by abundant glauconite. The Mn minerals are both oxides and carbonates. Up to this point, the Mn is clearly of terrigenous provenance and in terms of depositional environments, paleogeography and facies the Mn deposits are comparable with the Fe hydroxide and oxide cemented facies of the bedded ironstones, described earlier. The oolithic, typically shale-hosted ironstones do not have a textural counterpart among the Mn ores. Purely manganiferous ooliths are exceptional and the larger rounded concentric particles (pisoliths, microconcretions) characteristic of some Mn ores, formed both in weathering profiles and diagenetically.

Offshore shallow-marine Mn deposits are mostly hosted by shales and are composed of concretions (nodules) or of impure (pelitic) beds of a Mn(Fe) carbonate. Here, the Mn provenance need not always be terrigenous; in the Cretaceous Pierre Shale of the North American interior, the Mn nodule-rich sections correlate with tuffaceous (bentonitic) horizons, and at least a portion of the Mn has been derived from volcanic glass during diagenetic devitrification. The pyroclastic component came from an ash cloud, produced by a very distant explosive volcanism.

Quartz, glauconite sand(stone) and clay(stone) associated Mn ores: the Oligocene manganese ring around the Black Sea

The Oligocene nearshore sediments along the western, northern and eastern perimeter of the Black Sea Basin in Bulgaria, Rumania and the U.S.S.R., contain the world's largest (or second largest, after Kuruman) accumulation of manganese estimated at about 1.6 Bt Mn. The largest deposits are in the South Ukrainian Basin (Varentsov and Rakhmanov, 1974; Magak'yan, 1968; Strakhov et al., 1968; Fig. 19-6), located to the south and east of the extensive Precambrian banded iron formation region of Krivoi Rog.

In the oldest and best known Nikopol' deposit, the flat-lying to gently-dipping Mn ore horizon lies in the basal portion of the lower Oligocene nearshore poorly consolidated glauconite sandstones and mudstones, unconformably overlying a regolithic Precambrian crystalline basement. The ore bed ranges in thickness from several centimetres to 4.5 metres and covers about 250 km^2. The ore is mineralogically zoned, and a "primary" (depositional and diagenetic) oxide and carbonate facies can be distinguished. The oxide facies is dominant and represented by several varieties of ore. The concretionary ore has pyrolusite concretions several millimetres to 25 cm in size in a friable muddy sandstone, and manganite concretios in the mudstone. Hard, vuggy ores are composed of a porous mixture of

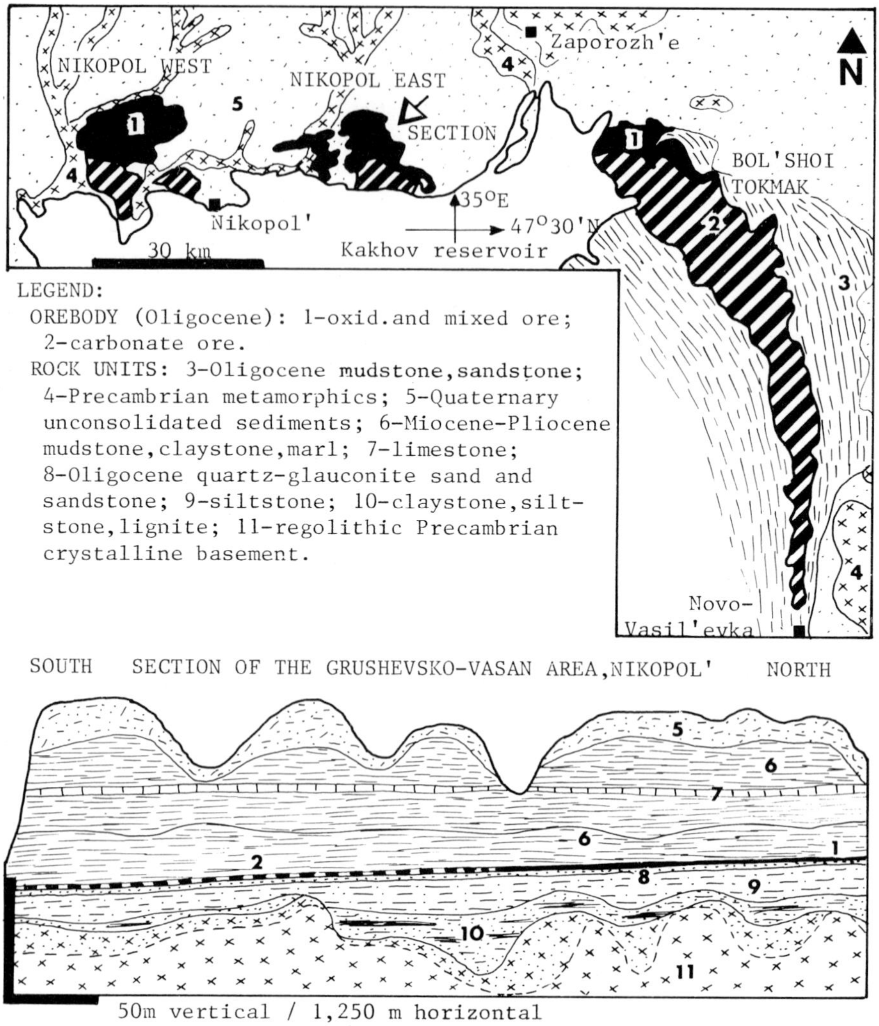

Fig. 19-6. South Ukrainian Mn Basin. Modified after Strakhov et al. (1968); scales added.

clastic sand grains, and Mn oxide pisoliths and granules, cemented by psilomelane, manganite and minor pyrolusite. Sooty Mn oxides impregnating sandstones represent a widespread, low-grade (10-15% Mn) material.

The carbonate facies appears in the former offshore region of the Oligocene Basin and is represented by massive bedded pelitic manganocalcite and rhodochrosite ores. This is only of minor importance in the Nikopol deposit but it accounts for the large underground ore reserves in the Bol'shoi Takmak field located in the eastern part of the Basin. Both the "primary" facies ores are weathering-altered and upgraded along the outcrop. Psilomelane infiltrations, crusts and botryoidal slabs are most common.

The Chiatura manganese deposit consists of a number of bedded Mn orebodies distributed along the fringe of the Dzhirula crystalline massif in the Caucasus Range in the Georgian Republic (Varentsov and Rakhmanov, 1974). The Oligocene ore unit rests unconformably on the eroded and karsted surface of the upper Cretaceous limestone. The stratiform ore horizon ranges from a few centimetres to 14 m in thickness (average 4.2 m), and consits of interlayered Mn oxide lenses and a sandy mudstone or muddy sandstone. The ore lenses are 1-50 cm thick and up to 200 m long, composed of both the oxide and carbonate facies as in Nikopol'. The ore facies transition is very gradational and secondary oxides are widespread along the ore outcrops exposed along the edges of a dissected plateau. The remaining Oligocene Mn deposits of the Black Sea-basin fringe are of substantially smaller size than the giant accumulations discussed above and are listed in Table 19-4.

In the Northern Urals Paleocene Basin, numerous Mn deposits are distributed along a 200 km long north-south – trending belt along the eastern slopes of the Urals (Varentsov and Rakhmanov, 1974). There, a Paleocene ore sequence, 20-30 m thick, unconformably overlies Cretaceous platformic sandstones and siltstones transgressive over the regolithic outcrop and subcrop of the Paleozoic "eugeosynclinal" volcanics. In the largest Ivdel' deposit, elongated ore lenses contain pelitic, concretionary and spherulitic calcic rhodochrosite grading into manganocalcite and oligonite, hosted by sandy mudstone.

Groote Eylandt, northern Australia

The Mn ores of the Groote Eylandt island east of the Arnhem Land, northern Australia (Ostwald, 1981) are the only major Mn accumulation in the western world comparable with the South Ukrainian deposits. There, the Mn ore occurs as a flat tabular sheet 2-20 m thick, and exposed over an area of 6 x 22 km. The Mn ore is a facies of a mantle of poorly consolidated – to-unconsolidated lower Cretaceous nearshore marine sandstone, sandy mudstone and marl, near its unconformable contact with the middle Proterozoic sandstone basement. The "primary" Mn ore consists of an almost continuous layer of cryptomelane and pyrolusite concretions and structureless masses, embedded in a sandy mudstone or mutually interconnected, grading upward into a loose, uncemented topping. There, gravel of loose pisoliths or concretions, blocks, infiltrations and encrustations of the same minerals, formed at the base of a Tertiary lateritic profile.

Ostwald (1981) argued in favour of microbial activity as one of the agents of Mn ore precipitation, supported by the presence of algal stromatolites, oncolites, and coccoid microfossils. A gray-green Cretaceous limy mudstone discovered by drilling in the western offshore region of Groote Eylandt, contains about 20% manganocalcite in the form of nodules and veinlets and also disseminated pyrite

Table 19-4. Shallow marine Mn ores in glauconitic sandstone - mudstone association

LOCALITY	AGE	GEOLOGY,MINERALIZATION	TONNAGE,GRADE	REFERENCES
South Ukrainian Basin (Nikopol',Bol. Takmak; USSR	Ol_1	pyrolusite and manganite,concretionary and pisolithic bedded ore in glauc.sandst.and mudstone overlying regolithic Precambrian crystalline basement,grade to pelitic manganocalcite and Ca-rhodochrosite in marl	1.65 Bt Mn/ 24-30%	Strakhov et al.(1968) Varentsov and Rakhmanov (1974)
Chiatura distr., Caucasus Mts., Georgia,USSR	Ol_1	4.5 m thick bed of oxide (pyrolusite,manganite) and carbonate (Mn-calcite, Ca-rhodochrosite) Mn ore is hosted by a sandy mudstone unconform.overlying Cr_3 limestone	600 Mt Mn/20%	
Laba R.Mn field,N. Caucasus,USSR	Ol_3	numerous short,discont.lenses of Ca-rhodochrosite,pyrolus. and manganite are hosted by friable sandst.,siltst.,mudstone.	2.91 Mt Mn/ 9.7%	Strakhov et al.(1968)
Mangyschlak Pen., Caspian region, Kazakhstan,USSR	Ol_{1-2}	35 km long N.W. - S.E. horiz.of sandy and silty mudst.,containing pyrolus.and manganite concr.,sooty Mn oxides,infiltr.,crusts	54 Mt Mn/6.5%	as above
Varna distr.,Black Sea coast,Bulgaria	Ol	140 km N.-S.belt,9 deposits of stratif. Ca-rhodochr.,Mn-calc.,psilom.,3-5 m thick,in shale. Assoc. glauconitic sandstone	5 Mt Mn/15.3-20.54%	Iovchev (1961)
N.Urals Paleocene Basin (Ivdel',Polunochnoe),USSR	Pc	Ca-rhodochr.,Mn-calcite,oligonite nodules, lumps,pelitic beds,form lenses in sandy mudst.horiz.transgr.over Cr sediments	32 Mt Mn/21.2%	Varentsov and Rakhmanov (1974)
Groote Eylandt Isl. N.Australia	Cr_1	pyrolus.and cryptomel.form pisoliths,nodules,concret.embedded in a flat-lying 2-20m thick horizon of sandy mudstone. The ore is loose on top of the horizon,below a Tertiary lateritic profile	142 Mt Mn/45% 80 Mt Mn/50%	Ostwald (1981)

crystals. The mudstone is considered to be older than the mined Mn ore.

Mn-Fe nodules in offshore, shallow marine shales: the Pierre Shale, North American interior

Pierre Shale is a very extensive, late Cretaceous Formation deposited in a 1,600 km wide north-south-trending shallow depositional trough across the North American interior in the Dakotas, Montana, Manitoba, Saskatchewan, and adjacent states and province (Schultz et al., 1980). In a cross-section, the formation is a westerly-derived wedge, thinning from several thousand metres in the west to less than 300 m in the east. The lithologic composition is extremely monotonous and poorly outcropping gray, greenish-gray fissile illitic shale-to massive claystone is the dominant rock. This monotony is interrupted by occasional conspicuous light (beige) bands of bentonite on one side, and of black organic shale on the other. A frothy-weathering shale distinctive in outcrop, contains montmorillonite derived from volcanic ash-fall, and an increased proportion of pelitic Fe-Mn carbonate nodules (Ca-rhodochrosite and Mn-siderite).

A 13 m thick nodule-rich zone in the Missouri Valley near Chamberlain, S.D. (Crittenden, 1956) was investigated during World War II as an emergency source of strategic manganese. There, the flat, discoid to irregular nodules of pale ocherous brown colour, weathering reddish-brown or brownish-black, range in size from 1 to 30 cm and comprise, at most, 5-6% of the shale bed. Nodule-rich lag deposits occur on the gentle valley slopes, and reworked nodule gravel and rubble is common along the river and creek banks. The "best brand" of the nodules contains only 15-16% Mn, hence the entire Mn shale grade is a mere 0.8-0.9% Mn. Given the enormous tonnage of the shale available in the 250 m long belt, the available Mn resource has been estimated as 10.7 Mt Mn.

Similar Mn carbonate nodules occur in the same unit in southern Manitoba (Miami-Morden area) and in numerous comparable shales of the world, but have never been investigated with the possibility of Mn prediction in mind.

RESEDIMENTED BAUXITE AND ALUMINOUS CLAYS IN MARINE SANDSTONE-SHALE ASSOCIATION

In the numerous classifications and interpretations of bauxite origin offered so far (some outdated; compare Bárdossy, 1982, for review), "geosynclinal", "marine", etc. classes have been distinguished. At the present state of knowledge, however, it appears that undisputably marine bauxites are confined to the carbonate association only (Chapter 20), and that the transported bauxites associated with shales (clays) and sandstones (sands) are continental to paralic (Chapter 25). Strakhov (1962) offered a plausible explanation for this in the context of his Al-Fe-Mn triad, within which the major ore accumulations of these metals are displaced successively basinward due to the mobility. The chances of finding a marine sedimentary bauxite deposit in a shale association are slim for another reason, which is the probability of a diagenetic restoration of Si (and K) and conversion of the fine aluminium hydroxides into alumosilicates in the marine environment.

METALLIFEROUS SHALLOW MARINE PHOSPHORITES

The Abu Tartur Plateau phosphorite deposit (west-central Egypt; Anonymous, 1976) is hosted by a Cretaceous shallow-marine dominantly detrital sequence overlying Jurassic-Cretaceous varicoloured claystones and siltstones. The phosphorite horizon, 17-57 m thick, has concretionary and pelletal phosphorite embedded in shale or impure carbonate. It is topped by glauconitic shale and sandstone and by a shale with reworked (gravel) phosphorite cemented by carbonate. A reserve of 1 Bt of 25.6% P_2O_5 and 2.8% F, has been outlined. If the normal trace content of 60 to 200 ppm U in marine phosphorites is assumed, some 100 Tt of U may be available.

METALLIFEROUS BLACK PHOSPHATIC AND BITUMINOUS SHALES: U (V) ASPECT

The upper Devonian Chattanooga Shale is the best-known and most thoroughly studied example of slightly metalliferous black shales, that will probably be used in the future for the complex production of uranium, hydrocarbons, phosphate and aluminium, plus the nonmetallic pulp. The Shale (Conant and Swanson, 1961; Fig. 19-7) is a relatively thin (about 10 m) portion of a blanket of black shale and other marine rocks, deposited in an extensive but shallow epicontinental sea located along the western side of the Appalachian orogenic belt. The average composition is 20-25% of quartz, 25-30% of clay minerals and fine mica, 15-20% of organic matter, 10% of feldspar, 10-15% pyrite, and a variable content of phosphorus. The shale is brownish-black to satin-black, fissile, and usually coated by yellow powdery jarosite in an outcrop. Several thin bentonite horizons and phosphorite nodules have been recorded. The shale has a high trace-uranium content ranging from 30 to 80 ppm, and the upper member of the shale averaging about 60 ppm has been estimated to contain about 6 Mt U over an area of 12,000 km^2 in east-central Tennessee. The uranium is "invisible", does not accumulate in discrete minerals, and is believed to have been adsorbed from seawater under euxinic conditions.

More North American shales are as uraniferous as the Chattanooga (or even more) but are poorly exposed, discontinuous or indistinct. The Cretaceous Pierre Shale already discussed under the "Mn aspect", contains several black shale members, among which the Sharon Springs Member, distributed over large tracts of territory in the Dakotas and Nebraska has the highest U content (250 ppm U) in several beds under 12 cm thick (Kepferle, 1959). As in the previous example, bentonitic

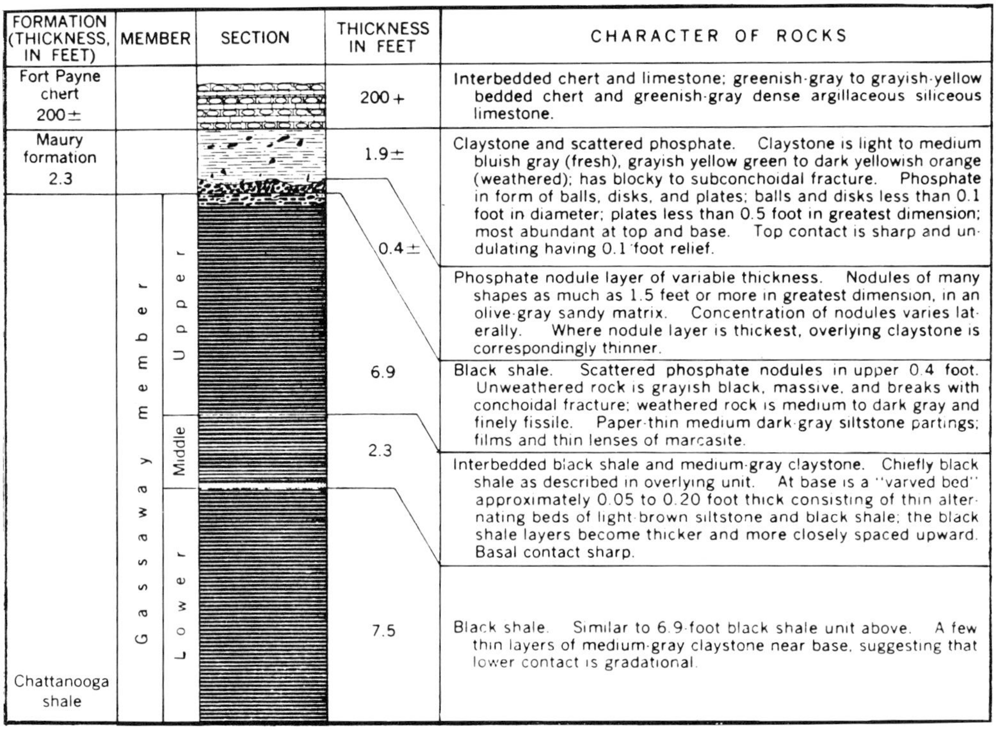

Fig. 19-7. Chattanooga Shale, a detailed section of the uraniferous Gassaway Member. From Conant and Swanson (1961).

layers intervene. The black shale interbedded with bentonite near Miami, Manitoba, contains 425 ppm V.

The 22-23 m thick middle-upper Cambrian epicontinental marine "Alum" or "Kolm" Shale of southern Sweden (Armands, 1973) is in many respects comparable with the American shales, except that it has already been mined for the recovery of various commodities for several centuries, and its complex utilization is now a reality. The shale is underlain by lower Cambrian sandstones resting on the Precambrian basement, and overlain by lower Ordovician shales and limestones. The highest uranium content (250-320 ppm U) is in the 2.5-4 m thick upper Cambrian Peltura Scarabaeoides biozone at Mt. Billingen near Ranstad, composed of bituminous shale containing 22% of organic matter (partly recoverable as kerogen) and 13% pyrite. About 90% of the uranium is evenly distributed in the seam and the rest is in the "kolm": nodular, carbon-rich material similar to anthracite, interpreted as being partly diagenetically remobilized sapropel. The separated "kolm" averages 0.45% U. The richest, currently mined segment of the Alum Shale in southern Sweden (Västergotland) is distributed over an area of at least 500 km^2 and estimated to contain some 300 Tt U (aver. grade 270 ppm U), 600 Tt V (650 ppm) and 320 Tt Mo (300 ppm).

Additional metalliferous black shale deposits occur in other areas of Sweden as well, the largest being along the eastern border of the Caledonides (e.g. Tasjo). Taken together, the Swedish metalliferous shales have been estimated (Min. Mag., December 1977) to contain a resource of some 600 Bt of shale with 50-150 ppm U, equivalent to 60 Mt U and ten times as much as in the Chattanooga Shale. Recent data on the Ordovician Dictyonema Shale in Estoni described as an oil shale are not available, but according to the scattered prewar data its U contents ranged from 50 to 250 ppm U.

METALLIFEROUS BLACK SHALES: Pb-Zn (Ag) ASPECT

Anomalous syngenetic and diagenetic concentrations of lead and zinc in shallow marine black shales are spatially separated from the U (V) shales discussed earlier. Although probably more abundant than the uraniferous shales, they have received substantially less publicity primarily for economic reasons: rocks with 0.1% Zn or Pb forming thin beds are of no economic interest at present, nor in the very near future.

Zn- or Pb-rich metalliferous shales occur in several sub-associations, some of which are rather distinct. Those identical with, gradational into, or comprising members of the facies lineage dominated by the "Kupferschiefer" (and its characteristic setting along the "red" and "gray" beds facies transition) are treated in one of the following Sections. Here, the emphasis is on shales that occur in the "black and white" suite of interbedded fine gray or black clastics and carbonates, formed in the epicontinental and marginal marine shelf environments.

Several examples of such shales have been described by Coveney (1979) in the central United States (Missouri, Kansas). In the Kansas City area, a thin but extensive shale horizon contains between 0.1 and 0.95% Zn. This horizon is of little economic interest in itself, but similar shales occur stratigraphically above the carbonate-hosted MVT zinc deposits in the Tri-State district, and may have been, at least in part, a source of the zinc carried by connate brines into the karsted carbonates below.

More Zn-rich shales have been described in the Interior Plains of north-western Canada. Several shales in the Great Slave Lake region laterally equivalent to the Pine Point dolomite (host of the large MVT deposit; Chapter 20) have been found to contain up to 0.32% Zn and 82 ppm Pb (Macqueen et al., 1975). The shales are black and highly organic (up to 40.8% organic matter) and there is a positive correlation between their metal content and the contents of organic carbon. There is also an excess of potassium, and up to 82 ppm U in some places.

BITUMINOUS SAND(STONES)

The lower Cretaceous Athabasca Tar Sands of northern Alberta (near Fort McMurray; Rose, 1973) underlie more than 2,500 km^2 and are up to 50 m thick. Lithologically, they are poorly consolidated sandstones, impregnated by a heavy bitumen, resting on Paleozoic carbonates. They contain a slightly elevated vanadium contents (360 ppm V_2O_5). The vanadium accumulates in the petroleum coke (up to 4% in ash), and plans have been made for the ultimate vanadium recovery in the petroleum production process. With the heavy oil content in the northern Alberta tar sands estimated at 750 Bbl, there may be some 60 Mt of vanadium available, thus the sands constitute a technological (by-product) V ore.

19.5.3. Well indurated but unmetamorphosed shale, sandstone, minor limestone association

IRON ORE ASPECT

Iron ores (Zitzmann and Neumann-Redlin, 1976) and in particular, sedimentary iron ores (Kimberley, 1978), have recently been classified from a compositional and environmental viewpoint, but neither classification appears to be entirely suitable for the purposes of this book. Iron ores occupy quite a wide range of sedimentary facies. Although the "Clinton type" of red "oolithic" hematite ore is generally considered to be an archetype of a shelf shale or sandstone-hosted ore, limestone is often much in evidence. In Birmingham, Alabama, it constitutes up to 50% of some types of the ore, and regionally it underlies the predominantly detrital ore-bearing sequence there.

The "red" (hematitic) and "gray" (chlorite, siderite, magnetite) ores form separate bodies at distinct stratigraphic horizons (as in the Barrandian "Basin" in western Czechoslovakia), or alternatively are intermixed within a single deposit.

Dominantly "red" (hematitic) particulate ironstones

Birmingham iron ore district, Alabama (Simpson and Gray, 1968; 700 Mt Fe/32%; Fig. 19-8) is the largest mining area in the Silurian "Clinton ore belt" of the Appalachians. The district is located along the northern flank of the Appalachian orogenic belt in the gently folded and thrust-faulted Valley and Ridge Province. The 100 to 170 m thick host sedimentary sequence rests on limestones and is composed of alternating crossbedded lenticular sublitharenite to quartz arenite (quartzite) and gray to greenish-gray shale. The iron ore forms several parallel lenticular stratiform beds (seams). In the Vulcan mine area, the basal Irondale Seam consists of about 1.7 m of ferruginous sandstone. It is overlain by 1.3 m of flat pebble ferruginous or limestone sandstone, which in turn is topped by the

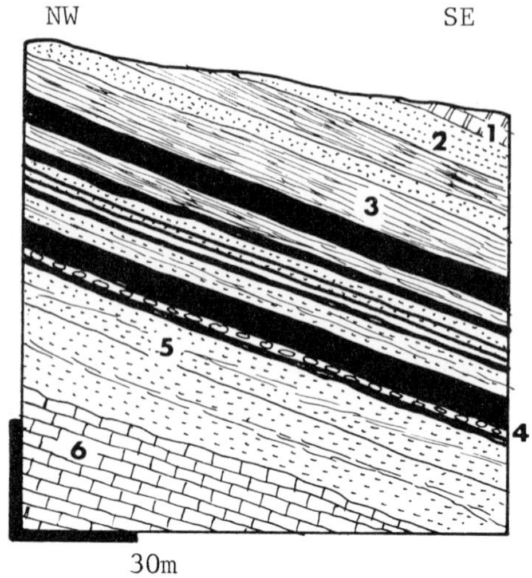

Fig. 19-8. Birmingham iron ore district, Alabama. From LITHOTHEQUE, data from Simpson and Gray (1968).

Big Seam, about 6 m thick. The Ida Seam, stratigraphically the highest, is subdivided by about 10 m of barren sandstone.

The mined Big Seam is internally composed of numerous lenticular beds of intermixed hematite, sandstone, shale and limestone which accounts for its relatively low average grade of 32% Fe. The single dominant iron ore mineral is a red, pelitic ("amorphous") hematite distributed as flat grains with rounded edges and generally a concentric microstructure; as matrix and cement and as fossil replacement. Four compositional and textural varieties of ore are usually distinguished: (1) fine-grained conglomerate or sandstone, with clasts coated by hematite and with calcite-hematite cement; (2) hematitized fossil fragments and hematite ooliths cemented by hematite (the highest grade ore); (3) hematite ooliths in a calcite matrix or cement; and (4) flattened concretions and pebbles of hematite with hematitized fossils cemented by hematite.

In the zone of weathering, the ores are friable because of the loss of the calcitic cement to leaching. The Birmingham ores are generally considered to be purely sedimentogenic, formed by iron oxide precipitation from solution and by diagenetic replacement. It seems likely that a lot of ferruginous mud was supplied from the continent and no coeval volcanism has been reported.

Several lesser iron ore deposits of the "Clinton type" are known in the Appalachian belt (Gadsden, Wolf Creek, Clinton; Table 19-5), but only the Wabana field (Bell Island, Newfoundland) is comparable with

Birmingham in size and production. There (Snelgrove and Baird, 1953), three workable seams and numerous thin layers and lenses of oolithic and granular iron ore occur within a 130 m thick unit of gently dipping Ordovician interlayered, crossbedded sublitharenite and micaceous shale. The ore beds have a maximum thickness of 5m, and are composed of dominant hematite (55-63% of the ore), chamosite (23%), siderite (3-13%), quartz (4-6%) and phosphatized carbonate skeletal remains. It is thus a transitional style between the red and gray iron ores. The Wabana tonnage is estimated to be at least 1.3 Bt Fe/51.3%, much of which is located under the seabed and thus difficult and costly to mine.

Important "red oolithic hematite" sedimentary iron ore districts are widespread worldwide (in Spain, France, Czechoslovakia, China; Table 19-5) and most are closely comparable with the Birmingham deposit in the general appearance of their ores and in lithologic association. In the Barrandian "Basin" of Czechoslovakia, Ordovician hematites are present at several levels,and they generally correlate with the local outbursts of basalt or spilite volcanism, so that their iron source may be at least partly volcanogenic (compare Chapter 10).

In the Zapla iron ore belt in northern Argentina (Bossi and Viramonte, 1978), pelitic-to-fine granular red hematite forms lenses and bands in a green micaceous sandstone, and is associated with thuringite. Paraconglomerates and siltstones with dropstones interpreted as having a glaciogenic origin, are present occasionally and indicate a transition to the iron ores treated in Chapter 25.

Dominantly "gray" (leptochlorite, siderite) particulate ironstones

In the Barrandian Basin of western Czechoslovakia, some of the shallow-marine Ordovician lithofacies composed of alternating, heavily burrowed and bioturbated sublitharenites and micaceous shales interpreted, sometimes, as flysch (e.g. the Nučice Group), contain several horizons of "gray" iron ores that supported the local steel industry for over a century. In the Nučice-Chrustenice iron ore field, a bed of ironstone up to 6 m thick is composed of dark gray chamosite ooliths in siderite, and locally of magnetite cement and matrix. It is situated on a base of dark gray micaceous shales topping the sandstones (Svoboda, ed., 1966). The ore deposition was coeval with the waning phase of a minor episode of submarine basaltic volcanism, but direct association of the iron ore with volcanics is nowhere apparent. Almost identical ores occur in the Ordovician of Thuringia (East Germany), Morocco, and elsewhere (Table 19-5).

INDURATED, UNMETAMORPHOSED MARINE SHALE, SANDSTONE, MINOR LIMESTONE: Mn ASPECT

Mn carbonate deposits in the form of stratiform beds or conformable lenses are moderately common in the shallow-marine sediments of orogenic belts and platforms. In the Avalon Peninsula (eastern Newfoundland; Snelgrove and Baird, 1953) a persistent, widely

Table 19-5. Particulate (oolithic, granular) sedimentary bedded ironstones hosted by Paleozoic indurated but unmetamorphosed shale, sandstone, minor limestone association

LOCALITY	AGE	GEOLOGY,MINERALIZATION	TONNAGE,GRADE	REFERENCES
Birmingham,Alabama	S_2	3 seams of granular to conglomeratic red hematite, grading to hematite - impregnated sandstone, interbedded with sublitharenite, gray to green shale,minor conglomerate, resting on limestone	700 Mt Fe/32% granular ore 260 Mt Fe/25% sandstone	Simpson and Gray (1968)
Gadsen district, Georgia,Alabama	S_2	as above	23 Mt Fe/30%	Wright et al. (1968)
Wolf Creek,Paris Mt. Virginia	S_2	hematite - cemented sandstone; equivalent to that in Birmingham	30 Mt Fe/20%	as above
Wabana,Bell Island, E.Newfoundland	Or_1	granular hematite/chamos.,siderite; several faulted seams interbedded with black micaceous shale, siltstone and quartzite	1.3 Bt Fe/ 51.3%	Snelgrove and Baird (1953)
Normandy iron ore district, S.E. of Caen, N.France	Or_2	oolithic hemat.,magnet.,chlorite ore forms folded,faulted and locally contact metamorphosed stratiform lenses and beds 1-6 m thick, interstratified with quartzite	370 Mt Fe/37%	Horon (1976)
Anjou-Bretagne Fe distr.(Ségre), N.W.France	Or_1	four 1-6 m thick oolithic magn.,hemat.,chlorite,siderite folded seams cropping out in W.N.W. synclines; feldspathic sandst. in fw, shale in hw.	400 Mt Fe/40%	Horon (1976)
Saalfeld distr., Thuringia,East Germany	Or_2	oolithic chamosite and siderite seams interstratified with dark shale,siltstone and graywacke	44 Mt Fe/40%	Zitzmann,ed., (1976)

Praha-Plzeň Fe ore zone (Barrandian Basin),W.Czechosl.	Or	granular,oolithic to massive hematite in several stratiform horizons; interbedded with black micaceous silty shale,litharenite and quartzite,basal horiz.rest on basalt flows and tuffs	130 Mt Fe/ 27%	Ilavský et al. (1976)
Nučice Fe horizon, Barrandian Basin, W.Czechoslovakia	Or	oolithic chamosite,siderite folded and faulted bed, at contact of black micaceous shale and sublitharenite; minor coeval submarine basaltic volcanism	8.4 Mt Fe/28%	as above
Gara Djebilet, S.W.Algeria	D_1	oolithic 6-10 m thick rich magnetite-siderite bed, and a low-grade chlorite-siderite bed, interbedded with shale and sandstone	1.3 Bt Fe/45%	Popov (1976)
Sierra Zapla,Sierra de Santa Barbara, N.Argentina	S_3	oolithic,granular hematite,chamosite,thuringite,goethite, 2 0.9-4 m thick stratif.lenses, interb.with quartz wacke, underlain by diamictite	175 Mt Fe/35%	Angelelli (1950) Haude and Weber (1975)
Ninghsiang "type" Fe ores (e.g.S.W.Hupeh, W.Kweichow,Hunan), China	D_{2-3}	bedded oolithic hematite in up to 3 1-2 m thick, extensive but discont.beds at base of sandstone and shale sequence	900 Mt Fe/50%	Ikonnikov (1975)
Fouling "type" Fe ores,Szechwan, Yunnan, China	Cb-Pe	bedded oolithic hematite interbedded with sandstone and shale	30 Mt Fe	as above

distributed but thin and low-grade (10% Mn over 1.7 m to 33% Mn over 0.5 m) Mn carbonate horizon is present near the base of a middle Cambrian unit. The host rocks are red and green, locally cleaved shales, in places containing interbeds of nodular limestone. The Mn horizon consists of gray-to-pink nodules of rhodochrosite and manganoan calcite, grading into thin solid Mn carbonate beds.

In the Oldakit Mn field located at the north-eastern end of the Angara Range, Siberia (Varentsov and Rakhmanov, 1974; about 200 Tt Mn/10%), several horizons of bedded, microspherulitic rhodochrosite and manganocalcite are hosted by lower Cambrian litharenite and slate.

METALLIFEROUS PHOSPHATES

The middle Cambrian phosphate deposits of the Georgina Basin, northwestern Queensland (e.g. the Duchess deposit; Russell and Trueman, 1971) are almost identical lithologically with the phosphorites of the Phosphoria Formation and of the Karatau Range, described in Chapter 17. They have been placed in the present section because of the environment of their deposition, located in a nearshore (some geologists even say lagoonal to intertidal) zone of an epicontinental sea. About fifteen phosphorite deposits have, so far, been located along the edge of the sedimentary basin and basement highs. The phosphatic horizons are almost always flat-lying, continuous sedimentary beds of pelletal phosphorite interbedded with phosphatic argillites, cherts and siltstones, and changing facies into limestones and dolomites. About 4 Bt of 18% P_2O_5 material have been outlined in the Duchess sector alone and with an average content of 94 ppm U they represent a resource of almost 400 Tt U.

The phosphorites of the upper Triassic-lower Jurassic Pucará Group in the Peruvian Andes (up to 20 m thick, 5-20% P_2O_5; Szekely and Grose, 1972), are similar.

19.6. PHYLLITE (SCHIST) - QUARTZITE ASSOCIATION

19.6.1. Metamorphosed equivalents of red and/or gray ironstones in orogenic belts

In the Ponferrada-Astorga zone in north-western Spain (Lunar and Amoros, 1979), magnetite, chlorite and siderite, conformable with the original bedding, are hosted by a middle Ordovician phyllite-schist association. The association includes chlorite and sericite phyllite, quartz phyllite and chloritic quartzite and is interpreted as being a regionally and contact-metamorphosed nearshore marine assemblage corresponding to those described earlier. The ore is magnetite-rich and a relict oolithic texture is often recognizable. Pyrrhotite, pyrite and arsenopyrite are widespread locally.

The iron ore deposits near Moncorvo, north-eastern Portugal (de Carvalho, 1976) are probably located in the western continuation of the same mineralized belt. The ore comprises several beds 1-8 m thick, interlayered with schist and with micaceous quartzite. The most abundant ore has a slatey appearance, enhanced by flakes and folia of specularite arranged in a mica-like fashion along the planes of schistosity. It is similar to the Proterozoic itabirite of Brazil. The iron ores near Vila Cova, about 100 km east of Oporto, are magnetite-rich and interbanded with chloritic schists and metaquartzites.

Greenschist-metamorphosed chamosite, siderite, magnetite ores that show surprisingly little mineralogical and textural change due to metamorphism occur in the Yugoslav Macedonia and have been described by Page (1958). At Tajmište, several tectonically dismembered stratiform lenses and sheets are hosted by Devonian graphitic phyllite.

Many of the Paleozoic "banded iron formations" summarized by O'Rourke (1961), such as those near Kathmandu, Nepal, associated with highly fossiliferous beds, have much in common with the typical Paleozoic ironstones reviewed above, and probably represent their local variation. The oolithic and granular hematites from the western Czechoslovak Ordovician that occur in the vicinity (or on top of) mafic volcanics, also contain abundant red jasper, which probably marks the transition into the Lahn-Dill "type" ores (Chapter 10). Some of the "itabiritic" (specularitic, schistose) deposits reviewed by O'Rourke (1961) are probably metamorphosed sandstones (quartzites) with a hematitic matrix, equivalent to the low-grade ores associated with the Clinton "type" in the Appalachians (e.g. Vila Cova). Alternatively, they could be magnetite paleoplacers (as in Talladega).

19.6.2. Metamorphosed equivalents of shallow marine Mn ores

This mineralization style is exemplified by several small occurrences of Fe-Mn ores in the Lienna Valley near Stavelot, Belgium (Ancion et al., 1956; about 130 Tt Mn/10%). There, rhodochrosite with hematite and supergene psilomelane forms cement and concordant lenses in upper Cambrian metaquartzites, interbedded with quartz phyllites and garnet micaschists. The ore layers are thin (0.2 to 1.1 m) and low-grade (7-16% Mn).

19.6.3. Monazite nodules in shallow-marine phyllites

Donnot et al. (1973) described an unusual occurrence of a gray nodular monazite, hosted by the Ordovician phyllites and schists of Britanny (north-western France). The millimetre-size "nodules" (or porphyroblasts ?) have a polycrystalline, radial or rimmed monocrystal internal structure and transgress the schistosity planes. The monazite-bearing metasediments overlie sedimentary ironstone and a detrital rutile, zircon and monazite-rich sandstone. They are interpreted as being metamorphosed, diagenetic products of rhabdophane transformation. The described occurrence is of mineralogical interest only (50-200 ppm monazite) but large tracts of weathered phyllite could presumably supply some released monazite into the placers.

19.7. INTERACTION AND "SPECIAL ASSOCIATION" METALLOGENESES INVOLVING SHALLOW MARINE (META)SHALE, SANDSTONE, MINOR LIMESTONE ASSOCIATION

19.7.1. Shale, quartzite, minor mafics or ultramafics, tectonism: Hg aspect

ALMADEN, SPAIN

Almaden is the world's most productive and highest-grade mercury ore field (487 Tt Hg/6%; Park and McDiarmid, 1975; Fig. 19-9), located on the northern slope of the Sierra Morena, 130 km S.W. of Madrid. The field is hosted by Silurian marine shales (some of them black) interstratified with quartzites which, in turn, are underlain and overlain by a monotonous terrigenous flysch-like association of alternating shales and sublitharenites. Mafic subaqueous volcanics form subordinate but interesting occurrences in the mining field, mostly in the form of mafic tuffs (reminescent of the Schallstein), flows and dikes. The region is folded and faulted (normal and thrust faults; many faults are wide, filled by breccia or mylonite gouge).

The ore is mineralogically simple: cinnabar with some pyrite and minor barite forms low-grade interstitial to high-grade, quartz-destructive replacements and veinlets in black and white sedimentary quartzite, interbedded with black argillite. In the largest Almaden mine, three east-west striking, 8-14 m thick, 330 m long subparallel "bonanza" (14-15% Hg in places !) orebodies are hosted exclusively by the quartzite so they are stratabound. The orebodies terminate sharply against the adjacent shale, or rarely diabase dikes, fault mylonite, or volcanic breccia. The genetic interpretation of Almaden remains controversial and has passed through all the fashionable models over the past fifty years. More recently, Saupé (1967) argued in favour of diagenetic origin of the Almaden ores on the basis of microscopic evidence that some cinnabar grains are enclosed in quartz overgrowths of the clastic quartz (or chert ?) grains, and the lack of mercury mineralization along faults, so common elsewhere. Whether true or not, the main problem of the Almaden origin remains the source (supply) of the enormous amounts of mercury that can hardly be explained by a simple diagenetic accumulation within a geochemically ordinary sedimentary pile. From the associational point of view, the most interesting rock in the Almaden field is the "Frailesca", a dark-gray carbonatized breccia that sometimes appears to be gradational into, or confused with, a carbonatized fault gauge but which is now mostly interpreted as an altered mafic tuff. Confusions are relatively easy in a vertically-dipping sequence. Some small Hg occurrences (e.g. the San Teodore mine) are hosted entirely by this rock. Although there are important differences, "Frailesca" nevertheless bears many points of

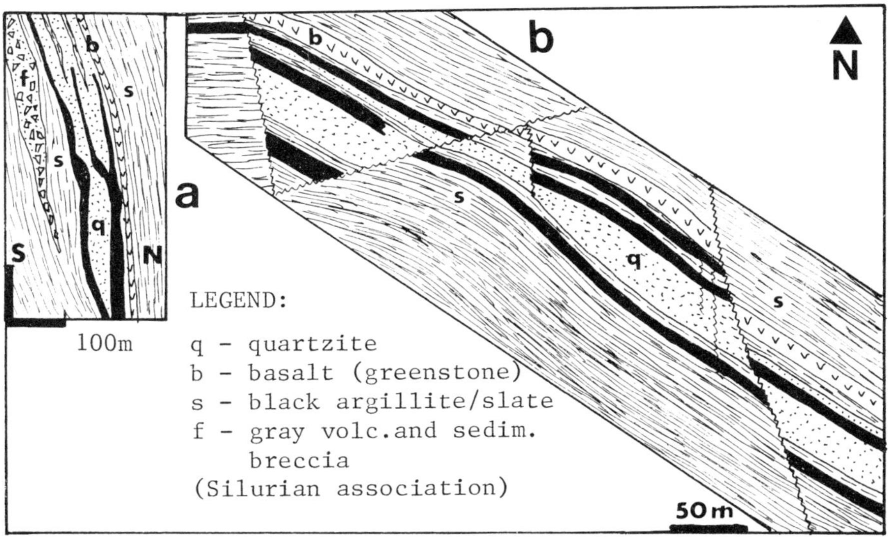

Fig. 19-9. Almaden mercury mine, Spain. (a) schematic section; (b) map, 14th level. From LITHOTHEQUE, modified after A. Almela S. and data in Park and McDiarmid (1975); Almaden Mine Staff, guided tour (1970).

similarity to the silica-carbonate rock formed by hydrothermal alteration of mafic or ultramafic rocks by CO_2-rich hot springs activity in Califormia (Chapter 8).

Several other mercury deposits and occurrences in shallow marine sediments display some points of similarity with Almaden in style and lithologic association but no close analogue is known. The Karareis deposit in western Turkey (Höll, 1966) comes closest. There, up to 5 m thick and 400 m long horizon of disseminated cinnabar, marcasite and arsenopyrite (average 2% Hg) is hosted by silicified middle Silurian graphitic slates interbedded with carbonate, cherty carbonate and sandstone. Höll (1966) argued for the stratiform nature of the orebody, and drew it as a conformable band. Yildiz and Bailey (1978), on the other hand, assumed it to be a fault-controlled epigenetic mineralization coeval with silicification. Subaqueous mafic volcanics are present in the area, but not in immediate contact with the orebody. In the nearby Kaleçik mine, cinnabar with pyrite and several rarer minerals, is hosted by millimetre-thin black argillite interbeds in rhythmically banded felsic tuffs, and also occurs as superimposed epigenetic veinlets or veins. Both deposits were interpreted by Höll (1966) as subaqueous-volcanogenic.

Epigenetic cinnabar deposits hosted by shallow marine shale or sandstone association are quite common and controlled by faults, crush zones, intervening volcanics or intrusive bodies, and by the porosity of the host rocks. Their predictability using the host sediments facies is at best applicable only locally, where cinnabar orebodies have already been found by other methods. In the Huancavelica mercury district, Peru (Yates et al., 1951; 51 Tt Hg), the most productive cinnabar orebodies in sandstone associated with pyrite, arsenopyrite, realgar and orpiment, occur mostly as a filling between and replacement of, quartz grains. The widespread presence of hydrocarbons (heavy black oil, tar) in all the types of Huancavelica mercury deposits is remarkable and it appears that in certain phases of their origin the migrating mercury and petroleum shared the same pathways and the same traps.

In Draževici, N.N.E. of Sarajevo, Yugoslavia (Ramović and Kulenović, 1964; 4.2 Tt Hg/1.4%), numerous small peneconcordant bodies of cinnabar, marcasite and calcite impregnations, stringers and veinlets, are hosted by a Triassic sandstone, shale, conglomerate and dark limestone near thrusts. Subaqueous mafic volcanics, tuffs and Mn oxide mineralized cherts and jaspers, are members of the same sequence.

19.7.2. Mineralization in areas of interaction between shallow-marine (meta)shale and arenite, and plutonic or subvolcanic intrusions

This is a fairly widespread type of interaction, in which mineralization is influenced by (1) the composition and differentiation history of the intrusive rocks; (2) the composition of the basement in which the intrusions were generated or through which they passed; and (3) facies or geochemical anomalies in the shallow marine sediments. In this section, the influence of (3) is emphasized.

Sb ASPECT

Mineralogically simple quartz-stibnite, lesser arsenopyrite, pyrite fissure veins and shear fillings, are frequently hosted by gray-to-black slates in the proximity of granitic intrusions. Example localities include the Silurian to lower Devonian sediments in the Melbourne Trough, Victoria (Australia), intruded by Devonian granites (e.g. Costerfield, 22.4 Tt Sb; Hill, 1975). The Lake George antimony mine 40 km west of Fredericton, New Brunswick (35 Tt Sb) is located in Silurian shallow-marine hornfelsed slates and siltstones near a Devonian quartz monzonite intrusion.

Au ASPECT

The stibnite deposits listed above are often lightly gold-bearing, and spatially associated with stibnite-free gold-quartz veins that are

are widespread in similar setting. In the upper Ordovician to Devonian shallow-marine (meta)shales and sublitharenites of the Melbourne Trough, Australia (a nonvolcanic successor basin association), numerous small gold-quartz veins and placers are associated with Devonian mafic and acid dikes intruding the sediments. In the richest field (Wood's Point), the majority of lodes are hosted by the dike rocks.

In the Cariboo (Barkerville) gold field, central British Columbia (Sutherland Brown, 1957; 108 t Au), placer and lode gold has been derived from hydrothermal gold-quartz fissure veins and from sulphide-rich replacements in a limestone band. The host rocks are lower Cambrian shallow-marine slates to phyllites, interbedded with abundant quartzite and rare bands of gray limestone. The veins formed mostly in the brittle quartzites and pyrite, minor galena, sphalerite, arsenopyrite, cosalite, bismuthinite and free gold form 15-25 % of the filling of the quartz-ankerite and minor sericite veins. Sericitization, carbonatization and bleaching are the dominant alterations. The replacement sulphide orebodies confined to a single marble band consist mostly of massive fine-grained auriferous pyrite, assaying as much as 140 ppm Au.

The ores are indisputably epigenetic and their structural control is both regional (flanks of a large anticline) and local (hundreds of transversal, diagonal and strike faults). The best gold grades are found in the short veins filling faults transversal to bedding, while most of the lodes parallel to the bedding schistosity are filled by the very abundant but barren milky ("secretion") quartz. The gold source, timing of mineralization, driving agent, etc., are largely unknown because of the complex geological history of the area. Of interest are the minor occurrences in highly ankeritized diorite sills. Despite the epigenetic and discordant nature of the ores, a degree of lithostratigraphic control has been acknowledged even in the papers written at the peak of preoccupation with the structural control in the 1930s-1940s. Most of the bedrock gold orebodies near Wells are hosted by the brittle metasediments of the Snowshoe Formation, but immediately at or close to a contact with the underlying Midas Formation. The latter is a conspicuous pyritic "black" metasedimentary unit composed of black siliceous phyllite, metasiltstone, black quartzite, limestone, and could have been the source bed to the gold.

The presence of black sediments, particularly the unusual black pyritic quartzite and conglomerate, often correlates with the presence of epigenetic gold-quartz veins in the shallow marine sedimentary associations (e.g. in the Beaconsfield goldfield, northern Tasmania).

Sn ASPECT

The Cordillera Real Tin Province, Bolivia

The Cordillera Real and Cordillera Quimsa Cruz ranges N.E. and S.E. of La Paz (Ahlfeld and Schneider-Scherbina, 1964), are dominated by a thick, sombre-looking sequence of Ordovician to Devonian dark gray terrigenous slate to phyllite, grading into mica-rich sublitharenite and having frequent interbeds of quartzite. These sediments are zeolite to greenschist metamorphosed, folded and thrusted. No carbonates or coeval volcanics are present, but the sequence has been intruded by numerous quartz monzonite batholiths of the Triassic to Miocene ages, and farther south by late Tertiary high-level subvolcanic bodies. The 1,000 km long N.W.-S.E.-trending belt of the Paleozoic shallow detrital sediments is believed to rest on (and be derived from) a Precambrian crystalline basement; a possible continuation of the Rondônia terrain from Brazil. The belt is also a major tin-tungsten province containing about 1000 known Sn occurrences and representing some 6 Mt of ore tin. This includes deposits in the plutonic and subvolcanic suites of rocks. The association or interaction of Paleozoic sediments with the tin-tungsten deposits can be treated in terms of three empirical categories (Fig. 19-10): (1) cassiterite manto mineralization in quartzite beds; (2) cross-cutting cassiterite bodies in "granite" exocontacts; and (3) cassiterite in and around the Tertiary subvolcanic complexes.

(1) Five groups of occurrences of "tin mantos" (cassiterite mineralization confined to quartzite beds enclosed in shales and slates) had been known and mined in the Cordillera Real before the publication of the Ahlfeld and Schneider-Scherbina (1964) book, in which they are summarized. Recently, these mantos have been reinterpreted as cassiterite paleoplacers and Schneider and Lehmann (1977) demonstrated their confinement to the Silurian Catavi and Llallagua Formations. The most readily accessible mantos are in the high mountains under Huayna Potosí, at altitudes of 4,600-5,200 m and about 30 km N.E. of La Paz. The Huallatani mine at present produces cassiterite from a single manto although more quartzite beds are known, some hosting cassiterite occurrences.

The Kellguani (Kellhuani) mine (Figures 19-11, 19-12) contains several mineralized quartzite mantos, and casual visitors expecting to see stratiform deposits are certainly impressed by lines of old workings (short adits) following the mantos. However, once one gets to the actual mine workings it becomes obvious that the mineable grade (0.5% Sn plus) is epigenetic, accumulated in fractures and as veins and associated with hydrothermal alteration. In an outcrop, the fractures usually occupy one or two directions. Both directions can have the same minerals, or the mineralogy may differ. Both, or only one system, can carry cassiterite. The density of fracturing and veinlets vary, from 1-2 to about 20-30 per metre. The thickness ranges from hair-thin fractures with flat mineral grains coating the walls, to discrete veins several centimetres thick. The following types of vein filling can be recognized: (a) simpliest association of milky quartz, dense or drusy inside. These veins are universally present in all the quartzites. Near Kellguani, they also occur in the "unmineralized" mantos, and contain sporadically scattered cassiterite crystals. (b) Tourmaline veinlets. The tourmaline is brownish-black, fibrous to needle-like, either arranged perpendicularly to the

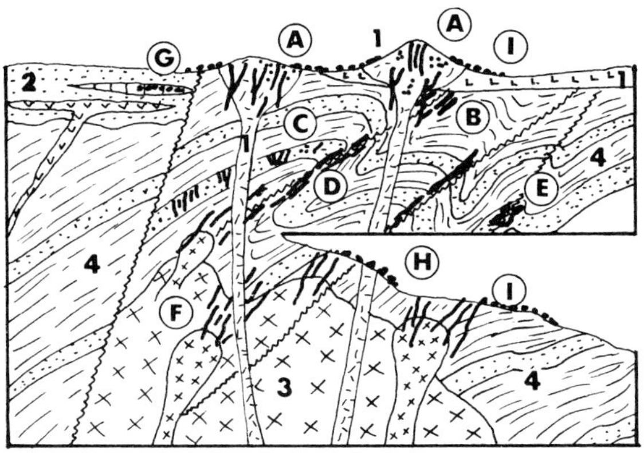

1 T_{2-3} felsic volcanic centres, tuff, volcanic conglomerate
2 Cr_1 red sandstone, mud-stone, basalt, minor limestone
3 Tr-Mi quartz monzonite granodiorite
4 Or-D shale to slate, quartzite, sub-litharenite

Fig. 19-10. Principal styles of the tin mineralization in Bolivia. (A) Pervasive disseminations, veins, stockworks, in Miocene-Pliocene vents and adjacent sediments (Llallagua, Potosí); (B) conformable pyrite lenses in Paleozoic black slates (Potosí); (C) mantos, cassiterite fracture fillings and disseminations confined to quartzite horizons (Kellguani); (D) persistent exocontact fissure veins (Milluni); (E) fissure veins, stockworks, at fault and quartzite intersection (Tanapaca); (F) endocontact veins or stockworks in granite (Caracoles); (G) replacement mantos in Cretaceous carbonates (Colavi). From LITHOTHEQUE.

fracture walls or having the form of flat, radiating aggregates. The tourmaline filling is either sharply outlined or it reaches into the alteration rim of the veinlet. In some instances the whole manto is fine-tourmalinized and cut by a younger generation of tourmaline veinlets. The tourmaline-only veinlets rarely carry cassiterite, but cassiterite is common in the quartz-tourmaline association. (c) Quartz, bladed white feldspar, pyrite, cassiterite veinlets, with centres usually filled by ankerite. These are important ore carriers in the ore mantos, and the same minerals could be found as fine, cloudy alteration assemblages in the quartzite. (d) Flaky chlorite fracture coating and chloritized quartzite. (e) Ankerite veinlets.

The associations (c) to (e) often mix and form composite veinlets in which cassiterite is scattered in the form of brown grains or needle-like crystals, and is common. All the veinlets range from massive to vuggy; open vugs are sometimes the result of leaching of the central soluble carbonate. Although all the quartzite beds in the Kellguani mine area carry some tin (under 0.1% Sn), the currently economically mineable ore must have at least 0.3% Sn; even so, recovery is rarely over 60%. All the economic ore shoots are confined to altered quartzites. The stratiform concept applied to the cassiterite mantos appears to have been oversold. Although accessory

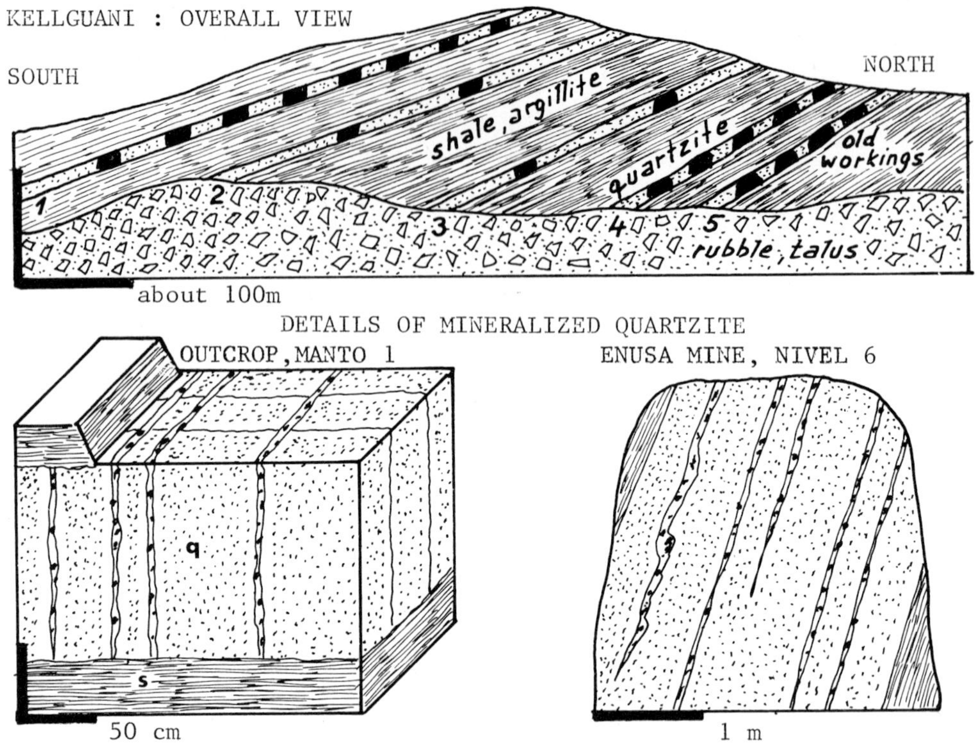

Fig. 19-11. Kellguani cassiterite mantos, Cordillera Real, Bolivia. Epigenetic quartz, tourmaline, adularia, chlorite, ankerite, cassiterite veinlets, confined to beds of gray, argillaceous and micaceous quartzite. These in turn are interbanded with a gray silty shale or a micaceous siltstone, containing local bands of black shale. Lightly thermally metamorphosed. From LITHOTHEQUE, based on 1983 visit.

zircon, ilmenite and other heavy resistates in the quartzite are supportive of the cassiterite paleoplacer interpretation, the economic ore, although stratigraphically confined, is also structurally controlled. The associated galena has been dated as Cretaceous and this supports its epigenetic origin, although the contemporary origin of cassiterite could be assumed only indirectly. Whether the ore tin has been provided by a buried granite cupola or whether it had been remobilized from originally detrital cassiterite by the hydrothermal front driven by the granite is uncertain. The black slate horizons are another possible source of tin. At the Enusa property alone, 2.67 Mt of a 0.6% Sn ore has been outlined in three mantos.

Fig. 19-12. Kellguani Sn mantos (a) overall view showing old shallow mine workings. (b) Distribution of cassiterite in quartz veinlets fringed by tourmaline and chlorite haloes. (c) Flat crystals and aggregates of quartz, feldspar and cassiterite on the surface of a fracture.

Some manto deposits are transitional into quartz-cassiterite stockworks. Serrano orebody in the Monte Blanco field (Turneaure and Welker, 1947) is a stockwork of an irregular outline hosted mostly by a quartzite unit about 30 m thick, but also reaching outside it. The stockwork is composed of numerous disconnected veinlets, the majority of which are parallel to the stratification. Twelve more mineralized seams occur within a width of 2.5 or 5 cm and convey an impression of a ribbon ore.

(2) Quartz-cassiterrite and lesser wolframite, sphalerite, galena, chalcopyrite, etc. hydrothermal veins concentrate (as in the rest of the world) in and above the apical portions of granite and quartz monzonite cupolas (compare Chapter 28). The majority of the veins lie in the granite exocontact, in hornfelsed slates and indurated quartzites. At a distance from the granite contact (outside the zone of contact hornfels) many fissure veins are strongly influenced by the host rock competency. Many veins (e.g. the Veta Principal, Viloco) are well developed when in quartzites but then ramify and rapidly lose the ore content upon reaching the slates. In some mines, orebodies are controlled by through-going faults, but occur exclusively along the fault intersection with quartzites, for example in the form of thick mineralized breccia pockets (bolsons).

Turneaure and Welker (1947) discussed the influence of sedimentary wallrocks on the epigenetic ore deposition in Bolivia and concluded that the influence had been largely structural. They admitted, however, the existence of a local chemical influence as well, supported by the observation that cassiterite veins in shales tend to be sulphide-rich. The composite Rothschild Vein in the Milluni field hosted by a thin-bedded black shale, is particularly high in sphalerite.

(3) The three large Neogene vein and "porphyry tin" fields Llallagua (3.5 Mt Sn), Cerro Rico, Potosí (520 Tt Sn, 52 Tt Ag) and Oruro (500 Tt Sn, 12 Tt Ag) (Sillitoe et al., 1975; Fig. 19-13) are all underlain or partly hosted by Ordovician and Silurian fine detrital sediments that include prominent pyritic black slates.

The Cerro Rico hill above Potosí is formed by an oval high-level intrusive body of thoroughly altered ? quartz latite porphyry, with an area of about 2 km^2. The intrusion, however, has a mushroom shape and at approximately the level of the Paleozoic / Cainozoic unconformity, it tapers down to a diameter of about 100 m. The stock is heavily mineralized by a variety of hydrothermal Sn and Ag bodies (pervasive disseminations of cassiterite in the porphyry; system of parallel Sn-Ag veins in a zone of sheating in the intrusion; fracture veins). At a depth, a significant proportion of the veins are hosted by the Paleozoic, low-grade metasediments. On the 9th level (at a depth of 800 m) of the Unificada mine on the Cerro Rico, siliceous black slates contain numerous bedding laminae of pyrite, coalescing into lenses of an almost massive pyrite. This pyrite appears to be contemporary with the slate. Superimposed hydrothermal fissure veins either cut sharply across the pyritic shales, or alternatively form

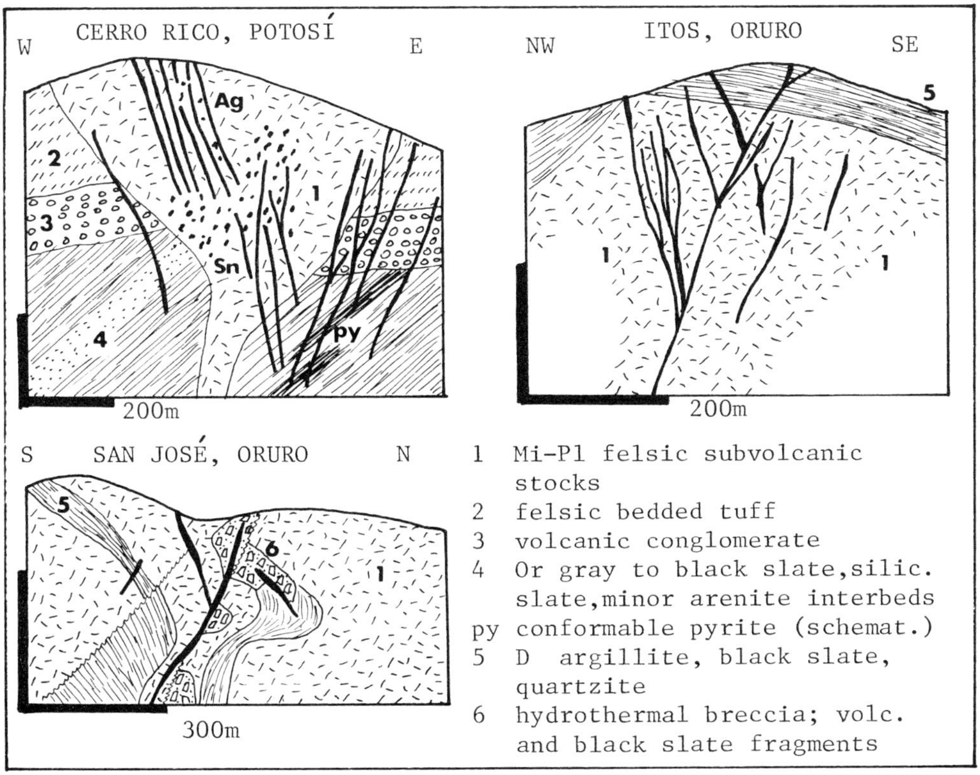

Fig. 19-13. Sn-Ag subvolcanic vein and disseminated mineralization associated with Paleozoic black argillite or slate, Bolivia. From LITHOTHEQUE, after Turneaure (1960); Ahlfeld and Schneider-Scherbina (1964); Chace (1948) and COMIBOL mine staff, guided tours (1977).

diffuse, loosely defined replacement bodies of an introduced pyrite, containing finely dispersed cassiterite. Pyritization and pyrite veining coeval with or shortly postdating the emplacement of the Cerro Rico stock, is widespread, and occurs in several generations.

Hydrothermal intrusion breccias are abundantly developed in Potosí, Llallagua and Oruro (Sillitoe et al., 1975). They are usually located along the contact of the intrusive stock with the surrounding sediments, as well as within the stock. The breccias have the form of circular pipes, narrow dikes and sheets and consist of irregular to rounded fragments of porphyry and the black Paleozoic sediments, in all proportions. The black slate fragments frequently contain pyrite laminae, and many are virtually unaltered. The matrix of the breccias consists of fine porphyry and a sedimentary rock flour, and is often drusy and hydrothermally tourmalinized (black). Any of those breccias affected by superimposed hydrothermal alteration and mineralization carry Sn and/or Ag values.

The contribution, if any, of the black Paleozoic sediments to the formation of the giant Sn and Ag accumulations has not been determined. but the presence of the black sediments is an interesting empirical feature to be considered in exploration.

Most tin districts of the world such as those in Cornwall (Dines, ed., 1956), Yakutia in Siberia (Flerov, 1976), eastern Tasmania, New South Wales, western Alaska, etc. (Chapter 28) are hosted by shallow-marine terrigenous sediments resting on, or adjacent to, an old crystalline basement, and intruded by granitic rocks. Some such associations have a more pronounced "flysch" character, so they have been treated in Chapter 17 (e.g. the Herberton field, Australia); others occur in geologically complex situations difficult to classify (e.g. some of the western Tasmanian tin deposits). Elsewhere (e.g. in Malaya, the Tin Islands of Indonesia), the deep tropical weathering lessened our awareness of the shallow marine sediments as tin ore hosts, and emphasized the role of granites. A visit to one of the few south-east Asian underground tin mines (e.g. Sungei Lembing, eastern Malaya) with veins hosted by shallow-marine sediments, puts one's ideas into a proper perspective. One might conclude this section with a quotation from the excellent review of tin distribution patterns by Professor Hosking (1979, p.3): "It seems to me reasonable to expect shales and other sedimentary rocks that are generated from the detritus of tin provinces to be, on occasion, distinctly stanniferous. Sediments that cannot claim to be called stanniferous placers must, on occasion, contain sufficient tin possibly for them to be converted by metamorphic process of one sort or another, or by magmatic invasion, to tin deposits of economic importance. I cannot believe that tin released into the drainage system by superficial process can never more take part in those geochemical syntheses which result in the development of new hardrock tin deposits. Only by postulating the mobilisation, both of released tin and of anomalous concentrations of tin in hard rocks can I find an acceptable explanation for the presence of primary tin deposits, of two or more widely differing ages, in a given metallogenic province".

W ASPECT

Tungsten overlaps with tin in many mineralized occurrences (e.g. in the deposits of the plutonic suite in Bolivia), but is conspicuously missing in other tin mineralization styles (such as the high-level "porphyry tin" deposits). In the shale, sandstone, granite mineralized system, tungsten (represented by wolframite) accumulates in the exocontact zone and coincides with the cassiterite-bearing quartz veins hosted by hornfelsed slates, arenites, or gray-to-black slates. The empirical affiliation of the tungsten accumulations with certain "black slate" units has been stressed earlier, in connection with the deeper-marine and volcanics-influenced sediments (compare Chapter 17). It is quite probable that certain "black slates" in the shallow-marine shale-quartzite association may also accumulate high trace tungsten and act as a source rock.

The Bolsa Negra mine located on the south side of the Mururala Batholith in the Cordillera Real zone, Bolivia (Turneaure and Welker, 1947), is an interesting example suggesting such a possibility. There, a 50 m thick ore horizon of dense, hard quartz, tourmaline and chlorite hornfels is enveloped by a thin-bedded black slate. The ore deposit confined to this horizon is over 330 m long and consists of numerous quartz-wolframite veins lying in an almost horizontal position and peneconcordant with the bedding. The individual quartz veins and lenses rarely exceed 25 cm in thickness, and are preferentially accumulated in three or four, 1.5-3 m thick, sections.

In the well-known tungsten deposit Panasqueira, Portugal (Kelly and Rye, 1979), a large number of parallel quartz-ferberite veins is also hosted by black hornfelsed slates to phyllites, above a greisenized granite cupola. There, however, the near-horizontal veins cut sharply across the steep bedding and foliation of the late Proterozoic or Cambrian pelitic metasediments. Quartz-wolframite veins hosted by hornfelsed dark or gray slates have a worldwide distribution, e.g. in the Aberfoyle and Storey's Creek fields in eastern Tasmania; Burnt Hill mine, New Brunswick; Kiangsi province in China and elsewhere.

Pb-Zn ASPECT

Pb-Zn veins (and replacement bodies in carbonates when these are present) are by far the most common style of metallic mineralization hosted by the shallow marine shale - sandstone sequences present in "granite" aureoles. Mineralized areas of global importance include the Great Basin of Utah, western Nevada and adjacent states; portions of the Mexican Mesa Central; portion of the Central Peruvian Andes; western European Variscan (Hercynian) belt; the Tian Shan belt of Soviet Central Asia; a portion of the Tasman belt of Australia and others. The metallogenetic relationships are comparable with those already discussed in Chapter 17, except that massive sulphide deposits like Rammelsberg are not normally expected to be found in this association. In many instances, however, the shallow marine sediments (shales, slates, arenites, quartzites, carbonates) that actually host the epigenetic orebodies, grade into, or are underlain by, an earlier "deeper marine" association often with submarine volcanics in which the presence of massive sulphides is possible and a metallogenetic overprinting (a lift of Pb,Zn,Ag into the top level rocks) quite likely. This has already been noted in Chapter 17, and it is reinforced here using the example of the Slocan, British Columbia, hydrothermal vein Pb,Zn,Ag district (Fig. 19-14 and 19-15), hosted by Triassic shallow marine sediments (Ross and Kellerhals, 1968).

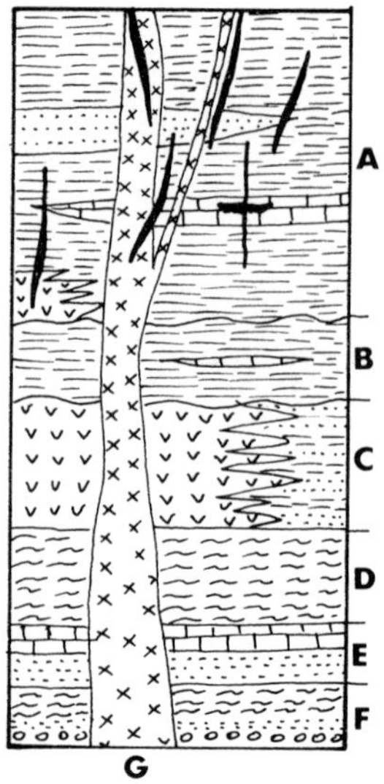

A Triassic shale,argillite,sandstone, quartzite,limestone; rare intermediate volcanics
B Mississippian-Permian argillite, limestone
C Silurian,Devonian graywacke,slate
D Middle Cambrian-Silurian schists
E Lower Cambrian marble,metaquartzite
F Late Proterozoic metaconglomerate, quartzite,schist
G Jurassic-Cretaceous quartz monzonite, granodiorite, quartz-feldspar porphyry, felsite
BLACK: Pb-Zn-Ag orebodies

restored,generalized

Fig. 19-14. Slocan Pb,Zn,Ag district, British Columbia; a generalized section showing setting of the granite-coeval, hydrothermal epigenetic vein deposits. From LITHOTHEQUE, based on Ross and Kellerhals (1968).

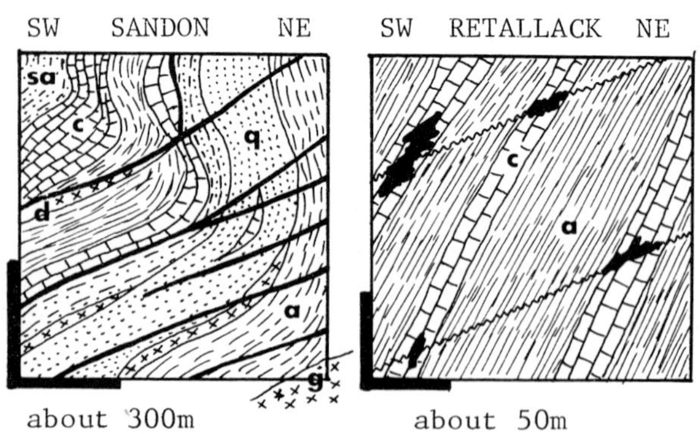

Triassic Slocan Group

sa sandstone/shale
c carbonate
q quartzite
a argillite

J-Cr Nelson Batholith

d intrusive dikes
g granodiorite

Fig. 19-15. Slocan Pb,Zn,Ag district, British Columbia; diagrammatic representation of the two principal mineralization styles: fissure veins (left) and carbonate replacements (right). From LITHOTHEQUE.

There, several tens of mostly north-east-trending quartz, siderite, sphalerite, pyrite, galena, minor tetrahedrite, Ag sulphides, etc. fissure veins and minor carbonate replacements are hosted by gray-to-black argillites interbedded with quartzite and subordinate limestone. Most of the veins are perpendicular or oblique to the host rocks strike, and sharply postdate emplacement of the Jurassic-Cretaceous Nelson Batholith. The Triassic shallow-marine sediments are underlain by a middle Paleozoic deeper-marine schist, slate, mafic volcanics assemblage known to host Zn-Pb mineralization outside the Slocan district.

In the Zeehan Pb,Zn,Ag district, western Tasmania (Blissett, 1962) as well as in parts of the Rheinische Schiefergebirge of Germany and elsewhere, the stratigraphy is reversed. The stratigraphically lower and older assemblage hosting the bulk of the Pb,Zn,Ag fissure veins, consists of shallow-marine sandstones (quartzites) and shales (slates), and the stratigraphically higher assemblage present within, or in the vicinity of the ore field, is developed in the deeper-water and/or volcanic facies. The emplacement of granite that was broadly coeval with the ore vein formation postdated the "deep" depositional period and was usually contemporary with the renewed shallow-marine sedimentation.

Although structural control (open faults, zones of axial cleavage, shears) and thermal aureole control are undisputably the immediate controls governing distribution of the ore veins, the "favourable" host or basement lithologies (in particular the abundance of black slates and/or minor contemporary mafic volcanics or subvolcanic bodies), is a more subtle but empirically equally important attribute probably governing the developing metal sources.

Attention to the "favourable" sediments (volcanics) may enhance the exploration strategy, or result in the discovery of a pre-vein mineralization coeval with deposition of the supracrustal rocks. Fuchs and Lang-Villemaire (1981) reported the discovery of peneconcordant lenses of disseminated sphalerite, galena and other minerals in lower Devonian shallow marine sediments, in the Carboniferous epigenetic fissure vein Pb-Zn Bensberg field (Rheinische Schiefergebirge, West Germany). Similar mineralization may exist in the large (620 Tt Zn, 320 Tt Pb) Ramsbeck field in the same region (Bauer et al., 1979). There, about 20 subparallel Pb-Zn veins of a late Carboniferous and partly younger age are hosted by Devonian shallow-marine quartzites and black slates in a complex structural setting on the flanks of an anticline. Diabase dikes and sills locally sheared and converted to greenstones are widely developed in the Ramsbeck field. A close analogy in the host lithology (black phyllites, quartzite, diabase to greenstone bodies) occurs in the important Mayo-Keno Hill Ag,Pb,Zn vein district, Yukon.

In the recently developed polygenetic Real de Angeles Ag-Pb-Zn deposit near Loreto, Zacatecas state, Mexico (Gonzales, 1984 and oral communication, 1985) a wide range of bedding-conformable to cross-cutting and "syngenetic" to "epigenetic" features is particularly well developed. This is an example of a large but low-grade "bulk" deposit (about 90 Mt ore with 30 to 150 ppm Ag; 1.3% Pb and 1.1% Zn in the "high-grade" zone only; this corresponds to at least 170 Tt Pb and 145 Tt Zn) hosted entirely by an interbedded sequence of Mesozoic nearshore-marine sediments. Laminated gray-to-black siltstone and argillite are interbedded with sublitharenite and the sedimentary unit is hard, indurated and lightly silicified suggesting setting at the fringe of an intrusive aureole. The host unit has been overthrusted by upper Cretaceous thin-bedded cherty limestone and locally capped by up to 40 m of Cainozoic calcrete. No inrusive rocks have been so far detected in the mine field, although remnants of a Tertiary ignimbrite sheet are preserved within several kilometers of the mine.

The ore zone has the form of a thick "manto" dipping 30° to the south, crossing stratification. Four major styles of distribution of ore minerals (iron-rich sphalerite, galena, pyrite, pyrrhotite, arsenopyrite, jamesonite, minor freibergite, proustite and stibnite) are shown on Figures 19-16 and 19-17. The mineralization is presently interpreted as partly remobilized syndepositional metal accumulation and a galena crystallization temperature of 270° C has been determined. There is no conspicuous widespread hydrothermal alteration superimposed on the host sediments, except for minor local haloes of axinite and hedenbergite or chlorite adjacent to veins. A small oxidation zone in argillized sediments contains argentojarosite, cerrusite and Ag-halides.

19.7.3. Shallow-marine sediments, thrust faults, regional volcanism and plutonism: "invisible" replacement gold deposits (Carlin "type")

The 1961 discovery of the Carlin, Nevada, gold deposit focussed attention on a previously unpublicized (although actually mined at several localities) style of gold mineralization, departing from the conventional models. The "Carlin-type" deposits were missed by the early prospectors whose principal tool was the panning dish, because the gold is extremely finely dispersed and invisible even under an electron microscope, so that it cannot be panned. It is believed that at Carlin the gold resides in arsenian pyrite or in carbonaceous material.

Carlin (Hausen and Kerr, 1968; Roberts et al., 1971; min. 120 t Au; Fig. 19-15) is located in the faulted Basin-and-Range Province and the gold is hosted by shallow-marine Silurian thin-bedded black argillaceous siltstone, dolomitic and limy siltstone, and banded argillaceous limestone locally containing chert nodules. An irregular, gently-dipping peneconcordant elongated mineralized zone up to 30 m thick and 300 m long, lies several tens of metres under the Roberts Mountain Thrust, along which a western "eugeoclinal" chert, siliceous carbonate and shale assemblage was tectonically transported over the eastern "miogeoclinal" suite. The ore is inconspicuous, because virtually no ore minerals or striking colour changes due to alteration are apparent. The auriferous pyrite is fine-grained and

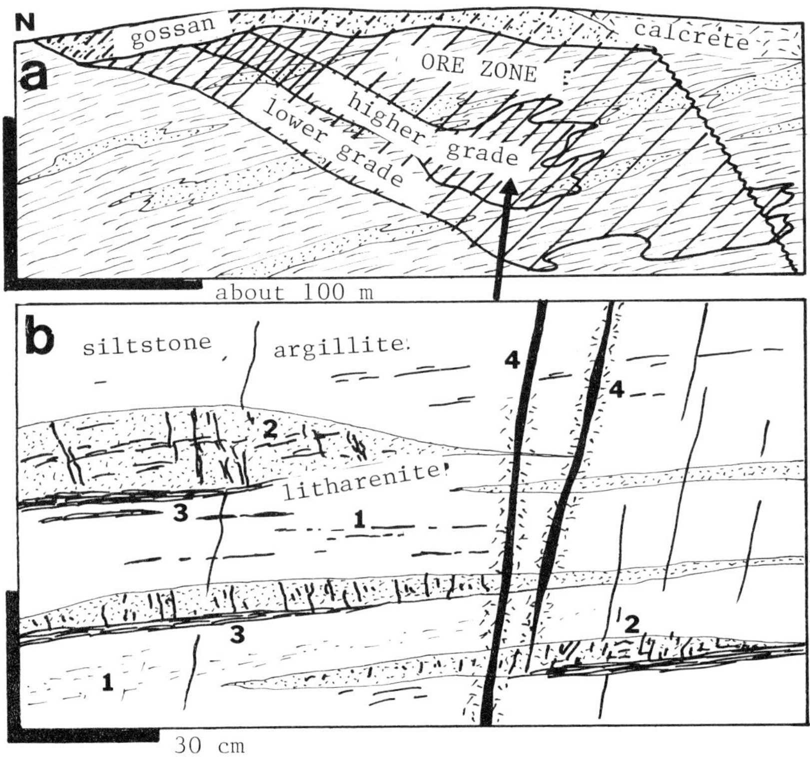

Fig. 19-16. Real de Angeles, Zac., Mexico. (a) Orebody cross-section. (b) Diagrammatic representation of the major styles of ore distribution in the "primary" (non-oxidized) ore zone: (1) disseminated sulphide (pyrite, pyrrhotite, sphalerite, galena, arsenopyrite) grains grading into thin lenses and partitions parallel with bedding in the black siltstone; (2) short sulphide veinlets and fracture coatings ("ramaleos") largely confined to sandstone bodies; (3) densely disseminated to massive sphalerite-rich sulphide lenses peneconcordant with bedding, most common along sandstone and siltstone contacts; (4) through-going sulphide fracture coatings, veinlets, grading to quartz, fluorite, calcite, sulphide fracture veins. From LITHOTHEQUE, (a) generalized and modified after Gonzales (1984), (b) original, based on 1985 visit.

sparse, the hydrothermal silicification very light, and the only visible sulphide occasionally found coating fractures is the orange-red realgar.

In the oxidation zone extending to a depth of about 50 m even the realgar is gone, and the ore is just an ordinary rusty spotted beige porous siltstone or argillite, from which the soluble carbonate has been leached out. The average grade at Carlin is about 9.2 ppm Au, and there are increased trace contents of As, Hg, Sb and Tl. Veins and sheet-like masses of black barite, carbon and fine gold having a grade of 5-69 ppm Au have been recorded.

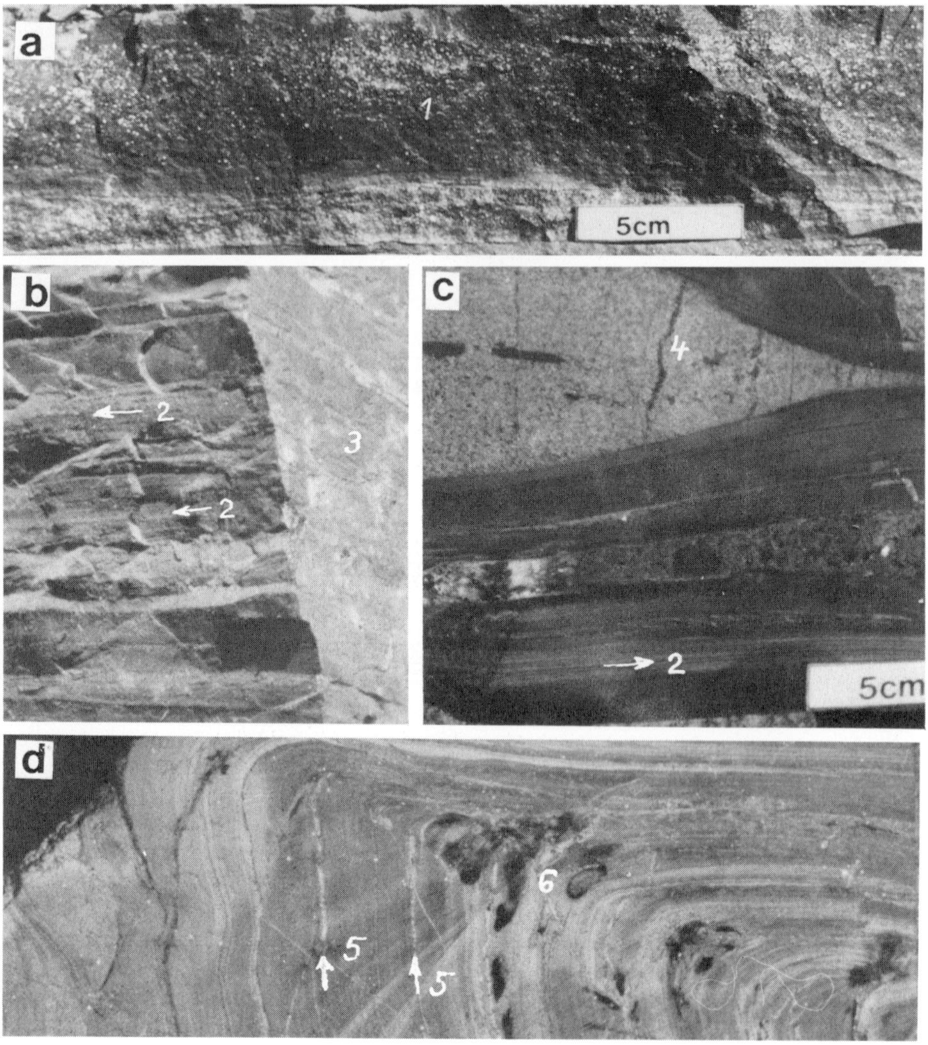

Fig. 19-17. Real de Angeles, Mexico. (a) Bedding-conformable, densely disseminated sulphide, 1. (b) Disseminated sulphide // with siltstone lamination, 2; sulphide fracture coatings, 3. (c) Ramaleos (veinlets, 4) in a sublitharenite lens. (d) Bedding-conformable sulphide strings, 5, and remobilizations in the axial zone of folds, 6.

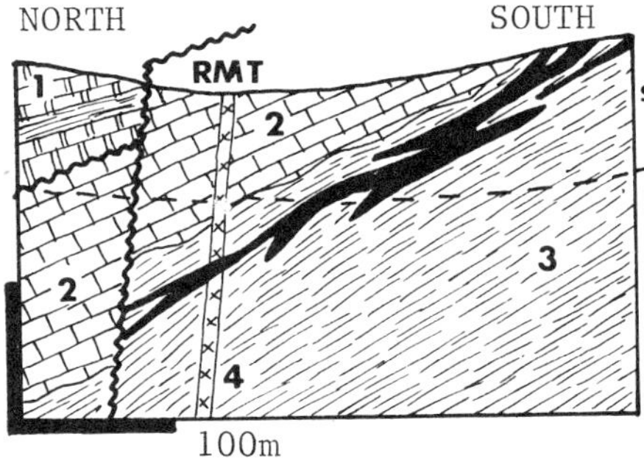

Fig. 19-18. Carlin, Nevada, deposit of finely dispersed gold in lightly altered black siltstones, under thrust. From LITHOTHEQUE, after Hausen and Kerr (1968) and Roberts et al. (1971).

The origin of the Carlin deposit is usually attributed to low-temperature hydrothermal fluids that rose along normal faults and replaced rock-forming carbonate under the confining floor of the thrust. The derivation of the fluids is either (1) attributed to the late Tertiary continental volcanism, and the hydrotherms are compared with the hot spring systems such as those that presently precipitate Hg, Sb, and Au near Steamboat Springs, Nevada (Chapter 5); or (2) associated with the granitic plutonism and the ores are believed to have formed at depths exceeding 700 m (compare Roberts et al., 1971, for review).

In the absence of clearly recognizable controls, exploration for the "Carlin-type" deposits is difficult. In the western United States, the footwall of the Roberts Mountain Thrust (and other thrusts) has been most thoroughly explored, and several discoveries made. This thrust control is quite reminescent of the conditions that control the Hg and Sb mineralization in the extensive South Ferghana Belt (Chapter 17). Elsewhere, any other faults or thrusts would probably do as well. The presence of the black sediments, contemporary volcanism or magmatism, and As, Hg, Sb, Tl geochemical anomalies are other helpful exploration leads.

Up to now, about 10 deposits said to be of the "Carlin-type" have been reported, most of them in the western United States (Table 19-6). They differ considerably in detail. In the Cortez deposit, Nevada (Wells et al., 1969) the fine gold is in a conspicuously bleached and silicified (jasperoid) thinly laminated carbonate-rich siltstone, along a front of a thick, sill-like mass of porphyry. The bleaching caused oxidation of the original pyrite in the siltstone into hematite, and removal of a portion of the organic pigment, so that the gold-bearing rocks have a light-gray to dark-red colour. In the Gold Acres mine west of Cortez, the fine gold has been superimposed over a

Table 19-6. Carlin - "type" deposits of finely dispersed gold, hosted by shallow marine sediments

LOCALITY	AGE OF HOSTS	AGE OF ORE	GEOLOGY, MINERALIZATION	TONNAGE, GRADE	REFERENCES
Carlin, Nevada	S	Ol ?	the orebody is an assay-bounded pene-conc.sheet of lightly silicif.gray li-my and dolom.siltst. and argill.,impr. by colloid.Au,pyr.,minor realgar, in footwall of a thrust	120 t Au/9.2 ppm 50 t Hg	Roberts et al. (1971) Hausen and Kerr (1968)
Cortez,Nevada (72 km S.W. of Carlin)	S	Ol	fine dissem.gold is in light gray to red,silicif.,bleached,hemat.pigmented calcar.siltst,at porph.cont.,under thrust	26 t Au/8.5 ppm 127 t Ag	Wells et al. (1969)
Gold Acres,Nevada 15 km W.of Cortez	Or,S	Cr_2	fine dissem.Au with pyrite in tactite, jasperoid,hornf.,greenst.within imbr. portion of thrust,under granit.massif	9 t Au	Wrucke and Armbrustma-cher (1975)
Prospect Ridge, Eureka district Nevada	Cm	Cr_1	low grade, invis.gold,pyr.,ars.,in roughly circular shoots in alter.do-lom.along N.E.faults.In qtz.diorite aureole	0.5 t Au	Nolan (1962)
Mercur,81 km S.W. of Salt Lake City Utah	Ms	T	bedded replacement of fine gold in silicif.black argill.limest.envel.by black shale. Py,realg.,orpim,cinnab., barite in fract.veins. High Tl cont.	31.7 t Au/12ppm	Gilluly (1932)
Specogna prosp., Graham I.,Queen Charlotte Islands Pacific Canada	Cr Mi-Pl	Mi-Pl	fine dissem.gold with pyrite, cinnabar chalcop.,sphalerite is in silicif. sandstone,congl.,argillite and rhyol. breccia in hangingwall of a thrust	180 t Au/1.8 ppm	Barr (1980)

variety of rocks along the locally thick and imbricate zone of the Roberts Mountain Thrust, which includes the "miogeoclinal" sediments of the eastern autochthon as well as the "eugeoclinal" rocks (Ordovician chert, argillite, greenstone) of the allochthon.

At the Mercur mine, Utah (Gilluly, 1932), finely dispersed gold bound in carbon was recovered from peneconcordant silicified bodies hosted by black argillaceous Mississippian limestones. Pyrite, realgar, orpiment and cinnabar were locally associated, or formed fracture veins with barite or calcite. The Mercur ores were unusually rich in thallium. The recently discovered isolated Specogna (Cinola) low-grade Au-Ag deposit, located on Graham Island, Queen Charlotte Islands (Pacific Canada; Barr, 1980; 180 t Au/1.8 ppm, 280 t Ag/2.8 pp m), has been assigned to the "Carlin-type". There, fine-grained disseminated gold is hosted by silicified late Tertiary or Cretaceous "molasse" sandstone and conglomerate, argillite and rhyolithic breccia in the hangingwall of a thrust fault. The gold is accompanied by sparsely disseminated iron sulphides, cinnabar, chalcopyrite and sphalerite.

19.8. MARINE DETRITAL, MINOR CARBONATE-EVAPORITE ASSOCIATION, OVERLYING CONTINENTAL "RED-BEDS"

19.8.1. Metalliferous marine shales (Kupferschiefer-style)

The facies progression of continental red—beds →(black) marine shale or slate → evaporite → carbonate correspond to one of the visually most striking, predictable and metallogenically fertile sedimentary sequence known. Some Recent depositional environments located along low-energy arid coasts, such as in the Persian Gulf (Chapter 6), provide modern analogues, unfortunately, lacking metallic mineralization. There is an extensive literature discussing the genetic and paleoenvironmental aspects of the sequence (e.g. Strakhov, 1962; Renfro, 1974; Gustafson and Williams, 1981). Most papers deal with the various aspects of sedimentogenesis and metallogenesis of the upper Permian sediments of the European Zechstein Sea and its Kupferschiefer horizon: the indisputable example and model area (Hirst and Dunham, 1963; Wedepohl, 1964; Oberc and Serkies, 1968). In this section, the basal metalliferous marine sediments will be reviewed preferentially. The continental red-beds are treated in Chapter 25, marine evaporites in Chapter 21 and carbonates in Chapter 20.

KUPFERSCHIEFER AND THE ZECHSTEIN BASIN (Cu,Zn,Pb,Ag,etc. aspect)

The Kupferschiefer (= copper shale) is a thin stratigraphic unit (a marker horizon), present predominantly in subcrops under Britain, the Netherlands, Germany and Poland, in a belt 1,500 km long with an estimated area of 60,000 km^2. The horizon is transgressive over a a peneplain, composed of a lower Permian red-bed sandstone with local bimodal volcanic occurrences (in the central and eastern portion), and over Carboniferous "gray" litharenites and chert in the north-west. Over most of its area, the Kupferschiefer is only 30-50 cm thick, and lithologically is a black, thinly-laminated and fissile bituminous marl, composed of alternating laminae of dolomitic carbonate and clay with bitumen, floored by a thin basal sandstone or conglomerate. Upward, it is gradational into a limestone. Figure 19-19 is a detailed section in the Richelsdorfer Hills, the most freely accessible occurrence located on the West German territory.

The Kupferschiefer is a significant repository of metals. Most of it contains elevated trace contents of Cu, Pb, Zn, Ag, Mo, V, U, Ni, Co, Mn (compare Table 19-2), but this is still far from constituting an ore (not considering the practical difficulties of mining !). Base-metal concentrations higher than 0.3% occur over a fraction of the total area underlain by the Kupferschiefer: 1% in the case of copper, 2-3% in the case of lead, 5% in the case of zinc (Wedepohl, 1971). The actual mining areas in Germany and Poland are in the range of 0.2 to 0.5% of the total extent of the Kupferschiefer. The "normal" shale is estimated to contain some 60 Mt Zn, 20 Mt Cu and 20 Mt Pb, and the quantity of metals recovered so far from the operating mines or present in reserves are, estimated conservatively, about 20 Mt Cu and 40 Tt Ag (compare Table 19-7). The amounts of Zn and Pb in the ore are even greater, but no reliable figures are available. An old wartime estimate assumed 100-150 Mt Pb and 200-250 Mt Zn just in

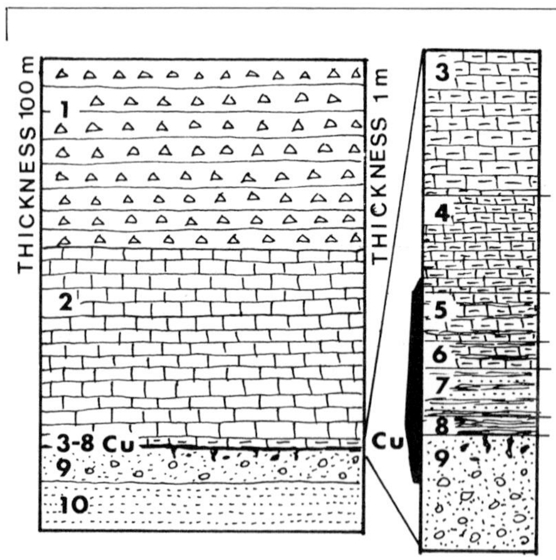

Pe_3 - ZECHSTEIN

1 Werra anhydrite,gypsum
2 Zechstein limestone
3 Dachberge marl
4 lightly bituminous marl
5 moder. to lightly bitum. marl and calcar.shale
6 bitum. marl to calcareous shale
7 Grosse Lette-bitum. sandy thin bedded shale
8 Feine Lette-impure to sandy, highly bitum.shale
9 Rote Fäule sandstone

Pe_1 - ROTHLIEGENDE

10 red sandstone

Fig. 19-19. Kupferschiefer, detailed section in the Richelsdorf field, West Germany. From LITHOTHEQUE, after Kautzsch (1942).

the area between Magdeburg and Richelsdorf in Germany and if the Polish portion is included, the figure might double.

In the Kupferschiefer, the dominant base metals are zonally distributed on a variety of scales. Copper is accumulated near the base, lead higher and zinc highest in the stratigraphic section. This is repeated worldwide. The copper ore in the basal part of the Kupferschiefer is represented by fine-to-extremely-fine-grained, evenly distributed chalcocite and bornite, grading locally into almost massive small sulphide nodules or lenses. In most cases the sulphides are practically invisible, revealed only by malachite or azurite coating on the surface. Alternatively, they may form megascopic grains in barite, calcite or quartz fracture veinlets. At most German and Polish localities, however, visible irregular disseminations, nodules, veinlets, carbonatized wood impregnations, fracture coatings, etc. of chalcocite, bornite or chalcopyrite, occur in the white arkosic sandstone-to-conglomerate, immediately underlying the black pelitic band. In places, particularly in the Harz Mts. Foreland, the continuity of the Kupferschiefer is interrupted by numerous unmineralized "highs" of the footwall sandstones, interpreted as being sand dunes (Kautzsch, 1942).

In south-western Poland, metalliferous Zechstein is developed in two mining areas west and N.N.W. of Legnica. In the more important Lubin-Połkowice-Sieroszowice district (Oberc and Serkies, 1968; Fig. 19-20), the mineralized horizon averages 4 m and the Kupferschiefer marker is in the centre. The marker is underlain by mineralized white sandstone, and topped by mineralized dolomite, but the ore grade rapidly decreases in both directions. The metal zoning is especially well developed, and the district can be subdivided into a north-westerly lateral "Copper Zone" (Cu, lesser Pb,Zn); central "Lead Zone" (Pb, lesser Cu,Zn) and easterly "Zinc Zone" (Zn, lesser Pb,Cu) in addition to the vertical zoning developed everywhere within the mineralized horizon.

The recent detailed studies carried out by the Polish geologists (Kucha, 1982), outlined widespread accumulations of precious (Au,Pt,Pd,Hg) and rare (U,Bi,Se) metals, hitherto unnoticed in comparable associations. These accumulations in some instances reach grades far surpassing those in "bonanza" deposits (up to 0.3% Au, 0.1% Pd, 0.034% Pt !). In most instances, accumulations of these metals are explained as due to organic matter (kerogen), in particular to a process of its auto-oxidation and desulfurization, aided by the Pt-group elements acting as catalysts. Accumulations of these metals are controlled by sedimentary microfacies and by metal zoning. Within the ore district, the largest noble metals occurrence is in the Lubin mine, along the Cu/Pb zone boundary. The parallel Ag and Hg-rich shale is situated within the Pb zone.

In the vertical zoning pattern, Au,Pd,Pt,Ag and Bi accumulated along the bottom part of the Kupferschiefer shale marker ("pitchy shale"), and there gold reaches 10 to 30 ppm over about a 30 cm thickness, probably present as chelates and thiocompounds. The "Noble Metal-Bearing Shale" of Kucha (1982) is a band up to 10 cm thick

Table 19-7. Cu (Zn,Pb,Ag) stratabound ores hosted by the "Kupferschiefer" and comparable units, in marine beds transgressive over continental red-beds.

LOCALITY	AGE	GEOLOGY,MINERALIZATION	TONNAGE,GRADE	REFERENCES
Richelsdorfer Hills Hessen, E.C. West Germany	Pe_3	finely dissem.chalcoc.,born.in 27 cm thick, thin bedded black bitum.calcar.argillite,resting on footw.white sandst.-conglom. Scattered cp,chalcoc.veinlets,impregn.in sandst.	1.075 Mt Cu/1% 700 Tt Zn/1.3% 2.7 Tt Ag/25ppm	Kautzsch (1942)
Mansfeld-Sangerhausen d.,S.E.Harz Foreland,E.Germany	Pe_3	as above; the Kupferschiefer horizon is 32-48 cm thick	2 Mt Cu/2.9% 12,797 t Ag/ 191 ppm	Jung and Knitzschke (1976)
Zajaczek-Grodzice, N.Sudetian Syncl. S.W.Poland	Pe_3	30-60 cm thick black bitum.claystone interlamin.with argill.carbonate, contains finely dissem.py,chalcoc.,born,cp,gal,sf	2 Mt Cu/1.2% 6.7 Tt Ag/40ppm	Wyzykowski (1971)
Lubin,Połkowice, Sieroszowice d., Fore Sudetian Monocline,S.W.Poland	Pe_3	4-6 m thick miner.band including 0.5 m thick black dolom.argillite in the centre (finely dissem.cc,born); gray sandst.in fw (cp,cc replace cement);dolomite in hw (gal,sf)	15 Mt Cu/1% 30 Tt Ag?	Oberc and Serkies (1968) Kucha (1982)
Niemstów-Tomice-Krepa (Pb-subdistrict in Fore Sudet.Monocline)	Pe_3	galena,sphalerite,pyrite,minor chalcopyrite dissem., fine grained, framboidal,conform. with lamin.of black bitum.argillac.dolomite; associated bedded barite	12 Mt Pb/5% ?	Harańczyk (1970)
Chapada do Araripe, (near Juazeiro do Norte),N. Brazil	Cr_1	crystals of galena and pyrite are consistently disseminated in an about 1 m thick bed of black lamin.bitum.marl, and overl.limestone conglomerate. Underl.by shale, yellow and red sandst.,congl.,near Precambr.crystall.basem.		Farina et al. (1978)

Lena and Angara R. Basin (N.W.of Irkutsk),Siberia, USSR	Cm_2-Or_1	extensive horizons of a lagoonal-deltaic gray shale grading into evaporite and carbonate, contain scattered stratabound occurrences of chalcoc.,chalcop.,and malachite in a variety of sedimentary hosts.Underl.by Cm red-beds	350 Tt Cu ?	Bogdanov et al. (1973)
Flowerpot Shale, S.W.Oklahoma (Creta,Mangum)	Pe	finely dissem.chalcocite in greyish-green mudstone under gypsum beds in a domin.red shale, siltst., sandst.sequence. Two Cu shale horizons, 22 and 36 cm thick	500 Tt Cu/2% 500 t Ag	Johnson (1974)
Jameson Land Basin,E.-C.Greenland	Pe_3-Tr	1-2 km thick sequence of interb.marine and contin.sediments contains many stratabound occur.of Cu,Pb and Zn in cont.and marine hosts. Black marine Pe_3 Posidonia Shale at Wegener Halvø is high in Pb and comparable with the Kupferschiefer		Thomassen et al.(1982)

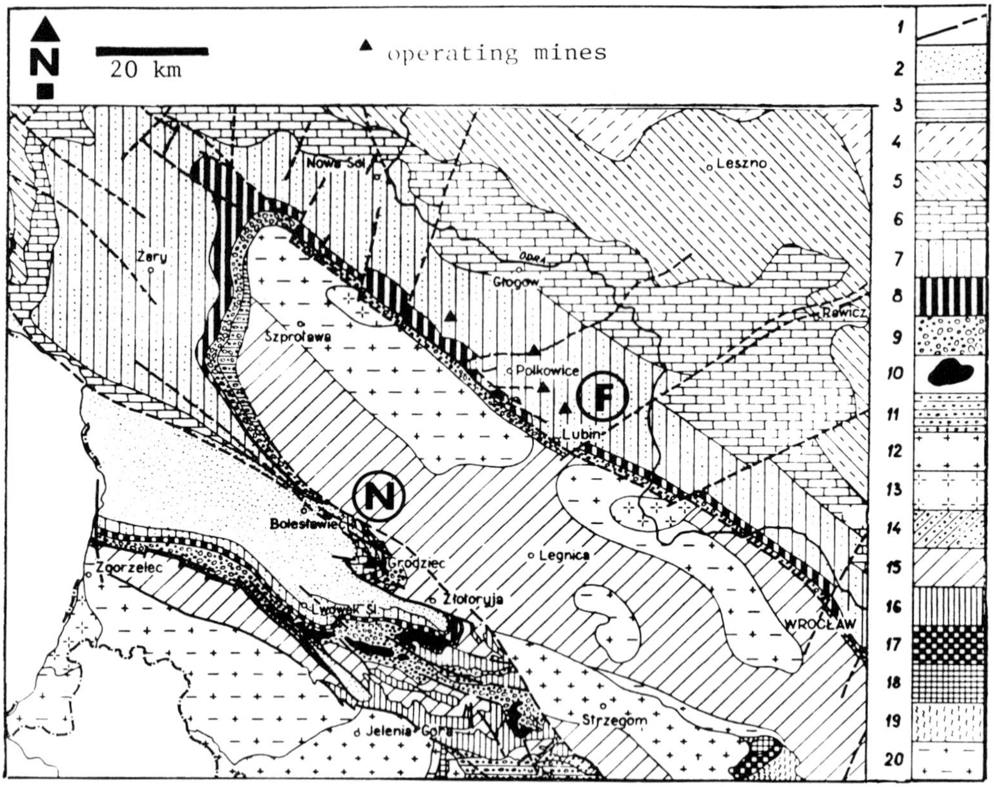

Fig. 19-20. North-western portion of Upper Silesia, Poland, showing the location of the Fore Sudetian Monocline (F) and Northern Sudetian Syncline (N) Cu and Pb-Zn districts. LEGEND: (1) faults; (2) Cr3 sediments; (3) J sediments; (4,5) $Tr_{2,3}$ sediments; (6) Tr_1 limestone (Muschelkalk); (7) Tr_1 mottled sandstone; (8) Pe_3 Zechstein sequence; (9) Pe_1 Rothliegende red beds; (10) Pe_1 rhyolites and basalts; (11) Cb coal-bearing sequence; (12) Cb granites; (13) Zawidów Granodiorite; (14) D_3 sediments; (15) lower Paleozoic rocks; (16) Cm greenstones; (17) Pt gabbro; (18) Pt amphibolites; (19) Pt serpentinite; (20) other Pt rocks. From Oberc and Serkies (1968), courtesy of Economic Geology.

sandwiched between the "pitchy shale" and the "boundary dolomite", a basal thin dolomite band overlying the basal white sandstone. In it, the gold content reaches up to 3,000 ppm in places, and there is also 700 ppm Pt, 400 ppm Pd, 0.1% Se. There, glauconite, phosphates and an organogenic limestone, are associated. The noble metals accumulate in detectable Cu-Fe sulphides mooihoekite and haycockite, which in turn are confined to red lenses of Fe-Ca phosphate in the black shale.

Where uraniferous (thucholithic) shale tops the Noble Metal-Bearing Shale, Pt and Ir are dispersed in the clay organic matrix but accompanying Cu, Co, Ni and Au are detectable by microprobe.

In the Rudna field of the Fore-Sudetic Monocline (Harańczyk, 1970) rich (10% Pb plus) galena accumulations hosted by black shale and carbonate rhythmites, are separated by an anhydritic sandstone barrier from the Cu-Pb-Zn zoned shales, and are interpreted as having formed in a restricted, hypersaline lagoon.

The British equivalent of the Kupferschiefer, the "Marl Slate" of eastern Durham in north-eastern England (Turner et al., 1978), is a black-to-light bituminous dolomitic laminite with discontinuous lenses of silt. It has an anomalous trace metal content with spot concentrations of up to 0.26% Pb and 3.1% Zn, but it nowhere does approach a potentially economic ore. Base metal sulphides chalcopyrite, galena, sphalerite and pyrite occur in discrete small lenses, intimately associated with hypersaline indicators and interpreted as produced during diagenesis by anhydrite replacement.

PERMIAN "COPPER SHALE" OF SOUTH-WESTERN OKLAHOMA

The Permian red-beds of western Oklahoma and north-eastern Texas (Johnson, 1974) are continental, tidal- to - lagoonal sediments frequently containing gypsum and numerous sandstone-hosted small copper occurrences. The first marine "copper shale" deposit, Creta, was discovered in 1962, and additional discoveries followed. The host unit, Flowerpot Shale, consists of reddish-brown shale to mudstone with thin interbeds of greenish-gray mudstone, gypsum, dolomite, siltstone and sandstone. It is about 50-60 m thick.

The greenish mudstone beds comprise about 15% of the sequence, are 3-60 cm thick, and usually underlie the gypsum beds. The copper mineralization at Creta is in a 3 m thick copper shale zone resting on a red sandstone, and there are two cupriferous beds of which only the stratigraphically lower bed (Prewitt copper shale) is economically mineable. The shale (or mudstone) is about 20 cm thick, and contains a very finely-disseminated chalcocite that is not megascopically apparent. Only minor malachite coating existed on discovery outcrop (Fig. 19-21). The Creta deposit has been recently mined and the shale developed over an area of about 100 km^2 may contain some 400-500 Tt Cu. In the nearby Mangum deposit, the copper horizon is stratigraphically higher and it lies in a laminated gray silty shale an average of 36 cm thick, and traceable in the outcrop for about 10 km. Chalcopyrite is the principal mineral. Additional copper shale occurrences have been detected in subcrops in central Kansas.

Although the Flowerpot Shale mineralization has frequently been compared with the Kupferschiefer in the literature, the field association of both the copper-bearing horizons and stratigraphic setting is substantially dissimilar and by visual impression only, the Oklahoma copper occurrences are more properly members of the red-beds association (Chapter 25).

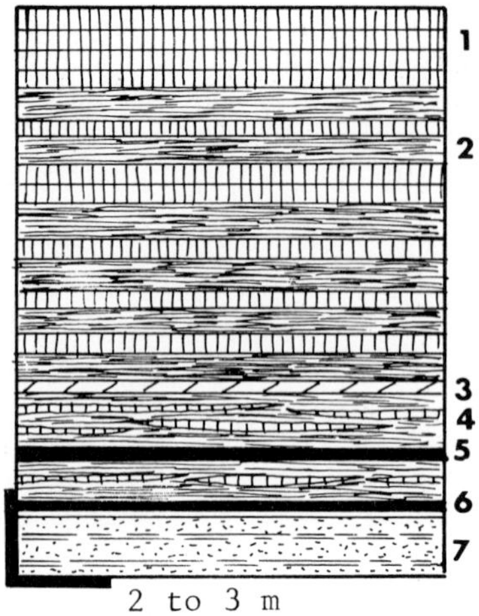

PERMIAN SEDIMENTS

1 massive and brecciated gypsum, anhydrite
2 alternating gypsum and red shale
3 marly dolomite
4 alternating gypsum and brownish-red shale
5 Meadows Cu Shale : greyish-green to black shale
6 Prewitt Cu Shale : laminated gray shale underlain by blocky, greenish-gray mudstone; finely dispersed chalcocite
7 red sandstone, shale

Fig. 19-21. Creta, Oklahoma, copper shale deposit. From LITHOTHEQUE, based on data in Johnson (1974).

ADDITIONAL OCCURRENCES

Numerous Cu, Zn and Pb occurrences in upper Permian and Triassic sediments in the Jameson Land Basin, east-central Greenland, are hosted by a sequence 1-2 km thick, of interbedded continental red-beds and marine evaporites, fine detritals, and carbonates (Thomassen et al., 1982). The upper Permian Posidonia Shale at Wegener Halvø, is a black marl comparable with the Kupferschiefer in age and in setting. It has a high-trace metal content (up to 0.13% Pb), but ore grade occurrences are restricted to a few horizons, 1-30 m thick, of a cherty, bioclastic limestone containing about 10% Pb and Zn. Copper occurrences have been reported from Triassic marine black mudstones at Pingel Dal.

Black calcareous or dolomitic marine shales associated with evaporites, carbonates, and underlain by red-beds, are one of the mineralization styles recorded in the extensive middle Cambrian to Ordovician Lena Basin of Siberia (Bogdanov et al., 1973).

At Chapada Araripe, north-eastern Brazil, a 1 m thick, but extensive, layer of lower Cretaceous black marl contains scattered galena crystals and subgrade lead contents (Farina et al., 1978). The layer has an intraformational setting within a marine shale unit, transgressive over red sandstones.

An interesting Zn-Pb sulphide mineralization has been reported from Miocene gypsiferous claystones and mudstones overlying several gypsum horizons, and underlying an evaporite sequence, in the Soviet Carpathians (Truskavets, Dzvinyach; Korenevskii, 1973; Fig. 19-19).

There, disseminations, veinlets, short massive lenses of galena, sphalerite and brunckite with some sulphur and gypsum, accumulate in thin stratabound lenses within a stratigraphic section 150 m thick. The grade ranges between 2-5% Zn and 0.8-2.5% Pb and the ore is now interpreted as being diagenetic. The presence of mudstone with broken sandstone blocks along with thinning and interruption of the gypsum horizons, suggests solution collapse.

Additional Cu-Zn-Pb-bearing shales, comparable in some respects with the Kupferschiefer, have been reported from the Permian of the Donetsk Basin, U.S.S.R.; from the Permo-Triassic of the Mangyschlak Peninsula, Caspian region, U.S.S.R. and from the Artemovsk Basin, U.S.S.R. Related deposits are also known from the Precambrian (Volume 2).

KUPFERSCHIEFER-STYLE CU,Zn,Pb TRANSITIONS AND ANALOGUES

Due to facies variation, the Kupferschiefer marl to shale is locally substituted by carbonate as the basal Zechstein unit. In other basins, shale and marl have not developed at all and the first

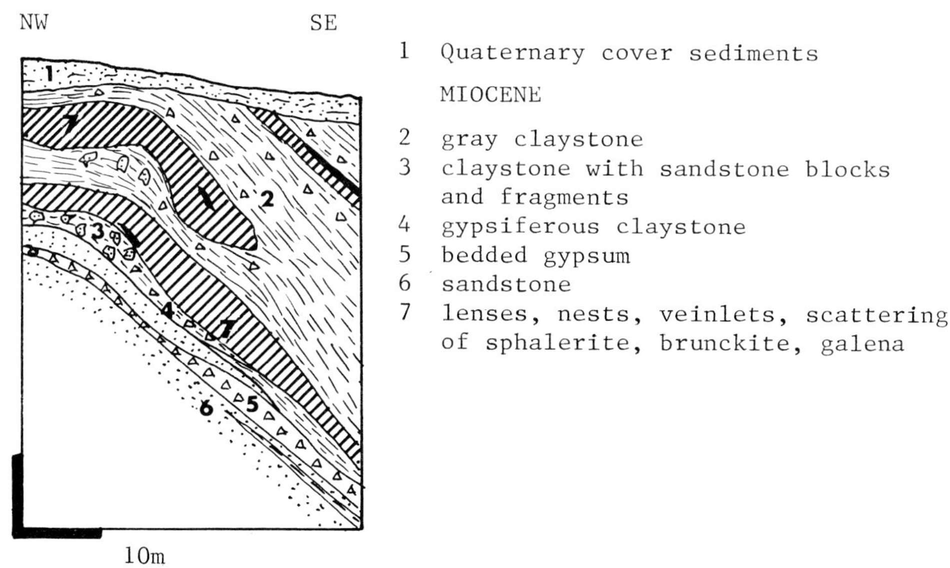

Fig. 19-22. Truskavets Zn-Pb deposit, Carpathians, U.S.S.R, a peneconcordant mineralization in a gypsiferous claystone. After Korenevskii (1973).

marine bed topping the red-bed basement is entirely carbonate (e.g. Dzhergala, U.S.S.R. and elsewhere). The accumulated metals, zoning and many other aspects of the Kupferschiefer origin remain comparable. Examples are treated in Chapter 20. The "Sandstone Pb-Zn deposits" treated in the following section, transgressively topping red-beds or crystalline basement, are also genetically comparable with the Kupferschiefer.

Mn (Cu) IN MARINE SHALE, LESSER CARBONATE, SANDSTONE OVERLYING RED-BEDS OR CRYSTALLINE BASEMENT

Wādi al Arabah (Dead Sea Rift).
In the southern portion of the Dead Sea Rift Valley, a thin Cambrian marine sequence consisting of dolomite, sandy dolomite and black shale transgresses over continental red-beds arkoses, which in turn top the Precambrian crystalline basement. Significant copper and manganese mineralization is hosted by interbedded dolomite and silty and sandy shale (including black shale), about 70 m stratigraphically above the base of the sequence.

Near Dana on the Jordanian side (Bigot, 1981), the principal mineralization style is confined to a lower-middle Cambrian horizon of black shale. This is about 3 m thick, encloses nodules and masses of pyrolusite and minor cryptomelane and ramsdellite with interspersed malachite and is traceable for several kilometres. Reserves of about 266 Tt Mn/33.2% and 13.6 Tt Cu/1.7% have been estimated. Infiltration, fracture coating and dispersed Mn oxides and oxidized copper minerals (chrysocolla, malachite, azurite, atacamite, brochantite) also occur in at least two stratigraphically higher and lower horizons of dolomite and sandstone.

Identical mineralization both of concretionary and infiltration manganese (252 Tt Mn/44%, 2.3 Mt Mn/23%; Bentor, 1956) and of copper (Bartura and Wurzbürger, 1974; 510 Tt Cu/1.89%) occur near Timna on the Israeli side, 25 km north of Elat. The copper mineralization is spectacular. It is entirely oxidic and the bright green and blue minerals (dominantly chrysocolla) fill cracks and impregnate or replace shale and sandstone along several stratigraphic horizons. The "primary" (depositional and diagenetic) mineralization is unknown. The andesitic and rhyolithic volcanics of the regolithic basement most probably provided both Cu and Mn.

The Wādi al Arabah Mn occurrences are closely related to the carbonate (Ulutelyak, Um Bogma) or evaporite- (Walton) -hosted Mn deposits overlying red-beds, described in Chapter 21.

CONTROLS TO THE "SHALE"-HOSTED METALLIC MINERALIZATIONS OVERLYING RED-BEDS

The single most important controls not only stressed in the literature, but also clearly and directly visible in the field, are the geochemical border surfaces: boundary planes between the oxic and anoxic conditions, the "red" and "green" or "black" sediments, which in turn have been influenced by the depositional environments and bathymetry. It has been well documented that the "ore grade" portions of the Kupferschiefer formed in restricted environments (? lagoons), while the bulk of the shale which is metal-enriched, but far from an ore, formed in the "open sea". In terms of geotectonics and the geotectonic cycle, the European Zechstein had been traditionally considered as an example of the "final stage" of geosynclinal development, overlying the Hercynian "molasse" and representing a transition into the platformic ("Germanotype") regime of development. Recently (Ziegler, 1978) it has been demonstrated that the Zechstein transgression coincided with the initiation of an aulacogene (aborted paleorift).

The crucial question of the origin of the Kupferschiefer mineralization discussed extensively but inconclusively (compare the reviews in Strakhov, 1962, and Wedepohl, 1971) remains the metal source. Although the actual precipitation of the metals may have been from seawater or alternatively from fresh groundwater assisted by the euxinic environment, it is logical to assume that the waters in turn were metal-anomalous, and that they probably obtained their metals from the land. The footwall red-beds contain numerous copper occurrences which is a worldwide phenomena. They also contain copper-rich amygdaloidal, lightly load-metamorphosed basalts (e.g. Kautzsch, 1957). The basalts seem to be particularly attractive as a metal source since there is a close analogy with the Keweenawan Cu province in Michigan (Volume 2). The source of zinc and particularly Pb (from feldspars or arkoses ?) is harder to explain.

Although the epigenetic theories of mineralization and subsurficial introduction of metals have been considerably discredited over the past 30 years as an agent of generation of the Kupferschiefer-like deposits, there has been a slow comeback recently (e.g. in the African Copperbelt; Volume 2). The incredibly high-grade, though patchy, accumulations of Au, Pt, Pd, etc. near Lubin are particularly difficult to explain and I do not believe that kerogen catalysis alone would have been powerful enough to achieve such concentration. It would be indeed tempting to assume the possibility of the existence of a buried endogenous mineralizing system under the ore-bearing segments of the Kupferschiefer, possibly in the form of differentiated mafic-ultramafic intrusions of the Noril'sk "type" (Chapter 32).

TECTONOMAGMATIC INTERACTION

The example localities of metalliferous shales discussed above are unmetamorphosed and virtually undisturbed tectonically. Small-to-medium-scale superimposed epigenetic features, such as barite, calcite, siderite, etc. fracture veins often containing ore minerals (Cu sulphides, Co-Ni arsenides, galena, sphalerite), are, however, common. The best-known examples often quoted in the older textbooks are the cobalt-rich fissure veins from Schweina and Kamsdorf, Germany (Beyschlag et al., 1914), that most probably formed by a selec tive lateral secretion from the Kupferschiefer. The low-temperature mobilization features are of little practical importance, but have an "indicator" value. They may help to reveal the presence of megascopically inconspicuous mineralized horizons.

Spatial coexistence of syn- to post-metalliferous shale intrusive bodies has not yet been demonstrated convincingly in the areas discussed above, except in eastern Greenland. There, the frequently metalliferous red-beds and shallow - marine Permo-Triassic sediments are intruded by, or adjacent to, the two Cretaceous to Tertiary hydrothermal Zn-Pb (Blyklippen) and Mo (Malmbjerget) ore deposits (Chapter 33).

19.8.2. "Sandstone lead" deposits in basal marine sandstones and quartzites, overlying sialic basement

Recently, Bjørlykke and Sangster (1981) outlined the "sandstone lead" deposits as a distinct and characteristic class of metallic ores. Their host sequences, however, are not uniform in terms of depositional environments and consequently lithologic associations. Most are hosted by continental sediments (particularly red-beds) and are discussed in Chapter 25. A minority of the lead deposits developed along the edge of the Caledonian orogenic belt in Sweden and Norway, are hosted by "white" quartzites believed to have been deposited in tidal channels and on beaches in a rather unique lithologic association of, at least initially, probably glaciogenic affiliation. The latter localities only are reviewed briefly in this section.

In Scandinavia, three major and seven minor disseminated Pb(Zn) deposits in quartzites occur discontinuously along a north-easterly trending belt approximately 2,000 km long, marking the border between the Caledonian thrust belt in the west, and the Precambrian crystalline basement of the Baltic Shield in the east (Bjørlykke and Sangster, 1981). The mineralized sediments are of uppermost Proterozoic and lower Cambrian age and were deposited along the margin of an epicontinental sea, outside the Caledonian domain and prior to the Caledonian tectonism. They are now part of an autochthonous sequence.

Laisvall in the Swedish Lappland (Rickard et al., 1979; Lilljequist, 1973; 3.2 Mt Pb/4%, 325 Tt Zn/0.44%, 72 t Ag/9 ppm; Fig. 19-23), is the largest and most extensively studied deposit. There, the basement consists of Proterozoic weathered and fractured pink granite, overlain by 20-25 m of basal conglomerate, arkose and pebbly shale and interpreted as being of glaciogenic origin. This is topped by a 40-50 m thick horizon of alternatively light and dark marine quartz arenite to quartzite which in turn is covered by marine siltstone and shale, including the pyritic "Alum" black shale. On top of the autochthonous unit rest two thrust sheets composed of

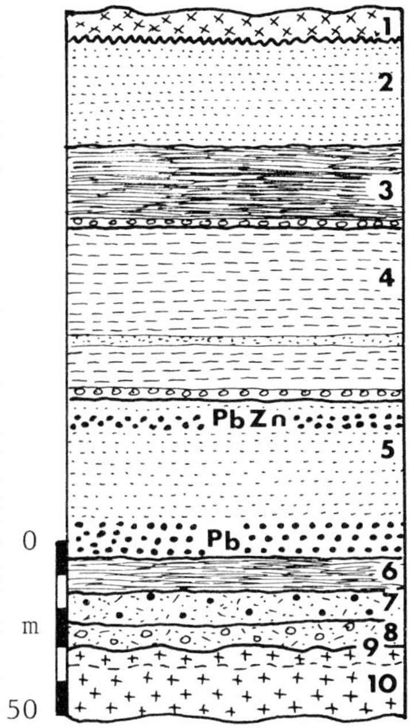

1 ALLOCHTHON ; Proterozoic gabbro, syenite, granite, porphyry

AUTOCHTHON, Lower-Middle Cambrian

2 litharenite ("graywacke")

3 graptolithic, pyritic "Alum Shale"

4 siltstone to shale

5 quartzite to hard cemented quartz-arenite, light to dark-coloured. Contains two horizons of disseminated galena; sphalerite is only in the top horizon

6 shale

7 basal arkose; contains minor, sparsely disseminated galena

8 mixtite (tillite ?) to polymictic conglomerate

9 regolithic Proterozoic granite basement

10 Proterozoic granite

Fig. 19-23. Laisvall Pb(Zn) field, Sweden; generalized section. From LITHOTHEQUE, after Bjørlykke and Sangster (1981) and Boliden Staff, guided tour (1975).

metamorphics, intrusive rocks and sediments.

The Pb(Zn) mineralization lies in eight blanket-like peneconcordant orebodies that occur along a strike length of 6.5 km, at two stratigraphic horizons. Galena only (in the basal horizon) and galena with minor sphalerite and pyrite in the upper horizon, form poikilotopic spots and knotten in the cement grading into schlieren. The mineralization is interpreted as being epigenetic, precipitated from 150° C warm brines ascending along permeable horizons in the original, poorly cemented sandstone (Rickard et al., 1979).

The remaining Scandinavian "sandstone-lead" deposits are variations of the Laisvall style. At Vassbo (275 Tt Pb/5.5%) the basement are Proterozoic "red-bed" sandstones, quartz porphyries and diabase dikes. Disseminated galena is hosted by a fine-grained calcite-cemented quartz arenite. At Dorotea (290 Tt Pb), the Precambrian basement is granitic and the galena disseminations are hosted by a coarse-grained lower Cambrian quartzite under a siltstone capping. The remaining Scandinavian localities are reviewed in Bjørlykke and Sangster (1981); the entire belt represents some 4 Mt of lead.

Somewhat similar to the Scandinavian deposits, but hosted by both sandstone and a black shale in alternating coastal marine units, are the lower Cretaceous deposits of Gabon (Kroussou) and Angola (Loeto). The first occurrence has been described by Caïa (1976), the second by Van Eden (1978). In Kroussou, the mineralized marine sediments were deposited on continental, poorly sorted breccias, conglomerates and channel arkoses and sandstones, filling depressions in the Precambrian basement. The ore is represented by disseminated, interstitial galena and pyrite with secondary minerals forming lenses and tabular bodies in carbonate-cemented gray sandstones, interbedded with black pyritic siltstones and shales. The Pb grade varies between 1 and 20% but no tonnage figure is available.

Loeto is primarily a zinc deposit, consisting of discrete thin layers of sphalerite and pyrite (2-2.5% Zn grade) in interbedded thin sandstones and black shales.

"A scientific fashion is actually a state of mind affecting investigators for a certain period of time during which they display characteristic associations of symptoms or syndroms... In the field of carbonate rocks the Bahama syndrome was followed by the excesses of supratidal dolomitization... At present the sabkha syndrome and its numerous subphases are taking their toll following recent investigations in the Persian Gulf".
A. Carozzi (1975, p.6).

CHAPTER 20

Shallow-Marine Carbonates

20.1. INTRODUCTION

Carbonates are very common sedimentary rocks, forming one fifth to one quarter of the stratigraphic record (Pettijohn, 1975) and are even more common in the Phanerozoic record considered alone. They have been summarized from a variety of points of view. Bathurst (1975), Frost et al. (1977), Milliman (1974), Wilson (1975) and others focussed on depositional environments and facies. Scholle (1978) and Pettijohn (1975) reviewed petrography and petrology. The recent excellent A.A.P.G. volume full of colour illustrations (Scholle et al., 1983) summarizes both aspects.

To an ordinary non-specialized field or exploration geologist, carbonate sequences are among the most monotonous ones. Their mineralogical composition is unspectacular. Pure carbonates are up to 98% either calcite or dolomite. Impure carbonates contain a detrital fraction corresponding to common clastic sediments (e.g. shale, argillite, sandstone) that is either dispersed in carbonate beds, or physically separated (interbedded, interlaminated) in a variety of mutual arrangements. Carbonates have been intensively studied scientifically for no more than 25 years and remain very much a "specialist's rock". One has to invest in a study of considerable depth in order to understand all their often subtle non-mineralogical attributes (components, particle sizes, fauna, facies), in order to break the thick carbonate piles of the Alps, Dalmatia, the Canadian Rocky Mountains, and elsewhere into manageable subunits and, ultimately, to deduce their depositional paleoenvironment and spatial connections. How much of this fine detail is really necessary for a prospector intent on finding an ore, rather than publishing a specialized paper ?

As it will be demonstrated later (and as already discussed in Chapter 5 dealing with the Recent carbonate depositional environments), carbonates act primarily as reservoir rocks (traps) to ores introduced from outside, rather than as source rocks in which the ores are generated simultaneously with the sedimentation and early (pre-lithification) diagenesis of their hosts. Thus, except for the few cases of syngenetic-diagenetic mineralizations, the pre-lithification history of carbonates largely has an indirect influence on the presence or absence of ore and need not be the first priority of a prospector. The reservoir (host) rock function of carbonates is controlled mostly by their non-compositional properties, such as porosity, fracturing, brecciation and plumbing systems, as well as by their post-lithification compositional changes, like dolomitization or silicification. The latter are equivalent to hydrothermal alterations in magmatogene systems.

The non-compositional properties of carbonates are attributes of considerable importance to the petroleum geologist, and consequently they have been studied intensively and are reasonably well understood. A petroleum geologist, however, deals with a single commodity migrating "under its own power" from a restricted selection of relatively proximal source rocks into a restricted selection of traps usually within a single basinal sequence. The reasoning of a metalliferous geologist is bound to be much more complex, because the sources of metals and the driving forces to the mineralization process may be (and usually are) located outside the carbonate association itself: adjacent to the carbonate "basin"; in the crystalline basement; in synchronous magmatism. Consequetly a multi-associational approach is paramount. A prospector has to look beyond the carbonate or sedimentary unit at large in searching for ores. Certain clues within the sedimentary unit, however, habitually overlooked by the carbonate specialist (e.g. tectonic structures, tuffaceous units, volcanic flows, intrusions) are of importance and have to be considered.

20.2. CARBONATE LITHOFACIES

On a global scale, carbonates are members of numerous distinct lithofacies some of which have been discussed earlier (e.g. Chapters 4, 5, 17). Others will be reviewed later (Chapters 21, 22, 25). The "most typical" shallow-marine carbonates are usually arranged into a facies progression based on the bathymetry of their depositional environment (from shore through shelf to slope). Such progression may be either simple or composite. "Reefs" (organic buildups) and evaporites intervene in the composite progressions.

Most ancient "reefs" are hardly continuous organic framework structures as some modern coral reefs, but rather mounds reinforced by and rich in large, attached bottom-dwelling carbonate secreting forms (corals, stromatoporoids, archaeocyaths, some algae), or composed of sediment-trapping and binding algae (algal mats, stromatolites). The main geological function of a "reef" is: (a) creating variable energy microenvironments (high-energy fore-reef, low energy back-reef lagoons); (b) restricting the depositional environment and (c) causing rapid local variations in bathymetry. The "reefs" occur in a variety of positions (fringing, barrier, patch) and the setting of some "reefs" indicates structural hinges or a position above growth faults.

Evaporites, of which the less soluble ones (gypsum, anhydrite) are the most common in ancient carbonate sequences, are important indicators of hypersalinity. Being more soluble than carbonates, they are usually leached first during emergence.

The caıbonate facies have been rewiewed by Wilson (1975; Fig. 20-1). Figure 20-2 taken from the Schneider (1964) chapter dealing with the distribution of Zn-Pb ores in the eastern Alps, treats carbonate facies as the major controlling agent of ore distribution. The Schneider (1964) interpretation goes further than many treatments afforded to facies in the specialized literature. It includes non-carbonate rocks (shales) and facies attributes generated by diagenetic (pre-lithification) and post-lithification causes, mainly dissolution. The "special facies" of the middle Triassic carbonates in the eastern Alps repeatedly associated with Zn-Pb ores, consists of products of both depositional and post-depositional diversification: thinly-layered (rhythmic) bituminous carbonates with slump structures (depositional diversification) as well as dolomitization, porosity, micro- and macro- brecciation and metallic mineralization (post-lithification diversification). Some sediments are controversial. The typically pale green marls or shales which were frequently interpreted in the past as being tuffs, are now increasingly considered to be residuals remaining after carbonate dissolution.

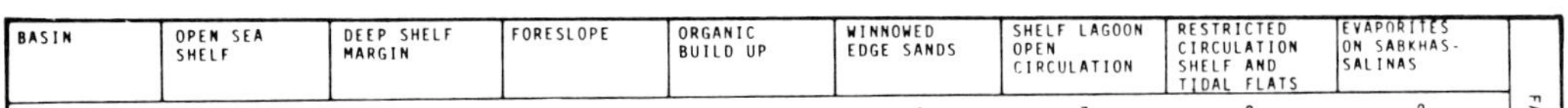

Fig. 20-1. Standard facies sequence of shallow marine ("platform") carbonates. From Wilson (1975), courtesy of Springer Verlag.

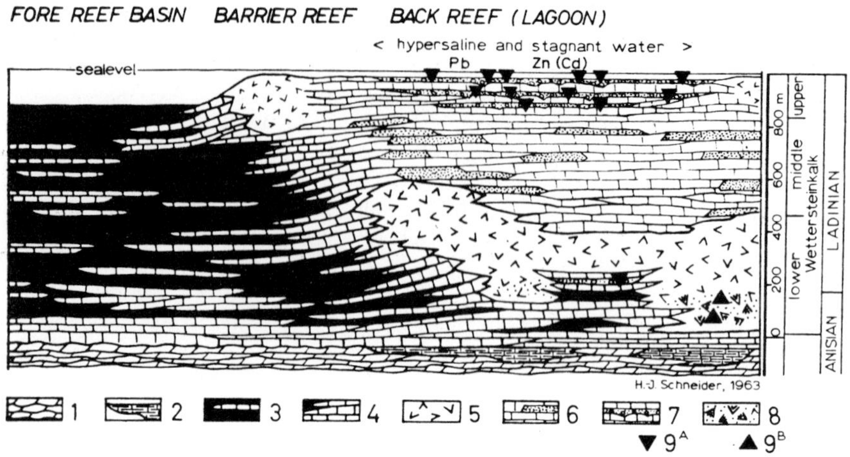

Fig. 20-2. APT mineralization of the Eastern Alps, as related to evolution and facies of a Ladinian "reef" platform. (1) uppermost "Alpine Muschelkalk"; wavy-clumpy, thinly bedded bituminous limestone with chert nodules; (2) andesitic green tuffs (ash and crystal tuffs, few lapilli) with thin layers of marl and limestone; (3) "Partnach Mergel", clayey marls, shales with lenses of layered limestone (like 4) (Ladinian basin facies); (4) "Partnachkalk", bituminous, marly, layered limestone units (Ladinian basin facies); (5-8) different types of "Wettersteinkalk" (Ladinian reef facies); (5) massive limestone and dolomite, partly cavernous or relict patterns of bioherms (often coral colonies); (6) well layered gray limestone (mainly calcarenite) with debris and colonies of algae (Dasycladacee), single algal patch reefs; (7) predominantly thinly layered limestone with intercalations of the "Sonderfazies" in special sequences (backreef units, tuffaceous marls, slump structures, ore sediments, etc.); (8) late diagenetic alteration of the cavernous reef body by recrystallization of dolomite, quartz and different Fe-dolomites; (9A) Pb-Zn sulphide ores with sedimentary fabrics; (9B) Pb-Zn sulphide ores primarily enriched in metasomatic replacement bodies, locally associated with small amounts of Cu-Sb-As minerals. From Schneider (1964). Courtesy of Professor Schneider.

20.3. TRACE METAL GEOCHEMISTRY AND MINERALIZATION

The ordinary, pure carbonate rocks are strongly depleted, rather than enriched, in major and trace metallic elements (average trace metal contents in high-purity limestones: Fe, 0.35%; Al, 0.22%; Mn, 0.14%; Co, 0.1 ppm; Ni, 20 ppm; Cu, 4 ppm; Mo, 0.4 ppm; Sn, 1 ppm; Cr, 11 ppm; etc., Rösler and Lange, 1972; see also the extensive summary in Wolf et al., 1967). This is consistent with the observation that carbonates act predominantly as reservoir rocks to ores. Compared with most other lithologic associations, carbonates are rendered unique by their virtual absence of "syngenetic" (syndepositional) metal accumulations (compare Chapter 5).

Detrital (clastic) metallic minerals are absent, except in transitional environments or associations, and all occurrences known so far are of mineralogical interest only. Examples include beaches along tropical coasts as in northern Jamaica, where detrital Ti-magnetite heavy mineral streaks and laminae often overlap with the carbonate skeletal sand, and many become preserved in carbonate beachrocks. Clastic chromite occurs in certain Ordovician limestones on the western coast of Newfoundland, derived from ophiolite thrust sheets.

Syndiagenetic accumulations of metals (Fe, Mn) and nonmetals (P) having a capacity to accumulate selectively some trace metals, such as uranium, form locally on the present seafloor as nodules, encrustations or dispersed components of oozes and muds (Chapter 5). Except for the phosphorite accumulations, however, they are uncommon in the fossil, shallow-marine carbonates. Despite the significant accumulation of some trace metals in living marine organisms including those that produce calcareous skeletons (e.g. oysters; compare Chapter 3), equivalent enrichment has not been proven in either Recent or ancient sediments composed of skeletal remains so it can be assumed that the many varieties of rock-forming calcium carbonate particles have no significant metal accumulating capacity. The organic matter, however, does have such a capacity and so do a variety of "carriers" (clay minerals, pyrite).

It is probable that certain "special facies" of impure carbonates formed in environments marked by an anomalous supply of certain trace metals, became enriched in such metals. This is most likely in case of the generally thinly bedded, gray-to-black carbonates that are members of the continental red-beds → evaporites → carbonates facies transition (also containing the Kupferschiefer metalliferous slates treated in Chapter 19). If so, however, the final metal fixing took place diagenetically following a burial.

Diagenesis is of paramount importance in the development of rock carbonates as seen in the present outcrop (Chilingar et al., 1979), as well as for the generation of a substantial proportion of the carbonate-hosted ores (e.g. Wolf, 1976). Due to the unique chemical properties of carbonates, such as the high solubility that distinguishes them from the other classes of common sedimentary rocks, the carbonate diagenesis is a much broader and more complex field of study than, for example, the diagenesis of a shale or sandstone. It not only involves processes of compaction, solution and reprecipitation, lithification, etc. taking place in the presence and under the influence of formational waters in the buried, subsurficial portion of the sedimentary pile, but also the near-surficial diagenesis in the freshwater phreatic and vadose zones. Consequently, there is no sharp break between the deep-seated diagenesis and shallow processes which, in other lithological associations, are usually treated under the heading "weathering" and its extension (e.g. infiltration, meteoric water circulation). Two characteristic categories of the near-surficial processes and products involve karsting and the generation of dissolution and reprecipitation profiles over carbonates ("caliche"). These are treated in some detail in Chapter 23 and the discussion will not be repeated here.

One of the earliest events of carbonate diagenesis that has demonstrable mineralogical and geochemical implications and possibly some metallogenetic implication as well, is the conversion of the unstable $CaCO_3$ polymorphs secreted (precipitated) in the original depositional environment (that is, aragonite and high-Mg calcite) into the stable low-Mg calcite. Although, as stated earlier, the skeletal remains contain only sub-clarke concentrations of trace metals, the polymorph conversion demonstrably releases and sets in motion the major and common trace elements (e.g. Mg and Sr), and probably releases some of the sub-clarke metals (such as Zn, Pb) as well. The released magnesium may form dolomite. The released lead and zinc could concentrate as galena and sphalerite when sulphur is available. Although Zn-Pb deposits are characteristically associated with dolomitization, a simple genetic link to the early polymorph conversion has not been established.

Carbonates are hosts to about 600 Mt Al in bauxite, 70 Mt Fe, 75 Mt Zn, 55 Mt Pb, 3 Mt Mn, 500 Tt Cu, 1 Mt U (low grade by-product) contained in ore deposits showing a close affiliation with at least one phase of carbonate rock development. An additional 40 Mt Zn, 25 Mt Pb, 3 Mt Cu, 100 Mt Fe, 20 Tt Hg, etc. are carbonate-hosted, but the ore emplacement is demonstrably due to interaction with a younger event, superimposed on the existing carbonate (e.g. igneous emplacement). Yet another, broad class of ore deposits resting on or hosted by karsted carbonates, is virtually unrelated to the carbonate lithogenesis and merely makes use of the structural control (provision of a depression or an opening to be filled), the carbonates offer. With the exception of some bauxites that are transitional, these ores will not be considered here and the reader is referred to Chapter 25 for further discussion.

Disregarding the ores associated with the "special facies" of carbonates (e.g. evaporite or red-beds affiliation) and the interaction categories, there are two major classes of carbonate-hosted deposits: (1) bauxites; (2) Zn-Pb deposits of the Mississippi Valley (MVT) and Alpine or Appalachian (APT) "types". They do not mix, and each category is governed by a rather distinct set of genetic attributes studied by specialists. These attributes and the overall metallogeny will be reviewed briefly later, in the corresponding sections.

20.4. MONOTONOUS CARBONATE, MINOR SHALE TERRAINS; SYNGENETIC-DIAGENETIC MINERALIZATIONS

In the "most normal" carbonate-dominated terrains, coeval sedimentogenic metal accumulations are exceedingly rare (Fig. 20-3; Table 20-1). Manganese ores, phosphates and barite have been described most frequently. These commodities are most widespread in the condensed pelagic units and at diastems and have already been treated in Chapter 15.

20.4.1. Mn deposits

In addition to the pelagic manganiferous beds of low-to-moderate Mn grade (7-15% Mn), very low-grade (0.X-3% Mn) manganese carbonate units that may, nevertheless, contain relatively significant aggregate Mn tonnage, are known to exist. The Cambrian Shady Dolomite of the Appalachians in Georgia and Tennessee is the best-known example. In northern Tennessee, this unit (King and Ferguson, 1960) contains from 0.15 to 1.24% Mn substituting for Ca and Mg in the lattice of calcite and dolomite and this content is megascopically undetectable until revealed by the presence of residual Mn oxide deposits that frequently form on the surface of similar units (e.g. in Cartersville, Georgia).

20.4.2. Uraniferous phosphorite

Thinly scattered phosphorite (collophanite) grains and nodules are relatively common in many carbonate units deposited on stable shelves or in warm epicontinental seas. Cathcart and Gulbrandsen (1973) listed numerous phosphate-containing limestones, marls and carbonatic shales in North America. These rocks, having the usual high trace U content (of the order of 60-100 ppm in the phosphate) are normally uneconomic, unless they are gradational into bedded diagenetic phosphorites (as in Morocco), or unless the scattered nodules have been residually enriched (as in Florida).

Bedded phosphorites hosted by shallow-marine carbonates are in most respects comparable to the "deeper" marine phoshorites (Chapter 17), and they form layers and lenses associated with limestone, marl and chert. Black slate is absent or rare, but glauconite-bearing sediments are common. Table 20-2 lists example localities.

RESIDUALLY ENRICHED NODULAR PHOSPHORITES: THE LAND-PEBBLE PHOSPHATE REGION, FLORIDA

The Land-Pebble Phosphate District (e.g. Cathcart, 1963; Fig. 20-4) is a shield-shaped area of about 7,000 km^2 located east of Tampa, from which some 80% of the United States' phosphate production comes at

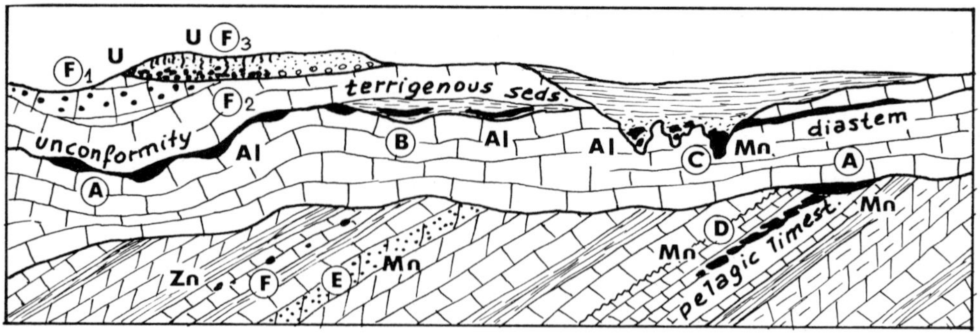

Fig. 20-3. Shallow marine carbonates, principal mineralization styles 1: syndepositional, diagenetic and unconformity—controlled (see Table 20-1 for an explanation of the letter codes).

Table 20-1. Shallow marine carbonates, mineralization styles 1: syndepositional, diagenetic and unconformity - related

(A)	residual, reworked or detrital ores at intracarbonate unconformities (buried paleokarst); dominantly ore components "Mediterranean - type" bauxite; X00 localities, 450 Mt Al uraniferous phosphorites; X localities, X00 t U Mn oxides; X localities, 1.5 Mt Mn Fe hydroxides; X0 localities, 30 Mt Fe
(B)	as above, ore is accompanied by terrigenous sedimentary "waste" sandwiched unconformably between two carbonate units; bauxite, X00 localities, 160 Mt Al
(C)	ore (bauxite) is on base of a detrital terrigenous association resting on karsted carbonate; treated in Chapter 24
(D)	Mn,Fe carbonate nodules, beds in pelagic limestones; treated in Chapter 15
(E)	metals / minerals finely dispersed in beds of "monotonous" carbonates; Mn carbonate, X0 localities, X0 Mt Mn (low grade) uraniferous phosphorite
(F)	diagenetic nodules, concretions, of uraniferous phosphorite;
(F1)	bedrock occurrences; X00 localities;
(F2)	physically reworked nodule gravels; X0 localities, 600 Tt U (low-grade resource);
(F3)	Al phosphates; infiltrations in tropical weathering profiles; X localities, 100 Tt U (low-grade)

Table 20-2. Bedded sedimentary uraniferous phosphate deposits in shallow marine carbonates

LOCALITY	AGE	GEOLOGY, MINERALIZATION	U TONNAGE, CONT.	REFERENCES
Central and Northern Florida and Atlantic Coastal Plain, USA	Mi Pl	in Florida, 4-40 m thick Mi biosparite contains diagenetic phosphorite nodules; these have been reworked and upgraded into the overlying Pl continental nodular gravels. This continues, discontinuously, further N.E.	600 Tt/110-160 ppm U (hypothet. resource)	Cathcart (1963) Riggs (1979)
Olinda, Recife, N.E.Brazil	Cr_3	diagenetic phosphorite, stratiform horizon assoc. with shallow marine shale, limestone	34 Tt U/170 ppm	Putzer (1976)
Mahattat al Hasa area, Jordan	Cr	several phosphate - rich shale and marl layers and lenses, up to 10 m thick	6 Tt U/100-125 ppm	Bender (1975)
Ar Rusayfah, 15 km N.E. of Amman, Jordan	Cr	4 layers of phosphorite in shallow marine sandstone, shale, limest. sequence, up to 30 m thick	4.2 Tt U/100-125 ppm	as above
Youssoufia and Khouribga fields, Morocco	Cr_3	granular and cementing collophanite, stratiform horizon in phosphatic limestone and marl	14 Tt U? /85 ppm	de Kun (1965)
Cabinda region, Angola	Cr	stratiform, undisturbed phosphorite horizon in shallow marine marl, limestone, sandstone association	9.8 Tt U/80 ppm	de Kun (1965)

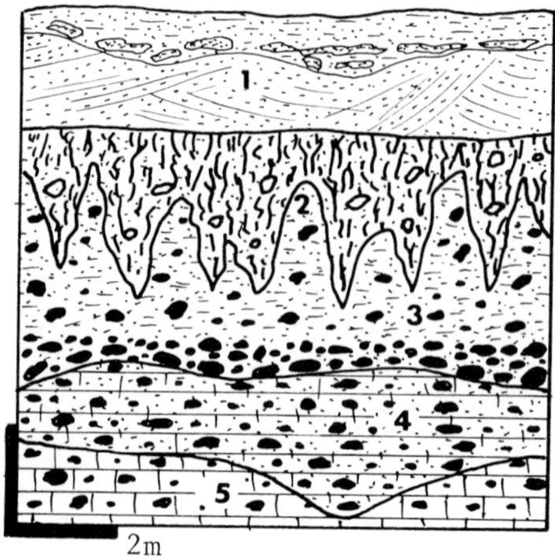

Fig. 20-4. Florida Land-Pebble Phosphate district, Fort Meade area, diagrammatic section. From LITHOTHEQUE.

present. The region is underlain by a thick section of marine sediments, of which the Miocene Hawthorn Formation is of greatest economic importance as the "primary" carrier of the phosphorite. The flat-lying Formation, 4 to 40 m thick, is composed of white biosparite limestone changing into a hard dolomite in the northerly direction, and grading into a thin, non-calcareous sandstone and claystone unit at the top. The unit is slightly phosphatic (2-5% P_2O_5) and the phosphorus is in microcrystalline F, CO_2 apatite that takes the form of a dispersed orthochemical "mud" as well as orthochemical pellets and nodules (Riggs, 1979).

The grains and nodules of variable size (1-30 mm) are structureless, yellowish-brown, with a smooth, shiny surface and have a low U content (about 50 ppm). The "primary" (diagenetic) phosphatic carbonate is uneconomic to mine and all the production (and calculated reserves) come from subaerially-enriched profiles ("hard rock phosphate", a secondary apatite replacing limestone, precipitated in karst cavities and cementing sand; developed only in northern Florida) and from continental sediments overlying the Hawthorn Formation and containing reworked phosphatic nodules. The Pliocene Bone Valley Formation ranges from 0 to 17 m in thickness of which the basal 2-4 m consists of unconsolidated sand, clay and gravel with a high content of phosphatic nodules representing the main ore horizon. In this, the nodules average 20-30% P_2O_5 and about 110 to 160 ppm U. The contained uranium is estimated to be of the order of 250 Tt U.

Tropical weathering and leaching superimposed on the Bone Valley Formation generated a zone of infiltrated and grain-cementing Al (and minor Fe) phosphates crandallite, millsite and wavellite that may constitute up to 20-30% of the sediment and contain 5-6% P_2O_5 and 90-300 ppm U. This could represent an additional 80 Tt of contained U. The Land-Pebble Phosphates in the Bone Valley Formation are only a portion (although the richest one) of a 1,000 km long belt of similar discontinuous occurrences in the Atlantic Coastal Plain in Florida, Georgia and the Carolinas (Riggs, 1979), that may represent a resource of the order of 600 Tt U. In addition to uranium, rare earths are enriched in the nodule apatite (586 ppm REE), and are potentially recoverable. Uraniferous aluminium phosphates also form large accumulations in the lateritically weathered phosphatic shallow marine marbles and limestones in the Thies area, Senegal (Patterson, 1967), and in Nigeria.

20.4.3. Probably diagenetic (stratiform) sphalerite, galena, fluorite in undisturbed "normal" carbonates

The overwhelming majority of Zn-Pb sulphide deposits and mineralogical occurrences in "ordinary" carbonates is epigenetic, and demonstrably diagenetic occurrences are quite rare. In Bruce Peninsula of southern Ontario (Sangster and Liberty, 1971), sphalerite and mixed sphalerite-dolomite concretions (nodules) are quite widespread, although in economically insignificant quantities. The concretions are disc or egg-shaped, up to 30 cm in diameter, fine crystalline, and have a characteristic light grayish-blue colour on the weathered surface of a white, bleached carbonate. Most nodules are confined to a thin-bedded dark gray-to-black finely crystalline bituminous dolomite of Silurian age. This is about 17 m thick.

Similar, but rare and small yellowish-brown sphalerite nodules occur, together with scattered single crystals of galena and yellow or purple fluorite, in black, thin-bedded siliceous Devonian limestones near Radotín, Czechoslovakia. Routhier (1963) mentioned widespread microconcretions or small concordant lenses of barite, sphalerite and galena, present along the base of the Lias in the western border of Cévennes, France. The demonstrably diagenetic occurrences of the Zn and Pb sulphides and fluorite are distinct from the coarse crystalline or metacolloform sphalerite located in solution cavities and along fractures in the same area as well as in others.

20.5. PURE CARBONATES, EVIDENCE OF SUBAERIAL EXPOSURE (UNCONFORMITIES, KARSTING), RESIDUAL OR PHYSICALLY DEPOSITED ORES

20.5.1. "Karst" (Mediterranean "type") bauxites

"Karst" bauxites (those associated in some way with carbonates; Bárdossy, 1982) are the most significant variety of "fossil" bauxite. Fossil bauxites formed by subaerial surficial processes of tropical weathering in the past in essentially the same way that the present lateritic bauxites form. After this they were rapidly buried by younger sediments. The burial as a mechanism of preservation is a factor almost as important as bauxite generation, bearing on the existence of pre-Cainozoic bauxites. Without the burial, the surficial bauxites would not be preserved and, in fact, probably 400 times the amount of fossil bauxite now preserved had once formed during the Phanerozoic period alone, but was subsequently eroded and dispersed. This figure has been calculated from the data presented by Bárdossy (1982: the world's bauxite resources, 45 Bt; of it, lateritic bauxites 0-10 m.y. old represent 39.5 Bt; fossil bauxites 10-600 m.y. old account for 5.5 Bt), taking into account the principle of actualism.

Carbonates combine a capacity to support the surficial bauxite-forming process when subaerially exposed, such as in the present-day Jamaica (Chapter 23), with a unique mechanism of preservation. The latter is a consequence of carbonate subaerial dissolution (karsting) and a rapid submarine carbonate buildup under conditions of low energy: that is, with a minimum of contemporaneous erosion. The karsting produces deep local dissection in the form of sinks and an uneven surface within which bauxite can form not only out of the insoluble residue in the parent limestone, but also out of the allogenic materials (such as transported clay or volcanic ash). Having once formed at the bottom of the sinks or washed into them, bauxite is then usually rapidly and easily buried by a continuous pile-up of continental sediments or by transgressing marine sediments. Bauxite located in deep karst sinks can survive the planar erosion that accompanies marine transgressions. The bauxite forming flat lenses in shallow depressions is no match for the clastics-depositing transgressions, but can and has been widely preserved under transgressive carbonates. As a consequence, fossil "karst" bauxite deposits now occur in carbonate-dominated to carbonate-containing lithologic associations displaying the evidence of subaerial exposure. Unconformities marking the exposure hiati are the single most important control for bauxite exploration (compare Laznicka, 1985f)

From the standpoint of regional lithologic associations (that differs markedly from the "type" classification offered by Bárdossy, 1982), carbonate-associated bauxites come in four distinct, although mutually transitional styles (Fig. 20-5).

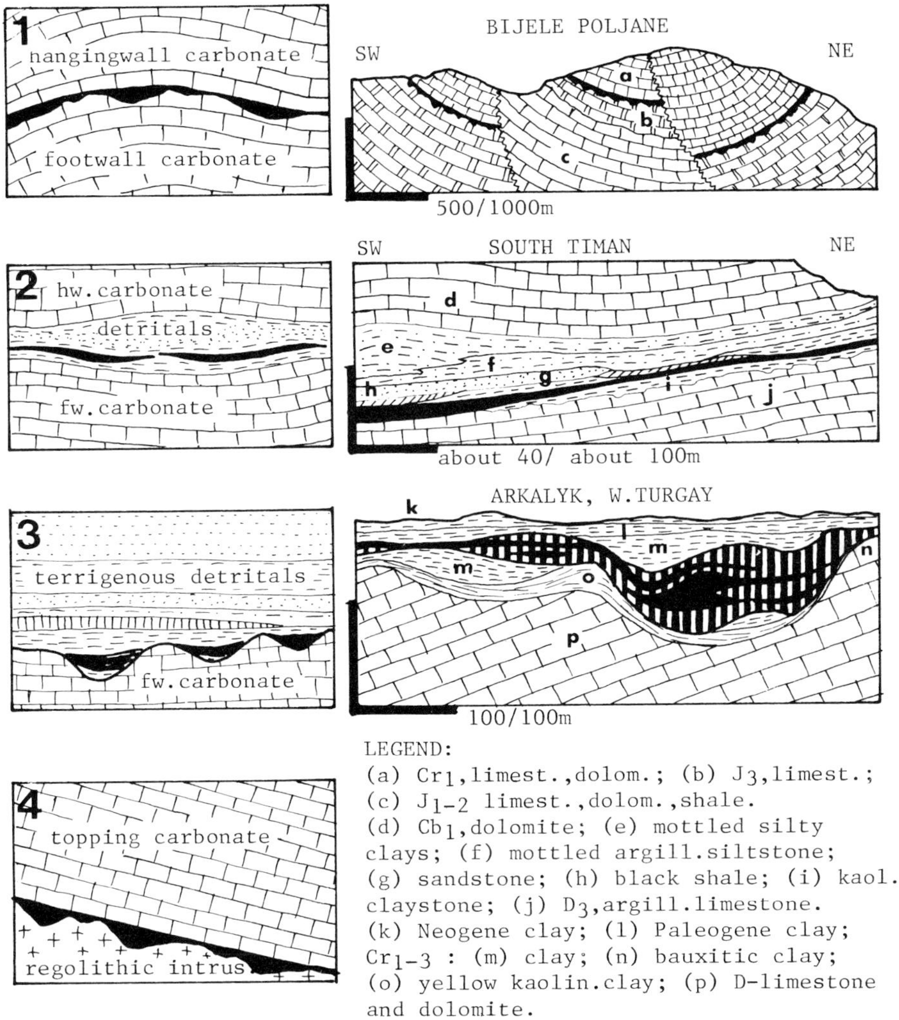

Fig. 20-5. Four principal styles of carbonate-associated fossil bauxites. Example localities: Bijele Poljane, Montenegro, Yugoslavia; South Timan, N.-C. Russian Platform, U.S.S.R.; Arkalyk, W. Turgay district, W. Kazakhstan. Modified after Burić (1966), Kirpal' and Tenyakov (1974) and Krivtsov (1973).

Style (1) is the textbook image of the "Mediterranean-type" deposits. There, high-purity bauxite lenses with the minimum of dilutants (that is, clay, sand, coal, etc.) occur along the surface of disconformity between two units dominated by chemically pure carbonates. A single such unconformity studded by hundreds of separate bauxite orebodies can be traced for tens to hundreds of kilometres, as in the carbonate terrains of the Dinarides and Hellenides in Yugoslavia and Greece (Fig. 20-6). The footwall carbonate (limestone, lesser dolomite) is karsted (Fig. 20-7) and the bottom contact of the bauxite body is uneven, often jagged and sometimes underlain by carbonate rubble. The bauxite hangingwall is smooth, planar and sometimes topped by a basal conglomerate derived from the lithified bauxite. The bauxite bodies may be residual (formed by lateritic weathering "in situ"), but in most cases the bauxite was brought in, although most probably from the immediate

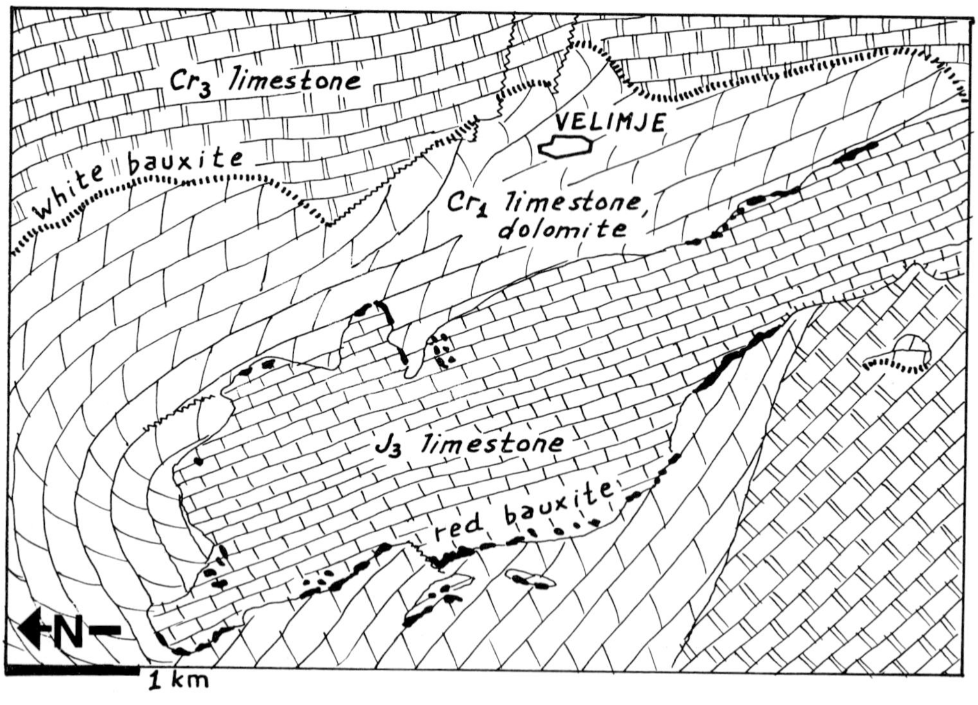

Fig. 20-6. Geological map near Velimje, Montenegro, Yugoslavia, showing two horizons of "karst" bauxites (style 1): (a) red bauxite at J_3/Cr_1 unconformity; (b) white bauxite at Cr_1/Cr_3 unconformity. Modified after Burić (1966).

LEGEND

10 Cr_3 limestone
9 coarse conglomeratic bauxite
8 yellow claystone
7 breccia/conglom.bauxite
6 spotted and red bauxite
5 white bauxite
4 coaly bauxite
3 pyritized bauxite and claystone
2 footwall carbonate breccia
1 Cr_1 limestone,dolomite

Fig. 20-7. White "karst" bauxite, generalized section, Montenegro, Yugoslavia. Based on Burić (1966).

vicinity. The bauxite bodies grade from planar peneconcordant sheets with an area of up to 7 km^2 ("stratiform" deposits of Bárdossy, 1982), through "blankets" (bauxite layers that copy the bedrock relief) and strip-like deposits to bauxite-filled grabens, canyons, sinkholes, nests and bags. The thickness of the bauxite bodies is of the order of several metres (up to about 35 m). The length and width are up to about a kilometre.

Petrographically, the bauxite ranges from massive, pelitic through pisolithic to concretionary and rubbly and the former is most typical. The colour is white to deep red and light pink is the most frequent variety. Mineralogically, most "Mediterranean" bauxites hosted by Mesozoic-Cainozoic carbonates are composed of boehmite, but diaspore is fairly common in some Paleozoic bauxites in the U.S.S.R. (e.g. Bokson) and China. Corundum appears in the rare partly metamorphosed bauxites (e.g. Salair Range, Siberia; Kirpal' and Tenyakov, 1974). In rare instances, bauxite grades into transported Fe or Ni oxide ores. In the Marmara deposit west of Athens, Greece (de Weisse, 1967), a bauxite lens located at an unconformity between a Triassic and Cretaceous limestone is topped by a claystone containing green nickel hydrosilicates and interpreted as being a reworked Ni laterite.

The "Mediterranean-type" deposits occur in fold belts (in the "miogeoclinal" megafacies) more often than the remaining styles, hence they are the archetype of the "geosynclinal" bauxites and frequently so named in the Russian literature. A selection of the "Style 1" locality examples is in Table 20-3.

Table 20-3. "Karst" bauxites, style 1: bauxite only, at unconformity, sandwiched between footwall and hangingwall carbonate. Selection of deposits / districts with over 10 Mt bauxite only, as summarized in Bárdossy (1982)

LOCALITY	FOOTWALL CARBONATE		BAUXITE*	HANGINGWALL CARBONATE		TONNAGE
	AGE	COMPOSITION	thickness	AGE	COMPOSITION	(Mt of bauxite)
Alpilles, Provence, France	J_3-Cr_1	neritic lim.	lenticular, sheets 10-30m	Cr_3	lacustr. lim.	160
Abruzzi, C. Italy	Cr_3	neritic lim.	lenticular, nests, 2-10 m	Cr_3	nearshore lim.	20
S. Giovanni Rotondo Apulia, Italy	Cr	neritic lim.	lenticular, 4-15 m	Cr_3	reef limestone	15
Padurea Craiului, Rumania	J_3	neritic lim.	lenticular, 4-20 m	Cr_1	freshwater lim.	50
Nagyvazsony, Hungary	Tr	neritic dolom.	sheet, lens, 10-35 m	Pl	freshwater lim.	30
Gánt, Iszkaszent-györgy, Hungary	Tr_3	limestone	sheets, lenses, 5-40 m	Eo_2	limestone	110
Centr. and Southern Istria, Yugoslavia	Cr_1	neritic lim.	blanket, 10-50 m	Eo_1	freshwater lim.	40
Lika, Bjelaj Lipa, Dalmatia, Yugosl.	Tr	neritic lim. marl	sheet, lens, 5-25 m	Tr	freshwater lim.	130
Bos. Krupa, Jajce, Yugoslavia	Cr_3	neritic lim.	blanket, lens, nest, 2-40 m	Cr_3	nearshore lim.	50
Drniš, Mostar; Herzegovina, Yugosl.	Cr_3-Eo_1	neritic lim.	lens, blanket, nest, 5-40 m	Eo_2	nearshore lim. marl	75
Nikšic, Montenegro Yugoslavia	Tr_3	neritic lim.	lenses, strips, 10-65 m	J_2	calcar. marl	120

Stara Crna Gora, Yu.	J_3	neritic lim.	lenses, nests, 2-10 m	Cr_3	neritic lim.	10
Petrovici, Velimlje, Yugoslavia	J_3	neritic lim.	lenses, nests, 4-40 m	Cr_3	neritic lim.	30
Bijele Poljane, Yug.	J_2	neritic lim.	lenticular, 3-15 m	Cr_3	marly limest.	70
Ulcinj, Buda; Yug.	Cr_1	neritic lim.	lenses, sheets, 2-40 m	Eo_2	neritic limest.	50
Helikon R., Greece	J_2	neritic lim.	lenses, nests, 2-10 m	J_2	neritic lim.	25
Kiona, Parnassos; Greece	$J_{2,3}$	neritic lim.	lenses, nests, 8-25 m	J_3	neritic limest.	140
Kiona, Parnassos; Gr.	Cr_3	neritic lim.	blankets, lenses, 10-25 m	Cr_3	neritic limest.	250
Euboea Isl., Greece	Cr_3	neritic lim.	nests	Cr_3	neritic limest.	10
Chalkidike Penins., Greece	Tr_3	neritic lim.	lenticular, 5-8 m	Cr_3	limestone	50
Menderes Massif, Turkey	Pe_3	marble	lenses, sheets, 2-10 m	Tr_1	marble	60
Akseki, Seydisehir W.Taurus, Turkey	Cr_3	neritic lim.	lenses, nests, 5-50 m	Cr_3	brackish-lacustr. limestone	60
Adama, Bolkardağ; E.Taurus, Turkey	Pe_1	neritic lim.	lenses, nests, 2-8 m	Pe_3-Te_1	dolomite, shale	10
Payas, Islahye; Turkey	Cr_3	neritic lim.	lenticular, 5-20 m	Cr_3	neritic lim.	30
Bukan, Iran	Pe_3	neritic lim.	sheet, 2-12 m	Tr_1	neritic limest.	10

*NOTE: the age of bauxite is assumed to be identical with, or shortly predating the age of the hangingwall carbonate; boehmitic composition.

Style (2) bauxite deposits are immediately hosted by a continental to marine terrigenous sedimentary association of moderate thickness (several tens of metres, out of which the bauxite body itself is several metres thick), sandwiched between lower and upper confining carbonate units. Typical stratigraphy of the bauxite-bearing sequence resting on karsted and/or rubbly footwall carbonate, starts with a kaolinitic claystone and is followed by bauxitic clay, bauxite, carbonaceous clay, coal, sandstone or siltstone, mottled clays and hangingwall carbonate. As in the previous style, red to white massive, pisolithic to rubbly boehmitic bauxite is the dominant variety.

In the Halimba bauxite field (southern Bakony Mts., Hungary; Morvai, 1982) several large almost flat-lying bauxite bodies ranging from 1 to 7 km^2 in size and from 8 to 10 m in thickness, rest on a brecciated to loose (leached) upper Triassic dolomite and limestone. The commercial bodies of red-brown, clayey or massive, detrital, pisolithic and nodular boehmite bauxite (aver. 50.5% Al_2O_3, 2.7% TiO_2, 24.3% Fe_2O_3), are enveloped by bauxitic clays, and overlain by an upper Cretaceous terrigenous sequence of conglomerate, claystone and lignite topped by a marine shale. This, in turn, is overlain by Eocene marls and argillaceous limestone. Table 20-4 shows a selection of example localities.

Style (3) has bauxite orebodies resting, as in the previous style, on karsted carbonate. It also has a similar stratigraphy but lacks the ultimate carbonate cover unit. Most of the example localities are shallow, near-surface, undisturbed – to-little– disturbed deposits on stable platforms, and a continental, terrigenous host sequence (kaolinitic clays to claystones, lignitic clays, lignites, sandstone, siltstone) is the unit that actually crops out on the surface. Because of this, it is treated in Chapter 25. Style (4) is rare and of little commercial importance.

20.5.2. Other metallic mineralizations

Accumulations of several metals in settings equivalent to those described for Styles (1) and (2) of bauxites have been recorded in the literature.

URANIFEROUS PHOSPHORITES AT UNCONFORMITIES

At Pécsely (Balaton Highlands, Hungary; Morvai, 1982), a rhythmically banded peneconcordant horizon of brownish gray or yellow phosphorite, 30 to 120 cm thick, is sandwiched between middle Triassic dolomite (footwall), and laminated bituminous siliceous limestone and marl. The surface of deposition is probably a brief emergence surface or a diastem and in places the phosphorite alternates with tuffaceous layers of alkaline basalt. The phosphorite is composed of

Table 20-4. "Karst" bauxites, style 2: bauxite associated with continental terrigenous (and organic) sediments between the footwall and hangingwall carbonates

LOCALITY	FOOTWALL CARBONATE		BAUXITE-BEARING UNIT		HANGINGWALL CARBONATE		BAUXITE TONNAGE (Mt)
	AGE	COMPOSITION	AGE	COMPOSITION	AGE	COMPOSITION	
Haute Var, France	J_2	limestone	Cr_3	baux., topped by sd, lign., clayst.	Cr_3	limestone	10
Brignoles, Provence France	J_2	limestone	Cr_3	baux. topped by sandstone	Cr_3	overthrusted limestone	150
Halimba, Hungary	Tr_3	dolomite, limestone	Cr_3	baux. topped by clay, lignite	Eo_{1-2}	limestone, marl	75
Szőc Basin, Hungary	Tr_3	dolomite	Eo	baux. topped by clay	Eo	limestone	25
Nyirád, Hungary	Tr_3	dolomite	Eo_1	baux. topped by multicoloured and lignitic clay	Eo_2	limestone	60
Fenyőfő, Hungary	Tr_3	dolomite	Eo_2	baux. topped by sandst. in fault graben	Eo_2	limestone	35
South Timan, USSR	D_3	limest., dolomite	Cb_1	baux. topped by clayst., shale, siltstone	Cb_1	dolomite	35

Data from Bárdossy (1982) except for South Timan (Kirpal' and Tenyakov, 1974)

fluorapatite, fluorite and calcite. The uranium is bound in apatite, and there is minor tyuyamunite coating fractures in brecciated phosphorite along faults.

Mn OXIDES

At Urkút (Bakony Mts., N.W. Hungary; Morvai, 1982), a locality already mentioned in Chapter 15 (Fig. 15-1) three varieties of Mn ore are hosted by Mesozoic carbonates: (a) carbonate Mn ore in Jurassic radiolarian marls (234 Tt Mn/18%) grading to a (b) Jurassic oxide facies. The latter one is further gradational into (c) a rubbly, concretionary, pisolithic manganite, psilomelane and pyrolusite ore, formed by redeposition of the former ores during the period of lower Cretaceous emergence, and now buried under the lower Cretaceous unconformity. About 1 Mt Mn/22.5% is contained in the Mn oxide styles (b) and (c). Urkút is closely equivalent in terms of setting to Style (1) of karst bauxites. At the nearby Eplény Mn deposit (Morvai, 1982) the Mn carbonate and minor oxide ore affiliated with a diastem within the Jurassic pelagic carbonate or marl sequence, has been partly oxidized and reworked during the Cretaceous-Eocene emergence. The redeposited ore now rests on the karsted surface of Jurassic limestone under the confining cover of lower Eocene claystone and sandstone, and it corresponds closely to Style (3) of bauxites.

In the Les Cabesses field (French Pyrenees; Lougnon, 1956; 76.5 Tt Mn/16.6%) pyrolusite and manganite concentrations occur at unconformity between the karsted basement upper Devonian limestones containing Mn carbonate and hausmannite diagenetic mineralization, and lower Cretaceous cover rocks.

Fe HYDROXIDES AND Ni,Cr,Co OCHERS

The four styles of carbonate-associated bauxites treated earlier, have an exact counterpart in goethitic iron ores and clay/goethite ochers. Such occurrences are always located near (or adjacent to) the source of the iron minerals, which in most cases had been lateritically weathered mafics or ultramafics. Such ores often have a significant content of detrital residual chromite, Ni and Co. The carbonate / mafic or ultramafic association is most common in the Mediterranean of Yugoslavia, Albania and Greece, as well as in the Urals of the U.S.S.R. and several example localities have been discussed in Chapters 7 and 25.

In the Alapaevo deposit, the Urals (equivalent to Style (3) of bauxites; Samonov and Pozharisky, 1974; 16 Mt Fe/38%), nodular, concretionary and granular goethite in a clay or chlorite matrix, forms 8-10 m thick lenses resting on karsted lower Carboniferous limestone. The orebodies traceable up to 10 km along the basal unconformity are gradational upward into, and contemporary with, an upper Cretaceous continental coaly mudstone and sandstone sequence.

The Wādi Husainiya, Iran (Skoček et al., 1971) occurrence of pisolithic, concretionary and clayey goethite ore bodies hosted by a sandstone-claystone unit unconformably resting on Triassic limestone and topped by Jurassic limestone, has a setting and lithologic association comparable with that of Style (2) bauxites. The authors, however, suggest a marine origin for the unit and the contained ironstone.

20.6. CARBONATE, LESSER SHALE, EVAPORITE ASSOCIATION TRANSGRESSIVE OVER RED-BEDS OR CRYSTALLINE BASEMENT

20.6.1. Pb-Zn (Cu) aspect

This is a fairly distinct and metallogenically fertile association transitional into or interchangeable with the Kupferschiefer (Chapter 19) and similar sedimentary sequences. The best example localities are to be found in the Soviet Central Asia and are summarized by Popov (1962).

In the Dzhergala field (northern Kirghizia; Fig. 20-8), a middle Carboniferous continental and marine sequence is transgressive over lower-middle Paleozoic granites. The basal sedimentary unit is up to 600 m thick and it comprises red gypsiferous feldspathic conglomerate, sandstone and siltstone topped by a stratabound green to gray shale and sandstone horizon about 2 m thick, mineralized by copper sulphides. The detrital sediments are overlain by a 40-80 m thick marine unit of a thin banded, black-to-gray limestone and dolomite interbedded with black shales. This unit hosts two rhythmically banded carbonate beds mineralized by disseminated grains and nodules of galena and sphalerite. The marine carbonate is topped by another red detrital-evaporitic unit.

In the Sumsar field (southern slopes of the Chatkal Range, Kirghizia; Popov et al., 1967; Fig. 20-9), discontinuous Pb-Zn mineralization is confined to a folded, thin stratigraphic horizon of middle Devonian dolomitized limestone, traceable for about 100 km. In the Sumsar mine, the main Pb-Zn ore unit has galena and sphalerite grains, laminae and lenses hosted by a gray rhythmically banded dolomitized limestone. This is immediately underlain by a thin unit of black slates, that in turn rest on red arkosic sandstones. In the hangingwall, the carbonates change into alternating limestones and gypsiferous siltstones and sandstones topped by red sandstones. In the footwall and hangingwall of the Pb-Zn bearing carbonate two Cu-Ag rich sulphide horizons have developed locally.

Additional, comparable Pb-Zn and Cu-mineralized marine carbonate units intimately associated with red-beds exist in the Moldotau Range in Kirghizia (middle Carboniferous); in the Arkalyk Pb deposit, central Kazakhstan (upper Devonian); and elsewhere. The Soviet localities discussed above have never been subjected to independent scrutiny by western geologists in order to establish their approximate equivalency with some of the international mineralization "types".

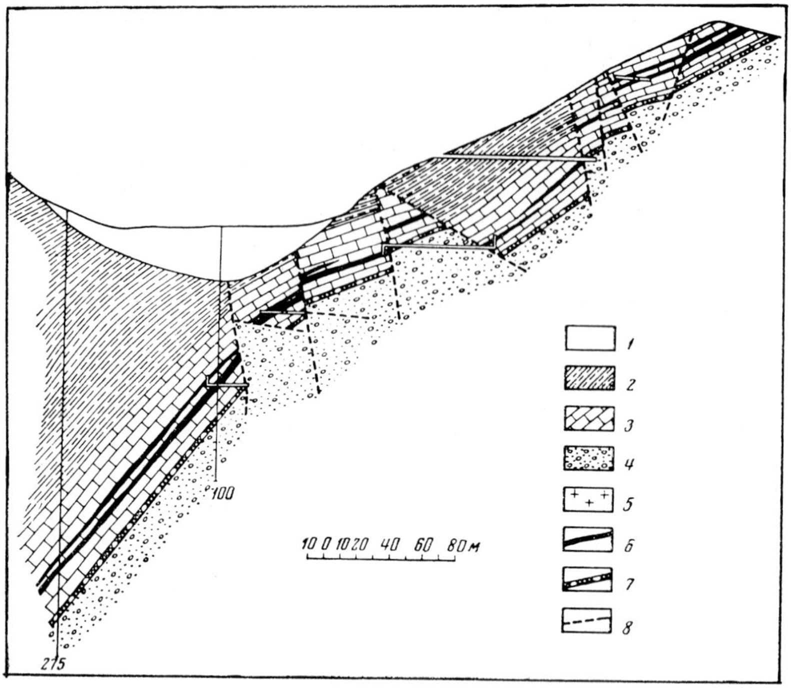

Fig. 20-8. Ikichat deposit, Dzhergala ore field, Soviet Central Asia: a stratabound Pb and Cu mineralization in carbonates overlying red beds. (1) Quaternary sediments; Permian rocks: (2) sandstone, conglomerate, shale, marl, limestone, gypsum, anhydrite, halite; (3) limestone interbedded with dolomite; (4) red conglomerate, sandstone, shale; (5) middle Paleozoic granitic rocks; (6) Pb-mineralized horizon; (7) Cu mineralized horizon; (8) faults. From Popov (1962).

Their genetic interpretation passed through the same maximum popularity stages that we have experienced here (granite plutonic - sedimentary diagenetic - groundwater epigenetic, etc.), so an armchair genetic assignment would have little credibility. The examples reviewed above represent a link between the Kupferschiefer and the MVT mineralization styles, being closer to the former because of the facies affiliation with the red-beds and evaporites. The associated copper orebodies suggest, as in the case of the Kupferschiefer, a strong oxidation/reduction boundary control.

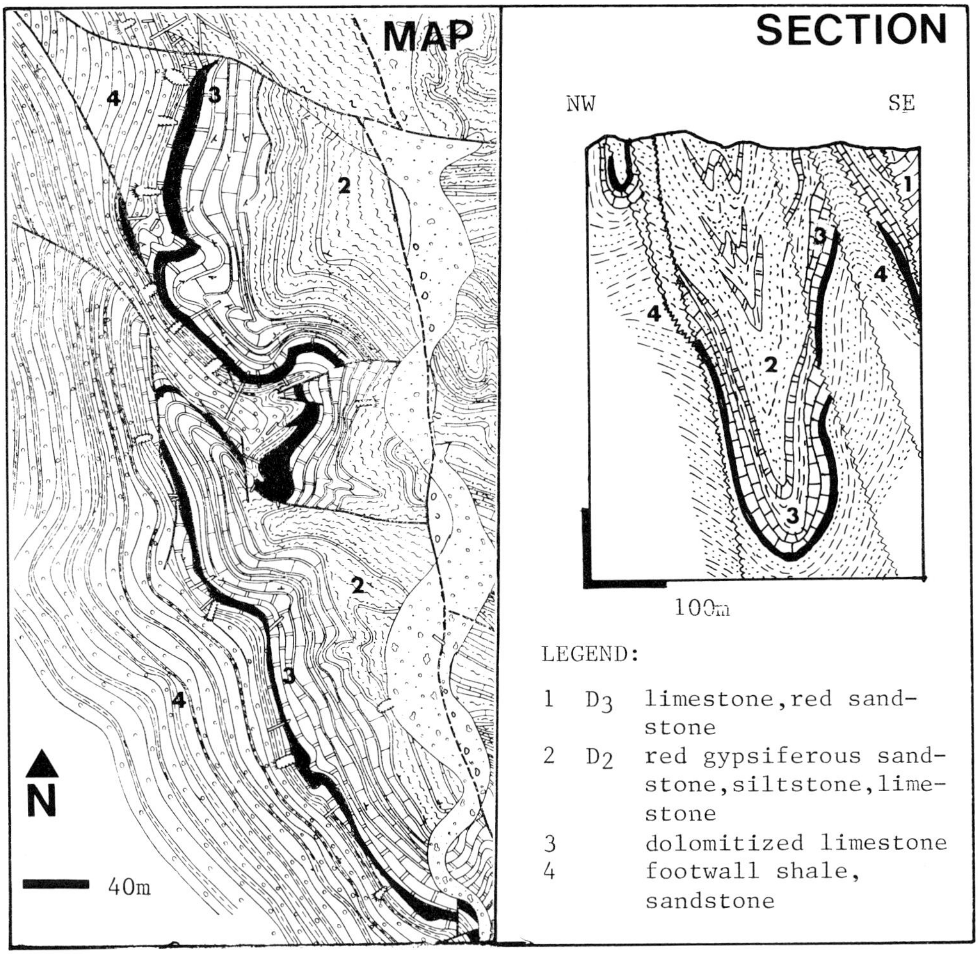

Fig. 20-9. Sumsar, stratabound Pb-Zn deposit, Kirghizia. From Popov (1962) and modified after Popov et al. (1967).

20.6.2. Mn aspect

Stratabound Mn deposits hosted by marine carbonates overlying continental red-beds or an "arid" continental crystalline basement, are analogous to the deposits in similar settings in which the host rock is shale (Chapter 19). The Um Bogma deposit in south-western Sinai, Egypt (Mart and Sass, 1972) is in a 20 m thick unit of lower Carboniferous shallow-marine dolomite, disconformably overlying, and also topped by, red sandstones. The orebodies are thick, discontinuous lenses resting on the footwall unconformity and hosted by a dolomite. The mineralogical composition is pyrolusite and psilomelane in the inner zone (45% Mn plus), grading laterally into an iron-rich fringe.

Several bedded Mn deposits in Morocco (Bouladon and Jouravsky, 1952) occur directly the contact of transgressive marine carbonates and continental red-beds, as well as within the carbonates and within the red-beds. At Imini, a thick, bedded pyrolusite body rests directly on the upper Cretaceous red – beds / dolomite contact. In Tashdremt, a Mn oxide ore lens lies in upper Cretaceous fine-grained dolomites and marls, overlying a red conglomerate, sandstone and mudstone unit. Sedimentary breccia with fragments of dolomite and Mn ore is commonly present along the footwall contact. At Bou Arfa, three Mn oxide beds are located in lower Jurassic calcareous dolomites and red clays, about 40 m stratigraphically above red-beds conglomerate and the crystalline basement. At M'Koussa, a lenticular psilomelane and pyrolusite orebody occurs in lower Jurassic Mn-stained sandy dolomites and red clays close to the Permotriassic basement basalts. The Mn accumulations are usually interpreted as being lagoonal precipitates derived from eroded and redeposited volcanogenic and vein Mn deposits in the basement.

20.7. CARBONATE-HOSTED Zn-Pb DEPOSITS: AN INTRODUCTION

Zinc and lead form more carbonate-hosted deposits than all the rest of metals taken together. Several sub-associations and interaction associations can be recognized, and some of them have already been reviewed (Fig. 20-10). By far the most widespread, productive and publicized, however, are the low temperature epigenetic deposits widely known as "Mississippi Valley-type" (MVT; Fig. 20-11) when on platforms, and "Alpine" or "Appalachian-type" (APT; Fig. 20-12) when in orogenic belts.

In the 1930's to 1950's these deposits, like most others, were considered hydrothermal, granite-affiliated and so premeditated exploration usually involved a search for "granites". Since few were found these deposits were placed into the "telethermal" and/or "cryptobatholithic" categories to justify it. When the stratiform and sedimentogenic concepts reached their peak of popularity in the 1960s-1970s, a complete turnaround took place and the MVT/APT deposits have been interpreted by means of a variety of purely sedimentogenic processes. The petroleum geologist's models of fluid migration and entrapment were modified to suit the more complex situation involving ores, resulting, for example, in the well-known Jackson and Beales (1967) model. In this, metals were believed to be released by migrating connate brines from basinal shales and reprecipitated in porous (usually carbonate) traps by reaction of the metalliferous brine with a precipitant.

Emphasis on karsting, reefs, carbonate facies, etc., typically one-sided and often with a tendency to exclude other possibilities,

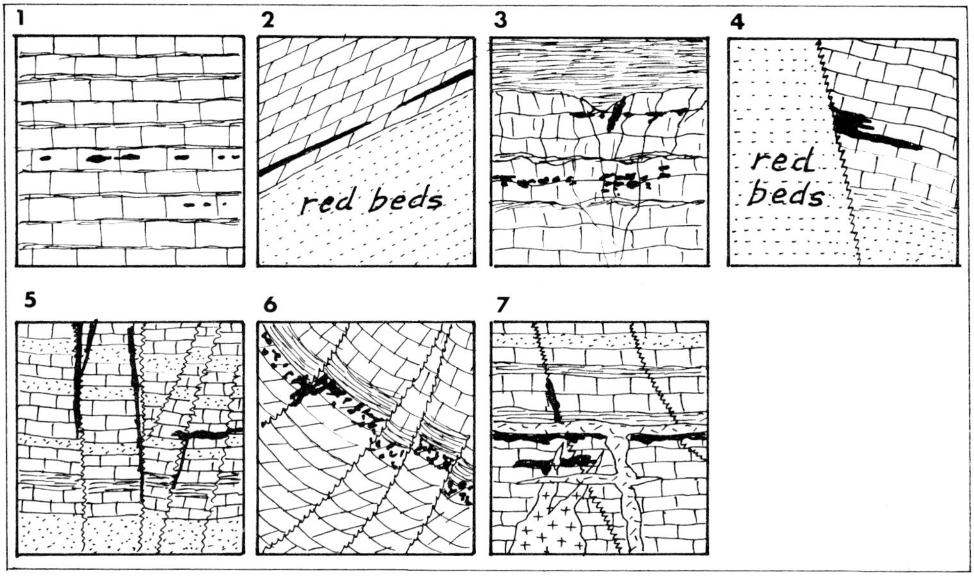

Fig. 20-10. Principal sub-associations and styles of shallow carbonate-hosted Zn-Pb deposits. (1) "Undisturbed" carbonates hosting diagenetic nodular or sparsely scattered sphalerite (e.g. Bruce Peninsula, Ontario); Section 20.4.3. (2) Basal carbonate units overlying red beds and hosting a bedded Zn-Pb ore (e.g. Sumsar, U.S.S.R.); Section 20.6.1. (3) Flat-lying carbonates hosting mostly peneconcordant epigenetic MVT Zn-Pb orebodies associated with solution, dolomitization, silicification. Section 20.8.1. (4) Fault contacts of red-beds/carbonates associated with carbonate-hosted Pb-Zn deposits partly contemporary with sedimentation (ex. the Irish Pb-Zn province); Section 20.9. (5) Flat-lying or folded carbonates mineralized by late, epigenetic fluorite, barite, sphalerite, galena cross-cutting veins controlled by normal faults and lacking conspicuously affiliated volcanics / plutonic rocks (e.g. Illinois - Kentucky district); Section 20.10. (6) Folded carbonates hosting peneconcordant or fault—controlled epigenetic APT Zn-Pb orebodies accompanied by solution, brecciation, dolomitization, silicification, etc. Section 20.9.1. (7) Vein or replacement Zn-Pb bodies in carbonates located in intrusive contact aureoles. Section 20.10.1.

also competed for attention. The main drawback of the numerous models was that they rarely told the whole story; for example, karsting can demonstrably redeposit and sometimes upgrade an earlier mineralization but does not itself generate the Pb-Zn ore out of "nothing". MVT karst models therefore, have also to consider pre-karsting metallogeny in sufficient depth to be credible.

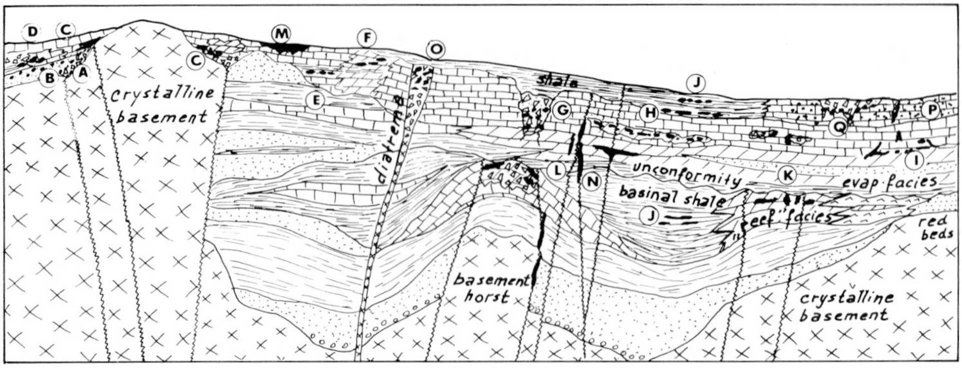

Fig. 20-11. MVT and associated Zn-Pb deposits hosted by unfolded shallow marine carbonates. (A) orebodies in fault graben, hosted by breccia, sandstone; (B) disseminated sulphides in basal sandstone, usually replacing calcite cement; (C) ore in carbonates that abutt basement knobs; (D) facies-controlled ores in general, especially in reef or back-reef facies; (E) ores controlled by shale, confining a porous carbonate; (F) ores associated with metasomatic dolomitization of porous brittle limestone (e.g. reef facies); (G) ores in a buried (fossil) vertical collapse structure; (H) ores in peneconcordant solution-brecciation bodies (e.g. pseudobreccia), usually under an unconformity; (I) ore in minor reverse, normal and bedding faults due to solution thinning and collapse (pitches and flats); (J) syngenetic-diagenetic high trace Zn-Pb content (or scattered nodules, crystals of sphalerite, galena) in continental or marine black shales; (K) Pb-Zn sulphides in solution-brecciated and dolomitized limestones at basinal shale / mound carbonate / evaporite facies change; (L) deposits in solution-collapse breccia under unconformity, controlled by faults; (M) oxidation zones, gossans and reworked gossans over MVT; (N) peneconcordant, replacement and discordant vein deposits associated with through-going faults; (O) ores associated with diatreme (cryptoexplosion ?) breccias; (P) low-grade, evenly disseminated Zn,Pb sulphides in carbonates; (Q) surficial karst; residual galena, sphalerite in karst sink fill, and/or replacements and infiltration of secondary Zn minerals in the underlying carbonates.

An exploration geologist, unfortunately, was trapped in the middle of the genetic debate; the only sure lead to MVT are its carbonate hosts. Except for this the orebodies can occur almost anywhere, most notably in both the platformic and orogenic settings (the latter represented by the APT style). Athough the MVT/APT genetic controversy is far from over and I have no new genetic model to offer as to their origin, a complex approach is recommended.

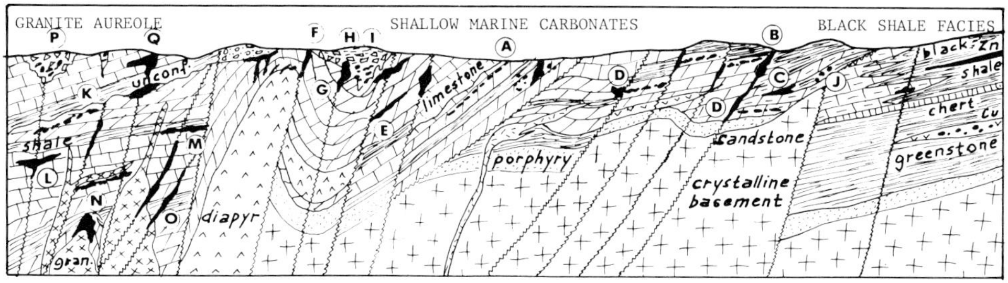

Fig. 20-12. APT and associated Zn-Pb deposits, hosted by folded shallow marine carbonates. (A) Pre- deformation, sub-unconformity, peneconcordant Pb-Zn orebodies; (B) peneconcordant, disseminated galena, sphalerite in dolomitized limestone under a shale (black shale) cover; (C) peneconcordant disseminated sphalerite, galena in carbonates at or near a facies transition to shales; (D) ore replacements controlled by faults in intrusive thermal aureoles; (E) bedded replacements, solutions were moving from faults; (F) ores associated with diapiric piercement structures and carbonate friction breccias; (G) finely disseminated galena, sphalerite, present in large volumes of carbonate; (H) secondary Zn(Pb) minerals in oxidation zones, and infiltrated into or replacing underlying carbonates along faults; (I) residual galena, sphalerite in near-surficial karst cavities fill; (J) stratiform disseminations or nodular galena, sphalerite, in black shale;

Interaction Zn-Pb orebodies hosted by sedimentary carbonates in intrusive aureoles: (K) ore mantos under unconformities; (L) ore mantos adjacent to faults; (M) ore mantos under impervious shale; (N) ore skarns in "granite" exocontacts; (O) endo- and exo-contact fissure veins; (P) solution-collapse breccias with mineralization that was probably depth-derived; (Q) Zn(Pb) carbonate or silicate ? descendent replacement bodies.

Although the oilman's explanation of the MVT/APT origin is fairly credible, it need not be exclusive. Almost all (if not all) MVT's located on platforms, correlate with zones or sites of disturbance. These zones are of the extensional type and often are not apparent on the surface but the subsurface disturbance may rival or exceed in intensity the deformation within orogenic belts. Mountain belts and folding, however, may be absent so that such terrains do not fit into the standard "oreogenic belt" pigeonhole. Some correspond to horst and graben systems (e.g. the Boothia Peninsula of the Arctic Canada). The Soviet school evolved the concept of "activated terrains", and in the West the link between the intracratonic disturbances and "rifts" has sometimes been stressed (Chapter 30).

Whatever the rôle of the above disturbances in the origin of the MVT, their presence is one of the empirical observations important in MVT prediction. Buried structural disturbances can generally be detected, if not by direct mapping, then geophysically. For MVT/APT ores the presence of carbonates is a necessity. Other favourable factors are the existence of Zn or Pb-rich black sediments (usually shales) or arkoses, granitic basement, unconformities, basement domes, karsting or even intrusive activity (not necessarily "granitic"; alkaline rocks are often present). In short, MVT/APT deposits are most likely to be found in complex areas that combine most of the favourable factors and depart, in most respects, from the "normal", orderly carbonate terrains.

MVT/APT deposits: empirical indicators of favourability

In the exploration for MVT/APT deposits, the geologist is advised not to stage everything on the latest genetic model in vogue. In 1982, there was still no agreement with regard to the origin of these deposits (compare Amstutz, ed., 1982); in fact, it appears that these deposits can form in a variety of ways each having the same result. The reader is invited to compare two recent papers interpeting two very similar ore districts: Pine Point and Kraków-Silesia, with either a sedimentogenic (Skall, 1975) or a magmatogene-hydrothermal (Sass-Gustkiewicz et al., 1982), bias. Although both districts have comparable characteristics they were given different meanings and different priority.

Regardless of genesis, MVT/APT deposits are accompanied by a variety of characteristic, observable field features that are useful in exploration but there is a considerable variation among individual ore deposits and districts. This is demonstrated in Fig. 20-13 that compares the empirical characteristics of the eighteen most prominent MVT/APT fields and districts. Two characteristics are dominant and strongly developed in all of them: (a) alteration dolomitization and (b) brecciation, mostly of solution-collapse origin. The presence of these features in the field, however, does not necessarily guarantee the presence of an ore.

20.8. CARBONATE, LESSER SHALE, SANDSTONE ASSOCIATION OF PLATFORMS AND STABLE BLOCKS

20.8.1. Epigenetic, peneconcordant Zn-Pb sulphide mineralization (MVT style)

NORTH AMERICAN PLATFORM

The North American Platform (Fig. 20-14) is the classical area of MVT mineralization, and several thousand such occurrences have been reported from its southern portion alone (compare Heyl, 1968). This region also contains two Pb or Zn districts of global importance (South-Eastern Missouri and Tri-State), and two significant districts

(Upper Mississippi Valley, and Central Tennessee). A third significant Illinois-Kentucky district lies in an intensively faulted segment of the crust and departs considerably from the expected style of predominantly horizontally arranged deposits. It is treated in Section 20.10.

The Platform extends far beyond the most densely mineralized region of the central-eastern United States to central, northern and Arctic Canada and, discontinuously, into the Canadian Atlantic provinces. There, only a few mineralized occurrences have been recorded, although two of the districts are of global importance (Pine Point and Little Cornwallis Island) and in operation. The scarcity of ore occurrences is, to a considerable degree although not exclusively, due to poor outcrop. The Paleozoic carbonates are buried under a thick pile of Mesozoic-Cainozoic sediments and/or glacially scoured and covered by Quaternary glaciogenic sediments.

The carbonate sequence of the North American Platform is facially very uniform, and was deposited between the Cambrian and Carboniferous periods in a warm epeiric sea. The basement is composed everywhere of Precambrian crystalline rocks and the carbonates either transgress directly over this basement, or more often, are underlain by an upward-fining (conglomerate – sandstone – shale) unit. The dominant structural style is a system of gently-dipping basins (Michigan and Williston Basins), separated by arches and uplifts (e.g. Cincinnati Arch). Some of the uplifts have a Precambrian basement cropping out in the centre (e.g. Ozark, Llano uplifts). Extensional faults and fault-bound structures (grabens, horsts, aulacogens) of greater than average mobility and deep reach, have increasingly come to light over the past 25 years. Most were established during the Precambrian age, but several have been repeatedly rejuvenated during the Phanerozoic; some may still be episodically active now, as indicated by seismic activity (e.g. the New Madrid earthquake). The normal tectonism appears to be an important controlling factor that has to be taken into account in interpretations of the carbonate-hosted deposits.

<u>South-Eastern Missouri lead district</u>

This district has been the principal producer of lead in the United States (20 Mt Pb; Fig. 20-15). It is located on the fringe of the Proterozoic granite and rhyolite outcrop area in the Ozark Uplift, about 100 km south-east of St.Louis. The Paleozoic sedimentary sequence starts with Cambrian arkose and quartz arenite (Lamotte Sandstone) that fills depressions in the relatively youthful crystalline basement and abutts the buried rhyolite and granite knobs (Snyder and Gerdemann, 1968; Thacker and Anderson, 1977). The sandstone is represented locally by basal rhyolitic or granitic conglomerate. The carbonate-cemented sandstone grades upward into the initially sandy but later pure limestones of the Bonneterre Formation, which in turn change vertically into a sequence of Cambrian and Ordovician limy shale, dolomite and impure limestone. The principal lead mineralization is hosted by the Bonneterre Formation which is up to 500 m thick and comprises a variety of carbonate lithofacies

	S.E.MISSOURI USA	TRI-STATE USA	UPPER MISSISSIPPI USA	CENTRAL TENNESSEE USA	PINE POINT CANADA	CORNWALLIS ISLANDS CANADA	GAYS RIVER,N.S. CANADA	DANIELS HARBOUR CANADA	AACHEN-MORESNET GERMANY-BELGIUM	CEVENNES BORDER (Les Malines)-FR.	SILESIA-KRAKOW POLAND	MASCOT-JEFFERSON CITY USA	AUSTINVILLE-IVANHOE USA	FRIEDENSVILLE,Pa. USA	BLEIBERG-KREUTH AUSTRIA	CAVE DI PREDIL (Raibl),ITALY	SALAFOSSA ITALY	ROBB LAKE CANADA
1 AGE	Cm3	Ms	Or2	Or1	D2	Or2	Ms	Or1	Cb1	Cb-J	Tr2	Or1	Cm1	Or1	Tr2	Tr2	Tr2	D
2 THICKN.	100	300	120	60	300	X00	X0	100	X00	X00	80	70	500	30	200	200	100	100
3 CARBNTE	L	LD	LD	LD	L	LD	LD	LD	LD	LD	LD	LD	LD	L	DL	DL	DL	LD
4 FACIES	■	X	X	O	#	O	#	X	////	////	////	O	////	X	////	////	X	////
5 "REEFS"	////	O	O		#		////		////	O	X			O	////	X	O	////
6 BLACK S.	X	O	O	O	O	O		O	X	O	X		X	////	////	////	O	////
7 EVAPOR.					O		X				O				O			O
8 CHERT		////	O	O										O				O
9 SANDST.	////	O	O	O								O						
10 CONGL.	O																	
11 VOLC.	O										O				O	O		
12 Zn/Pb	■	O	X		X	X	X	O	O	X	X		X		X	X	O	X
13 FeS2	O	////	O	O	O	O	O	O	////	O	O	O	O	////	////	////	O	O
14 Cu,Co	////	O	X				O		O		O				O	O		
15 FLUOR.	O	O		////	O		X		O	O	O	O	O		O	O		
16 BITUM.	O	O	O	O	O	O	O	O	O	O	O	O	O	O	O	O	O	O
17 GOSSAN	O	O	X						■	X	////	O	O	////	////	////	O	O
18 ALT.DL.	////	////	////	#	■	#	#	#	#	#	■	#	#	#	#	#	#	#
19 ALT.SI.	O	#	X	X				O		O		O	X	X	X	X	X	////
20 GRADE	6	3.5	7	4.3	9.6	18.4	7	7.7	11	7	6	4	4.3	6.5	6.5	16	6.9	7.3
21 ORE DS.	MXI	XCI	VMX	IXB	IBX	MX	MB	IB	MXC	MXI	IBM	IXB	IVM	XI	EXB	MXV	IXM	IXB
22 BRECCIAS	#	#	X	#	#	#	////	////	////	////	////	#	////	////	////	////	#	////
23 BR.INT.	DS	S	ST	S	S	ST	S	S	ST	SDT	ST	S	ST	ST	STD	ST	S	TSD
24 ORIENT.	O	O	#	O	////	X	X	O	X	////	X	X	////	X	////	////	////	////
25 FAULT	O	O	X	O	O	X	O	O	////	////	X	////	////	////	#	#	#	////
26 TIMING	A	A	C	A	B	BCA	B	A	BCA	BCA	BA	A	A	A	AB	BA	BA	A
27 UNCONF.	A	U	U	U	U	U	DA,A	U	U	DU,U	U	U	U	U	DU,U	DU,U	U	U
28 VOLC.S	O						O		O	O	O		O		O	O	O	
29 VOLC.M	O		O				O		O	O	O		O	O	O	O		
30 UNDERL.	AC	IS	ISA	ISA	IS	IS	AR	I	SA	IS	IS	I	A	I	I	I	IR	IS
31 BASEM.	0-100	X00	500	X00	X00	X00	X0	X00	X00	X00	X00	500	150	100	X00	X00	150	X00
32 BSM.FLT.	////	X	O		X	#	////	O	////	////	X		O	O	O	O	O	

INTENSITY SCALE: (blank) 1; O 2; X 3; //// 4; # 5; ■ 6

Fig. 20-13. Empirical characteristics of the major MVT and APT ore fields and districts. NOTES AND EXPLANATIONS.

Pre-mineralization characteristics of the host units:	
(1)	age
(2)	approximate thickness of the mineralized interval, in metres;
(3)	composition of unaltered carbonate in the mineralized interval L=limestone, D=dolomite
(4)	facies diversity within the ore interval: single facies throughout (0) to greatest variety of facies (5)
(5)	"reef" (buildup) facies: (0)=absent to (5)=comprises the entire unit
	Other pre-alteration rocks present in the host unit:
(6)	black shale (0-5)
(7)	evaporites (0-5)
(8)	chert, nodular or bedded (0-5)
(9)	sandstone (quartz arenite); 0-5
(10)	basal conglomerate (0-5)

(11)	tuff, volcanic sediments (0-5)
Mineralization and its controls	
(12)	sphalerite and galena are dominant minerals at all localities; (0) sphalerite only; sphalerite : galena = 1:1 (3); galena only (5)
(13)	pyrite or marcasite absent (0), exceeding galena and sphalerite (5)
(14)	Cu,Co,Ni minerals: absent (0) rare (1) widespread but not recoverable (2) locally recoverable (3)
(15)	fluorite, absent (0), dominant (5)
(16)	liquid bitumen or pyrobitumen: absent (0), present in minor quantities (1)
(17)	gossan (oxidation zone) Zn-Pb minerals: absent (0), dominant (5)
(18)	alteration dolomitization (typically coarse crystalline, sparkingly white or beige, vuggy dolomite) absent (0), dominant (5)
(19)	alteration silicification (chert or jasperoid): absent (0), dominant (5)
Structural control, shape of orebodies, distribution of ore	
(20)	approximate average ore grade
(21)	major style of ore distribution: irregularly scattered small grains (I); evenly disseminated fine grains (E); bunches (B) masses (M); breccia cement (X); veins, veinlets (V); cavity encrustations (C)
(22)	breccias as hosts to ore: absent (0), dominant (5)
(23)	prevalent breccia genetic interpretation: depositional (e.g. talus (D); solution collapse (S); tectonic (T)
(24)	orientation of orebodies to bedding: all peneconcordant (0); all discordant (5)
(25)	is the ore related to faulting ? no (0), entirely (5)
(26)	the ore emplacement took place: after faulting (A), before faulting (B), contemporaneously with faulting (C)
(27)	when the ore is related to unconformity, it is directly under (DU); directly above (DA); up to 300 m under (U) or above (A)
(28)	tuff beds, volcaniclastics, lava flows, contemporary with deposition of the host suite: absent (0), minor (1)
(29)	traces of igneous activity postdating the sediment deposition, but possibly coeval with the mineralization: absent (0), minor (2)
Miscellaneous aspects of the regional association:	
(30)	the unit immediately underlying the ore hosting unit: (I) identical with the host; (S) shale dominated; (A) white or gray arenite; (R) red-beds, evaporites; (C) crystalline basement
(31)	approximate depth of the crystalline basement (in metres)
(32)	extensive tectonism (major normal faults, grabens, horsts) in the basement, probably influencing the overlying units: absent (0); strong and dominant (5)

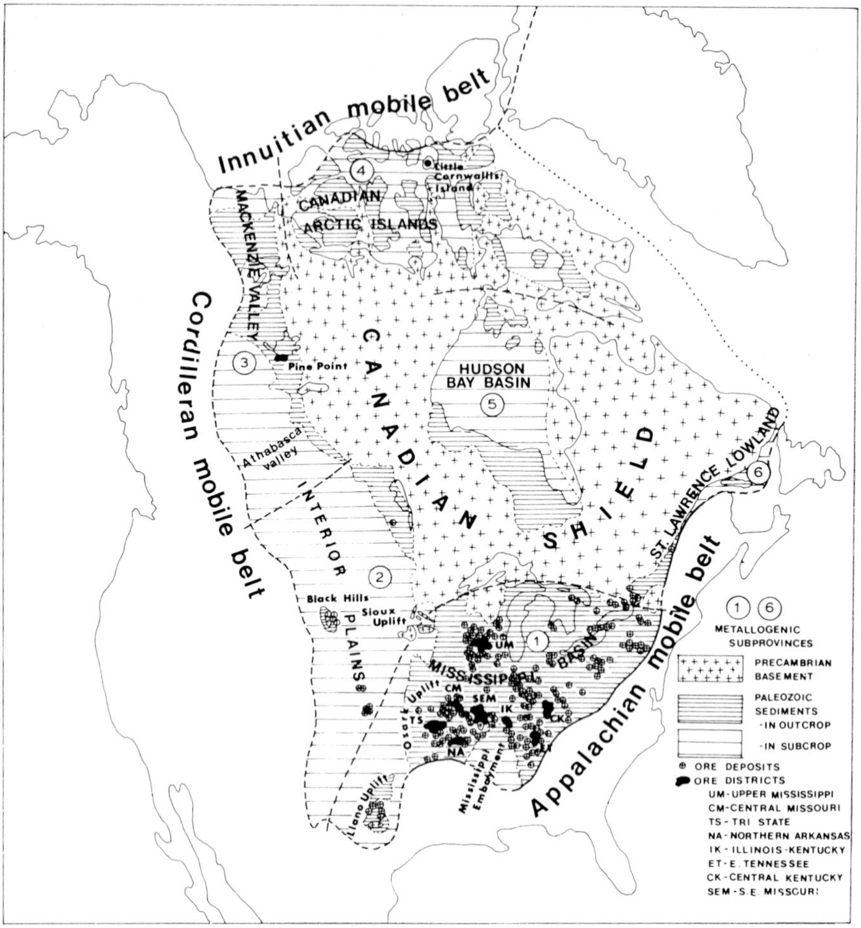

Fig. 20-14. North American Platform and its metallogenic subprovinces: (1) Mississippi Basin; (2) South-western Interior Plains; (3) Athabasca-Mackenzie Valleys area; (4) Canadian Arctic Islands; (5) Hudson Bay Basin; (6) St. Lawrence Lowland. From Laznicka (1981a).

corresponding to the forereef, backreef, reef complex and shelf depositional environments (Lyle, 1977). The only minor non-carbonate rocks are black shale, green shale and siltstone, and a thin tuff bed. Diagenetic dolomitization cuts across the facies boundaries and is relatively light. Conspicuous paleokarst is absent.

The mineralization is discontinuously distributed over an area of about 15,000 km^2, in irregular local ore clusters (ore-fields; about 100-200 km^2) and in a 50 km long north-south trending narrow belt (Viburnum Trend). The Mine La Motte near Fredericktown in the eastern part of the district (Buckley, 1909; Snyder and Gerdemann, 1968;

about 2 Mt Pb, 55 Tt Ni, 10 Tt Co; Fig. 20-16/3) is an example of a mineralized cluster. There, narrow linear orebodies are controlled by stratigraphy, lithology and facies of the basal portion of the Cambrian strata. The ore minerals: galena (over 90%), pyrite, chalcopyrite and siegenite, form rare massive nests or, more commonly, scattered single crystals in the lowermost 17 m of the Bonneterre Fm., in the ferruginous dolomites and in the uppermost, calcite (dolomite) cemented Lamotte Sandstone. The grade varies between 1 and about 10% Pb and the presence of siegenite contributes about 0.5% Ni and 0.15% Co to the ore in places. The basic trend of mineralization is roughly conformable with primary depositional structures (stratification, algal lamination), but mineralized cross-cutting fractures are very common particularly in the Lamotte. The most intensively mineralized zone lies immediately above or near the Lamotte pinchout and the orebodies have a narrow, linear and arcuate form, circling a knob of Precambrian granite.

The remaining ore clusters and belts are a variation on a theme, and have been recently described in detail in a series of papers published in Economic Geology (particularly in the special issue on the Viburnum Trend, vol. 72, 1977); see also Fig. 20-16/1,2,4. Because of the considerable grade variation among individual orebodies and their portions, average grades are rarely published. The reserve figure for the No.27 mine north of Viburnum (7.3 Mt ore with 2.9% Pb, 0.23% Zn, 0.17% Cu; Grundmann, 1977) underlies the fact that the south-eastern Missouri ore as a bulk is rather low grade, and lead-dominated. At other localities, zinc and copper are still lower grade to even absent.

As expected, genetic interpretation of the S.E. Missouri district kept pace with the changing ideas. The latest version (Davis, 1977) assumed ores to have formed at "favourable" (i.e. porous) sites in the Bonneterre carbonate, where metalliferous brines moving upward from the Lamotte Sandstone (which also supplied the metals, probably from decomposed Precambrian detrital feldspars) encountered locally produced sulphide (or sour gas).

The Cambrian Bonneterre level is not the only mineralized sequence in the district. Widespread barite (11.5 Mt produced) and minor quantities of Pb-Zn occur stratigraphically higher, in Cambrian dolomites along the Potosi / Eminence Formations contact about 170 m above the top of Bonneterre (Fig. 20-16/5). Most of the barite has been gathered from surficial residual soil in Washington County, in the north-western corner of the district, and little is known about the nature and controls of the unweathered ores. They are most probably epigenetic replacements and fracture fillings.

Tri-State zinc district

The Tri-State district (Hagni, 1976; 11 Mt Zn, 2.65 Mt Pb; Fig. 20-17) covers about 5,000 km^2 of discontinuously mineralized ground in the border regions of Kansas, Oklahoma and Missouri. The bulk of the production (7.3 Mt Zn, 1.8 Mt Pb) came from only 25 km^2 in the Picher Field (McKnight and Fischer, 1970).

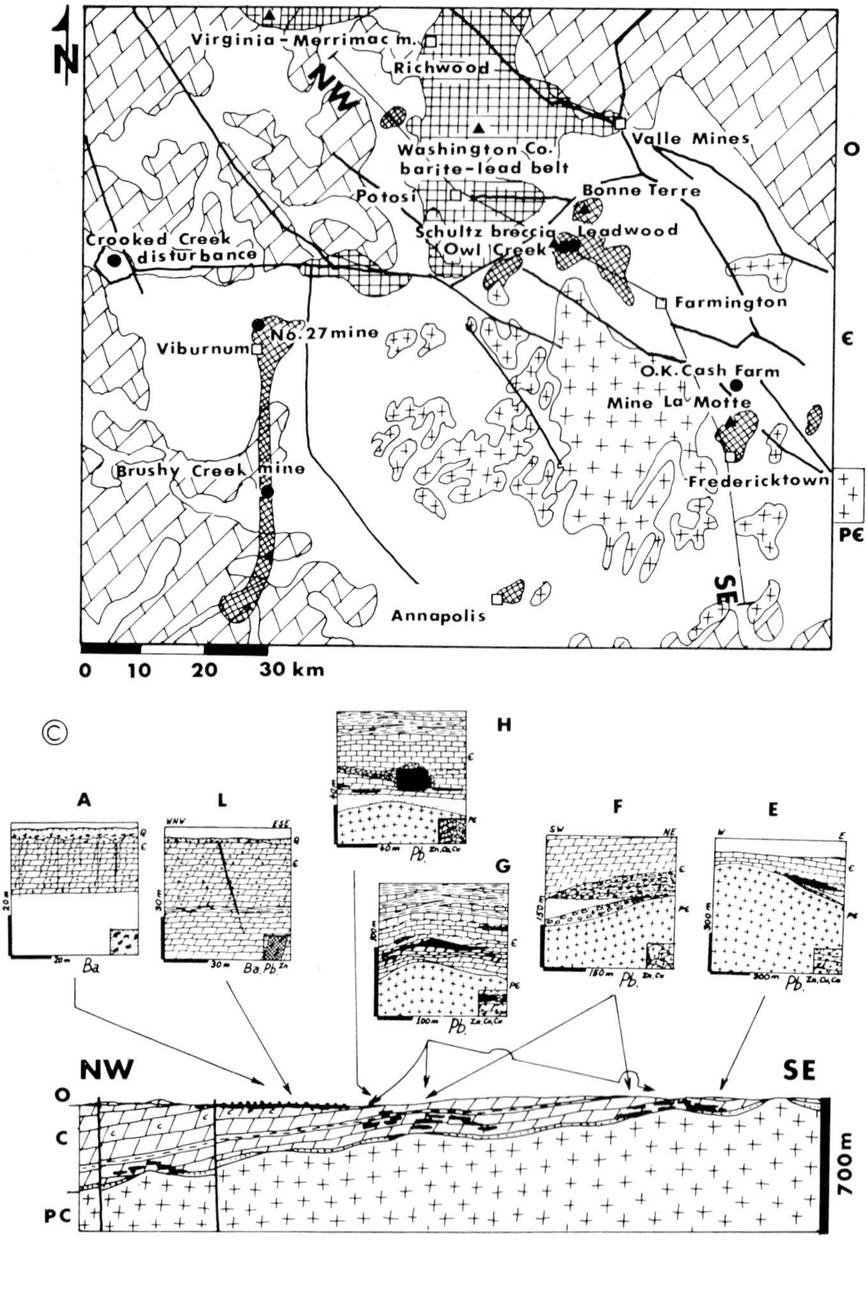

Fig. 20-15. South-East Missouri Lead district. From Laznicka (1981a).

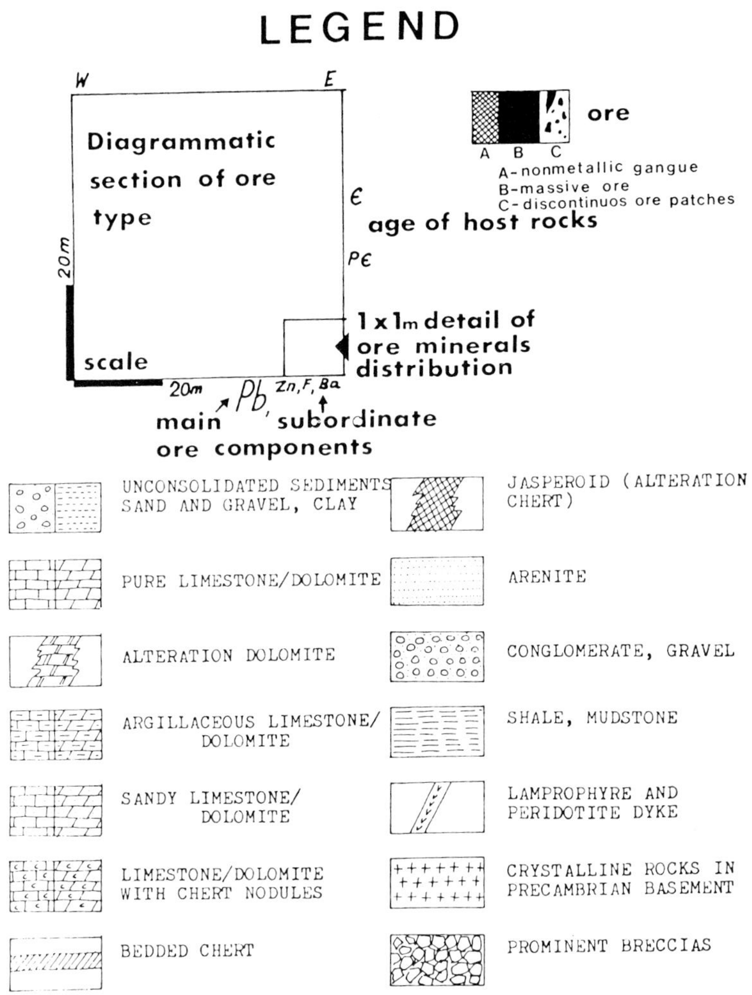

AGE OF HOST ROCKS CODES:

Q	Quaternary
Cr	Cretaceous
P	Pennsylvanian
M	Mississippian
D	Devonian
S	Silurian
O	Ordovician
Є	Cambrian
PЄ	Precambrian (Middle Proterozoic)

LEGEND for Figures 20-15, 20-17, 20-18 and 20-22.

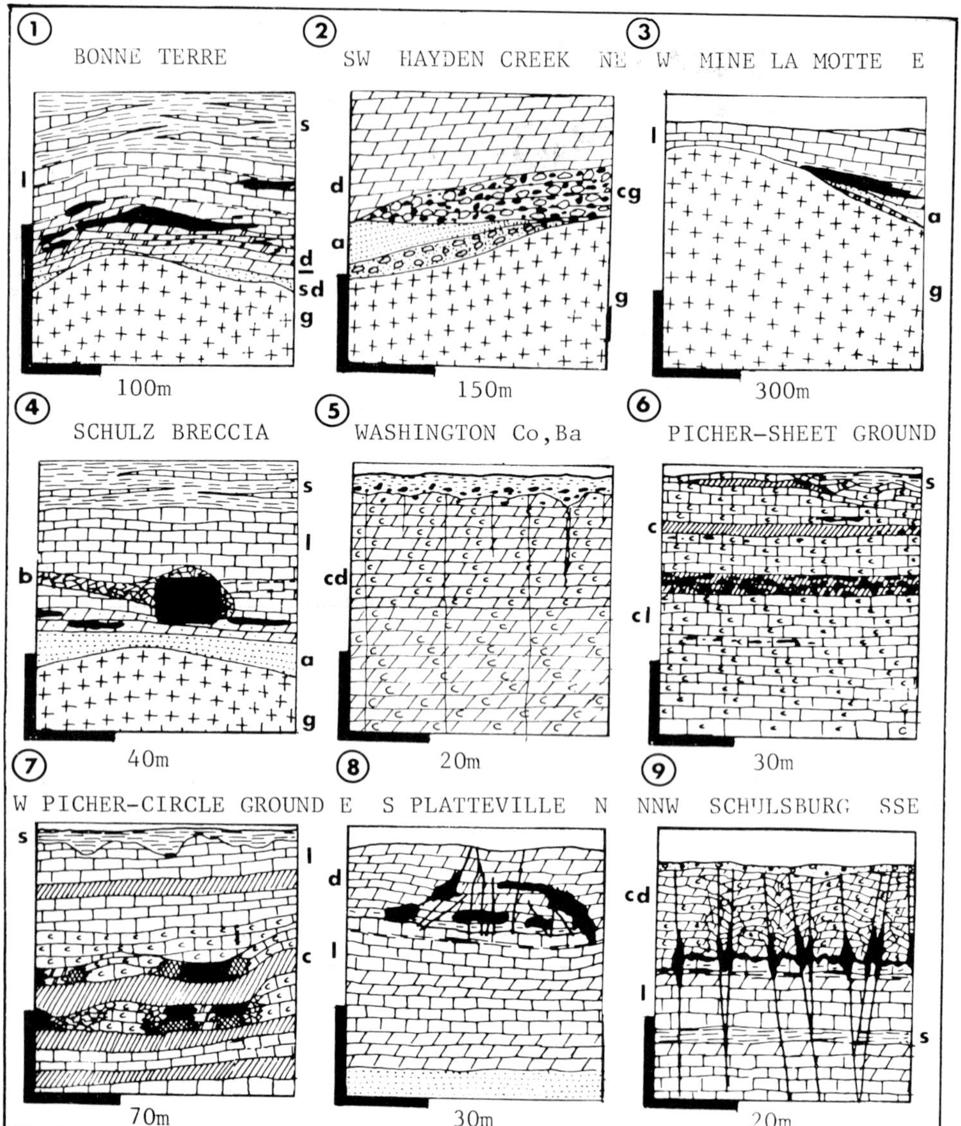

Fig. 20-16. Principal styles of the MVT mineralization, Missssippi Basin, U.S.A. Localities (1)-(5) in the S.E.Missouri district are hosted by Cambrian carbonates; (6,7) in the Tri State district, by Mississippian carbonates; (8,9) in the Upper Mississippi District, by Ordovician carbonates. LEGEND: (s) shale; (1) limestone; (d) dolomite; (a) quartz arenite; (cg) conglomerate; (g) Precambrian crystalline basement; (c) bedded chert; (ad),(al), argillaceous dolomite or limestone; (sl),(sd), sandy limestone or dolomite; (cl),(cd), limestone, dolomite with chert nodules; (b) breccia; (j) jasperoid; (d) alteration dolomite. Rearranged from Laznicka (1981b).

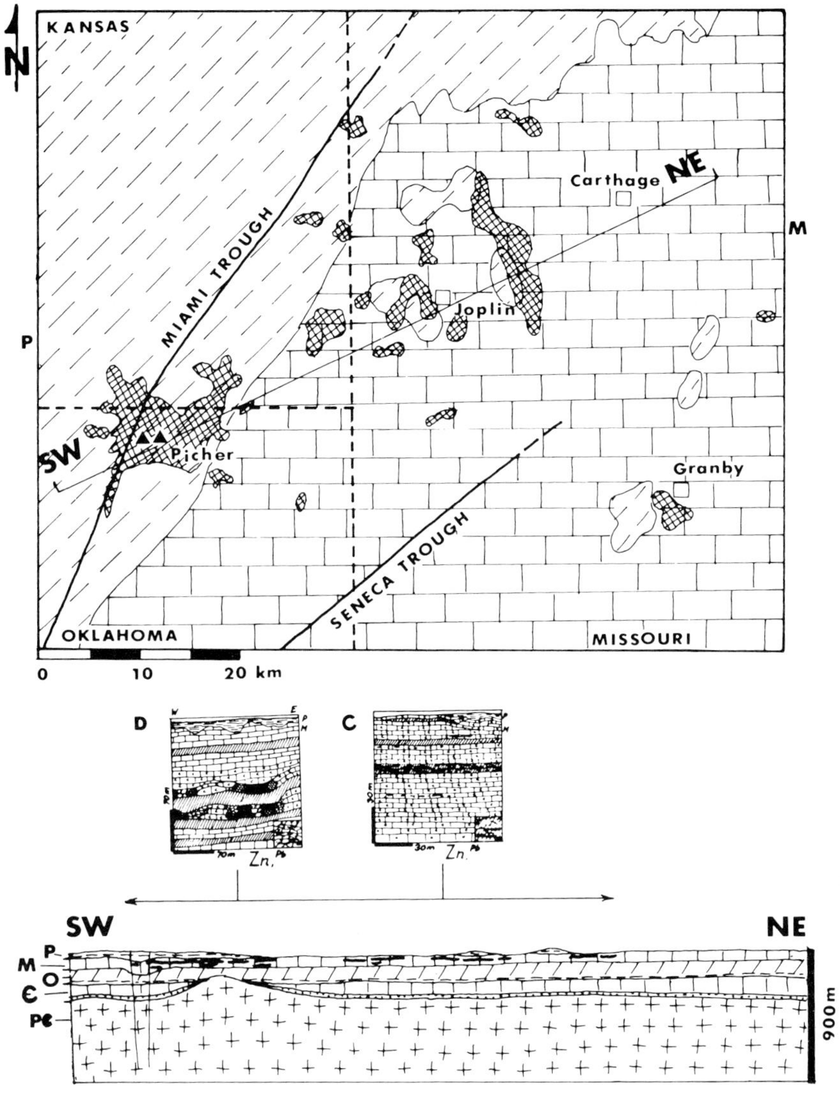

Fig. 20-17. Tri-State district, U.S.A. See legend following Fig. 20-15. From Laznicka (1981a).

Most orebodies are hosted by Mississippian shelf limestones set in the middle of a thick carbonate-detrital epicontinental sequence, several hundred metres above the Precambrian crystalline basement. The marine sequence is unconformably topped by a paralic Pennsylvanian association composed of black coaly shale, minor sandstone, limestone and coal. Carbonate buildups ("reefs") occur in the area but do not seem to contribute to the facies variation in the immediate mineralized zones to the extent that is reported in S.E. Missouri.

The bulk of the Zn-Pb ore in the district is located within an interval of the Mississippian sediments about 200 m thick, and there are about 15 "beds" that are more favourable than the rest but there are no sharp ore confinement limits. The host carbonates are mostly micritic, slightly argillaceous bioclastic limestones rich in chert nodules, interbedded with massive chert; spotty, partly silicified limestone; minor oolithic limestone and greenish or black shale. The ore deposits are bodies of scattered ore and gangue minerals, and the 1946 average grade of 1.99% Zn+Pb in the Picher field established a record low for a Zn ore mined in an industrialized country. The earlier grades, however, contained in the region of 4-6% Zn+Pb. Sphalerite is the dominant sulphide and it occurs as massive (but small) replacements, as scattered replacement grains and as loose crystals coating the numerous vugs. The crystals are typically ruby red in the centre, with a black outer zone. Galena is subordinate and chalcopyrite, enargite and luzonite are exceptional. Pyrite, marcasite, coarse calcite and dolomite are the associated or gangue minerals, while dolomitization and silicification (jasperoid) are the principal alterations.

The most typical orebodies occur within flat-lying stratigraphic intervals ("beds") several meters thick. Such bodies ("runs") have a curvilinear trend following the joint system. Two superjacent runs may coalesce locally and adjacent runs may unite to form blanket deposits with irregular outlines ("sheet grounds"; Fig. 20-16/6). The sheet grounds grade into "circle grounds" (Fig. 20-16/7; e.g. in the Joplin and Picher fields) that are roughly circular on plan, and controlled by solution collapse, brecciation, silicification and dolomitization. They are most common under the Pennsylvanian unconformity. The solution collapse "circles" are characteristically alteration zoned. The centre is composed of a crystalline dolomite replacing limestone. This is surrounded by a rim of jasperoid silicification which, in turn, is bordered by a zone in which the limestone has been completely removed leaving the original chert nodules and fragments as a mass of open rubble ("boulder ground"). All the three zones are mineralized and the greatest concentration of sphalerite is just within the edge of the dolomite with rest of galena just outside.

The exact age of mineralization in the district is not known but is believed to be Mesozoic. The metalliferous brines had temperatures of 115-135° C. The most influential controlling factors appear to have been the solution brecciation, silicification and dolomitization. The almost flat-lying and little disturbed rocks in the district are underlain by two linear north-east trending grabens and the ore formation is attributed to "formational brines mixing". It is interesting to note that the Pennsylvanian black shale unconformably overlying the ore district, has an anomalous Zn and Pb content.

Upper Mississippi Valley district

This district covers about 6,800 km^2 (Fig. 20-18) on each side of the upper Mississippi River valley in Illinois, Wisconsin and Iowa (centres: Dubuque, Platteville; Heyl et al., 1959). It is located on the south-western limb of the broad Wisconsin Arch, in upper Cambrian to Silurian sediments gently dipping to S.S.W. and warped into low broad undulations. Several thousand Zn, Zn-Pb and minor Cu, barite, Fe, occurrences are hosted by rocks in the entire sedimentary section but the bulk of Zn-Pb production came from an interval of middle Ordovician carbonates about 120 m thick, located about 500 m above the crystalline basement. "Basinal" biomicrite and fine dolomite rich in chert nodule prevail, unconformably topped by a gray to black shale.

The ore consists of a typical MVT assemblage of sphalerite (often metacolloform Schallenblende) galena, pyrite, marcasite, coarse calcite, dolomite, and local chalcopyrite and barite. Dolomitization and silicification (jasperoid) are the dominant alterations. The ores are vein fillings, cavity fillings in solution breccias, and disseminations and replacements. A visitor expecting to see orderly conformable orebodies will be disappointed because the numerous small ore deposits are mineralized reverse faults ("pitches"; Fig. 20-16/8), bedding planes and bedding faults ("flats"), joints (Fig. 20-16/9) and breccias. Within the rather broad mineralized stratigraphic interval, the immediate ore control is structural and most orebodies formed around the ends of synclines and anticlines. It appears that a substantial proportion of the faults, joints and breccias is the result of gravity tectonics, caused by solution thinning of carbonate units and collapse. Carbonate beds or thin units can be represented entirely by layers of their insoluble residue, typically light-green clay to claystone. This material is often sulphide-impregnated. Several major tensional faults detected geophysically in the basement coincide with, and control, the gentle folds and some of the faults in the Paleozoic sediments.

Central Tennessee district (Elmwood)

Elmwood (Kyle, 1976) represents the most recent major discovery of a MVT field in the United States. There, reddish-brown sphalerite associated with bornite, fluorite, calcite and galena is disseminated, and cements peneconcordant breccias, in a lower Ordovician limestone-dolomite sequence. There are several generations of breccia, the earliest having formed during a period of emergence under the mid-Ordovician unconformity.

Pine Point, Northwest Territories, Canada

Pine Point district (Skall, 1975; min. 3.15 Mt Zn/7%, 1.2 Mt Pb/2.6%; Fig. 20-19), is located on the southern shore of the Great Slave Lake far from major population centres, but it has recently been in production and has been extensively studied. There, over fourty discrete, mineralogically simple, crystalline and metacolloform sphalerite, galena, marcasite and calcite orebodies occur in an

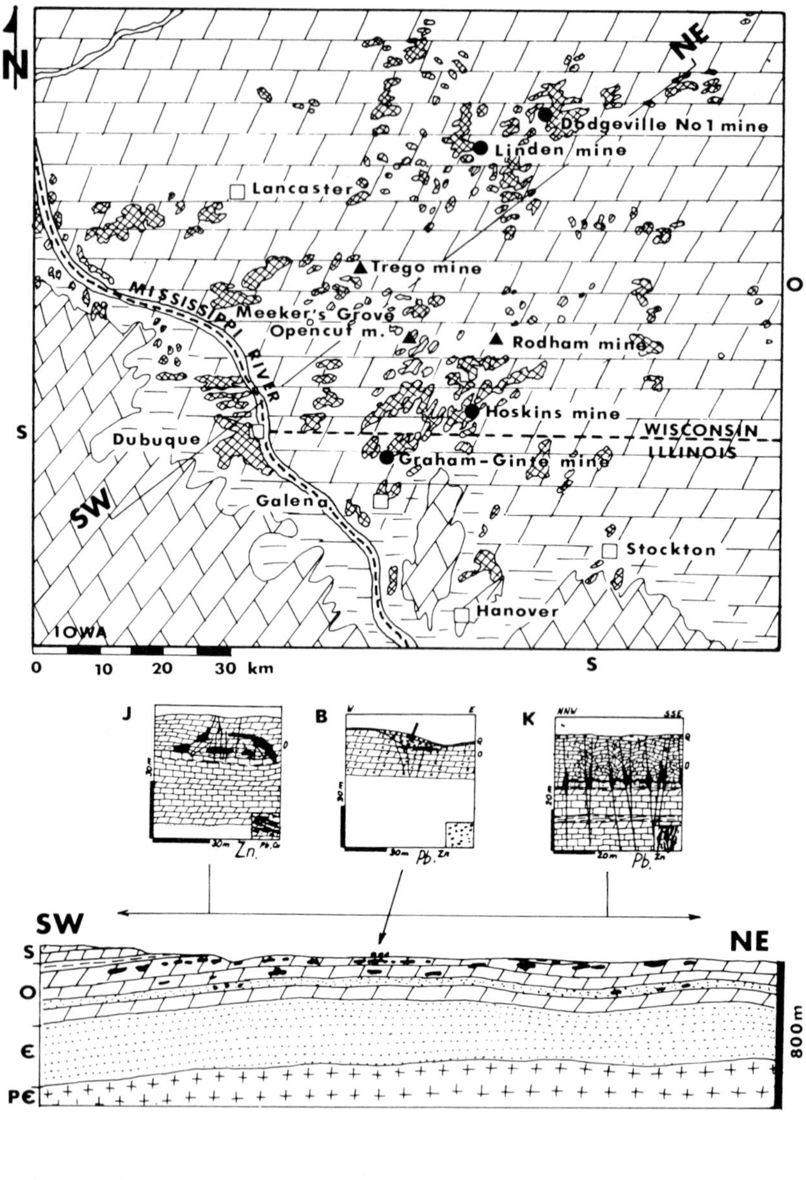

Fig. 20-18. Upper Mississippi Valley district, U.S.A. See legend following Fig. 20-15. From Laznicka (1981a).

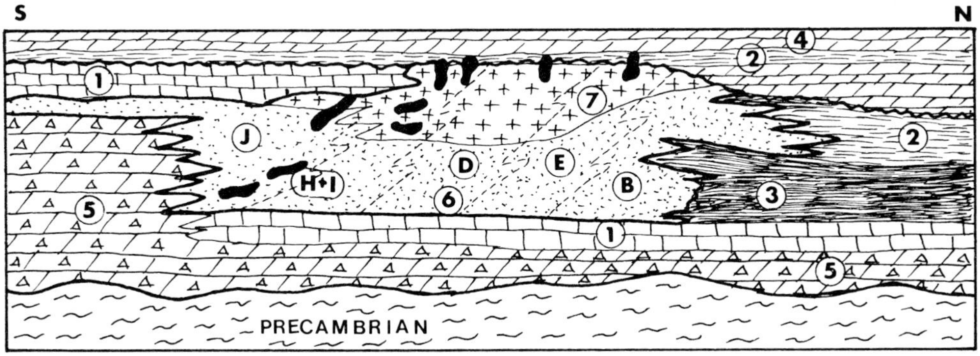

Fig. 20-19. Pine Point district, N.W.T., Canada. An example of a complex MVT mineralization in dolomitized limestones at, and below, an unconformity. Schematic section from LITHOTHEQUE, after Irvine (1972) and COMINCO mine staff, guided tour (1975). The length of section represents about 20 km. All rocks (except the Precambrian basement) are Devonian (Eiffelian to Givetian): (1) limestone; (2) shale; (3) black bituminous shale; (4) dolomite; (5) evaporitic dolomite with gypsum; (6) metasomatic, sandy dolomite; (7) coarse crystalline, porous, metasomatic dolomite (Presquille Facies); BLACK: projected Zn-Pb orebodies (not to scale). Interpreted paleoenvironments: (J) southern "reef" flank; (H+I) backreef to lagoon; (D) barrier reef; (E) forereef; (B) offreef (basin).

E.N.E.-trending belt some 100 km long and 50 km wide. The belt parallels a major lineament in the Precambrian basement present at a depth of several hundred metres beneath the middle Devonian ore-hosting sediments. The orebodies may be relatively extensive, but low-grade, peneconcordant-subhorizontal, tabular masses of metasomatic dolomite and dolomite-cemented breccia with ore minerals filling pores and vugs. Alternatively, they may be vertical "prisms" of a collapse carbonate breccia cemented by sulphide masses.

The ore bearing sequence is about 300 m thick and is characterized by a rapid facies change from an offshore shale and limestone through a "reef" (a former carbonate mud mound deposit with abundant organic remnants but lacking a continuous growth framework), to carbonate and evaporite. The original rocks have been intensively metasomatically dolomitized and converted into a central mass of a coarse crystalline vuggy dolomite associated with most (but not all) orebodies. In addition, there are numerous peneconcordant dolomite "pseudobreccia" bodies, but these are not mineralized in Pine Point.

The very characteristic environmental and facies setting of Pine Point contributed, more than any other deposit, to the development of the Jackson and Beales (1967) model, in which Zn and Pb derived from metal-rich basinal shales (up to 0.13% Zn found in the Pine Point basinal shales; Macqueen et al., 1975) met H_2S-rich hypersaline and petroliferous brines derived from the evaporitic facies within the locally most porous member (the metasomatic dolomite), causing mineralization. The role of the basement lineament, if any, has not been addressed in this hypothesis.

Arctic Archipelago, Canada

A major Zn-Pb district has been discovered in the past twenty years in the inaccessible Arctic desert of northern Canada and Polaris (the world's most northerly mine at 75° 25' N) has recently been brought into production (Kerr, 1977). About fourteen Zn-Pb occurrences on Cornwallis, Little Cornwallis, Truro, Bathurst and other islands are located in solution-collapse and possibly tectonic breccias hosted by middle Ordovician to Silurian carbonates and shales close to the Devonian unconformity. The sediments belong to a basinal facies, and crop out within the "Cornwallis Fold Belt", an intracratonic steep-sided anticlinorium of Proterozoic to Devonian sediments that overlie a horst of Precambrian crystalline basement.

The single most important Polaris deposit is in middle-upper Ordovician dolomitized biomicritic limestone, close to its contact with an overlying petroliferous basinal shale. Metacolloform and crystalline sphalerite occurs in vugs, fractures and cavities in quantities ranging from sparse disseminations to high-grade masses. Galena is rarer and coarse crystalline dolomite is the main gangue material. Liquid petroleum as well as pyrobitumen globules are common in the ore zone. The orebodies are irregular-to-peneconcordant breccia bodies and fault control is important.

North American Platform, remaining occurrences

It is estimated that at least 5,000 additional Pb-Zn or barite, fluorite occurrences, most of them minor, exist in the Platform, outside the large districts already reviewed. Two medium-sized localities (Daniel's Harbour and Gays River) are listed in Table 20-5; the remaining occurrences are too numerous to tabulate.

CENTRAL AND WESTERN EUROPE

Low-temperature zinc and lead deposits in carbonates were probably discovered and mined in Europe before the discovery of America (e.g. near Wiesloch, the Cévennes, South Pennines), but were not recognized as a special class distinct from the granite aureole-controlled deposits until recently. This is not surprising because in contrast to the North American Platform MVT localities, where the hiatus between the widespread intrusive activity at the time of the crystalline basement formation and platformic sediments deposition is of the order of 800 to 2,000 million years, in central and western Europe it is anywhere between zero and about 50 m.y. (This does not include the truly "platformic" Paleozoic carbonate occurrences resting on the Precambrian Baltic Shield e.g. on the Gotland Island, that are fully comparable with the North American Platform).

All the European MVT localities are in flat-lying sediments of very young platforms, formed shortly after stabilization of the Hercynian and Caledonian orogenic belts, or even during the latest stages of the belt consolidation (topping the molasse facies). Consequently the "platformic" sedimentogenesis and metallogenesis overlapped with the "orogenic" one, including the depth emplacement of granitic rocks. In addition to this, continental fragmentation ("rifting") regimes replaced the mountain-building (orogenic) ones almost immediately after the consolidation of the Caledonian and Hercynian orogens (e.g. Ziegler, 1978). The metallogenetic consequence is one of considerable complexity and also the existence of a transitional series of carbonate-hosted ore deposits ranging from granite-coeval hydrothermal veins and replacements (Chapter 28) through granite-unrelated (?) veins and replacements, to deposits resembling in many respects the American MVT. MVT may be present in independent and rather uniform ore districts, free of other major mineralization styles in the basement (e.g. the Upper Silesia region), or they may just be one of a variety of mineralization styles that are present locally. The European MVT districts and fields are characterized briefly in Table 20-5; most are small- to medium-sized, with the exception of the Kraków-Silesia district in Poland which is of global importance.

Kraków-Silesia (Upper Silesia) district, Poland

This district, located in southern Polands is economically the most important MVT region of the world due to its approximate content of 36 Mt Zn and 10 Mt Pb. It has been reviewed recently by Sass-Gustkiewicz et al.(1982), but there is an extensive literature available, most of it in Polish. The district is situated largely along the eastern and northeastern margin of the Carboniferous Upper Silesia Coal Basin, north and east of Katowice and north-west of Kraków. Its area is about 1,500 km^2 but concentrated mineralization is confined to five ore fields, the largest (Olkusz) with dimensions of about 15 x 10 km. The ores are hosted entirely by a unit of flat-lying Triassic carbonates approximately 200 m thick, resting in the south on paralic coal—bearing sediments and in the north on rocks of the Caledonian orogenic belt intruded by late Paleozoic plutons (Fig. 20-20).

In the Triassic shallow-water carbonate sequence, a better sorted, "clean" high-energy central bioclastic facies is enveloped by micritic and marly carbonates, located stratigraphically above and below. The central facies has been extensively converted into peneconcordant bodies of metasomatic dolomite, reminescent of the Presquille Dolomite at Pine Point. The bulk of the Zn-Pb orebodies are hosted by this dolomite. The dolomite contains irregularly scattered ore minerals throughout practically its entire extent and represents an enormous low- grade Zn-Pb resource. Typical orebodies are peneconcordant to irregular, tabular to nest-like accumulations of sphalerite (both metacolloform Schallenblende and the macrocrystalline variety), lesser galena, wurtzite and marcasite in concentrations greater than about 3% Zn+Pb. Three styles are usually distinguished: (1) replacement disseminations to masses in dolomite; (2) scattered crystals, nests

Table 20-5. MVT ("Mississippi Valley - type") Zn-Pb deposits in shallow marine carbonates of platforms: selected example localities

LOCALITY	AGE	HOST UNIT, MINERALIZATION	TONNAGE, GRADE	REFERENCES
S.E.Missouri dis., 100 km S.E. of St.Louis, USA	Cm_3	several thousands of orebodies over 50,000 km^2; subhorizontal gal//py,sf,cp,siegenite dissem.,nests,mineraliz.breccias,forming peneconc.blankets,runs. "Reef"facies lightly dolom.limest.resting on arkosic sandst. and Pt granite - rhyolite basement	20 Mt Pb 60 Tt Ni 12 Tt Co	Snyder and Gerdemann (1968);Thacker and Anderson (1977)
Barite - Lead Belt (Washington Co.), S.E.Missouri	Cm_3	replacem.masses,veins,nests of barite and minor gal,sf in micritic dolom.and argill. dolom.,mined entirely from residual soil	11.5 Mt barite 50 Tt Pb 20 Tt Zn	Buckley (1909)
Tri-State distr., Kansas,Oklahoma, Missouri,USA	Ms	sf//gal,py scattered crystals and metacolloform coating,vug filling,replac.in peneconc.sheets,runs,solution collapse brecc. in flat-lying dolom.and silic.biomicrites; topped by Ps Pb-rich coal shale	11 Mt Zn 2.65 Mt Pb	McKnight and Fischer (1970) Hagni (1976)
Upper Mississippi R.Valley distr., USA	Or_2	sf/gal,py,marc.,vein and breccia filling, impregn.,replac.along reverse and bedding faults,in gently warped basinal lim.and dolom.,500 m above Precambr.basement	1.6 Mt Zn 750 Tt Pb 2 Tt Cu	Heyl et al., (1959)
Centr.Tennessee d. (Elmwood),USA	Or_1	reddish sf,barite,fluor.,gal,calcite,scattered in peneconc.solution collapse brecc. under Or_2 unconf.,in fine cryst.dolom. and coarse dolomitiz.limest.;peneconcord.oreb.	3.2 Mt Zn/4.3%	Kyle (1976)
Pine Point,Great Slave Lake region N.W.Canada	D_2	cryst.and metacol.sfa/gal,py,marc.,open space fill.in peneconc.sheets and collapse brecc.in coarse dolomitiz."reef" limest.fac.	3.15 Mt Zn/7% 1.2 Mt Pb/2.6%	Skall (1975)

Gays River, Nova Scotia, Canada	Ms	sf/gal/marc,cp,fluor,bar.,mass. to dissem., vug—filling, peneconc.in "reef" limestone assoc.with anhydr.,gypsum,above contin.red and white clastics	490 Tt Zn/4.5% 272 Tt Pb/2.5%	MacLeod (1975)
Daniel's Harbour, N.W.Newfoundland	Or_1	sf//py,gal,dissem.and vug coating in pene-conc. dolomitiz.solution pseudobreccia under unconf.;basinal and intertidal lim., shale	416 Tt Zn/7.7%	Cumming (1968)
Polaris m.,Little Cornwallis Island, Arctic Canada	Or_{2-3}	sf/gal, open space fill.of solution and tect.brecc.control.by faults and under D un-conf. Dissem.to mass.ore; basinal biomicri-te overl.by shales and lim.,locally metas. dolomitized	3.6 Mt Zn/14.1% 1.1 Mt Pb/4.3%	Kerr (1977)
Mendip Hills, S.W.England	Cb_1-Tr_1	gal,sf,calc,barite,fissure veins and pene-conc.bodies along unconf.in basal dolom. congl.,limest.resting on S volc. and D red beds basement	160 Tt Pb	Ford (1976)
South Pennines, N.C.England	Cb_1 Pe_3	gal/barite,sf,fluorite miner.along bedding planes,miner.solution brecc.,faults; vari-ety of limest.facies,widespread contemp. basalt lavas and tuffs	506 Tt Pb 44 Tt Zn 2.63 Mt fluorite	Ford (1976)
Cévennes Border, S.E.France; e.g. Les Malines,Tréves	Tr,J	sf/gal replacem.and open space filling in J,Tr,Cm dolomitiz.limest.under many uncon-formities; veins,miner.solution collapse breccias, paleotalus	Les Malines 334 Tt Zn	Routhier (1963)
Kraków-Silesia (Upper Silesia) distr.,S.Poland	Tr_2	collom.sphalerite,wurtzite,gal,marc.,diss. to mass.,breccia cement,coating,scattering; peneconc.,discord.bodies in dolomitiz.lim.	36 Mt Zn 10 Mt Pb	Sass-Gustkie-wicz et al. (1982)

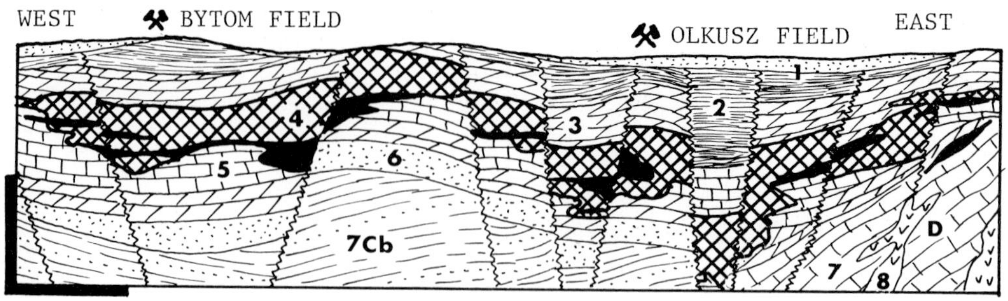

Fig. 20-20. Silesia-Kraków Zn-Pb district, Poland; schematic section. (1) Quaternary cover; (2) Tr_3 shale; (3) Tr_2 early diagenetic dolomite; (4) metasomatic "Ore Dolomite"; (5) limestone; (6) Tr_1 varicoloured sandstone, arkose, conglomerate, minor evaporite; (7) pre-Triassic basement: Cb, D_2; (8) late Paleozoic diabase and porphyry. Modified after Osika (1969).

and masses coating rock fragments or filling open spaces, in depth-generated solution collapse breccias; and (3) fragments, loose sulphide grains, stalagmites, etc. in loose fill (mostly granular, disaggregated dolomite) in the near-surficial karst sinks. Rich bodies of smithsonite and hemimorphite together with goethite were mined in the past from the oxidation zone. Sass-Gustkiewicz et al. (1982) argued that the mineralization is hydrothermal-epigenetic, contemporary with karsting ("hydrothermal karst") and derived from extrabasinal hot ore-bearing solutions that entered the Triassic carbonate aquifer.

THE REST OF THE WORLD

MVT equivalents are known in all the continents except Antarctica but in most cases they are small to medium-sized deposits comparable in setting and in the general style of mineralization to the examples treated earlier. The more important examples are summarized briefly in Table 20-5.

20.8.2. Deposits of other metals and some commodities equivalent in style to the "classical MVT"

FLUORITE

Fluorite is a minor constituent of several MVT deposits but it can also form independent peneconcordant carbonate-hosted orebodies.

Several large deposits are to be found in the Jurassic platformic limestones along the western margin of Morvan (north-eastern outlier of the Massif Central, France; Soulé de Lafont, 1967; about 3.5 Mt CaF_2). The fluorite is often colourless and so very inconspicuous.

BARITE

Barite free of Pb-Zn sulphides or with only accesory amounts of them is extremely common in platformic carbonates as nodules, fracture and fault fillings, veins, bedded replacements, breccia cement and fill in solution-collapse breccias. It is usually white, beige to light pink, and very inconspicuous when not crystallized.

IRON

Anomalous accumulations of iron sulphides (pyrite and marcasite) that normally accompany Zn and Pb sulphides in MVT deposits, may result in independent carbonate-hosted pyrite or marcasite orebodies. None has been of economic importance.

Several examples of industrial iron oxide deposits hosted by platformic carbonates form a heterogeneous, poorly understood group. In the Furness field (Cumberland, England; 37 Mt Fe/46%; Dunham et al., 1978), massive hematite forms inverted cone-shaped deposits up to 350 m wide and 180 m deep in Carboniferous limestones under the cover of Triassic sandstone. Quartz is the principal gangue mineral and there are small amounts of fluorite, barite and specularite. The most likely process of origin is attributed to 80-115° C warm hypersaline brines, travelling updip, leaching iron from the Permo-Triassic red sandstone, and depositing it chemically in the karsted limestones.

In the Bahariya iron ore deposit, western Egypt (Said, 1962; 68 Mt Fe/52.4%), peneconcordant goethite, minor hematite orebodies replace middle Eocene platformic limestone. The mineralization is interpreted as being "diagenetic replacement", or as being a replacement due to hydrothermal solutions, driven by the heat of plateau basalt volcanism.

20.9. TRANSITIONAL CARBONATE, DISTAL VOLCANISM, FAULTING, ASSOCIATION: Zn,Pb,Cu ASPECT

CENTRAL IRISH PLAIN MINERALIZED PROVINCE

The Central Irish Plain (Fig. 20-21), a broad, low-lying area encompassing about two-thirds of Ireland, coincides with an important province of polymetallic deposits all of which are hosted by lower Carboniferous carbonates (Morissey et al., 1971). The Province is now known to contain about 10.3 Mt Zn, 3.4 Mt Pb, 160 Tt Cu, 1,500 t Ag and 300 t Hg. It is transitional in its geotectonic setting (flat-lying to folded shallow-water carbonates); in its association of accumulated metals (significant Cu and Ag in addition to Pb-Zn) and

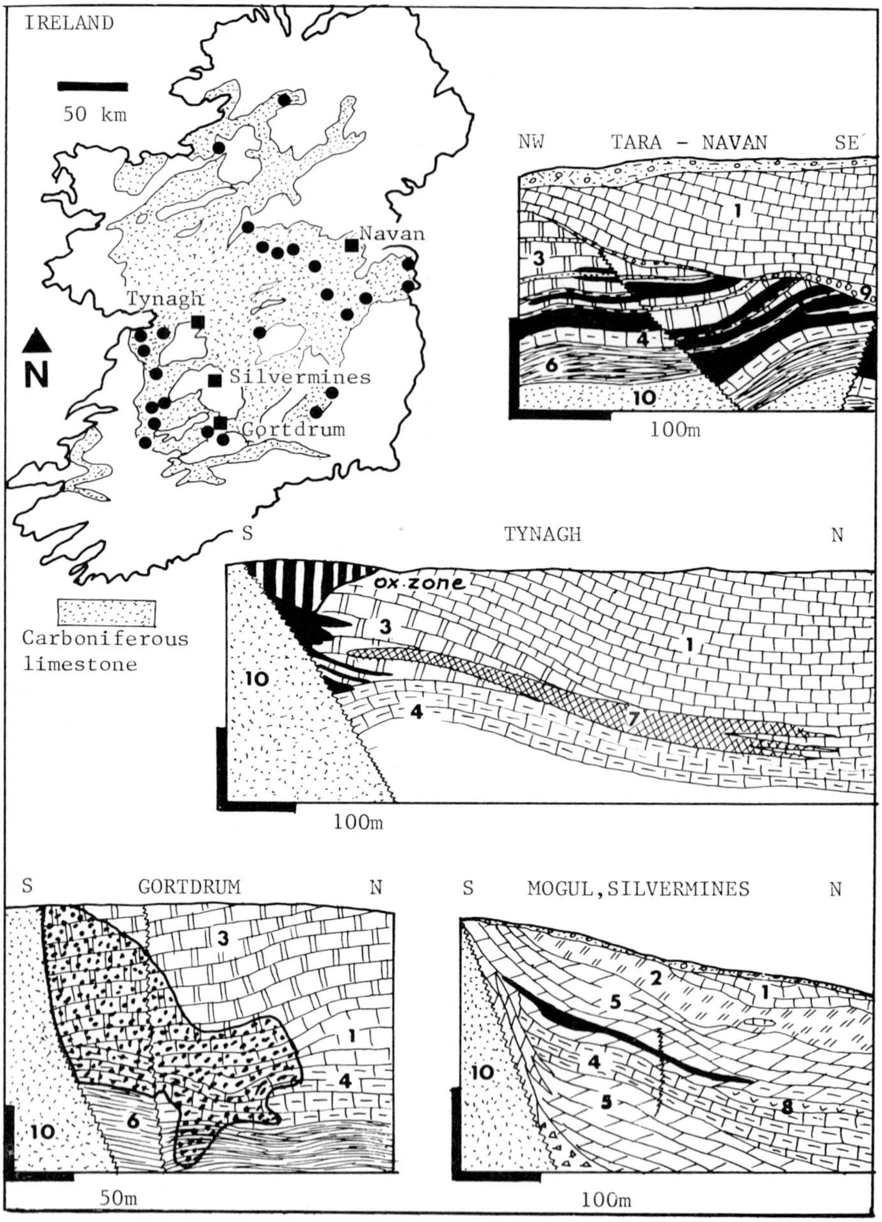

Fig. 20-21. The Irish Pb-Zn Province. LEGEND: Cb_1 rocks: (1) dark compact limestone; (2) chert; (3) bank limestone; (4) transitional, shaly (muddy) limestone; (5) dolomite, dolom. limestone; (6) siltstone, shale; (7) iron formation; (8) tuff; (9) conglomerate. Devonian Old Red Sandstone (10). After Hoppe (1977), Russell (1975), and Evans (1976).

in the presumed ore genesis (subaqueous hydrothermal metal supply along growth faults), into associations treated in this chapter as well as in Chapters 17, 19 and 25. There is a considerable degree of transition even within the province itself, and Morissey et al.(1971) have pointed out that "no two of the known deposits are identical, so that the search for new orebodies must take into account the probability that they will resemble one or more of the known deposits in certain respects, but differ from all the known deposits in others".

Briefly, the host Carboniferous sediments consist of sandy and shaly units near the base, grading upward into a thick sequence of dark, foliated limestone. These, in turn, interfinger with light limestones of the organogenic buildup ("reef") facies overlain by, and laterally transitional into, a dark compact limestone. The sequence is up to 1,000 m thick. The Carboniferous rocks rest conformably-to-unconformably on unmetamorphosed upper Devonian red-beds interpreted as being an intermontane "molasse" fill within the Caledonian orogenic belt. The sequence is terminated in the south by the northern edge of the Hercynian (late Paleozoic) orogenic belt, where it changes facies into shale and graywacke. Sporadic volcanicity synchronous with the carbonate sedimentation occurred at several places in central Ireland and trachyte, trachybasalt to olivine basalt lavas, pyroclastics and small intrusions, were produced.

The majority of ore deposits (one, Tara-Navan, of global importance; four significant deposits; several tens of small deposits to occurrences) are spatially associated with the carbonate buildup (Waulsortian) facies, but additional occurrences are also located in the underlying sediments (Schultz, 1971). The dominant control of ores is structural; all the major deposits with the exception of Navan, are located in the immediate vicinity of east and E.N.E.-striking faults. The orebodies tend to extend into and along the bedding of the carbonates adjacent to the faults, often without showing evidence of a typical post-lithification replacement. This is usually interpreted as due to synsedimentary faulting ("growth faults") concurrent with the influx of hydrotherms that were partly discharged on the seafloor, or injected into sediments undergoing lithification. The subsidiary control is stratigraphic or lithologic as the presence of local sulphide, barite and hematite-rich chemical (? volcanichemical) sediments suggests.

Tynagh (Fig. 20-21), the first important deposit discovered (Boast et al., 1981; Russell, 1975; 664 Tt Pb, 612 Tt Zn, 53 Tt Cu, 641 t Ag) is hosted by a lower Carboniferous dolomitized "mud bank" massive limestone along an east-west fault with a slip dip of at least 640 m, which brought the continental red-bed sandstone against the limestone. The epigenetic Pb,Zn,Cu orebody has a fine-grained pyrite, sphalerite, galena and minor chalcopyrite and tennantite as replacement and cavity fillings in a barite gangue, or directly in the host carbonate. A substantial secondary deposit consisting of a residual host rock rubble and mud mixed with residual (barite) and secondary minerals, marked the outcrop of the orebody. A conspicuous feature of Tynagh is the presence of a stratiform composite lens of finely laminated to massive iron-rich sediments (hematite, chert, stilpnomelane, chlorite, magnetite) interbedded with limestones, and apparently in facies transition of the "mud bank" facies. This "iron formation" is enriched in Zn,Pb and Mn and interpreted as being hydrothermal ("exhalative") sediment.

The Silvermines field (Larter et al., 1981; about 1.35 Mt Zn, 500 Tt Pb, 400 t Ag; Fig. 20-21) consisting of eight separate deposits or zones, is similar in setting to Tynagh. Most of the orebodies lie at the interface of a dolomitic breccia and the "mudbank" limestone and argillaceous limestone 200 m above the base of the Devonian red sandstone, but three zones lie in the sandstone. The largest "G" orebody (9.33 Mt 2.4% Pb, 9.2% Zn, 21 ppm Ag) is composed of fine-grained massive pyrite, sphalerite and galena. It is 4-26 m thick, associated with black shale, and interpreted as being synsedimentary on the evidence of abundant framboids, globules, slump structures, etc. In the Ballynoe deposit, a stratiform barite lens (2.5 Mt ore with 85% $BaSO_4$) is gradational into a pyritic Zn-Pb lens, underlain by a siderite horizon. Larter et al. (1981) demonstrated the presence of pyritic hydrothermal feeder channel in the mining field.

Gortdrum deposit (Schultz, 1971; 50 Tt Cu/1.2%, 88 t Ag/21 ppm, 300 t Hg; Fig. 20-21), is basically a copper mine. Disseminated bornite, tennantite-tetrahedrite and chalcopyrite are located along fractures and as rock replacements in partly dolomitized Carboniferous limestones and shales in a hangingwall of an E.N.E.-W.S.W. trending, 70° N. dipping fault. Red sandstones and shales are found in the footwall.

Navan (Tara), 50 km north-west of Dublin (Hoppe, 1977; 7.37 Mt Zn/11%, 1.61 Mt Pb/2.4%; Fig. 20-21), is the largest Irish deposit and the one that is closest in appearance to the "classical MVT". There, epigenetic sphalerite and galena with minor marcasite, pyrite, barite and calcite form disseminations, stringers, breccia cement and finely banded masses in lower Carboniferous massive dolomitized limestone. Argillaceous limestone, shale and sandstone are part of the unit, resting on Devonian red sandstone and conglomerate.

20.10. STRONGLY FAULTED SEGMENTS OF PLATFORMIC CARBONATES: VEIN AND REPLACEMENT FLUORITE, BARITE AND Zn-Pb SULPHIDES

ILLINOIS-KENTUCKY DISTRICT, U.S.A.

This district is a region of concentrated mineralization, covering about 1,400 km^2 in southern Illinois and western Kentucky (main centres: Rosiclare, Cave in Rock, Salem, Marion; Fig. 20-22). It is the principal producer of fluorite in the United States (Grogan and Bradbury, 1968; Pinckney, 1976; 9.5 Mt CaF_2, 122 Tt Zn, 54 Tt Pb). The district is underlain by over 4,000 m of Cambrian to Mississippian

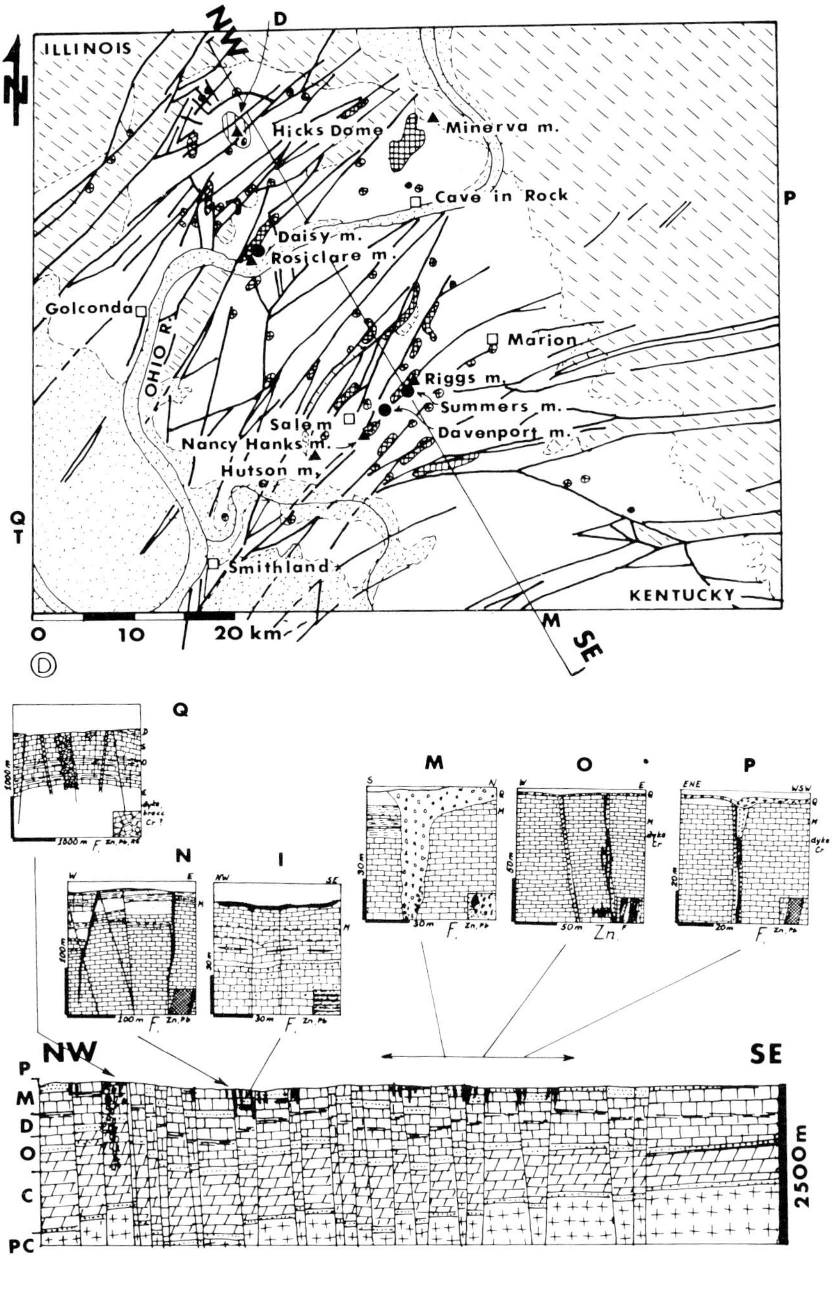

Fig. 20-22. Illinois-Kentucky fluorite district. See legend following Fig. 20-15. From Laznicka (1981a).

sediments of epeiric seas (carbonate, sandstone, shale) partly topped by Pennsylvanian paralic coal-bearing association. The sedimentation was followed by non-deposition, uplift and a crustal extension that culminated in Mesozoic. As a result, the region is now a broad, N.W.-trending arch, centered around the Hicks Dome (a diatreme), and a system of horsts and grabens that cross the arch and trend N.E. to east. A dense system of normal faults parallels the grabens, and a ring of short, radial and concentric faults surrounds the Hicks Dome. Vertical movement on the graben-bounding faults is about 200 m, and the maximum displacement is less than 900 m.

The mineralization identified so far is distinctly epigenetic (Fig. 20-23), and has been attributed by Brecke (1979) to heated connate fluids mobilized by pressure and heat associated with the upper Cretaceous igneous activity in the adjacent Mississippi Embayment. Because most orebodies occur within a stratigraphic interval approximately 100 m thick, the district is sometimes listed as an example of stratiform mineralization. The chief minerals are fluorite, quartz, calcite, lesser sphalerite, galena, minor barite, strontianite and chalcopyrite. Several hundred deposits and occurrences are known. These can be subdivided into four principal styles:

(1) Veins. Most veins are open-space fillings that occupy faults, chiefly the N.E. and E.N.E. horst and graben-bounding faults. The largest veins have been mined in Rosiclare (1,740 m along strike, 275 m downdip, 4 m wide; Fig. 20-23), but the majority of veins are several hundred metres long and around 1 m wide. There is little wallrock alteration. When the wallrock is carbonate, it is usually partly replaced.

(2) Bedded replacements. These orebodies take the form of peneconcordant lenses in limestones, or along limestone-sandstone contacts. Each subhorizontal orebody overlies one or more mineralized feeder fissures, and mineralogical zoning is common. Most known examples are in the Cave in Rock field (Fig. 20-23).

(3) Mineralized solution-slump breccias in carbonates. This is a large volume, low-grade mineralization style, associated with bedded replacements. The ore minerals fill the clay and sandstone breccia matrix or partly replace limestone blocks along the edges.

(4) Mineralized diatremes. Fluorite-mineralized diatreme breccias encountered within the Hicks Dome contained minor sphalerite and galena, as well as rare earths and thorium minerals (monazite and florencite).

PENNINE DISTRICT, CENTRAL ENGLAND (incorporates Derbyshire orefield)

The Pennines are a low north-south trending range of hills in northern England that contains widespread fluorite, barite, Pb-Zn mineralization. In the literature, the North Pennines (Alston and Askrigg blocks; Ineson, 1976) and South Pennines (Derbyshire field; Ford, 1976) are usually distinguished. The latter field has already

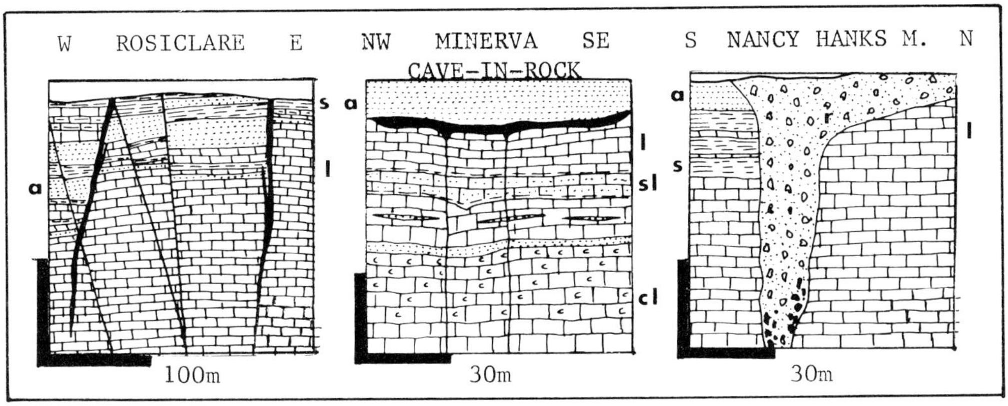

Fig. 20-23. Principal mineralization styles in the Kentucky-Illinois fluorite district, hosted by Mississippian sediments. (a) quartzite, quartz arenite; (s) shale; (l) limestone; (sl) argillaceous limestone; (cl) limestone with chert nodules; (r) rubble. From Laznicka (1981a), with minor modifications.

been mentioned briefly in the previous section.

The North Pennine orefield (20 Mt fluorite, 4 Mt Pb) is a fault block with an area of about 1,600 km^2. Numerous steep epigenetic fluorite, barite, galena, sphalerite fissure veins are hosted by a Carboniferous sedimentary unit about 900 m thick, composed of dominant marine limestone with minor shale, sandstone and paralic coal, resting on a Devonian granite basement. The mineralization and structural style are extremely similar to the Illinois-Kentucky district, except for the abundant diabase sills and dikes in the North Pennine field. Peneconcordant replacement orebodies in the form of "flats", "wings" or "mantos" occur in ankeritized limestone adjacent to feeder veins and fissures, particularly at sites where limestones are covered by an impermeable shale screen.

In the South Pennines (Derbyshire) the mineralization is similar, but peneconcordant orebodies in limestones showing many characteristics of the "classical MVT" are represented. The orebodies in the Golconda mine, Brassington, described by Ford and King (1965), are probably the most instructive. There, fluorite, barite, minor galena and calcite, occur in subhorizontal mineralized zones hosted by Carboniferous limestone, under a former emergence surface. The minerals line abundant solution cavities, line or cement voids in breccias, and fill "rakes" and "flats".

INTERPRETATION

Illinois-Kentucky and Pennine districts are just two examples of epigenetic fluorite-dominated mineralizations hosted by platformic sediments (dominantly carbonates in the above examples), but ultimately related to crystalline basement doming and extensional faulting (initial "rifting"). Similar (but not identical) localities of carbonate-hosted epigenetic fluorite include the north-eastern fringe of the Massif Central, France (Morvan), the Asturia Fluorite Belt in N.W. Spain (Förster, 1974) and other localities. Comparable fluorite mineralization is even more common at many crystalline basement localities devoid of cover sediments, where the ores exclusively form fissure fillings along high-level breccia, mylonite or gouge-filled faults (e.g. Schwarzwald, Nahe region, Erzgebirge, etc., in Germany; Chapter 28).

There appears to be a link between some of the latter ores in the crystalline basement and the "classical MVT" ores in the thick, platformic sedimentary cover. Such a connection is not apparent in the MVT region of the central United States primarilly because (with the exception of the S.E. Missouri district) such a basement is located at a considerable depth. It is, however, apparent elsewhere, for example at the fringe of the Schwarzwald. We are probably dealing with a composite mineralization system. Sawkins (1966) impressed by the compositional zoning of the North Pennine orebodies (fluorite and Pb-Zn in the core, barite on fringe) suggested, on the basis of fluid inclusions, fluorite and base metal sources to have been in "depth-derived" (hydrothermal, magmatogene) solutions. Barite, on the other hand, precipitated most likely from connate, sediment-derived brines moving updip, at sites of mixing with the former fluids.

20.9. SHALLOW-MARINE CARBONATE, MINOR SHALE, ARENITE, ETC. ASSOCIATION OF MOBILE BELTS (FOLDED, THRUSTED, FAULTED)

20.9.1. Zn,Pb (F,Ba) aspect : APT deposits

Alpine or Appalachian-"type" (APT) Zn-Pb deposits of mobile belts (Fig. 20-12) are generally considered to be the equivalent of the MVT of platforms, so that the basic difference between both "types" is that the first one is hosted by folded carbonates, the other by unfolded ones. As in other cases, however, there is an almost uninterrupted range of transitions. Alpine or Appalachian-"type" deposits have, indeed, many features identical with the MVT (mostly their mineralogical composition) but there are some important differences as well, mostly in the structural control. One aspect seems to be particularly important in interpreting and trying to understand the APT: a demonstration of whether the mineralization is pre-deformational (as, for example, the APT in the Mascot-Jefferson City district, Tennessee), or post-deformational (as many of the small occurrences in the Canadian Cordillera, discussed later). Each genetic category has different controls.

While the majority of the classical MVT is hosted by sediments formed in epicontinental seas, the majority of APT occurs in rocks of the nearshore facies of the complex facies progression developed across the Atlantic-type continental margins (Chapter 5), that also include the deeper-water sediments of the continental slope (Chapter 17). The facies transition between the shelf and slope sediments is particularly fertile metallogenically.

EASTERN CANADIAN CORDILLERA (ROCKY MOUNTAINS BELT)

Although not immortalized in an ore "type" name, the N.E. Cordillera (Douglas, ed., 1970), is a better example of a host to APT than the Alps or the Appalachians. It is a relatively orderly, extremely well exposed, 3,000 km long, N.N.W.-trending, "one-sided" orogen, thrust faulted in places but free of major nappes. A sequence of unmetamorphosed middle Proterozoic to middle Mesozoic limestones and dolomites and subordinate clastic sediments deposited on a "carbonate platform" (shelf and nearshore region) are exposed mostly in a Foreland Thrust and Fold Belt. The sequence is several thousand meters thick. The belt flanks a metamorphosed crystalline core in the west (Omineca Crystalline Belt) and merges with, or is thrusted over, the flat-lying sediments of the North American Platform in the east (Fig. 20-24). The direction of tectonic transport is consistently eastward.

In approximately the western third of the East Cordilleran Paleozoic sedimentary belt as shown on geological map, the shallow-water "shelf" facies changes into a fine clastic and chert, often "black", oceanward succession treated in Chapter 17. The transition is remarkably sharp, contained within a single stratigraphic unit (within some 5 km; the change is often visible in the walls of a single east-west valley), but because of deformation and shifts in the facies boundary, it now appears as a transitional zone several tens of kilometres wide. Dark (black) carbonates interfingering with shales are a particularly common rock type along the facies transition boundary, and so is a variety of breccias and Zn-Pb occurrences (Fig. 20-25).

About 500 Pb-Zn occurrences hosted by the shallow-water carbonates have been recorded by now in the N.E. Cordilleran belt. With the exception of the Gayna River deposit (2.5 Mt Pb+Zn) most occurrences are small, and only a few "camps" (districts) are cumulatively of "medium" size. Only two deposits (Monarch-Kicking Horse and Cadillac) have been mined or are ready for production. The total metal content produced and contained in reserves is of the order of 2.5 Mt Zn and 1.5 Mt Pb.

Macqueen (1976) and Macqueen and Thompson (1978) have recently reviewed some of the Zn-Pb deposits in the Cordilleran carbonates, and their controls. Briefly, the bulk of occurrences are clearly epigenetic, emplaced after lithification of their host carbonates and

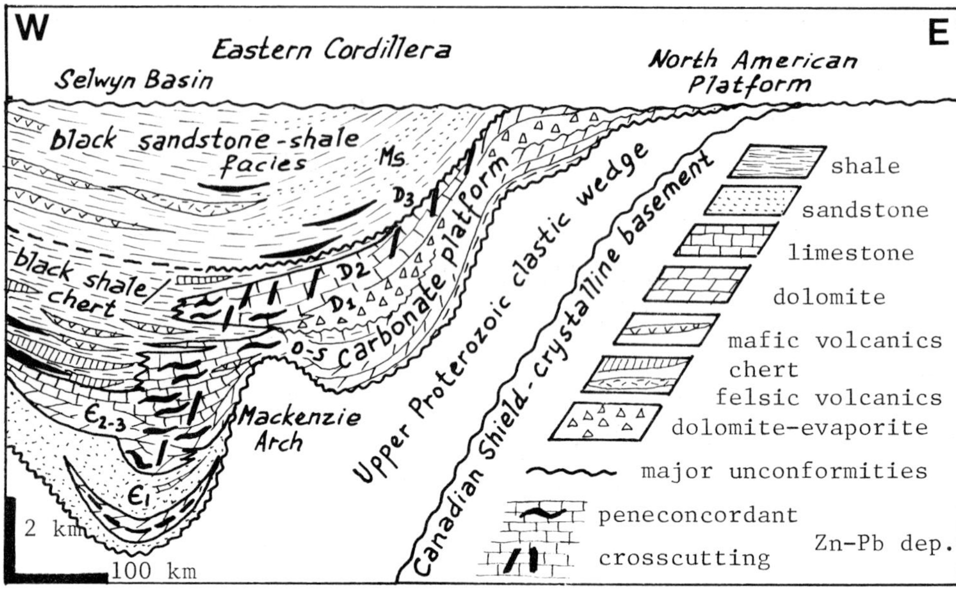

Fig. 20-24. Restored stratigraphic section across the north-eastern Cordillera, Canada, showing the projected position of the APT Zn-Pb deposits in the carbonate platform facies.

usually designated as "stratabound" in conference conversations. This implies confinement to a certain named lithostratigraphic unit but within such a unit the actual ore occurrences range from peneconcordant to discordant (the latter are more widespread). The principal characteristics of the epigenetic occurrences as summarized by Macqueen (1976) include: (1) simple mineralogy: sphalerite, galena, pyrite, marcasite, dolomite, calcite, rarely barite; (2) dominance of sphalerite over galena (10:1 or greater); (3) common presence of bitumen or its derivatives; (4) high porosity and permeability of the host rocks, preferentially dolomites. Further

Fig. 20-25. Examples of carbonate-hosted Zn-Pb APT mineralizations in the eastern (Rocky Mts.) belt of the Canadian Cordillera. The sections are diagrammatic, not accurate in detail. LEGEND: (1) limestone, argillaceous limestone; (2) dolomite, argillaceous dolomite; (3) shale; (4) quartz arenite, quartzite; (5) brecciated rock, mineralization. Post-lithification (alteration) dolomitization and/or silicification are not shown. From LITHOTHEQUE, (H) section only modified after Ney (1954).

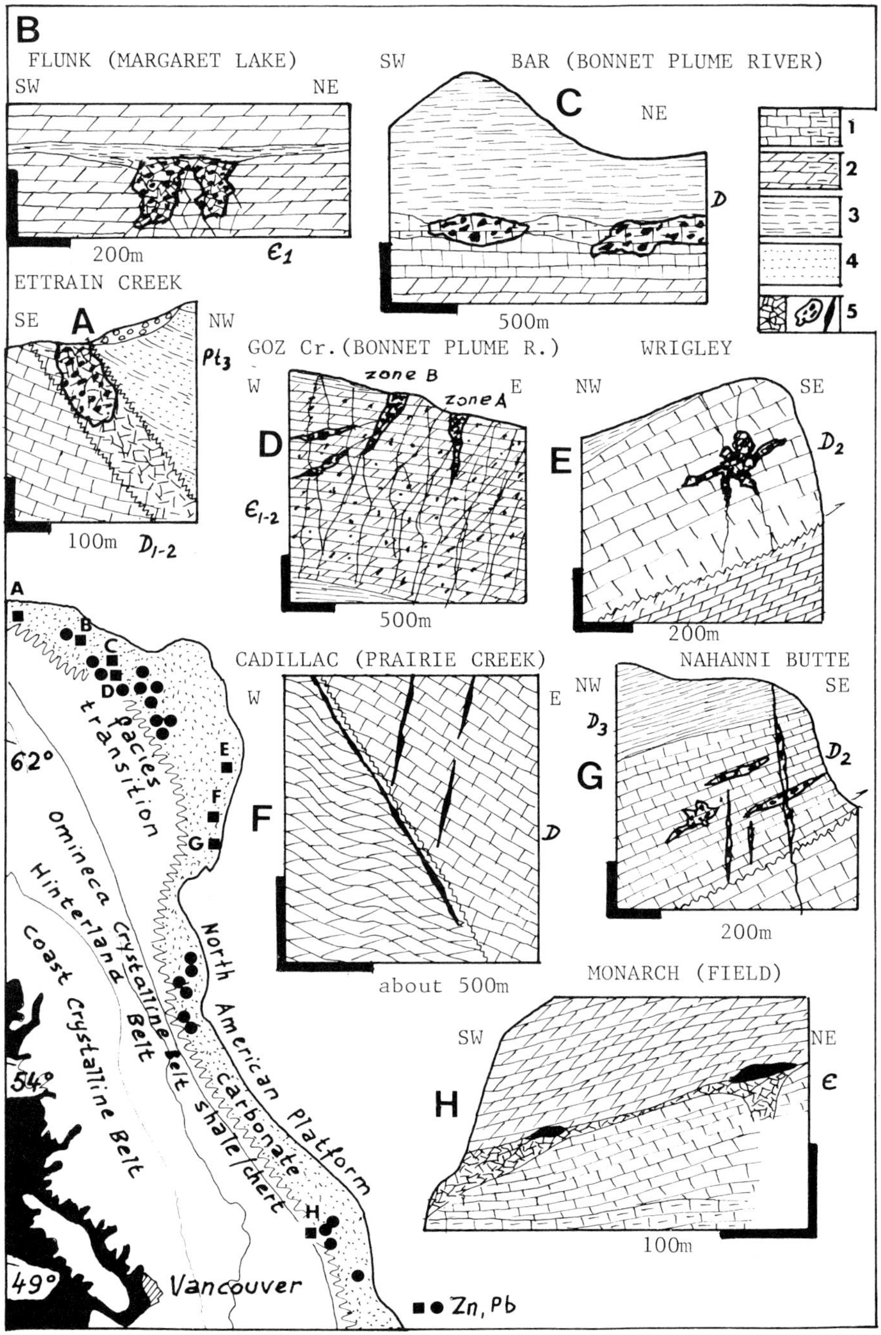
B
FLUNK (MARGARET LAKE)
SW
NE
200m
SW
BAR (BONNET PLUME RIVER)
C
NE
D
500m
1
2
3
4
5
ETTRAIN CREEK
SE
A
NW
100m
GOZ Cr. (BONNET PLUME R.)
W
zone B
zone A
E
D
500m
WRIGLEY
NW
SE
E
D2
200m
A
B
C
D
facies transition
E
F
G
62°
Omineca Crystalline Belt
Hinterland Belt
Coast Crystalline Belt
54°
49°
Vancouver
North American Platform
carbonate
shale/chert
H
Zn, Pb
CADILLAC (PRAIRIE CREEK)
W
E
F
D
about 500m
NAHANNI BUTTE
NW
SE
D3
G
D2
200m
MONARCH (FIELD)
SW
NE
H
100m

points made are (5) open space filling of pre-mineralization cavities by the ore minerals is dominant. (6) Most orebodies "stratabound" on large scale, are cross-cutting on an outcrop or hand-specimen scale. (7) Some deposits (e.g. Robb Lake) are close to the platform carbonate - basinal shale facies change, but others are well within the "platform". (8) Igneous rocks are virtually unknown and the host sediments are unmetamorphosed. (9) The degree of immediate structural control is highly variable. Some occurrences are not obviously associated with faults, while others are intimately fault-bound.

Both pre-deformational and post-deformational occurrences are known. The largest deposits seem to be largely pre-deformational (although subsequently modified), and are often compared with the Pine Point MVT deposit. The majority of the small cross-cutting occurrences seem to be post-deformational. The small-to medium-sized but high-grade Cadillac (Prairie Creek) deposit is a system of lenticular sphalerite, galena, lesser tetrahedrite and chalcopyrite orebodies in a quartz-carbonate gangue, parallel with a shear.

Truly stratiform (syngenetic-diagenetic) Pb-Zn deposits are extremely rare in the platform carbonates; e.g. several localities in the lower Cambrian Sekwi Formation, Mackenzie Mountains. These are hosted by basal carbonates deposited in hypersaline environment, overlying quartz arenite units and somewhat reminiscent of the style described in Section 20.6.1.

The Rocky Mountain Zn-Pb occurrences are distributed throughout the entire stratigraphic section, without consistent confinement to "time-bound" horizons (the lower Cambrian interval is richest). The genetic interpretaion is still imperfect due to the lack of data, but Macqueen (1976) favours the "oilman's" hypothesis of 70-160° C hot oil-field brines transporting Zn and Pb derived from the black basinal shales (these are Zn-anomalous in many places of the Cordillera), and precipitated in suitable traps on mixing with H_2S-rich fluids.

VALLEY AND RIDGE PROVINCE, THE APPALACHIAN BELT

Valley and Ridge Province is a north-easterly-trending thrust and fold belt about 2,000 km long but less than 80 km wide, located along the north-western margin of the Appalachian orogenic belt. The Province contains a sequence several kilometres thick, of Cambrian to Mississippian sediments comparable in age, lithology and facies distribution to the Eastern Cordilleran belt. Volcanics are exceptional (minor mafic lavas occur in the lower Cambrian units; several horizons of volcanic ash are located in the middle Ordovician rocks), and intrusive rocks are absent. Several hundreds of APT Zn-Pb occurrences have been recorded, most of them in three medium-tonnage districts.

The Appalachian Zn-Pb deposits have been reviewed recently by Hoagland (1976), with particular emphasis on ore genesis. Hoagland's major thesis is that mineralization in most of the economically important deposits is pre-orogenic and closely comparable to the contemporary MVT deposits in the adjacent North American Platform domain. There is, in particular, a high degree of correspondence between the APT Mascot-Jefferson City district, and the MVT East Tennessee district, both hosted by equivalent lower Ordovician carbonates. Regional unconformities marking paleosurfaces of emergence are particularly strong guides to the presence of ores. The middle Ordovician unconformity and extensive paleokarst phenomena related to it have been observed from Newfoundland to Alabama and the ores in eastern and central Tennessee and at Friedensville (Pennsylvania) are in dissolution breccias as much as 240 m below the unconformity,

Mascot-Jefferson City district 30 km N.E. of Knoxville (Crawford and Hoagland, 1968; min. 2 Mt Zn/4%) is the most productive and best-known example of APT deposits and also a model area of paleoaquifer metallogeny (this subject has been treated in a special issue of Economic Geology v. 66, No. 5, 1971). There, low-grade sphalerite and lesser pyrite are hosted by a 66 m thick section of lower Ordovician limestones and dolomites, under a middle Ordovician unconformity. A pale yellow, inconspicuous sphalerite is scattered in the matrix of dolomitized breccia in a variety of stratigraphically confined solution-collapse breccia bodies. Minor mineralized mantos are lateral projections from breccia bodies containing open space sphalerite fillings and replacements in coarse crystalline dolomite. The ores formed during the relatively short interval of emergence between the lower and middle Ordovician periods in two stages of dissolution and dolomitization (the earlier was barren, the second ore-forming). They are buried by unmineralized middle Ordovician carbonates. The entire sequence was later folded and thrust faulted.

Copper Ridge Zn(Pb) belt is located about 50 km north-east of the Mascot-Jefferson City district, at the same stratigraphic level, and is partly a fault offset continuation of the former. In the largest Flat Gap mine (Hill et al., 1971), two distinct styles of ore have been recognized: (1) older, reddish-brown sphalerite, galena ore; (2) pale-yellow sphalerite ore. The former is confined to a single faulted eastward-plunging narrow shoot at least 2,400 m long. A net-like pattern of numerous yellow sphalerite shoots radiates from the "dark" sphalerite ore. The remaining important Appalachian APT localities: Friedensville (Pennsylvania) and Austinville-Ivanhoe (Virginia) have been summarized by Hoagland (1976), and are listed in Table 20-6. The Friedensville ore is spectacularly plastically deformed in places.

THE ALPS

Carbonate-hosted Zn-Pb occurrences and small deposits (total of about 500) are scattered throughout the great Alpine arc in eastern France (Menglon, La Plagne), Switzerland (Binnenthal), the Bavaria-Austria border region (Northern Limestone Alps Zn-Pb belt, e.g. Lafatsch) and the Italy, Austria, Yugoslavia border region

Table 20-6. APT ("Appalachian-Alpine type") Zn-Pb deposits in shallow marine carbonates of orogenic belts: selected example localities

LOCALITY	AGE	HOST UNIT, MINERALIZATION	TONNAGE, GRADE	REFERENCES
Goz Creek, Bonnet Plume Valley, Yukon, Canada	Cm_1	extensive low-grade yellow sphalerite mineralization over 130 m strat.thickess of folded dolomitiz.limest.interb.with slate and qtz.sandst. Ore-grade sphalerite, minor py, boulang., is in silicif.breccia of vuggy dol.	202 Tt Zn/ 13.5%	Sinclair et al. (1976)
Robb Lake, N.E. Cordillera, British Columbia, Canada	D_{1-2}	peneconc.to crosscutting sf/gal, py, scattered void and vug-filling breccia oreb.in dolomitiz.micritic limest.near facies change into basinal shale	445 Tt Pb+Zn/ 7.3%	Macqueen and Thompson (1978)
Monarch-Kicking Horse near Field, British Columbia	Cm_2	scattered sf/gal, marcas.as void filling, several peneconc.bodies within 70 m interval in dolomitiz.limest.near shale facies change	84 Tt Zn/10% 30 t Ag/34 ppm 59 Tt Pb/7%	Ney (1954)
Friedensville, Pennsylvania, USA	Or_1	sphalerite, minor pyrite, repl.masses, veinlets, vug-fillings, in a 30 m thick breccia of dolomitiz.limest.under unconf.Signif.hemimorphite, smithsonite, goethite gossan	226 Tt Zn/6.5%	Hoagland (1976) Callahan (1968)
Austinville-Ivanhoe Virginia, USA	Cm_1	replac.and fract.fill.sf, gal, py in a breccia of dolom.limest.formed across a thrust-faulted anticline in laminated limest.interfingering with pyritic black shale	1 Mt Zn/3.7% 200 Tt Pb/0.7%	Brown and Weinberg (1968)
Mascot-Jefferson City distr., Tennessee	Or_1	yellow sf, minor py, scattered crystals in peneconc.and crosscutt.bodies of dolomitiz.limest.breccia, stratigr.confined under an Or_2 unconformity	min.2 Mt Zn/4%	Hoagland (1976) Crawford and Hoagland (1968)

Copper Ridge, E. Tennessee	Or_1	like Mascot-Jefferson City, but red and yellow sphalerite,pyrite,minor galena	min.600 Tt Zn 50 Tt Pb	Hill et al. (1971)
Reocin m.,Torrelavega,N.Spain	Cr_1	sf,py,gal,scattered and replac.peneconcord. lenses / masses in brecc.dolomitized limestone in oolithic and "reef" limest.,marl, shale association	1.67 Mt Zn/14% 298 Tt Pb/2.5%	Monseur (1962)
Aachen-Moresnet d., West Germany-Belgium	Cb_1	sf (metacollof.and cryst.),py,gal,open space filling in solution collapse breccia, veins; in dolomitiz."reef" and basinal lim., interb.with black argillite. Significant hemimorphite,smithsonite,goethite gossan	600 Tt Zn 180 Tt Pb	Beyschlag et al.(1916)
Salafossa m.,near Sappada,N. Italian Alps	Tr_2	sf,calcite/ga,py,fill voids and cement a solution collapse breccia up to 100 m thick, in dolomitiz.limest.under Tr_2 unconformity. $Pe+Tr_1$ sandy shale,limest.,red beds are in the basement	570 Tt Zn/6% 86 Tt Pb/0.9%	Lagny (1975)
Bleiberg-Kreuth, S.Austrian Alps	Tr_2	sf,gal,py,marcas.,scattered in peneconc.bodies in porous metasomat.dolom. interb.with shale; rhythm.banding,stromatolites,residual marls,peneconc.breccias; also fracture ore veins; oxidic ore in gossan	2.1 Mt Zn/5.37% 400 Tt Pb/1.2% 8.4 Tt Cd	Brigo et al. (1977) Holzer and Stumpfl (1980)
Cave di Predil (Raibl),Iulian Alps N.E.Italy	Tr_2	veins and columns of mass.to dissem.sf/gal, marcas.,py,in solution collapse and tect. breccias along N.-S. fault zones; in dolom. and dolomitiz.limest.,topped by black argill.limest.to limey argillite	505 Tt Zn/14% 76 Tt Pb/2%	Brigo et al. (1977)

Table 20-6 (continued). APT ("Appalachian-Alpine type") Zn-Pb deposits in shallow marine carbonates of orogenic belts: selected example localities

LOCALITY	AGE	HOST UNIT, MINERALIZATION	TONNAGE, GRADE	REFERENCES
Mežica, Karawanke Mts, N. Yugoslavia	Tr_2	sf, gal/py, marc, fluorite, dissem., void filling, masses in peneconc. and fault-controlled brecc. dolomitiz. limest. and in black resedim. breccias; gossan; "reef" facies	min. 310 Tt Zn/3.25% 240 Tt Pb	Brigo et al. (1977)
Sedmochislenitsi, Vracha distr., W. Bulgaria	Tr	massive to dissem. sf, gal, py, in dolomitized limest. breccia, peneconc. lenses at eight stratigr. horiz.; assoc. limest., marl, shale	930 t Ag/65 ppm 230 Tt Pb/1.41% 185 Tt Zn/1.11%	Iovchev (1961)
Achisay, Karatau Range, S. Kazakhstan USSR	D_2-Cb_1	py/sf, gal, scattered and masses, in fault-controlled mineraliz. solution collapse breccia in dolomitized limest. resting on continental red-beds	930 t Ag/65 ppm 454 Tt Zn/14%	Smirnov and Gorzhevsky (1974)
Mirgalimsay, Kararau Range, S. Kazakhstan, USSR	D_3	gal, sf, py, dissem., bands and nests in ribboned dolomitiz. bitum. limestone; concordant zone of (solution ?) breccia resting on tuffac. shale and D_2 red-beds	839 Tt Pb 504 Tt Zn	Smirnov and Gorzhevsky (1974)
Touissit-Bou Beker S.E. of Oudja, N.E. Morocco	J	gal, sf, py fill dissolution vugs, dissem., veinlets; several peneconc. lenses, runs in dolomitiz. limest., near crystall. basement	1.1 Mt Pb 950 Tt Zn	Bouladon (1952)
N.W. Tunisia Pb-Zn province	Tr-Mi	gal, sf, py $\pm$ barite, fluorite void filling, replac., veins in dolomitiz. limest. along faults and near diapirs of gypsum-shale brec.	960 Tt Pb 910 Tt Zn	Sainfeld (1956)
Shah Kuh Range, S. of Isfahan, Iran	J_1 Cr	hemimorphite, smithsonite, hydrozinckite, cerussite after sulphides in solution cavities along faults in limest., shales, marls	127 Tt Zn/25% 51 Tt Pb/10%	Geol. Surv. Iran (1964)
Kanchanaburi Prov. W. Thailand	Or-J	smiths., hemimorph., gal, sf, py, peneconc. lenses, in shallow marine massive "reef" limest. surrounded by dark limest., shale, slate	Song Tho dep.: 191 Tt Pb/8.7% 84 Tt Zn/3.8%	Diehl and Kern (1981)

(Southern Limestone Alps Zn-Pb region with the major deposits Salafossa, Cave di Predil, Bleiberg, Mežica). In terms of economic importance, the latter region, an east-west trending belt about 200 km long and 30 km wide, accounts for at least 80% of the Alpine zinc-lead production and reserves (about 3.5 Mt Zn and 1 Mt Pb) and has been most thoroughly studied. It will be reviewed briefly here. Some of the remaining mineralized areas are listed in Table 20-6.

In the Southern Limestone Alps belt (Fig. 20-26), the Zn-Pb ore fields are located in several mountain ranges (Gailtaler Alpen, Karawanken, Julian Alps), on both sides of an important W.N.W.-trending lineament (Periadriatic line). The mineralized belt is located south of the axial zone of exposed pre-Mesozoic crystalline basement, and is "time"- and strata- bound within an about 1,000 m thick middle-upper Triassic sequence. Within this, an interval of a massive Ladinian limestone (Wettersteinkalk) about 200 m thick under the base of the overlying Carnian Raibl Beds (thinly bedded argillaceous limestone and black argillite) is most consistently and intensively mineralized. The Wettersteinkalk was deposited on a widespread carbonate platform (broad lagoon ?), bound by a barrier "reef". Gray-to-black shales and marls with thin limestone lenses (Partnach Beds) formed contemporaneously in an off-reef basin (Schneider, 1964; compare Fig. 20-2).

The period of deposition of the mineralized carbonates coincided with local basaltic to andesitic volcanism, but no direct association of the Zn-Pb ores and volcanics has been demonstrated. Schneider (1964) pointed out a close temporal relation between the climax of Triassic "geosynclinal" volcanism, and the maximum supply of ore matter. The mineralogy of the APT deposits is, as is usual, simple. Sphalerite, both macrocrystalline and metacolloform, is five to ten times as abundant as galena. Pyrite is widespread. Calcite, dolomite, occasional fluorite, quartz, barite and celestite are the gangue minerals. The ores are consistently associated with a mappable "special facies" introduced by Schneider (1964), which is quite distinct if compared with the monotonous rest of the Wettersteinkalk. The facies formed in the back-reef (lagoonal) environment, was modified by dissolution during periods of emergence, and consists of rhythmically banded dolomite, black breccias, abundant stromatolites, oolithic limestones, green marls, etc. The 5-15 cm thick black breccia layers consistently maintain a fixed stratigraphic position 10, 25 and 50 m below the base of the Raibl Beds, so they serve as an important marker (Brigo et al., 1977).

The dominant styles of mineralization are peneconcordant (stratabound) sheets, lenses and "runs", located at stratigraphically well-defined horizons. Sometimes these are cyclic (4 cycles in the Bleiberg-Kreuth field). Within each orebody, the ore minerals are disseminated in dolomitized limestone, filling or coating pores. Rhythmically banded sulphides ("Bodenerz") are common. Discordant ore-bearing fissure-filling and replacement veins or irregular bodies are less common, although in places they have a significant vertical span (up to 600 m at Mežica). The fault-controlled mineralization in

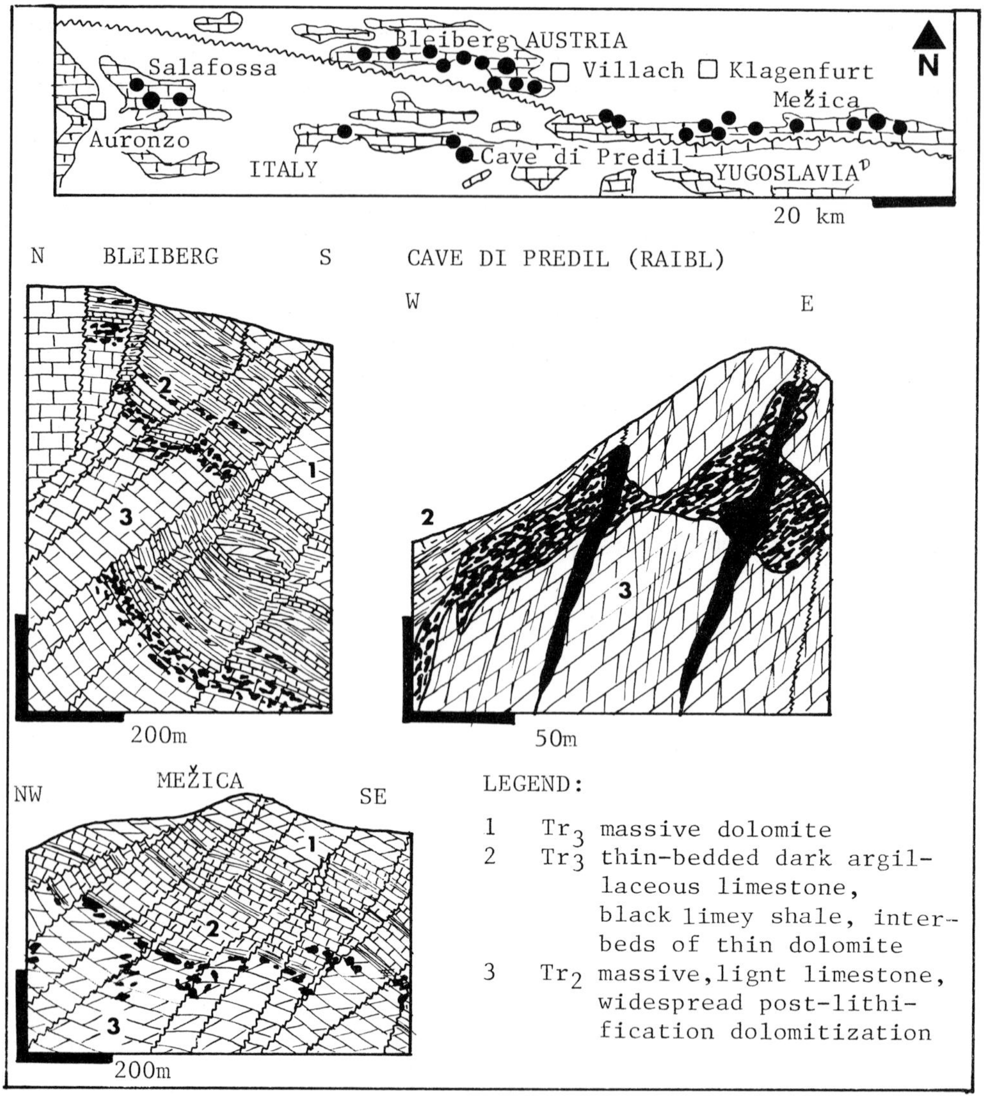

Fig. 20-26. Principal Zn-Pb deposits of the APT in the southern Limestone Alps. From LITHOTHEQUE, modified after Höller (1953), di Colbertaldo (1950) and Štrucl (1966).

Cave de Predil (Fig. 20-26) is considered to be syndepositional along a middle Triassic ?growth fault. There is a slight thermal aureole surrounding a Triassic porphyry intrusion in the ore field, but demonstrably post-lithification ores or mineral remobilization along faults are very widespread.

The Salafossa orebody is columnar in shape and elongated parallel to a N.N.E.-striking tectonic contact of the mid-Triassic host dolomite and the pre-Triassic basement (Brigo et al., 1977). There, the mineralization is attributed to karsting below a middle-upper Triassic unconformity.

A wide range of genetic hypotheses has been advanced to account for the origin of the Alpine Zn-Pb ores: the surficial, intra-karstic fluid-mixing and pore-filling hypothesis (Bechstädt, 1975) has gained ground recently. The widespread, but economically unimportant scattered galena, sphalerite with some tetrahedrite and chalcopyrite, found in the Permian Bellerophon Formation and the Tregiovo Beds in the Southern Alps (e.g. near Trento, Italy; Brusca et al., 1972), are different from the Triassic-hosted ores, discussed above. The host sequence is a gypsum, argillite, marl, dolomite and evaporite breccia unit grading upward into bituminous thin-bedded limestone and resting on Permian continental rhyolitic ignimbrites. The ore minerals appear to be largely stratiform, diagenetic disseminated grains or small concretions.

ATLAS OROGENIC BELT OF NORTH AFRICA

In the Atlas ranges of North Africa, several hundred Zn-Pb occurrences hosted by Mesozoic and Cainozoic carbonates are located in a E.N.E.-trending belt, 1,800 km long, between Marakech and the coast of Tunisia. In the important Touissit-Bou Beker field near Oujda (N.E. Morocco; Bouladon, 1952) galena, sphalerite and pyrite fill dissolution vugs and form disseminations and veinlets, in several peneconcordant lenses and runs in dolomitized Jurassic limestone. Secondary Zn and Pb minerals in the gossan are prominently developed.

The most dense accumulation of ore occurrences is in the north-western Tunisia Zn-Pb province (Sainfeld, 1956; Nicolini, 1970). This province with approximate dimensions of 200x200 km is in the strongly deformed region of northern Tunisia and follows the north-easterly structural grain. Of the several hundreds of ore occurrences, probably only about six fields contain in excess of 100,000 t Pb and Zn and the total Pb+Zn tonnage is of the order of 1 Mt Pb and 300 Tt Zn. Individual deposits occur in most structural zones between the Kabylian crystalline massif in the north and the African craton in the south, and are hosted by sediments ranging from Triassic to Miocene (with the maxima in Cretaceous and Miocene). A variety of ore styles is present, ranging from "stratiform" (more correctly peneconcordant) bodies through mineralized paleokarst systems to cross-cutting fissure veins, replacements and impregnations in tectonic breccias. The mineralogy is the typical MVT/APT suite but the majority of the sulphide deposits has been substantially to completely oxidized, so that the ore outcrops are now largely composed of a mixture of hemimorphite, smithsonite and hydrozinckite (calamine) with minor cerussite and remnants of galena.

Diapirism and tectonically controlled injections of the Triassic "basement" into the younger sediments along the north-eastern faults is peculiar to the Tunisian Pb-Zn region, and one of the frequently stressed empirical controls of ore presence. About two-thirds of the Pb-Zn occurrences are near (within 1 to several km) outcrops of the Triassic terrains, and all deposits (Nicolini, 1970) are located along axes of diapirism. Although they are not directly hosted by the Triassic assemblage itself, they appear to be controlled largely by long-lasting "living faults" (failles vivantes or growth faults).

The Triassic assemblage typically consists of varicoloured shale-gypsum breccia, containing blocks of dolomite and sandstone, and the adjacent pierced carbonates have often been converted into a friction breccia. The latter is a common host to Pb-Zn "contact deposits", particularly if "reef" limestones are involved. Dimitrov and Mankov (1973) observed that the proportion of conformable deposits increases with decreasing geological age of the host unit, so that most of the "stratiform" (peneconcordant) deposits are of Miocene age. Such deposits are not confined to the carbonate host alone but also occur in the basal terrigenous suite of the continental Miocene in coarse conglomerates and sandstones (e.g. Sidi-Dris). Some deposits combine a variety of styles: for example, in Umm et Tmim, fissure ore vein cutting Eocene limestones changes into a peneconcordant lens parallel to the basal unconformable contact of the Miocene conglomerate.

As expected, genetic interpretation of the Tunisian Pb-Zn deposits is in a state of flux and is shared almost equally between the proponents of the sedimentogenic and post-magmatic hydrothermal schools. Most deposits are Miocene, and Tunisia is the youngest geologically significant APT province of the world.

KARATAU RANGE, TIAN SHAN OROGEN, SOVIET CENTRAL ASIA

The north-westerly trending Karatau Range is a spur of the extensive Tian Shan orogenic belt, and an important province of carbonate-hosted Zn-Pb deposits (estim. 2.5 Mt Zn+Pb). The mineralization there combines the "typical APT" features with those of the transitional Irish Province. As in Ireland, middle-upper Devonian red continental ("molasse") sandstones, shales and conglomerates are overlain by upper Devonian to lower Carboniferous tuffaceous shales with interbeds of dolomitic limestones. These, in turn, grade upward into a sequence of limestones and dolomites approximately 1,000 m thick. Numerous intraformational breccia intervals probably coincide with periods of emergence and karsting.

In the best-known deposit Mirgalimsay (Smirnov and Gorzhevsky, 1974), the Pb-Zn ore is confined to a 2-28 m thick concordant horizon of a Fammenian ribboned dolomitized ("sparkling") bituminous limestone, located about 200 m above the base of the carbonate sequence. Within this horizon, pyrite, galena, sphalerite with dolomite, calcite and lesser barite, ankerite and quartz gangue, are scattered or form discontinuous veinlets and ore masses. The mineralization is considered to be a polyphase one, where an originally diagenetic mineralization within a "special facies" was later upgraded contemporaneously with the post-lithification dolomitization. The latter was largely controlled by faulting and fracturing. The 50 m deep gossan contained rich anglesite and smithsonite masses.

In the nearby Achisay field, sixty four discordant orebodies 0.5-3 m thick with sharp contacts appear to be associated mainly with block faults-controlled solution collapse breccias. Massive-to-streaky pyrite laced with sphalerite and galena veinlets, constitutes the main ore.

REMAINING AREAS

Major APT-style mineralization is also known in the Cretaceous of the Cantabrian Ranges of Spain (e.g. Reocin); in the Devonian and Carboniferous carbonates in the Aachen-Moresnet district, West Germany and Belgium; in Triassic carbonates in the Carpathians and in the Balkans (e.g. Poníky, Sedmochislenitsi); in the Ordovician and Silurian carbonates of Burma and Thailand; in the latest Precambrian-lower Paleozoic carbonates of south-eastern Yakutia, Siberia (Sardana and Urui); in the Devonian region of northern Australia; in Vietnam and China and elsewhere (compare Table 20-6). Numerous occurrences of lesser importance or unknown potential are widespread worldwide.

20.9.2. APT analogues other than Zn-Pb

FLUORITE AND BARITE

As in the MVT, fluorite and/or barite that sometimes accompanies Zn-Pb sulphides in the APT deposits, may also form independent accumulations. Schneider (1964) reported the frequent presence of fluorite in rhythmically laminated bituminous layers of his "special facies", in the Northern Limestone Alps of Bavaria and Austria. Peneconcordant and vein fluorite hosted by carbonates is mined in Asturia, N.W. Spain (Forster, 1974). Unless, however, such fluorite has its characteristic violet or light green colour and particularly when it is finely grained, it is very inconspicuous and often practically unrecognizable in the field.

Barite is even more common than fluorite and takes the form of white – to – lightly – pinkish pore-filling, replacements, veins and irregular masses in carbonates. In cavities, barite forms druses of thick tabular white or colourless crystals. In areas of tropical weathering, barite is enriched in tropical soils.

Cu ASPECT

Accessory quantities of copper minerals (usually tetrahedrite, less commonly chalcopyrite) are relatively common in the APT zinc-lead deposits. Schneider (1964) pointed out that increasing tetrahedrite contents in the Eastern Alpine APT deposits indicated the direction of flow of depth-derived hydrothermal fluids. It is also possible that the presence of tetrahedrite in APT indicates a higher temperature of the ore-forming solutions (above about 140° C), and metal-derivation from outside of the carbonate unit most probably from mafic flows, tuffs or dikes. Numerous small tetrahedrite occurrences in the carbonates hosting the APT are known worldwide (e.g. in the Mackenzie Mountains, N.W. Canada). The large Hg/Ag rich tetrahedrite deposit Schwaz (Austria) hosted by dolomites, has been reviewed earlier (Chapter 17).

Chalcopyrite or bornite accumulations in shallow marine carbonates are unusual except in small quantities along the basal contacts overlying the red-beds, in units containing minor basalt flows or in carbonates interacting with black shales which, in turn, show signs of interaction with mafic magmatites. Medium-to-large Cu sulphide deposits in carbonates sometimes considered comparable in style to the lead-zinc bearing APT (e.g. the Ruby Creek deposit, Alaska; Runnels, 1969), are most likely products of copper (re)mobilization from mafic source rocks.

Fe ASPECT

Certain siderite, ankerite and goethite deposits hosted by folded or faulted limestones, in many respects resemble the APT Zn-Pb mineralizations. The main point of dissimilarity is the dominant replacement origin (rather than open space filling) of the iron minerals. The ore genesis is usually interpreted as (1) replacement of the bedrock limestone or dolomite by descending, iron-carrying meteoric waters, that derived their iron from weathering profiles or from coal-depositing basins (e.g. Clifton Forge, Berezovo), or (2) hydrothermal metasomatism by metalliferous brines ascending along faults (e.g. Rudabánya). The (1) is transitional into some of the karst-hosted iron deposits (Chapter 24). The (2) is transitional to the deposits located in intrusive aureoles (Chapter 28).

In the Berezovo deposit, East Transbaikalia, U.S.S.R. (Sokolov and Grigor'ev, 1974; 175 Mt Fe/35%), siderite and goethite cement and replace carbonate rubble and breccia as well as the solid Jurassic limestone under an unconformity, topped by detritus-filled depressions and coal-bearing lacustrine sediments. The orebodies are irregular blankets. In the Clifton Forge deposit, Virginia (Wright et al., 1968; 6 Mt Fe/43%), the "Oriskany-type" iron ore consists of goethite forming tabular orebodies in a Devonian limestone overlain by sandstone. The goethite is believed to have been formed from descending solutions resulting from pyrite undergoing oxidation in stratigraphically higher pyritic shale beds.

At Rudabánya, the principal Hungarian iron ore deposit (Morvai, 1982; min. 20 Mt Fe/24%) crystalline siderite with barite replaces Triassic limestone and dolomite along a major fault zone. The orebodies have the form of mantos, irregular masses and veins. The temperature of the mineralizing solutions in the main siderite phase has been determined as 100-150° C, which corresponds closely to the temperature of the Pb-Zn carrying solutions in the APT deposits. A later, superimposed higher—temperature hydrothermal phase (200-300° C) deposited a minor fracture-filling chalcopyrite and quartz.

U ASPECT

The small but interesting Tyuya Muyun U-V deposit (Soviet Central Asia; Kazanskii, 1970) in karsted (more correctly paleo-karsted) limestones is comparable to some APT ores. There, solution cavities and fractures in upper Devonian-lower Carboniferous limestone are encrusted by a zonally arranged suite of precipitatd minerals. The innermost mineral zone adjacent to the karsted limestone consists of a fibrous calcite, often overlapping with residual marl. The following zone contains uranium minerals dominated by tyuyamunite which, in turn, is encrusted by a coarse platey barite and several generations of calcite-hematite, barite and calcite. The minerals crystallized from 150-210° C hot solutions, probably during the Tertiary period.

The small occurrences of tyuyamunite and metatyuyamunite associated with calcite, hematite, barite, opal, fluorite, celestite and pyrite in the Pryor Mt.-Little Mountain district, Montana (Hart, 1958) are similar. They are hosted by Mississippian limestones, repeatedly karsted and affected by solution-collapse between the Pennsylvanian and Tertiary periods.

Hg and Sb ASPECT

Cinnabar and/or stibnite deposits associated with shallow-marine limestones are relatively common, but rather poorly understood. In the literature, the origin of the "primary" orebodies replacing or filling carbonates is usually attributed to coeval intrusive activity (e.g. in Yugoslavia; Janković, 1960). Another ore style is represented by occurrences of loose particles of cinnabar present in residual clays filling young karst (compare Chapter 23). It appears, however, that a close empirical and genetic equivalent of MVT/APT does exist among the carbonate-hosted Hg-Sb deposits. In the Hunan-Kweichow Hg belt, China (Huang and Chu, 1945) cinnabar and minor stibnite in quartz, calcite, dolomite and bitumen gangue forms peneconcordant "bedded veins", stringers, disseminations and drusy cavity coatings developed along two well-defined stratigraphic horizons in Cambro-Ordovician limestones and dolomites. Cross-cutting veins and veinlets are often superimposed. Berger (1977) described peneconcordant, epigenetic Hg-Sb ores hosted by middle Cambrian dolomites in the Kelyanskoe deposit in north-western Transbaikalia. There, nests of cinnabar and stibnite are associated with fluorite and hosted by thin jasperoid lenses and a breccia superimposed on paleokarst, formed in massive black dolomites. He interpreted the ore as "telethermal", formed in Permian, and composed of metals derived from the surrounding carbonates.

Additional examples of probable Hg (Sb) APT equivalents include the Dzhizhikrut deposit in the Gissar Range, Tadzhik S.S.R.; Sizma-Ladik field, Turkey; and portions of the Mt. Amiata, Italy.

20.10. INTERACTION METALLOGENY OF SHALLOW-MARINE CARBONATES

20.10.1. Carbonates in "granitic" aureoles

Several important Pb-Zn producing districts or entire metallogenic belts of the world are hosted by shallow-marine carbonates subsequently intruded by granitic plutons (Table 20-7). Characteristically, the plutonism is substantially younger than the sedimentation in the craton-floored basin, and has been preceded or accompanied by block faulting. The typical mineralization styles are sphalerite, galena, and pyrite, pyrrhotite, minor tetrahedrite, chalcopyrite, etc. masses in metacarbonates that bear the shape of sheets or lenses peneconcordant with the carbonate bedding (mantos), or discordant pipes, columns or veins. The sulphides are either accompanied by a skarn silicate association (grossularite-andradite, diopside, hedenbergite, hornblende, wollastonite, ilvaite, etc.), or are surrounded by a metasomatic chert (jasperoid); iron-enriched marble; or just a recrystallized marble. The skarn orebodies are usually located along the immediate "granite" contact. The jasperoid and marble-associated orebodies are more distant from the contact, and the existence of an intrusion is suggested by the presence of dike rocks or just by the effects of the thermal metamorphism that are often subtle.

NORTH AMERICAN CORDILLERA

It is interesting to note that many single Zn-Pb orebodies (particularly the mantos) located in granitic aureoles, almost match in setting and geometry the single orebodies present in the MVT/APT fields in which igneous intrusions are (not yet ?) known. Even the immediate alteration effects in selected deposits from both the settings described above could be virtually identical. In the Lovering (1972) memoir on jasperoids, jasperoids from the Tri State district (a MVT example) are treated side by side with the petrographically corresponding jasperoids from Tintic, Goodsprings, and similar localities (plutonic affiliation examples). Obviously, there is a link (or a convergence) between both affiliation categories.

There is another interesting observation suggesting convergence. The Paleozoic carbonate-dominated continent→epicontinental basin→shelf→ offshore basin environmental and facies belts are remarkably well preserved in the northern Cordillera (in Canada and Alaska) and have been discussed earlier. They have their counterparts in the western United States and Mexico (Fig. 20-27). The latter, however, are interrupted, dismembered and partly removed as a consequence of extensional faulting (e.g. in the Basin and Range Province); major basement uplifts (e.g. in the Colorado Rocky Mountains) and a thick Mesozoic-Cainozoic volcanic and sedimentary fill. They are, moreover, intruded by a variety of granitic plutons.

The metallogeny of the northern and southern Cordilleran transects is an exercise in contrasts. Both are Zn-Pb dominated, but the northern mineralization has, at least since the 1960s, been attributed to the "sedimentogenic" causes, in contrast to the southern mineralizations attributed solely to "granites". The complete lack of "sedimentogenic" Pb-Zn examples in the South, although possibly modified by tectonism or plutonism, is against the geological logic and is almost certainly due to the interpretational, rather than objective, causes. Only recently Callahan (1977) pointed out the striking similarity of some of the western Cordilleran "replacement mantos" with some MVT/APT deposits, complete with evidence of a pre-granite karsting, dissolution, collapse, etc.

The Tintic district in Utah represents a prototype of the "replacement mantos" granite-coeval mineralization style. Here several manto orebodies are controlled by peneconcordant horizons of dissolution and brecciation in Cambrian limestones developed under disconformities. These structures were faithfully recorded in the comprehensive sections of Lindgren and Laughlin's (1919) memoir on Tintic but the public paid little attention. Other convincing evidence for a pre-intrusive control of Pb-Zn mantos exists in the Metaline district, Washington and in the adjacent Salmo district, British Columbia; in Goodsprings, Nevada; in Gilman and Leadville, Colorado; in Santa Eulalia, Chihuahua and elsewhere (Fig. 20-28). The localities involved are summarized in Table 20-8.

Callahan (1977) concluded his paper with a maxim that "wherever MVT sites are present ore could be there", and suggested the application of his paleophysiographic premises of MVT occurrence to exploration everywhere, regardless of whether or not "granites" are represented. This removes the "granitic" mental block from the prospector's mind and gives him another direction of ore search formerly neglected: away from granite intrusions and into the sedimentary basinal sequences.

Table 20-7. The variety of Zn-Pb mineralization styles in selected districts or ore fields of interacting shallow marine carbonates and subsequent "granitic" intrusions

LOCALITY	CARBONATE ASSOCIATION	PLUTONIC ASSOCIATION	MINERALIZATION STYLE					REFERENCES
			VI	VB	SC	MT	DR	
Salmo, British Columbia	Cm limest., dolom. interbedded with black phyll.	J_3-Cr_1 granodior., qtz monzonite	+	+	+++	++	+	Douglas, ed., (1970)
Metaline, Wash., USA	comparable to above	as above		+	+	+++	+	McConnell and Anderson (1968)
Bingham, Utah USA	Ps-Pe limest., quartzite, shale; cont. metam.	Ol quartz monzonite	+++		+	++	+	Hollister (1978)
Santa Rita, New Mexico, USA	Ms, Ps limest., dolom., shale, quartz. cont. met.	Cr_3-T_1 qtz monzon.	+++		++	+	+	Hollister (1978)
Tintic, Utah, USA	Cm limest., dol., shale	T_1 qtz. monzonite	++	+	+	+++	++	Lindgren and Laughlin (1919)
Leadville, Colorado, USA	Cm-Ms limest., dolom, shale, quartzite	Ol qtz. monzonite	+		+	+++	++	Tweto (1968)
Eureka, Nevada, USA	Cm-Pe limest., dolom, shale, quartzite	Cr_3 qtz diorite, qtz porphyry	+		+	++	++	Nolan (1962)
Santa Eulalia, Chih., Mexico	Cr_1 limest., dolom., shale, sandstone	Eo-Ol qtz monzon., porphyry, rhyolite	+		++	+++	+	Gonzáles-Reyna (1956)
Morococha, Peru	Tr, J limest, dolom, shale, sandst, anhydr.	T_2 qtz monzon., granodior., diorite	+++	+	+	++	+	Bellido and de Montreuil (1972)
Kurama Range, Uzbekistan, USSR	S-Cb_1 limest, dolom, shale, sandstone	Cb_3-Pe dior., granodior., qtz. monzon.	+++		++	++	++	Smirnov ed. (1968)

Table 20-7 (continued)

Tetyukhe, Sikhote Alin, USSR	Tr_3 limest, sandstone shale	Cr-Pc dior., granod, syen. porphyry		+	+++	++	++	Smirnov and Gorzhevsky (1974)
S.W. Sardinia district, Italy	Cm-S dolom, limetone, phyllite, sandstone	Cb granodiorite, diorite	+		++	+++	+	Moore (1969)

ABBREVIATIONS, MINERALIZATION STYLE: VI=veins, stockworks, disseminations in the intrusive bodies
VB=veins, etc. in the basement of the sedimentary suite
SC=ore skarn at carbonate / intrusion contacts
MT=peneconcordant ore mantos in carbonates
DR=discordant replacement bodies in carbonates

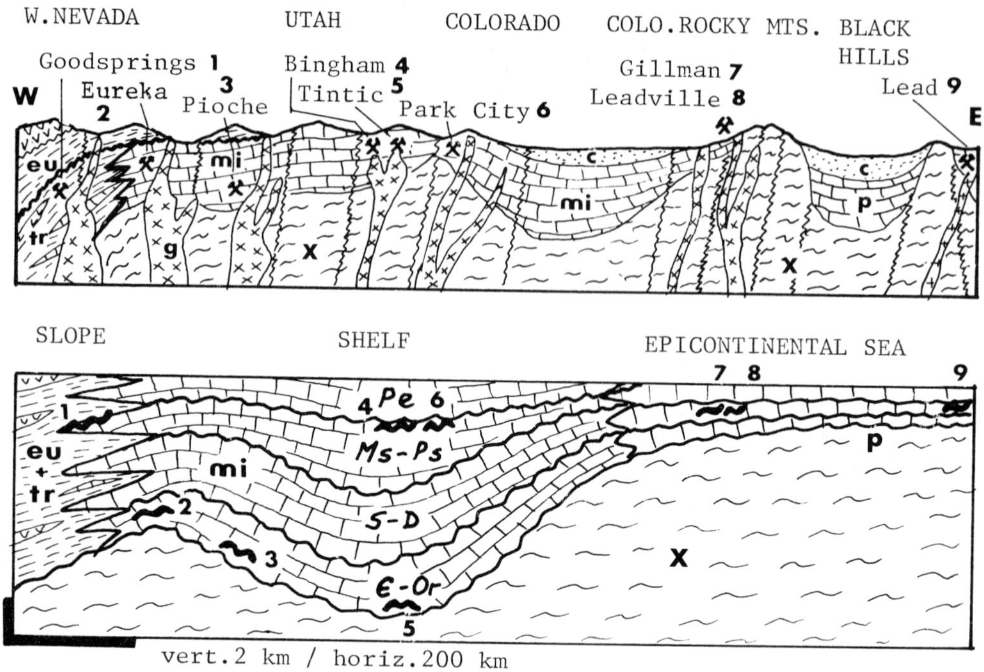

Fig. 20-27. Diagrammatic section across the western Cordillera at approximately 40° N latitude, with major deposits containing carbonate-hosted Pb-Zn mantos in intrusive aureole, projected on the line of section. Top: present situation; bottom: restored stratigraphic section (structure, intrusions removed; eroded sediments added). The Pb-Zn mantos, when plotted onto their host units (bottom), are in the "right" position, and their distribution resembles the distribution of the APT in the intrusions-deficient northern Cordillera in Canada (Fig. 20-25). LEGEND: (c) Mesozoic-Cainozoic cover rocks; (g) Mesozoic-Cainozoic plutons. Cambrian to Permian sediments: (eu) volcanic-sedimentary, deeper (slope ?) facies; (tr) transitional facies, (mi) shelf facies (carbonate platform plus detritals); (p) epicontinental (platform) facies. (X) Precambrian basement.

Reinterpretations similar to Callahan's were also made elsewhere. Kantor (1977) assumed syndepositional or diagenetic origin and later remobilization for the Pb-Zn ores in Triassic carbonates near Poníky, Slovakia. Brevart et al. (1982) used lead isotopes to determine the provenance of thirty-four Pb-Zn deposits situated along the southern margin of the Massif Central (France). A population of less radiogenic deposits there consists of (1) stratiform lenses of mostly disseminated galena and sphalerite in Cambrian carbonates, some possibly influenced by contemporary volcanism (e.g. Brusque,

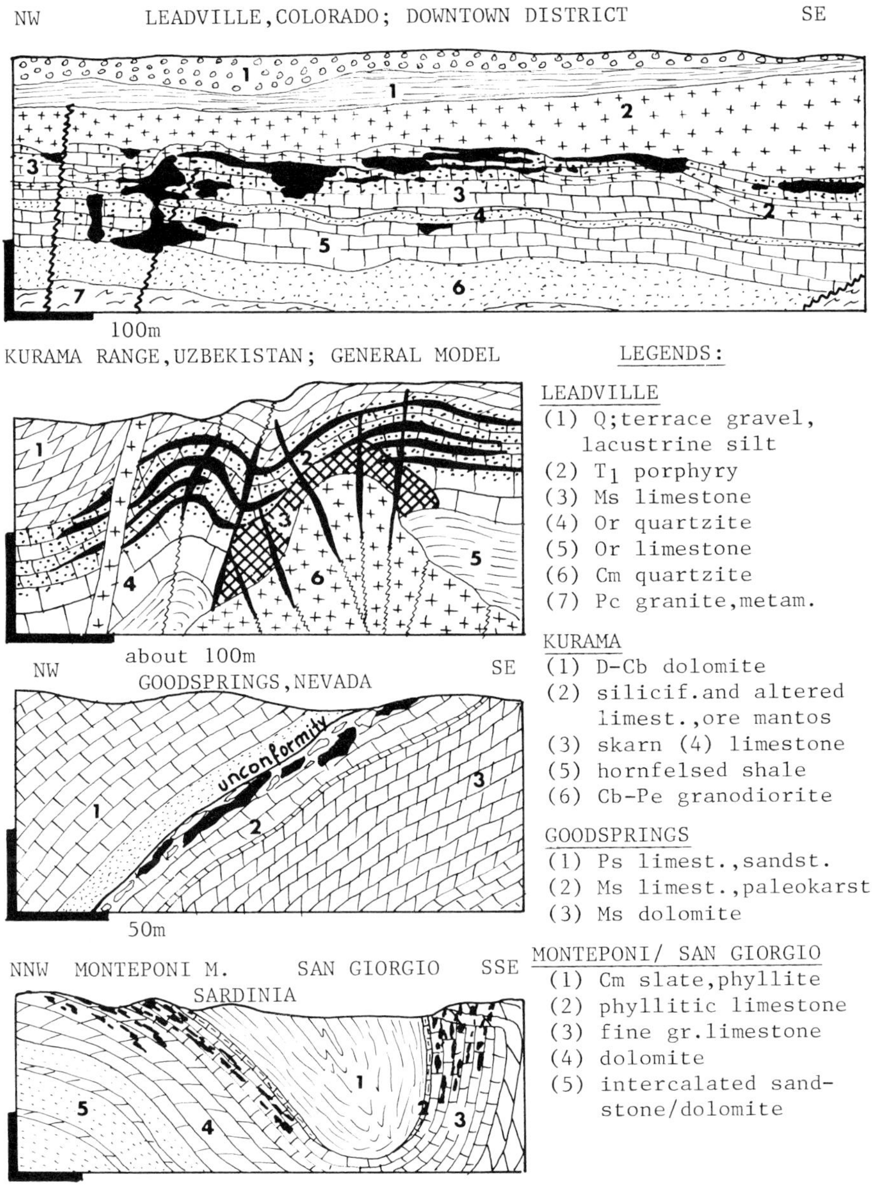

Fig. 20-28. Peneconcordant Zn-Pb sulphide mantos and other mineralization styles, shallow marine carbonate / granitic plutons interaction association. Formerly considered hydrothermal replacements, the idea that some orebodies are pre-granitic APT / MVT mineralizations is becoming increasingly acceptable. Modified after Emmons et al. (1927), Surgai (1970), Albritton et al. (1954) and Moore (1969); Leadville and Goodsprings are from LITHOTHEQUE.

Table 20-8. Zn-Pb sulphide "mantos" in shallow marine carbonate sequences located in intrusive aureoles, suggesting pre-intrusive origin of the ore or the ore-bearing structures; selected examples

LOCALITY	AGE OF HOST	AGE OF INTR.	GEOLOGY, MINERALIZATION	REFERENCES
Metaline, Washington, USA	Cm_2	J_3-Cr_2	stratabound py,sf,gal mantos in metasediments (Cm dolomitiz.limest.) under unconf.topped by black m-argillites; diorite,granodior.intrus. 400 Tt Zn/2.5%; 180 Tt Pb/1.2%	McConnel and Anderson (1968)
Leadville, Colorado, USA	Cm-Ms	T_1 Ol	blanket and manto replacements and veins of gal,sf, py,po,etc.,gossan; peneconc.mantos with dolomitiz. limest.most common under unconf.filled by porphyry sills	Tweto (1968)
Gilman, Colorado, USA	Ms	T_1	massive sf,gal,py,etc. manto in karsted,solution breccia containing dolomitized limest.resting on Cm quartzite, and overl.by Penns.shale. 786 Tt Zn/5.86%; 138 Tt Pb, 94 Tt Cu	Radabaugh et al.(1968)
Tintic, Utah USA	Cm	T_1	gal,sf,pyr,etc.massive and dissem.mantos in limestone,dolom.lim. and silicified lim. dominant in 50 m thick stratigr.interval marked by cave,solution breccias, under unconformity. 1.07 Mt Pb	Lindgren and Laughlin (1919)
Goodsprings, Nevada, USA	Ms_1	T	gal,sf,gossan miner. form peneconcordant mantos in dolomitiz.limest.with abundant solution cavities, sand-filled caves, under unconformity. 97 Tt Zn/14.5%, 42 Tt Pb/5.5%	Albritton et al.(1954)
Santa Eulalia, Chih., Mexico	Cr_1	Eo-Ol	py,po,sf,gal,mantos and chimney in limestone under volcanics-covered unconformity; 1.87 Mt Zn, 1.74 Mt Pb, 12 Tt Ag	Gonzáles-Reyna (1956)

La Loubatière, Bibaud); and (2) discordant late Hercynian galena, sphalerite, barite veins, hosted by Cambrian carbonates, shales and sandstones (Cusses, Camprafaud, Berlou). Villeneuve appears to combine both categories. Brevart et al. (1982) considered the latter to be products of remobilization of the former and lead isotopes are believed to be capable of distinguishing the remobilized deposits from those of similar morphology, but precipitated from hydrothermal fluids heated by the Hercynian "granites". The latter have more radiogenic leads. Although both populations of deposits had their original sources in the sialic basement, they had a different history of emplacement.

SOUTH-WEST SARDINIA Pb-Zn and BARITE DISTRICT, ITALY

The South-West Sardinia Pb-Zn district (also known as Iglesiente-Sulcis; Moore, 1969; 12 Mt Zn+Pb) has dimensions of about 50 x 30 km. The city of Iglesias lies in the centre. The district is underlain by a thick sequence of middle Cambrian shallow-marine sediments (basal sandstone with dolomite bands grading upward into carbonates) that, in turn, are topped by slates and sandstones. These are unconformably covered by Ordovician and Silurian sediments and intruded by a late Paleozoic granite batholith.

Three principal styles of mineralization have been recognized: (1) pyrite, sphalerite, galena "replacements" stratigraphically confined to limestones and dolomites of the Metallifero Formation; (2) zoned hydrothermal veins (e.g. Montevecchio); and (3) skarns at "granite" contacts. The possibility of a polygenetic mineralization in Sardinia and also the similarity of (1) to the MVT, has been considered for some time. The district is also a well-known type area of magmatogene-hydrothermal remobilization (e.g. Zuffardi, ed., 1969). Considered with particular emphasis on the pre-granite supracrustal hosts, style (1) is of paramount importance. Most deposits are confined to a fine-grained limestone member of the Cambrian Metallifero Formation, about 300 m thick. In it, the most common ore distribution consists of large but low-grade (3-5% Zn, 0.3-1% Pb) peneconcordant zones in limestone, containing disseminated and scattered small crystals of sphalerite, pyrite and minor galena (e.g. San Giovanni mine). Superimposed are numerous irregular, higher grade replacement bodies. The largest carbonate-hosted deposit Monteponi (min. 1.35 Mt Zn, 400 Tt Pb) is composed of many peneconcordant masses of pyrite, sphalerite and galena in metasomatically dolomitized limestone. There are numerous filled paleokarst cavities and solution breccias.

Substantial amounts of zinc have been recovered in the past from the oxidation zone (hemimorphite, smithsonite). Moore (1969) concluded that the rich Zn-Pb orebodies in the Sardinian carbonates were structurally (mostly by cleavage) controlled, and represent the loci of a late metal concentration superimposed on the original "syngenetic" mineralization in the limestone.

20.10.2. Zn(Pb) silicates, carbonates and oxides hosted by shallow-marine (meta)carbonates

Smithsonite, hemimorphite (calamine) and lesser cerussite are common minerals in oxidation zones over carbonate-hosted Zn-Pb deposits of all styles. Some MVT deposits (e.g. the La Calamina in the Moresnet field, Belgium) were almost entirely represented by the Zn silicates and carbonates, but the preserved remnants of the "primary" ores and the relict structures left no doubt about the ore genesis. Such doubts, however, arise when an orebody is composed entirely of the oxidic Zn minerals hosted by carbonates. In such a case, we may be dealing with (1) a complete, pseudomorphic conversion of a sulphidic ore into an oxidic one; (2) a proximal, selectively transported oxidation zone from an adjacent sulphide deposit that may no longer be preserved ("exotic" mineralization); and (3) other, unknown origin.

Example localities include the small smithsonite deposits Ardovo and Tiba, located in Mesozoic carbonates of southern Slovakia; the large and rich (3.5 Mt ore with 35% Zn) smithsonite, hemimorphite, hydrozinckite and Zn-bearing clays deposits in the Kanchanaburi Province, Thailand; a portion of the Vazante zone in Brazil; the Zn-Pb oxidic deposits in South Australia and others. When goethite, Mn oxides or Cu silicates or carbonates are absent, the oxidic Zn minerals may easily be missed because they form white, gray or variously pigmented masses or encrustations, not substantially different visually from the surrounding carbonates. The best indicator is the high specific weight of the Zn ore.

The South Australian oxidic zinc deposits located in the northern Flinders Ranges (Beltana and Aroona) have recently been mined and well documented (Muller, 1972; Horn, 1975; 368 Tt Zn/36.8%, 22 Tt Pb/2.2%). Their origin, however, remains controversial. The deposits are hosted by a lower Cambrian limestone overlying the Paleozoic sediments of the Adelaide "Geosyncline" in a region marked by open folding, numerous faults and a widespread presence of ? injection breccias, interpreted as diapirs. The largest Beltana (Puttapa) orebody is in contact with, and partly within, one such diapir and its host rocks are folded in a complex manner, dolomitized and recrystallized. The orebody is an irregular mass composed of massive, metacolloform, banded and brecciated red (hematite-pigmented) and white willemite, with subordinate smithsonite, hedyphane, coronadite and mimetesite. The orebodies are broadly conformable with the Tertiary erosion surface and the water table, and are interpreted as having formed by a weathering or pedogenic redistribution of a low grade Zn-Pb mineralization present in Cambrian black slates stratigraphically under the host limestone.

"When observed Phenomena are not readily explained by forms whose limitations are known to us, we naturally invoke other forms whose limitations are less well understood".
E. S. Shepard quoted by Augustus Locke (Econ. Geol. v.77, 1982, p.197).

CHAPTER 21

Marine Evaporites

21.1. Introduction

Marine evaporites (anhydrite, gypsum, halite, K and Mg salts) constitute an exciting field of geological study, covered by an extensive literature (Mattox, ed., 1968; Till, 1978; Braitsch, 1971). The geochemistry of evaporites has been summarized by Stewart (1963) and the economic geology by Korenevskii (1973). The worldwide distribution of salt resources was tabulated by Lefond (1969).

Unfortunately, pure evaporites contain almost no economic metallic accumulations and are strongly depleted in virtually all trace metals. The textbook and reference literature remarks referring to occurrences of the various deposits of Cu, Pb-Zn, U, and other metals "in evaporites", are inaccurate. Such deposits are invariably hosted by non-evaporitic rocks such as shale or carbonate, associated with evaporites. A portion of the dolomite, however, could have precipitated by evaporation. Alternatively, evaporites may have acted as an interacting agent in ore genesis, for example by producing structural traps and a precipitation ground for the metalliferous fluids or by supplying salts to the metal-transporting fluids. For this reason, evaporites receive a shallow treatment in this book, and are organized by characteristic, although transitional and empirical associations and styles apparent in the field. Figure 21-1 and Table 21-1 summarize the principal mineralization styles.

21.2. EVAPORITES IN THE RED-BEDS → GRAY-BEDS FACIES TRANSITION

Modern sebkhas (salt flats) have been reviewed in Chapter 5, and the black shale, carbonate and red-beds members of the sequence in Chapters 19, 20 and 25, respectively. Several types of lithofacies profiles have been described in the literature, their variations depending on the depositional regime; (transgressive or regressive); cycle (basal or intra-sequential) and basement composition. A "typical" red-beds-floored "paleo-sebkha", exemplified by the European Zechstein (Chapter 19) has a red sandstone in the footwall, passing laterally and vertically into a thin white sandstone → black marl (Kupferschiefer) → marly limestone or dolomite → anhydrite, → and a multiphase "saline giant".

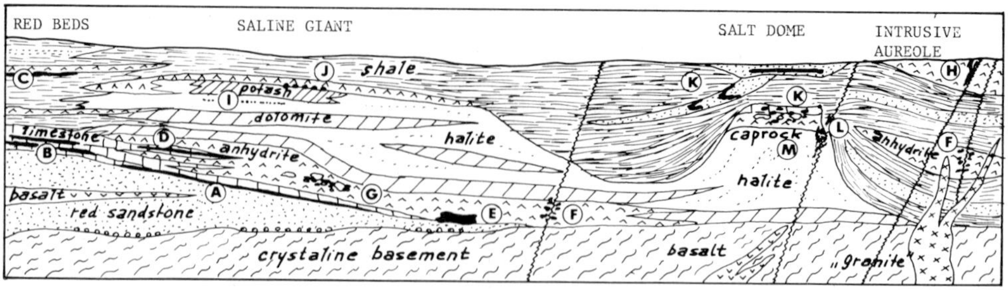

Fig. 21-1. Marine evaporites, principal mineralization styles (see Table 21-1 for an explanation of the letter codes).

Table 21-1. Marine evaporites, principal mineralization styles

(A)	black, gray, green, etc. metalliferous shale or marl at red beds / marine sediments interface (e.g. Kupferschiefer); Cu, Pb-Zn, Ag, Mo, Au, Pt, etc.; treated in Chapter 19
(B)	Pb-Zn ores in carbonate facies at red-beds / marine sediments contact; treated in Chapter 20
(C)	Cu shales in gray / green gypsiferous horizons, within red beds; treated in Chapter 19
(D)	Mn carbonate and oxide in marl tongues in anhydrites; X localities, 1 Mt Mn
(E)	barite, Pb,Zn,Ag,Cu sulphides, replacing anhydrite above red beds and sandstone basement; 1 locality (Walton)
(F)	Cu sulphide epigenetic replacements of anhydrite, gypsum, near faults and intrusions; X localities
(G)	trace metals (Se,As,Hg,etc.) associated and probably recoverable as by-product, with biogenic sulphur, in bedded Ca sulphates
(H)	Hg-Sb sulphides in anhydrite and dolomite breccia in an intrusive aureole; 1 locality (Huitzuco), 2.5 Tt Hg, 1 Tt Sb
(I)	locally high trace metal contents in saline giant deposits (Hg,Au)
(J)	rare earths - containing borate nodules at sylvite - anhydrite contact; 1 locality (Moab)
(K)	U,V minerals infiltrations in porous aquifers above salt domes X localities, X0 Tt U
(L)	cinnabar veinlets and impregnations in faulted salt domes; X localities, X00 t Hg
(M)	metalliferous bitumens impregnations in caprocks of salt domes; uranium (Panhandle field)

In the Permian "paleo-sebkha" of Oklahoma (Chapter 19), metalliferous shale interbedded with gypsum and minor dolomite is enveloped by red-beds above and below. The Jurassic "paleo-sebkha" in the counties of Sussex and Dorset in southern Britain (Till, 1978) rests on calcareous sandstones and oolithic carbonates. It displays an upward succession of black shale → laminated algal (stromatolitic) limestone with gypsum or anhydrite nodules or crystals → nodular and chicken wire anhydrite and gypsum interbedded with carbonate and clay.

In the above sequence the most critical units metallogenically are the black, gray or green beds (shales, marls, argillaceous thin bedded carbonates, exceptionally sandstones or conglomerates) immediately underlying (or interbedded with) the gypsum or anhydrite. These are associated with the major ore horizons in the European Zechstein, in the Cambrian cover sequence on the Siberian Platform, in the Ebro Basin, Spain, Donetsk Basin, Dzhezkazgan and elsewhere (Korenevskii, 1973). Pb-Zn are in the same setting in the Zechstein and elsewhere.

Metallic minerals - hosting evaporites in this association are exceptional and in many cases it appears that the "evaporite hosting" is just a matter of scale. The metallic minerals are in fact hosted by a thin shale, marl or dolomite band enveloped by a much thicker unit of anhydrite, so that in a section the anhydrite appears as the host. This is clearly exemplified in the Ulutelyak Mn deposit, reviewed below. Alternatively, metals hosted by evaporites are epigenetic, postdating the evaporite diagenesis and unrelated to the sedimentary facies. The Walton barite and polymetallic deposit, Nova Scotia, is an example.

The Ulutelyak deposit (Gribov, 1972; 1 Mt Mn/10%; Fig. 21-2) is located in the lower Permian red-beds in the Urals Foreland Basin. There, a 120 m thick lens of lagoonal anhydrite rests on limestone and changes facies into dolomite. Two conformable 6-10 m thick beds of manganoan marl are enclosed in the anhydrite and contain seams up to 1 m thick of rhythmically laminated, oolithic or lumpy manganocalcite. Residual, friable vernadite and psilomelane have formed in the outcrop area.

Walton, Nova Scotia, Canada, orebody (Boyle et al., 1976) lies in a Mississippian marine evaporite-carbonate sequence, resting on terrestrial red and white sandstones. A large, massive pipe-like to irregular body of cryptocrystalline barite contains a separate body of siderite, Ni-Co pyrite, marcasite, galena, sphalerite, chalcopyrite, tennantite and proustite near its base. The orebody is interpreted as having formed by the replacement of bedded and brecciated anhydrite and gypsum by deep circulating brines. Additional example localities where epigenetic metallic sulphides replace red-beds associated gypsum or anhydrite, include: Myrthengraben in the Semmering Pass, Austria (Tufar, 1982). There, enargite, tennantite, wurtzite, high-Se galena and other minerals replace Triassic gypsum-anhydrite; Špania Dolina, Gretl, and other localities in the Slovak Carpathians, where chalcopyrite and tetrahedrite replace lower Triassic anhydrites; and others.

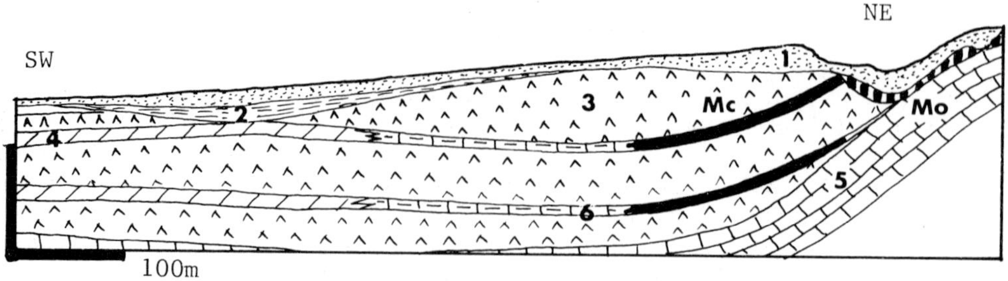

Fig. 21-2. Ulutelyak, Cisural, U.S.S.R.; a stratiform Mn carbonate deposit in Permian evaporite-carbonate association. (1) Quaternary cover; (2) Tertiary clay; Permian: (3) anhydrite; (4) dolomite; (5) limestone; (6) marl. (MC) = Mn carbonate ore; (MO) = residual Mn oxide ore. After Gribov (1972).

21.3. BEDDED GYPSUM OR ANHYDRITE BODIES (RED-BEDS ARE MISSING OR NOT STRONGLY DEVELOPED)

This is a relatively common, although a rather monotonous lithologic association. Occurrences and deposits of native sulphur now fashionably called "bioepigenetic" are present in some areas. These formed by bacteria reduction of the rock sulphate in the presence of bitumens. In the Miocene units of the Carpathian Foredeep in Poland and the western Ukraine (Pawlowski et al., 1979) sulphur bodies are found within gypsum horizons interbedded with sandstone, marl, clays and minor limestone. Bentonite beds occur locally. The sulphur is accompanied by calcite, barite and celestite and has a high trace content of some metallic elements (As, Se) that may possibly be economically recovered. Small occurrences of Zn-Pb sulphides are associated with gypsiferous carbonates and black mudstones or claystones in the Foredeep. The Truskavets deposit (Korenevskii, 1973) has already been reviewed in Chapter 19 (Fig. 19-19).

In the Burshtyn Mn deposit in the Carpathian Foredeep, U.S.S.R. (Korenevskii, 1973), Mn carbonate (manganocalcite, rhodochrosite) gradational into a Mn oxide (vernadite) horizon is located on top of a Miocene gypsum-anhydrite unit, covered by gypsiferous claystones. The Mn carbonate forms rhythmically banded lenses 0-26 m thick, and the Mn oxides mainly take the form of nodules and cement in the marly claystone. The Mn ore changes facies laterally into marl and algal limestone.

21.4. SEDIMENTARY VERSUS ALTERATION GYPSUM OR ANHYDRITE AND ASSOCIATED METALLIC DEPOSITS

Gypsum and anhydrite are important hydrothermal alteration minerals of some plutonic and subvolcanic ores (e.g. some porphyry coppers; massive sulphides; alunite Au-Ag ores; compare Chapters 26, 28). In the Hokuroku Miocene volcanogenic massive sulphide district (Chapter 14), both alteration and sedimentary gypsum form substantial rock bodies that occasionally host the metallic sulphides ("gypsum Kuroko"). At several localities, sulphide-hosting anhydrite and gypsum are of controversial origin.

In the Morococha porphyry copper, vein and replacement Cu, Mo, Ag, Pb, Zn deposit, Perú (Nagell, 1960), a major body of anhydrite and gypsum tops Permian volcanics and is either a sedimentary member of Permian red-beds or a product of alteration of Mesozoic carbonates. It is a minor host to sulphide orebodies.

In the Huitzuco Hg-Sb field, Guerrero, southern Mexico (McAllister and Hernandez, 1945; 2,500 t Hg, 738 t Sb; Fig. 21-3), an unusual Hg-Sb ore is represented by livingstonite ($HgSb_4S_8$) accompanied by stibnite and native sulphur. These minerals are scattered and form pockets in brecciated anhydrite, gypsum and dolomite along fractures, and also in semi-consolidated rubble, filling solution sinks. The Ca sulphates at Huitzuco are interpreted as being evaporites or alternatively carbonate replacements. The Hg-Sb sulphides probably precipitated from intrusive-coeval hydrothermal solutions, reacting with the sulphur that was probably released from the anhydrite.

21.5. SALINE GIANTS

It has recently become customary to refer to the thick, complete (or almost complete) massive evaporite sequences as "saline giants" (e.g. Till, 1978). In these, enormous accumulations of halite (of the order of hundreds of trillion tons), sylvite (tens of billion tons) and occasional carnallite, tachyhydrite and other minerals, have been identified. The question of origin of the saline giants, particularly in terms of their depositional paleoenvironment ("deep" or shallow water basin) is still unresolved and will not be dealt with here. It is interesting to note that ancient halite is several times more widespread than gypsum and anhydrite; about one third of the North American platformic sequences are underlain by halite-dominated evaporites. In contrast to the calcium sulphate that can be studied in surficial exposures, however, halite and K-Mg salts are known only from subsurface exposures, drilling and geophysical logs. As a consequence the broader lithologic association, particularly of the rocks below the evaporite unit, is poorly known unless it crops out, not far from the evaporite unit, along the edge of the basin. This is exceptional and is exemplified by the Zechstein assemblage of north-central Germany where the metalliferous Kupferschiefer marl

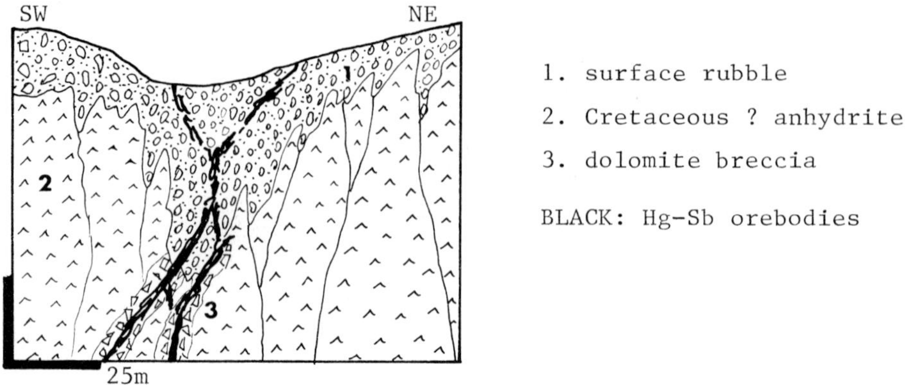

Fig. 21-3. Huitzuco, Mexico. Anhydrite and dolomite-hosted Hg-Sb (livingstonite) deposit. From LITHOTHEQUE, modified after McAllister and Hernandez (1945).

(Chapter 19) occurs on the base of the Permian evaporite. Should a similar metalliferous unit exist under the Devonian evaporites of the Williston Basin (under Saskatchewan and adjacent provinces or states); under Cretaceous evaporites of the Sergipe Trough, Brazil; in the Gabon and Congo Basins, Africa; on the Khorat Plateau, Thailand and Laos; or elsewhere, it has not yet been reported. The evaporite geology of some of the above basins has been reviewed recently in a special issue of Economic Geology (vol. 74, 1979, no.3) devoted to evaporites and phosphates.

As already stated, the saline giants themselves are dwarfs as far as the metal accumulations are concerned. There could, however, be some surprises and ultimately certain metals (e.g. Al from interbedded or overlying clays) could be recovered when complex, waste-free processing of materials becomes commonplace. One of the "surprises" is an anomalous gold content (23 ppm in blue halite, 4 ppm in blue anhydrite) reported by Stewart (1963). The figures appear erroneous (an error may have arisen through the substitution of "ppm" for "ppb"); if true, however, the blue halite would contain three times more gold than the Witwatersrand conglomerate.

Another "surprise" is a diagenetic, rare earths-containing borate, discovered in reddish-pink nodules abundantly distributed through a 15 cm thick layer in anhydrite that immediately overlies sylvite in the Pennsylvanian evaporites near Moab, Utah (Cane Creek mine; Raup et al., 1967). Further surprises may follow should the saline giants receive more interdisciplinary attention. Mercury is one metal to watch, judging from its relatively common occurrence in the Soviet salt domes. Korenevskii (1973) reported the presence of mercury in supra-clarke quantities even in some ordinary halites and anhydrites.

Interaction mineralization involving soluble evaporites and cases in which bedded halite is a host to an ore introduced from outside, are virtually unknown (a possible exception arises in the case of widespread tectonic collapse on sites of dissolved evaporites along the edge of evaporite bodies that may exert structural control on epigenetic ores). Another exception is in the magnetite province in the Angara Basin of Siberia, where Paleozoic salt partly hosts massive replacement magnetite related to a Permo-Triassic mafic magmatism (Chapter 31).

21.6. SALT DOMES, DIAPIRS AND ANTICLINES

Salt domes are marvels of structural geology and also important oil reservoirs, covered by a substantial specialized literature (e.g. Halbouty, 1979). Salt domes are diapiric or piercement structures, formed around a central, nearly equidimensional halite plug. The plug has risen through the enclosing sediments from a mother salt bed, located up to 10 km beneath the top of the plug. Salt anticline is a linear, usually a fault-controlled equivalent to a salt dome.

Many salt domes are topped by a cap rock, which is a material of variable composition and texture (usually a breccia) formed by a concentration of the less soluble minerals, insoluble particles (clays) and rocks (anhydrite, dolomite) originally associated with the salt. Many caprocks are compositionally zoned: a basal anhydrite directly topping the salt plug is followed by a transition zone and calcite (limestone) zone. The two latter zones usually contain the bio-epigenetic sulphur accumulations, liquid as well as solid bitumens, and sometimes masses of barite (e.g. in the Jefferson Island Dome, Louisiana; Halbouty, 1979).

Most salt plugs have almost vertical walls and some are overhung. The enclosing sediments are deformed, brecciated, turned up and faulted in a complex fashion. Collapse structures caused by leaching out of the soluble salts are complementary to domes and salt anticlines in many areas. Salt domes are widespread in the Gulf Coastal Plain and in the adjacent Gulf of Mexico in North America; in the North German Plain and North Sea in Europe; in the Azov and Caspian Sea regions, southern U.S.S.R; in Iran and elsewhere. The type area of salt anticlines lies in the Paradox Basin in south-western Colorado (Carter, 1970).

The metallogenetic function of salt domes and anticlines is manifold: (1) they provide structural control for metalliferous brine migration and epigenetic mineral deposition and (2) as reservoir rocks they confine gas, petroleum and formational water pools. The above fluids could be mildly metalliferous, and could precipitate a portion of their metals within the caprock or adjacent to it. Alternatively, (3) the fluids or gases could assist in the precipitation of metals from the extraneous fluids (including hydrotherms), largely by providing a reducing environment or by supplying sulphur.

Function (1) is the most widespread but is often misinterpreted, because salt accumulations that once caused halokinematic deformation structures could have been removed subsequently by dissolution. The most often quoted example of (1) probably combined with (3) is in connection with the formation of the U-V infiltration ores in the Uravan and La Sal districts of Utah. There, the ore minerals (carnotite, uraninite, etc.) are located in Jurassic channel sandstones cropping out on the flanks of salt anticlines, caused by Pennsylvanian evaporites (compare Chapter 25).

Factor (2) is responsible for the precipitation of metalliferous solid bitumens in the caprock or in its vicinity, and less commonly, for the precipitation of metallic minerals believed to have been present in oilfield brines (Pb-Zn, Hg). The first possibility is exemplified by the uraniferous caprocks in the Panhandle oil and gas field, Texas (Pierce et al., 1964). There, up to 1% U occurs locally as uraninite, smaltite, chloanthite and pyrite as well as urano-organic compounds, dispersed in asphaltite. The asphaltite is a brittle, lustrous, black organic residuum forming nodules, pods and infiltrations in a caprock composed of dolomite rhombs, anhydrite, chert, celestite, pyrite and containing brine and oil-filled pores (Fig. 21-4). The highest overall concentration of uranium in the caprock is between 10-20 ppm U, extending through a thickness of about 80 m. The asphaltite is interpreted as being a petroleum derivative and the trace metals are believed to be locally derived and selectively concentrated in petroleum compounds. The Panhandle field is well known because it contains the largest helium reserve in the United States and also anomalous concentration of radon.

Factors (2) and (3) have jointly been responsible for the formation of Pb-Zn and cinnabar occurrences in and near some salt domes in the north-western Donets Basin, U.S.S.R. (e.g. the Slavyansk deposit; Korenevskii, 1973; Kuznetsov, 1974; Fig. 21-5). There, pyrite, galena, marcasite, sphalerite and cinnabar are associated with solid bitumens as well as with sulphur and fluorite, and form veinlets and disseminations in a tectonic breccia along a north-west dipping fault transecting a dome of Devonian salt, gypsum and anhydrite emplaced into Permian sediments. The bulk of the ore is hosted by the caprock and by the overlying mottled Permian sandstones. The breccia contains a high proportion of fragments brought in from some depth and including Carboniferous argillites, carbonates and sandstones, in addition to some altered basalts.

Factor (3) is responsible for precipitation of a small U oxide orebody, hosted by Miocene-Pliocene beds directly above the Palangana salt dome, southern Texas (Weeks and Eargle, 1960; about 800 t U; Fig. 21-6). There, a tabular peneconcordant orebody composed of uraninite and pyrite interstitial infiltrations and grain coatings, formed in a flat-lying, friable sandstone. The uranium is believed to have derived by groundwater leaching of a zeolitized and silica-cemented horizon of an ash-fall tuff, located stratigraphically under the orebody. The orebody was probably a consequence of U and pyrite precipitation due to the reducing environment of the H_2S, emanating from the caprock.

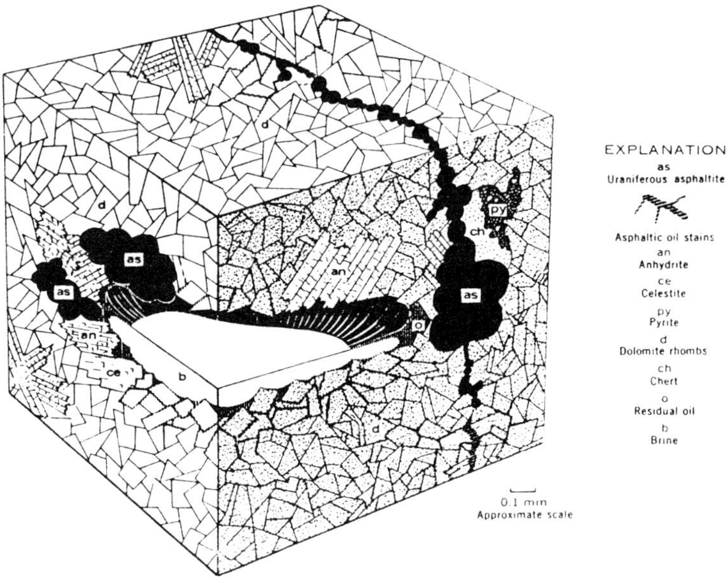

Fig. 21-4. Uraniferous asphaltite in porous "brown dolomite", Panhandle Gas Field, Texas. From Pierce et al. (1964).

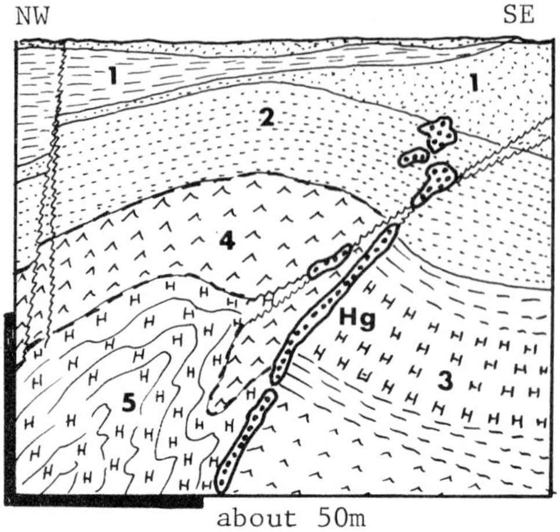

LEGEND

1 Triassic to Quaternary sediments
2 Pe_3 mottled sandstone
3 Cb_3-Pe shale and bedded gypsum, halite
4 Devonian salt dome : limestone / anhydrite caprock and tectonic breccia
5 halite

OUTLINED: Cinnabar veinlet, impregnation bodies

Fig. 21-5. Slavyansk Hg showing, Donetsk Basin, U.S.S.R. Fault-controlled cinnabar mineralization associated with a salt dome. Modified after Kirikilitsa and Laskov, in Kuznetsov (1974).

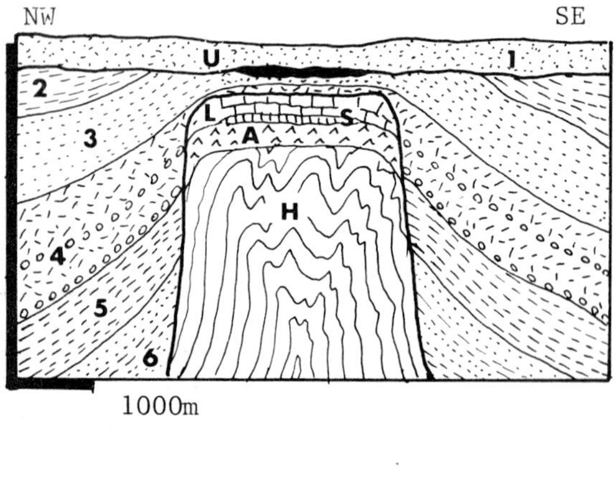

LEGEND

1 Pliocene sand
2 Mi clay
3 Mi sand, sandstone
4 Mi ash-fall tuff, conglomerate
5 Ol clay
6 Eo claystone, sandstone

Fig. 21-6. U-oxide infiltration orebody in roof of the Palangana salt dome, Texas. Salt dome rocks: (L) limestone; (S) sulphur; (A) anhydrite; (H) halite. After Weeks and Eargle (1960).

"Whatever the final answers may be to the questions on the genetic relationships between hydrocarbons and ores in sediments, there is little doubt that their terminal stage of migration and accumulation are controlled by similar factors".
K. H. Wolf in Ore Genesis Influenced by Compaction, 1976, p. 655.

CHAPTER 22

Petroleum, Natural Gas, Solid Bitumens

22.1. INTRODUCTION

Solid, liquid and gaseous hydrocarbons are a special class of earth materials. They are hardly "rocks" because like ores, they are minor components hosted by a variety of rock associations. The liquid and gaseous substances have a remarkable capacity for moving under their own power. Although the bulk of hydrocarbons are hosted by shallow-marine and continental sediments, there are important exceptions. It is not the purpose of this book to review the origin, occurrence and distribution of hydrocarbons: there are numerous reference publications doing so (e.g. Tissot and Welte, 1978; Hobson and Tiratsoo, 1981; Chilingarian and Yen, 1978; Halbouty, ed., 1970). The objective of this short chapter is to review briefly the metals and metallic deposits contained in hydrocarbons or associated in some way with the hydrocarbon deposits.

Hydrocarbons, more than any other category of natural materials (except perhaps coal and seawater), are capable of yielding a variety of mineral products (compounds and elements) during processing. Many such elements occur in unimpressive clarke – to – sub – clarke concentrations but are cheap to recover or have to be recovered to satisfy the government's anti-pollution regulations. Sulphur and mercury in natural gas and crude petroleum are examples of such "technological ores". The case of sulphur recovered in the process of purification of "sour" natural gas in scrubbing plants is particularly instructive. A single scrubbing plant in Alberta, Canada, recovers up to 4,000 t of sulphur daily. In the past 30 years, this cheap by-product sulphur gained a substatial share of the global sulphur market (in the 1970's, 25% of the world's sulphur production came from hydrocarbon processing plants), and put many deposits producing pyrite for sulphur recovery out of business.

22.2. NATURAL GAS

In addition to sulphur, mercury is another (and the only metallic) commodity recoverable from natural gas. It is probably already being produced in East Germany. Ozerova (1981) reviewed the mercury potential of the European gas fields located along the buried W.N.W.-trending Karpinskii lineament in Holland, Germany and Poland. The gas fields are largely associated with the Zechstein (upper Permian) metalliferous and evaporite province (Chapter 19), but the concentration of mercury in the gas is infinitessimally low (1.5 to 9 mg Hg/m^3), thus I would certainly be opposed to calling the string of the gas fields a "new mercury ore belt", as Ozerova (1981) did. There are, nevertheless, some 5,000 tons of a "technological" mercury contained in the known gas reserves, some 75% of which can be extracted. Most of the documented Hg resources are in the Groningen gas field in the Netherlands (gas reserves: 15 trillion m^3; average content 2 mg Hg/m^3; Hg resource 3,000 t).

22.3. LIQUID PETROLEUM

In the late 1950s-1960s, liquid hydrocarbons received attention as a possible source of by-product uranium (Erickson et al., 1954; Bell, 1960; Hyden, 1961). The U content in crude petroleum is low (of the order of 1 ppb, max. 5 ppb) but may increase substantially in the natural products of petroleum degradation (e.g. asphalts, pyrobitumens) or in the ash remaining after petroleum combustion. The latter, however, is virtually impossible to collect economically and consequently to utilize as an "ore".

Ni, V, Mo, Hg and to a lesser degree Ga, Co, Ti, Cr, Ag often show above-clarke, but erratic enrichment in petroleum. Boyle (1979) reported 0.438 ppb Au in an Alberta petroleum. Many trace-metal-enriched petroleum pools are hosted by lithologic associations themselves enriched in the same metal; one of the most notable examples are the southern California fields (e.g. the Cymric field) in, or on the contact with, ophiolites, all components of which (that is, gas, petroleum, waters) are Hg-anomalous.

22.4. OILFIELD WATERS (BRINES)

It is well known from the popular geological literature that in oil pools "petroleum floats on water", and that large quantities of this water are produced together with oil and have to be separated. The former practice of discharging this water directly into the streams has been halted long ago in the industrialized countries on grounds of pollution, and the water has to be treated before disposal. This treatment results in recovery of a variety of solids (NaCl, Br, I), so oilfield waters are one of the "technological ores" of our times.

The ordinary saline waters are connate (fossil) fluids diluted to various extents by meteoric waters, and their heavy metal content is low (of the order of 1-450 ppb Cu, 1-15 ppb Ni, 1-25 ppb Sn, 36-1,800 ppb Mn, up to 2.5 ppb V and up to 5 ppb Co; Collins, 1975). The trace metal content is of geochemical interest (tracing the petroleum source rocks and their migration paths), strongly influenced by geochemical provincionalism, but normally of little economic importance.

Several examples of "special", highly metalliferous oilfield brines have been described in the literature. The potassium-rich brines in the central Mississippi oilfields (Goshen Springs, Knoxo, Loreno; Carpenter et al., 1974) contain up to 360 mg/l Zn and 110 mg/l Pb and some brines precipitate rings of platey barite, scales of metallic lead and/or tiny crystals of galena on the walls of the well tubing. They are believed to have originated as interstitial fluids in halite evaporites, leached their metals from a shale, and were expelled upward during compaction. The brines were intersected in depths ranging from 2,600 to 4,300 m and were mildly thermal.

The oil brines known to deposit lead, galena, sphalerite, barite and other minerals in the wells of the Cheleken Peninsula (Caspian Sea, U.S.S.R.; Lebedev, 1973) are interstitial to Neogene red-beds containing interbedded evaporites, resting on a shale sequence. Their temperature is under 100° C and the brines contain up to 77 mg/l Pb and 5.4 mg/l Zn.

It is easy to imagine that the dominantly connate oilfield brines are transitional into, or mix with, magmatically or geothermally heated hot springs (e.g. in southern California). Such a mixed fluid parentage is reflected in many epigenetic mercury deposits containing accessory solid hydrocarbons (e.g. New Almaden, Wilbur Springs, Idrija), or even mixed hydrocarbon-mercury ores (e.g. the "kidney" ore of Idrija)

22.5. SOLID AND SEMI-SOLID HYDROCARBONS

Locally high trace-metal contents in the organic fraction of sediments, including the "oil shales" and "tar sands" have already been described. The "high-grade" accumulations of asphalt or asphaltite are commonly enriched in vanadium, and sometimes have a slight enrichment of uranium which, however, is rarely considered to be economically recoverable. Bell (1960) cited the example of the Edna asphalt deposit near San Louis Obispo, California, containing about 250 Mt of rock with 11% of bitumen. The bitumen contains 21 ppm U, its ash has 0.376% U, and the total uranium content is about 650 t U.

22.6. INTERACTION MINERALIZATION: BITUMENS / IGNEOUS INTRUSIONS

The enigmatic, now exhausted and rather inaccessible Minas Ragra (Quisque) deposit high in the Peruvian Andes (Hewett, 1909; 100 Tt V/6%; Fig. 22-1), is hosted by asphaltite and a bitumen coke, and may be the product of hydrocarbon infiltration and subsequent or concurrent interaction with intrusive dikes.

The orebody is a N 20° W-trending vertical to steeply dipping, downward tapering, discordant vein of solid anthracite or

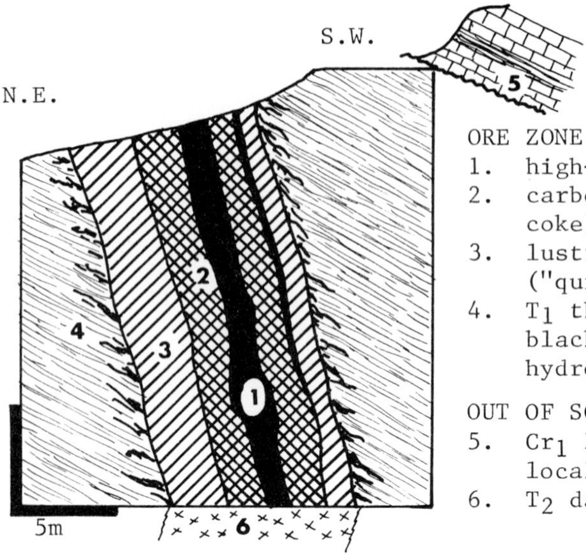

LEGEND:

ORE ZONE:
1. high-grade patronite ore
2. carbon-rich sinter (natural coke), vanadiferous
3. lustrous asphaltite ("quisqueite")
4. T_1 thin-bedded r[illegible] black shale, V minerals and hydrocarbon impregnations

OUT OF SCALE ROCK ASSOCIATIONS:
5. Cr_1 limestone and bituminous locally vanadiferous shales
6. T_2 dacite porphyry dike

Fig. 22-1. Minas Ragra (Quisque), Peru; a vanadiferous meta-asphaltite. From LITHOTHEQUE, modified after Hewett (1909).

coke-resembling asphaltite. It contains a central zone rich in vanadium sulphide patronite. The body fills a fissure in shale, which is gray (reduced) adjacent to the orebody. It contains a low proportion of hydrocarbons and dispersed vanadium minerals. Outwardly, the shales grade into brown, red and black shales, members of the Tertiary red-beds association. The oxidation zone is marked by red, blue, green and yellow secondary vanadates and vanadium oxides (vanadinite, descloizite, minasragraite). A narrow dike of andesite to dacite porphyry perpendicularly intruded the mineralized zone and caused a slight contact metamorphism in the adjacent shale.

Cretaceous limestones with interbeds of locally vanadiferous bituminous shales were thrusted from southwest over the red-beds and crop out near the orebody. McKinstry (1957) considered Minas Ragra, an oversized and modified vanadiferous asphaltite deposit. Driving off the volatiles, presumably by the heat of the intrusion, would account for coking of the material and thus accomplish some vanadium concentration.

"Analysis of the conditions of preservation of earlier formed economic concentrations under conditions of subsequent geological development is an important task of ore prognostication, so far little studied".
D. V. Rundkvist in Kriterii Prognoznoi Otsenki Teritorii, 1978.

CHAPTER 23

Weathering, Soil Profiles, Karst

23.1. INTRODUCTION

Under certain conditions (glaciation, very rapid stream erosion in mountains, human activity including that of geologists), virtually fresh bedrock can be broken and removed without altering (modifying) in any way its original appearance and composition. These are, however, a rather special set of conditions if considered in the context of the geological history. Under more normal and substantially more widespread natural conditions, the bedrocks outcropping on the surface undergo modification and usually a subsequent removal by the action of the exogenetic agents the first stage of which is weathering. The weathering (physical, chemical, biological and combined) is influenced by climate and physiography, so it is slightly different in different locations, even when the bedrock is identical. Fortunately, the effects of the contemporary weathering are directly observable and predictable with a high degree of confidence. The intensity of weathering usually reflected by the thickness of the decomposed mantle and the degree of chemical alteration of the original bedrock, is proportional to the magnitude of the two principal climatic agents: temperature and humidity. It increases from zero in the Arctic regions where above-freezing temperatures and liquid water are rare, to the maximum attained in the tropical rain forest environment. The "types of weathering" treated in textbooks and reference books (e.g. Strakhov, 1962; Figs. 23-1, 23-2) are closely related to the weathering intensity. The third important variable bearing on the intensity of weathering is physiography; only thin (or no) weathering crusts develop in rugged mountains regardless of their geographic location, whereas thick crusts are the rule in lowlands and depressions. The fourth important variable bearing on the intensity of weathering is time. At sites protected from erosion, or where the rate of erosion is less than the rate of weathering, the thickness of the weathered mantle is usually proportional to the duration of the weathering process and the weathering products are cumulative.

Weathering is usually associated with pedogenesis (soil formation), and the terms weathering profile and soil profile are often used interchangeably. In places such as in the Arctic, however, thin soil

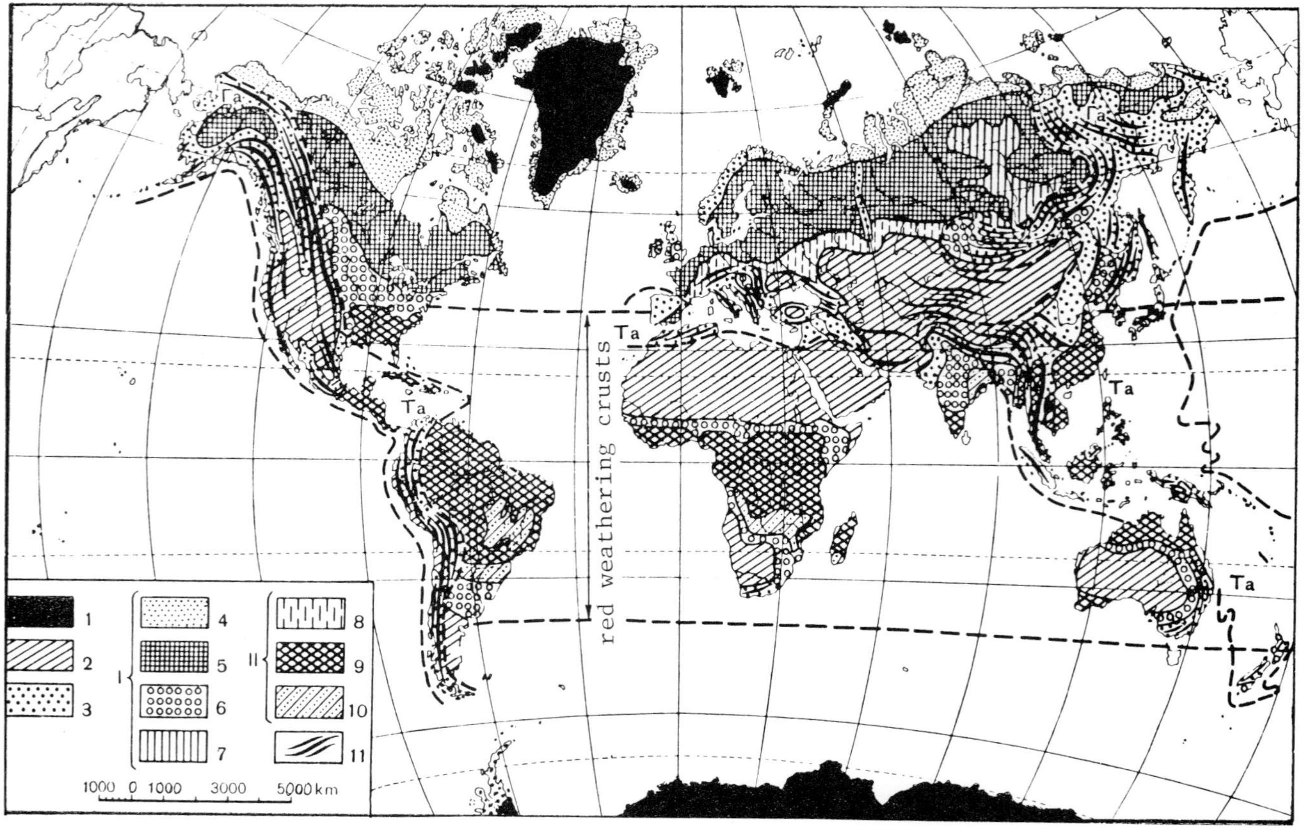

Fig. 23-1. "Types of weathering" as distinguished by Strakhov, worldwide, are closely related to the intensity of (mainly chemical) weathering. This, in turn, depends on temperature, humidity and physiography. Idealized global map from Strakhov (1962).

is usually found on a virtually unweathered bedrock. Many tropical weathering profiles, on the other hand, have no soil developed. Weathering and pedogenesis as a sediment-forming process, have two components: (1) subtractive and (2) additive. The first component causes selective removal of certain components (constituents) from the parent bedrock, so that the composition of the rock undergoing weathering changes continuously. The more mobile components are sequentially removed, the immobile ones residually enriched. This is important for both lithogenesis and metallogenesis, and empirically demonstrable as a law: all the cations present in the truly residual weathering profiles had earlier been present in the parent rock.

The second (additive) component is responsible for bringing in new elements from the outside (from the atmosphere and hydrosphere), and for incorporating them into the weathering profile. Addition of hydrogen and oxigen causes hydration of the originally anhydrous bedrock minerals (mostly silicates) and oxidation, pervasive throughout the weathering profile. The addition of carbon is more restricted, and it dominates the uppermost horizon of a weathering profile, soil. The soil supports vegetation and the decaying vegetation in turn generates the soil. Humous soils (e.g. the black chernozem) are visually striking and easily distinguishable from the

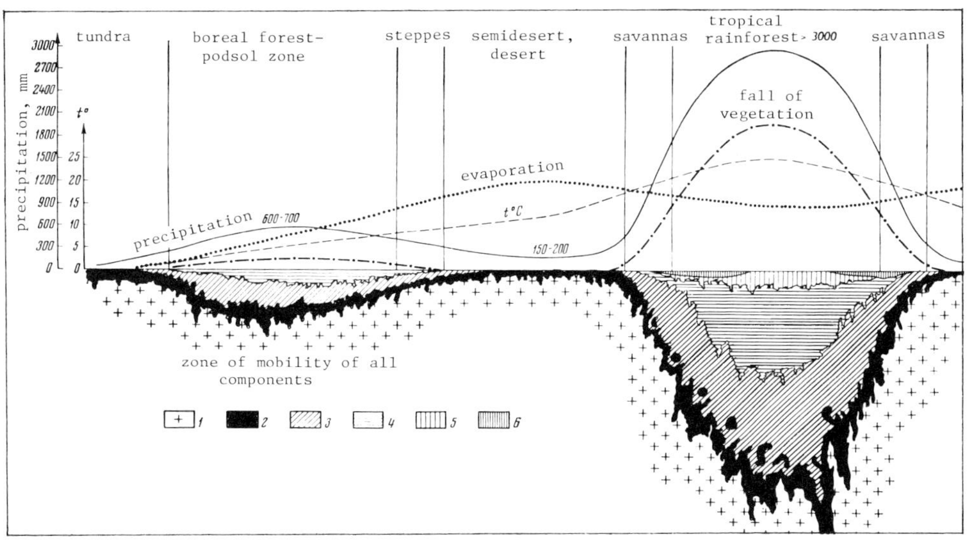

Fig. 23-2. Idealized "types of weathering profiles" in the various climatic zones of the world developed on a uniform bedrock, the thickness of which is related to the intensity of weathering. From Strakhov (1962).

humus-free lower zones of the weathering profile, on which they rest. Some tropical soils (latosols), however, need not contain any humus and there could be virtually no visual difference between the "soil", and the "weathered rock" (saprolite). This overprints into classifications and literature descriptions of ores, where it causes some misunderstanding. Laterite, for example, is technically a tropical soil, but in the "nickel laterite" (a mineralization style), the ore zones are mostly in the underlying saprolite. To be absolutely correct, one should refer to such mineralizations as "nickel deposits in tropical weathering and soil profiles" a term which, unfortunately, is too long. For the above reasons, it is convenient to use the term regolith, when referring to joint weathering and soil profiles.

Weathering and soil formation are important geochemical processes causing metal enrichment and impoverishment and sometimes generating ores. The efficiency of the metal concentration as expressed by the factor (clarke) of concentration, however, is low: within one order of magnitude for most metals and settings, and exceptionally within two orders of magnitude for some metals in "special" settings (e.g. over highly soluble, well drained rocks, or where infiltration is involved). This is important, because virtually all "weathering-generated" ores form from bedrock precursors already enriched in such metals. When such a precursor is "unusual" (as, for example an iron formation or a Mn-quartzite), the ore-forming process is often headlined in economic geology textbooks as a "secondary enrichment" of a pre-existing ore (or a protore), and treated as a minor transformation. When the precursor is a "rock" (e.g. nepheline syenite or shale to a bauxite; serpentinite to a nickel "laterite"), the mineralization weathering is given a greater emphasis, although both factors of concentration involved could be identical (e.g. 2 in the case of the "direct shipping Fe ores" formed over a banded iron formation; 3 to 8 for Mn oxides formed over a "gondite"; 2 for residual bauxites over nepheline syenites; 5-10 for Ni "laterites" over serpentinites).

To predict and evaluate properly a weathering-generated mineralization, one should free one's mind from the artificial distinction between the "primary ore" and "rock" precursors, and base the philosophy squarely on the geochemical elementary balances. Some metals tend to accumulate in tropical weathering profiles (e.g. Al, Fe, Ni, Mn); others tend to leave early, and become depleted (e.g. U, Cu, Zn, Pb). When the bedrock is anomalously high in even the most mobile metals, however, unusual metalliferous regoliths may form (e.g. U, V, Au, Zn, Pt, etc. "laterites").

Within the context of geological history, weathering profiles are short-lived and transient. In most cases, their destruction keeps pace with their formation, so many profiles maintain a constant thickness. Only a few profiles, however, have been preserved in ancient rock sequences, because a special delicate mechanism of preservation is required. Those metalliferous regoliths that survived are treated in Section 23-4.

Earlier, when discussing the balances of the "local" versus the introduced elements in a general way, a point was made that the cations were all "local", derived from the parent bedrock. Considered on a more detailed scale (within the individual zones of a thick weathering profiles), however, this is less obvious and element (metal) introduction is widespread and often substantial. Within a "typical" deep weathering profile where precipitation exceeds evaporation (and consequently the water moves by gravity from the top to the bottom), components leached from above are not always laterally removed but are often carried downwards, where, often, they accumulate. Consequently, the lower portions of weathering profiles are often enriched in the relatively mobile metals and it is usually not difficult to predict potentially accumulated metals in the deeper portions of a weathering profile, when the bedrock was the same throughout. In lithologically mixed profiles, however, the top rocks could be different from the bottom rocks; moreover, the top rocks may have gone by now (erosion removed). In such a case, the metals accumulated in the deeper reaches of a weathering profile, may be a surprise.

One can expect even more surprises, when the "weathering" (essentially the water movement) is reversed: that is, from the bottom up. This is in regions in which evaporation exceeds precipitation, and where the upward-seeping groundwater alters the rocks through which it passes, and occasionally deposits the elements (metals) it carries. There, surprises are commonplace, because one cannot usually see the buried metal source rock. To complicate matters even more, climates and physiographies change, and regoliths once formed in a rain-forest lowland swamp may now be exposed on top, or on flanks, of a desert mesa. There, incomplete (partially eroded) relict lateritic profiles undergo conversion into arid duricrust profiles, the consequence of which is a puzzling complexity. Australia, southern Africa, south-western United States, Mexico, Chile, Argentina, etc., are rich in such hybrid profiles, and their unravelling has only recently started. The potential for discoveries of ores there is high.

Because of the significant influence of the bedrock composition on the resulting metal accumulation in weathering profiles, the weathering has rather more the character of an interacting influence or an added factor than an agent to be emphasized separately as a mineralizer. For this reason, metallogenetic effects of weathering and the resulting ores have, to a considerable degree, already been covered in the preceding chapters (e.g. the Ni "laterites" have been reviewed in Chapter 7). The descriptions made earlier will not be repeated here, but references will be made. The ores, which for various reasons have not been described in the earlier or the following chapters, will be given greater attention.

FACTORS OR PROCESSES GOVERNING THE FORMATION OF WEATHERING AND SOIL PROFILES

Climate, vegetation zones, physiography, pedogenesis and other aspects treated in many branches of science, are of paramount importance in the understanding of weathering profiles, and prediction of their potential mineralization. Unfortunately, they are far too extensive to be covered meaningfully in this book, even in the form of a short abstract. Fortunately, excellent textbooks, reference books, and review articles on the subjects mentioned are available, and a short classified selection follows: (1) general, multidisciplinary textbooks: Strahler (1971), Holmes (1965); (2) climate: Köppen (1923), H.H.L. (1974); (3) soils: J.Pa. (1974), Hunt (1972); (4) pedogenesis and ore formation: Lelong et al. (1976); (5) weathering: Ollier (1975).

The excellent Physical Geographic Atlas of the World (Gerasimov, ed., 1964; in Russian) available in most reference libraries, contains colour maps of continents showing geology, geomorphology, soils, climate (including several hundreds locality climatograms), vegetation, etc. Soil maps of the world (1: 5 million, with 10 explanatory volumes), have been produced and published by FAO-UNESCO (1971).

23.2. HUMID BOREAL AND TEMPERATE CLIMATES: WEATHERING AND SOIL PROFILES

Most of us would not think of the temperate and boreal weathering profiles as sites of important ore accumulations, which they definitely are not. The profiles are important to an exploration geologist in a different capacity: as a bedrock cover from which, in the absence of an outcrop, geochemical samples are taken. Geochemical data on element mobility in weathering profiles are available mostly for the major elements (e.g. Lelong et al., 1976). The data for trace metals are fragmented and unsystematic.

Even with the low efficiency and incompleteness of leaching in the boreal and temperate profiles, iron frequently accumulates in the podzolic soil profiles at the base of the B horizon, in the form of the visually striking rusty "limonitic" clay, powder, encrustations or concretions. Often the "limonite" is manganiferous (e.g. Winters, 1938). In the past, such iron hydroxide accumulations often served as a resource base to backyard iron industries.

Soils developed on, and mixed with, metalliferous sediments (mostly placer sands) grade downward into the "ore" and may contain the same economic minerals in lesser quantities. The surficial portion of some fossil ilmenite, zircon, etc. placer sands has a high proportion of infiltrated humus, and can be classified as a sandy soil. Many organic-rich soils developed on alluvial gold placers had substantial gold contents. It is now widely recognized that many gold nuggets (particularly the compositionally zoned nuggets, that have a substantially higher purity than the bedrock gold) were formed by a partial or total accretion of gold in continental environment, particularly in placers (compare Boyle, 1979). It is conceivable (but not sufficiently well documented) that nuggets could also form in the soil profiles over auriferous deposits. In the literature, gold contents in soils reaching up to 6 ppm, and gold contents in ashes of recent plants of up to 600 ppm, have been reported. Boyle (1979, p.79-85) compiled a very detailed review including his own data. Although most anomalous gold values in soil and plants came from known areas of gold mineralization, some values came from regions where no gold occurrences are known, and some from areas where they are rather unlikely to be expected. This might indicate a significant gold concentrating ability of some plants (e.g. the horsetail). Although the time when industrial plantations of terrestrial plants may be harvested for gold and other metals extraction may still be several decades away, plants have already been used for geobotanical exploration with varying degree of success in several countries

ZONES OF WEATHERING ALTERATION OVER SULPHIDE DEPOSITS

Sulphide deposits are particularly strongly transformed by weathering. Gossans (iron hydroxide – cemented oxidation zones) and zones of secondary enrichment (of low-oxygen oxides and low-sulphur sulphides) are most typical. This subject is too extensive to be reviewed satisfactorily here, and the reader is referred to the classical (Emmons, 1917; Smirnov, 1951), as well as to the more recent (Blanchard, 1968; Blain and Andrew, 1977) reviews. Brief references are also made in chapters dealing with the "primary" deposits.

23.3. HUMID TROPICAL CLIMATES, WEATHERING AND SOIL PROFILES

23.3.1. Introduction

The "most typical" contemporary tropical weathering profiles free of superimposed modifications, form actively in the present rain-forest vegetation zone. There, however, they are masked by vegetation and so rarely exposed (Fig. 23-3). The available temporary exposures at slide scars on valley slopes, in road cuts and excavations, in agriculturally developed regions, etc., are usually incomplete and are rapidly modified by erosion. Less typical, to some degree modified, tropical weathering profiles that include many of those studied in detail and described in the literature, are now located in dissected savannas or even semi-deserts. The post-formational modifications include partial enduration by drying, erosion and dissection. The profiles display a considerable local variety and almost every outcrop is a combination of a relict

Fig. 23-3. Forari Mn deposit, Efate Island, Vanuatu (formerly New Hebrides). Surficial installations of the mine, densely overgrown by jungle six years after closure, indicating the intensity of humid tropical weathering and plant propagation.

weathering or soil profile formed in the past, and a currently forming profile that reflects the contemporary climatic conditions (Fig. 23-4).

A first-time geological visitor to a tropical country educated in the stereotype belief that, in the tropics, everything is completely, thickly and uniformly weathered and that solid rocks occur only in deep mines and in museums, is often pleasantly surprised to find locally fairly abundant bedrock outcrop. This is because the outcrop is more a matter of physiographic dissection than the intensity of weathering. In the tropical countryside of central Brazil (e.g. along the fringe of the Paraná Basin in central Goiás) underlain by Precambrian metamorphics intruded by Mesozoic alkaline rocks, bedrock outcrops are relatively plentiful. There, the depth of weathering ranges from several centimetres to several hundreds metres, and the depth of soil cover (lateritic soil = latosol) from zero to several tens of metres (Fig. 23-5). Scattered low knobs and hills marked by piles of exfoliation-rounded intrusive rock boulders grade rapidly into the intrusive bedrock. Samples of completely fresh rocks can be obtained from broken boulders, the surface of which is marked by weathering rinds, a few centimeters thick. Downslope from the knobs, the boulders grade into a sandy colluvium and ultimately into a red lateritic soil. The greatest thickness of the red soil underlain by saprolite is in valleys which, in turn, often follow faults or shear

Fig. 23-4. Two residual deposits established in thick relict tropical weathering profiles presently being modified by a climatically milder pedogenesis. (A): Araxá, Brazil, the world's richest Nb carbonatite, produces residual pyrochlore from a thick ocherous regolith over a carbonatite. (B) shows the ore appearance. (C): Cartersville, Georgia, produced Fe oxide ochers, Mn oxides, and residual barite from a Tertiary regolith topping lightly mineralized Cambrian metasediments.

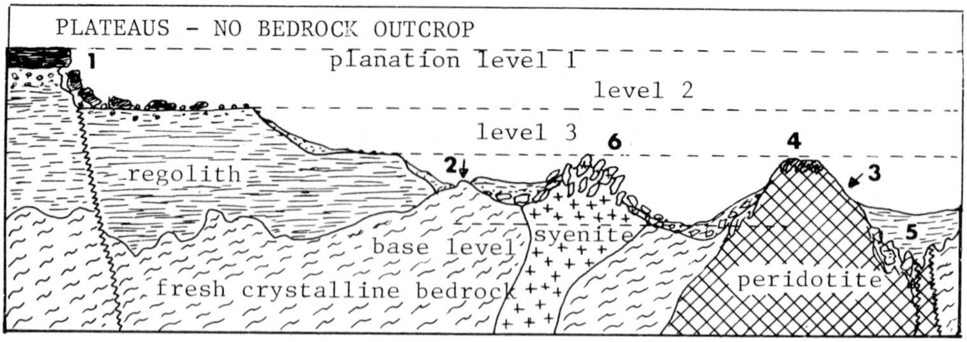

Fig. 23-5. Regoliths and bedrock outcrops in a dissected tropical landscape (as on the fringe of the Paraná Basin, central Goiás, Brazil). (1) Hardpan (ferricrete)-capped oldest planation surface. The basement crops out in road cuts (2) and on the flanks of an exhumed peridotite "morro" (3). (4) Remnants of saprolite under the third planation surface. (5) Colluvial and alluvial, transported lateritic clay. (6) Bouldery outcrop.

zones. Much of this soil is transported rather than residual and it is frequently possible to find, in road cuts and excavations, bouldery bedrock outcrops (or single boulders with fresh rock kernels) covered by, or suspended in, the lateritic soil.

The abrupt junction between the unweathered, massive rocks and their tropically weathered equivalents has been noted several times in the literature and named "basal surface of weathering" or "weathering front" (Ollier, 1975). This does not, however, develop in soft and/or porous rocks such as shales or sandstones. The area in central Brazil discussed above is located on a plateau which, in its early history, was as featureless as many present lowland basins. Its dissection was due to a slow but steady uplift and stream incission. During the long history of landscape development, several erosion surfaces formed separated from each other by scarps. The bedrock outcrops are on the lowermost surface currently undergoing development, while the older surfaces in the area are flat, featureless, generally free of bedrock outcrop and often duricrust-capped. Due to the variations in climate, successive landscape surfaces may have produced slightly or substantially different relict weathering crusts. When such crusts are metalliferous, understanding of the regional physiographic history is paramount in exploration.

23.3.2. Vertical zoning of tropical soil and weathering profiles

Compositional (chemical and mineralogical) and textural zoning is an integral part of every tropical weathering profile. The zoning is essentially a consequence of repeated hydration, breakdown and leaching of the original bedrock constituents, combined with removal of the more mobile elements. The physico-chemical principles of deep tropical weathering have been discussed frequently in the literature (e.g. Lelong et al., 1976; Ollier, 1975; Sivarajasingham et al.,1962) and will not be repeated here.

A deep tropical weathering is a great lithological equalizer: that is, the uppermost zones (those lacking the relict bedrock structures) developed on granite, gneiss, arkose, rhyolite, diorite, andesite, even gabbro, basalt, etc., would, under equal climatic and physiographic conditions, be equivalent. It would not be possible to distinguish what the original bedrock was, and distinct maxima of element accumulation (e.g. Fe at top, Al under it; etc.) would have the same position regardless of the bedrock composition. There are, however, limits. Iron and alumina-free rocks such as quartzites or pure limestone, will obviously not produce weathering profiles identical to those of granites, due to the lack of Fe or Al. Alternatively, the thickness of such profiles would be strongly reduced, because only the impurities in the quartzite or carbonate would take part. Rocks anomalous in iron (e.g. banded iron formations) or alumina (e.g. shales, nepheline syenite), on the other hand, would produce unusually strong metal accumulation peaks in the respective zones of the weathering profile. This has an important economic implication.

As one moves down the profile, the differences due to the bedrock composition will become more pronounced. Relict structures would appear, and the incompletely leached lower zones would increasingly display a strong bedrock heritage: for example, relict quartz would be found in the residual clay developed over a granite, but would be missing in a corresponding zone over a syenite. In a "most typical" lateritic profile developed on geochemically "average" silicate rocks (e.g. granitoids, gneisses, shales) and exposed in a present savanna climatic and vegetation zone, the following zones usually can be distinguished (e.g. Schuylenborgh, 1971; Zeegers and Leprun, 1979; Fig. 23-6); from surface downwards:

(1) Zone of a young (Recent) surficial soil; loose, with a little humus or without humus. Commonly, the top is sandy (A_1), the bottom clay-rich. The thickness ranges from several centimetres up to approximately 1 m.

(2) Iron crust (cuirasse, ferricrete), which is a reddish-brown to brown, very resistant sheet about 1-5 m thick, capping plateaus and mesas. Its structure and texture is variable; most commonly hematite or goethite concretions, pisoliths and fragments cemented by the same interstitial minerals. The rock ranges from porous to compact. Blocks and fragments of cuirasse litter the flanks of the plateaus topped, and spread onto the surface of the lower erosional levels and

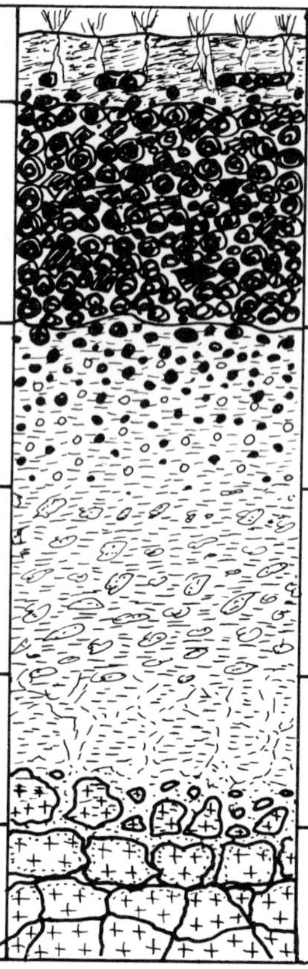

1	SURFICIAL ZONE; Recent soil of variable colour,may be humous
2	FERRICRETE (cuirasse,hardpan); hard,resistant,porous to compact crust (slab) composed of hematite,goethite concretions,pisoliths or fragments cemented by the same minerals. Minor clay pockets may be present.
3	SOFT FERRICRETE (carapace); concretions of goethite/hematite in ocher,gibbsite or kaolinite matrix. Sometimes gibbsite nodules. Red to pink.
4	SPOTTED (MOTTLED) CLAY; kaolinite $\pm$ gibbsite,goethite,hematite. Yellow or brown iron pigmented spots are surrounded by lighter clay matrix. Minor preserved residual bedrock minerals,e.g.quartz.
5	SAPROLITE (lithomarge); kaolinite $\pm$ montmorillonite residual clay with preserved bedrock texture. Widespread residual quartz above granite.
6	PARENT ROCK (BEDROCK); blocky,bouldery and bleached on the top.

Fig. 23-6. Usual zones of a common lateritic weathering profile over silicate rocks (e.g. granite). From Laznicka (1984).

into streams.

(3) The hard, resistant cuirasse passes downwards into a soft, plastic, non-endurated equivalent ("carapace"). This is of lighter colour than the cuirasse, made up of goethite and hematite pisoliths, concretions and fragments in red powdery Fe oxides and clay (gibbsite or kaolinite) matrix. The iron oxide microconcretions are often of uniform size ("buckshot"), and litter the lower erosional surface on which they spread by the action of running water.

(4) In the spotted (mottled) clay zone, irregular yellow, brown or red iron-pigmented spots formed on the background of a light-coloured kaolinite and gibbsite clay. Goethite or hematite pisoliths from the zone (3) are a common accessory. This zone usually formed by an "in situ" argillization, so it is technically a part of saprolite ("expanded saprolite"). The parent rock texture, however, is not preserved.

(5) The underlying zone ("pain d'epice", "lithomarge"), preserves the bedrock texture, and is also compositionally immature. It usually consists of kaolinite and/or montmorillonite, nontronite, chlorite (when formed over an intermediate or a basic rock).

(6) The base of the profile is the parent rock, usually blocky and partly bleached (discoloured) along joints, fractures and faults. The boundary of the zones (5) and (6) is rarely regular and sharp; partly leached to fresh, loose, bedrock blocks are usually abundant above the bedrock base

23.3.3. Variations of tropical weathering profiles over different bedrocks

Tropical weathering profiles over other than the "average" rocks may (1) have the same zones as reviewed earlier, but of different thickness (e.g. iron-deficient rocks have thin zones 2+3); (2) have some zones missing; (3) be topped by a completely different type of weathering crust (Fig. 23-7). Quartz-free, aluminous rocks yield bauxitic zones 3 and 4. Ultrabasics free of quartz and feldspar develop the very predictable classical lateritic profiles dominated by Fe-smectites (nontronite) in zones 4 and 5 and grading directly into prominent Fe oxide or hydroxide-rich toppings (compare Chapter 7).

Carbonate bedrocks yield a great variety of tropical soil profiles and the main peculiarity is that there is practically no equivalent to the saprolite in silicate rocks. Chemically pure carbonates do not "weather" (i.e. undergo hydration and oxidation); they dissolve and yield only their minor insoluble residue to enter the soil-forming process. Consequently, zones 4 and 5 are not developed in pure limestones and dolomites and fresh carbonate is in direct contact with the equivalents of zones 2 and 3. Massive ferricrete is, however, never developed. Sometimes, tropically weathered carbonates do contain saprolitic units, but these were formed from interbedded and topping silicate units. This seems to be of particular importance in the interpretation of bauxite origin, which is discussed later. The "ordinary" chemically pure limestones, for example those of young uplifted coral-floored platforms, produce a very thin residual soil mantle. Most of such soils are red or reddish-brown ("terra rossa"), and despite the visual similarity with the Jamaican bauxitic soils, few terra rossas contain substantial amounts of alumina hydroxides. Isphording (1973) studied the residual soils of the Yucatan Peninsula (Mexico) and found that both zoned rendzinas and unzoned terra rossas were represented. Their thicknesses ranged from less than 10 cm to 3 m; halloysite and chlorite were the major and talc, boehmite and montmorillonite the minor components. Al_2O_3 comprised over 33% of the total, but there was an average of 46% SiO_2 which makes such soils unsuitable as an alumina resource processable by means of present technology. Elsewhere as in the Mesozoic carbonate terrains of southern Slovakia and northern Hungary, the recent terra rossa is rich in iron oxide and hydroxide pisoliths, and contains 32-50% Fe_2O_3 and

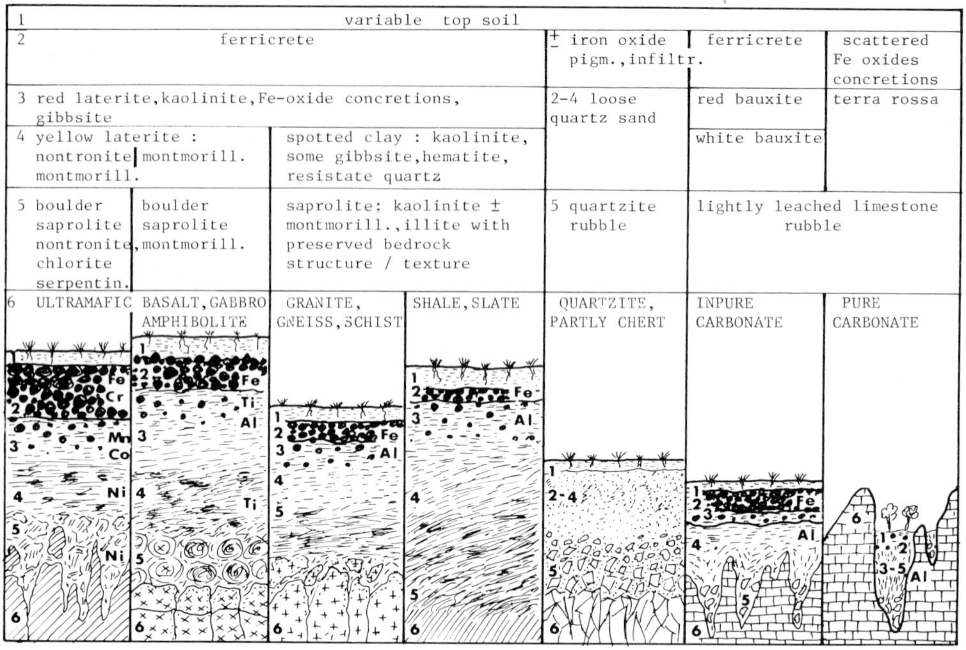

Fig. 23-7. Usual variations in tropical weathering/soil profiles over selected bedrocks. Chemical symbols indicate zones of maximum metal accumulations that could result in economic deposits. From Laznicka (1984).

23-26% Al_2O_3 (Ilavský, 1976). The clay fraction is represented by halloysite and minor gibbsite.

A variety of locally distinct weathering profiles has been recorded in impure carbonates, particularly those associated with "unusual" lithologic associations such as iron formations. Dorr (1969) described the "splash rock" widely formed on the Proterozoic manganoan and ferroan dolomite in the Quadrilatero Ferrífero, Brazil. This is a chocolate brown to black fine cellular boxwork of limonite with Mn oxides occurring locally, enclosing voids now filled with water. When struck with a hammer, the hammer sinks into the material with a splash; hence its name. Splash rock is an excellent guide to the underlying carbonates in deeply weathered terrains. Compositionally identical, but texturally different from splash rock and found in the same area, is a porous lateritic soil containing abundant pisoliths or microconcretions of Mn oxide or limonite. The Mn oxide accumulations reach economic proportions locally. High-alumina clay and structureless bauxite in the Quadrilatero Ferrífero forms from the aluminous residue after impure carbonates in zones 3 and 4, just under the ferricrete ("canga") capping.

23.3.4. Ironstones in weathering profiles and their variety

Ferricrete, ferruginous nodules, and other forms of iron oxides and hydroxides in tropical weathering profiles are certainly impressive. Unless they, themselves, are the target of exploration, they are a serious obstacle to exploration for other metals, because they (1) cover and mask the bedrock; (2) cause false anomalies (e.g. some concretionary "buckshot" iron oxides are magnetic (maghemite), so they cause magnetic anomalies; Fig. 23-8); and (3) they may be visually indistinguishable from gossans and other varieties of ironstone accumulations. This latter aspect has been addressed by Butt (1981), because it is of considerable importance in the interpretation of outcrop evolution in Australia. Butt distinguished a variety of types of ironstones, present in the partly erosion-stripped, deeply weathered Australian landscape: ferricrete or lateritic layered ironstone; rubbly and transported laterite; gossan and transported gossan ironstone; fault leakage and infiltration ironstone; "stratigraphic" ironstone (i.e. outcrop expression of ferruginous rocks) and drainage ironstone (one precipitated in stream beds). He also discussed the methods of discrimination. No single foolproof method was found, and a discriminant analysis of multiple factors was recommended.

Fig. 23-8. "Buckshot", residual iron oxide granules common on old planation surfaces, as in Australia.

23.3.5. Ore-grade weathering profiles (metalliferous "laterites")

As already noted in the introduction, the deep tropical weathering causes a significant geochemical differentiation in all the rocks over which it is superimposed, but only a fraction of such differentiates are called "ore". For example, zones 2 and 3 (Fig. 23-7) are always iron-enriched compared with the bedrock, but although at least 20 million km^2 of the present and former humid tropics are underlain by such zones, only a fraction has ever been considered to be an iron ore and treated as such in the economic geologic literature. Iron accumulations economic under present conditions do occur in the tropical belt in the "right" zones of the weathering profile, but have to be "unique": thicker, of higher grade, more extensive, suitably located, etc. To form, they require better than average conditions, many of which have been discussed earlier. The single most important condition is an anomalously iron rich bedrock precursor (e.g. a banded iron formation, ultramafics). The same conditions apply to other metals. The variety of ore concentrations in tropical weathering profiles is considerable, and they can be subdivided according to their mechanism of derivation of the ore substance that can be: (1) residual, unmodified, also designated as "elluvial placers" (usually resistate minerals such as chromite, rutile, cassiterite, gold, etc., originally present in the bedrock); (2) modified residual (the original resistate minerals were mineralogically or chemically altered during their residence in the weathering profile, for example ilmenite into leucoxene); and (3) authigenic (formed "in situ" within the lateritic profile; e.g. Fe oxide pisoliths, gibbsite).

BAUXITE AND ALUMINOUS LATERITE

The classification of bauxites is a highly subjective and evolving exercise and it will not be treated here; the reader is advised to consult Bárdossy (1982) or Lelong et al. (1976) for a recent review. Bauxite is an aggregate of hydrated alumina oxides, which many geologists associate with karsted carbonates. The terms "Al-laterite" or "lateritic bauxite" are often given to residual bauxites formed "in situ" by preferential lateritic weathering of a silicate rock. Both varieties are a part of tropical weathering profiles. Some of them still form actively, but the majority is "dead" (completed) and slowly undergoing dispersion. Many have been mechanically reworked and now appear as detrital sediment in continental sediments (Chapter 25) or, occasionally, in marine sediments.

Only a few ancient (pre-Tertiary) deposits remain "rooted" in the weathering profile in which they formed, protected from destruction by a superimposed rock cover. Most of the geologically old lateritic bauxites presently outcropping on the surface had been buried shortly after their formation, but recently exhumed. Most residual bauxites formed on silicate rocks form blankets, whereas most bauxites resting on carbonates fill a variety of karst depressions. There is, however, a complete gradation and overlap between both categories.

The origin of bauxite is not as straightforward and simple as it might appear, and many aspects of it still remain unanswered (compare Lelong et al., 1976; Kotschoubey and Truckenbrodt, 1981). First of all, the typical lateritic profile on a common rock (Fig. 23-7) need not be representative for many bauxite deposits developed on equivalent or on quartz-free (e.g. syenite) bedrocks. Bauxite often takes the form of a residual saprolite with a well developed relict texture (that is, it is in zone 5; e.g. the granitic—textured bauxitic clay in Arkansas, Fig. 23-9), or it could substitute for and be texturally equivalent to the spotted clay (zone 4). It is now generally accepted that most bauxites are polygenetic and form in response to several phases of uplift and subsidence, which in turn govern the successive cycles of selective leaching and induration. Some of these cycles may involve upward-seeping fluids, and the plateau bauxites of northeastern South America have recently been placed into the category of duricrusts by Aleva (1979).

The appearance of bauxite indeed varies significantly as a consequence of enduration, from a soft, structureless, massive, clayous, or powdery white to dark red material one would not notice unless brought to it by a local geologist (e.g. most of the residual bauxites in the Quadrilatero Ferrífero, Brazil) to hard, indurated, plateau-capping white to red pisolithic and concretionary bauxitic claystone sheets (e.g. the Guyanas, N.E. Brazil). The former tends to be prominently covered by a protective capping of ferricrete. The latter may often entirely lack ferruginous capping.

The process of bauxitization requires a well-drained setting usually along the edges of topographically high plateaus (remnants of ancient peneplains), or porous, karsted or fractured basement rocks.

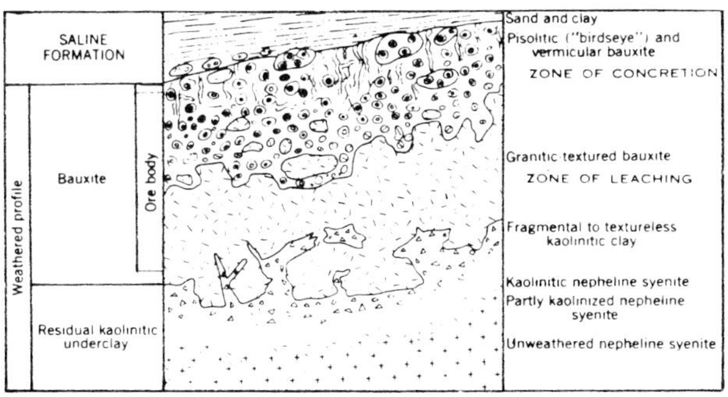

Fig. 23-9. Granitic-textured residual bauxitic clay in the Arkansas bauxite district, formed in situ over a Cretaceous nepheline syenite. From Gordon et al. (1958).

Two aspects are essential in the bauxite genesis: (1) removal of silica (quartz); and (2) removal of iron. In the common tropical weathering profiles over granite, residual quartz dilutes the gibbsite content and makes the product economically unusable. Quartz is ultimately soluble and removable under the conditions of tropical weathering, but it takes too much time and requires long periods of stability. The quartz removal has, however, often been accomplished by physical means, such as by physical reworking of a kaolinized or partly gibbsitized granitic or gneissic regolith, into separate layers of sand and clay. In the next phase, the pure clay underwent bauxitization. Some of the world's largest "plateau bauxites" (in Guyana, Suriname, Brazil, northern Queensland) formed in such a way.

Iron and alumina are intimately intermixed in ferralitic soils, the bauxite precursor, and the separation of both elements requires a repeated selective leaching. Obviously, such leaching has been less complete in bauxites formed over mafic or ferruginous precursors.

From the viewpoint of lithologic associations and characteristic settings, several bauxite styles can be distinguished and they are reviewed briefly below.

(a) "Peneplain" (or platform) residual bauxite blankets formed primarily by bauxitization of young clays or arkosic sandstones

This style represents some 15 Bt of high-grade bauxites in north-eastern Brazil, Guyana, Suriname, northern Australia, and elsewhere (Table 23-1). In all the areas mentioned, Precambrian crystalline basement composed of gneisses, granite gneisses, minor charnockites, etc., has been peneplained and covered by flat-lying Mesozoic and Cainozoic platformic sediments: mostly alluvial sand(stones) alternating with clay(stones). The bauxite deposits are located at distinct planation surfaces, currently subdivided into a series of erosional remnants (Fig. 23-10).

In the Paragominas district (north-eastern Brazil; Kotschoubey and Truckenbrodt, 1981; Fig. 23-10B), the bauxite formed by lower Tertiary alteration of upper Cretaceous continental claystones, interbedded with friable sandstone. In the basal zone of the weathering profile, the saprolitic sediments grade upward into a zone of spotted kaolinitic clay that contains abundant gibbsitic concretions. Above is the main ore zone of massive, compact, columnar, light coloured bauxite, locally alternating with red ferruginous bands. This is the main ore horizon about 1 m thick and underlain by 2 m of nodular gibbsite. The orebody is topped by a red, nodular to massive ferricrete which, in turn, is sometimes overlain by another, friable horizon of bauxitic concretions. The profile is topped by several metres of clay.

In Suriname (Aleva, 1979), the bauxite shows similar zoning, and it formed by lateritic weathering of lower Tertiary kaolinitic clays and sands during the Eocene-Oligocene hiatus that coincided with the period of peneplainization and intensive weathering. The bauxite bodies are exposed on flat topped hills (Moengo Hills), rising 5-80 m above the coastal plain, or buried under a cover up to 30 m thick of

Table 23-1. Cainozoic "peneplain" (platform) bauxite blanket deposits, formed on young continental claystones, siltstones and sandstones

REGION	SOURCE SEQUENCE		BAUXITE			TONNAGE	REFERENCE
	age	main rock types	age	compos.	textures		
Lower Amazon Basin, N.E. Brazil	Q Pl	kaolinitic clay feldspathic sand	Q Pl	gibbsite kaolin	massive concretions	3 Bt Al	Dennen and Norton (1977)
Paragominas district, N.E. Brazil	Cr_3	claystone sandstone	Eo-Ol	gibbsite kaolin	massive, concretions	319 Mt Al	Kotschoubey and Truckenbrodt (1981)
Guyana and Suriname Coastal Plain Linden, Onverwacht	Cr_3 T_1	kaolinitic clay sandstone over PCm basement	Eo-Ol	gibbsite kaolin	concretions massive low Fe	400 Mt Al/ 30.7%	Aleva (1979)
Cape York Penins. Weipa, Arukun N.E. Queensland	Cr_3 T	arkosic sand clay, silt	Cr_3 T	gibbsite boehmite kaolin	concretions loose to compact	1 Bt Al/ 28.6%	Evans (1975)
Gove (Nhulunbuy) Arnhem Land, N.Australia	Cr_1	mottled clay resting on PCm metamorphics	Cr_3 Tl	gibbsite hemat. goethite	as above	68 Mt Al/27%	Somm (1975)
Gulf Plain, U.S.A. Andersonville, Eufaula	Eo	clay, sand	Eo	gibbsite kaolin	clayous, concretions	7 Mt Al	Patterson (1967)

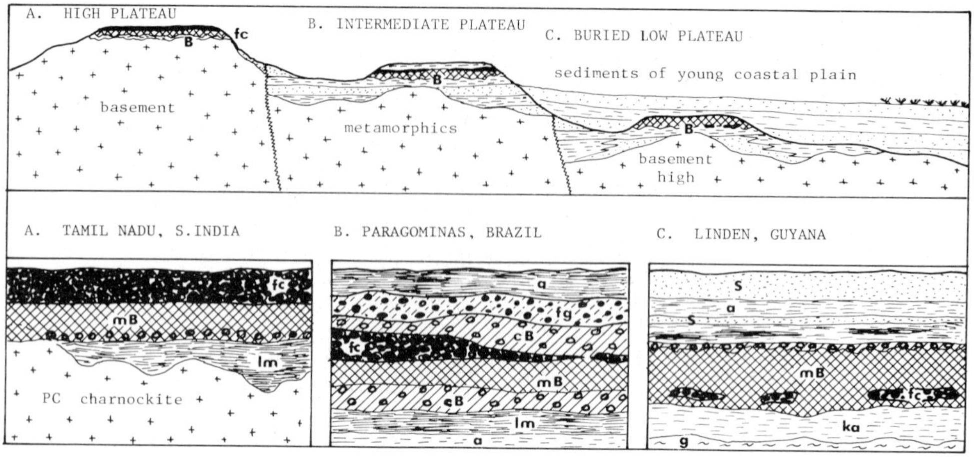

Fig. 23-10. Common types of "plateaus" (old planation surfaces) underlain by a bauxite regolith. fc=ferricrete; B=bauxite; mB=massive (structureless) bauxite. From Laznicka (1984).

young sediments (Onverdacht). Bauxites have virtually the same setting in the Demerara River district in neighbouring Guyana (Linden; Figures 23-10C, 23-11). It is interesting to note that no substantial bauxitization took place after Oligocene in the Guyanas. In contrast to this, the extensive bauxite deposits in the Lower Amazon basin (Rios Trombetas and Jari; Dennen and Norton, 1977) have been dated as of Pliocene or Plio-Pleistocene age, formed on Pliocene silty clays.

The north-eastern Queensland bauxite deposits of the Cape York Peninsula (Weipa, Arukun, Escape River; Evans, 1975) form an extensive blanket of unconsolidated to semiconsolidated gibbsite, boehmite and kaolinite pisoliths, to compact claystone. The ore blanket is, on average, 2.4 m thick and it overlies Tertiary arkosic sandstone, claystone and siltstone resting on peneplained metamorphics.

(b) Residual bauxite on nepheline syenite, syenite, phonolite and anorthosite

Bauxitization of alumina-rich, quartz-free parent rocks low in mafic minerals produces high quality bauxites. Unfortunately, such bauxites are rather rare because of the uncommon coincidence of the feldspathic or nephelinitic rocks, tropical climates and suitable physiography. Several examples of deposits (Arkansas, Iles de Los, Poços de Caldas) formed on nepheline syenite or phonolite, are also noted or briefly described in Chapter 33.

In the Iles de Los bauxite field (near Conakry, Guinea; Lelong et al., 1976), the weathering profile starts with a massive, light

Fig. 23-11. Bauxite (B) at Linden, Guyana. The ore layer is capped by lignitic clay and sand.

coloured, indurated bauxitic crust with a compact or vesicular structure, locally associated with bauxite breccia or a pisolitic bauxite. This grades downward into a friable to lightly indurated porous bauxite with intrusive relict texture, and a fresh nepheline syenite.

(c) Residual ferruginous bauxites on basalts, gabbros, greenstones and amphibolites

These bauxites, also discussed briefly in Chapters 7, 31 and 32, are usually dark-red, high in iron (23-30% Fe_2O_3; 40-50% Al_2O_3), and gradational to aluminous ironstones. They are also highly titaniferous (5-8% TiO_2). The deposits of ferruginous lateritic bauxites on the Kauai and Maui Islands, Hawaii (Patterson, 1971), are examples of Quaternary weathering profiles that are still forming actively in the rainforest environment. The profiles take the form of red, brown and brownish-gray mantle developed over Tertiary basalts, and are 20 to 42 m thick. Their greater portion is a saprolite in which the bedrock structures are well preserved. Remnants of spheroidally rounded residual boulders of hard basalt are common. The typical saprolite, however, is soft enough to be dug with a pick and it is protected by vegetation from erosion. When the vegetation is removed, destruction of the profile follows rapidly. In areas of good drainage, the saprolite is topped by a thin veneer of lateritic soil, and both are enriched in Al, Fe and Ti to such an extent that they constitute a poor quality ferruginous bauxite (e.g. on Maui,

33.4% Al_2O_3, 31.4% Fe_2O_3, 6.7% TiO_2). In poorly drained, swampy areas the thickness of the lateritic layer increases considerably (up to about 10 m), and a light-coloured plastic clay composed of halloysite, illite, quartz, anatase and gibbsite forms. Gibbsite accumulates mostly in impure nodules. These clays are low in iron, have a moderate alumina content (max. 35% Al_2O_3), and anomalously high titania (up to 30.6% TiO_2).

Bauxite deposits in the Serra dos Carajás, north-eastern Brazil (e.g. Clareira deposit; Assad and de Rui, 1978; 120 Mt bauxite), are an example of the "platform" (plateau) style, formed on peneplained mafic metavolcanics. The bauxite forms a continuous blanket, several metres thick, on the top of plateaus that stand approximately 100 m above the surrounding countryside, and are bordered by steep slopes. The weathering profile zones, from top to bottom, are as follows: (1) earthy, ferruginous, friable gibbsitic bauxite containing a small percentage of ferruginous nodules and pisoliths near the bottom; (2) porous and cavernous, sometimes pisolithic ferruginous laterite; (3) plastic reddish clay (lithomarge); (4) metabasalt saprolite; and (5) fresh rock.

(d) Residual bauxites on high-grade metamorphics

Gneisses, migmatites, granulites, charnockites, etc. often support tropical weathering profiles containing economic bauxite deposits. The "platform"-style bauxites topping the erosion-dissected peneplain plateaux are most common in India (e.g. Lohardaga in Bihar, 17 Mt bauxite on granite gneiss; the East Coast District in Orissa and Andhra Pradesh, 1.5 Bt bauxite on garnet-sillimanite gneiss; and Tamil Nadu State, southern India, e.g. Salem District, Nilgiri Hills, bauxite on charnockites; Krishnaswamy, 1972). The Tamil Nadu bauxite cappings are located on a high plateau (1,250-2,285 m altitude; Subramanian, 1978; Fig. 24-7A), and are irregular in shape and areal extent. Generally, the exposed portion has a brick red, scoriaceous and pitted ferricrete, passing downward into a fine grained, white to brown gibbsite, minor boehmite bauxite. Streaks and lenses of an almost pure, white gibbsite clay occur at some localities. The bauxite is underlain by kaolinitic clay and a weathered charnockite. The bauxites are both "primary" (i.e. in situ, charnockitic relict texture) and "secondary" (expanded, structureless).

Although the bauxite demonstrably formed by a direct weathering of the exposed crystalline bedrock, it appears that in some instances the bauxitization selectively and preferentially affected a pre-existing mantle of colluvial coarse feldspathic sediments, derived from the metamorphics. This has been demonstrated convincingly by Grubb (1971) in Jarrahdale, Western Australia. The latter case represents a transition into the blanket-style bauxites formed on the young continental clays and sands, described earlier. Aleva (1979) noted that in Suriname many bauxite blankets surround buried basement knobs.

(e) Residual bauxite on other silicate rocks

Residual bauxites formed extensively on the young volcanic ash and tephra blankets, as in Costa Rica; on a hornfelsed Triassic slate and volcanics situated in a granite contact aureole, as on the Bintan and Singkep Islands, Indonesia; on banded iron formations and associated impure (argillaceous) dolomites as in the Quadrilatero Ferrifero, Brazil; and on a variety of other rocks. It is unfortunate that the bedrock composition is frequently missing in publications reviewing bauxites.

(f) Cainozoic residual bauxites on carbonates (uncovered): Jamaica bauxite province

A considerable portion of the island of Jamaica, West Indies, is underlain by middle Eocene to lower Miocene, very pure limestone, formed on a carbonate platform. The surface of this limestone has been converted into a tropical karst, a system of rolling, densely vegetated, low hills and depressions. Finely divided, dark red, reddish brown to yellow, soft and loose bauxite resembling a silty soil in consistency, fills the karst depressions over a vast area of the island. The bauxite is moderately to highly ferruginous (about 20% Fe_2O_3), medium to high-grade in alumina (about 50% Al_2O_3), and has some titania (about 2% TiO_2). The orebodies are pockets and blankets governed by the shape of depressions in which they occur (Comer, 1974; Fig. 23-12). The overall reserves are of the order of 1.2 Bt bauxite.

The bauxite orebodies are simple. The ore starts at the very top of the soil layer and continues to a depth of between one and about 10 m. The thickest bauxite layers measured 33 m. Towards the base, the bauxite is diluted by carbonate rubble and when resting on dolomitized limestone, by residual dolomitic sand. In the footwall are karsted limestone pinnacles or scarps.

The fundamental problem in Jamaica and at a few other places where similar bauxite has developed (e.g. S.W. Haiti, Dominican Republic), is the alumina source. Although a purely residual origin of bauxites has been demonstrated beyond doubt when the parent rock was an impure (argillaceous) carbonate as in the Quadrilatero Ferrifero, Brazil, a similar source is unconvincing when the bedrock is a very pure limestone because of the enormous volume of $CaCO_3$ that would have to be dissolved and removed to yield the insoluble residue. Comer (1974) argued that the source of the Jamaican bauxite was an airborne Miocene volcanic ash accumulated as a blanket on the limestone surface, gradually washed down into the karst depressions, and subject to bauxitization. Sporadic occurrences of bentonitic clay layers in the middle-upper Miocene carbonates in Jamaica as well as the bauxite trace geochemistry, supports this interpretation. The bauxite is predominantly gibbsitic with some hematite, boehmite and kaolinite.

IRON ORES

Although ferricrete and the Fe oxide concretionary clays constitute a good iron ore, often having a grade exceeding 40% Fe, they have

Fig. 23-12. Bauxite (B), a small blanket on the bottom of a karst depression, northern Jamaica.

rarely been produced in the past. The main problem was the distance from the markets. The few exceptions of mined and shipped ores include the ferricrete from Mayarí, Cuba and Conakry, Guinea. Ferricrete represents an enormous resource of a medium quality iron ore and only a few regions have been thoroughly investigated so far and had their reserves estimated (Table 23-2). In New Caledonia alone, thick and high-grade ferricrete formed over ultramafics represents some 15 Bt of a 30-65% Fe ore (Fig. 23-13).

It has been noted earlier and indeed it is a central thesis of this book, that a geological process of metal accumulation superimposed on an originally metal-rich precursor might produce a still richer final product, ore. Applying this rule, tropical weathering superimposed on an earlier ironstone or an iron formation is, theoretically, likely to produce a richer and thicker equivalent of a ferricrete: a super-ferricrete. This has been empirically confirmed, and indeed the "hematite-enriched banded iron formation", the most important single category of iron ores of the present century, occurs predominantly in regions of coincidence of the B.I.F. occurrence and the present (or past but geologically young) weathering (e.g. in Brazil, Venezuela, S.Africa, W. Africa, Australia, India). Some exceptions to this rule such as the enriched hematitic ores of the Lake Superior region (e.g. the Mesabi Range) and the Labrador Trough, areas affected by Quaternary glaciation, have been satisfactorily explained. There, the

Table 23-2. Selected localities of ferricrete / Fe-oxide concretionary ores in tropical weathering profiles over silicate rocks, mined in the past or investigated as potential future Fe (Ni,Cr,Co) deposits

LOCALITY	PARENT ROCK age	PARENT ROCK lithology	AGE OF WEATH.	TONNAGE,GRADE	REFERENCES
Mayarí, N.E. Cuba	J-Cr_3	serpentinite	T,Q	49 Mt Fe/49%; 1.3 Mt Ni/1.3% 100 Tt Co/0.1%	Boldt (1967)
Donbrubari, etc., Suriname	MZ, T	basalt,diabase ?	T,Q	109 Mt Fe/45%	Putzer (1976)
Medellín to Belmira Antioquía, Colombia	Cr	ultramafics	T,Q	188 Mt Fe/47%; 1.4 Mt Ni/0.35% 5.4 Mt Cr/1.35%	Singewald (1950)
Kukusan Mts.,S.E. Kalimantan, Indones.	Cr	ultramafics, metabasalts	T,Q	76 Mt Fe/45%; 1.12 Mt Ni/0.66% 1.1 Mt Cr/0.64%	van Bemmelen (1949)
Sebuku Ialand, S.E. Kalimantan, Indones.	Cr	ultramafics, metabasalts	T,Q	135 Mt Fe/45%; 1.98 Mt Ni/ 0.66%; 1.92 Mt Cr/0.64%	van Bemmelen (1949)
Surigao del Norte N.E. Mindanao, Philippines	Eo	serpentinite	T_{2-3}	207 Mt Fe/47.8%; 5 Mt Ni/ 0.8 - 1.1%; 12.26 Mt Cr/2.85%	Bryner (1969)
New Caledonia Isl., S.W. Pacific	Eo	peridotite,minor serpentinite	Ol-Q	15 Bt Fe/30-65%	Paris (1981)
Conakry area,Guinea W. Africa	MZ	dunite	T,Q	650 Mt Fe/55%; 15 Mt Cr/1.25%	de Kun (1965)
Agbaja and Enugu areas, S.E. Nigeria	MZ?	basalt, diabase	T,Q	66 Mt Fe/51%; 21.5 Mt Fe/43%	de Kun (1965)

Fig. 23-13. Landscape in southern New Caledonia (A), showing dissected, rubbly iron oxide cuirasse (C) formed on tropically weathered peridotite. (B) shows an almost massive, dense, brecciated hematite.

hematites formed by an earlier (Proterozoic to Mesozoic) tropical weathering, and miraculously survived the scouring activity of the glaciers.

It has been demonstrated convincingly that tropical weathering forms residual enriched goethite and hematite iron ore caps over banded iron formations in the present rain-forest setting (e.g. in the Tonkolili District, Sierra Leone; Gaskin, 1975). It has, however, also been shown that the easily traceable late (Cainozoic) weathering cannot account for all, or even most, of the hematite - enriched ores in such giants as the Hamersley Range, Western Australia (Trendall, 1975). There, the main period of formation of the enriched ores is now placed in an interval of about 1.6-1.3 b.y., and the Cainozoic weathering is credited with a partial alteration (hydration, formation of goethite out of hematite, physical and chemical lateral "spreading" of the iron oxides and hydroxides over the surrounding countryside) of the earlier already enriched ores, rather than with generating the bulk of them. The processes and settings that had controlled the early ore enrichment are not known with certainty.

It seems obvious that, like many bauxites, most enriched iron ores are polygenetic: to an even a greater degree, and within a substantially longer time framework. There is very probably a wide degree of variation. Tropical weathering may be the only agent of iron enrichment; it can be one of several agents; and it could even be an agent of retrogressive destruction. In this chapter, the empirical features of enriched iron ores that are the obvious consequence of a geologically young tropical weathering, are reviewed briefly. A great variety of parent bedrocks is involved, of which the Precambrian metasediments and metavolcanics-hosted banded iron formations are dominant. Tropically enriched Phanerozoic ironstones are relatively rare. Many supergene iron oxide accumulations formed over pyritic (meta)sediments, so technically they are close to gossans. The latter, however, often overlap with the oxide, silicate and carbonate B.I.F.'s and cannot be easily separated.

North-Western Australia

The bulk of the Western Australian rich hematite ores are believed to have formed by a pre-Mesozoic process of enrichment. The Cainozoic weathering caused either a local partial to full hydration of the hematite into goethite, or it produced minor goethite orebodies. In the Mount Tom Price deposit, Hamersley Range (Gilhome, 1975), the principal hematite orebody is capped by a goethite layer reaching an average depth of 18 m (up to 50 m in synclinal troughs). Goethite is irregularly distributed along joints and fractures in hematite, or replaces shale bands. The goethite-rich ore is of lower grade than the hematite it has replaced (58.2% Fe versus 68.0% Fe), so here the weathering caused a mild impoverishment, not enrichment.

At the Mount Goldsworthy iron mine (Pilbara Province; Neale, 1975; Fig. 23-14) located in the Archean metavolcanics, the main orebodies are composed of a high grade (68% Fe), dense hematite adjacent to a fault and striking parallel with the synclinal axis. The steeply

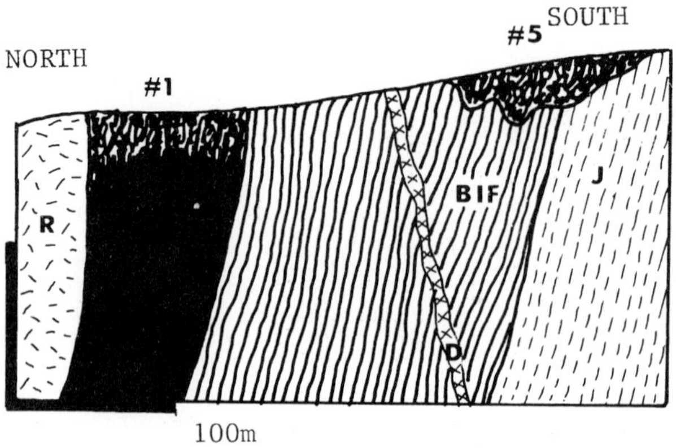

ARCHEAN:

J jaspillite
BIF banded iron formation
R brecciated, altered metarhyolite
D diabase dike
1, # 5 orebodies

Fig. 23-14. Two orebodies of contrasting origin and composition at the Mount Goldsworthy mine, Pilbara Province, W. Australia. The orebody #1 is a steep body of a dense hematite controlled by structure and formed by Proterozoic enrichment of an Archean banded iron formation. The orebody #5 is a thick goethitic blanket, produced by a Cainozoic weathering enrichment. From LITHOTHEQUE, modified after Neale (1975).

dipping, peneconcordant orebodies resemble sills or dikes. The thick, blanket-like No.5 orebody was composed of a residual goethite, formed by a Tertiary or later weathering enrichment of jaspilite and pyritic meta-argillite.

Quadrilatero Ferrífero, Minas Gerais, Brazil

This great iron ore province contains a large number of enriched iron orebodies, superimposed on a Proterozoic siliceous, dominantly non-volcanic, banded iron formation (Fig. 23-15). The primary B.I.F. ("itabirite") although relatively rich by world standards (30-40% Fe), is not considered "ore" in Brazil. Rich bodies of a high-grade specular hematite are superimposed over the B.I.F. in mostly structurally controlled locations, are pre-metamorphic, and unrelated to the young tropical weathering. The latter, however, combined with some transportation, generated a variety of ores usually described under the generic name "canga". Canga (Dorr, 1969) has been defined as a lithified surficial or a near-surface rock composed of variable quantities of detrital fragments (commonly a high-grade hematite or itabirite), cemented by goethite to form a blanket-like body (an unconsolidated equivalent of canga is called rolado in Brazil). The iron grade of canga depends on the composition of the fragments (the highest grade is achieved when the fragments are entirely of specular hematite) and porosity, and is usually between 30 and 64% Fe. The canga bodies immediately cap the iron ore outcrops, or they form extensive blankets on the high and intermediate planation surfaces where they merge with the ordinary ferricrete or with pisolithic

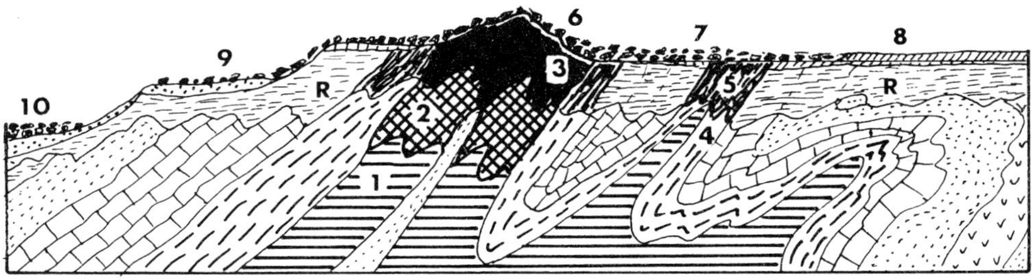

Fig. 23-15. Variety of "secondary" iron mineralizations derived from and superimposed on an earlier siliceous banded iron formation (1) in the Quadrilatero Ferrífero, Minas Gerais, Brazil. (2), "old" hematite-enriched ore; (3) goethite (hydration) superimposed on (2); (4) pyritic schist; (5) goethite ore formed by pyrite oxidation or schist replacement; (6) scree ore; (7) canga (Fe ore fragments cemented by goethite); (8) "normal" ferricrete; (9) colluvial, goethite-cemented ore; (10) stream Fe-ore gravel. R=regolith. The mineralizations 3-10 are broadly contemporary, Cainozoic. From LITHOTHEQUE, not to scale.

alluvial goethite. In addition to this, canga is abundant on the gentle slopes of valleys located below the B.I.F. outcrops, where it forms a cemented scree and a hill wash. There, in turn, it merges with the colluvial ferricrete and is ultimately transported into the valley streams and reworked into a goethitic gravel (and later cemented conglomerate).

Eluvial iron sands of Japan

Accessory magnetite, ilmenite and hemo-ilmenite present in a granite or gabbro saprolite, used to be recovered mechanically at several places in Japan. The Sanin district of S.W. Japan is the best known locality (Tsusue and Ishihara, 1975). These were "ores of opportunity", mined because of tradition, an established market and cheap labour, but there was nothing "special" about the parent rocks that contained only the ordinary amounts of the iron accessories (1-2% Fe_2O_3). Nevertheless, over a period of years, possibly several million tons of 48% Fe concentrates have been produced.

Residual Fe hydroxide over ferrous carbonates (siderite, ankerite)

Residual goethite ("limonite") accumulations over (a) mildly ferruginous limestone or dolomite; (b) moderately ferruginous ankerite; and (c) substantially ferruginous siderite carbonates, are frequent. (a) Is represented by Cartersville, Georgia (Kesler, 1950), where 5.5 Mt of about 50% Fe goethite ore with some residual specularite, was produced from residual clay pockets and blankets overlying Cambrian dolomite. The iron (and also Mn, barite) source is not known with certainty, and may include both hydrothermal replacements and the original (? diagenetic) iron component of the carbonate.

(b) Is represented by the Podrečany deposit of a porous ocher in Slovakia, that is a ferruginous residue after leaching out Ca and Mg from an ankerite lens. Similar ochers or "paint rocks" are often manganiferous, as in the Steep Rock iron field near Atikokan, Ontario.

(c) Is most common, and residual goethitic ores mark the outcrop of practically every siderite deposit of a variety of genetic styles, except those glacially scoured in the Quaternary.

MANGANESE ORES

Manganese, a relatively abundant metal (clarke about 800 ppm), responds wonderfully to the weathering and soil-forming processes and the conditions of its geochemical differentiation are reasonably well known (e.g. Lelong et al., 1976). In the rocks with near-clarke Mn contents undergoing tropical weathering, Mn is strongly impoverished in the weathering profiles of most rocks, except ultramafics. There, some lateritic soils may contain up to 5% Mn, usually distributed in the form of powdery patches, smears, fracture coatings of the black Mn oxides, and occasionally local accumulations of Mn oxide nodules. Small quantities of Mn oxides are recovered as a by-product of Ni laterite mining or exploration (e.g. Ora Banda, Western Australia). Over ultramafics, Mn oxides are usually cobaltiferous (asbolite).

Tropical weathering superimposed on Mn-anomalous rocks or on the "primary" Mn ores, caused a low factor of concentration enrichment which, nevertheless, marks a difference between an uneconomic "protore" and an economic "secondary" ore. The economy is, furthermore, helped by a (1) decomposition (softening) of the original hard Mn ore or a rock; and (2) geochemical separation of Mn from the formerly complex minerals. The weathering-generated Mn minerals are almost exclusively high oxides (psilomelane, cryptomelane, ramsdellite, nsutite, etc.). Residually-enriched Mn oxide ores in Cainozoic tropical weathering profiles are fairly important economically in themselves (about 20% of the global Mn production and reserves). Their contemporary or geologically older equivalents, now largely erosion removed, were important members of multistage genetic chains in which the nearshore sedimentary Mn ores (e.g. Nikopol'; Chapter 19) or the unconformity-controlled Proterozoic ores (Kuruman) formed.

Many of the "protores" to the residual Mn oxide deposits have been long misunderstood and their presence often unrecognized. This is quite understandable, because the mining had first to progress to sufficient depths that one could see the precursors to the secondary manganese. This applies particularly to the Precambrian manganiferous metamorphics that include spessartite or rhodonite-rich quartzites, schists or granulites (gondites); Mn carbonate marbles;

Ca-Mn silicate gneisses and hornfelses (queluzite); and iron formations. The Mn-rich metamorphics are metamorphosed equivalents of fairly well known unmetamorphosed manganiferous rocks and ores, known from the Phanerozoic record.

Conselheiro Lafaiete Mn district, Minas Gerais, Central Brazil

The Lafaiete district (Abreu, 1973; Fig. 23-16), is underlain by an Archean or lower Proterozoic medium-grade metamorphosed "schist belt" (a submarine volcanic-sedimentary association), composed of phyllite, micaschist, amphibolite, black schist, garnet quartzite, marble and Ca(Mn) silicate gneiss or hornfels, intruded by granite and pegmatite. The manganiferous members: spessartite quartzite (e.g. in the Cocoruto mine) and a rhodochrosite marble and rhodonite hornfels (Morro da Mina mine), underlie several tens of residual Mn oxide deposits. In the largest Morro da Mina deposit (8 Mt of 44% Mn oxides; Mt of 26% Mn primary ore), pyrolusite, cryptomelane and manganite composed a rubbly, botryoidal and powdery residual ore mass up to 85 m thick. This gradually changed into a saprolithic and ultimately fresh gray Mn metacarbonate and a Mn-silicate hornfels ("queluzite"). This rock, in turn, is enclosed in black and green schists.

Comparable overall geology, comparable Mn protores and comparable residual ores are also present in the Serra do Navio district in the Amapá Territory, N.E. Brazil (part of the Guyana Shield; Fig. 23-16). In the recently discovered Azul Mn deposit, Serra dos Carajás, N.E. Brazil (Bernardelli, 1982), a variety of enriched "in situ" as well as downslope transported Mn oxide ores formed by lateritic weathering superimposed on manganiferous black meta-marls. The manganiferous unit (top of the Río Fresco Fm.) is composed of rhythmically alternating black microcrystalline rhodochrosite bands, interbedded with a pyritic gray to black slate to siltstone. Two manganiferous horizons ranging in thickness from 18 to 54 m and having a grade of 15-26% Mn, are represented.

In the Quadrilatero Ferrífero (e.g. Conta História deposit), the protore is a slightly manganiferous iron formation with an associated Mn carbonate. In the coastal Mn belt of Bahía (e.g. Maraú), the protores are garnet quartzites and granulites. Table 23-3 lists additional example localities.

Ni (Co, Cr)-MINERALIZED WEATHERING PROFILES ("LATERITES")

Tropical weathering profiles over ultramafics represent the world's largest resource of nickel (in silicate or oxide form), and they also contain substantial quantities of resistate grains of chromite recoverable as a by-product. Cobalt accumulates in asbolite. The profiles have already been described in Chapter 7.

Compared with the Fe, Al and Mn weathering profiles treated above, the Ni, etc. profiles have a similar physiographic setting and are found in both the "high plateau" and "low" (buried) plateau settings.

NW MORRO DA MINA, CONS. LAFAIETE SE

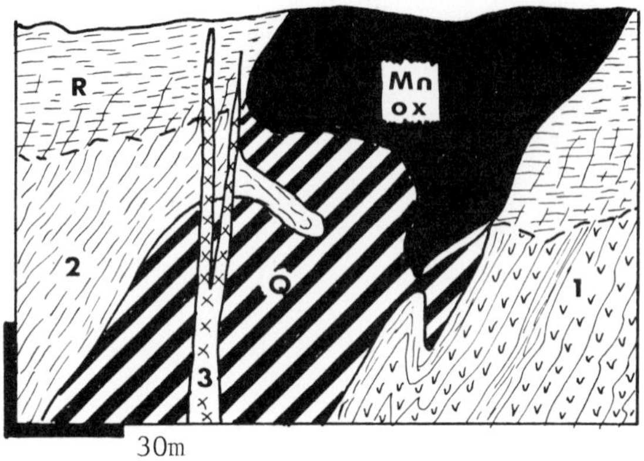

ARCHEAN:

1. amphibolite, hornblende schist, graphite schist
2. micaschist, graph. schist
3. Pt granite, diorite, pegmatite

Mn/ox secondary cryptomelane-pyrolusite ore

Q "queluzite"- gray massive Mn carbonate marble interbedded with black schist; spessartite-tephroite hornfels

R regolith

SW TEREZINHA, SERRA DO NAVIO, AM. NE

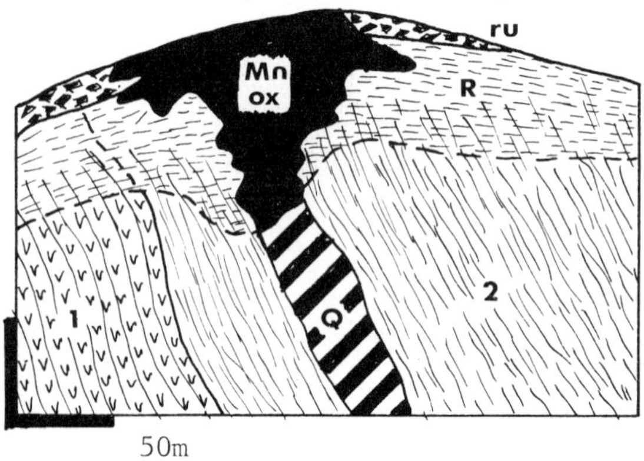

PRECAMBRIAN (pre- 1.7 b.y.)

1. amphibolite
2. garnet-biotite schist, spessartite quartzite

Q "queluzite" - Mn carbonate marble grading into black spessartite-rhodonite-tephroite schist

Mn/ox pyrolusite-cryptomelane secondary ore

R regolith

Fig. 23-16. Examples of two high-grade Mn oxide deposits in tropical weathering crusts over manganiferous carbonate or silicate metamorphics in Brazil. Diagrammatic, from LITHOTHEQUE, based on Abreu (1973).

In the former case, Ni accumulates in the lowermost zone of the profile, in coarse (bouldery) ultrabasic saprolite, and as infiltrations of silicates reaching deep into the fractured bedrock (e.g. New Caledonia; Riddle, Oregon). These profiles are usually capped by a well developed ferricrete. The "low plateau" setting is less distinctly zoned, and nickel resides mostly in goethitic ocher in the middle horizon of the "yellow laterite" (e.g. in Cuba; Fig. 23-17).

Table 23-3. Mn oxide deposits formed by residual enrichment of primary manganiferous rocks under conditions of Cainozoic tropical weathering

LOCALITY	AGE OF WEATH.	PARENT ROCK AGE	PARENT ROCK LITHOLOGY	OREBODIES	TONNAGE/GRADE	REFERENCES
Cartersville, Georgia, U.S.A.	T_3-Q	Cm_1	Mn dolomite	earthy blankets, pockets; gt,pr	200 Tt Mn	Kesler (1950)
Quemado district, Las Villas,Cuba	T_3-Q	J_3-Cr_1	Mn chert,shale silic.limest.	nodules,blocks, blanket in red residual soil	52 Tt Mn/9%	Simons and Straczek (1958)
Matthews Ridge, Guyana	T-Q	Pt_1 (Ar?)	Mn carb.marble spessart.qtzt. black schist greenstone	residual blanket pr,cm	5.28 Mt Mn/40%	Putzer (1976)
Maripa, Suriname	T-Q	Pt_1 (Ar?)	spessart.qtzt., chlor.,seric., graph.,graph. schist	blanket, cryptomelane	2.25 Mt Mn/25%	Putzer (1976)
Serra do Navio, Amapá, N.Brazil	T-Q	Ar-Pt_1	spessart.qtzt. rhodochr.marble Mn silic.hornf.	residual blanket pr,cm	25 Mt Mn/46%	Abreu (1973) Putzer (1976)
Azul deposit, Serra dos Carajás, Brazil	T-Q	Pt_1	rhythm.banded Mn carb.schist	pisol.,blocky, powdery,resid.and transport.blanket	23.68 Mt Mn/38%	Bernardelli (1982)
Maraú, Bahía coast Mn belt, Brazil	T-Q	Pt	spessartite quartzite	residual tabular blankets	2.66 Mt Mn/38%	Abreu (1973)

Table 23-3 (continued). Mn oxide deposits formed by residual enrichment of primary manganiferous rocks under conditions of Cainozoic tropical weathering

LOCALITY	AGE OF WEATH.	PARENT ROCK		OREBODIES	TONNAGE/GRADE	REFERENCES
		AGE	LITHOLOGY			
Miguel Burnier, Conta História, Quad.Ferrífero, M.G., Brazil	T-Q	Pt_1	Mn itabirite (BIF), Mn-Fe carbonate	blankets,psilomelane,pyrolusite	400 Tt Mn	Abreu (1973)
Cons.Lafaiete, Minas Gerais, Brazil	T-Q	Ar	rhodochr.marble, Mn silic. hornfels,black schist	resid.blankets, pockets; pr,cm	4.5 Mt Mn/44% secondary ore 1.3 Mt Mn/26% primary protore	Abreu (1973)
Tambao, Upper Volta	Q-T	Pt_1	rhodochr.marble Mn silic.hornf. amphibolite	concret.blanket cm,pr	2.25 Mt Mn/50%	Perseil and Grandin (1978)
Mokta, Ivory Coast	T-Q	Pt_1	spessart.qtzt. amphibolite	resid.blanket cm,nsutite,ramsdellite, manganite	1.5 Mt Mn/30%	Perseil and Grandin (1978)
Nsuta, Ghana	T-Q	Pt_1	rhodochr.marble quartzite, micaschist	concret.blanket cm,nsutite	20 Mt Mn/40%	Perseil and Grandin (1978)
Mrima Hill, Kenya	T-Q	MZ	carbonatite	nodular,boulders in saprolite	136 Tt Mn	de Kun (1965)
Peak Hill, W.Australia	T	Ar	banded iron fm.	botryoid.,pisol. blanket; pr,gr, psilomelane	238 Tt Mn/42%	Blockley (1975)

ABBREVIATIONS: gt = goethite; pr = pyrolusite; cm = cryptomelane

Fig. 23-17. Nicaro, Holguín Province, N.E. Cuba. Example of a large "low plateau" Ni laterite, developed on Mesozoic ultramafics.

Ti ACCUMULATIONS IN TROPICAL WEATHERING PROFILES

Titanium accumulations in tropical weathering profiles have two forms: (1) residual grains of resistate minerals (ilmenite, rutile) in saprolite and (2) authigenic minerals formed in the laterite from Ti released by silicate (e.g. biotite, pyroxene) breakdown.

(1) In the first case, no substantial increase of the ore substance has taken place (except for the weight loss and shrinkage of the

parent rock due to leaching); the resistate Ti minerals were merely loosened from the solid hard rock, and this substantially lowered the processing costs. As a consequence, weathering profiles over certain silicate rocks containing their "normal" complement of Ti resistates and having an overall Ti content only slightly higher than the Ti clarke, became Ti ores ("eluvial placers"). Example localities of titaniferous saprolites over metamorphics are reviewed briefly below.

At the Kundybaevo deposit in Kazakhstan (Borisenko, 1974), the weathering crusts over ultramafics contain 180 kg/m^3 ilmenite and 74 kg/m^3 rutile. The richer portions of the regolith are selectively quarried and the orebodies have assay boundaries. At Shooting Creek, North Carolina (Herz and Eilertsen, 1968), rutile occurs in thin lenses and stringers of secretion quartz in a garnet micaschist member of the Carolina Gneiss, along a belt 16 km long. The soft saprolite is occasionally quarried and the loose rutile recovered. The available reserves were estimated as 41,280 t Ti in 154 Tt of concentrate.

Residual ilmenite is recovered mostly from saprolites developed over anorthosites, and example localities include Roseland, Virginia (about 1.2 Mt Ti); Yadkin Valley, North Carolina (66 Tt Ti); Sherbo River, Sierra Leone (about 15 Mt Ti); Volyn Massif and the Korosten Complex, the Ukraine; and other localities (compare also Chapter 24).

(2) Authigenic titania concentrates mostly in anatase, which is a frequent constituent of residual clays including bauxites and kaolin. All bauxites developed over basalts or gabbros are high in TiO_2 (up to 8%), and the titania can be recovered as a by-product of aluminium recovery. High titania residual clays (up to 30.6% TiO_2 on Maui, Hawaii; Patterson, 1971, already discussed) form in poorly drained locations and have, so far, been insufficiently investigated.

OTHER METALS AND MINERALS

Saprolization (decomposition and softening by weathering) of an originally hard bedrock causing loosening of resistate minerals, is a process that generates "eluvial placers". Such placers together with colluvial and alluvial placers formed by a downslope transportation and hydrodynamic reworking of the former, account for a considerable portion of the cassiterite (e.g. in the Thailand, Malaya and Indonesia belt; in Manono, Zaïre; etc.), and a small quantity of gold, recovered worldwide. The South Asian cassiterite is usually explained as derived from "granite", but a considerable proportion of the residual cassiterite in fact comes from small hydrothermal orebodies: veins, greisens, veinlets, marble replacements, etc., located in granite as well as in a variety of other rocks in the granite exocontact. The Manono deposit is in a pegmatite saprolite.

Example localities of the "eluvial gold placers" have been reviewed by Boyle (1979). In Ambatolampy, Madagascar, gold occurs in what is interpreted as a saprolite over an auriferous band of a banded iron formation. It may be similar to the "jacutinga" ore of the

Quadrilatero Ferrífero, which is a disintegrated auriferous itabirite and a high-grade (specular) hematite. Some "canga" occurrences have also been found to be slightly auriferous. At Dahlonega and other localities in the Piedmont in Georgia, as well as in the greenstone belts of Guyana, in the Hargraves and Lucknow goldfields, N.S.W., Australia and elsewhere, auriferous saprolite formed over greenstone metamorphics hosting gold-quartz veins and stockworks. At Kuranakh, Yakutia (Siberia), gold has been derived from weathering-decomposed auriferous feldspathic and other metasomatites in carbonates.

A variety of frequently unusual residual mineralizations is associated with regoliths over alkaline rocks. Many are one of a kind, and they are reviewed in Chapter 33.

"Uranian laterites" have been mentioned briefly by Brown and Dey (1975) as present in small patches containing between 0.013 and 0.27% U_3O_8 in several states of India, presumably in association with the monazite-bearing high-grade metamorphics. "Vanadian laterite" from Coates Siding east of Perth, Western Australia, formed over a gabbro containing vanadiferous Ti-magnetite. A cassiterite-containing ferruginous laterite formed over hydrothermally mineralized Carboniferous sediments in the Nam Pathene district, Laos (Fontaine and Workman, 1978). Pb, Zn, Hg, Sb, W, and other metals "laterites" are special forms of weathering-modified outcrops of sulphide and oxide, usually hydrothermal deposits of considerable variety, that require individual treatment.

23.4. ANCIENT, BURIED WEATHERING PROFILES (NOW LOCATED AT OR BELOW AN UNCONFORMITY)

23.4.1. Introduction

Although preservation of an ancient regolith under a cover of younger sediments or volcanics is relatively uncommon, it does occur. Many such occurrences are also metallic ores of the same style as that described in the previous section, that is: bauxite, enriched iron ores, Mn oxides and others. The earliest buried regoliths have been recorded from the lower Proterozoic sequences and these are reviewed later. The Phanerozoic occurrences are substantially more common, and their number increases with the decreasing age. Naturally, there is a transition between the buried and exposed weathering and soil profiles, and an exact boundary is difficult to draw. The bauxite deposits near Linden, Guyana, mentioned earlier, are covered by a unit of unconsolidated lignitic clays and sands several meters thick, so that they are actually buried. A shallow burial like this, however, did little to modify the bauxites.

Substantially deeper burial subjects regoliths to initially diagenetic, and later metamorphic, modifications. The diagenesis includes compaction, partial dehydratation, and a mineralogical modification. This is most apparent in bauxites, in which the original gibbsite converts into boehmite and diaspore. All the fully boehmitic and diasporic bauxites have once been buried. Although it is likely that all the pre-Mesozoic regoliths had once been buried, they need not be so at present. Many have recently been exhumed so that they occur on the present surface. Their present exposure is, however, only temporary speaking in the dimensions of geological time. Surficially exposed ancient regoliths undergo a new phase of weathering and the relict weathering effects are usually difficult to distinguish from the effects of modern weathering. In general, the modern weathering tends to remove the effects of the burial modification.

Known ancient weathering profiles that continue to be buried are, naturally, rare because it requires extensive drilling to discover them. They seem to be most common on platforms topped by sediments deposited in epicontinental seas, because of the generally low-energy depositional environment that has a low scouring capacity. Most examples are known from the Russian Platform (e.g. Krivtsov, 1973), where several horizons of pre-Mesozoic regoliths are to be found in both the basal (crystalline basement/basal sediment) and intrasedimentary settings.

The basal regoliths are easiest to determine. Most of them are incomplete, represented by a thin saprolite zone several meters deep. There are few exceptions, however. A typical regolith on intermediate intrusive rocks intersected at a depth of about 1,300 m, had the following zones (from top to bottom): (1) chlorite, hydromica, kaolinite; 3.3 m thick; (2) chlorite, hydromica, accessory kaolinite; 3.7 to 5.75 m thick; (3) hydromica; 2.1 to 3 m thick; and (4) light initial alteration; 7.5 to 35 m thick.

In sequences of marine sediments, regoliths corresponding to periods of emergence are located at persistent unconformities that can usually be traced for long distances. The "Mediterranean-type" bauxites sandwiched between two carbonate units (Chapter 20) are typical representatives. In sequences of continental sediments, however, the intra-sequential unconformities are often difficult to recognize. With respect to a plane of unconformity, the saprolite portion of regoliths is always under the unconformity and if mineralized, we are dealing with sub-unconformity ores. The position of mineralized soils (laterites), however, is less clear. Sometimes they are gradational upward into transported sediments that are also the basal members of a sedimentary sequence. In such a case there was no time break between the pedogenesis and sedimentogenesis, and the unconformity between the crystalline basement and the topping sedimentary sequence becomes gradational. Button and Eriksson (1981) coined the term "graded unconformity", applicable to such situations. The role of unconformities in the formation and preservation of ore deposits is reviewed more extensively in Laznicka (1985f).

23.4.2. Ancient regolithic bauxites

Almost all early Cainozoic and older residual bauxites had once been covered and protected by younger sediments, but most have been recently exhumed. This has facilitated their discovery. Few members of this category are entirely residual; many are partly or fully reworked (resedimented). The Arkansas bauxite district described by Gordon et al. (1958) is one of the few examples of an in-situ preservation of a full spectrum of tropical weathering products, and a variety of redeposited bauxite. The in-situ regolith is the result of an early Tertiary lateritic weathering of a Cretaceous nepheline syenite and a complete spectrum of zones from a fresh nepheline syenite through kaolinized and bauxitized nepheline syenite (saprolite), covered by pockets of a residual bauxite laterite, is represented.

Elsewhere, for example in the North Onega bauxite region situated on the eastern slope of the Baltic Shield, U.S.S.R. (Krivtsov, 1973), the tropical weathering superimposed on the Precambrian basement metamorphics has been dated as lower Carboniferous. There, only the lower montmorillonite-hydromica and the upper kaolinite zones of the saprolite have been preserved. The bauxite in the district is entirely resedimented, although its original source area can be reasonably well reconstructed.

Perhaps the most complete and the best-preserved ancient residual bauxites have formed on the crystalline basement of the Voronezh Massif (a Precambrian "high" in the west-central part of the Russian Platform) and are buried under the thick cover of Phanerozoic platformic sediments. There, buried bauxite deposits have, so far, been found in the Belgorod and Novyi Oskol Districts. The former is more important and coincides with the productive portion of the Kursk Magnetic Anomaly. In this part of stable Europe, thick, tropical weathering crust has been forming locally since the upper Proterozoic to middle Devonian period. From the middle Devonian to Jurassic ages, block faulting related to the development of the Donetsk Trough (an aulacogene) caused interruption, erosion, and widespread reworking (Yermakov, 1977). Most of the bauxite deposits of the Russian Platform developed in the late Devonian and early Carboniferous periods by reworking of the earlier regoliths. The Belgorod and Oskol bauxite districts are examples of the few regolith relicts that survived.

The best studied and most fully described bauxite deposit in the Belgorod District is Vislovsk (or Vislovskoe; Kirpal' and Tenyakov, 1974; Krivtsov, 1973). There, lateritic bauxite is part of a 5-170 m thick weathering crust (weathering along fault zones reaches up to 700 m depths), selectively formed on "heads" (erosional highs) of steeply dipping, lower Proterozoic quartz, chlorite, sericite phyllites. These phyllites, in turn, are interbedded with banded iron ores of the Kursk Iron Formation, over which the weathering profiles contain enriched iron ores, discussed later. The bedrock lithologic control of the economic products of lateritic weathering in the Vislovskoe deposit is almost identical to the control apparent in the Quadrilatero Ferrífero, Brazil.

The residual bauxite horizon is, on the average, 5 m thick, composed of pisolithic and relict-structured boehmite, chamosite, kaolinite, diaspore and hematite. The colour is red to gray and the Al_2O_3 content is between 43.7 and 50%. The iron content is around 12-15%. Most of the bauxite is a saprolite; "in situ" bauxite laterite is subordinate, and a redeposited bauxite exceptional. The bauxite saprolite grades downwards into "allites" (mottled boehmite-kaolinite residual claystone), ferruginous kaolinite saprolite, hydromica-kaolinite saprolite, bleached and fresh bedrock. Smectites appear in the saprolite formed over amphibolite. The Vislovsk bauxite is topped by 450-600 m of lower Carboniferous to Recent epicontinental sediments.

The Vysokopol' bauxite field (Kirpal' and Tenyakov, 1974) located on the southern slopes of the Ukrainian Shield, is an example of residual bauxite deposits formed on the basement of the Russian Platform during the younger period of stability folowing the Mesozoic-Cainozoic block faulting. It rests on a Precambrian amphibolite, it formed in the middle Eocene, and is composed of a poor quality (38% Al_2O_3) gibbsite ferruginous bauxite.

Ancient counterparts of the important late Cainozoic residual ferruginous bauxites formed over basalts and intermediate volcanics, are represented by the Carboniferous bauxites of Central Timan (N.E. Russian Platform; Kirpal' and Tenyakov, 1974); Mesozoic bauxites formed on the plateau basalts of the Siberian Platform and other localities. Particularly interesting are the "synvolcanic" residual bauxites rapidly formed on the tops of continental basalt flows, and topped by succeeding flows (e.g. in the Inverell area, N.S.W., Australia; Oswego Lake, Oregon, etc.; see also Chapter 31).

23.4.3. Buried enriched iron ores

Compared with bauxites, the known deposits of completely buried iron ores are more common because they are much easier to find (geophysically; the Vislovsk bauxite was found accidentally during an exploration for iron ores). The Voronezh Massif with its Kursk Magnetic Anomaly (KMA) again represents an outstanding example, undisturbed by post-exhumation erosion and retrogressive weathering (Magak'yan, 1968; Sokolov and Grigor'ev, 1974). Two important iron ore fields have been developed there, so far.

The Yakovlevo field in the Belgorod District (130 km S.S.E. of Kursk; Fig. 23-18), is adjacent to, and overlaps, the Vislovsk bauxite bodies. The rich iron ores formed by the pre-Carboniferous tropical weathering, and are now buried under 490-550 m of platformic sediments. They rest on steeply dipping beds of a lower Proterozoic siliceous iron formation and interact or overlap with the specularite facies of rich iron ores formed, as in many other similar districts of the world, much earlier. The residual ores are typically red and

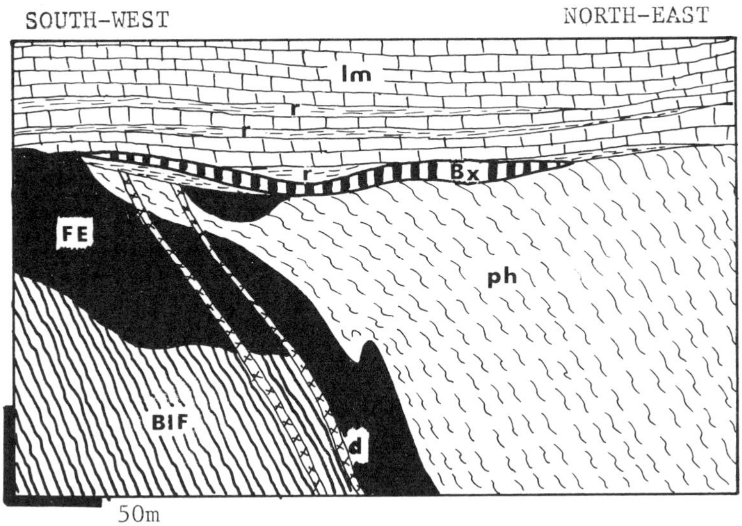

Fig. 23-18. Yakovlevo iron ore and bauxite field (Belgorod district, Kursk Magnetic Anomaly, U.S.S.R.), example of a pre-Carboniferous residual bauxite and enriched iron ore buried under 500m of Paleozoic and Mesozoic platformic sediments. lm=Cb_1 limestone; r=residual clay; Bx=pre-Cb_1 (D?) bauxite; FE= weathering-enriched hematite, goethite, martite. BIF=Pt_1 siliceous banded iron formation; ph=phyllite, slate; d=diabase. Modified after Krivtsov (1973) and Magak'yan (1968).

brown, composed of porous to dense goethite and hydrogoethite. Earlier magnetite caught in the zone of tropical weathering was converted into martite. Local sections of siderite probably formed in a paludal environment. The enriched ores are controlled by a north-westerly trending structural trough, traceable for 70 km, and the enriched ore mantle has a thickness ranging from 40-160 m (average 45 m). The reserves are 15-17 Bt of a 60% Fe ore.

The other developed field in KMA, Mikhailovskoe (Sokolov and Grigor'ev, 1974), is located 100 km N.W. of Kursk and its enriched iron ore mantle rests on a banded iron formation under the protective cover of platformic sediments only 35 to 114 m thick. Consequently, open pit operations are possible. In addition to the hematite-goethite ores, friable and fragmental specularite-martite ores reminescent of the "canga" are present. The rich iron ore reserves are substantially less than in the previous locality, 340 Mt of 45-58% Fe material.

23.4.4. Other metals

Selin (1981) described a cinnabar-bearing late Mesozoic residual clay in the Yeskongo deposit (Kurai zone, Gornyi Altai, U.S.S.R.) filling sinkholes in lightly and irregularly Hg-mineralized late Proterozoic marbles. It is probably a recently exhumed, Mesozoic regolith.

Collins and Loureiro (1971) described and interpreted the Pedra Verde copper deposit, Ceara State, north-eastern Brazil, as possibly being a buried mineralized regolith. There, a thin blanket of finely disseminated chalcocite, bornite, chalcopyrite and pyrite is hosted by a saprolitic but chemically reduced, Proterozoic phyllite, grading upward into a red oxidized phyllite. This, in turn, is unconformably topped by a Cambro-Ordovician basal conglomerate.

Many more examples can be extracted from the literature (e.g. native copper in buried Ordovician residual clays in the Mt.Lyell Cu field, Tasmania; residual clays impregnated by bismuth oxides and arsenates and mined as a small but rich deposit around the turn of century at San Gregorio near Cerro de Pasco, Peru; and others). However, all are small deposits and occurrences of uncertain genesis. Some have been reviewed by Laznicka (1985f).

23.5. WEATHERING AND SOIL PROFILES IN ARID CLIMATES AND THE PROBLEM OF DURICRUSTS

23.5.1. General

Arid climates are highly variable and transitional, and sand dune desert (erg) or a rocky desert (hamada) are their most extreme landform representatives. Indeed, there are no soils and weathering is confined to rinds, several centimetres thick, of bleaching and/or Fe,Mn oxide pigmentation in outcrops or on the surface of boulders. Semi-arid climates, however, do produce soil and support some vegetation. The most important quantitative indicator used to characterize arid regions is the amount of evaporation greatly exceeding that of evaporation. This has an important hydrogeological implication. When groundwater is available, it can move upward by capillary action. Consequently, it has a reversed vector of movement compared with the downward-moving meteoric water in the humid regions, and a reversed gradient of removal, transportation and precipitation of the dissolved components. Under ideal conditions, the equivalent of the (A) horizon of leaching in a "normal" soil profile, is to be found under the (B) horizon of deposition in the ascending water profile. This relationship is only exceptionally ideal, because descending and laterally flowing waters are also periodically involved in most arid regions of the world, but despite the imperfections duricrusts generated by the upward-seeping waters are a characteristic feature of the arid regions and one that appears to have an increasing metallogenetic significance.

The A.G.I. Glossary of Geology defines duricrusts as "a generic term for a hard crust on the surface of, or layer in the upper horizons of, a soil of a semiarid climate. It is formed by the accumulation of soluble minerals deposited by mineral-bearing waters that move upward by capillary action, and evaporate during the dry season". This, then, is the prevalent American usage of the term. In the rest of the world, particularly in Europe and in some large countries located in the humid tropics (most notably India), duricrusts include all the pedogenically formed cappings the most common representative of which is a "laterite" (mostly ferricrete, and sometimes bauxite; compare Goudie, 1973). The latter can hardly be considered products of an arid climate.

It is not the purpose of this book to engage in terminological debate, but the scope of coverage of this section has to be defined somehow. Bauxites and laterites will no longer be discussed despite the fact that climatic baking (permanent, as in arid climates containing relicts of humid weathering profiles; seasonal, as in savannas) undoubtedly does contribute to the hardening of the humid tropical soils. The remaining duricrusts have a distinctive mineralogical composition and include calcrete, silcrete, gypsicrete and several other rare "cretes", and these are indeed characteristic of the arid regions. Goudie (1973), however, presented many more genetic alternatives for the formation of these rocks than just precipitation from ascending groundwater, so the A.G.I. definition should be considered as reflecting one alternative only, although the most important one.

As in the humid landscapes, actively forming or geologically young duricrusts (and their embryonal equivalents) are invisible, because they form in the subsurface. They become highly obvious capping mesas only after an uplift and dissection at which time they are usually "dead" (inactive), undergoing degradation (dissolution, dispersal). As mentioned previously, later arid duricrusts may interact with earlier humid weathering profiles, should the climate change. The opposite situation has been observed less frequently.

23.5.2. Calcrete (calcicrust, calcicrete, kankar, caliche)

Calcrete, the most widespread duricrust composed of calcite, is also terminologically most confusing. This is no doubt because of its numerous variations in appearance and competing terminologies coined by specialists in several disciplines of earth sciences. Two emphases (research directions) are particularly strong and have to be taken into account because of their potential to disorient the readers.

(1) "Calcrete" and its many synonyms used throughout the world by the general and field geologists (e.g. caliche, kankar; compare Table 2 in Goudie, 1973), has a variety of definitions but most emphasize its precipitation in the arid regions from the infiltrating and ground waters. The formational (progressive) aspect of calcrete is emphasized.

(2) In North America, the term "caliche" has been adopted by the (mostly marine) carbonate sedimentologists to refer to the zones of post-depositional, subaerial modifications of marine carbonates. Esteban and Klappa (1983) considered caliche as one end-member of the subaerial diagenetic carbonate facies. Here, the emphasis is on what happens to an earlier marine carbonate exposed to the (mostly arid) subaerial modifications, that is on a transformation or destruction.

In this section, the emphasis is on (1) and the calcitic duricrust will be termed uniformly calcrete. Of the variety of calcrete types and genetic models listed in the literature (1) pedogenic models; (a) per descensum; (b) per ascensum; (2) groundwater precipitate in main reservoirs or conduits; (3) groundwater precipitates in capillary fringe; (4) fluvial precipitate; (5) lacustrine precipitate; (6) in situ modified earlier limestone; (7) detrital continental limestone and others. The first three are of greatest importance.

The pedogenic calcrete (1) is product of a special case of soil formation in climates averaging under 500 mm (some authors say 200 mm) of mean annual rainfall and potential evaporation in excess of 3,000 mm/year. It comes in a variety of forms and each form may occur alone or several forms may combine together in a zoned profile. The idealized caliche profile of Esteban and Klappa (1983) formed over a limestone bedrock, has a well indurated light laminated to massive limestone hardpan on the top. This changes downward into a "platy caliche", a thinly-bedded variety in which the beds are often substituted by loose limestone plates interbedded with soil. Often underneath there is a nodular caliche, changing further into a chalky, powdery caliche. This is followed by a rubbly, fractured to massive bedrock carbonate.

Although most pedogenic calcretes formed over carbonate bedrocks, many other bedrocks and $CaCO_3$ source rocks have been recorded (Goudie, 1973): calcareous shales; basalts (the calcite came from amygdules and fracture veinlets); various volcanics; plutonic rocks and metamorphics ($CaCO_3$ was derived mostly by breakdown of plagioclase); Fig. 23-19. When a non-carbonate bedrock underlies a calcrete, the lowermost zone usually consists of a chalky limestone fracture coating grading upward into a silicate rock rubble cemented by a chalky material, in turn grading into a nodular, platy and massive calcrete. The massive calcrete is usually free of bedrock fragments and it is in turn usually topped by a thin humous soil veneer or a blown sand or silt.

Calcretes frequently cement dry stream (arroyo) or residual plateau gravels, sand dunes, relict weathering crusts formed during past humid climates and volcanic ash falls. The pedogenic calcrete is widespread on the High Plains, Llano Estacado, Reynosa Plateaus, Basin and Range Province, etc. of the western United States and Mexico; in the Pampas of Argentina; in the Kalahari Basin and its surroundings in southern Africa; in parts of the East African Rift; in the Mediterranean; Central Asia; Australia and elsewhere.

The same lithogenetic activity forming calcrete may also accumulate minor metallic minerals, and occurrences of urano-vanadates have been

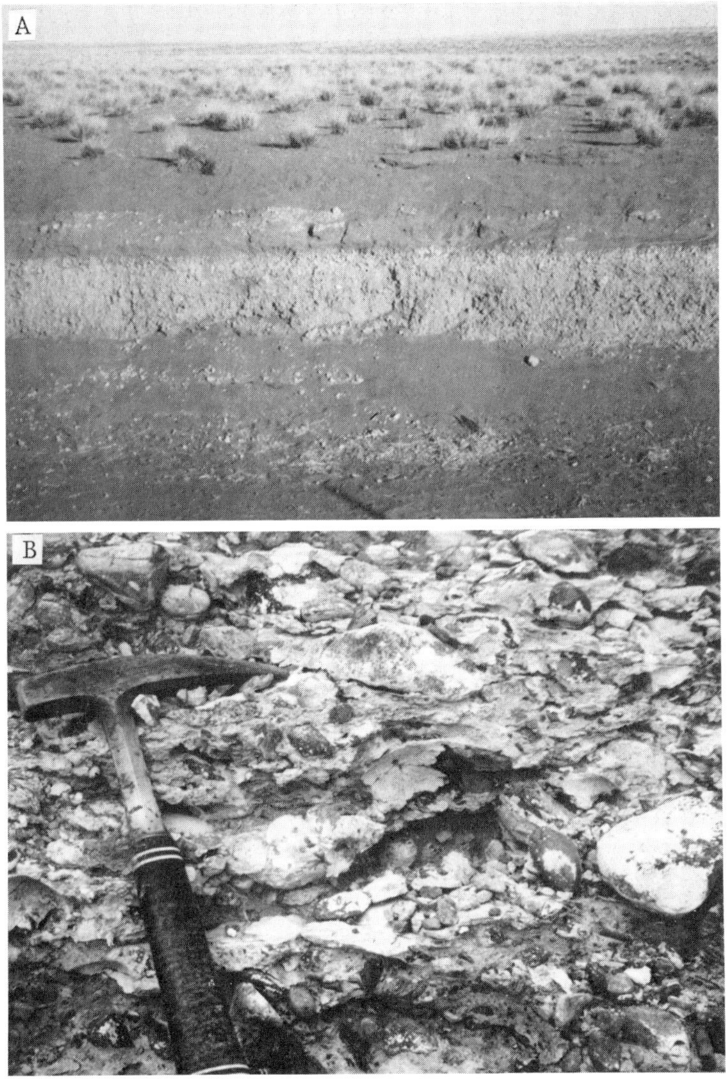

Fig. 23-19. Pedogenic calcrete. (A) in subsoil of the Kalahari Desert, S.W. Botswana. (B) Cementing metavolcanic gravel in an "arroyo" near El Arco, Baja California, Mexico.

reported most frequently In the Erie and Arden areas (south of Las Vegas, Nevada; Lovering, 1954), tyuyamunite occurs as small pellets and flakes disseminated through calcrete. Near Miller Hill, Wyoming, uranophane forms widespread coatings in a slightly silicified and brecciated calcrete, possibly dating from the Pliocene period.

In the Lost Creek area (Great Divide Basin, S.W. Wyoming; Sheridan et al., 1961), schroeckingerite occurs extensively in a calcrete developed over Tertiary and Quaternary continental sediments interbedded with a felsic volcanic ash, and themselves weakly uraniferous. The greenish-yellow mineral occurs as powdery coatings, small nodules and pellets, and shallow tabular orebodies about 45 cm thick are located 60 to 240 cm under the surface, above the water table. The uranium occurrences cover an area of about 1.3 km^2, and the grade varies between 0.001 and 0.26% U. The schroeckingerite is believed precipitated from upward-seeping groundwater.

Copper carbonates are relatively common in calcretes developed over copper mineralized bedrocks, for example near Arco, Baja California; Battle Mountain district, Nevada; Arizona; Atacama Desert in Chile (e.g. at Chuquicamata) and elsewhere. Their value is as indicator of a bedrock mineralization in exploration, rather than as ore in itself. Some of the "exotic" Cu silicate deposits reviewed later, may pass into calcrete-associated ores.

(2) Groundwater calcrete was named by Netterberg (1969) and is a non-pedogenic limestone precipitate located in broad, ancient drainage systems now filled by permeable sediments, and capped by unconsolidated sediments. Unless uplifted and dissected, this calcrete is not visible on the present surface except where it forms low mounds, protruding above the surrounding countryside. This calcrete is common in Western Australia north of latitude 30° S, and the discovery of a major U-V mineralization there (Yeelirrie) by the Western Mining Cororation in the late 1960s indicates the practical importance of these rocks to a prospector, and resulted in several publications (Langford, 1974; Mann and Horwitz, 1979).

Most of the Australian groundwater calcretes are linear, tabular limestone bodies about 10 m thick, traceable for distances of up to 100 km. Many terminate on the floor of ephemeral lakes, forming widened "chemical (precipitated) deltas". The appearance and the internal stratigraphy of a calcrete is highly variable. In contrast to the pedogenic calcrete, dense limestone varieties are dominant and the chalky coatings are usually substituted by solid encrustations.

The U-V deposits and occurrences in Western Australia (about 7 localities are now known, representing some 56.7 Tt U) are late-stage chemical precipitates and are represented almost exclusively by carnotite, often accompanied by cryptomelane and pyrolusite. Near the Yeelirrie cattle station (Gee et al., 1976; own visit, 1981; 39.1 Tt U/0.12%; Figures 23-20, 23-21), the old planation surface topped by silcrete has been rejuvenated during Miocene and Pliocene, exposing Archean granite bedrock under 10 to 30 m of tropically weathered regolith. The north-west trending Yeelirrie drainage channel formed as a consequence and it is now filled by a unit of a fluvial clay, 20 to 85 m thick with some quartz, feldspar and blown sand lenses. Groundwater calcrete has a spotty distribution in the channel and it reaches a thickness of about 8 m in the mineralized area that measures about 6 x 0.5 km. There, under several centimetres of a red, silty loam, isolated nodules of pedogenic calcrete start appearing and grade downward through laminated limestone crusts and foamy, earthy or fragmental calcrete into a hard, white, porcellaneous groundwater calcrete that is locally dolomitic. With increasing depth this calcrete becomes conglomeratic or brecciated and, ultimately, it changes into a basal, brown, silty carbonate transitional into a breccia of red siltstone fragments enclosed in a light-brown calcrete matrix. Underneath is an unconsolidated, non-calcareous, reddish-brown alluvial sandy clay to silt.

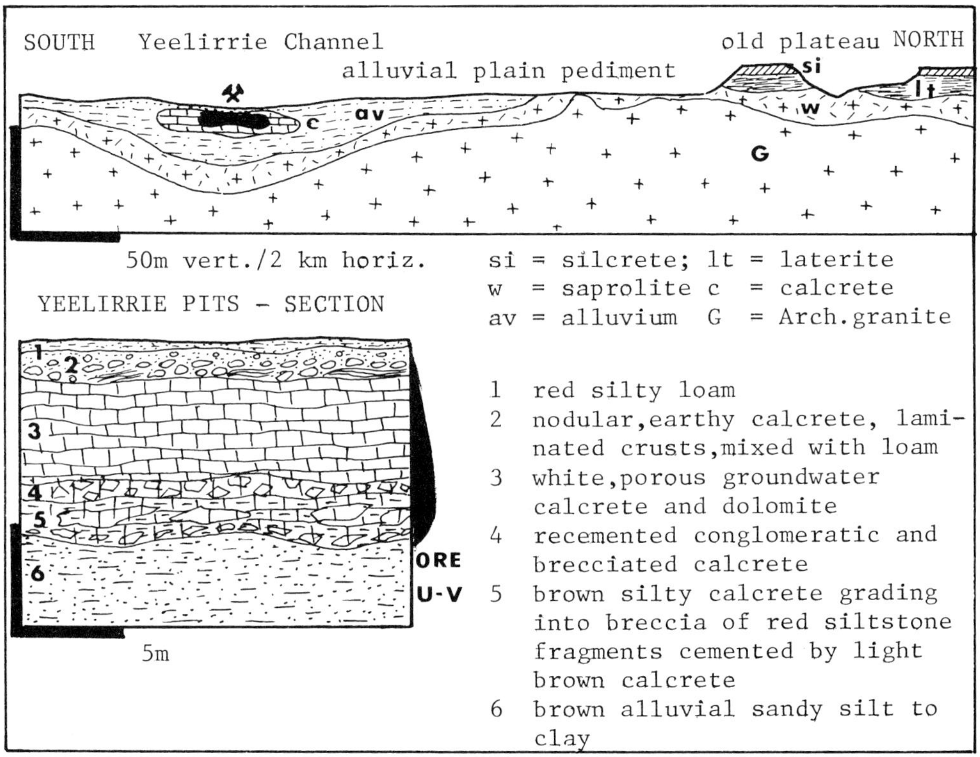

Fig. 23-20. Yeelirrie, Western Australia, carnotite deposit in Cainozoic groundwater calcrete. Regional section modified after Gee et al. (1976), local section from LITHOTHEQUE based on a 1981 visit.

Fig. 23-21. Yeelirrie U deposit, W. Australia. (A) General view of the buried groundwater channel. (B,C) Sections exposed in shallow pits. The massive, dense calcrete (in the lower portion of C, the main host to carnotite) grades upward into rubbly calcrete and loose calcrete nodules in loam.

The ore takes the form of bright yellow to greenish-yellow carnotite coatings and impregnations in most varieties of the groundwater calcrete, and carnotite is among the youngest minerals precipitated. The uranium is believed leached from the regolithic Archean granite in the catchment area; the granite is anomalously radioactive.

Additional occurrences of uraniferous calcretes are known from the Soviet Central Asia (Aktash River area); from Langer Heinrich, Namibia; from Botswana (Mokabesi occurrence; there, the calcrete overlies Karoo shales, is one metre thick, and is covered by one metre of overburden; 437 t U/350 ppm; O.E.C.D., 1979), but there are insufficient data to determine the genetic type of the host calcrete.

23.5.3. Silcrete

Silcrete is regarded by some geologists as being "as Australian as the kangaroo or the Vegemite", but it has a much wider distribution in southern Africa; in the Sahel; in east-central Brazil and elsewhere (Goudie, 1973). The recent Australian literature, however, dominates the field (e.g. Langford-Smith, ed., 1978).

Silcrete is usually a hard, indurated rock composed mainly of quartz or quartzite fragments of variable size, cemented by a microcrystalline quartz, chalcedony or opal. It is white, gray, yellow, red or brown weathering resistant material capping mesas, and widely distributed as rubble on mesa slopes and as relict blocks and gravel on the underlying planation surfaces (Fig. 23-22). Because of its resistance to weathering and abrasion, it is collected by streams and redeposited in alluvial gravels and ultimately in beaches. The formation of silcrete has rarely been observed and Hutton et al. (1978) pointed out that all the common Australian silcrete with microquartz cement is a relict deposit not forming at present, and a likely product of pedogenic processes that took place under rather humid conditions.

Silcrete is a substantial nuisance to an exploration geologist, constituting a "false bedrock" in regions where only deeply weathered rocks lacking outcrop are present. Where silcrete rests on an ultramafic regolith as locally in the Iporá belt, central Goiás, Brazil (e.g. Cerro Prieto), its basal portion grades into ferricrete and ferruginous clay. The resulting rocks then have very much the appearance of a gossan over sulphide mineralization complete with hydrothermal silicification. The positive aspect of pedogenic silicification (not always developed in the characteristic form of the granular quartzite or conglomerate) has been in preservation of the fine textural detail in some gossans, particularly over the Western Australian nickel sulphide deposits.

The ordinary silcrete in itself has not, so far, been of any importance as a host to metallic ores, although varieties anomalously rich in Ti and/or Zr and Nb have been described (Hutton et al., 1978). The silcrete from Beda Hill, South Australia, contains up to 15.2% Ti, 0.25% Zr, and 240 ppm Nb. At Mt. Keith, Western Australia, up to 0.5% Zr and 150 ppm Nb was found. Both localities are represented by the "skin silcrete": a thin (1-80 cm thick) yellowish-brown silica encrustations, coatings and sheets lining underlying sandstones and their fractures. Under the microscope, the skin silcrete is composed of angular quartz fragments embedded in a matrix of fine quartz, with authigenic rutile and zircon granules and cement. Uranium can occasionally accumulate in silcrete, particularly in areas containing contributions of vitric tuff. Bell (1956) described silcrete and occurrences of lag residual silica with up to 200 ppm U from the Colorado Plateau region, western United States.

Fig. 23-22. Australian silcrete scenery, west of Wiluna, Western Australia.

23.5.4. Gypsicrete (gypcrete)

Duricrusts composed of gypsum and transitional to evaporites are treated in Chapter 21. Locally, however, gypsum is associated with and gradational into calcrete (e.g. in Tunisia, in Namib Desert; Goudie, 1973), or into a silcrete with opaline matrix (Lake Eyre, South Australia; Hutton et al., 1978). A rare U-V mineralization of a transitional style associated with gypsicrete has been reported from the Mudugh Province, Somalia (O.E.C.D., 1979; 4.4 Tt U/0.06%). There, carnotite impregnates Miocene clays and fluvial sandstones and the contact with a massive gypsum crust. The ore zone is about 2.45 m thick and covered by 3-8 m of overburden.

23.5.5. Other duricrusts

Duricrusts of exotic composition form above geochemically anomalous bedrocks and above ore deposits. Nodular magnesite and magnesite crusts may substitute for calcrete above ultramafics (for example, along the edge of the Bushveld Complex; near Messina, South Africa; locally in Western Australia). No metallic mineralization has been reported, although some nickel enrichment could be expected.

23.6. CARBONATE KARST

23.6.1. General

Karst (Jennings, 1971; Herak and Stringfield, eds., 1972; Sweeting, 1973; Jakucs, 1977; Bogli, 1980) is a distinct surficial and subsurficial topography formed over water-soluble rocks, carbonates and evaporites. It is also a depositional environment in which new sediments contemporary with the active period of the karst (karsting) form. To a carbonate sedimentologist, karst is a facies of a subaerial "diagenetic" modification of earlier formed, usually marine, carbonates (Esteban and Klappa, 1983). This "diagenesis", however, is post-lithification, generally destructive, and many geologists would hesitate to use this term in its "stretched" form in the present context.

The term "karst" has been popular recently in economic geological interpretations and writings, particularly in connection with one variety of bauxites ("karst bauxites"; Bárdossy, 1982) and with the Zn-Pb, carbonate-hosted deposits of the Mississippi Valley "type" (e.g. Bernard et al., 1972). The entire subject of karst and economic mineral deposits has been summarized recently by Quinlan (1972) and Zuffardi (1976). The term "paleokarst" is used interchangeably with karst, and is usually applied to products (structures, rocks, associations) formed by the agents active in the karst environments, in the past.

Karst has both surficial and subsurficial expression and as in other geological interpretations, present and relict forms have to be differentiated. The relict forms were generated in the past and are being altered at present. It is now believed that many of the landforms exposed in the type area of karst in the Adriatic (Istria and Dalmatia) are relict, formed under humid tropical conditions.

In a typical profile, limestone-hosted karst has the following zones, from top to bottom (see also Fig. 23-23, that appears later):

(1) Infiltration zone (or upper vadose zone). Its surficial

landscapes are grooved exposed limestones (lapies), funnel-shaped sinkholes (dolines); flat-floored depressions (poljes) and mature karst plateaus. The landscape is partly barren, partly covered or filled by residual soil and rubble. In the subsurface, this zone is represented by solution-enlarged joints, fractures and faults gradational to vertical caves. The geochemical regime is oxidizing and marked by abundant red and brown soil or common goethite coatings and infiltrations. Precipitated carbonate is rare and mostly takes the form of crusts of chalky and friable consistency.

(2) Percolation (lower vadose) zone, marked by a system of sub-vertical, pipe-like cavities by which the water moves rapidly downward. There is little carbonate deposition except in the lowest part of the zone near the phreatic water table, where abundant speleothems form in the capillary water fringe.

(3) Zone of permanent circulation (upper phreatic zone) includes the water table located at the level of outlets, and is characterized by intensive hydraulic erosion, dissolution and precipitation resulting in abundance of mainly subhorizontal caves. Limestone precipitation and speleothem formation is intensive. Perched water tables located at various levels account for the frequently complex karst profiles. The zone is often modified by a partial or complete roof collapse, producing solution-collapse breccias. The waters are neutral, or slightly oxidizing.

(4) Zone of imbibition (lower phreatic zone) is located below the level of outlets, where the water circulation is restricted to non-existent. It is typical of the senile stage of karstification and is missing in many simple and shallow karst systems. The physico-chemical conditions in this zone are neutral to reducing; biogenic sulphurization is locally important, and sedimentation on the floor of depressions is widespread. The sediments include free detrital particles of residual clays; intrakarstic allochems (e.g. "cave pearls "; fragments of speleothems); mineral nodules, crystals, veinlets, etc., originally present in the bedrock limestone and released by dissolution; outside detritus brought in by the subterranean streams and other substances. The sediment deposition took place typically under low-energy conditions in the absence of burrowing organisms, so the sediment is frequently delicately laminated and usually of gray (reducing) shades. The imbibition zone need not occupy large open spaces, but could be largely filled by collapse rubble. In such a case, the fine sediment precipitates in the interstitial spaces of the rubble, filling and cementing a collapse breccia.

23.6.2. Metallogeny in the karst environment

The attention given to the karst-associated ore formation in the recent literature contrasts sharply with the virtual absence of observable metal deposition in the present karst areas. This led to occasional charges that the karst-controlled ores are a specialist's invention. Both sides of the dispute have raised some valid points, but both are equally guilty in oversimplifying and/or misdirecting the problem. The present stalactite caves correspond to the zone of permanent circulation, which happens to be the metallogenically least productive one. The "karst metallogeny" is, moreover, multistage and polycyclic, and in the majority of cases the karst provides nothing more than the open space for ore deposition. Filling of such open spaces can (and usually does) take place long after the karst formation, in greater depth, and by ore crystallization from heated brines.

The site and process of precipitation of the archetype of the "karst deposits", the MVT zinc-lead ores (Chapter 20), depends on 70-150° C hot connate or other brines at a depth of several kilometres under the surface, and this certainly bears little resemblance to the presently observable shallow karst activity, based on meteoric water movement in the vadose and phreatic zones. The specialists should point out very clearly the difference (probably by a constant reference to a paleokarst or a relict abyssal karst). There is also the "hydrothermal karst" due to the dissolution caused at any level within the crust (endokarst) by hydrothermal fluids, which usually lacks completely the characteristic surficial features and is unrelated to the karst zoning outlined above.

If only a direct metallogenesis taking place during the active period of karst development is considered, the number of examples of ore deposition shrinks drammatically and the bulk of the remainder will include cases of regeneration and/or entrapment of ores generated earlier. As in other environments, the importance of heritage is paramount. In the following paragraphs, the mineralization styles associated in a variety of ways with the active and near-surficial stage of karst development (that is, those within the reach of the surficially derived groundwater), are reviewed briefly (Figures 23-23 and 23-24; Table 23-4). Many of them have already been mentioned or described in other chapters.

(1) Surficial, karst-generated depressions (bedrock irregularities), physically filled by a sediment, genetically unrelated to the carbonate surroundings. This is a common case, exemplified by the poorly sorted, boulder, sand, clay, metalliferous colluvium near Guadalcázar (San Luis Potosí State, Mexico; 14.8 Tt Sn, 1,664 t Ag, 10.2 t Au, 2,640 t Hg; Foshag and Fries, 1942). There, the weathering residue derived from a small mineralized granodiorite stock and its aureole was washed into nearby karst depressions developed over Mesozoic limestones. Cassiterite, Ag sulphides, gold, cinnabar and scheelite are the most important heavy minerals recovered. The cinnabar came from impregnations in the carbonate interpred as telethermal. Similar alluvium or colluvium-filled karst depressions host some of the cassiterite placers in Malaya (e.g. in the Kinta Valley). In the Bakouma uranium field, Central African Republic (OECD, 1979; 16 Tt U/0.26%), uranium is associated with phosphate lenses interstratified with Eocene silt and siliceous horizons,

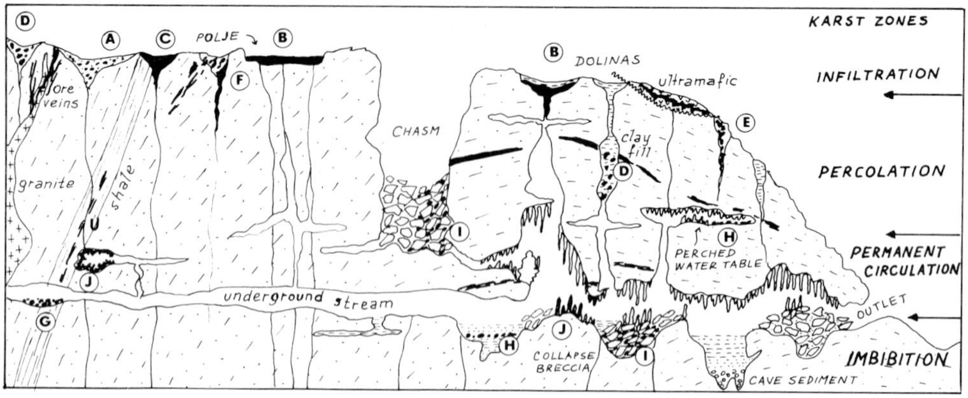

Fig. 23-23. Principal mineralization styles in a limestone karst coeval with karsting (vadose and phreatic water regime). See Table 23-4 for letter code explanation. The black unlettered mineralizations in limestone indicate the pre-karsting ores. From Laznicka (1984).

Table 23-4. Principal mineralization styles in a limestone karst

(A)	Placers filling surficial karst depressions (X0 localities, 30 Tt Sn, 5 Tt Hg, 15 t Au, 1.7 Tt Ag).
(B)	Bauxite in karst depressions, formed by tropical weathering of introduced material (rubble, volcanic ash; e.g. as in Jamaica). X0 localities, 500 Mt Al.
(C)	Karst depressions filled by residual "ore clay" (e.g. bauxite), or by metalliferous nodules resting in a clay (X0 localities, 100 Mt Fe, X0 Mt Al, X Mt Mn).
(D)	Ore blocks, lumps, grains resting in residual clay in karst depressions (X0 localities, X0 Tt Pb+Zn, 15 Mt barite, X Tt Hg, X0 Tt Sb, X t Au).
(E)	Ni oxides/hydrosilicates from weathering profiles over ultramafics washed into karst depressions (X localities, X0 Tt Ni).
(F)	Ore minerals formed by carbonate replacement by "per descensum" solutions (X localities, X0 Mt Fe).
(G)	Floor sediments in caves brought in from outside (placers, clay). X localities, X Tt Mn, X00 t Sn.
	EARLIER MINERALIZATIONS REDEPOSITED IN IMBIBITION ZONES OF RECENT KARSTS
(H)	in cave floor sediments;
(I)	in solution-collapse breccia and rubble;
(J)	as encrustations, impregnations, replacements or mineralized speleothems.

filling karst over Proterozoic dolomites.

(2) Surficial karst-generated depressions (bedrock irregularities) are filled by a non-carbonate rubble and/or volcanic ash, subsequently tropically weathered, together with the host carbonates. This corresponds to the genetic model for the Jamaican bauxite by Comer (1974).

(3) Surficial karst depressions are filled by residual clay, derived partly or fully from the host carbonates.

(a) The clay itself is an ore, as e.g. a gibbsite bauxite (alternative genetic model for the Jamaican bauxite); goethite-nontronite ocher, or Mn oxides (Cartersville, Georgia; Kesler, 1950). Alternatively, goethite or Mn oxide pisoliths or concretions formed in the lateritic profile over an impure carbonate, are embedded in a residual clay (kaolinite, nontronite) that, itself, is not an ore. The result is the "bean" or "lump" Fe or Mn ores mined in the past from many small deposits in Virginia (e.g. Reed Island, Fe); north-eastern Tennessee, Mn; West Germany and Switzerland (Franconian and Schwabian Alb, Jura Mts.; Fig. 23-24/3), Alapaevsk in the Urals, and other localities.

The Alapaevsk iron deposit (Sokolov and Grigor'ev, 1974; 65 Mt Fe/38.5%) has been placed by Zitzmann, ed. (1976) into the class of "elluvial limonite deposits on karst limestone". There, a karsted Carboniferous limestone is covered by varicoloured residual claystone that, in turn, grades upward into a deluvial-proluvial claystone breccia with silicified limestone, jasperoid and quartz fragments. The iron ore is represented by concretions, crusts and irregular masses of goethite and minor siderite within a green possibly residual clay. The ore horizon varies between 0.5 and 70 m in thickness, and can be followed for 5-10 km along the strike.

(b) The clay contains blocks, lumps, fragments, concretions, etc. of relatively resistant ore minerals released from the karsted carbonate. Examples are numerous:

Pb-Zn. Lag residual deposits of galena and sphalerite blocks are common in the debris-filled sinkholes over carbonates in Yunnan, southern China (Searls, 1952). The blocks are partly converted to smithsonite, hemimorphite and cerussite and derived from an APT or a plutonic-hydrothermal mineralization in the bedrock carbonates. Similar occurrences are known in Jebel bou Arhous and other localities in Tunisia (Nicolini, 1970; Fig. 23-24/6); from Sardinia; Thailand; Embreeville, Tennessee; Laurium, Greece.

Barite, minor Pb-Zn. This is one of the very common and industrially important styles of barite mineralization. Barite blocks, usually partly dissolved, have been produced from residual clays filling shallow sinks in Cambrian limestone and dolomite in the Washington County, Missouri (11.5 Mt barite produced; Fig. 23-24/2); Cartersville, Georgia; Sweetwater, Tennessee and elsewhere.

Cinnabar. Unique deposits of cinnabar in residual clays filling an extensive karst are known from southern China (e.g. Gunchen and

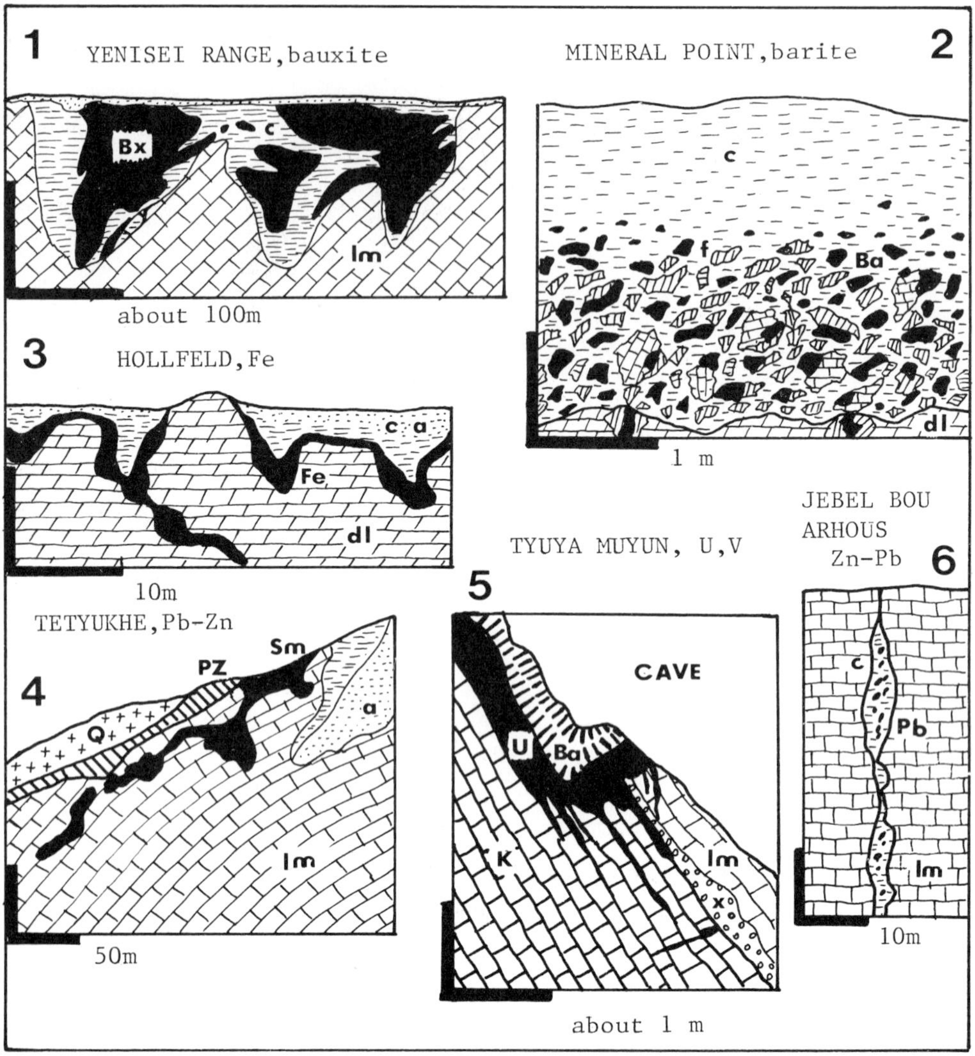

Fig. 23-24. Selection of karst-associated deposits (letters in brackets give the approximate correspondence to the mineral styles in Fig. 23-23). (1) Residual bauxite (Bx) filling karst depressions (B or C); (2) Barite (Ba) blocks in near-surface residual clay filling karst irregularities in dolomite (D); (3) Residual goethite nodules and friable "limonite" (Fe) blanketing karsted dolomite (C); (4) Secondary smithsonite (Sm) filling karst openings and replacing limestone in the oxidation zone of a Pb-Zn (PZ) sulphide replacement mineralization (F). (5) Tyuyamunite (U) and barite (Ba) encrusting coarse hydrothermal calcite (K) in a mineralized karst; interpreted as a hydrothermal ore superimposed on a paleokarst (J). (6) Galena blocks (Pb) scattered in residual clay in a vertical cave (D).

Linchuan deposits; Fedorchuk, 1960). There, angular fragments of cinnabar float are embedded in residual and transported red clay. The cinnabar was derived from hydrothermal impregnation and replacement deposits in the bedrock carbonates. Similar deposits have been described from Tashrabat (Tian Shan Range, U.S.S.R.; Surgai, 1980) and from the Kurai zone (Gornyi Altai, U.S.S.R.; Selin, 1981).

Antimony. Widespread concretions and fragments of Sb oxides (mostly kermesite and cervantite) occur in red residual clays formed over peneconcordant stibnite-jasperoid deposits in carbonates in the Kwangsi and Hunan Provinces, southern China (Siczyuan and Yuczyuan Rivers Basin; Fedorchuk, 1960). A portion of the Sb minerals has been reworked into proximal alluvial gravels.

Gold. The polygenetic deposit Kuranakh (Yakutia, U.S.S.R.; Pustovalov, ed., 1965) is an example of a karst superimposed on Mesozoic hydrothermal-replacement bodies in Cambrian carbonates. The residual clay filling karst depressions is mixed with fragmentary and powdery quartz, fragments of silicified carbonates and irregularly distributed gold.

Nickel. The "contact-karst type" of Ni silicate and oxide deposits (Chapter 7) is closely related to Ni "laterites", and is known from several localities in the Ural Mts., U.S.S.R. (e.g. Rezh, Ufalei; Pustovalov, ed., 1965). There, Paleozoic marbles are intruded by peridotite bodies and both are deeply weathered: the peridotites lateritized, the marbles karsted. A portion of the nickeliferous laterite and saprolite over peridotite have been reworked and redeposited in adjacent karst depressions.

(4) Carbonate in near-surface karst depressions, replaced by ore minerals (demonstrably precipitated from "per descensum" solutions). This style has many variations, the most common being transitions into (3). Most of the known examples represent variations of gossans (oxidation zones) over sulphide deposits, others probably correspond to the "exotic deposits" (transported oxidation zones).

Fe deposits. In the Berezovo deposit, East Transbaikalia, U.S.S.R. (Sokolov and Grigor'ev, 1974), limestone rubble and a massive limestone are replaced by siderite and ankerite in a karst, or along edges of detritus-filled depressions. The orebodies are situated close to a Cretaceous coal-bearing lacustrine complex, and the sideritization is explained as being due to descending iron-bearing meteoric swamp waters.

Zn carbonate and silicate deposits. Smithsonite and hemimorphite are common minerals in gossans over carbonate-hosted Zn-Pb sulphides,

---

Remaining abbreviations: lm=limestone; dl=dolomite; c=residual clay; a=sandstone; x=carbonate karst sediment; Q=quartz porphyry. Modified and simplified after Smirnov (1969); Buckley (1909); Magak'yan (1968); Kazanskii (1970); and Nicolini (1970).

in which they directly replace (pseudomorph) the sulphide aggregates. Impressive examples have been described, for example from the Aachen-Moresnet district (Belgium and West Germany). Because of high mobility, the zinc solutions released by sphalerite decomposition can migrate laterally and downward and replace the host rock carbonates along conduits. Such replacement can take place locally within the gossan and pose no interpretational problem. Alternatively, the zinc solutions can migrate far away from the site of the original sphalerite; the latter could in the meantime have been completely removed, so the genetic connection is obscured and an "exotic" Zn deposit results. The small deposit in Bertha, Virginia, and the large field Vazante, Brazil, are the frequently quoted examples.

In Vazante (Minas Gerais; 3.15 Mt Zn/9%, 175 Tt Pb/5%; Putzer, 1976), up to 200 m wide N.N.E.-trending zone in limestone and dolomite marbles of the Proterozoic Bambuí Group, is associated with Zn-Pb occurrences traceable for a distance of over 13 km. The hypogene minerals (sphalerite, galena, pyrite, chalcopyrite) are disseminated or form replacement masses in the tectonic breccia. Within the oxidation zone which is up to 70 m thick, hemimorphite with minor zincite, hydrozincite, willemite and cerussite are enclosed in a residual red clay or in a hematite-goethite gossan. Solution collapse breccias composed of partly dissolved carbonate and residual clay cemented by zinc oxides occur locally, for example at the Serra do Poço Verde locality (Abreu, 1973).

Vanadium. Vanadates (mostly descloizite) are commonly associated with oxidation and cementation zones of a distinct class of carbonate-hosted Zn,Pb,Cu deposits. Example localities include Berg Aukas and others in Namibia, Kabwe (Broken Hill) in Zambia, Itacarambí-Januaria zone in Brazil and other localities. Because of its superior mobility, vanadium migrates extensively and replaces carbonates along the walls or in blocks in debris-filled karst depressions, often at some distance from the residual gossan.

(5) Ore minerals accumulated in the karst subsurface, in the zone of permanent circulation. As already noted, the "tourist's karst", the system of stalactite and stalagmite decorated caves, is ill-suited for metallic deposition coeval with karsting. No metallic occurrences (with the exception of a few Fe and Mn oxide encrustations and wishing well coins) have been observed to have been formed there recently. No data are available on the trace metal content (if any) of the bat guano, the only nonmetallic commodity recorded as occasionally recovered from karst caves.

The subterranean streams flowing over some cave floors occasionally bring a non-carbonate detritus from the outside, and several insignificant metallic occurrences have been recorded in such sediment. In the eastern Cuban Mn province (Chapter 13; Simons and Straczek, 1958), soft Mn oxides have been found locally forming thin layers on limestone cave floors. In one cave, stalactites projected into the ore. Hosking (1974) reported cassiterite placers deposited by an underground stream in a limestone cave near Kaki Bukit, Malaysia.

(6) Ore minerals accumulated in the zone of imbibition. Although no recently formed ores are known from the presently accessible, active karst, there is widespread evidence indicating that mineralization, coeval with karsting, may have operated in this zone in the past. All the examples described in the literature (sphalerite and galena disseminations, veinlets, encrustation and cement in collapse breccias; authigenic crystals as well as clastic grains of galena or shalerite in vadose karst sediments; sphalerite pisoliths, encrustations, stalagmites; Harańczyk, 1974; Panek and Szuwarzynski, 1975; Zuffardi, 1976; Bernard et al., 1972; Quinlan, 1972) are from pre-Quaternary sequences. Consequently, they correspond to paleokarst.

The spectacularly mineralized "caves" of Tyuya-Muyun in the Soviet Central Asia (Kazanskii, 1970; Fig. 23-24/5) decorated by zoned encrustations of calcite, barite, tyuyamunite and other U, V and Cu minerals, are interpreted as an example of paleokarst. The karst openings formed earlier are believed to have been filled posthumously by low-temperature hydrothermal solutions.

Plate 9. Leduc glacier below the outcrop of the Granduc copper deposit near Stewart, British Columbia. Mineralized fragments are part of the rock debris carried by the glacier and deposited in the moraine.

The present is the key to the past
but don't jump to conclusions too fast.
Evolution's the thing
that can cause ideas to sink.
Without considering it interpretations may not last.

CHAPTER 24

Recent Continental Sedimentary Environments and Cainozoic Unconsolidated Continental Sediments

24.1. GLACIAL ENVIRONMENTS AND DEPOSITS

24.1.1. Introduction

At present, about 10% of the Earth's surface is covered by glacial ice and up to 30% of the surface may have been ice covered during the Cainozoic glaciations in the not very distant past. The bulk of the remaining great continental glaciers (in Greenland and in the Antarctica) is known on the surface only. Most of the data on which the glacial geology is based have been collected along the edges of the large continental glaciers, and in the mountains studded by remnants of numerous Alpine glaciers curently receding rapidly. In the above setting, the glacial sediments are fresh and unmodified.

Moving away from the existing glacier margins, we enter the extensive tracts of northern Eurasia and North America, discontinuously blanketed by the still unconsolidated and largely unburied Cainozoic glacial sediments. These are now all relict sediments modified, often beyond recognition, by the exogenic processes presently taking place in the boreal to temperate climatic zones.

Glacial geology has an extensive literature (e.g. Flint, 1971; Jopling and McDonald, eds., 1975; Legget, ed., 1976; Edwards, 1978), in which the substance of this scientific subject is treated in depth. Given that the purpose of this book is to look upon environments and rocks as concentrators and hosts to metallic deposits, however, one needs only a fraction of the detail. This is because the metallogenetic balance of the Cainozoic glacial environments and rocks is highly negative; that is, glaciation acts as an agent of destruction and dispersal of the earlier mineralizations without generating any new, substantial ore accumulations in the process. The infrequent ores hosted by the glacial sediments (Fig. 24-1, Table 24-1) are overwhelmingly the results of incomplete dispersion of the earlier deposits.

There is, however, still some challenge left and a promising direction for further research. One of the very practical problems the solution of which might provide a handsome economic reward, would be the interpretation of the fate of the pre-glacial auriferous surficial sediments and weathered crusts that once almost certainly

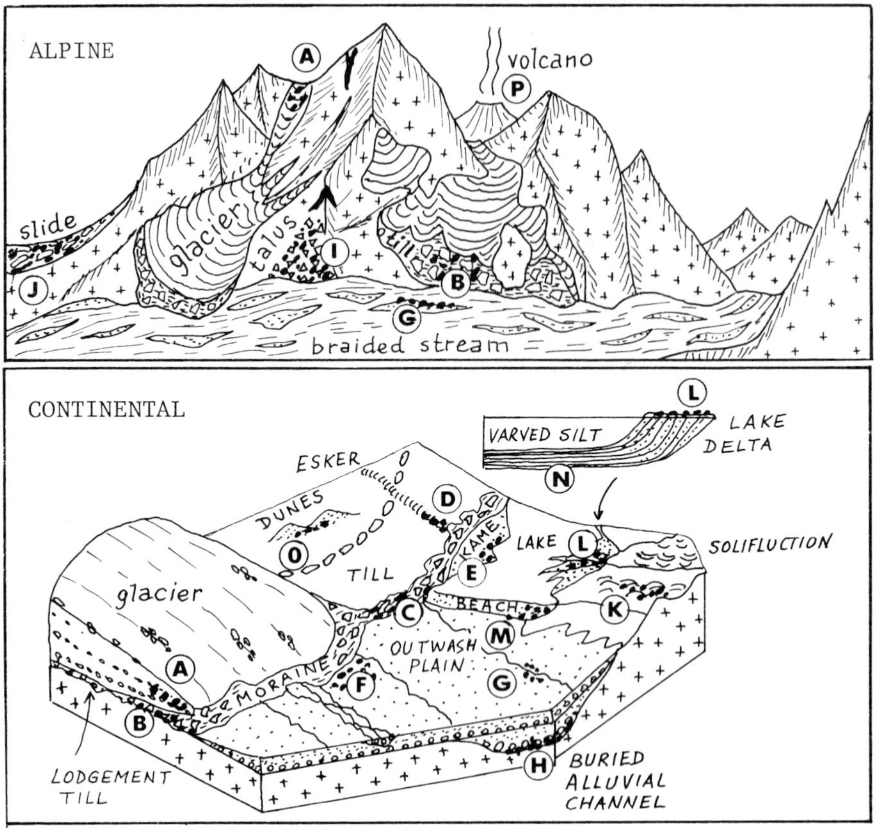

Fig. 24-1. Glacial depositional environments and sediments, principal mineralization styles. See Table 24-1 for letter symbol explanations.

existed over the great bedrock gold deposits in the recently glaciated areas such as the Canadian Shield. There, the heavily gold-mineralized greenstone belts, such as the Abitibi (Ontario and Quebec; production and reserves of lode gold over 2,500 t), have yielded virtually no "secondary" gold. This is in sharp contrast with the worldwide experience, where the lode : secondary gold production ratio ranges between 1:5 and 10:1. If even the lowest ratio applied to the Abitibi belt, there would have to be, somewhere, some 250 t of gold to be won.

In the meantime, however, glacial sediments are a nuisance as far as an exploration geologist is concerned. They mask the possibly mineralized bedrock and are relatively hard to interpret. Numerous prospection techniques have been developed to minimize this. On the plus side, however, glaciation has been beneficial in scouring the bedrock clean and fresh, so bedrocks in the recently glaciated areas

Table 24-1. Principal minealization styles in glacial environments and glaciogenic sediments

(A)	Ore fragments in glacier ice (e.g. Kennecott, Alaska); X localities, 50 Tt Cu.
(B)	Mineral or ore fragments scattered in till, moraine, drift. X0 localities, about 100 Tt Sn, 420 Mt Fe, 10 t Au.
(C)	Authigenic minerals in till (Mn; 1 occurrence, uneconomic)
(D,E,F)	Glaciofluvial deposits in eskers (D), kames (E) and outwash (F) X0 localities, X00 Tt Sn, about 10 t Au.
(G)	Stream-reworked till or glaciofluvial deposits, containing detrital ore minerals in alluvial sand or gravel (e.g. ilmenite, zircon, gold, etc.). X0 localities, 700 Tt Ti, about 100 t Au.
(H)	Detrital minerals in pre-glacial alluvial channels buried by glaciers or glacial sediments. X0 localities, X0 t Au, X Tt Sn.
(I)	Detrital minerals and mineralized blocks in talus and scree. X localities, about 100 Tt Sn, 20 Tt Cu.
(J)	Detrital ore minerals in debris flows and slides. X localities, about 2 Tt Sn, 700 t W.
(K)	Detrital minerals in solifluction mixture. X localities, 0.X t Au.
(L)	Detrital minerals in proglacial lake delta sand. X localities.
(M)	Detrital minerals in proglacial lake beaches (e.g. ilmenite, monazite, gold). X localities, 0.X t Au.
(N)	Authigenic minerals in varved lake sediments (Fe,Mn oxides). X localities, X0 Tt Fe,Mn.
(O)	Mineralization in or associated with dune sand and loess.
(P)	Glacial-volcanic association.

are of excellent quality. This, some say, far outweighs the benefits of tropical weathering with its thick mantle of decomposed rocks and soils which, despite having a positive metallogenetic balance (they generate new ores), make bedrock recognition and tracing difficult.

24.1.2. Glacial environments

Descriptions and interpretations of glacial environments are available from all text and reference books on geomorphology and sedimentology (e.g. Strahler, 1971; Reineck and Singh, 1975; Leeder, 1982) to which the reader is referred. Briefly, "cold" and "warm" (temperate) glaciers are two end-members based on the meltwater budget. In the former, meltwater is almost lacking because the temperature is well below the pressure melting point. In the latter, meltwater and the sediments it transports or deposits are abundant within, under and in front of the glacier.

The morphological classification of glaciers usually distinguishes (1) valley (Alpine) glaciers of the mountainous regions, that can be long, narrow, deep and comparable with river systems in their overall linear or dendritic distribution; (2) Piedmont glaciers, ice sheets developed by the coalescence of several valley glaciers in a lowland area; and (3) ice sheets (continental glaciers) or ice caps (remnants of continental glaciers) of a large areal extent, and thickness measured in thousands of metres. They are characteristic of lowlands and often extend into the sea as ice shelves. The active glaciers are centerpieces of a more extensive surrounding, shaped by meltwater or seawater (glaciofluvial, glaciolacustrine, glaciomarine environments).

24.1.3. Glaciogenic sediments and mineralization: drift, till and moraine

To a non-specialist, glacial drift represents all varieties of the rock debris deposited in close association with Quaternary ice sheets. It is often used interchangeably with the terms moraine and till. Leeder (1982) defined moraine as "essentially ridges produced by glacial dumping of both ice-deposited tills and water-deposited outwash", and summarized typical moraine stratigraphies. Terminal, recessional, interlobate and ground moraines are usually distinguishable. "Till" is both a petrographic term (boulder clay or mixton after Schermerhorn, 1966; its consolidated equivalent is tillite that corresponds to paraconglomerate or mixtite) and an environmental-associational term. Basal or lodgement till consists of a material dragged along and beneath the moving ice, and having a plastering effect. Such a till could be moulded into drumlins: streamlined, smooth-topped bedforms. Ablation (residual, melt-out) till consists of debris held within the ice and dumped as a result of seasonal melting. Flow till is generated by the release of the fragments from basal debris bands at the glacier snout. All the varieties of till could be subaqueously deposited and are either massive or banded. The massive tills are dominant; banded tillites are usually interpreted as being the result of mixing of two populations.

Petrographically, till is the worst sorted sediment known and an end member of the sorting sequence (Fig. 24-2). Compositionally, there are three main types of till sources (Edwards, 1978): (1) extrabasinal (from distant sources; "exotic"); (2) intrabasinal (locally derived); and (3) intraformational, re-transported glacial sediments, including the poorly sorted ones such as outwash and laminites. It has recently been repeatedly proven that the locally derived material in till is dominant (of the order of 90%), and the exotic material subordinate. The latter, however, is certainly more conspicuous and unusual, and has received prominence in the literature emphasizing the distance the material travelled (up to 2,000 km).

The till and moraine metallogenesis, insignificant as it may be, is a faithful reflection of their petrogenesis. A glacier advancing over a mineralized bedrock, will produce a mineralized drift. An outstanding, often quoted example has been the "glacier Cu ore" at

Fig. 24-2. Till, the worst sorted sediment known. Matanuska Glacier, S. Alaska.

Kennecott, Alaska (Bateman and McLaughlin, 1920). There, 220 Tt of high-grade chalcocite ore in the form of interglacial debris (moraine) have been mined from an active Alpine glacier, shortly below the bedrock ore outcrop in a crumbling cliff at the site of the Bonanza lode mine. This was a relatively rich "orebody", but the original ore source was still richer and larger; consequently, the Kennecott glacier ore was a case of an initial ore dispersion. Several additional examples of occurrences of scattered mineralized fragments in an active glacier have been recorded (e.g. near the Granduc copper mine, British Columbia), but the majority of "mineralized drifts" or moraines described are relict, fossil glacial sediments, and the parent glacier is generally out of sight.

El Rodeo tin deposit near Viloco in the Cordillera Quimsa Cruz (south-east of La Paz, Bolivia; Ahlfeld and Schneider-Scherbina, 1964; Fig. 24-3), is an example of a mineralized moraine in a glaciated alpine landscape. This is a virtually flat body of unsorted till, composed of a chaotic mixture of locally derived granite, metamorphics, vein quartz, cassiterite, magnetite and minor gold. It also contains minor pockets of intraformational, poorly sorted gravels. The body is interpreted as a ground moraine formed during the third Quaternary glaciation (equivalent of the North American Illinoian), and subsequently uplifted and dissected by the Rio Yaco valley. It has the shape of a 7 x 3 km large blanket, capping a small meseta. The metallic minerals, particularly cassiterite, have many

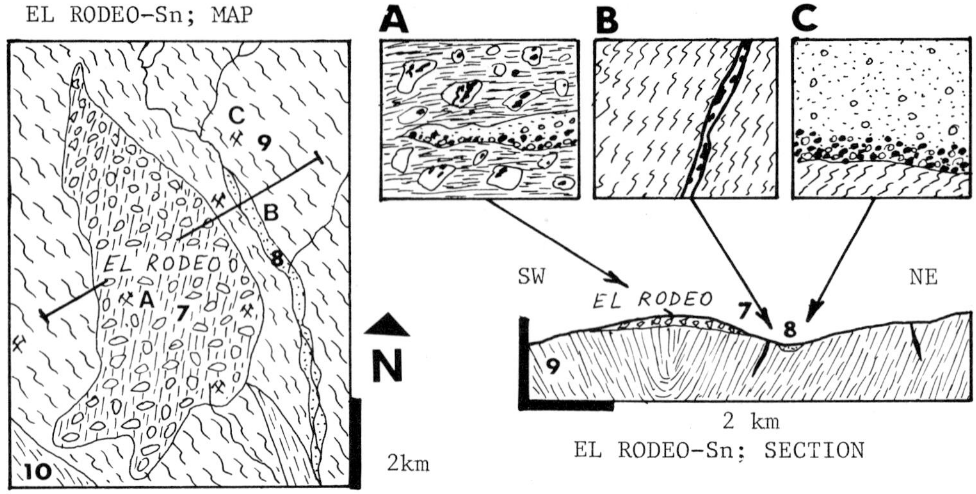

Fig. 24-3. El Rodeo, Bolivia, a cassiterite-bearing moraine. MAP: 7=mineralized Q (Illinoian) moraine; 8=Q gravels locally with placer cassiterite; 9=D schists; 10=Pe red — beds. Diagrammatic representations of mineralization styles: (A) moraine with blocks of rock/gangue with occasional cassiterite, containing a lens of a sorted gravel enriched in detrital cassiterite; (B) primary Sn hydrothermal vein; (C) creek bottom placers. Based on data in Ahlfeld and Schneider-Scherbina (1964).

forms of distribution: (1) veins, stockworks, disseminations, etc. in large blocks of the original gangue and/or the host rock; (2) bands or disseminations in fragments of hydrothermal quartz; (3) loose cassiterite grains that are a part of the sandy fraction of the till; and (4) clastic cassiterite grains slightly enriched in certain bands in the pockets of the weakly sorted sediment. The moraine body as a whole is believed to hold about 48 Tt of tin, but the average grade is only about 200 ppm. Putzer (1976) gave the reserve figure of 500 Mt "ore" with a grade of 0.02-0.1% Sn, that might correspond to some 300 Tt Sn. Up to now, only the richer pockets at El Rodeo have been selectively mined on a small scale, using primitive labour-intensive techniques. Even more extensive (20 x 1 km) stanniferous moraine having an average grade of 600 ppm Sn, has been reported by Rivas (1977) from the Apolabamba zone, Bolivia, where it tops a granite and slate basement.

In the Cobalt-Gowganda silver vein district, Ontario, recent (1983) mining of the glacial drift containing sparse mineralized boulders, has extended the life of this almost exhausted mining camp. This represents an unusual case when the mineralized drift was discovered and utilized long after the discovery of the primary (bedrock) orebodies. Usually, it is the other way round.

Several tens of references to "detrital gold in moraine" can be extracted from the literature, and Boyle (1979) gave a brief summary. In the Nome area, western Alaska (Nelson and Hopkins, 1972), glacial till contains consistently small amounts of gold, the average value being about 70 ppb (this corresponds to a factor of concentration of about 40). The gold was derived from numerous small vein occurrences in the slate and schist bedrock. Very similar but substantially more extensive (up to 200 km long) occurrences of auriferous till are found in the Allakh-Yun River valley in Siberia (Bilibin, 1955); at several places in the Juneau Gold Belt, Alaska (Spencer, 1906); near Stark and Libby, Montana (Schrader, 1911) and in the vicinity of the "greenstone belts"-hosted bedrock deposits in Ontario (Boyle, 1979). Only insignificant quantities of gold, however, have ever been recovered by sluicing.

Most of the gold recovered from a "moraine", in fact, came from outwash horizons or from a younger generation sediments produced by reworking (clay winnowing) and partial sorting of the moraine (till). Near Nome (Nelson and Hopkins, 1972), the highest gold contents (0.6-2.5 ppm) are in relict lag gravels that veneer the drift bodies, and these can be both subaerial as well as submarine ("offshore beaches"; compare Chapter 5). Elsewhere in the mountainous landscape, the reworked gold accumulated later in gulch placers in the upper reaches of creeks. In gently rolling landscapes, gold has been reworked in a series of alluvial terraces and river channels (e.g. Allakh Yun Valley, Siberia); see also the next section.

Glacial drift in the vicinity of bedrock iron formations in Canada and northern United States, contains abundant fragments of these rocks. Geologists working in the Yukon have long been familiar with the bright red jasper and hematite boulders and pebbles scattered in drift and redeposited in streams long before their source, the Snake River iron formation (Crest) has been discovered. The Canadian Charleston mine near Atikokan, Ontario, however, was probably the only locality where pebbles, cobbles and boulders of high-grade hematite and goethite present as a component of an end moraine, have been recovered commercially. The iron ore was derived from the enriched, Archean Steep Rock iron formation, and about 650 Tt concentrate was produced from 4.5 Mt of gravel.

The exotic (of distant derivation) till fraction has no significance as a mineable metal accumulation, but it provides important data for paleogeographic interpretations and for exploration by tracing "boulder trains". The latter technique has been reviewed recently by Ferguson and Freeman (1978), and also in Davenport, ed. (1982).

All the metallic mineralizations listed above as associated with till (moraine) were allogenic, detrital. Chew (1978) described an interesting occurrence of crystalline Mn oxides in a supraglacial flow till near Aberdeen, Scotland, which he interpreted as authigenic (chemically precipitated in situ). There, powdery and crystalline manganite and pyrolusite formed blocks and boulder coatings, some of which partially replaced granite fragments. The oxides were virtually iron-free, and the mineralization attributed to a groundwater leaching and precipitating activity under, or adjacent to, a peat bog.

24.1.4. Glaciofluvial environments and deposits: eskers, kames, outwash plains

Sand and gravel carried as bed load in meltwater streams can accumulate in a variety of settings. Eskers form in meltwater tunnels under the glacier and have a form of long, narrow ridges (shoestrings) of a stratified gravel and sand, roughly perpendicular to the glacier front. They are usually preserved following the glacier retreat.

Kames, on the other hand, are aligned parallel with the ice front and form long, low, steep-sided hills or short ridges, probably deposited as alluvial fans or deltas against the terminal margin of a melting glacier.

Glacial outwash has a typical form of smooth-surfaced gravel and sand plains (sandurs), sloping gently away from the terminal moraine and formed by coalescence of numerous alluvial fans.

Ideally, the glaciofluvial deposits are contemporary with unsorted till and were generated by the same parent glacier in time of its retreat. In regions covered by "fossil" glacial deposits now located in the humid boreal or temperate belt, however, the glaciofluvial deposits as well as till may have undergone partial or full reworking by the action of modern streams, and their original landforms may be no more recognizable. Because a degree of sorting has occurred, glaciofluvial sediments are in a slightly better position than till to accumulate detrital metallic minerals, provided that they are located in an area containing a bedrock mineralization.

In the literature on mineral deposits, mineralized glaciofluvial bodies are rarely listed. One of the rare examples is the Munro Esker in Ontario (Kirkland Lake area; Ferguson and Freeman, 1978). This esker is slightly gold-bearing in places and it also contains abundant pyrope garnet. A search for the garnet source resulted in the discovery of a kimberlite dike on the 840 m level of the Upper Canada gold mine. In his very detailed book, Boyle (1979) listed several examples of glaciofluvial gold placers. Several are in Park County, Colorado, in outwash aprons adjacent to the moraine. Elsewhere, glaciofluvial sediments are the initial metal-bearing association, over which and from which, a variety of superimposed alluvial placers formed. In the Tipuani placer field 100 km north of La Paz, Bolivia (Stoll, 1961; 17 t Au), gold occurs in Tertiary and Quaternary glaciofluvial gravels, interglacial alluvial terraces and in recent terraces and floodplain deposits.

The large alluvial placers in the Bear Valley, Idaho, that contain a variety of resistate heavy minerals (1.7 Mt ilmenite, 200 Tt monazite, 45 Tt zircon, 6,750 t euxenite, 1,500 t uranothorite in 485 cubic yards of coarse sand; Mackin and Schmidt, 1956), have also a combined origin. They are mostly post-glacial mountain stream sands, formed by a first-cycle reworking of Pleistocene till and glaciofluvial gravels, derived from granitic rocks of the Idaho batholith.

In Bolivia, the glaciofluvial tin placers are usually called veneros. In the Cerro Turqui area (Cuchu Jara and San Jose; Murillo et al., 1968), two horizons of sand and gravel deposited on a plain bordered by the Rio Pilcomayo, are selectively mineralized by detrital cassiterite to such a degree that they constitute a mineable orebody. One horizon, 60 cm thick, is located immediately at the bedrock of granite and metasediments in its aureole. The second horizon is 25 m stratigraphically above the base. Ahlfeld and Schneider-Scherbina (1964) listed all the important Bolivian tin placers adjacent to major bedrock deposits (e.g. the El Centenario placer at Llallagua, the Rio Huanuni placers, Anteguera-Avicaya and others), as glaciofluvial "veneros".

24.1.5. Buried sediment-filled valleys (channels)

Although this section is primarily concerned with the ores deposited during glaciation (and partly with those subsequently reworked), an exploration geologist has to be aware about the possible existence of a pre-glacial exogenic (or weathering modified or generated) mineralization, buried under the glacial sediments. Despite the tremendous scouring capacity of moving glaciers, pre-glacial sediment—filled valleys and/or secondary enriched bedrock deposits have survived here and there. Bilibin (1955) listed several examples of buried pre-glacial valley placers from the Allakh-Yun valley in Siberia, an area underlain by a relatively soft and fissile basement of gold-bearing shales and slates. The base of the moving glacier, choked and pasted-over by the slippery veneer of shale slivers and mud, had apparently lost much of its grinding ability, and it glided without causing significant harm, over several auriferous gravel-filled channels. Many such channels are located along faults or in narrow tectonic grabens. In some channels (e.g. in the Anmandykan Creek; Trushkov, 1971; Fig. 24-4), the auriferous buried gravel has been partly reworked into the base of the till.

Buried, sub-glacial channels are sometimes confused with alluvial pre-glacial detrital gold deposits located in regions which, usually due to their relative aridity, were not glaciated in the Cainozoic period. Strangely enough, some such placers occur in the present permafrost belt (e.g. the Klondike, Livengood, and other placers of east-central Alaska and western Yukon). The placers are permanently frozen and have to be melted for gold recovery, but have not been dynamically modified by the ice motion.

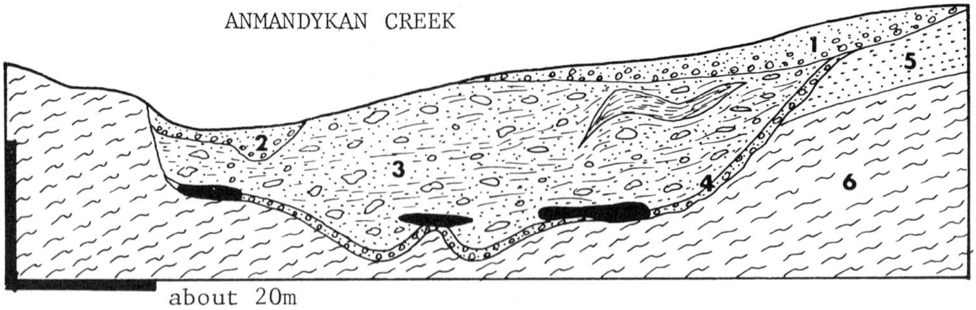

Fig. 24-4. Anmandykan Creek, Yakutia, U.S.S.R., an alluvial placer buried by, and partly reworked by, till. 1=proluvium; 2=present day creek sediments; 3=till; 4=pre-glacial (Tertiary) alluvial gravels; 5=Mesozoic quartzite; 6=shale. Based on data in Trushkov (1971).

24.1.6. Talus slopes and rock slides

In an Alpine landscape, glacial cover is often discontinuous, and glacier-filled cirques and valleys coexist with oversteepened, usually highly fragmented rock cliffs and slopes. Debris falls and landslides are common. A portion of the debris lands on the surface of a moving glacier into which it is subsequently incorporated as a moraine. The remainder accumulates as talus cones on and under mountain slopes, or as slide tongues in the valleys. The debris preserved in rock glaciers is less permanent. Orebodies hosted by the bedrock located uphill from the talus or slide bodies are naturally likely to be represented in the debris, in an unsorted form. The "glacier ore" at Kennecott mentioned earlier had its counterpart in the non-glaciated talus slopes, from which 90 Tt of a high-grade chalcocite was recovered (Bateman and McLaughlin, 1920). A significant quantity of scheelite ore has been recovered from talus under the Felberthal tungsten mine near Mittersil, Austria.

In Bolivia, the talus ores ("pallacas") locally contain important tin (and in places silver) resources. Cassiterite-mineralized blocks of quartzite are recovered from talus under the "tin manto" deposits at Kellguani. In Potosí, Fig. 24-5, an apron of talus ore, up to 50 m thick contours the Cerro Rico hill and contains an estimated 20 Mt ore having a grade between 0.3 and 0.6% Sn, and up to 300m ppm Ag (Murillo et al., 1968).

The Bolivian "llamperas" tin ores: a chaotic, unsorted debris of rock and ore mineral fragments cemented by goethite, are close to the water-saturated debris flows. The Guemul "llampera" under the cassiterite and wolframite mineralized Cerro Negro (Quebrada

Fig. 24-5. Cassiterite-bearing talus ("pallacas") on slopes of the Cerro Rico, Potosí, Bolivia.

Ajehulpaya; Ahlfeld and Schneider-Scherbina, 1964), contains about 1,760 t Sn and 960 t WO_3 in 4 million cubic metres of material.

24.1.7. Solifluction slopes

In periglacial environments with permafrost to a depth of 1-2 m the thin, surficial layer of embryonal soil mixed with bedrock fragments thaws briefly during the short Arctic Summers and slowly creeps downslope. Metalliferous solifluction components, most frequently muddy sand and gravel from high river terraces dissected by erosion, have been reported from gold placer goldfields of Tipuani (Bolivia) and from the goldfields and tinfields of Yakutia (eastern Siberia; Trushkov, 1971).

24.1.8. Glaciolacustrine environment

Proglacial lakes are typically associated with the "warm" retreating glaciers and can accumulate a variety of sediments. Small deltas with steeply inclined foresets are typically composed of a rhythmically alternating seasonal complex of ripple-laminated sand (Summer deposits), subdivided by thin clay laminae (Winter deposits); Edwards (1978). These pass downslope into fine grained, varved lake floor deposits of alternating fine sand to silt and clay. These are the most extensive and characteristic proglacial lake sediments ("lake beds"). Varved sediments of the Pleistocene lakes, dissected by modern streams, are now widely exposed in the formerly glaciated terrains of the Northern Hemisphere. Beaches, composed of a thin blanket of a relatively well sorted sand and gravel, may cover large areas. Subaqueous moraines and outwash formed in places where the glacier terminated in a standing body of water.

Bilibin (1955) discussed the possible presence of gold placers in deltas of proglacial lakes which, in contrast to most alluvial placers, would be relatively high in the stratigraphic sequence, in the subaerial (topset) portion of the delta. He, however, gave no actual locality examples. Several occurrences of resistate heavy minerals have been recorded in sands and gravels of the proglacial lake beaches. Bilibin (1955) reported minor gold placers from Yakutia. Folinsbee (1955) described a lenticular beach placer 15-33 m long, 1-3 m wide and 15 cm thick containing detrital monazite, magnetite and ilmenite, from Yamba Lake, Northwest Territories, Canada.

Metallic mineralization associated with the formation and early diagenesis of the offshore "lake beds" is rare and of little significance (the postglacial mineralization in lakes that may have started as proglacial, e.g. the Great Lakes of North America, is treated in Section 24-3). The exceptions include occasional layers of a dark organic-rich sediment that may be locally pyritic. Anthony (1977) described rhythmic laminae of authigenic siderite in varved sediments of the Lake of the Clouds, north-eastern Minnesota. Hewett et al. (1963) mentioned a bed of an earthy, brown, incoherent Mn oxides in a Pleistocene glacial lake bed 30 m below strand line, and assumed a possible coincidence with syndepositional hot springs. In the Leadville Pb,Zn,Ag district, Colorado, "lake beds" locally bury oxidized ore outcrops and occasionally contain infiltrated goethite and Zn-Pb minerals.

24.1.9. Glaciomarine environments

Glaciomarine sediments and associated ores have already been discussed in Chapter 5. One of the most popular aspects of glaciomarine sedimentation, the deposition of dropstones from floating icebergs, is a typical process of dispersion from the metallogenetic point of view and can have hardly any economic implications. Overstreet (1967, p.25) mentioned the occurrence of detrital monazite in unconsolidated sediments located in areas with volcanic bedrocks, which he atrributed to ice rafting from an original granitic source (that included a lake ice). Such monazite, however, lacked suficient concentration to be of practical interest.

In regions where auriferous moraines and outwash are dumped directly into the sea without the usual alluvial intermediary, wave reworking may accumulate detrital gold in both nearshore and offshore environments (e.g. off Nome, Alaska, and in the Bering Sea; Chapter 5).

The last relationship that could be of significance in the interpretation of ancient glaciogenic sequences (Section 25-5) but appears to be of little practical importance at present, is the geochemical change between the continental (potentially oxidized, iron enriched) and marine (reduced, iron depleted) diamictites. This is quite distinct along the sequence of the continental Antarctic glacial sediments grading into diamictites of the Ross Sea; Frakes and Crowell (1975) explained the iron depletion in the marine sediments by transportation of the reduced iron beyond the continental shelf.

24.1.10. Windblown sands and silts (loess)

Blown sand and silt accumulations are common in the "Arctic desert" environments, and large quantities of a windblown silt (loess) derived by winnowing of the fines from the exposed drift, lake beds and river terraces, now blanket countryside in central United States, China, western Europe and elsewhere. Recorded mineral occurrences in loess are rare and insignificant. Vennum (1980) reported green surficial encrustations of atacamite, antlerite, broachantite and other copper minerals on windblown sediments and on the bedrock, formed in the cold desert environment of the southern Antarctic Peninsula by water evaporation during the short summer. The metals were derived from the mineralized bedrock.

Interesting but uneconomic groundwater-metasomatic iron phosphate (destinezite, delvauxite) mineralization formed at the contact of Ordovician phosphatic and pyritic shales with the unconformable cover of Pleistocene loess in the northern outskirts of Prague, Czechoslovakia. The phosphates form abundant nodules along the base of the loess and are the result of replacement of its $CaCO_3$ fraction by Fe and SO_4 ions, carried by ascending groundwater from the shale.

24.1.11. Glacial-volcanic association and interaction

In several places around the world, active volcanism coincides or overlaps with glaciation (e.g. in Iceland, Alaska, Cascade Mts., Kamchatka, New Zealand, Mt. Erebus in Antarctica, etc.). In Iceland, subglacial basalt eruptions took place in pools of meltwater. Pillow lavas and hyaloclastites formed as a result (compare Chapter 4). On glacier or snow cap-topped high stratovolcanoes (Mt. Rainier, Mt. Baker, Mt. Ruapehu, Mt. Erebus, etc.), ice melting due to volcanic heat caused widespread and often catastrophic lahars, and provided water to recharge hot springs. In Alaska, snowfields and glaciers incorporate significant amounts of volcanic ash and tephra, later transferred into the drift. No metallic mineralizations or significant metal accumulation trends have been reported.

24.1.12. Miscellaneous influence of glaciation on metallic mineralization

Glaciers moving over soft or plastic floor (including its own sediments) cause deformation, and spectacular folds have often been reported (e.g. from proglacial lake sediments in Alaska; from Jurassic bedded gypsum near Gypsumville, Manitoba). The glacier load and movement also influences groundwater movement and discharge. McGinnis (1968) proposed a glacial-artesian model of origin of the MVT deposits, applicable to the Upper Mississippi Valley District, in which subsurface brines moving out of deep basins at abnormally high rates due to glacier loading, were heated and discharged beneath ice margins to produce ores.

24.2. ALLUVIAL ENVIRONMENTS AND UNCONSOLIDATED CAINOZOIC ALLUVIAL SEDIMENTS

24.2.1. Introduction

Alluvium is a "general term for clay, silt, sand, gravel or similar unconsolidated detrital material deposited by a stream or other body of running water as a sorted or unsorted sediment in the bed of the stream or on its flood plain or delta, or as a cone or fan at the base of a mountain slope"(A.G.I. Glossary of Geology). Alluvial sediments have been laid down in a variety of depositional environments or sub-environments (if alluvium is given the meaning of an environment), and it is quite characteristic that the recent review articles, chapters and books (Allen, 1965; Collinson, 1978a; Reineck and Singh, 1975) tend to treat such sub-environments in a rather non-uniform way.

In much of the sedimentogenic literature there is a tendency to emphasize several distinct "channel models" recently researched in detail (e.g. the point-bar model of meandering streams; low-sinuosity stream model; braided stream model; Collinson, 1978a) and assume that the bulk of streams are basically comparable. As has been the case before, theoretical sedimentologist and exploration geologist have different priorities, and the latter can easily skip over some of the fine detail and still reach his objective. Other factors (particularly the "primary" metal sources and bedrock peculiarities) and sub-environments rarely considered by the sedimentologists (e.g. gulches, mountain creeks) have to be added, so that the "prospector-friendly" selection of alluvial subenvironments, as treated here, has some peculiarities of its own.

Compared with the remainder of the continental environments and recent lithologic associations treated in this book, the alluvium is, economically, the most significant. Important quantities of gold and cassiterite and lesser, though still significant quantities of other metals (Ti, Zr, Th, U, Hg, W, etc.) came from recent alluvial environments and deposits.

METALLIFEROUS ALLUVIAL PLACERS

Most of the metals produced from alluvial deposits came from placers (Fig. 24-6), that is, particulate accumulations of heavy mineral grains in a variety of unconsolidated rocks, but mainly sands and gravels. The word placer came from the Spanish and basically denotes a mineralization style in which the economic substance is heavier than the rest and can be extracted (beneficated) by panning. Although alluvial placers are the most characteristic ones, metalliferous placers are common to a much greater variety of depositional environments or lithologic associations such as weathering crusts ("eluvial placers"; Chapter 23); marine beaches (Chapter 5 and others); aeolian sediments and some pyroclastics. Detrital metallic minerals accumulations in consolidated ancient sediments believed to be genetically equivalent to the modern placers, are called paleoplacers and are treated in several Chapters (19 and 25).

Metalliferous placers have a voluminous literature, but it consists of thousands of scattered articles and papers, many of them of vintage date. Very few comprehensive review papers and books are available in English (Hails, 1976; Harrison, 1954). The Russian literature is much more comprehensive, with several book volumes devoted solely to placers (Bilibin, 1955; Trushkov, 1971; Bykhovskii et al., 1981). Of the placer-concentrated commodities, gold occurs in the greatest variety of styles, sub-environments and sub-associations and Boyle (1979) gave a brief, but very complete introduction into their classification and geology. It is customary to refer to the economically mineable portion of a placer (an orebody) as a paystreak and this term is retained here.

Among the numerous properties and classification attributes of placers, some are objective and useful, others are subjective and to some extent incomprehensible. The latter have been avoided in this book. To an exploration geologist, the following aspects are of most practical importance:

(1) Derivation. "Eluvial placers" are heavy minerals essentially in situ, enclosed in weathering-softened parent rocks (saprolite) or soils. Colluvial and deluvial placers moved a short distance downslope, usually with most of the residual remains of their parent rocks. In alluvial placers, hydrodynamic sorting took place and the ore component is usually well separated from its former bedrock associates.

(2) Degree of consolidation. This ranges from a completely loose sediment (sand and gravel) through friable material to hard and brittle rocks.

(3) Grain size of the ore substance. Most ore minerals in placers have particle sizes in the sand to silt range (2 mm to 0.01 mm). When the grain size diminishes further, the placer dissapears because the very fine metalliferous particles are no longer mechanically recoverable.

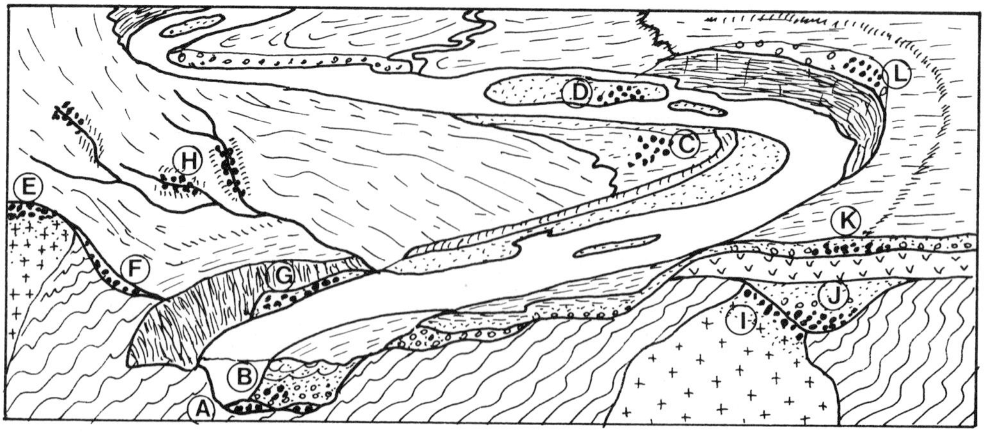

Fig. 24-6. Meandering stream system and associated placers. Principal mineralization styles. A=channel lag placers; B=point bar placers; C=flood plain placers; D=sand bar, islands placers; E=eluvial placers (saprolite); F=colluvial placers; G=proluvial placers; H=gulch and creek placers; I=buried eluvial placer; J=buried stream channel placer; K=terrace placer; L=bench.

recoverable.

(4) Environment of deposition or lithologic association.

(5) Present geomorphologic position: normal, elevated, depressed (with the placer usually buried).

Boyle (1979) integrated aspects 4 and 5 (Table 24-2). Further attributes are discussed later as required.

24.2.2. Alluvial fans, colluvium, proluvium

Alluvial fans (Collinson, 1978a; Fig. 24-7) are "localized deposits whose shape approximates to a segment of a cone. They develop in areas of high relief where there is an abundant supply of sediment". Coalescence of adjacent alluvial fans results in an almost continuous alluvial slope (bajada). Although most characteristic in semi-arid regions, alluvial fans are also present in the humid mountain environments where they are gradational into braided rivers. They are transitional into many other environments such as wadis, lakes, alluvial or deltaic plains, or marine environments.

The particle composition of an alluvial fan and the presence or absence of a metallic component, is directly related to the composition of the source (catchment) area above the fan. Texturally, the following varieties of sediments are usually recognized (Collinson, 1978a): (1) Debris flow deposits; an unsorted aggregate of rock blocks or fragments of variable size embedded in a sand or mud matrix; (2) sheet flood deposits, consisting of layers of fairly well

Table 24-2. Classification of alluvial placers according to their geological location and tectonic history (uplift or depression). From Boyle (1979).

PRESENT FORM EQUIVALENTS	ELEVATED EQUIVALENTS	DEPRESSED
1.Gulch, canyon and creek gravels	High-level gulch and creek gravels	Deeply buried gulch and creek gravels
2.River and bar gravels	Bench gravels High level river and bar gravels	Deeply buried river and bar gravels
3.River flood-plain gravels	High level flood-plain gravels	Depressed (deeply buried) flood-plain gravels
4.Deltaic gravels	Elevated deltaic gravels	Depressed deltaic gravels
5.Beach and shore line gravels	Elevated beach and shore line gravels	Depressed beach and shore line gravels

sorted sand and gravel; (3) stream channel deposits represented by lenticular, bedded, poorly sorted gravels and sands; and (4) sieve deposits, composed of clast-supported gravel lobes.

When ore minerals are present, they occur either as free particles, or enclosed in their gangue or wallrock materials. All the facies of alluvial fans can carry the mineralization. At most localities where ore minerals are in alluvial fans, they also occur in the "primary" bedrock outcrops, in creeks, gulches, alluvium, etc.

In the Basin and Range Province in the western United States and its extension into Mexico, alluvial fans and bajadas are an important component of the landscape and cover 20 to 80% of it. Typically, bedrock (lode) mines there are situated on slopes of mountain ranges above the fans, or in pediments. Although many such deposits were discovered by panning or tracing mineralized float in the alluvial fans, few fans have been mined for their ore content. Gold has been the most frequently recorded example, and the Hillsboro (3.4 t Au) and Jicarilla (7.2 t Au) deposits in New Mexico are often quoted in the literature as examples of gold placers in alluvial fans. In the Jicarilla Mts. (Segerstrom and Ryberg, 1972), the Pliocene auriferous fanglomerate is distributed over an area of 420 acres, is 2.3-3.3 m thick, and contains about 1.5 ppm Au. The gold source was in the mineralized aureole of an Eocene-Oligocene granodiorite laccolith emplaced into Permian limestones. Other, similar localities include a small proportion of the gold placers in the Battle Mountain district,

Nevada; Osgood Mts., Nevada and elsewhere.

An intriguing, but as yet poorly studied, aspect of alluvial fan metallogeny, is the potential presence of mineralizations precipitated within a fan from meteoric water solutions. Such solutions can be either (1) descending, bringing in dissolved metals from the fan source area or from the upper layers of the detritus, or (2) ascending, moving upward by capillarity and precipitating minerals in the same manner in which duricrusts form. The most favourable sites for precipitation are along the base of the fan and along its bedrock and sediment contact, but such sites are, unfortunately, rarely exposed. By analogy with the ancient "red-beds" association (Section 25.8), copper oxides, carbonates or silicates; uranium and manganese minerals are most likely to be found. Coatings of copper salts are usually abundant at localities where mine waters from producing porphyry copper deposits enter alluvial fans (e.g. Mineral Park, Arizona; 55 Tt Cu produced from chrysocolla infiltrations; Fig. 24-7C).

The large Exotica "exotic" (rootless) copper deposit in Chile (Roethe, 1975; 3.3 Mt Cu/1.65%; Fig. 24-8), is an outstanding example of Recent chemically precipitated copper orebodies in arid alluvial fans. The deposit is located in the Atacama desert, several kilometres downslope (south of) the large Chuquicamata porphyry copper deposit. It is hosted by a poorly sorted sand and gravel fanglomerate, containing interbeds and lenses of volcanic ash. Late Paleozoic amphibolites intruded by diorite and granodiorite are to be found in the basement. The mineralization has formed and continues to form through precipitation of chrysocolla (mainly as cement of the porous Quaternary sediments), atacamite and brochantite (as fracture coating in the crystalline basement), out of migrating solutions derived from the oxidation zone of the Chuquicamata porphyry deposit and also from the Chuquicamata ore detritus in the upper portion of the fan. The silica required for the chrysocolla precipitation is being produced by devitrification of the volcanic glass in the ash.

Although Exotica is unique in style and size, several tens of occurrences and deposits of "exotic" copper oxides, carbonates and silicates are known from regions containing bedrock sulphide deposits. There is a wide variation in style. Some deposits such as the Mineral Creek adjacent to the Ray porphyry copper, Arizona (5 Tt Cu/0.5%; Phillips et al., 1971), are surficial precipitates and they contain botryoidal goethite, jarosite, cuprite, tenorite, azurite and

---

⇦

Fig. 24-7. Alluvial fans. (A) A small northern alluvial fan, Mackenzie Mts. near Redstone, Canada. (B) Dissected arid alluvial fan, Quebrada del Toro near Salta, N. Argentina. (C) Alluvial fan (bajada) below the Mineral Park porphyry copper deposit, Arizona, that hosted a small infiltration chrysocolla deposit.

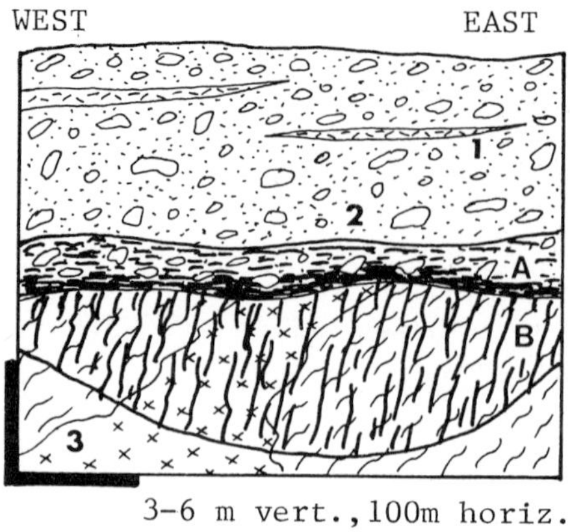

QUARTERNARY

1. lenses of volcanic ash
2. poorly sorted sandy gravel of an alluvial fan
3. D-Cb ? metamorphic complex amphibolite,partly granitized,intruded by diorite

A. ORE; chrysocolla coating and cementing gravels
B. fracture coating of atacamite and brochantite

Fig. 24-8. Exotica "exotic" (=rootless) copper orebody, Chuquicamata field, Chile. From LITHOTHEQUE.

malachite, cementing and coating gravel in an arroyo (ephemeral creek). The much larger Copper Butte deposit in the same area (1.17 Mt Cu/0.7%; Phillips, 1976), is hosted by an Oligocene conglomerate although the chrysocolla ore probably formed quite recently. In the Dundee-Arizona mine in the Jerome District, Arizona, adjacent to a Precambrian volcanogenic massive sulphide deposit, chrysocolla not only coats gravel clasts, but replaces the clasts of limestone as well as the footwall limestone bedrock along fractures (Anderson and Creasey, 1958).

Colluvium is defined in the A.G.I. Glossary of Geology as "alluvium deposited by unconcentrated surface runoff or sheet erosion, usually at the base of a slope", and proluvium is a "complex, friable deltaic sediment accumulated at the foot of a slope as a result of an occasional torrential washing of fragmental material". Consequently, these are rather close to and often overlap with, alluvial fans, or they substitute for alluvial fans where the climatic or physiographic conditions were unsuitable. In the literature, the term "slope-wash deposits" is frequently used and in the Russian literature "deluvial placers" correspond either to placers in colluvium, or in glacial sediments.

Colluvium and proluvium commonly contain placers very proximally derived. Such placers are poorly sorted, high in clay content, and the ore substance is irregularly distributed within a large mass of material. The Sn, Au, W, Hg, etc. placers near Guadalcázar, Mexico (filling karst topography and mentioned in Chapter 23; Fig. 24-9) are an example. A very similar mineralized slope-wash has been described by Hotz and Willden (1964) from the Dutch Flat district in the Osgood

Fig. 24-9. Poorly sorted colluvial gravel fills a karst depression near Guadalcázar, S.L.P., Mexico, and contains a detrital (placer) cassiterite, scheelite, cinnabar and gold.

Mountains, Nevada. There, unconsolidated sediment deposits 1.7 to 8.3 m thick, accumulated on the sides of canyons, topping a hydrothermally mineralized granodiorite intrusive into Paleozoic metasediments. They contain scattered gold, scheelite and cinnabar fragments, having an average value of $ 1.50/square yard (1954 prices). Colluvial and proluvial gold placers are common in the humid regions and merge or overlap with the gulch and creek placers, treated in the next section.

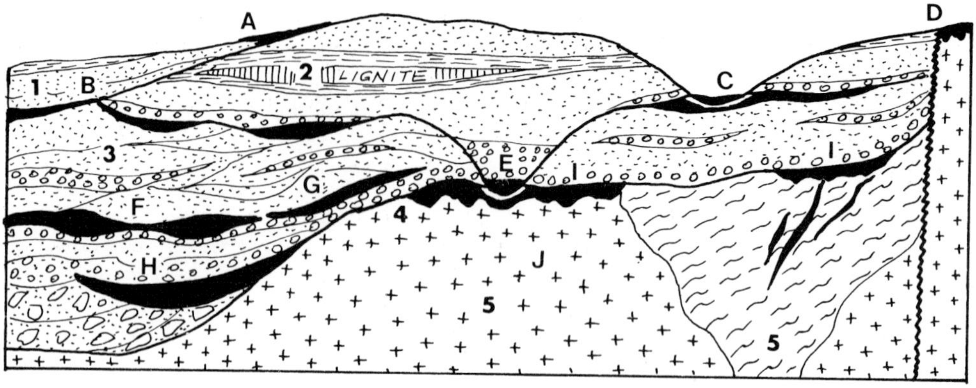

Fig. 24-10. Diagrammatic representation of the Sundaland (Thailand, Malaya, N.E. Indonesia) cassiterite placers. ROCK TYPES: 1=Q marine sediments; 2=Q younger alluvium; sand, gravel, clay, lignite; 3=Pl-Q, older alluvium; gravel, sand; 4=Miocene saprolite; 5=Pe-Tr tin granite, PZ metasediments. PLACER VARIETIES: A=Recent beach placers; B=buried beach placers; C=Recent stream channel placers; D=Recent alluvial placers; E=buried channel placers; F=buried braided stream placers; G=buried colluvial placers; H=buried alluvial fan and piedmont placers; I=buried alluvial placers; J=primary cassiterite in bedrock. From LITHOTHEQUE, based on data in Batchelor (1979) and Hosking (1969, 1974).

Although the various slope placers considered above are hardly impressive from the economic standpoint if considered alone and in isolation, they played a major role in the early stages of development of the world's largest detrital tin deposits in the Sundaland (Thailand, Malaya, Indonesia) region (Batchelor, 1979; Fig. 24-10). There, the detrital cassiterite accumulation started to develop in the middle Miocene period on the lateritically weathered Permo-Triassic tin granites and on their mineralized aureole. The colluvial cassiterite placers (kulit) formed first under savanna climatic conditions, and some are still preserved and mined as a superficial cover on stream divides, a few hundred metres from their source.

These were followed, during the Pliocene period, by the development of piedmont fan placers ("gugup"), represented by "debris flows of sodden stanniferous regolith including core boulders and talus". These occur as discontinuous massive wedges along the valley sides and they contain important cassiterite concentrations in the Kinta Valley and elsewhere. The cassiterite in both types of slope placers discussed above is coarse and angular.

Tectonic control to alluvial fans and slope deposits has not been emphasized so far, but it is of paramount importance, at least to the majority of orebodies. Fault-bounded sedimentary basins with periodic fault movements are not only important for fan deposition, but also for preservation due to subsidence and burial. A repeated step-like tensional backfaulting ("zony tektonicheskikh ustupov") as an important control of the shape, thickness and distribution of several Soviet cassiterite placers, has recently been summarized by Bykhovskii et al. (1981) and Lugov, ed., (1979); Fig. 24-11. The Tenkeli tinfield in northern Yakutia, U.S.S.R. (Fig. 24-12), is an example of a Tertiary alluvial-proluvial tin placer developed along a series of step-like faults that separate it from a primary, hydrothermal cassiterite stockwork. The close proximity of the placer source is a characteristic feature of the tectonically strongly controlled placer fields. The rapidity of placer development and burial facilitated preservation of semi-resistate to non-resistate ore minerals, and almost all the mineralogically unusual placers are adjacent to faults or are in tectonic grabens.

Bykhovskii et al. (1981) described the interesting Inkur River hübnerite placers, located in the Dzhida District, East Transbaikalia. There, three gravel horizons economically enriched in hübnerite are present in a sequence of muddy gravels up to 90 m thick, resting on quartz diorite adjacent to a fault. Near the fault, the mineralized zone is up to ten times as thick as the average thickness. The placer seems to have been preserved thanks to burial under a flow of Holocene basalt.

24.2.3. Hillside gulches and creeks

Gulches, gullies and ravines (Fig. 24-13) are small erosional valleys on hillsides, usually occupied by creeks that are tributaries of the main drainage system in the area. They are unspectacular to a sedimentologist and so rarely described as an environmental entity, but thousands of them have been worked, mostly for placer gold, typically in the initial stages of opening of goldfields and other metalliferous areas. The main characteristics of gulch placers as reviewed by Boyle (1979) are as follows: (1) most are confined to hilly country showing the signs of protracted weathering and denudation. Those in alpine landscapes have spotty paystreaks. (2) The stream gradients are moderate. (3) The sources of gold and other heavy minerals are very proximal, at the heads of gulches or along their sides. (4) The gold comes as coarse, free flakes, nuggets, wires and crystals of higher purity than the primary gold and much formed by nucleation and growth of initially finer gold in the exogenic setting. (5) The paystreaks are usually small but rich and generally on the bedrock or in the top metre or so of the bedrock. (6) In the absence of glacial deposits, the overburden is very shallow.

Dublin Gulch and Haggart Creek near Mayo, Yukon, the example locality of gulch placers described by Boyle (1979), is located in the non-glaciated portion of the Cordillera, in a low, hilly, forrested topography with numerous small streams tributary to the main course,

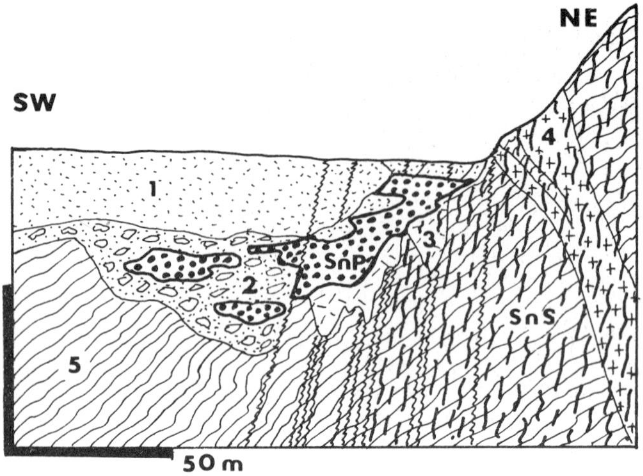

Fig. 24-11. Tenkeli tin field, northern Yakutia, N.E. U.S.S.R., an example of a Tertiary tin placer (SnP) developed over a period of time along a zone of repeated backfaulting. The placer is adjacent to a cassiterite stockwork orebody (SnS), probably located above a granite cupola. 1=Q alluvial sand and gravel; 2=T alluvial-proluvial gravels; 3=regolith; 4=Mesozoic granite porphyry; 5=hornfelsed metasediments. Based on data in Lugov, ed.,(1979) and Bykhovskii et al.(1981).

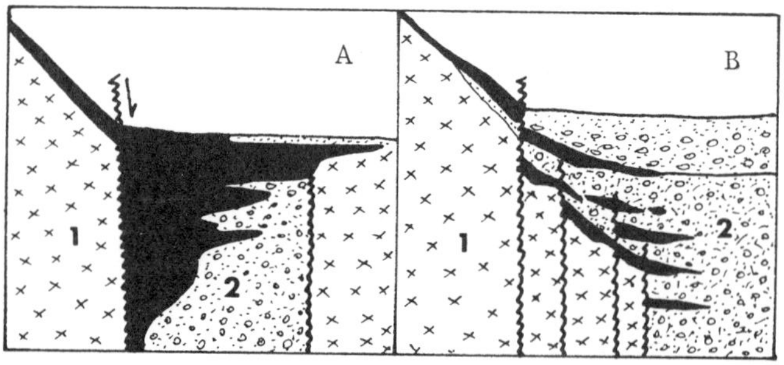

Fig. 24-12. Two varieties of alluvial tin placers in Yakutia (Siberia), controlled by syndepositional tensional backfaulting. (A) A thick wedge-like placer accumulated as a result of a slow, but steady downfaulting. (B) Step-like series of placers controlled by an interrupted tectonic subsidence. Modified after Bykhovskii et al. (1981).

Fig. 24-13. Gulches south-west of Helena, Montana, some of which contained gold placers derived from veins in the aureole of the Boulder Batholith.

indicating a deeply dissected upland. The primary mineralization consists of several auriferous quartz, arsenopyrite, pyrite veins formed in the aureole of a Cretaceous granite stock emplaced in micaschist, black schist and marble, and there is also one known tourmaline-cassiterite lode and several wolframite-scheelite occurrences. The primary ores are oxidized to depth of at least 6 m, and the oxidation zone is enriched in residual gold. The original stream profile in Dublin Gulch has by now been obliterated by eighty years of operations, but Boyle's (1979) reconstruction assumes 2-2.6 m of surficial slope wash on the surface, succeeded by 0.6-1 m of bluish clay and 1-1.3 m of limonite-coated gravels and weathered debris. This is underlain by the weathered bedrock. The paystreak is on the bedrock in the bottom 30 cm of the lower gravels and weathered debris, is about 33 m wide and several kilometers long. The average grade was about 1g Au/cubic yard and there was a minor scheelite and cassiterite by-product. The placer gold fineness was 900-925, compared with the 850 fine gossan gold.

Gold gulch placers have been widely distributed in Alaska, the Yukon, the Sierra Nevada of California, Idaho and Montana, the Black Hills, the southern Appalachians, Victoria Goldfields in Australia, New Zealand, the Urals, eastern Siberia and elsewhere. Gulch placers are often tributaries of the more extensive river placers. Few other metals, however, came from similar settings except as a by-product, because gulch deposits are generally too small and unprofitable for less valuable commodities such as cassiterite.

24.2.4. Braided streams and low-sinuosity sandy rivers

Braided streams have a distinct channel pattern, marked by a number of active shallow channels subdivided by longitudinal and transverse gravel and sand bars. They are characterized by rapid and continuous shifting of the sediments and repositioning of channels. Braided streams form on steeper slopes and, in a way, are transitional among alluvial fans, glacial outwash plains and the "true" valley streams. Consequently, braided streams are most widespread in semiarid, periglacial and mountainous setting. Braided rivers often change downstream into what Collinson (1978a) calls "sandy low-sinuosity rivers", the well-studied examples of which include the Platte River (Colorado and Nebraska), Brahmaputra, Hoangho and others.

Sediments of braided rivers consist of sand and gravel, mixed and interlensed in various proportions. Minor silt and clay lenses resting with sharp contacts upon gravels mark the channel fill remnants. The sedimentary pattern is a randomn interbedding of trough and tabular sand sets. Mud and clay are subordinate.

Contemporary braided and low-sinuosity streams have a remarkably low metallogenetic productivity, although when they drain a heavily mineralized or geochemically enriched bedrock region they act as a temporary settling environment for the ore minerals in transit (e.g. the Rio Huanuni, Bolivia, where cassiterite placers are located several km below the Huanuni bedrock tinfield and source area; 5,000 t Sn; Ocurí placers, Bolivia; 4,800 t Sn; Ahlfeld and Schneider-Scherbina, 1964).

One of the very few examples of the beneficial influence of braided streams on metal accumulation comes from the South-East Asian placer tinfields, reported by Batchelor (1979). There, early Pleistocene braided streams flowing over the earlier alluvial fan and piedmont placers, caused winnowing of fines and a widespread reworking. Under favourable conditions, the stanniferous gravels remained in situ as sieve deposits and additional concentration by elutriation formed the extremely rich kaksa and karang placers located as a thin layer on bedrock pediments. In situations in which the early placers were not gravelly, the question of preservation or dispersion of the cassiterite largely depended on gradients of the braided streams and bottom irregularities. Batchelor (1979) pointed out that gradients exceeding 1 in 150 resulted in total bed load transport and cassiterite dispersion, whereas gradients of around 1 in 300 allowed concentration. Braided stream placers formed locally up to 5 km from the source, especially on smooth clay bottoms. Although the Asian braided stream placers show a decrease in cassiterite grade and grain size downstream, this is balanced by their greater thickness and more uniform distribution, compared with the alluvial fan placers.

24.2.5. Meandering streams

The stream configuration in which a meandering channel lies within a much broader alluvial plain, forms the dominant pattern of the present rivers (Fig. 24-6). It is also one where the fluvial complex can be subdivided into a set of rather clear-cut sub-environments, and each considered and modelled separately. Some meandering streams are economically important, because the largest portion of metalliferous placers is associated with them. Briefly, fluvial deposits of a modern meandering stream can be treated in terms of three (sometimes only two) major associations (Reineck and Singh, 1975): (1) Channel deposits, those resulting from the activity of river channels. They include channel lag, point bar, channel bank and channel fill deposits. (2) Bank deposits, which are river-bank sediments produced during flood periods. They include natural levee and crevasse splay deposits. (3) Flood basin (plain) deposits, that comprise fine-grained sediments formed during heavy floods when water overflowed the levees into the flood basin. Flood basin and marsh deposits are included here.

All the fluvial deposits and their depositional sub-environments are important in exploring for metallic minerals, when such are present in the region (available from the bedrock in the drainage basin). Some sub-environments are the preferential ore hosts, others tend to be consistently barren.

Channel lag deposits consist, in the upper reaches of a stream, of gravels and pebbles that lagged behind while the sand moved farther downstream as a bed load. Typically they form discontinuous lenticular patches in the deeper parts of the channel, commonly resting on the bedrock. As the channel position keeps changing, the lag deposits are quickly covered by the finer sediments of point bars or channel fill and ultimately buried (and preserved) under the broad alluvial plain. In the lower reaches of rivers channel lag deposits may be missing, or substituted by driftwood; lumps of unconsolidated sediment and in the "civilized" environments by garbage.

Gravelly channel lag deposits are the most characteristic repositories of detrital heavy minerals, particularly gold, and a focal point of most metalliferous placers. The richest paystreaks tend to be on the bedrock bottom, trapped in cracks and other irregularities. The paystreaks may persist for several metres upward, the gold flakes being in the gravel matrix. The gold is relatively coarse. "False bottom" paystreaks (intraformational auriferous gravels usually resting on clay bottoms rather than on the bedrock) are less frequent, and difficult to predict.

Most channel lag placers differ to some extent, from the model examples described in the sedimentological literature and are, in many cases, a combination of or a transition between, gulch creeks and meandering streams. In short streams it is particularly difficult to differentiate between upper and lower reaches, and in the Klondike Goldfield as well as in many Siberian placers (e.g. those in the Kolyma River Basin), roots, leaves, tree trunks, animal bones and entire frozen animals such as mammoths, are intermixed with the gravel and boulders. The above examples are located in the zone of permafrost and the deeper reaches of channels (and former channels) are frozen, having been conserved motionless for several hundreds of thousands of years. This prevented further reworking. In "live" streams the sorting was longer and better and much of the organic remains removed.

In addition to gold, detrital platinum accumulated in channel lag deposits, as in the early Pleistocene "bench paystreak" in the Goodnews Bay, Alaska (Mertie, 1976). The lighter heavy minerals cassiterite, ilmenite, zircon, rutile, are less selective as to their site of deposition.

Point bar deposits are the convex sides of meanders. The lateral and downstream sedimentary accretion of point bars is volumetrically the most important mechanism for sedimentary entrapment in streams, and deposits so generated ultimately fill the greater part of the valleys. The thickness of point bar sediments is usually equivalent to the depth of the channel, and the upward-fining sequence of gravel, sand and silt-clay is usually marked by abundant large scale ripples and cross-stratification. The individual sediment bodies within the point bar sequence are, however, lenticular and discontinuous.

Point bar deposits may carry placer gold when it is locally available, but such gold tends to be fine and insufficiently localized so that the single-stage paystreaks are low grade, although relatively extensive. Repeated reworking of the point bar placers particularly during the successive stages of stream downcutting, however, can considerably upgrade the paystreak by winnowing out much of the light detrital constituents. Much of the present "terrace placers" are modified former point bar deposits.

Considering the "lighter" heavy minerals, the point bar deposits are much more suitable and more common sites of accumulation, than the channel lags. Much of the ilmenite-rutile placers near Roseland, Virginia, are in point bars, derived by reworking of the deeply weathered metamorphics and anorthosites in the region (Minard et al., 1976), and so are many of the fluvial tin placers of Yakutia (U.S.S.R.).

Flood basin and floodplain deposits. Flood basins are the lowest-lying part of a river floodplain (Reineck and Singh, 1975), separated from the channel by natural levees. Floodplain is a broader term applied by sedimentologists to deposits from overbank flows formed when rapidly shifting rivers lack well developed basins. The overwhelming majority of publications does not differentiate between floodplains and flood basins, so only the broader term is used here. Because of the poor drainage and general flatness, floodplains act as settling basins for the fine-grained suspended sediment from overbank flows. In addition to this, floodplains in the humid climatic zones are densely vegetated and often develop into backswamps. Floodplain sediments are mostly fine silt and clay, interrupted by sand and silt intercalations. Thicker sand beds may have formed as crevasse-splay deposits or natural levee deposits, but their interpretation is uncertain in the ancient fluvial sequences.

Detrital gold, at least gold coarse enough to be physically recovered, can hardly accumulate here. There is, however, a possibility that fine to very fine ("colloidal") gold can form a very low grade but relatively extensive accumulations in floodplains, and possibly represent a future resource. Minard (1971) reported up to 1.2 ppm gold in floodplains near Jefferson, South Carolina, in an area of known gold mineralization in a regolithic bedrock.

In contrast to gold, many of the lighter, rock-forming heavy minerals do form floodplain placers. Much of the fluvial monazite placers in the Inner Piedmont of Virginia, the Carolinas and Georgia (Overstreet et al., 1968) occur in this setting. There, the bedrock consists of deeply weathered high-grade metamorphics containing little more than accessory quantities of monazite. Erosion and redeposition of saprolite, residual soil and colluvium generates an uninterrupted flow of finer-grained detritus to the valleys, where it is being spread over the floodplains. The monazite placers are confined to floodplains 1.6 to 3.2 km downstream from headwaters and from there to the trunk stream, and they consist of fine sand, silt and clay with about 10% gravel. The placers are low-grade and a total of about 4,950 t of monazite concentrate was produced before 1917 in the Savannah-Catawba Rivers area. A large resource still remains (about 1.034 Mt monazite containing about 48 Tt ThO_2 and 520 Tt rare earths oxides).

Because of the floodplain capacity to accumulate clay and mud, it is conceivable that, under conditions of special supply, metalliferous mudrocks could accumulate in this setting. This applies in particular to gibbsite. A portion of the ancient continental redeposited bauxite deposits (such as those in the Tikhvinsk district, U.S.S.R.; Kirpal' and Tenyakov, 1974), may indeed be interpreted as floodplain deposits, the rest mostly being considered to be lacustrine. The optimal conditions for recent fluviatile bauxite accumulation exist several kilometres downstream from the original source (tropically weathered bedrock or an older bauxite deposit). Such deposits are to be expected in the basal portions of the valley fill and these may conceivably be masked by tropical vegetation: the source of coal and carbonaceous sediments known commonly to top or interfinger with the ancient bauxites.

25.2.6. Elevated fluvial deposits: terrace, bench and plateau gravels and sands

The majority of stream valleys are fringed by a system of terraces: long, narrow, relatively level or gently inclined surfaces, bounded along the edges by steeper ascending and descending slopes, respectively (Fig. 24-6). The terraces are covered by unconsolidated sediments (mostly sand and gravel) which are former valley deposits successively dissected by the stream downcutting activity. Most of the terrace sediments were laid down by the same river now flowing on the valley floor. They are compositionally almost equivalent, but cases where the earlier streams had different courses and consequently different detritus provenances, are also common. Some terraces carry lake sediments.

The word "bench" is commonly used by placer miners as a synonym for terrace, although the correct meaning is a level surface eroded in a solid rock. Sometimes this is the case and gold paystreaks may be developed in crevices in the top metre or so of the bedrock, a former stream bottom. More true benches are covered by a thin veneer of residual gravel, often gold-bearing.

Well-preserved fluvial terraces are typically composed of crossbedded gravel and sand, sometimes with pockets or layers of silt or clay. Occasionally, some of the distinct facies of a meandering stream discussed earlier, can be interpreted, but it has only very rarely been actually done. The literature usually mentions "terrace gravels" (or sands) or "elevated stream gravels", without giving more details. The bulk of terrace sediments are probably point bar deposits; floodplain sediments are rarer, channel lag deposits rarest. It should not, however, automatically be assumed that the terrace sediments formed in a meandering stream; many were deposited in low-sinuosity streams, braided streams, glacial outwash, or even rare alluvial fans.

It has already been mentioned that terrace gravels may contain gold paystreaks. In the important goldfield systems such as the Klondike, Bodaibo (Siberia), Kolyma, and others, several terrace levels plus the present valley floors are auriferous. The terrace gravels are either modified or unmodified. The modification includes the winnowing of fines and the light clasts, as well as a clast weathering. Intensive (tropical humid) weathering can eliminate selectively the unstable clasts and convert them into clay. Such clay either remains in the terrace forming a matrix to the resistate quartz and other minerals, or it is winnowed and a residual quartz only gravel deposits may form. Both modifications may be beneficial to gold enrichment.

Erosion of earlier terraces by the downcutting streams is an even more beneficial mechanism for gold concentration and accumulation. In the Klondike Goldfield (Boyle, 1979; Fig. 24-14), the Tertiary high-level White Channel terrace gravels contain low, but persistent gold values and are areally very extensive. They must once have contained an enormous gold accumulation. Since the close of their formation, further supply of new gold in the area was minimal and the 300 t of Klondike gold was won from numerous rich valley bottom placers 100 to 200 m below the White Channel gravels, into which it was supplied almost entirely by recycling and accumulation of the dispersed high-terrace gold. The prolonged, multistage generation of the great placer fields is a rule, not an exception, and understanding of the landscape development history is of paramount importance.

24.2.7. Buried alluvial deposits: deep-lead placers

Metalliferous placers are vulnerable to erosion, so burial under a cover of younger supracrustal rocks is an important mechanism for preservation. Burial is a transitional stage without sharp boundaries and in fact almost all placers are buried to some extent, even if only under several centimetres to metres of sediment of the same stream. In the literature, however, buried placers are considered those to be topped by at least several tens of metres of overburden and geologically significantly older than the modern surficial deposits. Particularly challenging are the placers unrelated to the present drainage, the majority of which are Tertiary or still older. Such placers are often called deep leads (Fig. 24-15).

Buried placers may either be unconsolidated, or partly cemented; some of the buried placers in the Sierra Nevada Foothills, California (Lindgren, 1911), are semi-friable. With increasing consolidation and age, buried placers grade into paleoplacers (treated in Chapter 25). The majority of buried placers, at least those that carry gold, are in erosional remnants of former stream valleys, in their deepest portions, so that in terms of depositional paleo-environments they are mainly channel lag deposits. The setting or morphology of the paystreaks as well as the detrital gold appearance are usually little different compared with the Recent placers, but a degree of diagenetic alteration could be apparent. The most common consequence of diagenesis is decomposition and disintegration of the unstable clasts and matrix constituents, and because there was no stream winnowing of the products of decomposition as was the case in many exposed

Fig. 24-14. Klondike goldfield, Yukon. TOP: Confluence of the Klondike and Yukon (right) rivers. The floor of the Klondike valley and its tributaries (e.g. Bonanza Creek, bottom picture) is covered by tailings remaining after the valley placers. The high-level terrace gravels are marked "w".

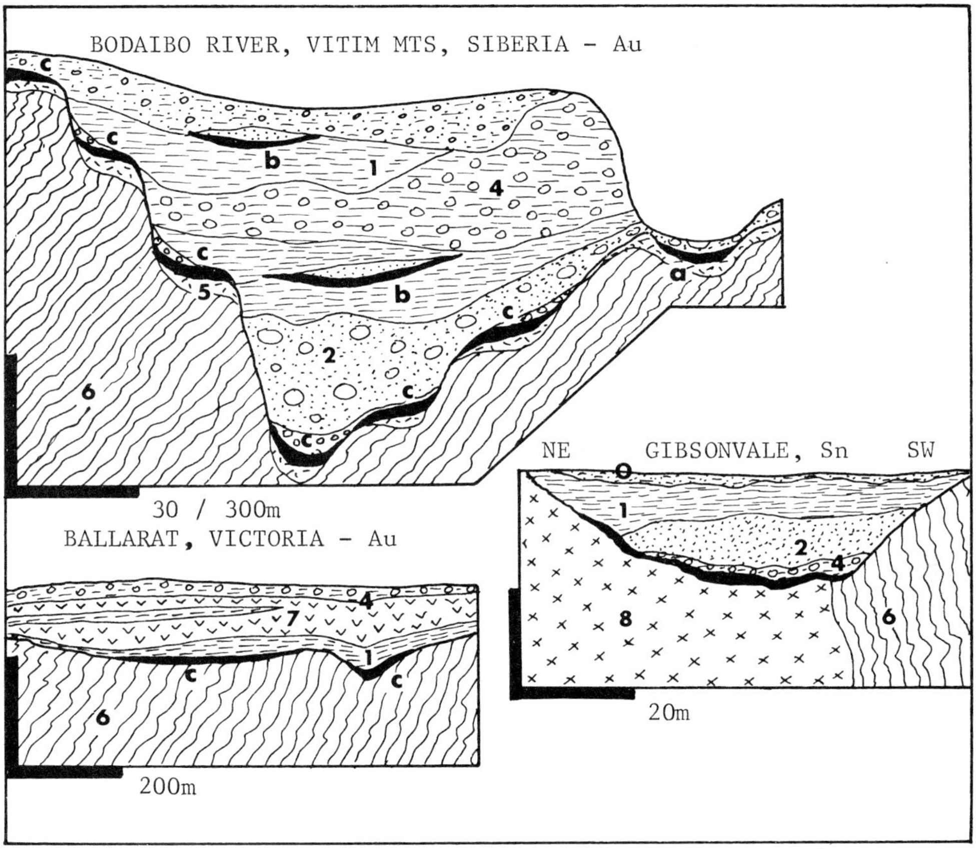

Fig. 24-15. "Deep lead" buried placers, selected examples. Bodaibo R. valley, Vitim Mts. (Lena Basin), Siberia. Pre-glacial and inter-glacial gold-bearing deep leads. Ballarat, Victoria Goldfields, Australia; Tertiary gold deep leads over quartz-gold lodes in Ordovician slates. Gibsonvale, N.S.W., Australia; cassiterite deep lead over a Devonian greisenized granite. EXPLANATIONS: 0=soil; 1=clay; 2=sand, 3=gravel; 4=clayous gravel; 5=regolithic bedrock; 6=bedrock slates, schists; 7=plateau basalt flow; 8=bedrock granite and greisen; a=Recent (modern) placer; b=interglacial placer; c=Tertiary (pre-glacial) deep lead.
Modified after McAndrew (1965), Belyaev in Vakhromeev (1961) and Campi et al. (1975).

terraces, the buried placers are often matrix (clay)-choked. They also tend to be "limonite"-rich, the iron having been derived by leaching from the decomposed mafic silicates and oxides. Under strongly reducing conditions, authigenic siderite and pyrite may have formed.

Deep leads are buried under a variety of rocks, including younger alluvial sediments; nearshore low energy marine sediments (e.g. Sundaland deep leads); plateau basalt flows (e.g. some fields in Victoria, Australia); andesitic tuffs and flows (Sierra Nevada Foothills, California), or glacial deposits (eastern Siberia). The latter are particularly intriguing because of the proven ability of glaciers to scour deeply and to remove, rather than to conserve, the ground over which they roll. Some glacial sub-environments, however, have a lower than normal scouring ability (e.g. the fine detrital bottom deposits of proglacial lakes). Alternatively a lack of scouring may have been due to special properties of the bedrock. This has already been discussed in Section 24.1. A special category of deep leads is in karst depressions. Selected example localities of metalliferous deep leads have been assembled in Table 24-4 that appears later.

24.2.8. Examples of representative alluvial placers

Metalliferous alluvial placers (this includes some eluvial placers) have produced, or contain in currently economic reserves, at least 12,608 t Au, 11.52 Mt Sn, 428 Tt REE, 5 Mt Ti, 3.2 Tt U, 52 Tt Cr, 136 Tt Th, 23.6 Tt W, 641 t Pt, some Fe, Zr, Nb, Ta, Be and other metals (data from Laznicka, 1985g). Most placer fields are polycyclic and composite; that is, more than one sub-association hosts the ore minerals and consequently selected localities have to be tabulated jointly (Tables 24-3 and 24-4).

SUNDALAND (THAILAND, MALAYA, ISLANDS OFF SUMATRA) TIN PLACER BELT

This 3,000 km long belt has produced the bulk of the world's tin, and 95% of the production came from placers (Fig 24-16). There are still significant resources left. The belt contains several thousands of individual tin deposits and there is a considerable variety of mineralization styles ranging from bedrock deposits through cassiterite-enriched saprolite (eluvial placers), alluvial fan and piedmont deposits, stream placers to marine beach placers. Some placer deposits are exposed, others are buried under continental or marine sediments. Reference to several styles has already been made and the belt has an extensive literature (Hosking, 1969, 1974; Batchelor, 1979). Hosking suggested that eustatic sea level fluctuations have been the major influence on the origin of the tin placers located in what is also an extensive bedrock tin geochemical province. Widespread and repeated reworking contributed to the formation of rich orebodies.

Table 24-3. Selected examples of metalliferous alluvial placers (all considered Quaternary, unless marked T = Tertiary)

METAL	LOCALITY	PLACER CHARACTERISTICS	BEDROCK SOURCE	TONNAGE	REFERENCES
Au	Livengood, 100 km N.W. of Fairbanks, Alaska	1x8 km, in permafrost channel gravel along tectonic trench	MZ hydrothermal veins	21.6 t Au (grade 0.54 ppm)	Hargreaves (1975)
Au	Klondike, Dawson, Yukon, Canada	low grade high terrace T gravels, frozen valley creek gravels under organic muck	numerous small qtz-Au veinlets in schist	300 t Au	Boyle (1979)
Au	Deadwood, Black Hills, S.Dakota	numerous gulch placers, free gold flakes, nuggets	Homestake deposit (qtz-arsenop. lodes in Pt_1 schists)	5.5 t Au	Koschmann and Bergendahl (1968)
Au	West slopes of Sierra Nevada, California	valley placers and T deep leads under volcanic and sedim. cover	gold-quartz veins in schist and greenst., in plutonic aureole	1,717 t Au	Lindgren (1911)
Au	Pato district, Antioquía, Colombia	valley, minor high terrace placers	gold-quartz veins in schists	108 t Au	Singewald (1950)
Au	Rio Tapajós, Amazonas, Pará St., Brazil	low valley placers over 220 km^2	qold-quartz veins along shears in PCm metamorphics	5,632 t Au	Damião (1980)
Au	Yana-Kolyma belt, Eastern Siberia U.S.S.R.	numerous valley, terrace, coluvial, reworked moraine placers along 750 km belt	veins in MZ cleaved slates, granite aureole	min. 650 t Au max. 5000 t	Borodaevskaya and Rozhkov (1974)
Au	Morobe Goldfield (Wau-Edie Creek) New Guinea Island	stream channel, valley placers, contain low fineness gold (530)	high-level T veins stockworks in qtz. porph. and metam.	97 t Au	McLeod, ed., (1965)

Table 24-3 (continued). Selected examples of metalliferous alluvial placers (all considered Quaternary, unless marked T = Tertiary)

METAL	LOCALITY	PLACER CHARACTERISTICS	BEDROCK SOURCE	TONNAGE	REFERENCES
Au	Westland, Nelson, Otago Prov., South Island, New Zealand	valley and terrace, minor glaciofluv. placers; access. scheelite	gold-quartz veins in PZ+MZ slate, schist, granite aureole	426 t Au	Williams (1965)
Pt Ir Os	Goodnews Bay, Alaska	valley, stream channel, bench, fluvioglacial placers; platinum, minor Os-Ir alloy	Cr_3 dunite and pyroxenite intrusions	226 t Pt met. 2 t Au	Mertie (1976)
Pt	Chocó Department, Colombia	channel, valley, bench placers resting on lateritized bedrock. Pt, Au, magnetite, ilmenite	post-J regolithic dunite, peridotite, picrite bodies	155 t Pt met. 6 t Au	Mertie (1976)
Pt Os Ir	Centr. and S. Urals (Nizhnyi Tagil) U.S.S.R.	channel, valley, T high terrace placers; Pt, Os-Ir, chromite grains, nuggets	PZ small circular differentiated dunite-pyroxenite complexes	min. 180 t Pt metals	Mertie (1976)
Sn	Rondônia State, N.W. Brazil	cassiterite in saprolite, alluvial valley placers	greisens and veins assoc. with Pt granite	1.85 Mt Sn	Abreu (1973)
Sn	Cornwall and Devon, S. England	valley placers, some deep leads, cassiterite	derived from Cb-Pe veins, greisens in granite aureole	550 Tt Sn	Dines, ed., (1956)
Sn	Sundaland (S.E. Asia) placer belt	great variety of placers along 3000 km belt; eluv., alluvial, deep leads, minor beach placers; cassiterite	Pe-Tr tin granites and their aureole	7.3 Mt Sn	Batchelor (1979)

Sn W	South Kiangsi, China	channel, valley, terrace placers, eluvial placers; cassiterite, wolframite, scheelite	MZ granites and their aureole, veins, greisens	96 Tt Sn 13.3 Tt W	Ikonnikov (1975)
Sn	N.E. Chukotka Pen. placers, E.Siberia	eluvial, alluvial fan, channel, valley, fluvio-glacial cassiterite placers	MZ granite and their aureole	min. 80 Tt Sn	Materikov (1974)
Sn	New England (Tingha, Tenterfield), N.S.W.,Australia	eluvial,valley and terrace placers, deep leads under basalt	Pe-Tr_1 granites, greisens, veins	210 Tt Sn	McLeod, ed., (1965)
Sn Ta Nb	Jos Plateau, central Nigeria	cassiterite, columbite in saprolite, laterite, alluv. placers; some deep leads	J peralkaline granites	79 Tt Sn 26.6 Tt Nb 1,640 t Ta	Williams et al., (1956)
W	Bainazar, Central Kazakhstan, U.S.S.R.	valley and alluvial fan placers, arid environment wolframite	veins in Pe granite aureole	estim. X00 t W	Bykhovskii et al.(1981)
W	Inkur placer, Dzhida distr., E. Transbaikalia	3 horiz.with detrital hübnerite in up to 90 m thick alluv. fan placer in a tectonic graben	hydr. veins in schist and MZ diorite	X00 t W estim.	Pustovalov et al. (1965)
Cr	Piedmont Upland Serpentinite Belt, Maryland	channel and valley placers, chromite	Pt_3 dunite and peridotite with podiform chromite	55 Tt chromite	Pearre and Heyl (1960)

Table 24-3 (continued). Selected examples of metalliferous alluvial placers (all considered Quaternary, unless marked T = Tertiary)

METAL	LOCALITY	PLACER CHARACTERISTICS	BEDROCK SOURCE	TONNAGE	REFERENCES
Nb	Bugaboo Creek, 72 km S.of Golden, British Columbia	black sands,alluv.of mountain streams; euxenite, polycrase, pyrochlore	two MZ peralkaline granite stocks	3,570 t Nb	Rowe (1958)
Nb Ta	Bear Valley, Idaho	mountain stream valley placers; euxenite, columbite, ilmenite	MZ quartz monzonite, qtz. diorite, pegmatite	5.6 Tt Nb 1.64 Tt Ta	Mackin and Schmidt (1956)
REE Th Zr	Broad Savannah and Catawba R.Basins, North Carolina	fine grained floodplain placers; ilmenite, monazite, zircon	katazonal granite, pegmatite, gneiss; saprolitic	1.03 Mt monazite	Overstreet et al. (1968)
REE Th NbTa	Kathaldih, Purulia; West Bengal, India	valley placers; monazite, columbite-tantalite	PCm saprolitic katazonal granite, pegmatite, gneiss, granulite	210 Tt REE 112 Tt Th 150 Tt Ta+Nb	Brown and Dey (1975)
REE Nb Ta	Kikune, 100 km N.W. of Seoul, Korea	valley placers topped by sand and peat; fergusonite, columbite, monazite, zircon	PCm gneiss, schist intr. by pegmatite and katazonal granite	55 Tt monaz. 32 Tt fergus. 23 Tt columb.	Miyazawa (1971)
Ti	Sherbro R., Mogbwemo, etc. area, S. Sierra Leone	poorly sorted sheet wash, fanglomerate, saprolite with rutile	PCm granulite pegmatite	15 Mt Ti/0.9%	Raufuss (1973)

Table 24-4. Selected example localities of metalliferous "deep lead" placers

METAL	LOCALITY	DEEP LEAD PLACER		COVER ROCKS		TONNAGE	REFERENCES
		AGE	GEOLOGY	AGE	COMPOSITION		
Au	Victoria Goldfields Australia (e.g. Ballarat)	T_1-MZ	valley, channel gravels up to 180 m deep; deriv.from lodes in slates	T_1	plateau basalts Eo alluvial sediments	440 t Au	Bowen and Whiting (1975)
Au	Lena Basin (e.g. Bodaibo R.),Siberia, U.S.S.R.	T	buried channel and terrace gravels in depth 20-150 m; from veins in Pt_3 schists	T,Q	alluvial and glacial sediments	minimum 1,200 t Au	Borodaevskaya and Rozhkov (1974)
Au	Yuba-American River, Sierra Nevada, California	Pg-Eo	channel gravels, gold derived from veins in slate, granodiorite	T,Q	andesite flows, pyrocl.,fluvial and glacial seds.	153 t Au	Yeend (1974)
Sn	Gibsonvale, N.S.W., Australia	T_2	cassiterite in wash and channel sedim. above tin granite	T_3, Q	fluvial gravel and sand, clay, loam, soil	6.5 Tt Sn	Campi et al., (1975)
Sn	Blue Tier,Pioneer, Ringarooma, etc., Tasmania	T	buried stream channels, cassiterite derived from tin granites	T,Q	plateau basalt, gravel, sand, clay	73 Tt Sn	Jennings (1975)
Pt	Novyi Log, the Urals Mts., U.S.S.R.	MZ	platinum in limonitized, partly cemented gravel and rubble, partly on karsted carbonate bedrock	T,Q	alluvial gravel, sand, clay	n.a.	Pustovalov, ed. (1965)

Fig. 24-16. Examples of dominantly alluvial (top) and, to a considerable extent, eluvial (saprolite) cassiterite placers in the Kuala Lumpur area, Malaysia.

CASSITERITE "DEEP LEADS", PIONEER MINE, N.E. TASMANIA

In the Pioneer mine, relatively shallow buried late Oligocene-Miocene deep lead cassiterite placer is mined in an open pit (K. Morrison, guided tour, 1981; Fig. 24-17). This is a small deposit (13 Tt SnO_2) but relatively uncomplicated thus suitable as an example. Detrital cassiterite of consistently medium sand size with fine accessory monazite and several other silicate resistates, is dispersed in a muddy and sandy gravel grading to sand, and interpreted as a braided stream deposit. The gravel rests on a Devonian tin granite saprolite which, in the mine area, is unmineralized, but the richest placer paystreaks are directly on top of the bedrock, progressively diminishing upward. The ore zone with an average grade 500-600g SnO_2/t is about 6m thick, and is overlain by finer grained and tin-poor sediments of a lacustrine delta. Spectacularly coloured leaching and soil profile zones are superimposed on all the rocks in the pit, but do not influence the cassiterite distribution. A portion of the deposit is covered by Miocene olivine basalt flow that has protected it from removal until the recent exhumation. The cassiterite has been derived from an apical portion of a tin granite and its aureole during the Tertiary.

SIERRA NEVADA FOOTHILLS DEEP LEAD GOLD PLACERS, CALIFORNIA

Sierra Nevada Foothills placers have yielded some 1,562t gold (Yeend, 1974). The recorded production came from both Tertiary fluvial gravels now preserved as deep leads in buried channels, and from Quaternary gulch and river placers. Both placer groups have derived their gold from the highly deformed Paleozoic and Mesozoic "geosynclinal" metasediments and metavolcanics intruded by granitic rocks. A proportion of the Quaternary placers formed by recycling of the Tertiary placers.

The Tertiary deep leads formed in valleys of rivers that flowed south-westward, and Lindgren (1911) pointed out the close correlation of the economic gold placers with the metasedimentary, metavolcanic and ultramafic rocks in the basement, located in the contact aureole of "granite". The gravels overlying the "granite", however, were barren. In the recently investigated buried gravels of the ancestral Yuba River (Yeend, 1974), 153 t of gold is known to be present, but its mining has been blocked by a court injunction since 1884. There, the Eocene gravel rests on Paleozoic phyllite containing minor fault-bound slivers of serpentinite and intruded by granodiorite. In the North Columbia diggings the auriferous gravel thickness ranged from 100 to 150 m, and the bottom 24-30 m of polymictic boulder and cobble, predominantly slate gravel, contained most of the gold. The overall grade is less than 1 ppm/cubic yard, and the richest 60 cm thick section contained about 6 ppm Au. The gravel is largely situated under the water table and is reduced ("blue"). Authigenic pyrite is abundant locally. The "upper gravel" contains abundant silt

Fig. 24-17. Partly exhumed Tertiary cassiterite "deep lead" at Pioneer, N.E. Tasmania. The section (B) shows a conspicuous colour change due to groundwater. (C) Main ore-bearing gravel. (D) Contact of granite saprolite (under the pick) and gravel; the richest cassiterite paystreaks are located immediately on the bedrock.

and clay beds, milky quartz and quartzite is dominant, and gold content is low. The gravels are buried under Tertiary andesite tuff, volcanic mudflows, and locally under a biotite-rich rhyolite tuff.

KLONDIKE PLACERS, YUKON

Klondike River is a tributary of the Yukon River near Dawson, N.W. Canada, and extensive placers distributed over about 2,500 km^2 in its basin have yielded 321 t of gold so far. The goldfield (Boyle, 1979) is in a thoroughly dissected upland marked by rounded hills and numerous small stream tributaries (gulches) to the main watercourses. The area was not glaciated, but is in the present permafrost zone, frozen up to a depth of 70 m. Consequently, the ground had to be thawed before dredging operations.

The bedrock is probably a Paleozoic pyritic graphitic schist with some greenstone and Boyle (1979) found up to 2 ppm Au in several pyrite samples. Otherwise, the primary gold occurs in small amounts in numerous small quartz veins and stringers, but no bedrock gold occurrence has ever been mined succesfully. Alluvial gravels occur at several levels; the high elevation, Tertiary White Channel gravels containing widespread but low-grade detrital gold have already been mentioned. The low level gravels on the bottoms of the present creeks and gulches are 1.2-3.3 m thick, and rest on a bedrock of decomposed and broken schist. They are topped by 0.5-10 m thick sheet of a black frozen organic-rich muck. Low level gravels located in the main stream channel are exceptional.

Additional example localities of gold and other metalliferous alluvial placers are summarized in Tables 24-3 and 24-4.

24.3. LAKES, BOGS, SWAMPS AND LATE CAINOZOIC UNCONSOLIDATED LACUSTRINE AND PALUDAL SEDIMENTS

24.3.1. General

Chapters dealing with lake deposition in sedimentologic texts (e.g. Reineck and Singh, 1975; Reading, ed., 1978; Leeder, 1982) usually start with an appreciation of the considerable variety of lakes and the wide selection of possible classifications. Collinson (1978b) commenced with a metaphor, comparing the status of lakes with a hole in a doughnut. There is a substantial amount of literature on lakes, of which Hutchinson (1957), Lerman, ed.,(1978) and Matter and Tucker, eds. (1978) offer the most comprehensive treatment under a single cover. Förstner (1981) reviewed the Recent heavy metal accumulation in limnic sediments.

The "most normal" lakes are standing bodies of water: freshwater to saline, shallow to deep, and they are gradational into the ephemeral lakes that do not usually display any surface water (ever or for most of the year) on one side, and into swamps and bogs that are water soaked, but the water is interspersed with and masked by vegetation. Continental lakes grade into a variety of brackish or marine basins.

Lakes have many geological features comparable with marine basins (e.g. the shoreline features such as beaches, estuaries and deltas) already treated earlier (Chapter 5). They also have, however, many peculiarities of their own and most writers emphasize the sensitivity of lakes to the influence of climate. This is mostly due to the lack of the homogenizing effects that influences the seawater.

As in the preceding sections, the strongest emphasis here is on those aspects of lakes and the related transitional environments that have the greatest significance in the accumulation of ores. The "most normal" lakes are sub-mediocre ore accumulators, now as well as in the geological past (Chapter 25). Their metallogenesis is more interesting theoretically than practically. Certain types of lakes (e.g. playas) and their sediments, however, are significant environments or associations to some non-metallic commodities (e.g. soda, sodium sulphate, borates) and some metals (Al, W) could be possibly recovered as by-products.

24.3.2. Bogs, marshes and swamps

Water-saturated, vegetated and poorly drained areas have been given a variety of names in the literature and the names are used interchangeably. The A.G.I. Glossary of Geology makes a distinction based on the kind of vegetation involved. A bog (or peat bog) is dominated by a spongy groundmass of mosses as well as sledges and heaths, that may develop into peat. Marsh has an aquatic and grasslike vegetation without peat. A swamp differs from a marsh in the presence of trees. Sometimes, local environmental equivalents are widely used, e.g. muskeg (in Canada).

Bog and swamp sediments (paludal) and ores have often been equated with lake (lacustrine) sediments and ores, particularly in the economic geological writing. Although both bogs (swamps) and lakes could exist side by side without a sharp boundary and generate a mixed (palustrine) sediment as in the small northern lakes underlain by glacial drift, both environments also occur independently.

BOREAL (NORTHERN) PEAT BOG AND SMALL LAKE ENVIRONMENT

Peat consists of partly decayed vegetable matter, inorganic minerals and water in varying proportion. Most frequently, it is about 10% solid matter and 90% water. Peat is a fibrous, woody to powdery light yellow to black material. In the northern latitudes, it consists of a variety of mosses (Sphagnum is the most common) and it requires very special climatic and topographic conditions to form. The climate has to be humid enough to support a rich vegetation, and the water table must be high enough to allow prolonged accumulation of plant debris in a reducing environment. Because the Sphagnum moss is capable of holding water even on a slope, peat accumulates not only in low flat bogs, but also forms upland peat areas (high moors or hilly peat).

Northern peat bogs and lakes are widespread in Canada, Alaska, Scandinavia and the northern parts of the U.S.S.R., in regions of chaotic drainage disrupted by the Quaternary glaciation. Both rest either directly on a glacially scoured bedrock, or on a glacial sediment. Fig. 24-18 shows the typical environmental setting of a northern bog and lake in Ontario.

The bogs and lakes are of interest to the coal geologist as a living model of coal formation (although very few ancient preserved coals actually formed in this environment; compare Section 25.4.). They have also been of minor importance as a source of iron and occasionally manganese ores to support local industries in the pre-industrial or early-industrial period. In times of emergency, some countries continued mining their bog and lake ores as late as the 1940s. In Finland (Marmo, 1952), 2.1 Mt of the bog or lake ores dredged up to 1947 contained 700 Tt Fe/30% and 252 Tt Mn/12%.

The bog or lake Fe and Mn ores are composed of goethite, amorphous dispersed or clayous iron hydroxide gel, manganite, birnessite, psilomelane and "wad" in variable proportions. Iron phosphates (vivianite) and siderite are early diagenetic products in peat. In a bog, the above minerals form encrustations, powdery or clayous impregnations, spheroidal, pea-sized concretions, or partly cemented fragmentary masses. On the lake floor encrustations, nodules, sand impregnations, ooliths and granules are the predominant form.

Descriptions of numerous localities with known bog or lake ores are widespread in the literature from the 1880s to the 1940s (e.g. Kindle, 1932, 1936; Trois Rivière, Quebec, Fe; Dawson Settlement, New Brunswick, Mn) and in vintage textbooks (e.g. Beyschlag et al., 1916), but few modern studies are available. The peat bog environment of Fe-Mn ores has been treated in Lukashev et al. (1971). At several localities of Quaternary bogs in Byelorussia, western Russia and the Ukraine (e.g. Rzhavets, Sinyukha), iron minerals are zonally distributed. In the Rzhavets hilly peat deposit, oxidized iron phosphates occur in the uppermost layer of the peat and change downward into hydrogoethite "beans". The slightly compacted peat under the water table contains concretions and thin lenses of limonite and siderite.

The lake environment and its Fe-Mn ores, is exemplified by the Punnus-Yarvi lake in Karelia (N.W. U.S.S.R., near the Finnish border), described in an unusually detailed study by Strakhov et al. (1968; Fig. 24-18b). This locality is a post-glacial, young lake with dimensions of 7 x 1.3 km and an average depth of 7 m. Its shores are sandy and gravelly so that water—clogged bogs are not typically developed. The sand grades basinward into a clay and mud. Embryonal Fe-Mn mineralization occurs as scattered nodules in clays, but extensive, medium-grade ore (19.48% Fe, 22.7% Mn) is present in the nearshore facies. The ore minerals start to appear at water depths of

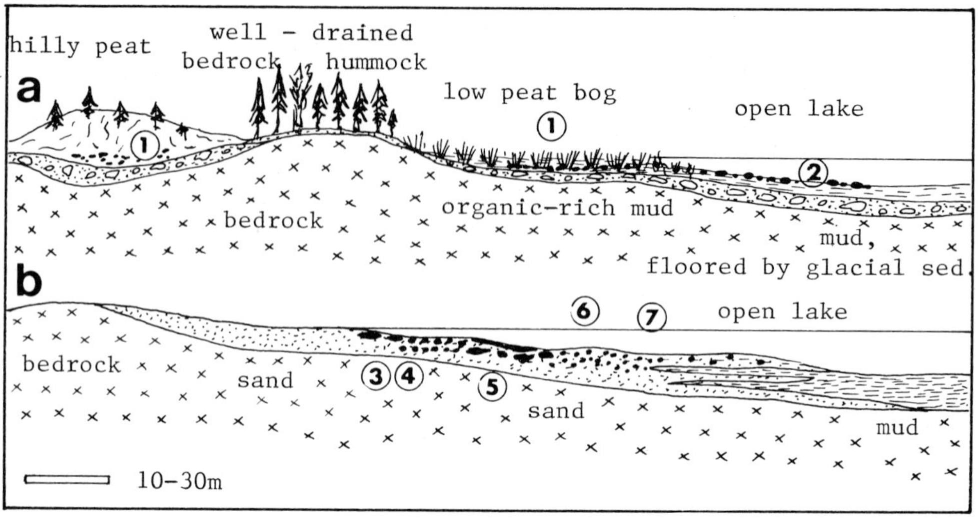

Fig. 24-18. Boreal peat bog-small lake association. (a) Bog/lake Fe and Mn ore distribution, generalized situation corresponding to N.W. Ontario, Canada. Muddy lakeshore, not to scale. 1=dispersed minerals in peat; 2=ore concretions on lake floor. (b) Mn oxide lake deposit, Punnus-Yarvi Lake, Karelia, U.S.S.R., with a sandy shore (in scale). 3=flat ore concretions; 4="bean" and "nut"-shaped small concretionary ore; 5=blocky and encrusting ore; 6=ore ooliths in sand; 7=scattered ore ooliths in clayous sand.

2 m and persist to about 10 m. Pisolithic and "bean" ore nearest the shore changes into coarse lumps and encrustations hosted by sand and resting on clay and this, in turn, is gradational into sand-size, granular or oolithic ore allochems scattered in a slightly clayous sand.

Although iron and manganese are the two most common metals accumulated in peat bogs in the northern lakes, Cu, U, Ni, Zn, Au, Ag and a few other metals accumulations are known and some have even been mined on a small scale in the past. Most of the metallic occurrences overlie, or are close to, a bedrock mineralization; consequently, the metals in bogs or lakes at present are more important for the geochemical exploration of bedrock ores, than as ores in themselves.

The Turf copper mine on the east side of Harlech Dome, North Wales (Dunham et al., 1978) is the best known example of a copper mineralized peat actually mined and processed. There, peat containing impregnations of copper carbonates and native copper was mined in the 19th century, burned on the spot, and the ash sent to Swansea for smelting. The Coed-y-Brenin disseminated copper deposit recently discovered in the vicinity supplied the copper. A very similar occurrence has recently been described by Lett and Fletcher (1980) from Whipsaw Creek near Princeton, British Columbia. There, a hillside bog, up to 3 m thick, contains over 1% Cu in chalcopyrite and covellite grains mixed with authigenic pyrite, and in irregularly distributed native copper grains. The peat is in a shallow, till-filled depression, overlying the copper-bearing Triassic Nicola Group intruded by a quartz-feldspar porphyry stock. Near Cooke City, Montana (Lovering, 1927), spongy masses of native copper are in thin beds of black muck enclosed in a peat bog. Near Jefferson City, Montana (Forrester, 1942), native copper in a peat bog is interspersed with precipitated limonite. It is situated downstream from the sulphide-bearing veins in the Boulder batholith.

Boyle (1979) listed several examples of auriferous peat bogs. Several bogs in New Brunswick contained up to 1.4 ppm Au and up to 2 ppm Ag, co-precipitated with Mn oxides. The Elk Park and Rudolphs Bog mountain peats in the Front Ranges of Colorado, contained 0.3 ppm Au.

Uraniferous peat bogs have been described from the U.S.S.R., Sweden, northern Manitoba, Sierra Nevada, and other places. The Sierra Nevada localities (Hoffman Meadow and Pettit Ranch Meadow; Vine, 1962) contain over 1,000 t of peat grading up to 0.7% U.

Additional occurrences of mineralized peat occurrences have been tabulated in Laznicka (1985g).

MARSHES AND SWAMPS IN WARM HUMID REGIONS

In warm climates, rates of evaporation and oxidation are higher than in the boreal belt, so that marsh vegetation growth and accumulation requires a nearly level water table (Wanless, 1975). Low moors may form as a result, and they are located in broad river valleys (e.g. the lower Mississippi River valley), large delta plains, and on coastal plains. These are more characteristic equivalents of the ancient coal depositing swamps than the boreal bogs. Geochemical studies made in some of these environments (e.g. Casagrande and Erchull, 1977, in the Okefenokee Swamp, Florida) found most trace metals in sub-clarke quantities, so this is hardly a promising metal-accumulating environment except for the mineralogical occurrences of "limonite" and siderite that are relatively common. The lack of metals is probably due to the monotony and lack of mineralization in the bedrock. In the metallogenically productive regions, the regionally enriched metals may spill-over into the marshes or swamps as well. Trescases and Guillon (1977) found up to 2.6% NiO in authigenic Ni-nontronite in peat in New Caledonia (Yaté map sheet). The peat formed in valleys and depressions in the vicinity of nickeliferous regoliths over ultramafics.

24.3.3. Permanent, non-saline lakes

The distribution and lithology of clastic sediments in large non-saline lakes is comparable with sediments of epicontinental seas, from which the ancient lacustrine sediments are often difficult to distinguish. The standard granulometric facies distribution (from shore to the basin centre: gravel - sandstone - mudstone - sometimes carbonate) is often disturbed due to the presence of relict sediments. This is most common in the northern lakes (e.g. the Great Lakes of North America), initiated as proglacial lakes. There, coarse to very coarse sediments representing reworked till or glacial outwash, may appear anywhere within the lake basin.

The "common" metallic mineral occurrences in the contemporary lakes occur as (1) detrital heavy minerals. These are most common in beaches, but some are in deltas and in submerged, partly reworked moraines. (2) Authigenic, syndiagenetic minerals usually present as concretions or encrustations on the surface and in the shallow subsurface of the lake floor. (3) Lake floor accumulations of ore allochems, such as iron oxide ooliths. The "uncommon" ore occurrences are associated with earlier proximal mineralizations or anomalous metal sources.

(1) Relatively few detrital heavy mineral occurrences have been reported from recent lakes and none have been mined on a significant scale. Beds of black sands, up to 30 cm thick, containing up to 11% TiO_2, 20% Fe, and 0.2% V, occur locally in the Lake Superior beaches south of Sault Ste Marie in Michigan. Lugov (1979) reported cassiterite placers in beaches of the small Lake Baky in Yakutia, Siberia. The lake is situated in a polygenetic tin province.

(2) The authigenic lake floor metallic accumulations are dominated by ferro-manganese oxide nodules that are, with the exception of some trace metals, comparable with the marine nodules already discussed (Chapters 4 and 5). The subject has recently been reviewed by Callender and Bowser (1976) and most examples came from the North American Great Lakes. The largest occurrences are in Lake Michigan (particularly in the Green Bay), where the nodules cover almost 500 km^2 and may contain 40-60 Tt Mn per km^2. Fe-Mn oxide coatings, crusts and dispersed oxide occur in addition to the nodules. The Mn contents range from 4.7 to 17.1%, Fe contents range from 20 to 36%, and of the base metal elements high Zn and Pb contents have been reported from the Lake Ontario (2,020 ppm Zn and 1,890 ppm Pb) and Lake Oneida (2,020 ppm Zn). Ni and Co contents, in turn, are much lower than in the marine nodules. Rare Mn carbonate (rhodochrosite) nodules have been reported from Lake Michigan.

Pyrite is the only other ore mineral relatively common in the recent lake sediments. Dell (1975) reported sand-size concretions composed of aggregates of pyrite spherules, associated with organic matter in a post-glacial lacustrine clay from South Bay, Lake Huron. Elsewhere in the Great Lakes sediments, the most common form of FeS_2 is a black amorphous material distributed in streaks in the post-glacial silty clays.

A special category of metalliferous lakes comprise small, undrained or poorly drained ponds located in the vicinity of "primary" (usually sulphide) ore deposits, undergoing oxidation, that collect water containing dissolved metals. A portion of the dissolved or suspended metals precipitates often as colloidal bottom sediments, while a portion remains in solution or suspension in the water, giving it a distinct colour. Most such "lakes" are artificial, formed by flooding of old quarries (e.g. the brown to red pigmented lakes located in abandoned pyrite pits, such as in the Rio Tinto, Spain, belt or near Hromnice, Czechoslovakia). Few localities are, however, natural, such as the Gayskoye Ozero close to the massive sulphide deposit Gai in the southern Urals, U.S.S.R. (Igoshin, 1966). This lake is colored blue by the dissolved and colloidal Cu sulphates and silicates, brought in by springs containing up to 38 ppm Cu.

(3) A unique example of an oolithic iron sediment in the process of formation has been described by Lemoalle and Dupont (1973) from Lake Chad (Africa), north-west of Fort Lamy. There, fine ooliths composed of iron hydroxides form a 5 cm thick layer of oolithic sand, distributed over 2,700 km^2 of the lake bottom. The sand contains up to 27% Fe, and the entire iron accumulation could represent some 30 Mt Fe. The iron was brought in solution and suspension by the river Chari and precipitated in the recent undrained lake situated in an arid location.

24.3.4. Saline lakes : arid, perennial

Perennial saline lakes are those that persist for many years without drying up (Collinson, 1978a,b; Hardie, Smoot and Eugster, 1978). The Great Salt Lake, Utah, and the Dead Sea in the Near East, serve as examples of shallow and deep saline lakes, respectively. The shallow saline lakes may turn into playa lakes with a slight climatic change, while the deep lakes are more permanent. Most perennial saline lakes have an appreciable influx of stream water, that balances the net evaporation.

As far as the metallogeny is concerned, these lakes are a transitional category between the non-saline lakes (Section 24.3.3.) and playa lakes (Section 24.3.5.) and as they have no unique features of their own, they will not be treated here.

24.3.5. Playa lakes (inland sebkhas, salars)

GENERAL

Playas are broad, shallow, dry depressions in arid regions, that may be filled seasonally by a thin sheet of water to become ephemeral playa lakes (Fig. 24-19). They are typical transitional lesser-order environments treated in the literature either under the heading "Lakes" (e.g. Leeder, 1982), or "Deserts" (Collinson, 1978c). Inland sebkhas and salars are equivalent terms used in the Middle East and the Andes, respectively.

Playa lakes are fed by water from two major sources: (1) surface runoff; this follows storms in deserts, comes from higher mountains

Fig. 24-19. Appearance of saline playas. TOP: Lake Magadi, East African Rift, S. Kenya. The salt is mostly trona. BOTTOM: Floor of the Death Valley, California. The pinnacled crust is mostly halite.

that have a greater precipitation than the valleys, follows a wet season, or results from melting of a winter snow cover; (2) groundwater supplied by springs or by the general rise of the water table. In exceptional cases the groundwater may be thermal. Water table fluctuation under playas has a far-reaching influence on the movement of fluid, and consequently it influences the precipitation of minerals from solutions.

Most playas form in closed basins that are commonly fault-bounded (grabens) and there they are adjacent to, or overlapping with, the local alluvial fans. The detrital fraction of the playa sediments depends granulometrically and compositionally on the local physiography and geology. In broad playas located in vast, essentially flat terrains (e.g. the Arabian, Western Australian, Saharan, Namibian, etc. playas) the detrital material is a fine clay to mud. Such depressions are sometimes called claypans and are often fringed by sand dunes. In narrow tectonic depressions, such as Death Valley or some of the Chilean salars, the proximal, coarse fragmental facies of alluvial fans, gravel of the dry creeks (arroyos) or landslides, are often interbedded with the playa clays and cemented by the chemically precipitated minerals. Playa deposits are often transitional into duricrusts (Chapter 23). Contemporary volcanism is important in many playas and is of substantial importance for metallogenetic interpretations. Volcanic ash added to the epiclastic and chemical playa sediments is the most common product, followed by coarse volcanic ejecta (lapilli, bombs), and occasionally, lava flows.

If coarse clastic or volcanic constituents are considered "special", a "typical" playa association consists of interbedded thin layers of clay or mud (sometimes sand) and evaporite minerals. Playa sediments are frequently cyclic (rhythmic) and consist of couplets or triplets of a coarser-grained sand or silt layer grading upward into a finer clay layer. A triplet forms when the clay layer is overlain or coated by evaporites. The origin of the rhythmites is often compared with the origin of the seasonally-laminated varved sediments in proglacial lakes.

The present surfaces of playas and the ancient playa sequences (Chapter 25) are rich in a variety of indicators of dessication, such as mud cracks and polygons; sediment-filled dessication wedges; lumps of peeled-off, often curled laminae, and others. The surficial oxidized layer of the recent playas is light brown to reddish-brown (unless there is a thin surficial coating of salts in which case the colour could be bright white), but light gray colours prevail in the subsurface, formed under reducing conditions. Bacterial reduction of evaporite sulphates results in the release of H_2S, and formation of black clays. Some clays contain pyrite and/or native sulphur and appear to be ideal sites for the capture of migrating trace metals and metallic ore deposition, provided the environment is metallogenically fertile.

The evaporites are the most interesting and complex association in the present playa environments and in the ancient associations. Texturally, they have a variety of forms ranging from thin powdery sprinklings to encrustations on the smooth surface of playas, through craggy pinnacled surfaces of partly dissolved salt to finely laminated subsurface layers. Compositionally, some 300 mineral species are involved. The common playa evaporites are dolomite, gypsum and halite, and they are frequently compositionally zoned with the more soluble minerals prevalent near the centre of the basin. The "non-standard" playa evaporites have a high degree of uniqueness and some are mined as important nonmetallic commodities. Figure 24-20 shows selected examples. The geological details of these occurrences,

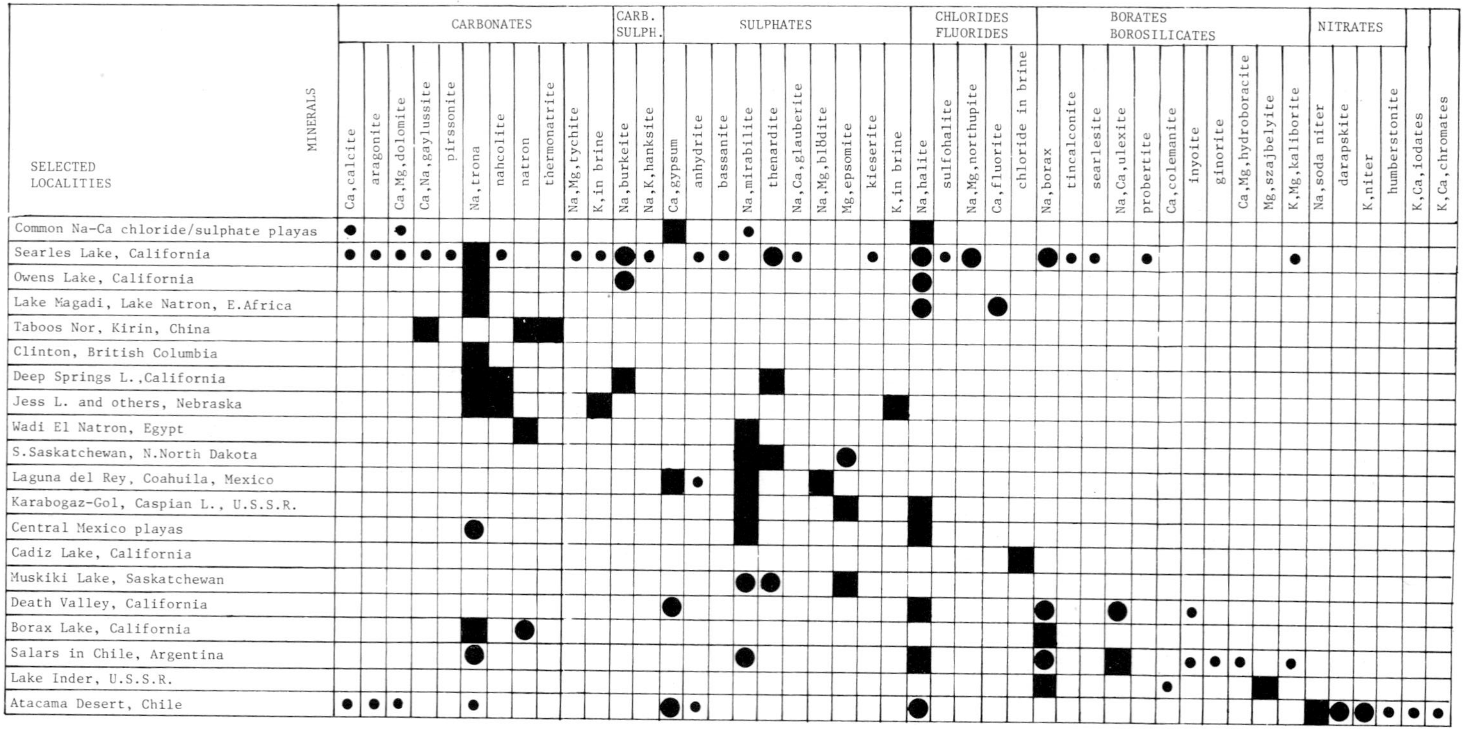

Fig. 24-20. Selected examples of evaporite mineral associations in the Recent playa lakes or in their brines (mostly pre-diagenetic salts). Based on the data in Lefond, ed.(1975); Tomkins (1954); Smith (1979); and Ericksen (1981).

unless they also contain associated metallic concentrations, cannot be reviewed here and the reader is advised to consult the chapters on boron, fluorspar, iodine, nitrogen compounds, salt, sodium carbonate, and sodium sulphate, in Lefond, ed. (1975).

METALLIC OCCURRENCES

Despite their uniqueness and importance as sources of certain nonmetallics, the Recent playas and playa associations are of limited practical importance as sources of metals (NOTE: lithium, although partly recovered as metallic Li, is not on the list of metals considered in this book; if it were, its resources in the playa environment would be of the order of 1.67 Mt Li_2O). The principal mineralization styles and tonnages (Fig. 24-21; Table 24-5) do contain some interesting figures, but in most cases the metals reside in materials that have to be considered as "technological ores" and have been reported, so far, from only a few isolated localities. Alternatively, the mineralizations are transitional. The metals dissolved in subsurface brines and recoverable as a by-product, appear to be of greatest practical importance. The Searles Lake, California, is an outstanding example at one time hailed as the largest tungsten resource of the United States because of some 68 Tt W, dissolved in its brine reserve (Carpenter and Garret, 1959). This locality will be briefly reviewed here.

Searles Lake (Smith, 1979; Figures 24-22 and 24-23) is a salt pan located in a fault-bounded graben, a part of the Basin and Range Province of eastern California. The active playa floor is underlain by several thousand metres of Quaternary sediments, of which the uppermost 270 metres of interbedded salts and mudstones have been drilled and studied in detail. The salts are composed of Na-Ca carbonates and sulphates, halite and borax. Interstitial brines occur in two evaporite horizons in the subsurface, and are pumped out and treated in two large chemical plants. The overall economic importance of Searles Lake arises from the cumulative value of mineral production that has been \$ 1 billion up to 1979. This figure includes the value of the Na, K and B salts produced, but no heavy metal recovery has been made so far. There is a significant salt and brine reserve left that contains some 50 Mt B_2O_3 and the following trace contents: 150 ppm Li_2O, 32 ppm W and 98-102 ppm As. The tungsten content corresponds to a factor of concentration of about 20, and its provenance is unclear. Searles Lake is located in a province containing numerous small bedrock scheelite occurrences within a radius of 100 km, and the tungsten may have been supplied to the lake floor at the time of deposition of the Lower Salt Bed (middle Wisconsinian), by subsurface springs.

The Clayton Valley playa lake in Nevada is another locality where a saline brine is pumped out and the constituents (NaCl, minor K salts, Mg salts, 650 ppm Li_2O) industrially recovered.

The accumulations of fine detrital or authigenic metallic minerals

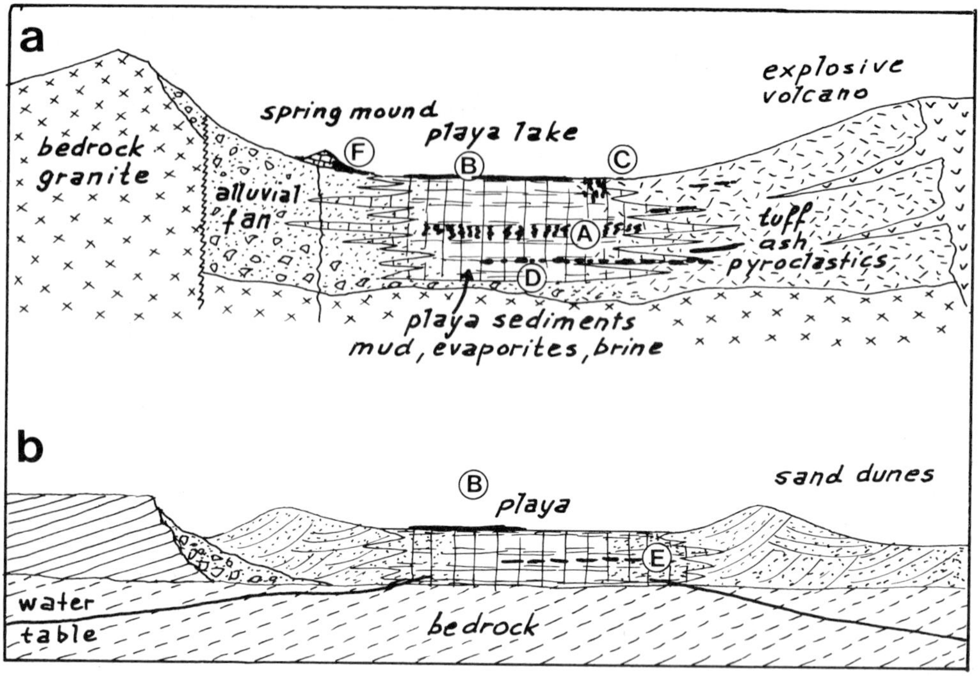

Fig. 24-21. Distribution of metallic mineralizations in playa lakes and their sediments. See Table 24-6 for the letter code explanations. (a) Playas in mountain valleys with a strong tectonic control. (b) Playas in flat deserts. Dimensions are grossly exaggerated. From Laznicka (1984).

Table 24-5. Existing and potential metallic mineralization styles in Recent playa lakes

(A)	Metals dissolved in subsurficial brine (X localities, 68 Tt W, X Mt Li).
(B)	Fine detrital and authigenic metallic minerals in (on) the surficial layer of playa sediments (alunite, fluorite, Mn oxides). X localities, 12 Mt alunite.
(C)	Infiltration minerals in the surficial detrital sediments of playas. 1 locality, estim. X00 t U.
(D)	Authigenic/infiltration minerals in the playa/pyroclastic association.
(E)	Dispersed metals in the playa evaporites (e.g. Zn).
(F)	Minerals in chemically precipitated aprons of hot or cold springs, extending into the playa lake association (e.g. Mn, Fe, W, etc.).

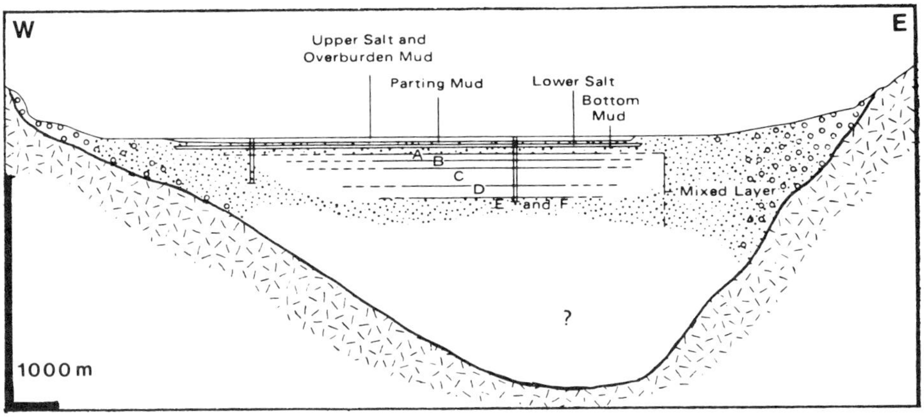

Fig. 24-22. Searles Lake, California, a cross-section showing named Quaternary stratigraphic units of interbedded mud and evaporites. The W and Li-bearing brine resides in the Upper Salt unit. From Smith (1979).

Fig. 24-23. Searles Lake, California, view of the southern end.

in the surficial layer of playa sediments (Style B, Fig. 24-21), is represented by several localities. In the Lake Campion near Chandler (300 km E.N.E. of Perth, Western Australia; Fitzgerald, 1945), a fine sediment containing up to 60% alunite mixed with detrital quartz, mica and Fe oxides, forms the top 6 m of the lake sediment. The lake contains a minimum of 12 Mt alunite, and its origin is uncertain. There has been no Cainozoic volcanism within a radius of 3,000 km.

The Pidinga alunite deposit in South Australia (King, 1953), consists of several orebodies located in Tertiary clay beds, deposited in a shallow playa lake. It is believed that the alunite formed by the reaction of sulphatic groundwaters released from oxidation of pyrite disseminated in the interbedded lignite, with the originally present kaolinitic clay.

Playa lakes concurrent with an intensive subaerial volcanic activity could become choked with ash and coarser pyroclastics. The post-depositional development of the pyroclastics (weathering, devitrification, diagenesis) overlaps with the playa sedimentogenesis and transitional lithologic associations result. There is an enhanced metallogenetic fertility mostly due to migration of the mobile elements leached from the fresh volcanics, or released during the breakdown of volcanic glass and the unstable silicates (compare also Chapter 5). Reported examples of mineralizations from this setting include fluorite, uranium and manganese. Fluorite is not a metallic mineral, but since it is a common associate of metallic ores, it is reported here.

The Pianciano fluorite deposit, 34 km north-west of Rome, Italy (Anonymous, 1978), is an example of a bedded fluorite in Pleistocene soft pyroclastics reworked in a playa lake setting. The fluorite is inconspicuous, extremely fine-grained (under 2 microns), and forms rhythmic bands alternating with clay (55.7% CaF_2), or cement to a volcanic arenite (20.74% CaF_2). Manganese oxide specks are common and there is an accessory barite, celestite and a locally increased trace U content. The host pyroclastics are members of the highly potassic Rome volcanic province. There are eight mineralized zones ranging from 0.16 to 3.37 m in thickness and the reserves are about 5.2 Mt CaF_2.

Near Roma, Oregon (Sheppard and Gude, 1969), authigenic fluorite forms submicroscopic, nearly spherical grains in a zeolithic tuff, tuffaceous mudstone and terrigenous mudstone in a Pliocene lacustrine sediment. The fluorite is non-uniformly distributed and the grade is low, between 5 and 16% CaF_2. The reserve is at least 12 Mt CaF_2.

Uranium can accumulate in the playa sediments when a suitable U source is located nearby. Such a source could be either in broadly contemporary continental volcanics or in the basement granites or metamorphics. The Antsirabe deposit in central Madagascar (O.E.C.D., 1979) probably represents an example of the first alternative. There, uranocircite impregnates volcanic clays and a sandstone in Pliocene-Pleistocene lacustrine sediments adjacent to the Ankaratra volcanic mountain chain. The second alternative is exemplified by the autunite occurrences in tuffaceous Neogene lake beds overlying granitic basement near Hallelujah Junction, north-eastern California (Gay, 1966). More examples are known amongst the geologically older "lake beds", treated in Chapter 25.

Manganese accumulates in the playa and volcanic association in the form of oxides (cryptomelane, psilomelane, pyrolusite) and is most common in devitrified and zeolitized, reworked vitric tuffs. Several examples (e.g. Huachipato, Chile; Lake Mead, Nevada) are treated in Chapter 26.

Accumulations of metals other than U and Mn in playa sediments have been recorded only rarely. Davidson (1965) reported up to 0.2% Zn in "modern evaporites" (presumably the black sulphurous and gypsiferous muds) in Lake Eyre, South Australia. Lake Eyre is adjacent to the fault-bounded Mt. Painter block well known for its bedrock uranium occurrences. Rare finds of highly oxidized chromium minerals are known among the Atacama, Chile, nitrate deposits (Ericksen, 1981). These constitute a transitional category treated in the next section. Metallogeny of the playa lakes associated with rifts and influenced by contemporary alkaline (especially carbonatite) volcanism such as Lake Natron and Lake Magadi in eastern Africa, is discussed in Chapter 33.

24.4. DESERT ENVIRONMENT

24.4.1. General

Deserts are usually defined as regions so devoid of vegetation that they are incapable of supporting any population, and regions where evaporation greatly exceeds precipitation. The upper limit for an average rainfall in a desert is usually given as 250 mm. Although the hot, tropical to subtropical deserts, such as the Sahara, are most popular and have been studied in detail, they are not the only ones. Deserts have formed at other latitudes as well. The middle latitude deserts are typified by the extensive Gobi Desert of Central Asia, and there are also the polar ice and snow deserts.

Deserts as depositional environments and the desert lithofacies receive coverage in all the text and reference books on sedimentology (Leeder, 1982; Reineck and Singh, 1975; Collinson, 1978c) and in the monograph by Glennie (1970). Wolf (1973) compiled a useful conceptual model of the desert environments.

Geologically, deserts are characterized by the dominance of physical weathering, wind activity resulting in both erosion and deposition, infrequent activity of running water and localized precipitation from groundwater. Reineck and Singh (1975) distinguished the following desert sub-environments (facies): (1) hamadas, arid rocky terrains; (2) serirs, stony deserts; (3) ergs and draas; aeolian sand deposits; (4) wadis, periodic streams; (5) desert lakes and inland sebkhas, and their surficially and sub-surficially precipitated sediments; and (6) dust or loess deposits. To this should be added coastal desert deposits that have some special features of their own, and (8) products of desert volcanism.

Modern deserts as environments of metal accumulation are of no significance and their metallogenetic balance is highly negative. The number of the pre-existing metallic deposits destroyed or buried by the geological processes operating in deserts greatly exceeds the number of deposits newly created in the same environment. To the student of economic geology, both metalliferous geology and the geology of petroleum, this may seem untrue, because the "desert environments" receive considerable publicity in the literature as creators of the frequently mineralized "red-bed" association (Chapter 25) and many of the aeolian sand bodies that constitute an important category of oil traps. Although both the facies mentioned above may have formed in paleodeserts, the metallic minerals and the petroleum were introduced subsequently, after burial and during diagenesis. Consequently, they could not be observed in the process of accumulation in the recent deserts.

Of the eight desert sub-environments and facies listed, some are areas of degradation where rocks are removed, not deposited (hamadas, partly serirs); others have already been treated elsewhere in this book (e.g. desert lakes; Section 24.3.5.), so they will not be repeated. The rest will be very briefly reviewed.

24.4.2. Serirs

Serirs are horizontal to sub-horizontal deflation lag surfaces, marked by the accumulation of coarse-grained sediments such as gravel and coarse sand. The fines (clay, silt, fine sand) have been selectively removed by wind action. Serirs blanket large areas of the present deserts and are often incomplete; under the coarse surficial layer are the finer sediments protected from deflation.

The metallogenic function of serirs and other desert lag deposits, is in their ability to cause relative residual concentration of the coarse (or fine but heavy) fraction, and when such fraction is metalliferous, ore accumulations may form. This, then, is a process entirely dependent on the presence of an earlier mineralization. Although there are few accurate published data, it seems logical that a proportion of the gold won in the Pharaohs' time in the Eastern Desert of Egypt may have come from the surficial lag accumulations. The deflation lag debris strewn surfaces over several Western Australian goldfields have been instrumental in their discovery. Cassiterite and ilmenite in the vicinity of regolithic bedrock Sn and Ti deposits are also probably enriched in serirs in the vicinity and so are rubbly iron ores in desert situated deposits. In Australia, in the sub-Saharan belt of Africa, and elsewhere where the present desert environment was superimposed over a previously lateritically weathered surface, serirs are marked by a conspicuous enrichment of the concretionary hematite (buckshot).

24.4.3. Aeolian sand deposits

Aeolian sands form a variety of dunes extensively treated in the literature (e.g. McKee, ed., 1979). They cover about 20% of the recent desert surface, forming extensive "sand seas" (ergs). Sand dunes, however, also occur abundantly adjacent to beaches in coastal humid areas, so they are not confined to deserts.

Recent sand dunes contain accessory amounts of heavy minerals, but on average, substantially less than the alluvial or beach sands contain. Although "dune placers" have been recorded in the literature (e.g. a proportion of the diamond placers near the Oranje River mouth, Namibia; ilmenite, rutile, zircon, etc. dune placers at the Stradbrook Island, Queensland, and elsewhere along the eastern coast of Australia; small and rare dune gold placers), they are almost always immediately adjacent to alluvial or beach placers and are in their extension. In the eastern Australian province of ilmenite and rutile sands, the metalliferous sands are most widespread in the upper beach and at the foot of the adjacent dune. There, the heavy mineral accumulation has been partly accomplished by winnowing the light fraction (mostly quartz). The above examples of mineralized dunes were almost exclusively from humid coastal areas. The ergs located within, or moving across, the desert interiors are virtually sterile, as far as metallic minerals are concerned.

24.4.4. Wadi deposits

Wadis are the dry courses of streams in deserts that occasionally fill with water following heavy rains. When in action, the rock debris is swept from higher ground by torrential flash floods and carried down the wadis by currents of such velocity that little sorting could take place. The floods quickly dissipate due to the high permeability of the desert sediments. Wadi sediments are poorly sorted and chaotic, and the process of their formation favours dispersal and dissipation rather than metal concentration, so placers are unlikely. Greenwood and Bleackley (1967) tested the heavy mineral concentrating capability of wadis by panning for heavy concentrates in wadi deposits accumulated at the foot of a vein carrying unoxidized galena in Southern Yemen. No traces of lead-bearing minerals were detected.

Although wadis are not very promising as hosts to placers, their buried channels have some potential for accumulating metallic mineralizations chemically precipitated from groundwater. Generally, this requires a major bedrock metal source, such as a bedrock exposed orebody undergoing oxidation and leaching upstream. A variety of "exotic" deposits of Cu, Mn, U and other metals can be expected.

The Chilean chrysocolla deposit Sagasca (Roethe, 1975; 372 Tt Cu/2%) is probably an example of such a mineralization. It is located on the fringe of the Atacama Desert and composed of colloform

Fig. 24-24. Wadi (dry stream) deposits, Death Valley, California.

chrysocolla cementing Pliocene gravel in a stream paleochannel resting on Cretaceous volcanics. The ore-bearing channel is capped by an ignimbritic rhyolite. The Huinquintipa deposit (72 Tt Cu/2.4%) is similar. The largest example of the "exotic" deposits is the Exotica orebody near Chuquicamata, treated above.

25.4.5. Desert volcanism

Active and inactive Quaternary volcanism in the desert environment (e.g. the Red Sea area, Tibesti, portion of the Andes) is not substantially different from the volcanism elsewhere, except for the excellent preservation due to the lack of weathering and erosion. The lithologic associations and geological features described by Greenwood and Bleackley (1967) from South Yemen, a young volcanic area in a desert environment, almost exactly correspond to certain "volcanic red beds" regions treated in Chapter 25. In South Yemen, contemporary desert sediments and two suites of late Cainozoic plateau basalts top Precambrian crystalline basement containing numerous copper occurrences, but no mineralization has been reported in the cover rocks. The unconsolidated sediments, in fact, have clarke to sub-clarke trace copper values. This appears to confirm the diagenetic or epigenetic character of copper and other ores hosted by the "volcanic red beds".

24.5. TRAVERTINE AND SPRINGS PRECIPITATES

24.5.1. "Cool" streams or springs

Travertine is an "accumulation of calcium carbonate in springs (karstic, hydrothermal), small rivers, and swamps, formed mainly by incrustation, cement precipitation and/or biochemical precipitation" (Julia, 1983). The "common" or "most normal" travertines precipitated from surficial waters and seen frequently in outcrop or as an ornamental stone, are delicately laminated to rubbly, usually porous limestones often of "Swiss chesee" appearance. They grade to semi-consolidated tufa or moss-, sponge- or popcorn-like appearance. They are most frequent in limestone terrains, forming aprons at the foot of limestone cliffs with hanging springs, terraces and mounds in creek and river valleys (Fig. 24-25). These travertines precipitated from surficial meteoric waters that carried dissolved $CaCO_3$, usually with biochemical assistance. There is no genetic controversy and no associated metallic occurrences are known.

Cool groundwater travertines precipitated from the mineral-rich waters of "common" springs: that is, ordinary fresh groundwater springs that are often locally tapped for drinking water supply. Despite the considerable hydrogeological variety, the common feature of these springs is that they leach soluble components from rock in the aquifer through which they pass, and deposit it at, or close to, the site of discharge. Such springs are most common along valley sides, forming horizontal lines of seepages and trickles. These springs are not confined to limestone terrains and their precipitate (if any) is not always limestone, although travertine is still the most common and its deposits most voluminous.

The metalliferous precipitates reported so far include iron and/or manganese hydroxides and oxides. Numerous occurrences are known, but they are of little economic importance. Near Roseisle, Manitoba, a spring located along the foot of a low escarpment composed of Cretaceous shales, deposits limestone tufa and travertine with which are locally interlayered patches of black, powdery Mn oxides often mixed with organic matter or with iron hydroxides. The manganese is probably leached from the shale that is slightly manganiferous, and contains scattered Mn-Fe carbonate nodules. Similar occurrences have been described by Hendricks and Laird (1943) in the Turtle Mountains, North Dakota. The Mn tonnage is negligible.

24.5.2. Thermal springs

Warm or hot springs have been touched on in this book under a variety of headings, depending upon the environment in which they occur or in what lithologic association. The environment is often suggestive of the water origin or source (magmatic, metamorphic, formational, meteoric, etc.); the agent of heating (volcanism, magmatism, geothermal heat); and metallogeny. Most springs are

Fig. 24-25. Travertine mound in a valley near Castrovirreyna, the Peruvian Andes, precipitated from an almost cool spring.

polygenetic, hard to interpret and difficult to classify. In this section only those springs situated outside of areas of volcanic and focussed tectonomagmatic activity are treated. Of course, it cannot be guaranteed that undetected intrusions do not indeed form in depth.)

Waring et al. (1965) prepared an extensive list and bibliography of thermal springs of the world (several thousand are listed !) and from maps in this work it is obvious that about half of the springs are in regions of Cainozoic volcanism, and most of the remainder are in regions of faulting (particularly block faulting) causing physiographic rejuvenation. Over 95% of springs have total dissolved solids contents lower than seawater (under 3.5%; most contain between 200 and 1,000 ppm), and calcium bicarbonate and sodium chloride are the main constituents of the high-solid springs. The Ca bicarbonate springs precipitate travertine. The remainder of the dissolved constituents (Ca, Na, Mg chlorides, sulphates and carbonates, H_2S, N_2, rare Si, K, B, Li, F, Ba salts) do not, as a rule, precipitate (except rare barite and fluorite as in the Teplice region, N.W. Czechoslovakia). In contrast to the springs in active volcanic areas, very little silica precipitates and conspicuous deposits of opal ("geysirite") do not form.

Metalliferous precipitates are rare and most reported examples involve small accumulations of Mn oxides, comparable with those precipitated from cold springs described earlier. Fe hydroxides are also common; frequently they pigment travertine, or thin laminae of "limonite" are interstratified with calcite in travertines.

The small polygenetic Mn and W deposit Golconda, Nevada, described by Kerr (1940; 15 Tt Mn/50%, 500 t W/2.2%) has already been mentioned here in connection with limnic ore deposition. There, flat-lying to gently inclined blanket-like bodies of porous tungsten-bearing

Mn oxides (psilomelane and hollandite with up to 6.5% of adsorbed WO_3) are interbedded with limonitic ocher and limestone tufa. The tufa is probably a deposit of a hot spring, discharging into the Pleistocene Lake Lahontan, at present dry. The locality is situated in the extension of the Osgood Mts. scheelite belt of Mesozoic age, which explains the possible tungsten source in the subsurface.

In the Warmwaterberg Springs field in the Cape Province, South Africa (White, 1981), hard and soft layers of travertine alternate with bands of limonite and massive, crystalline to earthy Mn oxides in hot spring mounds distributed along a fault. Most of the mineralization dates from the Pleistocene age, and it is estimated that there are some 600 Tt of 57.5% Fe_2O_3, 8.55% MnO_2, and 1.45% BaO material.

Many hot springs are radioactive, but uranium minerals have been found in spring sediments only rarely (e.g. rare sooty pitchblende in the Teplice, Czechoslovakia, springs). Cadigan and Felmlee (1975) reported briefly a radioactive travertine in the Gunnison River Valley, Colorado, and Popov (1979) reported 0.1% U in a spring sinter in Bad Gasstein, Austria.

Base metal precipitates have been recorded in mineralogical quantities in areas of hot springs, such as in the Teplice region, Czechoslovakia (sphalerite, galena), and from the Cheleken Peninsula, U.S.S.R. (Dvorov and Pavlov, 1978; see also the recent review in White, 1981). The former area is located in the Sub-Erzgebirge Graben that was the site of an extensive alkaline volcanism in the late Tertiary period (Chapter 33). The Cheleken Peninsula is part of an extensive Pliocene to Quaternary uplift in the eastern Caspian Sea region, completely lacking any recorded magmatic activity. The subsurface waters there form a vertically hydrochemically zoned system explored by drilling to a depth of 3,000 m and hosted by an aquifer in Neogene red-beds laterally and vertically confined by impervious Quaternary clays. Highly concentrated chloride-bicarbonate brines are located in collectors in the upper part of the aquifer and in a narrow zone adjacent to faults and crush zones they are metalliferous (brine in the E-85 well: 1.3 mg/l Cu, 0.2 mg/l Pb, 0.4 mg/l Zn, over 0.2 mg/l Cd; Dvorov and Pavlov, 1978). Pyrite, galena, sphalerite and native lead are actively depositing at well heads, in pipes and settling tanks, as well as along the brine conduits in the aquifer, and Lebedev (1973) calculated the yearly tonnage of base metals brought on the surface by the brines to be 300-360 t Pb, 48-50 t Zn, 24-25 t Cu, 18-24 t Cd and 6-8 t As. Dvorov and Pavlov (1978) argued convincingly that the metals have been locally derived, and the Cheleken brines are often quoted as a modern example of fluids that have generated the MVT deposits in the past. Additional examples are treated by White (1981).

Plate 10. Badlands topography developed in the Cretaceous tuffaceous coal-bearing sequence near Drumheller, Alberta, Canada.

"Ore deposits are rock masses in which occur one or more metalliferous minerals in sufficient quantity to make them useful as sources of metals".
T. Crook, Miner. Mag., 1914, p. 76.

"Ores are no more or no less than rocks".
R. L. Stanton, 1972.

CHAPTER 25

Pre-Quaternary Continental Sedimentary Association

25.1. INTRODUCTION

25.1.1. General

The continental (as opposed to marine) origin of many ancient sedimentary sequences has been demonstrated using the numerous sedimentogenic criteria listed in the literature (e.g. Reading, ed., 1978; Reineck and Singh, 1975; Rigby and Hamblin, eds., 1972). Compared with most marine sequences, however, ancient continental deposits display many more facies transitions and lithologic variations. They are also characterised by their frequently fragmentary and incomplete development, caused by repeated subaerial erosion shortly following most depositional events. Continental deposits change facies into marine sediments and in some associations (e.g. in cyclothems containing paralic coal seams), continental sediments deposited in the nearshore region contain so many marine beds that a regional paleo-environmental separation is virtually impossible. The marine / continental facies boundaries are favourite loci of metal accumulation.

Although the bulk of continental deposits are epiclastic detrital sediments derived from weathered and eroded subaerially exposed bedrock, subordinate chemical (evaporites) and biochemical (some limestones, coal) sediments also occur. The volcanic component is important locally, taking the form of intercalated lavas (mostly mafic, as in the basalt-red-beds association), tuff horizons, or reworked volcanic fragments of a variety of sizes. Volcanic detritus derived from ash falls often occurs far from the parent vents, in an otherwise typically stable, non-volcanic setting.

It is extremely difficult to subdivide continental sediments into rational but meaningful categories for description, helpful to a field geologist or a prospector. The environmental classification that is adequate for the Recent sediments (compare Chapter 24) is impractical for the entire ancient continental sedimentary association, although some distinctive sub-categories (such as the paleo-lacustrine sediments) can be so classified. The reasons are numerous. First, insufficient interpretations have, so far, been available in the literature and those areas recently studied in detail and used as examples in reference books (e.g. Reading, ed., 1978), tend to be largely unmineralized. Second, environmental interpretations are, to a considerable degree, subjective and change with time. A recent example includes many of the distinct, strikingly cross-bedded Mesozoic sandstones in the western United States (e.g. the Navajo Sandstone) formerly interpreted as aeolian, now considered littoral. Third, continental sediments have been more profoundly modified by diagenesis than any other group of rocks except, perhaps, the nearshore carbonates.

The diagenesis in particular often caused striking changes in rock colour (gray, green, etc., to red and vice-versa) and in many instances it is not certain whether the distinct colour of, for example, the "red-beds" is the product of a depositional or a diagenetic environment. This, naturally, makes a paleo-environmental interpretation difficult or impossible. Fourth, only a few metallic mineralizations hosted by the continental sediments (e.g. paleoplacers, bauxites, evaporites) are syndepositional and consequently genetically influenced by the original depositional environment of their host associations. More ores (e.g. the "sandstone" U-V deposits, Cu, Pb-Zn, etc.) were epigenetically emplaced into their present, porous hosts during diagenesis and after lithification, so they are not controlled by the depositional paleo-environment of the sedimentary association or at least not directly.

In the absence of a satisfactory rational classification of continental deposits suitable as a framework for a description of metallic mineralizations, a heterogeneous grouping based on various dominant attributes (paleo-environment, lithology, striking colour, distinct single rock or ore, etc.) recognizable in the field, had to be employed in organization of this chapter. The reader should be aware that these categories are not mutually exclusive classes and that they overlap considerably.

25.1.2. Geotectonic setting

Continental sediments formed (1) in continental margin orogenic belts; (2) on stable platforms; and (3) in intracratonic mobile belts.

In (1) they are most abundant in the foreland belts (molasse lithofacies), in which they usually top the earlier marine deposits. A proportion of the "successor" and intermontane basins has a continental sedimentary fill. A considerable proportion of continental sediments covering large, relatively stable crustal blocks (including median massifs) within broad, complex orogenic belts (such as the Colorado Plateau within the Cordilleran belt) are difficult to classify.

In setting (2), continental sediments most frequently occur in the basal and top portions and on fringe of marine sequences, formed in the extensive epicontinental marine basins. Areally and voluminously less important are the former glacial deposits, alluvial valley fills and windblown sediments. All deposits discussed above show a minimum of tectonic control.

The setting (3) is most intriguing and it contains sediments deposited in a variety of local basins controlled by normal faults. Sediment-filled grabens adjacent to bedrock hosts are most characteristic. Although (3) is most conspicuous when superimposed on large, continuous cratons for example in the form of aulacogenes (=aborted rifts), equivalent structures are equally widespread in composite mobile belts in which they are commonly superimposed over the blocks and units underlain by an ancient continental crust. The "foreland basins" of Wyoming known for their uranium deposits and trending in all directions are adjacent to (and largely derived from) uplifts of the Archean crystalline basement. They are substantially different from the linear foreland belts adjacent to and parallel with the usually thrust-faulted continental edges of orogens as in the Canadian Cordillera, the Appalachians, the Andes, etc.

25.2. MOST NORMAL ("GRAY", EPICLASTIC) CONTINENTAL DETRITAL ASSOCIATION

25.2.1. General

Most (at least two thirds by volume) of the ancient continental sediments preserved in the geological record, are contained in thick sequences (usually wedges) of alternating epiclastic conglomerate, sandstone and shale having generally a light colour ranging from beige to light tan or gray. The sediments are moderately to well sorted, submature to mature, and formed in humid climates largely in alluvial to deltaic environments. Coal seams are often present (these are described separately in Section 25.4.), as are broad tracts of intertongued marine sediments.

The molasse facies exposed in foreland belts between orogens and stable platforms usually resting on a depressed craton and composed of detritus derived from the rising orogenic front, is a most common and characteristic example worldwide. The Oligocene and Miocene Northern Alpine Molasse (van Houten, 1974) is a late orogenic to post-orogenic sedimentary suite formed in the Alpine Foredeep from detritus derived from the rising Alps in alluvial fan, floodplain with river channels, braided stream, lake, coal swamp, marine to brackish delta and tidal flat sub-environments (the fluvial sub-environment was dominant). The sedimentary wedges are up to 6 km thick, tens of kilometres wide and several hundred kilometres long. They rest partly on an autochthonous flysch topped by mudstone and transitional paralic sequence, and partly on crystalline rocks of the craton and are lithologically heterogeneous and lenticular. In the proximal facies adjacent to the Alpine orogenic front in the south (this source area was almost exclusively sedimentary: carbonates, some with chert nodules, lesser shale, sandstone, quartzite) fanglomerates and repetitive cycles of fining-upward fluvial sediments are present.

The conglomerates often exceed 1000 m in thickness, are composed of closely packed, well-rounded clasts of chert (from nodules in carbonates), quartzite, quartz, sandstone, and significant carbonate. The fluvial deposits comprise submature, lithic and feldspathic arenites and drab mudstones with occasional thin coal lenses and minor freshwater (lacustrine) limestone. The distal facies is dominated by sub-litharenite and sub-arkose sandstone and mudstone (shale) and marl with marine intercalations. Evaporites (gypsum, anhydrite, halite) formed in the transitional (lagoonal) environment. The Alpine Molasse and its continuation in the Carpathians posesses an anomalousy high carbonate content (in clasts and in the cement), reflecting the carbonate-dominated source area.

The late Jurassic to Oligocene molasse in the northern Cordilleran foreland belt in Canada (Eisbacher et al., 1974); the Cainozoic molasse of the eastern Andes (van Houten and Travis, 1968); the late Miocene to Recent molasse of the Himalayas (Gansser, 1974); and the Carboniferous to Permian molasse in the Appalachian foreland, are closely comparable with the Northern Alpine Molasse.

As far as metallic occurrences are concerned, all the thick and extensive foreland molasse belts listed above are virtually barren, and are almost on a par with the flysch belts as one lithologic association of least interest to the prospector. Small occurrences of iron ores (infiltrations of "limonite" in sandstones, pelosiderite concretions in shales) and Mn ores (ferroan rhodochrosite, manganoan pelosiderite, etc. nodules in shales) have no economic importance. The metallogenic fertility of the molasse facies and similar associations in other geotectonic settings, increases as a consequence of a variety of geological diversifications and "special" conditions. These are either concurrent with sedimentation, or superimposed and include:

(1) The presence of sub-environments favourable for chemical precipitation during syn- and anadiagenesis (e.g. muddy swamps, lake bottoms).

(2) Multiple reworking of the earlier sediments: sparse and uneconomic Fe or Mn concretions in shales could physically accumulate as gravelite deposits in stream valleys.

(3) Availability of metallogenically fertile units or earlier deposits in the source area. Generally, crystalline rocks are more fertile than the earlier sediments. Resulting ores (e.g. paleoplacers, reworked former weathering crusts e.g. Fe hydroxides, bauxite or Mn oxides, metalliferous infiltrations) tend to accumulate very proximally to the source. Eisbacher et al. (1974) in their study of the Canadian Cordilleran successor basins, made the point that in elongated basins the majority of the sedimentary fill was deposited by longitudinal paleo-currents, and the minority by short streams flowing perpendicularly to the mountain ranges. The former sediments usually had a distant source and an uncertain provenance, the latter were much more proximal, source relateable and have a much better opportunity for carrying ores.

(4) Volcanic supply concurrent with sedimentation or postdating it. In most instances this takes the form of ash falls. The ash represents a weathering-undepleted possible source of trace metals (e.g. U) that may, under favourable conditions, concentrate to form diagenetic or epigenetic deposits in the continental sedimentary unit.

(5) Tectonism, hydrothermal activity, magmatism.

In the "most normal" continental associations, the lithogenetic diversifications listed above are rare and of low intensity. In the "special" associations (Sections 25.3 to 25.7), they are widespread and forceful.

25.2.2. Examples of mineralizations

GOLD PALEOPLACERS

In the Laba River basin gold field on the northern slopes of the Caucasus, U.S.S.R. (Krendelev, 1974), detrital gold originally derived from hydrothermal veins in the Proterozoic and Paleozoic "eugeoclinal" metamorphics was detected in alluvial conglomerates and gravels of at least four age groups (Carboniferous, Permian red-beds, lower Jurassic basal conglomerates, Quaternary stream gravels). Repeated reworking of gold from the geologically older sediments into the younger ones has been demonstrated. The Jurassic occurrences are members of a foreland belt suite.

Near Jefferson, South Carolina (Minard, 1971), up to 1.2 ppm of gold occurs in small areas in the partly consolidated basal fluvial conglomerates of the Cretaceous Tuscaloosa Formation. The Formation is a member of the monotonous, largely unmineralized strata of the Gulf Coastal Plain and the detrital gold can be found over an area of about 55 km^2, where the basal sediments transgress over the regolithic lower Paleozoic metavolcanic and metasedimentary basement known to contain gold-quartz and gold-pyrite orebodies (Brewer and Haile mines).

Tertiary buried alluvial gold placers ("deep leads") known to exist in many goldfields of the world (e.g. Victoria, Australia; Sierra Nevada; Klondike; Yakutia) also belong into this category, but because they are largely unconsolidated and often overlap with (or are reworked into) Quaternary placers, both age groups have been reviewed simultaneously in Chapter 24. Exceptionally, the "deep lead" placers are consolidated, as at Forrest Hill, Sierra Nevada (Lindgren, 1911).

Ti-MAGNETITE, ILMENITE, ZIRCON, ETC. PALEOPLACERS

Numerous mineralogical occurrences that fit into this category are known to exist in continental sandstones or conglomerates, but economic or potentially economic deposits are exceptional. The Tobol River paleoplacers (Rundkvist, ed., 1978) are in Oligocene continental sediments in the Turgai Downwarp (N.W.Kazakhstan, U.S.S.R.). There, ilmenite, rutile and zircon (aver. grades 33.8, 4.19, 3.66 kg/m^3, respectively) occur in a N.N.W.—trending zone over an area of several tens of thousands km^2. The poorly consolidated sandstone, siltstone and mudstone unit blankets the Paleozoic crystalline basement and the highest ore grades are in lenses of quartz sandstone several centimetres to 30 m thick.

Near Almaşu Mare in the Zlatna Basin (Metaliferi Mts., Rumania; Ianovici and Borcoş, 1982), detrital vanadiferous Ti-magnetite derived from a mineralized ophiolite suite is contained in a zone, 100 m long and 2-3 m thick, in Tertiary molasse sandstones and conglomerates.

Fe OXIDE CONGLOMERATE (GRAVELITE) PHYSICALLY REWORKED AND CHEMICALLY PRECIPITATED ORES

Physically reworked boulders, pebbles, cobbles and granules of iron ore are common in many basal conglomerates of the continental sequences, transgressive over earlier units containing ironstones and iron formations. In the Santa Rita field, New Mexico, the Cainozoic Gila Conglomerate contains abundant cobbles of magnetite, derived from skarn deposits in the area. Several occurrences of hematite or goethite conglomerates unconformably overlying truncated siderite bodies in the basement, were described by Ilavský (1976) from the Spiš-Gemer region of Slovakia. Extensive and economically important deposits of comparable style hosted by continental sequences are, however, uncommon and the known examples (e.g. the later Proterozoic deposit Angara Pit, Siberia; Lisakovsk; deposits in the northern Aral Lake region; Sokolov and Grigor'yev, 1974) are all transitional.

In the large Lisakovsk deposit, 110 km S.E. of Kustanai (Turgai Downwarp, N.E. Kazakhstan), the middle Oligocene semi-consolidated fluvial channel deposits rest unconformably on eroded Eocene-lower Oligocene marine claystones. The iron ore is composed of goethite, siderite and leptochlorite pebbles, granules, pisoliths and irregular particles cemented by the same minerals. The orebodies are hosted by, and are gradational into, a conglomerate and take the form of layers, lenses and thin shoestring bodies controlled by the stream channels.

Oligocene Fe deposits in the northern Aral Lake region (e.g. Kutan-Bulak, Taldy Espeh; Strakhov, 1962; Fig. 25-1) are similar, in many respects, to Lisakovsk, except that, in addition to the river paleo-channel filling orebodies, additional reserves are found in sediments of the marine and deltaic facies. The latter consist of oolithic and granular chamosite, siderite and goethite and form stratiform beds and lenses. Taldy Espeh is credited with 35.3 Mt Fe/35.3% (Sokolov and Grigor'yev, 1974).

The Berezovsk deposit in the Argun Basin (E. Transbaikalia, Siberia; 170 Mt Fe/35.7% and 44.6% Fe; Sokolov and Grigor'yev, 1974) is situated at both sides of an unconformity that separates the upper Jurassic continental lacustrine sequence from the Cambro-Silurian phyllite, dolomite and limestone basement. There, siderite and goethite-replaced carbonate rubble and talus on the base of the Jurassic sequence above unconformity grades into a mineralized solution collapse breccia and has replaced carbonate in the sub-unconformity region.

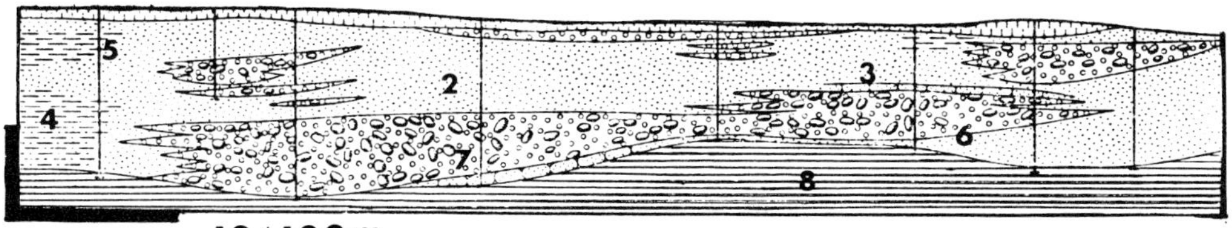

Fig. 25-1. Kutan Bulak iron deposit, Lake Aral district, Kazakhstan, U.S.S.R., an example of a combined ore conglomerate or oolithic iron ore deposit. 1=Q loam; 2=T_1 quartz siltstone; 3=ferruginous siltstone and sandstone; 5=silty claystone; 6=oolithic ore; 7=pisolithic, oolithic, conglomeratic ore; 8=marine claystone. From Strakhov (1962).

CHEMICALLY PRECIPITATED IRON ORES

Chemically precipitated iron ores typically developed along the base of continental sedimentary sequences transgressive over an older basement. They are gradational into the ores reviewed above. In the Amberg field, N.E. Bavaria, Germany (Gudden, 1975b), several small deposits of phosphatic concretionary goethite, siderite and chamosite occur along the base of an upper Cretaceous sandstone, overlying karsted Jurassic dolomite and limestone in several separate "ore troughs". The ores are believed to have formed as alluvial precipitates in stream valleys, fed by iron brought in solution from the sides. The total tonnage is of the order of 20 Mt Fe/31-44%.

The two small iron ore deposits Woodstock and Russelville in Alabama (Brown, 1948; 13 Mt Fe/40-57%) consist of an earthy and sandstone-cementing goethite. The orebodies are blankets and pockets along the base of the upper Cretaceous friable quartz conglomerate and sandstone, unconformably topping Mississippian limestone.

In the Paz del Rio field near Bogota, Colombia (Putzer, 1976; 50 Mt Fe/40%), particulate and cementing chamosite, chlorite, hematite, siderite and goethite form a single ore bed 4-6.5 m thick, in a fresh- to brackish-water deposited molasse sequence of conglomerate, quartz arenite and mudstone.

25.3. DETRITAL (REDEPOSITED) BAUXITE IN CONTINENTAL SEDIMENTS

The three most important associational or genetic categories of bauxite deposits are (a) "in situ" and transported bauxites along unconformities in marine limestones (Chapter 20); (b) "in situ" residual lateritic bauxites formed over a variety of rocks (Chapter 23); and (c) redeposited (transported) bauxites in ancient continental sediments (this chapter). There are frequent transitions and genetic links between the categories listed above; for example, a continental transported bauxite filling depressions in karsted limestone combines the characteristics of categories (a) and (c).

The latter has already been mentioned briefly in Section 20.5.1. and Fig. 20-5 as "Style 3" of carbonate-associated bauxite deposits.

Local combinations of bauxite styles (b) and (c) in non-carbonate hosts are less frequent and some are difficult to recognize. The Arkansas bauxite district described in the classical memoir by Gordon et al. (1958) is the most extensively studied example locality showing such a transition. In Arkansas, bauxite forms two groups of occurrences located about 8 km S. and 40 km S.W. of Little Rock. The deposits lie entirely in the Gulf Coastal Plain Cainozoic sedimentary province, a short distance S.E. of its boundary with the Ouachita Foldbelt. Highlights of the local geology include a late Cretaceous mass of nepheline syenite, erosion unroofed and tropically weathered shortly after its emplacement. This is overlapped by Paleocene marine sediments that, in turn, are unconformably topped by Eocene continental sediments. The basal Eocene consists of interbedded gray silty clay, greenish-gray sand and lignite, grading into a distinct facies of detrital kaolinitic clay adjacent to nepheline syenite hills under the unconformity. This facies includes many orebodies of reworked bauxite. Stratigraphically higher, silty to sandy claystone, red friable ferruginous sandstone, gravel, sand and lignite form erosional remnants of a formerly more extensive sequence, burying the bauxite.

Bauxite in the region is present as discontinuous, pocket and blanket-like "in situ" residual deposits in the nepheline syenite regolith (compare Chapter 23), and as lenticular stratiform resedimented bodies along the base, and within the Eocene continental sediments. Gordon et al. (1958; Fig. 25-2) subdivided the Arkansas bauxite bodies into four types: (1) Residual deposits on the upper slope of partly buried nepheline syenite knobs. These bodies are zoned, with a lower granitic textured saprolite and an upper concretionary bauxitic laterite zones; (2) deposits on the base of the Eocene continental association, in the proximal bauxite-kaolin facies. This bauxite has a basal zone with preserved fragmental clay textures, topped by a thick concretionary zone. There is, commonly, a siliceous hardcap upslope and the bauxite grades laterally and vertically into kaolin; (3) bedded, silicified alluvial deposits composed of sorted or crossbedded pebbles, reworked concretions, pisoliths and grains of bauxite in a clayous bauxite matrix. They truncate, overlie and fill channels in type 2 deposits and (4) bauxite conglomerate and rubble deposits composed of bauxite fragments derived from type 1, and located a short distance downslope from the source.

Types 1 and 2 are the most extensive and the most important commercially. The ore is composed of gibbsite with accessory kaolinite, siderite, hematite, goethite, magnetite, pyrite, anatase, etc. It contains anomalous trace quantities of Nb,Ga,Zr,REE,Se,etc., reflecting the alkaline bedrock heritage. The aggregate tonnage of the Arkansas bauxite province was of the order of 120 Mt of a 50% Al_2O_3 material.

The Arkansas bauxite province, although of moderate size only, illustrates perfectly the transitionality of the bauxites associated

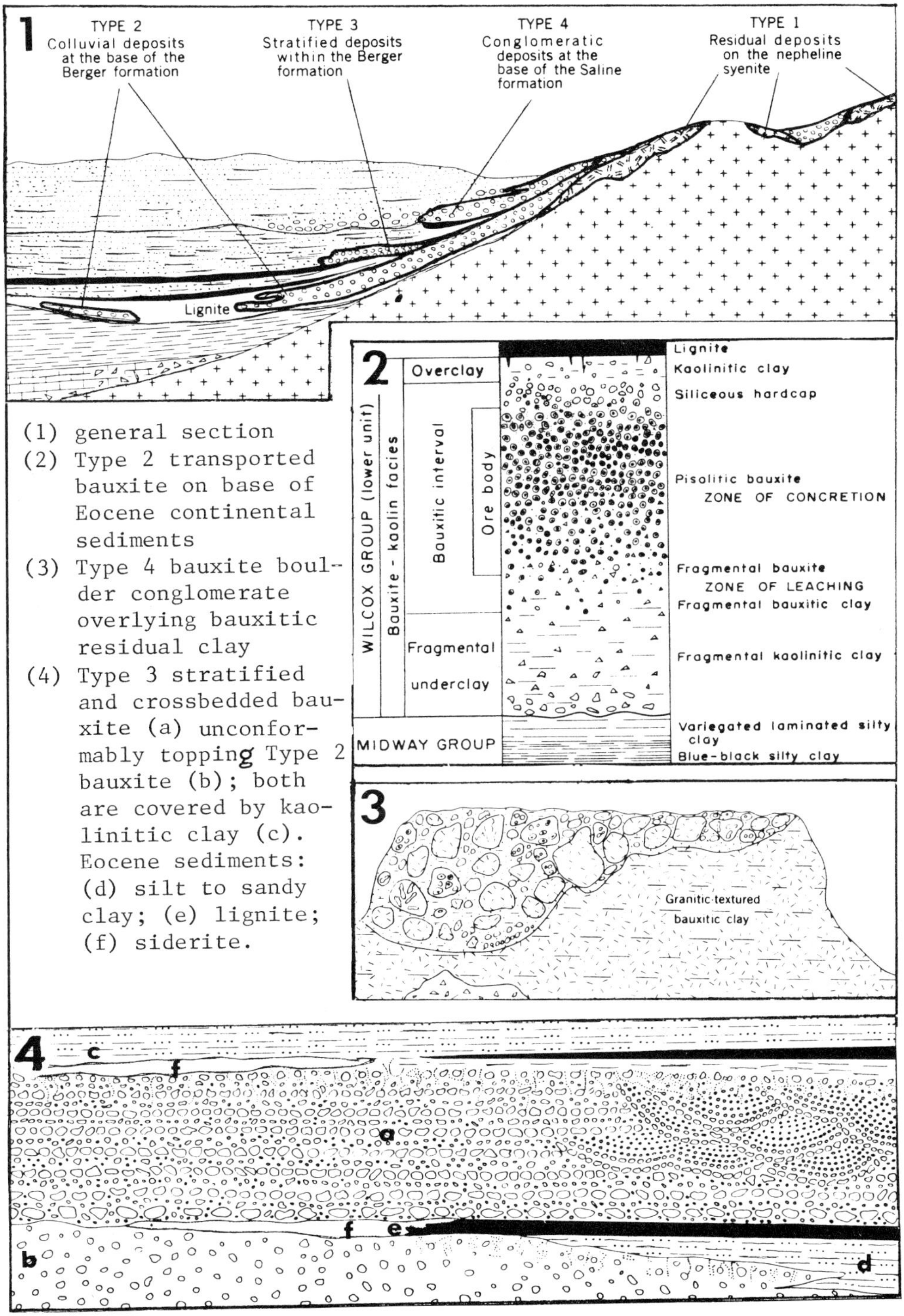

(1) general section
(2) Type 2 transported bauxite on base of Eocene continental sediments
(3) Type 4 bauxite boulder conglomerate overlying bauxitic residual clay
(4) Type 3 stratified and crossbedded bauxite (a) unconformably topping Type 2 bauxite (b); both are covered by kaolinitic clay (c). Eocene sediments: (d) silt to sandy clay; (e) lignite; (f) siderite.

Fig. 25-2. Four types of Arkansas bauxites of Gordon et al.(1958).

with continental sediments. It also illustrates the expected variations, which do not appear clearly in the global classifications offered so far. Krivtsov (1973) suggested a morphogenetic classification of the Soviet bauxite deposits in which four classes are applicable to the continental sediments-hosted ones. These comprise (1) "Valley" bauxites, e.g. Tikhvin. Numerous small orebodies are located in branching stream channels, eroded into the basement; (2) blanket-like, lacustrine-paludal deposits. These are most common in the Kursk Magnetic Anomaly region, where they are gradational into their "in situ" regolithic equivalents and Al sources (compare also Section 23.4.2.); (3) karst or transported bauxites, resting on karsted carbonates but gradational upward into a continental detrital suite and (4) slope (deluvial-proluvial, e.g. S. Timan) deposits, formed in shallow depressions on the slopes of basement highs, that acted as the sources of the bauxite detritus. The Arkansas bauxite types approximately match classes 1,2 and 4 of the Krivtsov categorization.

The brevity required in compiling this book precludes description of additional examples of bauxite deposits, but brief data on selected localities appear in Table 25-1. In this table a further convenient method for the grouping of bauxites has been applied, the advantage of which is that it is non-genetic and objectively recognizable in the field (Fig. 25-3). The groups are as follows: (1) bauxites at the base of continental sediments, floored by or adjacent to the source regolith; (2) bedded, stratiform bauxites entirely within a continental sedimentary suite; (3) transported (allochthonous) bauxites resting on karsted carbonate, but gradational upward into continental sediments and (4) bauxites intimately associated with kaolinitic clay(stones) of uncertain or dual origin. These are subdivided into (a) bauxites in clastic sediments brought in from the outside; (b) bauxite formed by an "in situ" alteration of kaolin brought in from outside; (c) kaolin formed by retrogressive silicification of bauxite. The category (4) overlaps with the "peneplain bauxites" (e.g. Guyana, N.Brazil), reviewed in Section 23.3.5.

BAUXITIC AND HIGH-ALUMINA CLAY(STONES) AND SHALES

Residual and transported bauxitic clays are associated with most bauxite deposits where they represent a waste. Similar sediments, however, also occur independently of the "classical" bauxites and they range in composition from rare, purely boehmite and/or diaspore claystones to shales, through mixed Al hydroxide-kaolinite rocks to purely alumosilicate (kaolinite) clays or shales. In contrast to common bauxites, these rocks tend to occur in gray (reduced) shades and many occur in the coal association with which they are treated in this book (Section 25.4).

Table 25-1. Transported bauxites within or on base of pre-Quaternary continental sediments

LOCALITY	STYLE (Fig. 25-3)	FOOTWALL UNIT	BAUXITE	TOPPING UNIT	REFERENCES
Saline and Pulaski Counties, C.Arkansas, U.S.A.	1,2, R	Cr_3 nepheline syenite Pc marine sediments	Eo,lenses and sheets, conglomeratic,pisolithic,massive; 20 Mt Al/28%	Eo claystone, kaolin,lignite	Gordon et al. (1958)
Andersonville, Georgia, U.S.A.	4	Cr_3,Pc,sandstone and mudstone	Eo_1 gibbsitic bauxite lenses in kaolin sheet 5.75 Mn Al/24%	Eo_3-Ol sand, clay	Patterson (1967)
Tikhvin (Boksitogorsk), N.W. Russia, U.S.S.R.	3	D_3 limestone	Cb_1 lenses of boehm. bauxite, bauxitic clay; est. 110 Mt Al/27%	Cb_1 sandstone, claystone	Krivtsov (1973)
North Onega district, N.W. Russia, U.S.S.R.	1,2	PCm metamorphics incl. ultramafics; Cm-D siltstone, shale	gibbs.-boehm.bauxite, pisol.to mas. lenses; estim. 27 Mt Al/27%	Cb_1 sandstone, shale	Kirpal' and Tenyakov (1974)
Amangel'da distr., E.Turgai, Kazakhstan, U.S.S.R.	3	D_3-Cb_1 limestone, minor Tr basalt	Cr_{1-2} gibbs. hard, friable, argillac. bauxite; 60 Mt Al	Eo claystone, lignite, siderite, sandstone, shale	Kirpal' and Tenyakov (1974)
West Turgai district, N.W. Kazakhstan, U.S.S.R.	3	PZ,Cb_1 karsted carbonates along fault zone	Cr, gibbs. concret. to pisolit. baux. lenses; 83 Mt Al	Cr_3-Eo clays, lignite grading to marine shale, sandstone	Kirpal' and Tenyakov (1974)

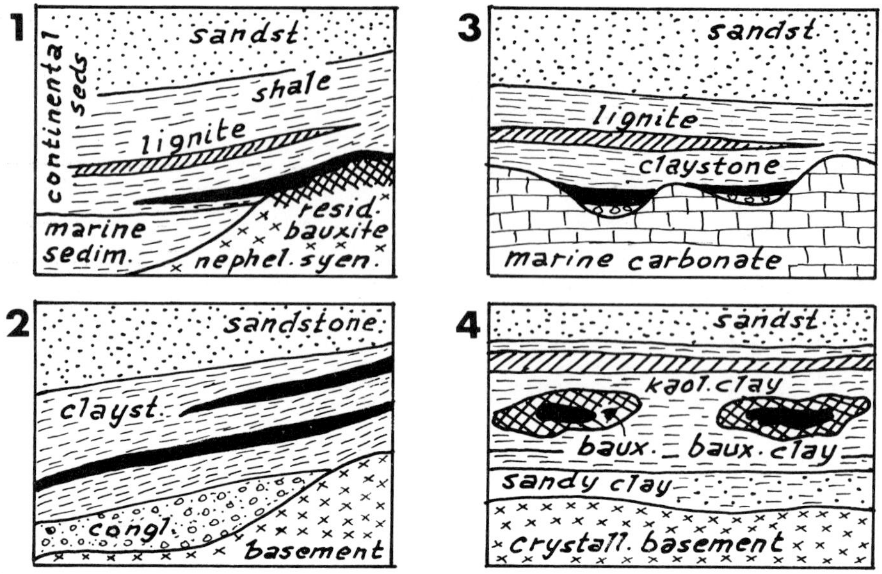

Fig. 25-3. The four principal styles of transported (allochthonous) bauxites in a continental sedimentary association.

25.4. COAL ASSOCIATION

25.4.1. General

Coal, an organic sedimentary rock, forms less than 0.01% of the sedimentary record, having an estimated tonnage of the order of 10^{13} tons (Fettweis, 1979). Despite this, about 4% of the total sedimentary volume is actually coal-bearing or at least potentially coal-bearing. There, the coal seams, because of their conspicuous appearance, are an important indicator of a restricted range of depositional paleoenvironments and lithologic associations.

Coal forms concordant (stratiform) seams that range in thickness from several centimetres to the recorded maximum of 233 m in the Tertiary lignite basin of southern Victoria, Australia. The more common average thickness is between 60 cm and 6 m; seams less than 60-70 cm thick are generally uneconomical to mine and consequently are rarely recorded or discussed. Lignite and brown coal seams are, on average, thicker than those of black coal and anthracite because of a lesser degree of compaction. In extent, coal bodies range from short, thin, rapidly splitting lenses (or single coalified logs) several square metres in area to giant, regular sheets that underlie territories in excess of 50,000 km^2 (e.g. the Pittsburgh Seam,

Pennsylvania which is of area 56,980 km^2, thickness 1.8-3 m and has a workable tonnage at least 100 Bt of coal; Pettijohn, 1975). In many coalfields, however, numerous coal seams occur (in the Ruhr-Westphalen Basin, West Germany, over 90 seams have an aggregate thickness of 84 m).

With the increase in ash content, coal grades into a black (carbonaceous, coaly) mudstone or shale. Another relative of coal is a sandstone with coalified logs or carbonatized wood trash. Most coal deposits are at least slightly faulted or folded (those accounting for 90% of world production; Galloway and Hobday, 1983), but a few have been metamorphosed (to form graphite ultimately, as in Sonora, Mexico). Interaction of coal with intrusive bodies (except diabase dikes) and volcanism is also rather uncommon.

This book is not concerned with coal as a major fuel and/or a nonmetallic economic commodity, but deals exclusively with coal as a carrier of ore metals. In addition to the coal itself, however, certain metals are hosted by other members of the "coal sedimentary association", in which their accumulation was either dependent or independent of generation of the coal itself. Most of the sediments associated with coal also occur independently (without coal) as members of the "most normal" gray, epiclastic continental detrital association (Section 25.2).

Although a "proto-coal" has been recorded from sequences in excess of 1.7 b.y. in age (Pettijohn, 1975) and a "coaly substance" from still older rocks, the first widespread coal generation in the Devonian and Carboniferous periods coincided with the appearance of terrestrial wood plants. Coal, therefore, has a restricted stratigraphic range.

The megascopic appearance, petrography and composition of coal on an intermediate scale (hand specimen to outcrop) depends primarily on the degree of coalification (Table 25-2) and to a lesser degree on the derivation (plants, microorganisms, pollen, etc.; the plant-derived coals, humites, are absolutely dominant). The large-scale features of coal basins and coal associations (facies) depend on the geotectonic setting and on depositional paleo-environments. The coal-associated metallic mineralization is very heterogeneous and is controlled by the above, as well as some additional conditions. There is an extensive review literature on the coal geology and petrology, that has a high proportion of relatively old but still valid, classical works (e.g. White and Thiessen, 1913; Murchison and Westall, 1968; Ivanov, 1967; Stach et al., 1975; Fettweis, 1979; Galloway and Hobday, 1983; etc.).

25.4.2. Geotectonic and paleo-environmental setting of coal deposits

Traditionally, geotectonic settings of coal associations have been attributed to either (1) tectonically controlled basins within and at the fringe of orogenic belts, or (2) atectonic basins on cratons or along the stable continental margins. In terms of depositional

Table 25-2. Coalification ranks

CLASS OF COAL	APPEARANCE	APPROX. LIMIT OF FIXED CARBON	DEPTH OF COALIFICATION	COMMON LITHOLOGY OF ASSOCIATED SEDIMENTS
anthracite, anthracitic coal	bright, almost submetallic-looking, jet black, massive and uniform coal, clean on touch	86–98%	4000 m plus	slate, argillite to phyllite, hard cemented sandstone to quartzite
bituminous (black) coals	black, distinctly banded coal (bands of glossy vitrain alternating with semi-gloss clarain and a matte durain); vertically jointed, low moisture content	53–86%	3000–4000 m	shale, well-cemented sandstone (wacke, sublitharenite, arkose) and conglomerate.
sub-bituminous (hard brown coals)	brown to brownish-black, massive to banded coal, dull, "dirty" on touch, but of homogenous appearance	45–55	1000–3000 m	mudstone, claystone, "soft" cemented sandstone, conglomerate, marl
lignite	brown, soft, disintegrating coal, high in moisture content, "dirty" touch, plant remnants are recognizable; logs, branches, etc. are physically separable	35–45%	10–1000 m	mudstone, claystone, friable sandstone to unconsolidated sand, gravel, clay
peat	brown, powdery matrix interspersed with loose woody components; extremely high moisture content	about 30%	10 m	sand, clay, gravel

mega-environments, (a) continental (entirely and permanently fresh-water swamps) and (b) paralic (fresh and/or brackish water swamps with periodic marine incursions) have been distinguished.

The simple geotectonic division is now outdated and is considered insensitive. The two contrasting environmental categories, however, are still applicable in general discussions and are particularly favoured by metalliferous geologists, because the major aspects of ore distribution are influenced by the presence or absence of marine sedimentation. Galloway and Hobday (1983) recently summarized the setting (Tables 25-3 and 25-4) and depositional environments of coal basins. Diessel (1970) discussed the climatic controls of coal deposition and pointed out that although the pre-Permian coals formed almost exclusively in humid tropical regions, a substantial proportion of geologically younger coals was laid down in temperate and even boreal regions. The 15 to 30° latitude belt was virtually free of coal, because of the general aridity.

25.4.3. Metal geochemistry and mineralization

This subject was reviewed recently by Laznicka (1985d) and this section has been summarized from this work making some additions and alterations. The coal geochemistry was also treated recently by Bouška (1981).

Coal is an outstanding anomaly of selective concentration and accumulation of carbon. The average carbon content of coal, about 65% C, contrasted with the carbon clarke of 200 ppm, represents a factor of concentration of 3,250. The tonnage of an average coal deposit (about 1 Bt coal, that is some 650 Mt C), represents a carbon accumulation equivalent to 3.25 x 10^{12} tons of the average crustal material. The C concentration factor in coal is of the same order as the factor in high-grade Pb or low-grade Au deposits, and greater than the Cu concentration factor in porphyry copper deposits.

Because of the geochemical uniqueness of coal, it might be assumed that certain metallic elements with an affinity for carbon or influenced by the numerous special environmental conditions that accompany the deposition and diagenesis of the organic matter resulting in coal, would co-precipitate with coal to produce metallic deposits. This has happened although on a substantially smaller scale than one would expect. About 15 Bt Fe, 200 Mt Al, 200 Tt U, 40 Tt Mo, 1 Tt Ni, etc. is estimated to have been produced or to remain in reserves of bona fide metallic deposits intimately associated with coal. Additional metals occur as only slightly enriched trace elements in coal and could be recovered from wastes remaining after coal combustion.

Coal and its wastes thus represent an outstanding "technological ore" that contains a minimum of 1 Mt Ge, 500 Tt Ga, unknown quantities of Be, Sc, and other metals possibly profitably recoverable if there is a demand. The coal-associated sedimentogenic ores (Fig. 25-4; Table 25-5) will be reviewed in the framework of the following

Table 25-3. Common geotectonic setting of ancient coal basins (based on data in Galloway and Hobday, 1983, with additions)

GEOTECTONIC ENVIRONMENT	CHARACTERISTICS	EXAMPLES OF COAL BASINS
subduction-related foredeep and backarc basins	dynamic systems, up to 13 km thick, rapid burial, usually large number of coal seams, strong deformation, minor syn-depositional volcanism	T, S.W. New Guinea Foredeep Pe, Sydney Basin, Australia
intracontinental rifts, aulacogenes, grabens, strike-slip pull-apart basins	generally thinner coal-bearing successions, bounded by normal faults. Limited number of seams, often narrow, elongated seam pattern ("shoestrings"). Frequent washouts, seams interrupted by sandstone and conglomerate. Rapid transitions to red-beds	Cb-Pe, Sillon Houiller, France; Donetsk Basin, U.S.S.R.; Asturias Basins, Spain; Bohemian Massif; T, La Trobe Valley, Victoria, Australia
epeirogenic, generally atectonic, stable cratonic basins	thin and uniform coal-bearing sequences with one or few seams; persistent, shallow, generally low degree of coalification	Cb, Sub-Moscow Basin, U.S.S.R.; T_1 lignites, Williston Basin, U.S.A. and Canada
foreland (molasse) belts and basins (collision-related)	generally assymetric elongated basins filled by detritus from nearby actively rising source areas, bordered by tectonically active terrains deformed by low-angle thrust faults. Cyclic deposition, frequent marine interbeds	Cr-T_1 coalfields in the Rocky Mountains and their foreland, Canada and U.S.A. Ps, Appalachian Foreland
coastal plain basins at Atlantic-type (passive) continental margins	atectonic to step-faults-bound wedges of coal-bearing continental sediments, intertonguing with marine sediments; frequently cyclic	Ps, basins in cyclothems in mid-United States

categories: (1) metallic ore substances present immediately in the coal seam (a) in dispersed state; (b) as a physically separable material; (2) widely distributed metallic substances accumulated in close proximity to coal seams; (3) ores occasionally present in the coal association and (4) metallic ores in facies equivalents of coal and/or the coal association. Additional mineralizations are the result of (5) non-sedimentogenic alterations of coal resulting from a variety of interactions.

1a. METALLIC ORE SUBSTANCES PRESENT DIRECTLY WITHIN THE COAL SEAM (OR COALY MUDROCK), IN A DISPERSED STATE

All the known elements have been identified in coal or in coal ash in trace quantities (Table 25-6), but only a few (Ge,Ga,As,V,B,Mo) tend to be present consistently in supra-clarke amounts; even so, their clarkes of concentration rarely exceed 10. Truly anomalous ("ore" and "protore") contents of the above-mentioned metals as well as of U,Ag,Pb,Se,Cu,Au,Sn,Ni,Hg,Be and others, may appear locally and indeed one of the most characteristic features of coal geochemistry is the considerable degree of local variation in the concentration of metals controlled by local heritage and the formational history of coal.

Trace metals in coal may occur as (a) constituents of detrital, mechanically separable, inorganic rock or mineral particles interlayered with or interspersed in the organic coal substance; (b) constituents of the mineral matter (ash), evenly dispersed in coal; (c) inorganic fracture or bedding plane coating and infilling in coal; (d) constituents of various inorganic compounds, etc. in coal; (e) organo-metallic compounds in coal; (f) finely dispersed metallic oxides or sulphides adsorbed on coal; (g) ions dissolved in the coal moisture. Because most trace metal determinations in coal have been based on analyses of coal ash, the sites in which the metals occur have only rarely been determined. The ash metal contents, moreover, preserve only the refractory elements (fortunately, they are the majority), while much of the volatile metals (As, Se, Hg, Sb) is lost.

Of greater interest to a prospector trying to predict the presence of metalliferous coals, is the source and timing of the trace metal enrichment. It now appears that only a minority of elements in coal accumulated by the metabolism of living plants. Most trace metals were clearly introduced posthumously, in part probably during the decay of the organic substance in the coal swamp, in part after burial during the diagenesis and even after lithification. Consequently, the bulk of the trace elements in coal appear to be epigenetic, introduced from outside, although in most cases from areas in very close proximity. Coals tend to be good indicators of the regional geochemical provincionalism, established before the deposition of the coal association. For example, the Ge-rich coals in the Plzeň Basin (Czechoslovakia) developed over a crystalline basement containing germanian sphalerite veins, and the Cu-rich coals on the fringe of the

Table 25-4. Common coal depositional (paleo)-environments. Based on data in Galloway and Hobday (1983), with additions

ENVIRONMENT	EXAMPLE AREA / SEQUENCE	CHARACTERISTICS
PARALIC ENVIRONMENTS		
stream-dominated deltas (birdfoot type or elongated); most common	Q, Mississippi River delta T, lignites of Gulf Coastal Plain, U.S.A.	coal seams form in blanket swamps, developed in abandoned lobate deltas, less frequently in inter-distributary tracts
wave-dominated deltas; common	Niger, Mekong, Ganges-Brahmaputra and other deltas	peat and coal are generated in sloughs and embayments between shore-parallel beach ridges
tide-dominated deltas; coal very rare	Cr, Horseshoe Canyon Fm., Alberta, Canada Q, Klang and Langat R., Malay	peat / coal form in freshwater swamps behind mangrove tracts
back-barrier basins and lagoons	Q, Shuffedy Swamp, S.C. Ps, Beckley Seam, West Virginia	marshes formed behind deltas or coastal barriers; thin, discontinuous coal seams. Coal tends to be highly pyritic
delta plain swamps	Q, Mississippi delta	marshes and swamps are located between beach ridges on strand-plains, on lower delta plains, etc. Thin but persistent coal seams
CONTINENTAL ENVIRONMENTS		
fluvial environments	Q, Mississippi R. alluvial plain; Pc, Eo Powder River Basin; Pe, Karoo Basin	coal seams are located on top of upward-fining sequences accumulated in backswamps adjacent to migrating channel tracts. Extensive, uniform and persistent coal seams, frequent washouts

limnic (lacustrine) environments	Q, Canadian Shield swamps (muskeg); Upper Nile, Amazon River, Okawango Swamps, etc. Pc, Powder River Basin	coals formed within widened, dammed, sluggish portions of rivers, in lacustrine deltas, in alluviated interior basins, in disrupted drainage systems. Thin coal seams, often accompanied by freshwater limestones
alluvial fan environments	T, Latrobe Valley, Victoria Pe, Sydney Basin Cb, N. Spain (Asturias)	coal-bearing sequences formed in intermontane and foredeep plain setting, along rifts, in pull-apart basins. Coal seams are often stacked and reach a considerable thickness. Abundant conglomerates, upward fining sequences. Coal seams are usually short, rapidly pinching out

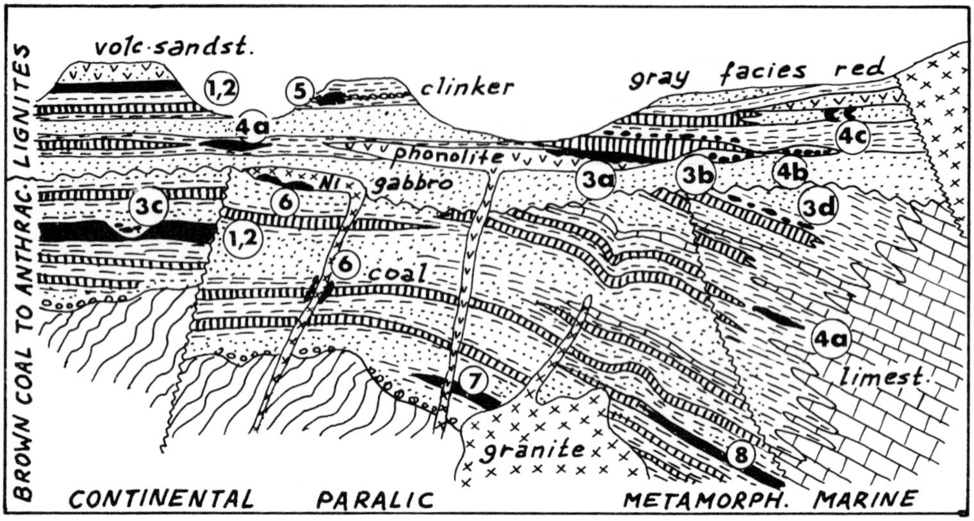

Fig. 25-4. Setting of coals and principal coal-associated mineralization styles. See Table 25-5 for explanations of the letter codes.

Table 25-5. Principal mineralization styles in the coal association.

(1)	Dispersed metals in coal in clarke or slightly supra-clarke concentrations. May be recovered from ashes as "technological ores". Estim. global resource 1 Mt Ge, 500 Tt Ga, 200 Tt U, 140 Tt Mo, 10 Tt Be, etc.
(2)	Physically separable ore substances within coal seams. Black-band iron ores (X0 localities, 3 Bt Fe); sphalerite (X localities, 50 Tt Zn); bauxite (X localities, 5 Mt Al).
(3)	Ores in immediate footwall or hangingwall of coals. Pelosiderite (X0,000 localities, 12 Bt Fe). Bauxite and Al, Ti clays (X0 localities, 195 Mt Al); Ti, Zr, Au, Sn, etc. paleoplacers.
(4)	Ores in facies equivalents of coal. Bauxite and Ti-Al clays (X0 localities); Sn, Au paleoplacers (X localities); infiltrated deposits in arenites (U-V, Cu, Pb-Zn; X0 localities).
(5)	Thermally metamorphosed rocks resulting from coal burning, and natural clinker. Mineralogical occurrences of Ge substances.
(6)	Thermal contacts of coal and mafic intrusions. Mineralogical occurrences of Cu, Ni sulphides in endocontact. Possible favourable influence on generation of the Noril'sk (Siberia) ores.
(7)	Graphite generated by contact metamorphism of coal.
(8)	Graphite generated by regional metamorphism of coal. Locally interesting trace metal contents only.

Table 25-6. Trace elements in coal ashes

	1	2	3	4	5
Ag	under 3	-	10	-	1-5
As	-	900	-	-	100-500
B	600	148	-	2200	-
Ba	1000+	300	2000	-	300-900
Be	8	14	30	-	1-19
Co	40	87	80	69	300
Cr	50	45	160	350	100-400
Cu	100	-	500	200	20-200
Ga	25	-	180	81	100
Ge	40	44	-	250	50-500
Li	-	-	350	-	-
Mn	1000	1200	350	-	-
Mo	10	60	110	87	100-200
Ni	150	131	400	320	50-800
Pb	100	572	450	-	5-50
Sn	20+	9	150	27	16-200
V	150	120	280	540	100-1000
Zn	300+	218	400	590	100-1000
Zr	800	-	200	-	100-500

EXPLANATIONS: Values are in ppm. Localities: (1) New South Wales, bituminous coals (Swaine, 1962); (2) Sydney Coalfield, Nova Scotia (Hawley, 1955); (3) West Virginia coals (Headlee and Hunter, 1953); (4) coals of the Eastern Interior Region, U.S.A. (Zubovic et al., 1964); (5) average trace metal content in coal ash (Krauskopf, 1955).

Lower Silesia Basin (Poland and Czechoslovakia) reflect well on what is a distinct, polygenetic, Permo-Carboniferous and older copper province.

Alternatively, metalliferous coals could have formed during periods of increased transfer of trace metals in groundwater, derived from above, from younger rocks burying the coal (as in the case of the North and South Dakota Paleocene uraniferous lignites mineralized by U derived from Oligocene-Miocene tuffs). In the majority of cases, coal bore the metallogenic role of a passive chemical or physicochemical trap, screen and reservoir rock for the migrating metals, rather than a source bed. This has a bearing on the exploration philosophy. In the regions of development of metalliferous coals, other suitable trap rocks (e.g. sandstones, carbonates, etc.) and the rocks along the paths of migrating fluids, could contain ores of the same metal(s) as well. Where, on the other hand, a known metallic ore coexists with coal, the coal deserves examination for possible enrichment in the same metals.

Germanium in coals

High Ge values in coal were first reported in the 1930s and the use of Ge in the early generation of transistors in the 1940s and 1950s triggered an intensive study of coal as a possible Ge source, worldwide. Numerous papers on trace-metal geochemistry of coal were published almost simultaneously in most industrialized countries (e.g. Aubrey, 1955; Stadnichenko et al., 1953) and many localities with Ge-rich coals were identified. Before a large-scale production could start, however, Ge in semiconductors was substituted by the cheaper silicon, and the demand for Ge returned to "normal" (the 1974 world production was 80 t Ge !).

The average Ge content in coals is about 7 ppm, and the high-Ge coals are rather erratically distributed in the coal sequence, being highest usually in the top and bottom seams, or in coal seams located above a thick parting of a porous rock (e.g. sandstone). This is indicative of predominantly infiltrational origin of Ge. Although spotty Ge contents in coal seams and particularly in isolated coalified logs enclosed in paralic or marine sediments may reach up to 2% Ge (up to 7.5% Ge in the ash of lignitic logs in the Atlantic Coastal Plain of the U.S.A.), the industrial Ge recovery, if it ever comes, would most probably be based on the waste products remaining after coal combustion rather than on fresh coal. Such products have the character of "technological ores", already mined, pre-processed and delivered to industrial complexes. The fine dust deposited in the flues of British coal-burning power plants contained up to 2% Ge (Aubrey, 1955) and it has been estimated that in the United Kingdom alone up to 2,000 t Ge can be recovered annually from the wastes based on an ordinary coal (Paone, 1970). The Ge resources in the British coal ashes have been estimated to be some 200 Tt Ge in materials with a Ge content over 0.1% Ge, and at least 1 Mt Ge in ashes with more than 0.01% Ge. The world resources probably exceed 50 Mt Ge.

Uraniferous lignites

The lower Tertiary and partly Cretaceous sedimentary basins of the western Great Plains and the Cordillera in South and North Dakota, Montana, Wyoming, Utah and Idaho, contain considerable resources of lignite (Denson and Gill, 1965; Vine, 1962). The bulk of these lignites and also the lignitic (carbonaceous) mudstones contain insignificant amounts of trace U; only a fraction of the total resource has from 0.005 to 0.002% U, so that it can be designated as uraniferous lignite. The U content in a single lignite seam is not uniform, and within an extensive low-grade body there are often small irregular and lenticular masses of coal which, in places, contains 0.1% U or more (up to 5% U_3O_8; Bailey and Childers, 1977). Some of the higher grade sectors were mined on a small scale in the 1950s and 1960s, but at present no mining is in progress and the known occurrences remain resources for the future.

In the S.W. Williston Basin (the Dakotas and Montana border region), several lignite beds are exposed on flanks of low erosional mesas (Fig. 25-5). The beds are members of the essentially

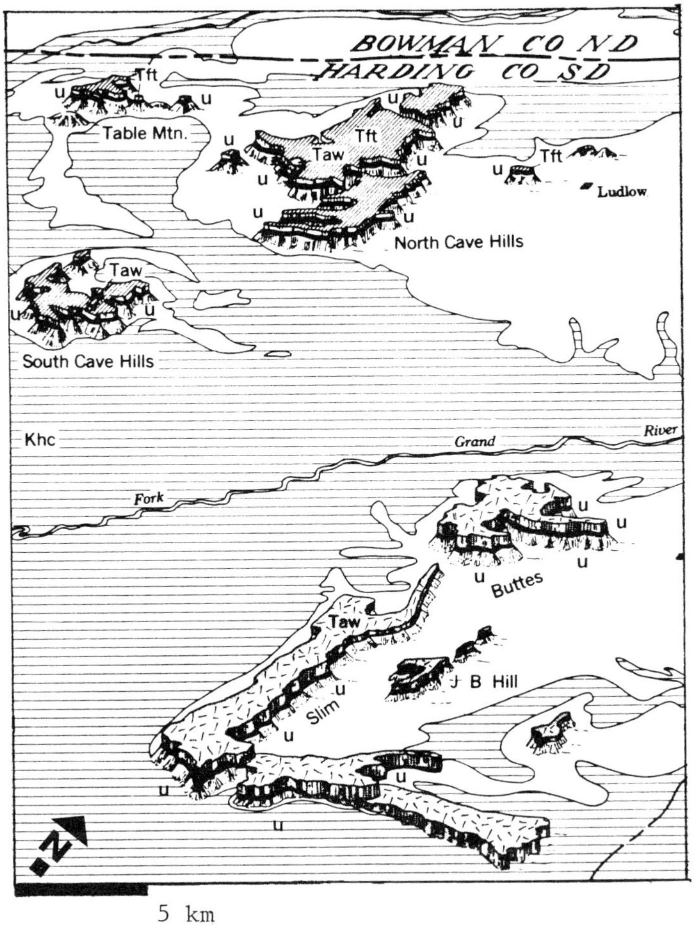

Fig. 25-5. Uraniferous lignites (U) in N.W. South Dakota. The uraniferous coals are located stratigraphically under the Oligocene and Miocene tuffaceous beds (Taw); these are still preserved in the Slim Buttes area, but are largely eroded in the remaining areas. Ttf, Tfc, Tfl: Pc continental sandstones and mudstones; Khc: Cretaceous marine shale. From Denson et al., (1959).

epiclastic, continental sublitharenite to feldspathic arenite, mudstone, lignite-bearing Paleocene Fort Union Formation, resting on Cretaceous marine shales. The mesa tops are formed by white Oligocene and Miocene tuffaceous sandstone, claystone, bentonite and basal conglomerate. The volcanic component has its source in the ash fall, derived from the Absaroka volcanic field, some 500 km W.S.W. Among the numerous lignite and coaly mudstone beds in the area, only the stratigraphically highest beds, or the erosion dissected intermediate beds exposed within about 70 m stratigraphically below the Oligocene unconformity, carry uranium. The lenticular uraniferous seams are only about 15 to 70 cm thick, the U is invisible, believed to be bound in an organo-uranium complex. Authigenic analcite is common in the adjacent sandstone, and visible secondary U-V minerals (carnotite, autunite, uranophane) have been recorded locally as bedding plane and fracture coatings. The vanadium content is very low, but locally the molybdenum content is high. The lignitic mudstones in the Little Missouri River escarpment contain up to 0.7% Mo.

The U in lignites is clearly epigenetic, leached by groundwater from the slightly uraniferous Oligocene and Miocene volcaniclastics, released during devitrification of the volcanic glass and diagenesis. Circulation of the U-bearing groundwater along and under the sub-Oligocene unconformity resulted in U precipitation in sites where the solutions came into contact with the coal. The stratigraphically lower lignite beds were not exposed to the uraniferous groundwater circulation or were penetrated by the depleted descending waters only, so that no U enrichment resulted. The U emplacement was intermittent. It was most intensive after the Miocene ash falls and it ceased about 250,000 years ago. The U content in the Williston Basin lignites is of the order of 60 Tt U/0.005-0.12% and the highest consistent grade (0.12% U) is near Belfield, North Dakota.

Elsewhere in western North America (Table 25-7), uraniferous lignites are comparable in age and style with the Williston Basin deposits, and account for approximately another 22 Tt U. The U source relationship is, however, less clear. In the Great Divide Basin, Wyoming, the U in Eocene lignites may have been derived from the Archean basement granites either directly, or via the adjacent lower Tertiary epiclastic arkose facies (Bailey and Childers, 1977).

Outside North America, uraniferous lignites appear to be fairly common (Table 25-7) and their importance increases with the increasing price of uranium. Unfortunately, only fragmented literature data are available here. The lignites of N.E. Spain (Ebro Basin); Miyagi Prefecture, Japan (Ouchi deposit); Bavaria and western Czechoslovakia, show a close source relationship to the basement granites rather than to younger volcanics. Characteristically, in the Ouchi lignite mine, Japan (Saito, ed., 1960), the U contents are highest and economic only in the lowest part of the lignite seam, indicating precipitation from ascending groundwaters.

Uraniferous brown, black coals and anthracites

The worldwide ratio of uraniferous to normal brown, black and anthracite coals is much lower than the ratio of uraniferous to barren lignites. Danchev and Strelyanov (1973) argued that advanced coalification destroys the complex humid compounds of lignites that have the capacity to adsorb U and replace them with much simplier compounds lacking such capacity. Consequently, U is not preferentially withdrawn from groundwaters percolating through coal after its advanced coalification, but the same coal can still retain the U accumulated in the earlier stages of development. The same coal

Table 25-7. Uraniferous lignites, data on selected example localities

LOCALITY	AGE OF LIGNITE	AGE OF MINERAL.	TONNAGE	REFERENCES
Little Missouri R. Escarpment, N.Dakota	Pc	01 to Q	X0 Tt U/0.12% also Mo	Denson and Gill (1965)
Sentinel Butte, N.Dakota	Pc	01 to Q	315 t U/0.007%	ditto
Chalky Buttes, N.Dakota	Pc	01 to Q	2,618 t U/0.017%	ditto
Ekalaka Hills, E.Montana	Pc	01 to Q	825 t U/0.005%	Gill (1959)
North and South Cave Hills, South Dakota	Pc	Mi_3-Q	424 t U/0.008%	Denson et al.(1959)
Slim Buttes, S.Dakota	Pc	Mi-Q	24,500 t U/0.007% 2,450 t U/0.005%	Denson and Gill (1965)
Great Divide Basin, Wyoming	Eo	Eo-Q	1,500 t U/0.005% 20,700 t U/0.003%	Masursky (1962)
Fall Creek, S.Idaho	Cr_1	Mi-Q	2,925 t U/0.045%	Vine (1962)
La Ventana, Sandoval Co, New Mexico	Cr_3	Q	132 t U/0.1%	Bachman et al. (1959)
Anonymous deposit (Far East ?), U.S.S.R.	T	T	X,000 t U also Be,Ge	Danchev and Strelyanov (1973)
Ebro Basin, N.E. Spain	01	01-Q	80,000 t U	O.E.C.D. (1979)

can also accomodate epigenetic U as an open space filling. Transitional deposits in which epigenetically infiltrated U is found in the coal joints and crush zones as well as in sandstones filling washouts in coal seams, are common (e.g. Radvanice, N. Czechoslovakia). In contrast to the uraniferous lignites, U in the higher grade coals is frequently represented by its own minerals (pitchblende, coffinite) accompanied by pyrite and base metal sulphides (galena, sphalerite, chalcopyrite). In the Carboniferous Ronchamp coal basin, southern Vosges, France (Fluck et al., 1975) the early, pre-coalification stratiform U in dispersed form is located in layers of coaly mudstone. The late epigenetic mineralizations of pitchblende, dolomite, barite, pyrite, galena and sphalerite, fill fault breccias.

Deposits of purely uraniferous black coals are few, low grade and of low tonnage (Table 25-8). Transitional deposits involving brown coal in which both the coal and associated sandstone are mineralized, are more significant (e.g. in the Mécsek Mts., Hungary and in several anonymous deposits in the U.S.S.R.). It is estimated that there are some 150 to 200 Tt U in the coals of the world that are either economic at present or likely to be in the near future. Vine (1962) gave an excellent, comprehensive review of the many aspects of U accumulation in coaly carbonaceous rocks.

Other metalliferous coals

Coal is capable of adsorbing and hosting almost any element locally available in an anomalous supply and may become an "ore" under suitable technological, economic and political conditions. Table 25-9 lists examples of the various metalliferous coals reported in the literature, but few "coal ore" deposits have actually been delineated, reserves calculated and sufficient data published. Many deposits are transitional and coal is just one host to the accumulated metal(s), other hosts being sandstones, conglomerates, tuffs, etc.

Of the metals recorded, Ga and Mo appear to be most consistently concentrated in coals. Both tend to be dispersed ("invisible"), although Mo sometimes enters jordisite or ilsemanite found as infiltrations or fracture coatings. Ga distribution in coals usually correlates with Ge, being some 50-66% less abundant. Be appears to be anomalously enriched in several Soviet coals, as well as in some localities on the Czech side of the Erzgebirge Mts. (Mezirolí and Kocourek in the Sokolov Basin: 0.2 and 0.3% Be, respectively; Obr, 1980). In both regions the occurrences are associated with a Be-rich granitic basement as well as with a contemporary continental volcanism. The genetically unusual anonymous deposit in Uzbekistan described by Petrov (1961), has a variety of ores developed in Triassic and Jurassic continental volcanics and sediments along a paleosurface. Mo,W,U,Pb,Zn,Be, etc. are all present in sulphide (pyrite, galena, sphalerite, molybdenite, scheelite) veins and stockworks, as well as dispersed in regolithic volcanics topped by Jurassic basal sediments that include coal seams.

Pb,Zn and Cu sulphides are relatively widespread in coal in

mineralogical quantities, particularly as coatings of joints in coal, or as accessory crystals in septarian pelosiderite concretions. In Lubná, W. Czechoslovakia, abundant galena in coal correlates with the local epigenetic hydrothermal Pb veining in Carboniferous porphyries and felsic volcanics. The coals high in copper in the Permo-Carboniferous Walbrzych and Žacléř Basins (Polish and Czech Sudetes) are members of a Cu-anomalous province containing a variety of volcanogenic (bimodal continental suite) and sedimentary copper deposits.

Ni-bearing lignites and brown coals ("kersinites"; 0.7-8% Ni) are one of the several varieties of metalliferous sediments that overlie the Mesozoic-Cainozoic ultramafic regolith at several localities in the central and southern Urals, U.S.S.R. (Glazkovsky et al., 1974). The nickel is mostly bound in dispersed millerite and its presence is indicated by the light-green crusts of the secondary morenosite. In the Cheremshan deposit in the Gornyi Ufalei district, middle Oligocene lignite contains a mere 2.3% of the total Ni reserves in the deposit (probably several hundred tons Ni).

Gold and silver have been detected in several coals (Boyle, 1979), but their distribution is unpredictable and erratic. Almost always there is a correlation with a local hydrothermal-epigenetic Au-Ag ore in silicate host rocks.

(1b) METALLIC ORE SUBSTANCES PRESENT DIRECTLY WITHIN THE COAL SEAM (OR COALY MUDROCK), AS A PHYSICALLY SEPARABLE COMPONENT

Blackband, pelosiderite and iron sulphides in coal

Blackband is a siderite or ankerite intermixed or intimately interstratified with coal and clay, sometimes to such a degree that the components cannot easily be separated. Pelosiderite is an impure (argillaceous) micritic siderite. Both materials form concretions, lenses and thin beds within or in the proximity of a coal seam, are diagenetic in origin, and were recovered in the past as a by-product of coal mining. Fig. 25-6 shows graphically the principal varieties of siderite associated with coal.

In the best studied area of the British blackbands, the North Staffordshire Coalfield seams, up to 2.5 m thick, at four stratigraphic horizons have been worked over an area of 70 km^2, replacing coal and interbedded with it. The raw blackband contained about 29% Fe, and it could be calcination upgraded to about 45% Fe. Some blackbands were almost perfect mixture of ingredients used in the primitive iron smelting furnaces. The tonnages of by-product blackband in any single coal mine were small (of the order of 10^5-10^6 t ore), but the aggregate tonnages of certain coalfields were better, and the remaining reserves substantial (e.g. North Staffordshire Coalfield, 1.3 Bt; South Wales Coalfield, 5 Bt; S.E. Ohio Coalfield, 1.5 Bt; Wright et al., 1968; Zitzmann, ed., 1976). The world's known production and reserves of blackband ores are about 10 Bt ore (about 3 Bt Fe), of which some 70-80% is in Britain. At present,

Table 25-8. Uraniferous brown, black and anthracite coals, selected example localities

LOCALITY	COAL		URANIUM MINERALIZATION	REFERENCES
	AGE	TYPE		
Saint-Hippolyte, Vosges, France	Cb_2	coaly black mudstone	fine, dispersed carburan in organic matter; remobilized As,Pb,Zn sulphides X00 t U, 0.02-0.1%, max. 0.7% U	Fluck et al. (1975)
Ronchamp, E. of Lure, S. Vosges, France	Cb_3-Pe	coaly mudstone black coal	fine, dispersed stratiform U in mudstone, superimposed epigenetic pitchblende, dolom., sulphides in fault breccia	ditto
Freital, S.W. of Dresden, East Germany	Pe_1	black coal to anthracite, shale	dispersed stratiform U and superimposed fissure filling coffinite, Pb,Zn,Cu sulphides; estim. X00 t U	Ruzicka (1971)
Königstein on Elbe, S.of Dresden, E.Germany	Cr_3	brown coal claystone qtz. arenite	dispersed carburane, pitchblende and coffinite; transitional to SUV infiltration deposits; estim. X00 t U	ditto
Teplice Basin, N.Bohemia, Czechoslovakia	Cr_3	as above	as above, estim. X00 t U	ditto
Hamr n.J.-Mimoň distr. N. Bohemia, Czechosl.	Cr_3	as above	as above, locally high Zr in baddeleyite; estim. min. 20 Tt U	ditto
Hroznětín, etc., Cheb and Sokolov Basins, N.W. Czechoslovakia	Mi	brown coal to lignite, sandstone	dispersed U-humates, pitchblende, pyrite; estim. X00 t U	ditto
Radvanice, Bečkov, N.W. Czechoslovakia	Cb_3	black coal	dispersed U in coal, fracture coating of pitchblende, galena, sphalerite, chalcopyrite, carbonate; est.X00 U	ditto

Mécsek Mts. (Uránváros), S.Hungary	Pe,J	brown coal	U dispersed in Jurassic coal overlying SUV-mineralized Permian red-beds deposited over a basement granite	Szalay (1954)
Anonymous deposit (probably in Turgai Downwarp),Kazakhstan, U.S.S.R.	J_3	brown coal, sandstone	pitchblende, coffinite disseminated in coal and sandstone; estimated XO Tt U, about 0.1% U	Kovalev (1970)
Anonymous deposit (probably in Argun Basin, E.Transbaikalia), U.S.S.R.	J_1	black coal, sandstone	U dispersed in coal, pitchblende, coffinite; also in sandstone, supergene enrichment; estim. X Tt U	Danchev and Strelyanov (1973)
Springbok Flats, N. Transvaal, South Africa	Pe	black coal	U dispersed in coal; X Tt U	O.E.C.D. (1979)
North of Fort Scott, Kansas, U.S.A.	Ps	black coal	finely dispersed U; min. 700 t U/0.007%	Vine (1962)
Rio de Peixe Coalfield, N. Paraná, Brazil	Cb	black coal sandstone	dispersed U and SUV infiltrations	Vine (1962)

Table 25-9. Metalliferous coals (other than U); selected example localities

LOCALITY	METAL	COAL		METALLIC MINERALIZATION	REFERENCES
		AGE	TYPE		
Great Britain, most coal basins except South Wales	Ge	Cb	black coal	trace Ge dispersed in coal, aver.7-10 ppm in coal, 50 ppm in ash, max. 2% in flue dust	Aubrey (1955)
Plzeň Basin (e.g.Vej-prnice), W. Czechosl.	Ge	Cb_3	black coal	5-40 ppm dispersed Ge in coal, up to 1% Ge in ash	Havlena (1964)
Slané Basin (e.g. Lo-touš), W. Czechoslov.	Ge	Cb_3	black coal	66-233 ppm Ge dispersed in coal	ditto
Mittweida, S. East Germany	Ge	Mi_1	lignite to brown coal	250 ppm Ge dispersed in ash	Rösler and Lange (1972)
Lower Kittaning Coal Bed, E.Ohio Coal-field, U.S.A.	Ge	Ps	black coal	100 ppm Ge in coal, up to 0.2% Ge in ash	Stadnichenko et al.(1953)
Meziroli, Kocourek; Sokolov Bs.,Czechosl.	Ge Be As	Mi	lignite to brown coal	up to 0.205% Ge dispersed in coal; also 0.2-0.3% Be, max. 0.6% As, 0.1% Cu, 0.1% Pb, 0.1% Y, U	Obr (1980)
West Virginia (some seams), U.S.A.	Ga	Ps	black coal	dispersed Ge in coal; max. 0.1% Ga in ash	Stamper (1970)
Argun Basin, Trans-baikalia, U.S.S.R.	Be	J	brown coal	dispersed Be in coal, estim. 0.1% in ash; epigenetic source in granitic basement	Danchev and Strelyanov (1973)
Cave Hills near Buf-falo, S.Dakota	Mo	Pc	lignite	dispersed Mo in coal, rare ilsemanite afflorescences; min. 15 Tt Mo/0.3%	Pipiringos et al.(1965)
Anonymous deposit, Uzbekistan, U.S.S.R.	Mo W,Ge	J	brown coal	dispersed and veinlet jordisite, Fe,Pb sulph.,in coal above miner. regolith	Petrov (1961)

Southern Sakhalin coalfields, Eastern U.S.S.R.	Re	MZ-T	lignite to brown coal	Re dispersed in coal, 0.39 ppm aver., max. 19.5 ppm	Kler et al. (1978)
Nikitovka, Donetsk Basin, Ukraine	Hg	Cb_3	black coal sandstone	Hg dispersed in coal, or in cinnabar infiltrations and veinlets. Coincides with fault-controlled Hg ore field	ditto
Pleasant Valley, Utah, U.S.A.	Au	T	brown coal to lignite	Au dispersed in coal, up to 0.5 ppm Au in ash	Boyle (1979)
Cambria, Wyoming, U.S.A.	Au Ag	T	lignite, brown coal	Au and Ag dispersed in coal; up to 4 ppm Au, 12 ppm Ag	ditto
Timok Basin, N.E. Serbia, Yugoslavia	Au Ag	MZ	brown coal to anthracite	0.117-0.573 ppm Au, 2.56-6.07 ppm Ag dispersed in coal	Popović (1954)
Brandov Basin, Erzgebirge, N. Czechosl.	Ag	Pe	anthracite	fracture and veinlet base metal miner. in coal	Havlena (1964)
Lubná near Rakovník, W. Czechoslovakia	Pb	Cb_3	black coal	dispersed, fract. coating, veinlet galena in coal and contemp. rhyolite	Kratochvíl (1961)
Chvaleč, Svatoňovice, Žacléř Bs.,Czechosl.	Cu	Cb	black coal	disseminated and joint coating chalcopyrite in coal and sandstone	Kudělásek (1959)
Milan County, Texas	Sn	Eo	lignite	Sn dispersed in coal, 0.1-1% Sn in ash	Deul and Annell (1956)
Cheremshan, Gor.Ufalei d.,Urals,U.S.S.R.	Ni	Ol_2	brown coal	Ni in dispersed millerite in coal above Ni laterite; estim. 600 t Ni	Glazkovsky et al. (1974)
Olbersdorf near Zittau, East Germany	V	Mi_1	lignite	up to 0.62% V in ash; V dispersed in coal	Rösler and Lange (1972)

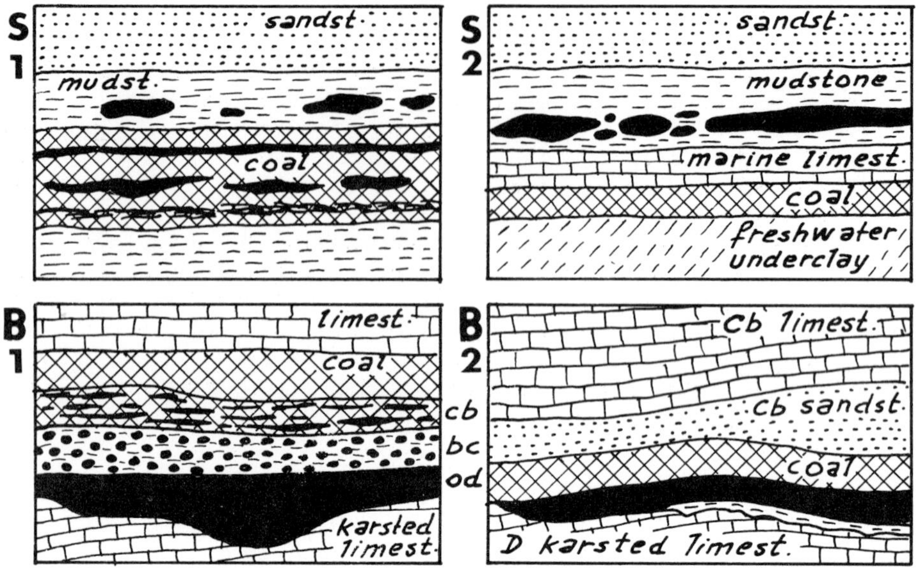

Fig. 25-6. Common varieties of diagenetic siderite and bauxite associated with coal. S1, blackband siderite (British Carboniferous); S2, pelosiderite in hangingwall mudstones (e.g. Ruhr Carboniferous); B1, coaly bauxite (cb), bauxitic clay (bc), oolithic or pisolithic chamosite and diaspore. Honan, China or Jammu, Kashmir. B2, bauxite and bauxitic clay in place of underclay, Timshersk, U.S.S.R. Approximate scale: coal seams are 2-5 m thick.

however, they are uneconomic to mine.

Iron bisulphide marcasite (dominant in young coals) and pyrite (dominant in mature coals) are widespread accessories in all coal seams. Here, they take the form of disseminated crystals, faceted nodules, laminae, joint coatings and vein fillings and are both diagenetic and post-coalification. Some coal seams have an anomalous FeS_2 content (50% plus) and were, mostly in the past, mined as a pyrite ore (e.g. Borovichi near Moscow, U.S.S.R.; Sokolov Basin, Czechoslovakia; several localities in the eastern United States' Pennsylvanian basins). In the Po-Shan coalfield in northern China (Onuki, 1967), two pyritic beds were mined. A 10 cm thick bed was interbedded with coal, and a 1.2 m thick bed with large reserves was found in the coal footwall. In contrast to the volcanogenic pyrites, the coal-associated pyrite contains no recoverable minor and trace metals (Cu,Au,Ag,etc.) and is of little practical value at present.

Sphalerite-rich coals and coaly carbonaceous rocks, U.S. interior

The coal-bearing paralic Pennsylvanian cyclic sedimentary association located on the stable North American craton, contains several coal seams or horizons of coaly carbonaceous shales anomalously enriched in zinc. In the Illinois coal basin (Hatch et al., 1976), Zn content in the coal is fairly uniform averaging 0.15% Zn and this may increase up to 1.5% Zn in areas where the coal is tectonically disturbed or in clastic dikes. In coal, sphalerite forms scattered small crystals and is most conspicuous in fracture veinlets and on joints. By-product recovery of the sphalerite during coal washing has been proposed.

Coveney (1979) reviewed the zinc distribution in black shales in the Kansas City area, which are brackish or marine sediments, the latter corresponding to the interval of marine incursion within the coal-bearing Pennsylvanian cyclothems. There, four horizons of shale distributed over areas of tens to hundreds km^2 contain from 0.07 to 0.25% Zn accumulated in sphalerite, enriched in thin, brown phosphatic laminae. The Zn enrichment is considered syngenetic to diagenetic and Coveney (1979) speculated that these shales could have acted as source beds to some of the stratigraphically lower Mississippi Valley "type" Zn-Pb deposits in the region.

(2) WIDELY OCCURRING METALLIC SUBSTANCES IN CLOSE PROXIMITY TO COAL SEAMS

Repetitive occurrence of several metalliferous commodities in close proximity to the coal suggest some causality in origin. Pelosiderites are the most common coal companions and although some may be hosted by the coal itself, the bulk is in the hangingwall and less frequently footwall sediments (claystones, mudstones, shale). The most persistent pelosiderite occurrence is in paralic cyclothems as in the Pennsylvanian of Ohio (Stout, 1944). There, siderite is most common in the marine intervals marking the marine transgression onto the former coal swamp, and it lies directly on the marine limestone. The British Carboniferous pelosiderites are similar. The Black Bed Ironstone of Yorkshire is a bedset of alternating iron-rich and iron-poor layers of black shale, 1.8-3.6 m thick. It is persistent over an area of several tens of square kilometres. The pelosiderite is mostly nodular, and the pelosideritic black shales were considered to be ore when they contained at least 20% of physically separable iron ore (Hemingway, 1968). The cutoff was about 7% Fe, i.e. the world's lowest grade iron ore ever mined! (if one disregards the Japanese iron sands).

In the continental coal association (e.g. in the Kladno Basin, Czechoslovakia), pelosiderite is found in an equivalent stratigraphic position, in lacustrine black, coaly mudstones and sandstones immediately above the coal seam, and less frequently in the underclays. Although the cumulative resources of pelosiderites are considerable (about 36 Bt of ore worldwide, i.e. some 10 Bt Fe, most of it in Great Britain), they are uneconomic to mine at present and the past production (mostly as a by-product) amounted to only several tens of million tons.

Some of the claystones (shales) closely associated with the coal seams (in particular the structureless underclays or "seatearths" containing coalified plant roots situated immediately under the coal seam), have been utilized for a long time by the ceramic industry. More recently, however, they have been investigated at some localities as possible ores of aluminium and titanium. These are reviewed in the next paragraph.

(3) METALLIC ORES OCCASIONALLY PRESENT IN THE COAL ASSOCIATION

Al or Ti-rich claystones and shales

The bulk of claystones (shales) associated with coal is kaolinitic in composition, but aluminium hydroxides (boehmite, diaspore) may appear in accessory to substantial quantities, so that some kaolinitic clays could be substituted by kaolinite-Al hydroxide to purely Al hydroxide (bauxite) clays or shales. The latter correspond to transported bauxites (Section 25-3) from which they differ mostly in appearance (displaying gray colours and usually being structureless), so that they can be easily overlooked.

Bauxitic clay deposits locally underlying the Carboniferous coal measures in the Midland Valley of Scotland formed by decomposition of Carboniferous felsic volcanics, and are currently mined for the manufacture of Al sulphate (Dunham et al., 1978). The late Eocene underclays under coal seams 30 km S.E. of Seattle, Washington (Patterson, 1967) consist of kaolinite and boehmite and contain 23.7 Mt material with 33% Al_2O_3. The Pennsylvanian "diaspore clays" of the eastern United States (Cheltenham Clay of the Ozark region, Missouri; Pottsville Formation in the Clearfield District, Pennsylvania) are recently exhumed or still buried diaspore, kaolinite (minor boehmite) continental underclays, characteristically resting on karsted carbonate basement (Keller, 1975). The Clearfield deposit contains 5-6 Mt of material with 40-70% Al_2O_3.

A substantial proportion of the Chinese bauxites and aluminous shales have a setting comparable with the American "diaspore clays" (Fig. 25-6), and represent a large aluminium resource (e.g. the Carboniferous Hsiuwen and Kueiyang districts, Kweichow, 93 Mt Al; Kunghsien and Tengfeng district, Honan; Tzupo and Laiwu districts, Shantung, 60 Mt Al; Penchi district, Liaoning, 21 Mt Al; Ikonnikov, 1975). These shales, composed of dominant diaspore with lesser quantities of kaolinite, boehmite, quartz, siderite and rutile are commonly (but not always) topped by coal seams. In some cases (in Honan) as well as in Jammu, Kashmir and Timshersk, U.S.S.R., coal and bauxite are physically intermixed to form coaly bauxite and a coal with bauxitic ash (Fig. 25-6).

Unusual deposits of allophane and gibbsite bauxite in the Moscow coal basin summarized by Bárdossy (1982; his "Tula type") formed entirely in the subsurface, by reaction of sulphuric acid released by decomposition of pyrite in coal and coaly claystones, with claystone and limestone in the footwall of Carboniferous coal measures.

Materials of high titania content are represented by the Miocene bentonitic clays forming footwall to the brown coals in the Chomůtov-Ústí nad Labem basin, Czechoslovakia. 4-15% TiO_2 is bound in submicroscopic anatase derived from phonolite, and an industrial recovery process based on material with min. 6% TiO_2 and max. 3% Fe_2O_3 has been designed and tested (Havlena, 1963).

Dawsonite

Dawsonite ($NaAl(CO_3)(OH)_2$), considered as one of the future aluminium sources, is widespread in several Soviet coal associations from where it can be recovered as a by-product of coal benefication. In the Zalomen Depression (N.E. Kuznetsk Basin, Siberia; Rundkvist, ed., 1978), thin interbeds in coal containing 30-40% dawsonite are distributed in a section of a Permo-Carboniferous coal sequence, 900 m thick.

Paleoplacers

A significant, locally economic content of ilmenite, zircon and rutile has been reported from several Soviet coal basins (e.g. from middle Carboniferous alluvial to deltaic sandstones, Donetsk Basin; Jurassic lacustrine-alluvial sandstones of the Irkutsk Basin; Lavrov, 1972). The widespread Cretaceous "black sandstones" in the western United States (Chapter 19) are also topped by coal in places. It is to be expected that some of the Recent gold (e.g. Klondike) or tin (Malaya) placers capped by peaty muck or a swamp, may have equivalents in the geological record.

(4) METALLIC ORES IN FACIES EQUIVALENTS OF COAL AND/OR COAL ASSOCIATION

In the literature, the humid coal-bearing continental facies is often contrasted with the arid red-beds facies. Given the broad and frequently uncertain status of the red-beds discussed later, there are probably numerous exceptions. Facies transition between the "red-beds" and the "gray-beds" associated with coal, has a high incidence of occurrence of the "sandstone" Cu,Pb and even U ores. In most cases, however, this appears to be the result of post-depositional infiltration rather than the facies control.

Rare examples of metallic mineralizations interpreted as proximal facies equivalents of coal, include (1) Miocene cassiterite paleoplacers on the southern slopes of the Erzgebirge Mts., N.W. Czechoslovakia (Sattran et al., 1966), considered facies equivalent of the coal seam Josef. (2) Marine particulate or massive, bedded ironstones, such as the Eocene ironstones at Paz del Rio, Colombia; some British Liassic ironstones; Karoo ironstones of South Africa; middle Tertiary siderite-leptochlorite ironstones in the Transurals; Eocene ironstones in the Western Siberian Iron Ore Basin.

(5) NON-SEDIMENTOGENIC ALTERATION OF COAL (AND THE COAL ASSOCIATION), RESULTING FROM A VARIETY OF INTERACTIONS

Spontaneous coal burning and thermal metamorphism

Coal seams in the shallow subsurface ignited by natural causes (sulphide oxidation, lightning, volcanism, etc.) are capable of slow burning lasting thousands of years. The end-product of coal burning is clinker, a porous, fused ash residue marking the former position of the coal seam, surrounded (mainly topped) by an aureole of thermally metamorphosed adjacent rocks ranging from dense porcellanite to orange baked shales. The orange shales are relatively weathering resistant, and a conspicuous component of scenery in the North American Badlands (e.g. Drumheller, Alberta; Little Missouri R. Badlands, North Dakota) and elsewhere.

No significant metallic mineralization is known associated with the burned-out coal, although the process obviously has a direct, as well as an indirect, metallogenic potential (thermal mobilization of metals, generation of porosity and local collapse, etc.). Ge,Ga,U,Mo and base metals remain in the clinker following the coal burning, so they may be slightly enriched. Thermally metamorphosed (naturally calcined) pelosiderites are slightly upgraded, and mineralogical quantities of native iron have been found in a few localities. Žák (1951) described an interesting occurrence of germaniferous opal, filling fractures in porcellanites above burned-out Miocene seam near Březno (N.W. Czechoslovakia). This represents a minor epigenetic redeposition of trace Ge by descending water.

Spontaneously burning waste dumps near collieries often produce an interesting assemblage of sublimed minerals on the surface or in the shallow subsurface, reminescent of fumarole minerals (sulphur, salmiak, realgar, selenium, Pb oxides, etc.).

Thermal metamorphism or metasomatism at coal / igneous contacts

Rather voluminous literature is now available on the effects of metamorphism and metasomatism at intrusive contacts of usually mafic dikes and sills with coal (e.g. Kisch and Taylor, 1966). There, coal is typically converted to natural coke or even graphite, coaly mudrocks to a variety of porcellanites, and intrusive rocks in the endocontact to albite-, chlorite-, etc.-rich assemblages. No significant, directly associated mineralization has ever been reported, although the coincidence of coal seams in the sedimentary sequence forming the footwall of the Ni-Cu mineralized differentiated mafic intrusions in the Noril'sk-Talnakh district, Siberia (Chapter 32), could have been a contributary factor in the ore genesis (Sobolev et al., 1978).

Granite / coal contacts have been reported only occasionally, for example from Sonora State, Mexico, where a broad range of contact metamorphic products terminating with "amorphous" graphite is represented. A graphite sample from an unknown locality in Mexico (Janda and Schroll, 1960) contained 20 ppm Ag, a grade equivalent to low-grade silver deposits.

Regional metamorphism of coal

Quinn and Glass (1958) demonstrated, using the example of the metamorphosed coals in the Narragansett Basin of Rhode Island, U.S.A., that anthracite persists throughout the lower greenschist facies, and meta-anthracite through the higher greenschist facies, without being converted to graphite. Graphite, however, is the coal equivalent at higher metamorphic grades. Janda and Schroll (1960) studied the trace element contents of many graphites and recorded some interesting local high trace-metal contents. Probably only a minority of such graphites, however, were meta-coals.

25.5. DIAMICTITE (PARACONGLOMERATE)-FEATURING ASSOCIATION

25.5.1. Introduction

Diamictite is "any nonsorted or poorly sorted terrigenous sediment that consists of sand and/or larger particles in a muddy matrix" (Flint et al., 1960). Paraconglomerate and conglomeratic mudstone are substitute terms. It is important to note that the above terms do not indicate genesis. Pettijohn (1975) summarized this class of rocks, which he subdivided by interpreted origin into tillite (lithified till, a glacial sediment) and tilloid (Geröllton, non-glacial origin). Because of the widespread availability of excellent contemporary continental glacial environments, the continental glacial model has been well understood for a long time and during the first half of this century, applied to virtually all known ancient lithologic associations containing paraconglomerates, throughout the world.

Some such interpretations withstood the test of time, although it is now believed that glacio-marine, rather than glacio-continental deposits are dominant (e.g. most of the Permo-Carboniferous diamictites of the Southern Hemisphere), whereas others have been reinterpreted as subaerial to subaqueous mudflows, landslides, slumps and turbidite flows (e.g. Crowell, 1957). Diamictite-featuring associations are still not well understood and remain enigmatic, despite the considerable progress made in the past decades. There is a substantial difference in emphasis and interest between the sedimentologist and a metalliferous geologist, dealing with this lithologic association. Sedimentologists have emphasized and preferentially studied diamictite terrains that have the greatest degree of correspondence with the recent glacial model. Like their modern equivalents, however, such terrains are virtually unmineralized. The mineralized diamictite-featuring terrains, such as the Rapitan Group in N.W. Canada and the Jacadigo Group in South America, on the other hand, depart in many respects from the modern glacial model and include many "special features" such as pre- and syn-depositional faulting; volcanism and magmatism; prominent chemical sediments as well as contemporary, earlier or later carbonates; metallogenically fertile provenance associations involving greenstones, gabbros and granites; etc. These have been barely touched upon by modern geological research.

In the tillite (glacial diamictite) model, Edwards (1978) distinguished (1) massive tillite (mixtite), an equivalent of subaerial or subaqueous till; (2) stratified, partly sorted sandstone and conglomerate (consolidated continental glaciofluvial sediments), and (3) laminated mudstone or siltstone with or without dropstones (the principal glaciomarine sediment), as the three principal lithofacies. To these should be added the products of superimposed hydrodynamic reworking, such as sandstone or conglomerate-filled channels and beaches. Frequently described examples of ancient tillite terrains include large tracts of the Dwyka Tillite of southern Africa, Ipanema Tillite of Brazil, the late Paleozoic glacial beds of eastern Australia and Tasmania and others. The sharp contrast between the richly mineralized pre-Carboniferous lithologic associations and the completely barren Permo-Carboniferous tillites, is a striking feature of the ore distribution in Tasmania (e.g. in the Zeehan ore field).

25.5.2. Mineralization

DETRITAL ORES IN TILLITES (PALEO-GLACIOGENIC ASSOCIATIONS)

Mineralized occurrences compatible with the modern glaciogenic model are extremely rare. Putzer (1962) described mineralogical occurrences of monazite in syenite clasts from Cerro Itapopo in Paraguay. The clasts are components of an upper Carboniferous tillite.

The economically important paleoplacers in the Karoo (Triassic) sandstones near Bothaville and Wolmaransstad in South Africa, are in regressive beach sediments immediately topping stratified and unstratified Dwyka Tillites. The latter, in turn, blanket the lower Proterozoic Ventersdorp mafic lavas and granitic rocks of the Archean basement. Near Bothaville (Behr, 1965; 26.8 Mt of heavy mineral content, including 18 Mt ilmenite and 326 Tt monazite), the gray to black, heavy minerals-rich sandstone, forms several parallel north-west trending stratiform sheets in the ordinary quartz-rich sandstones. Ilmenite and leucoxene are dominant minerals, followed by anatase, zircon, monazite, garnet, chromite and other minerals. The deposits appear to be genetically comparable with the modern auriferous beach sediments near Nome, Alaska (Chapter 5), formed by reworking of gold from both till and bedrock occurrences.

CHEMICALLY PRECIPITATED ORES IN DIAMICTITE-FEATURING ASSOCIATIONS

Rapitan Group, north-eastern Cordillera, Canada

The late Proterozoic Rapitan Group (Eisbacher, 1976) is a distinct diamictite-dominated unit, exposed in the core of the Mackenzie Mountains. It is up to 300 m thick. The diamictites immediately overlie a unit composed of graded rhythmic maroon and green siltstones with sandstone interbeds, and grade upward into a varicoloured, upward-coarsening argillite, siltstone and sandstone unit. The diamictite unit consists of alternating beds and lenses of green or maroon paraconglomerate, composed of greenstone, gabbro, minor dolomite and limestone pebble to boulder-size well-rounded clasts, resting in rhythmically laminated to homogeneous siltstone matrix. The clasts distribution ranges from densely packed pockets or clusters to sporadic clast occurrences in siltstone. Several surficial striations of clasts reported appear to be due to tectonic shearing. Crossbedded, quartz-rich lenses are locally interstratified with the diamictites and siltstones.

The large Crest iron deposit in the Snake River area, Yukon (Thorpe, 1969; 11.62 Bt Fe/46%; Fig. 25-7) is a band composed of lenses of interbanded massive hematite and bright red jasper, hosted by and interbedded with the diamictite. The band is 51 km in length and up to 110 m thick. Orthoconglomerate, quartz-rich litharenite and shale form minor interbeds, partings and channels interrupting the iron ore beds. The Crest ore zone is folded and faulted, but virtually unmetamorphosed.

In the Jasper Valley Fe-Cu prospect, the basal portion of the Rapitan Group is composed of a greenish arenite to siltstone and diamictite, containing several horizons of a low-grade cherty iron formation (alternating bands of gray hematite and red jasper or grayish chert). Within a stratigraphic thickness of about 70 m there is an anomalous copper content, and visible disseminated and fracture filling chalcopyrite and bornite with rare chalcocite occur locally.

The origin and controls of the Rapitan Group-hosted mineralizations are unclear, but syndepositional volcanism and tectonism are believed to have been involved.

Jacadigo Group, Corumbá area, Brazil and Bolivia

Several isolated hogbacks and low plateaus rising above a flat, swampy lowland in the border region of western Brazil and eastern Bolivia in the Mutún-Urucúm district, contain an enormous reserve of rich Fe and Mn ores (22 Bt Fe/54%, 50-100 Mt Mn/49%; Dorr, 1945; Ahlfeld and Schneider-Scherbina 1964; Walde et al., 1981; Fig. 25-8). The host unit, Jacadigo Group, consists of 430 m of clastics and 350 m of chemical sediments and is now considered to be late Proterozoic, about 1 b.y. old (it was formerly considered Cambro-Ordovician). It is interpreted as a continental sequence with marine incursions composed of calcareous and ferruginous arkose, polymictic conglomerate including diamictite, laminated siltstone with dropstone pebbles to blocks, chert, jasper, and bedded siliceous Fe and Mn formation. The Group rests on a granite some 1.2 b.y. old which, in turn, intruded an older Precambrian crystalline basement.

The iron ores are of the "jaspilite type" (alternating laminae and beds of fine crystalline steely hematite and a bright red jasper), and several horizons are present. Dorr (1945) estimated the original tonnage as exceeding 500 Bt of ore (some 250 Bt Fe) over 800 km^2, the

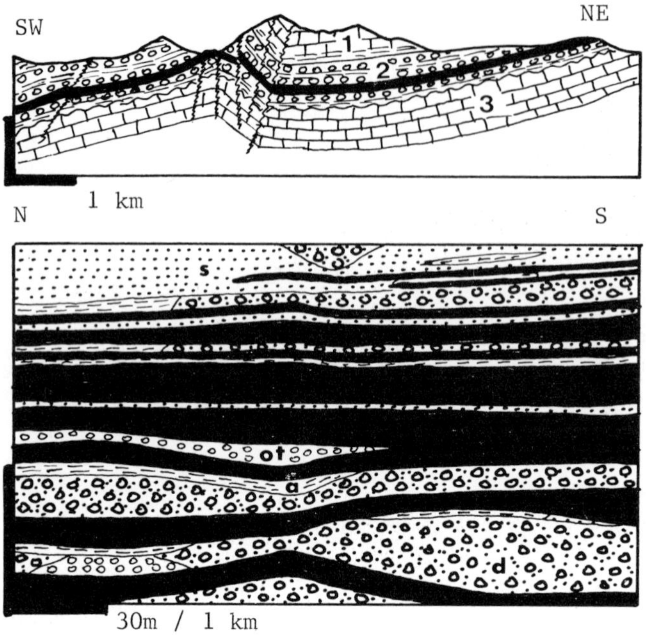

Fig. 25-7. Snake River (northern Canadian Cordillera) iron formation, cross-section and a detail of the ore zone. 1=Pt_3 limestone, dolomite, argillite; 2=Pt_3 Rapitan Group, diamictite, iron ore, sandstone, argillite; 3=Pt_3 dolomite. s=sandstone; a=argillite; ot=orthoquartzite; d=diamictite. BLACK: red jasper interbedded with specular hematite. From LITHOTHEQUE, modified after unpublished materials prepared by Crest Exploration, Ltd., 1964.

present deposits being just a small erosional remnant.

The manganese ores consist of lenses and seams up to 6m thick of fine laminated, rhythmically banded and concretionary cryptomelane and minor pyrolusite, closely associated with conglomeratic and dropstone-containing siltstones. These interrupt the continuity of the siliceous iron formation. One to four Mn ore horizons have developed in the area. In addition to the unenriched and slightly enriched "primary" ore, a large quantity of colluvial (scree) ore is available. As in the previous example, the genetic controls are largely unknown, except that there is no evidence of syndepositional volcanism in the Jacadigo Group.

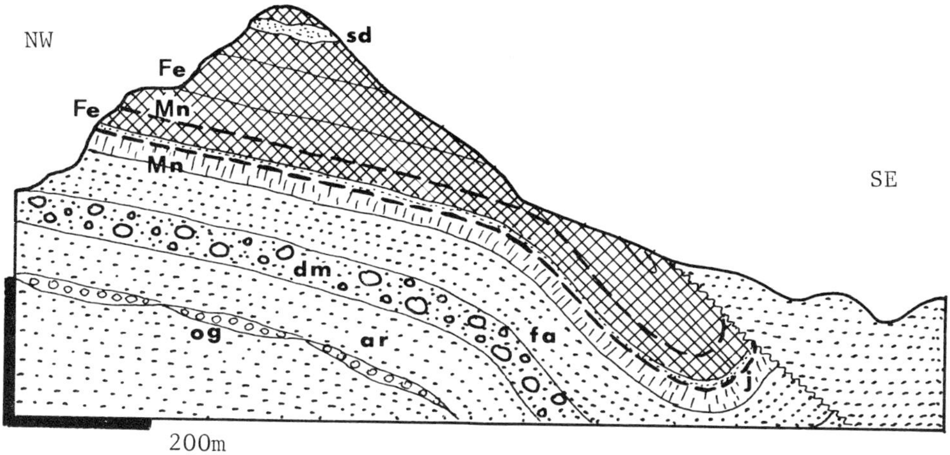

Fig. 25-8. Morro do Urucún (S. of Corumbá, Mato Grosso, Brazil) banded hematite-jasper (Fe) orebody with two manganiferous beds (Mn) near the base. All rocks are Pt_3 Jacadigo Group. sd=sandstone; j=jasper; fa=ferruginous arkose; dm=diamictite; ar=arkose; og=orthoconglomerate. Modified after Dorr (1945).

OTHER OCCURRENCES

Diamictites occur in the vicinity of various metallic ores but their contribution to ore formation (controlling aspect) if any, remains unknown. In the Zapla iron ore field, northern Argentina, diamictites are associated with Silurian hematite-chlorite particulate ironstones (Chapter 19). In the Bolivian tin province, minor diamictite interbeds are members of the Paleozoic Llallagua and Catavi shale and quartzite Formations, which are hosts to a variety of tin deposits. In Laisvall, Sweden (Bjørlykke and Sangster, 1981; Chapter 19), a thin veneer of mixtite overlain by feldspathic sandstone forms the base of the galena-mineralized Eocambrian sequence, resting on a Precambrian basement. Sporadic galena is found in the feldspathic sandstone, topped by a green to gray shale with pebbles, interpreted as a glacial sequence. The main Laisvall Pb orebody is stratigraphically higher, in marine quartzites.

25.6. LAMINATED MUDSTONE, CARBONATE, EVAPORITE (PALEO-LACUSTRINE) ASSOCIATION

Sediments deposited in ancient lakes appear in many other continental associations treated in this chapter, in which they received no special attention. There, they commonly overlap with alluvial sediments often beyond recognition. Because of the great variety and insufficient study, no general models are available. Several ancient lacustrine associations, however, have recently been studied in greater detail. Despite the variations, mostly due to climatic conditions and setting in which the lakes formed, the common features of paleo-lacustrine associations include the presence of delicately laminated chemical and authigenic sediments interbedded with reduced (gray, black) fine clastics and frequent marls. This differentiates the lacustrine association from the rest of the generally coarser, oxidized continental sediments and supports a distinct metallogeny. Ancient lake deposits and the criteria for their recognition have recently been reviewed by Collinson (1978b); Matter and Tucker, eds. (1978); Picard and High (1972) and others (see also Section 24.3).

GREEN RIVER FORMATION, WESTERN UNITED STATES

The Eocene Green River Formation discontinuously distributed in four "basins" in Wyoming, Colorado and Utah, is one of the most extensively studied and also economically important lacustrine associations of the world (contained reserves are over 1 trillion barrels of hydrocarbons in oil shale; at least 75 Bt of trona; 24.5 Bt dawsonite, containing 3.5 Bt aluminium; 30 Bt of nahcolite, etc.; Bradley, 1964; Bradley and Eugster, 1969). The Formation was laid down in a shallow and arid, often ephemeral Lake Gosiute, situated on a relatively stable thick sialic basement of the Colorado Plateau at the time when the less stable fringe of the Plateau was undergoing normal and thrust faulting, folding and magmatic activity. The lacustrine facies is immediately floored and topped by, and laterally interfingers with, fluviatile sediments.

Green River Formation has been subdivided into three Members, of which the lower and upper members are mostly composed of monotonous mudstones, highly tuffaceous near the top of the sequence. The middle (Wilkins Peak) Member is most interesting, comprising alternating facies of (1) dolomitic mudstone-composed flat pebble conglomerate; (2) calcarenite and mud cracked dolomitic mudstone; (3) laminated and thin bedded dolomitic mudstone; (4) oil shale (kerogene-containing dolomitic marl laminite); (5) trona, halite, minor nahcolite, shortite, northupite facies; (6) well-sorted channel sandstone and (7) felsic tuff. The facies distribution is cyclic and the fine grained rocks have the character of thin persistent rhythmic laminites. The saline facies (5) contains such an enormous quantity of sodium salts that a "normal" sedimentogenesis alone is considered to have been incapable of generating them and unusual sources of sodium have been sought. Most authors invoke hot springs, known to have been active along the southern flanks of the Rock Springs uplift.

Metallic minerals accumulated in the Green River Formation are, by comparison with the huge accumulation of hydrocarbons and salts, almost insignificant. The only "ore mineral" for which economic assesments have been made and which will probably be recovered as a by-product commodity in the future, is dawsonite (Smith and Milton, 1966). This Na-Al carbonate contains 35% acid-extractable Al_2O_3 (about 18.5% Al) and it is present throughout a 230m thick section of "oil shale", mostly in the Piceance Basin in western Colorado (N.W. of Rifle; Fig. 25-9). Most of the dawsonite takes the form of an inconspicuous, microscopic admixture in the dolomitic marl (10-15.5% by volume), except for a rich zone 15m thick and 2.4 km long in the Lower Piceance Creek area that has a slightly more massive appearance and up to 50% or more dawsonite. Dawsonite is considered to be a product of diagenetic reconstitution of analcite, that in turn formed by devitrification of the tuffaceous component.

Although an impressive list of minerals including rare B,Ba,Sr,REE,Zr,Nb,etc.-containing species has been compiled, the minerals are rare, dispersed and without a practical significance. The minerals are present mostly in the evaporite facies. The only locally accumulated metal is uranium, preferentially enriched in phosphatic radioactive zones (Love, 1964). Twenty five uraniferous zones (aver. 0.005% U, max. 0.15% U; aver. 2.2%, max. 18.2% P_2O_5) ranging in thickness from 7 cm to 2 m, have been recorded in the Wilkins Peak Member and are most intimately associated with the trona horizons.

OTHER LACUSTRINE SEQUENCES

The additional examples of ancient lacustrine associations reviewed in Collinson (1978b) are mudstone-dominated (with abundant rhythmites), contain horizons of impure carbonates, and evaporites are rare except for gypsum. Pyrite is widespread in the "black shale" horizons, but no appreciable base metal content was found. The Taboca occurrence in the Ceara State, N.E. Brazil (Cassedane, 1965), may be an exception. There, rhythmically banded bituminous shale interbedded with marl, gray limestone, sandstone, gypsum and interpreted as lacustrine, contains disseminations of framboidal pyrite, galena, sphalerite and marcasite concordant with lamination.

Uranium occurrences have frequently been reported although with the exception of the Lodève Basin, France, few have ever been mined. Love (1964) reported more occurrences of uraniferous phosphates from the western United States in addition to those in the Green River Formation: in Eocene sediments in the Pine Mountain area, Wyoming (aver. 0.06% U, max. 0.29% U); on the Beaver Divide area, S.C. Wyoming; in the Uinta basin, Utah (numerous zones, comparable with those in the Green River area) and others. Occasional layers of U phosphate have been found in the Tertiary "lake beds" of Nevada, e.g. near Tonopah. Although the uranium in all the occurrences was probably derived from volcanic ash, its fixing has most probably been diagenetic.

Lakes located in areas of contemporaneous intrusive explosive volcanism may become overwhelmed by the volcanic supply and their

Fig. 25-9. Eocene "oil shale" containing also dawsonite, Piceance Basin N.W. of Rifle, W. Colorado.

sediments lose much of the "typical" lacustrine character. Mineralized volcanic-lacustrine associations that include the Spor Mt. Be deposit, Utah; Date Creek Basin U deposit, Arizona; the California borate deposits containing As and other localities, are reviewed in Chapter 26.

Lodève Basin, France

Lower Permian sediments in the Lodève Basin, southern France (20 Tt U/0.04-0.09% U) are transitional, variously interpreted as lacustrine or lagoonal, and on the larger scale members of the red-beds association. The actual lithology and (except where remobilized) probably diagenetic fixing of the uranium in bituminous sediments, however, justifies its description in the context of lacustrine deposits. The reduced, lacustrine unit in the Lodève Basin (Herbosch, 1974) is composed of cyclic alternations of finely laminated dolomitic and bituminous shales, high in dispersed organic matter and authigenic minerals. It forms a faulted monocline resting on a red conglomerate and on a Cambrian basement. The ore consists of microscopic pitchblende and coffinite accompanied by arsenopyrite, ilsemanite, sphalerite and galena dispersed in bituminous shale in the form of stratiform to peneconcordant lenses, 0.3-2 m thick, in the gray shales.

25.7. TRANSITIONAL DETRITAL ASSOCIATION, DOMINATED BY URANIUM AND VANADIUM ORES IN SANDSTONES (SUV)

25.7.1. General (Figures 25-10, 25-11; Table 25-10)

Despite the best efforts to the contrary, this lithologic association had to be artificially created to accomodate the "sandstone" (or "Western States-type", Colorado Plateau-Wyoming "type", etc.) peneconcordant and roll uranium-vanadium deposits (subsequently SUV). These deposits, despite their many varieties, are members of a distinct mineralization style formed under a set of quite characteristic conditions. Because these deposits are epigenetic, emplaced after lithification of their hosts, their geological controls had to be treated in terms of two, unfortunately overlapping and often blurred, sets of considerations: (1) origin and characteristics of the ore-hosting rocks (older event) and (2) origin of the ores (younger event).

The considerations involved in (1) are reflected in the lithologic association, and in the past the variable rock appearance (particularly the colour) was automatically considered as established

Fig. 25-10. Mitten Butte, Monument Valley, N. Arizona, a remnant of a red Permian and Triassic sandstone and siltstone plateau. The butte is topped by a thin veneer of a white channel conglomerate (arrow), the regional U-V hosting unit.

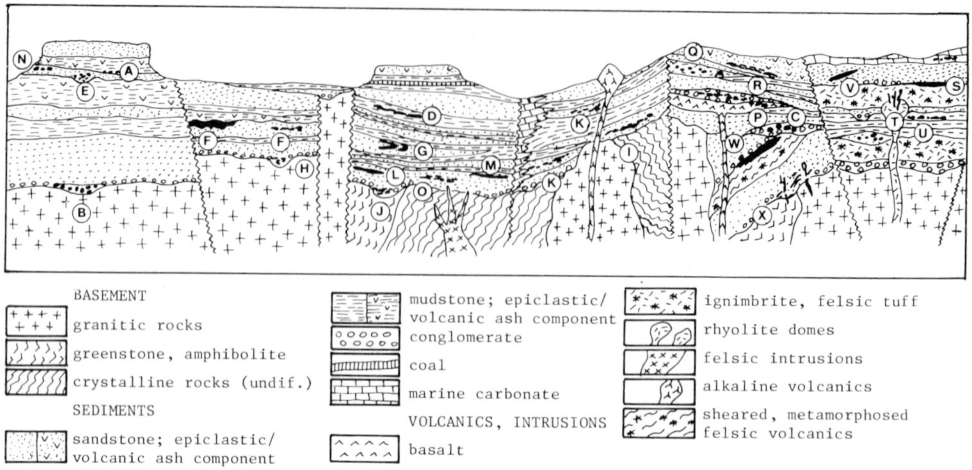

Fig. 25-11. Diagrammatic setting and the main mineralization styles in transitional, red beds and volcanic red-beds ancient continental associations. See Table 25-10 for explanation and letter codes.

Table 25-10. Principal mineralization styles in the transitional, red-beds and volcanic red beds ancient continental associations

(A,B,C)	Detrital heavy minerals (paleoplacers), gold (A; X localities, 0.01 t Au); ilmenite, zircon, monazite (B; X localities, 900 Tt Ti); reworked iron ores (C; X0 localities).
(D)	Diagenetic nodular, disseminated mineralizations in lacustrine, alluvial, etc. mudrocks (Cu,Pb,Zn sulphides). X0 localities, about 1 Mt Cu.
	INFILTRATION, EPIGENETIC ORES IN SANDSTONES, USUALLY PRECIPITATED AT REDOX INTERFACES; VARIETIES:
(E)	U-V ores in intraformational channels, enveloped by tuffaceous sediments (Colorado Plateau).
(F)	Sub-unconformity U-V infiltrations controlled by organic matter.
(G)	Intraformational mineralized "rolls" in tuffaceous sediments near granitic basement highs (Wyoming).
(H)	U (V,Mo) ores in sandstones filling channels incised into granitic basement (as in Japan).
(I)	U-Zr infiltrations in sandstones in areas of alkaline volcanism (Hamr-Mimoň, Czechoslovakia).
	Total U-V infiltration deposits: X00 localities, 1.06 Mt U, 150 Tt V.

(J)	Cu sulphide and oxide infiltration deposits in sandstones; X00 localities, about 13 Mt Cu.
(K)	Pb (Zn) sulphide impregnations in continental sandstones; X0 localities, 7 Mt Pb, 1 Mt Zn.
(L)	Mn oxide impregnations in sandstones; X00 localities, about 2 Mt Mn.
(M)	Impregnation cassiterite mantos (Colavi, Bolivia); about 100 Tt Sn.
(N)	Silver halide (cerargyrite) and sulphide impregnations in sandstones (Silver Reef, Utah); 204 t Ag.
(O)	Hydrothermal infiltrations in sandstones (U,Mo); X localities.
	VOLCANIC RED-BEDS
(P)	Cu-mineralized basalts, tuffs; X0 localities, 10 Tt Cu.
(Q)	Ilmenite-rich basalt tuffs (Yastrebovsk, Russian Platform); est. X00 Tt Ti.
(R)	Cu sulphide impregnations in conglomerate and sandstone close to basalts. X0 localities, X00 Tt Cu.
(S)	Mn oxide bodies in felsic tuffs, ignimbrites, volcaniclastics in the red-beds association. X0 localities, X00 Tt Mn.
(T)	U, Pb-Zn, Cu-Ag, etc. veins in altered felsic volcanic/subvolcanic rocks in the red-beds association. X0 localities.
(U)	Diagenetic, disseminated or nodular U, Cu, Pb minerals in intervolcanic limnic and lacustrine sediments (gray). X0 localities.
(V)	Infiltration (redox) U (Mo,V,Cu) ores in volcanics - associated sandstones. X localities, 3 Tt U.
	DEFORMED, MODIFIED, ENIGMATIC VOLCANIC RED-BEDS
(W)	Banded siliceous iron formations on base of red-beds topping ignimbrites (X localities, 1 Bt Fe).
(X)	Dynamometamorphosed, schistose red-beds sediments, volcanics and crystalline basement with a variety of vein, stringer and remobilization Cu,Sb,Ag sulphides (Špania Dolina). X localities, 200 Tt Cu, 1 Tt Ag.

early (during sedimentation or early diagenesis) and not modified, at least on a large scale, at the time of epigenetic mineralization. This resulted in fragmentation of uranium deposits in some classifications based on host associations. The Colorado Plateau ores located in a regional setting that is dominantly and strikingly "red" (the orebodies themselves are in minor "gray" beds) were placed into the "red-beds U" class. The Wyoming deposits surrounded by mostly beige, yellow, gray, buff, etc. rock scenery, were placed into the "ordinary sandstone" class or similar classes. Because the rock colour was believed to indicate depositional environment, the "red" association and its ores were also described by a variety of synonyms such as "arid", "desert", etc., the "gray" or "white" host rock association was "humid", "alluvial", etc.

Although colour may indicate aridity or humidity during deposition or diagenesis and most of the "red beds" treated in the next section are probably true arid sediments, colour alone has proved to be highly misleading in evaluation of the SUV ores. Rock colour changes due to alteration in the immediate ore zone were interpreted relatively early, but in the shallow deposits mined in the 1950s they were rather subtle, light gray to beige. A far-reaching and striking regional colour change has been recognized only relatively recently, for example by Shawe (1976) who demonstrated convincingly in the Slick Rock District of Colorado that many "white" rock units were earlier "red". Discoloration was the result of alteration by reducing fluids expelled downward (into porous rocks) as well as upward during compaction of the reduced, marine Mancos Shale. More mechanisms of colour change, however, have been proposed. The alteration discoloration is most striking and convincing when the red and gray (and yellow) colour boundary cuts across bedding, as in the Antimony Canyon, Utah. Although the change from "red" originally to "white" subsequently is most common, reverse cases are known. In the Highland uranium mine, Powder River basin, Wyoming, the alteration-oxidized and uranium-depleted side of alteration rolls has a bright pink to orange-red colour. Colour variation is an important lead in the exploration for SUV orebodies.

The rock colour variation is most striking in the Mesozoic continental sediments of the Colorado Plateau, where the majority of rocks were "red" shortly after lithification. There, the altered sediments now have the appearance of varicoloured, often mottled lithofacies, composed of red, green and gray components. The two principal U(V)-bearing lithostratigraphic units there are the upper Triassic Chinle Formation and the Jurassic Morrison Formation. The 300-450 m thick Chinle (Stewart et al., 1972) overlies unconformably the red and white lower Triassic and Permian platformic sediments and is a unit deposited in fluvial (meandering stream) and lacustrine environments, receiving, at times, significant quantities of felsic volcanic ash fall. The stratigraphically lower members comprise variegated reddish-purple and light greenish-gray mottled bentonitic claystone, clayey siltstone and sandstone. The top units are dominantly epiclastic, composed of a reddish-orange blocky cross-bedded siltstone and sandstone. The Shinarump and Moss Back Members are light coloured, ledge-forming, cross-bedded, sandy conglomerates. They are quartz-rich and usually interpreted as point bar deposits filling paleochannels of variable thickness. They contain frequent patches of coalified wood trash and are the most important host to SUV ores.

The less strikingly coloured U-hosting sediments in Wyoming are exemplified by the Puddle Springs Arkose Member of the lower Eocene Wind River Formation. This unit interpreted as a wet alluvial fan system, hosts SUV ore in the important Gas Hills District. It was described by Soister (1968) and is summarized in Galloway and Hobday (1983). The Formation was deposited on an erosion surface cut into late Cretaceous to Cambrian sediments on the north flank of the ancestral Granite Mountains (an uplifted Archean granite). Most of the flat-lying to very gently dipping sediments consist of variegated to drab claystone and siltstone with interbedded lenticular to sheetlike sandstone beds, except in the Gas Hills U field where the bulk of it is represented by the 130 to 260 m thick arkose member. This member comprises massive, coarse grained conglomeratic arkosic sandstone and granite conglomerate ranging from granule to boulder, with thin beds of finer grained feldspathic to arkosic sandstone, siltstone, claystone and rare thin-bedded carbonaceous shale. Carbonized and silicified wood fragments are locally abundant. On the surface, the arkose is oxidized and grayish-yellow, but below the water table it is light bluish- or greenish-gray and gray. The upper portion of the Wind River Formation has numerous beds of drab tuffaceous and bentonitic mudstone beds, alternating with the arkose.

In the rest of the world, the hosts to SUV deposits are comparable except for minor local variations. These may include calcite-cemented arenites, limestones (both lacustrine and marine), lignite, brown coal and gypsum. The most common actual host rock to the U(V) minerals is a light-coloured sandstone to conglomerate, typically enriched in plant debris (detrital organic substance). Alternatively, interstitial organic substance may have been brought in by descending organic-rich meteoric waters under swamps (Figures 25-12, 25-13). The organic matter has an important function as an intrinsic reductant, causing precipitation of U from oxidizing solutions (Fig. 25-14). In rare instances, other compounds may act as intrinsic reductants (some zeolites, calcite, clay minerals, pyrite) and cause U precipitation. In such cases, organic matter need not be present.

Alternatively, U may be precipitated by an extrinsic (outside) reductant, which is the basis of the two-fluids genetic model (in which oxidizing U-bearing fluid that is usually descending meets the reducing fluid which is usually ascending). Other models assume that upward-streaming natural gas serves as reducing agent or that bacteriogenic reduction is acting. (Bacteriogenic reduction, however, requires organic matter to operate). Other reduction mechanisms are also possible. The Texas Coastal Plain U Province has sediments with very little visible organic matter, so extrinsic reductants must have been particularly powerful.

Mudstone or siltstone interbedded with sandstone are important as confining screens to the flow of mineralizing solutions, and the ore-bearing sandstone units are sandwiched between (as in the Highland Mine, Wyoming), or topped by, the mudrocks. Sandstone provenances in an older basement granite; presence of tuffaceous (bentonitic) interbeds or a topping unit or both, are essential as the most typical U source. Like petroleum, the SUV deposits are controlled essentially by porosity and permeability, to which have to be added the chemical agents discussed above. The facies control is of paramount importance but there is not necessarily a consistent stratigraphic control.

Fig. 25-12. Common intrinsic reductants in SUV ores. TOP: coalified wood trash. BOTTOM: asphaltite. Both examples are from the Temple Mountain field, Utah.

Fig. 25-13. A visually striking combination of infiltrations of a black sapropelic substance under a Quaternary bog (black), and embryonic geochemical cells (light) in a road cut near Zeehan, Tasmania. Slightly elevated trace U contents are in the "right" places of the roll.

25.7.2. Geotectonic setting

All the known occurrences of SUV deposits are in regions underlain by an ancient, sialic crust covered by a veneer of continental to shallow marine sediments that are either uniformly thin or change thickness rapidly. This setting, however, is usually not identical with that of the stable platforms, quiescent for a long period of time, but with block-faulted continental segments within (1) platforms (e.g. S. Texas Coastal Plain), (2) stable blocks (e.g. median massifs) within broad, composite orogenic belts (e.g. Bohemian Massif, Pannonian Basin) and (3) "foreland belts" situated along a craton / orogenic belt transition (e.g. Wyoming, Argentina SUV provinces). The Colorado Plateau SUV Province is transitional between (2) and (3).

Block faulting generated steep local gradients that facilitated a rapid erosion of the uplifted blocks and a rapid, proximal sediment deposition in the depressed local basins. Sialic or alkaline explosive volcanism was locally affiliated, or active in the proximity of the depositional basins within the reach of airborne volcanic ash. All the three settings listed above are members of a transitional geotectonic megadomain recognized by the Soviet school as "regions of tectonomagmatic activation" (e.g. Shcheglov, 1967; see also review in Laznicka, 1981b).

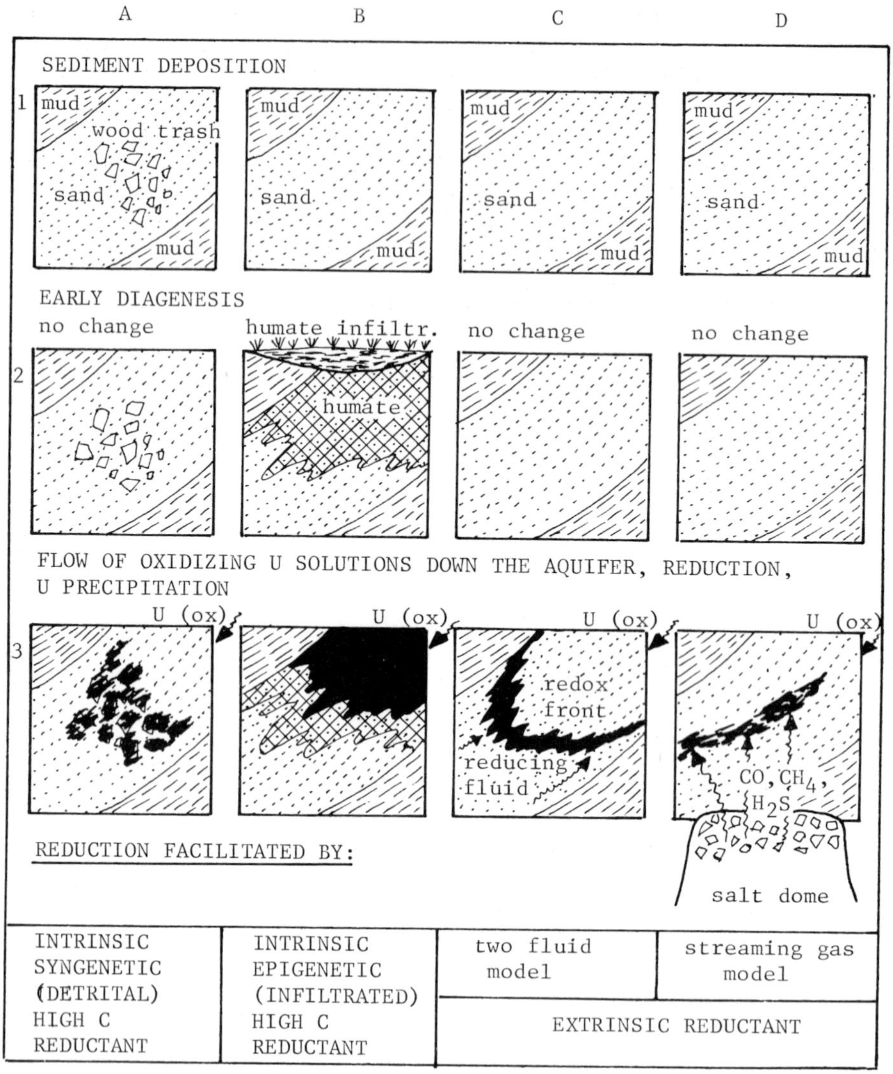

Fig. 25-14. Common varieties of reductants causing precipitation of uranium from percolating oxidizing solutions. Four alternatives showing ideal cases (see description in the text). From Laznicka (1984).

The Wyoming SUV Province (Fig. 25-15) is one of the most instructive activated areas, illustrating remarkably well the interplay of basement tectonism and occasional volcanism, sedimentogenesis, evolving U-sources and groundwater regimes, that together created what is now a significant area of U accumulation. The province is a part of the "Cordilleran Foreland" south of the Lewis and Clark lineament, a series of anticlinal mountains (uplifts) and broad assymetric basins (Osterwald and Dean, 1961). This contrasts markedly with the foreland along the more northerly segment of the Cordillera in Canada, adjacent to an uniform and extensive linear fold and overthrust belt.

The mountain ranges in Wyoming trend in almost all directions

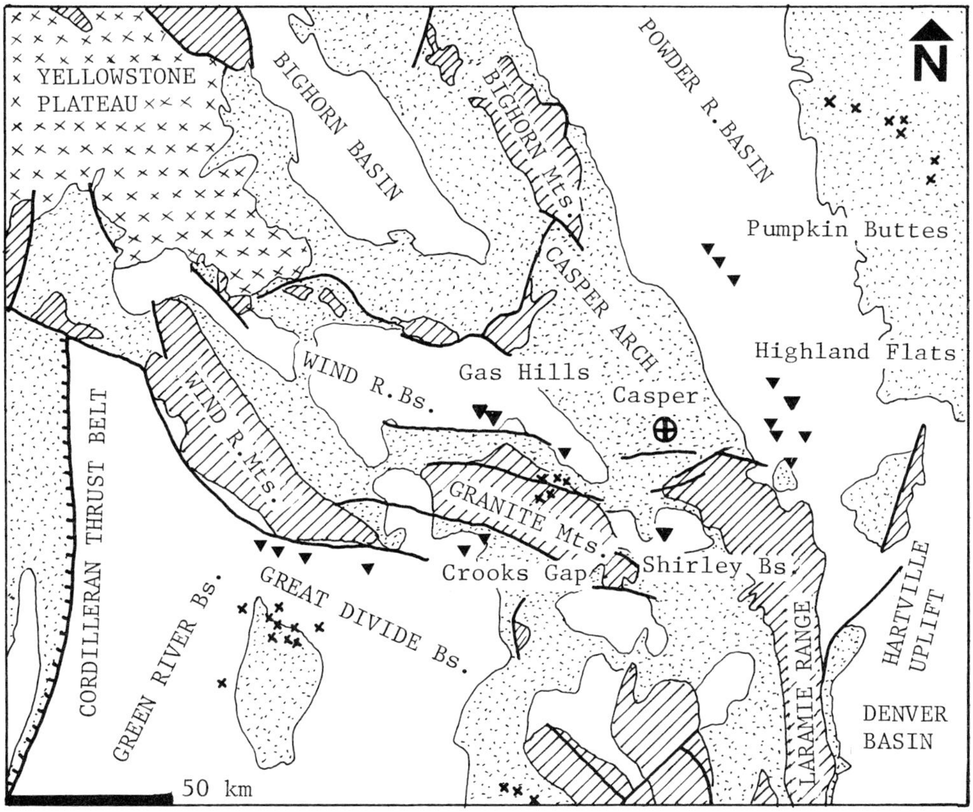

Fig. 25-15. Wyoming Foreland uranium province, U.S.A. Diagonally ruled are Precambrian basement metamorphics in uplifts; dots are PZ and MZ sediments; X=Tertiary volcanics; triangles="sandstone" U deposits; blank=T sedimentary basins.

(N.N.W. and W.S.W. are dominant), reach altitudes of over 4,000 m and characteristically have a core of Precambrian granite or granite gneiss flanked by Precambrian metamorphics and Paleozoic and Mesozoic sediments. To every uplifted block there is a complementary basin, filled by Tertiary terrigenous detritus and subordinate volcaniclastics. The basement granites as well as the Tertiary volcaniclastics have slightly elevated U contents. Galloway and Hobday (1983, p.317-320) summarized the geological history of the Wyoming Foreland, resulting in the SUV deposition. The history and SUV mineralization in the Pre-Cordillera and the Pampean Ranges in Argentina are very similar.

25.7.3. Uranium (vanadium) mineralization

"Sandstone U-V" deposits (Rackley, 1976; Bailey and Childers, 1977; Nash et al., 1981) represent some 1.06 Mt U economically recoverable at present (grades 0.1-0.35% U_3O_8), plus some 150 Tt V. There is also an unknown quantity of Cu,Mo and Se, possibly recoverable as a by-product. The "primary" deposits formed by epigenetic precipitation in the subsurface under reducing conditions, are represented by black, sooty, grain-coating or interstitial uraninite sometimes accompanied by coffinite, roscoelite, montroseite and widespread pyrite in sandstone. The sooty U oxides are visually inconspicuous and indistinguishable from coaly powder or asphaltite with which they tend to be affiliated. Occasionally, the ore entirely lacks identifiable minerals and the U is bound in the organic component ("carburan"), in zeolites or in clay minerals.

The "secondary" deposits formed by degradation of the "primary" ores by oxidation are, on the other hand, visually conspicuous. A variety of hexavalent U(V) minerals (carnotite is dominant, tyuyamunite, autunite, torbernite, schroeckingerite, etc. are subordinate) form bright yellow to yellowish-green impregnations and coatings, often around a residual core of the black "primary" minerals.

The ore impregnations in sandstones congregate to form mineable deposits and a variety of classifications of the SUV orebodies have been offered. Childers and Bailey (1979) distinguished five morphological styles (Fig. 25-16). (1) "Trend deposits" are large tabular bodies, (1a) peneconcordant or (1b) discordant with the sandstone bedding and paleo-drainage (channels). The peneconcordant bodies are dominant and they range from thin, small lenses to bodies up to 9m thick, 240 m wide and 2 km long in the Grants (New Mexico) region. Individual trend bodies could aggregate into discontinuous mineralized belts such as the Uravan (Utah) belt, 115 km long. (2) Roll-front (or roll) deposits have the characteristic "C" shaped (in cross section) and sinuous to elongate (in plan view) form of a single or multiple impregnation orebody. Such bodies developed, in "typical" form, perpendicularly to the trend and stratification of the porous host.

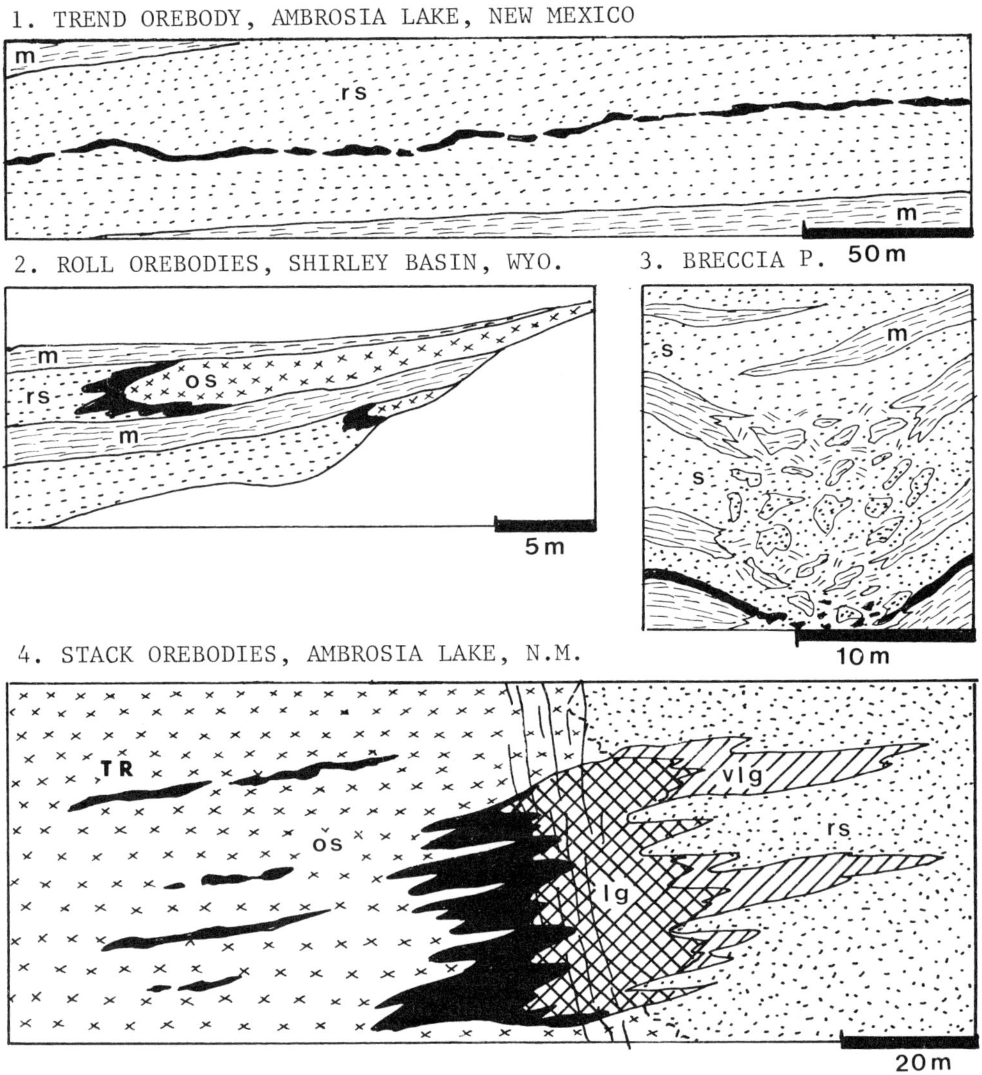

Fig. 25-16. Principal mineralization styles of the SUV ("sandstone U-V") deposits. os=oxidized (altered) sandstone and conglomerate; rs=reduced sandstone and conglomerate; s=sandstone; m=mudstone, siltstone; black=ordinary or high-grade U ore; lg=low-grade U; vlg=very low-grade U; TR=remnants of uranium "trend" orebodies. From LITHOTHEQUE, based on data from Childers and Bailey (1979); Hilpert (1969); Moench and Schlee (1967).

The (2a) "bifacies roll fronts with both oxidized and reduced facies" are distinctly geochemically and mineralogically zoned and the zoning is attributed to the alteration and mineralization activity of the groundwater. Under conditions corresponding to the most common model, the oxidizing groundwaters were descending and moving in the same direction as the original stream that had deposited the detrital sediment down the aquifer. In the upstream direction, the aquifer sandstone underwent oxidation that may have modified its originally light-gray colour into red, yellow, beige or brown. In the downstream direction, unaltered (and unmineralized, "normal") gray, reduced and often carbonaceous sandstone prevails.

The ore roll is usually darker gray than the original sandstone and it is pyrite- and interstitial calcite-rich, in addition to bearing the U(V) minerals. The ore and trace metal accumulation maxima within a geochemically well zoned roll are (in the downstream direction): (Cr,U)-Se-U(V)-Mo (Fig. 25-17). At the concave roll side (interface of the oxidation and reduction zones), limonite or hematite impregnations are conspicuous and sometimes they contain patches of hexavalent U minerals. The (2b) monofacies rolls of Childers and Bailey (1979) occur in uniformly reduced sandstones, beneath ancient erosion surfaces (unconformities).

(3) "Stack deposits" are post-depositional modifications of the trend deposits, common in the Grants region and usually associated with faults or crush zones. Mineralized pipe deposits (e.g. the Woodrow deposit, Laguna field; Moench and Schlee, 1967) are rare and

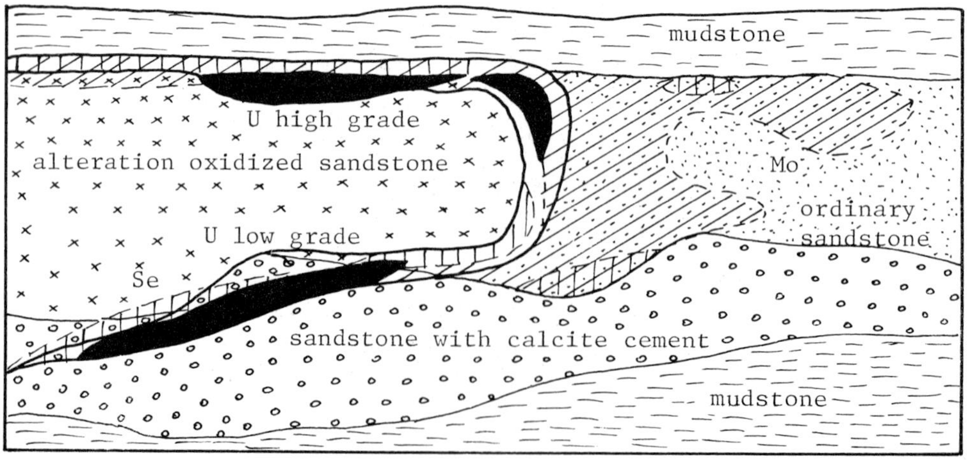

Fig. 25-17. Overlapping U (black and thick outline), Mo (oblique lines) and Se (vertical lines) zoning in a solution front (roll). After Kashirtseva and Sidel'nikova (1973).

of little importance economically. Disseminated to massive pyrite and marcasite with coffinite and several accessory minerals impregnate and replace sandstone and carbonaceous matter within a nearly vertical sandstone collapse (?) bodies ranging from 2 cm to 66 m in diameter and 2 to 100 m in height. To the four classes recognized above must be added (5) fracture-controlled impregnation and replacement U(V) bodies in limestone, as in the Grants district, New Mexico (1.9 Tt U/0.19%).

The bulk of the American and overseas SUV deposits appear to have precipitated from cool and probably largely descending solutions. Orebodies of similar morphology and in similar settings (terrigenous sediments mantling a block-faulted, activated granitic basement) but accompanied by higher P/T alteration, have been described in an anonymous locality in the U.S.S.R. (probably in the East Transbaikalia; Garbuzova et al., 1970; Fig. 25-18). There, peneconcordant impregnation and partly replacement bodies of massive and sooty uraninite, coffinite and U-asphaltite are stacked at several

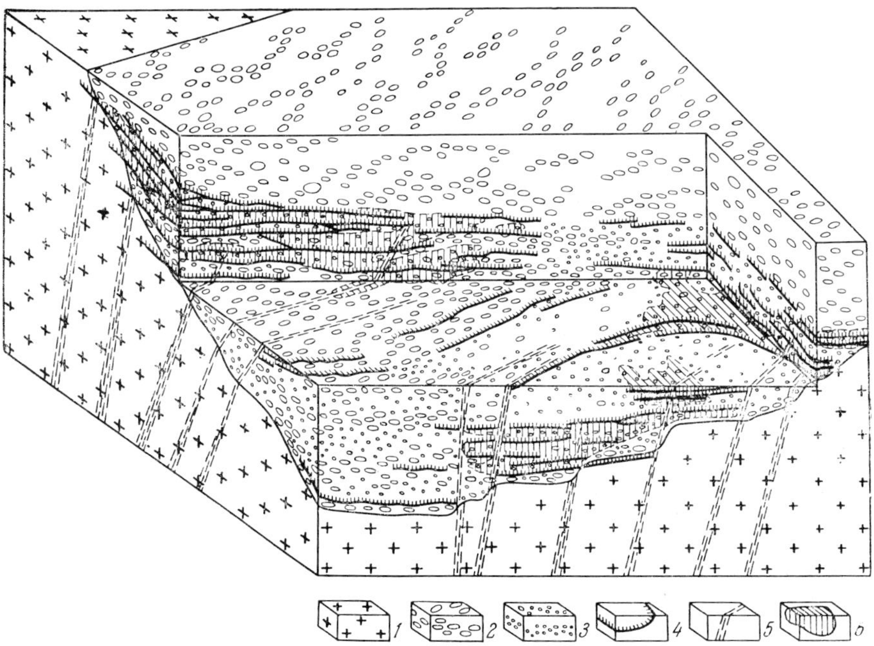

Fig. 25-18. Low-temperature hydrothermal peneconcordant SUV equivalent in propylitic and quartz-sericite altered sediments. 1=basement granites; 2=conglomerates; 3=gravelites; 4=flat fractures and faults; 5=steep faults; 6=orebodies. Unknown locality in the U.S.S.R., from Garbuzova et al. (1970).

levels controlled by growth faults and bedding in conglomerate interbedded with tuff, resting on granite basement. The orebodies are enveloped by sericite, chlorite, epidote, zeolites alteration assemblage, grading into zones of cherty silicification, carbonization and argillization.

25.7.4. Examples of SUV deposits and provinces

COLORADO PLATEAU

The Colorado Plateau is a large, oval outlier of the North American Platform in Utah, Colorado, Arizona and New Mexico, subdivided from the rest of the Platform by the eastern (intracratonic) branch of the Cordilleran orogenic belt. There, an older Precambrian crystalline basement is topped by middle Proterozoic to Cainozoic sequence of only lightly and locally deformed sediments with local volcanic or intrusive centres. The Triassic and Jurassic interval of dominantly red, locally tuffaceous continental sediments hosts the bulk of the SUV deposits (total about 480 Tt U, 150 Tt V). The deposits congregate into several densely mineralized belts and clusters, are epigenetic, and three channel sandstone or conglomerate units contain most of them. These units represent less than 100 m out of the overall thickness of 600-1000m of the Triassic-Jurassic strata.

Grants district, New Mexico

The Grants district contains three U-V mineralized belts or clusters (Ambrosia Lake, Laguna, Gallup; Fig. 25-19) and it has the largest U tonnage in the entire Plateau (reported 120 Tt U, 50 Tt V, estimated 390 Tt U, 160 Tt V; Moench and Schlee,1967; Kelley et al., 1968). The dominant peneconcordant "trend" impregnation orebodies are in Jurassic volcanogenic feldspathic sandstone, intimately affiliated with organic matter precipitated from groundwater shortly after deposition. The ore has the appearance of a friable, black to brown, tar-like sandstone. The reduced ore minerals (coffinite, roscoelite, jordisite, ferroselite) accompanied by pyrite and marcasite are finely dispersed, "invisible", but a short period of surficial exposure is usually sufficient to "reveal" the U presence in the form of bright yellow patches of secondary minerals. The two unusual mineralization styles in the district (U in collapse sandstone pipes and in Todilto Limestone) were of minor interest only.

Uravan V-U zone

The Uravan zone (Motica, 1968) is one of the oldest Colorado Plateau U-V localities, mined before the war for radium and vanadium. It is a zone trending approximately east-west containing numerous peneconcordant tabular and roll bodies, hosted by the Jurassic Salt Wash channel sandstone and conglomerate. Sooty uraninite, coffinite, vanadian clays, montroseite, and secondary U-V minerals impregnate sandstone and replace carbonaceous trash. Uravan is notable for the

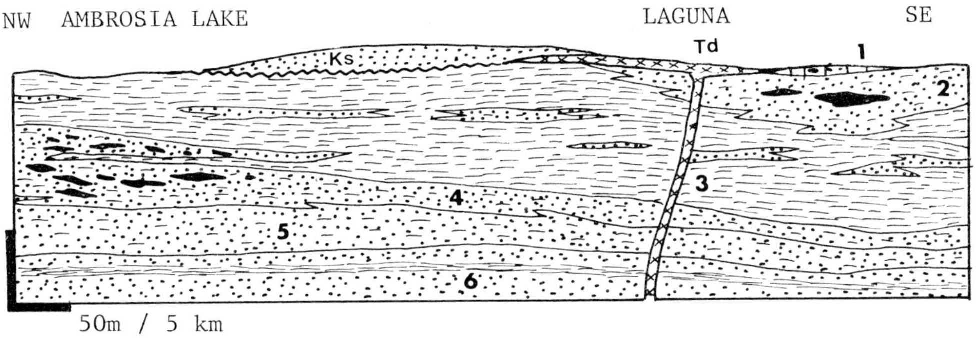

Fig. 25-19. Eastern portion of the Grants region, New Mexico, showing diagrammatically the lithostratigraphic units and setting of U deposits. Ks=Cretaceous sandstone; Td=Tertiary diabase; Jurassic rocks: 1=Todilto Limestone; 2=Entrada white and red sandstone; 3=Brushy basin, tuffaceous mudstone with sandstone lenses; 4=Westwater Canyon, feldspathic and montmorillonitic sandstone; 5=Recapture sandstone; 6=Bluff red sandstone and siltstone. From LITHOTHEQUE, modified after Hilpert (1969).

dominance of vanadium over uranium (U:V=about 1:3).

Monument Valley, Arizona and Utah

Monument Valley mines (Lewis and Trimble, 1959) now exhausted, were located along cliffs just outside the boundary of one of the most striking scenic areas of the world. Although small, this field has become well-known for its paleo-drainage control. All the small tabular to lenticular orebodies ranging in thickness from a few centimetres to about 3 m, are composed to ancient fluvial channels. The channels are above an unconformity and scour the underlying Moenkopi red mudstone and siltstone (Fig. 25-20). The ore is composed mainly of the oxidized minerals (tyuyamunite, uranophane, autunite, corvusite, hewettite), replacing calcareous cement in Triassic sandstone and conglomerate and impregnating the coaly trash. The ores are notable for their substantial copper content mostly occurring in the form of malachite.

San Rafael Swell, Utah

San Rafael Swell is an erosion-dissected broad, uplift dome with a distinct badlands topography. The Temple Mountain U-V field (Hawley et al., 1965; Figures 25-21, 25-22) has been a small producer (min. 600 t U, 1.7 Tt V), with interesting geology. Temple Mountain, a local landmark, is an erosional remnant of Triassic red-beds unconformably topping Permian limestone and sandstone, and containing a collapse structure. The collapse is associated with alteration

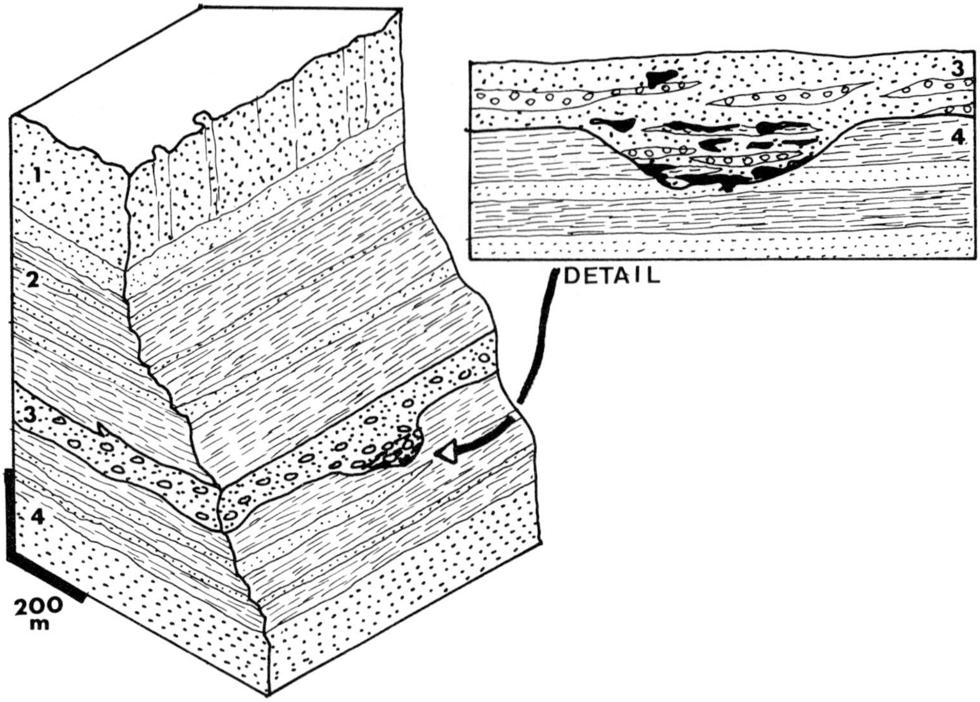

Fig. 25-20. Monument Valley, Arizona, U-V area, a diagrammatic section showing the confinement of orebodies to fluvial paleochannels. The detail shows the Happy Jack mine. 1=J, thick-bedded, light Navajo Sandstone; 2=Tr_3 Chinle Fm., red siltstone, mudstone with sandstone interbeds; 3=white, resistant Shinarump conglomerate and sandstone; 4=Tr_{1-2}, Moenkopi red mudstone on top of massive Pe sandstone. From LITHOTHEQUE, modified after Lewis and Trimble (1959).

dolomitization, bleaching and Fe oxide infiltration, and contains small patches of uranian asphaltite along the unconformity. A belt of more "normal" small roll and tabular SUV orebodies lies in Triassic channel sandstones and conglomerates. The ore minerals (uraninite, coffinite, carburan, pyrite, minor Zn,Pb and Cu sulphides) are associated with asphaltite and introduced humate as well as with detrital coaly trash. The ore has a high Se content (aver. 0.024% Se) bound in ferroselite.

Rifle Creek, W. Colorado.

Rifle (Fischer, 1960) is an isolated SUV locality hosted by altered Jurassic quartz and feldspathic sandstone to conglomerate, sandwiched between "gray" and "red" beds. The only metal recovered was vanadium (2,350 t V/7%), accumulated in bluish-gray—weathering montroseite and

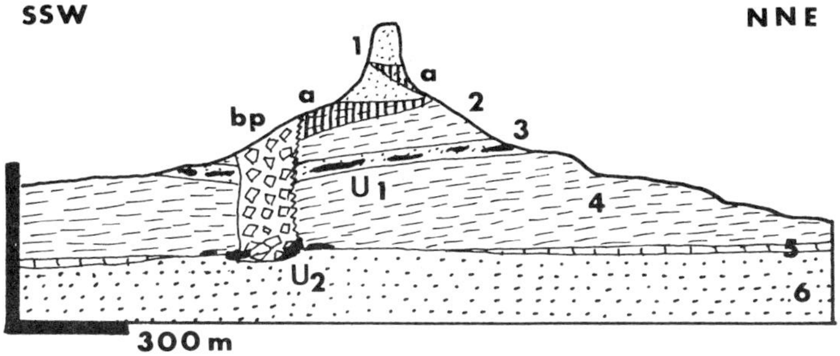

Fig. 25-21. Temple Mountain, San Rafael Swell, Utah, showing collapse breccia and (diagrammatically projected) the two settings of U-V mineralization. 1=Tr_3 Wingate sandstone; 2=Tr_3 Chinle Fm., red limey siltstone; 3=white channel sandstone; 4=Tr_{1-2} Moenkopi Fm., red mudstone to sandstone; 5=Pe Kaibab Limestone, solution removed under breccia pipe; 6=Pe Coconino Sandstone. bp=solution collapse breccia; a=dolomitization, bleaching, Fe addition; U1=peneconcordant U-V orebodies in asphaltites; U2=irregular U-V orebodies related to collapse. From LITHOTHEQUE, after Hawley et al. (1965).

Fig. 25-22. Temple Mountain, Utah (see description in the text).

vanadian micas (roscoelite) that fill pores and replace the host sandstone. Peneconcordant tabular bodies were followed for a distance of 2.5 km. The slightly larger Placerville field on the flank of the San Juan Mts. is similar to Rifle, except that there are several vanadiferous horizons.

WYOMING AND BLACK HILLS, SOUTH DAKOTA

Gas Hills field

This field is located in a barren, gently rolling landscape west of Casper. Three N-S-trending belts of orebodies are found in Eocene feldspathic arkose to conglomerate aquifers, derived from an Archean basement uplift, and screened by volcanic mudstones (Soister, 1968). Finely dispersed uraninite, coffinite, pyrite and jordisite coat clasts and impregnate sandstone along margins of altered, tabular sandstone beds, forming crescentic and "C"-shaped rolls. The ore varies from very inconspicuous to megascopically unrecognizable.

Shirley Basin

Shirley Basin, about 50 km south of Casper, contains significant U reserves, and is considered an example locality of solution roll orebodies and associated alteration, documented in the now classical memoir by Harshman (1972). The host sediments are comparable with those in the Gas Hills field. The orebodies composed of uraninite and pyrite coating grains and impregnating sandstone, occur as rolls along margins of sandstone tongues and as tabular layers along the top and bottom sandstone surfaces. The average grade varies between 0.1 and 0.7% U_3O_8 and much of the ore is visually unrecognizable, unless oxidized.

Powder River Basin

This basin filled by Paleocene and Eocene epiclastic and volcaniclastic alluvial sediments is about 100 km N.E. of Casper. Uranium was discovered in 1951 and in the late 1950s, several surficial operations (Pumpkin Buttes and other localities) mined small oxidized U-V deposits. The ore consisted of tyuyamunite, carnotite and paramontroseite dispersed in limonite and Mn oxides-impregnated sandstone. 872 t U were produced.

A drilling program in the late 1960s based on the solution front alteration model, resulted in the discovery of an important unoxidized ore in the buried Highland Zone (EXXON Corporation; Dahl and Hagmaier, 1974; at least 20 Tt U/0.17%; Figures 25-23, 25-24). There, three vertically stacked feldspathic sandstone-filled paleochannels are enclosed in tuffaceous mudrocks. Finely dispersed, invisible grain coating uraninite and coffinite form multiple ore rolls as well as connecting tabular bodies in sandstone channels and along their edge. The alteration colours at this locality are extremely bright, ranging from brick-red to orange-red in the oxidized roll centre through rusty-yellow just behind the roll, to

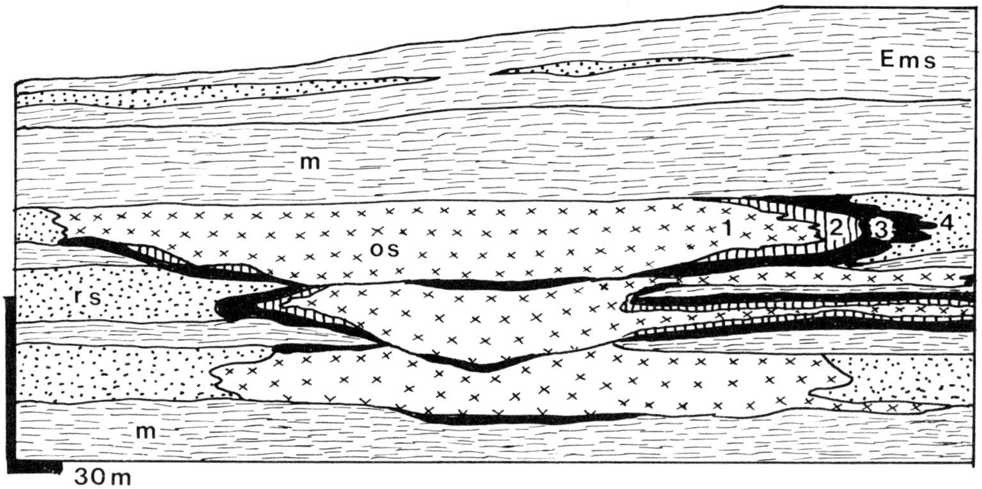

Fig. 25-23. Highland U zone, Powder River Basin, Wyoming. Roll and tabular orebodies controlled by three vertically stacked sandstone-filled fluvial paleochannels. Ems=Eo mudstone, sandstone Paleocene rocks: m=tuffaceous gray mudstone; rs=gray (reduced) arkose; os=red (oxidized, altered) arkose. Alteration zoning across a roll: 1=red, hematite pigment; 2=limonite pigment in arkose, high Se, U appears; 3=greenish-gray, pyrite and U oxide rich ore zone; 4=calcite cemented gray arkose. From LITHOTHEQUE, based on Dahl and Hagmaier (1974); Langen and Kidwell (1974); and Walter C. Duke, guided tour, 1979.

greenish-gray at the leading edge of the roll, to gray in the unaltered sandstone.

Black Hills

Small and mostly oxidized tabular and podiform U-V impregnation deposits are scattered along the perimeter of the Black Hills Uplift (e.g. Edgemont, Carlile, Hullet Creek; Hart, 1968), hosted by fluvial lower Cretaceous sandstone with much coaly debris. The ores have an anomalously high Mo content, dispersed as well as bound in ilsemanite and jordisite. In the Badlands of South Dakota, a purely molybdenum occurrence equivalent in style to the SUV deposits has been mentioned briefly in the U.S.Geol. Survey Professional Paper 750-A. At this locality jordisite, ilsemanite and yellow Mo oxide coat quartz grains in a basal Oligocene volcaniclastic arenite, filling a scour in the underlying marine Cretaceous shale. The ore lenses are 2.5 cm to 3.3 m thick, and the grade ranges from several hundreds ppm to over 10% Mo.

Fig. 25-24. Highland U mine, Powder River basin. A photo of the pit wall approximately showing the situation in the right-hand portion of Fig. 25-23. m=mudstone; r-red altered sandstone tongue; U=mineralized roll front.

SOUTH TEXAS COASTAL PLAIN U PROVINCE

Uranium deposits in the South Texas Coastal Plain (about 80 km S.E. of San Antonio; Galloway and Hobday, 1983; Bunker and MacKallor, 1973) are of minor economic importance (8,160 t U), but unusual in two respects: (1) the SUV impregnations are hosted by fluvial as well as littoral and deltaic sediments; (2) there is hardly any carbonaceous matter associated with the ore and the reduction necessary to precipitate the U has been entirely extrinsic, attributed to the reaction of sulphide-rich thermobaric waters with oxidizing meteoric waters bringing in uranium. The orebodies are visually very inconspicuous lenses, sheets and rolls with diffuse boundaries (Fig. 25-25), aligned along margins of fluvial axes. The source of U is sought in Oligocene-Miocene windblown ash component.

OTHER NORTH AMERICAN LOCALITIES

SUV occurrences are known in the southern British Columbia (Kettle River area; mostly autunite coatings of young, semi-consolidated gravels capped by plateau basalts); in northern Washington (Northwest U mine, Sherwood mine; uraninite impregnations in tuffaceous

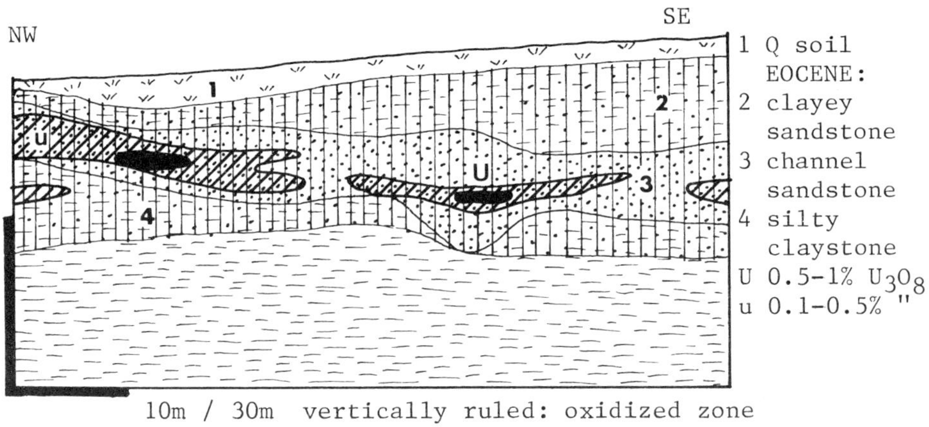

Fig. 25-25. Korzekwa deposit, 15 km S.W. of Falls City. Example of a shallow, peneconcordant oxidized U deposit in Eo channel sandstone, S.E. Texas Coastal Plain. From LITHOTHEQUE, after Bunker and MacKallor (1973).

Oligocene channel sandstone overlying granite regolith; 5,000 t U); in the Permo-Carboniferous basins of the Appalachians (Magdalen Islands, Prince Edward Island; Carbon County, Pennsylvania) and from the Nuevo Leon State, N.Mexico (La Sierrita; 2.2 Tt U in Oligocene sandstone).

SOUTH AMERICA

Several tens of mostly small U-V occurrences are located in the mountain ranges situated in the Andean Foreland of Argentina (Sierras Pampeñas, Sierra de Cordóba, Precordillera). Their setting is stratigraphically heterogeneous (Permian to Eocene) and in many fields peneconcordant impregnation deposits are accompanied by fault-controlled occurrences in the Precambrian crystalline basement.

The best-known Brazilian Phanerozoic Cambuí-Figueira SUV field is affiliated with the Permian paludal-fluvial Río Bonito Formation of the Paraná Basin (Daemon et al., 1982). There, two mineralization styles are present: (1) diagenetically-fixed U (0.02-0.09% U_3O_8) in black coaly shale marked, at its outcrop, by coatings of uranocircite; (2) epigenetic impregnations of interstitial uraninite, pyrite and jordisite in a cross-bedded channel sandstone with coaly trash, forming peneconcordant lenses 1.5 to 3.5 m thick (0.2-0.5% U_3O_8, 0.12-0.42% MoO_3) in the hangingwall of a coal seam.

HERCYNIAN EUROPE

Scattered SUV occurrences are distributed in Permian to Miocene sediments topping granites of the Hercynian (late Paleozoic) orogen in western and central Europe (Mazarete, Iberian Ranges of Spain, 850 t U in Triassic sandstone; Rodez, France, 750 t in Permo-Carboniferous sandstone; Cantal, France, 1,160 t U in Miocene sandstone near Neogene volcanics; several localities in upper Cretaceous sandstones of N.W. Czechoslovakia and southern East Germany, representing at least 30 Tt U).

The largest U accumulations are in the freshwater sandstones generally rich in coal trash, interlensed with mudstone at the base of the Cretaceous sedimentary sequence in the Labe (Elbe) River Basin near Hamr and Mimoň (Ruzicka, 1971). There, the block faulted Cretaceous sediments are an important artesian aquifer, resting on a variety of basement rocks (Proterozoic and lower Paleozoic metamorphics intruded by Carboniferous granites and topped by Permian ignimbrites and rhyolites; Carboniferous coal association), some of which have a slightly increased U content. The Cretaceous is also intruded by numerous dikes and necks of late Tertiary alkaline volcanics. The orebodies are lenses and sheets of sandstone containing finely dispersed sooty pitchblende and coffinite. Čadek et al. (1975) described an unusual local accumulation of zirconium (in authigenic baddeleyite), associated with U.

ALPINE EUROPE

In the Alps, small U deposits and occurrences are scattered in a belt 700 km long, consistently hosted by often metamorphosed Permian red-bed sandstones and conglomerates. The largest concentrations are in the Trento region of northern Italy and near Ljubljana, N.W. Yugoslavia (Žirovsky Vrh). The Mécsek Mts. uranium field near Pécs, Hungary (Uránváros), is in an outlier of the Alpine Permian belt in the Pannonian basin. Additional U occurrences are located in eastern Slovakia (Novoveska Huta) and in the Banat district of Rumania.

The facies development and the mineralization style is remarkably similar in all the occurrences. Sooty to disseminated uraninite with pyrite and frequent Cu,Pb and Zn sulphides, lie in peneconcordant bodies in gray or light-green (reduced) arkoses to conglomerates, enveloped by red-beds. Continental ignimbritic rhyolites are contemporary with the ore-bearing sandstones or are stratigraphically below them, and are considered to be the source of sandstone detritus and possibly also the uranium. In places, the felsic volcanics themselves are mineralized.

The Alpine U deposits vary from mildly to significantly sheared (dynamometamorphosed) and partly remobilized. The most intensively modified deposits (Förstau, N. Austria, in greenschist facies sericite quartzite; Iserábles, Switzerland, in chlorite-sericite and sericite-albite schist; Hügi, 1967) are of little importance.

JAPAN

Several tens of U occurrences precipitated from groundwater have been discovered in Japan, most often in block-faulted areas underlain by Cretaceous-lower Tertiary regolithic granite, and topped by Miocene to Pliocene epiclastic and volcanic continental sediments (Doi et al., 1975). The occurrences and small deposits (Ningyotoge and Tono fields) are comparable in most respects with the average SUV model, except for the U mineralogy which is frequently different. In the Tono field, uranian zeolite (clinoptilolite) and U-calcite are important ore carriers. Elsewhere, U-apatite is present. In Ningyotoge, autunite, uraninite and ningyotogite are the principal ore minerals, accompanied by pyrite and gypsum. The Japanese orebodies have a very low vanadium content.

NIGER

Although the U province of Niger (Gangloff, 1970) in sedimentary basins adjacent to the western flank of the Aïr Massif contains the second largest known resource of uranium in the SUV orebodies (around 190 Tt U), the literature coverage is patchy and hard to obtain. Ccffinite and uraninite infiltration ores have a locally high Cu content and occur at several stratigraphic levels (Carboniferous, Triassic, lower Cretaceous) in fluvial-deltaic sandstones and conglomerates. The host rocks have, as elsewhere in the world, a dual provenance (1) in the granitic and crystalline basement and (2) in airborne pyroclastic detritus. The orebodies are peneconcordant sheets, trends and rolls.

AUSTRALIA

In Australia, SUV occurrences have been reported from Mount Eclipse, 300 km N.W. of Alice Springs (uraninite in Carboniferous feldspathic and coaly sandstone), and from the playa Lake Frome Basin, South Australia. The Lake Frome deposits (Beverley, 13,430 t U; Haynes, 1975) are sand lenses interbedded with black clay, filling young incised fluvial channels. The orebodies are irregular masses, pods and rolls with finely dispersed uraninite in sandstone.

REST OF THE WORLD (Table 25-11)

Additional SUV deposits and occurrences are in Oligocene sandstones in the Western Desert of Egypt; in the Triassic Karoo sandstone of South Africa (Beaufort West and Fraserburg); in Jurassic sandstones, Khorat Plateau, Thailand and in Neogene sediments of Turkey (Salihli-Koprubasi, 2,125 t U). Localities in the U.S.S.R., Mongolia and China remain anonymous and no data are available concerning their size and tonnage.

Table 25-11. Important SUV (="sandstone uranium-vanadium") regions hosted by continental sandstones and conglomerates

LOCALITY	HOST UNIT	MINERALIZATION	REFERENCES
Grants region, New Mexico, U.S.A.	J_3, volcanogenic (sanidine-bearing) sandst. interbedded with bentonitic mudstone, minor gypsum, limestone	J_3-Cr_1 peneconc., tabular "trend", less "roll" coffinite etc. impregnation deposits, assoc. with infiltrated org. matter. Min.120 Tt U/0.19%, 50 Tt V; est. 390 Tt U.	Kelley et al. (1968); Moench and Schlee (1967)
Uravan field, W. Colorado, U.S.A.	J_3, red to gray qtz. sandstone with carb. wood trash, fill channels in mudstone	sooty uraninite, coffinite, V-clays, montroseite, carnotite; rolls, trends, sheets. 18 Tt U/0.23% (38 Tt whole belt), 65 Tt V.	Motica (1968)
Lisbon Valley, 60 km S.E. of Moab, Utah	Tr, qtz. and feldsp. sandst. with calcar.cement, fills channels above unconformities	uraninite, less coffin., roscoelite, impregn. sandst. with wood trash. Tabular, amoeba orebodies. 22 Tt U/0.3%, 8 Tt V/0.1%	Wood (1968)
Monument Valley and White Canyon, S.Utah and N.Arizona	Tr, qtz. and feldsp. sandst. and congl. with coal trash, fill channels in siltst. and mudstone	zoned uraninite, coffinite, montroseite, corvusite, etc. infiltr. rods, patches, sheets in paleochannels. 9 Tt U/0.25%, 7.2 Tt V, 25 Tt Cu	Lewis and Trimble (1959)
Rifle Creek, W. Colorado, U.S.A.	J, light gray to red qtz. or feldsp. sandstone interbedded with mudstone	roscoelite and montroseite pore filings in altered sandst., tabular peneconc. layers along a 2.5 km long trend. 2,350 t V/7%	Fischer (1960)

Gas Hills, E. of Riverton, Wyoming, U.S.A.	Eo_1,lenticular alluvial coarse granite derived congl. to arkose fan, intertongued with tuffaceous mudstone	22-24 m.y.,uraninite,coffinite, pyr.,jordisite, etc.impregnations, in roll and blanket orebodies. Min. 30 Tt U/0.17%	Soister (1968)
Shirley Basin, S.of Casper, Wyoming, U.S.A.	Eo_1, arkosic sandst.separated by siltst. and bentonitic mudstone	18 m.y.,uraninite,pyrite,etc.as impregn. along edges of roll solution fronts. 120 Tt U/0.34%	Harshman (1972)
Highland mine, Powder R. Basin, Wyoming, U.S.A.	Pc,Eo, feldspathic sandst.,3 stacked channel fills enveloped by tuffaceous mudstone	Ol-Pl,finely dispersed uraninite, coffinite in multiple rolls in sandst. Striking colour of alteration. 20 Tt U/0.17%	Dahl and Hagmaier (1974)
Black Hills fringe, Wyoming and South Dakota, U.S.A.	Cr_1 fluvial channel sandstone interb. with mudstone; coal trash	dominantly second. U-V minerals (carnotite,tyuyamunite,etc.) and relicts of uranin.,coffin.,impregn. pods,patches. 2.75 Tt U	Hart (1968)
Crow Butte near Crawford, Nebraska	Ol, channel fill tuffac. and feldsp. sandst. interbedded with tuffac. mudstone	uraninite grain coatings, impregnations in channel sandstone	Chenoweth (1982)
Karnes, Live Oak, Duval Counties, S.Texas Coastal Plain, U.S.A.	Eo-Pl,friable fluvial,deltaic to beach, qtz. and tuffac. sandst., mudst., with marine interbeds	Mi-Pl,uraninite, coffinite grain coatings and impregnations in roll and tabular bodies. 8,160 t U/0.14%	Galloway and Hobday (1983)
La Poma-San Carlos area, 140 km S.of Salta, N.Argentina	Cr_2 sandst. with coaly trash, mudstone, marl, dolomitic limestone	tyuyamunite,carnotite,autunite uraninite, penecon. impregnations in sandstone	Putzer (1976)

Table 25-11 (continued). Important SUV ("sandstone uranium-vanadium") regions hosted by continental sandstones and conglomerates

LOCALITY	HOST UNIT	MINERALIZATION	REFERENCES
Valle de Punilla, 60 km N.W. of Cordoba, Argentina	Eo, sandst., mudstone resting on Precambrian metamorphics	carnotite and tyuyamunite, peneconcordant impregnations in sandstone; 2,550 t U	Haude and Weber (1975)
Sierra Pintada, Mendoza, W.Argentina	Pe,Tr crossbedded sandstone, conglomerate, with tuffac. mudstone	uraninite impregnations in 10m thick peneconcordant orebody 16 Tt U	Putzer (1976)
Malargüe district, Mendoza, W.Argentina	Cr_3 faulted sandst., conglom., topped by Oligocene volcanics	pitchbl., chalcop., born., U,V,Cu oxides impregnate asphaltic pods in sandstone. 850 t U, 800 t V	Haude and Weber (1975)
Cambuí-Figueira, N. Paraná, Brazil	Pe black coaly shale, coal, siltstone, channel sandstone	(1) diagenetic carburan in shale; (2) impregn. uraninite, pyrite, jordisite, peneconc. bodies in channel sandstone. 4 Tt U, 2.5 Tt Mo	Daemon et al. (1982)
Hamr-Mimoň distr., N.W. Czechoslovakia	Cr_3 freshwater sandst. with coal trash interb. with mudst. on block-faulted cryst.basem. intr. by Mi alkal. volcanics	sooty uraninite, coffinite impregnations form peneconc. bodies in sandst., locally high Zr content. Estim. min. 25 Tt U	Ruzicka (1971)
Trento U region, Northern Italy	Pe_2, gray and green volc. arkose and congl. below a red-beds assoc. Resting on and derived from rhyolite	pyr., pitchbl. in sandst., local im- impregn. in coal trash; peneconcordant oreb. in gray sandst. under red-beds. Min. 2 Tt U	Mittempergher (1972)

Žirovski Vrh and Brodovlje, 30 km W. of Ljubljana, N. Yugoslavia	Pe_2 gray and red sandstone, conglom., siltst., slate. Thrusted allochthon, deformed	17 km long ore zone, impregn. of pitchblende as peneconc. lenses. 5 mineraliz. horizons, 20 m thick ore zone. 5 Tt U/0.05-0.2%	Lukacs and Florjančić (1974)
Mécsek Mts. (Uránváros), W. of Pécs, S. Hungary	Pe_2 volcanic arkose, green and gray sandst., siltst., congl., overl. Pe quartz porphyry (rhyolite) and PZ metamorphics	sooty uraninite, coffinite, galena, chalcop., fine dissem. in sandst. Peneconc. bodies in sections rich in vegetal remains. Min. 5,000 t U	Barabas and Kiss (1958)
Agadés Basin (Arlit, Akouta, Azelick), Niger	Cb-Cr_1 fluvial-deltaic sandstone, conglom. with coaly trash	uraninite, coffinite, Cu sulph. and oxides, form peneconcordant and roll impregn. in analcime-rich sandstone. 190 Tt U	Gangloff (1970)
Tono, Gifu Pref., Honshu Isl., Japan	Mi, tuffac. zeolitized arkose, sandst., congl., topping Cr-T_1 granite. Epicl. and pyroclastic sediments	U-calcite, U-zeolite, autunite, carburan, coffinite, uraninite, pyrite, etc. impregn. bodies in sandst. 4,500 t U	Doi et al. (1975)
Ningyo-Toge, Honshu S.W. Japan	Mi-Pl, epiclastic and volcanic sandst., arkose, conglom. fill channels in Cr-T_1 basement granite	autunite, ningyotogite, uraninite, pyrite form 12 peneconc. impregnation bodies within a 20x20 km area. 3,060 t U/0.05%	Doi et al., (1975)
Lake Frome U district (Beverley Prosp.), S. Australia	Mi, coarse lacustrine sand, carbonaceous clay, silt	finely dispersed uraninite impregnates lentic. sand bodies enclosed in carbonaceous clay. 13,430 t U/0.2%	Haynes (1975)

25.8. RED AND VARICOLOURED BEDS ASSOCIATION

25.8.1. General

"Red-beds" is a traditional and popular lithofacies term that has been reviewed comprehensively and is covered by an extensive reference literature (Anatolyeva, 1972; Van Houten, 1973a; P. Turner, 1980). "Typical" red-beds are hematite-pigmented continental (alluvial, aeolian, less deltaic, lacustrine) sediments in which the presence of detrital feldspar indicates incomplete weathering in the source area. This distinguishes the "true" red-beds from other red coloured sediments, such as red pelagic shales, slates and limestones; jaspers and a variety of hematitic ironstones; lateritic weathering crusts; thermally metamorphosed sediments in the vicinity of burning coal beds and others.

Few "red-beds" terrains are uniformly red. The content of the hematitic pigment varies and is usually proportional to the clay mineral content. Consequently, mudstones, siltstones and fine grained sandstones with a clay matrix are the darkest, whereas well sorted, usually thick-bedded quartz arenites are light red or mottled, yellow or white. In addition to the colour variation due to the abundance of pigment in uniformly oxidized rocks, the other reason for colour variation is the oxidation state of the pigment. Light green, gray, beige and mottled rocks are usually the consequence of diagenetic or post-lithification reduction. This has been reviewed in the previous section. Light-brown (limonite pigmented) rocks are usually recently surficially oxidized, formerly reduced rocks, because the red, hematite pigmented sediments are remarkably resistant to hydration in outcrops.

The origin of red-beds has been a much discussed controversy, reviewed in detail in the references quoted above as well as in Glennie (1970), Walker (1975) and elsewhere. At present, at least in North America, the opinion of the majority seems to favour the Walker (1975) interpretation of authigenic hematite pigment formation by intrastratal breakdown of the rock-forming ferrous minerals during diagenesis in arid basins. Although in this interpretation the emphasis is on the sedimentary basin rather than the climate and source area as in the other models, an arid source area such as the present Sonoran rocky desert is implied. It is most compatible with the observed facts (e.g. inhibited chemical weathering in the source area enabling the unstable silicates to accumulate in the depositional basin; inhibited plant growth, reflected in the lack of organic matter; oxidizing interstitial waters).

Lithologically, the red-beds include a broad spectrum of sediments. The first cycle (i.e. derived from crystalline bedrocks) epiclastic rocks, particularly the source-proximal ones (e.g. the "wash", a proximally reworked profile of weathering and bedrock disintegration) are lithic or arkosic in composition, poorly sorted (matrix-rich) conglomerates to sandstones, reflecting the bedrock composition. Sorting and maturity improves and quartz content increases with the length of transportation. Red-beds laid down some distance from their source are mostly sublitharenites or subarkoses alternating with mica-rich siltstones and mudstones. Recycled sediments derived from earlier sedimentary bedrocks as well as highly mobile sediments with very distant sources (e.g. dune sands) are quartz-rich, up to quartz arenites. Most red-bed sequences also contain thin units of continental carbonates and units of evaporites. Volcanic components, mostly present as airborne ash, are very common in red-beds because the arid climate inhibits rapid ash decomposition. In addition to this, felsic or mafic terrestrial volcanism contemporary with the red-beds deposition is common and is reviewed in some detail in the following section.

25.8.2. Tectonic setting

Red-beds occur (1) in the sedimentary cover of stable, atectonic platforms; (2) in continental margin orogenic belts; (3) in regions of block faulting, crustal extension and rifting, superimposed on (1) or (2). The last setting seems to be particularly characteristic and also the best mineralized. It is this setting in which are located many of the present environments considered to be modern counterparts of the ancient red-beds (e.g. Baja California, Dead Sea, Gulf of Suez, Red Sea region; compare also Chapter 30). Active tectonic regimes are considered necessary to maintain a steady supply of the fresh, first cycle detritus (in contrast to peneplains where reworking with little new detritus supply is the rule); to support rapid interfingering of facies; to facilitate burial and preservation; to channel groundwaters; to control magmatism, etc. Narkelyun et al. (1962) investigated the tectonic position of the cupriferous sandstones and shales that included members of the red-beds association, and concluded that virtually all were associated with extensional faulting.

25.8.3. Trace metal geochemistry and metallogeny

Despite the vivid colours, red-beds are either not enriched in iron at all or are only slightly so. Some tend to have slightly increased trace Mn,Cu,Pb,U, etc. contents. This, however, appears to be strongly influenced by heritage of the source rock composition. Others are depleted in these metals with respect to the detrital source and there is usually clear evidence of diagenetic and later intrastratal migration of elements in groundwaters moving through permeable aquifers.

Red-bed sequences contain about 13 Mt Cu, 7 Mt Pb, 3 Mt Mn, 1 Mt Zn, 900 Tt Ti, 1 Tt Ag and lesser quantities of U, V, Sn and other metals in presently economic ore deposits. With the exception of Dzhezkazgan (Cu,Pb,Zn) which is a "world class" metal accumulation, the rest of the deposits are mostly small. Genetically, virtually all the mineralized occurrences are members of a single category of epigenetic impregnation deposits formed in porous sediments at "redox" or "solution front" interfaces. Consequently, there is a high degree of analogy between the Cu,Pb,Zn, etc. and the U,V deposits described in the previous section. All the base metal deposits occur exclusively within the "gray" or "green" reduced horizons or patches. The red rocks are uniformly barren except for minor Mn accumulations, so that they can be by-passed during exploration. The red rocks, however, may still contain detrital metallic resistates. Several Cu,Ag,Pb, etc. deposits mimic the setting of U(V) mineralizations and occur in the same units; for example, the Nacimiento, New Mexico copper deposit and the Silver Reef, Utah, silver deposit are both hosted by a coaly trash-rich section of reduced Triassic sandstones.

The major difference between the U(V) and Cu infiltration deposits is the additional need for sulphur as precipitant of Cu sulphides; different metal sources (mafic rocks); different composition of the metal-bearing brines (sulphate rich, Gustafson and Williams, 1981) and possibly prevalent upward direction of movement of the copper bearing brines, as compared with the downward movement of uraniferous fluids.

Marine facies sediments immediately topping red-beds or grading into them laterally are important hosts to bedded Cu,Pb,Zn,Ag, etc. deposits such as the "Kupferschiefer". These have been treated in Chapter 19.

25.8.4. Mineralization styles

DETRITAL GOLD (PALEOPLACERS)

Several small gold placers are hosted by basal Mississippian red beds conglomerates, transgressing over the Meguma Group basement gold province, Nova Scotia (Van de Poll, 1978). Similar occurrences are located in the Permian red—beds, Laba River Basin, N. Caucasus, U.S.S.R. Near Paria, Utah (Colorado Plateau) Butler et al. (1920) recorded an extremely fine-grained gold in varicoloured Triassic claystones, topping the pebbly channel sandstones of the Shinarump Member (a common host to the SUV ores). The gold content ranged from about 0.1 to 0.15 ppm and the gold was very pure. The source and origin are unknown.

DETRITAL OR PRECIPITATED Fe AND Ti OCCURRENCES

Basal conglomerates of the low-grade metamorphosed Permian Verrucano facies near Rudňany, Slovakia (Ilavský, 1976) contain small hematite-clast sandstone and conglomerate bodies derived from hydrothermal siderite deposits in the basement. Van de Poll (1978) reported small lenticular and tabular bodies of specularite and red hematite from the base of the Mississippian fluvial sequence in the Cape Breton Island, Nova Scotia.

Ikonnikov (1975) reported 20 Mt of detrital ilmenite in alluvial paleoplacers in Tertiary red-beds in Luanping district (S.Hupeh, China).

DIAGENETIC Cu SULPHIDES IN BLACK, GRAY AND GREEN MUDSTONES AND CARBONATES

Reduced lacustrine and paludal mudstones, marls and marly limestones (dolomites) are sometimes interbedded with red-beds. They may contain diagenetic copper sulphides in the form of scattered nodules ("Vyatka type" of Lur'ye, 1973). This association and its ores are often indistinguishable from the marine "copper shales". In the type area in the western Cisural, U.S.S.R., covellite, digenite, chalcocite and pyrite nodules are sparsely scattered (6-8 kg nodules per 1 m^2, i.e. about 0.12% Cu) in Permian mudstones along the facies transition from dark-gray to reddish-brown sediments.

Virtually identical but marginally higher-grade zoned chalcocite, bornite and pyrite nodules were mined from beds of Permian greenish-gray calcareous claystone enveloping a conglomerate marker bed near Horní Vernéřovice, N.W. Czechoslovakia (Čadková, 1971). A variety of rare Cu,Ag,Hg,V,Ge and other minerals have been described from mineralogical occurrences in a similar setting, worldwide (e.g. Cu, native amalgam and roscoelite, S.W. New Brunswick, Van de Poll, 1978; native silver, vésignièite, cuprite; Horní Kalná, Czechoslovakia). Considered separately, the lacustrine intercalations appear to be a style of little importance economically.

The enigmatic occurrence of nodular stibnite in sandy, gypsiferous mudstone intercalated with Tertiary red-beds in Antimony Canyon, Utah (Fig. 25-26), may be of similar origin.

EPIGENETIC INFILTRATIONS OF Cu MINERALS IN POROUS SANDSTONES AND CONGLOMERATES ("Cu sandstone-type")

This is the best-known mineralization style hosted by the red-beds, credited with some 13.2 Mt Cu. The common, transitional style can be illustrated by the small, economically unimportant Red Gulch occurrence near Cotopaxi, Colorado (Emmons, 1905; Fig. 25-27). There, diagenetic chalcocite nodules and impregnations are found in gray Pennsylvanian carbonaceous shale with abundant coal chips and small lenses (a paludal sediment) immediately topping a red, polymictic, poorly sorted conglomerate "wash". The latter is gradational into a Proterozoic granite and amphibolite regolith on one side, and an alluvial channel sandstone and conglomerate on the other. The more porous sections of the red conglomerate contain infiltration veinlets, disseminations and cement of chalcocite surrounded by light colour reduced haloes. The Cu source is in the basement which is known to contain several small Cu-Zn massive sulphide deposits.

The economically more important Nacimiento deposit near Cuba, New

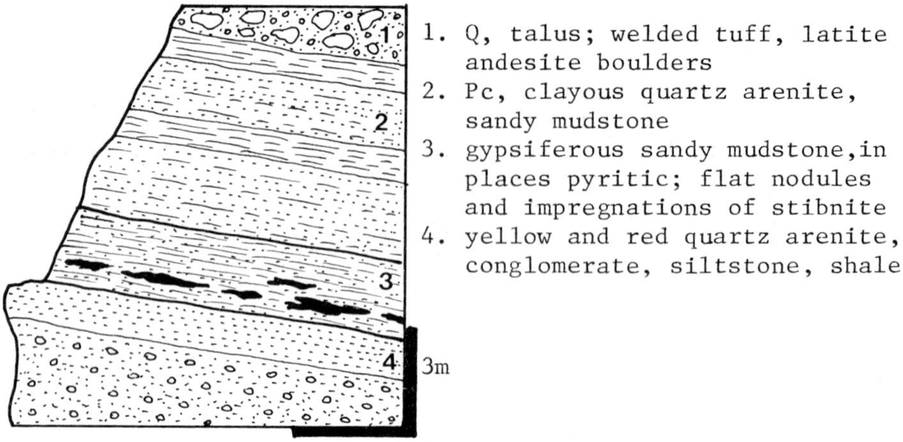

Fig. 25-26. Antimony Canyon, Utah, a stratiform deposit of nodular stibnite. From LITHOTHEQUE.

Mexico (Woodward et al., 1974; 71.5 Tt Cu/0.65%; Fig. 25-27), has chalcocite and minor pyrite partly replacing and partly cementing carbonaceous trash and coalified logs, resting in a permeable white channel sandstone sandwiched between red siltstone and argillite. The host unit is the Triassic Chinle Formation.

The Corocoro copper field hosted by Neogene red—beds on the Altiplano, N.W. Bolivia (Ljunggren and Meyer, 1964; 774 Tt Cu/1.3-5%; Figures 25-27, 25-28), contains two mineralogically different styles subdivided by a fault. The older, Oligocene Totora Formation east of the Corocoro fault, is composed of fine, maroon sandstones alternating with shales and local bedded gypsum. There are some reworked tuff beds, and native copper forms several parallel peneconcordant impregnation bodies in the sandstone. The younger Caquiaviri Formation (late Oligocene-lower Miocene) starts with a mudstone with some gypsum nodules, topped by a reduced, light-coloured arkose, lithic arenite, shale (ore bearing unit) that, in turn, are topped by red (oxidized) mudstone to conglomerate. The orebodies are numerous peneconcordant lenses of interstitial rhombic chalcocite with minor galena, sphalerite and domeykite in the gray arkose. The ore emplacement predates faulting and probably formed by a proximal redeposition of copper from the sedimentary detritus that had a provenance in the hypogene Cu-rich source area (a porphyry copper province) along the western fringe of the Altiplano in Chile. In Chacarilla (95 km S.E. of Corocoro; Ahlfeld, 1967; 1.2 Mt Cu/3.4%), a "Corocoro-type" Cu mineralization occurs in several mantos on both sides of a steep anticline, considered by some to flank a diapir.

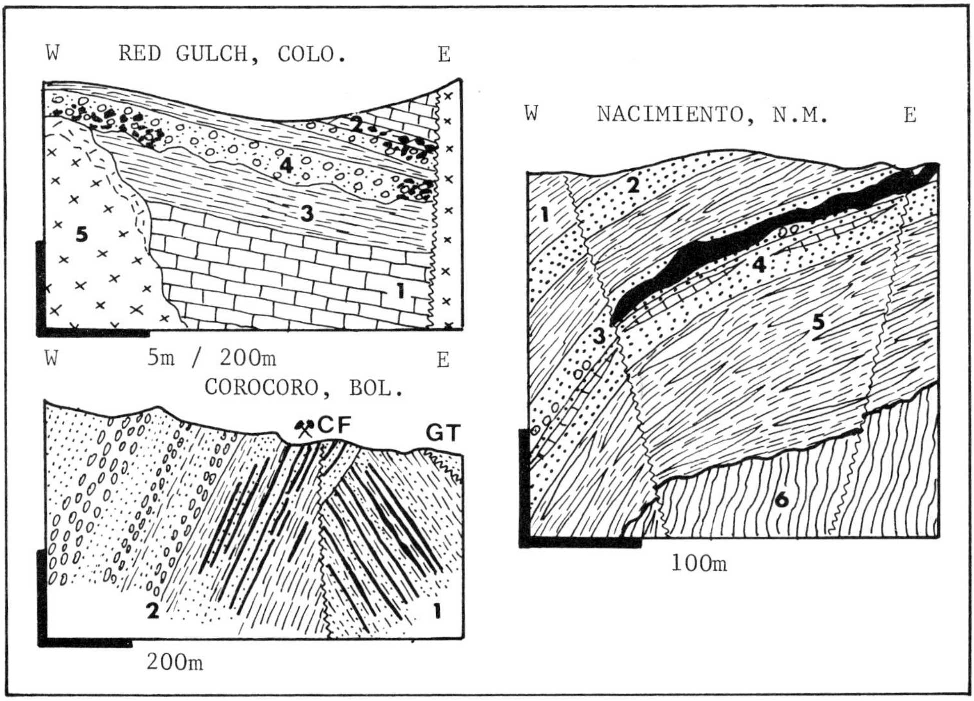

Fig. 25-27. Epigenetic infiltration deposits of Cu minerals in porous sandstone or conglomerate.

Red Gulch, Colorado (diagrammatic); Ps sediments: 1=limestone; 2=dolomitic shale and sandstone, local chalcocite nodules and impregnations; 3=red and green shale; 4=poorly sorted lithic conglomerate to sandstone with disseminations and bunches of chalcocite, malachite coatings; 5=Proterozoic granite, granite gneiss, greenstone rafts.

Nacimiento near Cuba, New Mexico. Triassic 1=red and green shale; 2=sandstone with coaly trash; 3=white to buff sandstone and channel conglomerate with coaly trash. Permian: 4=fine red sandstone, limestone lenses; 5=red mudstone, arkose lenses. 6=PCm metamorphics.

Corocoro, Bolivia. 1=Ol Totora Fm., fine maroon sandstone alternating with shales, minor gypsum; contains native Cu orebodies. 2=Ol_3-Mi_1 alternating red and green mudstone, gray, white and red arkose and litharenite, conglomerate. Contain chalcocite orebodies. CF=Corocoro Fault; GT=Guallatiri Thrust.

From LITHOTHEQUE, based on Emmons (1905); Woodward et al. (1974); Ljunggren and Meyer (1964); COMIBOL Corocoro Staff, guided tour and maps, 1977.

Fig. 25-28. Corocoro Cu field, Altiplano, Bolivia. TOP: general view looking N.W. towards the snow-covered Quaternary stratovolcanoes and parallel with the Corocoro fault. Cu-sulphide orebodies ("Vetas") are on the left side of the valley. The native copper orebodies ("Ramos") are on the right. BOTTOM: oxidized ore zone at Viscachani.

The Dzhezkazgan district (Fig. 25-29), a giant metal accumulation in central Kazakhstan, U.S.S.R. (estimated content are 10 Mt Cu in a 70 km long belt having an area of 600 km^2; Magak'yan, 1968; Bogdanov et al., 1973; Samonov and Pozharisky, 1974), is the largest Soviet copper producer. It is hosted by a 700m thick, transitional (marine to continental), middle-upper Carboniferous gray- and red-beds sequence and it is often quoted as the closest Phanerozoic equivalent to the Proterozoic Zambian Copperbelt. The ore-bearing unit is a

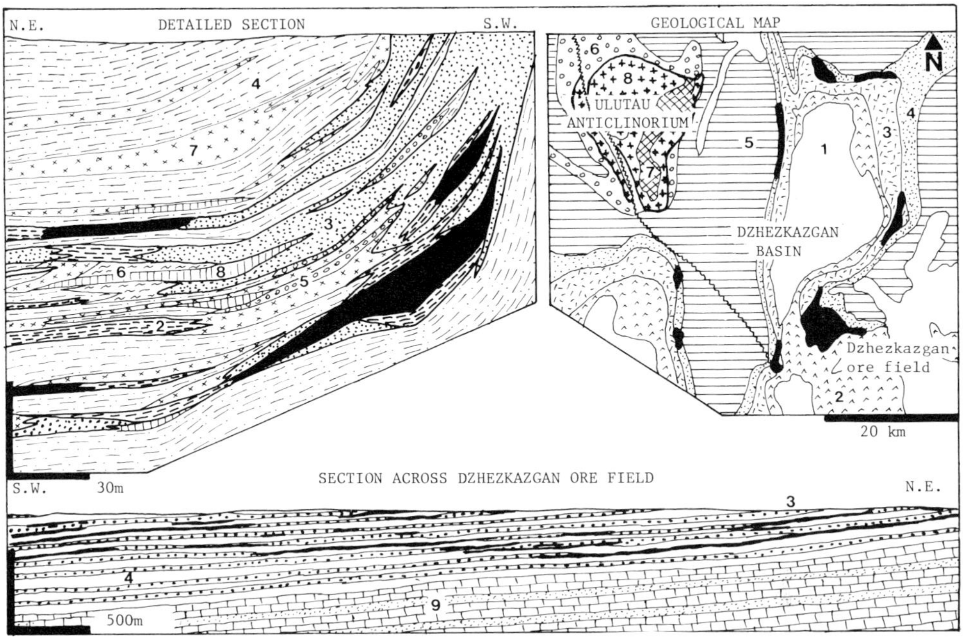

Fig. 25-29. Dzhezkazgan Cu(Pb,Zn) ore field and mineralized region, central Kazakhstan, U.S.S.R. DETAILED SECTION (all rocks Cb_{2-3} Dzhezkazgan Suite): Black=high-grade ore in shale and siltstone; 2=low-grade ore; 3=low-grade disseminated ore in sandstone; 4=red sandy shale; 5=conglomerate; 6=green shale, mudstone; 7=arkose; 8=chert. GEOLOGICAL MAP: 1=T platformic sediments; 2=Pe red-beds; 3=Cb_{2-3} Dzhezkazgan Suite, gray and red-beds; black=ore-bearing sections; 4=Cb_{2-3}, gray and red-beds; 5=Cb_1 terrigenous and carbonate deposits; 6=D_{2-3} red-beds; 7=pre-Or ultramafics; 8=PCm and Cm-Or crystalline basement. LONG SECTION ACROSS DZHEZKAZGAN FIELD (Cb_{2-3} Dzhezkazhan Suite). Black=high-grade orebodies in shale and siltstone; 3=gray sandstone with low-grade disseminated Cu ore horizons; 4=red sandy shale; 9=Cb_1, sandstone and limestone.

Modified after Magak'yan (1968), Pustovalov, ed. (1965) and Bogdanov et al. (1973).

member of a Devonian to Permian "molasse" situated on the flanks of a Precambrian crystalline block uplift, and partly covered by Permian sediments of an interior basin. The setting is, in its tectonic style, reminescent of the Wyoming "foreland".

The productive Dzhezkazgan Suite consists of rhythmically alternating gray and red lithic sandstone unit, grading into siltstones and sandstones with minor conglomerate channels. There are at least 51 rhythms, each 2-42 m thick. Druzhinin (1973) interpreted the paleo-environment as an alluvial coastal plain (red members) grading to paralic deltas and lagoons and shallow marine offshore (gray-beds). He attributed most of the deposits to the regressive phase. The detritus appears to be mostly epiclastic, with a minor felsic pyroclastic contribution. The sediments are lightly open folded and block faulted. The orebodies are hosted by the "gray" units, and there are 26 stratabound ore horizons consisting of up to 2 km long and up to 18 km thick lenses. Zonally arranged chalcopyrite, bornite and chalcocite are disseminated in the gray litharenite and shale, and there are separate galena and light-coloured sphalerite mineralized bodies stratigraphically above the copper orebodies and in terms of the paleo-environment, on the "offshore" side. The Cu ores have a high rhenium content, the rhenium residing in bornite and dzhezkazganite.

The origin of Dzhezkazgan is controversial and interpretations range from syndepositional-diagenetic through groundwater-epigenetic (infiltrational) to hydrothermal-epigenetic. Some cross-cutting veins are present, now mostly attributed to post-ore remobilization. It is notable that Dzhezkazgan is contemporary in origin with the porphyry copper deposits in the Lake Balkhash region, several hundred kilometres to the east.

Additional examples of the "copper sandstone" deposits have been assembled in Table 25-12.

EPIGENETIC DISSEMINATED Pb-Zn MINERALS IN POROUS SANDSTONE OR CONGLOMERATE ("SANDSTONE Pb OR Zn")

Disseminated deposits of galena and minor sphalerite in marine sandstone and quartzite (e.g. Laisvall) have already been treated in Chapter 19. Similar deposits also occur in continental sandstones that are direct or transitional members of the red-beds association. Although of low to very low grade, these deposits summarized by Bjørlykke and Sangster (1981) together represent some 6.5 Mt Pb, 800 Tt Zn and 700 t Ag.

The L'Argentière deposit located on the south-eastern fringe of the Massif Central, France (303 Tt Pb/3.7%, 83 Tt Zn/0.6%, 164t Ag/20 ppm; Fig. 25-30) has recently been interpreted genetically by Samama (1968). There, galena, sphalerite and minor pyrite, bornite and chalcopyrite occur, unevenly distributed in the cement of Triassic continental feldspathic arenites, and as small masses, breccia cement and fracture veinlets in the overlying dolomites. Five broadly stratabound horizons are developed, and Samama (1968) stressed their position at the interface of a Triassic alluvial fan (or bajada) flanking the trace Pb-enriched Hercynian crystalline basement, and the carbonate-evaporite lagoonal basin. The ore zone is accompanied by a fringe of silicification of the arenite cement.

By far the largest lead accumulation in continental sandstones is hosted by the lower Triassic white sandstone topped by red sandstone and mudstone along the northern fringe of the Eifel uplift in West Germany. There, the larger, 10 km long S.W.-trending

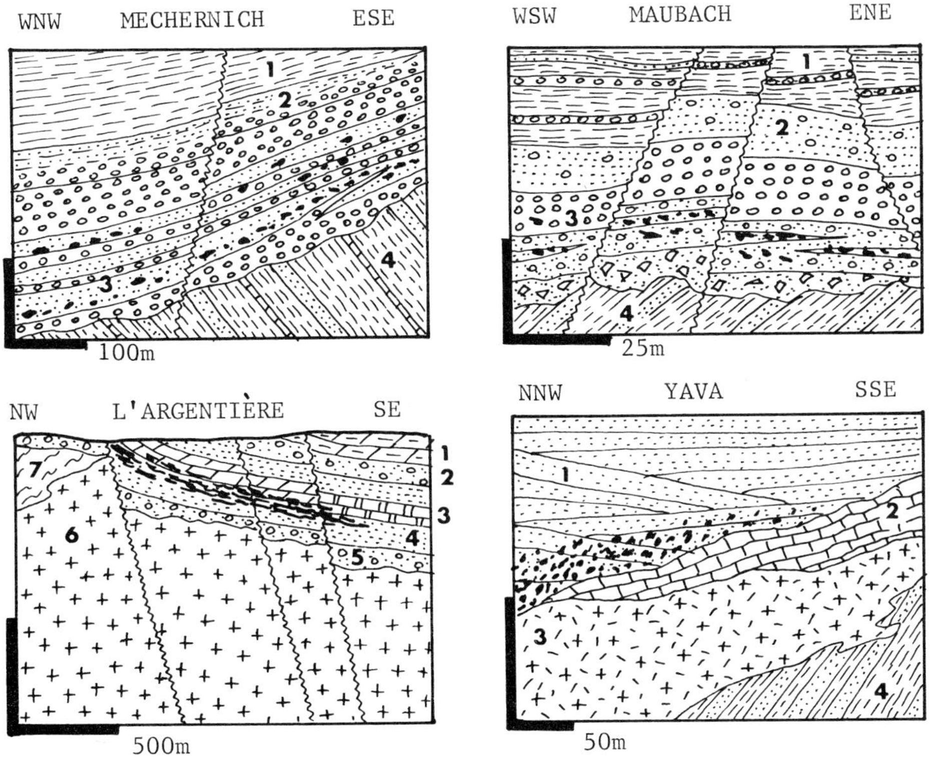

Fig. 25-30. Examples of arenite-hosted lead deposits. (1) Mechernich, West Germany: 1=Tr_{1-2} red mudstone, minor sandstone; 2=Tr_1 pink to red conglomerate and argillaceous quartz arenite; 3=white ore-bearing conglomerate and sandstone; 4=D_2 slate, sandstone, limestone. (2) Maubach, West Germany (legend identical with Mechernich). (3) L'Argentière, France: 1=J_1 marly limestone resting on sandstone and conglomerate; 2=Tr arkose, sandstone; 3=dolomite, limestone, shale change facies to anhydrite and gypsum; 4=conglomerate to mica sandstone topped by sandy shale; 5=Pe, red basal conglomerate to siltstone; 6=Cb_3 red porphyritic granite; 7=PZ gneiss, micaschist. (4) Yava (Silvermine), Solomon River, Nova Scotia. 1=Cb_{2-3} conglomerate, sandstone, siltstone, coaly trash; 2=Cb_1 limestone, minor anhydrite and gypsum; 3=D_2 pink felsite to marginal facies of biot. quartz monzonite; 4=sandstone, siltstone, quartzite.

From LITHOTHEQUE, based on data in von Behrend (1950), Schröder (1954), Samama (1968) and Bjørlykke and Sangster (1981).

Table 25-12. Peneconcordant and stratabound Cu-sulphide (oxide) impregnated sandstones ("Cu-sandstone" style); selected example localities

LOCALITY	HOST UNIT	MINERALIZATION	REFERENCES
Nacimiento mine near Cuba, New Mexico, U.S.A.	Tr_3 contin. gray arkosic sandstone rich in coaly trash, sandwiched between red siltstone and argillite	chalcocite, minor pyr., impregn. in sandst. and organic detritus. Oreb. is a block faulted peneconc. lens. 71.5 Tt Cu/0.65%	Woodward et al. (1974)
Corocoro, Altiplano N.W. Bolivia	Ol-Mi_1, brown and green-gray lithic sandst. interb. with reworked tuff, shale and gypsum; topped by red-beds, fault.	2 sets of multiple peneconc. ore lenses subdiv. by a fault. 1. Nat. copper impregn., 2. chalcocite impregn. 774 Tt Cu/1.3-5%	Ljunggren and Meyer (1964) Ahlfeld (1967)
Chacarilla, Altiplano, Bolivia	Ol-Mi, as above. Folded, faulted	3 mantos with dissem. native Cu, chalcoc. in litharenite, on both flanks of an anticline. 1.2 Mt Cu/3.4%	Ahlfeld (1967)
Škofje, Cerkno, etc. Slovenia, N.W. Yugoslavia	Pe_2 interbedded red and gray quartz and feldsp. sandst., congl., siltst., shale	bornite, chalcoc., chalcop. form stratabound impregn. bodies in gray sandst., up to 750 m long. 23.3 Tt Cu/0.97%	Drovenik et al. (1972)
Western Cisural (Kargalin Group), U.S.S.R.	Pe_3 fluvial channel sandst. to conglom. incised into red marls	chalcoc., bornite, malachite impregn. sandst. and replace coaly trash. 7,500 occurrences, old prod. 600 Tt Cu, 150 t Ag	Lur'ye (1973)
Donetsk Basin (e.g. Kartamysh deposit) U.S.S.R.	Pe_1 rhythmic red argillite, siltst. and sandst. interbed. with gray sandstone	chalcoc., chalcop., born., pyr., dissem. in sandst. or replace coaly trash in gray sandstone. 31 showings, estim. 50 Tt Cu	Lur'ye (1973)

Pri-Dniester dist. Russian Platform, U.S.S.R.	D_1 interbedded red and gray sandstone, siltstone, argillite	chalcoc., lesser chalcop., born. impregn. in channel sandstone. Up to 6.5 m thick, 1 km long lenses	Bogdanov et al. (1973)
Dzhezkazgan distr. W.C. Kazakhstan, U.S.S.R.	Cb_{2-3} rhythmic, 700 m thick assembl. of marine-deltaic-lagoonal gray litharenite and shale, and red continental sandstone and shale	26 stratigr. horizons contain ore beds and lenses of gray shale and litharenite with dissem. chalcocite, bornite, chalcop. Zonally arranged Pb-Zn sulphides. Estim. 10 Mt Cu, also Pb, Zn, Ag, Re	Bogdanov et al. (1973) Druzhinin (1973)
Naukat district, N.Ferghana Depres. Ctr.Asia, U.S.S.R.	T_{2-3} red marls and litharenites altern. with gray arkose, minor contin. limestone	nat.copper, cuprite, concret. and impregn. in dark feldsp. sandst. with calcar. cement. Zone 20x3 km, 200m thick, 7-24 ore horizonts	Bogdanov et al. (1973)
Minussinsk Depres. (Bazinskoe, etc.) S.Siberia, U.S.S.R.	D_{1-2} red sandst., arkose, congl. limnic shale, marl, overlie explosive volcanic basement	detrital and diagenetic chalcop. and born. in sandstones; 11-14 ore horiz., each 0.5-3 m thick, 1.3 km long. Est.min. 100 Tt Cu	Bogdanov et al. (1973)
Merija, Tansrift, Aïn-Sefra, Morocco	Cr_1 lagoonal-deltaic congl., sandst., siltst., shale	chalcoc. dissem. in shale with plant remains, impregn. in sandst.	Caïa (1976)
E. Antiatlas (e.g. Talate n'Ouamane) Morocco	Pt_3-Cm volc. sandst. and congl. with andes. and rhyol., rest on Pt_3 greenstone and granite	chalcop., chalcoc., born. dissem. and concretions in sedim. fill of paleochannels; 18 Tt Cu/1.8%	Caïa (1976)
Cachoeiras do Binga, Angola	Cr_1 red sandst., white sandst. with vegetal remains, overl. by evaporite, rest on PCm basem.	chalcop., born., digenite, covellite impregn. sandst. in peneconc. lenses 1-2 m thick, in paleochannels. 140 Tt Cu/2%	Caïa (1976)

Mechernich-Kommern ore zone (von Behrend, 1950; Schröder, 1954; 3.51 Mt Pb/1.56%, 675 Tt Zn/0.3%, 428 t Ag; Fig. 25-30) contains scattered galena grains and groups of grains ("Knotten") replacing cement and filling intersticies in a white, slightly argillaceous, quartz arenite. In the nearby Maubach (von Behrend, 1950; 240 Tt Pb/2%, 96 Tt Zn/0.8%, 24 Tt Co/0.2%, 18t Ag; Fig. 25-30) the mineralization is more irregular and groups of fine galena, minor sphalerite and chalcopyrite grains are to be found in several indistinct horizons of light quartz arenite. The host fluvial sandstones are located near the base of Triassic succession, unconformably overlying Devonian slate, litharenite and impure limestone in what is a distinct polygenetic Zn-Pb province also containing basement vein deposits and MVT ores in Devonian-Carboniferous carbonates.

The other German region of "sandstone lead" occurrences lies in Triassic sandstones in the Oberpfalz area of Bavaria (Freyhung; Gudden, 1975a; 200 Tt Pb/2%). There, elevated trace Pb content and scattered groups of galena, cerrusite or pyromorphite are distributed over an area of 1,700 km^2, confined to a stratigraphic thickness of 150m. The mineable occurrences are confined to a fracture zone and are probably a product of remobilization.

The Yava or Silvermine deposit, Salmon River, Nova Scotia (1.43 Mt Pb/2.09%, 60t Ag/0.81 ppm; Bjørlykke and Sangster, 1981; Fig. 25-30) is the best-known Canadian locality. There, a very fine, indistinct galena impregnates a Pennsylvanian white to brown feldspathic and micaceous quartz sandstone to conglomerate rich in plant trash, overlying a thin Mississippian carbonate-evaporite unit or Cambrian felsite, granite and metasedimentary basement.

The last significant "sandstone Pb" locality, the Zeida-Bou Mia field near Midelt, Morocco (Schmitt and Thiry, 1977; 713 Tt Pb/3.2%) has disseminated galena and cerussite hosted by a Triassic arkose and conglomerate, filling paleochannels in a regolithic Hercynian granite basement.

Additional occurrences are known among the lower Cretaceous sandstones in Morocco and Gabon (e.g. Bou-Sellam, Kroussou; Caïa, 1976); from Triassic sandstones along the N.W. tip of Madagascar; from the Bakhmut depression, Donetsk Basin, U.S.S.R.; in Permian sandstone at Rio Bavaro near Bolzano, Italian Alps (Brondi et al., 1973); in Cerro Sayari, Bolivia; Salina, Utah and elsewhere. The ores hosted by the basal Lamotte Sandstone in the S.E. Missouri district transitional upward into the carbonate-hosted MVT deposits (Chapter 20), are closely comparable with the "sandstone Pb" deposits in composition and style.

DISSEMINATED SILVER DEPOSITS IN CONTINENTAL SANDSTONE ("SANDSTONE Ag")

This rare mineralization style is represented by a single locality, the Silver Reef field N.E. of St. George, S. Utah (204 t Ag/500 ppm; Proctor, 1953). There, irregular elongate peneconcordant orebodies are hosted by the fluvial, upper Triassic Chinle feldspathic quartz arenite rich in plant debris, and to a lesser degree by interbedded conglomerate and mudstone. Cerargyrite and rare argentite coat sandstone grains, accompanied by malachite, azurite, carnotite and other rare minerals. The ore probably precipitated along a "redox" front, and the silver source is unknown.

DISSEMINATED CASSITERITE IN CONTINENTAL SANDSTONES

The enigmatic Colavi tin deposit 40 km N.N.E. of Potosí, Bolivia (Ahlfeld and Schneider-Scherbina, 1964) contains cassiterite accompanied by pyrite, ankerite, siderite and barite replacing calcite cement in a lower Cretaceous sandstone, forming a conformable manto 1-1.5 m thick and 2 km long. The manto is located in a 6m thick basal unit of a white massive crossbedded calcite-cemented sandstone and conglomerate, grading upward into red sandstones and shales. The Cretaceous sediments unconformably top Ordovician slates and are locally intruded by Tertiary rhyodacite sills. The genetic interpretations offered range from a modified paleoplacer to a hydrothermal replacement.

25.9. "VOLCANIC RED-BEDS" : CONTINENTAL SEDIMENTS INTIMATELY ASSOCIATED WITH CONTEMPORARY VOLCANICS

25.9.1. Introduction

Red-beds lithofacies is commonly associated with, or overlaps, products of contemporary continental volcanism. In the two previous sections, the involvement of the late orogenic ("subsequent") explosive andesite, dacite, rhyolite volcanism was frequently mentioned, mostly as a supplier of the ash diagenetically converted into bentonite or bentonitic shale. Red-beds depositing basins formed in extensional (taphrogenic) systems superimposed on orogenic belts or on previously stabilized platforms and grading into "rifts" (Chapter 30), are usually filled by sediments overlapping with mafic (basalts), felsic (rhyolites) or mixed (bimodal) continental volcanics. The volcanics alone and their metallogeny are reviewed in Chapter 31, but the volcanics and sediments are often intimately intermixed and cannot be separated. These are treated here.

25.9.2. Mafic volcanics and red-beds

Interaction of contemporary continental basalts and sediments (as well as the initiation, mixing and ultimate merging of red-bed sediments having a dual provenance (1) epiclastic, in granitic and metamorphic basement; (2) volcaniclastic, in contemporary basalts), can be observed clerly in the superb exposures of the Cretaceous Salta Group in the Quebrada Cafayete and its environs in northern Argentina

(Reyes and Salfity, 1973; Figures 25-31, 25-32). There, the basal (Subgroup Pirquo) epiclastic lineage rests on crystalline regolith, and starts with a matrix-rich arkose and conglomerate ("granite wash") changing facies into a better—sorted red feldspathic arenite and siltstone, quartz arenite and quartzite, mudstone and shale.

The mafic volcanics are represented by numerous olivine to locally picritic basalt massive, amygdaloidal to scoriaceous flows and sparse feeder dikes. Massive flows and dikes are dark green, scoriaceous flowtops brown. Some flows were emplaced under water (lakes, streams), where they expanded to form fractured flows and autoclastic breccias, changing into granulates, pépérites, proximal conglomerates and volcanic arenites. The volcanic sediments mix with the contemporary epoiclastic sediments and rapidly lose identity several tens to hundreds of metres from their source. Basalts and the proximal volcanogenic sediments are cemented and veined by hydrothermal alteration minerals, mostly calcite, minor prehnite and quartz. Patches of "green" (reduced) and "brown" (oxidized) rocks cut

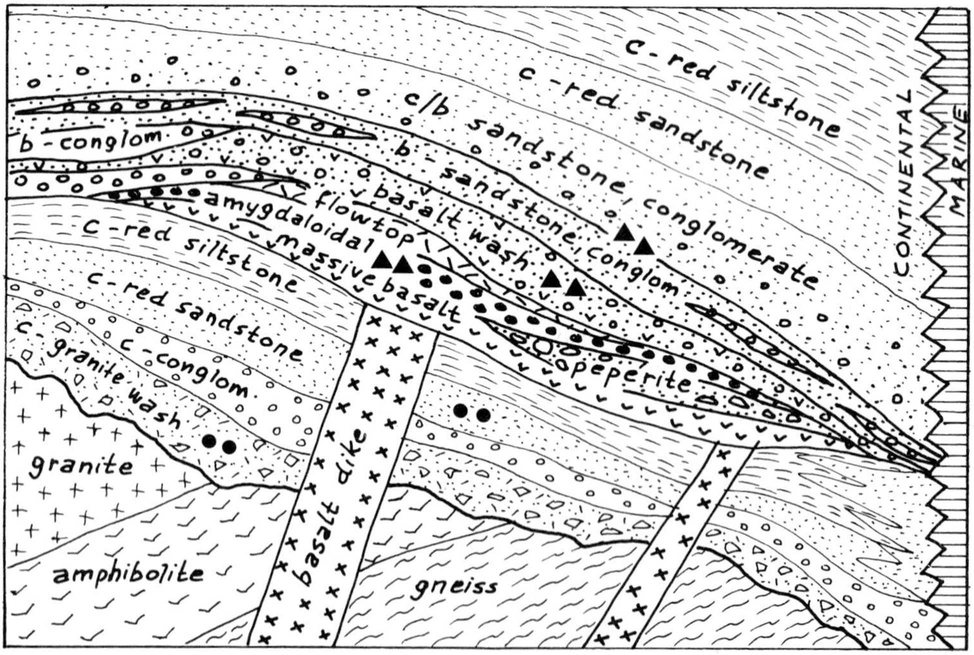

Fig. 25-31. Representation of continental lithofacies derived from the crystalline basement (c) and a synsedimentary extrusion of basalt (b). Diagrammatic, based on the Cr Salta Group in the Quebrada Cafayete, N. Argentina. Triangles: common setting of Cu occurrences; dots, common setting of U occurrences.

across lithologic boundaries in both sediments and volcanics.

The Salta Group hosts mineralogical occurrences and several small deposits of Cu and U. The copper occurrences are mostly of the "Cu sandstone" style, having a form of Cu sulphide and carbonate infiltrations in the basal sediments, showing provenance links to basalts. Minor occurrences of native copper are in the amygdaloidal and scoriaceous basalts themselves. Uranium, on the other hand, is stratigraphically higher, in sandstones of the epiclastic suite, controlled by solution fronts and most likely source-related to the granitic basement.

Similar conditions, less-strikingly exposed (or inaccessible) but better mineralized, exist in the Bolivian Altiplano; in the Triassic Newark Trough, N.E. Appalachians; in the Jurassic and Cretaceous red-beds of central and eastern Yunnan, China; in Permian of the Intra-Sudetic Basin, Czechoslovakia and elsewhere.

TERTIARY BASALT / CONTINENTAL SEDIMENTS MINERALIZED PROVINCE OF THE ALTIPLANO, BOLIVIA

The Altiplano is underlain by a thick wedge of Tertiary epiclastic and pyroclastic largely continental sediments interbedded, in its upper part, with several flows of continental basalts. There, scattered copper occurrences are known over a length of 1000 km, making up a distinct copper province (Ahlfeld, 1967). Economically the most important copper deposits at Corocoro and Chacarilla are found in the nonvolcanic red-beds. These were reviewed earlier. The small deposits Tambillo and Azurita, however, bear a close relationship to Miocene basalts. At Tambillo, native copper and cuprite are disseminated in an amygdaloidal and scoriaceous, prehnite-pumpellyite altered basaltic flow enveloped by sandstone. At Azurita, native copper dissemination mantos and cuprite veins occur in a 120 m thick bed of argillized polymictic conglomerate, near its footwall contact with a basalt body. In the Cuprita mine, the same conglomerate horizon is irregularly mineralized by cuprite and malachite, but no basalt is in sight.

VORONEZH MASSIF, U.S.S.R.

An unusual, nickel-bearing variety of the "Cu-sandstone" was described in the vicinity of Nizhne Mamon in the Voronezh Massif, Russian Platform (Ageykin et al., 1968). There, bravoite, pyrite, chalcopyrite and minor arsenopyrite cement a 0.2-0.3 m thick horizon of upper Devonian arkose, immediately overlying a unit of Devonian continental basalts and amygdaloids. The sediments are interpreted as lagoonal-deltaic and the source of Ni (and probably As) was probably located in the Proterozoic basement ultramafics of the Mamon Complex, known to contain pyrrhotite-pentlandite orebodies. The sulphides account for up to 30-35% of the rock volume.

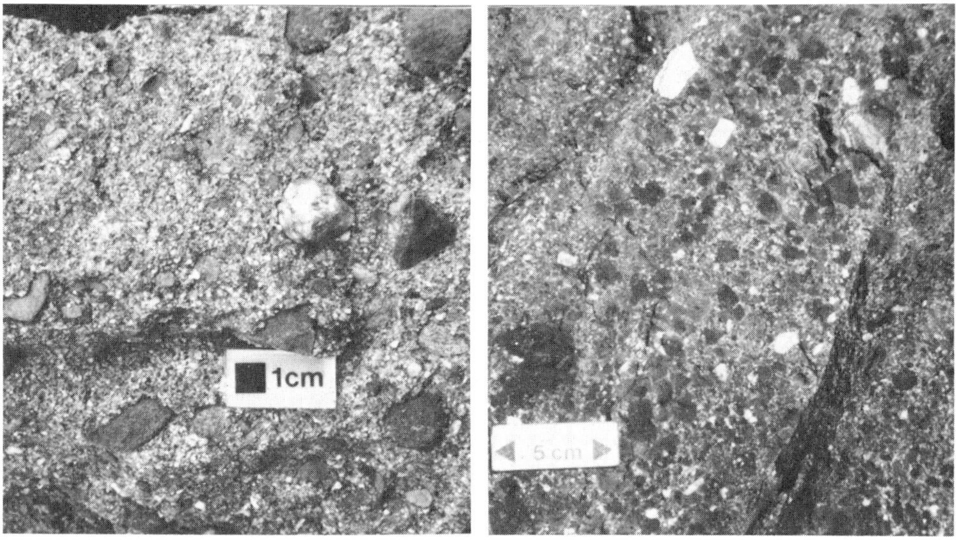

Fig. 25-32. Quebrada Cafayete, Cr volcanic red-beds. sd=red sandstone; bs=basalt. Pictures C and D show fractured and quartz, calcite, silicate—veined basalt and basalt breccia. E is a proximal granitic "wash", F is a pépérite transitional into a subaqueous basaltic breccia.

25.9.3. Felsic and bimodal volcanics / red-beds

PERMIAN RED-BEDS OF THE ALPS AND CARPATHIANS

The Permian volcanic-sedimentary sequence in the Alps and Carpathians (popularly known as "Verrucano" facies, a term applicable to the basal polymictic conglomerate in the southern Alps, but often used in a broader sense to include the entire Permian red-beds) has a remarkable continuity and sequential persistence covering a distance of over 1000 km. Over most of its outcrop area, the sequence is considerably deformed and its metamorphic grade ranges from zeolite to greenschist facies.

In the Bolzano-Trento-Merano area (N. Italy; Brondi et al., 1973), the sequence starts with (1) basal "Verrucano" conglomerate, unconformably resting on Carboniferous metamorphics. This is succeeded by (2) felsic volcanic complex up to 2000 m thick, consisting of (a) lower intertonguing dacite to latite and minor andesite; (b) middle ignimbritic rhyodacite and rhyolite layers with intercalated sediments; (c) upper rhyolite layers interbedded with, or topped by, a thick continental to marine sedimentary unit. Above this is (3) a post-volcanic, quartz-rich sandstone unit (Val Gardena

Sandstone) that is the host to the majority of SUV uranium deposits mentioned earlier. On the top is a (4) carbonate-evaporite unit.

In the western Carpathians of Slovakia (south of Spišská Nová Ves; Mahel' and Buday, eds., 1968), the overall stratigraphy and facies development is comparable with the Alps except for the absence of a distinct thick sandstone unit. The volcanic-sedimentary unit (equivalent of 2) hosts numerous Pb-Zn, Cu-Ag and U-Mo showings, but only a few occurrences have been mined or contain a substantial ore reserve economic at present.

In northern Italy, minor stockwork and vein galena, sphalerite, tetrahedrite, fluorite, barite, etc. showings are found in propylitized and sericitized rhyolites and quartz porphyries. More common Pb-Zn occurrences, however, appear to be diagenetic, confined to lacustrine or lagoonal-deltaic, reduced, carbon-rich sediments, deposited in volcanic-tectonic depressions and floored by and/or intercalated with rhyolites. Near Tregiovo (Mostler, 1966; Fig. 25-33), 160 m thick, thin-bedded sequence of alternating volcanic arkose, argillite, black shale, black limestone, chert and quartz porphyry conglomerate contains numerous thin horizons enriched in scattered crystals of diagenetic galena and sphalerite. Flat sulphide nodules conformable with bedding are located in the shales. Hydrothermal galena, sphalerite and fluorite occur in altered quartz porphyry clasts in the conglomerate, and tetrahedrite with chalcopyrite form sporadic post-lithification fracture veinlets.

The Permian syn-volcanic radioactive mineralization is represented by Boarezzo, northern Italy, and by Novoveska Huta, Slovakia. At Boarezzo (Frenzel et al., 1975) a polymictic conglomerate overlies a gneissic basement and grades upward into rhyolite to rhyodacite lapilli tuff transitional into a volcanic conglomerate to arenite. The latter contains a 3 m thick horizon mineralized by uraninite, coffinite and lesser uranoan rutile and brannerite scattered in the matrix. Pyrite, marcasite, arsenopyrite, galena are associated and organic matter is also present. The mineralization is interpreted as groundwater-precipitated and the U derived from the volcanics during devitrification of the glass.

At Novoveská Huta, Slovakia (Ilavský, 1976), massive to disseminated uraninite, chalcopyrite, tetrahedrite and molybdenite form patches and lenses gradational into low-grade pyrite-chalcopyrite disseminations within a shale-, sandstone-, felsic tuff- and anhydrite-hosted, several metres thick and up to 700 m long horizon. The 400 m thick host unit is a locally expanded, sediment-rich segment (probably a volcanic-tectonic depression) of a more extensive belt of rhyolite pyroclastics. Relatively massive portions alternate with sheared sections along which the volcanics were converted to quartz-sericite schists (porphyroids). The mineralization is polygenetic, most often interpreted as a groundwater precipitate of the locally leached trace metals, partly remobilized during the alpine deformation.

At two predominantly Cu-Ag deposits in the Banská Bystrica area, central Slovakia, mined in the past, the geology is substantially more

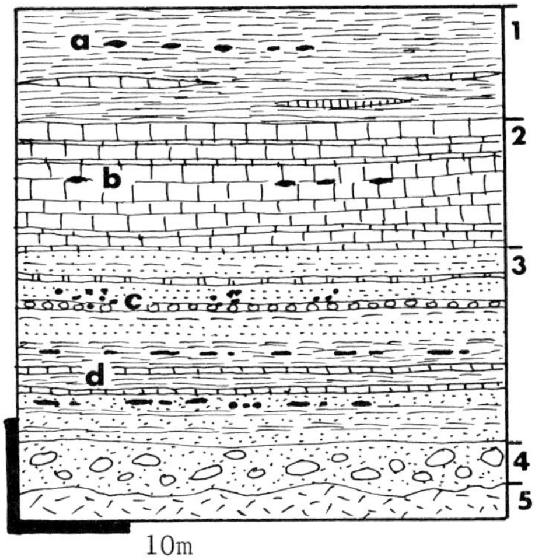

1. Pe_2 thin bedded shale, silic. shale, minor impure limestone, chert
2. thin to medium bedded gray marl, locally siliceous, grading to chert
3. alternating arkose and thin bedded silic. or black shale, thin porphyry conglomerate
4. partly reworked porphyry regolith (wash)
5. Pe_1 red, compact rhyolite flows and tuffs

Fig. 25-33. Tregiovo, 30 km W.S.W. of Bolzano, N. Italy. Lightly Pb-Zn and Cu mineralized Pe volcanic-sedimentary sequence overlying continental rhyolite. Mineralization styles: a=sphalerite, minor fluorite in concretions; b=galena concretions in marl; c=disseminated sphalerite, galena, hematite, fluorite in sandstone and conglomerate; d=disseminated and concretionary galena in a thin-bedded shale. From LITHOTHEQUE, based on data in Mostler (1966).

complex and the mineralization polygenetic and difficult to decipher. At Špania Dolina (Herrengrund; Ilavský, 1976; Laznicka, fieldwork 1965; Fig. 25-34), pre-Permian cataclastic gneisses and granodiorites are unconformably overlain by a thick unit of the Permian "Verrucano" conglomerate, grading upward into arkose. The arkose contains intercalations of felsic tuffs and anhydrite, and is interpreted as a volcanic-sedimentary unit deposited in a hypersaline lake or a lagoon. The Permian rocks are greenschist-metamorphosed and sheared. Chalcopyrite, Ag-tetrahedrite, pyrite and several other minerals in siderite, ankerite and quartz gangue, form (1) veins along faults, shears and the basal Permian unconformity; (2) probably originally conformable, but remobilization-modified impregnation horizons in the Permian sediments and (3) post-deformational infiltrations, impregnations and veinlets. In the similar but smaller chalcopyrite and tetrahedrite deposit L'ubietová (Libethen), the shearing is so intense that the original felsic volcanics and arkosic sediments are represented by an augen quartz-sericite schist, and are mutually indistinguishable.

Fig. 25-34. Špania Dolina (Herrengrund), Slovakia, a pictoresque small mining town. In the 16th century, managed by the Fugger family, it was one of the most significant Cu and Ag producers in Europe.

SOUTH-EAST ALTAI IRON PROVINCE, U.S.S.R. AND MONGOLIA

This is an important (1 Bt Fe/30-40%) but transitional and enigmatic iron ore province, hosted by Devonian rhyolitic ignimbrite and a sedimentary association about 230 km S.E. of Kyzyl, southern Siberia and adjacent Mongolia (Kalugin, 1972; Fig. 25-35). There, an extensive iron ore horizon 50 km long, up to 65 m thick and 20,000 km^2 in area immediately tops ignimbrites and non-welded tuffs interbedded with minor rhyolite and quartz keratophyre flows, volcanic conglomerates and red siltstones. The iron ore unit is composed of rhythmically interbanded chert, jasper, specular hematite and minor magnetite, floored and intercalated with coarse volcaniclastics, some containing terrestrial plant remains. The ores are very similar in style to the Precambrian iron formations and are topped, and often channelled, by members of the coarse detrital, terrestrial, varicoloured suite containing abundant iron ore clasts at the base.

The paleo-environmental setting and origin of these ores is not clear. The sediments are alternatively interpreted as continental or nearshore marine and the ores most frequently as subaqueous-hydrothermal. Although the proposed origin would correspond to the Lahn-Dill "type" (Chapter 11), the subaerial and shallow setting and the distinct highly potassic, basalt-poor lithologic association indicates that this is a different category.

OTHER AREAS

Permian porphyries and rhyolites in the bimodal volcanic red-beds association in the Intra-Sudetic basin, N. Czechoslovakia, contain several mineralogical occurrences of vein, stockwork and disseminated copper sulphides (e.g. Běloves). Contemporary association in the Rheinpfalz area, West Germany (Schneiderhöhn, 1941) contains several small epigenetic cinnabar deposits.

25.10. CONTINENTAL CARBONATES AND SILICEOUS SEDIMENTS

Continental carbonates form in arid soil profiles (calcrete duricrusts, Chapter 23); in karst (Chapter 23); in lakes (Chapter 24) and by precipitation from stream, hot or cold spring systems. They may also form by subsurficial hydrothermal alteration. Chemically precipitated silica sediments form in soil and weathering profiles (silcrete, Chapter 23); as lake deposits (diatomite); as hot springs precipitates (Chapter 5) and as products of hydrothermal and other silicification. Both carbonate and silica sediments preserved in the geological record are inconspicuous, because lithification, recrystallization, deformation, metamorphism, etc. usually alters their frequently delicate fabric and obliterates their identity.

These sediments as a separate group are rare, and insignificant as hosts to mineralization so they are rarely emphasized in the

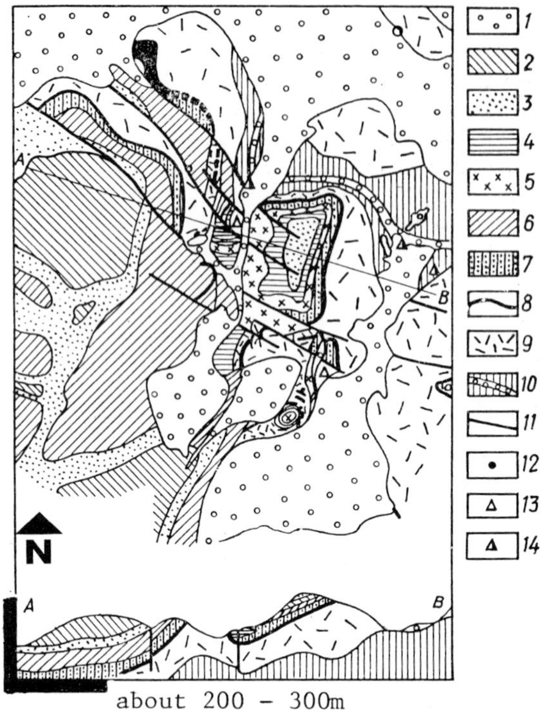

LEGEND:

1. Q,cover rocks
2-6. D_2 sandstone, siltstone, red conglomerate, volcanics, breccia with ore fragments
7. D_2 volcanic sandstone and conglomerate with ore clasts
8. rhythmic hematite ore with remains of continental plants
9. D_1 felsic ignimbrite
10. S_1 marine siltstone
11. faults
12. D_1 flora
13. D_2 fauna
14. S_1 fauna

Fig. 25-35. Vodopadnoe deposit, S.E. Altai iron ore province, U.S.S.R. From Kalugin (1972), approximate scale added.

literature. There are some exceptions, however. A small cinnabar deposit 24 km E. of Weiser, Idaho (Idaho Almaden mine; Ross, 1956; 136t Hg) was discovered in 1936 by chance. It is located in an agricultural area, underlain by semiconsolidated Miocene stream, alluvial fan and pediment deposits (arkose, siltstone, mudstone) with minor intercalated plateau basalt flows. Similar rocks cover extensive areas in the western United States and elsewhere, completely devoid of metallic occurrences. The Weiser deposit consists of cinnabar, pyrite, opal, chalcedony and clay, replacing the porous sediments along fractures and to a lesser degree along bedding planes. The gangue opal and chalcedony (opalite) are hot springs precipitates, probably deposited under an impervious mudstone cover. They form a flat peneconcordant lens much more extensive then the cinnabar-bearing portion, and constitute an important exploration lead when recognized. The ore is probably related genetically to bimodal volcanism.

"De 1545 a 1600 Potosí habría producido la mitad de la producción mundial de plata de aquel tiempo". F. Ahlfeld and A. Schneider-Scherbina, 1964, p.203.

"The immortal Potosí is a hill full of legends but empty of minerals. It is said that the noble metal extracted from the hill during the colonial time would provide a silver bridge from Potosí to Spain". S. Rivas in Geol. Soc. Malaysia Bull., 1979.

CHAPTER 26

Continental, Pre-Quaternary, Calc-Alkaline Volcanic and Subvolcanic Association

26.1. INTRODUCTION

The term "continental" setting as used here has two meanings: (1) subaerial, and (2) formed upon and within the continental crust. Minor departures from the above definition include volcanic rocks emplaced fully or partly into the water of continental lakes, streams, glaciers, etc., and volcanics emplaced into originally non-continental (oceanic, transitional) associations such as ophiolites, present in time of emplacement as minor varieties of the composite continental lithosphere.

"Volcanic rocks" were emplaced on the surface or close to it. The depth limit is hard to establish accurately because interiors of thick lava flows as well as volcanic necks and domes solidified entirely in the subsurface. The term "subvolcanic" was first used by economic geologists (e.g. Schneiderhöhn, 1941) to refer to hydrothermal deposits emplaced at shallow depths (300 to 1,000 m) in terrains of "subsequent" (post-orogenic, continental) volcanism. "Plutonic" deposits formed at progressively deeper levels. More recently volcanologists (e.g. Williams and McBirney, 1979, p.68) have also adopted the term "subvolcanic" for rocks intermediate in setting and appearance between lava flows and epizonal plutonic stocks. For example, in intermediate volcanic-plutonic association, lavas have the composition of andesite; subvolcanic intrusions (dikes, plugs, domes) are of andesite porphyry; and plutonic stocks are represented by diorite porphyry. In this book the term "subvolcanic" is used in a general sense to indicate concern with lithogenesis and metallogeny that have taken place in the relatively shallow subsurface (about 200 to 2,000 m) in paleo-volcanic terrains, above the "plutonic" level that is treated in Chapter 28. Most hydrothermal deposits (e.g. the epithermal veins) emplaced in the subvolcanic level and treated in this chapter are genetically related to, and contemporary with, the evolving plutonic stocks at greater depth so that there is a considerable overlap between the matter treated here and that covered in the next chapter. Placement of certain mineralizations into either chapter is highly subjective. In searching for certain mineralization styles (e.g. porphyry coppers), the reader is therefore advised to peruse both chapters.

The modern (Quaternary) equivalents of the present association and its surficial lithogenesis and metallogeny have already been treated in Chapter 5, so these aspects will not be repeated here. In contrast to the clastic sediments, "typical" volcanics are "ready made" right on the surface, virtually within sight and touch of an observer and do not require, as in the case of the sedimentary rocks, a burial with diagenesis to change from a loose material into a solid rock. This makes the study of paleo-volcanics and their metallogeny easier; the "dry-emplaced" volcanics remain, after shallow to moderate burial, unchanged and identical with their modern counterparts. Brecciated and fragmented volcanics, porous tuffs, volcaniclastics, however, do undergo diagenesis in a water-saturated environment after burial. This causes some compositional convergence and tends to dimimish the contrast between volcanics and sediments.

Deep burial, hydration and dynamometamorphism, on the other hand, produce significant compositional and appearance changes in the volcanics. Slight change (incipient hydration) in geologically old volcanics is apparent in a colour change. The black to gray recent basalt, for example, changes into a green paleo-basalt. In some European literatures, paleo-volcanics have different names (e.g. paleobasalt = diabase; paleoandesite = porphyrite; paleorhyolite = quartz porphyry). This can cause confusion of which the reader should be aware. Subaerial calc-alkaline volcanics are always associated with some sediments. The transitional and sediment-dominated continental lithologic associations are treated in Chapter 25. The proportion of continental sediments is, however, very small when compared with sedimentary volumes present in the marine volcanic-sedimentary associations (Chapters 10-14).

The "most typical" continental calc-alkaline volcanics correspond to the sequentially differentiated basalt, andesite, dacite, rhyolite suite, in which andesite is usually dominant. Its marine counterpart is treated in Chapter 14. Locally, however, basalts, rhyolites or andesites may be the single dominant rocks and their marine counterparts are treated in Chapters 10, 11 and 13, respectively. The "late orogenic" or "final" bimodal (mafic/felsic) and mafic (plateau basalt) continental suites commonly appear during the closing stages of continental volcanism and mineralization associated with them overlaps with the ores related to the sequentially differentiated volcanics. Separation of such ores by affiliation is rather difficult, as experienced, in particular, by geologists working in the basin-and-range terrains of the western Cordillera. The mafic and bimodal continental volcanics are treated in Chapter 31, and marine bimodal suite is reviewed in Chapter 12.

26.2. GEOTECTONIC SETTING, DEVELOPMENT, AND PETROCHEMICAL AFFILIATION

Belts of calc-alkaline volcanics on the surface of continental margin magmatic arcs are one of the pillars of the present plate tectonic model (e.g. Mitchell and Garson, 1976, 1981; compare also Chapter 5). These belts are believed to have formed above active Benioff zones, hidden at depths of about 150 to 300 km (or more). Based on the principle of uniformitarianism, it is believed that the ancient calc-alkaline volcanic associations formed in an equivalent manner, although they are no longer floored by geophysically detectable Benioff zones. Former location of the zone has to be interpreted from megafacies distribution and from rock petrochemistry.

Along modern subductive continental margins, however, several petrochemical series of magmatic rocks are being generated (Chapter 5), the calc-alkaline series being just one of them. It has been demonstrated empirically that the calc-alkaline rocks start forming only when the top lithospheric plate reaches a certain stage of evolutionary maturity so that its composition approaches that of an average continental crust and it achieves a substantial thickness. Most calc-alkaline suites are located in former "Andean-type" magmatic belts or in evolutionarily mature island arcs. Magmatic melts generated (or their generative sequence initiated) along a Benioff zone at some depth, have to pass through the earlier lithosphere and acquire a strong continental heritage (Hamilton, 1979). In addition to this, much magma is generated within the continental crust itself (compare Chapter 28).

Recent studies correlated generation of the calc-alkaline suite and affiliated ore deposits with active subduction, and attributed the appearance of minority petrochemical suites or time-bound changes into petrochemically different assemblages, to standstills, cessations or reversals of the geotectonic regime. Clark et al. (1982) in their interpretation of the Cainozoic mineralized magmatic arc in Mexico pointed out that the calcic and calc-alkaline magmatites and related ores correlate with the 140 to 16 m.y. period of eastern subduction (Cu,Mo,W ores are related to the increasing plate convergence between 80-42 m.y.; Pb,Zn,Ag deposits correlate with the decreasing convergence between 49-26 m.y.). They pointed out also that the minor alkaline rocks affiliated with fluorite and Mo occurrences in the eastern portion of the belt, correlate with several stillstands above the deepest subduction depths.

Similar correlations have been made farther north, in the western United States, and elsewhere (in the Andes, Japan, etc.; compare Sillitoe, 1977 and Mitchell and Garson, 1981, for a review). In the western United States, the eastward subduction generating calc-alkaline magmas and affiliated ores now preserved at a variety of erosional levels, diachronically ceased over most of the region between 31 and 5 m.y. (e.g. at 31-29 m.y. in S.W. New Mexico; at 20-16 m.y. in central Nevada; at 5 m.y. in western Nevada), and has been substituted by an extensional tectonic regime such as the basin-and-range faulting; initiation and filling of the Snake River

Downwarp; and establishment of the Rio Grande "Rift". This was accompanied by a change into a bimodal (basalt-rhyolite) or purely basalt magmatism. The bimodal magmatism is affiliated with most of the Miocene and younger ores in the region (Au-Ag, Sb, As, Hg, Sn), including the present hot springs activity (e.g. Steamboat Springs; Chapter 5). Because most of these ore deposits are relatively distal in regard to their parent magmatic bodies (in most cases probably buried evolving intrusions) and are actually hosted by the syn-subduction andesite-dominated rocks, the bimodal affiliation of such ores is not immediately apparent. Equivalent relationships are probably common worldwide, but they have just started to be unravelled.

It has not yet been demonstrated convincingly that calc-alkaline magmas cannot be related to processes other than subduction and form in settings other than those above the converging continental margins (e.g. along the intracontinental or collisional activation terrains). The subduction affiliation is particularly doubtful in areas located over a thousand kilometres from the nearest identifiable paleo-trench as in the case of the Tertiary magmatites in Colorado. Christiansen and Lipman (1972) went to great lengths to find a subduction mechanism for these rocks, but their ideas are not universally accepted.

26.3. GEOLOGY AND LITHOLOGY

The impressive, youthful volcanic scenery as seen in the Quaternary calc-alkaline magmatic belts is as time passes by, rapidly modified by erosion and/or burial under a pile of younger rocks, so that only the deeply dissected and denuded relicts of the former volcanic edifices remain. Lithofacies distribution, however, generally makes it possible to interpret approximately the paleo-volcanology of most ancient volcanic suites, as long as the subaerially emplaced rocks (lavas, pyroclastics, sediments) are preserved. More intensive or longer erosion, however, tends ultimately to remove all the subaerially formed rocks so that only the former basement intruded by plutonic equivalents remains. Neogene subaerial volcanic terrains have reached their "prime time" at present. Remnants of their former volcanic edifices are often still preserved, yet local erosion cut sufficiently deep to expose the "roots" of the former volcanoes or calderas; the basement on which the volcanics formed; the plutonic equivalents and the whole range of metallic deposits from the surficially-generated to the plutonic ones. The subvolcanic (mostly epithermal) deposits are dominant and most "strongly" developed. With increasing geological age, volcanic associations formed along former "Andean-type" margins disappear quickly (subaerial equivalents in the island arc setting tend to be preserved longer) thus becoming an exception in the Paleozoic setting, and rarity in the Precambrian setting.

The continental calc-alkaline association is usually reviewed in petrography and petrology text and reference books (e.g. Carmichael et al., 1974; Hughes, 1982) under the headings of "orogenic basalt, andesite, rhyolite volcanism"; "subsequent continental volcanic association"; "igneous rocks above Benioff seismic zones" and similar. The rock types have a compositional range extending from basalts to rhyolites. Andesites (hypersthene and amphibole andesites more evolved than arc andesites and basaltic andesites) are the dominant species in most regions, followed by dacites, rhyodacites, latites and rhyolites. Basalts are most common in the "final stages" of continental volcanism, as well as in the earliest stages of volcanism where they form gentle lava shields on which the later explosive andesitic volcanoes formed. The proportions of volcanic rocks listed for the various regions are highly subjective, particularly when the continental volcanic suite rests on an older, less mature volcanic association formed in the island arc setting and both associations are treated jointly.

In settings in which the continental volcanics rest on an earlier nonvolcanic continental basement (e.g. San Juan Mts., Colorado; central Andes; inner Carpathians), the Tertiary volcanic piles are several kilometres thick and composed of numerous overlapping phases. In contrast to the "maturation trend" apparent in the polycyclic oceanic island arcs (Chapter 5), Andean-style volcanic piles show little evolutionary consistency as a whole and andesites, dacites and rhyolites overlap repeatedly. The volcanics have a high explosive index so that fragmental rocks dominate the active volcanic phase. Coarse fragmental volcanogenic rocks include (Williams and McBirney, 1979; Fisher and Schmincke, 1984): agglomerates (deposits composed of bombs of intermediate or basaltic composition in monolithic scoriaceous matrix); agglutinates (welded basaltic spatter); volcanic breccias (pyroclastic rocks composed of angular blocks); tuff breccias (volcanic breccias with abundant ash-size matrix); a variety of air-fall and ash-fall tuffs; welded felsic tuffs (ignimbrites); lahars (boulder mudflows, partly stratified and sorted volcaniclastic beds formed by downslope flow) and a variety of water-deposited conglomerates. These coarse fragmental rocks tend to be the most impressive visually when exposed, for example, on valley walls or in road cuts.

Massive lavas, dikes, domes and small subvolcanic intrusions are substantially rarer. Fine airfall tuffs and medium to fine volcanic sediments (sandstones, mudstones) increase towards the fringe of the volcanic regions (or gather in the internal or marginal depressions), so that a typical elevated composite subaerial volcanic pile has the characteristics of a huge lag deposit from which much of the fines have been winnowed out.

There is a broad colour variation caused by rock basicity (from black to dark-green basalts to white or pink rhyolites) and also by the syn-depositional oxidation state. Andesite flows and volcaniclastics range from dark-green to bright purple, depending on the degree of pigment oxidation established during emplacement or diagenetically (an analogy with sediment pigmentation, e.g. in the "red-beds" suite). Another colour effect is due to hydrothermal alteration which may be local or regional. The latter covers uniformly large territories. Regional propylitization that impairs dark greenish-gray hues to the rocks is most common in dominantly andesitic terrains. Regional argillization and "solfataric" alteration is common and most striking, for example in the unvegetated terrain of the Peruvian Andes between Castroverreina and Ayacucho (Fig. 26-1). It consists of a patchy distribution of white, beige hues due to clays, bright-yellow hues due to jarosite and brown hues caused by goethite. It is due to attack by the sulphuric acid released by pyritic weathering.

Paleovolcanic (environmental) interpretation of ancient volcanic associations is useful, because many industrially important metallic deposits have a strong paleo-environmental control. In general, former eruptive centres (relicts of stratovolcanoes, calderas, vents and domes) marked by a variety of rapidly changing facies, are substantially more interesting to a prospector than monotonous tracts of subhorizontal to gently-dipping, facially more persistent volcanics and related sediments, formed on volcanic flanks, between adjacent volcanic centres, or on broad plateauxs. Several examples of the ore controlling volcanic centres are reviewed later.

26.4. MINERALIZATION STYLES AND METALLOGENY

26.4.1. General

Pre-Quaternary continental calc-alkaline volcanic and subvolcanic terrains as summarized in this chapter represent about 700 Mt Fe, 260 Mt Al, 7 Mt Mn, 10 Mt Cu (plus 40 Mt Cu in volcanics-hosted porphyry coppers from Table 26-6), 20 Mt Zn, 15 Mt Pb, 110 Tt Mo (plus 720 Tt Mo in volcanics-hosted porphyry Cu-Mo), 6 Mt Sn, 120 Tt W, 7 Tt Au, 420 Tt Ag, 60 Tt Hg, 30 Tt U, 600 Tt Sb, 600 Tt As, 100 Tt Bi, 10 Tt Te, 35 Tt Be and minor quantities of other metals, in ores economic either at present or in the near future. Considering the rather limited occurrence of this association, the mineralization density per outcrop area is high, although the occurrences tend to be very unevenly distributed. Most are clustered in or near volcanic centres.

Metallic deposits hosted by the association are either closely or loosely genetically related to it (the bulk), or they are unrelated having been formed long after its completion by outside agents (e.g. by younger magmatic events). The latter are not considered here. Geochemically, the bulk of continental volcanics tend to be rather "neutral", having their trace metal values close to the averages for the respective compositional groups listed in the literature. Anomalous trace metal contents, at least of the basaltophile elements, are less common than in the "marine" volcanics (Chapters 10-14). This seems to reflect the multiple recycling evolutionary history of these rocks. However, lithophile metals (Sn,Be) as well as Au,Ag, etc., may form local geochemical anomalies. In most cases, these can be attributed to the local crustal heritage.

Fig. 26-1. Areal argillic and pyritic alteration in the Andes between Castrovirreyna and Ayacucho, Peru.

Within a thick, compositionally average volcanic pile, the main ore category to be expected, the epithermal veins, is compositionally predictable and the high-silver Pb-Zn and/or Au deposits have a truly global distribution. This suggests a high degree of uniformity in systems evolution and the ore forming process, recently interpreted mostly as convective hydrothermal leaching of volcanic piles. The Pb-Zn-Ag and Au(Ag) deposits have a fairly "average" host assemblage (andesite to rhyolite lavas and pyroclastics), showing little preference for any particular rock type. Some Pb-Zn deposits are members of zoned systems gradational downward (or centre-ward) into plutonic Cu or Mo accumulations.

The anomalously rich or anomalously abundant Pb-Zn, Au(Ag) or Ag deposits or clusters may have required more than the "ordinary" conditions to form, and almost invariably they correlate with a "metallogenically fertile" basement composed of one of the lithologic associations treated elsewhere in this book (e.g. marine "black shales", ophiolites, tholeiitic basalts). This suggests higher than usual, relatively proximal basement contribution to the evolving lithogenetic and metallogenetic system. The metals less commonly accumulated in the continental volcanics in mineable quantities (Sn,Cu,Mo,Hg,etc.) appear to have required a "special" setting and conditions. The volcanics-hosted tin deposits are consistently affiliated to rhyolite plugs, and the Bolivian tin giants (Llallagua,

Potosí) are in the middle of a polygenetic "tin province" the unifying feature of which is the basement composed of gray to black shallow marine slates.

The volcanic and subvolcanic metallogeny was dynamic and rather rapid. There are virtually no intramagmatic or synmagmatic ores attributable to the classical process of magmatic differentiation, with the possible exception of the magnetite lavas that are controversial anyway. The bulk of ore deposits are post-magmatic, composed of metals largely extracted from a relatively sizeable rock "pile" and precipitated at relatively narrow intervals of physico-chemical change. It has to be emphasized that the subvolcanic ore-forming complex reaches far outside the preserved volcanic piles and their deep roots, scavenging metals from or forming orebodies in, the basement rocks. Orebodies in the basement may remain preserved even if the volcanics are completely stripped away by erosion and only relicts remain. The spatial affiliation of orebodies to continental magmatism may thus be lost or confused. The "bonanza" silver veins of Guanajuato, broadly contemporary with a period of local subaerial volcanism, now fill fissures in older conglomerate, black slate and granite.

The principal mineralization styles in this association are assembled in Table 26-1 and shown in Fig. 26-2. It can be seen that the surficial volcanic styles are relics of those already tabulated in Table 5-4 for the regions of contemporary volcanism. The remainder, however, are new, sub-surficial styles.

26.4.2. Ore phenocrysts

Ore phenocrysts, dominantly (Ti)magnetite already treated in Section 5.7.1., are omnipresent in a variety of volcanics, but in a dispersed state. Although they supply heavy minerals to metalliferous placers, they are without economic importance in their original state.

26.4.3. Mineralized miaroles, amygdules, etc. in volcanics

Miaroles (small angular cavities in magmatic rocks into which project small crystals of the rock-forming or accessory minerals) and lithophysae (bubble-like spherulites) are relatively widespread in rhyolites, quartz latites and similar felsic rocks. In the Thomas Range, Utah (Staatz and Carr, 1964) porphyritic rhyolites locally contain topaz, beryl and fluorite as phenocrysts or interstitial accessories and these minerals are most conspicuous and megascopically apparent in miaroles and in lithophysae. Although of little economic importance in themselves, the cavity minerals may indicate a geochemical enrichment or metallogenic "specialization" of their host suite, important in exploration strategy. Near the Spor Mountain "invisible" volcanic sediments-hosted bertrandite deposits, rhyolites are enriched in Be to about four times the background concentration.

In the Taylor Creek region, New Mexico, minor cassiterite occurs with topaz in miaroles in a Tertiary rhyolite.

Mineralized amygdules are confined to basalts and basaltic andesites and most commonly contain minor native copper and Cu sulphide occurrences associated with prehnite, pumpellyite, epidote and other minerals. These are common to other lithologic associations with which they were described in greater detail (Chapter 10).

26.4.4. Finely dispersed metals in volcanics and volcaniclastics ("metalliferous volcanics")

All freshly deposited subaerial pyroclastics release small, but cumulatively significant quantities of leacheable metals (Cu,U,Pb,Zn, etc.) when percolated by meteoric waters, shortly after deposition (Section 5.7.4.), so they act as slightly metalliferous source rocks. Another batch of trace metals is released during devitrification and trapped, for example, in some metalliferous coals. Few volcanics have an ore-approaching metal content at the time of deposition, but some may become mineralized after burial or in the postmagmatic stage. Some volcanics with economically leacheable slightly elevated trace metals may become the technological ores of the future. There is only a small number of recorded examples, and all are controversial and transitional.

Koeberlin (1937) reported the presence of minute quantities (0.01-0.05 ppm Au) of a "syngenetic" gold locked in the glassy matrix of pyroclastics at several locations in the Valle Central, Chile. The content was still too low to be recoverable using the techniques currently available. Fine, invisible "disseminated silver" in Neogene volcanics probably mineable in an open pit, is known to exist in the San Cristóbal field near Uyuni, Bolivia. Lovering (1954) reported extensive but low-grade deposits of uraniferous opalite in Virgin Valley, Nevada. The opalite, diagenetically silicified porous pyroclastics or lacustrine sediments, form discontinuous layers up to 2m thick and 800 m long, conformable with bedding of the host felsic ash and tuff. The grade ranges from 0.002 to 0.12% U, and there are sparse carnotite coatings.

26.4.5. "Contact sweeps", mineralizations in basal volcanics above basement unconformities

The Hart and Getchel occurrences in the Ivanpah Quadrangle, S.E. California (Hewett, 1956) are example localities of this style. There, small veinlets and veins of chert and quartz with scattered megascopic gold occur in late Tertiary rhyolite breccias, flows and tuffs at or only slightly above the unconformity with Precambrian schists and gneisses. Hewett offered no genetic explanation, but a "sweep-out" and remobilization of an auriferous regolith by the basal volcanics appears to be likely.

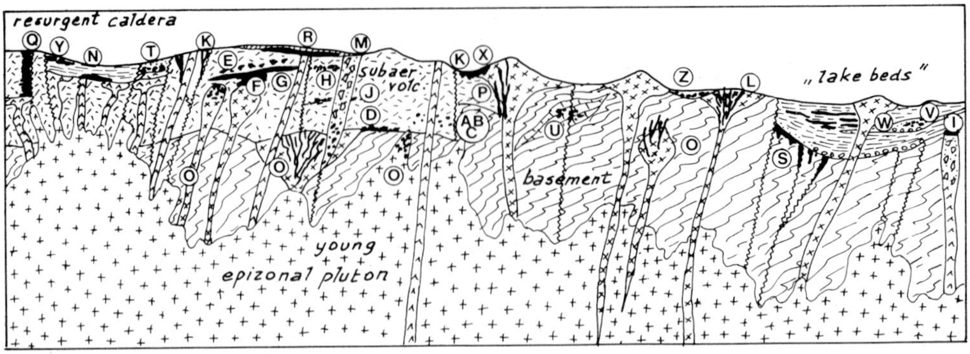

Fig. 26-2. Principal mineralization styles in pre-Quaternary, continental volcanic and subvolcanic association. See Table 26-1 for an explanation of the letter codes.

Table 26-1. Principal mineralization styles in the pre-Quaternary, continental calc-alkaline volcanic and subvolcanic association.

(A)	Ore phenocrysts (e.g. magnetite) in volcanics and subvolcanics.
(B)	Mineralized miaroles, amygdules (e.g. beryl, cassiterite).
(C)	Finely dispersed supra-clarke quantities of metals in lavas, tuffs, volcaniclastics ("metalliferous volcanics"); X0 localities, X0 t subgrade Au, X00 t Ag, X00 t U.
(D)	"Contact sweep" mineralization in basal volcanics above unconformities. X localities, 0.X t Au.
(E)	Mineralized ejecta (e.g. magnetite, hematite lapilli, ash). X localities.
(F)	Solid ore lava flows (magnetite). X localities, 600 Mt Fe.
(G)	High-grade metalliferous ash and tuff (e.g. magnetite, hematite). X localities, included with (F).
(H)	Fumarolic impregnations, coatings, veinlets in lavas and pyroclastics. X0 localities, about 30 Tt Sn, 150 Tt Cu, 300 t Ag.
(I)	Subaqueous-hydrothermal ("exhalative") deposits (e.g. in flooded craters).
(J)	Volcanic-diagenetic or low-grade metamorphosed disseminated deposits (e.g. U, Cu sulphides). X00 localities, 6 Tt U; Cu listed in Chapters 12-14.
(K-N)	Epithermal deposits
(K)	fissure veins and their swarms;
(L)	stockworks;
(M)	mineral disseminations and cements in breccia pipes;
(N)	replacements.
	Epithermal mineralizations are in (a) the coeval volcanics/intrusions, or (b) in older basement. X00 localities, 6 Tt Au; 350 Tt Ag, 5 Mt Zn, 3 Mt Pb, 500 Tt Cu, 30 Tt Bi, 10 Tt Te, 500 Tt Sb, 500 Tt As, 3 Tt U, 2 Tt Mo, 12 Tt Hg, 2 Tt Be.

(O)	Mesothermal (deeper-seated) deposits. Fissure and replacement veins, stockworks, disseminations, massive to disseminated replacements. X00 localities, about (min.) 10 Mt Zn, 6 Mt Pb, 6 Mt Cu, 100 Tt Mo, 60 Tt Ag, 200 t Au.
(P)	"Xenothermal" deposits. X0 localities, 1 Mt Cu, 600 Tt Zn, 100 Tt Sn, 10 Tt W.
(Q)	Mineralized alteration zones (e.g. alunite). X localities, 250 Mt Al.
(R)	Ancient hot springs deposits and replacements (e.g. opalite). X0 localities, 15 Tt Hg, 10 Tt Sb,As.
(S)	Infiltration deposits, descendent veins, stockworks, etc. under unconformities. X0 localities, 200 Tt Mn, X Tt U.
(T)	Infiltration/impregnation ores in epiclastic sediments interbedded with or topping underlying volcanics. X0 localities, 10 Tt U.
(U)	Volcanic-associated karst in carbonates (e.g. cinnabar). X localities, 30 Tt Hg (transitional).
(V)	Volcano-fluvial deposits, paleoplacers and epigenetic infiltrations. X0 localities, 200 t U.
(W)	Volcano-lacustrine deposits. X0 localities, 3 Mt Mn, 10 Tt U, 30 Tt Be, 50 Tt As, 50 t Au.
(X)	Mineralized humid tropical weathering profiles over volcanics (bauxitic laterite). X0 localities, 10 Mt Al.
(Y)	Arid regoliths over volcanics; traces of U, Cu.
(Z)	Unconsolidated Cainozoic placers derived from continental volcanics. X0 localities, 200 Tt Sn.

26.4.6. Mineralized ejecta

Minor pyroclastic fragments and blocks of iron ore have been reported from the explosive volcanic units shortly postdating formation of the Cerro Mercado iron deposit near Durango, Mexico. Forster and Knittel (1979) mentioned magnetite lapilli in the late Proterozoic Mishdovan iron deposit, Iran. Although of little direct economic importance in themselves, mineralized ejecta may be valuable in an exploration for hidden subsurface ores.

26.4.7. Solid ore lava flows

The important Quaternary deposit of magnetite lavas El Laco, Chile (Chapter 5) has several probable ancient counterparts. At Cerro de Mercado, Durango (Lyons and Clabaugh, 1973) a thick, wedge-shaped layers of massive martitized magnetite top a minor intravolcanic unconformity and are interpreted as subaerial lava flows at the margin of a large Tertiary caldera. Several much older (late Proterozoic) iron ore deposits in the Bafq-Saghand district, Iran (Forster and Knittel, 1979) are interpreted in a similar fashion. Both regions are reviewed in greater detail later.

26.4.8. High-grade metalliferous ash and tuff

A tuff formed by lithification of a metalliferous ash (rather than mineralized after burial by infiltration or replacement) is a theoretical, though a remote possibility. Lyons and Clabaugh (1973) interpreted a thin layer of hematite powder as possibly being an Oligocene ash blanket, deposited on an erosional surface several hundred km^2 in extent and adjacent to the Cerro de Mercado (Mexico) iron ore deposit. Although of no economic significance, this is an excellent marker and a possible guide to concealed orebodies closer to the volcanic centre.

26.4.9. Fumarolic deposits

Fumaroles are vents and conduits that emit volcanic gases and vapours as well as, by implication, the gases and vapours themselves. They have already been reviewed in Section 5.7.4. As agents of ore deposition, fumaroles are notable for their virtual contemporaneity with the active phase of volcanism (in contrast with the epithermal deposits that postdate the volcanic cataclysms). They are also notable for their very proximal relation to their host rocks that, in most cases, also supplied the heat, metals and ore transporting fluids. The fumarolic ore occurrences are believed to have precipitated from high temperature gases (e.g. Sn) or hydrotherms. In active volcanic terrains they mostly take the form of surficial encrustations, while in the ancient volcanic associations they occur as (1) vertical to subvertical joint and fracture coatings and veinlets (e.g. cassiterite-specularite coatings and veinlets, Cerro Remedios, Durango), often grading into stringers and irregularly distributed disseminations (Black Range, New Mexico); (2) fine impregnations and replacements of permeable tuff horizons (e.g. of hematite and minor magnetite in the vicinity of Cerro de Mercado, Durango and (3) as laterally persistent ("stratabound") disseminations, breccia cement, pipes, stringers in the upper parts of rhyolite flows (e.g. Amolanos Cu deposit, Chile).

The host rocks to fumarolic mineralizations are almost exclusively rhyolite, rhyodacite, occasionally dacite flows, ignimbrite sheets and domes. The ores may be in completely fresh unaltered rocks, or alternatively there may be a relatively light, low-pressure alteration, e.g. silicification, argillization. In the Amolanos mine near Copiapó, Chile, interpreted as fumarolic (Lortie and Clark, 1974; about 80 Tt Cu/2.5-3%) chalcocite, bornite and chalcopyrite occur as sub-spherical "nodules" rimmed by hematite, and dispersed through the upper part of an extensive Paleocene rhyolite flow. Quartz, sericite and minor barite and fluorite are irregularly present as gangue minerals. The host rhyolite is brecciated and grades upward into a breccia cemented by quartz, calcite and sericite. The mineralization is interpreted as having formed during the cooling of the rhyolite sheet.

The Jardín Cu-Ag deposit in the same area (some 25 Tt Cu/1.75; 210 t Ag/140 ppm) is another locality interpreted as fumarolic by Lortie and Clark (1974). There, disseminations and veinlets of chalcocite, bornite and tennantite occur in the uppermost part of a thick, welded Paleocene ignimbrite sheet, as well as in an overlying breccia, black shale, red sandstone and mudstone. The lithologic association corresponds to the "volcanic red-beds", as treated in Chapter 25. The main reason for interpretation of this deposit as fumarolic is the presence of thin vertical pipes suggesting an upward penetration of mineralizing gases or solutions through still wet sediments, during the period of continuing cooling and degassing of the ignimbrite sheet.

26.4.10. Subaqueous-hydrothermal deposits

Although water-filled reservoirs (crater lakes, playa lakes, stream floodplains, glacier subfloors) are occasionally present in active continental volcanic areas, their shallow depth constitutes a considerable restraint on the formation of the "exhalative" massive sulphide deposits so characteristic of the submarine volcanic regions. Several Quaternary examples of pyrite, lesser Hg,As,Sb,Au, etc. and Fe-Mn oxide occurrences formed recently on crater lake floors (e.g. the Mendeleev Volcano, the Kuriles); in sea-flooded, breached calderas (Rabaul); in shallow, unconsolidated sediments at a foot of a volcano (Vulcano Islands, Japan; Chapter 5), however, may have more ancient equivalents but none have been convincingly and unequivocally interpreted in this way. The bedded "exhalative-sedimentary" Mn and Fe deposit San Francisco (Jalisco, Mexico) is in Tertiary lake sediments and transitional into or identical with, the low-temperature Mn and Fe deposits in the volcanic-lacustrine association reviewed later.

26.4.11. Hydrothermal subvolcanic deposits: epithermal and mesothermal

Epithermal deposits are high-level and relatively low temperature (100-200° C) mineralizations, precipitated in waning stages of volcanism (or post-volcanically) from convecting hydrothermal waters. They are the "flagship" of the subvolcanic metallogeny and occur as a variety of veins, stockworks, disseminations and replacements. Because of their uniqueness and economic importance, they are more thoroughly reviewed in a separate section.

The classical mesothermal deposits were hydrothermal-plutonic orebodies with a range of filling temperatures between 300 and 600° C (Schneiderhöhn, 1955). In the present association these deposits are always closely spatially and genetically associated with intrusive plutons, but they may overlap with the epithermal deposits (the latter being usually of a younger phase) or be hosted by volcanics. In the great Cerro de Pasco orebody (Einaudi, 1977), the earlier mesothermal mineralization phase (up to 500°C) produced quartz, pyrite and pyrrhotite, chalcopyrite, sphalerite, galena, replacements in carbonate adjacent to a quartz monzonite stock, in a former vent. Later, epithermal (around 300° C and less) enargite, luzonite, pyrite veins formed.

Mesothermal deposits can be expected at a depth under continental volcanic centres. Several examples of orebodies showing a close relation to such centres will be reviewed briefly, but a more detailed treatment of mesothermal deposits as a class appears in Chapter 28. When epithermal, near-surface and mesothermal, moderate-depth deposits coexist in a single volcanic-intrusive system, metal zoning is commonly developed. Typically, Au-Ag or Pb-Zn-Ag epithermal veins occur above Cu-rich mesothermal orebodies.

26.4.12. Mineralized alteration zones

This heading is almost a misnomer, because every ordinary epithermal vein consists of a metal-bearing portion (the vein) enveloped by an altered wallrock. In a simplified economic approach the vein is the "valuable component" (except its gangue), the alteration zone is the "waste". In this paragraph we are concerned with alteration envelopes associated with ore veins or present alone as independent zones, that have a substantial content of a recoverable metal and so may possibly be considered as metallic ores in their own right. If altered wallrocks containing lower grade haloes of the same metals that are present in ore veins are excluded, alunite and high-alumina clays, both considered as one of the possible future sources of aluminium, remain as the most important example.

Alunite (K-Al sulphate) is a white clay-like mineral containing about 35% Al_2O_3 (18.6% Al). Although it is inferior to bauxite in its Al content, the possible by-products (sulphuric acid and K-compounds) could make the recovery profitable. In the U.S.S.R. (Dashkesan), one alunite-based aluminium smelter is in operation and feasibility studies for a similar smelter were, at one time, under way in the U.S.A.

Hall (1978) recently reviewed the world alunite resources and although not all deposits are associated with continental volcanics, the bulk of alunite is. In the Zaglik deposit in the Soviet Azerbaijan (Kirpal' and Tenyakov, 1974) that supplies the Dashkesan smelter, gently dipping body of alunite (grade 40-60%) mixed with microquartz lies in a series of concordant replacement (alteration) layers in upper Jurassic felsic tuffs and volcaniclastics, overlying a limestone. The reserve was about 100 Mt ore with 50% alunite, containing a mere 8 Mt Al.

The largest American low-grade alunite deposit in the Wah Wah Range, Utah (Hall, 1978; 634 Mt ore with 28-33% alunite, that is about 33.6 Mt Al), is in a thick pile of Tertiary rhyolitic ash-flow tuffs and ignimbrites. Alunite, as in Zaglik, is intermixed with silica and grades into an almost pure silicified core ("secondary quartzite"). The Fanshan deposits 450 km S.S.W. of Shanghai in Cretaceous ? rhyolites and tuffs intruded by a quartz monzonite stock are almost equivalent genetically. They are reported to contain 2 Bt of 50% alunite ore (about 180 Mt Al). In the Alunite Ridge, Marysvale District, Utah (Hall, 1978), high-grade (88% alunite) fissure veins in Oligocene rhyolite, tuff and breccia, are probably open-space precipitates from a hydrothermal fluid. They are fringed by a lower-grade, replacement alunite.

26.4.13. Hot springs deposits

The surficial and shallow sub-surficial hot springs deposits described in Chapter 5 are rarely preserved in the ancient rocks. The recorded examples are rarely older than Mesozoic. Cinnabar and/or stibnite, realgar and orpiment, are the most frequent minerals and interaction (e.g. with ophiolitic ultramafics) was probably necessary to generate economically significant deposits (compare Section 8.3.).

26.4.14. Volcanic-diagenetic, groundwater-epigenetic, metamorphogenic, hydrothermal, etc. ore disseminations in porous volcanics and volcaniclastics

Most examples of this style are "mantos": elongated lenses to sheets, concordant to peneconcordant with bedding, composed of disseminated, stringer or cement-forming ore minerals hosted by a volcanic rock (e.g. amygdaloidal or fractured lava, porous tuff) or by units of alternating flows, tuffs and sediments. Copper is the most frequently accumulated metal, followed by manganese, tin, gold and occasionally other metals.

These deposits are transitional, common to volcanics in both continental and nearshore marine settings. Most copper-bearing mantos (as in Chile) are located on both sides of the marine / continental facies transition in Jurassic to lower Tertiary andesites and the exact environmental setting is usually impossible to determine accurately. The mantos hosted by Jurassic basalts to andesites at Buena Esperanza; mantos in porphyritic andesite at Lo Aguirre; and mantos in a laharic conglomerate at Sustut, have already been reviewed in Chapters 10 and 13, respectively.

Sillitoe (1977) attributed the origin of the Chilean "manto-type copper deposits" to the hydrothermal systems active during cooling of the spatially related volcanics, so they would overlap with the "fumarolic deposits" reviewed earlier in this section. Because most of the sulphides in the mantos are accompanied by the "critical" prehnite-pumpellyite and/or epidote silicate association, they could equally well be interpreted as ores formed at depth as a result of redistribution of the trace copper locked originally in high-temperature magmatic silicates, during diagenesis and metamorphism.

Mn oxide mantos (Forari, Coquimbo Mn belt, eastern Cuba) are reviewed in Section 13.4.; Au mantos (Gold Ridge) in Section 13.6.

Subvertical bodies of disseminated ore minerals in a similar setting and in similar host rocks as the mantos, occur in greater numbers but few are economically significant bodies. They overlap with other mineralization styles listed here and a sharp separation is impossible. Several small uranium deposits and showings fit into this category. In the Villa Aldama U region (54 km N. of Chihuahua, Mexico; O.E.C.D., 1979; about 5,600 t U/0.08-0.25%), uraninite and secondary oxides are disseminated in fractured rhyolite and tuff and also in cavities and fractures in limestone. At the Rosamond prospect in the Mojave Desert, S. California (Fig. 26-3), autunite is erratically distributed over about 2.5 km^2 as thin coatings along fault planes and fracture surfaces in tuffaceous sandstones and felsic lithic tuffs overlying a basalt flow and a granitic basement. Lesser quantities of autunite impregnate the tuff.

26.4.15. Descendent (infiltration) veins, stockworks, impregnations under unconformities

The subject of unconformity-controlled mineralization in general has been reviewed recently by Laznicka (1985f). Among such mineralizations in continental volcanic terrains one could expect to find a variety of orebodies formed from descendent solutions. Such solutions could have originated on the paleosurface in weathering profiles long since erosion removed. Alternatively, they may have originated within the volcanic pile where the plane of unconformity could have acted as a surface of permeability to the laterally travelling fluids. Except in a few localities, the direction of movement of the mineralizing solution (up or down) has not been proven, so the assumptions made here are hypothetical.

Certain Mn oxide deposits represent a class for which a per descensum origin has frequently been proposed. Some of the small Mexican vein, chimney or breccia deposits of psilomelane, cryptomelane, hollandite, coronadite and similar minerals in gangue of chalcedony, quartz, calcite or barite hosted by silicified rhyolite along faults (e.g. Talamantes, Manganita, La Negra; Trask et al., 1946), may belong here. The small uranium deposits listed in the previous paragraph may overlap with the present style.

26.4.16. Volcano-fluvial and volcano-lacustrine deposits

Terrestrial sediments having a minor to significant volcanogenic (pyroclastic) component, have been reviewed in Chapter 25. There is a complete transition between epiclastic (=derived by fragmentation on the older land surface) and entirely volcaniclastic sediments; in this paragraph we are concerned with predominantly volcaniclastic sediments containing, at most, only accessory quantities of epiclasts.

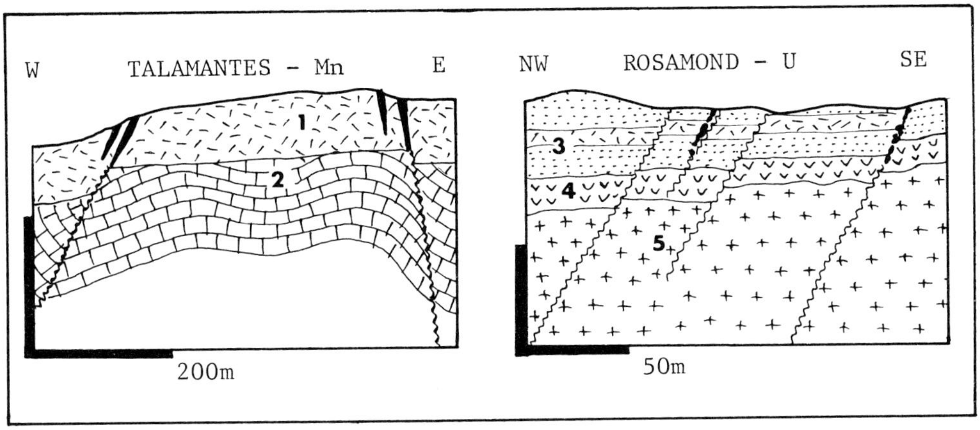

Fig. 26-3. An example of probable "per descensum" veins (Talamantes-Mn, Mexico), and infiltrations under intravolcanic unconformities (Rosamund-U, California). 1=T, rhyolite, volc. breccia, tuff; 2=Cr? limestone; 3=T,felsic tuff, tuffac. sandstone; 4=T, basalt flow; 5=Cr_3-T_1 granite. From LITHOTHEQUE, based on data in Trask et al. (1946).

In the European literature, such rocks are usually called tuffites. Here, entirely pyroclastic sedimentary rocks are usually called reworked tuff, water-laid tuff or similar. Pettijohn (1975) included even the "in situ" tuffs among "volcaniclastic sediments".

The most distinct volcaniclastic sediments formed in streams and lakes contemporary with explosive volcanism, at times virtually choked and paralysed by a sudden influx of the airborne detritus. This interfered with the orderly development of the usual facies as described in the literature from the "normal" (nonvolcanic) fluvial and lacustrine environments. It also interfered with the expected metallogeny. Volcano-fluvial sediments are texturally and compositionally immature in many respects. They contain a high proportion of unstable clasts, are poorly sorted and rounded and the granulometry is anomalous. In regions of quartz-free ancient andesitic volcanism, the only texturally recognizable sediments may be conglomerates composed of unstable volcanic clasts (e.g. andesite porphyry) in a mudstone matrix, and mudstones. Arenites may be missing, because their unstable clasts could have been diagenetically argillized or converted to chlorite, sericite, silica, etc. epimatrix. The general and pervasive immaturity virtually eliminates metalliferous paleoplacers (except occasional magnetite). The few mineralization styles occasionaly present include iron oxide lenses precipitated in valleys from hot springs (equivalents of the style "K", Table 5-4, Chapter 5), as well as occasional post-depositional metalliferous infiltrations (e.g. Mn or U oxides, Cu silicates, etc.).

None is of much importance. The small Good Will (Yellow Chief) U deposit in the Thomas Range, Utah (Staatz and Carr, 1964; 180 t U/0.18%; Fig. 26-4), is an imperfect example of metalliferous infiltrations in fluvial sediments. There, uranophane and weeksite coat volcanic and minor limestone pebbles in a late Tertiary, poorly consolidated sandstone and conglomerate sandwiched between a bentonitic (=argillized) felsic crystal and vitric tuff and reworked tuff. The uranium was probably leached from the pyroclastics during devitrification.

Volcano-lacustrine sediments bear a greater facies resemblance to sediments of the non-volcanic lakes both perenial and playas (Chapters 24, 25), except that the sandstone-dominated delta and beach facies is very rare. The most characteristic Tertiary volcano-lacustrine sediments ("lake beds") in the western United States are thin-bedded units of alternating bentonitic (montmorillonitic) and zeolithic claystones, volcanic siltstones, lithic arenites, marls, rare freshwater limestones and occasional lignites, gypsum and a variety of evaporites. Tuffs and lavas are commonly interbedded and channel conglomerates or sandstones are common. In this setting, Mn,U,Be,As,Au, etc., mineralization is moderately common and of minor importance except for Be, whose largest described accumulation occurs in this setting. Of the non-metallics, fluorite deposits are found increasingly in this setting.

Most metallic accumulations are concordant to peneconcordant sheets (mantos) formed: (1) by precipitation from hot springs on the lake floor during the active period of sedimentation; (2) by redistribution of trace metals released chiefly by devitrification of

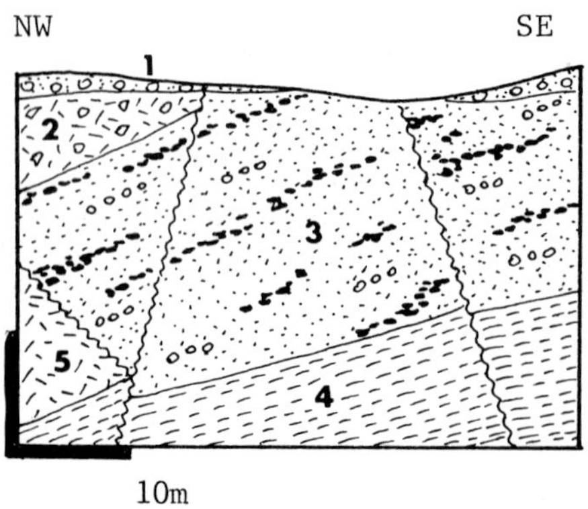

1. Q, alluvium
2. T_2, felsic bentonitic tuff
3. fluvial, poorly consolidated sandstone and conglomerate, tuffaceous, local limestone clasts
4. felsic water - laid tuff
5. crystal tuff

Fig. 26-4. Good Will mine, Utah, example of U infiltrations in volcano-fluvial setting. From LITHOTHEQUE, based on data in Staatz and Carr (1964).

the volcanic glass during the early or late diagenesis; (3) by syn-diagenetic or post-lithification infiltration from groundwater or from hot springs or (4) by a combination of (1)-(3) and by other mechanisms, e.g. by hydrothermal introduction from depth. Several example localities (San Francisco and lake Mead, Mn; Date Creek and Roma Province, U; Spor Mountain, Be and Kramer borate with As-Sb) are reviewed briefly below. Some very low grade, large tonnage "resources for the future" are found in the "lake beds". Koeberlin (1937) reported about 400 Mt of material containing 0.2 ppm Au (=80 t Au content) in Tertiary lake beds near Hartsel, Colorado (Cripple Creek region). Similar material is to be found in parts of the Death Valley sequence.

The presently exposed late Tertiary volcano-lacustrine association is in its "prime time" of development. Its geologically older equivalents are substantially less distinct, and rarely recognized or described.

26.4.17. "Volcanic karst"

Volcanics (mostly airborne pyroclastics) filling karst depressions (1), or subvolcanic bodies intruded into a previously karsted limestone (2), are frequently associated with a transitional, interaction mineralization. In the situation corresponding to case (1), residual bauxite may ultimately result following an episode of tropical weathering and leaching. Known example deposits are located on islands distant from the source of the volcanic detritus (e.g. Jamaica; Chapter 23), or in transitional continental-marine sedimentary environment of volcanic island arcs (Solomon Islands, Chapter 5). No important deposits have, so far, been reported from a typical continental volcanic setting. Leaching of karst-filling pyroclastics could also result in precipitation of U minerals in the carbonate, as in the small La Domitila deposit (Villa Aldama district, Chihuahua, Mexico; O.E.C.D., 1979).

In the situation corresponding to alternative (2), the volcanic heat drives hot springs systems, transporting the most volatile metals (Hg,Sb,As). As a consequence, deposits of cinnabar (Terlingua, Texas; Monte Amiata, Italy; Guadalcázar, Mexico; partly Huancavelica, Peru), stibnite and realgar-orpiment (Tajov, Slovakia) contemporary with local volcanism or hot springs activity result, commonly hosted by paleo-karst.

26.4.18. Humid tropical weathering profiles

Economically significant weathering profiles (e.g. metalliferous laterites, Chapter 23) are rare in this setting and are of limited importance. The best example includes residual bauxites formed over the late Paleozoic to Triassic Pahang Volcanics (rhyolite, trachyte, andesite) in southern Malaya, and on the Indonesian Bintan and Singkep

Islands (Patterson, 1967). The bauxite consists mostly of hard gibbsite concretions dispersed in soft, mottled residual clay and the available tonnage is of the order of 10 Mt Al.

26.4.19. Arid regoliths

Occurrences of uranium and vanadium oxides have been found in calcrete overlying young continental volcanics as in southern Nevada (compare Section 23.5.2.) and uranium mineralization comparable with Yeelirrie (Australia) in groundwater calcretes can be expected in settings where felsic vitric volcaniclastics and calcrete coincide.

26.4.20. Metalliferous placers

Economically important placers are virtually unknown in the areas underlain by ancient continental calc-alkaline volcanics although detrital magnetite derived from phenocrysts is strongly in evidence, in mineralogical quantities, in the stream sands. The magnetite beach sands in Japan, New Zealand and elsewhere (Chapter 5) contain a portion of magnetite derived from pre-Quaternary volcanics.

26.5. EPITHERMAL DEPOSITS

26.5.1. Introduction

The term "epithermal" originally assigned to high-level hydrothermal vein deposits formed within the temperature range 100-200° C, peaked in the 1950's (Bateman, 1951; Schneiderhöhn, 1955) and has been resurrected and slightly modified in the past decade. Sillitoe (1977) presented a detailed modern review of epithermal deposits and summarized their general characteristics in the six points paraphrased, with modifications, below. (1) Epithermal deposits are emplaced in andesite to rhyolite associations of the mid- to late-Cainozoic age. Observable relationship to intrusive rocks may or need not be evident. (2) Fissure and fault veins are the principal style of orebodies. These are increasingly complex (branching) upward, are more persistent laterally than vertically and often contain exceptionally high-grade ore shoots called, in the case of Ag and Au, "bonanza" orebodies. (3) Volcanics in the vicinity of veins are usually propylitized or, alternatively, argillized. Sericitization or silicification, often with adularia or advanced argillic alteration occasionally with alunite, pyrophyllite and diaspore, are evident in the more restricted alteration haloes, immediately adjacent to the veins. (4) The deposits are most characteristic for their silver (in argentite, native silver, As/Sb sulphosalts, tetrahedrite) and gold (in electrum, tellurides, rare selenides) contents. Precious metals-carrying veins are gradational into Pb-Zn veins. Copper is less common in the epithermal association (there, it is usually present in enargite, tennantite, luzonite or famatinite), but may form separate orebodies at greater depth. Stibnite, cinnabar, Sn,Bi and other minerals may also be present. The main gangue is a low-temperature quartz (often chalcedony, cherty quartz). Calcite, dolomite, rhodochrosite, barite, adularia, rhodonite, fluorite and alunite are also frequent. (5) Drusy cavities, open space filling, comb, colloform, and crustification textures as well as inter-mineral breccias are frequent, suggesting low confining pressure. Gangue minerals are typically fine-grained, but the ore minerals have a variable grain size and large crystals (e.g. of stibinte) are often found. (6) The shallow emplacement depth beneath the surface (100 to 1,000 m) is emphasized by the abrupt downward bottoming of mineralization and the downward increase of Pb,Zn and Cu sulphides at the expense of Au and Ag.

In Tertiary volcanic areas, epithermal deposits typically formed in the waning stages of volcanism (after the cataclysmatic period), often as a response to emplacement and consolidation of high-level intrusive stocks in depth. Such stocks may be located directly under the volcanic centre or in other locations and in many ore fields they are unknown, although commonly inferred. Sillitoe (1977) generalized that epithermal ore deposition commonly follows less than 1 m.y. to 3 m.y. after accumulation of the host volcanic rocks, but in areas of recurrent magmatism and multiple mineralization ages this can occur earlier or later (up to about 15 m.y. in the San Juan Mts.; Lipman et al., 1976).

Epithermal deposits hosted by the pre-volcanic basement rocks out of which the volcanics had been removed by erosion, are particularly difficult to assign to a parent intrusion. These deposits have some additional characteristics and variations unaccounted for in the six points listed above, which are mostly due to the variable nature of the host rock. In carbonates, for example, replacement Pb-Zn sulphide chimneys and mantos form, enveloped by alteration jasperoid.

Recent fluid inclusion and stable isotope studies demonstrated that epithermal deposits formed from about 100-370° C hot (the most common filling temperatures were 200-300° C) low salinity, slightly acid fluids, dominated by meteoric waters. "Magmatic" and other waters, however, were also identified locally. In the thoroughly studied Creede, Colorado, ore depositing system (Barton et al., 1977), the fluids were part of a freely convecting hydrothermal system leaching metals and sulphur from the wallrocks and redepositing them according to solubilities. Minerals with normal (=more soluble in hot solutions) solubilities were precipitated near the top of the convecting cell, as a direct consequence of boiling and the loss of acid components. They are capped by sericite-altered wallrocks. Minerals with retrograde solubilities (e.g. anhydrite and molybdenite) were experimentally determined to have concentrated in the hottest part of the system (that is, at a substantially greater depth). A stockwork molybdenite mineralization was predicted possibly to be present at some depth. Both sets of ores would be separated by a wide gap between the shallow and deep facies of mineralization.

The beauty of the Barton et al. (1977) model lies in its predictive value, suggesting that at some depth under an epithermal deposit (or a corresponding alteration cap even if no actual economic mineralization had formed) one can, with a great deal of luck, find a stockwork Mo or a porphyry Cu deposit. Even if one multiplies this assumption with some healthy scepticism realizing that geological systems were rarely perfect, this is a very important piece of exploration strategy. Since the publication of the Barton et al. (1977) pioneering paper, several examples of coexisting bi-level epithermal/"porphyry" mineralized systems have been recorded in the literature. Berbeleac and David (1982) reported one from the Musaria ore zone near Brad (Apuseni Mts., W. Rumania). There, late Tertiary Ag and Au tellurides with native Te, pyrite, chalcopyrite, alabandite, etc. in quartz-clay gangue are hosted by propylitized and argillized subvolcanic dacite and quartz diorite in the upper part of a mineralized system. Porphyry copper showing in zonally altered identical host rocks is located at a greater depth.

26.5.2. Paleovolcanic setting and structural control of epithermal deposits

Virtually every epithermal vein is controlled by a discernible fault or a fracture, so that immediate structural control of these orebodies is beyond doubt and has been extensively publicised in the "age of structural control" in the 1930s to 1950s (e.g. Newhouse, ed., 1942). A recognition that on the ore field or district-scale groups of veins (or other orebodies) are also controlled by paleovolcanic centres, such as calderas, stratovolcanoes or volcanic-tectonic depressions, came later (e.g. Steven and Ratté, 1960). Brief syntheses appeared almost simultaneously in the United States (Albers and Kleinhampl, 1970) and in the U.S.S.R. (Smirnov et al., 1968; e.g. his Fig. 216 on p. 506). Continuing work mostly by members of the U.S. Geological Survey in the San Juan Mts., Colorado (Lipman et al., 1976; Steven and Lipman, 1976) and elsewhere in the western United States (e.g. Byers et al., 1976) provided a wealth of data especially on the development of calderas and cauldrons and associated mineralization. The publicity thus generated had some negative side-effects in monopolizing the issue and giving an impression that calderas are the only significant paleovolcanic features to control ore.

Elsewhere, however, as in the Andes (Sillitoe, 1977), the Rumanian Carpathians (Ianovici and Borcoş, 1982) and in other areas, former stratovolcanoes were the more important controls to ores. Central and linear volcanic-tectonic depressions, linear fault zones and all types of "ordinary" faults, apparently unrelated to any particular centre, also control epithermal orebodies.

CALDERAS AND CAULDRONS

A caldera "is a large volcanic collapse depression, more or less circular or cirquelike in form, the diameter of which is many times greater than that of any included vent". "The name cauldron is often applied to calderas formed, at least in part, by passive subsidence into a broad, shallow magmatic reservoir; there is no sharp demarcation between calderas and cauldrons" (Williams and McBirney, 1979). Calderas and cauldrons are the result of foundering due to volume loss in the subsurface. Such a loss was accomplished either by a subsurficial withdrawal (drainage) of magma (e.g. Katmai, Hawaiian, etc. types) or by explosive eruptions of siliceous pumice (Krakatoa, Valles types). The "Valles type" (after the Pleistocene Valles caldera near Los Alamos, New Mexico, in the bimodal Rio Grande Rift) is close to the "Nevada type" (an informal name not listed by Williams and McBirney, 1979) named after another well studied locality, the 16-9 m.y. old Timber Mountain-Oasis Valley resurgent caldera and cauldron complex in southern Nevada (Byers et al., 1976). Most of the mineralized paleo-calderas in Colorado, Nevada and California are variations of the "Nevada type", although the Timber Mountain complex itself contains almost no ores.

To a prospector searching for ores, modern (active) or very young (e.g. 1-3 m.y., like the Valles Caldera) calderas are of little consequence, because they have not reached the ore producing stage that tends to come later. They are, however, a good model for the interpretation of the paleo-calderas, now preserved as relicts in various stages of erosional or tectonic dismemberement. Many caldera complexes and particularly the mineralized ones, have had a long and turbulent, multistage history marked in particular by numerous events of resurgent volcanism and intrusive emplacement. Understanding of this history in order to interpret the former position and function of preserved paleo-caldera relicts, is important in the ore search. Fig. 26-5 reprinted from Byers et al. (1976) shows in detail the seven million years of active history of the Timber Mountain-Oasis Valley complex. It started with the initial volcanic eruption and minor local collapse (1,2); passed through several collapse and resurgence stages (3-7); and was completed (stabilized) by crystallization of the magma chamber to form a granitic stock (stages 8-10).

In most paleo-calderas, epithermal ores formed only during the stages 8-10 or shortly after the completion of stage 10 in the above-intrusion region and laterally to the intrusion, particularly along fault intersections. In the San Juan Mountains, the richest ore deposits were emplaced 5-15 m.y. after caldera formation, genetically related to minor intrusions of quartz-bearing silicic porphyry (Lipman et al., 1976) so there are many exceptions to the rule. Intra-caldera, rim and ring and outer caldera faults, acted as important conduits of the mineralizing solutions and sites of ore deposition at the time of active hydrothermal circulation, regardless of the actual age of faulting. In calderas filled by a thick internal moat, the buried faults often remained open and active, propagating

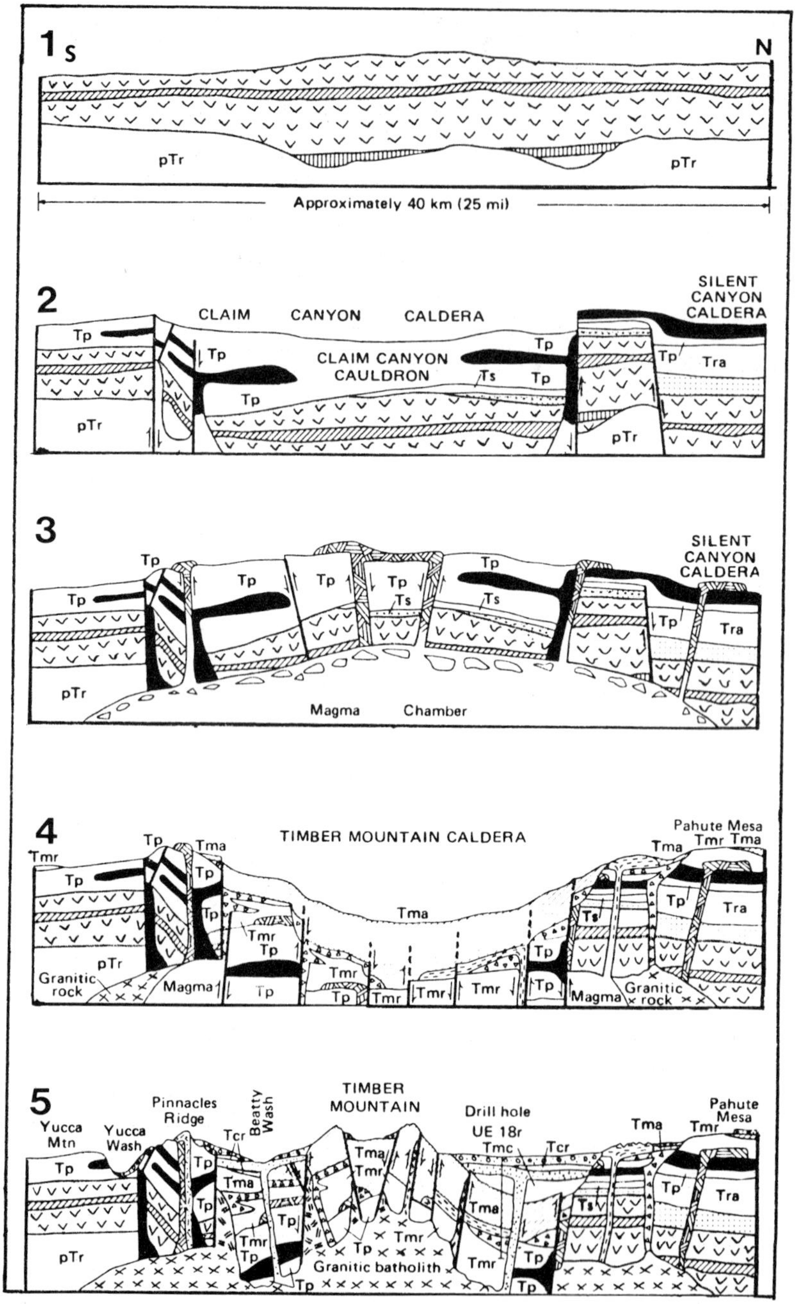
1
S
N
pTr
pTr
Approximately 40 km (25 mi)
2
CLAIM CANYON CALDERA
SILENT CANYON CALDERA
CLAIM CANYON CAULDRON
Tp
Ts
Tra
pTr
3
SILENT CANYON CALDERA
Tp
Ts
Tra
pTr
Magma Chamber
4
TIMBER MOUNTAIN CALDERA
Pahute Mesa
Tmr
Tma
Tp
pTr
Granitic rock
Magma
Tra
Ts
5
Yucca Mtn
Yucca Wash
Pinnacles Ridge
Beatty Wash
TIMBER MOUNTAIN
Drill hole UE 18r
Pahute Mesa
Tcr
Tmc
Tma
Tmr
Tp
Ts
Tra
pTr
Granitic batholith

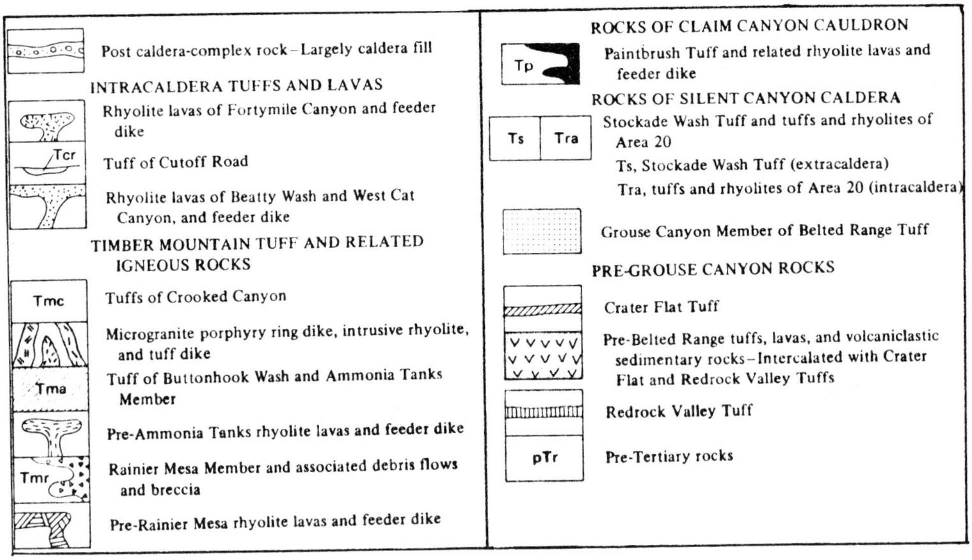

Fig. 26-5. Active period of the tectono-magmatic development of the Timber Mountain-Oasis Valley caldera complex, S. Nevada, illustrated by a sequence of interpretative diagrams. Condensed from Byers et al. (1976).

into their cover. Many control vein systems (as in the Cripple Creek goldfield, Colorado). Several local examples of mineralized calderas are cited in the regional review that follows.

COMPOSITE VOLCANOES (STRATOVOLCANOES)

Modern, predominantly andesitic stratovolcanoes (Fig. 26-6) give numerous surficial manifestations of a possible ore depositing activity taking place at some depth (e.g. fumarolic encrustations, hot springs, sulphur deposits, etc.; Chapter 5). Sillitoe (1975) presented a model of the probable depth zoning of metallic mineralizations under stratovolcanoes, supported by ore occurrences actually recorded from a glacially dissected Pleistocene volcano El Cueva in northern Argentina. There, as well as elsewhere, epithermal veins are found in the upper, central part of a volcanic structure, within or in the vicinity of, former vents.

Levelled paleo-volcanic edifices are often marked by outward-dipping alternating lava and pyroclastic strata and by coarsening of the volcaniclastics. An almost unlimited wealth of ore setting examples in dissected paleo-stratovolcanoes can be gleaned from the recent publications describing the classical, although very

Fig. 26-6. Glaciation-dissected Pl to Q stratovolcanoes on the Altiplano between Arica and La Paz, South America.

little known western Rumanian (Transylvanian) Au-Ag and polymetallic province (e.g. Ianovici and Borcoş, 1982; Lang, 1979 in English; Ianovici et al., 1976, in Rumanian, with excellent sections of paleovolcanoes). Additional examples have been described located in the western United States, in the Andes, in Japan, Kamchatka, New Zealand and elsewhere.

Several distinct styles are apparent in regions of dominantly mafic andesitic subaerial volcanism with a moderate explosive index and rigid (crystalline) basement as in the Miocene and Pliocene volcanic terrain in north-western Rumania. Mineralized mushroom-shaped bodies of dense andesite, andesite porphyry or andesite breccia are particularly characteristic (Figs. 26-7,8). In a typical "mushroom", the greatest density of epithermal, vertically-zoned fissure veins is contained within a 300-1000m vertical interval centered around the level of the rapid widening of a narrow andesite stem. This level, in turn, usually corresponds to the former paleosurface on which the volcanism started, or to the top of a crystalline basement. The andesitic "mushrooms" have usually been interpreted in a general way only as "cores of stratovolcanoes", but at least four alternative interpretations are possible (compare also Williams and McBirney, 1979; Fig. 26-9): (1) small shield volcanoes; (2) lava plugs filling central conduits of large composite volcanic cones and solidified entirely in the subsurface; (3) upheaved domes, formed by congellation of extruded lava in a volcanic crater and (4) intrusive domes. The second alternative appears to be the most common and similar plugs may also have solidified outside the central vents. Much of the rock composing the plugs is an ordinary dense and jointed mafic andesite to andesite porphyry. The former is unrecognizable in a hand specimen from surficially extruded lavas. It is a matter of semantics whether the plugs should or should not be named subvolcanic

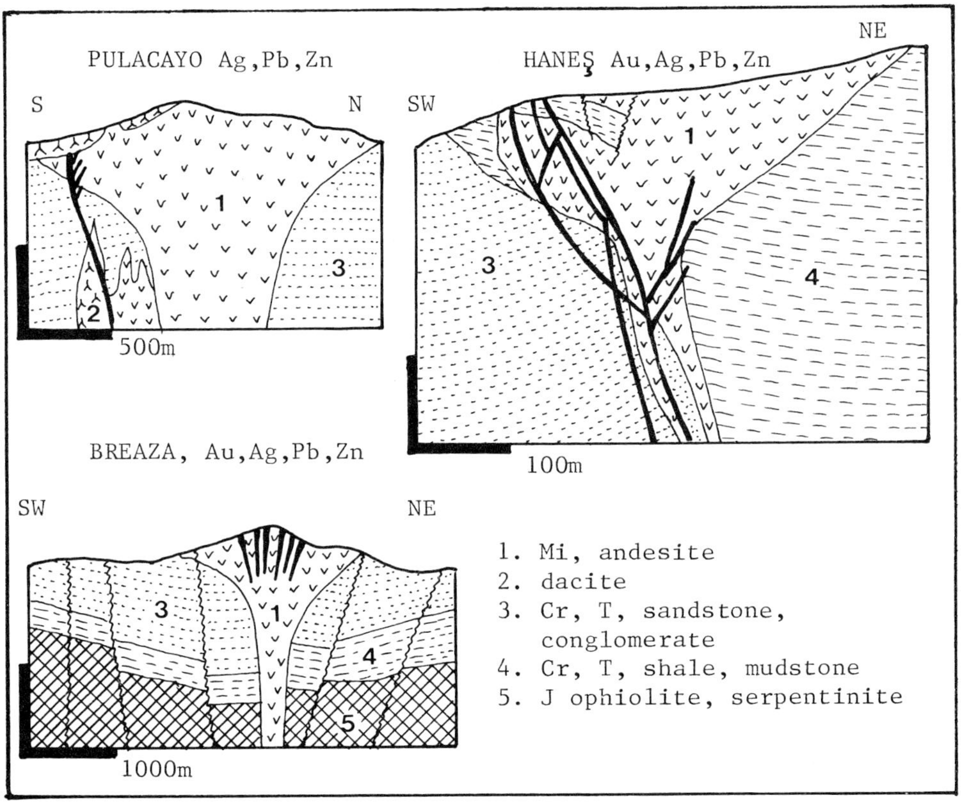

Fig. 26-7. Single mushroom-shaped bodies of dense andesite, mineralized by epithermal Au-Ag and Pb,Zn,Cu veins. Selected examples from Bolivia and from western Rumania. Modified after Lyons (1963), Ianovici et al. (1976) and Ianovici and Borcoş (1982).

intrusions. Williams and McBirney (1979) considered subvolcanic intrusions to be the holocrystalline stocks at a depth of 2 km or more beneath crater floors (1-2 km in Tahiti). Others (including this author) would call these epizonal intrusions, intrusive stocks, plutons, etc. In any case, the holocrystalline stocks are generally responsible for the epithermal mineralization above them and hypabyssal ores (such as porphyry coppers) may appear when their tops are reached.

The andesite plugs are sometimes gradational with depth into microdiorites without a sharp boundary, and both rock types host epithermal veins (e.g. in Herja, Rumania; Petrulian, 1973; Fig. 26-10). Not all andesite plugs need be mushroom-shaped and cylindrical. Pipe-like and dike-like varieties are also common (Fig. 26-11). It is likely that most of the mineralized volcanic plugs

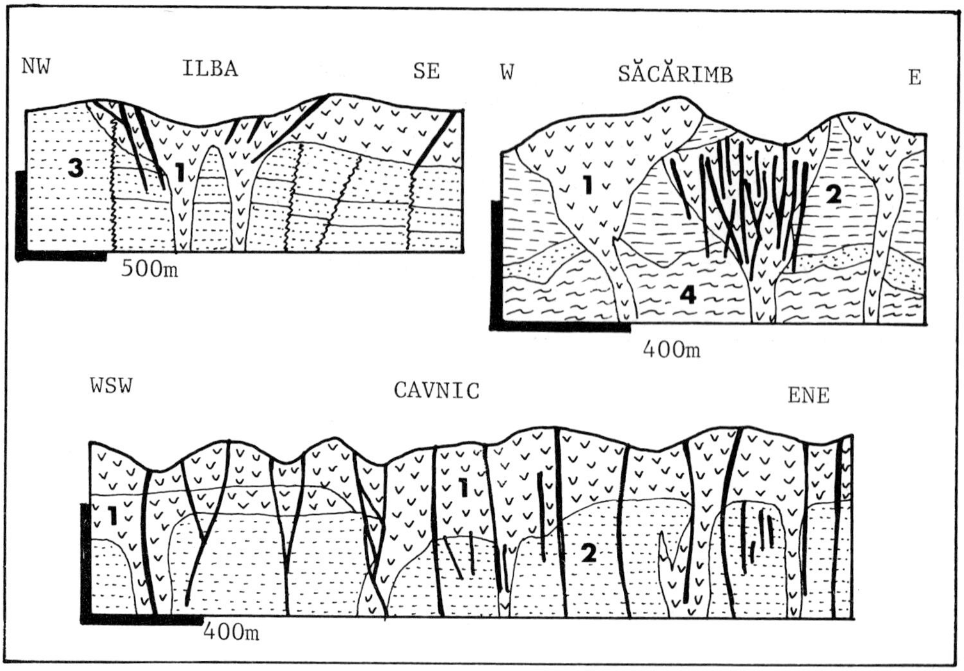

Fig. 26-8. Composite, mushroom-shaped bodies of dense andesite, mineralized by epithermal Au-Ag and Pb,Zn,Cu veins. Selected examples from the Apuseni and Gutăiu Mts., Rumania. Generalized after Ianovici and Borcoş (1982) and Petrulian (1973). 1=Mi andesite plugs; 2=T marls, shales; 3=T, Cr_1 sandstone, conglomerate; 4=PZ phyllite.

represent the terminal local volcanic products, and that they did not communicate with the surface.

Breccia pipes and breccia-filled vents, some of them mineralized, are another control to mineralization found in volcanic centres (Fig. 26-12). In centres in which breccias are associated with andesitic plugs, they are usually younger and the ore they may contain could either be an older, relict fragmental ore, or a post-breccia, hydrothermally introduced ore (the relict fragmental ore may occur as mega-fragments or part of the rock flour in a breccia matrix. Part of such an ore could be redistributed post-depositionally).

Mineralized felsic volcanic centres (Fig. 26-13) such as those containing disseminated and stockwork cassiterite in Bolivia (Potosí, Llallagua, Chorolque) or chalcopyrite veins and stockworks (Ashio, Japan), show some similarity in shape to the "mushroom" andesite bodies but pervasive alteration has often obliterated the distinction between subvolcanic and plutonic bodies and sometimes even between magmatic intrusions and breccias.

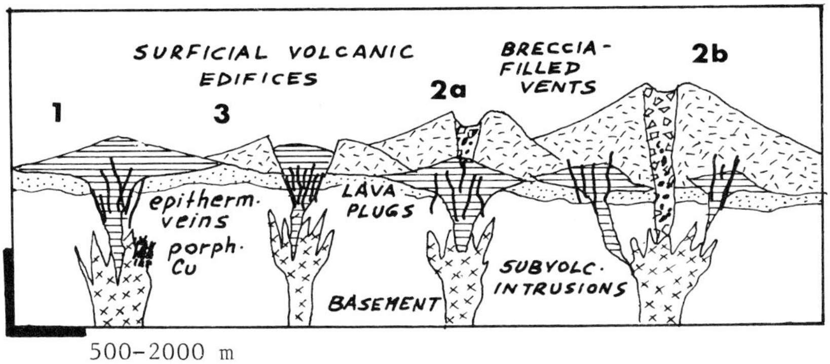

Fig. 26-9. Alternative models for the formation of mineralized, mushroom-like andesite bodies. 1=shield volcano; 2a=lava plugs filling central conduits of volcanic cones; 2b=lava plugs outside central vents; 3=upheaved domes in craters. From Laznicka (1984).

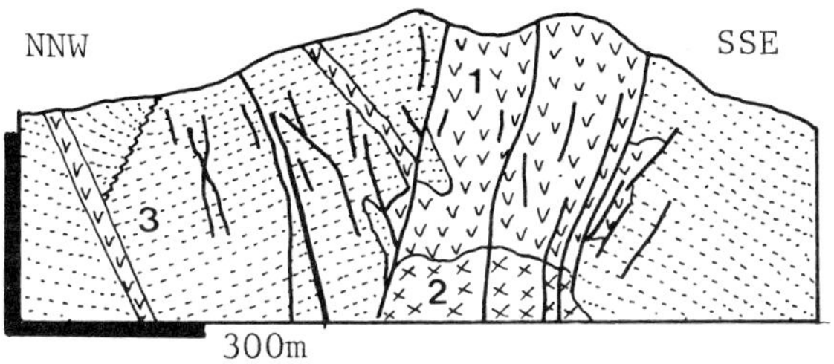

1. Pl, pyroxene-amphibole andesite
2. porphyritic microdiorite
3. Eo and Mi_3-Pl_1 sediments

Fig. 26-10. Dense Miocene andesite lavas filling a plug are gradational into a microdiorite at some depth. Both rocks host epithermal Au,Ag,Pb,Zn,Cu veins. Herja deposit, Gutăiu Mts., N.W. Rumania. Simplified and generalized after E.M. Herja and R. Jude in Petrulian (1973).

VOLCANO-TECTONIC DEPRESSIONS

Volcano-tectonic depressions are approximately transitional between calderas and small rifts or grabens and in terms of dimensions they are usually larger than calderas. They are collapse structures framed by pre-existing faults, so they tend to be rectangular rather than circular. Spectacular collapses as in the Pleistocene Lake Toba depression, Sumatra (van Bemmelen, 1949) were accompanied by large

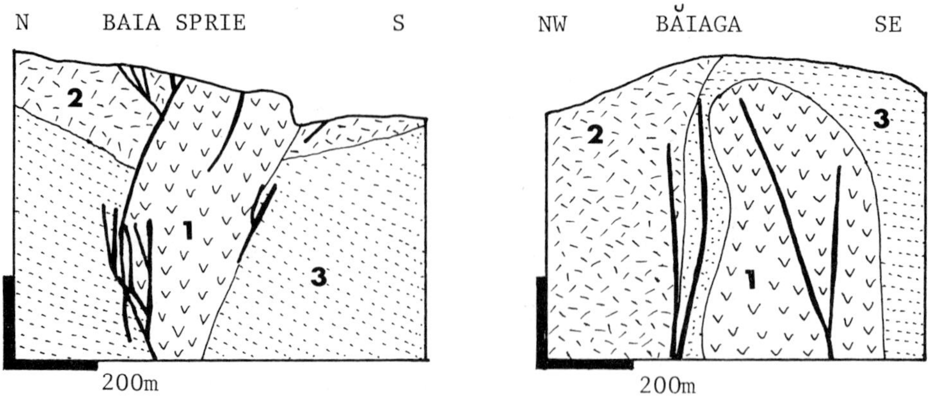

Fig. 26-11. Cylindrical, pipe-like, dike-like, etc. subsurface plugs of mineralized dense andesite: Baia Sprie, Gutăiu Mts., Băiaga, Apuseni Mts, both in Rumania. 1=Mi-Pl andesite plugs; 2=Mi, older andesite lavas; 3=Cr+T sediments. Simplified after Ianovici and Borcoş (1982) and Ianovici et al.(1976).

pumice and felsic tuff eruptions. The Taupo Volcanic Zone, New Zealand (Chapter 5) is a frequently quoted recent example of a volcano-tectonic depression, complete with evidence that epithermal ores are presently forming there at a depth. The Tertiary Hauraki Goldfield, New Zealand; Elkhorn Mts.and their Au,Ag and Pb,Zn province, Montana, are either partly or entirely former volcano-tectonic depressions.

OTHER STRUCTURES

Despite the elegance of caldera, stratovolcano and volcano-tectonic depression ore control, the majority of epithermal deposits cannot easily be fitted into any of them. Of the 80 major Tertiary mineralized volcanic centres in Nevada evaluated by Albers and Kleinhampl (1970), only 17 are certainly or even probably associated with paleo-calderas and none with distinct former stratovolcanoes. Among the rest, the most frequent interpretation of volcanic centres are "intrusions with possible uplift" (e.g. Bodie, Comstock Lode, Tonopah).

Fig. 26-12. Examples of mineralized breccia pipes and breccia-filled vents in andesitic volcanic centres. Roşia Montană, Baia de Arieş and Fericeana-Stanija are in the Apuseni Mts., Şuior is in the Gutăiu Mts., Rumania. Bassick is in the Wet Mts., Colorado, El Teniente is

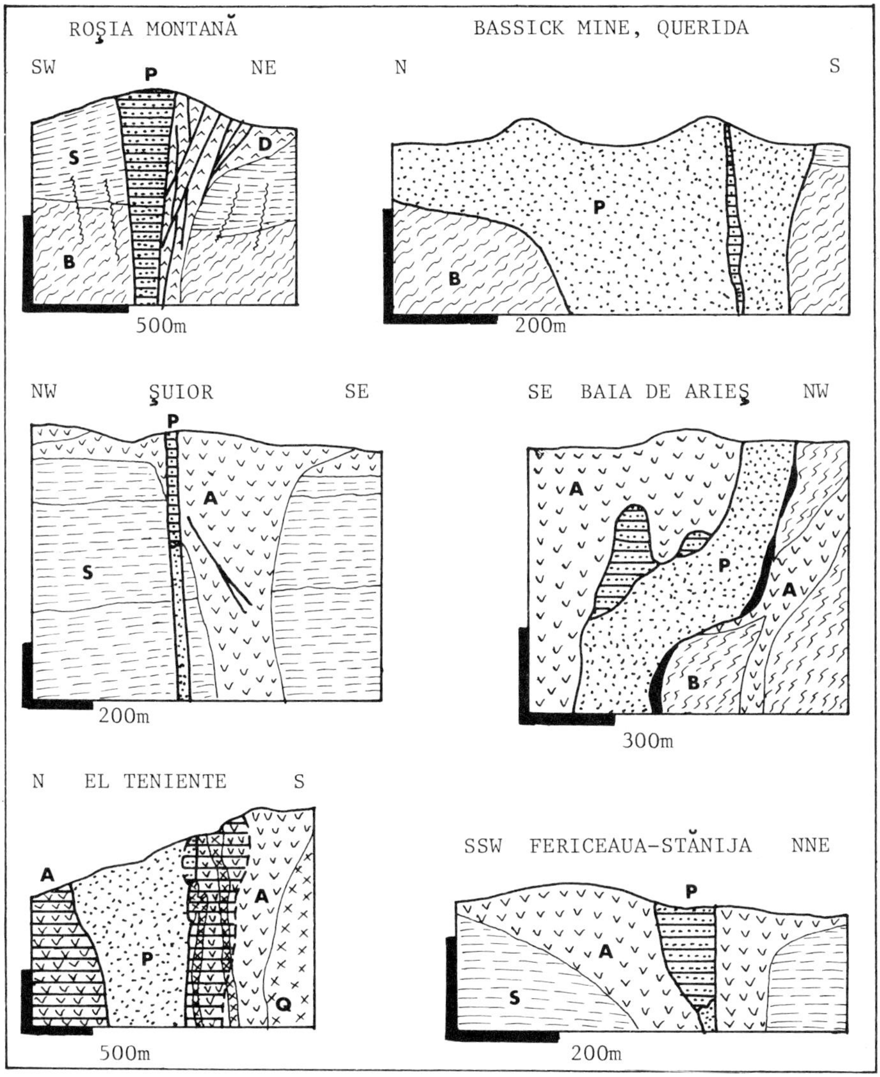

in Chile. All deposits except El Teniente have disseminated stockwork or vein Au,Ag(Pb,Zn) mineralization shortly postdating the breccia pipe formation. The El Teniente pipe is post-ore, largely unmineralized, while the porphyry Cu-Mo ore is in the wallrock andesite. P=Mi-Pl breccia pipe; A=andesite; D=dacite; S=Cr,T sediments; Q=Pl quartz diorite; B=pre-Tertiary basement; black=Pb-Zn replacements and skarns; horizontally ruled=disseminated and stockwork Au,Ag(Pb,Zn).

Simplified and modified after Ianovici et al. (1976); Ianovici and Borcoş (1982) and Camus (1975).

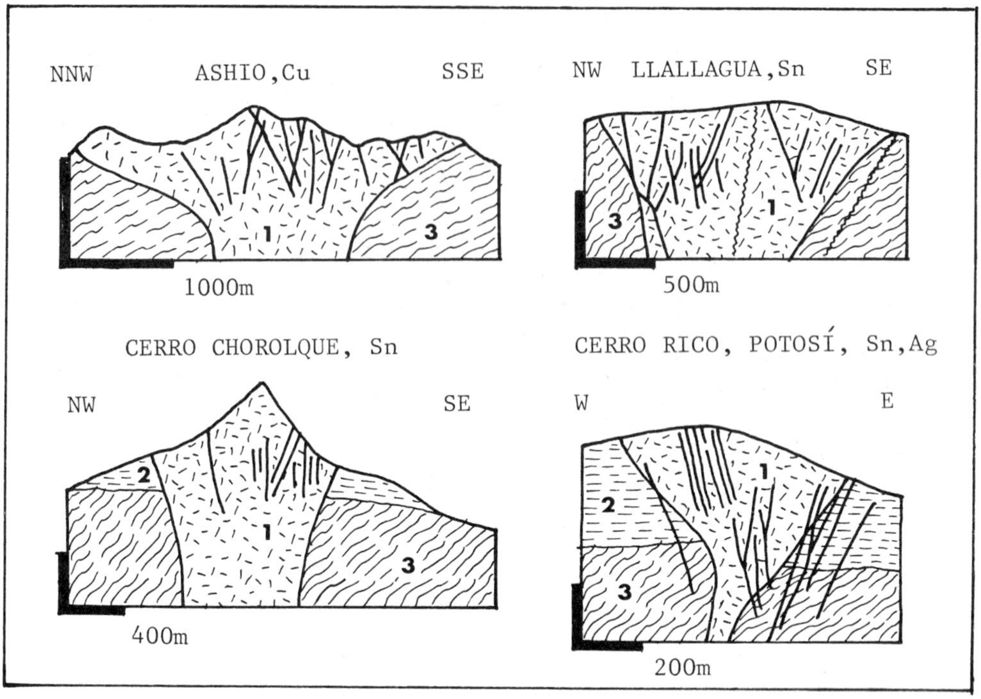

Fig. 26-13. Mineralized aggradational felsic volcanic centres. 1=Cr-Mi hydrothermally altered felsic body; 2=surficial volcanics, volcaniclastics; 3=Cb-Pe (Ashio), Or-S (the rest) marine slate, arenite. From LITHOTHEQUE, simplified after Nakamura (1971); Sillitoe et al. (1975); and Grant et al. (1977).

26.6. MINERALIZED EXAMPLE REGIONS

26.6.1. Late Cretaceous and Tertiary epithermal Au-Ag and Pb-Zn provinces

SAN JUAN VOLCANIC FIELD, S. COLORADO

San Juan field (Steven and Lipman, 1976; Lipman et al., 1976; Fig. 26-14) is an erosional relict of an Oligocene-Miocene volcanic pile situated in the southern extension of the Colorado Mineral Belt. Volcanic activity started in the lower Oligocene period when a coalescing assemblage of andesitic stratovolcanoes formed over a variety of basement rocks (Proterozoic metamorphics and granites, Cambrian to Cretaceous platformic sediments). Thick and areally dominant sequences of acid calc-alkaline ash-flows and related lavas formed in the middle and late Oligocene times as a response to large volume recurrent eruptions from 17 identified or postulated calderas.

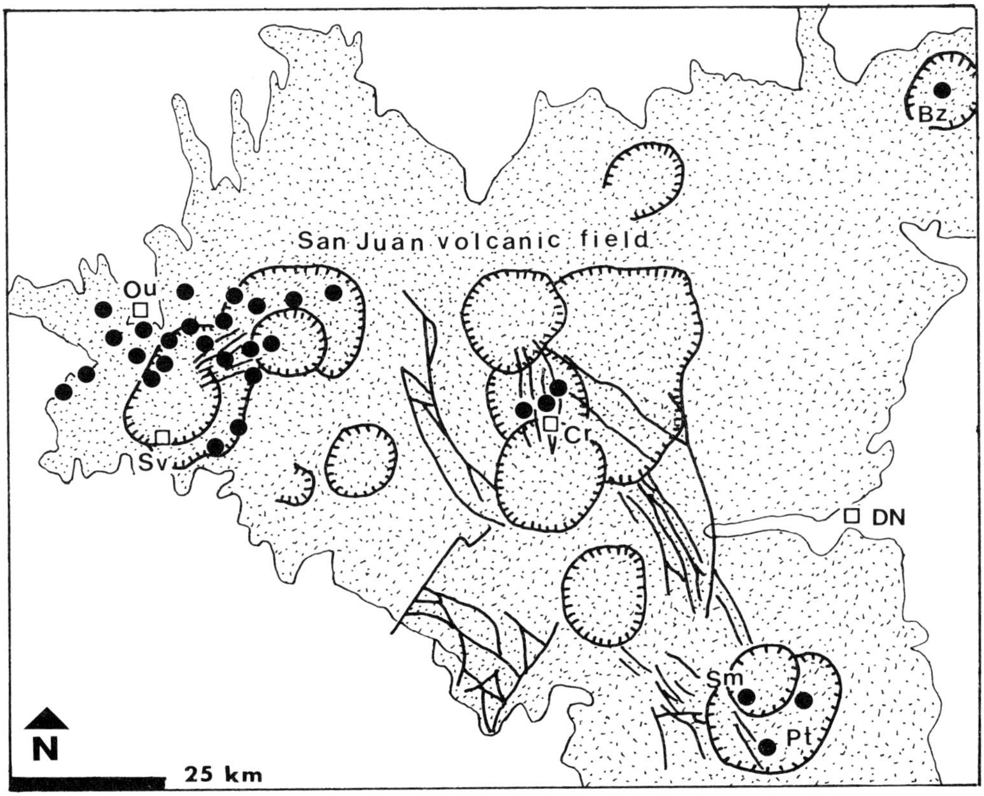

Fig. 26-14. The San Juan volcanic field (stippled), Colorado, showing major paleo-calderas, faults and Pb,Zn,Cu and Au,Ag orebodies (black dots). Abbreviations: Bz=Bonanza; DN=Del Norte; Cr=Creede; Ou=Ouray; Sm=Summitville; Sv=Silverton; Pt=Platoro. Modified after Steven and Lipman (1976).

By the late Miocene period, the cataclysmatic explosive volcanism was superseded by sporadic occurrences of bimodal basalt-quartz latite suite, increasing in magnitude towards the south-east (Rio Grande Rift).

Although the ash-flow eruptions and caldera subsidence are attributed to the development of a batholithic magma chamber at some depth, the intrusive rocks presently exposed in the field are mostly dikes and small stocks. Most of them are exposed on the flanks of the volcanic field. The region contains two mutually independent suites of metallic deposits: (1) the "Laramide" (Cretaceous to Eocene) suite associated with granitic plutonism (e.g. Rico, La Plate, Ouray) and (2) the epithermal subvolcanic suite, formed within a time span of 35-10 m.y. by largely meteoric water convection heated by evolving plutons and by minor intrusions of quartz-bearing silicic porphyries. In this chapter we are concerned only with suite (2).

Several hundred vein and lesser replacement and breccia pipe epithermal orebodies represent some 700 Tt Pb, 420 Tt Zn, 70 Tt Cu, 7 Tt Ag and 260 t Au. They are irregularly and very unevenly distributed. The eastern portion of the volcanic field has only three small ore clusters (Bonanza, Summitville, Platoro). The central portion hosts the important Creede field. The western portion is by far most extensively mineralized containing numerous deposits surrounding the Silverton caldera.

In the Creede ore field (Steven and Ratté, 1965; Barton et al., 1977; Fig. 26-15), several N.N.W. and N.W. fissure vein systems are confined to a complex graben that extends N.N.W. from the Creede caldera and connects it with the San Luis caldera (Fig. 26-16). The systems are up to 3 km long. The mineralization has been dated at 24.6 m.y., an age corresponding to emplacement of a minor local felsic intrusion and 1-2 m.y. later than the age of the youngest volcanics in the area. The most common host rocks are devitrified welded tuffs. The crustification or breccia-cementing vein filling carries galena,

Fig. 26-15. Entrance to the Creede gorge, San Juan Mts., Colorado, showing workings on the Amethyst vein system.

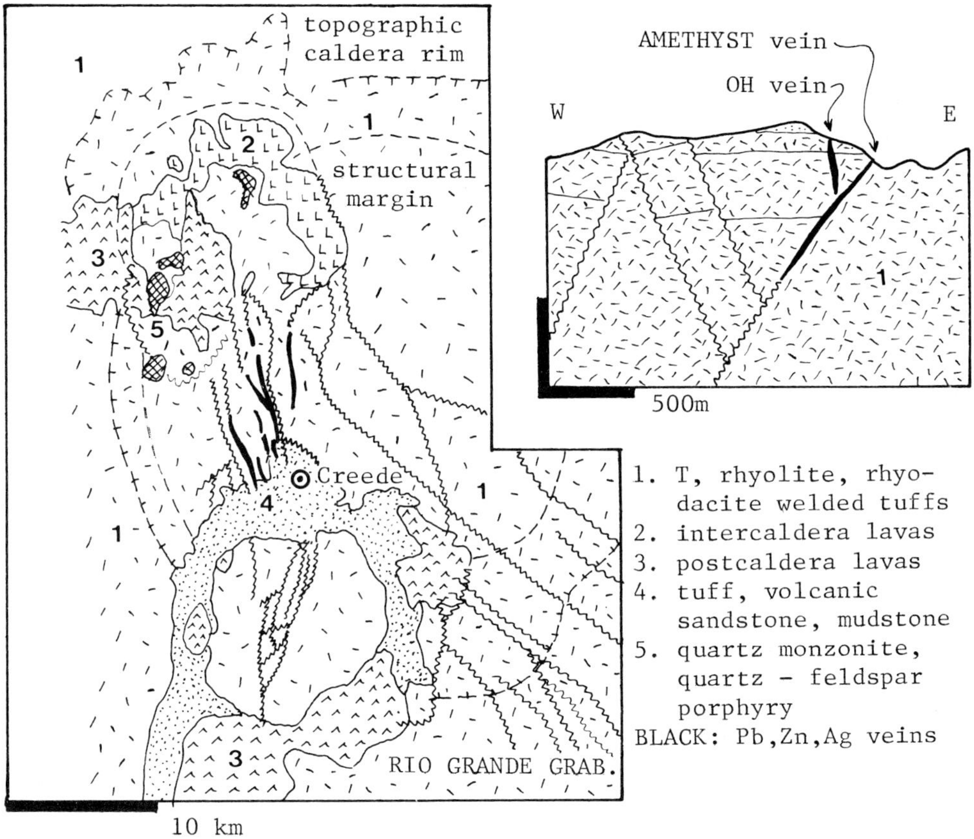

Fig. 26-16. Tertiary calderas and epithermal veins in the Creede field, San Juan Mts., Colorado. After Steven and Lipman (1976), section from LITHOTHEQUE.

sphalerite, pyrite, minor chalcopyrite, rare Ag sulphosalts in quartz, rhodochrosite, adularia, chlorite and siderite gangue. There is a minor development of replacement jasperoid, mostly in the fault gouge. Wallrock alteration (silicification, carbonate impregnation, chloritization, sericitization, argillization) is generally light and confined to the immediate vicinity of the veins.

Silverton district (Lipman et al., 1976; Varnes, 1963; Fig. 26-17) has several ore fields and clusters (Eureka, Red Mountain, Telluride, South Silverton) along the fringe of the Silverton caldera. The richest orebodies are concentrated in the north-east and north-west. In the north-east, veins fill faults and fractures along the northern side of the N.E. trending Eureka Graben. This is another connecting graben system located between two calderas (Silverton and

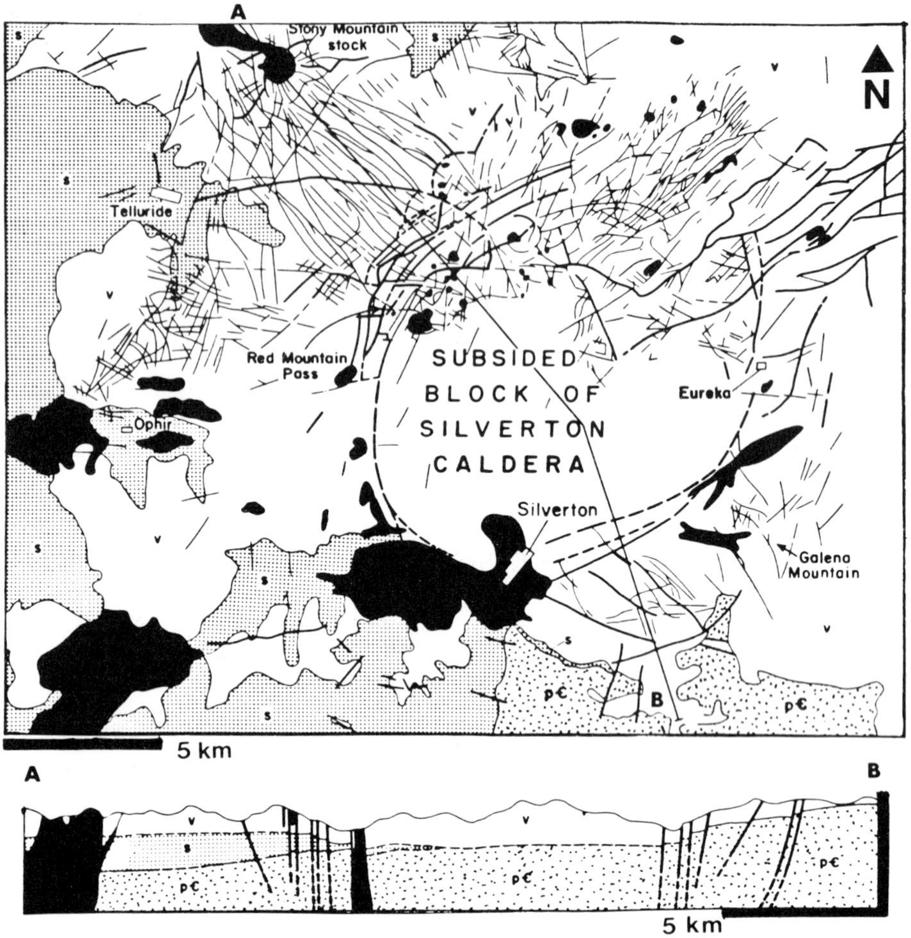

Fig. 26-17. Pb,Zn,Ag,Au veins around the Silverton Caldera, Colorado. Black=Ol-Mi intrusive rocks; v=Ol andesite to rhyolite volcanics; s=MZ sediments; pC=PCm metamorphics; thick black lines=faults; thin black lines=hydrothermal veins. From Varnes (1963).

Lake City) formed during resurgent doming of the latter in the late Oligocene period. There, the Sunnyside mine (Casadevall and Ohmoto, 1977) is the largest producer. In the north-west, a system of sub-parallel N.W. fissure veins is situated perpendicular to the ring fault system of the Silverton caldera (Idarado and Camp Bird mines). In addition to the veins, mineralized breccia pipes occur within the ring faults themselves (Red Mountain field).

The vein deposits such as the Sunnyside and Idorado mines (Table 26-2) contain banded to massive galena, sphalerite, pyrite, lesser chalcopyrite, bornite, hematite, gold telluride filling in quartz, rhodochrosite and minor fluorite gangue. These veins are spatially and genetically associated with silicic intrusions (coarse quartz rhyolite porphyry), substantially younger than the surrounding areal volcanics. The intrusions have strong silicification haloes. Replacement massive sphalerite, galena, chalcopyrite and pyrite bodies also occur in Eocene conglomerates in the Idarado mine. At Red Mountain, numerous small chimney-shaped massive sulphide replacement deposits occur in silicified and advanced argillized (clays, alunite, diaspore) breccia pipes related to stocks and plugs of quartz latite porphyry.

GREAT BASIN IN NEVADA (MINOR OCCURRENCES IN E.CALIFORNIA, UTAH AND S.W. IDAHO)

Over 100 clusters of epithermal deposits affiliated with a 38-20 m.y. period of calc-alkaline continental volcanism and a 16-10 m.y. period of bimodal volcanism, represent an aggregate tonnage of some 14 Tt Ag and 600 t Au, plus minor quantities of Pb,Zn,Cu. In contrast to the San Juan Mountains, gold-only (e.g. Goldfield) and gold-silver (e.g. Comstock Lode, Tonopah) fields with only minor base metal sulphides, are much more important economically than the rest of the small Pb-Zn-Ag veins.

Goldfield, Nevada (Ashley and Keith, 1976; 145 t Au, 39 t Ag; Fig. 26-18) is a Miocene volcanic centre composed of quartz latite, andesite and dacite, unconformably overlying a Cambrian black argillite (compare also Chapter 17 and Fig. 17-10). The volcanics are extensively kaolinized, alunitized and silicified over an area of more than 40 km^2, but significant gold deposits have so far been found only in a 1,200 m long, broad, N.-S.—trending, 35° E-dipping zone in the western part of the centre, near the town. There, the brownish-yellow ground is dotted by shafts that followed several productive, gently dipping sheet-like alteration zones ("ledges"), several metres wide. These zones contained shoots of cryptocrystalline quartz with specks and lenses of auriferous famatinite, enargite, bismuthinite, goldfieldite, electrum and Au-Ag tellurides. The age of mineralization is 20-21 m.y.

Comstock Lode in Virginia City, Nevada (241 t Au, 5.8 Tt Ag; Whitebread, 1976; Albers and Kleinhampl, 1970; Fig. 26-19) is a N.N.E.-trending mineralized fault transecting middle Miocene explosive continental volcanic pile (rhyodacite to andesite flows, breccias) intruded by a high-level granodiorite. The lode fault dips east away from the intrusion and it contains several ore shoots (bonanzas) of chalcedonic quartz interspersed with bunches of argillized and silicified wallrocks in places phenomenally rich in electrum, argentite and native silver. The rocks immediately adjacent to the shoots display advanced argillic alteration and silicification and there is a visually conspicuous regional bleaching and propylitization

Table 26-2. Epithermal Au-Ag and Pb-Zn provinces, a selection of examples

PROVINCE	BASEMENT	CONTIN. VOLCANICS	MINERALIZATION	REFERENCES
San Juan Mts., S. Colorado, U.S.A.	1.7 b.y. PCm metam. and granites; Cm-Cr platform. sediments; Eo, granitic plutons, volc., sedim.	Ol-Mi andes. strato-volc. changing to felsic ash-flow tuffs from calderas. Termin. by bimodal volcanics, intrus.	Ol-Mi epith. Pb-Zn, less Ag, Au fiss. veins // faults, minor mass. sulph. brecc. pipes; rel. to small post-caldera intrus. 700 Tt Pb, 450 Tt Zn, 6.4 Tt Ag 260 t Au.	Lipman et al. (1976); Steven and Lipman (1976)
Great Basin, Nevada and Utah, U.S.A.	Pt_1-Pt_3 metam., PZ, MZ marine metased., metavolc., MZ dior.-granodior. batholiths	Mi and., dac., qtz. latite, rhyol. pyrocl. and lavas; small stocks of granod., qtz. porph. Latest bimodal assoc.	over 100 scattered groups of Mi-Pl epith. deposits. Au and Ag-Au most important, Pb-Zn numer. but small. About 600 t Au, 14 Tt Ag	Albers and Kleinhampl (1970)
Central-Western Mexico, 2400 km long N.W.-S.E. belt	Pt-MZ marine metased. and metavolc., intr. by Cr_3-T_1 plutons; widespr. "black shale" horiz. T_1 red beds	Eo-Q andes., lesser dac., rhyol., basalt. Pl volc. locally bimodal or alk., small dior. to qtz. monz. and qtz. porphyry stocks	X000 occur. of Pb-Zn-Ag, Ag and Ag-Au epith. veins, widespread but small Sn in rhyolites. Minim. 140 Tt Ag, 1 Tt Au	Clark et al. (1982)
Gulf of Fonseca to Costa Rica, Centr. America	Pt metam., topped by PZ-MZ sedim. and volc. intr. by Cr_3-T_1 plutons	T_1-Q, andes. to rhyol. contin. volc., rare subvolc. qtz. latite dikes	Mi, small scattered Pb-Zn-Ag and Au-Ag veins (e.g. Aguacate). 5.4 Tt Ag, 260 t Au	Sillitoe (1977)

Central Perú mining region	PZ,MZ "miogeoclinal" sedim. (abund.carbonate,black shale) topped by T_1 red beds	$T_{2,3}$ andes.,dac., rhyol. centres, deeply eroded, abundant plutonic equivalents	Widespr. epitherm.veins and carb.replac.,lesser porphyry Cu. About 10 signif. epithermal Pb-Zn-Cu-Ag-Au dep. contains 11 Tt Ag, 700 Tt Zn, 600 Tt Pb	Putzer (1976) Petersen (1965)
Southern and Central Bolivia min. region	Or to D "miogeos." gray to black slates, quartzites; Cr-T red beds; Tr-Mi felsic plutons	scattered remn.of andes.,dac.,rhyol. volc.centres,intr.by high-level rhyol.to qtz.latite stocks, dikes	Ol-Pl,large high-level dissem.,stockw.,vein Sn miner.stocks (Lllallagua) and Sn-Ag deposits (Potosi) repres. 4.5 Mt Sn, 70 Tt Ag. Lesser Pb-Zn-Ag veins (Pulacayo) repr.600 Tt Zn, 3.5 Tt Ag, 100 Tt Pb	Ahlfeld and Schneider-Scherbina (1964)
Sierras Pampeñas, N.W. Argentina	Block faulted Pt cryst.basem.overl.by PZ+MZ "miogeos." sediments	T-Q, several small andes.,dac.,rhyol. volc.piles; deeply eroded,plutonic equiv exposed	several small groups of epith. Au-Ag-Pb-Zn veins (Capillitas), high-level porph.Cu (Mi Vida)	Sillitoe (1977) Angelelli (1950)
Betic Cordillera, S.E. Spain	Pt-Tr amphib.to greensch.metam.,Pe-Tr meta red beds, Mi mudstone and conglomerate	Mi-Pl, scattered and., rhyol.,trachyandes. small volc.centres, minor brecc.pipes, small intrusive stocks	Cartagena d.,minor epith. Pb-Zn-Ag veins. Major repl.mantos in carb.with greenalite and magnetite,possibly remobiliz. PZ. Rodalquilar,epith.Au-Ag; 4.15 Mt Pb,1.35 Mt Zn	Oen et al. (1975)

Table 26-2 (continued). Epithermal Au-Ag and Pb-Zn provinces, a selection of examples

PROVINCE	BASEMENT	CONTIN. VOLCANICS	MINERALIZATION	REFERENCES
Kremnica-Štiavnica Mts.,Slovak Carpathians, Czechoslovakia	PZ? crystall. rocks, overl. by Tr carbonates, shales	Mi-Pl andes.-domin. centre, intr.by subvolc. rhyol.,dacite porph.,granod. stocks	Mi_3-Pl Štiavnica center is a paleo-caldera,epitherm.Au-Ag-Pb-Zn-Cu veins,buried porph.Cu. Kremnica-epith. Au Estim. 90 t Au,2 Tt Ag	Mahel' and Buday, eds. (1968)
Apuseni Mts., Transylvania, W. Rumania	PZ_1 metam., PZ_{2-3} spilite-keratophyre, J ophiolite, Cr_3-T_1 dior.,qtz.monz., flysch	Mi-Pl volc.pile,andes. domin.,minor subvolc. dac.,rhyol.porph.and qtz.diorite stocks	Mi_3-Pl,epith.veins, stockw.,breccia pipe, Au with Te domin.,lesser Ag,Pb,Zn. Miner. centra under stratovol. Est.400 t Au, 3 Tt Ag	Ianovici and Borcoş (1982)
Japan, Mi-Pl continental volcanic province	variable; ranges from 1.8 b.y. metam.,Cb-Pe maf.marine volc.-sed. to Cr-T granodior.plutons	Mi-Pl,marine to cont. andes.,dac.,rhyol. assoc.,centres, subv. intrus. Regional propylitization	Mi_3-Pl Zn-Pb veins (Hosokura, Toyoha) and Au-Ag bonanza veins (Sado, Mochikoshi) locally coexist with submarine "kuroko". 1.6 Mt Zn, 600 Tt Pb,200 t Au	Saito, ed., (1960)
N. New Zealand neovolcanic prov.	MZ graywacke-argillite (turbidite) association	Eo-Mi andes.,lesser dacite. Pl-Q rhyol. sheets and domes, hot springs activity	Hauraki Goldfield: extensive, epitherm. Au-Ag veins, minor Pb-Zn. 465 t Au, 860 t Ag. Small cinnabar occur. of hot springs origin	Williams (1965)

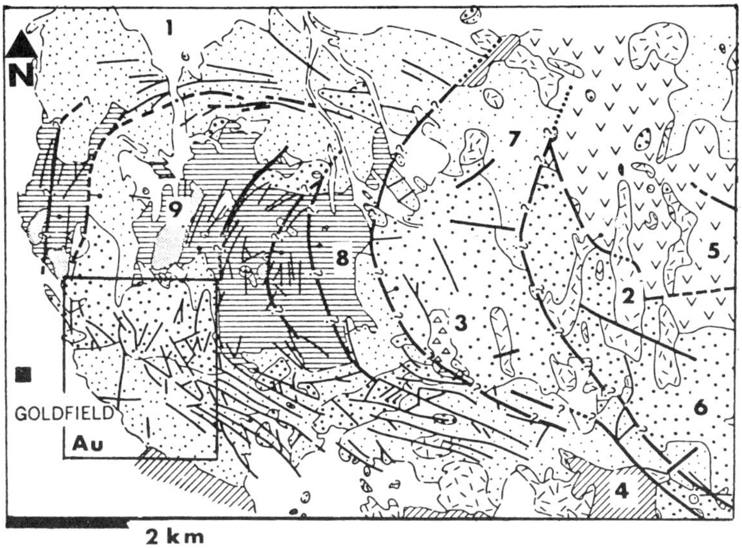

Fig. 26-18. Goldfield, Nevada, epithermal gold field. 1=Q alluvium; younger (Mi) volcanics: 2=andesite, minor basalt; 3=volcanic breccia; 4=dacite vitrophyre; 5=rhyolite flows; 6=dacite; 7=Milltown Andesite; 8=older T volcanics; 9=MZ,PZ alaskite, metasediments. From Albers and Kleinhampl (1970).

on the hills above the town (Virginia City is the type locality of propylitization).

Tonopah, Nevada (Spurr, 1905; Albers and Kleinhampl, 1970; 64 t Au, 4,920 t Ag) is the last significant Au-Ag field in the Great Basin. There, a system of E.-W. and N.E.-S.W.-striking fissure and replacement veins filled by quartz, carbonates, barite, galena, sphalerite, chalcopyrite, electrum, argentite and polybasite, occurs along faults and fractures. The orebodies are situated on the N.W. side of a composite Miocene volcanic centre containing numerous small subvolcanic and plutonic stocks. The mineralized area is affected by widespread albitization and quartz, sericite, adularia alteration. This is surrounded by regional propylitization. There are about 20 major veins, up to 500 m long and 0.3-13 m wide.

Additional localities with production of the order of several hundred tons of silver include: Calico, California (rich pockets of cerargyrite, embolite and native silver were recovered from the secondary enriched zone over low-grade mineralized faults and shattered zones in Miocene volcanics and lake beds); Bodie, California (7.2-8.6 m.y. gold- and silver-bearing quartz, adularia, pyrite "reefs" cut Miocene propylitized and silicified andesites); Silver Peak, Nevada (barite, calcite, argentite veins cut 5.9 m.y. old trachyandesite); Bullfrog near Beatty, Nevada (numerous Au-Ag veins

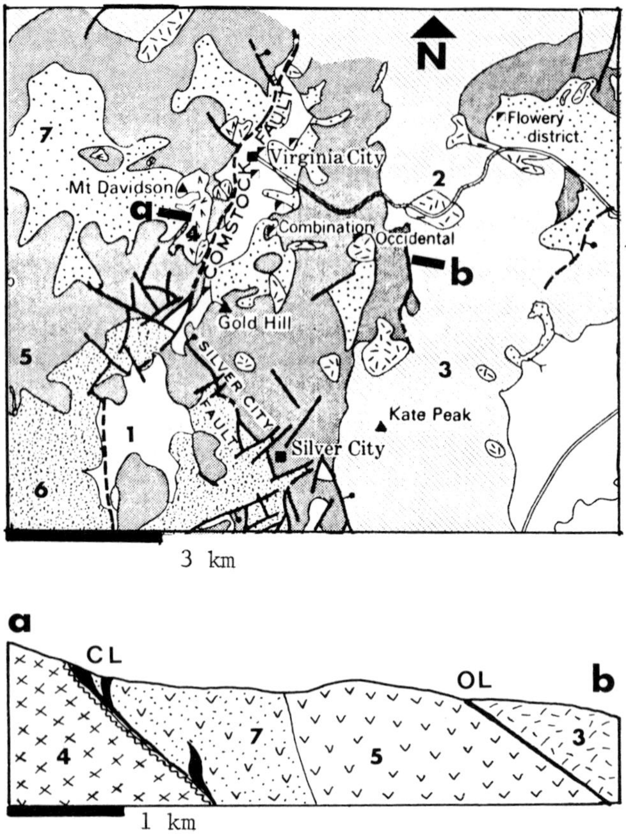

1. Q, unconsolidated sediments
2. Mi, rhyodacite to andesite subvolcanic intrusions
3. Mi, rhyodacite to andesite flows, breccias
4. Mi, granodiorite
5. andesite flows, breccias
6. Mi, rhyolite, volc. sediments, granodiorite
7. alteration bleached units 2,3,5

Fig. 26-19. Virginia City, Nevada (Comstock Lode), an epithermal Au-Ag field. CL=Comstock Lode; OL=Occidental Lode. From Albers and Kleinhampl (1970) and LITHOTHEQUE.

within and alongside the rim of a Miocene caldera); Mojave, Rosamond, Rand, California (late Miocene Ag-rich gold-bearing veins in Precambrian schist basement and silicified andesites and rhyolites); Delamar, S. Idaho (disseminated naumannite in quartz veins hosted by Miocene silicified rhyolite and tuff; 1,200 t Ag produced).

Black (manganiferous) calcite-silver veins

Several rich occurrences of near-surface cerargyrite and native silver mined in the past in the western United States (e.g. Hamilton-Treasury Hill, Nevada, 600 t Ag; Lake Valley field, Nevada, 140 t Ag; Escalante field near Modena, Utah), were recently explained as products of residual enrichment of hypogene bodies of "black calcite" by Smith (1970). This is a coarse crystalline calcite finely intergrown and pigmented by aurorite, an argentian manganate. At Treasure Hill, it occurs in lenses and irregular pods and veins ranging from 0 to 1m in thickness and up to 10 m in length, hosted by limestone and breccia along low-angle bedding plane faults.

OTHER EPITHERMAL LOCALITIES, WESTERN UNITED STATES

Outside the Great Basin, epithermal gold and silver deposits exist in the young volcanic centre of Cripple Creek, Colorado (the largest gold accumulation, 576 t Au; this is partly in alkaline volcanics and treated in Chapter 33); in the Republic District, N.E. Washington; in the Mogollon District, New Mexico; in the Silver Cliff and Rosita Oligocene-Miocene volcanic centres, Colorado and elsewhere. The latter two localities (Sharp, 1978) are interesting since they are hosted by a combination of composite stratovolcanoes and calderas, and for the occurrence of Au, Ag and base metals in deep-reaching breccia pipes.

The Bassick Pipe in the former town of Querida (Fig. 26-20) is a subvertical column about 400 m in diameter emplaced into andesitic flows, breccias and tuffs during subsidence of the Rosita stratovolcano. The pipe is composed of an agglomerate-like mass of subrounded andesite blocks in a rock flour and only its central part is mineralized. The ore minerals encrust the larger fragments and the encrustations are zoned (the innermost shell is mostly sphalerite, lesser galena, changing into Ag sulphosalts and electrum-rich outer shell). Tetrahedrite and Au-Ag tellurides are also disseminated in the argillized matrix. The pipe was followed to a depth of over 600 m and yielded about 1.5 t Au and 40 t Ag. Similar breccia pipes in the Silver Cliff centre reach into the Precambrian basement and contain a large amount of altered basement fragments.

Silver Cliff is of further interest for the fairly extensive deposit of argentian cryptomelane, coating fractures and cementing breccias in Oligocene rhyolite (Hildebrand and Mosier, 1974; Fig. 26-21). This ordinary looking "Mn oxide" assayed as much as 24% Ag, but the silver was unrecoverable by the last century cold processing techniques. The primary source of the silver is not known.

Several large to medium size Pb,Zn,Ag replacement deposits in carbonates reviewed in the following chapter, are also high-level mineralizations coeval with Tertiary volcanism and shallow intrusive activity. The important Tintic, Utah, district dating from the Oligocene age, is situated within and on the edge of a large caldera, and quartz latite flows and tuffs are evident locally. The majority of replacement orebodies are related to small quartz monzonite stocks and display little connection with the surficial volcanics. Some fissure and replacement veins in the district, however, are found in argillized and propylitized volcanics.

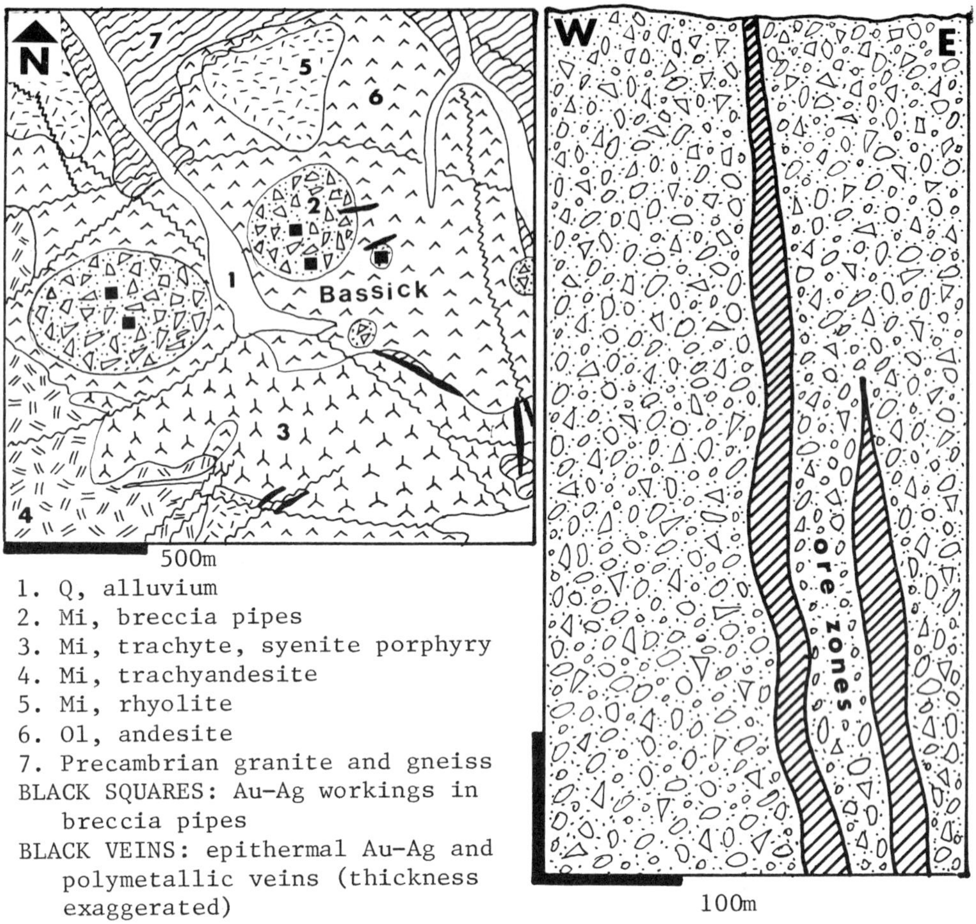

Fig. 26-20. Bassick Au-Ag bearing breccia pipes, Querida, Colorado. Map and a diagrammatic section. From LITHOTHEQUE, after Sharp (1978).

MEXICO

A 2,500 km long belt of scattered clusters of ore deposits is traceable from the American border in the north to Puebla state in the south-east. This belt is particularly well known as the world's richest and most extensive silver province. The Cainozoic metallogeny of Mexico has recently been reviewed by Clark et al. (1982) and as in the United States, it consists of an older "plutonic" suite and a younger "subvolcanic" suite of mostly epithermal Oligocene to Pliocene vein deposits. Only the latter suite is considered here.

Fig. 26-21. Silver Cliff, Colorado, old pit that has produced Ag-halides and Ag-cryptomelane (c) from fracture coatings and infiltrations in rhyolite. The Sangre de Cristo Range is in the background.

Epithermal deposits are the dominant style in the Oligocene to Recent continental volcanic terrain of Mexico, almost continuously exposed in the Sierra Madre Oriental and in the Trans-Mexican volcanic belt. The latter, however, is almost unmineralized. "Bonanza" silver deposits are the most unique type, but much of the silver produced in more recent times came from complex Pb,Zn,Ag vein and replacement deposits, such as Santa Eulalia, considered in the following chapter.

Pachuca-Real del Monte, Hidalgo (Geyne et al., 1963; 44 Tt Ag, 230 t Au; Fig. 26-22) is the largest representative of the "bonanza" Ag-Au style. This ore field 100 km N.N.E. of Mexico City is hosted by a thick Oligocene to Pliocene largely andesitic and dacitic volcanic pile, resting on Cretaceous sediments and positioned farther north-east with respect to the main Mexican "silver belt". The volcanics are intruded by a variety of andesite, dacite, rhyolite dikes and small subvolcanic stocks. The mineralization is considered to date from the middle Pliocene period, shortly postdating an episode of block-faulting and it takes the form of N.W.- and N.-trending fissure quartz veins. The veins are both parallel and oblique to the dike trend, and over a hundred veins have developed in an area of 12 x 10 km. The vein filling is a product of several mineralization stages and the bonanza ore shoots (late stage quartz, calcite, acanthite, argentite, lesser polybasite, miargyrite, pyrargyrite, proustite, stephanite, native silver, sternbergite) fill openings along the earlier-stage quartz, pyrite, calcite, rhodonite, albite, sphalerite, galena and chalcopyrite veins. The wallrock alteration is a near-vein silicification, sericitization and albitization, fringed by the more regional propylitization and carbonatization.

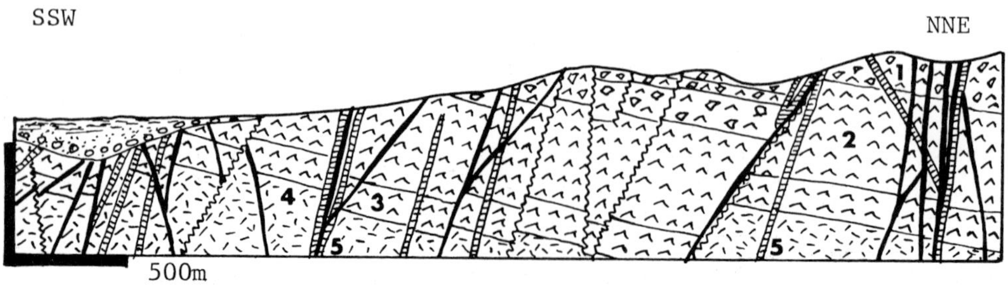

Fig. 26-22. Pachuca-Real del Monte, Mexico, an epithermal "bonanza" Au,Ag deposit. 1=Mi andesite, dacite, volc. breccia, basalt tuff; 2=Ol-Mi andesite, dacite lavas and breccias; 3=Ol andesite flows; 4=Ol rhyolite and dacite flows; 5=Ol dacite porphyry and other dikes. Modified after Geyne et al. (1963).

Guanajuato (Gross, 1975; 30 Tt Ag, 115 t Au; see Fig. 17-9 and 26-23) is the second-richest Mexican silver field. It has veins similar in character to those in Pachuca, but located in basement black slates (considered as the possible Ag source) and red conglomerates. It was briefly reviewed in Section 17.7.

Zacatecas (26 Tt Ag; Fig. 26-24) has a variety of epithermal veins hosted by Triassic and older phyllites, slates and greenstones, intruded and topped by Tertiary andesite, rhyolite and diorite porphyry. The banded, vuggy and breccia chalcedonic quartz, jasper and amethyst veins carry Fe,Zn,Pb and Cu sulphides as well as bonanza shoots with argentite, polybasite and pyrargyrite.

Taxco (Fig. 26-24, a famous mining town that produced considerable amounts of silver in the 16th century, now produces massive to banded complex siliceous Pb-Zn ores only with a moderate silver content. The recent reserves (Busch, 1980) are 525 Tt Zn/6.6%, 200 Tt Pb/2.8% and 1,100 t Ag/145 ppm. These ores form fissure veins and peneconcordant mantos in Proterozoic black and sericite phyllite, greenstone and diorite association, and in Cretaceous marine sediments. The silver produced in the past probably came from quartz or carbonate "bonanza" silver veins with very little base metals. These are now completely exhausted and their geology imperfectly known.

Tayoltita, Durango (Randall, 1972; 15 Tt Ag/425 ppm, 210 t Au/6 ppm) is one of the westernmost Mexican "bonanzas" and also a locality where the bulk of production came from recent, rather than antique, mining. This field which also has several plutonic Sn,Mo and W occurrences, is in a thick Eocene-Oligocene rhyolite and andesite pile intruded by small Oligocene granodiorite and granite stocks and is probably situated on the margin of a caldera. Most of the Ag-Au veins are hosted by a 500 m thick section of dense andesite porphyry flows, and they consist of Cu-rich argentite and electrum in quartz, lesser albite gangue. Molybdenite and Mn silicates are common in some veins. The alteration is represented by intense albitization in the immediate vicinity of the veins, grading outward to propylitization.

Parral ore field, Chihuahua (Koch, 1956) is an example of a vein deposit producing mainly Zn-Pb rich ore (average grade in the Frisco mine was 5% Pb, 8% Zn, 0.6% Cu, 150 ppm Ag, 0.5 ppm Au). At Frisco mine, 32 fissure veins are hosted by Cretaceous ? gray and black calcareous shales to limestones, and consist of sphalerite, galena, chalcopyrite, pyrite and arsenopyrite replacements of shale fragments in the vein breccia structure, cemented by drusy quartz, fluorite and calcite. The wallrocks have a thin calc-silicate (diopside, actinolite, epidote, ilvaite) alteration fringe and are silicified.

THE ANDES

The Andes are a classical region of epithermal deposits affiliated to the widespread Tertiary volcanics. As elsewhere, the ore deposits are in discontinuous groups, clusters and mineralized regions that start to appear in the north near the Colombia-Ecuador border (Titiribí, Colombia, Au-Ag), followed by scattered deposits in Ecuador; a belt in northern (Cordillera Negra) and central Peru;

Fig. 26-23. Mexican "Silver Cities", Guanajuato (TOP). BOTTOM: the citadel-like Guanajuato silver mines (e.g. the Valenciana) from past centuries, established on the Veta Madre vein system.

Fig. 26-24. Mexican "Silver Cities" Zacatecas (TOP) and Taxco (BOTTOM).

scattered occurrences over the vast region of the Altiplano-Puno plateau in Bolivia, Argentina and Chile and extending to the adjacent Cordilleran and pre-Cordilleran ranges. The entire region has been reviewed by Putzer (1976) and additional data on individual deposits can be found in the national syntheses (Bellido and de Montreuil, 1972; Ahlfeld and Schneider-Scherbina, 1964; Angelelli, 1950; Ruíz, 1965).

Central Peru

The central Peruvian metalliferous belt (centre: La Oroya, 150 km N.E. of Lima; Petersen, 1965), is one of the great mineralized areas of the world that contains the giant Cerro de Pasco deposit. The belt is underlain by a considerable thickness of Paleozoic and Mesozoic largely "miogeoclinal" sediments, including a considerable proportion of carbonates and black shales to slates. Erosional remnants of Tertiary andesite, dacite, rhyolite flows and pyroclastics attest to a formerly extensive continental volcanism and also the rapid rate of denudation compensating for the Andean uplift. As a consequence many of the Tertiary Pb-Zn-Ag deposits in central Peru (e.g. Morococha) are hydrothermal systems closely affiliated to small quartz monzonite to diorite plutons and as such treated in Chapter 28. Several Miocene-Pliocene deposits and groups of deposits are still hosted by the continental volcanics and have the undisputable character of epithermal veins (e.g. Casapalca, Colquí, Julcani, Huachocolpa, Castrovirreyna). The Cerro de Pasco orebody is a transitional high-level replacement treated in Chapter 17 (compare Fig. 17-7).

In the Casapalca ore field (Wu and Petersen, 1977; Fig. 26-25), andesite porphyry flows and volcaniclastics possibly dating from the Miocene period overlie Paleocene-Eocene red-beds, and are intruded by small stocks of late Tertiary andesite porphyry and granodiorite. A system of branching fissure veins up to 5 km long and having a vertical interval of over 2,000 m, carry zonally arranged sphalerite, galena, Ag-tetrahedrite, chalcopyrite and minor argentite, stibnite, realgar, orpiment in quartz, pyrite and rhodochrosite gangue. Alteration ranges from near-vein silicification through sericitization and pyritization into propylitization and carbonatization.

The small Colquí field (60 t Ag, 1.12 t Au, 32 Tt Zn, 19 Tt Pb, 1 Tt Cu) is notable for the variety of coexisting ore associations, including epithermal Pb,Zn,Cu deposits and separate Ag-Au "bonanza" shoots. These were recently studied and described in a very detailed paper by Kamilli and Ohmoto (1977). The field and its principal Finlandia vein are located in late Tertiary continental volcanics, intruded by a rather unusual variety of intrusive stocks (granodiorite, quartz diorite, quartz norite). The Finlandia vein dated at 10.3 m.y. is unusual in that the "bonanza" Ag-Au assemblages (quartz, tetrahedrite, Ag sulphosalts, electrum, etc.) are earlier and formed at higher temperature than the base metal sulphides (quartz, barite, galena, sphalerite).

The Caylloma Ag field (Putzer, 1976) located farther south in the Arequipa Department, has a typical "bonanza" Ag-Au ore in about 30

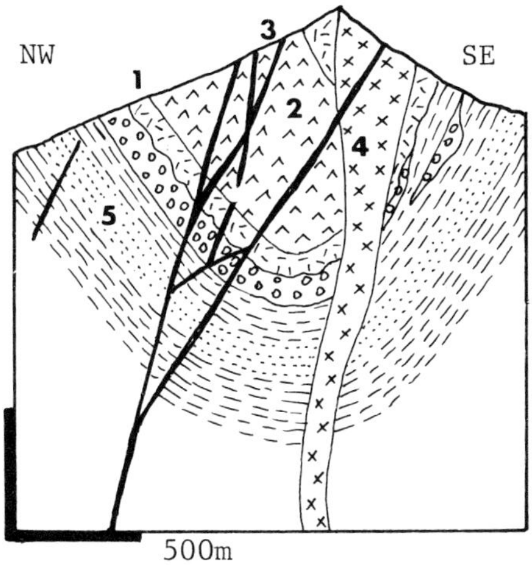

Fig. 26-25. Casapalca Ag,Pb,Zn,Cu veins, central Peru. Diagrammatic, from LITHOTHEQUE.

fissure veins rich in polybasite, pyrargyrite, native silver and other minerals in quartz, rhodochrosite and calcite gangue. Pb,Zn and Cu sulphides are present only in subeconomic quantities.

Bolivia

In central and southern Bolivia, Tertiary calc-alkaline continental volcanics are widely scattered and rather deeply eroded, so that often only the subvolcanic and plutonic equivalents remain. The important high-level cassiterite stockworks (Llallagua) are discussed later in this chapter and so is the unique Sn-Ag partly epithermal association (Potosí, Oruro). Significant epithermal deposits that are not members of the above associations are rare in Bolivia. They include the Miocene-Pliocene Pb,Zn,Ag veins in Pulacayo (Fig. 26-26), hosted by propylitized quartz latite and dacite; Laurani and La Joya stockworks of auriferous pyrite, enargite, bornite, tennantite, etc. in subvolcanic dacite stocks; Carangas argentiferous stockworks with Pb-Zn sulphides and pyrargyrite in barite, rhodochrosite and calcite gangue, hosted by silicified late Tertiary subvolcanic rhyolite stocks. There are also minor, economically marginal but geological unusual ore occurrences, such as those in the Berenguela field in southern Bolivia, that carry rhythmically banded metacolloform sphalerite and greenockite veins containing up to 16% Cd. These thin, parallel veins are hosted by a Pliocene subvolcanic dacite stock. All the localities mentioned are described in Ahlfeld and Schneider-Scherbina (1964).

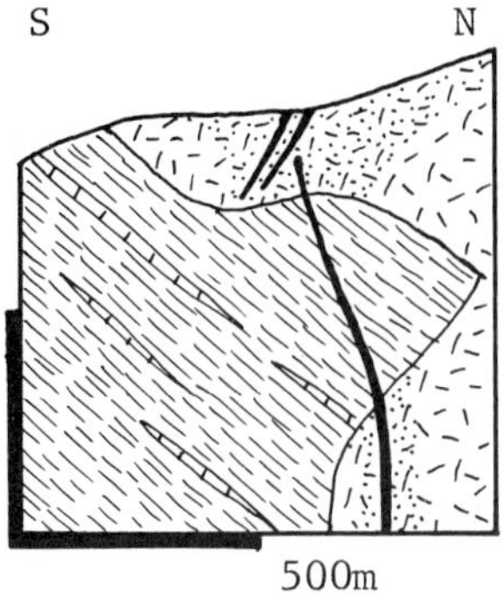

1. Mi-Pl quartz latite to dacite flows and a stock, in places propylitized (dots)
2. Cr_3 - T_1 red marl, shale, siltstone

Fig. 26-26. Pulacayo Pb,Zn,Ag vein deposit, Bolivia. From LITHOTHEQUE, data from Ahlfeld and Schneider-Scherbina (1964).

Puna and Sierras Pampeñas, N.W. Argentina

This extensive region (e.g. Sillitoe, 1977) is underlain by a Precambrian crystalline basement, covered by Paleozoic and Mesozoic "miogeoclinal" sediments and locally patched by Cainozoic volcanic piles. Rapid dissection of stratovolcanoes often exposed young (under 10 m.y.) plutonic equivalents at some depth, so that plutonic porphyry Cu often coexist with epithermal Ag-Au and Pb-Zn deposits in one district (e.g. Farallon Negro epithermal rhodochrosite, Mn-calcite, quartz veins with polybasite, argentite, gold and minor base metal veins coexist with the Mi Vida porphyry copper).

BETIC CORDILLERA, S.E. SPAIN

Betic Cordillera is a Cainozoic orogen trending N.E. along the south-eastern coast of Spain in the Murcia Province. Within its relatively small territory, three ore districts or fields genetically associated with Tertiary continental magmatism occur: Cartagena-La Unión, Mazarron and Rodalquilar. The first two are polymetallic (Pb,Zn,Ag; high Fe), the latter is a "bonanza" Au-Ag.

The Cartagena-La Unión district has been in discontinuos production for over 2,000 years, and for 15 years it has been a battleground for the proponents of remobilization and multistage metallogenic heritage (Pavillon, 1969) and advocates of a hydrothermal-postmagmatic mineralization related to a single or multiple intrusive centres (Oen et al., 1975). This short review is based on the Oen et al.(1975) results mainly because they are the more recent and more accessible ones concerning the district that are available in English. This, however, does not necessarily imply endorsement of all the published ideas - quite the contrary. The ores in the La Unión field in particular are extremely complex and unique and the uniqueness is almost certainly due to the peculiarities of the pre-Tertiary local geological history rather than to any abnormality of the Tertiary high-level magmatic system. To the New World readers, La Union may serve as an introduction to the metallogeny of the entire Alpine-Mediterranean orogen because some of the aspects discussed here are also found elsewhere e.g. in Rumania, Yugoslavia, Greece, Turkey and Iran.

In the La Unión field, the basement is represented by several units of Paleozoic to Triassic carbonates alternating with meta-pelites and minor meta-basites, variously interpreted as autochthon plus cover, or as allochthonous slices. Metamorphic grade decreases upward from amphibolite through greenschist facies to virtually unmetamorphosed rocks. The lithologic association of the basal units is reminescent of the one treated in Chapter 17 (compare Fig. 17-1), particularly the one having a "Fe aspect" (Section 17.3.). The units discussed above are topped by an epi-metamorphosed red-beds and evaporite unit comparable with the Alpine "Verrucano" facies, and by Triassic dolomitic limestones. These units are unconformably topped locally by Miocene marls and conglomerates and by Miocene-Pliocene rhyodacite to trachyandesite pyroclastics. All units are intruded by subvolcanic necks, dikes and breccia pipes.

The ore deposits of La Unión have produced Pb,Zn,Ag and some iron. A long list of metals (including Sn, Bi) is present in supra-clarke quantities. Of the ten or so mineralization styles distinguished by Oen et al. (1975), only those directly hosted or immediately associated with the young volcanics and subvolcanics in the central zone, are undisputably epithermal. These are predominantly N.N.W.-trending veins, exposed almost entirely in the oxidation zone, and filled by a mixture of Fe and Mn oxides and hydroxides, cerussite, relict cassiterite, jarosite, alunite, etc. They are of little economic importance. The bulk of production came from low-grade replacement mantos, replacement veins and pods in carbonate units. The most productive and also the most unusual are the greenalite, magnetite, siderite, lesser pyrite, galena, sphalerite mantos replacing Triassic or Paleozoic limestone (e.g. in the Tomasa pit; Fig. 26-27). The replacements appear to be fault controlled, but there are always altered greenstones (possibly meta-diabases) in the vicinity. Swarms of metasomatic veinlets, albite-chlorite altered and bleached schists and a high degree of stratigraphic control, are prominent.

INNER CARPATHIAN ARC

About ten large and several hundred small Miocene-Pliocene continental calc-alkaline volcanic piles and/or centres occur on the inner (southern) side of the great Carpathian folded arc, along and outside its border with the Pannonian Basin (a median massif filled by young sediments; Fig. 26-28). Politically, the volcanics are located

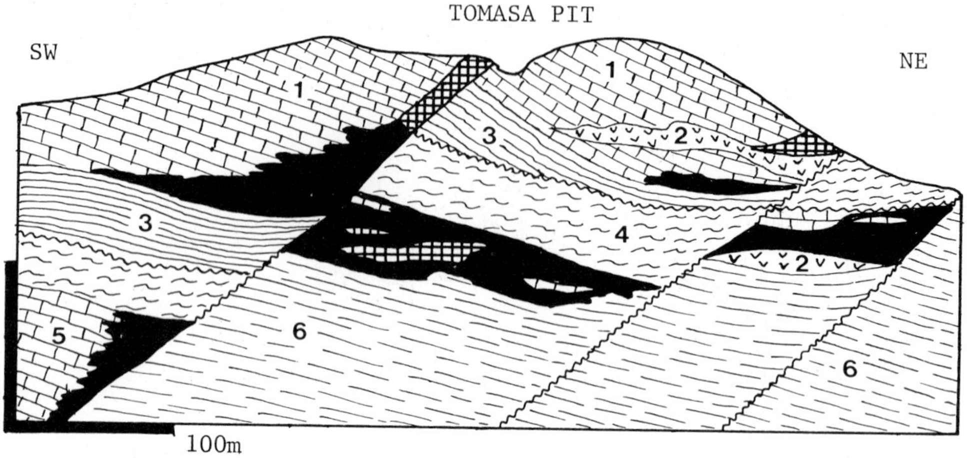

Fig. 26-27. La Unión ore field, Sierra de Cartagena, Spain. An example of a magnetite, greenalite, low-grade Zn,Pb mantos interpreted as Tertiary hydrothermal replacements. 1=Tr limestone; 2=Tr greenstone, metabasalt; 3=phyllite; 4=Tr or older, sericite and micaschist, quartzite, albite schist; 5=siliceous marble; 6=gray and black micaceous chert, marble; black="primary" orebodies (greenalite, magnetite, galena, sphalerite, pyrite, quartz, siderite); grid pattern=oxidation zone (Fe-Mn oxides and hydroxides, Pb-Zn carbonates and sulphates, gypsum, jarosite). From LITHOTHEQUE, simplified after Oen et al. (1975).

on the territories of Czechoslovakia, Hungary, U.S.S.R. and Rumania. Three volcanic areas (Kremnica-Štiavnica Mts. in Slovakia, Gutăiu Mts. in northern Rumania and Apuseni Mts. in western Rumania) contain important and classical Au-Ag and Pb,Zn,Ag fields, widely treated in the pre-1900 textbooks of economic geology (before the American examples started to dominate the field). The two Rumanian regions located in Transylvania, until 1918 a province of Hungary, were particularly famous. There, most localities are known by two (Rumanian and Hungarian) or even three (plus German) local names. This adds to the confusion. The mining in the region is still going on and some of the Rumanian localities are among the world's most significant gold producing fields. Unfortunately, modern quantitative data are virtually unavailable and only a trickle of the modern Rumanian re-interpretations have reached the West (e.g. in Ianovici and Borcoş, 1982). Outside the three principal mineralized regions, Pb-Zn and Au-Ag deposits of local importance also occur in the Mátra Mts., Hungary (Gyöngyösöroszi; about 70 Tt Zn, 50 Tt Pb, 3t Au), Tokai Mts. (Telkibánya)and Börzsöny Mts. (Nagybörzsöny). The latter have been reviewed recently by Morvai (1982).

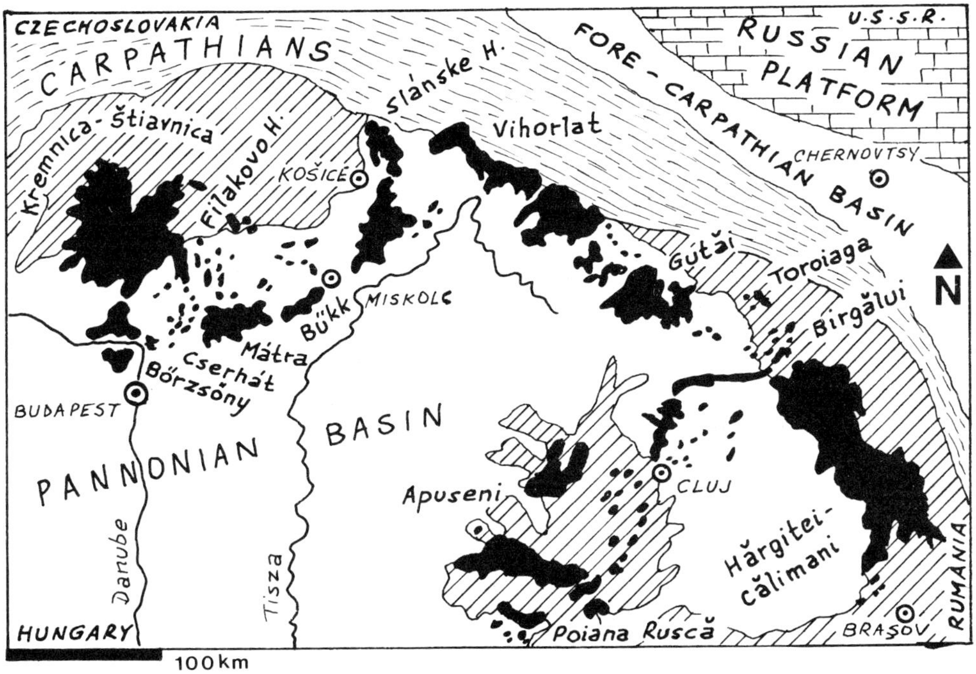

Fig. 26-28. Tertiary continental volcanic complexes on the inner side of the Carpathian arc (shown in black). Diagonally ruled: Internal zone of the Carpathians (crystalline core enveloped by Paleozoic and Mesozoic sediments); dashed, Cr to lower Tertiary flysch lithofacies; white, Neogene sediments-filled basins.

Kremnica-Štiavnica Mts. (Rudohorie), Slovakia

The Banska Štiavnica volcanic centre coincides with an old Au-Ag and polymetallic district (listed as Schemnitz in the old literature; at least 65 t Au, 1.8 Tt Ag), that was recently interpreted as a resurgent, middle to late Miocene 17-11 m.y. caldera (Burian et al., 1974). The caldera collapse followed a constructional phase, during which a dominantly pyroxene andesite stratovolcano was built on the basement of Triassic limestone and shale, and Eocene conglomerate. During doming and caldera resurgence, subvolcanic bodies of dacite porphyry and high-level granodiorite stocks were emplaced. The mineralization related to the intrusions is multistage, and incorporates minor magnetite skarns at granodiorite/marble contacts (Vyhne); subeconomic porphyry-Cu showings in granodiorite (Hodruša); galena-sphalerite replacements in silicified limestone; and vertically zoned fissure veins (top: Au-Ag bonanzas, bottom: massive and banded galena, sphalerite, chalcopyrite in quartz gangue). There are about thirty major veins.

Gutăiu (or Gutii, Gutin) Mts., N.W. Rumania

The Gutăiu Mts. are a short mountain range and a Miocene-Pliocene volcanic pile in the north-western corner of Rumania. The range practically coincides with the important Baia Mare Pb,Zn,Cu and Au-Ag epithermal vein district (Lang, 1979). The district is underlain by Eocene flysch, Neogene continental sediments and Neogene continental volcanics. The latter are dominated by pyroxene andesite, lesser dacite, rhyodacite, rhyolite and their pyroclastics. Epithermal deposits are represented by generally steeply to vertically dipping systems of subparallel fissure veins and rarer stockwork and impregnation zones, concentrated in an E.-W.-trending belt 60 km long and 5 km wide. Ore fields coincide with local subvolcanic andesite intrusions which in turn are mostly controlled by N.W. and E.-W. faults and their intersections. Hydrothermal alteration is represented by an almost regional propylitization, with locally superimposed wallrock low-temperature K-feldspathization (adularia). Sericitization and argillization are also important.

Ten ore fields are known in the belt of which at least three contain mines operating at present. The famous localities include: Baia Mare (city) field ("bonanza" Au-Ag veins and stockworks productive up to a depth of 800m); Herja (Fig. 26-8, a system of parallel Pb,Zn,Ag veins in Neogene sediments and subvolcanic pyroxene andesite porphyry, gradational into microdiorite); Baia Sprie, formerly Felsöbanya (Fig. 26-29; the principal vein, 2.4 km long, 1000 m deep and 15-20 m thick, is well zoned with an Au-Ag "bonanza" at top, Pb-Zn sulphides in the middle and Cu near the bottom. 100 mineral species were recorded) and Cavnic, formerly Kapnik (14 sub-parallel epithermal veins in Neogene sediments and a pyroxene andesite laccolith).

Apuseni Mts. (including Munţii Metaliferi), W. Rumania

The Apuseni Mountains (Ianovici at al., 1976; Ianovici and Borcoş, 1982; Fig. 26-30) are in western Rumania (in Transylvania, north of Deva. The region is also known as Siebenbürgen in the German and older literature). They coincide with an unusually complex, polygenetic mineralized region. The basement is composed of Proterozoic and lower Paleozoic metavolcanics and metasediments, intruded by old granitic plutons. This is topped by Devonian spilite-keratophyre association and Jurassic ophiolites, each in turn intruded by "Laramide" (late Cretaceous-lower Tertiary) diorite, granodiorite, and quartz monzonite suite. As expected, the basement mineralization includes Cu and Cu-Zn volcanics-hosted massive sulphide deposits, vanadiferous Ti-magnetite in gabbro, Cu-Ni and Co showings in ophiolites and two porphyry Cu-Mo deposits.

A first-order epithermal, predominantly Au,Ag,Te province (district) is located in the S.E. corner of the Apuseni Mts., in the vicinity of Brad and Baiţa, and in the literature it usually goes under the name Metalliferous Mountains (Munţii Metaliferi). This province coincides with a Miocene-Pliocene volcanic pile dominated by andesite and lesser rhyolite and rhyodacite and formed in three

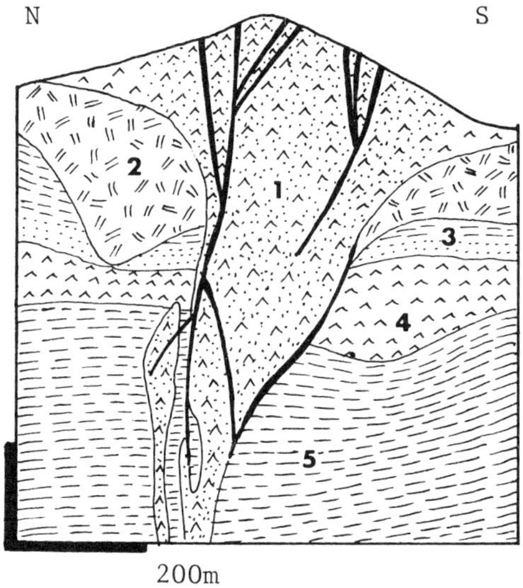

1. Mi-Pl subvolcanic pyroxene-amphibole andesite, propylitized
2. quartz-andesite flows and tuffs
3. volcanic arenite, marl
4. pyroxene andesite flows
5. shale, marl

Fig. 26-29. Baia Sprie (formerly Felsöbánya), N.W. Rumania, Pb,Zn,Au,Ag,Cu epithermal veins. Modified after Socolescu (1972).

successive stages. Individual volcanic centres are controlled by N.W. and E.-W., N.E. faults and their intersections. Most appear to have been stratovolcanoes and simple vents rather than calderas. The volcanic centres coincide with ore fields and have been deeply eroded and dissected shortly after the formation. Many, however, were preserved from a still deeper erosion by the cover of Neogene "molasse" from beneath which many have only recently been exhumed.

The Neogene subvolcanic orebodies take the form of fissure veins, multiple clusters of short veins in the vent region and above high-level intrusive stocks, grading into breccia pipes and chimneys or replacements in limestones and marls. Many examples have been depicted in Figures 26-7, 8, 11 and 12. Gold as electrum and in tellurides (this is the classical area of Au-Ag tellurides, from which many have been described in the last century, e.g. sylvanite named after Transylvania; nagyagite named after Săcărîmb) and also native tellurium are dominant, silver has been of lesser importance and Pb,Zn,Cu have mostly been recovered only as by-products. The production and reserves in the region mined since Roman days have not been published, but they are conservatively estimated to be at least 600t Au. The dominant wallrock alteration is adularization and also biotite, sericite, advanced and "normal" argillization including alunite. The filling temperatures were slightly higher than usual for the epithermal range (185 - 360° C). It is interesting to note that the richest Au-Te centres (Brad-Săcărîmb, Baiţa) coincide with the

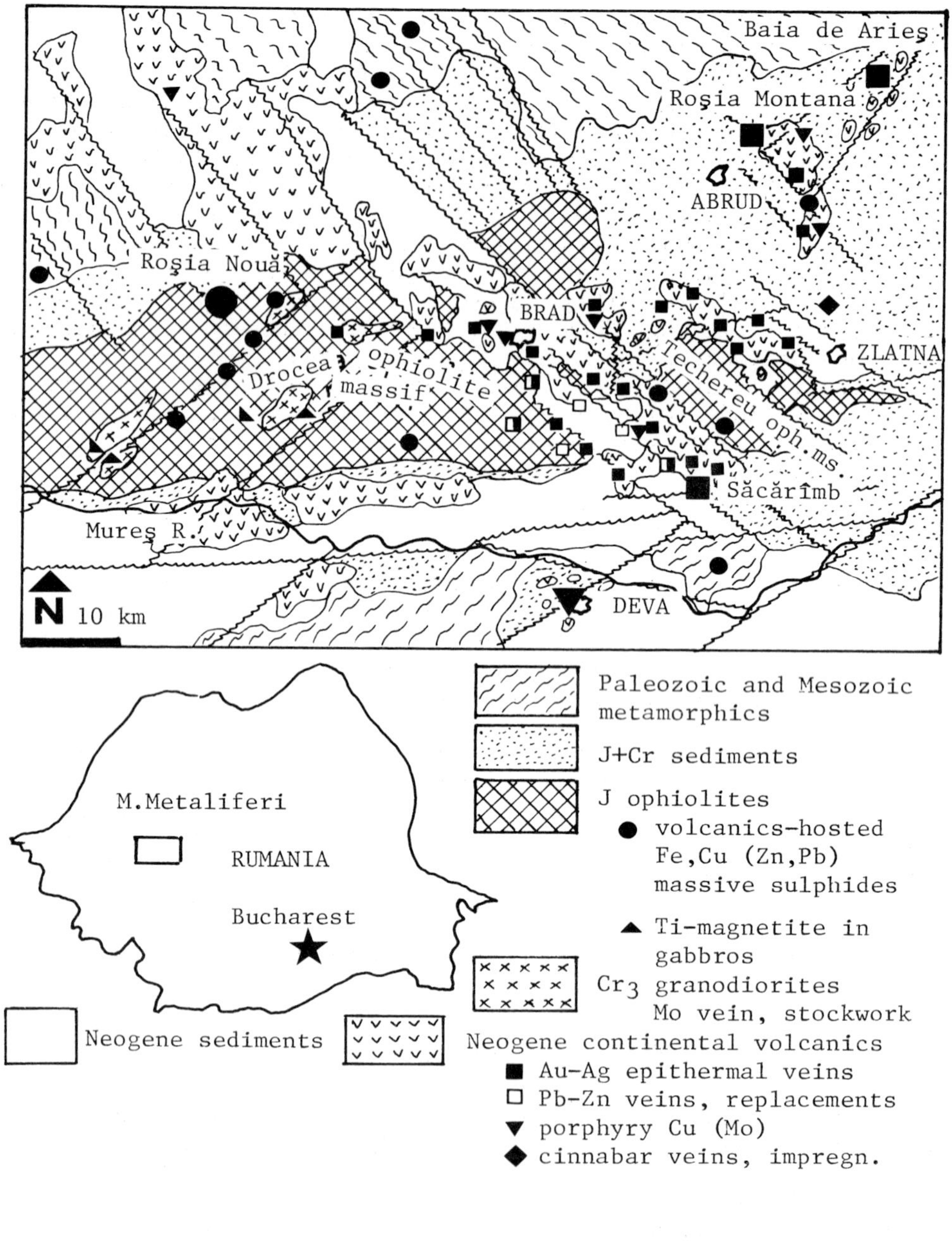

Fig. 26-30. Geology and principal metallic deposits in the Metaliferi Mountains (southern portion of Apuseni Mts., Rumania). Based on data in Ianovici et al. (1976) and Ianovici and Borcoş (1982).

Jurassic ophiolite basement, a relationship also noted in other regions of the world (compare also Section 7.10.7.).

The most important N.W.-trending Brad-Săcărîmb volcanic and mineralized zone consists of about fourteen circular volcanic centres, in the same time ore fields. In the best-known and most important field Săcărîmb (formerly Nagyág; briefly summarized in Boyle, 1979, p.264) a paleo-volcanic centre including also a caldera stage evolved along an intersection of N.W. and E.-W. faults. The ore is located in four groups of veins (about 220 veins) intimately controlled by the shape of an upward — widening plugs of a biotite-enriched quartz-amphibole andesite (Fig. 26-8). The highest density of veins is found in the widened, funnel-like andesite bodies surrounded by soft Neogene sediments, near the basement unconformity. Paleozoic metamorphics are located in the immediate basement. The veins contain electrum, nagyagite, altaite, krennerite, sylvanite, hessite, petzite, native tellurium and sub-economic amounts of the common Pb,Zn,Cu sulphides scattered in quartz-carbonate gangue. Some veins grade into low-grade disseminated gold deposits in K-feldspathized (adularia) volcanics. The present grades in the Săcărîmb mine are 5-8 ppm Au in the veins, 2-3 ppm Au in the disseminations (Wagner, 1977).

The only other epithermal deposit currently mined in the Metalliferous Mountains is Roşia Montană in the N.E. corner of the volcanic province. There, a Miocene volcanic centre dominated by biotite-amphibole subvolcanic andesite and dacite formed over Cretaceous flysch and Paleozoic or older metamorphics. The deposit mined since the Roman days (estim. 350 t Au, 1,500 t Ag) is in a subvolcanic dacite plug interrupted by several breccia pipes. The pipes are cemented by quartz with lesser rhodochrosite, pyrite, Cu,Zn,Pb sulphides and electrum, and grade into stockworks and discrete fissure veins. Pervasive alteration has proximal K-feldspar (adularia) and quartz-sericite - pyrite zones, grading to argillic and propylitic zones. Wagner (1977) quoted the low-grade reserves of disseminated Au ore at 100 Mt containing 2 ppm Au (=200 t Au). The currently mined veins and breccias grade between 2-20 ppm Au and 60-150 ppm Ag.

REMAINDER OF THE ALPS - MEDITERRANEAN - HIMALAYA OROGEN

Scattered occurrences of Tertiary volcanic, subvolcanic and intrusive rocks indicating former calc-alkaline continental volcanic piles, are widespread in Yugoslavia, Bulgaria, Greece, Turkey, Iran, Afghanistan and Pakistan.

In Yugoslavia (Janković, 1982), these are most widespread in the Serbo-Macedonian Massif. Epithermal Au-Ag and Pb-Zn veins closely affiliated with dacites and andesites are in the Lece, Zletovo, Blagodat-Osogovo and other fields. Elsewhere (Kopaonik, Trepča) Pb,Zn,Ag replacement bodies are located in the basement metacarbonates related to high-level Tertiary granodiorite coeval with the continental volcanics.

In Bulgaria and northern Greece the important Pb,Zn,Ag vein and carbonate replacement districts in the Rhodopen crystalline massif are genetically associated with subvolcanic rhyolite dikes and small quartz monzonite stocks mostly dating from the Oligocene age (Bogdanov, 1982). Similar, small deposits are scattered on the Greek Aegean Islands (Marinos, 1982).

In Turkey, the most productive subvolcanic field has been the Balya (W. Turkey; 400 Tt Pb; Ryan, 1957). There, Permo-Carboniferous limestone is covered by Triassic conglomerate, sandstone and shale, folded and topped by Eocene dacite and rhyolite flows and dikes. A younger, Miocene period of volcanism produced hornblende andesite flows, pyroclastics and dikes. The ores are related to Eocene subvolcanic and high-level plutonic bodies and take the form of (1) disseminated and fracture veinlets of galena, sphalerite and pyrite in fluorite-adularia gangue or alteration envelope, in fractured rhyolite and dacite porphyry ("porphyry Pb-Zn"); (2) epithermal veins along contacts of porphyries and metasediments; (3) veins and replacements in marble and (4) bedding replacements in metasediments. A substantial proportion of the past production came from the oxidation and cementation zones.

INDONESIA

In Indonesia (van Bemmelen, 1949), hydrothermal ore deposits associated with Cainozoic subaerial calc-alkaline volcanic piles are widespread on the larger islands (Sumatra, Sulawesi, Java), particularly in sectors with a relatively thick continental crust due to the geologically old basement. The most important epithermal Au-Ag deposits occur near the western coast of central Sumatra, in the Padang region. The largest and best-known deposit Redjang Lebong (89 t Au, 843 t Ag) contains five subparallel quartz, chalcedony, calcite and truscottite breccia veins with scattered pyrite, chalcopyrite, sphalerite, aguilarite, berzelianite and electrum. The veins are located in a S.W.-dipping lower Miocene andesitic fault breccia and hornblende andesite, emplaced in a Miocene Globigerina marl. Subvolcanic andesite and dacite bodies are found in the vicinity and the ore zone and its environs are heavily feldspathized to resemble a pseudo-rhyolite or pseudo-trachyte. The Lebong fault zone is probably situated on the edge of a volcanic-tectonic depression. Additional Au-Ag epithermal deposits in the Padang region are Mangani (5.5 t Au, 237 t Ag, high Se); Simau (28 t Au, 300 t Ag) and others.

In Java, the best-known epithermal field is Cikotok (100 km S.W. of Jakarta; 2.5 t Au, 160 t Ag). There, quartz, calcite, rhodochrosite, pyrite, galena, sphalerite and chalcopyrite veins are hosted by an Eocene to Miocene andesite-rhyolite suite.

PHILIPPINES

In the Philippines, the most productive and most thoroughly studied epithermal Au-Ag region is the Baguio district in north-central Luzon (Fernandes and Damasco, 1979; 300 t Au). There, numerous (96 veins in the Balatoc Plug only) quartz, rhodochrosite, calcite, pyrite, minor sphalerite, chalcopyrite, gold, Au-Ag tellurides containing N.W. and E.-W. epithermal veins grade into mineralized breccias. The most common host rocks are chloritized and locally anhydrite-impregnated Miocene andesites and quartz dacites. The mineralization, however, is Pliocene to Pleistocene, temporary affiliated to high-level granodiorite and dacite intrusions. The pre-volcanic basement in the district is metallogenically fertile and consists of Cretaceous and earlier ophiolites and metamorphics, followed by late Cretaceous to Oligocene keratophyres and andesites.

NEW GUINEA AND ADJACENT ISLANDS

In Niugini, the most productive mineralization associated with young continental volcanism is found in the Morobe Goldfield (around Wau and Edie Creek; Rebek, 1975; 14 t Au from lodes, 101 t Au from placers). The Pliocene (3.1-3.8 m.y.) bedrock ore is only developed at Edie Creek and Golden Ridges, where it bears the character of (1) thick, persistent and sharply defined N.W. belt of quartz, Ca-Mn carbonate, pyrite and minor galena, sphalerite, stephanite, electrum fissure veins; (2) groups of tabular, poorly defined quartz-gold veins and veinlets in shear zones and (3) Ca-Mn carbonate, quartz, pyrite veins, veinlets and disseminations in a Pliocene breccia of uncertain origin (possibly diatreme or a scree breccia), resulting in a "bulk" orebody with an average grade of 4.5 ppm Au/t. The mineralization is genetically affiliated to Pliocene quartz-biotite porphyry intrusions, emplaced into the basement of Miocene volcanics and granodiorites and Mesozoic slate and chloritoid schist. The prevalent alterations include silicification, sericitization and argillization but this is modified or even obliterated by the deep lateritic weathering in the area.

FIJI

The largest ore field in the country near Vatukoula (northern Viti Levu Island; Colley and Greenbaum, 1980; Denholm, 1967; 115 t Au, 40 t Ag; Fig. 26-31) produces Au,Ag and Te and is associated with a Pliocene (4-5 m.y.) caldera ("Tavua Basin"). The caldera formed by the collapse of an olivine basalt and trachyte stratovolcano and provides a unique example of continental paleo-volcanism barely "out of water" in an immature island arc formed on a "quasi-continental" crust. The caldera shortly after its collapse, was intruded by augite andesite cone sheets and dikes, repeatedly sheared and faulted

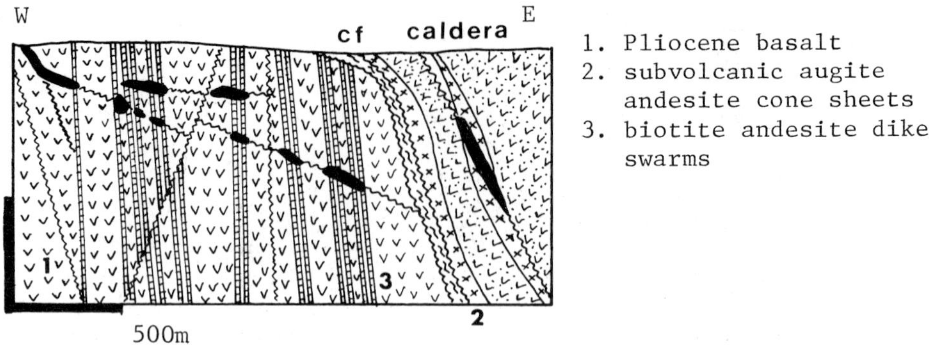

Fig. 26-31. Vatukoula, Fiji, an epithermal Au-Te ore field in a Pliocene island arc caldera and along its rim. Ore shoots (black) are at the intersection of "flatmakes" (flat lode structures) and dikes. Cf=caldera fault zone. From LITHOTHEQUE, modified after Denholm (1967).

and intruded by biotite andesite dikes and a monzonite (shoshonite) plugs. This was followed by the Au,Ag,Te epithermal mineralization.

There is a great variety of orebodies mostly formed along shears parallel with earlier andesite dikes, along flat tensional structures and intersections of the former with the young generation of dikes and along shatter structures. The orebodies lie both within and outside the caldera and the richest ones in the Emperor mine are found just west of the caldera rim fault. The gold-bearing orebodies are distinct open-space filling quartz, calcite, pyrite, sylvanite, krennerite, electrum, native Te, Ag sulphosalts, etc. shoots, as well as low-grade disseminations and veinlets in the shears. The alteration is largely propylitic.

The much less productive Yanawai Goldfield on the Vanua Levu Island (2t Au, 0.15 t Ag) comprises quartz-barite veins hosted by propylitized Miocene-Pliocene andesitic volcanics and subvolcanics.

NEW ZEALAND

The Hauraki Goldfield (Williams, 1965) is a 160 x 30 km long, N.N.W.-trending belt of Miocene-Pliocene subvolcanic Au-Ag and, to a lesser degree, Pb,Zn,Cu mineralization. It is located in the North Island about 100 km east of Auckland. The basement rocks are Mesozoic argillites and volcanic graywackes, unconformably topped by Eocene to Miocene andesite and dacite breccias, lavas and pyroclastics intruded by small quartz diorite to granodiorite stocks. These, in turn, are postdated by Pliocene-Pleistocene rhyolite sheets and domes and, locally, by hot spring sinter deposits. The numerous ore fields (Waihi, Thames, Coromandel, Te Aroha) are tectonically controlled. The largest Waihi field (208 t Au, 1,115 t Ag) has a conjugate system of fissure veins, hosted by a block-faulted Miocene andesite topped by a post-ore rhyolite. The largest vein, the N.E.-striking Martha Vein has been traced for 1,600 m. The ore shoots consist of lamellar and chalcedonic quartz with orthoclase, calcite, rhodochrosite, pyrite, Zn,Pb,Cu sulphides, pyrargyrite, acanthite and electrum.

The Thames field was famous for its rich bonanzas, discovered about 170 m under the surface. The Tui mine in the Te Aroha field, recently described by Robinson (1974) is locally unusual due to its dominance of base metals (average grade 17% Zn, 7% Pb, 37 ppm Ag). It is hosted by hydrothermally propylitized Miocene andesite and the alteration, mineralization is Pliocene (2.5-7 m.y.).

JAPAN

The transition between continental and submarine (near-marine) mineralized volcanic associations noted briefly in Fiji, constitutes the general rule in Japan. The Miocene volcanics occur not only in different paleo-environmental settings in various parts of the country (submarine, as in the "Green Tuff" assemblage in S.W. Hokkaido and N.E. Honshu; subaerial in Kyushu) and carry different mineralizations, but the ores formed under volcanic-marine and continental conditions often overlap (Saito, ed., 1960; Ishihara, ed., 1974).

The principal mineralization styles in the subaqueous "Green Tuff" association (Section 14.3.1.) are the "kuroko-type" volcanogenic massive sulphide deposits, whereas the subaerial volcanic terrains carry epithermal veins. Veins, however, are also present in the "kuroko"-dominated districts (e.g. in the Hokuroku district), where they overlap with the massive sulphides. The veins are interpreted partly as being younger than the "kurokos", formed in a later period when their environment was subaerial. Alternatively they have been interpreted as being endogenous deposits contemporary with the "kuroko" or older and formed as endogenous ores within a volcanic pile (regardless of whether submarine or subaerial), at sites that had no substantial communication with the seawater. Both alternatives are probably valid in some locations. Both polymetallic (Pb,Zn,Ag) and Au-Ag epithermal deposits are widespread and economically important in Japan.

The Toyoha mine in S.W. Hokkaido (40 km S.W. of Sapporo; Shikazono, 1975; 650 Tt Zn, 260 Tt Pb, 1.1 Tt Ag; Fig. 26-32) has been the largest Japanese Pb-Zn producer for some time. It is hosted by the Miocene "Green Tuff" assemblage consisting of a lower marine suite of sandstone, mudstone, shale, conglomerate, andesite to dacite lavas, breccias and tuffs, topped by a non-marine suite of conglomerate, lignite and subaerial volcanics. Several subsequent phases of magmatic activity accompanied by faulting and local

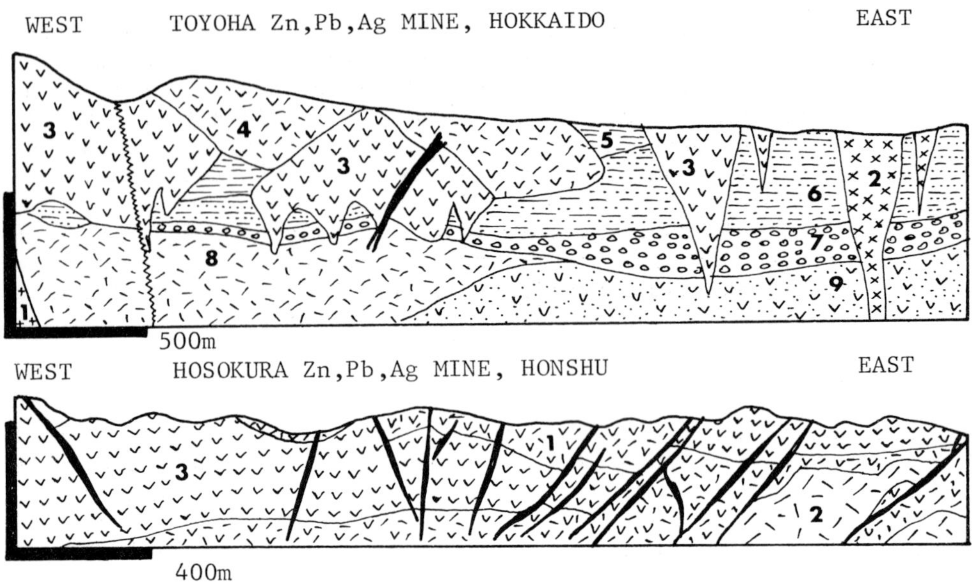

Fig. 26-32. Toyoha and Hosokura Pb,Zn,Ag mines, Japan, examples of Tertiary epithermal veins. TOYOHA (Mi): 1=quartz porphyry dikes; 2=basalt porphyry dikes; 3=propylitized andesite porphyry; 4=andesite tuff, tuff breccia; 5=volcanic mudstone, felsic tuff; 6=litharenite, mudstone; 7=conglomerate; 8=rhyolite welded tuff; 9=propylitized andesite. HOSOKURA (Mi): 1=propylitized andesite tuff; 2=rhyolite; 3=propylitized andesite flows. From LITHOTHEQUE, simplified after materials received from Nippon and Mitsubishi mining companies, 1972.

subsidence, followed. The principal veins are contained in an E.-W. swarm extending for a distance of over 2.6 km. They contain banded to massive filling of sphalerite, galena, pyrite in quartz and Mn-calcite gangue. Silicification is the prevalent wallrock alteration when the host rock is a mudstone, and silicification and propylitization are most frequent in andesite. The veins are probably late Miocene, believed to have been affiliated to subvolcanic rhyolite stocks and dikes.

Hosokura mine in northern Honshu (60 km N. of Sendai; Imai, 1978; 640 Tt Zn, 235 Tt Pb; Fig. 26-32) is the second largest epithermal Pb-Zn deposit in Japan. There, relatively simple N.E. and N.W. fissure veins fill wrench faults in Miocene propylitized andesite, tuff and domal rhyolite.

Sado mine, located on Sado Island in the Sea of Japan (Imai and Bunno, 1978; 77 t Au, 2.3 Tt Ag, 5.4 Tt Cu) is the largest Japanese representative of the epithermal Au-Ag category. There, fissure veins composed of milky quartz, adularia, calcite, montmorillonite, rhodochrosite, electrum, native silver, stromeyerite and other minerals are hosted by Miocene and Pliocene submarine to continental volcanics (andesite tuffs and breccias, rhyolite, rhyolite tuff, black shale). Silicification, argillization and propylitization are the main wallrock alterations. Chitose mine in Hokkaido (16.5 t Au, 81.5 t Ag), Mochikoshi mine on the Izu Peninsula, Honshu (18 t Au, 500 t Ag), etc., are other epithermal Au-Ag deposits.

Several epithermal rhodochrosite veins in Hokkaido bearing only low Au,Ag,Pb,Zn grades (Ohe mine) or entirely lacking recoverable base and precious metals (Inakuraishi mine), are mined for their Mn content.

SOVIET FAR EAST

Numerous subvolcanic deposits of base metals and Au-Ag occur in several extensive continental volcanic belts developed along the Pacific coast (the Primor'ye belt N.E. of Vladivostok; Okhotsk-Chukotka belt centered on Magadan) and in the East Transbaikalia (centre: Chita). Four example Au-Ag localities (Mnogovershinnoe, Karamken, Khakandzha and Balei) were reviewed recently in the English translation of the "Gold" chapter, in the Ore Deposits of the U. S. S. R. (Borodaevskaya and Rozhkov, 1974). Economically the most important Balei field, East Transbaikalia (Magak'yan, 1968; Lozovskii et al., 1960; estim. min. 125 t Au; Fig. 26-33) is located in a tectonic graben floored and flanked by Paleozoic granitic rocks, and filled by upper Jurassic continental andesites and tuffs, and lower Jurassic "molasse" conglomerate, sandstone and mudstone. These have been intruded by small stocks of Cretacoeus dacite porphyry. Several hundred fissure veins grading into stockworks of short, 10-20 cm wide veins, are tectonically controlled and most consistently present at the intersections of flat ore zones and steep faults. The vein filling consists of a banded chalcedonic quartz, adularia, kaolinite, pyrite, arsenopyrite, chalcopyrite, tetrahedrite, pyrargyrite and electrum. The wallrock alteration is sericitization, K-feldspathization, pyritization and argillitization. The ores are interpreted as having formed from 150-270° C hot fluids, at a depth of only 150-200 m under the paleosurface.

REST OF THE WORLD AND SUMMARY

The most important example localities of the epithermal Pb, Zn and Au, Ag deposits are briefly reviewed in Tables 26-3 to 26-5.

Table 26-3. Epithermal Pb-Zn ⟩⟩ Au,Ag deposits, selected examples

LOCALITY	GEOLOGY	MINERALIZATION	REFERENCES
Creede, C. San Juan Mts.,S. Colorado, U.S.A.	Ol_{2-3} volc.pile,domin.rhyol.and Mi rhyodacite, on 1.75 b.y. metam. basement. Ore zones in N.N.W. graben between two calderas	(24.6 m.y.) N.N.W. persist.veins along faults; gal.,sphal.,lesser pyr.,chalcop. Qtz,rhodochr., chlorite,etc.gangue.Light alter. 100 Tt Pb,20 Tt Zn,1.6 Tt Ag	Steven and Ratté (1965) Barton et al. (1977)
Sunnyside m.,Silverton d.,San Juan Mts.,Colorado	Ol, qtz.latite lavas,tuffs,laharic breccias,pyrox. andesites. Tect. graben between two calderas, regional propylitization	Mi_2 (13-16.6 m.y.),N.E. and N.W. persistent fiss.veins (Sunnyside vein 2.1 km long);qtz,sphal.,gal. pyroxmangite,pur.,rhodochr.,Au-Ag tellur.,tetrahedr.,etc. 313 Tt Zn, 212 Tt Pb, 488 t Ag, 20.4 t Au	Casadevall and Ohmoto, (1977)
Idorado,Telluride orebodies, N.W. Silverton distr., San Juan Mts.	Pt basem.overl.by PZ-Eo "miogeos." sediments. Ol tuff breccia, and., qtz.latite, rhyol.cont.volc. Small rhyol. and qtz.porph. stocks	Several // epith.N.W. fiss.veins, qtz.,sphal.,gal.,pyr. Some contain shoots of mass.sulph. (as above) in qtz.,epid.,chlor.,rhodochr. gangue in Eo conglom. 97 t Au, 1.45 Tt Ag, 115 Tt Pb	Lipman et al. (1976)
Parral, Chihuahua, N. Mexico	Cr ? marine calcar. shale, topped by T andesite volc. and pyroclast. and intr. by rhyol.porph. and qtz. monzonite	Mi, sets of // N.-S. fissure breccia and calcar.shale replacing veins. Qtz.,calc.,fluor., sphal.,gal.,pyr.,chalcop.,ars. 400 Tt Zn, 260 Tt Pb, 600 t Ag	Koch (1956)
Casapalca, Central Peru	T_1 red bed sedim.topped by Mi-Pl andes.porph. and volcaniclastics. Intruded by T_3 andes.porph. and granodiorite	Mi-Pl,long, persist.,zoned fiss. veins. Qtz,pyr.,Mn-calc.,rhodochr.,sphal.,gal.,Ag-tetr.,minor argent.Silic.,seric.,propylit. 350 Tt Zn,250 Tt Pb,2 Tt Ag	Wu and Petersen (1977); Petersen (1965)

Cerro de Pasco, C.Peru, late veins	D,black phyllite; Tr_3-J_1 limest. and dolom.,intr. by Mi diatreme, quartz monzonite stock	Late stage veins, masses: Mi-Pl, enargite,pyr.,sphal.,tennantite, luzonite, pyrargyrite. 7.3 Tt Ag, 216 Tt Pb, 198 Tt Zn, 16 Tt Cu	Einaudi (1977)
Pulacayo,Bolivia	Mi ? qtz.latite to dacite subvolc. stock intrus. into propylit.dac. and andesite and Cr_3-T_1 red bed sediments	Mi-Pl, 4 fissure veins,banded and drusy. Qtz.,pyr.,bar.,sphal., gal.,Ag-tetrah. Qtz. seric. and propyl. volc., bleached sedim. 586 Tt Zn/3.5%, 3,050 t Ag	Ahlfeld and Schneider-Scherbina (1964)
Banska Štiavnica, E. Czechoslovakia	Mi_2 andes.,dac.,rhyol. volcani-clastics, minor lake beds, intrus. dacite and granod.porph.assoc. with paleo-caldera built on Tr-Eo sediments	Mi_3, about 30 epith.vertic.zoned N.N.E. fiss.veins; top,Au-Ag "bo-nanza",bottom gal.,sphal.,chalcop. tetr. Min.65 t Au, 2.7 Tt Ag	Burian et al. (1974)
Baia Mare distr., Gutâiu Mts., N.W. Rumania	Mi pyrox. andes.,lavas,pyrocl., subvolc. bodies, T_2 sedim.,res-ting on Eo terrig.flysch; in depth Pt and PZ metamorphics	Mi_3-Pl E.-W. belt 60x5 km cont.10 Au-Ag "bonanza" and Pb-Zn-Ag epi-thermal vein groups controlled by subvolc. andes. plugs. Vertic.zo-ned,min.120 t Au,5.5 Tt Ag,250 Tt Zn, 200 Tt Pb.	Lang (1979)
Toyoha mine, S. Hokkaido, Japan	Mi, marine to contin. mudst., sandst.,andes.to rhyol.lavas and pyroclastics, propylitized	Mi_3, E.-W. zone of // fiss.veins. Qtz.,Mn-calcite,sphal.,gal.,pyr. Wallrock silic. and propyl. 650 Tt Zn, 260 Tt Pb, 1.1 Tt Ag	Shikazono (1975)
Hosokura mine, N.Honshu, Japan	Mi, propylit.andes. and andes. tuff, intruded by rhyolite domes	Mi, N.E. and N.W. fissure veins along faults. Gal.,sphal.,pyr.in qtz. gangue, wallr. silic.,propyl. 650 Tt Zn/4%, 235 Tt Pb/1.47%	Imai (1978)

Table 26-4. Epithermal Au ≫ Ag,Pb,Zn deposits, selected example localities

LOCALITY	GEOLOGY	MINERALIZATION	REFERENCES
Goldfield, Nevada, U.S.A.	Mi volcanic centre,qtz.latite, andesite, dacite unconf.overl. Cambrian black slate	numerous chalcedonic quartz shoots with famatinite,enargite,gold, Au-tellurides, in silicif. and alunitized alteration sheets. 145 t Au	Ashley and Keith (1976)
Cripple Creek, Colorado	Mi, ellipt. fault-bound basin surr. by Pt cryst.rocks. Filled by over 1000m of water-laid,mixed proven.sedim.intr.by phonolite,syenite,basalt plugs,dikes	simple to complex fiss.veins,stockworks,impregn.,brecc.pipes,cemented by vein chalced.and cherty qtz. Pyr., adular.,fluor.,carb.,barite,gold, Au-Ag tellur.;light propyl.,argilliz. 696 t Au, 60 t Ag	Koschmann (1949)
Săcărîmb deposit Apuseni Mts., W.Rumania	Mi biot.-hornbl.subvolc. andes. emplac.into Neogene molasse sediments and PZ-MZ crystalline basement, near J ophiolites	Mi_3-Pl,220 fiss.veins in 4 groups, some grading to dissem.Au in feldspathic metasomatites.Qtz.-carb.gangue, electrum,Au-Ag tellurides,access. Pb,Zn,Cu sulph. Est.min. 200 t Au	Ianovici and Borcoş (1982)
Baguio district, Luzon,Philippines	Mi qtz.dior. and andes. formed over T_1 keratoph.-andes.,and Cr and older ophiol. and metam. Intr. by Pl granod. and dacite porphyry	Pl_3,Q_1 open and vughy qtz.,calcite, rhodochr.,pyr.,gold,Au-Ag tellur, fiss. veins and brecc.filling in the vicinity of a volc. plug. 300 t Au/7 ppm	Bryner (1969) Fernandez and Damasco (1979)
Vatukoula, Tavua caldera, Fiji	Pl,oliv.basalt and trachybasalt postdated by caldera collapse, aug. and biot. andes.dikes,cone sheets, monzonite stocks	Pl,qtz.,calcite,pyr.,electrum, Au-tellurides ore shoots and low-grade veinlet/dissem.oreb. along shears and shear/dike intersections. 115 t Au, 40 t Ag, estim. 80 t Te	Denholm (1967)

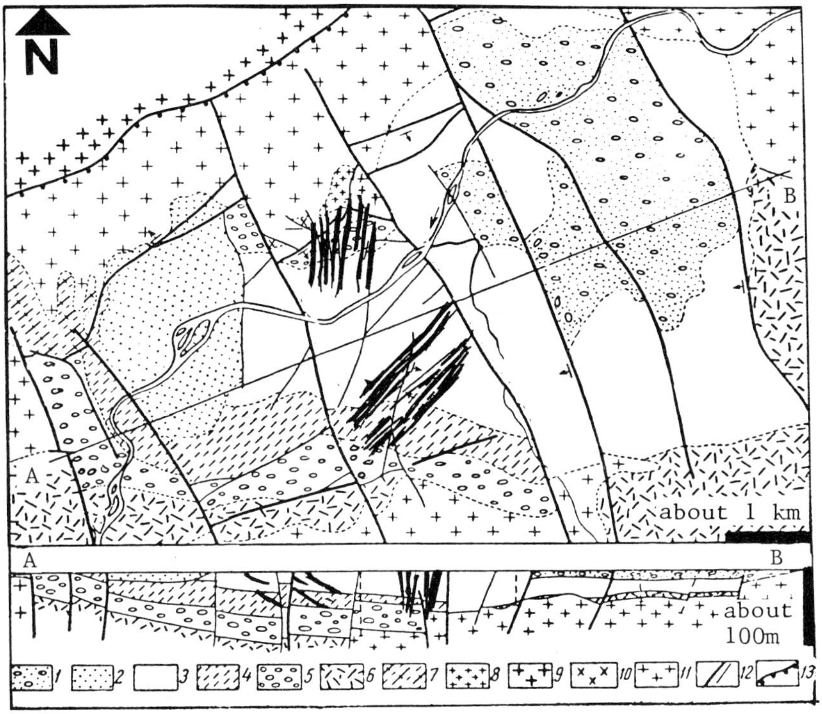

Fig. 26-33. Balei Au-Ag epithermal field, East Transbaikalia, U.S.S.R. 1=Tr conglomerate; 2=Cr_1 sandstone; 3=Cr_1 sandstone and conglomerate; 4=Cr_1 litharenite; 5=Cr_1 conglomerate; 6=J_3 andesites and tuffs; 7=Pt metamorphics; 7=Cr diorite porphyrites; 9=Cr granite; 10=PZ leucogranites; 11=PZ granodiorites; 12=faults; 13=Borschovochnyi Fault. From Lozovskii et al. (1960).

26.6.2. Pre-Cretaceous epithermal Au,Ag and Pb,Zn deposits

Although probable equivalents of subvolcanic epithermal deposits have been preserved in continental volcanics that date as far as the Proterozoic age, pre-Cretaceous examples are quite rare and are usually indistinguishable from the dominant "plutonic" deposits. This is because the parent volcanic piles, the feature most indicative of the possible presence of subvolcanic deposits, have been largely removed by erosion or sheared, metamorphosed, etc. It should be realized by prospectors that the mere presence of paleo-volcanics is not sufficient evidence that the orebodies such volcanics host are coeval and subvolcanic. In most cases such ores are younger, bearing no direct genetic relationship to their host assemblage, and controlled largely by structure and by the development of another mineralizing system unrelated to the volcanics. This is usually a

Table 26-5. Epithermal Ag $\rangle$ Au ("bonanza") deposits

LOCALITY	GEOLOGY	MINERALIZATION	REFERENCES
Comstock Lode, Virginia City, Nevada, U.S.A.	Mi andes.to rhyodac.volc.pile intr. by granodior. In basement Cr granit.pluton,metamorphics	qtz.,electrum,argentite,etc. ore shoots in N.N.E. trending fault zone. Advanced argillic alter.,regional propyl. 242 t Au, 5.8 Tt Ag	Whitebread (1976)
Tonopah, Nevada, U.S.A.	Mi andes.,dac.,rhyol.volcanic centre intruded by felsic stocks	over 20 E.-W. and N.E.-S.W. fissure and replac.qtz.,carbon.,adularia, electrum,argent.,polybasite,Cu,Pb, Zn assoc.sulph.,silic.,albitiz.,seric.alter. 4,920 t Ag, 64 t Au	Spurr (1905) Albers and Kleinhampl (1970)
Republic, N.E. Washington, U.S.A.	Eo-Mi andes.,dac.,rhyol. overl. Cr_3 granod.bathol., Pe mafic metavolcanics	chalced.quartz,pyr.,electrum,etc. fissure veins; silic. and argillic alteration. 26 t Au/10 ppm, 255 t Ag	Muessig (1967)
Batopilas,Chihuahua, N.Mexico	T, fine grained diorite intruding andes. brecc. and flow near basement unconformity	thin (under 1m) calc.,quartz,argentite, access.Co,Ni arsenides, nat.Ag veins in propylitized hosts. 1,300 t Ag	Wisser (1966)
Tayoltita, Durango, N.W. Mexico	Eo-Ol, rhyol.-andes.pile on margin of caldera, intruded by Ol granod. and dior. stock	qtz,alb.,argent.,electrum fiss.veins in albite altered halo,grading to propylitiz. 15 Tt Ag/425 ppm, 210 t Au/6 ppm	Randall (1972)
Zacatecas, Zac., Mexico	Tr and older phyll.,seric.sch., limest.,congl. T porph.andes., rhyol. and dior. porphyry	qtz (chalc. and drusy),pyr.,sfal., gal.,tetr.,argent.,polybas.,pyrarg., gold, calcite E.-W. and N.W. fissure veins and miner.breccias; silicif., propyl. 26 Tt Ag, 7 t Au, 700 Tt Zn, 450 Tt Pb	Mapes Vazques and Stone (1956)

Guanajuato, Gto, Centr. Mexico	MZ folded black slate, minor sandst., limest. intr. by Cr_3 diorite, granod. T_1 red polymictic congl. T_2 contin. volcanics	Ol_3, persistent (up to 23 km) qtz. lode zones (miner. faults) have local Ag "bonanza" shoots (acanthite, aguilarite, naumann., electrum). 30 Tt Ag/450 ppm, 115 t Au/2.5 ppm	Gross (1975)
Pachuca-Real del Monte, Hidalgo, C. Mexico	Ol-Pl andes., dac., rhyol. pile resting on Cr marine seds., intruded by rhyol. porph. dikes, small stocks	Pl_2 N.W. and N.S. // fract. veins filled by early qtz. and minor Pb, Zn, Cu sulph., and late qtz., calc., argent., acanth. shoots. 44 Tt Ag/231 ppm, 400 t Au/1.35 ppm, 80 Tt Pb	Geyne et al. (1963)
Roşia Montană, Apuseni Mts., W. Rumania	Mi subvolc. hornbl. biot. andes. and dac. intruding Cr_3 flysch sandstones	Mi_3-Pl, qtz., carb., pyr., electrum, Au-Ag tellur., veins and stockw. grade to low-grade Au dissem. in adularia-argillic altered brecc. pipe in a dacitic plug. Est. 350 t Au	Ianovici and Borcoş (1982)
Sado mine, Sado Isl., Japan Sea, Japan	Mi-Pl andes. tuff and flows interb. with black mudst., underlain by rhyolite	Pl, qtz., adularia, calc., rhodochr., electrum, nat. Ag, Ag sulfosalts fiss. veins in silic. and propylitiz. rocks 77 t Au/4.7 ppm; 2.3 Tt Ag/153 ppm	Imai and Bunno (1978)
Edie Creek, Wau, New Guinea Island	MZ chloritoid sch. and slate; Mi volc., granod.; Pl breccia, qtz. biot. porphyry	Pl, discrete Ca-Mn carb., qtz., pyr., electrum veins, veinl. in shears, veinl. and dissem. in breccia. Silicif., seric., argilliz. 115 Tt Au	Rebek (1975)
Waihi, Hauraki Goldf., N. Island, New Zealand	Mi propylitized andesite, topped by post-ore rhyolite	Mi-Pl syst. of conjug. fiss. veins of chalced. qtz., orthocl., calcite, Zn, Pb, Cu sulph., pyrarg., acanthite, electrum. 240 t Au, 754 t Ag	Williams (1965)

"granitic" intrusion. If this is the case, laborious paleo-volcanic interpretations are of little consequence because both the volcanic centre and flanks rocks have an approximately equal chance of hosting orebodies.

The Cracow gold deposit, Queensland (Ransom and Knight, 1975; 17.5 t Au, 18.1 t Ag) is an example of a mineralization of uncertain genetic affiliation that is probably epithermal. There, the gold-bearing ore shoots (quartz, adularia, carbonate, sphalerite, chalcopyrite, pyrite, bornite, hessite, electrum) occur within a steeply dipping east-trending zone linking two N.W. fault zones. The ore zone is about 1 km long and 240 m deep. The orebodies are either crustification fissure veins or quartz-filled breccias or stockworks. The host rocks are lower Permian trachytes, dacites, rhyodacites and trachyandesites, intruded by rhyolite dikes. The wallrock alteration is light (silicification and minor sericitization, pyritization, carbonatization). The ore is clearly structurally controlled and emplaced sometimes between Permian and middle Triassic.

26.6.3. "Xenothermal" (telescoped) deposits

The term "xenothermal" was introduced in 1935, to refer to hydrothermal deposits formed at high temperatures but at shallow or moderate depths. This was in the classical period of economic geology when the experimental methods of determination of the actual filling temperatures of hydrothermal veins were in their infancy. Sn,W,Bi,Mo and several other elements were automatically assumed to be "high temperature" and Pb,Zn,Cu,Ag, etc. to be "moderate to low temperature" products. As a consequence, apparently subvolcanic deposits combining Sn,W,Cu,Zn,Pb and Ag such as Ikuno and Akenobe, or Sn,Ag such as Potosi or Oruro, were placed into this category.

Recently, Imai et al. (1975) investigated fluid inclusions in ores from the Japanese localities, and found that their filling temperatures ranged 155 and 355° C, as is found in ordinary epithermal deposits. Although at Akenobe the decrepitation temperatures had a tendency to increase with depth, they did not show any distinction between the Sn-W and Cu-Zn suites.

I believe that the unusual congregation of metals at Akenobe and Ikuno is due to two or more contrasting lithologic associations hosting the orebodies, and also, probably, having a source relationship with the ores. The ophiolitic to mafic volcanic-sedimentary on one side, and terrigenous-sedimentary association on the other, appear also to influence the petrochemistry of the later intrusions, one dominantly dioritic having the characteristics of "I granites", the other quartz porphyritic and granitic more closely matching the "S granite" family.

The Akenobe ore field in Ohya Town, Hyogo Prefecture, W.Honshu (Imai et al., 1975; Mitsubishi Akenobe Staff, guided tour, 1972; 750 Tt Cu/1.26-2.5%, 330 Tt Zn/2.2%, 7.5 Tt W; Fig. 26-34), probably holds a world record for the number of metallic elements accumulated

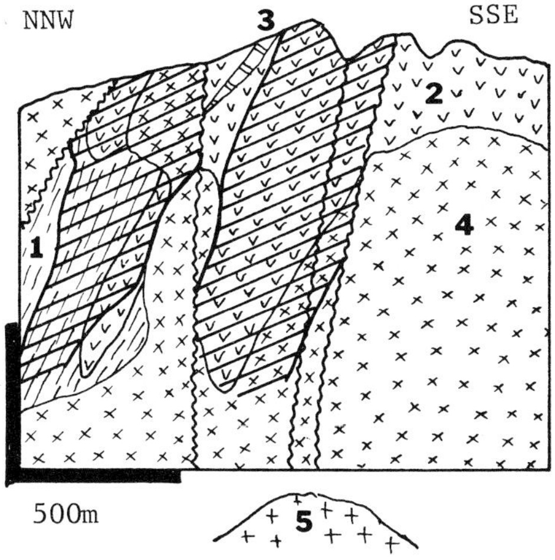

1. Cb-Pe phyllite
2. mafic metavolcanics and pyroclastics ("Schallstein")
3. chert
4. Cr_3-T_1 diorite
5. T_1 postulated granite cryptobatholith (not to scale)

diagonally ruled: projections of N.W.-S.E. veins

Fig. 26-34. Akenobe Cu,Zn,Sn,W mine, Ohya, Honshu. From LITHOTHEQUE, based on data in Imai et al. (1975) and Mitsubishi Akenobe Staff, guided tour (1972).

in supra-clarke quantities in its ores (Zn,Cu,Pb, Sn,W, Ag,Au, Mo,Bi, As,Sb,Se,Te, In,Ge,Ga,Tl, etc.). The field is situated in a Permian marine mafic metavolcanic and metasedimentary terrain (in the Maizuru orogenic belt) dominated by slate to phyllite, greenstone (incl. the varicoloured "Schallstein") and chert. Ophiolite slices are also present in the area and small deposits of "Kieslager-type" massive pyrite-chalcopyrite (Minamitani) and pyrrhotite, Ni-Co arsenides (Natsume) are hosted by these rocks. In the late Cretaceous, continental andesite, basalt and rhyolite suite associated with probably comagmatic andesite porphyry, diorite, felsite and granite intrusions, invaded the older rocks. Following a faulting episode a swarm (about 100) of mostly N.W.-trending fissure veins was emplaced on both sides of the N.-S. Akenobe Fault. The veins are usually banded, with multistage filling (typically: quartz, calcite, chlorite, chalcopyrite, sphalerite, cassiterite, bornite, galena, wolframite, scheelite) and the ore field is famous for the overlapping metal zoning (from centre outward: Sn,W - Sn,Cu - Cu,Zn - Zn,Pb - Ag). The outermost zone has several "bonanza" epithermal veins with argentite, polybasite, stephanite, native silver, etc. in quartz, adularia, calcite and rhodochrosite gangue. The mineralization is presumably situated above a buried "Sanin-type" "I granite" cupola.

Ikuno ore field (17 km S.E. of Akenobe; 350 Tt Zn, 165 Tt Cu, 8.5 Tt Sn, 2.3 Tt Pb, 310 t Ag, 1.5 t Au; Imai et al., 1975) has a system of zoned N.W. fissure veins, comparable with Akenobe. The veins are, however, hosted almost entirely by the late Cretaceous felsic and intermediate volcanics, quartz porphyry and granite.

26.6.4. Copper-mineralized veins in continental volcanic association

Copper sulphides in accesory quantities are almost always present in the Au-Ag and Pb-Zn epithermal deposits, but purely coper-bearing epithermal veins are less common. The known examples carry usually Cu sulphosalts (e.g. enargite, luzonite, tennantite) or hypogene chalcocite, bornite, covellite, etc. Many formed in the late stages of development of plutonic porphyry copper systems (e.g. in Butte, Morococha, partly probably in Récsk). The more frequently occurring styles of Cu mineralization in the young, unmetamorphosed subaerial volcanic piles are shown diagrammatically in Fig. 26-35.

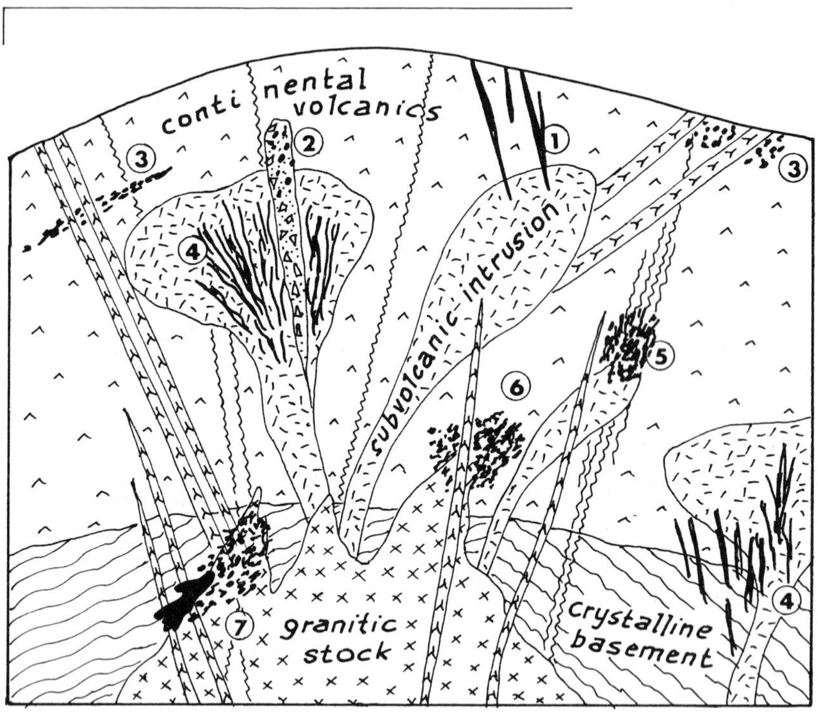

Fig. 26-35. Common Cu mineralization styles in young, unmetamorphosed subaerial volcanic piles. 1=enargite, luzonite, etc. epithermal veins (Laurani); 2=high — level disseminations in breccia pipes; 3=disseminations, veinlets in volcanics at dike or sill contacts; 4=stockworks, disseminations or veins in subvolcanic bodies (Ashio); 5=massive to disseminated replacements in altered volcanics and in subvolcanic "porphyry" bodies (volcanic and subvolcanic "porphyry Cu"); 6=plutonic stockwork and disseminated porphyry Cu systems in volcanic and intrusive hosts; 7=porphyry Cu and sometimes skarn systems entirely within plutonics and basement rocks. From Laznicka (1984).

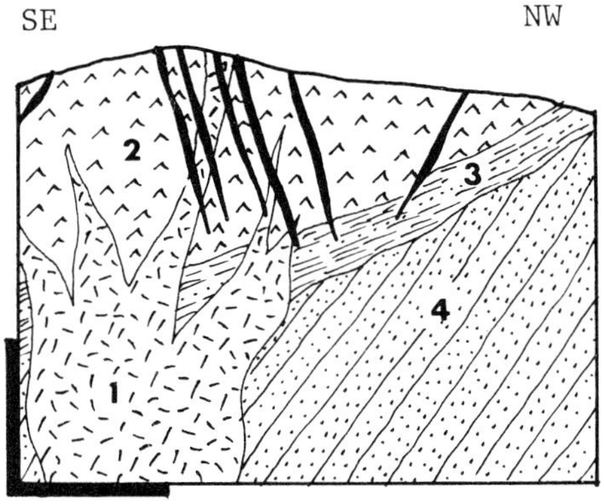

Fig. 26-36. Cerro Laurani, Bolivia. From LITHOTHEQUE, diagrammatic, based on data in Ahlfeld and Schneider-Scherbina (1964).

At Cerro Laurani near Sica Sica, Bolivia, an example of "style 1" (Ahlfeld and Schneider-Scherbina, 1964; Fig. 26-36), a system of banded and drusy epi- to mesothermal fissure veins and minor replacements is hosted by "Laramide" andesites and dacites, above the apical portion of a Miocene-Pliocene dacite and rhyolite porphyry stock. The veins are composed of quartz, pyrite, enargite, bornite, tetrahedrite, lesser luzonite and famatinite in the zone nearest to the stock. At a greater distance carbonates, lesser barite, galena, sphalerite, tetrahedrite, jordanite and Ag sulphosalts-bearing veins are dominant. The wallrock alteration consists largely of quartz-sericite and silicification.

Ashio field in central Honshu, Japan (Imai et al., 1975; 700 Tt Cu; Fig. 26-37) is an example of "style 4". There, about 1,400 discrete veins are located within a conical mass of a late Cretaceous to early Tertiary "rhyolite" (variously interpreted as welded tuff, flow or a subvolcanic intrusion) emplaced in a Permian marine bimodal volcanic-sedimentary association, and intruded at some depth by a granodiorite porphyry stock. The veins are composed of quartz, pyrite, chalcopyrite, pyrophyllite, clay minerals, minor arsenopyrite, sphalerite and galena. The felsic host is pervasively altered by quartz, sericite, pyrite and by an advanced argillic assemblage. There are many variations of this style throughout the world, many of which are conspicuous for their intensive silicification often creating completely silicified rocks ("secondary quartzites"; style 5).

High-level mineralizations transitional between porphyry coppers and various forms of subvolcanic deposits, are known in several regions of the world, but few have been described in detail. Several localities exist in the Capillitas district near Andalgalá, Catamarca,

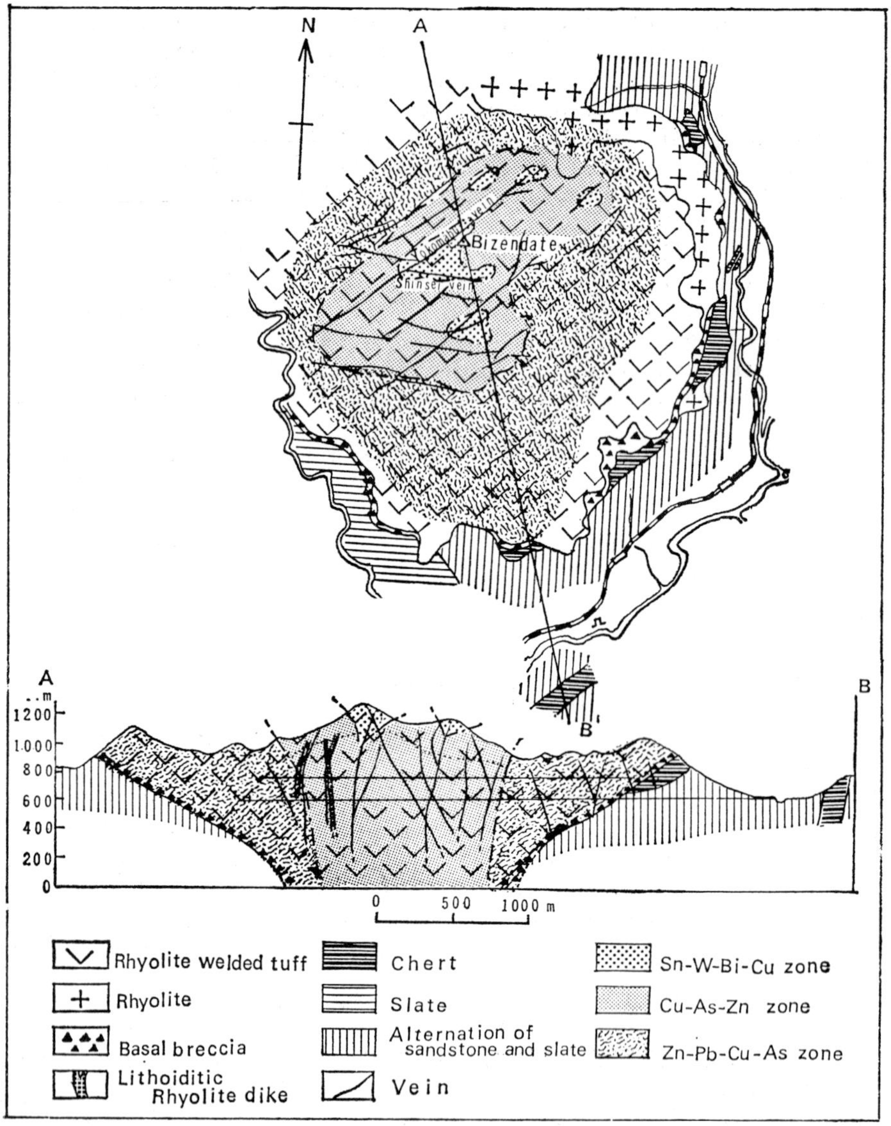

Fig. 26-37. Ashio Cu mineralized structure (Cr), Honshu, Japan. From Nakamura (1971), courtesy of Professor Takeshi Nakamura.

N.W. Argentina, already mentioned in connection with the epithermal Au-Ag veins at Farallon Negro. There, Pliocene continental basalt, andesite, dacite, rhyolite suite with comagmatic syenodiorite, monzonite and granodiorite bodies, top and intrude lower Paleozoic and Proterozoic basement. At the Mi Vida prospect (Koukharsky and Mirre, 1976), disseminated, veinlet and vein pyrite, covellite, bornite, enargite, molybdenite, sphalerite, galena, rhodochrosite, and other minerals are found in a Pliocene rhyolite breccia pipe. The wallrocks are sericitized and silicified and these alterations grade into argillized rocks with alunite and zunyite. In the Bajo de la Alumbrera showing, about 1.2 Mt Cu, 100 t Au and 750 t Ag are located in a Pliocene quartz-sericite and argillic-altered quartz monzonite porphyry stock, situated at the top of the plutonic level under epithermal Au,Ag, Pb,Zn,Mn veins. The Cu grade is between 0.3 and 0.55%.

In the Valea Morii Nouă deposit in the Apuseni Mts., Rumania (Ianovici and Borcoş, 1972), a subvolcanic body of a Miocene andesite porphyry intrudes a Mesozoic ophiolite basement and Miocene hornblende-quartz andesite lava flows. Disseminated and veinlet pyrite-chalcopyrite mineralization ("porphyry Cu") is present largely in sub-economic quantities throughout the entire subvolcanic stock. Sequentially later quartz, calcite, rhodochrosite, sphalerite, galena, chalcopyrite and native gold epithermal veins are superimposed along faults.

The Musariu Nou deposit in the same area (Ianovici et al., 1976; Fig. 26-38) is associated with a cylindrical subvolcanic stock of a Miocene-Pliocene andesite and quartz andesite, emplaced into Jurassic ophiolites. In the top levels (down to about 500 m beneath the original surface) a system of epithermal quartz, calcite, barite, Cu,Pb,Zn sulphides, gold and native arsenic veins mined for gold, comprise the principal mineralization style. Recent drilling to a depth of about 1,500 m established the presence of "porphyry Cu" ore in fractured and altered microdiorite facies of the andesite, rich in magnetite.

In the Récsk deposit, Mátra Mts., Hungary (Morvai, 1982; 2.8 Mt Cu/0.4%, 50 Tt Mo), an Eocene andesite subvolcanic complex intrusive into Triassic sediments carries enargite-luzonite veins and stockworks in the near-surface region. These have been mined on small scale in the past. These orebodies are in quartz-sericite altered andesites and are enveloped by pyrite and kaolinite-altered wallrocks. A large tonnage (700 Mt) but low grade (0.4% Cu) "porphyry" orebody was discovered more recently at some depth, hosted by altered biotite-hornblende andesites and by diorite.

Lepanto (Philippines) and Chinkuashih (N. Taiwan) Cu,Au,Ag deposits are examples of enargite-dominated mineralizations affiliated to continental or transitional (into marine volcanics) magmatic and sedimentary suites of young island arcs. The ores are, in many respects, transitional into volcanics-hosted massive sulphides, particularly those of the "kuroko-type" (compare Chapter 14).

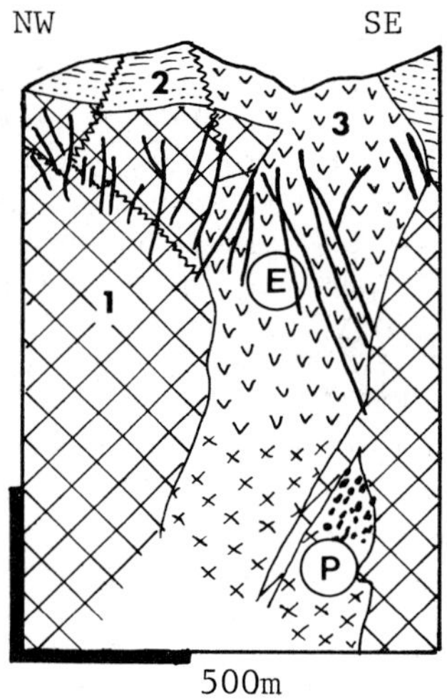

1. Jurassic ophiolites
2. Miocene mudstone, sandstone
3. Miocene - Pliocene cylindrical intrusion of quartz - hornblende or hypersthene andesite porphyry grading to microdiorite

E epithermal gold veins

P porphyry copper - style mineralization (diagrammatic only)

Fig. 26-38. Association of epithermal Au,Ag veins (E) and "porphyry copper" (P) at greater depth, illustrated by the example of the Musariu Nou deposit, Apuseni Mts., Rumania. Modified after T.P. Ghitulescu in Ianovici et al. (1976).

Au/3.7 ppm, 160 t Ag), enargite with luzonite, tennantite, Au-Ag tellurides and other minerals, is found in a 2 km long, pencil-shaped replacement orebody along a N.-S shear. Chalcedonic quartz constitutes the gangue and also the main wallrock-altering mineral. The hosts are Cretaceous to Paleogene marine meta-basalts and meta-andesites and the orebody is located close under the unconformable capping of Miocene-Pliocene dacitic pyroclastics. Miocene trondhjemite pluton and Pliocene subvolcanic stocks are in the area.

Chinkuashih (Folinsbee et al., 1972; min. 47 t Au/3.1 ppm, 150 t Ag, 30 Tt Cu/0.9%) contains a system of replacement veins grading to ore masses and disseminations, some filling breccia pipes. The host is a Pleistocene dacitic mushroom-shaped subvolcanic intrusion, emplaced in Miocene marine sandstone and shale. The ore consists of pyrite, enargite, luzonite, wurtzite, electrum, etc. in quartz and lesser barite, alunite, clay gangue. It is accompanied by strong silicification, argillic alteration and chloritization.

26.6.5. Volcanics and subvolcanics-hosted disseminated ("porphyry") Cu-Mo deposits

Styles 6 and 7 in Fig. 26-35 are bona fide porphyry coppers in which the major feature of "porphyry" deposits, association with intrusive rocks, is uncontested. Style 5 and sometimes Style 4 are often referred to as "subvolcanic" or "volcanic" porphyry coppers. There is little unity in the terminology and the meaning of "volcanic porphyry Cu" as applied by Sutherland Brown (1976) has already been discussed in Section 13.7.4. (compare also Fig. 13-9). In Chapters 13 and 14, the volcanics surrounding the copper orebodies were members of a fundamentally marine association. In this section we deal with largely continental andesitic (also basaltic, dacitic, rhyolitic) volcanic piles pervasively altered and mineralized by stockwork and disseminated chalcopyrite in company with other common minerals such as pyrite, bornite or minor molybdenite. Conspicuous "granitic rocks" are missing at the present level of erosion but a variety of "porphyry" dikes and small stocks is present.

From what we have learned so far about the epithermal ore system and from what is known about the anatomy and setting of the porphyry copper deposits, there is little doubt that the volcanics-hosted disseminated and stockwork Cu deposits are not an independent style, but merely the topmost expression of a mineralization system driven and partly supplied, by a buried, evolving intrusion. The fragmentation of the porphyry Cu family in this book has been carried out for the convenience of the prospector, who can immediately see and evaluate the orebodies in their volcanic and subvolcanic setting. This section should be read concurrently with Section 28.5.

Table 26-6 lists selected examples of continental volcanics-hosted porphyry coppers lacking conspicuous "granitic" rocks. These are transitional into similar deposits in which the "granites" are apparent. The latter are listed in Table 28-3. On global scale, volcanics-hosted "porphyries" are included in the porphyry copper belts and provinces shown on Fig. 28-41 and are briefly characterized in Table 28-2.

Schaft Creek, British Columbia (Fox et al., 1976; 1.32 Mt Cu, 72 Tt Mo) is hosted by upper Triassic andesitic and basaltic flows, pyroclastics and a volcanic breccia, intruded by felsic porphyry dikes. Bornite, chalcopyrite and molybdenite occur as disseminations and veinlets in a N.-S. system of elongated zones of fracturing and brecciation. The confusing mixture of various alteration minerals is interpreted as a former broad prograde biotite alteration zone bordered by an epidote-rich propylitic fringe. Later, the biotite was largely chloritized.

Safford deposit, Arizona (Robinson and Cook, 1966; 2.5 Mt Cu/0.5%, 47 Tt Mo; Fig. 26-39) is hosted by Cretaceous subaerial andesites and volcanic breccias, intruded by a swarm of subparallel dikes and small plugs of rhyolite, quartz latite and dacite in a probable vent region of an Eocene volcano. The rocks are pervasively altered. Hydrothermal biotite is most common in andesite, quartz, sericite and pyrite in felsic dikes. Disseminated, stockwork and vein chalcopyrite occur in a large orebody having an area of 8 km2 (dimensions 4x2 km). The hypogene zone is topped by the remnants of a Tertiary chalcocite blanket now substantially converted into chrysocolla and brochantite, by leached capping and is buried, in places, under younger

Table 26-6. Continental volcanics and sub-volcanics hosted disseminated ("porphyry") Cu(Mo) deposits, lacking conspicuously associated "granites" at the present level of erosion

LOCALITY	M*	HOST VOLCANICS	MINER. AGE	CONTROLS	ALTER./MINER.	TONNAGE	REFERENCES
Schaft Creek, Stikine area, Br.Columbia, Canada	5	Tr_3 andes.,basalt flow,volcaniclast. fels.porph.dikes	182 m.y.	steep dipping irreg.z.of intrus.and tect.brecc.	veinl.,diss. KF,bio,sil bn,cp,py	1.32 Mt Cu/ 0.4%; 73 Tt Mo,495 t Ag 10.5 t Au	Fox et al. (1976)
Safford, Arizona	6	Cr_3 and.,and.brc., intr.by 58 m.y. stocks,dikes of qtz.latite,rhyol. qtz.monz.porph.	Eo 58 m.y.	veinl.,diss. in former volc.centre contr. by faults	veinl.,veins shears; KB,ph; py,cp bn,mag	8 Mt Cu/ 0.4%	Robinson and Cook (1966)
Cerro Colorado, W.Panama	7	Ol,subaer.andes.fl. intr.by Mi_3-Pl qtz. latite porph.,rhyol. rhyodac.,qtz-feldsp. porphyries	4.2-7.2 m.y.	irreg.oreb. contr.by faults and intrus.cont.	diss.,veinl. phyllic alt. py,cp,bn,mo 25% ore in enrich.z.	18 Mt Cu/ 0.6%; 450 Tt Mo,168 t Au,10 Tt Ag	Kesler et al. (1977)
Majdanpek, Timok Massif, Serbia, Yugoslavia	11	Pt_3 schist,marble PZ phyll.,J limest. Cr_3 contin.andes. T_1 minor stocks of dior.,qtz.dior.p.	T_1	wide N.-S. fault and fract.zone stockw.,vein limest.repl.	sil,KB,less ph; cp,bn,py rare tetr. mass.pyrite	3 Mt Cu/ 0.6%; 30 Tt Mo; 1.8 Tt Ag; 190 t Au/0.37%	Starostin (1970) Janković (1982)
Bor, Timok Massif E.Serbia, Yugoslavia	11	Cr_3 andes.volc.and pyrocl.,congl.,intruded by dior. porph.pluton	T_1	broad fault zone,volc.facies,above intrusion setting	irreg.mass. py,cp,enarg. bodies in arg.volc., porph.Cu in depth	2.75 Mt Cu/ 1%; 190 t Au; 1 Tt Ag	as above

Veliki Krivelj, Timok Massif, Yugoslavia	11	Cr_3 andes.volc. and pyrocl.,minor dior. porph. intrusions	T_1	tect.contr. N.W.-S.E. zone 1.5 x 0.4 km	stockwork, KB,less ph, argill.; py, cp,mo; about 50% superg. enrichment	3.2 Mt Cu/ 0.4%; 120 Tt Mo,240 t Ag, 64 t Au	Janković (1982)
Santo Niño, Tub-lay, N.Luzon, Philippines	14	$Cr-T_1$ m-andes.,ba-salt,dac.,rhyol. gabbro,dior.intr.	T	fault control	stockwork, bn,cp,py	175 Tt Cu/ 0.371%; 14.6 t Au	Wolff (1978) Bryner (1969)
Waisoi dep.,Namo-si area, S.E. Viti Levu, Fiji	15	Mi_3 cont.andes.fl., agglom.,over $Eo-Mi_2$ maf./interm.volc. Qtz.-feldsp.porph. stocks	Mi_3-Pl	several over-lap.zones of miner. and alter.,fault control	veinl.,diss. KB,ph,low gr.in prop. py,cp,bn,mo	2.29 Mt Cu/ 0.48%	Colley and Greenbaum (1980)
Goonumbla, Parkes dist.,N.S.W., Australia	22	Or,marine m-andes., phyllite,intr.by D? qtz.-feldsp. porphyry	D?	fault control	stockw., veinl.,veins sil,ph; py, cp	2.28 Mt Cu/ 0.7%; 84.5 t Au; 1,015 t Ag	E/MJ January (1982)

ABBREVIATIONS: M*=number of the host "porphyry Cu belt" in Fig. 29-X and Table 29-X. py=pyrite; cp=chalcopyrite; bn=bornite; mo=molybdenite; mag=magnetite KF=potash feldspar (alteration); KB=biotite; ph=sericite (quartz) sil=silicification; arg=argillization

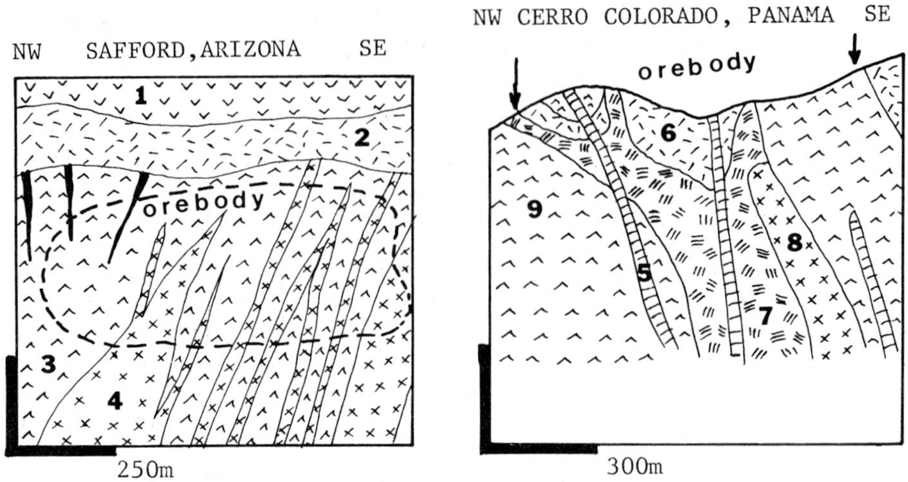

Fig. 26-39. Volcanics-hosted Cu-Mo deposits, two American examples (diagrammatic). 1=Mi ? basalt flow; 2=Eo-Mi intermediate continental volcanics; 3=Cr3 andesite, andesite breccia; 4=Pg quartz latite, rhyolite, dacite, quartz monzonite porphyry; 5=Pl rhyodacite porphyry; 6=Mi_3-Pl_1 quartz latite porphyry; 7=Mi_3-Pl_1 rhyolite to feldspar porphyry; 8=Mi_3-Pl granodiorite porphyry; 9=Ol subaerial andesite flows. From LITHOTHEQUE, based on data in Robinson and Cook (1966), Clark et al. (1977), Compañía de Cobre Cerro Colorado Staff, oral comm., 1977.

unmineralized Tertiary volcanics.

Cerro Colorado in the Chiriquí Province of western Panama (Kesler et al., 1977; Fig. 26-39) is one of the largest porphyry copper deposits discovered in the late 1960s (18 Mt Cu or more) in an area without any significant earlier mining tradition. There, pyrite, chalcopyrite, bornite and molybdenite form pervasive, fine-grained disseminations and veinlets in mainly quartz-sericite altered and silicified late Miocene and Pliocene subvolcanic quartz latite and rhyodacite porphyries, and in exocontact Oligocene subaerial andesite flows. About one quarter of the reserves lies in sooty chalcocite in the zone of secondary enrichment, covered by leached capping. The ore is considered to be Pliocene, and the ore zone has a roughly circular outline with a diameter of about 4 km.

The three large copper deposits in the Timok volcanic massif (Eastern Serbia, Yugoslavia): Bor, Majdanpek and Veliki Krivelj (approximate tonnage about 10 Mt Cu) are largely volcanics-hosted disseminated Cu deposits although replacement massive sulphide bodies are important in Bor, and Cu skarns and replacements in Majdanpek. The Timok Massif is about 70 km long and 20 km wide. It is a

N.-S.-trending late Cretaceous andesitic volcanic pile, intruded by small "Laramide" quartz diorite and granodiorite pluton. At the present level of erosion, the plutons represent less than 5% of the area.

The northernmost Majdanpek deposit (Starostin, 1970; Janković, 1982) is associated with a N.-S.-trending fault zone at the northern extremity of the Timok Massif. There, Cretaceous andesites are located in a narrow graben, floored and surrounded by Jurassic limestones, conglomerates and sandstones and by Paleozoic or Proterozoic phyllites, gneisses and amphibolites. Small hypabyssal quartz diorite porphyries occur in the zone. Most of the ore reserves take the form of very fine pyrite and chalcopyrite, evenly disseminated in biotite and quartz-sericite–altered andesite. The average grade is low, but the orebody is persistent for several kilometres. Disseminated chalcopyrite is also hosted by altered gneisses. Small massive pyrite-chalcopyrite bodies are found in fractured and silicified limestone and have been mined in the past.

Bor (Fig. 26-40) has been in operation for 80 years and the ore mined so far has come from a series of massive to stockwork pyrite, lesser bornite, enargite, chalcopyrite, chalcocite, covellite etc. orebodies in silicified and quartz, sericite, dickite and kaolinite-altered andesites. The andesites are both submarine lavas and breccias and small bodies of subvolcanic andesite porphyries, difficult to distinguish in the highly altered ore zone. In some respects the mineralization style resembles Mount Morgan, Queensland (Chapter 11) or Kedabeg, Kafan and Allaverdy in the Soviet Transcaucasia. Recently, a large disseminated porphyry copper body in diorite porphyries has been discovered at a depth of 600 m under the surface (Janković, 1982).

In the Philippines, the Santo Niño deposit in the Mountain Province of northern Luzon (Baguio district; Bryner, 1969), is an example of a volcanics-hosted disseminated copper deposit. There, chalcopyrite and minor bornite, molybdenite, with quartz and calcite form veinlets in biotite altered, sheared and faulted Tertiary hypabyssal basalt, andesite, dacite and albite porphyry. The basement rocks are Cretaceous to Paleogene mafic metavolcanics and metasediments intruded by gabbro.

In Fiji, the Waisoi deposit near Namosi (Colley and Greenbaum, 1980; 2.29 Mt Cu) has pyrite, chalcopyrite, bornite and minor molybdenite disseminations and stockwork veinlets in biotite-sericite and propylitic–altered Miocene-Pliocene quartz-feldspar porphyry stocks, and in adjacent late Miocene andesites. The ore appears to be controlled by a paleo-volcanic centre (stratovolcano).

In the Tasman orogenic belt of eastern Australia, several Paleozoic copper occurrences (e.g. Copper Hill, Peak Hill, Goonumbla in the Lachlan orogen, N.S.W.) correspond to volcanics-hosted "porphyry" deposits. The largest orebody discovered recently is near Goonumbla in the Parkes-Forbes gold zone (2.28 Mt Cu), and it is a gold-rich pyrite, chalcopyrite stockwork in sheared and altered Ordovician marine metasediments intruded by subvolcanic porphyry stocks probably of Devonian age.

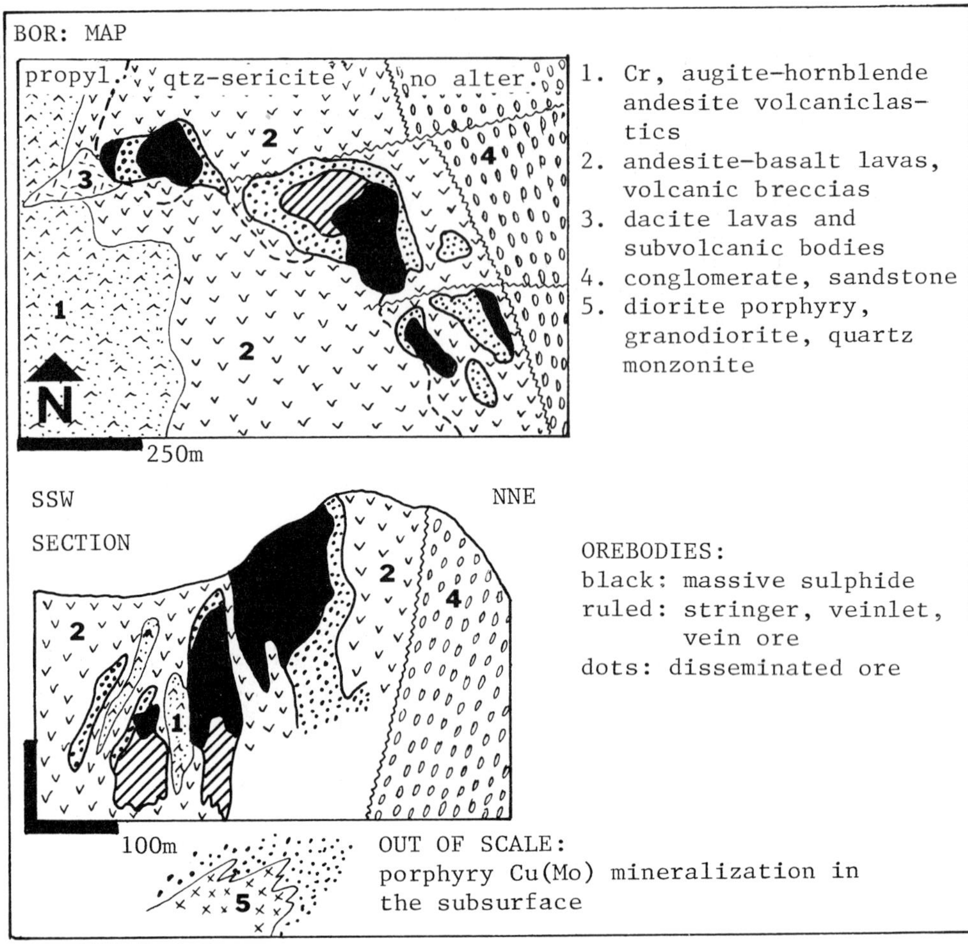

Fig. 26-40. Bor, Yugoslavia, massive and disseminated Cu deposit in andesites. Modified after Starostin (1970).

26.6.6. Sn (W,Mo) deposits in continental volcanic association

WESTERN UNITED STATES

Several occurrences of cassiterite (particularly the metacolloform variety "wood tin") are associated with Tertiary rhyolitic plugs and breccias. No important deposit has, so far, been discovered and the small production of several hundreds of tons of cassiterite concentrate came largely from placers. The largest bedrock deposit Majuba Hill in western Nevada (Albers and Kleinhampl, 1970;

2.7 Tt Cu, 350 t Sn) is a complex rhyolite plug emplaced into Triassic slates. The rhyolite is disturbed by several varieties of silicified, sericitized and tourmalinized breccias containing scattered grains of chalcopyrite, pyrite, arsenopyrite, cassiterite and a variety of secondary minerals. Near Izenhood, Nevada, cassiterite occurs as veinlets in rhyolite flows and placer "wood tin".

In the Black Range, S.W. New Mexico, cassiterite with specularite, quartz, bixbyite, pseudobrookite, topaz and monazite form scattered crystals in miaroles; encrustations on fractures; stringers and disseminations, all in a flow-banded, porphyritic rhyolite of the late Oligocene age. Lufkin (1972) considered these occurrences to be cogenetic with the host rhyolite, probably fumarolic.

MEXICO

More than 1,000 cassiterite showings are known in the Tertiary rhyolites in a N.W.-trending belt over 1,000 km long, situated in the western portion of the Mesa Central (Foshag and Fries, 1942). The cumulative production until the 1960s has been about 3,000 t Sn, but the largest field (Sierra de Chapultepec) contained only 600 t Sn/0.5%. The bulk of the occurrences are sparsely scattered cassiterite grains in miarolitic rhyolite, fracture coatings, minor disseminations in felsic tuffs, talus breccias, rhyolite flow breccias, tectonic breccias and alluvial placers.

More recently, a re-evaluation of the Mexican volcanics-hosted tin deposits (e.g. Ypma and Simons, 1969) demonstrated that possibly economic low-grade, low-temperature (about 150° C) Sn mineralization is most likely to be found in intrusive rhyolites (marked by steeply dipping flow banding) and in particular in altered felsic breccias. In the Zaragoza area N.E. of Madero, Durango (80 Tt Sn/0.2%), tin values occur in a shallow, funnel-shaped rhyolitic breccia crisscrossed by a stockwork of veinlets containing cassiterite, specularite, wood tin, varlamoffite and other minerals. In some cases the ore minerals are submicroscopic (invisible). The gangue is quartz, tridymite, chalcedony, opal, zeolites and minor fluorite.

BOLIVIA AND NORTHERN ARGENTINA

The famous Bolivian Tin Belt contains a variety of tin deposits in three distinct bedrock associations: (1) Paleozoic shallow marine sediments (Chapter 19); (2) Tertiary volcanic-subvolcanic complexes; (3) Mesozoic and Cainozoic granitic plutons and their aureoles (Chapter 28). The association of type (2) contains a variety of mineralization styles. The most important are (a) diseminated, stockwork and vein deposits in high-level altered central stocks. The less important are (b) fracture coating, miarole, etc. occurrences of cassiterite in rhyolites comparable with the Mexican occurrences. The (a) style has already been reviewed briefly in Chapter 19, where the emphasis was placed on the persistently affiliated Paleozoic bedrock sediments. Additional information will be provided here. Several examples of the very young (Pliocene to Pleistocene) style (b) have been treated briefly in Chapter 5, and will not be discussed any further here.

The Tertiary "porphyry tin" (Sillitoe et al., 1975) or "sub-volcanic" (Grant et al., 1977) deposits that include the world's largest single tin deposit Llallagua and two or three other large deposits, are situated in southern part of the Cordillera Oriental. There, the basement is a folded marine association of gray to black slates interbedded with quartzite and litharenite (quartzite). The basement is unconformably covered and intruded by erosional remnants of Tertiary continental felsic volcanics and intrusions. These include small scattered andesite, dacite, rhyolite centres and extensive Pliocene ignimbrite sheets. The tin mineralization is confined to the former. Grant et al. (1977) distinguished eruptive centres in which only intrusive rocks are preserved (Llallagua, Morococala); centres where both intrusions and coeval volcanics coexist (Chorolque, Oruro, Potosí) as well as partially eroded volcanic superstructures lacking any intrusive rocks (Chocaya, Tatasi).

In Llallagua (3.5 Mt Sn/0.4%; Figs. 26-41 and 26-42), the mineralized Salvadora intrusive stock is a downward-tapering isolated body about 1 km in diameter, intruding Silurian and Devonian sediments, and interpreted as a former volcanic vent. The stock is composed of intensively altered porphyry (possibly quartz latite), converted to a mixture of sericite, tourmaline, quartz and clay minerals. Megascopic as well as microscopic cassiterite, stannite, pyrite, pyrrhotite and arsenopyrite are pervasively disseminated throughout the stock (usually replacing feldspar phenocrysts in the company of tourmaline, sericite and quartz) or fill hairthin veinlets. Almost the entire stock is a single large low grade (0.2-0.5% Sn) orebody having a tonnage approaching 1 Bt ore. Over this are superimposed patches, pipes and dikes of cassiterite-bearing hydrothermal breccia and several sets of mostly N.E.-trending veins (30-80 m wide), vein zones and zones of stringers. These veins often reach beyond the stock outline into the exocontact. The vein filling is quartz, cassiterite, pyrite, pyrrhotite, stannite, chalcopyrite. Silver minerals are rare and silver is not recovered. The age of mineralization is Miocene (21 m.y.).

The Cerro Rico stock above Potosí (52 Tt Ag/300 ppm, 520 Tt Sn/0.8%; Figures 19-12 and 26-43) represents the world's largest silver accumulation and is also a major tin deposit. The mineralization is related to a funnel-shaped, downward narrowing stock of intensively altered quartz latite porphyry intruding an anticlinal structure in Ordovician shales and arenites, and also to a lesser extent the capping Miocene volcaniclastics. Hydrothermal brecciation and intensive alteration: quartz-sericite in the main intrusive mass; silicification in the cap; bleaching, silicification and pyritization in the Ordovician shales in the exocontact, is usually interpreted as being a product of a late stage autometasomatic event. Portions of the stock and the adjacent metasediments contain pervasively disseminated pyrite and cassiterite, or veinlet quartz, pyrite, cassiterite stockwork resulting in a "porphyry tin" deposit containing a large tonnage of 0.3 to 0.6% Sn material. Over 35 predominantly N.E.-trending fissure veins are superimposed over the disseminated mineralization in the endocontact and over the exocontact sediments. Their multistage and frequently telescoped filling consists of (from the earlier to the later): quartz, pyrite, cassiterite, arsenopyrite, lesser wolframite, bismuthinite, tetrahedrite, sphalerite, andorrite, matildite, pyrargyrite, jamesonite, boulangerite, cherty quartz, alunite and stibnite. The Cerro Rico mineralization is known to extend to a depth of 1,150 m and dates from late Miocene age (13.8 m.y.).

Similar but less productive complex deposits are to be found in Oruro (Fig. 19-12), Chorolque, Morococala and Carguaicollo. Sillitoe et al. (1975) pointed out that these ores formed at temperatures ranging from about 300° C to about 530° C and at pressures between 200 and 500 bars. They speculated that this took place at depths ranging from 1,500 m to 3,000 m under the surface, most probably under stratovolcanoes. The Chocaya deposit, however, located within a thick volcanic complex, lacking intrusive bodies and disseminated tin, formed closer to the surface. It has quartz, pyrite, arsenopyrite, stannite, cassiterite, bismuthinite, wolframite, freibergite and pyrargyrite filling of fissure veins. The alteration is argillization and pyritization.

The Huanuni tin field between Oruro and Llallagua (Ahlfeld and Schneider-Scherbina, 1964; COMIBOL Huanuni Staff, guided tour, 1977; 125 Tt Sn/2.31%) produces from a group of approximately twenty irregular veins in Silurian litharenites. Fracture filling and replacement veins as well as minor mineralized breccias occur. The hypogene vein minerals are quartz, tourmaline, cassiterite, siderite, pyrite, pyrrhotite and minor teallite, and the gossan contains residual cassiterite in a matrix of iron hydroxides, quartz and clay minerals. The host arenites are silicified to form hard quartzites and in places they are tourmalinized. Except for several late Tertiary altered quartz porphyry dikes, there is no major intrusion exposed, but an intrusive stock is postulated to exist at some depth.

KHINGAN-OLONOY TROUGH TIN DISTRICT, SOVIET FAR EAST

The Khingan-Olonoy Trough is located in eastern Siberia, north of the Chinese border. It is a structural graben, formed over a Proterozoic to Paleozoic basement and filled by a lower Cretaceous rhyolite-dominated volcanic-sedimentary suite. Several shallow cassiterite deposits of controversial origin are hosted by the felsic volcanics. In the Dzhalinda deposit (Grushkin and Vedernikov, 1977; Fig. 26-44) often considered to be of the "Mexican-type", the ore zone is controlled by a subhorizontal unit of fine rhyolite crystal tuff

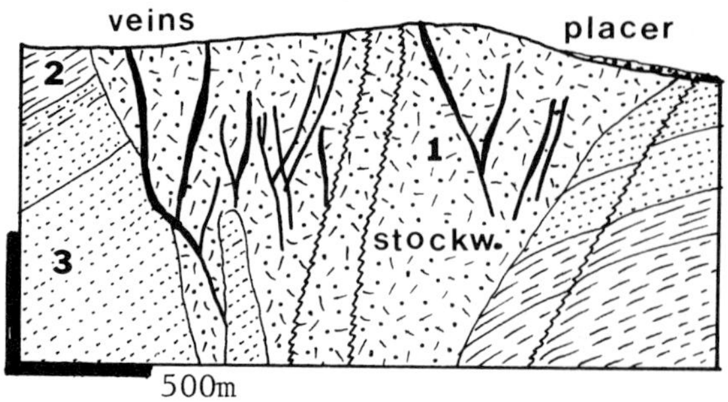

1. Mi-Pl Salvadora Stock pervasively altered qtz. latite porphyry gradational to intrusive breccia
2. D_1 thin bedded shale
3. $S_{2,3}$ coarse to fine sublitharenite

Fig. 26-41. Llallagua, Bolivia, high-level Sn mineralized stock. From LITHOTHEQUE, based on data in Ahlfeld and Schneider-Scherbina (1964); Sillitoe et al. (1975) and COMIBOL Llallagua Staff, mine tour, 1977.

Fig. 26-42. Llallagua, Bolivia, a giant tin deposit.

Fig. 26-43. The ancient mining city Potosí, Bolivia and its "richest hill on earth", Cerro Rico, a prominent landmark. Bottom: pyrite-cassiterite fracture veinlets, part of the Cerro Rico Sn stockwork.

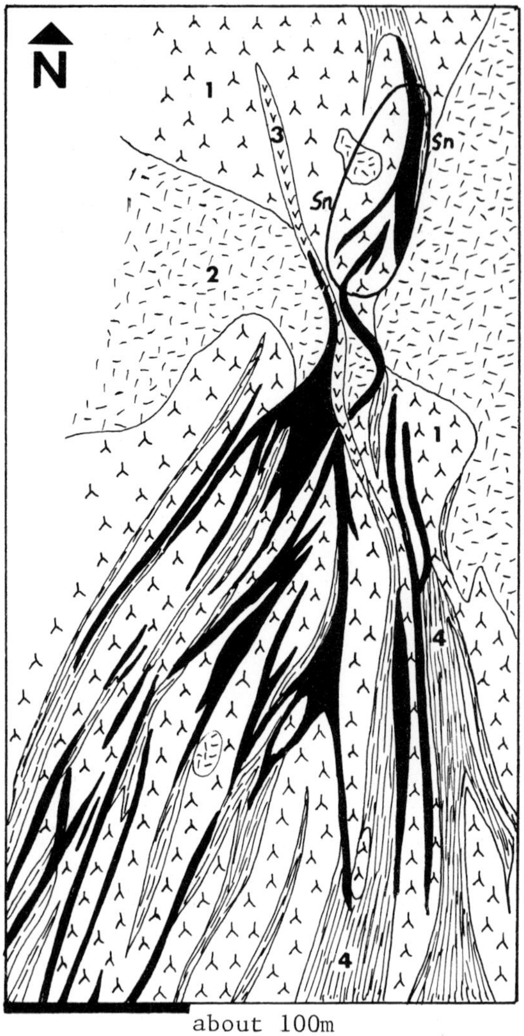

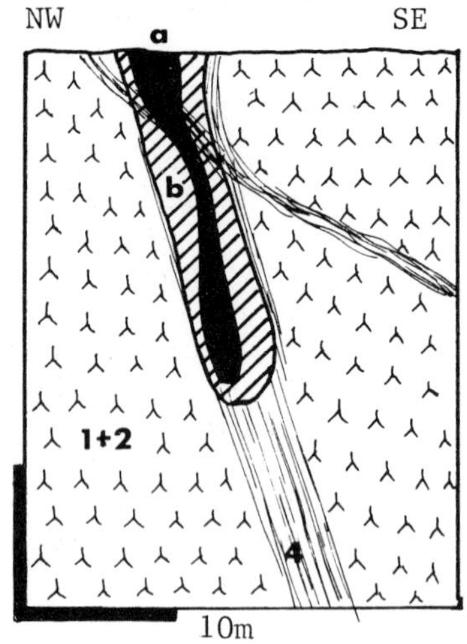

1. Cr rhyolite
2. rhyolite tuff
3. diabase dike
4. argillized fracture zones

BLACK: quartz - sericite and quartz alteration zones with cassiterite

Sn main economic deposit with wood tin

a Sn orebody

b silicified and sericitized rhyolite

Fig. 26-44. Dzhalinda Sn deposit, Malyi Khingan, U.S.S.R.; an example of a "wood tin" deposit in felsic volcanics, a map and a detailed section. After Itsikson et al. (1959).

approximately 40 m thick, resting unconformably on earlier felsic flows and ignimbrites. Superimposed on this is a fan-like series of branching fractures 1-1.2 km long, along which the tuff was completely silicified into jasperoid. The jasperoid zones extend no deeper than 120 m under the present surface, and each contains a 1-2m thick zone mineralized by scattered "wood tin" cassiterite and by quartz-cassiterite veinlets also containing minor arsenopyrite, galena, sphalerite and indite. The ore grade is about 0.8% Sn and 0.14% Pb. The reserve probably does not exceed 1 million tons of ore. The mineralization formed at temperatures around 260° C, at depths of between 80 and 200 m below the paleosurface and it is considered to be volcanogenic (possibly fumarolic).

In the Karadub field (Itsikson et al., 1959), disseminated cassiterite occurs in shallow quartz-sericite altered bodies of Cretaceous ignimbrite, felsic tuffs, rhyolite and andesite flow breccia along steep faults. Several tens of N.W.-trending lens-like bodies are present, only two of which were considered to be economic.

The Khingan deposit is a stockwork of quartz, chlorite, fluorite, cassiterite and minor As,Zn,Cu,Pb sulphides, hosted by a breccia pipe in rhyolite, intruded by subvolcanic bodies of granite porphyry.

MOUNT PLEASANT, NEW BRUNSWICK, CANADA

Mount Pleasant paleo-volcanic complex dating from the Mississippian age (Ruitenberg and Fyffe, 1982; Parrish and Tully, 1971; 60 Tt W/0.2%, 30 Tt Mo/0.1%, 24 Tt Bi/0.08%, 10 Tt Sn/0.42%, 83t Ag; Fig. 26-45) is an isolated occurrence emplaced in late Silurian folded slates and phyllites. Erosional remnants of subaerial quartz latite ash-flow tuffs and quartz feldspar porphyries were intruded by two plugs (possibly former necks) of fine-grained quartz feldspar porphyry or rhyolite gradational as the depth increases into microgranite. These became shortly after emplacement, centres of hydrothermal alteration and mineralization.

Two orebodies are present. (1) A large disseminated and stockwork wolframite, molybdenite, bismuth, arsenopyrite body in the south and (2) a discontinuous, small cassiterite, arsenopyrite, sphalerite, chalcopyrite zone in the north. Both orebodies are hosted by silicified and locally feldspathized or greisenized topaz, fluorite, chlorite and sericite-altered porphyries and the minerals are very fine and erratically distributed so that the visual identification of the ore is often impossible. In the lowermost levels of the northern ore zone, cassiterite exists in chloritized microgranite and porphyry.

CHU-ILI OROGEN, KAZAKHSTAN

Several high-level molybdenite showings in Devonian paleo-volcanics were described by Shcherba et al. (1972) in the southern part of the Chu-Ili orogenic belt, Kazakhstan. In the Shaitan-Simess field,

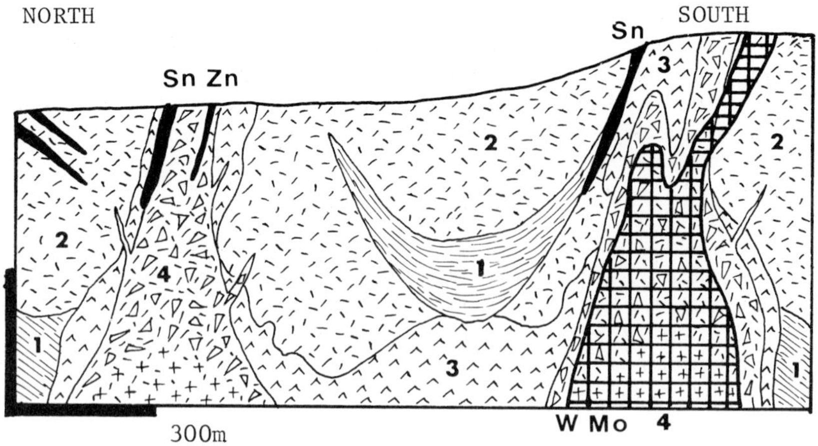

Fig. 26-45. Mt. Pleasant, New Brunswick, an example of Cb_1 subvolcanic stockwork and disseminated Mo,W,Bi,Sn,Zn, etc. mineralization. 1=S_3D_1 hornfelsed phyllite, slate, litharenite; 2=Ms altered latite porphyry with metasedimentary inclusions and rafts; 3=Ms rhyolite (quartz feldspar) porphyry; 4=Ms vent facies: fragmental rhyolite breccia gradational to microgranite in depth; black=Sn-Zn orebodies; cross ruled=W,Mo,Bi orebody. From LITHOTHEQUE, modified after Parrish and Tully (1971).

finely disseminated femolite (low-temperature metacolloform MoS_2) and jordisite are present in propylitized trachyte, dacite and rhyolite in exocontacts of andesite, diorite and syenite porphyry stocks. With increasing depth, the disseminations change into a stockwork of molybdenite veinlets in quartz-sericite altered intrusions.

The class of the large stockwork molybdenite deposits ("Climax-type") is reviewed in Chapter 28.

26.6.7. High-level tungsten deposits

High-level tungsten deposits carry either ferberite or hübnerite, or scheelite, in a gangue of chalcedonic or cherty quartz, carbonate, barite and fluorite. Sulphides are often associated. Known occurrences hosted by black slate or phyllite are often considered to have been (re)mobilized.

Gümüsler mine in the Niğde district, Turkey (Yildiz and Bailey, 1978) contains patches of scheelite in stibnite accompanied by cinnabar and realgar, in a 0.25-3 m thick E.S.E.-trending fault zone in Paleozoic marbles and metamorphics. The ore is interpreted as polygenetic; the scheelite is attributed to Paleozoic granites, the Sb,Hg and As sulphides to hot springs activity triggered by late

Cainozoic volcanism.

The Conde Auque field 45 km N.E. of Oruro, Bolivia (Ahlfeld and Schneider-Scherbina, 1964) has scheelite, ferberite, hübnerite, stibnite, fluorite and calcite filling fissure veins in Devonian slates and sandstones, and cementing tectonic breccia.

At Atolia, California (Hulin, 1925; 6.6 Tt W), scheelite is found in steeply-dipping, high-grade quartz-carbonate fissure veins in Jurassic quartz monzonite, genetically affiliated to Miocene subvolcanic intrusions in the area. Stibnite and cinnabar are present in accessory amounts, and the portion of the scheelite recovered came from detrital accumulations in a Quaternary alluvial fan.

High-level cherty quartz-ferberite veins hosted by crystalline basement rocks in the Boulder County (Netherland) field, Colorado, are described in Section 28.5.8.

26.6.8. Iron deposits in continental volcanic association

The Cerro de Mercado deposit located on the outskirts of Durango City, Mexico, is currently the largest Mexican iron ore producer (min. 45 Mt Fe/60%). It is located in a part of the extensive Sierra Madre Occidental volcanic field dating from the Oligocene, hosted by a voluminous rhyolitic ash-flow tuff sequence resting unconformably on older andesites. The deposit has recently been interpreted (Swanson et al., 1978) as situated in the moat of a resurgent caldera, above a rhyolite, miarolitic syenite and pyroxene porphyry flow dome and plug facies (Fig. 26-46). The orebody is composed of massive martite and magnetite forming a thick (up to 100 m) peneconcordant lens, resting on a temporary synvolcanic erosional paleosurface. The ore is interpreted as being a lava flow interrupted by several discordant feeder dikes of magnetite. The massive ore grades into a breccia ore in syenite. Also present are numerous chalcedonic quartz, magnetite, apatite veins, lenses and nests. The Cerro de Mercado orebody changes laterally into a thin, widespread, hematite layer that blankets the intravolcanic unconformity over an area of about 100 km^2 and is interpreted as being composed of Fe oxide pyroclastic.

Another ancient synvolcanic iron ore has recently been interpreted as existing in the Bafq iron ore district, central Iran (Förster and Knittel, 1979). There, an unmetamorphosed Infracambrian felsic volcanic unit unconformably overlies Precambrian metamorphics, and is itself overlain by lithic arenite and dolomite. The stratigraphy is thus strongly reminescent of the "volcanic red-beds" (e.g. in the S.E. Altai iron ore district; Chapter 25). The magnetite ore in the Bafq region (min. 600 Mt Fe/53-65%) is stratabound and always located at the top of the rhyolite sequence. Individual ore beds range in thickness from 20 cm to 6 m and are traced for more than 1,000 m along the strike. The immediate host to ore is a green felsic agglomerate and the ore is composed of small magnetite crystals lumped together without being intergrown. This is interpreted as being a paleo-volcanic ash. Magnetite lapilli and ash, furthermore, occur in

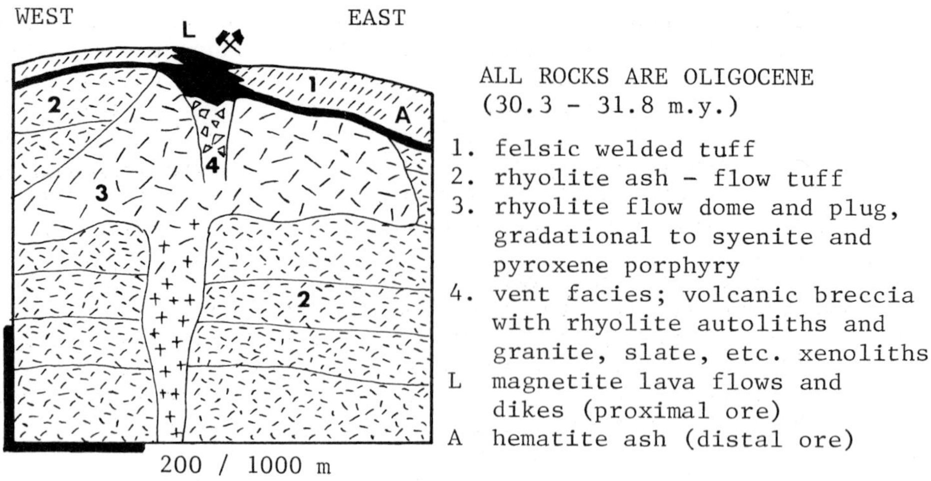

Fig. 26-46. Cerro de Mercado iron deposit and environs, Durango, Mexico, interpreted as being a magnetite lava flow and dikes (L) above a vent, laterally changing into a hematite ash (A) topping an intravolcanic unconformity. Diagrammatic from LITHOTHEQUE, based on data in Swanson et al. (1978).

the agglomerate. The ore petrography and setting are strongly reminescent of the Precambrian Kiruna orebodies, Sweden (Volume 2).

The last example of what is probably a synvolcanic major iron ore deposit in young continental volcanics is the Pliocene Abovyan deposit in Armenia, U.S.S.R. (Mezhlumyan, 1977; 65 Mt Fe/27%). There, magnetite and lesser hematite and apatite form veins, lenses and masses in brecciated and altered andesite porphyry.

26.6.9. Mn deposits

Occurrences and small deposits of Mn oxides in continental volcanic sequences are exceedingly common. They take one of two basic forms: (1) discordant veins, breccia fillings, infiltrations in porous volcaniclastics or (2) bedded lenses and sheets in pyroclastics or volcanic sediments (in particular in the "lake beds").

VEINS

Hewett (1964) reviewed the Mn oxide veins in the western United States. Of the several hundred veins, 80% are in Tertiary continental volcanics and individual orebodies rarely contain in excess of several hundred tons of ore. Typical veins are located in simple or complex, linear or slightly sinuous, steeply-dipping fractures, and are shallow. The deepest vein terminated about 160 m under the surface. Some veins are filled by bands or encrustations of psilomelane, cryptomelane, hollandite, coronadite or pyrolusite. Others are volcanic or fault breccias with Mn oxides in the matrix. Gangue minerals, where present, are usually quartz (often chalcedonic or jasper), barite or, occasionally, fluorite and calcite. All types of volcanics and volcaniclastics ranging from basalts to rhyolites can host Mn occurrences.

The Mn oxides clearly precipitated from solutions, both descending (meteoric) and ascending (hydrothermal, low temperature) and a genetic interpretation is often difficult. Wallrock alteration is usually absent or very light, although Hewett (1964) reported several examples of adularia feldspathization. Most veins carry increased trace contents of W,Pb,Zn,Cu and As and tungsten is particularly common in quantities ranging from 0.1 to 1.5% WO_3.

In the Talamantes Mn field, northern Mexico (Wilson and Rocha, 1948; 50 Tt Mn/25%, 1.5 Tt W/0.75% WO_3; see Fig. 26-3), sooty to massive psilomelane, cryptomelane and hollandite cement breccia along normal faults in Tertiary subhorizontal rhyolite flows and pyroclastics. Barite veins occur along the same structures, sometimes intermixed with or impregnated by chert and jasper and in several cases they overlap with the Mn oxide-filled veins. The most persistent Mn oxide vein was 550 m long and 2 m thick and the greatest depth of exploration was a mere 42 m. The veins are interpreted as being of descendent origin, formed under an intravolcanic unconformity.

BEDDED Mn OXIDE OREBODIES

Bedded or peneconcordant Mn oxide deposits in tuffs formed by epigenetic infiltration or by replacement in a porous aquifer from cool or hot solutions. In most cases, Mn oxides are associated with chert or jasper. In the Lucifer deposit N.W. of Santa Rosalía (Baja California, Mexico; Trask et al., 1946; 141 Tt Mn/45%), there exists a gently dipping lens of soft cryptomelane and pyrolusite gradational into a silica-impregnated ore. This orebody lies in Pliocene felsic tuff and volcanic conglomerate unconformably overlying Miocene basalts and is 1-6 m thick.

Concordant Mn orebodies hosted by water-reworked pyroclastics believed to have formed in shallow lakes contemporary with a local explosive volcanism, are represented by several deposits in the western United States and Mexico. The San Francisco (Autlán) deposit in Jalisco State, Mexico (Zantop, 1978; 1.6 Mt Mn/40%) is a conformable and thinly bedded lens of braunite, calcite and pyrolusite. This measures 1.6 x 1 km and is up to 3 m thick. The Mn orebody changes facies into a banded chert-hematite body. The Tertiary ore horizon is immediately enveloped by a red mudstone, which in turn, rests on a volcanic conglomerate and sandstone and is topped by a volcanic siltstone and sandstone locally containing trash of carbonized wood. In one place, a travertine plug interfingers with the Mn ore. Zantop (1981) proposed a syn-depositional "sedimentary-exhalative" origin; that is, that hot springs or hydrothermal solutions coming from below and spreading laterally, brought in and precipitated the Mn and Fe.

The largest United States' Mn district of comparable style and origin dating from the Pliocene age is located in the Lake Mead region, 15-50 km E.S.E. of Las Vegas, Nevada (McKelvey et al., 1949; about 500 Tt Mn/20% or better; additional 400 Tt Mn of low-grade ore). In the largest and only producing deposit Three Kids (Fig. 26-47) a thick unit of basalt and basaltic breccia is overlain by an interbedded sequence of felsic volcanic sandstone, another basalt flow, siltstone, gypsiferous sandstone and zeolitized tuff. The manganese is in a "wad", a soft, powdery black material probably composed of cryptomelane and hollandite and also containing some psilomelane and neotocite. The manganiferous beds consist of a wide range of host rocks in which the "wad" is a minor impurity, through soft "wad" blebs or irregular nodules, to black horizons composed of reworked tuff or volcanic sand densely intermixed with the "wad". The Mn beds may locally be partially or fully silicified. The genetic interpretation ranges from a "syngenetic" to an "epigenetic" supply of Mn from hot springs, to descendent infiltrations of Mn released from the volcanic pile during devitrification. The Mn oxide at Three Kids is anomalously rich in lead (2% Pb in coronadite), thus the ore reserve also contains some 40 Tt Pb.

The manganiferous beds in the Artillery Mountains, Arizona in Pliocene alluvial fans and lake beds (6.125 Mt Mn/3.5%) are comparable with the Lake Mead occurrences, except that they are of still lower grade.

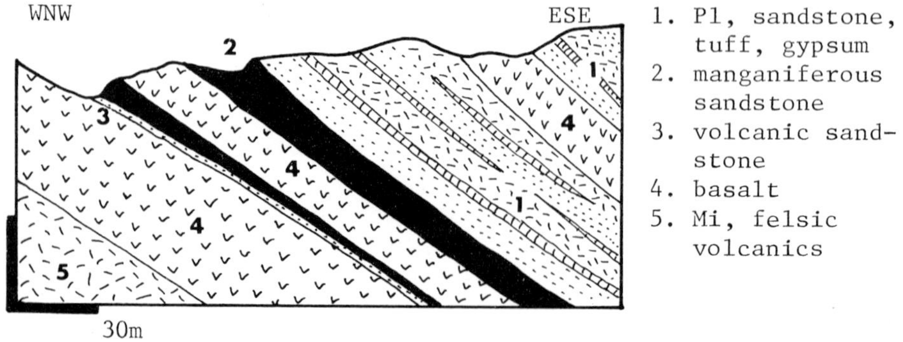

Fig. 26-47. Three Kids Mn mine, Nevada. From LITHOTHEQUE, after McKelvey et al. (1949).

26.6.10. Continental volcanics-hosted uranium deposits

Leaching of acid volcanics and the pyroclastic component of sediments is an important mechanism for the release of trace uranium. Because of the high U mobility in an oxidizing environment, such U is then transported in solution until it is trapped on a reducing surface. Important examples of U deposits formed in such a way are the "sandstone U-V", uraniferous lignites, and others (Chapter 25). The continuous "export" of the trace uranium from the volcanic piles causes impoverishment in the mobile U available for local accumulation and, as a consequence, volcanics-hosted U deposits are infrequent and of only marginal importance. As in the case of Mn, volcanics and volcanic sediments-hosted U deposits are (a) cross-cutting, and (b) concordant to peneconcordant bodies, veins, stockworks and disseminations.

CROSS-CUTTING U OREBODIES

Marysvale district, Utah, is the most comprehensively studied and the most frequently quoted example of subvolcanic to high-level plutonic vein uranium deposits (Cunningham et al., 1982; Fig. 26-48) and yet a mere 550 t U have been produced. There, numerous but very narrow (1-10 cm) fissure veins and mineralized fault breccias are filled by fluorite, chalcedonic silica and pyrite, as well as by spherulites, encrustations and veinlets of pitchblende and coffinite. Jordisite and molybdenite are associated in some veins and their quantity increases with depth. Silicification and hematite pigmentation are the principal alterations. The mineralization is Miocene (18-19 m.y.), hosted by alkali rhyolite domes, tuffs and ash-flow tuffs as well as by fine grained granite and Oligocene-Miocene quartz monzonite. The youngest (late Miocene) rocks are members of the bimodal continental suite. The U(Mo) orebodies are considered to be products of hydrothermal activity related to hypabyssal felsic stocks and dikes. A disseminated U mineralization ("porphyry U") or a stockwork molybdenite is postulated possibly to be present at some depth. If so, the presently known deposits might be only minor near-surface indications of a more important buried wealth.

The Ben Lomond U-Mo deposit discovered recently about 70 km W. of Townsville, Queensland (Valsardieu et al., 1980; 1,500 t U) is a close Carboniferous counterpart to Marysvale. There, a swarm of mineralized subvertical E.-W.- trending veins is confined to a zone of intense shearing in welded rhyolite and rhyodacite tuffs that are members of the felsic volcanic-plutonic sequence unconformably topping Precambrian crystalline rocks of the Georgetown Inlier. The most common near-surface minerals are the complex yellow urano-molybdates umohoite and iriginite, grading downward into a primary coffinite, pitchblende, molybdenite and jordisite. The gangue minerals are pyrite, chlorite, minor sericite and green tourmaline. Silicification

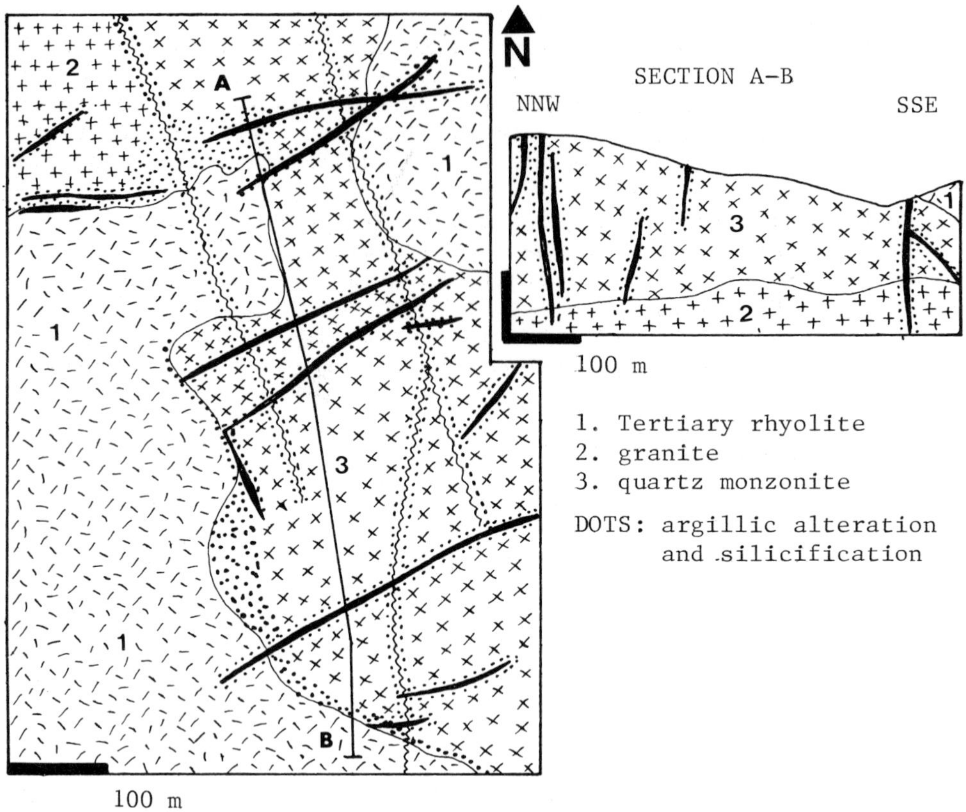

Fig. 26-48. Marysvale field, Utah, epithermal U(Mo) veins. From LITHOTHEQUE, after Kerr et al. (1957).

and hematitization of the wallrocks results in the characteristic pink to brick-red alteration jasperoid developed up to two metres away from the vein. Outwardly, this changes into dolomitization. The mineralization is considered epigenetic, shortly postdating emplacement of the host volcanics.

At the Zletovska Reka showing in S.E. Yugoslavia (Janković, 1982), pitchblende with minor sphalerite, pyrite, galena and tetrahedrite occur in fracture veins in argillized, silicified, alunitized and propylitized Tertiary brecciated dacite and andesite pyroclastics.

The relatively large U accumulation in the Aurora orebody in the McDermitt caldera complex located along the Oregon-Nevada state line (6.5 Tt U/0.0425%; Rytuba and Glanzman, 1979) is unusual. Uranium occurs with pyrite and leucoxene in faulted and altered Neogene basaltic flows along the northern rim of the caldera, in an ore field that also produced cinnabar from opalite deposits.

Several Mo-U vein deposits hosted by ignimbrites, dikes and rocks of the neck facies of felsic volcanic centres, have been reported from the U.S.S.R. (probably mainly from the East Transbaikalia and central Kazakhstan). Unfortunately the localities and the host units are anonymous, and no quantitative data (including dimensions and map scales) are available.

Very minor fracture coatings and infiltrations of U minerals in porous volcaniclastics explored in the Mojave Desert, California (e.g. Rosamond; compare Fig. 26-2); in the Challis Volcanics, Idaho; in Sonora, Mexico (El Rayon); in Nevada and elsewhere, are of little importance. The small volcanics-hosted U deposits in Chihuahua (Villa Aldama district) have already been mentioned earlier in this chapter.

CONCORDANT TO PENECONCORDANT U BODIES

Infiltrations of U and Mo minerals in porous pyroclastics within felsic explosive volcanic piles forming peneconcordant lenses, have been described in the Soviet literature. The localities, unfortunately, are anonymous and quantitative data are lacking.

In the West, several varieties of peneconcordant U bodies are known in volcanic association. The first variety represented by the Northwest Uranium Mine near Spokane, Washington and by the Maureen U-Mo prospect N.N.W. of Georgetown, Queensland, includes deposits hosted by basal epiclastic sediments filling channels on a granitic or metamorphic basement and subsequently capped by continental volcanics. There, the volcanics were important for the preservation of the deposits, but have probably contributed little to the origin of the ores. They are therefore comparable with the SUV deposits treated in Chapter 25.

In the Maureen prospect (O'Rourke, 1975; 3,120 t U/0.1-0.15%) uraninite, pyrite, molybdenite, renardite, fluorite and secondary minerals impregnate basal Permo-Carboniferous feldspathic and micaceous continental sandstone, locally rich in carbon-rich matter. The orebodies are gently-dipping, peneconcordant lenses locally grading into fluorite-rich pipes. The basement Archean metamorphics are micaschists, granite gneiss, pegmatite, etc., and the mineralized unit is topped by Permo-Carboniferous bimodal continental volcanics (rhyolite ignimbrite, agglomerate, basalt).

Another variety of uraniferous bodies includes mineralization in the volcanic-lacustrine facies, transitional and in fact little different, from other paleo-lacustrine U deposits (e.g. those in the Lodève Basin, France; Chapter 25), except for the presence of widespread continental volcanics in the vicinity, as interbeds among the "lake beds" and as the possible source of U. An outstanding example of the volcanic-lacustrine U(Mo) mineralization in Arizona is known as Anderson Mine or Date Creek Basin (Sherborne et al., 1979; min. 6.3 Tt U/0.07%; Fig. 26-49). This is located 72 km N.W. of Wickenburg, in the Basin and Range Province. There, gently dipping

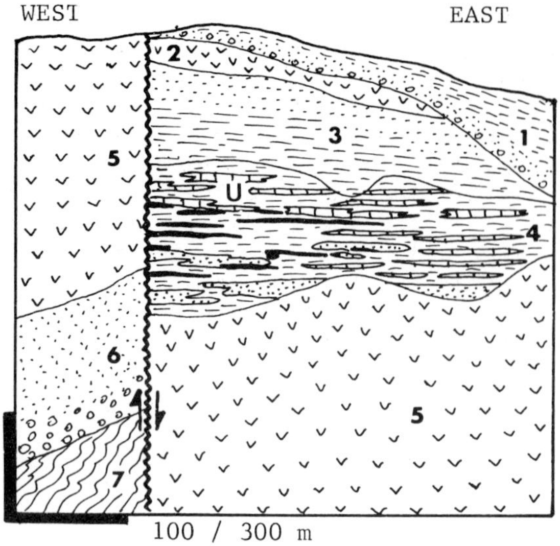

1. Q, alluvium
2. Mi, basalt and agglomerate
3. siltstone, sandstone, conglomerate
4. arkose, tuffaceous and carbonaceous mudstone, marl, conglomerate
5. Eo-Ol, basalt, andesite, tuff, conglomerate
6. Eo?, arkose fanglomerate, ignimbrite
7. PC?, crystalline basement

Fig. 26-49. Anderson Mine, Arizona, an U-bearing volcanic-lacustrine association. Diagrammatic, modified after Sherborne et al. (1979).

Miocene lacustrine carbonaceous mudstones and siltstones interbedded with fine tuffaceous sediments, freshwater marlstones, limestones and sandstones, rest on top of Eocene andesites, covered by Miocene sediments and basalts. The deposits have a tabular, blanket and lens-like configuration, extend for 1 to 1.5 km, and consist of several stacked mineralized beds often reaching an aggregate thickness of more than 15m. The major mineral is a dispersed coffinite, and there are high trace contents of Mo,V,Mn,Li and F. The orebodies are believed to have formed during compaction and dewatering of the U-rich lake sediments, by precipitation and fixation of U in contact with the paludal sediments.

26.6.11. Beryllium deposits

Trace Be is slightly or strongly enriched in felsic volcanics, particularly when a compositional heritage from an older basement known to contain Be pegmatites is suspected (as in the western Utah Be province). Rare occurrences of beryl (e.g. in the Thomas Range, Utah), chrysoberyl or phenakite phenocrysts have been recorded in rhyolite porphyries, particularly the topaz-bearing, miarolitic varieties, but until the discovery of the large Spor Mountain bertrandite deposit in 1959, no economically important volcanogenic Be deposits were known.

The Western Utah Be Province constitutes a portion of the Great Basin, measuring about 100 x 60 km. It is underlain by Precambrian granites and metamorphics covered by Paleozoic "miogeoclinal" sediments, intruded by Mesozoic-Cainozoic "granites" and studded by several discontinuous remnants of Eocene to Pliocene volcanic piles. In addition to the beryl phenocrysts in Tertiary rhyolites and Be in the basement pegmatite and granite, two volcanogenic Be styles have been reported (Lindsey, 1977): (1) discordant, epithermal bertrandite-bearing quartz, adularia, calcite veins and disseminations in silicified and altered quartz monzonite S.E. of Gold Hill (1,750 t Be/0.175%) and (2) peneconcordant bodies of disseminated bertrandite in water-laid tuff and breccia with carbonate clasts, as at Spor Mountain and Honeycomb Hills.

The Spor Mountain Be field (Lindsey et al., 1973; Lindsey, 1977; 24 Tt Be/0.3%; Figures 26-50, 51), contains six or more N.E.-S.W.-elongated tabular bodies of Be-bearing volcaniclastics, up to 5 km long. Each body is a horizon of altered water-laid Pliocene crystal vitric tuff (tuffite), deposited in an ash-choked lake, resting on basement of Ordovician and Silurian marine dolomite. The tufite in the mineralized zones is argillized (montmorillonite, hectorite) and adularia-feldspathized, in contrast to its non-mineralized equivalents that remain vitric or zeolitized vitric tuffs. In the ore zone, the principal Be mineral bertrandite (Be hydrosilicate) is white, finely dispersed, and megascopically unrecognizable. Bertrandite is associated with abundant, dispersed fluorite and with subspherical zoned calcite, silica, fluorite nodules. These nodules are conspicuous because of the light purple colour of the fluorite and they are interpreted as replaced dolomite clasts from the carbonate basement. Some nodules contain up to 2% Be, 1% Zn and 0.15% U. Infiltrations of Mn oxides are also abundant. Lindsey (1977) pointed out that the Be deposits are associated with a Pliocene topaz-bearing rhyolite and that they are located above projections of basement faults that probably channelled the mineralizing solutions. If so, the orebodies are epigenetic and the possible presence of Be-mineralized veins, tactite or greisen at greater depth has been suggested.

Small Be occurrences in continental volcanics are known in the vicinity of Truth or Consequences, New Mexico.

26.6.12. Volcanic-lacustrine borates and associated As,Sb accumulations

Quaternary playas and Tertiary "lake beds" contain the bulk of borates, worldwide. This book is not concerned with nonmetallic commodities as such, but the Kramer borate deposit near Boron, California (Barnard and Kistler, 1956), the world's largest, contains a persistent interval of borates 0.6 to 7 m thick. These, in turn, contain relatively large amounts of As and Sb. The Kramer borate body is hosted by Miocene tuffaceous lacustrine mudstone, claystone and

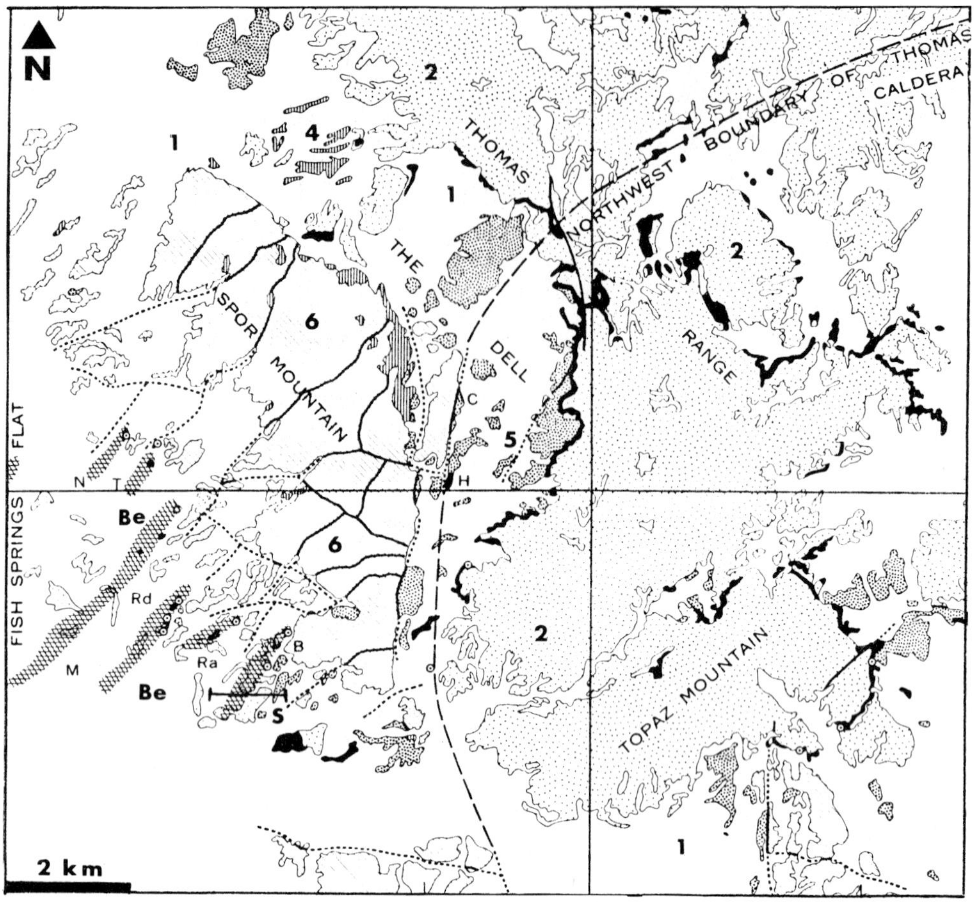

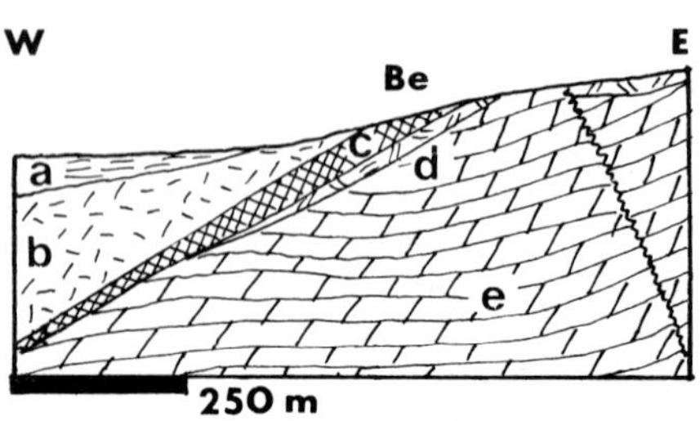

a Q, lake clay and silt
b Mi,Pl, topaz - bearing rhyolite
c Mi,Pl, water-laid felsic tuff with altered dolomite clasts, Be orebody
d latite flows
e Or, S, marine dolomite

Fig. 26-50. Spor Mountain bertrandite deposits, Utah. Map explanations: 1=Q alluvium and lake sediments; 2=T, younger volcanics; 3=T, water-laid tuffs, locally Be-bearing; 4=T, intrusive breccia; 5=T, older volcanics; 6=PZ marine carbonates and clastics; Be=beryllium deposit; S=location of the section. The map is from Lindsey et al. (1973), the diagrammatic section is from LITHOTHEQUE.

Fig. 26-51. Spor Mountain bertrandite deposit (top), and the "spotted leopard" variety of ore, argillized and adularized felsic tuffite with bertrandite, fluorite and Mn oxide infiltrations.

tuff, resting on a continental basalt flow and topped by arkosic sandstone. The borate orebody consists of a core composed of borax and kernite crystals (Na-borates), enveloped by an ulexite and colemanite (Ca-borate) facies. The borates are believed to have precipitated from lake brines, fed by thermal springs coeval with the local bimodal continental volcanism, and later diagenetically modified. The As-Sb stratabound horizon is located in the Upper Ore Zone. It consists of disseminated, abundant realgar and minor stibnite rosettes and it serves as a stratigraphic marker. As and Sb are not recovered at present.

26.6.13. Hg, Sb and As deposits

As and Sb accumulate in sulphoarsenides and sulphoantimonides in epithermal deposits and Hg may reach substantial concentration in tetrahedrite. In addition to this, Hg, Sb and As form their own deposits affiliated to continental volcanism, practically devoid of the base metal sulphides and of precious metals. Because of the high volatility of these elements and their capability to precipitate from low-temperature solutions (hot springs), the Hg,Sb,As orebodies often substantially postdate the periods of volcanism manifested by surficial volcanic products, and/or occur at a distance from the volcanic centres so that the volcanic affiliation could be rather loose or even not obvious at all. Hg,Sb and As occurrences associated with existing hot springs have been reviewed in Chapter 5. Deposits formed by interaction of hot springs with distinct lithologic associations (e.g. mélanges, ophiolites, karst) in the past, have been treated in Chapters 7, 8 and 20, and there some overlap in coverage is unavoidable. On the grounds of lithologic associations, one can distinguish three categories: (1) Hg,Sb,As deposits hosted by young continental volcanics and presumably formed largely at temperatures over 100° C, so they are the equivalents of the base and precious metal epithermal deposits treated earlier. (2) Low-temperature epigenetic deposits hosted mostly by a variety of "basement rocks" in regions of young volcanism. (3) Deposits hosted by demonstrable hot springs precipitates (siliceous sinter, travertine) in a variety of settings.

VOLCANICS-HOSTED EPITHERMAL DEPOSITS (1)

An extensive "province" of scattered epithermal mercury deposits has formed in the Carpathian bend in eastern Slovakia, the Soviet Transcarpathia and northern Rumania, associated with Miocene-Pliocene continental volcanics. In Merník, Slovakia (920 t Hg), cinnabar and metacinnabar in chalcedony, quartz and calcite gangue are hosted by an altered obsidian dike. In the Vyshkovo field, U.S.S.R. (Kuznetsov, 1974) cinnabar, pyrite, metacinnabar, barite and quartz veins and replacements form shallow orebodies along a fault zone cutting

Miocene-Pliocene granodiorite porphyries, tuffs and volcanic sediments along a major lineament. At the nearby Olenevsk, cinnabar and metacinnabar with some galena, sphalerite, pyrite form fine disseminations in pervasively argillized and carbonatized Pliocene explosive breccia in an andesite and basalt vent intruding Paleogene sediments. The Sintîmbru Hg deposit in the Harghita Mts., N.E. Rumania said to be "significant" (Ianovici and Borcoş, 1982), has cinnabar and metacinnabar associated with pyrite and marcasite in a stockwork of quartz veinlets in argillized and silicified pyroclastics overlying pyroxene andesite lava flows. The temperature of deposition was between 137 and 215° C. Tonnage figures are unavailable on the Carpathian Hg Province, but it is estimated that it represents about 10 Tt Hg. Small volcanics-hosted stibnite deposits are also present, e.g. in Zlatá Baňa, Slovakia.

Outside the Carpathians, Hg deposits hosted by young continental volcanics are known in eastern Siberia (in the Kamchatka and Chukotka Peninsulas), in Mexico, in Turkey, in Japan and in other regions. In the significant Plamennoe deposit, Chukotka Peninsula, U.S.S.R. (Kuznetsov, 1974), a peneconcordant replacement and impregnation body of cinnabar, stibnite, minor Fe bisulphides and realgar, occurs in adularia and kaolinite-altered lower Cretaceous rhyolite pyroclastics intruded by a lower Cretaceous subvolcanic andesite and high-level stocks of diorite and granodiorite. In the Chempura Hg deposit, Kamchatka, sheet-like bodies of cinnabar, realgar, orpiment, stibnite, freibergite, etc. in quartz, sericite, calcite and barite gangue, were superimposed on a complex alteration pattern (silicification, advanced argillization, sericitization, propylitization). The alteration formed along fault zones in a Tertiary andesite-dacite association intruded by quartz diorite.

In the Itomuka Hg mine, Hokkaido (Saito, ed., 1960; about 1,000 t Hg/0.35-0.4%) cinnabar and native mercury in quartz-chalcedony gangue formed stockworks and veins in a Miocene propylitized andesite and overlying tuffaceous sandstone.

LOW-TEMPERATURE EPIGENETIC DEPOSITS IN "BASEMENT" ROCKS (2)

A large number of Hg deposits and a lesser number of Sb and As deposits are located in or near deformed ophiolites interacting with young continental volcanics. Several examples (e.g. the Clear Lake area, California) have already been discussed in Chapter 8. In the Rajičeva Gora Sb deposit, Serbia, Yugoslavia (Janković, 1982; 10 Mt ore), stibnite and also galena and sphalerite form a N.E.-S.E.-trending vein and several irregular orebodies in faulted silicified serpentinite and limestone in the contact aureole of a Tertiary dacite and trachyte subvolcanic body. In the Lojane As-Sb deposit in the Yugoslav Macedonia (Janković, 1982; 40 Tt Sb/4%, 50 Tt As/5%, some Ni) realgar, stibnite, minor orpiment, bravoite, etc. in quartz-dolomite gangue fill a 1-2 m wide vein. The vein is controlled by a fault zone that follows the contact between a Paleozoic serpentinite and Tertiary andesite and rhyodacite.

In Alšar, S. Macedonia, Yugoslavia (Janković, 1982; min. 12 Tt Sb, 2 Tt As, 500 t Tl), veins, lenses, nests and impregnations of stibnite, realgar, orpiment and local vrbaite and lorrandite, occur discontinuously along a 24 km long, N.-S.-trending fault zone. The host rocks are Triassic limestones and dolomites and Paleozoic mica- and chlorite-schists, intruded by Pliocene trachyte and andesite.

Terlingua, Texas (Fig. 26-52); Huancavelica, Peru; and Monte Amiata, Italy, comprise three major Hg districts containing a multi-style mineralization. The main orebodies are hosted by karsted or hydrothermally replaced limestones (Chapter 20) or by cinnabar impregnations in sandstone (Huancavelica), but there are also minor orebodies hosted by the young continental volcanic and subvolcanic rocks. The continental volcanism and subvolcanism is generally considered to have been the driving force for the mineralization.

DEPOSITS IN PRE-QUATERNARY HOT SPRINGS PRECIPITATES (3)

Hg and lesser Sb and As deposits hosted by extensive subhorizontal bodies of silica (opal, chalcedony or quartz) are usually designated as being of "opalite-type". The silica could be a direct hot spring precipitate (geysirite). Alternatively, it could be a cemented and recrystallized lacustrine diatomite or a product of hydrothermal replacement and impregnation of volcanics or sediments. Transitional opalites are common.

The McDermitt (Opalite) district straddling the Nevada-Oregon border (Yates, 1942; min. 20 Tt Hg; Fig. 26-53) is the most commonly quoted example locality. There, a flat-lying Miocene basalt and rhyolite are overlain by Miocene tuffaceous lake beds considered by some (Albers and Kleinhampl, 1970) to fill a caldera. Cinnabar forms irregular ipregnation and replacement orebodies in opalized and chalcedony-replaced or impregnated tuffs and lake sediments, as well as in the non-opalized but argillized wallrocks along faults. The orebodies in opalite are of a lower grade (0.3% Hg) than those in the non-silicified hosts (0.8-0.9% Hg).

➡▷

Fig. 26-52. Terlingua, Texas. A variety of cinnabar mineralization styles that are probably epithermal. (1) Chisos mine, fracture filling cinnabar; (2) Fresno mine area, cinnabar in solution breccia in fractured limestone; (3) 248 Mine, cinnabar in diatreme ? breccia; (4) Mariposa mine, epithermal ? cinnabar in karst and cave fill near an intrusive dike. 1=Q, alluvium; 2=Eo rhyolite, trachyte; 3=Cr_3 claystone, sandstone; 4=Cr_3 limestone; 5=Cr_3 calcareous claystone; 6=Cr_1 limestone; 7=Eo, breccia pipes; black=Hg occurrences (projected); 8=Eo, syenogabbro; 9=Eo, soda latite. From LITHOTHEQUE, after Yates and Thompson (1959).

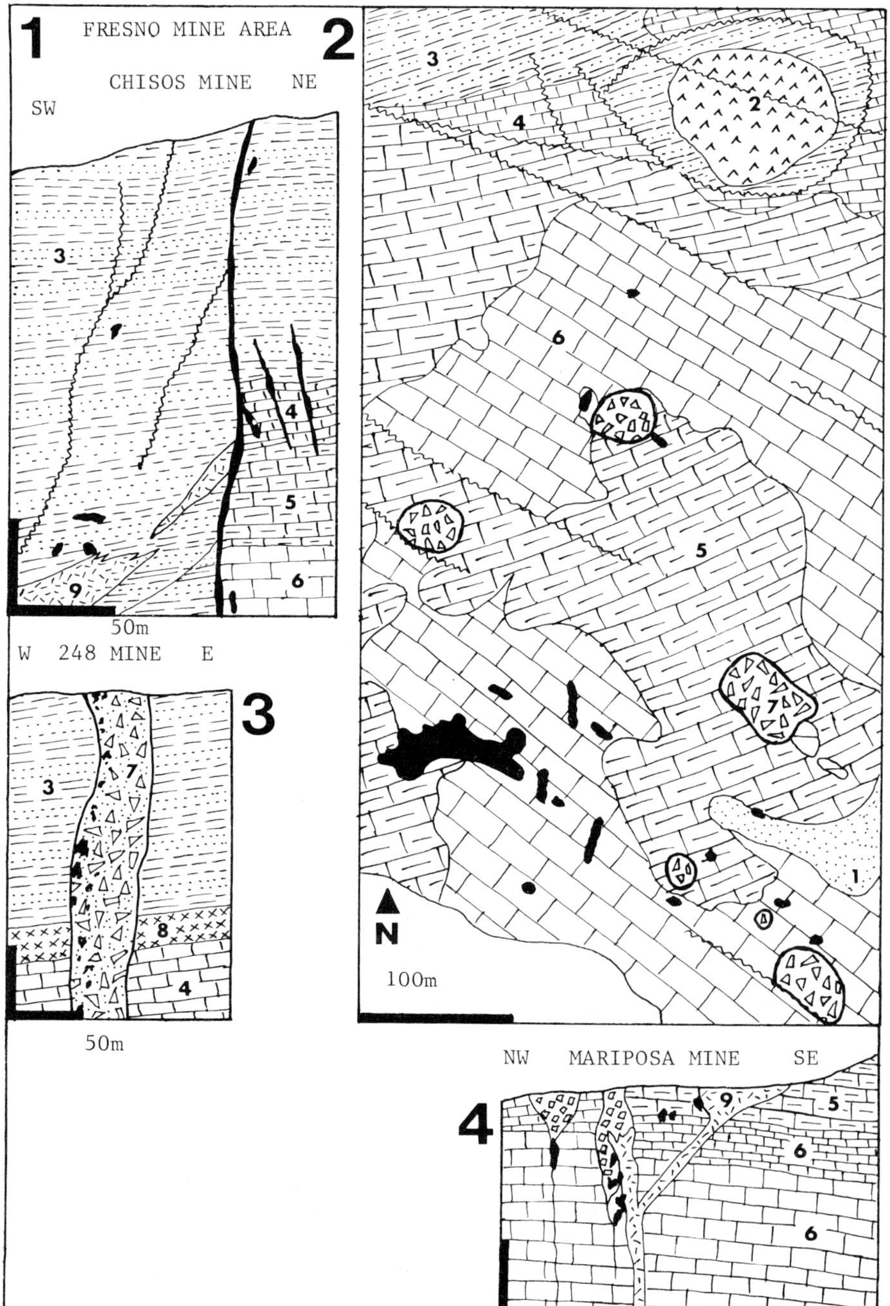
1
FRESNO MINE AREA
CHISOS MINE NE
SW
3
4
5
6
9
50m
W 248 MINE E
3
3
7
8
4
50m
2
3
2
4
6
5
7
1
N
100m
NW MARIPOSA MINE SE
4
9
5
6
6

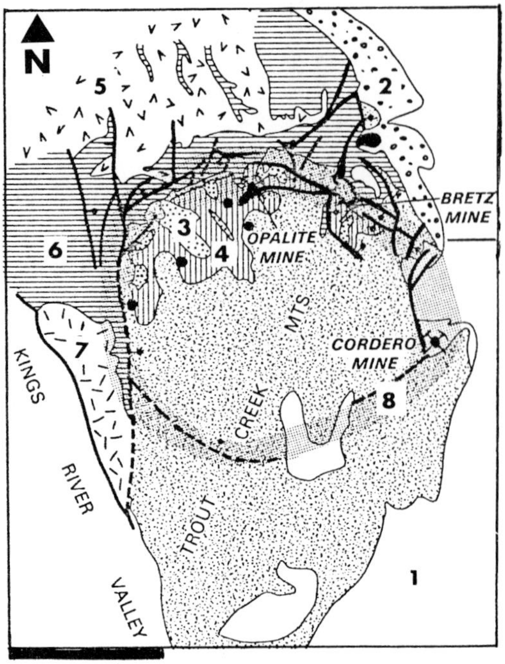

15 km

1. Q, unconsolidated sediments
2. pediment gravels
3. Pl, pediment deposits
4. Mi, continental volcanics and volcaniclastics
5. welded tuff
6. andesite and basalt flows
7. Cr_3-T_1 granodiorite
8. approximate limit of caldera collapse

BLACK: silicic vent rocks, silicified rocks

Fig. 26-53. Opalite mercury district, Oregon and Nevada. From Albers and Kleinhampl (1970).

Another "opalite-type" cinnabar deposit near Weiser, Idaho shows little signs of volcanic affiliation and has been reviewed in Chapter 25. Additional small Hg opalite localities are found in Turkey (Murat Daği area), the U.S.S.R. and elsewhere.

Small deposits of realgar or orpiment often associated with stibnite are common on fringes of volcanic piles, hosted by a variety of rocks (e.g. marls and sandstones in Dzhul'fa, Causasus, U.S.S.R.; karsted limestones near Tajov, Slovakia and metamorphics in the Kutahya area, W.Turkey). They are interesting scientifically but are of little economic importance except as indicators of the possible presence of the "invisible" (Carlin-"type") gold orebodies.

Developments in Economic Geology, 19

empirical metallogeny

Depositional Environments, Lithologic Associations and Metallic Ores

Vol. 1: Phanerozoic Environments, Associations and Deposits

PART B

Further titles in this series

1. *I.L. ELLIOTT and W.K. FLETCHER (Editors)*
GEOCHEMICAL EXPLORATION 1974

2. *P.M.D. BRADSHAW (Editor)*
CONCEPTUAL MODELS IN EXPLORATION GEOCHEMISTRY
The Canadian Cordillera and Canadian Shield

3. *G.J.S. GOVETT and M.H. GOVETT (Editors)*
WORLD MINERAL SUPPLIES
Assessment and Perspective

4. *R.T. SHUEY*
SEMICONDUCTING ORE MINERALS

5. *J.S. SUMNER*
PRINCIPLES OF INDUCED POLARIZATION FOR GEOPHYSICAL EXPLORATION

6. *R.A. RICH, H.D. HOLLAND and U. PETERSEN*
HYDROTHERMAL URANIUM DEPOSITS

7. *J.G. MORSE (Editor)*
NUCLEAR METHODS IN MINERAL EXPLORATION AND PRODUCTION

8. *M. KUŽVART and M. BÖHMER*
PROSPECTING AND EXPLORATION FOR MINERAL DEPOSITS

9. *C.R.M. BUTT and I.G.P. WILDING (Editors)*
GEOCHEMICAL EXPLORATION 1976

10. *G.B. FETTWEIS*
WORLD COAL RESOURCES
Methods of Assessment and Results

11. *R.G. TAYLOR*
GEOLOGY OF TIN DEPOSITS

12. *H.K. GUPTA*
GEOTHERMAL RESOURCES
An Energy Alternative

13. *C.R.M. BUTT and R.E. SMITH (Editors)*
CONCEPTUAL MODELS IN EXPLORATION GEOCHEMISTRY, 4
Australia

14. *G.BÁRDOSSY*
KARSTIC BAUXITES

15. *A.W. ROSE and H. GUNDLACH (Editors)*
GEOCHEMICAL EXPLORATION 1980

16. *R.W. BOYLE*
GEOCHEMICAL PROSPECTING FOR THORIUM AND URANIUM DEPOSITS

17. *G.R. PARSLOW (Editor)*
GEOCHEMICAL EXPLORATION 1982

18. *M. KUŽVART*
INDUSTRIAL MINERALS AND ROCKS

DEVELOPMENTS IN ECONOMIC GEOLOGY, 19

empirical metallogeny

Depositional Environments, Lithologic Associations and Metallic Ores

Vol. 1 : Phanerozoic Environments, Associations and Deposits

PART B

Peter Laznicka

Department of Earth Sciences, University of Manitoba, Winnipeg, Man. R3T 2N2, Canada

ELSEVIER, Amsterdam — Oxford — New York — Tokyo 1985

ELSEVIER SCIENCE PUBLISHERS B.V.
Sara Burgerhartstraat 25
P.O. Box 211, 1000 AE Amsterdam, The Netherlands

Distributors for the United States and Canada:

ELSEVIER SCIENCE PUBLISHING COMPANY INC.
52, Vanderbilt Avenue
New York, N.Y. 10017, U.S.A.

ISBN 0-444-42530-6 (Vol. 19A)
ISBN 0-444-42553-5 (Vol. 19B)
ISBN 0-444-42554-3 (Set)
ISBN 0-444-41250-6 (Series)

Printed in The Netherlands

"Broadly regarded, the depth to which a metalliferous region has been eroded affects its present characteristics as a mineral province, because of the exposure of the various depth zones in which the ores were formed".
E. Sherbon Hills in Geology of Australian Ore Deposits, 1953, p. 57.

CHAPTER 27

Intracrustal and Subcrustal Environments (Introduction to Chapters 28-32)

The depositional environments treated so far, and the lithologic associations generated in them, have either been wholly surficial (supracrustal) or they have had a significant supracrustal component. In contrast, the environmemts in which the rocks treated in Chapters 9,28,29,32 formed, were intracrustal. The lithologic associations described in Chapters 4,5,7, 11-14, 31,33, are partly intracrustal.

Because the active intracrustal environments are not directly accessible and consequently they cannot be observed and summarized in empirical terms, they will not be discussed further in this book (but their products, lithologic associations formed at a greater depth, are). The reader is referred to the theoretical literature based on geophysical, geochemical and other data (e.g. Condie, 1982; Ringwood, 1975; Turner, 1980). Figures 27-1 to 27-5 show diagrammatically the depth setting of intracrustal lithologic associations and orebodies, at their time of origin.

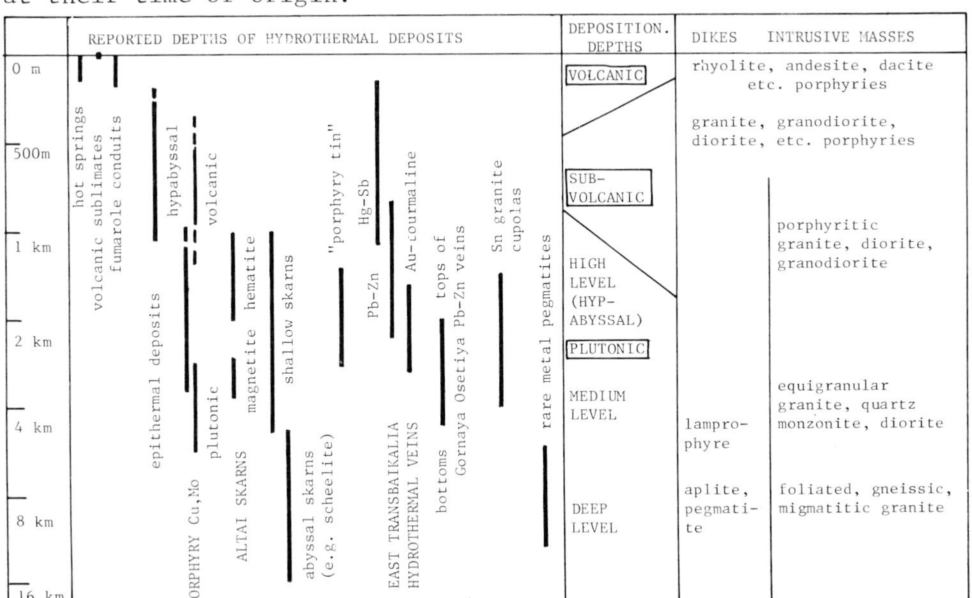

Fig. 27-1. Depths of formation of hydrothermal deposits and associated intrusives.

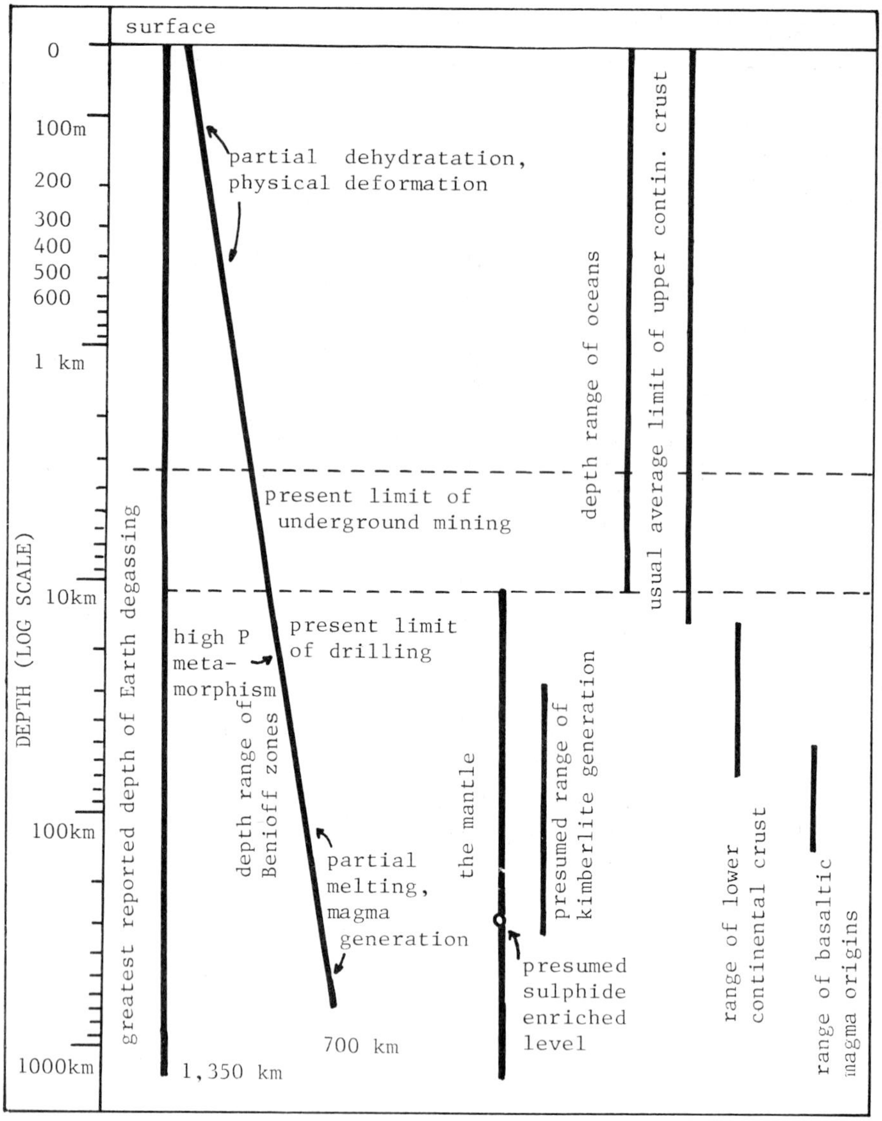

Fig. 27-2. Depth ranges of selected rock and ore-forming zones, planes and processes.

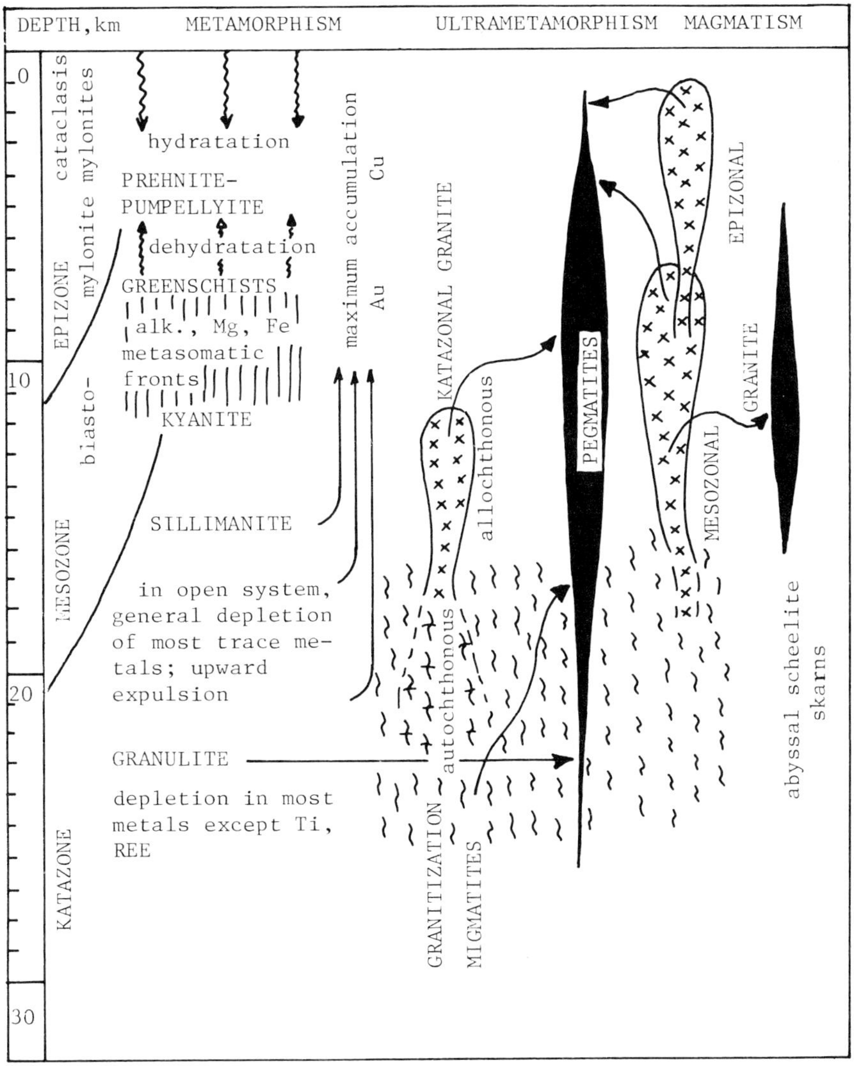

Fig. 27-3. Interpreted depth levels of metamorphism, ultrametamorphism and granite magmatism.

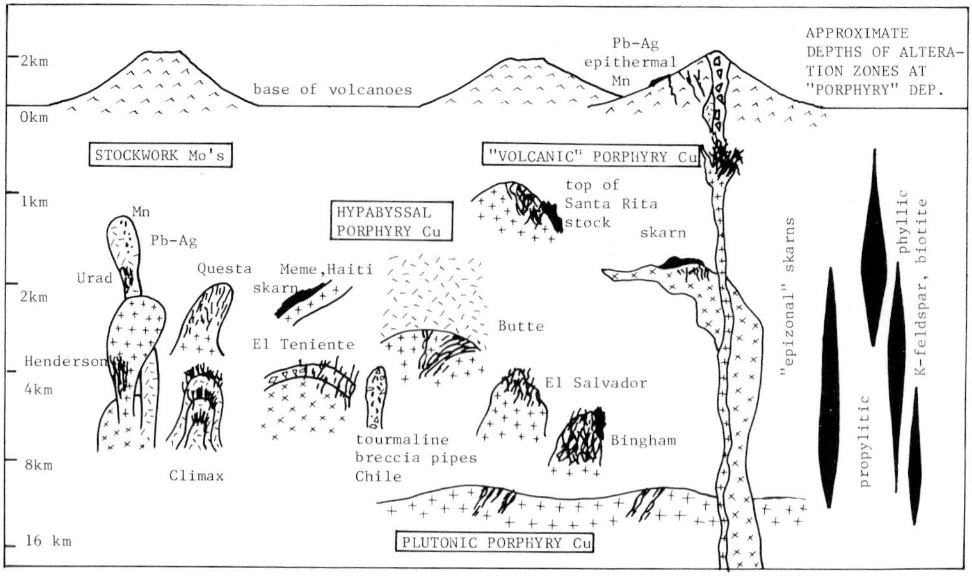

Fig. 27-4. Porphyry Cu and Mo deposits, interpreted depths of origin.

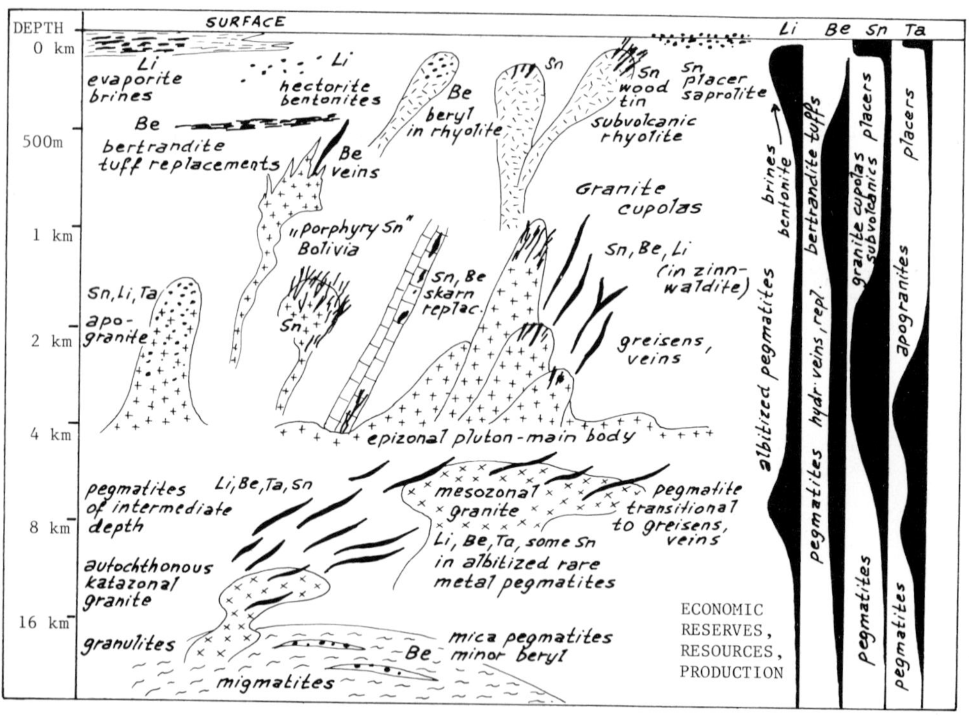

Fig. 27-5. Depth variation in genetic association and mineralization style of rare element deposits.

"Granite is the balance of what was there originally, plus what has migrated in, minus what has been driven out".
Arthur Holmes, 1937.

"Porphyry copper deposits are not found, they are made".
Donald K. Mustard, 1976.

"The economic geology of the (S.E. Asian) tin belt is still a riddle wrapped in a mystery inside an enigma despite certain naïve utterances to the contrary".
K. F. G. Hosking (paraphrasing Sir Winston Churchill) in 2nd Tech. Conf. on Tin, 1970.

CHAPTER 28

Plutonic Granite, Diorite, (Gabbro) Association (GDG) and its Aureole

28.1. INTRODUCTION

Granite, diorite, (gabbro) associations abbreviated GDG are also known as "granites", granites sensu lato, granitic rocks or granitoids because of the granitic or hypidiomorphic-granular texture characteristic of these rocks regardless of their colour index. This is, without doubt, the most complex lithologic association to be handled from the point of affiliated orebodies. It is so because few of the affiliated ores are directly hosted by these rocks, and even when this is the case such ores are almost always "epigenetic", that is, younger than the adjacent rock and generally emplaced when the host was a "dead" (solidified, lithified, fractured) recipient only. With the exception of some gabbros, syn-magmatic and orthomagmatic metal accumulations generated during the magmatic crystallization and hosted by the parent rock in the same manner in which detrital heavy minerals in placers are enclosed in the sediments with which they formed, are rare and of little importance, economically. The bulk of ores is post-magmatic, that is, formed from highly mobile fluids released in the latest development stages of a magmatic body and migrating generally upward, or even generated outside of the intrusion, for example, from the pore waters in the rocks present in the GDG roof or from downward percolating meteoric waters.

In the latter case, the evolving GDG intrusion could have supplied heat only, the principal driving energy. Yet, without the heat, the ore depositing process would not have taken place so the presence of a GDG system is genetically important, causal and indispensable for most hydrothermal mineralizations, even if the orebodies themselves (e.g. hydrothermal veins) are actually located outside the intrusions, often far outside, in an endless variety of host lithologic associations.

No parent (driving, generating) intrusive body is known for many hydrothermal veins at the present level of exposure. Even when veins do occur within or in the proximity of a "granite", the genetic affiliation of both should not automatically be assumed without conclusive evidence, as was the case in the past. To bypass the problem of uncertain affiliation, Stanton (1972) in his book in which he pioneered the treatment of ore deposits in the framework of lithologic associations, coined a category, "ores in vein association". Such association is neutral, truly independent of the nature of the host rocks and one does not need a "granite" in sight to search for such orebodies. But veins themselves are extremely rare rock bodies seldom cropping over more than 0.001% of an area surface, so the "vein association" is of little assistane to a prospector who needs a regionally widespread lithologic association to start looking for the veins. For this purpose the GDG association is useful, because most postmagmatic hydrothermal deposits do occur in the proximity, in the aureole of igneous intrusions, and this aureole is more or less definable by the effects of thermal metamorphism.

When exploring in GDG terrains and their aureoles, one cannot avoid genetic speculation and reasoning no matter how imperfect. In doing so, however, an exploration geologist should always try to differentiate the recipient and generator rocks or units. The former actually host the orebodies, but they did not set into motion the processes that produced or drove the ore-bearing fluids that came from the outside (however, they caused the ores to precipitate and in some instances even supplied the metals). The latter provided the heat (energy) to drive the hydrothermal system and in many cases are still considered to have supplied the ore-forming fluids and metal as well. In some instances a rock may be both reservoir and generator rock (e.g. gabbros containing accumulations of ilmenite crystallized from an immiscible liquid). In others, both functions are close to each other in space and time so that they overlap and cannot be separated (e.g. certain cassiterite disseminations in endocontacts of granite cupolas). The simple empirical observation that an ore is hosted by an intrusion, however, does not necessarily indicate that the host rock is both reservoir and concentrator medium. In most cases (as in the porphyry copper, stockwork molybdenum or tin cupola complexes) we are dealing with multiphase, magmatic systems and typically the older phases, more visible and volumetrically abundant, provide reservoirs to fluids generated by younger (often the youngest) phases that are usually volumetrically insignificant (stocks, dikes, sills), if they are exposed in outcrop or in the mine workings at all. The Climax Mo complex is the best known example. Here three generations of orebodies appear above their generator intrusions hosted by sequentially earlier intrusions forming the roof in the time of ore deposition. The plutonic GDG association treated in this chapter includes the "most typical granites" formed at intermediate (mesozonal) and shallow (epizonal) crustal levels by solidification of introduced melts. There is a gradual upward transition and overlap between this association and the continental volcanic and subvolcanic terrains treated in Chapters 5 (Recent) and 26 (pre-Quaternary). The transition into the "marine" volcanic-sedimentary associations

(Chapters 10-14) is less typical and more controversial. With increasing depth, the mesozonal granites treated here change into katazonal granites, that are set in or coincide with the level of high-grade metamorphism and ultrametamorphism (partial melting), treated in Chapter 29. The latter is a zone of generation of at least some of the granites treated here, so some of the petrogenetic principles applicable to the generation of granitic magmas are discussed there. The GDG-affiliated metallogeny depends to a considerable degree on interaction with the outside rocks, so it is transitional with the metallogeneses of almost all the ancient lithologic associations treated in this book.

28.2. PETROGRAPHY, ORIGIN AND SETTING OF GDG PLUTONIC ROCKS

28.2.1. General

Examples of exposed "granitic" batholiths or plutons are known from almost the entire range of ages measured in the outcropping rocks (about 3.6 b.y. to 1.2 m.y.) and are among the most widespread rock types. Granites, granodiorites and diorites account for 21.6% of the total volume of the crust (Ronov and Yaroshevsky, 1969). Tonalite (light quartz diorite) "is the dominant rock type in the continental crust and the average composition of the continents is virtually tonalite" (Brown, 1979). The above statement, however, is applicable in the compositional rather than textural sense because the bulk of Brown's "tonalites" are the Precambrian "granite gneisses" treated in Volume 2 of this book.

GDG rocks occur as composite batholiths (gabbro, diorite, tonalite, granodiorite, quartz monzonite, granite) having a total silica content ranging from 45 to 78% (as in the Andean Batholith, Peru), or as homogeneous bodies composed virtually of a single rock type. "Granites" are perhaps most famous for the periodic genetic controversies they generate, the most notable having been one in the 1950s (magmatic versus metasomatic origin of "granites"). The controversies are the inevitable consequence of the fact that nobody has actually observed and measured a granite in the process of formation so that our interpretation is based on hypotheses gradually improving by the addition of constraints obtained in the field and in the laboratory. The granite controversy is not yet settled, as comparison of two recent statements indicates: (1) "Production of melts of the granite class is a natural culmination of metamorphic processes in the crust" (Fyfe, 1973); and (2) "Calc-alkaline batholiths are furnished by between a 70-100% contribution from melts derived from the upper mantle above subduction zones" (Brown, 1979).

To an exploration geologist, the origin of "granite" is of some importance, because it is at the core of a predictive philosophy of where to find the "granite"-related ore deposits. In view of "the simplistic (and one-sided; my remark) models in which petrologists are prone to indulge" (Hughes, 1982), the existing models should be used with utmost care, opposing opinions compared and the empirically observable facts given the highest priority. The basic aspects of petrology, origin and setting of GDG rocks relevant to the associated metallic ores are reviewed briefly in the following paragraphs. In-depth coverage can be found in the petrographic and petrologic literature, e.g. Marmo (1971), Carmichael et al. (1974) and Hughes (1982).

An explorationist trying to find ores has to go farther than a petrologist in regional geological interpretations, and take into consideration the lithologic associations surrounding the "granite" as well as the regional structure and alterations, as clues to mineralization.

The bulk of "granites" occur in moderately to deeply eroded orogenic belts marking the former continental margins. The present is a "prime time" of exposure of the lower Tertiary and late Mesozoic high-level (epizonal) granites, because the plutons have been sufficiently erosion dissected yet remnants of their locally comagmatic ejecta roofs are still preserved. It is also a "prime time" of exposure of the late Paleozoic to early Mesozoic medium-level (mesozonal) granites. These terrains are characteristically calc-alkaline and subsurficial equivalents of the continental volcanics (Chapters 5, 26) when such were developed. Granitoids with tholeiitic affinities (the gabbro-plagiogranite suite) occur in orogenic belts formed in evolutionarily primitive island arcs, e.g. in the Izu Peninsula, Japan and elsewhere; compare Chapters 7, 10 and 12. Alkaline granitic rocks within orogenic belts (compare Brown, 1979) usually formed along intraplate reactivated terrains, in collision belts and along intermontane or interarc extensional structures.

A proportion of "granites" constituting a very small volume is affiliated with intra- and inter-continental ("rift") structures and systems and this association is dominated by alkaline members although calc-alkaline rocks (such as biotite granites) are also present. This association is largely treated in Chapter 33.

Intrusive rocks are conventionally subdivided into hypabyssal and plutonic (Hughes, 1982). The hypabyssal rocks include small near-surface dikes, sills and plugs that have cooled quickly and are fine-grained. Plutonic rocks are coarse-grained, because they have cooled slowly at generally deeper levels. The most common large bodies of intrusive rocks are almost equidimensional or elongated plutons, batholiths or massifs. These terms are usually used interchangeably although the prevalent application is as follows (based on the A.G.I. Glossary of Geology): (1) Batholiths are large, generally discordant plutonic masses, having a surface exposure of over 100 km^2, and lacking a visible floor. (2) Pluton is a general term, used as a synonym for "igneous intrusion", as well as for a body of rocks formed by metasomatic replacement. (3) Massif is an alternative general term for a plutonic rock body and its underlying area. Small stocks and plugs are also common and largely self-explanatory. Additional varieties of intrusive bodies will be discussed later.

In a "most normal" GDG terrain (within a core of an orogen, eroded to a depth of 5 km), the dominant rock type is an equigranular, hypidiomorphic granodiorite composed of white, creamy or greenish plagioclase; creamy to pink or also white potash feldspar; gray translucent quartz and biotite and/or hornblende. Changing proportions of the five rock-forming minerals result in granite (sensu stricto) containing at least twice the volume of K-feldspar and above 20% of quartz, diorite, gabbro and syenite. At present, the nomenclature of Streckeisen (1976) is most widely used (Fig. 28-1).

The equigranular rocks grade readily into porphyritic equivalents, in which feldspar tablets are the most common phenocrysts. The homogeneity of the plutonic rocks is often interrupted by the presence of autoliths, xenoliths, schlieren, mineral nodules and intrusive dikes, and many granitic bodies are lightly to strongly foliated. Autoliths (or cognate xenoliths) are usually rounded inclusions or fragments of older intrusive rocks genetically related to the host. They tend to be more melanocratic, but mineralogically equilibrated with the surrounding intrusion. The rest of the xenoliths are fragments of country rocks ranging from sharply-outlined to almost-digested end members.

Didier (1973) prepared a definitive work on the enclaves in granites and their variety. The term schlieren is used to describe usually streaky, irregular masses or trains of inhomogeneities (minerals or inclusions) in granites. Mineral nodules are defined by Didier (1973) as more or less diffuse, heterogeneous patches not as sharply outlined as xenoliths, nor linear as the schlieren. Tourmaline and cordierite nodules tend to be the most common.

Frequent dikes in "granites" often occur in several mutually crosscutting generations and are controlled by joints. In the meso- and epizonal granites virtually all dikes have sharp boundaries. Compositionally, the dikes are: (1) porphyritic equivalents of the plutonites they transect (e.g. granodiorite porphyry); (2) leucocratic equivalents, polymineralic (e.g. aplite, pegmatite) or monomineralic (vein quartz); and (3) lamprophyres (minette, kersantite, spessartite or vogesite).

The field relations in plutonic terrains and in particular their interaction with the country rocks are, to a considerable degree, influenced by the depth level of emplacement. Buddington (1959) in his classical paper on the North American granites, proposed a three-fold depth subdivision of plutons and batholiths into epizonal, mesozonal and katazonal varieties. This division and terminology is "alive" and widely used in igneous petrology, although the same terms applied to metamorphic assemblages are now generally considered obsolete.

28.2.2. Epizonal plutons

Buddington (1959) characterized epizonal (epizone) "granites" as emplaced at depths of 1-6 km; having sharp, discordant contacts with

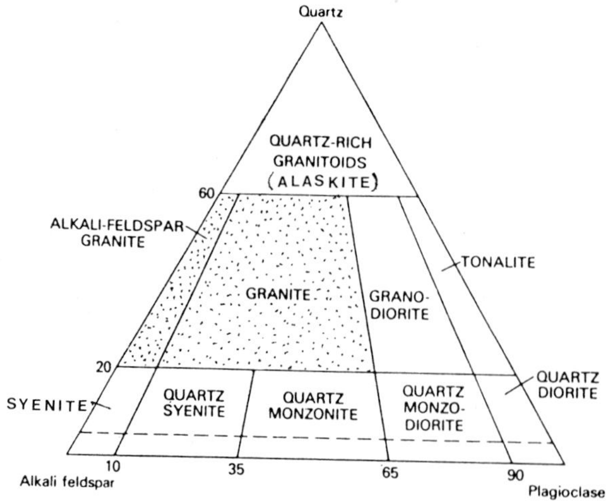

Fig. 28-1. Nomenclature of granite, diorite, syenite association, proposed by Streckeisen (1976).

their surroundings; composed of nonfoliated rocks; emplaced by large-scale magmatic stoping and as ring dikes or cone sheets. Additional characteristics contributed by Hutchison (1977) and Hughes (1982) include: extreme temperature difference between "granite" and its country rocks expressed in the well-developed contact metamorphic aureole (hornfelsing); frequently documented relationship to comagmatic volcanics and sub-volcanics; structurally unstable feldspars (e.g. mesoperthite) due to rapid loss of water; "post orogenic" and "orogenic" setting; predominance of equigranular to micro-porphyritic rocks; virtual absence of "typical" pegmatites; general concordance of Rb:Sr and K:Ar dates and other properties. Excluding alkaline and peralkaline complexes, the most frequent epizonal plutonites are granodiorites, quartz monzonites and granites commonly emplaced into their slightly earlier dacite, rhyodacite, quartz latite and rhyolite equivalents.

Both vented (=communicating with the surface) and unvented varieties of epizonal "granites" are known. Cunningham (1976) tabulated criteria indicative of former vents: (1) clastic texture; (2) partly aphanitic groundmass; (3) vertical pipe-like form located at the centre of the youngest plutonic rock; (4) pervasive alteration and disseminated Fe oxides; (5) evidence of a high content of volatile material as indicated by miarolic cavities; (6) fluid inclusions: zonality and increasing proportion of gas-rich inclusions. Most of the mineralized volcanic-subvolcanic terrains are underlain by epizonal "granites" some of them in the initial stages of unroofing, others still buried. There, caldera collapse, ring dikes and stratovolcanoes, are considered as surface expressions of near-surface plutons.

Examples of epizonal plutons associated with mineralization are numerous. Two of them, the Coastal (Andean) Batholith of Peru and the

Boulder Batholith, Montana, have recently been studied in some detail and have been included here as two representative examples.

COASTAL (ANDEAN) BATHOLITH OF PERU

This intrusive body has an extensive literature summarized in the recent memoir by Cobbing et al. (1981) and reviewed briefly in Hughes (1982); Fig. 28-2. The batholith, in its uninterrupted and dominantly plutonic portion, forms an elongated belt over 1,600 km long and 50 km wide. Minor isolated intrusive bodies believed to be connected with the main batholith at a depth, however, occur within an additional width of 150-200 km and are up to 15 km thick. The entire complex is composed of over 800 mappable plutons and small intrusions emplaced in two periods (105 to 55 m.y. and 33 to 12 m.y.) separated by a 20 m.y. hiatus. The batholith includes phases formed at both the epizone and mesozone levels and sharp demarcation is impossible. The intrusive rocks comprise 58% tonalite and granodiorite, 25.5% quartz monzonite, 16% gabbro and diorite and 0.5% granite. The plutons were emplaced by stoping or cauldron subsidence into a passive upper crust comprising (1) marine "eugeoclinal" volcanic-sedimentary association dominated by pyroxene andesite; (2) Precambrian metamorphics; (3) Paleozoic and Mesozoic (meta)sediments including abundant "black slates" and carbonates and (4) its own ejecta of subaerial andesites, dacites and rhyolites. Although the magmatism was contemporary with, and presumably related to, subduction the structural history and intrusive emplacement in western Perú was dominated by vertical tectonics. The batholith, together with its surrounding supracrustals, is transversely segmented and this segmentation appears to control the metallogeny to a considerable degree.

The excellent outcrops along the arid Peruvian coast and in the numerous quebradas (dry canyons) illustrate convincingly, in three dimensions, the intrusive morphology and composition which makes a fairly detailed interpretation possible. Myers (1975) and Cobbing et al., (1981); Fig. 28-3, interpreted four distinct mechanisms of intrusion in the Lima segment, marked by mostly permissive emplacement and a lack of metasomatism:

(1) Piecemeal stoping. This is illustrated by pluton apophyses "frozen" in the process of prying-off slabs of andesite from the roof, blocks of andesite "showering down from a roof of volcano" and trains of angular wallrock blocks enclosed in "granite", showing a gradual reduction in size and progressive rounding, recrystallization and dispersion. Despite the overwhelming evidence for stoping along the intrusive margins, the interiors of most plutons are clear of inclusions indicating that stoping was only a secondary process in creating space for magma.

(2) Cauldron subsidence is demonstrated by the "bell-jar" shape of the majority of plutons in which flat-lying roofs pass into steep walls. Some such intrusions are multiphase, nested, and include a subsidence of the central block and the downward passage of tabular plutons into

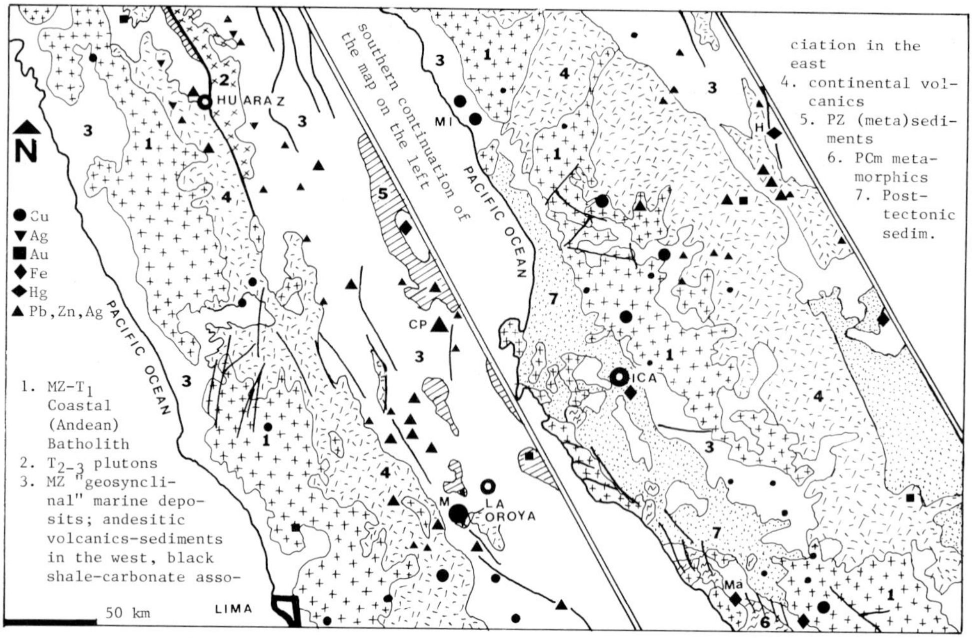

Fig. 28-2. Coastal (Andean) Batholith of Perú, map and distribution of ore deposits in the central sector. Based on the Mapa Metalogénico del Perú (Bellido E., Girard D., and Paredes J.) in Bellido and de Montreuil (1972). Abbreviations: CP=Cerro de Pasco; M=Morococha; Ml=Mala, Cu; Ma=Marcona, Fe; H=Huancavelica, Hg.

ring dikes.
(3) Fluidization and entrainment resulted in intrusive breccias that form some ring dikes. Such breccias comprise large blocks of volcanics, gabbros and earlier plutonic phases, enclosed in a "fluidized" matrix. Tuffisitic breccia dikes are considered to represent fluidized magmas rising ahead of a pluton.
(4) Uplift of the roof.

The distribution of ore deposits within and around the Andean Batholith of Peru (Bellido and de Montreuil, 1972; Putzer, 1976; Hollister, 1978) is quite characteristic of the overall batholithic metallogeny worldwide. The most striking feature is that the intrusive mass itself is virtually barren, almost devoid of economic orebodies (Fig. 28-2). This also correlates with the scarcity of metasomatism and hydrothermal alteration. The ores, however, start to appear in force along the contact. Some conformable orebodies in the hornfelsed exocontacts (e.g. the Cu "mantos" near Mala; Chapter 13) are probably thermally modified, volcanic-sedimentary relicts. Other

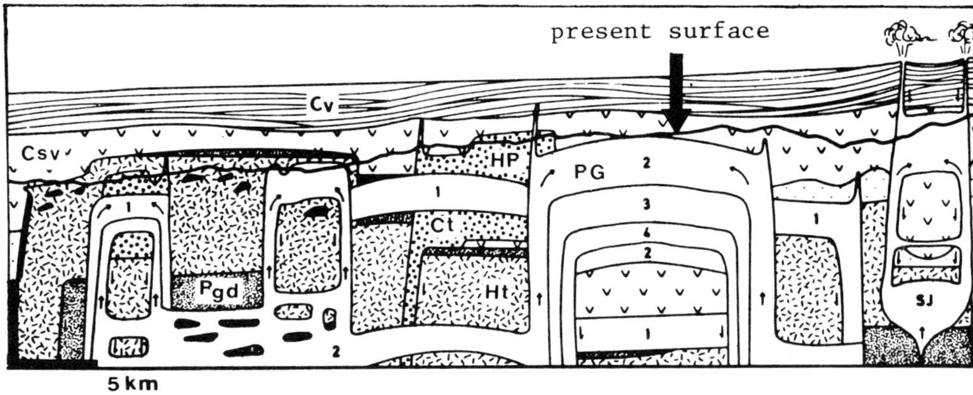

Fig. 28-3. Generalized section showing the major structures of the Coastal (Andean) Batholith of Peru during the latest intrusive episode about 50 m.y. ago. The batholith is composed of numerous plutons emplaced by repeated cauldron subsidence into their own volcanic ejecta. Abbreviations: SJ=San Jeronimo Granite; PG=Puscae Granite; half arrow=movement of block; arrow=movement of magma, during intrusion of SJ and PG; HP=Huampi Piruroc Granodiorite; Ct=Corralillo Tonalite; Ht=Huaricanga Tonalite; Pgd=Patap Gabbrodiorite; CV=Calipuy Volcanics; Csv=Casma Volcanics. From Myers (1975), reprinted with permission of the author.

localities (e.g. the large Marcona zone of magnetite bodies replacing Jurassic marine limestones at contact with an earlier Jurassic deeper-seated granodiorite batholith in an area underlain by Precambrian metamorphics; Putzer, 1976) are also relicts, partly modified. At Marcona, subvolcanic andesite and dacite dikes, probably related to the Coastal Batholith and intersecting the magnetite orebodies, produced magnetite, actinolite, minor calcite, quartz, chalcopyrite, talc, etc. possibly mobilization veins.

The major porphyry copper deposits that have so far been recognized in three areas in Peru (in the south-eastern corner of the Coastal Batholith near Arequipa, Toquepala-Cerro Verde district; in the centre, Morococha and in the north, Michiquillay) straddle the intrusive contacts and are related to "Laramide", Pliocene and Miocene subvolcanic dacite and high-level quartz monzonite porphyry intrusions, respectively.

The bulk of the Peruvian base and precious metal deposits occur along the fringe of the subaerial, late Tertiary volcanic pile topping the "miogeoclinal" and transitional basement megafacies. Both are intruded by sparse (under 20% by area) high-level plutonic bodies. The ores include major Pb,Zn,Ag replacements and veins (e.g. Cerro de Pasco, Casapalca; Chapter 26), all marked by intensive hydrothermal alteration. A strong basement metal source heritage is suspected (Chapter 19). These high-level orebodies, however, postdate emplacement of the major phase of the orogenic belt granites as in many other regions of the world (Colorado Mineral Belt, Great Basin, W. Rumania, etc.). The regional metal zoning pattern across the batholith based on the metals most conspicuously accumulated in ore deposits is (from west to east) approximately Fe,Cu-Pb,Zn,Ag and to a lesser extent Cu,W-Zn,Pb.

BOULDER BATHOLITH, S.W. MONTANA

This is a composite batholith 120 km long and 50 km wide, composed of coalesced epizonal plutons compositionally ranging from gabbro to alaskite (Hamilton and Myers, 1967; Klepper et al., 1971; Miller, ed.,1973; Fig. 28-4). The batholith was emplaced syn- and post-tectonically during an interval from 78-68 m.y. ago (late Cretaceous) in a basement of steeply dipping Proterozoic metamorphics covered by "miogeoclinal" Paleozoic and Mesozoic sediments, and into and under the roof of its own ejecta, the late Cretaceous Elkhorn Volcanics.

These volcanics are entirely continental and consist of dacite, rhyodacite and quartz latite flows and especially felsic welded tuffs. The "granite" / volcanics contacts range from horizontal (predominant) to steep faulted ones (in the east), and the intrusive emplacement was predominantly permissive but forceful in the east where wallrocks were locally sheared and vertically stretched. Hamilton and Myers (1967, Fig. 28-4) envisage emplacement of the batholith as a "gigantic mantled lava flow across a broad basin whose subsidence may have been due to the withdrawal of magma from depth". Portions of the batholith solidified within 1,600 m of the surface.

The constituent plutons formed over a 10 m.y. period, generally in order of increasing SiO_2 content, and can be placed into one of the four major compositional groups resulting from mixing of two different magma series ("normal" and sodic). These comprise (1) early mafic rocks (syenogabbro, syenodiorite, monzonite), (2) granodiorite, (3) Butte quartz monzonite (predominant phase) and (4) late leucocratic quartz monzonite, granodiorite, granite, quartz feldspar porphyry, aplite, alaskite, etc. The batholith is notable for the presence of a relatively widespread pegmatite, an unusual member of epizonal plutons.

Boulder Batholith and its aureole are densely mineralized. Several hundred of vein deposits, mostly small, produced Pb,Zn,Ag,Au and small quantities of U. There are small skarn orebodies, stockwork Mo occurrence, and the great Butte ore field in the southern part of the batholith. In contrast to the Peruvian Coast batholith, most of the deposits are within the granitic intrusive rocks, although relatively close to its former roof.

Hydrothermal epigenetic veins occur along joints, fractures and faults, often in swarms, trending E.-W. in the north-eastern portion of the batholith. They are parallel with the local alaskite, porphyry and quartz latite dikes replacing fault gouge or fractured wallrocks,

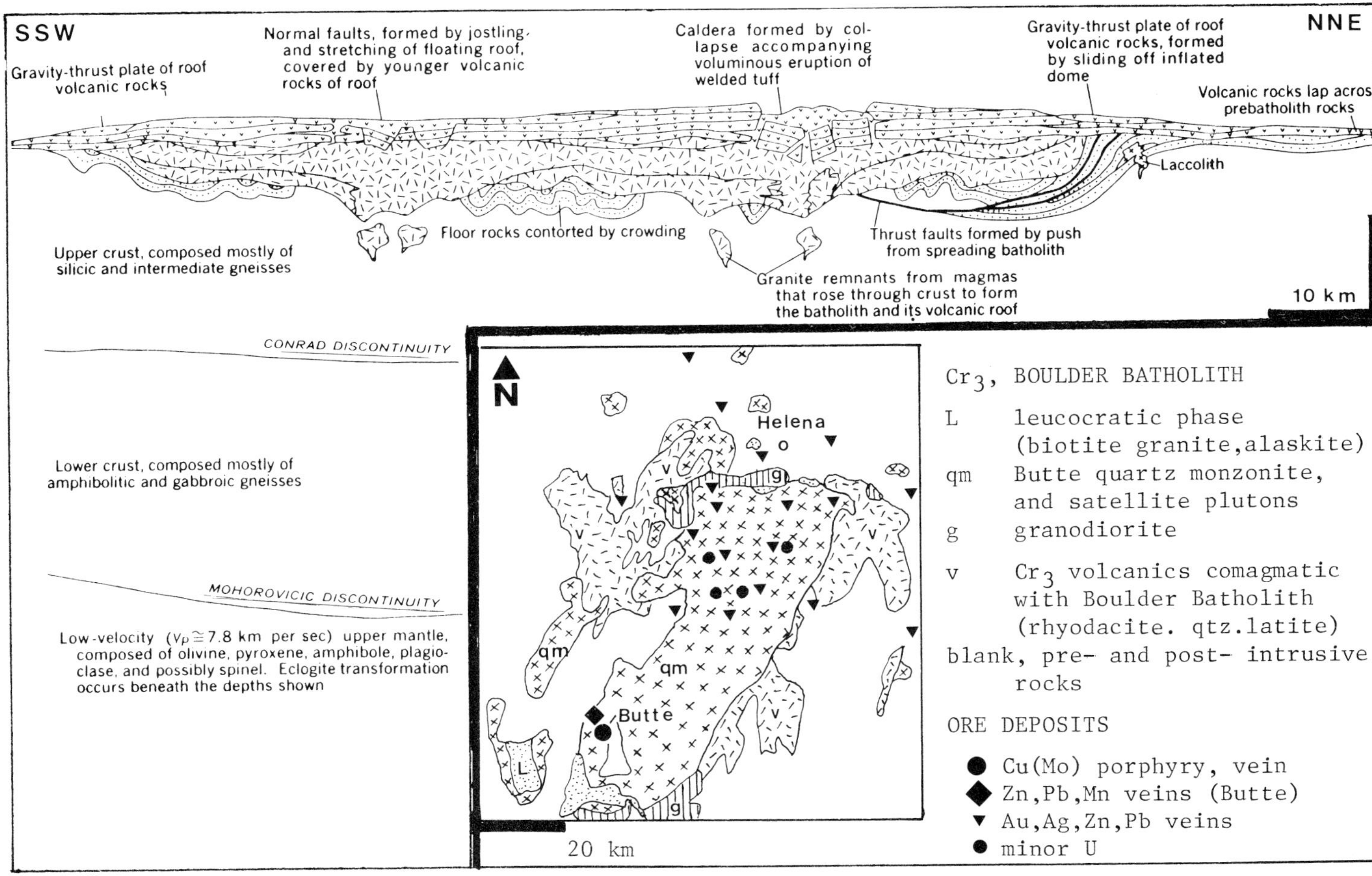

Fig. 28-4. Boulder Batholith, western Montana. TOP LEFT: Diagrammatic longitudinal geologic and crustal section, from Hamilton and Myers (1967). BOTTOM RIGHT: Simplified geological map and major ore deposits, from LITHOTHEQUE.

and some veins were repeatedly sheared and then healed. The gangue is almost exclusively silica, either gray quartz or chalcedony. The mineralization consists of disseminated , scattered to banded pyrite, sphalerite, galena, arsenopyrite, minor tetrahedrite, frequently accompanied by Ag-sulphosalts, gold or pitchblende. The total production has been about 120 Tt Pb, 476 t Au (much of it from gulch placers) and 2.5 Tt Ag, from about 15 ore fields or districts. Rimini was the most productive field. The wallrock alteration ranges from light to strong, and quartz, sericite, pyrite assemblage is most common.

Several mineralized breccia pipes and "pebble dikes" are known (e.g. the Montana Tunnels deposit, Obelisk Mine). The former locality contains disseminated pyrite and minor sphalerite and galena coating bleached rounded cobbles and pebbles resting in a tuff matrix, probably formed at base of the Elkhorn Volcanics. Most recently, a large tonnage of a very low-grade material has been established there (58.5 Mt ore with 0.93 ppm Au, 8.5 ppm Ag, 0.7% Zn, 0.3% Pb; Engineering and Mining Journal, November 1984).

The Butte ore field contains a multistage mineralization hosted by the Butte quartz monzonite in an area conspicuous for the abundance of late stage quartz-feldspar rocks (aplite, alaskite, pegmatite) apparently transitional into hydrothermal metasomatites and hydrothermal open space fillings. The recent results quoted repeatedly in the guidebook of Miller, ed., (1973), however, suggest a discontinuity between the process of batholith crystallization and ore deposits, representing a hiatus of some 10-15 m.y. This hiatus may have been accompanied by a partial de-roofing and initiate groundwater circulation through the completely crystallized and fractured quartz monzonite.

Butte (Brimhall, 1977; Meyer et al., 1968; Fig. 28-5) has been famous for almost a century as an example of numerous high-grade Cu-Ag veins, metal zoning, chalcocite supergene blankets and the overall richness. The veins formed during the Main Stage (58-57 m.y.) show an ore field-wide zoning of its mostly E.-W, N.E. and N.W.-striking fissure veins and branching vein sets from a central copper zone (with a small segment rich in quartz-molybdenite veins at its outer edge) to Zn and Mn zone on the periphery. The veins in the Cu zone can be subdivided into several varieties that differ by the degree of sulphurization of the copper minerals. The high-sulphur assemblage also rich in silver has mainly hypogene chalcocite and enargite, lesser tetrahedrite in quartz-pyrite gangue, and it is gradational into covellite, digenite, colusite and bornite-chalcopyrite veins. The wallrock alteration is mainly sericitization, quartz-sericitization or locally advanced argillization (dickite-kaolinite). The Zn-Mn zone has mostly sphalerite veins with rhodochrosite gangue.

An earlier (62 to 63 m.y.; Fig. 28-6) Pre-Main Stage porphyry Cu-Mo mineralization was recognized and economically utilized later. It consists of a stockwork, containing veinlets less than 3 cm wide and composed of quartz, K-feldspar, chalcopyrite, lesser anhydrite,

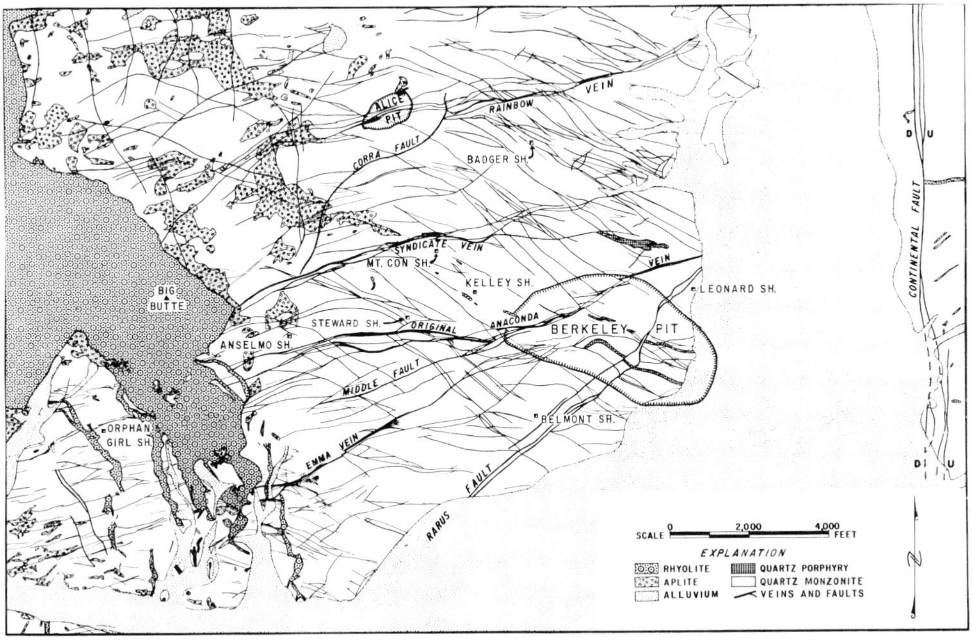

Fig. 28-5. Butte ore field, Montana. From Meyer et al. (1968), courtesy of the Society of Mining Engineers of A.I.M.E.

molybdenite, magnetite and biotite, surrounded by alteration envelopes of overlapping potassic (K-feldspar, biotite, anhydrite) and advanced argillic (andalusite, corundum, muscovite) assemblage. This represents a considerable resource of low-grade ore (min. 3 Mt Cu) and appears shortly to postdate (or possibly overlap with) emplacement of quartz porphyry and a biotite igneous breccia. In this breccia, fragments of quartz monzonite, aplite, vein quartz, sulphides, etc. are enclosed in biotite and K-feldspar matrix.

Although the Butte field has recently been considered to be located in the vicinity of a "root zone" of the Boulder Batholith, the reason for its existence and its metal sources remains as enigmatic as ever. Following ore deposition and partial unroofing, Boulder Batholith and portions of the Butte field have been intruded by post-ore rhyolite and rhyodacite dikes and blanketed by Eocene quartz latite continental volcanics. These rocks are virtually unmineralized.

NORTH LAKE BALKHASH (PRIBALKHASH) REGION, KAZAKHSTAN, U.S.S.R.

This is an outstanding region of Permo-Carboniferous continental volcanism and comagmatic plutonism so little eroded that several hundred paleovolcanic remnants (stratovolcanoes, calderas, cauldrons, volcano-tectonic depressions) are still recognizable. These are

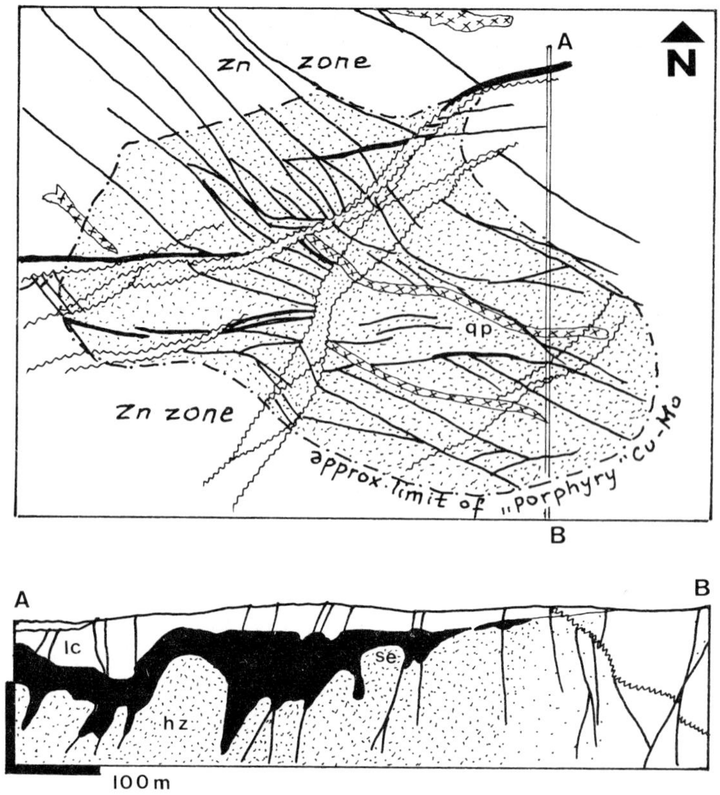

Fig. 28-6. Butte, Berkeley pit, the early stage "porphyry" Cu-Mo mineralization intersected by "main stage" chalcocite and enargite veins. Butte Quartz Monzonite is the host rock throughout. lc=leached capping; qp=quartz porphyry; hz=hypogene zone; se=secondary sulphides zone. From LITHOTHEQUE, based on data in Miller, ed. (1973)

interspersed with numerous epizonal plutons (Yesenov, ed., 1972; Nakovnik, 1968; Fig. 28-7).

Central Kazakhstan is extremely complex in geological terms. It remains an enigmatic accumulation of several crustal megablocks, comprising blocks of thickened (50-55 km) Proterozoic to Permian transitional to continental crust forming structural highs, alternating with thinner (40-45 km) synclinoriums. In the Permo-Carboniferous period, this was probably an "Andean-type" continental margin with a northward polarity which later collided. Basalt, andesite, rhyolite volcanism started in the middle-upper Devonian and lavas, pyroclastics and volcaniclastics were deposited over the immediate basement of Silurian marine sediments, in both subaqueous and subaerial environments. Of the several phases of "orogenic" volcanism, the middle-upper Carboniferous was most intensive and increasingly subaerial. In the Tokrau-Bakanass volcanic belt a number of eruptive centres produced andesites and basalts in downwarps adjacent to deep faults, and dacites-rhyolites along uplifts.

The upper Carboniferous to Permian interval was dominated by "subsequent" (post-orogenic) continental volcanism, and andesite, dacite, rhyolite lavas and pyroclastics issued from numerous centres controlled by block faults. The Permo-Carboniferous volcanism was cyclic and closely comparable with the Tertiary cycles reviewed (as in the San Juan Mts.) in Chapter 26. Each cycle (duration 20-25 m.y. in the earlier cycles, 10-15 m.y. in the later cycles) was initiated by (1) deposition of coarse dacite tuffs containing basement rock fragments on an eroded fundament. This was followed by (2) andesite and dacite lavas and subvolcanic andesite bodies, sometimes postdated by granodiorite intrusions. (3) After a break marked by deposition of conglomerate and sandstone, widespread deposition of rhyolite, dacite and latite lavas, ignimbrites and ash-flow tuffs was contemporary with caldera and volcano-tectonic depression development. This was shortly followed by consolidation, in depth, of granodiorite-quartz monzonite massifs. The closing stages (4) produced mostly felsic dike complexes and a variety of hydrothermal metasomatites dominated by "secondary quartzites" (relict-free zones of silicification; compare Fig. 28-7a). The late magmatism frequently had alkaline tendencies (trachybasalts, trachytes, monzonites, syenites).

In the North Balkhash area, intrusive Permo-Carboniferous plutons are now exposed over about 30% of the territory. 40% is occupied by the comagmatic volcanics, and the rest are rocks of the basement, Permo-Carboniferous sediments and Mesozoic-Cainozoic sediments of the platformic and para-platformic covers. The intrusions range from usually elongated large plutons of irregular outline to subrounded small massifs and rounded to irregular stocks, dikes, vent breccias and ring dikes, scattered within and on flanks of a former caldera. The small massifs are often interpreted as marking the cores of stratovolcanoes.

The region is densely mineralized and contains the principal Soviet porphyry Cu-Mo province estimated to represent some 15 Mt Cu and 200 Tt Mo. Most of these metals are in the large Kounrad (Cu-Mo) and East Kounrad (Mo-W) fields. There are two or three additional large porphyry Cu and skarn/porphyry deposits (Karatas, Sayak, Borly), one large Zn-Pb replacement in Paleozoic carbonates (Kyzyl-Espeh), and 116 Cu, 112 Zn,Pb,Ag, 75 Mo, W, Sn and 17 Au medium-to-small deposits and occurrences. Unfortunately, no size or tonnage figures are available. Neither, in most cases, are the exact locations of the deposits.

The absolute majority of the North Balkhash ore deposits (and all the economically important ones) are hydrothermal-epigenetic, postmagmatic mineralizations formed during stage (4) of the magmatic development mentioned above. All appear to be high-level occurrences, formed at the subvolcanic to uppermost plutonic levels, in the roof

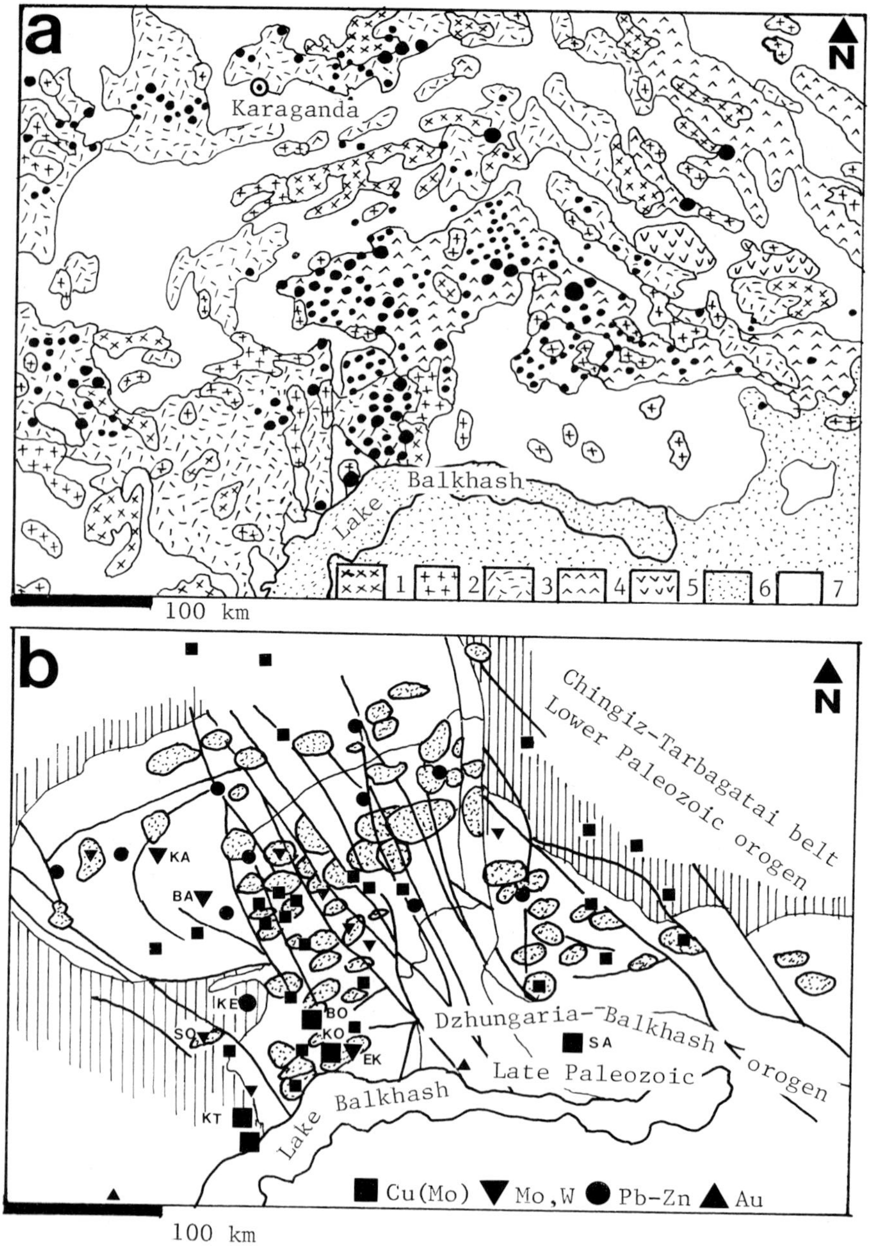

Fig. 28-7. North Lake Balkhash (Kazakhstan) late Paleozoic mineralized region.
a) Simplified geological map, showing the distribution of massifs of "secondary quartzites" (black dots; a hydrothermal silicification usually associated with aluminous products of sericitic and advanced

and above-roof zones of the intrusive bodies formed under volcanic centres. Because of the magmatic cyclicity (at least 3 major cycles), virtually identical ores are associated with upper Devonian to upper Permian (or even lower Triassic) magmatism, although the importance with regard to Mo, W and Sn increases with the decreasing age (maximum in upper Permian), while Cu had reached its peak of accumulation during the middle Carboniferous period.

The best guide to the location of ores on the gently rolling surface of the semiarid Kazakh Steppe are the topographically positive zones of intensive silicification ("secondary quartzites"), often gradational into quartz-sericite and advanced argillic (andalusite, diaspore, corundum) alterites. The "secondary quartzites" and even more the adjacent silicate alterites host most of the orebodies: porphyry Cu-Mo stockworks and pervasive disseminations; gold-quartz veins grading to "linear stockworks" (some reminescent of the Butte "horsetail" veins); quartz, molybdenite, wolframite veins to stockworks (East Kounrad) and a variety of vein Pb-Zn deposits. Yesenov, ed., (1972) reported 500 major occurrences of "secondary quartzites" in the North Balkhash region, of which 60 are associated with porphyry Cu deposits or occurrences. Additional Cu porphyries (and Sn,W,Mo "cupola" stockworks, veins and disseminations) have been discovered in the more recent phase of exploration, affiliated with K-silicate-altered and greisenized granitoids lacking the conspicuous silicification.

The Sary-Oba ore field described by Laumulin et al. (1973; Fig. 28-8) is a rather typical example of a mineralized North Balkhash volcanic-intrusive centre. It is located 110 km W.N.W. of the city of Balkhash, in the Tasaral-Kyzyl Espeh Anticlinorium. There, the substratum is represented by lower-middle Carboniferous continental dacites, pyroxene-hornblende andesites, and the rhyolite ignimbrites that are most important in terms of volume. The felsic volcanics eruption overlapped with intrusion of a biotite leucogranite massif in depth. Remnants of small circular vents surrounded by a radial system of fractures filled by small bodies of quartz-syenite porphyries, are still recognizable. The latter, as well as a circular granite stock, are probably a product of cauldron subsidence. A variety of dikes (felsite, diabase, quartz-syenite porphyry) intrude all the earlier rocks.

⟵

argillic alteration such as andalusite, diaspore). 1=late Paleozoic granitic rocks; 2=early Paleozoic "granites"; Paleozoic felsic=3, intermediate and mixed=4 and mafic=5 volcanics. 6=sediments of post-orogenic basins; 7=Paleozoic (meta)sediments. After Nakovnik (1968).

b) Major hydrothermal deposits related to late Paleozoic granites, and their correlation with granite cupolas, cauldrons and Cb-Pe volcanic centres (stippled). Abbreviations: KA=Kairakty; BA=Bainazar; Bo=Borly; KE=Kyzyl-Espeh; SO=Sary Oba; KT=Karatas; Ko=Kounrad; EK=East Kounrad; SA=Sayak. Base map simplified after Yesenov, ed., (1972); ore deposits added.

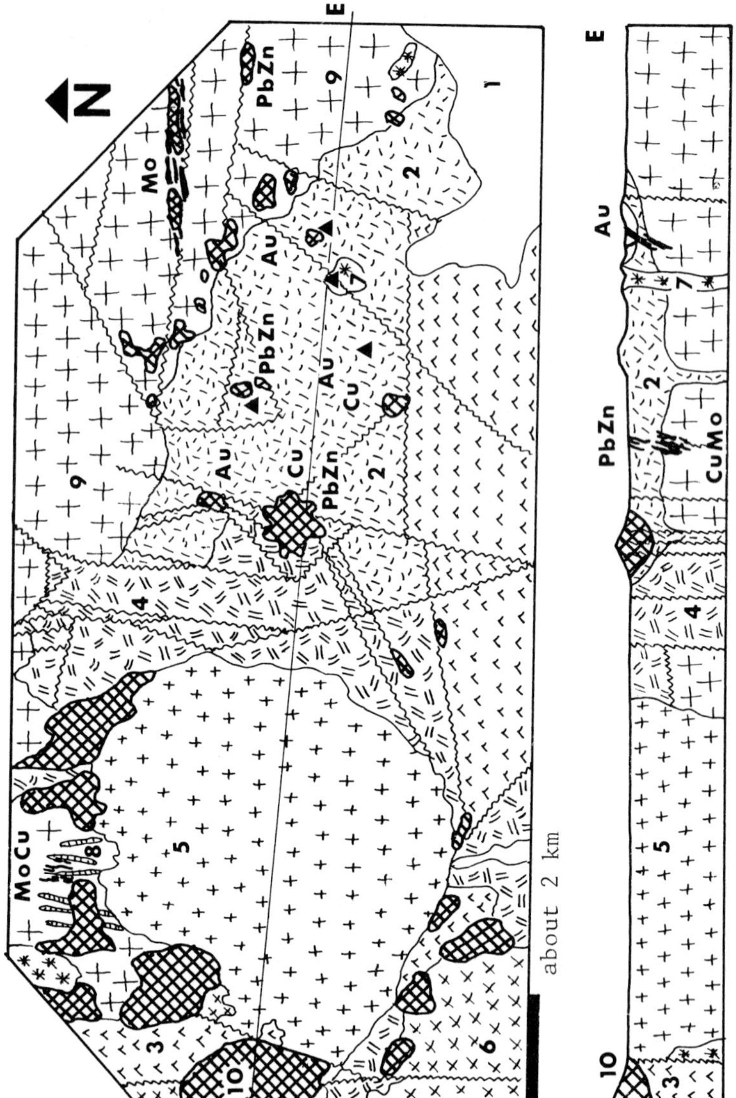

Fig. 28-8. Sary-Oba (Kazakhstan), a high-level mineralized volcanic–plutonic complex. Most ore occurrences are associated with fault-controlled silicified massifs ("secondary quartzites"). 1=Q, alluvium; 2=Cb_2 ignimbrite, rhyolite flows, tuffs; 3=$Cb_{1,2}$ andesite; 4=Cb_1 rhyolite flows and tuffs; 5=$Cb_{2,3}$ leucocratic granite; 6=granodiorite; 7=quartz syenite porphyry; 8=felsite porphyry dikes; 9=Cb_1 biotite leucocratic granite; 10=secondary quartzites; triangles= volcanic vents; Mo=molybdenite stockworks, veins; Cu=disseminated Cu sulphides assoc. with feldspar porphyry dikes. Modified after Laumulin et al. (1973).

Quartz-dominated post-magmatic hydrothermal metasomatites (secondary quartzites) are widespread, (1) fringing the leucogranite and controlled by the cauldron rim faults; (2) marking former vents and (3) situated along faults and fault intersections. The pure "secondary quartzite" is a fine-grained, granoblastic rock replacing both volcanics and intrusions, and it contains grains of ilmenite and rutile. It is gradational into quartz-sericite, quartz-kaolinite, quartz-alunite and alunite-diaspore alteration assemblages. It is quite possible that a portion of the "quartzite" formed by supergene modification of other alterations, because the largest masses do not continue to depth. A variety of ore showings has been recorded, but no information is available concerning their size and economic importance. Gold occurrences are situated along fault zones accompanied by strong pyritization. Most promising are the simple lens-like quartz-pyrite veins grading to stockworks. Persistent quartz-barite veins also carry low gold values in depth. Minor Pb,Zn,Cu-bearing veins hosted by the ignimbrites along faults, grade downward into poor "porphyry Cu-Mo" stockworks in granite. The Mo showings (e.g. the Birek showing) appear to be the most interesting. Molybdenite occurs as disseminated flakes in quartz veins and in silicified, greisenized and albitized biotite granite along a S.W.-N.E. fracture zone, up to 2 km long and 100-250 m wide.

Kounrad porphyry Cu-Mo deposit (estim. 10 Mt Cu, 70 Tt Mo; Gazizova, 1957; Samonov and Pozharisky, 1974; Fig. 28-9) is the largest Soviet "porphyry" located 17 km from Balkhash (city). It cropped-out in a low hill of heavily silicified rocks surrounded by Cainozoic alluvium. Lower Carboniferous volcanics (andesite, basalt, rhyolite lavas and tuffs and ignimbrite locally albitized to keratophyre) have been forcibly intruded by a high-level granodiorite porphyry body, fractured, invaded by felsic dikes, altered and mineralized. The dominant alteration is pervasive silicification grading to quartz-andalusite, quartz-sericite and propylitic alterations. Lesser diaspore, dickite and alunite are present and andalusite is recovered as a by-product. This phase of alteration is considered pre-ore, affecting all rocks, but only the fractured granodiorite porphyry, silicified or quartz-sericite altered, contains the ore. The hypogene mineralization consists of disseminations, veinlets and discrete veins filled by quartz, pyrite, chalcopyrite and minor molybdenite. Late-stage rich quartz, pyrite, enargite, bornite veins are rare and occur locally. The Kounrad orebody is pipe-like with dimensions of about 1.7x1.3 km. It is distinctly supergene zoned. Leached capping (about 27 m thick) is followed by an oxidation zone (22m) with an overall subeconomic grade of 0.2% Cu, but containing local rich patches. A thick (135m) zone of secondary chalcocite enrichment with grades between 0.7 and 1.2% Cu has provided most of the ore mined so far. The hypogene zone is said to average between 0.3 and 0.4% Cu.

NORTH-EASTERN QUEENSLAND

In North-Eastern Queensland (Georgetown Inlier and the adjacent Tasman Foldbelt), Carboniferous epizonal granites and comagmatic continental volcanics interact in numerous cauldron subsidence, ring and neck complexes described in detail by Branch (1966; Fig. 28-10). In contrast to the earlier treated examples affiliated with Cu, Pb-Zn and Au mineralization this is a distinct Sn(W) province.

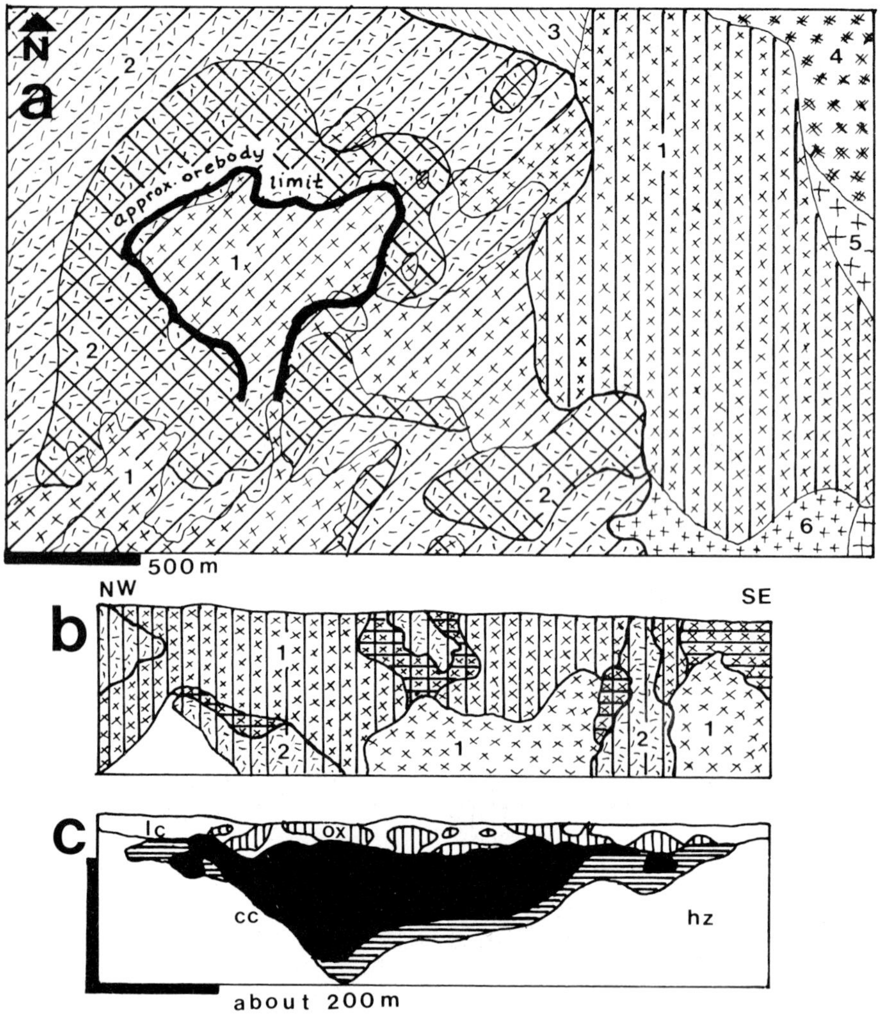

Fig. 28-9. Kounrad "porphyry" Cu-Mo deposit, North Lake Balkhash area, Kazakhstan. (a) Geological map. Rock types: 1= Cb_1 granodiorite porphyry; 2= Cb_1 rhyolite; 3=D sandstone, shale, contact hornfels; 4= Cb_1 granodiorite; 5=Cb-Pe biotite granite; 6=Pe aplite granite. Alterations: (shown by ruling superimposed on geology); N.E.-S.W ruling: quartz-sericite (secondary quartzite); N.W.-S.E. ruling: andalusite-sericite, grading in depth into diaspore-corundum; vertical ruling: supergene argillization. (b) Cross-section, showing alterations; vertical ruling: early silicification and quartz-sericite ("secondary quartzite"); horizontal ruling: late argillization. (c) Cross-section, mineralization; cc=chalcocite blanket; ox=oxidized ores; lc=leached capping; hz=hypogene zone. Modified after Nakovnik (1968).

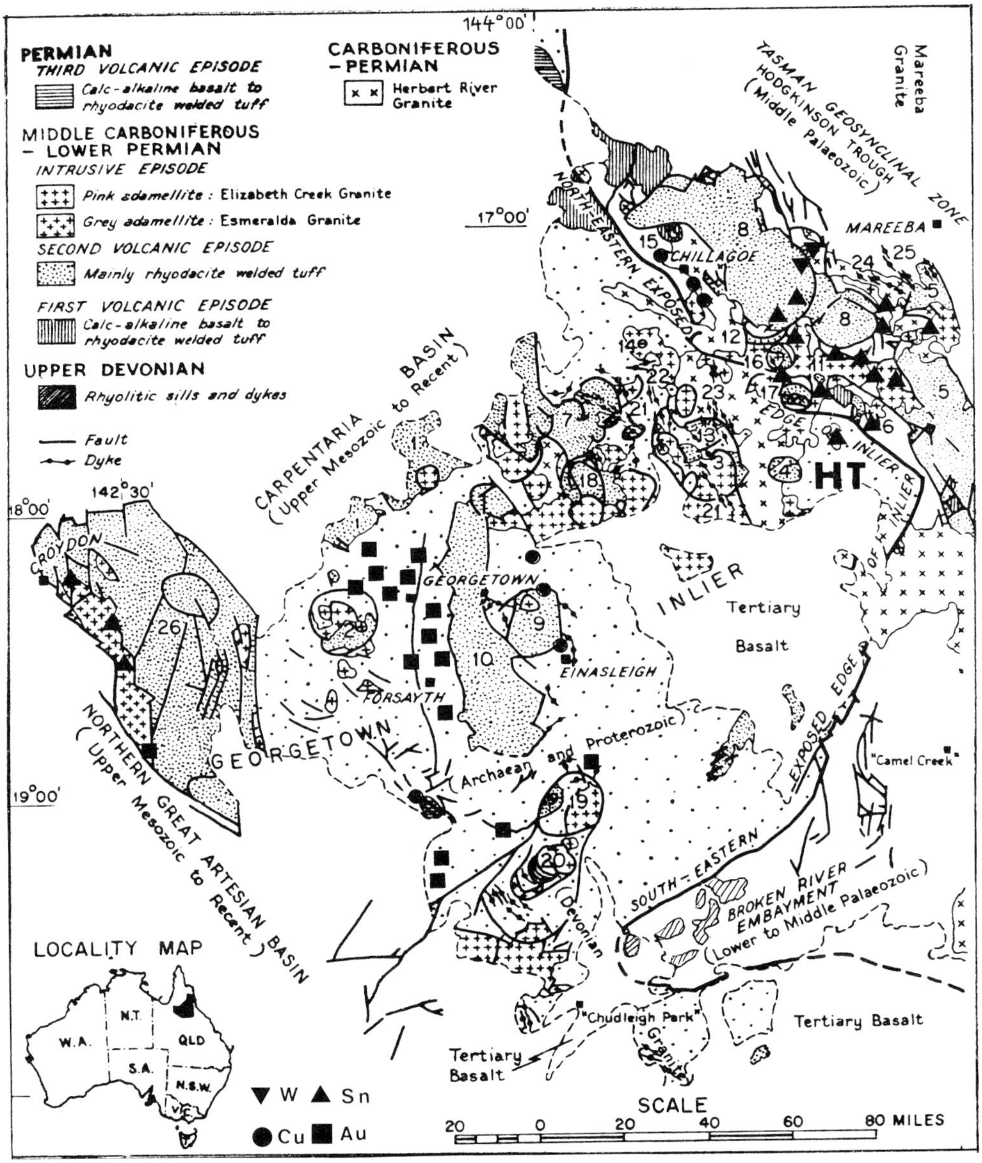

Fig. 28-10. Permo-Carboniferous high level plutonic province of N.E. Queensland and its metallic mineralization (granite-related, interaction and relict). HT=Herberton Tinfield; numbers indicate named complexes, 11=Gurrumba volcanic neck (Fig. 28-11); 17=Claret Creek ring complex (Fig. 28-12). The gold deposits in the Georgetown district are of Proterozoic age. From Branch (1966), with additions. Courtesy of the Bureau of Mineral Resources, Geology, Geophysics, Canberra.

In the Georgetown Inlier (western) portion, the basement rocks are Precambrian metamorphics and granitic rocks. In the east, the old crystalline basement is topped by a prism up to 13 km thick, of detrital and very minor carbonate sediments of a stable shelf facies, ranging in age from upper Ordovician to Carboniferous. The late Paleozoic continental magmatic activity followed a folding in the "geosyncline" and fragmentation (activation, rejuvenation) in the basement, and it proceeded in several cycles. The initial extrusion of small amounts of calc-alkaline acid, intermediate and mafic volcanics in the Carboniferous period was followed by a vast outpouring of compositionally monotonous rhyodacite ignimbrite and ash-flow tuffs. Ring complexes and cauldron subsidence centres formed at intersections of major basement fractures. After this, acid magma stoped through the upper crust and intruded the base of the felsic pyroclastic cover. This was followed by dike emplacement and terminated, in the Permian period, by another weak outburst of sequentially differentiated volcanics.

The volcanic-intrusive central complexes treated in Branch (1966) are interesting geologically, but they carry little mineralization contemporary with their formation. Small Pb-Zn veins (Greenhills), Cu veins (Mt. Jardine), Mo,Sn,W veins (Lochaber), Au and Sn veins (Croydon), Au veinlets associated with microdiorite in the Gurrumba volcanic neck (Fig. 29-11) and others, have been reported. The U-Mo veins in Carboniferous welded tuff (Ben Lomond) have been reviewed in Chapter 26. The bulk of the Northern Queensland ore metals (over 100 Tt Sn, 10 Tt W, some Bi, Pb-Zn, Cu, magnetite) is, or was, contained in over 2,500 lode deposits and several hundred placers, all genetically related to late Carboniferous granite plutons, postdating the circular centres. This is apparent from Fig. 28-12, showing the minor Sn and Cu-Fe occurrences along the fringe of the Claret Creek ring complex.

The Elizabeth Creek Granite, a high-level pink leucocratic biotite quartz monzonite, is responsible for most of the ore occurrences. It was emplaced as little as 170 m below the surface, under its comagmatic ignimbrites. The Esmeralda Granite and its hornblende-biotite granodiorite phase, accounts for the rest (e.g. magnetite and/or chalcopyrite skarns in the Chillagoe area, small porphyry Cu, Ruddygore and Zn-Pb replacements in carbonate).

Most of the Sn(W) orebodies are small greisen or quartz, chlorite, cassiterite pipes or fissure veins located in both endo- and exocontact in the granite roof. Taylor and Steveson (1972) argued convincingly that the known mineralization accumulated preferentially in a number of small, circular to elliptical, metal-zoned patches, is controlled by a series of emanative centres in the Elizabeth Creek Granite that in turn overlie the "highs" (cusps, ridges and cupolas) on the generally flat surface of the pluton. The "highs", in turn, are fault-controlled and some of the structures are cauldron subsidence rims or branching faults.

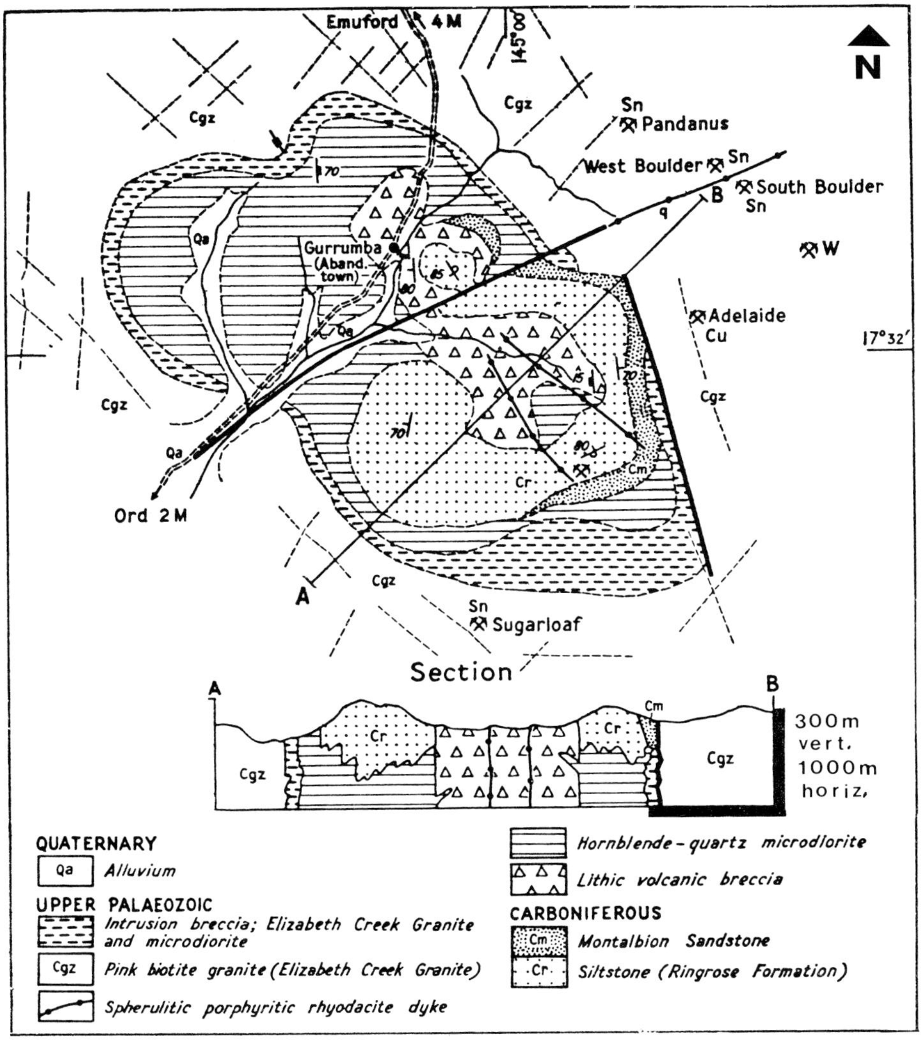

Fig. 28-11. Gurrumba (paleo)volcanic neck, 35 km W.S.W. of Herberton, N. Queensland. The Sn and Cu mineralization postdates the neck emplacement and is related to the postmagmatic phase of the Elizabeth Creek Granite. From Branch (1966), courtesy of the Bureau of Mineral Resources, Geology, Geophysics, Canberra.

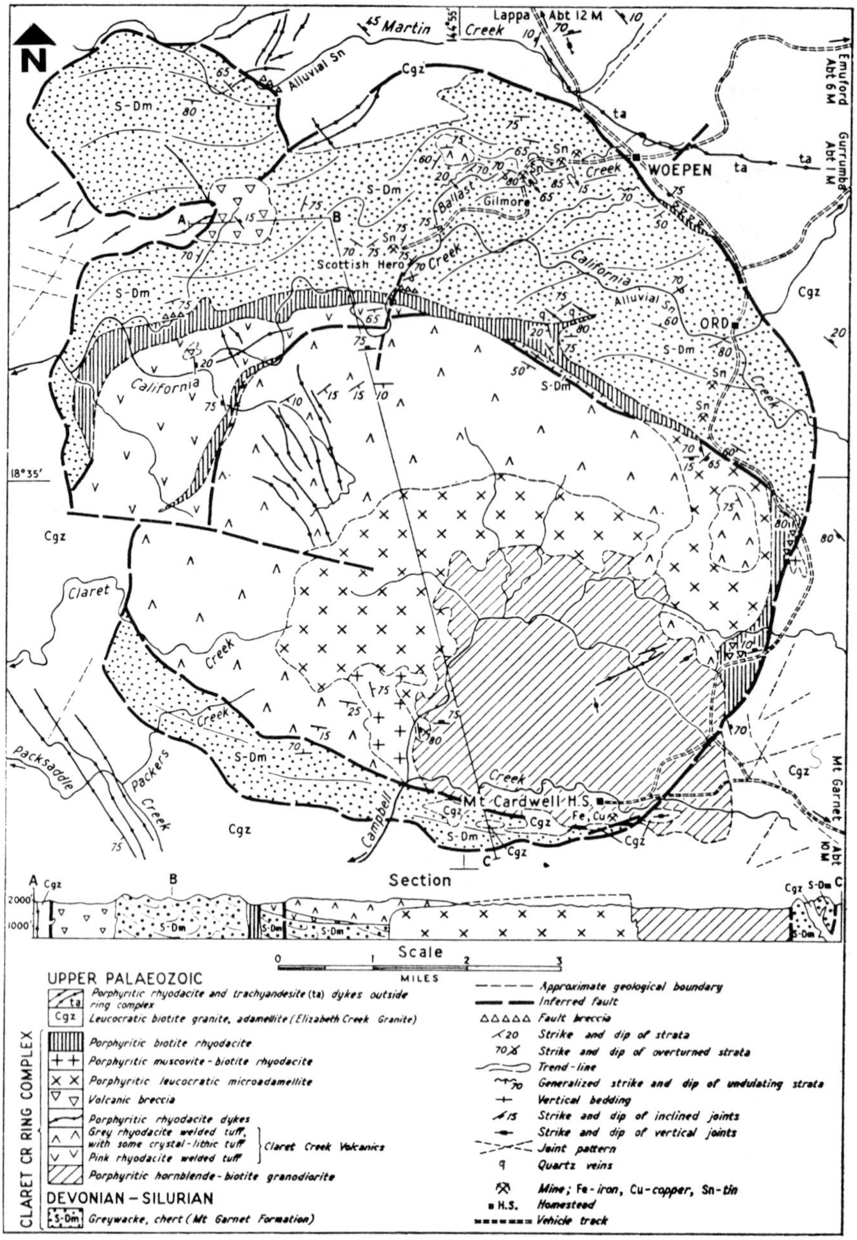

Fig. 28-12. The Claret Creek late Paleozoic ring complex 70 km W.S.W. of Herberton, N. Queensland. The minor mineralization (placer cassiterite and magnetite, chalcopyrite disseminations and replacements), are related to the young intrusion of biotite leucogranite (Cgz). From Branch (1966), courtesy of the Bureau of Mineral Resources, Geology, Geophysics, Canberra.

COLORADO ROCKY MOUNTAINS MINERAL BELT

The Colorado Rocky Mountains are a N.N.E.-trending young uplift deformed during the Laramide Orogeny (late Cretaceous-early Tertiary period), and it separates the little deformed Colorado Plateau in the west from the North American Platform in the east. The Rockies are dominated by deeply eroded lower Proterozoic metamorphics (gneisses, migmatites, schists) and kata- to mesozonal granites. There are lesser Paleozoic and Mesozoic platformic sediments preserved on flanks of the basement domes and ridges. The Laramide Orogeny was followed by magmatic activity, and two age groups of magmatic rocks are present: (1) pre-Oligocene and (2) Oligocene to Pliocene. The magmatic belt is in the N.E. extension of the San Juan Mts. volcanic field (Chapter 26).

The "Laramide" magmatites (Lovering and Goddard, 1950) are mostly porphyritic high-level intrusions of calc-alkaline diorite to alaskite forming small, often inconspicuous stocks, dike swarms and single dikes, "porphyry" laccoliths, etc. Most of these intrusions congregate in a narrow belt crossing obliquely the Front Ranges, extending in the S.W. direction from Boulder to Breckenridge. Most of the classical Pb,Zn,Ag and Au,Ag vein districts or fields occur in the aureole, forming the Colorado Mineral Belt (Fig. 28-13). Additional intrusions are scattered outside this belt and some of them (e.g. Climax, Red Mountain) are affiliated to giant Mo stockworks.

Although extrusive equivalents of the Laramide intrusions are known locally on the flanks of the Rocky Mountain belt (e.g. in the Denver basin), the intrusions themselves generally lack affiliated volcanics and were emplaced mainly in the crystalline basement along faults and crush zones. The high emplacement levels are demonstrated by intrusive textures and many wide, open, breccia or mylonite-filled faults ("breccia reefs"). It is likely that the Colorado Rockies' small intrusions are apophyses of a mesozonal batholith situated at a greater depth.

The Colorado Mineral Belt is one of the classical regions of metalliferous geology and of genetic affiliation of "granites", faults and postmagmatic hydrothermal ores. It contains several thousand showings. Several hundred small to medium-size Pb,Zn,Ag; Au; few W, U deposits; two significant Zn,Pb,Ag fields (Leadville and Gilman); two or three large Mo stockworks (Mt. Emmons, Mt. Tolman) and two Mo giants (Climax and Red Mountain). This represents some 4 Mt Mo, 1.8 Mt Pb, 1.9 Mt Zn, 15 Tt Ag and 975 t Au (portion of Au came from placers). Cu,W,Sn,U,Bi,Re and other minor metals form lesser accumulations.

Metallogenically, this is a distinct "continental" association and the deficiency in copper is conspicuous. Disregarding young placers and old relict deposits in the intruded basement (e.g. pegmatites, katazonal skarns and greisens, hydrothermal veins, minor concordant pyrite and Zn-Pb sulphide bodies or metalliferous horizons), three ore styles are the most prominent: (1) fissure veins in intrusive exocontacts; (2) replacements in exocontact carbonates and (3)

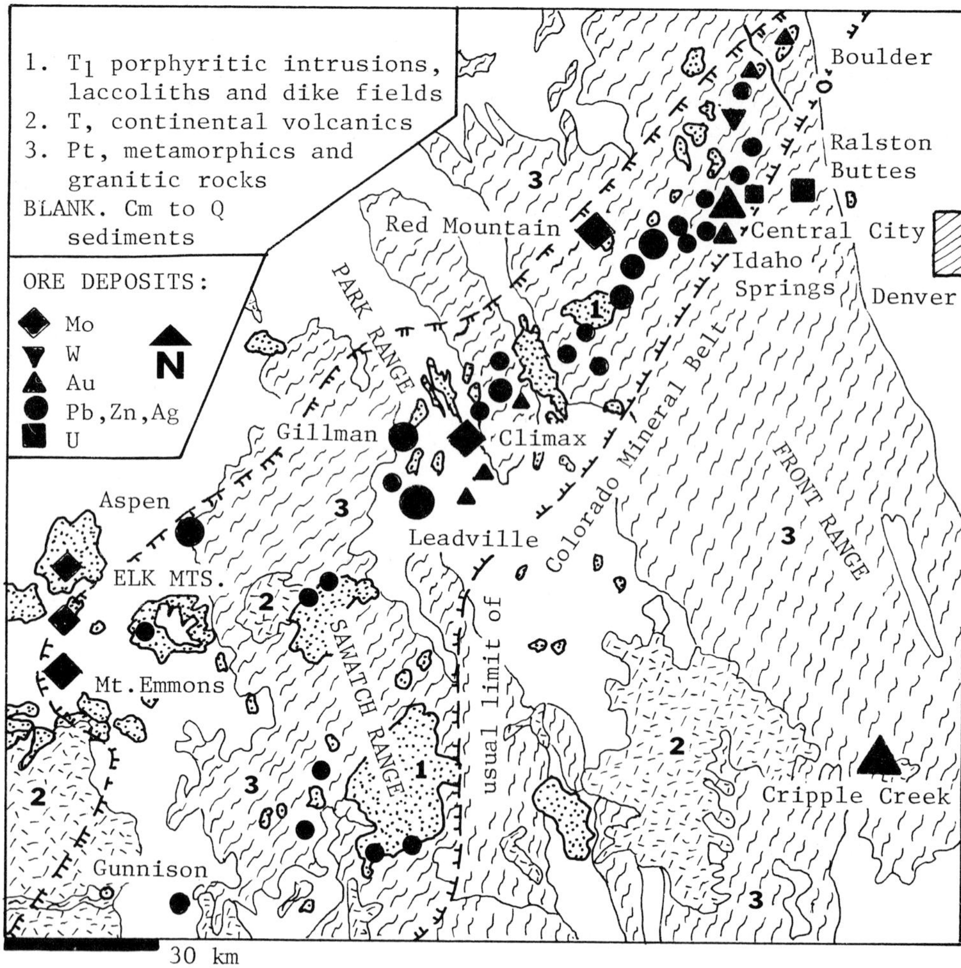

Fig. 28-13. Colorado Mineral Belt. From LITHOTHEQUE.

sulphide stockworks in altered "porphyries".

Category (1) is most common but represented only by small to medium-sized deposits. The veins either singly or in groups, are exercising a variety of structural controls for which they have been extensively treated in the literature from the first half of this century. The brittle rocks (Proterozoic granites, metaquartzites, amphibolites, etc.) are favourite hosts whereas sheared and faulted schists and shales produce tight structures and inpenetrable gouge so they rarely contain veins. Most mineralized structures are approximately parallel with the regional trend of the "porphyry" belt and many such fissures carry ore when they intersect the otherwise barren N.W.-trending "breccia reefs". The veins usually have quartz gangue and a banded-to-scattered filling of pyrite, sphalerite and galena. Lesser chalcopyrite, arsenopyrite. Ag-sulphosalts, gold, Au tellurides, occur locally. Pb-Zn only; Pb,Zn,Au; and Au-only orebodies have been identified in almost all districts. Individual ore shoots tend to be short with length several hundred metres, but the mineralized depth range may exceed 700m.

Category (2) is best developed in the Leadville field, already mentioned briefly in Section 20.10.1. (see also Fig. 20-28). There, both fissure filling and discordant replacement Zn,Pb and Fe sulphide veins grade into peneconcordant mantos in Mississippian limestones, directly at the contact or in the vicinity of quartz-feldspar porphyry sills and dikes.

Category (3) is primarily represented by the great molybdenite stockworks at Climax and Henderson (Red Mountain), described later. There, quartz, pyrite and molybdenite veinlets and fracture coatings are situated in a large mass of an altered cylindrical composite intrusive body. Recently, Mutschler et al. (1981) argued that the Colorado Mo deposits are affiliated to the petrochemically true granites and occur in cupolas on tops of epizonal batholiths coincident with, or shortly predating, a regional extensional tectonism. Metals other than Mo rarely accumulated in endocontact stockworks. The small Jessie mine in the Breckenridge field (about 2 t Au; Lovering and Goddard, 1950; Fig. 28-14) is an example of a gold-bearing stockwork in sericite-altered Tertiary tongue of quartz monzonite porphyry.

OTHER MINERALIZED EPIZONAL PLUTONS

These have a worldwide distribution and are affiliated to the bulk of porphyry Cu-Mo deposits (in the south-western United States and adjacent Mexico, in Chile and Argentina, in Yugoslavia, Rumania and Bulgaria, in Iran, the Philippines, Melanesia and elsewhere), to tin deposits (Sundaland, Cornwall, Sikhote Alin, eastern Australia), to Pb,Zn,Ag provinces (e.g. western United States and Mexico, Rhodopen Mts. in Bulgaria), to Au, W, Mo, Sb and other metals. Example deposits will be discussed below.

HIGH-LEVEL "GRANITIC" LACCOLITHS

Laccoliths are concordant injected masses that lifted their roofs by arching, so they appear not to be located immediately by major regional tectonic structures. A major factor in their emplacement has been the local lithostratigraphy, in which incompetent rocks were favourable for a laccolith generation. In the La Sal Mountains, Colorado, laccoliths rose on salt anticlines and spread mostly in the salt horizons.

Laccoliths typically developed in platformic cover sequences, and

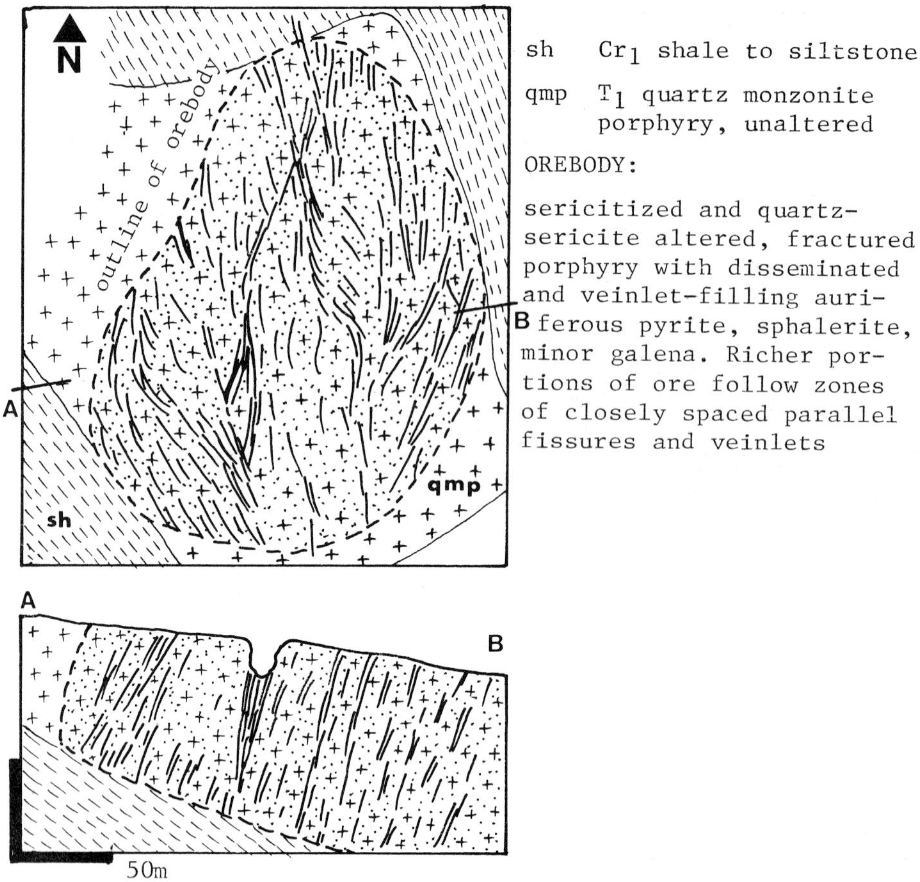

Fig. 28-14. Jessie gold mine, Breckenridge, Colorado. From LITHOTHEQUE, after Ransome in Lovering and Goddard (1950).

classical terrains of their distribution are the Colorado Plateau (e.g. Henry, Navajo, La Sal, Carrizo, La Plata, Rico, etc. Mountains; Hunt, 1956) and the Cordilleran Foreland in Montana. Compositionally, many laccoliths consist of alkaline or highly undersaturated rocks (Chapter 33), but GDG intrusions are widespread. Given the nature of their emplacement, laccoliths are a special form of epizonal intrusions and deserve a brief treatment.

In the Colorado Plateau province, Hunt (1956) and Hunt et al. (1953) recognized two major phases of laccolithic emplacement. In the earlier phase diorite, monzonite and syenite porphyries formed by physical injection of viscous magma of low temperature and a low content of volatiles (Fig. 28-15). In the later phase, compositionally similar rocks formed in part by assimilation or

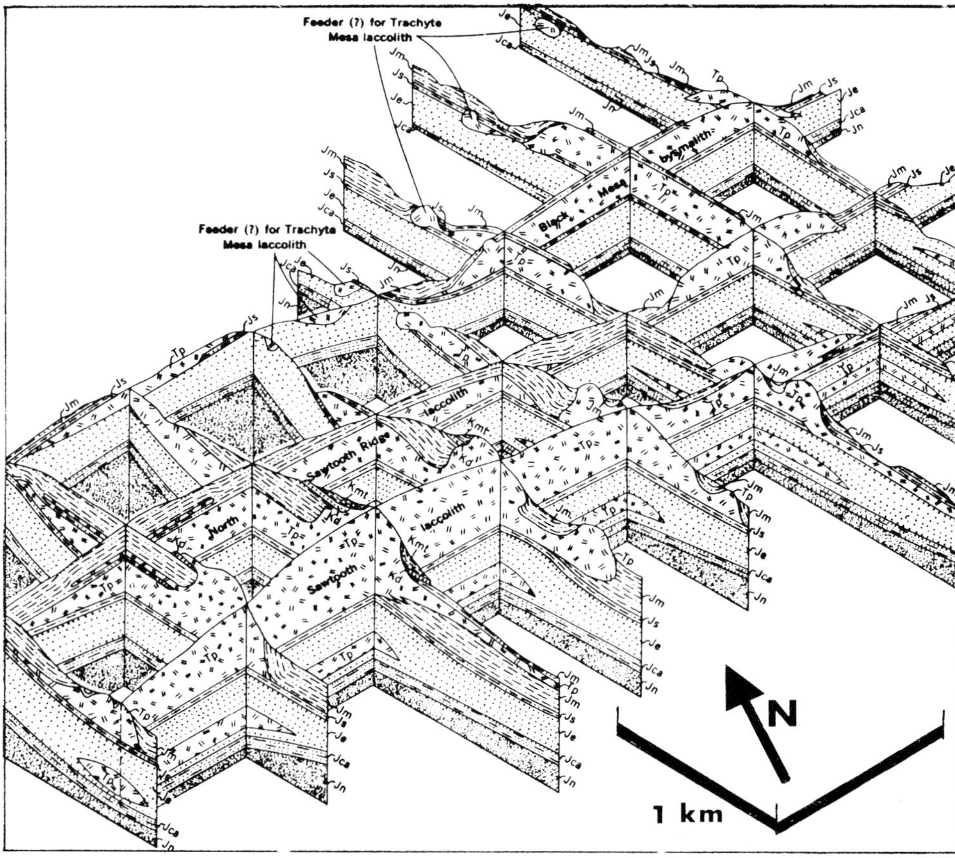

Fig. 28-15. Tertiary diorite porphyry laccoliths and bysmaliths (Tp) emplaced in Jurassic platformic sediments and shales, Sawtooth Ridge, Henry Mts., Utah. From Hunt et al. (1953).

replacement of the earlier intrusion. Their magmas, however, were more fluid, of higher temperature and with higher volatile content. The emplacement was usually forceful. The magmas solidified as holocrystalline stocks and dikes (Fig. 28-16). The small gold occurrences recorded by Hunt et al. (1953) from the Henry Mts. (Mt. Ellen, Mt. Pennell) are related to the late stage stocks, and carry auriferous pyrite in fissure veins and in shear zones along intrusive contacts. The succesful processing of the very low grade vein and stockwork gold deposits at several localities in the Cordilleran Foreland in Montana (Little Rocky Mts., Judith Mts., etc. described

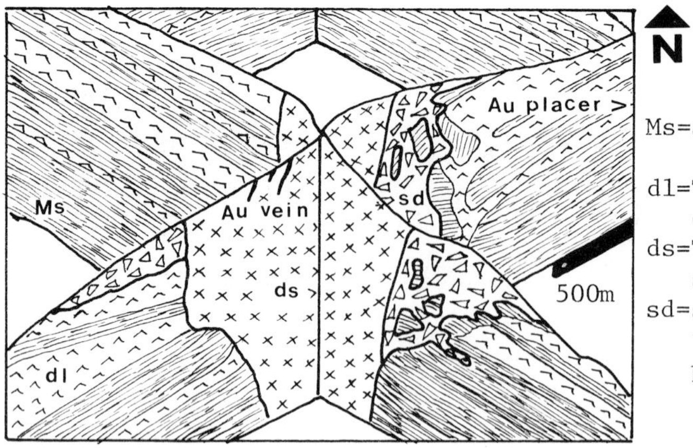

Ms=J and Cr sediments (sandstone, shale)
dl=T, diorite porphyry of the laccoliths
ds=T, diorite porphyry stock
sd=shattered endocontact of diorite porphyry stocks

Fig. 28-16. Mount Ellen stock (ds) and adjacent laccoliths (dl), Henry Mountains, Utah, showing diagrammatically the minor ore occurrences. Modified after Hunt et al. (1953).

later), indicates a possible potential of the laccolith/stock systems for gold.

The majority of the Colorado Plateau laccoliths are unmineralized, but there are minor exceptions. In the La Plata district N.W. of Durango, Colorado (Eckel et al., 1949; Fig. 28-17), the Phase 1 diorite porphyries are unmineralized but Phase 2 diorite, monzonite and syenite stocks are genetically associated with exocontact fissure veins of quartz, carbonate, galena, sphalerite, gold and Au tellurides (6.2 t Au, 57 t Ag, 322 t Pb). The Copper Hill showing contains disseminated and stockwork hematite, magnetite, chalcopyrite, quartz, ankerite and garnet in the endocontact of a syenite stock.

In the Iron Springs iron field, Utah (Mackin, 1968; min. 200 Mt Fe/47.5% Fe; Fig. 28-18), replacement exocontact bodies of magnetite and hematite in Jurassic limestones and minor endocontact magnetite veins, occur at or near the contact with a 19-24 m.y. old laccolithic quartz monzonite intrusion. The most unusual feature of the Iron Springs geology is the general concordancy of the intrusion with the sedimentary bedding as well as with Laramide bedding thrusts. The intrusive emplacement has probably been controlled by a graben structure. The replacement iron orebodies are tabular and lens-like, high in fluorine and immediately adjacent to (or hosted, as in the case of low-grade ores) by hornfelsed metasedimentary or intrusive breccias. The low-intensity contact alteration rarely reached the skarn stage, but the endocontact has been altered, possibly deuterically (K- and Na-feldspathization, biotitization, silicification). Mackin (1968) believed that the endocontact metasomatism was a sufficiently powerful process to release the metals from the parent intrusion and accumulate the iron along the contact.

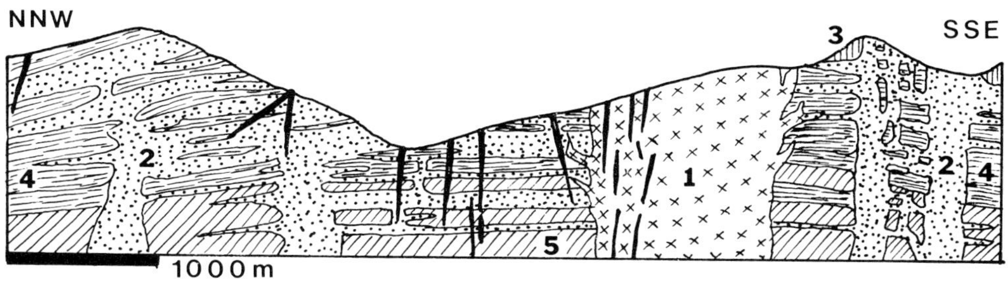

Fig. 28-17. La Plata Au,Ag,Pb,Zn district, S.W. Colorado, showing distribution of ore veins (black) in relation to Cr-T syenite stocks (1) and slightly earlier diorite-monzonite porphyry laccoliths (2). 3=J-Tr red shale, sandstone; 4=Pe red shale, sandstone; 5=Ps gray and red shale, sandstone, limestone. From LITHOTHEQUE, after Eckel et al. (1949).

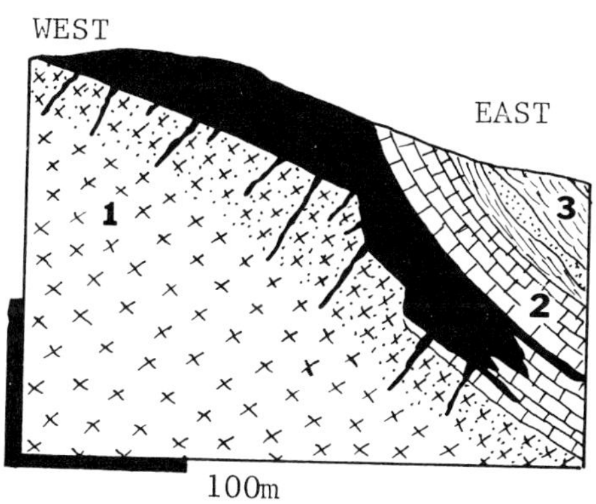

1. Ol_3 quartz monzonite laccolithic intrusion
2. J siltstone,argillite,limestone, thrust faulted, brecciated and altered in the ore zone
3. J_3 maroon and gray shale, interbedded with arkosic sandstone

BLACK: replacement magnetite-hematite body

Fig. 28-18. Iron Springs, Utah, replacement iron orebody related to laccolithic intrusion of quartz monzonite. From LITHOTHEQUE, based on data in Mackin (1968).

28.2.3. Mesozonal plutons

Mesozonal plutons, the most widespread and "typical" bodies of the "granites" (Buddington, 1959; Hutchison, 1977; Hughes, 1982), formed at depths of between 6-12 (or 4-16) km by forceful diapiric emplacement. They generally display a primary igneous foliation (alignment of feldspar tablets, biotite or hornblende crystals), and have a sharp to gradational boundary against the wallrocks. In the higher levels of a pluton, the wallrocks are usually regionally greenschist metamorphosed and over this metamorphism is superimposed a thermal aureole marked by biotite, andalusite, cordierite, etc. hornfelses. In their lower levels, mesozonal batholiths may grade into katazonal zones of granitization or, when they terminate at depth, be floored or flanked by "contact" migmatites. At considerable depths there was only a minor thermal contrast between the "granite" and the wallrocks, so thermal aureoles are inconspicuous or missing completely.

Mesozonal batholiths frequently contain contact metamorphosed roof pendants, septa and rafts of the country rocks, often partly assimilated. Pegmatites and aplites are moderately common. The K-feldspars are pure or perthitic stable polymorphs, orthoclase or microcline. The rocks range from gabbro to granite, and tonalite or granodiorite compositions are the most common. The majority of rocks are equigranular and medium crystalline, but coarse porphyritic varieties are also common. Large, syn- to post-kinematic elongated composite batholiths in cores of orogens are characteristic. There is no visible presence of comagmatic volcanics, but sometimes (as in the Sierra Nevada Batholith/Great Valley Sequence couple) sedimentary petrofacies faithfully record an early pyroclastic supply attributed to continental volcanism, gradually changing into an epiclastic detritus of "granitic" provenance derived from an unroofed pluton undergoing erosion. Many (perhaps most) mesozonal batholiths, however, probably never communicated with the surface.

Late Paleozoic and early Mesozoic plutons are in their "prime time" of exposure, but mesozonal "granites" range from Archean to middle Tertiary. Granitization and assimilation effects are frequently apparent, particularly when the granitic magma came into contact with compositionally contrasting wallrocks (e.g. mafic metavolcanics, carbonates). Some writers (e.g. Lee and van Loenen, 1971) even maintain that, locally, all the equivalents of the classic differentiation sequence formed through assimilation of chemically distinct and contrasting rocks, such as quartzite, shale, limestone and amphibolite.

Mesozonal granites sometimes generated and drove far-reaching mafic and acid "fronts". These fronts are believed to have caused metasomatic alteration of certain metamorphics so now, megascopically, they resemble plutonic rocks such as quartz diorite or tonalite. In the gneissic (pseudo)quartz diorite and tonalite near Orofino (Idaho batholith, Hietanen, 1962), sedimentary relic textures and structures are apparent. Fe,Mg,Ca,Al and Na were the principal elements introduced into, and Si, K removed from, the altered metamorphics. Metasomatic fronts are often associated with increased proportions of accessory ilmenite, zircon, apatite and allanite.

SIERRA NEVADA BATHOLITH, CALIFORNIA

The late Triassic to Cretaceous Sierra Nevada Batholith of California is 55-110 km wide. Its exposed length is 650 km, and it has been the most frequently studied mesozonal batholith in the New World (Hamilton and Myers, 1967; Bateman et al., 1963; Peck and Wones, 1980; Fig. 28-19). The batholith straddles the "eugeoclinal" and "miogeoclinal" megafacies boundary and occupies the axial part of a complex N.N.W.-trending synclinorium. It is composed of several nested plutons which frequently become progressively younger, less mafic and more potassic towards the centre, and are interpreted as products of fusion events in the lower crust. The intrusion took place in five major epochs within a time span of about 131 m.y.

The plutons comprise quartz-bearing intrusive rocks having a compositional range from quartz diorite to alaskite, but the batholith also includes small scattered masses of diorites and gabbros ("forerunners"), remnants of metasediments and metavolcanics and local ultramafic inclusions. The discrete plutons have either sharp, steep contacts, or are separated by septa of metamorphics, mafic igneous rocks or later aplitic dikes. The plutons along the western ("eugeoclinal") side are older, more mafic, and indicate greater involvement of a more primitive, possibly partly oceanic, basement in their origin. Assimilation of wallrocks along contacts is widespread, and the flow structures in plutons indicate that they have risen past their wallrocks. The plutonic intrusion was mostly forcible, although local contact breccias indicate stoping during late stages of intrusion. Hamilton and Myers (1967) assumed, on geophysical evidence, that the batholith is relatively thin (15-20 km) and floored by more mafic and heavier rocks of the lower crust.

The mineralization and metallogeny of the Sierra Nevada Batholith (Fig. 28-20) has recently been reviewed by Dodge and Bateman (1977) and by Albers (1981). It is quite representative of many of the mesozonal batholiths of the world. If one plots all the metallic deposits of California and Nevada onto a map, the Sierra Nevada "granitic" core will appear as a distinct minimum in the density and size of ore occurrences, surrounded by a heavily mineralized fringe. Of the sparse ore occurrences within the batholith , at least 90% is scheelite, all of which is in skarns hosted by metasedimentary roof pendants and rafts, in the granite exocontact. The only two mineralization styles of proven or potential importance, hosted by the plutonic rocks themselves, are the gold-quartz veins in the Grass Valley-Nevada City field and the Lights Creek porphyry copper deposit. It is worthy of note that both the above deposits occur in small satelite plutons outside the main batholithic mass. The small fracture infiltrations of U minerals (mainly autunite) in granodiorite explored

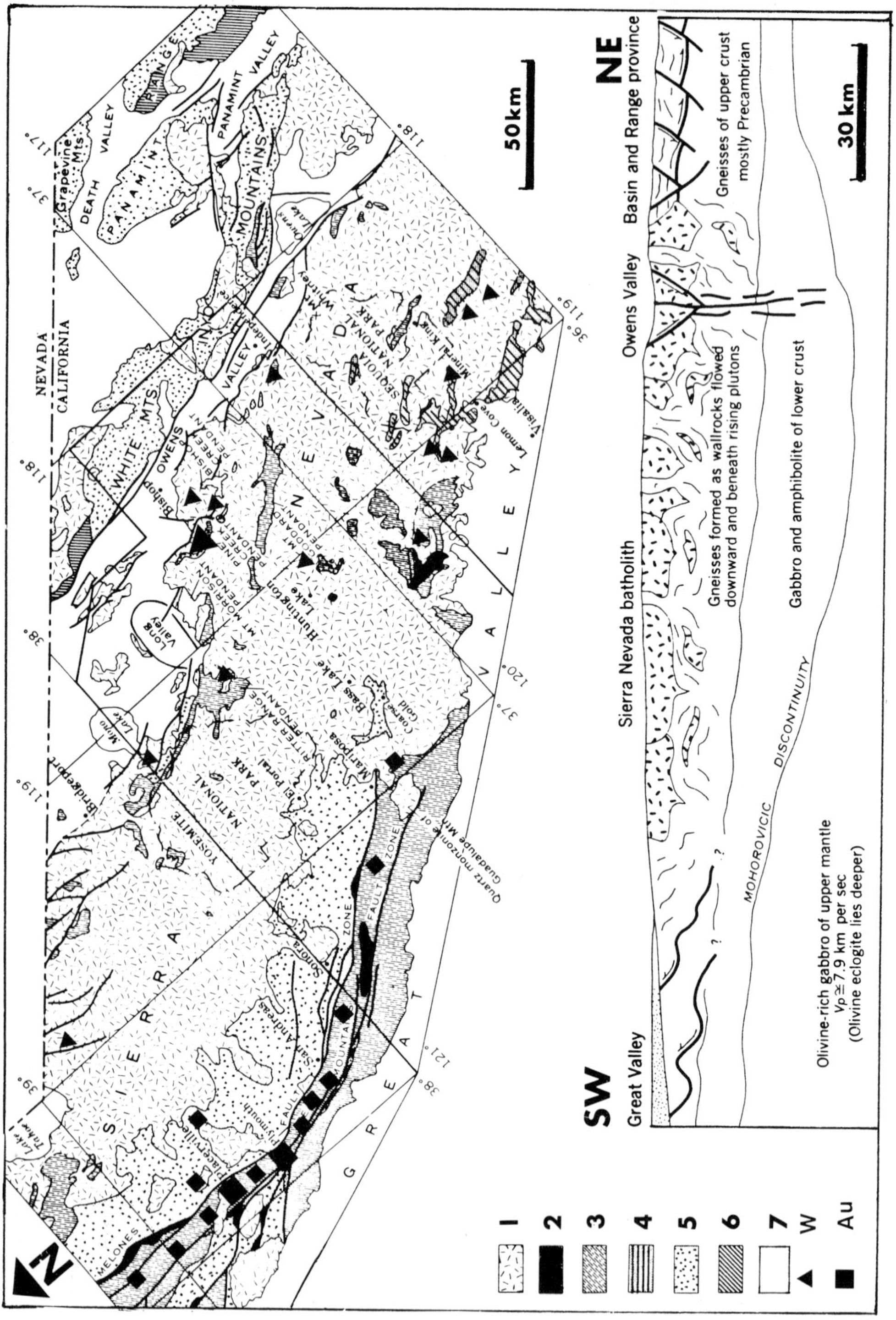

NEVADA
CALIFORNIA
DEATH VALLEY
PANAMINT VALLEY
PANAMINT MOUNTAINS
Grapevine Mts
WHITE MTS
OWENS VALLEY
INYO
Bishop
Independence
Mt Whitney
SEQUOIA NATIONAL PARK
Mineral King
Lemon Cove
Visalia
Long Valley
Mono Lake
Bridgeport
YOSEMITE NATIONAL PARK
El Portal
Mariposa
Bass Lake
Huntington Lake
Sonora
San Andreas
Placerville
Plymouth
Lake Tahoe
SIERRA NEVADA
GREAT VALLEY
MELONES FAULT ZONE
Quartz monzonite of Guadalupe Mtn
50 km
SW
NE
Great Valley
Sierra Nevada batholith
Owens Valley
Basin and Range province
Gneisses of upper crust mostly Precambrian
Gneisses formed as wallrocks flowed downward and beneath rising plutons
Gabbro and amphibolite of lower crust
MOHOROVICIC DISCONTINUITY
Olivine-rich gabbro of upper mantle
Vp ≥ 7.9 km per sec
(Olivine eclogite lies deeper)
30 km
1
2
3
4
5
6
7
W
Au

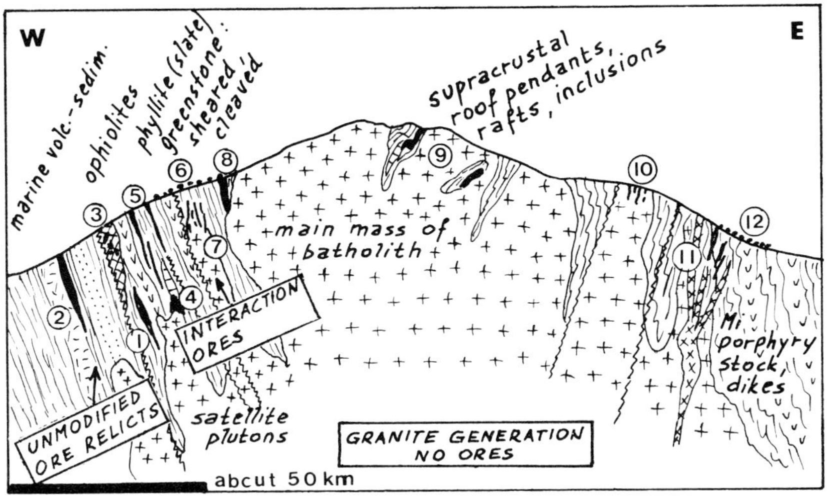

Fig. 28-20. Diagrammatic section of the Sierra Nevada Batholith, showing ore styles: 1=stratiform Mn in chert; 2=unmodified metavolcanics – hosted Cu-Zn massive sulphides; 3=podiform chromite; 4=magnetite skarns; 5=gold-quartz lodes; 6=gold placers; 7=porphyry Cu; 8=Ba silicate, barite; 9=scheelite skarns; 10=U oxide infiltrations (Kern Co.); 11=scheelite veins (Atolia); 12=scheelite placers. From LITHOTHEQUE.

in Kerr County have no practical importance.

The metallic deposits along the western fringe of the Batholith in the Sierra Nevada Foothills are located in late Paleozoic and Mesozoic greenschist-metamorphosed "eugeoclinal" metasediments (gray, green and black phyllites, metagraywackes), greenstones (metabasalts), serpentinites and mélanges. Most are gold, arsenopyrite, quartz, ankerite shear and fissure lodes and low-grade disseminations or stockworks in altered greenstones, in the Mother Lode system (already reviewed in Chapter 10). There are also several stratiform lenses of Mn silicates and oxides in jasper gangue, and occurrences of podiform chromite in serpentinite. The two latter styles even if, in places, they occur in the thermal aureole of the plutons, are clearly relics contemporary with the formation of their hostrocks.

◁━

Fig. 28-19. Sierra Nevada Batholith, California, geological map showing major scheelite (W), gold deposits, and crustal section. 1=J-Cr granitic rocks; 2=PZ-MZ serpentinite; 3-6=PZ-MZ metamorphics; 7=MZ-CZ flanking sediments. From Bateman et al. (1963) and Hamilton and Myers (1967), ore deposits added.

The gold deposits in supracrustals, on the other hand, are largely epigenetic, and show genetic or associational links with both the phyllites or greenstones and with "granites". Satellite plutons are widespread in or near the Mother Lode system so at least a portion of the gold probably accumulated as a consequence of interaction.

No recognizable metal zoning of the classical (Emmonsian) type is evident around the batholith (Albers, 1981). The only zoning is in the style and genetic interpretation of the ores, from the (sparse) (1) intra-plutonic ores outward into (2) granite-supracrustals interaction ores, to (3) granite (thermally, hydrothermally) modified relic ores, to (4) unmodified relic ores. The latter, outside the granite influence, are represented by the volcanics-hosted massive Fe,Cu,Zn sulphides in the Foothills Copper Belt (Chapter 12). Dodge and Bateman (1977) concluded that the source of most metals in the Batholith and its vicinity may have been in the country rocks adjacent to the deposits, rather than in end-stage hydrothermal solutions supplied by the plutons.

Geological and metallogenic conditions, similar to those around the Sierra Nevada Batholith, prevail in the plutons of the Klamath Mountains, except that the latter are more mafic (quartz diorite is the most common variety). The economic mineralization is almost entirely gold in exocontact lodes. Significant scheelite has not, so far, been discovered.

COAST BATHOLITH OF BRITISH COLUMBIA

This batholith (Douglas, ed., 1970) is, with its length of 1,800 km, one of the world's longest continuous intrusive bodies. It is a composite assembly of plutons ranging in age from upper Triassic to Miocene; the bulk is Jurassic to early Eocene. The batholith is surrounded by a high-grade gneiss complex from which probably at least a portion of the "granitic" magma was derived, and transitions from katazonal autochthonous migmatitic masses into mesozonal allochthonous intrusive plutons are widely apparent and have recently been described by Hutchison (1970). Unmetamorphosed and weakly metamorphosed rocks contemporary with the Coast plutonism and earlier occur on flanks of the pluton. They are most obvious in the east (the western margin of the batholith is, to a considerable degree, hidden under the Pacific ocean). Quartz diorite and granodiorite are the most common plutonic rocks, followed by diorite and quartz monzonite. Gabbro and granite are rare.

In terms of metallic mineralization, the Coast Batholith is the weakest mineralized belt in the Canadian Cordillera (with the exception of the Rocky Mountains thrust belt). This is particularly apparent in southern British Columbia north of the U.S. border, where both the adjacent facies belts (Insular Belt in the west, e.g. on Vancouver Island and Interior Belt in the east) are densely mineralized. The ore occurrences within the batholith fall into the following categories (Fig. 28-21): (1) relic concordant massive sulphide lenses: (a) in high-grade gneisses at the level of katazonal

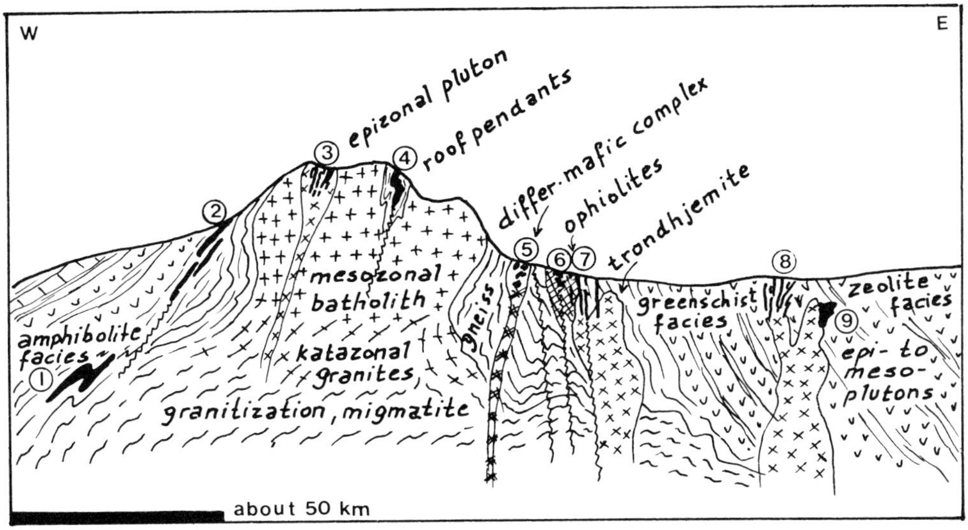

Fig. 28-21. Coastal Batholith of British Columbia, mineralization styles (diagrammatic): 1=relic massive sulphides (Ecstall); 2=Mother Lode-style Au; 3=stockwork Mo, wolframite vein; 4=massive sulphides (Britannia) in roof pendants, screens; 5=Ni-Cu sulphides; 6=disseminated cinnabar; 7=Au veins in shears and fissures (Bralorne); 8=porphyry Cu and aureole Au veins; 9=Cu skarn. From LITHOTHEQUE.

plutons, e.g. Ecstall and (b) in moderately regionally metamorphosed but contact hornfelsed metavolcanics or metasediments (Anyox, Granduc, Britannia). The Britannia deposit (Chapter 12) is particularly interesting, being hosted by a small roof pendant completely surrounded by granodiorite yet only slightly remobilized. (2) Magmatogene pyrrhotite, pentlandite, chalcopyrite in small gabbro-ultramafic stocks and dikes (Hope, Wellgreen, etc.; Chapter 9). (3) Endo- and exocontact quartz-gold lodes such as Bralorne (in soda "granite" intrusive into ophiolite and chert assemblage), Surf Inlet. (4) Endocontact wolframite, scheelite and minor Pb,Zn,Cu veins, e.g. Red Rose near Hazelton. (5) Mo-stockworks in altered hypabyssal intrusions as in the Alice Arm district, Adanac, Salal Creek. All the deposits of the latter category and probably also those in the group (4) are affiliated to epizonal granites of Miocene high-level centres, emplaced into older mesozonal plutons and surrounding metamorphics. In contrast to Sierra Nevada, scheelite skarns are virtually missing.

The above review reinforces the observation made earlier that the metallogenic balance of the deep mesozonal granite terrains is highly negative. They generate hardly any new ores while the deep erosion (of the order of 16 km) has removed much of the former wealth presumably present. As a consequence, preservation potential becomes an important indicator of exploration predictions and conformable massive sulphides that have one of the highest preservation potentials tend to survive until actually stoped away or removed by erosion. The mineralized epizonal "granites" are a random, superimposed feature. They may be suspected but are generally impossible to establish until radiometric ages become available. Once, however, one leaves the batholithic core and moves into the flanking assemblage of volcanic and sedimentary low-grade supracrustal metamorphics intruded by small meso- to epizonal satelite plutons, ore occurrences (porphyry Cu; Cu, magnetite, Zn-Pb skarns; stockwork Mo; Pb,Zn,Ag and Au veins, etc.) become plentiful.

CENTRAL BOHEMIAN PLUTON, W. CZECHOSLOVAKIA

The Central Bohemian Pluton (Svoboda, ed., 1966; Fig. 28-22), is a N.E.-elongated batholith, 150 km long and 30 km wide. It is composed of about 20 named intrusive bodies of variable composition and appearance. The Pluton occupies a former complex fault zone along the boundary between the high-grade metamorphosed block of Moldanubicum in the south and south-east, and the little metamorphosed late Proterozoic ("Algonkian") slate and spilite association in the north-west. It is interpreted as a late syn-kinematic to post-kinematic product of the late Paleozoic Variscan orogeny (middle Devonian to Permian; mostly lower Carboniferous).

Katazonal granites, interpreted by some as feldspathized metasomatites, occur along the south-eastern margin of the Pluton, represented mostly by porphyritic melanocratic biotite granite and syenodiorite (durbachite) and melanocratic biotite-pyroxene syenite. The central and north-western body of the Pluton has, mostly, truly mesozonal equigranular hornblende-biotite tonalites and granodiorites, lesser biotite quartz monzonites and small bodies of biotite granite. Minor bodies of gabbro and gabbrodiorite are known. There are abundant dikes of porphyries, aplites and lamprophyres and pegmatites are rare (except in the south-eastern katazonal domain). The intrusive rocks range from cataclastic to undisturbed (massive) and the large number of stone quarries provides excellent outcrops of fresh rocks. The area is easily accessible and has been studied in considerable detail. Virtually every outcrop has been recorded and documented which makes the Pluton and its aureole very suitable for metallogenic study. In particular, a wealth of data is available for a comparative study contrasting the "mineralogical" (minor) and "depositional" (economically significant) metallic occurrences.

The Pluton and its surroundings contain over a thousand individual metallic occurrences (Fig. 28-22), amongst which is one large Pb,Zn,Ag and U district (Příbram; compare also Chapter 17 and Fig. 17-8); one medium-size Au ore field (Jílové) and close to one hundred small mines, formerly producing Au, Au-Sb, Pb,Zn,Ag and U. There are virtually no Cu,Sn,and W accumulations. The total economic metal content is estimated to be of the order of 1.2 Mt Pb, 1 Mt Zn, 6 Tt

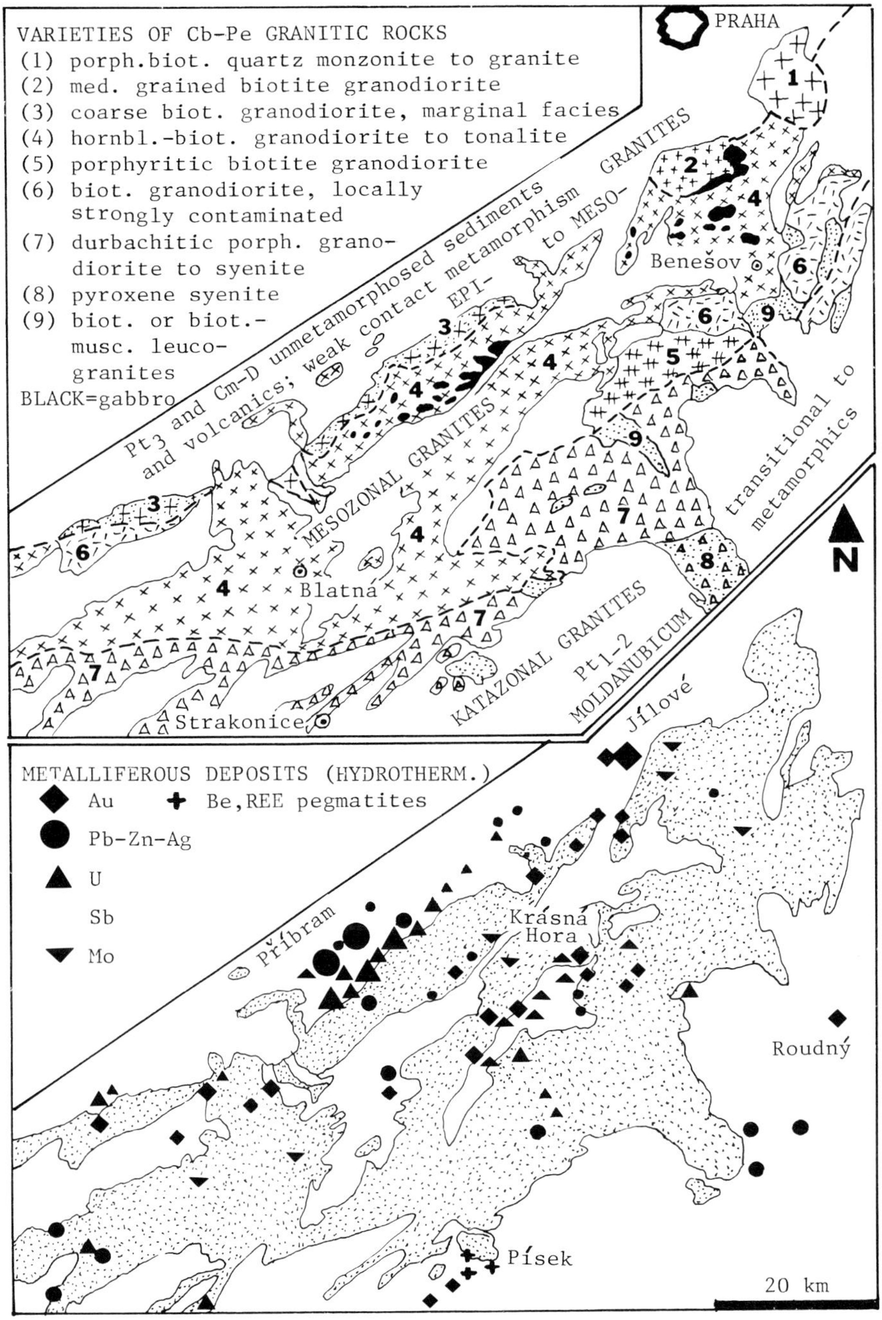

Fig. 28-22. Central Bohemian Pluton, Czechoslovakia, intrusive varieties and postmagmatic mineralizations. From LITHOTHEQUE.

Ag, 100 t Au (most in ancient placers), 40 Tt Sb and 90 Tt U. The metallogeny of the Central Bohemian Pluton has been discussed in numerous publications (Koutek, 1963; Sattran et al., 1966; Bernard and Klomínský, 1974) and much of the field observations assembled below is based on this writer's fieldwork in the period from 1957 to 1967.

Relic mineralization in the exocontact of the intrusive massifs (this includes several inliers, "islands" of supracrustals preserved within the Pluton) is extremely rare and represented by stratiform pyritic horizons in hornfelsed slates (Plíškovice) and probably Lahn-Dill "type" occurrences near Vranov. Small pegmatite masses formed at two separate depth levels. Those associated with katazonal granites near Písek and Sušice carry mineralogical occurrences of ilmenorutile, strüverite and monazite with minor volumes of REE,Th,Nb, etc. or Li-minerals and beryl. A single pegmatite dike near Skalsko is adjacent to a mesozonal granite and it contained a small orebody of scattered molybdenite.

Hydrothermal veins are the dominant style associated with the Pluton and they have been placed into the following major paragenetic associations: (1) quartz, albite, carbonate, pyrite, arsenopyrite, gold; e.g. Jílové, Libčice, Kasejovice, N. Knín; (2) quartz, siderite, galena, sphalerite, Pb,Zn sulphides, Pb,Cu,Ag sulphosalts; Příbram, Bohutín, Vrančice, Velhartice, etc.; (3) quartz, stibnite, gold or frequently stibnite-only forming stringers in fault gouge and mylonite; Krásná Hora and Milešov and (4) dolomite, calcite, pitchblende, lesser Co-Ni arsenides, native silver; Kamenná-Bytíz zone in the Příbram district, Újezdec.

Two age maxima for the veins formation have been distinguished: upper Devonian for the gold veins, and Permian for the polymetallic veins (and possibly also the uranium veins). All the vein deposits are within several hundred meters of igneous/supracrustal contacts, in both exo- and endocontacts. All display a strong structural control by open faults, and most veins are interchangeable with (or occur within) dikes of intrusive porphyries (diabases in Příbram, aplites near Bytíz, kersantites near Krásná Hora) filling fairly open, mylonite-filled or slickensided faults. This would indicate a relatively high level of mineralization. The main intrusive host masses are mesozonal, but the veins may have formed in the epizone. Bernard and Klomínský (1974) pointed out that there was a time gap of up to 100 m.y. between the vein and intrusive emplacements.

There is a characteristic correlation between the accumulated metals and the lithologic associations in the exocontact hosts. The Jílové gold-quartz veins are situated in cleaved slates and greenstones as well as in tonalite and they resemble the Mother Lode-Grass Valley (California) styles. The Pb,Zn,Ag and U veins show close spatial association with black slates, members of the slate-spilite association.

Major disseminated, stockwork, breccia, etc. deposits are absent, a fact which is easy to attribute to the depth of erosion. There are, however, widespread "mini-showings" (mineralogical occurrences that mimic, on small scale, the aspects of industrially important styles) corresponding to the porphyry (stockwork) Cu and Mo, apparent on joints in granites and tonalites exposed in stone quarries. There, in particular, the Q joints (open joints parallel with the stress vector) contain thin veinlets of quartz, K-feldspar, chalcopyrite, lesser molybdenite or quartz-molybdenite, fringed by a narrow pink feldspathized stripe. The presence of Cu minerals in the Permian "molasse" arkoses and mudrocks preserved in a small remnant near Český Brod (adjacent to the N.E. termination of the Pluton) indicates such a possibility.

The Pluton and its environs are transected by a system of relatively open faults, broad open zones of brecciation and mylonite or breccia-filled "reefs", some of which frame narrow grabens. The structures are believed to be ancient (pre-Variscan), repeatedly rejuvenated. Some contain fracture coatings of torbernite and autunite, grading downward into simple dolomite, calcite, pyrite, pitchblende, sometimes fluorite lens and stringer-like veins in the mylonite (e.g. Heřmaničky, Kovářov, Kvasejovice). There is a distinct wallrock "reddening" (bleaching and hematite pigmentation) and the origin is uncertain, but there is an almost exact correspondence with some of the U occurrences in the Massif Central, France, interpreted by Barbier (1974) as descendent infiltration veins.

THE MAIN RANGE BATHOLITH, MALAYSIA

The great Sundaland tin belt (Burma, Thailand, Malaya, Indonesia) can be subdivided into several parallel facies zones as well as several transversal segments. By far the most important is the western belt in the Malaya Segment, corresponding to the Main Range Batholith (MRB). This granite belt is interpreted by Hutchison (1977) as mesozonal in contrast to the Eastern Granite Belt which is epizonal. MRB is emplaced into penetratively deformed lower Paleozoic metasediments (phyllite, some marbles), and less deformed and metamorphosed upper Paleozoic slates. In the east, this belt is in fault contact with sediments of the Central Triassic-Jurassic Graben, and there are some ophiolite slices along the boundary lineament.

The majority of the MRB granites are Permian to Triassic, but Hosking (1973) also reported occurrences of upper Carboniferous and late Cretaceous to Tertiary granites. The bulk of the granites are coarse porphyritic biotite quartz monzonites to muscovite-biotite granites containing perthitic microcline. Contact metamorphic aureoles against the phyllites and slates are generally narrow to indistinct. It should be noted that fresh outcrops are extremely rare in the humid tropics and this could affect some geological conclusions.

As already noted, the plutonic mineralization is overwhelmingly dominated by tin, and the bulk of Sn comes from a variety of placers (eluvial, colluvial, alluvial, beach) although the primary cassiterite source is in granites and their aureole. The Kinta Valley tinfield (near Ipoh) and Kuala Lumpur tinfield, are among the richest in the world. Disregarding the secondary mineralization, the MRB and its aureole-hosted metallic ores can be subdivided into the following styles: (1) relic orebodies in the wallrocks. These are rare, represented by the small occurrences of disseminated and podiform chromite in ultramafics along the eastern tectonic contact of MRB. (2) "Magmatic" disseminated cassiterite in granites has been mentioned several times in the literature, but the occurrences examined by Hosking (1973) were all epigenetic, postmagmatic. (3) Pegmatites. These contributed cassiterite and columbite-tantalite to placers,and small regolithic occurrences (Gunong Kedah, Bakri) were mined in the deepest eroded Carboniferous granites. (4) Magnetite, magnetite-cassiterite, cassiterite (Ampang) and scheelite (Kramat Pulai) skarns and replacements at granite/carbonate contacts. (5) Hydrothermal veins and pipes of quartz-cassiterite and minor quartz-wolframite in granite endo- and exocontacts, mostly formed during the postmagmatic phase of the Triassic mesozonal granites, and associated with their apical cusps. Hosking (1973) believed that several "xenothermal" deposits in the Kinta Valley and elsewhere are near-surface mineralizations that would require a younger (possibly Tertiary) high-level intrusion.

28.2.4. Katazonal plutons

Generation of katazonal plutons is usually placed in the depth interval exceeding 12 or 16 km. Because most katazonal granites are autochthonous (in situ), this is also the zone of granite generation. There, the granites cannot easily be separated from the high-grade metamorphics so both rocks will be discussed in Chapter 29.

28.2.5. Petrogenesis and setting of GDG association

In the voluminous literature from the past 15 years (summarized in Hughes, 1982) the sites of magma generation that ultimately produce the GDG association, have been sought (1) above Benioff (subduction) zones, (2) under collisional fronts and (3) under intercontinental and occasionally under interoceanic taphrogenic (extensional) systems ("anorogenic continental terrains") e.g., in the initial stages of rifting, which, in turn, were attributed to the heat supplied by mantle plumes (hot spots). A site of type (1) is now favoured asa craddle of the bulk of the calc-alkaline "granitic" magmas (e.g. Hughes, 1982, treats the batholiths in orogenic belts under the heading "Igneous rocks above Benioff seismic zones"). The site of type (2) is generally ignored by the petrologists although frequently mentioned or reviewed in geotectonic and metallogenic compilations (Burke et al., 1976; Mitchell and Garson, 1976; Sawkins, 1984). Rocks developed along sites of type (3) are jointly reviewed in Chapters 30 and 33. As a metalliferous geologist sensitive to transitions and anomalies, I find the above division too simplistic. In particular, it accomodates poorly the transitional situations among the supra-Benioff belts and the intracontinental melting events as well as magmas generated as a consequence of a small-scale taphrogenesis along consuming plate margins.

Wyllie (1981) prepared a recent review on magma generation concurrent with subduction but, significantly, did not impose a numerical limit on the proximity of the magma-generating site to the Benioff zone. Such an explanation thus allows the magma melting to take place fairly high in the continental crust in the ultrametamorphic environment, which makes it possible to bypass the former awkward dogma that magma had to form directly along the subduction plane and then progress, somehow, to the near-surface region through a slab of mantle and crust 150-700 km thick. Wyllie (1981) listed three "end-member" variations of magma generation by partial fusion above subduction zones roofed by continental margins or mature island arcs: (1) fusion of mantle peridotite modified by siliceous melts or hydrothermal fluids produces a magma ranging in composition from quartz-normative basalt to andesite; (2) fusion of oceanic crust produces intermediate magmas (andesites) and (3) fusion of a deep-seated continental crust approximately composed of granite and amphibolite equivalents, produces tonalitic magmas (or a granitic/rhyolitic magma when the crust is entirely "granitic"). It should, however, be realized that the magma generated by partial melting undergoes further differentiation and modification within the continental crust, so one can finally obtain petrographically identical differentiates (e.g. a granodiorite) out of the three parental magma series (and their mixes).

It seems logical that provenance has some bearing on the trace metal distribution in the parental magma and this, in turn, influences the metallogeny. It follows from the earlier review of mineralization styles associated with mesozonal and epizonal granites that petrographically comparable rock types (e.g. a quartz monzonite) could be associated with a very different suite of ore minerals (e.g. Cu,Mo as in Butte; Mo in Smithers; Au in the Colorado Mineral Belt; Sn-W as in N.E. Queensland). Disregarding, for the time being, the possibility of a proximal metal extraction from the country rocks, it appears natural that the site of magma derivation (and by implication the geographic location and geotectonic setting of the magmatic system) influenced the metal selection and metallogeny. This possibility was not considered in the classical (1930s to 1950s) models of postmagmatic metallogeny associated with granites and it resulted in several over-generalizations e.g. in interpretation of the metal zoning patterns. The latter subject is reviewed more fully in Laznicka (1985c).

PETROMETALLOGENETIC SERIES; I AND S GRANITES; ILMENITE AND MAGNETITE GRANITES

Abdullaev (1964), Sattran et al. (1970) and others applied the concept of petrometallogenic series of intrusive rocks to explain and predict the affiliated mineralization. The concept relied on the combination of magma series (calc-alkaline, K-alkaline, Na-alkaline) and basicity of the differentiate. For example, within the GDG association the strongest contrast was found between the "intermediate" (dioritic) series affiliated with Au deposits, and "granitic" series associated with tin deposits. Out of the regional examples discussed earlier, the Sierra Nevada and Central Bohemian plutons are members of the "intermediate series" with mostly Au deposits. The N.E. Queensland and western Malaya massifs belong into the "granite" series with tin deposits.

Later on, Chappell and White (1974) working on a different premise, distinguished two contrasting "types" of granitic rocks (I and S) which, in terms of affiliated mineralizations, broadly correlate with the two petrometallogenic series distinguished above (I granites carry mainly Au, S granites Sn). The above writers assumed that a granitoid magma is derived by partial melting and consists of a mixture of granitic liquid and a residuum (restite). The restite are inclusions as well as much of the crystal content of plutonic rocks, while the former granitic liquid is now represented by the interstitial phases.

The I granites are believed to be derived from igneous, the S granites from (meta)sedimentary precursors (essentially shales). Both granite varieties have distinct linear geochemical variation trends when plotted on the Harker diagram. Despite the elegance and present popularity of the model, there are many areas where it does not work (e.g. along continental margins where volcanogenic graywackes were partially melted as in the Coast Batholith of British Columbia or in the Precambrian shields). Such sediment-derived granites usually have an I signature. Most recently, several authors added "A" (for "anorogenic") and "M" (sodic granites of immature island arcs) granites.

The latest contribution to the classification of the granitoid series from metallogenic viewpoint came from Ishihara (1981). This author distinguished "magnetite series" and "ilmenite series" of granitic rocks, based on the accessory content of either mineral. The magnetite series has a positive $\delta^{34}S$ value and a low $\delta^{18}O$ value, is depleted in lithophile elements and is commonly associated with sulphide mineralization. It more or less correlates with the I granite. The ilmenite series contains less than 0.1% magnetite, has a low magnetic susceptibility, negative $\delta^{34}S$ values, high $\delta^{18}O$ value, and enrichment in lithophile elements. It is commonly associated with Sn,W,Be and fluorite, and includes equivalents of both I and S granites. Ishihara (1981) believes that the two series of granitoids resulted from the prevalence of different oxygen fugacities during evolution of the granitic magmas. The dissociation of water in hydrous magmas was the main oxidizing agent in the magnetite series, and incorporation of crustal carbon was the most essential reducing medium for the ilmenite series magmas.

So now we have at least three methods for the interpretation of the granite metallogeny using rock geochemistry. None is perfect. There are numerous exceptions, but at least a rough correlation between rock geochemistry and prevalent metallic mineralization is now available to justify the earlier empirical observations. When both I and S granites form facies belts along a continental margin, I tends to be more internal (closer to the paleo-ocean), S more external (closer to the paleo-craton). The former have relatively low Rb, Rb/Sr, Sr 87/86 and high K/Rb and Sr signatures. The latter have higher Rb, Rb/Sr, Sr 87/86 and low K/Rb and Sr values and tend to be strongly peraluminous.

A down-to-earth exploration geologist is cautioned against placing excessive faith in the above generalizations, because a small metasedimentary microcontinent accreted along with dominant metavolcanics to a continental margin, may produce a "S" enclave capable of, for example, accumulating tin "where it should not have been". Faults, thrusts, "suspect terrains", multiphase terrains, etc. can do the same and have been found to do so.

28.2.6. The magmatic/metasomatic controversy and hydrothermal-metasomatic "granites"

The high point of the classical controversy of the 1940s-1950s regarding the derivation of granitic magmas, has now been more or less settled with the realization that the magma first formed by partial melting of a solid (that is, anatexis when in the continental crust), after which the melt travelled into the higher levels of the crust in the form of a "granitic magma". The neo-controversy partially outlined earlier is concerned with the site of the melting. Marmo (1971) gave a good review but old problems still remain although often under different names. One problem of fundamental importance to metalliferous geologists is the possibility of hydrothermal origin of "granites" by means of metasomatic fronts. The concept of granitization or de-granitization as summarized by Marmo (1971, p.72,73) is as follows: in the upper levels of the SIAL, extraction of potassium from sediments undergoing metamorphism takes place in loaded areas (granodioritization); K and Si move to areas of low energy in anticlinoria, causing granitization. When the granitized segments are exposed to further load due to the advancing orogeny, potassium removal will take place and potassium will move to areas with a still lower free energy. The formation of igneous-looking metasomatites interpreted by Hietanen (1962) was reviewed earlier.

Approached from a different angle, the formation of igneous-looking rocks virtually identical compositionally and texturally with orthomagmatites, by a high-level hydrothermal metasomatism and mineral crystallization in open spaces, is now a well-documented reality. Potash feldspar and biotite neoformation in porphyry copper systems tend to produce hard, fresh-looking rocks, visually often unrecognizable from magmatic "granites", and often containing disseminated sulphides. They are a far cry from the popular image of an altered wallrock as being a spotty, dirty, disintegrating tortured rock. In the North American literature the rocks produced by alteration (alterites) do not have a status of "independent rocks". In the Russian literature they are called "metasomatites" (e.g. K-feldspathite, biotitite, secondary quartzite). Some old descriptions of disseminated chalcopyrite in "unaltered granodiorites" were later identified as K-alterites.

An equivalent situation, though more obscure, is a part of the "magmatic cassiterite" controversy. Is the cassiterite sometimes found in a hard, fresh-looking, feldspathic rock (granite, aplite, pegmatite) really a product of magmatic crystallization, or has it been introduced in a highly mobile fluid from outside, deposited by replacement, and the associated alteration feldspars erroneously interpreted as melt crystallizates ? The second interpretation is now prevalent. Apogranites (Smirnov, ed., 1968; hydrothermal metasomatites in apexes of granite cupolas) include albitites, quartz-albitites, microclinites, and other rocks. These hard, fresh-looking feldspathic rocks often carry disseminated columbite-tantalite, cassiterite, Be-minerals and grade into greisens. Feldspathization (particularly albitization) may obscure petrochemical groups of intrusive rocks , in particular place an altered calc-alkaline rock into the alkaline family.

Many albite "syenites": pink, often miarolitic syenite-looking rocks within granite batholiths as well as outside them, are hydrothermal metasomatites. Laznicka and Edwards (1979) interpreted "syenites" associated with gabbro and diorite dikes at Dolores Creek, Yukon, as albitites formed in dilations and in porous breccias by metasomatic albitization and open space filling. They contained minor disseminated chalcopyrite, and resembled "porphyry coppers". Many pegmatite-looking veins are in fact hydrothermal feldspathites.

28.3. INTRODUCTION TO GDG METALLOGENY

28.3.1. General

Average trace metal contents in granitic rocks are available from the literature. Given the wide basicity range and the various magmatic series involved, such contents are inaccurate in detail. They are usually used for approximating the average composition of the upper continental crust and, contrasted with the trace metals in basalts, give some appreciation of the evolutionary trends. Average trace metal data for narrower rock categories are in Table 5-3.

There is an enormous number of factors governing "granite" metallogeny, most of them poorly understood. Some are related to the "granite" petrogenesis and petrochemistry as discussed earlier and others are to be sought primarily in the exocontact lithologic association, in structure, thermal and hydrological regimes during cooling, etc. The classical, all-embracing model of "granite"-affiliated postmagmatic hydrothermal metallogeny as presented in text and reference books from the 1930-1950 period (Lindgren, 1933; Bateman, 1951; Schneiderhöhn, 1955) is now largely outdated, although several restricted subjects still remain valid.

A substitute overall granite metallogeny model is not yet available because, to a considerable degree, it depends on the conclusion of the problem of GDG derivation and petrogenesis which, despite the considerable progress made in the past fifteen years, is still not yet here. In the meantime, modern models have been developed for several styles and categories of ore deposits (e.g. porphyry coppers, stockwork molybdenums, skarns, granite cupola tin, etc.) and these are briefly reviewed later. There is, however, little cross-pollenation among these models and large gaps abound. "Fringe" ore styles lack modern coverage altogether.

Modern interpretation of hydrothermal systems, the core of the "granite" metallogeny, is now a highly sophisticated and specialized field of research. The reader is referred to the summaries and reviews available in Stanton, 1972; Holland, 1972; Barnes, ed., 1967 and 1979; Cathles, 1981 and others. In the following paragraphs, several problems directly influencing the field occurrence of the GDG-affiliated ores, will be touched upon briefly.

28.3.2. Hydrothermal systems

CONVECTIVE SYSTEMS

At present, the best documented hydrothermal system is a convective system utilizing meteoric water, seawater or connate water (that is, non-magmatic, exo-granitic fluids), driven by a cooling intrusion (Cathles, 1981). The active periods of such systems are considered short in geological terms, ten thousand to hundred thousands years. Much depends on permeability. Longer active times can be expected if the top of a geothermal system is impermeable. Hydrothermal convection usually takes place in the above-intrusion zone largely in supracrustal exocontact rocks, and the area affected gradually shrinks and drops to involve, subsequently, the top of the cooling magmatic body itself. Recognition of paleo-convective systems facilitates the interpretation of metal sources to ore-bearing fluids and ultimately to metallic deposits. Such sources tend to be relatively proximal and often in the same unit that hosts the orebodies.

Certain convective systems could have been driven by the heat from radioactive decay. Such heat sources could have been unusually persistent, and convection initiated within and in the vicinity of an U-enriched pluton every time fracturing took place (Fehn et al., 1978). This model seems to be supported by empirical evidence of the frequent pitchblende deposits formed in relatively open mylonite or breccia-filled fractures and faults, the filling of which considerably postdated the host granite emplacement and cooling (e.g. in the Hesperian and Central Massifs, Spain and France; Bohemian Massif). Alternative genetic explanations , however, are available for the same deposits.

MAGMATIC VOLATILES

The mass of magmatic volatiles vented by an intrusion is much less than the mass of the outside water likely to circulate convectively in response to such intrusion (Cathles, 1981). The volatiles stream upward and tend to accumulate in granite cupolas or in small elevated stocks, under the confinement of impervious strata. This is facilitated by a permissive intrusive emplacement in which the roof is not shattered. In settings having a shattered or permeable roof, venting and volatile escape may take place. Explosive outbursts of volatiles, however, may produce mineralized breccia pipes. Particularly complex but effective mineralization systems may develop when magmatic volatiles interact with convecting groundwaters, as interpreted for several porphyry copper systems.

Once a hydrothermal system is in operation, ore precipitation is governed by the physical and chemical principles many of which had already been satisfactorily interpreted in the classical period (e.g. the "second boiling"), while the rest benefited from the recent advances in fluid inclusion and stable isotope studies. To form an orebody, however, an actively convecting system requires a metal source. Many recent theoretical studies assumed "normal" (=close to clarke values) metal sources as the basis for calculation. This may, under favourable conditions, suffice for ore deposition, but the system has to operate at top efficiency, something that is fairly rare in nature. Empirical observations and probabilistic deductions based on field data, however, indicate that in most cases ore formation probably took place in systems interacting with "better than normal" metal sources, so interpretation of and search for such sources are an important ingredient of exploration strategy.

28.3.3. Metal sources to ore-bearing fluids

In the classical model of the 1930-1950s the "granitic magma" (usually interpreted as being a differentiate of basaltic magma) was automatically believed to carry along the entire complement of ore metals, sequentially released and deposited in the course of cooling of the intrusion. A classical (Emmonsian) metal zoning pattern around and above the intrusion (from centre outward: Sn-W-Au-Cu-Pb,Zn-Ag-Sb-Hg) resulted (Bateman, 1951; Fig. 28-23A). Concurrently and also later, alternative models have been introduced, mostly in response to the changing interpretation of granite origins. The "granitization model" in which the granitic magma obtained its components including metals from crustal sources during anatexis

(Fig. 28-23B) became the chief alternative of the 1950s. By now, many more alternatives have been proposed (Fig. 28-23C). The subject of metal sources to ore deposits and zoning has recently been reviewed by Laznicka (1985c,e).

Very briefly, the ore metals may enter the "granitic" magma (become dissolved in it or carried as solid phase): (a) at the site of the dominant magma generation (within the mantle, along and above Benioff zone; within the continental crust); (b) above the magma passage that is chiefly in the lower and upper continental crust or (c) in magma reservoirs within the upper continental crust and at sites of emplacement. These metals are considered to be relatively mobile within the developing magma, and capable of accumulation during the cooling history at sites located both inside (endocontact) and outside (exocontact) of the intrusion. With respect to the intrusion, these are active metal sources.

Ore metals, however, need not actually enter the magma and pass through the stage of dissolution (dispersion) in the magma first, followed by release, migration and accumulation in orebodies. The metals could be present in the rocks surrounding the generator intrusion, leached, transported and accumulated by convecting hydrotherms at sites asociated with a rapid drop in solubility. This has been discussed earlier. Alternatively, solidified and fractured "granite" itself may be the site of hydrothermal convection and metal leaching. Both the above alternatives have been termed "passive metal sources" in Laznicka (1985e).

Recognition of the difference between active and passive metal sources is a useful ingredient in speculations on mineralization potential of a "granitic" intrusion and ore prediction. The distant metal sources (in the mantle, in the oceanic crust and other rocks melted along subduction zone, etc.) are impossible to observe or to deduce from field observation, so they are highly hypothetical and changing with fashion and the progress of science. Their interpretation is made within the limits of geochemical and petrochemical constraints (e.g. stable isotopes, trace metals such as the REE), and the degree of controversy is high. Academic debates about the metal sources at depths of hundreds of kilometres are of dubious merit to a prospector.

Numerous down-to-earth speculations involving milieus much closer to the present sites of ore deposits and supported, as much as possible by actual field data can, on the other hand, aid considerably in exploration. Several examples of such speculations appear later in this chapter, in sections dealing with specific metals. The problem of trace metal and rock-forming mineral compatibility is reviewed briefly here as an example of reasoning that is of practical help.

During magmatic differentiation trace base, rare and precious metals can behave as compatible or incompatible in regard to the major rock-forming minerals (Stanton, 1972). The compatible elements are removed from the melt by crystallizing major minerals, while the incompatible elements remain in the rest liquid and ultimately accumulate in the intergranular spaces, or they migrate with

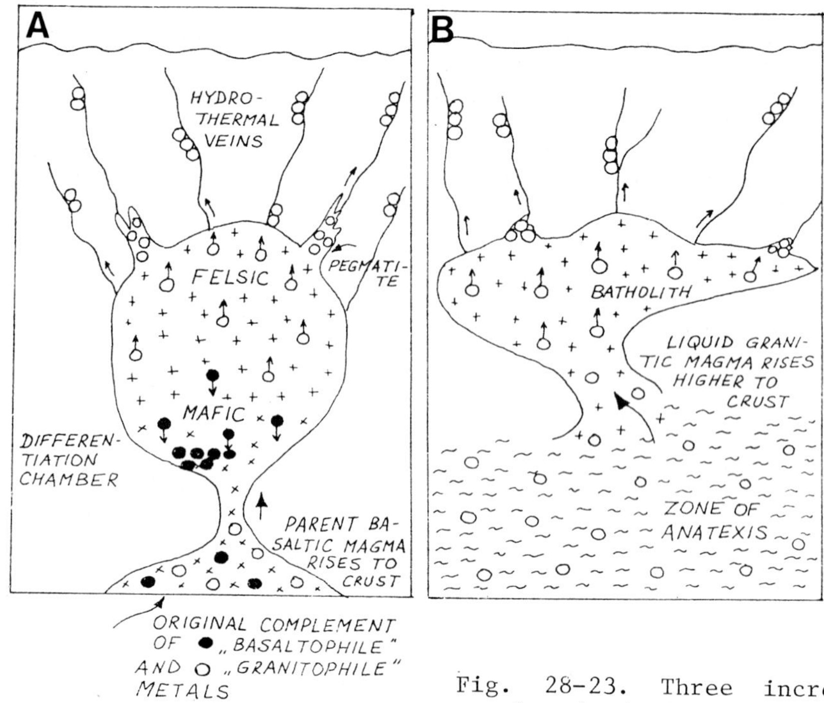

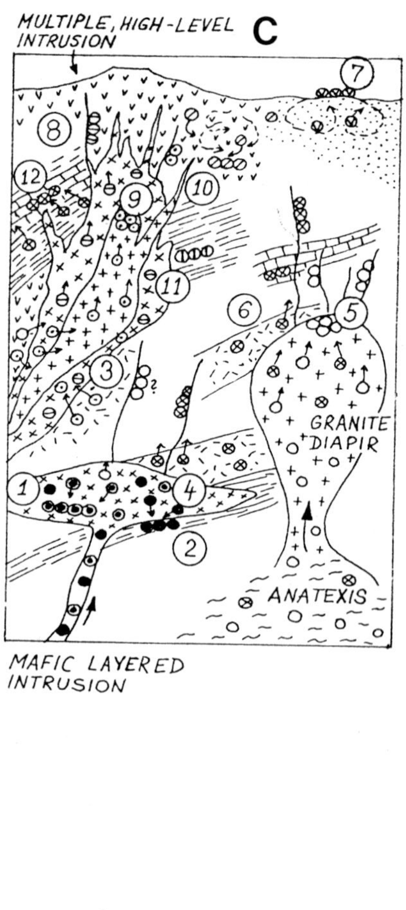

Fig. 28-23. Three increasingly more complex (and also more recent) models of "granite" affiliated mineralizations. A: Classical, closed system model of differentiation and crystallization of a "juvenile" mafic parent magma. B: Crustal granitization or anatexis, ascent of granitic magma, followed by its fractionation and crystallization. C: Selected contemporary alternatives in which a wide range of interactions of magmatic rocks with their surroundings is stressed.
EXPLANATIONS: Scattered small circles indicate dispersed trace metals. Small circles with arrows indicate directions of trace metals movements. Groups of small circles indicate orebodies; not to scale.
(continued on right)

postmagmatic fluids. It is known that various rock-forming minerals preferentially accumulate certain trace metals. Olivine and Mg-pyroxenes accumulate Ni,Cr,Hg; pyroxenes and amphiboles accumulate Cu,Zn; biotite accumulates Sn,U,Th; magnetite accumulates Cu,Zn; K-feldspars accumulate Pb. The variable proportions of these minerals in rocks and their preservation or destruction, can thus have a substantial influence on the ore-generating potential of an intrusion. The general rule is that in a progressive magmatic differentiation process reaching completion, the metals that remain most incompatible throughout the entire process have the greatest chance of accumulating in magmatic or post-magmatic deposits so they constitute an important active metal source.

Eilenberg and Carr (1981) studied the anomalously copper-rich andesitic and basaltic volcanics of the recently active San Miguel Volcano (El Salvador), and explained the high Cu content in the volcanics as being due to the lack of early augite and magnetite in the rocks. They reasoned that, since these minerals tend to

---

Fig. 28-23 explanation (continued)

INDEXES USED IN SECTION C: 1=cumulus settling (e.g. chromite); 2=immiscible liquid, settling of droplets, squeeze-up, down or laterally beyond the limit of intrusion; also possible contact sulphurization and precipitation of non-sulphide trace Ni content; 3=gaseous or hydrothermal transfer of fractionated metals (Sn) to veins; 4=magmatic heating expels high trace metals (with their carriers) in surrounding rocks to veins (e.g. Cobalt Ag,Co; Ontario); 5=mobile metals and volatiles gather in apical portions of a granitic diapir; 6=granite heat and hydrothermal activity triggers metal remobilization from metalliferous contact rocks; 7=hydrotherms (heated groundwater) circulate in a metal- enriched aquifer dissolving, concentrating, reprecipitating metals; 8=trace metals which arrived as a component of the original magma, ascend and fill veins; 9=trace metals picked-up from intruded (or replaced) crustal rocks, enriched in an older intrusive phase; from there they are overtaken by a younger intrusive phase, further concentrated and deposited as ores; 10=trace metal enriched in porous overlying sediments was picked-up by circulating heated groundwaters, carried downward, and redeposited stratigraphically below the source; 11=an existing orebody in intruded wallrocks was modified by an intrusion; 12=geochemically enriched metal in argillite was remobilized by magmatic heat or solutions and redeposited in adjacent, reactive carbonate (e.g. as scheelite skarns).
Black symbols: "basaltophile" metals; white symbols: "granitophile" metals.

In the complex diagram (C) several internal patterns have been added to both categories of metal symbols, and these serve no purpose other than to help the reader to trace the hypothetical metal movement from the source to the site of deposition. From Laznicka (1985e).

accumulate copper, their lack prevented the early depletion of the magma in trace Cu and allowed Cu to reach its maximum trace content in the relatively dispersed form easily leacheable, not only by high pressure/temperature hydrotherms, but even by rainwater (compare Chapter 5). Eilenberg and Carr (1981) concluded that porphyry copper deposits can form preferentially in the roots of a volcano whose mafic to intermediate lavas are relatively poor in magnetite and augite, so that the abundance of these two minerals could be used as one of the exploration leads for porphyry Cu source areas.

The above reasoning, although probably correct in regard to the surficially leached copper, can be reversed (Fig. 28-24). If magnetite and augite do exist in abundance and consequently the content of incompatible (dispersed) copper is low, the rocks may still be high in their trace metal content but it would be a compatible Cu stored in magnetite and augite. Indeed, the Cu-anomalous (around 200 ppm Cu) andesitic and basaltic associations hosting some porphyry coppers (e.g. the Nicola, Nikolai and Karmutsen Groups, North American Cordillera; the Farellones Formation of Chile) are conspicuous for having abundant porphyritic augite andesites with high magnetite contents. These rocks could have provided Cu to the porphyry coppers, but they would have been passive sources, requiring a thermodynamically intensive pyroxene and magnetite-destructive process to release the Cu and set it in motion. Intensive hydrothermal metasomatism around and under porphyry copper systems attests to a possible existence of such a process. Similar reasoning can be applied to tin deposits associated with young, altered muscovite-biotite granites usually superimposed on older biotite quartz monzonites possibly storing Sn in biotite. In nature, most metalliferous magmatic systems do not become productive and parental to a distinct suite of associated metals suddenly, but as a result of a lenghty, multistage development history progressing in the direction of "increased favourability".

28.3.4. The rocks above and outside of the "granitic" intrusion

The rocks outside any "granitic" intrusion under consideration can be (1) older than the "granite" and have intrusive contacts with it. In high-level plutonic systems such rocks show a distinct thermal metamorphism, which can be indistinct in the more abyssal emplacement levels. Alternatively, they may be (2) older or younger than the "granite", having a mutual tectonic contact or (3) younger than the "granite", having a non-conformable contact and deposited transgressively over the eroded granite, and composed either of the granite-derived detritus, or independent of it.

In terrains undeformed and unmetamorphosed after the "granite" emplacement, the "granite"/outside rock relationship is usually easy to recognize. In repeatedly deformed and metamorphosed terrains this could cause some problems. In this short section, we will be concerned with the type (1) relationship. Virtually all the known

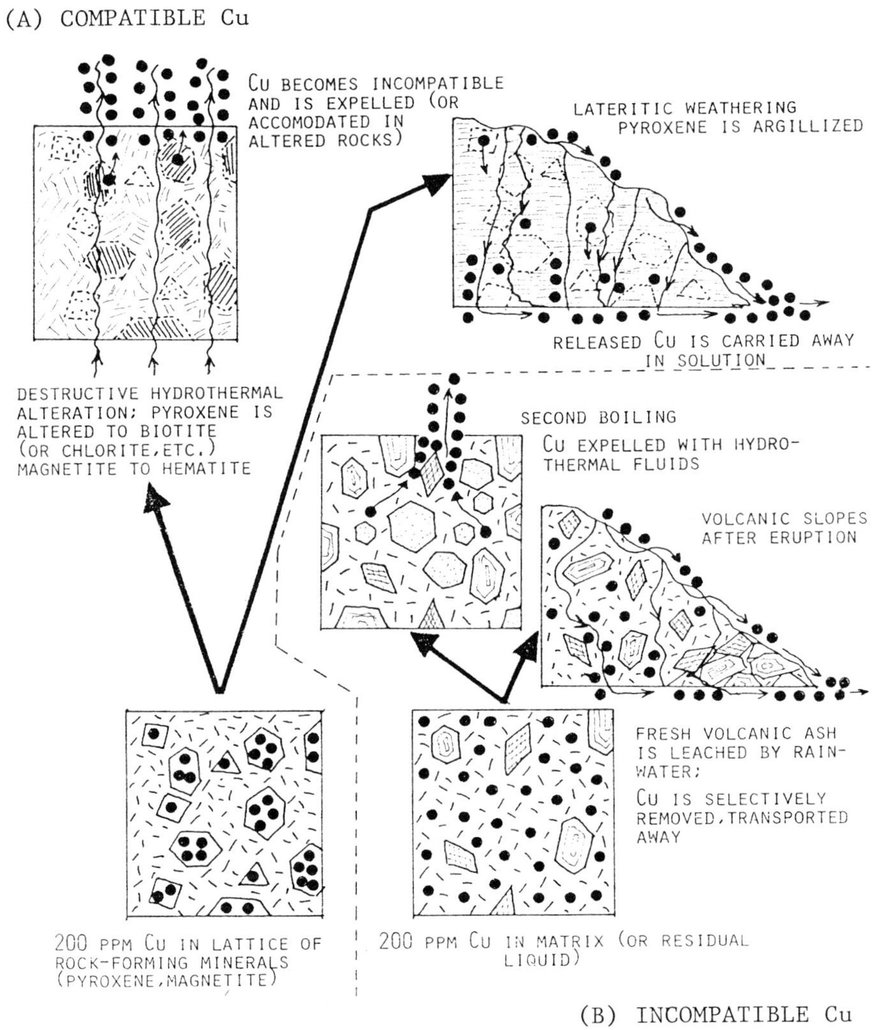

Fig. 28-24. The contrasting behaviour of compatible (in regard to the common rock-forming minerals during the magmatic crystallization) and incompatible copper in the post-magmatic history of a basalt or andesite. Black dots denote schematic representation of the trace copper, arrows indicate copper movement and the direction of concentration. From Laznicka (1985e).

pre-Quaternary lithologic associations can be intruded by "granites", including older generations of "granites" themselves. The contacts range from knife-sharp ones to gradational fronts of hybrid rocks on both sides of the contact (but more in the exocontact).

In regard to orebodies, the exocontact rocks have several functions that have to be evaluated carefully when interpreting granite aureoles. (a) The exocontact rocks host orebodies and some 40% of Cu orebodies, 30% Sn orebodies, 20-40% Mo orebodies, 90% Pb-Zn orebodies and over 95% of scheelite orebodies, are in the exocontact. There, they fill open spaces or replace solid rocks. The orebodies have a significant structural as well as lithological control, and the variety of mineralized structures has been adequately described in the classical literature (e.g. Bateman, 1950; Newhouse, ed., 1942). When orebodies (e.g. veins) fill open spaces, brittle rocks tend to be preferentially mineralized whereas the more ductile ones remain barren.

(b) Exocontact rocks exert an important physical influence on ore deposition, in particular by providing impervious screens. When a flat-lying or gentle dipping unit of a ductile shale tops a brittle fractured quartzite, replaceable marble, etc., ascending mineralizing solutions are trapped under the impervious horizon and often forced to precipitate their load and form orebodies. Such orebodies often take the form of peneconcordant lenses, frequently confused with stratiform orebodies.

(c) Chemical and physicochemical influences that cause metals to precipitate are often exerted by a variety of rocks placed in the way of ascending mineralizing solutions. The most widely quoted examples are horizons of graphitic and pyritic rocks.

(d) Exocontact rocks often constitute the metal sources for ores. Conforming to the "principle of minimum effort" (it states that metallic deposits will preferentially form on sites where the requirement of energy for their generation will be the lowest, and anomalous metal content the nearest; Laznicka, 1985e), lithologic associations anomalously enriched in certain elements or even containing earlier orebodies, have the greatest potential for influencing the selection of metals in hydrothermal orebodies to form in the vicinity. Some lithologic associations or rock types are predictably (systematically) enriched in certain trace metals. Fig. 28-25 shows the clarke values of selected lithologic associations contrasted with average ore grades. Empirical experience indicates that many metallic deposits are preferentially affiliated to units (rocks) having the highest clarke values, and "granite"-related metallic deposits are similarly associated with interaction complexes involving such rocks. In the latter case, sufficient metal mobility in hydrothermal fluids is a necessary prerequisite and immobile (e.g. Ti) or insufficiently mobile (Cr) metals rarely form depositional accumulations precipitated from hydrothermal fluids, even if their environment is anomalously rich in such metals.

Zn,Pb,Sb,Cu,W,Bi, Mo,U,Ag and Au have the systematically highest clarkes in a variety of mostly black mudrocks and indeed a large

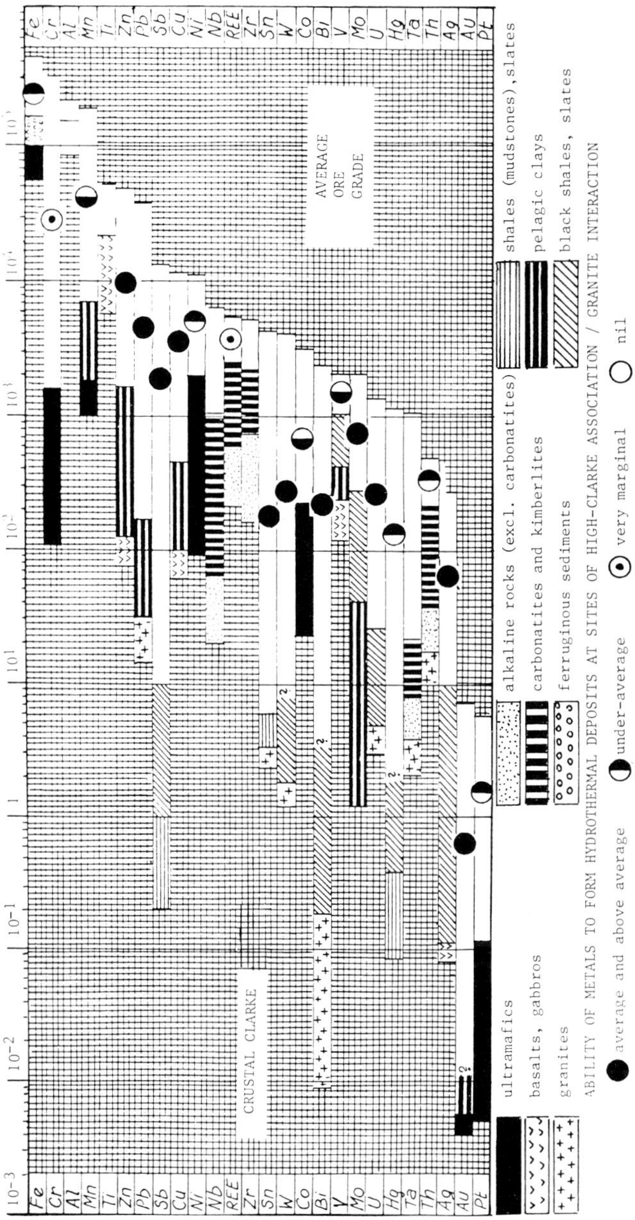

Fig. 28-25. Ability of metals to form hydrothermal deposits at sites of high clarke lithologic association and granite interaction. Rock groups with the highest average trace metal contents are contrasted with the ore grades of respective metals. Log scale. The grid pattern represents the crustal clarke (left) and the average ore grade (right). The gap between the highest trace metal content in a rock category and average ore grade corresponds to the factor of concentration (when it can be demonstrated that the ore and the rock category have a source relationship). Expanded, after Laznicka (1985e).

number of postmagmatic hydrothermal deposits of these metals does occur within or near these rocks (compare Chapters 7,15,17 and 19). Jensen (1971) postulated that most of the hydrothermal polymetallic deposits in the western United States (mostly in the Great Basin) had their metal sources in the organic-rich (black) "shale" horizons penetrated and flushed by "granite"-heated convective systems.

Cu and Zn have also high clarkes in (meta)basalts, andesites and gabbros, and these rocks are repeatedly associated with postmagmatic plutonic deposits of these metals (e.g. porphyry coppers). Pb,Sn,W,Bi and U are significantly enriched in felsic granites, and indeed many or even most of their hydrothermal deposits are proximal to older generation of granites or are part of multiphase granite intrusive systems. Non-systematic metal enrichment in a variety of rocks is more difficult to predict. Often it is regional, crossing lithologic boundaries and suggesting a heredity.

Availability of metals for ore generation in passive source rocks is not only a function of abundance, but also a function of setting (bonding) of the metal. Such settings can be arranged by the ease of release of the metal from the host mineral or rock as follows (from easiest to most difficult to release): (1) metals forming their own sulphide (or other) phase, located in fracture veinlets or in pores; (2) the same minerals, located on surface of rock-forming minerals and interstitially; (3) the same minerals present as inclusions in rock-forming minerals; (4) trace metals carried by common sulphides (e.g. pyrite); (5) native metals and metallic oxides forming their own phases; (6) trace metals present in easily soluble and degradeable rock-forming minerals (carbonates, organic compounds, volcanic glass); (7) trace metals forming own accessory silicates or (8) trace metals substituting in the lattice of rock-forming silicates and oxides.

28.3.5. Mineralized versus barren "granites"

The problem of why some "granites" (plutons, batholiths, massifs) are mineralized and others barren, is from time to time, discussed in the literature, but at present there is not a single, universally valid answer. All the aspects of "granite" origin and emplacement reviewed earlier, seem to be of importance. Probably of greatest importance is the depth of erosion. Epizonal and upper mesozonal plutons are often mineralized, while the deeper plutons are uniformly barren except for pegmatites or modified relic deposits.

The second most important feature appears to be the pluton morphology. In the tin-bearing regions in particular, elevated cupolas, cusps and ridges "sitting on the back" of a more or less flat batholith, are foci of mineralization while the rest is barren. Porphyry Cu and stockwork Mo also favour central, upward-evolving intrusive systems.

Third in importance is the state and composition of the over-intrusion lithologic association, in which convection took place. In regions where extensive "granitic" batholiths are roofed by a variety of rocks, segments composed of the more variegated and "fertile" assemblages (black metasediments, marbles, amphibolites, a variety of metavolcanics), tend to be more commonly mineralized than monotonous gneiss, schist, sandstone or common shale terrains.

A significant number of geologists (Noble, 1970; Routhier, 1980;

the Soviet school) interpret richly mineralized segments of the crust (including those intruded by granites) as being predetermined by local metal anomalies in the mantle, or simply by the relative proximity to the mantle. When a mantle inhomogeneity and a "granitic" intrusion overlap, the granites are believed to have acted as "revelateurs", "developing" the anomaly and lifting a portion of its metals into the near-surface region.

28.4. MINERALIZATION STYLES ASSOCIATED WITH PHANEROZOIC (HIGHER-LEVEL) GRANITE, DIORITE, (GABBRO) ASSOCIATION (Fig. 28-26; Table 28-1)

28.4.1. Ore minerals disseminated or scattered in "normal" (unaltered) "granites"

The ordinary rock-forming and accessory metallic minerals in "granites" (magnetite, ilmenite, zircon, rutile, monazite) are of little importance as sources of metals, unless additional conditioning, sorting and accumulation of the valuable substance takes place. This happens by the action of tropical weathering and by a hydrodynamic transportation and sorting of the detritus. Weathering slightly enriched accessory magnetite in "granites" used to be recovered in the past in Japan as an iron ore (Chapter 23). A portion of the detrital zircon, monazite, ilmenite and rutile in alluvial and beach placers described earlier came from granitic rocks.

Gabbros are particularly high in ilmenite, occasionally enriched anomalously in schlieren and inhomogeneities. Such ilmenite after weathering release could be added to placers, but no "primary" ilmenite or Ti-magnetite accumulations in "granite"-affiliated gabbros are known.

Rare and "special" accessory minerals in "granites" include orthite (allanite), xenotime, euxenite, columbite-tantalite, beryl, uranothorite, uraninite, fergusonite, betafite, molybdenite and cassiterite. Most of these accessories are fine crystalline and megascopically unrecognizable. They may accumulate in proximal placers. Granites unusually rich in some of the accessory minerals listed above may develop into "metalliferous granites", that might be economic to mine in bulk in the future. Most examples of such granites are members of the alkaline or peralkaline association (Chapter 33), katazonal granites (Chapter 29), Precambrian granites (Volume 2), or both.

Of the reminder one can mention: uranothorite-rich "granite" in the Khantau massif, southern Kazakhstan, and several orthite-rich granites in the Susamyr Batholith, Tian Shan (both in the U.S.S.R.; Lyakhovich, 1973), both containing up to 3% Th. Elberton Granite in Georgia, rich in orthite, contains, as a whole, about 45 ppm Th, and feasibility tests have been made to extract this thorium directly by chemical leaching in situ. In the You Yangs Granite in Victoria, Australia (Baker, 1937), allanite there is a rare accessory in the granite, but it accumulates in mafic schlieren in which it constitutes

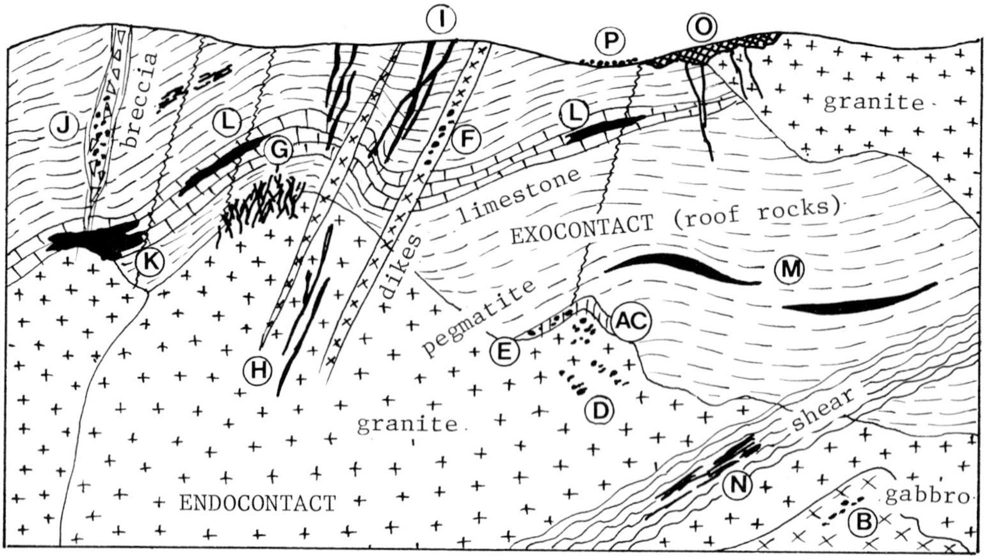

Fig. 28-26. Principal mineralization styles associated with Phanerozoic GDG ("granitic") intrusions. See Table 28-1 for explanation of letter codes. From Laznicka (1984)

Table 28-1. Principal mineralization styles associated with Phanerozoic "granitic" intrusions

(A)	physically separable ore minerals disseminated (scattered) in unaltered "granites" (zircon, monazite, U,Th,REE minerals, magnetite, beryl, rutile
(B)	higher than accessory disseminations, schlieren, etc. of ilmenite, magnetite or chalcopyrite in gabbros
(C)	dispersed metals in granites in supra-clarke concentrations (U,Th)
(D)	mineralized inhomogenities in "granites"
(E)	mineralized high-level pegmatites (Sn,W,Mo,Be)
(F)	mineralized intrusive dikes
(G)	stockworks and pervasive ore disseminations along altered contacts in non-carbonate hosts (Au,W)
(H)	veins in intrusive endocontacts
(I)	veins in intrusive exocontacts
(J)	mineralized breccias, breccia pipes, replacement pipes
(K)	skarns
(L)	non-skarn carbonate replacements
(M)	relic and remobilized ores in granite aureoles
(N)	mineralized meta-granite and metamorphic ores in granites
(O)	ore accumulations resulting from weathering of "granites" and their mineralized aureoles
(P)	ores formed by sedimentogenic reworking of "granite" - associated mineralizations

3.5% of the rock. Pink perthitic post-kinematic granites forming small stocks in the Sawtooth Massif S.W. of Stanley, Idaho (Killsgaard et al., 1970) and in the Bugaboo and adjacent creeks near Spillimacheen, S.E. British Columbia, contain disseminated accessory euxenite-polycrase. In both areas, placer accumulations have formed. Fergusonite is said to be abundant in the marginal zone of a granodiorite in the Nanling Range, S. China.

Accessory uraninite has been found in almost all granites known to host vein or "unconformity-type" uranium deposits, or interpreted as uranium sources to groundwater-epigenetic deposits in the vicinity. Many such localities (N. Australia, Wyoming) are Precambrian, but the French uraniferous granites (particularly those in Vendée, N.W. France; Barbier, 1972) are late Paleozoic. There, the discovery of one U mineralization style (e.g. U veins) can serve as a starting point for a search for other styles, including uraniferous granites.

Beryl-bearing granites have been discussed by Burke et al. (1964). Several occurrences of scattered, fine-grained beryl mostly in high-level granites of ring complexes in Ireland, Scotland and elsewhere, have been documented. Such beryl is very inconspicuous, but detectable by portable berylometers. At most localities, disseminated beryl in granite is a minor variety located on fringe of Be greisens or pegmatites. It is economically unimportant

Cassiterite-, molybdenite-, wolframite-, scheelite-, etc.-bearing granites are genetically enigmatic, and discussions as to whether such occurrences are "orthomagmatic" or "postmagmatic" appear periodically in the literature (see Hosking, 1973 and Taylor, 1979, for discussion). In most cases this is a matter of misunderstanding or carelessness in writing, in which authors equated "granite-hosted" ores with orthomagmatic ones. It appears that emplacement of the economically interesting quantities of cassiterite and some other minerals postdated the main crystallization of granites and accumulated in late stage pegmatites (in deep-seated granitic terrains) or in hydrothermal replacements, stockworks and veins accompanied by alteration. As mentioned earlier, the high-temperature hydrothermal feldspathization or biotitization often produces minerals virtually indistinguishable from their magmatic counterparts and gives the rocks a fresh, "primary" look. The largest accumulations of cassiterite in "unaltered granite" as in the Čistá deposit, Czechoslovakia (Ďurišová et al., 1969) are always situated along fringes of greisenized or otherwise altered zones so their orthomagmatic emplacement has a low credibility.

Molybdenite is a much more common (and conspicuous) "accessory" in "granites". Close examination, however, almost always indicates that the mineral is, in fact, in fracture veinlets, coats joints, or replaces biotite flakes within an alteration band. The same applies to wolframite and scheelite (e.g. in the Atolia field, California).

28.4.2. Granites with low-grade dispersed metallic component ("metalliferous granites")

Earlier we discussed metals accumulated in discrete ore minerals, but some granites could have an anomalous trace content of various metals perhaps recoverable in the near future by cheap, bulk leaching methods. The biotite phase of the Jurassic Conway Granite, New Hampshire (Finch et al., 1973) has been studied with this in mind. The granite averages 12 ppm U and 56 ppm Th (this corresponds to clarke of concentration of 4 and about 10, respectively), but this represents about 675 Tt U and 3 Mt Th available within an area of 768 km^2 to a depth of 330m.

The increased trace metal contents in granites are either evenly distributed throughout the rock, or restricted to certain major or accessory minerals. There is a possibility that accumulation of such "metalliferous rock-forming minerals" may evolve into future ore deposits. Most examples involve tin. Hosking (1973, 1974) reported up to 2% Sn in some zircons, up to 10% in sphene, up to 2% Sn in andradite and up to 1.6% Sn in axinite. High Sn contents are in högbömite, an inconspicuous mineral resembling biotite. Furthermore, a high content of Sn (X00 ppm) is found in some amphiboles, biotites, and staurolites. Lead, on the other hand, accumulates in feldspars (up to 3% Pb), particularly in the green microcline (amazonite).

28.4.3. Mineralized inhomogeneities in granites

Inclusions (xenoliths) and schlieren may contain ore minerals that their host granites lack. The common mafic (gabbroic) xenoliths in tonalites and granodiorites almost always contain scattered grains of chalcopyrite (together with pyrrhotite), probably released during conversion of pyroxene into amphibole and amphibole into biotite. Didier (1973) described "mineral nodules" from leucogranites in Britanny (N.W. France) containing minor beryl.

Tsypukov and Vladykin (1976) described occurrences of native gold (up to 2.1 ppm) associated with gabbro rafts in some "granites" in Mongolia. The gold was located in strongly hybrid sections accompanied by pegmatite and fluorite. Relics of former metalliferous deposits in supracrustals invaded by "granites" often survived, remarkably little modified, in xenoliths and rafts. One example of a "granite"-dismembered conformable massive sulphide lens has already been reviewed and illustrated (Chapter 12, Fig. 12-11).

Xenoliths and rafts of reactive rocks (e.g. carbonates) may carry contact metasomatic metalliferous replacements such as magnetite or scheelite skarns. Many scheelite skarns in the Sierra Nevada, California, are in such rafts. The rafts grade down in size into centimetres-sized xenoliths and into even smaller particles. It is possible that several occurrences of disseminated scheelite occurrences in "granites", in fact, contain small metacarbonate inclusions completely replaced by scheelite.

The Minto copper deposit, Yukon (Pearson and Clark, 1979; 160 Tt Cu/1.86%, 55 t Ag, 3.6 t Au) is one of the recently described enigmatic deposits associated with inhomogeneities in "granite". The main orebody is a flat-lying lens of foliated biotite granodiorite and gneiss with dimensions of 335x247 m and 30 m average thickness, enclosed in a generally unfoliated late Triassic mesozonal granodiorite. The ore confined to the foliated matrix consists of disseminations and veinlets of chalcopyrite, bornite, lesser magnetite and pyrite, often accompanied by clusters of biotite. After deposition, the ore and its surroundings suffered deformation, light metamorphism and dismemberment by faults and intrusive dikes. Although Pearson and Clark (1979) considered the mineralization to be pre-granite, the original protolith and ore style remain uncertain.

28.4.4. Mineralized high-level pegmatites

Granitic pegmatites are most widely associated with katazonal granites (and sillimanite-grade metamorphics), so their main review is in Chapter 29 (also in Volume 2, because the majority of economic pegmatites are Precambrian). Here, the small and relatively uncommon pegmatites affiliated with mesozonal, and even epizonal, granites are reviewed briefly. In the classification of pegmatite associations by depth prepared by Ginzburg et al. (1979), two classes fall into the present field of coverage. They are (1) miarolitic pegmatites forming pods in the upper parts of granitic intrusions and containing miaroles filled by quartz, fluorite, topaz and beryl and (2) intermediate-depth rare-element (Li,Rb,Be,Ta,Sn,Nb) pegmatites. The type locality of the miarolitic pegmatites is in the Precambrian Korosten Massif, hence they are considered in Volume 2. The rare element pegmatites are all associated with apical portions of "S granites" in Sn,W,Mo,Be,etc. mineralized provinces where they represent a minor mineralization style of limited economic importance. They could be lithian, but Li is mostly in zinnwaldite or lithionite. These pegmatites should not be confused with the "typical" Li-pegmatites (with lepidolite, spodumene, etc.) treated in the next chapter.

The rare-element pegmatites as outlined above occur (1) as "syngenetic" schlieren, nests and small lenses in acid granites, or as "linings" of granite stocks along contacts with the outside rocks ("Stockscheiders") or (2) as independent pegmatite bodies such as dikes, sills, lenses, filling dilations in both endo- and exo-contacts in granites. Both varieties may carry irregularly scattered crystals of wolframite and scheelite (Soktui and Adun-Chalon Massif, Transbaikalia; several massifs in the Kolyma belt in Siberia; Shcheglov and Butkevich, 1974; Modoto and Tsagan-Daba deposits, Mongolia; Marinov et al., 1977); cassiterite (Kalba district, Kazakhstan; Baga-Gazryn Sn deposit, Mongolia); stannite (Vernéřov near Aš, W. Czechoslovakia, in amblygonite pegmatite); molybdenite (Condor Huta deposit near Milipaya, Bolivia, where an apophysis of a coarse granodiorite changes into a pegmatitic granite and ultimately pegmatite with scattered flakes of molybdenite; Ahlfeld and Schneider-Scherbina, 1964. Similar deposits have also been found near Skalsko, Czechoslovakia); columbite-tantalite, beryl and other minerals (Zhanchublin and Khukh-Del-Ula occurrences, Mongolia; Umana, Bolivia). Although of little value considered searately, the occurrences listed above could be gradational into quartz, feldspar, muscovite, tourmaline, cassiteite veins (e.g. Zigei-Khundei showing, Mongolia); to quartz, wolframite or cassiterite veins and to molybdenite veins. Sn-W greisens could be superimposed on pegmatites and their surrounding. The pegmatite presence can serve as an exploration lead for the more productive styles of Sn,W(Mo,Ta-Nb) deposits, and widespread occurrences of stanniferous pegmatites may contribute detrital tin to placers.

Probably the best known representative of a small mesozonal Sn pegmatite deposit is the Mina Fabulosa in the Sorata Batholith, Cordillera Real, Bolivia (Ahlfeld and Schneider-Scherbina, 1964). There, numerous dikes of microcline-quartz pegmatite are found at both sides of contact of a coarse Mesozoic muscovite granite and Paleozoic spotted andalusite phyllites and biotite contact hornfelses. Coarse cassiterite in zoned crystals, up to 15 cm in diameter, is haphazardly scattered in the dikes, and there is also minor lazulite, triplite, trifyline-lithiophyllite, molybdenite, pyrrhotite, arsenopyrite and stannite. The average grade is 1% Sn and the pegmatite grades to quartz-muscovite veins and is sometimes partly converted into greisen along sahlbands.

Several additional varieties of "pegmatites" are actually hydrothermal metasomatites or open space fillings. This is probably where the "amazonite" (microcline) pegmatites belong, that are associated with albitized "apogranites"; pegmatitic nests and portions of K-feldspar and biotite metasomatites in porphyry copper systems; etc. Many metallic ores hosted by pegmatites are epigenetic, of hydrothermal origin, and introduced after completion of the pegmatite stage. When the hydrothermal gangue minerals are feldspar, quartz and mica, the superimposed origin of the ore is commonly missed. In this category, for example, belong several localities of the "auriferous pegmatites" listed by Boyle (1979, p. 248,249).

28.4.5. Mineralized porphyry, lamprophyre, diabase, etc. dikes

Intrusive dikes are a typical feature of most hydrothermal vein and stockwork fields, where they often occupy the same fissures filled by veins. They are often hydrothermally mineralized in preference to the other, more voluminous hostrocks. In the Březové Hory field, Příbram district, Czechoslovakia, the majority of Pb,Zn,Ag veins are located in narrow diabase dikes. In the Berezovsk goldfield in the Urals (Magak'yan, 1968) granite, plagiogranite and syenite porphyry dikes believed to be derived from a late Paleozoic granitic pluton, intersect Devonian mafic metavolcanics, metasediments and serpentinites. The dikes are 2-40 m thick and up to 8 km long. Within the 64 km^2 of the field, the dikes are intersected by more than 70,000 thin quartz, pyrite, tourmaline, gold veins. About 40% of the total length of dikes is productive and on average, each 3 m of a dike contains one ore vein. In the Kochkar goldfield (the Urals, U.S.S.R.) over 1,500 mafic to felsic dikes are biotite, amphibole, feldspar-altered and carbonatized in the aureole of a small microcline granite stock. The altered dikes were then preferentially mineralized by quartz, arsenopyrite, pyrite, tetrahedrite, tellurides, gold, etc. veins.

Elsewhere, dikes often overlap with the ore vein deposition and pre-, syn- and post-ore dikes have been reported. Dikes are considered to be a favourable indication of the possible presence of ore veins, mainly because dike swarms are one of the indicators of an over-intrusive setting. Abdullaev (1957) devoted a whole book to the ore and dike association. The many varieties of vein arrangement in dikes have been thoroughly treated in the classical literature on structural control of ore deposits. In addition to the predictable varieties of mineralization hosted by dikes, there are several unique examples of metalliferous dikes. Boyle (1979, p.248) listed several examples of auriferous dikes. In the Central City district, Colorado, a set of quartz-bostonite porphyry dikes is highly radioactive (max. 0.013% equivalent U) and intimately associated with pitchblende-bearing hydrothermal veins in the district (Wells, 1960).

28.4.6. Stockworks and pervasive disseminations in altered endo- and exocontacts

The bulk of the large volume, low grade, hydrothermal-plutonic deposits of Cu(Mo), Mo and Sn occur in distinct stockworks and disseminations-dominated alteration-mineralization systems (porphyry coppers, stockwork molybdenums, cupola tin deposits), that are reviewed as units later. The remaining metals may occasionally accumulate in stockwork or dissemination-style orebodies, but these deposits are of substantially lesser importance than alternative ore styles, particularly veins and replacements (Fig. 28-27). There is an intermediate style of orebodies distribution between stockworks and fissure veins, taking the form of swarms of closely-spaced, parallel veins (e.g. Endako, British Columbia, Bainazar, Kazakhstan and Mt. Carbine, Queensland, Mo and/or W deposits).

GOLD DEPOSITS

The small gold stockwork at Jessie mine, Colorado, has already been mentioned. Boyle (1979) reviewed additional localities of auriferous stockworks, many of which are in volcanics and subvolcanics typically associated with explosive volcanic centres (Chapter 26) or with Precambrian intrusions (Volume 2). Further examples of plutonic Au

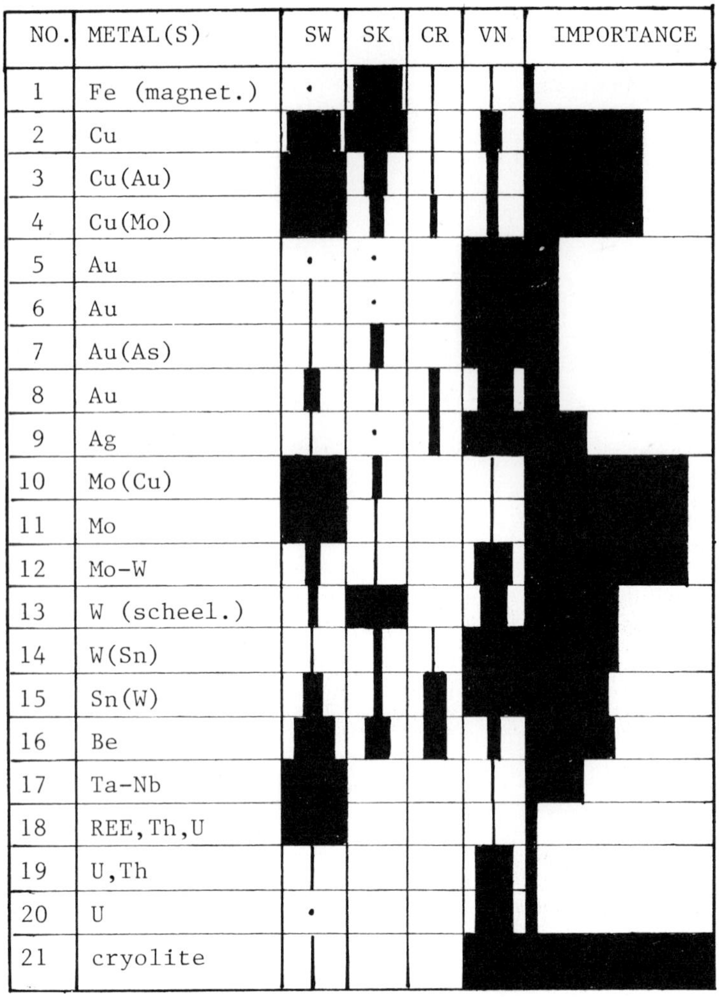

Fig. 28-27. Relative importance of stockwork or disseminated, skarn, carbonate replacement, and vein deposits of metals. Abbreviations: SW=stockworks; SK=skarns; CR=carbonate replacements; VN=veins. Column IMPORTANCE denotes the importance of granite-associated deposits of the respective metal among all deposits of such a metal.

Table below lists examples of associated intrusives and endocontact stockwork, disseminated and exocontact skarn ore deposits:

NO.	METAL(S)	AFFILIATED INTRUSIVE ROCKS	STOCKWORKS/ DISSEMINATIONS	SKARNS
1	Fe (magnet.)	quartz diorite, quartz monzonite	Iron Springs, Utah	Iron Springs, Utah

2	Cu	soda granite (trondhjemite) "primitive" assoc.	Tsessovka, Voznessenskoe	Tur'ya (partly)
3	Cu(Au)	quartz-free (diorite, syenite) association	British Columbia, Boshchekul	Princeton, Brit. Columbia
4	Cu(Mo)	high-quartz qtz.monzonite, granodiorite	widespread worldwide	Ely, Bingham, Tucson South
5	Au	soda granite (trondhjemite) "primitive" assoc.	Bralorne,B.C.	
6	Au	quartz-free (diorite, syenite) association	Vunda, Fiji	
7	Au(As)	quartz monzonite, granodiorite, qtz. diorite	Jessie,The Patch, Colo.	Hedley, B.C.
8	Au	syenite, granosyenite to qtz.monzonite, granite	Zartman-Landuski, Zarmitan	Judith Mts.
9	Ag	quartz monzonite, granodiorite porphyries	Nenzel Hill, Hahn's Peak	
10	Mo(Cu)	quartz monzonite, granodiorite, porphyries	Boss Mountain, Shakhtama	Cannivan Gulch, Rossland, B.C.
11	Mo	high K calc-alkali and alkali-calcic granite	Climax, Henderson, Colorado	
12	Mo-W	high-K leucogranite and related porphyries	Akchatau, Koktenkol'	
13	W (high lv.sclt.)	granodiorite, quartz monzonite, granite	Gabbs, Boguty, Haut Auxelles	Tyrny Auz Xi Zhoyouang
14	W(Sn)	biotite and musc.-biot. granite, leucogranite	Hemerdon, S. England	
15	Sn(W)	as above	Altenberg,Címovec, Krásno	Doradilla,Moina, Lost River
16	Be	as above	minor in Ireland,Scotland	Xi Zhoyouang Iron Mountain
17	Ta-Nb	peralkaline to alkaline granite	E.Transbaikalia Kaffo Valley,	
18	REE,Th,U	as above	Khantau, Susamyr, Bugaboo	
19	U,Th	peralkaline granite	Bokan Mt., Sawtooth Massif	
20	U	syenite, bostonite	Central City, Colorado	
21	cryolite (Al)	peralkaline granite	Ivigtut, Greenland (PCm)	

stockworks include the Zarmitan Complex of Permian granosyenites in Uzbekistan (Garkovets et al., 1979). There, gold-bearing stockworks and discrete quartz veins with auriferous pyrite and arsenopyrite are enveloped in microcline and albite alteration metasomatites.

In the Klyuchi gold field (E. Transbaikalia; Borodaevskaya and Rozhkov, 1974), networks of quartz, tourmaline, pyrite, chalcopyrite, arsenopyrite and gold are hosted by strongly tourmalinized Paleozoic and Mesozoic granitoids. In the Zartman-Landusky field in the Little Rocky Mts., Montana, numerous, very low-grade auriferous veinlet stockworks and veins in Tertiary quartz monzonite and syenite are mined on large scale, and locally heap leached.

SCHEELITE DEPOSITS

Disseminated and stockwork ("porphyry") scheelite deposits are known, both in the form of local sections of composite ore zones, and as independent orebodies. Examples include the Victory Mine near Gabbs, Nevada (scheelite disseminated in granodiorite); Boguty stockwork, S. Kazakhstan (quartz-scheelite veinlets enveloped in greisenized and quartz, sericite, chlorite-altered Carboniferous granite and Ordovician sandstone); Haut Auxelles, Vosges, France (quartz-scheelite veinlet stockwork in a silicified dike of microgranite) and other localities. In the Yellow Pine ore zone, Idaho (Cooper, 1951), disseminations and fracture coatings of scheelite, stibnite, pyrite and gold are found in altered quartz monzonite along an important shear.

28.4.7. Veins in intrusive endocontacts

A great variety of hydrothermal veins formed in the brittle rocks of granite endocontacts, is discussed in the classical literature on structural control and in textbooks. The more typical veins emplaced contemporaneously with, or shortly after, the cooling of a post-tectonic granite body in a stress field, fill the regularly developed perpendicular joints in granites. Of the three regular joint directions, the open Q joints (parallel to stress) and the "flat" (low angle, L) joints tend to be preferentially filled by veins. In granite cupolas such veins often occur in parallel systems and may grade into stockworks. In Cínovec, Czechoslovakia, a system of parallel flat quartz, cassiterite, wolframite, etc. veins is arranged in an onion-skin fashion parallel with the outer perimeter of a granite cupola.

In forcefully emplaced plutons, in late tectonic plutons, in plutons deformed by post-solidification faulting, etc., regular and predictable vein patterns are rare. The veins fill fissures and faults often healed earlier by dikes, striking in all directions and mostly dipping steeply. Conjugate vein systems may form, as for example in the Grass Valley field, California. Several vein examples have been recorded located in thrusts. Lodes filling shears in "granite" plutons are also known. Open-space filling veins are slightly more abundant than replacement veins.

28.4.8. Veins in granite exocontacts

Exocontact veins show even greater variety than the veins hosted by granites (Fig. 28-28), because of the wide range of rock competency (e.g. in alternating slates and quartzites), reactivity with hydrothermal fluids (carbonates versus silicate rocks) and a variety of "topomineralic" influences and proximal metal source relationships.

In aureoles of high mesozonal to epizonal "granites" emplaced in a slate unit (as is common in many tinfields), the contact-proximal hard and brittle quartz, biotite, plagioclase hornfels "welded" to the granite is mechanically comparable with the "granite" and often hosts stockworks of numerous short fracture veins or veinlets. Outside the hornfelsed rim, discrete, more persistent and usually richer veins form along faults. In terrains of alternating slates and quartzites (as in the Paleozoic sediments of Bolivia), through-going fissures are tight and virtually unfilled in the slates, but widen to form "bolsons" (ore shoots) in the brittle quartzites. The latter often take the form of mineralized breccias or stockworks.

This general relationship extended to include the endocontact veins, has been supported statistically by evaluation of the average hydrothermal vein dimension, prepared by Smirnov, ed. (1968; Fig. 28-29). There, the veins of metals preferentially situated at the greatest distance from granite contacts (Pb-Zn, Au, Cu) are long and deep. Those present close to the contact (W, Sn, Mo, fluorite) are short and shallow.

Probably the greatest regularity and predictability of vein occurrence exists above granite cupolas, in compositionally homogeneous metasediments. There, veins, in most cases, are subvertical and subparallel with cleavage (e.g. Burnt Hill, New Brunswick), but there are important exceptions. At Panasqueira, Portugal (Kelly and Rye, 1979), a subhorizontal swarm of quartz-ferberite veins cuts sharply across the steep bedding and foliation of pelitic schists, above and in the exocontact (and partly endocontact) of a granite.

In regions of alternating silicate rocks and carbonates, veins in the former and bedded replacements in the latter often coexist. Under special conditions, ore veins are repetitively associated with a "favourable" lithologic and stratigraphic unit (horizon) and may occur within it (// with bedding and schistosity; e.g. wolframite veins in Bolsa Negra, Bolivia). Alternatively barren fissures or veins may become mineralized in the intervals cutting such horizons (e.g. Kti Teberda W deposit, the Caucasus). This strongly suggests (re)mobilization of metal from an earlier metalliferous horizon.

There are numerous variations of fissure veins, rated by the regularity and persistence of filling. Some are continuously

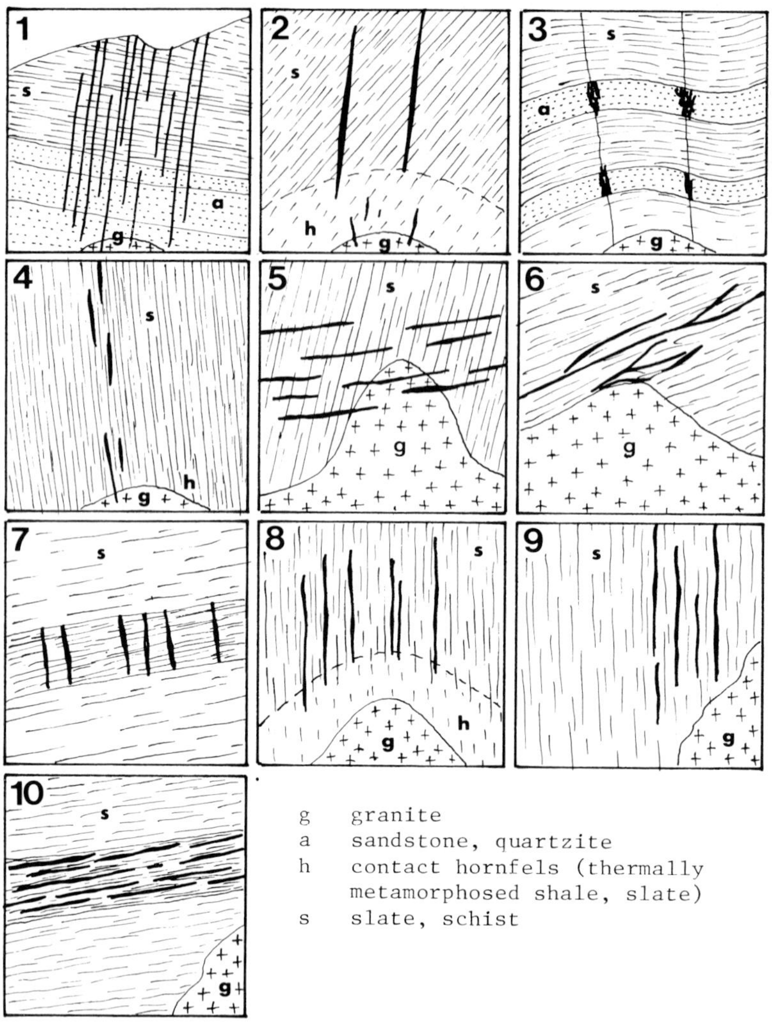

Fig. 28-28. Diagrammatic representation of the variety of hydrothermal Sn and W veins in non-carbonate exocontact rocks: 1=closely spaced veins almost normal to bedding (e.g. Tachishan); 2=persistent veins along faults in slates (Bolivia); 3=thin veins or barren fissures (leaders) in slates widen and become economic in quartzite (Bolivia); 4=veins parallel with slate cleavage above granite cupola (e.g. Burnt Hill); 5=subhorizontal fissure veins crossing bedding above, and partly within, granite cupolas (e.g. Panasqueira); 6=an inclined vein swarm in hornfelsed slates, above granite (Storeys Creek); 7=productive ore shoots confined to a "favourable" sedimentary unit (Kti-Teberda); 8,9=veins parallel with foliation (e.g. Boriana); 10=veins and veinlets confined to and conformable with, bedding of a "favourable unit" (e.g. black slates; Bolsa Negra). From Laznicka (1984).

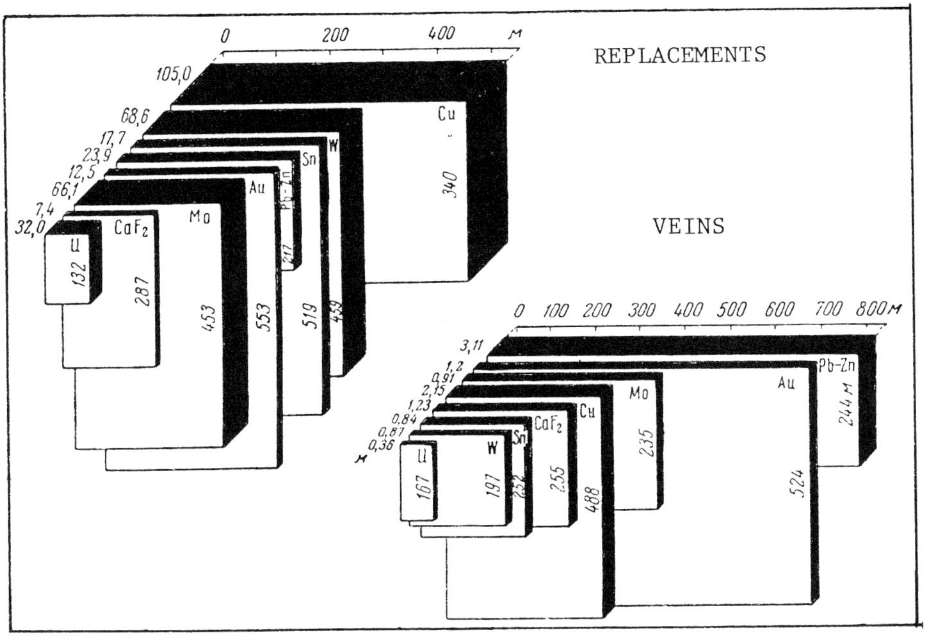

Fig. 28-29. Mean dimensions of postmagmatic hydrothermal veins and replacements. From Smirnov, ed.,(1968).

mineralized, for example by ore bands interlayered with gangue minerals; gangue filling containing scattered ore minerals; or rarely massive ore fill. Other veins change zonally with depth (the proportion of ore substance usually decreases), so in the deeper portions the vein often becomes barren although still continuing. Other "veins" are essentially tectonic structures mineralized at intervals, or filled by gangue material in which ore-bearing shoots appear from time to time.

28.4.9. Zoning in veins and hydrothermal mineralized structures

Metal (e.g. Sn-W-Mo), mineralogical (e.g. cassiterite → wolframite → molybdenite), ore style (e.g. greisen → endocontact stockwork → exocontact vein), alteration (e.g. greisenization → tourmalinization → sericitization) and other zonings are common features of a "granite"-affiliated mineralization system. These have been well known for over 80 years. Understanding and application of local zoning patterns contributed to ore discoveries but, on the other hand, orebodies have been missed and ore discoveries delayed for decades by insensitive application of the few popular zoning patterns highlighted in the literature. The extensive subject of zoning has recently been reviewed by Laznicka (1985c) and local zoning patterns as well as the repetitive zonings pertinent to specific mineralization styles are reviewed later in this chapter.

Given the variety of the petrochemical types of "granites" and their settings, surrounding rocks, variety of mineralization processes, etc., it is not surprising that there is not a single, master zoning pattern applicable to all granitic intrusions and their aureoles. The classical Emmonsian metal zoning pattern and its variations and modifications that adorn classical textbooks had already been termed "a wonderful oversimplification" by several authors at the time of its introduction in the 1920s, so it is not treated here. Instead, a diagram (Fig. 28-30) showing the frequency of occurrence of the various hydrothermal epigenetic ores at various distances from a "granite", is provided. The diagram is entirely empirical, based on a population of several hundred orebodies, and free of undocumented speculation. It shows how far from a genetically related igneous intrusion various hydrothermal orebodies most frequently occur worldwide, but it in no way suggests that all such orebodies (or their majority) are to be expected in the same area and that the pattern of the graph might appear as an actual vein, field, district or even mineralized belt zonality. There seems to be little doubt that mineralizations that have a narrow interval of occurrence and are constantly proximal to the "granite" contact (such as Sn greisens, skarns), are also genetically closely related to it. Ores showing a wide spread (e.g. quartz-hematite veins) or those that are systematically located in areas showing few signs of "granite" presence (e.g siderite veins), need not be genetically related to "granites" at all, at least not directly.

28.4.10. Breccias, breccia pipes and replacement pipes

Mineralized and unmineralized breccias are fairly common in the plutonic association, being particularly widespread in and near epizonal plutons. The ore-bearing breccias were favourable loci for ore deposition by virtue of their porosity and also because they were often the products of the dynamic interplay that generated ores regionally. Finally, breccias may contain ore fragments derived from older mineralizations present below, above or laterally, so that they often constitute an excellent exploration guide for hidden orebodies.

There is a bewildering variety of breccias and brecciated rocks in general, matched by an extensive list of genetic explanations proposed to account for their origin. These will be reviewed in a separate chapter in Volume 2. Here, only those breccias located in calc-alkaline intrusive terrains are considered. A characteristic feature of such terrains is a variety of breccia bodies having a circular or elliptical outline, and vertical to steep attitude. Some may be eroded volcanic vents (e.g. the Braden "Formation" at El Teniente, Chile; Camus, 1975), while the remainder are genetically uncertain. Gilmour (1977) termed breccias believed to be associated

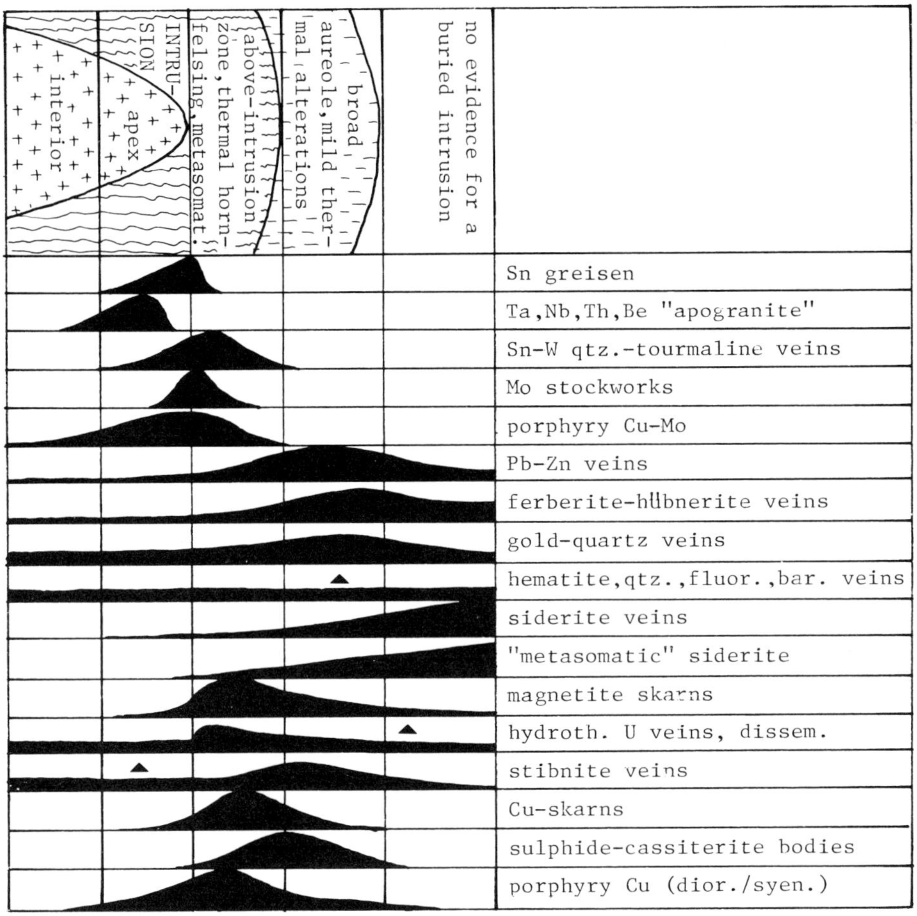

Fig. 28-30. Empirical frequency of occurrences of metallic mineralizations with respect to granitic intrusions (the ores need not be contemporary with, or shortly postdating, the granite emplacement). Triangles: major mineralizations postdating granite emplacement. From Laznicka (1984)

in some way with intrusive activity "intrusive breccias". Bryant (1968) who contributed a classical study of breccias present in the Bisbee porphyry copper field, Arizona, on the other hand, used a different terminology. His "intrusive breccias" were intruded (that is, forcefully emplaced) into their present position, not necessarily as a consequence of an igneous intrusion. At Bisbee, such breccias take the form of heterolitic irregular sheets and masses controlled by bedding planes of metasediments and by faults. Intrusion breccias, on the other hand, developed within an igneous intrusion. Many geologists find these terms too confusing.

The breccias vary in appearance and composition. Some are porous, almost loose and uncemented. The rest are hydrothermally altered or cemented and/or replaced by alteration minerals and sometimes ores. Where there is strong evidence for emplacement as a suspension in a hydrothermal fluid, the term "hydrothermal breccia" is sometimes used.

Most breccia pipes are downward-tapering, shallow to deep bodies, the depth ranging from several metres to over 800 m. Their diameter range from about one metre to about 2 km and most occur in clusters, usually controlled by fault zones. Their fragments range from angular to rounded (rounded fragments are most characteristic in "pebble dikes") and there is a variable degree of fragment rotation. Almost unrotated fragments are generally considered local. The fragment and matrix composition is also variable although volcanic and intrusive fragments are dominant. The genetic interpretations offered include: (1) fluidization, mostly by upward-streaming volatiles released from a late stage intrusive stock; (2) hydrothermal dissolution that created void subsequently filled by collapse; (3) hydrothermal or late magmatic deposits filling void created by magmatic water trapped in the apical region of an intrusion (Norton and Cathles, 1973); (4) meteoritic impact and (5) other mechanisms.

Because of the considerable heterogeneity and interpretational uncertainity, breccia pipes cannot be classified satisfactorily at present either on the morphological and compositional basis, or by their origin. The safest way is to treat them individually (by region) or by associated mineralization. Proportionally, most breccia pipes are associated with porphyry copper systems (e.g. Cananea), including copper-bearing tourmaline breccia pipes (Rio Blanco-Disputada); with stockwork Mo; with apical portions of Sn,W,Mo,Bi mineralized granites; with gold disseminations and stockworks; etc. Fig. 28-31 shows several examples diagrammatically. Additional pipes at subvolcanic level are treated in Chapter 26. More detailed description of several pipes appears under the heading of metals they contain.

In terms of exploration, the most important species are the breccia pipes suggesting the presence of concealed mineralized plutonic systems, especially porphyry coppers. They are of two possibly end-member styles: (1) conical, deep-reaching subvertical pipes and (2) thin to wide, lens to cap-like breccia hoods topping intrusive cupolas. Examples and case histories from Arizona have been reviewed by Gilmour (1977).

Certain pipes in granite endocontacts are dominated by hydrothermal replacements to such an extent that the original breccia fragments may have been obliterated entirely. Alternatively, the hydrothermal minerals (usually quartz)-filled pipe need not originally have been a breccia, but merely a densely fractured zone. Short pipes taking the form of downward (into the granite) tapering cones are most common in some Sn,W,Mo,Bi regions. The pipes in the Wolfram Camp, N. Queensland (Plimer, 1975; about 3.1 Tt W, 900 t Mo, 190 t Bi) are hosted by sericitized Carboniferous quartz monzonite. They are elliptical in cross-section and often bulbous or branching. They range from several centimetres to 6m wide and reach a depth of about 200 m.

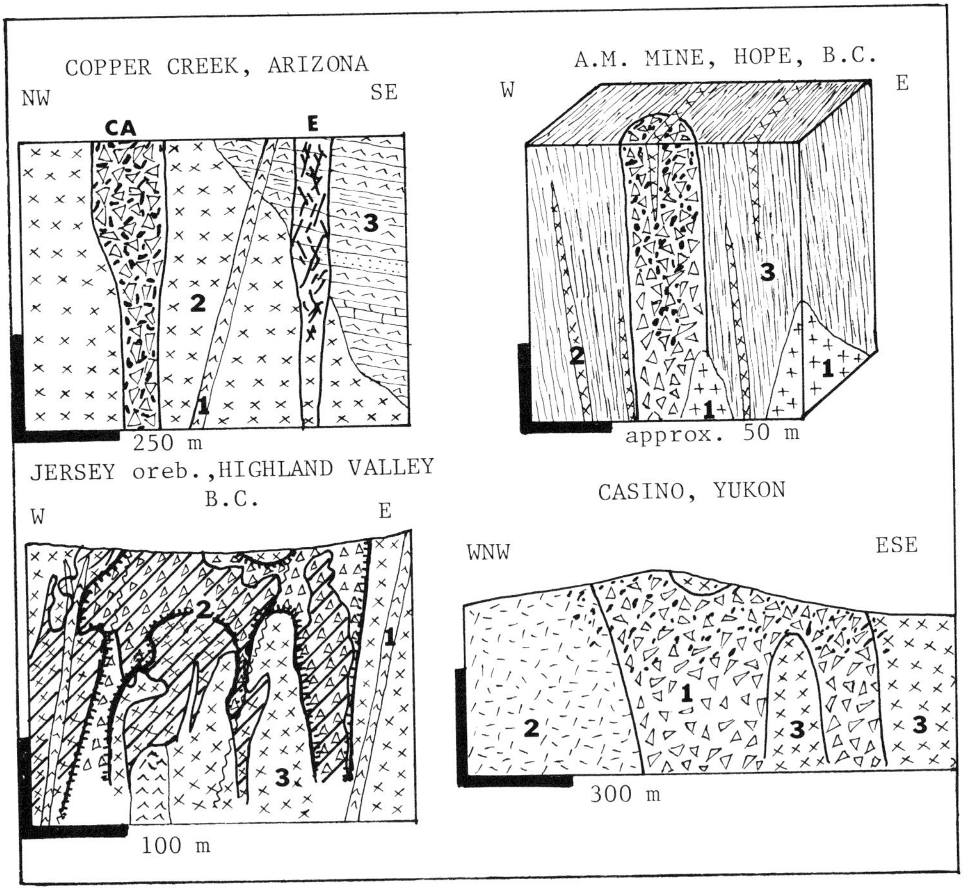

Fig. 28-31. Mineralized breccias in granitic intrusions and non-carbonate exocontacts.

Copper Creek Cu,Mo, Arizona; CA=Childs Aldwinkle altered and mineralized breccia pipe; E=Eagle deposit, fracture Cu stockwork in andesite; 1=T_1 latite and andesite dike; 2=Cr_3 granodiorite; 3=Cr andesite, tuff, minor sediments.

A.M. Mine, cp,po,py disseminations in the matrix of a breccia pipe. 1=Cr_3 ? granodiorite, quartz diorite; 2=Cr ? diorite to gabbro sills; 3=J_{1-2} hornfelsed distal turbidites.

Jersey mine, mo,cp disseminations and stockworks in granodiorite and breccia; 1=Tr dacite, latite, porphyry dikes; 2=breccia; 3=Tr granodiorite; ruled: orebody.

Casino, stockwork cp,mo in breccia, porphyry, quartz monzonite; 1=Cr_3 breccia; 2=Cr_3 quartz-feldspar porphyry; 3=Cr_2 quartz monzonite.

From LITHOTHEQUE, generalized after Kuhn (1941), Gilmour (1977), Briskey and Bellamy (1976) and Godwin (1976).

Compositionally they consist of quartz in a greisen envelope, and carry patches of wolframite, molybdenite, Bi,As,Cu,Zn and Pb sulphides.

An extensive but relatively unproductive cluster of Mo-Bi-bearing replacement pipes is found near Kingsgate (New England, Australia), hosted by a Permian leucogranite. The pipes are 1 to 20 m in diameter and up to 170 m long. They have been traced down to a depth of about 80 m. The core is composed of vuggy quartz grading into brecciated and silicified granite. Flakes of molybdenite and grains of bismuth and bismuthinite are irregularly scattered in all the zones.

28.4.11. Skarns

The term skarn originated in the Precambrian mineralized terrain of central Sweden (Bergslagen, Volume 2), and it referred to a mineralogical association of Ca,Fe,Mg silicates (Ca-Fe garnet, Fe-rich pyroxene) associated with iron ores. Elsewhere, similar silicates may be associated with ores of other metals (Cu,W,Sn,Au, etc.) and in such association the garnets and pyroxenes need not be highly ferruginous (e.g. grossularite, diopside). Some authors coined the term "tactite" for the iron-low assemblage.

The skarn association may have a variety of origins: (1) localized metasomatic replacement of limestone or dolomite in "granite" exocontacts (Fe and Si have to be supplied from outside, Ca and Mg are local); (2) isochemical metamorphism of a supracrustal rock of suitable composition (e.g. pelosiderite, ferruginous marl); (3) local exchange of components between adjacent rocks of contrasting composition during a high-grade metamorphism ("reaction" or "bimetasomatic" skarn); (4) replacement or exchange of components between broad mobile fluid "fronts" triggered by metamorphism or plutonism, and solid rocks or (5) multistage origin (e.g. initial formation of volcanic-hydrothermal or hydrothermal-replacement siderite or ankerite, followed by an essentially isochemical metamorphism; e.g. Rushiţa, Rumania).

Possibility (1) is most characteristic of the meso- and epizonal granite aureoles to such an extent that most authors writing about skarns consider only the most "typical" variety of them, that is the "granite"/carbonate contact metasomatic variety. After a period of conceptual uncertainty, skarns, s.s., have now been organized satisfactorily and modelled (Einaudi et al., 1981; Economic Geology Special Issue on Skarns, v.77, No. 4, 1982). Consequently, only a short summary is provided here and the reader is referred to the sources quoted above.

Skarn profiles are considered most typical when the garnet, pyroxene, etc. assemblage has formed on the (meta)sedimentary side of the "granite"/sediments contact, that is in the exocontact. The Milford Cu and magnetite skarn, Utah (Fig. 28-32) is a typical example. In exceptional cases the mineralogical skarn association formed on the intrusive side of the contact (endocontact). An example of this is found at Christmas, Arizona. In these cases, however, a less intensive alteration is always present. Many skarns formed along minor dike contacts or along conduits entirely within the sedimentary host, in which case an efficient structural control was important to provide plumbing to the mineralizing solutions. The composition of the skarn silicate assemblage depends, to a considerable degree, on the composition of the replaced precursors, and calcic (limestone-replacing) and magnesian (dolomite-replacing) skarns are the two most common varieties. Most skarns are evolutionarily complex and were formed over the course of at least three successive stages (Einaudi et al., 1981; Fig. 28-33):

(1) Contact metamorphic stage: formation of a zoned contact metamorphic aureole of generally iron-poor silicates (diopside-plagioclase, wollastonite, grossularite, etc. hornfelses in impure carbonates; biotite-plagioclase, lesser quartz hornfelses in pelites) and carbonate marbles. This stage is usually devoid of ore minerals, unless they are relic ores (including siderite and ankerite precursors).

(2) Metasomatic skarn growth. The Stage (1) was usually followed by brittle fracturing, and late-stage magmatic fluids rose along the outer contact of the pluton, along dikes, fissures, sedimentary contacts, etc., and infiltrated the wallrocks. As a consequence a second generation of zoned anhydrous skarns formed and these skarns, as a rule, possess a greater mineralogical variety and coarser grain.

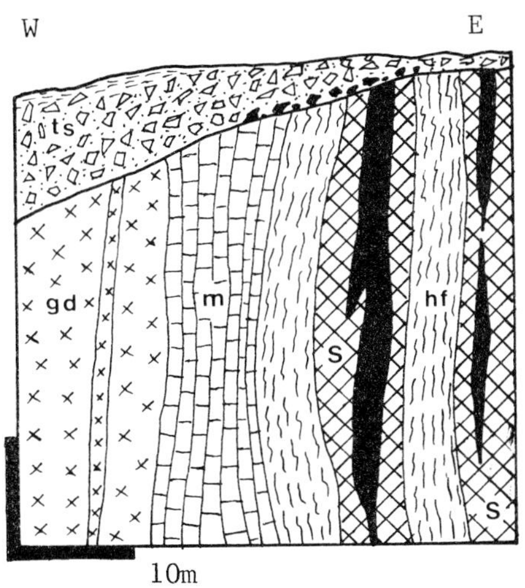

ts Q talus

EXOCONTACT

m PZ, white to beige marble

hf thin-bedded plagioclase-diopside-quartz hornfels

S grossularite, minor epidote skarn with lenses of magnetite (black) and scattered blebs of chalcopyrite

ENDOCONTACT

gd T_1 granodiorite to quartz monzonite with aplite dikes

Fig. 28-32. Simple magnetite chalcopyrite skarn, Milford, S.W. Utah. From LITHOTHEQUE.

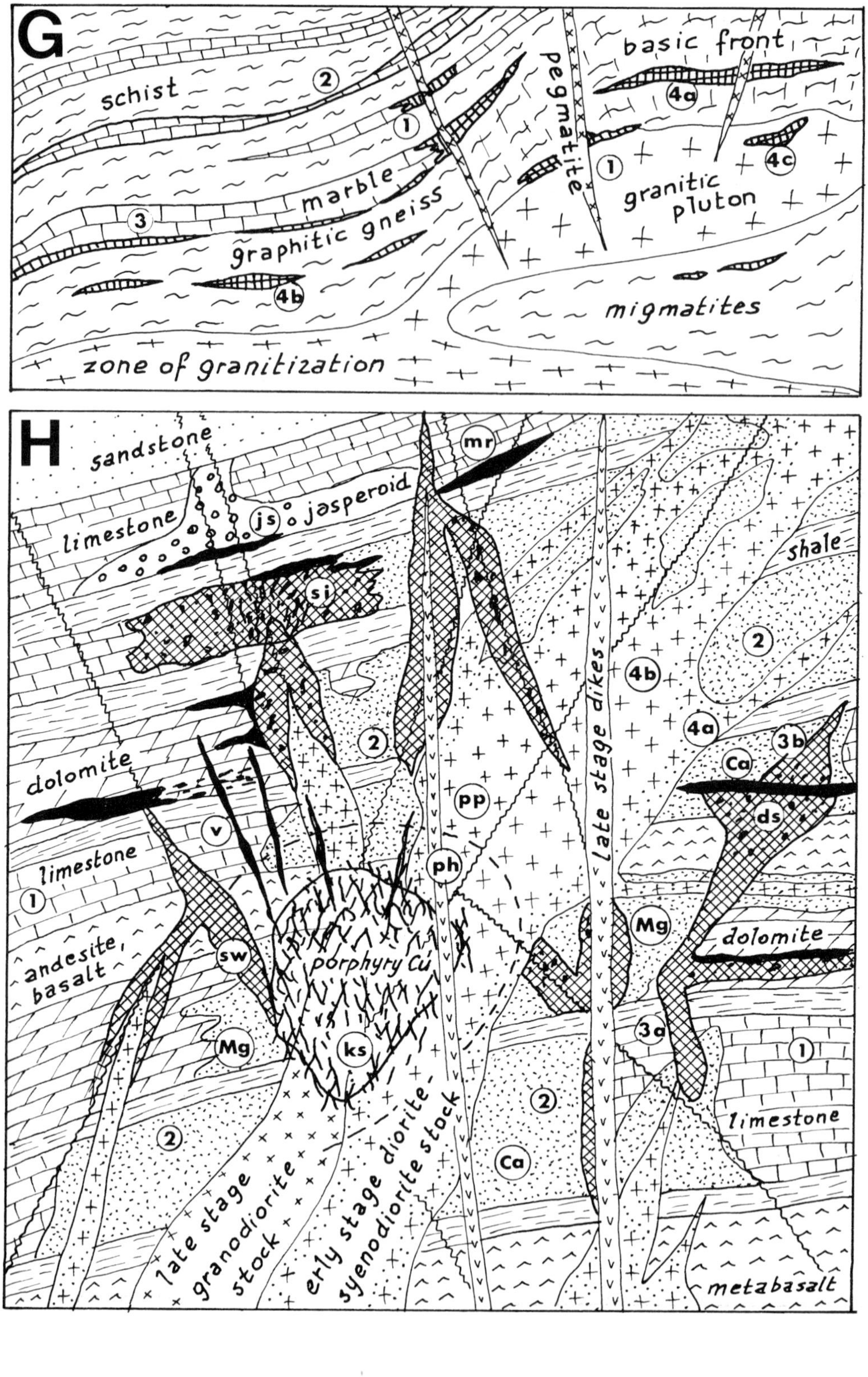
G
schist
2
basic front
pegmatite
4a
1
1
4c
marble
granitic pluton
3
graphitic gneiss
4b
migmatites
zone of granitization
H
sandstone
mr
limestone
js
jasperoid
si
shale
2
late stage dikes
4b
4a
3b
Ca
dolomite
2
pp
ds
v
ph
limestone
1
Mg
dolomite
andesite, basalt
sw
porphyry Cu
3a
Mg
ks
1
2
limestone
2
Ca
late stage granodiorite stock
erly stage diorite-syenodiorite stock
metabasalt

Fig. 28-33. General skarn model.

(G) General concept and deep level skarns (katazone to lower mesozone). 1="Normal" (contact magmatic or hydrothermal metasomatic) skarn; 2=skarnoid (pseudoskarn); skarn association resulting from regional isochemical metamorphism of a rock of suitable composition (e.g pelosiderite, pelitic ankerite); 3=bimetasomatic (reaction) skarn, generated by a local exchange between unlike lithologies in thermal or metamorphic aureoles; 4=multistage, genetically uncertain or relic skarns: a=skarn produced in a "basic front" of magmatism or metamorphism; b=relic skarn in metamorphic terrains; c=relic skarn in rafts and xenoliths in plutonic rocks.

(H) Skarns (and affiliated mineralizations) associated with high level "wet" plutonic systems emplaced in carbonate containing supracrustal rocks. All the skarns here correspond to the "normal" (magmatic and hydrothermal metasomatic) skarn (1) in (G) above. Based on Cu skarn-porphyry copper system.

1=Isochemical contact hornfelses and recrystallized carbonates in place of all sediments present in the thermal aureole (the rock names used are the original compositions); 2=early ("primary") metasomatic skarn resulting from local exchange at high temperatures between magmatic and wallrock components. Garnet, pyroxene, forsterite, etc. + magnetite. No sulphides formed; Ca=calcic skarn (in place of limestone); Mg=magnesian skarn (in place of dolomite). 3=Late ("postmagmatic hydrothermal", infiltrated) skarn, formed by a large scale transfer and exchange of components between hydrothermal fluids and carbonates. Sulphides may replace skarn (2) or any other rocks (including the intrusive rocks) along faults and fractures. 3a=Prograde skarn, invades rocks not previously converted to (2) skarn; 3b=retrograde, retrogressively altered (e.g. by hydration) skarn (2), along fractures.

Common mineralizations associated with (3): ds=disseminations of sulphides in (3a) or (3b) skarns; sw=stockworks superimposed on skarn (actinolite, chlorite, quartz, calcite, etc. gangue); si=stockworks, impregnations superimposed on non-skarn hornfelses; v=fracture veins; mr=massive replacement, typically at "marble line" (contact of skarn and unreplaced marble); js=massive or disseminated sulphide replacements in silicified carbonate (jasperoid).

4=Endoskarn (within the intrusion); 4a=proximal (garnet, pyroxene, epidote); 4b=distal (biotite and hornblende converted to diopside, epidotized or argillized plagioclase, etc.). The same mineralizations as in (3) may be present. 5=Late stage intrusive stock (more siliceous) associated with porphyry Cu stockwork in a K-silicate alteration envelope (ks); ph=phyllic alteration; pp=propylitic alteration. From Laznicka (1984).

Simultaneously deposited ore minerals include magnetite, borates (e.g. ludwigite) and scheelite.
(3) Hydrothermal stage (retrograde alteration and mineralization). The cessation of the skarn growth (2) was followed by an influx of hydrothermal fluids of magmatic derivation as well as of groundwater, connate and similar water origin. The fluids caused retrogressive alteration of the earlier skarn minerals, as well as alteration of any other silicate rocks present, and dissolution of the rock carbonate remains. The skarn alteration products include hornblende, biotite, epidote, chlorite, albite, plagioclase, tremolite, actinolite, serpentinite, talc, etc. The ore minerals, mostly sulphides (pyrrhotite, pyrite, sphalerite, galena, etc.) as well as a late-stage scheelite, cassiterite, etc., accumulated as open space fillings (in breccias, stockworks, fractures) or as replacements of carbonates. The replacement of small carbonate relics in the skarn zone produced a scattered mineralization. Replacement of the marble along the skarn/marble contact ("marble line") produced the largest and richest hydrothermal-metasomatic orebodies, particularly of Zn-Pb. This was caused by the sudden neutralization effect of carbonate on the mineralizing fluids.

Realisation that the economic mineralization of most base, rare and precious metals in skarns is a late-stage hydrothermal one is of a considerable practical importance to the prospector. It places several constraints on the ore search in the skarn association. Since the hydrothermal stage was, to some extent, independent of formation of the earlier skarn zones, hydrothermal deposits other then skarn could have formed simultaneously in the area, in both the fractured and altered intrusive (e.g. porphyry Cu) as well as in the supracrustals (e.g. Pb-Zn replacements in carbonates; compare the next section; metalliferous fissure veins and stockworks).

Skarns are widespread and have a range in age from Proterozoic to Pliocene, although Mesozoic and early Tertiary skarns have now reached their "prime time" of exposure and are most numerous and productive. Skarns are common members of complex local mineralization systems (e.g. porphyry Cu stockwork + skarn + veins; scheelite skarn + Mo stockwork). Examples of metalliferous skarns are reviewed later in the framework of economically accumulated metals.

28.4.12. Replacements other than skarns

Replacements of carbonates in granite aureoles are the most common. These are frequently present without the associated skarn silicates at some distance from major intrusive contacts, or adjacent to intrusive dikes and sills. The most common affiliated alteration is silicification by which the carbonate was converted into a gray alteration chert (jasperoid of Lovering, 1972). Lovering described the petrography and field occurrence of jasperoid in considerable detail and concluded that, in mineralized regions, jasperoid silicification is usually of an early stage and was followed by a later sulphide emplacement. In galena-sphalerite, jasperoid-hosted deposits (e.g. East Tintic, Pioche, Leadville, Gilman, Eureka) the sulphides typically fill open spaces (fractured, brecciated portions) in jasperoid and also replace carbonate relics.

Silicification in carbonates is common and not every jasperoid indicates the presence of ores. Lovering (1972) pointed out that phaneritic texture, abundant vugs, brown colour, highly variable grain size, reticulated microtexture, the presence of goethite, jarosite, pyrite and high trace metal contents, are characteristic features of the ore-associated jasperoids. In some cases discussed in greater detail in Chapter 23 (under karst), jasperoid and hydrothermal sulphides were superimposed on paleokarst, and some Pb-Zn (and other) mantos could be modified relic Mississippi Valley-"type" ores.

Replacements or jasperoid-associated open space fillings in carbonates usually take the form of mantos (peneconcordant sheets and lenses), chimneys (subvertical tubular bodies) or irregular masses. Fault control is important. Faults are often hard to interpret due to the "healing effect" of carbonate recrystallization and redeposition.

Metals other than Pb-Zn and Sb are relatively rare in jasperoids, unless one includes the copper ores in silicified carbonates in porphyry copper systems such as Clifton-Morenci, Bisbee, Ely and others. Gold associated with pyrite has been reported in jasperoid replacing Cambrian carbonates in the Drum Mountains, Utah.

Despite the jasperoid abundance, it is not necessarily a dominant gangue in all non-skarn replacements. The pyrrhotite-cassiterite mantos in dolomite in north-western Tasmania (Renison Bell, Mt.Bischoff, Mt. Cleveland) are contained in recrystallized and lightly talc-altered dolomites with little quartz. Many Pb-Zn replacements are also located here.

Ore replacements in non-carbonate rocks are substantially rarer and usually enigmatic. Often they formed originally as open space fillings (e.g. stockworks or breccias) and later healed to such a degree that the former avenues of passage of the ore-bearing hydrothermal fluids were obliterated. Examples are numerous. In the Midnite uranium mine near Spokane, Washington (Nash et al., 1981), pitchblende and coffinite form replacements as well as disseminations parallel with foliation and fracturing in Proterozoic schists and metacarbonates, about one metre from the margin of a late Cretaceous porphyritic granite. A special case in which a carbonate (dolomite) was first generated by hydrothermal alteration of ultramafics and was later replaced by Pb-Zn and Sn ores (in the Dundas field, Tasmania), was reviewed in Chapter 7.

28.4.13. Relic and remobilized orebodies in "granite" aureoles

Every metallic mineralization present in a "granite" aureole and older than the postmagmatic suite of ores is a relic from the past, genetically unrelated to the "granite". Some such relics, mostly veins, may have formed shortly before the granite emplacement at the present (higher) crustal level, where they were later approached by the rising and stoping intrusion. Other orebodies formed during older periods of magmatic activity, during deposition of the supracrustal rocks e.g. from hydrothermal fluids discharged on the seafloor or during continental sedimentation (paleoplacers), etc. Concordant and peneconcordant deposits and metal-enriched horizons are most characteristic.

A relic orebody may suffer a contact metamorphism, but remain otherwise unmodified and generally recognizable. Examples of this include the metavolcanics-hosted massive Cu-Zn sulphide orebodies in the Akenobe field, Japan, mentioned in Chapter 26. In remobilized orebodies, earlier metallic accumulations were thoroughly reconstituted (texturally as well as, usually, mineralogically) and redeposited within the newly generated pattern of plutonic deposits, from which they are often difficult to distinguish. Ore remobilizations (reconstitutions involving no substantial increase in grade or tonnage) and mobilizations (accumulation of ore substance from a previously more dispersed precursor that required concentration) have recently been reviewed by Laznicka (1985b).

28.4.14. Mineralized metamorphosed "granites" and metamorphogenic ores in meta-granites

Regionally metamorphosed granitic rocks often act as passive, genetically unrelated hosts to younger intrusions and affiliated orebodies (e.g. the Precambrian "granites" in the Tertiaty Colorado Mineral Belt), but metamorphosed old relic hydrothermal-plutonic ores, particularly veins, are quite rare. This is because the deep portions of "granite" batholiths that form the bulk of the meta-granites originally lacked associated ores, and because thin ore veins involved in penetrative deformation and metamorphism most probably underwent dispersion. The recorded examples of metamorphosed deposits of plutonic affiliation are all controversial.

The Gibraltar (Granite Mountain) copper deposit, British Columbia (Sutherland Brown, 1974b; 1.33 Mt Cu/0.371%, 37 Tt Mo; Fig.28-34), is interpreted as an example of a dynamometamorphosed porphyry copper, with a gneissic fabric. This deposit is hosted by a foliated Triassic tonalite and granodiorite pluton, intruded into a Permian "oceanic" volcanic-sedimentary assemblage. The plutonic rocks are marked by a primary foliation over which is superimposed a largely parallel secondary foliation. The most intensively foliated "granite" is now a phyllonitic schist entirely lacking feldspar. The Cu-Mo mineralization is in four elliptical, W.N.W.-elongated orebodies, that have well developed leached capping and oxidation, cementation zones. The hypogene ores are in stockworks of quartz, pyrite and chalcopyrite veinlets and veins, with sericite and chlorite selvedges parallel with foliation. Superimposed on these are irregular and lensoid, thicker veins, reminescent of the "secretion veins" in metamorphic terrains. They contain large nests of pyrite, chalcopyrite and molybdenite and

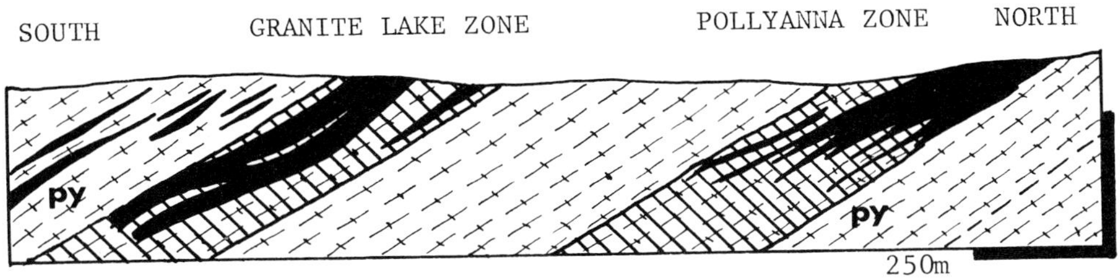

Fig. 28-34. Gibraltar field, British Columbia, disseminated Cu mineralization in Mesozoic foliated and metamorphosed quartz diorite. Diagonally ruled section denotes the ore zone; black=orebodies; py=pyrite halo. From LITHOTHEQUE, after Drummond et al. (1976).

may be drusy. Visually, the ore specimens are reminescent of ores in deformed and altered marine metavolcanics, such as those described in Chapter 11 (e.g. Mt. Lyell).

Another example, the Tungsten Queen hübnerite deposit in the Hamme District, North Carolina, was in a vein, 3.5 km long and up to 10 m thick, traceable through a slate belt and a granitic pluton. Eight en-echelon ore shoots are in the vein. Foose et al. (1980) demonstrated recently that the vein and its surroundings have been deformed, lightly metamorphosed and remobilized during at least two episodes of folding and shearing. This produced a pronounced alignment of ore and gangue minerals and may have caused the en-echelon arrangement of the ore shoots.

28.4.15. Ore accumulations resulting from weathering of "granites" and their mineralized aureoles

Tropically weathered "granites" are of little importance economically, although the accessory magnetite used to be recovered from regoliths in Japan (Chapter 23). Gibbsite nodules may form in lateritic profiles over "granites", but bauxite deposits formed only on sorted clays winnowed from such profiles. Weathering of hydrothermally mineralized regions results in a variety of leached cappings (Blanchard, 1968), gossans and secondary enriched zones, too complex to be satisfactorily reviewed here but treated in textbooks and several reference books (e.g. Smirnov, 1951).

Weathering and residual enrichment superimposed on zones that are only lightly metal enriched (for example in Mn) and are uneconomic in the fresh state, may produce economically valuable orebodies. Sizeable Mn-oxide bodies formed over altered carbonates containing replacement and vein Zn-Pb ores, as in Philipsburg, Montana or Novo Brdo, Yugoslavia.

28.4.16. Ores formed by sedimentogenic reworking of "granites"

Ordinary "granites" contribute too little of a valuable heavy mineral fraction to form significant placer deposits. Gabbros are more fertile and may supply significant quantities of ilmenite. Low-grade and widely dispersed occurrences of resistate minerals in mineralized granites and their aureoles (e.g. cassiterite, columbite-tantalite, wolframite, scheelite) can result in valuable placer accumulations (as in the Sundaland, S.E. Asia). Trace metals from granites released in solution (particularly U), often find their way into proximal sediments in which they can precipitate (e.g. the SUV deposits). To a lesser extent, U accumulation can take place already within the granite by infiltration and precipitation from descending solutions along fractures.

28.5. "PORPHYRY" (STOCKWORK, DISSEMINATED) Cu-Mo DEPOSITS

28.5.1. Introduction

"Porphyry" Cu(Mo) deposits constitute a broad and complex category the members of which have only two common characteristics: (1) they represent relatively low-grade but large-tonnage ore accumulations almost always outlined by assay boundaries and (2) they are genetically related to calc-alkaline (exceptionally other) intrusive activity and tend to have a simple hypogene mineralogy (pyrite, chalcopyrite, lesser molybdenite). Their economic importance is extraordinary. The deposits around the "Pacific Basin" alone (Titley and Beane, 1981) represent 365 Mt Cu in 56.7 Bt ore, and 7.02 Mt Mo in 24.2 Bt ore. The "porphyry" deposits outside this probably represent a further 53 Mt Cu and 1 Mt Mo, yielding a grand total of about 420 Mt Cu and 8 Mt Mo.

The reservoir and generator rocks to "porphyry Cu" need not be porphyritic (although in the typical localities they are), so the term is not completely representative. Considered from the viewpoint of lithologic associations, "porphyry Cu" in this book have been broken down into deposits in volcanic association (Chapters 10, 13, 14, and 26), deposits in plutonic association (this chapter) and Precambrian deposits (Volume 2). The plutonic "porphyry" association is the largest, most important and most "typical".

Mesozoic and early Cainozoic "porphyry" Cu(Mo) deposits are currently enjoying their prime time of exposure, but the age range of this class is from Archean to about 1.2 m.y. "Porphyry" Cu(Mo) are the most intensively studied category of hydrothermal deposits and the one on which the most detailed and numerous experimental and physicochemical data (many of which provide genetic constraints) have accumulated. Their treatment, however, is far beyond the scope of this book and the reader is referred to the recent review of Titley and Beane (1981). There are few monographies dedicated to porphyry coppers in general (e.g. Pavlova, 1978), but excellent summaries can be found in several regional reviews and special volumes (e.g. Sutherland Brown, ed., 1976; Hollister, 1978).

28.5.2. Porphyry Cu terrains: the hypabyssal and plutonic levels

Porphyry coppers occur in former (some in still existing) island arcs and Andean-type continental margins (Chapter 5), and much has been written about their position above former Benioff zones. They are characteristically associated with calc-alkaline, I-type, multi-stage plutonic systems and the bulk of orebodies formed within the depth range between about 500m and 5 km (most between 2-3 km; compare Fig. 27-3). As a consequence, porphyry Cu are most abundant in continental margin plutonic belts eroded to depths of around 2-3 km. In less eroded regions marked by andesite to rhyolite volcanics, the "porphyries" are still largely hidden in depth. In deeper-eroded regions they have disappeared irreversibly. The depth variations of porphyry coppers have been integrated in the often quoted model by Sillitoe (1973), and two (transitional) depth levels: hypabyssal and plutonic, are repeatedly emphasized in the literature.

In the higher level of exposure, hypabyssal, "phallic", or "stock" porphyry copper systems are developed (Nielsen, 1976; Sutherland Brown, 1976; Hollister, 1978; Fig. 28-35). There, the orebodies are associated with high-level (epizonal) composite intrusions dominated by porphyritic textures. These are usually emplaced in a comagmatic volcanic suite typically deposited within a former subaerial volcanic centre and composed of andesite and rhyolite lavas and pyroclastics. In many cases, the volcanics are completely missing and a great variety of basement rocks genetically unrelated to the magmatic complex surround the intrusions and interact with them. The intrusive stocks usually consist of a variety of small plugs, intrusive breccias and dikes having a substantial vertical extent but a small area. Petrographically, quartz-feldspar porphyry, latite and quartz monzonite porphyry, granodiorite porphyry, are the most common in the "continental" setting. Quartz diorite and diorite are dominant in island arc settings, where the surrounding rocks range from basalts to pyroxene andesites.

The copper (molybdenum) mineralization is usually associated with a specific intrusive phase that tends to be among the youngest, but often not the youngest because, almost always, the mineralized phases are intersected by minor post-ore dikes. The economic orebodies are most commonly situated at the periphery of the generator intrusion in both endo- and exocontact, and the alteration-mineralization pattern is broadly concentric and centered on the intrusion. When reactive rocks (carbonates, calcareous tuff) form the wallrocks, the stockwork or disseminated Cu-Mo ore in the intrusive endocontact usually grades into a copper skarn in the exocontact.

At the lower level of exposure corresponding to the "plutonic" or "batholithic" porphyry copper systems (Fig. 28-36), a fairly deeply eroded mesozonal pluton is the dominant unit in the ore zone, and the

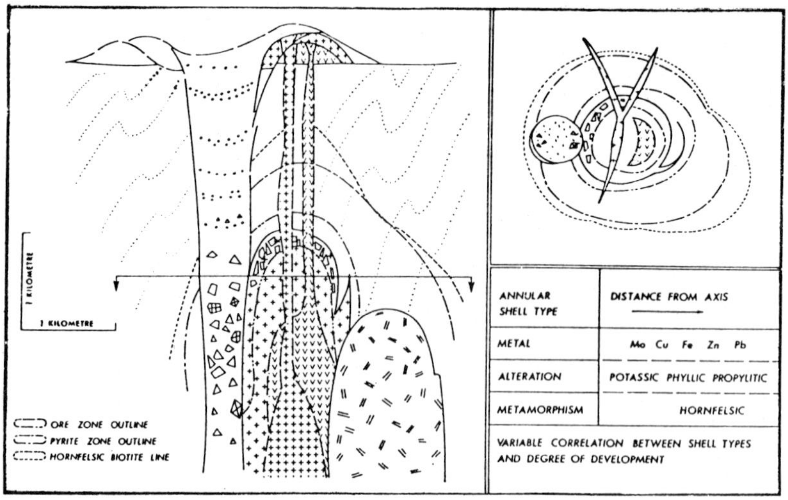

Fig. 28-35. Hypabyssal ("phallic") porphyry copper model, plan and section. From Sutherland Brown (1976), courtesy of the Canadadian Institute of Mining and Metallurgy and the author.

former roof rocks occur mainly as relics outside the ore field. The mineralized plutons are usually multiphase, and porphyry Cu tends to be associated with the youngest phases and occur in the interior or at margins of plutons. Structural control is of paramount importance and mineralization is often contemporaneous with active deformation, commonly associated with more felsic minor higher-level intrusions (stock, dike swarm) emplaced into a more mafic reservoir rock. The reservoir rocks are entirely intrusive when ore zones are situated in plutonic interiors, and differentiation between generator and reservoir rocks may prove to be problematic. Marginal orebodies can be partly hosted by the basement rocks in plutonic exocontacts (including carbonates), but in general the variation of mineralization styles in plutonic deposits is less than in the hypabyssal systems. This statement is, however, invalid when multistage mineralizations involving leaching and redeposition of an earlier ore took place, as in Butte, Montana. The ore zones are more commonly tabular rather than equidimensional. They have a linear surficial expression and are controlled by the intensity of fracturing. The alteration-mineralization zones consist most commonly of broad areas showing weak alteration and low-grade ore mineral distribution, over which are superimposed more localized, higher-grade ore-bearing fracture zones or discrete veins.

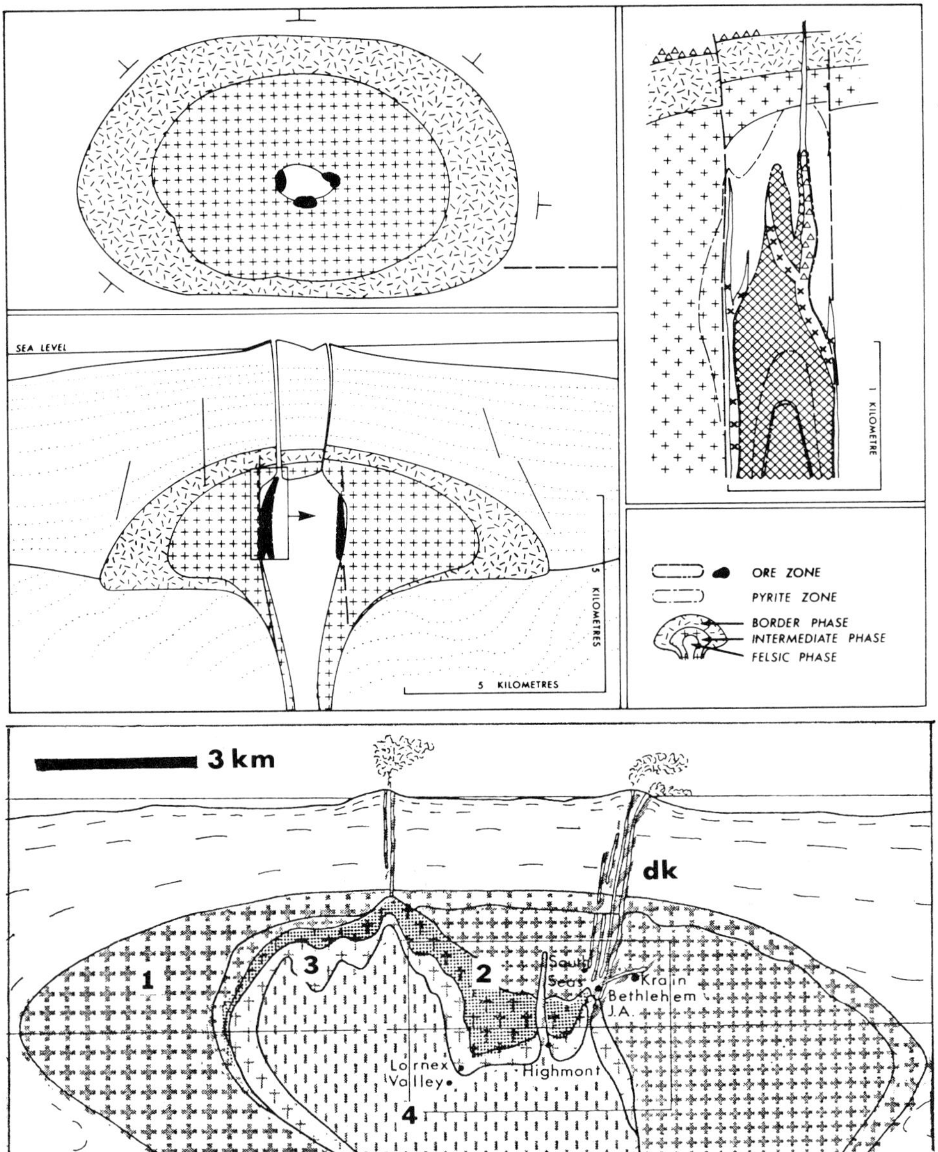

Fig. 28-36. Plutonic porphyry copper model: general (top) from Sutherland Brown (1976), and representing the Highland Valley plutonic porphyry copper field (bottom), from McMillan (1976). Highland Valley is hosted by the late Triassic Guichon Creek Batholith. 1=Border phase, diorite to quartz diorite; 2=quartz diorite to granodiorite; 3=leucocratic granodiorite; 4=quartz monzonite to granodiorite; pd=porphyry dikes. Courtesy of the Canadian Institute of Mining and Metallurgy and the authors.

28.5.3. Intrusive associations

Regardless of the level of emplacement, there are two or three distinct intrusive associations affiliated with porphyry coppers. These have been contrasted in the literature (e.g. Hollister, 1978). They comprise (1) high-quartz, granodiorite or quartz monzonite plutons, marked by low $Na_2O:K_2O$ ratios (calc-alkaline, "normal", etc. association); (2) quartz-free plutons (diorite, syenite, alkalic suite, etc. association) and (3) low-quartz quartz diorite and soda granitoid plutons of former island arcs have a high $Na_2O:K_2O$ ratios and are transitional in many respects between (1) and (2). In the literature, the latter are most commonly considered with an association of type (1).

The intrusive association has a considerable influence on the alteration pattern: (1) usually has a "full" (model) range of alteration zones, from potassic through phyllic to propylitic and (2) generally lacks the phyllic zone and potassic alterites grade directly into the propylitized rocks. The metallogeny differs substantially too. Copper in chalcopyrite is the dominant metal in both associations, but in (1) Mo in molybdenite is the principal by-product in the disseminated or stockwork ores (the Cu:Mo ratios have a range between 20:1 and 100:1) and Pb,Zn,Ag veins or replacements commonly occur along the fringe of a porphyry stock. In (2), Mo is virtually absent and so are Pb-Zn occurrences, but the ores are usually high in Au and small gold showings are sometimes associated zonally. The type (1) and (3) associations carry at least 90% of porphyry copper occurrences. The (2) affiliated "porphyries" have been documented in only a few places in the world, being best developed in the mafic volcanic facies belts in British Columbia.

In a recent paper dealing with petrochemical affiliation of the stockwork molybdenite deposits, Mutschler et al. (1981) recognized "granodiorite" and "granite" molybdenite systems. The former have affiliation identical to the "normal" porphyry coppers into which they are gradational (e.g. Brenda or Boss Mountain, British Columbia). The latter Mo stockworks ("Climax-type") are spatially separated from porphyry and other copper deposits (as in the Colorado Mineral Belt). The affiliation of several Mo-Cu deposits in eastern Mongolia and East Transbaikalia (e.g. Zhireken) said to be in "granites", is uncertain.

28.5.4. Hypogene alteration and mineralization patterns

Although it had long been known that porphyry Cu deposits were associated with intensive alteration, it was not until the 1970s that the prevalent alteration / mineralization pattern had been incorporated into several all-embracing models. The model of Lowell and Guilbert (1970) based on the San Manuel-Kalamazoo deposit in Arizona and believed to be close to an average situation in the western United States, is essentially representative of the hypabyssal porphyry system emplaced into silicate rocks. In a later paper

(Guilbert and Lowell, 1974), additional alteration pattern variation due to the presence of carbonate wallrocks or deeper levels of exposure (plutonic porphyry Cu) have been incorporated. Gustafson and Hunt (1975) in their meticulous paper on the El Salvador deposit, Chile, stressed the dynamism of the porphyry Cu systems complete with numerous progressive and retrogressive events creating new alterations and mineralizations as well as destroying or modifying the existing ones. Hollister (1978) reviewed the "diorite" model of lithologic association and alteration. At present, alterations in porphyry deposits almost possess the status of a separate science and there is a broad range of variation (Figures 28-37 and 28-38).

Briefly, as in the skarn systems, the first consequence of igneous intrusion into the wallrock is contact metamorphism. This is most apparent in pelites or in marine volcanic associations, that change into brownish biotite-plagioclase or greenish amphibole-plagioclase hornfelses. There is virtually no obvious thermal metamorphism, when intrusions were emplaced into older "granites", gneisses or migmatites. Brittle fracturing in contact hornfelses as well as in the intrusive rocks provided openings for the movement of hydrothermal fluids.

In silicate hostrocks, the earliest K-silicate progressive alteration resulted either in the growth of neoformed minerals (K-feldspar and biotite), or in reconstitution of existing potassic constituents. Quartz and minor albite, sericite, anhydrite and apatite may be associated. The K-silicate alteration assemblage forms within or close to the generator intrusion and because it involves minerals that are also common constituents of unaltered granitic rocks, it may be difficult to recognise. Fresh-looking fracture veinlets filled by K-feldspar and scattered fresh biotite flakes cutting across fractured and generally altered (decomposed plagioclase, amphibole, biotite) "granite" or other rocks, are most diagnostic. The K-silicate zone is of greatest importance economically and it usually carries the bulk of chalcopyrite and molybdenite in the form of scattered grains on the walls of dense fractures, in quartz or silicates-filled veinlets or pervasively disseminated in an alteration partially healed hostrock. Hand specimens of low-grade hypogene ores are very inconspicuous. Many are virtually unrecognizable from a "rock" granite.

Phyllic (sericite, quartz-sericite) alteration is substantially more conspicuous than the anhydrous K-silicate alteration. The altered rocks have a mottled or monotonous beige to yellowish-green appearance when fresh, contain abundant scattered pyrite and are often laced by fractures covered by slickensided coatings of greasy-looking sericite. Chalcopyrite may be present and several porphyry coppers have their ore in the phyllic zone. In weathered outcrops, sericite zones create a mottled beige, yellow and brown scenery, a consequence of pyritic weathering. It has been suggested that K-silicate alteration results largely from the magmatically-derived fluids, while the phyllic (and argillic, propylitic) alterations are the result of convecting meteoric waters.

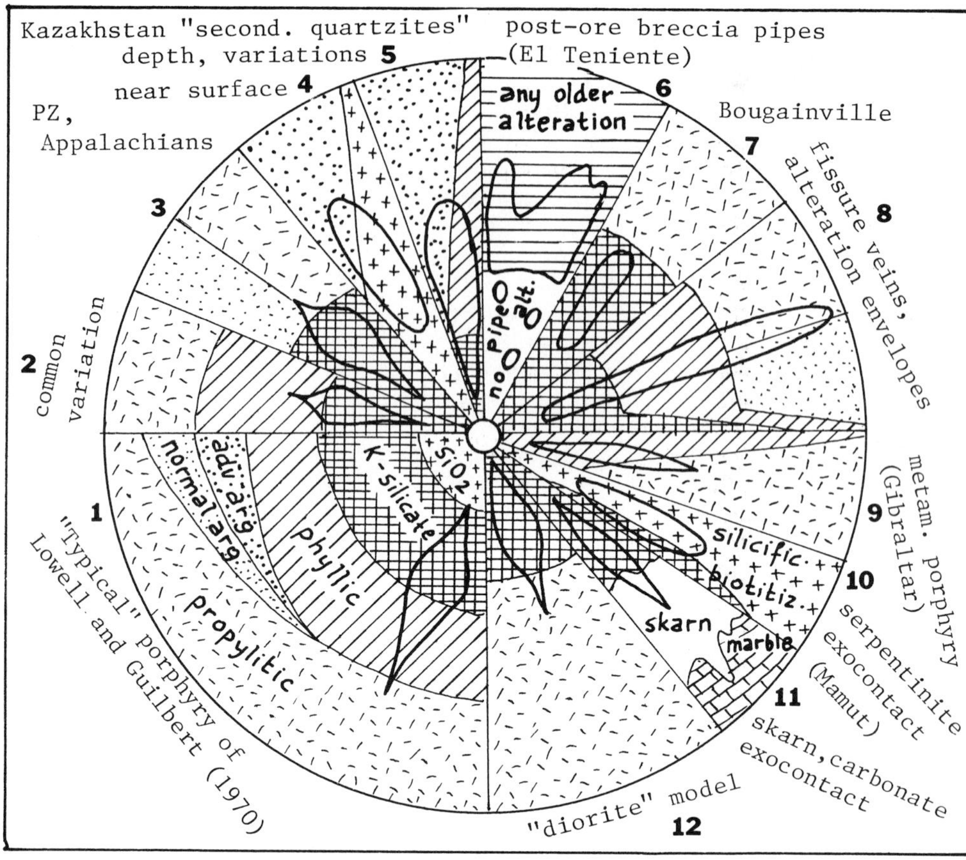

Fig. 28-37. Examples of the variations of porphyry copper alteration patterns. The position of economic Cu mineralizations is outlined by a thick black line. From Laznicka (1984).

Argillic alteration ("normal": mostly kaolinite or advanced: dickite, diaspore, andalusite, corundum, alunite, dumortierite) resulted from acid leaching. A portion of it is attributed to supergene processes caused by descending meteoric waters containing high concentrations of sulphuric acid released by sulphide breakdown. This alteration is sulphide-destructive and it causes impoverishment. Cu mineralization in this zone largely takes the form of relics of chalcopyrite in fracture quartz veinlets or veins, or chalcocite-coated pyrite. Hypogene, late-stage sericitization and argillization are often associated with the process of hypogene leaching, short-distance metal transfer and generation of localized, high-level fissure veins filled by enargite, chalcocite, bornite and other minerals which are gradational into epithermal veins (e.g. in

Butte, Morococha, Récsk).

Propylitic alteration is closely equivalent mineralogically to the greenschist metamorphic facies (chlorite, epidote, calcite, albite). In completely zoned alteration profiles propylites are generally devoid of ore minerals, but they may contain orebodies in several "volcanic" porphyry coppers hosted by metabasalts or mafic metaandesites; by quartz-free intrusions of the "diorite" model; by pre-ore diabase dikes; by retrograded skarn profiles, etc. (e.g. Afton, B.C., El Arco, Sierrita, Esperanza, Bozschekul', etc.).

A variety of additional alterations involve tourmalinization and rare greisenization. Thorough silicification often results in the generation of monomineralic quartz metasomatites ("secondary quartzites"). Silicification is the most widespread and genetically the most controversial. Following a literature review of several hundred porphyry coper occurrences and a brief examination of at least one hundred of them in the field, one gains the impression that there are at least three major genetic categories of silicification which are often transitional. These are as follows: (1) a high-temperature, early "magmatic" silicification, localised in the core of zoned hypabyssal porphyry systems, flanked by, and often transitional into, the zone of K-silicate alteration. It may carry scattered flakes of molybdenite. (2) "Convective" ascending silicification, associated with phyllic and advanced argillic envelopes. Its position is variable. It reaches into moderate depths where it wedges out. Underneath are often K-silicate altered intrusive rocks. (3) Supergene silicification in some respects reminescent of silcretes, associated with leached cappings and zones of descendent argillization.

28.5.5. Weathering modification and supergene ore zoning

Dense fracturing and widespread sulphides (Cu and Mo sulphides as well as the more widespread pyrite) combined with humid climates, cause a rapid and deep weathering of exposed porphyry copper deposits. In mountainous, rapidly eroded and well-drained humid tropics (e.g. in the Caribbean, Panama and western Colombia, Indonesia and the Philippines) a porphyry copper deposit may be exposed, weathered and completely removed within several hundred thousand years. From the discussion above it may appear that weathering of porphyry coppers is a completely destructive process. This is not always the case. In fact, the early generation of "porphyries" in the western United States developed in the early 1900s, were all producing ores from the secondary supergene enriched zones that had the form of a locally almost solid chalcocite blanket. At that time primary (hypogene) "porphyries" were uneconomic. El Teniente, Chile, was one of the first primary porphyry Cu-Mo deposits to start production in the 1930s, but there the ore was unusually rich (1.5-2% Cu). A wave of post-World War II discoveries and development of "porphyries" in the 1960s consisted of a majority of deposits not substantially enriched, having a grade as low as 0.284% Cu.

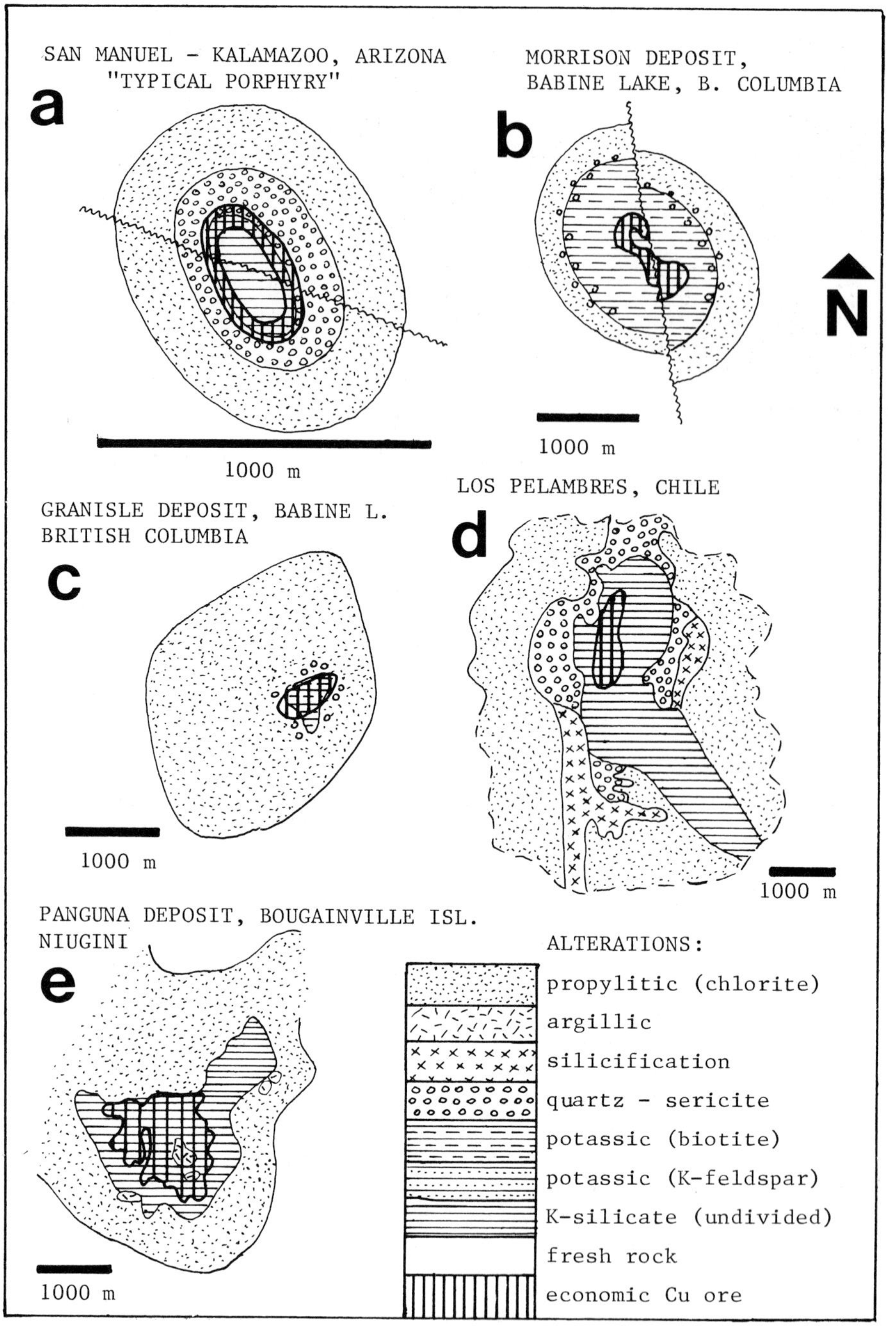
SAN MANUEL - KALAMAZOO, ARIZONA
"TYPICAL PORPHYRY"
a
1000 m
MORRISON DEPOSIT,
BABINE LAKE, B. COLUMBIA
b
1000 m
N
GRANISLE DEPOSIT, BABINE L.
BRITISH COLUMBIA
c
1000 m
LOS PELAMBRES, CHILE
d
1000 m
PANGUNA DEPOSIT, BOUGAINVILLE ISL.
NIUGINI
e
1000 m
ALTERATIONS:
propylitic (chlorite)
argillic
silicification
quartz - sericite
potassic (biotite)
potassic (K-feldspar)
K-silicate (undivided)
fresh rock
economic Cu ore

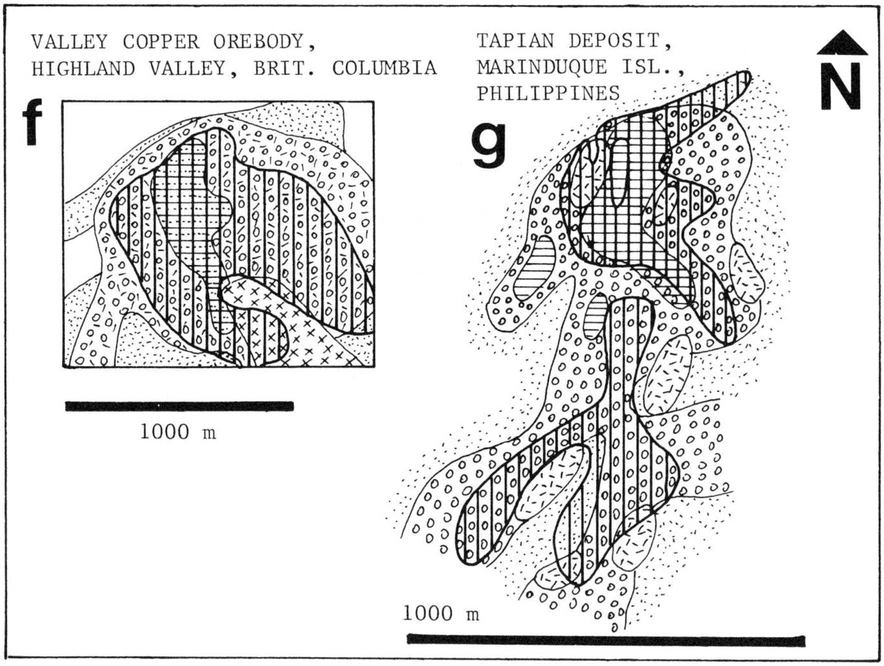

Fig. 28-38. Alteration patterns (in plan) of selected porphyry copper deposits; a-e=hypabyssal deposits, f,g=plutonic deposits. From Laznicka (1984), based on data in Lowell (1968), Carson and Jambor (1974), Sillitoe (1973b), Baumer and Fraser (1975), Osatenko and Jones (1976) and Loudon (1976).

At present, a few deposits still produce exclusively from secondary enriched zones (e.g. Clifton-Morenci, Arizona). A large number of deposits started in the enriched zone and now produce both primary and secondary ore (e.g. most Arizona "porphyries"). The majority of mines produce from the hypogene zone (e.g. the Canadian "porphyries") or from non-enriched ores in the various stages of weathering modification (most of those in the humid tropical belt).

From the economic point of view, it is important to make a distinction between weathering modification and secondary enrichment. All porphyry deposits are weathering modified at the outcrop, the style and intensity of which is influenced by the local climatic zone. In the recently glaciated terrain of Alaska and Canada, the primary (hypogene) zone often crops out as a hard rock on the surface often in glacially scoured hummocks, and is virtually fresh except for thin superficial coatings of malachite or limonite on sulphide grains (e.g. the discovery outcrop of the Brenda mine). Elsewhere, yellowish-brown "limonite" coating marks the mineralized rock (e.g. the Orange Hill near Nabesna, Alaska), but this coloration tends to be better

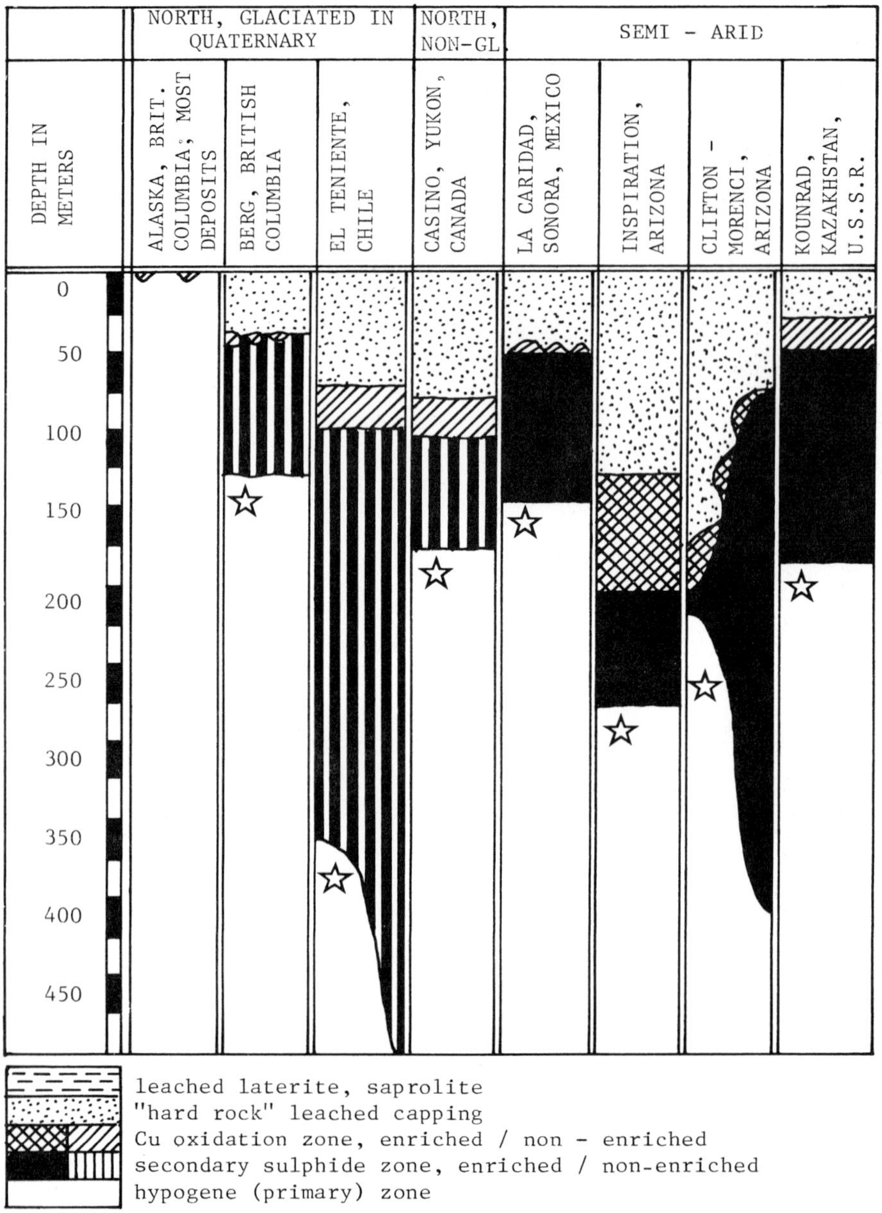

Fig. 28-39. Supergene zoning at selected porphyry copper deposits in different climatic zones. Stars indicate full or partial relic (pre-Quaternary) zoning. From Laznicka (1984).

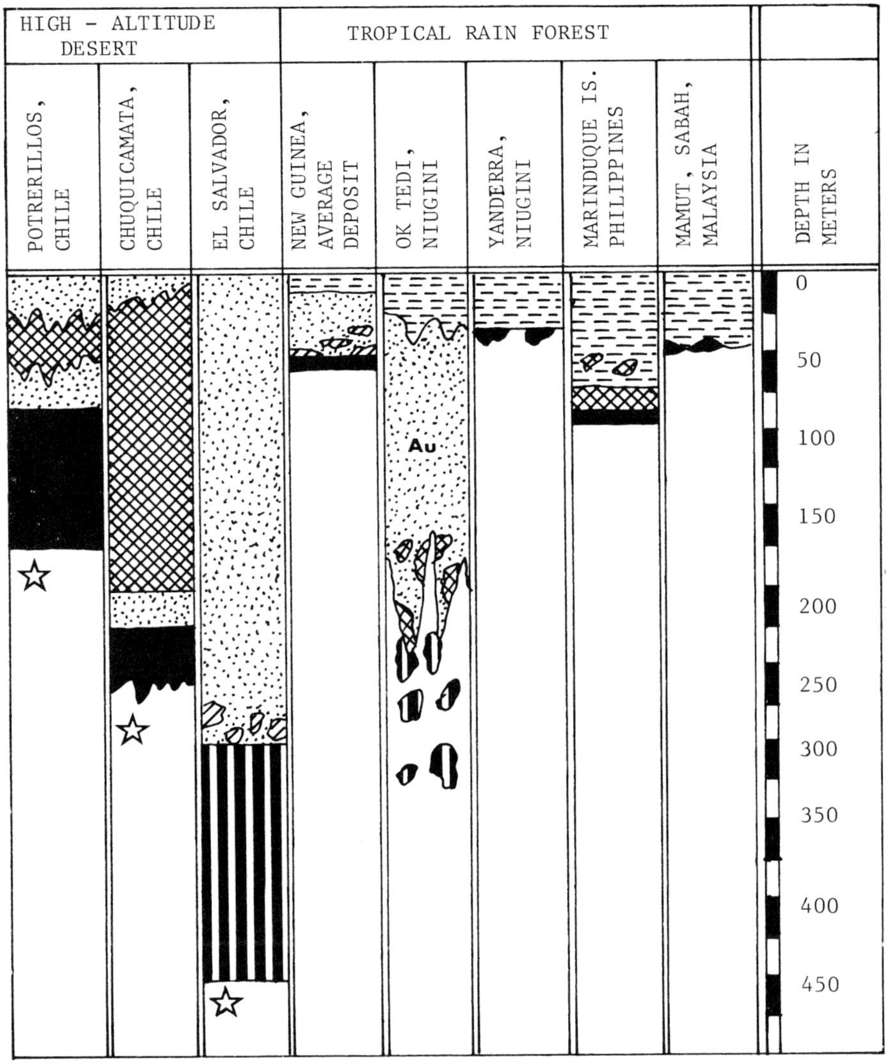

developed over the sericite-pyrite altered fringe than over the Cu-Mo bearing core (when this is in the K-silicate alteration zone). There are, however, exceptions and some recently glaciated deposits are weathering modified (e.g. Casino, Yukon, to a depth of 170 m; El Teniente, Chile, to a depth of up to 600 m; Fig. 28-39).

In the rugged humid tropics (as in New Guinea, Melanesia, Philippines), considerable differences in the thickness and nature of the weathered zones exist among deposits situated on high ridges (up to 330 m thick weathered zone, Ok Tedi, Niugini), steep flanks (thin weathered zone) and floors of mountain streams (fresh rock). In the ridge setting as well as over "porphyry" deposits located in a rolling hills countryside, the top several tens of metres above a porphyry orebody are represented by a soft lateritic soil devoid of visual traces of mineralization as well as being significantly impoverished in the trace Cu content. Trace Mo, however, may occur in higher concentration and gold is often preserved in the residuum. There, ore discoveries have been made chiefly by stream geochemistry. Sometimes, oxidized but not lateritized alteration and ore zones were sighted in stream valleys and on landslide and erosional scars on mountainsides.

In high-pyrite ore systems the next zone under the laterite is a silica-rich leached capping, coated and impregnated by goethite, jarosite and hematite. Again, visible copper minerals are virtually absent and sporadic chrysocolla, cuprite, etc. may appear towards the bottom of this zone. Gold, however, remains and at Ok Tedi a deposit of 30 Mt of 3 ppm Au material (=90 t Au) remains in the leached capping. Zones of secondary sulphides (chalcocite, covellite) are usually patchy and poorly developed in the tropics as well as in the immature recently glaciated profiles. In most cases the Cu sulphides coat or pseudomorph relic hypogene pyrite or chalcopyrite. This is underlain by the primary zone.

In low-pyrite systems or where the acidity is neutralized by wallrock carbonates, serpentinites, etc., brown "limonitic" gossans tend to develope on the surface (e.g. at Mamut, N. Borneo), and malachite, azurite or chrysocolla stains and pseudomorphs after sulphide grains are common in the underlying oxidation zone. In the climatic zone of northern and mountain forests, malachite-stained cliffs were sometimes the discovery sites (e.g. Catface, Lorraine, deposits, British Columbia). At some localities (e.g. Viví Arriba, Puerto Rico) the mineralized intrusion has the form of a tropical saprolite with a well preserved relic texture and feldspars pigmented green by infiltrated Cu oxides.

In the arid to semi-arid zone as found at the Chuquicamata deposit located in the Atacama Desert, Chile (Ruíz, 1965), a thin, siliceous, leached capping is underlain by a rich and conspicuously green oxidation zone composed of a variety of minerals (antlerite with lesser atacamite, brochantite, krohnkite and other). Underneath is a barren (leached) gap floored by an enriched blanket of secondary sulphides (chalcocite, covellite, bornite), followed by the hypogene zone. This local departure from an orderly zoning pattern is interpreted as being a consequence of a fluctuating water table. The rich oxidation zone at Chuquicamata formed by destructive replacement of an ancient zone of secondary sulphides. The present chalcocite zone is the product of more recent supergene activity. A similar situation exists at Potrerillos.

At the El Salvador deposit (Gustafson and Hunt, 1975) located in the same arid setting, no secondary enrichment took place. The uppermost leached zone has a jarosite and goethite coating gradational downward into a zone of hematite-goethite coating. This, in turn, changes into a zone where chalcocite coats relics of the primary sulphides. Little enrichment took place in the latter zone and the grade is only slightly higher than in the primary ore. The effects of supergene processes, however, are traceable to a depth of 450 m.

In the classical area of secondary enriched ores in the western United States (Butte, Bingham, Globe-Miami, Santa Rita, Bisbee, etc.), a beige to rusty, siliceous leached capping with little clay, is underlain by a zone of supergene argillization containing small areas and pockets of jarosite and goethite-limonite in its upper part; sporadic Cu oxides (chrysocolla, malachite, azurite, cuprite) in the centre, and chalcocite with lesser covellite at the bottom. The chalcocite is black, sooty, fracture-coating or pseudomorphing pyrite or chalcopyrite, but locally it can accumulate into almost pure, homogeneous masses of steely gray ores. The chalcocite blankets were up to 200 m thick and were significantly enriched (1-2% Cu with local 30% Cu plus "bonanzas"), compared with the lean hypogene ore (0.1-0.6% Cu). The enrichment is a relic, formed in places as early as the Mesozoic and, at most localities, in the lower Tertiary age. Enriched supergene profiles over the Arizona porphyry coppers were already buried under a thick pile of middle Tertiary volcanics, thus preserved from removal, and only recently exhumed. Livingston et al. (1968) pointed out that all the western U.S. enriched porphyry coppers crop out within few kilometres of the edge of the cover volcanics.

In the Globe-Miami field, Arizona (Peterson, 1962; Fig. 28-40), the different effects of two periods of supergene modifications of porphyry coppers are well apparent. The Tertiary period produced chalcocite enrichment blankets, while the Quaternary period caused oxidation of a portion of the Tertiary chalcocite and also of the erosion-exposed hypogene ore. Substantial amounts of copper are present in "in situ" as well as exotic (transported) chrysocolla and malachite ores, recovered by heap as well as in-situ leaching at the Inspiration mine.

Supergene alterations of porphyry coppers are a fascinating, as well as a practically important, subject. Surprisingly, little new

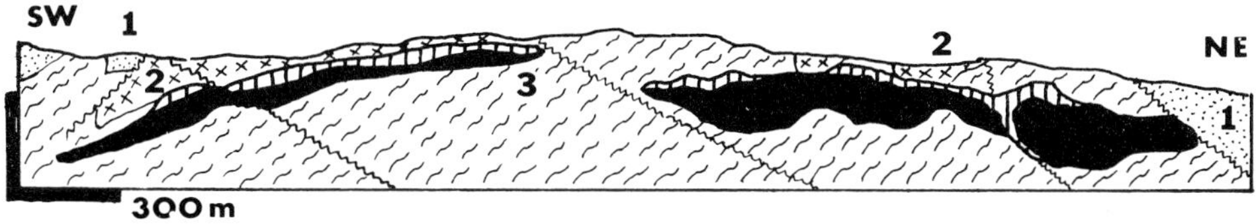

Fig. 28-40. Miami-Inspiration mine area, Arizona, showing an oxide and sulphide enriched porphyry copper. 1=Q-T continental gravels (fanglomerate), volcanics; 2=Cr_3-T_1 quartz monzonite porphyry; 3=Pt 1-2 schist. BLACK: sulphide enriched (chalcocite) orebodies; VERTICALLY RULED AREA: oxide (chrysocolla, malachite) and mixed oxide-sulphide ores. From LITHOTHEQUE, after Peterson (1962).

research has been carried out in the past 15 years, in contrast to the progress in the interpretation of the hypogene "porphyry" systems. The recent comprehensive reviews of porphyry coppers barely mention the subject at all. The success or failure of a prospecting venture for porphyry deposits in terrains other than the recently glaciated ones really depends on the correct interpretation of their surficial expressions, particularly of leached cappings. There are few other mineralization styles known in which their outcrop area is as completely devoid of any conspicuous traces of copper minerals as the high-pyrite "porphyries" of the tropics and the temperate zones. Blanchard (1968) reviewed the art of the interpretation of leached outcrops in considerable depth.

28.5.6. Porphyry copper regions of the world (Fig. 28-41, Tables 28-2, 28-3)

N.W. PACIFIC PROVINCE: THE ALEUTIANS AND THE CORDILLERA IN CANADA AND IN THE UNITED STATES

This extensive province has recently been reviewed by Hollister (1978) and by a number of authors in Sutherland Brown, ed. (1976). Its most consistent unifying characteristic is the general lack of deep supergene modification which is the consequence of glacial scouring of ancient weathering profiles (exceptions: Casino, Krain, Berg, Afton, Cariboo Bell, Taurus, Butte, deposits) and a generally cold climate that accounts for the low rate of present weathering. For that reason, "fresh" porphyry Cu complexes generally crop out at the very surface.

The Aleutian Islands

The western portion of the Aleutian Islands is a Cainozoic active island arc, developed on the oceanic crust. Ore occurrences are rare and porphyry coppers are represented by a showing at the Attu Island affiliated with a 5.9 m.y. old stock of a sodic quartz diorite porphyry, emplaced into Cretaceous mafic volcanics (Hollister, 1978). The eastern Aleutians as well as the Pacific continental margin of Alaska are dominated by Mesozoic and Cainozoic low-grade metasediments and metavolcanics, believed to have been formed in an arc-trench gap setting. Scattered porphyry copper showings are affiliated with Tertiary quartz diorite to quartz-monzonite porphyries and no important deposits have, so far, been discovered.

Pacific coastal belt

This belt follows approximately and discontinuously the Pacific coast west of the great Coast Ranges Batholith from eastern Alaska to Oregon and it includes the islands in the Alaska Panhandle, Queen Charlotte Islands and Vancouver Island. It is marked by a great variety (and also antiquity) of basement rocks (Paleozoic metamorphics, Carboniferous and Permian marine volcanics and sediments

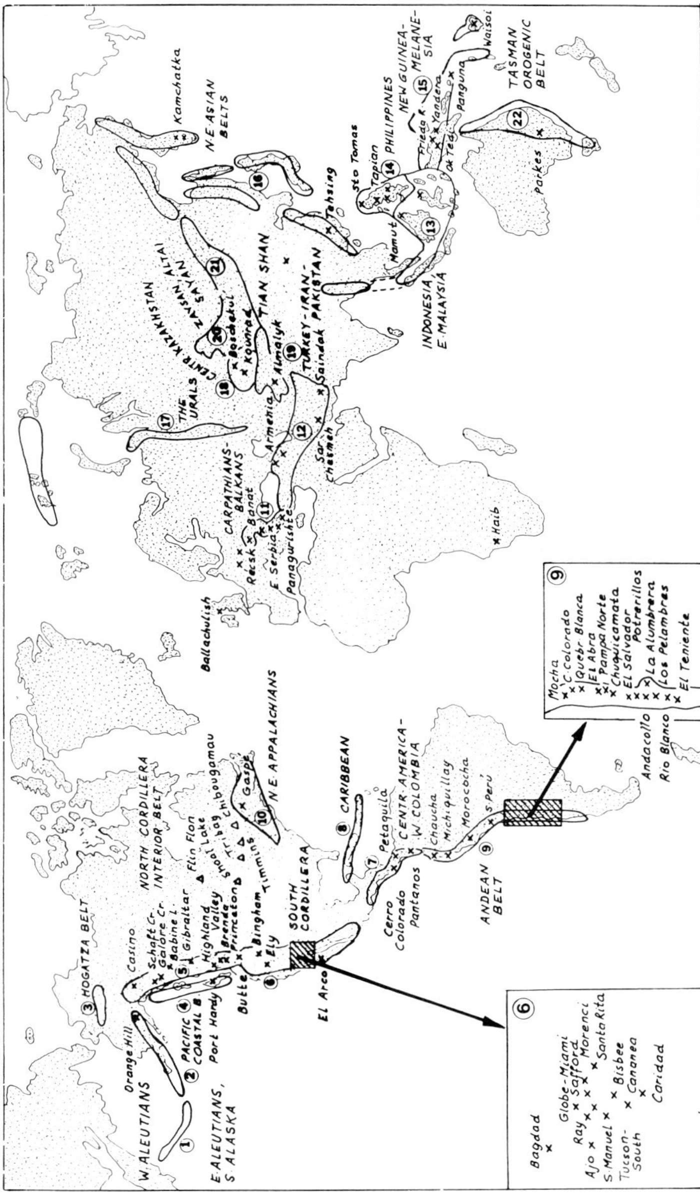

Fig. 28-41. Major porphyry copper provinces of the world (compare also Table 28-2).

including ophiolites, Triassic to Cretaceous island arc and continental margin volcanics and sediments). It is intruded by Permian to late Tertiary granitic rocks.

Of the cluster of ore occurrences in eastern Alaska (south of the Denali Fault), the most important is Orange Hill near Nabesna (Table 28-3).

On the Queen Charlotte Islands, Vancouver Island and on the

Table 28-2. Major porphyry copper provinces and sub-provinces of the world

PROVINCE, SUBPROVINCE	BASEMENT GEOLOGY	MINERALIZ.INTRUSIONS	PORPHYRY COPPERS	REFERENCES
Western Aleutian Isl. (Attu showing), U.S.A. **1**	Cr_3-Q immature isl. arc developed on oceanic crust	T, sodic quartz diorite intrusion	stockwork, most ore in phyllic zone	Hollister (1978)
E.Aleutians and Alaska continental margin U.S.A. **2**	J-T low-grade metagrayw., slate, mafic to bimodal volc.	T(3.3-59 m.y.) quartz dior. and qtz.monzon. porphyry	stockw., brecc., ph, less KF alter., hypog. ore only; 11 show.	Hollister (1978)
Hogatza Plutonic Belt, W.C. Alaska U.S.A. **3**	D, Pe metased., metavolc., Cr marine andes., sedim.	Cr_3 qtz.monzonite porph. and D? granodiorite	4 stockw., 1 breccia, 1 skarn; ph, some ppl alter., 6 showings	Hollister (1978)
N.E.Pacific coastal belt (E.Alaska, Brit. Columbia, Washington, Oregon) **4**	$PZ_{1,2}$ metavolc., metased.; Cb-Pe ophiolites, marine seds., volc. Tr-Cr island arc, cont.mg.andes.	Pe qtz.monzon.pluton Cr granod.(5 dep.) T granod., qtz.monz., qu.-feldsp.porph.(the bulk)	stockworks, less brec. ph, KF alter., 36 prsp. 2 dep., about 1.5 Mt Cu	Hollister (1978)
N.Cordillera Interior Belt; Alaska, Yukon, Br.Columb., Washingt. Idaho, Montana **5**	Pt_1, PZ metam.; Cb, Pe marine sd., ophiol. Tr_3-Cr marine and. and seds.	Tr_3-J dior., qtz.dior., granod., syenite J_2-T_1 qtz.diorite to qtz.monzonite	300 occurr., 2 giants (Butte, Highland Val.) most calc-alk.stockw. in KF, ph alter.; lesser "diorite" md. 66% lack superg.alt. 36 Mt Cu, 1.3 Mt Mo	Sutherland Brown, ed. (1976) Hollister (1978)
S.Cordillera, U.S.A. and Mexico (W.of Rocky Mts., E.of Sierra Nevada) **6**	Pt_1 schist, grenst., granite; $Pt_{2,3}$ diabase, quartzite; PZ, carb., shale; MZ marine sedim., andesitic volc.	J-Mi qtz.diori.to qtz. monzon.intrusions; most "Laramide" (Cr_3-T_1)	mostly endocont.stwks import.exoct.skarns Signif.superg.enrich. Local breccias. X00 occur., 103 Mt Cu, 3 Mt Mo	Lowell and Guilbert (1970)

Central America (Panama,Costa Rica) **7**	Cr_3-Q tholeitic to andes.volc.,pyrocl. early qtz.diorite	34-4 m.y. potass.grano-dior. and qtz.-feldspar porphyries	veinl.and dissem. stockw.in endo and exocont. Cerro Colo-rado mostly in volc. 25.2 Mt Cu,510 Tt Mo	Kesler et al. (1977)
Caribbean (Cuba,His-paniola,Pto Rico, Virg.Islands,Jamaica) **8**	J-Eo arc basalts, andes.,graywacke, shale,minor carb.	63-31 m.y.qtz.diorite, less granodior.porph.	stockworks in KB or ph.alter.,low Mo, high Au; about 30 occur.,1 Mt Cu	Hollister (1978)
Andean belt, 4,800 km long, South America **9**	PZ sedim.,MZ-T ma-rine to cont.basalt andes.,rhyol.volc. local PCm metam.	Cb to 4.3 m.y.(most Cr_3-T_2) quartz diorite to quartz monzonite intr.	mostly Cu-Mo stockw. assoc.with hypabys. intrusions; some important tourm. brecc.pipes (Rio Blanco); X00 occur., 182 Mt Cu,3.25 Mt Mo	Oyarzún and Frutos (1980) Hollister (1978)
N.E. Appalachian belt, Canada and U.S.A. **10**	PCm to D basement metam.,overl.by Cm to D marine volc.-sedim. and sedim.	Cm to D qtz.monzon.equi-granul.to porph.; D most common	mostly stockw.,py and cp.,minor moly.KF and KB alter.,ph.mostly absent. Grade to skarns. 20 occur.,1 deposit; 1.9 Mt Cu	Hollister (1978)
Carpathians,Balkans, Hellenides in Hunga-ry,Rumania,Yugosla-via, Bulgaria,Greece **11**	PCm metam.basem. topped by PZ,MZ ophiol.,volc.-sed., sedim. association	Cr_3-Pl qtz.dior. to qtz. monzon., contemp. with subaerial andesite-dacite volcanism	largest dep.in Timok Mass.are in andes., rest in endo-,exo-cont.stockw.grading to skarns; about 20 occur.,12.5 Mt Cu	Ianovici and Borcoş (1982) Janković (1982)

Table 28-2 (continued). Major porphyry copper provinces and sub-provinces of the world

PROVINCE, SUBPROVINCE	BASEMENT GEOLOGY	MINERALIZ. INTRUSIONS	PORPHYRY COPPERS	REFERENCES
Turkey, Caucasus, Iran, Pakistan belt **12**	MZ marine volc.-sedim. mobile belt encl. Pt-PZ metam. blocks; Cr_1-Q cont. volcanism	Eo-Pl qtz. dior. to qtz. monzon. stocks and plutons	zoned stockworks in endo and exocont., KB, KF, minor ph. alt. 12.5 Mt Cu	Pidzhyan (1975)
Burma, Indonesia, E. Malaysia (Sumatra, Sulawesi, E. Borneo) **13**	mostly MZ-CZ island arc volc. and sedim.	T hypabyssal stocks and plutons, quartz dior. to qtz. monzon.	stokworks, about 30 showings, 1 mine; 1 Mt Cu	Hamilton (1979)
Philippines (Luzon, Marinduque, Negros, Cebu Islands) **14**	Cr-T_1 marine sedim. and mafic metavolc. (basalt, spilite, andesite)	60-2 m.y. dior. and qtz. dior., rarely granodior. and qtz. feldsp. porph. plutons and stocks	dissem., fault controlled prism. to elong. oreb., irreg. zoning, silicif., KB, KF alter. X00 occur., 80 depos. 25-30 Mt Cu, low Mo	Bryner (1969) Wolff (1978)
New Guinea-Melanesia belt (the Solomons, Vanuatu, Fiji) **15**	NG: J-Q sedim. cover PCm sialic block in S. Tr-Q seds, volc., in N. and islands	Ol-Pl dior. to quartz dior., granod. and qtz. monz. rare and latest	stockw. high in Au, low in Mo, in KB, KF alter.; miner. 01 to 1.2 m.y. 35 occurr., 15 Mt Cu, 900 t Au 4 Tt Ag	Davies (1978) Titley (1975)
N.E. Asian discontin. porphyry Cu occurrences (S.E. China, Okhotsk-Chaun belt, Japan, Kamchatka) **16**	MAINLAND: MZ-T_1 cont. marg. andes. to rhyol. volc. belt ISLANDS: T-Q isl. arc volc., sedim.	MZ-T_1 qtz. dior. to qtz. monzonite T diorite to granodior.	mostly endocont. stockw. in KB, KF alt. granodior., qtz. dior. about 20 occurrences	Laznicka (1976a)
The Urals, U.S.S.R. **17**	PCm to Cb deform. metam. maf. volc. sd.	D-Cb dior. to qtz. monz. plutons	"granod." and "dior." stockw., skarn; 500Tt Cu	Laznicka (1976a)

North Lake Balkhash distr.,Kazakhstan, U.S.S.R. **18**	Pt metam.,Cm-D marine volc.-sed. D_2-Pe marine to cont.and.-rhyol.	Cb-Pe high level granodior. to qtz.monzon. qtz.-feldsp.porphyry	calc.-alk.porph. Cu-Mo stockw. in KF, KB,ph,silicific. alt. superg. enrichment; about 100 occur., 15 Mt Cu, 500 Tt Mo	Laznicka (1976a)
Tian Shan orogen, Soviet Centr.Asia, Kurama Range distr. **19**	Pt-Pe metam.,sed., andes.-rhyol.volc.	Cb-Pe syenodior.,granodior.,qtz.monzonite	stockw. in KB,KF,ph and silicif.zones, superg. enr.,about 20 occur.,6 Mt Cu	Laznicka (1976a)
Zaysan,Altai, Sayan orogens,S.Siberia, Mongolia, China **20**	Pt to Pe metam., volc., sedim. on S. fringe of Siberian Platform	Cm dior.,syen.complexes Or-Pe granod.,qtz.monz. granosyenite	"diorite" model calc-alk.Cu stockw. grad.to subvolc.Cu-Mo and Mo-Cu; about 50 occur., estim. 4 Mt Cu	Laznicka (1976a)
E.Mongolia, East Transbaikal, Amur Belt, U.S.S.R. **21**	PCm basement, PZ_3 deeply eroded dior.,gran.plut.	J-Cr epizonal plutons, qtz. monzon.,granite	Sn,W,Mo stockw. in KF alter.grade to Mo-Cu stockworks; about 5 Cu occur.	Marinov et al. (1977) Laznicka (1976a)
Tasman Orogenic Belt Eastern Australia **22**	Cm-Cb fold.sedim., andesitic volc.	Cm to Cr_1 dior. to qtz. monzon. plutons	fract.controlled plutonic porph.Cu, KF,KB alter. 120 occur.,2.5 Mt Cu	Horton (1978)

ABBREVIATIONS: KF=K-feldspar; KB=biotite; ph=phyllic; ppl=propylitic alteration
BOLD NUMBERS=provinces as shown of Fig. 28-45

Table 28-3. Hypabyssal and plutonic porphyry Cu-Mo deposits, selected example localities

1. LOCALITY	2,NO	3,4. SETTING (3=BEDROCK, 4=INTRUSION)
Orange Hill,Nabesna, central Alaska	4	3) Pe mafic metavolc.,metasediments 4) 105m.y. quartz dior.porph.intruding quartz-diorite pluton
Casino,Yukon, Canada	4	3) PZ or PCm? schists intr.by Cr_2 bathol. 4) Cr_3 (70 m.y.) subvolc.compl.,qtz-feldspar porphyry, tuff breccia
Babine Lake (Granisle,Bell,Morrison) C. Brit. Columbia	5	3) J_1-Cr siltst.,limest.,andes.and volcanocl.,hornfelsed 4) Eo, small stocks,plugs, dikes; quartz dior.,qtz.monz.,biot.-feldsp.porphyry
Highland Valley, S. Brit.Columbia Canada	5	3) Cb-Pe argill.,chert,congl.; Tr_3 m-bas., andes.,argill.,volcaniclastics 4) Tr_3 semiconc.,domal compos.bathol., dior. to qtz.monzon.,abundant dikes
Butte,Montana U.S.A.	5	3) PCm-PZ metam.,Cr_3 qtz.latite cont.volc. 4) Cr_3-T_1 (70-72 m.y.) qtz.monzon.pluton followed by aplite,qtz.porph.,biotite breccia; miner. 57-63 m.y.
Bingham, Utah, U.S.A.	6	3) Ps calcar.quartzite,sandst.,limestone 4) 39-32 m.y. qtz.monzon.stock, porphyry, porphyry dikes; miner. 35.8 m.y.
Ely (Robinson), Nevada, U.S.A.	6	3) D-Pe hornfelsed platf.shale,limestone 4) Cr_1 qtz.monzon.porphyry (115-103 m.y.) Eo-O1 (41-37 m.y.)post-ore dikes
Globe-Miami field, Arizona, U.S.A.	6	3) Pt_1 schist, $Pt_{2,3}$ granite,diabase; PZ limest.,shale 4) Cr_3-T_1 qtz.monzonite,qtz.monzon.porph.
Ray, Arizona, U.S.A.	6	3) Pt_{2-3} quartzite,diabase; D-Cb limest. 4) Cr_3 (70-60 m.y.) qtz.dior.,granod.porp.
Bagdad, Arizona, U.S.A.	6	3) Pt_1 greenst.,schist,granite 4) Cr_3 (71 m.y.) granod.,qtz.dior.porph.
Ajo (New Cornelia), Arizona, U.S.A.	6	3) PCm gneiss; Cr andes.to rhyol.flows, brecc.,tuff,continental; 4) Cr_3 (63 m.y.) qtz.dior.,qtz.monzon.
San Manuel,Arizona U.S.A.	6	3) Pt_1 quartz monzonite pluton 4) Cr_3 (67 m.y.) qtz.monzon.porph.dikes and stocks

5.STYLE	6.ALTER.	7.ORES	8.ENR.	9.TONN./GRADE	REFERENCES
H/S I,W	CA; KF,ph *KB	distant AuV	0	800Tt Cu/0.4%	Hollister (1978)
S,H,B I,W	CA; KF,ore in KB	AuV Pb-ZnV	?70%	570Tt Cu/0.37% 30Tt Mo/0.002%	Godwin (1976)
S,H I,W	CA; KB *ppl	none	?10%	940Tt Cu/0.42-0.51%	Carson and Jambor (1974)
P/D,Fr I	CA; KF,KB *ph, ppl	CuV,CuSk AuV	2%	9 Mt Cu/0.4% 150Tt Mo	McMillan (1976)
P,*S D,Fr I	CA; KF,KB ph	++CuAsV ++ZnMnV	30-60%	8.3Mt Cu 2.2Mt Zn 19.3Tt Ag 84 t Au	Miller,ed. (1973)
H I,*W	CA; KF,KB *ph	++PbZnR ++PbZnV Au plac.	?50%	21.5Mt Cu 820Tt Mo	Lanier et al. (1978)
H skarn	CA; KF, *ph,Si	++CuSk CuZnPbAg veins	?30%	4.3Mt Cu/0.8%	James (1976)
P,H,Fr I,*W	CA; KF ph,arg,at second.ore	++CuV ZnPbV AuV	60-70%	5 Mt Cu	Peterson (1962)
H/Fr W,*I	CA; KB,KF ph,ppl	exotic Cu	? 25%	1.5Mt Cu 1.2t Au, 120tAg	Phillips et al. (1974)
H I,*W	CA; KF,ph	PCm mass. sulphides CuZn;WV	?50%	+760Tt Cu/0.76%	Lowell and Guilbert (1970)
H I,*W	CA; KF,ph *Si		?10%	3.8Mt Cu	Wadsworth (1968)
H I,W	CA; KF.*ph	PbZnMoV	?10%	4.33Mt Cu/0.75%	Lowell (1968)

Table 28-3 (continued). Hypabyssal and plutonic porphyry Cu-Mo deposits, selected example localities

1. LOCALITY	2.NO	3,4. SETTING (3=BEDROCK, 4=INTRUSION)
Clifton-Morenci, Arizona, U.S.A.	6	3) Pt_1 schist, granite, granod.; PZ,MZ shale, limestone 4) Cr qtz.dior.laccol.,dikes,intr.by Cr_3-T_1 qtz.monzonite porphyry stock
Santa Rita (Chino) New Mexico, U.S.A.	6	3) Cm-Pe carb.,shale; Cr shale 4) Cr (63 m.y.) granod. and qtz.monzonite
Bisbee (Warren) Arizona, U.S.A.	6	3) Pt_1 schist, PZ limest.,shale; Cr shale 4) 163 m.y. gran.and qtz.felsp.porph. stock and dikes
Tucson-south (Pima, Mission,Twin Buttes) Arizona, U.S.A.	6	3) PZ siltst.,sandst.; Tr-Cr contin.andesite, rhyol., ignimbrite, conglom. 4) Cr_3-T_1 qtz.monzon.,granodior.,andes. rhyolite
Cananea, Sonora, Mexico	6	3) Pt_2 (1.44 b.y.) qtz.monzonite; Cm-Ps limest.,dolom.,shale,quartzite; MZ, continental volcanics 4) 69-64 m.y. quartz monzonite porphyry
La Caridad,Sonora, N.Mexico	6	3) T_1 contin.volc.,dior.to granod.pluton 4) T_1 younger phase qtz.monzon.porphyry, intrusive breccia
Pegadorcito and Pantanos, N.Colombia	7	3) T volcanics, Ol tonalite bathol. 4) T_2 qtz.dior.porph.,intrus.breccia, qtz.monzon.,dacite porphyry
Chaucha, Ecuador	9	3) T andesite, granodiorite 4) Mi (9-12 m.y.) qtz.monzon.and granodiorite porphyry
Michiquillay, N.Peru	9	3) Cr quartzite, limestone 4) 20.6 m.y. qtz. monzonite porphyry
Morococha (Toromocho) central Peru	9	3) Pe contin.andes.and dac.volcanics Tr_3-J_1 limest.,shale,dolom.,basalt,anh. 4) 7 m.y. quartz monzonite porphyry
Cerro Verde and Santa Rosa, southern Peru	9	3) MZ granodiorite 4) 58.8 m.y. granodior.,qtz.monzon.porph.
Cuajone, southern Peru	9	3) Cr_3-T_1 volcanics,basalt,rhyol.,andes.

5.STYLE	6.ALTER.	7.ORES	8.ENR.	9.TONN./GRADE	REFERENCES
H I,*W	CA; KF,ph	CuV	100%	13Mt Cu/0.8%	Moolick and Durek (1966)
H W,*I	CA; KF,ph Si,skarn	+CuSk +FeCuSk ZnPbR	?66%	6.28Mt Cu/0.78%	Hernon and Jones (1968)
H,D,Fe,Br W,*I	CA; ph,*Si skarn	CuSk,CuR PbZnR	?50%	4.4Mt Cu	Bryant (1968)
H W,*I	CA; KF,KB ph,skarn		?30%	5.92Mt Cu	Lowell and Guilbert (1970)
P,Br I,W	CA; KF,KB ph,skarn	CuSk CuZnPbSk, replac.	?50%	15.255Mt Cu 135Tt Mo 360Tt Zn	Meinert (1982)
P,H I,W	CA; ph,*KF		100%	5.3Mt Cu/0.8% 66Tt Mo/0.01%	Saegart et al.(1974)
H,Br I,W	CA; ph			9.7Mt Cu/0.5%	Hollister (1978)
H I,W	CA; ph			1.6Mt Cu/0.7%	Hollister (1978)
H I,W	CA; ph,arg.		?20%	4.14Mt Cu/0.72%	Hollister and Sirvas (1974)
H,S I	CA; KB,ph	CuAsAgV CuAsR PbZnR;AgV	?75%	2.7Mt Cu/0.76%	Velazquez, oral comm. (1977)
Br I,W	CA; KF,KB, ph,tourm.	AgPbZnV		8.04Mt Cu/0.67%	Bellido and de Montreuil (1972)
H,S,Br I,*W	CA;KF,ph tourmaline	PbZnV	16%	13Mt Cu/1%	Manrique and Plazalles (1975)

Table 28-3 (continued). Hypabyssal and plutonic porphyry Cu–Mo deposits, selected example localities

1. LOCALITY	2.NO	3,4. SETTING (3=BEDROCK, 4=INTRUSION)
Toquepala, southern Perú	9	3) Cr volcanics, T_1 diorite intrusion 4) 58.7 m.y. dacite porphyry
Quelleveco, southern Perú	9	3) Cr volcanics 4) T_1 quartz monzonite porphyry
Mocha, Tarapacá, northern Chile	9	3) T volcanics; 4) 56.4 m.y. quartz diorite porphyry
Cerro Colorado, Tarapacá, N.Chile	9	3) J marine sedim.,Cr volcanics 4) T quartz diorite porphyry
Quebrada Blanca, northern Chile	9	3) MZ volcanics 4) T granodiorite, dacite porphyry
El Abra, northern Chile	9	3) J_3 shale,sandst.,limest.,granodiorite 4) 33.2 m.y. dior.to syen.complex,dacite
Pampa Norte, northern Chile	9	3) T granodiorite 4) 29 m.y. quartz monzonite porphyry
Chuquicamata, northern Chile	9	3) D-Cb schist,gneiss,amphib.,gabbro Cb-granite; J_3 shale,siltst.,limest., andes.,granodiorite 4) 29.2 m.y. qtz.monzonite porph.stock
El Salvador, Chile	9	3) Cr_3 andesite flows,volcaniclastics T_1 rhyolite,ignimbr.,granod.porphyry 4) 39.1 m.y. granodiorite porphyry
Potrerillos, Chile	9	3) J limest.,shale,sandstone 4) 34.1 m.y. quartz diorite porph.stock
Andacollo, Chile	9	3) J-Cr andesite volcanics, diorite 4) 90 m.y. granodiorite porphyry
La Alumbrera, N.W. Argentina	9	3) T continental andesite-rhyolite volc. 4) 7.5 m.y. qtz.monzonite porphyry
El Pachón (Argentina) and Los Pelambres, Chile	9	3) J_3-Cr contin.andesite volc.and volcaniclastics 4) 9.8 m.y. tonalite, diorite porphyry
Rio Blanco-Disputada central Chile	9	3) Eo contin.basaltic andes.flows and pyroclastics; T_2 qtz.monzonite, granodiorite plutons 4) 4.3 m.y. dacite porphyry dike

5.STYLE	6.ALTER.	7.ORES	8.ENR.	9.TONN./GRADE	REFERENCES
H,Br I,*W	CA; ph, tourmal.		?20%	4Mt Cu/1%	Hollister (1978)
H I,W	CA; ph.	PbZnV		1.9Mt Cu/0.95%	Bellido and de Montreu-il (1972)
H I,W	CA; ph, arg.			1.07Mt Cu/1%	Hollister (1978)
H I,W				1.2Mt Cu/1.3%	Hollister (1978)
H I,W	CA; KF,ph.	PbZnCuV		2Mt Cu/1%	Hollister (1978)
H to P Br I,*W	CA to D KB,ph	CuV	20% oxid.	16.35Mt Cu/1.09%	Ambrus (1977)
H I,W				1.82Mt Cu/0.7%	Hollister (1978)
H to P I,*W	CA; ph, *KF,KB	exotic Cu,Mn	70%	43Mt Cu/1.3% 1.1 Mt Mo/0.04%	H.Soto P., pers.comm. (1977)
H I,*W	CA; KF *ph		?60%	5.5Mt Cu/1.34% 200Tt Mo	Gustafson and Hunt (1975)
H,Br I,W	CA;ph,KF	CuSk	?80%	3Mt Cu/1.43%	Ruiz (1965)
H I,W	CA; KF *KB,ph	AuV PbZnV		2.45Mt Cu/0.7%	Hollister (1978)
H I,W	CA; KF			2.5Mt Cu	Hollister (1978)
H,D I,W	CA; KF,KB		under 5%	8 Mt Cu 127 Tt Mo	Sillitoe (1973b)
H to S,Br I,W	CA;Si,ph, tourmal.		none	17Mt Cu	Ruiz (1965)

Table 28-3 (continued). Hypabyssal and plutonic porphyry Cu-Mo deposits, selected example localities

1. LOCALITY	2.NO	3,4. SETTING (3=BEDROCK, 4=INTRUSION)
El Teniente (formerly Braden) near Rancagua central Chile	9	3) Eo contin.andesite flows, Ol-Mi quartz diorite 4) 4.3 m.y. dacite porphyry stock,breccia
Gaspé Copper, Murdochville,Québec, Canada	10	3) D_1 calcar.siltst.to limestone 4) D_3-Ms rhyodac.porph.sills, plugs
Moldova Nouǎ, Banat, S.W. Rumania	11	3) MZ limest.,hornfels, skarn 4) Cr_3-T_1 qtz. diorite porph.stock
Radovište (Bučim, Borova Glava), S.E. Yugoslavia	11	3) Pt gneiss, schist, amphibolite 4) andesite porphyry stock
Panagurishte field (Medet),Bulgaria	11	3) PZ granodiorite,diorite 4) Cr_3 granod.,qtz.monzon.porphyry
Ankavan,Armenia, U.S.S.R.	12	3) PCm-PZ_1 schist, marble interbeds 4) Ol granodiorite,qtz.monzonite
Kadzharan, Armenia, U.S.S.R.	12	3) Eo hornfelsed ands.,dac.,rhyol. 4) Ol or Mi quartz-feldspar porphyry
Agarak, Armenia, U.S.S.R.	12	3) Eo hornfelsed andes.,dacite,rhyolite volcanics and volcaniclastics 4) Ol qtz.monzonite and granosyenite
Sar Chesmeh, Kerman, southern Iran	12	3) T_1 andes.lavas and pyrocl., granodior.porphyry 4) T_3 granod.,qtz.monzon.porph.;breccia laced by late-stage dikes
Saindak, Baluchistan, N.W.Pakistan	12	3) Cr_3-Ol andes.volc.-sedim. association 4) 19-20.3 m.y. tonalite porph.,granodior. porph.,andesite,diorite dikes
Kalmakkyr dep.,Almalyk distr.,Uzbekistan, U.S.S.R.	19	3) Or-S sandst.,shale; D-Cb limest.,sand., shale,basalt,andes.,dac.,rhyol.,horfld. Cb_{1-2} syenodiorite, diorite 4) Cb_{2-3} granodiorite porphyry
Kounrad, N.Lake Balkhash distr.,Kazakhstan, U.S.S.R.	18	3) Cb_1 contin.andes.,basalt lavas,tuffs 4) Cb_{2-3} granod.porph.,qtz.-feldsp.porph.
Boshchekul', N.E. Kazakhstan, U.S.S.R.	18	3) Pt_3-Cb_1 m-chert, spilite, m-basalt 4) Cb_2 diorite, monz.porph. dikes
Erdentuin field, N.E. Mongolia	20	3) Cm metavolc.,Pe contin.basalt, andes., rhyol.; Pe granosyen.,monz.,diorite 4) Pe gran.,granosyen.,granod.porphyry

5. STYLE	6.ALTER.	7.ORES	8.ENR.	9.TONN./GRADE	REFERENCES
H to S I,W	CA; KF,KB *ph; post- ore breccia	CuV	?10%	62Mt Cu/1.05% 1.6Mt Mo	Camus (1975)
H I,W	CA; KF,KB, ph,arg., anhydrite	CuR	?10%	1,848Mt Cu	Allcock (1982)
H I,W	CA; KB, skarn	CuFeSk	?20%	230Tt Cu/0.23 43Tt Mo	Ianovici and Borcoş (1982)
S,H W,*I	CA; KF	AuAgTeV		459Tt Cu	Janković (1982)
H I,W	CA; KB *ph		?10%	525Tt Cu/0.43 13Tt Mo	Iovchev (1961)
H I,*W	CA; KB	CuSk		min. 150Tt Cu	Pidzhyan (1975)
H I	CA; KB			min.500Tt Cu 30Tt Mo	Pidzhyan (1975)
H I	CA; KB			min.700Tt Cu 17Tt Mo	Pidzhyan (1975)
H I,*W	CA; KB,KF ph		20%	9.04Mt Cu/ 1.13% 240Tt Mo	Hakim (1980)
I,W	CA; KF,KB			1.47Mt Cu/0.4	Sillitoe and Khan (1977)
P to S I,W	CA; KF,KB ph	CuFeSk,R AuV PbZnR;Be	?80%	2.1Mt Cu	Musin (1970)
P to H Br,Fr I	CA; Si, adv.arg. ph	CuAsV	90%	10Mt Cu/0.95%	Gazizova (1957)
H,D I,W	D; ppl, *KF,KB		?80%	2.16Mt Cu/0.4 to 0.81%; Au	Trofimof (1973)
P to S I,*W	CA to D ph,Si	AuV Au plac.	100%	3Mt Cu/1%	Marinov et al.(1977)

Table 28-3 (continued). Hypabyssal and plutonic porphyry Cu-Mo deposits, selected example localities

1. LOCALITY	2.NO	3,4. SETTING (3=BEDROCK, 4=INTRUSION)
Mamut, Mt.Kinabalu, Sabah, N.E. Borneo, Malaysia	13	3) Cr_3-T_1 flysch,ultramaf.tectonite,tuff, black shale, sandst.,deformed 4) 9 m.y. qtz.monzon.porphyry,granod.porph.
Sto Tomas II, Tuba, Luzon, Philippines	14	3) Eo-Ol slate,phyllite,chert,limest., volcanics 4) T diorite to andesite
Tapian, Marinduque Isl.,Philippines	14	3) Eo-Ol slate,phyllite,chert,limest.volc. 4) Mi_2 qtz.dior.,granodior.stock
Sipalay, W. Negros Island, Philippines	14	3) Cr_3-T_1 mafic metavolcanics 4) T, quartz diorite stock
Basay (Muhong Creek) S.W. Negros Isl., Philippines	14	3) Cr_3-T_1 mafic tuff, metasedim. 4) T, diorite stock
Lutopan group, Cebu Isl.,Philippines	14	3) Cr_3-T_1 metavolc.(spilite,basalt) 4) 59.7 m.y. biot.diorite porphyry
Ok Tedi (Mt.Fubilan) Niugini	15	3) MZ-T fine clastics; Ol-Mi limestone 4) 4-1.2 m.y. monzonite stock,dior.pluton
Frieda Prospect, Sepik R. area, Niugini	15	3) Cr-Eo shale,phyll.,seric.sch.,marble mafic metavolc.,ultramafics; Mi clast. 4) Mi_2 porph. microdior.plut.incl. hornblende, hornbl.-pyrox.porphyries
Yandera, N.W.of Goroka, Niugini	15	3) 12-14 m.y. synorog, granod.batholith 4) 6.5 m.y. qtz.monz.porph.,qtz.diorite, dacite
Panguna, Bougainville Isl.,Niugini	15	3) Ol, andesite volc.,4-5 m.y. qtz.diorite pluton 4) 3.4 m.y. biot.dior.,granod.,qtz.dior.

NOTES AND EXPLANATIONS:

Column 2: number of porphyry Cu province from Table 28-2.

Column 5: style of ore-bearing intrusion and mineralization; H=hypabyssal; P=plutonic; S=subvolcanic; D=dikes; Fr=fault control (elongated); Br=breccia I=ore in intrusion (could be of older phase); W=ore in wallrocks

5.STYLE	6.SLTER.	7.ORES	8.ENR.	9.TONN./GRADE	REFERENCES
H I,W	CA; Si,KB tremolite actinolite	CuV	?5%	852Tt Cu/0.48% 90t Au/0.5ppm	Kosaka and Wakita (1978)
H,S I,W	CA-D; ph,Si anhydrite		?20%	645Tt Cu/0.43% 133t Au/0.89ppm 150t Ag	Bryner (1969)
H,P I,W	CA; KF,Si		?20%	1.6Mt Cu/0.57%	Loudon (1976)
H,Br I,W	D-CA; KF,KB			8.71Mt Cu 65.1Tt Mo 1.6Tt Ag; 48t Au	Wolff (1978)
H I,W	D, KF,KB			960Tt Cu/0.48%	Wolff (1978)
H,F	D,KB,*Si			6 Mt Cu/0.46% 121t Au/0.124 ppm; 600t Ag	Bryner (1969)
H I,W	D-CA; KF skarn	FeCuAuSk CuR	Au in leach cap.	2.33Mt Cu/0.88 262t Au 29Tt Mo	Bamford (1972)
H I,W	D-CA; ppl arg.			2.5Mt Cu/0.5% 100t Au/0.2ppm	Whalen and Britten (1982)
P,H,Br I,W	CA; KF,KB			1.43Mt Cu/0.42	Grant and Nielsen (1975)
H,Br I,W	CA; KB	AuV,CuV Au plac.		4.32Mt Cu/0.48 450t Au/0.5ppm 2.7Tt Ag/3 ppm	Baumer and Fraser (1975)

Column 6: alterations; CA=calc-alkaline parentage; D="diorite" model; KF=K feldspar; KB=biotite; ph=phyllic; ppl=propylitic; Si=silicification; arg=argillic

Column 7: other ore styles present in the ore field; Sk=skarn; R=replacements and V=veins

Column 8: approximate percentage of secondary enriched ore.

adjacent British Columbia coast, magnetite-chalcopyrite skarns are the most widespread plutonic copper mineralization style (Chapter 10). The largest "porphyry" (Island Copper near Port Hardy) is largely volcanics-hosted, and has been described in Chapter 14. The plutonic "porphyry" prospects (O.K., Catface) are of limited importance (Carson, 1969).

In the Cascade Range in Washington and Oregon, Hollister (1978) recorded fourteen "porphyry" occurrences, associated with Pliocene (6.2 m.y.) to Oligocene quartz diorite, granodiorite and quartz-monzonite porphyries (also dacite porphyries), intruded into Paleozoic metamorphics, Cretaceous and Tertiary sediments and Tertiary volcanics. Most are stockworks in potassic alteration zones. No economic deposit has been found, so far.

The Northern Cordilleran Interior porphyry Cu-Mo Belt which is over 3,000 km long and including the western Yukon, British Columbia, eastern Washington, Idaho and Montana (Sutherland Brown, ed., 1976; Hollister, 1978) contains almost 300 "porphyry" occurrences. This includes two giant fields (Butte, Montana and Highland Valley, British Columbia) and almost 50 major deposits, some of them in production (Table 28-3). The eastern and southern portion of this belt is underlain by Proterozoic and lower Paleozoic metamorphics along the western continental margin of North America. Also included are some of the "exotic" terrains added to this continent by collisions and displaced along strike-slip faults. Late Paleozoic "oceanic" association (ophiolites, deep marine sediments; Chapters 7, 10) crops out intermittently in the central part of this belt. The two sequences mentioned above form a "passive basement" to the porphyry copper complexes (Ney and Hollister, 1976), all of which are affiliated with Triassic to lower Jurassic (215 to 180 m.y.) and upper Jurassic to lower Tertiary (150 m.y. to 42 m.y.) plutonic rocks.

In the earlier lithogenetic and metallogenetic epoch, widespread mafic andesitic volcanics and associated sediments (e.g. Nicola Group, Chapter 13) formed in an island arc setting. "Volcanic" porphyry deposits corresponding to the "diorite" association-alteration model of Hollister (1978; e.g. Cariboo Bell, Galore Creek, described in Chapter 13) as well as "normal" plutonic porphyry coppers (Highland Valley, Lorraine, etc.) are the main examples of affiliated ores.

In the later epoch, marine to continental sedimentation and basalt, andesite, rhyolite volcanism were active in a series of intercontinental and marginal basins, situated to the east of the rising magmatic arc. The deeply eroded Coastal Batholith (Section 28.2.3.) is virtually devoid of ores, except for a few relicts. Porphyry copper deposits, however, are associated with smaller epizonal and upper mesozonal plutons located east of this batholith. Most of them are in terrains underlain by upper Paleozoic or Triassic volcanics.

The Casino deposit in western Yukon, described by Godwin (1976), is in an upper Cretaceous (70 m.y.) subvolcanic complex of quartz-feldspar porphyry plugs and breccia pipes, emplaced in a mid-Cretaceous granitic batholith. This batholith, in turn, intruded a Paleozoic or earlier metamorphic complex. Chalcopyrite and molybdenite are concentrated in a phyllic alteration zone surrounding a K-silicate altered breccia pipe. This deposit situated in the present zone of permafrost (average yearly temperature -5° C) is notable for the presence of up to 170 m deep supergene alteration zones (Fig. 28-39), interpreted as a relict of Paleogene subtropical weathering and a recent sporadic activity of groundwater heated by pyrite oxidation.

In the Babine Lake area, central British Columbia (Carson and Jambor, 1974), a north-west trending string of medium-size porphyry Cu deposits is associated with Eocene hypabyssal intrusions (stocks, dikes, dike swarms) of quartz diorite, quartz monzonite and in particular biotite-feldspar porphyry (Fig. 28-38). These intrude Jurassic to Cretaceous marine to continental andesite, litharenite, shale association as well as Jurassic mesozonal plutons. Chalcopyrite is the dominant mineral, whereas molybdenite occurs in mineralogical quantities and is not recovered. In the largest Granisle orebody, chalcopyrite and minor bornite fills fractures, 1-5 cm thick, in a stockwork hosted by biotite-altered porphyry and hornfelsed metavolcanics, and there is a gradation into irregular veins up to 30 cm wide, filled by coarse ("pegmatitic") chalcopyrite, bornite, quartz, biotite and apatite. The Granisle orebody is fresh to the very top, whereas at the nearby Bell deposit a supergene chalcocite zone extends to a depth of as much as 100 m.

The Highland Valley porphyry Cu(Mo) field (McMillan, 1976; Fig. 28-38) is located between Ashcroft and Merritt, southern British Columbia, and is the largest Canadian porphyry copper complex (over 9 Mt Cu; Table 28-3). Five major and several minor orebodies are located in the central part of a composite upper Triassic Guichon Creek Batholith that has an area of about 1,000 km^2. All orebodies are of the "plutonic" style. The batholith intruded broadly comagmatic basalt and andesite volcanics, pyroclastics and volcaniclastics of the Nicola Group (Chapter 13), and probably also Permo-Carboniferous marine sediments in the basement. The early plutonic phases are hybrid and largely of diorite-quartz diorite composition, but the younger phases are granodiorite and quartz monzonite. The low-grade stockwork chalcopyrite-lesser molybdenite mineralization is in broadly equidimensional to elongated orebodies largely controlled by N.-S., E.-W., and N.W.-S.E.-trending fault zones and grabens. The orebodies are affiliated to one of the youngest granodiorite to quartz monzonite phases (Bethsaida Phase), with swarms of porphyry dikes or with intrusive breccias. The model alteration and mineralization zoning is developed, but its regularity is considerably modified by the intensity of fracturing and faulting.

The Lorraine deposit (Duckling Creek; N.C. British Columbia; Wilkinson et al., 1976; Fig. 28-42) is insignificant by modern porphyry copper standards (67 Tt Cu/0.6-0.75% Cu, 2.03 t Au), but it has been included as an example of a "diorite" (syenite)-model mineralization set deeper than the "volcanic" equivalents reviewed in Chapter 13. The ore zone is associated with an upper Triassic-Lower

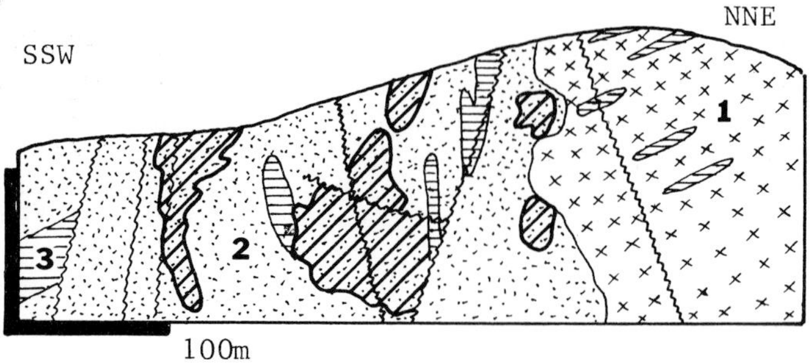

Fig. 28-42. Lorraine (Duckling Creek) porphyry Cu orebody, British Columbia. After Wilkinson et al. (1976).

Jurassic syenite-monzonite complex, a N.W. elongated, intrusive body emplaced into amphibolitic basement metamorphics and upper Triassic island arc basalts, andesites and their volcaniclastics. This measures 32x5 km. Erratically disseminated and veinlet chalcopyrite and bornite with local pyrite and magnetite are associated with high biotite and chlorite contents, potash feldspathization, pervasive sericitization, and the presence of accessory epidote along faults. The immediate host rock is an enigmatic "syenite migmatite", a hybrid rock, and rare bodies (possibly rafts or resisters) of biotite pyroxenite. The ore was emplaced at about 175 m.y. and it gives an impression of a K-silicate metasomatized mafic metavolcanics along a fault or shear zone.

Of the porphyry deposits on the American side of the border, the largest and most famous is Butte, briefly described earlier.

SOUTHERN CORDILLERA, U.S.A. AND MEXICO

Scattered porphyry copper occurrences are distributed over a broad region 3,300 km long and up to 800 km wide, reaching from southern Idaho through Utah, Nevada, Arizona and New Mexico to the latitude of Mexico City. The western United States' segment of this belt is the craddle of porphyry coppers, where this type of deposit was first utilized and interpreted. It is also the most thoroughly investigated porphyry copper region and the second richest (after the Andean belt) of the world. It represents some 103 Mt Cu contained in at least 500 occurrences that include 3 giant fields with more than 10 Mt Cu (Bingham, Clifton-Morenci, Cananea), 12 large fields with more than 1 Mt Cu and at least 50 deposits with more than 100 Tt Cu.

The region is bordered in the west by the Sierra Nevada Batholith in the United States and the Pacific Ocean in Mexico. In the east, no porphyry coppers are known at present, either within or to the east of the Laramide thrust belt and the Precambrian uplift of the Rocky

Mountains (but there are important stockwork Mo deposits). The geology as might be expected is very heterogeneous. The greater portion of the area is underlain by Precambrian metamorphics, topped in the east by Paleozoic and Mesozoic platformic and "miogeoclinal" sediments. These change facies westward into an "eugeoclinal" facies represented by deeper marine, siliceous sediments, mafic volcanics and minor ophiolites. As in the northern Cordillera, the porphyry deposits started to form only in the Mesozoic time (the oldest are Triassic) and the formation terminated at the onset of the Basin-and Range block faulting and continental rifting in the Miocene age. The bulk of the deposits, as in Canada, are "Laramide" (Cretaceous to Oligocene), associated with calc-alkaline plutons sometimes accompanied by subaerial volcanic equivalents. Most porphyry coppers are associated with small hypabyssal stocks and "plutonic" occurrences are rare.

The lonely Bingham, Utah (Lanier et al., 1978; 21.5 Mt Cu, min. 820 Tt Mo; Fig. 28-43) is the largest porphyry copper deposit in North America mined from what may be the world's largest (certainly the most pictoresque) open pit. The deposit is associated with an Oligocene (39-32 m.y.) multiphase (six equigranular to porphyritic phases) quartz monzonite stock emplaced into Pennsylvanian calcareous siltstones and sandstones, against which it is bordered by a zone of hybridization and recrystallization. The stock is cut by porphyry dikes. The porphyry Cu orebody is in the centre of a very regularly concentrically zoned metal distribution pattern, underlain in depth by a dominantly stockwork molybdenite body and flanked laterally by a ring of galena-sphalerite hydrothermal veins and limestone replacements. Over 80% of the Cu-Mo ore is in the K-silicate and phyllic-altered intrusive stock, the rest is in the metasediments. The pear-shaped copper orebody has dimensions of 1.7 x 2.2 km, and is composed of bornite, chalcopyrite and minor primary chalcocite veinlets and disseminations. A rich chalcocite zone topped the hypogene ore.

Ely, Nevada (James, 1976) is a west-trending zone of alteration and mineralization, 13 km long and 460 m wide, associated with six bodies of lower Cretaceous quartz monzonite porphyry. It is localised by a thrust fault separating Pennsylvanian limestone and Mississippian black shale. Chalcopyrite and pyrite are the main hypogene minerals and they occur as disseminations and fracture fillings in K-silicate altered porphyry, hornfelsed shale, and skarn. A quartz-sericite alteration zone in silicate rocks has silicification as its counterpart in limestones. There is widespread pyrite and numerous small Cu, Au and Pb-Zn bearing fissure veins were mined in the past.

Globe-Miami field in Arizona (Peterson, 1962) is notable for the variety in its rock units (Lower Paleozoic schist and amphibolite, middle Proterozoic granite, late Proterozoic diabase, Paleozoic sediments, "Laramide" granodiorite and quartz monzonite and Tertiary volcanics and sediments. The porphyry mineralization is associated with the upper Cretaceous-lower Tertiary quartz monzonite phase of a composite pluton, and it has chalcopyrite, pyrite, minor molybdenite

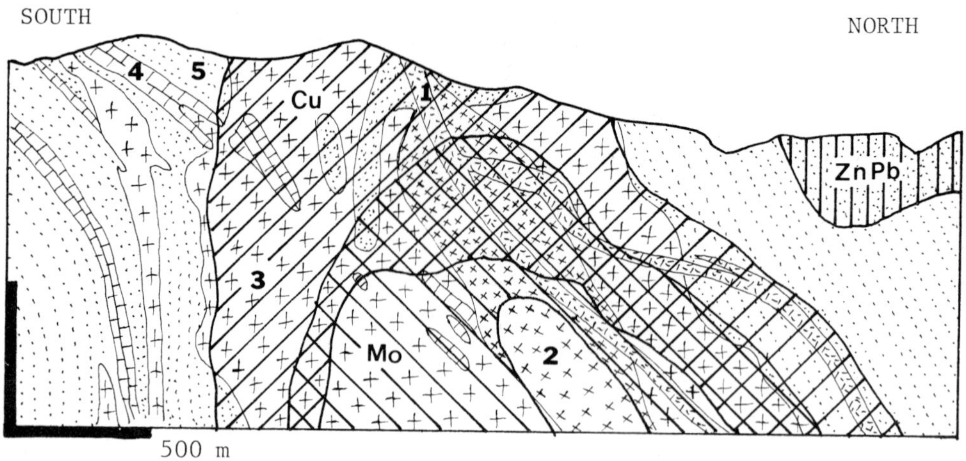

Fig. 28-43. Bingham porphyry Cu-Mo deposit, Utah, showing overlapping Cu and Mo stockwork zones; 1=Eo latite porphyry; 2=Eo quartz monzonite porphyry; 3=Eo monzonite and quartz monzonite; 4=Ps limestone; 5=Ps quartzite. From LITHOTHEQUE, after John (1978).

disseminations and veinlets in both endocontact and exocontact. There is a strong fault control. The porphyry orebodies are fringed by a system of fissure and breccia veins filled by quartz, pyrite, chalcopyrite and specularite, that contributed about 450 Tt Cu to the production. The "porphyry" mineralization is both hypogene and supergene. In addition to chalcocite blankets formed chiefly during the Tertiary period and recently exhumed, Cu oxides (chrysocolla, malachite, azurite) pseudomorphically replacing sulphide grains or infiltrated along fractures, have been recovered by leaching. The Inspiration Company has pioneered the leaching methods since 1926. Some of the secondary enriched orebodies are outside the original K-silicate altered hypogene mineralized zone, in the phyllic zone or in argillized schist.

The Ray deposit, Arizona (Phillips et al., 1974; Fig. 28-44) has a similar setting as Globe-Miami. Most of the disseminated and veinlet high chalcopyrite and bornite, low pyrite ore, is in biotite and chlorite-epidote altered Proterozoic diabase and quartzite near "Laramide" quartz diorite and granodiorite porphyry stocks. The alteration type and zoning is strongly lithologically controlled. Potash feldspar missing in the diabase is developed in the porphyry and quartzite.

Bagdad ore field, Arizona (Anderson et al., 1955), is interesting because of the coexistence of a stockwork and disseminated pyrite, chalcopyrite and molybdenite orebodies mostly in the endocontact of a late Cretaceous quartz monzonite porphyry stock, and several small

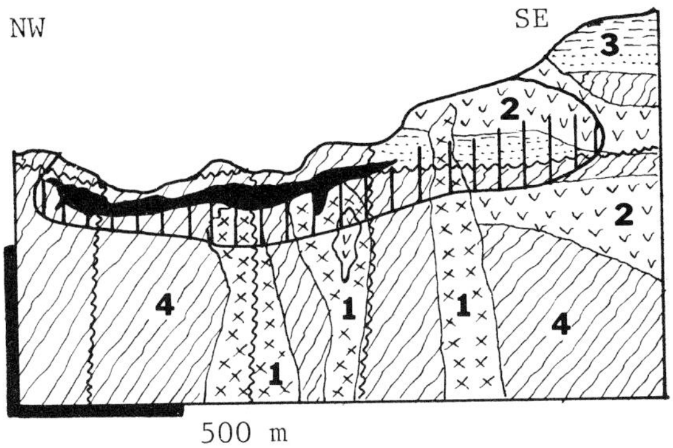

1. Cr_3 quartz monzonite porphyry
2. Pt_3 diabase and amygdaloidal basalt
3. Pt_3 quartzite, shale, conglomerate
4. Pt_2 Pinal Schist

OREBODY:
black=enriched zone
vertically ruled: hypogene zone

Fig. 28-44. Ray porphyry copper deposit, Arizona. From LITHOTHEQUE, after Metz and Rose (1966).

pyrite, sphalerite and chalcopyrite massive sulphide bodies. The latter are relics, hosted by the lower Proterozoic greenstone basement. Minor quartz-wolframite fissure veins are further associated with a Proterozoic granite.

San Manuel and Kalamazoo deposits, Arizona (Lowell, 1968) are two segments of a tilted and fault-displaced porphyry copper orebody. The locality has been made famous as an "average" western U.S. deposit and a model locality of the hypogene alteration-mineralization zoning of Lowell and Guilbert (1970). The orebodies are hosted by a coarse-grained Precambrian deep mesozonal quartz monzonite, intruded by late Cretaceous swarms of quartz monzonite porphyry dikes and irregular stocks. Two broken ring-shaped orebodies (Fig. 28-38) hosted by both the Precambrian and Cretaceous quartz monzonites, have very gradational assay boundaries, and consist of hypogene pyrite, chalcopyrite, minor molybdenite and bornite disseminations and veinlets. The ore shell has an average thickness of 200 m and is situated close to the overlapping border of the K-silicate and quartz-sericite alteration zones. The relative uniformity of the alteration zones and the mineral distribution is attributed to the compositional and physical near-uniformity of the host "granites", regardless of their significant age difference.

Clifton-Morenci field, Arizona (Moolick and Durek, 1966; 13 Mt Cu) has been in operation since 1872, and is most notable for the fact that 100% of its ore is secondary enriched. The hypogene material with its 0.1-0.15% Cu remains a protore even today. The hypogene mineralization which is virtually invisible, is hosted by and genetically associated with a stock of Laramide quartz monzonite porphyry, intruded into a variety of wallrocks ranging from Proterozoic schist, granite, granodiorite to Paleozoic and Mesozoic shale and limestone and an earlier Cretaceous diorite porphyry laccolith and sills. Three small breccia pipes are known. The principal orebody has been a chalcocite blanket up to 330 m thick, averaging 0.8% Cu, formed in the Tertiary time under 70-200 m of leached capping and recently exhumed.

Santa Rita (or Central Mining District), New Mexico (Hernon and Jones, 1968; Fig. 28-45), is one of the "complete" concentrically zoned districts. There, an apex of a "Laramide" quartz monzonite and granodiorite stock (Chino mine) has a porphyry copper deposit, a stockwork pyrite-chalcopyrite topped by a supergene chalcocite blanket. The basement is a thick sequence of Cambrian to Permian shallow marine carbonates interbedded with shale, topped by Cretaceous shale. These rocks host skarn and replacement chalcopyrite bodies in the exocontact of the Chino stock. Magnetite and chalcopyrite skarns at Fierro, and sphalerite skarn and sphalerite-galena replacement bodies in silicified limestone near Hanover and Bayard, are in the exocontact of the Chino and Hanover-Fiero stocks.

Bisbee (Warren) field, Arizona (Bryant and Metz, 1966), has a wide variety of ore styles, most of which depart considerably from the image of a "typical" porphyry copper. In fact, Bisbee ores are closer in appearance to those in Cerro de Pasco, Peru or Burra,

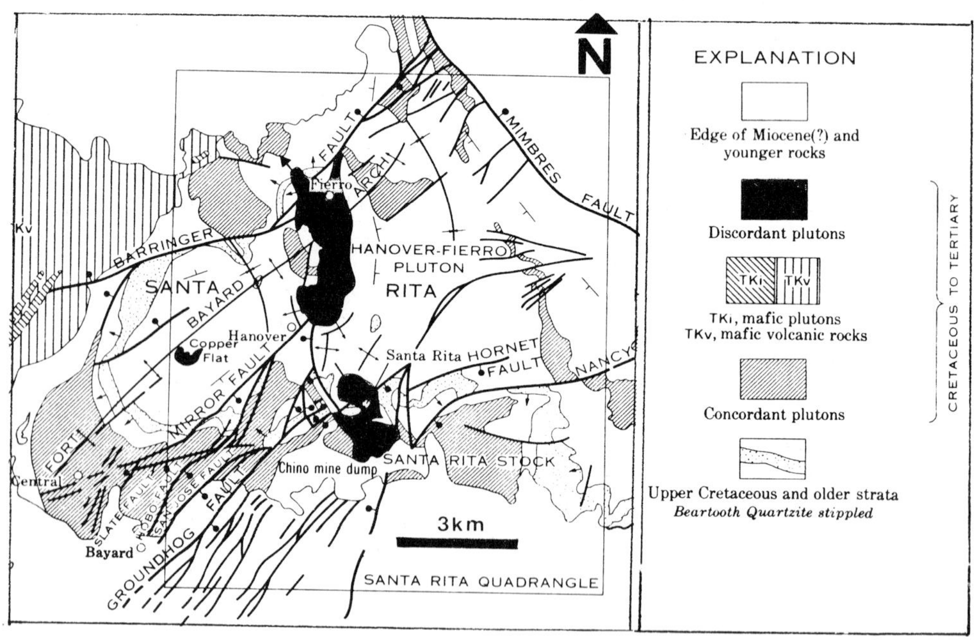

Fig. 28-45. Santa Rita (Central) district, New Mexico. The Chino porphyry Cu deposit is in the Santa Rita stock. From Jones et al. (1967).

South Australia, than to other Arizona porphyry coppers such as San Manuel. The basement at Bisbee is composed of a lower Proterozoic schist, overlain by Paleozoic limestone and shale, and intruded by a Jurassic quartz-feldspar porphyry stock. The area is intensively faulted, and ores occur in three ways: (1) as bedded (mantos) and discordant (pipes, sheets) massive pyrite, chalcopyrite, lesser galena, sphalerite replacements in silicified limestone (rarely skarns) in porphyry exocontacts, along faults and along bodies of intrusive breccia; (2) as low-grade disseminations of chalcopyrite and pyrite in the matrix of intrusion breccia or (3) as disseminations and veinlets in sheared and altered quartz-feldspar porphyry and its brecciated segments. Supergene alteration is extremely intensive and thick in the ore field, and it masks much of the primary alteration and mineralization. Silicification and argillization are the most conspicuous alterations, probably superimposed on hypogene sericitization. Many of the orebodies in carbonates were converted to rich and colourful Cu-oxides (cuprite, malachite, azurite, chrysocolla), whereas a blanket of sooty chalcocite has been the major ore style in the breccia and in porphyries, produced from the Lavender Pit. The intrusive breccias described by Bryant (1968) are heterolithic unsorted masses controlled by faults, and emplaced largely as sills peneconcordant with sedimentary bedding, dikes, pipes and irregular bodies, rather than as vertical columns as elsewhere. Bryant assumed that they had been transported in fluidized state by hydrotherms.

The Cananea ore field in Sonora, northern Mexico (Velasco, 1966; Meinert, 1982) is a giant copper accumulation (15.255 Mt Cu; Fig. 28-46). It has many points of similarity with Bisbee in that a substantial portion of the ore is in skarns, replacement mantos in carbonates and in breccia pipes. The rest is in "Laramide" quartz monzonite and quartz porphyries. The basement in Cananea is represented by Cambrian to Carboniferous limestones and basal quartzite, unconformably resting on middle Proterozoic porphyritic quartz monzonite. The most important primary orebodies at Cananea are associated with three major breccia pipes. The Cananea-Duluth pipe is a steep, oval-shaped vertical structure, 80 m wide and 400 m long. It cuts through gently dipping Tertiary felsic volcanics. It is composed of altered and brecciated weakly mineralized volcanic interior with a strongly mineralized peripheral ring. In the ring chalcopyrite, sphalerite, quartz and minor adularia cement the rock fragments. The La Colorada pipe is related to an intrusive plug of quartz porphyry and it carries a massive and brecciated body of chalcopyrite, bornite and molybdenite in its core. This grades into a pyrite-rich outer ring.

Meinert (1982) established a time framework for the altération-mineralization, particularly in the exocontact carbonates where it follows the general model of skarn origin discussed earlier. After hornfelsing and subsequent metasomatic skarnization, emplacement of a series of quartz monzonite porphyry stocks, followed by

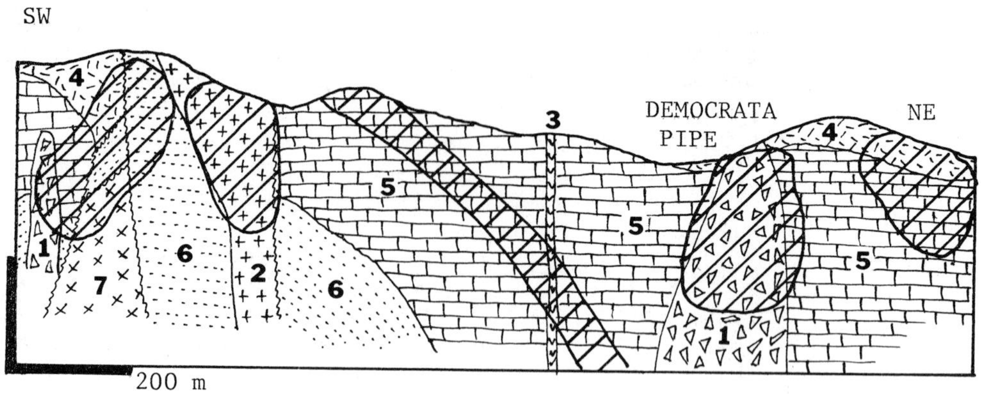

Fig. 28-46. A portion of the Cananea (Sonora, Mexico) Cu field (generalized). 1=T breccia pipes; 2=Cr_3-T_1 quartz porphyry and quartz monzonite porphyry stocks; 3=Cr_3-T_1 diabase dikes; 4=Cr_3 continental rhyolite, andesite, dacite, and tuffs; 5=Cm,D,Ms-Ps limestone, dolomite, minor shale, siltstone; contact metamorphosed; 6=Cm quartzite; 7=1.44 b.y. porphyritic quartz monzonite. Diagonally ruled are disseminated, skarn, fracture filling, breccia replacing Cu-Mo orebodies. From LITHOTHEQUE, based on data in Velasco (1966).

sericitization, produced stockwork porphyry Cu mineralization in the endocontact and disrupted the skarn and hornfels. This was followed by formation of Fe,Cu and Zn sulphide mantos in the carbonates.

La Caridad, Sonora, Mexico (Saegart et al., 1974) is another major porphyry deposit, discovered in the 1960s. Stockwork and disseminated pyrite, chalcopyrite and molybdenite are mostly in quartz, sericite, pyrite-altered Tertiary quartz monzonite porphyry stock and intrusive breccia, emplaced in an older diorite and granodiorite. Most of the economic ore comprising fossil supergene blanket, up to 250 m thick, mineralized by sooty chalcocite, buried under a thick leached capping.

Scattered small to medium-sized porphyry Cu deposits (Santo Tómas, Sinaloa; 750 Tt Cu/0.5%; La Verde, Michoacan; 770 Tt Cu/0.7%; Inguarán, Michoacan) have been discovered in the past 15 years. Disseminated copper deposits hosted largely by volcanics and subvolcanic bodies (Safford, Arizona; El Arco, Baja California, 3.9 Mt Cu/0.6%, were reviewed in Chapter 26).

CENTRAL AMERICA

A new porphyry copper province in Panama continuing westward into Costa Rica and eastward into Colombia, has been discovered in the past 15 years. The mineralization (Kesler et al., 1977) is affiliated with high-level potassic granodiorite intrusions 33-4 m.y. old. These represent the latest stage of igneous activity that took place originally in an island arc setting 60-70 m.y. ago. The largest deposit Cerro Colorado with a potential of 18 Mt Cu is largely hosted by volcanics and by subvolcanic bodies and was reviewed in Chapter 26. The deposits in N.W. Colombia (Pantanos, Pegadorcito) are located in the extension of the mineralized belt in eastern Panama.

THE CARIBBEAN

In the Caribbean (Hollister, 1978), porphyry copper occurrences have been reported on all the major islands of the Greater Antilles and Jamaica. The largest and most extenisively studied occurrences that include two small to medium-size deposits are in Puerto Rico (Fig. 28-47). The oldest dated rocks in the Greater Antilles are Jurassic, and they are marine "eugeoclinal" sediments and volcanics. Between the Jurassic and Eocene periods the region was an active island arc and a considerable thickness of largely andesitic volcanics, derived sediments and comagmatic quartz diorite-granodiorite intrusions, accumulated. The porphyry Cu occurrences are stockworks, almost invariably associated with "Laramide" quartz diorite porphyries. The ore zones are most commonly situated in biotite-altered endo- and exocontacts. They are low in Mo, but relatively high in Au.

ANDEAN BELT, SOUTH AMERICA

If the Colombian porphyry Cu occurrences are not considered here, the Andean porphyry Cu occur in a N.N.W.-S.S.E. trending belt 4,500 km long between Chaucha, Ecuador, and Campana Mahuida, Argentina (Hollister, 1978; Putzer, 1976). Several hundred occurrences are known and these include 5 giant deposits (10 Mt Cu plus; see Table 28-3); 16 large deposits (between 1 Mt and 10 Mt Cu) and several tens of medium to small-sized deposits. As in the Cordillera, the porphyry Cu belt is situated between the chain of Mesozoic coastal batholiths in the west, and the eastern limit of the Precambrian craton and the "miogeoclinal" thrust and molasse belt of the Eastern Cordillera of the Andes. The intrusions with which the "porphyries" are associated range in age from 4.32 m.y. (El Teniente) to lower Carboniferous (Oyarzún and Frutos, 1980) and they have been emplaced in a continental to transitional crust environment along faults. Further generalization on the setting of the deposits is impossible, and each has to be considered individually.

The isolated northern Peruvian deposit Michiquillay (Hollister and Sirvas, 1974) is a stockwork of chalcopyrite, pyrite, minor bornite, chalcocite, molybdenite and sphalerite veinlets in a phyllic and argillic alteration zone of a Miocene hornblende-biotite quartz monzonite stock. The stock intruded Cretaceous quartzite and limestone without causing much contact metamorphism.

Morococha, 110 km E. of Lima, Peru (Petersen, 1965; M. Velasquez R., guided tour, 1977; Fig. 28-48) is a zoned Cu-Mo, Cu-As, Pb-Zn,

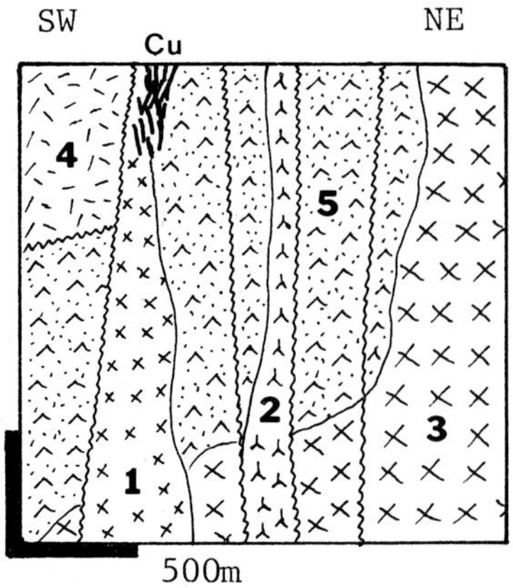

1. Eo, qtz. diorite stocks
2. Eo, hornblende quartz diorite and diorite
3. Cr_3–T_1 quartz diorite and granodiorite pluton
4. Eo, marine to contin. andesite, dacite, tuff, volc. sediments
5. Cr, mafic andesite, volc. arenite, shale, limestone

Fig. 28-47. Diagrammatic representation of the setting of the Puerto Rican porphyry coppers (Viví Arriba, Piedra Hueca) S.S.E. of Utuado. From LITHOTHEQUE, based on data in Cox et al. (1973).

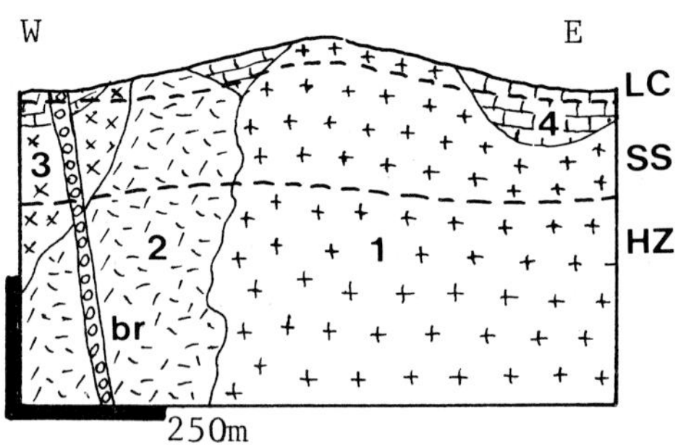

supergene zones:
LC=leached capping
SS=secondary sulphides
HZ=hypogene zone

geology:
1=Mi_{2-3} granodiorite stock
2=Mi quartz porphyry
3=8 m.y. porphyritic diorite
4=Tr_3–J_1 cont.metasom.carbonate
br=Mi breccia pipe

Fig. 28-48. Toromocho porphyry Cu-Mo deposit, Morococha, Peru. From LITHOTHEQUE, diagrammatic, after M. Velazquez R., guided tour, 1977.

Ag, ore field. The mineralization is associated with Pliocene quartz monzonite porphyry stocks intruded into a petrographically variable sequence of Triassic-Jurassic limestone, black shale, anhydrite, gypsum and metabasalt suite. This rests unconformably on Permian continental andesite and dacite lavas, tuffs and breccia within a symmetrical, N30°W-trending anticline. As in Santa Rita, the most central mineralization is a porphyry Cu-Mo stockwork (Toromocho deposit). The hypogene ore consists of pervasively disseminated and fracture coating pyrite, chalcopyrite, molybdenite with quartz, in an intensely altered quartz monzonite. In places, biotite forms an almost monomineralic matrix to sulphide grains. The largest tonnage of economic ore is in the zone of secondary sulphide enrichment where quartz, pyrite, chalcocite coatings are in quartz-sericite altered and in places argillized hostrocks.

Above and laterally with respect to the porphyry copper, rich pyrite, chalcopyrite and enargite "contact pipes" formed along the intrusion-carbonate contact, in diopside, tremolite or serpentinite, chlorite or talc-altered carbonate. Mantos and irregular replacement bodies in lightly altered carbonates further from the contact carry rhodochrosite, ankerite, pyrite, enargite and bornite, or galena and sphalerite. Several tens of E.-W.-trending epithermal fissure veins filled by quartz, pyrite and enargite are hosted by another quartz monzonite stock. Silver-rich quartz, pyrite, ankerite, tennantite veins are situated mostly in the Permian basement volcanics.

The important string of porphyry copper deposits in the Arequipa area, S.W. Peru (Cerro Verde, Cuajone, Quelleveco and Toquepala), together represent some 27 Mt Cu. The largest deposit, Cuajone (Manrique and Plazalles, 1975) has a stockwork ore hosted by a K-silicate and sericite-altered Tertiary quartz monzonite porphyry, emplaced into a late Cretaceous-early Tertiary continental basalt, andesite and rhyolite suite. Chalcopyrite-tourmaline mineralized breccia pipes also occur, but they are best developed at Toquepala (Hollister, 1978). There, a breccia pipe affiliated with and emplaced into a dacite, has a diameter of about 600 m and is concentrically zoned. Its central zone is a "pebble breccia". It postdates alteration and ore emplacement and it changes into a mixed fragment breccia and a monolithic (dacite) breccia. The outer breccia zones are quartz-sericite altered and cemented by tourmaline containing scattered pyrite, chalcopyrite and molybdenite. A stockwork of quartz, tourmaline and chalcopyrite veinlets encircles the pipe.

The large and interesting Chilean porphyry coppers start, in the north, with Chuquicamata (Ruíz, 1965; Sillitoe, 1973; 43 Mt Cu, 1.1 Mt Mo; Fig. 28-49). This deposit has dimensions of about 4x2 km and has been explored to a depth of 600 m. Hypogene pyrite, chalcopyrite, lesser bornite, enargite and molybdenite occurs as stockwork, pervasive disseminations and discrete veins in a zone of quartz-sericite, lesser K-feldspar and biotite alteration, within and around a composite Oligocene granodiorite to quartz-monzonite stock. The Chuquicamata orebody is one of several bodies located in a N.-S. trending, fault-controlled belt. The ore mined at present and in the past has come from a zone of secondary enrichment, in which chalcocite and minor covellite are in supergene argillized wallrocks. Formerly, the oxidation zone with antlerite, atacamite, brochantite, chrysocolla, and other rare minerals contributed to copper production (and excitement of the visiting mineralogists). The Exotica chrysocolla and atacamite, broachantite deposit, still actively forming by precipitation in the alluvial fan below Chuquicamata, has been described in Chapter 23.

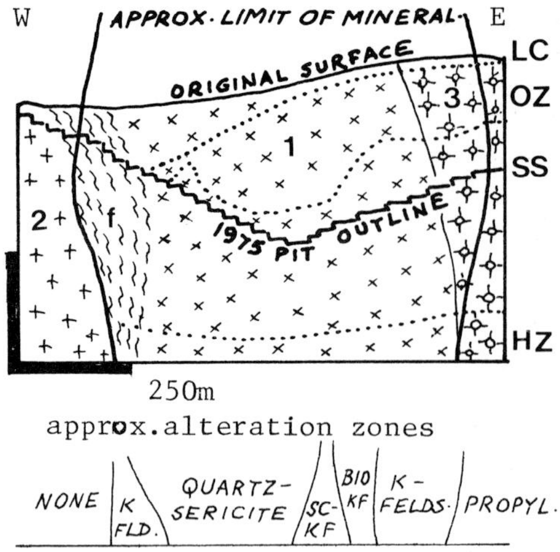

supergene zones:
LC=leached capping
OZ=oxidation zone
SS=secondary sulphides
HZ=hypogene zone
geology:
f=Falla Oeste fault zone
1=46-32 m.y. altered quartz monzonite to granodiorite stock
2=35 m.y. granodiorite
3=J granodiorite

Fig. 28-49. Chuquicamata porphyry Cu-Mo deposit, Chile. From LITHOTHEQUE, after H. Soto, guided tour, 1977 and CODELCO materials for visitors.

El Abra porphyry copper deposit (Ambrus, 1977; 16.35 Mt Cu), is 42 km north of Chuquicamata and controlled by the same tectonic structure (Falla Oeste). There, a multiple alteration-mineralization related to a diorite-syenite complex was later intruded and remobilized by several dacite necks, capped by biotite breccia. The low pyrite, chalcopyrite-molybdenite ore lies in structurally controlled bodies. The hypogene ore is capped by a chrysocolla-mineralized oxidation zone.

El Salvador, a "minor" deposit by Chilean standards (5.5 Mt Cu; Fig. 28-50), is best known for its excellent modern description and interpretation by Gustafson and Hunt (1975). The ore is associated with a system of high-level granodiorite porphyry intrusions (39.1 m.y.), emplaced into comagmatic and slightly earlier andesite-rhyolite volcanic pile. Economically, the most important earlier mineralization phase lies in endo- and exocontact quartz veins, fracture coatings and pervasive disseminations, and has chalcopyrite, bornite and pyrite in K-feldspar and biotite—altered rocks. In the later stage pyrite-rich quartz, anhydrite, bornite, chalcopyrite, enargite, tennantite, sphalerite and galena veins and veinlets formed, surrounded by feldspar-destructive sericite and sericite-chlorite alteration haloes. The economic mineralization originally outlined coincides with a supergene enriched blanket containing chalcocite, pyrite and bornite. This is confined under a leached capping.

Los Pelambres, Chile and the adjacent El Pachón, Argentina (Sillitoe, 1973; about 8 Mt Cu), are related to a late Miocene

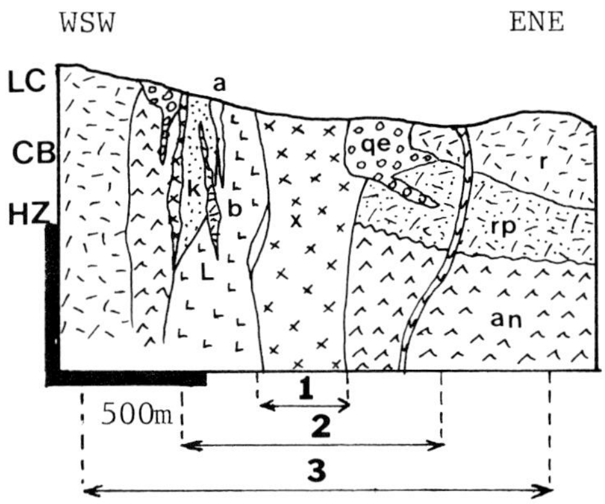

41 m.y. old rocks
b=igneous breccia
k,L,a,x=varieties of granodiorite porphyry
qe=45 m.y. quartz-eye porphyry
r=50 m.y. rhyolite dome
rp=T_1 ignimbrite, rhyolite pyroclastics
an=Cr_3 andesite flows, pyroclastics
supergene zones:
LC=leached capping
CB=chalcocite blanket
HZ=hypogene zone

alteration-mineralization zones
1=low sulphide zone
2=K-silicate alteration, chalcopyrite-bornite
3=qtz,pyrite,sericite alter., chalcopyrite-pyrite miner.

Fig. 28-50. El Salvador Cu-Mo deposit, Chile. From LITHOTHEQUE, simplified after Gustafson and Hunt (1975).

tonalite and diorite porphyry stock and dike. The intrusive rocks are emplaced in Mesozoic continental andesites and volcaniclastics. Chalcopyrite, bornite, pyrite and minor molybdenite stockwork is hosted by biotite and K-feldspar altered rocks in the centre of a concentrically zoned alteration complex.

Rio Blanco-Disputada. This giant field is located in glaciated alpine country 50 km N.E. of Santiago, within sight of Aconcagua (Ruíz, 1965; 17.6 Mt Cu). The area is underlain by early Tertiary andesitic lavas of the Farellones Formation (the same unit that hosts El Teniente), intruded by a granodiorite-quartz monzonite pluton. These "basement" units are intruded by subvolcanic bodies of dacite porphyry and capped by several erosional remnants of rhyolite. Cu and minor Mo,Au,Ag, Pb-Zn mineralization occurs within an area of about 11 x 5 km, and is subject to significant control by N.-S. fault zones.

The Disputada (Los Bronces) orebody is hosted by granodiorite and it consists of a breccia composed of angular intrusive clasts in a matrix of tourmaline, pyrite and chalcopyrite. The larger Rio Blanco (Andina) group of orebodies is mostly in andesite at contact with dacite porphyry, and also in a subvertical subvolcanic porphyry plugs. The andesite-hosted orebodies are mostly tabular breccias and brecciated metavolcanics, silica, biotite, sericite, tourmaline and chlorite – altered. Pyrite, chalcopyrite, specularite, lesser molybdenite, bornite and enargite are disseminated or form veinlets in the breccia matrix. Secondary enrichment is insignificant.

The El Teniente (formerly Braden) deposit east of Rancagua, central Chile (Camus, 1975; 62 Mt Cu/above 1%, 1.6 Mt Mo; Fig. 28-51), is a giant metal accumulation and the world's largest porphyry copper. The orebody having more then 0.5% Cu is a N.W. elongated zone measuring 6x3 km of fractured and altered volcanic and intrusive rocks, surrounding the almost barren Braden breccia pipe. The mineralization correlates with the emplacement of a 4.3 m.y. dacite porphyry into the bedrock composed of Paleocene to Eocene subaerial, massive homogeneous andesite and minor conglomerate (Farellones Formation), and an Oligocene-Miocene quartz diorite. The Farellones andesites are load metamorphosed and thermally contact hornfelsed in the ore field, and they have a high trace Cu content. They host about 20% of the stockwork ore, the rest of which is in the quartz diorite and dacite porphyry. Hypogene chalcopyrite, pyrite, bornite, molybdenite, tennantite-tetrahedrite, lesser magnetite, specularite, enargite, etc., occur without gangue (or with quartz and anhydrite) in a K-silicate and less quartz-sericite and propylitic alteration envelope. The K-silicate zone has a biotite, K-feldspar, lesser anhydrite and quartz assemblage. Tourmaline is locally widespread. Argillic alteration is supergene and it contained patchy chalcocite and small amounts of oxidation zone minerals such as chrysocolla, malachite, brochantite and antlerite, reaching along faults to a depth of 600 m.

NORTH-EASTERN APPALACHIANS, CANADA AND THE UNITED STATES

About 20 occurrences of "porphyry Cu-style" deposits ranging in age from Cambrian to Devonian, have been recorded in the Appalachian Orogen between Newfoundland and Maine (Hollister, 1978). With the

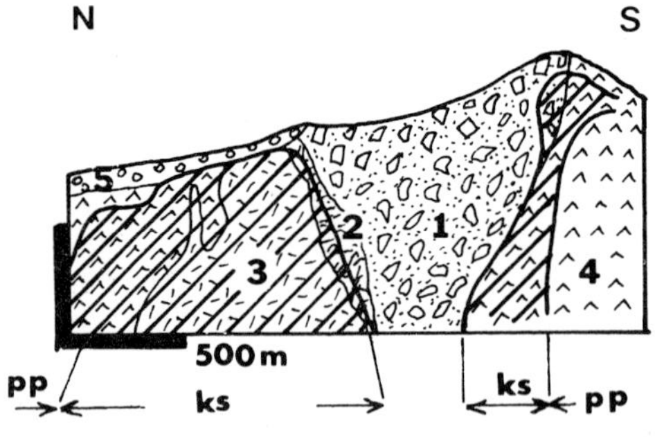

1. Pl, breccia pipe (fragments of andesite, dacite porphyry, quartz diorite)
2. brecciated rocks
3. Pl, dacite porphyry
4. Pc-Eo, subaerial metaandesite flows, agglomerates; contact hornfelsed
5. Q, talus, glacial sediments

ruled: orebodies above 0.5% Cu cutoff

Fig. 28-51. El Teniente porphyry Cu-Mo, Chile. From LITHOTHEQUE, after Camus (1975).

sole exception of the Gaspé (Murdochville) deposit, none is of economic importance. As expected, the occurrences are deeply eroded so that most appear to represent only the "roots of the porphyry system". Among the peculiarities of these occurrences as a set, Hollister (1978) noted the general lack of a well developed phyllic zone (most mineralizations are in the potassic zone); most intrusions having a fresh, unaltered core; many intrusions being equigranular instead of porphyritic, and some containing a sparse, possibly magmatic chalcopyrite. There is no clear connection between intrusions and extrusive volcanics; missing argillic zones; and pyrite is rarer than in the Meso-Cainozoic systems. The "porphyry" deposits are members of an ensialic orogen, formed during periods of distension immediately following a compressive event.

The Gaspé Cu field (near Murdochville, Gaspé Peninsula, E. Quebec; Allcock, 1982; Fig. 28-52), contains several disseminated copper orebodies genetically associated with a late Devonian to Mississippian intrusion of rhyodacite porphyry sills, dikes and plugs into lower Devonian calcareous sediments. The orebodies are of two styles: (1) replacement bodies in metasediments at Needle Mountain (736 Tt Cu/0.7-1.35%). These consist of disseminated to massive chalcopyrite and pyrrhotite, lesser bornite, and they are peneconcordant, lithologically controlled or discordant, controlled by faults and (2) disseminations and veins in and around fractured Copper Mountain porphyry plug (1.112 Mt Cu/0.4%). There, the ore is composed of pyrite, chalcopyrite and minor molybdenite and is of five generations. The Stage 2 quartz, anhydrite, K-feldspar veinlets with traces of chalcopyrite and magnetite are coeval with K-feldspar and biotite alteration. Stage 3 which is economically the most important contains quartz, lesser anhydrite, chalcopyrite, pyrite and molybdenite veins associated with kaolinization, sericitization and minor anhydrite alteration in the endocontact, and with the formation of tremolite, epidote and chlorite in the exocontact.

CALEDONIAN AND HERCYNIAN OROGENS OF EUROPE

Scattered occurrences of "porphyry copper-style" mineralization are known in several places in Caledonian and Hercynian Europe (e.g. Ballachulish, Scotland; Miedzianka, Polish Silesia), but are presently of little economic significance. The only exceptions are the chalcopyrite stockworks in the subvolcanic felsic intrusions underlying some massive sulphide systems, as in the Rio Tinto belt (e.g. Cerro Colorado; compare Chapter 12).

CARPATHIANS, BALKANS, THE HELLENIDES OROGENS

An economically significant porphyry Cu province has been outlined only recently in the Cainozoic ("Alpine") orogen of the Carpathians and in its continuation into the Balkans and Hellenides systems. The

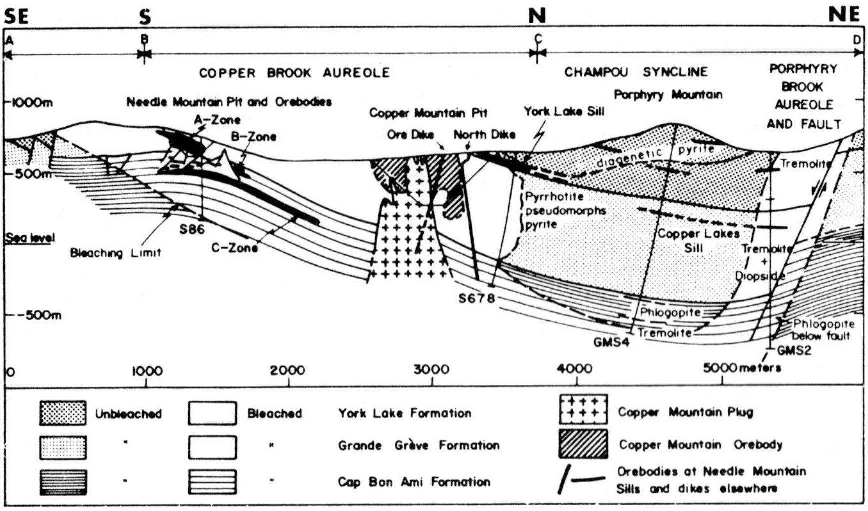

Fig. 28-52. Gaspé Cu field, Quebec, Canada, showing Needle Mountain skarn and Copper Mountain porphyry deposits. From Allcock (1982), courtesy of Economic Geology.

known deposits in Hungary (Récsk); Rumania (Moldova Nouă, Ciclova, Deva, etc.); Yugoslavia (Bor, Majdanpek, Veliki Krivelj, Bučim); Bulgaria (Panagurishte) and Greece (Skouries) are associated with later Cretaceous and Tertiary intrusive activity, and represent some 12.5 Mt Cu.

The Récsk deposit is mostly volcanics-hosted, so it has been reviewed in Chapter 26 as was the Musariu Nou occurrence in the Rumanian Apuseni Mts., and the Yugoslav deposits Bor and Majdanpek. In the Banat district of S.W. Rumania (S.E. of Timişoara; Ianovici and Borcoş, 1982), three parallel N.-S. lineaments are marked by late Cretaceous quartz diorite to quartz monzonite plutons, emplaced in the geologically complex basement of Devonian bimodal volcanic-sedimentary association, Jurassic ophiolites and Cretaceous sediments including limestones, and andesitic volcanics. Several tens of Fe, Cu, Pb-Zn, Ag, Au deposits mostly of vein, skarn and limestone-replacement styles occur in the hydrothermal aureole of the plutons, but endocontact porphyry Cu mineralization has been outlined only recently in about 10 localities.

In the largest exploited deposit, Suvorov in the Moldova Nouă field (230 Tt Cu/0.23%, 43 Tt Mo/0.0043%) disseminated pyrite, chalcopyrite, minor molybdenite in a biotite-altered quartz diorite porphyry grade into magnetite, pyrite, chalcopyrite skarn. The skarn is in the immediate exocontact as well as rafts and inclusions in the intrusive body. Additional porphyry copper localities in the Banat are Ciclova, Sasca Montană, Oraviţa, Lapusnic, and others.

In Bulgaria, the producing porphyry copper deposit Medet is situated in the Panagurishte district in the Sredna Gora Range (Iovchev, 1961). There, late Cretaceous continental explosive andesite, dacite, rhyolite association and comagmatic intrusions were emplaced in a Proterozoic crystalline basement intruded by late Paleozoic tonalite-granodiorite plutons, along a system of W.N.W.-trending fault zones. Medet is a low-grade stockwork of pyrite, chalcopyrite and molybdenite veinlets in a biotite-altered granodiorite and quartz monzonite porphyry endocontact, and in the Paleozoic tonalite exocontact. Several small massive pyrite-chalcopyrite deposits hosted by quartz-sericite altered dacites (e.g. Elshitsa), are probably the subvolcanic equivalents of the "porphyry" ores.

The Serbian-Macedonian Massif, a large block of Paleozoic crystalline basement extending from S.E. Yugoslavia to Greece and Bulgaria, contains several porphyry copper occurrences associated with Tertiary volcanics and intrusive stocks. The ore zone near Radovište, Yugoslavia (Janković, 1982; 459 Tt Cu) is mostly situated in K-silicate altered Proterozoic metamorphics in the exocontact of subvolcanic andesite and hypabyssal quartz diorite intrusions. The occurrences in the Eastern Chalkidiki Peninsula in Greece are associated with granodiorite and quartz diorite porphyries, and they contain one possibly economic orebody (Skouries).

TURKEY, CAUCASUS, IRAN, PAKISTAN

This 3,000 km long portion of the Alps-Himalayas orogenic system is in the eastern continuation of the Balkan-Hellenides belt, and terminates on the Himalayan collisional structures. Throughout almost its entire course, it is marked by numerous rigid blocks of older (probably Precambrian) continental crust, most of them in the position of median massifs. These are surrounded by Mesozoic-Cainozoic foldbelts composed of "eugeoclinal" (marine volcanic-sedimentary) megafacies rocks. Porphyry copper deposits (about 9) and occurrences (about 100) represent some 12.5 Mt Cu and they occur either isolated or in groups. Most are associated with Tertiary hypabyssal intrusions emplaced in the transitional crust along margins of the old continental blocks.

In Turkey, the important copper deposits Lahanos and Murgul (Chapter 26) represent a transition between deposits that are probably volcanogenic or subvolcanic and porphyry coppers, and are sometimes treated as members of the latter category. The Bakircay porphyry Cu showing in northern Turkey is associated with a 42 m.y. old granodiorite porphyry.

The economically important porphyry copper province in the Lesser Caucasus Range of the Soviet Armenia (Pidzhyan, 1975), forms three mineralized clusters within a N.W.-trending Pambak-Zangezur belt. This belt is parallel with the Iranian border and ultimately crosses into Iran. It represents at least 2 Mt Cu and 80 Tt Mo. The most important deposits Dastakert, Kadzharan and Agarak are within the 50 km long southern extremity of the belt, west of the town of Kafan (Zangezur). The porphyry Cu belt formed along the margin of and within an old median massif in the south, south of an ophiolite belt. There, a Precambrian and lower Paleozoic crystalline basement is covered by Devonian to Cretaceous carbonates and detrital sediments. This is topped by an Eocene to Miocene andesite, dacite, rhyolite volcanic pile, and intruded by late Eocene to Miocene high-level granodiorite, quartz monzonite, granite and quartz-feldspar porphyry stocks. The porphyry Cu-Mo deposits are endo- and exocontact (in hornfelsed lower Eocene andesite-rhyolite volcanics) zoned stockworks in K-silicate alteration zones, and of all the Soviet "porphyries" they are most reminescent of the "typical porphyries" of the American Southwest.

The Tertiary volcanic-intrusive belt continues south-eastward across Iran to Baluchistan, and contains the important porphyry Cu deposit Sar Chesmeh (Hakim, 1980; 9.04 Mt Cu). There, disseminated and veinlet chalcopyrite, pyrite and lesser molybdenite lie in a K-silicate altered late Tertiary granodiorite to quartz monzonite stock emplaced in andesites, and laced by several generations of porphyry dikes.

In Saindak, Pakistan (Sillitoe and Khan, 1977), stockwork porphyry copper is in Miocene K-silicate altered tonalite and granodiorite stock intruded into a late Cretaceous to Oligocene volcanic-sedimentary sequence.

BURMA, INDONESIA, EASTERN MALAYSIA REGION

Porphyry copper occurrences have been discovered in the Indo-Burman Ranges, in Sumatra (Lake Singkarak, Tangse), northern Sulawesi (Tapadaa, Tombuililato) and in Borneo, associated with Tertiary high-level intrusives generally emplaced into andesitic volcanics.

The Mamut Mine on the slopes of Mount Kinabalu in Sabah (N.E. Borneo; Kósaka and Wakita, 1978) is, at present, the only deposit being exploited. This deposit has an unusual setting which corresponds to the somewhat unusual alteration style and mineralogy. The basement is part of a broad Cretaceous to possibly Miocene mélange zone (Hamilton, 1979) composed of imbricated flysch sandstone and mudstone intruded by slices of tectonite ultramafics. This has been further intruded by several phases of the late Miocene Kinabalu Batholith, out of which small hypabyssal quartz monzonite porphyry and granodiorite porphyry bodies are responsible for the mineralization. Most of the ore is in pyrrhotite, pyrite and chalcopyrite endocontact disseminations, accompanied by silicification and biotite alteration. The orebody is situated in an outer shell of a quartz monzonite stock, and in a hornfelsed siltstone and serpentinite exocontact. Inward (into the stock), the biotitization changes into tremolite-actinolite alteration and in the core, hornblende alteration is a most likely consequence of hybridization by the ultramafics.

THE PHILIPPINES

The Philippines are a prominent porphyry copper province. Several hundred occurrences are known, and Wolff (1978) listed 75 medium to small deposits and 6 large deposits (over 1 Mt Cu) for which tonnage figures were reported. The overall copper content in the Philippine "porphyries" is of the order of 25-30 Mt Cu. The geology of the Philippines (an active as well as a relic island arc system since the Cretaceous period) and also the ore distribution, were reviewed by Bryner (1969). Wolfe (1973) commented on the special characteristics of the Philippine porphyry coppers that set them apart from the rest of the world, in particular from the "typical porphyries" of Lowell and Guilbert (1970).

The porphyry copper occurrences range in age from late Cretaceous to almost Pleistocene (2 m.y. at Lepanto). They are typically low in Mo and high in Au, associated with quartz-deficient rocks (diorite to quartz diorite) often emplaced in or intermingled with andesite, so that sharp separation of both rocks is impossible ("intimate" andesite-diorite association; compare Chapter 13). The most acidic common rock associated with mineralization is trondhjemite, although quartz porphyries may form dikes or small stocks. In the "Visayan Type" of porphyry Cu deposits introduced by Wolfe (1973) no hypabyssal rocks are associated and the ore is in basic andesites (Chapter 13).

The majority of the Philippine porphyry Cu deposits are related to intrusions emplaced in the vicinity of Tertiary geanticlines into late Cretaceous or early Tertiary metavolcanics. The orebodies are stockworks controlled by faults, so the orebodies tend to be prismatic or tabular and are rarely concentric. Secondary enrichment is minor and usually erratic. Occurrences are known on Luzon, Marinduque, Negros, Cebu and Mindanao Islands.

The Tapian (Marcopper) ore field situated on the small Marinduque Island (Loudon, 1976; 600 Tt Cu/0.58%; Fig. 28-53) has been in production since 1969. There, middle Miocene elongated quartz diorite pluton intrudes folded and low-grade metamorphosed Eocene-Oligocene marine mafic metavolcanics and metasediments that include also reef limestones. Granodiorite, granodiorite porphyry and quartz diorite porphyry in the vicinity of orebodies probably represent a locus of a multiple intrusion. The intrusives and the hornfelsed rocks in their aureole are highly fractured and disrupted by a N.W.-S.E.-trending swarm of faults. Hypogene chalcopyrite, pyrite and magnetite fill fractures in a K-altered and silicified endocontact and to a lesser extent exocontact rocks. There is an irregular secondary enriched zone extending to a depth of 90 m.

NEW GUINEA-MELANESIA BELT

This belt extends from the western tip of New Guinea to Fiji, a distance of almost 6,000 km. About 35 porphyry copper occurrences are known, four of which have reserves over 1 Mt Cu. Geology and

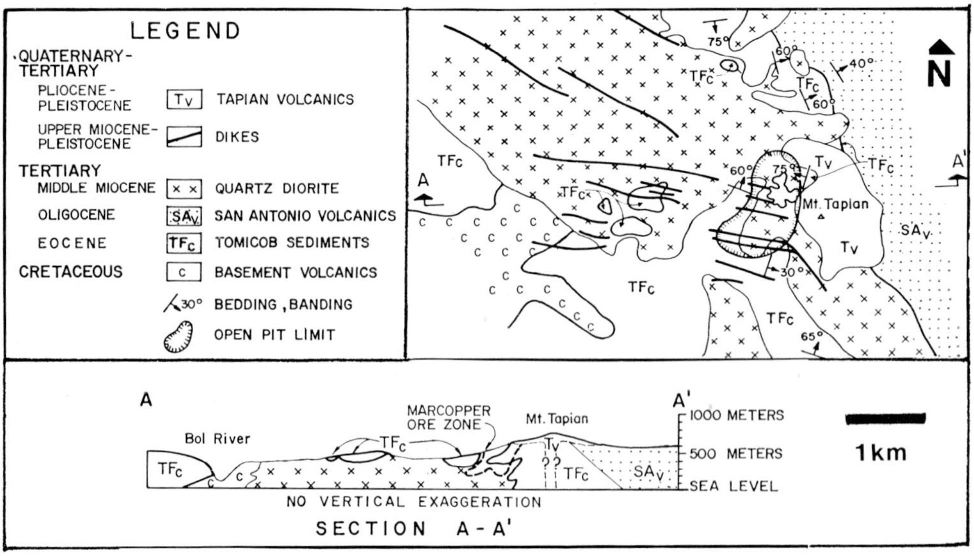

Fig. 28-53. Marcopper porphyry copper, Marinduque Island, Philippines. From Loudon (1976), courtesy of Economic Geology.

"porphyry" characteristics in this belt have been summarized recently by Titley (1975).

The island of New Guinea is underlain in the south by the stable continental crust of the Australian block, covered by Jurassic to Quaternary sediments. These change northward into an increasingly more structurally complex and metamorphosed sequence of an orogenic belt. This belt consists of Triassic to Eocene (meta)sediments and (meta)volcanics including, in the south-east, an ophiolite belt. The deformed rocks are intruded by an Oligocene to Pleistocene suite of diorite, granodiorite, syenite and granite. The large, low-grade porphyry copper deposits (Ok Tedi, Frieda River, Yandera) are associated with diorite or quartz diorite plutons and their late-stage hypabyssal differentiates, and the mineralization is in geological terms very young (middle Miocene to 1.2 m.y. old). Despite the difficult access, the large "porphyries" have recently been described in detail in a series of papers in Economic Geology.

The Ok Tedi (Mount Fubilan) deposit located near the Indonesian border (Bamford, 1972; 2.33 Mt Cu, 262 t Au) has the distinction of being the world's youngest porphyry copper deposit (1.2 m.y.). It is a part of a W.N.W.-trending tectonic belt of unmetamorphosed Tertiary and older detrital and carbonate sediments, intruded by a number of small Plio-Pleistocene diorite plutons. The ore deposit consists of a central, distinctive orthoclase-rich cylindrical intrusion of quartz latite porphyry, mineralized by disseminated chalcopyrite, bornite and marcasite in a potassic alteration zone. The intrusion is surrounded by a number of magnetite and chalcopyrite skarn bodies and a pyrite-pyrrhotite massive replacement body in the exocontact limestones. Supergene alteration is intense and chalcocite replaces the hypogene minerals to a depth of 330 m. This is topped by an oxidized and partly leached capping that includes a residual gold orebody. Data on the Frieda River and Yanderra deposits, two remaining large porphyry coppers on the New Guinea (Niugini) mainland, appear in Table 28-3.

The Melanesian island arc chain includes the territory of Niugini (New Britain, New Ireland, Bougainville, and other islands), Solomon Islands, New Hebrides (Vanuatu) and Fiji. These islands, already mentioned in Chapter 5 (Davies, 1978; Arthurs, 1979; Colley and Greenbaum, 1980), consist of Tertiary or latest Cretaceous marine to continental volcanics with volcanogenic and biogenic (coral limestone) sediments, intruded by contemporary plutonic rocks. These rocks formed on or within an oceanic or a "quasi-oceanic" crust. Major porphyry prospects are on New Britain (Uasilau, Kulu, Pleysumi, Pelapuna), New Ireland (Legusulum), Manus Island (Mt. Kren), Guadalcanal (Koloula), Viti Levu (Namosi area) and other islands. The Panguna deposit on the Bougainville Island is the largest deposit, in production since 1972. The large Waisoi Prospect (2.29 Mt Cu) in Fiji is volcanics-dominated and was reviewed in Chapter 26.

Panguna deposit (Baumer and Fraser, 1975; 4.32 Mt Cu, 450 t Au, 2.7 Tt Ag; Fig. 28-54) is a young (3.4 m.y.) system of chalcopyrite, bornite, pyrite, lesser magnetite, molybdenite veins, stockworks and disseminations. It is situated on the edge of a Pliocene high-level intrusive complex of granodiorite and quartz diorite porphyries, emplaced in a slightly older quartz-diorite pluton and Oligocene

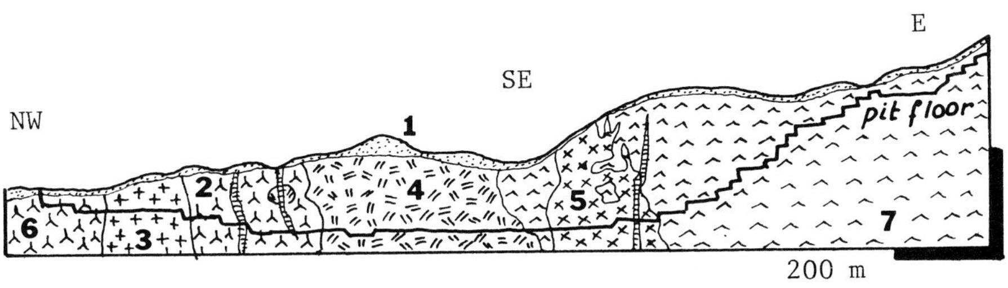

Fig. 28-54. Panguna porphyry copper deposit, Bougainville Island, Niugini. 1=Q volcanic ash, landslide, alluvium; 2=Pl pebble dikes; 3=Pl, granodiorite; 4=Pl, biotite granodiorite (concentrator intrusion); 5=Pl leucocratic quartz diorite; 6=biotite diorite; 7=Ol_1Mi_1 andesite and volcanic breccia. The orebody is approximately indicated by the pit outline. From LITHOTHEQUE, simplified after Baldwin et al. (1978).

hornfelsed andesite. The ore is in both endo- and exocontact, best developed in the zone of biotite alteration. Supergene alteration caused argillization and leaching, but no regular enrichment zone formed.

NORTH-EASTERN ASIAN BELT

This is an enormous region extending from Indochina to the Koryak Range of Siberia, unified by our ignorance rather than by geology. Included in this region are segments of a great Mesozoic to lower Tertiary andesite, dacite, rhyolite continental margin volcanic belt (in S.E. China and in the Okhotsk-Chaun volcanic belt), as well as segments of the presently active island arc (Taiwan, Japan, Kuriles, Kamchatka). It is puzzling that, given the many points of similarity with the "porphyry belts" reviewed so far and the general geological favourability, there are so few porphyry copper showings known; the best described ones being on the Kamachatka Peninsula (Tumannoye, Krasnogorskoe) and economically, the most important ones in China (e.g. the Tehsing district in Kiangsi, estimated size 8 Mt Cu). Basic information about the "porphyry" occurrences in the north-eastern U.S.S.R. can be found in Laznicka (1976a).

THE URALS BELT, U.S.S.R.

The Urals form a N.-S.-trending Paleozoic orogenic belt 2,000 km long, separating the Russian and Siberian platforms. Their complex geology is dominated by a Cambrian to Carboniferous, mostly mafic marine volcanic-sedimentary association, that hosts several hundred volcanogenic massive sulphide deposits (compare Chapters 10-14). Plutonic rocks were emplaced in several periods, but only Devonian to Carboniferous "granites" have a significant aureole mineralization, mostly of gold. The thirteen "porphyry Cu" occurrences usually listed (Laznicka, 1976a) are in the massive sulphides-mineralized terrains. Some "porphyry coppers" are closely associated with massive sulphides mostly as deeply eroded equivalents of felsic domes underlying and feeding seafloor-deposited conformable sulphide lenses. One deposit (Tsessovka) has been reviewed in Chapter 14. Others are similar in style to the "diorite-model" porphyry coppers of Hollister (1978).

The Voznessenskoe deposit (Fig. 28-55), the only one for which a pre-war reserve figure is available (250 Tt Cu), has a low-sulphur chalcopyrite, bornite, pyrite and magnetite stockwork and veins controlled by faults in Devonian gabbros and syenites, and exocontact greenstones. The orebodies are linear to irregular and are accompanied by diffuse albitization, epidotization and chloritization. In the Turya (Tur'inskoe) and other magnetite and chalcopyrite skarn fields of the Urals, minor Cu veins and stockworks have occasionally developed in the granodiorite or quartz diorite endocontacts. The Ural-Tobol anticlinorium east of the central "eugeoclinal" belt

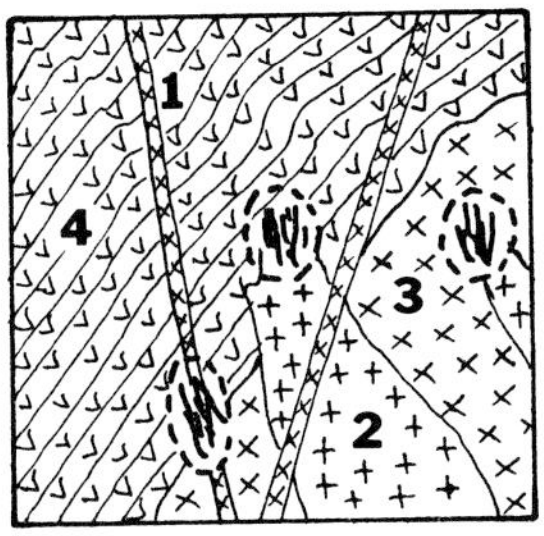

1. porphyry dikes
2. D_{1-2} quartz monzonite, granosyenite
3. quartz diorite, gabbro
4. D_2 metabasalt, spilite-keratophyre, metasediments

Fig. 28-55. Voznessenskoe stockwork Cu deposit, the Urals; diagrammatic representation. After Laznicka (1976a).

intruded by Carboniferous "granites", is a setting where one would expect to find porphyry deposits of the S.W. United States style. Unfortunately, this zone is deeply eroded and the only better known "porphyry" occurrence is the Yelenovo quartz, tourmaline, chalcopyrite stockwork in Silurian and Devonian volcanics intruded by a granosyenite.

CENTRAL KAZAKHSTAN, U.S.S.R.

This strongly block-faulted region centered on Lake Balkhash is located near the junction of four major cratonic blocks. It includes orogenic units ranging from lower Paleozoic to Permian in age of which metallogenically the most fertile had been the Devonian to Permian period of widespread marine to continental basalt, andesite, rhyolite volcanism and intrusive activity. The North Balkhash (Cisbalkhash) porphyry copper region represents some 15 Mt Cu, most of it in the giant Kounrad deposit. The porphyry coppers are associated with silicification and sericitization of Carboniferous granodiorites, and mineralization in the K-silicate alteration assemblage is less common (Kaskyrkazgan and Ken'kuduk occurrences). This region has already been discussed (Section 28.2.2.).

The relatively small Boshchekul' orogenic belt located in the N.E. part of Central Kazakhstan has a late Paleozoic metamorphosed chert-spilite association at its base, overlain by Cambrian and Ordovician "arc" flysch and volcanic assemblage. The large Boshchekul' porphyry copper deposit (2.16 Mt Cu, 0.4-0.81% Cu) is a member of the "diorite-model", affiliated to diorite and monzonite porphyry dikes that are probably of middle Carboniferous age, intruding lower Cambrian propylitic and amygdaloidal metabasalt flows, tuffs and volcanic arenite. The veinlet and disseminated chalcopyrite lies in both the endocontact and exocontact of the hypabyssal intrusions, in a zone of propylitic alteration (Trofimov, 1973).

It appears that patchy K-feldspar metasomatism is also present. The bulk of the economic ore at Boshchekul' comes from the secondary chalcocite zone and to a lesser extent from the oxidation zone with malachite.

TIAN-SHAN OROGEN, SOVIET CENTRAL ASIA

Tian-Shan is a complex Hercynian (late Paleozoic) orogen located south of the Central Kazakh block and continuing into the Sinkiang Province of China. This is essentially a Pb-Zn, W and Au province and copper is rare, yet the important Almalyk porphyry Cu district with several producing mines is located in the Kurama Range, S.E. of Tashkent. The Almalyk district (Musin, 1970) contains 338 copper occurrences distributed over an area of approximaletly 300 km^2. Two large deposits (Kalmakkyr, Fig. 28-56, and Sary-Cheku) are in production. Veinlet and disseminated chalcopyrite-molybdenite ore occurs in K-feldspar, biotite and quartz-sericite alteration envelopes (1) in middle-upper Carboniferous granodiorite porphyry stocks and (2) in large rafts and blocks of lower-middle Carboniferous syenodiorites and quartz porphyries of the Karamazar Batholith. Most of the ore comes from chalcocite blankets and the overlying oxidation zone. The adjacent and basement Devonian and lower Carboniferous carbonates frequently contain skarn and replacement magnetite and chalcopyrite occurrences, replacement Zn-Pb, vein Au, helvite (Be) metasomatites and Pt-Pd-bearing massive replacement pyrite pipes. The Kalmakkyr deposit has dimensions of 3x1.3x0.5 km, the ore grade was 0.8% Cu for the oxide ore and 0.72% for the sulphide ore, and the pre-war reserves were 2.1 Mt Cu.

ZAYSAN, ALTAI, SAYAN OROGENS (S. SIBERIA, N. CHINA, MONGOLIA)

This is a complicated assemblage of N.W. and E.-W.-trending Paleozoic orogenic belts wrapping around the southern tip of the Siberian Platform between Tian-Shan and Lake Baikal (Laznicka, 1976a). The geology is too complex to summarize briefly. Several clusters of porphyry copper showings are known to occur in the Zaysan (Rudnyi Altai) belt (Bukhtarma, Novonikolaevskoe), in the Kuznetsk Alatau Range and its eastern extremities (Batenevski Range; e.g. Kiyalykh-Uzen and Sora fields) and in the northern Mongolia (Erdentuin, Tsagan Suburga districts). The "porphyries" are associated with Cambrian diorites and syenites and grade into skarns (Kuznetsk Alatau); with Devonian granite porphyries, granosyenites and alaskites (Batenevski Range); with Carboniferous subvolcanic quartz-feldspar porphyries (Rudnyi Altai) and with Permian granite, granodiorite and granosyenite porphyry (Erdentuin). The last locality is a major industrial deposit (about 3 Mt Cu/1%).

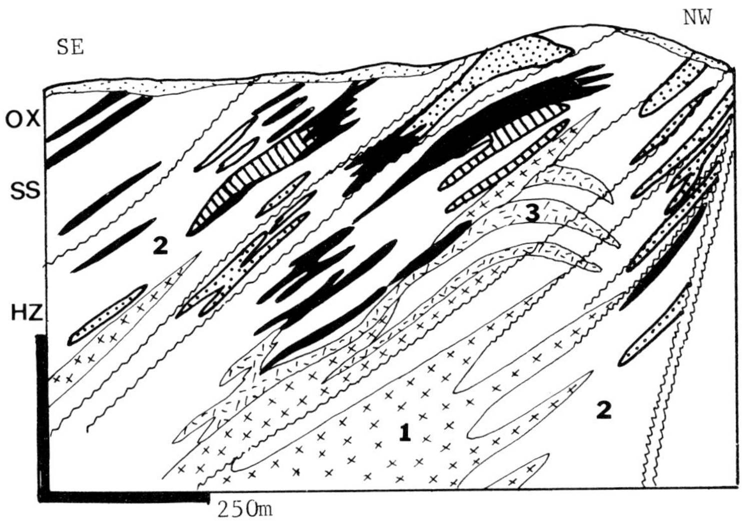

Fig. 28-56. Kalmakkyr porphyry Cu-Mo deposit, Almalyk district, Uzbekistan. Cb_{1-2} intrusive rocks: 1=granodiorite porphyry; 2=syenodiorite, diorite; 3=quartz porphyry. Zones of complete alteration (metasomatites): black: silicification; diagonally ruled: K-silicates; dots: sericitization. Supergene zones: OX=oxidation; SS=chalcocite; HZ=hypogene zone. After Musin (1970), scale and orientation added.

EASTERN MONGOLIA, EAST TRANSBAIKALIA, AMUR BELT

This belt, located in the eastern extension of the territory described earlier, is marked by a largely continental faulted old basement intruded by several generations of granitic plutons during the late Paleozoic orogeny and Jurassic-Cretaceous "activation". This is largely a Sn-W-Mo province affiliated with the Mesozoic epizonal plutons, and copper is recovered as a by-product of several vein and stockwork molybdenite deposits (e.g. Shakhtama). The Aryn-Nur showing in Mongolia (Marinov et al., 1977) has a transitional porphyry Cu-Mo mineralization associated with Jurassic granites emplaced in a Permian batholithic environment. The K-feldspar alteration is gradational into greisenization.

TASMAN OROGENIC BELT, EASTERN AUSTRALIA

The Tasman Belt traceable from Cape York to Tasmania is a composite Paleozoic orogen marginal to the Australian craton in the west. It is usually described in terms of three segments separated by post-orogenic basins. Almost 120 porphyry Cu occurrences are known. Of these, 85 are in the northern Queensland segment (Horton, 1978). In the entire orogen, there is, so far, only one large deposit (Goonumbla, N.S.W.; 2.28 Mt Cu) which has been described under the heading "volcanic porphyries" in Chapter 26, and 3 to 5 small deposits, not exceeding 100 Tt Cu (Moonmera and Coalstoun, Queensland; Copper Hill, N.S.W.). The Queensland "porphyries" (Horton, 1978) are associated with Silurian to early Cretaceous granodiorites and quartz monzonites, emplaced in a variety of basements out of which andesitic volcanics are the most common. Fracture-controlled mineralization, usually patchy, is controlled by a weak potassic alteration, but phyllic and propylitic assemblages are present in most cases. Porphyry coppers in the southern (Lachlan) segment resemble most closely the Uralian deposits and occur in settings that also contain volcanogenic massive sulphide deposits.

28.6. COPPER SKARNS AND CARBONATE REPLACEMENTS

About 20 Mt Cu are contained in skarns areally associated with porphyry copper systems, and possibly another 6 Mt are to be found in skarns or replacements hosted by carbonates in the contact region of "barren" intrusions. This represents less than 3% of the world's economic copper. Despite this, at least 1,000 individual copper skarn localities have been recorded, ranging in size from several tons of Cu to 4.73 Mt Cu (at Twin Buttes, Arizona). Accessory amounts of copper are present in and locally recovered from, skarns and replacements mined for other dominant metals (Fe,Pb-Zn,W,Sn).

Copper as well as other skarns have been reviewed recently by Einaudi et al. (1981), who summarized their characteristics as follows: (1) Cu skarns are associated with syn- to late-orogenic continental margin calc-alkaline granodiorite to monzonite stocks. The stocks are porphyry-textured, hypabyssal, locally converted to pyroxene-garnet endoskarn in a narrow zone along the contact, and frequently pervasively altered (K-silicate and phyllic alteration). (2) Compositionally, they have high garnet to pyroxene ratios, relatively oxidized assemblages (andraditic garnet, diopsidic pyroxene, magnetite and hematite) and moderate to high sulphide contents. (3) Sulphides occur as disseminations, as massive streaks, as veins and stockworks in fractured and brecciated skarns, and as massive marble replacements at the skarn front metacarbonates. In calcic skarns the pyrite, chalcopyrite, magnetite assemblage is characteristic of the more proximal contact garnet zones, bornite and chalcopyrite association is most common in the outer wollastonite zone. (4) Most Cu orebodies are in calcic, rather than magnesian, skarns and they have the form of mantos, pipes, stockworks, or irregular bodies. (5) Some pyrite, chalcopyrite, bornite replacements and open space fillings occur within or outside the skarn zone, associated with retrograde hydrous silicates, most characteristically actinolite and chlorite. Retrogression often "spreads" the mineralization outside the skarn assemblage, into fractured marbles.

Skarns are typical interaction deposits and in searching for them the characteristics of both the supracrustal association that hosts them, and of the plutonic association that generates them, are important. The most "natural" associations hosting Cu skarns are the "eugeoclinal" suites dominated by trace copper-rich marine or continental mafic volcanics (basalts and andesites), interbedded with or topped by carbonates. Examples of such associations including short descriptions of several Cu-Fe skarns have been given in Chapters 10, 12 and 13. Cu skarns are common in belts mineralized by volcanogenic massive sulphides (e.g. in the Urals), where they are related to either an early, evolutionarily immature phase of plutonism (sodic plagiogranite), or to later calc-alkaline intrusions. In this setting it seems probable that, at least a portion of the Cu and Fe, was derived from the mafic volcanics.

Important skarns, however, are also hosted by platformic or "miogeoclinal" carbonates as in the western United States (Bingham, Santa Rita, Ely), completely lacking any volcanics. There, Cu and Fe were clearly brought in from some depth with the intrusive system and the sedimentary association itself gives no clues as to the possible Cu skarn presence, except for the universally valid indicators such as the intensity of fracturing and signs of contact and hydrothermal metamorphism.

As for the intrusion favourability, a general rule is that whichever intrusion hosts (or is considered suitable to host) porphyry coppers, it could also generate skarns when carbonates (or calcic tuffs, shales, etc.) are locally available. The range of intrusive favourability for Cu skarns is, however, broader than for the porphyry Cu. In the Urals where porphyry Cu are rare and of little industrial importance, skarns are widespread and economically important.

Brief data on Cu skarns are summarized in Tables 28-4 and 5. The Christmas deposit, Arizona, is described in greater detail below as a typical example. The reader is also referred to the description of several British Columbia Cu skarns in Chapter 10, and the Tur'ya field, the Urals, in Chapter 12.

Christmas mine (Eastlick, 1968;, 700 Tt Cu/0.7%; Fig. 28-57) is 12 km N.E. of Hayden, Arizona. The orebodies are located in three Devonian to Pennsylvanian limestone, dolomite and shale units, overlying Cambrian quartzite. These are topped by late Cretaceous continental andesite flows and pyroclastics. The mine area is

Table 28-4. Chalcopyrite (bornite)-bearing skarns, 1: associated with porphyry Cu mineralizations in the endocontact

LOCALITY	HOST ASSOCIATION,INTRUSION	MINERALIZATION	REFERENCES
Craigmont mine, Merritt, British Columbia, Canada	Tr_3 m-graywacke,limest.,mylonite; Tr_3-J_1 andesite, diorite,quartz diorite. Minor access.dissem. cp. in qtz.dior.,Highland Valley porph.Cu field adjacent	1) dissem. magn.,cp.,specularite in exoskarn; 2) late stage fract.veins of coarse pink feldspar with cp.,mag.,specularite. 515 Tt Cu/1.77%	Chrismas et al. (1969)
Bingham, Utah, U.S.A.	Ps calcar.sandst.,quartzite,limestone; 39-32 m.y. qtz.monzon. porph.,qtz.feldsp.porph.,contain major porph.Cu-Mo orebody	widespread garn.,pyrox.,pyr., cp.,born. contact rim skarn; 3.8 Mt Cu/1.2-2%	Lanier et al. (1978)
Ely, Nevada, U.S.A.	D-Pe hornfelsed shale, limestone; Cr_1 qtz.monzonite porphyry; major porphyry Cu-Mo orebodies	quartz,calcite,magn.,pyr.,cp., veins, veinlets, replacements superimp. on earlier garn.-pyrox skarn and in marble; 1 Mt Cu/1%	James (1976)
Christmas, Arizona, U.S.A.	Cm, sandst.; D-Ps limest.,dolom., hornfelsed shale; 62 m.y. biotite diorite to qtz.diorite; minor endocont. dissem. cp.	calcic and magnes.exoskarn mineraliz. by magn.,cp.,born.; mantos, contact masses, fract. lodes; 700Tt Cu/0.7%	Eastlick (1968)
Twin Buttes, Arizona U.S.A.	PZ limest., hornf. shale; Tr-Cr contin. volcanics; Cr_3-T_1 quartz monzon.,andes.,rhyol.dikes; major porph. Cu (1.18 Mt Cu)	widespread masses of garnet, pyrox.,quartz, magn.,pyr.,cp., skarn; 4.73Mt Cu/0.78%	Einaudi et al. (1981)
Mission, Arizona U.S.A.	PZ limestone, hornfelsed shale; 60 m.y. qtz.monzon.porphyry with major porph.Cu miner.(268 Tt Cu)	garnet,pyrox.,cp.,pyr.skarn; high-grade bornite replacements in marble; 1.072Mt Cu/0.8%	Einaudi et al. (1981)

Santa Rita, New Mexico, U.S.A.	Cm-Pe hornf. shale, carbonate 63 m.y. granodior.,qtz.monzonite stock with major porph.Cu-Mo	garnet, epid.,magm.,cp. and late stage pyr.,qtz.,actinolite, skarn; 900Tt Cu/0.9%	Hernon and Jones (1968)
Yerrington, Nevada, U.S.A.	Tr_2-J metaandes.,rhyol.,hornf. sandst.,shale,black shale,limest. J, granod.,qtz.monzon.,porph. Cu	many small endo- and exoskarn deposits: magn.,ludwigite, cp., consid.second.enr. and oxide minerals (chrysoc.,malach.); 2.5 Mt Cu/0.54% (with porph.Cu)	Harris and Einaudi (1982)
Cananea (Capote Basin,Puertecitos), Sonora, Mexico	Cm-Cb limest., dolom., minor quartzite; MZ volcanics; T_1 qtz. monzon.porph.,major porph.Cu-Mo	Capote Basin:cp.,sphaler.,calc., hemat. manto superimp. on brecc. older skarn; 350Tt Cu, 400Tt Zn.	Meinert (1982)
Gaspé field, Murdochville, Québec	D_1 hornfelsed calcar.siltstone to limest. D_3-Ms rhyodac.porph. sill, plugs; porph.Cu miner.	dissem. cp. in calc-silic.hornfels, less in fine grained skarn in exocont.,often parall. with bedding; 1.2Mt Cu	Allcock (1982)
Banat (e.g.Moldova Nouǎ), S.W. Rumania	J-Cr limest., hornfelsed shale Cr_3-T_1 granod.,dior.,qtz.diorite; low-grade porph. Cu-Mo	garn.-magnet.skarn,rims and rafts in intrusion, contains inclus.,dissem.,masses of cp., pyr. Estim. 100Tt Cu	Ianovici and Borcoş (1982)
Ok Tedi (Mt.Fubilan), Niugini	MZ-T fine clastics, Ol-Mi limest. 4-1.2 m.y. dior.,monzon.stock wit major porphyry Cu orebody	massive garnet,pyrox. skarn and massive pyr.,pyrrhot.,cp., replacements in limest. 750Tt Cu	Bamford (1972)

Table 28-5. Chalcopyrite or bornite-bearing skarns, 2: associated with "barren" intrusive stocks

LOCALITY	HOST ASSOCIATION, INTRUSION	MINERALIZATION	REFERENCES
Méme, Haiti	Cr_3 limestone, andesite flows and tuffs; Cr_3 (66 m.y.) syenodiorite, granod., dikes and stocks	5-70 m wide zone of skarn surrounds marble pendant; garnet-epid., diops. skarn, mass.+dissem. magn., cp, pyr., moly. 30Tt Cu/2.5%	Kesler (1968)
Cobriza, Cordillera Oriental, ea-tern Peru	PZ slate, sandst., limest., congl. T_1 quartz monzonite stock	magn., lesser cp., ars., pyr., in banded hornbl., diops., quartz skarn; 600Tt Cu/1.3%, 650 t Ag	Petersen (1965)
Tintaya, Cordillera Occidental, southern Peru	Cr, limestone, shale T, granodiorite, qtz. monzonite	s-shaped, narrow skarn zone (magn., cp, born., chalcoc.) topped by 80m deep oxid. zone; 996Tt Cu/2.12% (hypog.z.)	Eng. Min. Journ. Nov. 1984
Sayak, central Kazakhstan, U.S.S.R.	Cb_{1-2} marine andesite flows, volcanicl., arenite, shale, limestone; Cb_{2-3} granod., qtz. monz., stocks, dikes	N.-S. zone of faults miner. by magnetite skarn; cp, born., moly., pyrrh. superimp., assoc. with epid., actinolite along fract. Estim. 600Tt Cu	Samonov and Pozharisky (1974)
Tongling distr., central China	PZ_1-J shelf sedim. (shale, sandst., limest.) incl. thick Tr limestone. J qtz. diorite, granodiorite	several hundred magn., cp., born., oreb. in garnet-pyrox. exocont. skarns; est. 400Tt Cu/1.18%	Min. Magazine Aug. 1980
Kamaishi mine, N. Honshu, C. Japan	Cb andes. m-pyrocl., slate, limest. Cr_1 diorite, diorite porphyry	linear, N.-S. line of magn., cp., pyrrhot., cubanite in garn.-pyroxene exoskarn; min. 80Tt Cu	Uchida and Iiyama (1982)
Ertsberg, W. Irian, Indonesia	PZ sandst., phyll., limest., volc., coal; T, limestone; 3.1 m.y. granodiorite intrusion	plug-like magn., cp., born. oreb. in garn., diops., actin., chlorite, epid. skarn in a m-carb. block engulfed by granod. 825Tt Cu/2.5	Adams (1973)

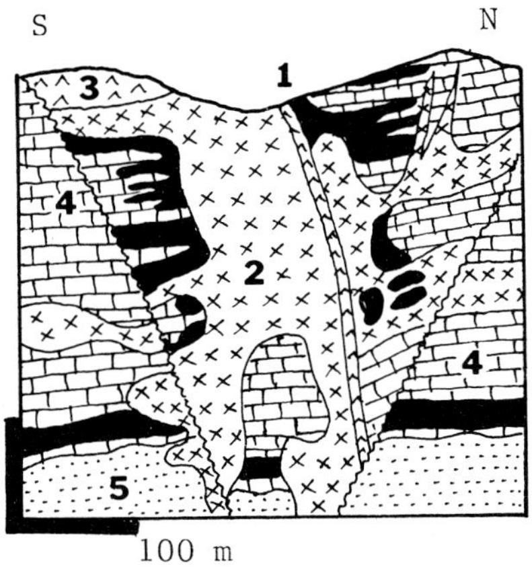

1. T, andesite dikes
2. Cr_3-T_1 quartz-mica diorite
3. Cr_3 subaerial andesite flows and pyroclastics
4. D-Ps limestone and dolomite
5. Cm sediments, mainly quartzite

BLACK: exo- and lesser endoskarn with disseminated chalcopyrite

Fig. 28-57. Christmas mine, Arizona, a Cu-skarn deposit. From LITHOTHEQUE, based on data in Eastlick (1968).

intensively block faulted and intruded by a "Laramide" quartz-mica diorite porphyry stock and by a swarm of lower Tertiary dikes. In the central part of the intrusive stock, magmatic stoping caused fragmentation of the gently dipping sediments and their engulfement in the diorite to form "floating" blocks, rafts, vertical sheets and xenoliths. The sediments are hornfelsed and recrystallized in the intrusive aureole and the carbonates in proximity of the intrusion are converted to skarn.

Calcic skarn (andradite, diopside) formed in the stratigraphically higher limestones. Magnesian skarn (forsterite-magnetite) is contained in the lower dolomites. Christmas mine is well known for the strong postmagmatic hydrothermal endoskarn alteration in the diorite. At some distance from the orebody, hornblende and biotite are replaced by clinopyroxene, and plagioclase is epidotized or argillized. Brown crystals of sphene appear as a common accessory. Close to the contact, diorite is non-selectively replaced by garnet which, in turn, may be selectively substituted by diopside. The endoskarn is not substantially mineralized, but small disseminated pyrite-chalcopyrite porphyry copper orebodies occur at several places in the ore field, accompanied by potassic alteration.

The principal orebodies occur in both the calcic and magnesian endoskarns and they are tabular bodies that selectively replaced carbonate beds interstratified with the silicate beds, irregular masses along the subvertical intrusive-sedimentary contact, subvertical mineralized skarn sheets along faults and irregular replaced carbonate rafts in the diorite. The largest and most continuous orebody in the lowermost (Devonian) limestone has, overall, the shape of a ring, surrounding the diorite stock. Disseminated chalcopyrite and bornite are interstitial to the silicates in skarn, or form veinlets or swirly layers interbanded with magnetite and silicates.

The enigmatic Ruby Creek Cu orebody in the Brooks Range, Alaska (Runnels, 1969; 1 Mt Cu/1%) has epigenetic replacements of chalcopyrite, pyrite, bornite, tennantite and chalcopyrite in Devonian limestone. Cymrite, a barium silicate, and some jasper are the principal alteration minerals and there are no skarn silicates. The Cu orebodies are peneconcordant lenses (mantos), and there is widespread diagenetic pyrite in the associated black phyllites. Cretaceous albitic gneissic granite and soda-aplite are present in the vicinity. The general context of this mineralization suggests a proximal hydrothermal remobilization of a copper probably geochemically enriched in the sedimentary sequence or in a gabbro sill complex.

Large replacement mantos of copper sulphides in silicified (jasperoid) carbonates far outside skarn zones are rare. This contrasts with the abundant Pb-Zn sulphides in equivalent settings.

28.7. COPPER VEINS

Copper sulphides-dominated fissure veins can best be thought of as an end-member of complex and gradational copper mineralized systems. They occur in the company of porphyry copper stockwork complexes (the late stage veins in Butte), as minor remobilization or feeder veins in and around volcanics-hosted massive sulphide deposits, as remobilized veins in mineralized sediments such as the Kupferschiefer and in association with skarns. Where fissure veins occur independently of disseminated, massive or stratabound copper sulphide accumulations, only a few are dominantly copper-bearing. Copper is common in most polymetallic (Pb-Zn), "silver" (e.g. tetrahedrite veins), cassiterite, uranium, siderite, and other veins in lesser quantities, and may be recovered as a by-product. Most veins have probably been filled by metals precipitated from convecting hydrotherms. Although most such hydrotherms were probably driven by the heat from "granitic" intrusions, other alternatives have also been recognized. Of these, gabbro and diabase intrusions related to crustal extension have also affiliated Cu veins (e.g. Churchill River field, British Columbia). These are treated in Chapters 31 and 32.

The number of examples of hydrothermal-plutonic dominantly Cu veins listed in classical textbooks shrunk considerably as a consequence of genetic reinterpretation. It is interesting to note how many localities have been excluded from the category of "Mesothermal Cu veins" listed in Schneiderhöhn (1955). Kitzbühel, Britannia mine, Hidden Creek and others are now considered to be deformed volcanics-hosted massive sulphides and Copper Mountain is now an example of the "diorite-model" porphyry copper. The remainder of "true" granite-affiliated Cu veins, however, still represent some 3 Mt Cu and although this is less than 0.5% of the world's copper contained in ores, it could be of significance locally. The largest Austrian copper deposit Mitterberg, for example, is a vein.

Table 28-6 lists selected examples of dominantly Cu veins, and it is apparent that each locality has a high degree of individuality and modern typification is virtually impossible. The common features of these veins include: (1) pronounced structural control and (2) occurrence in a "granite" exo- or endocontact, in regions that are distinct "copper provinces", that is containing other styles of Cu accumulations as well, most of them economically more important than the veins. Cu veins are thus often useful exploration leads. (3) Proximity to a unit rich in trace Cu, a postulated Cu source (e.g. greenstone, gabbro, diabase, red beds sediments, some black "shales").

Naturally, there are numerous exceptions. The copper-bearing lodes near Redruth, Cornwall, very productive in the past, constitute a link between Sn and Cu provinces. Copper is abundant when "greenstones" are present in the wallrock suite. The chalcopyrite-bearing siderite or quartz, siderite, barite veins that have a wide distribution in the Paleozoic phyllite and schist cores of the Alps and Carpathians (e.g. Mitterberg, Slovinky) do not show clear association with "granites", but buried plutons are now postulated under the Spiš-Gemer region of Slovakia.

Most copper veins contain chalcopyrite, sometimes with bornite, as the main ore minerals. The common gangue associations are quartz (e.g. Tocopilla); siderite, lesser quartz, specularite, barite, ankerite (Mitterberg, Slovinky); quartz-tourmaline; quartz-chlorite; amphibole, chlorite, tourmaline and other silicates (Carrizal Alto); axinite, etc. Some veins are dominated by tetrahedrite and these are mined mostly for silver (e.g. Sunshine, Idaho). Fissure veins may grade into stockworks and replacements in carbonates (e.g. Schwaz, Austria). The wallrock alterations include either or all alterations known from porphyry coppers (K-feldspathization, biotitization, quartz-sericite alteration, propylitization) and in addition to this silicification, tourmalinization, carbonatization and light bleaching. Alteration usually forms narrow selvedges in the vein wallrocks. Where systems of closely spaced veins formed, the alteration can affect large blocks of rocks.

The Magma mine in Superior, Arizona (Hammer and Peterson, 1968; 657 Tt Cu) is a frequently quoted example of a hydrothermal Cu vein. It is hosted by a Proterozoic and Paleozoic sequence of folded and faulted schist, diabase, quartzite, shale and limestone, intruded by Cretaceous to Tertiary stocks and dikes of quartz diorite, diorite porphyry and quartz monzonite porphyry. The Magma vein follows an east-west fault and has been traced to a depth of 1,500 m. It is a branching and splitting sheet composed of high-grade quartz, pyrite, chalcopyrite, bornite, enargite, tennantite, chalcocite and other minerals that have replaced fault gouge and the sheared wallrocks. Up to a depth of 300 m, supergene leaching and sulphide enrichment took place. At an intersection with Devonian limestone, a replacement manto formed.

The Rosen ore field (Burgas district, E. Bulgaria; Iovchev, 1961; 77 Tt Cu; Fig. 28-58) is an example of Cu veins that have a setting almost equivalent to the "diorite model" porphyry coppers. There, 230

Table 28-6. Hydrothermal Cu sulphide veins, selected examples

LOCALITY	GEOLOGY	MINERALIZATION	REFERENCES
Magma mine, Superior, Arizona, U.S.A.	Pt_1 schist, Pt_{2-3} quartz., diabase; Cm-Ps limest., dolom., shale; Cr_3-T_1 qtz.dior., qtz. monzon.stocks, dikes	Cr_3-T_1 E.-W./65°N composite replac. vein in shear; qtz., pyr., cp., born., chalcoc., sphal., etc. filling. Manto formed at inters.of vein with lim. Ox.and sulph.enrichment; 675Tt Cu/5.69%; 716t Ag; 12.4t Au	Hammer and Peterson (1968)
Tocopilla, northern Chile	J, mesozon. tonalite pluton intr.into contin.to marine J metabasalts, metaandes., diab.	6 en echelon N.E. fract. veins; qtz., specul., magnet., pyr., cp., minor moly, U, Co, Ni. Oxid.and sulph. enr.zone; 110Tt Cu/3.1%	Ruíz (1965)
Carrizal Alto, Atacama, Chile	MZ diorite, empl.into PZ_1 ? quartzite, phyll., micaschist; diabase dikes	N.E. fiss. veins, hornbl., tourm., chlor., qtz., calc., apatite, pyr., cp., tetr. Chlorite alter., 150 Tt Cu/5%	Ruíz (1965)
Cornwall and Devon Cu veins, S.W. England, U.K.	D_1-Cb_3 slate, shale, sandst., limest., greenst., intr.by Cb_3-Pe high-level granites	fissure or bedding quartz, chlorite, tourm., cp., born., veins with or without cassiterite (e.g.Devon Great Consols, Dolcoath); 2 Mt Cu	Dunham et al. (1978)
Mitterberg mine, Austria	S, black and green phyllite Cb_3 quartzite, congl., chlor. schist (metam.red-beds)	E.-W./40-90°S, 4.5 km long fissure vein; qtz., siderite, pyr., cp., specularite; 180Tt Cu/1.4%	Bernhard (1965)
Rosen, Burgas district, E.Bulgaria	Cr_{2-3} andes.and latite lavas and tuffs, intr.by T_1 gabbro, monzon., syen.pluton, stocks	230 N.E. fiss.veins; qtz., pyr., specul., chlor., epid., cp.; wallrock silicif., chloritiz.; ox. and enrich. zones; 77Tt Cu/1.44%; 161t Mo	Iovchev (1961)
Vrli Bryag, Burgas distr., E. Bulgaria	same setting as Rosen	53 N.E. striking fiss.veins over 1 km long in volc. along intrusive cont. Fill as above, 48Tt Cu/1.42%	Iovchev (1961)

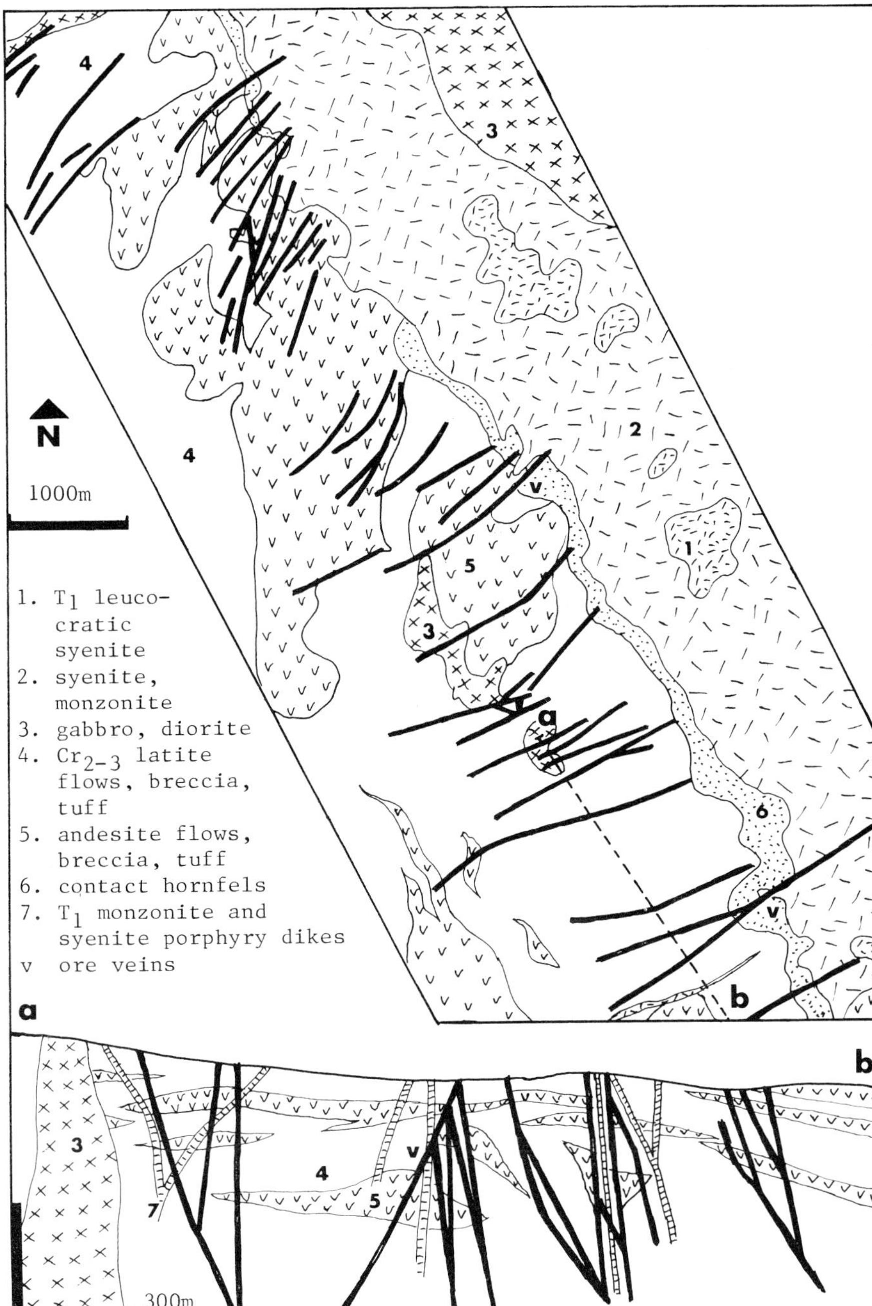

Fig. 28-58. Rosen, Burgas district, E. Bulgaria, an example of a postmagmatic hydrothermal copper vein field. Simplified after Iovchev (1961).

N.E.-striking subparallel fissure veins with an average length of 500 m and 0.1-10 m thick, are hosted by middle Cretaceous andesite and latite flows and tuffs within a 5x1 km block. The vein field is adjacent to a Paleogene gabbro, gabbrosyenite, monzonite, and syenite pluton, the small apophyses of which also include the volcanics in the ore field and host about 20% of the veins. The vein filling is quartz, pyrite, specularite, chlorite, epidote, chalcopyrite and other minerals deposited during four successive periods. The wallrock alteration is silicification, chloritization and pyritization.

28.8. Sn (W,Bi,Mo,Be,Ta-Nb) MINERALIZATIONS ASSOCIATED WITH GRANITE PLUTONS

28.8.1. Introduction

Most intermediate level (i.e. those associated with epizonal or upper mesozonal plutons) hydrothermal and possibly pneumatolytic deposits of lithophile metals are genetically and spatially associated with distinct elevations (cupolas, cusps, protuberances, ridges) that "sit on the back" of a larger buried "S-granite" pluton. A variety of actual mineralization styles (disseminations, stockworks, veins,replacements), alterations, host rocks, mineral associations, etc. can be recognized (Fig. 28-59). The most popular generalization, "tin cupolas" are similar in many respects to porphyry copper systems. They supply the bulk of a major metal, but individually there are endless "variations on a theme". The many styles of actual mineralizations that form tin orebodies are usually causally linked and one may lead to another. It is rather foolish to initiate an exploration program for, e.g., "tin skarns" and follow it blindly without taking into account that once a hydrothermal Sn-depositing system was set in motion, tin accumulated in many styles, skarns being just one of them, easily interchangeable for another style as the local conditions changed slightly.

Despite the many similarities in style and conceptual approach between "tin cupolas" and porphyry coppers, both systems are mutually exclusive and overlaps are extremely rare although examples have been recorded (e.g. in Central Kazakhstan and East Transbaikalia). This is clearly due to the geological setting and related petrogenesis of the parent "granites". While the bulk of the porphyry Cu affiliated "granites" are I- or "magnetite"-types, generated within or emplaced into the "transitional" crust of continental margins, the "tin granites" are overwhelmingly S- or "ilmenite"-types, with provenance within the continental crust. This is supported by the worldwide empirical experience, although competing ideas about mantle sources of tin (and tin granites) have also been suggested. This problem is reviewed in greater depth by Taylor (1979), and briefly touched upon in the earlier sections of this Chapter.

The tin granite mineralized systems have recently been reviewed by Hosking (1973, 1974), Štemprok et al., (1978), Taylor (1979) and by

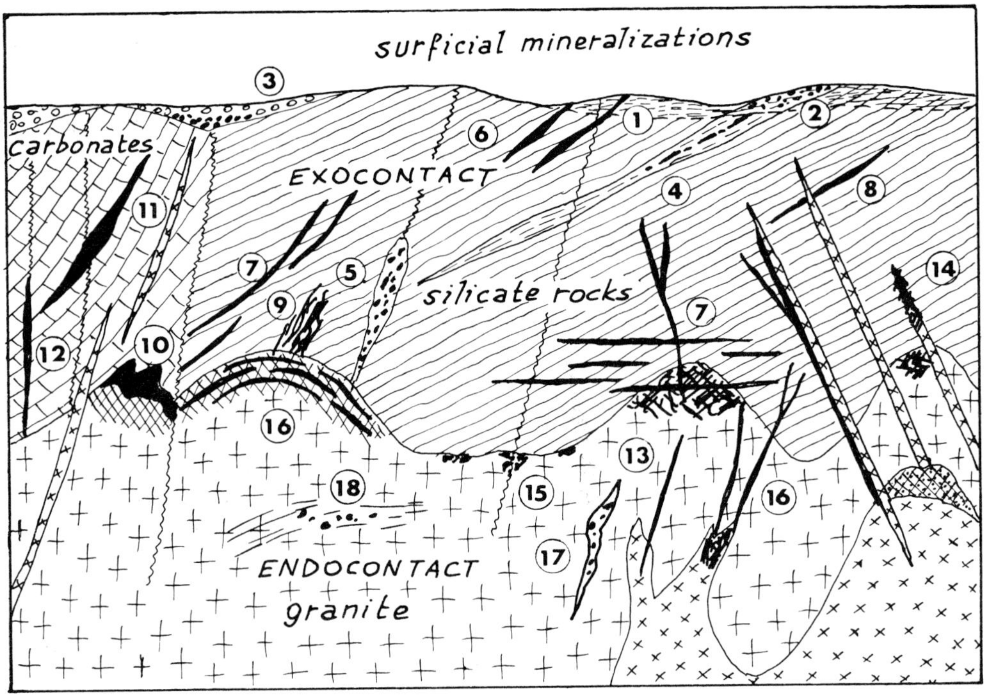

Fig. 28-59. The principal varieties of Sn(W,Bi,Mo,Be) mineralizations associated with high- to medium-level granitic plutons.

Surficial mineralizations: 1=gossans and oxidation zones; 2=regoliths; 3=transported placers (alluvial, beach).

Exocontact ores in silicate rocks: 4=relic mineralizations (unrelated to granite emplacement), e.g. paleoplacers, subaqueous hydrothermal stratiform deposits; 5=pipes, mineralized breccias; 6="distal" veins (no apparent connection with the "granite"); 7=proximal veins; 8=replacements in silicate rocks; 9=exocontact stockworks, disseminations.

Exocontact mineralizations in carbonate rocks: 10=exoskarns; 11=replacements in carbonates, mantos, pipes, etc.; 12=replacement veins.

Endocontact mineralizations: 13=stockworks, disseminations in cupolas; 14=ditto, in dikes; 15=small ore occurrences along contacts outside cupolas; 16=endocontact veins (sheeted flat veins); 17=pegmatites; 18=cassiterite disseminations in schlieren and inhomogenities. From Laznicka (1984).

the Soviet authors (Smirnov, ed., 1968; Smirnov 1969). The Russian-language authors brought many fresh ideas and contributed valuable descriptions of the Soviet and Mongolian tinfields, yet they used several conceptual approaches that differ from those common here. This led to several recent misinterpretations, the most usual one being consideration of the "apogranites" (essentially albitized granite apexes) as a petrochemically distinct magmatic rock.

It has to be emphasized that the concept of "tin granites" and "tin cupolas" is, to a considerable degree, an enigmatic one, and is confusing in many respects due to the variety of scales of consideration. The "main body" of the batholiths or plutons regionally associated with tin mineralization is usually an equigranular or porphyritic, monotonous biotite quartz monzonite (adamellite) with only slightly elevated tin contents (of the order of 3-5 ppm Sn) and virtually no associated ores (except when it was rejuvenated after intrusion). Such a precursor or "grandparent" granite (e.g. the Gebirge Granite in the Erzgebirge Tin Province) is then intruded or topped by a later stage, lesser volume more siliceous and leucocratic ("leucogranite") muscovite-biotite granite (e.g. the Erzgebirge Granite). This has the form of small epizonal massifs, stocks, dikes, etc. permissively formed just under the confining roof assemblage. The latter are rich in volatiles and the relatively impervious roof confinement prevented venting and volatile loss. Consequently, the volatiles including the metals they carried, accumulated in the apical regions. In some instances the "cupola granites" are not younger stage intrusions, but hydrothermally autometamorphosed apices of a single intrusive phase.

It appears that the depth modelling of magmatic systems applied to porphyry coppers ("volcanic" or subvolcanic, hypabyssal, plutonic) is immediately applicable to the tin (and some W,Be,Mo) deposits as well, and the absolute depth levels are also comparable (see Fig. 27-4). Table 28-7 gives the approximate balance of economic tin as related to mineralization styles and their depth levels. Also applicable to tin deposits is the analogy in multistage mineralization separated by a "barren interval" (as in Butte), where low-grade but more extensive disseminated or stockwork ore was partly upgraded into more restricted and higher-grade veins by a superimposed event (e.g. in the Cornwall; Panasqueira, Portugal; etc.). As in the Cu-mineralized fields, Sn ores occur, in relation to a granite body, in the endocontact or proximal exocontact (there, the association with a granite is intimate and immediately obvious), as well as in the distant exocontact where the granite is often not in sight but is usually postulated.

Two profound differences in comparison of porphyry Cu and Sn granite systems, however, stem from the fact that the ore components in the former are sulphides and in the latter are mostly oxides. The chemistry of the ore-bearing solutions and consequently the wallrock alterations are, therefore, to some extent different (high-sulphide Sn deposits are, however, transitional). A further difference is in the response of ores to weathering. The Cu sulphides are rapidly weathering degraded, while cassiterite is highly weathering resistant

Table 28-7. Approximate balance of tin in Phanerozoic deposits affiliated with mesozonal-epizonal granitic plutons

STYLE	TONNAGE IN:		
	1. BEDROCK DEPOSITS	2. PLACERS PROPORTIONALLY RE-ASSIGNED TO BEDROCK SOURCES	3. TOTAL
Early endocontact ores			
in feldspathized apogranites	20Tt Sn	-	20Tt Sn
greisen, stockworks, veins	1,250	3,825Tt Sn	5,075
pegmatites	6	5	11
postmagmatic exocontact ores			
veins, stockworks	5,000	4,250	9,250
replacements in skarns	570	400	400
replacements in non-skarn carbonates	1,900		1,900
placers (considered independently)	8,500		
		GRAND TOTAL	17,226Tt Sn

so, in contrast to the Cu-depleted leached capping over porphyry coppers, similar weathering profiles over tin granites may preserve their hypogene Sn content, or may even be slightly residually enriched in tin.

28.8.2. Sn (W,Bi,Mo,Be, Ta-Nb) disseminations, stockworks and veins in altered granite cupolas and stocks.

This is the most distinct ("textbook") style of tin mineralization, but it is very unevenly distributed on a world-wide scale. It is the flagship of tin ores in the Erzgebirge, but elsewhere economic deposits of this style may be quite uncommon. This style, however, has the potential to supply low and very low grade tin ores in the future. In the literature, the content of this section is often treated under headings "greisen Sn deposits", "Sn apogranites", etc. The meaning of apogranites has been defined earlier. Greisens are hydrothermally altered felsic wallrocks (mostly granites), from which feldspar has been completely removed or altered to mica and/or topaz. The typical greisen assemblage is quartz, muscovite or zinnwaldite, topaz, frequently fluorite and apatite. Greisens commonly host disseminated or stockwork cassiterite or form alteration envelopes to ore veins.

In a "typical" tin-bearing granite cupola, the apogranitic feldspathic metasomatites and greisens are members of a generally low-sulphide alteration-mineralization system superimposed on leucogranites. In some sulphide-rich systems affiliated with quartz monzonites and granodiorites (as in the Sikhote Alin region, U.S.S.R.), the alteration pattern is different, resembling, in many respects, that in porphyry coppers.

"TYPICAL" (K/Na FELDSPAR, GREISEN) ALTERED GRANITE CUPOLAS

Cínovec (formerly Zinnwald) deposit in the Erzgebirge, N.W. Czechoslovakia (Štemprok and Šulcek, 1969; min. 150 Tt Sn/0.3%, 30 Tt W; Fig. 28-60) is an example of an almost ideal mineralized granite cupola that combines Li-enriched zones of feldspathization ("apogranite") with greisen alteration enveloping flat Sn,W,Li veins. The regularity is probably due to the fact that there is very little exocontact mineralization resulting from "leakage" of the mineralizing fluids into the roof.

The unaltered parent rock of the Cínovec cupola is a medium-grained porphyritic biotite granite, that is interrupted, in places, by microgranite zones generally parallel with the outer outline of the cupola. The intrusion dates from the Permian age and is generally considered to be one of the protuberances on the back of a much larger batholith that underlies the entire Erzgebirge region. The immediate roof to the Cínovec granite is a remnant of a formerly extensive pink rhyolite flow and ignimbrite sheet, locally intruded by granite porphyry dikes and stocks. The rhyolite is considered to be broadly comagmatic and only slightly older than the Cínovec tin granite.

The granite cupola has an elliptical cross-section, dipping gently (30°) in the north and south, steeply (80°) in the west. Hydrothermal alteration starts imperceptibly at a depth of about 730 m under the roof of the cupola, and is marked by the increase of initially intergranular, later K-feldspar-destructive albite, and by the formation of protolithionite-type micas. Higher in the profile the intensity of metasomatism increases and the albitization is locally complete so that almost pure albitites form as lenses and pipes, associated with zinnwaldite mica. The most common rock is albitized lithian granite composed of quartz, K-feldspar, zinnwaldite, sericite and kaolinite with accessory fluorite, topaz and rare disseminated cassiterite, zircon, columbite and other minerals. Approximately in the middle of the albitized profile, several bodies of K-feldspar metasomatites (metasomatic syenites) appear, and the entire contact of the cupola against the roof rhyolite is represented by a K-feldspar pegmatite lining ("Stockscheider"). There is no visible structural control to this early metasomatism other than the surfaces of concentric sheeting, and the alteration appears shortly to postdate

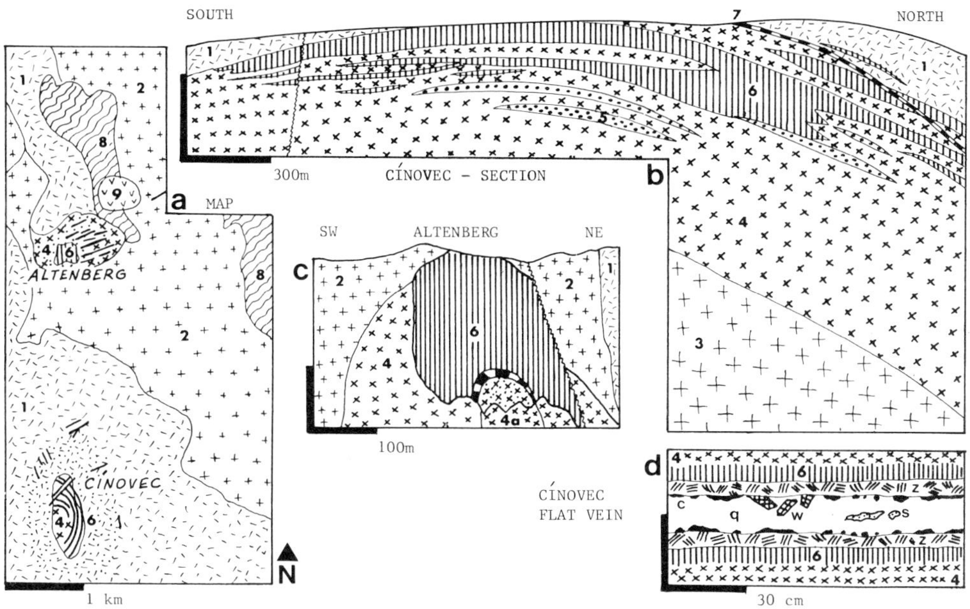

Fig. 28-60. Cínovec (Czechoslovakia) and Altenberg (East Germany) Sn(W) deposits hosted by greisen zones in feldspathized granite cupolas.
1=Fe rhyolite; 2=granite porphyry; 3=biotite granite (fine to coarse); 4=albitized zinnwaldite granite; 4a=younger phase at Altenberg; 5=metasomatic K-feldspar syenite; 6=greisen with flat ore veins; 7=border pegmatite (Stockscheider); 8=gneiss; 9=T_3 flood basalt.
Legend to Cinovec flat vein: c=cassiterite; q=quartz; w=wolframite; s=scheelite; z=zinnwaldite. Modified after Štemprok and Šulcek (1969), Chrt and Bolduan (1966); Rösler et al. (1968) and authors fieldwork, 1965.

the emplacement of the latest phase of the composite intrusion. The alteration regime was alkaline and is, in many respects, comparable with the K-feldspar alteration in porphyry coppers.

This was followed shortly afterwards by alteration due to acid fluids that was accompanied by formation of several parallel flat vein (lode) systems subparallel with the cupola surface, enveloped or interconnected by a quartz, zinnwaldite, lesser topaz and fluorite greisen. The veins filled and partly replaced dilations probably resulting from contraction of the intrusion during cooling. Minor, localized sericitization and argillization followed. Compared with porphyry coppers, greisenization is an "inserted" alteration phase between feldspathization and sericitization and is often gradational into silicification.

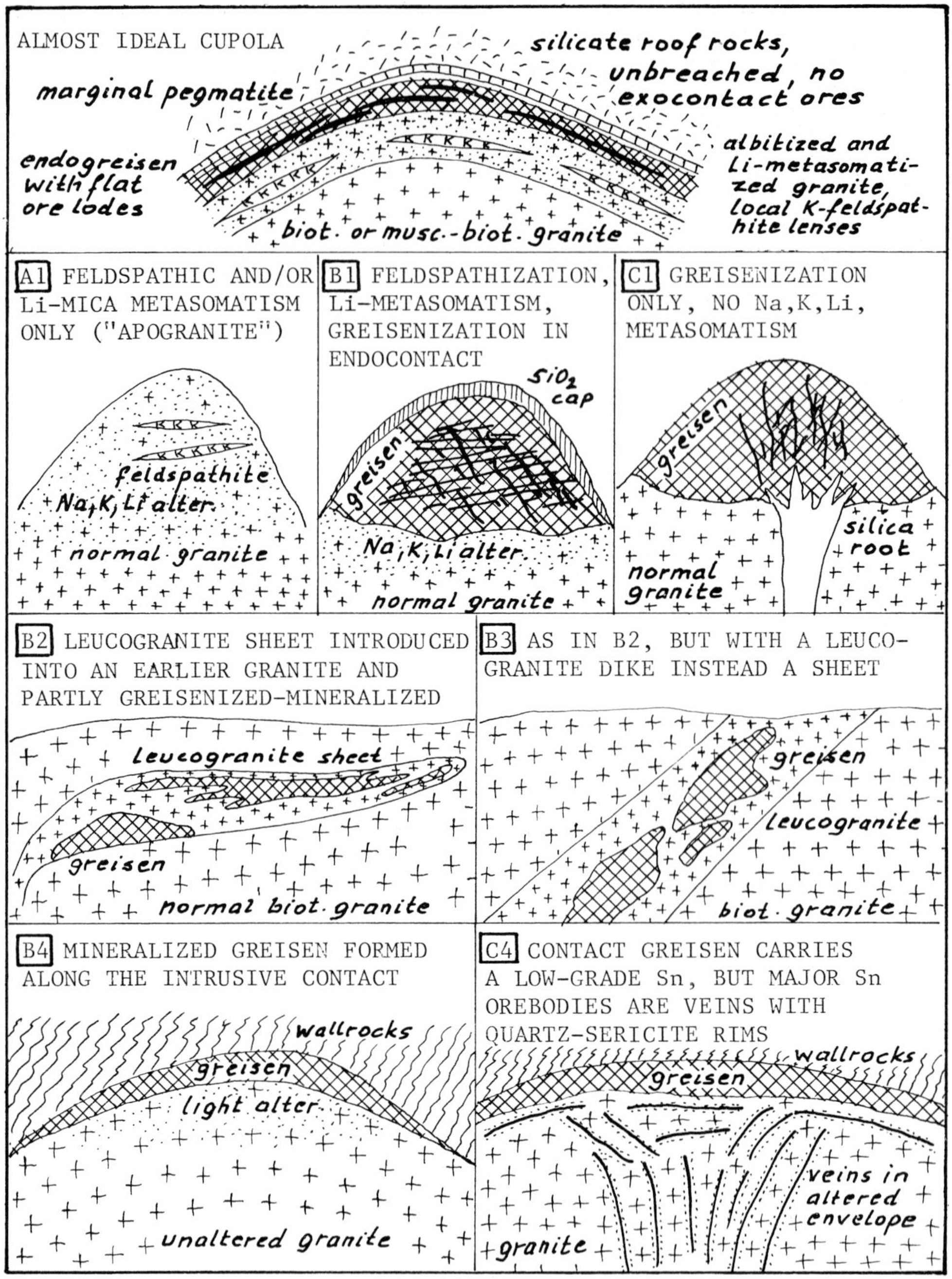

Fig. 28-61. Sn(W,Mo,Bi) mineralized altered granite cupolas, variations on a theme (see explanations in the text); from Laznicka (1984).

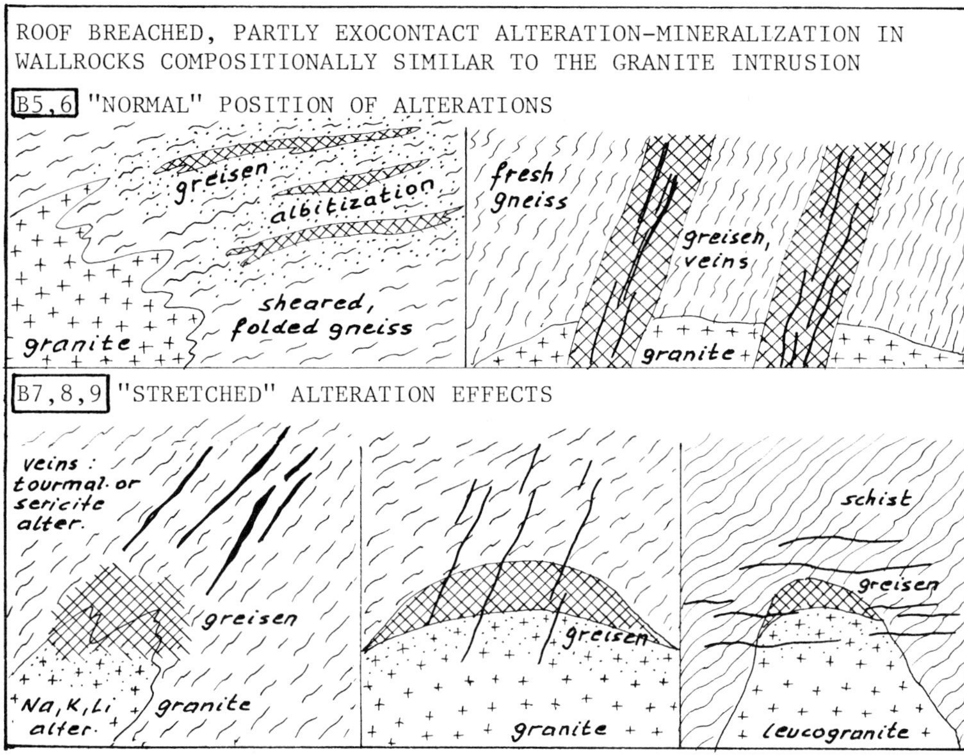

The ore minerals cassiterite, wolframite and minor molybdenite, bismuth, etc., may occur in three ways. These are as follows: (1) fine-grained disseminations in greisens. This constitutes a low-grade (0.1-0.2% Sn) but large tonnage ore; (2) coarse crystals of ore minerals scattered in zinnwaldite selvedges along the lode/greisen contacts or (3) large to very large crystals scattered in the flat quartz veins (Fig. 28-60d). The veins are up to 1 m thick, often symmetrical, and lesser quantities of scheelite, arsenopyrite, galena, sphalerite, stolzite, etc. are associated with wolframite and cassiterite. The thickness of the productive zone (veins and greisens) in Cínovec is about 200-300 m, but only a small proportion of this is an actual ore.

Other Sn(W,Bi,Mo) deposits in granite cupolas represent variations on a theme and many of the literature "subtypes" can be derived from the "ideal" example by adding and subtracting various attributes (Fig. 28-61). When only feldspathization and Li-metasomatism took place, sometimes mineralized "apogranites s.s." (A1) formed. (B1) is represented by a well developed endocontact greisen, crisscrossed by quartz veinlets and veins with scattered cassiterite, wolframite, aresenopyrite, and molybdenite (e.g. Hub Stock, Krásno and Altenberg). Greisenization grades into silicification and portions of the greisen are, in fact, masses of grayish-white vuggy and crystal-rich quartz with haphazardly scattered flakes of zinnwaldite and Sn,W,Mo,Bi,As ore minerals. The greisen is superimposed on the previously feldspathized granite, but this is often not apparent when mining or drilling was not carried out at a depth greater than the bottom of the productive zone. Occasionally, the greisen body changes downward into a quartz root (pipe).

(C1) shows greisen developed in a granite not previously albitized. Situations (B2) and (B3) show a leucogranite sheet or a late stage dike emplaced in an earlier granite, in which the dike suffered an autometasomatism terminating by the formation of endogreisen and/or a mineralized stockwork in it. The pre-greisen feldspathization was often light and inconspicuous and Li addition to the micas recognizable only analytically. At (B4), greisen is developed along the intrusive contact and at (C4) (as in Bukuka, Transbaikalia) a contact greisen with a low-grade Sn mineralization is developed, but the bulk of cassiterite is in a stockwork of endocontact quartz veins that have narrow quartz-sericite alteration rims, but are free of greisenization.

Increased variety of mineralization styles is the result of breach (rupture) of the roof to the granite cupola, that allows the fluids to migrate into the exocontact rocks of considerable lithological variety and uneven structural preparation. Fig. 28-61 cannot show more than a few "highlights". When the tin granite is emplaced in a wallrock compositionally similar to it (an older granite, gneiss; B5,6), the endocontact alteration can cross the contact and be developed on both sides of it (e.g. Čistá). In the remaining instances (B7-9) the alteration-mineralization system "stretches" and one may find feldspathization deep in the parent granite, greisenization at the very top of the granite or in the exocontact (exogreisen) and quartz-cassitetite veins in no further altered (or sericitized, tourmalinized) hornfelsed metasediments at some distance from the contact.

GRANITES (APOGRANITES, "A" in Fig. 28-61), FELDSPATHIZED ONLY, MINERALIZED BY Ta(Nb,Be,Sn,REE,Th)

Feldspathization modifies granite composition and petrochemistry and makes it more alkaline than it was at the end of the magmatic stage. When the existence of such post-magmatic alteration is not recognized, a calc-alkaline granite can be mistakenly placed into the alkaline or peralkaline family. The latter family, however, also contains apogranites, for example in Nigeria (Chapter 33). Apogranites of both petrochemical affiliations have been thoroughly described in the Soviet literature (e.g. Smirnov, ed., 1968), but the quality of description suffers from anonymous localities and a complete lack of quantitative data (witheld for state security reasons).

Most occurrences of apogranites that have been described correspond compositionally to the feldspathized zone at Cínovec and the pervasively but incompletely feldspathized bulk of the original granite is interrupted by fracture-controlled pods, pipes, lenses, sheets, fracture veinlets or stockworks of almost pure albitites (resembling aplite) or microcline-syenite metasomatites. The latter are commonly composed of the conspicuous light-green (the colour is particularly well developed when lightly weathered) microcline variety amazonite. These advanced feldspathites and their lithian mica (zinnwaldite, exceptionally lepidolite)-containing varieties, frequently contain scattered crystals of beryl, columbite-tantalite, ilmenorutile, monazite and other Be, Ta-Nb, REE and Th minerals. This suggests a convergence to and correspondence with the "rare metal pegmatites" that exist at greater depths. The ore minerals are too dispersed in most cases to constitute a currently economic ore, but they may have contributed much of the columbite and monazite recovered in many placer cassiterite fields of the world. Beryl, however, is rarely reported from tin placers. This is due to its low stability in the zone of weathering. Beryl accompanied by columbite, monazite, scheelite and other minerals, however, is present in several placers formed over apogranites in arid zones, as in Mongolia (e.g. Zhanchublin district; Marinov et al., 1977).

At some Mongolian apogranite localities (e.g. Buren-Tsogto tungsten deposit; Marinov et al., 1977), amazonite forms veinlets and veins in apogranite resembling pegmatite and it carries disseminated wolframite, beryl, fluorite, cassiterite and sulphides.

Of particular scientific interest (and readily accessible to the western visitors) are the apogranite cupolas at Montebras and Echassières in the French Massif Central (Aubert, 1968). The Montebras cupola is a ring dike measuring 350x200 m of a porphyritic muscovite-biotite granite emplaced by magmatic stoping into an older granite. It is surrounded by a pegmatitic (Stockscheider) and silicified envelope and a mica-greisen. It is heavily albitized. The albitized cupola and an adjacent separate intrusion of albite granite contain disseminated cassiterite associated with amblygonite-montebrasite and niobotantalite, but the Sn and Ta contents are subeconomic. Echassières is a biotite granite cupola 3 km in diameter, faulted after emplacement, and intruded by several volatile-rich younger intrusive phases including the Beauvoir albite-lepidolite apogranite. Lepidolite constitutes 10-20% of the granite and there are disseminated cassiterite, herderite, montebrasite, microlite and topaz. The disseminated ore represents 12 Mt of material with 0.14% Sn (16.8 Tt Sn), 0.97% Li_2O, 219 ppm Be and 117 ppm W. A wolframite stockwork in the exocontact schist represents 6.24 Tt W.

It appears that large masses of feldspathized and Li-metasomatized granites in the Sn and W provinces of the world may become important future low-grade, large tonnage resources of Ta and Be in their own right, or in combination with epigenetic vein or carbonate replacement bodies in the exocontact (e.g. Lost River, Alaska). Already at present, Proterozoic apogranites (in Egypt and Saudi Arabia) hold the largest known Ta resources of the world. These are treated in Volume 2.

An anonymous apogranite cupola located in the Sn-W-Mo province of East Transbaikalia and described in Smirnov, ed. (1968; p. 324, 325), appears to contain a Ta orebody economic at present. There, scattered crystals of columbite-tantalite, Mn-tantalite and rare microlite are hosted by a zone of lepidolite, amazonite, quartz, albite metasomatite 20-100 m thick, formed within a cupola of a Jurassic porphyritic biotite and muscovite-biotite granite, emplaced into a schist roof. The complex alteration zoning is of the Cínovec style, but the massif suffered an early post-magmatic faulting and fracturing that facilitated the movement of the metasomatizing fluids. The Ta_2O_5 grades in the uppermost zone (up to a depth of 20-70 m from the contact) range from 0.04 to 0.01%. If the zone is persistent, the deposit can easily contain 50 Mt ore with an average of 0.02% Ta (=10 Tt Ta), a significant accumulation.

DISSEMINATED AND STOCKWORK Sn (W,Be,Mo) ORES IN GREISEN-ALTERED GRANITE CUPOLAS ("B" and "C" in Fig. 28-61; Table 28-8)

Erzgebirge and Slavkovský Les tin province (Czechoslovakia and East Germany)

This N.E.-elongated province that straddles the Czech and German border is about 150 km long and 60 km wide (Chrt and Bolduan, 1966; Tischendorf et al., 1978). In the southwest, it is separated by the Ohře River graben from a smaller satellite Sn-W and U, polymetallic Slavkovský Les district. Cínovec, described earliers, is situated in this province. The province coincides with a fold- and metamorphic belt of late Proterozoic and Paleozoic metasediments and metavolcanics which are believed to be almost completely underlain by a Carboniferous to Permian granitic batholith dated 305 and 240 m.y. The older intrusive phase is mostly a porphyritic granite to quartz monzonite, the younger phase is the cupola-forming volatile-rich granite, believed to be responsible for most of the ores.

The granites crop out in several plutons over 20% of the territory and are in shallow subsurface elsewhere. The present level of erosion intersects probably the most optimal level of hydrothermal mineralization and almost a thousand occurrences of Sn,W, U, Pb,Zn, Ag, Cu, Fe and fluorite, barite are known in the granite aureole. Erzgebirge has been a classical area of economic geologic studies in past centuries (Agricola, v.Cotta, Stelzner, and others). About 200-300 Sn(W) occurrences within apical granites are known. This includes at least 3 deposits and ore fields containing over 100 Tt Sn (Table 28-8). The cumulative Sn tonnage is of the order of 1 Mt Sn (more when the very low-grade material around 0.1% Sn is included).

The largest single tin deposit in the Erzgebirge is Altenberg (Chrt and Bolduan, 1966; Rösler et al., 1968; Fig. 28-60). This deposit is only 5 km from Cínovec on the German side of the border and although both deposits are members of the same system, each orebody is separate. In Altenberg the ore is located in a mass of greisen formed within a cupola of a Permian granite emplaced into a comagmatic

Table 28-8. Disseminated and stockwork tin deposits and fields in greisen-altered granite cupolas. Selected examples

LOCALITY	WALL-ROCKS AND TIN GRANITE	MINERALIZATION	REFERENCES
Cínovec, Erzgebirge Mts., N.W. Czechoslovakia	Pe rhyolite, granite porphyry Pe biotite granite (apogranite) cupola	system of flat qtz., zinnwaldite, cassit., wolfr., etc. lodes // with the cupola contour envel. by greisen; floored by alb., Li-mica apogr. Min. 150Tt Sn/0.3%, 30Tt W	Štemprok and Šulcek (1969)
Altenberg, Erzgebirge East Germany (adjacent to Cínovec)	Pe rhyolite and quartz porph. Pe biot. granite, multiphase double stacked cupola	homogenous greisen body around younger apex; cassit., wolfr., moly., Bi, arsenop., fill hairthin fract. 200Tt Sn/0.3% (est. 600Tt Sn/0.1%)	Rösler et al. (1968)
Horní Slavkov, Krásno, Čistá zone, N.W. Czechoslovakia	Pt_3-PZ migmatite, gneiss, meta-quartz., amphibolite; Pe biot. gran., albitized, seric. altered	5 greiseniz. cupolas and sheets in endocont. have dissem. and stockw. cassit., wolfr., ars., moly. etc. Minor Sn-W, import. U, Ag, Pb veins in exocont. Est. 300Tt Sn	Chrt and Bolduan, (1969)
East Kemptville, S.W. Nova Scotia, Canada	Cm-Or quartzite, slate; D? biot. granite and qtz. monzon.	cassit., pyr., pyrrh., arsenop., pods in mass. greisen and greis. selvadges // qtz. veins in gran. Minor fluor. veins; 76Tt Sn/0.2%	Richardson et al. (1982)
Ardlethan, N.S.W., Australia	S, quartzite, micasch., slate D, biot. to musc.-biot. granite	numerous zones of qtz.-tourm. alter. and topaz greis. along joints; pipes with dissem. cassiter. grade to pipes, stockw., fiss. veins; 10.2Tt Sn/0.17%	Garretty (1953)
Anchor M., Lottah, N.E. Tasmania, Austr.	S schist, phyllite; D older bio. gran., sills of musc.-biot. gran.	700x160 m greis. zone with fine dissem. cassit., cp.; 7Tt Sn/0.3%	Groves and Taylor (1973)

granite porphyry and rhyolite. This cupola was thoroughly greisenized shortly after emplacement, and once more following the emplacement of another, younger phase granite stock (the "Inner Granite"). The latter is capped by a thin pegmatite body (Stockscheider) and the volatiles produced by the Inner Stock caused an autometasomatic alteration of its apical portion (including the pegmatite that was converted into an almost pure topaz), as well as additional greisenization and mineralization veining in the above-intrusion region of the older cupola. The earlier greisen is a dark topaz, quartz and Li-biotite rock (also called "Zwitter"), the younger greisen is light, bleached, hematite-pigmented rock. Cassiterite, wolframite, molybdenite, bismuth, bismuthinite, arsenopyrite and rarer minerals occur as fillings of hairline (0.06-0.1 mm) fractures in the greisen, giving the whole mass an average grade of 0.3% Sn.

Horní Slavkov, Čistá, Krásno (Ďurišová et al., 1969; Chrt and Bolduan, 1966) is a zone about 10 km long, of greisen stocks and sheets, developed within and over a Permian lithian albitized leucogranite (apogranite) emplaced into an earlier multiphase pluton, under the confining cover of a Proterozoic biotite gneiss (Fig. 28-62). Two adjacent mineralized stocks mined since the Middle Ages (Huberstock and Schnödenstock) had a massive quartz, topaz, zinnwaldite greisen bodies mineralized by stockwork and microscopic disseminated cassiterite, wolframite, arsenopyrite, chalcopyrite, molybdenite, etc. The intensity of greisenization diminishes downwards. The solid greisen changes into a system of flat greisenized lodes alternating with kaolinized and chloritized granite. The Huberstock alone is estimated to have contained originally about 150 Tt Sn and 70 Tt W.

The Čistá deposit at the western end of the zone is in relatively lightly albite feldspathized and Li-altered biotite granite (K-feldspars and even plagioclase are largely preserved, biotite is replaced by zinnwaldite and lithionite, topaz was added) under a migmatite roof. Several diffuse, elongated mica-rich greisen bodies subparallel with the intrusive grain are present, and they grade into a sericitized granite. Fine-grained cassiterite and minor wolframite and molybdenite impregnations form seven elongated zones up to 20m thick and 400 m long, subparallel with the contact. The ore zones have diffuse (assay) boundaries and are largely in the most intensively greisenized and sericitized wallrocks although their fringe is in megascopically virtually unaltered rocks. The orebodies, however, cross lithologic boundaries and follow faithfully the alteration zones. The Sn content is between 0.2 and 0.3%, W and Mo contents are less than 0.05%, but there is a high Ta content (up to 0.13%) locally. Minor stockworks and veins are superimposed.

The remaining Erzgebirge tin stockworks (Sadisdorf, Ehrenfriedersdorf, Tannenberg, Geyer, Schneckenstein, Krupka, Přebuz) are less important economically and are considered exhausted after more than eight centuries of mining.

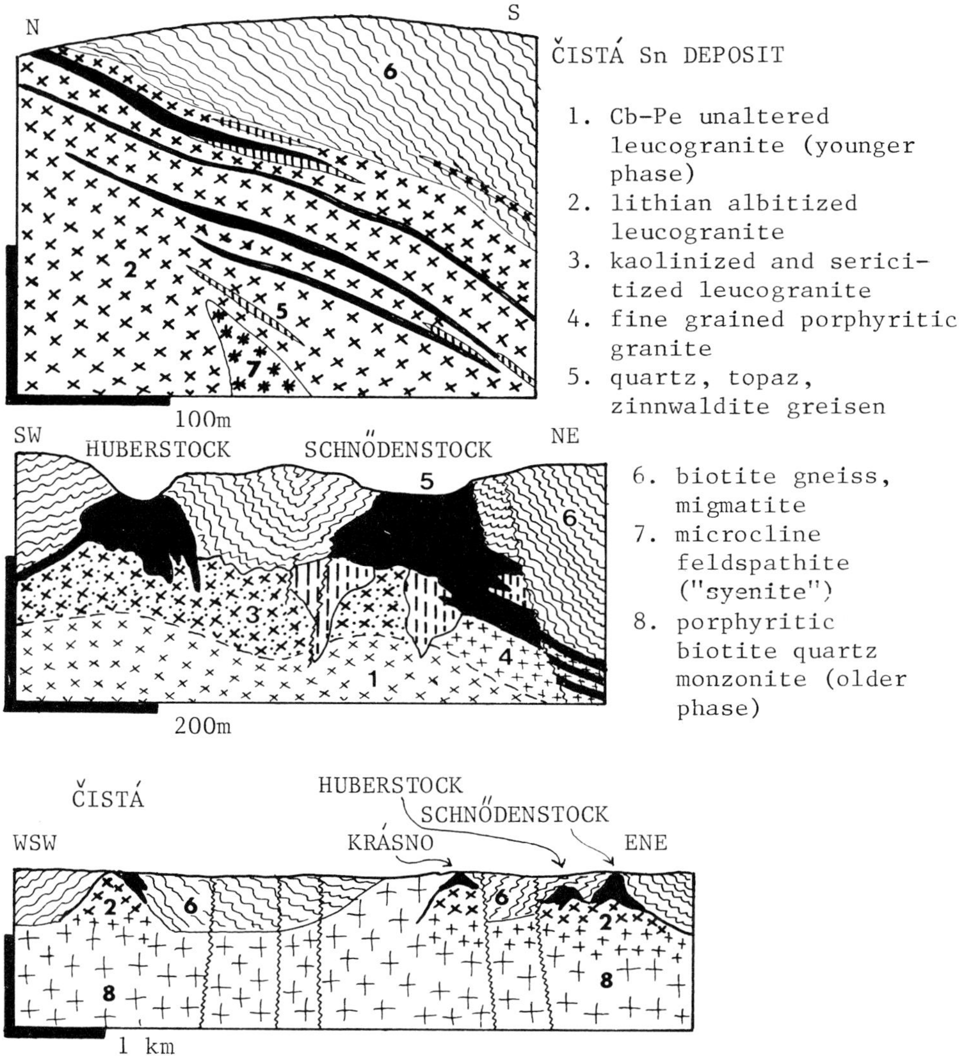

Fig. 28-62. Cassiterite and wolframite-bearing greisenized granite stocks in the Horní Slavkov-Čistá area, W. Czechoslovakia. Modified after Ďurišová et al. (1969) and Chrt and Bolduan (1966). Bottom section is diagrammatic from LITHOTHEQUE.

CORNWALL AND DEVON, S.W. ENGLAND

South-west England (Cornwall in particular) is a well-known classical tin mining area that has produced more then 2 Mt Sn. The bulk of the tin, however, came from exocontact veins (and from placers), so a more detailed locality description is provided in a subsequent Section. Mineralized cupolas, however, have recently become a favourite exploration target because of their supposed potential for containing large, low-grade deposits. Hemerdon, the largest such deposit so far discovered, is a tungsten mine with minor tin by-product (45 Mt ore with 0.136% W, 0.029% Sn) so is discussed later.

FRANCE, SPAIN, PORTUGAL

Massifs of Permo-Carboniferous "tin granites" appear throughout the western European Hercynian foldbelt and in the Hesperian Massif. The past production came mostly from veins and placers, but several endocontact stockworks have been discovered or evaluated recently. In western Spain, the large but very low-grade Baltar cassiterite and wolframite stockwork (50 Mt ore with 0.12% W, 0.08% Sn) is primarily a tungsten mine and so is the Panasqueira vein deposit in Portugal.

EASTERN SIBERIA AND N.E. MONGOLIA

This vast area contains several Sn,W,Mo provinces (Table 28-9 and Figure 28-63) genetically associated with Jurassic to lower Tertiary granites. Most orebodies are in the granite exocontact. Those having the form of stockwork in greisenized granite cupolas include the Kester deposit in the Ege Khaya field, Yakutia, described by Materikov (1974); Bukuka in East Transbaikalia (Magak'yan, 1968); Baga-Gazryn and Khara-Moritu in Mongolia (Marinov et al., 1977). The Butugychag Sn deposit in the upper Kolyma River (Materikov, 1974) departs from the usual style in that it has stockworks and parallel systems of quartz, albite, K-feldspar, cassiterite, wolframite and molybdenite veins filling what appear to be zones of torsion fracturing in a granite cupola. The host rock is an albitized apogranite, locally greisenized.

NOVA SCOTIA, CANADA

Recent geochemical exploration resulted in the discovery of a large but low-grade (38 Mt ore with 0.2% Sn) deposit of disseminated cassiterite in Devonian granite near East Kemptville (Richardson et al., 1982). This deposit is contained in a greisenized and sericitized granite just under the quartzite roof. It is notable that the ore occurs beneath an inflection of the roof, rather than in a

Fig. 28-63. Principal tin provinces of the world.

typical cupola. Albite alteration of the host granite outside the greisen has not been emphasized in the preliminary paper, but the presence of quartz-phosphate (triplite, apatite) suggests apogranite association.

SUNDALAND (S.E. ASIA) TIN BELT

Over 90% of the tin production and reserves in this extensive belt come from placers and the rest from exocontact veins, replacements and stockworks. Because of the humid tropical environment, bedrock outcrops are scarce. Several small endocontact cassiterite occurrences are known in Malaya (Hosking, 1973, 1974). On the Indonesian "Tin Islands", the Pemali deposit on Bangka (Schmidt, 1976; 12.22 Tt Sn/0.12%), is the only significant though small bedrock deposit. Cassiterite is in a greisen-hosted stockwork in the contact zone of a Mesozoic granite.

EASTERN AUSTRALIA

The Tasman orogenic belt of eastern Australia contains a series of tinfields distributed from northern Queensland to Tasmania, but in order of importance endocontact tin stockworks are in fourth place (after carbonate replacements, placers and exocontact veins). Over 100 occurrences of disseminated and stockwork cassiterite in small

Table 28-9. Phanerozoic plutonic tin provinces of the world, brief data

PROVINCE	GEOLOGY	MINERALIZATION	REFERENCES
Cornwall and Devon, S.W. England, U.K.	D-Cb slate, silts., minor greenstone (metabasalt), limest., intr. by 295 m.y. qtz.monzon. pluton, 280-275 m.y. qtz. feldsp.porphyry dikes	X000 main stage (postdating dikes) fiss.and repl.cassit., cp., arsenp., pyrrh., qtz., tourm., chlor.veins; minor wolfr.stockw.in greisenized cupolas, replac.shears. 2.5Mt Sn (portion from placers); 2 Mt Cu; 250Tt Pb	Halliday (1980) Dunham et al. (1978)
Tin granite massifs in the Hercynian belt of France, Spain, Portugal	Pt_3-PZ high-grade to greensch. metased.and metavolc., intr.by several phases of Cb-Pe granitic plutons; typical cupolas rare (e.g. Echassières)	mostly qtz-cassit. or wolfr. fissure veins in exo-and endocont.; rare stockw. in greisens (Baltar) and feldspathiz. Li apogranites (Montebras); about 70Tt Sn, 70Tt W	Taylor (1979)
Erzgebirge and Slavkovský Les Mts., W. Czechoslovakia and East Germany	Pt_3-D greensch.to amphib.fac. maf.metavolc., metased. in a D-Cb deformed N.E. orog.belt; 305 m.y.mesozon.porph.granite to qtz.monzon., 240 m.y. epiz. "tin granites"	the bulk of econ.Sn comes from several large endocont.greisens with cassit.stockw., dissem. and Li mica. X00 of qtz veins with cassit.or wolframite are less productive; placers. 1 Mt Sn, 3 dep.100Tt Sn+	Tischendorf et al.(1978) Chrt and Bolduan (1966)
Central Kazakhstan Sn Province (S.W. of Karaganda)	Or-S siltst., shale, sandst. intr. by Cb_3-Pe qtz.monzonite,	numerous miner.centres in aureole of granite stocks and plutons; apogr., endocont.greisens and veins, lesser exocont.veins, domin. by W-Mo; Sn is rare. Estim. 5Tt Sn, 50-100Tt W	Borukaev and Shcherba eds. (1967)
Kal'ba-Narym Ranges Sn Province, E.Kazakhstan, U.S.S.R.	D_1-Cb_1 shale, sandst.; Cb_3-Pe large gran.and qtz.monz.pluton, cont.dorsal ridges and cupolas of Sn granite; Pe-Te subaerial volcanism	endocontact greisens and stockworks with cassit., minor apogranites with Ta, Be, pegmatites, placers. Estim. 30Tt Sn	Smirnov ed.(1968)

N.E. Mongolia, East Transbaikalia, U.S.S.R. province	PCm and PZ folded basement includes a PZ_3 granitic bathol. J-Cr meso-epizon.diorite to alaskite plutons, andes.-rhyol. continental volcanism	rare endocont.stockw. and veins (Bukuka); many exocont.veins,replac. and stockw.(Khapcheranga, Sherlovaya Gora). Qtz.,tourm.,chlor.,cassit. and sulphide-cassit. Est. 500Tt Sn	Materikov (1974)
N.E. Yakutia and Kolyma Basin Province, Siberia, U.S.S.R.	Tr flyschoid sandst.,slate,on margin of a median massif, intr.by J_3-Cr_1 granodiorite, granite	rare endocont.veins,stockw.(Kester, Butugychag);widespr.exocont.qtz., tourm.,chlor.,cassit.veins and stockw.,minor cassit.-sulph.bodies. Estim.min.300Tt Sn	Materikov (1974) Magak'yan (1968)
N.E.Chukotka, U.S.S.R., and west. Alaska, U.S.A.	Tr-Cr_1 litharen.,slate; J_2-T_1 dior.to alaskite. Sn related to Cr qtz.monz.,gran. and leucogr.,30-50 intrus.	in Lost River,Alaska,cassit.is in cupola and dike greis.,exocont. veins,skarn (with Be,fluorite,Li micas); similar in Siberia,where most import.qtz.-cassitt.,less sulph.-cassit.veins; placers. Est. min.250Tt Sn	Taylor (1979) Materikov (1974)
Komsomol'sk (Miao-Chai) distr.,Amur area, S.E. U.S.S.R.	J terrig.sedim.,basalt,diab., gabbro; Cr andes.,dac.,rhyol. volc.-sedim.; Cr dior.to granite plut.; Sn assoc.with diorite porphyries	qtz.,tourm.,cassit.replac.bodies along faults to qtz.-chlorite repl. cut by qtz.-cassit. and cassit.-sulph.veins. At Solnechnoe,ore zone is up to 8km long, 115m wide. Estim. 100Tt Sn	Materikov (1974)
Kavalerovo distr., Sikhote Alin, Soviet Maritime Province (Primor'e)	PZ_3-MZ sandst.,siltst.,minor lim.,chert,spil.,gabbro; Cr_1 terrig.sedim.; Cr_3-Eo buried bath.with gran.cupolas; Cr-T_1 contin. andes.-rhyol. volcanism	Eo, silicate (tourm.,chlorite) vein, lesser replac. cassiterite oreb. Biot.,seric.alteration envel. around granod.to granite stocks; ore veins, stockworks. Estim. 250Tt Sn, by-product Zn,Pb,Ag	Materikov (1974)

Table 28-9 (continued). Phanerozoic plutonic tin provinces of the world, brief data

PROVINCE	GEOLOGY	MINERALIZATION	REFERENCES
Southern Kiangsi (parts of Kwangsi, Hunan), China	Pt, Cm-S sandst.,shale,quartz. D,Cb limestone, intruded by J_{2-3} granite	numerous endocont.greisen stockworks mostly feeding placers; exocont.veins; recent large Sn,W, Be,Bi,etc.skarns (Xi Zhoyouang); min. 850Tt Sn, 650Tt W	Ikonnikov (1975) Juan (1946)
Yunnan, S.W. China (Kochiu)	Tr limest.,shale 85 m.y. granite intrusions	hematite,cassiter.,arsenop.,pyr., sphal.,gal.,massive replacem.in limest.;also residual and placer Sn 1.65 Mt Sn/3.65%	Juan (1946)
E.Thailand,Laos (Cholburi-Chantaburi; Nam Pathene)	PZ quartzite,limest.,shale Cb,Tr-J granite plutons	probably greisen,vein and skarn, but all prod.estim.to max.10Tt Sn comes from saprolite,alluv.and beach placers	Fontaine and Workman (1978)
Peninsular Burma, Thailand,Malaya tin belt	D_2-Cb_1 marine shale,sandstone (qtz.rich turbidite);Pe thin sandst.,shale,minor limest. Cb_1-T_1 biot.gran.,coarse two mica gran.with access.tourmal.	few bedrock mines (e.g.Sungei Lembing); complex cassit.exocont.veins greisens,skarns; thousands of small regolithic occurr.contrib.cassiter. to a variety of eluv.,alluv.,beach placers; 4.9 Mt Sn	Hosking (1973,1974)
Indonesian Tin Islands (Bangka,Belitung, Singkep)	Cb-Pe_1 sandst.,shale,siltst., minor tuff.seds.,hornfelsed; Tr_3 (213 m.y.) biot.and musc. biot.granite,tropic.weathered	hydroth.miner.198 m.y.,small greisen,skarns,veins; almost entire prod.and res.comes from onshore and offshore placers; 2.6 Mt Sn	Hosking (1973,1974)
N.E. Queensland Tinfields,Australia (Herberton,Cooktown, etc.)	S_2-Cb_1 sublitharen.,shale,minor carb.; Cb_1-Pe_3 acid cont. rhyodac.volc.,cauldrons,qtz. monz.to gran.bath.,cupolas	20-30 separ.miner.centres above cupolas above cryptobath. in Herberton field, 10 more elsewhere; greisens,pipes,veins,placers;163 Tt Sn	Taylor (1979) McLeod,ed. (1965)

New England tin-fields, N.S.W., Australia	PZ_1-Pe_1 folded terrigen.seds. Pe_3 rhyodacitic contin.volc., large qtz.monzon. and granite batholiths, minor cupolas	cassit.mostly as endo-and exocont. veins, greisen pipes, fract.stock-works in hornfelsed exocontact seds. The bulk of cassit.came from placers; 210Tt Sn	Taylor (1979) McLeod, ed. (1965)
Lachlan belt tin fields, N.S.W. and Victoria, Australia	Or-D metased., terrig.slate, qtz-rich litharen., minor carb. S_3-D qtz.monzon.to granite batholiths	cassit.in small greisen bodies, less exocont.veins; the bulk came from placers; 53Tt Sn	McLeod, ed. (1965)
N.E.Tasmania (Rossarden, Blue Tier, Ringarooma, etc.)	PZ_1 (S) slate, qtz.-rich litharenite; D_3 qtz.monzonite, granite	cassit.in greisen (Blue Tier) and exocont.fiss.veins (Rossarden); the rest is from placers, mostly alluv. deep leads; 95Tt Sn	Williams (1978)
N.W. Tasmania (Renison Bell, Mt.Cleveland, Bischoff)	Pt_3-Cm_1 siltst., quartzite, minor lim., dolom., greenst., ophiolites; D qtz.monzonite, leucogranite, qtz. feldsp.por.	3 large cassiterite-pyrrhotite repl. in carbonates; about 20 small greisen, vein and placer occurrences; 400Tt Sn	Williams (1978)
Nova Scotia batholith, Canada	PZ_1 quartzite, slate intr.by D granod., qtz.monzon.bathol., 350 m.y. alaskite dikes	East Kemptville, low-grade cassit. dissem.in greisen and sericitiz. granite-hosted stockwork	Richardson et al.(1982)
Cordillera Real plutonic Sn province, Bolivia	Or-D gray slates interb.with qtz.-rich litharen. and quartzites; Tr to Mi qtz.monzon.to granite batholiths, stocks	exocont. qtz., tourm., chlor., cassit., wolfr., sphal., etc. veins are most common, followed by peneconc.mantos in quartzite and placers. 2Mt Sn	Ahlfeld and Schneider-Scherbina (1964)

endocontact greisen pipes are known but most are small (of the order of 10-100 t Sn; Fig. 28-64). They have, however, been of greater importance as suppliers of cassiterite to placers (e.g. at Gibsonvale, Cooktown, Tingha, Ringarooma, etc.).

The two largest deposits mined on their own have been Ardlethan, N.S.W. (10.2 Tt Sn/0.17%) and Anchor mine in the Blue Tier Batholith, Tasmania (7 Tt Sn/0.2-0.5% Sn). The latter deposit has been studied in detail (Groves and Taylor, 1973) and it consists of a very fine (practically invisible) disseminated cassiterite and some chalcopyrite in a light greenish-gray, quartz, mica, chlorite greisen. The greisen forms irregular, sheet-like bodies in the apical region of a thick sill-like intrusion of biotite-muscovite granite, emplaced in an unmineralized Devonian porphyritic biotite granite to quartz monzonite. The mineralized zone is about 700 m long, 160 m wide, and positioned within the upper 40 m of the granite sill.

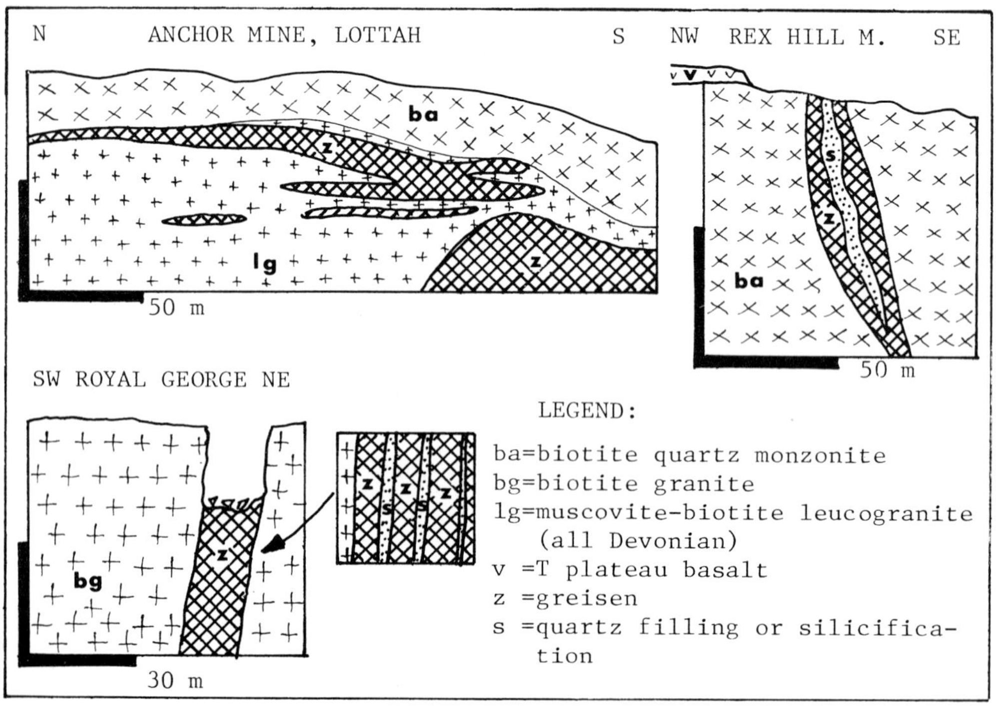

Fig. 28-64. Examples of small endocontact greisen Sn deposits in N.E. Tasmania. From LITHOTHEQUE, after Groves and Taylor (1973), Urquhart (1967) and authors fieldwork, 1971.

28.8.3. Postmagmatic Sn (W,Bi,Mo) veins, stockworks and replacements in granite aureoles (non-carbonate hosts; Table 28-10)

This is the mainstay of the plutonic tin association. Thousands of occurrences are known worldwide and although individual deposits are small to medium in size, the cumulative importance is significant because exocontact veins and similar bodies occur in every known plutonic tin province. The classical, narrow and small tonnage fissure veins have recently been joined by several "modern" styles: mineralized breccias and fracture (joint) stockworks in hornfelsed supracrustal rocks.

CORNWALL AND DEVON, S.W. ENGLAND

This N.E.-trending metalliferous area occupies a peninsula in the S.W. tip of England. It measures about 150x60 km (the same dimensions as those of the Erzgebirge) and has produced about 2.5 Mt Sn and 2 Mt Cu mainly from veins and placers. Cornwall is one of the sacred places of metalliferous geology (e.g. a craddle of the zoning theory), and it has recently been re-interpreted in a series of publications (Dunham et al., 1978 ; Jackson, 1979; Halliday, 1980). As in the Erzgebirge, the greater part of S.E. England is underlain by a major Carboniferous batholith exposed in five cupolas. Deformed, greenschist and contact-metamorphosed Devonian and Carboniferous "molasse" slates, litharenites, lesser greenstones and carbonates occur in the roof. Several thousand hydrothermal ore occurrences congregate in the immediate vicinity of the cupolas, at both sides of the contact (but more in the exocontact).

Halliday (1980) demonstrated that the ores kept forming for about 75 million additional years after the granite emplacement, according to the following time pattern:

295 m.y.: emplacement of major plutons;
285-280 m.y.: endocontact greisens with sheeted veins and Sn-W stockworks (rare);
280-275 m.y.: after fracturing, emplacement of granite porphyry dikes and shortly afterwards the majority of Sn, W and Cu lodes. Main Stage of mineralization;
270 m.y.: formation of most Pb,Zn,Sb,Ag, etc. veins (minor) and
220-230 m.y.: U and remaining polymetallic mineralization.

In S.W. England both the intrusive dikes and ore veins (lodes) follow the same fracture sets and shear zones, both in the endo- and exocontacts of the early granites (but in exocontacts of the younger generation of granitic stocks when such are present). When the "late" (postmagmatic) Main Stage veins are hosted by the older granite, they are recognizable from the earlier endocontact greisen apogranite association by their lack of intimacy with their host, by lack of control by the granite cupola morphology and contacts and by different mineral filling and alterations. Most reviews of the Cornish metallogeny recognize the following styles of the Main Stage and later

Table 28-10. Postmagmatic Sn (W,Bi,Mo,etc.) veins, stockworks, replacements, in granite aureoles (non-carbonate hosts); selected examples

LOCALITY	HOST ROCKS	MINERALIZATION	REFERENCES
Wheal Jane mine, Truro, Cornwall, Great Britain	D_1 slate, siltstone intr.by 280 m.y. qtz.feldsp.porph. dikes	up to 23 m thick complex N.E. striking replac. vein in shear along dike/wallrock contact. Cassit.,ars., local mass.pyr.,sphal.lenses. 62.5Tt Sn/1.25%	Rayment et al. (1971)
Dolcoath Main Lode, Camborne, Cornwall, Great Britain	D-Cb slate, siltst.,greenstone intr. by Cb granite	1.5 km long, 1 km deep, 0.3-12 m thick N.E. striking fracture filled by brecc. cementing qtz.,tourm., cassit. 93.5Tt Sn, 355Tt Cu	Jackson (1979)
Khapcheranga, East Transbaikalia, U.S.S.R.	Pe hornfelsed shale,quartzite; J,porph.biot.granite; Cr,T, gran.porph.,rhyol., dacite	over 20 fiss.filling and fault gauge replac. veins along N.W. tect.zone; pyrrh., fine cassit.,less arsenop.-cassit. and qtz.,calc.,chlor.,cassit. veins; ox.zone to 50m; 240Tt Sn/3.78%	Magak'yan (1968)
Ege-Khaya field, Yakutia, N.E. Siberia, U.S.S.R.	Tr folded terrig.litharen., slate, intr. by J_3-Cr_1 granodiorite	miner.fault brecc.and fiss.veins along N.E. crush zones; qtz.,tourm., chlor.,pyrrhot.,cassit.,arsenop., sphal.,cp.,stannite; zoned; min. 70Tt Sn, partly from placers	Magak'yan (1968)
Sungei Lembing mine Pahang, E.Malaya	PZ hornfelsed slate, quartzite intr. by Cb_3 granite	600 m wide band of hornfelsed sedim. // with granite contact, cut by complexly zoned veins with cassit., chlor.,arsenop.,cp.,sphal.,pyr., locally quartz	Hosking (1973)

Sn(W,Cu,etc.) ores (essentially the same styles are universal worldwide):

(1) Fissure veins

Veins supplied most of the metal production. They are usually tabular, several centimetres to about 2 m wide, traceable along a strike for 500-1,000 m (exceptionally up to 6 km), and followed to depths of up to 900 m. There is no persistent relationship between the granite form and vein trends. Some veins occupy the same fissures as dikes, but dike-cutting veins are more common. Most fissure veins have polyphase filling and many show metal zoning with depth. Cassiterite usually accompanied by chalcopyrite, chalcocite, specularite, sphalerite, wolframite and arsenopyrite is scattered in quartz, chlorite, tourmaline, feldspar, hematite or mica. Some major veins were high in Cu and low in Sn (e.g. the Dolcoath Vein produced 355 Tt Cu and 93.5 Tt Sn), others carried exclusively chalcopyrite and chalcocite (e.g. Devon Great Consols, 776 Tt Cu). There is no satisfactory explanation for the widely variable presence of Cu in the tin veins. The high-copper areas appear to correlate with abundant greenstone wallrocks (spilites, metabasalts), the most likely Cu sources.

The Dolcoath Main Lode near Camborne (Fig. 28-65), the most productive vein in the district, is a N.E.-trending, 50-90° S.W. dipping structure continuing for 1.5 km and tin-mineralized for over 1 km downdip. Its thickness varies from 0.3 to 12 m and the vein is filled by brecciated quartz, tourmaline and cassiterite at a depth where granite is the wallrock, and comb quartz, pyrite and chalcopyrite near the surface, in slate and greenstone wallrocks. The wallrock alterations include silicification, sericitization, tourmalinization and chloritization.

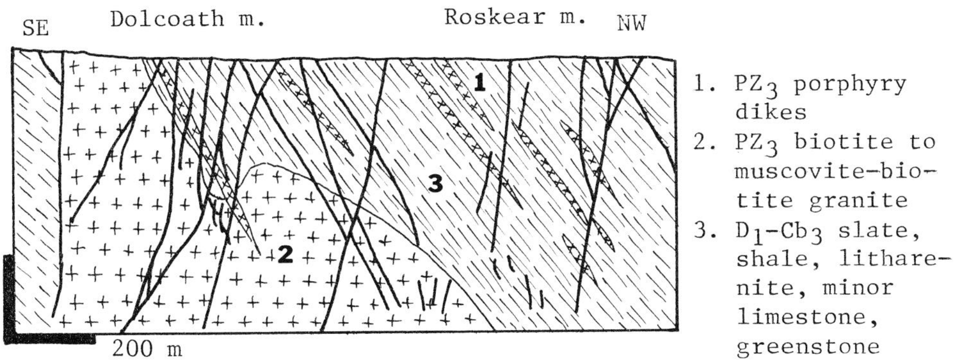

Fig. 28-65. Sn and Sn-Cu veins in the Camborne area, Cornwall. Simplified after Dines, ed., (1956).

(2) Replacement veins

Dunham et al. (1978) distinguished three varieties: (a) replaced wallrocks adjacent to a narrow fissure, (b) completely replaced selected horizons (e.g. carbonates) within the metasediments and (c) replacements of mylonite and gouge along faults and shears. The Great Flat Lode is an example of (a). It has been traced for 6 km and it consists of tourmalinized and silicified granite mineralized by finely disseminated cassiterite and minor sulphides, adjacent to quartz, tourmaline and cassiterite leaders. The lode width ranged from 1.3 to 4.5 m.

In the Wheal Jane mine near Truro, an example of (c) and one of the few recently producing Cornish deposits (Rayment et al., 1971; 62.5 Tt Sn/1.25%), unusually thick (up to 23 m) and complex lodes follow shears along the contact of gently N.W.-dipping quartz-feldspar porphyry dikes, emplaced into Devonian slates and siltstones. The thickest B lode is a shear zone replaced and impregnated by cassiterite, tourmaline, arsenopyrite, lesser chalcopyrite and pyrrhotite. In places it contains lenses of massive pyrite and sphalerite. The wallrock black slates are sericitized.

(3) Replacement pipes.

These are uncommon and of little importance. The largest pipe in East Wheal Lovell mine is roughly cylindrical, 130 m deep and 4 m in diameter, consisting of a 1 cm thick central stringer of quartz surrounded by silicified and sericitized granite with some disseminated cassiterite, sulphides and fluorite.

ERZGEBIRGE MOUNTAINS, CZECHOSLOVAKIA AND EAST GERMANY

In the Erzgebirge, veins and late postmagmatic tin stockworks are considerably subordinate in importance to mineralized greisens. Sn-W and W lodes were mined in the Pechtelsgrün field, Germany, Horní Slavkov (Gellnauer Vein), Rolava and Přebuz, Czechoslovakia and elsewhere. Tourmalinized stanniferous hornfelses formed by thermal metamorphism of phyllites at the contact with the Horní Blatná Stock in Podlesí (Roos, 1966), constitute a potential (0.6% Sn) tin ore. In the past, two small orebodies (2 m thic and 40 m long) marked by sporadic visible cassiterite, pyrite and arsenopyrite on occasional fracture veinlets, were mined. In the absence of associated fracture minerals the cassiterite is finely dispersed and megascopically unrecognizable. Similar hornfelses may form economically important deposits in the future.

EASTERN CORDILLERA TIN BELT, BOLIVIA

The "plutonic" share of the Bolivian Tin Province probably represents about 2 Mt Sn, most of it in exocontact veins, stockworks and replacements affiliated with Triassic to Miocene high-level granites. The Province has already been reviewed briefly in this book

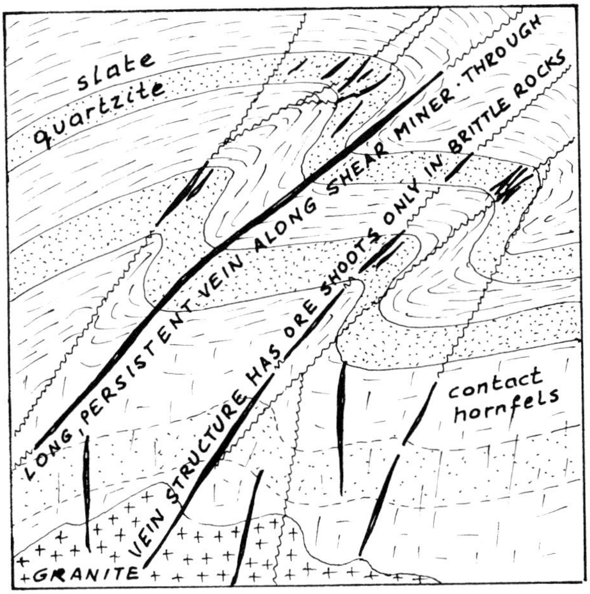

Fig. 28-66.
The usual control of the Bolivian exocontact veins by bedrock competency.

in connection with its distinct bedrock, composed of alternating shallow marine shales and quartzites (Section 19.7.1., Fig. 19-10). The vein morphology is strongly influenced by the competency of wallrocks (Fig. 28-66).

EAST TRANSBAIKALIA, U.S.S.R.

In East Transbaikalia where Sn,W,Mo ores are associated with Jurassic and Cretaceous granites, several stages of mineralization have been recognized just as they have in the Cornwall (Tomson et al., 1970). The early stage (165-170 m.y. at deposits in Khapcheranga and Bukuka) produced endocontact greisens with disseminated and stockwork cassiterite and wolframite. A later stage (148 m.y.) generated postmagmatic hydrothermal cassiterite-sulphide veins.

The Khapcheranga deposit (Magak'yan, 1968; min. 240 Tt Sn/3.78%) is in a Permian shale and quartzite intruded by Jurassic biotite granite stocks. The mineralization is located in the exocontact, along a set of parallel N.W.-striking fault zones over 2 km long. Several varieties of ore are present, the most important being massive pyrrhotite-cassiterite forming lenses replacing fault gauge. The cassiterite is very fine grained and invisible. Quartz, calcite, chlorite, cassiterite veins and cassiterite-arsenopyrite veins are alternative fillings. The oxidation zone is well developed to a depth of 50 m and it consists of a goethite-scorodite mass with relic cassiterite crystals.

In the Sherlovaya Gora deposit a quartz, chlorite, hematite, cassiterite, ferberite stockwork lies in a breccia, filled by upper Jurassic rhyolite fragmental material. The mineralization is fault controlled and about 600 m from a stock of lower Cretaceous granite porphyry.

SUNDALAND (S.E. ASIA) TIN BELT

This N.-S.-trending belt, approximately 3,000 km long, represents some 7.5 Mt Sn, most of it recovered from placers (Chapters 23, 24). The secondary cassiterite, however, comes entirely from a belt of Carboniferous to lower Tertiary granitic plutons and their aureoles. The Main Range Batholith has already been described as an example of mesozonal batholiths. Although several hundred bedrock tin occurrences are known and they span the entire range of styles discussed in this Chapter (Hosking, 1973, 1974), only about 5 major "bedrock" deposits are known and have been in production. One of them, the Sungei Lembing mine near Kuantan, eastern Malaya (Table 28-10), is of greatest importance. It contains a system of "Cornish-style" chlorite, cassiterite, arsenopyrite, chalcopyrite, etc. veins in hornfelsed sediments along a granite contact.

NEW ENGLAND, AUSTRALIA, THE MOLE GRANITE AUREOLE

The Mole Granite (Weber, 1974; Fig. 28-67) is a late Permian "tin granite", and one of the youngest intrusions of the composite New England Batholith. It is an elliptical body with an area of 800 km^2, emplaced in folded and hornfelsed Permo-Carboniferous sediments (originally shales, siltstones, sandstones) and in lower Permian continental volcanics. The batholith is in an early stage of unroofing as is borne out by the presence of a small roof pendant. About 150 mostly small Sn,W,Bi and base metal occurrences are scattered on top of the unroofed granite (these are small chlorite-cassiterite or quartz-wolframite pipes), in the pendant and on its flanks. A local speciality are ore occurrences in "silexite", resembling, in some respects, the "secondary quartzites" and formed largely by postmagmatic silicification of the granite along fractures and in the immediate exocontact.

At the Bismuth mine (Fig. 28-68A), the "silex" is a fracture-filling and wallrock-replacing white quartz, grading to coarse crystalline quartz, fluorite, topaz masses and a fine-grained greisen. It contains scattered crystals of wolframite and lesser molybdenite, bismuth, safflorite and skutterudite.

Of greater practical importance are large exocontact cassiterite stockworks, two of which are being mined near Emmaville, about 10 km south from the Mole Granite margin. The earlier known stockwork at the site of the Great Britain mine (Figs. 28-68, 69) is located in brittle, hornfelsed former Permian shales and felsic volcanics in the

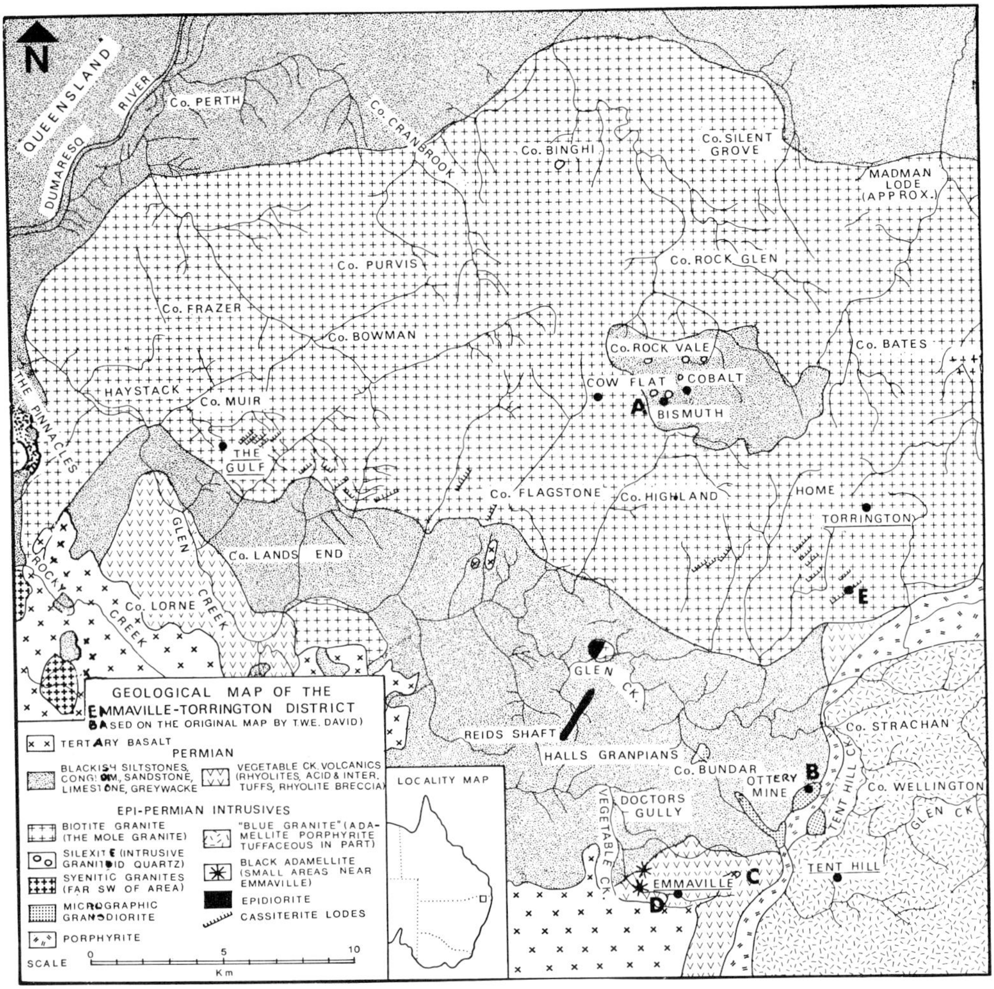

Fig. 28-67. Geological map of the Mole Granite massif and its aureole, N.S.W., from Lawrence (1975); courtesy of the Australian Institute of Mining and Metallurgy. Localities assembled in Figure 28-68 are marked by letters A to E.

immediate roof of a hornblende-biotite granodiorite. A system of parallel N.E.-trending, steeply dipping fractures extends over several square kilometres, having a density of 5 to 50 fractures per metre. The fractures are filled by thin white vuggy quartz veinlets, some of which contain scattered cassiterite crystals. Some fractures are dotted by flat cassiterite crystals without quartz. Every fracture is fringed by a bleached rim, 1-2 cm wide and some contain fine radial or fibrous black tourmaline. The overall appearance is very similar to the Bolivian mineralized quartzite mantos (e.g. Kellguani;

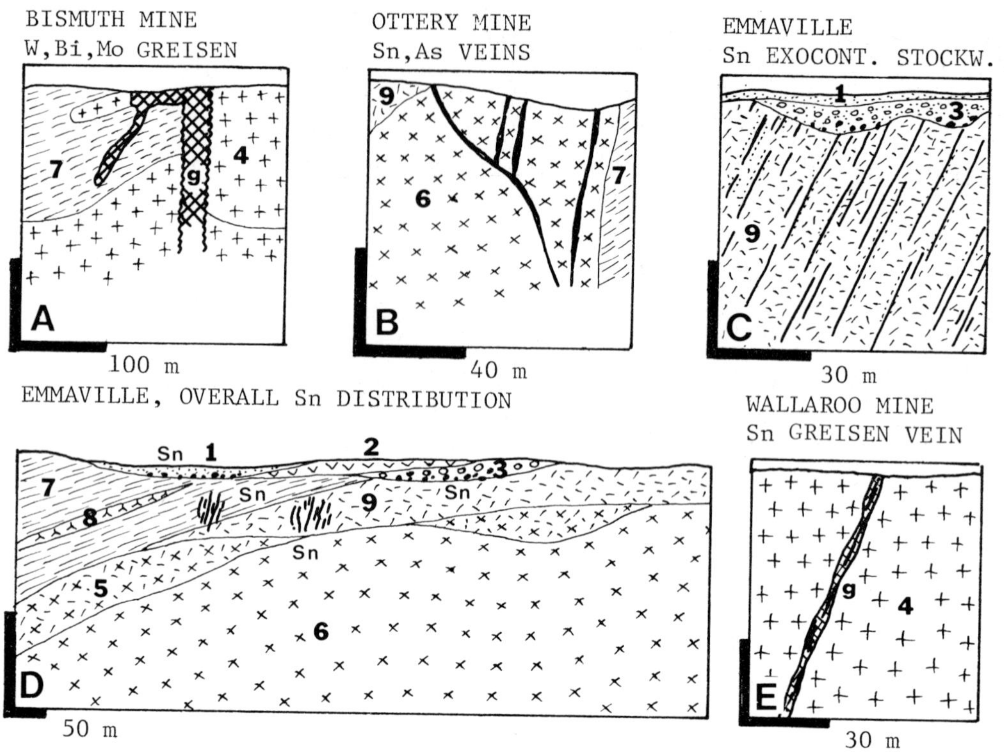

Fig. 28-68. Diagrammatic representation of the principal mineralization styles associated with the Mole Granite and its aureole, New England, N.S.W.
1=Q, Recent alluvium; 2=T plateau basalt, partly lateritized; 3=T, sub-basalt gravels; 4=Pe Mole Granite, coarse porphyritic biotite granite with a finer grained marginal facies; 5=black epidotized feldspar hornblende porphyry; 6=xenolithic hornblende biotite granodiorite; 7=Pe hornfelsed siltstone, conglomerate, sandstone; 8=greenstone; 9=rhyolite to rhyodacite flows, tuffs, breccia. From LITHOTHEQUE.

Chapter 19). The payable portions of the stockwork are irregularly distributed. 520 t Sn was produced between 1970 and 1973, and the overall recoverable tin content is probably of the order of several thousand tons.

In the nearby Grampians (Tarongo) deposit, the reserves of comparable ore are 17 Mt of material with 0.17% Sn, 0.05% Cu and 4.4 ppm Ag (=29 Tt Sn, 75 t Ag). Other deposits in the Mole Granite aureole (Fig. 28-68) include quartz, arsenopyrite, cassiterite fissure veins (Ottery mine, 2.4 Tt Sn); short quartz, chlorite, cassiterite

Fig. 28-69. Emmaville, New England, Australia, a quartz, cassiterite, lesser sulphides exocontact fracture stockwork and sets of parallel veins in hornfelsed Permian metasediments and metavolcanics. The ore minerals are located in thin fractures emphasized by narrow alteration rims.

veins in granite (e.g. Wallaroo mine); a Pb,Zn,Ag sulphide replacement in a dolomitic bed (Collisons) and others.

SIKHOTE-ALIN DISTRICT (U.S.S.R.) AND THE TIN-SULPHIDE DEPOSITS IN BIOTITE AND SERICITE ALTERATION HALOES

Sikhote Alin (or Kavalerovo) tin district (Materikov, 1974; Magak'yan, 1968) is near the Pacific coast of Siberia, facing Hokkaido over the Sea of Japan. It contains a variety of hydrothermal Sn and

Pb-Zn deposits probably associated with Eocene high-level quartz diorite, granodiorite and granite stocks at a depth. The existing orebodies, however, are situated in the roof where only intrusive dikes are in evidence and they are controlled by alteration-zoned structural domes. A buried pluton is postulated to be present at a depth of 1.5-3.5 km.

The mineralized region is a synclinorium filled by a monotonous, folded upper Paleozoic and Mesozoic shale, siltstone, minor limestone, chert, spilite sequence (reminescent of the Cornish lithology), flanked on the west by an uplifted and deeply eroded late Cretaceous granitic batholith, and in the east by a Cretaceous continental margin andesite-rhyolite belt. The latter volcanics overlap into the ore-bearing sedimentary sequence.

The ore fields and deposits (Rudnoe, Smirnovskoe, Lifudzin, Khrustal'noe, Dalnee, Bol'shaya Sinancha) are associated with N.E. and N.S.-trending fault zones, and ore distribution is controlled by distinct alteration envelopes described and interpreted by Razmakhnin et al. (1974); Fig. 28-70. A typical domal alteration-mineralization complex in pelitic sediments has a core of an early intensive prograde biotitization, grading outward into a propylitic (mainly chlorite) zone. The biotites have an increased trace Sn content. On this are superimposed products of a later biotite-destructive acid leaching, consisting of sericite, quartz-sericite, quartz, chlorite and other assemblages containing biotitized relics. The best economic tin mineralization accumulated above the buried biotitized cores, and it has a vertical span of 200-600 m. Most commercial ores do not descend more than 200 m into the biotitized cores and those orebodies that do so have a strong envelope of sericitization.

The principal ore style comprises long (2-7 km) sets of thin, parallel fracture veins grading into stockworks and replacement veins. The tin ores are sulphide rich (arsenopyrite, chalcopyrite, galena, sphalerite, stannite), and cassiterite is usually finely crystallized and "invisible" except for the early stage quartz-tourmaline or chlorite, cassiterite veins and stockworks. The veins with high stannite content are difficult to process and they are of marginal economic importance.

As with other Soviet localities, the impossibility of an on-site inspection makes it difficult to ascertain the degree of uniqueness of the Kavalerovo district. Biotite formation is an integral part of the thermal metamorphism in the above intrusive zone everywhere where the granite was emplaced into a shale sequence, which is the most common case. Later-stage hydrothermal veins and stockworks hosted by the biotite hornfelses often are in the bleached and sericite-altered hornfelses. Razmakhnin et al. (1974), however, emphasized hydrothermal biotitization resulting in almost pure biotitites, so there is likely to be a degree of, at least quantitative, uniqueness.

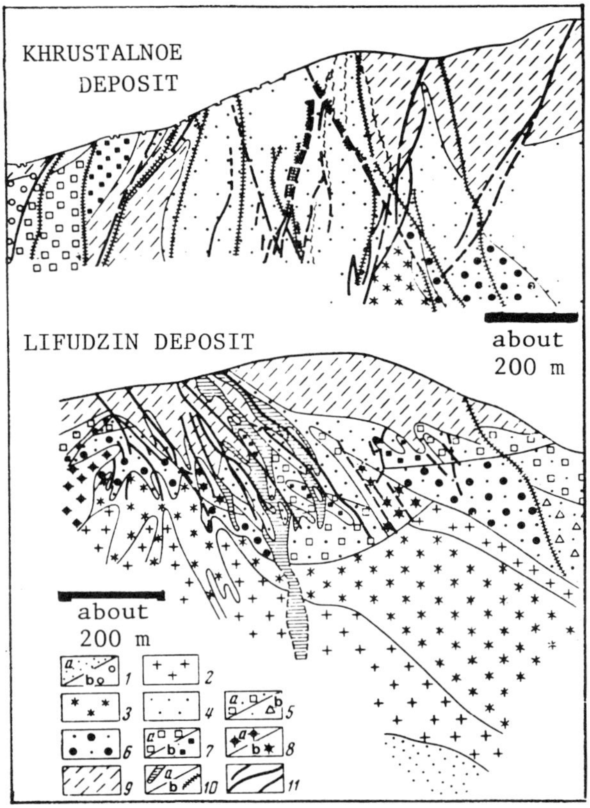

1. biotite metasomatites and biotitized rocks
2,3. zone of hydrothermal leaching
2. light
3. strong
4-6. zone of simultaneous leaching and deposition represented by white silica metasom.
4. after propylites
5. after cryptic biotitization
6. after biotitites
7. cryptic biotitization
8. zones of "gathering" biotitization
9. propylitic rocks
10. dacites, andesites, diorite porphyries
11. orebodies

Fig. 28-70. The variety of hydrothermally altered rocks and metasomatites present at two deposits in the Kavalerovo district, Sikhote Alin, S.E. Soviet Union. From Razmakhnin and Razmakhnina (1973), approximate scale added.

28.8.4. Postmagmatic Sn (W,Bi,Mo) ores in skarn and altered carbonate

The development and success of the Renison Bell deposit in Tasmania focussed attention of geologists on the carbonate-hosted tin deposits, a previously little publicized style of mineralization. About 2.55 Mt Sn is believed to be present in metacarbonate-hosted orebodies. In the popular literature, all the tin ores in metacarbonates are often headlined as being in skarn or tactite and considered, implicitly or explicitly, to be contemporary with the formation of the "main stage" silicates. This is an oversimplification. Approximately half of the deposits are located in skarns or tactites but the rest are in recrystallized, dolomitized or silicified carbonates lacking the skarn silicates. In the former association the timing and characteristics of the tin mineralization conform to the general skarn model, reviewed earlier.

TIN-BEARING SKARNS

"Tin skarns" have recently been reviewed in Einaudi et al. (1981). In skarns tin can be (a) bound in the lattice of silicates, (b) present as fine but "invisible" cassiterite surrounded by a more common "carrier mineral" such as magnetite or pyrrhotite or (c) megascopically apparent in fracture fillings and veins. The (a) group is, at present, of questionable economic value, because the silicate tin cannot be economically recovered but if a large deposit of stanniferous silicates were ever discovered, a recovery process would certainly be found. The Doradilla deposit near Bourke (N.S.W., Australia) is said to contain 10-12 Mt of material with 0.3% Sn (=30 to 36 Tt Sn) largely in grossularite but also in malayite in skarn. The rest of the currently known occurrences of Sn silicates are small and have a patchy distribution. Most are located in the "gangue" of bodies of a later cassiterite. The silicates include Sn garnets (grossularite or andradite, up to 5.8% SnO_2); Sn-amphibole (up to 3% SnO_2); axinite, idocrase and few other minerals. Magnetite may contain up to 0.4% SnO_2. These minerals form in the earliest stage of skarn growth.

Stanniferous sphene, malayite ($CaSnSiO_5$) as well as nordenskioldine, högbömite and some other rare minerals, form in the later stages of skarn formation. Malayite is most widespread in wollastonite skarns and is an inconspicuous light-brown mineral of nonmetallic appearance resembling common sphene, grossularite or idocrase. Like scheelite, malayite is fluorescent under ultraviolet light. Because of its high tin content, malayite accumulations could be of practical importance.

The latest, hydrothermal stage of skarn development is the most important for tin deposition, and cassiterite accompanied by fluorite, amphibole, phlogopite, hematite, pyrrhotite and sometimes by scheelite, bismuthinite and molybdenite accumulates as stockwork veinlets and veins within and outside the skarn. Commonly, the late-stage stockworks in exocontact skarns are transitional into greisens in the endocontact, or into fluorite, quartz, tourmaline, axinite, datolite and other veins anywhere in the contact aureole.

In Moina, Tasmania (Kwak and Askins, 1981; Fig. 28-71) cassiterite with malayite, scheelite and bismuthinite occurs in a rhythmically banded idocrase, magnetite, fluorite skarn on the periphery of a pyroxene, idocrase, magnetite skarn. The skarn formed by replacement of an Ordovician limestone, 200 m above an altered cupola of tin granite. The tin-bearing skarn (45 Tt Sn/0.15%) was recognized only recently in what had been a small ore field producing Sn,W and Bi (689, 438, 90 t, respectively) from five E.-W.-trending fracture quartz, fluorite, topaz, muscovite, chlorite veins hosted mostly by quartzite.

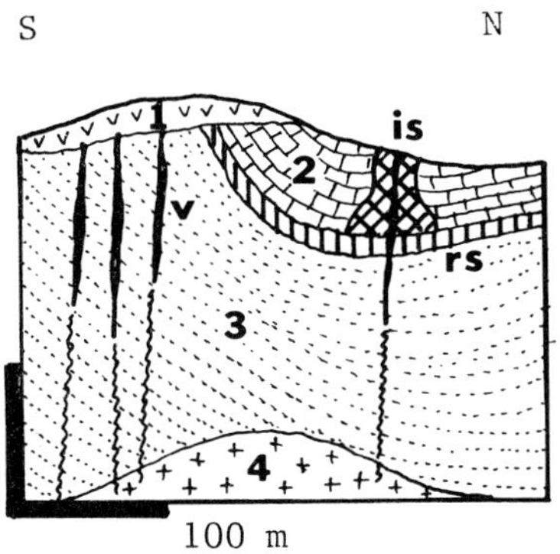

1. Mi olivine basalt capping
2. Or, recrystallized limestone (marble)
3. Or, sandstone to contact meta-quartzite
4. D, altered granite cupola

rs=reaction garnet,pyroxene,magnetite skarn

is=rhythmic infiltrational fluorite, magnetite, vesuvianite, cassiterite, scheelite, sphene

v=quartz,fluorite,cassiterite veins

Fig. 28-71. Moina Sn-W skarn and vein deposit, Tasmania (diagrammatic, from LITHOTHEQUE).

The Lost River ore field in the York Mts., Seward Peninsula, Alaska (Dobson, 1982; Fig. 28-72) has close geological affinities with the Chukotka belt in Siberia rather than with the rest of Alaska. Ordovician limestone resting on earlier sediments was intruded by a late Cretaceous "tin granite" stock and a complex fluorite, Sn,W,Be mineralization formed in both the endo- and exocontact. The interesting fluorite-Be veins predate the granite emplacement and are reviewed in a later section. The granite-hosted greisen is of the "ordinary" type, composed of fine-grained quartz with topaz, tourmaline, cassiterite and sulphides.

The skarn is composed of an early andradite garnetite (up to 6% SnO_2 in andradite) which is cut by a layered fluorite, magnetite, idocrase vein skarn. The latter carries some helvite and malayite. The layered skarn, in turn, is often intersected by idocrase and garnet veins containing abundant sulphides (galena, sphalerite, chalcopyrite), some scheelite, and light-brown cassiterite. The latest abundant hydrothermal assemblage connected with the skarn development has fluorite, green biotite, lesser hornblende and white mica, and it forms veins and replacement veins in the earlier skarn as well as in the exocontact limestones. Sulphides (dominant sphalerite and pyrite) occur as massive zones and pods often intergrown with cassiterite. The late stages of skarn development were contemporary with the greisenization in granite. The published Sn and W reserves in the Lost River area are 40.5 Tt Sn/0.15%, 8.1 Tt W/0.03%, and there is some 4.4 Mt of fluorite.

Xi Zhoyouang mine in Hunan, S.E. China (Jovanović and Ramović, 1981) is truly a giant deposit if the published figures are correct (420 Tt Sn, 600 Tt W, 40 Mt CaF_2, 200 Tt of beryl, 230 Tt Bi,

Table 28-11. Skarn-hosted hydrothermal Sn (W,Mo,Bi,Be) deposits; selected examples

LOCALITY	HOST AND INTRUSIVE UNITS	MINERALIZATION	REFERENCES
Lost River, Seward Peninsula, Alaska, U.S.A.	Pt or Cm, slate,argill.limest., gabbro; Or, argill.limest. Cr_3 (71 m.y.) granite stock, porph. dikes, rhyolite	complex miner.system; Sn is high in andradite and malayite in skarn; cassit.is in endocont.greisen, late stage stockw. and veins in skarn. 40.5Tt Sn/0.15%; 8.1Tt W	Dobson (1982)
San Antonio mine, S.Eulalia, Chih., N. Mexico	Cr limest.converted to skarn; T_1 granite, rhyolite dikes	vertic.chimney of garn.,epid.,actinol. skarn has late stage qtz.,fluorite,topaz,cp.,cassit.veins in 2 km N.-S. zone; 8Tt Sn	Hewitt (1943)
Campiglia Marittima, central Italy	J limest.,shale; 5.7-4.7 m.y. qtz.monz.,tourm.gran.,porphyry	cassit.with sphal.,gal. dissem.or in late-stage veins in Mn-ilvaite, hedenb.skarns and marble; 4Tt Sn	Corsini et al. (1980)
Beatrice Pipe, Selibin,Perak,Malaysia	Pe limestone; Tr? tin granite	cassit.,ars.,cp.,in fluor.and barite gangue, late stage repl.masses in tremol.skarn; a pipe under a tin placer; 8Tt Sn	Hosking (1973)
Xi Zhoyouang dep., S.of Changsha, Hunan, China	PZ sandst.,shale; D-Cb limest. J (172, 139 m.y.) biot.granite	900x600x300 m complex skarn,superimp. stockw.of cassit.,scheel.,Bi, moly.,etc.;cassit.also dissem.in late stage skarn and greisen; 420Tt Sn,600Tt W, 230Tt Bi,40Mt fl.	Jovanović and Ramović (1981)
Doradilla near Bourke, N.S.W., Australia	Cm-Or slate,chert,marine metavolc.,rare limest.; D granite, greis.porph.dikes,leucogranite	15 km long, 3m wide horiz.of grossul.,wollast.,idocr.skarn with Sn-diops.,garnet,malayite; late stage Sn-magn.,cassit.,sulphides. 36Tt Sn/0.3%, partly in silicates	Plimer (1980)

Moina, N.W. Tasmania, Australia	Or,quartzite,limest.,hornfels; D, autometasom. tin granite at a depth of 200 m	cassit.,malayite, Sn-garnet in late stage assoc.in idocr.,magnet., fluor.skarn; also Sn,W,Bi veins outside of the skarn; 45Tt Sn/ 0.15%, plus 690 t vein tin	Kwak and Askins (1981)

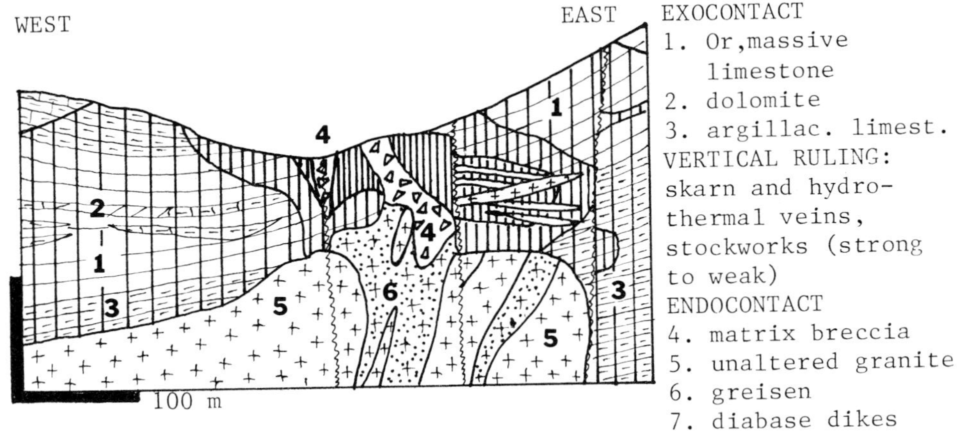

Fig. 28-72. Lost River mine, Alaska, hydrothermal plutonic cassiterite superimposed on greisen and skarn. Slightly modified after Dobson (1982).

110 Tt Mo), and it is a "variation on a theme" of lithophile metal skarn/greisen systems. The ores are contained in both the endo- and exocontact of a Jurassic biotite "tin granite" (two phases, 172 and 139 m.y.) emplaced in Devonian and Carboniferous limestones in a faulted anticline. The earliest skarn has a garnet, diopside, idocrase and wollastonite assemblage, followed by a hornblende-epidote skarn. These are only lightly mineralized by sporadic veinlets and nests with sulphides, cassiterite, chrysoberyl, tafeiite, helvite and beryl. The economic orebodies associated with the early skarn contain 0.238% W, 0.124% Sn, 0.105% Bi and 0.023% Mo.

The most important orebodies, as in the Lost River field, are contained in the transitional greisen-skarn zone dominated, in the exocontact, by late-stage veins and replacement masses of fluorite, garnet, diopside, hornblende, feldspar, quartz, scheelite, wolframite, molybdenite and bismuthinite. In this zone tungsten (0.465% W) is the dominant metal, and the tin content is 0.105%.

The remainder of the known tin skarns are all small deposits (Table 28-11); e.g. Campiglia Marittima, Italy, 4 Tt Sn; Santa Eulalia, Mexico, about 6 Tt Sn; Beatrice Pipe, Malaya, 8 Tt Sn).

TIN IN NON-SKARN CARBONATES (Table 28-12)

Cassiterite occurrences in altered carbonates lacking extensive skarn zones are most extensively developed in the trio of important deposits (Renison Bell, Mt. Bischoff and Mt. Cleveland) in north-western Tasmania (Figs. 28-73, 28-74). There, cassiterite is always associated with predominant massive pyrrhotite, to the point of being completely "invisible". All three major deposits and additional small occurrences (e.g. Mt. Razorback; Chapter 7) are hosted by a Cambrian and latest Proterozoic sedimentary and volcanic marine sequence of quartzite, slate, minor pillowed metabasalt greenstone, litharenite, bedded chert, limestone and dolomite. These low-grade metamorphosed but locally strongly deformed and tectonite serpentinite intruded rocks, occur in a broad N.E.-trending synclinorium (Dundas Trough), confined between two Precambrian "highs" (Williams, 1978). A series of late Devonian to lower Carboniferous quartz monzonite and granite plutons, small stocks and dikes with thermal aureoles up to 2.5 km broad, have been emplaced in the supracrustal rocks. About 20 small cassiterite occurrences of the common styles (endocontact greisens, exocontact veins) are associated with the potassic leucogranite phase, and there is little doubt that the tin accumulations in carbonates have the same affiliation, despite occasional statements to the contrary.

Renison Bell, the largest deposit (Patterson et al., 1981; 282 Tt Sn/1%; Fig. 28-74) contains a series of massive pyrrhotite, lesser chalcopyrite and cassiterite sheet (manto)-like orebodies hosted by three horizons of impure dolomite interbedded with Cambrian quartzite, shale and minor volcaniclastics. Lesser discordant orebodies of disseminated, veinlet and bleb pyrrhotite and cassiterite are situated

Table 28-12. Carbonate-hosted non-skarn hydrothermal tin deposits; selected examples

LOCALITY	HOST AND INTRUSIVE UNITS	MINERALIZATION	REFERENCES
Renison Bell, N.W. Tasmania, Australia	Cm_1 shale,siltst.,sandst.,minor tuff,3 impure dolomitic beds; D qtz.monzon.,granite,qtz. porphyry (not in cont.with ore)	set of dolom.-replacing pyrrh., cp.,cassit. manto oreb.,extending from faults; minor discord.orebod. 282 Tt Sn/1%	Patterson et al. (1981)
Mt.Bischoff, Waratah, N.W. Tasmania Australia	Pt_3 or Cm_1 quartzite,slate,minor dolom.interbeds; D, qtz.feldsp.porph. dikes	massive pyrrh.,qtz.,talc,cassit. replace dolom.; dissem.cassit.,pyr. sphal. in altered porphyry dikes; minor qtz.-cassit.veinl.; 80Tt Sn	Groves (1972)
Mt.Cleveland, Luina, N.W. Tasmania, Australia	Cm, low-grade met.argill.,lithàren.,chert,spilite,maf.tuff, limest.; intr.by gabbro,diab. D, granite plut. 4 km away; altered,W,Mo miner.granite dike, at lower mine levels	system of parallel dissem.to mass. pyrrh.,cp.,cassit. lenses in calcsilic.hornf.("chert") and marble, underl. by qtz.,fluor.,wolfr.,Bi, molybd. stockw. in altered porphyry 31.2Tt Sn, 10Tt Cu	Collins (1981)
Kochiu, Yunnan, southern China	Tr limest., shale, underl. by PCm gneiss, schist, phyllite; Cr_3 (85 m.y.) granite cupolas of biot.-tourm.granite, believed floored by a major cryptobatholith	replacem.bodies of hemat.,cassit., ars.,pyr.,sphal.,gal. in limestone; in endocont. apogranite and greisen cont. feldspathite veins, lepidolite, lithionite, qtz. veins; minor vesuv.-wollast. skarns; 1.6 Mt Sn/3.65%	Meng et al. (1937)

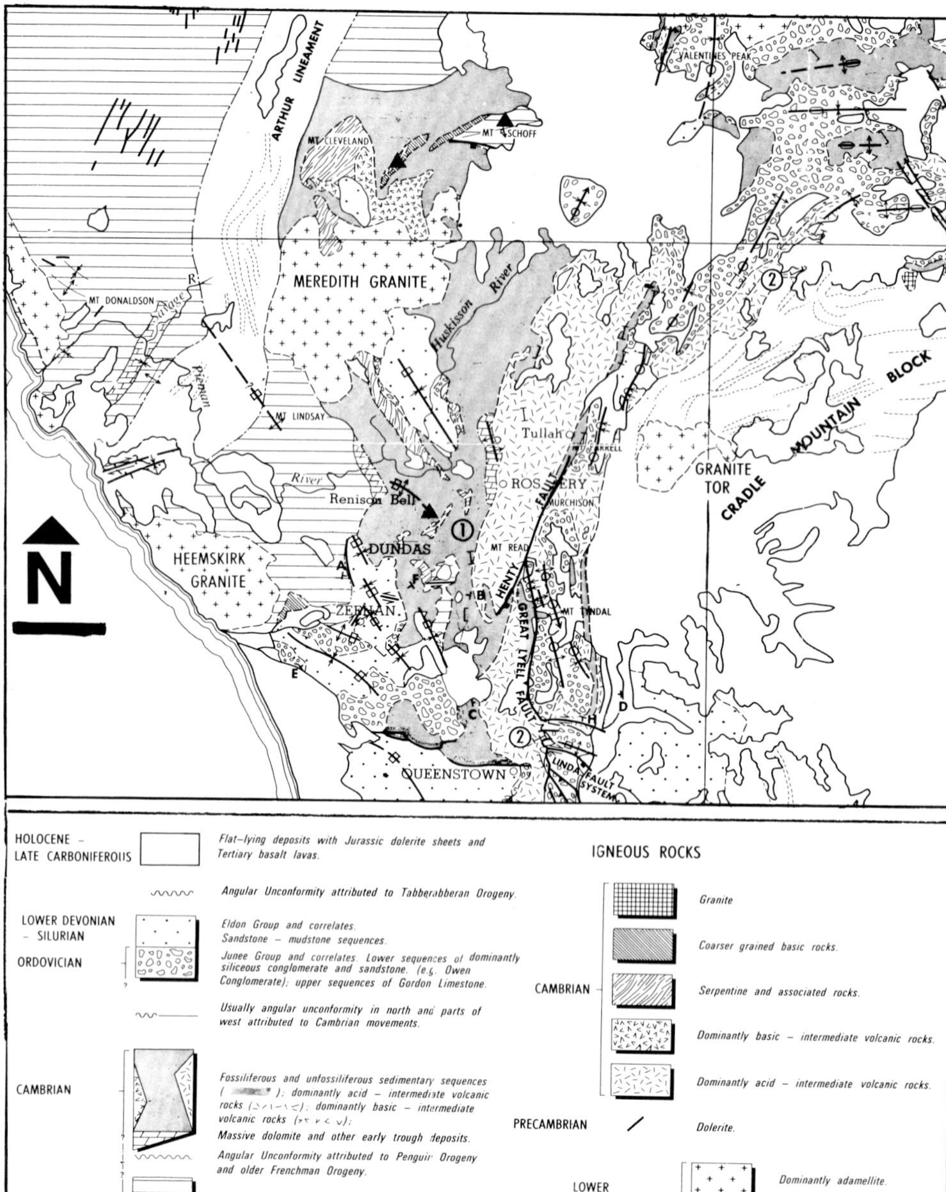

Fig. 28-73. The metalliferous area of N.W. Tasmania, showing the location (by triangles) of the three important sulphide cassiterite replacement deposits in carbonates. From Williams (1978).

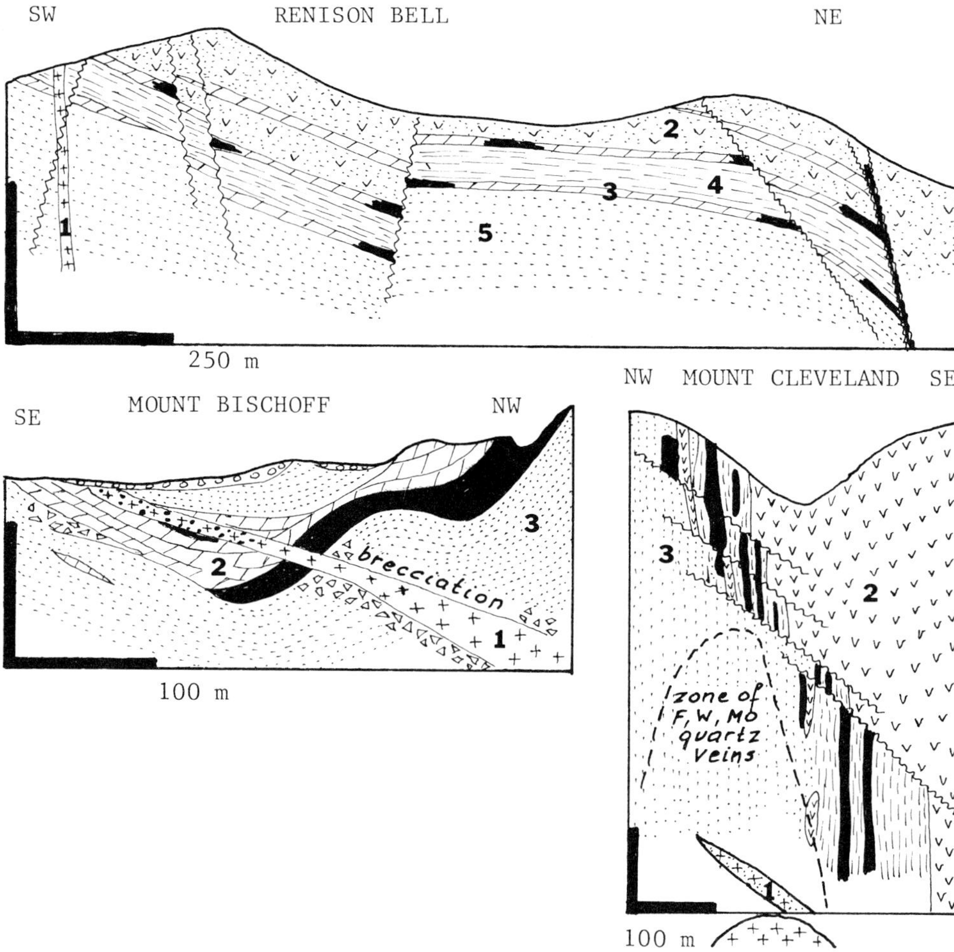

Fig. 28-74. Replacement pyrrhotite-cassiterite deposits in carbonates, N.W. Tasmania.

Mount Cleveland: 1=D altered granite porphyry body (diagrammatic only); 2=Cm greenstone (meta-spilite); 3=Cm meta-argillite, siltstone, shale, graywacke.

Mount Bischoff: 1=D quartz feldspar porphyry dike; 2=Pt_3 or Cm_1 dolomite; 3=quartzite and slate.

Renison Bell: 1=D quartz feldspar porphyry; 2=Cm basaltic volcarenite, siltstone; 3=ferroan dolomite layers; 4=thinly bedded quartzite, siltstone, shale, sandstone; 5=quartzite, sandstone.

From LITHOTHEQUE, after Newnham (1976), Groves (1972), Collins (1981) and 1980-81 property tours.

in talc-altered and coarse-recrystallized dolomite along a fault zone. When a quartzite bed is present in the mineralized sequence, it is often replaced by alternating bands of tourmaline, quartz and pyrrhotite. Small stocks of Devonian biotite granite and quartz monzonite, and several dikes of quartz porphyry including one that is greisenized and contains minor cassiterite, occur in the vicinity and probably represent the concentrator rock suite to Renison Bell. The ore was interpreted as having precipitated from hot stannous chloride complexes at 350° C.

Mount Bischoff, Waratah (Groves, 1972; Figures 28-74, 75) was discovered in 1871 and it produced 60 Tt Sn. Several tens of separate orebodies are arranged at both sides of the contact of a probably Eocambrian dolomite and Devonian altered quartz-feldspar porphyry dikes. The most important orebodies were irregular pyrrhotite, quartz, talc, fine cassiterite masses , replacing dolomite. Next in importance were patches of pervasively disseminated cassiterite with pyrite and marmatite, in argillized and topaz-tourmaline altered quartz-feldspar porphyry endocontacts.

Mount Cleveland in Luina (Collins, 1981; 31.2 Tt Sn; Fig. 28-74) is in Cambrian deep marine ("oceanic") mafic volcanic-sedimentary association (spilitic greenstone, mafic pyroclastics, argillite, chert, litharenite, limestone) of the type described in Chapter 10. The lower greenschist-metamorphosed host unit is in fault contact with ophiolites, and is intruded by diabase and gabbro dikes. A Devonian granite pluton crops out 4 km south of the mine and probably underlies the mine field in depth. The ore consists of a series of vertical to steeply dipping lenses of massive to disseminated pyrrhotite and fine-grained (invisible) cassiterite (and chalcopyrite), peneconcordant with the relic bedding of a calc-silicate hornfels. The ore grades to replacement sulphide masses in a relatively pure former limestone. In the footwall of the peneconcordant sulphide lenses is a stockwork of quartz, fluorite, wolframite, molybdenite, bismuthinite and minor cassiterite veins hosted by a litharenite. In the deepest mine levels, the stockwork enters an altered (silicified) Devonian quartz porphyry dike. The origin of Mt. Cleveland is enigmatic but much less so now than it was in the past, when the presence of a mineralized granite in depth was unknown. The lithophile metals accumulation is hardly "stratiform" or "exhalative" as suggested by some, but the Fe and Cu sulphide orebodies are in the "right" lithological association to make such an origin possible and could be pre-granite.

Fig. 28-75. Mount Bischoff tin deposit, Waratah, Tasmania. Top and centre: general view and one of the workings. The white material is weathering–argillized altered porphyry. The dark rock is altered dolomite with remnants of pyrrhotite–cassiterite replacements.

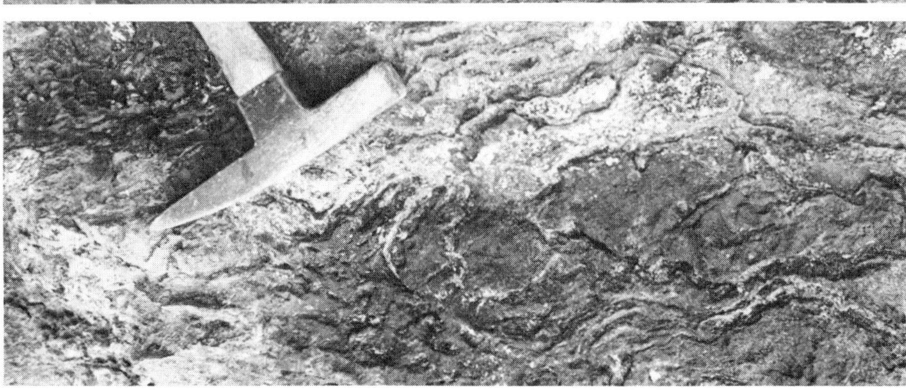

The Kochiu district, Yunnan, southern China, by far the largest carbonate-hosted tin accumulation (1.6-2 Mt Sn/3.65%) has been mined for over 100 years, but there are few modern published data available in European languages. The fragmentary information (Meng et al., 1937; Ikonnikov, 1975) indicates that the district contains complex ore near a late Cretaceous apogranite, complete with albite, lepidolite and protolithionite zones. Known orebodies occur in greisen, in quartz, cassiterite, wolframite veins in the granite and in Precambrian basement metamorphics and in Triassic carbonate. Monotonous Triassic limestones are much in evidence in the area, and they are mineralized at and close to a contact with granite cupolas. A portion of the cassiterite (and also probably Sn-silicates that are not recovered) is situated in a late-stage idocrase-wollastonite skarn, but the bulk come from "sausage-like" and manto replacement orebodies peneconcordant with the limestone bedding.

28.8.5. Relic (pre-granite) Sn(W,Mo) orebodies in granite aureoles

In plutons-affiliated ore fields, the most frequently reported relic mineralization is in the supracrustal rocks, with which it is more or less conformable ("stratiform"). Such mineralization can be preserved almost intact (only isochemically metamorphosed), or suffer remobilization of variable intensity. The cassiterite mineralized quartzites at several localities in Bolivia which are interpreted as being paleoplacers, have already been treated in Chapter 19. More reported examples of relic Sn ores or anomalous enrichments in metamorphosed volcanic-sedimentary supracrustal associations are currently interpreted as former "exhalites" (=hydrothermal, seafloor sediments).

The Gierczyn-Nové Město pod Smrkem (Poland and Czechoslovakia) and Halsbrücke-Bräunsdorf (Erzgebirge) zones are treated in Chapter 29. More examples are known and Plimer (1980) reviewed the problem of "exhalative" precursors to the granite-associated Sn and W deposits. In several instances when cassiterite is present in a conformable massive sulphide or in iron formation horizons, it is epigenetic and superimposed (e.g. Zlatý Potok, Erzgebirge). There is a close analogy with the setting of cassiterite in skarns.

28.9. TUNGSTEN (WOLFRAMITE AND SCHEELITE) VEINS, STOCKWORKS AND DISSEMINATIONS IN NON-CARBONATE ROCKS IN GRANITE AUREOLES

28.9.1. Introduction

The economic tungsten supply comes approximately equally from (1) wolframite and lesser scheelite veins and stockworks in silicate rocks and (2) scheelite skarns in carbonates. The geological difference between tungsten veins and scheelite skarns is more than just the variety of host rocks, mineralogy and style of deposition. In contrast to Cu skarns which, in most cases, are just a "special" wallrock variety of porphyry coppers and Sn-skarns that are often an exocontact extension of greisens, the bulk of W veins and scheelite skarns are mutually exclusive and do not occur together (there are, however, exceptions as, for example, at the Xi Zhoyouang deposit).

The bulk of wolframite and scheelite veins in provinces that also carry tin, are associated with potassic leucogranites rooted in large quartz-monzonite to granite batholiths. The granites are of the "S" (ilmenite) type and the regional geological history is marked by events of fragmentation and reconstitution of an earlier continental crust (doming, initial rifting, collisions). Scheelite skarns, on the other hand, are genetically associated with quartz diorite to quartz monzonite (mostly granodiorite) plutons of continental margin belts, and their parent intrusions show I (magnetite) type tendencies, although this is not always conclusive. No appreciable quantities of tin are associated. There is also a significant contrast in the level of emplacement. Scheelite skarns belong to the "deep" (depth of formation around 12 km) skarns, W veins formed at epizonal and upper mesozonal intrusive levels.

Ferberite and hübnerite (end members of wolframite) vein deposits are not, with some exceptions (e.g. Dzhida, Transbaikalia; Rwanda) members of the Sn-W vein/stockwork provinces, and have a transitional setting.

28.9.2. Wolframite-bearing greisens and veins in W-Sn provinces

This mineralization style is a complete analogue of the cassiterite deposits in greisens described earlier. Cassiterite is usually present as an accessory component in wolframite greisens and complete transitions exist between both. Most tin deposits listed in Table 28-8 produced some tungsten as well. In some systems of granite cupolas with a greisen and exocontact roof sediments with veins, the former preferentially carry Sn, the latter W. In Cornwall this zonality is reversed and most endocontact greisens (e.g. Hemerdon) carry W deposits, whereas the bulk of the tin occurs in veins. Examples of interesting tungsten deposits in Sn-W provinces are briefly reviewed here and in Table 28-13, but the reader should consult the previous sections for data on the provinces.

Hemerdon, located just 11 km N.E. of Plymouth, (Anonymous, 1979b) is the largest W deposit in the metalliferous region of S.W. England. There, a dike-like body of Devonian biotite granite intrudes hornfelsed siltstone, mudstone and mafic volcanics. The granite is partly greisenized, silicified and tourmaline-altered, and pervaded by a stockwork of thin quartz, K-feldspar, hematite veinlets and veins. The veins range in thickness from several millimetres to several centimetres. Wolframite, arsenopyrite and rare cassiterite are scattered in the veins in a most irregular fashion ("few nuggets in some bands, then nothing for some distance") that makes exploration difficult. The outcrop is marked by a kaolinized saprolite down to a

Table 28-13. Wolframite-bearing greisens grading to veins; selected examples

LOCALITY	GEOLOGY	MINERALIZATION	REFERENCES
Hemerdon, Devon, S.W. England	Cb-Pe granite intruded into D hornfelsed siltst.,mudst. and marine metavolcanics	qtz.,microcline,hemat.,wolfr.,arseno-pyr.,cassit. stockwork grading into veins in greiseniz.granite; 61.2Tt W/0.136; 13.05Tt Sn/0.029	Anonymous (1979b)
Baltar deposit, western Spain	Cb-Pe stocks of greiseniz. leucogran.intrus.into coarse biot.and musc.granite	large,low-grade stockw. of quartz, wolfr.,cassit.veinlets; 60Tt W/0.12% 40Tt Sn/0.08%	Min.Magaz. June 1980
Panasqueira, S. Portugal	Cb-Pe gran.to qtz.monzon. greiseniz.cupola,intr.into Pt_4 or Cm hornf.argillite and litharenite	290 m.y.; 200-300 m thick horiz. swarm of qtz.,wolfr.,lesser cassit., cp.,veins at a level of cupola; W miner.postdate granite and greisen; 42Tt W, by-product Cu,Sn	Kelly and Rye (1979)
Akchatau, central Kazakhstan,U.S.S.R.	S_1-Cb_1 sandst.,siltst.,argill.,andes.,rhyol.tuff; Pe biot.qtz.monzon. intr. by leucogran.bodies	clusters of qtz.,musc.,topaz greisen and vein bodies in 4 belts; 300 separate bodies with scatter.wolfr., fluor.,tourm.,molybd.	Shcheglov and Butkevich (1974)
Antonova Gora, East Transbaikal, USSR	J sandst.,shale; J med.grained musc.granite	series of // qtz.,musc.,fluor.,wolf. veins 0.6-0.7 m thick, up to 1 km long, in apical part of gran. stock	ditto
Bukuka, East Transbaikal, U.S.S.R.	J_2 sandst.,shale; J_3 biot.granite stock,dikes	80 qtz.,wolfr.,lesser sphal.,cp.,pyr. gal.,bismuth veins and stockw. in greiseniz.,silicif.,seric.granite	ditto
Dzhida, East Transbaikalia, U.S.S.R.	Cm schist,metavolc.,ophiolites; PZ qtz.dior.,gabbro, granite; Tr syen. Tr-J_1 leucocr.gran.,gran.porph.	140 m.y. qtz.,hübnerite,lesser scheelite and sulphide veins.	ditto

Yugodzyr, S.E. Mongolia	PZ, sandst.,slate; PZ_3-J_1 granite, dikes; 210-220 m. y. aplite,tourm.pegm., greiseniz.gran.porphyry	1)gently dipping qtz.-wolfr. veins in exocont.; 2) moly. dissem.and stockw. in endocont.greisen; dissem. moly, wolfr.,lesser scheel.,helvite,pyr., ars. Est.min. 30Tt W, 50Tt Mo	Marinov et al. (1977)
Iul'tin, Chukotka Pen.,Siberia, U.S.S.R.	Cb_1-Tr_2 schist,marble,gabbro overl.by shale,siltst., sandst.; Cr_1 granod.,qtz. dior.porph.,biot.granite	104 short,en-echelon veins, up to 1.2 km long ore zones; qtz.,musc., lesser albite,fluor.,wolfr.,cassit., sulph. in hornfels. exocontact	Shcheglov and Butkevich (1974)
South Kiangsi W distr.,S.E. China; SELECTED DEPOSITS:	PZ, N.E. trending sandst. and hornf.slate,intr.by J_2 and Cr granitic bathol.	160-180 m.y. mineraliz. assoc. with granites, in endo- and exocontacts. 5.5 Mt W (minor quant. from placers)	Ke-Chin Hsu (1943) Kazanskii (1972)
Sihuashan, near Tayu	biot., musc.-biot. granite, alaskite, intruded to hornfelsed pelites	qtz.,wolfr.,feldsp.,musc.,moly.,cp., in about 200 // fiss. veins N80°W grad.to stockw.,dissem.,in endocont. 891Tt W/0.64%	
Pankushan	D, thick-bedded quartzite, S.W. dipping	E.-W. vertic.fissure veins of qtz., wolfr.,bismuth,arsenop.,zinnwald., scheel. in exo- and endocontact; 112Tt W/1.2%	
Yachishan		fissure qtz.,wolfr.,ars.,bismuth. veins in endocont.; 118Tt W	
Kweimeishan	Cm-Or dark gray quartzite, phyllite interbeds, near granite cupola	closely spaced N20°E qtz.,wolfr. veins grading to sheeted miner.zone; 1,500x500 m; 107Tt W	
Tachishan	Or-S phyllite, argillite, quartzite, diorite dikes	N65-80°W // veins to sheeted zone, qtz.,wolfr.,zinnwald.veins filling joints in tourmaliniz.hornfelses; 52Tt W/2%	

depth of 40m.

In the Erzgebirge, Pechtelsgrün (East Germany; Rösler et al., 1968) is an often quoted cupola wolframite deposit. There, a coarse-grained porphyritic Carboniferous biotite granite is intruded at a depth of 340 m by a younger "inner" leucogranite stock, the apical portion of which is capped by the marginal pegmatite, and greisenized. The tungsten mineralization is located in a N.W.-trending fracture zone up to 8 m thick and 1,200 m long, containing a large amount of narrow quartz, muscovite, feldspar, wolframite, arsenopyrite, etc. veinlets forming a stockwork. At some depth, these veinlets gradually unite to form several discrete veins. The ore zone is contained in the older granite and dissapears upon reaching the inner granite cupola. In the greisen, pyrite is abundant but the W and Mo contents are unimportant.

Panasqueira, Portugal (Kelly and Rye, 1979) has been the most productive western European tungsten mine for some time. There, a large number of near-horizontal parallel fissure veins occur in hornfelsed phyllite at a level of a greisenized granite cupola topped by a silica cap. The veins are, however, younger than the granite and composed of quartz with scattered wolframite and lesser cassiterite and chalcopyrite. The ore originated from NaCl-rich brines well below their critical temperature. The veins cut sharply across the steep foliation of their hosts and there is only an inconspicuous alteration.

In the W(Sn,Mo) province of central Kazakhstan, U.S.S.R. (Shcherba, 1968), wolframite is in the "conventional" style of stockworks and veins in granite endo- and exocontacts (e.g. Akchatau, Bainazar, Boguty). The Bainazar deposit is comparable, in many respectss, with Pechtelsgrün, only the ore band is substantially wider and is hosted by hornfelsed sandstones and slates in the exocontact above a greisenized leucogranite cupola. The parallel but discrete veins in the stockwork (quartz, molybdenite, wolframite) loose identity once they reach the granite cupola, and change into a mineralized fracture stockwork in the zinnwaldite, topaz, quartz greisen. There is an increase in the proportion of molybdenite with depth.

Eastern Transbaikalia and Mongolia contain a large number of W, W-Sn, and W-Mo deposits associated with Jurassic and Cretaceous granites. Yugodzyr, the largest tungsten deposit in Mongolia (Marinov et al., 1977), is in the Nukut-Daban Range in the south-eastern part of the country. There, a stock of lower Jurassic quartz porphyry topped by a small cupola is emplaced in lower Paleozoic hornfelsed slates and sandstones. The cupola as well as a younger phase of granite porphyry dikes are greisenized. The ore comprises two types: (1) disseminated and stockwork molybdenite, lesser wolframite, scheelite, arsenopyrite, etc. assemblage in greisens and (2) several systems of discrete parallel quartz, muscovite, wolframite, beryl, molybdenite (in the exocontact) and quartz, molybdenite, fluorite, ankerite, pyrite, lesser wolframite (in the endocontact) fracture veins. The veins are horizontal to gently dipping and the deposit morphology is strongly reminescent of Panasqueira.

Tumen-Tsogto deposit near Bayan-Oba, E. Mongolia (Marinov et al., 1977) is associated with a middle-upper Jurassic granitic complex intruded into Permian and Triassic continental felsic volcanics, Devonian terrigenous sandstones and limestones, and late Paleozoic metamorphics. Several apogranite cupolas with early albitites and pegmatites (or amazonitic K-metasomatic syenites) contain small quantities of beryl and wolframite. The most important ore from an economic point of view, however, is contained in a lens up to 30 m thick, 150 m long, of faintly banded ? replacement quartz, rimmed by greisen and situated near the contact of fine and coarse granites. Wolframite is the main ore mineral, present in the form of scattered, relatively large, tabular crystals. Additional W and Mo occurrences are situated in greisen zones. In the Buren-Tsogto W deposit, Mongolia, amazonite and albite apogranites are cut by quartz, wolframite, muscovite, beryl, fluorite, scheelite, cassiterite and sulphide veins.

On the Soviet side of the border in Transbaikalia, the Dzhida tungsten field is best known (Shcheglov and Butkevich, 1974; Fig. 28-76). There, 140 m.y. old quartz, hübnerite, scheelite and sulphide veins and stockworks intersect Paleozoic quartz diorite and granodiorite, Triassic syenite and lower Jurassic leucocratic granite and dikes. There is a considerable time gap between the termination of the intrusive activity and the hydrothermal veining. At the Kholtoson deposit, 140 subparallel veins have been recorded, 70 of them economic.

In the Bolivian plutonic tin belt, most cassiterite deposits produced lesser quantities of tungsten, the largest producers having been Chojlla (8 Tt W; Fig. 28-77) and Viloco (35 Tt W, 60 Tt Sn). There is, however, no large wolframite deposit.

The situation in the Tasman orogen of eastern Australia is similar. Mt Carbine, 50 km W.S.W. of Cairns, Queensland, has recently been explored as a large but very low grade deposit (15-25 Mt ore with 0.072% W = 14.4 Tt W; Plumridge, 1975) that gives it the distinction of being probably the world's lowest-grade hydrothermal tungsten deposit considered for production. There, a swarm of nearly vertical parallel quartz, K-feldspar, wolframite, lesser scheelite and minor sulphide veins striking 135°, is hosted by Devonian to Carboniferous hornfelsed slates with minor volcanics, 1 km from a Carboniferous granite contact. The veins represent 9.8% by volume of the ore to be mined in bulk.

28.9.3. Wolframite-bearing greisens and veins, outside of major tin provinces

NAN-LING RANGE (S. KIANGSI) TUNGSTEN PROVINCE, S.E. CHINA

The N. E.-trending Nan-Ling Range in the Kiangsi, Hunan and Kwangtung Provinces (about 200-400 km north from Canton) is estimated to contain about 5.5 Mt of tungsten in several hundred plutonic vein

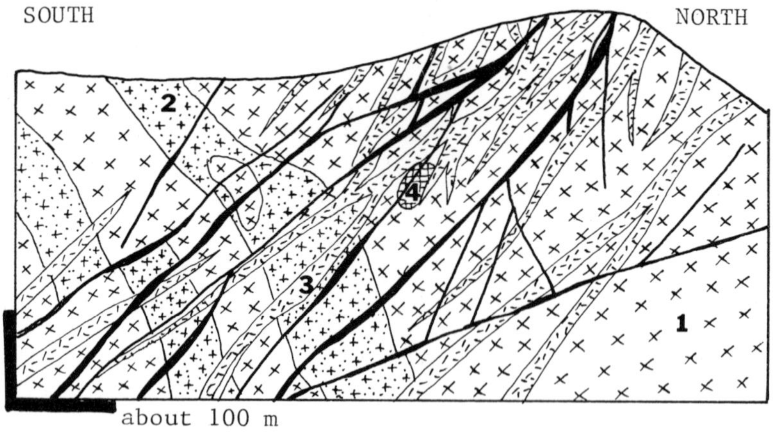

Fig. 28-76. Kholtoson W deposit, Dzhida field, East Transbaikalia, U.S.S.R. 1=PZ_3 quartz diorite; 2=J_1 granite porphyry; 3=J_1 lamprophyre; 4=PZ_1 hornblendite relics; 5=quartz–hübnerite veins. After N. K. Nefedov in Vakhromeev (1961).

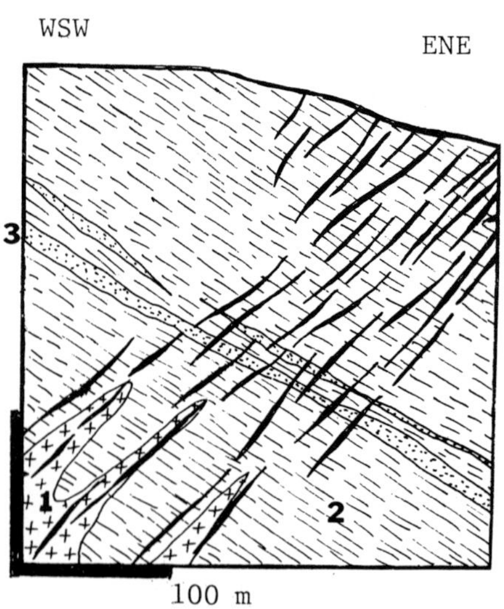

1. 190 m.y. altered leucogranite (partly greisen) in apical portion of a granitic pluton
2. Or, gray slate transferred to knotten schist and biotite hornfels in granite aureole
3. Or quartzite interbeds

BLACK: lenticular quartz veins with scattered wolframite and cassiterite

Fig. 28-77. Chojlla Sn-W mine, Cordillera Real, Bolivia. From LITHOTHEQUE, based on data in Ahlfeld and Schneider-Scherbina (1964) and mine tour, 1977.

occurrences distributed over an area of about 350x150 km (Kazanskii, 1972). Over 90% of this wealth is in the Southern Kiangsi region (36,000 km^2; Fig. 28-78). Geologically, this is a part of the Kuei-Hsiang-Kan orogenic belt. This belt consists of lightly metamorphosed lower Paleozoic marine sandstones and shales marginal to and resting on a Precambrian crystalline basement, continuing here from the Cathaysian Platform in the west. The folded rocks are partly topped by Devonian to Triassic platformic sediments including quartz arenite, shale and limestone and intruded by Jurassic and Cretaceous granitic rocks. Most of the region is probably floored by a large quartz monzonite and biotite granite batholith with a large number of cupolas. Numerous wolframite vein deposits are mostly hosted by the lower Paleozoic detrital sediments in the aureole of Jurassic granites. Further west, in the thick platformic carbonate sequence in Hunan, there are numerous skarn and carbonate replacement Pb-Zn and occasional Sn-W deposits (e.g. Xi Zhoyouang). As expected, modern non-Chinese literature on this province is scarce, and the classical paper by Ke-Chin Hsu (1943) is still widely quoted.

The tungsten deposits are mostly veins in granite (44% of old production) and in the sedimentary exocontact (56% of production), near the apex of granite cupolas. As elsewhere, greisens are common in the endocontact, while tourmalinization is the most common alteration of the sedimentary wallrocks. The veins typically fill systems of parallel E.-W. fissures filling tension joints perpendicularly or at an angle to the metasedimentary bedding. The vein persistence and regularity is one of the peculiarities of the enormous cumulative tungsten concentration. The veins are composed mainly of massive quartz, muscovite, tourmaline, K-feldspar, wolframite, lesser molybdenite, chalcopyrite, arsenopyrite and Bi-minerals. At Chiulungnau, a quartz lens, up to 50 m wide, was emplaced along the granite-phyllite contact and is lined by a greisen on the endocontact side. The quartz contains swarms of small wolframite crystals. A small proportion of tungsten came from "pegmatites", more probably hydrothermal quartz, K-feldspar, fluorite, lithian mica, topaz, wolframite and cassiterite veins that accompany the quartz veins. The richest deposit, Sihuashan (891 Tt W) has dimensions of 2 x 1.2 km and is entirely contained within a small granite stock.

NORTH AMERICA

In the Americas (except Bolivia) granite and greisen associated wolframite deposits are conspicuously scarce. This contrasts with the large and abundant scheelite skarns in the Cordillera. In the Appalachian belt, wolframite veins are represented by the small Burnt Hill deposit in New Brunswick (over 30 short quartz, muscovite, topaz, wolframite veins in hornfelsed lower Paleozoic argillite and quartzite above a Devonian granite cupola), and by the Hamme mineralized zone in the Carolina Slate Belt. The latter locality (Foose et al., 1980;

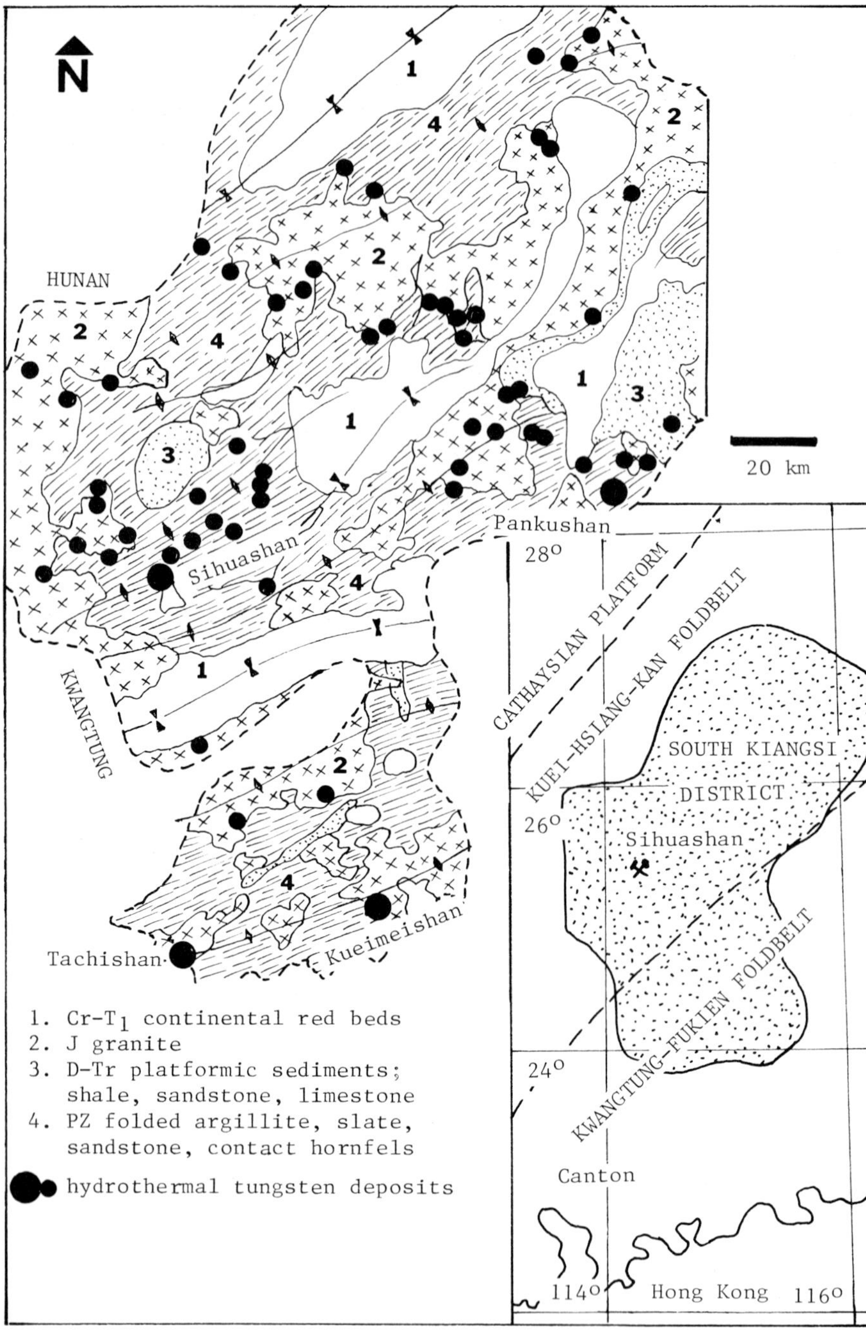

Fig. 28-78. The wolframite vein and stockwork district of South Kiangsi, southern China. After Ke-Chin-Hsu (1943).

7.2 Tt W) has a series of steeply dipping quartz-hübnerite veins, concentrated along a lower Paleozoic granite-slate contact. The N35°E trending vein band is about 3.5 km long and it contains eight en-echelon lodes fringed by a thin alteration envelope of quartz-sericite greisen. The orebody has been folded and locally penetratively deformed along a series of shears.

In the western Cordillera, the Red Rose mine near Hazelton, British Columbia (Sutherland Brown, 1955; 850 t W) is located at the eastern edge of the Coast Batholith (Section 28.2.3.), hosted by a Mesozoic diorite near a lower Tertiary porphyritic granodiorite and felsite dike swarm. The orebody is a vein, 1.2-13 m thick, filling a N35°W striking, 65° S.W. dipping shear, and is composed of a massive, drusy quartz with lesser amounts of feldspar, biotite, hornblende, tourmaline, ferberite, scheelite and chalcopyrite.

The Boulder County (Netherland) tungsten district in Colorado (Lovering and Tweto, 1953; 12 Tt W) has orebodies in Proterozoic gneissic quartz monzonite as well as minor schist and gneiss, associated with middle Tertiary felsite, aplite and "pegmatite" dikes. The orebodies are N.E. and E.N.E.-trending fissure, shear and breccia filling low-temperature (200-300° C, i.e. epithermal) veins. They are composed of ferberite and lesser pyrite, sphalerite, tetrahedrite and adularia scattered in drusy and cherty horn quartz to hematite-pigmented jasper. The veins grade into barren silica-cemented "reefs" or into gold-telluride veins. The wallrock alteration consists of a thin sericitized or silicified selvedge, rapidly passing out into an argillized margin. This is a high-level mineralization probably related to an intrusive stock in depth and substantially different from the "greisen-style" lodes reviewed so far.

SOUTH AMERICA

The Pasto Bueno deposit in the northern Andes of Peru (Landis and Rye, 1974) is more interesting for its geology and geochemistry than for its economic importance. It is a tungsten-polymetallic deposit (monthly concentrate production in the 1970s: 20 t W, 40 t Cu, 45 t Pb, 189 kg Ag), surrounded by a halo of K-silicate (feldspar and biotite), greisen, phyllic, argillic and propylitic alteration. It is associated with a 9.5 m.y. old quartz monzonite stock having I-type characteristics. The ores are in a near-vertical quartz, fluorite, sericite, carbonate, wolframite, tetrahedrite, sphalerite and galena vein system in the endo- and exocontact of the stock. The Pasto Bueno Stock is a separate intrusion located about 50 km N.E. of the main body of the Andean batholith, reviewed in Section 28.2.2.

In Argentina, several hundred small quartz, wolframite, scheelite, bismuthinite, etc. orebodies occur in uplifted blocks of the old crystalline basement east of the main Andean ranges (in Cordillera Frontal, N. Patagonian Massif), but their aggregate tonnage is only about 7.5 Tt W. The largest deposit, the Los Condores mine in the

Sierra de San Luis (Haude and Weber, 1975; 4,592 t W, 173 t Bi) contains three parallel veins, 0.5-2.5 m thick, in phyllite and schist, parallel with pegmatite injections.

28.9.4. Scheelite veins and stockworks in non-carbonate rocks

Accessory scheelite is a common mineral in tin and tungsten veins as well as in some molybdenite and gold deposits. Hydrothermal veins and stockworks in which scheelite is the only or dominant mineral are, however, uncommon and also inconspicuous (Fig. 28-79).

(1) Disseminated and stockwork scheelite in granite endocontacts (in cupolas and dikes) which is reminescent of porphyry coppers, has been reported in several small occurrences such as the Victory mine near Gabbs, Nevada, or Haut-Auxelles in the Vosges Mountains, N.E. France (Fluck et al., 1975). At the latter locality, scheelite is located in an intensely silicified and less sericitized, tourmalinized dike of Carboniferous microgranite, intruded into lower Carboniferous continental volcanics and sediments. A stockwork of quartz or quartz-feldspar veinlets occupies 80% of the rock and a very fine, inconspicuous scheelite is scattered throughout. Up to 30% of sulphides (pyrite, pyrrhotite, arsenopyrite, molybdenite) are present locally.

Stockworks of crosscutting veinlets are gradational into swarms of parallel quartz-scheelite veins, or into few "strong" veins in both endocontact (2) and exocontact (3). This style is represented by the very low grade (0.05-0.07% W) vein swarms in the Permo-Carboniferous granite near Barruecopardo, W. Spain, as well as in the Boguty deposit, S. Kazakhstan.

Boguty (Shcheglov and Butkevich, 1974) is in production and it is probably the largest currently operating scheelite vein/stockwork deposit known. There, a flysch-like Ordovician sequence of dominant sublitharenites and minor slates and limestones was folded and intruded by a composite pluton which is probably of Permo-Carboniferous age. The earlier intrusive phase produced biotite-hornblende quartz monzonite and the later phase albitized and greisenized leucogranites. Early K-feldspar, biotite and tremolite alteration surrounds the intrusion. Later stage scheelite is situated in an N.E.-trending altered and mineralized fracture zone, up to 2.5 km long and 200 m wide, located along the granite/metasedimentary contact, largely on the exocontact side. In the stockwork, several large and relatively persistent veins are accompanied by a complex network of fine veinlets. The veins and veinlets are composed of quartz, muscovite, purite and scheelite. Wolframite, molybdenite, chalcopyrite, galena, pyrite, tourmaline, fluorite and K-feldspar occur in small quantities. Most of the scheelite is located in selvedges, whereas the central portions of thicker veins are visually unmineralized. Within the stockwork, quartz, sericite, chlorite alteration is most common. The Verkhnye Kairakty deposit in Kazakhstan is similar.

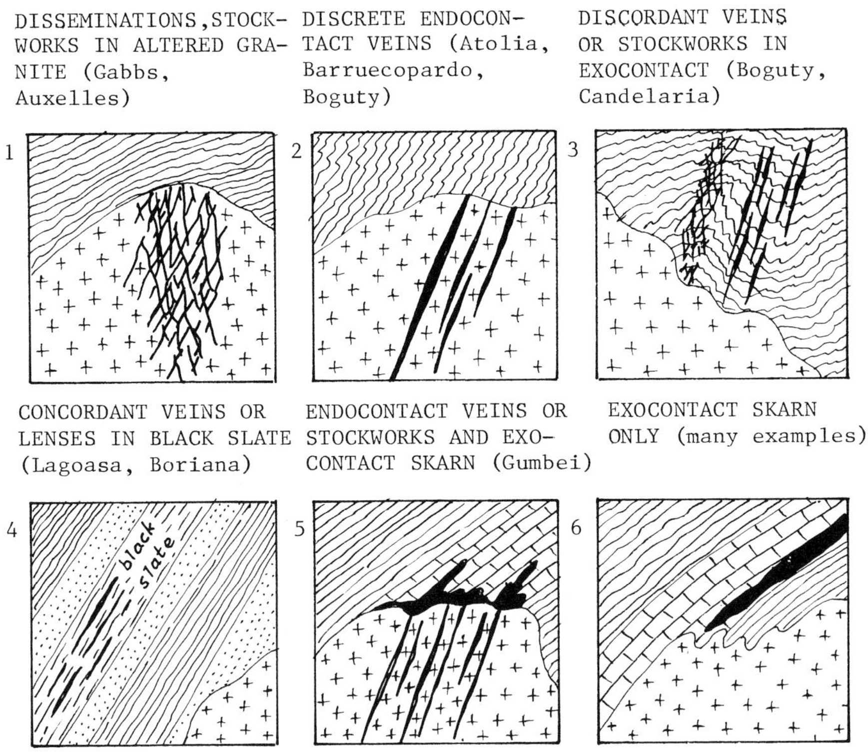

Fig. 28-79. Diagrammatic representation of the varieties of epigenetic hydrothermal scheelite deposits associated with granitic plutons.

The recently discovered Logtung property at Logjam Creek (close to the Yukon-British Columbia border, Canada; Tempelman-Kluit, 1981) is a large, low-grade "porphyry scheelite-Mo" deposit. It contains geological reserves of 162 Mt ore with 0.0952% W and 0,032% Mo (i.e. 152.3 Tt W and 50 Tt Mo). The deposit consists of scheelite and molybdenite disseminated in a stockwork of quartz veins in a Cretaceous quartz monzonite stock, a member of the Seagull group of intrusions.

Orebodies consisting of quartz-scheelite lenses and stringers running parallel to foliation and/or bedding of black metasediments (slates, phyllites) in granite aureoles (style 4) are known in several localities, for example, Lagoasa in Portugal. Such occurrences are generally considered (re)mobilized stratiform tungsten enrichments and these were recently reviewed by Denisenko and Rundkvist (1977). The low-temperature (epithermal) scheelite veins in quartz monzonite near Atolia, California, were reviewed in Chapter 26. When carbonates are available in the exocontact of scheelite-mineralized granites (style 5; e.g. Gumbei deposit in the Urals), both endocontact veins and exocontact scheelite skarns result. In the large scheelite skarn deposits reviewed in the next section, the granite endocontacts frequently contain mineralogical quantities of scheelite, but this is rarely of economic significance.

28.10. SCHEELITE SKARNS

Phanerozoic scheelite skarns contain about 950 Tt W, a significant increase from virtually nothing before World War II. The potential for new discoveries is good, mainly because of the relatively deep-seated origin of this mineralization style. Scheelite skarns have been reviewed recently by Einaudi et al. (1981). From the associational point of view, scheelite skarns (SS) can be subdivided into (1) high-level skarns (hypabyssal or aposkarns, associated with epizonal and upper mesozonal plutons) and (2) deep-seated (abyssal) skarns. Skarns of type (1) are associated with greisenized granite cupolas and scheelite is almost always accompanied by substantially more widespread cassiterite or sulphides. These have been reviewed sufficiently earlier and will not be considered again here.

Skarns of type (2) are the "typical" scheelite skarns, believed to have formed at depths ranging from 5 to 15 km, at the level of lower mesozonal or even katazonal granites. Because of their deep origin, SS now exposed are almost entirely hosted by Triassic and older carbonates and related to Cretaceous and older intrusions. In complex orogenic belts, scheelite skarns often survive in the deepest eroded cores of major batholiths, as in the Sierra Nevada, California. The SS are rarely associated with other metalliferous skarns and hydrothermal deposits except, sometimes, the "hypothermal" gold-quartz lodes and later-stage, superimposed ores. As summarized by Einaudi et al. (1981), the properties of scheelite skarns are as follows:

(a) They are associated with coarse-grained, K-feldspar megacrysts containing porphyritic granodiorite to quartz monzonite stocks and batholiths. Aplites and pegmatites are common; (b) the plutons are largely unfractured, permissively emplaced; (c) endoskarns are narrow and unimpressive, containing diopside, plagioclase, epidote; (d) SS are hosted either by a "black association" of argillaceous carbonates converted to calc-silicate hornfelses alternating with thermally hornfelsed pelite, or by pure marbles; (e) the above rocks are overprinted by infiltration and diffusion metasomatic skarns that are typically zoned. A barren pyroxene, plagioclase, lesser epidote skarn fringed by an outer amphibole zone forms in a hornfels, mineralized garnet-pyroxene skarn with an outer wollastonite-idocrase zone, forms in the marble; (f) SS are often manto-shaped, peneconcordant ("stratiform"), and tend to follow specific stratigraphic units (usually the stratigraphically lowest marble bed if more beds are present); (g) SS are either "reduced", formed in a "black" exocontact association and at a greater depth, or "oxidized", formed in light or hematitic hosts. Reduced skarns contain hedenbergite, almandine-rich garnet, biotite and hornblende. Oxidized skarns contain andradite and epidote; (h) early, anhydrous phases of skarns are usually overprinted by subsequent retrograde hydrous phases with biotite, hornblende, actinolite, epidote, sphene and apatite; (i) early anhydrous skarns are more persistent, and contain low-grade but consistent mineralization of fine grained, high-Mo disseminated scheelite, accompanied by disseminated pyrrhotite and accessory molybdenite. Molybdenite is sometimes a helpful megascopically visible indicator of scheelite presence. Retrograde skarns are patchy, coarser-grained, and contain low-Mo, coarse scheelite with masses of pyrrhotite and chalcopyrite. Rich ore masses are often located at the outer skarn-marble contact, replacing marble.

Fig. 28-80 is an idealized example of a composite "reduced" SS developed in a "black shale"-impure limestone and minor greenstone association, intruded by granodiorite. In this example, the supracrustal association is predominant in terms of volume. In many cases, however, SS crop out as thin screens on the surface of an exposed eroded pluton or form isolated rafts or xenoliths within the "granite" (as in Sierra Nevada, California). In the latter case, the "granite" is predominant and scheelite-mineralized remnants can be expected anywhere within the batholithic body. Commonly, SS occurrences within a deeply eroded batholith form long, discontinuous belts.

Mactung deposit located in the Macmillan Pass region of the N.E. Canadian Cordillera (Dick and Hodgson, 1982; Fig. 28-81) is a major tungsten accumulation (reserve figure is given as either 216 or 479 Tt W/0.9% WO_3). It is hosted by a lower Cambrian to Devonian association of gently dipping hornfelsed micaceous phyllite, slate and limestone, intruded by Cretaceous quartz monzonite stocks. The SS are peneconcordant with the bedding, replacing relatively pure limestone beds interlayered with non-calcareous, commonly graphitic hornfels. Scheelite is concentrated in a hedenbergite-almandine rich garnet skarn, locally retrograded to actinolite-biotite and clinozoisite-plagioclase assemblages. It is accompanied by abundant pyrrhotite, lesser chalcopyrite and rare sphalerite and ferberite. Minor quantities of scheelite are located in quartz veinlets in hornfels and in the quartz monzonite endocontact.

The Cantung deposit (81 Tt W/1.28%; Fig. 28-82) located farther south in a comparable setting is smaller and has three separate orebodies, two of them high-grade. The Pit orebody is in a coarse garnet, diopside, epidote skarn. The East orebody is essentially a banded pyrrhotite mass in marble with much tremolite but few other dark silicates in the matrix. Scheelite and lesser chalcopyrite are disseminated in the pyrrhotite.

The tungsten skarns of the Sierra Nevada, California, a classical region, have recently been summarized by Newberry (1982). There, scheelite occurrences are contained within a N.W.-S.E.-trending belt measuring 250 x 80 km, at or near contacts of late Triassic to middle Cretaceous granodiorite and quartz monzonite plutons and lower

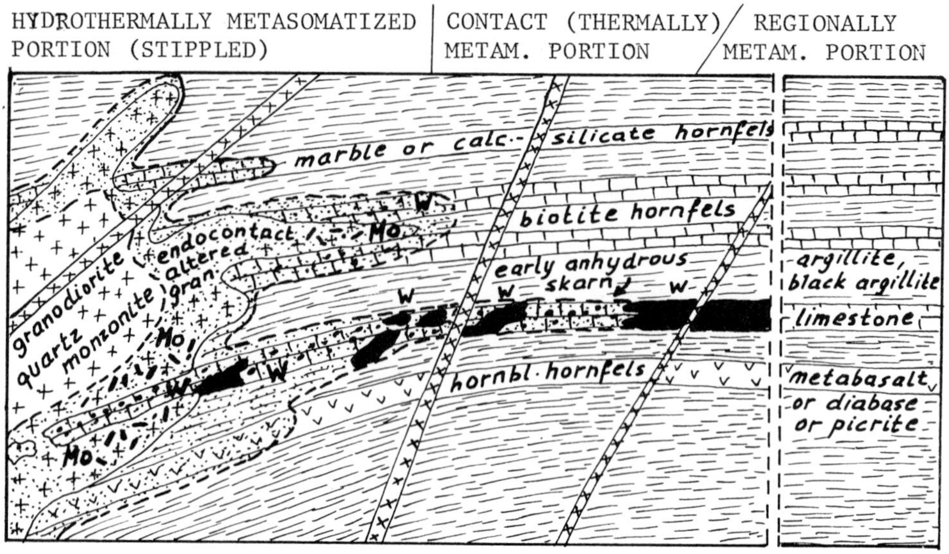

Fig. 28-80. Usual setting of a deep-seated "reduced" scheelite skarn in a "black" carbonate, pelite, lesser greenstone exocontact association. BLACK: late retrogressively hydrated skarn with patches of rich scheelite; W and Mo: erratic, subeconomic disseminations of scheelite and molybdenite. From Laznicka (1984).

Cambrian to upper Triassic carbonate metasediments. The largest deposit, Pine Creek mine near Bishop (27 Tt W/0.4%) is contained in a 1,200 m long septum of Paleozoic metasediments forming rafts and roof pendants within the plutonic rocks. Comparable skarns mined in the Osgood Mountains, one of the Basin and Range blocks in western Nevada, were described in detail in a memoir by Hotz and Willden (1964; Fig. 28-83). There, SS are on the contact of Cambrian to Pennsylvanian metacarbonates with lower Cretaceous granodiorites.

Salau in the French Pyrenees (Derre et al, 1980) is a zone of discontinuous skarn lenses in Ordovician marble, about 1,600 m long, in the roof of what is probably a Carboniferous granodiorite. The scheelite is disseminated in the central portion of pyrrhotite masses enveloped by a scheelite-free pyrrhotite. Pyrrhotite, in turn, is hosted by a diopside-idocrase skarn, particularly by pyroxene-biotite bands within it. Stratabound Pb-Zn occurrences are situated in the same sedimentary horizon at a greater distance from the contact.

Tyrny Auz, located at the northern slopes of the Great Caucasus Range, U.S.S.R. (Pek et al., 1970; Pokalov, 1974; Fig. 28-84), has been the most significant Soviet W and Mo deposit of the "early generation" of discoveries. This is actually a geologically complex polygenetic and multistage field, that departs considerably from the

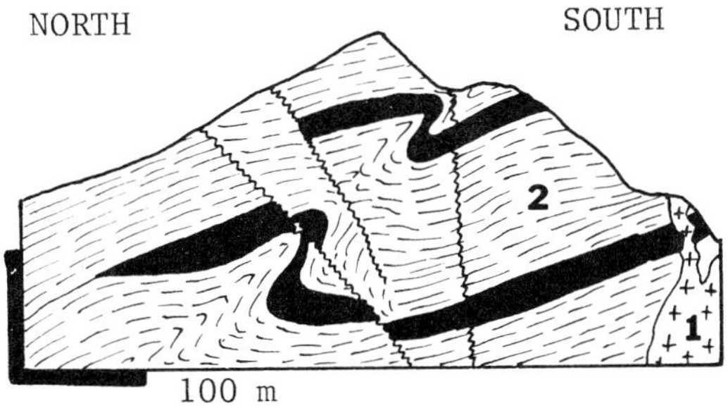

1. Cr (89 m.y.) biotite quartz monzonite
2. Cm-D hornfelsed metasediments (biotite hornfels, black graphitic hornfels, minor calc-silicate hornfels

BLACK: scheelite skarn

Fig. 28-81. Mactung mine, Macmillan Pass, N.W.T., Canada. From LITHOTHEQUE, slightly modified after Dick and Hodgson (1982).

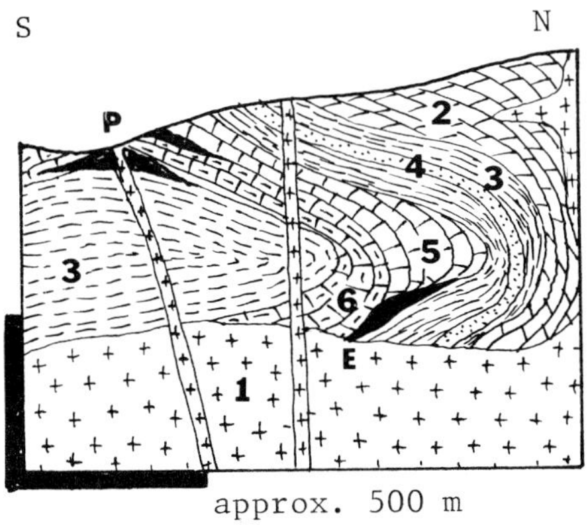

1. Cr granodiorite, qtz.monzonite, dikes and a stock
2. Cm dolomite
3. contact hornfelsed argillite
4. quartzite
5. gray and white banded limestone marble
6. spotted calc-silicate hornfels

BLACK: scheelite skarn orebodies P=Pit ore body; E=E orebody

Fig. 28-82. Cantung scheelite deposit, Tungsten, N.W.T., Canada. From LITHOTHEQUE, based on data in Dick and Hodgson (1982) and N. Cawthorne, guided tour, 1975.

"typical" model of SS shown earlier. The stratigraphic sequence there starts with a Proterozoic or a lower Paleozoic basement, overlain successively by early Devonian shale and limestone, lower Carboniferous ophiolites and bimodal (spilite-keratophyre) volcanic-sedimentary association. This is topped by late Carboniferous and Jurassic continental sediments. These were intruded by a series of granitic intrusions ranging in age from Jurassic to Pliocene, and emplaced at various levels ranging from a probably

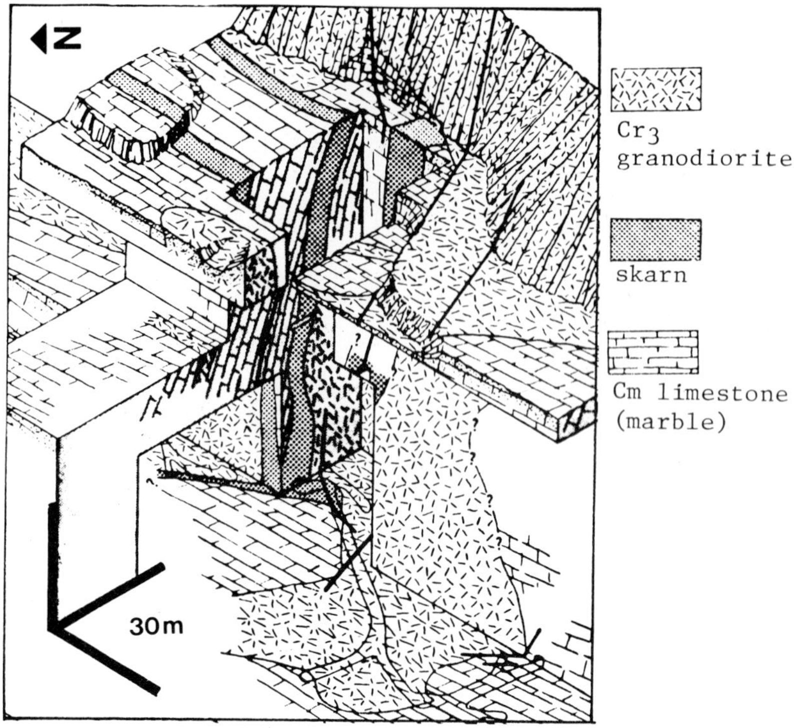

Fig. 28-83. Pacific mine scheelite skarn, Osgood Mts., Nevada. From Hotz and Willden (1964).

middle mesozone to surficial volcanic extrusions.

The two styles that are economically the most important at Tyrny Auz are scheelite skarns and molybdenite stockworks. The latter are younger, superimposed on the former. The principal scheelite orebodies are in the roof of a probably Cretaceous porphyritic biotite granite intruded by a younger leucogranite. Scheelite skarns occur along the contact of hornfelsed pelites and marbles. Scheelite and minor molybdenite are disseminated in garnet, idocrase and wollastonite skarns formed from relatively pure limestones. Molybdenite is concentrated in quartz-molybdenite veins and veinlets, forming a dense stockwork superimposed on the biotite hornfels adjacent to the SS, and partly on the skarn itself. A variety of late-stage veins with sulphides of base metals carrying Au,Ag, and Sn appear to be contemporary with a late-stage fracture-controlled retrogressive skarn assemblage, containing stanniferoue ilvaite.

Sangdong, the best known ore deposit of Korea (Farrar et al., 1978; 105 Tt W/0.56%) is in the popular literature headlined as a major SS, but the mineralization is much more complex and controversial.

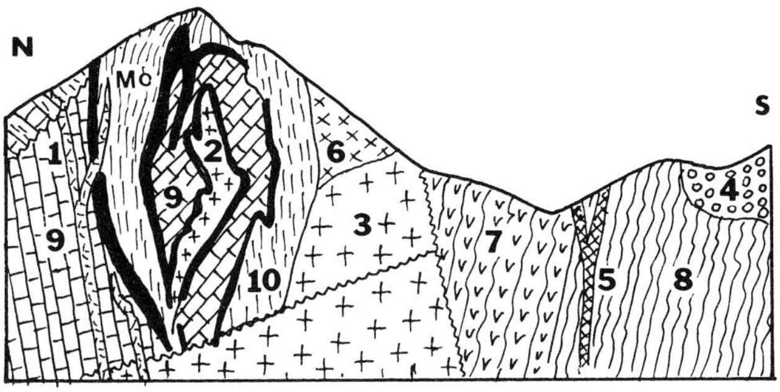

1. Cr-P1 (80-2 m.y.) rhyolite dikes
2. leucogranite
3. porphyritic biot. granite
4. J_1 conglomerate
5. J_1 peridotite
6. Cb? qtz.diorite
7. Cb_1 ophiolites, spilite-keratoph.
8. Cb_1 phyllite
9. D, marble; D, biotite hornfels

BLACK: scheelite-Mo skarn; Mo=molybdenite stockwork

Fig. 28-84. Tyrny-Auz, N. Caucasus, U.S.S.R., scheelite-Mo skarn and Mo stockwork. After Pek et al. (1970).

The principal orebody is a narrow, persistent zoned stratabound horizon 3.5-5 m thick and 1.5 km long, a member of a lower Cambrian alternating shale, sandstone and impure limestone unit. The central zone in this horizon has a quartz, biotite, muscovite assemblage, grading outward to hornblende-quartz, minor biotite and ultimately to diopside-garnet skarn. The ore consists of closely spaced quartz-scheelite and quartz, scheelite, sulphide veinlets in the quartz-mica core, that has a grade of 1.5-2.5% WO_3. This drops to 0.3-1.5% WO_3 in the hornblende-quartz zone, and to sub-economic contents in the skarn. Farrar et al. (1978) interpreted the mineralization as upper Cretaceous, epigenetic, introduced from outside but 20 m.y. later than the age that corresponds to the closest known granitic stock. Alternative interpretations consider the deposit as stratiform (metamorphosed and partly remobilized hydrothermal sediment); epigenetically mineralized shear zone and other. Discordant tension fractures filled by earlier quartz, scheelite, molybdenite and later quartz, bismuthinite, tetradymite, chalcopyrite, etc., are widespread in the mine field and contribute important quantities of the by-product bismuth.

King Island tungsten field off Tasmania is the most important of the current Australian W producers (71.5 Tt W), and an example of a "completely zoned" SS recently described by Kwak and Tan (1981). The largest No.1 and Dolphin orebodies form a system of mineralized lenses, broadly conformable within two stratigraphic horizons in a contact metasomatized assemblage of Cambrian mafic volcanics, argillites, carbonates and what is possibly a tillite, in the contact aureole of a lower Carboniferous quartz monzonite intrusion. The total length of the E.-W.-trending zone is about 1.5 km, the width is about 50 m. The grade distribution varies and within the ore zone several barren "horses" occur. The main ore mineral, Mo-rich scheelite, is associated with abundant pyrrhotite and minor molybdenite and chalcopyrite. The scheelite has the form of fine to coarse disseminated and erratically scattered crystals in andradite, andradite-hedenbergite, and grossularite-calcite skarn. Less frequently, coarse crystalline scheelite forms masses and grains along crosscutting calcite veinlets.

Although alternative genetic interpretations of the King Island SS have been offered (e.g. remobilized stratiform "exhalite"), the epigenetic, granite-related postmagmatic nature of the ore is demonstrated beyond doubt by the crosscutting relationships and sharp contacts within the host marble beds. The lithologic peculiarities of the exocontact association have already been discussed in Chapter 17.

Table 28-14 gives a brief summary of the major SS deposits of the world.

28.10. STOCKWORK MOLYBDENITE DEPOSITS

28.10.1. General

Stockwork molybdenite deposits (SMD) in high-level intrusions represent some 7 Mt of contained economic (or nearly economic) Mo at present. They stand about half way between the porphyry Cu-Mo deposits (that represent some 8 Mt of by-product Mo) and stockwork W-Sn(Mo) deposits (some 200 Tt of by-product Mo), in many aspects of geotectonic setting, petrochemical affiliation, ore mineralogy, etc. The North American Cordillera and, in particular, a small segment of the Colorado Rocky Mountains is a "home" for the majority of stockwork Mo deposits, so it is no wonder that most recent descriptions, systematics and genetic models regarding these deposits originated there (Clark, 1972; Soregaroli and Sutherland Brown, 1976; White et al., 1981; Mutschler et al., 1981; Westra and Keith, 1981).

In North America at least, it is possible to subdivide the SMD into two subcategories, showing a good deal of mutual independence in character and setting. The "granite molybdenite systems" (Mutschler et al., 1981) or "Climax-type" deposits (White et al., 1981) are on one side, genetically affiliated to "true" granitic magma. The "granodiorite (or quartz monzonite) systems" are on the other. The former, although a unique class, have generally high trace or accessory contents of W and Sn (recovered in Climax), so they are on the more "sialic" side of the porphyry Cu-stockwork Mo-cupola Sn sequence. The "granodiorite" Mo stockworks, on the other hand, are gradational into porphyry coppers. Some Mo deposits, particularly those outside North America and of pre-Tertiary age, appear transitional and there is not enough data to determine clearly their affinities. In multiphase intrusive complexes the older phases may be granodiorites to quartz monzonites and the youngest phases granites, so the classifier has a choice of placing the Mo occurrences into either category. The B.C. Moly deposit in the Alice Arm district placed, in the literature, in the granodiorite-quartz monzonite class, is associated with true granite.

28.10.2. Granite-associated stockwork molybdenite deposits

These deposits (Table 28-15) contain a greater share of Mo than the "granodiorite molybdenums" (5 Mt against 2 Mt). Westra and Keith (1981) characterized the "granite Mo stockworks" as associated with alkali-calcic, metaluminous to peraluminous granitic differentiates. Their source plutons contain from 25 to over 250 ppm Nb, 200 to 800 ppm Rb, less than 125 ppm Sr, less than 0.2% TiO_2, and are enriched in F,Sn,W and Mo. Molybdenite and pyrite are the main ore minerals, minor W is present as wolframite or hübnerite, tin is in cassiterite, and monazite is relatively common. Quartz and sometimes K-feldspar are the dominant gangue minerals, but fluorite and topaz occur occasionally. K-feldspar metasomatism and silicification are intense in the core of mineralization systems. The further proposed subdivision of the "granite Mo stockworks" into a "Climax subtype" and "Questa subtype" based on the degree of differentiation of the parent magma, is not recommended for global empirical purposes, because of the transitionality. Molybdenite stockworks hosted by peralkaline or alkaline (alkalic) granites and syenites (or set in alkaline provinces) are treated separately in Chapter 33.

"Granite molybdenite deposits" are most common in tectono-magmatically rejuvenated old sialic blocks ("terrains of autonomous activation" of the Soviet writers). These may be positioned on the continental fringe of broad marginal orogens (e.g. the Colorado Rocky Mountains), or in an intracratonic setting having a long history of doming, block faulting and rifting (East Transbaikalia). The difference between the two above settings is, at best, quantitative anyway. Comparable regions in the Fore-Cordilleran Ranges in Argentina, the "diwa" (activated) terrains in S.E. China and elsewhere, are the probable sites of future stockwork Mo discoveries.

Most of the Colorado stockwork Mo deposits are hosted by high-level intrusions of high silica rhyolite and granite porphyry, having the form of intrusive domes, cylindrical stocks and less frequently of ring dikes, radial dikes, breccias, etc. Many are positioned as protuberances on the "back" of a buried regional batholith, much the same as in the cassiterite-bearing systems. Multiple intrusions are even more common here. The host (roof) rocks into which the intrusions were emplaced, are mostly old high- to medium-grade metamorphics (gneisses, migmatites, amphibolites, as in Colorado). Some are interrupted by tectonic grabens filled by continental felsic volcanics and volcaniclastics, broadly coeval with the mineralized intrusions. The contact metamorphic alteration affecting the metamorphics can be virtually undetectable. Tectonic control (grabens, horsts, lineaments, block faults and their intersections) is important. The "prime time" of exposure of the hypabyssal mineralized complexes is Miocene to Cretaceous.

Table 28-14. Scheelite skarns in Phanerozoic carbonates, selected example localities

LOCALITY	HOST AND INTRUSIVE UNIT	MINERALIZATION	REFERENCES
Macmillan Pass (Mactung), N.W.T., Canada	Cm_1-D altern.marble and hornf. often black argill.,loc.brecc. 80-87 m.y. qtz. monzonite stock	scheel.,pyrrh.,lesser cp.,sphal., ferber. forming lenses in a penecon. horiz. of hedenb.,garnet skarn; 479Tt W/0.76%	Dick and Hodgson (1982)
Cantung, Tungsten, N.W.T., Canada	Cm_1 interb.hornf.argill.,lim., dolom.,minor quartzite, greens. 92 m.y. granod. and qtz. monz. stock	3 ore lenses cont. dissem.scheel. two are in garn.,diops.,epidote skarn, one is mass.pyrrhot. in tremol. skarn and marble away from cont.; 80.9Tt W/1.28%	ditto
Salmo, S.British Columbia, Canada	Cm hornfelsed black argill., gray limestone; J_3 qtz.monzonite, aplitic granite	hedenb.,garn.,biot.,hornbl.skarn lenses with pyrrh.,cp.,scheel., gal.-sphal. replac. in marble outside of contact; 2,439Tt W/0.47%	Little (1959)
Osgood Mts., W. Nevada, U.S.A.	Cm-Pe hornf. limest.,shale, chert, quartzite, interm. to mafic volc.; 69 m.y. granod. with qtz.dior.border facies	dissem.scheel.in garn.-epid.skarn at immed.granod./marble contact; irreg.to tabul.oreb.,minor scheel. in granod. 5,040t W	Hotz and Willden (1964)
Pine Creek mine, Sierra Nevada, California,U.S.A.	PZ_3 hornf.argill.,limest., quartzite, rafts and roof pendants in 92 m.y. porph. qtz. monzon. with pegmatite pods	contact lenses and sheets of garnet,idocr.,wollast.,hornbl.skarn with dissem. scheel.,cp.,moly., magn.,pyr. 24Tt W/0.4%	Newberry (1982)
Salau, Pyrenees, France	Or, hornf.marble, shale, sandstone; Cb? granod. to qtz.diorite at margin	lenses and columns of mass.pyrrh. with dissem.scheel.in diops.-idocrase skarn and pyrox.-biot.bands 10.15Tt W/1-1.6%	Derré et al. (1980)

Uludağ mine, Bursa distr., N.W. Turkey	Pt_3? schist, marble, amphibolite; T? granodiorite	scheel. with pyrrhot. in diops.-garnet skarn, peneconc. lenses along strat. horiz.; minor quartz-wolfr. veins; 28Tt W/0.28%	Schumacher (1956)
Tyrny Auz, N. Caucasus, U.S.S.R.	D hornf. shale, limest., quartzite; Cr porph. biot. gran. plut., leucogran. stocks, followed by Cr-Pl rhyol. porph. dikes	garn., idocr., wollast. skarn at marble cont. carry scheel. and lesser moly.; superimp. qtz.-moly. stockw. and late stage Pb, Zn, Sn, Au, Ag veins. Est. 70Tt W, 50Tt Mo	Pokalov (1974) Pek et al. (1970)
Chorukh-Dairon, Tadzhikistan, U.S.S.R.	D_3-Cb_1 limest. xenol. and rafts in Cb qtz. dior., granod., gran. porph., aplite, pegm. pluton	2 km ore zone, vein and layer-like retrogr. scapol. skarns repl. garn., pyrox., amphib., epid., scheel. endo- and exoskarn; Est. 20Tt W	Shcheglov and Butkevich (1974)
Sangdong, South Korea	Cm, hornf. shale, sandst., impure limest.; Cr qtz. monzon. intr. 4 km away	peneconc. zoned horiz. in metased. miner. by qtz., scheel., sulph. veinlets in its centr. zone (qtz., biot., musc.), fringed by skarn. Interpr. as hydrotherm. replacem. 105Tt W/0.56%, 9Tt Bi, 6.5Tt Mo	Farrar et al. (1978)
King Island off Tasmania, Australia	Cm, hornf. argill., limest., mafic metavolc., breccia (tillite ?); Cb_1 qtz. monzonite	Mo-rich scheel. dissem. in several lenses of andrad.-pyrox. and grossul.-calcite skarn along two stratigr. horizons; 71.5Mt W	Kwak and Tan (1981)

Table 28-15. Stockwork molybdenum deposits, 1: granite affiliation ("Climax-type")

LOCALITY	HOST AND INTRUSIVE UNIT	MINERALIZATION	REFERENCES
Climax, Colorado, U.S.A.	Pt_{1-2} biot.schist to gneiss, qtz.monzon.; 33-18 m.y. multiple intrus.,coaxially stacked conical gran.to rhyol.porph.	3 stacked annular stockw.oreb.of qtz. moly.,pyr.,topped by a tungsten (hübnerite) zone, in exocontact above intrus.apex; 2Mt (or 4.8Mt)Mo/0.33%	Wallace et al.(1968)
Red Mountain (Henderson and Urad deposits), Colorado,U.S.A.	Pt_{1-2} qtz.monzonite; 28-23 m.y. compos.high-level multiph.intrus.suite of granite + rhyol.porph.,minor intrus.breccia	small Urad oreb.is in an early stage subvolc.porph.plug at surface; Henderson is a large Climax-like blind stockw.in depth, in K-feldsp.altered porphyries; 1.5Mt Mo/0.3%	White et al. (1981)
Mt.Emmons-Redwell Basin,Colorado, U.S.A.	MZ sandst.,shale; T, hornf.sedim.,rhyolite; 17 m.y. rhyol. porph.,felsite, ign.breccia	2 oreb.; circular,inverted teacup-shaped oreb.,760 m diam.,90 m thick (Mt.Emmons); moly. stockw.below an intrus.brecc.pipe (Redwell Bs.) Mt.Em. only: 409Tt Mo/0.264%	White et al. (1981)
Questa, New Mexico, U.S.A.	Pt gneiss, gran.,pegm.,quartz. T_2 rhyol.,andes.,latite,sedim. 23 m.y. hornbl.and biot.gran., biot. leucogranite	pyr.,moly.,qtz.veinlets,fract.coat., fiss.veins in K-feldsp. to biot.-seric.alter.arcuate and linear shell along granite-andes.cont. 250Tt Mo/ 0.11%	Clark (1972)
East Kounrad, N. Lake Balkhash, U.S.S.R.	PZ_1 sedim.,volcs.;D_{1-2} silic. volc.,granod.,qtz.diorite Pe,coarse leucogran.,aplite, pre-ore apogranite	W.N.W. system of parallel veins grading into stockw.;qtz.,moly.; qtz., wolfr.; qtz. mica greisen. Estim. min. 70Tt Mo	Pokalov (1974)
Bugdaya, East Transbaikalia, U.S.S.R.	PZ granite, granosyen.; J_{1-2} shale, sandst.; J_3 continent. volc.,granod.,qtz.dior.; J_3 leucogran.,qtz.porph.,apl.	circular stockw.of qtz.,moly.,pyr., Mo-scheelite, fluorite in sericitiz. granite enveloping barren silicif. core. Est.min. 40Tt Mo	Pokalov (1974)

Deeper-seated (plutonic) equivalents of the Colorado cylindrical mineralized intrusions probably exist in the East Transbaikalia and central Kazakhstan Mo provinces, but are often difficult to distinguish from the granodiorite-affiliated Mo stockworks. The plutonic Mo deposits tend to be linear or rectangular in form, and densely veined stockworks are usually substituted by bands of closely spaced veins. Mo stockworks are, as a rule, very proximal to their concentrator intrusion, being in their apical portions or, more frequently, immediately above the apex in the above-intrusion roof. In single phase intrusions, the above-intrusion orebodies are in metamorphics, in basement granites or in felsic volcanics. In multiple intrusion/mineralization systems such as Climax, the top orebody is in the roof rocks and the lower orebodies are in the earlier intrusions. The very intensive silicification and K-feldspar alteration associated with orebodies often obliterates the former intrusion/roof contact, so that it may appear that the entire orebody is in the intrusive endocontact.

Most of the known Mo stockwork deposits occur alone (no other ores are directly associated), although there are exceptions. At Red Mountain, Colorado, embryonal epithermal Mn,Pb,Zn,Ag mineralization is to be found in the late stage porphyries above the Henderson and Urad stockworks. In Tyrny-Auz (Greater Caucasus; Section 28.10), Mo stockworks are superimposed on an earlier scheelite skarn and adjacent hornfelses, in the exocontact of a granite stock. A stockwork Mo deposit is postulated (although not yet proven) under the U(Mo) veins near Marysvale, Utah. Pb-Zn veins and replacements as well as gold deposits may be broadly contemporary with stockwork molybdenum deposits and be present in the same metallogenic belts, as in the Colorado Rocky Mts.

Supergene alteration and its zoning over stockwork Mo deposits is poorly developed and imperfectly known. To a considerable degree this is the result of the setting of the major deposits which is in recently glaciated terrains, from which most of the fossil weathering crusts have been stripped away. The orebody outcrops at Climax, Questa, etc. are marked by fracture coatings of the bright yellow ferrimolybdite, and portion of the released Mo is in jarosite and goethite coatings and infiltrations. No equivalents to the zone of sulphide enrichment over porphyry coppers have been reported.

<u>Climax, Colorado,</u> is the largest, richest and oldest known representative of this clas (Wallace et al., 1968; 2 Mt Mo/0.33%; Fig. 28-85). It is a complex of three coaxial Oligocene intrusions of granite and rhyolite porphyry, emplaced into Proterozoic biotite schist and quartz monzonite. The intrusions are arranged in such a way that each progressively younger intrusion is emplaced at a lower level into the roof of an older intrusion. Intrusive dikes corresponding to each phase as well as the mineralization and alteration are in the roof above each intrusive phase. In Climax, there are two separate, arcuate, concave downward annular ore shells capping the apex of the two lower (younger) intrusive stocks, and an erosional remnant of a third (upper) shell. Taken together, these

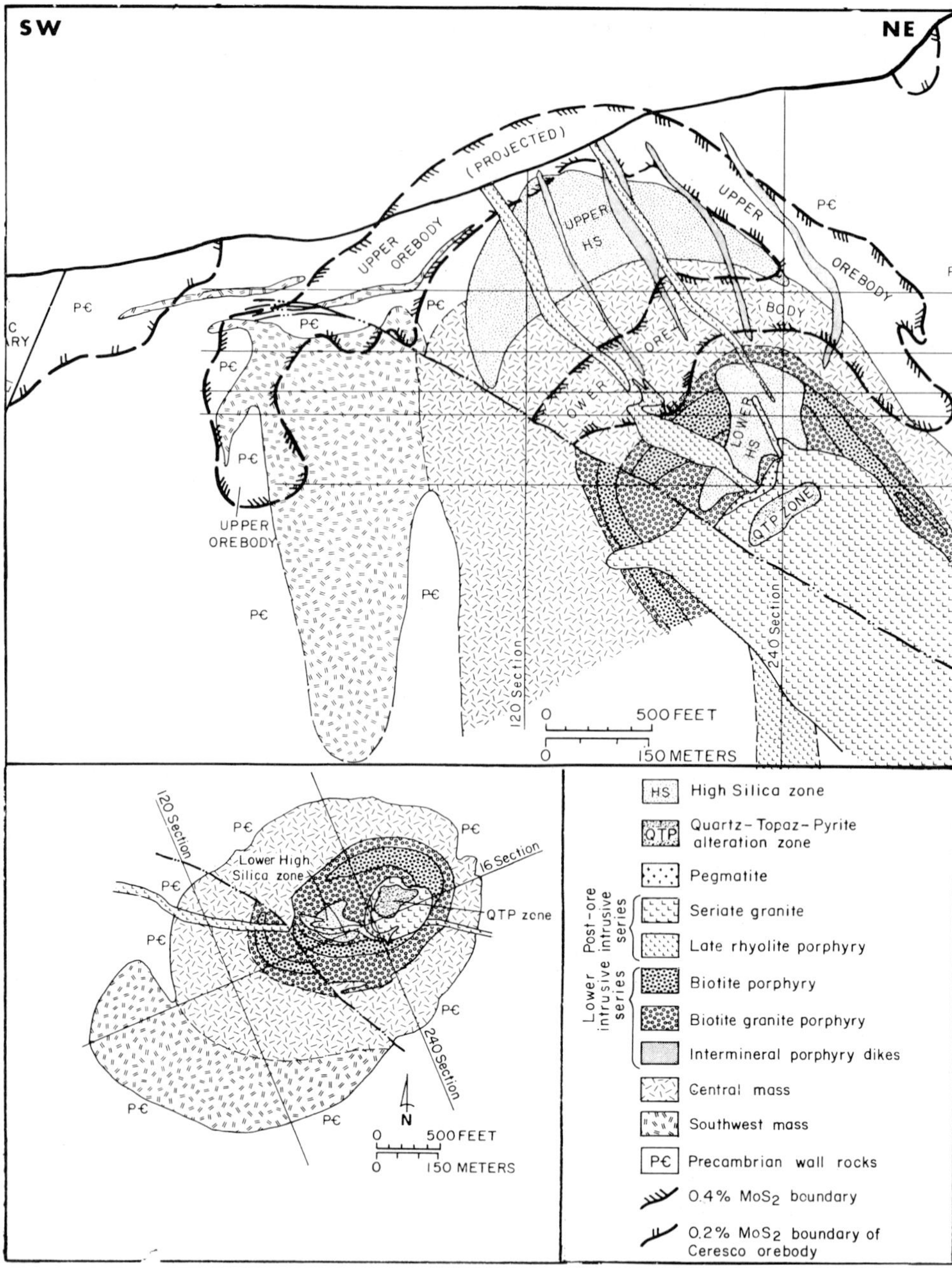

Fig. 28-85. Climax Mo deposit, Colorado; generalized geology and ore zones in section and a map. From White et al. (1981), courtesy of Economic Geology.

orebodies form an inverted hollow cone with upper and lower diameters of 1,400 and 750 m, respectively, and a height of about 450 m.

An orebody is composed of a stockwork of crisscrossing quartz, molybdenite, pyrite veinlets less than 3 mm thick, hosted by a silicified and K-feldspar altered beige granite or rhyolite porphyry in the centre. Quartz, sericite, pyrite alteration marks the outer fringe. Each molybdenite orebody is accompanied by a tungsten zone located above, from which the by-product hübnerite and cassiterite are recovered. In the pre-Quaternary relict oxidation zone at Climax, ferrimolybdite, Mo-"limonite" and Mo-jarosite are present in place of molybdenite, without a marked enrichment or impoverishment in Mo.

The Urad-Henderson twin deposits (White et al., 1981; Fig. 28-86) are located on and under Red Mountain, a short distance N.E. of Climax. Urad is a small (12 Mt ore with 0.38% MoS_2) orebody discovered in an outcrop. Henderson is a large (272 Mt ore with 0.49% MoS_2), blind orebody discovered in depth, 900 m under the surface. Both deposits are members of a multistage late Oligocene probably vented system (probably a stratovolcano) emplaced into the wallrock of a Proterozoic quartz monzonite. The system consists of a downward-widening composite stock of rhyolite, granite and aplite porphyries, many of which crop out at the Red Mountain summit. The Urad orebody is located in an early subvolcanic porphyry plug, dike swarm and intrusive breccia, and it is a relic of an initially larger body destroyed by intrusion of the Red Mountain porphyry. The latter is altered and impregnated or veined by rhodochrosite.

The deep-seated Henderson Mo deposit is similar in shape to the lower Climax orebodies, is younger than Urad and draped over a four-phase mass of a granite porphyry (Primos porphyry) as well as a cupola of Henderson Granite. The Mo stockwork contains much pure molybdenite coating, without quartz, hairline fractures, and also molybdenite flakes in quartz, pyrite and fluorite veinlets. Greater portion of the orebody coincides with K-feldspar alteration zone, but occurs also in silicified rocks. Among the unusual alteration zones at Henderson are topaz-magnetite, greisen, and spessartite garnet zones. The latter is not a skarn, but an associate of a late-stage hydrothermal galena, sphalerite and rhodochrosite suite.

Two Permian "granite Mo stockworks" in central Kazakhstan, Koktenkol' and East Kounrad (Pokalov, 1974), differ from the Colorado deposits by their predictably deeper erosional level and widespread greisenization. Koktenkol' is above a large granite cupola rooted in a more extensive buried pluton. The quartz-molybdenite stockwork is elongated parallel with a N.N.W.-trending shear zone, and wolframite is locally abundant in the quartz-molybdenite veins. The wallrock is K-feldspathized and there are greisen patches and vein rims.

The East Kounrad deposit is unusual because being set in what is a porphyry copper province, associated with Carboniferous granodiorites (e.g. Kounrad; Section 28.2.2.). Despite its proximity (about 10 km) to the porphyry copper field, East Kounrad is related to a younger, Permian, petrochemically different suite of biotite granite and leucogranite. This is a plutonic level deposit, located in a

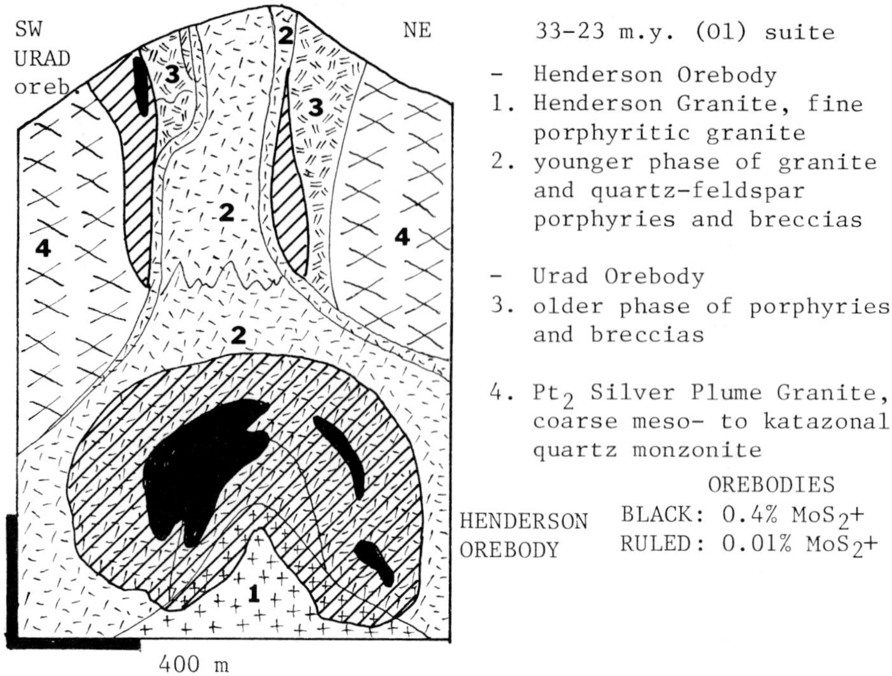

Fig. 28-86. Red Mountain stockwork Mo mineralized complex, Colorado. From LITHOTHEQUE, simplified after Wallace et al. (1978).

considerably eroded granite cupola, and most orebodies are swarms of parallel, discrete quartz, molybdenite, lesser wolframite veins in K-feldspar altered, greisenized and sericitized granites.

28.10.3. Granodiorite (quartz monzonite) associated stockwork molybdenum deposits

This ore category (Table 28-16) is associated with calc-alkalic and high-potassium calc-alkalic magma series and peraluminous mesozonal to epizonal plutons. This is the same association that produced the majority of porphyry Cu-Mo deposits. Geochemically, these rocks have less than 20 ppm Ni, 100-800 ppm Sr and 100-350 ppm Rb. They are most common in a continental margin setting in which the majority of rocks is believed emplaced under compressive conditions (Westra and Keith, 1981). There, SMD favour small, late-stage stocks emplaced into older, polyphase and deeply eroded major batholiths (such as the Coast Batholith of British Columbia; Section 28.2.3.) at a much higher crustal level. Such stocks are easily overlooked in the "sea" of the relatively monotonous "granite" mass. Similar stocks occur on flanks

Table 28-16. Stockwork molybdenite deposits 2: granodiorite-quartz monzonite affiliation

LOCALITY	HOST AND INTRUSIVE UNIT	MINERALIZATION	REFERENCES
Quartz Hill, S.E. Alaska, U.S.A.	J-Cr foliated granod.,tonal., migmat.,schist, sillim.gneiss; 27-30 m.y. compos.stock,biot. gran.to porph.,aplite	hypabyssal miner.,all in endocont. molybd.in fract.coatings,qtz.-moly. veinlets; silicif.,K-silicate alt. 1.2Mt Mo/0.08%	Hudson et al. (1979)
Alice Arm field, N.W. British Columbia, Canada	J_3-Cr_1 argill.,siltst.,grayw., hornfelsed; 48-54 m.y. qtz. monzon.,granod.porph.,alaskite plugs, stocks	cylindrical stockw.,qtz.,moly.,pyr.; K-feldsp.,biot. alter.,phyllic outer zone; cluster of 13 dep.and occurr.; late stage Zn,Pb,Cu veins; 225Tt Mo	Woodcock and Carter (1976)
Adanac, 40 km N.E. of Atlin, northern British Columbia	Cr plutonic rocks,deep-seated; 62 m.y. qtz.monzon.,qtz. monz. porph.,alaskite	elliptical, lens-shaped stockw.of qtz.veins, veinlets,with moly.,pyr. Weak qtz.,seric.,pyr.alteration; 91Tt Mo/0.1%	Soregaroli and Sutherland Brown (1976)
Hudson Bay Mt. near Smithers, central Brit. Columbia	$J_{1,2}$ hornfelsed marine shale, andes.metavolc.,granodiorite; 70-84 m.y. qtz.monz.porph., quartz latite, cylindr.stock	extens.tabular stockw.of qtz.,moly., pyr.veins and veinl. in biot.,silica, seric.alteration. 109Tt Mo/0.12% or 6Mt Mo/0.05%	Bright and Jonson (1976)
Endako, central British Columbia Canada	137-141 m.y. granodior.bathol. 141 m.y. elong. qtz.monz.body	stockw.and set of subparallel qtz., moly.,pyr. veins in K-feldsp.,biot., seric. and argill. alter.envelope; 248Tt Mo/0.09%	Kimura et al. (1976)
Mačkatica-Surdulica, S.E. Yugoslavia	PZ gneiss,schist,migmatite; T subvolc.dacite,nearby granodiorite pluton	several small pipe-like qtz.,moly., pyr.stockw. in silicif. and seric. altered dacites; 20Tt Mo/0.1%	Janković (1982)
Zhireken, East Transbaikalia, U.S.S.R.	PZ granite bathol.,J_{2-3} biot. hornbl.granod.pluton; J_3 gran. porph.stock,dikes;dacite porp. dikes	qtz.,moly.,pyr.veinlets in K-feldsp. stockw.in apical zone of granite endocont.; access.cp.,qtz.,tourm., wolfr.veins; est.min. 50Tt Mo	Pokalov (1974)

of major batholiths, emplaced into supracrustal volcanics or into sedimentary associations. Stockwork molybdenum deposits can be found within or on the fringe of porphyry copper provinces, or alternatively in terrains noted for scheelite skarn occurrences. This is reflected in the accessory minerals. Chalcopyrite is common and scheelite occasional, but there is no wolframite or cassiterite. Some deposits such as Brenda, British Columbia, stand, in terms of geochemical concentration, about halfway between porphyry coppers and stockwork Mo's.

In the Canadian Cordillera (Soregaroli and Sutherland Brown, 1976), hypabyssal stockwork Mo's are associated with composite, silicic, leucocratic quartz monzonite stocks (e.g. Boss Mountain, Alice Arm district). Plutonic stockworks occur in later stages of composite quartz monzonite batholiths. Their ages range from 140 m.y. to 8 m.y. Most stocks were emplaced in marine pelitic rocks and are surrounded by biotite hornfels. The orebodies occur in both endo- and exocontact, are largely fracture-controlled, and range from stockworks of crisscrossing veinlets in the hypabyssal bodies to zones of closely spaced parallel veins in the plutonic deposits (Endako). K-feldspar and biotite alteration is associated with ore in the former, sericite alteration in the latter, styles. Silicification is widespread.

Endako deposit, 160 km W. of Prince George, British Columbia (Kimura et al., 1976; 248 Tt Mo/0.09%; Fig. 28-87), is the largest operating Mo mine in Canada. The ore is located in a late Jurassic quartz monzonite, a phase of a large composite batholith, emplaced in Triassic marine andesitic volcanics and sediments. The orebody is an irregularly elongated, N60^{o}W-trending mass of densely fractured and diked quartz monzonite, containing a series of en-echelon N.E.-striking ore bands. It is 3.4 km long. The bands consist of discrete, 15-100 cm wide ribboned quartz-molybdenite veins, and stockworks of thin quartz-molybdenite and molybdenite veinlets. Magnetite, pyrite and lesser chalcopyrite are associated. The large veins are bordered by a narrow (3-5 cm) pink K-feldspar altered envelope. The thin veinlets and the stockworks as a mass are quartz, sericite, pyrite altered.

Hudson Bay Mountain (also Glacier Gulch, Yorke-Hardy) deposit near Smithers, British Columbia (Bright and Jonson, 1976) is under a local landmark (Fig. 28-88). It is under and adjacent to a glacier. It may contain as much Mo as Climax or more, but in about 10 times lower concentration (2 Mt Mo/0.07%). The deposit is located above a flank of a late Cretaceous quartz monzonite stock or cupola, intrusive into domed and hornfelsed Jurassic marine metavolcanics and metasediments. The roof rocks were earlier intruded by a thick granodiorite sheet and a composite subvolcanic plug composed of rhyolite porphyry, intrusive breccia and dikes. The plug is truncated at depth by the quartz monzonite stock and a block of various rocks in the exocontact of the stock contains erratically distributed molybdenite on hairline fractures and in two generations of quartz-molybdenite veinlets and veins. The dimensions of the mineralized block are 1.5 x 2.5 x 2.1 km. There is no striking, localized alteration within the system

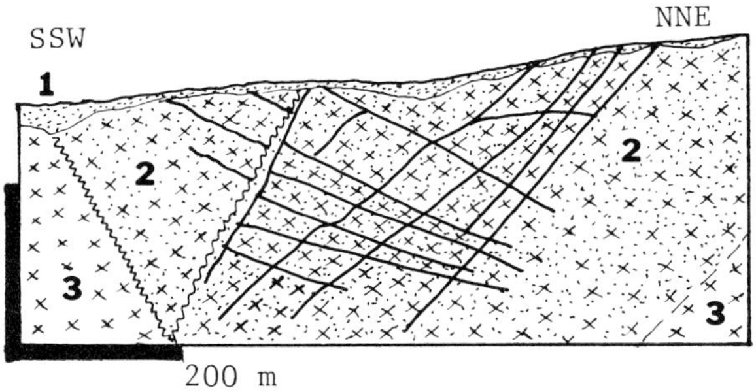

1. Q,glacial drift
2. J quartz monzonite, lightly pervasively kaolinized
3. fresh quartz monzonite

BLACK: major veins and stockworks, K-feldspar and quartz-sericite-pyrite envelopes

Fig. 28-87. Endako, British Columbia, multiple vein to stockwork Mo deposit. From LITHOTHEQUE, simplified after Kimura et al. (1976).

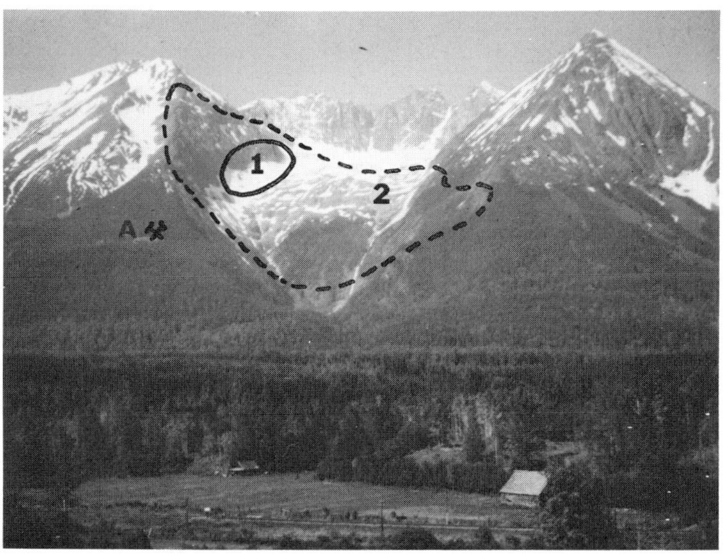

Fig. 28-88. Hudson Bay Mountain near Smithers, British Columbia, from the north-east. 1=Higher-grade molybdenite stockwork projected to surface; 2=approximate outline of a weak surface mineralization. A=adit.

(except for contact hornfelsing, bleaching and intense silicification within the early breccia plug), and no focussed and dense network of fractures. Bright and Jonson (1976) pointed out that if the entire 8.5 km^2 block is considered to be a low-grade ore (with an arbitrarily assigned grade of 0.05% MoS_2), then there are some 6 Mt Mo in the system. The potentially economic orebody is a U-shaped sheet of rocks with a greater density of quartz, molybdenite, lesser scheelite, chalcopyrite, pyrite, carbonate and K-feldspar veinlets with a grade better than 0.12% Mo.

Alice Arm district in the S.W. corner of British Columbia (Woodcock and Carter, 1976) is unusual in having a cluster of some thirteen stockwork occurrences. Normally, these deposits tend to occur in isolation. The Mo stockworks are associated with small oval and elongated high-level stocks of quartz monzonite to granite porphyry, most of which are multiple intrusions. These were emplaced in Jurassic to Cretaceous hornfelsed metasediments. As a rule, quartz monzonite is the oldest and most widespread phase. It is bordered by granodiorite and quartz diorite, and intruded by younger dikes and irregular masses of alaskite or, occasionally, intrusive breccia. Quartz, pyrite, molybdenite stockworks and veins are in the zone of potassic alteration and small, late stage quartz, carbonate, pyrite, galena, sphalerite, tetrahedrite, etc. veins are associated with most occurrences.

Additional Cordilleran localities are summarized in Table 28-16, and many more deposits and showings are known there (e.g. Thompson Creek, Idaho, 136 Tt Mo/0.15%; Cannivan, Montana, 178 Tt Mo/0.096%; Opodepe, Sonora, Mexico, 109 Tt Mo/0/08), but not enough data are available so far.

In Europe, the best-known stockwork deposits are near Mačkatica in the Serbo-Macedonian massif of Yugoslavia (Janković, 1982). There, quartz, pyrite, molybdenite veinlets comprise numerous irregular, pipe-shaped orebodies in a silicified and sericitized Tertiary subvolcanic dacite intrusion, emplaced in Paleozoic basement metamorphics. The dacite is located near the Surdulica granodiorite complex and could be a surficial manifestation of a more complex intrusive system at depth.

East Transbaikalia, U.S.S.R., contains several deposits that appear transitional between the "granite" and "granodiorite" stockwork molybdenums. At Zhireken (Pokalov, 1974), a quartz-molybdenite stockwork is located in an apical portion of a small, late Jurassic, composite stock of a fine grained granite porphyry as well as in a dacite dike swarm emplaced in an early Jurassic granite batholith. The orebody is situated in a K-feldspar alteration zone overprinted by argillization. This deposit contains a relatively high concentration of accesory chalcopyrite (recovered as a by-product) as well as tungsten (associated with quartz-tourmaline veinlets).

In the rest of the world, stockwork Mo deposits occur in the Okcheon orogenic belt, South Korea (e.g. Yongwol, 192 Tt Mo/0.24%), in south-eastern China and elsewhere, but the information available here is patchy.

28.11. MOLYBDENITE (WULFENITE) VEINS AND SMALL PIPES

Separate and discrete quartz and quartz, feldspar, molybdenite fissure veins associated with granitic plutons were mined on a small scale before the World War II, at the onset of the wave of discoveries of stockwork molybdenum deposits in the Cordillera. Not a single vein produced more than 5,000 t Mo (most produced several tens to hundreds t Mo), despite the frequently high grade (1-3% Mo). Locality examples include Olalla, Lost Creek and other small deposits in British Columbia, Daito-Yamasa area in Japan, Pupio in Chile and others. At present the only practical importance of small molybdenite veins in mesozonal to epizonal granite terrains is as an indicator of a possible "leak" from a more extensive stockwork system in the subsurface, so they should be examined as a possible exploration guide. In Endako and Alice Arm fields such veins were known to exist before the discovery of the large "bulk" deposits.

Wulfenite ($PbMoO_4$) forms minor accumulations in some replacement Pb-Zn deposits in carbonates (e.g. Mapimi, Mexico) and rare small vein deposits in silicate rocks. The best known American locality is the San Anthony vein near Mammoth, Arizona (1,700 t Mo). It was a late-stage quartz, adularia, galena, sphalerite, wulfenite, etc. fissure vein in the roof of the large San Manuel-Kalamazoo porphyry copper. The main economic importance of the few small known wulfenite occurrences is as a source of crystallized mineral specimens to collectors, although some may be indicative of possibly buried Mo or Cu-Mo stockwork, or Pb-Zn vein or replacement deposits.

Small quartz-molybdenite replacement and breccia-cementing pipes in granites near their roof contacts, are best known in eastern Australia and were mentioned earlier. Their importance is minimal.

28.12. MOLYBDENUM SKARNS

Skarns mined for molybdenum only are rare, yet Einaudi et al. (1981) managed to list eleven example localities worldwide. Cumulatively, they represent some 30 Tt of Mo, to which should be added the larger quantity of Mo present in skarns that are minor exocontact variations of mineralized intrusive stocks (e.g. Little Boulder Creek, Idaho, about 90 Tt Mo/0.09%; Cannivan Gulch, Montana). By-product Mo is also recovered from scheelite, Sn-W-Mo-Be skarns and Pb-Zn skarns. Molybdenum skarns are not a unique mineralization style (as, e.g., are the scheelite skarns) and can be anticipated whenever a Mo-mineralized complex interacts with exocontact carbonates. Molybdenite is the only major Mo mineral and pyrrhotite, pyrite, chalcopyrite, scheelite, bismuthinite, are often associated. The igneous intrusion may range from granodiorite to granite.

At the Yangchiachangtzu mine (Liaonong, China; Hsieh, 1950; 21 Tt Mo/0.3%), a Mo-skarn was discovered by chance at depth under a Pb-Zn replacement deposit in carbonates. The Mo orebody consists of veinlets, disseminations and lumps of molybdenite with magnetite, pyrite, sphalerite and galena in a garnet-dioside and later stage skarn lens, peneconcordant with bedding in a Cambro-Ordovician "black" limestone-shale sequence.

The Coxey mine near Rossland, British Columbia (Eastwood, 1967b), produced 1,653 t Mo from a small mineralized system involving a Tertiary brecciated and altered granodiorite dike emplaced in a brecciated diopside skarn. Molybdenite stockwork was superimposed over both rocks across the contact.

28.13. POSTMAGMATIC Be DEPOSITS

28.13.1. General

Hydrothermal-plutonic beryllium deposits are an entirely post-Lindgren and post-Schneiderhöhn class of ores, missing from the classical textbooks. Before and shortly after World War II the only beryllium resource was beryl from pegmatites, and the world Be reserves were estimated to be about 7.1 Tt Be in 1952 (of this, 525 t Be were contained in U.S. pegmatites). In the past thirty years, a variety of non-pegmatitic Be deposits has been found (Fig. 28-89; Table 28-17), of which the subvolcanic bertrandite deposits at Spor Mountain, Utah (Chapter 26, 24 Tt Be) are of greatest importance.

Beryllium deposits associated with granitic rocks for which some tonnage estimates have been made (Warner et al., 1959), represented 6,530 t Be in the United States only (when epithermal deposits are included; 4,780 t Be when the latter are excluded). Beryllium reserves in the Xi-Zhoyouang skarn deposit, Hunan, China (Section 28.8.4.) correspond to 7,7 Tt Be. There are no quantitative data on the Soviet Be reserves but a guestimate based on the six or so unnamed major plutonic Be deposits repeatedly described in the literature, is about 15 Tt Be. Altogether, non-pegmatitic Be deposits associated with Phanerozoic granitic rocks may represent some 30 Tt Be. Be affiliated to alkaline and peralkaline intrusions will be treated in Chapter 33, and is not included in the above figures.

Classification of granite-hydrothermal Be deposits is highly subjective and mostly based on mineral associations (parageneses; e.g. Zabolotnaya, 1974). Shcherba (1968b) proposed a classification of Be deposits affiliated with greisens that included also the exocontact metasomatites such as skarns. On the grounds of intrusive petrochemistry, the bulk of Be deposits tend to be associated with "tin granites" or similar alkali-calcic, highly siliceous, peraluminous granites. In the distal accumulations (subvolcanic and epithermal), this cannot be determined easily. It is possible that Be deposits have, like the Mo stockworks (Section 28.10) dual affiliation and in addition to true granites, they may be also related to quartz monzonites as in the Basin and Range province, western United States and its extension into Mexico. In the latter case, it is hard to imagine the Be accumulation to be the result of magmatic differentiation alone. The western U.S. "berrylium belts" (Warner et

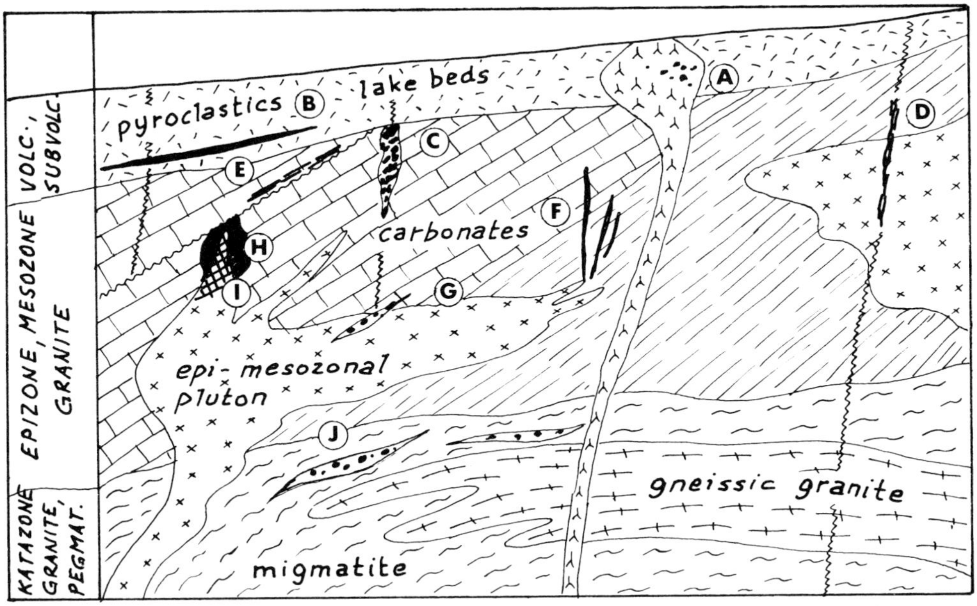

Fig. 28-89. Principal styles of Be deposits associated with calc-alkaline magmatism (see Table 28-17 for explanation of letter codes).

Table 28-17. Principal styles of Be deposits affiliated with calc-alkaline magmatism (emphasis on independent postmagmatic deosits related to epi- and mesozonal granites and quartz monzonites; compare also Fig. 28-89)

VOLCANIC-SUBVOLCANIC LEVEL (described in Section 26.6.11)	
(A)	disseminated beryl in subvolcanic rhyolite
(B)	disseminated bertrandite in water-laid tuff (e.g. Spor Mountain; about 30 Tt Be)
(C)	fluorite-bertrandite pipes, veins, replacements
(D)	epithermal quartz, adularia, bertrandite veins (e.g. Gold Hill; about 1,750 t Be)
EPI- AND MESOZONAL PLUTONIC LEVEL	
(E)	quartz, fluorite, scheelite, phenacite, bertrandite fault-filling veins and replacements (Mt. Wheeler, 260 t Be)
(F)	quartz, tourmaline, fluorite, wolframite, beryl, helvite, etc. veins (300 t Be)
(G)	high-level pegmatites with minor beryl
(H)	skarns with Be-andradite, Be-idocrase, helvite
(I)	greisens with accessory beryl
KATAZONAL SYSTEMS	
(J)	Be pegmatites (10 Tt Be)

al., 1959) seem to correlate with a Be-anomalous Precambrian crystalline basement.

From the associational point of view, the largest and most complex group of Be deposits occurs in the intrusion and over-intrusion complexes of "tin granite" cupolas, involving apogranites, greisens, skarns and veins. The rest are scattered, lonely localities, giving few clues as to their grand system affiliation, if any. Each category will be reviewed separately.

28.13.2. Be deposits in granite cupola endocontacts and exocontacts

The anatomy of "tin granite" cupolas has already been discussed. Here, only the setting of Be deposits will be considered (Fig. 28-90, Table 28-18). The "normal" (unaltered) granite (marked A on Fig. 28-90) associated with Be ores is likely to have a high trace Be content (of the order of 20 ppm Be), and may contain sparsely scattered beryl grains within its principal mass, or in pegmatite bodies.

The post-magmatically feldspathized portion of the cupola (apogranite), may reach up to 200-300 ppm Be in its partly but evenly albitized main body (B), and small crystals of beryl and rare bertrandite may be present. Greater accumulation of coarser beryl crystals can be found in metasomatic veins and pods of almost pure albitite ("albite aplite", C); microclinite ("amazonite syenite, D), or in mineralogically variable fracture veins (E). No currently economic Be accumulations, however, have been reported from this setting, nor from the marginal pegmatite (Stockscheider, F) of the cupola perimeter.

The feldspathized granite is terminated by a greisenized capping, and among the greisen facies distinguished by Shcherba (1968b), muscovite-quartz (G); topaz-quartz (H); mica (muscovite) only, (I) and quartz-tourmaline (J) greisens are most characteristic. They could be zonally arranged, but more often they form patches. Almost completely silicified masses (K) are sometimes considered to be a special facies of greisens. Beryl may be present as an accesory mineral in all the greisen facies, always accompanied by wolframite, cassiterite, molybdenite and other metallic minerals. Exclusively beryl-bearing greisens are not known. Shcherba (1968b) listed a late-stage bertrandite, sericite, fluorite association superimposed on greisen or formed along its fringe. As a source of Be, greisens are of limited importance.

Economically the most important accumulations of Be in granite intrusive and over-intrusive systems occur in exocontacts in carbonate rocks. About 80-90% of the postmagmatic plutonic Be comes from them. In the early skarns (L), Be usually accumulates in idocrase (Be-idocrase, up to 1.09% BeO) as in the Snake Range, Nevada and Iron Mountain, New Mexico, or in grossularite. Helvite group minerals $R_8(BeSiO_4)_6S_2$; R in helvite is Mn, in danalite Fe, in genthelvite Zn) are common in late stage skarns (M), e.g. in ribboned magnetite, fluorite, helvite skarn at Iron Mountain, New Mexico. Local variations of Be associations in skarn have been described, almost all in the late stage skarn or in retrograde veins and replacements superimposed on skarn.

In the large polymetallic skarn deposit Xi-Zhoyouang, Be accumulated in the beryl-epidote zone of late skarn, and in feldspar, helvite, chrysoberyl, tafeiite, bertrandite, cassiterite veinlets superimposed on skarn. Shcherba (1968b) listed a variety of mineral assemblages of late stage veins and stockworks superimposed on skarn in Kazakhstan and the Soviet Far East: (1) phenacite, chrysoberyl, fluorite, muscovite, tourmaline, calcite, scheelite, euclase, beryl, sulphides; (2) beryl, mica, fluorite, phenacite, scheelite, cassiterite, albite and (3) helvite, vesuvianite, fluorite, beryl, danalite, chrysoberyl, fluorite. Simple fluorite, scheelite, beryl and quartz, beryl veins are superimposed over skarn in the Victoria mine, New Mexico (Warner et al., 1959) and fluorite-helvite veins are found in the Grandview mine, New Mexico.

Exocontacts in alumosilicate rocks (N) are less exciting as sites of Be accumulations. Disseminated beryl rarely occurs in fracture veinlets superimposed on biotite or tourmaline hornfelses. Exocontacts in mafics (amphibolite, gabbro, serpentinite; O, P) are getting much attention in the Soviet literature (Shcherba, 1968b; Zabolotnaya, 1974) because they are well developed at several localities in the Urals (e.g. Takovaya). There, however, they are a source of a gem quality chrysoberyl (alexandrite) and emerald, rather than a bulk Be ore. Usually, two endocontact zones may be distinguished along mafic or ultramafic contacts (O): inner, more proximal, phlogopite and/or amphibole (actinolite, tremolite, anthophyllite) zone; (P) outer zone of chloritization and talc alteration. Shcherba (1968b) listed two Be associations, most likely to be present in the late-stage veins superimposed over both zones: (1) beryl, chrysoberyl, phenacite, bavenite, phlogopite, biotite, fluorite, feldspar, quartz and (2) phenacite, bavenite, beryl, albite, phlogopite, etc.

Above-intrusion veins and/or replacement bodies in carbonates (S, T) that are more or less independent of the skarn/greisen profile along the immediate contact, are common and economically important. At Lost River, Alaska (Sainsbury, 1969; this deposit contains the largest calculated reserve of granite-affiliated hydrothermal Be in the United States), the fluorite, white mica, diaspore, tourmaline, chrysoberyl veins date from an earlier period than the skarns and greisens. Fluorite, phenacite, bertrandite; and fluorite, mica, beryl associations, are most common in carbonate replacement veins and masses in the U.S.S.R.

In alumosilicate exocontacts, quartz, K-feldspar, molybdenite, wolframite, etc. veins and stockworks carry beryl as an accesory mineral, but no exclusively Be-mineralized veins have been reported.

The Lost River ore field, western Alaska (Sainsbury, 1969; Dobson, 1982; 4,050 t Be/0.2%) is also an important tin locality, already mentioned earlier. A granite cupola below the Lost River tin deposit contains some phenakite in greisenized and argillized granite. The exposed greisen bodies are virtually Be-free, but a fluorite, tourmaline, cassiterite-bearing rock in selvedges along the greisen/carbonate contact contains up to 3.3% BeO largely bound in chrysoberyl. The early stage andradite, magnetite, idocrase skarn is generally Be-poor, except for the local contents of up to 0.45% BeO in idocrase. Banded fluorite-magnetite skarn at the Tin Creek locality carries Be in helvite. The bulk of the economic Be in the Lost River area is contained in fracture filling and replacement veins in Ordovician carbonates, relatively distant from the granite. The veins are controlled by thrust faults, normal faults and zones of brecciation, and their usual mineralogy consists of dominant fluorite with lesser quantities of white to lilac mica, Be-diaspore, tourmaline

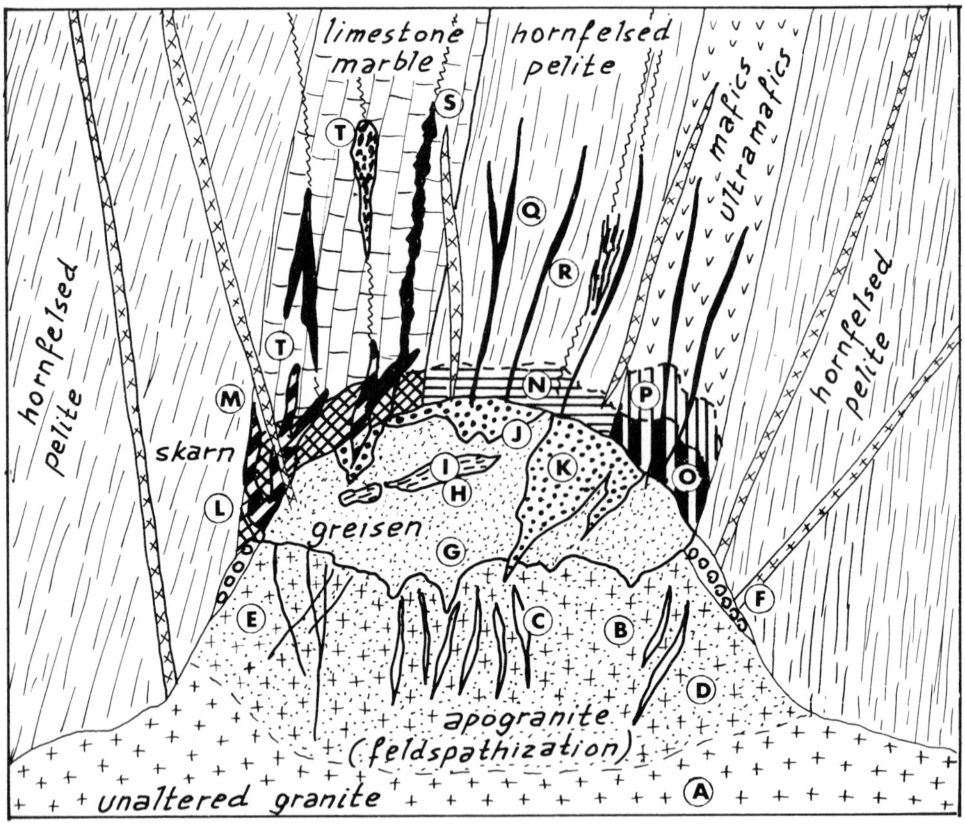

Fig. 28-90. The setting of Be deposits in intrusive and overintrusive region of granite cupolas. See Table 28-18 for an explanation of letter codes. From Laznicka (1984)

Table 28-18. The setting of Be deposits in the intrusive and over-intrusive regions of granite cupolas

(A)	NORMAL GRANITE; high trace Be or beryl disseminations in weathered granite
	FELDSPATHIZED GRANITE (APOGRANITE)
(B)	small crystals of beryl or bertrandite disseminated in evenly albitized apogranite
(C)	"albite aplite"-style replacement veins containing disseminated beryl
(D)	"amazonite syenite" replacement bodies with beryl
(E)	fissure veins
	MINERALIZED GREISENS (X00 localities, X00 t Be)
(F)	occasional disseminated beryl in all varieties of marginal pegmatite (Stockscheider)
(G)	muscovite-quartz greisen
(H)	topaz-quartz greisen
(I)	monomineralic mica (muscovite, lithionite) greisen
(J)	tourmaline-quartz greisen
(K)	silicified masses, pipes
	IMMEDIATE EXOCONTACT IN CARBONATES
(L)	early stage skarn; Be-andradite, Be-idocrase; X0 localities, X00 t presently uneconomic Be
(M)	late-stage skarn, superimposed veins (fluorite, magnetite, helvite; beryl, epidote; etc.); X0 localities, about 20 Tt Be
(N)	IMMEDIATE EXOCONTACT IN ALUMOSILICATE ROCKS: biotite, tourmaline, etc., hornfels
	IMMEDIATE EXOCONTACT IN MAFICS, ULTRAMAFICS
(O)	amphibole, phlogopite, sphene, etc. metasomatites, beryl (also emerald), chrysoberyl; several localities, X00 t Be
(P)	chlorite, talc-altered rocks; Be minerals in superimposed quartz, tourmaline, etc. veins
	ABOVE-INTRUSION VEINS, REPLACEMENTS; Be occurs in a variety of associations, of which fluorite, quartz, tourmaline, crysoberyl, phenakite, beryl, etc. orebodies are most common; X localities, about 10 Tt Be
(Q)	fissure veins
(R)	stockworks
(S)	replacement veins
(T)	replacement masses, pipes, mantos

and chrysoberyl. Euclase, bertrandite, phenakite and beryl may be present in minor quantities. Fluorite is colourless to light-green and inconspicuous. The largest lode system at Camp Creek is a tabular body up to 800 m long and 70 m thick, gently dipping south beneath a thrust fault. Numerous ore veins and veinlets are located within the fractured and brecciated limestone and dolomite (Fig. 28-91).

Iron Mountain near Socorro, New Mexico (Jahns and Glass, 1944; 300 t Be/0.15%) is a frequently quoted, readily accessible locality. There, a Paleozoic limestone and calcarous shale were intruded by a Tertiary rhyolite, granite and aplite complex, and a small skarn zone formed in the exocontact. The early massive magnetite-garnet skarn grades into a marginal, banded "ribbon rock" composed of fluorite, magnetite and helvite replacing carbonate. Helvite is yellowish-brown or reddish-brown. It is inconspicuous and megascopically difficult to tell from garnet or idocrase. The higher-grade "ribbon rock" contains about 0.7% BeO, the lower-grade material has 0.2% BeO. Lesser quantities of Be occur in the same assemblage superimposed along shears and breccia zones on the earlier skarn.

The Soviet Be deposits are considered to be of great strategic importance, so their names, locations, tonnages, etc. have not been released. Descriptions and illustrations of anonymous localities lacking even such elementary detail as scales, orientations, names of stratigraphic units, etc., are pitiful and will not be reviewed here. The reader is referred to the original source (Zabolotnaya, 1974; Shcherba, 1968b). Only the general setting of the Soviet localities can be gleaned, after an exhaustive search for clues. Several occurrences are in the Chukotka Tin Belt of eastermost Siberia (western extension of the Seward Peninsula of Alaska), and these are comparable in many respects with the Lost River field. Beryllium albitites superimposed on Precambrian metamorphics are probably located in the Aldan Shield. The beryl-containing greisens are found in the East Transbaikalia (also at the Chulun-Khuriete showing in adjacent Mongolia), and in the Central Kazakhstan W-Mo-Sn belt.

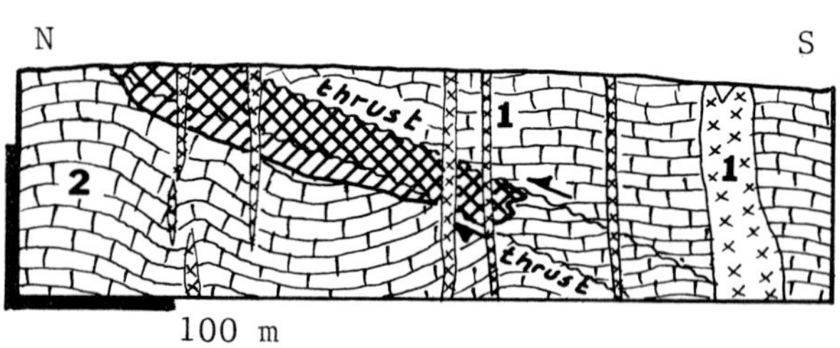

1. Cr lamprophyre, porphyry dikes, stocks
2. Or limestone, dolomite

Be. fluorite, diaspore, mica, tourmaline, chrysoberyl veins, veinlets

Fig. 28-91. Rapid River Be deposit, York Mountains, W. Alaska. After Sainsbury (1969).

The Be occurrences at granite/ultramafic contacts are located in the central Urals.

28.13.3. Independent Be occurrences

Many small occurrences of hydrothermal Be minerals are isolated, one-of-a-kind entities that cannot be fitted into the granite cupola model. Most are exocontact veins or replacements.

The California Vein at Mt. Antero (Sawatch Range, Colorado; Dings and Robinson, 1957) is a quartz, beryl, pyrite vein 215 m long and at most 1 m wide. The beryl is light-green to bluish and there is an accesory molybdenite. Several quartz, hübnerite, molybdenite veins are known in the vicinity and stockwork molybdenum deposits have been discovered in the area recently.

The Mount Wheeler deposit, Nevada (Stager, 1960) probably contains a minimum of 260 t Be/0.26%. There, Cambrian black limestone and interbedded shale are faulted, silicified and some of the E. or N.E.-trending fault zones are filled by "linear stockworks of quartz veinlets and parallel veins mineralized by scheelite. Quartz monzonite to granodiorite body is postulated to be present at a depth of less than 300 m. A pale blue beryl was detected in veinlets and small isolated crystals in hornfelsed shale below the main mineralized zone originally mined for scheelite, but the presence of the two most widespread minerals phenacite and bertrandite, was discovered much later. Both minerals are colourless to white and resemble quartz. They are accompanied by fluorite, pyrite, Mn-siderite and scheelite. The Be mineralization was traced for over 800 m downdip, and 1,300 m along the strike. The ore shoots are up to 3.3 m thick, extend vertically for 5 to 7 m and average about 1% BeO.

Additional Be localities, many listed by Shawe (1966) include quartz, beryl, wolframite and fluorite veins mostly associated with quartz monzonite stocks (e.g. Little Dragon Mts., Arizona; Sheeprock Mts., Utah; Victoria Mine, New Mexico; etc.), fluorite-helvite veins (Grandview Mine, New Mexico) and others. Volcanics and subvolcanics-hosted Be accumulations have been described in Chapter 26.

28.14. Pb-Zn(Ag) DEPOSITS IN "GRANITE" AUREOLES

28.14.1. General

Pb-Zn(Ag) ("polymetallic") deposits and occurrences are the most common variety of ores to be found in "granite" aureoles. At least 100 districts and large ore fields carry polymetallic deposits, worldwide. Considering the frequently large number of individual veins or orebodies present in a single field or district (e.g. 1100 veins in the Freiberg district, Erzgebirge, about 100 of them producing), the figure of over 10,000 producing Pb-Zn(Ag) orebodies in the world is probably a conservative estimate. These orebodies represent some 80 Mt Pb, 70 Mt Zn, 200 Tt Ag plus a considerable quantity of by-product Cd,Sb,Cu, As,Bi,Au, Sn,W,Mo, Hg,Te,Se, In,Ge,Ga, etc.

Although several colleagues would nominate a symmetrically filled quartz, galena, sphalerite vein as a "typical postmagmatic-hydrothermal orebody", not all veins are alike. There is a great variety of geometries, mineral associations and settings of the polymetallic orebodies and their origin continues to be controversial. No single model can accomodate this variety, and almost every summary statement made subject to scrutiny can be proven to include numerous exceptions. This is the price one pays for generalization and for the convenience of having a 20 page summary representing an ore category treated in no less than 50,000 pieces of literature.

Hydrothermal Pb-Zn(Ag) deposits have been traditionally (e.g. Schneiderhöhn, 1955; Park and McDiarmid, 1975) placed into the mesothermal, epithermal and telethermal categories on the grounds of assumed or demonstrated temperature of the ore-forming solutions, and into the subvolcanic and plutonic associations by their presumed level of emplacement and derivation. The high-level epithermal deposits hosted, by or spatially associated with, continental volcanic terrains have been thoroughly reviewed in Chapter 26. Most epithermal deposits studied recently by modern methods have been proven to have formed in the waning stage of volcanism or post-volcanically from convecting fluids mostly heated by an evolving intrusive body in the subsurface. Consequently, there is no qualitative difference between the "subvolcanic" and "plutonic" deposits and both categories overlap.

The recent advances in ore petrology, particularly the progress in the study of strata-related deposits (these have been extensively reviewed in the 14 volumes of the Handbook of Stratiform and Strata-bound Ore Deposits, edited by Wolf, 1976, 1981 and 1985), further demonstrated the considerable variety and transitionality of the Pb-Zn deposits. Not long ago, the peneconcordant massive sulphide ores in volcanic-sedimentary associations (e.g. Chapter 14) as well as the Mississippi Valley and Appalachian "types" (Chapter 20) were considered granite-related, postmagmatic. Now we know better...(do we really ?).

In many cases the style and origin of Zn-Pb ores correlates with the host lithologic association (e.g. sedimentary ores are in sediments), the "granites" being the most notable exception. Because of the distal position of postmagmatic-plutonic Pb-Zn orebodies with respect to their "emanative centre", such orebodies are hosted by older (pre-granite) rocks rather than by the concentrator intrusion. When a "granite" hosts orebodies (mostly veins), it is an old granite acting as a passive host, genetically unrelated to the mineralization. Several minor, economically unimportant or marginal exceptions, are reviewed in the following paragraphs. Parent/daughter relationships between plutonic intrusions and hydrothermal Pb-Zn deposits located in their aureole are difficult to demonstrate exactly. Numerous modern determinations based on isotopic age data indicate considerable gaps between the ages of igneous crystallization and ore emplacement (of the order of tens to hundreds of millions of years; about 100 m.y. in the Stříbro district, Czechoslovakia). Pb-Zn deposits occur particularly frequently in the tectono-magmatically "activated" (block faulted, stretched, rifted) terrains the formation of many of which shortly postdated termination of the continental granitic plutonism. In such terrains, metalliferous veins may have resulted from hydrothermal convection driven by other than granite-provided heat sources (geothermal heat, mafic dikes, small alkaline intrusions, younger buried granitic intrusions) and in most cases such a source is impossible to determine. A prospector should realize that although many Pb-Zn(Ag) deposits are undisputably members of a postmagmatic suite related to an evolution of an igneous intrusion and occur orderly in its aureole, there are many exceptions and departures from such a scenario some of which may, by their setting, mimic the former. In addition to this, relic, remobilized and mobilized Pb-Zn(Ag) deposits are unusually frequent. One should bear in mind the limitations of the aureole ore distribution model and be ready to see beyond it. The uncertainities and limitations mentioned in general above naturally overprint into the contents of the present book section as well. Indeed it cannot be demonstrated that all the examples treated here in the framework of plutonic deposits, are beyond doubt "granite"-affiliated.

Regarding geotectonic setting and host association, all Pb-Zn districts and "belts" formed in regions underlain by a relatively mature continental crust at the time of intrusive emplacement and ore generation. The most favourable terrains are "activated" portions of former platforms or "miogeoclines" such as the eastern portion of the Great Basin in Eastern Nevada and Utah; uplifts bordering "eugeoclinal" and "miogeoclinal" domains in orogenic belts (e.g. the Omineca belt, British Columbia); uplifts within the "ensialic" portions of continental margin orogenic belts and median massifs (e.g. the Bohemian Massif, Massif Central); ancient sialic continental blocks within island arcs (e.g. N. and C. Honshu, Japan); intracontinental uplifts, grabens, aulacogenes, rifts in their early stage of development (before the appearance of widespread basalts) and others.

The supracrustal lithologic associations treated in Chapters 11,12,14, 17,19,20, 26,28 and 30, typically host postmagmatic Pb-Zn deposits. Small isolated Zn-Pb occurrences may be present in the remaining associations. Some Zn-rich skarns and veins containing virtually no Pb are transitional into Cu skarns and veins and are hosted by "granite"-intruded intermediate to mafic marine volcanic-sedimentary suites of continental margins.

28.14.2. Pb-Zn sulphide deposits in non-carbonate hosts

STOCKWORKS, DISSEMINATIONS

In contrast to the disseminated or stockwork Sn,Mo,Cu,W, and other deposits, Pb-Zn ores normally do not accumulate in a similar way in the immediate contact area of their concentrator intrusion. Several minor examples of "porphyry" (or stockwork) Pb-Zn(Ag) occurrences are largely due to unusual interactions, are very local and, to date, of no economic importance. Most are high-level (subvolcanic; e.g. disseminated sphalerite in porphyry dikes, Mt.Bischoff; disseminated galena and sphalerite in felsic subvolcanic stocks and dikes, locally in the Cartagena and Mazarron districts, Spain).

The disseminated, very low grade occurrence at Hahns Peak, Colorado (Young and Segerstrom, 1973; Fig. 28-92) is often quoted as an example of this type of mineralization. There, pyrite, galena, sphalerite, lesser tetrahedrite, chalcopyrite and proustite are sporadically scattered in certain zones of an altered (silicified, sericitized, kaolinized) breccia pipe intruded by an 11 m.y. old porphyry stock and dikes. This deposit is estimated to contain 6.39 Tt Ag, 639 Tt Pb and 387 Tt Zn at grades of 9.7 ppm, 0.1 and 0.06%, respectively. Similar occurrences although not of much economic importance in themselves at present, may indicate concealed mineralized apical intrusions (e.g. stockwork Mo).

FISSURE FILLING VEINS

Fissure veins are the most characteristic style of Pb-Zn deposits in non-carbonate hosts. All varieties of vein filling structures and textures treated in economic geology textbooks (symmetrical, banded, breccia cementing, cockade, etc.) are represented. The veins occur at some distance from the concentrator intrusion. When such an intrusion is a granite cupola intruded into pelitic sediments, the bulk of Pb-Zn veins tend to be located outside the thermally hornfelsed fringe, in rocks showing only minor effects of thermal metamorphism. A considerable variety of vein-filling mineral associations has been recorded (Tables 28-19, 20). Quartz, carbonate (calcite, dolomite, siderite) and barite gangue; galena, sphalerite, pyrite sulphides; and alteration sericitization, silicification, carbonatization, argillization and bleaching are most common. The alteration envelopes to veins tend to be narrow and developed in the immediate vein vicinity.

Some veins show mineralogical or geochemical zoning, most commonly showing an increasing Zn:Pb ratio and increase in the Cu and Au contents with depth (in many subvolcanic veins, on the other hand, gold is at its highest concentration near the top). Pb-Zn veins commonly follow dikes and are often preferentially hosted by them (e.g. in Příbram, Czechoslovakia). The economic importance of Pb-Zn(Ag) fissure veins has diminished considerably in the past four

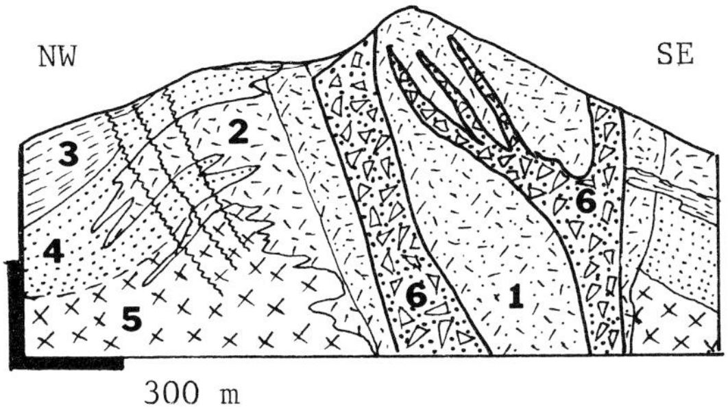

1. T, bleached, silicified rhyolite porphyry
2. dacite, quartz latite. rhyolite porphyry
3. Cr shale, sandstone
4. J_3-Pe red beds
5. Pt_1 crystalline rocks
6. mineralized and altered breccia zones

Fig. 28-92. Hahn's Peak, Colorado, a low-grade disseminated Ag,Pb,Zn sulphide deposit. After Young and Segerstrom (1973).

or five decades mostly due to exhaustion and a labour-intensive method of mining. Individual veins tend to be small or, at best, medium-sized orebodies (several tens to hundred thousand tons of ore), but an aggregate tonnage of multiple vein fields could be significant. Compared with the many "modern" orebodies, veins tend to have high Pb and Zn grades and high silver contents. In the Oberharz district, West Germany, the Grund vein deposit was more profitable than the massive sulphide Rammelsberg, in the 1970's (H. Sperling, pers. commun.,1975).

REPLACEMENT AND RELIC OREBODIES IN NON-CARBONATE HOSTS

Replacement Pb-Zn(Ag) orebodies in silicate wallrocks occur at only a few localities but at least one of them (the Coeur d'Alene district, Idaho) is of substantial economic importance. There, steeply dipping narrow veins composed of fine crystalline galena, sphalerite, pyrite, pyrrhotite, ankerite, siderite, quartz, tetrahedrite and other minerals, are located in bleached siliceous Proterozoic slate, phyllite and sericite quartzite. At Coeur d'Alene and elsewhere, fault gouge to mylonite were most commonly replaced. Sometimes, replacement orebodies enclosed in silicate wallrocks formed by complete substitution of carbonate lenses, or by selective replacement of the carbonate component in, for example, a limy shale or a limy tuff.

Many Pb-Zn sulphide orebodies formerly interpreted as hydrothermal replacements in the "granite" aureole have subsequently been re-interpreted as relic stratiform deposits. Examples include several localities in Mexico (Fresnillo, San Martín, Plomosas and others; De Cserna, 1976a), Peru (Lehne and Amstutz, 1982) and elsewhere. This subject has been reviewed in Section 17.6.2. In addition to little modified earlier relic orebodies, one can expect a

Table 28-19. Pb,Zn,Ag multiple vein deposits, fields and districts; selected examples

LOCALITY	HOST AND INTRUSIVE UNITS	MINERALIZATION	REFERENCES
Elsa-Keno Hill field, Yukon, Canada	PZ_3 or MZ micasch.,gray to black phyll.,m-quartz., m-diabase sills; Cr_3 qtz. monzonite	28 vein zones N.E./S.E., in 1.6-20m wide gouge and brecc.filled zones; sider.,gal.,sphal.,pyr.,freiberg., pyrarg.; light (bleaching) to none alt. 190Tt Pb,123Tt Zn, 4.5Tt Ag	Douglas, ed., (1970)
Slocan-Kaslo distr. S. British Columbia Canada	Tr black slate, m-quartzite litharen.,minor limest., greenst.; J_1 granod.,qtz. dior., qtz. monzonite	many fiss. veins, rare limest.replac. sider.,qtz.,calc.,sphal.,gal.,pyr., tetr. 198Tt Zn; 164Tt Pb; 1,077 t Ag	Douglas, ed., (1970)
Coeur d'Alene distr. Idaho, U.S.A.	Pt_2 phyllite, siltst., m-quartz.; Cr_3 qtz. monzon. and dior. stocks, dikes	9 N.N.W. trending, miner.fault contr. "belts"; replac.lodes, sider.,anker., gal.,sphal.,pyrrh.; qtz.,sider.,Ag-tetrahedr. Main miner. period Cr_3-T_1, early Pt_2 miner.; possibil.of remobil. 6.9 Mt Pb; 3Mt Zn; 140Tt Cu; 31Tt Ag	Fryklund (1964)
Colorado Mineral Belt (Rocky Mts., U.S.A.)	Pt_1 gneiss,migmat.,amphib. Pt_2 qtz. monzon.,granite Cr_3-T_2 qtz. monz. stocks, dikes;granod.,lamproph.	almost contin.belt of fissure veins; qtz.,pyr.,gold; qtz.,carb.,pyr.,sphal. gal. Short, impersistent ore shoots. 250Tt Pb; 130Tt Zn; 2.8Tt Ag; 210 t Au (signif. portion of Au from placer)	Lovering and Goddard (1950)
Park City field, Utah, U.S.A.	Ps-Tr limest.,dolom.,shale sandst.,folded,faulted, thrusted; T_1 dior.,diorite porphyries	replac.mantos in jasperoid // with bedd.planes; fiss.and replac.veins; qtz.,rhodochr.,rhodon.,pyr.,sphal., gal.in veins; Ag-bonanzas. 1.15Mt Pb, 600Tt Zn, 50Tt Cu, 6.76Tt Ag, 28t Au	Barnes and Simos (1968)

Santa Barbara, Chih., Mexico	Cr_1 limest., Eo, contin. andesites; Eo-Ol rhyol., qtz. monzonite in depth	pyr., gal., sphal., qtz., fissure veins in sedim. and andesites. 1.1 Mt Pb/3% 1.7 Mt Zn/4.8%; 266Tt Cu/0.75%; 5.2 Tt Ag	Busch (1980)
Fresnillo, Zacat., Mexico	Tr_3 andes. pillow lava, greenst.; Cr_1 limest., shale; Cr_3 flysch sediments; Eo_3? granod., qtz. monzon., rhyolite	bedded replac. or relic ? mantos; N.W. fissure veins, stockw.; py, sph., gal., cp., lesser pyrrh., ars., tetr., pyrarg. Veins have qtz., calc., sider. gangue. 980Tt Zn, 920Tt Pb; 7.75 Tt Ag	DeCserna (1976b)
San Martin, Zacat., Mexico	MZ blue limest., hornfelsed shale; T granod. porphyry	N.E./45-70°W up to 1.5 m thick veins; qtz., calc., fluor., sphal., cp., gal., argent., Ag; 669Tt Zn/5.75%; 180Tt Pb; 162 Tt Cu/1.35%; 4.2Tt Ag/150 ppm	Busch (1980)
Mina Matilde, northern Bolivia	D gray folded slate, sandstone; no intrus. known.	2 fiss. veins N23°E/90° up to 1.2-7 m thick, banded; sider., gal., sphal. 540 Tt Zn, 180Tt Pb, 12 Tt Cu, 120t Ag	Ahlfeld and Schneider-Scherbina (1964)
Oberharz district (Clausthal), Harz, West Germany	D slate, limest., sandst.; Cb_1 litharenite, slate; Cb_3-Pe_1 and later: buried granite ?	16 N.N.W. discont. miner. fault zones; qtz., sider., calc., gal., sphal., lesser bourn., tetr., Ag-sulphosalts. 1.8Mt Pb, 700Tt Zn, 4.7Tt Ag	Buschendorf et al. (1971)
Freiberg, Erzgebirge, East Germany	Pt_3 biot. gneiss, migmatite; Cb-Pe cryptobatholith	1,100 N.-S. and W.-E. fissure veins. 2 major miner. stages (PZ_3 and Tr-T); qtz., sider., gal., sphal., Ag-sulphosalts (see Table 28-20); 1.7Mt Pb, 7Tt Ag	Baumann (1976)
Rheinische Schiefergebirge, W. Germany (Ramsbeck, Enns, Holzappel, Bensberg)	D qtz. rich grayw., slate, quartzite; deformed, cleaved, but no known granitic intrusions are present	fissure, shear, bedding veins, some may be remobiliz. relict oreb.; barite, carb., sphal., pyr., gal., lesser tetr., ars. 2.74Mt Zn; 1.22 Mt Pb; 1.1Tt Ag	Schneiderhöhn (1941)

Table 28-19 (continued). Pb,Zn,Ag multiple vein deposits, fields and districts; selected examples

LOCALITY	HOST AND INTRUSIVE UNITS	MINERALIZATION	REFERENCES
Příbram district, Czechoslovakia	Pt_3 slate,graywacke,greenstone (spilite); Cm litharenite; Cb-Pe buried pluton,qtz.diorite,diabase	2 parallel N.E. trending fault zones containing N.N.E. fiss.veins up to 1.7 km deep in fract.feathering from master faults; qtz.,sider.,carb.,gal., sphal.,boulang.,Ag-sulphosalts; estim. 1.2Mt Pb, 700Tt Zn, 4.5Tt Ag	Kutina (1963)
Sierra Morena district, Spain (Linares,San Quintin,La Carolina,etc.)	Pt_3 schists; Or, quartz., litharen.,slate,schists; Cb_1 grayw.,slate; Cb-Pe granod.,qtz. monzonite	subvert.fiss.veins in vein zones up to 12 km long; veins 1-2 m thick, qtz., anker.,carb.,gal.,sphal.,boulanger., Ag-sulphosalts; in metased.and granite 2.9Mt Pb, 670Tt Zn, 4.2Tt Ag	Schneiderhöhn (1941)
Leadhills-Wanlockhead d.,Scotland	Or grayw.,black slate, volcaniclastics; Cb-Pe buried plut., dikes	Pe_3; 70 W.N.W. to N.W. banded fissure veins up to 5 m thick; dolom.,calc., gal.,sphal.,pyr.,chalcop. 300Tt Pb	Dunham et al. (1978)
English Lake District.,N.W. England	Or, shale, grayw.,marine andes.,rhyol.; D_1 granite pluton; Cb-Pe cryptobath.?	fissure veins (e.g. Greenside vein); qtz.,bar.,sphal.,gal.,cp.; 230Tt Pb	Dunham et al. (1978)
Central Wales district, Great Britain	Or,Si folded slate, litharenite; unknown intrus.	356 m.y. miner.,low-temp.fissure veins 275 occurrences; qtz.,pyr.,sphal.,gal. cp. 382 Tt Pb, 75Tt Zn, 109t Ag	Dunham et al. (1978)
Rhodopen district, S.Bulgaria (e.g. Madan)	Pt gneiss,amphib.,marble; PZ deep-seated granite; Ol qtz.monz.,granod., syenite, rhyolite	swarms of fissure veins in 43 groups, minor marble replac.; qtz.,carb.,gal., sphal.,pyr.; 2.77Mt Pb/2-3%; 2.18Mt Zn; 240Tt Cu; 4Tt Ag	Iovchev (1961)
Zeehan-Dundas d., S.W. Tasmania Australia	Pt_3-Cm slate, quartzite; Cm-D slate,greenst.,sd., limest.;D tin granite	D; about 30 N.N.W. and N.N.E. fissure veins of qtz.,sider.,gal.,sphal.,pyr.; minor stann.,cassit; 324 Tt Pb 900t Ag	Blissett (1962)

Table 28-20. Selected examples of single Pb-Zn-Ag fissure veins

VEIN	HOST ROCKS	AGE, FILLING, MINERALOGY, ALTERATIONS, MAGNITUDE, TONNAGE, ETC., OF THE VEIN	REFERENCES
Neue Hoffnung Gottes Freiberg, E.Germany (noble quartz assoc.)	Pt_3 gray gneiss, biotite gneiss	Cb-Pe; qtz.,sphal.,ars.,pyr.,pyrrh.,lesser cassit.,pyrarg.,miarg.,stephanite,polybasite. Seric.,silicif.; N.N.E.,2 km long, up to 2 m wide, 460 m deep; X00 t Ag	Schneiderhöhn (1941,1955)
Himmelfahrt, Freiberg East Germany (Edle Braunspathformation)	Pt_3 gray gneiss biotite gneiss	Cb-Pe; sider.,lesser calc.,gal.,sphal.,qtz., tetr.,freibergite,pyrarg.,proustite; N.N.E., 600-1,000 m long, 0.75-1.5 m wide; X00 t Ag	ditto
Silbernaler Gang, Grund, Harz Mts., West Germany	Cb_1 litharenite, slate	Cb_3-Pe_1; qtz.,lesser calc.,dolom,sider.,bar., gal.,sphal.,lesser cp.,pyr.,tetr.,pyrargyrite; banded and brecciated; W.N.W., 6 km long, up to 50 m wide; 11.5 Mt ore	Buschendorf et al. (1971)
Holzappel Hauptgang, Rhein. Schiefergebir-ge, West Germany	D litharen.,sla-te,diabase sill, porphyroid	Cb?; qtz.,sider.,sphal.,tetr.,gal.,lesser pyr.,cp.; qtz.-seric.alter.; 3.4 km long, 0.6-7 m thick, 1,017 m deep;	Schneiderhöhn (1941,1955)
Main Vojtěch Vein, (Hlavní Vojtěšská) Příbram, Czechoslov.	Cb diabase dikes in Cm litharenite; in depth Pt_3 slate,greenst.	Cb?; sider.,gal.,sphal.,bournon.,Pb-Sb sulphosalts; in depth: fine qtz.,sider.,gal.,sphal., cp.; pyrarg.,steph.,polybasite throughout. Seric.,carbonatiz.,bleaching alter.; N.N.E./70-85°E; 2 km long, depth to 1,700 m	Kutina (1963)
Greenside Vein, English Lake Distr. Great Britain	Or, marine andesite,rhyol.,qtz. porph.stocks	PZ_3; barite,qtz.,sphal.,cp.,gal.; single N.-S. east dipping fault filling vein, 1.3 km long, up to 10 m wide, 600 m deep; 167Tt Pb, 65 t Ag	Dunham et al. (1978)
Mayflower Vein, Park City field, Utah, U.S.A.	Ps-Pe limestone, dolomite; T_1 diorite	T_1; quartz,lesser calc.,rhodochr.,rhodon.,pyr., sphal.,gal.,small shoots of enarg.,tetr.,cp.; silicif.,qtz.-seric.alter.; N70°E/65°N, 670 m depth.	Barnes and Simos (1968)

variety of (re)mobilized mineralizations, always controversial and difficult to interpret. The Coeur d'Alene district (Gott and Cathrall, 1980) described in greater detail later, illustrates a case where (re)mobilization has been frequently assumed, but not yet convincingly proven.

28.14.3. Zn-Pb sulphides in carbonates: skarns

Zn-Pb skarns (Table 28-21) represent some 12 Mt Zn and 9 Mt Pb, worldwide. Their Zn:Pb ratio varies widely. As a rule, the skarns near intrusive contacts are high in Fe-sphalerite (marmatite) and low in galena, but they may contain significant amounts of chalcopyrite. Distal skarns tend to be Pb-rich, particularly in orebodies formed in carbonates outside the skarn envelope.

Zn-Pb skarns were briefly reviewed by Einaudi et al. (1981), and these authors summarized their most distinctive features as follows: (1) frequent occurrence of Mn-rich skarn silicates (johannsenite, bustamite); (2) common occurrence in distal position to intrusive contacts; (3) location along structural or lithologic contacts; (4) preferential association of sulphides with pyroxene, rather than with garnet; (5) absence of significant metamorphic aureoles around the skarn and (6) retrograde minerals represented by Mn-rich ilvaite, pyroxenoids, subcalcic amphibole and chlorite.

The skarns are mostly calcic, although few examples of magnesian skarns are known. They are all related to high-level intrusions (upper mesozone to subvolcanic) ranging from granodiorite through quartz monzonite to leucogranite, and from holocrystalline plutons through small stocks and porphyry dikes to breccia pipes (Trepča, San Antonio Mine, Chih.). Some skarns lack known associated intrusive rocks, but are located along structurally or lithologically controlled solution pathways. The skarns vary in shape from irregular masses through chimneys and pipes to peneconcordant mantos. Selective replacement of thin carbonate lenses or beds sandwiched between silicate rocks (e.g. shales, quartzites) results in orebodies that may be confused with stratiform deposits. Some isochemically metamorphosed Zn-Pb deposits in supracrustal rocks, on the other hand, may correspond compositionally to skarns. Relic supracrustal Zn-Pb orebodies overprinted by skarn and partly remobilized, are virtually indistinguishable from true metasomatic skarns. Fig. 28-93 shows selected examples of Zn-Pb skarns and

28.14.4. Zn-Pb sulphides hosted by carbonates: no skarn assemblage

Disseminated (scattered) to massive galena, sphalerite, pyrite, pyrrhotite and other sulphides lacking the skarn silicates, form large and rich orebodies in many mining fields throughout the world (Table 28-22), in which high-level plutons and subvolcanic bodies intruded sedimentary carbonates. They represent some 25 Mt Zn and 27 Mt Pb.

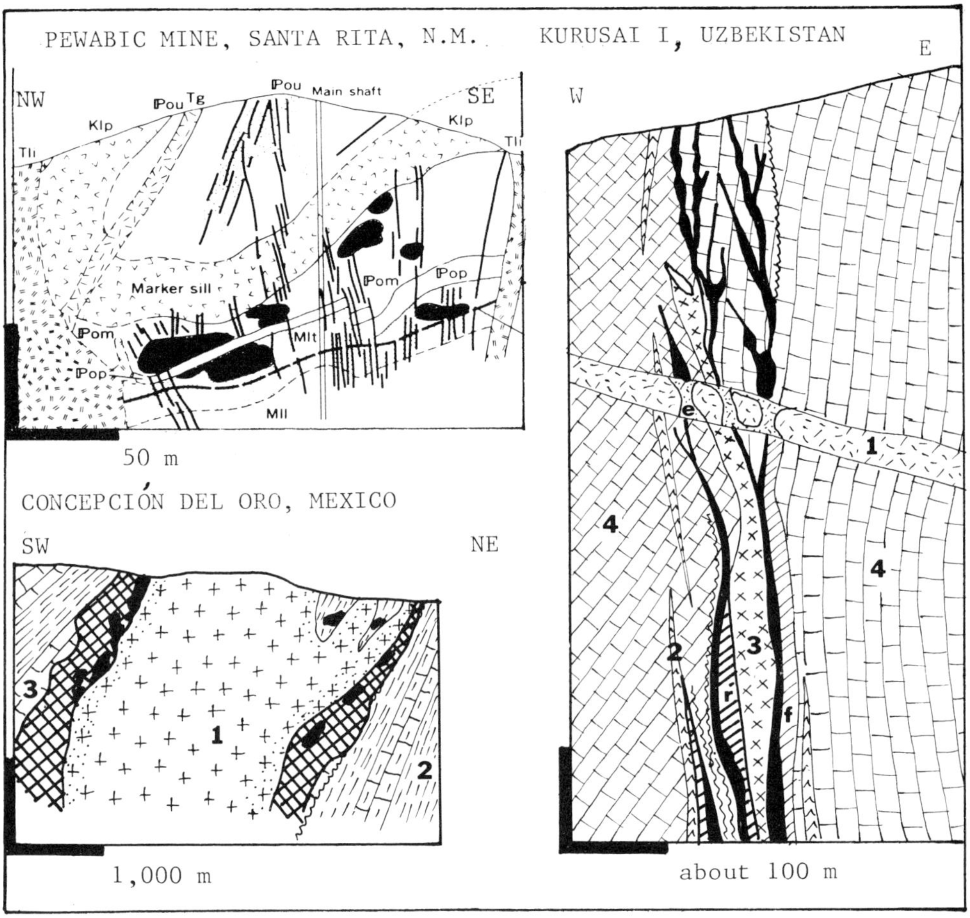

Fig. 28-93. Examples of Zn-Pb and Zn-Cu skarns.

Pewabic mine: Tli=T_1 rhyodacite porphyry; Tg=epidotized granodiorite porphyry; Klp=Cr_3 hornblende quartz diorite; Pou, Pom=Ps limestones; Pop=Ps shale; Mlt, Mll=Ms limestone.

Concepción del Oro, Mexico: 1=Eo-Ol granodiorite; 2=Cr_3 thin bedded limestone and black shale; 3=J limestone. Black area: mineralized grossularite skarn; grid: wollastonite skarn and marble.

Kurusai I (Kurama Range, Uzbekistan): e=epidote endoskarn; r=reaction skarns; f=forsterite calcite hornfels; l=Pe syenite porphyry; 2=quartz porphyry and diabase; 3=diorite, syenodiorite; 4=Cb_1 marble; 5=D3 marble; black: vein skarn with sulphides. From Jones et al. (1967), LITHOTHEQUE (diagrammatic) and Lukin et al. (1968).

Table 28-21. Zn-Pb skarns, selected examples of deposits and ore fields

LOCALITY	HOST UNITS AND INTRUSIONS	MINERALIZATION	REFERENCES
Hanover, Santa Rita distr.,New Mexico, U.S.A.	Cb limestone; T_1 granod.stock, granod.porphyry dikes	sphal.,lesser gal.,cp.,pyr.,moly., in garn.,pyrox.,amphibole skarn; 920Tt Zn/14%; 132Tt Pb/0.3%; 70Tt Cu/1%; 658t Ag/112 ppm	Einaudi et al. (1981)
Naica, Chihuahua, N.Mexico	Cr_1 folded limest.,shale; T rhyol. dikes, qtz.-feldspar porphyry	42 chimney and manto oreb.of dissem. pyr.,gal.,sphal.,lesser cp.,arsenp., pyrrh.,scheel. in wollast.,garnet, idocr.,hedenb. skarn; fault control. 620Tt Pb/4.4%; 550Tt Zn/3.9%; 47Tt Cu/0.37%; 1,875t Ag/133 ppm;	Stone (1959)
Concepción del Oro, N. Mexico	J-Cr limest.,shale,sandstone; hornf.,skarnized; 40-38 m.y. porph.granod., qtz.monzonite	dissem.and mass.magn.,specul.,pyr., cp.,sphal.,gal. in garn.,epid.. diops.,wollast. rim skarn. 300Tt Cu, 30t Au; 633 Tt Pb; 546 Tt Zn	Buseck (1966)
El Mochito mine, Honduras	Cr limest.,red-beds clastics, shale, skarn; block-faulted, graben; T diab.dikes,granod. stock postul.in depth	fault-control.distal skarn; pipe-like bodies of magn.,sphal.,gal.,pyr., cp.,pyrrh. in Mn-hedenb.,andradite, ilvaite,bustam.skarn; 857Tt Zn/8.2%, 508Tt Pb/4.84%; 1,617t Ag/154 ppm	Eng.and Min. November 1977
Aguilar, northern Argentina	Cm quartzite, limestone; T_1 quartz monzonite stock	sphal.,gal.,pyr.,ars.,pyrrh. mantos and columns // intr.cont.and faults in hedenb.,garn.,wollast.,rhodonite, amphib.,chlorite skarn; 1.8 Mt Zn/16%; 1.38Mt Pb/11%; 2.7Tt Ag	Angelelli (1950)
Kurama Range, Uzbekistan, Soviet Central Asia	D_2-Cb_1 limest.,dolom.,andes., tuff,rhyol.porph.; Cb_2 granodior.,gran.porph.,stocks,dks.	sphal.,gal.,pyrrh.,magn.,cp.,in mantos and pipes fringing porph. dikes, in skarn and marble. Est.2Mt Zn,Pb	Smirnov and Gorzhevsky (1974)

Tienpaoshan, Sikang, China	Or, limestone; MZ quartz monzon. stock	sphal.,gal.,cp.,pyr. masses and mantos in hedenb.,epid.,fluor.skarn; min.180Tt Zn, 150Tt Pb, 56Tt Cu	Hsieh (1950)
Shuikoushan, Hunan, China	Pe limestone; MZ-T quartz monzonite stock	sphal.,gal.,pyr.,cp.,irregular masses in contact garnet,pyrox.,epid. skarn; 750Tt Zn, 690Tt Pb, 900t Ag	Hsieh (1950)
Yeonhwa, Ulchin distr., South Korea	Pt_1 basem.granite; Cm-Or folded carbonate,quartzite,shale; 94-50 m.y. granod.,qtz.monzon. porphyry	several skarn oreb.in a 25 km long E.-W. belt; pyrrh.,sphal.,lesser gal.,cp.,in high-Mn pipe-like bodies of grandite-andrad.,rhodon.,bustam., Mn-pyrox. 723Tt Zn, 84Tt Pb	Yun and Einaudi (1982)
Tetyukhe, Sikhote Alin, south-eastern U.S.S.R.	Cb to Cr_1 shale,sandst.,limestone incl.silic.karst brecc., Cr_3-T_1 contin.andes.,dac.,rh. T_1 qtz.porph.,dior.,granod.	sphal.,gal.,pyrrh.,dissem.and masses in skarn mantos and pipes (Mn-hedenb.,ilvaite skarn) and replac. in marble; min.2Mt Zn,1.6Mt Pb, 2 Tt Ag	Magak'yan (1968)
Kamioka, N.C. Honshu, Japan	Pt_1 hornbl.and biot.gneiss, marble,calc-silic.gneiss; Cb-Pe amphibolite,m-gabbro, m-quartzite; J? granitization; Cr,qtz. and gran.porph.	80 small sphal.,gal.,cp.,pyr.,graphite,pyrrh.oreb. in hedenb.skarn and partly in marble; 3 major clusters; 4.44Mt Zn/5.7%; 624Tt Pb; 2.7Tt Ag; 6,240 t Bi; 338t Te	Nishiwaki et al. (1970)
Chichibu mine, C. Honshu, Japan	Cb-Pe litharen.,slate,chert, limest.,maf.tuff,m-basalt; hornfelsed,cont.skarn; Mi, qtz.dior.porph.,qtz.porph.dk.	pyr.,pyrrh.,magn.,cp.,in skarn; pyr.,sphal.,gal.,cp.,stibn.,bismuthin.,Au, etc.in brecc.superimp.on skarn and limest.; 53Tt Zn, 9Tt Cu	Miyazawa (1970)
Nakatatsu mine, C. Honshu, Japan	Cb_2 limest.,black slate, diabase; J_2-Cr_1 sandst.,shale; Cr_3-T_1 (60 m.y.) gran. porph.	sphal.,lesser gal.,cp.,in a penecon-cord. manto in clinopyrox. skarn; 385Tt Zn/5.5%; 35Tt Pb/0.5%; 21Tt Cu/0.3%; 210 t Ag/30 ppm	Shimizu and Iiyama (1982)

Table 28-22. Pb-Zn replacement sulphide deposits in carbonates, selected examples

LOCALITY	HOST AND INTRUSIVE UNITS	MINERALIZATION	REFERENCES
Bluebell mine, Riondel, British Columbia, Canada	Cm_1 calcite to dolom.marble, m-quartzite, micasch.,graph. sch.,calc-silic.hornfels, tactite; 170-30 m.y. qtz. monz.plut.and stocks in area	tabular replaced ore mantos in 3 zones, not in direct cont.with intrus. Coarse pyyrh.,sphal.,gal.,knebelite, quartz, calcite; 280Tt Pb	Ohmoto and Rye (1970)
La Encantada, N.W. Coahuila, Mexico	Cr, limest.,shale; recryst., partly skarn; T qtz.monzon. stock in depth	high-grade ore chimney; gal.,sphal., magn.,pyr.,proustite in limest.,underlain by sulph. skarn. Almost all prod. comes from oxid.zone with cerussite, argentojaros.,angles. 769Tt Pb/12%; 2.52Tt Ag/400 ppm	Lozej and Beales (1977)
Providencia, Zacatecas, Mexico	J_3-Cr limest.,shale; folded, hornfelsed; Ol (40 m.y.) granodior. stock	elong. replac. pipes; sphal.,gal.,pyr. calc.,quartz; high proport.of oxidiz. ore; 633Tt Pb/6.5%; 546Tt Zn/13%; 3.24Tt Ag/200 ppm	Sawkins (1964)
Cerro de Pasco, central Peru	Tr_3-J_1 black limest.,dolom., silic.shale resting on D black phyllite; Mi_2 (14.4-14.2 m.y.) vent breccia, qtz. monzonite dikes	early stage mass.repl.bodies of pyrrh. sphal.,arsenop. and pyr.,silica in carbonate, followed by late stage qtz. pyr.,enarg.,luzonite veins and Ag-bonanza formed by hydroth. leaching; 100Mt pyr.,7 Mt Zn; 3Mt Pb; 48Tt Ag	Einaudi (1977)
Trepča, Kosovska Mitrovica, Yugoslavia	Or-S marble, phyll.,qtz.-seric. sch.,m-quartzite; Mi vent fac.brecc.pipe comp. of qtz.-seric.alter.trach., dacite fragm.,carbonate; dacite-granod. plugs	massive, coarse sphal.,gal.,pyrrh., arsenop.,cp.,etc. in pipe-like or irreg.bodies replace marble or early skarn near vent; 3Mt Pb/6%; 2Mt Zn/4%; 5Tt Ag/100 ppm	Janković (1982)

Olympias mine, Chalkidiki Pen., Greece	Pt,PZ micasch.,gneiss, amphibolite, marble; T aplit.granite,granod.	peneconc.tabular replac.body 400x 1,500x12 m; pyrrh.,pyr.,sphal.,gal., ars.,cp.; Mn-oxide rich gossan; 450Tt Zn; 360Tt Pb; 240Tt As; 1.3Tt Ag; 70 t Au	Marinos (1982)
Laurium, Attica, Greece	PZ marble, micasch.; MZ slate, congl.,limest.,greenstone metavolcanics; Mi? granod.porph.sills,buried stock; minor skarn	gal.,sphal.,pyr. replac. of carbonate along faults and bedd.planes,less superimp.on skarn; karsting,oxid. zone rich in Ag; 497Tt Zn, 1.21Mt Pb, 8,010 t Ag	Marinos (1982)
Argun R.belt (Nerchinsk), East Transbaikalia,USSR	Cm_1 black argill. dolom., limest.,shale; J_{2-3} qtz. porph. dikes, stocks	pipe,vein,lens replac. in jasperoid and dolomitiz. limest.; gal.,sphal., pyr.,arsenop.,stannite, tetr.; 2.3Mt Pb; 1.5Mt Zn, 800t Ag	Smirnov and Gorzhevsky (1974)

NOTE: North American fields Metaline, Leadville, Gilman,Tintic, Santa Eulalia, are treated in Table 20-8.

Virtually all the orebodies are distal with respect to intrusive plutons or stocks, but could be in immediate contact with minor porphyry dikes or sills (as in Leadville).

The orebodies are strongly structurally and lithologically controlled, and the most common alteration associated with the ore is pervasive or partial silicification resulting in jasperoid (metasomatic chert). When jasperoid is developed, the sulphides either occur as fine disseminations in jasperoid giving it a gray to black colour (e.g. in the Taylor ore field, Nevada), or they form rich, often coarse crystalline masses, outside the jasperoid on the site of replaced marble masses or as cavity fillings. Jasperoid and its characteristics have been treated in detail by Lovering (1972). In a later paper (Lovering and Heyl, 1974) the difference between a barren and mineralized jasperoid and application of jasperoid as a guide to concealed ore, have been discussed.

The characteristic shapes of Pb-Zn sulphide orebodies are peneconcordant sheets to lenses (mantos) controlled by lithology (stratigraphy) and also by bedding plane dilations and paleokarst; subvertical chimneys and pipes and a variety of mineralized breccias, veins and stockworks (Fig. 28-94). Until recently, it has been automatically assumed that all the carbonate-hosted orebodies are replacements, but this is no longer the only possible explanation. Callahan (1977) suggested that many of the well-known "mantos" in the western Cordillera are hydrothermally modified and reconstituted earlier deposits of the Mississipi Valley-"type". The most important piece of evidence applied by Callahan was the widespread presence of karst features such as open caves, solution collapse breccias, etc. at many localities, recorded in earlier publications (e.g. in Leadville, Tintic, Goodsprings). This genetic alternative together with some description and cross-sections of the example localities is treated in Section 20.10.1. and will not be repeated here. Even if the Zn-Pb ores were not earlier and were introduced by granite-related hydrotherms, they could have still precipitated in a paleokarst so this represents an important structural control.

The carbonate-hosted Zn-Pb orebodies may occur alone (no skarns present), or alternatively may coexist with skarns. In the latter case, skarns occur in closer proximity to the intrusion. Transitional orebodies between true skarns and carbonate replacements are also known. In the Bluebell mine, British Columbia (Ohmoto and Rye, 1970), coarsely crystalline pyrrhotite-marmatite rich massive to disseminated

---

Fig. 28-94. Selected examples of Zn-Pb sulphide manto and chimney deposits in carbonates in granite aureoles.

Gilman: 1=Cr_3-T_1 quartz biotite porphyry; 2=Ps shale; 3=Ms_2 dolomitized limestone; 4=Cm_3-Ms_2 quartzite, sandstone, shale, dolomite; 5=Pt_2 metamorphics.

Ophir: 1=01 rhyolite and quartz monzonite porphyry; 2=Cm_{2-3} limestone, dolomite; 3=Cm_1 hornfelsed shale, shaly limestone; 4=Cm_1 quartzite.

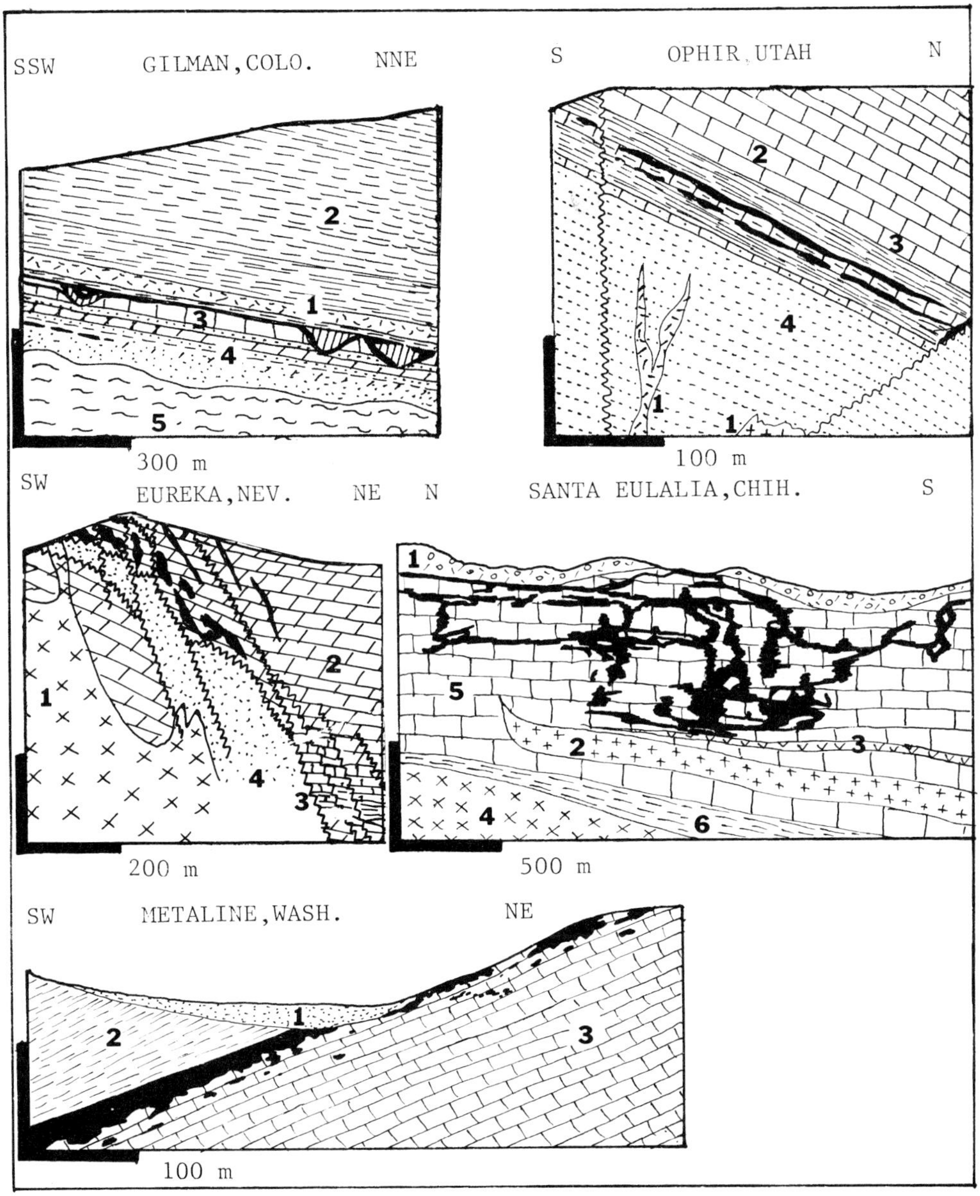

Eureka: 1=Cr_3 quartz diorite; 2=Cm_2 massive dolomite and limestone; 3=Cm_2 black limestone and shale; 4=Cm_1 quartzite.

St. Eulalia: 1=T,Q conglomerate, tuff, volcanic flows; 2=T_2 rhyolite, microgranite; 3=T_2 diabase, microdiorite; 4=granodiorite; 5=Cr thick-bedded limestone; 6=MZ carbonate evaporite association.

Metaline: 1=Q cover; 2=Cm slate and argillite; 3=Cm gray limestone.

From LITHOTHEQUE, after Radabaugh et al. (1968), Gilluly (1932), Nolan (1962), Hewitt (1943), Dings and Whitebread (1965).

sulphides replaced Cambrian limey to magnesian marble beds enveloped by metaquartzite and micaschist. Quartz, calcite and knebelite (Fe-Mn olivine) occur in the ore, and tremolite, diopside and talc have a patchy distribution in the marble and form beds of a calc-silicate hornfels.

28.14.5. Examples of Pb-Zn(Ag) deposits and districts (Tables 28-19 to 22, Fig. 28-95)

FREIBERG, SAXONY (EAST GERMANY); A STRUCTURALLY CONTROLLED VEIN FIELD IN THE ROOF OF A BURIED PLUTON

Freiberg ore field, Erzgebirge (Baumann, 1976; Figs. 28-95 and 96) is hosted by monotonous and lithologically uniform biotite gneisses and migmatites. These metamorphics form a gentle dome, believed to be underlain in depth by a late Paleozoic granitic pluton. The field contains about 1100 discrete mineralized fissure veins, of which about 100 were productive. The bulk of the veins occur within a 5x12 km N.N.E.-trending belt situated approximately at the centre of the dome. The remainder are located in a less distinct E.-W. system along the dome fringe.

Baumann (1976) distinguished two types of fissures which he interpreted as shear joints and feather joints. The shear joints (fissures) have a great strike extension (up to 20 km) and dip vertically. Their width ranges up to 6m, and they are filled by breccia, rock powder and mylonite. The ore minerals they carry have often a form of impregnations. The feather joints trend N.W.-S.E., dip 30-70° S.W., are shorter (average 2km), and filled by massive to banded ore veins.

The hydrothermal veins belong to two major depositional cycles: (1) Late Paleozoic (Permo-Carboniferous) post-magmatic cycle, associated with the "granite" and (2) Triassic to Tertiary cycle of a high-level "activation" (Fig. 28-97). Two major vein associations formed during the first cycle: (a) quartz, arsenopyrite, pyrite, sphalerite, galena and (b) siderite, pyrite, galena, freibergite, jamesonite, Ag-sulphosalts and native silver. The wallrock alteration was a comparatively light sericitization, silicification and carbonatization. The second cycle produced the following association: (c) fluorite-barite, hematite, lesser galena, sphalerite and marcasite; (d) cherty quartz, carbonate, skutterudite, niccolite, proustite, argyrodite and (e) quartz, hematite, Mn-oxides. The alteration was gentle bleaching, argillization and quartz impregnation. The zoning pattern is indistinct, expressed in terms of the maxima of metal concentration, from the centre outwards: ZnPb-PbAg-AgSb.

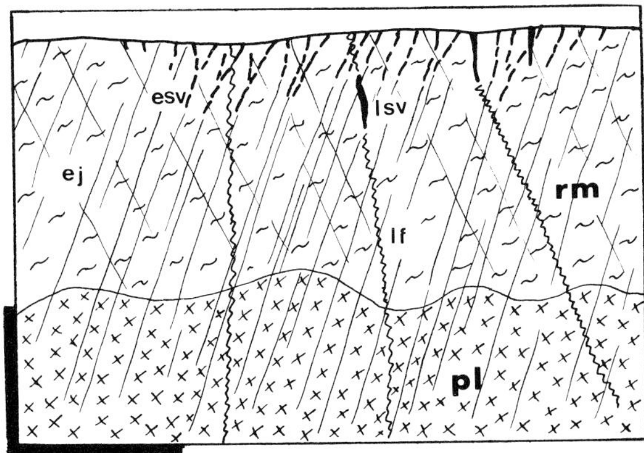

vert. 1 km, horiz. 5 km

a Mineralization in two regular systems of joints filled by early stage veins (esv) above buried granite (pl). Some veins were rejuvenated (lsv) along faults (lf). rm=roof metamorphics (Pt). Freiberg, Saxony, East Germany. Permo-Carboniferous and some Mesozoic Pb,Zn,Ag minor Cu,Sn ores.

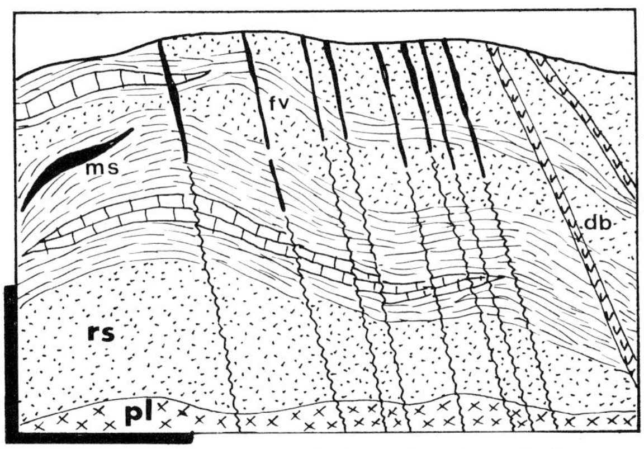

vert. 1 km, horiz. 5 km

b Mineralization in fissure veins (fv) filling faults and fault zones in Cb_1 "graywacke" and slate. Stratiform massive sulphide orebody (ms; Rammelsberg) is in D_2 slates. db=diabase rs= D_1-Cb_1 roof sediments; pl=postulated buried pluton. Oberharz district, West Germany; Cb_{2-3} ores

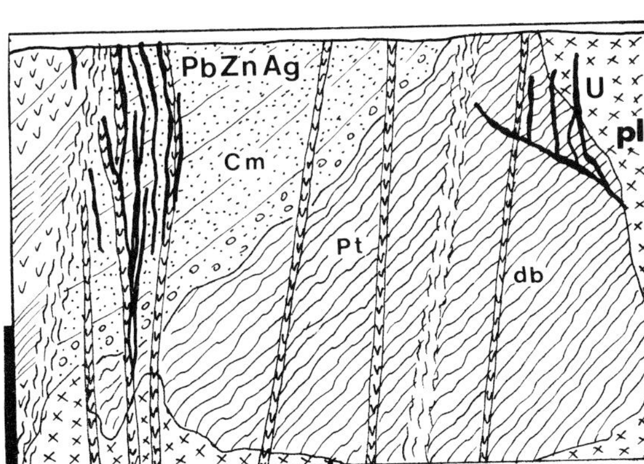

1 km

c Fissure Pb-Zn-Ag veins (on the left) are closely affiliated with diabase dikes along fissures, branching from a major N.E. fault. U veins (on right) with lesser Pb,Zn,Ag,Cu are along faults and shears near a plutonic (pl) margin. db=diabase. Příbram district, Czechoslovakia; Cb to MZ? ores

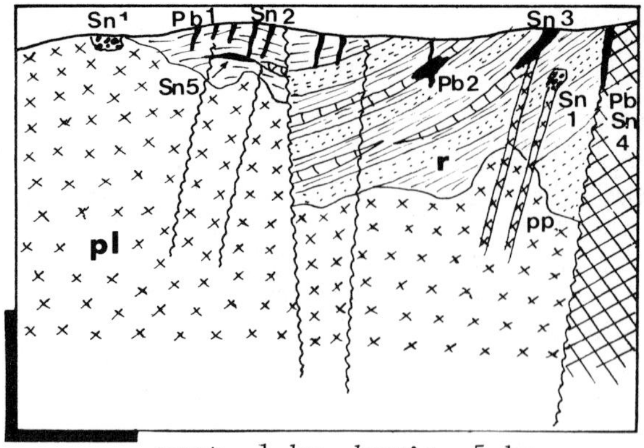

vert. 1 km, horiz. 5 km

d A variety of Pb-Zn-Ag (=Pb) and Sn ores hosted by Pt_3-D sediments and minor volcanics, intruded by D quartz monzonite and granite (pl), Zeehan-Renison Bell area, W.Tasmania. pp=porphyry r=roof rocks; op=ophiolite Pb 1=fissure veins; Pb 2= carbonate replacements; Sn 1=greisens; Sn 2=stannite veins; Sn 3-cassiterite-pyrrhotite replacements; PbSn 4=replacemens in altered ultramaf.

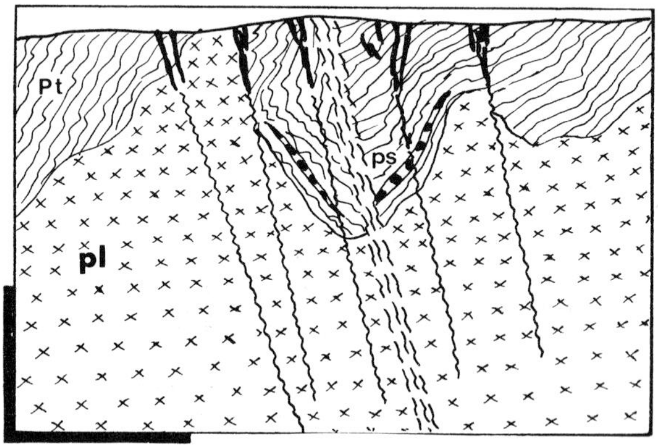

vert. 1 km, horiz. 5 km

e Galena-sphalerite and siderite-tetrahedrite replacement veins in bleached Pt_2 fine detrital sediments, along faults above Cr_3-T_1 pluton (pl). ps=concordant pyritic slate. Coeur d'Alene district, Idaho

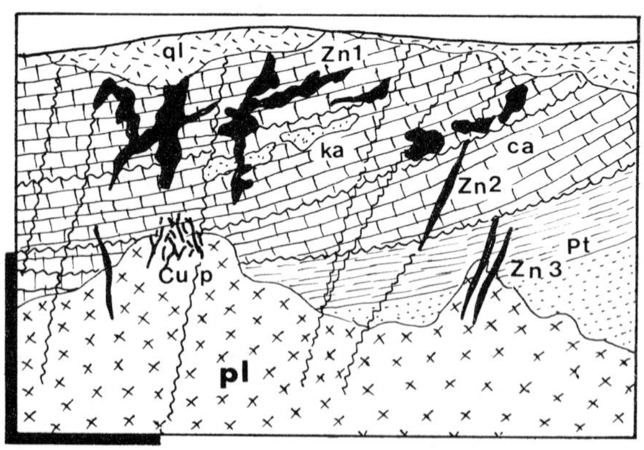

vert. 1 km, horiz. 2 km

f Zn-Pb replacement bodies in silicified (jasperoid) Cm-Cb carbonate (Zn 1); replacement veins (Zn 2); fissure veins in altered PCm basement (Zn 3); Cu p=porphyry Cu in Ol qtz. monzonite pluton (pl); ka=karst cavities,solution collapse breccias; ca=carbonate. Tintic district, Utah

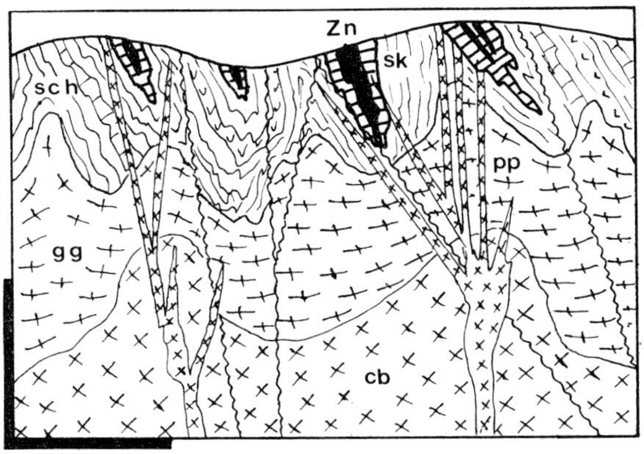

vert. 1 km, horiz. 2 km

g Replacement sphalerite, galena, pyrrhotite, pyrite etc. masses in skarn (sk) in Pt_2-PZ high-grade metamorphics (sch) at and near contact with Cr porphyry dikes (pp) believed related to a cryptobatholith (cb). gg=J ? katazonal granite, granitization. Kamioka district, Honshu, Japan.

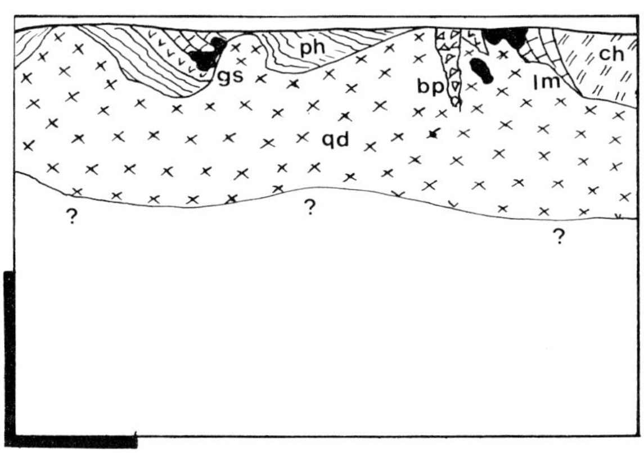

vert. 1 km, horiz. 2 km

h Zn–Cu, minor Pb, skarn orebodies at contact of Cb-Pe marbles (lm), members of phyllite (ph), greenstone (gs), chert (ch) association with Mi ? quartz diorite (qd). Chichibu Mine, Honshu, Japan.

Fig. 28-95. Diagrammatic representation of the major styles of postmagmatic Zn-Pb ore fields, genetically associated with buried and exposed granitic plutons. Not accurate in detail. The dimensions of important features (orebodies, fault zones) are exaggerated and telescoped. The approximate scale represents the overall dimension of the area considered only. From LITHOTHEQUE, based on references listed in the text and field reconnaisance.

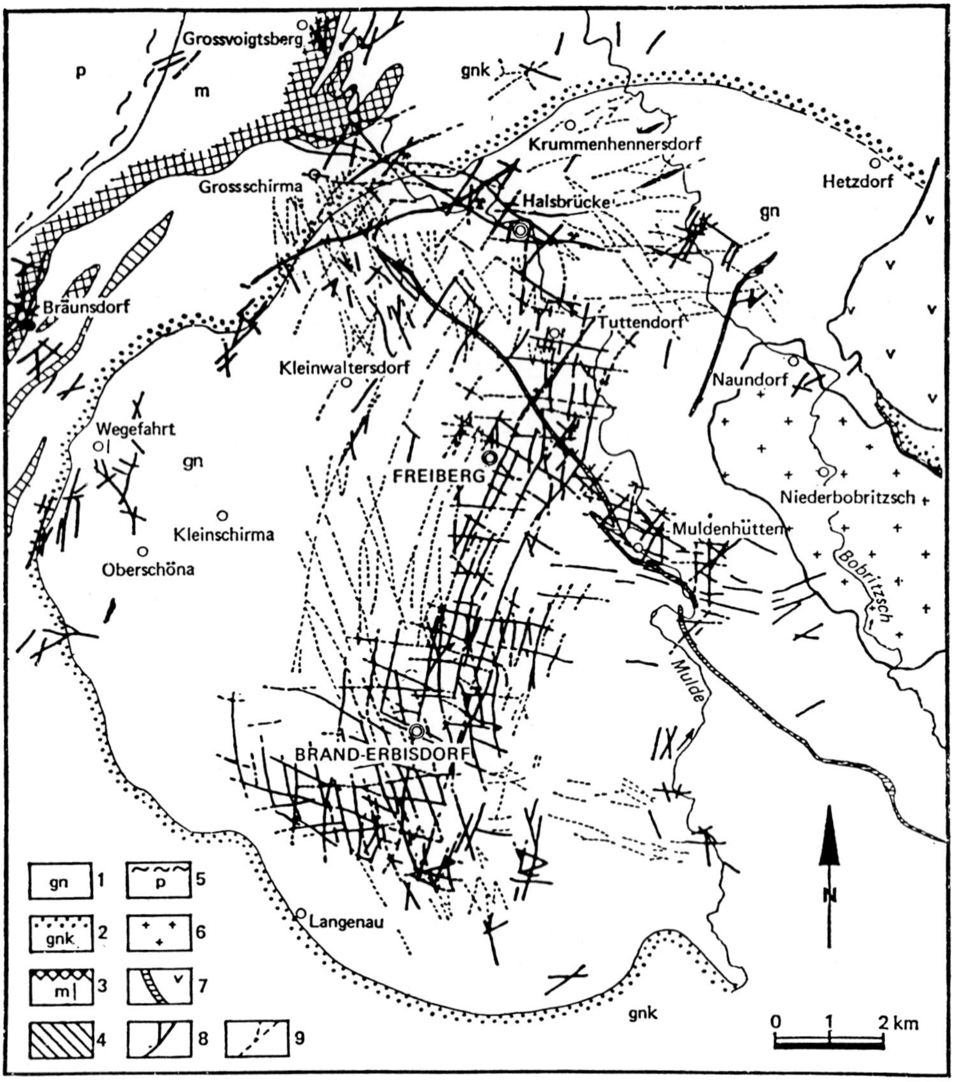

Fig. 28-96. Central part of the Freiberg (Erzgebirge) polymetallic vein ore district. 1=Biotite gneiss; 2=mica gneiss; 3=micaschist; 4=red gneiss; 5=phyllite; 6=PZ_3 granite; 7=porphyry; 8=ore veins; 9=tectonic structures. From Baumann (1976), courtesy of the Scottish Academic Press.

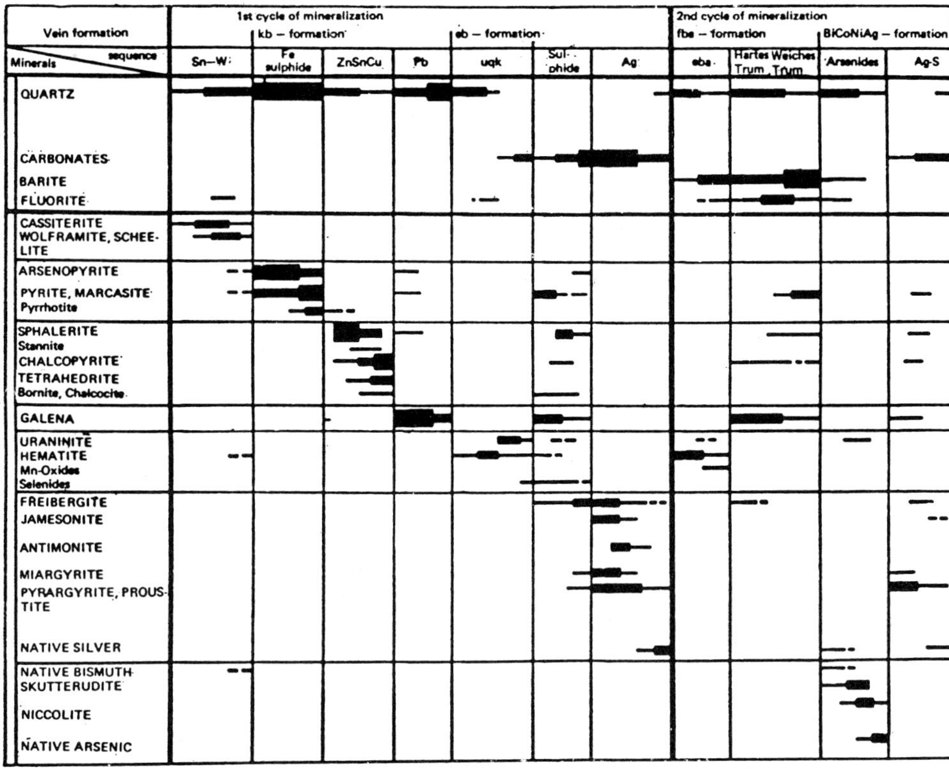

Fig. 28-97. Mineral associations and sequence of mineralization in the Freiberg (Erzgebirge) polymetallic ore veins. kb=pyritic sphalerite–galena association; eb=precious metals and siderite association; uqk=U, quartz, calcite association; eba=Fe-barite association; fba=fluorite-barite association. From Baumann (1976), courtesy of the Scottish Academic Press.

OBERHARZ DISTRICT, WEST GERMANY: A SYSTEM OF PERSISTENT, SUBPARALLEL MINERALIZED STRIKE-SLIP FAULTS (Fig. 28-95b)

Oberharz (Upper Harz, or Clausthal-Zellerfeld district) is a mineralized area, measuring 22x25 km in the northwestern extremity of the Harz Mountains (Buschendorf et al., 1971). This is part of the late Paleozoic foldbelt consisting of a sequence, several thousand metres thick, of Devonian marine sandstones, slates and minor limestones, unconformably topped by lower Carboniferous siliceous slate and quartz-rich litharenite interbedded with slate. The middle Devonian black slates contain the important stratabound massive sulphide deposit Rammelsberg near Goslar, described in Chapter 17. At the end of Devonian the supracrustal sequence was folded, faulted and intruded by a bimodal association of diabase and keratophyre.

In the upper Carboniferous period, the complex was intruded by granite, disrupted by a system of subparallel W.N.W.-trending strike-slip faults with a variable vertical downthrow component, and mineralized. The ore deposition took place in four or more phases of open space filling and intra-vein metasomatism. The vein structures are up to 20 km long and up to 50 m wide, filled by breccia, rock flour and gouge, but the economic ore was situated in more restricted sectors (up to 6 km long). Nine major fields exploiting one or more veins, have produced 1.8 Mt Pb, 700 Tt Zn, 4.7 Tt Ag and about 32 Tt Sb, till 1980.

The Grund field alone, producing from three vein structures, accounts for about one half of the production (Sperling, 1973; 950 Tt Pb, 400 Tt Zn, 2 Tt Ag). There, the ore shoots consisted of mineralized fault breccia, banded veins and linear fault zone stockworks. Quartz is the most common gangue, followed by siderite, calcite, barite. Sulphides are represented by several generations of sphalerite and galena, lesser bournonite, boulangerite, tetrahedrite and Ag-sulphosalts. The wallrock alteration is light: sericitization, local light silicification, ankerite or siderite impregnation or veining and bleaching.

COEUR D'ALENE DISTRICT, IDAHO: Pb,Zn,Ag REPLACEMENT VEINS WITH A 1.4 B.Y. HISTORY OF MINERALIZATION

Coeur d'Alene district (near Kellogg and Wallace, N.Idaho; Fryklund, 1964; Gott and Cathrall, 1980; Figs. 28-95e, 28-98) represents 6.9 Mt Pb, 3 Mt Zn, 140 Tt Cu, 31 Tt Ag and at least 70 Tt Sb in terms of metal quantities accumulated in its ores. The core of the district is a system of subparallel W.N.W.-trending "mineral belts", developed on both sides of the major Osburn Fault. The latter is a strike-slip and normal fault with a lateral displacement of up to 26 km and some 5,000 m of vertical offset. The district has approximate dimensions of 40x20 km and covers an area of 780 km^2. The bulk of orebodies are hosted by a sequence of monotonous middle Proterozoic marine detrital sediments (slate, argillite, siltstone, quartzite) at least 7,000 m thick, deposited on a stable continental margin basement ("miogeocline"), and later metamorphosed in the lower greenschist facies. The sediments have been strongly deformed during the "Laramide orogeny" and intruded by several small stocks and dikes of quartz monzonite, diabase and lamprophyre of Cretaceous to lower Tertiary age that may be a part of a larger buried pluton.

The eight or so Coeur d'Alene "ore belts" are W.N.W.-trending, south dipping second-order fault structures, parallel with the Osburn Fault. Both shears and tensional fractures are represented. The principal faults as well as branching third order fissures are mineralized. Most characteristic are the steeply dipping veins replacing fault gouge, mylonite and breccia. The veins tend to have been narrow at outcrop, but many widened with increasing depth. The productive veins shortly postdate emplacement of the "Laramide"

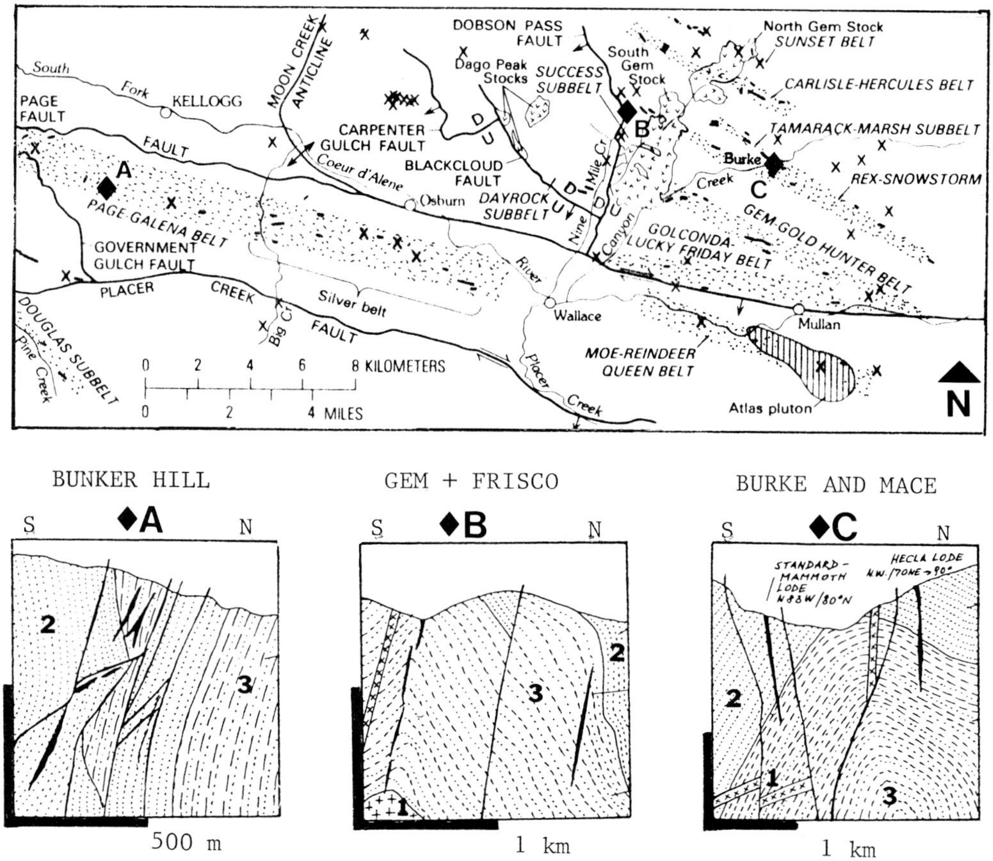

Fig. 28-98. Coeur d'Alene Pb,Zn,Ag district, Idaho. TOP: location map, from Gott and Cathrall (1980). BOTTOM: examples of deposits, from LITHOTHEQUE: 1=Cr_3-T_1 quartz monzonite; 2=Pt_{2-3} quartzite; 3=Pt_{2-3} argillite, siltite, quartzite interbeds.

intrusive stocks, and two mineral associations are of the greatest importance. These are as follows: (1) siderite, ankerite, galena, sphalerite with occasional biotite, garnet and grunerite gangue, and pyrrhotite, magnetite, tetrahedrite (e.g. the Bunker Hill mine) and (2) quartz, siderite, Ag-tetrahedrite veins (e.g. Sunshine, Galena mines).

The former are thick and account for all the lead and zinc produced plus some silver. The latter are thin, but they produced the bulk of the silver. Gold-quartz, stibnite and uraninite veins have been of little importance. The most widespread and conspicuous wallrock alteration in the district is a bleaching, ranging from slight sericitization to a mere pigment removal, as well as impregnation and thin veining of the wallrocks by siderite or ankerite. The bleaching is a consequence of the prevalent wallrock composition, that already includes a rock-forming sericite that remained stable in the presence of mineralizing fluids. In the few examples of "granite"-hosted veins, the alteration is a feldspar-destructive sericitization. There have been six dated periods of vein mineralization in the district, ranging from Precambrian (uraninite veins) to Tertiary.

Widespread occurrences of stratabound copper mineralization are known in the host sequence outside of the Coeur d'Alene district, and the basal sedimentary unit in the district (Prichard Formation) contains several horizons of pyritic phyllites and it correlates stratigraphically with the unit hosting the important Sullivan orebody, British Columbia (Volume 2). This, as well as the "ancient" isotopic lead composition and the low temperature of the ore fluids (130-175° C) makes it tempting to assume some form of remobilization of metals initially brought in during the Proterozoic, by the "Laramide" intrusions.

ZEEHAN, TASMANIA; Pb-Zn-Ag VEINS IN THE ROOF OF A "TIN GRANITE"

Zeehan ore field is in western Tasmania and has dimensions of about 15x15 km (Blissett, 1962; 190 Tt Pb, 780 t Ag; Fig. 28-95d). There, block-faulted late Proterozoic and Cambrian quartzite, slate, siltstone and minor greenstone and Ordovician to Devonian sandstone, limestone and slate, are believed to be located in the roof of a Devonian (Mt. Heemskirk) granite pluton. About 30 fissure veins and an isolated carbonate replacement orebody formed along N.N.W. and N.N.E. faults. The veins carry relatively short (X0-330 m) and narrow but rich ore shoots composed of quartz-siderite or siderite and galena, brown sphalerite, pyrite, lesser chalcopyrite and tetrahedrite. A small replacement galena, pyrite and sphalerite orebody in Mn-siderite envelope is hosted by Ordovician limestone in the Oceana mine.

The Heemskirk Granite west of the Zeehan field contains small relics of cassiterite greisen and vein orebodies. Near Zeehan, in the granite roof, tin accumulated in the Stannite Lode, a thin quartz-pyrite fissure vein containing scattered stannite, arsenopyrite, minor cassiterite, tetrahedrite, molybdenite, bismuthinite and galena. The calculated content of metals in this vein was 775 t Sn, 941 t Cu and 60 t Bi, although not all these metals have actually been produced. In the same group of mines, several overlapping thick lenses of pyrite closely associated with greenstones (meta-spilite) contained 0.3-0.4% Sn, mostly in the form of dispersed cassiterite. About 1 Mt of 1% Sn ore has recently been discovered in the vicinity. The Renison Bell replacement tin deposit in carbonates (Section 28.8.4.) is located about 8 km N.E. from the margin of the Zeehan field, probably above a buried Devonian granite cupola.

TINTIC ORE FIELD, UTAH; CARBONATE REPLACEMENTS IN THE AUREOLE OF AN INTRUSIVE STOCK WITHIN AN ERODED PALEOVOLCANO.

Tintic (Morris, 1968; Morris and Lovering, 1979; 1.01 Mt Pb, 167 Tt Zn, 127 Tt Cu, 8.3 Tt Ag, 82.5 t Au; Fig. 28-95f) is in a north-trending fault block range near the eastern margin of the Great Basin, central Utah. The host rocks range from Precambrian quartzite and argillite through a 3.3 km thick section of Cambrian to Carboniferous sediments, over 60% of which are limestone and dolomite. The remainder consists of quartzite and sandstone. The sediments have been extensively thrust-faulted during the late Cretaceous period, and carried eastward for as much as 160 km. The thrusts were subsequently folded and faulted. During the Oligocene period, a composite volcano including a caldera stage, buried the sedimentary range under a thick pile of latite ash flow tuffs, flows and agglomerates.

Several quartz monzonite stocks, plugs and dikes intruded the sediments and a portion of the volcanic pile. A widespread hydrothermal activity followed emplacement of the Silver City quartz monzonite. The earliest, pre-intrusion solutions converted many faulted limestone bodies to hydrothermal dolomite and caused chloritization and propylitization of the volcanics. This may have produced numerous cavity openings ("hydrothermal karst"), although the possibility of an earlier karst complete with Mississippi Valley-"type" Zn-Pb deposits was considered by Callahan (1977).

The early postmagmatic solutions were acidic, and they caused widespread advanced argillic alteration of silicate rocks along the faults. Slightly later fluids caused widespread pyritization, sericitization and silicification (jasperoid) in the carbonates. Metallic mineralization was associated with the latest, potassic alteration in the alumosilicate rocks (adularia, zunyite) and Mn introduction into the carbonates.

Over 90% of the metals produced came from large, irregular orebodies that have replaced the Paleozoic carbonates along faults above the quartz monzonite stocks and breccia pipes. The replacement bodies are composed of a cherty jasperoid containing barite, rhodochrosite, and scattered to massive sphalerite, galena, pyrite, lesser enargite and argentite. Economically the most important are large columnar masses (chimneys), followed by mantos and a variety of irregular bodies. Replacement veins in carbonates and fissure veins in the alumosilicate rocks are rare. The existence of a porphyry copper mineralization in one of the intrusive stocks has recently been revealed at some depth.

The Pb-Zn-Ag ores resulted from solutions that rose from numerous centres in the field, and spread in a manner resembling the trunk of a tree. The orebodies at higher elevations branch and divide, extending along bedding faults. In the downward direction, several of the "ore trunks" terminate in breccia masses, invaded by intrusive plugs and dikes.

CERRO DE PASCO, PERU; MASSIVE PYRITE, Zn,Pb,Cu REPLACEMENT IN CARBONATES ALONG A MARGIN OF A MIOCENE VENT

Cerro de Pasco is a giant deposit located in the metalliferous belt of central Peru (Einaudi, 1977; 100 Mt pyrite, 7 Mt Zn, 3 Mt Pb, 1.5 Mt As, 48 Tt Ag; see Fig. 17-7). It is hosted by an upper Triassic-lower Jurassic limestone and dolomite with interbeds of siliceous shale, unconformably resting on Devonian black phyllites. These sediments were block faulted and intruded by a Miocene (14.4 m.y.) vent facies breccia-conglomerate, followed by quartz monzonite porphyry dikes (14.2 m.y.). The dikes probably make contact with a buried intrusive stock. A large, N.-S. elongated funnel-shaped orebody, 1,800 x 300 m long, composed of almost massive pyrite mixed with chert, chalcedony and quartz, formed by replacement of Mesozoic limestone and minor Tertiary volcanics along a margin of the Cerro vent.

Einaudi (1977) interpreted the massive sulphide body as being an initially pyrrhotite, marmatite, arsenopyrite mass, altered by late fluids to pyrite-marcasite and FeS-poor sphalerite. The silicification and pyrite deposition in carbonates was contemporary with quartz-sericite, pyrite alteration of quartz monzonite and the volcanics. The sulphide mass is mineralogically zoned and the greatest accumulation of sphalerite and galena is on the fringe and above massive pyrrhotite pipes. The massive sulphide deposition was followed by a second stage of hydrothermal deposition, that produced quartz, sericite, pyrite, enargite and luzonite veins fringed by quartz-sericite alteration envelopes. Late-stage, low temperature hydrothermal leaching resulted in the formation of collapse breccias and the open spaces were mineralized by pyrite, hematite, gratonite, baumhauerite, aramayite, argentite, freibergite and light (low Fe) sphalerite and galena. This assemblage contains silver bonanza orebodies. Supergene secondary enrichment generated mantos of chalcocite-covellite, and a silver-rich limonitic gossan.

Trepča (Stari Trg) deposit near Kosovska Mitrovica, (Janković, 1982) is the largest Pb-Zn deposit of Yugoslavia (3 Mt Pb, 2 Mt Zn, 5 Tt Ag) and is somewhat reminescent of Cerro de Pasco in terms of association with a breccia-filled vent. The orebodies are irregular to pipe-shaped, coarse sphalerite, galena, pyrite, pyrrhotite, lesser chalcopyrite, magnetite and arsenopyrite masses. They replace Ordovician to Silurian limestone marble and an earlier hedenbergite, andradite, ilvaite, epidote, etc. skarn. The orebodies are situated immediately at the contact of a Miocene breccia pipe and dacite-granodiorite plug, or along a limestone-phyllite contact. The pipe fill is intensively silicified, sericitized, pyritized and carbonatized, and hosts minor occurrences of chalcopyrite, enargite and arsenopyrite.

CHICHIBU MINE, JAPAN: Zn,Cu,Fe SKARN IN A MAFIC VOLCANICS-CARBONATE ASSOCIATION

Deeper marine basalt (spilite), shale, litharenite, chert, limestone and often keratophyre association and its mineralization have been treated in Chapters 10 and 12. There, stratabound iron deposits (magnetite, hematite, siderite), Mn carbonates and Fe,Cu,Zn massive sulphides, constituted the distinct mineralization styles. When similar associations containing abundant limestone was later intruded by "granitic" plutons, skarn and carbonate replacement orebodies formed. Postmagmatic, "granite"-related ore-bearing complexes often display a remarkable degree of metallogenic heritage with respect to their supracrustal hosts (e.g. coexistence of Fe, Zn-Cu and Mn orebodies), the causes of which range from direct, in-situ remobilization of the earlier volcanic-sedimentary orebodies, to proximal mobilization of anomalous metal contents from the supracrustal unit.

The Chichibu ore field in central Honshu, Japan (Miyazawa, 1970; 4.7 Mt ore with 35.5% Fe, 1.12 % Zn, 0.19 % Cu, 0.07% Pb; Fig. 28-95h) is located in a Permo-Carboniferous suite of litharenite, slate, chert, lesser limestone, varicoloured tuff ("Schallstein") and metabasalt flows. These rocks have been intruded by a Miocene quartz diorite pluton and by quartz-porphyry dikes. All the supracrustals in the contact aureole have been extensively hornfelsed and a two-stage skarn (1: garnet-hedenbergite; 2: idocrase, wollastonite, ilvaite) formed along carbonate contacts. Tourmalinization, axinite alteration and silicification followed by sericitization and propylitization are widespread. About 12 small orebodies are located along the quartz diorite-carbonate contact, and they consist of an earlier pyrite, pyrrhotite, magnetite and chalcopyrite masses in skarn, and superimposed pyrite, sphalerite, galena, chalcopyrite, stibnite, bismuthinite and gold fillings in brecciated skarn and also as limestone replacements outside the skarn limit. The Mn ores are represented by rhodochrosite, Mn-ankerite and Mn-siderite replacing limestone, and by secondary oxides.

KAMIOKA, JAPAN: Zn-Pb SKARN IN A PRECAMBRIAN HIGH-GRADE METAMORPHIC TERRAIN, RELATED TO GRANITE PORPHYRY DIKES AND A PROBABLE CRYPTOBATHOLITH

The Kamioka ore field (Nishiwaki et al., 1970; 4.44 Mt Zn, 624 Tt Pb, 2,707 t Ag; Fig. 28-95g) is the largest zinc producer in Japan. About 100 small to medium-sized orebodies are hosted by hedenbergite, lesser epidote, garnet, actinolite, diopside and ilvaite skarn, and also by marble, in three major clusters. The ore field, about 8x4 km in cross-section, is situated in the crystalline Hida Block, composed of a lower Proterozoic core of tonalitic gneiss, migmatite and marble, with infolded Permo-Carboniferous mafic metavolcanics (now amphibolites), metacherts (quartzites) and metagabbros. The foliation strikes N.N.W., but turns almost 300° to N.N.E. owing to a composite anticline.

The metamorphics have been intruded and partly granitized by a foliated katazonal granite, probably dating from the Jurassic period and an augen gneiss. Following a rapid uplift and high-level open faulting in Cretaceous or Tertiary, quartz porphyry and granite porphyry dikes and small stocks probably related to a cryptobatholith penetrated along faults and produced skarn metasomatism in the marble. The skarn forms numerous horizons interfoliated with the alumosilicate rocks, and adjacent horizons often coalesce along anticlinal and synclinal axes and along drag folds. About 10% of the skarns carry ore, in the form of scattered sulphide and magnetite grains grading to almost massive pods. Much of the skarn contains relic graphite, recovered as a by-product. The porphyry dikes contain widespread scattered molybdenite flakes, and small localized stockworks. There is a strong possibility that there exists a Mo-stockwork mineralized cupola under the present ore field.

28.15. HYDROTHERMAL-PLUTONIC SILVER DEPOSITS

In plutonic association, silver deposits are substantially less conspicuous as a class than the epithermal "bonanza" Ag(Au) veins in the subvolcanic setting, treated in Chapter 26. The following styles of silver mineralizations can be recognized:

(1) Epithermal argentite, silver, Ag-sulphosalts identical with those described in Section 26.6.1., but hosted by the "basement rocks" (including older "granites") with few continental volcanics penecontemporary with the mineralization in evidence (e.g. Austin, Nevada; Fig. 28-99).

(2) Epithermal enargite-luzonite or tennantite-tetrahedrite veins hosted by the "basement" rocks (Cerro de Pasco, partly Butte).

(3) Ag-tetrahedrite (freibergite) veins, fracture stockworks and carbonate replacements in non-volcanic areas (e.g. Sunshine and Galena mines, Coeur d'Alene, Fig. 28-100).

(4) Silver-rich tops of Pb-Zn vein or replacement systems formed by supergene enrichment or a late-stage hydrothermal leaching and precipitation (e.g. Chañarcillo, Fig. 28-101).

(5) Silver-rich "Main Stage" and "Late Stage" veins or shoots in essentially Pb-Zn fields (Příbram, Kutná Hora, Czechoslovakia; Freiberg, Erzgebirge; Keno Hill, Yukon; Beaverlodge and Silbak Premier mine near Stewart, British Columbia, Fig. 28-102).

(6) Silver-only (no base metals) low-temperature veins, stockworks, replacements; Batopilas, Mexico, partly St. Andreasberg, Harz Mts.

(7) Disseminated ("porphyry") silver deposits, e.g. Rochester, Nevada, Fig. 28-103; partly Hahns Peak, Colorado.

(8) Ag,Co,Ni,Bi,U,As association veins (treated separately in Section 28.16.).

An ore deposit has to contain a substantial amount of silver minerals undiluted by substantial amounts of base metal sulphides, or

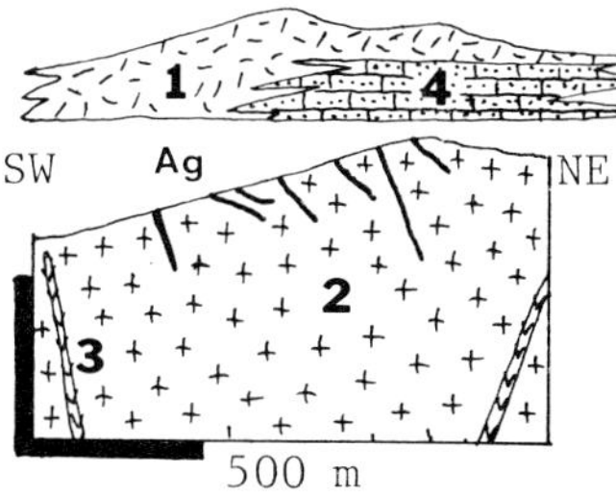

(top out of scale)

1. T, felsic lavas and pyroclastics
2. J ? quartz monzonite
3. J-T lamprophyre, aplite, pegmatite dikes
4. Cm? carbonatic arenite to sandy limestone

Fig. 28-99. Austin (Reese River) Ag vein field. Diagrammatic, from LITHOTHEQUE.

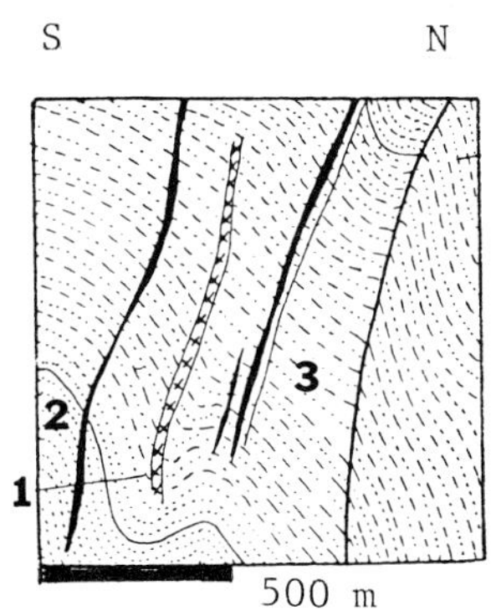

1. T_1 diabase and diorite dikes
2. Pt_{2-3} meta-quartzite
3. thin-bedded quartzite gradational into and interbedded with argillite, siltite and dolomitic, calcitic quartzite

Fig. 28-100. Sunshine Ag mine, Big Creek, Coeur d'Alene district, Idaho. From LITHOTHEQUE, based on data in Fryklund (1964).

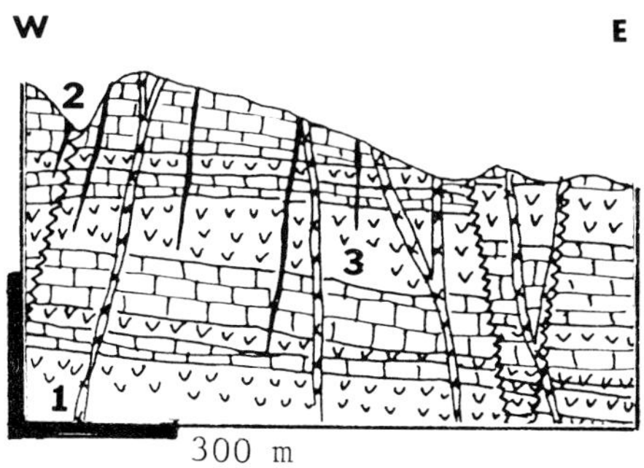

1. Cr_{1-2} (100 m.y.) granodiorite and diorite stocks and dikes
2. Cr_1 black, gray and white limestone, minor hornfelsed shale
3. Cr_3 andesite to keratophyre pyroclastics

Fig. 28-101. Chañarcillo Ag field, Atacama, Chile. From LITHOTHEQUE, after Whitehead (1919).

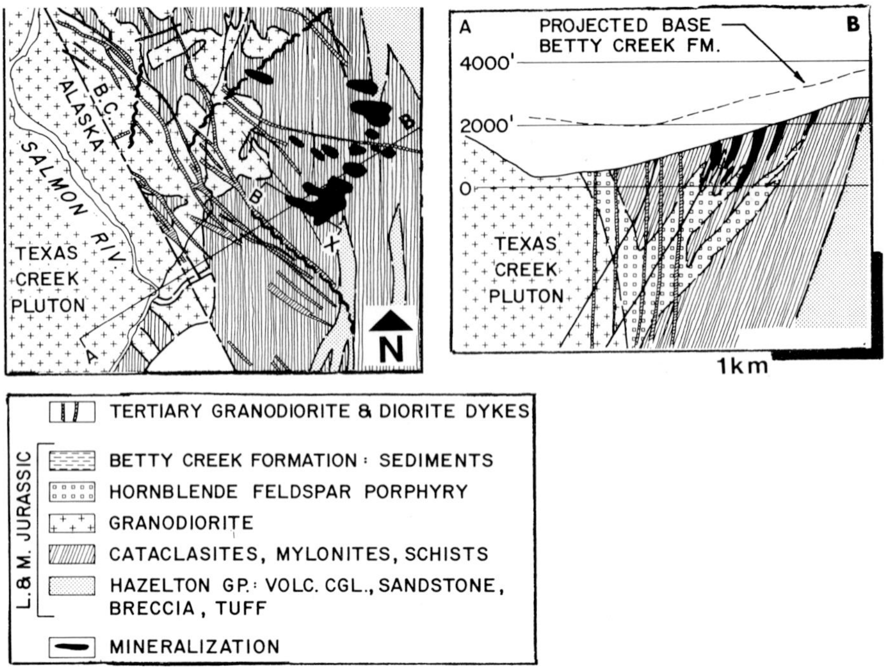

Fig. 28-102. Silbak Premier Au-Ag mine near Stewart, N.W. British Columbia. From Barr (1980), courtesy of the Canadian Institute of Mining and Metallurgy and the author.

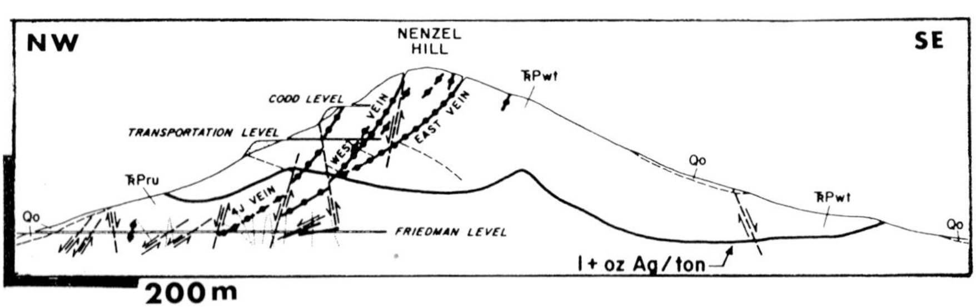

Fig. 28-103. Low-grade disseminated Ag mineralization, Rochester field, Nevada. Qo=Q alluvium; TrPwt=Tr-Pe rhyolite ash-flow tuffs and volcaniclastic rocks; TrPwf=rhyolite flows; TrPru=rhyolite flows and tuffs, undivided. From Vikre (1981), courtesy of Economic Geology.

must grade about 280 ppm Ag (=10 oz/t) or more, to qualify as a silver deposit. Many silver deposits of the above listed styles have been reviewed already in other sections, so selected examples only are considered briefly below.

Chañarcillo, 50 km S. of Copiapó, central Chile (Whitehead, 1919), produced 2,300 t of silver from a 40-150 m thick zone of supergene sulphide enrichment (argentite, native silver, dyscrasite, stephanite, amalgam), topped by a mineralized oxidation zone (cerargyrite, iodobromite, bromyrite, embolite, iodyrite). The hypogene ore is situated in a system of nearly vertical, parallel N. to N.E.-trending fracture or replacement veins containing only a low-grade, scattered galena, sphalerite, pyrite, chalcopyrite, proustite and pyrargyrite in calcite and barite gangue. The veins are located in a zone of facies transition between lower Cretaceous marine black limestone and partly continental andesite tuff, intruded by 100 m.y. old granodiorite and diorite stocks and dikes.

Austin, Nevada (Ross, 1953) had about 100 narrow, discontinuous N20-40°W striking, east-dipping fracture and shear veins, and minor silicified (replacement) ribs, hosted by a quartz monzonite massif (? mesozonal pluton) of probably Jurassic age. Quartz and pyrite are the most common vein minerals, and rhodochrosite, sericite, calcite, dolomite, galena, sphalerite, tetrahedrite, argentite, pyrargyrite, stephanite and polybasite are rare. Quartz, sericite and pyrite alteration is limited to the immediate vicinity of the veins. Most of the past production (600 t Ag, 540 t Pb, 60 t Cu, 0.5 t Au) came from proustite and tetrahedrite-rich shoots. The ores have epithermal characteristics and are probably related to high-level Tertiary intrusions, rather than to the postmagmatic phase of the host "granite".

Nenzel Hill, Rochester field, Nevada. This is a large, low-grade ("bulk") disseminated Ag(Au) deposit, described by Vikre (1981; Fig. 28-103) and estimated to contain about 100 Mt of 28 to 57 ppm Ag ore. It is hosted by Permo-Triassic rhyolite flows, ash-flow tuffs and volcaniclastics intruded by quartz monzonite to granodiorite stocks, related to the upper Cretaceous fringe of the Sierra Nevada Batholith. The intrusive exocontact is intensively quartz, sericite, pyrite or tourmaline altered, and there are sectors with dumortierite-andalusite (advanced argillic) alteration. The mineralization is located in a few quartz, K-feldspar, sericite, pyrite, sphalerite, Ag-tetrahedrite, stromeyerite, pyrargyrite, etc. higher grade and persistent veins mined in the past, as well as in thin stockwork veinlets of similar composition, in wallrock pyrite and in supergene "limonite". The latter is estimated to contain 80% of the available reserves.

St. Andreasberg, Oberharz, Germany. This is an ancient deposit (Schneiderhöhn, 1941; 313 t Ag) hosted by lower Carboniferous "graywacke" and siliceous slate, unconformably resting on Devonian slate, black slate, metabasalt and carbonate. The deposit lies in a contact aureole of a Permo-Carboniferous "granite". The ore veins fill N.W. or E.-W. fissures. They are under 1 m thick, and are filled by gray calcite with fine disseminated sulphides, arsenides and native elements (stibnite, breithauptite, arsenic, antimony, niccolite, smaltite, löllingite, sphalerite, galena). Three superimposed mineralization phases produced drusy, coarse crystalline aggregate of quartz, fluorite, Ag-tetrahedrite, chalcopyrite, galena, sphalerite, millerite, native silver, Ag-sulphosalts, realgar and crystallized zeolites. The wallrock alteration is very light (chloritization, calcite impregnations).

Batopilas (Chihuahua, Mexico; Wisser, 1966; 1.550 t Ag) has calcite, quartz, barite, arsenopyrite, sphalerite, galena, native silver, Ag-sulphosalts, safflorite and rammelsbergite veins, hosted by a fine-grained Tertiary diorite and quartz porphyry dikes, intruding andesitic volcanic breccias and flows. The veins were under 50 cm thick, close to the basement and low in base metal sulphides and in gold.

28.16. THE Ni,Co,Bi,Ag,U,(As) ASSOCIATION

The "Five Elements" association accumulated in hydrothermal veins has been popular mostly with the German-language geologists (e.g. Schneiderhöhn, 1941 and earlier works of the Freiberg Mining Academy Staff), because of its numerous classical occurrences in the Erzgebirge Mts. Close to fifty Phanerozoic localities of this association are now known worldwide, and it is estimated that they account for about 10 Tt Ag, 50 Tt Ni, 70 Tt Co, 3 Tt Bi and 50-100 Tt U. These ores have always been a genetic enigma, particularly because they bring together metals that possess a contrasting geochemical and environmental affiliation like the basaltophile and ultramafics-related Ni and Co, and granitophile U. Alternatively, all five metals plus As could conceivably accumulate together in certain marine sediments e.g. black shales and phosphorites. The fahlbånds (metalliferous schists) near Schladming, Austria, contain stratabound and remobilized niccolite, glaucodote, smaltite, bismuth, arsenic, Ag-tetrahedrite, etc. More examples are known in the Precambrian terrains (Cobalt, Ontario; Modum, Norway; Volume 2).

The known Ni,Co,Bi,Ag,U,As occurrences and incomplete, related metal combinations (e.g. Co-As, Ni-As, Co-Ni-As, U-Ag, etc.), although locally unique, appear to be variations on a theme. All are low-temperature, multistage mineralizations having a carbonate (calcite, dolomite), jasper, chalcedonic quartz, fluorite or barite gangue. All occur in block-faulted terrains, and the faults are high-level, filled by graphite-rich gangue, mylonite or breccia. The host terrains are invariably marked by an early stage of development involving mafic marine volcanics, black metalliferous pelites, minor carbonates, ultramafics, gabbros, and diabase dikes. This was followed, usually after a considerable time gap, by emplacement of "granites". The "granites" are of several age generations and include synorogenic tonalites and post-orogenic granodiorites to biotite granites. Skarns are frequently present. Although the "Five Element" veins usually occur in "granite" aureoles, they tend to be substantially younger than the hydrothermal phase of the "visible" granites, and usually superimposed. In Freiberg (Baumann, 1976) and Jáchymov (Chrt and Bolduan, 1966), the "Five Element" veins are Mesozoic to Tertiary, in contrast to the spatially associated late Paleozoic Pb-Zn and Sn postmagmatic veins and granite cupola stockworks. Post-granite diabase, lamprophyre or porphyry dikes often overlap with the Ni,Co,Bi,U,Ag veins.

The Jáchymov-Abertamy ore field on the Czech side of the Erzgebirge (Mrňa and Pavlů, 1967; Figs. 28-104, 105) is a historically important locality quite apart from its geological importance. Formerly known under the German name Joachimsthal, it was the home town of Georg Bauer (Agricola), craddle of the dollar (the local 16th century coinage was known as "Joachimsthaller", abbreviated to thaller, taller, dollar) and the first known uranium deposit of the world, in whose ores uranium was first discovered by Klaproth and Ra and Po by the Curies. The field started as a medieval silver-mining camp and subsequently passed through Co (for porcelain paints), Ra and U mining stages. The metal quantities in about fifty veins located within an area of 12x6 km are estimated to have been between 30-50 Tt U, about 2 Tt Ag, 5 Tt Co and 300 t Bi. 240 grams of Ra was recovered in the pre-war period.

The ore field is located in a Cambro-Ordovician micaschist-dominated series with interbeds of graphitic schist, metachert, metacarbonate, acid and mafic volcanics, situated in the roof of a Permo-Carboniferous granitic pluton. The bedrock is intersected by a system of N.W. and E.-W. faults, many of which are followed by granite porphyry and lamprophyre dikes. Erosional remnants of late Tertiary continental basalts and their feeder dikes are preserved over less than 10% of the area.

Postmagmatic mineralization along granite contacts and in their thermal aureole is represented by several swarms of quartz, cassiterite, wolframite veins and greisens, outside the limits of the Jáchymov-Abertamy field. Several bodies of magnetite, sphalerite, pyrite, chalcopyrite mineralized skarns are known within the field. The numerous post-granite hydrothermal ore veins are located in portions of gouge-filled E.-W., N.-S., and N.W. fractures. The veins are up to 1 km long, up to 1 m wide, and congregate in six clusters ("ore knots") separated by barren ground. Ore shoots are most common at fault intersections, and have often the form of columns composed of en-echelon ore lenses. The ore and gangue lenses are often separated from the wallrocks by altered fault clay, and in many cases the fault gouge is dispersed in the vein filling. The wallrock alteration ranges from silicification and sericitization around veins filled by the early quartz-sulphide stage to chloritization, argillization, hematitization and pyritization.

The hydrothermal ores are the result of six mineralization stages, from oldest to youngest: (1) quartz, arsenopyrite, pyrite, galena, sphalerite, chalcopyrite; (2) quartz-hematite; (3) dolomite, pitchblende, lesser pyrite, fluorite; (4) quartz, silver, skutterudite, rammelsbergite, niccolite, safflorite, bismuth;

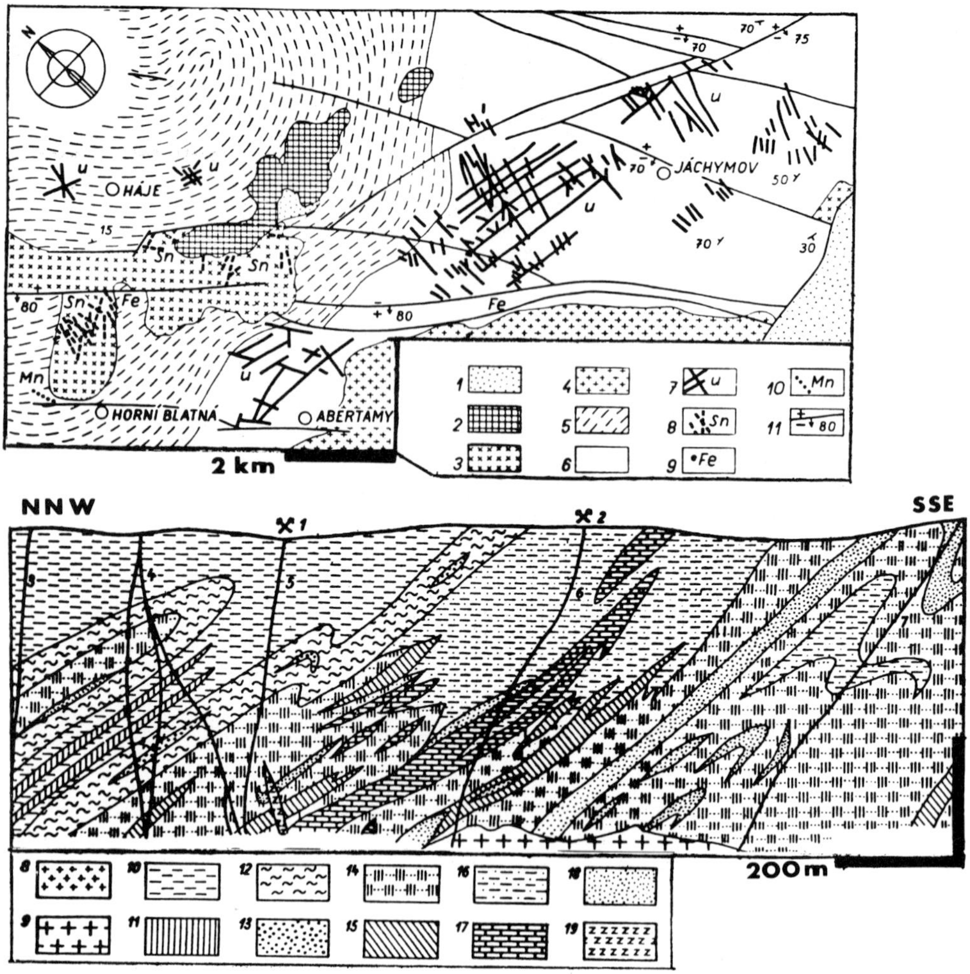

Fig. 28-104. Jáchymov-Abertamy U and Ag,Ni,Co,Bi vein field, N.W. Czechoslovakia (top) and a section through the Svornost (Einigkeit) mine in Jáchymov. After Mrňa from Mrňa and Pavlů (1967) and Mrňa (1963), courtesy Ústřední Ústav Geologický, Praha. Map legend: 1=Q,T sediments; 2=T plateau basalts; 3=Cb-Pe porphyritic granite (older mesozonal phase); 4=Pe epizonal leucogranite; 5=PZ_1 phyllites; 6=PZ_1 micaschists; 7=veins of Ag,Bi,Co,Ni,U association; 8=greisens with cassiterite; 9=quartz hematite veins; 10=Mn oxide veins; 11=fault zones. Section legend: 1=Svornost mine; 2=Josef mine; 3=Dorotea vein; 4=T basalt; 5=Kraví vein; 6=Geier vein; 7=Šindler vein; 8=granite porphyry; 9=granite. 10=PZ_1 (top unit) biotite-muscovite schist; 11=biotite schist; 12=PZ_1 (middle unit) biotite-muscovite schist; 13=biotite schist; 14=PZ_1 or Pt_3 (lowest unit) biotite muscovite schist; 15=biotite schist; 16=phyllite and micaschist; 17=calc-silicate hornfels; 18=metaquartzite; 19=lamprophyres.

Fig. 28-105. Jáchymov (formerly Joachimsthal), "the town that gave you the dollar" (and also U,Ra,Po isolated for the first time from its pitchblende). City centre with the shaft Svornost on the right.

(5) dolomite, arsenic, proustite, argentite, sternbergite, stephanite and (6) calcite, pyrite, galena, sphalerite, chalcopyrite, realgar. Brecciation and intravein metasomatism are widespread.

Jáchymov-style Ni,Co,Bi,Ag,U,As veins having a comparable geological as well as cultural history, have been mined at numerous localities on both the Czech (Potůčky, Přísečnice, Vejprty, Horní Slavkov) and East German (Schneeberg, Marienberg, Johanngeorgenstadt, Oberschlemma, Annaberg, Buchholz, Aue, Freiberg) sides of the Erzgebirge and Slavkovský Les. Similar deposits are known in the Příbram district in central Bohemia, in the Czech and Polish Sudeten (Kowary, Zálesí), in the Schwarzwald, Germany (Wittichen), in the French Alps (Chalanches), Pyrenees (Gistain), New Mexico (Black Hawk) and elsewhere.

The Khovu-Aksy field in the Tuva region, S. Siberia (U.S.S.R., close to the Mongolia border; Krutov, 1974) has Co-Ni arsenide veins superimposed on skarn and is reportedly a site of a large mining-metallurgical complex. The ore field has dimensions of about

4x2 km and is located in a geologically extremely complex block situated in a zone of deep N.-S. faults. The stratigraphically lowest units are lower Cambrian marine basalts, andesites, limestones and dolomites, intruded by plagiogranites and ophiolitic ultramafics. This is topped by Silurian to Devonian conglomerates, sandstones, siltstones and limestones. A 4 km long horizon of Silurian carbonate is almost completely converted into skarn in the exocontact of a lower Devonian granitic pluton.

The Co,Ni,Bi,Cu (and possibly U) orebodies are almost entirely contained in a series of N. and N.E.-trending fissure veins, superimposed on skarn and on Silurian siltstone. The veins are several hundred metres long and composed of chalcopyrite, smaltite-chloanthite, safflorite, lesser niccolite, rammelsbergite, löllingite, bismuth and emplectite in dolomite, ankerite and chlorite gangue. Most of the veins are compositionally banded, and chloritization is a common wallrock alteration.

Other localities with Ni-Co arsenides

Smaltite, chloanthite, cobaltite, gersdorffite, rammelsbergite and similar arsenides are present in small quantities in at least 100 hydrothermal vein and replacement deposits, worldwide. They are relatively most common in siderite orebodies and in the past, small quantities of Ni-Co concentrates have been produced as a by-product of siderite benefication. The Siegerland siderite district of West Germany and the Spiš-Gemer area of Slovakia, were the best known producers. A small siderite-gersdorffite deposit near Dobšiná (Dobschau), Slovakia, yielded about 1,000 t Ni.

28.17. GOLD DEPOSITS

28.17.1. GENERAL

Many (perhaps 40-50%) of the postmagmatic hydrothermal deposits treated earlier (e.g. porphyry coppers, Pb-Zn veins) contain small quantities of gold, which is recovered as a by-product. Such deposits may grade into mineralizations in which the Au:base metal ratio (expressed in units of metal value, not in weight units !) is about one. Ultimately they may grade into deposits in which gold is the dominant or the only commodity recovered. "Gold deposits" thus cannot be differentiated sharply and in this section we concentrate on ores in which gold is the dominant commodity under an average economic climate.

Gold is frequently and characteristically associated spatially with "granitic" plutons, but this does not automatically guarantee a direct genetic connection. Its setting is a most elusive one. Of all metals, gold has the broadest range of intrusive rock types with which it is (presumably) genetically associated, and the broadest range of possible host rocks. Almost all alteration assemblages have been found at least once with gold. Gold accumulations have the greatest vertical range of formation and occurrence of all metalliferous deposits ranging from the deep katazone to the subvolcanic and even volcanic level and, curiously, at least one style of a significant gold accumulation can be distinguished at either level. The old prospector's adage "gold is where you find it" perhaps best expresses the heterogeneous nature of gold metallogeny and furnishes a plausible excuse for the lack of universal workable models of field gold association. "You could write books about gold deposits alone", a colleague remarked, and this is precisely what Boyle (1979) and before him Emmons (1937) and other writers have done. The reader is advised to consult the above compilations and the original literature for detail. The material in this section is just a brief summary.

Gold deposits spatially associated with Phanerozoic granitic plutons represent some 12,000 t Au and this includes placer gold presumably derived from corresponding bedrock sources. This figure overlaps with those given for "interaction gold" in some earlier chapters. In the Sierra Nevada Foothills, for example, gold is associated with both the characteristic slate-greenstone suite and tonalite-granodiorite intrusions, and the share of either association on the gold accumulation cannot be accurately expressed quantitatively.

Hydrothermal gold deposits can be classified and subdivided in many ways and the classical way based on the presumed temperature of hydrothermal solutions and mineral association (e.g. Schneiderhöhn, 1955; Bateman, 1951; Park and McDiarmid, 1975) is probably the most popular (at least in university courses). It is, however, one not particularly suitable for exploration and one that is outright misleading and contradicts the research results from the past twenty years (e.g. some greenstone-hosted peneconcordant deposits now interpreted as reconstituted hydrothermal sediments or metamorphogenic mobilizates, are placed into the "hypothermal" class. Here the temperature is the temperature of metamorphism, not precipitation from freely moving postmagmatic fluids). Simple substitute classification, however, does not exist. One can distinguish a variety of gold orebodies identical to orebodies of the metals treated so far (disseminations or stockworks, veins, carbonate replacements, etc.). One can subdivide gold deposits on mineralogical grounds (there is a two end-member sequence between Au hosted by entirely sulphide carriers and free Au only in quartz; Table 28-23 lists sample mineralogical associations of plutonic gold deposits). One can arrange gold deposits by membership in the regional "metal belts"; Table 28-24) and one can arrange gold deposits by lithologic associations as has been done in this book. The brief treatment of gold deposits in this section is arranged by mineralization styles within and outside a plutonic aureole, regardless of the supracrustal rock association.

In each association, gold in ore can be present as free-milling, visible or invisible, or as a gold in the lattice of sulphides (mainly arsenopyrite and pyrite). The latter is more expensive to extract. The gold bullion has a variable purity, ranging from about 995 (per mil) to around 500.

Table 28-23. Sample mineral associations of hydrothermal gold deposits associated with Phanerozoic granitic plutons

quartz, lesser pyrite, chalcopyrite, pyrargyrite, gold (Balei, epithermal)
quartz, stibnite, lesser siderite, pyrite, sphalerite, gold (Zlatá Idka, Slovakia)
quartz, carbonates, pyrite, lesser galena, sphalerite, chalcopyrite, tennantite, pitchblende (Central City, Colorado)
quartz, ankerite, calcite, lesser pyrite, arsenopyrite, galena,sphalerite, chalcopyrite, gold (Grass Valley, California)
quartz, ankerite, fluorite, lesser galena, sphalerite, gold tellurides, gold (Jamestown, Colorado)
quartz, tourmaline, pyrite, lesser chalcopyrite, arsenopyrite, freibergite (Klyuchi, East Transbaikalia)
quartz, pyrite, arsenopyrite, scheelite, bismuthinite (Au is located in pyrite and arsenopyrite); Zarmitan, Uzbekistan
quartz, calcite, lesser pyrite, arsenopyrite, pyrrhotite, scheelite, chalcopyrite, gold (Jílové, Czechoslovakia)
quartz, ankerite, calcite, albite, lesser arsenopyrite, pyrite, pyrrhotite, gold (Bendigo, Victoria)
quartz, carbonate, pyrite, lesser arsenopyrite, galena, sphalerite, stibnite, gold (Ballarat, Victoria)
pyrite, chalcopyrite, pyrrhotite, lesser quartz, calcite, gold (Rossland, British Columbia)
pyrrhotite, pyrite, chalcopyrite, molybdenite, bismuth, tetradymite, Au tellurides (Ol'khovo, Sayan Range, Siberia)

Table 28-24. Hydrothermal-plutonic gold deposits arranged by association with "belts" and "provinces" of other metals, sometimes resulting in mixed provinces (e.g. Au-Sn; Au-Mo; etc.)

Pb,Zn,Ag	Colorado Front Ranges, e.g. Georgetown, Silver Plume, Breckenridge; quartz, carbonate, pyrite, galena, sphalerite, arsenopyrite, gold veins
Mo	E. Transbaikalia Au-Mo belt, e.g. Klyuchi, Darasun; quartz, tourmaline, pyrite, molybdenite, sulphides, gold stockworks and veins
Cu	porphyry copper fields with zonally associated gold-quartz veins; e.g. Panguna, Bougainville; Almalyk, Uzbekistan
Sb	auriferous quartz, stibnite, pyrite, arsenopyrite veins; La Lucette, France; Milešov, Magurka, Dúbrava, Czechoslovakia
W (scheelite)	quartz, scheelite, gold veins, e.g. Moose River, Nova Scotia; Zarmitan, Uzbekistan
Sn	Yana-Kolyma belt, Siberia; gold-quartz lodes overlap with quartz-cassiterite veins and greisens
Hg	complex quartz, carbonate, tetrahedrite, amalgams, gold veins; Los Mantos mine, Punitaquí, Chile
Ag	common epithermal, less common plutonic association

28.17.2. Disseminated and stockwork gold in granitic stocks and breccia pipes or dikes

This ore style is most common in Tertiary volcanic or subvolcanic settings (Chapter 26) and also in the Precambrian greenstone belts (Volume 2). Boyle (1979) mentioned only two Phanerozoic plutonic-associated deposits. More localities, however, are known and with the increasing gold price and application of new techniques of low-grade gold extraction (e.g. in-situ and heap leaching), interest in low-grade "disseminated" gold deposits increases.

The small gold stockwork in quartz monzonite at the Jessie Mine, Colorado (2 t Au) has already been described briefly (Section 28.2.2.). In the Central City gold vein field in the same region, a large pipe-like stockwork ("The Patch"; Sims, Drake and Tooker, 1963) composed of quartz, pyrite, sphalerite, chalcopyrite, tennantite, galena and gold filling thin fracture veinlets, is located in a pyrite-sericite-altered quartz monzonite.

The Klyuchi deposit (East Transbaikalia Au-Mo belt, U.S.S.R.; Borodaevskaya and Rozhkov, 1974) is a network (stockwork) of auriferous veinlets and larger veins in Jurassic granite and granodiorite porphyry. The host high-level intrusion is emplaced into Paleozoic granite gneisses. The veinlet composition is quartz, tourmaline, pyrite, chalcopyrite, arsenopyrite, freibergite, and other minerals. Gold is located mostly in pyrite, and the wallrocks are intensively sericite altered. The large, low-grade Kidston deposit in northern Queensland, Australia (20 Mt ore with 2.5 ppm Au=50 t Au) is similar.

In the Permian Zarmitan granosyenite intrusive complex, Uzbekistan (Garkovets et al., 1979), gold is situated in quartz, pyrite, arsenopyrite, scheelite, bismuthinite veins and stockworks. Most of the gold is carried by pyrite and arsenopyrite, and the stockwork is located in a zone of pervasive microcline and albite metasomatism.

28.17.3. Gold vein fields around small central hypabyssal intrusions

Gold deposits equivalent in overall geometry to the granite cupola-controlled Sn deposits or Mo-mineralized intrusive stocks, are relatively rare. The high-level Darasun field in the East Transbaikalia Au-Mo belt, is an example (Borodaevskaya and Rozhkov, 1974; min. 90 t Au). This deposit is located within a Paleozoic crystalline basement (gabbro, amphibolite, intruded by late Paleozoic granodiorite, granite and syenite), block-faulted and activated in the Mesozoic. The gold is to be found in about 200 thin but persistent N.E. and N.W.-trending fissure veins, hosted by gabbros and other Paleozoic rocks in the exocontact of a small middle-upper Jurassic plagiogranite stock and intrusive breccia. The veins are filled by quartz, carbonate, tourmaline, chlorite, sericite, pyrite, arsenopyrite, sphalerite, galena, tetrahedrite, Pb-sulphoantimonides, bismuthinite, Au-tellurides, etc. The wallrock alteration is quartz-sericite and propylitic and there is a distinct metal and mineral zoning apparent on the surface as a series of overlapping concentric zones (from centre outward the series runs as follows: quartz-tourmaline, pyrite-arsenopyrite, galena-sphalerite, chalcopyrite-bournonite, sulphoantimonides).

In the Jamestown ore field near Boulder in the Colorado Mineral Belt (Lovering and Goddard, 1950; 29 t Au, 235 t Ag), swarms of mineralized fissure veins surround an almost circular small stock of Oligocene sodic granite (1 km in diameter). The host rocks to the veins are a small pluton of lower Tertiary granodiorite and a Proterozoic granite. Gold occurs in two types of veins: (1) quartz, fluorite, gold tellurides, gold and (2) pyrite, chalcopyrite, fine gold.

The Central City field in the same belt (Sims, Drake and Tooker, 1963; 116.4 t Au; Fig. 28-106) contains over a hundred veins. The bedrock is a Proterozoic crystalline terrain composed of biotite gneiss, migmatite and pegmatite intruded by many small dikes and irregular high-level plutons of an early Tertiary intrusive suite. The latter includes granodiorite, quartz monzonite and bostonite porphyries. Abundant, closely spaced intersecting faults produce a meshlike network. The veins are largely fault fillings and range from simple, tabular bodies to complex branching lodes in subparallel fractures. Most veins are 0.3-1 m wide, filled by quartz and chert with minor carbonate, fluorite and barite.

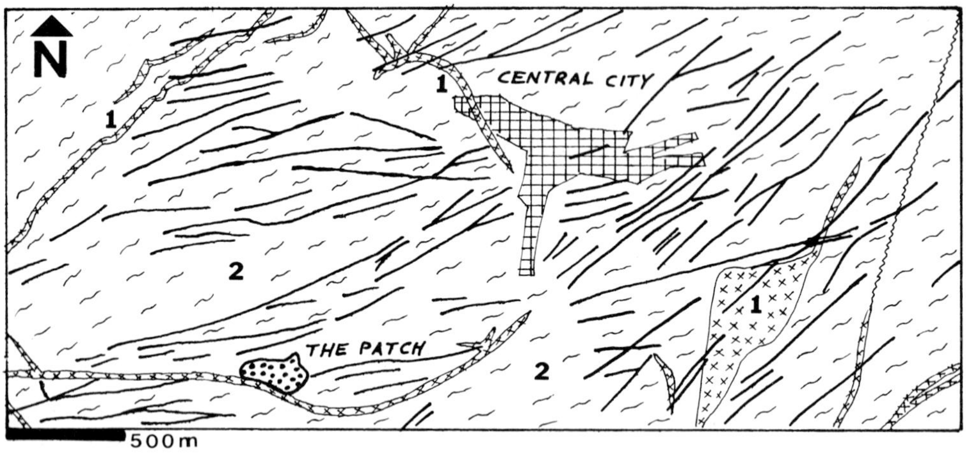

Fig. 28-106. Central City field, Rocky Mountains, Colorado, Au(Pb-Zn,U) veins. 1=T_1 granodiorite, quartz monzonite, granosyenite, bostonite, trachyte dikes and stocks; 2=Pt_{1-2} biotite sillimanite gneiss, migmatite. From LITHOTHEQUE, after Sims, Drake and Tooker (1963).

Pyrite is the dominant metallic mineral, and there are variable quantities of galena, sphalerite, chalcopyrite, tennantite, enargite and pitchblende. Gold occurs as free metal, or in pyrite. Wallrocks adjacent to the veins are usually sericitized, and sericitization is succeeded outward by argillization. A faint mineralogical zoning in the district is represented by the dominance of pyritic veins in the core, surrounded by a peripheral zone rich in galena and sphalerite.

28.17.4. Gold veins in and around moderately eroded mesozonal plutons

It has frequently been observed that gold-quartz veins of simple mineralogy often occur in moderately to deeply eroded "granitic" (mostly quartz diorite to granodiorite) plutons, in a setting in which other metals (except scheelite skarns) rarely accumulate. In the classical model of the 1930s, such veins were placed into the hypothermal or katathermal categories but more recent studies (e.g. Coveney, 1981) have demonstrated that the filling temperature range of similar veins was much broader (from about 200° to 491° C).

Grass Valley-Nevada City gold field, California (Johnston, 1940; Albers, 1981; 330 t Au/7-14 ppm, 90 t Ag; Fig. 28-107) is a classical locality now exhausted (but one mine has been preserved as a state park). This field is in the N.W. extension of the Mother Lode Belt, hosted by and probably associated genetically with a Mesozoic granodiorite pluton emplaced in late Paleozoic and Mesozoic slate, litharenite, greenstone and serpentinite terrain. The gold-bearing veins form a conjugate N. and N.W.-striking system. The veins in granodiorite have gentle dips (aver. 35°), those in the supracrustals as a rule are steep. The veins fill both normal and reverse faults and fissures, marked by shattered wallrock. Quartz is the principal vein mineral, associated with carbonates (ankerite, calcite) and pyrite, arsenopyrite, lesser sphalerite, chalcopyrite, galena and gold.

Gold deposition was contemporary with galena, and both minerals are closely associated. The veins were remarkably persistent and no change in mineralogy was observed over a vertical interval of more than 1,300 m. The wallrocks are strongly altered. Sericite, pyrite and ankerite alteration assemblage adjacent to veins grades away into propylitization (chlorite-epidote). Aplite, porphyry and lamprophyre dikes, both pre-ore and post-ore, are widespread in the field.

Deposits very similar in style and environment to those in Grass Valley include the Stepnyak field in northern Kazakhstan (gold-quartz veins in a diorite pluton, intruded into mafic volcanic meta-arenites); Dorozhnyi in the Yana-Kolyma belt; Milešov-Krásná Hora, Czechoslovakia; Bralorne, British Columbia; Segovia (Antioquía, Colombia) and others (Table 28-25).

The important <u>Kochkar ore field</u> in the S.E. Urals, U.S.S.R. (Magak'yan, 1968; min. 120 t Au; Fig. 28-108) is in a large Carboniferous plagiogranite pluton, emplaced in a lower Devonian to middle Carboniferous greenstone (mafic volcanic-sedimentary) suite.

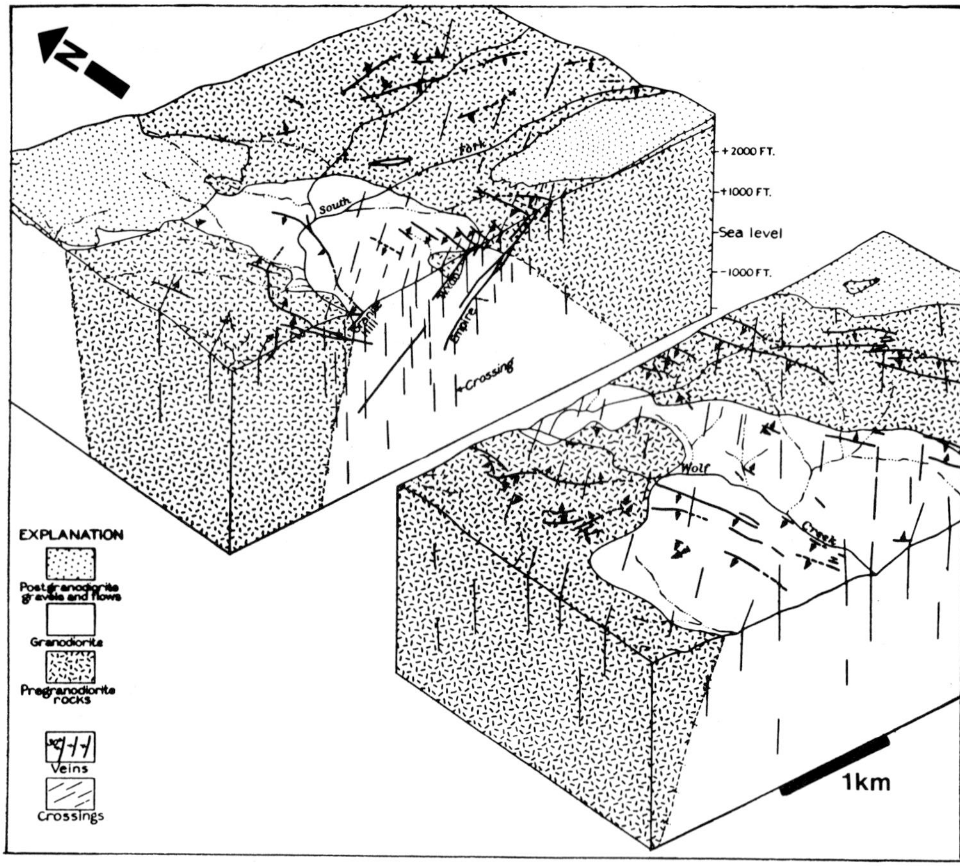

Fig. 28-107. Gold quartz veins in the Grass Valley, Nevada City field, California. From Johnston (1940).

A younger intrusive phase of microcline granite (quartz monzonite) is believed affiliated with the gold mineralization. The field contains over 1,500 veins (200 of them economic) in the microcline granite exocontact, that are controlled by two conjugate systems of E.N.E. shear fractures. The bulk of the ore veins are hosted by carbonatized dikes of diorite, diorite porphyry, plagiogranite porphyry, syenite porphyry, felsite and diabase. The ore veins are 0.4 to 2 m wide and tens to hundreds of metres long. The filling is quartz, lesser tourmaline, ankerite, calcite, pyrite, galena, chalcopyrite, tetradymite. The gold occurs in both free milling form and bound in pyrite and arsenopyrite. Portion of it is contained in tellurides. The gold bullion contains up to 10% Pt.

Rossland ore field, British Columbia (Barr, 1980; 80.1 t Au/13.35 ppm, 106 t Ag/17 ppm, 62 Tt Cu/1%) is an example of sulphide-gold

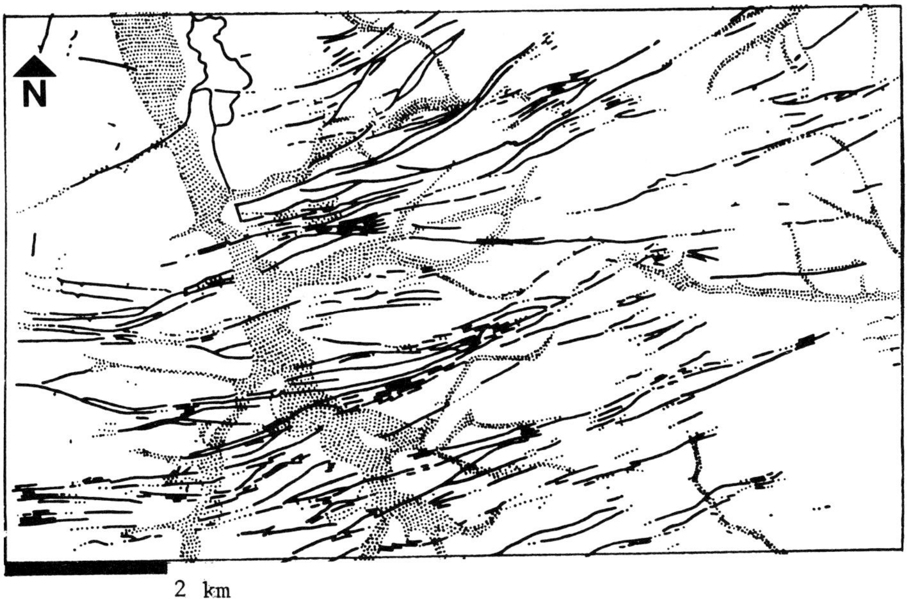

Fig. 28-108. A portion of the Kochkar goldfield, the Urals, U.S.S.R. Black area: Au-bearing veins; stippled: Q valley sediments, containing alluvial placers; blank: Cb_1 plagiogranite massif intersected by E.N.E. altered dikes. From Magak'yan (1968).

veins within and along the contact of an eroded intrusive pluton. The basement is formed by Carboniferous siltstone, sandstone, conglomerate and minor limestone, topped by Jurassic intermediate to mafic volcanic pile composed of andesite volcanic breccia, ash-flow tuff, augite porphyrite and volcaniclastics. The supracrustal rocks were intruded by several plutons and stocks between the Jurassic and lower Tertiary periods , and the intrusive rocks range from quartz diorite through monzonite, granodiorite, granite to syenite. The Rossland orebodies cross the intrusive contact between the monzonite and augite porphyrite, and are believed to be related genetically to a minor stock of Eocene (48 m.y.) quartz diorite and a swarm of diorite porphyry and lamprophyre dikes.

The ore veins are controlled by E.N.E. and S.E. sets of fractures and ore shoots in the veins commonly terminate against the northerly set of dikes. Although the vein zones extend for 1.3 km or more, the ore shoots are short and narrow and composed of almost massive pyrrhotite, chalcopyrite, lesser pyrite and native gold. The vein width is very variable, ranging from several centimetres up to 43 m, and it would be tempting to assume remobilization of an earlier volcanogenic or subvolcanic relic mineralization in the origin of these veins. No sufficient evidence, however, is available.

28.17.5. Gold veins in distal plutonic exocontacts, or predating the "granite" emplacement

These are by far the most common and productive variety of auriferous veins. Both gold-quartz and sulphide-gold end-members of filling occur, although the former is the dominant one. In this setting, the veins display a considerable lithological (associational) and structural-metamorphic control, and several distinct lithologic associations treated in this book carry the bulk of the veins. They are: (1) greenstone-phyllite or slate gold belts (e.g. Juneau, Klondike, Mother Lode, Piedmont Gold Belt, Jílove, and others; Section 10.4.3., Figs. 10-6 and 10-11); (2) felsic metavolcanic belts (e.g. Howie, Haile, Hay Mt. mines, S. Piedmont; Section 12.5.) and (3) slate belts of "flyschoid" character (Meguma Group of Nova Scotia; Victoria Goldfields, Australia; Yana-Kolyma belt; Chapters 16, 19).

In all the regions listed above, penetratively deformed and low-grade metamorphosed supracrustal rocks host the ore veins that fill (or replace) shears, faults, fractures as well as dilations conformable with foliation. "Granites" are invariably present within the mineralized fields or adjacent to them. In fact, it appears that in most of the supracrustal belts a sizeable pluton underlies the metasediments and metavolcanics. The genetic relationships between the "granites" and the veins, however, remain controversial and in some instances (in Nova Scotia, Yana-Kolyma, etc.) it has been demonstrated that some of the gold veins are older than the local "granites", which dismember and contact metamorphose them.

The gold deposits that have already been reviewed will not be discussed again here, but basic data on several example localities are summarized briefly in Table 28-25.

28.17.6. Gold-bearing skarns and carbonate replacements

Many Cu, Pb-Zn, Bi, and other skarns yield gold as a by-product, but only a few skarn deposits have ever been mined for gold as the sole, or the main, product. Gold in skarns is always associated with sulphides (arsenopyrite, pyrite, rarely pyrrhotite) that accumulate in the late-stage infiltrational skarn or are superimposed on the earlier garnet, hedenbergite, epidote, amphibole, etc. skarn in the form of veins or fracture stockworks.

<u>The Hedley ore field,</u> British Columbia; (Barr, 1980; 49.2 t Au/12.4 ppm; Fig. 28-109) is underlain by late Triassic marine andesite tuff, breccia, volcanic argillite, arenite, and impure limestone. The supracrustal rocks were folded, intruded by middle to upper Jurassic and Cretaceous quartz diorite and granodiorite stock and dikes as well as by augite diorite and quartz gabbro. The latter rocks could, at least in part, be hybrids, created by mingling of the supracrustal basaltic andesite and intermediate intrusions. The granitic intrusions produced contact hornfelses out of the silicate sediments, and converted the limestone into garnet, hedenbergite,

axinite and epidote skarn.

In the principal Nickel Plate zone, the orebodies are sheet-like, overlapping en-echelon masses of skarn within 80 m of the skarn/marble contact, mineralized by disseminations, scattered blebs and masses of arsenopyrite and minor pyrite and sphalerite. The arsenopyrite carries the gold, and its content in the ore was most commonly between 10 and 50% by volume. Arsenic was not recovered and is not included in the statistics, but if the average As grade were 5% As (a conservative estimate), there would have been about 200 Tt As in the ore.

Boyle (1979, p. 250-251) briefly reviewed more localities of auriferous skarns in the western United States (Cable, Montana; Ouray, Colorado; Battle Mountain, Nevada), Nicaragua (Rosita, La Luz), southern Siberia (Sinyukha, Lebedskoe, Natal'evskoe, Tardanskoe), Sarawak (Bau) and North Korea (Suian). At most localities, copper sulphides were abundant. Gold tellurides are important in the Soviet deposits. In Bau stibnite is associated with

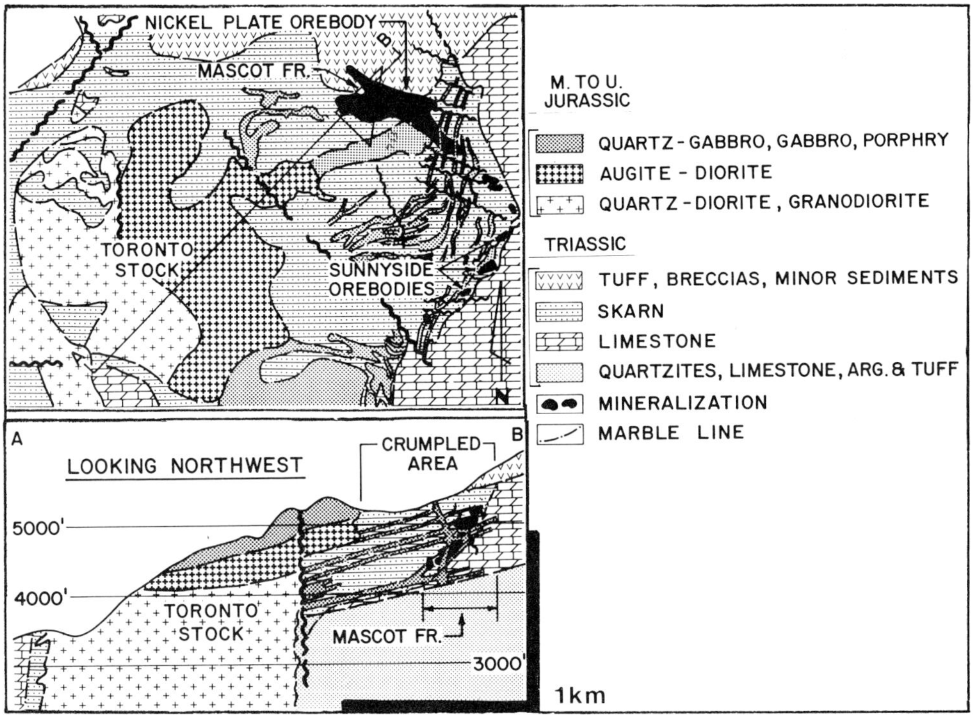

Fig. 28-109. Hedley field gold skarns, southern British Columbia. From Barr (1980), courtesy of the Canadian Institute of Mining and Metallurgy and the author.

Table 28-25. Essential data on the more important plutonic and interaction-plutonic Phanerozoic gold deposits, districts and belts of the world. Abstracted from MANIFILE (Laznicka, 1973)

1. LOCALITY	2. CL.	3. ST.	HOST UNIT			
			4. AGE	5. MT.	6. CH.	7. HOST ROCKS
Willow Creek, Alaska, U.S.A.	OF	FV,SV	PZ,MZ	A	29	gneiss,schist
Nome, Alaska, U.S.A.	DT	PL,FV	PZ	G,A	17,29	schist,gneiss
Fairbanks, Alaska	DT	PL,FV	Pt,PZ	A	29	gneiss,schist,qtzt.
Chicagoff Island, Alaska, U.S.A.	OF	SV,FV	Cr	1G	13,16	andes.tuff,grayw., slate
Juneau, Alaska	OF	RV,SV	Tr_3	Z-1G	10,16	slate, graywacke, greenstone
Premier near Stewart British Columbia	DP	SV,FV	J_{1-2}	Z-1G	13	andes.tuff, slate, argillite
Zeballos,Vanc.Isl. British Columbia	OF	SV,FV	Tr_3	Z-1G	13	andesite tuff
Bralorne, British Columbia, Canada	OF	SV,FV, SW	Pe,Tr	Z-1G	10,7 13	chert,slate,serpen. greenst.,andesite
Hedley, British Columbia, Canada	OF	SK	Tr_3	Z-1G CT	13,16	limest.,quartz., argill.,andesite
Rossland, British Columbia, Canada	OF	RV,FV SV	J_2	Z-1G	13	andes.tuff,argill.
Wells (Cariboo dist) lodes, Br.Columbia	OF	FV,SV RV	Cm	G	19,17	schist,m-quartzite argill.,limestone
Salmo-Ymir, southern British Columbia	OF	FV	Cm_1	1G	19	quartzite,phyllite, limestone
Klamath Mts.,Oregon and Calif.,U.S.A.	DT	PL,SV FV	PZ	G,A	12,29 7	gneiss,greenst., micasch.,limest.
Blue Mts.,Oregon, U.S.A.	DT	PL,SV FV	PZ	G,A	12,10	schist, greenst., slate
Boise Basin, S.W. Idaho, U.S.A.	DT	PL,FV	$J-Cr_1$		28	granodiorite, quartz monzonite
Marysville, Montana U.S.A.	OF	PL,FV	Cr_3		28	qtz.dior.,granod., granite
Rimini, Montana U.S.A.	OF	FV,PL	Cr_3		28	qtz.dior.,qtz.monzon.,granodior.
Elliston, Montana U.S.A.	OF	PL,FV SV	Cr_3		28	qtz. monzonite

CONCENTRATOR INTRUSION			OREBODIES	
8.ST	9.AGE	10.PETROGRAPHY	11.MINERALOGY	12.TONNAGE
1	Cr_3-T_1	qtz.diorite	qtz,py,ar,sf,tetr,ga,Au	17.24t Au
2	Cr_3-T_1	granod.,granite	qtz,py,ar,Au	100.7t Au
2	Cr_3-T_1	qtz.dior.,gran.	qtz,py,ar,sf,sb,sch	87.3t Au
2	Cr	dior.,qtz.dior., granodiorite	qtz,py,Au	22t Au
3-4	Tr_{2-3}	qtz.dior., granodiorite	qtz,py,ar,ga,sf,cp,Au	187.2t Au
3-4	J_2?	qtz.diorite porphyry	qtz,py,ga,sf,cp,tetr, pyrargyrite,Au	50.54t Au
2	J_2	granod.,qtz. diorite	qtz,py,sf,cp,ga,po,Au	12t Au
1-2	J_1	plagiogranite, dior.,qtz.dior.	qtz,carb,py,ar,Au	95.2t Au
3-4	J	granodiorite	ar,po,cp,py,sf,Au	44.3t Au
3-4	J-T_1	dior.,monzon., granodiorite	qtz,po,cp,py	82.5t Au
5	J-Cr?		qtz,ank,py,ar,sch,sf, ga,Au	74.8t Au
3-4	J_1	qtz.dior.,gran., granodiorite	qtz,carb,py,ga,sf,cp,Au	35t Au
3	J-Cr_1	dior.,qtz.dior., granodiorite	qtz,py,ar,cp,sf,ga,Au	162t Au
3	Cr_2	qtz.diorite, diorite	qtz,py,cp,ar,sf,ga,sb, Au	110t Au
1	Cr_1	qtz.monzon., granod.,porph.dk	qtz,py,ar,sf,tetr,sb,Au	79.24t Au
1	Cr_3	qtz.dior.,dike rocks	qtz,py,tetr,cp,ga,sf,Au	42t Au
1	Cr_3	qtz.monzonite	qtz,ga,sf,py,ar,Au	112.6t Au
1	Cr_3	qtz.monzonite, porph. dikes	qtz,ga,sf,py,Au	33t Au

Table 28-25 (continued).

1.LOCALITY	2.CL.	3.ST.	HOST UNIT 4.AGE	5.MT.	6.CH.	7.HOST ROCKS
Basin and Boulder, Montana, U.S.A.	OF	BR,FV	Cr_3		28	qtz.monzonite
Winston, Montana U.S.A.	OF	PL,FV	Cr_3		28	qtz.dior.,granod., qtz.monzonite
Virginia City, Mont. U.S.A.	OF	PL,SV	Pt	A	29	gneiss,schist
Grass Valley-Nevada City,Calif., U.S.A.	OF	SV,FV RV	Cb-J	G-1A CT	10,12 7	micasch.,greenst., serpentinite
Mother Lode, Sierra Nevada, California	MB	SV,FV RV,PL	Cb-J	G-1A	10,12 7	slate,schist,greenstone,serpentinite
Central City, Colorado, U.S.A.	OF	FV,SV	Pt_2	A,UM	29	gneiss,migmatite, granite,pegmatite
Idaho Springs, Colorado, U.S.A.	OF	FV,SV	Pt_2	A,UM	29	as above
Antioquía Gold Province, Colombia	MA	PL,FV SV	PZ PZ-Cr	G,1A	10,12 16,17	slate,greenstone, ophiol.,tonalite
Segovia mine (incl. in above)	DP	FV	Cr_3		28	qtz.diorite, dacite porphyry
Berlin mine (incl. in Antioquía)	DP	FV	PZ	G,1A	12	greenschist, chlorite schist
Pataz-Parcay area, eastern Peru	DT	SV,FV PL	Pt,PZ	G,A	16,17 29	phyllite,schist, granite
Nova Scotia goldfields, Canada	MA	SR,SV FV	Cm	1G	16	slate,sublitharen. quartzite
Southern Piedmont gold belt, U.S.A.	MB	SR,SV FV,PL	Cm	1G	16	slate,graywacke
Jílové, western Czechoslovakia	OF	SV,FV SW	Pt_3	Z-1G CT	12	m-keratoph., greenst.,plagiogr.
Roudný mine, Zvěstov W. Czechoslovakia	DP	RV,SW	Pt_2	A,UM	29	biot.-sillim.gneiss calc-sil.gn.,amphib
Berezovsk, Ural Mts. U.S.S.R.	OF	FV,RV SW	Cm_1-Cb	1g,A	10,12 7,28	serp.,greenst,qtz-seric.sch.,plagiog.
Kochkar, Ural Mts., U.S.S.R.	OF	FV,RV	Cb_1 Or-D_1	G,A	28,10 12	plagiogr.,m-basalt m-andesite
Muruntau, S.Tian-Shan, U.S.S.R.	OF	SW,FV SV	Pt? S?	G	12,16 17	slate,phyllite, sandstone (quartzt)
Kommunard, Altai-Sayan, U.S.S.R.	OF	SW,SK SV,FV	Cm	G	10,12 28	m-basalt,m-keratop. diorite,gabbro

CONCENTRATOR INTRUSION			OREBODIES	
8.ST	9.AGE	10.PETROGRAPHY	11.MINERALOGY	12.TONNAGE
1	Cr_3	qtz.monzonite, porph.dikes	qtz,py,ga,sf,cp,tetr,Au	75.6t Au
1	Cr_3	qtz.monzonite, porph.dikes	qtz,py,ga,sf,ar,tetr,Au	56.6t Au
2-3	Cr_3-T_1	granod.,quartz monzonite	qtz,py,ga,sf,tetr,sb,Au	74t Au
1	Cr_3	granodiorite	qtz,ank,alb,sct,py,ar,cp, Au	330t Au
3-4	J-Cr	qtz.dior., granod.,qtz.mnz.	qtz,ank,alb,py,ar,sch,Au	440t Au
2-3	T_1	qtz.monzon.porp. bostonite	qtz,carb,py,sf,cp,tetr, ga,Au	116.4t Au
2-3	T_1	as above	qtz,py,cp,sf,ga,ar,Au	51.1t Au
2	Cr_3	tonalite	qtz,ga,sf,py,ar,Au	1,450t Au
5	post Cr_3		qtz,carb,py,sf,ga,Au	45t Au
4	Cr_3?	tonalite	qtz,py,ga,sf,po,ar,cp	14.6t Au
3-4	PZ?	granodior.,qtz. diorite	qtz,py,ar,tetr,sf,ga,Au	280t Au
3,4	D	granod.,qtz. monzon.,granite	qtz,py,ar,sch,Au	26.3t Au
3,4	D	granod.,quartz monzon.,gran.	qtz,py,src,chl,ar,Au	80t Au
3	D-Cb	granodiorite, qtz.diorite	qtz,carb,py,ar,po,Au	est.12t Au
5	PZ_3		qtz,dolom,ar,py,Au	10t Au
2	Cb	qtz.porph., granod.dikes	qtz,tourm,py,tetr,ar,cp, sch,Au	est.200t Au
2,3	Pe	microcl.granite	qtz,ar,py,cp,ga,sf,Au	est.120t Au
3-4	MZ?	plagiogr.porph. dk.,granod.	qtz,py,ar,Kf,biot,carb, tourm,sch,Au	est.250t Au
2	PZ_1	diorite	qtz,calc,py,po,ar,cp, tetr,magn,Au	est. 50t

Table 28-25 (continued)

1.LOCALITY	2.CL.	3.ST.	HOST UNIT 4.AGE	5.MT.	6.CH.	7.HOST ROCKS
Darasun, East Trans-baikalia, U.S.S.R.	OF	FV	PZ	G-A	28,29	granod.,gabro, gr. syen.,amphib.
Yana-Kolyma belt, Siberia, U.S.S.R.	MB	Pl,FV	Cb_3-Cr_1	Z-G	16	litharenite,slate
Croydon, N.W. Qld., Australia	OF	FV,SW	PCm	A	29	gneiss,amphibolite
Charters Towers, Qld.,Australia	OF	FV,SV	PZ_1? Pt?	A	29	micasch.,chlorit. sch.,quartzite
Gympie,southern Qld.,Australia	OF	FV	PZ	1G	16?	slate,argillite
Beechworth-Myrtle-ford, Victoria,Aus.	OF	PL,FV	Or_{2-3}	G	16?	phyllite,shist, hornfels
Walhalla, Victoria Australia	OF	SV,FV	D_1	1G	16	shale,sandstone, graywacke
Wood's Point, Victoria, Australia	OF	FV,SV	D_1	1G	16	slate,sandstone, graywacke
Victoria Goldfields, Australia	DT	SV,FV SR,PL	Or_1	1G	16	slate,black slate, qtz-rich litharen.
Beaconsfield, N.Tasmania, Austral.	OF	FV	Cm,Or	Z	16?	litharen.,conglom., greenst.,ophiolite
Reefton, Westland, New Zealand	OF	SV,FV	Pt?	G	16	graywacke,argillite phyllite

EXPLANATIONS:

Column 2, CLASS; DP=deposit; OF=ore field; DT=district; MA=metalliferous area; MB=metalliferous belt

Column 3, STYLE; BR=breccia; FV=fissure vein; MR=massive replacement; RV=replacement vein; SR=saddle reef; SV=shear vein; SW=stockwork; PL=placer

Column 4, AGE=geological age

Column 5, METAMORPHIC INTENSITY; Z=zeolite (and slates); 1G=lower greenschist; G=greenschist; 1A=lower amphibolite; A=amphibolite; UM=ultrametamorphism (granitization); CT=contact

Column 6, CH=Chapter in which the lithologic association is treated in this book

CONCENTRATOR INTRUSION			OREBODIES	
8.ST	9.AGE	10.PETROGRAPHY	11.MINERALOGY	12.TONNAGE
2	J_{2-3}	plagiogr.,granod porph.dikes	qtz,carb,tourm,chl,sct,py, ar,sf,ga,cp,tetr,Au	est. 80t Au
2-4	J_3	dior.,albitite, porph.dikes	qtz,py,ar,sb,sch,Au	min.1,800 t Au
3,4	Pe_3	granite dikes	qtz,carb,py,ar,Au	21.7t Au
1	PZ_1	granod.dikes	qtz,calc,py,ga,sf,cp,ar, tellurides,Au	192t Au
4	Pe_3	granod.,gabbro diorite	qtz,py,Au	96.2t Au
2	D_3	granite,qtz. monzonite	qtz,py,Au	118.2t Au
2	D_3	diorite dikes	qtz,ank,py,ar,born,Au	41.85t Au
1	D_3	diorite dikes	qtz,ank,py,ar,cp,Au	16t Au
4	D	granodiorite	qtz,ank,alb,py,ar,Au	2,081t Au
3,4	D	lamprophyre dikes	qtz,py,cp,ga,sf,Au	24.2t Au
3-4	MZ		qtz,cp,py,ar,sb,cp,ga	57t Au

Column 8, STATUS; intrusive unit 1=hosts all or most orebodies; 2=hosts some ore; 3=present in ore endocontact; 4=distant from orebody; 5=not established

Column 11,MINERALOGY; qtz=quartz; py=pyrite; ga=galena; sf=sphalerite; ar=arsenopyrite; alb=albite; sct=sericite; sb=stibnite; cp=chalcopyrite; sch=scheelite; cl=chlorite; ank=ankerite; carb=carbonate; Au=gold or electrum the gold, and in Suian the skarn is particularly high in boron minerals such as kotoite and ludwigite.

Złoty Stok (formerly Reichenstein; min. 12 t Au) in the Polish Silesia (Schneiderhöhn, 1941) is an enigmatic mineralized skarn in which gold is carried by löllingite as much as by arsenopyrite. The orebodies are contained in magnesian (diopside, tremolite, chondrodite, forsterite) skarn formed by replacement of a dolomite, considerably retrogressively hydrated into a meta-serpentinite. Masses of löllingite, arsenopyrite, pyrrhotite, magnetite and chalcopyrite are situated in a compact ore lens measuring 40x35 m and the grade was between 5 and 35 ppm Au. The skarn formed along the contact of Proterozoic or lower Paleozoic schist, amphibolite and marble sequence, intruded by late Paleozoic syenite and diorite.

The small "skarn" gold deposit at Junction Reefs near Mandurama, N.S.W., Australia (Section 13.6.) is contained in peneconcordant layers interbedded with meta-andesites and volcaniclastics. This could be a thermally metamorphosed volcanogenic mineralization.

Auriferous sulphide replacements in carbonates lacking the skarn association also occur. Like their Pb-Zn counterparts, they may be associated with widespread silicification (jasperoid), or they may be accompanied by another kind of alteration such as recrystallization, Fe or Mn introduction, hydrothermal dolomitization, etc. Purely gold occurrences in metasomatic jasperoid are uncommon. In the Drum Mts. (N.W. of Delta, Utah), gold occurs with pyrite in jasperoid, replacing Cambrian carbonates near a small Tertiary diorite, monzonite, andesite and quartz latite stocks and dikes. Spot contents of 56-85 ppm Au have been reported.

Gold-bearing massive pyrrhotite (visually indistinguishable from a massive pyrrhotite associated with replacement deposits of cassiterite, scheelite, etc.) with lesser pyrite, chalcopyrite, molybdenite, bismuth, tetradymite and gold tellurides, is mined in the Olkhovo deposit in E. Sayans, Siberia (Shakhov, 1972; Borodaevskaya and Rozhkov, 1974; Fig. 28-110). There, replacement sulphide masses containing a variable proportion of quartz, carbonate, sericite, chlorite and serpentinite gangue, are located in lower-middle Cambrian limestone and dolomite (members of a volcanic-sedimentary suite), at the contact with Ordovician plagiogranite and granodiorite. Skarn envelopes may be present locally.

Similar auriferous massive replacements carry gold, in addition to fissure veins, at the Salsigne gold mine in southern France (Reynolds, 1965; 49.5 t Au/11 ppm, 149 t Ag/33 ppm, 149 Tt As/3%, 1,800 t Bi/0.04%). The host rocks are Cambrian to Devonian phyllites, limestones and sandstones but there is no exposed intrusive body in the vicinity.

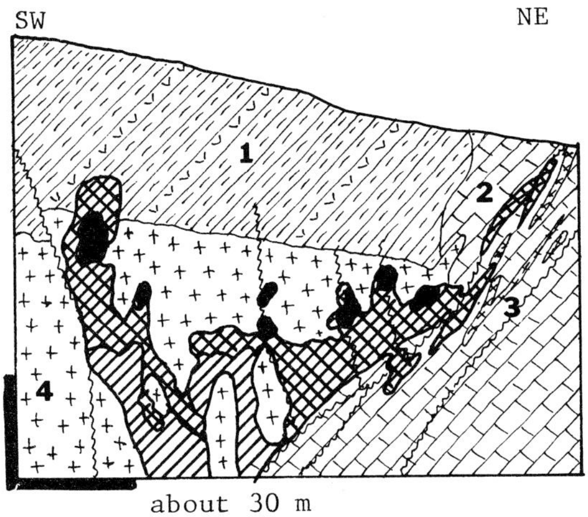

1. $Cm_{1,2}$ hornfelsed slate, andesite, pyroclastics
2. limestone marble
3. diabase porphyry dikes
4. Or, granitic rocks

BLACK: high grade auriferous massive sulphide ore
GRID, DIAGONALLY RULED: lower grade, siliceous disseminated ore

Fig. 28-110. Olkhovskoe Au deposit, E. Sayans, Siberia, auriferous sulphide replacements. Modified after Shakhov (1972), scale and orientation added.

28.17.7. Fine, "invisible" gold disseminated in (replacing) sediments, with or without a visible association with plutonic rocks ("Carlin-type")

The "Carlin-type" gold deposits (after Carlin, Nevada) contain a fine (almost colloidal) gold typically disseminated in calcareous, little metamorphosed or unmetamorphosed sediments, usually accompanied by finely dispersed pyrite, arsenopyrite, realgar or orpiment. Such mineralizations are inconspicuous, because (1) the content of visible sulphides rarely exceeds quantities common as accessories in ordinary sediments; (2) the gold is invisible and cannot be panned and (3) the alteration is inconspicuous (particularly in the weathered outcrop), limited to light bleaching, silicification, argillization or Fe, Mn introduction. The orebodies found so far have in common a fault (often thrust) control and a low-temperature hydrothermal origin, but the hydrotherms (hot springs) could have been a consequence of contemporary volcanism, high geothermal gradient or evolving igneous intrusion in depth. The proximity of plutons, therefore, hardly constitutes a helpful guide. With increasing intensity of alteration, "Carlin-type" deposits may presumably pass into jasperoid-hosted bodies and carbonate replacements.

Deposits of this "type" have already been reviewed briefly in Sections 17.8. (Getchell); 19.7.3. (Carlin, Cortez, Mercur) and 28.21. (Stibnite-Yellow Pine).

28.18. HYDROTHERMAL IRON ORES (EXCEPT SKARNS)

28.18.1. General

Phanerozoic deposits of iron that are still considered to be hydrothermal (other than subaqueous hydrothermal-sedimentary and skarns), represent some 2.8 Bt Fe contained in the past production and reserves. Of this, about 130 Mt Fe come from veins and the remainder from carbonate replacements. The bulk of iron is contained in siderite or in products of its oxidation (goethite, hematite; Fig. 28-111) and only a fraction is found in "primary" hematite veins. Metallogenic statistics involving iron are inaccurate because of genetic uncertainty (particularly in the past) and uneven economic parameters. The genetic link between hydrothermal iron ores and "granites" is the weakest one to demonstrate, of all the metals. Some siderite deposits formerly considered to be postmagmatic, "granite"-affiliated, are now interpreted by some authors as submarine "exhalative" or even groundwater replacements.

As far as the economic factors are concerned, a vein containing 95% siderite and 5% tetrahedrite as in the Coeur d'Alene district is mined for silver and the siderite is a waste, consequently not included in production statistics. The same style of veins formerly mined in Siegerland, West Germany, has a virtually complete production record, because the veins were mined as an iron ore and the minor chalcopyrite, tetrahedrite, etc. they contained recovered as a by-product.

28.18.2. Hematite fissure veins

Fissure veins filled by massive, often fibrous aggregates of hematite are relatively common in the Erzgebirge Mountains. There, they are preferentially associated with S.W. and east-trending high-level fault zones, that indiscriminately cut across the late Paleozoic granites and also rocks of the crystalline basement. Some such zones are up to 10 km long and up to 8 m wide. The hematite filling often changes into quartz, jasper, quartz-barite, fluorite, minor chalcopyrite or uraninite. Mn-oxides (pyrolusite, manganite) are locally abundant. The wallrocks are silicified or argillized.

These veins are among the youngest members of the Erzgebirge vein association, now considered to be a product of "activation" and substantially postdating the hydrothermal phase of the granites they intersect. In the past, these veins were mined at several localities (Bludná, Pernink in N.W. Czechoslovakia; Suhl, Eisenbach, Lauterberg in Germany) as a high-grade and high-quality, but small tonnage iron ore (X0-X00 Tt ore per vein). Comparable veins were mined in the Harz Mountains (e.g. Wieda-Zorge area), West Germany.

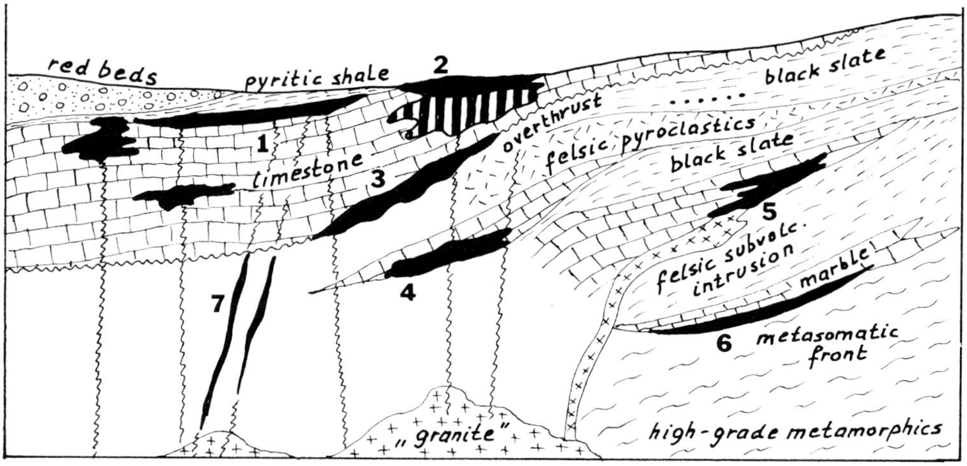

Fig. 28-111. Non-sedimentary siderite deposits: styles, settings, probable interpretations. 1=Replacements from descending waters; 2=low grade replacement siderite (ruled) topped by limonitic gossan (black); 3=hydrothermal siderite replacements along overthrusts; 4=ditto, in granite roof; 5=possibly remobilized subaqueous hydrothermal siderite; 6=possibly siderite replaced limestone in Fe metasomatic front; 7=fissure and shear veins.

28.18.3. Siderite fissure veins

Siderite veins were substantially more widespread and economically important than the hematite veins. Those in Siegerland have been the most important source of iron ore in Germany until 1885. At present, however, this style of ore is considered to be of local importance only.

In the Siegerland-Wied siderite vein district of Germany (Neumann-Redlin et al., 1976; 64.5 Mt Fe/30%, 11.4 Mt Mn/5.3%), several hundred veins occur within an area of about 3,000 km^2. The veins are hosted by a thick, monotonous sequence of Devonian shale and litharenite and there is no evidence of a proximal association with "granites". A large, probably mafic plutonic body has been detected geophysically at a depth greater than 10 km. The veins are lenticular, up to 12 km long, 2-6 m wide, and filled by coarse crystalline manganoan siderite with variable quantities of quartz, calcite and scattered chalcopyrite. Wallrock alteration consists of a slight bleaching in slates, and argillization and siderite impregnation in the host arenites.

An extensive siderite district of local importance is found in the Paleozoic core of the West Carpathians, particularly in the Spiš-Gemer region of Slovakia (Mahel' and Buday, eds., 1968; about 80 Mt Fe/30%,

5.75 Mt Mn/2.5%). This region and its siderite deposits have already been mentioned briefly in this book (Sections 10.4.1. and 11.2.). There, siderite occurrences are unusually widespread (over 2,000 known) and hosted by low-grade metavolcanics and metasediments ranging in age from Silurian to Permian. A portion of the deposits are peneconcordant lenses and sheets, interchangeable with ankerite, magnesite, dolomite and graphitic limestone. Formerly interpreted as hydrothermal replacements probably dating from the Cretaceous period and related to buried granite plutons (Bernard and Hanuš, 1961), many deposits were recently re-interpreted as being "stratiform" Paleozoic volcanic-sedimentary deposits, deformed, metamorphosed and partly remobilized during the Mesozoic alpinotype deformation (Ilavský, 1976). The remainder of the Slovak siderite deposits are bona fide structurally controlled fissure veins and, as in Siegerland, there is no granite is evidence. The veins contain high trace Ni,Cr,Co and Hg as well as accessory fuchsite, gersdorffite, Hg-tetrahedrite, chalcopyrite and cinnabar, indicating a possible genetic association with mafic-ultramafic magmatic rocks.

The largest vein deposit Rudňany (31 Mt Fe; Section 10.4.1.) could be the largest single vein siderite deposit of the world.

28.18.4. Siderite lenses and masses in (or interchangeable with) carbonates (Table 28-26)

This category of deposits used to be called "metasomatic siderites" (it still is in Zitzmann, ed., 1976), and was interpreted as limestone or dolomite replaced by siderite, by the action of hydrothermal fluids. In the past decade, however (compare Wolf, ed., 1976, 1981), there has been a tendency to re-interpret many of these orebodies (e.g. Eisenerz, Austria, the Carpathian deposits of Slovakia and Rumania) as "exhalative" and stratiform. As on other occasions in the history of geology, a sweeping global re-interpretation is harmful and it is virtually certain that subaqueous-hydrothermal, hydrothermal (solid host)-replacement and late diagenetic-replacement siderite deposits are possible and can exist side by side. An association with "granites", however, is rarely apparent.

It is beyond the purpose of this book to "bring order" into the siderite genetic problem so this will not be considered further. Siderite deposits are most common in the following lithologic associations: (1) low-grade metamorphosed mafic, felsic and bimodal volcanic-sedimentary (Chapters 10, 11, 12, respectively); (2) deeper-marine "black shale" association and its transition into carbonates (Chapter 17) and (3) shallow marine carbonate, evaporite and shale associations transitional into continental "red beds" (Chapter 19); shallow-marine platformic and "miogeoclinal" carbonate association (Chapter 20).

Zitzmann and Neumann-Redlin (1976) distinguished two "types" among the "metasomatic siderites": (a) Bilbao "type", in which the principal economic commodity is the weathering-enriched hematite-goethite zone and (b) Eisenerz "type", producing unenriched siderite or ankerite.

In the northern Spain iron ore province centred around Bilbao (Zitzmann, ed., 1976; about 216 Mt Fe), the siderite is hosted by lower Cretaceous folded and faulted limestones. The orebodies are always fault controlled and have the form of irregular replacement siderite masses to mantos that are thickest near the fault. The siderite contains, on average, 38% Fe and 0.5% Mn, but the bulk of the commercial ore was a massive, rubbly to earthy residual hematite and goethite in the oxidation zone (51-57% Fe; 0.4-0.7% Mn).

The Kremikovtsi deposit near Sofia, Bulgaria (Iovchev, 1961; 76 Mt Fe, 15.3 Mt Mn, 46 Mt barite, 1 Mt Pb; Fig. 28-112) is unusual in the variety of mineral commodities it contains. The orebody is hosted by a middle Triassic dolomitic limestone at the base of a nappe, floored by upper Jurassic marls and argillaceous siltstones. The orebody is an erosional remnant of a composite lens over 1,000 m long and up to 267 m thick, the bulk of which is now residual hematite and limonite enclosing lenses of relic barite. The "primary" ore is a massive gray siderite, laced by a second generation of siderite veinlets and it is transitional into barite and hematite. Hematite occurs in strongly brecciated zones of the original dolomitic limestone as well as the siderite, and it replaces both carbonates. Pyrite, galena, tetrahedrite, chalcopyrite and sphalerite are widespread at fractures and in remobilization veinlets within the siderite and barite bodies. The average Fe ore contains 0.4% Pb, 0.14% Cu and up to 10 ppm Ag. Additional disseminated galena occurs in Triassic dolomitized limestone on the fringe of the iron orebody.

The original, detailed description of the Kremikovtsi deposit given in the Bulgarian language by Iovchev (1961) makes a good case for the epigenetic, replacement origin of the orebody, shortly postdating the late Mesozoic thrusting. In conformity with the latest fashion, however, an "exhalational" origin has also been proposed recently.

Additional example localities of members of the former class of "metasomatic siderites" are summarized briefly in Table 28-26.

Jebel Ouenza siderite field in Algeria (Dubourdieu, 1952; Fig. 28-113) is an interesting case of replacement siderite in a lower Cretaceous carbonate unit, strongly tectonized in the Eocene period and dismembered by diapirs of Triassic sediments. Siderite replaces limestone in the core of a large N.E.-oriented anticline and is oxidized into "limonite" to a depth of 250 m.

28.19. MAGNETITE SKARN AND REPLACEMENT DEPOSITS

28.19.1. General

Magnetite skarns and replacement deposits (this category may include some subaqueous hydrothermal-sedimentary deposits subsequently modified in an intrusive aureole), represent some 6 Bt of economic iron. If the world's economic content of iron is given as 300 Bt Fe,

Table 28-26. Massive siderite deposits and districts ("metasomatic siderites"), selected examples

LOCALITY	HOST UNIT	MINERALIZATION	REFERENCES
N.Spain iron ore province (Bilbao)	Cr_1 folded and faulted limestone	irreg.to manto orebodies of siderite replacing limestone along faults (35% Fe); capped by residual hematite-goethite 51-57% direct shipping ore;216 Mt Fe	Zitzmann, ed. (1976)
Sierra Menera Fe ore region, east-central Spain	Or dolomite, pyritic shale, quartzite; Cb_3 dolomite	goethite and hemat. formed by oxidation of PZ_3 siderite and ankerite, in an up to 80 m thick ore lens replacing dolomite near fault; probably groundwater replacem. 60 Mt Fe/60%	Zitzmann, ed., (1976)
Marquesado deposit east of Granada, S. Spain	Tr limestone	irreg. siderite masses to mantos replace carbonates along tect. zones; siderite was converted into goethite and hematite to form layers and pockets in karsted carbonates; 90 Mt Fe/55%	Zitzmann, ed. (1976)
Erzberg near Eisenerz, Austria	S-quartz seric.schist (meta qtz.keratoph.); D-limest., black phyll.	large stratabound siderite, ankerite zone with small branching veins, interpeted as syndepositional or post-Tr replacement 214 Mt Fe/33% Fe; 13.75 Mt Mn/2.5%	Holzer and Stumpfl (1980)
Ljubija, Bosnia, Yugoslavia	PZ limest., schist, sandst., Cb_2 metagraywacke altern.with slate Cb mafic volc., tuff	76 struct.controlled siderite and goethite deposits, replacement lenses, veins. 500 Mt Fe/38-46.6%, 1.82-2% Mn	Čičić (1980)
Spiš-Gemer region, E.Slovakia, Czechoslovakia	S-Cb mafic and felsic metavolc., black and green phyll., lenses of limest., anker., magnes., sider. Cr cryptobath.	over 2,000 sider.occurrences, both veins and conformable to irregular lenses and mantos (Železník, Nižná Slaná, Dobšiná). Interpr.at either "exhalational" or "teletherm.replac.". 50 Mt Fe/30%, 4.5 Mt Mn	Mahel' and Buday (1968) Ilavský (1976)

Rudabánya near Miskolc, northern Hungary	PZ and Tr limestone	irregular masses of siderite, ankerite, minor chalcopyrite, galena; thick goethite gossan; PZ mineralization remobilized in Tr_{2-3}	Morvai (1982)
Teliuc-Ghelar, Poiana Ruscă Mts, W. Rumania	D_3 greensch.(mafic metavolc.),black,seric., chlor.,phyll.,carbonate, felsic metatuff	lenses of massive to banded siderite assoc.with quartz,magnetite,stilpnomel., thuringite,pyrite,ankerite interpreted as "exhalational", partly mobilized; goethitic gossan; 22 Mt Fe	Kräutner (1977)
Ouenza Bou Khadra, N. Algeria Fe ore region	J_1,Cr,Eo limestone, dolomitized limestone	4 mineralized zones controlled by faults and diapirism. Irregular masses to mantos. Siderite, goethite in oxid. zone; minor barite, access.chalcop.,chalcoc., pyr., galena; controversial origin; 135 Mt Fe/54%	Popov (1976)
Djerissa, Tunisia	Cr_1 "reef" limestone exposed in a faulted dome (diapir ?)	6 irregular to stratabound siderite orebodies oxidized to goethite and hematite; 25.12 Mt Fe/54%	Massin (1976)

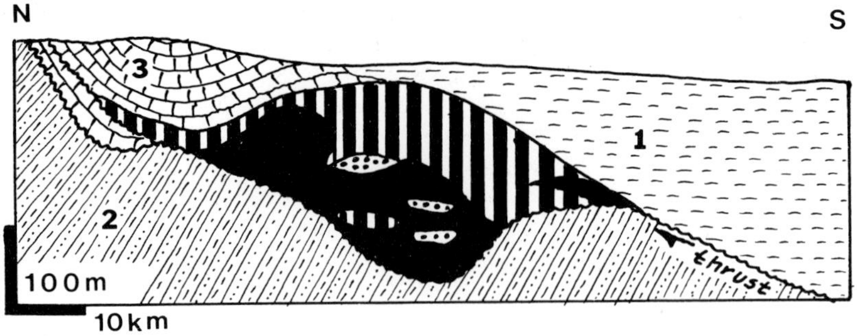

Fig. 28-112. Siderite, limonite, barite deposit Kremikovtsi, western Bulgaria. 1=T_3 sediments; 2=J flysch shale, sandstone; 3=Tr_2 sediments. Tr_2 orebodies; ruled area: dominant limonite; black area: siderite; dots: barite. Modified after Iovchev (1961).

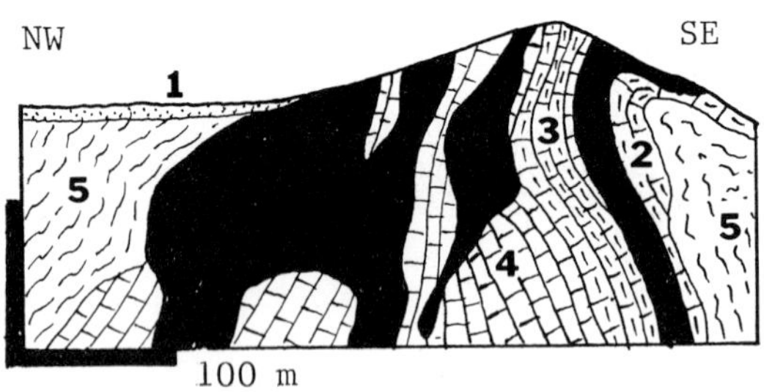

1. Q cover sediments
2. Cr_3 marls
3. Cr_2 argillaceous limestone
4. massive limestone
5. Tr red beds / evaporite association in a diapir

BLACK: metasomatic limonite (goethite) gradational into siderite in depth

Fig. 28-113. Metasomatic limonite siderite deposit in limestone, Jebel Ouenza, Algeria. Modified after Dubourdieu (1952).

this is just about 2%. If the iron resources are considered in trillions of tons, the role of skarns fades into insignificance. Skarn magnetite deposits range in size from local showings containing several tens of tons of ore, to giant ore fields with over 2 Bt Fe (Turgai district, U.S.S.R.; Marcona field, Peru). Magnetite skarns formed in both limestones (calcic skarns) and dolomites (magnesian skarns) and several occurrences of replaced ankerite or siderite may also be present.

The bulk of skarns mined for magnetite only occur in mafic volcanic-sedimentary associations believed to have been generated in immature island arcs (Chapters 10,12,13) and containing minor lenses and interbeds of carbonates. The intrusions responsible for the skarn generation are I- ("magnetite")-type granitoids, ranging from gabbros to granodiorites. Quartz diorites are the most common. Lesser number of magnetite skarn deposits formed within former Andean-type margins. There, magnesian skarns are important, probably because dolomite is a rare rock in the previous setting. The concentrator intrusions here are more felsic and alkalic, represented by granodiorite, monzonite, syenite. Magnetite also occurs, and is recovered locally, from Cu, Pb-Zn, Sn, and other skarns. The skarns associated with gabbro or diabase intrusions in rifted continental margins (e.g. Cornwall, Pennsylvania) are reviewed in Chapter 32.

28.19.2. Calcic skarns in mafic volcanic-sedimentary association

Einaudi et al. (1981) recently reviewed the magnetite skarns, and pointed out the following basic characteristics of the "island arc calcic magnetite skarns": (1) the frequent presence of endoskarns, some of them mineralized by disseminated magnetite; (2) widespread Na-metasomatism in the endo- and exocontact (albitization, scapolitization); (3) anomalous trace concentrations of Ni and Co; cobalt is sometimes recovered as a by-product; (4) the most widespread skarn silicates are andradite, hedenbergite and epidote, but skarn zoning is poorly developed because of the large variety of rock types in the exocontact and (5) magnetite orebodies are closely associated with garnet skarns, or are hosted by limestone outside the skarn zone. The magnetite bodies are usually irregular masses close to the contact, but some magnetite orebodies are "stratabound" sheets or lenses hosted by a favourable carbonate horizon at some distance from an intrusion. The latter are difficult to distinguish from pre-skarn bedded ores.

The Gora Magnitnaya magnetite skarn deposit near Magnitogorsk, southern Urals, U.S.S.R. (280 Mt Fe/56-65%) is usually quoted as a type locality for calcic magnetite skarns (e.g. in Zitzmann, ed., 1976) and there is copious literature covering it, some of it in English translation (Magak'yan, 1968; Baklaev, 1973; Fig. 28-114). The deposit is hosted by a Devonian to lower Carboniferous sequence of marine basalt to basaltic andesite flows, tuffs and volcaniclastics with interbeds of limestone. These rocks are folded, faulted and intruded by Carboniferous quartz monzonite, granodiorite and diorite stocks. There are two peneconcordant ore sheets (1.8x2.5 km and 1.4x1.2 km in size, and, on average, 40-44 m thick). The orebodies are disrupted by a swarm of diorite dikes. Magnetite, the principal mineral with lesser pyrite, hematite, chalcopyrite and quartz, is contained in an earlier garnet, hedenbergite, magnetite skarn, and in a later epidote, calcite, magnetite, quartz skarn.

The Turgai iron ore province (Sokolov and Grigor'ev, 1974; Baklaev, 1973; 3 Bt Fe) is located in north-western Kazakhstan, in a transitional zone of the Urals into the Kazakh Block. This is a prominent polygenetic mineralized region with iron deposits in both the pre-Mesozoic basement and in the Mesozoic-Cainozoic platformic cover. The skarns in the basement are hosted by Carboniferous shallow

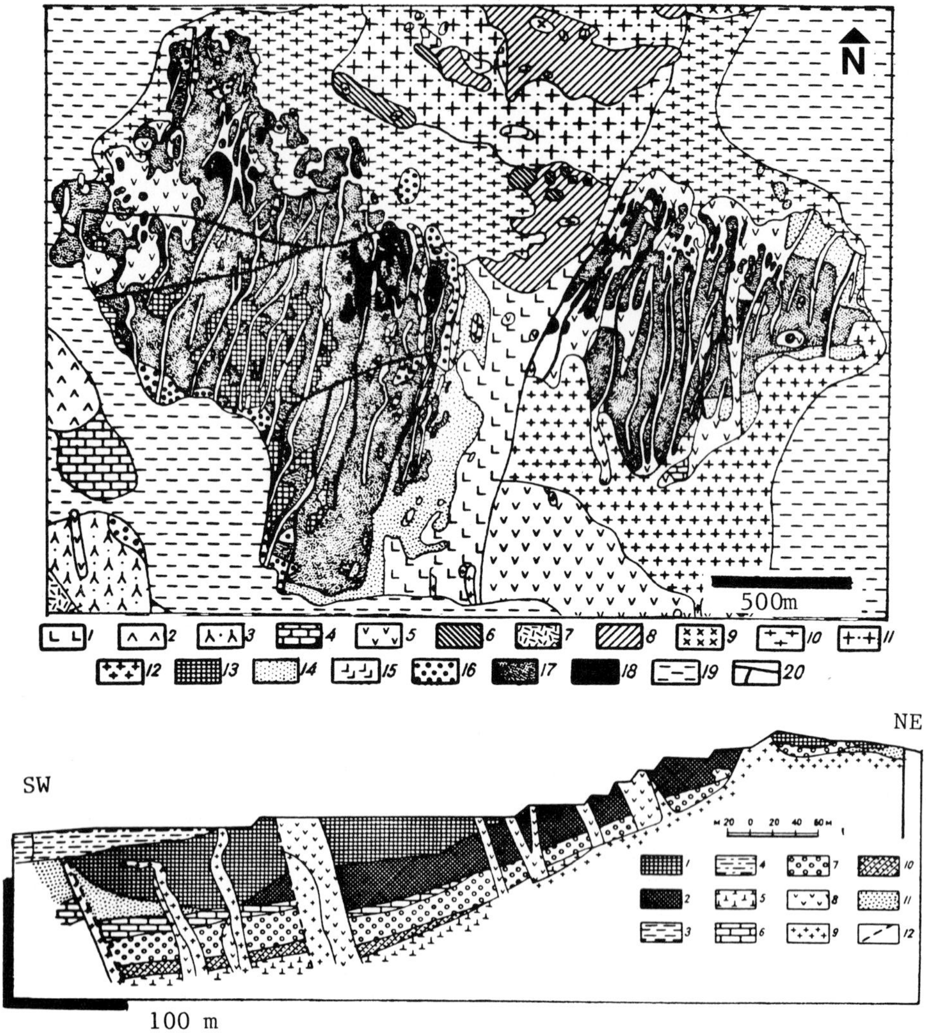

Fig. 28-114. Gora Magnitnaya magnetite skarn deposit, southern Urals, U.S.S.R. From Magak'yan (1968). MAP LEGEND: 1=$D-Cb_1$ basalt porphyry; 2=extrusive rhyolite porphyry; 3=rhyolite pyroclastics; 4=limestone; 5=diorite; 6=gabbrodiorite; 7=breccia with a diorite matrix; 8=granodiorite; 9=hornblende granite; 10=leucogranite; 11=low quartz granite and syenite; 12=granite porphyry, microgranite, albitophyre; 13=hornfels; 14=andalusite, sillimanite, cordierite hornfels; 15=metaporphyries; 16=garnet skarn; 17=martite; 18=magnetite; 19=talus; 20=faults. SECTION LEGEND: 1=oxidized ores; 2=primary ores; 3=barren talus; 4=ore-bearing talus; 5=porphyries; 6=limestone and marble; 7=skarns; 8=diorites; 9=granites; 10=andalusite, sillimanite, cordierite hornfels; 11=rubble ; 12=faults.

marine to continental andesites, their pyroclastics, volcaniclastics and lesser impure and bituminous limestones and minor evaporites, exposed in a N.-S.-trending anticline. The supracrustals are intruded by Carboniferous diorite porphyry, gabbrodiorite, plagiogranite and andesite porphyry. Granite and quartz-feldspar porphyries are also represented, e.g. in the Kachar deposit. The intrusives are epizonal to subvolcanic bodies, partly comagmatic with the volcanic-sedimentary suite and their emplacement has been strongly controlled by structure and lithology. Elongated plutons, dikes and sills are the most common and the intrusions overlap in time with the mineralization.

The igneous emplacement into the probably still "wet" volcanic-sedimentary pile is believed to have caused widespread sodic metasomatism and hydration (scapolite, albite, chlorite, actinolite, apatite) of the intrusive rocks, earlier hornfelsed silicate rocks and the early garnet-hedenbergite skarn along the intrusive contacts. The hydrous metasomatites were, in turn, often progressively converted to anhydrous skarn assemblage as the intrusive front advanced. Scapolite and albite metasomatism in the Turgai district is a high-level feature, rapidly diminishing with depth.

Magnetite deposits in the district are typically sheet-like, relatively persistent bodies (mantos), controlled to a considerable degree by "favourable lithology" of replaceable carbonate layers and bedded calcareous pyroclastics and volcaniclastics. In the large Kachar deposit (Sokolov and Grigor'ev, 1974; 630 Mt Fe/45%) the following petrographic types of iron ores are represented: (1) Massive, homogeneous magnetite; (2) martite after magnetite, in the oxidation zone; (3) scapolite-magnetite; (4) pyroxene, garnet, magnetite skarn; (5) albite-magnetite and (6) actinolite, chlorite, zeolites, calcite, magnetite. The above assemblages are transitional and the proportion of the silicate gangue minerals naturally controls the ore grade that ranges from under 30% Fe to 65% Fe (average about 45% Fe).

The Dashkesan magnetite deposit in Azerbaidzhan, U.S.S.R. (Smirnov, 1969) is notable for its anomalous cobalt content. Cobalt resides in accessory cobaltite and glaucodot, which, in turn, are present in a garnet, pyroxene, amphibole, magnetite, ilvaite skarn replacing late Jurassic andesitic volcaniclastics, near the contact with a lower Cretaceous gabbro-granodiorite intrusion. Separate accumulations of massive pyrite, arsenopyrite, cobaltite and other sulphides, moreover, occur in the hangingwall of the magnetite skarn lens, enveloped by a Cl-rich (up to 7% Cl) hornfels ("dashkesanite").

Additional examples of magnetite skarns (in British Columbia) have been reviewed briefly in Chapter 10.

28.19.3. Magnesian skarns

Einaudi et al. (1981) explained the case of magnetite formation in magnesian (dolomite-hosted) skarns by the fact that forsterite, talc and serpentine do not accept much iron in solid solution under conditions of skarn formation. Consequently, such iron accumulates in magnetite rather than in Fe-garnet or pyroxene, as at limestone contacts. The high-temperature skarns may include a diopside-spinel assemblage close to the igneous contact, and forsterite-calcite near the dolomite marble line. A garnet-pyroxene calcic skarn may form as a later-stage overprint.

The Teia (Teiskoe) iron deposit in Khakassia (S. Siberia, U.S.S.R.; Sokolov and Grigor'ev, 1974; Ivankin, ed., 1973; 48 Mt Fe/33%; Fig. 28-115), is commonly quoted as a typical producing example of magnesian magnetite skarns. There, a pipe-like body of skarn containing magnetite masses is underlain by limestone marble and topped by a dolomite marble. These rocks are a part of a Cambrian bimodal volcanic-sedimentary metamorphosed complex that also includes amphibolite, metagabbro and meta-keratophyre. The Cambrian supracrustals were intruded by Cambro-Ordovician or Devonian granosyenite, syenite, intrusive breccias and aphanitic rocks designated as felsites and keratophyres. The latter are probably metasomatic albitites. The ore zone is over 1.5 km long and up to 300 m thick. It contains 12 lens-like masses of magnetite-serpentinite; dolomite, magnetite, lesser phlogopite and magnetite-hematite composition. The magnetite is very fine grained and there are minor amounts of accessory sulphides and Ni-Co arsenides. The skarn envelope consists of forsterite, chondrodite, clinohumite, diopside and spinel skarn on the dolomite side, and garnet-pyroxene skarn on the limestone side. There is a considerable variety of late-stage metasomatites including phlogopite, scapolite, cordierite, sphene, tourmaline, etc. in the exocontact, and Na- and K-feldspathites in the intrusive breccia.

Table 28-27 gives basic data on selected localities of magnetite skarns.

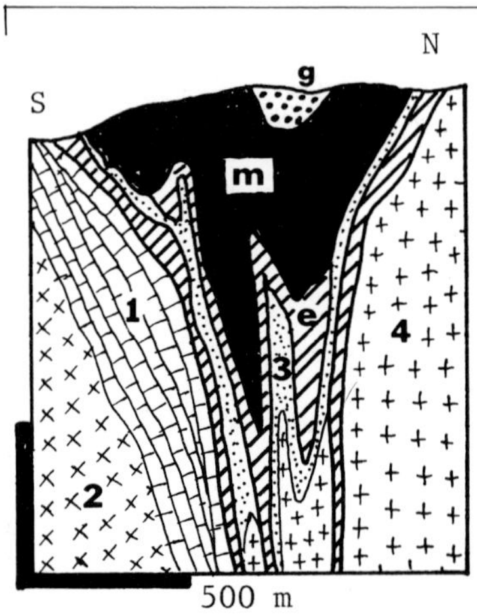

1. Cm, limestone, dolomite
2. diorite, amphibolite
3. D felsite porphyry, albitophyre, albite metasomatites
4. granosyenite

SKARN ZONES:

e=zone of poor magnetite ores and serpentine-phlogopite skarn

m=zone of rich and thick magnetite ore in garnet-pyroxene skarn

g=zone of thin magnetite ore layers in epidote-chlorite metasomatite

Fig. 28-115. Teia (Teiskoe) Fe ore field, U.S.S.R., an example of a magnesian magnetite skarn. Modified after Ivankin, ed. (1973).

Table 28-27. Phanerozoic magnetite skarn and replacement deposits, selected examples

LOCALITY	SUPRACRUSTAL HOST UNIT	INTRUSION	MINERALIZATION	REFERENCES
Insular Belt of British Columbia Canada	Tr_3 basalt to andesite topped by limestone minor black slate	Tr_3-Ol qtz. diorite, granodior.	irregular, mostly small magnetite lenses, common chalcopyrite, in skarn 25 Mt Fe, 120 Tt Cu	Sutherland Brown (1968)
Tasu Sound, Queen Charlotte Islands, British Columbia (incl.above)	Tr massive metabasalt topped by limestone	J qtz.diorite,porphyry dikes	irregular sheets to masses of magnetite in skarn,thickened along faults and porphyry dikes; widespread chalcopyrite. 16 Mt Fe/37%	Sutherland Brown (1968)
Las Truchas-Ferrotepec, Michoacan, Mexico	J ? andesites Cr_2 calcareous marine sediments	T_1 granodiorite, aplite	several irreg.orebodies of magnetite in garnet,epidote skarn; 40 Mt Fe/49-55%	Min.Magaz. Oct. 1976
Marcona, southern Peru	PCm gneiss; PZ metasedim.incl.limest. and dolom.; J limestone,andesite,shale	Cr ? diorite granodiorite	numerous magnet.,pyr.,quartz actinol. up to 100 m thick bodies in a 20x8 km E.-W. zone; 2.65 Bt Fe/60%	Bellido and de Montreuil (1972)
El Pedroso,Jerez de los Caballeros,S.W. Spain	PZ limestone,marble	PZ_3 qtz.dior. granodiorite	magnetite bodies,interpr.as possibly volcanic-sediment., metamorphosed; 250 Mt Fe	Zitzmann, ed. (1976)
Elba Island, Italy	PZ andalus.schist Tr limest.,meta-keratophyre	Mi, quartz monzonite	magnetite-ilvaite masses near granite converted to goeth. gossan; interpreted as therm.modified "exhalative" ores; 35 Mt Fe	Zitzmann, ed. (1976)

Table 28-27 (continued). Phanerozoic magnetite skarn and replacement deposits, selected examples

LOCALITY	SUPRACRUSTAL HOST UNIT	INTRUSION	MINERALIZATION	REFERENCES
Sokolovka dep., Turgai distr., W. Kazakhstan, U.S.S.R.	Cb_2 andesite, tuff, limestone	Cb_3 gabbro, diorite granodior.	7.5 km long ore zone, peneconc. to discord., mass. to dissemin. magnet. in pyrox., actinolite skarn, envel. by albitization; 396 Mt Fe/41% Fe	Baklaev (1973)
Cürek and Divriği dep., E.C. Anatolia, Turkey	Eo limestone, serpentinite	Eo syenite, monzonite	2 orebodies, a contact magnet. skarn, and distal magnetite oreb. along limest.-serpentin. thrust contact; 56 Mt Fe/55%	Zitzmann, ed. (1976)
Dashkesan, Azerbaidzhan, U.S.S.R.	J_{2-3} andes., rhyol. tuff siltst., sandst., marl, slate	Cr_1 gabbro-granodior. gabbro-syen.	peneconc. magnetite skarn body up to 4 km long, garnet-pyrox., local access. cobaltite, glaucodote; 84 Mt Fe/42%, estim. 50 Tt Co	Sokolov and Grigor'yev (1974)
Teia (Teiskoe) dep., Kuznetsk Alatau, Siberia, U.S.S.R.	Pt_3-Cm_1 dolomite, basalt, andesite	Cm-Or syen., granosyen., diorite	magnesian skarn, forster., humite, spinel, diops., amphib., serpent., magnet., minor sulphides; a lens 1,500x300 m; 48 Mt Fe/33%	Sokolov and Grigor'yev (1974)
Tashtagol field, Kuznetsk Alatau Siberia, U.S.S.R.	Cm_2 greenschist, meta-andes., basalt, albitophyre, limestone	Cm-Or syenite, quartz syenite, syen. porph.	layer to lens, fault dismember. up to 1 km long, 40-70 m thick garnet, epid., magnet. skarn oreb. 152 Mt Fe/45%	Sokolov and Grigor'yev (1974)

Banat distr.(Dognacea,Ruschiţa, Ocna de Fier), Rumania	D,bimodal volc.-sed. assoc.,carbonates; Cr-J limest.,dolom.	Pc granod.	irreg.lenses of magnet.,ilvaite,hedenb.,garnet,ludwigite skarn, variable content of Cu,Zn,Pb sulph. Est.20 Mt Fe	Ianovici and Borcoş (1982)
Gora Vysokaya near Nizhn.Tagil,Ural Mts. U.S.S.R.	S_3 limest.,basalt, andes.,pyroclast., volcaniclastics	D ? syenite	4 km^2 peneconc.replac. magn. oreb. in garnet-pyrox. skarn martitized,rubble zone; 31 Mt Fe/48.8%, 12 Tt Co	Baklaev (1973)
Peshchansk dep., Serov distr.,central Urals, U.S.S.R.	D_1 limest.,andes.and basalt volc.,sedim.	D-Cb diorite	up to 60 m thick magnet.,garnet, pyrox.,epid. skarn; 87 Mt Fe/51%	Sokolov and Grigor'yev (1974)
Gora Blagodat, central Urals, U.S.S.R.	S, volc.-sedim.assoc. basalt,tuff,limest.	D-Cb diorite, syenite	tabular, lentic.garnet,epid., magnet.skarn; 50 Mt Fe/35% 22 Tt Co/0.022%	Sokolov and Grigor'yev (1974)
Magnitnaya Gora, Magnitogorsk, S.Ural Mts.,U.S.S.R.	S_1-D_3 metabasalt, andes.,phyllite, limest.over ophiolite	D-Cb gabbro to qtz.dior., granodiorite	2 peneconc. garn.,pyrox.,magnet.skarn lenses up to 115 m thick; 280 Mt Fe/56-65%	Baklaev (1973)
Kachar deposit,Turgai distr.,W.Kazakhstan, U.S.S.R.	Cb_2 andes.,andes.tuff, bituminous limest., slate	Cb_3 gabbro-dior.,granodior.,dior. porphyry	peneconc. magnet.bodies in scapol.-magn.;pyrox.,garnet, magnet.; albite-magnet.assoc. 630 Mt to 2.23 Bt Fe/45%	Baklaev (1973)
Sarbai dep.,Turgai distr.,W.Kazakhstan U.S.S.R.	Cb_2 limest.,andes. flows,tuffs,volcaniclastics	Cb_3 diorite porphyry	massive to dissem.conformable pyrox.,scapol.,albite,epid., actinol.,magnetite lens up to 185 m thick, 2 km long; 684 Mt Fe/45.6%	Sokolov and Grigor'yev (1974)

28.19. MANGANESE DEPOSITS

Independent manganese deposits associated in some way with granitic plutons or forming hydrothermal orebodies are of limited importance and represent some 4 Mt Mn. In contrast, the Mn content of the siderite veins and replacements alone is at least 50 Mt Mn. Manganese mineralization styles are close equivalents of the iron deposits treated in the earlier paragraphs, and the following varieties can be recognized:

(1) Hypogene pyrolusite or manganite-filled veins in fault zones, greatly postdating the emplacement of granites which they may intersect. In the Erzgebirge, these veins are gradational into or interchangeable with the fibrous hematite-filled veins. Additional localities include the Harz Mts. and Schwarzwald in Germany; Sierra del Norte (part of Sierras Pampeñas), Argentina and others.

(2) Rhodochrosite, kutnahorite, manganocalcite or oligonite-filled veins. These are rare as independent orebodies, but common as gangue to Pb-Zn ores (e.g. at Philipsburg and Butte, Montana).

(3) Mn carbonates (the same as under 2) replacing limestone, dolomite or jasperoid. At Leadville, Pioche and other localities, Mn carbonates formed an alteration envelope around galena-sphalerite replacements.

(4) Mn-rich skarns (Mn-ilvaite, Mn-hedenbergite, rhodonite, dannemorite, etc.) are commonly associated with Pb-Zn sulphide orebodies, but there has been no substantial manganese production reported.

Most of the Mn production from "granite" aureoles was derived from psilomelane and cryptomelane accumulations in oxidation zones over Zn-Pb vein or replacement deposits with manganiferous gangue or alteration (e.g. Leadville, 550 Tt Mn; Pioche, Nevada, 330 Tt Mn; Philipsburg, Montana, 174 Tt Mn).

28.20. URANIUM DEPOSITS

28.20.1. General

Phanerozoic hydrothermal uranium vein deposits represent some 250 Tt U, and U minerals impregnations and fracture coatings in granite exocontacts acount for another 25 Tt U or so. This makes up a total of about 275 Tt U, a rather insignificant quantity compared with the 4 Mt U in currently recoverable ores, and additional tens of million tons of U in low-grade uraniferous sediments. Curiously, this style of U deposits is almost exclusively European and all but some 25 Tt U included in the 275 Tt total occur in a discontinuous belt of sialic, Proterozoic basement-floored Hercynian (=late Paleozoic) median massifs and blocks. These are intruded by Permo-Carboniferous granites and traceable, discontinuously, from Spain and Portugal through Cornwall, France and West Germany to the Bohemian Massif and Sudeten (Fig. 28-116). The bulk of the uranium came from the French

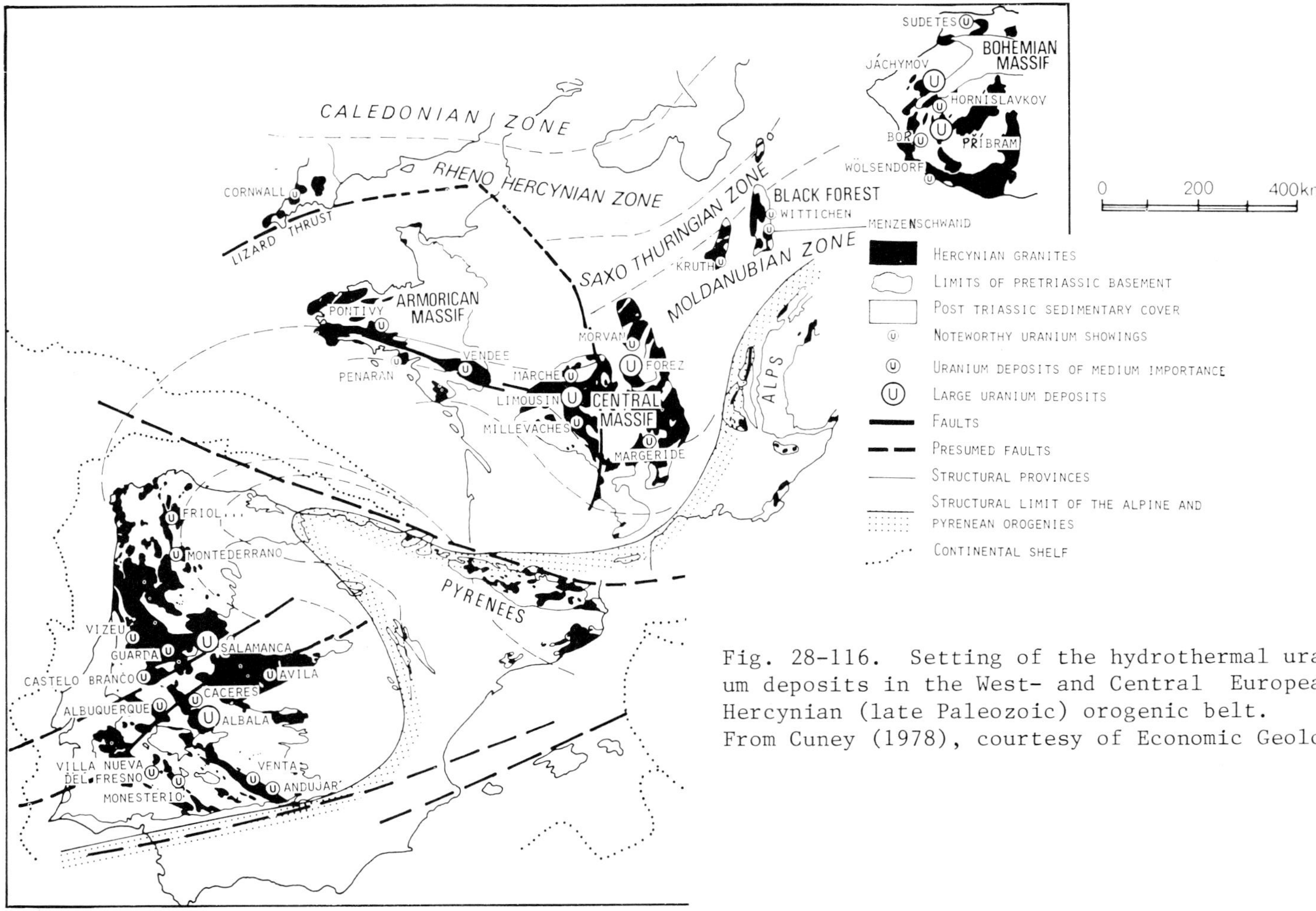

Fig. 28-116. Setting of the hydrothermal uranium deposits in the West- and Central European Hercynian (late Paleozoic) orogenic belt. From Cuney (1978), courtesy of Economic Geology.

(65 Tt U) and Czechoslovakian (estimated 150 Tt U) deposits.

Uranium deposits including those associated with granites have been reviewed recently by Nash et al. (1981) and Ruzicka (1971), and these authors distinguished three major "types" (or rather end-members) of granite-affiliated (or hydrothermal with uncertain affiliation) mineralizations: (1) mineralogically simple veins; (2) mineralogically complex veins and (3) exocontact fracture coating or disseminated ores. To this could be added (4) disseminated or stockwork U concentrations in granite endocontacts. Fig. 28-117 shows, diagrammatically, the U mineralization styles in the present setting.

28.20.2. Mineralogically simple uranium veins (Table 28-28)

These ores are contained in granite endocontacts (most of the French deposits), in metamorphics in granite exocontacts (most of the Czech deposits, Schwartzwalder mine in Colorado), or in both.

A "typical" small and simple intragranite orebody, such as the La Faye deposit near Grury, France (Roubault, 1956 and own visit, 1978; 2 Tt U/0.2%; Fig. 28-118), is hosted by an upper mesozonal, late Carboniferous coarse porphyritic biotite granite to quartz monzonite pluton. The main intrusive phase is cut by fine grained biotite microgranite, aplite and kersantite dikes. The intrusive rocks show evidence of high-level brittle fracturing with numerous thin zones of brecciation and mylonitization. Many fractures and faults have distinct and conspicuous red or pink alteration haloes, and several breccia or fracture zones have been converted into "episyenites", feldspathic (K-feldspar or albite) rocks formed by (a) dissolution and removal of quartz, sericitization of plagioclase and biotite, feldspathization of muscovite, preservation of orthoclase or (b) wholesale hydrothermal feldspathization of the wallrock. The variety of type (a) tends to be porous and reminescent of miarolitic structure. It is sometimes called "sponge rock". The (b) variety is usually hard, homogeneous and fresh-looking.

The mineralization at La Faye is represented by open space vein filling of fleshy red microcrystalline quartz, gradational into zones of silicification (jasper). The vein has numerous voids and vugs, and is about 1 m thick. The adjacent zone of brecciation and alteration is up to 10 m thick. Veinlets and nests of brown smoky quartz, dark violet fluorite and white or pink barite cut the microcrystalline quartz. Masses, veinlets, blebs of black pitchblende accompanied by pyrite and rare Ni-Co arsenides are irregularly distributed in the vein quartz and the adjacent breccia, and are accompanied by inconspicuous black sooty uraninite coatings. The oxidation zone has impregnations and fracture coatings of torbernite, autunite and kasolite.

The contrasting variety of uranium minerals apparent in this style of mineralization (the colourful, hexavalent U compounds; sooty pitchblende; massive, metacolloform pitchblende) triggered a

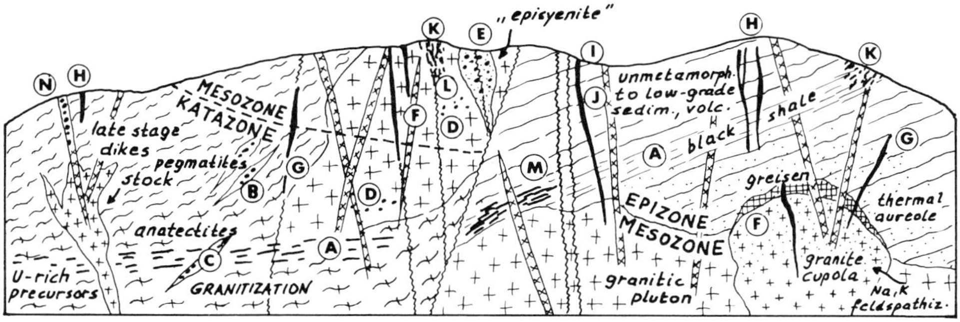

Fig. 28-117. Common styles of U mineralization spatially associated with granitic plutons (styles A-C are treated in Chapter 29). From Laznicka (1984). A=High trace U layered (foliated) supracrustal rocks (e.g. graphitic slate, schist); B=uraniferous anatectites, aplite and pegmatite veined katazonal granites (e.g. Rössing); C=discrete U pegmatites; D=minor (non-ore) accessory uraninite in mesozonal and katazonal S granites; E=disseminated "primary" and "secondary" U^{+4} oxides in "episyenite" (=desilicified and/or feldspathized granite); simple vein composition.

Discrete "main stage" quartz or carbonate, pitchblende, sulphides, etc. veins; F=wholly within granite; G=in granite and its roof; H=in the roof only. Supergene altered tops of ascending veins: $I=U^{+6}$ oxides (torbernite, autunite, etc.) in the oxidation zone; $J=U^{+4}$ sooty pitchblende in cementation zone. Shallow, descending infiltration veins, impregnations, stockworks of: $l=U^{+6}$ oxides; L=sooty pitchblende. $M=U^{+4}$ oxide stockworks, impregnations in granite-schist hornfelsed exocontacts; N=radioactive high level dikes.

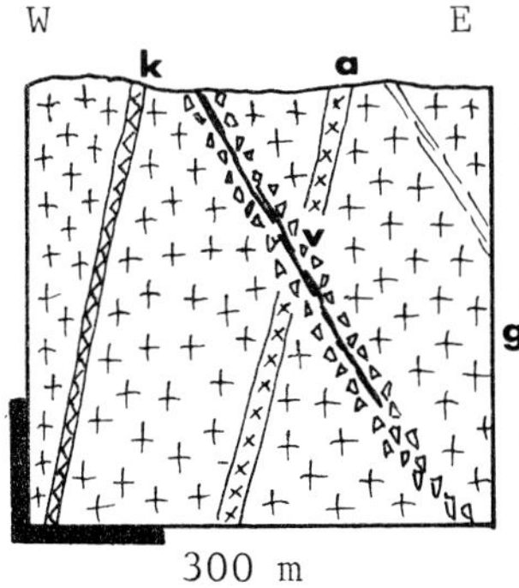

k Cb-Pe kersantite dikes

a fine grained biotite leucogranite to aplite dikes

g Cb_3 coarse porphyritic biotite granite to quartz monzonite

v vein structure in brecciated and altered granite

Fig. 28-118. La Faye vein, Grury field, east-central France. Diagrammatic from LITHOTHEQUE, based on data in Roubault (1956).

Table 28-28. Phanerozoic hydrothermal uranium vein deposits, selected examples

LOCALITY	HOST UNIT	MINERALIZATION	REFERENCES
La Crouzille, Limousin, France	350-360 m.y. muscovitized, albitized two mica katazon. granite cut by lamprophyre dikes	275 m.y.,pitchbl.,pyr.,cherty hematitic quartz,musc.,fluorite in brecc.veins in granite. 33 Tt U/ 0.16%	Leroy (1978)
Bois Noirs-Limouzat, Forez, France	335 m.y. biot. granite, microgranite, 270 m.y. quartz porph.,lamprophyre	W.N.W.,N.W. vein structures, miner. porous gran.brecc.,pitchbl.,marc., pyr.,coffin.,01 miner. 6.4 Tt U	Cuney (1978)
Grury, Morvan, France	Cb_3 porph.biot.gran. to qtz.monzon.,microgranite, lamprophyre	fiss.veins,brecc.+ mylonitiz.zones in silicif.,argilliz.,hemat.alter. gran. 1 m thick, red microcr.qtz., fluor.,barite,pyr.,pitchbl. 2 Tt U	Roubault (1956)
Les Pierre Plantées, Le Cellier,Margeride, France	Cb, granite massif, kersantite and lamprophyre dikes	LPP: patchy pitchbl.dissem.in min. episyenite granite; LC: pitchbl., coffinite vein in kersantite; 4 Tt U.	Geffroy (1971)
Saint-Sulpice des Feuilles, Basse-Marche, France	Cb granite, leucogranite	small qtz.,pitchbl.,coffinite, fluorite veins and masses in "episyenite"-altered granite	Geffroy (1971)
Urgeiriça, E.-C. Portugal	Cm schist, intruded by Cb granite, lamproph.dikes; N.E. faults, shears, mylon. zones	80-100 m.y. quartz,jasper,pyr., metacollof. and sooty pitchbl., gal.,sphal.,etc. veins in seric., silic.,chloritiz.,tourmaliniz. granite. 13.6 Tt U	Rich et al. (1971)
Menzenschwand, Schwarzwald, West Germany	granite gneiss, gneiss, schist, Cb granite	botryoid. and sooty pitchbl.fiss. qtz.,fluor.,barite, pitchbl., autunite veins; 60 t U.	Rein (1960)

Jáchymov-Abertamy field, N.W. Czechoslovakia	Pt_3-PZ_1 micasch.,amphibolite, calc-silic.gneiss,chl., biot.,graph.schist; Cb-Pe_1 biot.granite,gabbro	E.-W.,N.-S. veins of quartz,carbonate,pitchbl.,lesser fluor.,pyr., Ni,Co arsenides,Ag,Bi; estim. 40 Tt U	Ruzicka (1971) Mrňa and Pavlů (1967)
Johanngeorgenstadt-Potůčky, E.Germany and N.W. Czechoslovakia	Cm-S phyl.,andalus.micasch., biot.,graph.schists; Cb-Pe_1 granite, leucogranite	qtz.,dolom.,anker.,calc.,pyrite, pitchblende, Ni-Co arsenides, Ag sulphosalts veins; estim. 8 Tt U, 500 t Bi,Co	Ruzicka (1971)
Horní Slavkov, N.W. Czechoslovakia	Pt_3-S gneiss,schist,amphibolite,skarn. Cb_{2-3} biot.granite, leucogranite,greisen cupolas	multistage fissure veins in gran. exocont.,qtz.,dolom.,calc.,sider., pitchbl.,pyr.Hematitiz.wallrocks, some fluorite; N.W.,N.veins; estim. 10 Tt U	Ruzicka (1971)
Vítkov near Bor, N.W. Czechoslovakia	Pt or PZ gneiss, intrud.by Cb granodior.,aplite dikes	qtz.,carbon.,pitchbl.,pyr.impregn. in N.E. trending sericitiz.fault zone; estim. 6 Tt U	Ruzicka (1971)
Zadní Chodov,N.W. Czechoslovakia	Pt_3,PZ,biot.gneiss near contact with PZ_3 granodior., qtz. monzonite pluton	fine,sooty pitchbl. in mylonitiz., graphitiz.,carbonatized,chloritized shear; estim. 6 Tt U	Ruzicka (1971)
Bytíz-Kamenná zone, Příbram district W.Czechoslovakia	Pt_3 slate, grayw.,spilite; Cb qtz.diorite,granodior. pluton; diabase,lamprophyre dikes	25 km long N.E. fault lineament, 11 vein groups; lenses,chimneys, stockw.,calc.,dolom.,pitchbl.,pyr., gal.,sphal.,chalcop.,minor Ag, Ni-Co arsenides. Estim. 70 Tt U	Kutina (1963) Ruzicka (1971)
Rožínka-Olší zone, Centr.Czechoslovakia	Pt biot.gneiss,amphib.,marble,graph.gneiss	W.N.W.miner.shear zones,carbonate, chlor.,pitchbl.,pyr.;est.15 Tt U	Ruzicka (1971)
Schwartzwalder M., Ralston Creek,Colorado, U.S.A.	Pt_1 garnet-biot.gneiss,biot. schist,calc-silic.gneiss, amphibolite,quartzite	62 m.y.flat or horsetail pitchbl., coffin.,adular.,pyr.,hemat.veins in subsidiary faults	Young (1979)

controversy in interpretation. The genetic debate has essentially been between the proponents of descending origin and supergene U release (e.g. Barbier, 1974) and ascending origin with hypogene U sources, of the uranium deposits. Although many small occurrences can probably form by either mechanism, it appears that at sizeable and relatively deep mineralization systems (1) the massive pitchblende is the "primary" ore, most probably precipitated from ascending fluids in depth; (2) sooty pitchblende is the equivalent of the zone of secondary sulphides at base metal suphide deposits (e.g. porphyry coppers) and (3) the hexavalent U salts mark the oxidation zone. Leached capping usually marked by barren quartz, porous silicification quartz and clay fault gouge residue, is sometimes developed.

Cuney (1978) recently interpreted, in detail, the geological environment and the activity resulting in the formation of the Bois Noirs-Limouzat uranium vein system (Forez, N.E. Massif Central, France; 6.4 Tt U). The deposit is hosted by a mesozonal Carboniferous (335 m.y.) granite interpreted as anatectite, emplaced into a Proterozoic gneiss and amphibolite terrain. The early phase granite was partly deuterically altered (to quartz, microcline, albite and chlorite). Accessory uraninite associated with quartz-muscovite alteration crystallized subsequently from a residual fluid phase. The economic uranium ore consists of pinch-and-swell ore shoots composed of comb and microcrystalline quartz, pitchblende, marcasite, pyrite with minor fluorite, chalcopyrite, hematite, galena, etc., filling N.W.-trending fractures and breccia zones. The main stage mineralization is interpreted as being a product of low-temperature (300-70° C; 100-70° C for pitchblende) uranium mobilization througout an interval of about 65 m.y. duration following the granite crystallization. A post main-stage process modified portion of the earlier pitchblende into coffinite, and a "per descensum" remobilization during the Oligocene age produced sooty pitchblende and hexavalent U minerals.

The small deposit Les Pierre Plantées in the Margeride ore field, S. Massif Central, France (Geffroy, 1971) is an example of a mineralized "episyenite" alteration pipe in granite. The pipe contains local patches of pitchblende impregnations.

The Schwartzwalder mine near Golden, Colorado (Young, 1979; 8.5 Tt U; Fig. 28-119) is an example of a mineralogically simple uranium vein system in metamorphics, and also the largest Phanerozoic uranium vein deposit in the Americas. The host rocks are lower Proterozoic garnet-biotite gneiss, calc-silicate gneiss and metaquartzite. "Laramide" (late Cretaceous-lower Tertiary) quartz monzonite and granite porphyry dikes are known within 2 km of the deposit. The veins are structurally controlled by an intersection of a fold nose with a N.-S.-trending, steeply west dipping fault zone, and the ore is persistent vertically for over 900 m without change in grade or mineralogy. The ore shoots are both within the main fault, and in branch and feathering-away fractures. The vein filling is ankerite, pyrite, adularia, hematite, pitchblende, coffinite. The ore deposition has been dated between 52 and 73 m.y., which is broadly

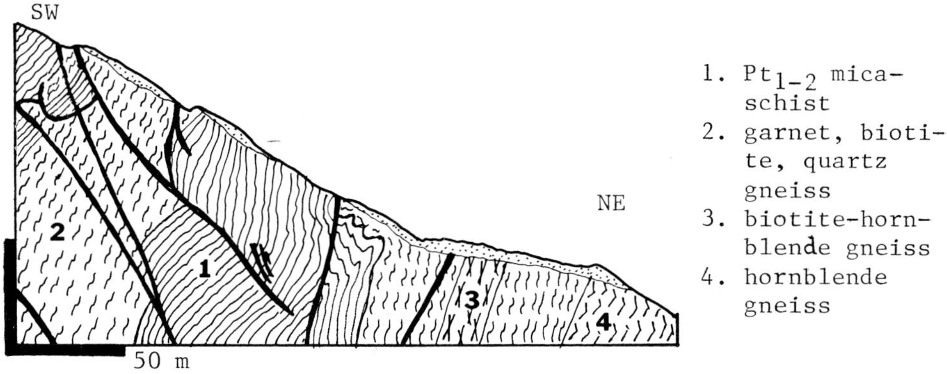

Fig. 28-119. Schwartzwalder U deposit, Ralston Creek, Colorado. After Sheridan et al. (1967).

contemporary with the Laramide phase of intrusive activity in the Colorado Front Ranges.

28.20.3. Mineralogically complex uranium veins

The hydrothermal Ni,Co,Bi,Ag,As association described in Section 28.16., frequently combines with uranium to form mineralogically complex veins. The Jáchymov-Abertamy field described earlier is the best known example which, in its 500 years of mining history, evolved from a silver-mining camp through cobalt, radium production into a post-World War II uranium boom town, before its final exhaustion in the 1960s. The geology of the field was reviewed earlier (compare Fig. 28-104). Pitchblende formed in the third mineralization stage only, accompanied by dolomite, lesser pyrite and fluorite. It is interesting to note that in all the mineralogically complex U vein deposits, the actual U-depositing stage is compositionally identical to the "simple U veins" treated earlier. The additional metals were introduced either earlier or later.

In the Eastern (uraniferous) ore zone in the Příbram district, western Czechoslovakia (Ruzicka, 1971; see also Fig. 17-8 and short description in Section 17.6.3.), a N.E.-trending fault system, 25 km long, contains eleven or more discontinuous uranium vein groups (segments). The tonnage figures have not been released, but it is estimated that the zone represents a minimum of 70 Tt U. The faults are developed in, and the ore hosted by, late Proterozoic gray to black pyritic slates and litharenites with minor tuffaceous and mafic metavolcanic horizons. It parallels the contact of the composite, mainly Carboniferous tonalite and granodiorite-dominated Central Bohemian Pluton in the east (see description of the Pluton in Section

28.2.3.). A portion of the metasediments is thermally hornfelsed and intruded by diabase, granite porphyry and lamprophyre dikes.

Ore shoots within the mineralized segment, many of them blind, have the form of short lens-like veins, chimneys, ore bunches or stockworks. They are locally crowded by, or gradational into, fault gouge and mylonite. About 12% of the vein-filled faults contain economic uranium shoots. The ore deposition took place in three stages which are as follows: (1) siderite, quartz, sphalerite, galena, dolomite, arsenopyrite, lesser Ni-Co arsenides, native silver. This phase is comparable with the vein filling in the western ("Old") Příbram Pb,Zn,Ag zone. (2) calcite, dolomite, ankerite, pitchblende, rare U-pyrobitumen, pyrite and (3) barren calcite, lesser pyrite. The carbonate-U stage took place at low temperature and alteration was slight, mainly chloritization, sericitization, hematitization and carbonate impregnation and veining of the fault gouge.

28.20.4. Exocontact fracture coating and disseminated ores (Table 28-29)

The Midnite mine, 50 km N.W. of Spokane, Washington (Nash et al., 1981; 4.6 Tt U/0.18%; Fig. 28-120), is frequently quoted as a type locality for a "contact" uranium mineralization. There, several tabular orebodies are hosted by a middle Proterozoic thermally hornfelsed siliceous pyritic phyllite and calc-silicate schist, at contact with late Cretaceous porphyritic biotite granite. Sooty pitchblende, coffinite, lesser pyrite, marcasite, arsenopyrite, chalcopyrite and sphalerite occur as disseminations along foliation and stockworks of fracture fillings in phyllite, and replacements in metacarbonate bands. Autunite and metaautunite occur in the oxidation zone. The Fé and Nisa deposits in Spain and Portugal are similar in many respects.

28.20.5. Disseminated and stockwork uranium in granite endocontacts

Under a tempting headline of "resources of the future", Armstrong (1974) popularized the concept of "porphyry uranium deposits". Unfortunately, most of the examples he used are mineralized deep-seated anatectites, katazonal granites and pegmatites far away from the mineralized "porphyries" treated earlier as hosts to Cu,Mo,Sn and other ores. The deep-seated ore systems will be reviewed in Chapter 29.

Curiously, high-level calc-alkaline intrusions (with the exception of several examples associated with continental volcanics and discussed in Chapter 26) do not seem to accumulate uranium substantially in a form comparable with the Cu,Mo or Sn stockworks or pervasive accumulations. The highly mobile uranium tends to leave the system and accumulate preferentially in the derived sediments. Exceptions occur when U (together with Th,REE,Ta,Nb, etc.) is bound in

Table 28-29. Phanerozoic "granite contact" uranium deposits

LOCALITY	HOST UNIT	GRANITIC INTRUSION	MINERALIZATION	REFERENCES
Midnite mine near Spokane, Washington U.S.A.	Pt_2 partly graphit. schist, calc-silicate hornfels, marble	Cr_3 (75 m.y.) porph. granite	380x210x50 m tabular oreb. of dissem., replac. pitchbl. and coffinite in exocontact partly // with foliation; miner. age 50-52 m.y. 4.6 Tt U/0.18%	Nash et al. (1981)
Les Bondons dep. near Mt.Lozère, Cévennes, France	Cm-Or graph.schist micasch., quartzite andalus.-biot.met.	290-280 m.y. granite locally altered to episyenite	impregn. along fract. in schists, up to 200m from gran.contact; pitchblende, coffinite, pyrite; Tr_3 supergene alteration	Eulry and Vargas (1980)
Fé deposit near Ciudad Rodrigo, Spain	Pt_3, Cm, hornfelsed meta-graywacke, shale	Cb, biot.granite	pitchbl. with pyr., chlorite in fract.veinlets in hornfelsed metased. 11.9 Tt U	O.E.C.D. (1979)
Nisa deposit, Alto Alentejo, Portugal	pre-Or schists	PZ_3 granite pluton	autunite, saleeite, bassetite phosphouranylite, impregn. along faults, fractures, foliation, breccia zones	O.E.C.D. (1979)

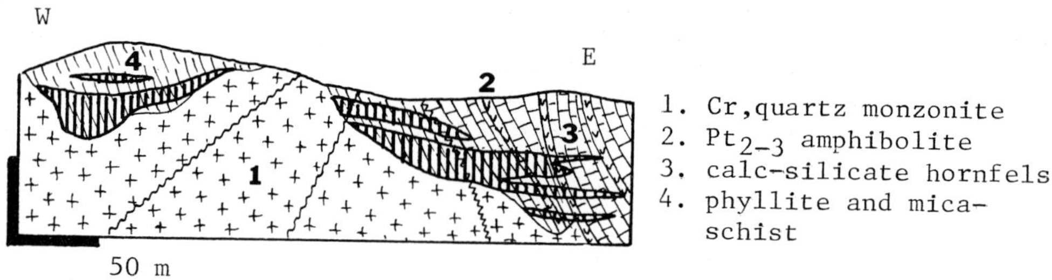

Fig. 28-120. Midnite U mine, Washington, U.S.A. Generalized after Nash et al. (1981).

accessory heavy minerals such as fergusonite, euxenite or monazite (e.g. Bugaboo, British Columbia), or is present in dispersed or cation-substituting form as in the bostonite dikes in the Central City gold field, Colorado (up to 100 ppm U in some dikes). Several stockwork and disseminated U accumulations in the alkaline association will be reviewed in Chapter 33.

28.21. ANTIMONY DEPOSITS

Antimony deposits are a very interdisciplinary group, difficult to pigeonhole. Of the total of some 6.5 Mt Sb in ores, worldwide, about 50% is in silicified carbonates (jasperoid) under a shale or schist screen (or possibly a Sb source), controlled by faults and lithology. Almost 33% of the total is contained in a single field, Hsikuangshan in China. The above low-temperature mineralizations are, in regard to granitic intrusions, distal, and their formerly automatically considered association with "granites" is no longer considered to be universal. For that reason, these deposits have been reviewed under the headings of their common host rocks, carbonates and "black" argillites (Section 17.10.4.).

Of the remainder, about 10% Sb occurs in the Precambrian deposits, reviewed in Volume 2; 6% is in subvolcanic deposits (Chapter 26) and 4% in (pene)concordant lenses in (meta)volcanics and sedimentary associations. The latter are presently largely considered to be stratiform, although often subsequently modified by tectonism (Stadt Schlaining, Austria) or by intervening later granitic intrusion (Pezinok, Pernek, Slovakia). About 30% of Sb deposits (Table 28-30) appear to be "classical" epigenetically emplaced hydrothermal bodies (veins, stockworks, replacements), some of them undisputably affiliated with "granites".

Antimony is widespread in lesser quantities in polymetallic deposits in which the main (or most valuable product) is Pb-Zn, Ag, Au, W, Hg and a few other metals. It is, however, rarely recovered in the metallic form and sold hence the presence of antimony is rarely recorded in the statistics. This class of deposits is gradational into another class of mixed ores, where the Sb content is substantial and is recovered as a by-product, together with other metals. Some example localities are listed in Table 28-31. The last, most distinct category, consists of stibnite-only (or quartz, stibnite, pyrite) orebodies. Some of these are gold-bearing.

Antimony is a cheap metal (in the region of $ 3-4 per kilogram), so only very high-grade orebodies (3% Sb plus, or around 30-40% Sb in the case of hand-cobbed concentrates) have been economic to mine until recently. Since the Sb clarke (mean crust content) is very low (0.2 ppm), the common antimony orebodies have enormous factors of concentration (150,000 times the clarke). With the exception of Hsikuangshan, most Sb deposits and fields are small to medium-sized. Low-grade "bulk" Sb deposits are little known, unless they contain other, usually more valuable commodities (e.g. gold and scheelite as in the Yellow Pine mine, Idaho).

Stibnite deposits are hosted both by "granites" and by supracrustal rocks. Stibnite in "granites" is always substantially younger than its host, and appears to be related to a later-stage hydrothermal activity frequently corresponding to an "activation" regime. Black slate to phyllite or schist, flyschoid, and marine volcanic-sedimentary associations are the most common hosts to Sb ores. Some such occurrences have been interpreted, with increasing frequency, as modified "stratiform" accumulations (e.g. Höll, 1966, 1977; Höll and Maucher, 1976), or products of lateral secretion (Roger, 1972).

The Lake George deposit, 40 km W.S.W. of Fredericton, N.B., Canada (Abbott and Watson, 1975; about 35 Tt Sb; Fig. 28-121) is an example of the simple stibnite veins, probably genetically associated with mesozonal granites. The ore is hosted by a folded, cleaved and hornfelsed Silurian slate, calcareous slate, siltstone and litharenite. Devonian porphyritic biotite quartz monzonite is anticipated in depth, and there are kersantite dikes in the mine field. Several E.-W. fracture veins filled by quartz and carbonate contain discontinuous ore bunches and lens-like ore shoots. Stibnite forms more than 90% of the ore minerals, the rest being arsenopyrite, pyrite and rare tetrahedrite. The grade ranges from 5.22 to 12.7% Sb and Au and Ag are so low and irregularly distributed that they are not recovered.

The Hillgrove field 30 km E. of Armidale, N.S.W., Australia (Harrison, 1953; own visit, 1980; over 16 Tt Sb, 15 t Au, 1.4 Tt W; Fig. 28-122), is an example of an Sb,Au,W-mineralized discrete veins. The field is underlain by cleaved, sheared and thermally metamorphosed lower Permian "flyschoid" gray slate to schist, containing thin quartzite and litharenite interbeds. This has been permissively intruded by a later Permian biotite quartz monzonite and sheared, fractured shortly after the intrusion cooling. A system of numerous N.W.-trending shears and fractures in both endo- and exocontact is filled by granodiorite porphyry dikes and steep quartz,

Table 28-30. Hydrothermal Sb deposits or districts hosted by or affiliated with "granites"; selected examples. NOTE: distal Sb replacements or "stratiform" deposits accompanied by jasperoid are reviewed in Table 17-6.

LOCALITY	HOST ROCKS	MINERALIZATION	REFERENCES
Lake George, N.B., Canada	S,contact hornfelsed slate, siltst.,calcar.siltst.; D,porph.biot.qtz.monzonite, kersantite dikes	several E.-W. striking quartz,stibnite,lesser arsenopyrite,pyrite fissure veins; 35 Tt Sb/3%	Abbott and Watson (1975)
Yellow Pine, Idaho, U.S.A.	J-Cr,slightly foliated biot. qtz.monzon. and aplite,intr. Or? carb.and detrit.metased. T, qtz.latite, rhyol.dikes	Large irreg.complex oreb. hosted by shattered and alterd qtz.monzon.in hangingw.of a shear. Low-grade gold bearing dissem.pyr.+arsenop.is followed by stockw.scheel.and stibnite 79 Tt Sb/0.81%,206 t Au,6,365 t W	Cooper (1951)
Tupiza region, S. Bolivia	Or,D,slate with black slate interbeds,thin arenite partings; minor qtz.monz.intr.	simple discord.qtz.,stibn.,pyrite minor ferber.veins,along shears and faults; 279 Tt Sb/8.78%	Ahlfeld and Schneider-Scherbina (1964)
Brioude-Massiac, Massif Central, France	Cm-Or biot.schist,gneiss, migmatite; Cb sandstone; Cb biot.granite,microgranite	Cb_3 veins along N.E. faults. Qtz., carbonate,stibn.,pyr.,lesser berthierite,arsenop.filling; 40 Tt Sb	Roger (1972)
La Lucette, Armoricain Massif, N. France	S-D metaquartzite, schist	Cb, quartz,stibnite,arsenop.,pyrite lesser galena,gold, fissure veins	Routhier (1963)
Dúbrava, Magurka, Nízké Tatry Mts., E. Czechoslovakia	PZ_3 cataclast.meso-to katazonal granite,granodior., quartz monzonite	E.-W. and N.-S. quartz,stibnite,pyrite,minor jameson.,berthier.,gold fissure and shear veins,impregnations in graphitic mylonite; estim. 65 Tt Sb, 3 t Au	Mahel' and Buday, eds. (1968)

Helcmanovce, Čučma, Spiš-Gemer Mts., E. Czechoslovakia	S-Cb black phyllite, chert, carbonate, bimodal metavolc. Cr granite cryptobathol.?	quartz veins with massive lenses to dissem. of stibnite, minor pyr., ars., sphal., gal.; estim. 35 Tt Sb	Mahel' and Buday, eds. (1968)
Krupanj-Zajača, Drina district, Yugoslavia	Cb-Tr litharenite, slate, limest., dolom.; folded, thrusted; intr. by Mi dacite, andesite, granodiorite	peneconc. replac. bodies of stibnite with jasperoid in carbonates; breccia cementing stibnite; quartz, calcite gangue, minor galena, boulangerite; 120 Tt Sb	Janković (1982)
Sarylakh, N.E. Yakutia, U.S.S.R.	Tr_3 sandst., siltst., shale; J_3 qtz. diorite porphyry stocks	N.W. striking miner. crush zones, irregular quartz, stibnite, pyrite, arsenop. veins; silicif., sericitiz. wallrocks; estim. 20 Tt Sb	Zharikov (1974)
Bau district, Sarawak, Malaysia	J-Cr_1 polymictic subduction melange rich in sheared shale, grayw., greenst., chert, ultramaf. Cr_3 granitic pl.	auriferous stibnite in silicif. limest. and in veins; Au in eluvial placers; about 55 Tt Sb, 28 t Au	Kho (1976)
Hillgrove-Metz, New England, N.S.W., Australia	Pe_1 hornfelsed, cleaved slate to schist; $Pe_{2,3}$ quartz monzon., sheared; granodior. dikes, diorite stocks	Pe_3 stibnite or qtz.-stibnite, scheel., gold shear and fissure veins, N.W. striking, in hornfels and granite; 16 Tt Sb, 1,370 t W, 15 t Au	Harrison (1953)
Costerfield, Victoria, Australia	S-D_1 gray to black slate, litharenite; D, biotite granite	D, quartz-stibnite simple and branching fissure veins up to 6 m thick; 22.4 Tt Sb	Hill (1975)

Table 28-31. Mineral associations of selected polymetallic vein deposits containing significant proportion of antimony

METAL ASSOC.	MINERALOGY	EXAMPLE LOCALITIES
Sb-Au	quartz, calcite, ankerite, stibnite, arsenopyrite, gold	Magurka, Czech.
Sb-Hg	dolomite, livingstonite, stibnite, anhydrite, native sulphur	Huitzuco, Mexico
Sb-As	quartz, stibnite, arsenopyrite, pyrite, galena, gold	La Lucette, France
Sb,As, Ni,Co,U	quartz, dolomite, stibnite, realgar, pyrite, marcasite, bravoite, pitchblende, vaesite	Lojane, Yugosl.
Sb,Ag, Pb,Zn	siderite, quartz, ankerite, galena, sphalerite, chalcopyrite, tetrahedrite	Coeur d'Alene, Idaho
Sb,Cu	siderite, tetrahedrite, quartz	Schwaz, Austria
(Ag,Hg)	siderite, quartz, tetrahedrite, pyrite	Sunshine m., Idaho
Sb,W	quartz, wolframite, pyrite, scheelite	Osanica, Yugosl.
	quartz, dolomite, calcite, pyrite, stibnite, arsenopyrite, scheelite	Yellow Pine, Idaho
	quartz, scheelite, pyrite, arsenopyrite, stibnite, calcite, gold	Hillgrove, N.S.W.
Sb,As Tl	quartz, dolomite, stibnite, realgar, orpiment, arsenopyrite, marcasite, pyrite, sulphur, lorandite, vrbaite	Alšar, Yugoslavia
Sb only	quartz, stibnite, pyrite	Zajača, Yugosl.
	quartz, stibnite, pyrite, arsenopyrite, berthierite	Brioude-Massiac, Fr.

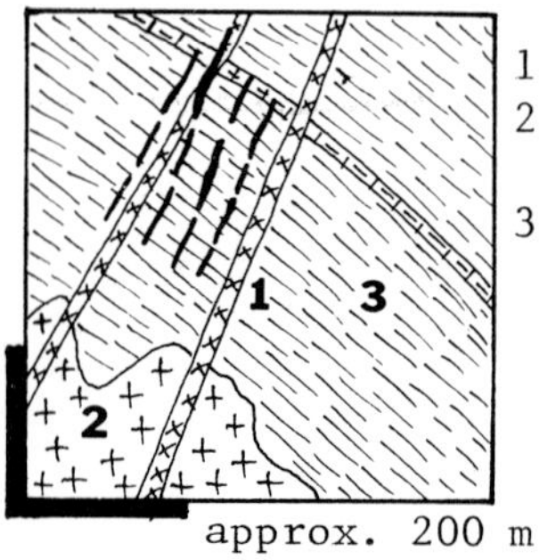

1. D, biotite kersantite dikes
2. D, porphyritic biotite quartz monzonite
3. S slate, calcareous slate, siltstone

Fig. 28-121. Lake George Sb deposit, New Brunswick, Canada (diagrammatic). From LITHOTHEQUE.

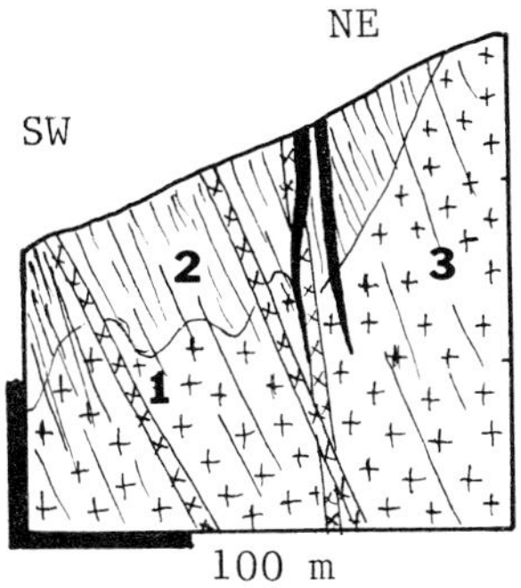

1. Pe_3 qtz.monzonite and granodiorite porphyry dikes
2. Pe_1 gray cleaved slate to schist, minor quartzite and litharenite
3. Pe_2 medium crystalline to porphyritic protoclastic biotite quartz monzonite sheared and altered in the ore zone

Fig. 28-122. Hillgrove vein stibnite and scheelite field, N.S.W., Australia (generalized). From LITHOTHEQUE.

quartz-stibnite, quartz-gold and quartz-scheelite veins. The Sb-mineralized structures contain fine crystalline ("steely") or fibrous stibnite as bunches, lenses, fracture fillings or as disseminations in slightly bleached or unaltered hornfels, or in a slightly sericitized cataclastic quartz monzonite. Brecciated and sheared vein stibnite and stibnite-coated slickensided fractures are common.

The Stibnite (a ghost town) ore field in the Yellow Pine district, central Idaho (Cooper, 1951; 79 Tt Sb, 206 t Au, 6,365 t W, 170 t Ag; Fig. 28-123) has an interesting combination of metals and mineralization styles probably emplaced at a variety of levels along and within a N.E.-trending broad shear and fracture zone. The host rock is a mesozonal Jurassic to Cretaceous slightly foliated biotite quartz monzonite, intersected by numerous aplite dikes and emplaced into carbonates and detrital metasediments of probably Ordovician age. These rocks have been faulted and sheared probably in the early Tertiary period and the Meadow Creek Fault zone is a fault gouge-filled structure up to 50 m wide. Fracturing and shearing is intense in quartz monzonite in the hangingwall (west of) the main shear in the Yellow Pine mine field, and an irregular block 660 m long, 230 m wide and more than 130 m deep, contains the mineralization.

The ores formed in three successive hydrothermal phases probably in the middle or late Tertiary age, and broadly coincide with the emplacement of felsic (quartz latite, rhyolite porphyry) and mafic dikes. The earliest phase produced the broadest zone of sparsely disseminated auriferous pyrite and arsenopyrite. This zone contains gold values ranging from 1.4 ppm to 6.53 ppm Au and is currently mined as a large ("bulk"), low-grade gold orebody reminescent of the "Carlin-type". The gold is invisible, bound in the sulphides. The associated alteration is a broad fracture carbonatization,

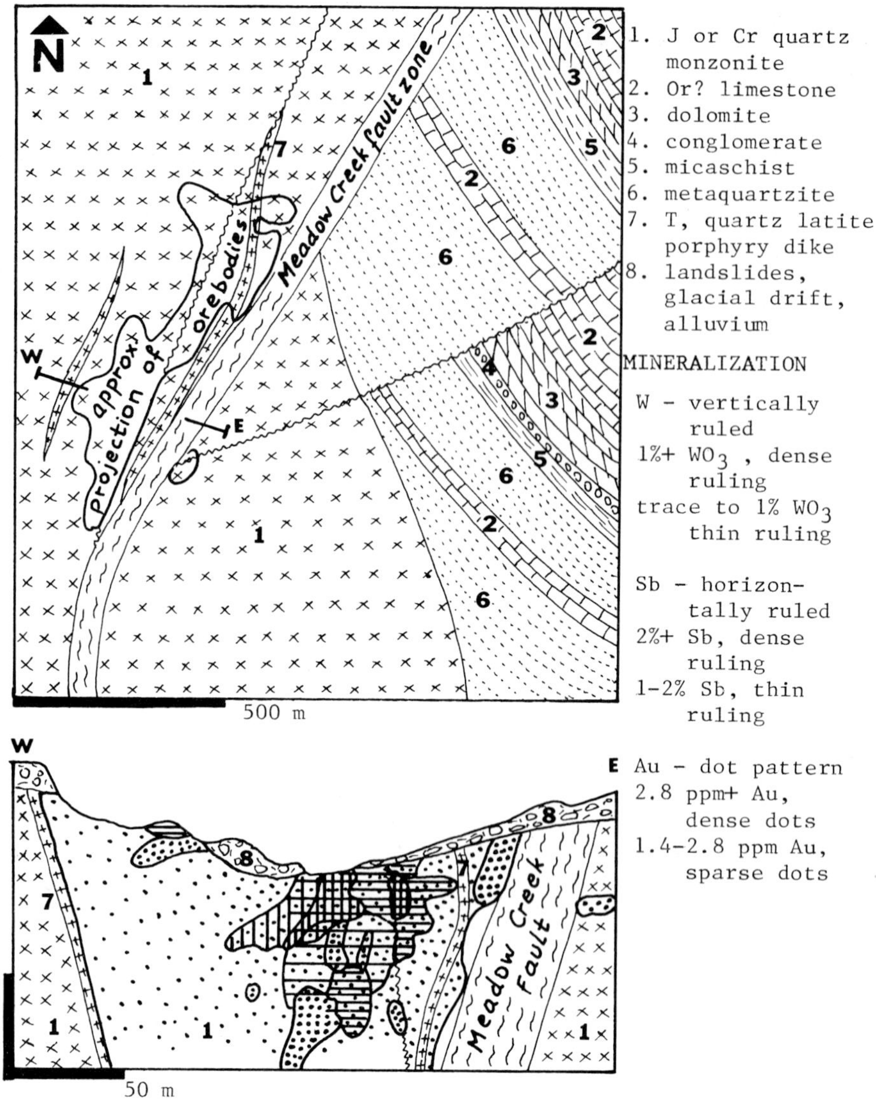

Fig. 28-123. Yellow Pine Sb-W-Au mine, Idaho, map and section. From LITHOTHEQUE, modified after Cooper (1951).

sericitization, silicification and biotite bleaching.

A relatively high-grade (1.7% W) but small scheelite orebody formed during the second phase, in the centre of the mineralized zone. Most of the scheelite is fine grained, disseminated in brecciated gold ore, or filling small veinlets and stringers. A portion of the scheelite appears to have replaced an earlier generation of calcite fracture veinlets. Stibnite orebody envelopes and overlaps the scheelite body, and consists of disseminations, thin veinlets, stockworks, massive lenses, small veins, and scattered crystals on fractures. The antimony grade in the period of wartime mining was 4% Sb, but there is still a large amount of ore with 0.81% Sb left. The stibnite is silver-bearing. Several cinnabar occurrences (the largest was Hermes mine) are known that are located within 5 km from the Yellow Pine Au,W,Sb orebody.

28.22. MERCURY DEPOSITS

In the classical model of the 1940s, cinnabar deposits were placed into the "telethermal" category in terms of their association with granitic intrusions. These days there is a tendency to dismiss the "granite"/cinnabar affiliation altogether, and attribute Hg deposition to hot springs, "exhalations", volcanism, etc. Some of the late Cainozoic low-temperature cinnabar depositing convecting spring systems such as those in the Clear Lake area, California (McLaughlin, 1981; compare Section 5.7.5. and Fig. 5-5 for discussion), however, have been interpreted as having been heated and controlled by an evolving silicic intrusion 7 km under the surface (hence a mesozonal granite according to the depth zoning as shown in Chapter 27!).

Consequently, the "granite"/cinnabar genetic affiliation cannot be dismissed altogether but, given the highly disparate depth levels of formation of both, the presence of "granite" on the surface is indeed of little assistance in exploration for Hg orebodies. Epigenetic, low-temperature hydrothermal cinnabar concentrations hosted by ophiolites, subduction mélanges, "black" sedimentary association, shallow-marine and continental sediments, have been treated in Chapters 7,8,16, 17,19,20 and 26, respectively.

Several small Hg occurrences are known which are hosted by older, genetically probably unrelated "granites". At the Yamato mine, Japan (Saito, ed., 1960), late Paleozoic biotite granite hosts irregular cinnabar mineralized fissure veins. The ore emplacement is probably Miocene or Pliocene.

At several localities in the Gordonbrook serpentinite belt, New England, Australia (Yulgilbar, Pulganabar, Lionsville; Carne, 1913), cinnabar was deposited in small quantities at sites of interaction of ultramafics and Permian "granites". In Yulgilbar, cinnabar forms disseminations and narrow siliceous replacement veins in hornblende granodiorite near the contact with diorite dikes and about 1-2 km from the edge of the serpentinite belt. Only about 2.3 t Hg were produced. At Lionsville, sparsely disseminated cinnabar is located on both sides of a contact of felsite porphyry dike in serpentinite. At Pulganabar, quartz, calcite, specularite, Hg-tetrahedrite and cinnabar veins are situated in the endo- and exocontact of granodiorite and tourmaline granite.

28.23. BISMUTH DEPOSITS

Most depositional accumulations of bismuth are associated with "granites", but usually Bi is a by-product or unrecovered companion metal in Sn-W, Pb-Zn, Au, Cu, Ni,Co,Ag, and U deposits. Orebodies mined for bimuth only are exceptional (e.g. Tasna, Bolivia). Data on minor and trace Bi distribution in ore deposits are fragmentary and often unreliable. A substantial proportion of Bi is recovered in smelters that process ores from a broad region (e.g. the La Oroya smelter, Peru, to which the single largest ore supplier is Cerro de Pasco). Of the 350 Tt Bi in major ore accumulations that could be accounted for, 275 Tt Bi are in Sn-W-Mo skarns (Xi Zhoyouang, China, 230 Tt Bi; Gejiu, China, 30 Tt Bi ?; Sandong, Korea, 9 Tt Bi; Brichmula-Ustarasai, Soviet Central Asia, no data). 8 Tt Bi is in Zn-Pb skarns (Kamioka, Japan, 6.24 Tt Bi; Baiţa Bihor, Rumania). 2 Tt Bi is in magnetite skarns (e.g. Biggenden, Queensland; Chokadam-Bulak, Uzbekistan). 28 Tt Bi is in subvolcanic and high-level plutonic W,Mo,Sn veins and stockworks (Mt. Pleasant, Canada; 24 Tt Bi; Tasna, Bolivia, min. 1.2 Tt Bi). 10 Tt Bi is in massive pyritic replacements and superimposed veins (Cerro de Pasco). 7 Tt Bi is in quartz, gold, arsenopyrite veins (Salsigne, 5 Tt Bi). About 5 Tt Bi is in Ni,Co,Bi,Ag,U veins (Schneeberg, Johanngeorgenstadt, Jáchymov). About 10 Tt Bi is in Precambrian deposits (Tenant Creek, Boliden) and 5 Tt Bi in miscellaneous hydrothermal ores. Most of the Bi-containing mineralization styles have already been described earlier, and bismuthinite with native bismuth are the two principal Bi minerals.

In the Biggenden ore field, S.E. Queensland (Clarke, 1969), lower Permian nearshore marine to continental andesite and basalt flows, pyroclastics, volcaniclastics, argillite and limestone, are intruded by lower Triassic porphyry dikes and biotite granodiorite. In the exocontact, basic hornfels formed from the volcanics and hornblende, epidote, scapolite, magnetite, sphene skarns were generated from the impure carbonate. Bismuthinite and rare native bismuth with lesser chalcopyrite, sphalerite, tetrahedrite, cobaltite and molybdenite, are disseminated in magnetite. The total production has been about 470 t Bi and 1.28 t Au.

28.24. ARSENIC DEPOSITS

Arsenic is extremely widespread in many styles of hydrothermal-plutonic ores, but it is rarely recovered and sold due to lack of demand. Consequently, quantitative data on As accumulations are scarce. About 3 Mt As can be accounted for as having been present in Phanerozoic hydrothermal-plutonic deposits (largest accumulations for which data are available: Cerro de Pasco, 1.5 Mt As; Olympias mine, Chalkidiki, Greece, 240 Tt As/2.4%, in Pb-Zn skarn and replacement; Hedley, British Columbia, about 200 Tt As/5%, gold-bearing arsenopyrite skarn; Salsigne, France, 150 Tt As/3%, complex gold vein deposit).

Few major deposits have ever been mined for As alone. When arsenic was recovered, it was almost always a by-product, usually of gold. This applies even to the largest-known massive arsenopyrite accumulation in the Boliden deposit, Sweden (544 Tt As/6.8%; this is a Precambrian orebody treated in Volume 2). Magak'yan (1968) discussed two Soviet localities (Mosrif and Takeli, both in Tadzhik S.S.R.) as examples of As deposits, but both are complex high-As carbonate replacements containing also Pb,Zn,Cu, Au,Ag, and other metals.

Arsenic deposits have traditionally been subdivided into three broad classes: (1) high-temperature deposits developed in the proximal intrusive exocontact, such as As and complex skarns and carbonate replacements, and quartz-arsenopyrite veins. In the endocontact, arsenopyrite is significantly distributed in many greisens (e.g. Krásno, Czechoslovakia). Magak'yan (1968) mentioned a "porphyry arsenopyrite" occurrence in granodiorite from Takeli, having a grade of 2% As; (2) medium-temperature (mesothermal) deposits, in which As is partly contained in arsenopyrite, partly in enargite or tennantite (Butte, Cerro de Pasco, Morococha); in Ni-Co arsenides (Jáchymov) or in Ag-sulphoarsenides such as proustite and (3) low-temperature realgar-orpiment deposits, limited to high-level accumulations usually attributed to the activity of hot springs (Getchell, Nevada; Matra, Corsica; Tajov, Slovakia). Most styles of As-containing mineralizations have already been described.

28.25. MINOR METALS AND METALLOIDS: Se,Te,Tl, In,Ge,Ga, Cd

The above metals (except Cd) are produced in small quantities from "technological ores" such as the residues after electrolytic Cu and Zn recovery; from slimes that accumulate during sulphuric acid production from pyrite or from similar materials. The deposits from which the rare metals were derived need not demonstrate any anomalous local enrichment of such metals and need not contain their discrete minerals (e.g. Butte is credited with the total production of 107 t Te and 143 t Se; the smelter in La Oroya, Peru, produced about 50 t In, 20 t Te and 20 t Se in 1982; Eng. and Min. Journ., November 1984, derived from 6 major and about 200 small deposits in the region).

In contrast to the "technological ores", certain deposits and occurrences contain widely quoted minerals of rare metals, but in quantities so small that industrial production of the contained metals would be uneconomic. Such localities are, nevertheless, listed in the literature as "deposits" of such metals.

Selenium

The largest recorded Se accumulation in metallic ores (2.4 Tt Se/0.03%) has been reported in the Precambrian Boliden deposit, treated in Volume 2. Of the rest, Se minerals have been recorded from: (1) pitchblende vein deposits (clausthallite, berzelianite, tiemannite, e.g. in Olší, Slavkovice and others localities, Czechoslovakia); (2) Ag deposits with naumannite; (3) Au,Se,Pb occurrences, e.g. Corbach, West Germany and (4) carbonate-polymetallic veins, e.g. Tilkerode, Germany; Sierra de Umango, Argentina; etc.

Tellurium

Tellurium is most conspicuous (but rarely recovered) in Au and Ag tellurides (sylvanite, hessite, petzite, krennerite, nagyagite) and as native tellurium, in the Au-Te association. This could be either subvolcanic (e.g. Apuseni Mts., Rumania; Cripple Creek, Colorado) or plutonic (e.g. Bohuliby, Czechoslovakia; Darasun, Siberia; Jamestown, Colorado). Non-precious metals tellurides (e.g. tetradymite) are common at vein gold deposits as well as in some polymetallic deposits.

Thallium

Thallium is enriched in some "Carlin-type" deposits of "invisible" dispersed gold replacements in carbonaceous sediments (e.g. Mercur, Utah; 1.8 Tt Tl) and in some probably subvolcanic As-Sb deposits (e.g. Alšar, Macedonia, Yugoslavia; probably 2 Tt Tl).

Indium, germanium, gallium

These metals form rare minerals of their own. Indium forms indite, germanium forms germanite and argyrodite and gallium forms gallite, respectively. Among the common sulphides, the highest enrichment of these metals is in sphalerite.

Cadmium

Cadmium is different from the above metals, because it is produced in relatively large quantities and its recoverable economic metal content is measured in millions of tons. Except for a few interesting but uneconomic small Cd-dominated occurrences (e.g. Berenguela, Bolivia), Cd is always present in sphalerite, substituting for Zn. The light-coloured (yellow, light brown) sphalerites are usually richest in Cd (up to 1.5% Cd) and the high-Fe black marmatites poorest (up to 0.05% Cd). Cadmium is not confined to hydrothermal-plutonic sphalerite deposits only, and is equally widespread in the MVT, APT, volcanic-sedimentary and other deposits.

There does not seem to be a systematic difference in Cd contents in sphalerites of different origin. Sphalerites from thirteen North American deposits or regions listed by Wedow (1973) contained from 0.11 to 1.34% Cd (MVT localities: Pine Point, 0.11%; Joplin, 0.358%; APT localities: E. Tennessee, 0.32% Cd; S.E. United States, 0.52% Cd; hydrothermal veins and replacements: Illinois-Kentucky, 0.69%;

Central Kentucky, 1.34%; Colorado Rocky Mts., 0.28%; Butte, 0.58%; Coeur d'Alene, 0.4%). Wedow (1973) also calculated the world's average Cd content in 55% Zn concentrates, which is 0.24% Cd. This corresponds to 4.36 kg Cd per ton of zinc. Using this ratio, the Zn(Pb) deposits in "granite" aureole represent some 305 Tt Cd.

The Berenguela field in S.W. Bolivia (Ahlfeld and Schneider-Scherbina, 1964) contains unique mineralization dominated by cadmium. There, probably subvolcanic or hot-spring impregnations, veins and replacements of rhythmic greenockite, "Schallenblende" (metacolloform sphalerite), freieslebenite and other minerals containing up to 11% Cd, are hosted by Meso-Cainozoic arenites and quartzites and late Tertiary andesite pyroclastics.

28.26. SUMMARY GRAPHS OF ORE DISTRIBUTION PATTERNS

This chapter reviewing the granite-diorite (gabbro) association and associated metallic mineralization is concluded with a set of graphs, showing semiquantitatively: (1) the preferential association of the various ores with the different host rock assemblages and with the compositional variety of their concentrator intrusions (Fig. 28-124); (2) the preferred mineralization and alteration combinations (Fig. 28-125) and (3) the frequent evolutionary and timing pattern of intrusions, host rocks and mineralizations (Fig. 28-126).

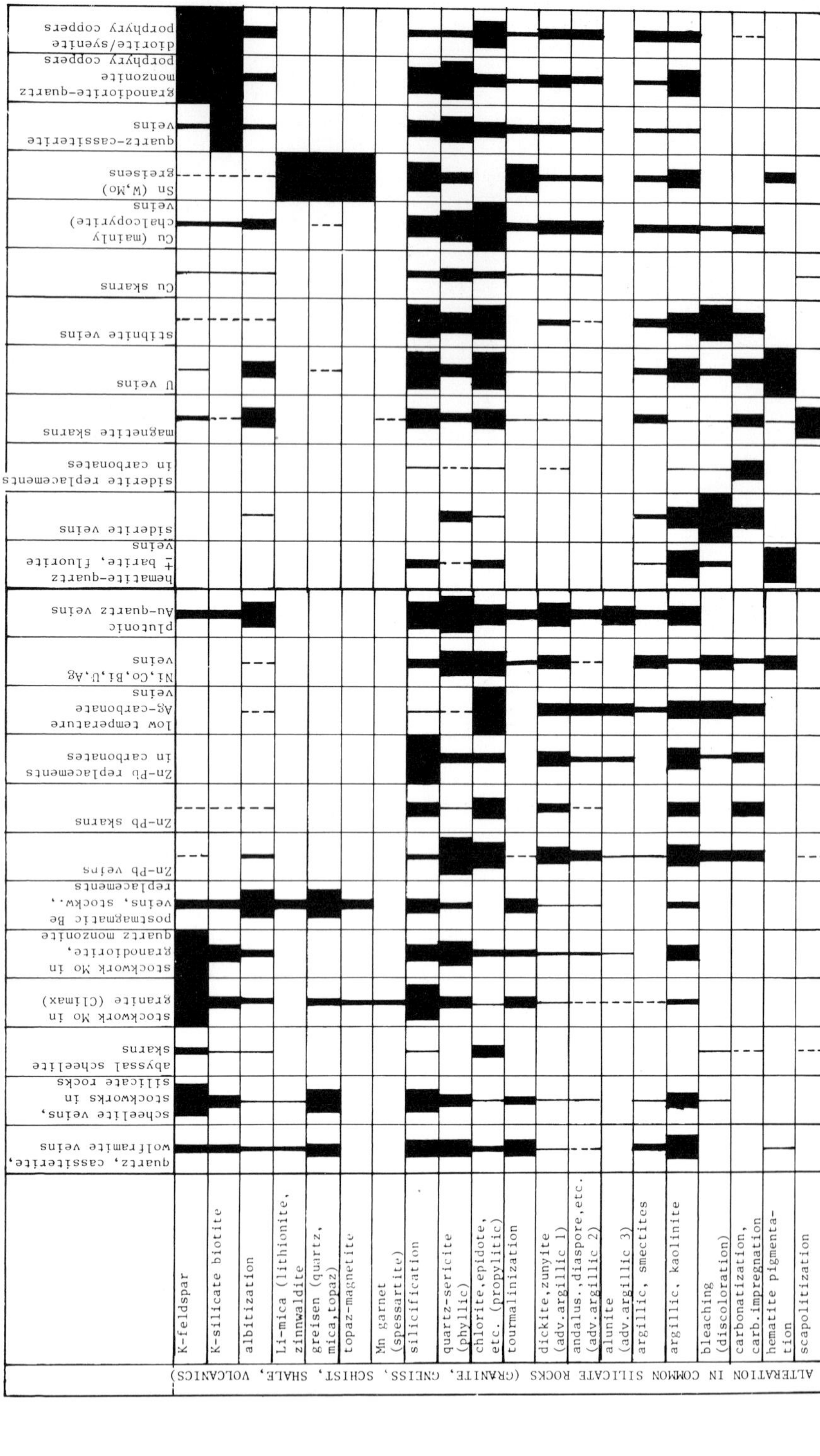
ALTERATION IN COMMON SILICATE ROCKS (GRANITE, GNEISS, SCHIST, SHALE, VOLCANICS)
K-feldspar
K-silicate biotite
albitization
Li-mica (lithionite, zinnwaldite
greisen (quartz, mica,topaz)
topaz-magnetite
Mn garnet (spessartite)
silicification
quartz-sericite (phyllic)
chlorite,epidote, etc. (propylitic)
tourmalinization
dickite,zunyite (adv.argillic 1)
andalus.,diaspore,etc. (adv.argillic 2)
alunite (adv.argillic 3)
argillic, smectites
argillic, kaolinite
bleaching (discoloration)
carbonatization, carb.impregnation
hematite pigmentation
scapolitization
quartz, cassiterite, wolframite veins
scheelite veins, stockworks in silicate rocks
abyssal scheelite skarns
stockwork Mo in granite (Climax)
stockwork Mo in granodiorite, quartz monzonite
postmagmatic Be veins, stockw., replacements
Zn-Pb veins
Zn-Pb skarns
Zn-Pb replacements in carbonates
low temperature Ag-carbonate veins
Ni,Co,Bi,U,Ag veins
plutonic Au-quartz veins
hematite-quartz ± barite, fluorite veins
siderite veins
siderite replacements in carbonates
magnetite skarns
U veins
stibnite veins
Cu skarns
Cu (mainly chalcopyrite) veins
Sn (W,Mo) greisens
quartz-cassiterite veins
granodiorite-quartz monzonite porphyry coppers
diorite/syenite porphyry coppers

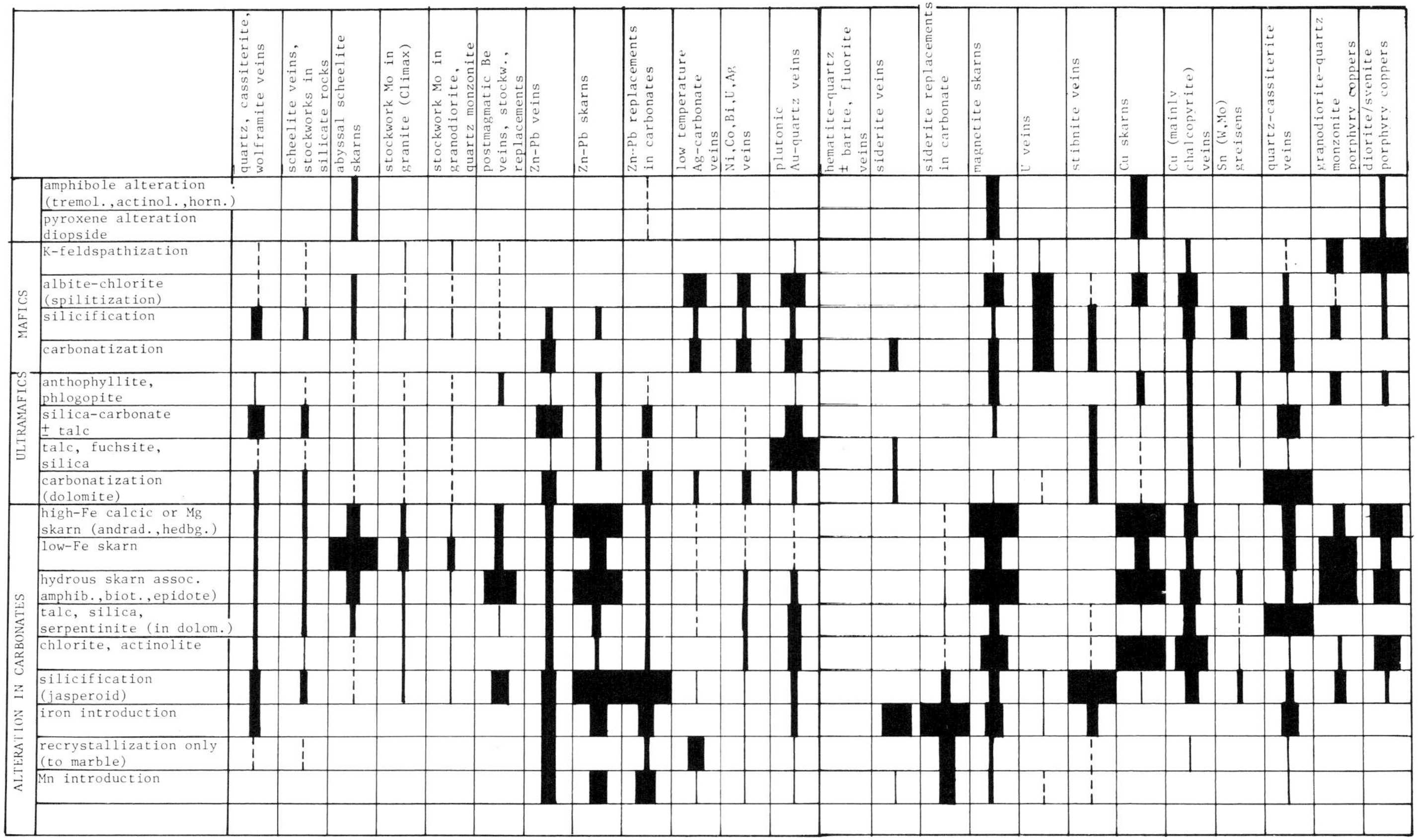

Fig. 28-124. "Granite"-associated (hydrothermal, epigenetic, postmagmatic) metallic ores as related to wallrock alterations. From Laznicka (1984).

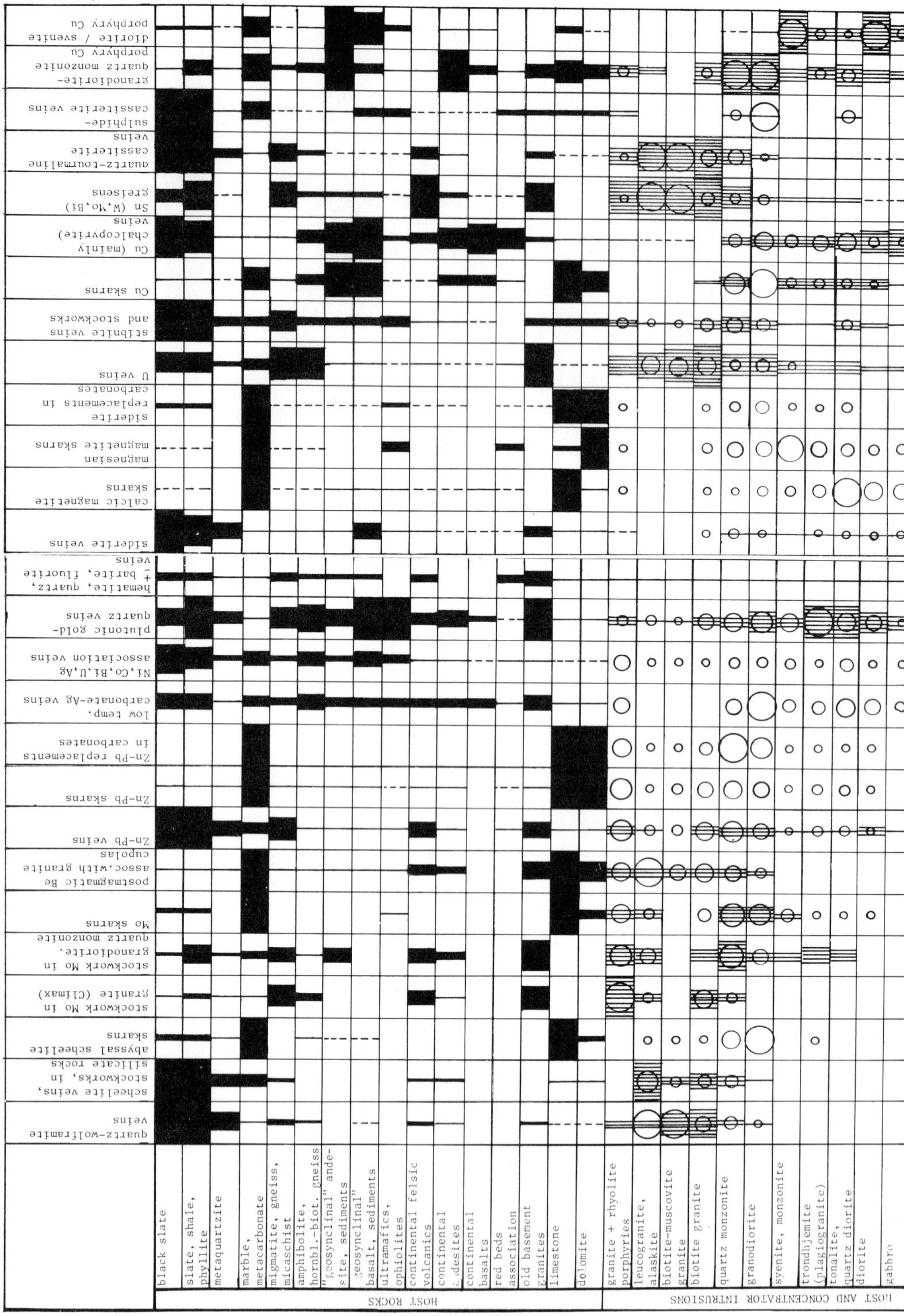
diorite / syenite porphyry Cu
granodiorite-quartz monzonite porphyry Cu
sulphide-cassiterite veins
quartz-tourmaline cassiterite veins
Sn (W,Mo,Bi) greisens
Cu (mainly chalcopyrite) veins
Cu skarns
stibnite veins and stockworks
U veins
siderite replacements in carbonates
magnesian magnetite skarns
calcic magnetite skarns
siderite veins
hematite, quartz, ± barite, fluorite veins
plutonic gold-quartz veins
Ni,Co,Bi,U,Ag association veins
Low temp. carbonate-Ag veins
Zn-Pb replacements in carbonates
Zn-Pb skarns
Zn-Pb veins
postmagmatic Be assoc. with granite cupolas
Mo skarns
stockwork Mo in granodiorite, quartz monzonite
stockwork Mo in granite (Climax)
abyssal scheelite skarns
scheelite veins, stockworks, in silicate rocks
quartz-wolframite veins
black slate
slate, shale, phyllite
metaquartzite
marble, metacarbonate
migmatite, gneiss, micaschist
amphibolite, hornbl.-biot. gneiss
"geosynclinal" andesite, sediments
geosynclinal" basalt, sediments
ultramafics, ophiolites
continental felsic volcanics
continental andesites
continental basalts
red beds association
old basement granites
limestone
dolomite
granite + rhyolite porphyries
leucogranite, alaskite
biotite-muscovite granite
biotite granite
quartz monzonite
granodiorite
syenite, monzonite
trondhjemite (plagiogranite)
tonalite, quartz diorite
diorite
gabbro
HOST ROCKS
HOST AND CONCENTRATOR INTRUSIONS

Fig. 28-125. "Granite"-associated (hydrothermal, epigenetic, postmagmatic) metallic mineralizations indicating, semiquantitatively, the frequency of their host rock and concentrator intrusion compositions. From Laznicka (1984).

HOSTS TO ORES:

"Passive" host rock units (those genetically unrelated to the magmatic and postmagmatic heat sources and processes, driving the hydrothermal systems; mostly (meta)sediments and (meta)volcanics)

Intrusive units acting as "passive" hosts to ores (some may be, however, sequentially earlier members of complexes that include "concentrator" intrusion coeval with the ores)

"Concentrator" intrusions (those driving the hydrothermal systems)

FREQUENCY:

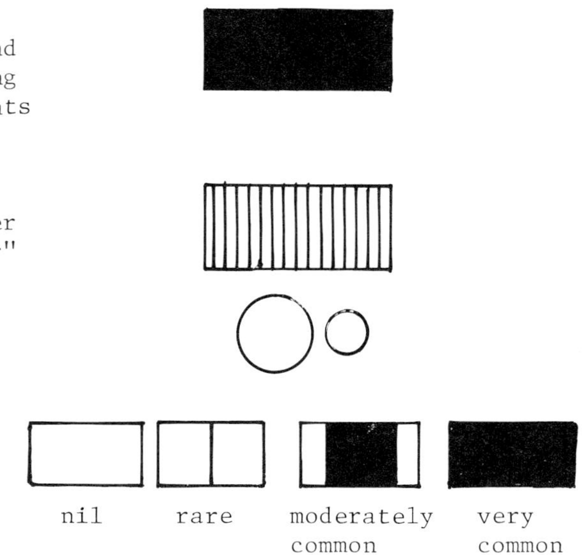

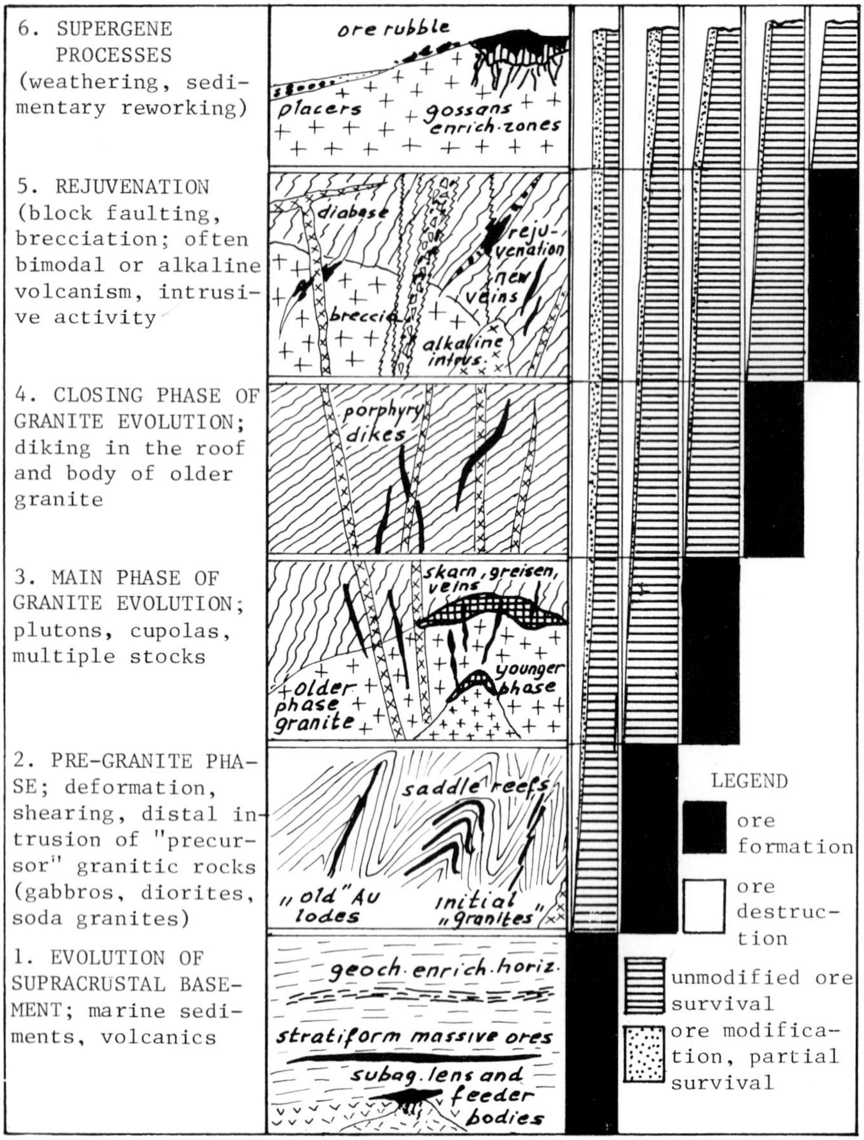

Fig. 28-126. The usual evolutionary phases in a multiphase polygenetic mineralized region, showing a real as well as an apparent genetic association with granitic intrusions. From Laznicka (1984).

"I believe that it will be discovered that a great number, if not the majority, of ore deposits, are not the result of a single segregation, but are the accumulated fruits of a great interrupted process of segregation, a part of the metals for the deposits having been worked over many times by the metamorphic process".
C. R. Van Hise in Treatise on Metamorphism, 1904.

CHAPTER 29

High- to Medium-Grade Metamorphosed Terrains, Katazonal Granites, Pegmatites

29.1. INTRODUCTION

Metamorphism starts gradually as a process superimposed on preexisting rocks. Initial, mild metamorphic effects (e.g. of zeolite facies) are barely recognizable megascopically during the general fieldwork, and rarely emphasized on geological maps and in field reports. With increasing metamorphism, the appearance and mineralogical composition of rocks keeps changing, until a point is reached beyond which the metamorphism-generated features far exceed the relic features of the original rock. The original rocks become no more directly recognizable megascopically and have to be interpreted. The interpretation of many metamorphic rocks is still in its infancy. Finally, ultrametamorphism that involves partial or total melting of older rocks or metasomatic crystalloblastesis, produces completely "new" rocks some of which remain "in situ", while others are found in new locations into which they moved in the form of plastic or fluid magma or migma.

There is a tendency amongst field and exploration geologists to disregard the effects of metamorphism (or dismiss them as a nuisance modification only), as long as the original rock type and depositional style can be easily recognized and interpreted. In the Canadian Shield, amphibolites composed of an entirely metamorphic assemblage, but still showing well-preserved pillow structures in outcrop, are designated as "basalts". Quartz-sericite schists are "rhyolites", and a variety of schists are "graywackes" or "shales". Although petrographers have serious reservations about the above practice (I require my students at least to use the prefix meta- in naming such rocks; e.g. meta-basalt, meta-shale), it is undeniable that metallogenic analysis of low- to moderately-metamorphosed associations appears to be more convenient and faster when done in the pre-metamorphic framework. This, however, requires an assumption (which most geologists accept as an established fact) that most orebodies in low-grade metamorphic terrains formed before the metamorphism which they survived as easily recognizable relics (e.g. of volcanics-hosted massive sulphides). Alternatively, ores were generated by epigenetic introduction during or after metamorphism (e.g. granite-affiliated postmagmatic hydrothermal deposits), usually at higher crustal levels than those in which the metamorphism took place. Metamorphism as a new ore-forming process had not been seriously considered, until very recently.

In this book, so far, the ore-hosting lithologic associations and their ores have been arranged according to their pre-metamorphic state, and although the modifications due to metamorphism have been noted and recorded, the intensity of metamorphism was not applied to draw first-order division between the categories. The time has now come to do so.

In this chapter, the emphasis is on high-grade metamorphic and ultrametamorphic (=granitization) terrains regardless of their pre-metamorphic composition, and on their metallogeny. Katazonal granites and pegmatites are generally present at the same level and overlapping, so they are also considered. As for the other lithologic associations, this category has gradational boundaries and overlaps with the subjects treated in Chapters 7-17, 19, 20-22, 24, 26, 28, 32-36. The overwhelming majority of high-grade metamorphic terrains are of Precambrian age and reside in Precambrian shields, in the basement of platforms and in large (e.g. median) massifs within orogenic belts. These are treated in Volume 2. A substantially smaller proportion of high-grade metamorphics are exposed in cores of Phanerozoic orogenic belts. Although many of these rocks are of Precambrian age anyway, Phanerozoic units are usually also involved. Post-Precambrian metamorphism, ultrametamorphism and interaction with Phanerozoic surroundings and roof assemblages, justifies treatment of such terrains in this chapter.

29.2. PETROGRAPHY, ORIGIN AND SETTING OF HIGH-GRADE METAMORPHIC TERRAINS

There is not sufficient space in this book to provide an adequate introduction into the metamorphic petrology and petrography, and the reader is urged to consult some of the recently published text and reference books (e.g. Winkler, 1974; Turner, 1980). The excellent book by Suk (1983) is particularly suitable for use as a basis for metallogenic interpretation of metamorphic terrains. Table 29-1 summarizes briefly some of the classifications, aspects, contrasts, etc. of metamorphism and its products of the greatest practical importance, largely abstracted from Suk (1983) and Turner (1980).

The heterogeneous nature of metamorphics and the numerous directions from which they can be studied, makes their elegant organization virtually impossible. Even the theoretically-minded textbooks (e.g. Turner, 1980) had to devote considerable space to brief descriptions of numerous local areas to cover the field adequately, and this book is no exception. High-grade metamorphism overlaps with "granitic" magmatism, so it is advantageous to borrow some of the recent concepts bearing on the petrochemistry and setting of "granites" (reviewed in Chapter 28) for the purpose of organization of the present subject.

All Phanerozoic progressive high-grade metamorphics occur in orogenic belts. There, the micaschist, gneiss, migmatite, granite assemblage is most common, and it occurs in thermal domes, ridges and anticlinal belts. Two end-member varieties of the above terrains can be distinguished: (1) the more internal (i.e., near-oceanic) ones, characterized by associated I ("magnetite")-type granitic rocks and evolutionarily less mature (e.g. the Coast Range metamorphic complex, British Columbia) and (2) the more external (i.e., near-continental) ones, marked by S ("ilmenite")-type granites and a long period of residence within or at a margin of a thick sialic craton (e.g. the Shuswap Complex, British Columbia).

Suk (1983) pointed out that metamorphic equivalents of all rocks are known. This is summarized briefly in Table 29-2.

29.3. TRACE METAL GEOCHEMISTRY AND ORE GENESIS

Trace metal geochemistry of high-grade metamorphics is a function of the pre-metamorphic heritage combined with the metamorphic differentiation. In relatively open systems that permit upward migration of metamorphically displaced mobile elements, trace-metal migration starts and, in fact, is probably most intensive, under the conditions of low-grade metamorphism (particularly along the intervals of accelerated dehydration, such as chlorite→amphibole and illite→muscovite). Quantitative evidence for syn-metamorphic metal migration is, however, difficult to obtain because it is impossible to sample a lithologically and geochemically uniform rock sequence over a considerable vertical span. Trace-metal contents of originally similar rock associations affected by prograde metamorphism to form zones of increasing metamorphic intensity, indicate gradual depletion in trace Cu,Au,Zn, U,Ni,Cr,V and enrichment in Ba,Sr,Zr, Ce and La, with the increasing metamorphic grade (e.g. Belevtsev, 1976, 1979). The majority of studies have been performed in the Precambrian shields, and their results are further treated in greater detail in Volume 2.

Boyle (1979; p. 116, 117) stated the case for gold deposits as follows: "Nearly all types of epigenetic gold deposits are restricted to rocks that exhibit a low to moderate degree of metamorphism (greenschist and amphibolite facies). One does not find epigenetic gold deposits in rocks that do show the effects of considerable recrystallization and alteration such as regional propylitization. From this fact the logical conclusion seems to follow that epigenetic gold deposits are more closely related to metamorphism than to magmatic phenomena. In the grand scheme of orogenesis the occurrence of epigenetic gold deposits in the vicinity of intrusive granitic rocks is not fortuitous. The reason for this, however, may not be that the granitic bodies provided the gold, but that these bodies are simply one in the series of products of intensive metamorphism. These

Table 29-1. Selection of practically important classifications, definitions and concepts bearing on metamorphism and its products. Summarized and slightly modified from Suk (1983) and Turner (1980)

VARIETIES OF METAMORPHISM BY SCALE, SETTING AND PROCESS

Regional metamorphism (RM; generates metamorphic rocks over extensive areas); varieties:

- RM of continental (Precambrian) shields
- RM of central parts of thermal domes
- RM of orogenic areas
- RM in the zone along lithospheric plate edges
- RM of ocean floor

Local metamorphism

- contact metamorphism
- dislocation metamorphism
- shock metamorphism
- combustion metamorphism (surficial or shallow subsurficial coal or bitumen burning)

VARIETIES OF METAMORPHISM BY PRESSURE/TEMPERATURE RELATIONSHIP

(1) high temperature, almost no pressure (e.g. baking, hornfelsing at surficial volcanics and sediments contacts

(2) high temperature, moderate pressure ("normal" metamorphism as e.g. in orogenic belts)

(3) high pressure, moderate temperature ("blueschist-type")

(4) high pressure, almost ordinary temperature (e.g. cataclasis and mylonitization along shallow faults)

SUBDIVISION OF THE "NORMAL" METAMORPHISM BY THE DEGREE OF INVOLVEMENT OF INTRUSIVE PLUTONS

(1) Enorogenic (Barrovian, Dalradian, "typical regional", kyanite-sillimanite) metamorphism, predominantly pressure activated

(2) Periplutonic (Buchan, Abukuma, andalusite-sillimanite) metamorphism; heat activated in the proximity of intrusive plutons

VARIETIES OF METAMORPHISM BY ITS CHEMICAL CHARACTER

(1) Chemically conservative metamorphism (isochemical); no elements added or removed

(2) Metasomatism (metasomatic, allochemical, etc. metamorphism); element exchange along rock contacts (bimetasomatism, reaction metasomatism), or element addition and removal to and from the site under consideration

(3) Ultrametamorphism; rock transformations under conditions near to the melting of the rock system involved (migmatitization, formation of anatectites)

METAMORPHIC DIFFERENTIATION: a process that brings about mineral heterogeneity in an originally homogenous rock, by migration of substances within the rock (e.g. quartz or pegmatite secretions, concretions, boudins, etc.)

METAMORPHIC HOMOGENIZATION: the opposite, largely a consequence of equilibration of the chemistry of the fluid phase

METAMORPHIC FACIES AND FACIES GROUPS

"A metamorphic facies is a set of metamorphic mineral assemblages repeatedly associated in space and time, such that there is a constant and therefore predictable relation between mineral composition and chemical composition" (Turner, 1980, p.54). Den Tex (1971) distinguished four facies groups of regional metamorphism:

(1) Very low grade metamorphism. Zeolite, prehnite-pumpellyite, laumontite-albite, portion of lawsonite-glaucophane
(2) Low-grade metamorphism. Greenschist facies, albite epidote amphibolite facies
(3) Medium grade metamorphism, amphibolite facies
(4) High-grade metamorphism, granulite facies

PROGRESSIVE AND RETROGRESSIVE METAMORPHISM

Progressive metamorphism is recorded in a sequence of rocks showing a rising grade of metamorphism out of previously unmetamorphosed rocks. Retrogressive metamorphism produces a metamorphic facies equivalent lower in the sequence than the modified precursor rock, and it is accomplished mostly by localized tectonic deformation and action of hydrothermal fluids

METAMORPHIC CONVERGENCE AND DIVERGENCE

Convergence: formation of identical metamorphics out of a variety of precursors (e.g. biotite gneiss out of granite, arkose, litharenite, shale)

Divergence: formation of different metamorphics out of a single precursor as a consequence of a different metamorphic grade or style (e.g. phyllite, micaschist, gneiss out of shale)

bodies also include the gold deposits, the gold being derived from piles of sedimentary and volcanic rocks and concentrated as a result of granitization and later metamorphic processes that continued long after the emplacement, crystallization and consolidation of batholiths, stocks and dikes of granite".

Overstreet (1967) in his exhaustive review of the world monazite occurrences, argued that most monazite accumulations in high-grade metamorphic terrains and in associated magmatic rocks (pegmatites, granites, charnockites) are metamorphogenic, rather than a detrital relic. The global distribution of bedrock monazite and proximal monazite placers, supports this thesis beautifully. Moreover, the chemical composition of monazite, particularly its Th/REE ratio, appears to increase with the metamorphic grade and the Sri Lanka monazite with 10% ThO_2 derived from charnockite association, is among the highest Th monazites commercially produced. Although scattered monazite in high-grade metamorphics (granites, charnockites) may

Table 29-2. High-grade metamorphic equivalents of pre-metamorphic rocks, a brief summary.

HIGH TO MEDIUM GRADE METAMORPHIC	PROTOLITH
muscovite, biotite, staurolite, kyanite, garnet, sillimanite, etc. schist and gneiss; partly granulite	shale, siltstone, graywacke, arkose, conglomerate
graphitic schist, marble, quartzite	bituminous or coaly pelite, carbonate, chert
graphite (amorphous)	humitic coal, rarely sapropelite
metaquartzite	quartz arenite, chert
hornblende-biotite schist, gneiss	intermediate to mafic tuffs, volcaniclastics
amphibolite (schistose)	basalt, andesite flows, pyroclastics, volcaniclastics, marls
gabbro-amphibolite (massive)	diabase, gabbro
limestone or dolomitic marble	sedimentary carbonates; rarely igneous carbonatite
calc-silicate gneiss, hornfels	impure sedimentary carbonate, marl
magnetite-quartz rock	ironstone or iron formation
corundum and corundum-magnetite	bauxite and laterite
tourmaline schist (some)	sedimentary borates
albitites (some)	halite with clay layers
apatite-rich metamorphics	sedimentary phosphorite
scapolite metamorphics	sedim. gypsum, anhydrite, halite
meta-anhydrite	sedimentary gypsum, anhydrite
leucocratic orthogneiss, leptynite, partly granulite	rhyolite and felsic pyroclastics, granite
nepheline gneiss (partly)	nepheline syenite, phonolite
prasinite (chromian barroisite hornblendite)	high-Mg basalts, picrites

glaucophanite (blueschist)	spilite or any metabasalt of graywacke metam. under high P/low T metamorphic conditions
eclogite	gabbro, basalt, troctolite, spilite
concordant olivinite	ultramafics (peridotite, serpentine)
gedrite, anthophyllite, actinolite schists	ditto
skarnoids	pelosiderite, peloankerite, ironstone, iron formation

represent a case of metamorphogenic mineralization, the bedrocks (even regoliths !) never approach an economic grade and the monazite can be profitably recovered only when reworked into placers.

It appears that the bulk of significant metamorphogenic and metamorphics-hosted (other than relic) deposits formed by precipitation from a mobile phase (magmatic, hydrothermal) displaced during ultrametamorphism or during metamorphogenic hydration or dehydration. Such ores and their host rocks fill dilations or replace earlier rocks usually above (rarely within) the metal source regions and often merge with the "granite"-affiliated ores. In the recent classification of metamorphogenic ore deposits by Belevtsev (1976, 1979; Table 29-3), the former (in situ) deposits are designated as autochthonous, the latter (transported) as allochthonous.

29.4. MINERALIZATION STYLES IN METAMORPHOSED NON-CARBONATE SUPRACRUSTALS (SCHISTS, GNEISSES; Fig. 29-1)

29.4.1. "Metalliferous schists", containing dispersed to disseminated metals conformable with foliation (Fig. 29-1a)

These are thin, short lenses to extensive horizons of fine-grained schistose metamorphics usually differing compositionally from their environment and anomalously enriched in trace metals. "Black" (graphitic) metasediments are most common, combined with (or followed by) pyritic metasediments, impure metacarbonates, amphibolites, etc. The best-known examples are the Proterozoic "Fahlbånds" of Norway (Volume 2), which are closely comparable with some of the graphitic and pyritic siliceous schists in the late Proterozoic and Paleozoic of the Erzgebirge Mountains (e.g. in the Jáchymov area).

The "metalliferous schists" are rarely economic in themselves, but they often appear to have been the source rocks of metalliferous fissure veins or stockworks, such as veins with Ni,Co,Bi,U,Ag

Table 29-3. Classification of metamorphogenic ore deposits of Belevtsev (1976, 1979)

		TEMPERATURE OF METAMORPHISM		
		LOW	MEDIUM	HIGH
METAMORPHOSED DEPOSITS (pre-metamorphic, later metamorph.)	META-SEDIMENTARY	chlorite-magnetite b.i.f., U-conglomerates (Elliot Lake)	amphibole-magnetite b.i.f.	pyroxene-magnetite b.i.f.
	META-VOLCANIC	volcanogenic massive sulphides; Cu in meta-basalts (Michigan)	volcanogenic massive sulphides; Mn gondites	PCm iron quartzites; Mn khondurites (India)
	META-MAGMATOGENE	Ni-Cu sulphides in mafic magmatites (Pechenga)	Ni-Cu sulphides (Voronezh Massif)	ilmenite-rutile ores, the Urals
METAMORPHIC DEPOSITS (formed during regional metamorphism)	AUTOCHTHONOUS (generated in situ by concentration of ore matter present in the hosts)	rich hematite-magnetite ores of the PCm b.i.f.; massive sulphides in the Urals; Pb-Zn deposits Zhairem, Mt. Isa	rich amphibole-magnetite b.i.f.; "Alpine-type" veins	rich magnetite-pyroxene ores (Azov Sea); Mary Kathleen U-REE skarns
	ALLOCHTHONOUS (generated from transported components released during metamorphism)	Cu-Ni sulphides in schists (Baltic Shield); pyrite-polymetallic deposits, Lake Baikal region	Fe metasomatites, Gora Vysokaya, Blagodat', the Urals	corundum, phlogopite, saphirine, wollastonite materials

<table>
<tr><td rowspan="2">ULTRAMETAMORPHIC DEPOSITS (formed under the influence of ore-bearing fluids, released in zones of polygenetic and metasomatic granitization)</td><td>PALINGENETIC-METASOMATIC resulted from metasomatic granitization; also affiliated with rheomorphic granites and pegmatites</td><td></td><td>ore-bearing granites
rare elements pegmatites and metasomatites</td><td>Sn-W muscovite pegmatites
Sn-Nb granites (in Nigeria)</td></tr>
<tr><td>POST-GRANITIZATION DEPOSITS formed in waning stage of ultrametamorphism at the same structural level by precipitation of metals migrating from depth along faults</td><td>U,Au,Hg,etc. mineralized albitites, berezites, listvenites</td><td>rare elements albitites; U vein deposits (Beaverlodge); Cu, Singbhum</td><td>rare elements metasomatites (Siberia)</td></tr>
</table>

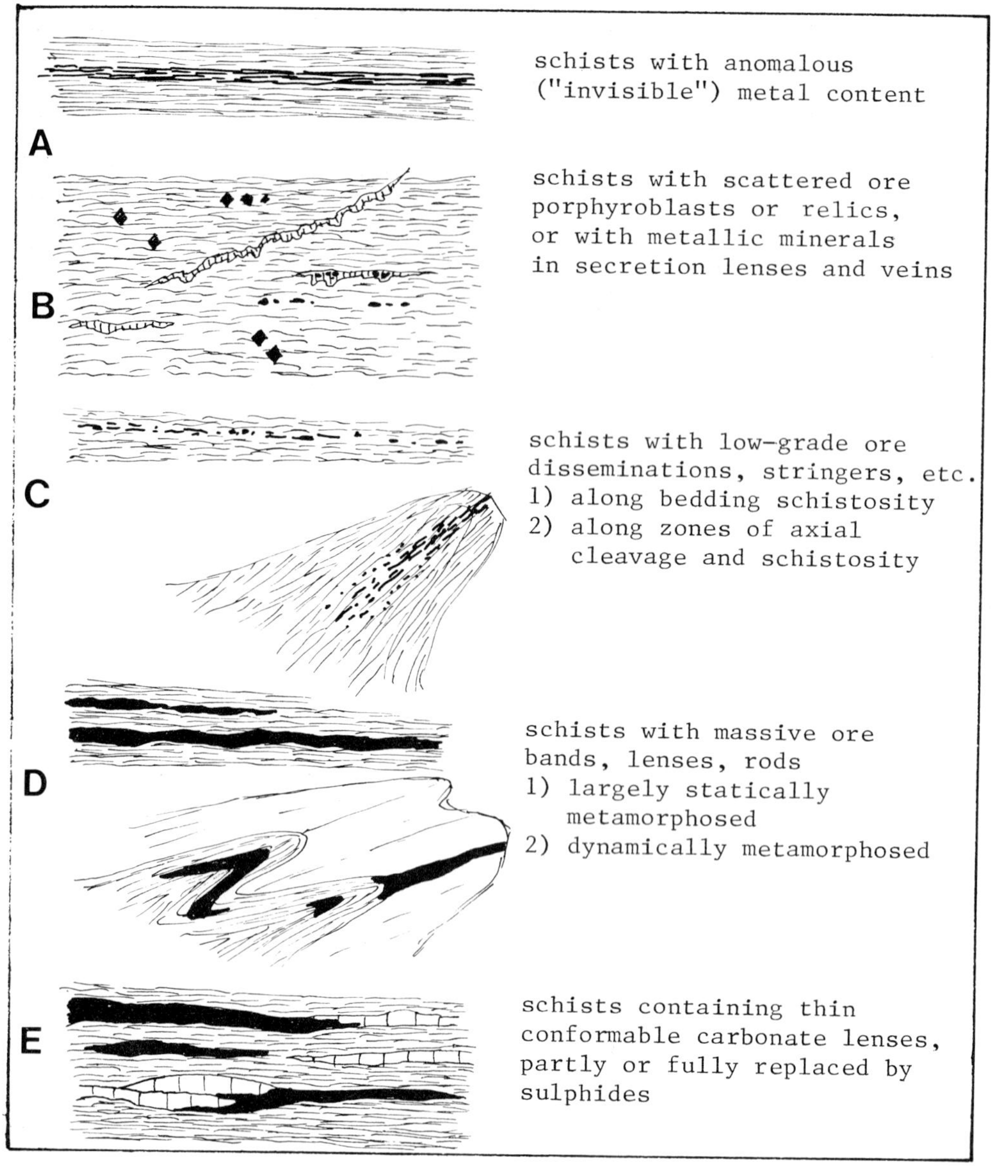

Fig. 29-1. Diagrammatic representation of mineralization styles in non-carbonate, progressively metamorphosed supracrustal schists. From Laznicka (1984).

association. The most commonly expected enriched metals are U,Ag,Au, Ni,Co,Bi, Mo,V,Hg,Sb, or their combinations. Visible ore minerals in schists occasionally appear as scattered porphyroblasts (most common are pyrite or arsenopyrite), or as components of secretion quartz, carbonate or silicate lenses; fillings of gash veinlets; etc.

The economically important Rönneburg uranium field, East Germany (Ruzicka, 1971) is situated in black schists that are metamorphosed equivalents of lower Paleozoic graptolithic shales. Visible U minerals (pitchblende) appear sporadically in secretion veins. The mineable ore comes in irregular lenses grading between 0.07 and 0.14% U.

29.4.2. Schists with scattered ore mineral porphyroblasts or relics (Fig. 29-1b)

This style is a close equivalent of the volcanic flows with ore phenocrysts, paleoplacer beds or sedimentary beds with ore nodules. Lenses and intervals of a variety of so mineralized metamorphics are conformable with schistosity and usually also with bedding. The ore mineral particles may be haphazardly scattered throughout the entire thickness of the ore horizon (the thickness usually ranges from several centimetres to several metres). Alternatively they may be arranged into trains parallel with schistosity. Genetically, the ore particles are either metamorphically modified relics (e.g. heavy mineral grains in metasediments), or porphyroblasts produced by metamorphic growth. Sometimes, their origin is uncertain. A variety of inhomogeneities superimposed on the ordinary mineralized horizon such as secretion quartz or quartz, feldspar, mica, etc. lenses, carry the same minerals in the form of larger, more conspicuous grains. Currently economic ore deposits of this style are rare, but the ore particles can be reworked into placers.

RUTILE OR ILMENITE-BEARING SCHISTS

At Shooting Creek, North Carolina (Williams, 1964; 37 Tt Ti), rutile and minor ilmenite are associated with sugary white quartz secretions in a Proterozoic garnet mica schist, discontinuously along a 16 km long belt. The mineralization is interpreted as being metamorphogenic, and small quantities of concentrate have been produced from regolith and small placers.

In the Yadkin Valley Ti deposit near Richlands, North Carolina (Bryant and Reed, 1966; 66 Tt Ti), ilmenite with magnetite are disseminated in a talcose layer conformable with enclosing cataclastic micaschist, quartzite and gneiss. The orebody consists of a series of closely spaced mineralized lenses, 110 m long and 6-17 m thick, conformable with foliation. Most of the production came from saprolite.

MONAZITE, XENOTIME, ETC.-BEARING SCHISTS

Lower Proterozoic biotite schists and gneisses that are part of the early Tertiary Colorado Rocky Mountains uplift, contain localized intervals enriched in monazite and xenotime. These minerals are megascopically invisible or inconspicuous, and have been discovered by a chance microscopic examination or by increased radioactivity. In the Central City district (Young and Sims, 1961), lenses up to 1.6 m thick and several tens of metres long, contain up to 5% by volume of xenotime and monazite. The authors favour the concept of metamorphogenic origin of the heavy minerals, over that of paleoplacers.

TIN-BEARING SCHISTS

The frequently mentioned Gierczyn (Göhren), Przecznica, Nové Město pod Smrkem stanniferous zone (Polish and Czech Sudeten; Jaskołski, 1960; Chrt and Bolduan, 1966) has been listed several times under the heading "Tin paleoplacer". This is an E.-W.-trending band, about 20 km long, of sericite, chlorite, garnet micaschist of late Proterozoic or Cambro-Silurian age, interfoliated with amphibolites, and periplutonically metamorphosed in the mantle of a late Paleozoic granite gneiss core. The tin is contained in inconspicuous, light-coloured small scattered grains of cassiterite, accompanied by scheelite, pyrrhotite, sphalerite, chalcopyrite, bismuthinite, niccolite and cobaltite. The ore minerals accumulated locally in lenses, 1-8 m thick and 15-100 m long, rich in two varieties ("white" and "blue") of "secretion" quartz, peneconcordant with schistosity (Fig. 29-2). The tin values are very irregularly distributed and range from 0.02 to 5% Sn. The paleoplacer origin has not been demonstrated convincingly and an alternative explanation assumes a (re)mobilized "fahlbånd-style" metalliferous horizon, with a portion of the Sn possibly being supplied from the "granite".

29.4.3. Schists with low-grade disseminations, stringers, veinlets, etc. of sulphides, gold, scheelite and other metallic minerals (Fig. 29-1c)

This style is transitional into the previous one. The ore minerals are distributed as disseminated flat grains, coatings, stringers, fracture veinlets, bunches, etc. in a schist, and there appear to be two fundamental end-member ore arrangements: (1) along bedding schistosity, where the orebodies are usually flat, stratigraphically controlled (meta-stratabound) lenses or (2) along zones of axial cleavage and schistosity, in which the orebodies usually cut across the original bedding. Type (1) cases are usually interpreted as metamorphosed sulphidic sediments or volcanics and tend to be of lower grade but greater persistence. Several examples have been reviewed in

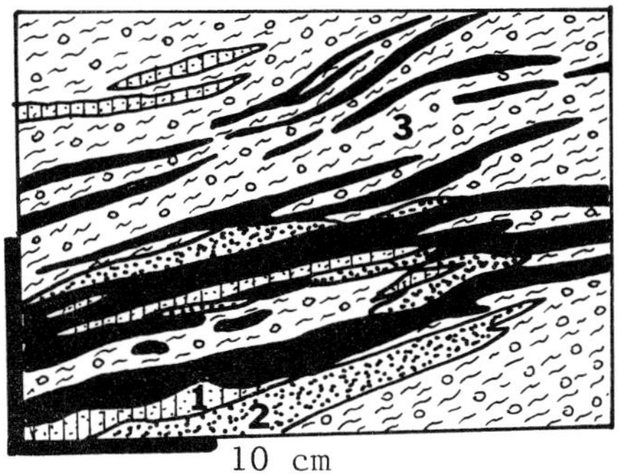

1. milky (secretion ?) quartz lenses
2. gray, transparent quartz
3. Pt_3-PZ_1 garnet-chlorite schist

BLACK: magnetite and sulphide lenses, scattered masses, grains, stringers, hosted by the gray quartz; contain microscopic cassiterite

Fig. 29-2. Detailed section of the cassiterite, sulphide, magnetite mineralization near Gierczyn, S.W. Poland. Modified after Jaskolski (1960).

Chapters 10-17. Type (2) cases are clearly metamorphogenic or metamorphically (re)mobilized orebodies.

KANMANTOO COPPER DEPOSIT, SOUTH AUSTRALIA

This locality (Verwoerd and Cleghorn, 1975; 122.5 Tt Cu/1%; Fig. 29-3) is an example of (2). There, the orebody is a system of flat ore lenses parallel with the axial plane schistosity, which has an overall shape of a 80° N-plunging elongated pipe. The host rock is a lower-middle Cambrian coarse grained porphyroblastic garnet-chlorite schist, enveloped by garnet-andalusite and quartz-mica schist. Pyrite, chalcopyrite and minor arsenopyrite with or without quartz, form parallel veinlets, stringers and flat grains along the schistosity, fine-grained disseminations and local massive pockets and bands. The final act of ore emplacement was contemporary with the last, third phase of deformation which dates from the early Ordovician period.

COPPER DEPOSITS IN THE CABO ORTEGAL COMPLEX, N.W. SPAIN

A group of copper deposits in high-grade metamorphosed lower Paleozoic rocks interpreted as original ophiolites near Santiago de Compostela (200 Tt Cu/0.6-0.6% Cu), has recently been studied by Badham and Williams (1981). The mafic complex is composed of granulite, meta-ultramafics, eclogites and amphibolites, and is situated in thrust-fault contact with a Silurian mélange. Disseminated, stringer, veinlet and occasionally massive pyrite, pyrrhotite and chalcopyrite form lenses, up to 10 m thick, in a

Fig. 29-3. Kanmantoo sulphide orebody, South Australia. From LITHOTHEQUE, modified after Verwoerd and Cleghorn (1975).

distinct stratigraphic horizon of quartz, pink garnet, gedrite and cummingtonite containing amphibolite, enclosed in a fine garnet amphibolite. The ores are also structurally controlled and confined to masses of isoclinal folds. Two medium-sized deposits (Arinteiro, 11 Mt ore with 0.67% Cu and Bama, 20 Mt ore with 0.5-0.6% Cu) have been mined, and additional small deposits and prospects are known in the area.

THE SCHEELITE DEPOSIT, FELBERTAL, AUSTRIA

The interesting scheelite deposit Felbertal near Mittersil in northern Austria which is hosted by a lower Paleozoic metamorphosed bimodal volcanic-sedimentary association, has already been discussed briefly earlier (Chapter 12). There, scheelite is scattered in metaquartzite bodies, confined to a "prasinite" (=chromian hornblendite interpreted as being a metamorphosed high-Mg basalt or picrite), amphibolite, albite gneiss and porphyroid gneiss unit. Höll (1977) emphasized the stratabound nature of the scheelite orebodies, confined to a 400 m thick sequence traceable for several kilometres, and to a prevalent concordancy of the "primary" (fine crystalline, metamorphosed) disseminated scheelite with the relic pre-metamorphic fabrics of its hosts, implying common and contemporary origin (by "exhalation").

"Secondary" (metamorphogenic) scheelite occurs as large scheelite-powellite porphyroblasts with dimensions of up to 1 cm accompanied by pyrrhotite, chalcopyrite, molybdenite, galenobismuthite, etc. in quartz stockworks. The Felbertal deposit is in the mantling complex of the Hohe Tauern ultrametamorphic core, deformed and metamorphosed in the Oligocene. No allochthonous granites are known so far, but the scheelite-bearing sequence is situated in such a way that it may have acted as a confining screen to metasomatic fronts. The main scheelite carrier, metaquartzite, occurs as peneconcordant intercalations, but also as crosscutting bodies. Holzer and Stumpfl (1980) pointed out the probable genetic association of quartz and ore. The considerable thickness and variety of the scheelite-bearing rocks (readily apparent in the paper of Höll, 1977, Fig. 2) can be attributed equally well or possibly better to metamorphic (re)distribution, as to an isochemical metamorphism of a syndepositional mineralization.

29.4.4. Schists with massive ore bands, lenses and rods (Fig. 29-1d)

GENERAL

Massive Fe and Mn oxide or silicate ores, massive sulphides and other commodities are commonly located in strongly schistose silicate metamorphics. There, they have the form of thin, persistent bands or laminae alternating with a barren rock, or of more localized, thin to thick massive ore lenses. As in the previous section, two end-members can be distinguished. They are as follows: (1) largely statically metamorphosed but little deformed orebodies, in which the ore bands are obviously relics inherited from the time of supracrustal deposition. Although complete or partial metamorphic recrystallization and corresponding mineralogical changes of the ore bands took place, the overall pre-metamorphic ore distribution pattern has been well preserved; and (2) penetratively (dynamically) metamorphosed orebodies, usually massive ore lenses, preferentially accumulated along shears, zones of axial cleavage, pressure shades, etc. These orebodies are concordant to peneconcordant with schistosity, but concordant to discordant with the original bedding. Although the ore can be banded, the banding is not necessarily a relic of a depositional structure. More commonly, it is a metamorphically formed banding.

The orebodies, in addition to having undergone the mineralogical and textural changes due to metamorphism, could have been reshaped, plastically moved into a new location, cataclastically comminuted and subsequently recrystallized to form ore blastomylonites, remobilized through a fluid phase, etc. In many cases, the orebodies need not have been present before the deformation and metamorphism at all, and are truly metamorphogenic: that is, precipitated from metamorphic-hydrothermal fluids during, or shortly after, the metamorphism.

The origin of similar orebodies has to be interpreted painstakingly. It is not apparent on first sight, and even less amenable to an armchair re-interpretation by a distant reader. It is becoming increasingly apparent that many of the "stratiform" massive sulphide orebodies in metamorphics were not in their present position at the onset of deformation and metamorphism, so they are metamorphogenic. This matter is further discussed later in this chapter. It is also discussed in Volume 2, and it has been reviewed recently in Laznicka (1985b).

Many of the massive ore deposits in metamorphics have already been reviewed earlier, in the lithologic framework of the pre-metamorphic host associations. The earlier descriptions will not be repeated and the emphasis in this section is on the particularly highly modified deposits not treated before.

Statically metamorphosed iron ores (limited mobilization)

The most common statically metamorphosed iron ores are the Precambrian iron formations, treated in Volume 2. Examples in Phanerozoic orogenic belts are much rarer. In the Desná Gneiss Dome, Jeseník Mts., N.C. Czechoslovakia (Gruzczyk and Pouba, 1968; Svoboda, ed., 1966; Fig. 29-4), delicately banded quartz-magnetite horizons in late Proterozoic gneisses are traceable for over 100 km. The dynamometamorphic overprint is apparent mainly on a local scale (boudinage, interruption of bands, thinning and swelling).

Statically metamorphosed manganese ores

In the eastern Carpathians, Rumania (Sebeş and Semenic Mts.; Ianovici and Borcoş, 1982), the Proterozoic basement contains numerous lenses and bedding schistosity-concordant Mn orebodies. The host rocks are metaquartzite, gneiss and amphibolite, and the Mn ores are composed of spessartite, rhodonite, pyroxmangite, dannemorite, knebelite, rhodochrosite, piemontite, and jacobsite. The Delinesti field has produced about 100 Tt Mn.

Statically metamorphosed bedded sulphide ores

Probably the best-known examples are the "vasskis" (or Leksdal-"type" orebodies) of the Norwegian Caledonides, described in numerous publications (e.g. Holtedahl et al., 1960; Vokes, 1976, 1978). These are persistent, regular, horizons of pyrite and pyrrhotite-impregnated schist, metaquartzite or oxide-silicate iron formation 1-2 m thick, grading to massive sulphide lenses. The sulphides are most frequently hosted by amphibolite (greenstone), black schist, or occur at lithological contacts such as between an ultramafic body and graphitic schist. Their lithofacies, relic structures and geochemistry indicate, quite convincingly, the pre-metamorphic origin of these ores. Although at several Scandinavian localities the "vasskis" are spatially associated with Fe,Zn,Cu massive sulphide bodies, both styles may occur independently.

Skarnoids and skarn enveloped by non-carbonate rocks (Fig. 29-1e)

Skarnoids are metamorphosed rocks composed of "skarn mineralogical assemblage" (essentially Ca-garnet and diopside-hedenbergite; compare Chapter 28), formed by metamorphism of rocks of appropriate composition (marls, mafic tuffs, sedimentary iron ores, some

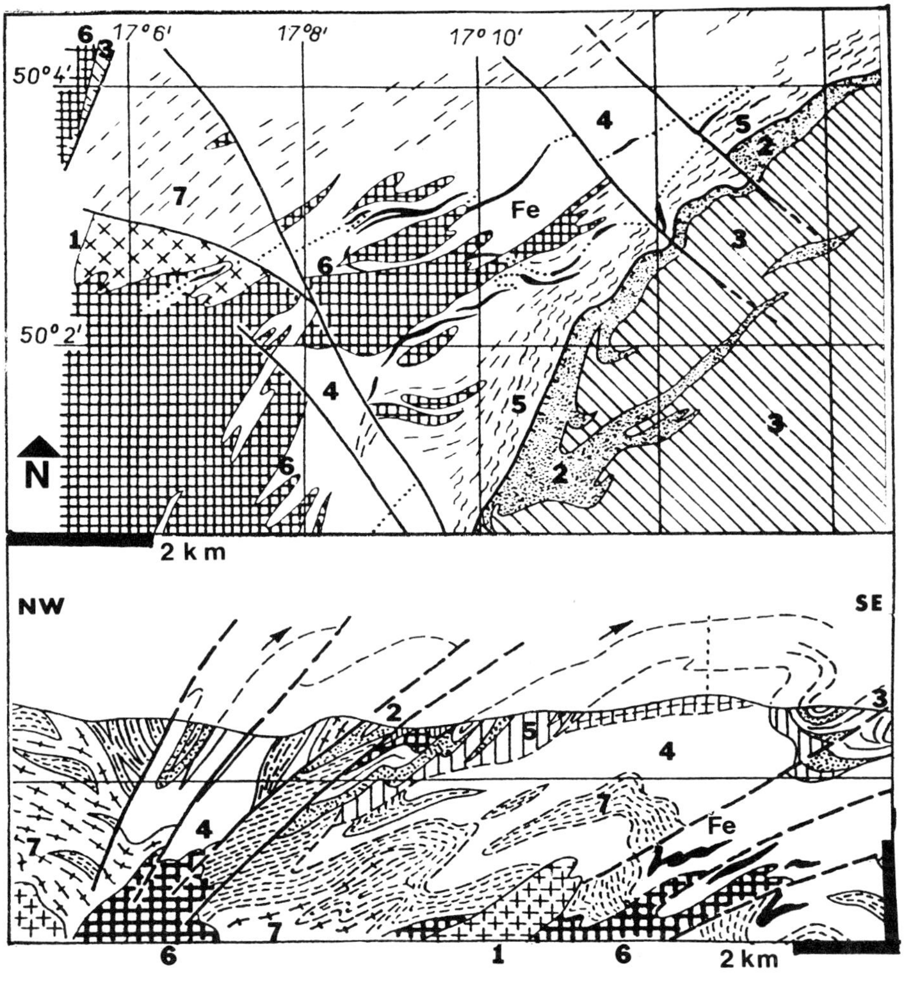

Fig. 29-4. Banded siliceous magnetite ores considered of "Sydvaranger type" in late Proterozoic or Devonian amphibolites in the core of the Desná (thermal, gneiss) Dome (Jeseník Mts., N.C. Czechoslovakia). After Pouba and Mísař from Svoboda, ed.(1966), courtesy of the Ústřední Ústav Geologický, Praha. The magnetite bands shown on the map are in correct position; those in the section are diagrammatic projection only, in the correct stratigraphic position. 1=PZ_3 granitic plutons; 2=D quartzite; 3=D phyllite; 4=D and/or Pt_3 biotite paragneiss; 5=chloritized biotite gneiss; 6=amphibolite; 7=Pt_3 migmatite to orthogneiss; Fe=banded magnetite ore.

metamorphics, some regoliths; Suk, 1983). In high grade metamorphic and ultrametamorphic terrains the skarnoids are difficult to distinguish from metasomatically completely skarnized limestone lenses sandwiched between silicate wallrocks, and also regionally metamorphosed earlier periplutonic metasomatic skarns. Suk (1983) coined the term "metaferrolite" which is applicable to the magnetite skarnoids formed by metamorphism of bedded iron ores.

Numerous small lenses of polymetamorphic skarns or skarnoids, some containing magnetite and many mined on a small scale in the past (e.g. Vlastějovice, Budeč, Županovice), occur in the Moldanubian metamorphic terrain of central Czechoslovakia. Their origin is uncertain. Elsewhere, skarns or skarnoids in high-grade metamorphics carry rare or base metal minerals. This could be the case of the stanniferous "Felsite Horizon" on the N.W. rim of the Freiberg Dome (near Halsbrücke and Bräunsdorf, Erzgebirge; Baumann, 1970). This is a 600 m thick, N.E.-trending metamorphic unit of interbedded gneiss, micaschist, amphibolite, dolomitic limestone and "metarhyolite", containing irregularly disseminated cassiterite with minor chalcopyrite. These minerals are confined to a magnetite and hematite-bearing, retrogressive chlorite schist and "skarn". Four mineralized horizons have been proven so far.

Dynamically metamorphosed iron ores

Probably the most extensive horizon of Phanerozoic (possibly Cambro-Silurian) industrial-grade metamorphosed ironstones is situated in the Dunderland Group of northern Norway, traceable for over 550 km strike length (Bugge, 1978). In the Mo-i-Rana region (Dunderlandsdalen iron field; 170 Mt Fe/34%; Fig. 29-5), the ore-bearing sequence contains interbedded dolomite and calcite marble, quartz-mica schist, and calcareous micaschist. Two iron ore horizons have developed. The upper horizon hosted by the quartz-mica schist, consists of specularite, granular hematite and magnetite in quartz, calcite, epidote, biotite gangue. The lower horizon in calcareous micaschist has magnetite and minor apatite in quartz, calcite, biotite, hornblende and grünerite gangue. The orebodies are penetratively deformed, 20-50 m thick and swelling in the axial zones up to a combined thickness of 70 m. The ores are interpreted as being metamorphosed chemical sediments. Metavolcanics are rare in the host sequence.

The Fosdalen deposit N.E. of Trondheim, central Norway (Bugge, 1978; 10.5 Mt Fe/35%) occurs along a E.N.E.-trending stratigraphic horizon 150 km long, located in Cambro-Ordovician metavolcanics and metasediments (schists, marbles, quartzite, amphibolite), interpreted as being a bimodal spilite-keratophyre suite. The orebodies are steeply dipping, ruler-shaped and composed of magnetite with accessory quantities of pyrite and chalcopyrite and hornblende, biotite, epidote, calcite and quartz gangue. They are interpreted as being former subaqueous-hydrothermal (Lahn-Dill "type") deposits.

At Kowary, S.W. Poland (formerly Schmiedeberg; Banaś and Mochnacka, 1974), a metamorphosed and repeatedly deformed lens of magnetite hosted by late Proterozoic gneiss, metacarbonate and amphibolite, is interpreted as being a pre-metamorphic orebody. Over it is superimposed a late Paleozoic periplutonic skarnization of the metacarbonate, and still later ("activation") emplacement of pitchblende, and Ni,Co,Bi, Cu,As,Hg,Se minerals veins and stockworks.

Dynamically metamorphosed manganese ores

The most common representatives are the Precambrian "gondites" treated in Volume 2. Phanerozoic examples are few, low-grade, and economically of little importance. Example localities include Mn-silicate horizons in Cambro-Silurian schists in the Mo-i-Rana area, Norway (Bugge, 1978). The most important horizon located stratigraphically above the Dunderlandsdal iron ore unit is fine grained, several metres thick and composed of quartz, Mn-almandine, accessory spessartite and dannemorite. The rock has a brownish-red colour and an average content of about 3.5-7% Mn (max. 14% Mn).

Dynamically metamorphosed massive sulphides

Large-scale physical transformations of ores are the result of differences in the physicochemical properties of orebodies, as compared with their host rocks. Stretching and flattening of sulphide lenses is most common, followed by boudinage and dismemberement. Brecciation and "Durchbewegung" (i.e., kneading; thorough deformation caused by inhomogeneous rotational strain) are also common. In many massive sulphide bodies mechanically more resistant fragments (of pyrite, arsenopyrite, garnet, quartz, feldspar) rotate, and are surrounded by a flow-banded matrix of pyrrhotite, chalcopyrite, galena, sphalerite and other minerals.

At most massive sulphide localities, the ore had lower strength than the wallrocks, so it was more mobile and flowed plastically. This is expressed almost universally in a variable thickness of the originally

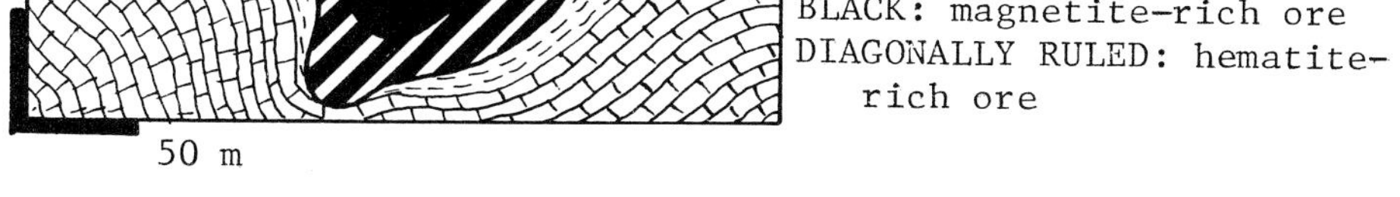

Fig. 29-5. Dunderlandsdal iron district, N. Norway; Ørtvann pit. From LITHOTHEQUE, after Mining Magazine, November 1980.

uniform ore beds and lenses, thickening in fold hinges and thinning along limbs. Many sulphide lenses in metamorphics which have been traditionally interpreted as "stratiform" have recently been re-interpreted as being structurally composite products of intrafolial folding (e.g. the Caribou deposit, New Brunswick; Davis, 1972). Rod and ruler-shaped orebodies, strongly elongated parallel to fold hinges of intensively compressed fold structures, constitute another extreme example of the dynamometamorphic control of orebodies. An ore rod can form by (1) multiple refolding of an earlier concordant sheet, or (2) replacement or dilation-filling in the axial area of folds, by fluid-transported regionally mobilized metals. Several deposits in the Scandinavian Caledonides (e.g. Killingdal; Rui, 1973) and the Sanbagawa metamorphic terrain in Japan (Chichara and Shingu mines; Kanehira and Tatsumi, 1970), contain rod-shaped orebodies.

Dynamometamorphism interferes with the banding or lamination in massive sulphides, which can either be destroyed, or newly created. The former case is a common result of metamorphic homogenization, in which originally heterogeneously banded sulphides revert partly or wholly to a massive, non-banded, monosulphide solution. The latter case results from metamorphic differentiation. The banding is oriented in regard to the stress vector, and need not be parallel to the wallrocks.

The last effect of dynamometamorphism on massive sulphide bodies that will be mentioned, is that of remobilization through a fluid phase. Stringers, veinlets, fracture networks, schlieren, disseminations of paragenetically younger minerals (in particular chalcopyrite) overprinting a usually pyritic main ore lens and also injected into the wallrocks, are commonplace.

Mineralogical and chemical compositions of dynamometamorphosed massive sulphide deposits in high-grade metamorphic terrains are almost identical with those in the lightly metamorphosed terrains, except for the greater representation of pyrrhotite at the expense of pyrite. The principal base metals ratios (Zn,Cu,Pb) also correspond to the latter, thus giving some indication as to the environmental and geotectonic setting. As expected, Cu-dominated orebodies are in amphibolite (mafic metavolcanic) terrains, e.g. Fornas, Cabo Ortegal Complex, Spain; Sulitjelma, Norway; Cajla, S.W. Slovakia (Table 29-4).

Zn-Cu orebodies are contained in mafic and felsic metavolcanics and associated metasediments, apparently corresponding to bimodal mafic-felsic volcanic-sedimentary terrains (e.g. Ducktown, Ore Knob, Gossan Lead and other deposits in the southern Appalachians; Tezuitlan, Mexico; Table 29-5). Zn-Pb orebodies are situated in metasedimentary (typically graphitic); metasedimentary-mafic metavolcanic (taphrogenic) and sequentially differentiated metavolcanic-metasedimentary terrains (e.g. Bleikvassli, Shuswap Complex, Pflersch, Achik-Tash; Table 29-6).

Nickel and Ni-Cu sulphides spatially associated with the metamorphosed ultramafics are common and economically important in the Precambrian terrains (e.g. Thompson, Manitoba; Selebi-Phikwe,

Table 29-4. Massive to disseminated, Cu-dominated schistosity-peneconcordant sulphide bodies in high-grade metamorphic terrains (mostly amphibolites)

LOCALITY	HOST UNIT	MINERALIZATION	REFERENCES
Sulitjelma, N. Norway	S, amphibolite, retrograde chlorite schist, meta-keratophyre, minor ultramafics; in allochthon	up to 1.5 km x 400 x 8 m oreb. at several stratigr. horizons near or within amphibolite; massive to disseminated pyrite, pyrrhot.,chalcop.,minor sphalerite, cubanite, valeriite; 370 Tt Cu	Bugge (1978)
Arinteiro and Bama deposits, Galicia, Spain	S ? garnet amphibolite (meta-ophiolite ?), granulite, gneiss	chalcop.,pyrrhot.,pyrite dissem., veins and veinlets assoc. with garnet porphyroblasts in amphibolite; elongated ore lenses in noses of isoclinal folds, up to 10m thick; 200 Tt Cu/0.5-0.67%	Badham and Williams (1981)
Cajla, Male Karpaty Mts.,Czechoslovakia	S ? micasch., graph.schist, amphibolite, intr. by late to post-orogenic Cb granite	pyrrhot.,pyrite, minor chalcopyr., dissem. to locally massive in amphibolite	Mahel' and Buday, eds.(1968)

Table 29-5. Massive to disseminated, Zn-Cu dominated schistosity-peneconcordant sulphide bodies in high-grade metamorphic terrains

LOCALITY	HOST UNIT	MINERALIZATION	REFERENCES
Ducktown, Blue Ridge, S.E. Tennessee, U.S.A.	Pt_3 chlorite-garnet gneiss, seric.-biot.schist, biot. quartzite, amphibolite	string of massive pyrrh.,pyr.,mag., chalcop.,sphal.,in calcite,dolom., qtz.,tremol.,chlor.,biot.gangue in schists, peneconc.with foliation; 8 tabular oreb., 9 km length 500 Tt Cu/0.7-1.6%, 18 Tt Co	Magee (1968)
Ore Knob, North Carolina, U.S.A.	Pt_3 biot.-sillimanite gneiss, amphibolite	pyrrhot., pyr.,chalcop.,quartz, biot.,amphibole peneconc.massive to dissem.bodies, 1.3 km long zone 31.5 Tt Cu/2.22%; 4.1 t Ag	Kinkel (1967)
Gossan Lead, Blue Ridge, Virginia, U.S.A.	Pt_3 gneiss, quartz-mica schist, minor amphibolite	N.E.-trending zone of discontin. massive pyrrhot.,lesser sphal., chalcop.,galena, 28 km long, en-echelon lenses. 100 Mt pyrrhot., estim. 2 Mt Zn/2%	Kinkel (1967)
Mofjell, Northern Norway	Cm-S, feldspar gneiss	sphalerite,galena,chalcopyrite in 3 ruler-shaped orebodies, plunging 8° east; 147 Tt Zn/3.5%; 35.7 Tt Pb/ 0.85%; 11.8 Tt Cu/0.28%	Bugge (1978)
Goldstream deposit Shuswap Complex, Brit. Columbia	Pt_3-PZ_1 graphit.schist, calc-schist,marble,amphibolite (metabasalt, metatuff)	massive pyrrhot.,sphal.,chalcop., 3-4 m thick, 1.2 km long lens in quartz-rich schist interb.with garn.-chlor. and graph. schist. 160 Tt Cu/3.7%; 117 Tt Zn/2.2% 69 t Ag/16 ppm	Höy (1979)

Table 29-6. Massive to disseminated, Zn-Pb schistosity-peneconcordant sulphide bodies in high-grade metamorphosed schist, gneiss, and/or minor intercalated carbonate horizons

LOCALITY	HOST UNIT	MINERALIZATION	REFERENCES
Mt.Copeland, Shuswap Complex, British Columbia	Pt_3 or PZ sillim.-biot.gneiss, micasch., metaquartzite, calc.-silic. gneiss, amphib.	up to 3m thick fine sphal., gal., pyr., pyrrhot. massive repl. in a thin marble horiz.; min.161 Tt Zn/6.2%, 146 Tt Pb/5.6%; 88 t Ag	Fyles (1970)
Ruddock Creek, Shuswap Complex, Brit. Columbia	micasch., calc-silic.gneiss, marble, veined by pegmatite and katazonal granite	lenses of mass. sphal., pyrrhot., gal., pyr., barite, fluorite up to 1.6 m thick; discontin. deformed horizons. Estim.min. 400 Tt Zn	COMINCO Staff oral commun. (1976)
Groundhog Basin, S.E. Alaska U.S.A.	PZ? schist, gneiss, calc-silicate gneiss, minor marble	massive pyrrhot., sphaler., lesser pyrite, galena; tabular oreb. max. 3m thick, 1,070 m long; 20 Tt Zn	Buddington and Chapin (1929)
Tracy Arm and Sumdum, S.E. Alaska U.S.A.	PZ? muscovite schist with a marble band	mass. to dissem. pyrrh., chalcop., sphalerite, lesser galena; 144 Tt Cu/0.6%, 96 Tt Zn/0.4%	as above
Bayindir, Menderes Massif, Turkey	Or-S, garnet-chlorite schist, calc-silic.gneiss, graphite schist, marble, metaquartzite	3 up to 3 km long gal., sphal., pyr., dissem. to massive ore layers	Dora (1977)
Bleikvassli, N. Norway	Pt_3 or Cm-S gneiss, schist, metaquartzite, graph.quartz., garnet schist, marble	series of interconn. and branching mass. and dissem. pyr., sphal., galena, pyrrhot., chalcopyr. lenses 163 Tt Zn/6.81%; 88 Tt Pb/3.67% 10 Tt Cu/0.42%	Vokes (1963)

Botswana; Western Australia; Volume 2), but insignificant in the Phanerozoic terrains.

The Zn-Pb dominated massive sulphide deposits in high-grade metamorphics (sometimes designated as Broken Hill-"type") are a special class of orebodies showing some links with carbonate hosts and with migmatite terrains, so they are reviewed in an independent section. Deformation and metamorphism of massive sulphide deposits not only produced the profound textural and structural changes discussed earlier, but also obliterated, to a considerable extent, traces of the pre-metamorphic depositional systems and environments (provided the orebodies are indeed pre-metamorphic), such as the characteristic "stratiform" lens and discordant footwall-feeder configuration (compare Chapter 6).

Sundblad (1980) convincingly demonstrated the disappearance of the anticipated feeder stockwork, following a twelve-fold flattening of an originally rooted massive sulphide lens along a shear, in Ankarvattnet, Sweden. The original vertical zoning common in massive sulphide deposits, moreover, had been tectonically transposed so, at present, it shows a lateral gradient.

Ducktown ore field, Tennessee, is one of the best-studied examples of Zn-Cu bearing massive sulphides in medium-grade metamorphics (Magee, 1968; Addy and Ypma, 1977; over 500 Tt Cu, 700 Tt Zn, 18 Tt Co; Fig. 29-6). There, eight folded and penetratively deformed, originally tabular orebodies exist with a composition: 60% pyrrhotite, 30% pyrite, 4% chalcopyrite, 4% sphalerite and 2% magnetite. They occupy a system of subparallel N.E.-trending shear zones in late Paleozoic staurolite, chlorite, garnet schists and biotite-rich metamorphics. These rocks are interpreted as being partly retrogressively metamorphosed metagraywackes, shales and conglomerates, cut by gabbro dikes. The orebodies appear to be at least partly conformable with the original bedding, and are interpreted as being the result of polystage mineralization. This involved syn-sedimentary hydrothermal seafloor precipitation, followed by shearing, partial metamorphogenic remobilization, accretion of additional sulphides, and retrogressive sericitization and chloritization.

In the early days of mining at Ducktown, 11 Tt Cu were produced from a chalcocite secondary enrichment blanket, and about 1 Mt Fe came from a thick limonitic gossan.

Ore Knob, North Carolina (Kinkel, 1967; 31.5 Tt Cu/2.22%; Fig. 29-7), is a much smaller massive pyrrhotite, pyrite, chalcopyrite, quartz, biotite and amphibole deposit vein-like to lenticular in shape, hosted by a late Proterozoic gneiss. As in Ducktown, the ore zone is contained in a narrow shear to breccia zone and is peneconcordant to foliation, slightly crossing the gneissic banding. The ore shoot is at least 1,300 m long and aligned along the plunge of lineation. The ore is polygenetic and surrounded by a fringe of retrogressive alteration that is up to 1.6 m thick. Most of the remaining deposits of this style are variations on a theme.

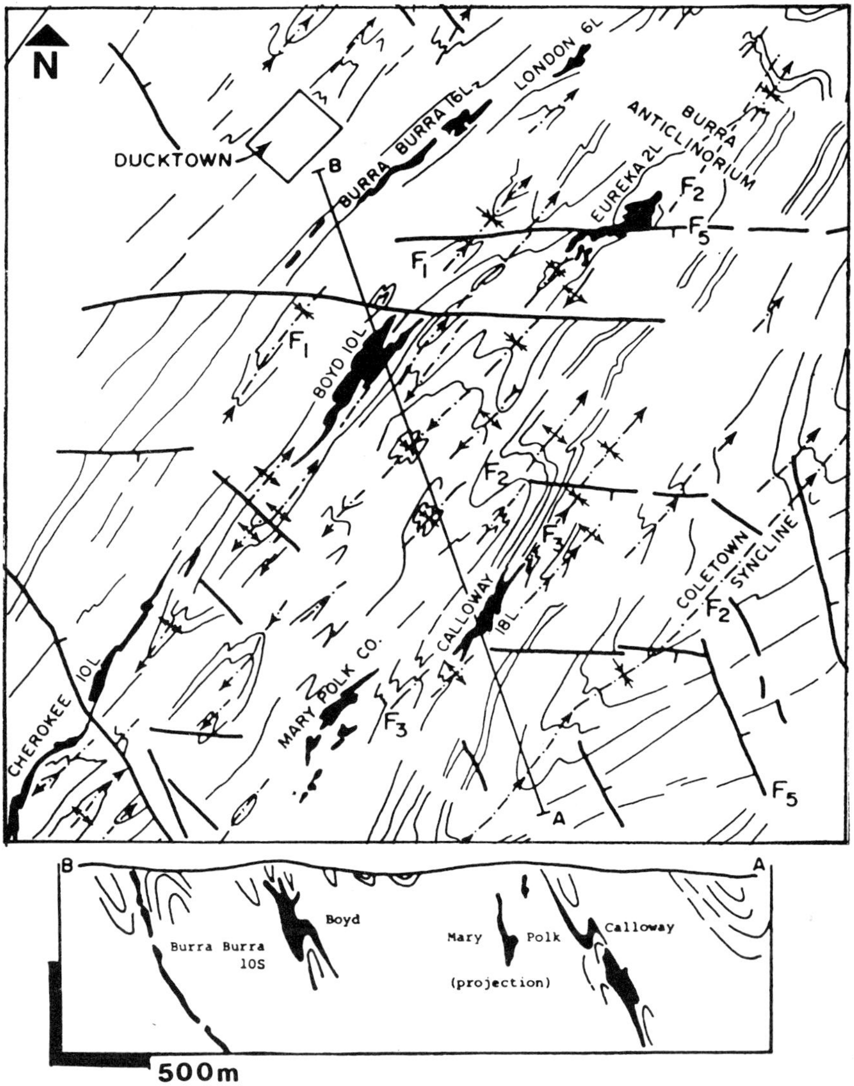

Fig. 29-6. A portion of a structural map of the Ducktown field, Tennessee, sulphide orebodies conformable with foliation in the host metamorphics. From Addy and Ypma (1977), courtesy of Economic Geology.

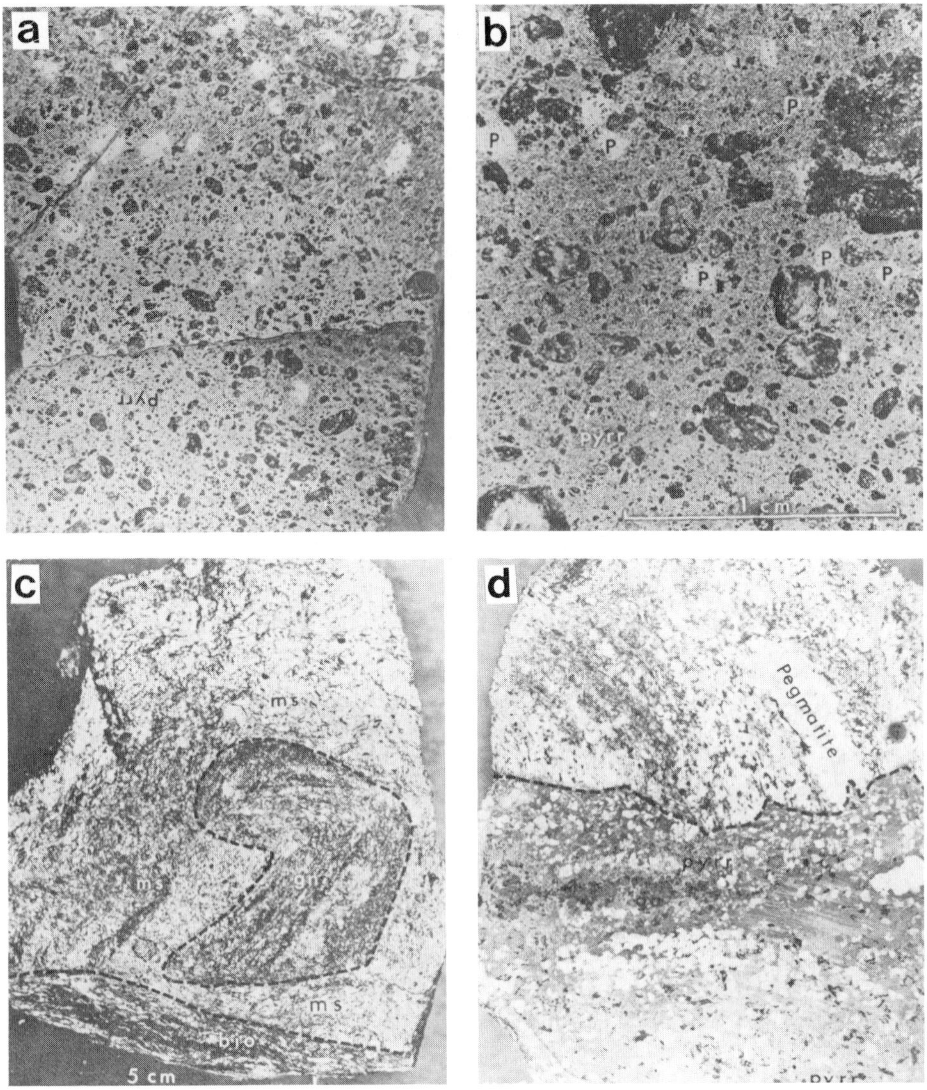

Fig. 29-7. Fabrics of high-grade metamorphosed massive sulphides and their host rocks in the Ore Knob deposit, North Carolina. From Kinkel (1967). (a) Rounded rock and mineral fragments (black) in massive, fine grained pyrrhotite (pyrr). White areas are pyrite porphyroblasts. The sharp contact between ore and sheared gneiss (gn) is a common feature. (b) Rounded rock and mineral fragments in massive sulphide ore. Silicate and quartz fragments have partial or complete biotite rims. The gray matrix (pyrr) is massive pyrrhotite. The small black minerals are biotite and hornblende crystals;

29.4.5. Disseminated to massive Zn-Pb sulphide deposits in high-grade metamorphics, peneconcordant with schistosity: the Broken Hill "type"

Broken Hill, N.S.W., Australia, is a composite ore zone, made up of six rich orebodies of coarse crystalline galena, sphalerite, pyrrhotite, bustamite and other minerals. The zone is conformable with the regional schistosity in a high-grade metamorphosed assemblage of gneiss, amphibolite, iron formation, carbonate and synorogenic granite. Broken Hill is lower Proterozoic in age and situated in the Willyama Precambrian block. It is therefore described in Volume 2. Zn-Pb deposits showing many (but not all !) Broken Hill characteristics, however, are known in many Phanerozoic orogens and these are reviewed briefly here.

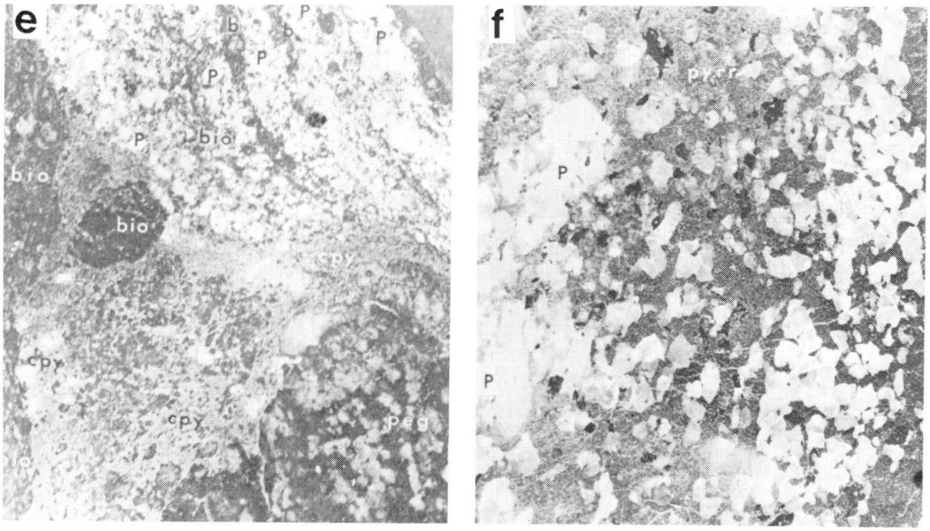

P=pyrite. (c) Twisted partly recrystallized fragments of gneiss (gn) in massive sulphide (ms). Coarse-grained black biotite (bio) at bottom of specimen is aligned parallel to vein wall. (d) Massive pyrrhotite ore (pyrr) cutting gneissic banding in wallrock. The vein contains much coarse granular white calcite and garnet; the wallrock contains no calcite or garnet. The black crystals in pyrrhotite are biotite and hornblende. (e) Pyrite bands (P) in gneiss that has been altered to biotite but retains planar structure. Banded biotite gneiss (bio) bends around a large fragment of unoriented, recrystallized pegmatitelike gneiss (peg), which is altered to coarse grained biotite, quartz and plagioclase. Chalcopyrite (cpy) enclosing quartz and rock fragments was formed in the pressure shadow and as rims on rounded pyrite grains. (f) Coarse calcite grains in massive pyrrhotite (pyrr) and minor pyrite (P).

The Mount Copeland Zn-Pb orebody (also known as River Jordan, True Fissure; Fyles, 1970; Riley, 1961; Fig. 29-8), is situated in the Shuswap Metamorphic Complex of British Columbia. There, a remobilized granite core is enveloped by late Proterozoic or Paleozoic mantling assemblage of a gray gneiss, biotite-sillimanite and biotite gneiss, quartz-mica schist, quartzite, amphibolite, marble and calc-silicate gneiss. The age of metamorphism is Jurassic or Cretaceous. The Zn-Pb mineralization is confined to a discrete, stratigraphically-controlled (stratabound) horizon of a calc-silicate gneiss to marble, less than 10 m thick, which is folded and traceable with interruptions for several kilometres. The ore reserve available in a single outcropping segment of the ore zone is 2.6 Mt with 6.2% Zn, 5.6% Pb and 34 ppm Ag, but it is estimated that the entire zone, prior to erosion, contained at least 60 Mt ore.

The orebody itself is up to 3 m thick, and consists of fine grained pyrrhotite, sphalerite, galena, pyrite, containing "eyes" and bunches of quartz and pegmatitic feldspar. When the ore is located at the contact with the footwall sillimanite, biotite, garnet schist or gneiss or the hangingwall light-gray quartzitic micaschist, its boundary is sharp. When the entire width of the host horizon is represented by sulphide, the orebody has the form of sharply outlined, concordant band enveloped by the silicate metamorphics. When the ore minerals constitute less than 100% of the ore horizon, the carbonate background shows up. The ore then displays the character of replacement pockets, blebs, veinlets and disseminations in the metacarbonate, often along contacts with short, impersistent lenses and veins of pegmatite. Although the ore appears to be metamorphosed in a grade identical to that of the wallrocks, it is by no means certain that this was originally a seafloor-deposited sedimentary ore layer as is popularly believed.

The remaining schistosity-peneconcordant Zn-Pb zones in the Shuswap Complex (Big Ledge, Ruddock Creek; Fig. 29-9; Wigwam, Cottonwood; Table 29-6), are all immediately hosted by a thin, often boudinaged or completely sulphide-replaced horizon of metacarbonate, sandwiched between gneiss or schist, so that they correspond closely to the situation at Mount Copeland. Even at Broken Hill, N.S.W., calcite, Mn carbonates and Ca or Mn silicates probably produced by contact metasomatism are widespread in the ore zone and although there are no conspicuous relics of unreplaced marble, a complete pseudomorphic replacement of an earlier carbonate body is one of the possible alternatives of the Broken Hill origin.

In the Shuswap Complex, graphitic metasediments are neither widespread nor typical, but elsewhere they may be abundant and prominently associated with metamorphosed Zn-Pb sulphide masses. The Bleikvassli deposit in the northern part of the Norwegian Caledonides (Vokes, 1963; 163 Tt Zn/6.81%, 88 Tt Pb/3.67%, 10 Tt Cu/0.42%; Fig. 29-10) is in Proterozoic or Cambro-Silurian micaschist and gneiss at or near contact with a unit of graphitic and muscovitic quartzite and schist. A series of interconnected and branching, schistosity-concordant lenses of massive and disseminated pyrite,

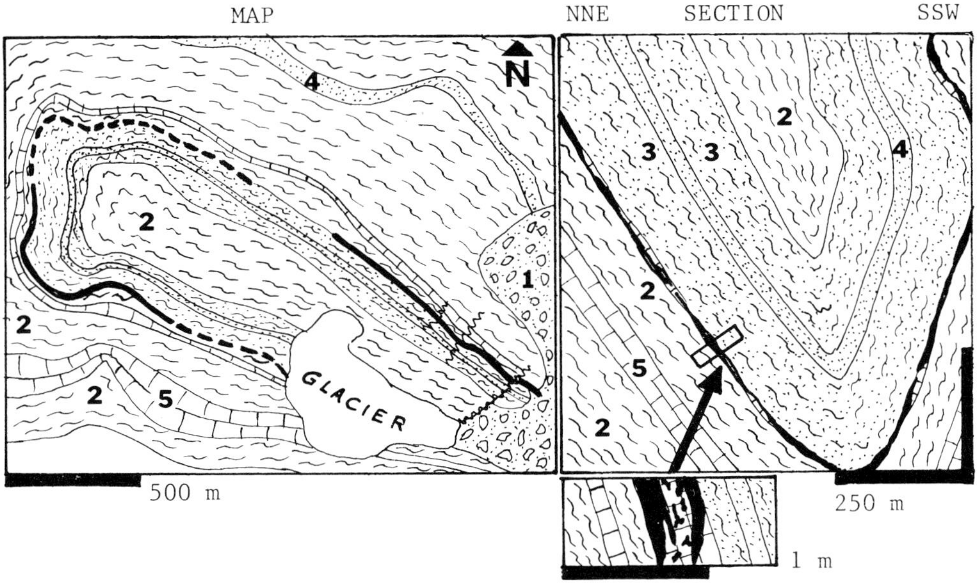

Fig. 29-8. Mount Copeland near Revelstoke, British Columbia, a schistosity–concordant Zn-Pb massive orebody in amphibolite grade metamorphics. 1=Q glacial sediments; 2=$^{Pt}3$ or PZ mantling assemblage of the Shuswap Complex (J metamorphism), biotite or biotite sillimanite gneiss and schist; 3=mica quartzite to quartz-mica schist; 4=metaquartzite; 5=calc – silicate gneiss to marble; black: Zn-Pb ore horizon. From LITHOTHEQUE, after Fyles (1970), Riley (1961) and own fieldwork.

sphalerite, galena, pyrrhotite, lesser chalcopyrite, stannite and cassiterite, show typical metamorphic ore textures and has the same metamorphic grade as the enclosing rocks, so it is pre-metamorphic. Although there are no traces of carbonate in the orebody itself, lenses of marble do occur in the host series, an allochthonous unit. The associated metamorphics suggest that Bleikvassli could have been a member of the "black slate" lithologic association which was reviewed, in its pre-metamorphic state, in Section 17.5.

29.4.6. Granulite, charnockite, eclogite

Granulite facies consists of completely dehydrated rocks, such as ordinary (quartz, feldspar or garnet) granulite, hypersthene granulite and charnockite. Eclogites are composed of omphacite pyroxene and garnet. In the recent literature (see review in Suk, 1983, p.213), two varieties of granulite terrains have usually been distinguished:

Fig. 29-9. Ruddock Creek massive Zn-Pb sulphide zone, British Columbia. (A) General view. (B) Gneiss of the host sequence, showing initial partial melting and metasome segregation. (C) Contact of massive sphalerite, galena, pyrrhotite band (s) and marble (m).

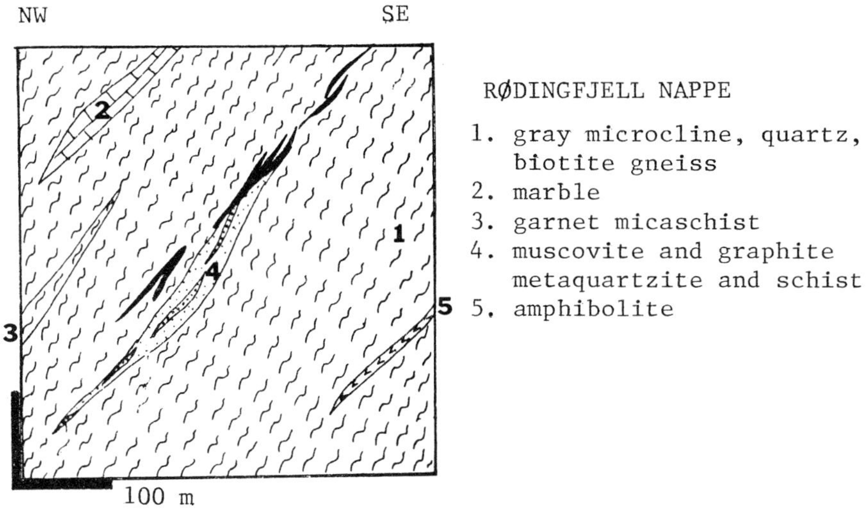

Fig. 29-10. Bleikvassli, N. Norway, Zn-Pb massive sulphide deposit. From LITHOTHEQUE, based on data in Vokes (1963).

(1) those in Precambrian shields, Vol. 2, considered "typical" granulite facies metamorphics and (2) those in Phanerozoic orogenes, often interpreted as originally anhydrous rocks of suitable composition such as continental rhyolite or andesite, that achieved their granulitic character under conditions of lower, amphibolite facies metamorphism. Those of type (2) are relatively uncommon, although nine areas of distribution are known from the Bohemian Massif alone. No metallic occurrences in the granulites proper are known, with the exception of small quantities of accessory pyrite in some of the hypersthene varieties.

The category of the "deepest-seated" pegmatites of Ginsburg et al. (1979) is spatially associated with granulites, being gradational into pegmatites in migmatite terrains or a product of retrogression. Monazite and/or allanite have been reported in mineralogical quantities from some of the "deepest-seated" pegmatites, and Be (taafeite) and complex Th-U minerals have been found in exceptional cases. Small amounts of monazite recovered from terrace and channel sands of some of the rivers draining the southern Bohemian granulites (Blanice, upper Vltava), may have been derived from such pegmatites.

Eclogites had probably formed under similar conditions of pressure and temperature as granulites, although from different precursors (spilites, gabbros, basalts). They occur in association with amphibolites and ultramafics in Barrovian metamorphic belts, as well as in high-pressure blueschist associations. No metallic deposits are known from the former. A rutile deposit hosted by the high-pressure eclogites in the Urals, has been described briefly in Chapter 8.

29.5. MINERALIZATION STYLES IN METACARBONATES

29.5.1. Meta-bauxites

Meta-bauxites associated with metacarbonates (former "Mediterranean-type" or "karst" bauxites; Chapter 20) are the least controversial example of metamorphosed relic mineralization in carbonates. Unfortunately, thoroughly studied meta-bauxite occurrences are extremely rare, and the best examples can be found on the Greek Cyclade Islands, particularly on Naxos. There (Jansen and Schuiling, 1976; Fig. 29-10), a migmatitic gneiss dome having an Oligocene age of metamorphism, is enveloped by metasediments (marble, micaschist) and mafic-ultramafic members of ophiolites (amphibolite, serpentinite) with concentrically decreasing metamorphic grade (away from the central dome). A Miocene high-level granodiorite intrusion is to be found in the western part of the island.

The dominant rock, the metacarbonate, contains numerous lenses and pockets of meta-bauxite. The close proximity of the mafic-ultramafic rocks to the meta-bauxite, its highly ferruginous nature and the presence of unusual minerals (e.g. Ni-fuchsite) suggest that the ophiolites could have been the source of the Fe and Al-oxides rich residuum, washed into the carbonate karst. The meta-bauxites are present in a wide range of metamorphic zones and change their mineralogical composition accordingly. In the lowest metamorphic zone (greenschist facies), the meta-bauxite has an assemblage of diaspore, chloritoid and hematite. With increasing metamorphism, dehydration resulted in the formation of corundum and magnetite, and emery deposits formed. In the biotite-chloritoid zone, the meta-bauxites contain corundum, chloritoid, kyanite, magnetite assemblage in the centre, and corundum, calcite, margarite, chloritoid along the contact. In the kyanite zone, the emery consists of kyanite, corundum, staurolite, magnetite, grading into a marginal margarite, anorthite, corundum assemblage. In the sillimanite zone, the emery is composed of corundum, staurolite and magnetite.

29.5.2. Skarns

In the previous section, the peculiarity of skarns and skarnoids in the predominantly non-carbonate high-grade metamorphic terrains were discussed briefly. This section is concerned with skarns hosted by high-grade metamorphosed, widespread carbonates (marbles). There, the skarn bodies could have formed in several ways. These are as follows: (1) by isochemical metamorphism of a pre-metamorphic compositionally suitable assemblage (pelosiderite, marl, ankerite); (2) by bimetasomatism or local infiltration metasomatism at a marble-contrasting rock (e.g. mafic or ultramafic body, ironstone) contact, in the metamorphic or ultrametamorphic environment; (3) by pre-metamorphic infiltration metasomatism at the marble-"granite" contact and subsequent regional metamorphism that generated a

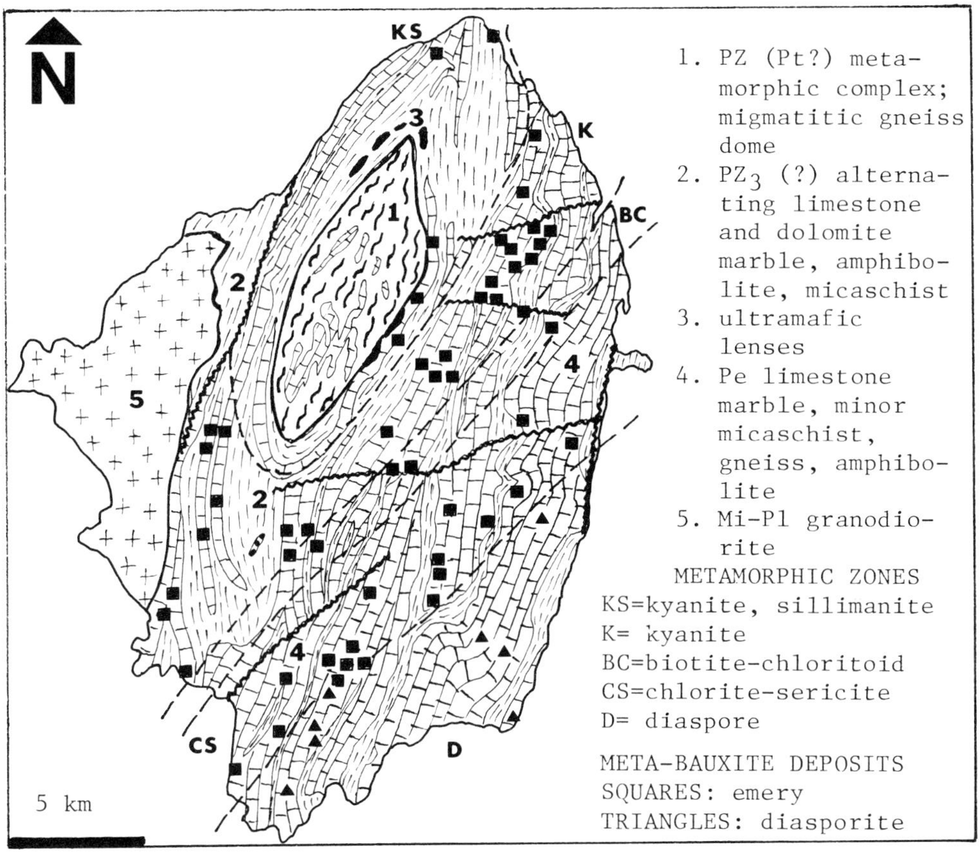

Fig. 29-11. Setting of meta-bauxite occurrences on Naxos Island, Greece. Slightly simplified after Jansen and Schuiling (1976).
Redrawn with permission from the American Journal of Science and the authors.

meta-skarn and converted the granite into an orthogneiss or (4) by post-metamorphic infiltration metasomatism at marble-"granite" contact. The latter possibility was treated in Chapter 28 and will not be considered further here. Categories (1-3) are virtually indistinguishable in the field and the literature interpretations are usually the consequence of the interpreter's bias.

In the Erzgebirge Mountains (East Germany and Czechoslovakia; Svoboda, ed., 1966; Chrt and Bolduan, 1966; Fig. 29-12), skarns which are interpreted as belonging to categories (3) and (4) are widespread and locally of economic importance. The (3) skarns are hosted by minor late Proterozoic carbonate lenses in a complex gneiss, micaschist, amphibolite, metacarbonate and ultramafics sequence, have the form of lenses concordant with schistosity, and the skarnization is attributed to synorogenic early Paleozoic "granites" (now a "red orthogneiss"). Magnetite is the main commodity produced from these skarns (e.g. at Měděnec) and Cu,Pb,Zn sulphides are minor. The average Fe content is 37% Fe. The Měděnec skarn deposit is a lens, over 100 m thick and 500 m long consisting of hedenbergite-andradite skarn with local portions of amphibole, epidote, idocrase and biotite skarn, enveloped by garnet-muscovite micaschist. A medium- to fine-grained muscovite or two-mica orthogneiss is located in the vicinity. Banded magnetite-rich lenses up to 60 m thick with minor and irregularly distributed pyrite, pyrrhotite and chalcopyrite are being mined.

The type (4) skarns in the Erzgebirge are more variable in composition, and formed by replacement of Proterozoic and lower Paleozoic carbonates, mafic calcic tuffs and partly carbonatic metabasalts in the contact aureole of the post-orogenic Permo-Carboniferous "granites". Polymetallic skarns (Zn,Pb,Cu), some with superimposed fracture filling cassiterite (Zlatý Kopec, Breitenbrunn), are common.

→

Fig. 29-12. Geological map and section of a polymetamorphic terrain in the Erzgebirge Mountains near Kovářská, N.W. Czechoslovakia, showing ore occurrences.

1=Q peat bogs; 2=quartzitic micaschist and metaquartzite; 3=garnet micaschist; 4=two mica gneiss; 5=gneiss with feldspar porphyroblasts; 6=two mica gneiss and paragneiss; 7=massive graywacke gneiss; 8=two mica gneiss with layers of graywacke gneiss; 9=marble; 10=amphibolite; 11=skarn; 12=serpentinite; 13=migmatitic orthogneiss; 14=coarse augengneiss; 15=granite porphyry; 16=lamprophyre; 17=T volcanics; 18=tuffs. ORES: Fe=magnetite skarns (circled); Ag=Ag,Co,Ni,As,Pb veins; F,Ba=T? fluorite and barite veins along faults. After Sattran and Škvor in Svoboda, ed. (1966); courtesy of the Ústřední Ústav Geologický, Praha. Ore occurrences added.

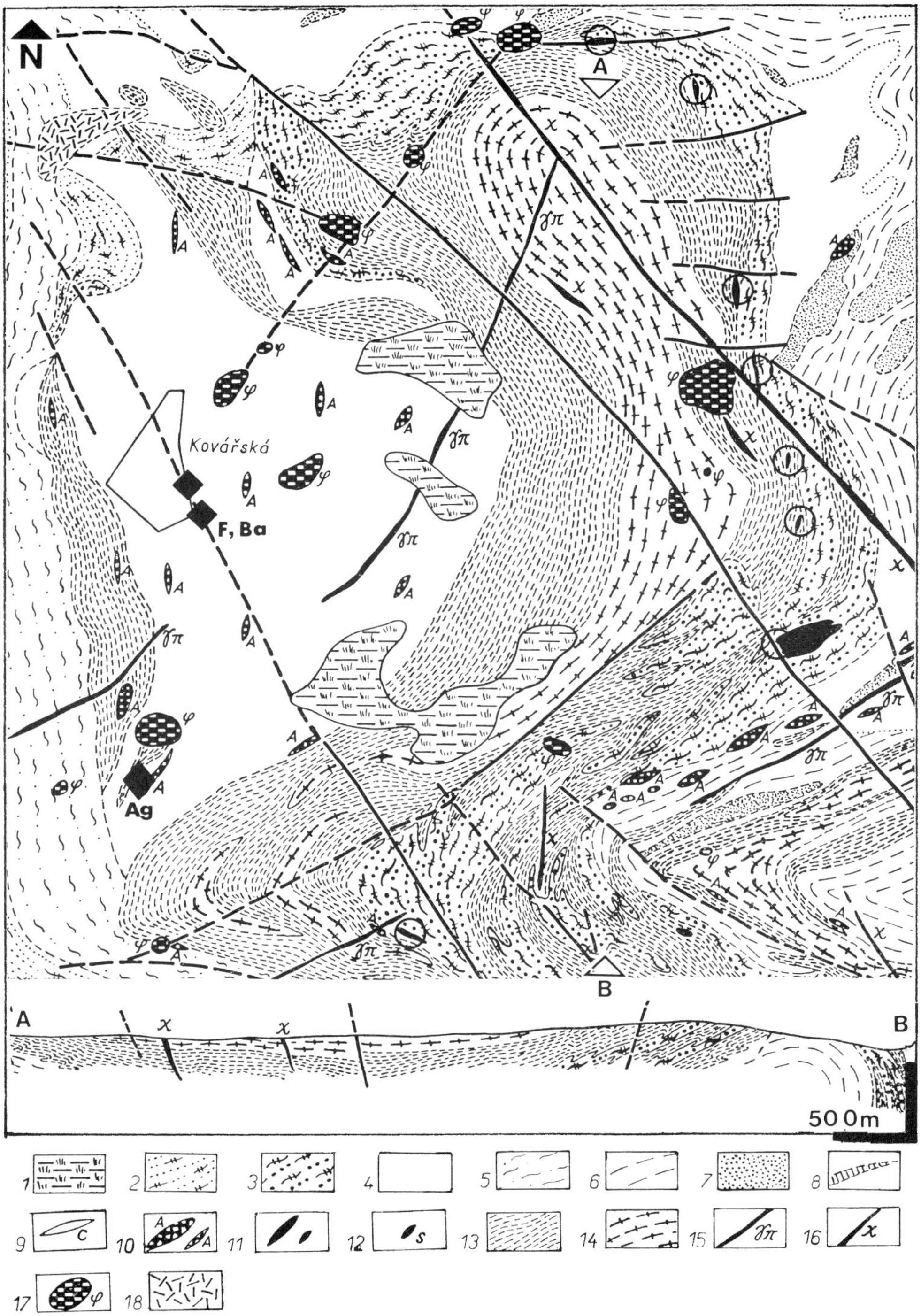
N
A
B
Kovářská
F, Ba
Ag
500m

29.5.3. Manganese ores

Phanerozoic equivalents of the Precambrian "queluzite" (i.e., rhodochrosite marble and rhodonite hornfels) acting as a protore to rich residual Mn-oxide deposits (e.g. Cons. Lafaiete and Serra do Navio, Brazil; Section 23.3.5. and Fig. 23-10), are rare. One example is the Takhta-Karacha zone in the Zeravshan Range, Soviet Central Asia, where a Mn-member in a Silurian metacarbonate sequence is traceable for over 250 km (Varentsov and Rakhmanov, 1974). The Mn-member is a dark gray to black finely banded to massive limestone marble, interbedded with minor amphibolite. Two seams of rhodochrosite and manganocalcite with minor "floating" crystals of spessartite, 6-8 m thick, crop out in numerous small deposits, some of which contain secondary vernadite, psilomelane and pyrolusite in the oxidation zone.

In the small James River-Roanoke River Mn field in Virginia (Espenshade, 1954; 16 Tt Mn/43.3%), lumps and masses of hard psilomelane occur in residual clay, formed over late Proterozoic to lower Cambrian white marbles that carry a mere 0.11 to 0.85% MnO. The marble is interbedded with schist and meta-quartzite. Some Mn deposits in the lower Cambrian Tulgheş Series, N.E. Rumanian Carpathians (Ianovici and Borcoş, 1982), are hosted by marbles.

29.5.4. Zn-Pb ores

Metamorphosed galena-sphalerite bodies hosted by widespread marbles are usually designated Balmat "type" in the literature, and are common in several Precambrian metamorphic terrains (e.g. Balmat, New York; Marmorilik, Greenland; Vol.2). Only a few minor occurrences are known dating from the Phanerozoic time.

In the Rodna Mountains (northern Rumanian Carpathians; Ianovici and Borcoş, 1982), late Proterozoic marbles interbedded with micaschists in a thrust, contain tabular peneconcordant metamorphosed bodies of shpalerite, galena and barite. These are interpreted as being partly remobilized, metamorphosed Mississippi Valley-"type" deposits.

29.6. MINERALIZATION STYLES IN AND RELATED TO THE ZONE OF ULTRAMETAMORPHISM AND GRANITIZATION

29.6.1. General

Ultrametamorphism causes transformations of rocks under conditions near to the melting point of the rock system involved (Suk, 1983). This can take place in regions of high-grade regional metamorphism, but also at intrusive contacts. It has been demonstrated that ultrametamorphism is sometimes independent of isometamorphic zones, but requires a high water pressure. Ultrametamorphism starts with partial melting of rocks. That triggers a chain reaction: melting→ metamorphic differentiation→ metasomatism by the mobile substance and cation exchange→ formation of magmatic melts and their local injection, or rise into the higher crust in the form of granitic magma (Mehnert, 1968).

The most common rocks in the zone of ultrametamorphism are migmatites, rocks composed of the usually darker immobile residue of the partially melted material (paleosome, melanosome, substrate, restite), and the lighter mobile melt (neosome, leucosome, metatect, mobilizate). Granite gneiss is "a gneiss derived from a sedimentary or igneous rock and having a granite mineralogy" (A.G.I. Glossary of Geology) and is either an evolutionarily more advanced product of granitization that also generated migmatites (a product of metamorphic homogenization rather than differentiation like the migmatite), or an equivalent of migmatite in a suitable environment.

The structural varieties of migmatites and granite gneisses are treated in most textbooks and will not be considered here. Compositionally, the bulk of migmatites have a substrate of biotite, feldspar, quartz, lesser sillimanite, garnet gneiss, and a quartz-feldspar metatect. The second most common variety are the migmatites with a green amphibole-rich (amphibolitic) metatect. The remainder of high-grade metamorphics rarely form regularly banded migmatites and the metatect tends to form fracture or bedding veins, replacements, breccia matrix, etc., as well as a variety of reaction and infiltration metasomatites.

For the purposes of metallogenic interpretation and description (Fig. 29-13), it is advantageous to treat separately (I) the zone of actual granitization (migmatite and granite-gneiss terrain); (II) the high- to medium-grade metamorphosed mantle; (III) the katazonal granites, and (IV) pegmatites.

29.6.2. Migmatite and granite-gneiss terrains

These are among the least attractive terrains to a prospector, except in cases where the deep-generated migmatite terrains were uplifted and subsequently intruded by high level "granitic" (e.g. the Colorado Mineral Belt) or other, e.g. alkaline, intrusions. In the latter case, however, there is a substantial time difference between the ultrametamorphism and high-level intrusion. The granitization processes have a highly negative metallogenetic balance, that is, more earlier orebodies are destroyed than are newly created.

Although reliable analytical data are scarce, there is an abundant circumstantial evidence that in a reasonably open system (having avenues of communication with the shallower zones of the crust via ascending granitic magmas, chains of metasomatic fronts, deep lineaments, etc.), a considerable proportion of the transition metals leave the system and move upward. There is no clear evidence that any metals accumulated residually in the granitized rocks. The rarely encountered mineralizations in this zone correspond to four styles:

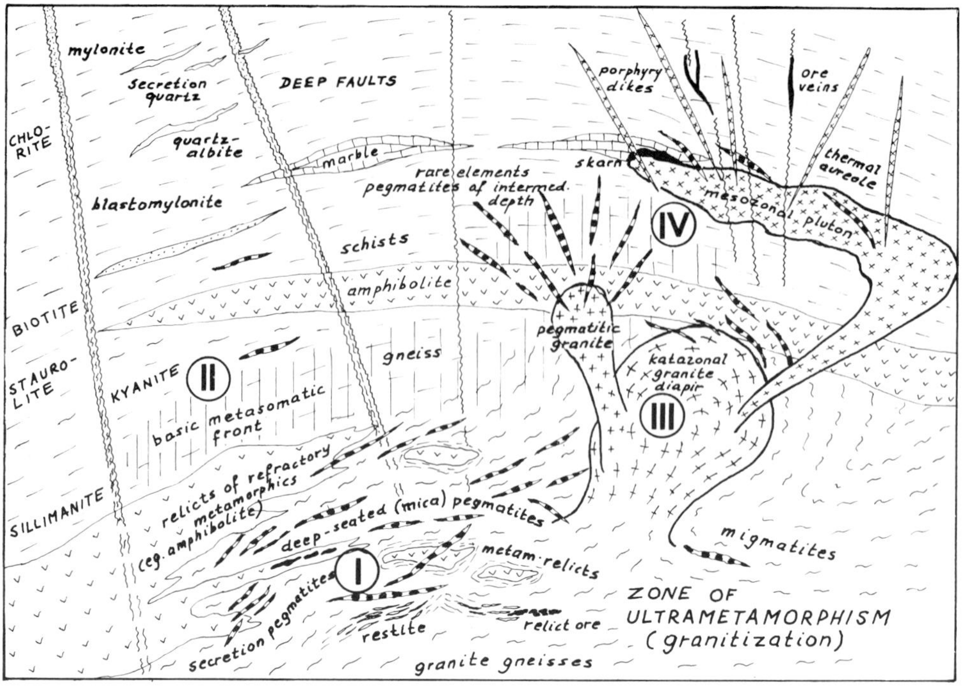

Fig. 29-13. The zone of ultrametamorphism and adjacent zones. Diagrammatic only, dimensions distorted and telescoped. Roman numerals: zones of metallogenic interest as described in the text: I=core zone of granitization; II=metamorphism in mantling assemblage; III=katazonal granites; IV=metalliferous pegmatites of intermediate depth. From Laznicka (1984).

(1) resisters (relics); (2) semi-resisters; (3) metal accumulations in secretions within the granitization zone and (4) metals intercepted during their upward escape and forced to precipitate at or under physical or chemical screens.

(1) RESISTER (RELIC) ORES

Relic mineralizations occur either hosted by refractory rocks preserved as rafts, inclusions or non-granitized segments of earlier rocks, surrounded by or in contact with migmatites, or as "denuded" bodies in the migmatites themselves. The former category includes small magnetite skarns enveloped by migmatites. Several examples (Budeč, Županovice, Vlastějovice) are known from the Moldanubian terrains of the Bohemian Massif.

The Vlastějovice deposit (Koutek, 1963; Fig. 29-14) is confined to

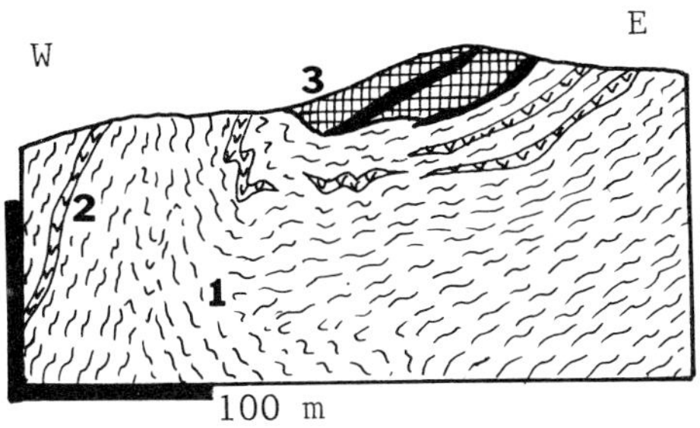

Pt, Moldanubicum Complex metamorphics

1. biotite gneiss to migmatite
2. amphibolite
3. garnet-pyroxene polymetamorphosed skarn with two lenses of magnetite

Fig. 29-14. Polymetamorphosed magnetite skarn in Vlastějovice, central Czechoslovakia. Modified after Koutek (1963).

two isolated synclinal lenses of andradite-hedenbergite and hedenbergite-amphibole skarn, surrounded by Proterozoic migmatites. The skarn is intersected by numerous hybrid pegmatite veins. The magnetite-rich skarn occurs in several lenses up to 10 m thick, and several million tons of iron ore have been produced.

Bands, schlieren, disseminations and small lenses of magnetite are common in rafts and blocks of amphibolite resisters in migmatites and granitic gneisses. There are few significant Phanerozoic examples, and Precambrian localities (e.g. Dover, New Jersey; Labrador Iron Belt) are reviewed in Volume 2 as well as in Laznicka (1985b).

"Denuded orebodies" in terrains of granitization (Fig. 29-15) are represented by several examples of massive Zn,Pb(Cu) orebodies in the Proterozoic Moldanubian Complex in Bavaria and western Czechoslovakia, granitized in the late Paleozoic. The historical Bodenmais deposit in the Bavarian Forest (Teuscher, 1982) is the largest sulphide orebody along a 30 km long W.N.W.-trending mineralized zone, running parallel with the remnants of schistosity in migmatites. The orebody, nearly 1,000 m long and 150 m deep, is a system of boudins over 10 m thick, thinning to about 1 m in the connecting necks. The ore is coarse grained and composed of pyrrhotite, pyrite, sphalerite, chalcopyrite , magnetite and additional ore minerals. The immediate host rock is a garnet, cordierite, sillimanite restite, and pegmatitic mobilizate, pegmatite. Monazite is anomalously concentrated in the migmatite in a separate layer adjacent to the sulphide orebody. The Pohled deposit in central Czechoslovakia is very similar.

(2) SEMI-RESISTER OREBODIES

Semi-resisters, an informal term, is applied here to former metal concentrations modified by granitization often beyond recognition, and to a large degree digested (dissipated) so that only minor remnants of the former orebody remain (Fig. 29-15). Of particular interest are

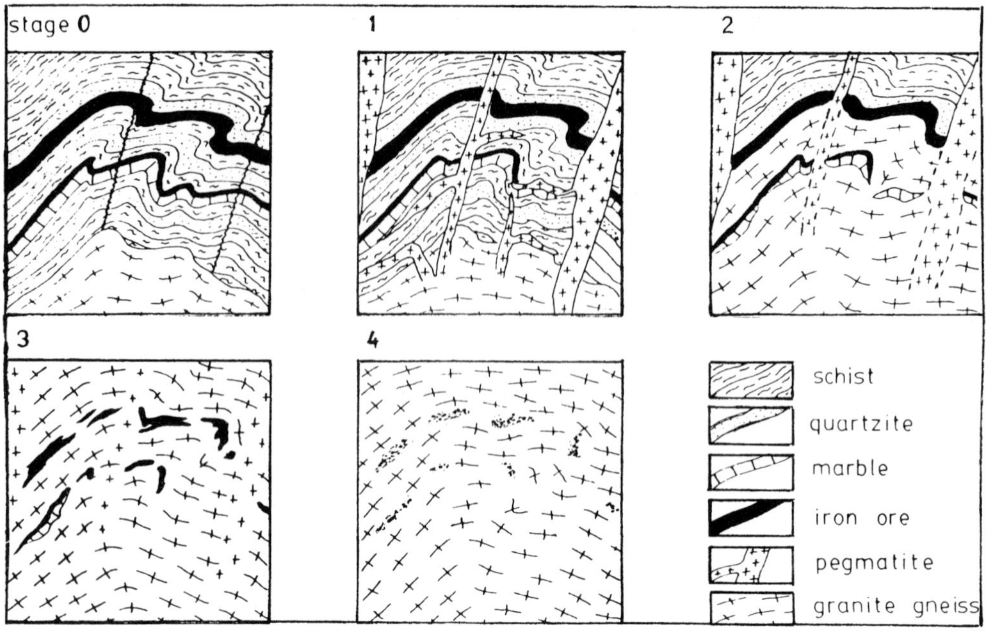

Fig. 29-15. Development of "denuded" resister and semi-resister orebodies in terrains of granitization. From Laznicka (1985b).

the cases of digestion of metalliferous horizons in which the base metals (or gangue elements such as Ba), were accomodated as inconspicuous silicates or oxides. Zinc frequently accumulates in dark green crystals of gahnite (Zn spinel), Pb in the green (amazonite) microcline, Ba in barian feldspars, Co in amphibole, Sn in sphene, staurolite, garnet, etc. Most examples are located in the Precambrian shields (e.g. in the Broken Hill lode zone, gahnite and hyalophane).

Höll and Maucher (1976) described occurrences of scheelite semi-resisters in the Central Gneiss (a migmatite) in the Hohe Tauern Range of Austria. They interpreted the scheelite as being a product of palingenetic regeneration and partly anatectic granitization of the Paleozoic scheelite-bearing strata, during the Variscan orogeny.

The observation made about metallogenically negative processes earlier (e.g. about glaciation) that even when a previously metal-enriched environment is subject to destruction transitional mineralization may appear before such destruction is complete, is applicable here as well. As an example one may quote semi-resister style local uranium enrichments in granitized terrains, formed at the site of uraniferous black sediments.

(3) METAL ACCUMULATIONS IN SECRETIONS AND CONCRETIONS WITHIN THE GRANITIZATION ZONE

Secretions and concretions are nest, lens, vein-like inhomogeneities composed by the most mobile minerals (usually pegmatitic, or of quartz, tourmaline, etc.). They frequently contain porphyroblasts of metalliferous minerals such as spessartite, monazite, allanite and others, but only in mineralogical quantities. Deeply weathered ultrametamorphic terrains may supply these minerals into placers, but only monazite is sufficiently resistant chemically to be able to persist.

(4) METALS INTERCEPTED AND PRECIPITATED DURING THEIR UPWARD MIGRATION

Granitization and anatexis of slightly to strongly uraniferous sediments causes U mobilization and ultimate removal, unless a portion of the U on the move is trapped. The most popular and economically important example of this is the Rössing U deposit in Namibia (and Charlebois Lake, partly Bancroft, Canada; Crocker's Well, South Australia). At Rössing (Berning et al., 1976), uranium concentrated in residual melt of alaskitic composition during syntexis formed low-grade U-oxide disseminations now occurring in a wide variety of rocks including pegmatite, aplite, migmatite, gneiss, and several types of resisters. All the above deposits, often incorrectly designated as "porphyry uraniums", are Precambrian, thus treated in Volume 2. The known Phanerozoic examples, so far, have been insignificant. Recently, Rogers et al. (1978) reviewed this style of mineralization, and offered a prognosis of its possible occurrences in the eastern United States.

29.6.3. Granite-gneiss domes, the core and the mantling complex

GENERAL

This subject overlaps with the subject matter treated earlier in Sections 29.4. and 29.5. The rock types involved are the same, but the emphasis, at present, is on the mutual relation of the metamorphic mantle assemblage and the ultrametamorphic core. In this context, the fundamental metallogenic function of the mantle to be considered is its likely ability to provide a reservoir for the entrapment of metalliferous volatiles (melts, fluids, gasses), supposedly moving upward from the core. This emphasis is different from the emphasis placed on pre-metamorphic relic ores earlier. In many respects the present exercise involves processes and conditions parallel with those taking place during the interaction of medium to high-level "granitic" plutons with their roof, treated in Chapter 28.

The mantled gneiss (thermal) domes recognized by Eskola and studied later by Wegmann, de Sitter, Zwart and other investigators (compare

Suk, 1983, for review), are a characteristic fixture of Precambrian shield as well as Phanerozoic orogenic belt geology. In some geologically young domes, such as in the Shuswap Complex, British Columbia (Jurassic to Cretaceous metamorphism), the geologically younger mantle (late Proterozoic to Paleoozic) formerly resting unconformably on the early Proterozoic basement, is now largely converted into granite gneiss. The contrast in style and degree of metamorphism between the core and the mantle is substantial. In older, deeper eroded domes, such as the late Paleozoic Central Moldanubian Dome in central Czechoslovakia (Fig. 29-16), migmatites are widespread and dominant and the core/mantle relationship less distinct. In both the above-mentioned, as well as other domes, katazonal to mesozonal granites appear near the thermal centres.

Ever since Reid and Reynolds, the idea of metasomatic fronts (acid, basic, alkalic) initially triggered by granitization and gradually propagating upward, has from time to time been discussed or applied, so far with inconclusive general results (e.g. Hietanen, 1962; Velikoslavinsky et al., 1968; see also review in Suk, 1983). The "basic front" supposed immediately to top the ultrametamorphic or periplutonic dome, could be of interest in metallogeny. The alkalic fronts, interchangeably K or Na specialized, appear, in many respects, to be larger scale and substantially more heterogeneous counterparts of the system that produced the apogranite alteration assemblages (Chapter 28). Regional metasomatic fronts have occasionally been suspected in the generation of the regionally distributed ankerite, siderite and magnesite deposits in the Alps and Carpathians, and possibly as a cause of several major gold-mineralized provinces.

GEOLOGY

The characteristic anatomy of a multiphase mantled granite-gneiss dome is shown in Fig. 29-17. The bulk of the core consists of migmatite (A) and synorogenic katazonal granite (B). The granite has numerous subvarieties, such as anatectite formed by a complete in-situ remelting and homogenization of the original rock; anatectite transported for short distances, during which it could have acquired a different character due to assimilation of various host rocks. Alternatively, metasomatic granite formed by feldspathization and feldspar porphyroblastesis of paragneisses (C). The latter is, for example, represented by the "durbachitic" granodiorite, syenodiorite and syenite with large, tabular, feldspar porphyroblasts resting in a foliated dark biotite-rich groundmass, described from the Schwarzwald Range (Germany) and Central Bohemian Pluton border (Czechoslovakia; Svoboda, ed., 1966). Numerous but minor, non-persistent bodies of compositionally simple (quartz, feldspar, mica) pegmatites and aplites are abundant (D).

Emplacement of sub-autochthonous to allochthonous leucocratic plutons formed from palingenetic magmas (E) took place during the closing stages of synkinematic granitization and shortly afterwards.

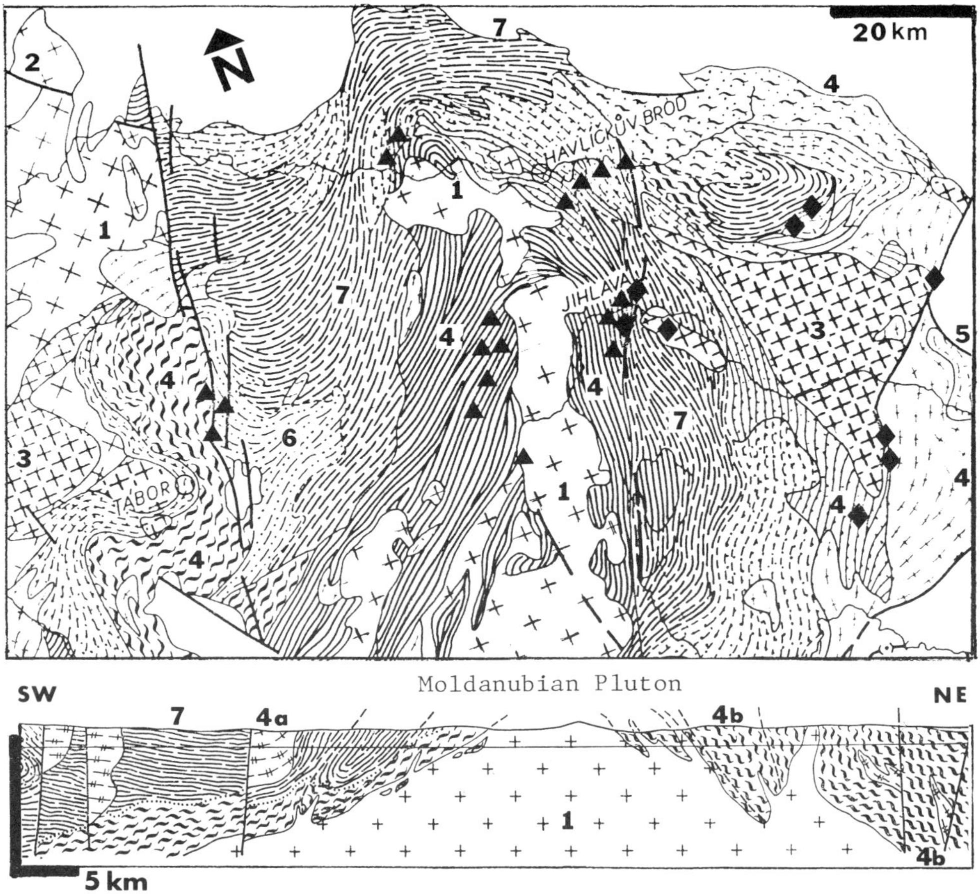

Fig. 29-15. The axial portion of the Moldanubian Complex, Bohemian Massif, intruded by a mesozonal Moldanubian Pluton. After Kodym and Suk from Svoboda, ed. (1966), courtesy of the Ústřední Ústav Geologický, Praha. 1= PZ_3 biotite granite and quartz monzonite; 2=granodiorite, diorite, gabbro; 3=melanocratic granite, syenite, syenodiorite. Pt metasediments metamorphosed in PZ_3. 4=orthogneiss (a) and migmatite (b); 5=granulite; 6=micaschist, two mica gneiss; 7=paragneiss. Triangles: Pb-Zn veins; diamonds: Li pegmatites.

These granites reached some way into the sillimanite or at most kyanite zones in the mantling complex. Andalusite and locally cordierite-anthophyllite assemblages occur frequently in the exocontact and sometimes, in lesser quantities, in the granite endocontact as well. The plutons exhibit both gradational and sharp contacts and many appear to narrow downward and to be rooted in the core granite gneiss. Lamprophyre (minette, kersantite) (F) and more evolved, more persistent aplite and pegmatite dikes and sills (G) are

LITHOLOGIC ASSOCIATIONS

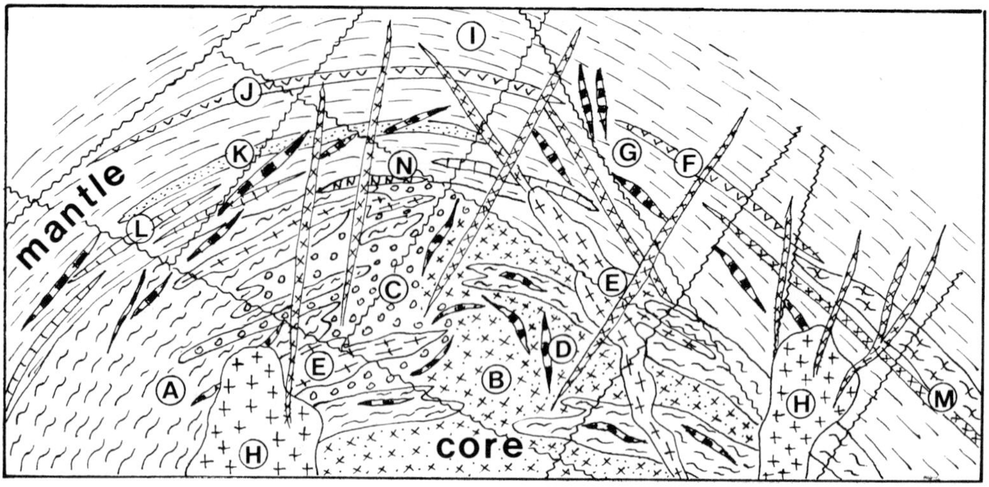

ORES

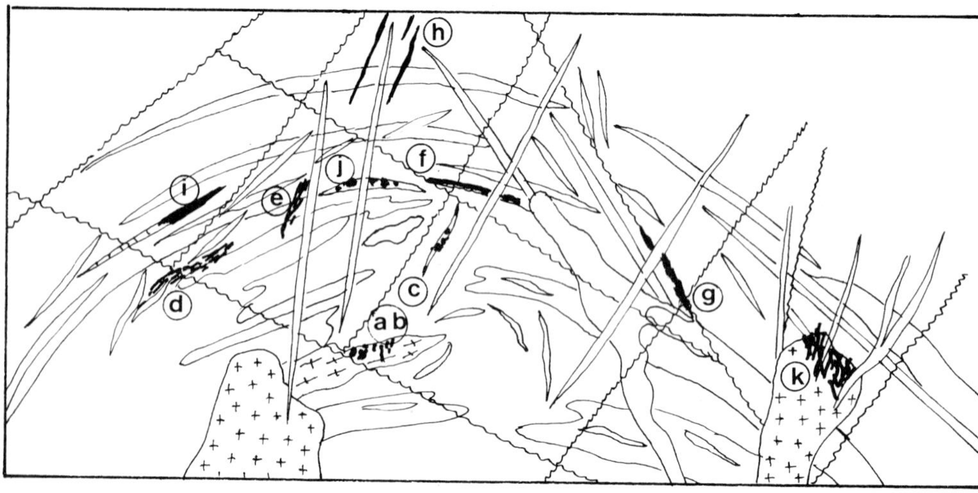

Fig. 29-17. Multiphase mantled granite gneiss domes and their mineralization styles.

TOP: rock types. A=migmatite; B=synorogenic katazonal granite; C=metasomatic (porphyroblastic) granite; D=simple (mica) pegmatite; E=late synorogenic, early postorogenic kata- to mesozonal plutons; F=lamprophyre dikes; G=evolved pegmatite dikes, sills; H=high-level postorogenic intrusive stocks; I=micaschist, biotite gneiss; J=amphibolite; K=metaquartzite; L=marble, calc-silicate gneiss; M=orthogneiss; N=nepheline syenite gneiss. (continues on right) contemporary with or shortly postdate the leucocratic plutons.

High-level granodiorite, monzonite or leucogranite stocks (H) may have formed during a period of tectonomagmatic activation following an uplift of the dome into the shallow crustal region, coupled with considerable erosion (e.g. in the Coast Range Complex of British Columbia or the Colorado Rocky Mountains).

The metasedimentary and metavolcanic mantle is dominated by biotite (sillimanite) paragneiss grading to micaschists at most locations (I). It commonly contains units of amphibolite (J), metaquartzite (K), marble or calc-silicate gneiss (L) and sometimes an "orthogneiss" (M) that could have formed from felsic metavolcanics, arkoses, granitic plutons, etc. Less common rocks include graphitic schists or gneisses, meta-ultramafics, eclogite, nepheline syenite gneiss (N), meta-anhydrite, etc.

The nepheline syenite gneiss in the Shuswap Complex, British Columbia (McMillan and Moore, 1974) is a controversial peneconcordant body, enveloped by a calc-silicate gneiss and marble. Its interpretation ranges from a synkinematic alkaline metasomatite to pre-metamorphic alkaline intrusion. The Ditrău alkaline massif in the crystalline core of the East Carpathians in Rumania (Ianovici and Borcoş, 1982) is similar in some respects, but the recent interpretation is that it is a postkinematic, superimposed intrusion fringed by an aureole of gneissic fenite.

MINERALIZATION STYLES

Late synorogenic to postorogenic plutons (E on Fig. 29-17)

These plutons can, under favourable conditions, reach up to the lower mesozone level so that they merge and overlap with the plutons discussed in Chapter 28. The "typical" deep-seated plutons are rarely mineralized. Near their roof, they contain numerous inhomogeneities in the form of semi-digested xenoliths or rafts (most commonly of biotite migmatite composition or amphibole-rich), or aplite-pegmatite secretions. Endocontact mineralization is rare. Scattered black allanite is relatively common in the pegmatite or aplite schlieren, or in deuterically altered granite (e.g. Red Rock, Nevada, in aplite pods in quartz diorite; Elberton, Georgia). Some granites emplaced into a

---

←

BOTTOM: mineralization styles. (a) Minerals scattered in endocontact inhomogenities (e.g. allanite, uraninite); (b) redistributed secondary U minerals on fractures of slightly uraniferous granites; (c) "simple" pegmatites, rare Th,U,Ti complex minerals; (d) remobilized massive sulphide orebodies in granite aureole; (e) sulphide and quartz sulphide veins, probably remobilized from pre-metamorphic ore accumulations; (f) bedding lodes; (g) shear lodes; (h) high level fissure veins; (i) carbonate replacements; (j) molybdenite in nepheline syenite gneiss; (k) apical stockworks. From Laznicka (1984).

uranium-rich environment contain minor accessory uraninite. So far, this does not constitute an ore in itself, but it provides an easily leacheable source of uranium for a later hydrothermal or groundwater deposition.

The Precambrian katazonal granites in the Darwin region of Australia (Vol.2) provided at least some of the U to the significant "unconformity uranium" deposits there. Some Archean granites in the Wyoming crystalline basement supplied U to the Cainozoic "sandstone uraniums" (Chapter 25), and the older phases of the French Hercynian granites lightly enriched in uraninite supplied U to the later generation of "episyenites" and veins (Chapter 28). In the oxidation zone, the sub-economic and usually haphazard uraninite occurrences produce often conspicuous autunite, torbernite, uranophane, etc. fracture infiltrations. Many such occurrences were investigated in the Massif Central of France and the Central Moldanubian Massif of Czechoslovakia.

Zn-Pb or Cu sulphides in cordierite or cordierite-anthophyllite gneisses

This is a quite distinct ore style most common in the Precambrian shields (e.g. Manitouwadge, Sherridon, etc. in Canada), and known also in the Phanerozoic orogens. The small but interesting Gull Pond (Gullbridge) deposit in N.W. Newfoundland was recently studied by Bachinski (1978). There, pyrrhotite is intergrown with pyrite and chalcopyrite in a lens-like massive to disseminated sulphide orebody, situated along a shear zone in cordierite-anthophyllite and andalusite schist probably dating from the Ordovician period. These schists grade into metabasalt, siliceous iron formation, metarhyolite (porphyroid) and quartz-sericite phyllite. The mineralization is interpreted as being an Ordovician synvolcanic subaqueous-hydrothermal stratabound orebody, remobilized in the thermal aureole of Devonian kata- to mesozonal granite.

In the Jihlava and Havlíčkův Brod districts located in the aureole of the late Paleozoic Central Moldanubian Pluton (Svoboda, ed., 1966), Pb,Zn,Ag association is most common. At the Pohled and Bartoušov deposits (Němec, 1965), massive pyrrhotite, marmatite, galena and pyrite grading to quartz-sulphide veins, are to be found along shears and fractures in migmatite, cordierite gneiss and lamprophyre near "granite" contact. The ores are clearly epigenetic, post-granite and post-lamprophyre, and remobilization of an earlier orebody is a possibility.

In the Hitachi massive sulphide deposit, Japan, cordierite, anthophyllite and andalusite form wallrock to the Permo-Carboniferous pre-metamorphic massive sulphides, where these were intruded by a Mesozoic "granite".

Scheelite skarns

Scheelite skarns have been reported from several mantling complexes where they occur at or near contact with the core. In the Fairbanks district, Alaska (e.g. Stepovich mine; Byers, 1957) thin, boudinaged marble horizons interbedded with a Proterozoic micaschist, quartzite and amphibolite envelope a gentle thermal dome intruded by Mesozoic synorogenic porphyritic biotite granite, and Tertiary post-orogenic granodiorite. Scheelite occurs in skarn-replaced marble and the richest ore shoots are located at intersections of skarn with pegmatite. In the Pedro Dome area, scheelite exists in quartz veins along shears.

"Katathermal" gold-quartz lodes

In the Fairbanks area, Alaska (Byers, 1957) numerous but impersistent bedding, shear and fracture gold-quartz lodes are located in the metamorphics of the mantling complex, close to granite contact. In the Liberty Bell mine, a nearly horizontal lode, 2-10 m thick, contains arsenopyrite, lesser chalcopyrite, pyrite, löllingite, bismuthinite and gold with or without quartz. It consists of small ore lenses and stringers parallel with foliation. Most of the gold in the Fairbanks district (220 t Au) came from placers.

In the gold region of southern Piedmont, U.S.A. (Pardee and Park, 1948), several gold deposits are situated along or near the contact of granite gneiss and a mantling gneiss and schist complex. In the Barlow mine S.E. of Dahlonega, the country rock is hornblende gneiss intruded by irregular masses of granitoid gneiss. Gold occurs in many quartz stringers and lenses most of which are conformable with foliation, but some are crosscutting. In the quartz, gold occurs as free metal as well as in the minerals pyrite and arsenopyrite, but its distribution is, to a considerable degree, masked by deep weathering. At the Barlow mine, small quantities of gold are widespread in the metamorphics.

Molybdenite in nepheline syenite gneiss, Mt. Copeland, British Columbia

At Mt. Copeland near Revelstoke (McMillan and Moore, 1974; 2,075 t Mo/1.1%; Fig. 29-18), nepheline syenite gneiss occurs as a thick lens along a stratigraphic horizon of a calc-silicate gneiss. The latter is member of the mantling complex, close to the migmatite and granite gneiss core of the Frenchman's Cap mantled gneiss dome, Shuswap Complex. The emplacement age of the nepheline syenite and associated minor carbonatites is not known, but the rocks suffered multiple deformation and partial anatectic melting. The melting produced aplitic, syenitic and pegmatitic schlieren and dikes. Scattered molybdenite in small quantities occurs as an accesory mineral in the main body of the nepheline syenite gneiss, but the commercial orebodies were located along its northern contact with the calc-silicate gneiss, hosted by aplite-pegmatite bodies. Most such bodies lie parallel with the foliation and are up to 3 m thick. They are composed of K-feldspar with exsolved albite. Molybdenite with some pyrite, pyrrhotite and minor chalcopyrite fills thin fractures. There is a light kaolinite, sericite, calcite alteration and the mineralization has been dated 44 m.y., contemporary with the emplacement of lamprophyre dikes. Molybdenite also occurs in veins in

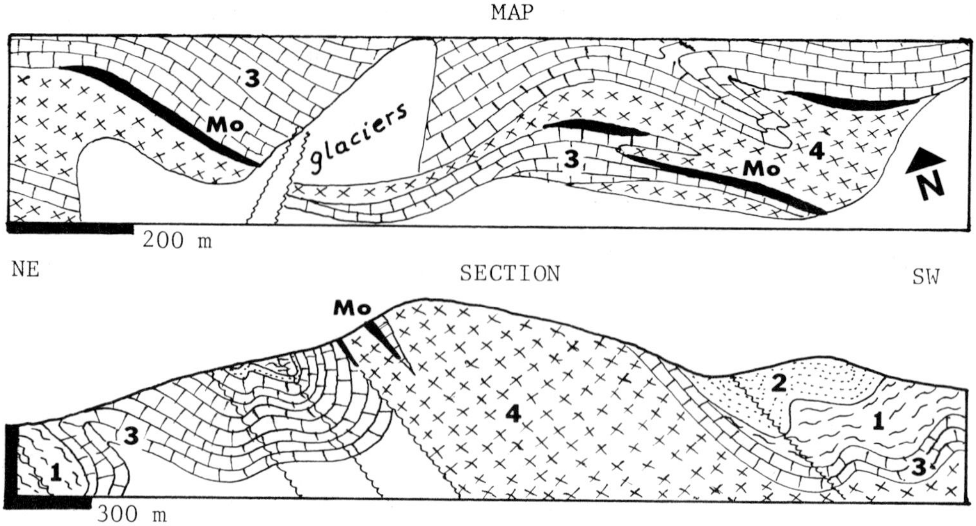

Fig. 29-18. Mount Copeland molybdenite deposit in nepheline syenite and calc-silicate gneiss, British Columbia. Pt_3 or PZ Shuswap Complex (J metamorphism). 1=biotite gneiss and schist; 2=quartzite; 3=calc-silicate gneiss and marble; 4=nepheline syenite gneiss. From LITHOTHEQUE, after Fyles (1970).

alkaline diorites along the northern margin of the Ditrău alkaline complex (Rumania; Ianovici and Borcoş, 1982). Monazite, orthite, yttrocalcite, calcite, siderite, etc. veins are hosted by albite metasomatites.

Monazite, allanite, etc. in marble, Mineral Hill district

In the Mineral Hill district, N.E. Idaho and W. Montana (Anderson, 1958), monazite, allanite, ilmenorutile and apatite occur in thin marble lenses within a belt of occurrences 40 km long and 4 km wide, in the schist, amphibolite, pegmatite, etc. association mantling the Idaho Batholith. This could be a true metasedimentary marble possibly mineralized within the reach of a "metasomatic front", or alternatively a series of meta-carbonatite sills.

Postmagmatic hydrothermal fissure veins, stockworks, etc. in granite-gneiss domes and their roofs

Fissure veins bearing a variety of metals (Pb,Zn,Ag, Au,Cu, etc.) located in granite-gneiss domes, but emplaced later at relatively shallow levels following the dome uplift, are common mostly because of the favourable conjugate or grid-like fracture patterns. The veins or stockworks are genetically related to high-level intrusive stocks either visibly affiliated (e.g. Climax-Mo, Colorado), or postulated to exist in depth (e.g. Freiberg Pb,Zn,Ag, East Germany).

Examples of mineralized Phanerozoic gneissic domes are listed in Table 29-7.

29.7. GRANITIC PEGMATITES

29.7.1. General

The occurrence of the bulk of pegmatites (excluding the high-level miarolitic pegmatites and those associated with tin granite cupolas, Chapter 28) are restricted to zones of medium to high-grade metamorphism. In enorogenic metamorphic terrains pegmatites usually correlate with the kyanite to sillimanite assemblage (almandine amphibolite facies) of their host rocks. In periplutonic terrains, pegmatites are most common in the metamorphics of the andalusite, cordierite, muscovite zone. In the classification of pegmatites drawn up by Ginsburg et al. (1979) according to their depth of formation and occurrence, the former (enorogenic) terrains host the "pegmatites of great depths", the latter (periplutonic) terrains host the "pegmatites of intermediate depths". Both seem to have a slightly different derivation and different setting, although both pegmatite varieties appear to be members of the "S granite" suite. It is quite remarkable that the large masses of "I granites" and associated metamorphics generated along the former Pacific- and Andean-type continental margins, are virtually devoid of pegmatites.

Pegmatites have an extensive literature and a large following of investigators. Numerous and comprehensive reviews are available in several languages (Fersman, 1960; Jahns, 1955; Schneiderhöhn, 1961; Ginsburg et al., 1979; Černý, ed., 1982). This is because of the mineralogical uniqueness of these rocks, rather than their economic importance. Pegmatites are not impressive as a source of metals, having accounted for less than 0.05% of the metals production value in the 1980s, in the entire world (they are more important as a source of nonmetallic commodities). Even so, the majority of metalliferous pegmatites are Precambrian and so treated in Volume 2. The Phanerozoic pegmatites are estimated to represent some 20 Tt Be, 20 Tt Sn, 5 Tt Ta and 5 Tt Nb. The figures quoted above include minerals accumulated in placers demonstrably and proximally derived from pegmatites. Some components of placers of mixed provenance or "distal placers" (such as monazite, zircon), may have come partly from pegmatites.

29.7.2. Pegmatites as metallic ores

When a large volume of literature is reviewed, it becomes obvious that virtually all metals have been found, at least once, in some sort of affiliation with pegmatites, and variable but mostly small quantities of Be,Sn,Ta, Nb,U,Th, Au,Mo,W, Bi,Cu,Sb, Pb,Zn,Ag, As,Ni,Co have actually been produced from this setting. The above list of

Table 29-7. Selected examples of mineralized gneiss domes in Phanerozoic orogenic belts

GNEISS DOME	AGE OF METAM.	GEOLOGY	MINERALIZATION	REFERENCES
Central Gneiss Complex, Coast Plutonic Belt, N.W. Canada and S.E. Alaska	MZ	C: migmatite to gran. gneiss; M: PZ to J gneiss, schist, amphibolite; I: J-Cr diorite to quartz monzonite	Ecstall, mass.py, po, in marble; Surf Inlet, Au, qtz.-pyr. vein in diorite; Britannia, massive Cu-Zn sulph., relict in mantle	Douglas, ed. (1970)
Fairbanks area, Alaska, several domes	MZ	C: PCm? migmatite, granitic gneiss; M: PCm? micaschist, gneiss, lesser amphibolite, marble; I: MZ porph.granite, T_1 post-orog.granodiorite	Au, qtz. veins in mantle along foliation, shears, fissures; scheelite skarn in mantle; Au placers	Byers (1957)
Shuswap Complex, Brit.Columbia, several domes	J	C: 1.7 b.y. anatectic granite, migmatite; M: Pt_3-PZ gneiss, amphibolite, quartz., calc-sil. gneiss, marble, neph.syen.gn. I: MZ-T_1 granite, qtz.dior., granodiorite	massive Zn-Pb sulphides conform. with foliation; dissem.molybd. in neph.syen. assoc.; Pb-Zn veins, replac.	Fyles (1970) Douglas, ed. (1970)
Hohe Tauern Mts., Austria	MZ	C: PZ_1 gran.gneiss, migmatite; M: PZ_1 gneiss, schist, marble, amphibolite, prasinite, quartzite; I: not known	scheelite dispersed in core migmatite; Felbertal scheelite in mantling metamorphics	Höll and Maucher (1976)
Menderes Massif, W. Anatolia, Turkey	PZ-J	C: PCm augen gneiss; M: Or-S qtz.-musc.schist, amphibolite, calc-silic.gneiss, graphitic schist, metaquartzite; I: MZ ophiolites, T contin.volc.	Pb-Zn Bayindir, mass.sulph. peneconc. with schistosity; Hg, in retrogress.fault zone, Haliköy	Dora (1977)

metals can be subdivided into those that are systematically associated with the deep and moderately deep-seated pegmatites treated in this chapter (Be,Ta,Nb, partly Sn, REE, Th, U) and those due to one-of-a-kind, chance associations usually involving higher-level pegmatites interacting with unique environments (the rest). The latter will not be treated here. The systematically associated metals have the following distribution in pegmatite bodies and their contacts (in nature, usually two or more distribution styles combine):

(1) Lump (blocky) ore. Discrete, large to small crystals (e.g. of beryl) are irregularly scattered in pegmatite, often confined to a particular compositional and textural zone. It is virtually impossible to estimate the average grade and tonnage of the metallic mineral in such bodies. A single orebody (pegmatite lens) having a tonnage of several tens to hundred thousands tons of pegmatite, may contain several tons or tens of tons of beryl (beryl itself contains only 3.85% Be). Few such deposits have ever been mined for the rare metal content itself, but a minor (X0-X00 t of beryl) cumulative production has often been achieved over a period of several years or decades by handpicking and stockpiling the beryl as a by-product of mining of a bulkier nonmetallic commodity, such as feldspar, mica or quartz. Few other minerals occur in lumps amenable for handpicking. Those that occasionally do are usually rapidly carted away by the collectors.

(1a) In humid tropics, handpicking of beryl and other ore lumps from deeply tropically weathered pegmatites has occasionally been carried out, sometimes with the aid of hydraulicing. Most such occurrences (in Africa, India) date from the Precambrian age.

(2) Scattered (disseminated) ore. The ore minerals in small (0.1-10 mm) but physically separable and recoverable crystals are usually haphazardly scattered throughout the pegmatite mass, usually confined to a specific compositional and textural zone. In the past, production was achieved by complete or selective mining and milling of the pegmatite and the metallic component (beryl, cassiterite, columbite-tantalite, etc.) were the only products obtained. At present, complex recovery (several products including feldspar, quartz, spodumene, muscovite, lepidolite) is more common.

(2a) Deeply weathered pegmatites, and

(2b) proximal colluvial or alluvial placers, are economically more profitable varieties of (2).

(3) Metals present as invisible and physically unrecoverable trace elements in the bulk of a pegmatite, or more commonly in particular mineral carriers. The range of trace metal contents is considerable even within single pegmatite bodies. Kuzmenko, ed. (1976) gave the following ranges for the Soviet pegmatites: BeO, 9-4,700 ppm; Ta_2O_5, 2-5,000 ppm; Nb_2O_5, 10-1,400 ppm; SnO_2, 13-35,500 ppm. The usual background values of most medium-level Li pegmatites are in the region of 10-20 ppm Ta_2O_5; 20-60 ppm Nb_2O_5.

(4) Metals accumulated in the wallrocks and in the sahlbands of pegmatites. Although interesting Li (in holmquistite) and Cs (in caesian biotite) contents are known in several pegmatite exocontacts,

no significant accumulations of Ta,Nb,Sn and other rare metals have been reported.

Pegmatites (including the Precambrian ones), contain about 20% of Be (around 40-60 Tt Be); 7.7% of Ta (about 20 Tt Ta); about 3-4% of Sn and less than 0.1% of the Nb present in economic ores, worldwide. 75% of the Ta share and about 66% of the Be share is said to be present in Precambrian pegmatites. Table 29-8 gives the usual grade and expected tonnage of Be,Ta and Nb in the five most common systematically mineralized compositional varieties of pegmatites. Even though the magnitude ranges appear excessive (up to 100 Tt Ta_2O_5 is in the "subalcalic" lithian pegmatite. The largest Ta accumulation of the world for which published data are available, the Abu Dhabbab in Egypt, is credited with 131.5 Tt Ta, and it is probably a late Proterozoic apogranite), the relative importance of the pegmatite varieties is well recognized in the table.

Pegmatites in the higher crustal levels are often transitional into quartz, feldspar, cassiterite, wolframite, scheelite, molybdenite, beryl, etc. veins and into mineralized greisens. These are treated in Chapter 28.

29.7.3. Internal features of pegmatite bodies

Most small pegmatite bodies are simple, rather homogeneous aggregates of feldspar, quartz, mica and other minerals, lacking any systematic internal variation. Larger pegmatites usually show some internal variations and these have been subdivided by members of the wartime U.S. Geological Survey pegmatite evaluation project (e.g. Cameron et al., 1949; Jahns et al., 1952) into three fundamental categories: (1) zones; (2) replacement bodies an (3) fracture fillings. Although minor changes have been suggested in the past 30 years (compare Černý, ed., 1982), the U.S.G.S. subdivision has proven to be a remarkably suitable one for fieldwork.

(1) Pegmatite zones represent textural and compositional variations within a pegmatite body, established at the time of the primary (main stage) crystallization. They form concentric to irregular, complete to incomplete successive shells between the wallrock and the core. Four zones can be recognized in many pegmatites, and the examples of zones offered here are based on those recognised by Jahns et al.(1952) in the S.E. Piedmont pegmatite belt (Fig. 29-19):

(a) Border zone (the outermost one) is a fine grained selvedge, 5-7 cm thick, that separates the pegmatite from its wallrock. In the Piedmont pegmatites, the zone consists of quartz-rich, sugary or granitoid rock with sharp or gradational contacts against the host rock.

(b) Wall zone, the successively inner zone, is coarser grained and thicker (up to 11 m). The most widespread constituent is sodic plagioclase with lesser amounts of quartz, muscovite, occasional biotite, perthite, garnet, apatite and beryl. Wall zones most commonly form complete envelopes along the inner units.

Table 29-8. Usual grade and tonnage range in five compositional varieties of Be,Ta and Nb-mineralized pegmatites (all ages). After Rundkvist, ed. (1978)

PEGMATITE VARIETY	USUAL RANGE OF GRADE	USUAL RANGE OF TONNAGE OF ORE
"Standard pegmatite" (mica, ceramic); microcline, beryl, columbite	BeO 50-100 ppm Ta_2O_5 30-70 ppm Nb_2O_5 70-100 ppm	1-100 t 1-10 t 1-10 t
microcline-albite with tantalite, wodginite, beryl	BeO 400-500 ppm Ta_2O_5 130-250 ppm Nb_2O_5 100-150 ppm	100-1,000 t 100-10 Tt 100-10 Tt
subalcaline lithian pegmatite; spodumene, microcline, albite, lepidolite, pollucite, tantalite, beryl	BeO 400-600 ppm Ta_2O_5 150-300 ppm Nb_2O_5 70-150 ppm	100-10 Tt 1 Tt-100 Tt 1 Tt-100 Tt
albite, tantalite, beryl, ± spodumene	BeO 800-1,500 ppm Ta_2O_5 150-250 ppm Nb_2O_5 100-150 ppm	1-50 Tt 100-1 Tt 100-1 Tt
albite-spodumene ± beryl, columbite, fergusonite	BeO 300-500 ppm Ta_2O_5 40-100 ppm Nb_2O_5 60-120 ppm	2-10 Tt 1-10 Tt 1-10 Tt

(c) Intermediate zones are the least regular and discontinuous of all the pegmatite zones. More than one such zone may be present, each possessing a slightly different mineral assemblage. The composition is highly variable. Coarse to almost massive perthite is most common, followed by quartz, plagioclase, coarse muscovite and spodumene in lithian pegmatites.

(d) Core zones tend to be irregular, ranging from thin ribs to cigar-shaped and ellipsoidal masses. The most common mineral is massive quartz, or quartz with scattered perthite or plagioclase crystals. An alternative filling is a coarse, blocky feldspar.

(2) Replacement bodies postdate the primary crystallization and can be superimposed over any (rarely all) of the pegmatite zones, although the internal zones (core, intermediate) are most often replaced. The replacement bodies are fracture-controlled stockworks and veinlets, dikelets and schlieren, irregular masses, sheets along pegmatite and wallrock contacts or along zone boundaries, etc. The replacement can be complete or partial, and a sugary or bladed ("cleavelandite") albite is by far the most characteristic mineral.

In geochemically complex pegmatites (lithian, rare earths, Ta-Nb, etc.) the bulk of the rare element minerals have been introduced with the metasomatic albite and sometimes K-feldspar (often green microcline, amazonite). There is a considerable degree of

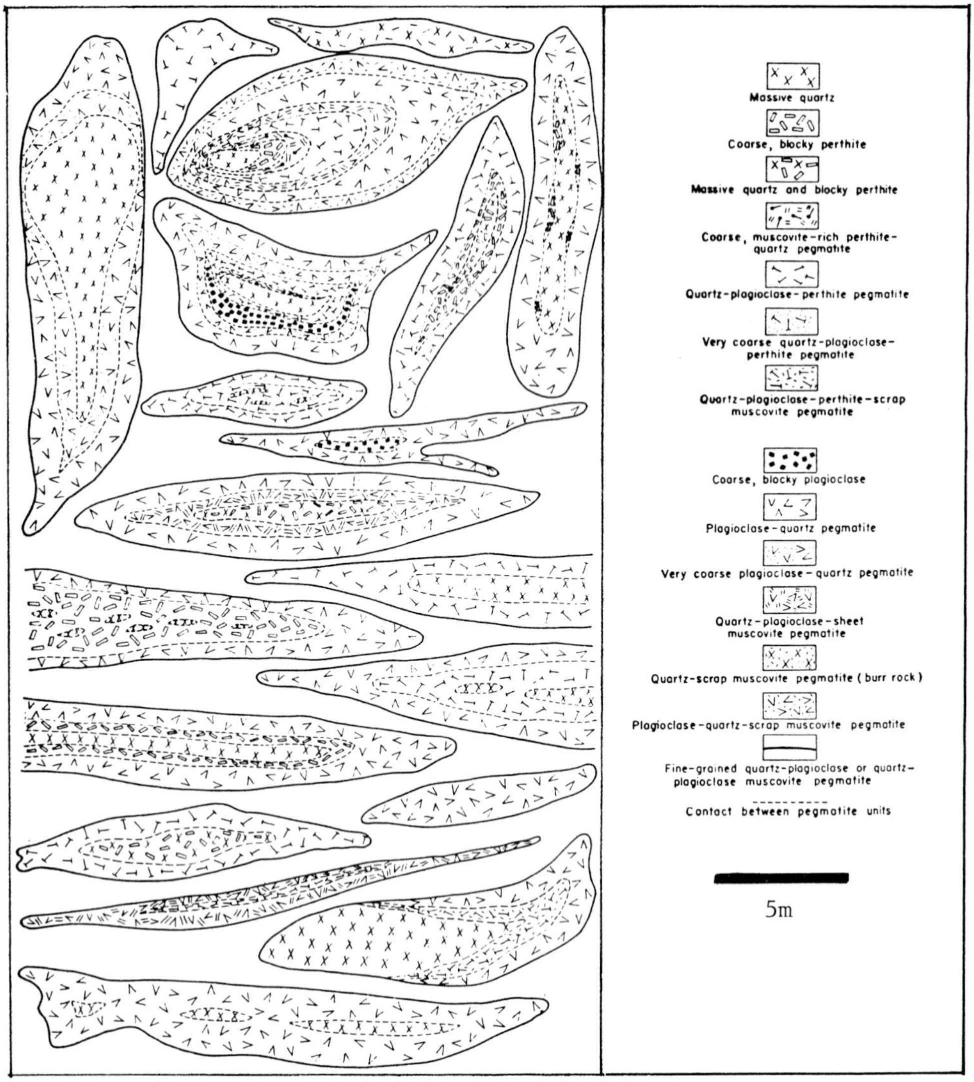

Fig. 29-19. Idealized maps, showing pegmatite zoning in the S.E. Piedmont mica pegmatite region, U.S.A. From Jahns et al. (1952).

compositional, although not necessarily textural, similarity between the Na and K metasomatites and apogranites, described in Chapter 28. Apogranites are thus most probably epizonal equivalents of the mesozonal to katazonal replacement pegmatite bodies.

(3) Fracture fillings may postdate the "primary" or the replacement pegmatite, or they may be equivalent in time to the replacement pegmatite. They have the form of veins, veinlets, stringers, and may

be composed of a great variety of hydrothermal minerals. Quartz, carbonates and sulphides are particularly characteristic.

"Pocket pegmatite" (Fig. 29-20) is an additional, minor feature of some pegmatites. It has the form of porous or clay-filled patches, marked by an euhedral form of constituent minerals or by terminated crystals extending into a cavity. Fersman described "walk-in" pockets located in the Murzinka pegmatite field, southern Urals. Quartz (often smoky quartz or amethyst), K-feldspar, albite, muscovite are most common in pocket pegmatites, augmented by a variety of rare minerals. Their origin is attributed to a late stage hydrothermal leaching and precipitation.

29.7.4. Pegmatites of "great depths" in migmatite and sillimanite- or kyanite-grade metamorphic terrains

Pegmatites of "great depths" are closely associated with granite-gneiss and migmatite domes and ridges, and with biotite-sillimanite schists in the immediate vicinity. In terms of their body shape and setting, two end-members can be distinguished (Fig. 29-21): (a) tabular, lens-like, trough-shaped, bulbous, pinching and swelling bodies that are peneconcordant with foliation of the metamorphics, or schlieren subparallel with foliation in granite gneiss and (b) dikes cutting across the structural grain.

The (a) variety has a very proximal ultrametamorphic origin, corresponding to local accumulations of the leucosome in migmatites. The pegmatites bear no clear relationship to granite masses when they are present in the area, and occur in the form of broadly subparallel bodies together making up pegmatite fields and belts. Their thickness ranges from centimetres to several metres and the length of individual pegmatites is rarely over 20 m.

The pegmatite dikes corresponding to type (b) are more persistent (often for several hundred metres). They are tabular crosscutting bodies usually with sharp boundaries. They may be younger than the (a) bodies which they often intersect, and are most common within and in the proximity of katazonal granites.

Pegmatites in enorogenic terrains tend to be simple in composition. They usually comprise feldspar, quartz, muscovite or biotite, occasionally tourmaline, garnet, hornblende, although they could still be zoned. Most have been mined for mica, some for blocky feldspar and quartz. Accessory metallic minerals are rare and dominated by beryl. Monazite, although rare in a megascopically visible form, is probably present and it accumulates in proximal placers where the pegmatites have been tropically weathered.

The pegmatite belt of southern Blue Ridge and Piedmont (Appalachian Mts.), described in considerable detail in the U.S. Geological Survey Professional Paper 248 (Jahns et al., 1952, and following papers; Fig. 29-22), is a classical region of Phanerozoic "mica pegmatites". Over 1,600 mica deposits have been mined within a N.E.-trending belt, some 1,000 km long. The pegmatites in the western portion of the belt

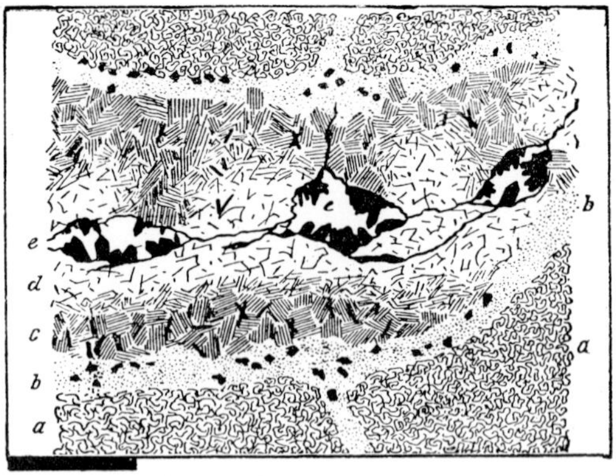

a granite
b aplite zone
c dense graphic pegmatite with scattered mica
d blocky quartz-feldspar pegmatite
e crystal-lined cavities

Fig. 29-20. "Pocket pegmatite" from Murzinka, the Urals. a=Granite; b=aplite zone; c=dense graphic pegmatite with mica flakes; d=blocky quartz, K-feldspar; e=crystal lined cavities. From Fersman (1960).

are hosted by late Proterozoic to lower Paleozoic mica, sillimanite and hornblende schists, gneisses and migmatites with numerous bodies of late Paleozoic synorogenic anatectic granites. Small quantities of accessory beryl are present throughout the area, but most pegmatites contain less than 0.3% beryl (that is, under 0.01% Be). Monazite is a widespread accessory in some pegmatites in the north (South Carolina border region), where it was reworked into placers mined in the past and still containing significant reserves (e.g. Catawba River area; Chapter 24).

Within the Piedmont pegmatite belt, the metamorphic grade decreases generally eastwards and patches of periplutonic (Buchan, Abukuma) metamorphics surround the late Paleozoic still deep-seated, but at least partly allochthonous granites in places along the eastern boundary of the belt. There, "medium depth" pegmatites with a granite affiliation (e.g. in the Carolinas "Tin-Spodumene Belt"; near Coosa, Alabama) and having a very limited area of distribution, account for the entire production and reserves of Li minerals, cassiterite (about 500 t Sn) and Ta-Nb (about 50 t). These are treated in the next section.

The "great depth" pegmatites are the most widespread (probably over 80% of pegmatites) but least interesting pegmatite variety to a metalliferous geologist. They are present in all the ultrametamorphic terrains worldwide, and no additional localities need to be reviewed.

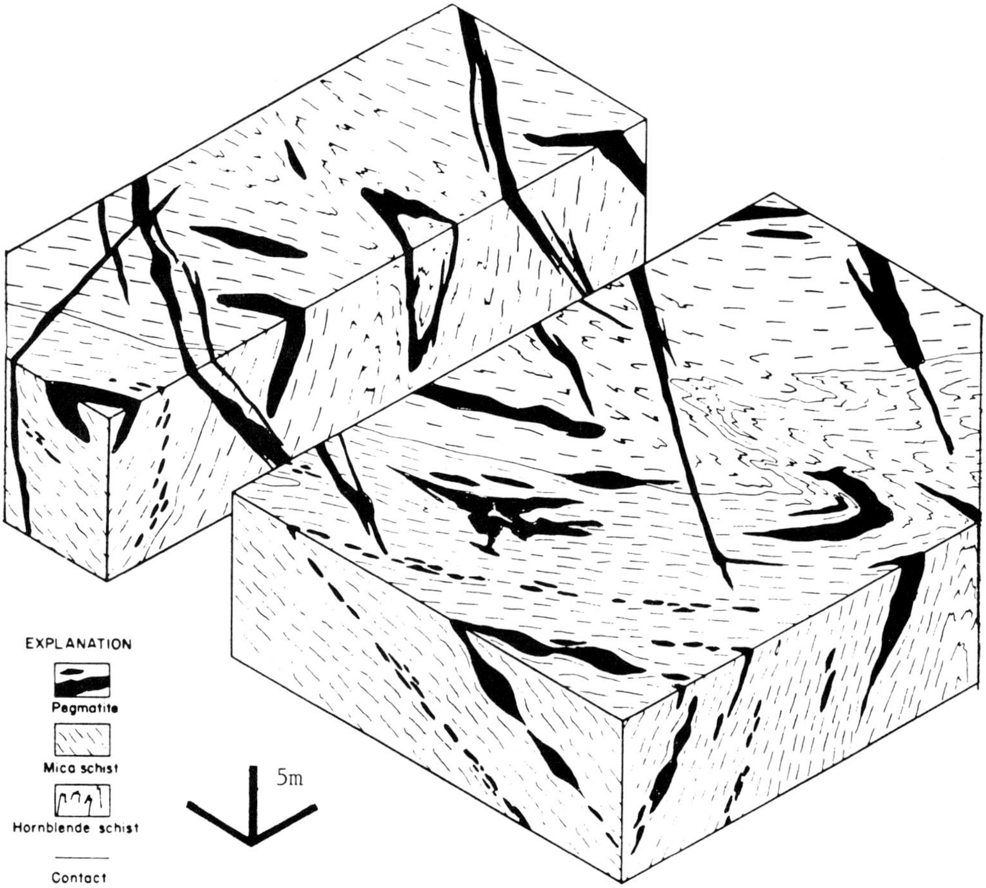

Fig. 29-21. Forms of mica pegmatite bodies and their relations to wallrock metamorphics, S.E. Piedmont, U.S.A. From Jahns et al. (1952).

29.7.5. Pegmatites of "intermediate depths" in periplutonic andalusite, cordierite, muscovite or staurolite schist metamorphic terrains

This category of pegmatites is substantially more evolved than the earlier one. Much of the pegmatite bodies are identical in shape to those already described, but extremely persistent dikes (up to 4 km long in Nuristan, Afghanistan) and thick lenses (spodumene pegmatite up to 100 m thick at Long Creek, North Carolina) are a specialty. Probably the most characteristic feature is the well-documented association with late orogenic to early post-orogenic, katazonal to lower mesozonal granites. Most of the Li and rare metal-bearing pegmatites occur in exocontact metamorphics typically 1 to 3 km from

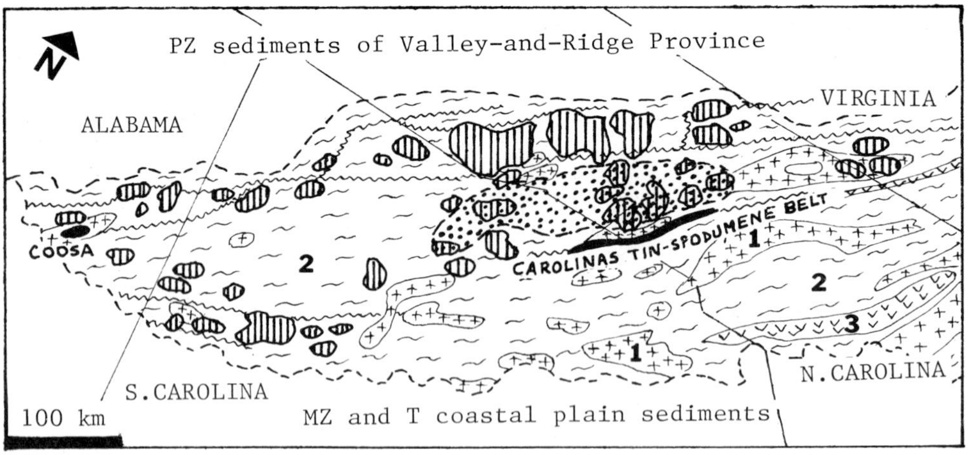

Fig. 29-22. Southern Appalachian pegmatite province, showing the fields of mica pegmatite (vertically ruled); spodumene cassiterite pegmatite (black) and detrital monazite placers (dots). $1=PZ_3$ granite; $2=Pt_3$-PZ high grade gneiss, schist and greenstone; 3=Tr basalt and sediments in grabens. From LITHOTHEQUE.

the contact, but endocontact pegmatites are also known. Persistent pegmatite bodies initiated in the granite and continuous into its roof have been described from the Nilau-Kulam and Alingar pegmatite fields, Nuristan, Afghanistan (Rossovskiy and Chmyrev, 1976); from the Aksu-Pushtiru field in the Soviet Central Asia (Beus, 1956) and from elsewhere.

The parent granite is most often a leucocratic biotite-muscovite coarse-grained pegmatitic "S" variety. It is usually a sequentially young (or the youngest) phase of the local orogenic magmatism (third phase in Nuristan). Partly digested sillimanite gneiss inclusions or rafts, or accessory andalusite, are common. The pegmatite bodies are zoned to unzoned, and it is interesting that many of the largest Li and Sn, Ta-Nb, Be orebodies such as those in the Carolina Tin-Spodumene Belt and some in Nuristan, are unzoned. The mineralogical composition is highly variable, but the presence of metasomatic albite and usually also of Li-minerals (spodumene, lepidolite, amblygonite, petalite) is an essential prerequisite of the rare metal (Be,Sn,Ta-Nb) mineralization. In some pegmatite fields, rare metal and mineralogical zoning is apparent (from the granite outward: barren zone → beryl → Be,Ta,Nb → Li,Be,Ta-Nb → Li,Cs,Be, Ta-Nb, Sn,U).

Rundkvist, ed.(1978) pointed out that 89% of the rare metals producing pegmatites are of Precambrian age and of the remainder 10% are Paleozoic. The pegmatite province of Eastern Afghanistan of lower Tertiary age is exceptional and exposed only as a result of an unusually rapid uplift and erosion.

In the Nuristan pegmatite belt, E.Afghanistan (Rossovskiy and Chmyrev, 1976), late Triassic graphitic micaschists and quartz, biotite, garnet, staurolite schists with minor amphibolites and marbles, mantle upper Paleozoic sillimanite-garnet biotite gneiss and migmatite basement. Both sequences are intruded by multiphase katazonal to lower mesozonal batholiths dating from the Cretaceous to lower Tertiary age that include early granodiorite, norite and quartz diorite followed by granodiorite, biotite-hornblende quartz monzonite and a two-mica granite. The youngest biotite-muscovite granites form large elongated massifs on the fringe or in places within the composite batholith. The granites are foliated, elongated layer- or lens-like bodies peneconcordant with the surrounding metamorphics or arranged along major N.E. faults. They are nearly equivalent compositionally to the Proterozoic ultrametamorphics from which they formed by anatexis, and they range from fine and medium grained to pegmatoid varieties. Garnet and sillimanite are typical minor minerals and there are accessory zircon, monazite, allanite, cassiterite and beryl.

The rare-metal pegmatites are peneconcordant sheets, swelling veins, dikes and lenses, ranging in thickness from 1 to 60 m and in length from several tens of metres to 5 km. The bulk of them are hosted by schists exposed in "graben synclinoriums", and most pegmatites congregate in swarms (pegmatite fields) that range in area from 10 to 1,500 km^2. Internally, the pegmatites are massive, banded or indistinctly zoned. The following mineralogical assemblages have been recognized in the Afghan pegmatites: (1) oligoclase-microcline and biotite-muscovite, with both tourmaline and rare beryl; (2) albitized microcline, black tourmaline, muscovite, beryl; (3) albitized microcline (nests of blue cleavelandite), lepidolite, spodumene, Li-tourmaline; (4) albite with nests of lepidolite, spodumene, pollucite and tantalite; (5) spodumene, microcline, albite and apodumene, albite and (6) lepidolite, spodumene, albite, Li-tourmaline, pollucite, tantalite. The pegmatites with an assemblage of type (2) are the most important source of coarse (hand-cobbed) beryl. Pegmatites with assemblages of type (4) or (6) produce tantalum (e.g. the Nilau deposit).

The Tin-Spodumene Belt of the Carolinas (Kesler, 1942) is a 40x3 km N.E.-trending zone of many discontinuous pegmatite occurrences hosted by Proterozoic and lower Paleozoic gneiss, schist, minor marble, quartz schist, intruded by kata- to mesozonal granitic plutons. Cassiterite was detected in several hundred small pegmatite occurrences that form tabular, lenticular to irregular bodies having a maximum length of 90 m and a thickness of around 30 cm. Cassiterite has the form of sparsely scattered, small dark brown crystals hosted by fine albite-quartz unzoned pegmatite. The Ross mine near Gaffney accounted for over 50% of the recorded tin and contained material with a highest grade of 2% Sn. At about 92 localities, cassiterite also occurs in a quartz-muscovite transitional rock between pegmatite and feldspar. The sum of the known and estimated Sn potential of the belt (including placers) hardly exceeds 500 t Sn.

Economically the most important pegmatites in the belt are near Kings Mountain, where eight bodies of spodumene, quartz, muscovite, albite and microcline pegmatite are mined primarily for lithium (30 Mt material with 1.47% Li_2O). Beryl, columbite-tantalite and other rare minerals occur in small quantities and are not recovered.

Additional regions of Phanerozoic rare metal pegmatites are summarized in Table 29-9. Table 29-10 contains a brief list of metallic commodities present in pegmatites.

29.7.6. Hybrid pegmatites

Hybrid pegmatites are contaminated as a consequence of exchange of elements with their wallrocks. Exchange between pegmatite and biotite, cordierite, kyanite or staurolite schists generates large crystals or masses of these minerals in the pegmatite (e.g. giant cordierite or sekaninaite in the Dolní Bory pegmatite, Czechoslovakia). Exchange with mafic and ultramafic rocks (amphibolite, gabbro, serpentinite) causes desilicification of the pegmatite, and the appearance of accessory hornblende, diopside, sphene, corundum, etc. Distinct anthophyllite or anthophyllite-biotite rims form along exocontacts of pegmatite bodies in the ultramafics. In carbonates, pegmatites commonly contain scapolite, wollastonite, tremolite, diopside, phlogopite, graphite, etc. and may have skarn envelopes.

Beryllian hybrid pegmatites in ultramafics are rare but economically important as a source of gem emerald and chrysoberyl, as well as metallic Be. Their usual composition is oligoclase, albite, phlogopite, margarite and the Be minerals are represented by beryl, chrysoberyl, phenacite, bavenite, bertrandite, etc. A characteristic feature of these pegmatites is their tendency to split into systems of subparallel or branching dikes or veins of considerable persistency.

The Izumrudnye Kopy field 80 km N.E. of Sverdlovsk in the Urals, U.S.S.R. (Fig. 29-23) is a former gem emerald and alexandrite locality and more recently one of the important Soviet Be deposits described, without being named, in Beus (1956) and Zabolotnaya (1974). There, the country rock is a strongly schistose Devonian-Carboniferous metamorphosed mafic-ultramafic complex (amphibolite, serpentinite and talc schist), intruded by hornblende diorite and dunite-pyroxenite massifs. This was further intruded by a deep-seated Carboniferous granitic batholith, the muscovite-biotite granite phase of which is associated with the pegmatite dike swarm. The Be pegmatites occur in a N.N.W. trending, S.W. dipping band of numerous subparallel and anastomosing veins (or dikes) 2-3 m thick, each of them surrounded by a distinct alteration envelope. The band is about 100 m wide and 700 m in length.

Three varieties of Be-bearing veins (desilicified pegmatites) are present: (1) muscovite-plagioclase and muscovite veins with beryl; (2) beryl, muscovite, fluorite veins and (3) beryl, muscovite, quartz

Table 29-9. Rare elements pegmatites of "intermediate depths", selected Phanerozoic examples

LOCALITY	ENCLOSING ROCKS	PARENT INTRUSION	PEGMATITE	REFERENCES
Carolinas Tin-Spodumene Belt, U.S.A. (91 depos.)	PZ_1 amphibolite, schist, gneiss, deform.,metam.Dev.	D, quartz monzon. biot.-musc. granite	several groups of unzoned microcl.,alb.,qtz.,spodumene beryl,phosph.,columb.-tantal., cassit.pegm. 500 t Sn, X0 t Be	Kesler (1942)
Harding pegmatite, Taos Co.,New Mexico, U.S.A.	PCm amphibolite, schist	granite containing sphene	pegm.lens 305x183x18 m,zoned, spodum.,albite, musc.,lepidol. replac.pegmat.,microlite,tantalite,beryl. 85 t Ta/0.082%	Berliner (1949)
Moldanubian Massif W. Czechoslovakia Rožná,Jeclov,Dolní Bory,Písek,etc.	PCm biot.gneiss, schist,amphibolite,marble; PZ_3 plutons	PZ_3 musc.-biot. and biot.kata-to Mesoz.granites about 300 m.y.	widespread "common" pegmatites (e.g.Dol.Bory); Albite,lepidolite,spodum.,Li-tourmal. pegm.,minor beryl,columb.-tantalite, cassiterite	Svoboda, ed., (1966)
Central Urals pegm. region (Murzinka, Izumrud.Kopy,etc.) U.S.S.R.	Pt,PZ gneiss,ophiolites,bimodal metavolcanics	Cb,two mica gran. younger phase of composite batholiths	"normal" pegmatites; albite, Li-pegmatite contains columb.-tantalite; beryl most in hybrid pegm. in ultramafics and mafics	Fersman (1960)
Kalba-Naryn Ranges E.Kazakhstan, U.S.S.R.	PZ schist,quartz-rich litharenite	PZ_3 granite	"normal" graphic pegm.with tourm.cont.rarely scheel.and cassit.,grade to greisens, veins; albite-spodum.pegm. have rare dissem.cassiterite	Fersman (1960)
Nuristan and Hindukush Mts.,Eastern Afghanistan	Pe-Tr schists, quartzite,marble, unconf.on PCm gneiss and migmat.	Cr_1-T_1 two mica granite, qtz.diorite,granodior.	2 // pegm.belts,N.E.,large number of albite,spodum.,lepidol.,beryl,tantalite pegm. 5,580 t Be, also Ta	Rossovskiy and Chmyrev (1976)

Table 29-10. Brief list of metallic mineralizations in pegmatites

Be (beryl); low-alkalies, yellow, light green beryl is common in the wall, intermediate and core zones; widespread, e.g. Nuristan high-alkalies beryl, present in albitized and Li pegmatites; white, bluish; widespread
Be (chrysoberyl); in desilicified pegmatites intruding ultramafics, amphibolites; rare; Izumrudnye Kopy, Urals; Maršíkov, Czechosl.
Sn (cassiterite); in Li-pegmatites with spodumene; Kings Mountain, U.S.A.; Rožná, Czechoslovakia in K-feldspar pegmatites, transitional to greisens; Fabulosa, Bolivia; Kalba, Kazakhstan
Sn (stannite); rare, in quartz, amblygonite, microcline, cassiterite, phosphates pegmatite; Vernéřov, Czechoslovakia
Ta-Nb; columbite-tantalite, wodginite, microlite in Li pegmatites; widespread in small quantities, economic in Nuristan
REE; monazite, xenotime, samarskite, fergusonite, gadolinite, ilmenorutile, etc. in albitized pegmatites (Nuevo, California); monazite, accessory in small quantities in ordinary mica pegmatites, concentrated in placers (Catawba, Piedmont)
REE, orthite; sometimes with cerite, parisite, etc. in hybrid pegmatites in ultramafics; Borzovka, S. Urals; orthite (allanite) is common in "normal" pegmatites, particularly when emplaced into marbles, skarns, amphibolites
Zr; zircon, common accessory mineral in most pegmatites; reworked into placers; disseminated zircon in wall and border zones of "normal" pegmatite, Zirconia distr., N.C.; many "Zr pegmatites" are of alkaline affiliation (Chapter 33)
Th, thorianite, rare accessory in normal and albitized pegmatites, e.g. in Soafia, Betroka, Madagascar
U; uraninite, occasionally in K-feldspar or albitized pegmatite, e.g. Bemasoandhro, Madagascar; New Hampshire
Mo; molybdenite, in red blocky orthoclase pegmatite gradational into quartz veins, Skalsko, Czechoslovakia
Al, corundum; alone in desilicified pegmatite in marble, Pokojovice, Czechoslovakia; with vermiculite, chlorite, plagioclase at Corundum Hill, N.C., in southern Urals
As, Bi, Sb, etc. sulphides; occasionally present in arsenopyrite, löllingite, bismuthinite, e.g. Dolní Bory, Czechoslovakia

Fig. 29-23. Izumrudnye Kopy, the Urals, U.S.S.R. beryllium ore field. Cb desilicified pegmatite bodies in PZ_3 ultramafics.

Legend to (C): 1a=phlogopite glimmerite; 1b=non-oriented biotite; 2=pegmatite; 2b=kaolinized pegmatite; 3a=gray quartz; 3b=ferruginous quartz; 4=schists; 5=light beryl; 6=emerald; 7=talc schist; 8=actinolite schist; 9=granodiorite, diorite dikes; 10=aplite dikes; 11=tourmaline; 12=apatite. From Beus (1956) and Fersman (1960).

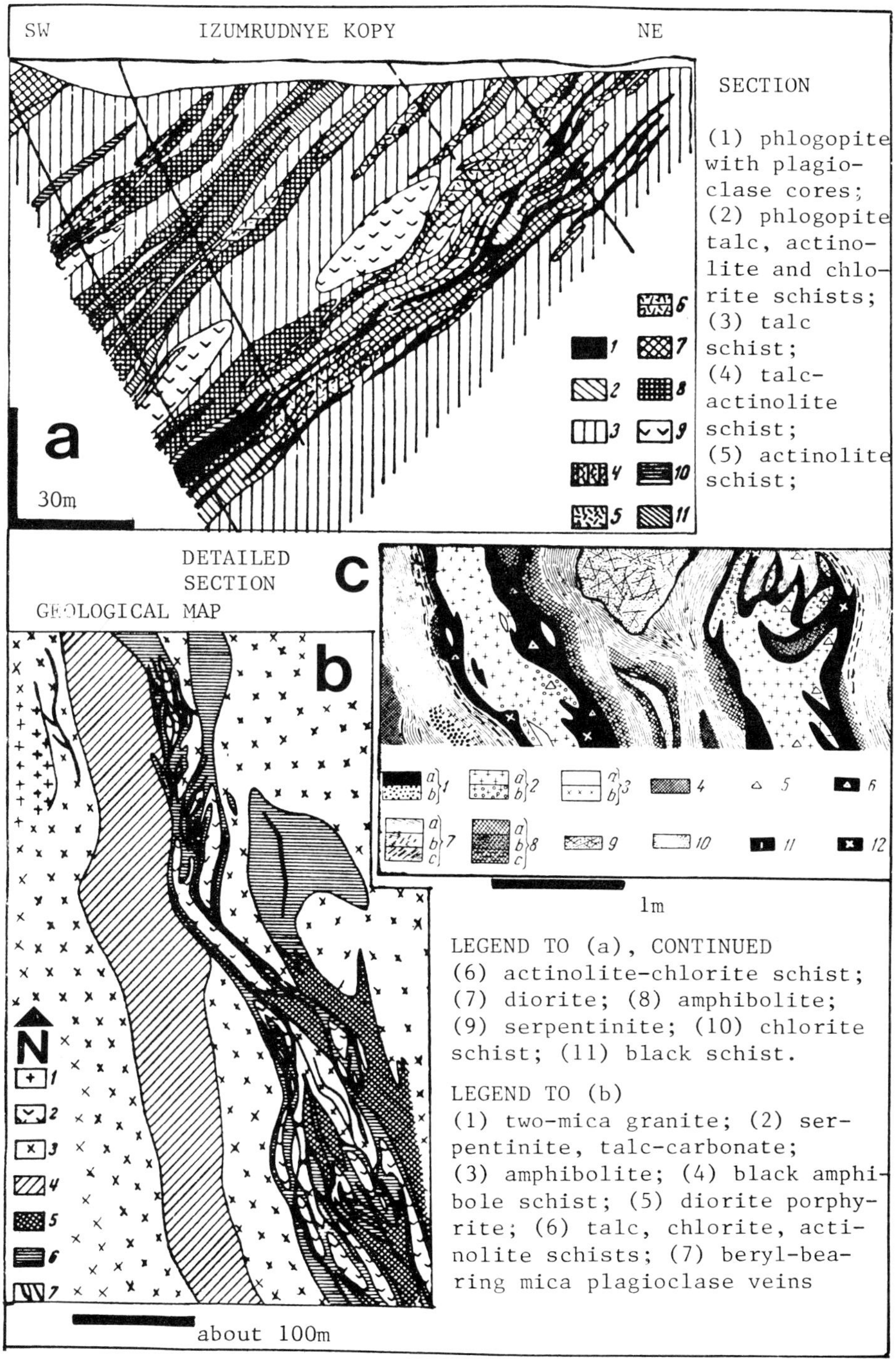
SW
IZUMRUDNYE KOPY
NE
SECTION
a
30m
(1) phlogopite with plagioclase cores; (2) phlogopite talc, actinolite and chlorite schists; (3) talc schist; (4) talc-actinolite schist; (5) actinolite schist;
DETAILED SECTION
c
GEOLOGICAL MAP
b
N
1m
about 100m
LEGEND TO (a), CONTINUED
(6) actinolite-chlorite schist; (7) diorite; (8) amphibolite; (9) serpentinite; (10) chlorite schist; (11) black schist.
LEGEND TO (b)
(1) two-mica granite; (2) serpentinite, talc-carbonate; (3) amphibolite; (4) black amphibole schist; (5) diorite porphyrite; (6) talc, chlorite, actinolite schists; (7) beryl-bearing mica plagioclase veins

veins. The (1) veins are most common (Fig. 29-23b). They consist of a plagioclase core kaolinized near surface. The core is zonally surrounded by a rim of phlogopite, up to 6 m thick, grading outward into an actinolite zone, chlorite zone, talc zone and unaltered serpentinite. The bulk of the beryl is contained in the phlogopite, where a small proportion of it is in the form of gem quality emerald. Additional minerals in this zone are phenacite, chrysoberyl, margarite, tourmaline, bavenite, fuchsite, molybdenite and apatite. Lesser quantities of scattered beryl crystals are present in the pegmatite, accompanied by rarer columbite and molybdenite. The chlorite zone contains phenacite, chrysoberyl and lesser beryl.

The type (2) veins are hosted by diorite. They are rare and less persistent than the veins of type (1). Typically they are 1.5 m thick and 20 m long. They contain up to 50-60% beryl in large grains 1-5 cm across, intergrown with fluorite. Small quantities of apatite, molybdenite and native bismuth are present. The Be-rich cores are rimmed by muscovitic glimmerite and plagioclasite (probably a wall pegmatite zone). No quantitative data are available for the Izumrudnye Kopy field, but a conservative estimate is about 2,000 t Be.

29.8. RETROGRADE METAMORPHICS, MYLONITES, CATACLASITES

29.8.1. General

Retrograde (retrogressive) metamorphism (or diaphtoresis) is a lower pressure and temperature rank metamorphism superimposed on a rock assemblage produced under higher pressure-temperature conditions. Its results are extremely relative in time, space, cause and effect and overlap with many of the mineralization systems already desribed earlier or with those that will be treated later (e.g. hydrothermal deposits hosted by high-grade metamorphics, Chapter 28; ores associated with "rifting", Chapter 30; etc.).

In order to produce retrograde effects in a higher-grade metamorphosed protolith, uplift into higher crustal levels is a first prerequisite. There, however, the bulk of the metamorphics remain as unmodified relics for unlimited periods of time, as the presently outcropping metamorphic terrains indicate. Retrogression, thus, is not automatic but a selective, localized process controlled by focussed tectonism along shear zones, lineaments and faults. The intensity of the retrogression is proportional to the depth at which it took place and pressure under which it took place. Its products range from cataclastic rocks, breccias and fault gouges near the surface through mylonites and ultramylonites into blastomylonites, phyllonites and retrogressed schists in depth. Higgins (1971) presented a comprehensive review of cataclastic rocks together with examples of their field distribution along major fault zones (Fig. 29-24).

The retrogression could be either "dry" or "wet". Retrogressed schists generated from completely dehydrated progenitors, such as

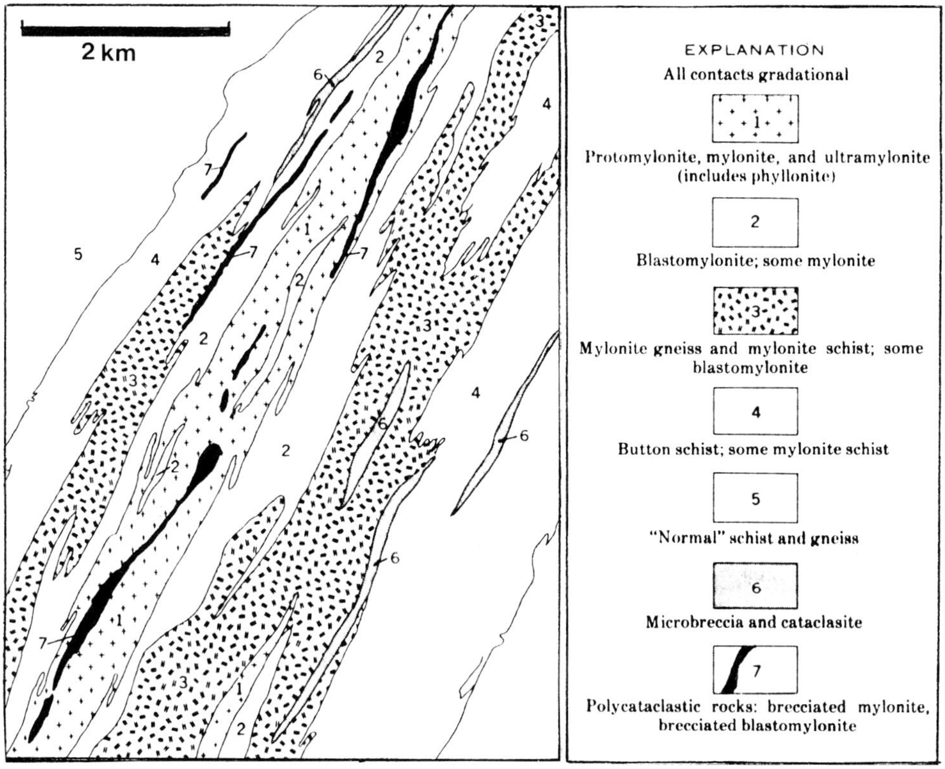

Fig. 29-24. Distribution of rocks resulting from high to medium level of retrogressive dynamometamorphism along major fault and shear zones. A hypothetical situation, strongly influenced by conditions along the Breward Fault, southern Appalachians, U.S.A. From Higgins (1971).

granulites required hydration, so they must have formed under "wet" conditions. Dry retrogression has a negative metallogenic balance because it results in destruction (dispersion) of earlier orebodies (when it coincides with an orebody), without creating any new mineralization. Wet retrogression, on the other hand, may produce new orebodies from metals introduced by the migrating fluids, in addition to the partial or complete remobilization of earlier orebodies. As a consequence, retrograde-metamorphogenic (i.e., metamorphism created) mineralization is virtually always hydrothermal and converges, overlaps and merges with structurally and lithologically controlled ores in intrusive aureoles or zones of fluid convection driven by any heat source.

The truly metamorphogenic mineralization (the one unrelated to igneous intrusions) depends on two fundamental genetic varieties of fluids: (1) depth-derived fluids, mostly components of the metamorphic fronts moving as a result of metamorphic dehydration in depth or (2) convecting groundwaters heated by geothermal gradient and/or radioactive decay. Belevtsev (1979) devoted an extensive coverage to the retrogressive-metamorphogenic hydrothermal process causing enrichment of lean iron formations and concentration of uranium in albitites, considered the two most common and rather convincingly interpreted manifestations of the process. Upward-migrating albitizing and NaCl-containing metamorphic fluids are probably connected with K-feldspar blastesis at greater depth (Suk, 1983). Albite blastesis at the expense of K-feldspar or muscovite, in turn, releases potassium into solution to cause sericitization, biotitization, or higher-level K-feldspathization. Other varieties of fluids carry silica, fluorine (probably displaced from earlier micas or added from subcrustal sources), barium, iron, carbonates, and other components. In anomalously metal enriched or previously mineralized terrains any metals can be redeposited. The metamorphogenic and/or convective fluids either react with their wallrocks to produce metasomatites (alterites), or they fill available dilations (rock pores, fissures, breccia zones) resulting in impregnations, veins and breccia cements. There is a convergence among the metasomatites and hydrothermal feldspathites and pegmatites.

The products of retrograde metamorphism at a variety of levels are best apparent in old high-grade crystalline massifs ("tortured massifs"), repeatedly subjected to several phases of shallow or intermediate level reactivation (rejuvenation) under alternating compressional and tensional regimes. Median massifs in divergent orogens (e.g. Hesperian, French Central, Bohemian massifs in the Hercynian belt of Europe; the Alpine massifs such as the St. Gotthard, Aar, Belledonne, etc.), as well as portions of Precambrian shields and platforms (such as the Aldan Shield, Azov Block, etc.), are the most frequently discussed examples. In areas where block-faulted, reactivated or retrogressively metamorphosed crystalline blocks are partly or fully covered by a thin veneer of young sediments and volcanics, the mineralization in the basement often overprints into the cover sequence. This establishes a link with some of the ores treated in Chapter 18. In this section, the coverage is restricted to the regions retrogressively metamorphosed and mineralized after the Precambrian.

29.8.2. Pitchblende deposits in mylonite-filled reactivation faults, Bohemian Massif

The Bohemian Massif (Svoboda, ed., 1966; Škvor, 1979) centered in western Czechoslovakia, but also including portions of German, Austrian and Polish territories, is a rhomb-shaped, periodically rejuvenated Proterozoic rigid block of approximate dimensions 300 x 300 km. It is fault-bound and surrounded by Hercynian and Alpine mobile belts, or sediments of the Mesozoic-Cainozoic platformic cover. The middle Proterozoic metamorphic basement (Moldanubicum) is interrupted by several superimposed late Proterozoic to Devonian "graben-style" mobile belts, and transformed by Carboniferous periplutonic metamorphism.

The late Palozoic granitic plutonism is responsible for the numerous Au, Pb-Zn, Sn-W and U deposits, mostly contained in veins. Several deposits, however, including to some extent, the important Příbram and Jáchymov (Chapter 28), significantly postdate the assumed cooling period of the plutons. The Bohemian Massif is disturbed by a network of repeatedly rejuvenated N.E., N.W., and N.-S. regional faults and weakened zones and this system controls the post-Carboniferous epigenetic mineralization as well as (in northern Bohemia) the emplacement of Tertiary alkaline volcanics (Chapter 33). The post-lower Permian (post-intrusive) hydrothermal mineralization is represented by pitchblende in quartz or carbonate-filled or unfilled shears and mylonitized zones, by fluorite-barite deposits, by minor occurrences of Pb-Zn and Au and by the Ni,Co,Bi,Ag,U arsenide association.

A group of uranium deposits hosted by Moldanubian biotite-sillimanite gneisses in the western part of the Bohemian Massif (Zadní Chodov, Damětice, Ústaleč, Altransberg, Tirschenreuth, Wölsendorf, etc.; Ruzicka, 1971) is typically associated with a system of persistent N.W. and N.E. zones of silicification and quartz veining. The quartz is fringed or interrupted by graphite-rich mylonite and blastomylonite. In Altransberg (Bavaria; Bultemann and Hofmann, 1982), pitchblende with pyrite and rare coffinite fills thin fractures in the extension of feather-joint quartz veins. The mineralized zone of upper Permian age is 3 m wide and averages 0.1% UO_2. Torbernite, metatorbernite, autunite, phosphuranylite, parsonite and kasolite are present in the oxidation zone.

In the Damětice deposit (Janout, 1972; Fig. 29-25), Moldanubian biotite migmatite intruded by granodiorite porphyry is intersected by a series of steeply N.W.-dipping faults with mylonite filling. The most persistent fault is filled by at least three generations of quartz grading into silicified gneisses and mylonites, up to 30 m wide. The U ore is located in several mylonite, graphitic gouge, quartz and calcite-filled faults postdating the main generation of quartz. Pitchblende is the only hypogene uranium mineral, and it exists mostly in the form of thin veinlets and lenses in mylonite and the carbonate gangue.

Similar probably late Paleozoic hypogene pitchblende, galena, sphalerite and chalcopyrite vein filling in Moldanubian gneisses near Okrouhlá Radouň, S. Bohemia (Mrázek, 1972) was reactivated during the period of Tertiary block faulting, and the U dispersed along faults and in mylonite in the form of sooty pitchblende.

In addition to high-level, open faults, the vein U mineralization in high-grade metamorphics of the Bohemian Massif is controlled by the host rock lithology. Graphitic schists yielding graphite-rich mylonites, are preferentially mineralized. In the Rožínka-Olší U belt, central Czechoslovakia (Jürgenson and Hájek, 1980), mineralized faults are developed in biotite-hornblende gneiss alternating with amphibolite, calc-silicate gneiss and marble. The ore is absent where

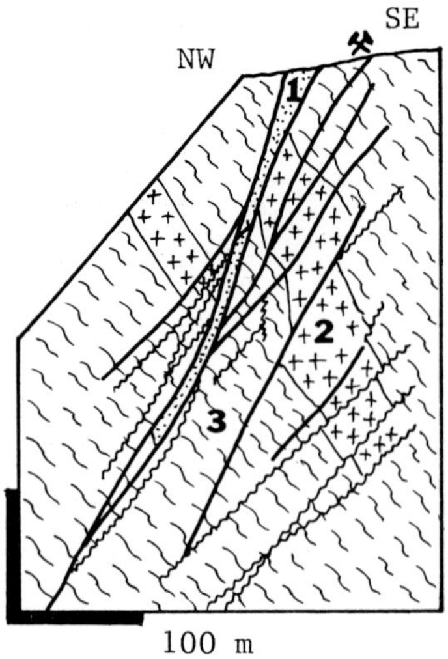

1. "Pfalz" quartz filling a fault zone, grading to silicified mylonite and cataclastic gneiss
2. PZ_3 granodiorite porphyry
3. Pt_{2-3} biotite gneiss, migmatite

Fig. 29-25. Damětice vein uranium deposit, western Czechoslovakia. Simplified after Janout (1972).

the fault intersects with advanced migmatites.

Other regions

Comparable U ores hosted by brecciated and mylonitized fault and shear zones are known in many other regions and overlap with the granite-hosted localities reviewed in Section 28.20). At Urgeiriça, N.Portugal (Rich et al., 1977) veinlets, coatings and sooty dispersions of pitchblende with or without a cherty quartz gangue of Cretaceous age (80-100 m.y.) are located in reactivated structures superimposed on late Paleozoic granites and Cambrian metamorphics. At Val Vedello in the Central Alps of northern Italy (O.E.C.D., 1979; 1,000 t U/0.1%), uranium minerals are found in a mylonite zone along a tectonic contact between the crystalline basement and overlying Permo-Carboniferous continental sediments.

29.8.3. Wet Mountains, Colorado, Th and REE province

The Wet Mountains are a small portion of the Colorado Rocky Mountains S.W. of Pueblo (Christman et al., 1959; Armbrustmacher, 1979). The bedrock is composed of middle Proterozoic (1.72-1.45 b.y.) granite gneiss, migmatite, amphibolite, metagabbro, metapyroxenite, metaquartzite, calc-silicate gneiss, etc. The high-grade metamorphics to ultrametamorphics are locally retrogressed, and intersected by numerous long shear, mylonite and breccia zones. Some of these zones are filled or fringed by local patches of pink to red K-feldspar or albite and quartz, barite, carbonate, galena and sphalerite. Thorogummite, thorite and other Th minerals occur sporadically in the above association and are in most cases decomposed and megascopically invisible, masked by "limonite". About 800 small Th occurrences have been recorded in the Wet Mountains (Fig. 29-26), having an aggregate tonnage of about 3,564 t Th (grade better than 0.088% Th).

In the early stages of research, the Th mineralization was believed to be genetically affiliated to a Precambrian syenite body and to a lower Cambrian albite syenite stock and associated breccias, dikes and quartz-feldspar veins. These were considered by some to be products of retrograde metamorphism, comparable with the uranian and thorian albitites and K-feldspathites discussed by Belevtsev (1979), mostly from the Precambrian terrains of Siberia (Aldan Shield) and the Ukraine (Zheltye Vody). The subsequent discovery of Cambrian alkaline syenite, nepheline syenite, gabbro, pyroxenite, carbonatite and alkaline lamprophyre bodies in the Wet Mountains, caused a re-interpretation. The Th orebodies are now considered to be related to the alkaline magmatism, or to the interaction of the alkaline fluids with their retrogressed environment. There is a remarkable convergence between the effects of alkaline contact metasomatic feldspathization as found in zones of fenitization (Chapter 33), and the retrogressive metamorphic feldspathization as well as carbonatization.

29.8.4. Pb,Zn,Ag provinces

Vein or replacement deposits of galena, sphalerite and associated minerals in quartz, carbonate, barite or fluorite gangue are very common in partially retrograded, rejuvenated and deep-faulted high-grade metamorphic terrains (e.g. the Colorado Mineral belt, Massif Central, Vosges and Schwarzwald Mts, Bohemian Massif, Rhodopen, Menderes Massif, etc.). In many regions mineralized in this way the Pb-Zn bodies show spatial and temporal association with large plutons, with small intrusive stocks and dikes, with geophysically buried plutons, with surficial andesite to rhyolite volcanism or with alkaline magmatism. Abundant deposit examples have been treated in the corresponding chapters. The above affiliations, however, need not exert the complete control and genetic over-generalization could be harmful in exploration. It can be assumed that orebodies equivalent to the above can also accumulate from hydrotherms driven by the more obscure agents (geothermal or radioactive decay heat, metasomatic fronts, etc.).

In the Bohemian Massif, a belt of small Pb,Zn,Ag deposits controlled by the Blanice Furrow, a distinct narrow N.N.E.-trending fault graben in the Moldanubian metamorphics, completely lacks any apparent magmatic agents of mineralization. The Furrow is bordered by a discontinuous, pinching and swelling system of mylonite-filled faults, to which are locally adjacent short, diagonal feather faults. Quartz, barite, ankerite, sphalerite, minor arsenopyrite, tetrahedrite

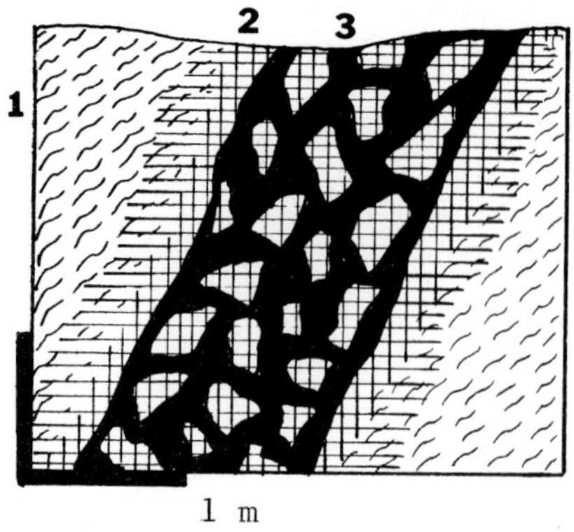

1. Pt migmatite, amphibolite, pegmatite
2. red altered (bleached, partly feldspathized) rocks
3. fragments of red altered and feldspathized ("albite syenite") wallrocks in a matrix of drusy gray to smoky quartz. The quartz contains patches of barite and calcite with scattered thorite. At surface, calcite is leached and substituted by Th-containing residual ocher

Fig. 29-26. Wet Mountains, Colorado, detailed section of a Th-bearing lode near Rosita. From LITHOTHEQUE.

and Ag sulphosalts-filled veins in the Stříbrná Skalice, Ratibořské Hory, Rudolfov and other fields (cumulative production about 170 t Ag), preferentially fill the feather faults. The prevalent wallrock alteration is a slight sericitization and chloritization.

In the Pb,Zn,Ag fields in the Rhodopen Massif (Bulgaria and Greece, e.g. Madan; Iovchev, 1961), the ore shoots fills are often breccia and mylonite-filled faults in the crystalline basement, tens of kilometres long. The mineralization is believed to be coeval with the Eocene-Oligocene continental volcanism and at some deposits (e.g. Madzharovo), epithermal orebodies are hosted by the volcanics. This, however, need not be valid universally.

29.8.5. Gold deposits

Gold deposits hosted by high-grade metamorphics are usually closely associated with "granites" or with subvolcanic intrusions, and localities lacking this association are rare. One example is the Roudný deposit in the Blanice Furrow, Bohemian Massif, a structure mentioned in the previous paragraph (Koutek, 1963). This deposit is hosted by Proterozoic biotite-sillimanite gneisses and migmatites containing numerous bodies of aplite, pegmatite and orthogneiss. The orebody has the form of a westerly plunging fault-bound triangular prism, framed by a massive to brecciated quartz lode 10-150 cm thick and gradational into quartz stockwork in the adjacent silicified gneiss. The ore minerals are scattered pyrite, arsenopyrite, lesser sphalerite, tetrahedrite and galena. The grade ranges between 4 and 25 ppm Au, and the gold has a low purity (0.675). Some 6.6 t Au was produced or remains in the tailings.

29.8.6. Antimony and mercury deposits

Small stibnite deposits are occasionally associated with faults of activation superimposed on high-grade metamorphic complexes. Shcheglov (1967) mentioned the Boguchan Sb deposit in the Bureia median massif, E. Siberia (chalcedony, fluorite, stibnite veins and mineralized zones in Cretaceous sandstones and tuffs above crystalline basement) and the Ribkovo Sb deposit in southern Bulgaria.

A series of cinnabar deposits associated with open faults and mylonite zones in high-grade metamorphics of the Menderes Massif, W. Turkey, has recently been described by Yildiz and Bailey (1978). At the Haliköy mine (3,509 t Hg/0.254%; Fig. 29-27), cinnabar, metacinnabar, quartz, pyrite and marcasite form veinlets and disseminations along the footwall in a clayey gouge-filled zone up to 35 m wide. The zone is a E.-W.-trending, north-dipping thrust fault separating micaschist from a granitic gneiss. In the Turkonü deposit, cinnabar and pyrite are disseminated in a steep E.-W. fault. The ore zone 1-5 m thick and 500 m long is contained in a granitic augen gneiss and the production has been 242 t Hg. Similar but smaller deposits are found in the Tire area.

29.8.7. Fluorite and barite veins

Fluorite, barite and usually quartz, carbonate veins (or replacements in carbonates) are by far the most characteristic mineralizations associated with shallow retrogression, faults, fault breccias and mylonite zones. Their description has been delayed because these are nonmetallic commodities. Metallic minerals are, however, often associated in small quantities and the veins grade to one of the metallic fillings discussed earlier. In the review by Shcheglov (1967), fluorite and barite deposits are always the youngest products of "activation", significantly postdating plutonic intrusive activity and the older generations of hydrothermal deposits. Important fluorite belts are particularly common in the crystalline basement of median massifs and their covers.

One of the largest examples is the Eastern Mongolian fluorite belt which is 1,000 km long and 300 km wide. Lesser occurrences are found in the Hesperian, French Central, Morvan, Schwarzwald (Wieden), Bohemian (Moldava) and Rhodopen (Slavianka) Massifs, as well as in the Colorado Rocky Mountains (Poncha Springs), in the Santa Catarina crystalline complex, S.E. Brazil (Morro da Fumaca) and elsewhere. Of the minor metallic minerals, galena and sphalerite are the most common. Small production of Pb,Zn and Ag was realized from the Schwarzwald (e.g. from Wieden and Badenweiler; the latter is a Tertiary silicified mylonitized zone), from the Massif Central (e.g. Ussel field, upper Triassic-Jurassic fluorite, galena, sphalerite veins) and other localities.

Uranium minerals are known in small quantities in several fluorite deposits, and the fluorite itself is often radioactive. At Kletno in

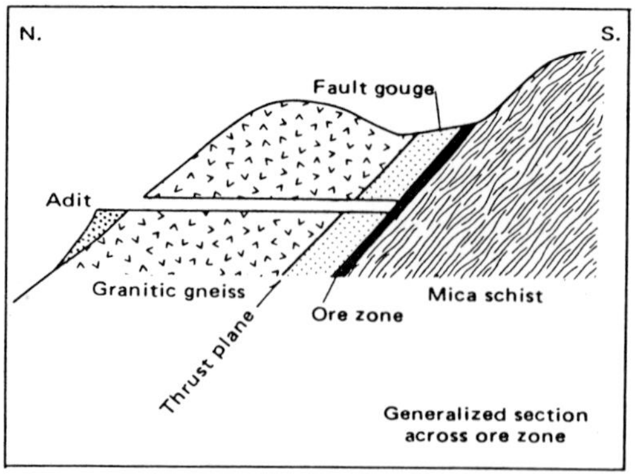

Figure 29-27

Haliköy mine, Turkey, an example of a cinnabar deposit along fault in high-grade metamorphics. From Yildiz and Bailey (1978)

the Polish Sudeten Mts. (Banaś and Mochnacka, 1974), quartz-fluorite veins located in a N.N.W.-trending overthrust of augen gneisses and marbles over micaschists, contain nest-like accumulations of U, Cu and Se minerals. Chalcopyrite, chalcocite and pyrite are most common, pitchblende, uranophane, torbernite, gummite, clausthallite and umangite occur in lesser quantities.

29.8.8. Fe and Mn deposits

These are of minor importance. The Jurassic Mn veins near Romanèche (eastern tectonic margin of the Massif Central, France; Lougnon, 1956; about 160 Tt Mn, up to 1 Tt W content) consist of quartz, barite, fluorite and psilomelane. The veins are up to 20 m thick, and controlled by fault zones in the basement metamorphics and Carboniferous granite. Locally, the veins are gradational into psilomelane replacements in Jurassic limestones of the platformic cover. Similar veins are found in the Schwarzwald (Eisenbach).

29.8.9. "Alpine" veins

"Alpine veins" are mineralized fissures (often just groups of mineral crystals sitting on walls of empty clefts or fissures), particularly well developed in the crystalline massifs of the Alps (Parker, 1923). The mineral association, prized by collectors, consists of a variety of nonmetallic silicates and oxides (quartz crystal, epidote, adularia, sphene, zeolites, chlorite) with only small quantities of Ti and REE metallic minerals (rutile, anatase, brookite, aeschynite, synchisite, gadolinite, etc.). The minerals crystallized from hydrothermal fluids at temperatures ranging from 450 to 100° C, during the retrogressive phase of gneisses and schists. The vein components have been leached from the wallrocks and are often accompanied by quartz gangue. Except for specimen minerals, the only economic product of the "Alpine veins" has been piezoelectric quartz.

"Hebung, Spaltung, Vulkanismus". Hans Cloos in Geol. Rundschau v. 30, p. 401-527, 1939.

CHAPTER 30

Continental Fragmentation, Rifts and Paleo-Rifts

30.1. INTRODUCTION

Rift zones are "regions of stretching where the crust is somewhat thinned and the mantle has a lower than normal density" (Milanovskii, 1972). The word "rift" is used as a synonym for rift zones, or as a summary term for a broad tectonic environment involving crustal fragmentation and the affiliated lithologic associations. "Rift" is usually used in the general (time unconfined) sense, but many authors using the term imply contemporary geotectonic environment: in other words, the processes and forces generating rifts that are at work at present and can be observed and measured.

Remnants of tectonic structures and characteristic lithologic associations interpreted as being a result of "rifting" but with the tectonic activity that ceased a long time ago, are usually termed paleo-rifts. A sharp division between contemporary, active rifts and paleo-rifts does not exist and many regions of Mesozoic (e.g. St. Lawrence Graben, Canada) or even Paleozoic rifting are still seismically active at present.

An extensive literature on "rifts" and "rifting" has been generated in the past 30 years (e.g. Illies and Mueller, eds., 1970; Girdler, ed., 1972; Milanovskii, 1976; Grachev, 1977; Neumann and Ramberg, eds., 1978; Morgan and Baker, eds., 1983) and it offers numerous, often conflicting definitions and interpretations of rifts and associated features (Fig. 30-1). Particularly uncertain are the limits of rifts, that is, where they start and terminate. "True rifts" are gradational into "aulacogenes" and into a variety of graben and horst-and-graben systems and extensional (taphrogenic) lineaments. Aulacogenes (=failed rifts, failed tripple arms of rifts) are wedging-out graben-like depressions, often forming re-entrants oriented at an angle to rifted continental margins or mobile belts. They contain a thick fill of terrigenous sediments, basalts, and are often folded; e.g. Salop and Scheinmann (1969), Burke (1977).

Although "typical" continental rifts are established within "cratons" (platforms), comparable, although usually lesser-scale, taphrogenic systems can be found in mobile belts as well. There they are either (1) relatively late, superimposed on an assemblage that had passed earlier through phases of intensive marine sedimentation and volcanism ("geosynclinal stage"), metamorphism, deformation and batholithic emplacement ("orogenic or batholithic stage", e.g. the Rio Grande Rift, New Mexico) or (2) relatively early, controlling the initial stages and facies of the "geosynclinal" deposition.

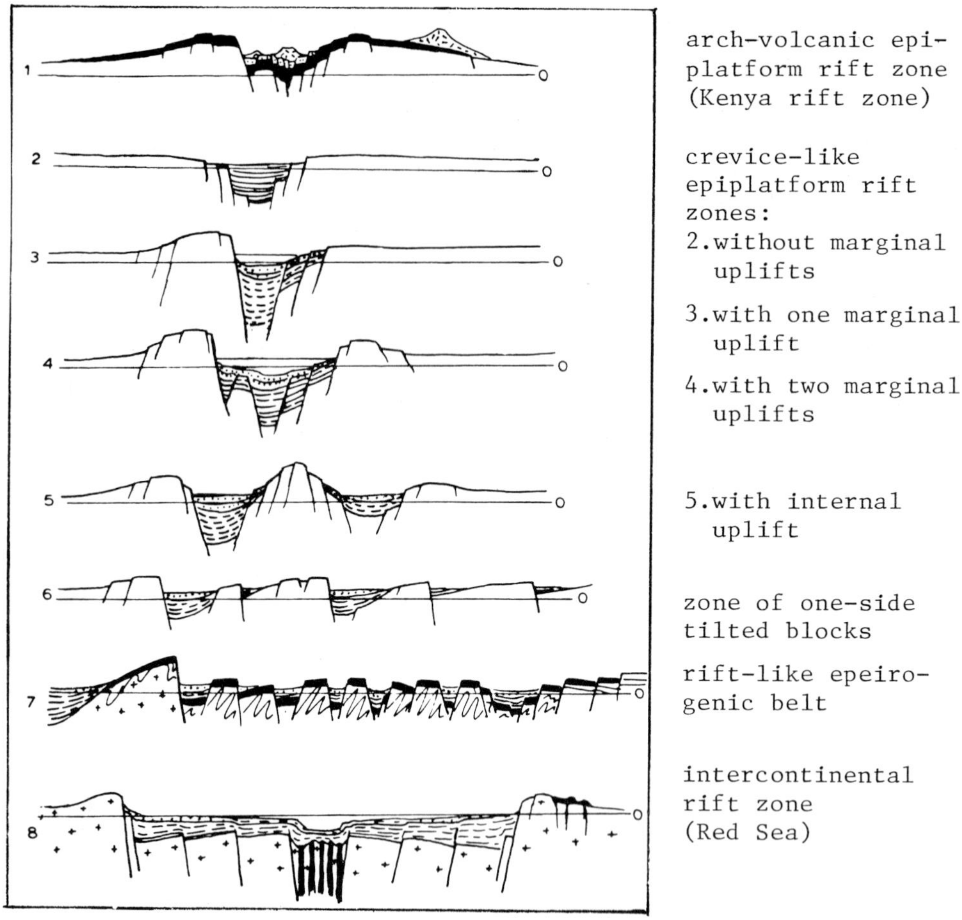

Fig. 30-1. Characteristic structural types of rift zones. From Milanovsky (1972).

Mineralization associated with the "rifting" episodes in orogenic belts is treated in Chapters 6, 26, 28 and 29.

Milanovskii (1972) distinguished three relatively objective, empirical categories of the rift zones of the earth, differentiated by the type of crust involved: (1) continental (intracontinental) rifts, where both the rift floor and shoulders are composed of thinned continental crust (e.g. East African Rift); (2) intercontinental rifts, narrow waterways where the rift (its axial part) has an oceanic crust (the Red Sea, Gulf of California) and (3) oceanic (intraoceanic) rifts, where the axial graben is bordered by oceanic crust (e.g. the Mid-Atlantic Ridge). The latter category was treated in Chapter 4.

"Rifts" and "rifting" have variable roles in the generation of ore deposits (Fig. 30-2; Table 30-1). Certain mineralizations are so intimately affiliated to the process of active rifting that they can be designated with much credibility as "rifting-generated" (e.g. the metalliferous brines and young ore precipitates in the axial trough of the Red Sea). The latter are treated in the present chapter. Other ores are immediately genetically associated with characteristic parent rocks (e.g. alkaline rocks such as nepheline syenites, carbonatites, etc.) which, in turn, are one of the characteristic lithogenetic consequences of rifting, so occurrences of such ores are also, to some extent, (indirectly) controlled by rifting. In many cases, however, the rift structures are obscure or not obvious at all, so only the lithologic association remains. Consequently, ore occurrences associated with rifting-generated lithologic associations are treated with the latter, in the following chapters: 24 (recent sediments of rift lakes); 31 (plateau basalts) and 33 (alkaline rocks).

It is easy to understand why the most convincingly interpreted "rifting-generated" ores are Recent, or geologically very young (Tertiary). It is often because their origin can still be directly observed (e.g. the Red Sea metalliferous brines and sediments). Alternatively, in some instances, even when the actual ore formation has already ceased, the paleo-environmental configuration is still so well preserved that a credible genetic interpretation can be suggested. Although the model for the origin of the Red Sea metalliferous muds has been applied to several ancient ore deposits such as Sullivan, Canada; Mount Isa, Lady Loretta, Broken Hill, etc., Australia; Bawdwin, Burma; Rammelsberg, Germany and others (Blissenbach and Fellerer, 1973; Sawkins, 1976, 1978, 1984; Robbins, 1983; etc.), the credibility of interpretation is low and numerous alternative rift-unrelated models have been offered to explain the origin of the above deposits. As a rule, only those interpretations of "rifting-generated" ore deposits that offer good empirical evidence of rifting still preserved in the area and applicable in actual exploration, are treated in this chapter.

30.2. STAGES OF RIFTING AND RECENT EXAMPLES

Intracontinental rifting is a long-lasting, dynamic and additive process (Fig. 30-3). It is initiated in the earlier stabilized continental crust and terminates when, at a site under consideration, the continental crust is completely destroyed (most probably removed by spreading and/or replacement) and substituted by oceanic crust. The same process then usually continues, but has a character of intraoceanic spreading. Although intracontinental and intraoceanic rifting and spreading systems may converge (become identical) in the subcrustal region, they differ in terms of interaction with the crustal environment.

The rift stage marked by the formation of conspicuous laterally extensive fault, graben and horst systems usually accompanied by dominantly mafic or bimodal alkaline or tholeiitic volcanism, is preceded by uplift, block faulting and predominantly felsic

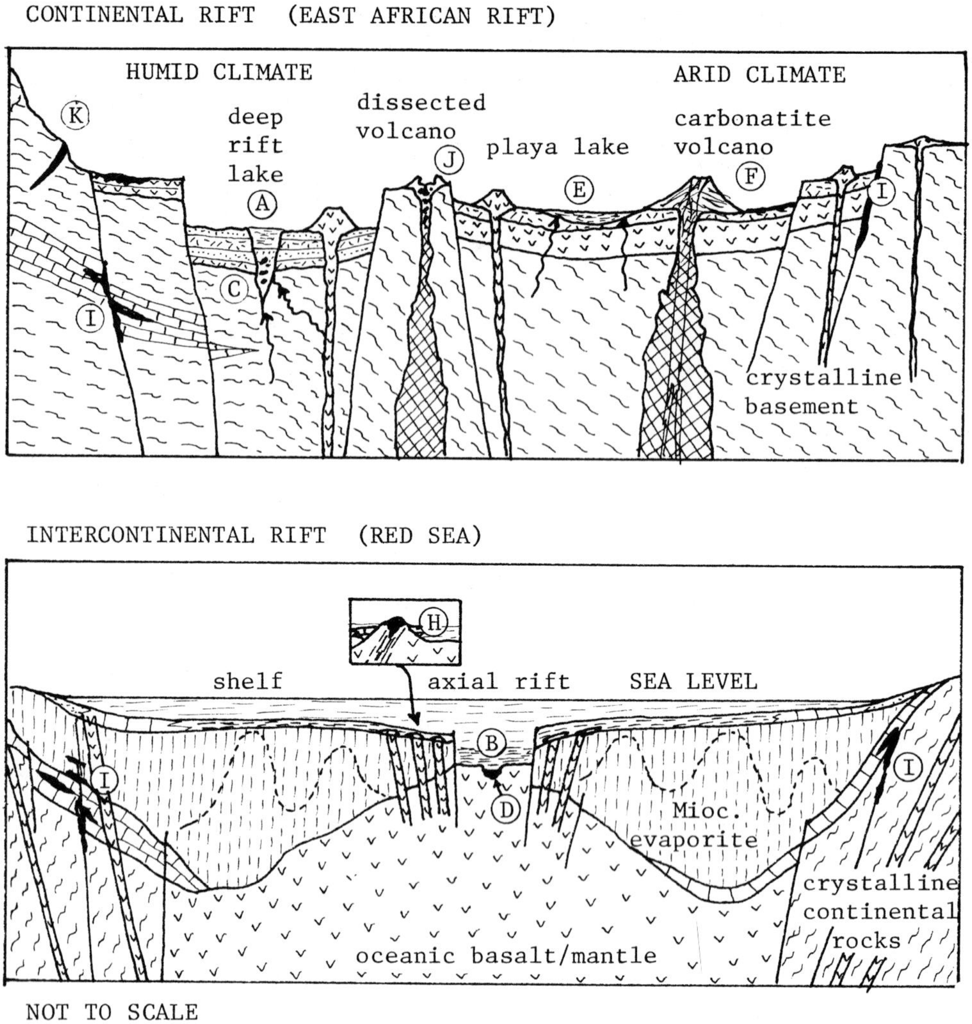

Fig. 30-2. The major sub-environments, lithologic associations and mineralization styles in the Recent continental and intercontinental rift environment (see Table 30-1 for explanation of letter codes). From Laznicka (1984).

Table 30-1. Principal mineralization styles in the Recent continental and intercontinental rift environments

(A)	Anomalous metal (Zn,Pb,Cu,etc.) and F content in the water of rift lakes (e.g. L. Kivu) and feeder springs
(B)	Anomalous metal contents in highly saline, thermal seawater and in the feeder springs (Red Sea; Mn,Fe,Zn,Cu,Pb,Ag,etc.)
(C)	Anomalous base metal contents (Mn,Zn,Mo,etc.) and diagenetic metallic minerals (Mn-Fe oxides, carbonates) in bottom sediments of rift lakes
(D)	Metalliferous sediments (muds, oozes) filling depressions in the axial zone of rift seas (Red Sea); about 10 Mt Mn, 2 Mt Zn, 500 Tt Cu, 150 Tt Pb, 6 Tt Ag
(E)	F, trace metals (U,Zn) and minor metallic minerals in recent evaporites of rift playa lakes
(F)	U and other metals in calcretes over alkaline volcanics in arid climatic zones
(G)	Residual bauxites over alkaline volcanics in humid tropics
(H)	Ni-silicate infiltrations in saprolitic mineralized serpentinite ("oceanic" islands in the Red Sea); 250 t Ni
(I)	Fissure veins, epigenetic replacements and ? stratiform bodies of fluorite, barite, galena, sphalerite, Mn-ores, etc. in pre-rift rocks (mostly young carbonates)
(J)	Ores at deeper levels of the alkaline igneous association (e.g. pyrochlore); see Chapter 33.
(K)	Various pre-rift mineralizations exposed in the uplifted basement blocks on rift shoulders

calc-alkaline (to peralkaline) volcanism and intrusive activity (Cloos, 1939). The latter activity and its rock products have been treated in the Soviet literature as "tectonomagmatic activation", a term now used with increasing frequency in the western literature as well. Some authors (e.g. Sawkins, 1978) consider rifting as a part of activation, whereas the Russian school usually treats "activation" and "rifting" separately.

There is indeed continuity and overlap between both regimes and it is impossible to make a sharp distinction, but there is a different emphasis. Lithogenesis and mineralization influenced by tectonomagmatic activation of the continental crust are treated in Chapters 24,25,26, 28,29,31-33. The recent "rifts" (i.e., regions undergoing the rifting stage of development) are represented by two outstanding example areas, that serve as models for interpretation of the less perfectly developed areas elsewhere: (1) the East African Rift System represents the less advanced stage of rifting developed fully (at the present level of exposure) within the continental crust and (2) the Red Sea, a representative of the advanced stage of rifting in which oceanic lithosphere already floors the axial portion of the rift, but the continental lithosphere on both flanks is still well in sight and involved in the lithogenesis and metallogeny.

1 Incipient gentle doming within sialic cratons (may influence facies of platformic sediments)	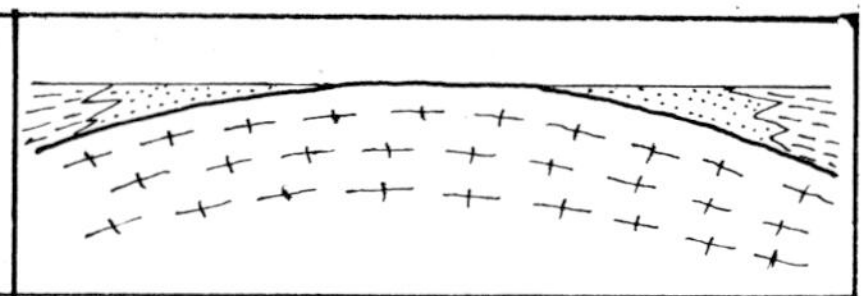
2 Mantle upwelling and modest thinning of the continental crust; "anorogenic" granite, rhyolite, syenite, trachyte, etc. magmatism; may grade into or overlap with (3) to form bimodal associations	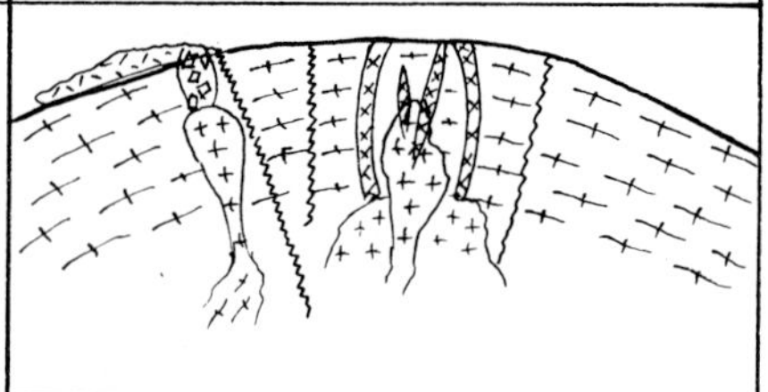
3 Fracturing, light crustal extension, mantle rise, plateau basalt volcanism; diabase diking, formation of high level mafic magma chambers in the continental crust	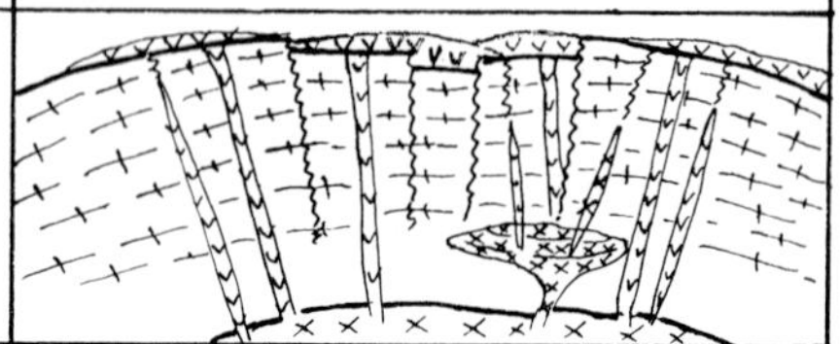
4 More intensive fracturing, horst and graben formation (rifting), accompanied by alkaline magmatism (as in the East African Rift)	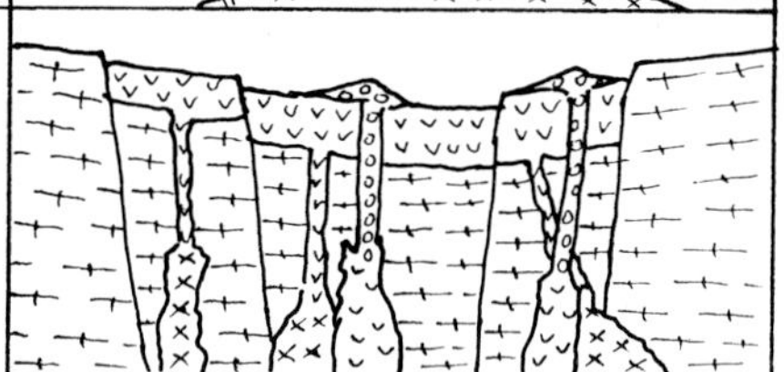
5 Increased crustal spreading, subsidence of axial troughs, rise of mafic magmatic systems, attenuation and ultimately interruption of the continental crust; oceanic basalt flows, diabase dikes, gabbro intrusions; evaporites accumulate in restricted basins (as in the Red Sea)	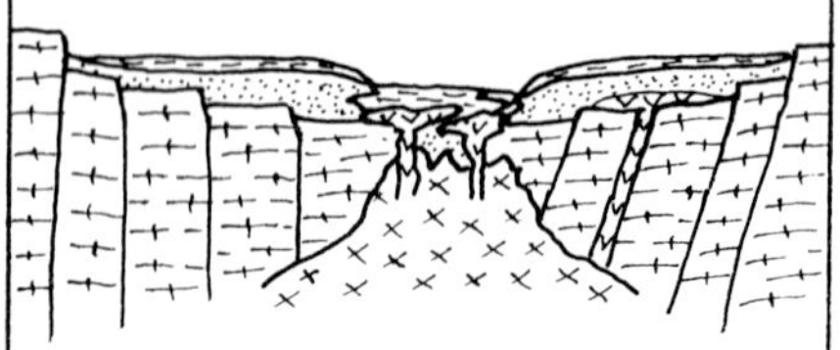
6 Growth of oceans	

Fig. 30-3. Idealized stages of rifting. From Laznicka (1984).

30.3. EXAMPLES OF MODERN RIFT AND TAPHROGENIC SYSTEMS

30.3.1. East African Rift (EAR)

The classical East African Rift system (Dixey, 1956; McConnell, 1967; Khain, 1971; Fig. 30-4) is an epiplatformic orogenic system of block uplifts interrupted by a system of anastomosing grabens and faults. It is more than 4,000 km long. The system can be traced from the Zambezi River valley in the south through Malawi and Zambia to southern Tanzania. From there, the western branch follows lakes Tanganyika, Kivu and Albert in the border region of Zaïre, Burundi, Rwanda and Uganda. The eastern branch (Gregory Rift) continues to Kenya and Ethiopia. In Ethiopia, the central graben widens to form the Afar Triangle, a site of a triple junction. There, the EAR loses its identity and the Red Sea and Gulf of Aden extend in the form of a N.W. and N.E. arms, respectively.

McConnell (1967) argued that much of the EAR has been superimposed in the Cainozoic time on an ancient network of mylonite and blastomylonite-filled shears and lineaments in the Precambrian basement. The individual EAR structures are true rift valleys (elongated sunken areas descended between parallel faults, 30-60 km wide and several hundred km long), as well as assymetrical narrow fault and tilted block, and fault and flexure, structures. The floors of both are intensively faulted. In its present configuration, EAR has been most active tectonically during the past 15 m.y., and the seismic activity continues. The Precambrian uplifted blocks have predominantly been regions of erosion except for local sections, where they are covered by Cainozoic volcanics or intruded by igneous rocks. The rift valleys have been areas of continental volcanism and sedimentation, both subaerial and subaqueous-lacustrine. The deep rift lakes remain important depositional basins.

The EAR volcanism (King, 1970) operated in several stages, and generally displayed a change from a pre-rift fissure areal volcanism (basalt and phonolite flows and shield volcanoes) to rift-controlled central volcanoes (basalt, trachyte, phonolite, nephelinite, peralkaline rhyolite, minor carbonatite). The volcanism is strongly to mildly alkaline and although the volcanic centres are most common within rift valleys and along rift faults, they are widespread on the rift flanks as well up to a distance of about 300 km from the rift axis (Fig. 30-5). The metallogeny of the rift-controlled alkaline volcanic and intrusive suite is treated in Chapter 33. The young volcanics are almost devoid of mineralization, whereas subvolcanic to intrusive stock carbonatites exposed in several dissected centres, carry the expected accumulations of Nb, REE, apatite, Sr, etc.

The rift valley lakes (Fig. 30-6) form a string extending over 3,000 km along the rift trace. Those located in the high rainfall area are typically narrow and deep. They accumulated a considerable thickness of sediments (up to several kilometres; Degens and Ross, 1976). The lake sediments are mostly mud, silt, reworked volcanic ash or tuff, minor diatomite, aragonitic carbonate and minor evaporites.

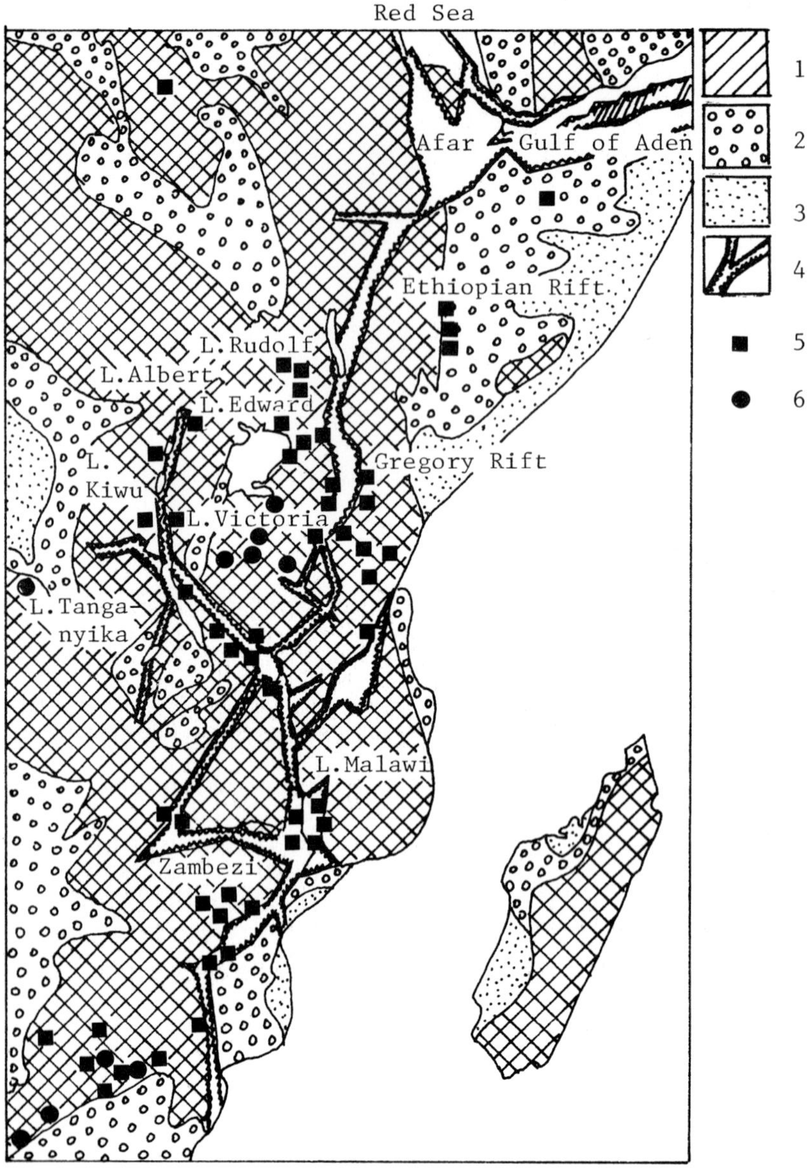

Fig. 30-4. The East African Rift. 1=PCm crystalline basement; 2=Pt_3 to PZ platformic cover; 3=MZ-T platformic cover; 4=rifts; 5=alkaline magmatic centra including carbonatites; 6=kimberlites. From LITHOTHEQUE.

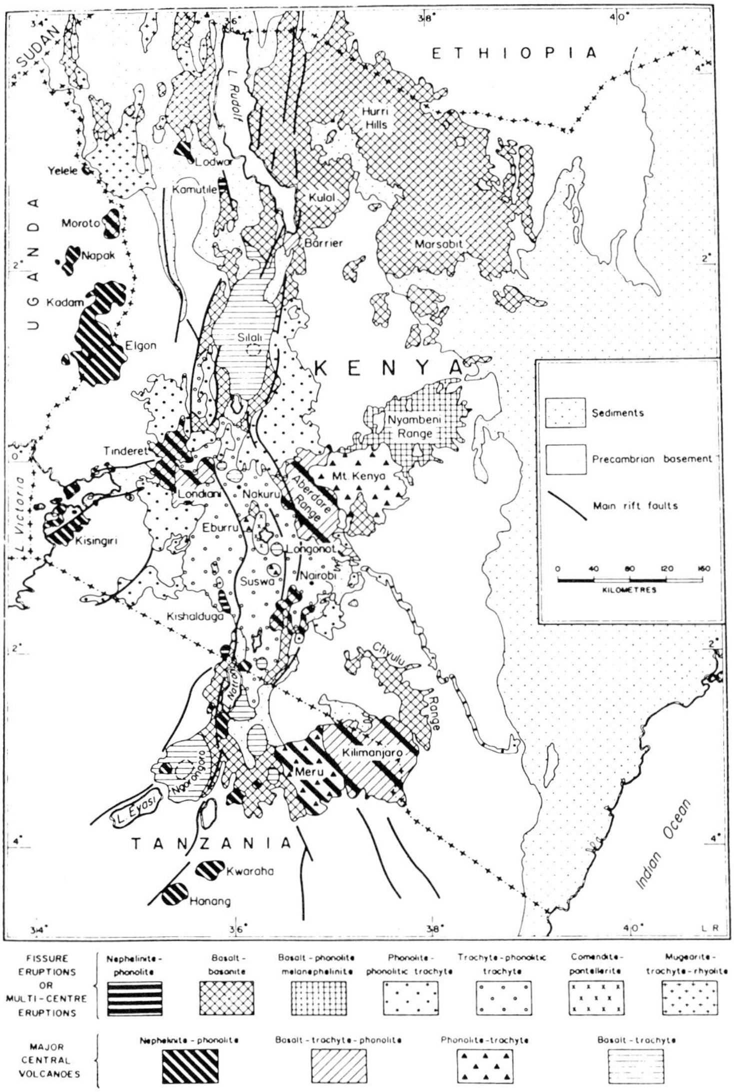

Fig. 30-5. East African Rift volcanism along the Gregory Rift in southern Kenya and northern Tanzania. From Williams (1969). Reprinted from Nature, with permission; © 1969, Macmillan Journals.

Fig. 30-6. East African Rift, lakes of the semi-arid zone. Top: Lake Manyara, Tanzania (the rift escarpment is on the right); bottom: soda Lake Magadi, Kenya.

Evaporites are more characteristic of the shallow ephemeral lakes such as Natron and Magadi, located in the dry grassland climatic belt. The African rift lakes, at present, yield only nonmetallics (trona, possibly fluorine), and no industrial metallic occurrence has been discovered so far. Several interesting geochemical metal anomalies, however, have been recorded and studied and they contribute considerably to our understanding of the rift lacustrine metallogeny.

A portion of such metallogeny is comparable with ore occurrences recorded in lakes in general, treated in Chapter 24 (e.g. heavy mineral placers in lacustrine beach sands such as monazite in Monkey Bay, Lake Nyasa, McNaughton, 1958, authigenic manganosiderite or rhodochrosite nodules in terrigenous mud or clay, Lake Kivu; Degens and Ross, 1976). The remainder is unique, being the result of interaction of the rift volcanism (particularly hot springs) and pluvial lake sedimentation. Most of the available information comes from the research done by Degens and co-workers (Degens et al., 1972; Degens and Kulbicki, 1973; Degens and Ross, 1976) in the Lake Kivu.

Lake Kivu located in the border region of Zaïre and Rwanda, is situated about 1,500 m above sea level, surrounded by active volcanoes of the Virunga Range. It extends to a depth of 500 m and reducing conditions prevail below about 50 m. The lake is fed by both rainwater and salty hydrothermal springs. The springs are the major source of water in the restricted Kabuno Bay. The Lake Kivu water itself has an anomalous zinc content (2 ppm Zn, that corresponds to 1 Mt dissolved or suspended Zn in the entire lake), much of which is bound in micron-size spheres of sphalerite enclosed in resin globules, suspended in the water. The lake floor sediments present in the reducing zone contain chemically precipitated nodules of Mn-Fe carbonates and high-Ni pyrite, as well as anomalous trace contents of Pb,Zn,Mo and Cu. The heavy metals are probably bound in organic complexes.

Müller and Förstner (1973) described an occurrence of nontronite, limonite, opal and vivianite, currently forming in aerated shallow parts of Lake Nyasa (Malawi) near Nkhota. The minerals form pellets with 20-45% Fe, accumulated in a layer, over 80 cm thick, topping diatomite or detrital sediments.

Harris (1961) recorded an unusual occurrence of thick phosphatic beds, developed around a small basement outcrop of Archean gneisses projecting through recent sediments of Lake Manyara, Tanzania. The phosphate is radioactive and is possibly a result of guano accumulation on an island. Mineral occurrences in buried Cainozoic sediments of the African lakes have rarely been reported probably because of the lack of drilling core. Harris (1961) recorded a bed of uraniferous strontianite at a depth of about 70 m in the Bahi depression near Dodoma, Tanzania.

Fluorine is a widespread element in EAR, clearly associated with rifting. Up to 22% villiaumite (NaF) and 6% fluorite is present in trona in the playa Lake Magadi, Kenya (van Alstine and Schruben, 1980). Fluorite is also frequent in tuffaceous lake beds. A substantial (9 Mt ore with 50% CaF_2) fluorite mineralization has been described from Kerio Valley in W.C. Kenya (Nyambok and Gaciri, 1975). There, fluorite lodes replace marbles of the Mozambique System along rift faults. The mineralization is interpreted as being of low-temperature hydrothermal origin and appears to be associated with basalt dikes.

30.3.2. The Salton Sea, California, hydrothermal brines

Salton Sea is a recently formed lake located in Imperial Valley (California and Baja California), a complex rift graben situated in the northern extension of the Gulf of California. Drilling for geothermal energy has outlined areas, measuring several square kilometres, underlain by a highly saline brine with an extraordinarily high content of metals (600 ppm Zn, 85 ppm Pb, 2.7 ppm Ag, 2,000 ppm Mn, 6 ppm Cu; White, 1967). Gray travertine deposited at the surface and in drillhole casings contained 10% Mn, 0.7% As, 0.07% Zn, 0.0015% Cu, in the form of Mn oxide and Mn calcite with finely dispersed sulphides. Megascopic sphalerite, chalcopyrite, tennantite and other minerals were discovered in well cuttings at depth. The Salton Sea active hydrothermal system serves as a model for fracture and paleoaquifer-controlled fossil hydrothermal mineralizations in paleorifts.

30.3.3. The Red Sea, summary of geology and mineralization

The Red Sea occupies an elongated N.N.W. depression over 2,000 km long, between north-eastern Africa and Arabia (Fig. 30-7). Its width ranges from 28 km in the southern narrows to 360 km at the latitude of Asmara. The geology of the Red Sea has been summarized by Coleman (1974). A pre-Miocene downwarp between the Arabian and African Precambrian basement swells was filled by a considerable thickness (over 3 km) of Miocene evaporites and clastic sediments with minor interbedded plateau basalts. These rocks, covered by Pliocene and Quaternary coral reef limestones, marginal clastics and marine oozes, underlie a broad shelf along both shores of the Red Sea. The shelf continuity is interrupted by an axial trough, 48-74 km wide, formed in the Pliocene time and related to a major rift. The trough is a steep-walled depression with basaltic floor covered by a thin vener of sediments. The basalt is interpreted as being oceanic tholeiite to alkali basalt and is believed to represent a newly formed oceanic crust along a spreading center.

Intermittently distributed on the seafloor of the axial trough are local depressions marked by anomalous heat flow, highly saline brine pools, hydrothermal sediments and sometimes metal accumulations. Thirteen brine pools had been discovered by the early 1970s (Schoell et al., 1974), five of which carried appreciable mineralization. The first discovered and largest metallic accumulation, the Atlantis II Deep, has been most extensively studied and is described in a large volume of literature (e.g. Degens and Ross, eds., 1969; Hackett and Bischoff, 1973; Backer and Richter, 1973 and others). It has already been test-mined. Its reserves are now estimated to be 1.7 Mt Zn, 400 Tt Cu and 5 Tt Ag in a 4-8 m thick layer distributed over 4.5 km^2 of the seafloor (Mining Magazine, 1981).

The Atlantis II mineralization (Fig. 30-8) is covered by about 2,000 m of normal seawater, and about 200 m of warm brine at 60-65° C

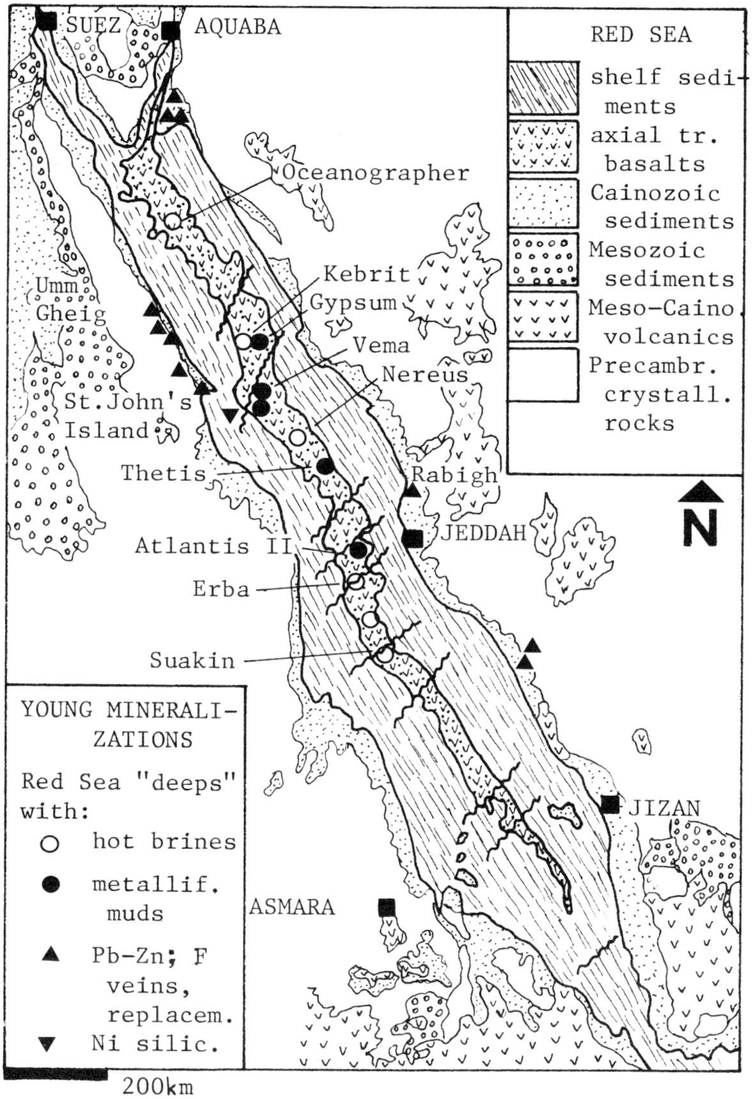

Fig. 30-7. Red Sea, geology and Cainozoic mineralization; geology after Coleman (1974).

with 44 to 56° salinity, stratified by temperature. The metals are present as suspended matter in the brine, from which they gradually settle to form the seafloor metalliferous sediments. The 12 to 30 m thick layer of hydrothermal sediments rests on a basaltic bedrock and is compositionally zoned (from top to bottom: amorphous hydrosilicates → smectites with sphalerite, pyrite, manganosiderite, chalcopyrite → Fe-hydroxides, manganite → lower sulphide zone →

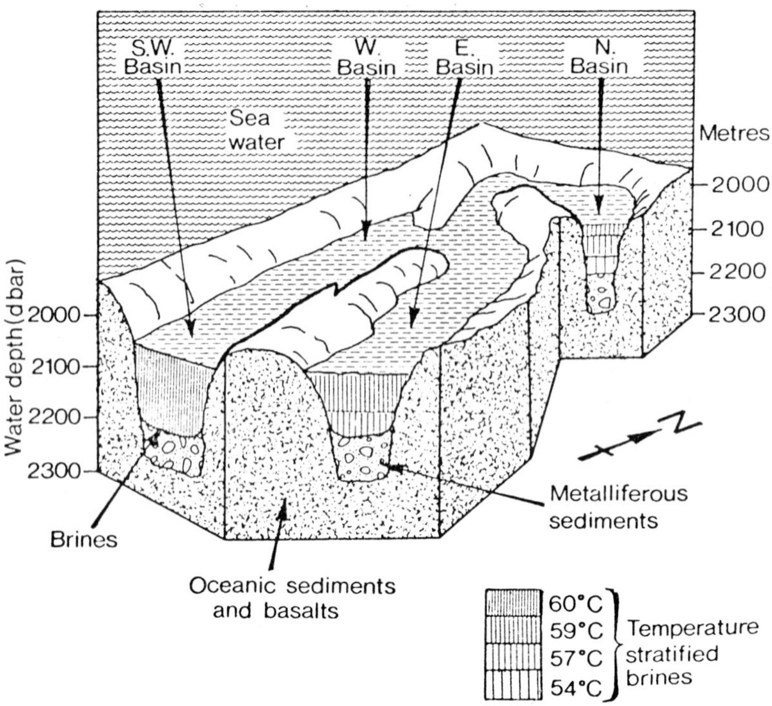

Fig. 30-8. Atlantis II Deep and its metalliferous brines and sediments. After Schoell, reprinted with permission from the Mining Magazine, August 1981.

---

coquina, limonite, minor sulphides).

The "ore" is black, fine grained and of paste-like consistency. It is usually interpreted as being a product of chemical precipitation from a subsurface brine. The brine is: (1) heated and driven by basaltic volcanic heat; (2) deriving its high salt content from the subsurface Miocene evaporites and (3) leaching its constituent metals from rocks en route, in particular from the Precambrian metamorphics. The Atlantis II Deep sulphide facies contains from 15.74 to 20.31% Fe, 1.11 to 2.81% Mn, 5.31 to 10.93% Zn, 1.34 to 2.22% Cu, 0.11 to 0.18% Pb and increased contents of Ag,Hg,Co. The same elements are anomalously concentrated in the oxide and silicate facies of sediments, but in sub-grade amounts (Bignell et al., 1976).

The young (Pliocene to Recent) Red Sea basalts occassionally crop out under subaerial conditions. One well-known occurrence is situated on the small St. John's Island (about 75 km S.E. of Berenice on the Egyptian coast). In addition to basalt and coral reefs, this island consists of serpentinized peridotite (Garson and Shalaby, 1976) and it is not known whether the peridotite is a tectonically emplaced raft or a member of the Cainozoic orogenic assemblage. In the peridotite, narrow and short, but rich infiltration zones of limonite and garnierite formed (5-6,000 t ore with 4.86% Ni, 0.93 ppm Pt), probably derived from "primary" disseminated or massive sulphides.

30.3.4. Afar Triangle

The Afar Triangle of eastern Ethiopia (Tazieff et al., 1972) is considered to be a displaced segment of the axial (rifted) portion of the Red Sea, recently uplifted so that the ocean ridge basalts and thick evaporites recently deposited subaqueously are now subaerally exposed. Bonatti et al. (1972b) described the En Kafala Fe-Mn deposit, interpreted as having formed some 200,000 years ago by submarine hydrothermal activity in a Red Sea-style brine pool. The mineralization is similar to the manganese deposits of Cuba, Efate (New Hebrides) and other localities, generated in an island arc setting. A basalt foundation is overlain by a basalt conglomerate locally cemented by Mn-oxides. These, in turn, are topped by a goethite-rich and manganite (birnessite) and pyrolusite layer. The Fe-Mn oxide layers are around 2 m thick and are capped by a reef limestone. Tooms (1976) mentioned occurrences of Cu-rich brines, copper deposits topping gypsum beds and manganiferous horizons in the Northern Ethiopia Rift Valley.

30.4. EXPOSED PALEO-RIFTS

Paleo-rifts, the sites of former rift systems that are no longer active, can be identified using a set of criteria (Table 30-2): (1) relic morphology, such as fault and relic graben pattern comparable with the EAR (e.g. the St. Lawrence Valley, Kumarapeli and Saull, 1966); (2) characteristic rock association and rock body configuration (e.g. alkaline ring complexes) or (3) both. Compared with the recent rifts, paleorifts carry substantially more widespread and important metallic mineralization than the former. This is clearly due to the greater depth of erosion exposing subvolcanic and plutonic levels of the rift magmatites. Since obvious rift grabens have rarely been preserved and faults may have been partly healed (e.g. by intrusive dikes) or draped by younger sediments, establishment of a paleorift requires a painstaking reconstruction and the degree of uncertainty is often high.

In some areas (e.g. in the Irish Zn-Pb province hosted by Carboniferous carbonates) where orebodies that usually occur associated with intrusive rocks lack such association, tentative suggestions have been made that orebodies might be related to rifting-generated fault systems, in particular by fault intersections (e.g. Russell, 1968; Vokes, 1973). The ore, faults, rifting relationship may be only conjectural, but it is widely used. A genetic connection between rifting and an epigenetic Pb-Zn

Table 30-2. Some criteria for the interpretation of paleorifts

GEOPHYSICAL
-thinned continental crust (shallow MOHO)
-shallow earthquakes usually associated with Phanerozoic paleorifts
-large negative Buguer anomaly over the broad uplifted arch
-sometimes, narrow zones of gravity maxima over axial grabens
-commonly increased heat flow
PETROGRAPHIC
-intimately associated alkaline intrusions (Chapter 33)
-loosely associated tholeiitic (exceptionally oceanic) basalts
-regionally associated bimodal (basalt-rhyolite) association, high-level granite plutons
-thick sequence of evaporites
STRUCTURAL
-relic rift valleys, about 50 km wide
-bifurcating tectonic pattern, short angular deflections, dimensions comparable with the East African Rift system
-dense network of longitudinal (parallel) faults
GEOMORPHOLOGIC
-remnants of fault scarps and fault-line scarps
-tilted blocks and horsts
-remnants of grabens, often interrupted by rivers
-frequent hot springs

mineralization has been proposed to explain the origin of the Proterozoic Nanisivik Zn-Pb deposit, Baffin Island, Canada (Jackson and Iannelli, 1981); the Zurak-Abakaliki Pb-Zn belt in the Benue Trough, W. Africa (Burke et al., 1972); the Pb-Zn replacements in Miocene carbonates along the western shore of the Red Sea, Egypt (e.g. Umm Gheig) and eastern shore in Saudi Arabia (Motti et al., 1981); Pb-Zn and barite-fluorite veins south of the Ottawa Valley, Ontario (Kumarapeli, 1976) and others.

Sawkins and Rye (1979) offered an interesting paleorift-related interpretation of the origin of the important Messina (Transvaal) copper field, hosted by the Archean metamorphics of the Limpopo mobile belt. The Messina orebodies are Cu sulphide mineralized breccia pipes, disseminated replacements and fissure filling veins dating from post-Karoo (Mesozoic) age, aligned along a N.E. trend for a distance of 20 km. Supposedly they formed by precipitation from heated saline brines controlled by a Mesozoic paleorift. The brines acquired their Cu content from the rift-floor basalts believed formerly to have been topping the hydrothermal convection system.

30.5. ASSYMETRICAL PORTIONS OF PALEORIFTS PRESERVED UNDER "ATLANTIC-TYPE" CONTINENTAL MARGINS; PALEORIFTS WITHIN SHELVES COVERED BY YOUNG SEDIMENTS

Association of terrigenous sediments (particularly of the red-beds facies), evaporites, mafic volcanics and intrusions, is frequently present as a block-faulted fill of fault-bound troughs and grabens, now buried under a thick pile of continental shelf sediments under the Atlantic-type continental margins. These rocks represent the remnants of intercontinental and intracontinental rifts dismembered by continental drift and are particularly common under the Atlantic shelf of North America and under the North Sea sediments off western Europe (e.g. Sheridan, 1974; Ballard and Uchupi, 1975). Salt doming is widespread. Most of these buried occurrences are known from geophysical work and confirmed by oil drilling, and no economic mineralization has been reported so far, although metals (particularly copper) could be expected by analogy.

30.6. HYPOTHETICAL, METAMORPHOSED PALEORIFTS

Increasingly often, the paleorift interpretation is offered in the literature to interpret metamorphosed associations of amhibolite, marble, schists, and strata-related ores mostly Cu,Zn or Pb. This is particularly true in the instances where such associations form narrow lithologic belts supposedly generated with a significant taphrogenic component (e.g. Wopmay Orogen in northern Canada, Easton, 1981; "Amphibolite Belt" of Namibia; Goldberg, 1976; Mount Isa shale belt, Queensland; Sawkins, 1976, see Volume 2). Continental rift models have also recently been proposed to explain the formation of the Precambrian greenstone belts (e.g. Anhaeusser, Windley, Hunter, and others; see summary in Condie, 1981, p. 349-354).

Selected examples of mineralized rifts and paleorifts are listed in Table 30-3 and their location is marked in Fig. 30-9.

Table 30-3. Selected examples of mineralized rifts and Phanerozoic paleorifts

RIFTS ACTIVE IN THE CAINOZOIC
East African Rift system A system of block uplifts over 4,000 km long, interrupted by anastomosing grabens and faults, established on Precambrian crystalline basement. The central grabens are filled by arid to humid continental sediments. Deep lakes in the humid zone have terrigenous and volcanogenic sand to mud (high proportion of euxinic mud). Widespread areal and central tholeiitic, bimodal and alkaline

volcanism and intrusive activity in Neogene and Quaternary. ORES: rare Mn nodules in lake sediments; slightly metalliferous (mainly Zn) lake muds and waters; high F in playa evaporites and brines; apatite, minor pyrochlore in high-level Cainozoic carbonatites. McConnell, (1967); Khain, (1971); Mitchell and Garson (1981).

Red Sea

A N.N.W. depression, over 2,000 km long and 28-360 km wide. Over 3 km of Miocene evaporites and clastics fill a Precambrian basement swell, topped by Pl-Q tropical shelf sediments. These are interrupted by a Pl-Q steep-walled axial trough 18-74 km wide, with a basalt floor, covered by a thin veneer of euxinic sediments. ORES: Hot brine pools and metalliferous muds are known from several sites in the axial zone (largest: Atlantis II deep, 1.7 Mt Zn). Numerous fault-controlled fluorite, minor Pb-Zn vein and Pb-Zn carbonate replacement deposits, are along the margin of the Sea, in Neogene sediments or PCm basement. Coleman (1974); Degens and Ross, eds. (1969); Mitchell and Garson (1981).

Bekaa, Dead Sea, Wadi Arabah rift valley

A N.N.E.-trending continental extensional graben, 360 km long and 5-20 km wide, filled by up to 10 km of Oligocene to Recent sediments (fluvial, lacustrine) and some basalt flows. Up to 4 km of evaporites. ORES: small infiltration occurrences of Cu silicates and carbonates and Mn oxides. Bender (1975)

Mongolia, Baikal, Yakutia rift system, Siberia

A chain of assymetrical grabens and horsts, 2,500 km long and up to 300 km wide. The grabens are filled by up to 5 km of Ol-Q alluvial, paludal and lacustrine sediments and minor olivine basalt and trachybasalt flows. ORES: Minor Mn oxide nodules on Lake Baikal floor. Florensov (1966).

Rhine Graben

An Eocene to Quaternary seismically still active graben, 300 km long and 36 km wide, flanked and underlain by Hercynian crystalline blocks. Filled by continental sediments, minor flood basalts, centres of alkaline volcanism and intrusive activity including rare carbonatite (Kaiserstuhl). ORES: Small pyrochlore deposit in Kaiserstuhl carbonatite. Illies and Mueller, eds. (1970).

Gulf of California, Imperial Valley

(Transitional to oceanic spreading ridges and transform fault systems).

Succession of closed marine basins separated by fault scarps in continuation of the East Pacific spreading ridge. Deltaic to normal marine sediments, euxinic muds, diatomites. Arid continental sediments on land, minor rhyolite plugs and obsidian. ORES: Salton Sea, metalliferous hot brines in depth, geothermal area. Subaqueous present hydrothermal activity in Guaymas Basin with Fe,Cu,Zn sulphide mounds. The bedded Neogene Boleo (Santa Rosalia) deposit is considered by some to be related to the initial stage of the rifting. Byrne and Emery (1960), Sawkins (1984).

PRE-TERTIARY PALEORIFTS (INCLUDING AULACOGENES)

Newark Trough, eastern North America

A Triassic system of N.E. faults and grabens up to 100 km wide, traceable from Nova Scotia to the subsurface of Florida. Filled by red-beds continental fluvial and piedmont fan sediments, minor flood basalts, diabase and gabbro dikes. ORES: Numerous small occurrences of "red beds" Cu infiltrations in sandstones, traces of U, Mn. Significant skarn and replacement magnetite and Co-pyrite at limestone/diabase contacts (e.g. Cornwall). Burke (1977).

Oslo Graben, Norway

Remnants of a Permian graben formed within a Proterozoic basement and lower Paleozoic platformic sediments. The structure is almost entirely filled by Permian flood basalts and syenite, monzonite, granite subvolcanic and epizonal intrusive complexes (typical are circular cauldrons). Minor alkaline intrusive rocks. ORES: large number of small ore occurrences associated with the granites and syenites (Bordvika, Drammen: stockwork Mo; small magnetite skarns and Zn-Pb replacements in limestones; Pb-Zn veins) and with marginal faults and diabase or gabbro dikes (Kongsberg Ag veins). Vokes (1973).

Pripyat'-Donetsk aulacogene (south-western U.S.S.R.)

A graben system over 1,000 km long, up to 100 km wide, established in Devonian, active till Mesozoic. Filled by a sequence of Pe-Cb red beds and coal association 5-12 km thick and intruded by minor Jurassic intrusive stocks, alkaline dikes and lamprophyres. ORES: Minor "red beds Cu" deposits in reduced sandstones, lesser Pb-Zn in shales and sandstones. Cu,Pb,Zn,Ag veins (Nagolnyi Kryazh) affiliated with Mesozoic intrusive activity and so is probably the fault-controlled Nikitovka cinnabar deposit (cinnabar hosted by Pe-Cb sandstone, shale and coal. Novikova (1964).

Pachelma aulacogen, southern Russia

This is located between Voronezh and Volga-Uralian Massifs. Pt_3 to PZ_1 red-beds sequence including continental sandstones, red dolomite, marls, black argillites, minor basaltic tuffs. Buried under younger sediments of the Russian Platform. ORES: Substantial stratiform accumulations of Ti-magnetite in tuffs and reworked mafic tuffs in the Yastrebovsk Horizon (near Nizhnyi Mamon); Cu-Ni "sandstone" impregnations. Novikova (1964).

St.Lawrence River Rift and Ottawa-Bonnecherre Graben, E.Canada

E.-W.-trending broad area of updoming, interrupted by several graben systems in the Precambrian basement. Mainly Cretaceous, but still seismically active. Little Mesozoic sedimentation, but scattered Cretaceous small alkaline intrusive centers; largest: Monterregian Hills) dominated by nepheline syenite varieties. Minor carbonatites. ORES: Oka and St.Honoré pyrochlore carbonatite mined; probable Pb-Zn and barite, fluorite veins. Kumarapeli and Saull (1966).

Benue-Abakaliki Trough, Niger Delta, Africa

N.E.-trending graben system developed on PCm crystalline basement. The basal Cretaceous suite up to 7 km thick formed in an "aulacogene" and consists of a folded shale and arkose fill. This is topped by up to 4 km of Cr3-Eo ordinary marine quartz arenite and shale. Minor mafic volcanics. ORES: Pb-Zn veins and replacements in the Zurak-Abakaliki belt. Burke (1977).

Mississippi Embayment, south-central United States

N.E.-trending complex of subparallel narrow horsts and grabens within Paleozoic platformic sediments and their Precambrian basement. Transgressively overlain by Cretaceous and younger nearshore detrital sediments of the Gulf Plain. Minor alkaline intrusions (dikes, diatremes; e.g. Hicks Dome). ORES: Fluorite, lesser barite, Zn-Pb veins in the Illinois-Kentucky district; minor REE occurrences in diatremes.

Fig. 30-9. Rifts and related structures of the earth, showing principal localities treated in Table 30-3.
1=epiplatform arch, volcanic rift zones; 2=epiplatform crevice-like rift zones; 3=epeirogenic rift zones and belts; 4=intercontinental rift zones; 5=mid oceanic ridges with axial rift valleys; 6=as 5 but without rift valleys; 7=mid oceanic ridges with important volcanism; 8=large scale faults active in T. Paleorifts: 9=MZ_3 and T; 10=MZ_1; 11=PZ; 12=Pt_3; 13=areas of T volcanism outside of alpine orogenic belts; 14=zones of T epiplatform orogenesis; 15=Q deep sea troughs; 16=Q "geosynclines"; 17=zones of T folding; 18=zones of MZ folding; 19=zones of PZ folding; 20=zones of Pt_3 folding and regeneration; 21=pre-Pt_3 platforms; 22=oceanic floor with sub-oceanic crust; 23=deep sea depressions with sub-continental crust; 24=oceanic basins with oceanic type crust.

LOCALITIES: BR=Baikal Rift; BT=Benue Trough; DA=Pripyat'-Donetsk Aulacogene; DS=Bekaa, Dead Sea, Wadi Arabah; EAR=East African Rift; GA=Gulf of Aden; MV=Midland Valley; NT=Newark Trough; OG=Oslo Graben; OK=Orsha Krestzov Aulacogene; P=Pachelma Aulacogene; RG=Rhine Graben; RS=Red Sea; SL=St. Lawrence Rift; RG=Rhine Graben; RS=Red Sea; SL=St. Lawrence Rift. From Milanovskii (1972), localities added.

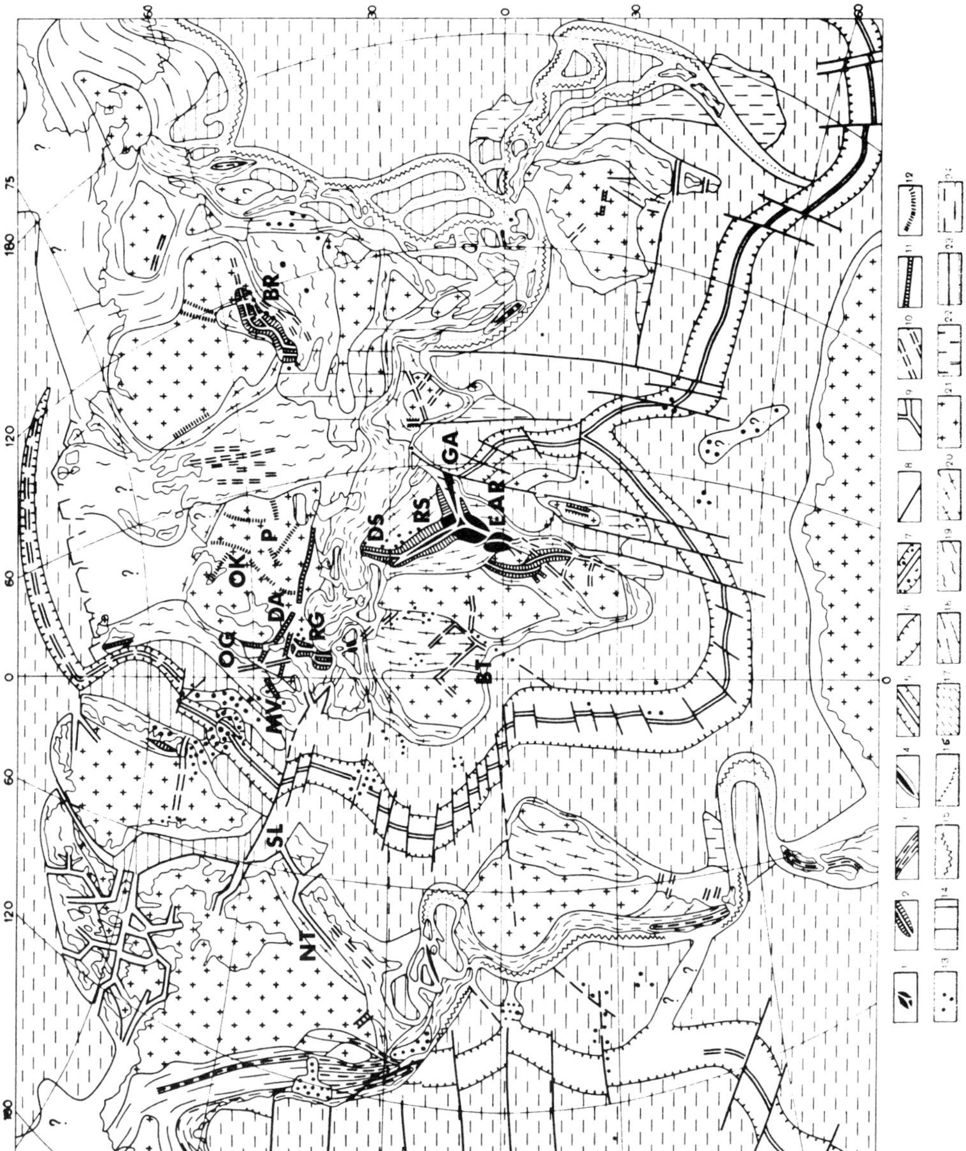

SL
NT
MV
OG
OK
DA
RG
P
DS
RS
GA
EAR
BT
BR

Plate 11. Quaternary continental basalt lava and cinder cones at the Sunset Crater near Flagstaff, Arizona.

"The arrangement of many transitional types (of ore deposits) into the (genetic) system is basically a matter of choice of the individual". Paul Ramdohr in The Ore Minerals and Their Intergrowths, 1969.

CHAPTER 31

Continental Plateau Basalt and Bimodal Volcanic Association

31.1. INTRODUCTION

Basalts are the greatest continental lavas in terms of volume and the majority are tholeiitic (that is, predominantly plagioclase-clinopyroxene, minor olivine, spinel, ilmenite, magnetite assemblages (Carmichael et al., 1974). Gradations to alkali basalts, however, are widespread and sharp distinction between these two series cannot be made in the field. Consequently, alkali basalt provinces are included here, whereas alkaline association (with feldspathoids) is excluded and treated in a separate Chapter 33.

Areally the most widespread and covering areas of up to 1.2 million km^2, are the massive accumulations of subhorizontal, fissure-fed lavas. The thickness of their central portions is typically 1-3 km, but in exceptional cases it reaches 5 or 6 km. These are usually designated as being plateau or flood basalts in the literature. The synonymous term "trap" (or trapp, trap-rock) is widely used in the European, Soviet and Indian literatures. The petrology and petrogenesis of plateau basalts is treated in Carmichael et al. (1974), the volcanology in Williams and McBirney (1979). The Basaltic Volcanism Study Project (1981) provides the most comprehensive summary covering these and other (e.g. geotectonics) subjects. The metallogeny of plateau basalts has mostly been treated in the framework of local areas, and the work of Oleinikov (1979) with the emphasis on metallogeny of the platformic mafic suite of Siberia, is among the most comprehensive. Two major contrasting types of volcanicity are recognized in most plateau basalt regions: (1) fissure eruptions that generate flood-basalt sheets of regional extent, underlain or fringed by feeder dikes or sills and (2) localized eruptions emitted from major central volcanoes (Williams and McBirney, 1979).

The basaltic sheets are composed of individual lava flows which have thickness in metres to tens of metres, and each flow usually shows distinct variation throughout, from massive base and centre through increasingly amygdaloidal upper third to scoriaceous, brecciated and often oxidized (brown in contrast to the green remainder) flowtop. Tops of individual flows often contain lenses or horizons of interflow continental sediments, such as fine brown argillite, siltstone, sandstone and conglomerate. The volcanic sequence near its margin and at the top frequently interfingers with fluvial or lacustrine sediments. Although most plateau basalts have erupted subaerially (and some flowtops are regolithic), subaqueous flows often occur: in subglacial environments, in continental lakes, river valleys, etc., where pillow lavas and/or hyaloclastites may occur. Basalts that have invaded water-saturated sediments (invasive flows of Swanson and Wright, 1981) have a sill-like appearance with glassy upper selvedges, and often grade to pépérites (mixed granular basalt and sediment). Plateau basalts may also grade into submarine flows, which are often pillowed. In large plateau basalt provinces, the lava deposition has often been cyclic (four cycles in the Paraná Basin).

In the subsurface, plateau basalts are represented by mafic dikes, sills, sheets and layered intrusions. Dikes tend to form swarms, usually oriented at right angles to the direction of crustal extension. Sills, where present, may be the immediate precursors of surface eruptions (Baragar, 1977). In Mesozoic plateau basalt provinces, such as the Paraná Basin, a zone of abundant diabase dikes or sills occurs on the flanks of the continuous basalt lava sheet. In many deeply eroded Precambrian regions of the world such as in the crystalline basement of the Canadian Shield, swarms of diabase dikes are present without spatially associated surficial basalts. The regional ratio of diabase sills to basalt flows is, according to Baragar (1977), governed by the quantity and availability of easily intruded sediments at the site of eruption. Where basalt overlies a thick sedimentary sequence, sills emplaced along the bedding planes of sediments are dominant. Where only a thin sediment covers a consolidated basement, flows are abundant and sills negligible.

Layered intrusions may occur at subvolcanic levels of platau basalt provinces and are usually interpreted as being shallow magma chambers. Compared with flood basalts and diabase dikes, such plutons are relatively rare, and only in exceptional instances are consanguinous lavas, dikes and plutons preserved in a single province (such as the Proterozoic Keweenawan basalts and Duluth Complex pair preserved in the Lake Superior region, or the Coppermine basalts and Muskox intrusion pair in northern Canada). More frequently, layered mafic plutons occur separately and the largest recorded example, the Bushveld complex of South Africa, is a classical locality. Even so, virtually all the layered intrusions exposed at present are of Precambrian age, and so have been treated in Volume 2.

The localized (central) eruptions in plateau basalt associations mostly display the characteristics of small cinder cones and low stratovolcanoes that emit lavas and pyroclastics compositionally identical with the flood basalts. The two styles of volcanicity usually overlap. Mafic pyroclastics are particularly common in the Siberian Platform and Deccan Plateau basalt provinces. Contrast differentiated (bimodal, i.e., mafic-felsic) central complexes are rarer, but petrographically and structurally highly variable lavas, pyroclastics and neck facies coexist and some complexes have a compositional range that extends from ultrabasics to rhyolites.

Usually, contrast-differentiated complexes are younger than the main bodies of plateau lava sheets over which they may be superimposed. Alternatively, such complexes may crop out along the plateau basalt fringe. Deep dissection has often exposed the subvolcanic (mafic and felsic dikes) and plutonic (small mafic and felsic stocks and plutons) levels. In deeply eroded ancient terrains the volcanic level is usually missing, and exposed plutons can only occasionally be assigned to a particular plateau basalt cycle.

In many plateau basalt provinces, the latest stages of continental magmatism are alkaline (see Chapter 33). Compared with magmatic complexes of mobile belts, there was very little interaction between the bulk of tholeiitic magmatic rocks and the earlier rock suites into which they were emplaced, as well as among volcanic flows and the contemporary overlapping sediments (when the latter are present at all). There are, however, exceptions many of which are associated with important metallic mineralizations.

Continental tholeiitic association emplaced into continental crust ranges in age from Archean to Recent (Baragar, 1977), and although the orthomagmatic conditions and characteristics of the association changed only slightly due to evolution, the colour and general appearance of the lavas changed as a consequence of ageing (diagenesis) and burial metamorphism, generally from gray in the recent basalts to green in the ancient ones. Because the present field appearance is of fundamental importance in the practical fieldwork, and because several important mineralizations in the supracrustal members appear to be controlled by post-depositional events, the colour of the basalts was taken into account in the subdivision that follows. The colour is usually proportional to the geological age.

31.2. GEOTECTONIC SETTING AND ORIGIN

Continental tholeiites are said to be the characteristic volcanics of the stable crust (cratons, platforms; Baragar, 1977). However, all occurrences of plateau basalts and their deeper-seated feeders or equivalents, are also associated with distention of the continental crust (taphrogenesis). In this way fissures were generated along which magma from the subcrustal region ascended. The degree of distention varies greatly, from tenths to several tens of percent of the original area. Large plateau basalt or diabase dike provinces correlate with major events of continental break-up. Generally, one can distinguish intracratonic provinces (such as the Paraná and Karoo Basins), and continental margin provinces (the Brito-Arctic Province, Atlantic coast of Brazil). Small plateau-basalt provinces are frequently linear and often discontinuous. They are controlled by grabens or "aulacogens" (i.e., failed rifts), superimposed on cratons. Comparable provinces also formed by taphrogenesis during the closing stages of orogenic belts development. These are mentioned briefly in Chapter 26.

Tholeiitic magma is believed to have been derived from the upper mantle, by partial melting above mantle plumes and plume-ridges (Baragar, 1977). This was followed in many cases by fractional crystallization in small magmatic chambers within the continental crust. Most continental tholeiites show signs of crustal contamination (Carmichael et al., 1974), and they are generally distinguishable from the "geosynclinal" volcanics by a higher content of "incompatible" elements.

31.3. ORE DISTRIBUTION AND ECONOMIC IMPORTANCE

The continental plateau basalt association has a typically "platformic" characteristic of distribution of metallic deposits. The bulk (over 90%) of the territory composed of subhorizontal lavas and sills is completely devoid of any metallic mineralization (except for the low factor of concentration ores of Al, Fe and Ti generated by tropical weathering). There is, however, significant mineralization confined to small local areas both inside and outside the monotonous lava or sills terrain. The bulk of this mineralization is (1) the product of magmatic differentiation in the feeder or magmatic chamber apparatus in depth (e.g. the Bushveld complex); (2) the product of interaction of intrusions with their environment (e.g.the Noril'sk district, Duluth Complex) and (3) a product of metal-concentrating mobilizations superimposed on the lavas (Keweenaw Peninsula-Cu, bauxites, Fe-ores).

The members of the continental plateau basalt association treated in this chapter contain about 20 Mt Cu, 7-10 Mt Ni, 6 Bt Fe, 30-40 Mt Al (in bauxites) and about 10 Tt Pd+Pt metals, that are distributed in the very few deposits that are economic at present, or are likely to be in the near future. There seems to be a good potential for future, very low-grade metal sources. Figure 31-1 and Table 31-1 summarize the major known mineralization styles.

Continental tholeiites are transitional into oceanic tholeiites (Wilkinson, 1981) and to basalts of the continental margin mobile belts, with which they share many lithogenetic and metallogenetic characteristics (see Chapters 4, 10, 12). Alkaline volcanics and intrusions (Chapter 33) frequently occur in terrains underlain by plateau basalts (Paraná, Deccan, Siberian Platform), in which they are usually geologically younger. Depth equivalents of plateau basalts, diabase or gabbro dikes emplaced into the continental crust and layered complexes of the Bushveld style, are treated in Chapters 32 and in Volume 2. Tholeiitic basalts located in narrow, linear, tectonic grabens within cratons (Chapter 26) are treated separately and so are the minor occurrences of basalts in the predominantly sedimentary sequences, for example the "red-beds" (Chapter 25).

31.4. MAJOR SUB-ASSOCIATIONS AND THEIR METALLIC ORES

31.4.1. Cainozoic (dominantly "gray") plateau basalts

The archetype of the young plateau basalts is a medium to dark gray, fine crystalline to aphanitic massive basalt, that occurs as subhorizontal flows which display characteristic columnar jointing. Small olivine phenocrysts are sometimes megascopically recognizable. At least 75% of the outcrop in plateau basalt provinces is composed of this extremely monotonous rock, ignored by exploration geologists for its apparently complete lack of mineralization potential. The most common other structural or textural variation of plateau basalts encountered in the field are the vuggy, scoriaceous and brecciated basalts of lava flowtops, and the cinders, scoria and spatter of the central vents. Many of the porous rocks listed above are oxidized (red). The pores remain unfilled. Some flows are covered by, or interbedded with, continental sediments.

Deep (several km) burial caused diagenetic or metamorphogenic migration of the mobile components and healing of the pores by calcite, zeolites, or silica. Walker (1960) demonstrated the correlation between pore filling minerals in basalts and the depth of burial, in Iceland.

METAL GEOCHEMISTRY AND METALLOGENY

Common tholeiitic lavas (and magmas) are enriched in the "basaltophile" trace metals to an extent, several orders of magnitude greater than the crustal average (Cr, 200-500 ppm, Ni: 70-300 ppm, Co: 30-50 ppm, Cu: 70-200 ppm). In olivine-rich rocks (olivine basalts that are most common near the base of thick sequences of tholeiitic flows, picritic basalts and picrites), the Cr, Ni and Co content increases further, to reach values of 0.18-0.22% Cr_2O_3 and 0.1-0.12% NiO, as in the Tertiary olivine tholeiites of Baffin Island (Canada) and Swartenhuk (W.Greenland; Clarke, 1977). The unusually high magnesium content and low proportion of incompatible elements-containing lavas in the Baffin-Western Greenland Province are exceptional among the Cainozoic basalts. They are considered by Clarke (1977) to be examples of unmodified (primitive) partial melts of the mantle, comparable with komatiites.

Alkali basalt lavas have appreciably lower trace Cr,Ni,Co and Cu contents (clarke to sub-clarke values). They have, however, slightly super-clarke contents of Al (12-20% Al_2O_3), Fe (10-15% Fe oxides) and Ti (1.0-3.5% TiO_2), but are low in sulphur. At the volcanic (near-surface) level, and during the magmatic stage, processes or agents of further concentration of the geochemically enriched metals are almost lacking, and consequently metallic deposits do not form. Rare, uneconomic metallic occurrences in basalts such as the emery (ie., a physical mixture of granular corundum and magnetite with minor hercynite and rutile) showing near Quirindi, N.S.W., Australia, are

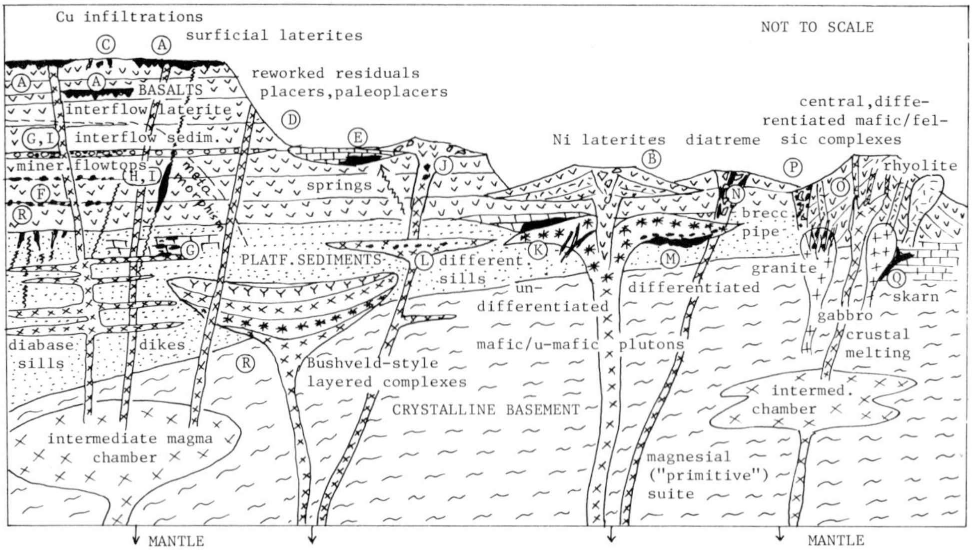

Fig. 31-1. Principal lithologic sub-associations and mineralization styles in the continental plateau basalt and bimodal associations (see Table 31-1 for an explanation of the letter codes). Some of the Precambrian associations and styles treated in Volume 2 are shown.

Table 31-1. Principal mineralization styles in the continental plateau basalt and bimodal associations
NOTE: Precambrian paleo-basalts described in Volume 2 are partly included in this table

	WEATHERING CRUSTS
(A)	residual bauxite and Fe oxides (50 Mt Al, 10 Mt Fe)
(B)	Ni laterites and saprolites formed on picrites, high-Ni olivine gabbros, norites, ultramafics
(C)	Cu silicates, carbonates, oxides, infiltrations in regolithic amygdaloidal metabasalts
	PRODUCTS OF SEDIMENTOGENIC REWORKING
(D)	ilmenite, Ti-magnetite, zircon, etc. in placers and paleoplacers (partly 10 Mt Ti)
(E)	PRECIPITATES FROM HOT SPRINGS DRIVEN BY BASALT HEAT; HYDROTHERMAL REPLACEMENTS (e.g. El Bahariya; controversial)
	GREEN (ALTERED) METABASALTS
(F)	native Cu, peneconcordant orebodies in prehnite-pumpellyite alteration assemblage (e.g. Keweenaw Pen.; 3 Mt Cu)

(G)	native Cu in interflow conglomerate, sandstone (Hecla, Calumet; 2.5 Mt Cu)
(H)	native Cu in discordant fissure veins in metabasalts, near flowtops (150 Tt Cu)
(I)	Cu sulphides in veins, breccias, hosted by altered amygdaloidal basalts (100 Tt Cu)
	DIABASE, BASALT, GABBRO DIKES, SILLS, CLOSELY AFFILIATED WITH PLATEAU BASALTS
(J)	native iron, disseminations and lumps in basalt dike, W. Greenland
(K)	magnetite disseminations, veins, stocks in diabase dikes and sills and their contacts (100 Mt Fe)
(L)	disseminated pyrrhotite, pentlandite, chalcopyrite in differentiated mafic sills (50 Tt Ni, Cu)
	DIFFERENTIATED MAFIC INTRUSIONS
(M)	disseminated and massive Ni-Cu sulphides in picritic gabbro-diabases (Noril'sk); about 10 Mt Ni, 15 Mt Cu, Pt,Pd,Au
(N)	magnetite veins, masses in diatremes and vent breccias (e.g. Angara-Ilim); 4 Bt Fe
	FELSIC INTRUSIONS
(O)	porphyry Cu-Mo occurrences in porphyries, granites, granophyres (e.g. Tribag; 200 Tt Cu, Mo; subeconomic)
(P)	sulphides in breccia pipes, breccias (Tribag; 400 Tt Cu)
	INDIRECTLY ASSOCIATED ORES, MISCELLANEOUS
(Q)	magnetite skarns on contacts of granite and limestone in bimodal volcanic-intrusive centres (10 Tt Fe)
(R)	Pb,Zn,Cu,Ag,U, etc., veins, replacements. disseminations in exocontacts of basalt, diabase sills, dikes

exceptional and of controversial origin.

Continental basalts have practically no hydrothermal phase, but they can heat groundwater in the intruded sediments to produce hot springs. These, in turn, are capable of metal leaching and redeposition, if suitable metal sources are locally available. The economically important but genetically controversial Fe-Mn deposit Bahariya in Egypt (hematite, magnetite and Mn oxides replace Tertiary limestone near Miocene plateau basalts; Basta and Amer, 1969), is interpreted in such a manner. Said (1962) gave more examples of Pb-Zn and U ores in Egypt, presumably deposited from hot springs driven by the heat of basaltic volcanism.

Conditions of ore deposition in plateau basalt terrains were presumably more favourable at the subvolcanic and plutonic levels and in load metamorphosed ("green") lavas, but the deeper levels are rarely exposed in the Cainozoic plateau basalts. Minor disseminated native copper occurrences in zeolitized and prehnitized flowtops of Tertiary plateau basalts have been recorded in the Faroe Islands (Cornu in Butler and Burbank, 1929) and malachite fracture coatings in amygdaloidal basalts have been recorded from the Deccan "traps"(Sahasrabudha, 1978).

Supergene processes superimposed on the volcanics remain the only agent responsible for the formation of economic metallic deposits known from Cainozoic tholeiitic and alkali-basalt volcanic terrains. Lateritic bauxites generated by Eocene tropical weathering on the Deccan "traps" in north-western India (states Gujarat and Maharashtra), represent a reserve of some 60-70 Mt. of ore containing from 38% to 63% Al_2O_3. The bauxites, both nodular and massive (blocky) with the relic texture of the source volcanics, favour basaltic pyroclastics (Sahasrabudha, 1978; Fig.31-2).

Over one hundred separate but small deposits of lateritic bauxites with an aggregate tonnage of about 22 Mt. of low-grade ore (35-52% Al_2O_3), are known from eastern Australia (Owen, 1954; Townsend, 1965). The deposits are the residual product of early- to mid-Tertiary humid tropical weathering, superimposed on Paleocene to Eocene alkali basalts. The major gibbsite and minor boehmite bauxites range from a hard, pisolithic variety to earthy bauxites, and most deposits have only recently been exhumed from under the cover of younger Tertiary sediments including coal (Gippsland in Victoria), or from under younger basaltic flows (Tamborine Mt., Inverell-Emmaville, etc.). Comparable fossil residual ferruginous and titaniferous bauxite deposits formed on plateau basalts in Oregon and in northern Ireland (Antrim County). There, gibbsitic bauxite occurs in two lateritic horizons sandwiched between 66-61 m.y. old tholeiitic flows. It is topped locally by an iron oxide-rich crust, mined in the past as an iron ore (Glenravel, 2 Mt. Fe produced Dunham et al., 1978). Several bauxite deposits in northern Ireland formed by residual weathering of rhyolite, obsidian and felsic tuff. 30 Mt of 20% Al bauxitic laterite and saprolite formed over Tertiary basalts on the Jan Mayen Island in the northern Atlantic (Patterson, 1967). The extremely low-grade (10-20% Fe) remnants of Tertiary limonitic pisolithic laterites topping Neogene basalts in the German Vogelsberg, were mined during the World War II emergency.

Weathering resistant and heavy accessory minerals of basalts (ilmenite, Ti-magnetite, zircon) released by weathering and transported by streams, contributed the economic substance to some alluvial and beach heavy mineral placers (e.g. those along the Pacific coast in Queensland and New South Wales).

INDIRECT INFLUENCE OF YOUNG PLATEAU BASALTS ON MELALLOGENY AND EXPLORATION

Cainozoic basalts are usually considered to be a nuisance, because they cover many areas with promising basement mineralizations and inhibit exploration (e.g. in the Victoria Goldfields, Australia; in

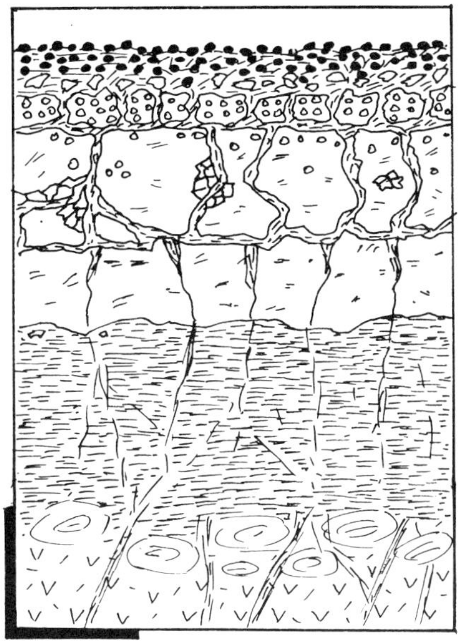

Fig. 31-2. Schematic profile of residual Eocene bauxite developed over plateau basalt tuff, Deccan Plateau, N.W. India. Based on data in Sahasrabudha (1978).

Tasmania; in New England, N.S.W.; in Oregon and Idaho; in central British Columbia and elsewhere). The positive role of the basaltic cover, however, is that of preserving the earlier mineralizations. Many Tertiary and older surficial ores nonresistant to erosion, such as placers (also laterites, gossans and supergene enriched zones over sulphide deposits, infiltration deposits), have been preserved until now thanks only to their burial under the protective cover of Cainozoic volcanics, including plateau basalts. The buried deposits have only been recently exhumed. Examples include many of the "deep lead" gold placers in Victoria; "deep lead" cassiterite placers near Emmaville and Torrington, N.S.W. (Weber, 1974), in the Jos Plateau, Nigeria and elsewhere; infiltration uranium deposits in soft Tertiary conglomerates under basalts near Beaverdell, British Columbia and others (compare Laznicka, 1985f). It may be assumed that many metallic deposits are still buried under Cainozoic basalts.

31.4.2. Pre-Cainozoic (dominantly "green") plateau basalts

Some Cretaceous and Jurassic basalts still remain "gray", but the proportion of gray basalts decreases in geologically older sequences, where they are substituted by the "green" basalt variety (this is often called "melaphyre" or "diabase" in the European literature). The green colour is caused by the partial or full development of metamorphic actinolite (uralite), pumpellyite or chlorite at the expense of pyroxene. The calcic plagioclase tends to be zeolitized, analcitized or albitized. The above changes, essentially hydration at elevated temperatures, can be the consequence of various processes: (1) hydrothermal autometasomatism at the time of the magmatic rock emplacement; (2) hydrothermal metasomatism triggered by the heat of a younger, often felsic, intrusion or (3) hydration and metasomatic recrystallization under the conditions of zeolite to greenschist facies burial (load, regional) or dynamic metamorphism. It is usually difficult to determine accurately the cause of the hydrothermal metasomatism, and several causes may combine.

A hydrothermal metasomatism superimposed on basalts not only caused pseudomorphic changes in the original minerals, but also triggered migration of the mobile components ($CaCO_3$, SiO_2, hydrous silicates), that were gradually filling available spaces: vugs, pores in breccia and scoria or fractures. As a consequence, amygdaloidal basalts, hydrothermal minerals-cemented scorias and breccias, or completely replaced rocks (e.g. Jolly and Smith, 1972), substitute for the initially porous varieties of mafic volcanics. In the flow top region of the dominantly "green" basalts, patches of brown, hematite-pigmented amygdaloids, breccias and scoriaceous basalts frequently occur.

Large Mesozoic intracratonic plateau basalt sheets (Paraná, Karoo, Siberian Platform) are usually located near the top of thick sequences of epiclastic or (bio)chemical sediments-filled basins. There, the volcanics are often interbedded with mature sediments such as quartz arenites, or with carbonates. Interaction of sediments with "green" plateau basalts and associated rocks is more intensive than the similar interaction involving the "gray" basalts, and has important metallogenic implications.

Because of the deeper erosional dissection achieved, at least locally, by the pre-Cainozoic plateau basalts, subvolcanic and plutonic equivalents are more widespread in recent outcrops. In the Siberian Platform, Permo-Triassic deep equivalents of plateau basalts, such as gabbro and dolerite sills, laccoliths and shallow intrusions, are densely distributed in an area surrounding the basalt-filled Tunguzka Basin, and are more voluminous than the basalts themselves (Vilenskii,1967). In Tasmania, Jurassic diabase sills and dikes are the only representatives of the continental tholeiite suite present (McDougall,1962).

METAL GEOCHEMISTRY AND METALLOGENY

The pre-burial metal geochemistry and metallogeny of the "green" basalts are identical with those of the "gray" basalts, described earlier. The hydration and metasomatism of the rock-forming minerals of basalts and particularly the short- to long-distance migration of the mobile components are important for metal concentration. Many trace metals migrate along with the major rock-forming elements. The Proterozoic native copper province of the Keweenaw Peninsula, Michigan (Volume 2), is considered to be a type area of such processes. The native copper deposits are located in altered metabasalt flowtops and in thin sedimentary interbeds. Amstutz (1977) placed the hydration and metal migration event responsible for the formation of Keweenaw ores into the deuteric hydrothermal stage of the host lavas, on the basis of a supposed congruency between mineralogic distribution and primary textures. Smith (1974), working in the same region, associated the hydration and metal migration event with burial metamorphism. Jolly (1974) and Jolly and Smith (1972), presented a rather convincing model of metamorphogenic mobilization of the trace coper in basalts, as a mechanism for the formation of the native copper ores. They demonstrated that the 5,000 m thick pile of tholeiitic lava flows is metamorphically zoned. At the top chlorite (12% H_2O) and pumpellyite (6% H_2O) metadomains formed by hydration of the volcanics, whereas the epidote metadomain at the base (2% H_2O) formed by dehydration. Copper, averaging 70 ppm in the hydrated basalts, was leached and expelled upward from the epidote zone undergoing dehydration. It was then added to the hydrated zone containing pumpellyite, prehnite, laumontite and chlorite. The coper-bearing, sulphur-deficient fluids may have contained up to 2,000 ppm Cu, and a portion of this copper precipitated in "stratabound" permeable zones (scoriaceous and amygdaloidal lava flow tops and interflow conglomerates). The precipitation was facilitated by the presence of impermeable barriers, for example diabase dikes. Numerous variations of the process described above are plausible. Copper ores in sediments associated with "green" basalts could have received their copper from Cu-bearing fluids, released from hydrated basalts. The widespread occurrences of native copper and Cu sulphides in brecciated interflow limestones, adjacent to altered plateau basalts and tuffs in the Arylakh and Kharaelakh areas, Siberian Platform (Dyuzhikov et al., 1976), probably formed in such a way.

The supergene metallogeny of "green" basalts and diabases is comparable with that of the "gray" basalts, at least in the case of bauxites and iron ores. Small residual ferruginous bauxite occurrences formed on basalts of the Paraná Basin and on Devonian basalts in the Central Timan bauxite region, U.S.S.R. The Miocene lateritic ironstone on Paraná basalts in the Misiones district, eastern Argentina (Haude and Weber, 1975), represents 20 Mt ore with 28-35% Fe and up to 4% TiO_2. Numerous but small occurrences of Cu oxides chrysocolla, tenorite, malachite, etc.) hosted by amygdaloidal basalts (as in the Alto Uruguai area, Brazil; Szubert et al., 1981) and interflow felsic conglomerates (e.g.Allouez Mine, Michigan) may, however, have a different origin. They could be the products either of supergene degradation of earlier sulphide or native copper accumulations, or they could be the equivalent of "exotic deposits" in which copper dissolved in groundwater was precipitated by the reaction with silica, possibly released from a volcanic glass undergoing devitrification.

NATIVE COPPER DEPOSITS IN PREHNITE-PUMPELLYITE ALTERED AMYGDALOIDAL AND SCORIACEOUS THOLEIITIC FLOWS AND INTERFLOW SEDIMENTS ("Michigan type")

In the classical copper-producing area of the Keweenaw Peninsula (total production 4.9 m.t.Cu), several hundred copper deposits and occurrences have been mined and described in the monograph by Butler and Burbank (1929). All of them are hosted by the Proterozoic Portage Lake Lava Series, which is a sequence over 5,000 m thick, of several hundred discrete continental flows of low-olivine meta-tholeiite. Between the flows are sandwiched several thin lenses of rhyolite conglomerates. Three fundamental styles of native copper deposits can be distinguished and all have a tabular form: (a) concordant to peneconcordant orebodies in amygdaloidal and scoriaceous lava flowtops (about 58% of the total production; W. S. White, 1968); (b) concordant to peneconcordant orebodies in thin lenses of interflow conglomerates and (c) fissure veins, discordant with the strike of the host amygdaloids. These are described in Volume 2.

Small native copper deposits and occurrences in plateau basalts, comparable with the "Michigan-type", have been recorded from various areas of the world (Table 31-2). In southern Brazil and Uruguay frequent showings are associated with the deposits of amethyst and agate (Jacques and Cassedanne, 1975), but the copper mineralization is always erratic and unpredictable. This tends to support the opinion of many geologists that the "Michigan type" is a poor exploration target. A sharp drop of the cut-off grade of copper ores in the future (to 0.1-0.2% Cu), however, could possibly make economic certain regions (or discrete basalt flows) with a large number of small Cu occurrences, as in the Itapiranga area, Brazil (large reserves of 0.145% Cu ore; Szubert et al., 1981). Comparable native copper occurrences are found in the amygdaloidal and scoriaceous metabasalts of orogenic belts (e.g. Novaya Zemlya, Kamchatka, Yukon, etc; Chapter 10), and basalts that are members of the "volcanic red-beds" association (Chapter 25).

COPPER SULPHIDE DEPOSITS IN LOW-GRADE ALTERED THOLEIITIC AMYGDALOIDS

Minor calcite-chalcocite seams and stringers that crosscut the native copper-bearing amygdaloids, are relatively common in the Keweenaw Peninsula district (Volume 2). A more significant mineralization that is predominantly sulphide in nature (chalcocite and minor digenite, bornite, chalcopyrite and pyrite in quartz, chlorite, epidote and calcice gangue), was recently described from Mt.Bohemia (M. Robertson, 1975). The ore is hosted by concordant amygdaloids, as well as by discordant andesite dikes. Rich, massive

Table 31-2. Selected localities of native copper mineralization in plateau (meta)basalts*

LOCALITY	AGE	MINERALIZATION	REFERENCES
Kharaelakh Mts., Sukhariki Basin, N.W. Siberia,U.S.S.R.	Pe-Tr	XO occurrences of nat.copper in amygd.basalts, lava breccias, tuffs, and tuffac.sedimentary interbeds (particularly limestones). Peneconc. bodies to veins; 0.2-4.45% Cu, up to 163 ppm Ag	Dyuzhikov et al. (1976)
Serra Geral Fm., Paraná Basin, S.Brazil (e.g. Itapiranga)	J-Cr	pod-like patches of chrysocolla, malachite, nat.copper, chalcocite in amygdaloidal and brecciated basalt flowtops	Szubert et al., (1981); Jacques and Cassedanne (1975)
Orozimbo Fm., Maranhão Basin, N.E.Brazil (e.g. Grajau)	J-Cr	native Cu or Cu-sulphide occurrences in amygdaloidal basalts, assoc.with zeolites, chlorite, prehnite, amethyst	Abreu (1973) Nunes et al. (1973)

* Precambrian native copper occurrences are treated in Volume 2; occurrences in mobile belt basalts (e.g. Buena Esperanza, Chile; the Philippines; etc.) are in Chapters 10,12,13.

chalcocite fissure veins and low-grade disseminations of chalcocite in a meta-basalt dike, have been mined near Mamainse Point on the Ontario (northern) shore of Lake Superior. The host rocks and ores are of the same age as the native copper deposits in Michigan (Keweenawan). Widespread showings of chalcocite and secondary Cu minerals have been recorded in the Cambrian Antrim Plateau amygdaloidal basalts of northern Australia (North. Territ. Geol. Surv., 1975).

The presence of felsic subvolcanic intrusions and plutons postdating the plateau basalts is common to both the areas mentioned above, and there is probably a genetic connection. The intrusions could have provided the heat and possibly sulphur, while the basalts provided the copper. These mineralizations are thus transitional into those more intimately related to the felsic intrusions often located along the fringe of plateau basalts and are described later.

31.4.3. Diabase feeders to plateau basalt

Diabase (dolerite) dikes and sills represent the subvolcanic equivalent of plateau basalts, to which they are compositionally identical. The dikes fill tensional fractures, and swarms of hundreds of dikes (most are several tens of metres wide, but some extend 100 to 300 m) are a convincing indication to the degree of crustal extension achieved in the magmatic period. The bulk of diabase (dolerite) dikes and sills are petrographically extremely monotonous and homogeneous, even more so than the plateau basalts. Thermal metamorphism and metasomatism at dikes and sills contacts is usually limited and inconspicuous, restricted to zones of baking and hornfelsing a few centimetres wide. They are apparent chiefly in sedimentary exocontacts.

The pre-Cainozoic diabases are green, containing partly or fully uralitized pyroxene. Mobile components released during deuteric or metamorphogenic mobilization fill fractures that are fairly common. White, coarse calcite is an almost universal fracture filling mineral. White or pink laumontite (zeolite) fracture coating, quartz, pyrite and prehnite veins also occur frequently. Calcite veins with asphalt-filled vugs are abundant near Rio Claro (São Paulo, Brazil), near diabase contacts with Permian oil shales.

The following examples of economically unimportant metallic occurrences have been recorded from diabase dikes intimately associated with plateau basalts:

(a) local pegmatitic schlieren that could contain coarse bunches of ilmenite;

(b) small-size inhomogeneities (schlieren, veinlets) within the diabase, or thin quartz or calcite veinlets reaching into the wallrocks, that could carry minor chalcopyrite in the form of scattered grains;

(c) small concentrations of magnetite, for example in fractures. These can grade into magnetite stockworks or thick veins in faulted or sheared diabase or in the adjacent supracrustal rocks, as in the

Tunguska Basin, Siberia (Pavlov,1960) and
(d) rare occurrences of native iron in disseminated grains and occassional boulders of controversial origin, recorded from western Greenland (Disko Island, Kitdlit; Fundal, 1975).

Massive replacement bodies of magnetite in reactive sediments (carbonates) adjacent to diabase sills as in the Severnaya River field, N.W.Siberia (Pavlov, 1960; Fig.21-3) is one of the few economic ore styles, directly associated with diabases. Tropically weathered diabases, like basalts, can yield ferruginous lateritic bauxites (e.g. at Ouse, Tasmania, where 15 small deposits of clayey bauxite contain 703,500 t ore with 40.4% Al_2O_3; Owen, 1954). Residual ilmenite or Ti magnetite reworked from argillized or lateritized diabases could accumulate in alluvial and beach placers (as along the Atlantic coast of south-eastern Brazil) and in paleoplacers.

31.4.4. Differentiated tholeiitic sills

Carmichael et al. (1974, p.437-451) provided an adequate description of tholeiitic diabase and gabbro sills from the petrological point of view. Many individual sills are over 300 m thick and traceable for tens of kilometres. They formed by injection of tholeiitic magma at liquidus temperatures into undisturbed sediments, or along unconformities. The thick sills tend to be gravity-differentiated, but the differentiation is chiefly expressed by variation of the mafic index only. Few sills have granophyric tops (e.g.Hamilton, 1965), and the olivine-rich (picritic, troctolitic) basal members are even less common. The latter, however, are the only sills that interest the exploration geologist directly, because some of them may carry small quantities of nickel sulphides and platinoids.

Differentiated sills are transitional into Bushveld-style layered mafic intrusions, and have some points in common with the Noril'sk-style differentiated mafic plutons described later. The most frequently quoted example of nickel-mineralized tholeiitic sill (intrusion) is the Mesozoic Insizwa sill (Nolagoni Massif, Mt.Ayliff district, Transkei, Southern Africa; Dowsett and Reid, 1967; Scholtz, 1936). It is a subvolcanic equivalent of the Karoo plateau basalts, represented by an undulating gabbroic sheet, up to 600 m thick, injected into Karoo sediments. Three main petrological zones can be recognized: (a) basal zone, consisting largely of picrite, troctolite and olivine hyperite; (b) central zone, composed of gabbro and (c) roof zone, that contains the felsic phase (granophyre) of the intrusion. Subeconomic, massive and disseminated Ni-Cu sulphide mineralization (pyrrhotite, pentlandite, chalcopyrite and cubanite with small quantities of Au,Ag and Pt metals) at Waterfall Gorge, is confined to the ultramafic members in the Basal zone and to the hybrid footwall contact. This locality has been considered to represent a classic example of magmatic ore formation by differential segregation and gravitational settling.

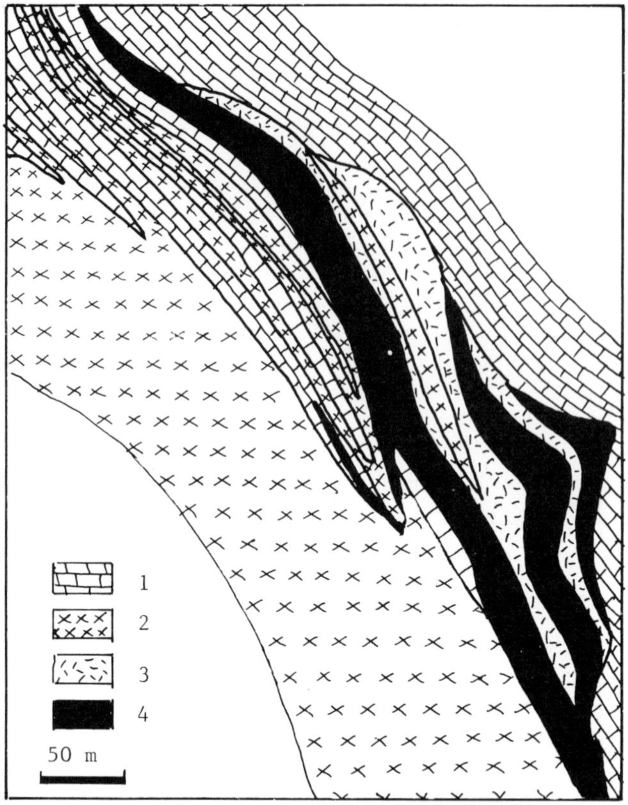

Fig. 31-3. Severnaya River iron field, N.W. Siberia, a contact metasomatic magnetite mineralization related to Pe-Tr gabbro-diabase sills. 1=Or marbles; 2=Pe-Tr sills of uralitized gabbrodiabase; 3=skarn; 4=magnetite. After Pavlov (1960).

Ni-Cu sulphide mineralization of the "Kureika-type" associated with diabase-troctolite sills (intrusions), is a member of the Permo-Triassic plateau basalt and diabase association of the Siberian Platform (Rundkvist, ed., 1978). The ore minerals in the Kureika showings (pyrrhotite, chalcopyrite, pentlandite, minor millerite, pyrite) are disseminated in olivine-rich layers distributed throughout the intrusion, but are most extensive in the middle and upper layers. Rich sulphide schlieren occur sporadically in the central parts of the disseminated ore horizons.

31.4.5. Central, explosive volcanic-intrusive complexes and the mafic-felsic bimodal suite

Tertiary central volcanic-intrusive complexes associated with the continental plateau basalt activity and usually younger than the earlier phase of fissure basalt eruptions, are known to exist in various parts of the world. The Brito-Arctic Province is considered to be a classical area, and differentiated central complexes have been described in the Scottish Hebrides (Mull, Skye, Ardnamurchan; Richey et al., 1961), in northern Ireland (Slieve Gullion Complex, Carlingford Volcano; Preston, 1981) and from Iceland (Breiddalur Volcano, Walker, 1963; Thingmuli Volcano, Carmichael, 1964). Other Tertiary occurrences are situated in South Yemen (Greenwood and Bleackley, 1967) and elsewhere. Pre-Tertiary central complexes suffer from fragmentary preservation.

A petrologically characteristic feature of most central complexes is the presence of abundant felsic members associated with basalts. The lack of intermediate members results in a distinct bimodal mafic-felsic suite. At the volcanic level, tholeiitic to alkali basalt lavas and pyroclastics represent the mafic mode, whereas trachyte and rhyolite lava domes, breccias, agglomerates and tuffs are members of the felsic mode. The latter have a tendency to accumulate in the core of volcanoes. At the subvolcanic and plutonic levels, diabase and gabbro dikes, stocks and plutons are dominant, and felsites, granophyre, quartz feldspar and feldspar porphyry, syenite and minor granite, are subordinate. Highly hybrid sections have occassionally resulted from the mixing of acid and mafic members in intrusive stocks under some central volcanoes (e.g.in Asturhorn, S.E. Iceland).

The lithologic complexity, the presence of vents and the frequent interaction of mafic and felsic members of central complexes (such as in the cone-sheet dominated Central Mull and Ardnamurchan, Scotland; compare the maps and sections in Richey et al., 1961) makes these centres favourable for ore deposition, but very few ore occurrences have actually been found. The small magnetite orebody located in skarn in the Skye (Scotland) volcanic centre is one of them. Small pipe and ribbon-like bodies of magnetite with minor Cu,Zn and Pb sulphides developed on contact of a small granite stock, and Cambro-Ordovician limestone. Only a few thousand tons of about 40% Fe ore have been mined (Dunham et al., 1978).

The scarcity of mineralization is probably due to the general lack of hydrothermal alteration. Propylitic alteration has been described by Walker (1963) from the core of the Breiddalur paleovolcano in S.E. Iceland, and although no mineralization has been reported from this particular locality, numerous small Cu and Cu-Mo showings and geochemical anomalies have been reported from S.E.Iceland, particularly from the Höfn area. The best-known occurrences described by Janković (1972) are those in the Ossura River zone, near Svinhöler farm (Fig. 31-4). There, chalcopyrite and minor galena and sphalerite occur in blebs and disseminations in a breccia pipe composed of fragments of rhyolite and basalt. The pipe was emplaced into Tertiary plateau basalt flows. Small copper sulphide occurrences are relatively common in felsic volcanics associated with pre-Tertiary plateau basalts (for example, in the Indian Mine, Michigan; Butler and Burbank, 1929), but economically important occurrences are, so far, lacking.

31.4.6. Small differentiated mafic plutons : Noril'sk-style

The discovery and development of the Noril'sk-Talnakh district of Siberia (the second largest accumulation of sulphide Ni and third largest accumulation of platinoids of the world), focussed attention on its setting and the nature of its host rocks. Over one thousand citations dealing with Noril'sk are now available, both in the Russian original (e.g.Zolotukhin and Vilenskii,1978; Sobolev et al., 1978; Urvantsev et al., 1975; Dobin et al., 1971) and in English translations (e.g.Glazkovsky et al., 1974; Ivankin et al., 1971; Tarasov, 1968).

The Noril'sk-Talnakh ore district and several tens of showings (about 40 known Ni-Cu sulphide occurrences) are located near the

Fig. 31-4. Chalcopyrite-mineralized breccia pipe in Tertiary rhyolite (left, X) at Ossura River, S.E. Iceland. Chalcopyrite with quartz is located in the matrix and in vugs interstitial to rhyolite fragments (right).

north-western margin of the Siberian Platform, in a region of numerous N.N.E. and N.E. deep faults and lineaments (see the description of the Siberian Platform later). There, the Proterozoic basement is overlain by an Ordovician to Permian sequence of platformic sediments, that includes Devonian evaporite unit (anhydrite, gypsum, halite, carbonates) and a Permo-Carboniferous coal-bearing unit. These units occur in the immediate vicinity of the sulphide mineralization.

The plateau basalt volcanism commenced in the Permian period and continued through the Triassic period. The second, Triassic phase of central mafic explosive volcanism, produced several hundred paleovolcanoes and their subvolcanic-plutonic equivalents. In the Noril'sk area, the volcanics fill series of elliptical basins, subdivided by basement highs (Fig. 31-5). Most of the mafic plutons crop out in the "highs" and intrude the Paleozoic sediments. The majority have a simple gabbro-diabase composition and are undifferentiated. Only about 1% of the intrusions are differentiated, and even here the differentiation is, at most, within the gabbro-diorite to troctolite range. Felsic differentiates are rare and separate bodies of ultramafics are not known. The ore-bearing intrusions are usually elongated, some 100-300 m thick, peneconcordant with the enclosing sediments or mafic lavas. They are members of the high-magnesium, evolutionarily primitive suite and the olivine-rich members have a high trace Ni content (0.11-0.2% Ni).

In the ore-bearing intrusions (Figures 31-5 and 31-6), the ore minerals pyrrhotite, pentlandite, chalcopyrite, bornite, cubanite and numerous rare Ni,Cu, Pt-Pd, Bi,Sb,Te, etc., minerals occur as: (1) low-grade (0.4-0.8% Ni) disseminations in troctolites and gabbro-diabases or norites ("taxites") in the basal endocontact of the intrusion; (2) massive and brecia-cementing ores along the footwall contact or (3) veinlet and stringer ores in altered Paleozoic footwall metasediments (Fig. 31-7). The orebodies, particularly those in the exocontact, are accompanied by intense hydrothermal and contact metamorphic alteration. The dominant and most widespread alteration is albitization (in sandstones and gabbros), skarnization in carbonates, hornfelsing in shales, and local biotitization in gabbros. Scapolitization occurs locally. Chloritization occurs on the fringe and along faults.

The Noril'sk-Talnakh district supplies 67% of the Soviet nickel production, and is credited with a minimum production and reserves of abut 5.5 million tons of Ni in ores with an average grade 0.5% Ni and 0.8% Cu. The richer ores have 1.23% Ni, 2.16% Cu, 0.1% Co, 11 ppm Pd+Pt and some Au,Ag and Te (Wagner and Berthold, 1979). The actual regional potential is probably several times higher, of the order of 10-15 Mt Ni.

The recent Soviet monographs quoted earlier present lists of criteria used to estimate the nickel potential of the Siberian "differentiated trap intrusions", which include: (1) association of intrusions with deep faults; (2) abundance of endo- and exocontact metasomatites, including skarns; (3) abundance of picritic basalts and subalkaline traps in associated volcanics; (4) presence of

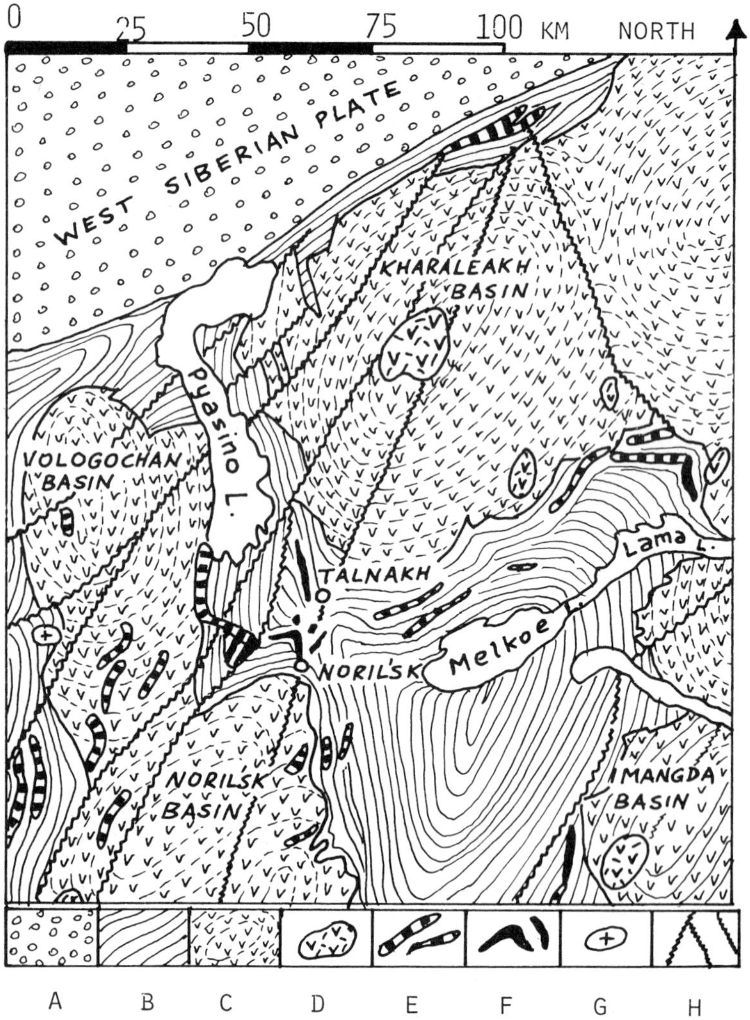

Fig. 31-5. Noril'sk-Talnakh Ni,Cu,Pt district, N.W. margin of Tunguska Basin, Siberian Platform, U.S.S.R.

A=MZ sediments of the West Siberian Plate; B=Pt_3 to Pe platformic sediments; C=Pe_3-Tr_1 plateau basalts – filled basins, isopachs are shown; D=volcanic centres; E=Pe-Tr undifferentiated mafic intrusions (diabases); F=Pe-Tr differentiated and Ni-Cu bearing mafic intrusions; G=granite porphyry intrusion; H=faults.

Modified after Akad. Nauk S.S.S.R., "Sul'fidnye medno-nikelevye rudy Noril'skikh mestorozhdenii", Nauka, Moscow, 1981.

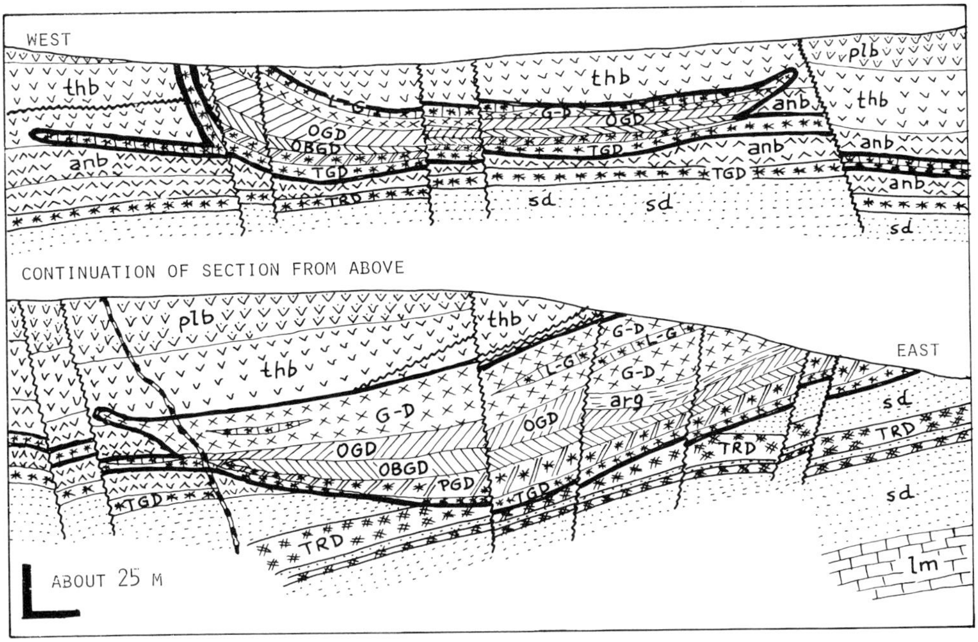

Fig. 31-6. Geological cross section of the Noril'sk I differentiated mafic intrusion (outlined by solid black line). D contact metamorphosed and altered sediments: lm=limestone; sd=sandstone; Tr_1 basaltic volcanics: plb=plagioclase basalt; thb=tholeiitic basalt; anb=andesitic basalt; Tr_1 plutonic members of the Noril'sk intrusion: TRD=trachydolerite; G-D=gabbrodiorite; LG=leucocratic gabbro; PGD, PD=picritic gabbrodiabase and diabase; OGD=olivine gabbrodiabase; OBGD=olivine biotite gabbrodolerite; TGD=taxitic gabbrodolerite. After Dobin et al. (1971).

anhydrite and/or gypsum in the exocontact sediments; (5) presence of taxitic (i.e., those with inhomogeneous texture) gabbro-dolerites; (6) presence of felsic hybrid rocks in the roof of intrusions as well as a series of petrochemical criteria.

So far, the unique "Noril'sk-style" of mineralized mafic intrusions has not been found outside the Siberian Platform. Most points of similarity with Noril'sk outside the U.S.S.R. can be found at several localities in the Thulean and Baffin Island-West Greenland Provinces, where some central volcanic-intrusive complexes are associated with high-nickel picritic lavas. Differentiated ultramafic to felsic intrusions on the fringe of the Paraná Basin in Brazil (e.g. the Cretaceous Iporá Group intrusions in Goiás; Guimarães et al., 1968), resemble Noril'sk in some respects. They have a silicate-based nickel-rich early ultramafic phase (dunite), followed by a felsic or mafic to felsic alkaline suite. In this regard, the Iporá magmatites

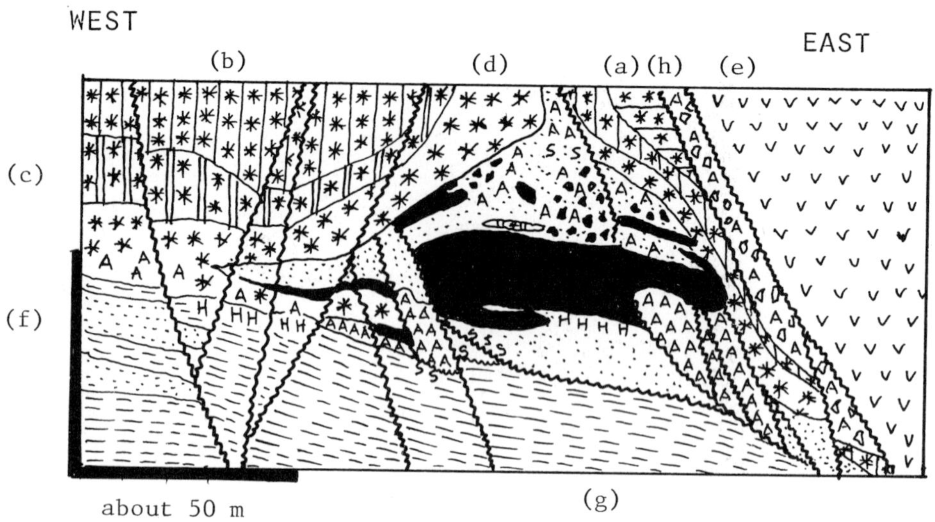

Fig. 31-7. Detailed section across the zone of exocontact Ni-Cu sulphide mineralization; northern ore shoot in the S.W. portion of the Talnakh ore deposit. a=olivine gabbrodiabase; b=picritic gabbrodiabase; c=troctolitic gabbrodiabase; d=taxitic and contact gabbrodiabase; a-d are members of Tr_1 Talnakh Intrusion. e=Pe-Tr plateau basalt; f=Cb sandstone; g=argillite, siltstone; h=fault breccia to mylonite. Black area: massive sulphide ore. Alterations: A=albitization; S=skarnization; H=hornfelsing. Modified after Tarasov (1974).

seem to bear a closer relationship to members of the "ultramafic-alkaline" association, developed separately in the Siberian Platform, for example in the Meimecha area (Bilibina et al., 1976).

Whereas the "Noril'sk-type" Ni-Cu ores are associated with highly magnesial members of the "trap" association, unique and economically important iron ores (reserves of 4-5 billion tonnes of iron in high-grade ores) are associated with the Triassic iron-enriched "trap" intrusions of the Siberian Platform. Vein and irregular replacement bodies of magnesian magnetite in the most productive Angara-Ilim region, are located in breccia pipes (diatremes) emplaced into Paleozoic platformic sediments (shales, sandstones, carbonates; Fig. 31-8). The magnetite replacements in carbonates are enveloped by pyroxene-actinolite and scapolite skarns, and some diatremes are surrounded by radial and concentric fractures filled by gabbro-diabase dikes (Pavlov, 1960). There is also a unique deposit where magnetite replaced sedimentary halite.

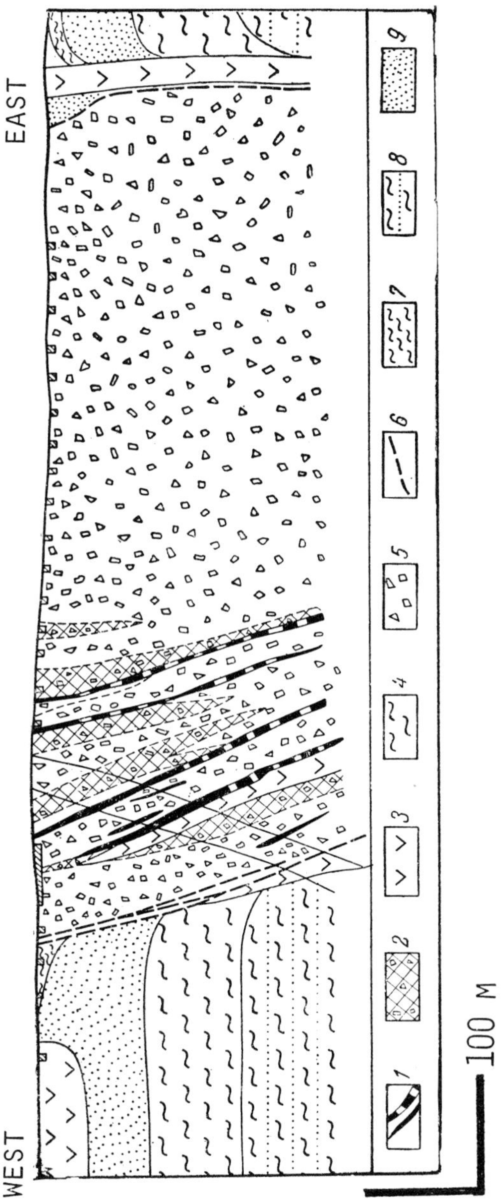

Fig. 31-8. Krasnoyarovskoe magnetite deposit, Angara-Ilim district, Siberian Platform, a Triassic mineralized explosive pipe (diatreme). 1=Magnetite veins; 2=disseminated magnetite; 3=Pe_3-Tr_1 dikes and sills of diabase; 4=S red argillite; 5=Tr_1 volcaniclastics and tuff breccias; 6=pre-ore faults; 7=Or argillites and calcareous shales; 8=sandstones and shales; 9=gray and yellow sandstones. From Pavlov (1960).

31.4.7. Miscellaneous lithologic associations and mineralizations indirectly associated with plateau basalts

Plateau basalt magmatism, particularly at the deep levels, introduces a considerable amount of heat into the continental crust. This heat can produce hydrothermal metal leaching and redepositing systems, and may generate palingenetic felsic magmas complete with a hydrothermal phase of their own, out of the continental lithosphere.

The hydrotherms or secondary felsic magmas can interact with members of the plateau basalt association to produce interference mineralization. Alternatively, they can evolve completely outside the plateau basalt complex, so that the relationship between the resulting mineralization and the plateau basalt magmatism is indirect, usually controversial and sometimes lost. The selected examples of mineralizations that follow have been included to focus the attention of an exploration geologist on ores that might be expected to be spatially associated with plateau basalt regions, although there may be little component relationship.

HYDROTHERMAL VEINS AND REPLACEMENTS IN SEDIMENTS.

Quartz-carbonate or barite, fluorite, fissure veins or replacements in carbonates mineralized by chalcopyrite, galena or sphalerite, occur locally in platformic sediments topped by, or interlayered with, plateau basalt flows or diabase sills. Staritskii and Tuganova (1971) recorded about 50 Pb-Zn occurrences in the Paleozoic sediments, genetically related to the plateau basalts, scattered over the Siberian Platform. They compared them with the American "Mississippi Valley-type".

MINERALIZED BRECCIA PIPES

The economically important Messina ore field, Transvaal (production 300 Tt Cu; Sawkins, 1977), contains several brecia pipes emplaced into Archean granulite facies metamorphics, and mineralized by copper sulphides. The mineralization dates from the Karoo (210-170 m.y.)age and is located about 50 km from the present edge of the Karoo plateau basalts. It is considered to be associated genetically with the late Karoo peralkaline magmatites. Further examples of hydrothermal metallic mineralizations associated with peralkaline to alkaline felsic intrusions, postdating the nearby plateau basalts, are reviewed in Chapter 33.

31.5. EXAMPLES OF PLATEAU BASALT PROVINCES

The two examples of plateau basalt provinces described below, i.e. the Brito-Arctic Province and the Siberian Platform Province, illustrate two principal age groups, the Cainozoic and the Paleozoic-Mesozoic. Additional examples are tabulated in Table 31-3 and their geographic distribution is shown on Figure 31-9. Proterozoic and Archean plateau basalt provinces are treated in Volume 2.

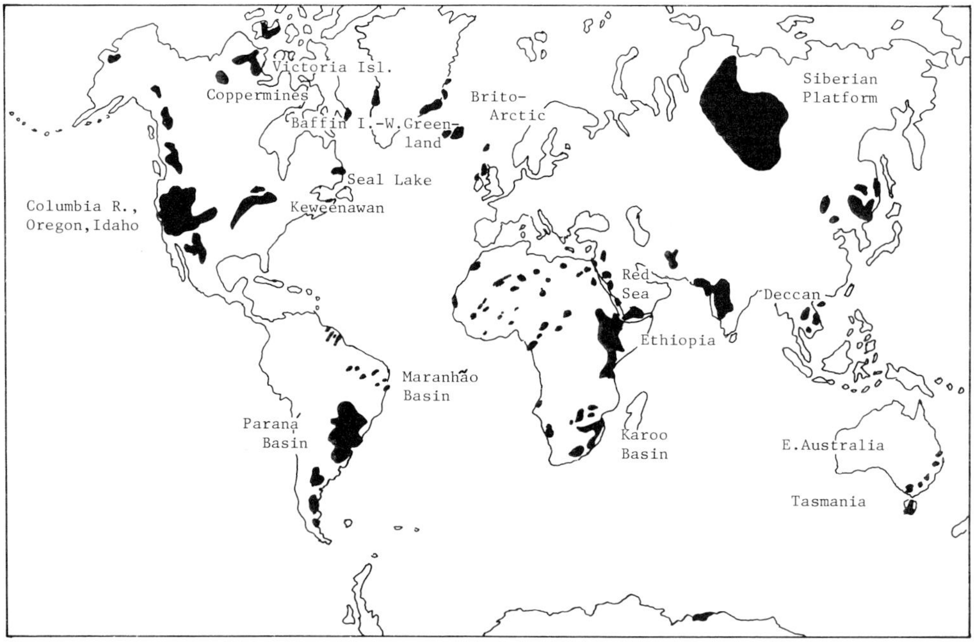

Fig. 31-9. Location map of the principal plateau basalt provinces of the world (see Table 31-4 for a brief listing).

THE BRITO-ARCTIC (THULEAN) PROVINCE

The Brito-Arctic Province (Fig. 31-10) comprises a transitional tholeiitic to alkali basalt association which has an area of at least 163,000 km^2 and reaching a thickness of up to 12 km in eastern Iceland. Still more of it is covered by the sea and the Greenish ice cap. The basalts erupted between 66 m.y. (upper Cretaceous) and 50 m.y. (Eocene) ago, in the initial stage of rifting, continental fragmentation and the North Atlantic opening. The province is dominated by fissure-errupted plateau basalts (dominantly tholeiites), and diabase (dolerite) dike swarms. Rare picrites, andesites, dacites and rhyolites also occur. Central, erosion dissected volcanic-plutonic complexes comprise the bimodal mafic-felsic suite (mildly alkaline basalts, such as hawaiites and mugearites; trachytes, rhyolitic pitchstones, rhyolites, felsites; gabbro, granophyre and granite in depth). The islands of Skye and Mull in the Inner Hebrides are type localities of cone sheets. Central plutonic intrusions of granophyre and granite are known in Skye and Rhum in the Hebrides, in Mourne in Ireland and in S.E. Iceland. Several layered, differentiated mafic to ultramafic complexes have been unroofed by erosion (e.g. Cuillin Hills on Skye; Rhum; Skaergaard in eastern Greenland; Richey, 1961; Preston, 1981; Carmichael et al., 1974).

Figure 31-3. Major Phanerozoic plateau basalt (diabase) provinces of the world

PROVINCE	AREA OR VOLUME	AGE	GEOLOGY	MINERALIZATION	REFERENCES
Brito-Arctic (Thulean); Scotland, Ireland, Iceland, E.Greenland	600T km^2 most under water, ice	Cr_3-Eo	plateau basalt flows, dolerite dikes, differ.central felsic compl., different. mafic/ultramaf.complexes	bauxite (Antrim) rare Cu in basalts, Cu in rhyol.breccia magnetite skarn	Richey et al. (1961) Preston (1981)
Red Sea Flanks (Egypt, Soudan, Arabia), Ethiopia	about 200T km^2	Eo-Mi	dominantly alkali basalt flows in desert envir., grade to submarine	Fe-Mn oxide replac. in sedim.near basalt (hot springs?)	Said (1962) Basta and Amer (1969)
Baffin Island (Canada)-Western Greenland	45T km^2	Pc-Eo	tholeiitic flows incl.local picrites; minor trachyte, interm.and acid pyrocl. Doler.sills, intrus.	up to 0.12% Ni in picrites; native Fe in basalt dikes (Disko Island)	Clarke (1977)
Columbia R., Wash. Oregon, Idaho, centr.Br.Columbia	500T km^2	Mi-Pl	highly uniform tholeiitic basalt lavas; local pillows, vents, minor dikes	ferruginous bauxite in Oregon	McDougall (1976)
Deccan Plateau, India	518T km^2	Cr_3-Eo	tholeiitic flows domin., locally abundant tuffs; minor picrite, rhyolite	variety of Eo bauxites; 60-70Mt ore	Poldervaart and Sukheswala (1958)
Paraná Basin, Brazil, Uruguay, Argentina	1.2M km^2	J_3-Cr_1	basalt flows interb.with aeolian sedim.; diabase sills and dikes along edge	nat.Cu in amygdal. basalt, minor Cu at dike contacts	Leinz et al. (1968); Abreu (1973)
Maranhão Basin, N.E. Brazil	50T km^2	Tr_3-Cr_1	3 units of tholei.basalt flows, diabase and gabbro sills, rare mafic intrus.	native Cu in amygdaloidal basalts	Nunes et al. (1973); Abreu (1973)

Karoo Basin, S.Africa, Lesotho, Namibia, Zambia	140T km^2	Tr_3-J_2	tholeit. lavas, diabase sills and dikes, differentiated mafic sills; minor granoph., followed by alkaline suite	Ni-Cu sulphides in different. gabbro sill; minor native Cu in basalts; Cu in breccia pipes	Du Toit and Haughton (1953)
Tasmania dolerite sills province, Australia	15T km^2	J_1	contin. tholeiitic suite; exclusively sills, dikes; minor granophyre	minor lateritic bauxite in regolith	McDougall (1962)
Eastern Australian contin. margin province	discont. 3,000x300 km belt	Cr_3-Q	alkali basalts, minor tholeiite flows, volc. centres rhyolite, trachyte	lateritic bauxite in over 100 small deposits	Ewart (1981) Owen (1954)
Siberian Platform, U.S.S.R.	1.5M Km^2	Pe_3-Tr_1	4 major eruptive cycles; doler. dikes, tholeiite flows; primitive picrites; mafic volc. centres, differentiated mafic intrusions	major Ni-Cu sulphides in different. intrusions; Cu in amygdaloids; magnetite in diatremes and at diabase contacts	Vilenskii (1967) Oleinikov (1979)

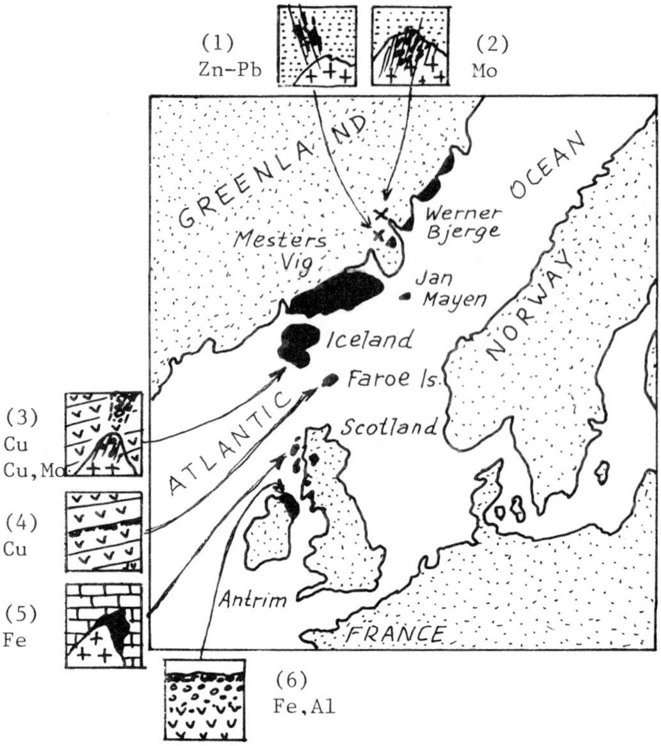

Fig. 31-10. Brito-Arctic (Thulean) plateau basalt province (in black), prior to rifting 60 m.y. ago. Mineralizations: (1) Mesters Vig, hydrothermal Pb-Zn in sediments; (2) Werner Bjerge, stockwork Mo in felsic intrusion; (3) S.E. Iceland, Cu in rhyolite breccia pipe, Cu-Mo in felsic intrusions; (4) Farøe Islands, native Cu in basalt amygdaloids; (5) Skye, magnetite skarn; (6) Antrim, residual iron oxides and bauxite.

Considering the petrographic variety and excellent outcrop along the rocky shores, it is surprising that very little metallic mineralization directly connected with the plateau basalt association is known. Only two metallic commodities have ever been mined: iron (in the small Broadford skarn deposit on Skye, located in Paleozoic limestones at the contact with a Tertiary granite member of a central compled and in ferruginous laterites formed on plateau basalts near Glenravel, Antrim, northern Ireland), and aluminium (in bauxites in Antrim County). Small copper showings have been reported in basaltic amygdaloids in the Farøe Islands and in a felsic breccia pipe probably related to granophyre in the Ossura River area, S.E. Iceland (Janković, 1972). Several molybdenite occurrences are associated with the felsic intrusions in the same area. Economically important Zn-Pb

(Mesters Vig) and molybdenite (Werner Bjerge) deposits located near the edge of plateau basalts in eastern Greenland, are related to the post-basalt peralkaline intrusions, treated in Chapter 33.

PERMO-TRIASSIC PLATEAU BASALT PROVINCE OF THE SIBERIAN PLATFORM

The Siberian Platform contains the world's most extensive plateau basalt province ("trap" in the Soviet literature). The area continuously covered by the Permo-Triassic lavas and densely intruded by contemporary sills and dikes, is estimated to exceed 1.5 million km^2 (Bilibina et al., 1976; Zolotukhin, 1964; Figure 31-11). The Platform has a consolidated Precambrian crystalline basement that crops out along its fringe (Yenisei Range, Aldan Shield), and in the centre (Anabar Shield). The basement is overlain by Ordovician platformic shallow-marine to continental sediments. The "trap" magmatism was initiated in the late Permian period (250 m.y. ago) and continued until the early Triassic period (210 m.y. ago). Subsequently, until the Jurassic age, several small alkaline basalt (Del'kan Suite) and alkaline-ultrabasic complexes (including kimberlite; Maymecha Suite), formed (Makarenko, 1976).

The core of the "trap" province is the Tunguska Basin, a broad depression filled by subaerial basalt lava flows which, locally, have a high proportion of basaltic tuffs. The subaerial volcanism reached its maximum intensity around the Permian-Triassic boundary and the volcanics range from tholeiites to alkali olivine basalts. Along the fringe of the Tunguska Basin, and over an extensive area south of it, diabase dike and sill fields predominate. The younger, lower Triassic phase of magmatism was dominated by explosive mafic volcanism, and numerous relics of paleovolcanoes are preserved, showing a transition from the central neck facies into the outer lava and tuff sheets. Diatremes are common. Compared with the Brito-Arctic Province, felsic differentiates are rare among the Siberian "traps" but, on the other hand, evolutionarily primitive magnesian magmas were prominent along the north-western margin (Noril'sk and Kureika areas).

The early stage "trap" intrusions are mostly undifferentiated diabases, that form sills to basin-like concordant intrusions, controlled by bedding or formational contacts of the enclosing Paleozoic sediments (Zolotukhin,1964). In the middle and closing stages of the "trap" magmatism in the lower Triassic phase, crosscutting, steep plutons became more common. About 1% of such plutons in the Noril'sk-Talnakh area are differentiated. Dominant gabbro-diabases grade into picrites, troctolites, teschenites or analcite dolerites, and there are prominent metasomatites formed in both endo- and exocontacts.

The Siberian "trap" province has the most complete metallogeny of all the plateau basalt provinces in the world, with the exception of metalliferous weathering crusts most of which were removed by the Quaternary glaciation. Economically outstanding is the Ni-Cu sulphide and platinoid mineralization associated with the differentiated

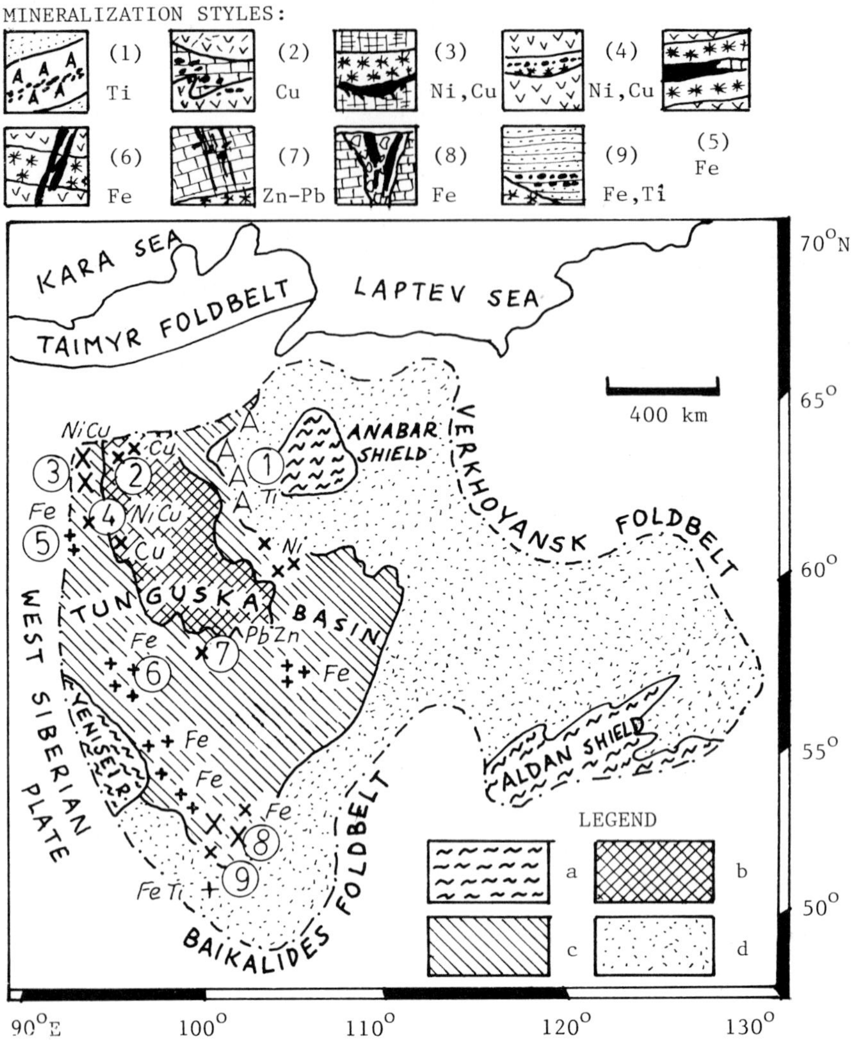
MINERALIZATION STYLES:
(1) Ti
(2) Cu
(3) Ni,Cu
(4) Ni,Cu
(5) Fe
(6) Fe
(7) Zn-Pb
(8) Fe
(9) Fe,Ti
KARA SEA
TAIMYR FOLDBELT
LAPTEV SEA
ANABAR SHIELD
VERKHOYANSK FOLDBELT
TUNGUSKA BASIN
WEST SIBERIAN PLATE
YENISEI R.
ALDAN SHIELD
BAIKALIDES FOLDBELT
400 km
LEGEND
a
b
c
d
70°N
65°
60°
55°
50°
90°E
100°
110°
120°
130°

◁▬ Fig. 31-11. Siberian Platform plateau basalt province (patterned). (a) PCm basement outcrops (shields); (b) area continuously covered by Pe-Tr plateau basalts; (c) PZ sediments densely intruded by Pe-Tr diabase dikes, sills and mafic intrusions; (d) PZ to T sedimentary cover, occasional Pe-Tr mafic dikes and intrusions. AAA=alkaline provinces.

Major mineralizations: (1) Meimecha-Kotui ilmenite and Ti-magnetite layers and disseminations in alkaline intrusions; (2) Arylakh, native Cu along plateau basalt and carbonate contact; (3) Noril'sk and Talnakh Ni-Cu sulphides with Pt+Pd at footwall of differentiated mafic intrusions; (4) Kureika, Ni-Cu sulphides disseminated in lightly differentiated gabbro sills; (5) Tunguska Basin, magnetite layers in or near contacts with gabbro and diabase sills; (6) Podkamennaya Tunguska, Bakhta Fe district; magnetite veins and sheets in or near dolerite sills; (7) Tunguska Basin, sphalerite, galena veins or replacements in PZ carbonates near diabase sills; (8) Angara Ilim district, magnetite veins in diatreme breccias; (9) southern Siberian Platform: magnetite, ilmenite, zircon paleoplacers in J sandstone.

---

gabbro-dibase plutons (Noril'sk-Talnakh), described earlier. In the Kureika region, Ni-Cu sulphides which, so far, have been uneconomic, are disseminated in differentiated gabbroic sills. Significant magnetite reserves are present in a variety of mineralization styles associated with diabase sills and intrusions (schlieren within sills, contact replacements in carbonates, veins and replacements in diatremes; Pavlov,1960). Economically, the most important are the mineralized breccia pipes in the Angara-Ilim region in the south.

Hydrothermal mobilization events superimposed on plateau basalts and dolerites generated widely distributed, but so far uneconomic, native copper occurrences in prehnite-pumpellyite to zeolite altered zones. They probably also contributed copper to the vein and replacement Cu sulphide deposits in the adjacent sediments (Dyuzhikov et al., 1976). In the Arylakh field, native copper mineralization is hosted by several stratigraphic horizons of brecciated carbonates, carbonate-rich tuffs and basalts. Scattered hydrothermal vein and replacement sphalerite and galena occurrences in Paleozoic sediments in the vicinity of diabase dikes and sills, probably formed by the mobilization of Pb+Zn from the sediments with the assistance of magmatic heat (Staritskii and Tuganova, 1971).

Ilmenite, rutile and zircon paleoplacers in lower Jurassic arenites in the southern part of the Platform, formed by reworking of the heavy resistates from weathered Triassic "trap" intrusions.

Plate 12. Old silver workings near Kongsberg, southern Norway. Calcite-native silver veins are closely associated with diabase dikes, transecting metalliferous schists (fahlbånds).

"L'utilité des mines ne se borne pas seulement à extraire du sein de la Terre des substances utiles: semblables en cela à la navigation, elles ont contribué à faire naitre et à étendre les sciences qui leur servent de guides".
Élie de Beaumont

"Study of geology sprang originally from the empirical observations of those engaged in mining for the useful minerals".
S. F. Emmons, 1886.

CHAPTER 32

Diabase, Gabbro and Similar Dikes and Sills

32.1. INTRODUCTION

In American usage, "diabase" is an intrusive rock composed of labradorite and pyroxene and characterized by ophitic texture (A.G.I. Glossary of Geology). It is the hypabyssal equivalent of basalt and gabbro. Dolerite, a term widespread in the British, South African, Australian and European literature is mostly a synonym for diabase, although it is sometimes used for rocks of diabasic composition with a doleritic texture. Diabase grades into gabbro and the terms diabase, dolerite, gabbro and sometimes diorite, are used interchangeably in the literature and need not indicate any substantial difference between rocks so named. One should be aware, however, that in the older European literature "diabase" means a "green" (i.e., pyroxenes partly altered to amphibole or chlorite) paleo-basalt, as well as the dike or sill rock.

Diabase dikes and sills are extremely widespread, being most conspicuous in old fractured and extended cratons, and common in young mobile belts as well. In the oceanic domain, dense swarms and sheets of diabase dikes underlie basalt flows. The overwhelming majority of diabase bodies formed as hypabyssal equivalents of surficially erupted basalts, filling feeder fissures (dikes); Fig. 32-1, or available lateral dilations (sills, sheets, laccoliths). They are members of the same petrochemical kindreds applicable to basalts (compare Basaltic Volcanism Study Project, 1981). When the connection of diabase with surficial basalts was obvious, both rocks were treated in the same chapter (4, 7, 10, 12, 25, 31). In deeply eroded regions, however, most of the superficial volcanics have been removed, yet diabase dikes and sills remained, often prominently exposed. Such an occurrence can be designated "independent diabase", and is treated in this chapter.

The overwhelming majority of "independent diabase" bodies occur in stabilized segments of the continental crust. The majority are related to distinct taphrogenic events, have the composition of

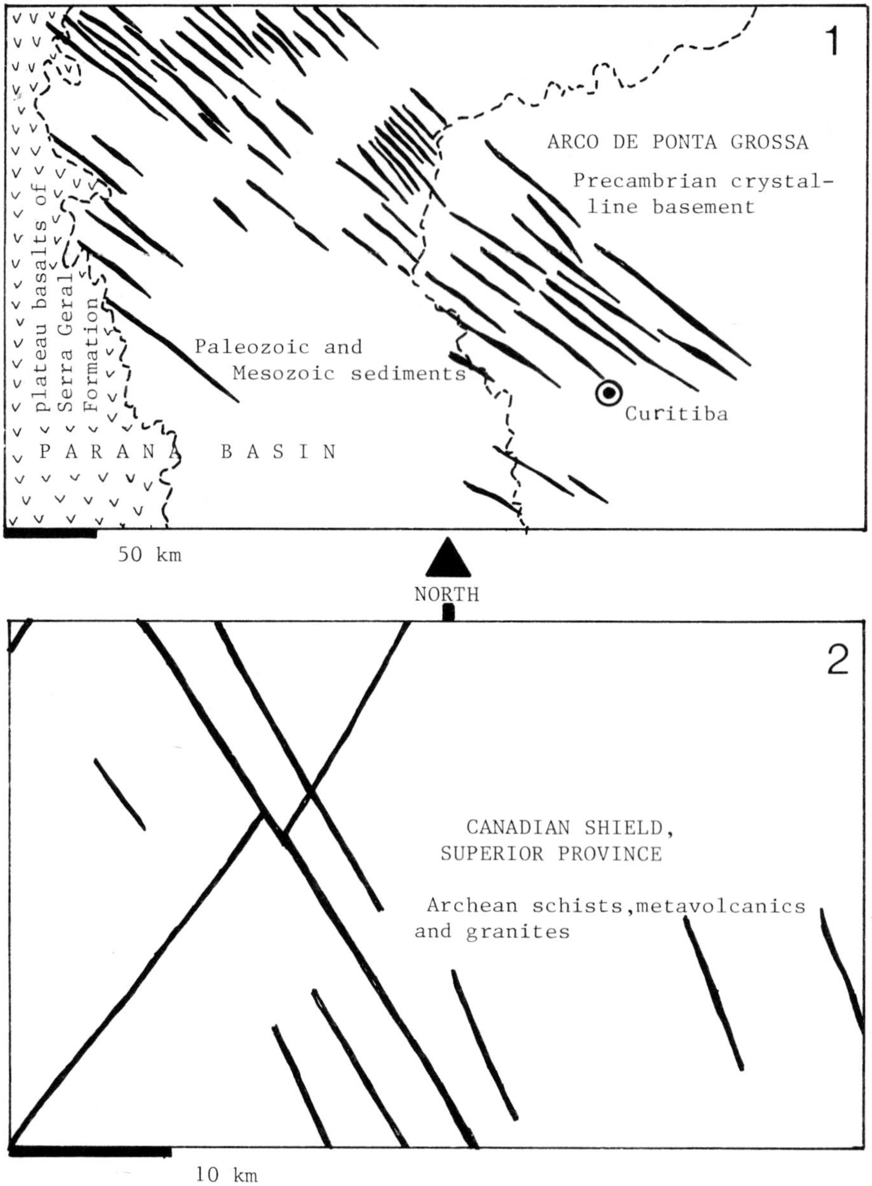

Fig. 32-1. The two characteristic regional distribution patterns of diabase dikes (shown in black). (1) Swarm of J-Cr diabase dikes in the Paraná State, S.E. Brazil, feeders to the plateau basalts of the Serra Geral Formation (after Carta Geol. do Brasil 1:1 million, sheet Curitiba, 1974). (2) The two major systems of Proterozoic diabase dikes in the Canadian Shield metamorphics south of Hearst, Ontario (after Ontario Geol. Surv. Map 2166, Hearst-Kapuskasing Sheet).

continental tholeiite and represent feeders to plateau basalts. Such diabases tend to be remarkably homogeneous and monotonous both compositionally and texturally. They are rarely associated with contemporary igneous rocks of contrasting composition. When they are, such rocks are usually felsic (granophyre, syenite porphyry, felsite or granite), resulting in a bimodal mafic-felsic association.

The minority of continental diabase bodies are spatially associated with granitic plutonism (Chapter 28) where they are most common in the earliest (pre-granite) and latest (post-granite) intrusive phases. Such diabases display a far greater petrographic diversity and are often accompanied or substituted by a variety of hypabyssal equivalents of diorite, granodiorite or granite.

The petrology of diabase is described in most textbooks (e.g. Carmichael et al., 1974). Frankel (1967) reviewed the diabase forms and structures. Both subjects are further treated in a variety of publications dealing with local areas (Walker and Poldervaart, 1949, Karoo, S.Africa; Blackadar, 1956; Fahrig and Wanless, 1964, Canadian Shield; McDougall, 1962, Tasmania; Hamilton, 1965, Antarctica and others).

Diabase dikes and sills are one of the rock bodies that show least evolutionary change with geological time, so there is no substantial difference between, say, lower Proterozoic and Cretaceous diabases except for their host association. For that reason, Phanerozoic and Precambrian diabases cannot be clearly distinguished and a complete ignorance of the latter in this volume would deprive the reader of a significant number of mineralization examples that can be found associated with the Phanerozoic rocks as well. For that reason, Precambrian diabases receive a "light" coverage in this chapter. For a more detailed description, the reader is referred to Volume 2.

32.2. METALLOGENY AND ORES ASSOCIATED WITH DIABASE DIKES AND SILLS

32.2.1. General

Abdullaev (1957), Lewis (1955) and others reviewed the relationship between metallic ores and dikes or sills in general and this included diabase dikes. They distinguished two fundamental relationships: (a) genetic and (b) structural. The (a) relationship is largely geochemical. Metal migration that parallels the various stages of the petrologic development of diabase, or the migration triggered by the interaction of diabase (mostly its heat) with the surrounding rocks, could have resulted in a local ore accumulation. In the (b) relationship, ores genetically independent of the diabase merely filled the same type of spaces (fractures, bedding planes, etc.) occupied by diabase dikes or sills, or post-solidification fractures in diabase. The fractures formed preferentially due to the higher competency (brittleness) of diabase and its baked hosts compared with the commonly incompetent wallrocks, e.g. sediments. The two relationships often overlap.

Metallic mineralization genetically associated with diabase (Fig. 32-2 and Table 32-1) is rare and, so far, of little importance economically although there are some exceptions (e.g. Cobalt, Ontario; a major silver camp). This contrasts with the wide distribution of diabase, most of which is completely devoid even of traces of metallic accumulations. Since it appears that most of the diabase-associated ores are a product of remobilization of earlier metal concentrations

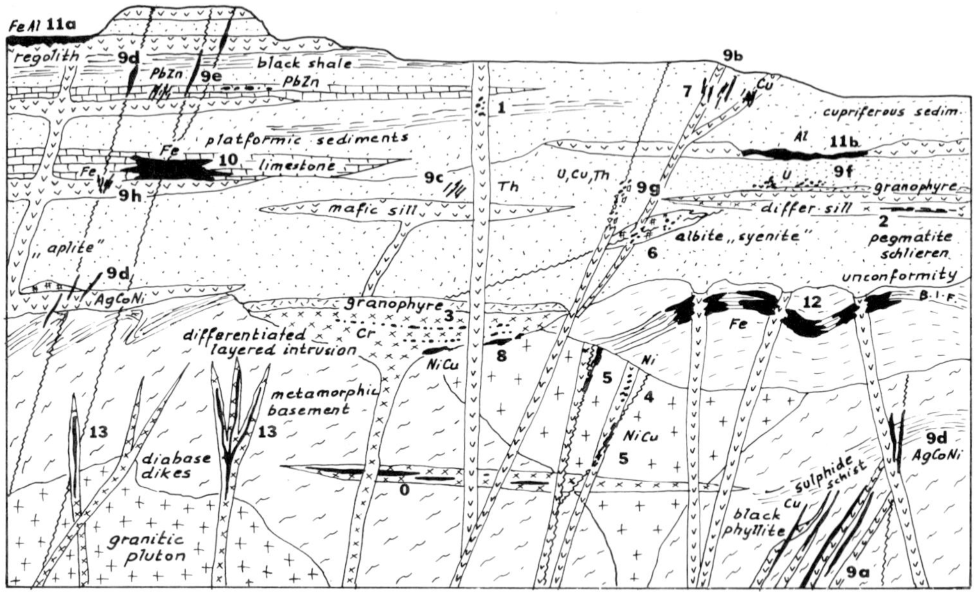

Fig. 32-2. Diagrammatic representation of the more common metallic mineralizations associated with independent diabase or gabbro dikes and sills (see Table 32-1 for explanation of the numerical codes).

Table 32-1. Principal mineralization styles associated with independent diabase and gabbro dikes and sills. NOTE: Some Precambrian mineralizations described in Volume 2 are included in this table

	ORES IN DIKES AND SILLS
(0)	Lenses of Ti-magnetite in gabbro-anorthosite sills, W. Australia (Precambrian); 130 Mt Fe, 45 Mt Ti, 1.96 Mt V
(1)	Sparsely disseminated sulphides (chalcopyrite, pyrrhotite, pentlandite, sphalerite, chalcocite) or platinoids, scattered grains and xenoliths of native iron (W. Greenland)

(2)	Pegmatitic schlieren containing oversized grains of Ti-magnetite, ilmenite; disseminated chalcopyrite
(3)	Disseminated ore minerals (chromite, pyrrhotite, pentlandite, chalcopyrite) in differentiated sills grading to layered intrusions (see Volume 2)
(4)	Disseminated to massive pyrrhotite, pentlandite, chalcopyrite in xenoliths-rich zones near gabbro or diabase contacts (about 15 Tt Ni, 18 Tt Cu)
(5)	Disseminated to massive pyrrhotite, pentlandite, chalcopyrite along shears and faults cutting thin diabase dikes (about 8 Tt Ni, 10 Tt Cu)
(6)	Disseminated specularite, chalcopyrite; uraninite, brannerite, Ti,Th minerals; etc., in light-coloured "syenites", "aplites" and breccia affiliated with mafic dikes (50 Tt Cu, 2 Tt Th, 500 t U)
(7)	Hydrothermal Cu (Ni) veins, veinlets, stockworks transecting fractured gabbro or diabase (about 100 Tt Cu)
	ORES AT OR NEAR CONTACTS OF MAFIC DIKES OR SILLS
(8)	Ni-Cu massive, breccia, stringer sulphides on the base of differentiated sills or layered complexes (see Volume 2)
	Hydrothermal veins in (meta)sediments and (meta)volcanics intruded by gabbro or diabase dikes:
(9a)	Cu veins hosted by lustrous slates (about 200 Tt Cu)
(9b)	Cu veins hosted by cupriferous sandstones, siltstones (about 100 Tt Cu)
(9c)	Th (Cu) veins in quartzite (about 500 t Th)
(9d)	Ag (Co,Ni,As,Bi,U) veins, Cobalt "type" (about 20 Tt Ag, 35 Tt Co)
(9e)	Pb,Zn (fluorite, barite) veins (about 50 Tt Pb+Zn)
(9f)	U veins, disseminations, replacements (about 200 t U)
(9g)	U, Th, specularite, chalcopyrite, etc. in albitized, hematitized, carbonatized breccia near or in the extension of gabbro, diabase dikes (1 Tt U+Th)
(9h)	Magnetite veins, stockworks, disseminations in altered gabbro or diabase (about 5 Mt Fe)
(10)	Magnetite or pyrite replacements in carbonates near gabbro-diabase sills, dikes (Cornwall, Pa.); 60 Mt Fe, 40 Tt Co
	WEATHERING CRUSTS
(11)	Residual bauxites and Fe-oxide laterites; (a) surficial, (b) buried at unconformities; 100 Mt Al, 10 Mt Fe
(12)	High-grade Fe oxides formed by enrichment of banded iron formation along mafic dikes
	MISCELLANEOUS
(13)	Ores in mafic dikes associated with differentiated granitic terrains

in the diabase intruded rocks, the exocontact association deserves as much or possibly more attention as the intrusions themselves when interpreting the diabase metallogeny.

32.2.2. Ores in diabase dikes or sills

Most diabase, gabbro and diorite dikes and sills are compositionally uniform and carry an ordinary complement of accessory metallic minerals (Ti-magnetite, ilmenite) which is too low (1-5% by volume) to be of practical importance.

The group of occurrences of native iron in the Tertiary plateau basalt province on western Greenland (Disko Island, Nugssuaq, Jakobshavn, etc.) has already been mentioned in Chapter 31. The iron, sometimes accompanied by pyrrhotite, forms scattered grains and inclusions of variable size and weight (from milligrams to 22.7 ton blocks) in basalt flows as well as in basalt and diabase dikes. The host rocks to the irons at the best known localities Uivfaq and Kitdlit are mafic dikes to breccias. Bird and Weathers (1977) reviewed the existing genetic interpretations of the native iron occurrences and concluded that the iron was derived from the sub-lithosphere mantle and brought to the surface as xenoliths.

Sulphides in diabase occur occasionally, in very insignificant quantities. Pyrite and pyrrhotite are the most common, followed by chalcopyrite. In the Coppermine area, northern Canada, the Willow Lake gabbro dike, traceable for over 8 km, contains sparsely disseminated pyrite and chalcopyrite and it grades around 0.1% Cu (Kindle, 1972). Similar occurrences are numerous in the nearby Bathurst Inlet area. Pentlandite can sometimes be microscopically determined in the minor pyrrhotite in diabase, and sphalerite in 0.015 mm large grains, was found by Desborough (1963) in the mafic intrusions in the Ozark Mountains, Missouri. Rare platinoids have occasionally been reported as disseminated accessories in diabase dikes, e.g. from Tilkerode in the Harz (East Germany).

Chilled margins of some mafic dikes and sills often show substantial variation in their trace metal content. Granger and Raup (1969) recorded an almost twofold increase in titania in the chilled zone (4.26% TiO_2), compared with the rest of the intrusion (2-3% TiO_2), in the Sierra Ancha sheet in Arizona. Local pegmatitic facies occur as small schlieren in some diabase bodies and often contain scattered mineralogical quantities of visible sulphides (pyrrhotite, chalcopyrite) as well as large phenocrysts or nests of the accesory ilmenite or Ti-magnetite.

Some mafic sills or dikes are multiple, composite or differentiated, showing a variety of rock types. One such variation is towards ultramafic members (picrite, troctolite), the other towards felsic members and alkali enrichment (syenite, granophyre, granite). There is much controversy as to whether such non-uniform rock bodies originated by an in-situ differentiation from a single surge of magma; whether the parent magma was pure, mantle-derived partial melt or a mantle melt modified by assimilation of the crust or whether the magma was an entirely crust-derived melt (compare Blackadar, 1956 and Basaltic Volcanism Study Project, 1981, for further discussion).

There is further controversy as to whether certain igneous-looking rocks associated with diabase, particularly the albite-rich felsic members (albite "syenites" and "aplites", albitites) are magmatic differentiates or hydrothermally metasomatized earlier rocks. If they are metasomatites, what was the timing of their conversion: deuteric and autometasomatic (of and by the diabase), or post-consolidation, unrelated to the diabase magmatism? The problem of albitic rocks runs parallel with the spilite problem, already reviewed in Chapters 10-12.

The hydrothermal-alteration origin of a red or pink albitic "aplite" in the Cobalt, Ontario, district, was proposed relatively early by Bastin (1935) and has been reported several times since. In the Dolores Creek Cu (Fe,U,Th) camp in the Yukon (Laznicka and Edwards, 1979), numerous gabbro dikes intrude Proterozoic fine clastic sediments and impure carbonates. Bodies of pink, coarse to fine albite-chlorite rock of syenitic appearance occur in the extension and near contacts of the gabbro dikes and appear to have formed by hydrothermal albitization of the gabbro as well as breccias. Specularite and local chalcopyrite are disseminated, producing low-grade orebodies of a sodic alteration equivalent to "porphyry coppers". There is evidence indicating that the trace copper has been released from the gabbro in the process of pyroxene destruction.

Copper showings of similar style modified by local conditions are relatively common, e.g. in the Luiri Hill area, Zambia (Phillips, 1959). Even more common is a variety of Cu-sulphide (usually chalcopyrite) mineralized hydrothermal veins, veinlets and stockworks in faults and fractures transecting the diabase (e.g. Kumarina mine, Bangemall Basin, Western Australia; Gee, 1975; numerous occurrences in the vicinity of the early Proterozoic Nipissing diabase sill, Lake Huron area, Ontario; Pearson, 1978). In the Lewis Range, southern Canadian Cordillera, gabbro dikes intrusive into Proterozoic sediments cause local remobilization of a weak stratabound mineralization (Morton et al., 1973). Altered mafic sills near Yarrow Creek and Spion Kop Creek contain abundant interstitial copper sulphides. Other sills intruding Proterozoic dolomites, carry disseminated sphalerite and galena, in addition to the Cu sulphides.

Diabase dikes transecting Cu-mineralized continental basalt flows as in the Coppermine area, Northern Canada (Kindle, 1972), often contain disseminated chalcocite. In the VIC showing, the zone of a diabase body contains 0.46% Cu in an area over 400 m long and 12 m wide. The gray chalcocite is very inconspicuous and when secondary minerals (malachite) are not developed, it is easy to overlook it .

Thick diabase sills grading to layered intrusions containing olivine-rich members and centrally situated or basal Ni-Cu sulphides (Kureika, Insizwa), have already been described in Chapter 31. Examples of Ni-Cu sulphide accumulations in relatively thin diabase dikes are much rarer, and most of them appear to be the result of post-solidification, hydrothermal alteration and mineralization along shears or faults intersecting the mafic dikes, or their proximal exocontacts. In the small Sohland-Rožany ore field near Šluknov, N.W. Czechoslovakia and S.E. East Germany (Beck, 1909; Kopecký et al., 1963), thin diabase dikes, measuring 5-20 m, intrude a late Proterozoic cataclastic granite. Massive lenses, veins and impregnations of pyrrhotite and pyrite with minor chalcopyrite, pentlandite and ilmenite form ore shoots, up to 400 m long, within the diabase. In the Thomson River Cu,Ni,Au, Pt-Pd mine east of Melbourne, Australia (Keays and Kirkland, 1972), a bulging Devonian diorite-pyroxenite dike contains massive to disseminated chalcopyrite, pyrite and minor pyrrhotite, pentlandite and Pt-Pd minerals along "contact shears".

32.2.3. Ores at or near contacts of gabbro or diabase dikes, sills

The contact effects of most diabase or gabbro dikes and sills are slight and usually limited to rims a few centimetres wide. Thermal metamorphism is most pronounced in mudstone hosts, where narrow zones of biotite and sometimes cordierite hornfels may develop. In the Middelburg, Ermelo, Wakkerstroom area, South Africa (Hammerback, ed., 1976), originally hematitic ironstones interstratified with sandstone and shale are converted by Karoo diabase sills into a hard, compact magnetite. This represents a mere 230 Tt ore with 67-69% Fe.

Arenites and crystalline rocks are usually unaffected, and dolomites transected by diabase dikes are often de-dolomitized (converted to calcitic marbles). The most unusual effects of thermal maetamorphism can be seen at diabase-coal contacts, as in the Natal Coalfield, South Africa. There, anthracite, natural coke or graphite formed in the exocontact. Under certain conditions, the hot mafic magma may transfer suitable country rocks into a plastic state capable of intrusion, to form rheomorphic dikelets and veins filling dilations in both endo- and exocontact of a diabase body (Frankel, 1967).

Thermal metamorphism alone, however, seems to be unimportant as an ore-generating process. This requires a widespread hydrothermal transfer. Because basaltic magmas are relatively "dry", hydrothermal systems in mafic dike or sill fields contemporary with the mafic magmatism are almost always the result of heating of the water, contained in the enclosed rocks. Hydrothermal systems younger than the mafic magmatism and superimposed on the diabase bodies can have diverse origins, e.g., emplacement of felsic plutons in depth. Hydrothermal veins were reported on both sides of diabase contacts, although the exocontact veins are substantially more common.

COPPER VEINS

Copper accounts by far for the greatest number of mineralized occurrences in the contact area of gabbro and diabase dikes and sills. In the Racing River-Gataga River area, N.E. British Columbia (Carr,

1971; Preto, 1972; 115Tt Cu/3-4%; Fig. 32-3), numerous swarms of thin but high-grade ankerite, quartz, chalcopyrite fissure veins are developed in Proterozoic black lustrous slates, densely transected by diabase dikes. Both veins and dikes fill the same dilations and overlap in time. The majority of the diabase dikes are slightly younger than the ore veins and many veins have been destroyed by diabase engulfment. The ore veins as well as some diabase dikes are fringed by a narrow halo of decarbonization.

In the Inyati mine, Zimbabwe (Roberts, 1973), chalcopyrite fissure veins are associated with basic dikes, subsequently intruded by granites.

MAGNETITE VEINS AND REPLACEMENTS

Magnetite forms common small occurrences in mafic dikes and sills. Puffer and Peters (1974) recorded a small locality near Laurel Hill,

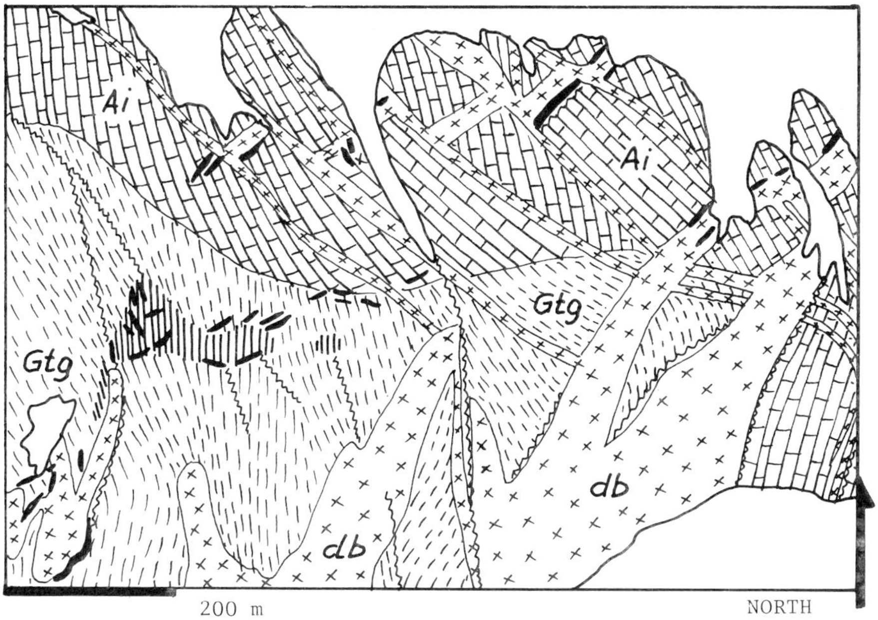

Fig. 32-3. Bronson Mountain group of Cu-mineralized veins (black) in the Racing River-Gataga River area, N.E. British Columbia, showing their close association with diabase dikes. Ai=$^{Pt}3$ carbonate, minor slate; Gtg=slate and phyllite; db=diabase. Vertical ruling: bleached phyllite with chalcopyrite mineralized quartz veinlets. White: glaciers and snowfields. From LITHOTHEQUE, modified after Preto (1972).

New Jersey, where parallel magnetite veins fill joints surrounded by bleached diabase. They interpret the magnetite as being a product of local lateral secretion of iron by deuteric fluids.

Much more extensive migration and entrapment of locally ? deuterically released iron, resulted in the economically important magnetite deposits of the "Cornwall-type" in S.E. Pennsylvania (Eugster and I-Ming Chou, 1979; Lapham, 1968; Fig. 32-4). The largest Cornwall deposit contains two major orebodies, replacing a Cambrian limestone directly above and below a saucer-shaped sheet of Triassic diabase. The ore is either massive, homogeneous, or it incorporates the relic banding of the host limestone. It consists of dominant magnetite, actinolite and subordinate hematite, pyrite, chalcopyrite and chlorite. The alteration and mineralization was multistage. The early-formed diopside, tremolite, phlogopite hornfels was faulted and modified by fissure-controlled localized K-feldspar metasomatism, chloritization, actinolitization and magnetite emplacement. Cornwall has produced over 100 Mt ore with 39.4% Fe, 0.29% Cu and 0.02-0.056% Co. Cobalt is concentrated in pyrite. The Grace mine located in the same area (Sims, 1968) is comparable, but smaller.

Numerous showings of mineralized "skarns" in the Lake Huron area, S.C. Ontario, developed in lower Proterozoic limestones at the contact with the Nipissing diabase sill, have been recorded by Pearson (1978). In addition to magnetite and chalcopyrite, one occurrence (Cobden River) caries scheelite in idocrase, diopside and garnet skarn.

Ag (Pb,Zn,Cu; Co,Ni,As,Bi) VEINS

Several well-known high-grade silver deposits and fields (Cobalt-Gowganda, Great Bear Lake, Kongsberg, etc.) are associated both in space and time with diabase dikes or sills. The Cobalt-Gowganda district, Ontario, has been a major metal producer (about 14.5 Tt Ag, 12 Tt Co, 1.3 Tt Ni), with the metals derived from several hundred thin and short steeply dipping fissure veins in the vicinity of a thick Proterozoic Nipissing Sill. The sill intrudes Archean greenstone terrain containing a horizon of metalliferous "exhalites", and Proterozoic detrital sediments locally pyritic and arsenopyrite-bearing. Cobalt-Gowganda and Great Bear Lake districts are treated in Volume 2.

The historical silver mines in Kongsberg, S. Norway (Ihlen and Vokes, 1978; Fig. 32-5), produced 1.4 Tt Ag from about 130 workings. Native silver accompanied by small quantities of pyrite, sphalerite, chalcopyrite, argentite, Ni-Co arsenides, etc., occurs in steep, narrow (5-10 cm) and short (max. 100 m) white calcite veins. The veins shortly postdate emplacement of diabase dikes in the area, and most are enclosed in Proterozoic metamorphic schists and gneisses containing bands of disseminated chalcopyrite, sphalerite and galena conformable with schistosity ("fahlbånds"). The silver occurrence is confined to (or is richer in) the vein segments intersecting the

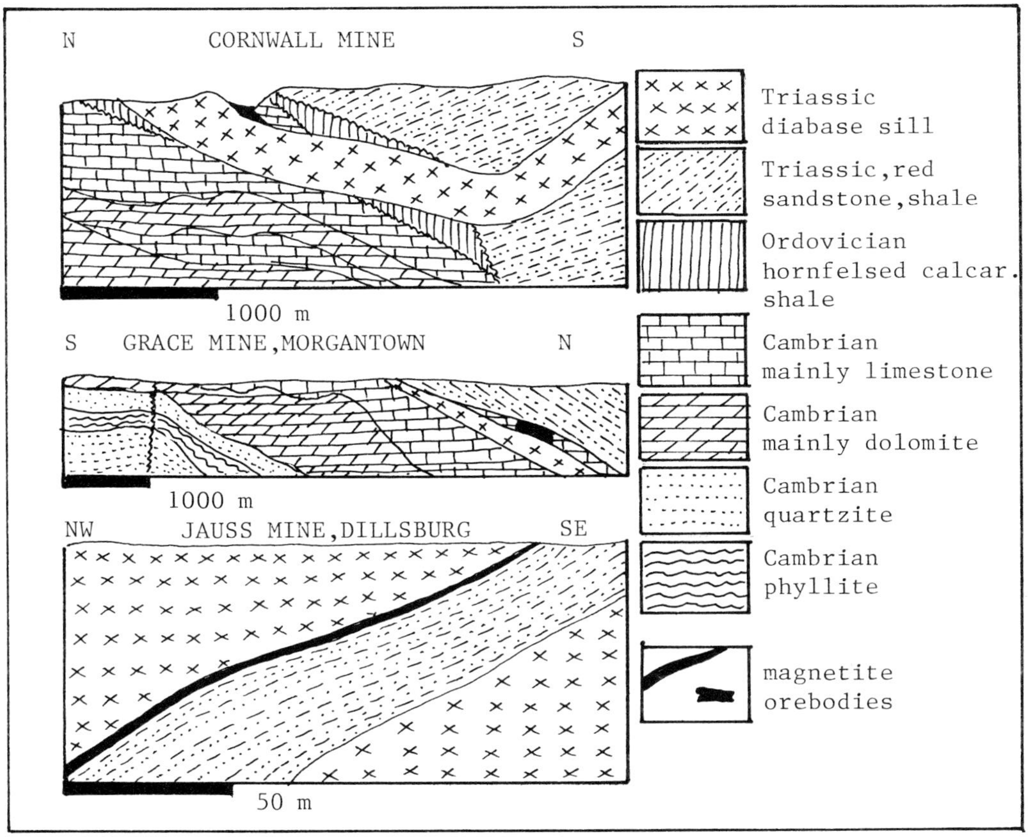

Fig. 32-4. Examples of replacement magnetite deposits of the "Cornwall (Pennsylvania) type", adjacent to Triassic diabase sills. From LITHOTHEQUE, modified after Lapham (1968), Sims (1968) and Spencer (1908).

sulphide schists, from which it is generally believed to have been extracted during the period of Permian rifting in the adjacent Oslo Graben.

Small vein, veinlet, dissemination, replacement, etc. occurrences of galena and sphalerite in quartz, calcite, fluorite or barite gangue are quite common in sediments adjacent to diabase dikes (at the Argent Station near Pretoria, South Africa, Hammerback, ed., 1976; in the Igarapé Xituba region in the Amazon Basin, Brazil, Rezende and João, 1980 and elsewhere).

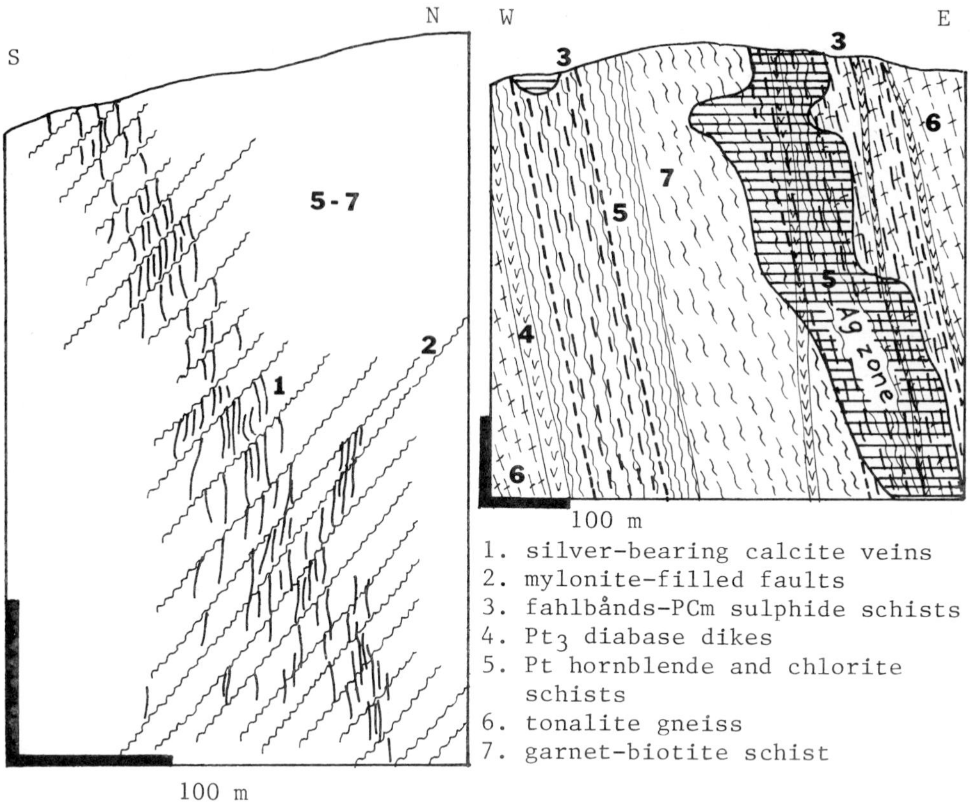

Fig. 32-5. Silver veins hosted by the Overberget fahlbånd, Kongsberg, southern Norway. From LITHOTHEQUE, after Bugge (1917) and Neumann (1944).

U and Th VEINS, MINERALIZED BRECCIAS, IMPREGNATIONS

Spatial association of radioactive mineralization and gabbro or diabase intrusions has been recorded from various localities (e.g. Nabarlek, N. Australia; Port Radium, N. Canada; Montreal River area, Ontario; Paukkajanvaara, Finland), but insufficient work has been done to determine the metallogenic function of the mafic bodies, if any. The small U deposits hosted by the Dripping Springs Quartzite (Gila County, N. Arizona; Granger and Raup, 1969; 21.3 Tt ore with 0.205% U_3O_8), probably give the best insight into the problems involved.

The Quartzite is a formation of the middle-upper Proterozoic Apache Group, a sequence of red to gray feldspathic sandstone, siltstone and shale resting on lower Proterozoic crystalline basement and topped by limestone. Intercalated minor flows of plateau basalts, and widespread diabase sills and dike feeders, are also present.

The U mineralization is represented by veinlets, blebs and disseminations of uraninite and small quantities of Cu,Pb,Zn and Mo sulphides in quartz, carbonate and chlorite gangue. The U occurrences occupy poorly outlined structures, usually gradational into the altered quartzite wallrock. The best-outlined veins are steeply dipping tabular bodies of disseminated uraninite, the central part of which is marked by a core of breccia or a narrow fissure filling. Most of the U orebodies show dual control: (1) confinement to the gray unit of the upper Member of the Dripping Springs Quartzite (=stratigraphic control) and (2) proximity to diabase sills, particularly to a suite of felsic rocks of intrusive appearance (granophyre, syenite and aplite) formed on both sides of the diabase-feldspathic sandstone contact. Smith and Silver (1975) argued convincingly that the felsic rocks were derived from the Dripping Springs Quartzite by an extensive interaction between the mafic magma and the feldspathic country rock, facilitated by the influx of groundwater into the magmatic roof region. The detrital sediments, often carbonaceous, appear to have been the source of U of infiltrational origin, comparable with the "sandstone uranium-vanadium" mineralization style (Chapter 25). The U is thus the product of remobilization of a stratabound U enrichment by secondary magmatic-hydrothermal processes triggered by the magmatic heat.

Similar effects of remobilization can be observed on small scale in, for example, the Laguna mine in New Mexico, but there the peneconcordant U ore in arenites is the dominant ore style, so that the local modification by diabase dikes has never been emphasized. The U occurrences discussed above have many common links with the widespread but so far uneconomic U (in uraninite) and U-Th (in brannerite) occurrences affiliated with unusual altered breccias and locally intervening diabase and gabbro dikes and metasomatic "syenites" in the Wernecke and Ogilvie Mountains in the Yukon. The Dolores Creek camp mentioned earlier represents an example locality where Cu is prominent, but other occurrences in the area (e.g. Quartet Lakes; Bell and Delaney, 1972) are Th-U dominated.

The thorium-bearing veins at Hall Mountain near Porthill, Idaho (Staatz, 1972; Fig. 32-6) are hosted by a Proterozoic quartzite in the upper contact of a quartz diorite (or gabbro) sill. The sill is a member of a middle Proterozoic swarm of "Purcell Sills". Quartz and calcite are the principal gangue minerals, and chlorite, magnetite, pyrite and biotite occur in lesser quantities. Brown crystals of thorite sometimes accompanied by allanite, chalcopyrite, specularite and other minerals are very erratically distributed in the gangue (the grade varies between 95 ppm to 21% Th). Although Hall Mountain appears to be of little economic importance at present, it constitutes a possible link with the more important Th accumulations such as the Lemhi Pass (Idaho and Montana), where mafic dikes are much less prominent.

32.2.4. Mineralizations in mafic dikes affiliated with differentiated granitoid terrains

This subject is treated in Chapter 28. The reader, however, should be aware that spatial affiliation of granitic rocks and diabase dikes does not necessarily imply a genetic relationship between the two, so that the possibility of a link with the subjects and problems treated in this chapter cannot be excluded. Preferential emplacement of ore veins into dikes of diabase and lamprophyre is particularly common in some deeply eroded block faulted plutonic aureoles, as in the Bohemian Massif, Czechoslovakia. There, the Příbram Pb,Zn,Ag field is by now almost a classical example of the close ore-diabase association, an association applied in exploration (Kutina and Tělupil, 1966; Fig. 32-7). Additional examples include the Krásná Hora stibnite-gold lodes hosted by kersantite dikes (Bernard et al., 1968); Premier Au-Ag mine, N.W. British Columbia and several localities listed by Lewis (1955).

32.2.5. Weathering crusts over diabase dikes and sills

Tropical weathering superimposed on gabbro and diabase intrusions could produce residual lateritic bauxite or iron ores. Small deposits

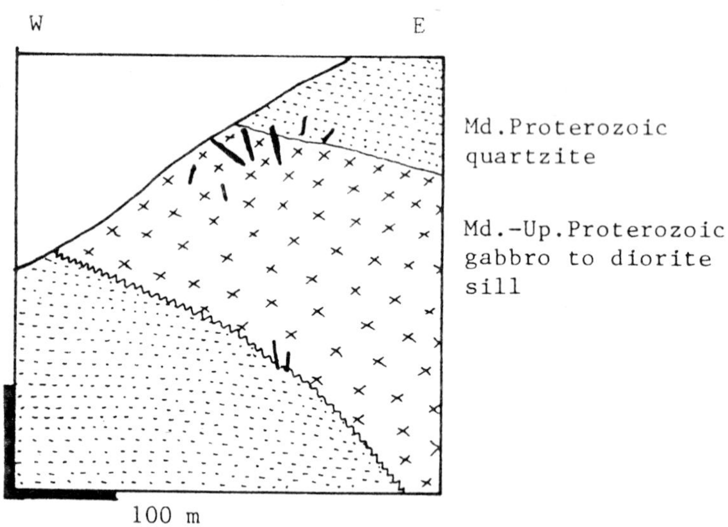

Fig. 32-6. Hall Mountain, Idaho, Th-bearing veins associated with a diorite sill. Diagrammatic section, ore veins (in black) projected on the plane of section. From LITHOTHEQUE, data from Staatz (1972).

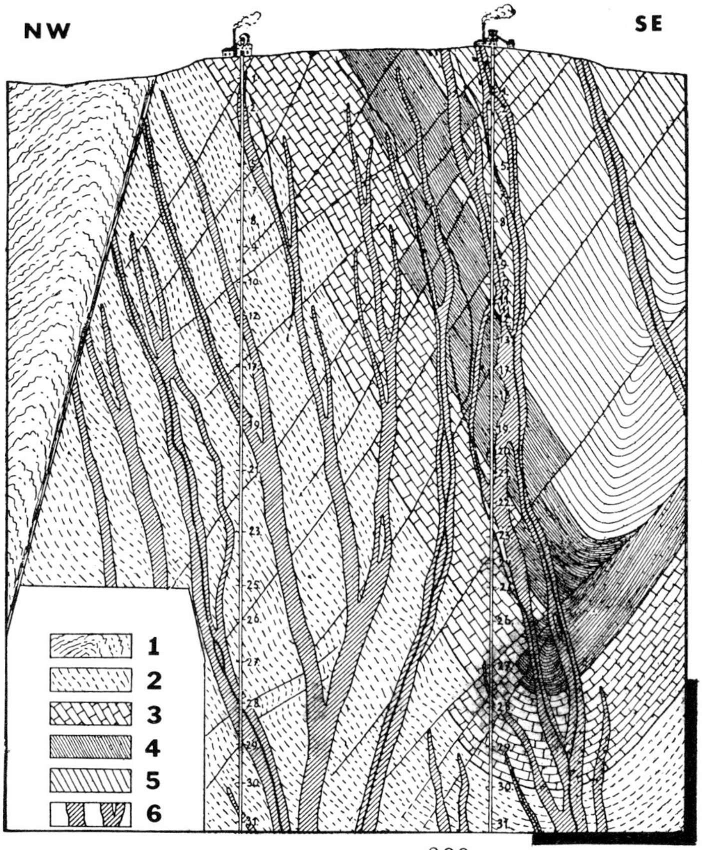

Fig. 32-7. Březové Hory Pb,Zn,Ag field, Příbram district, Czechoslovakia. Ore veins are controlled by late Paleozoic diabase dikes, postdating emplacement of a tonalite pluton. After Kettner (1918).

of this sort found in Tasmania have already been mentioned in Chapter 31, and numerous but ususally poorly documented occurrences exist in the humid tropical belt of western Africa. In the bauxite districts of Guinea (Tougue-Dabola, Boke, Fria, Fatala, etc.; de Kun, 1965), diabase sills and dikes are consistently listed as one of the parent rocks (perhaps the most important one) to the lateritic bauxite.

In some Lake Superior Proterozoic iron ore districts, diabase dikes and sills are closely associated with high-grade iron oxide orebodies formed by residual enrichment (leaching out of SiO_2, CO_2, Ca, etc.) of the low-grade banded siliceous iron formation. In the Marquette Fe ore district (Gair and Tsu-Ming Han, 1975; Fig. 32-8), the diabase bodies acted mainly as impermeable barriers, impounding and confining the groundwater flow through the iron formation so that their

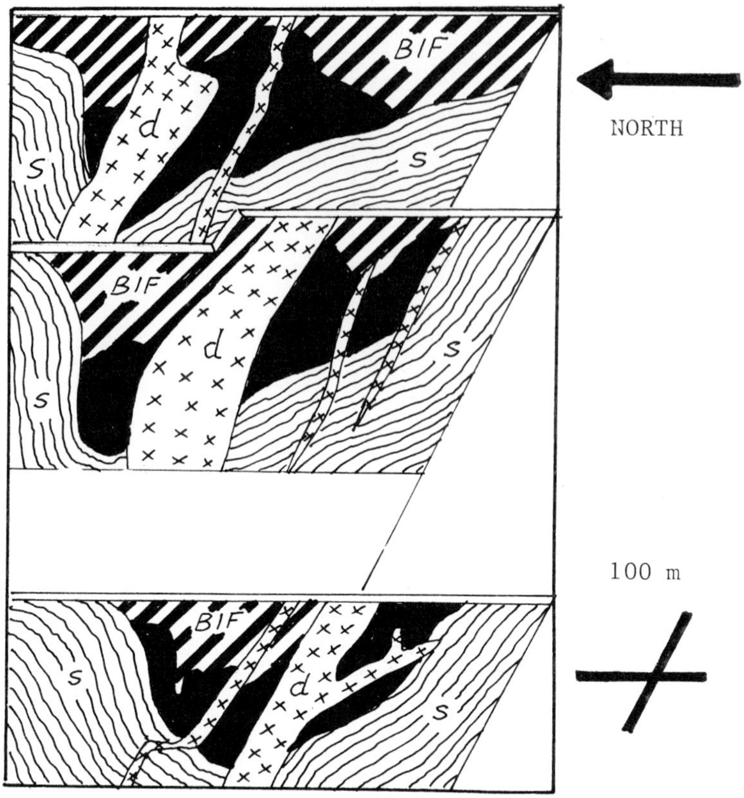

Fig. 32-8. Athens-Bunker Hill orebody, Marquette Fe district, Michigan; a set of isometric cross-sections showing the close association of diabase dikes (d) and "soft" Fe oxide ore (black), formed by residual enrichment of a lean Proterozoic oxide and carbonate banded iron formation; s=slate. After Gair and Tsu-Ming Han (1975).

contribution to the rich iron ore genesis was largely structural.

32.2.6. Placers derived from diabase dikes and sills

Diabase dikes and sills contributed ilmenite and Ti-magnetite to many placers and paleoplacers, but proximal occurrences of heavy minerals closely affiliated to parent diabase are rare. Nielsen (1973) reported the existence of sands containing 37-74% ilmenite in raised beaches in the Thule region, N.W. Greenland, adjacent to mafic dikes intruding Precambrian metamorphics.

"There is now general agreement that no single parent magma is responsible for the formation of alkaline rocks, since alkaline rocks occur in many petrological associations".
H. Sørensen, 1974.

CHAPTER 33

Alkaline Igneous Association

33.1. INTRODUCTION

33.1.1. General

The terms alkaline and alkalic rocks have different meanings to different investigators, and "the very definition of an alkaline rock is fraught with difficulty and controversy" (Currie, 1974; see also the historical review in the introductory chapters in Sørensen, ed., 1974). The brief definition stating that "alkaline igneous rocks are characterized by the presence of feldspathoids and/or alkali pyroxenes and amphiboles", either in the rock itself or in the chemical analysis recalculated into the C.I.P.W. norm, will be followed approximately here. To this have been added two categories of rocks transitional between the alkaline and calc-alkaline families that may lack the feldspathoids (peralkaline rhyolite/granite and feldspar trachyte/syenite associations).

The field of alkaline igneous petrography is burdened by a long list of rock names, many of them unnecessary (at least for the purpose of metal exploration). Fundamental rock class names (such as syenite, gabbro, phonolite) preceded by a qualifier, which in most cases is the name of a prominent mineral (e.g. nepheline syenite, analcite phonolite), have been used whenever possible here. The numerous rock names reflecting minor variations in accessory minerals, texture, petrology and local preference (such as tinguaite, foyaite, juvite, etc.) have been avoided and used only sparingly, mostly in cases where a certain variety controlled a particular mineralization. Whenever possible, the alkaline rock terminology outlined in Sørensen, ed. (1974) has been followed.

Strictly alkaline rocks (in the petrographic sense) are quite rare and their outcrop occupies no more than a few tenths of a percent of the present Earth surface. In most regions of concentrated occurrence (alkaline petrographic provinces) the alkaline rocks, often in the minority, are associated with alkali rocks (e.g. alkali basalts, those that contain 0-10% of feldspathoids), and also with tholeiitic and calc-alkaline rocks. The coexisting petrochemical kindreds cannot be clearly distinguished and many initially tholeiitic or calc-alkaline provinces often develop into alkaline provinces with the passage of time (e.g. Carmichael et al., 1974).

Compared with the plateau basalt association (Chapter 31), alkaline rocks are more localized and rarely form extensive bodies of uniform composition. In Cainozoic alkaline provinces, lava flows and pyroclastics are of limited areal extent and can usually be attributed to a discrete central volcano. Most of the alkaline volcanic provinces that have been described are wholly subaerial. Subaqueous equivalents have rarely been recognized and described although they do exist and could have interesting metallogenic implications.

The subsurface equivalents to alkaline volcanics occur as dikes, laccoliths, stocks and plutons. Simple, undifferentiated sills are rare, but ring dikes and composite ring complexes are common and characteristic. Layered alkaline intrusions comparable in terms of structure with the Bushveld (Lovozero, Ilímaussaq) are rare and controversial. Interaction of alkaline rocks with the environments into which they were emplaced is of variable intensity, ranging from zero to extensive alkaline metasomatism (fenitization) in the exocontact. Certain alkaline complexes are interpreted as being entirely of metasomatic origin, (for example those of the Urals Mts.). The influence of the environment on alkaline rocks in the endocontact is still imperfectly known and highly controversial, as are the effects of "ageing" (diagenesis and metamorphism), which may involve de-alkalization. Metamorphosed equivalents of the alkaline association are poorly known and rare.

Alkaline rocks are known that date from the Proterozoic age, although the majority are Phanerozoic. To prevent fragmentation, the Precambrian occurrences received only brief coverage in this chapter when warranted, and their more detailed description appears in Volume 2.

33.1.2. Geotectonic setting and origin

Like plateau basalts, alkaline igneous rocks are characteristic of (but not confined to) the highly stabilized continental crust (platforms, cratons). They appeared later in the geological record, when a sufficient quantity of such a crust had been generated. The actual emplacement of alkaline magmas into the continental crust, however, coincided with periods of extensional faulting (taphrogenesis, especially rifting) of variable intensity. Some peralkaline silicic rocks (e.g. comendites and granites) were emplaced as a consequence of the pre-rifting "epeirogenic doming" (Bowden, 1974). Certain bodies of alkaline rocks (e.g. kimberlites) initially required nothing more than a very minor crustal dilation, usually at an intersection of two prominent joint systems in the host rocks, to reach the surface. Others, such as the alkaline intrusions east of the Paraná Basin in Brazil (Fig. 33-1), were controlled by faults and narrow grabens. The most prominent alkaline provinces in the world are localized by major graben and horst systems (e.g. the East African Rift, Oslo Graben), which are members of the global rift system.

The belief that alkaline magmas originate at considerable subcrustal depth (between 40 and 100 km), in the mantle, is often

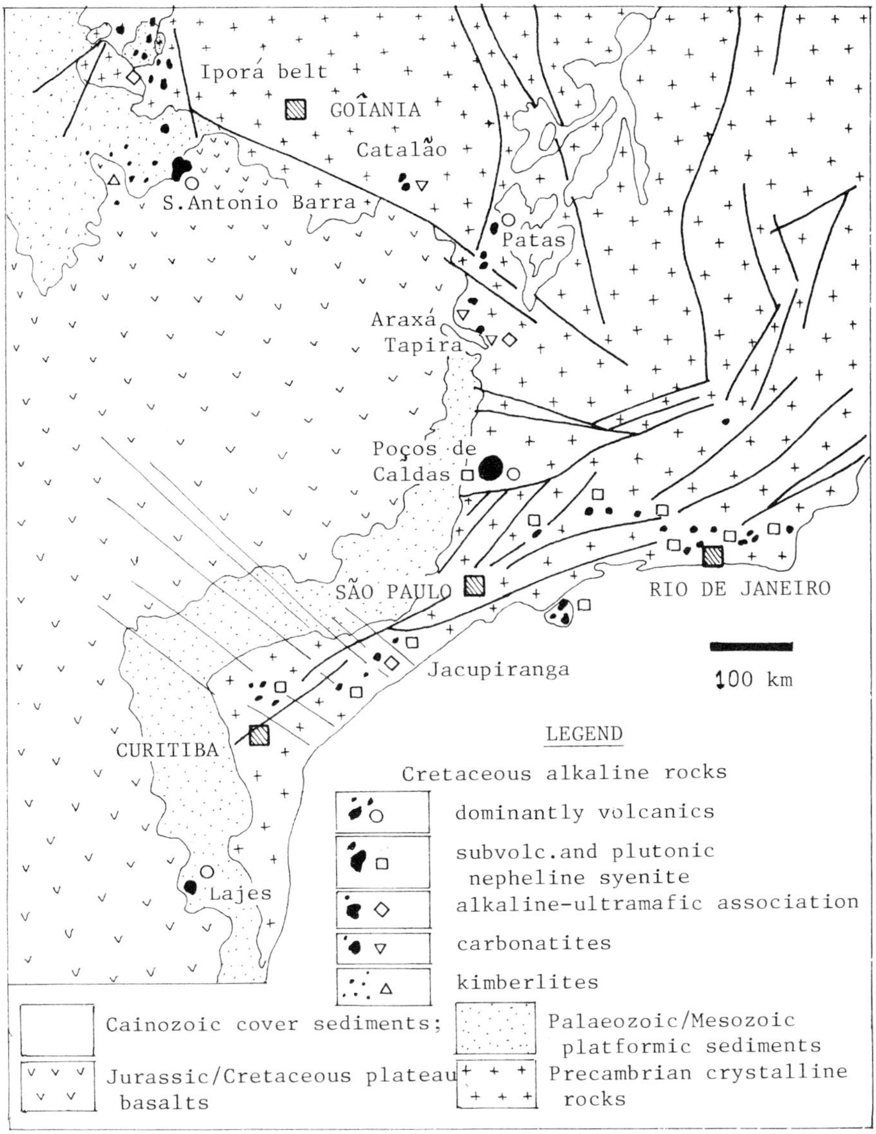

Fig. 33-1. The alkaline province of south-eastern Brazil (fringe of the Paraná Basin). From LITHOTHEQUE.

quoted in the literature (e.g. Carmichael et al., 1974; their Chapter 10). Bailey (in Harris and Bailey, 1974) concluded that felsic alkaline magmas are generated in the deep continental crust. The depth of derivation of alkaline rocks usually quoted is greater than the depth of derivation of tholeiitic magmas. The evidence of deep derivation is particularly compelling for kimberlites, rocks that often carry distinctive high-pressure minerals (e.g. diamond) as well as distinctive abyssal inclusions (garnet peridotite and eclogite).

The actual process of alkaline magma generation is now usually considered to be a partial melting of the mantle and/or the continental crust above "hot spots" (i.e., mantle plumes); compare Burke et al. (1981) for a review. Carmichael et al. (1974) suggested a considerable diversity of magmatic parentages, a variety of differentiation models and crustal contamination as agents of the alkaline lithogenesis.

33.1.3. Ore distribution and economic importance

The economic geology of all the alkaline rocks has been summarized briefly by Semenov (1974), and that of carbonatites alone by Deans (1966), Ginzburg et al. (1958), and others. Despite their very limited occurrence, the alkaline rocks as outlined in this chapter hold a virtual monopoly (over 90%) on the occurrence of niobium and the rare earths (about 35 Mt Nb; 42 Mt REE) and furthermore contain significant quantities of Zr,Ti,Ta and Th (about 200 Mt Ti, 250 Mt Zr, 1.5 Mt Th, 300 Tt Ta). The above tonnages are very unevenly distributed, the bulk being in few giant ore accumulations (e.g. Araxá, Lovozero, Ilímaussaq). Smaller but still significant quantities of Cu,Fe,Al, U,Be,Sn, Mo,Au,Ag, Pb,Zn, etc. are produced from, or are known to exist in, deposits affiliated with the alkaline rocks.

As a class, alkaline rocks appear to be highly mineralized, but the ore occurrence is selective. Carbonatites, that account for less than 0.01% of the area occupied by the alkaline association, contain the bulk (some 80%) of the economic metals listed above (that excludes the presently uneconomic, low-grade resource of Zr and some Ti). The intrusive alkaline equivalents are frequently ore-bearing, while the alkaline volcanics are rarely so. The predominantly volcanic alkaline provinces, such as the České Středohoří-Doupovské Hory region of western Czechoslovakia and the province of the East African Rift have, until recently, been dismissed as "barren" by the economic geologists. This notion is now slowly changing, thanks to new discoveries of metallic accumulations in sedimentary basins derived from the volcanics (e.g. the U-Zr infiltrational ores in northern Bohemia, Chapter 25); the utilization of new, low-grade materials for metal extraction (e.g. high-Ti and Al residual and reworked clays tested as a possible substitute of bauxite and a source of titanium) and the new, interesting data on metallogeny (e.g. the heavy metals dissolved or suspended in rift lakes) that may result in the discovery of new economic ore accumulations.

33.1.4. Major alkaline sub-associations and their mineralization

Occurrence of alkaline igneous rocks is highly nonuniform, ranging from single kimberlite pipes emplaced into platformic sediments and showing a high degree of individuality and separation from the host environmemt, to extensive volcanic-intrusive complexes composed of a wide variety of intermixed rocks that are not only alkaline. As a consequence, subdivision of the alkaline igneous association into sharply outlined sub-associations having a regional uniformity is difficult if not impossible. The petrochemical classifications of alkaline rocks (e.g. Currie, 1974) are beneficial mainly when used with small, homogeneous objects, ranging in size from hand specimens to small, uniform magmatic bodies. When exact classifications are applied to actual field areas, it is appreciated in most cases that although some local rocks qualify as members of a particular selected petrochemical class, other rocks (often in volumes exceeding the former) are members of different classes, many of which are not alkaline at all. This is well illustrated by the silicic peralkaline sub-association, popular with exploration geologists because of its affiliation with the "Nigeria-type" Sn provinces. The truly peralkaline members of this association in Nigeria (e.g. riebeckite granites), however, are in the minority in the field and rarely carry tin. Tin is associated with biotite granites that are little different from the "tin granites" of the plutonic association of orogenic belts, treated in Chapter 28.

The confusion resulting from "scale jumping" and the simplified designation of petrographically diverse provinces by names of certain rock groups alone is typical in geology and impossible to fully eliminate. The reader should bear this in mind. The nine subdivisions into which the alkaline association has been broken down are just convenient categories believed most suitable for a prospector. They overlap and are mutually transitional.

33.1.5. Alkaline provinces

This term is being frequently used in the petrographic literature (e.g. Carmichael et al., 1974; Sørensen, ed., 1974), but the "provinces" treated are not uniformly outlined and named. This is directly due to the virtual impossibility of outlining them accurately. The worldwide distribution of alkaline rocks ranges from relatively large, densely occupied regional occurrences of uniform geological age in which the ratio of alkaline to non-alkaline rocks approaches 50:50 (such as the Tertiary alkaline province of the Massif Central, France), to areally insignificant (0.X km^2) occurrences of kimberlite diatremes or dikes distributed over entire continents. Fig. 33-2 and Table 33-1 use a much looser form to indicate briefly

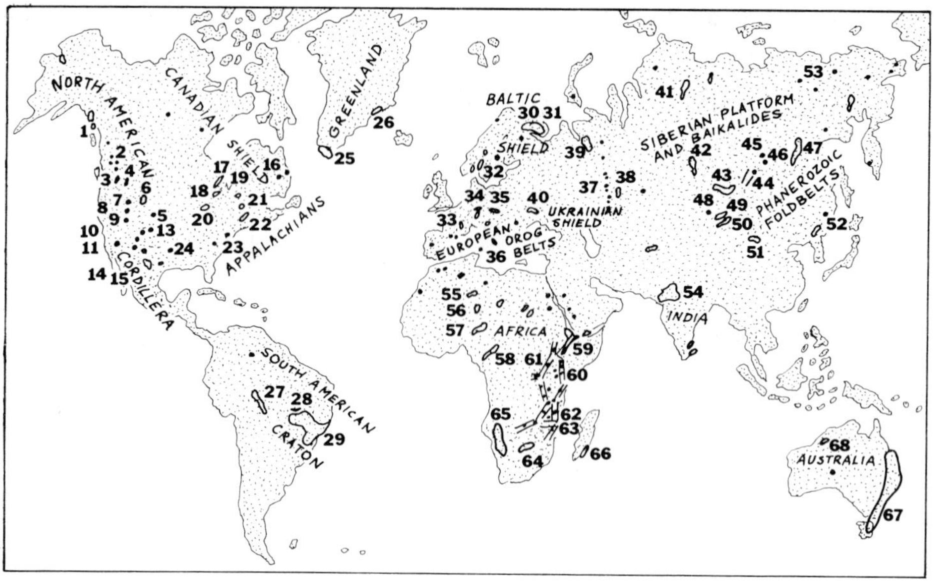

Fig. 33-2. Index map of mineralized alkaline occurrences and areas of the world, listed in Table 33-1.

Table 33-1. Principal areas of distribution of mineralized alkaline occurrences of the world (Figure 33-2)

NO.	AREA OR LOCALITY	AGE	ROCK ASSOCIATION EMPLACEMENT LEVEL	METALS IN ORE DEPOSITS
NORTH AMERICAN CORDILLERA				
1.	Bokan Mt., S.E. Alaska	Cr	A,s,p	U,Th
2.	Lonnie, C.Brit. Columbia	?	C,G,p,m	Nb
3.	Osoyoos-Penticton area, southern Brit. Columbia	Eo	B,D,v,s,p	U,Ag
4.	Crowsnest Volc.,Alberta	Cr	E,B,v,s	Cu,Mo
5.	Black Hills	T	C,A,v,s	Au,Pb,W
6.	Highwood Prov.,Belt Mts., etc., Montana	Eo	E,H,C,v,s,p	U,Pb-Ag,Au
7.	Libby, Montana	Cr?	F,p	Fe,V,Ti
8.	Lemhi Pass, Idaho, Montana	T	I,p	Th,REE
9.	Sawtooth Range, Idaho	T?	A,p	U,Ta,Nb
10.	Wet Mts., Colorado	Cm	D,G,p	Nb,REE,Th
11.	Powderhorn, Colorado	Cm	F,D,G,p	Ti,Fe,Nb, REE,Th
13.	Cripple Creek, Colorado	Mi	C,B,v,s	Au,Ag,Te
14.	Hopi reserve, Ariz.,N.Mex.	Pl	C,H,s	U,Cu

15.	Terlingua, Texas	T	C,D,A,v,s	Hg
CANADIAN SHIELD				
16.	Red Wine-Letitia, Labrador	1.5-1.6	D,G,p,m	Zr,U,Be
17.	Kapuskasing zone, Ontario	1.0-1.1 1.5-1.7	D,G,F,p	Nb,Fe,REE
18.	Port Coldwell, Ontario	1.1	B,D,p	Fe,Cu
19.	Nipissing Lake, Ontario	Cm	D,G,p,s	Nb,U
20.	Wausau, Wisconsin	Cm	B,D,p	Th,Zr
APPALACHIANS, ST.LAWRENCE GRABEN				
21.	Monteregian Hills and continuation, Quebec	Cr	D,F,G,s,p	Nb,REE
22.	White Mountains, N. Hampshire	D,Tr,J	D,B,A,p	U,Pb-Zn
23.	Elberton, Georgia	Or?	B,p	Mo,Cu
24.	Magnet Cove etc.,Arkansas	Cr	D,F,G,H,s,p	Ti,V,Nb, Al,REE
GREENLAND				
25.	Gardar Prov.(Ilímaussaq, Ivigtut)	1.25	D,A,G,v,s,p	Zr,REE,Nb Th,U,Be
26.	Werner Bjerge	Cr3-T1	B,A,p	Mo,Pb-Zn
SOUTH AMERICAN CRATON				
27.	Rondônia-Velasco Province	0.94	A,p	Sn,Au
28.	Iporá belt, Goiás, Brazil	Cr	F,H,v,s,p	Ni
29.	Eastern fringe of Paraná Basin (Araxá,Poços de Caldas)	Cr	D,F,G,H,C,p,s,v	Nb,REE,U, Zr,Th,Al
BALTIC SHIELD				
30.	Kola P., Caledonian suite (Kovdora, Afrikanda)	Cm-Or	F,G,p	Fe,Ti,Th REE
31.	Kola P.,Hercynian suite (Lovozero, Khibiny)	D	D,p	Zr,REE,Ti
32.	Oslo Graben, Norway	Pe	B,A,C,v,s,p	Fe,Mo,Pb
EUROPEAN OROGENIC BELTS				
33.	Auvergne, Massif Central, Fr.	Mi-Q	C,v	F
34.	Rhine Gr., Vogelsberg, Pfalz	Mi	C,D,G,v, rare s	Nb,F
35.	Bohemia, Silesia; Czechosl.	Mi-Q	C,D,E,H,v, rare s	Al,Ti,Pb
36.	Tuscany, Rome Prov., Pantelleria; Italy	Pl-Q	C,A,v,s	Hg,U
37.	Ural Mts., U.S.S.R.	D	D,I,p,m	Zr,Nb,REE
38.	Turgai Downwarp, Kazakhstan	D	F,D,G	Fe,Ti,U Nb,REE
39.	Timan Range, U.S.S.R.	D?	F,p	Fe,Ti
40.	UKRAINIAN SHIELD (Mariupol)	1.7-1.9	D,F,p	Ti,Zr,REE
SIBERIAN SHIELD AND BAIKALIDES				
41.	Meimecha-Kotui area	Pe-Tr	F,G,D,H,p,s	Ti,Fe,REE

Table 33-1 (continued). Principal areas of distribution of mineralized alkaline occurrences of the world (Figure 33-2)

NO.	AREA OR LOCALITY	AGE	ROCK ASSOCIATION EMPLACEMENT LEVEL	METALS IN ORE DEPOSITS
42.	Yenisei Range	PZ	D,F,G,p	Al,Nb,REE
43.	Eastern Sayan (e.g. Aksug)	PZ	D,A,B,p	Ta,Be,REE, Nb,U,Th
44.	N.E. Lake Baikal (Synuyr)	Cm,MZ	D,E,G,p	Fe,U,Nb
45.	Inagly, Aldan Shield		F,p	Fe,Pt
46.	Central Aldan district	MZ	F,D,p	Au,U,Pb
47.	Sette-Daban		G,F	Nb,REE
PHANER. FOLDB., SIBERIA, MONGOLIA				
48.	Kuznetsk Alatau, Minussinsk	Pe,MZ	D,A,p	Th,U,REE
49.	Eastern Tuva (e.g. Sangilen)	D-T	D,A,F,G,p,m	Nb,REE,Ta Th,U
50.	Northern Mongolia	D	D,p	Be,Zr,REE
51.	Southern Gobi, Mongolia	J	A,g	Ta,Nb,REE
52.	Sikhote Alin, Bureia Massif	J	F,D,G	Nb,REE
53.	E. Chukotka-W. Alaska	Cr?	D,A,C,p, less v,s	Ta,Be,Nb
54.	INDIA, Deccan Plateau	T	D	Al
AFRICA				
55.	Hogar (Ahaggar)	Cr-Q	C,B,v,s	Sn,U
56.	Aïr Massif, Niger	295	A,s,p,v	Sn,U
57.	Nigeria (Jos-Bauchi) etc.	J	A,C,B,p,s,v	Sn,Nb
58.	Benue, Cameroon	T,Q	A,C	Sn,Au
59.	Aden, Afar, Ethiopia	Mi,Q	C,G,H,D,v,s	Nb,REE
60.	Kenya, Uganda, Tanzania	Mi,Q	C,G,H,D,v,s	Nb,REE
61.	Virunga, Toro-Ankole, Kivu	Mi-Q	E,v	F
62.	N.W. Malawi (Ilomba)	650	D,p	Nb
63.	S. Malawi (e.g. Lake Chilwa)	J-Cr	A,D,G	Nb,REE
64.	Pilansberg, Transvaal	Pt	D,p	REE,Th
65.	Erongo, etc.,Namibia	J,Cr	A,B,D,G,p,s	Sn
66.	Southern Madagascar	Or	A	Ta
AUSTRALIA, OCEANIA				
67.	E. Australian volc. prov.	Eo-Mi	C,B,D,v,s,p	Fe,Ti
68.	W. and E. Kimberleys		E,H	

ABBREVIATIONS:

DOMINANT ROCK ASSOCIATION: A=peralk. granite or rhyolite; B=syenite, trachyte; C=nephelinite, basalt, phonolite, tephrite, trachyte; D=nephel. syenite, alkal. gabbro; E=potassic alk. rocks; F=alkaline ultramafics; G=carbonatites; H=kimberlites; I=subtle alkaline rocks

DOMINANT EMPLACEMENT LEVEL: v=volcanic; s=subvolcanic; p=plutonic; m=metamorphosed and tabulate the distribution of the important areas containing alkaline rocks in the world.

33.2. DOMINANTLY VOLCANIC ALKALINE PROVINCES AND OCCURRENCES

33.2.1. General

Alkaline volcanics are covered by a voluminous petrological literature (e.g. Carmichael et al., 1974) but they are rarely mentioned in economic geology textbooks and compilations because, in comparison with the intrusive alkaline equivalents, they appear virtually to be devoid of metallic deposits. Most alkaline volcanic provinces are located on old "stable" cratons and controlled by extensional tectonism, particularly rifting. Most of the volcanic piles, however, lie on rift shoulders (particularly on uplifted horst blocks) or near principal faults, rather than in the rift graben itself (e.g. in central Europe; Wimmenauer, 1974). Isolated alkaline volcanic centres have been recorded in mobile belts (e.g. the North American Cordillera), where they often merge with the dominant calc-alkaline volcanics (e.g. Cripple Creek, Colorado). The occasional alkaline volcanic occurrences in the oceanic domain were treated in Chapter 4.

Intracratonic alkaline volcanics have been almost exclusively emplaced subaerially and attacked by erosion shortly after their emplacement. The erosion exposed the subvolcanic and plutonic levels and removed a considerable portion of the volcanics in the geologically older terrains, so that in the pre-Tertiary alkaline provinces the intrusive equivalents are dominant and are treated separately in the following sections. As a consequence, most representatives of alkaline volcanics are late Cainozoic and correspond in their style of occurrence, freshness, etc. to the "gray basalts" reviewed in Chapter 31.

Alkaline volcanics rarely occur in isolation. In most extensive continental volcanic provinces the older, "main phase" of volcanism starts with alkali (subalkaline) basalts, the mafic members of the tholeiitic and calc-alkali magma series (Wilkinson, 1974). Clinopyroxene, plagioclase ("normal" basalts) and clinopyroxene, plagioclase, olivine (olivine basalts) are the most common rocks. Alkaline basalts (those with feldspathoids: basanite, tephrite, nephelinite, leucitite, etc.) appear later, preferentially arranged aroud discrete centres. The felsic rocks show a similar succession. Earlier, "normal" (without feldspathoids) trachytes are succeeded by light-coloured feldspathoidal rocks, mostly phonolites. Most petrographic provinces containing abundant alkaline volcanics are essentially bimodal, rich in mafic and felsic phases but lacking significant quantities of intermediate andesites, dacites and equivalents.

In the recent alkaline volcanic terrains, lavas and pyroclastics are approximately equally common. In the geologically older terrains, pyroclastics have been described substantially less frequently than alkaline lavas, but this could, to some degree, be due to the chemical instability of feldspathoids. Feldspathoids in the porous pyroclastics are soon converted into zeolites or clay minerals by weathering and diagenesis and disappear. Feldspar persists much longer. Many trachytes and phonolites, moreover, have been emplaced under a thin cover of usually soft platformic sediments acting as domes with "endogenic" structures (Sørensen, ed., 1974), without reaching the surface. From there they have been exhumed over a period, short in geological terms and are often indistinguishable from true surfical lavas.

Breccia-filled volcanic necks, diatremes, maars, and similar types of eruptions are common. Subvolcanic bodies (dikes, sills, laccoliths) and rare intrusive stocks are also exposed in Tertiary alkaline volcanic terrains. Buried intrusive equivalents are to be expected in the subsurface of all volcanic terrains, and such bodies seem to have maintained hydrothermal activity resulting in zones of alteration and mineralization, sometimes hosted by rocks of the volcanic level or by older, non-alkaline rocks. Even if no exposures of the subvolcanic and plutonic rocks are present among the alkaline volcanics, samples of these together with samples of the basement, can be found as xenoliths in lavas or as "exotic" components in pyroclastics.

33.2.2. Trace metal geochemistry and mineralization

Certain alkaline volcanics have a high alumina content (e.g. the phonolite from Špičák near Most, N.W. Czechoslovakia, 23.01% Al_2O_3, Kopecký in Svoboda, ed., 1966; phonolitic pumice tuff from the Laacher See district, West Germany, up to 23.09% Al_2O_3, Wimmenauer, 1974), which makes these rocks potential sources of aluminium of the future. In the meantime, the high Al_2O_3 content of phonolites and trachytes together with the lack of quartz, facilitates the formation of high-alumina residual clays and bauxites, themselves used for aluminium extraction. Titania is also highly enriched in some alkaline volcanics. There is 5.72% TiO_2 in the nepheline leucitite from Stráž nad Ohří, N.W. Czechoslovakia (Kopecký in Svoboda, ed., 1966). This content, however, although substantially higher than the TiO_2 content in many heavy mineral beach placers, cannot be recovered economically using the existing technology. Further TiO_2 enrichment took place in saprolitic phonolites and in residual clays, for example, in the Chomůtov region of Czechoslovakia and an industrial process to recover this titania could become available in the near future.

Alkaline volcanics are enriched in the same suite of trace elements as their deep-seated equivalents (Zr, REE, Nb, Th, etc.), but the enrichment appears, on average, to be about one order of magnitude lower (e.g. 0.02% ZrO_2 in phonolite, Špičák near Most) and discrete accessory minerals other than zircon and melanite (Ti-garnet) are rarely recognizable. The coarse crystalline, olivine rich protoenclaves common in olivine basalts and usually interpreted as being samples of the undepleted mantle, are high in trace Cr (1-5% Cr_2O_3). The Cr is bound in chrom-diopside and Cr-spinels, but no cases of additional Cr enrichment have been reported.

As mentioned earlier alkaline volcanics have a reputation, among exploration geologists, of being poor targets for metalliferous exploration. Although this is generally true, there are exceptions (e.g. Cripple Creek, Colorado; over 600 t of gold produced). Most of these exceptions are hydrothermal ores generated by convective systems driven by the heat of the subsurface intrusions. Cripple Creek, despite being hosted by, and affiliated to, alkaline and subalkaline volcanics and subvolcanic intrusions (phonolite, syenite), is comparable in all other respects with the calc-alkaline volcanic-intrusive complexes of the orogenic belts hosting the epithermal vein and stockwork systems (as in Nevada, W. Rumania, etc.), treated in Chapter 26.

33.2.3. Principal mineralization styles (Fig. 33-3; Table 33-2)

METALS OR MINERALS DISPERSED OR DISSEMINATED IN ALKALINE VOLCANICS AND PYROCLASTICS

Known examples of this style are few and heterogeneous. Barbosa (1967) described several occurrences of Cretaceous platiniferous (0.8-4 ppm Pt metals) melteigitic tuffs and re-sedimented volcaniclastics, in the Mata da Corda Plateau, M.G., Brazil. The platinum is probably finely disseminated together with perowskite. The host volcaniclastics are mildly phosphatic and carbonatic, suggesting carbonatite affiliation.

Some alkaline volcanics and pyroclastics are radioactive, for example in the Agur Lake area near Penticton, British Columbia (Anonymous, 1979a). The highest U and Th values (over 100 ppm U, 500 ppm Th) are associated with biotite and clay-altered fault zones in Tertiary crystal tuffs and syenites which are probably comagmatic suggesting epigenetic hydrothermal mineralization. In the Kaiserstuhl (West Germany) phonolites, opal veinlets and coatings filling fractures, in places contain 0.2-0.9% U (Wimmenauer, 1966).

HOT SPRINGS PRECIPITATES

Hot springs, the highly diluted, near-surface hydrothermal solutions, are commonly discharged along faults in both currently active and formerly active (e.g. Tertiary) volcanic areas. Within the East African Rift system, numerous hot springs leach trace elements from the young volcanics (as well as from their host rocks). The springs that discharge on the bottom of rift lakes, contribute to the high metal content of some lake waters and bottom sediments (e.g. Lake

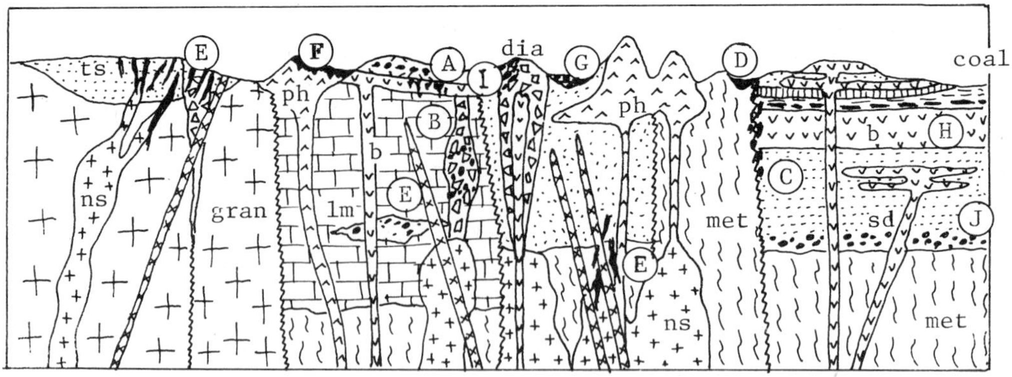

Fig. 33-3. Principal mineralization styles in continental alkaline volcanic terrains (see Table 33-2 for an explanation of the letter codes). Abbreviations: b=basalt; dia=diabase; gran=granite; lm=limestone; met=metamorphics; ns=nepheline syenite; ph=phonolite; sd=sandstone; ts=tuffaceous sandstone.

Table 33-2. Principal mineralization styles in continental alkaline volcanic terrains

(A)	Metals or minerals dispersed or disseminated in volcanics and pyroclastics (e.g. Pt)
(B)	Metallic minerals on fractures in volcanics (e.g. U)
(C)	Hot springs precipitates on the surface or along faults (U,F,Ba)
(D)	Continental lakes with anomalous dissolved metal content in waters or in bottom sediments (F,Fe,Mn,Zn)
(E)	Hydrothermal veins, stockworks, mineralized breccias, replacements (e.g. Cripple Creek); X0 Tt Pb,Zn; 620 t Au; 50 t Ag; 4.5 Tt Hg
(F)	Metalliferous weathering crusts; bauxite (20 Mt Al), Ti-clays
(G)	Coluvial or alluvial sands or gravel with detrital ore minerals reworked from alkaline volcanics (X Mt Ti, X0 Tt Zr, low-grade)
(H)	Reworked clays, claystones, shales (Al, Ti potential)
(I)	U infiltrations in sediments and volcanics in diatremes
(J)	Infiltration U-Zr deposits in sandstones intruded by melilitic volcanics (N. Bohemia)

Malawi, Fe; Lake Kiwu, Zn; see Chapter 30).

In the Děčín, Teplice, Duchcov region in the Sub-Erzgebirge Graben, N.W. Czechoslovakia (Sattran et al., 1966), recent hot springs have generated numerous small veins and veinlets of quartz, fluorite and barite, filling fissures in Permian rhyolites and Cretaceous sandstones. Some veins contain small quantities of pyrite, galena, sphalerite and sooty U oxides. The dark purple fluorite has a high U content, and also carries 0.01-0.03% Be and REE (Bernard et al.,

1969). The small veins and impregnations of Mn oxides running along fractures in Tertiary trachyte at Špičák near Teplá (N.W. Czechoslovakia), are also attributed to hot springs activity.

HYDROTHERMAL VEINS, STOCKWORKS, REPLACEMENTS

Hydrothermal ores in alkaline volcanic terrains mostly formed at subvolcanic or plutonic levels of burial, so a substantial depth of erosion was needed to expose such mineralizations. By analogy with the calc-alkaline volcanic provinces (e.g. the San Juan Mountains, Colorado; Chapter 26), the intrusive equivalents rather than the surficial volcanics are credited with driving the metal leaching, transporting and depositing hydrothermal systems. The hydrothermal activity was most productive in the vaning stages of the magmatic cycles.

In the classical, intracratonic provinces of alkaline volcanism (e.g. Auvergne, Vogelsberg, České Středohoří, East African Rift), metalliferous veins are exceedingly rare. In Roztoky nad Labem, České Středohoří Mts. (N.W. Czechoslovakia), inhomogeneous late Tertiary syenodiorite stock is emplaced into Cretaceous sediments and cut by numerous alkaline dikes (bostonite, monchiquite, etc.). The intrusion is enveloped by high temperature metasomatites (wollastonite, epidote, grossularite skarns) and intersected by three parallel N.W.-trending fissure veins. The veins are filled by quartz, barite, pyrite, Ag-rich galena, and high-Ga and In sphalerite (Kopecký, Chrt and Losert in Krutský, ed., 1964). The available ore tonnage probably does not exceed a few tens of thousands of tons of ore.

Scattered mineralized Au,Ag,Hg, etc. occurrences associated with Tertiary transitional to alkaline volcanics, subvolcanic and plutonic intrusions are located in the Cordilleran foreland (e.g. Black Hills, Little Rocky Mountains, Judith Mountains, Little Belt Mountains, etc., in Montana and South Dakota, reviewed later) and in the Cordilleran frontal belt (e.g. Cripple Creek, Colorado; Terlingua, Texas).

The Cripple Creek district (Koschmann, 1949; Lovering and Goddard, 1950; Fig. 33-4), has produced about 596 (or 620) t Au and 6 t Ag from a large number of high-level, high-grade hydrothermal bodies. It is located in an approximately elliptical Miocene fault-bound basin surrounded by Paleozoic crystalline rocks. The basin is filled by more than 1,000 m of water-laid sediments of mixed provenance (epiclastic detritus derived from the Proterozoic basement and volcaniclastics produced by explosive volcanism in the area contemporary with the basin subsidence) and intruded by dikes, sills, small plugs and necks of phonolite, phonolite with plagioclase phenocrysts and syenite, as well as alkali dike rocks. Undersaturated basalt cements a mineralized elliptical heterogeneous breccia pipe in the Cresson mine.

The Cripple Creek ore deposits are (1) simple veins or fissure veins; (2) irregular stockwork and impregnation deposits in shattered rocks and (3) pipe-like rubble or corroded fragments coated by ore

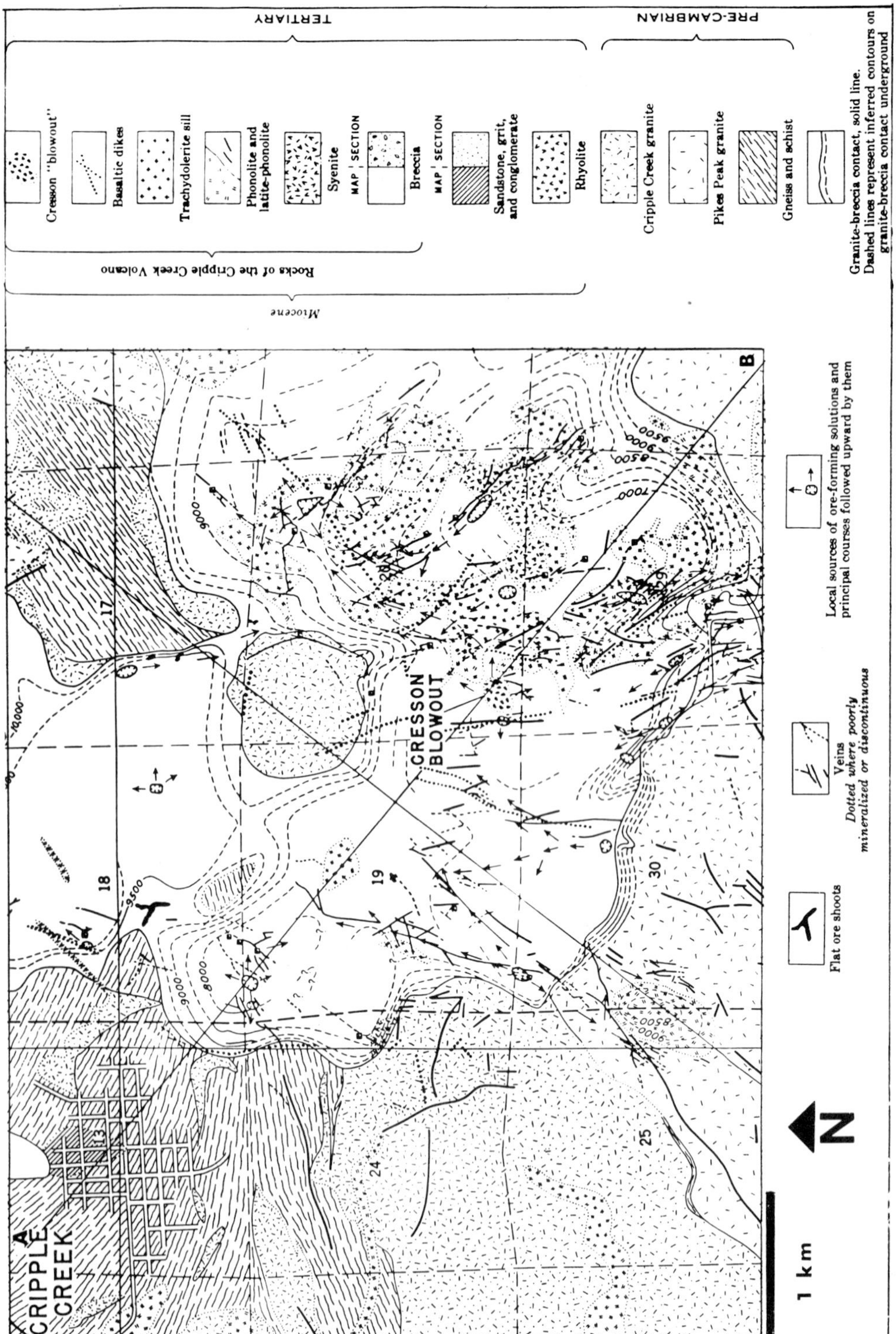
TERTIARY
PRE-CAMBRIAN
Miocene
Rocks of the Cripple Creek Volcano
Cresson "blowout"
Basaltic dikes
Trachydolerite sill
Phonolite and latite-phonolite
Syenite
MAP SECTION
Breccia
MAP SECTION
Sandstone, grit, and conglomerate
Rhyolite
Cripple Creek granite
Pikes Peak granite
Gneiss and schist
Granite-breccia contact, solid line. Dashed lines represent inferred contours on granite-breccia contact underground
Local sources of ore-forming solutions and principal courses followed upward by them
Veins
Dotted where poorly mineralized or discontinuous
Flat ore shoots
CRIPPLE CREEK
A
B
CRESSON BLOWOUT
17
18
19
30
13
24
25
10,000
9500
9000
8500
8000
7000
N
1 km

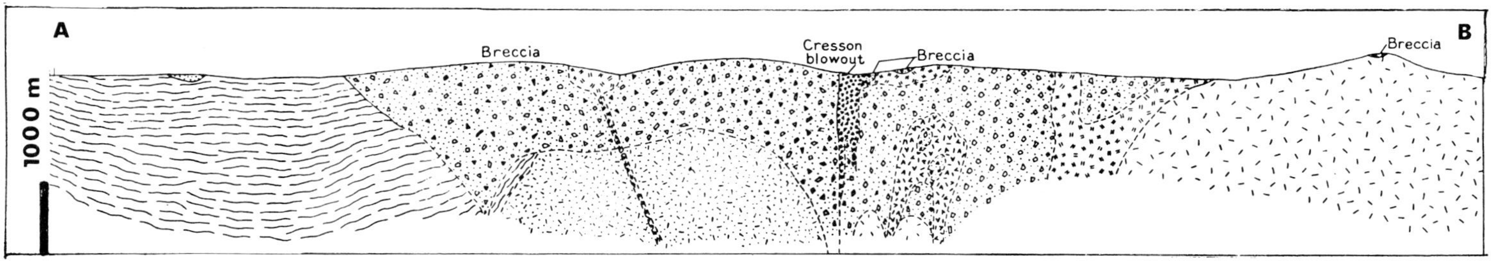

Figure 33-4. Map and section across the Cripple Creek, Colorado, sedimentary basin and volcanic-intrusive centre, showing the distribution of hydrothermal gold mineralization. The Cresson Blowout is a heterogeneous breccia of phonolite and dike rock fragments in a dense matrix of alkaline basalt and basaltic tuff. It is sericite-pyrite altered and hosts several hydrothermal veins. From Lovering and Goddard (1950).

minerals. The principal gangue and ore minerals are chalcedonic and cherty quartz, minor fluorite, pyrite, adularia, carbonates, barite, galena, sphalerite, native gold, Au-Ag tellurides and cinnabar. The wallrock alteration is slight. It is predominantly propylitization, argillization, carbonatization and bleaching.

In the Terlingua field, Texas (4.9 Tt Hg; Yates and Thompson, 1959; compare also Chapter 26), a variety of mercury ore styles is hosted by Mesozoic carbonates capped by continental volcanics and intruded by post-Oligocene dikes, sills and plugs of analcite gabbro and syenite, nephelinite, aegirine-riebeckite rhyolite and trachyte. The most common cinnabar orebodies are located in limestone and/or clay residue after carbonate dissolution, and in carbonate-clay breccia pipes. Only one locality is hosted by an intrusive rock. The ore is epigenetic and is attributed to the hydrothermal phase of the sodic alkaline magmatism.

WEATHERING CRUSTS

Alkaline volcanics lacking quartz (e.g. phonolite, trachyte) are very favourable substrata for the generation of low-iron residual bauxites (e.g. in the Poços de Caldas complex, Brazil). The relative scarcity of occurrences that have actually been recorded is due to the general scarcity of phonolites in the humid tropical belt.

Bauxitic residuals formed over alkaline volcanics, however, are not confined to the present tropical belt. Strnad (in Krutský, ed., 1964) described an 8 m thick regolith of middle Miocene age, formed over sodalite trachytes at Mariánská Hora near Ústí nad Labem, N.W. Czechoslovakia. The basal unit of the weathering profile is a grayish-green saprolite with a trachytic relic texture, some 2-3 m thick. It is abruptly topped by a zone of alternating white and red laterite which is 8 m thick. The laterite contains 43.47% Al_2O_3 and corresponds to a bauxitic claystone in which about one third of the Al_2O_3 is bound in gibbsite and the rest in kaolinite.

At other localities in N.W. Czechoslovakia (e.g. Braňany in the Chomůtov coal basin), Tertiary saprolite over kaolinized pohonolite in the footwall of a brown coal, contains up to 30% Al_2O_3 and has a a high TiO_2 content. There, kaolinite is the dominant clay mineral and gibbsite occurs only as an accessory. The material was tested as a possible aluminium source.

In the tropical regolith topping the occurrence of platiniferous tuffs near Mata da Cordo, Brazil (Barbosa, 1967), small quantities of Pt have accumulated in the limonitic "canga" (=ferricrete).

In arid regions, calcrete is often developed over young volcanics and it may cement pyroclastics (e.g. in the Lashain Hill, Ngorongoro crater and elsewhere in Tanzania). The true, groundwater-precipitated calcretes may host U deposits. Some "calcretes", however, may be carbonatite-related tuffs.

ORES IN SEDIMENTS FORMED BY REWORKING OF THE METALS OR MINERALS IN ALKALINE VOLCANICS

Detrital zircon derived from a variety of alkaline volcanic and subvolcanic bodies is present in small quantities (around 1%) in the Pleistocene pyrope-bearing colluvial gravels in N.W. Czechoslovakia (near Třebenice). The alkaline volcanic centres of N.E. New South Wales and S.E. Queensland, contributed a portion of the ilmenite, rutile and zircon recovered from the eastern Australian beach sands.

The Miocene sedimentary claystones in the footwall of brown coal in N.W. Czechoslovakia (Most-Chomůtov Basin; Sattran et al., 1966) contain 30-40% Al_2O_3, 4-15% TiO_2 and 0.2-0.4% V, and are expected to be mined and processed as an ore of these metals in the near future. The claystones formed by reworking of phonolite regoliths, followed by an additional metal enrichment during sedimentation and by diagenesis in freshwater lakes.

In the Hopi Buttes area, N.E. Arizona (Shoemaker, 1976; Figures 33-5 and 33-6) numerous, partly eroded maar-type Tertiary paleovolcanoes occur, associated with alkaline basaltic flows and tuffs. They were partly emplaced in a shallow Pliocene lake and a chaotic mixture of volcanic ejecta and lake sediments (siltstone, claystone, limestone) formed within and in the vicinity of the volcanic vents. In this setting small infiltrated uranium occurrences are located.

The Quaternary Rome volcanic province, central Italy (trachyte, latite, tephrite lavas, ignimbrites, volcanic fluviolacustrine deposits), was reviewed recently by Kimberley (1978a). The volcanics are enriched in U and Th by a factor of about ten above the clarke, and both metals are being actively redistributed by the action of hot springs, by CO_2 and H_2S-rich gas seepages and by groundwater. At the Sabatini deposit, dispersed uranium (about 500 ppm U) occurs in pyrite and marcasite lenses peneconcordant with bedding of kaolinized and silicified felsic pyroclastics. The mineralized layers have a total thickness of 5-10 m over an area of approximately 1 km^2.

Syka et al. (1978) described an unusual variety of a "sandstone U deposit" (Chapter 25) from the Stráž pod Ralskem area, Czechoslovakia, enriched in Zr (in baddeleyite). The zirconium was probably contributed by Miocene Zr-rich melilithic rocks in the area.

33.2.4. České Středohoří Mts., N.W. Czechoslovakia, as an example of a Tertiary alkaline volcanic province and its mineralization

České Středohoří (formerly Böhmische Mittelgebirge; Kopecký in Svoboda, ed., 1966; Wimmenauer, 1974; Fig. 33-6) is a N.E.-trending belt of Tertiary subalkaline and alkaline volcanics and minor intrusive bodies located in the northern part of the Bohemian Massif. The bulk of the Teriary magmatic rocks are controlled by the Sub-Erzgebirge Graben which is approximately 30 km wide, now reversed into an automorphic horst. The volcanics intrude and overlie upper

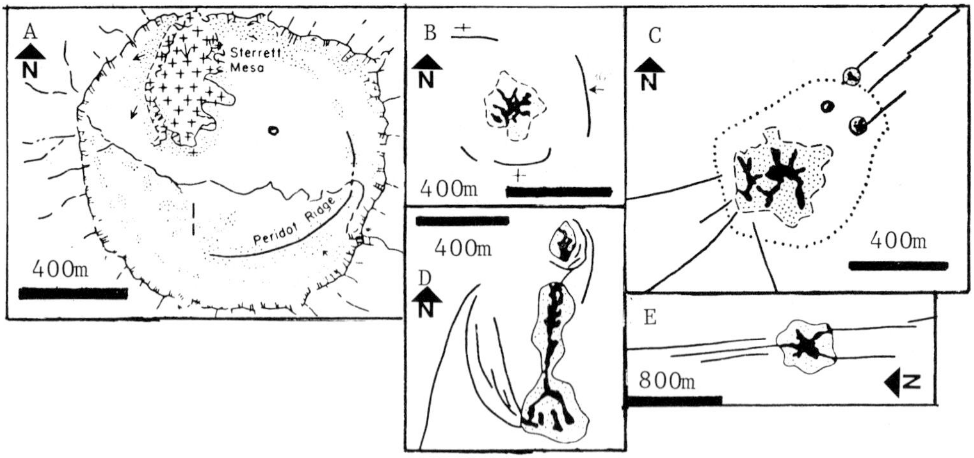

Fig. 33-5. Selected volcanic necks in the Navajo-Hopi fields, Arizona and New Mexico, U.S.A. (A) Buell Park, arcuate dike of leucite minette; other dikes of minette; lava cap on Sterrett Mesa ends southward in a depression among pyroclastic rocks; tuff-breccias, stippled. (B) Tuff-breccia neck, 8 km N. of Flying Butte, Hopi country, rises through horizontal Tr shales and is bordered by thin dikes of monchiquite; diagrammatic. (C) Shiprock; neck of minette tuff-breccia and dike of minette. (D) Boundary Butte, similar neck bordered by arcuate, vertical dikes. (E) Wildcat Peak, tuff-breccia necks associated with dikes of alnöite, monchiquite and related rocks. From Hunt (1956).

Cretaceous platformic sediments (quartz arenite, marl, claystone) and overlap a Tertiary brown coal-bearing sequence along the N.W. margin of the horst. The physiographically distinct core of the volcanic province (rounded and conical hills, exhumed necks, dikes) extends for about 80 km, but scattered magmatic occurrences including one major large composite paleo-volcano (Doupovské Hory), cover the entire northern half of the Bohemian Massif.

The volcanics belong to three eruptive phases (lower Miocene, middle Miocene and late Pliocene-Pleistocene), of which the first phase generated the bulk. The rocks are extremely variable compositionally, ranging from kimberlite to alkaline trachyte. Olivine basalts, trachytes and phonolites are the most widespread and conspicuous. The rock outcrops are abundant and generally fresh, and the region has been studied in considerable detail petrographicaly.

Metallic deposits directly hosted by the volcanics and their intrusive counterparts are extremely rare and only one locality (Roztoky, Pb,Zn,Ag veins) has ever been mined or explored. The residual and transported high Al and Ti clays derived from phonolites represent a large alternative resource of these metals for the future.

Indirectly associated ores are represented by the economically

Fig. 33-6. Shiprock Butte, northern Arizona; eroded neck of minette tuff-breccia and dikes.

important SUV infiltration deposits formed in an aquifer at the base of the freshwater Cretaceous sandstones. In the east (Hamr, Mimoň) several alkaline plugs, necks and dikes pierce the uranium field but contribution, if any, of the Tertiary magmatism to formation of the U deposits, is unclear. The volcanics, however, probably supplied the Zr to form local baddeleyite and zircon-rich bodies. In the north-west (near Teplice), modern hot springs and precipitates of Tertiary hot springs (fluorite, barite, minor base metal sulphides) have formed contemporaneously with the infiltration uranium ores.

33.3. PERALKALINE GRANITE-RHYOLITE ASSOCIATION

33.3.1. General

Peralkaline silicic rocks (granites and rhyolites) have a molecular excess of alkalies over alumina, which sets them slightly apart from their more widespread calc-alkaline equivalents. In terms of petrochemistry, geotectonic setting and metallogeny, this class of rocks bridges the gap between alkaline and calc-alkaline felsic associations. Bowden (1974) has prepared a brief summary of peralkaline silicic rocks, and Murthy and Venkataraman (1964) discussed their global distribution. Peralkaline granites and peralkaline rhyolites (named comendite and pantellerite in the petrologic literature) are the most common members of the association.

Most of the granites are leucocratic rocks containing K-feldspar (usually perthite), albite, quartz, aegirine and

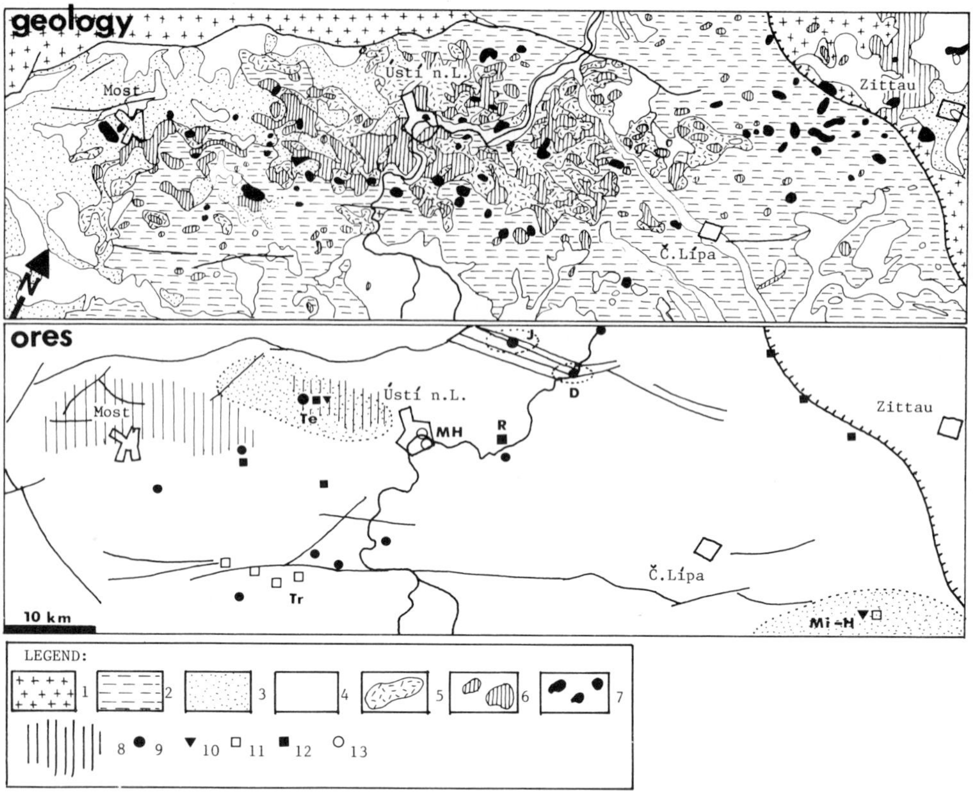

Fig. 33-7. České Středohoří Mountains, N.W. Czechoslovakia, geology and ore distribution. GEOLOGY: 1=Pe and older crystalline basement; 2=Cr_3 sediments; 3=T sediments; 4=Q unconsolidated sediments; 5=Mi volcanics, dominantly pyroclastics; 6=mafic lavas (basalt, nephelinite, tephrite, etc.); 7=felsic lavas (trachyte, phonolite). ORES: 8=High Al,Ti claystones, Most and Chomůtov areas; 9=fluorite veins, replacements; Te-Teplice; J=Jílové; D=Děčín; 10=infiltration U deposits in Cr sandstones; Mi-H=Mimoň, Hamr; 11=Zr ores, baddeleyite associated with U; Tr=Třebívlice, zircon in pyrope gravels; 12=Pb,Zn,Ag occurrences; R=Roztoky; 13=gibbsite, kaolinite residual clay; MH=Mariánská Hora. Simplified after Geological Map of Czechoslovakia 1:500,000 and Chrt and Bolduan (1966).

arfvedsonite-riebeckite as the principal minerals. The granites crop out in subvolcanic ring structures or as small stocks and massifs. The peralkaline rhyolites form lavas, tuffs or glass (obsidian) bodies, megascopically indistinguishable from calc-alkaline rhyolites. Quartz syenites, syenogabbros, gabbros and alkali basalts are sometimes associated, as are calc-alkaline biotite granites and muscovite-biotite leucogranites.

Most of the peralkaline silicic occurrences treated in this section are located in cratons, typically along structural domes of "epeirogenic" origin, marking the initial stage of tectonomagmatic activation (e.g. Jos, Aïr Mountains, etc., in W. Africa). Other occurrences are in or near intracratonic rifts (e.g. in Kenya, Ethiopia), or along extensionally faulted and attenuated continental margins underlain by a thick continental crust (e.g. Great Basin, Nevada). The rest is in orogenic belts, typically underlain by ancient sialic blocks.

Bowden (1974) discussed the origin of the peralkaline silicic magmas and concluded that while these rocks have their place in the evolutionary model of continental rifts, there is no unequivocal evidence to determine whether they formed by differentiation of a mantle-derived basaltic parent magma or by partial melting within the continental crust.

33.3.2. Trace metal geochemistry and metallogeny (Table 33-3, Figure 33-8)

Peralkaline silicic rocks are moderately to strongly enriched in some lithophile metals (Be, 6-56 ppm; Ga, 28-45 ppm; La, 95-580 ppm; Nb, 69-670 ppm; Sn, 5.7-52 ppm; Th, 17-23 ppm; Y, 73-490 ppm; data for peralkaline glasses, believed to correspond closely to the composition of the parent magmatic melt; Bowden, 1974). They are also high in fluorine (0.18-1.30% F). The Nigerian riebeckite granites contain 45 ppm Ga, 145 ppm La+Yb, 350 ppm Nb, 1,330 ppm Zr, 30-150 ppm Sn, 3-30 ppm Be and 0.76% F (MacLeod et al., 1971). These metals are contained either in the lattice of silicates, such as amphibole in the Nigerian albite-riebeckite granites (Olade, 1980) and biotite in biotite granites, or they form their own accessory minerals: pyrochlore and zircon in riebeckite granite or columbite and cassiterite in biotite granite.

The trace Sn and Nb contents in "ordinary" granites, however elevated compared with the clarke, do not make ore, not even in the deeply weathered regions such as northern Nigeria, where resistate cassiterite and columbite accumulate in eluvial and alluvial placers. The dispersed accesory minerals are too fine grained to be physically recoverable. A more substantial accumulation is needed.

Locally, cassiterite or columbite grains accumulate in granite endocontacts to such a degree that they constitute low-grade disseminated orebodies the economy of which has been further improved by weathering and sedimentogenic reworking (the bulk of the Nigerian tin comes from placers). Strong (1980) explained the origin of the "primary" cassiterite deposits in which cassiterite is evenly disseminated throughout the intrusive rock as a consequence of retention of the late magmatic aqueous phase into which Sn (and Ta and Nb) normally partition. Escaping aqueous, Cl, F and alkali-rich fluids, on the other hand, may leach, transport, concentrate and deposit the above metals epigenetically in the apical portions of

Table 33-3. Principal mineralization styles in the intracratonic peralkaline granite-rhyolite association (Fig. 33-7)

WEATHERING CRUSTS	
(A)	Cassiterite, columbite, pyrochlore, etc. enriched in saprolite over a granite containing disseminated ore minerals
PRODUCTS OF SEDIMENTOGENIC REWORKING	
(B)	Exposed cassiterite and/or columbite alluvial placers
(C)	Alluvial placers buried under plateau basalt flows
(D)	U infiltrations in swamps or bogs
(E)	U infiltrations in sandstone hosts
ANOMALOUS TRACE METAL CONTENTS OR DISSEMINATED MINERALS IN GRANITES	
(F)	Anomalous trace U content
(G)	Disseminated cassiterite and/or columbite in (albitized) biotite granite
(H)	Disseminated pyrochlore in albite-riebeckite granite
VEINS, REPLACEMENTS, MINERALIZED BRECCIAS, PEGMATITES, ETC. IN GRANITE ENDOCONTACTS	
(I)	Cassiterite-bearing greisen pipes and lodes, fissure veins
(J)	Schlieren of cassiterite pegmatites
(K)	Cryolite + Be minerals (bavenite, phenacite) in altered zones in riebeckite granites
(L)	Cryolite-siderite masses (Ivigtut)
(M)	Zircon, xenotime, Th and U minerals, pyrochlore with fluorite in albitized, Li-mica replaced, etc. zones superimposed on riebeckite granites
(N)	Stockwork molybdenite deposits in sericite (and/or K-feldspar altered) peralkaline granites, hypabyssal level
(O)	Discrete molybdenite-quartz veins in granites, plutonic level
VEINS, REPLACEMENTS, ETC. IN GRANITE EXOCONTACTS	
(P)	Mineralizations in skarns at granite/carbonate contacts (Mo, scheelite, magnetite, helvite, etc.)
(Q)	Fluorite replacements in carbonates containing Be minerals (bertrandite, leucophane, etc.)
(R)	Mineralized (Zr,Nb,REE,Th, etc.) feldspathite (albite, microcline) replacement veins and other bodies
(S)	Sulphide-gold replacement bodies in limestone and lodes in blastomylonite near granite dikes (Central Aldan)
(T)	Fissure replacement veins, disseminations, mineralized breccias, etc., of base metals (Pb,Zn,Ag,Cu, etc.) in contact aureole of granites

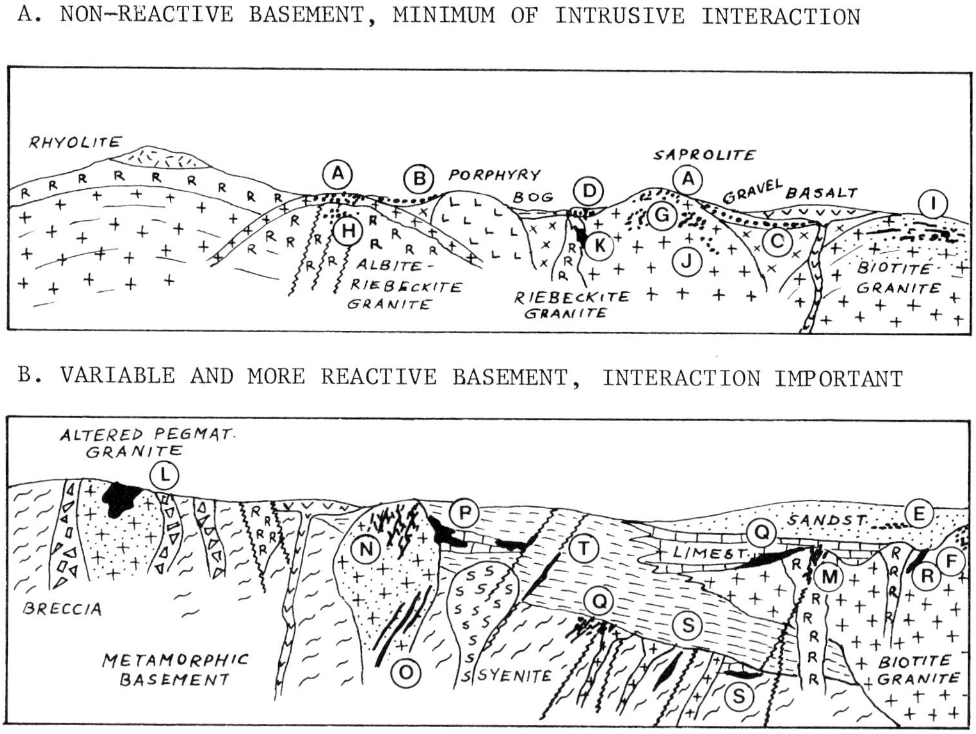

Fig. 33-8. Principal mineralization styles in the intracratonic peralkaline granite – rhyolite association; see Table 33-3 for an explanation of the letter codes.

granitic stocks and in their exocontacts. This is usually accompanied by a profound, often zonally arranged alteration (Na and K feldspathization, greisenization, etc.), most apparent in the "apogranites". Apogranites and the problems of granite-affiliated tin origin have already been discussed in Chapter 28 and will not be considered further here. The earlier conclusions are fully applicable to peralkaline granites as well.

33.3.3. Examples of major mineralization styles

DISSEMINATED Nb,Ta and Sn OXIDES IN PERALKALINE ASSOCIATION OF THE WEST AFRICAN "YOUNGER GRANITES"

The "Younger Granite Province" is a belt of over 60 magmatic complexes of Permian to Tertiary age, discontinuously distributed along an extensive N.-S.-trending strip in western Africa that measures 1,500 x 200 km (Black and Girod, 1970; see also Table 33-4). The most extensive and also economically the most important are the occurrences in northern Nigeria, particularly the Jos-Bukuru Complex. The latter is considered to be the type area for mineralized peralkaline granite complexes of the Western World.

The Jos-Bukuru Complex (MacLeod et al., 1971; Wright, 1970; Olade, 1980; Fig. 33-9), contains over 50 separate small occurrences of Jurassic high-level magmatic rocks, 90% of which are granites and rhyolites. The remainder are syenites, anorthosites, olivine gabbros and diabase. They intrude Precambrian high-grade metamorphics (migmatites, gneisses, local charnockites) and granites, and are in turn partly covered by Tertiary to recent basalt flows and alluvial sediments.

The felsic magmatites are well outlined topographically as hills, and they are circular or elliptical ring complexes, cone sheets, ring dikes, small stocks and massifs flanked by remnants of comagmatic felsic volcanics and pyroclastics. The dominant intrusive rock is biotite granite, followed by hornblende, pyroxene, fayalite granite; riebeckite, biotite granite and riebeckite aplite.

Cassiterite, columbite and pyrochlore are the three major minerals that locally form disseminated orebodies in the Nigerian granites. Cassiterite and columbite are associated exclusively with the altered apical portions (apogranites) of biotite granites. There, columbite forms true disseminations in the alteration feldspathized intrusion, but disseminated cassiterite in the same setting is exceptional and most of the Nigerian cassiterite came from numerous but small greisen lodes and small veins superimposed on the altered granite.

Three zones of disseminated columbite in the Nigerian apogranites grading close to 0.45% columbite (two in the Jos-Bukuru Complex, one in the Afu Granite in Benue Valley; MacLeod et al., 1971) contain about 70 Tt Nb_2O_5 in the soft surficial saprolite. Pyrochlore in greater than accessory concentrations (around 0.3% Nb_2O_5) is disseminated in albite-riebeckite granite in the Kaffo Valley, Liruei, and there are large reserves of this mineral in saprolite (360 Tt Nb_2O_5). Numerous additional pyrochlore occurrences are known in the riebeckite granite of the Jos-Bukuru Complex. Small showings of Be minerals (helvite, phenacite) and cryolite, have also been reported.

→

Fig. 33-9. Geological map and section showing a portion of the Jos-Bukuru tin region, Nigeria. bg=Biotite granite; arg=albite riebeckite granite; rbg=riebeckite–biotite granite; rag=riebeckite aegirine granite; hpfg=hornblende, pyroxene, fayalite granite; bg=biotite granite; qp=quartz porphyry; gp=granite porphyry; r=rhyolite. Simplified after MacLeod et al. (1971).

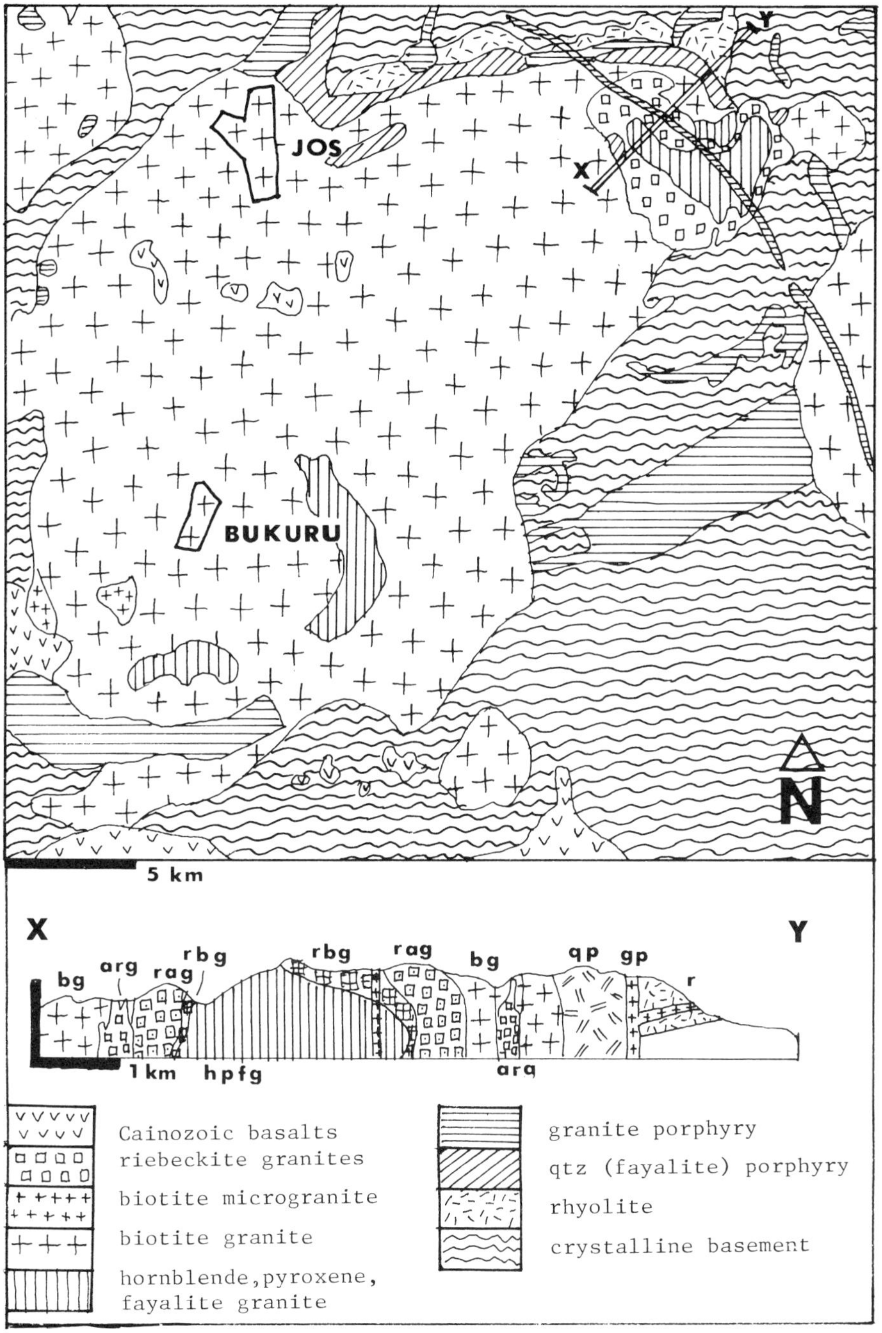
JOS
BUKURU
X
Y
N
5 km
X
Y
bg
arg
rag
rbg
rbg
rag
bg
qp
gp
r
1 km
hpfg
arg
Cainozoic basalts
riebeckite granites
biotite microgranite
biotite granite
hornblende, pyroxene, fayalite granite
granite porphyry
qtz (fayalite) porphyry
rhyolite
crystalline basement

Table 33-4. Phanerozoic peralkaline granite-rhyolite association, selected examples of mineralized provinces or occurrences

LOCALITY	GEOLOGY	MINERALIZATION	REFERENCES
Mesters Vig area (Werner Bjerge), E. Greenland	Cr_3-T_1 pyrox.,alk.granite,syenite,syen.porph.,intr.into Cb sedim.; diabase,gabbro,subvolcan.	stockw. of moly.,pyr.,qtz.,gal., sphal.,cp.fract.coatings,veinl. in gran.endocont.;K-feldsp.,sericite alter.; 1 m thick border zone of topaz greis.with wolfr. 175.5Tt Mo/0.15%	Nielsen (1973)
Bokan Mt.,Prince of Wales Island, S.E. Alaska, U.S.A.	8 km^2 boss of peralk. granite emplac. into D? metamorphics; Cr intrusion	minor zirc.,fluor.,uranothorite dissem.in gran.; uranoth. and uranin. veinl. in albitiz.zone // shear; prod.1957 102t U/0.64%	MacKevett (1963) Staatz (1978)
Topsails intrusion St.Lawrence,Newfoudland,Canada	minor peralkaline granite intrusion	fluorite veins, minor Pb,Zn,U occurrences	Strong (1980)
White Mts.Province Conway Granite New Hampshire, U.S.A.	Ms, Tr-J (185 m.y.);post-tect. peralk.plut.,gran.,syen.,monzon.,dior.,gabbro; aegirine-riebeckite granite, emplaced into D gneisses	granite trace U enriched Iron Mt.(Bartlett): repl.magnet., hemat.,danalite (up to 1.6% BeO) in granite; Silver Lake: gal., sphal.,qtz.,chlor.,veins in brec. near granite contact	Barker (1974) Cox (1970)
Sawtooth Massif S.W. of Stanley, Idaho, U.S.A.	T? pink peralkaline perthitic granite	10-2,000 ppm U, bound mainly in dissem.euxenite; euxenite accumulates in placers, U in bog	Killsgaard et al.(1970)
Velasco alkal.prov. eastern Bolivia (S.Ignacio de Vel.)	6 ring compl.Pt_3-PZ_1 of phonol., qtz.syen.,biot.and riebeckite gran.intr. PCm basem.gneiss	Cerro Manomo, cassit.,pyrochl., monaz.,topaz,fluor. dissem. in albitiz. granites	Fletcher et al. (1977)

Hoggar (Ahaggar) Mts., S. Algeria	Cr, several isolated intrus.of biot. and peralk. granite into PCm basem. gneiss // N.-S. shears and grabens	Ouan Rechla, dissem.cassit.,Li-musc.,minor monaz.,wolfr.,in albitiz.granite; Timgaouine distr., pitchbl. veins and stockw. in altered granite; 14Tt U; X00 t Sn	Black and Girod (1970) O.E.C.D. (1979)
Aïr Mts., Niger	295 m.y. biot. and peralk.gran. complex intr. PCm basement	cassit.dissem.and in stockw. in gran.,reworked to placers N.of Agades; 1,410 t Sn;	de Kun (1965)
Sabaloka ring com. Jebel Qeili, S. Soudan	porph.microgr.,biot. and musc.-biot. gran.,riebeckite-egirine syen.,acid lavas, empl. to PCm basement complex	biot.-musc. gran. contain greisens and qtz. stockw. with cassiterite, wolframite	Whiteman (1971)
Jos-Bukuru compl., northern Nigeria	160 m.y. ring and circul. compl. of biot.,riebeckite, etc. granite intrude PCm cryst. basement	cassit.and lesser columb.widespr.in albitized (apogranites) apical biot.gran.,in narrow greisen veins and stringers; entire prod.came from placers; 590Tt Sn, 49Tt Nb,1.6Tt Ta, 500t W	McLeoe et al. (1971)
Kaffo Valley, Liruei, Nigeria	as above	dissem.pyrochlor in localized albitized zones in granite; 235.5Tt Nb/0.21%	ditto
Afu Granite,Benue Valley, Nigeria	J peralk. and biot. granite, ring and apical complexes	dissem. columbite in albitized zones of granite	ditto
Cameroon and Chad (Mayo Darle, Tapare, Poli)	T, 20 small peralk.intrus.plutons along a N.E. tect. zone	cassit. in placers probably derived from stockworks in greisenized apogranites	de Kun (1965)

Table 33-4 (continued). Phanerozoic peralkaline granite-rhyolite association, selected examples of mineralized provinces or occurrences

LOCALITY	GEOLOGY	MINERALIZATION	REFERENCES
Central Lake Malawi province, Malawi	450-600 m.y. ring compl., small stocks, dikes, empl. into PCm basem. complex; alkal. and peralkal. granites	pyrochlore is a common access., also allanite, zircon; U and Th occur in pegmatite	Bloomfield (1970)
Chilwa alkal. prov. southern Malawi	J_3-Cr, ring compl.; qtz. syenite, perthosite, alkal. granite, riebeckite granite	minor accessory pyrochlore in peralkaline granites	Woolley and Garson (1970)
Brandberg and Spitzkoppe, Namibia	J, 20 km diam. circular stock of granite empl. into Karoo sedim. and lavas and PCm gran. basem.; hornbl., biot., tourm., arfvedson. granite facies	fissure qtz., cassit., minor wolfr. and scheel. veins; dissem. cassiterite suppl. to placers; 3Tt Sn	Hodgson (1973) Martin et al. (1960)
Erongo, Namibia	J, circul. gran. stock and ring intrusive into Karoo lavas and sedim.	small pegmatite lenses and schlieren cont. qtz., cassit., scheel., molybdenite; 236t Sn	Martin et al. (1960)
S. Madagascar thorianite province	485 m.y. post-orog. peralk. granites intrude (and replace ?) diops., spinel, phlogop. pyroxenite of the PCm basement	thorianite, phlogop., zircon, etc. form banded dissem.; reworked into placers	Besairie (1959)
"Kazakhstan A" complex (Khorgos Complex ?), USSR	PZ_3?, two cupolas of riebeck., microcl., qtz., albite apogranite superimp. on biot. granite massif intr. Cb_1 sandst., slate assoc. Riebeckite albitites formed in exocont. sandstones	main Nb, REE, Zr, Be, Th and U mineraliz. is in fissure veins of albitite, lesser microclinite, with dissem. zircon, thorite, pyrochlo., gagarinite, gadolin., etc.	Beus (1968)

"Siberia A" alk. province (East Tuva ?),U.S.S.R.	PZ_{2-3} metasom. zoned riebeckite, microcl., albite massif emplac. into PZ metacarb. and diorite-granodiorite	zircon, columb.,pyrochl.,etc. are in fissure-controlled endo-cont.metasom.; fluorite, bertrandite,phenacite repl. in the exo-contact marbles	Ginzburg and Fel'dman (1974) Beus (1968)
Numurgin and Udzhigin complexes, Mongolian Tuva	PZ_3 granosyen.,riebeckite, egirine apogranite, etc. emplaced into Cm-D volc. and sedim.,PZ_2 granite pluton	acces. fergusonite dissem. in granosyenite	Marinov et al. (1977)
"Siberia B" alk. prov. (central Aldan Shield ?), U.S.S.R.	J-Cr albite-riebeckite granite intr. Cm carb.,shale along major faults; several massifs are in Arch. basement; skarns and Na,K feldspathites along contact	Zr,Th,REE in endocont. assoc. with albite and qtz.-albite bodies; Be(leucophane) miner. is in fluorite-rich exocont. metasom.	Zabolotnaya (1974)
Central Gobi alk. province,eastern Mongolia	J, calc-alk. and peralk.gran., peralk. rhyol.,carbonatite	Ta,Li,Cs,F miner.in albitized apogranites; assoc. agpaitic intrus. carry REE,Zr,Nb	Kovalenko et al. (1977)

DISSEMINATED ORE MINERALS IN PERALKALINE GRANITES OUTSIDE WESTERN AFRICA

Simple, sub-economic disseminations of pyrochlore and also zircon and allanite in unaltered or albitized peralkaline granites comparable with those in Nigeria, have been recorded by Bloomfield (1970) and Woolley and Garson (1970) in Malawi. They also occur in the Siberian and Kazakh apogranites (Beus, 1968) and elsewhere, but have rarely been reported.

Accessory cryolite disseminated in peralkaline granite occurs in Ivigtut, Greenland (Blaxland, 1976) and minor disseminated cassiterite has been reported from the Rondônia, Brazil, granites (Kloosterman, 1968). Some peralkaline granites have anomalously high trace uranium content (100 ppm U plus in the Conway Granite, New Hampshire; up to 200 ppm in granite of the Sawtooth Massif, Idaho), so that they have been considered a possible future resource of low-grade uranium. Disseminated zircon, xenotime, fluorite and uranothorite in albitized riebeckite-acmite granitoids is one of the mineralization styles described in the Bokan Mountain stock, Alaska (MacKevett, 1963).

Overall, however, it appears that "simple" disseminations of metallic minerals in "unaltered" (or feldspathized-only) peralkaline granites can rarely produce an economic deposit, particularly in areas free of the deep tropical weathering and cheap labour as in Nigeria. An added metal concentrating process is obviously needed to super-concentrate the broadly distributed trace metal content of the granite into a few smaller size, but rich orebodies. Some such processes were initiated before the complete solidification of the intrusion, so they did not markedly alter the host rock (e.g. pegmatites and monomineralic masses). Other were superimposed on the fresh or feldspathized granite or on the rocks of the exocontact, accompanied by hydrothermal alteration.

MINERALIZED PEGMATITES AND MONOMINERALIC ORE MASSES IN PERALKALINE GRANITES

Small pegmatitic schlieren and vein-like bodies are relatively common in peralkaline granites emplaced at the plutonic level, and they frequently carry scattered megascopic ore minerals. Such minerals are often identical to the fine-grained "invisible" accessories in the host granite, or at least carry the same elements in which the granite is enriched. U and Th minerals, for example, have been observed in the peralkaline granites of central Malawi (Bloomfield, 1970) and minor pegmatites contribute cassiterite to the Rondônia placers. The numerous but small drusy pegmatites in the Jurassic granites of Spitzkoppe, Brandberg and Erongo, Namibia (Martin et al., 1960), contain fluorite, beryl, cassiterite, chalcopyrite and arsenopyrite. Some 236 t Sn were recovered at Erongo, partly from placers. No large mineralized pegmatite occurrences, however, are known in this association.

Rich, almost monomineralic, ore masses hosted by what appears to be a hydrothermally little-altered granite are unique, and are exemplified by the Ivigtut cryolite deposit in S.W. Greenland. This is a Proterozoic mineralization and has been described in Volume 2.

HYDROTHERMAL-EPIGENETIC ORES IN ALTERED APOGRANITES AND ALONG THEIR CONTACT

The Soviet school (Beus, 1968) differentiated between two apogranite series: (1) the "normal" series developed over "orogenic" calc-alkaline granitic massifs (Chapter 28) and (2) the alkaline series, associated with peralkaline granites emplaced in "anorogenic" cratons. The fundamental alteration-mineralization and zoning pattern of both series is comparable, but the alkaline series tends to have higher proportion of Ta,Nb,Zr, Th,REE, and U. Sn and Be are of approximately equal importance in both series. Only the alkaline apogranites will be discussed here.

Beus (1968) further subdivided the alkaline apogranites into (a) biotite, microcline, quartz, albite apogranites with Nb and Zr mineralization, represented by an unnamed massif in Kazakhstan ("Kazakhstan-A") and (b) riebeckite, microcline, quartz, albite apogranite with Zr,Nb,REE,Th and U mineralization. The latter is represented by an unknown locality in the Caledonides of Siberia (probably in the Tuva-N. Mongolia alkaline province), designated here as "Siberia-A". Both apogranite types correspond closely to the apogranites developed over the biotite (a) and riebeckite (b) granites of Nigeria.

Hydrothermal-epigenetic mineralization in apogranites (exclusive of the disseminated minerals such as columbite and pyrochlore which were described earlier and which some authors interpret as being hydrothermal-epigenetic as well) is controlled by alteration-mineralization zones formed along fractures, faults, shears, porous or replaceable rocks at both sides of the granite contact. Economic metallic mineralization in this setting is known in less than 20 regions of the world and each region has its own special characteristics. Because generalizations are difficult to make, a few example localities will be reviewed briefly (see also Table 33-4).

Tin-mineralized apogranites in the Jos-Bukuru region, Nigeria.

Here, cassiterite is confined to albitized cupolas of biotite granite, in which it is disseminated in narrow, usually subhorizontal, fracture-controlled zones of patchy and diffuse greisenization (reminescent of Cínovec, Chapter 28) and also as scattered crystals in thin, joint-controlled quartz veinlets. The greisen lodes are sometimes enveloped by zones of K-feldspar, sericite and hematite alteration (MacLeod et al., 1971). The richest cassiterite concentrations correlate with the recently unroofed intrusions. Disseminated cassiterite in greisenized apogranite and in endo- and exocontact quartz veins, also occurs in the Proterozoic Rondônia tin province, Brazil (Kloosterman, 1968; see Volume 2).

Nb (Zr,Be,Th, REE)-mineralized apogranites.

The "Siberia-A" locality (Beus, 1968; Ginzburg and Fel'dman, 1974; Fig. 33-10) is a small (2 km^2) composite apogranite massif emplaced into marble and granodiorite, diorite environment. It is distinctly zoned compositionally. The core is composed of Li-mica, microcline, quartz, albite granite, and there is a locally intensive silicification that increases with depth. The following zone has riebeckite, microcline, quartz, albite apogranite, and the endocontact is dominated by microcline metasomatites. The exocontact in carbonates has a 10 to 50 m wide zone of intensive fluoritization that carries bertrandite and lesser phenacite. The main Nb orebodies contain hematite, magnetite, zircon, columbite, pyrochlore, fluorite, and several additional rare minerals and they are patchy, fracture-controlled albitites, microclinites and quartz-microcline replacement bodies best developed in the Li-mica altered core zone.

In the somewhat similar "Kazakhstan-A" locality (Table 33-4), the economic Zr,Nb,REE, Th,Be, etc. mineralization is in fissure-controlled replacement feldspathite veins developed mainly in the exocontact of an albite-riebeckite metasomatite.

Th,Zr,Be, etc., mineralized apogranites.

The example locality "Siberia-B" (Zabolotnaya, 1974), is probably located in the Central Aldan ore field (near the Lebedinskoe gold mine), where a strongly tectonized Archean crystalline basement is topped by Cambrian platformic carbonates and intruded by members of a Mesozoic alkaline complex. Peralkaline members are represented by riebeckite-albite granites, fringed by an exocontact aureole of alkaline skarn. The metallic minerals were superimposed in several phases. In the earlier phase, metasomatic albitites and quartz-albite veins containing Th and Zr minerals formed in the endocontact. In the second, main stage, a variety of replacement albite or microcline-fluorite bodies (veins, masses, lenses), formed along the granite-sediment contact. The principal ore mineral leucophane $(CaNa(BeSi_2O_6)F)$ is accompanied by lesser phenacite, danalite and milarite. Th and REE mineralization (mainly gadolinite) formed at the same time in the endocontact.

U-Th mineralized apogranites.

These are best represented by the Cretaceous Bokan Mt. stock of mineralized peralkaline granite (S.E. Alaska; MacKevett, 1963; Staatz, 1978), emplaced into metamorphics possibly dating from the Devonian period. The intrusive outcrop is less than 7.5 km^2 and it is composed of a leucocratic quartz, albite, microcline, riebeckite, acmite granite, fringed by an aureole of albitization some 2.4 km wide. The economic orebodies formerly exploited by the Ross-Adams mine, are narrow veinlets of uranothorite and uraninite, developed in a north-trending zone of albitization along a shear.

The recently discovered Timgaouine uranium district in the Hoggar

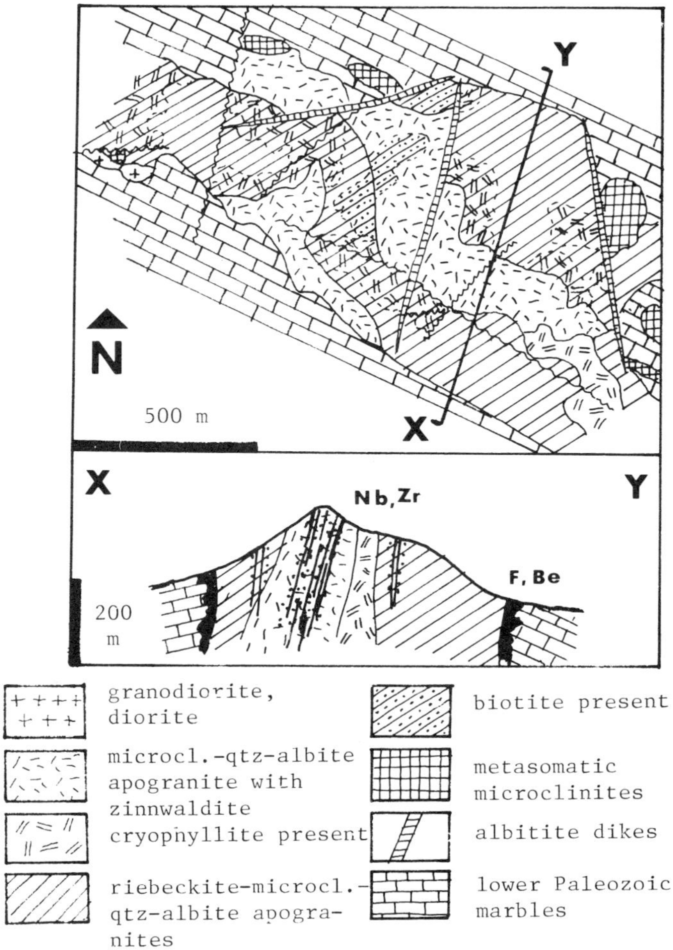

Fig. 33-10. A mineralized massif of albitic apogranites of the "alkaline series", located in Siberia (probably eastern Tuva). After Beus (1968), approximate scale and diagrammatic position of mineralization have been added (Nb,Zr: pyrochlore, columbite and zircon; Fe,Be: fluorite, bertrandite, phenakite exocontact replacements).

Mountains, S. Algeria (O.E.C.D., 1979), contains about 14 Tt U in three vein and stockwork deposits associated with muscovite-biotite "younger granite" emplaced along faults in a Precambrian metamorphic basement.

HYDROTHERMAL-EPIGENETIC ORES IN PERALKALINE GRANITES LACKING WIDESPREAD ALBITIZATION

Molybdenite-bearing peralkaline granite intrusions generally lack a widespread albitization. Instead, quartz, pyrite, molybdenite stockwork veinlets (at a shallow level, e.g. Bordvika in the Glitrevann Cauldron) or quartz-molybdenite veins (at a deeper level, near Drammen; both localities are in the Oslo Graben, Norway), are enveloped by quartz-sericite altered host rocks (Geyti and Scønwandt, 1979). Similar mineralization on a much larger scale (175 Tt Mo) formed in a lower Tertiary peralkaline granite stock at Werner Bjerge, eastern Greenland (Nielsen, 1973). The molybdenite localities mentioned above have many features in common with the "Climax-type" (White et al., 1981) stockwork Mo deposits of orogenic belts, treated in Chapter 28. The major contrast between the two settings of Mo deposits formerly considered to be important (the former in a "rift", the latter in a "mobile belt" environments) has been considerably weakened by the realization that the "Climax-type" deposits are also related to a graben formation. Petrochemically, the bulk of the "Climax-type" alkalies-rich granite porphyries are just below the treshold of the peralkaline class.

HYDROTHERMAL-EPIGENETIC ORES IN THE AUREOLE OF PERALKALINE INTRUSIONS

The exocontact mineralization adjacent to peralkaline granites is not substantially different from the same setting related to the calc-alkaline stocks (Chapter 28). The peralkaline parentage is sometimes reflected by high Be,Nb,Th, REE,Zr, etc. trace metal contents or accessory minerals in predominantly polymetallic, gold, iron, etc. veins or replacements. In an aureole of the Conway Granite (New Hampshire; Cox, 1970), small veins of galena and sphalerite also contain danalite $(Fe,Mn,Zn)_4Be_3(Si_4)_3S$. A larger accumulation of this inconspicuous, yellow to red-brown mineral, was recorded in a small skarn magnetite deposit in Bartlett. There, the Be content in the iron ore reaches as much as 1.6% BeO.

In the Central Aldan ore district, Siberia, two major gold fields (Lebedinskoe and Kuranakh; gold-sulphide veins and replacements in Cambrian carbonates) and numerous gold occurrences, are situated in a contact aureole of small Mesozoic alkaline and peralkaline intrusions. An unusual style of a gold ore, an auriferous quartz-orthoclase metasomatite in blastomylonite-filled lower Proterozoic faults, appears to be closely associated with dikes of Mesozoic peralkaline granite.

In the Oslo Graben, Norway (Bugge, 1978), several small magnetite skarn and Pb-Zn replacement bodies are contained in Silurian limestones at the contact with the Permian Drammen Granite. Pb,Zn,Ag fissure veins also formed in the granite endocontact.

RESIDUALLY ENRICHED ORE MINERALS IN WEATHERING PROFILES AND MINERALS, METALS REWORKED INTO SEDIMENTS

Those metallic minerals associated with peralkaline granites that are also heavy resistates, accumulate in a variety of placers (see Chapter 24). Over 95% of cassiterite produced from the peralkaline association (Nigeria, Rondônia) comes from placers.

Redeposition of the soluble metals leached from mineralized peralkaline granites in sediments has rarely been recorded. One example is the low-grade accumulations of U in the recent organic-rich bog sediments in Idaho (Killsgaard et al., 1970).

33.4. FELDSPAR SYENITE-TRACHYTE TRANSITIONAL ASSOCIATION

33.4.1. General

Small "anorogenic" or late orogenic syenite bodies are quite common in settings that are the same or similar to those in which the peralkaline granites are located, namely in terrains "frozen" in the initial stages of activation and "rifting", in the forelands of ensialic orogenic belts and within sialic blocks in marginal orogenic belts. Such syenites occur either alone, or as sub-members in (or a link to) other igneous associations such as the calc-alkaline diorite-quartz monzonite, peralkaline granite, bimodal basalt or gabbro-trachyte or syenite, agpaitic syenite, etc.

In the broad, discontinuous zone of small, high-level late Cretaceous to Tertiary intrusions emplaced into the little deformed platformic sediments in the eastern foreland of the North American Cordillera (Montana, Wyoming, South Dakota, Colorado, New Mexico, Mexico), syenite is a common member of a broad congregation of intrusive rocks that include quartz diorite to quartz monzonite, syenite, shonkinite and phonolite. Although often designated as an "alkaline province", only a proportion of the magmatic rocks is truly alkaline. Early laccoliths usually breached by later stage stocks and dikes are the most characteristic style of magmatic occurrence in the Cordilleran Foreland. The style is closely comparable with the laccolithic complexes of the Colorado Plateau, reviewed in Section 28.2.2.

In the Oslo Graben, Norway, Permian syenites together with remmants of their volcanic equivalents, are the most common rock type forming ring complexes and composite cauldron-filling bodies in an intrusive assemblage ranging from biotite granite through syenite, monzonite to nepheline syenite.

A variety of postmagmatic-hydrothermal deposits is associated with syenites. The low-grade Mo and Au stockworks, veins and replacement deposits appear to have the greatest potential, whereas Pb,Zn,Ag, W, Be, REE, etc. concentrations have, so far, been of limited importance.

One of the fundamental genetic problems involving feldspathic rocks of syenitic appearance is the uncertainty as to whether a particular rock is a magmatic crystallizate from a melt, or a metasomatite formed by essentially hydrothermal feldspathization of solid precursor rocks. Improved understanding of syenites is of considerable help for metallogenic interpretation, and facilitates exploration.

33.4.2. Example localities of mineralized syenites

Little Rocky Mountains, north-central Montana

The Little Rocky Mountains (Knechtel, 1944; Figures 33-11 and 33-12) form a subcircular dome about 30 km in diameter, composed of a core of unroofed late Cretaceous-early Tertiary intrusive sills, laccoliths and stocks, fringed by more than fifty faulted subordinate sedimentary domes probably floored by intrusive plugs in depth. The sediments are Cambrian to Cretaceous members of the carbonate and detrital platformic association of the North American Platform. The intrusions range from quartz monzonite to syenite and trachyte.

Gold placers and hydrothermal vein, stockwork and breccia pipes in the southern part of the igneous core (Zortman and Landusky fields) have been exploited since 1884, and since 1903 the district has been a pioneer in the application of cyanide leaching for the recovery of a very fine, low-grade gold; a process greatly improved and expanded in 1979 and in the years since (Fig. 33-13). Almost 14.8 t Au had been produced up to 1982.

The gold orebodies in the Little Rocky Mountains (Rogers and Enders, 1982) are structurally controlled, carry almost no base metals and are hosted by and related to the hydrothermal phase of the latest members (dikes) of the syenitic magmatism. Minor mineralization is contained in the Archean basement amphibolite and basal Paleozoic sediments in the intrusive exocontact. The orebodies are unpredictable funnel-shaped bodies marked by low to very low Au values (from cutoff of 1 ppm Au through average grade of about 3 ppm to maximum 60 ppm), lacking conspicuous alteration or a systematic preference for a particular intrusive or a sulphide enrichment. The ore cannot be recognized visually from waste and the selective mining is based entirely on assay data. In the Gold Bug orebody, for example, pyrite-veined wallrocks and a syenite breccia pipe are barren, whereas a fractured syenite along faults with an unimpressive limonite coating is the ore. The mine profitability is achieved by cheap recovery (heap leaching of blasted but uncrushed rock).

In the nearby Judith Mountains (Kendall), low-grade gold replaces sedimentary carbonate in the exocontact of a syenite similar to the one described above.

Little Belt Mountains, Montana

This is another mineralized Eocene laccolithic, volcanic and intrusive group of hills in central Montana, situated in the southern extension of the Highwood Mountains potassic province (Section 33.7.). It was recently described in detail and interpreted by Witkind (1973;

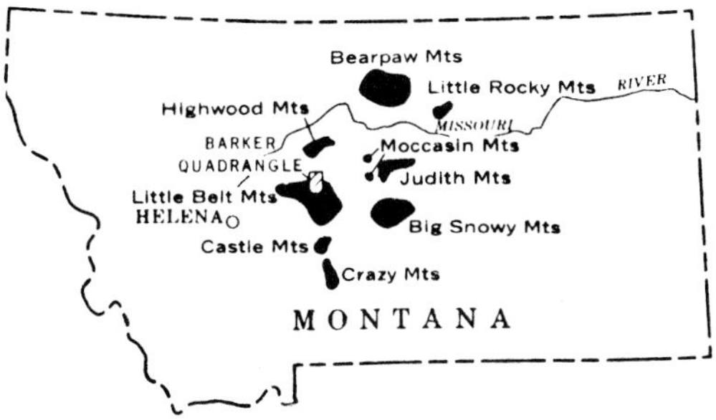

Fig. 33-11. Central Montana igneous province in the Cordilleran Foreland. From Witkind (1973).

Fig. 33-14) and it is representative of the wide variation in quartz saturation of the coexisting central Montana magmas. The prevalent intrusive and volcanic phases are quartz monzonite and quartz rhyolite porphyry, whereas syenite and shonkinite are minor. Eleven major intrusive bodies including one stock, eight laccoliths, a bysmalith and a buried ovoid pluton, are contained within an area measuring less than 30x20 km.

High-grade Pb,Zn,Ag vein and replacement deposits and scattered molybdenite and scheelite occurrence which were mined in the past occur in two ore fields (Barker and San Miguel), located within and on the margin of an intensively fractured Hughesville quartz monzonite stock and probably underlain by a buried pluton. Although the orebodies are younger than the stock, Witkind (1973) attributed the vein control by the stock to a long-lasting master conduit, repeatedly guiding the various magmas and mineralizing fluids towards the same site (Fig. 33-15).

Gallinas Mountains, New Mexico

In the Red Cloud District about 60 km N. of Carrizozo (Perhac and Heinrich, 1964), upper Cretaceous to lower Tertiary porphyritic trachyte, peralkaline leucorhyolite and various riebeckite and aegirine-containing high-level stocks, dikes and laccoliths intrude arched Permian sandstone and Precambrian granitic basement. Small breccia zones and mineralized faults in the sandstone close to trachyte bodies contained veinlets and cement of fluorite, galena, chalcocite, bornite, pyrite, barite and bastnaesite. A mere 4 Tt Cu and 60 t bastnaesite were produced.

Permian magmatic province, Oslo Graben

In this well-known igneous province, syenite and monzonite massifs together with remnants of their volcanic equivalents are one of the most common rock types. Numerous small replacement Pb-Zn sulphide deposits formed in skarns and marbles along the contact of syenite and Ordovician limey sediments, e.g. in the Grua area (Ineson et al.,

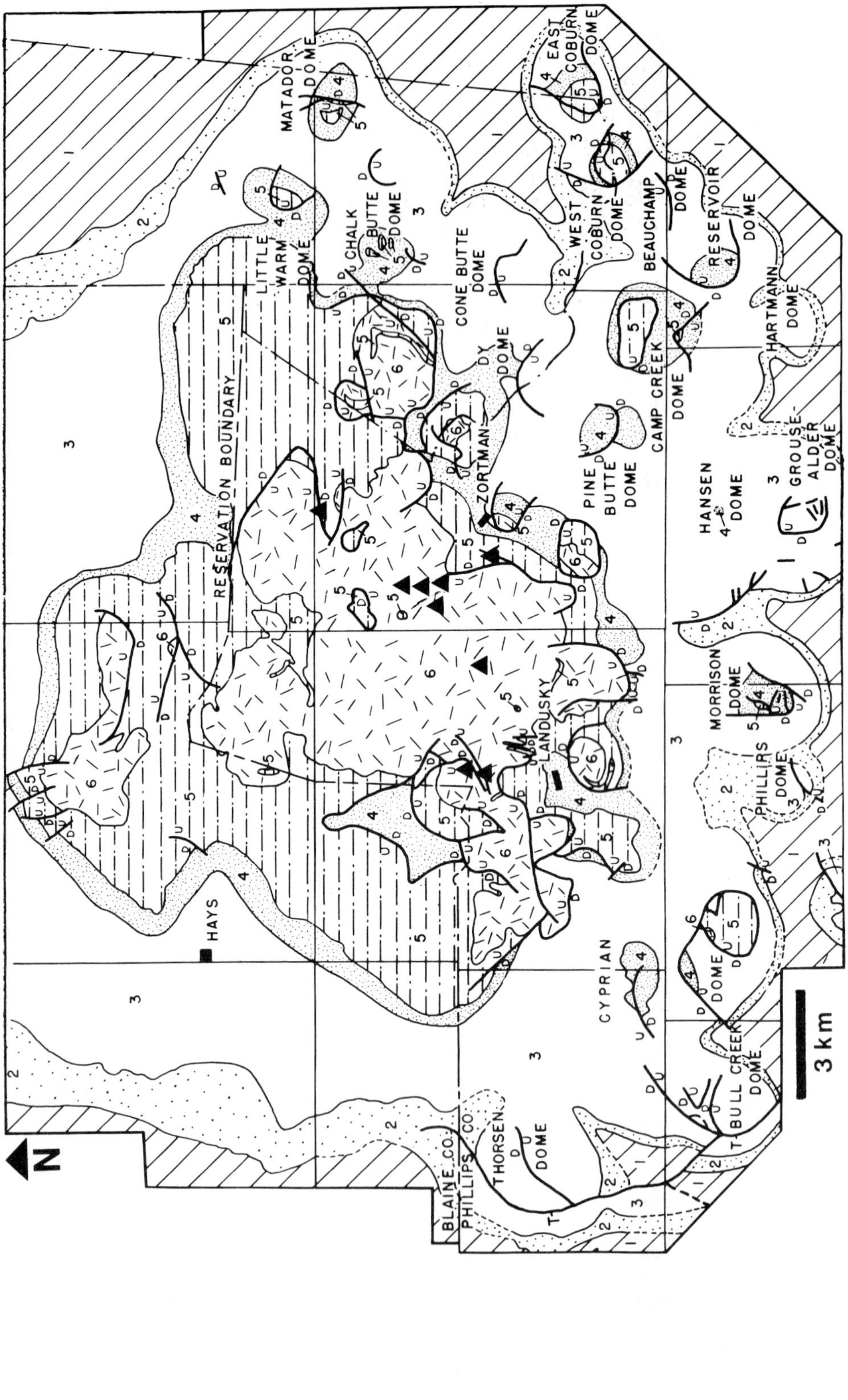
N
HAYS
RESERVATION BOUNDARY
LITTLE WARM DOME
MATADOR DOME
CHALK BUTTE DOME
CONE BUTTE DOME
DY DOME
ZORTMAN
LANDUSKY
PINE BUTTE DOME
EAST COBURN DOME
WEST COBURN DOME
BEAUCHAMP DOME
RESERVOIR DOME
HARTMANN DOME
CAMP CREEK DOME
HANSEN DOME
GROUSE-ALDER DOME
MORRISON DOME
PHILLIPS DOME
CYPRIAN
DOME
BULL CREEK DOME
THORSEN DOME
BLAINE CO.
PHILLIPS CO.
3 km

Fig. 33-13. Zortman, Little Rocky Mountains, Montana. Heap-leaching of blasted (but not crushed) rock made possible a profitable gold recovery from very low-grade auriferous stockworks and veins in syenite and quartz monzonite.

1975). Compared with the younger granite intrusive phase in the same area, the number and size of ore occurrences associated with syenite is substantially less. Fissure fillings of hematite and fluorite considered to be of fumarolic origin (that is, directly precipitated from magmatic gases) were mined near Baerums Verk (Holtedahl et al., 1960).

Mesters Vig, eastern Greenland

Several hydrothermal Pb-Zn deposits occur in the aureole of lower Tertiary alkaline syenite intrusives (Bondam and Brown, 1955). The largest deposit Blyklippen has been mined and produced 40 Tt Pb and 40 Tt Zn. There, galena and sphalerite in quartz gangue filled N.N.W.-striking veins, hosted by faulted upper Carboniferous sandstone, black slate and arkose.

<—

Fig. 33-12. Little Rocky Mountains, N.-C. Montana, showing a largely syenite core hosting gold mineralization, and a fringe of small faulted domes. 1-3=Cr_3 shale and sandstone; 4=J-Cr_1 shale and sandstone; 5=Cm-Cb limestone, dolomite, basal sandstone; 6=T intrusions and minor PCm basement metamorphics. Triangles:principal Au-Ag orebodies. After Knechtel (1944); ore occurrences added.

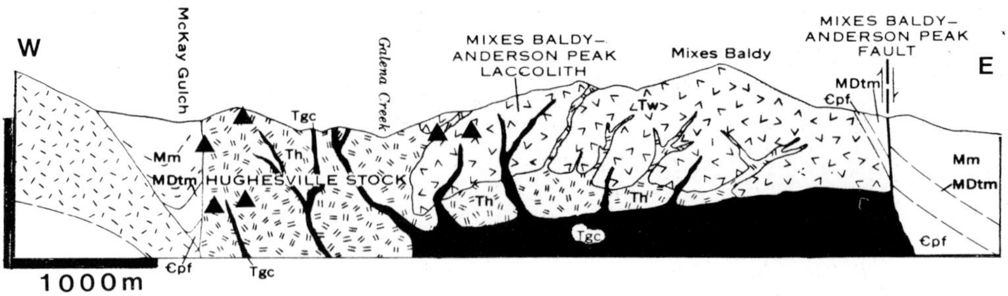

Fig. 33-14. Hughesville stock and adjacent intrusions, Little Belt Mountains, Montana, showing diagrammatically (by triangles) the usual setting of the Au and Pb-Zn mineralization. Tgc=Eo porphyry; Th=quartz monzonite; Tw=Eo porphyry; Mm, MDtm=D-Ms limestone and dolomite; Cpf=Cm limestone, shale, sandstone. From Witkind (1973), ore symbols added.

Wausau Complex, Wisconsin

A 560 m.y. old stock of medium to coarse syenite and minor nepheline syenite and pegmatite, intrudes an early Proterozoic basement of granite, greenstone, metaquartzite and metaargillite. Thorogummite, Th-zircon, zircon and allanite associated with hematite alteration, have been discovered in residual soil and in saprolite in the S.W. exocontact of the intrusion (Vickers, 1956). They have probably been derived from fissure veins, comparable with Th-bearing veins known in the western United States (e.g. Lemhi Pass).

Sivrihisar Kizilcaoren Th-REE deposit, Turkey

The O.E.C.D. (1979) reported the discovery of a significant deposit of thorium (min. 334 Tt Th/0.185%) and rare earths in bastnaesite and other minerals filling fractures and cementing breccia in the vicinity of a small granosyenite massif.

33.5. NEPHELINE SYENITE AND ALKALINE GABBRO-DOMINATED INTRUSIVE COMPLEXES

33.5.1. General

This category represents lithologic associations considered by an average geologist to be the "typical" alkaline rocks. Such rocks are undersaturated in silica and are members of the miaskitic (Na+K:Al=lesser than 1) and agpaitic (Na+K:Al=greater than 1) petrochemical classes. The miaskitic and agpaitic intrusives usually form equant, circular or elliptical central complexes, emplaced in a basement composed of a mature, stabilized continental crust (usually of crystalline rocks). Partly differentiated intrusive sills, dikes

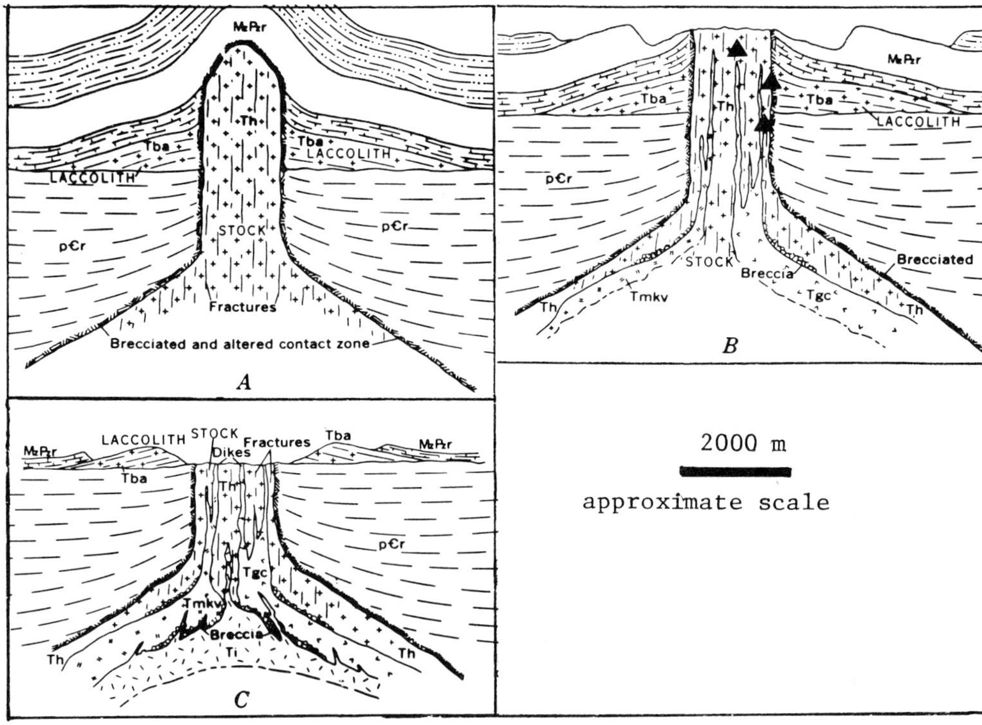

Fig. 33-15. Interpretation of magmatic and mineralization history of the Hughesville multiphase intrusive stock, Little Belt Mountains, Montana. (A): Magmas and ore solutions were channelled repeatedly into an old master conduit, presumably resulting in the radial pattern of laccoliths about the stock. (B): Intermediate magma concealed in the throat resulting in the Hughesville stock. The stock was later fractured, invaded by dikes and hydrothermal orebodies. (C): Subsequently, another pluton guided into the master conduit reactivated some fractures. From Witkind (1973).

and laccoliths, plugs, stocks, ring dikes and small homogeneous plutons also occur (Sørensen, 1974). Most complexes are considered to be "anorogenic" to postorogenic, controlled by crustal distention in cratons, and there is usually a substantial age difference between the alkaline intrusions and the basement.

In exceptional cases, minor occurrences of nepheline syenite and equivalent rocks have been reported in orogenic belts, where some may be broadly contemporary with the "regular" products of calc-alkaline magmatism. The Kruger alkaline syenite complex in the Similkameen Valley, British Columbia (Currie, 1974) is an example. It occurs along the margin of the Similkameen Pluton, a typical granitoid pluton in the Cordillera. It is of similar age.

Nepheline syenite and its varieties are the most widespread members of the present category, followed by alkaline gabbros. Carbonatites are commonly present, but in subordinate quantities. Complexes dominated by carbonatite are treated in Section 33.8. Alkaline complexes with more than one rock type are characteristically zoned, usually ranging from mafic on the outside to salic in the interior. Alkaline metasomatic aureoles (fenites) are usually developed, and some alkali metasomatites (e.g. feldspathites) are compositionally equivalent to rocks crystallized from magmatic melts (e.g. syenites) and hard to distinguish.

Syenitic rocks (including nepheline syenites) may also crystallize from an immiscible syenite magma produced from rheomorphically mobilized metasomatic rocks and crystallization of these melts may overlap with crystallization of the depth-derived alkaline magmas (Currie, 1974). This may have important metallogenic implications because crustally derived magmas or in-crust generated metasomatites would reflect the metallogenic character of the intruded environment rather than the expected subcrustal suite of metals known to be systematically associated with alkaline magmas.

33.5.2. Metal geochemistry and metallogeny

Uncontaminated, depth-derived alkaline magmas are strongly enriched in a distinct suite of trace metals (Gerasimovsky, 1974; Table 33-5). The enrichment factors of Zr and Nb, measured against the mean crustal content, exceed 20; that of REE,Ta, exceed 10; and that of Be,U,Th exceed 5. Other trace metals such as Cu,Cr,Ni are depleted in alkaline magmas. Of the major metals, the Al content of the light-coloured alkaline rocks is moderately enriched (up to 23% Al_2O_3 in some syenites) compared with the clarke, and the absence of quartz makes such rocks very suitable for an undiluted residual aluminium enrichment during lateritization. High-grade bauxites, therefore, form frequently on such rocks.

As a rule, the highest trace contents of the rare elements mentioned above are in rocks formed from strongly fractionated melts which are the end product of differentiation of alkaline basaltic magma. Petrographically comparable rocks interpreted as being of palingenetic origin (e.g. the nepheline syenites in the Mongol-Tuva alkaline province, Siberia; Gerasimovsky, 1974), have substantially lower contents of the same metals.

In intrusive rocks, the rare elements (Nb,Ta,Th,U,Be,Tl, etc.) (1) substitute in the lattice of the rock-forming minerals or of common accessories (e.g. in alkaline pyroxenes and pyriboles, in Ti-magnetite or ilmenite, in zircon); (2) are bound in complex accessory minerals (e.g. eudialyte, loparite, eucolite, lovozerite, catapleite, complex Ti-silicates and Zr-silicates) that are rare in the bulk of the alkaline rocks, but common or even dominant in rare alkaline rock types and (3) form their own rare minerals (e.g columbite-tantalite, pyrochlore, thorite, sørensenite, chalcothallite, etc.). Certain

Table 33-5. Trace elements highly enriched in alkaline intrusive rocks compared with the crustal average (in ppm). Data from Gerasimovsky (1974), Taylor (1964) and *estimated

	nephel.syenite Ilímaussaq	nephel.syenite Lovozero	nephel.syenite Khibiny	average nepheline syenite	average crust	enrichment factor
Li	330	55	20	30	20	1.5
Be	30	8.7	6.1	6	2.8	2.14
F	2,100	1,400	1,230	1,400*	625	2.34
REE	3,680	2,050	480	800*	174	4.6
Zr	4,735	3,480	625	550	165	3.3
Nb	525	696	152	200	20	10
Sn	115	10	6.6	12*	2	6
Ta	32	60	14	18	2	9
Th	38	35	14	20*	9.6	2.1
U	62	16.1	4.2	15*	2.7	5.5

magmatic differentiates become so anomalously enriched in either the category (2) or (3) minerals that they become low-grade ores having a relatively even distribution of the ore substance.

Ore occurrences having metal concentrations above the level of the "metalliferous rocks", however, require an additional mechanism of upgrading. Pegmatites produced from the volatile-enriched residual magmatic fluids do concentrate the rare elements and many (particularly those related to agpaitic rocks) become virtual "treasure troves" of rare minerals to collectors. The rare minerals, however, are scattered and few pegmatite bodies are persistent and extensive enough to constitute valuable industrial deposits. More valuable appear to be the local accumulations of some rare minerals with a minimum of gangue, that texturally and probably genetically resemble pegmatites. Such may be some of the bastnaesite veins (e.g. Ifasina in Madagascar; Chantraine and Radelli, 1970).

Hydrothermal processes associated with or superimposed on alkaline complexes as well as supergene leaching and residual enrichment are important agents of localized metal concentration that may generate important orebodies. Compared with the calc-alkaline "granitic" hydrothermal systems, however, the frequency and efficiency of the hydrothermal systems associated with alkaline complexes is very much subdued.

33.5.3. Major mineralization styles (Fig. 33-16, Table 33-6)

1. MAJOR MAGMATIC ROCK-FORMING MINERALS AS A METALLIC ORE

Nepheline syenite is frequently quarried as a nonmetallic mineral commodity for use in the ceramic and glass industries. In the U.S.S.R., a process has been developed for extracting metallic aluminium from raw nepheline syenite, or from nepheline concentrate (Rundkvist, ed., 1978). Three plants have, so far, been put into operation and one of them (in Leningrad) uses nepheline concentrate with 29-29.5% Al_2O_3 obtained as the by-product of processing of the Khibiny apatite ores.

2A. MAGMATIC ROCKS WITH ACCESSORY METALLIC MINERALS OR WITH HIGH TRACE METAL CONTENTS AS ORES, (A): LAYERED INTRUSIONS

Some alkaline layered intrusions, such as Lovozero or Ilímaussaq, have higher trace contents of certain elements (e.g. Zr) than "ores" of such elements currently mined on a commercial basis elsewhere (e.g. 0.47% Zr in the Ilímaussaq Complex versus 0.1-0.05% Zr in Florida heavy mineral sand deposits). In addition to Zr, Lovozero and Ilímaussaq contain high trace contents of REE,Nb,Ta,Th,U, etc., and their separate rock-forming minerals (e.g. nepheline, alkali feldspar) constitute important nonmetallic commodities. As a consequence, both massifs may be mined in the future and the ore processed as a complex raw material. Before this occurs, however, selective mining of discrete rock layers with substantially higher metal grades will probably have taken place, or is now taking place. Fig. 33-17 shows the grade-tonnage relationship of zirconium source materials in Lovozero and Ilímaussaq. Examples of metalliferous magmatic layers:

(a) Ilímauusaq Zr, Nb-bearing kakortokites (Nielsen, 1973; Bohse et al., 1971). Kakortokites are late stage agpaitic layered arfvedsonite, eudialyte, nepheline, feldspar syenites that are banded in a remarkably rhythmical manner. Within one rhythm, black (arfvedsonite-rich), red (eudialyte-rich) and white (feldspar and nepheline-rich) bands (in ascending order) are usually developed. Kakortokites contain 1.14% ZrO_2 in the black bands, 7.09% ZrO_2 in the red bands and 1.09% ZrO_2 in the white bands. The corresponding Nb_2O_5 contents are 0.05, 0.56 and 0.1%. The average content of the entire kakortokite is 1.2-1.4% ZrO_2 and about 0.13% Nb_2O_5 and they are estimated by Bohse et al. (1971) to contain about 51.6 Mt ZrO_2 and 5.4 Mt Nb_2O_5. This does not include the metal contents in the border pegmatite that must be considerable.

In Ilímaussaq, sphalerite is a widely distributed accessory mineral in the lujavrite layers, and 0.1% Zn contents have been recorded over large areas.

(b) Lovozero eudialyte lujavrites and eudialytites (Vlasov et al., 1959). The layered eudialytic lujavrites (mesocratic agpaitic nepheline syenites) constitute 18% of the Lovozero intrusion and form a rhythmic sequence 150 to 500 m thick, with an average content of 1.36% ZrO_2. Each rhythm consists of successive bands of leucocratic, mesocratic and melanocratic varieties, with gradual transition from one to the other. Eudialytites, eudialyte highly enriched conspicuously dark-red bands (up to 75% of eudialyte cumulus crystals in nepheline matrix), form lenticular intercalations ranging in thickness from a few millimetres to about 20-40 cm. In the Chivruái Valley (Fig. 33-18), 13 eudialytite bands form about 40% of a 3 m thick section of eudialytic lujavrite.

Eudialytites contain 6.76 to 8.68% ZrO_2, 0.39 to 0.93% $(Ta,Nb)_2O_5$ and 1.01 to 1.56% REE_2O_3 and they represent a resource of 10 to 100 Tt Ta, 100 to 1,000 Tt Nb, 100 to 1,000 Tt REE and some 4 to 40 Mt Zr (Rundkvist, ed., 1978).

(c) Lovozero loparite-bearing urtites and juvites (Vlasov et al., 1959). Loparite $(Ce,Na,Ca)_2(Ti,Nb)_2O_6$ in the form of black xenomorphic grains, is a widespread accesory mineral in the Lovozero layered intrusive rocks. Pure loparite contains about 8-10% Nb_2O_5, 0.65-0.75% Ta_2O_5, 0.62-0.76% ThO_2 and 16-17% REE_2O_3 When loparite content in an intrusive layer approaches about 10%, the layers can be selectively mined as a low-grade complex ore of REE,Nb,Ta,Th and U (Fig. 33-19). The loparite-rich layers are most common in urtites of the differentiated complex.

Rundkvist, ed. (1978) hinted that such layers containing 0.02-0.04%

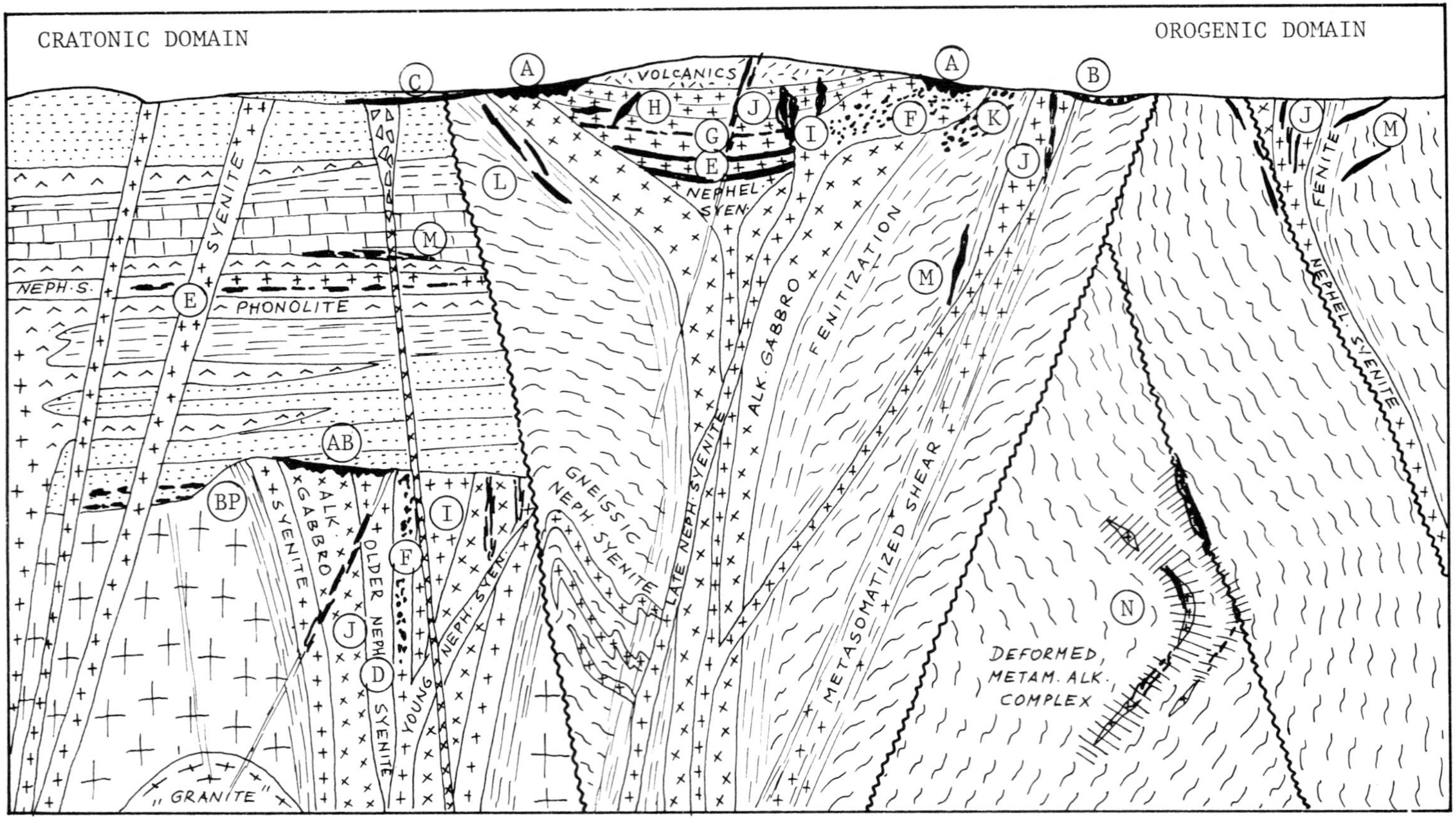

Fig. 33-16. Principal lithologic sub-associations and mineralization styles in the nepheline syenite-dominated alkaline complexes (see Table 33-6 for explanation of letter codes).

Table 33-6. Principal mineralization styles in nepheline syenite and alkaline gabbro-dominated alkaline complexes

	WEATHERING CRUSTS
(A)	Residual resistate minerals (Nb,Ta,REE,Zr); minerals formed by precipitation from leached components (Zr,U); residual bauxite over nepheline syenite and phonolite (60 Mt Al); residual Mn-oxides (500 Tt Mn) (AB) buried mineralized regoliths
	PRODUCTS OF SEDIMENTOGENIC REWORKING
(B)	Alluvial placers (caldasite, pyrochlore, zircon);(Zr,Nb,REE) (BP) paleoplacers
(C)	Resedimented bauxite (or sandy clay); Al (included in A)
	SYN-MAGMATIC ORES IN INTRUSIVE ROCKS
(D)	Major magmatic rocks and constituent minerals as metallic ores (e.g. nepheline syenite, nepheline, as future sources of Al)
(E)	Metalliferous magmatic layers: dominantly eudialyte-bearing syenites (min. 100 Mt Zr, 5 Mt Nb, 10 Mt REE, X0 Tt Ta, low-grade) pyrochlore, loparite, steenstrupine, lovozerite, etc. bearing syenites (min. 5 Mt Nb, 5 Mt REE, 50 Tt Ta, 500 Tt Th, low-grade)
(F)	Metalliferous non-layered magmatic rocks (massifs, sills, dikes); e.g. nepheline syenite with steenstrupine, monazite, thorite, etc. (40 Tt U, 85 Tt Th, 100 Tt REE)
	LATE MAGMATIC AND POSTMAGMATIC ORES IN INTRUSIVE ENDOCONTACT
(G)	Layered alkaline pegmatites, eudialyte rich (Ilimaussaq); possible 20 Mt Zr, 1 Mt Nb, low-grade
(H)	Patchy and dike alkaline pegmatites with murmanite, lovchorrite, etc. (about 20 Tt Nb,Ta, 10 Tt REE, 500 t Th)
(I)	Metasomatites (Be,REE,Nb,Th)
(J)	Hydrothermal veins genetically closely associated with alkaline intrusions (about 200 t Be, 280 Tt REE, 150 Tt Zr, 30 Tt Th, 1 Tt U, 160 Tt Ti, 17 Tt Mo); unaffiliated veins (Pb,Zn,Ag,Cu)
	HYDROTHERMAL ALTERATION AND MINERALIZATION IN INTRUSIVE EXOCONTACTS
(K)	Scattered, disseminated, patchy minerals in fenites (Fe,Ti,Zr,Nb, U, 50 Tt V)
(L)	Hydrothermal veins in fenitized aureole (U,Th,Ti,REE)
(M)	Ores in non-fenitized contact aureoles (Fe,Cu,Pb,Zn,Ag)
(N)	Syntectonic ores in deformed and metamorphosed alkaline complexes (Zr,Nb,Ta,U,Th,REE,Be)

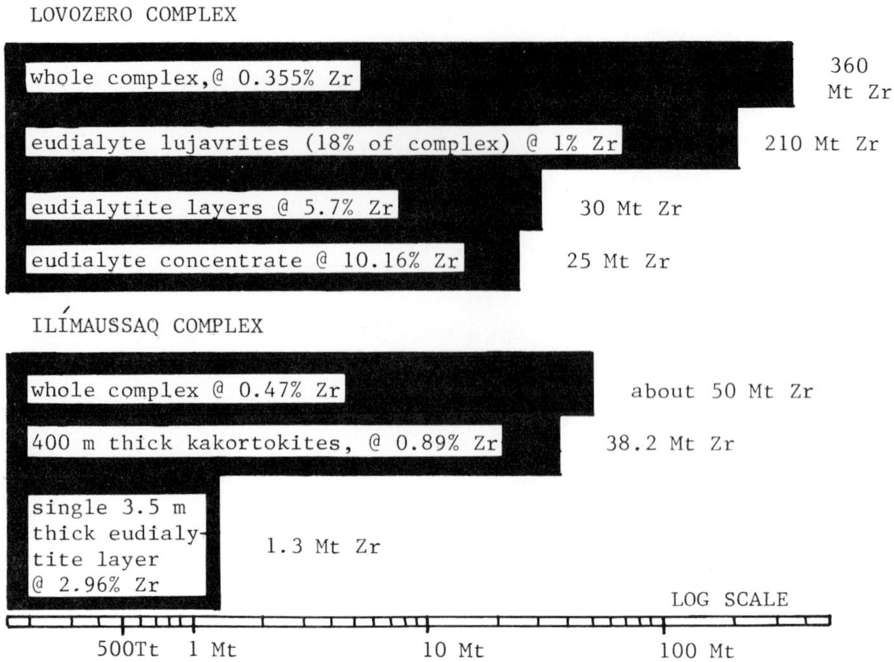

Fig. 33-17. Grade-tonnage relationship of zirconium source materials in Lovozero and Ilímaussaq alkaline complexes. Based largely on data in Vlasov et al. (1959) and Bohse et al. (1971), with gaps filled by estimates.

Ta, 0.2-0.4% Nb, 1-1.5% REE and probably 0.03% Th, may contain a mineable resource of some 1-10 Mt rare earths as well as 25% of the world's reserves of Ta (some 5-60 Tt Ta) and 12% of the world's reserves of Nb (some 5-8 Mt).

(d) Lovozero steenstrupine and lovozerite-bearing lujavrites. Steenstrupine $(Ce,La,Th,Ca,Na)_2(Mn,Fe)(SiO_3)_4 \cdot 5H_2O$, contains 14% REE_2O_3, 10.23% $ThO2$, 4.37% Nb_2O_5, 1.28% Ta_2O_5. Lovozerite $(Na,Ca)_2(Zr,Ti)Si_6O_{13}(OH)_6 \cdot 3H_2O$, contains 16.54% ZrO_2 and 0.56% REE_2O_3 (Vlasov et al., 1959). Both minerals occur in well-developed crystals and mineralogical quantities in some Lovozero pegmatites. They are much rarer accessories in the layered rocks, but recently several layers of steenstrupine and lovozerite-enriched lujavrites (mesocratic nepheline syenites) have been outlined. Beacause of the high REE,Th and Ta contents of steenstrupine, a rock with 1-2% of this mineral could become an industrial ore with 0.015-0.025% Ta, 0.2-0.3% Nb, 0.5% REE (Rundkvist, ed., 1978) and 0.X% Th. The potential Ta,Nb and REE resources in the steenstrupine and lovozerite lujavrites are only slightly less than the resources of the same metals in the

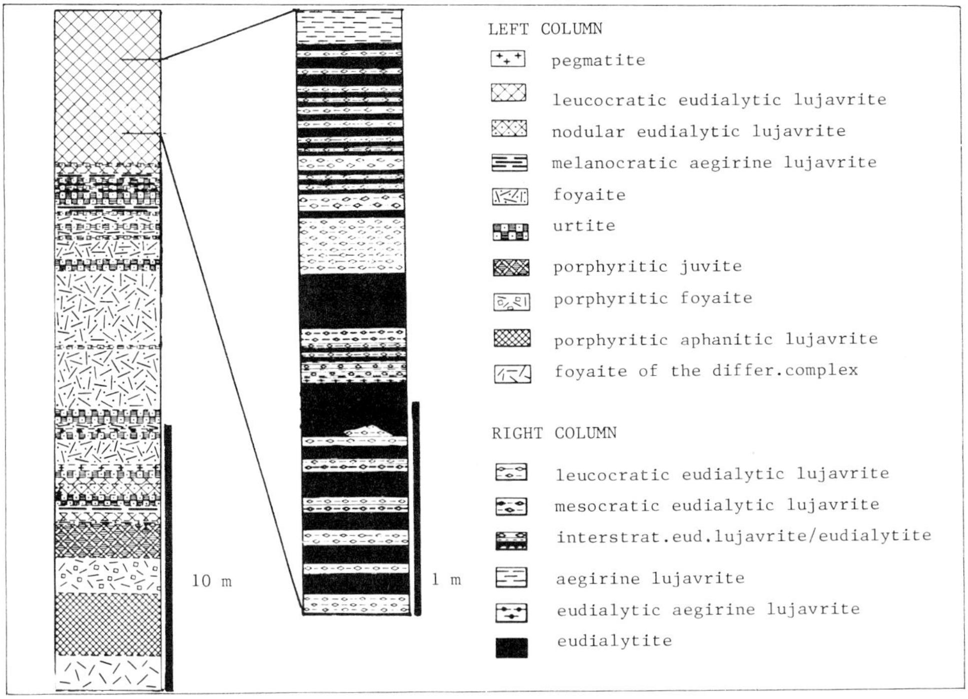

Fig. 33-18. Lovozero intrusion, columnar section of the basal portion of the eudialytic lujavrite complex (left) and detail of the eudialyte-rich 3 m thick interval in the Chivruai Valley (right). After Yeskova and Yeliseev, from Vlasov et al. (1959).

loparite ores, reported above.

(e) Metallic by-products of mining and processing of the Khibiny apatite. In the Khibiny Complex, a zone of brecciated aegirine nepheline syenite conformable with igneous layering, cemented and replaced by apatite, contains up to 65% of apatite and lesser quantities of Ti-magnetite, sphene and eudialyte. This zone is 2.4 km long and 160 m thick. The latter three minerals are recovered as by-products (Ivanova, 1963; Herz, 1976; Fig. 33-20). Sphene-rich layers in the hangingwall of the apatite orebody, moreover, contain 8-11% TiO_2 over a thickness of 5 to 30 m, and are periodically mined as a titania ore. The sphene concentrate contains 26% TiO_2.

The apatite in Khibiny and in other alkaline layered complexes usually contains a high trace content of the rare earth elements (almost 5% REE_2O_3), which is recoverable during the chemical processing of the apatite. Given the published apatite reserves of 2.7 Bt ore with 18% P_2O_5, this represents some 35 Mt REE_2O_3.

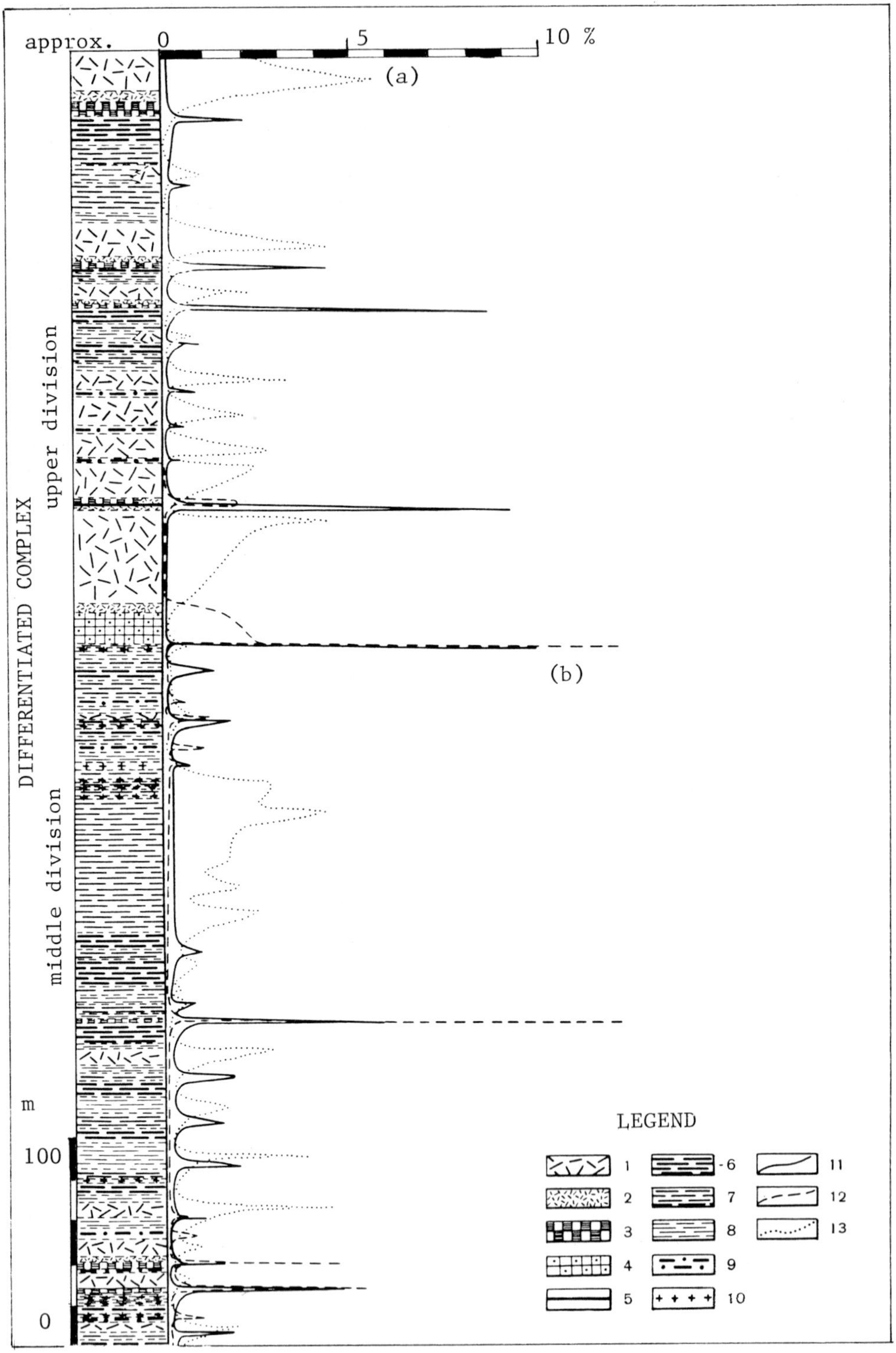
approx.
0
5
10 %
(a)
(b)
DIFFERENTIATED COMPLEX
upper division
middle division
m
100
0
LEGEND
1
2
3
4
5
6
7
8
9
10
11
12
13

2B. MAGMATIC ROCKS WITH ACCESSORY METALLIC MINERALS OR HIGH TRACE METAL CONTENT, (B): NON-LAYERED INTRUSIONS

In this style, the ore substance distribution is comparable with the style (2A) reviewed above, but the host intrusive unit does not show rhythmic layering. Consequently, the ore minerals are distributed either evenly throughout the entire intrusive body (e.g. dike, sill, small plug), or form irregular ore-grade patches.

In the Kvanefjeld area of the Ilímaussaq intrusion, a late nepheline syenite (lujavrite) dike intrudes earlier brecciated alkaline intrusive rocks, and reaches into the exocontact of the layered intrusion (Nielsen, 1973). The dike is enriched in steenstrupine and lesser amounts of monazite and thorite. The whole rock contains 100-800 ppm U, 200-2,000 ppm Th, 1.21% REE and the reserves are currently estimated to be at least 18 Mt of ore, containing 43 Tt U and 86 Tt Th.

Carbonatite dikes occur frequently in nepheline syenite complexes or in their fenitized aureole. They may carry economic quantities of pyrochlore.

(3) PEGMATITES OF ALKALINE SYENITES

Pegmatites are widespread in nepheline syenite complexes, particularly the agpaitic ones. They are often zoned and carry a variety of rare minerals that are megascopically developed. From the standpoint of a field geologist, two sub-categories of pegmatites can be distinguished: (a) patchy pegmatites and (b) horizon (layer) pegmatites. In the Lovozero Massif (Vlasov et al., 1959), patchy pegmatites are abundant (particularly in the eudialyte lujavrite complex). Their dimensions vary between X and X00 m and they have no preferred orientation. They carry the same minerals as their enclosing rocks and are of no economic importance.

The horizon pegmatites are concordant or peneconcordant with the magmatic units, being most widespread at boundaries of petrographic

---

Fig. 33-19. The character of REE, Nb and Ta mineralization in the layered differentiated complex, Lovozero intrusion, illustrated by distribution curves for loparite and lomonosovite-murmanite. The lomonosovite-murmanite peak (a) of 5.1% corresponds to a grade of about 0.27% $(Nb,Ta)_2O_5$. The loparite peak (b), about 11%, has an approximate grade of 1% Nb_2O_5; 0.07% Ta_2O_5; 0.08% ThO_2; and 3.5% REE_2O_3. The loparite-rich horizons are presumably mined selectively. 1=Foyaite; 2=juvite; 3=urtite; 4=ijolitic urtite; 5=malignite; 6=melanocratic lujavrite; 7=mesocratic lujavrite; 8=leucocratic lujavrite; 9=hornblende lujavrite; 10=pegmatite; 11=loparite distribution curve; 12=apatite curve; 13=lomonosovite-murmanite curve. Modified after Yeskova, from Vlasov et al. (1959). Approximate scale added.

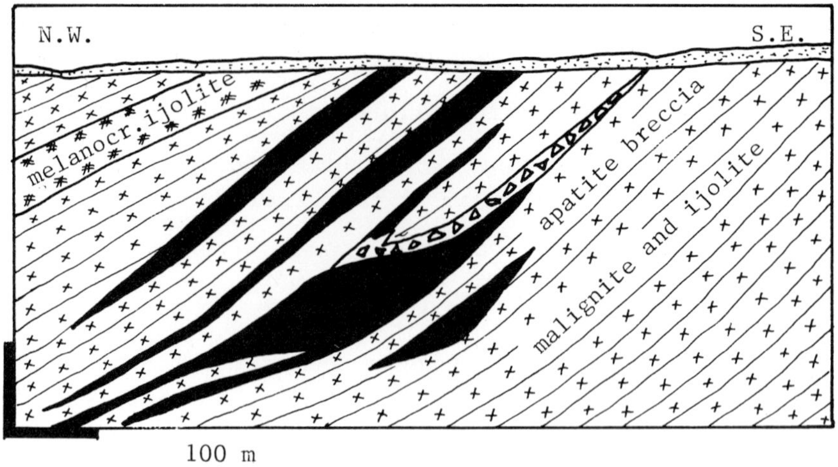

Fig. 33-20. Khibiny alkaline complex, Kola Peninsula, U.S.S.R. Cross-section of an open pit apatite mine near Kirovsk. Black areas: bodies of rare earths-rich apatite ore. After Pogrebitskii et al. (1968).

units ("marginal pegmatites"). The pegmatitic horizon beneath the ijolitic urtites in the Lovozero Complex consists of up to six subparallel pegmatitic layers ranging in thickness up to 2.5 m and traceable for several kilometres. The pegmatites are usually symmetrically zoned and typomorphic minerals are present in each zone. This facilitates selective mining and separation of the valuable minerals. Murmanite $Na_2(Ta,Ti,Nb)_2Si_2O_9$) content, for example, may reach up to 10% in the intermediate pegmatite zone. The greatest proportion and greatest thickness of the Lovozero pegmatites is in the poikilitic syenite complex. There, they contain the greatest diversity of rare minerals, and have an increased proportion of Th,REE and Be.

In the Ilímaussaq intrusion, the marginal pegmatite bordering the layered kakortokite unit against the marginal augite-syenite is one of the most impressive alkaline pegmatites known (Bohse et al., 1971). It is about 50 m thick and contains the same minerals as the kakortokites. The mineral distribution, however, is substantially more heterogeneous, yet the mean ZrO_2 content in the pegmatite is 2%, which is higher than the kakortokite average (1.2% ZrO_2).

In the Khibiny Complex (Ivanova, 1963), rare minerals-bearing pegmatites are widespread. At the Yukspor locality (near Kirovsk), a W.N.W.-trending swarm of several hundred K-feldspar, aegirine, lesser nepheline, arfvedsonite, apatite and lovchorrite pegmatite dikes up to 200 m long, was mined for its rare earths content (Afanasyev, 1937). The principal ore mineral was a dark yellow to brown lovchorrite $(Na_2Ca_4(Ce,La)(Ti,Nb)(Si_2O_7)_2(F,OH)_4$.

Outside the three important layered alkaline complexes mentioned above, nepheline syenite pegmatites are widespread but of little economic importance. Eudialyte pegmatites have frequently been reported (e.g. at Kipawa River, W. Quebec, Currie, 1974; Poços de Caldas, Brazil and numerous localities listed by Pavlenko, 1974, in the Mongol-Tuva province, Siberia). Langensundfiord in southern Norway (Brøgger, 1890) and Mount St.-Hilaire, Quebec, are famous mineral collecting localities in alkaline pegmatites.

4. ALKALINE METASOMATITES IN INTRUSIVE ENDOCONTACTS

Metasomatism is widespread in all phases of development of an alkaline intrusion. The products of intramagmatic and early postmagmatic metasomatism in alkaline complexes which presumably formed as a result of reaction of the volatile-rich residual fluids with the earlier crystallized consanguinous magmatites, are particularly distinct. Usually they form gradational patches, schlieren or veins controlled by fracture systems, zones of brecciation, contacts, etc. The composition, appearance and grain size of the metasomatites is variable and most striking are the coarse crystalline ones, that resemble pegmatites.

The majority of metasomatites are sodic, and albite is a particularly widespread constituent. Albite metasomatites often carry scattered minerals of Nb,Ta,REE,Zr,Th,Be (pyrochlore, thorite, zircon, sphene, apatite, allanite, etc.). K-feldspar (microcline)-rich metasomatites are most abundant where an alkaline intrusion is hosted by terrigenous (meta)sediments or granites (Omel'yanenko, 1974) and could be the result of remobilization of the earlier rock-forming potassium.

A variety of unusual mineralized metasomatites have been described in the Ilímaussaq Complex (e.g. by Engell et al., 1971). They include: (a) "lujavritized syenite", a metasomatite composed of albite and arfvedsonite, replacing nepheline syenite; (b) aegirine-replaced naujaite; metasomatic aegirine, partially or completely replacing a sodalite-rich agpaitic syenite. A portion of it carries scattered crystals of chkalovite $Na_2(BeSi_2O_6)$. Analcite-replaced naujaite formed in the same setting, and it is a relatively porous light-coloured rock with chkalovite, epistolite and Li-mica in crystal-lined miaroles. (c) an albitized arfvedsonite nepheline syenite dike offshoot in the Taseq area contains great number of Be minerals of several generations. The earliest chkalovite was later altered to tugtupite, which in turn changed to eudidymite and epididymite. This reflects the long-lasting, continuing history of alteration and mineralization in the area. The metasomatized intrusive rocks grade, without interruption, into the shear and fissure-filling mineral veins, treated in the next paragraph.

5. HYDROTHERMAL VEINS IN INTRUSIVE ENDOCONTACT

Hydrothermal veins cutting alkaline intrusions have frequently been recorded and some are important economically. Two end-member types are present: (a) veins that appear to be closely associated with the development of the host intrusion ("affiliated", "consanguinous", etc. veins). Such veins are often gradational into the metasomatites or pegmatites described earlier and when mineralized, the veins carry the alkaline suite of trace metals; (b) veins that do not reflect the mineralogical and geochemical characteristics of the host intrusion ("unaffiliated", "strange", etc. veins), and often correspond to or resemble the mineralizations known to occur outside the alkaline intrusion (for example, associated with the hydrothermal phase of calc-alkaline batholiths).

(a) Affiliated veins

In the Ilímaussaq complex, Be-mineralized hydrothermal veins formed by extension of the same process that produced some of the metasomatites reviewed earlier. In the Kvanefjeld area (Engell et al., 1971), albite and natrolite veins and veinlets carry scattered Be minerals chkalovite, tugtupite, beryllite and bertrandite, and several additional Nb minerals. Some of the veins persist into the exocontact formed by sheared mafic lavas, anorthosites and augite syenites. There, analcite is the principal gangue mineral, and the Nb_2O_5 and Ta_2O_5 contents of 0.6% and 0.008%, respectively, are in pyrochlore, epistolite $(Na,Ca)(Nb,Ti)(OH)SiO_4$ and murmanite $(Na(Ti,Nb)(OH)SiO_4$; Hansen (1968).

The largest Be accumulation in the Ilímaussaq and one that might become economic in the future (estim. reserves 180 Tt ore with 0.1% BeO; Engell et al., 1971), is contained in the Taseq slope area. There, a zone of hydrothermal veins and veinlets is superimposed on a steenstrupine-bearing arfvedsonite nepheline syenite dike. The main Be mineral is chkalovite, contained in albite-fluorite gangue. Analcite veins hosted by coarse nepheline syenite in the same area carry sodalite, natrolite, sørensenite, pyrochlore, neptunite, monazite, chkalovite and beryllite.

In the Poços de Caldas complex, a variety of hydrothermal veins, breccia fillings, stockworks, impregnations, etc. carry economic quantities of Th,REE,Zr,U and Mo. At Morro do Ferro (Cavalcante et al., 1979; Wedow, 1967; 300 Tt REE_2O_3, 35 Tt ThO_2, by-product Ti-magnetite), the mineralization is in a steeply dipping magnetite stockwork to massive vein, cutting altered phonolite and nepheline syenite. Th and REE are finely dispersed in colloidal form in the clay, and a lesser part of them is hosted by the magnetite. At present, the orebody is interpreted as being initially hydrothermal (magnetite, minor allanite, bastnaesite, cerianite and thorogummite fringed by a K-feldspar alteration envelope) and subsequently decomposed and further remobilized during lateritization.

The numerous though small "caldasite" (=mixture of zircon and baddeleyite) veins in the Poços de Caldas plateau follow fractures and small faults in phonolite and nepheline syenite. With rare exceptions all the veins are extremely shallow and developed in saprolite, so that the character of "caldasite" veins in the unweathered rocks is not well understood. Guimarães (1948) reported several thin veins of zircon in almost unaltered rocks and abundant accessory eudialyte in nepheline syenite at a depth of 85 m under the Serrote "caldasite" mine (Fig. 33-21). The "caldasite" is a very inconspicuous (except for its considerable specific weight) gray or hematite-pigmented material, the fragments of which are embedded in residual clay filling the vein structures. Crystalline, light-green zircon occasionally coats the fractures. The deposits of caldasite are generally interpreted (e.g. Ellert, 1978) as due either to the hydrothermal remobilization of Zr (and trace U) from zircon, eudialyte, astrophyllite and other minerals that are accessory in the host alkaline rocks, or to an exclusively supergene remobilization during lateritization. A combination of both processes is most likely to be the case.

In the presently exploited Cercado and Agostinho camps, Poços de Caldas complex, a low-grade U-Mo mineralization is superimposed on nepheline syenite and phonolite as well as the volcaniclastics and sediments in their roof. Maroon fluorite, zircon, uranothorianite, pyrite, coffinite, pitchblende, torbernite, jordisite and ilsemanite are disseminated or form impregnations in hydrothermal breccias, stockworks and veins, fringed by argillized and silicified hosts (Cavalcante et al., 1979).

In the Magnet Cove alkaline complex, Arkansas, thompsonite (zeolite) veins enriched in Be (up to 0.02% Be) have been reported (Ericson and Blade, 1963). At the Magnet Cove rutile deposit,

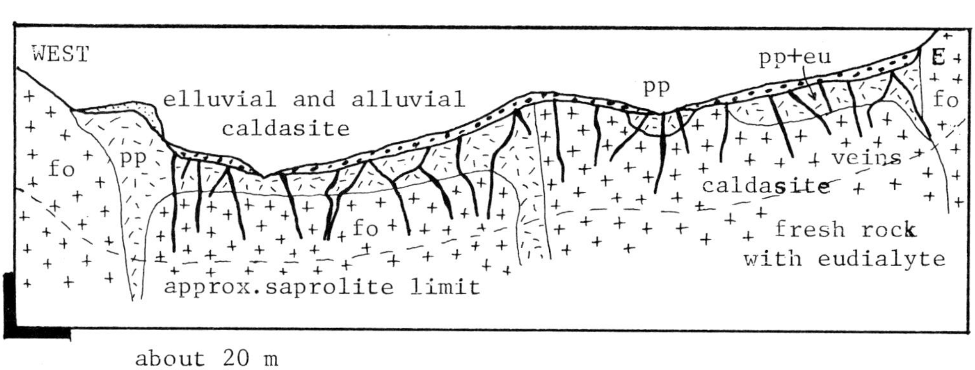

Fig. 33-21. Cross-section of the Serrote caldasite mine, Poços de Caldas Plateau, M.G., Brazil. pp=porphyritic phonolite; fo=foyaite (fine crystalline nepheline syenite); pp+eu=phonolite with eudialyte. Modified after Guimarães (1948).

feldspar, carbonate, rutile veins and masses occur in brecciated and altered phonolite. There is about 8 Mt of 3% TiO_2 ore (Fryklund and Holbrook, 1950), too low-grade to be profitable to mine, except in the soft saprolite. The rutile concentrate contains 2.2% Nb and 0.6% V. Another small Ti-oxide and Mo occurrence in the Magnet Cove complex is in microcline, albite, apatite, brookite, sphene, pyrite and molybdenite veins. The veins are up to 170 cm wide and 13 m long, N.W.-trending and hosted by a jacupirangite dike (Holbrook, 1948).

(b) Unaffiliated veins

The high-grade but thin Ag,Au, minor Pb,Zn fissure veins in the Horn Silver mine in southern British Columbia (Currie, 1974), are hosted by the alkaline intrusive rocks of the Kruger Mountain Pluton. The character of the veins, however, suggests that they are members of the hydrothermal aureole of the adjacent calc-alkaline batholiths.

6. MINERALIZATIONS IN THE EXOCONTACT OF ALKALINE INTRUSIONS

There is a considerable variety in the alkaline exocontacts and metallic mineralizations they contain. Empirically, it seems appropriate to subdivide the contact aureoles into (a) those that show distinct effects of the alkaline alteration or metasomatism (fenitization) and (b) those lacking fenites and showing no contact effects at all, or containing a variety of products due to isochemical contact (thermal) metamorphism.

(a) Fenitized exocontacts

Fenitization is an alkaline metasomatism, during which substantial amounts of Na or K and often Al and Fe are added to the original wallrocks from the alkaline intrusion, and silica is removed. Most varieties of alkaline intrusions (described in Sections 33-5,6,8,10) cause fenitization. The dominantly carbonatic Fen alkaline complex in southern Norway is the type area, and the memoir by von Eckermann (1948) describing the Swedish Alnö Island alkaline complex, is considered a classical paper on the subject. Von Eckermann demonstrated the zoned nature of fenites in Alnö, and his zones overprinted on Precambrian metamorphics are frequently quoted, mostly in the European literature (from outside towards the alkaline intrusion): (1) thermal shock zone, a shattered zone showing the effects of thermal metamorphism and minor metasomatism and comprised of feldspar, alkaline amphibole and pyroxene, carbonate, etc. fracture and breccia fillings; (2) a zone marked by spreading of the same minerals inward from fractures into their wallrocks and breccia fragments. Relic textures and quartz in the partly replaced rocks may still persist; (3) inner zone form, composed of microcline-aegirine, albite-riebeckite, or similar assemblage devoid of quartz. Original rock structures are often obliterated, and a massive rock of syenitic appearance forms in their place. The latter metasomatites are often indistinguishable from magmatic alkaline rocks.

The petrography of a fenite is strongly influenced by the character of the rocks intruded by the alkaline complex. The fenitization envelope around the Callander Bay alkaline complex, Ontario (Currie, 1974) superimposed on early Precambrian granitic rocks, starts to be noticeable about 650 m from the complex, and its zoning is close to the Alnö model. Gabbroic host rocks intruded by nepheline-dominated complexes are usually altered into rocks of ijolitic (=nepheline and pyroxene) composition, as around the Lackner Lake complex, Ontario; Parsons (1961). A considerable amount of iron seems to have been set in motion during fenitization of mafic precursors and Ti-magnetite is commonly found as disseminations, anastomosing veins, masses, etc. in ijolites (e.g. near Nemegos, Ontario).

Intensive fenitization tends to obscure the boundary between the truly magmatic crystallizates of the core alkaline suite, and fenitic metasomatites. Some practically indistinguishable syenites occur on both sides of the contact. Thin screens or slices of alkaline magmatites, moreover, frequently occur in many fenitized exocontacts. Fenitized remnants of former roofs to alkaline complexes may also occur within circular intrusions, where their metasomatic nature is often unrecognizable.

Although the majority of fenites described in the literature are situated around a known intrusion in the centre, several occurrences of fenites where such intrusion is missing have been recently described (e.g. from the Sudbury area, Ontario; Siemiatkowska and Martin, 1975). Presumably, the source of the alkaline fluids and probably the intrusion is at a depth and not yet unroofed. Metallic mineralizations in fenitized aureoles have the form of (i) anomalous metal enrichment or disseminated (scattered) ore minerals or (ii) hydrothermal ore veins in fenitized aureoles. Examples of mineralizations:

Zr enrichment

Ordovician shales at the contact with agpaitic syenite and phonolite at Mount St.-Hilaire, Quebec (Currie, 1974) have been sericitized in the inner contact zone, and converted to a Zr and Ti-enriched granular aggregate of albite, arfvedsonite and narssarsukite in the outer zone. A portion of the fenite aureole of the Lovozero complex is enriched in Zr, accumulated in accessory vlasovite (Vlasov et al., 1959).

Ti-enrichment

Many fenites are enriched in titanium, usually in the form of disseminations, lenses, veins and veinlets of rutile (e.g. Siemiatkowska and Martin, 1975). The analytically determined whole rock TiO_2 content is usually low, far from approaching the economically interesting threshold. In tropical areas, however, such rutile can be residually enriched in saprolite or reworked into placers. Rutile from fenites has usually high Nb and V trace content.

V enrichment

In Wilson Springs, Arkansas (Hollingsworth, 1967; D.R. Owens, guided tour, 1978), a major mine is producing vanadium from the fenitized exocontact of the Cretaceous Potash Sulfur Springs nepheline syenite intrusion. The Ordovician host rocks originally of chert, sandstone and impure carbonate composition, have been extensively fractured and converted into diopside, hedenbergite, lesser aegirine contact pyroxenite (alkaline skarn), to wollastonite tactite and metaquartzite. The latter is locally potash feldspathized. Argillic alteration is widespread on the fringe. Several alnöite dikes (alkaline lamprophyre) intrude the exocontact.

The vanadium ore (about 4.5 Mt ore with 1% V_2O_5) contains no megascopically recognizable minerals. In the unoxidized ore, vanadium is bound in rutile and in a variety of submicroscopic V minerals. There is a high scandium content. White, granular and veinlet apatite serves as a megascopic indicator of the V ore, that is richest in the calc-silicate hosts. A considerable proportion of V is contained in saprolite, in which it is believed to be bound in montmorillonite and goethite (also possibly montroseite).

Nb-U enrichment

In the Manitou Island alkaline complex, Ontario (Lumbers, 1971), a 568 m.y. alkaline syenite and pyroxenite intrusion is emplaced in fenitized Precambrian granitic gneisses. The inner zone of fenitization is composed of coarse K-feldspar and aegirine rock, intruded by thin crescentic carbonatite seams, and containing abundant thin feldspar-calcite veinlets. The carbonatite and in places the adjacent granite contain disseminated uranian pyrochlore.

Hydrothermal ore veins in fenitized aureoles

Thin but persistent fissure veins with a variety of mineral fillings (albite, hematite, aegirine, K-feldspar, calcite, quartz, bastnaesite, monazite, etc.) are common in the exocontact along the N.E. side of the Ilímaussaq intrusion (Hansen, 1968). Some veins carry from 60 to 4,500 ppm Th and 17 to 1,500 ppm U. U and Th-bearing veins have also been recorded along the western contact of the Poços de Caldas complex. In the aureole of the Magnet Cove complex, numerous but small quartz, brookite, lesser rutile and taenolite veins are well known, particularly to the mineral collectors.

Siemiatkowska and Martin (1975) described an uneconomic ankerite vein with scattered sphalerite from the fenitized breccia at Kusk Lake near Sudbury.

(b) Exocontacts lacking fenitization

Cordierite-andalusite, mica, hornblende-diopside, and biotite hornfelses, are developed along the contact of Devonian sediments and volcanics and the Lovozero alkaline pluton (Semenov, 1974). Small bodies of massive pyrrhotite were mined in a similar association along the Khibiny massif contacts.

In exocontacts of alkaline complexes rimmed by gabbro and emplaced into "greenstone belts", veins, veinlets and stringers of chalcopyrite are fairly common (e.g. Herman Lake, Ontario; Currie, 1974). Near Chipman Lake, Ontario, chalcopyrite is relatively common in gabbros rimming a syenite core. The copper has most probably been derived from the mafics. Veins of massive pyrite and pyrrhotite with erratic galena, sphalerite, chalcopyrite and arsenopyrite, are abundant in the aureole of the deformed and metamorphosed Ice River Complex, British Columbia (Currie, 1974).

7. DEFORMED AND METAMORPHOSED ALKALINE COMPLEXES

Several such complexes have been described in the literature (the nepheline syenite gneisses in high-grade metamorphic terrains have been reviewed in Chapter 29 and Volume 2, and are not consiedered here). The Paleozoic Ice River alkaline complex in the Rocky Mountains, British Columbia, was deformed by several events between 392 and 220 m.y. ago. The complex, however, acted as a rigid kernel, around which were wrapped much more intensively deformed metasediments (Currie, 1974).

In the Red Wine-Letitia alkaline association in Labrador described in greater detail in Volume 2 (Currie, 1974), remnants of a 1.5 b.y. old central complex comparable with Ilímaussaq, remnants of smaller intrusions, dikes and fenites, have been dismembered, penetratively deformed and metamorphosed. Patches of ore minerals common in undisturbed alkaline complexes (e.g. eudialyte, neptunite, pyrochlore, Be-silicates) occur throughout the area. The most striking feature appears to be a post-magmatic, syntectonic and syn-metamorphic recurrence of alkaline metasomatism, that overprint and/or replace earlier alkaline and non-alkaline rocks.

Syenitic metasomatites in the North Red Wine complex contain eudialyte and often larger amounts of joaquinite $Ba_2Na,Ce_2Fe(Ti,Nb)_2Si_8O_{26}(OH,F)_2$. At the Mann-1 locality, a probable meta-fenite lacking an associated intrusion, Be minerals occur in metasomatic pegmatites and in their vicinity (Evans and Dujardin, 1961).

8. WEATHERING-GENERATED RESIDUAL METAL AND MINERAL CONCENTRATIONS

Deep tropical weathering superimposed on bedrock mineralizations in alkaline complexes containing relatively stable ore minerals, results in their separation and isolation in the soft saprolite or laterite, from which they can be easily recovered. The alluvial caldasite deposits of Brazil are perhaps the best known example. Elsewhere, rhabdophane and bastnaesite have been detected in small amounts in a relic regolith over an alkaline massif located in the Yenisei Range, Siberia (Semenov, 1974). The regolith probably dates from the Cretaceous period.

Residual bauxite deposits frequently form over nepheline syenite and phonolite, and examples include the bauxites topping the Los alkaline complex near Conakry, Guinea; bauxites formed on phonolites and nepheline syenites in the Poços de Caldas complex, Brazil (reserves 200 Mt ore with 56% Al_2O_3; Fig. 33-22); bauxites resting on Cretaceous nepheline syenite, Lavrinhas municipio, Brazil (450 Tt ore with 56-58% Al_2O_3; Fonseca et al., 1979) and others. The Eocene bauxites over Cretaceous nepheline syenite in Arkansas (Gordon et al., 1958) have a high trace Nb content.

Small accumulations of earthy Mn-oxides sometimes cut by infiltration psilomelane veins, formed locally (e.g. near Cercado in the Pocos de Caldas complex; 2 Mt of 35-40% MnO_2 ore; Cavalcante et al.,1979).

9. MINERALIZATIONS RESULTING FROM SEDIMENTOGENIC REWORKING

All the residual commodities listed above also have proximally redeposited equivalents. The residual caldasite is reworked into proximal colluvial and alluvial placers, and the fragments become rounded ("favas"). Dissolution and reprecipitation of caldasite in soil or alluvial profiles results in "refinement". Almost pure, microcrystalline baddeleyite nuggets form.

Other resistate minerals from alkaline rocks sometimes accumulate in alluvial placers. Semenov (1974) mentioned pyrochlore and zircon, contributed to several Uralian streams by the Miask and Vishnevogorsk alkaline complexes. Resedimented bauxite is prominent in the Eocene sediments of Arkansas.

33.5.4. Examples of nepheline syenite-dominated complexes

The three examples of complexes described below, the Ilímaussaq, Lovozero and Poços de Caldas, illustrate two principal varieties, the layered and the non-layered ones. Table 33-7 contains a brief listing of additional examples of mineralized complexes.

ILÍMAUSSAQ, SOUTHERN GREENLAND

The Ilímaussaq intrusive complex is a member of the Gardar alkaline province and was emplaced 1,020 m.y. ago in early Precambrian basement granites under a roof of late Proterozoic lavas and sandstones. Although Proterozoic, it is described in this volume because of its unprecedented example value. The intrusion (Ferguson, 1964; Figures 33-23, 33-24) covers an area of approximately 150 km^2. A dicontinuous margin of chilled augite syenite was the earliest igneous member to crystallize, and it probably represents an example of rocks that compositionally are the closest to the parent magma for the rest of the intrusion.

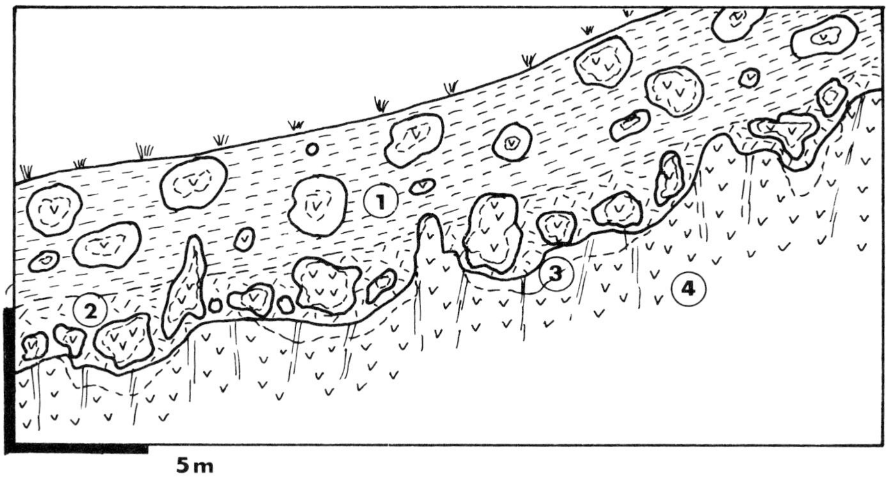

Fig. 33-22. "In situ" and colluvial lateritic bauxite on steep slopes N.W. of Poços de Caldas (city), formed over Cr phonolite. (1) Downslope transported red laterite with saprolitic and fresh allochthonous phonolite boulders; (2) in situ red and yellow mottled bauxite, local patches and coatings of black Mn oxides. In situ phonolite blocks; (3) leached phonolite; (4) fresh phonolite. From LITHOTHEQUE.

The augite syenite was followed by a central stratified series, composed of several saucer-shaped units. These include naujaite (= a poikilitic sodalite-nepheline syenite), sodalite foyaite (=intergranular-textured nepheline syenite) and a heterogeneous syenite believed to have formed by crystallization from the roof downward (Engell et al., 1971). Kakortokite which have already been discussed, have formed contemporaneously with the naujaite by bottom crystal accumulation. Lujavrites (trachytoid nepheline syenites) were subsequently intruded between the naujaite and the kakortokite, and their emplacement caused widespread metasomatism, hydrothermal alteration and epigenetic vein formation in the earlier rocks. The complex is cut by two N.E.-trending normal faults, that expose the intrusion at three progressively deeper crustal levels (Upton, 1974).

The Ilímaussaq complex is an outstanding geochemical and metallogenic anomaly, greatly enriched in Zr,REE,Nb,Ta,Th,U and Be. To call some of these anomalies "ore deposits", however, would require a considerable increase in the demand for Zr,REE and Nb, the most abundant metals there, to offset the logistic problems resulting from their remote location. In the economic geologic literature, Ilímaussaq is noted mainly for its explored low-grade uranium potential which, on the global scale, is only a moderate accumulation (43 Tt U). The spectacularly banded sequence of kakortokites which is

Table 33-7. Phanerozoic alkaline igneous provinces and complexes dominated by nepheline syenite

LOCALITY	GEOLOGY	MINERALIZATION	REFERENCES
Lake Nipissing Prov. Ontario, Canada	560 m.y. 5 nearly circular centr. compl. 3 km diam.; nephel.syenite minor carbonatite screens, lamprophyre dikes; empl.into fenitized PCm granite gneiss	U-pyrochlore dissem. in carbonatite; Manitou Island 25,540 t Nb/0.48-0.6%, 1,050 t U/0.035%	Lumbers (1971)
Monteregian Hills Prov., Quebec, Canada	190 km E.-W. string of small intrus., 110 m.y.; nephelinite, gabbro, nephel.syen., minor carbonatite	acces.pyrochl., enigmatite and rare minerals dissem.in neph. syen. and gabbros; economic only in carb.(see Oka)	Currie (1974)
Magnet Cove and Potash Sulfur Springs compl., Arkansas	Cr (95 m.y.); ring-dike compl.emplaced into PZ sedim.(shale, chert); 800 m wide fenitiz. aureole; ijolite and carb.core envel. by trach., phonol., neph.syen., neph. pyroxenite; dike rocks	qtz.-brookite + Mo veins in exocont.; feldsp., carb., fluor. Nb-rutile veins in endocont. (54Tt Ti, 2.2Tt Nb, 600t V); V at contact (Wilson Springs) parent rocks to Ark.bauxite	Erickson and Blade (1963)
Wind Mountain, Otero Co., S.E. New Mexico U.S.A.	circul.dome, laccol.of neph.syen. intr. into Pe limest., shale; dikes of melanosyen., nephel.syen. pegmatite	up to 20% eudial.in nephel. syen.dikes; Zr silic.with egirine, riebeckite, analc.at dike contacts; erratic Be enrichement	Warner et al. (1959)
Khibiny Massif, Kola Peninsula, N.W. U.S.S.R.	1,327 km^2, 290 m.y., 40 km diam. ring, 7 distinct intrus.phases; system of overlapping rings, domin. by aegirine and nephel. syen.empl.into Arch.gneisses; minor ijolite, alkal.dikes	large apatite and sphene dep. peneconc.with ign.layer., has high eudial.content; 2.7Bt of mater.with 18% P_2O_5, 0.4% REE_2O_3 (=935Tt REE); Yukspor Mt., W.N.W., up to 200 m long REE, Ta-Nb, Th lovchor.vein set	Gerasimovsky et al.(1974) Afanasyev (1937)

Lovozero Massif, Kola Peninsula, N.W. U.S.S.R.	659 km^2, 266 m.y. circular composite intrusion emplaced into Arch. granite-gneiss; layered, concentrically zoned, dominated by varieties of neph.syenite; dike and vein rocks	1)layered eudialyte neph.syen. 150-500m thick,aver.1.36% ZrO_2; 2)eudialytite layers cont.aver.5.8% Zr;0.5% Ta+Nb; 1.1% REE and estim.resource of 50Tt Ta,500Tt Nb,500Tt REE, 20Mt Zr; 3)layered loparite urtite cont.about 0.03% Ta, 0.3% Nb, 1.25% REE,0.03% Th (est.res.5Mt REE,30Tt Ta, 6.5Mt Nb; additional REE,Nb, Th,Ta is in steenstrup.rocks	Vlasov et al. (1959)
The Urals alkaline prov.(e.g.Vishnevogorsk,Miask) U.S.S.R.	several massifs usually conformable with struct.grain of enclosing PZ metam.;268 m.y.; neph. syen.,broad hybrid aureoles	widespr.zircon,pyrochl.,sphene,britholite in albitized and carbonatized neph.syen.and fenite; Nb,REE,Th produced	Semenov (1974)
Eastern (Soviet) and Mongolian Tuva alk.prov.(e.g.Sangilen,Botogol',Numurgin,Udzhigin complexes)	D_3-MZ, several tens of small intrus.:neph.syen.,urtite,phonol., alk.syen.,peralk.granite,rare carbonatite; fenite aureoles	Zr,REE,Nb,Ta,Th dissem.in eudialyte rocks; minor pyrochlore in carbonatite; Be asoc.with peralk.granites	Kuznetsov (1967)
Poços de Caldas Complex, M.G., Brazil	almost circular complex domin.by phonol. and subvolc.nephel.syen. (Cr) empl.into PCm gneiss basem. Several intrus.phases but compos. monotonous; local potassic rocks	Zr "caldasite" veins,resid. and placers cont.about 100Tt Zr and 391t U; Th and REE in Ti-magn.stockw.have 255Tt REE and 30Tt Th; U-Mo veins and stockw.in altered complex roof cont.about 8Tt U,1Tt Mo, 1.5Tt Th; residual bauxite and Mn ox. on neph.s.,phonol.	Cavalcante et al.(1979)

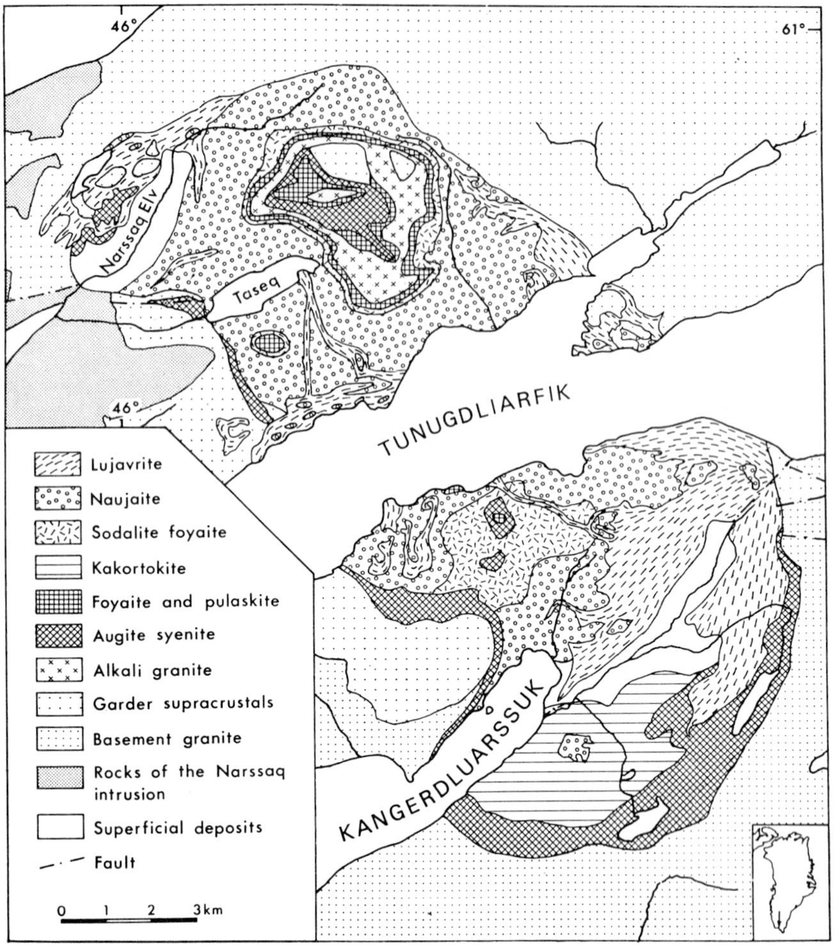

Fig. 33-23. Simplified geological map of the 1.02-1.18 b.y. Ilímaussaq alkaline complex, southern Greenland. After Ferguson, from Nielsen (1973). Courtesy of the Geological Survey of Greenland.

over 400 m thick, represents the most outstanding metal-bearing unit containing some 38.2 Mt Zr and 3.78 Mt Nb, that are uneconomic to work at the present time. Additional Zr+Nb resources exist in a thick layer of a marginal pegmatite.

The lujavrites, as a whole, are enriched in REE,Nb,Ta,Th and U concentrated in a variety of complex accessory minerals (pyrochlore, epistolite, steenstrupine, monazite, thorite, etc.) and host the uranium ore zone at Kvanefjeld. Additional concentration of the same metals is associated with the late stage hydrothermal activity that produced numerous albite, analcite, and zeolites-dominated

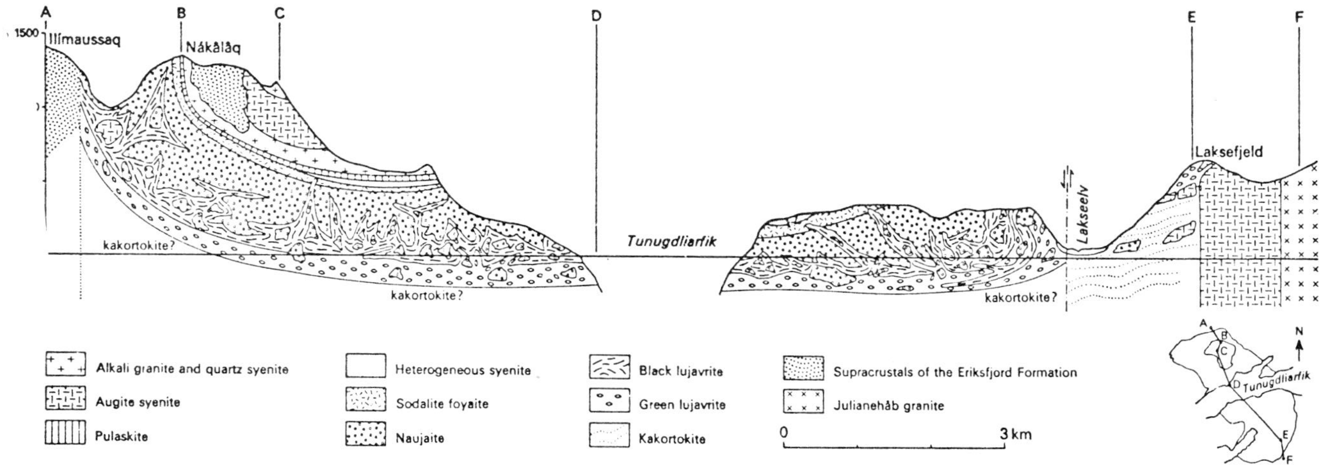

Fig. 33-24. N.N.W-S.E. section across the Ilímaussaq alkaline complex, southern Greenland. After Ferguson, from Allaart (1973).
Courtesy of the Geological Survey of Greenland.

metasomatites and veins (Alaart, 1973). Beryllium is irregularly accumulated in several types of hydrothermal veins and metasomatites containing chkalovite, tugtupite, epididymite and other minerals. A mere 180 t BeO is contained in the only outlined deposit (Taseq slope) in the N.W. portion of the complex (Engell et al., 1971).

LOVOZERO ALKALINE COMPLEX

Lovozero is a member of the Kola Peninsula alkaline province, north-western U.S.S.R. It is a large (650 km^2) composite intrusion dating from the Carboniferous age (266 m.y.; Fig. 33-25), emplaced into the surrounding Archean granite-gneiss terrain (Vlasov et al., 1959; Gerasimovsky et al., 1974). The intrusion has sharp contacts, a stock-like shape in depth and it is concentrically zoned. The upper part of the complex consists of a laccolith-like differentiated body.

The complex is a product of four successive magmatic phases. Rocks of the first phase (metamorphosed and metasomatized nepheline syenites) are preserved mostly as rafts and xenoliths. Phase 2 is represented by a rhythmically layered sequence with repetitive stratigraphy (from bottom to top: urtite, foyaite, lujavrite). The rocks of Phase 3 comprise coarsely crystalline layered lujavrites with a prominent zone of eudialyte lujavrite in the central part of the massif. Phase 4 consists of alkaline dikes (monchiquite, camptonite, tinguaite).

Rocks of the Lovozero complex are extremely agpaitic (the $Na_2O+K_2O:Al_2O_3$ ratio is 1.4), dominantly sodic, and anomalously enriched in Zr (0.48% ZrO_2 is the whole complex average). The typical rare-metal minerals in the Lovozero complex are members of the eudialyte and lomonosovite groups, loparite, REE-apatite and minor lovozerite, ramsayite, lamprophyllite, murmanite. They are present either as regularly distributed accessories in the layered rocks (0.01-0.5%), or as locally accumulated minor minerals (up to 5%). Eudialyte forms almost monomineralic rocks locally.

Although Rundkvist, ed. (1978) suggested that the Complex may contain some 25% of the world's reserves (or resources) of Ta and 12% of the world's Nb, it is not clear what type of ore is actually being mined or prepared for production. The most likely industrial candidates are sections with a high content of eudialyte, loparite-rich urtite horizons or steenstrupine and lovozerite-bearing lujavrites. Lovozero appears to be a prime candidate for a large-scale, complex production of multiple commodities.

→

Fig. 33-25. Lovozero alkaline complex, geology. 1=Q sediments; 2=pegmatites; 3=poikilitic syenites; 4=rocks of the differentiated complex; 5=rocks of the eudialytic and porphyritic lujavrite complex; 6=nepheline syenite fenites; 7=D, augite porphyries, tuff, quartzite, sandstone; 8=gneiss and granite gneiss. From Vlasov et al. (1959). Courtesy of Oliver and Boyd, Edinburgh.

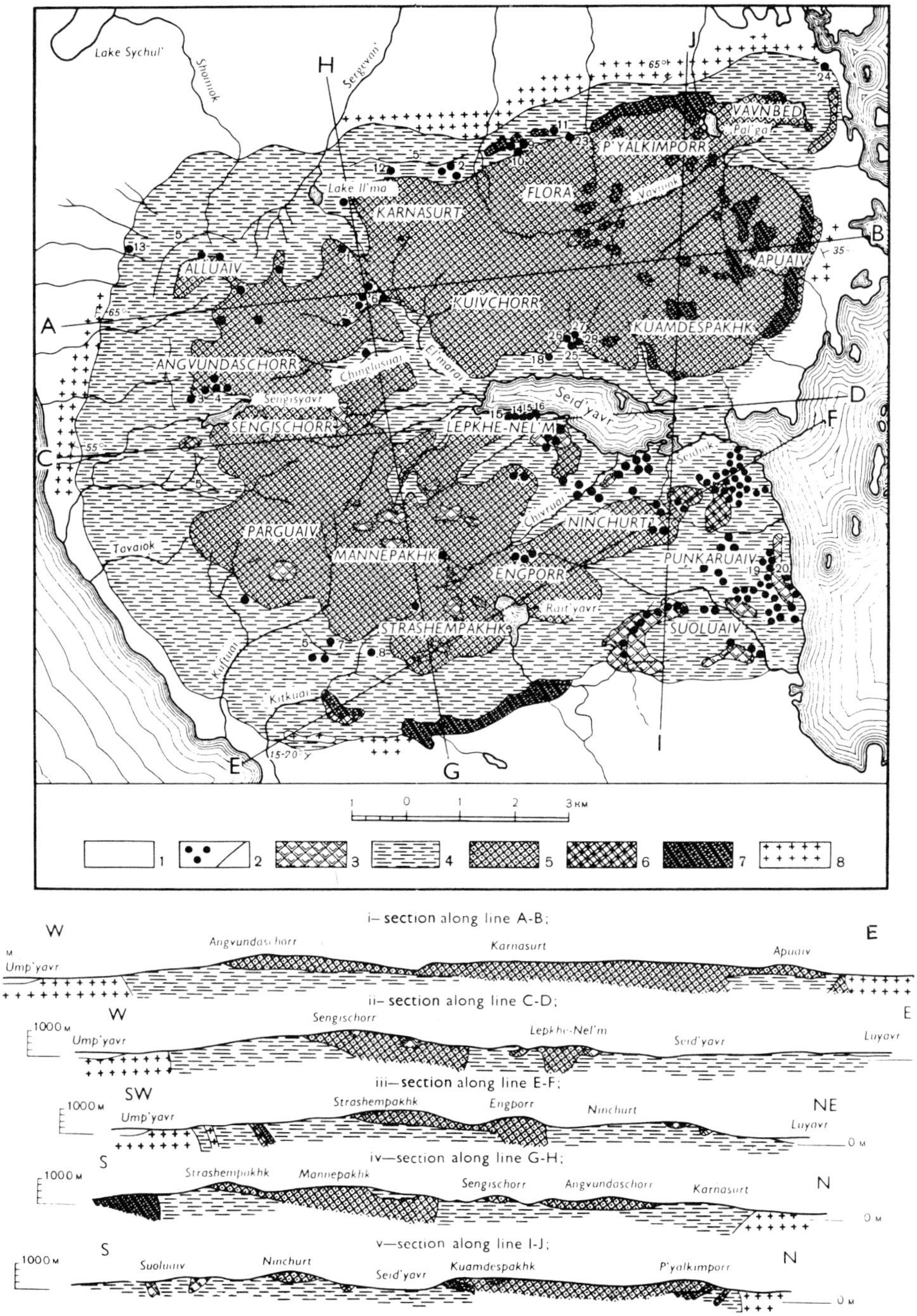
Lake Sychul'
Sergevan'
Lake Il'ma
KARNASURT
FLORA
P'YALKIMPORR
VAVNBED
ALLUAIV
KUIVCHORR
APUAIV
KUAMDESPAKHK
ANGVUNDASCHORR
Sengisyavr
SENGISCHORR
LEPKHE-NEL'M
Seid'yavr
NINCHURT
PARGUAIV
MANNEPAKHK
ENGPORR
PUNKARUAIV
STRASHEMPAKHK
SUOLUAIV
Tavaiok
Kitkuai
1 0 1 2 3 км
i– section along line A-B;
ii– section along line C-D;
iii– section along line E-F;
iv– section along line G-H;
v– section along line I-J;

Agpaitic pegmatites, hydrothermal veins, and fenite-hosted mineralization, although of considerable variety and scientific importance, treated in detail in the book of Vlasov et al. (1959), appear to have little practical value at present.

POÇOS DE CALDAS COMPLEX, MINAS GERAIS, BRAZIL

This complex (Fig. 33-26) forms, geomorphologically, a dissected plateau with a medium altitude of about 1,200 m, surrounded by an almost circular range of hills reaching 1,500-1,600 m. It is situated 250 km from São Paulo in the Cretaceous alkaline province on the fringe of the Paraná Basin, and has an area of about 800 km^2 (Cavalcante et al., 1979). The complex was emplaced in a Precambrian gneiss basement presumably under a thin roof of its own ejecta (tuffs, breccias, ankaratritic lavas) and associated sediments some 87-60 m.y. ago. The earliest phase produced ankaratrites, the following phases generated phonolites and intrusive hypabyssal nepheline syenites (foyaites, tinguaites).

Compositionally, the Complex is quite monotonous and nepheline syenite-phonolite account for 90% of the rocks. No distinct intrusive layering is present, and the phonolites are locally potassic. A pseudoleucite-rich facies occurs in places. Zircon and eudialyte appear to be widespread accessory minerals, but are mostly "invisible" because of the fine grain of most rocks and a deep weathering crust. Coarse eudialyte syenites and pegmatites are locally developed along the northern rim of the complex.

The Poços de Caldas district contains quite a variety of Zr,Th,U,REE,Al and Mn ores. These have all arisen as a result of local accumulation of the originally rock-forming trace and major elements by a variety of superimposed agents. Probably the best known example and a unique case is the "caldasite" vein, residual and placer mineralization. It is widespread (containing close to 100 separate workings) and represents some 230 Tt of 50-75% ZrO_2 and 0.2% U_3O_8 concentrate mined in the past and present in known reserves (Tolbert, 1966; Cavalcante et al., 1979). The Morro de Ferro, a magnetite vein and stockwork with superimposed Th and REE, contains some 30 Tt Th and 255 Tt REE.

The U-Mo veins, stockworks, and disseminations in hydrothermally altered nepheline syenites and roof rocks at Agostinho and Cercado, represent the first and only exploited and complexly treated Brazilian uranium deposit existing at present. The measured and inferred reserves are about 7,939 t U, 1,500 t Th, 1,000 t Mo and 64 Tt Zr (Ramos and Fraenkel, 1974).

Good quality bauxite is widespread, occurring in numerous small deposits formed, in situ, over nepheline syenite and phonolite or transported clays. The reserves are estimated at between 70 and 200 Mt and they support an operating smelter. Manganese oxides are sometimes recovered from lateritic profiles as a by-product of bauxite mining. The Mn deposit near Cercado contains about 480 Tt Mn in a

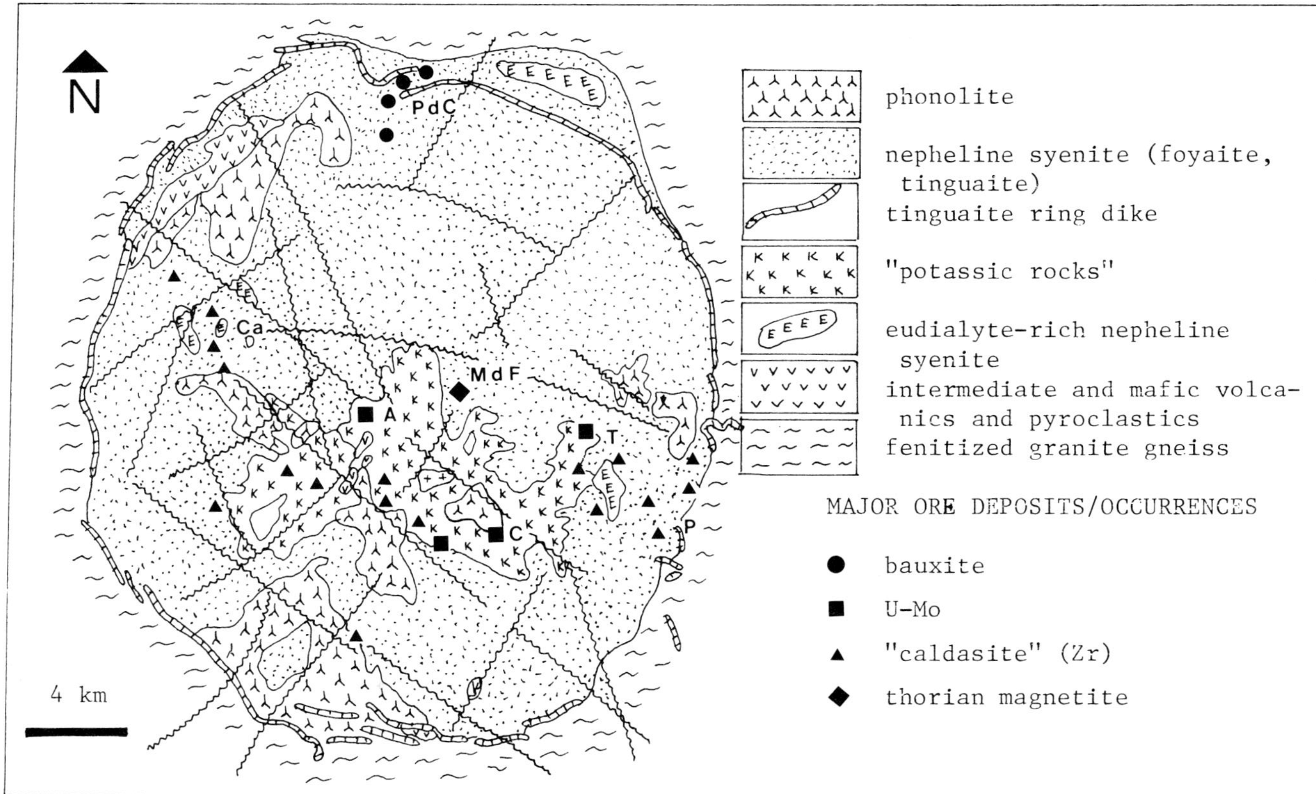

Fig. 33-26. Poços de Caldas alkaline complex, Minas Gerais, Brazil, showing locations of major ore occurrences. PdC=Poços de Caldas; Ca=Cascata; MdF=Morro de Ferro; T=Taquarí; P=Pocinhos; A=Agostinho; C=Cercado. The "potassic rocks" are phonolites with pseudoleucite and also probably remnants of fenitized roof rocks. From LITHOTHEQUE, compiled from published maps.

horizon 0.5-1 m thick, grading around 24% Mn.

33.6. ALKALINE-ULTRAMAFIC ASSOCIATION

33.6.1. General

Alkaline-ultramafic association is widely recognized in the U.S.S.R. (e.g. Rundkvist, ed.,1978) and it usually includes peridotite, pyroxenite, alkaline gabbros, alkaline syenites, carbonatite and kimberlite. In this section, only those complexes actually composed of major peridotite or pyroxenite units, are treated. Carbonatites or kimberlites may be present, but the dominantly carbonatite complexes are treated in Section 33.8., and the kimberlites in Section 33.9.

The mode of occurrence of alkaline-ultramafic intrusions is comparable with that of the nepheline syenite-dominated complexes, except that layered intrusions comparable with Ilímaussaq or Lovozero are not known. Circular or elliptical zoned intrusions are most characteristic and the fully developed ones have usually a dunite or peridotite core, fringed by pyroxenite, alkaline gabbroids and alkaline syenites. The fringe syenites and a variety of associated or substituting rock types (norites, "aplites", etc.) are genetically controversial and are currently interpreted as orthomagmatic differentiates by some and metasomatites by others. In the latter interpretation, the syenites would coincide with the most proximal zone of fenitization.

33.6.2. Metal geochemistry and metallogeny

Dunites and peridotites lacking alkalies are petrochemically neutral and common to alkaline as well as tholeiitic and calc-alkaline families. As a consequence, the ultrabasics considered alone display predictable trace metal anomalies (Cr, Ni, platinoids) and ore occurrences influenced by such anomalies. The nickel laterites represent an example. However, pyroxenites (e.g. jacupirangite; Upton, 1967) and in particular gabbros, show distinctly alkaline tendencies, although non-alkaline pyroxenites and gabbros also occur. The alkaline mafic rocks are enriched in the rare metals characteristic of alkaline complexes (REE,Zr,Nb,Th), but are depleted in Ni,Cu and S, the elements that normally combine to form a plethora of mineralizations in tholeiitic and calc-alkali mafic suites. Cu and Ni occurrences are consequently rare in alkaline gabbros.

Iron and titanium, however, had been mobile in several phases of development of alkaline pyroxenites and gabbros, and these rocks host a variety of Ti-magnetite, Ti-oxides, ilmenite, perowskite, etc. mineralizations that sometimes constitute important mineralized provinces (e.g. at least 1.4 Bt Fe, 15 Mt Ti, by-product apatite, vermiculite, REE,Ta,etc. in the Caledonian alkaline province of the Kola Peninsula).

33.6.3. Major mineralization styles (Fig. 33-27, Tables 33-8, 33-9)

EARLY MAGMATIC MINERALIZATIONS

These are hosted by magmatic rocks and believed to have formed during the process of original magmatic crystallization as cumulate crystals or as a solidified intercumulus phase. Minerals accumulated in the form of schlieren, disseminations, etc. in inclusion-rich zones may have formed as the result of magma contamination. In the overall balance, magmatic "syngenetic" ores in alkaline-ultramafic rocks (except carbonatites) are of limited economic importance.

Dunites often have high trace Ni content (e.g. 0.29% NiO in the fresh dunite from the Santa Fé massif, Iporá Group, Brazil), but this is a silicate-bound nickel and Ni sulphides have rarely been reported. The dunite core of the Inagly Massif in Siberia (Rundkvist, ed., 1978; Fig. 33-28) contains Pt metals in Pt-Fe and Pt-Ir alloys, cooperite and laurite, associated with chrome spinelides. These minerals are enriched in nests, lenses and schlieren in the central part of the dunite core and, so far, only the secondary placer deposits seem to have been of interest from an economic point of view.

Pyroxenites usually carry high accessory contents of Ti-magnetite, perowskite and sometimes (as in Africanda, Kola Peninsula) knopite (cerian perowskite). They are also the most common hosts to the large Ti-magnetite and apatite deposits, but most such deposits are metasomatites or epigenetic bodies emplaced after the pyroxenite solidification. The most notable exception where an industrial Fe-Ti deposit appears to have formed by a "syngenetic" magmatic crystallization from an immiscible liquid, is the mineralized jacupirangite (=alkaline pyroxenite) dike in Kodal, Norway (Bergstøl, 1972). Here, the Permian jacupirangite forms a dike 2 km long and 20 m wide emplaced in augite monzonite. Portions of the dike consist of dense intergrowths of Ti-magnetite (40%), apatite (18%) and ilmenite (8%) with aegirine, minor biotite, and amphibole. The textures are typical of magmatic crystallization.

Gabbros. Low-grade schlieren and disseminations of Ti-magnetite and ilmenite are common in the middle Proterozoic (1.8 b.y.) alkaline-ultramafic intrusions in Kola and Karelia (north-western U.S.S.R.; Yudin and Zak, 1970). In addition to these lean ores, high-grade densely disseminated to massive ilmenite and Ti-magnetite ores form layers and lens-like seams up to 100 m long, conformable with igneous layering. In addition discordant vein-like bodies of massive ores are formed. Such ore is interpreted as being late magmatic (hysteromagmatic) injection, but the possibility of metamorphogenic upgrading of lower-grade magmatic ores cannot be excluded.

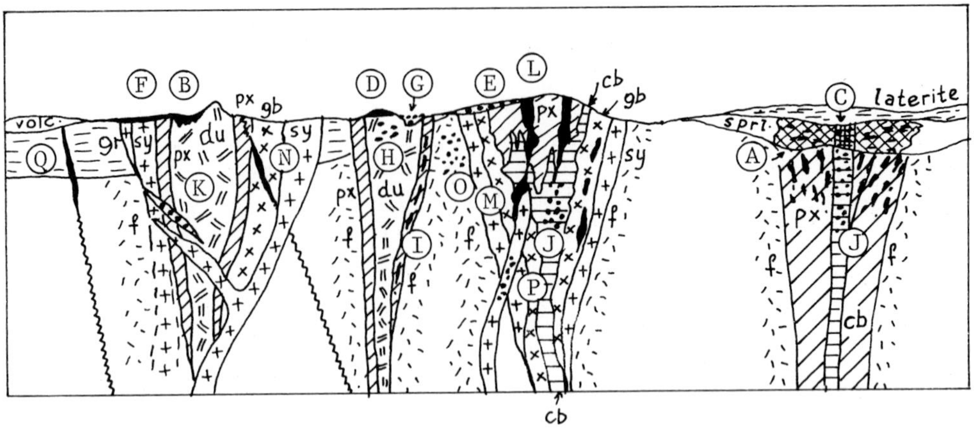

Fig. 33-27. Principal lithologic sub-associations, rock types and mineralization styles in alkaline ultramafic complexes (see Table 33-8 for an explanation of letter codes). gb=Gabbro (alkaline); px=alkaline pyroxenite; du=dunite; sy=syenite (magmatic and metasomatic); gr=granite; cb=carbonatite; f=fenitization; sprl=saprolite.

Pegmatites. Alkaline pyroxenites and gabbros commonly contain coarse ("pegmatitic") facies that are mineralogically identical with the rest of the host rock. Light (syenitic) pegmatites, usually corresponding in composition to nepheline syenites, often transect the mafic magmatic rocks. In the Inagli intrusion and elsewhere, such dikes carry small anounts of Zr,Nb, and REE minerals. Many pegmatites react with the mafic or ultramafic wallrocks and phlogopite or vermiculite is usually the only economically important product (as in Afrikanda, Kola Peninsula). Ti-magnetite, perowskite, chalcopyrite and other minerals may be present in small quantities and recovered as a by-product.

EPIGENETIC METASOMATITES AND VEINS SUPERIMPOSED ON ALKALINE PYROXENITES AND GABBROS

Most alkaline-ultrabasic complexes are polyphase, and the most volatile components (residual liquids, hydrothermal fluids) of the younger magmatic phases reacted with the older, solidified and usually fractured phases to produce a series of hybrids, metasomatites and hydrothermally-filled veins. The latter often carry important ore accumulations. The most common hosts to such ores are alkaline pyroxenites and the usual phases contemporary with the mineralization are nepheline syenites and carbonatites. The most common mappable hybrid rocks correspond petrographically to ijolites and urtites (nepheline-pyroxene rocks), to carbonate-pyroxene and to other

Table 33-8. Principal mineralization styles in alkaline-ultramafic intrusive complexes

	WEATHERING CRUSTS
(A)	Partly modified residual minerals (magnetite, ilmenite, anatase, leucoxene); about 20 Mt Ti (Tapira)
(B)	Residual and infiltration oxide and silicate Ni in laterite, saprolite (about 500 Tt Ni)
(C)	Residual apatite, francolite, enriched in REE, Th, etc.
(D)	Residual minerals in silicified ultramafics (birbirite); Pt
	PRODUCTS OF SEDIMENTOGENIC REWORKING
(E)	Fragmental ore and ore boulders in glacial drift (X Mt Fe)
(F)	Redeposited ferruginous (locally bauxitic) laterite
(G)	Alluvial placers (X0 t Pt)
	MAGMATIC "SYNGENETIC" MINERALIZATION
(H)	Scattered Pt metals in dunite
(I)	Magmatic "syngenetic" disseminations of Ti-magnetite, ilmenite, perovskite, apatite, etc. in pyroxenites and alkaline gabbros (X0 Mt Fe, X Mt Ti)
(J)	Disseminated Nb,REE,Th, etc. minerals in minor carbonatite bodies and screens
(K)	Mineralized pegmatites and their contacts (Zr,Nb,REE)
	EPIGENETIC METASOMATITES AND VEINS SUPERIMPOSED ON ALKALINE PYROXENITES AND GABBROS
(L)	Ti-magnetite, perowskite, apatite, etc. masses, veins, lenses in pyroxenite and alk. gabbro (Kovdora; 1.4 Bt Fe, 15 Mt Ti)
(M)	High-temperature silicate metasomatites (skarns)
(N)	Fracture filling hydrothermal veins
	ORES IN INTRUSIVE EXOCONTACTS
(O)	Fe,Ti,Nb, apatite, etc. disseminations, replacements in fenites
(P)	Radioactive syenite porphyry dikes (X Tt Th)
(Q)	Hydrothermal fissure veins (X0 Tt Th)

silicates mixtures. The common ore minerals are Ti-magnetite, ilmenite and perowskite. Apatite, phlogopite and vermiculite are the usual nonmetallic commodities.

In the Kovdora massif, Kola Peninsula, steeply dipping replacement veins or lenses of massive Ti-magnetite enveloped by an apatite-forsterite rock were emplaced in ijolite-urtite and pyroxenite (Borovikov and L'vova, 1962; Fig. 33-29). These, in turn, were intruded by a younger carbonatite, resulting in a mineralization strongly reminescent of Palabora, South Africa. By-product betafite, baddeleyite and minor copper, is recovered at Kovdora.

Similar mineralization but spreading over a smaller area is hosted

Table 33-9. Alkaline-ultramafic complexes, selected examples

LOCALITY	GEOLOGY	MINERALIZATION	REFERENCES
Libby, Montana U.S.A.	Cr? composite intrusion into Pt_{2-3} sediments; core of biotite pyroxenite with coarse-grained biot. and xenoliths of ultram.surround.by magnet.pyrox. ring dike;neph.syen.,phon.,dks.	perowskite conc. in vermiculite; magnet.pyrox. contains 15-21% Fe 0.13% V_2O_5; locally scattered accessory sphalerite	Boettcher (1967)
Powderhorn Complex S.W. Colorado U.S.A.	Cm; about 30 km^2 pear-shaped, multiphase compl. Pyrox.intrud. or replac.by nephel. or melil. rocks and carbonatite; intrud. into PCm gran.,fenitiz.aureole	vermicul.,Ti-magn.,perowsk. and apatite form dikes and irreg. lenses in pyrox.; 100 Mt ore with 6.5% TiO_2 and 11.7% Fe. Nb in carbonatite; Th widespr.in carb.,pyrox.,altered dikes, exocont. veins	Temple and Grogan (1965)
Gem Park complex Wind Mts.,Colorado U.S.A.	Cm; small funnel-shaped compos. intrus.;pyrox.,gabbro,minor lamproph.,neph.syen.,carbonatite,intr.PCm metam. Fenite envel. Believed underl.by large carb.	vermicul.,magn.;pyrochl.in carb. dikes; Nb,REE,Th miner.dissem. in fenite and anastomosing carbonatite dikelets	Parker and Sharp (1970)
Iporá Group, Paraná Basin fringe,S.W. Goiás, Brazil	Cr_3; 73 small intrus.,10 larger massifs; circul.to ellipt.intr. comp.of dunite nuclei surr.by pyrox.,alk.gabbro,neph.syen., syen. Some have Na-amphib.rich fenitiz.margin; intr.to PCm metam. and D_1 sandstones	Ni later.on dunite (100Mt ore with 1.5% Ni; largest Santa Fe massif, 21.14 Mt ore with 1.5% Ni	Danni (1976) Pena and Figueiredo (1972)

Tapira, Paraná Basin fringe, M.G. Brazil	Cr, elipt.intr.6 km diam., pyroxenite and perid.with small carbonatite, empl.into fenitiz. PCm metamorphics	dissem.primary apat., anatase, magn., phlogop.and perowsk., irregul.bodies and veins in pyrox. Actual prod.is from up to 100 m thick saprolite; 200Mt ore with 22% TiO_2 in pyrox., 113Mt ore with 1% Nb_2O_5 in carbonatite	Herz (1976) de Tarso, oral comm.(1980)
Kola Penins.early PZ alkal.ultram.pr. N.W. U.S.S.R.	14 plutons, 340-590 m.y., early dunite, pyrox., foll.by ijolite, melteigite, nephel.rocks	stock-shaped bodies of dissem. perowsk. and Ti-magn., high rare elements content	Yudin and Zak (1970)
Kovdora massif (member of above province)	D_2; concentr.zoned intr.with pyrox. and dunite core surr.by melil.rocks, outer ijol.zone and fenites; intr.by silic.carbont. Widespr.hybrid rocks	steep vertic.replac.veins and lenses of Ti-magn., apat., phlog. 1,300 m long in ijolite and pyroxenite; 708Mt ore with 35% Fe; 6.6% P_2O_5; 0.3% Zr; about 0.2% Nb and 0.01% Ta	Borovikov and L'vova (1962) Rundkvist, ed., (1978)
Afrikanda massif (member of above province)	pipe-like 400 m long body, peridot.core with erupt.dunite brec. transected by nephel-pyrox.pegm.	Ti-magnet. and Ce-perowskite, phlogop.veins, veinl.and metas. in pyrox.near pegmat.dikes; 626Mt ore with 13% TiO_2	Sokolov and Grigor'yev (1974)
Kodal, Oslo Igneous Province, Norway	Pe, 1.9 km long, 10-35m wide jacupirangite (alkal.pyrox.) dike empl. into augite monzon.	densely intergrown apat., Ti-mag. ilm.in pyrox., dissem.in monzon. 70Mt ore with 46% Fe oxides, 8.48% TiO_2, 7.98% P_2O_5	Notholt (1979)
Maimecha-Kotui Prov., N.Siberian Shield, U.S.S.R. largest: Gulya plut. (2,000 km^2)	Pe_2-T_1 central intrus.complexes; dunite, perid., pyrox.core envel. by ijolite-melteigite, carbonat., nephel.syen., syenite	several apatite and Ti-magnet. deposits cont. several billion t Fe; baddeleyite, pyrochlore, francolite in some pipes.	Rundkvist, ed. (1978) Epshteyn and Anikeyeva (1963)

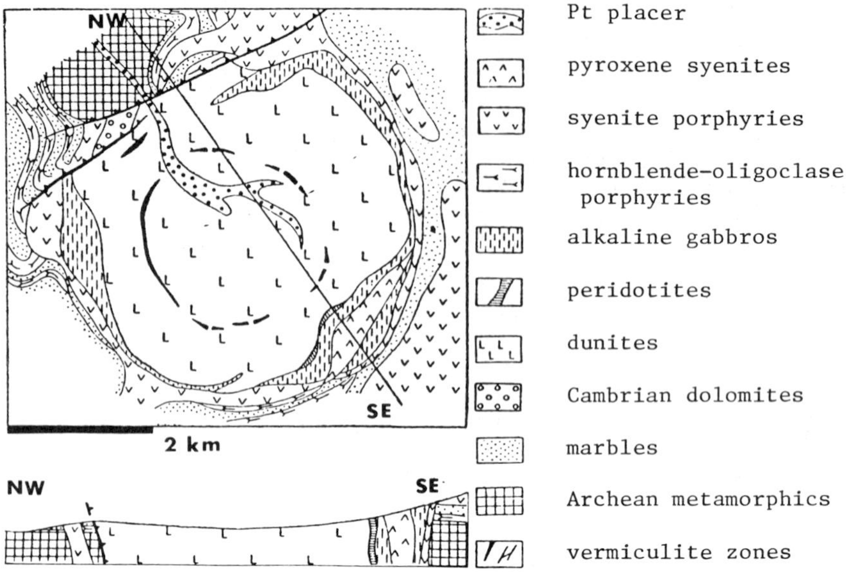

Fig. 33-28. Inagly alkaline ultramafic massif, N. Aldan Shield, Siberia. After Rozhkov and others, from Kazanskii (1972).

by pyroxenite and ijolite at Iron Hill, in the Powderhorn complex, Colorado (Temple and Grogan, 1965). There, the perowskite is locally rich in thorium.

In Tapira, M.G., Brazil, irregular bodies, veins, veinlets and scattered magnetite, perowskite, anatase together with apatite, phlogopite and vermiculite, replace pyroxenite (A. de Tarso, oral commun., 1980). All the minerals and particularly perowskite and anatase, have a high Nb content. The bulk of the ore mined in Tapira comes from the deeply weathered surficial portion.

High-temperature metasomatites comparable with skarns (garnet, alkaline pyroxenes, etc. rocks) are members of a hybrid rock suite formed along the centres of carbonate-rich pyroxenites, silicate carbonatites and urtite-melteigites in the Turiy Peninsula complex, Kola (Samoylov and Afanas'yev, 1978). Magnetite replacements occur in the vicinity, but are not directly hosted by the skarns.

Fracture-filling hydrothermal veins composed of analcite, natrolite, carbonates, etc. are widespread in the alkaline-ultramafic complexes, but no significant associated mineralization has been reported.

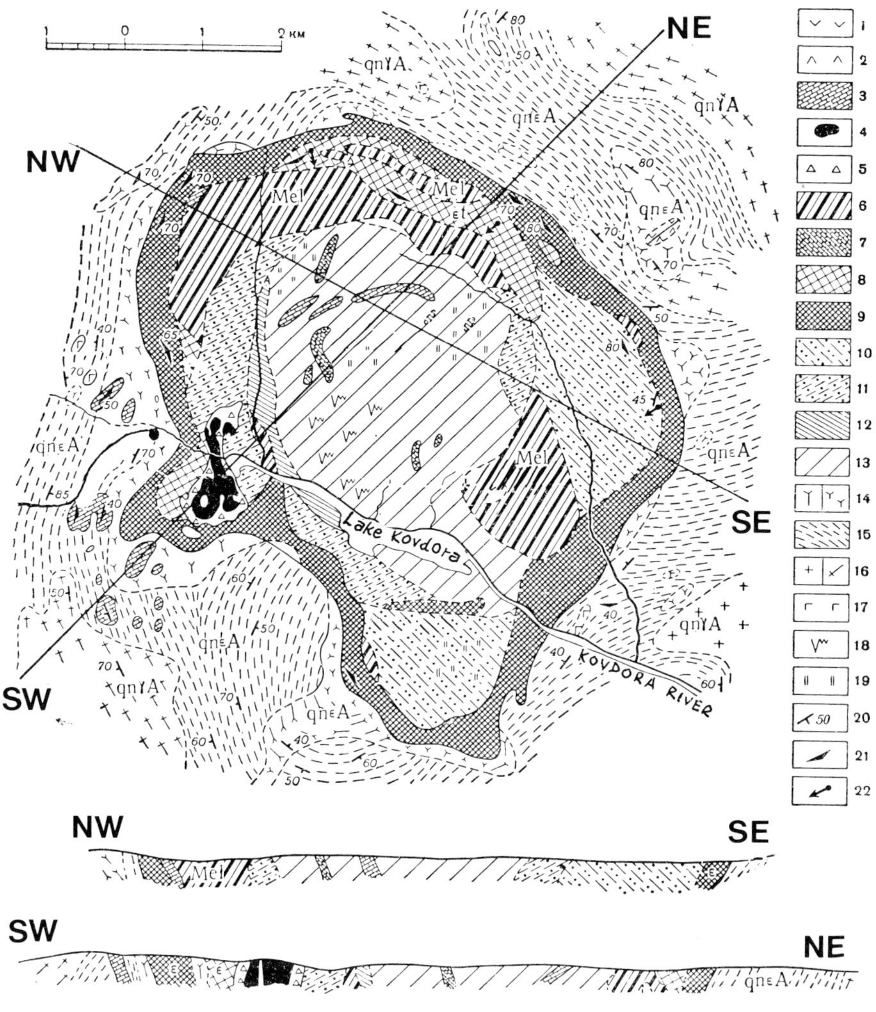

Fig. 33-29. Kovdora alkaline ultramafic massif, Kola Peninsula, U.S.S.R. PZ intrusive complex: 1=poikilitic and trachytic nepheline syenites; 2=trachytoid ijolites; 3=carbonatites; 4=magnetite; 5=apatite, forsterite, phlogopite rocks; 6=turyaites, melilitites, monticellite-melilite rocks; 7=vein ijolites; 8=coarse ijolites; 9=fine ijolites to melteigites; 10=nepheline pyroxenites, partly melteigites; 11=pyroxenites; 12=peridotites; 13=olivinites. Archean crystalline rocks: 14=fenitized gneisses, granite gneisses; 15=ditto, weakly fenitized; 17=amphibolites; 18=zones of sungulitized olivinites; 19=intensive phlogopite and vermiculite metasomatism; 20=schistosity; 21=banding in alkaline rocks; 22=lineation. From Borovikov and L'vova (1962).

MINERALIZATION IN INTRUSIVE EXOCONTACTS

Exocontacts of alkaline-ultramafic complexes are usually fenitized, and the fenitized zones are the same as those described in the previous sections. Certain ores normally formed in the intrusive endocontacts are, in some cases, found in fenitized exocontacts as well. Rundkvist, ed. (1978) reported the existence of apatite-bearing metasomatites which formed in altered quartzites in the exocontact of the Magan Intrusion (Meimecha-Kotui alkaline province, N. Siberia). Apatites normally have high trace contents of rare earth elements.

Parker and Sharp (1970) described a body of fenitized alkaline gabbro located within an alkaline-ultramafic intrusion in the the Gem Park Complex, Colorado. The fenite carries high Nb and V values which the authors attribute to a buried carbonatite body.

In the contact aureole of the Powderhorn complex, Colorado (within an alkaline metasomatic rim several hundred meters wide and beyond this, Th mineralization is widely distributed (Hedlund and Olson, 1961). Pink to red, fine grained trachyte porphyry dikes are radioactive, although no visible radioactive minerals have been detected. About 100 dikes have been mapped and the radioactivity is attributed to fine residual thorogummite, dispersed in earthy hematite pseudomorphs after pyrite. In addition to the dikes, 217 Th-bearing radioactive veins emplaced in Precambrian basement metamorphics have been recorded. The thin veins are most abundant in the proximity of outlying (metasomatic) syenite bodies. They follow shears and breccia zones, and carry scattered thorite crystals in quartz, orthoclase, barite, specularite, carbonate, fluorite, biotite, etc. gangue. The wallrocks are often feldspathized (microcline) and this results in conspicuous pink or red rims.

METAMORPHOSED ALKALINE-ULTRAMAFIC COMPLEXES

Schorscher et al. (1982) interpreted as being meta-jacupirangite several Proterozoic mafic meta-intrusives, located 120 km N. of Morro do Pilar, M.G., Brazil. This rock forms an elongated body, several hundred metres long and emplaced in the metamorphics of the basement complex and itself deformed and metamorphosed in the upper greenschist facies. Hedenbergite, sphene, apatite, carbonate and quartz are the major constituents, and there are anomalous contents of Th (0.21%), Y (0.13%) and other metals. Thorium is bound in orangite and yttrium in apatite.

WEATHERING-MODIFIED OR GENERATED ORES

In Tapira, Brazil (Herz, 1976), a weathering crust which formed in situ over the primary magnetite, ilmenite, perowskite and anatase ore in pyroxenites, is up to 100 m. thick. Within it, ilmenite and perowskite have been converted to "leucoxene" and the anatase was partly leached. The residual ore is slightly enriched in Ti due to removal of Ca,Fe and Nb and it is cheap to mine. The reserves are 132 Mt ore with 21.6% TiO_2. Residual, powdery niobian apatite ore (15% of apatite) has been preserved in the saprolite, but diluted or even dissolved in the topping red laterite.

Francolite (carbonate apatite) is a common alteration product of apatite hosted by or close to carbonatite. It is an inconspicuous, whitish mineral widespread, for example, in the apatite-ijolite and apatite-carbonatite breccias in the Magan complex (N. Siberia; Rundkvist, ed., 1978). Francolite usually has a high trace content of rare earths.

Lateritic weathering superimposed on nickel-rich dunites and peridotites may result in the formation of the usual style of Ni laterite and saprolite, already reviewed in Chapters 7 and 23. Numerous occurrences are known in the Cretaceous Iporá Group belt of alkaline-ultramafic intrusions, located on the fringe of the Paraná Basin, Brazil. The largest and best studied deposit is located in the Santa Fé massif (de Oliveira and Trescases, 1980). There, the following zones of the lateritic profile have been recognized. Running from top to bottom these are as follows: red laterite, yellow laterite, fine saprolite, coarse saprolite, bleached dunite, fresh dunite. The Ni-enrichment is most significant in the yellow laterite and saprolite (max. 7.25% NiO), and it consists of dispersions and, lower in the profile, infiltration veinlets (boxworks) of Ni-kerolite, pimelite and silica. The lateritization is Tertiary and the ore reserves at Santa Fe are estimated to be about 21.14 Mt of ore with 1.5% Ni.

Silcrete present in tropical regions is considered to be a negative development, because it masks the bedrock (compare Chapter 23). It is rarely mineralized. In the Yubdo ultramafic body, Ethiopia (considered by Rundkvist, ed., 1978, to be comparable with the Inagli massif, Siberia), silicified surficial portions of ultramafics ("birbirites") locally contain residual particles of platinum. Relict, residual jasper (ferruginous silica chert) locally present in the surficial portion of the Th-bearing veins near the Powderhorn complex, is radioactive (Hedlund and Olson, 1961). The radioactivity is probably due to finely dispersed secondary Th minerals.

ORES RESULTING FROM SEDIMENTARY REWORKING

The numerous Ti-magnetite deposits (Kovdor, Afrikanda) that crop-out in the recently glaciated regions, are often enveloped by glacial drift rich in ore boulders. Some such boulders have been recovered as an ore in the initial stages of mining of the bedrock deposits.

The Pt scattered in the Inagly and Yubdo peridotites has been reworked into alluvial placers. Those in Inagly are very proximal, directly overlying the source dunite.

33.7. POTASSIC ALKALINE ROCKS

The "most normal" alkaline rocks, such as nepheline syenites and phonolites, contain an excess of sodium over potassium (e.g. 8.46% Na_2O and 5.77% K_2O in an average nepheline syenite; Table 3 in Sørensen, 1974). Exceptions, however, are common and the nepheline syenites from Stjernøy and Vishnevogorsk (same table) have a slight excess of K_2O over Na_2O. Some alkaline rocks contain a large excess of K_2O (e.g. phonolite from Shonkin Sag, Montana; same table; 6.45% K_2O, 1.85% Na_2O) and have been included in a separate class of potassium-rich alkaline rocks. In addition to several distinctly potassic alkaline igneous provinces (e.g. the Highwood Mountains of Montana; Hurlbut et al., 1941), high-K rocks usually marked by abundant pseudoleucite have a patchy distribution within or along the margin of sodic alkaline complexes (e.g. Poços de Caldas). Many are probably metasomatites (fenites).

The class of potassic alkaline rocks is, however, considered unique by petrologists and it has a specialized literature (e.g. Sahama, 1974; Gupta and Yagi, 1980), but its metallogeny is indistinct. This may be due to the general rarity of this association. Geochemical enrichment in U and Th has been reported most commonly and it is interesting that isolated occurrences of K-alkaline rocks coincide spatially (although not genetically) with "granitic" tin provinces (e.g. Loučná in the Erzgebirge; Karagwe-Ankolean Province, E. Africa). The U-Mo mineralization in the Poços de Caldas complex (Section 33.5) is contained partly in potash-rich rocks and so are the hydrothermal veins with brannerite and thorite in an unnamed leucite and kaliophilite-rich late Proterozoic massif in Siberia (Semenov, 1974).

In the Rome and Tuscany Pleistocene volcanic provinces of Italy (von Backström, 1974), potassic alkaline rocks have high trace U and Th contents (25-50 ppm U, 130-240 ppm Th). A variety of low-grade reprecipitated orebodies in volcaniclastics have already been mentioned in Section 33.2.

In the sanidine-rich Cretaceous Crowsnest volcanics, S. Alberta, Canada (Currie, 1974), small Cu and Mo showings are hosted by their high-level intrusive equivalents.

33.8. CARBONATITES

33.8.1. Introduction

Carbonatites are very rare rocks that constitute no more than 0.00X% (by volume) of the alkaline rock suite, yet they are blessed with a copious volume of literature. Every known occurrence of carbonatite has probably been recorded in print and described and a high proportion of such descriptions have been abstracted in the international literature. New occurrences of carbonatites are being added as the result of continuing discoveries as well as genetic reinterpretation of known carbonate occurrences. There is, also, an abundance of petrologic and ore petrologic literature on carbonatites, and two widely available reference volumes exist in English (Heinrich, 1966; Tuttle and Gittins, eds., 1966). This makes carbonatites probably the best covered category of mineralized rocks in the English literature.

To some extent this is economically justifiable. Carbonatites are decidedly the most consistently and most frequently mineralized alkaline rocks, and they hold a disproportionate share of the world's production and reserves of niobium (about 30 Mt Nb) and rare earths (about 35 Mt REE). They also represent a case history of a fundamental scientific recognition and elegant re-interpretation of a formerly controversial rock assemblage. Once an attractive model of the magmatic origin of carbonatites from deep subcrustal sources had been generated, everyone rushed to contribute his or hers data.

While the genetic affiliation of carbonatites to some form of alkaline magmatic activity is not in dispute, not all occurrences are simple stocks or flows of a solidified carbonate magma. It is becoming increasingly evident that, as in the case of feldspathites, carbonatites could form as orthomagmatic crystallizates, as products of hybridization (mixing) of earlier magmatic rocks with carbonatitic magmas, as products of solidification of magmatically remobilized non-magmatic carbonates or as products of alteration metasomatism of earlier rocks by a variety of hydrothermal fluids and gases. Other alternatives are also possible. All these possibilities may host the characteristic rare metal mineralization.

Most "typical" carbonatite complexes have, by now, probably been discovered, described and evaluated economically. Not all are mined, however. There is a limited market for both Nb and REE and two single carbonatite localities alone (Araxá, Brazil and Mountain Pass, California) are capable of supplying most of the Western World's needs with respect to these metals. "Atypical" carbonatites and situations marginally linked to carbonatite activity and its interaction with a variety of environments, however, present a continuous challenge to an exploration geologist. It is likely that unknown "types" of mineral deposits (not only of Nb and REE) will be found in the future.

This section departs from the practice followed in the earlier sections, and treats carbonatites at all levels of emplacement: volcanic, subvolcanic and plutonic. Because of the easily available and comprehensive compilation literature, locality descriptions have been reduced and substituted by an abbreviated data summary in Table 33-10 and Figure 33-30.

Terminological remarks

The name carbonatite implies magmatogene (orthomagmatic) origin, particularly to non-specialists. This need not always be true. Smith (1956) defined carbonatites more sensibly, stressing their common lithologic association: "Carbonatites are carbonates that appear to behave as intrusive rocks and are closely associated with alkaline igneous rocks". This definition, on the other hand, neglected the existence of the empirically proven volcanic equivalents. The logic

Table 33-10. Selected occurrences of mineralized Phanerozoic carbonatites

LOCALITY	GEOLOGY	MINERALIZATION	REFERENCES
Oka complex, W.of Montréal, Québec	Cr_1 double Ca-carb.ring dike and Si-carb. stocks,surr.by melillite rocks,alk.pyrox.,melteig.,urtite; empl.into Pt_{2-3} cryst.basement	magn.,pyrochl.,perowsk.,niocalite in streaky disseminations; 735Tt Nb/0.35%; 267Tt REE/0.127%	Gold (1969)
Lake Nipissing Ontario,Canada	560 m.y. plut.ring dike compl., Ca-carb. lenses in fenite envel. adj.to alk.pyrox.,gabbro in Archean metamorphics	dissem. pyrochlore in carbonatite; 20.3Tt Nb/0.371%	Currie (1974)
Iron Hill, Powderhorn compl.,Colorado, U.S.A.	Cm pyrox.,ijol.,neph.syen.compl., late dolom.carb.stocks,dikes empl.into fenitiz. PCm metam. basement	high trace Th; dissem.Ti-magn., pyrochl.,bastnaes.; endocont. martite,apatite veins; 2.435Mt REE/0.34%; 288Tt Nb/0.04%; 26.2Tt Th/36 ppm	Armbrustmacher (1980)
Seis Lagoas,Guiana Shield,Amazonas, Brazil	circular carbonatite ring dikes and stocks empl. into Archean metamorphics; fenite; lateritically weathered	residual pyrochlore (pandaite) in saprolite; 110Tt REE/1.27%; 14.7Tt Nb; 5.5Tt Th	Bonow and Issler (1980)
Catalão, Goiás, Brazil	Cr_1 central Ca-carb.stock in fenitized PCm gneiss	dissem. pandaite, gorceixite, apat.,goeth.,magnet.,anatase in saprolite; 1.33Mt REE/1.7%; 863Tt Nb/0.6%; 102Mt Ti; 128Tt Th; 45Tt U/aver.180 ppm	Loureiro (1980)
Araxá (Barreiro), M.G., Brazil	Cr_1 Ca- and biot.carbonatite stock in fenitiz. Pt_2 m-quartzite and schist; 200 m thick regolith	pyrochl.,monaz.,apatite dissem. in biot.-rich xenolithic and brecc.carb.cem.by calc. All ore mined comes from saprolite; 18Mt Nb;34Mt REE;660Tt Th;	Grossi Sad and Torres (1978)

Jacupiranga, S.P., Brazil	J_3-Cr_1 Ca-carb.stock in centre of a compos.ellipt.massif (pyrox., perid.,alk.gabbro,neph.syen.) empl.to PCm micasch.,granodior. Lateritically weathered	residual apatite,magnetite, ilmenite with access.pyrochlore, baddeleyite; 26% Fe_2O_3 in residual soil	Melcher (1966)
Fen near Ulefoss, Telemark, southern Norway	600-413 m.y. multiphase Ca and dol.carb.empl.with alk.gabbro and pyrox. into fenitized Pt gneiss	1) dissem.magn.,pyrochl.,koppite,columb.,apatite in dike-shaped bodies of brecc.carb.; 2) Th and REE-bearing N.-S. hemat.veins; 56Tt Nb/0.35%	Vokes ed., (1960)
Alnö near Sundsvall,N.E. Sweden	563 m.y.cone sheet and ring dike compl.of Ca- and dol.,ank.-carb., syen.,kimberlite empl.to fenitized Arch. granitic gneiss	dissem. Th-rich pyrochl. and dysanalyte in Ca-carb.;fluorite, perowskite; miner. occurr. of Nb,REE,Th,U minerals	v.Eckermann (1966)
Sokli, northern Finland	334-378 m.y. zoned multist.silic., Ca-, dolom. carb. stock 18 km^2 emplac. into fenitiz. Archean gneiss and amphibolite; regolith	apatite/francolite relict regolith with minor dissem. pyrochl. baddel.,rhabdophane; 100Mt 19% P_2O_5	Vartiainen and Paarma (1979)
Turiy Peninsula, Kola Pen.,U.S.S.R.	PZ_2 small carb. core envel. by pyroxenite,skarn,analc.-calcite lamproph.,empl.into fenitiz. PCm granodiorite	steeply dipp.small magnet.lodes in phlog.,diops.,amphib.carbont. lens at cont.of pyrox. and ijol.	Samoylov and Afanas'yev (1978)
Kaiserstuhl N.W. of Freiburg, West Germany	Mi_2 subvolc.Ca and dol.,ank.carb. flanked by phonol.and alkal.basalt lavas, empl.into T,MZ sedim. Light fenitization	dissem.pyrochlore, koppite, dysanalite, magnet.,apatite; 4.9Tt Nb/0.23%	Wimmenauer (1966)
Odikhinch Massif, Meimecha-Kotui prov.,N. Siberia	Pe-J concentric zoned intrus., Ca-,dolom.,anker.carb.,alk.pyrox., neph.syenite in fenitized PZ seds.	pyrochlore dissem. in Ca-carb.; monaz. in dolom. carb.	Heinrich (1966)

Table 33-10 (continued). Selected occurrences of mineralized Phanerozoic carbonatites

LOCALITY	GEOLOGY	MINERALIZATION	REFERENCES
Mushugay-Kuduk, southern Mongolia	J_3 volc.-subv.alk.complex; trachyte,nephelin.lavas,pyrocl.,syenite,vent breccias,nephel.syen. carbonatite	stockw.of apat.,Ti-magnet.veins in trachyrhyol.vent breccia; dissem.magn.,fluor.,apat.,celestite in fine-gr.explos.breccia	Vakrushev and Vladykin (1979)
Khanneshin carbonatite volcano, S. Afghanistan	Pl-Q anker.carb.,carb.breccias, lavas,tuffs;dissected compos. volcano resting on T sedim.	bastnaes. and burbankite in hydr.alter.zones; U oxides in exocont.veins;dissem.and repl. magnetite in breccia	Alkhazov et al.(1977)
Yin-Shan Range, Inner Mongolia, China	probable vein/dike carbonatites empl. into Pt and PZ limestone, chrt,biot.schist,iron formation as large tabular bodies (Wula-Shan and Paiyunopo)	dissem.magn.,hemat.,fluor.,aegirine,aeschynite, beiyinite; 40-65% Fe, 2-7.2% REE_2O_3	Lee (1970)
Mrima Hill near Mombasa, Kenya	MZ-T high level Ca- and dolom. carb.plug and breccia; deeply tropically weathered	residual dissem.pandaite,Ba-gorceixite,earthy monaz.; resid. psilom.,pyrolus.,hausmann., ocher; 400Tt Nb/0.5%; 255Tt REE 1.275%; 100Tt Mn	Deans (1966)
Ruri,E.shore Lake Victoria, Kenya	5-11 m.y. circ. intrus.of Ca-carb. accomp.by phonol.,nephel. syen. into fenitiz. PCm meta-basalt; regolithic	magnet.,monaz.,bastnaesite in dike-like late ferrug.carb.; minim. 21Tt REE, 16.5Tt Th	Jaffé and Collins (1969)
Mbeya (Panda Hill) S.W. Tanzania	MZ Ca- and Ca-dolom.carbon. stock empl.to fenitiz. PCm metamorphics	pyrochlore dissem.in Ca-carb. and biot. fenite; 283Tt Nb/ 0.553%;	Deans (1966)

S.E.Uganda carb. group (Tororo,Bukusu,Sukulu)	T_2, volc.-subvolc. Ca-,dolom., anker.,silic.carb. with phonol.,nephelinite,neph.syen., pyrox.,etc. empl.to fenitiz. PCm basement	dissem.pyrochl.,baddeleyite, apatite, Ti-magnetite; Sukulu: 200Mt 13% P_2O_5, 30Mt Fe, 182Tt Nb/0.175%	Deans (1966)
Lueshe, Kivu Prov., Zaïre	zoned ellipt. Ca-,silic.,dolom. carb. compl. assoc. with cancrin. syenite, in fenitiz. PCm basem. schists	dissem. pyrochl. in Ca-carb. saprolite; 282Tt Nb/0.94%	Deans (1966) Gittins (1966)
Longonjo, Benguela Plateau, Angola	J carb.and syenite ring complex in fenitized PCm basement; regolithic	dissem.pyrochlore in saprolite 10.8Tt Nb; 36.8Tt Th	de Kun (1965)
Ondurakoruwe, Namibia	J? Ca- and dolom.carb.,circular plug 1.2 km diam.,in fenitized Pt schists; carbonatite dikes	strontian.,monaz.,pyrochlore, cerianite,columbite,zircon in dolom. carb. dikes; 2.55% REE, 0.002% U_3O_8	Verwoerd (1967)
Kalkfeld, C.Namibia	J? Ca-carb.pipe grading to volc. brecc.in fenitiz.Pt metasedim.	finely divided Th in replac.masses of Mn-rich hematite	Verwoerd (1967)
Chilwa Island, S.Malawi	J_3-Cr_1 4 compos.centres in a ring compl.;Ca-,dol.,ank.,sider. carb. with phonol.,nephel.,alnöite in feldsp.and fenitiz.brecc. Pt basement metamorphics	pyrochl. dissem. in Ca-carb.; Th,REE,gal.,fluor.,in late qtz.-calcite veins; Th dispers. in hydroth.metas.sider.carb. 4,550 t Nb/0.7%; 300Tt Mn	Garson (1966)
Kangankunde, southern Malawi	J_3-Cr_1 elong.stock of strontianite-rich carb.,ank.,sider.carb. carb.agglom. and brecc. in fenitiz.,feldsp..brecc.PCm gneiss	impregn.and rich veins of monaz. florencite,bastnäs. in carb.; replac.masses of stront.-monaz. min.54Tt REE; Th;	Garson (1966)
Tundulu complex, southern Malawi	J_3-Cr_1 Ca- and sider.carb.vents, ring,agglom.in Pt syen.,gran.,gn.	sider.,bastnaes. carb. dikes; apatite, pyrochl. in Ca-carb.	Garson (1966)

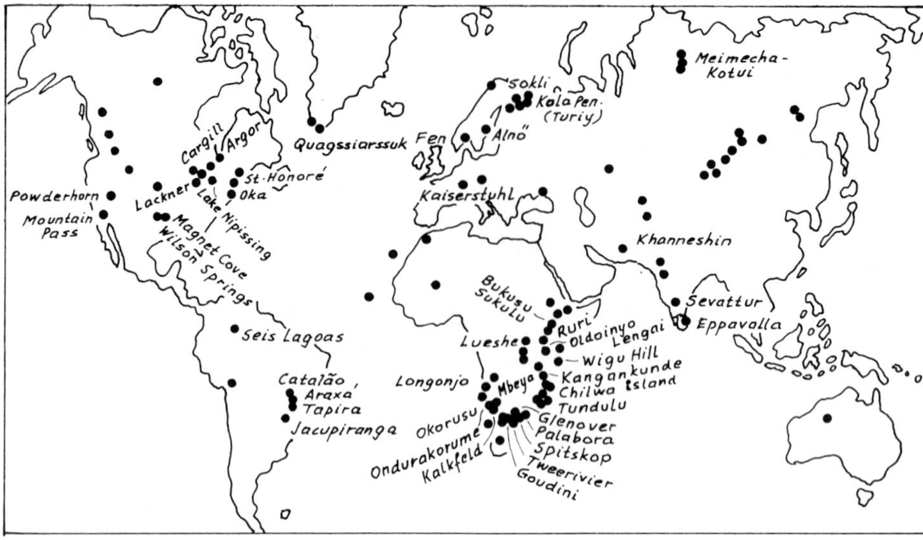

Fig. 33-30. Index map of major occurrences of the carbonatites listed in Table 33-10.

of carbonatite definitions was further reviewed by Moore (1973).

Rather unfortunately, we have inherited a list of petrographic variety names of carbonatites introduced into the literature throughout 50 years of study. Of these, the original Brøgger's terms for the two most widespread mineralogical varieties of intrusive carbonatites: "sövite" (calcite carbonatite) and "rauhaugite" (dolomite carbonatite) are widely used, and so is the von Eckermann's term "beforsite" for a hypabyssal, usually dike-forming dolomite carbonatite. In this book, carbonatites are uniformly identified with the help of a compositional prefix (calcite carbonatite or CC and dolomite carbonatite or DC). Other terms frequently used in the literature are silicate carbonatite or silicocarbonatite (SC) for rocks containing considerable proportions of silicate minerals (biotite or phlogopite, wollastonite, diopside, etc.).

Verwoerd (1967) proposed the use of the term "metacarbonatite" for all "secondary" carbonatites, formed by a hydrothermal reconstitution of earlier "magmatic" carbonatites. This term has been avoided here because of its possible confusion with regionally metamorphosed carbonatites.

33.8.2. Carbonatite complexes and emplacement levels

The outcrop appearance of a carbonatite complex is essentially governed by the erosional level, by the style (pattern) of emplacement and by the nature of the associated non-carbonate rocks. Idealized sections (models) showing the way in which the nature of a carbonatite complex changes with depth, have been presented by Garson (1966), Verwoerd (1967; his Fig. 19), Smirnov, ed. (1968; their Fig. 86) and Moore (1973; his Fig. 1). These models differ in detail, and all apply to zoned central complexes in which the carbonatite is situated in the centre ("normally zoned" carbonatite complexes; Smirnov, ed., 1968; Fig. 33-31). The "normally zoned" complexes are believed to have projected on the surface as central volcanoes (some, however, never vented) and the following levels can be recognized (Fig. 33-32):

(a) Volcanic cone level. This is the topmost level, preserved only in recently active volcanoes (e.g. Oldoinyo Lengai and Kerimasi in northern Tanzania). Oldoinyo Lengai (Dawson, 1966; Fig. 33-33) is a steep, cone-shaped composite volcano approximately 8 km in diameter, standing about 2,100 m above the surrounding plains. The summit is occupied by two craters, and the cone is dissected by radial gullies and interrupted by small parasitic cones, explosion craters and tuff rings. The majority of its lavas and pyroclastics are nephelinitic in composition. Some are ijolitic. Carbonatite volcanism is marked by periodic (approximately every 7 years) extrusions of Na-Ca carbonatite lavas, initially black but rapidly turning gray (composed of nyerereite) and explosions of soda ash. Soda is soluble in water so it is rapidly leached out by rains and carried into the adjacent Lake Natron (a playa lake), where it precipitates as an evaporite. The deeper levels of the volcano are not accessible for observation, but can be interpreted thanks to the exotic ejecta present in the pyroclastics. These include fenitized gneiss, urtite, ijolite, alkaline pyroxenes and alkaline gabbros.

(b) Intermediate volcanic level has been exposed in the late Tertiary alkaline volcanoes in the south-eastern corner of Uganda (Elgon, Napak, Toror, and others; King and Sutherland, 1966). Napak is a remnant of a volcanic cone with an original diameter of 32 km, deeply eroded in its central part. Remnants of the stratovolcano are composed predominantly of pyroclastics (chiefly agglomerates) and minor nephelinite lavas. The central intrusive complex contains a variety of alkaline pyroxenites, gabbros, nepheline syenites and members transitional to carbonatites. Ijolites, frequently present in a very coarse, pegmatitic form, are particularly abundant. Carbonatite constitutes a small hill in the centre (400 m diameter) and also forms a number of dikes intersecting the ijolite. The main mass shows a concentric vertical banding. The central intrusive is surrounded by shattered, sheared and fenitized metamorphics of the Precambrian basement complex.

(c) Subvolcanic level is well illustrated by the late Jurassic-early Cretaceous carbonatite complexes in southern Malawi (Tundulu, Chilwa Island, Kangankunde) described by Garson (1966). Here, the centre of the complex is filled by a variety of carbonatites (calcite, calcite-ankerite, siderite carbonatites), emplaced sequentially. The

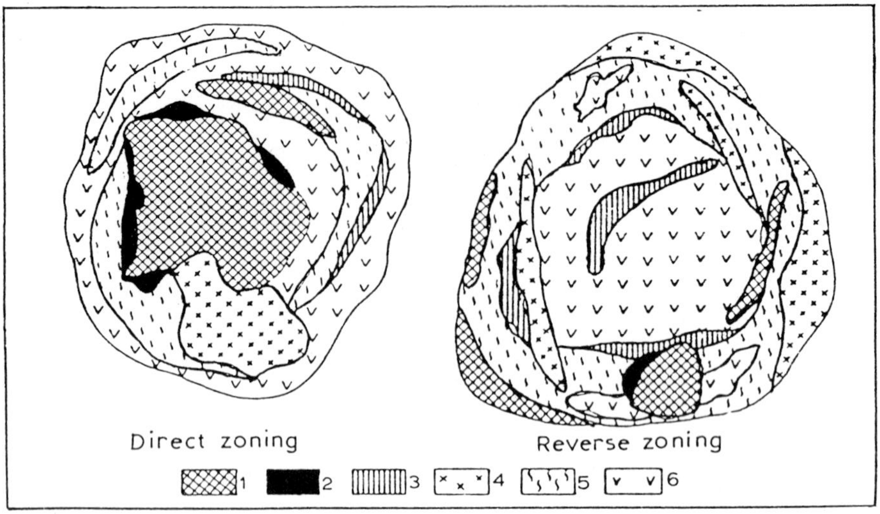

Fig. 33-31. "Normal" and "reverse" zoning in alkaline ultramafic complexes. 1=carbonatites; 2=forsterite, apatite, magnetite rock; 3=silicate metasomatites; 4=syenites; 5=ijolite melteigites; 6=ultramafics. From Smirnov, ed. (1968).

centre is surrounded by feldspathic breccia and agglomerate saturated by a carbonate matrix along the inner contact, and grading into a brecciated and fenitized Precambrian gneiss. The fenitization is predominantly potassic. Trachytic lava and pyroclastics form the remnants of a volcanic cone and there are dikes, plugs, and/or cone sheets of nepheline syenite rocks.

(d) Upper plutonic level of zoned central complexes is exemplified by numerous examples of Mesozoic and Paleozoic carbonatites in Africa (Dorowa, Shawa in Zimbabwe; Spitskop in South Africa; Kalkfeld in Namibia) and elsewhere. The central core of carbonatite is enveloped by alkaline pyroxenites, gabbros and syenites, grading to predominantly sodic fenites. Dikes are moderate to common.

(e) Deep plutonic level is most commonly exposed in Paleozoic and Precambrian carbonatite complexes, e.g. Powderhorn, Palabora (Fig. 33-34), Nemegos, and others. Ultramafics (pyroxenites), alkaline gabbros, syenites, are usually abundant. There is a broad halo of fenitization and widespread metasomatism. Garson (1966) believed that many alkaline pyroxenites are reaction products at the contacts of carbonatite with syenitic fenites.

Heinrich (1966, his Table 10-2) in his classification of carbonatites according to depth levels of emplacement, listed "plutonic catazone" as the deepest level. There, the most conspicuous

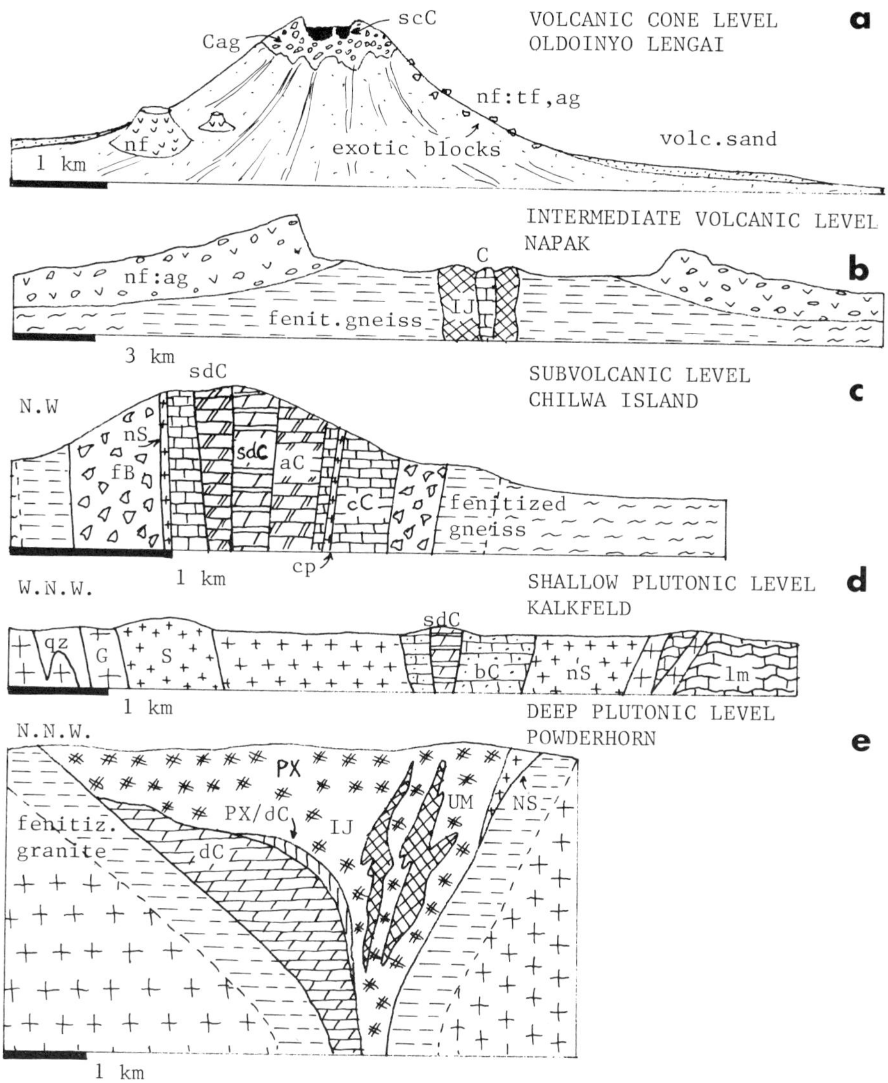

Fig. 33-32. Examples of "normally" zoned carbonatite occurrences arranged according to increasing depth of the erosional level. C=carbonatite; scC=Na-Ca carbonatite; sdC=siderite carbonatite; aC=ankerite carbonatite; cC=calcite carbonatite; bC=biotite containing carbonatite; Cag=carbonatite agglomerate; dC=dolomite carbonatite; G=granite; nf=nephelinite; nf:tf,ag=tuff, agglomerate; IJ=ijolite; UM=uncompaghrite; cp=camptonite; nS=nepheline syenite; fB=feldspathic breccia; lm=crystalline limestone; qz=quartzite; px=pyroxenite. (a) from LITHOTHEQUE; (b) after King and Sutherland (1966); (c) from LITHOTHEQUE, data from Garson (1966); (d) after Martin et al. (1960); (e) from LITHOTHEQUE, modified after Temple and Grogan (1965).

Fig. 33-33. Oldoinyo Lengai nephelinite, carbonatite volcano, northern Tanzania. TOP: general view; BOTTOM: crater; note the white soda and limestone afflorescences on top of the lavas and pyroclastics.

characteristics of the five depth levels treated above, the ring or circular expression and piercement tectonic features, are either not fully developed or are inconspicuous. Instead, "intimate mingling of carbonatitic fluids and various country rocks (including marble) akin to granitic migmatitization" is widespread. Most occurrences, moreover, are genetically controversial.

Although the "normally zoned" alkaline complexes that involve carbonatite are the most conspicuous, they are not the only form of carbonatite occurrence. Their symmetry is often disturbed and the carbonatite occurs eccentrically (e.g. at Iron Hill in the Powderhorn complex), or is substituted by diffuse carbonatite screens alternating with syenite in the zone of fenitization (e.g. Manitou Island, Ontario; Currie, 1974). In the same area (Lake Nipissing, Ontario), carbonatite forms rims at lamprophyre dikes or occurs as dikes in fenites. Carbonatite dikes, ring dikes and cone sheets on the fringe of alkaline ring complexes, occur frequently (Nemegos, Spitskop, Kovdozero, etc.).

Many carbonatite bodies associated with alkaline complexes have a thick sheet-like or irregular form (e.g. Mountain Pass, California)or form dikes, dike swarms, stockworks or sills (e.g. James Bay, Canada; McClure Mt., Colorado). Similar carbonatite bodies sometimes occur in terrains lacking directly affiliated alkaline complexes (Wigu Hill, Tanzania; Turyi Peninsula, U.S.S.R.). Some carbonatites, moreover, lack conspicuous fenite aureoles. The latter are difficult to identify and appear to be more common in mobile belts than within cratons.

In alkaline complexes, carbonatites are associated (1) with ultramafic rocks (dunites, peridotites, pyroxenites); (2) with alkaline pyroxenite and gabbroids, some with a variety of feldspathoids or minerals like melilite, cancrinite (e.g. ijolite, urtite or melteigite) and (3) nepheline syenite and syenite. Verwoerd (1967) pointed out that the above order of decreasing basicity also corresponds to the usual intrusive sequence and the succession of zones from the central carbonatite outward. Calcitic carbonatites are by far the most common, followed by dolomitic, ankeritic and sideritic carbonatites. Carbonatites containing abundant accessory silicates (biotite, alkaline pyroxenes and amphiboles, melilite, wollastonite, etc.) are usually transitional to predominantly silicate rocks, e.g. ijolites.

33.8.3. Multistage development

Few carbonate occurrences formed as a product of the simple crystallization of a single batch of magma. Most are the result of a long-lasting activity, during which carbonatitic (and other) magmas and fluids issued from several centres (or from a single, migrating centre), in several stages. The products issued from adjacent centres reacted with each other and sequentially younger products interfered with, and modified the older products. The result is a great diversity of magmatic, metasomatic and hydrothermal phases in differentiated carbonatite complexes, that exerts an important influence on the mineralization. Table 33-11 documents the sequential development of the Chilwa Island carbonatite ring complex in Malawi (Garson, 1966). Five development stages have been recognized in the Sokli (Finland) carbonatite complex (Vartiainen and Paarma, 1979).

In the Loolekop carbonatite, Palabora, South Africa (Verwoerd,

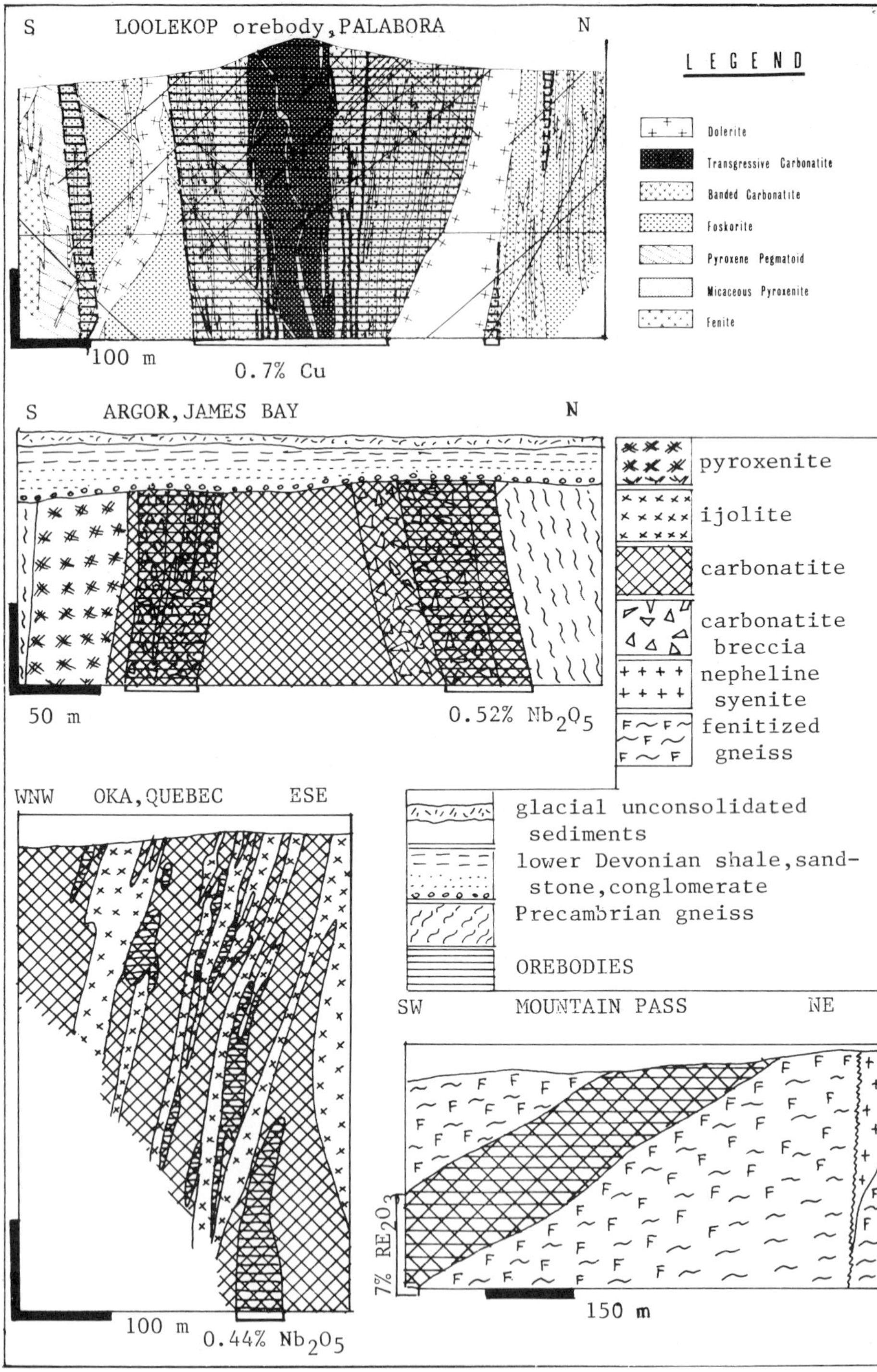
S LOOLEKOP orebody, PALABORA N
LEGEND
Dolerite
Transgressive Carbonatite
Banded Carbonatite
Foskorite
Pyroxene Pegmatoid
Micaceous Pyroxenite
Fenite
100 m
0.7% Cu
S ARGOR, JAMES BAY N
pyroxenite
ijolite
carbonatite
carbonatite breccia
nepheline syenite
fenitized gneiss
50 m
0.52% Nb2O5
WNW OKA, QUEBEC ESE
glacial unconsolidated sediments
lower Devonian shale, sandstone, conglomerate
Precambrian gneiss
OREBODIES
SW MOUNTAIN PASS NE
7% RE2O3
150 m
100 m
0.44% Nb2O5

Table 33-11. The sequential development of carbonatites in the Chilwa Island ring complex, Malawi. Slightly modified after Garson (1966)

LATE JURASSIC-EARLY CRETACEOUS CHILWA ALKALINE PROVINCE
CENTRE 4
Faulting and minor overthrusting
Intrusion of lamprophyric plug into Summit Plateau CC
Intrusion of radial dikes of syenite, olivine nephelinite, diabase
Hydrothermal introduction of BaO, ThO_2, Ce_2O_3, PbS, F, late quartz, calcite
Intrusion of manganiferous and sideritic C
Intrusion of ankeritic C and partial replacement of CC by ankerite
Apatite-veining of pressure cracks
CENTRE 3
Intrusion of nepheline syenitic and trachytic dikes and formation of silicate CC
Intrusion of CC with siderite
Intrusion of mobilized feldspathic breccias, brecciation
CENTRE 2
Intrusion of olivine nephelinite, alnöite and pyrochlore-bearing CC into shear planes
Intrusion of cone sheets of phonolite
Intrusion of aegirine-biotite CC into shear planes
Brecciation
CENTRE 1
Intrusion of Marongwe acmite CC
Feldspathization of contact breccia
Brecciation above carbonatite pluton
Intrusion of carbonatite in depth producing aureole of fenitization
UPPER PRECAMBRIAN TO LOWER CAMBRIAN
Quartz syenite and granite intruded into Basement Complex granulite, gneiss, limestone
ABBREVIATION: CC=calcite carbonatite (sövite)

Fig. 33-34. Cross-sections of four mineralized plutonic carbonatites (see text and Table 33-10 for description). Loolekop, after Palabora Staff (1976); approximate outline of the Cu orebody above 0.3% Cu grade added. Argor, after Stockford (1972). Oka, from LITHOTHEQUE, after Gold et al. (1967). Mountain Pass, from LITHOTHEQUE, based on data in Olson et al. (1954) and Molycorp Staff, guided tour, 1982.

1967), three major phases of carbonatite emplacement have been distinguished. There, the dominant "banded carbonatite" is the product of Phase 1. It consists of a relatively coarse calcite carbonatite, conformable with the walls of the feeder conduit. The banding is caused by the parallel arrangement of magnetite-rich layers and parallel lines of crystals of other minerals. Phase 2 produced the "main younger carbonatite", which is a transgressive, irregular body grading into a series of dikes. It was emplaced into the central part of the older carbonatite plug. The younger carbonatite tends to be finer grained than the older one, banding is absent, and the slightly higher magnetite and silicate content is preferentially concentrated in localized lenses. Minor copper mineralization is associated. Phase 3 is represented by the youngest transgressive carbonatite veins and thin sheets, traversing the central mass of Phase 2 carbonatite. It crystallized from hydrothermal solutions and carries the bulk of the Palabora copper ores.

Heinrich (1966), Verwoerd (1967) and other authors described the effects of sequential alterations of older ("syngenetic") carbonatites by younger carbonatitic fluids, resulting in the generation of a "metacarbonatite". Zhabin (1978) discussed the textural features, in particular the crystal habits of calcite, applicable in the interpretation of the metamorphic evolution of carbonatites.

33.8.4. Origin and recognition of carbonatites

Despite the diversity in style, composition and association, carbonatites are a remarkably distinct category of rocks. Their strong tectonic control and association with major continental extensional fault systems (rifts and paleorifts) has been demonstrated many times (see also the review in Heinrich, 1978, also Chapter 30). This together with the experimental evidence points to the subcrustal sources, or at least a subcrustal "initiation" of the carbonatite generation at higher crustal levels. Kapustin (1976) considered the origin of carbonatite independently of that of other alkaline rocks, and placed the site of carbonatite generation into the deep mantle, under the zone of generation of tholeiitic magmas. From there, carbon and hydrogen are carried into a higher chamber, where oxidation takes place to form CO_2. This, then, reaches the crust. Other authors prefer to interpret carbonatite as being a differentiate of a parent alkaline magma. Nephelinite, nepheline-picrite, melilite-basalt, pyroxenite, peridotite, kimberlite, ijolite, etc., have been considered to act as such magmas (compare reviews in Heinrich, 1966, 1978; Sørensen, ed., 1974).

From our practical point of view, however, the criteria for carbonatite recognition in the field and laboratory are more important than the latest genetic story because most carbonatites in hand specimens are megascopically virtually indistinguishable from the much more abundant crystalline carbonates of a less exotic derivation (calcic and dolomitic marbles, hydrothermal veins and replacements, fault impregnations, etc., of any affiliation). The common presence of alkaline silicates (aegirine, melilite, monticellite) or characteristic accessory minerals (apatite, pyrochlore) can help, provided that they can be recognized by the geologist. Biotite, diopside, wollastonite, magnetite, etc. that are common in carbonatites are not diagnostic, because the same minerals occur frequently in many contact- and regionally metamorphosed marbles. By far the most important field criterion is the association with other alkaline rocks (although metasedimentary marbles may be common in the basement as well). The presence of fenite aureole and the intrusive, transgressive nature of most carbonatites are also important criteria.

Geochemically, members of the carbonatite association (that includes not only an orthomagmatic carbonatite but also almost any rock altered, filled or modified by the most mobile carbonatite-related fluids) are characterized by a strong enrichment in one or more of the following elements: Sr,Ba,Nb and REE. These elements (Verwoerd, 1967) are markedly enriched in carbonatites even by comparison with the alkaline silicate rocks, while others (P,F,Zr,Hf,Ti,Th) accumulate in both. The isotopic ratios of $^{13}C/^{12}C$; $^{87}Sr/^{86}Sr$; $^{16}O/^{18}O$ differ from those in limestones and marbles, but are similar to those of basaltic rocks (Heinrich, 1966).

33.8.5. Metal geochemistry and metallogeny

The extraordinary enrichment of carbonatites in certain minor elements has already been mentioned above, and it is further apparent from the comparison of average abundances of trace elements in carbonatites and other rocks (Table 33-12). The minor elements can either be camouflaged in major rock-forming or accessory minerals (e.g. Ba in calcite, Nb in perowskite) or form their own minerals (e.g. barite, pyrochlore). The camouflaged elements show a tendency to accumulate in the early, main (magmatic) stage of carbonatite development, whereas late, usually superimposed processes, tend to collect the camouflaged elements to form their own species (Heinrich, 1966).

Even where the original carbonatite magma was so enriched in the minor elements that they had already formed their own minerals in the early magmatic carbonate (e.g. Nb,REE in pyrochlore, Zr in zircon, REE in monazite), such minerals are widely dispersed and may form only low-grade (but often large tonnage) orebodies. Late-stage carbonatite emplacement and a variety of mobilization processes, on the other hand, tend to redistribute metals bound in the early minerals and concentrate them locally to form usually smaller, but higher-grade orebodies. The mineral species in the late-stage mineralizations may either remain the same (e.g. pyrochlore), or new species may appear (e.g. thorite, bastnaesite, baddeleyite, etc.). Kapustin (1971) prepared a detailed summary of carbonatite mineralogy.

Weathering processes superimposed on mineralized carbonatites are unusually effective metal concentrating agents. Due to the high

Table 33-12. Comparison of average abundance of trace elements in carbonatites and other rocks (in ppm; after Gold, 1963)

	CARBONATITE	AVERAGE IGNEOUS ROCK	LIMESTONE
Sc	10	13	1
Co	17	18	0.1
Ni	8	100	20
Cu	2.5	70	4
Ga	1	26	4
Y	96	20	30
Zr	1,120	170	19
Nb	1,951	20	0.3
Mo	42	1.7	0.4
Sn	4	32	under 1
La	516	40	under 1
Ce	1,505	40	11.5
Cr	48	117	11

solubility of carbonate, relatively small amounts of the insoluble residuum remain after the solution removal of a considerable thickness of the original rock, so that the rare minerals present in insignificant quantities in carbonatite can accumulate in economically significant amounts.

33.8.6. Major mineralization styles (Fig. 33-35, Table 33-13)

MINERALIZED VOLCANIC CARBONATITES

Continental volcanics of any composition rarely contain economic metallic mineralization and the volcanic carbonatites are no exception. Ti-magnetite is an abundant accessory mineral in carbonatite lavas and pyroclastics and it can accumulate in small quantities in streams draining such volcanics (e.g. in the braided ephemeral streams along the base of Oldoinyo Lengai). Perowskite and baddeleyite have been reported as accessory minerals in Kerimasi (Tanzania) lavas and pyroclastics (Gittins, 1966) and they are slightly enriched in the red-brown soil filling the crater floor. Accessory pyrochlore, monazite, allanite and other minerals can usually be detected in small quantities in exotic blocks of carbonatite and silicate alkaline rocks (e.g. ijolite), that occur frequently in vent breccias and in agglomerates (e.g. at Oldoinyo Lengai, in the Laacher Lake area, Germany; Heinrich, 1966).

Carbonatite volcanics have a high trace content of rare elements, for example 800 ppm Nb, 400 ppm La and 34.5 ppm U at Oldoinyo Lengai (Dawson, 1966) and a high but unspecified content of the same metals at Khanneshin (Alkhazov et al., 1977). No information is available regarding the fate of these metals during weathering and sedimentogenic reworking at the first locality, but numerous occurrences of yellow uranium oxides coating the stockworks of fractures in Neogene red sandstones, forming veinlets, sinters and crusts in sandstone and leached carbonatite and associated with chalcedonic crusts in residual clays, have been recorded at Khanneshin. The uranium has presumably been leached from the carbonatite volcanics and redeposited by the action of hot springs as well as supergene agents. REE, Nb, barite, fluorite, magnetite, etc. occurrences at this interesting locality are controlled by hydrothermally altered zones, believed to have formed at the subvolcanic level, postdating the surficial volcanic emplacement of carbonatite.

Mobile metals leached from volcanics (including the areally widespread fine ash) and carried away in solution can accumulate, under favourable conditions, in the sediments of interior basins such as playa lakes (e.g. Lakes Natron and Magadi in the Gregory Rift, Tanzania and Kenya).

ENDOCONTACT MINERALIZATION IN SUBVOLCANIC AND PLUTONIC CARBONATITES

(a) Main mass of carbonatite as a metallic ore.

Siderite (e.g. Chilwa Island, Kalkfeld) or rhodochrosite (Muambe Hill, Mozambique; Gittins, 1966) carbonatite bodies might appear to constitute, respectively, a readily accessible Fe or Mn ore. The highest Fe and Mn contents in the Chilwa siderite carbonatite were found to be only 11.1% Fe_2O_3 and 6.7% MnO_2 (Garson, 1966). The 60% FeO+MnO, mostly hematitic iron ore hosted by the siderite carbonatite of Kalkfeld, Namibia (Verwoerd, 1967), probably formed as a result of a superimposed hydrothermal event.

The Fe- and Mn-rich carbonatites, however, constitute a suitable protore, over which economically important Fe and Mn residual ores can form by tropical weathering (e.g. Mrima Hill, Kenya).

(b) Highly dispersed trace metal content in carbonatite

In the Iron Hill carbonatite stock, Powderhorn complex (Armbrustmacher, 1980), the average content of Th is 36.2 ppm, and it is bound in rather regularly dispersed fine accessory minerals (e.g. pyrochlore). The Th content alone is too low to be economic, but modern complex processing of the entire mass also recovering also REE (0.4% REE_2O_3), Nb (570 ppm Nb_2O_5), accessory apatite, rutile, as well as the rock dolomite, could be profitable. As a consequence, the available Th tonnage figures in Iron Hill are quoted in the literature and this sets this locality apart from numerous similarly (or better) endowed carbonatite localities for which equivalent considerations have not been made or published.

(c) Disseminated (scattered) minerals in carbonatite

Thinly disseminated pyrochlore is present almost universally

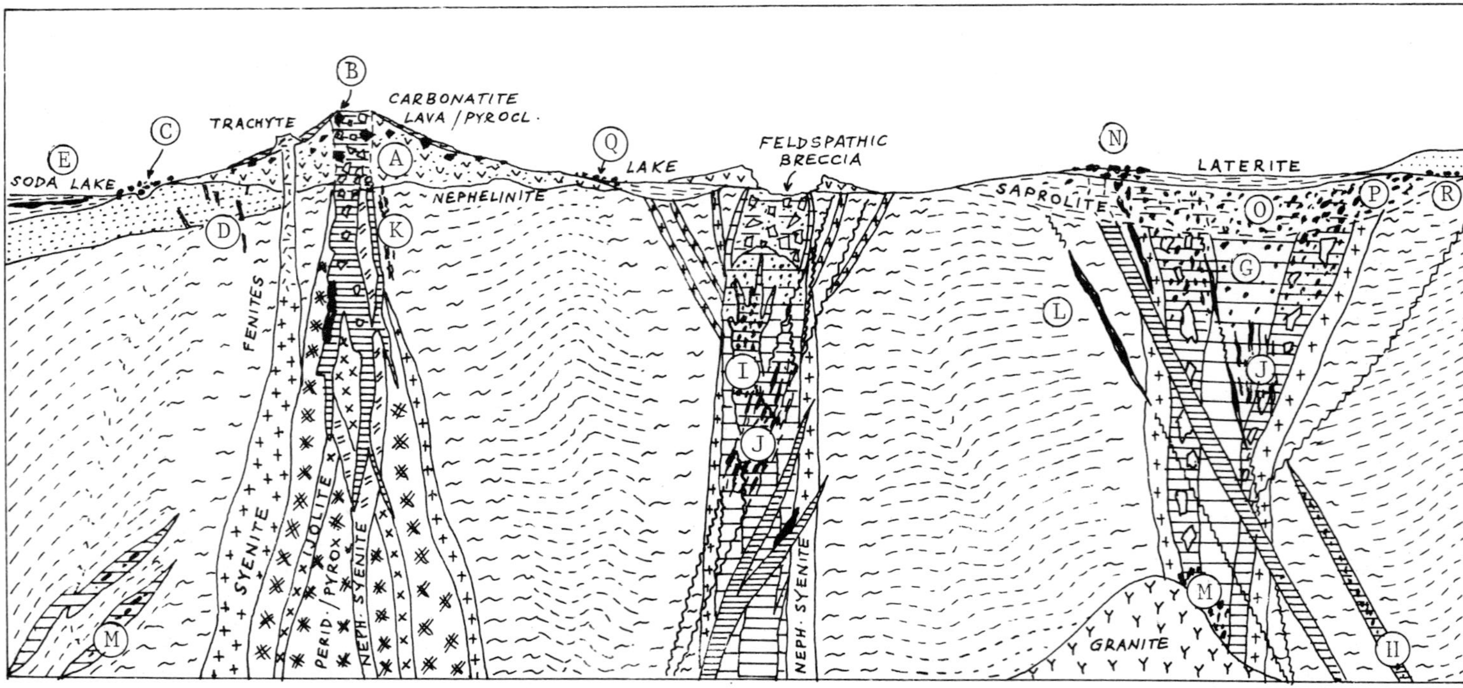

Figure 33-35. Principal patterns, rock types and mineralization styles in carbonatite complexes (see Table 33-13 for explanations of the letter codes). Carbonatites are shown by a horizontal line pattern. The closer the line spacing, the younger the carbonatite phase. From Laznicka (1984).

Table 33-13. Principal mineralization styles in carbonatites

	VOLCANIC CARBONATITES
(A)	High trace metals or accessory metallic minerals (Fe,Ti,U,Nb, REE,etc.)
(B)	Metallic minerals in subvolcanic or plutonic carbonatite inclusions brought from some depth
(C)	Placer Ti-magnetite, etc., in volcanics-derived sand
(D)	Infiltration veinlets of minerals (e.g. U) in volcanics and young associated sediments (X Tt U)
(E)	Metals (e.g. U,Th,REE,etc.) and fluorite in chemical sediments (e.g. playa evaporites)
	SUBVOLCANIC AND PLUTONIC CARBONATITES Recoverable dispersed anomalous trace metal contents (e.g. Nb,REE, Th,U) and disseminated minerals (e.g. pyrochlore), in:
(G)	Early stage (magmatic) carbonatite stocks;
(H)	Carbonatite dikes;
(I)	Late stage (metasomatic, vein) carbonatite;
(J)	Late stage hydrothermal metasomatites, veins and stockworks in carbonatites
(K)	Disseminated and replacement Fe,Ti,REE,Nb,Th ores in fenites
(L)	Mineralized fissure veins in fenites
(M)	Mineralized regionally or contact metamorphosed carbonatites
	WEATHERING CRUSTS
(N)	Surficial ore rubble (magnetite, hematite)
(O)	Residual resistate minerals; minerals formed by alteration of original ores or precipitated from leached components in laterite and saprolite
(P)	Buried mineralized regoliths
	PRODUCTS OF SEDIMENTOGENIC REWORKING
(Q)	Alluvial placers
(R)	Paleoplacers

present in carbonatites (in 0.X to 0.0X vol.% quantities) and so are magnetite, apatite (usually rich in REE), Nb-rutile, baddeleyite, bastnaesite and other minerals. Such disseminations become potentially economic ores when the content of pyrochlore reaches several percent, and that of apatite, magnetite, etc., several tens of percent. This is most common in the late-stage carbonatites. Recoverable amounts of disseminated magnetite have been outlined in the Clay-Howells carbonatite, Ontario (Currie, 1974).

In the mineralized portion of carbonatite in depth at Araxá, Brazil (Fig. 33-36), the ore minerals (pyrochlore, apatite, minor monazite) are hosted by a brown, biotite-rich xenolithic silica carbonatite (glimmerite) in the vicinity of younger phase veins and stringers of

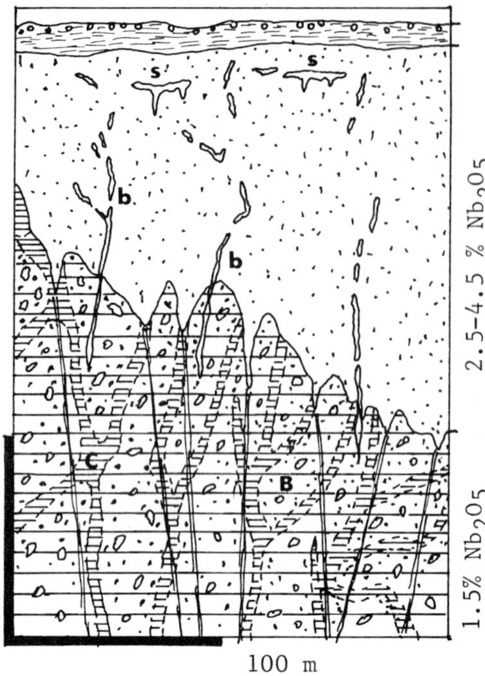

Fig. 33-36. Diagrammatic section across the pyrochlore orebody Barreiro near Araxá, M.G., Brazil. From LITHOTHEQUE, based on W. Betz, oral communication and tour, 1980.

white calcitic carbonatite (W. Betz, oral commun., 1980). In the Kangankunde, Malawi, carbonatite (Garson, 1966; Fig. 33-37) green monazite, pink florencite and pearly flakes of bastnaesite are present in carbonatites of several generations, but the spectacular concentration of monazite with strontianite there is located at the margins of late intrusive carbonatite dikes (Lonrho Staff, oral commun. 1980).

Carbonatite dikes themselves are commonly mineralized, for example at Ondurakorume, Namibia (von Backström, 1974; bastnaesite, strontianite, riebeckite, minor monazite, pyrochlore, cerianite, etc. in a dolomite carbonatite dike) and Mountain Pass, California (high-grade bastnaesite, lesser monazite, parisite, zircon mineralization in dolomite-barite and calcite carbonatite dikes; Molycorp Staff, guided tour, 1982).

Iron and base metal sulphides are relatively rarely disseminated in carbonatite. Widespread pyrrhotite, chalcopyrite and molybdenite are located in carbonatite dikes in the Callander Bay complex, Ontario (Currie, 1974), and minor chalcopyrite usually appears in carbonatites emplaced in mafic basement rocks.

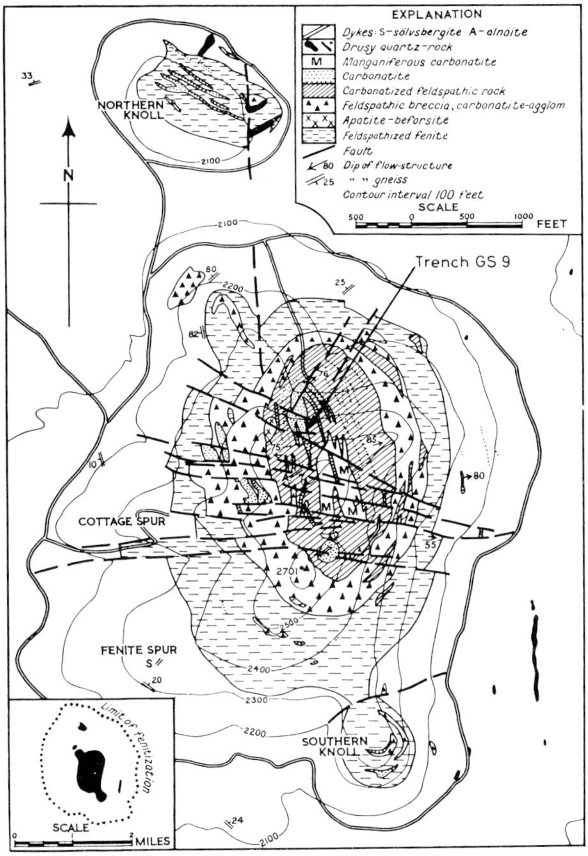

Fig. 33-37. Geology of the Mesozoic Kangankunde carbonatite complex, Malawi. From Garson (1966), courtesy of the Institute of Mining and Metallurgy, London, and the author.

(d) Late stage metasomatites and veins in carbonatites

In multiphase alkaline complexes, the sequentially younger phases, usually rich in volatiles, fill dilations in, or replace older phases. A variety of mineralized metasomatites and open space-filling orebodies (veins, stockworks, breccias) may result. The mineralizations incorporate both newly introduced metals as much as remobilized, earlier introduced metals.

Introduction of silica into earlier carbonatites may result in the formation of skarn-like metasomatites. In the Turiy Peninsula complex, Kola, U.S.S.R., such skarns are composed of diopside, monticellite, phlogopite and Na-pyroxenes. Lenses of replacement Ti-magnetite occur in the adjacent carbonatites (Samoylov and

Afanas'yev, 1978). Incomplete carbonatization of earlier silicate rocks, on the other hand, results in the generation of rocks that correspond compositionally to silica-carbonatites. The "apatite beforsite" dikes at Kangankunde, Malawi (Garson, 1966) contain brucite, portlandite, monticellite, melilite, gehlenite and apatite in a dolomite matrix, and are interpreted as being carbonatized apatite-rich olivine melilite nephelinites. The ore minerals are represented by baddeleyite.

Apatite-dominated breccias in carbonatites are quite common, and they usually have a high rare earths content (e.g. Eppawala, Sri Lanka), or carry a variety of accessory minerals (e.g. monazite in Glenover, South Africa). Magnetite, olivine, apatite rocks ("phoscorites") are frequently developed at the contacts of a carbonatite emplaced in an older pyroxenite, as in Palabora (Transvaal). Some carbonatized breccia zones and faults resemble dikes. The Th-rich "dikes" in Chilwa Island are brecciated zones in sideritic carbonate altered to goethite with interstratified quartz, fluorite and barite. Thoria (up to 2% ThO_2) is dispersed in goethite (Garson, 1966).

Ore-bearing hydrothermal veins and stockworks are common in carbonatites, and Palabora (Transvaal) is certainly the most prominent example. There (Verwoerd, 1967; Palabora Staff, 1976; Fig. 33-34,) a dike-like body of transgressive carbonatite was emplaced in an earlier, banded carbonatite along the intersection of two prominent fracture zones. A stockwork of transgressive carbonatite veinlets crosscut the older rocks along structural trends. This was further fractured and later stage hydrothermal copper sulphides introduced. The result is a large tonnage, low-grade stockwork and disseminated chalcopyrite, bornite, lesser chalcocite, valeriite, cubanite, pyrrhotite, etc. mineralization in carbonatite. In terms of ore distribution this is reminescent of the porphyry copper deposits and almost independent of the early development of the Palabora alkaline complex. Magnetite is widespread and the variety of by-product metals recovered in Palabora comes from two sources: (1) subsidiary metals introduced with the copper ores (Au,Ag,Pt) and (2) metals present in the early stage minerals scattered in carbonatite such as baddeleyite and U-thorianite (Zr,Th,U). A late-stage sulphide mineralization rich in copper, associated with metasomatic carbonatites and reminescent of Palabora, has been described at Bukusu, Uganda (Baldock, 1969).

A few additional examples taken from the large variety of ore-bearing hydrothermal veins in carbonatites, include the following: (a) high-grade (over 10% REE_2O_3) carbonate, bastnaesite, monazite veins from Wigu Hill, E. Tanzania (Deans, 1966) and (b) hematite-calcite veins in red coloured carbonatites, formed along N.W. fissures in the Fen Complex, Norway (Vokes, ed., 1960). These veins have been mined for iron in the past, and they contain high ThO_2 and REE contents (0.2 and 1%, respectively). (c) Veins and replacements of drusy quartz and purple fluorite in the Chilwa Island, Malawi carbonatites (Garson, 1966) carry small quantities of bornite and pyrochlore. (d) At Kangankunde, Malawi (Garson, 1966; Fig. 33-37), ramifying monazite-rich veinlets accompanied by strontianite, ankerite, quartz and barite, grade into spectacular, rounded bodies composed of radiating crystals of barite, strontianite, green monazite and other minerals.

ORES IN FENITIZED AUREOLES AROUND CARBONATITES

In multiphase carbonatite complexes, fenitization is not confined to exocontacts, but often affects its own members (earlier carbonatite and alkaline silicate rocks). In the Mountain Pass REE deposit, California, fenitization impairs spectacular lavender blue colours caused by pervasively distributed riebeckite both to a portion of the carbonatite and to the surrounding granite and syenite. It is related to the latest hydrothermal phase of alkaline magmatism, and not only to the carbonatite.

At some localities, early fenites are extensively carbonatized and so gradually acquire the character of metasomatic carbonatites (e.g. in Kangankunde, Malawi; Garson, 1966). Some such carbonatites are rheomorphically re-emplaced. Carbonatized fenites may carry accessory minerals, identical to those in magmatic carbonatites (apatite, pyrochlore, etc.). In addition to this, they host a variety of mineralized hydrothermal veins, such as: (a) vanadiferous Ti-magnetite veins (Tweerivier, South Africa; Verwoerd, 1967); (b) tabular bodies of Ti-magnetite in K-feldspar fenitized pyroxenite, Okorusu, Namibia (Verwoerd, 1967); (c) calcite, magnetite, pyrrhotite, galena, bastnaesite, apatite, etc. veins with finely dispersed ThO_2 and REE (0.67% and 9.1%, respectively), Itapirapua, Brazil (Gittins, 1966) and (d) quartz veins and stockworks in fenitized quartzite, containing synchisite, Glenover, South Africa (Verwoerd, 1967).

METAMORPHOSED CARBONATITES

Carbonatites affected by regional metamorphism as well as by thermal metamorphism at genetically unrelated contacts (e.g. with granites), have rarely been described. Heinrich (1966, p.376-385) reviewed the "metamorphosed carbonatites" of the Haliburton-Bancroft area, Ontario, located in the high-grade metamorphics of the Grenville province. The origin and interpretaion of these rocks is controversial. They were reviewed in greater detail in Volume 2.

Moore (1973) described greenschist to amphibolite facies metamorphosed carbonatites in the Strangways Range, 100 km N.N.E. of Alice Springs, Central Australia. Several occurrences in the Canadian Cordillera (Ice River, Verity, Lonnie) have been reviewed by Currie (1974). The metacarbonatites are usually inconcpicuous and difficult to distinguish from marbles. Some are discussed in Section 33.10.

WEATHERING-GENERATED RESIDUAL METAL OR MINERAL CONCENTRATIONS OVER CARBONATITES

The fully-developed regoliths over deep tropically weathered mineralized carbonates as in Araxá, contain the following zones running from top to bottom: (a) in situ or transported lateritic residual soil that may contain buckshot concretions, blocks or irregular horizons of ferricrete, or occasionally relict magnetite rubble; (b) powdery (clayous) saprolite, usually of brown or yellowish-brown colour; (c) irregular (pinnacled) top of karsted carbonatite and (d) fresh carbonatite.

Among the ore minerals present in regoliths one can distinguish (1) unaltered relic resistates having the same properties as in the fresh carbonatite (e.g. magnetite, partly pyrochlore, monazite, baddeleyite); (2) altered resistates (e.g. "pandaite", a pyrochlore in which the original Na and Ca were replaced by Ba and Sr in the weathering profile; francolite, etc.) and (3) newly formed compounds pseudomorphing the original minerals, or authigenically precipitated in the weathering profile (e.g. goethite, gibbsite, etc.). Some minerals (e.g. apatite) can occur in all the three modes.

In some apatite-rich carbonatites (e.g. Sokli, Finland; Vartiainen and Paarma, 1979), hard francolithic phosphorite breccia in the upper zones of a relict weathering profile, substitute for ferricrete blocks in the "normal" weathering profile. In Araxá and similarly zoned profiles (e.g. Catalão and Seis Lagoas, Brazil; Mrima Hill, Kenya), the upper lateritic zone is of little economic importance. The main orebody is contained in the brown saprolite. Such saprolite is composed of authigenic goethite, residual magnetite, variable quantities of apatite, residual and secondary pyrochlore ("pandaite") and monazite. Abundant barite forms drusy veins and veinlets.

At Araxá (Fig. 33-36), the reserves of the ochre are estimated to be 450 Mt, having a grade of 2.5% Nb_2O_5, 4% REE_2O_3, 30-40% Fe, 10-20% barite, and a variable grade of ThO_2. The calcite dissolution during weathering caused an improvement of about 160% in the Nb and REE ore grades, and resulted in an orebody that was cheap to mine (W. Betz, oral commun., 1980). This orebody supports the largest and highest grade Nb producer of the world.

At Mrima Hill, Kenya, a residue over carbonatite that is over 200 m deep, contains 40-70% goethite, 10% Mn oxides, barite, gorceixite and "pandaite", that represents an average grade of 30% Fe, 5% Mn, 0.7% Nb_2O_5, and under 5% REE_2O_3 (Deans, 1966). The Mn reserve is 136 Tt Mn.

In Kangankunde, Malawi (Garson and Morgan, 1978; Fig. 33-37), supergene leaching of carbonate from a complex ankerite, strontianite, monazite, bastnaesite carbonatite, resulted in surficial crusts greatly enriched in strontianite. Average grades range up to about 30-40%. The monazite content has also been enhanced, in contrast to bastnaesite which was largely destroyed and removed in solution. In addition to this, there are estimated to be about 100 Tt loose monazite in soils (Deans, 1966). In Glenover, South Africa (Verwoerd,

1967), the residual earth with hematite blocks contains yellow secondary U-Th patches and up to 1.16% Nb_2O_5.

Certain ore-bearing regoliths have not developed directly over carbonatites, but over mineralized fenites. In the Mata Creek section in the Araxá field, a saprolite formed over feldspathized fenite has a high rare earths content, based mostly on goyazite. Monazite, apatite and pandaite are subordinate. The reserves of this material are about 1.3 Mt with 12% REE_2O_3, 2.21% Nb_2O_5, 0.6% ThO_2, 0,05% U and 0.65% PbO (Grossi Sad and Torres, 1978).

Erdosh (1979) described a karst that pobably dates from the lower Cretaceous age, formed over the Proterozoic Cargill carbonatite in Ontario. The sinkholes are now mostly filled by apatite, goethite and clay residue and topped by a crandallite-rich blanket. The rare earths content in crandallite may have some economic significance.

Abundant Ti-magnetite rubble ore rests on the surface of the Bukusu, Uganda, carbonatite (Deans, 1966).

ORES RESULTING FROM PHYSICAL REDEPOSITION

Erdosh (1979) described significant quantities of an almost pure well-sorted apatite sand in Quaternary glacial lake sediments in the vicinity of the Cargill complex, Ontario, formed by reworking of residual apatite capping the carbonatite. The apatite has an enhanced REE content.

ORES RESULTING FROM CHEMICAL REDEPOSITION OF METALS BY GROUND OR METEORIC WATER

Alkhazov et al. (1977) mentioned radioactive stratiform mineralization formed in Quaternary soda lake beds as a result of leaching of uranium from the Quaternary carbonatite volcano Khanneshin in Afghanistan.

33.9. KIMBERLITES AND KIMBERLITIC DIATREMES

33.9.1. General

On a hand specimen scale, kimberlite is defined as "a porphyritic alkalic peridotite containing abundant phenocrysta of olivine and phlogopote and possibly geikielite and chromian pyrope in a fine grained groundmass of calcite and second generation olivine and phlogopite. It has accessory ilmenite, serpentine, chlorite, magnetite and perowskite" (A.G.I. Glossary of Geology, 1981).

When the term "kimberlite" is used to designate a lithologic field unit (for example, a host to diamonds), the term becomes much more complex. With the exception of the relatively deep-seated kimberlite dikes and sills, the "typical" (pipe-like) kimberlite occurrences are rather heterogeneous systems, strongly influenced by the depth of emplacement. A "kimberlite pipe" (or diatreme) as in northern Montana or South Africa (Hearn, 1968; Hawthorne, 1975; Fig. 33-38), starts in depth as a narrow sheet-like peridotite dike. With decreasing depth, it gradually changes into intrusive breccias. Together with the lithologic change, the shape of the intrusion also changes from an initially linear cross-section to an elliptical and ultimately a circular one. The middle and upper part of the kimberlite pipe is filled by a heterogeneous breccia of abundant subrounded to angular rock fragments in a finely crystalline matrix of serpentinite, calcite or dolomite, phlogopite, chlorite, etc.

The fragments are either (1) of subcrustal (mantle) origin and composed of garnet peridotite, lherzolite, eclogite, etc., or (2) of crustal origin, representing wallrocks brought in from great depth, or collected from above the present site and emplaced by gravity slumping and a free fall. The fragments corresponding to category (1) are considered as consanguinous with the kimberlite and are of considerable petrogenetic significance. The fragments coresponding to type (2) frequently represent samples of rock units in which the kimberlite was emplaced, but which have since been eroded away. Usually, the proportion of crustal rocks increases upward.

In the least eroded kimberlite pipes (e.g. Kasama, Mali), the pipe is fringed by a ring of marginal breccia and filled with kimberlitic ejecta and fragments of the wallrock that appear to have been washed into the pipe from the surface. Hawthorne (1975) coined the term "epiclastic kimberlite" for such an association. In outcrop, the kimberlite breccia bears the characteristic appearance of a yellowish or olive-green soft soily saprolite, enclosing less intensively weathered nodular fragments of ultrabasics, as well as hydrated flakes of phlogopite. Further downward, the "yellow soil" changes into a "blue soil".

A characteristic feature of a kimberlite pipe is the lack of thermal and hydrothermal metamorphism and metasomatism in both the endocontact (material within the pipe) and exocontact (the wallrock). Undisturbed fragments of coal and bitumen have been recorded in some African kimberlites.

Lorenz (1975) explained the origin of kimberlite diatremes by a cold emplacement, that followed a hot magmatic emplacement of the deep root dike. The gases and vapours causing fluidization and upward particle transport were geneated either by unmixing of the juvenile gas phase from the ultrabasic magma, or were vapourized local meteoric waters driven by the magmatic heat in the manner described for the phreatomagmatic eruptions. Kimberlites and diatremes (and also maars) may have a causal relationship, but not all diatremes are kimberlites.

Kimberlites are extremely rare rocks and in Africa, where they are most common, they cover an area of less than 100 km^2. There is, however, an extensive literature on the subject (e.g. Vol. 9, 1975, of the Physics and Chemistry of the Earth; several contributions in Wyllie, ed., 1967; Mitchell, 1970; Sobolev, 1977) because of their key scientific significance for providing material for the study of

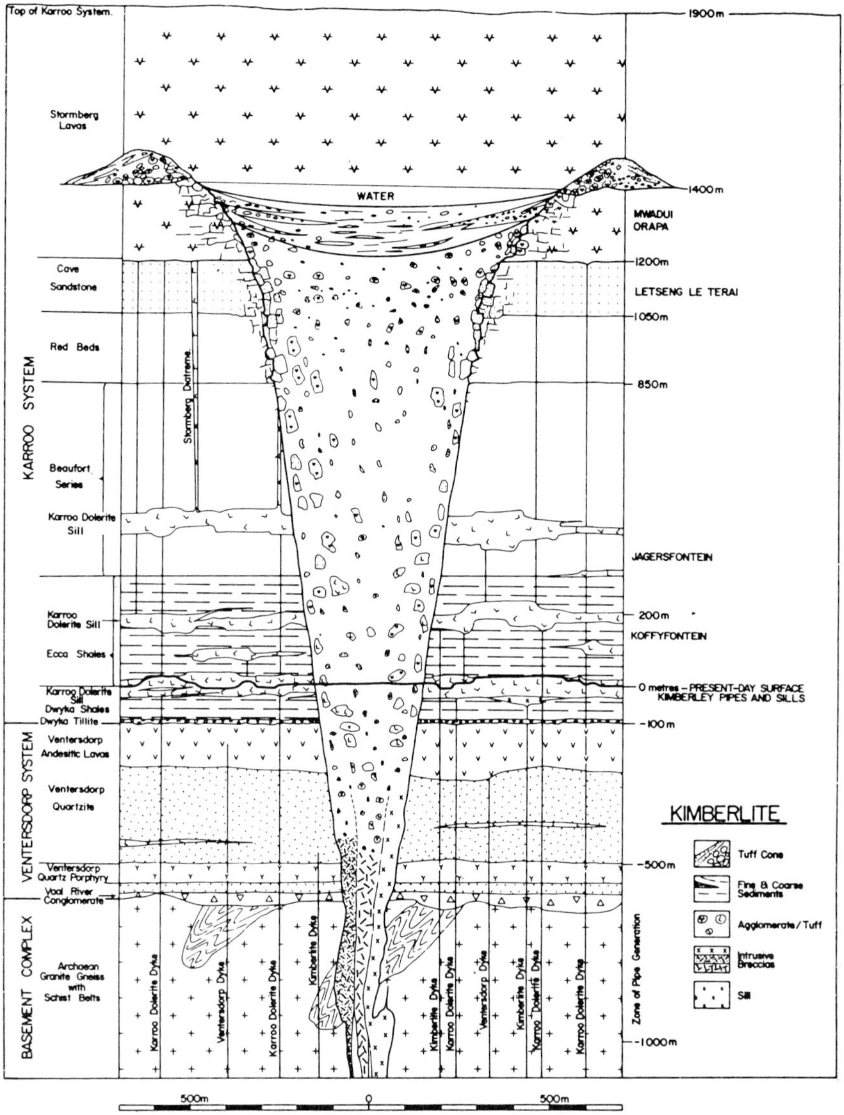

Fig. 33-38. Model of an African kimberlite pipe. From Hawthorne (1975), courtesy of the Pergamon Press.

the deep subcrustal layers of the earth, and because they are the only primary source of diamonds. This book, unfortunately, does not treat diamonds. Although kimberlites do contain a significant trace content of a variety of metals, they hardly represent a viable target for metalliferous exploration. They are too small. For that reason, kimberlites are entitled to only a brief review here.

33.9.2. Metal geochemistry and metallic mineralization

Like other members of the ultramafic-alkaline association, kimberlites are geochemically enriched in both the ultramafic (Ni, 0.13% in the Premier mine kimberlite; Cr, aver. 0.1%; Pt) and alkaline (REE, 300-800 ppm; Zr, 150-300 ppm; Nb, 100-300 ppm) suites of trace elements (Mitchell and Brunfelt, 1974). Kimberlitic breccias contain accessory Ti-magnetite, ilmenite, perowskite and chromite. In the Premier mine in Transvaal, kimberlite contains 2.5% ilmenite with 46-54% TiO_2, recoverable as a by-product of diamond mining. Wagner (1914) referred to the presence of small amounts of Pt metals in some South African kimberlite pipes.

At the Premier mine, the kimberlite breccia is cut by several intrusive magnetite, serpentine, calcite bodies, causing wallrock carbonatization (Robinson, 1975) and formerly interpreted as being carbonatite. Their metal content, however, is low (10-15% Fe oxides, 0.93-1.66% TiO_2 and 0.14% Ni).

Kashtanov (1967) reported several occurrences of redeposited alluvial bauxite, filling kimberlitic diatremes in eastern Siberia.

33.10. MINERALIZATIONS LINKED TO SUBTLE AND QUESTIONABLE ALKALINE INTRUSIVE PARENTS

This is a "convenience type", included in order to notify the exploration geologist of the possible presence of ores (e.g. of Nb,Th,REE,U,Zr) that are normally associated with alkaline complexes in terrains, where such "typical" complexes are not known. Most such terrains are situated in polyphase metamorphosed portions of mobile belts, usually in segments with an old, thick, formerly consolidated continental crust, later affected by retrograde metamorphism, fragmentation, "granitic" plutonism, etc. There is an overlap with the examples treated in Section 29.8.

The majority of the ore occurrences treated are fracture-filling or replacement veins as well as apparently conformable metamorphosed and mineralized rock bands. Some are dominated by feldspar gangue (albite or microcline) so that they resemble pegmatites. Others have quartz-carbonate or barite gangue. Some carbonate "veins" or "marble beds" may be thin carbonatite dikes or sills or metasomatically carbonatitized zones, lacking conspicuous wallrock fenitization, or with fenitization modified or obliterated by metamorphism. In most cases, the evidence for alkaline magmatic affiliation is insufficient or ambiguous. A considerable number of examples comes from the eastern portion of the North American Cordillera. Thorium-bearing veins, the most characteristic mineralization style there, have been reviewed by Staatz (1974).

Lemhi Pass District, Idaho and Montana

Lemhi Pass straddling the border of Idaho and Montana (Staatz, 1979) is underlain by low-grade metamorphosed middle to late Proterozoic schist and quartzite and is adjacent to an outcrop area of mid-Tertiary calc-alkaline volcanics (Challis Volcanics). About 250 Th,REE,Cu and sometimes Au-bearing fissure veins have been identified over a territory of about 130 km^2. The veins are composed of pink microcline, quartz, carbonate (calcite, ankerite, siderite) and sometimes barite. Thorite, in the form of small dark-red to reddish-brown crystals as well as a rarer monazite, are usually scattered within the carbonate. There is no obvious wallrock alteration.

In outcrop, the carbonates are completely leached out, and substituted by a radioactive limonitic residue. The veins are very inconspicuous, resembling pegmatites, and the presence of Th and REE is not megascopically apparent. The largest Last Chance vein is a tabular body, 1.4 km long and up to 13 m thick. The mineralization is mid-Tertiary, postdating the Challis Volcanics. There is a single occurrence of an inconspicuous (and questionable) carbonatite dike in the district, but more similar dikes have been reported by Heinrich (1966) in the nearby Ravalli County in Montana. Lemhi Pass contains a substantial share of the U.S. thorium reserves (242 Tt Th in ores with a grade better than 0.1% ThO_2), as well as 306 Tt REE. Similar veins without known alkaline rocks in the Cordillera have been recorded by Staatz (1974) from Diamond Creek, Idaho; Gold Hill, New Mexico; Quartzsite and Cottonwood, Arizona and Monroe Canyon, Utah.

Monumental Summit, Idaho

Carter (1973) described an occurrence of an earthy, brown to yellowish-brown rhabdophane ($(Ce,La)PO_4.H_2O$) and rare pyrite, rutile, zircon, etc. in association with Fe and Mn oxides and residual clays along a shear in a silicified marble near the contact with Proterozoic metaquartzite. This occurrence is located along the western paleo-caldera rim of the mid-Tertiary Challis Volcanics, and appears to be broadly contemporary with the Lemhi Pass ores. There are about 85 Tt of ore with 0.35% REE.

Iron Hill near Silver Cliff, Colorado

Hildebrand and Conklin (1974) described an interesting breccia dike composed of massive magnetite, apatite, molybdenite, and a variety of other minerals, cementing and replacing fragments of a Tertiary trachyandesite. The apatite contains 3.4-4% rare earths and the occurrence is, to some extent, reminescent of the Morro de Ferro deposit in Brazil. The ore occurrence is of mid-Tertiary age and is located in what is a typical calc-alkaline, epithermal Au-Ag and polymetallic district. Older (Cambrian) alkaline intrusions, however, are known in the Wet Mountains region in this vicinity.

Lonnie Prospect, Manson Creek, British Columbia

The small and uneconomic Lonnie Nb,Th,REE,U showing (Holland, 1955) has been included as an example of a marble band of uncertain origin conformable with foliation of adjacent metamorphics, that carry a low-grade mineralization corresponding compositionally to the alkaline suite. Sparsely scattered ore minerals are contained in a feldspathic marble grading into syenitic to pegmatitic rocks containing conspicuous acmite (columbite, ilmenorutile), and in massive to foliated marble (U-pyrochlore). The gneiss adjacent to the mineralized band contains abundant crossite and acmite, gradually dissapearing with increasing distance from the contact. The most likely genetic interpretation assumes a regionally metamorphosed carbonatite sill partly modified by interaction with a syn-metamorphic pegmatite.

Karonge, Burundi

Van Wambeke (1977) described the small deposit Karonge, consisting of veins and stockworks of monazite, bastnaesite with minor rhabdophane and cerianite in quartz, barite and goethite gangue. The veins are fault-controlled and hosted by Precambrian gneisses and metasediments, cut by pegmatites. About 2,300 t bastnaesite have been produced. Somewhat similar "pegmatite-like" veins with bastnaesite, parisite, chevkinite and monazite near Ifasina, Madagascar (Chantraine and Radelli, 1970) appear to be associated with 550 m.y. old intrusive syenite bodies.

An apatite-uranium mineralized field, U.S.S.R.

This unknown ore field (probably in the Kuznetsk Alatau or Minussinsk Basin region, Siberia) described by Kazansky and Laverov (1974) consists of replacement ore lenses, veins and pipes hosted by argillaceous and tuffaceous Ordovician limestone accompanied by andesite tuffs and porphyritic metabasalts. The orebodies consist of a fine crystalline fluorapatite, arshinovite, pitchblende, and a variety of base metal sulphides and Th,REE minerals. Silicification and albitization are the dominant alterations. The ore is Devonian or later, postdating the main orogeny in the region.

REFERENCES

Abbott, D. and Watson, D., 1975. Mineralogy of the Lake George antimony deposit, New Brunswick. C.I.M. Bulletin, July 1975, 11-113.
Abdullaev, Kh.M., 1957. Daiki i orudeneniye. Gosgeoltekhizdat, Moscow, 232 pp.
Abdullaev, Kh.M., 1964. Rudno-petrograficheskie provintsii. Nedra, Moscow, 135 pp.
Abdullaev, Kh.M. and Borisov, O.M., 1963. Evolution of central massifs. Int. Geol. Review, 7, 1361-1367.
Abele, G.C., 1977. Geologia del distrito minero de Cerro de Pasco. Centromin, Lima, unpublished, 40 pp.
Abreu, S.F., 1973. Recursos Minerais do Brasil, v. 2. Blücher, São Paulo, 754 pp.
Adamides, N.G., 1980. The form and environment of formation of the Kalavasos ore deposit, Cyprus. In A. Panayiotou, ed., Ophiolites. Cyprus Geol. Surv. Dept., Nicosia, 117-178.
Adams, B.W., 1973. The Ertsberg Project. Min. Mag., October 1973, 310-315.
Adams, J.E. and Rhodes, M.L., 1960. Dolomitization by seepage refluction. Amer. Assoc. Petr. Geol. Bull., 44, 1912-1920.
Adams, R.L., Burton, C.C.J., Druett, J.G., Hanson, N.H. and McNaught, I.S., 1976. The Rosebery and Hercules zinc-lead deposits. In Solomon, M. and Green, G.R., eds., Ore Deposits of Western Tasmania, 25th Int. Geol. Congr., Exc. Guide No. 31 AC, 31-49.
Adams, R.L. and Schmidt, B.L., 1980. Geology of the Elura Zn-Pb-Ag deposit. Bull. Austr. Soc. Explor. Geophys., 11, 143-146.
Addy, S.K. and Ypma, P.J.M., 1977. Origin of massive sulfide deposits at Ducktown, Tennessee; an oxygen, carbon and hydrogen isotope study. Econ. Geol., 72, 1245-1268.
Adyshev, M.M. and Kalmurzayev, K.E., 1968. The sedimentary-diagenetic origin of uranium mineralization in a carbonaceous cherty slate formation. Econ. Geol. 63 (abs.), 705-711.
Afanasyev, M.S., 1937. The Yukspor lovchorrite deposit. 17th Int. Geol. Congr., U.S.S.R., The Northern Excursion, Kola, Guidebook, Leningrad-Moscow, 115-119.
Ageykin, A.S., Chernyshov, N.M., Molotkov, S.P., and Bukovshin, V.V., 1968. Copper and nickel mineralization in upper Devonian terrigenous sediments of south-eastern Voronezh region. Intern. Geol. Rev., 11, 132-134.
Aguirre, L. and Mehech, S., 1964. Stratigraphy and mineralogy of the manganese deposits of Coquimbo Province, Chile. Econ. Geol., 59, 428-442.
Ahlfeld, F., 1949. La terminación meridional de la faja estañífera Boliviana. Mineria Boliviana, VI, No. 46, 5-9.
Ahlfeld, F., 1967. Metallogenetic epochs and provinces of Bolivia. Miner. Deposita 2, 291-311.
Ahlfeld, F. and Schneider-Scherbina, A., 1964. Los Yacimientos Minerales y de Hidrocarburos de Bolivia. Dept. Nac. Geol.

(Bolivia), Bol. No. 5, La Paz, 388 pp.

Aho, A.E., 1956. Geology and genesis of ultrabasic nickel-copper-pyrrhotite deposits at the Pacific Nickel Property, Southwestern British Columbia. Econ. Geol. 51, 444-481.

Ainemer, A.I. and Konshin, G.I., 1982. Rossypi Shelfovykh Zon Mirovogo Okeana. Nedra, Leningrad, 264 pp.

Akademia Nauk, S.S.S.R., 1981. Sul'fidnye Medno-Nikelevye Rudy Noril'skikh Mestorozhdenii. Nauka, Moscow, 481 pp.

Alabaster, T., Pearce, J.A., Mallick, D.I.J. and Elboushi, I.M., 1980. The volcanic stratigraphy and location of massive sulphide deposits in the Oman ophiolite. In A. Panayiotou, ed., Ophiolites; Cyprus Geol. Surv. Dept., Nicosia, 751-757.

Alaimo, R., Calderone. S. and Leone, M., 1970. Mineralogia e caratteri genetici degli ossidi di ferro a di manganese nelle concrezioni del flysch numidico siciliano. Atti. Acad. Sci. Lett., Palermo, Ser. 4, 30, 3-19.

Albandakis, N., 1981. The nickel-bearing iron ores in Greece. In UNESCO, Intern. Symp. on Metall. of Mafic and Ultramafic Complexes, Athens, 1980, 194-213.

Albers, J.P., 1961. Gold deposits in the French Gulch-Deadwood district, Shasta and Trinity Counties, California. U.S. Geol. Survey Profess. Paper 424-C, C1-C4.

Albers, J.P., 1966. Economic deposits of the Klamath Mountains. In E.H. Bailey, ed., Geology of Northern California; Calif. Divis. of Min. Geol. Bulletin 190, 51-62.

Albers, J.P., 1981. A lithologic-tectonic framework for the metallogenic provinces of California. Econ. Geol., 76, 765-790.

Albers, J.P. and Kleinhampl, F.J., 1970. Spatial relation of mineral deposits to Tertiary volcanic centres in Nevada. U.S. Geol. Survey Profess. Paper 700-C, C1-C10.

Albers, J.P. and Robertson, J.F., 1961. Geology and ore deposits of East Shasta copper-zinc district, Shasta County, California. U.S. Geol. Survey Profess. Paper 338, 107 pp.

Albritton, C.C., Jr., Richards, A., Brokaw, A.L. and Reinemund, J.A. 1954. Geologic controls of lead and zinc deposits in the Goodsprings (Yellow Pine) district, Nevada. U.S. Geol. Survey Bulletin 1010, 111 pp.

Aleva, G.J.J., 1979. Bauxitic and other duricrusts in Suriname: a review. Geol. en Mijnbouw, 58, 321-336.

Alkhazov, V.Yu., Atakishiyev, Z.M. and Azimi, N.A., 1977. Geology and mineral resources of the early Quaternary Khanneshin carbonatite volcano (southern Afghanistan). Int. Geol. Rev. 20, 281-285.

Allaart, J.H., 1973. Geological map of Greenland 1:100,000 Julianehåb, 60V.2 NORD. Meddel. om Grønland, Bd. 192, 4, 42 pp.

Allcock, J.B., 1982. Skarn and porphyry copper mineralization at Mines Gaspé, Murdochville, Québec. Econ. Geol. 77, 971-999.

Alleman, F. and Peters, T., 1972. The ophiolite-radiolarite belt of the North-Oman Mountains. Eclog. Geol. Helvet. 65, 3, 657-697.

Allen, D.G., Panteleyev, A. and Armstrong, A.T., 1976. Galore Creek.

Canad. Inst. Min. Metall. Spec. Volume 15, Montreal, 402-414.

Allen, E.T. and Zies, E.G., 1923. A chemical study of the fumaroles of the Katmai region. Nat. Geogr. Soc. Contr. Techn. Papers, Katmai Ser., No. 2, 75 pp.

Allen, J.R.L., 1965. A review of the origin and characteristics of Recent alluvial sediments. Sedimentology, 5, 89-191.

Altschuler, Z.S., 1980. The geochemistry of trace elements in marine phosphorites, Part I: characteristic abundances and enrichement. Soc. Econ. Paleont. Miner. Spec. Public. 29, 19-30.

Ambrus, J., 1977. Geology of the El Abra porphyry copper deposit, Chile. Econ. Geol. 72, 1062-1085.

Amstutz, G.C., 1959. Syngenese und Epigenese in Petrographie und Lagerstättenkunde. Schweiz. Miner. Petr. Mitteil. 39, 1-84.

Amstutz, G.C., 1968. Spilites and spilitic rocks. In H.H. Hess and A. Poldervaart, eds., Basalts, v. 2, Interscience, 737-753.

Amstutz, G.C., ed., 1974. Spilites and Spilitic Rocks. Springer Verlag, Berlin, 482 pp.

Amstutz, G.C., 1977. Time- and strata-bound features of the Michigan copper deposits (U.S.A.). In D.D. Klemm and H.-J. Schneider, eds., Time and Strata-Bound Ore Deposits. Springer Verlag, Berlin, 123-140.

Amstutz, G.C., ed., 1982. Ore Genesis, the State of the Art. Springer Verlag, Berlin, 804 pp.

Anatolyeva, A.I., 1972. Domezozoiskie krasnotsvetnye formatsii. Akad. Nauk S.S.S.R., Sibir. Otdel., Novosibirsk, Vyp. 190, 347 pp.

Ancion, Ch., Calembert, L. and Macar, P., 1956. Les ressources minerai de manganèse du sous-sol de la Belgique. Sympos. del Manganeso, 20th Int. Geol. Congr. Mexico, v.5, 9-17.

Anderson, A.L., 1958. Uranium, thorium, columbium and rare earth deposits in the Salmon region, Lemhi County, Idaho. Idaho Bur. Min. and Geol. Pamphlet 115, 81 pp.

Anderson, C.A. and Creasey, S.C., 1958. Geology and ore deposits of the Jerome area, Yavapai County, Arizona. U.S. Geol. Survey Prof. Paper 308, 185 pp.

Anderson, C.A., Scholz, E.A. and Strobell, J.D., Jr., 1955. Geology and ore deposits of the Bagdad area, Yavapai County, Arizona. U.S. Geol. Survey Profess. Paper 278, 103 pp.

Andrews, A.J. and Fyfe, W.S., 1976. Metamorphism and massive sulphide generation in oceanic crust. Geoscience Canada, 3, 84-87.

Angelelli, V., 1950. Recursos minerales de la Republica Argentina, v. 1., Yacimientos Metalíferos. CONI Edit., Buenos Aires, 535 pp.

Anonymous, 1976. Egypt's Abu Tartur phosphorite deposit. Min. Magaz. July 1976, 31-36.

Anonymous, 1978. Pianciano fluorite: development, appraisal. Min. Magaz. September 1978, 203-209.

Anonymous, 1979a. Exploration in British Columbia, 1978. Brit. Columb. Min. of Energy, Mines, Petr. Res., Victoria, p. E35, E36.

Anonymous, 1979b. Hemerdon-Britain's largest tungsten deposit. Min. Magaz. October 1979, 342-351.

Anonymous, 1981a. Mantos Blancos. Min. Magaz. December 1981, 458-469.

Anonymous, 1981b. Gravity surveys, the key at Neves-Corvo, Portugal. Min. Magaz. November 1981, p. 345.
Anthony, R.S., 1977. Iron-rich rhythmically laminated sediments in Lake of the Clouds, northeastern Minnesota. Limnol. and Oceanogr., 22, 45-54.
Antun, P., El Goresy, A. and Ramdohr, P., 1966. Ein neuartiger Typ "hydrothermaler" Cu-Ni Lagerstätten. Miner. Depos. 2, 113-132.
Antweiler, J.C. and Love, J.D., 1967. Gold-bearing sedimentary rocks in Northwest Wyoming-a preliminary report. U.S. Geol. Survey Circular 541, 12 pp.
Aramaki, S. and Ui, T., 1982. Japan. In R.S. Thorpe, ed., Andesites, J. Wiley, Chichester, 258-292.
Arkhangelskii, N.I., 1962. O tektonicheskikh zakonomernostyakh razmeshcheniya poleznykh iskopaemykh v mezozoe na vostochnom sklone Urala i v Zaural'e. Zakon. Razm. Polez. Iskop. 5, 464-482.
Armands, G., 1973. Geochemical studies of uranium, molybdenum and vanadium in a Swedish Alum shale. Stockh. Contrib. Geol. 27, 1-143.
Armbrustmacher, T.J., 1979. Replacement and primary magmatic carbonatites from the Wet Mountains area, Fremont and Custer Counties, Colorado. Econ. Geol. 74, 888-901.
Armbrustmacher, T.J., 1980. Abundance and distribution of thorium in the carbonatite stock at Iron Hill, Powderhorn District, Gunnison County, Colorado. U.S. Geol. Surv. Prof. Paper 1049-B, B1-B12.
Armstrong, F.C., 1974. Uranium resources of the future-"porphyry" uranium deposits. In Formation of Uranium Ore Deposits, Intern. Atomic Energy Agency, Vienna, 625-635.
Armstrong, J.E., 1949. Fort St.James map-area. Geol. Survey of Canada Memoir 252, 210 pp.
Arnould, A. and Routhier, P., 1956. Les gîtes de manganèse de Nouvelle-Calédonie. Un "type" de gisement de manganèse mèconnu: le type volcano-sédimentaire. Sympos. del Manganeso, v. IV, 20th Int. Geol. Congr. Mexico, 313-329.
Arrhenius, G., 1963. Pelagic sediments. In M.N. Hill, ed., The Sea, v.3., Interscience, New York, 655-727.
Arrhenius, G. and Bonatti, E., 1965. Neptunism and vulcanism in the ocean. In M. Sears, ed., Progress in Oceanography, v. 3., Pergamon Press, 7-22.
Arthurs, J.W., 1979. Mineral occurrences in the Solomon Islands. Solomon Isl. Geol. Surv. Bull. 13, 55 pp.
Ashley, P.M., 1969. The petrology and mineralization of the Coolac serpentine belt east of Brungle, New South Wales. Proc. Austral. Inst. Min. Metall. No. 230, 19-27.
Ashley, P.M., 1973. Petrogenesis of sulphide-bearing reaction zones in the Coolac ultramafic belt, New South Wales, Australia. Miner. Deposita, 8, 370-378.
Ashley, P.M., 1980. Geology of the Ban Ban zinc deposit, a sulfide-bearing skarn, Southeast Queensland, Australia. Econ. Geol. 75, 15-29.
Ashley, R.P. and Keith, W.J., 1976. Distribution of gold and other metals in silicified rocks of the Goldfield mining district,

Nevada. U.S. Geol. Surv. Profess. Paper 843-B, p. 17.
Asmus, H.E. and Porto, R., 1980. Differenças nos estágios iniciais da evolucão da margem continental brasileira: possíveis causas e implicações. Anais 31 Congr. Brasil. Geol., 1, 225-239.
Assad, R. and V. de Rui B., 1978. Deposito de bauxita na Serra dos Carajás. Annais do 30 Congr. Brasil. de Geol., Recife, 4, 1385-1390.
Aubert, G., 1968. Contribution à l'étude des granite à albite et mica blanc, riches en fluor, lithium, ètain, beryllium, niobium, tantale, etc. Les gisements de Montebras et d'Echassières. 23th Intern. Geol. Congr., Prague, v. 7, 215-232.
Auboin, J., 1965. Geosynclines. Elsevier, Amsterdam, 335 pp.
Aubrey, K.V., 1955. Germanium in some of the waste products from coal. Nature, 176, 128-129.
Auden, J.B., 1974. Afghanistan-West Pakistan. In A.M. Spencer, ed., Mesozoic-Cenozoic Orogenic Belts, Geol. Soc. London, 235-253.
Audley-Charles, M.G., 1972. Cretaceous deep-sea manganese nodules in Timor-implications for tectonics and olistostrome development. Nature, 240, 137-139.
Aver'yanov, I.P., Zelepukhin, L.P. and Chernov, A.A., 1968. O proyavlenii molibdenovoi mineralizatsii na vulkane Burevestnik. Geol. i Geofyz., 1968, No. 2., 116-120.
Avias, J. and Gonord, H., 1975. New Caledonia. In R.W. Fairbridge, ed., The Encyclopedia of World Regional Geology, Part 1; Dowden, Hutchinson, Ross, Stroudsburg, Pa., 365-372.
Bachman, G.O., Vine, J.D., Read, C.B. and Moore, G.W., 1959. Uranium-bearing coal and carbonaceous shale in the La Ventana mesa area, Sandoval County, New Mexico. U.S. Geol. Survey Bulletin 1055-J, 295-307.
Bäcker, H. and Richter, H., 1973. Die Rezente hydrothermal-sedimentäre Lagerstätte Atlantis-II Tief in Roten Meer. Geol. Rundschau, 62, 697-737.
Badham, J.P.N. and Williams, P.J., 1981. Genetic and exploration models for sulfide ores in metaophiolites, Northwest Spain. Econ. Geol. 76, 2118-2127.
Baibulatov, E.V., 1964. Achiktashskoe sernokolchedannoe mestorozhdenie i ego genezis. Akad. Nauk Kirghiz Rep., Frunze, 195 pp.
Bailey, E.H., Blake, M.C., Jr. and Jones, D.L., 1970. On-land Mesozoic oceanic crust in California Coast Ranges. U.S. Geol. Surv. Prof. Paper 700-C, C70-C81.
Bailey, E.H. and Everhart, D.L., 1964. Geology and quicksilver deposits of New Almaden district, Santa Clara County, California. U.S. Geol. Surv. Profess. Paper 360, 206 pp.
Bailey, R.V. and Childers, M.O., 1977. Applied mineral exploration with special reference to uranium. Westview Press, Boulder, 542 pp.
Baker, B.H., 1967. The Precambrian of the Seychelles Archipelago. In K. Rankama, ed., The Precambrian, v.3, Interscience, 122-132.
Baker, G., 1937. Orthite in some Victorian granites. Proc. Royal Soc. Victoria, 1, 50 pp.
Baklaev, Ya.P., 1972. Glavneishie zakonomernosti razmeshcheniya

zhelezorudnykh kontaktovo-metasomaticheskikh mestorozhdenii na Urale. Geol. Rud. Mestor. No.4, 1972, 3-16.

Baklaev, Ya.P., 1973. Kontaktovo-metasomaticheskie mestorozhdeniya zheleza i medi na Urale. Nauka, Moscow, 310 pp.

Baldock, J.W., 1969. Geochemical dispersion of copper and other elements at the Bukusu carbonatite complex, Uganda. Trans. Inst. Min. Metall., London, Sect. B., 78, B12-B18.

Baldwin, J.T., Swain, H.D. and Clark, G.H., 1978. Geology and grade distribution of the Panguna porphyry copper deposit, Bougainville, Papua New Guinea. Econ. Geol. 73, 690-702.

Ballard, R.D. and Uchupi, E., 1975. Triassic rift structure in Gulf of Maine. Amer. Assoc. Petrol. Geol. Bull. 59, 1041-1072.

Bamford, R.W., 1972. The Mount Fubilan (Ok Tedi) porphyry copper deposit, Territory of Papua and New Guinea. Econ. Geol. 67, 1019-1033.

Banaś, M. and Mochnacka, K., 1974. The vertical extent of ore minerals in some polymetallic deposits in Poland. Problems of Ore Deposition, Fourth IAGOD Sympos., Varna, 1, 45-52.

Banerji, P.K., Halder, D. and Ghosh, S., 1981. Metallogenesis in the Cretaceous embryonic ophiolites of Andaman-Nicobar island arc and its northern prolongation into the mountains along the Indo-Burman border. In Intern. Sympos. on Metal. of Mafic and Ultramaf. Compl., Athens, 1980, v.2, 21-38.

Banno, S., Takeda, H. and Sato, H., 1970. Geology and ore deposits in the Besshi Mining District. IAGOD, Tokyo-Kyoto meeting, Guidebook 9, Excursion B5, 29 pp.

Barabas, A. and Kiss, J., 1958. The genesis and sedimentary petrographic character of the enrichment of the uranium ore in Mécsek Mountains. Proc. 2nd U.N. Intern. Congr. on Peaceful Uses of Atom. Energy, Geneva, v.2, 388-401.

Baragar, W.R.A., 1977. Volcanism of the stable crust. Geol. Assoc. Canada Spec. Paper 16, 377-405.

Barbier, J., 1970. Zonalités géochimiques et métallogeniques dans le massif de Saint-Sylvestre (Limousin, France). Miner. Deposita, 5, 145-156.

Barbier, J., 1972. L'abondance des boxworks d'uraninite dans les granites: un guide stratégique possible pour les gisements d'uranium. Bull. B.R.G.M., Sect. II, No.1, 1-9.

Barbier, J., 1974. Continental weathering as a possible origin of vein-type uranium deposits. Miner. Deposita, 9, 271-288.

Barbosa, O., 1967. Projeto Chaminés. Geologia do região do Tríangulo Minéiro. Brasil, Dep. Nac. Prod. Min., 116 pp., unpublished.

Bardin, D., 1971. Les amas pyriteux de Sain-Bel (Rhône) liés au groupe spilites-kératophyres de la Brévenne. Bull. B.R.G.M., Sect. II, No.6, 17-41.

Bárdossy, G., 1982. Karst Bauxites. Bauxite Deposits on Carbonate Rocks. Devel. in Econ. Geol. 14, Elsevier, Amsterdam, 441 pp.

Barker, D.S., 1974. Alkaline rocks of North America. In H. Sørensen, ed., The Alkaline Rocks. Wiley, London, 160-171.

Barker, F., 1981. Introduction to special issue on granites and rhyo-

lites: a commentary for the nonspecialist. Journ. Geophys. Res. v. 86, No. B11, 10131-10135.
Barnabás, K., ed., 1966. Ásványtelepeink Földtana. Müszaki ed., Budapest, 316 pp.
Barnard, R.M. and Kistler, R.B., 1956. Stratigraphic and structural evolution of the Kramer sodium borate body, Boron, California. In J.L. Rau, ed., 2nd Sympos. on Salt, N. Ohio Geol. Soc., 133-150.
Barnes, H.L., ed., 1967. Geochemistry of hydrothermal ore deposits. Holt, Rinehart, Winston, New York, 670 pp.
Barnes, H.L., ed., 1979. Geochemistry of hydrothermal ore deposits. 2nd edition, Wiley, New York, 798 pp.
Barnes, M.P. and Simos, J.G., 1968. Ore deposits of the Park City District with a contribution on the Mayflower Lode. In J.D. Ridge, ed., Ore Deposits of the United States 1933-1967, A.I.M.E., New York, 1102-1126.
Barr, D.A., 1980. Gold in the Canadian Cordillera. Canad. Inst. Min. Metall. Bull., 73, 59-76.
Barr, D.A., Fox, P.E., Northcote, K.E. and Preto, V.A., 1976. The alkaline suite porphyry deposits-a summary. Canad. Inst. Min. Metall. Spec. Vol. 15, Montreal, 359-367.
Barron, B.J., Scheibner, E. and Slansky, E., 1976. A dismembered ophiolite suite at Port Macquarie, New South Wales. Rec. Geol. Surv. New South Wales, Sydney, 18, 69-102.
Barton, P.B., Jr., Bethke, P.M. and Roedder, E., 1977. Environment of ore deposition in the Creede mining district, San Juan Mountains, Colorado. Part III. Progress toward interpretation of the chemistry of the ore-forming fluid for the OH vein. Econ. Geol., 72, 1-24.
Bartura, Y. and Würzburger, V., 1974. The Timna copper deposit. In Sympos. Stratif. et Provinces Cupriferes, Liege, 277-285.
Basaltic Volcanism Study Project, 1981. Basaltic Volcanism on the Terrestrial Planets. Pergamon Press, 1286 pp.
Basta, E.Z. and Amer, H.I., 1969. El-Gidida iron ores and their origin, Bahariya oases, Western Desert, U.A.R. Econ. Geol. 64, 424-444.
Bastin, E.S., 1935. "Aplites" of hydrothermal origin associated with Canadian cobalt-silver ores. Econ. Geol. 30, 715-734.
Batchelor, B.C., 1979. Geological characteristics of certain coastal and offshore placers as essential guides for tin exploration in Sundaland, Southeast Asia. Geol. Soc. Malaysia Bull. 11, 283-313.
Bateman, A.M., 1950. Economic Mineral Deposits, 2nd ed. Wiley, New York, 916 pp.
Bateman, A., 1951. The Formation of Mineral Deposits. Wiley, New York, 371 pp.
Bateman, A.M. and McLaughlin, D.H., 1920. Geology of the ore deposits of Kennecott, Alaska. Econ. Geol., 15, 1-80.
Bateman, P.C., Clark, L.D., Huber, N.K., Moore, J.G. and Rinehart, C.D., 1963. The Sierra Nevada Batholith-a synthesis of recent work across the central part. U.S. Geol. Surv. Profess. Paper 414-D, 46 pp.

Bathurst, R.G.C., 1975. Carbonate Sediments and their Diagenesis. 2nd ed., Elsevier, New York, 658 pp.
Baturin, G.N., 1982. Phosphorites on the sea floor. Devel. in Sedim. 33, Elsevier, Amsterdam, 343 pp.
Bauer, G., Ebert, A., Hesemann, J., von Kamp, H., Müller, D., Pietzner, H., Podufal, P., Scherp, A. and Wellmer, F.W., 1979. Die Blei-Zink-Erzlagerstätten von Ramsbeck und Umgebung. Geol. Jahrbuch, D 33, Hannover, 377 pp.
Baumann, L., 1970. Tin deposits of the Erzgebirge. Trans. Inst. Min. Metall., London, Sect. B, 79, B68-B75.
Baumann, L., 1976. Introduction to Ore Deposits. Scottish Acad. Press, Edinburgh and London, 131 pp.
Baumer, A. and Fraser, R.B., 1975. Panguna porphyry copper deposit, Bougainville. In C.L. Knight, ed., Economic Geology of Australia and Papua New Guinea, 1, Metals. Austr. Inst. Min. Metall. Monograph 5, 855-866.
Bechstädt, T., 1975. Lead-zinc ores dependent on cyclic sedimentation. Miner. Deposita, 10, 234-248.
Beck, R., 1909. The Nature of Ore Deposits. McGraw Hill, New York, 685 pp.
Behr, S.H., 1965. Heavy-mineral beach deposits in the Karoo System. South Africa Geol. Surv. Memoir 56, 116 pp.
Béland, J., 1957. St. Magloire and Rosaire-St. Pamphile area, Quebec. Québec Geol. Surv., Geol. Rept. 76, 49 pp.
Béland, J., Marleau, R., Perusse, J. and Duquette, G., 1962. Metallic mineralization in the Appalachians of Southern Québec. Canad. Min. Journ., April 1962, 97-101.
Belevtsev, Ya.N., 1976. Istochniki rudnogo veshchestva pri metamorfogennom rudoobrazovanii. In Istochniki Rudnogo Veshchestva Endogennykh Mestorozhdenii, Nauka, Moscow, 66-84.
Belevtsev, Ya.N., 1979. Metamorfogennoe Rudoobrazovaniye. Nedra, Moscow, 275 pp.
Bell, K.G., 1956. Uranium in precipitates and evaporites. U.S. Geol. Surv. Prof. Paper 300, 381-386.
Bell, H., 1982. Strata-bound sulfide deposits, wall-rock alteration, and associated tin-bearing minerals in the Carolina Slate Belt, South Carolina and Georgia. Econ. Geol. 77, 294-311.
Bell, K.G., 1960. Uranium and other trace elements in petroleums and rock asphalts. U.S. Geol. Surv. Prof. Paper 356-B, B45-B65.
Bell, R.Y. and Delaney, G.D., 1977. Geology of some uranium occurrences in Yukon Territory. Geol. Surv. Canada Paper 77-1A, 33-37.
Bellido, E.B. and de Montreuil, L.D., 1972. Aspectos generales de la metalogenía del Perú. Serv. de Geol. y Min., Geol. Económica No. 1, Lima, 149 pp.
Beloussov, V.V., 1962. Basic Problems in Geotectonics. McGraw-Hill, New York, 809 pp.
Bender, F., 1975. Geology of Arabian Peninsula, Jordan. U.S. Geol. Surv. Profess. Paper 560-I, 36 pp.
Beneš, K. and Hanuš, V., 1967. Structural control and history of origin of hydrothermal metallogeny in western Cuba. Miner. Deposita,

2, 318-333.
Beneš, K. and Palas, M., 1964. Projevy radioaktivní metamorfózy uhlí z Č.S. části Dolnoslezské Pánve. Věstník Ú.Ú.G., 39, 201-204.
Benson, W.N., 1913-1918. The geology and petrology of the Great Serpentine Belt of New South Wales, Parts I-VII. Proc. Linn. Soc. N.S.W., 38:490-517, 569-596, 662-724; 40:121-173, 540-624; 42:223-245; 43:320-384.
Bentor, Y.K., 1956. The manganese occurrences at Timna (southern Israel), a lagoonal deposit. Symposium del Manganeso, 20th Int. Geol. Congr., Mexico, v. IV, 159-172.
Bentor, Y.K., ed., 1980. Marine phosphorites-geochemistry, occurrence, genesis. Soc. Econ. Paleont. Miner. Spec. Publ. 29, 249 pp.
Berbeleac, I. and David, M., 1982. Native tellurium from Musariu, Brad region, Metaliferi (Metalici) Mountains, Romania. In G.C. Amstutz, ed., Ore Genesis, the State of the Art. Springer Verlag, Berlin, 283-295.
Berce, B., 1958. Geologija živosrebrenega rudišča Idrija. Geologija, Ljubljana, v.4, 5-62.
Berg, H.C. and Cobb, E.H., 1967. Metalliferous lode deposits of Alaska. U.S. Geol. Surv. Bulletin 1246, 254 pp.
Berger, W.H., 1974. Deep-sea sedimentation. In C.A. Burk and C.L. Drake, eds., The Geology of Continental Margins. Springer Verlag, New York, 213-241.
Berger, V.I., 1977. Structure and genetic features of the Kelyansk antimony-mercury deposit. Int. Geol. Rev. 20, 295-307.
Bergstøl, S., 1972. The jacupirangite at Kodal, Vestfold, Norway. Miner. Deposita, 7, 233-246.
Berkman, D.A., 1975. Magnetite beach sand deposits of south-east Papua. In L.C. Knight, ed., Economic Geology of Australia and Papua New Guinea, v.1, Metals. Austr. Inst. Min. Metall., 1088-1092.
Berliner, M.H., 1949. Investigation of the Harding tantalum-lithium deposit, Taos County, New Mexico. U.S. Bureau of Mines Rept. Inv. 4607, 7 pp.
Bernard, A., Lagnay, P., and Leleu, M., 1972. A propos du rôle métallogénique du karst. 24th Int. Geol. Congr., Montréal, Sect. 4, 411-422.
Bernard, J.H., ed., 1969. Mineralogie Československa. Academia, Prague, 396 pp.
Bernard, J.H. and Hanuš, V., 1961. O časovém vztahu gemeridních granitů a turmalinisace k hydrothermálnímu zrudnění ve Spišsko-Gemerském Rudohoří. Věstník Ú.Ú.G., Prague, 36, 361-363.
Bernard, J.H. and Klomínský, J., 1974. Two groups of hydrothermal mineral associations related to variscan plutonic rocks in the Bohemian Massif. Problems of Ore Deposition, 4th IAGOD Symposium, Varna, v.2, 254-260.
Bernard, J.H., Rösler, H.J. and Baumann, L., 1968. Hydrothermal ore deposits of the Bohemian Massif. 23th Int. Geol. Congr., Prague, Guide to Excursion 22AC, 51 pp.
Bernardelli, A.L., 1982. Azul manganese deposit. Intern. Sympos.

on Archean and Early Proter. Geol. Evol. and Metallogenesis, Sept. 1982, Salvador, Brazil, 62-66.

Bernhard, J., 1966. Die Mitterberger Kupferkieslagerstätte Erzführung und Tektonik. Jb. Geol. Bundesanst., Wien, 109, 1-90.

Berning, J., Cooke, R., Hiemstra, S.A. and Hoffman, U., 1976. The Rössing uranium deposit, South-West Africa. Econ. Geol. 71, 351-368.

Bernouilli, D., Laubscher, H.P., Trümpy, R. and Wenk, 1974. Central Alps and Jura Mountains. In A.M. Spencer, ed., Mesozoic-Cenozoic Orogenic Belts. Geol. Soc. London, 85-108.

Berthold, G., 1980. Rohstoffwirtschaftlicher Länderbericht, XXV-Jugoslawien. Bundesanst. f. Geowiss., Hannover, 186 pp.

Besairie, H., 1959. Le socle crystallin de Madagascar. 20th Int. Geol. Congr. Mexico, Assoc. de Serv. Geol. Afric., 47-62.

Bespalov, V.F., 1971. Geologicheskoe stroeniye Kazakhskoi SSR. Nauka Kazakh S.S.R., Alma-Ata, 362 pp.

Beus, A.A., 1956. Berillii-otsenka mestorozhdenii pri poiskakh i razvedkakh. Gosgeoltekhizdat, Moscow. Engl. transl. "Beryllium", Freeman, San Francicso, 161 pp.

Beus, A.A., 1968. Albititovye Mestorozhdeniya. In V.I. Smirnov, ed., Genezis Endogennykh Rudnykh Mestorozhdenii. Nedra, Moscow, 303-377.

Beyschlag, F., Vogt, J.H.L. and Krusch, P., 1916. The Deposits of the Useful Minerals and Rocks, 2 vols., Macmillan, London, 1262 pp.

Bezrukov, P.L., Andrushchenko, P.F., Murdmaa, I.O. and Skornyakova, N.S., 1969. Fosfority na dne tsentral'noi chasti Tikhogo Okeana. Doklady Akad. Nauk S.S.S.R., v.185, 913-916.

Bezrukov, P.L., Petelin, V.P. and Skornyakova, N.S., 1970. Mineral'-nye resursy okeanov. In V.G. Kort, ed., Tikhii Okean, vol. Osadkoobrazovaniye. Nauka, Moscow, 322-340.

Bignell, R.D., Cronan, D.S. and Tooms, J.S., 1976. Red Sea metalliferous brine precipitates. Geol. Assoc. Canada Spec. Pap. 14, 147-179.

Bigot, M., 1981. Quelques données sur l'environement géologique et la gîtologie des occurrences cupro-manganésifères du Wādí Araba (Royaume Hashemite de Jordanie). Bull. de B.R.G.M., Sect.2, 1980-81, 153-163.

Bigotte, G. and Obelianne, J.M., 1968. Découverte de minéralisations uranifères au Niger. Miner. Deposita, 3, 317-333.

Bilibin, Yu.A., 1955. Metallogenic provinces and metallogenic epochs. Geol. Bull., Dept. Geol., Queens College, Flushing, 1968, 35 pp.

Bilibina, T.V., Afanas'eva, M.A. and Barkanov, I.V., 1976. Geologiya i metallogeniya shchitov drevnikh platform SSSR. Nedra, Leningrad, 339 pp.

Bird, J.M. and Weathers, M.S., 1977. Native iron occurrences of Disko Island, Greenland. Jour. Geol., 85, 359-371.

Bischoff, J.L. and Piper, D.Z., eds., 1979. Marine geology and oceanography of the Pacific manganese nodules province. Plenum Press, New York, 842 pp.

Bittencourt, A.C.da S.P., and Boas, G.de S.V., 1977. Ocorrência de

chamosita nos sedimentos recentes da Bahía de Aratu (BA). Revista Brasil. de Geocienc., 7, 230-238.
Bjørlykke, A. and Sangster, D.F., 1981. An overview of sandstone lead deposits and their relationship to red-bed copper and carbonate-hosted lead-zinc deposits. Econ. Geol. 75th Anniv. Volume, 179-213.
Black, R. and Girod, M., 1970. Late Paleozoic to Recent igneous activity in West Africa and its relationship to basement structure. In T.N. Clifford and I.G. Gass, eds., African Magmatism and Tectonics. Oliver and Boyd, Edinburgh, 185-210.
Blackadar, R.G., 1956. Differentiation and assimilation in the Logan Sills, Lake Superior District, Ontario. Amer. Journ. Science, 254, 623-645.
Blain, C.F. and Andrew, R.L., 1977. Sulphide weathering and the evolution of gossans in mineral exploration. Minerals Sci. Engin. 9, 3, 119-150.
Blake, D., 1972. Regional and economic geology of the Herberton/ Mount Garnet area, Herberton Tinfield, North Queensland. Bur. Min. Res. Geol. Geoph., Canberra, No. 124, 265 pp.
Blake, M.C., Jr. and Morgan, B.A., 1976. Rutile and sphene in blueschist and related high-pressure facies rocks. U.S. Geol. Surv. Prof. Paper 959-C, 6 pp.
Blanchard, R., 1968. Interpretation of leached outcrops. Nevada Bur. of Mines Bull. 66, 196 pp.
Blaxland, A.B., 1976. Rb-Sr isotopic evidence for the age and origin of the Ivigtut granite and associated cryolite body, South Greenland. Econ. Geol., 71, 864-869.
Blecha, M., 1974. Batchawana area-a possible Precambrian porphyry copper district. C.I.M. Bulletin, August 1974, 71-76.
Blissenbach, E. and Fellerer, R., 1973. Continental drift and the origin of certain mineral deposits. Geol. Rundschau, 62, 812-840.
Blissett, A.H., 1962. Geological Survey explanatory report, one mile geol. map series, K 55-5-50-Zeehan. Tasmania Dept. Mines, 272 pp.
Blockley, J.G., 1975. Peak Hill manganese deposit, W.A. In C.L. Knight, ed., Economic Geology of Australia and Papua New Guinea. Vol. 1, Metals. Austr. Inst. Min. Metall., 1020-1023.
Blondel, F. and Raguin, E., 1935. Les gisements de cuivre de la France et des possessions françaises. 16th Int. Geol. Congr., Washington, D.C., Copper Resources of the World, 2, 565-580.
Bloomer, R.O. and de Witt, W., Jr., 1941. Titaniferous sandstone near Buena Vista, Virginia. Econ. Geol., v. 36, 745-747.
Bloomfield, K., 1970. Orogenic and post-orogenic plutonism in Malawi. In T.N. Clifford and I.G. Gass, eds., African magmatism and Tectonics. Oliver and Boyd, Edinburgh, 119-156.
Boast, A.M., Coleman, M.L. and Halls, C., 1981. Textural and stable isotopic evidence for the genesis of the Tynagh base metal deposit, Ireland. Econ. Geol., 76, 27-55.
Boettcher, A.L., 1967. The Rainy Creek alkaline-ultramafic igneous complex near Libby, Montana, I: ultramafic rocks and fenite. Journ. Geology, 75, 526-553.

Bogdanov, B., 1982. Bulgaria. In F.W. Dunning, W. Mykura and D. Slater, eds., Mineral Deposits of Europe, 2, Southeast Europe. Miner. Soc./Inst. Min. Metall., London, 215-232.
Bogdanov, N.A., 1969. Thalasso-geosynclines of the Circum-Pacific belt. Geotectonics, 3, 141-147.
Bogdanov, Yu.V., Gur'yanova, V.N. and Mirayes, M., 1965. Outline of metallogeny of copper deposits of Cuba. Intern. Geol. Review 8, 1218-1225.
Bogdanov, Yu.V., and others, 1973. Stratifitsirovannye Mestorozhdeniya Medi SSSR. Nauka, Leningrad, 312 pp.
Bøggild, O.B., 1953. The mineralogy of Greenland. Medd. Grønland v. 149, No.3, 442 pp.
Bogli, A., 1980. Karst Hydrology and Physical Speleology. Springer Verlag, Berlin, 284 pp.
Bohdanowicz, K., 1952. Surowce Mineralne Świata, v. 1. Panstw. Inst. Geol., Warsaw, 476 pp.
Bohse, H., Brooks, C.K. and Kunzendorf, H., 1971. Field observations on the kakortokites of the Ilímaussaq intrusion, South Greenland, including mapping and analyses by portable X-ray fluorescence equipment for zirconium and niobium. Grønl. Geol. Unders. Rapport Nr. 38, 43 pp.
Bois, J.P., 1972. Carboire-un nouveau type de minéralisation stratiform en zinc-plomb dans les Pyrenees françaises. 24th Int. Geol. Congr. Montréal, v.4, 363-372.
Boldt, J.R., 1967. The Winning of Nickel. Longman, Toronto, 487 pp.
Bonatti, E., 1975. Metallogenesis at oceanic spreading centres. Ann. Rev. of Earth and Planet. Sciences, 146-166.
Bonatti, E., Fisher, D.E., Joensun, O., Rydell, H.S. and Beyth, M., 1972. Iron-manganese-barium deposit from the Northern Afar Rift (Ethiopia). Econ. Geol. 67, 717-730.
Bonatti, E., Guerstein-Honnorez, B.-M. and Honnorez, J., 1976. Copper-iron sulfide mineralizations from the Equatorial Mid-Atlantic Ridge. Econ. Geol. 71, 1515-1525.
Bonatti, E., Kraemer, T. and Rydell, H., 1972. Classification and genesis of submarine iron-manganese deposits. In D. Horn, ed., Ferromanganese Deposits on the Ocean Floor. Natl. Sci. Foundation, Washington, D.C., 149-165.
Bondam, J. and Brown, H., 1955. The geology and mineralization of the Mesters Vig area, East Greenland. Meddel. om Grønland, Bd. 135, Nr.7, 40 pp.
Bonnichsen, B., 1972. Southern part of Duluth Complex. In P.K. Sims and G.B. Morey, eds., Geology of Minnesota; Minn. Geol. Survey, St. Paul, 361-387.
Bonow, C.d.W. and Issler, R.S., 1980. Reavaliação e aspectos económicos do jazimento de terras raras e ferro-ligas do Lago Esperanca complexo carbonatítico dos Seis Lagoas-Amazonas-Brazil. Anais do 31. Congr. Brasil. de Geol., v.3, 1431-1443.
Bookstrom, A.A., 1977. The magnetite deposits of El Romeral, Chile. Econ. Geol. 72, 1101-1130.
Borisenko, L.F., 1974. Deposits of titanium. In V.I. Smirnov, ed.,

Ore Deposits of the U.S.S.R., v.1, Engl. Transl., Pitman, London, 237-255.

Borley, G.D., 1974. Oceanic islands. In H. Sørensen, ed., The Alkaline Rocks. Wiley, London, 311-330.

Borodaevskaya, M.B., 1962. Strukturnye usloviya lokalizatsii rudnykh tel v kolchedannykh mestorozhdeniyakh Yuznogo Urala. Zakon. Razm. Polez. Iskop. V, 304-320.

Borodaevskaya, M.B. and Rozhkov, I.S., 1974. Deposits of gold. In V.I. Smirnov, ed., Ore Deposits of the U.S.S.R., v. 3, Engl. Transl., Pitman, London, 3-81.

Borovikov, P.P. and L'vova, I.A., 1962. Tipy mestorozhdenii vermikulita, ikh promyshlennoe znacheniye i napravleniya dalneishikh geologorazvedochnykh rabot. Zakon. Razm. Polez. Iskop. VI, 470-488.

Borukaev, R.A. and Shcherba, G.N., eds., 1967. Geologiya i metallogeniya Uspenskoi Tektonicheskoi Zony. Nauka, Alma-Ata, 4. vols.

Bossi, G.E. and Viramonte, J.G., 1977. Contribución al conocimiento de la petrología de los yacimientos ferríferos sedimentarios de Zapla y Unchimé (provincias de Jujuy y Salta, Rep. Argentina). II. Congr. Ibero-Amer. de Geol. Econ., V.5, Buenos Aires, 181-202.

Bostrom, K. and Peterson, M.N.A., 1966. Precipitates from hydrothermal exhalations on the East Pacific Rise. Econ. Geol. 61, 1258-1265.

Bouladon, J., 1952. Plomb et zinc. 19th Int. Geol. Congr., Alger, Geologie des Gites Mineraux Marocains, 179-221.

Bouladon, J. and de Lapparent, A.F., 1975. Le minerai de fer d'Hajigak (Afghanistan). Position stratigraphique, cadre geologique et type du gisement. Miner. Deposita, 10, 13-25.

Bouladon, J. and Jouravsky, G., 1952. Manganese. In Geologie des gîtes minéraux Marocains, 19th Int. Geol. Congr., Alger. Monogr. Reg. Ser. Maroc, 1, 45-80.

Bouška, V., 1981. Geochemistry of Coal. Elsevier, Amsterdam, 284 pp.

Bowden, P., 1974. Oversaturated alkaline rocks: granites, pantellerites and comendites. In H. Sørensen, ed., The Alkaline Rocks; Wiley, London, 109-123.

Bowen, H.J.M., 1966. Trace elements in biochemistry. Academic Press, London, 241 pp.

Bowen, K.G. and Whiting, R.G., 1975. Gold in the Tasman Geosyncline, Victoria. In C.L. Knight, ed., Economic Geology of Australia and Papua New Guinea. Austr. Inst. Min. Metall., v.1, 647-658.

Boyle, R.W., 1951. An occurrence of native gold in an ice lens, Giant Yellowknife Gold Mines, Yellowknife, Northwest Territories. Econ. Geol. 46, 223-227.

Boyle, R.W., 1979. The geochemistry of gold and its deposits. Geol. Survey Canada, Bulletin 280, 584 pp.

Boyle, R.W. and Dass, A.S., 1971. Origin of the native silver veins at Cobalt, Ontario. Canad. Mineralogist, 11, 414-417.

Boyle, R.W. and Jambor, J.L., 1966. Mineralogy, geochemistry and origin of the Magnet Cove barite-sulphide deposit, Walton, N.S. Canad. Inst. Min. Metall. Transact., 69, 394-413.

Boyle, R.W., Wanless, R.K. and Stevens, R.D., 1976. Sulfur isotope investigation of the barite, manganese and lead-zinc-copper-silver deposits of the Walton-Cheverie area, Nova Scotia, Canada. Econ. Geol., 71, 749-762.

Bradley, W.H., 1964. Geology of the Green River Formation and associated Eocene rocks in southwestern Wyoming and adjacent parts of Colorado and Utah. U.S. Geol. Surv. Prof. Paper 496-A, 86 pp.

Bradley, W.H. and Eugster, H.P., 1969. Geochemistry and paleolimnology of the trona deposits and associated authigenic minerals of the Green River Formation of Wyoming. U.S. Geol. Surv. Profes. Paper 496-B, 71 pp.

Braitsch, O., 1971. Salt Deposits. Their Origin and Composition. Springer Verlag, New York, 297 pp.

Branch, C.D., 1966. Volcanic cauldrons, ring complexes and associated granites of the Georgetown Inlier, Queensland. Bur. Miner. Res. Geol. and Geoph., Canberra, Bulletin No. 76, 158 pp.

Brathwaite, R.L., 1974. The geology and origin of the Rosebery ore deposit, Tasmania. Econ. Geol., 69, 1086-1101.

Brecke, E.A., 1979. A hydrothermal system in the Midcontinent region. Econ. Geol. 74, 1372-1335.

Brevart, O., Dupré, B. and Allègre, C.J., 1982. Metallogenic provinces and the remobilization process studied by lead isotopes: lead zinc ore deposits from the southern Massif Central, France. Econ. Geol., 77, 564-575.

Bright, M.J. and Jonson, D.C., 1976. Glacier Gulch (Yorke-Hardy). Canad. Inst. Min. Metall. Spec. Vol. 15, 455-461.

Brigo, L., Kostelka, L., Omenetto, P., Schneider, H.-J., Schroll, E., Schulz, O. and Štrucl, I., 1977. Comparative reflections on four Alpine Pb-Zn deposits. In Klemm, D.D. and H.-J. Schneider, Time and Stratabound Ore Deposits. Springer Verlag, Berlin, 273-293.

Brimhall, G.H., Jr., 1977. Early fracture-controlled disseminated mineralization at Butte, Montana. Econ. Geol., 72, 37-59.

Brinckmann, J. and Hinze, C., 1981. On the geology of the Bawdwin lead-zinc mine, northern Shan State, Burma. Geol. Jahrbuch, Hannover, D43, 7-45.

Brinkmann, R., 1976. Geology of Turkey. Elsevier, Amsterdam, 153 pp.

Briskey, J.A., Jr. and Bellamy, J.R., 1976. Bethlehem Copper's Jersey, East Jersey, Huestis and Iona deposits. Canad. Inst. Min. Metall. Spec. Volume 125, 105-119.

Brøgger, W.C., 1980. Die Mineralien der Syenit-pegmatitgänge der Sudnorwegischen Augit- und Nephelinsyenite. Z. Kristallogr. Miner., 16, 1-663.

Brondi, A., Carrara, C. and Polizzano, C., 1973. Uranium and heavy metals in Permian sandstones near Bolzano (northern Italy). In G.C. Amstutz and A. Bernard, eds., Ores in Sediments; Springer Verlag, Berlin, 65-77.

Brooke, W.J.L., 1975. Cobar mining field. In C.L. Knight, ed., Economic Geology of Australia and Papua New Guinea, v. 1, Metals; Aust. Inst. Min. Metall., Monogr. 5, 683-694.

Brown, A., 1948. North Alabama brown iron ores. U.S. Bureau of Mi-

nes Rept. Inv. RI 4229, 82 pp.
Brown, E.H., Bradshaw, J.Y. and Mustoe, G.E., 1979. Plagiogranite and keratophyre in ophiolite on Fidalgo Island, Washington. Geol. Soc. Amer. Bull. Pt.1, 90, 493--507.
Brown, G.C., 1979. The changing pattern of batholith emplacement during earth history. In M.P. Atherton and J. Tarney, eds., Origin of Granite Batholiths, Geochemical Evidence. Shiva Publ., Orpington, 106-115.
Brown, J.C. and Dey, A.K., 1975. The Mineral and Nuclear Fuels of the Indian Subcontinent and Burma. Oxford Univ. Press, Delhi, 517 pp.
Brown, A.V., Rubenach, M.J. and Varne, R., 1980. Geological environment, petrology and tectonic significance of the Tasmanian Cambrian ophiolitic and ultramafic complexes. In A. Panayiotou, ed., Ophiolites; Cyprus Geol. Surv. Dept. Nicosia, 649-659.
Brown, P.R.L., 1969. Sulfide mineralization in a Broadlands geothermal drill hole, Taupo volcanic zone, New Zealand. Econ. Geol. 64, 156-159.
Brown, W.H. and Weinberg, E.L., 1968. Geology of the Austinville-Ivanhoe District, Virginia. In J.D. Ridge, ed., Ore Deposits in the Unites States 1933-1967, A.I.M.E., New York, v.1, 169-186.
Brusca, C., Dessau, G., Leroy Jensen, M., and Perna, G., 1972. The deposits of argentigerous galena within the Bellerophon Formation (upper Permian) of the Southern Alps. Geologija, Ljubljana, v. 15, 159-176.
Bryant, D.G., 1968. Intrusive breccias associated with ore, Warren (Bisbee) mining district, Arizona. Econ. Geol. 63, 1-12.
Bryant, D.G. and Metz, H.E., 1966. Geology and ore deposits of the Warren mining district. In S.R. Titley and C.L. Hicks, eds., Geology of the Porphyry Coppoer Deposits, Southwestern North America. Univ. of Arizona Press, Tucson, 189-203.
Bryant, B. and Reed, J.C., Jr., 1966. Mineral resources of the Grandfather Mountain window and vicinity, North Carolina. U.S. Geol. Survey Circular 521, 13 pp.
Bryant, B. and Reed, J.C., Jr., 1970. Geology of the Grandfather Mountain Window and vicinity, North Carolina and Tennessee. U.S. Geol. Surv. Prof. Paper 615, 190 pp.
Bryner, L., 1969. Ore deposits of the Philippines-an introduction to their geology. Econ. Geol., 64, 644-666.
Bublichenko, N.L., Vorob'ev, Yu.Yu., Ivankin, P.F., Inshin, P.V., Kuzebnyi, V.S., Lyubetskii, V.N., Popov, V.V. and Stuchevskii, N.I., 1972. Printsipy i metody prognozirovaniya mednokolchedannogo i polimetalicheskogo orudeneniya. Nedra, Moscow, 256 pp.
Buckland, F.C., 1959. Germanium in British Columbia. West. Miner. and Oil Rev., v.32, 9, 30-34.
Buckley, E.R., 1909. Geology of the disseminated lead deposits of St. Francois and Washington Counties. Missouri Bur. Geol., Mines, 9, pt. 1, 259 pp.
Buday, T., Cambel, B. and Mahel', M., 1962. Vysvetlivky k prehl'adnej geologickej mape ČSSR 1:200,000, Wien-Bratislava. Geofond, Bratislava, 249 pp.

Buddington, A.F., 1959. Granite emplacement with special reference to North America. Geol. Soc. Amer. Bull. 70, 671-747.

Buddington, A.F. and Chapin, T., 1929. Geology and mineral deposits of South-Eastern Alaska. U.S. Geol. Surv. Bull. 800, 398 pp.

Bugge, C., 1917. Kongsbergfeltets geologi. Norg. Geol. Unders. No. 82, 284 pp.

Bugge, J.A.W., 1978. Norway. In S.H.U Bowie, A. Kvalheim and Haslam, H.W., eds., Mineral Deposits of Europe, v. 1. Inst. Min. Metall. London, 199-249.

Bultemann, H.-W. and Hofmann, R., 1982. Uranium occurrences of Bavaria, Germany (with emphasis on the Bavarian Pfahl of the Altransberg district). In C.G. Amstutz, ed., Ore Genesis, the State of the Art. Springer Verlag, Berlin, 426-433.

Bunker, C.M. and MacKallor, J.A., 1973. Geology of the oxidized uranium deposits of the Tordilla Hill-Deweesville area, Karnes County, Texas; a study of a district before mining. U.S. Geol. Survey Prof. Paper 765, 37 pp.

Burckhardt, C.E. and Falini, F., 1956. Memoria sur giacimenti Italiani di manganese. Symposium del Manganeso, 20th Int. Geol. Congr., Mexico, v.5, 221-272.

Burian, J., Konečný, V. and Stohl, J., 1974. Banska Štiavnica ore deposit and its position within caldera (Czechoslovakia). 4th IAGOD Symposium, Varna, v.1, 233-239.

Burić, P., 1966. Geologija ležišta boksita Crne Gore. Geol. Glasnik, Sarajevo, Spec. Volume 8, 240 pp.

Burk, C.A. and Drake, C.L., eds., 1974. The Geology of Continental Margins. Springer Verlag, Berlin, 1009 pp.

Burke, K., 1977. Aulacogens and continental breakup. Ann. Rev. Earth Planet. Sci. 5, 371-396.

Burke, K.C., Cunningham, M.A., Gallagher, M.J. and Hawkes, J.R., 1964. Beryl in the Rosses Granite, North-West Ireland. Econ. Geol. 59, 1539-1550.

Burke, K., Dewey, J.F. and Kidd, W.S.F., 1976. Dominance of horizontal movements, arc and microcontinent collisions during the later permobile regime. In B.F. Windley, ed., The Early History of the Earth; Wiley, London, 113-129.

Burke, K.C., Dessauvagie, T.F.J. and Whiteman, A.J., 1972. Geological history of the Benue Valley and adjacent areas. Proc. of the Confer. on African Geology, Ibadan, 1970, 187-205.

Burke, K.C., Kidd, W.S.F., Turcotte, D.L., Dewey, J.F., Mouginis-Mark, P.J., Parmentier, E.M., Sengor, A.M.C. and Tapponnier, P. E., 1981. Tectonics of basaltic volcanism. In Basaltic Volcanism on the Terrestrial Planets, Pergamon Press, New York, 803-898.

Burnett, W.C., Veeh, H.H. and Soutar, A., 1980. U-series, oceanographic and sedimentary evidence in support of recent formation of phosphate nodules off Peru. Soc. Econ. Paleont. Miner. Spec. Publ. 29, 61-71.

Buroš, J., Grebe, W.H. and Wagner, H., 1977. Rohstoffwirtschaftliche Länderberichte, XIV-UdSSR-Mangan. Bundesanst. f. Geowiss., Hannover, 107 pp.

Busch, K., 1980. Rohstoffwirtschaftlicher Länderbericht, XXIV-Mexiko. Bundesanst. f. Geowiss., Hannover, 165 pp.
Busch, P.R., 1970. Chloride-rich brines from sabkha sediments and their possible role in ore formation. Trans. Inst. Min. Metall., London, Sect. B, v.79, B137-B144.
Buschendorf, F., Dennert, H., Hannak, W., Huttenhain, H., Mohr, K., Sperling, H. and Stoppel, D., 1971. Geologie des Erzgang-Reviers, mineralogie des Ganginhalts und Geschichte des Bergbaus im Oberharz. Geol. Jahrb., Beih., Hannover, No. 118, 212 pp.
Buseck, P.R., 1966. Contact metasomatism and ore deposition, Concepcion del Oro, Mexico. Econ. Geol., 61, 97-136.
Butler, B.S. and Burbank, W.S., 1929. The copper deposits of Michigan. U.S. Geol. Surv. Prof. Paper 144, 238 pp.
Butler, B.S., Loughlin, G.F., Heikes, V.C. an others, 1920. The ore deposits of Utah. U.S. Geol. Surv. Prof. Paper 111, 672 pp.
Butt, C.R.M., 1981. Some aspects of geochemical exploration in lateritic terrains in Australia. In Lateritization Processes, Balkena, Rotterdam, 369-380.
Button, A. and Eriksson, K., 1981. Precambrian paleoweathering and paleo-environments: controls on mineralization. Course notes, Dept. of Geology, Univ. of Western Australia, Nedlands, 290 pp.
Button, A. and Tyler, N., 1981. The character and economic significance of Precambrian paleoweathering and erosion surfaces in Southern Africa. Econ. Geol. Anniv. Vol., 686-709.
Butuzova, G.Yu., 1969. Sovremennyi vulkanogenno-osadochnyi zhelezorudnyi protsess v Kaldere Santorin (Egeiskoe More) i yego vliyanie na geokhimiyu osadkov. Nauka, Moscow, 110 pp.
Byers, F.M., Jr., 1957. Tungsten deposits in the Fairbanks district, Alaska. U.S. Geol. Surv. Bull. 1024-I, 179-216.
Byers, F.M., Jr., 1959. Geology of Umnak and Bogoslov Islands, Aleutian Islands, Alaska. U.S. Geol. Surv. Bull. 1028-L, 267-369.
Byers, F.M., Jr., Carr, W.J., Orkild, P.P., Quinlivan, W.D. and Sargent, K.A., 1976. Volcanic suites and related cauldrons of Timber Mountain-Oasis Valley caldera complex, southern Nevada. U.S. Geol. Surv. Prof. Paper 919, 70 pp.
Bykhovskii, L.Z., Gurvich, S.I., Patyk-Kara, N.G. and Flerov, I.B., 1981. Geologicheskii kriterii poiskov rossypei. Nedra, Moscow, 253p
Byrne, J.V. and Emery, K.O., 1960. Sediments of the Gulf of California. Bull. Geol. Soc. Am., 71, 983-1010.
Čadek, J., Mirovský, J., Novák, F., and others, 1975. Association of uranium and zirconium in the sandstone-type uranium deposits in northern Bohemia. Časop. Miner. Geol.(Prague), 20, 131-140.
Cadigan, R.A. and Felmlee, J.K., 1975. Radioactive mineral springs in Delta County, Colorado (abs.). Econ. Geol., 70, p. 1318.
Čadková, Z., 1971. Genesis of the Permian stratiform Cu deposit at Horní Vernéřovice. Sbor. Geol. Věd, LG, 14, 65-85.
Cady, W.M., Wallace, R.E., Hoare, J.M. and Webber, E.J., 1955. The Central Kuskokwim region, Alaska. U.S. Geol. Surv. Profess. Paper 268, 132 pp.
Caïa, J., 1976. Paleogeographical and sedimentological controls of

copper, lead and zinc mineralizations in the lower Cretaceous sandstones of Africa. Econ. Geol., 71, 409-422.
Cairnes, D.D., 1915. Upper White River District, Yukon. Geol. Surv. Canada, Memoir 50, 191 pp.
Callahan, W.H., 1968. Geology of the Friedensville zinc mine, Lehigh County, Pennsylvania. In J.D. Ridge, ed., Ore Deposits in the United States 1933-1967, A.I.M.E., New York, 95-107.
Callahan, W.H., 1977. Some thoughts regarding premises and procedures for prospecting for base metal ores in carbonate rocks in the North American Cordillera. Econ. Geol., 72, 71-81.
Callender, E. and Bowser, C.J., 1976. Freshwater ferromanganese deposits. In K.H. Wolf, ed., Handbook of Stratiform and Strata-Bound Ore Deposits, v. 7, Elsevier, Amsterdam, 341-394.
Calvert, S.E. and Price, N.B., 1970. Composition of manganese nodules and manganese carbonates from Loch Fyne, Scotland. Contrib. Min. and Petrol., 29, 215-233.
Calvert, S.E. and Price, N.B., 1977. Shallow water, continental margin and lacustrine nodules. In G.P. Glasby, ed., Marine Manganese Deposits. Elsevier, Amsterdam, 45-86.
Cameron, G.W. and Hood, P.J., 1975. Residial aeromagnetic anomalies associated with the Meguma Group of Nova Scotia and their relationship to gold mineralization. Geol. Surv. Canada Paper 75-1C, 197-211.
Cameron, E.N., Jahns, R.H., McNair, A. and Page, L.R., 1949. Internal structure of granitic pegmatites. Econ. Geol. Monogr. 2, 115 pp.
Campbell, I.H., Franklin, J.M., Gorton, M.P., Hart, T.R. and Scott, S.D., 1982. The role of subvolcanic sills in the generation of massive sulphide deposits. Econ. Geol., 77, 2248-2253.
Campi, D., McGain, A. and Ellem, C., 1975. Gibsonvale alluvial tin deposit, N.S.W. In C.L. Knight, ed., Economic Geology of Australia and Papua New Guinea, v.1, Metals. Austr. Inst. Min. Metall., Monogr. 5, 1049-1053.
Camus, F., 1975. Geology of the El Teniente orebody with emphasis on wall-rock alteration. Econ. Geol., 70, 1341-1372.
Cann, J.R., Winter, C.K. and Pritchard, R.G., 1977. A hydrothermal deposit from the floor of the Gulf of Aden. Miner. Magazine, No. 318, 193-199.
Carey, S.W., 1976. The Expanding Earth. Elsevier, Amsterdam, 488 pp.
Cargill, D.G., Lamb, J., Young, M.J. and Rugg, E.S., 1976. Island Copper. Canad. Inst. Min. Metall. Spec. Vol. 15, 206-218.
Carlisle, D. and Susuki, T., 1974. Emergent basalt and submergent carbonate-clastic sequences including the upper Triassic Dilleri and Welleri zones on Vancouver Island. Canad. J. Earth Sciences, 11, 254-279.
Carlson, G.G., 1977. Geology of the Bailadores, Venezuela, massive sulfide deposit. Econ. Geol., 72, 1131-1141.
Carmichael, I.S.E., 1964. The petrology of Thingmuli, a Tertiary volcano in eastern Iceland. Journ. Petrology, 5, 1-3.
Carmichael, I.S.E., Turner, F.J. and Verhoogen, J., 1974. Igneous Petrology. McGraw-Hill, New York, 739 pp.

Carne, J.E., 1913. Mercury or "Quicksilver" in New South Wales, with notes on its occurrence in other colonies and countries. Miner. Resources, Geol. Surv. N.S.W., 7, 53 pp.
Carne, R.C., 1976. The Tea barite deposit. Dept. Indian and North. Affairs, Canada, Open File Rept., unpubl., 20-31.
Carne, R.C. and Cathro, R.J., 1980. Metallogeny and national significance of "sedimentary exhalative" zinc-lead-silver deposits of the Selwyn Basin. Canad. Inst. Min. Metall. Ann. General Meeting, April 23, 1980, Toronto.
Carozzi, A.V., 1975. Sedimentary rocks: concepts and history. Benchmark Papers in Geology, 15, Dowden, Hutchinson and Ross, Stroudsburg, Pennsylvania, 468 pp.
Carpenter, A.B., Trout, M.L. and Pickett, E.E., 1974. Preliminary report on the origin and chemical evolution of lead- and zinc-rich oil field brines in Central Mississippi. Econ. Geol., 69, 1191-1206.
Carpenter, L.G. and Garret, D.E., 1959. Tungsten in Searles Lake. Min. Engin., March 1959, 11, No.3, 310-313.
Carr, J.M., 1971. Geology of the Churchill copper deposit. Canad. Inst. Min. Metall. Transact. 74, 152-156.
Carr, J.M. and Reed, A.J., 1976. Afton: a supergene copper deposit. Canad. Inst. Min. Miner. Spec. Vol. 15, 376-387.
Carr, M.J., Rose, W.I. and Stoiber, R.E., 1982. Central America. In R.S. Thorpe, ed., Andesites. Wiley, Chichester, 149-166.
Carson, D.J.T., 1969. Tertiary mineral deposits of Vancouver Island. Canad. Inst. Min. Metall. Bull., May 1969, 511-521.
Carson, D.J.T. and Jambor, J.L., 1974. Mineralogy, zonal relationship and ecomomic significance of hydrothermal alteration at porphyry copper deposits, Babine Lake area, British Columbia. Canad. Inst. Min. Metall. Bull., Febr. 1974, 110-133.
Carter, F.W., 1970. Geology of the Salt Anticline region in southwestern Colorado. U.S. Geol. Surv. Prof. Paper 637.
Carter, F.W., 1973. Mineral resources of the Idaho Primitive area and vicinity, Idaho. U.S. Geol. Surv. Bull. 1304, 431 pp.
Carter, L., 1980. Iron sand in continental shelf sediments off western New Zealand-a synopsis. N. Zealand Journ. Geol. Geoph.,23, 455-468.
Carter, W.D. and Aliste, N.T., 1964. Paleo-channels at the Guayacán copper mine, Cabildo district, Aconcagua Province, Chile. Econ. Geol., 59, 1283-1292.
Casadevall, T. and Ohmoto, H., 1977. Sunnyside mine, Eureka mining district, San Juan County, Colorado: geochemistry of gold and base metal ore deposition in a volcanic environment. Econ. Geol. 72, 1285-1320.
Casagrande, D.J. and Erchull, L.D., 1977. Metals in plants and waters in the Okefenokee swamp and their relationship to constituents found in coal. Geoch. Cosmoch. Acta, 41, 1391-1394.
Cassard, D., Nicolas, A., Rabinovitch, M., Moutte, J., Leblanc, M. and Prinzhofer, A., 1981. Structural classification of chromite pods in southern New Caledonia. Econ. Geol., 76, 805-831.

Cassedanne, J., 1965. Indice de sulfures sédimentaires de Taboca (Municipe de Crato, État de Ceara, Brésil). Bull. Soc. Geol. de France, 7, 177-186.
Cathcart, J.B., 1978. Uranium in phosphate rock. U.S. Geol. Surv. Prof. Paper 988-A, 6 pp.
Cathcart, J.B., 1963. Economic geology of the Keysville Quadrangle, Florida. U.S. Geol. Surv. Bull. 1128, 82 pp.
Cathcart, J.B. and Gulbrandsen, R.A., 1973. Phosphate deposits. U.S. Geol. Surv. Prof. Paper 820, 515-525.
Cathless, L.M., 1981. Fluid flow and genesis of hydrothermal ore deposits. Econ. Geol. 75th Anniv. Vol., 424-457.
Cavalcante, J. C. and others, 1979. Projeto Sapucaí-relatório final de geología. Dep. Nac. Prod. Min., Sér. Geol., No. 4, Geol. Basica, No.2, Brasília, 299 pp.
Čechovič, V., 1937. Beiträge zur Geologie und Genese der Kupferlagerstätte bei Rehova, Albanien. Zeitschr. f. Prakt. Geol., 45, 76-82.
Černý, P., ed., 1982. Short course in granitic pegmatites in science and industry. Miner. Assoc. Canada, Winnipeg, 555 pp.
Chace, F.M., 1948. Tin-silver veins of Oruro, Bolivia. Econ. Geol. vol. 43, 333-383 and 435-470.
Chace, F.M., Cumberlidge, J.T., Cameron, W.L. and van Nort, S.D., 1969. Applied geology at the Nickel Mountain mine, Riddle, Oregon Econ. Geol. 64, 1-16.
Chaikovskii, V.K.M., 1978. Tipy rudonosnykh fatsii. Nauka, Moscow, 222 pp.
Chantraine, J. and Radelli, L., 1970. Tectono-minerogenetic units of the basement of Madagascar. Econ. Geol. 65, 690-699.
Chappell, B.W. and White, A.J.R., 1974. Two contrasting granite types. Pacific Geology, 8, 173-174.
Chenoweth, W.L., 1982. Developments in uranium in 1981. Amer. Assoc. Petrol. Geol. Bull. 66, 2500-2508.
Chernyshev, V.F., 1960. Osobennosti geologicheskovo stroyeniya Tur'inskovo skarnovorudnovo pol'ya na Urale. In Osnovnye Voprosy i Metody Izucheniya Struktur Rudnykh Polei i Mestorozhdenii. Gosgeoltekhizdat', Moscow, 469-503.
Cherpasov, B.L., Pokrovskaya, I.V. and Kovrigo, O.A., 1972. O poligennom kharaktere orudeneniya Ridder-Sokol'nogo mestorozhdeniya. Geol. Rud. Mestor., 1972, No. 6, 30-45.
Chesnokov, B.V., 1960. Rutile-bearing eclogites from the Shubino village in the southern Urals. Int. Geol. Rev., 2, 936-945.
Chew, K.J., 1978. Crystalline manganese oxides in till from Bridge of Don, Aberdeen. Scottish Jour. Geol. 14, 329-334.
Childers, M.O. and Bailey, R.V., 1979. Classification of uranium deposits. Contrib. to Geology, Univ. of Wyoming, Laramie, v.17, 187-199.
Chilingar, G.V., Bissell, H.J. and Wolf, K.H., 1979. Diagenesis in sediments and sedimentary rocks. Devel. in Sedimentology 25A, Elsevier, Amsterdam, 247-424.
Chilingarian, G.V. and Yen, T.F., 1978. Bitumens, asphalts and tar

sands . Devel. in Petr. Sci., Elsevier, Amsterdam, 332 pp.
Chrismas, L., Baadsgaard, H., Folinsbee, R.E., Fritz, P., Krouse, H.R. and Sasaki, A., 1969. Rb/Sr, S and O isotopic analyses indicating source and date of contact metasomatic copper deposits, Craigmont, British Columbia, Canada. Econ. Geol. 64, 479-488.
Christiansen, R.L. and Lipman, P.W., 1972. Cenozoic volcanism and plate tectonic evolution of the western United States, II, Late Cenozoic. Phil. Trans. R. Soc., A271, 249-284.
Christie, R.L., 1979. Phosphorite in sedimentary basins of western Canada. Geol. Surv. Canada, Paper 79-1B, 253-258.
Christman, R.A., Brock, M.R., Pearson, R.C. and Singewald, Q.D., 1959. Geology and thorium deposits of the Wet Mountains, Colorado; a progress report. U.S. Geol. Surv. Bull. 1072-H, 491-535.
Chrt, J. and Bolduan, H., 1966. Die postmagmatische Mineralisation des Westteils der Böhmischen Masse. Sbor. Geol. Věd, Prague, LG, 8, 113-192.
Chrt, J., Neumann, J., Hoffman, V. and Trdlička, Z., 1972. Měděnokyzové ložisko Tisová u Kraslic. Sbor. Geol. Věd, Prague, LG, 15, 7-28.
Church, B.N., 1971. SG (Sam Goosly). In Geol., Explor., Miner. in Brit. Columbia, 1970. Brit. Columb. Dept. Min., Victoria, 126-128.
Church, B.N., 1974. Sustut Copper. In Geol., Explor. and Mining in Brit. Columbia. Brit. Col. Dept. Min., Victoria, 417-432.
Church, W.R., 1972. Ophiolite: its definition, origin as oceanic crust, and mode of emplacement in orogenic belts, with special reference to the Appalachians. Canad. Contrib. No.6 to the Geodynamic Project, 71-85.
Churkin, M., Jr., 1973. Paleozoic and Precambrian rocks of Alaska and their role in its structural evolution. U.S. Geol. Surv. Profes. Paper 740, 1-64.
Churkin, M., Jr., 1974. Paleozoic marginal ocean basin-volcanic arc systems in the Cordilleran foldbelt. Soc. Econ. Pal., Petr., Miner., Spec. Publ. 19, 174-192.
Čičić, S., 1980. Prilog poznavanju geolosko-ekonomskih odlika rezervi, proizvodnje i potrosnje zeljeznih ruda u svijetu i Jugoslaviji. Geoloski Glasnik, Sarajevo, 1980, 143-166.
Cissarz, A., 1957. Lagerstätten des Geosynklinalvulkanismus in den Dinariden und ihre Bedeutung für die geosynklinale Lagerstättenbildung. Neues Jb. Miner., Abh., 91, 485-540.
Clark, A.H., Farrar, E. and Kents, P., 1977. Potassium-argon age of the Cerro Colorado porphyry copper deposit, Panama. Econ. Geol. 72, 1154-1158.
Clark, K.F., 1972. Stockwork molybdenum deposits in the western Cordillera of North America. Econ. Geol. 67, 731-758.
Clark, K.F., Foster, C.T. and Damon, P.E., 1982. Cenozoic mineral deposits and subduction-related magmatic arcs in Mexico. Geol. Soc. Amer. Bull., 93, 533-544.
Clark, L.D., 1964. Stratigraphy and structure of part of the western Sierra Nevada metamorphic belt, California. U.S. Geol. Surv. Prof. Paper, 410, 70 pp.

Clark, W.B., 1970. Gold districts of California. Calif. Div. of Mines and Geol., Bull. 193, 186 pp.
Clark, W.E., 1969. Giant Mascot Mines Ltd., geology and ore control. Western Miner, No. 42, 40-46.
Clarke, D.E., 1969. Geology of the Mount Biggenden gold and bismuth mine and environs. Geol. Surv. Queensland, Rept. 32, 16 pp.
Clarke, D.B., 1977. The Tertiary volcanic province of Baffin Bay. Geol. Assoc. Canada Spec. Paper 16, 445-460.
Cloos, H., 1939. Hebung, Spaltung, Vulkanismus. Geol. Rundschau, 30, 401-439.
Cloud, P.E., Jr., Schmidt, R.G. and Burke, H.W., 1956. Geology of Saipan, Mariana Islands. U.S. Geol. Surv. Prof. Paper 280-A, 126pp.
Coats, R.R. and Stephens, E.C., 1968. Mountain City copper mine, Elko County, Nevada. In J.D. Ridge, ed., Ore Deposits of the United States 1933-1967; A.I.M.E., New York, v.2, 1074-1101.
Cobbing, E.J., Pitcher, W.S., Wilson, J.J., Baldock, J.W., Taylor, W.P., McCourt, W. and Snelling, N.J., 1981. The geology of the western Cordillera of northern Peru. Inst. Geol. Sc. Overs. Mem. 5, 144 pp.
Cole, J.W., 1979. Structure, petrology and genesis of Cenozoic volcanism, Taupo volcanic zone, New Zealand-a review. New Zealand Journ. Geol. and Geophys., 22, 631-657.
Coleman, P.J. and Hackman, B.D., 1974. Solomon Islands. In A.M. Spencer, ed., Mesozoic-Cenozoic Orogenic Belts. Geol. Soc. London, 453-461.
Coleman, R.G., 1971. Plate tectonic emplacement of upper mantle peridotites along continental edges. Journ. Geophys. Res. 76, 1212-1222.
Coleman, R.G., 1974. Geologic background of the Red Sea. In C.F. Burk and C.L. Drake, eds., The Geology of Continental margins; Springer Verlag, Berlin, 743-751.
Coleman, R.G., 1977. Ophiolites. Springer Verlag, Berlin, 229 pp.
Coleman, R.G. and Donato, M.M., 1979. Oceanic plagiogranite revisited. In F. Barker, ed., Trondhjemites, Dacites abd Related Rocks. Devel. in Petrol. 6, Elsevier, Amsterdam, 149-168.
Coleman, R.G. and Irwin, W.P., 1974. Ophiolites and ancient continental margins. In C.F. Burk and C.L. Drake, eds., The Geology of Continental Margins. Springer Verlag, Berlin, 921-931.
Colley, H. and Greenbaum, D., 1980. The mineral deposits and metallogenesis of the Fiji Platform. Econ. Geol. 75, 807-830.
Collins, A.G., 1975. Geochemistry of oilfield waters. Devel. in Petrol. Sc. 1, 496 pp.
Collins, P.L.F., 1981. The geology and genesis of the Cleveland tin deposit, western Tasmania: fluid inclusion and stable isotope studies. Econ. Geol., 76, 365-392.
Collins, J.J. and Loureiro, A.R., 1971. A metamorphosed deposit of Precambrian supergene copper. Econ. Geol. 66, 192-199.
Collinson, J.D., 1978a. Alluvial sediments. In H.G. Reading, ed., Sedimentary Environments and Facies; Elsevier, New York, 15-60.
Collinson, J.D., 1978b. Lakes. In H.G. Reading, ed., Sedimentary

Environments and Facies. Elsevier, New York, 61-79.
Collinson, J.D., 1978c. Deserts. In H.G. Reading, ed., Sedimentary Environments and Facies. Elsevier, New York, 80-96.
Colony, W.E. and Nordlie, B.E., 1973. Liquid sulfur at Volcán Azufre, Galapagos Islands. Econ. Geol. 68, 371-380.
Comer, J.B., 1974. Genesis of Jamaican bauxite. Econ. Geol., 69, 1251-1264.
Compagnoni, R., Elter, G., Fiora, L., Natale, P. and Zucchetti, S., 1981. Magnetite deposits in serpentinized lherzolites from the ophiolite belt of the western Alps, with special reference to the Cogne deposit (Aosta Valley). UNESCO Intern. Sympos. on Metall. of Mafic and Ultramaf. Compl., Athens, v.3, 376-394.
Conant, L.C. and Swanson, V.E., 1961. Chattanooga Shale and related rocks of central Tennessee and nearby areas. U.S. Geol. Survey Prof. Paper 357, 91 pp.
Condie, K.C., 1981. Archean Greenstone Belts. Devel. in Precambr. Geol. 3; Elsevier, Amsterdam, 434 pp.
Condie, K.C., 1982. Plate tectonics and crustal evolution. Pergamon Press, New York, 310 pp.
Coney, P.J., 1970. The geotectonic cycle and the New Global Tectonics. Geol. Soc. Am. Bull., 81, 739-748.
Constantinou, G., 1980. Metallogenesis associated with Troodos Ophiolite. In A. Panayiotou, ed., Ophiolites. Cyprus Geol. Surv. Dept. Nicosia, 663-674.
Constantinou, G. and Govett, G.J.S., 1973. Geology, geochemistry and genesis of Cyprus sulfide deposits. Econ. Geol. 68, 843-858.
Conybeare, C.E.B., 1979. Lithostratigraphic Analysis of Sedimentary Basins. Academic Press, New York, 555 pp.
Cooper, J.R., 1951. Geology of the tungsten, antimony and gold deposits near Stibnite, Idaho. U.S. Geol. Surv. Bull. 969-F, 151-197.
Corbett, K.D., 1981. Stratigraphy and mineralization in the Mt. Read volcanics, western Tasmania. Econ. Geol., 76, 209-230.
Cornwall, H.R., 1973. Nickel. U.S. Geol. Surv. Prof. Paper 820, 437-442.
Corsini, F., Cortecci, G., Leone, G. and Tanelli, G., 1980. Sulfur isotope study of the skarn (Cu-Pb-Zn) sulfide deposit of Valle del Temperino, Campiglia Marittima, Tuscany, Italy. Econ. Geol., 75, 83-96.
Coveney, R.M., Jr., 1979. Sphalerite concretions in mid-continent Pennsylvanian black shales of Missouri and Kansas. Econ. Geol. 74, 131-140.
Coveney, R.M., Jr., 1981. Gold quartz veins and auriferous granite at the Oriental mine, Alleghany district, California. Econ. Geol., 76, 2176-2199.
Cox, A., 1973. Plate Tectonics and Geomagnetic Reversals. Freeman, San Francisco, 702 pp.
Cox, D.P., 1970. Lead-zinc-silver deposits related to the White Mountain plutonic series in New Hampshire and Maine. U.S. Geol. Surv. Bull. 1312-D.
Cox, R., 1975a. Magnet silver-lead-zinc orebody. In C.L. Knight,

ed., Economic Geology of Australia and Papua New Guinea, v.1, Metals. Austr. Inst. Min. Metall. Monogr. 5, 628-631.
Cox, R., 1975b. Peak Downs copper mine, Clermont. In C.L. Knight, ed., Economic Geology of Australia and Papua New Guinea, v.1, Metals. Austr. Inst. Min. Metall. Monogr. 5, 771-773.
Cox, D.P., Larson, R.R. and Tripp, R.B., 1973. Hydrothermal alteration in Puerto Rican porphyry copper deposits. Econ. Geol., 68, 1329-1334.
Crawford, J. and Hoagland, A.D., 1968. The Mascot-Jefferson City zinc district, Tennessee. In J.D. Ridge, ed., Ore Deposits of the United States 1933-1967, A.I.M.E., New York, 242-256.
Crawley, R.H.A., 1955. Sources of germanium in Great Britain. Nature, v. 175, 291-292.
Crerar, D.A. and Barnes, H.L., 1974. Deposition of deep-sea manganese nodules. Geochim. Cosmochim. Acta, 38, 279-300.
Crerar, D.A., Namson, J., So Chyi, M., Williams, L. and Feigenson, M.D., 1982. Manganiferous cherts of the Franciscan assemblage, I. General geology, ancient and modern analogues and implications for hydrothermal convection at oceanic spreading centres. Econ. Geol. 77, 519-540.
Cressman, E.R. and Swanson, R.W., 1964. Stratigraphy and petrology of the Permian rocks of southwestern Montana . U.S. Geol. Survey Prof. Paper, 313-C, 275-569.
Crittenden, M.D., Jr., 1956. Syngenetic deposits. Symposium del Manganese, 20th Int. Geol. Congr., Mexico, v.3, 177-193.
Cronan, D.S., 1977. Deep-sea nodules: distribution and geochemistry. In G.P. Glasby, ed., Marine Manganese Deposits. Elsevier, 11-44.
Crook, K.A.W., 1972. Geotectonic significance of compositional variation of graywackes. In R.H. Dott, Jr., ed., Modern and Ancient Geosyncl. Sedimentation; Conf. Abstracts, Nov. 1972, Madison, 52-53.
Crowell, J.C., 1957. Origin of pebbly mudstones. Bull. Geol. Soc. Amer., 68, 993-1010.
Cruickshank, M.J., 1974. Mineral resource potential of continental margins. In C.A. Burk and C.L. Drake, eds., The Geology of Continental Margins. Springer Verlag, New York, 965-1000.
Cruzat, A.O., 1970. Genesis of manganese deposits in northern Chile. Econ. Geol. 65, 681-689.
Cumming, L.M., 1968. St.George-Table Head disconformity and zinc mineralization, western Newfoundland. Canad. Inst. Min. Metall. Bull., June 1968, 721-725.
Cuney, M., 1978. Geologic environment, mineralogy and fluid inclusions of the Bois Noirs-Limouzat uranium vein, Forez, France. Econ. Geol., 73, 1567-1610.
Cunningham, C.G., Jr., 1976. Petrogenesis and postmagmatic geochemistry of the Italian Mountain intrusive complex, eastern Elk Mts., Colorado. Geol. Soc. Am. Bull. 86, 897-908.
Cunningham, C.G., and others, 1982. Geochronology of hydrothermal uranium deposits and associated igneous rocks in the eastern source area of the Mount Belknap volcanics, Marysvale, Utah. Econ.

Geol. 77, 453-463.
Curray, J.R. and Moore, D.G., 1974. Sedimentary and tectonic processes in the Bengal deep-sea fan and geosyncline. In C.F. Burk and C.L. Drake, eds., The Geology of Continental margins, Springer Verlag, New York, 617-628.
Currie, K.L., 1974. The alkaline rocks of Canada. Geol. Surv. Canada Bull. 239, 228 pp.
Daemon, R.F., Paiva, I.B., Tavares, J.R.P. and Marquezan, R.G., 1982. Jazida de Figueira, aspectos da mineralização de uránio. Anais de 32. Congr. Brasil. de Geol., Salvador, v.5, 2099-2109.
Dahl, A.R. and Hagmaier, J.L., 1974. Genesis and characteristics of the southern Powder River Basin uranium deposits, Wyoming, U.S.A. In Formation of Uranium Deposits, Int. Atom. Energy Agency, Vienna, 201-216.
Daly, R.A., 1925. Relation of mountain building to igneous action. Proc. Amer. Phil. Soc., 64, 283-307.
Damião, R.N., 1980. Reservas geológico e potencial de ouro secundário do Brasil-metodologia de aviliação. Anais do 31. Congr. Brasil. de Geol., Camboriú, v.3, 1473-1481.
Danchev, V.I. and Strel'yanov, N.P., 1973. Uranougol'nye mestorozhdeniya i ikh glavneishie geneticheskie tipy. Geol. Rud. Mestor. 1973, No. 3, 66-81.
Danni, J.C.M., 1976. Magmatic differentiation of the alkaline-ultrabasic intrusions of the Iporá region, South West Goiás, Brazil. Proc. of the 1st Intern. Symp. on Carbonatites, Poços de Caldas, 1976, 149-167.
Davenport, P.H., ed., 1982. Prospecting in areas of glaciated terrain, 1982. Canad. Inst. Min. Metall., Montreal, 340 pp.
Davidson, C.F., 1965. A possible mode of origin of stratabound copper ores. Econ. Geol., 60, 942-954.
Davidson, D.F. and Lakin, H.W., 1961. Metal content of some black shales of the western United States. U.S. Geol. Surv. Profess. Paper, 424-C, C329-C331.
Davies, H.L., 1971. Peridotite-gabbro-basalt complex in eastern Papua: an overthrust plate of oceanic mantle and crust. Bur. Min. Res.,Geol., Geoph., Canberra, Bull. 128, 48 pp.
Davies, H.L., 1978. Geology and mineral resources of Papua New Guinea. Third Region. Conf. on Geol. and Miner. Res. of S.E. Asia, Bangkok, Nov. 1978, 685-699.
Davies, T.A. and Gorsline, D.S., 1976. Oceanic sediments and sedimentary processes. In J.P. Riley and R. Chester, eds., Chemical Oceanography, 2nd ed., Academic Press, London, 1-80.
Davis, F.F., 1966. Economic mineral deposits in the Coast Ranges.In E.H. Bailey, ed., Geology of Northern California. Calif. Dept. Nat. Res. Bull. 190, 315-321.
Davis, G.H., 1972. Deformational history of the Caribou strata-bound sulfide deposit, Bathurst, New Brunswick, Canada. Econ. Geol. 67, 634-655.
Davis, J.H., 1977. Genesis of the Southeast Missouri lead deposits. Econ. Geol., 72, 443-450.

Dawson, J.B., 1966. Oldoinyo Lengai-an active volcano with sodium carbonatite lava flows. In O.F. Tuttle and J. Gittins, eds., Carbonatites; Interscience, New York, 155-168.
Deans, T., 1966. Economic mineralogy of African carbonatites. In O.F. Tuttle and J. Gittins, eds., Carbonatites. Interscience, New York, 385-413.
de Carvalho, D., 1976. Les gisements de fer du Portugal. In A. Zitzmann, ed., Iron Ore Deposits of Europe. Bundesanst. f. Geowiss., Hannover, 255-260.
De Cserna, Z., 1976a. Mexico-geotectonics and mineral deposits. New Mexico Geol. Soc. Spec. Publ. No.6, 18-24.
De Cserna, Z., 1976b. Geology of the Fresnillo area, Zacatecas, Mexico. Geol. Soc. Am. Bull., 87, 1191-1199.
Degens, E.T. and Kulbicki, G., 1973. Hydrothermal origin of metals in some East African rift lakes. Miner. Deposita, 8, 388-404.
Degens, E.T., Okada, H., Honjo, S. and Hathaway, J.C., 1972. Microcrystalline sphalerite in resin globules suspended in Lake Kivu, East Africa. Miner. Deposita, 7, 1-12.
Degens, E.T. and Ross, D.A., eds., 1969. Hot brines and recent heavy metal deposits in the Red Sea. A geochemical and geophysical account. Springer Verlag, Berlin, 600 pp.
Degens, E.T. and Ross, D.A., eds., 1974. The Black Sea-geology, chemistry and biology. Amer. Assoc. Petr. Geol. Memoir 20, 640 pp.
Degens, E.T. and Ross, D.A., 1976. Strata-bound metalliferous deposits found in or near active rifts. In K.H. Wolf, ed., Handbook of Stratiform and Strata-bound Ore Deposits, v.4, Elsevier, 165-202.
De Geoffroy, J. and Wignall, T.K., 1972. A statistical study of geological characteristics of porphyry copper-molybdenum deposits in the Cordilleran belt-application to the rating of porphyry deposits. Econ. Geol. 67, 656-668.
De Kun, N., 1965. The Mineral Resources of Africa. Elsevier, Amsterdam, 740 pp.
De Launay, L., 1913. Traité de Métallogenie, 3 vols., Béranger, Paris.
Dell, C.I., 1975. Pyrite concretions in sediment from South Bay, Lake Huron. Canad. J. Earth Sc., 12, 1077-1083.
Denholm, L.S., 1967. Lode structures and ore shoots at Vatukoula, Fiji. Proc. Austr. Inst. Min. Metall. No. 222, 78-83.
Denisenko, V.K. and Rundkvist, D.V., 1977. New prospective types of stratiform tungsten mineralization. Int. Geol. Rev., 20, 575-586.
Denisov, S.V., 1979. Gold occurrences in marine beaches as an exploration feature in the ore and placer deposit of a nearshore land area. Intern. Geol. Rev., 22, 1194-1198.
Dennen, W.H. and Norton, H.A., 1977. Geology and geochemistry of bauxite deposits in the lower Amazon basin. Econ. Geol. 72, 82-89.
Denson, N.M., Bachman, G.O. and Zeller, H.D., 1959. Uranium-bearing lignite in northwestern South Dakota and adjacent states. U.S. Geol. Surv. Bull. 1055, 11-58.
Denson, N.M. and Gill, J.R., 1965. Uranium-bearing lignite and car-

bonaceous shale in the southwestern part of the Williston Basin-a regional study. U.S. Geol. Surv. Prof. Paper 463, 75 pp.
den Tex, E., 1971a. The facies groups and facies series of metamorphism, and their relation to physical conditions in the Earth's crust. Lithos, Oslo, 4, 23-41.
den Tex, E., 1971b. Age, origin and emplacement of some Alpidic peridotites in the light of recent petrofabric researches. Fortschr. Mineral., 48, 69-74.
de Oliveira, S.M.B. and Trescases, J.-J., 1980. Geoquímica da alteração supérgene dos rochas ultramáficos de Santa Fé (Goiás, Brasil). Revista Brasil. de Geocienc., 10, 243-257.
de Quervain, F. and Zitzmann, A., 1976. The iron ores of Switzerland. In A. Zitzmann, ed., The Iron Ore Deposits of Europe and Adjacent Areas. Bundesanst. f. Geowiss., Hannover, 295-297.
Derré, C., Fonteilles, M. and Nansot, L.Y., 1980. Le gisement de scheelite de Salau, Ariège, Pyrenees. 26th Int. Geol. Congr., France, Gisements Français, Fasc. E9, 42 pp.
Desborough, G.A., 1963. Magmatic sphalerite in Missouri basic rocks. Econ. Geol. 58, 971-977,
Desborough, G.A., Anderson, A.T. and Wright, T.L., 1968. Mineralogy of sulfides from certain Hawaiian basalts. Econ. Geol., 63, 634-644.
De Sitter, L.U., 1956. Structural Geology. McGraw-Hill, New York, 552 pp.
Dessau, G., 1977. Die Quecksilber- und Antimonlagerstätten den Toskana. Freib. Forschungsh., C, 328, 47-71.
Deul, M. and Annell, C.S., 1956. The occurrence of minor elements in ash of low-rank coal from Texas, Colorado, North Dakota and South Dakota. U.S. Geol. Surv. Bull. 1036-H, 155-172.
de Weisse, G., 1967. Sur la présence de nickel dans un gisement de bauxite prés de Mégare. Miner. Depos. 2, 349-356.
Dewey, J.F., 1974. Continental margins and ophiolite obduction: Appalachian-Caledonian system. In C.F. Burk and C.L. Drake, eds., The Geology of Continental Margins, Springer Cerlag, New York, 933-947.
Dewey, J.D. and Bird, J.M., 1970. Mountain belts and the new global tectonics. J. Geophys. Res., 75, 2625-2647.
Dewey, J.F. and Bird, J.M., 1971. Origin and emplacement of the ophiolite suite: Appalachian ophiolites in Newfoundland. Journ. Geophys. Res. 76, 3179-3206.
Dick, L.A. and Hodgson, C.J., 1982. The Mactung W-Cu(Zn) contact metasomatic and related deposits of the Northeastern Canadian Cordillera. Econ. Geol., 77, 845-867.
Dickinson, W.R., 1969. Evolution of calc-alkaline rocks in the gosynclinal system of California and Oregon. Proc. of the Andesite Conf., A.R. McBirney, ed., Oregon Dept. Geol. Min. Industr. Bull. 65, Portland, 151-156.
Dickinson, W.R., 1973. Reconstruction of past arc-trench systems from petrotectonic assemblages in the island arcs of the western Pacific. In P.J. Coleman, ed., The Western Pacific. Western

Austr. Univ. Press, Nedlands, 569-601.
Dickinson, W.R., 1974a. Sedimentation within and beside ancient and modern magmatic arcs. Soc. Econ. Paleont. Miner. Spec. Publ. 19, 230-239.
Dickinson, W.R., 1974b. Plate tectonics and sedimentation. Soc. Econ. Paleon. Miner. Spec. Publ. 22, 1-27.
Dickson, F.W. and Tunell, G., 1968. Mercury and antimony deposits associated with active hot springs in the western United States. In J.D. Ridge, ed., Ore Deposits of the United States 1933-1967, v.2, A.I.M.E., New York, v.2, 1675-1701.
Di Colbertaldo, D., 1950. The lead and zinc deposit at Raible in Friuli. 18th Int. Geol. Congr., London, Pt.7, 277-289.
Didier, J., 1973. Granites and Their Enclaves. Devel. in Petrol. 3, Elsevier, Amsterdam, 393 pp.
Diehl, P. and Kern, H., 1981. Geology, mineralogy and geochemistry of some carbonate-hosted lead-zinc deposits in Kanchanaburi Province, Thailand. Econ. Geol., 76, 2128-2146.
Diessel, C.F.F., 1970. Paralic coal seam formation. In The assessment of our fuel and energy requirements. Inst. of Fuel Conf., Brisbane, 1-22.
Dietrich, V., 1972. Ilvait, Ferroantigorit und Greenalith als Begleiter oxidish-sulfidischer Vererzungen in den Oberhalbsteiner Serpentinen. Schweiz. Min. Petr. Bull., 52, 57-74.
Dimitrov, R., 1977. Nekotorye osobennosti metallogenii Rodopskovo sredinnovo massiva. Geol. Rud. Mestor., 19, 32-42.
Dimitrov, D.K. and Mankov, S., 1973. Stratiformnye svintsovo-tsinkovye mestorozhdeniya severnovo Tunisa. Geol. Rud. Mestorozhd., No.3, 38-51.
Dimroth, E., 1977. Facies Models 5. Models of physical sedimentation of iron formations. Geoscience Canada, 4, 1, 23-30.
Dines, H.G., ed., 1956. The metalliferous mining region of South-West England. Mem. Geol. Surv. London, 2 volumes, 792 pp.
Dings, M.G. and Robinson, C.S., 1957. Geology and ore deposits of the Garfield Quadrangle, Colorado. U.S. Geol. Surv. Prof. Paper 289, 110 pp.
Dings, M.G. and Whitebread, D.H., 1965. Geology and ore deposits of the Metaline zinc-lead district, Pend Oreille County, Washington. U.S. Geol. Surv. Prof. Paper 489, 109 pp.
Dixey, F., 1956. The East African Rift system. Geol. Soc. London, Pr. no. 1533, 33-40.
Dobin, D.A., Batuev, B.N., Mitenkov, G.A. and Izoitko, V.M., 1971. Atlas porod i rud noril'skikh medno-nikelevykh mestorozhdenii. Nedra, Leningrad, 560 pp.
Dobson, D.C., 1982. Geology and alteration of the Lost River tin-tungsten-fluorine deposit, Alaska. Econ. Geol., 77, 1033-1052.
Dodge, F.C.W. and Bateman, P.C., 1977. The Sierra Nevada batholith, California, U.S.A., and spatially related mineral deposits. Geol. Soc. Malaysia Bull. 9, 17-29.
Doi, K., Hirono, S. and Sakamaki, Y., 1975. Uranium mineralization by groundwater in sedimentary rocks, Japan. Econ. Geol. 70,

628-646.

Donnot, M., Guigues, J., Lulzac, Y., Magnien, A., Parfenoff, A. and Picot, P., 1973. Un nouveau type de gisement d'europium: la monazite grise à europium en nodules dans les schistes paléozoïques de Bretagne. Miner. Deposita, 8, 7-18.

Dora, O.O., 1977. The strata-bound lead-zinc deposits from Menderes Massif in Bayindir (West Anatolia). In D.D. Klemm and H.-J. Schneider, eds., Time and Strata-Bound Ore Deposits. Springer Verlag, Berlin, 220-231.

Dorr, J.V.N., 2nd, 1945. Manganese and iron deposits of Morro do Urucúm, Mato Grosso, Brazil. U.S. Geol. Surv. Bull. 946-A, 47 pp.

Dorr, J.V.N., 2nd, 1969. Physiographic, stratigraphic and structural development of the Quadrilatero Ferrífero, Minas Gerais, Brazil. U.S. Geol. Surv. Prof. Paper 641-A, 110 pp.

Dorr, J.V.N., 2nd, Crittenden, M.D., Jr. and Worl, R.G., 1973. Manganese. U.S. Geol. Surv., Prof. Pap. 820, 385-399.

Douglas, R.J.W., ed., 1970. Geology and Economic Minerals of Canada. Geol. Surv. Canada, Econ. Geol. Rept. 1, 838 pp.

Dowsett, J.S. and Reid, T.N., 1967. An exploration programme for nickel and copper in the differentiated intrusives of East Griqualand and Pondoland. Trans. Geol. Soc. S. Africa, 70, 67-80.

Drake, C.L. and Burk, C.A., 1974. Geological significance of continental margins. In C.A. Burk and C.L. Drake, eds., The Geology of Continental Margins. Springer Verlag, Berlin, 3-10.

Drewes, H., Fraser, G.D., Snyder, G.L. and Barnett, H.F., Jr., 1961. Geology of Unalaska Island and adjacent insular shelf, Aleutian Islands, Alaska. U.S. Geol. Surv. Bull. 1028-S, 583-676.

Drovenik, F., Drovenik, M. and Grad, K., 1972. Kupferführende Grodener Schichten Sloweniens. Geologija, Ljubljana, 15, 95-107.

Drubina-Szabó, M., 1959. Manganese deposits of Hungary. Econ. Geol. 54, 1078-1093.

Drummond, A.D., Sutherland-Brown, A., Young, R.J. and Tennant, S.J., 1976. Gibraltar-regional metamorphism, mineralization, hydrothermal alteration and structural development. Canad. Inst. Min. Metall., Special Volume 15, 195-205.

Druzhinin, I.P., 1973. Litologiya karbonovykh otlozhenii Dzhezkazganskoi vpadiny i genezis plastovykh sul'fidnykh rud. Trudy A.N. U.S.S.R., Vyp. 222, Nauka, Moscow, 187 pp.

Dubois, J., Dugas, F., Lapouille, A. and Louat, R., 1978. The troughs at the rear of the New Hebrides island arc: possible mechanisms of formation. Canad. J. Earth Sc., 15, 351-360.

Dubourdieu, G., 1952. Les mines de fer de l'Ouenza et du Bou-Khadra. 19th Int. Geol. Congr., Alger, Sympos. Gisem. Fer du Monde, 1, Alger, 70-82.

Dudas, B.M. and Grove, E.W., 1970. Granduc mine. Geol., Explor., Mining in Brit. Columbia, Victoria, 67-73.

Duffield, W.A. and Sharp, R.V., 1975. Geology of the Sierra Foothills melange and adjacent areas, Amador County, California. U.S. Geol. Surv. Prof. Paper 827, 30 pp.

Dugas, J., Assad, R. and Marleau, R., 1969. Metallogenic concepts in Gaspé. Canad. Inst. Min. Metall. Trans., 72, 248-255.

Dunham, K., Beer, K.E., Ellis, R.A., Gallagher, M.J., Nutt, M.J.C. and Webb, B.C., 1978. United Kingdom. In Mineral Deposits of Europe, v. 1, Inst. Min. Metall./Miner. Soc. London, 263-317.

Ďurišová, J., Just, J. and Kušnír, I., 1969. Cínové ložisko Jeroným u Čisté ve Slavkovském (Císařském) Lese. Sbor. Geol. Věd, Prague, LG, 10, 25-53.

DuToit, A.L. and Haughton, S.H., 1953. The Geology of South Africa. Hafner, New York, 611 pp.

Dvorov, V.I. and Pavlov, D.I., 1978. Near-fault zones of sulfide enrichment and a possible mechanism for formation of metalliferous brines. Int. Geol. Rev., 21, 806-814.

Dyuzhikov, O.A., Distler, V.V. and Fedorenko, V.A., 1976. Stratiformnye proyavleniya samorodnoi medi v vulkanogennykh porodakh severa Sibirskoi Platformy. Geol. Rud. Mestor., 2, 62-75.

Dzułynski, S. and Walton, E.K., 1965. Sedimentary features of flysch and graywackes. Elsevier, Amsterdam, 274 pp.

Earney, F.C.F., 1980. Petroleum and Hard Minerals from the Sea. Wiley, New York, 291 pp.

Eastlick, J.T., 1968. Geology of the Christmas Mine and vicinity, Banner Mining District, Arizona. In J.D. Ridge, ed., Ore Deposits of the United States 1933-1967, A.I.M.E., New York, 2, 1191-1210.

Easton, R.M., 1981. Stratigraphy of the Akaitcho Group and the development of an early Proterozoic continental margin, Wopmay Orogen, Northwest Territories. Geol. Surv. Canada Paper 81-10, 79-95.

Eastwood, G.E.P., 1967a. A.M. and Invermay (Giant Explorations Limited). Brit. Columbia Minister of Min. and Petr. Res. Rept. for 1965, Victoria, 206-213.

Eastwood, G.E.P., 1967b. Geology of the Coxey-Giant area. Brit. Columbia Minister of Mines and Petr. Res. Rept. for 1966, 200-207.

Eckel, E.B., Williams, J.S. and Galbraith, F.W., 1949. Geology and ore deposits of the La Plata district. U.S. Geol. Survey Profess. Paper 219, 179 pp.

Edmond, J.M. et al., 1979. On the formation of metal-rich deposits at ridge crests. Earth Planet. Sci. Letters, 46, 19-30.

Edwards, M.B., 1978. Glacial environments. In H.G. Reading, ed., Sedimentary Environments and Facies. Elsevier, New York, 416-438.

Ehrenberg, H., Pilger, A. and Schröder, F., 1954. Das Schwefelkies-Zinkblende-Schwerspatlager von Meggen (Westfalen). Beih. Geol. Jb. 12, Hannover, 352 pp.

Eichelberger, J.C., 1975. Origin of andesite and dacite: evidence of mixing at Glass Mountain in California and at other circum-Pacific volcanoes. Geol. Soc. Amer. Bull., 86, 1381-1391.

Eilenberg, S. and Carr, M.J., 1981. Copper contents of lavas from active volcanoes in El Salvador and adjacent regions in Central America. Econ. Geol., 76, 2240-2256.

Einaudi, M.T., 1977. Environment of ore deposition at Cerro de Pasco, Peru. Econ. Geol., 72, 893-924.

Einaudi, M.T., Meinert, L.D. and Newberry, R.J., 1981. Skarn depo-

sits. Econ. Geol. 75th Anniv. Volume, 317-391.
Eisbacher, G.H., 1976. Proterozoic Rapitan Group and related rocks, Redstone River area, District of Mackenzie. Geol. Surv. Canada, Paper 76-1A, 117-125.
Eisbacher, G.H., Carrigy, M.A. and Campbell, R.B., 1974. Paleodrainage pattern and late orogenic basins of the Canadian Cordillera. Soc. Econ. Paleont., Miner., Spec. Public. 20, 143-166.
El-Hinnawi, E.E., 1965. Contributions to the study of Egyptian (U.A.R.) iron ores. Econ. Geol., 60, 1497-1509.
Ellert, R., 1978. The Poços de Caldas alkaline massif. Proc. of the First Intern. Sympos. on Carbonatites, Brazil. D.N.P.M., 305-313.
Elliott, T., 1978a. Deltas. In H.G. Reading, ed., Sedimentary Environments and Facies. Elsevier, New York, 97-142.
Elliott, T., 1978b. Clastic shorelines. In H.G. Reading, ed., Sedimentary Environments and Facies. Elsevier, New York, 143-177.
Ellis, A.J., 1977. Chemistry and Geothermal Systems. Academic Press, New York, 392 pp.
Elliston, J., 1953. Platinoids in Tasmania. In A.B. Edwards, ed., Geology of Australian Ore Deposits, 1250-1254.
Emery, K.O., 1965. Some potential mineral resources of the Atlantic continental margin. U.S. Geol. Survey Prof. Paper 525-C, C157-C160.
Emery, K.O., 1968. Relict sediments on continental shelves of the world. Amer. Assoc. Petr. Geol. Bull. 52, 445-464.
Emery, K.O. and Skinner, B.J., 1977. Mineral deposits of the deep ocean floor. Mar. Mining, 1, 1-71.
Emmons, S.F., 1886. Geology and Mining Industry of Leadville, Colorado. U.S. Geol. Survey Monograph 12, 770 pp.
Emmons, S.F., 1905. Copper in the red beds of the Colorado Plateau region. U.S. Geol. Surv. Bull. 260, 221-232.
Emmons, S.E., Irving, J.D. and Laughlin, G.F., 1927. Geology and ore deposits of the Leadville mining district, Colorado. U.S. Geol. Surv. Prof. Paper 148, 368 pp.
Emmons, W.H., 1917. The enrichment of ore deposits. U.S. Geol. Surv. Bulletin 625, 530 pp.
Emmons, W.H., 1937. Gold Deposits of the World. McGraw-Hill, New York, 562 pp.
Engell, J., Hansen, J., Jensen, M., Kunzendorf, H. and Lovborg, L. 1971. Beryllium mineralization in the Ilímaussaq intrusion, South Greenland, with description of a field berylometer and chemical methods. Grønl. Geol. Unders., Rapport Nr. 33, 40 pp.
Epshteyn, Ye.M. and Anikeyeva, L.I., 1963. Problems in geology and petrology of ultrabasic alkalic intrusive rock complexes. Int. Geol. Rev. 7, 307-324.
Erdosh, G., 1979. The Ontario carbonatite province and its phosphate potential. Econ. Geol. 74, 331-338.
Ericksen, G.E., 1981. Geology and origin of the Chilean nitrate deposits. U.S. Geol. Surv. Profess. Paper 1188, 37 pp.
Erickson, R.L. and Blade, L.V., 1963. Geochemistry and petrology of the alkalic igneous complex at magnet Cove, Arkansas. U.S. Geol.

Survey Prof. Paper 425, 95 pp.
Erickson, R.L., Myers, A.T. and Horr, C.A., 1954. Association of uranium and other metals with crude oil, asphalt and petroliferous rock. Am. Assoc. Petrol. Geol. Bull., 38, 2200-2218.
Espenshade, G.H., 1954. Geology and mineral deposits of the James River-Roanoke River manganese district, Virginia. U.S. Geol. Surv. Bulletin 1008, 155 pp.
Esteban, M. and Klappa, C.F., 1983. Subaerial exposure environment. In P.A. Scholle, D.G. Bebout and C.H. Moore, eds., Carbonate Depositional environments. Amer. Assoc. Petrol. Geol., Tulsa, 1-54.
Eugster, H.P. and I-Ming Chou, 1979. A model for the deposition of Cornwall-type magnetite deposits. Econ. Geol., 74, 763-774.
Eulry, M. and Vargas, J.M., 1980. Le rôle des altérations anteliasiques dans la métallogenése de la péripherie du Mont Lozère (France). Cas de l'uranium. Mém. B.R.G.M., No.104, 15-175.
Evans, A.M., 1976. Genesis of Irish base-metal deposits. In K.H. Wolf, ed., Handbook of Stratiform and Strata-Bound Ore Deposits, vol. 5, Elsevier, Amsterdam, 231-256.
Evans, E.L. and Dujardin, R.A., 1961. A unique beryllium deposit in the vicinity of Ten Mile Lake, Seal Lake area, Labrador. Proc. Geol. Assoc. Canada, 13, 45-51.
Evans, H.J., 1975. Weipa bauxite deposit, Q. In C.L. Knight, ed., Economic Geology of Australia and Papua New Guinea, v.1, Metals. Austr. Inst. Min. Metall., Monogr. 5, 959-963.
Ewart, A., 1981. Petrological aspects of the Tertiary anorogenic volcanism of southern and central Queensland, Australia. Geol. Survey of India, Mem. No. 3, 377-393.
Ewart, A., 1982. The mineralogy and petrology of Tertiary-Recent orogenic volcanic rocks: with special reference to the andesitic-basaltic compositional range. In R.S. Thorpe, ed., Andesites; Wiley, Chichester, 25-95.
Ewers, G.R. and Keays, R.R., 1977. Volatile and precious metal zoning in the Broadlands geothermal field, New Zealand. Econ. Geol. 72, 1337-1354.
Ewing, M. and Heezen, B.C., 1956. Antarctica in the international geophysical year. Amer. Geoph. Union Monogr. 1, 75-81.
Fahrig, W.F. and Wanless, R.K., 1964. Age and significance of diabase dyke swarms of the Canadian Shield. Nature, 200, 934-937.
Fahrni, K.C., Macauley, T.N. and Preto, V.A.G., 1976. Copper Mountain and Ingerbelle. Canad. Inst. Min. Metall. Spec. Vol. 15, 368-375.
Fairbridge, R.W., 1967. Phases of diagenesis and authigenesis. In G. Larsen and G.V. Chillingar, eds., Diagenesis in Sediments. Elsevier, Amsterdam, 19-89.
Falcon, N.L., 1974. Southern Iran: Zagros Mountains. In A.M. Spencer, ed., Mesozoic-Cenozoic Orogenic Belts. Geol. Soc. London, 199-211.
FAO-UNESCO, 1971. Soil map of the world 1:5,000,000. 10 explanatory volumes and maps.
Farina, M., Lins, C.A.C. and Scheid, C., 1978. Excursão No.02 Sulfetos sedimentogénicos de região do Araripe-Ceará. 30 Congr. Brasil. de Geol., Bol. No.02, Roteiro des Excursões; Recife, 27-44.

Farrar, E., Clark, A.H. and Ok Joon Kim, 1978. Age of the Sangdong tungsten deposit, Republic of Korea, and its bearing on the metallogeny of the southern Korean peninsula. Econ. Geol. 73, 547-566.
Fedorchuk, V.P., 1960. O vtorichnykh mestorozhdeniyakh rtuti i surmy. Izv. A.N. Kirghiz S.S.R., ser. estestv. i tekhn. nauk, 2, 9.
Fehn, U., Cathles, L.M. and Holland, H.D., 1978. Hydrothermal convection and uranium deposits in abnormally radioactive plutons. Econ. Geol. v. 73, 1556-1566.
Féraud, J., Fornari, M., Geffroy, J. and Lenck, P.-P., 1977. Minéralisations arséniées et ophiolites: le filon à réalgar et stibine de Matra et sa place dans le district à Sb-As-Hg de la Corse alpine. Mém. du B.R.G.M., Sect. 2, 2, 91-112.
Ferguson, J., 1964. Geology of the Ilímaussaq alkaline intrusion, South Greenland. Bull. Grønl. Geol. Unders. 39, 82 pp.
Ferguson, J. and Burne, R.V., 1981. Coastal interactions of saline continental groundwaters and marine carbonate sediments, Spencer Gulf (Abs.). Baas Becking Geobiol. Lab Research Sympos., Progr., Canberra, p. 14.
Ferguson, J. and Lambert, I.B., 1972. Volcanic exhalations and metal enrichments at Matupi Harbour, New Britain, NPNG. Econ. Geol. 67, 25-37.
Ferguson, J., Lambert, I.B. and Jones, H.E., 1974. Iron sulphide formation in an exhalative-sedimentary environment, Talasea, New Britain, P.N.G. Miner. Depos. 9, 33-47.
Ferguson, S.A. and Freeman, E.B., 1978. Ontario occurrences of float, placer gold, and other heavy minerals. Ont. Geol. Surv. Miner. Dep. Circular 17, Toronto, 214 pp.
Fernandez, H.E. and Damasco, F.V., 1979. Gold deposition in the Baguio gold district and its relationship to regional geology. Econ. Geol., 74, 1852-1868.
Ferrario, A. and Garuti, G., 1980. Copper deposits in the basal breccias and volcano-sedimentary sequences of the eastern Ligurian ophiolites (Italy). Miner. Deposita, 15, 291-303.
Ferrario, A., Garuti, G. and Sighinolfi, G.P., 1982. Platinum and palladium in the Ivrea-Verbano basic complex, Western Alps, Italy. Econ. Geol. 77, 1548-1555.
Fersman, A.E., 1960. Pegmatity. Akad. A.E. Fersman Izbrannye Trudy, vol. VI, Akad. Nauk S.S.S.R., Moscow, 742 pp.
Fettweis, G.B., 1979. World Coal Resources. Methods of Assessment and Results. Devel. in Econ. Geol. 10, Elsevier, Amsterdam, 415 pp.
Field, C.W., Dymond, J.R., Corliss, J.B., Dasch, E.J., Heath, G.R., Senechal, R.G. and Veeh, H.H., 1976. Metallogenesis in Southeast Pacific Ocean: Nazca Plate Project. Amer. Assoc. Petrol. Geol. Memoir 25, 539-551.
Finch, W.I. and others, 1973. Nuclear fuels. U.S. Geol. Surv. Profess. Paper 820, 455-476.
Firsov, L.V., 1966. Gold-quartz ore deposits of Yana-Kolyma belt. Intern. Geol. Rev. 9, 1544-1552.
Fischer, R.P., 1960. Vanadium-uranium deposits of the Rifle Creek

area, Garfield County, Colorado. U.S. Geol. Surv. Bull. 1101, 52pp.
Fisher, R.V. and Schmincke, H.-U., 1984. Pyroclastic Rocks. Springer Verlag, Berlin, 472 pp.
Fitzgerald, M.L., 1945. Production of potash from alunite, Lake Campion, Western Australia. Chem. Eng. and Mining Rev., v.37, No. 440, 241-248.
Fitzpatrick, K.R., 1975. Woolomin-Texas Block, Woolomin Beds and associated sediments. In N.L. Markham and H. Basden, eds., The Mineral Deposits of New South Wales. Geol. Surv. N.S.W., Sydney, 339-349.
Flerov, B.L., 1976. Olovorudnye mestorozhdeniya Yano-Kolymskoi skladchatoi oblasti. Nauka, Novosibirsk, 288 pp.
Fletcher, C.J.N., Aguilera, E.A., Appleton, J.D., Bloomfield, K., Llanos, A.Ll. and Roberts, J.L., 1977. The Velasco Alkaline Province. Simpos. Intern. del Estaño, La Paz, No.17, 19 pp.
Fletcher, K. and Couper, J., 1975. Greenvale nickel laterite, North Queensland. In C.L. Knight, ed., Economic Geology of Australia and Papua New Guinea, v.1, Metals. Austr. Inst. Min. Metal., 995-1000.
Flint, R.F., 1971. Glacial and Quaternary Geology. Wiley, New York, 892 pp.
Flint, R.F., Sanders, J.E. and Rodgers, J., 1960. Diamictite, a substitute term for synmictite. Bull. Geol. Soc. Am., 71, p. 1809.
Florensov, N.A., 1966. The Baikal rift zone. Geol. Surv. Canada Paper 66-14, 173-180.
Fluck, P., Weil, R. and Wimmenauer, W., 1975. Gîtes minéraux de la France, v. II: Géologie des gîtes minéraux des Vosges. Mém. du B.R.G.M., No. 87, 189 pp.
Folinsbee, R.E., 1955. Archean monazite in beach concentrates, Yellowknife geologic province, Northwest Territories, Canada. Royal Soc. Canada Trans., 3rd ser., v. 49, sect. 4, 7-24.
Folinsbee, R.E., Kirkland, K., Nekolaichuk, A. and Smejkal, V., 1972. Chinkuashih, a gold-pyrite-enargite-barite hydrothermal deposit in Taiwan. Geol. Soc. Am. Memoir 135, 323-335.
Folk, R.L., 1957. Petrology of Sedimentary Rocks. Hemphill's, Austin, Texas, 180 pp.
Fonseca, M.J.G., et al., 1979. Carta Geol. do Brasil ao milionésimo, Folhas Rio de Janeiro (SF.23) Vitoria (SF.24) e Iguape (SG.23). Dep. Nac. Prod. Min., Brasília, 240 pp.
Fontaine, H. and Workman, D.R., 1978. Review of the geology and mineral resources of Kampuchea, Laos and Vietnam. Proc. of the Third Reg. Conf. on Geol. and Miner. Res. of S.E. Asia, Bangkok, 540-603.
Foose, M.P., Slack, J.F. and Casadevall, T., 1980. Textural and structural evidence for a predeformation hydrothermal origin of the Tungsten Queen deposit, Hamme district, North Carolina. Econ. Geol., 75, 515-522.
Ford, T.D., 1976. The ores of the South Pennines and Mendip Hills, England-a comparative study. In K.H. Wolf, ed., Handbook of Strataform and Strata-Bound Ore Deposits, v.2, Elsevier, Amsterdam, 161-195.

Ford, T.D. and King, R.J., 1965. Layered epigenetic galena-barite deposits in the Golconda Mine, Brassington, Derbyshire, England. Econ. Geol., 60, 1686-1701.
Forrester, J.D., 1942. A native copper deposit near Jefferson City, Montana. Econ. Geol., 37, 126-135.
Förster, A., 1974. Die Flusspatlagerstätten Asturiens/Nordspanien und deren Genese. Geol. Rundschau, 63, 212-263.
Förster, H. and Knittel, U., 1979. Petrographic observations on a magnetite deposit at Mishdovan, Central Iran. Econ. Geol., 74, 1485-1510.
Förstner, U., 1981. Recent heavy-metal accumulation in limnic sediments. In K.H. Wolf, ed., Handbook of Stratiform and Strata-Bound Ore Deposits, v.9, Elsevier, Amsterdam, 179-270.
Foshag, W.F. and Fries, C., Jr., 1942. Tin deposits of the Republic of Mexico. U.S. Geol. Surv. Bull. 935-C, 99-180.
Fox, K.F., Jr., 1983. Melanges and their bearing on late Mesozoic and Tertiary subduction and interplate translation at the west edge of the North American Plate. U.S. Geol. Surv. Profess. Paper 1198, 40 pp.
Fox, P.E., Grove, E.W., Seraphim, R.H. and Sutherland Brown, A., 1976. Schaft Creek. Canad. Inst. Min. Metall. Spec. Vol. 15, 219-226.
Frakes, L.A. and Crowell, J.C., 1975. Characteristics of modern glacial marine sediments: application to Gondwana glacials. Third Sympos. on Gondwana Strat., Canberra, 1973; A.N.U. Press, 373-380.
Francheteau, J. et al., 1979. Massive deep-sea sulphide ore deposits discovered on the East Pacific Rise. Nature, 277, 523-528.
Frank, M., Groschopf, P., Sauer, K., Simon, P. and Wild, H., 1975. Sedimentäre Eisenerze in Süddeutschland. Geol. Jahrb., Hannover, D 10, 280 pp.
Frankel, J.J., 1967. Forms and structures of intrusive basaltic rocks. In H.H. Hess and A. Poldervaart, eds., Basalts, v.1, Interscience, New York, 63-102.
Franklin, J.M., Lydon, J.W. and Sangster, D.F., 1981. Volcanic-associated massive sulphide deposits. Econ. Geol. 75th Anniv. Volume, 485-627.
Fraser, D.C., 1961. A syngenetic copper deposit of recent age. Econ. Geol. 56, 951-962.
Frenzel, G., Ottemann, J. and Kurtze, W., 1975. Uran-vererzungen un uranhaltige Rutile in einem Perm vulkanischen Tuffit von Boarezzo (Valganna, Varese). N. Jb. Min. Abh., 124, 75-102.
Frets, D.C. and Balde, R., 1975. Mount Morgan copper-gold deposit. C.L. Knight, ed., Economic Geology of Australia and Papua New Guinea, v.1, Austr. Inst. Min. Metall., Monogr. 5, 779-785.
Friedrich, G., 1982. Genesis of low-grade chromite ore deposits in lateritic soils from the Philippines. In G.C. Amstutz, ed., Ore Genesis, the State of the Art. Springer Verlag, Berlin, 240-250.
Frost, S.H., Weiss, M.P. and Saunders, eds., 1977. Reefs and related carbonates-ecology and sedimentology. Am. Assoc. Petr. Geol. Stud. in Geol. No. 4, 421 pp.

Frutos, J.J. and Oyarzun, J.M., 1975. Tectonic and geochemical evidence concerning the genesis of El Laco magnetite lava flow deposit, Chile. Econ. Geol., 70, 898-990.

Fryklund, V.C., Jr., 1964. Ore deposits of the Coeur d'Alene district, Shoshone County, Idaho. U.S. Geol. Survey Profess. Paper, 445, 103 pp.

Fryklund, V.C., Jr., and Holbrook, D.F., 1950. Titanium ore deposits of Hot Springs County, Ark. Arkansas Resources and Devel. Comm., Div. Geol., Bull. 16, 173 pp.

Fuchs, Y. and Lang-Villemaire, C., 1981. Sur quelques concentrations plombo-zincifères du Devonian inferieur du Massif schisteux Rhenan. Miner. Deposita, 16, 339-355.

Fuller, A.O., 1979. Phosphate occurrences on the western and southern coastal areas and continental shelves of southern Africa. Econ. Geol., 74, 221-231.

Fundal, E., 1975. The Uivfaq dike and related hybrid dikes from southern Disko, West Greenland. Meddel. om Grønland Bd. 195, 28 pp.

Furnes, H., Fridleifsson, I.B. and Atkins, F.B., 1980. Subglacial volcanics-on the formation of acid hyaloclastites. J. Volcanol. Geotherm. Res., 8, 95-110.

Fyfe, W.S., 1973. Granites past and present. Geol. Soc. S. Africa Spec. Publ. 3, 13-16.

Fyfe, W.S. and McBirney, A.R., 1975. Subduction and the structure of andesitic volcanic belts. Am. J. Science, 275-A, 285-297.

Fyles, J.T., 1964. Geology of the Duncan Lake area, Lardeau district, British Columbia. Brit. Columb. Dept. Min. Bull. 49, 88 pp.

Fyles, J.T., 1970. Jordan River area near Revelstoke, British Columbia. Brit. Columb. Dept. Min., Bull. 57, 64 pp.

Gabelman, J.W., 1976. Strata-bound ore deposits and metallotectonics. In K.H. Wolf, ed., Handbook of Stratiform and Strata-Bound Ore Deposits, v.4, Elsevier, Amsterdam, 75-163.

Gair, J.E. and Tsu-Ming-Han, 1975. Bedrock geology and ore deposits of the Palmer Quadrangle, Marquette County, Michigan. U.S. Geol. Surv. Profess. Paper 769, 159 pp.

Galloway, W.E., 1974. Deposition and diagenetic alteration of sandstone in Northeast Pacific arc-related basins: implications for graywacke genesis. Geol. Soc. Amer. Bull. 85, 379-390.

Galloway, W.E. and Hobday, D.K., 1983. Terrigenous Clastic Depositional Systems. Springer Verlag, New York, 423 pp.

Gangloff, A., 1970. Notes sommaires sur la géologie des principaux districts uranifères etudies par la CEA. In Uranium Exploration Geology, Int. Atom. Energy Agency, Vienna, 77-105.

Gansser, A., 1974. Himalaya. In A.M. Spencer, ed., Mesozoic-Cenozoic Orogenic Belts. Geol. Soc. London, 267-278.

Garbuzova, V.F., Dankovtsev, R.F. and Kislyakov, Ya.M., 1970. Usloviya lokalizatsii endogennovo uranovovovo orudeneniya v kontinental'nykh terrigenykh formatsiyakh verkhnevo strukturnovo etazha aktivizirovannykh oblastei. In Ocherki po Geol. i Geokh. Rud. Mestor., Nauka, Moscow, 91-112.

Gardner, D.E., 1955. Beach-sand heavy mineral deposits of eastern

Australia. Austral. Bur. Min. Res. Geol. Geoph., Bull. 28, 103 pp.
Garkovets, V.G., Mushkin, I.V., Titova, A.P., Arapov, V.A., Ignat'eva, L.P. and Golovanov, I.M., 1979. Osnovnye cherty metallogenii Uzbekistana. A.N. Uzbek S.S.R., Tashkent, 272 pp.
Garretty, M.D., 1953. The Ardlethan Tinfield. In A.B. Edwards, ed., Geology of Australian Ore Deposits, Melbourne, 955-961.
Garson, M.S., 1966. Carbonatites in Malawi. In O.F. Tuttle and J. Gittins, eds., Carbonatites. Interscience, New York, 33-71.
Garson, M.S. and Morgan, D.J., 1978. Secondary strontianite at Kangankunde carbonatite complex, Malawi. Inst. Min. Metall., London, Transactions, Ser. B., 87, B70-B73.
Garson, M.S. and Shalaby, I.M., 1976. Precambrian-lower Paleozoic plate tectonics and metallogenesis in the Red Sea region. Geol. Assoc. Canada Spec. Paper No. 14, 573-596.
Gaskin, A.R.J., 1975. Investigation of the residual iron ores of Tonkolili district, Sierra Leone. Transact. Inst. Min. Metall. London, Sect. B, 84, B89-B97.
Gass, I.G., 1968. Is the Troodos massif of Cyprus a fragment of Mesozoic ocean floor? Nature, London, 220, 39-42.
Gass, I.G., 1977. Origin and emplacement of ophiolites. In Volcanic Processes in Ore Genesis. Inst. Min. Metall. London, Geol. Soc. London, 72-76.
Gass, I.G., 1980. The Troodos massif: its role in the unraveling of the ophiolite problem and its significance in the understanding of constructive plate margin processes. In A. Panayiotou, ed., Ophiolites. Cyprus Geol. Surv. Dept., Nicosia, 23-35.
Gay, T.E., Jr., 1966. Economic mineral deposits of the Cascade Range, Modoc Plateau and Great Basin region in northeastern California. In E.H. Bailey, ed., Geology of Northern California. Calif. Dept. of Natural Res. Bull. 190, 97-101.
Gazizova, K.S., 1957. Geologo-strukturnye i geneticheskie osobennosti mednovo mestorozhdeniya Kounrad. Gosgeoltekhizdat' Moscow, 130 pp.
Gealey, W.K., 1980. Ophiolite obduction mechanism. In A. Panayiotou, ed., Ophiolites. Cyprus Geol. Surv. Dept., Nicosia, 228-243.
Gee, R.D., 1975. Bangemall Basin-mineralization. In C.L. Knight, ed., Economic Geology of Australia and Papua New Guinea, v. 1, Metals. Austr. Inst. Min. Metal. Monogr. 5, 528-529.
Gee, R.D., Groves, D.I. and Fletcher, C.I., 1976. Archean geology and mineral deposits of the Eastern Goldfields. 25th Int. Geol. Congress, Australia, Excursion Guide No. 42A, 55 pp.
Geffroy, J., 1971. Les gîtes uranifères dans le Massif Central. In Symposium J. Jung, Plein Air Serv. Ed., Clermont-Ferrand, 541-579.
Genshaft, Yu.S. et al., 1975. Possible origins of the andesitic magmas of island arcs. Int. Geol. Rev. 19, 57-65.
Geol. Survey of Iran, 1964. The geology and structure of the Shah Kuh zinc mines. In CENTO Sympos. on Min. Geol. and the Base Metals, Ankara, 117-126.
Geotimes, 1972. Ophiolites, Penrose Field Conference. Geotimes, v. 17, 12, 24-25.

Gerasimov, I.P., ed., 1964. Fiziko-geologicheskii Atlas Mira. U.S.S.R. Acad. of Sciences, Moscow, 298 pp.
Gerasimovsky, V.I., 1974. Trace elements in selected groups of alkaline rocks. In H. Sørensen, ed., The Alkaline Rocks, Wiley, London, 402-412.
Gerasimovsky, V.I., Volkov, V.P., Kogarko, L.N. and Polyakov, A.I., 1974. Kola Peninsula. In H. Sørensen, ed., The Alkaline Rocks, Wiley, London, 206-221.
Germann, K., 1978. Vulcanism and manganese ore deposition in the Liassic of the Northern Calcareous Alps. In H. Cloos, ed., Alps, Apennines, Hellenides. Schweizerbart, Stuttgart, 96-98.
Geyne, A.R., Fries, C., Jr., Segerstrom, K., Black, R.F., Wilson, I.F. and Probert, A., 1963. Geologia y yacimientos minerales del distrito de Pachuca-Real del Monte, Estado de Hidalgo, Mexico. Publ. Cons. Rec. Nat. no Renov., Mexico, 5E, 222 pp.
Geyti, A. and Schønwandt, H.K., 1979. Bordvika-a possible porphyry molybdenum occurrence within the Oslo Rift, Norway. Econ. Geol. 74, 1211-1220.
Gilhome, W.R., 1975. Mount Tom Price iron orebody, Hamersley iron province. In C.L. Knight, ed., Economic Geology of Australia and Papua New Guinea, v.1, Metals; Austr. Inst. Min. Metall.,892-897.
Gill, J.B., 1981. Orogenic Andesites and Plate Tectonics. Springer Verlag, Berlin, 390 pp.
Gill, J.R., 1959. Reconnaisance for uranium in the Ekalaka lignite field, Carter County, Montana. U.S. Geol. Surv. Bull. 1055, 167-180.
Gilligan, L.B., Felton, E.A. and Olgers, F., 1979. The regional setting of the Woodlawn deposit. Journ. Geol. Soc. Australia, 26, 135-140.
Gilluly, J., 1932. Geology and ore deposits of the Stockton and Fairfield quadrangles, Utah. U.S. Geol. Survey Prof. Paper 173, 171 pp.
Gilmour, P., 1976. Transitional deposits. In K.H. Wolf, ed., Handbook of Stratiform and Strata-Bound Ore Deposits, v.1, Elsevier, Amsterdam, 111-160.
Gilmour, P., 1977. Mineralized intrusive breccias as guides to concealed porphyry copper systems. Econ. Geol. 72, 290-298.
Gimmel'farb, B.M., 1958. Regularities in tectonic distribution of phosphate deposits in the U.S.S.R. Int. Geol. Rev. 6, 317-343.
Ginsburg, R.N., ed., 1975. Tidal Deposits. A Casebook of Recent Examples and Fossil Counterparts. Springer Verlag, Berlin, 428 pp.
Ginsburg, A.I., ed, 1958. Rare Metal carbonatites (in Russian). Gosgeoltekhizdat', Moscow, 127 pp.
Ginsburg, A.I., Timofeyev, I.N. and Fel'dman, L.G., 1979. Printsipy Geologii Granitnykh Pegmatitov. Nedra, Moscow, 296 pp.
Ginsburg, A.I. and Fel'dman, L.G., 1974. Deposits of tantalum and niobium. In V.I. Smirnov, ed., Ore Deposits of the U.S.S.R., v.3, Engl. Transl. Pitman, London, 372-424.
Ginsburg, R.N. and James, N.P., 1974. Holocene carbonate sediments of continental shelves. In C.F. Burk and C.L. Drake, The Geology of Continental Margins. Springer Verlag, New York, 137-155.

Girard, P., 1969. Geology of the Madeleine Mines deposit. Canad. Inst. Min. Metall. Bull., August 1969, 837-845.

Girdler, R.W., ed., 1972. East African rifts. Tectonophysics, 15, Special Issue Nos.1+2, 353 pp.

Gittins, J., 1966. Summaries and bibliographies of carbonatite complexes. In O.F. Tuttle and J. Gittins, eds., Carbonatites. Interscience, New York, 417-570.

Glaçon, J., 1971. Le gisement de Cavallo (El Aouana, Algérie). Bull. B.R.G.M., Sér. 2, No. 6, 69-78.

Glasby, G.P. and Read, A.J., 1976. Deep-sea manganese nodules. In K.H. Wolf, ed., Handbook of Stratiform and Strata-Bound Ore Deposits, v.7, Elsevier, Amsterdam, 295-345.

Glasson, M.J. and Keays, R.R., 1978. Gold mineralization during cleavage development in sedimentary rocks from the auriferous slate belt of central Victoria, Australia: some important boundary conditions. Econ. Geol. 73, 496-511.

Glazkovsky, A.A., Gorbunov, G.I. and Sysoev, F.A., 1974. Deposits of nickel. In V.I. Smirnov, ed., Ore Deposots of the U.S.S.R., v.2, Engl. transl. Pitman, London, 1977, 1-79.

Glennie, K.W., 1970. Desert Sedimentary Environments. Devel. in Sedim. 14, Elsevier, Amsterdam, 222 pp.

Glennie, K.W., Boeuf, M.G.A., Hughes-Clarke, M.W., Moody-Stuart M., Pilaar, W.F.H. and Reinhardt, B.M., 1974. Geology of the Oman Mountains. Verh. Konink. Nederlands Geol. Mijnb. Geootschap, 31, 1974, 423 pp.

Glover, L., III, 1971. Geology of the Coamo area, Puerto Rico, and its relation to the volcanic arc-trench association. U.S. Geol. Surv. Prof. Paper 636, 102 pp.

Godwin, C.I., 1976. Casino. Canad. Inst. Min. Metall. Spec. Volume 15, 344-354.

Gold, D.P., 1963. Average chemical composition of carbonatites. Econ. Geol. 58, 988-991.

Gold, D.P., 1969. The Oka carbonatite and alkaline complex. Geol. Assoc. Canada, Miner. Assoc. Canada, Guidebook, Montreal, 43-62.

Goldberg, E.D. and Arrhenius, G., 1958. Chemistry of Pacific pelagic sediments. Geochim. Cosmoch. Acta 13, 153-212.

Goldberg, I., 1976. A preliminary account of the Otjihase copper deposit, South West Africa. Econ. Geol. 71, 384-390.

Golightly, J.P., 1981. Nickeliferous laterite deposits. Econ. Geol. 75th Anniv. Volume, 710-735.

Gonzáles, O.S., 1984. Basic geological aspects about Real de Angeles Ag-Pb-Zn deposit in central Mexico. Placer, Ltd., in-house exploration workshop, 1984, unpublished, 20 pp.

Gonzáles-Reyna, J., 1956. Riqueza Minera y Yacimientos Minerales de Mexico. Banco de Mexico, S.A., Mexico D.F., 497 pp.

Goossens, P.J., 1972. An exhalative volcanic iron sulfide stratabound deposit near San Fernando, Azuay Province, Ecuador. Econ. Geol. 67, 469-480.

Goossens, P.J., 1978. The metallogenic provinces of Burma: their definitions, geologic relationships and extension into China, India

and Thailand. Third Region. Conf. on Geol. and Miner. Res. S.E. Asia, Bangkok, 1978, 431-492.
Gordon, M.,Jr., Tracey, J.I., Jr. and Ellis, M.W., 1958. Geology of the Arkansas bauxite region. U.S. Geol. Surv. Prof. Paper 299.
Gott, G.B. and Cathrall, J.B., 1980. Geochemical-exploration studies in the Coeur d'Alene district, Idaho and Montana. U.S. Geol. Survey Prof. Paper 1116, 63 pp.
Goudarzi, G.H., 1970. Geology and mineral resources of Libya- a reconnaisance. U.S. Geol. Surv. Prof. Pap. 660, 104 pp.
Goudie, A., 1973. Duricrusts in Tropical and Subtropical Landscapes. Clarendon Press, Oxford, 174 pp.
Grachev, A.F., 1977. Riftovye Zony Zemli. Nedra, Leningrad, 247 pp.
Granger, H.C. and Raup, R.B., 1969. Geology of uranium deposits in the Dripping Springs Quartzite, Gila County, Arizona. U.S. Geol. Surv. Prof. Paper 595, 108 pp.
Grant, J.N., Halls, C., Avila, W. and Avila, G., 1977. Igneous geology and the evolution of hydrothermal systems in some subvolcanic tin deposits of Bolivia. In Volcanic Processes in Ore Genesis, Inst. Min. Metall./ Geol. Soc. London, 117-126.
Grant, J.N. and Nielsen, R.L., 1975. Geology and geochronology of the Yandera porphyry copper deposit, Papua New Guinea. Econ. Geol. 70, 1157-1174.
Green, D. and Cullen, D.J., 1973. The tectonic development of the Fiji region. In P.J. Coleman, ed., The Western Pacific: island arcs, marginal seas and geochemistry. Univ. W. Australia Press, Nedlands, 127-145.
Green, J.C., 1977. Keweenawan plateau volcanism in the Lake Superior region. Geol. Assoc. Canada Spec. Pap. 16, 407-422.
Green, L.H., 1972. Geology of Nash Creek, Larsen Creek and Dawson map-areas, Yukon Territory. Geol. Surv. Canada Memoir 364, 157 pp.
Greenwood, J.E.G.W. and Bleackley, D., 1967. Geology of the Arabian peninsula, Aden Protectorate. U.S. Geol. Surv. Profess. Paper 560-C, 73-75.
Gribov, Ye.M., 1972. Ulutel'yakskoe margantsevoe mestorozhdeniye (Bashkirskoe Priural'e). Geol. Rud. Mestor. No.6, 1972, 95-101.
Griffitts, W.R., Albers, J.P. and Ömer Öner, 1972. Massive sulfide copper deposits of the Ergani Maden area, Southeastern Turkey. Econ. Geol., 67, 701-716.
Griggs, A.B., 1945. Chromite-bearing sands of the southern part of the coast of Oregon. U.S. Geol. Surv. Bull. 945-E, 113-150.
Grogan, R.M. and Bradbury, J.C., 1968. Fluorite zinc-lead deposits of the Illinois-Kentucky mining district. In J.D. Ridge, ed., Ore Deposits of the United States, 1933-1967. A.I.M.E., New York, vol. 1, 370-399.
Gross, G.A., 1967. Geology of iron deposits in Canada, vol. 2. Geol. Surv. Canada Econ. Geol. Rept. 22, 111 pp.
Gross, W.H., 1975. New ore discovery and source of silver-gold veins, Guanajuato, Mexico. Econ. Geol. 70, 1175-1189.
Grossi Sad, J.H. and Torres, N., 1978. Geology and mineral resources of the Barreiro Complex, Araxá, Minas Gerais. Proc. of the First

Intern. Sympos. on Carbonatites, Brazil. D.N.P.M., 307-312.
Groves, D.I., 1972. Section 4: Geology. In A Century of Tin Mining at Mount Bischoff 1871-1971. Tasmania Dept. Min. Geol. Survey Bulletin 54, 165-258.
Groves, D.I. and Taylor, R.G., 1973. Greisenization and mineralization at Achor tin mine, northeast Tasmania. Transact. Inst. Min. Metall., London, Sect. B, 82, B135-B146.
Grubb, P.L.C., 1971. Mineralogical anomalies in the Darling Ranges bauxites at Jarrahdale, Western Australia. Econ. Geol. 66, 1005-1016.
Grundmann, W.H., Jr., 1977. Geology of the Viburnum No.27 mine, Viburnum Trend, Southeast Missouri. Econ. Geol. 72, 349-364.
Grushkin, G.G. and Vedernikov, P.G., 1977. The "rhyolite" association of tin-ore deposits (as in the Dzhalinda deposit). Intern. Geol. Rev., 20, 1059-1066.
Gruszczyk, H. and Pouba, Z., 1968. Stratiform ore deposits of the Bohemian Massif and of the Silesia-Cracow area. 23th Int. Geol. Congress, Prague, Guide to Excursion 23AC, 48 pp.
Gudden, H., 1975a. Zur Bleierz-Führung in Trias-Sedimenten der nördlichen Oberpfaltz. Geol. Bavarica, 1975, No. 74, 33-55.
Gudden, H., 1975b. Die Kreide-Eisenerzlagerstätten in Nordost-Bayern. Geol. Jb., Hannover, D 10, 201-238.
Guilbert, J.M. and Lowell, J.D., 1974. Variations in zoning patterns in porphyry ore deposits. Canad. Inst. Min. Metall. Bulletin v. 67, No. 742, 99-133.
Guillon, J.H., 1974. New Caledonia. In A.M. Spencer, ed., Mesozoic-Cenozoic Orogenic Belts. Geol. Soc. London, 445-452.
Guillon, J.H. and Lawrence, L.J., 1973. The opaque minerals of the ultramafic rocks of New Caledonia. Miner. Deposita, 8, 115-126.
Guimarães, D., 1948. The zirconium ore depoits of the Poços de Caldas Plateau, Brazil, and zirconium geochemistry. Minas Gerais Inst. Tecnol. Industr. Boletim 6, Belo Horizonte, 45-79.
Guimarães, G., Glaser, I. and Margues, V.J., 1968. Sobre a ocorrencía de rochas alcalinas na região de Iporá, Goiás. Min. e Metal., vol. 48, 283, Rio de Janeiro, 11-15.
Guiza, R., Jr., 1956. El distrito minero de Guanajuato. 20th Int. Geol. Congr., Mexico, Excursion A-2 and A-5, 141-152.
Gulbrandsen, R.A. and Krier, D.J., 1980. Large and rich phosphorus resources in the Phosphoria Formation in the Soda Springs area, Southeastern Idaho. U.S. Geol. Surv. Bulletin 1496, 25 pp.
Gümüs, A., 1964. Genesis of some cuprous pyrite deposits of Turkey. CENTO Sympos. on Min. Geol. and the Base Metals, Ankara, 147-154.
Gupta, A.K. and Yagi, K., 1980. Petrology and genesis of leucite-bearing rocks. Springer Verlag, Berlin, 152 pp.
Gustafson, L.B. and Hunt, J.P., 1975. Porphyry copper deposit at El Salvador, Chile. Econ. Geol. 70, 857-912.
Gustafson, L.B. and Williams, N., 1981. Sediment-hosted stratiform deposits of copper, lead and zinc. Econ. Geol. 75th Anniv. Volume, 139-178.

Hackett, J.P., Jr. and Bischoff, J.L., 1973. New data on the stratigraphy, extent and geologic history of the Red Sea hydrothermal deposits. Econ. Geol., 68, 553-564.
Hadjistavrinou, Y. and Constantinou, G., 1982. Cyprus. In F.W. Dunning, W. Mykura and D. Slater, eds. Mineral Deposits of Europe, 2, Southeast Europe. Miner. Soc./Inst. Min. Metall., 255-277.
Hagni, R.D., 1976. Tri-State ore deposits: the character of their host rocks and their genesis. In K.H. Wolf, Handbook of Stratiform and Strata-Bound Ore Deposits, v.6, Elsevier, Amsterdam, 457-494.
Hails, J.R., 1976. Placer deposits. In K.H. Wolf, ed., Handbook of Stratiform and Strata-Bound Ore Deposits, v. 3, Elsevier, 213-244.
Hakim, M., 1980. Petrographish-lagerstättenkundliche Untersuchung der Kupferlagerstätte Sar Chesmeh (Süd-Iran). Geol. Rundschau, 69, 2, 384-396.
Halbach, P., 1972. Lagerstättenkundliche charakterisierung des Ni- und Cr-haltigen Eisenerzvorkommen bei Trsteník (Yugoslawien). Miner. Deposita, 7, 141-153.
Halbouty, M.T., ed., 1970. Geology of Giant Petroleum Fields. Amer. Assoc. Petrol. Geol. Memoir 14, Tulsa, 575 pp.
Halbouty, M.T., 1979. Salt Domes (Gulf Region, United States and Mexico). 2nd ed. Gulf Publ. Comp., Houston, 584 pp.
Hall, J.M. and Robinson, P.T., 1979. Deep crustal drilling in the North Atlantic ocean. Science, 204, 573-586.
Hall, R., 1976. Ophiolite emplacement and the evolution of the Taurus suture zone, southeastern Turkey. Geol. Soc. Am. Bull. 87, 1078-1088.
Hall, R.B., 1978. World nonbauxite aluminium resources: alunite. U.S. Geol. Surv. Prof. Paper 1076-A, 35 pp.
Halliday, A.W., 1980. The timing of early and main stage ore mineralization in southeast Cornwall. Econ. Geol. 75, 752-759.
Hamachi, T., 1962. The uraniferous pelitic sediments closely related to manganese ore deposits in Japan. Japan. Journ. Geol. Geoph., 33, 53-72.
Hamilton, W., 1965. Diabase sheets of the Taylor Glacier region, Victoria Land, Antarctica. U.S. Geol. Surv. Prof. Pap., 456-B, 71 pp.
Hamilton, W., 1979. Tectonics of the Indonesian region. U.S. Geol. Surv. Prof. Paper 1078, 345 pp.
Hamilton, W. and Myers, W.B., 1967. The nature of batholiths. U.S. Geol. Surv. Prof. Paper 554-C, C1-C30.
Hammer, D.F. and Peterson, D.W., 1968. Geology of the Magma mine Arizona. In J.D. Ridge, ed., Ore Deposits of the United States 1933-1967. A.I.M.E., New York, 1282-1310.
Hammerbeck, E.C.I., ed., 1976. Mineral Resoures of the Republic of South Africa, 5th ed. S. Afr. Geol. Surv., Pretoria, 720 pp.
Hansen, J., 1968. Niobium mineralization in the Ilímaussaq alkaline complex, South-West Greenland. 23th Int. Geol. Congr., Prague, v. 7, 263-273.
Harańczyk, C., 1970. Zechstein lead-bearing shales in the Fore-Sudetian Monocline in Poland. Econ. Geol., 65, 481-495.

Harańczyk, C., 1974. Sulfur isotopes and karst features of the Zn-Pb ores (Kraków-Silesia Zn-Pb deposits). In Problems of Ore Deposition, 4th IAGOD Symposium, Varna, v.2, 77-85.

Hardie, L.A., Smoot, J.P. and Eugster, H.P., 1978. Saline lakes and their deposits, a sedimentological approach. In A. Matter and M. Tucker, eds., Modern and Ancient Lake Sediments. Intern. Assoc. Sedim., Spec. Publ. 2, Blackwell, 7-41.

Hargreaves, D., 1975. The Livengood placer deposit-a developing Alaskan gold mine. Min. Magaz., May 1975, 363-365.

Harley, D.N., 1979. A mineralized Ordovician resurgent caldera complex in the Bathurst-Newcastle mining district, New Brunswick, Canada. Econ. Geol. 74, 786-796.

Harper, L.F., 1920. The Lucknow Goldfield. N.S.W. Geol. Survey, Sydney, Mineral Resources No. 30, 1-40.

Harris, I.M., 1975. Sedimentological study of the Goldenville Formation, Nova Scotia. Geol. Surv. Canada Paper 75-1A, 171-174.

Harris, J.F., 1961. Summary of the geology of Tanganyika. Mem. No.1, Tanganyika Geol. Surv., Dodoma, 143 pp.

Harris, N.B. and Einaudi, M.T., 1982. Skarn deposits in the Yerrington District, Nevada: metasomatic skarn evolution near Ludwig. Econ. Geol. 77, 877-898.

Harris, P.G. and Bailey, D.K., 1974. Origin of alkaline magmas as a result of anatexis. In H. Sørensen, ed., The Alkaline Rocks. Wiley, London, 427-442.

Harrison, E.J., 1953. Scheelite-gold-antimony deposits of the Hillgrove district. In A.B. Edwards, ed., Geology of Australian Ore Deposits. Melbourne, 930-934.

Harrison, H.L.H., 1954. Valuation of Alluvial Deposits. Mining Publ. Ltd., London, 308 pp.

Harshman, E.N., 1972. Geology and uranium deposits, Shirley Basin area, Wyoming. U.S. Geol. Surv. Prof. Paper 745, 82 pp.

Hart, O.M., 1958. Uranium deposits in the Pryor-Bighorn Mountains, Carbon County, Montana and Big Horn County, Wyoming. 2nd U.N. Intern. Conf. on the Peaceful Uses of Atomic Energy, Geneva, Proceedings v.2, 523-526.

Hart, O.M., 1968. Uranium in the Black Hills. In J.D. Ridge, ed., Ore Deposits of the United States 1933-1967, A.I.M.E., New York, v.1, 832-837.

Hassan, M.A. and Al-Sulaimi, J.S., 1979. Copper mineralization in the northern part of Oman Mountains near Al Fujairah, United Arab Emirates. Econ. Geol. 74, 919-946.

Hatch, J.R., Gluskoter, H.J. and Lindahl, P.C., 1976. Sphalerite in coals from the Illinois Basin. Econ. Geol. 71, 613-624.

Hatherton, T. and Dickinson, W.R., 1969. The relationship between andesitic volcanism and seismicity in Indonesia, the Lesser Antilles and other island arcs. J. Geoph. Res., 5301-5310.

Haude, H. and Weber, R., 1975. Argentinien. Rohstoffwirtschaftliche Länderberichte, v.VI. Bundesanst. f. Geowiss., Hannover, 121 pp.

Hausen, D.M. and Kerr, P.F., 1968. Fine gold occurrence at Carlin, Nevada. In J.D. Ridge, ed., Ore Deposits of the United States,

1933-1967. A.I.M.E., New York, v.2, 908-940.
Havlena, V., 1963. Geologie Uhelných Ložisek, v.1. Akad. Sci., Prague, 342 pp.
Havlena, V., 1964. Geologie Uhelných Ložisek, v.2. Akad. Sci., Prague, 437 pp.
Hawkins, J.W., Jr., 1980. Petrology of back-arc basins and island arcs: their possible role in the origin of ophiolites. In A. Panayiotou, ed., Ophiolites. Cyprus Geol. Surv. Dept., 244-254.
Hawley, J.E., 1955. Germanium content of some Nova Scotian coals. Econ. Geol. 50, 517-532.
Hawley, C.C., Wyant, D.G. and Brooks, D.B., 1965. Geology and uranium deposits of the Temple Mountain district, Emery County, Utah. U.S. Geol. Surv. Bulletin 1192, 154 pp.
Hawthorne, J.B., 1975. Model of a kimberlite pipe. Physics and Chemistry of the Earth, v. 9, Pergamon Press, Oxford, 1-15.
Haynes, R.W., 1975. Beverley sedimentary uranium orebody, Frome Embayment, South Australia. In C.L. Knight, ed., Economic Geology of Australia and Papua New Guinea, v.1., Austr. Inst. Min. Metall., Monogr. 5, 808-814.
Haynes, S.J. and McQuillan, H., 1974. Evolution of the Zagros Suture Zone, southern Iran. Geol. Soc. Am. Bull., 85, 739-744.
Headlee, A.J.W. and Hunter, K.G., 1953. Elements in coal ash and their industrial significance. Industr. and Engin. Chem., 45, 3, 548-551.
Hearn, B.C., Jr., 1968. Diatremes with kimberlitic affinities in North-Central Montana. Science, v.159, 622-625.
Heath, G.R. and Dymond, J., 1977. Genesis and transformation of metalliferous sediments from the East Pacific Rise, Bauer Deep, and Central basin, northwest Nazca plate. Geol. Soc. Am. Bull., 88, 723-733.
Heckel, P.H., 1972. Recognition of ancient shallow marine environments. Soc. Econ. Paleont. Miner. Spec. Publ. 16, 226-286.
Hedlund, D.C. and Olson, J.C., 1961. Four environments of thorium-, niobium- and rare earths-bearing minerals in the Powderhorn district of southwestern Colorado. U.S. Geol. Surv. Prof. Paper 424-B, B283-B286.
Heezen, B.C., 1960. The rift in the ocean floor. Scient. American 203, 98-100.
Heezen, B.C., 1974. Atlantic-type continental margins. In C.F. Burk and C.L. Drake, Geology of Continental Margins. Springer Verlag, New York, 13-24.
Heezen, B.C. and Hollister, C.D., 1971. The Face of the Deep. Oxford Univ. Press, 659 pp.
Heinrich, E.W., 1966. The Geology of carbonatites. Rand McNally, Chicago, 555 pp.
Heinrich, E.W., 1978. Contrasts in the anatomy of neighbouring alkalic carbonatite complexes. Proc. 1st Intern. Sympos. on Carbonatites, Brazil. D.N.P.M., Brasilia, 23-36.
Hékinian, R., 1982. Petrology of the Ocean Floor. Elsevier, 393 pp.
Helby, R. and Morgan, R., 1979. Palynomorphs in Mesozoic volcanoes

of the Sydney basin. Quart. Notes, Geol. Surv. N.S.W., April 1979, 1-15.

Hemingway, J.E., 1968. Sedimentology of coal-bearing strata. In D. Murchison and T.S. Westall, eds., Coal and Coal-Bearing Strata. Oliver and Boyd, Edinburgh, 43-69.

Henderson, F.B., III, 1969. Hydrothermal alteration and ore deposition in serpentinite-type mercury deposits. Econ. Geol. 64, 489-499.

Hendricks, T.A. and Laird, W.M., 1943. The manganese deposits of the Turtle Mountains, North Dakota. Econ. Geol., 38, 591-602.

Herak, M. and Stringfield, V.T., eds., 1972. Karst. Important Karst Regions of the Northern Hemisphere. Elsevier, Amsterdam, 551 pp.

Herbosch, A., 1974. Facteurs controlant la distribution des elements dans les shales uranifères du basin Permien de Lodéve (Hérault, France). In Formation of Uranium Deposits, Intern. Atomic Energy Agency, Vienna, 299-312.

Hernon, R.M. and Jones, W.R., 1968. Ore deposits of the Central Mining District, Grant County, New Mexico. In J.D. Ridge, ed., Ore Deposits in the United States 1933-1967, A.I.M.E., New York, 1211-1238.

Herz, N., 1976. Titanium deposits in alkalic igneous rocks. U.S. Geol. Surv. Prof. Paper 959-E, E1-E6.

Herz, N. and Eilertsen, N.A., 1968. Titanium. U.S. Geol. Surv. Prof. Paper 580, 437-443.

Hewett, D.F., 1909. Vanadium deposits in Peru. Trans. Am. Inst. Min. Eng., 40, 274-299.

Hewett, D.F., 1956. Geology and mineral resources of the Ivanpah Quadrangle, California and Nevada. U.S. Geol. Surv. Profess. Paper 275, 172 pp.

Hewett, D.F., 1964. Veins of hypogene manganese oxide minerals in the southwestern United States. Econ. Geol., 59, 1429-1472.

Hewett, D.F., Fleischer, M. and Conklin, N., 1963. Deposits of the manganese oxides: supplement. Econ. Geol., 58, 1-51.

Hewitt, W.P., 1943. Geology and mineralization of the San Antonio mine, Santa Eulalia district, Chihuahua, Mexico. Geol. Soc. Am. Bull. 54, 173-204.

Heyl, A.V., 1968. Minor epigenetic, diagenetic and syngenetic sulfide, fluorite and barite occurrences in the central United States. Econ. Geol., 63, 585-594.

Heyl, A.V., Agnew, A.F., Lyons, E.I., Behre, C.H., Jr. and Flint, A.E., 1959. The geology of the zinc and lead deposits of the Upper Mississippi Valley District. U.S. Geol. Survey Profess. Paper 309, 310 pp.

H.H.L., 1974. Climate. Encyclopedia Britannica, v.4, 714-730.

Hietanen, A., 1962. Metasomatic metamorphism in western Clearwater County, Idaho. U.S. Geol. Surv. Profes. Paper 244-A, 116 pp.

Higazy, R.A. and Naguib, A.G., 1958. Study of Egyptian monazite-bearing black sands. 2nd Intern. Conf. on Peaceful Uses of Atom. Energy, Geneva 1958, Proceedings, vol. 2, 658-662.

Higgins, M.W., 1971. Cataclastic Rocks. U.S. Geol. Surv. Profess. Paper 687, 97 pp.

Hildebrand, F.A. and Conklin, N.M., 1974. A breccia dike containing rare-earths bearing apatite, molybdenite and magnetite at Iron Hill, Custer County, Colorado. Econ. Geol., 69, 508-515.

Hildebrand, F.A. and Mosier, E.L., 1974. Argentian cryptomelane and bromargyrite in volcanic rocks near Silver Cliff, Colorado. U.S. Geol. Surv. Bulletin 1382-C, C1-C24.

Hill, M., 1975. Sundry mineralization in Victoria. In C.L. Knight, ed., Economic Geology of Australia and Papua New Guinea, vol. 1; Austr. Inst. Min. Metall. Monogr. 5, 659-666.

Hill, M.N., ed., 1963. The Sea, vol. 3. Wiley, New York, 963 pp.

Hill, W.T., Morris, R.G. and Hagegeorge, C.G., 1971. Ore controls and related sedimentary features of the Flat Gap mine, Treadway, Tennessee. Econ. Geol., 66, 748-765.

Hilpert, L.S., 1969. Uranium resources of northwestern New Mexico. U.S. Geol. Survey Prof. Paper 603, 166 pp.

Himmelberg, G.R. and Loney, R.A., 1981. Petrology of the ultramafic and gabbroic rocks of the Brady Glacier nickel-copper deposit, Fairweather Range, southeastern Alaska. U.S. Geol. Surv. Profess. Paper 1195, 26 pp.

Hirst, D.M. and Dunham, K.C., 1963. Chemistry and petrography of the Marl Slate of S.E. Durham, England. Econ. Geol., 58, 912-940.

Hoagland, A.D., 1976. Appalachian zinc-lead deposits. In K.H. Wolf, ed., Handbook of Stratiform and Strata-Bound Ore Deposits, v. 6, Elsevier, Amsterdam, 495-534.

Hobson, G.D. and Tiratsoo, E.N., 1981. Introduction to Petroleum Geology. Gulf Publ., Houston, 352 pp.

Hodgson, C.J., Bailes, R.J. and Verzosa, R.S., 1976. Cariboo-Bell. Canad. Instit. Min. Metall. Spec. Volume 15, 388-397.

Hodgson, F.D.I., 1973. Petrography and evolution of the Brandberg intrusion, South West Africa. Spec. Publ. Geol. Soc. S. Africa, 3, 339-343.

Holbrook, D.F., 1948. Molybdenum in Magnet Cove, Ark. Arkansas Res. and Devel. Comm., Div. Geology, Bull. 12, 16 pp.

Höll, R., 1966. Genese und Alterstellung von Vorkommen der Sb-W-Hg Formation in der Turkei und auf Chios (Griechenland). Bayer. Akad. Wiss., Munchen, 118 pp.

Höll, R., 1977. Early Paleozoic ore deposits of the Sb-W-Hg Formation in the Eastern Alps and their genetic interpretation. In D.D. Klemm and J.-H. Schneider, eds., Time and Strata-Bound Ore Deposits. Springer Verlag, Berlin,170-198.

Höll, R. and Maucher, A., 1968. Genese und Alter der Scheelit-Magnesit-Lagerstätte Tux. Bayer. Akad. Wiss. Math. Naturw. Kl., 1967, 1-11.

Höll, R. and Maucher, A., 1976. The strata-bound ore deposits in the Eastern Alps. In K.H. Wolf, ed., Handbook of Stratiform and Strata-Bound Ore Deposits, v.2, Elsevier, Amsterdam, 1-36.

Holland, H.D., 1972. Granites, solutions and base metal deposits. Econ. Geol., 67, 281-301.

Holland, S.S., 1955. Lonnie. Brit. Columbia Minister of Mines Rept. for 1955, 29-30.

Höller, H., 1953. Der Blei- Zinkerzbergbau Bleiberg, seine Entwick-

lung, Geologie und Tektonik. Zeitschr. Carinthia, Klagenfurt, II, 143 pp.
Hollingsworth, J.S., 1967. Geology of the Wilson Springs vanadium deposits, Garland County, Arkansas. Geol. Soc. Amer. Field Conf. Central Arkansas. Arkansas Geol. Comm., Little Rock, 22-28.
Hollister, V.F., 1978. Geology of the Porphyry Copper Deposits of the Western Hemisphere. A.I.M.E., New York, 219 pp.
Hollister, V.F. and Sirvas, E.B., 1974. The Michiquillay porphyry copper deposit. Miner. Deposita 9, 261-269.
Holmes, A., 1965. Principles of Physical Geology. Ronald Press, New York, 1288 pp.
Holtedahl, O., and others, 1960. Geology of Norway. Norg. Geol. Unders., Nr. 208, Oslo, 560 pp.
Holzer, H.H. and Stumpfl, E.F., 1980. Mineral deposits of the Eastern Alps. Abhandl. Geol. Bundesanst., Wien, 34, 171-196.
Honnorez, J. and von Herzen, R.P., et al., 1981. Hydrothermal mounds and young ocean crust of the Galapagos: preliminary deep sea drilling results, Leg 70. Geol. Soc. Am. Bull., Pt.1, 92, 457-472.
Honnorez, J., Honnorez-Guerstein, B., Valette, J. and Wauschkuhn, A., 1973. Present day formation of an exhalative sulfide deposit at Vulcano (Tyrrhenian Sea), Pt. II: Active crystallization of fumarolic sulphides in the volcanic sediments of the Baia di Levante. In G.C. Amstutz and A.J. Bernard, Ores in Sediments. Springer Verlag, Berlin, 139-166.
Hoppe, R., 1977. Tara-portend of the future for Irish mines ? Engin. and Min. Journal, September 1977, 83-90.
Hopwood, T.P., 1976. "Quartz-eye"-bearing porphyroidal rocks and volcanogenic massive sulfide deposits. Econ. Geol., 71, 589-612.
Horon, O., 1976. Les gisements de fer de la France. In A. Zitzmann, ed., The Iron Ore Deposits of Europe and Adjacent Areas. Bundesanst. f. Geowiss., Hannover, 143-159.
Horn, R.A., 1975. Beltana and Aroona willemite deposits. In C.L. Knight, ed., Economic Geology of Australia and Papua New Guinea, v.1, Metals. Austr. Inst. Min. Metall.,Memoir 5, 548-554.
Horton, D.J., 1978. Porphyry-type copper-molybdenum mineralization belts in eastern Queensland, Australia. Econ. Geol., 73, 904-921.
Hosking, K.F.G., 1969. Aspect of the geology of the tin fields of Southeast Asia. In 2nd Techn. Confer. on Tin, Bangkok, 1969. Intern. Tin Council, London, 39-80.
Hosking, K.F.G., 1973. Primary mineral deposits. In Gobbett, D.J. and Hutchison, C.S., eds., Geology of the Malay Peninsula. Wiley, New York, 335-390.
Hosking, K.F.G., 1974. The Search for Tin Deposits. 4th World Council on Tin, Bath, 55 pp.
Hosking, K.F.G., 1979. Tin distribution patterns. Geol. Soc. Malaysia Bull., 11, 1-70.
Hotz, P.E., 1964. Nickeliferous laterites in southwestern Oregon and northwestern California. Econ. Geol. 69, 355-396.
Hotz, P.E. and Willden, R., 1964. Geology and mineral deposits of

the Osgood Mountains Quadrangle, Humboldt County, Nevada. U.S. Geol. Surv. Profes. Paper 431, 128 pp.

Houston, R.S. and Murphy, J.F., 1977. Depositional environment of upper Cretaceous black sandstones of the western interior. U.S. Geol. Surv. Prof. Paper 994-A, 29 pp.

Howard, P.F., 1959. Structure and rock alteration at the Elizabeth mine, Vermont. Econ. Geol. 54, 1214-1249.

Hower, H., et al., 1976. Mechanism of burial metamorphism of argillaceous sediments, 1: mineralogical and chemical evidence. Geol. Soc. Amer. Bull. 87, 725-737.

Höy, T., 1977. Big Ledge (82L/8E). Geology, Mining, Exploration in British Columbia, 1975, G 12-G-18.

Höy, T., 1979. Geology of the Goldstream area. Brit. Columbia Min. of Energy, Mines, Petrol. Res. Bull. 71, 85 pp.

Hsieh, C.Y., 1950. Note on the lead, zinc and silver deposits in China. 18th Int. Geol. Congr. London, Pt. 7, 380-399.

Hsü, K.J., 1968. Principles of mélange and their bearing on the Franciscan-Knoxville paradox. Geol. Soc. Am. Bull. 79, 1063-1074.

Hsü, K.J., 1970. The meaning of the word flysch-a short historical search. Geol. Assoc. Canada, Spec. Paper 7, 1-11.

Hsü, K.J. and Jenkyns, H.C., eds., 1974. Pelagic Sediments: on Land and Under the Sea. Int. Assoc. Sedim. Spec. Publ. 1, 447 pp.

Huang, Y.S. and Chu, H., 1945. Some remarks on the Quicksilver deposits in the Hunan-Kweichow border region. Bull. Geol. Soc. China, 25, 283-293.

Hudson, T., Smith, J.G. and Elliott, R.L., 1979. Petrology, composition and age of intrusive rocks associated with the Quartz Hill molybdenite deposit, South-Eastern Alaska. Canad. J. Earth Sc., 16, 1805-1822.

Hughes, C.J., 1973. Spilites, keratophyres and the igneous spectrum. Geol. Magazine, 109, 513-527.

Hughes, C.J., 1982. Igneous Petrology. Devel. in Petrology 7, Elsevier, Amsterdam, 551 pp.

Hügi, Th., et al., 1967. Die Uranvererzung bei Iserábles (Wallis). Beitr. Geol. Schweiz, Geotechnische Ser., 42.

Hulin, C.D., 1925. Geology and ore deposits of the Randsburg Quadrangle, California. Calif. Min. Bur. Bull. 95-1-52.

Hunt, Ch. B., 1956. Cenozoic geology of the Colorado Plateau. U.S. Geol. Surv. Profess. Paper 279, 99 pp.

Hunt, C.B., 1972. Geology of Soils. Freeman, San Francisco, 344 pp.

Hunt, C.B., Averitt, P. and Miller, R.L., 1953. Geology and geography of the Henry Mountains Region, Utah. U.S. Geol. Surv. Profes. Paper 228, 234 pp.

Hurlbut, C.S., Jr., Larsen, E.S., Burgess, C.H. and Buie, B.F., 1941. Igneous rocks of the Highwood Mountains, Montana. Geol. Soc. Amer. Bull., 52, 1857-1868.

Hutchinson, G.E., 1957. A Treatise on Limnology, v.1. Wiley, New York, 1015 pp.

Hutchinson, R.W., 1973. Volcanogenic sulfide deposits and their metallogenic significance. Econ. Geol. 68, 1223-1246.

Hutchison, C.S., 1977. Granite emplacement and tectonic subdivision of Peninsular Malaysia. Geol. Soc. Malaysia, Bull. 9, 187-207.
Hutchison, C.S., 1983. Economic Deposits and their Tectonic Setting. Wiley-Interscience, New York, 365 pp.
Hutchison, C.S. and Jezek, P.A., 1978. Banda Arc of eastern Indonesia: petrography, mineralogy and chemistry of the volcanic rocks. Third Region. Conf. on Geol. and Miner. Res. of S.E. Asia, Bangkok, 607-619.
Hutchison W.W., 1970. Metamorphic framework and plutonic styles in the Prince Rupert region of the central Coast Mountains, British Columbia. Canad. J. Earth Sc., 7, 376-405.
Hutton, J.T., Twidale, C.R. and Milnes, A.R., 1978. Characteristics and origin of some Australian silcretes. In Silcrete in Australia, Univ. New England, Armidale, 19-39.
Hyden, H.J., 1961. Distribution of uranium and other metals in crude oils. U.S. Geol. Surv. Bulletin 1100, 17-99.
Ianovici, V. and Borcoş, M., 1982. Romania. In F.W. Dunning, W. Mykura and D. Slater, eds., Mineral Deposits of Europe, 2, Southeast Europe. Miner. Soc./Inst. Min. Metall. London, 55-142.
Ianovici, V., Borcoş, M., Bleahu, M., Patrulius, D., Lupu, M., Dimitrescu, R. and Savu, H., 1976. Geologia Munţilor Apuseni. Edit. Academiei, Bucharest, 580 pp.
Ianovici, V., Giuscă, D., Mutihac, V., Mirauta, O., and Chiriac, M., 1961. Aperçu général sur la géologie de la Dobrogea. Asoc. Geol. Carpato-Balkan, Guide des Excurs. D-Dobrogea, Bucharest, 92 pp.
Igoshin, B.A., 1966. K voprosu o prirode obrazovaniya Gayskovo kuporonosnovo ozera na Yuzhnom Urale. Sovet. Geologiya, 1966, no.10, 121-130.
Ihlen, P.M. and Vokes, F.M., 1978. Metallogeny. In E.R. Neumann, ed., The Oslo Paleorift. Norg. Geol. Unders. 337, 75-90.
Ikonnikov, A.B., 1975. Mineral Resources of China. Geol. Soc. Am. Microform Public. No.2, 6 microfiche.
Ilavský, J., 1974. Príspevok k paleogeografii gelnickej série Gemeríd na zaklade rozšírenia stratiformných zrudnení. Zapadné Karpaty, ser. 1, Bratislava, 51-97.
Ilavský, J., 1976. An stratigraphische Hiate gebundene Vererzungen in den Tschechoslovakischen Karpaten. Miner. Deposita, 11, 93-110.
Ilavský, J., ed., 1979. Metallogenèse de l'Europe Alpine Centrale et du Sud-Est. Geol. Úst. D. Štúra, Bratislava.
Ilavský, J., Malkovský, M. and Odehnal, L., 1976. The iron ore deposits in the Czechoslovak Socialist Republic. In A. Zitzmann, ed., Iron Ore Deposits of Europe. Bundesanst. f. Geowiss., Hannover, 111-124.
Illies, J.H. and Mueller, S., eds., 1970. Graben Problems. Schweitzerbart, Stuttgart, 310 pp.
Imai, H., 1960. Geology of the Okuki mine and other related cupriferous pyrite deposits in southwestern Japan. N. Jb. Miner., Abhandl., 94, 352-389.
Imai, H., 1966. Formation of fissures and their mineralization in the vein-type deposits of Japan. Univ. of Tokyo Fac. of Engin.

Journal, v.28, 255-302.
Imai, H., 1978. Geologic studies of the mineral deposits in Japan and East Asia. Univ. of Tokyo Press.
Imai, H. and Bunno, M., 1978. Sado Mine, Niigata Prefecture. In Geologic studies of the mineral deposits in Japan and East Asia, Univ. of Tokyo Press, 54-56.
Imai, H., Min Sung Lee, Iida, K., Fujiki, Y. and Takenouchi, S.,1975. Geologic structure and mineralization of the xenothermal vein-type deposits in Japan. Econ. Geol. 70, 647-676.
INAL Staff, 1975. Nickeliferous laterite deposits of the Rockhampton area, Q. In C.L. Knight, ed., Economic Geology of Australia and Papua New Guinea, v.1, Metals. Austr. Inst. Min. Metall., Monogr. 5, 1001-1005.
Inden, R.F. and Moore, C.H., 1983. Beach. In P.A. Scholle, D.G. Bebout and C.H. Moore, eds., Carbonate Depositional Environments. Amer. Assoc. Petr. Geol., Tulsa, 211-266.
Ineson, P.R., 1976. Ores of the Northern Pennines, the Lake District and North Wales. In K.H. Wolf, ed., Handbook of Stratiform and Strata-Bound Ore Deposits, v.5, Elsevier, Amsterdam, 197-230.
Ineson, P.R., Mitchell, J.G. and Vokes, F.M., 1975. K-Ar dating of epigenetic mineral deposits: an investigation of the Permian metallogenic province of the Oslo region, southern Norway. Econ. Geol. 70, 1426-1436.
Innocenti, F., Manetti, P., Mazzuoli, R., Pasquare, G. and Villari, L., 1982. Anatolia and north-western Iran. In R.S. Thorpe, ed., Andesites. Wiley, Chichester, 326-349.
Iovchev, I.S., 1961. Polezni Iskopaemi na NR B'lgaria-Tsvetni Metali. Tekhnika, Sofia, 132 pp.
Irvine, T.N., 1967. The Duke Island ultramafic complex, southeastern Alaska. In P.J. Wyllie, ed., Ultramafic and related Rocks. Wiley, New York, 84-97.
Irvine, T.N., 1974. Ultramafic and gabbroic rocks in the Aiken Lake and McConnell Creek map-areas, British Columbia. Geol. Surv. Canada Paper 74-1A, 149-152.
Irvine, W.T., 1948. Britannia Mine. In G.M. Brownell, ed., Structural Geology of the Canadian Ore Deposits, Montreal, 105-109.
Irvine, W.T., 1972. Geological setting and mineralization of the Pine Point lead-zinc deposits. 24th Int. Geol. Congr., Montreal, Excursion A24-C24, Guidebook, 3-18.
Ishihara, S., ed., 1974. Geology of Kuroko deposits. Mining Geol. Spec. Issue 6, Tokyo, 435 pp.
Ishihara, S., 1981. The granitoid series and mineralization. Econ. Geol. 75th Anniv. Volume, 458-484.
Isphording, W.C., 1973. Origin, mineralogy and economic potential of the residual terra rosa soils of the Yucatan Peninsula. Econ. Geol. 68, p. 1215.
Itsikson, G.V., et al., 1959. Olovorudnye mestorozhdeniya Malovo Khingana. Trudy VSEGEI, v.27, Leningrad, 344 pp.
Ivankin, P.F., ed., 1973. Atlas Morfostruktur Rudnykh Polei. Nedra, Leningrad, 164 pp.

Ivankin, P.F., Lyul'ko, V.A. and Rempel, G.G., 1971. Morfogeneticheskie osobennosti rudnykh polei Noril'skovo rayona. Doklady A.N. SSSR, 1971, v. 199, 3, 674-676.
Ivanov, G.A., 1967. Uglenosnye Formatsii. Nauka, Leningrad.
Ivanov, S.N., Loginov, V.P., Necheukhin, V.M., Prokin, V.A. and Shteinberg, D.S., 1972. Nauchnye osnovy prognozirovaniya endogennykh mestorozhdenii mednykh rud na Urale. In Rudnaya Baza Urala, Nauka, Moscow, 148-165.
Ivanov, S.N., Prokin, V.A. and Dolmatov, G.K., 1962. Nature of ore-bearing brachyanticlinal uplifts in Ural. Intern. Geol. Rev. 8, 234-252.
Ivanova, T.N., 1963. Mestorozhdeniya apatita Khibinskoi Tundry. Gosgeoltekhizdat', Moscow, 288 pp.
Jackson, G.D. and Iannelli, T.R., 1981. Rift-related cyclic sedimentation in the Neohelikian Borden Basin, northern Baffin Island. Geol. Surv. Canada, Paper 81-10, 269-302.
Jackson, N.J., 1979. Geology of the Cornubian tin field, a review. Geol. Soc. Malaysia Bulletin, 11, 209-237.
Jackson, S.A. and Beales, F.W., 1967. An aspect of sedimentary evolution: the concentration of Mississippi-Vally type ores during late stages of diagenesis. Bull. Canad. Petrol. Geol. 15, 383-433.
Jacques, P.and Cassedanne, J.O., 1975. As jazidas brasileiras do ametisto. Mineracao/Metallurgia, 39, 183-194.
Jaffe, F.C. and Collins, B., 1969. Rare-earths concentrations in the South Ruri carbonatite in western Kenya. Inst. Min. Metall. Transact., Sect. B, v.78, B161-B163.
Jahns, R.H., 1955. The study of pegmatites. Econ. Geol. 50th Anniv. Volume, 1025-1130.
Jahns, R.H. and Glass, J.J., 1944. Beryllium and tungsten deposits of the Iron Mountain district, Sierra and Socorro Counties, New Mexico. U.S. Geol. Surv. Bull. 945-C, 45-79.
Jahns, R.H., Griffitts, W.R. and Heinrich, E.W., 1952. Mica deposits of the southeastern Piedmont, Pt. 1, general features. U.S. Geol. Surv. Prof. Paper 248-A, 101 pp.
Jakeš, P. and White, A.J.R., 1972. Major and trace element abundances in volcanic rocks of orogenic areas. Geol. Soc. Am. Bull., 83, 29-40.
Jakucs, L., 1977. Morphogenetics of karst regions. Wiley, New York, 284 pp.
James, H.L., 1966. Chemistry of the iron-rich sedimentary rocks. U.S. Geol. Surv. Profess. Paper 440-W, 61 pp.
James, L.P., 1976. Zoned alteration in limestone at porphyry copper deposits, Ely, Nevada. Econ. Geol. 71, 488-512.
James, N.P., 1983. Reef. In P.A. Scholle, D.G. Bebout and C.H. Moore, eds., Carbonate Depositional Environments. Amer. Assoc. Petr. Geol., Tulsa, 345-440.
Janda, I. and Schroll, E., 1960. Geochemische Untersuchungen an Graphit-Gesteinen. 21st Int. Geol. Congr., Norden, 1, 40-53.
Janković, S., 1960. Allgemeine Charakteristik der Antimon-Erzlagerstätten Jugoslawiens. N.Jb. Miner., Abh., 94, 506-538.

Janković, S., 1972. The origin of base-metal mineralization on the Mid-Atlantic Ridge. 24th Int. Geol. Congr. Montreal, 4, 326-334.
Janković, S., 1982. Yugoslavia. In F.W. Dunning, W. Mykura and D. Slater, eds., Mineral Deposits of Europe, vol. 2, Southeast Europe. Miner. Soc./Inst. Min. Metall., London, 143-202.
Janout, T., 1972. Uranové ložisko Damětice jižně od Horažďovic. Sbor. Geol. Věd, Prague, LG, 15, 67-75.
Jansen, J.B.H. and Schuiling, R.D., 1976. Metamorphism on Naxos: petrology and geothermal gradients. Amer. J. Sc., 276, 1225-1253.
Jaritz, W., Ruder, J. and Schlenker, B., 1977. Das Quartär im Küstengebiet von Moçambique und seine Schwermineralführung. Geol. Jb., Hannover, B26, 3-93.
Jaskółski, S., 1960. Beitrag zur Kenntnis über die Herkunft der Zinnlagerstätten von Gierczyn (Giehren) im Iser-Gebirge, Niederschlesien. N. Jb. Miner., Abh., 94, 181-190.
Jeffery, W.G., 1966. Iskut River; E and L. Brit. Columbia Min. of Mines and Petr. Res. Rept. for 1966, Victoria, 31-34.
Jenkyns, H.C., 1978. Pelagic environments. In H.G. Redaing, ed., Sedimentary Environments and Facies. Elsevier, New York, 314-371.
Jennings, D.J., 1975. Alluvial tin deposits of Tasmania. In C.L. Knight, ed., Economic Geology of Australia and Papua New Guinea, v. 1. Austr. Inst. Min. Metall., Monogr. 5, 1053-1055.
Jennings, J.N., 1971. Karsts. M.I.T. Press, Cambridge, Mass., 252 pp.
Jensen, M.L., 1971. Provenance of Cordilleran intrusives and associated metals. Econ. Geol., 66, 34-42.
Jensen, M.L. and Bateman, A.M., 1979. Economic Mineral Deposits, 3rd edition. Wiley, New York, 593 pp.
John, E.C., 1978. Mineral zones in the Utah copper orebody. Econ. Geol., 73, 1250-1259.
John, T.U., 1963. Geology and mineral deposits of east-central Balabac Island, Palawan Province, Philippines. Econ. Geol., 58, 107-130.
Johns, R.K., 1975. Adelaide geosyncline and Stuart Shelf-mineralization. In C.L. Knight, ed., Economic Geology of Australia and Papua New Guinea, v. 1., Austr. Inst. Min. Metal., 542-547.
Johnson, K.S., 1974. Permian copper shales of southwestern United States, In Gisements Stratiformes et Provinces Cupriferes, Liege, 383-394.
Johnson, M.G., 1972. Placer gold deposits of New Mexico. U.S. Geol. Survey Bull. 1348, 46 pp.
Johnson, R.W., 1982. Papua New Guinea. In R.S. Thorpe, ed., Andesites. Wiley, Chichester, 225-244.
Johnston, W.D., Jr., 1940. The gold quartz veins of Grass Valley, California. U.S. Geol. Surv. Prof. Pap. 194, 101 pp.
Joklik, G.F., Jackson, W.D. and Zani, J.A., 1975. Kimberley bauxite deposits, W.A. In C.L. Knight, ed., Economic Geology of Australia and Papua New Guinea, v.1, Austr. Inst. Min. Met., 968-986.
Jolly, W.T., 1974. Behavior of Cu, Zn and Ni during prehnite-pumpellyite rock metamorphism of the Keweenawan basalts, Northern Michigan. Econ. Geol., 69, 1118-1125.
Jolly, W.T. and Smith, R.E., 1972. Degradation and metamorphic

differentiation of the Keweenawan tholeiitic lavas of Northern Michigan, U.S.A. Journ. Petrology, 13, 2, 273-309.
Jones, D.L., Silberling, N.J. and Hillhouse, J., 1977. Wrangellia-a displaced terrane in northwestern North America.. Canad. Journ. Earth Sci., 14, 2565-2577.
Jones, W.R., Hernon, R.M. and Moore, S.L., 1967. General geology of Santa Rita Quadrangle, Grant County, New Mexico. U.S. Geol. Surv. Prof. Paper 555, 144 pp.
Jopling, A.V. and McDonald, B.C., eds., 1975. Glaciofluvial and Glaciolacustrine Sedimentation. Soc. Econ. Paleont. Miner.,Tulsa, Spec. Public. 23, 320 pp.
Joralemon, I.B., 1952. Age cannot wither or varieties of geological experience. Econ. Geol. 47, 243-259.
Joubin, F.R., 1948. Bralorne and Pioneer mines. In G.M. Brownell, ed., Structural Geology of Canadian Ore Deposits, 168-177.
Jovanović, R. and Ramović, E., 1981. Geološko-ekonomske karakteristike nekih ležišta obojenih i rejetkih metala u NR Kini. Geološki Glasnik, Sarajevo, 299-313.
Juan, V.C., 1964. Mineral Resources of China. Econ. Geol. 41, 399-474.
Julia, R., 1983. Travertines. In P.A. Scholle, D.G. Bebout and C.H. Moore, eds., Carbonate Depositional Environments. Amer. Assoc. Petrol. Geol., Tulsa, 64-72.
Jung, W. and Knitzschke, G., 1976. Kupferschiefer in the German Demokratic Republic (GDR) with special reference to the Kupferschiefer deposit in the southeastern Harz Foreland. In K.H. Wolf, ed., Handbook of Stratiform and Strata-Bound Ore Deposits, v.6., Elsevier, Amsterdam, 353-406.
Jürgenson, B.P. and Hájek, A., 1980. Influence of the migmatitization process on the mineralization of the uranium deposit Rožná-Olší. Věstník Ústř. Úst. Geol., Prague, 55, 223-227.
Juteau, T. and Whitechurch, H., 1980. The magmatic cumulates of Antalya (Turkey): evidence of multiple intrusions in an ophiolite magma chamber. In A. Panayiotou, ed., Ophiolites. Cyprus Geol. Surv. Dept., Nicosia, 377-391.
Kalugin, A.C., 1972. Geologiya i genezis rud tipa zhelezistykh kvartsitov v Devonskikh otlozheniyakh Altaya. In Geologiya i Genezis Dokembriiskikh Zhelezisto-kremnistykh i Margantsevykh Formatsii Mira. Nauk. Dumka, Kiev, 175-187.
Kamilli, R.J. and Ohmoto, H., 1977. Paragenesis, zoning, fluid inclusion and isotopic studies of the Finlandia vein, Colqui district, central Peru. Econ. Geol. 72, 950-982.
Kanehira, K., 1970. Conformable copper-pyrite deposits in the Iimori mining district. In T. Tatsumi, ed., Volcanism and Ore Genesis. Tokyo, 93-104.
Kanehira, K. and Bachinski, D.J., 1968. Mineralogy and textural relationships of ores from the Whalesback mine, northeastern Newfoundland. Canad. J. Earth Sc., 5, 1387-1395.
Kanehira, K. and Tatsumi, T., 1970. Bedded cupriferous sulfide deposits in Japan, a review. In T. Tatsumi, ed., Volcanism and Ore Genesis. Univ. of Tokyo Press, 51-76.

Kantor, J., 1977. Pb-Zn ores of the West Carpathian Triassic and the distribution of their sulphur isotopes. In D.D. Klemm and H.-J. Schneider, eds., Time and Strata-Bound Ore Deposits. Springer Verlag, Berlin, 294-304.

Kapustin, Yu.L., 1971. Mineralogiya Karbonatitov. Nauka, Moscow, 288 pp.

Kapustin, Yu.L., 1976. On the origin of carbonatites. Intern. Geol. Rev., 19, 997-1008.

Karig, D.E., 1970. Ridges and basins of the Tonga-Kermadec island arc system. Journ. Geophys. Res., 75, 239-254.

Karig, D.E., 1971. Origin and development of marginal basins in the western Pacific. Journ. Geophys. Res., 76, 2542-2561.

Karns, A.W., 1976. Submarine phosphorite deposits of Chatham Rise near New Zealand-summary. Amer. Assoc. Petr. Geol. Memoir 25, 395-398.

Kashirtseva, M.F. and Sidel'nikova, V.D., 1973. Raspredeleniye selena, urana, molibdena pri infiltratsionnom rudoobrazovanii. Geol. Rud. Mestor. No.3, 1973, 82-92.

Kashtanov, M.S., 1967. Alluvial bauxite in diatremes in eastern Siberia. Doklady A.N. U.S.S.R., 174, 145-147.

Katz, H.R., 1974. Margins of the Southwest Pacific. In C.F. Burk and C.L. Drake, eds., The Geology of Continental Margins. Springer Verlag, New York, 549-565.

Kautzsch, E., 1942. Untersuchungsergebnisse über die Metallverteilung im Kupferschiefer. Arch. Lagerst. Forsch., 74, 42 pp.

Kautzsch, E., 1957. Die Metallführung des Melaphyrs von Grossorner und ihre Beziehungen zur Metallführung im Kupferschiefer. Neues Jb. Miner., Abhandl., 91, 441-454.

Kay, M., 1951. North American Geosynclines. Geol. Soc. Am. Memoir 48, 140 pp.

Kay, M., 1975. Campbellton sequence: manganiferous beds adjoining Dunnage melange, northeastern Newfoundland. Bull. Geol. Soc. Amer. 86, 105-108.

Kaye, C.A., 1959. Shoreline features and Quaternary shoreline changes, Puerto Rico. U.S. Geol. Surv. Prof. Pap. 317-B, 140 pp.

Kazanskii, V.I., 1970. Geologo-strukturnye osobennosti mestorozhdeniya Tyuya-Muyun. In Ocherki po Geologii i Geokhimii Rudnykh Mestorozhdenii, 34-57.

Kazanskii, V.I., 1972. Rudonosnye tektonicheskie struktury aktivizirovannykh oblastei. Nedra, Moscow, 240 pp.

Kazansky, V.I. and Laverov, N.P., 1974. Deposits of uranium. In V.I. Smirnov, ed., Ore Deposits of the U.S.S.R., v.2, Engl. Transl. Pitman, London, 349-424.

Keays, R.R. and Kirkland, M.C., 1972. Hydrothermal mobilization of gold from copper-nickel sulphides and ore genesis at the Thompson River Copper Mine, Victoria, Australia. Econ. Geol. 67, 1263-1275.

Keays, R.R. and Scott, R.B., 1976. Precious metals in ocean-ridge basalts: implications for basalts as source rocks for gold mineralization. Econ. Geol. 71, 705-720.

Ke-Chin Hsu, 1943. Tungsten deposits of southern Kiangsi, China. Econ. Geol. 38, 431-474.

Keller, J., 1982. Mediterranean island arcs. In R.S. Thorpe, ed., Andesites. Wiley, Chichester, 307-325.

Keller, W.D., 1975. Refractory clays in the lower part of the Pennsylvanian System. U.S. Geol. Surv. Prof. Pap. 853, 65-71.

Kelley, V.C., Kittel, D.F. and Melancon, P.E., 1968. Uranium deposits in the Grants region. In J.D. Ridge, ed., Ore Deposits in the United States 1933-1967. A.I.M.E., New York, 747-769.

Kelly, W.C. and Rye, R.O., 1979. Geologic, fluid inclusion and stable isotope studies of the tin-tungsten deposits of Panasqueira, Portugal. Econ. Geol. 74, 1721-1822.

Kennedy, G.C. and Walton, M.S., Jr., 1946. Nickel investigations in southeastern Alaska. U.S. Geol. Surv. Bull. 947-C, 39-64.

Kennett, J.P., 1982. Marine Geology. Prentice-Hall, Englewood Cliffs, 813 pp.

Kepferle, R.C., 1959. Uranium in Sharon Springs Member of Pierre Shale, South Dakota and northeastern Nebraska. U.S. Geol. Surv. Bulletin 1046-R, 577-604.

Kerr, J.W., 1977. Cornwallis lead-zinc district; Mississippi Valley-type deposits controlled by stratigraphy and tectonics. Canad. J. Earth Sc., 14, 1402-1426.

Kerr, P.F., 1940. Tungsten-bearing manganese deposit at Golconda, Nevada. Geol. Soc. Amer. Bull. 51, 1359-1390.

Kerr, P.F., Brophy, G.P., Dahl, H.M., Green, J. and Woolard, L.E., 1957. Marysvale, Utah, uranium area. Geol. Soc. Amer. Spec. Paper 64, 212 pp.

Kesler, S.E., 1968. Contact-localized ore formation at the Meme mine, Haiti. Econ. Geol. 63, 541-552.

Kesler, S.E., Russell, N., Seaward, M., Rivera, J., McCurdy, K., Cumming, G.L. and Sutter, J.F., 1981. Geology and geochemistry of sulfide mineralization underlying the Pueblo Viejo gold-silver oxide deposit, Dominican Republic. Econ. Geol. 76, 1096-1117.

Kesler, S.E., Sutter, J.F., Issigonis, M.J., Jones, L.M. and Walker, R.L., 1977. Evolution of porphyry copper mineralization in an oceanic island arc: Panama. Econ. Geol. 72, 1142-1153.

Kesler, T.L., 1942. The tin-spodumene belt of the Carolinas. U.S. Geol. Surv. Bulletin 936-J, 245-269.

Kesler, T.L., 1950. Geology and mineral deposits of the Cartersville district, Georgia. U.S. Geol. Surv. Prof. Paper 224, 97 pp.

Kettner, R., 1918. Nový geologický profil příbramskými doly a příbramským okolím. Sbor. Čes. Spol. Zeměvěd., Prague, 24, 1-9.

Khadem, N., 1964. Types of copper ore deposits in Iran. In CENTO Symposium, Mining Geol. and the Base Metals, Ankara, 101-115.

Khain, V.Ye, 1971. Regional'naya Geotektonika. Nedra, Moscow, 548 pp.

Kharkevich, D.C., ed., 1968. Karta magmaticheskikh formatsii SSSR 1:2,500,000. VSEGEI, Moscow, 16 sheets.

Kho, C.H., 1976. Regional geology: West Sarawak. Geol. Surv. Malaysia Ann. Rept. 1975, 86-88.

Kholodov, V.N., 1973a. Trace-element distribution in the Kurumsak-

Chulaktau deposits of Karatau. Geoch. Intern., 7, 795-803.
Kholodov, V.N., 1973b. Osadochnyi Rudogenez i Metallogeniya Vanadia. Nauka, Moscow, 262 pp.
Killsgaard, T.H., Freeman, V.L. and Coffman, J.S., 1970. Mineral resources of the Sawtooth Primitive area, Idaho. U.S. Geol. Surv. Bulletin 1319-D, 174 pp.
Kimberley, M.M., 1978a. Origin of stratiform uranium deposits in sandstone, conglomerate, and pyroclastic rock. In Miner. Assoc. Canada, Short Course, Geology of Uranium Deposits, Univ. of Toronto, 339-381.
Kimberley, M.M., 1978b. Paleoenvironmental classification of iron formations. Econ. Geol. 73, 215-229.
Kimura, E.T., Bysouth, G.D. and Drummond, A.D. Endako. Canad. Inst. Min. Metall. Spec. Vol. 15, Montreal, 444-454.
Kindle, E.M., 1932. Lacustrine concretions of manganese. Amer. J. Science, 24, 496-504.
Kindle, E.M., 1936. The occurrence of lake bottom manganiferous deposits in Canadian Lakes. Econ. Geol. 31, 755-760.
Kindle, E.D., 1972. Classification and description of copper deposits, Coppermine River area, District of Mackenzie. Geol. Surv. Canada Bull. 214, 109 pp.
King, B.C., 1970. Vulcanicity and rift tectonics in East Africa. In T.N. Clifford and I.G. Gass, eds., African Magmatism and Tectonics. Oliver and Boyd, Edinburgh, 263-283.
King, B.C. and Sutherland, D.S., 1966. The carbonatite complexes of eastern Uganda. In O.F. Tuttle and J. Gittins, Carbonatites. Interscience, New York, 73-126.
King, D., 1953. Origin of alunite deposits at Pidinga, South Australia. Econ. Geol. 48, 689-703.
King, F.B. and Ferguson, H.W., 1960. Geology of Northeastermost Tennessee. U.S. Geol. Surv. Prof. Paper 311, 136 pp.
Kinkel, A.R., Jr., 1967. The Ore Knob copper deposit, North Carolina, and other massive sulfide deposits of the Appalachians. U.S. Geol. Survey Prof. Paper 558, 58 pp.
Kinkel, A.R., Jr., Hall, W.E. and Albers, J.P., 1956. Geology and base metal deposits of West Shasta copper-zinc district, Shasta County, California. U.S. Geol. Surv. Prof. Pap. 285, 156 pp.
Kirpal', G.R. and Tenyakov, V.A., 1974. Deposits of aluminium. In V.I. Smirnov, ed., Ore Deposits of the U.S.S.R., v.1., Engl. Transl. Pitman, London, 1977, 273-348.
Kisch, H.J. and Taylor, G.H., 1966. Metamorphism and alteration near an intrusive-coal contact. Econ. Geol. 61, 343-361.
Klemm, D.D. and Schneider, H.J., eds., 1977. Time and Strata-Bound Ore Deposits. Springer Verlag, Berlin.
Klepper, M.R., Robinson, G.D. and Smedes, H.W., 1971. On the nature of Boulder Batholith. Geol. Soc. Am. Bull. 82, 1563-1580.
Kler, V.R., Nenakhova, V.F. and Stroganova, L.I., 1978. Obrazovaniye poleznykh iskopaemykh uglenosnykh formatsii. In Tipy Rudonosnykh Formatsii, Nauka, Moscow, 139-182.
Kloosterman, J.B., 1968. Uma provincia do tipo Nigeriano no sul

da Amazonia. Miner./Met. (Brazil), 47, 59-64 and 167-168.
Knechtel, M.M., 1944. Plains adjacent to Little Rocky Mts., Montana. Oil and Gas Invest. Prelim. Map 4, U.S. Geol. Survey.
Knipper, A.L., 1980. The tectonic position of ophiolites of the Lesser Caucasus. In A. Panayiotou, ed., Ophiolites. Cyprus Geol. Surv. Dept., Nicosia, 372-376.
Knopf, A., 1929. The Mother Lode system of California. U.S. Geol. Survey Prof. Paper 157, 88 pp.
Kobayashi, K. and Nomura, M., 1972. Iron sulfides in the sediment cores from the Sea of Japan and their geophysical implications. Earth Planet. Sc. Letters, 16, 200-208.
Kobe, H.W., 1982. A strata-bound Ni-Co arsenide/sulfide mineralization in the Paleozoic of the Yauli Dome, central Peru. In G.C. Amstutz, ed., Ore Genesis-the State of the Art. Springer Verlag, Berlin, 150-160.
Koch, G.S., Jr., 1956. The Fricso mine, Chihuahua, Mexico. Econ. Geol., 51, 1-40.
Koeberlin, F.R., 1937. Gold in volcanic ash. Eng. and Mining Journ., 135, p. 394.
Kolesnikov, V.V., 1975. Mednorudnye formatsii severnoi chasti Dzhungaro-Balkhashskoi skladchatoi sistemy, usloviya ikh lokalizatsii i zakonomernosti razmeshcheniya. In Tsvetnye i Blagorodnye Metaly Kazakhstana. Nauka Kazakh SSR, Alma Ata, 67-78.
Komar, P.D., 1976. Beach Processes and Sedimentation. Prentice-Hall, N.J., 429 pp.
Kononov, V.I. and Polyak, B.G., 1976. Present geothermal activity in Iceland. Geotectonics, 9+10, 267-272.
Konstantinov, R.M. and Sirotinskaya, S.V., 1974. Logiko-informatsionnye issledovaniya endogennykh rudnykh formatsii i variatsionnye ryady rudnykh mestorozhdenii. In Problemy Endogennovo Rudoobrazovaniya. Nauka, Moscow, 68-82.
Kopecký, L., et al., 1963. Vysvětlivky k přehledné geologické mapě ČSSR, 1:200,000, M-33-IX Děčín. Ústř. Úst. Geol. Prague, 176 pp.
Köppen, W., 1923. Die Klimate der Erde. Walter de Gruyter, Berlin, 369 pp.
Korenevskii, S.M., 1973. Kompleks poleznykh iskopaemykh galogennykh formatsii. Nedra, Moscow, 299 pp.
Kort, V.G., 1970. Tikhii Okean, v.2. Nauka, Moscow, 420 pp.
Korzhnev, V.N., 1980. Raspredeleniye zhelezoorudeneniya v Kholzunskom rudnom polye v Gornom Altae. Trudy A.N. SSSR, Sibir. Otdel., Novosibirsk, vol. 465, 76-80.
Kosaka, H. and Wakita, K., 1978. Some geologic features of the Mamut porphyry copper deposit, Sabah, Malaysia. Econ. Geol., 73, 618-627.
Koschmann, A.H., 1949. Structural control of the gold deposits of the Cripple Creek district, Teller County, Colorado. U.S. Geol. Surv. Bull. 955-B, 19-60.
Koschmann, A.H. and Bergendahl, M.H., 1968. Principal gold-producing districts of the United States. U.S. Geol. Surv. Prof. Paper 610, 283 pp.
Koski, R.A., Goodfellow, R. and Bouse, R.M., 1982. Preliminary des-

cription of massive sulphide samples from the southern Juan de Fuca Ridge. U.S. Geol. Surv., Menlo Park, Open File Report 82-200B, 21 pp.
Kotschoubey, B. and Truckenbrodt, W., 1981. Evolucão poligenético das bauxitas do distrito de Paragominas-Acailandía (estados do Pará e Maranhão). Revista Brazil de Geocienc., 11, São Paulo, 193-202.
Koukharsky, M. and Mirré, J.C., 1976. Mi Vida prospect: a porphyry copper-type deposit in northwestern Argentina. Econ. Geol. 71, 849-863.
Koutek, J., 1963. Geologie Československých Rudních Ložisek, I. S.P.N. edit., Prague, 120 pp.
Kovalenko, V.I., Vladykin, N.V. and Goreglyad, A.V., 1977. Vostochnaya Mongolia-novaya provintsiya redkometal'noi mineralizatsii. In Trudy Sovmest. Sov.-Mongol. n. i Geol. Expedicii, v.22, 189-205.
Kovalev, A.A., 1970. Polygenetic character of uranium mineralization in coal-bearing deposits. Intern. Geol. Rev. 14, 345-353.
Krasil'nikova, N.A., Gurevich, B.G. and Bliskovskii, V.Y., 1965. Fosfority Altae-Sayanskoi skladchatoi oblasti. Litologiya i Poleznye Iskop., No.4, 161-181.
Kratochvíl, J., 1961. Topografická Mineralogie Čech, v. IV. Czech. Akad. Sc. Prague, 384 pp.
Kraume, E., Dahlgrun, F., Ramdohr, P. and Wilke, A., 1955. Die Erzlager des Rammelsberges bei Goslar. Beih. Geol. Jahrb., 18, 394 pp.
Krauskopf, K.B., 1955. Sedimentary deposits of rare metals. Econ. Geol. 50th Anniv. Vol., 411-463.
Kräutner, H.G., 1977. Hydrothermal-sedimentary iron ores related to submarine volcanic rises: the Teliuc-Ghelar Type as a carbonatic equivalent of the Lahn-Dill Type. In Klemm, D.D. and Schneider, H.-J., eds., Time and Strata-Bound Ore Deposits. Springer Verlag, Berlin, 232-253.
Krebs, W., 1966. Der Bau des Oberdevonischen Langenaubach-Breitscheider Riffes un seine weitere Entwicklung im Unterkarbon (Rhein. Schiefergebirge). Abh. d. Senckenbg. Nat. Gess., No. 511, Frankfurt/Main, 105 pp.
Krebs, W., 1972. Facies and development of the Meggen Reef (Devonian, West Germany). Geol. Rundschau, 61, 647-671.
Krebs, W., 1975. Formation of Southwest Pacific island arc-trench and mountain systems: plate or global vertical tectonics ? Amer. Assoc. Petrol. Geol. Bull. v.59, 1639-1666.
Krebs, W., 1976. Geology of European stratabound lead-zinc-copper deposits. Canad. Assoc. Petrol. Geol. Seminar, Univ. Calgary, unpublished, 28-54.
Krebs, W., 1979. Devonian basinal facies. Spec. Pap. Paleontol., London, 23, 125-139.
Krendelev, F.P., 1974. Metallonosnye Konglomeraty Mira. Nauka, Novosibirsk, 238 pp.
Krishnaswamy, S., 1972. India's Mineral Resources. Oxford Publ., Delhi, 503 pp.
Krivtsov, A.I., 1973. Domezozoiskie Boksity SSSR. Nedra, Leningrad,

383 pp.
Kropachev, S.M. et al., 1973. Blyavinsko-Kinderlinskaya rudonosnaya zona. In Osnovnye Printsipy i Metodika Sostavleniya Prognozno-Metallogenicheskikh Kart Rudnykh Raionov v Paleovulkanicheskikh Oblastiakh. Nedra, Moscow, 199-219.
Krutov, G.A., 1974. Deposits of cobalt. In V.I. Smirnov, ed., Ore Deposits of the U.S.S.R., v. 2, Engl. Transl., Pitman, London, 80-105.
Krutský, N., ed., 1964. Sborník k XV. Sjezdu s Exkursním Průvodcem. ČS. Spol. pro Miner., Geol., Teplice, 264 pp.
Kucha, H., 1982. Platinum-group metals in the Zechstein copper deposits, Poland. Econ. Geol. 77, 1578-1591.
Kudělásek, V., 1959. Stopové prvky Dolnoslezské Pánve II. Sborník Věd. Prací V.Š.B. Ostrava, 5, 457-481.
Kuhn, T.H., 1941. Pipe deposits of the Copper Creek area, Arizona. Econ. Geol., 36, 512-538.
Kumarapeli, P.S., 1976. The St. Lawrence rift system, related metallogeny, and plate tectonic models of Appalachian evolution. Geol. Assoc. Canada Spec. Paper 14, 301-320.
Kumarapeli, P.S. and Saull, V.A., 1966. The St. Lawrence Valley system: a North American equivalent of the East African Rift valley system. Canad. J. Earth Sc., 3, 639-658.
Kutina, J., 1963. Pb-Zn ore veins in the Příbram ore field. In Some Ore Deposits of the Bohemian massif, Guide to Excursion. Sympos. Probl. of Postm.Ore Depos., Prague, 55-85.
Kutina, J. and Telupil, A., 1966. Prospection for ore veins along the Clay Fault (Příbram ore field) with application of the principle of equidistances. Věstník Ústř. Úst. Geol., 41, 431-443.
Kuÿpers, E.P. and Denyer, P.Ch., 1979. Volcanic exhalative manganese deposits of the Nicoya ophiolite complex, Costa Rica. Econ. Geol. 74, 672-692.
Kuzmenko, M.V., ed., 1976. Polya Redkometal'nykh Granitnykh Pegmatitov. Nauka, Moscow, 332 pp.
Kuznetsov, V.A., 1967. Altae-Sayanskaya metallogenicheskaya provintsia i nekotorye voprosy metallogenii politsiklichnykh skladchatykh oblastei. Zakon. Razm. Polez. Iskop. 8, 275-303.
Kuznetsov, V.A., 1972. Ore Formations. Application of formations analysis in the study of ore deposits. Int. Geol. Rev. 15, 57-65.
Kuznetsov, V.A., 1974. Deposits of mercury. In V.I. Smirnov, ed., Ore Deposits of the U.S.S.R., v.2, Engl. Transl. Pitman, London, 298-348.
Kuznetsov, Yu.A., 1964. Glavnye Tipy Magmaticheskikh Formatsii. Nedra, Moscow.
Kwak, T.A.P. and Askins, P.W., 1981. Geology and genesis of the laminar F-Sn-W(Be-Zn) skarn at Moina, Tasmania, Australia. Econ. Geol. 76, 439-467.
Kyle, J.R., 1976. Brecciation, alteration and mineralization in the Central Tennessee zinc district. Econ. Geol. 71, 892-903.

Laboratoire de Géologie Appliquée, Univ. de Paris, 1973. Some major concepts of metallogeny. Miner. Depos., 8, 237-258.
Ladd, H.S., Tracey, J.I., Jr. and Gross, M.G., 1970. Deep drilling on Midway atoll. U.S. Geol. Surv. Prof. Pap. 680-A, 22 pp.
Lagny, P., 1975. Le gisement plombo-zincifère de Salafossa (Alpes italiennes orientales): Remplissage d'un paleokarst triasique par des sediments sulfurés. Min. Deposita, 10, 345-361.
Lambert, I.B. and Sato, T., 1974. The Kuroko and associated ore deposits of Japan: a review of their features and metallogenesis. Econ. Geol. 69, 1215-1236.
Lamey, C.A., 1950. The Blewett iron-nickel deposit, Chelan County, Washington. U.S. Geol. Surv. Bull. 969-D, 87-103.
Lamey, C.A. and Hotz, P.E., 1952. The Cle Elum River nickeliferous iron deposits, Kittitas County, Washington. U.S. Geol. Surv. Bull. 978-B, 27-67.
Landis, G.P. and Rye, R.O., 1974. Geologic, fluid inclusion and stable isotope studies of the Pasto Bueno tungsten-base metal ore deposit, Northern Peru. Econ. Geol., 69, 1025-1059.
Lang, B., 1979. The base metals-gold hydrothermal ore deposits of Baia Mare, Romania. Econ. Geol. 74, 1336-1351.
Langen, R.E. and Kidwell, A.L., 1974. Geology and geochemistry of the Highland uranium deposit, Converse County, Wyoming. The Mountain Geologist, Denver, 11, 2, 85-93.
Langford, F.F., 1974. A supergene origin for vein-type uranium ores in the light of the Western Australian calcrete-carnotite deposits. Econ. Geol. 69, 516-526.
Langford-Smith, T., ed., 1978. Silcrete in Australia. Univ. of New England, Armidale, 304 pp.
Lanier, G., John, E.C., Swenesen, A.J., Reid, J., Bard, C.E., Caddey, S.W. and Wilson, J.C., 1978. General geology of the Bingham mine, Bingham Canyon, Utah. Econ. Geol., 73, 1228-1241.
Lapham, D.M., 1968. Triassic magnetite and diabase at Cornwall, Pennsylvania. In J.D. Ridge, ed., Ore Deposits of the United States 1933-1967, A.I.M.E., New York, 72-94.
Lapin, B.N., 1977. Geologiya Terligkhaiskovo rudnovo polya (Tuva) i vopros o vozraste rtutnovo orudeneniya. Trudy Inst. Geol. Geogr. A.N. SSSR, Vyp. 364, 114-143.
Lapukhov, A.S., 1975. Zonal'nost kolchedanno-polymetallicheskikh mestorozhdenii. A.N. SSSR, Sibir. Otd., Vyp. 247, 264 pp.
Large, R.R., 1977. Chemical evolution and zonation of massive sulfide deposits in volcanic terrains. Econ. Geol. 72, 549-572.
Large, R.R. and Both, R.A., 1980. The volcanogenic sulfide ores at Mount Chalmers, eastern Queensland. Econ. Geol., 992-1009.
Larter, R.C.L., Boyce, A.J. and Russell, M.J., 1981. Hydrothermal pyrite chimneys from the Ballynoe barite deposit, Silvermines, County Tiperary, Ireland. Miner. Depos. 16, 309-318.
Laumulin, T.M., Tilepov, Z.T., Trubnikov, L.M. and Bukurov, G.S., 1973. Geologicheskie osobennosti i metallogeniya Tasaral-Kyzylespinskovo antiklinoriya v severo-zapadnom Pribalkhash'ye. Nauka, Kazakh SSR, Alma-Ata, 184 pp.

Laurent, R., 1975. Occurrences and origin of the ophiolites of southern Quebec, northern Appalachians. Canad. J. Earth. Sc., 12, 443-455.
Laville-Timsit, L. and Wilhelm, E., 1979. Comportement supergène des métaux autour du gîte sulfuré de Porte-aux-Moines (Côtes-du-Nord). Application à la prospection géochimique. Bull. B.R.G.M., Sect. 2, 2-3, 195-228.
Lavrov, V.V., 1972. Tipy uglenosnykh formatsii i parageneticheskie kompleksy ikh poleznykh iskopaemykh. Trudy VSEGEI, v.176, 180-189.
Lawrence, L.J., 1975. Emmaville-Torrington tin-tungsten base metal mineralization. In C.L. Knight, ed., Economic Geology of Australia and Papua New Guinea, v.1. Austr. Inst. Min. Metall., 725-729.
Lawrence, L.J., 1977. The syngenetic-epigenetic transition, an Australian example. In Klemm, D.D. and Schneider, H.-J., eds., Time and Strata-Bound Ore Deposits. Springer Verlag, 46-54.
Lawrence, L.J., 1978. Porphyry type gold mineralization in shoshonite at Vunda, Fiji. Proc. Austr. Inst. Min. Metall., 268, 21-31.
Laznicka, P., 1965. Regional mineralogical conditions in the western and central parts of the Bohemian Pluton of central Czechia and in the massifs of Stod and Stenovice. Acta Mus. Nat. Pragae, B 21, No.3, 93-156.
Laznicka, P., 1970. Quantitative aspects in the distribution of base and precious metal deposits of the world. Ph.D. Thesis, Univ. of Manitoba, Winnipeg, unpublished, 725 pp.
Laznicka, P., 1973. MANIFILE, the University of Manitoba file of nonferrous metal deposits of the world. Publ. No.2, Centre for Precambr. Studies, Univ. of Manitoba, Winnipeg, 533 and 767 pp.
Laznicka, P., 1974. LITHOTHEQUE, a system of rock and mineral specimens arrangement in geological education, documentation and exploration. Publ. No. 5, Centre for Prec. Stud., Univ. of Manitoba, Winnipeg, 32 pp.
Laznicka, P., 1975. Exploring with lithotheque. Western Miner, Vancouver, Febr. 1975, 32-39.
Laznicka, P., 1976a. Porphyry copper and molybdenum deposits of the U.S.S.R. and their plate tectonic setting. Transact. Inst. Min. and Metall., London, Sect.B, v. 85, B13-B32.
Laznicka, P., 1976b. Barite nodules of possibly late diagenetic origin from Twitya River area. Mackenzie Mountains, Northwest Territories. Canad. J. Earth Sc., 13, 1446-1455.
Laznicka, P., 1981a. The concept of ore types-summary, suggestions and a practical test. In K.H. Wolf, ed., Handbook of Stratiform and Strata-Bound Ore Deposits, v.8, Elsevier, Amsterdam, 449-511.
Laznicka, P., 1981b. Platform versus mobile belt-type stratabound/stratiform ore deposits: summary, comparisons, models. Ibid., v.8, 513-592.
Laznicka, P., 1981c. Data on the worldwide distribution of stratiform and stratabound ore deposits. Ibid., v. 10, 79-389.
Laznicka, 1981d. A rock encyclopedia that includes rock samples. Jour. Geol. Educ., 29, 56-59.
Laznicka, P., 1983a. Giant ore deposits: a quantitative approach.

Global Tect. nad Metall., Washington, 2, 41-63.
Laznicka, P., 1983b. The search for a more realistic metallogenic map format with reference to the Pine Creek "Geosyncline". B.M.R. Journal of Australian Geology, Canberra, v.8, 4, 293-306.
Laznicka, P., 1984. Mineral Deposits/Economic Geology. Course and lab scripta and notes, University of Manitoba, Winnipeg.
Laznicka, P., 1985a. Data on ore deposits: a critical review of their sources, acquisition, organization and presentation. In K.H. Wolf, ed., Handbook of Stratiform and Strata-Bound Ore Deposits, Elsevier, Amsterdam, v.11, 1-118.
Laznicka, P., 1985b. Concordant versus discordant ore deposits and ore transformations. Ibid, v.11, 119-316.
Laznicka, P., 1985c. Zoning and ores. Ibid, v.11, 317-524.
Laznicka, P., 1985d. The geological association of coal and metallic ores, a review. Ibid, v.13, 1-71.
Laznicka, P., 1985e. Metal sources of ore deposits. Ibid, v.12, 109-218.
Laznicka, P., 1985f. Unconformities and ores. Ibid, v.12, 219-360.
Laznicka, P., 1985g. Strata-related ore deposits classified by metals, lithologic associations, and some quantitative relationships. Ibid, v.12, 1-107.
Laznicka, P. and Edwards, R.J., 1979. Dolores Creek, Yukon-a disseminated copper mineralization in sodic metasomatites. Econ. Geol. 74, 1352-1320.
Laznicka, P. and Wilson, H.D.B., 1972. The significance of a copper-lead line in metallogeny. 24th Intern. Geol. Congr., Montreal, Sect. 4, 25-36.
Lebedev, L.M., 1970. Novye dannye po mineralogii kolchedannykh rud vulkana Mendeleeva. Doklady A.N. SSSR, v.191, 1130-1133.
Lebedev, L.M., 1973. Minerals of contemporary hydrotherms of Cheleken. Geochemistry Internat., 9, 485-504.
Leblanc, M. and Billaud, P., 1978. A volcano-sedimentary copper deposit on a continental margin of upper Precambrian age: Bleida (Anti-Atlas, Morocco). Econ. Geol., 73, 1101-1111.
Lee, D.E. and van Loenen, R.E., 1971. Hybrid granitoid rocks of the Southern Snake Range, Nevada. U.S. Geol. Surv. Prof. Paper 668, 48 pp.
Lee, K.Y., 1970. Some rare-element mineral deposits in Mainland China. U.S. Geol. Survey Bulletin 1312-N, 34 pp.
Leeder, M.R., 1982. Sedimentology, Process and Product. Allen and Unwin, London, 344 pp.
Lefond, S.J., 1969. Handbook of World Salt Resources. Plenum Press, New York, 384 pp.
Lefond, S.J., ed., 1975. Industrial Minerals and Rocks. Am. Inst. Min. Metall. Petr. Eng., New York, 1360 pp.
Le Fur, Y., 1971. Les indices de cuivre du groupe volcano-sédimentaire de Poli (Cameroun). Bull. B.R.G.M., Sect.2, 6, 79-91.
Leggett, R.F., ed., 1976. Glacial till: an inter-disciplinary study. Royal Soc. Canada, Spec. Publ. 12, 412 pp.
Lehne, R.W. and Amstutz, G.C., 1982. Sedimentary and diagenetic fabrics in the Cu-Pb-Zn-Ag deposit of Colquijirca, central Peru.

In G.C. Amstutz, ed., Ore Genesis, the State of the Art. Springer Verlag, Berlin, 161-166.
Leinz, V., Bartorelli, A. and Isotta, C.A., 1968. Contribução ao Estudo de magmatismo basaltico mesozoico da Bacia do Paraná. An. Acad. Brasil. Cienc., 40, 167-189.
Leitch, E.C., 1980. The Great Serpentine Belt of New South Wales: diverse mafic-ultramafic complexes set in a Paleozoic arc. In A. Panayiotou, ed., Ophiolites. Cyprus Geol. Surv. Dept., 637-648.
Leitch, C.H.B., 1981. Mineralogy and textures of the Lahanos and Kizilkaya massive sulphide deposits, Northeastern Turkey, and their similarity to Kuroko ores. Miner. Deposita, 16, 241-257.
Lelong, F., Tardy, Y., Grandin, G., Trescases, J.J. and Boulange, B., 1976. Pedogenesis, chemical weathering and processes of formation of some supergene ore deposits. In K.H. Wolf, ed., Handbook of Stratiform and Strata-Bound Ore Deposits, v.3, Elsevier, Amsterdam, 93-174.
Lemoalle, J. and Dupont, B., 1973. Iron-bearing oolites and the present conditions of iron sedimentation in Lake Chad (Africa). In G.C. Amstutz and A.J. Bernard, eds., Ores in Sediments. Springer Verlag, Berlin, 167-178.
Leonardos, O.H., Jr., 1974. Origin and provenance of fossil and recent monazite deposits in Brazil. Econ. Geol. 69, 1126-1128.
Lerman, A., ed., 1978. Lakes: Physics, Chemistry and Geology. Springer, New York.
Leroy, J., 1978. The Margnac and Fanay uranium deposits of the La Crouzille district (western Massif Central, France). Geologic and fluid inclusion studies. Econ. Geol. 73, 1611-1634.
Lett, R.E.W. and Fletcher, W.K., 1980. Syngenetic sulphide minerals in a copper-rich bog. Miner. Depos., 15, 61-67.
Levi, B., 1969. Burial metamorphism of a Cretaceous volcanic sequence west from Santiago, Chile. Contrib. Miner. Petrol. 24, 30-49.
Lewis, D.V., 1955. Relationships of ore bodies to dikes and sills. Econ. Geol., 50, 495-516.
Lewis, R.Q., Sr. and Trimble, D.E., 1959. Geology and uranium deposits of Monument Valley, San Juan County, Utah. U.S. Geol. Surv. Bulletin, 1087-D, 105-130.
Lilljequist, R., 1973. Caledonian geology of the Laisvall area, southern Norbotten, Swedish Lappland. Sveriges Geol. Unders., Ser. C, No. 691, 43 pp.
Lindgren, W., 1911. The Tertiary gravels of the Sierra Nevada of California. U.S. Geol. Surv. Prof. Paper 73, 226 pp.
Lindgren, W., 1933. Mineral Deposits. 4th ed., McGraw-Hill, New York, 930 pp.
Lindgren, W. and Laughlin, G.F., 1919. Geology and ore deposits of the Tintic district, Utah. U.S. Geol. Surv. Prof. Pap. 107, 282 pp.
Lindsey, D.A., 1977. Epithermal beryllium deposits in water-laid tuff, western Utah. Econ. Geol. 72, 219-232.
Lindsey, D.A., Ganow, H. and Mountjoy, W., 1973. Hydrothermal alteration associated with beryllium deposits at Spor Mountain, Utah. U.S. Geol. Surv. Prof. Paper 818-A, A1-A20.

Linn, R.K., 1968. New Idria mining district. In J.D. Ridge, ed., Ore Deposits of the United States 1933-1967. A.I.M.E., New York, v.2, 1623-1649.
Lipman, P.W., Fisher, F.S., Mehnert, H.H., Naeser, C.W., Luedke, R.G. and Steven, T.A., 1976. Multiple ages of mid-Tertiary mineralization and alteration in the western San Juan Mountains, Colorado. Econ. Geol., 71, 571-588.
Lisitzin, A.P., 1971. Sedimentation in the World Ocean. Soc. Econ. Paleont., Miner., Tulsa, Spec. Publ. 17, 218 pp.
Lissiman, J.C. and Oxenford, R.J., 1975. Eneabba rutile-zircon-ilmenite sand deposit, W.A. In C.L. Knight, ed., Economic Geology of Australia and Papua New Guinea, v. 1., Austr. Inst. Min. Metall. Monogr. 5, 1062-1069.
Little, H.W., 1959. Tungsten deposits of Canada. Geol. Surv. Canada Econ. Geol. Ser. 17, 251 pp.
Livingston, D.E., Mauger, R.L. and Damon, P.E., 1968. Geochronology of the emplacement, enrichment and preservation of Arizona porphyry copper deposits. Econ. Geol. 63, 30-36.
Ljunggren, P. and Meyer, H.C., 1964. The copper mineralization in the Corocoro basin, Bolivia. Econ. Geol. 59, 110-125.
Lorenz, V., 1975. Formation of phreatomagmatic maar-diatreme volcanoes and its relevance to kimberlitic diatremes. Physics and Chemistry of the Earth, 9, Pergamon Press, Oxford, 17-27.
Lorinci, G.I. and Miranda, V.J.C., 1978. Geology of the massive sulfide deposits of Campo Morado, Guerrero, Mexico. Econ. Geol., 73, 180-191.
Lortie, R.B. and Clark, A.H., 1974. Stratabound fumarolic copper deposits in rhyolitic lavas and ash-flow tuffs, Copiapó district, Atacama, Chile. Problems of Ore Deposition, IV. IAGOD Sympos., Varna, 1974, v.1, 256-264.
Loudon, A.G., 1976. Marcopper porphyry copper deposit, Philippines. Econ. Geol. 71, 721-732.
Loudon, A.G. and Bowman, H.N., 1967. Karamah copper prospect, Bingara. N.S.W. Geol. Surv. Rept. No. 55, Sydney, 8-12.
Lougnon, J., 1956. Rapport general sur les gisements de manganèse en France. 20th Int. Geol. Congr., Mexico, Sympos. Manganeso, v.5, 63-171.
Loureiro, F.E. de V.L., 1980. Uranio associado a rochas fosfaticas de origem ignea, o exemplo de Catalão. Anais do 31 Congr. Brasil. de Geol., v.3, 1635-1648.
Love, J.D., 1964. Uraniferous phosphatic lake beds of Eocene age in intermontane basins of Wyoming and Utah. U.S. Geol. Surv. Prof. Paper 474-E, 66 pp.
Lovering, T.G., 1954. Radioactive deposits of Nevada. U.S. Geol. Surv. Bull. 1009-C, 63-106.
Lovering, T.G., 1972. Jasperoid in the United States: its characteristics, origin and economic significance. U.S. Geol. Surv. Prof. Paper 710, 164 pp.
Lovering, T.S., 1927. Organic precipitation of metallic copper. U.S. Geol. Surv. Bulletin 795-C, 45-52.

Lovering, T.S. and Goddard, E.N., 1950. Geology and ore deposits of the Front Range, Colorado. U.S. Geol. Surv. Profess. Paper, 223, 319 pp.

Lovering, T.G. and Heyl, A.V., 1974. Jasperoid as a guide to mineralization in the Taylor mining district and vicinity near Ely, Nevada. Econ. Geol. 69, 46-58.

Lovering, T.S. and Tweto, O., 1953. Geology and ore deposits of the Boulder County tungsten district, Colorado. U.S. Geol. Surv. Profess. Paper 245, 199 pp.

Lowell, J.D., 1968. Geology of the Kalamazoo orebody, San Manuel District, Arizona. Econ. Geol. 63, 645-654.

Lowell, J.D. and Guilbert, J.M., 1970. Lateral and vertical alteration-mineralization zoning in porphyry ore deposits. Econ. Geol. 65, 378-408.

Lozej, G.P. and Beales, F., 1977. Stratigraphy and structure of La Encantada mine area, Coahuila, Mexico. Geol. Soc. Am. Bull. 88, 1793-1807.

Lozes, J., Guérangé, B., Doumenge, J.P. and Schmid, M., 1977. Notice explicative sur la feuille Ponérihouen. Territ. N. Calédonie/ B.R.M.G., Carte Géol. a l'échelle du 1:50,000. Nouméa, 35 pp.

Lozovskii, V.N., Cheglokov, S.V. and Sidorenko, A.V., 1960. Osnovnye cherty struktury Baleiskovo zolotorudnovo polya. In Ye.T. Shatalov, ed., Osnovnye Voprosy i Metody Izucheniya Struktur Rudnykh Polei i Metsorozhdenii. Gosgeoltekhizdat, Moscow, 608-621.

Ludwig, G., Müller, H. and Streif, H., 1979. Neuere Daten zur holozänen Meeresspiegelanstieg im Bereich der Deutschen Bucht. Geol. Jahrb., D 32, Hannover, 3-22.

Lufkin, J.L., 1972. Tin mineralization within rhyolite flow-domes, Black Range, New Mexico (abs.). Econ. Geol., 67, p. 1008.

Lugov, S.F., ed., 1979. Geologiya Olovyannykh Rossypei SSSR, ikh Poiski i Otsenka. Nedra, Moscow, 296 pp.

Lukacs, E. and Florjančić, A.P., 1974. Uranium ore deposits in the Permian sediments of Northwest Yugoslavia. In Formation of Uranium Deposits, Intern. Atom. Energy Agency, Vienna, 313-327.

Lukas, W., 1971. Tektonisch-genetische Untersuchung der Fahlerz-Lagerstätte am Falkenstein bei Schwaz/Tirol. N. Jb. Geol. Paleont., Mh., 1971, 47-63.

Lukashev, K.I., Kovalev, V.A., Zhukhovitskaya, A.L., Khomich, A.A. and Generalova, V.A., 1971. Geokhimiya ozerno-bolotnovo litogeneza. Nauka i Tekhnika, Minsk, 283 pp.

Lukin, L.I., et al., 1968. Osobennosti struktur gidrotermal'nykh rudnykh mestorozhdenii v rozlichnykh strukturnykh etazhakh i yarusakh. Nauka, Moscow, 295 pp.

Lumbers, S.B., 1971. Geology of the North Bay area. Ontario Dept. Mines, Geol. Rept. 94, 104 pp.

Lunar, R. and Amoros, J.L., 1979. Mineralogy of the oolithic iron deposits of the Ponferrada-Astorga zone, Northwestern Spain. Econ. Geol. 74, 251-762.

Lut'ye, A.M., 1973. Tipy mestorozhdenii medi v krasnotsvetnykh formatsiyakh Russkoi Platformy. Geol. Rud. Mestor. No.6, 79-88.

Lusk, J., 1969. Base metal zoning in the Heath Steele B-1 orebody, New Brunswick, Canada. Econ. Geol., 64, 509-518.
Lyakhovich, V.V., 1973. O biotite kak indikatore rudonosnosti granitoidov. Geol. Rud. Mestor., 1973, 41-51.
Lyle, J.R., 1977. Petrography and carbonate diagenesis of the Bonneterre Formation in the Viburnum Trend area, Southeast Missouri. Econ. Geol. 72, 420-434.
Lyons, W.A., 1963. Structural geology of Pulacayo mine, Bolivia. Econ. Geol., 58, 978-987.
Lyons, J.I., Jr. and Clabaugh, S.E., 1973. Pyroclastic and extrusive iron ore at Durango, Mexico (abs.). Econ. Geol. 68, p. 1216.
Maaløe, S. and Petersen, T.S., 1981. Petrogenesis of oceanic andesites. Jour. Geophys. Res., 86, 10273-10286.
Macdonald, G.A., 1949. Hawaiian petrographic province. Geol. Soc. Amer. Bull., 60, 1541-1596.
Machado, F., Quintino, J. and Monteiro, J.H., 1972. Geology of the Azores and the Mid-Atlantic Rift. 24th Intern. Geol. Congr., Montreal, Sect. 3, 134-142.
MacKevett, E.M., Jr., 1963. Geology and ore deposits of the Bokan Mountain uranium-thorium area, southeastern Alaska. U.S. Geol. Surv. Bull. 1154, 125 pp.
Mackin, J.H., 1968. Iron ore deposits of the Iron Springs district, southwestern Utah. In J.D. Ridge, ed., Ore Deposits of the United States 1933-1967, A.I.M.E., New York, 992-1019.
Mackin, J.H. and Schmidt, D.L., 1956. Uranium and thorium-bearing minerals in placer deposits in Idaho. U.S. Geol. Surv. Prof. Paper 300, 375-380.
MacLeod, J.L., 1975. Diagenesis and sulphide mineralization of a section of lower Carboniferous carbonates at Gays River, Nova Scotia; a progress report (abs.). Geol. Soc. Amer. Progr. with Abstr. v.7, 6, p. 812.
MacLeod, W.N., Turner, D.C. and Wright, E.P., 1971. The geology of the Jos Plateau. Geol. Surv. of Nigeria, Bull. 32, 269 pp.
Macqueen, R.W., 1976. Sediments, zinc and lead, Rocky Mountains Belt, Canadian Cordillera. Geoscience Canada, 3, 71-80.
Macqueen, R.W. and Thompson, R.I., 1978. Carbonate-hosted lead-zinc occurrences in northeastern British Columbia with emphasis on the Robb Lake deposit. Canad. J. Earth Sci., 15, 1737-1762.
Macqueen, R.W., Williams, G.K., Barefoot, R.R. and Foscolos, A.E., 1975. Devonian metalliferous shales, Pine Point region, District of Mackenzie. Geol. Surv. Canada Paper 75-1A, 553-556.
Magak'yan, I.G., 1968. Ore Deposits. Int. Geol. Rev., 10, 202 pp.
Magee, M., 1968. Geology and ore deposits of the Ducktown district, Tenn. In J.D. Ridge, ed., Ore Deposits of the United States 1933-1967, A.I.M.E., New York, 207-241.
Mahel', M. and Buday, T., eds., 1968. Regional Geology of Czechoslovakia, pt.II, The West Carpathians. Academia, Prague, 723 pp.
Makarenko, G.F., 1976. The epoch of Triassic trap magmatism in Siberia. Int. Geol. Rev. 19, 1089-1100.
Makarov, V.G. and Bordon, V.E., 1970. Perspektivy osadochnykh otlo-

zhenii BSSR na metallicheskie poleznye iskopaemye. In Tverdye Poleznye Iskopaemye BSSR, Minsk, 132-138.
Malcolm, W. and Faribault, E.R., 1929. Gold Fields of Nova Scotia. Geol. Surv. Canada Memoir 156, 253 pp.
Manheim, F.T., 1961. A geochemical profile in the Baltic Sea. Geoch. Cosmoch. Acta, 25, 52-70.
Manheim, F.T., 1965. Manganese-iron accumulations in the shallow marine environment. Occas. Publ. Narangansett Marine Lab. Univ. of Rhode Isl., 3, 217-276.
Mann, A.W. and Horwitz, R.C., 1979. Groundwater calcrete deposits in Australia: some observations from Western Australia. Journ. Geol. Soc. Austral., 26, 293-303.
Manrique, C.J. and Plazolles, V.A., 1975. Geologia de Cuajone. Bol. Soc. Geol. Perú, 46, 137-150.
Marinos, G., 1982. Greece. In F.W. Dunning, W. Mykura and D. Slater, eds., Mineral Deposits of Europe, 2, Southeast Europe. Miner. Soc./Inst. Min. Metall., London, 233-253.
Marinov, N.A., Khasin, R.A. and Khurtz, Ch., eds., 1977. Geologiya Mongolskoi Narodnoi Respubliki, v. 3-Poleznye Iskopaemye. Nedra, Moscow, 697 pp.
Marmo, V., 1952. Iron Ores of Finland. 19th Int. Geol. Congr., Alger, Symposium sur les gisements de fer du Monde, v.2, 45-57.
Marmo, V., 1971. Granite Petrology and the Granite Problem. Devel. in Petrol. 2, Elsevier, 244 pp.
Marsh, B.D., 1982. The Aleutians. In R.S. Thorpe, ed., Andesites. Wiley, Chichester, 99-114.
Mart, J. and Sass, E., 1972. Geology and origin of the manganese ore of Um Bogma, Sinai. Econ. Geol. 67, 145-156.
Martin, H., Mathias, M. and Simpson, E.S.W., 1960. The Damaraland sub-volcanic ring complexes in South-West Africa. 21st Int. Geol. Congr., Norden, Pt. 13, 156-174.
Massin, J.-M., 1976. Les gisements de fer de Tunisie. In A. Zitzmann, ed., Iron Ore Deposits of Europe. Bundesanst. f. Geowiss., Hannover, 303-308.
Master, J.M., 1956. Manganese ores of Pakistan. Symposio del Manganeso, 20th Int. Geol. Congr., Mexico, v. IV, 237-243.
Masursky, H., 1962. Uranium-bearing coal in the eastern part of the Red Desert area, Wyoming. U.S. Geol. Surv. Bull. 1099-B, 152 pp.
Materikov, M.P., 1974. Deposits of tin. In V.I. Smirnov, ed., Ore Deposits of the U.S.S.R., v. 3, Engl. Transl. Pitman, London, 229-294.
Matsuda, T. and Kitamura, N., 1974. North-East Japan. In A.M. Spencer, ed., Mesozoic-Cenozoic Orogenic Belts. Geol. Soc. London, 543-552.
Matter, A. and Tucker, M., eds., 1978. Modern and Ancient Lake Sediments. Intern. Assoc. Sedim. Spec. Publ. No. 2.
Mattox, R.B., ed., 1968. Saline Deposits. Geol. Soc. Amer. Spec. Paper No. 88, 701 pp.
Maucher, A. and Höll, R., 1968. Die Bedeutung geochemisch-stratigraphischer Bezugshorizonte für die Alterstellung der Antimonit

Lagerstätte von Schlaining in Burgenland, Österreich. Miner. Deposita, 3, 272-285.
Maxwell, J.C., 1970. The Mediterranean, ophiolites and continental drift. In H. Johnson and B.L. Smith, The Megatectonics of Continents and Oceans. Ruthers Univ. Press, New Brunswick, 167-193.
Maxwell, J.C., 1974. Early western margin of the United States. In C.F. Burk and C.L. Drake, eds., The Geology of Continental Margins. Springer Verlag, New York, 831-852.
McAllister, J.F., Flores, H.W. and Ruíz, C.F., 1950. Quicksilver deposits of Chile. U.S. Geol. Surv. Bull. 964-E, 361-400.
McAllister, J.F. and Hernandéz, D.O., 1945. Quicksilver-antimony deposits of Huitzuco, Guerrero, Mexico. U.S. Geol. Surv. Bull. 946-B, 49-71.
McAllister, A.L. and Lamarche, R.Y., 1972. Mineral deposits of southern Quebec and New Brunswick. 24th Intern. Geol. Congress Montreal, Guidebook, Excursion A58-C58, 95 pp.
McAndrew, J., 1965. Gold deposits of Victoria. In J. McAndrew, ed., Geology of Australian Ore Deposits. Melbourne, 450-456.
McBirney, A.R. and White, C.M., 1982. The Cascade Province. In R.S. Thorpe, ed., Andesites. Wiley, Chichester, 115-136.
McConnell, R.B., 1967. The East African Rift system. Nature, London, 215, 578-581.
McConnel, R.H. and Anderson, R.A., 1968. The Metaline district, Washington. In J.D. Ridge, ed., Ore Deposits of the United States 1933-1967. A.I.M.E., New York, 1460-1480.
McDougall, I., 1962. Differentiation of the Tasmanian dolerites-Red Hill dolerite-granophyre association. Geol. Soc. Amer. Bull. 73, 279-316.
McDougall, I., 1976. Geochemistry and origin of basalt of the Columbia River Group, Oregon and Washington. Geol. Soc. Amer. Bull. 87, 777-792.
McGinnis, L.D., 1968. Glaciation as a possible cause of mineral deposition. Econ. Geol. 63, 390-400.
McKee, E.D., ed., 1979. Study of Global Sand Seas. U.S. Geol. Surv. Prof. Paper 1052, 429 pp.
McKee, E.D. and Ward, W.C., 1983. Eolian. In P.A. Scholle, D.G. Bebout and C.H. Moore, eds., Carbonate Depositional Environmemts. Amer. Assoc. Petr. Geol., Tulsa, 131-170.
McKellar, J.B., 1975. The Eastern Australian rutile province. In C.L. Knight, ed., Economic Geology of Australia and Papua New Guinea, 1-Metals. Aust. Inst. Min. Metal., 1055-1062.
McKelvey, V.E., Wiese, J.H. and Johnson, V.H., 1949. Preliminary report on the bedded manganese of the Lake Mead region, Nevada and Arizona. U.S. Geol. Surv. Bull. 948-D, 101 pp.
McKelvey, V.E., Williams, J.S., Sheldon, R.P., Cressman, E.R., Cheney, T.M. and Swanson, R.W., 1959. The Phosphoria, Park City and Shedhorn Formations in the Western Phosphate Field. U.S. Geol. Surv. Prof. Paper 313-A, 47 pp.
McKinney, A.A. and Horst, H.W., 1953. Deadwood conglomerate monazite, Bald Mountain area, Sheridan and Big Horn Counties, Wyoming.

U.S. Atom. Energy Comm. RME-3128, 3-39.
McKinstry, H., 1957. Review of articles by H. Aguije S., "El vanadio en el Peru". Econ. Geol. 52, 324-325.
McKnight, E.T. and Fischer, R.P., 1970. Geology and ore deposits of the Picher field, Oklahoma and Kansas. U.S. Geol. Surv. Prof. Paper 588, 165 pp.
McLaughlin, R.J., 1981. Tectonic setting of pre-Tertiary rocks and its relation to geothermal resources in the Geysers-Clear Lake area. U.S. Geol. Surv. Prof. Paper 1141, 3-23.
McLeod, I.R., ed., 1965. Australian Mineral Industry: The Mineral Deposits. Bur. Min. Res., Geol., Geoph., Canberra, Bull. 72, 690 pp.
McMillan, W.J., 1976. Geology and genesis of the Highland Valley Ore Deposits and the Guichon Creek Batholith. Canad. Inst. Min. Metall. Spec. Volume 15, 85-104.
McMillan, W.J. and Moore, J.M., Jr., 1974. Gneissic alkalic rocks and carbonatites in the Frenchman's Cap gneiss dome, Shuswap Complex, British Columbia. Canad. J. Earth Sc., 11, 304-318.
McNaughton, J.H.M., 1958. Notes on economic minerals. Nyasaland Protect. Geol. Surv. Dept., Ann. Rept. 1957, 28-31.
McTaggart, K.C., 1971. On the origin of ultramafic rocks. Geol. Soc. Amer. Bull., 82, 23-42.
Mehnert, K.R., 1968. Migmatites and the Origin of Granitic Rocks. Elsevier, Amsterdam, 393 pp.
Meijer, A., 1982. Mariana-Volcano Islands. In R.S. Thorpe, ed., Andesites. Wiley, Chichester, 293-306.
Meillon, J.J., 1978. Economic geology and tropical weathering. Canad. Inst. Min. Metall. Bull., July 1978, 61-90.
Meinert, L., 1982. Skarn, manto and breccia pipe formation in sedimentary rocks of the Cananea mining district, Sonora, Mexico. Econ. Geol., 77, 919-949.
Melankholina, Ye. N. and Kovylin, V.M., 1977. The tectonics of the Sea of Japan. Geotectonics, No.4, 1976, 273-282.
Melcher, G.C., 1966. The carbonatites of Jacupiranga, São Paulo, Brazil. In O.F. Tuttle and J. Gittins, eds., Carbonatites. Wiley, New York, 169-182.
Mellon, G.B., 1961. Sedimentary magnetite deposits of the Crowsnest Pass region, Soithwestern Alberta. Res. Counc. Alberta Bulletin 9, 98 pp.
Meng, H.M., Chern, K. and Ho, T., 1937. Geology of the Kochiu tin field, Yunnan, a preliminary sketch. Bull. Geol. Soc. China, 16, 421-437.
Mero, J.L., 1965. The Mineral Resources of the Sea. Elsevier, Amsterdam, 312 pp.
Mero, J.L., 1977. Economic aspects of nodule mining. In G.P. Glasby, ed., Marine Manganese Deposits. Elsevier, Amsterdam, 327-356.
Merriam, C.W., 1963. Geology of the Cerro Gordo mining district, Inyo County, California. U.S. Geol. Surv. Prof. Paper 408, 83 pp.
Mertie, J.B., Jr., 1976. Platinum deposits of the Goodnews Bay district, Alaska. U.S. Geol. Surv. Prof. Paper 938, 42 pp.
Metz, R.A. and Rose, A.W., 1966. Geology of the Ray copper deposit,

Ray, Arizona. In S.R. Titley and C.L. Hicks, ed., Porphyry Coppers in the S.E. United States. Univ. of Arizona Press, 177-188.

Meyer, C., Shea, E.P., Goddard, C.C., Jr. and Staff, 1968. Ore deposits at Butte, Montana. In J.D. Ridge, ed., Ore Deposits of the United States 1933-1967. A.I.M.E., New York, 1373-1416.

Meylan, M.A., Glasby, G.P., Knedler, K.E. and Johnson, J.H., 1981. Metalliferous deep-sea sediments. In K.H. Wolf, ed., Handbook of Stratiform and Strata-Bound Ore Deposits, v.9, Elsevier, 77-178.

Mezhlumyan, G.B., 1977. Sostoyanie izuchennosti perspektivy zhelezorudnykh mestorozhdenii Armyanskoi SSR. Izv. AN Arm. SSR, Nauki o Zemle, Yerevan, 30, 113-121.

Milanovskii, Ye.Ye., 1972. Continental rift zones: their arrangement and development. Tectonophysics, 15, 65-70.

Milanovskii, Ye.Ye., 1976. Riftovye Zony Kontinentov. Nedra, Moscow, 280 pp.

Miller, R.L., 1945. Geology of the Katahdin pyrrhotite deposit and vicinity, Piscataquis County, Maine. Maine Geol. Surv. Bull. 2, 21 pp.

Miller, R.N., ed., 1973. Guidebook for the Butte field meeting of Society of Economic Geologists, Aug. 1973. Butte, 350 pp.

Milliman, J.D., 1974. Marine Carbonates. Springer Verlag, New York, 375 pp.

Milliman, J.D., Barretto, H.T., Barreto, L.A., Costa, M.P.A. and Francisconi, O., 1972. Surficial sediments of the Brazilian continental margin. Anais do 26 Congr. Brasil. de Geol., 2, 29-44.

Mills, J.W., 1971. Bedded barite deposits of Stevens County, Washington. Econ. Geol. 66, 1157-1163.

Minard, J.P., 1971. Gold occurrences near Jefferson, South Carolina. U.S. Geol. Surv. Bulletin 1334, 20 pp.

Minard, J.P., et al., 1976. Alluvial ilmenite placer deposits, Central Virginia. U.S. Geol. Surv. Prof. Paper 959-H.

Mining Magazine, 1973. Lateritic nickel mining in Greece. Min. Mag. July 1973, 12-19.

Mining Magazine, 1981. Two thousand meters under the sea. Min. Mag. August 1981, 114-122.

Mitchell, A.H.G. and Bell, J.D., 1973. Island arc evolution and related mineral deposits. Journ. Geol., 81, 381-405.

Mitchell, A.H.G. and Garson, M.S., 1976. Mineralization at plate boundaries. Minerals Sci. Eng., 8, 129-169.

Mitchell, A.H.G. and Garson, M.S., 1981. Mineral Deposits and Global Tectonic Setting. Academic Press, London, 405 pp.

Mitchell, A.H.G. and Reading, H.G., 1978. Sedimentation and tectonics. In H.G. Reading, ed., Sedimentary Environments and Facies. Elsevier, New York, 438-476.

Mitchell, R.H., 1970. Kimberlite and related rocks- a critical reappraisal. Jour. Geol., 78, 686-704.

Mitchell, R.H. and Brunfelt, A.O., 1974. Rare earth element geochemistry of kimberlite. Phys. and Chem. of the Earth, 9, 671-686.

Mitchell-Thomé, R.C., 1970. Geology of the South Atlantic Islands. Bornträger, Heidelberg, 376 pp.

Mittempergher, M., 1972. The paleogeographical, lithological and structural controls of uranium occurrences in the Alps. Geologija, Ljubljana, 15, 63-74.

Miyazawa, T., 1970. Geology and ore deposits of the Chichibu Mine. IMA-IAGOD 7th Gener. Meeting, Guidebook 5, Tokyo, 8-49.

Miyazawa, T., 1971. Heavy sand deposits of fergusonite and columbite in the Kikune (Kukkun) mine, Yonbaek-Kun, Hwanghae-do, Korea. In T. Ogura, ed., Geology and Mineral Resources of the Far East. Univ. of Tokyo Press, 85-99.

Mlakar, I. and Drovenik, M., 1972. Geologie und Vererzung der Quecksilberlagerstätte Idrija. Proc. 2nd Intern. Sympos. on the Mineral Deposits of the Alps. Geol. Transact. and Repts. 15, Ljubljana, 47-62.

Moench, R.H. and Schlee, J.S., 1967. Geology and uranium deposits of the Laguna district, New Mexico. U.S. Geol. Surv. Prof. Paper 519, 117 pp.

Moffit, F.H., 1913. Geology of the Nome and Grand Central Quadrangles, Alaska. U.S. Geol. Surv. Bulletin 533, 140 pp.

Moffit, F.H., 1954. Geology of the Prince William Sound region, Alaska. U.S. Geol. Surv. Bull. 989-E, 225-310.

Moiseyev, A.N., 1968. The Wilbur Springs Quicksilver district (California), example of a study of hydrothermal processes by combining field geology and theoretical geochemistry. Econ. Geol., 63, 169-181.

Monger, J.W.H., 1977. Upper Paleozoic rocks of the western Canadian Cordillera and their bearing on Cordilleran evolution. Canad. J. Earth Sc., 14, 1832-1859.

Monger, J.W.H., Souther, J.G. and Gabrielse, H., 1972. Evolution of the Canadian Cordillera: a plate tectonic model. Amer. J. Science, 272, 577-602.

Monseur, G., 1962. Étude metallogénique du secteur central de gisement de zinc de Reocin (Province de Santander, Espagne). Ann. Soc. Geol. de Belg., Liège, 85, Mem. No.1, 70 pp.

Moolick, R.T. and Durek, J.J., 1966. The Morenci district. In S.R. Titley and C.L. Hicks, eds., Geology of the Porphyry Copper Deposits, Southwestern North America. Univ. of Arizona Press, 221-231.

Moore, A.C., 1973. Carbonatites and kimberlites in Australia: a review of the evidence. Miner. Sc. Engn., 5, 2, 81-91.

Moore, G.W. and Silver, E.A., 1968. Gold distribution on the sea floor off the Klamath Mountains, California. U.S. Geol. Survey Circular 605, 9 pp.

Moore, J. McM., 1969. Influence of structure on the base metal deposits of Southwest Sardinia, Italy. Transact. Inst. Min. Metall. London, Sect. B., 78, B135-B147.

Morávek, P., 1964. Tafilalet-perspektivní oblast rudného hornictví Maroka. Geol. Průzkum, 7, Prague, 201-202.

Morgan, J.P. and Shaver, R.H., eds., 1970. Deltaic sedimentation, modern and ancient. Soc. Econ. Paleont. Miner., Tulsa, Spec. Publ. 15, 312 pp.

Morgan, P. and Baker, B.H., eds., 1983. Processes of Continental

Rifting. Elsevier, Amsterdam, 680 pp.
Morganti, J.M., 1981. Ore deposit models-4: Sedimentary-type stratiform ore deposits. Some models and a new classification. Geocs. Canada, 8, 2, 65-75.
Morin, J.A., 1976. The MM zinc-lead-silver deposit, a volcanogenic origin. A paper presented at the 4th Geosc. Forum, Whitehorse, 1976. Unpublished, 6 pp.
Morris, H.T., 1968. The Main Tintic mining district, Utah. In J.D. Ridge, ed., Ore Deposits of the United States 1933-1967. A.I.M.E., New York, 1043-1073.
Morris, H.T. and Lovering, T.S., 1979. General geology and mines of the East Tintic mining district, Utah and Juab Counties, Utah. U.S. Geol. Surv. Profess. Paper 1024, 203 pp.
Morris, R.C., 1974. Sedimentary and tectonic history of the Ouachita Mountains. In W.R. Dickinson, ed., Tectonics and Sedimentation. Soc. Econ. Paleont. Miner., Tulsa, Spec. Publ. 22, 120-142.
Morrissey, C.J., Davis, R.R. and Steed, G.M., 1971. Mineralization in the lower Carboniferous of central Ireland. Trans. Inst. Min. Metall., London, Sect. B, 80, B174-B185.
Morrow, D.W., Krouse, H.R., Ghent, E.D., Taylor, G.C. and Dawson, K.R., 1978. A hypothesis concerning the origin of barite in Devonian carbonate rocks of northeastern British Columbia. Canad. J. Earth Sc., 15, 1391-1406.
Morton, R., Goble, E. and Goble, R.J., 1973. Sulfide deposits associated with Precambrian Belt-Purcell strata in Alberta and Brit. Columbia, Canada. In Belt Symposium Proc., Moscow, Idaho, 159-179.
Morton, R.D., Aubut, A. and Gandhi, S.S., 1978. Fluid inclusion studies and genesis of the Rexspar uranium-fluorite deposit, Birch Island, British Columbia. Geol. Surv. Canada Paper 78-1B, 137-140.
Morvai, G., 1982. Hungary. In F.W. Dunning, W. Mykura and D. Slater, eds. Mineral Deposits of Europe, 2, Southeast Europe. Miner. Soc./ Inst. Min. Metall., London, 13-53.
Mostler, H., 1966. Sedimentäre Blei-Zink Vererzung in den mittelpermischen "Schichten von Tregiovo". Miner. Deposita 2, 89-103.
Motica, J.E., 1968. Geology and uranium-vanadium deposits in the Uravan mineral belt, southwestern Colorado. In J.D. Ridge, ed., Ore Deposits of the United States 1933-1967. A.I.M.E., New York, 805-813.
Motti, E., Vasquez-Lopez, R. and Bigot, M., 1981. Les minéralisations Zn-Pb-Ba et Cu de la marge du rift de la mer etre Yanbu Al Bahr et le Golfe d'Aquaba (Arabie Saoudite). Bull. du B.R.G.M., Sect. II, 1-2, 113-134.
Mrázek, P., 1972. Geologické poměry v okolí ložiska U-rud Okrouhlá Radouň u Jindřichova Hradce. Sbor. Geol. Věd, LG, 15, 83-93.
Mrňa, F., 1963. Jáchymov, in Some Ore Deposits of the Bohemian Massif, Guide to Excursion. Sympos. Problems of Postmagm. Ore Deposits, Prague, 1963, 46 pp.
Mrňa, F. and Pavlů, D., 1967. Ložiska Ag-Bi-Co-Ni-As formace v Českém Masívu. Sbor. Geol. Věd, LG, Prague, 9, 7-104.
Muessig, S., 1967. Geology of the Republic Quadrangle and a part

of the Aeneas Quadrangle, Ferry County, Washington. U.S. Geol. Surv. Bull. 1216, 135 pp.
Mukanov, K.M., Baishev, K.S. and Makhmutov, A.T., 1966. Opyt geokhimicheskovo kartirovaniya nekotorykh rudnykh polei i mestorozhdenii Tsentral'novo Kazakhstana. Trudy Inst. Geol. Nauk Im. Satpaeva, v.15, Alma-Ata.
Mulina, C., 1973. Hidrotermalno-egzogena ležišta silikatnih ruda nikla i kobalta Glavice i Cikatova. Rud. Metal., Tehnika, 28, 1919-1925.
Muller, D.W., 1972. The geology of the Beltana willemite deposits. Econ. Geol. 67, 1146-1167.
Müller, G. and Förstner, U., 1973. Recent iron ore formation in Lake Malawi, Africa. Miner. Deposita, 8, 278-290.
Murata, K.J., 1966. An acid fumarolic gas from Kilauea Iki, Hawaii. U.S. Geol. Surv. Prof. Paper 537-C, 6 pp.
Murchison, D. and Westall, T.S., eds., 1968. Coal and Coal-bearing Strata. Oliver and Boyd, Edinburgh, 418 pp.
Murillo, J., Cordero, G. and Bustos, A., 1968. Geología y yacimientos minerales de la region de Potosí, v.2,. Serv. Geol. Boliv. Boletin 11, La Paz, 188 pp.
Murray, C.G., 1975. Tasman Geosyncline in Queensland-mineralization. In C.L. Knight, ed., Economic Geology of Australia and Papua New Guinea, v.1. Austr. Inst. Min. Metall., 738-754.
Murray, J. and Renard, A.F., 1891. Rport on deep-sea deposits based on the specimens collected during the voyage of H.M.S. Challenger in the years 1872-1876. Rept. Voyage "Challenger", Longmans, London, 525 pp.
Murthy, M.V.N. and Venkataraman, P.K., 1964. Petrogenetic significance of certain platform peralkaline granites of the world. The Upper Mantle Sympos., New Delhi, 1964, 127-149.
Musin, R.A., 1970. Formatsii rudnykh metasomatitov i metallogenicheskie osobennosti Almalyksovo raiona. FAN, Uzbek SSR, Tashkent, 211 pp.
Mutschler, F.E., Wright, E.G., Ludington, S. and Abbott, J.T., 1981. Granite molybdenite systems. Econ. Geol., 76, 874-897.
Myers, J.S., 1975. Cauldron subsidence and fluidization: mechanisms of intrusion of the Coastal Batholith of Perú into its own volcanic ejecta. Geol. Soc. Am. Bull., 86, 1209-1220.
Nagell, R.H., 1960. Ore controls in the Morococha district, Perú. Econ. Geol., 55, 962-984.
Nagle, F., Fink, L.K., Bostrom, K. and Stipp, J.J., 1973. Copper in pillow basalts from La Desirade, Lesser Antilles Island Arc. Earth Planet. Sci. Letters, 19, 193-197.
Nakamura, T., 1970. Mineral zoning and characteristic minerals in the polymetallic veins of the Ashio copper mine. In T. Tatsumi, ed., Volcanism and Ore Genesis. Univ. of Tokyo Press, 231-146.
Nakamura, T., 1971. Mineral zoning as related to intersecting structures to fractures in the subvolcanic hydrothermal polymetallic vein-type deposits in Japan. Soc. Min. Geol. Japan, Spec. Issue 3, 47-51.

Nakhla, F.M., 1958. Mineralogy of the Egyptian black sands and its applications. Egypt. Journ. Geol., 2, 1-22.

Nakovnik, K.I., 1968. Vtorichnye kvartsity SSSR. Nedra, Moscow.

Narkelyun, L.F., 1962. Geologiya i orudeneniye Dzhezkazganskovo mestorozhdeniya. Trudy IGEM, 87, 130 pp.

Nash, J.T., Granger, H.C. and Adams, S.S., 1981. Geology and concepts of genesis of important types of uranium deposits. Econ. Geol. 75th Anniv. Volume, 63-116.

Naylor, D. and Mounteney, S.N., 1975. Geology of the North-West European Continental Shelf, 2 vols. Graham, Trotman and Dudley, London, 162 pp.

Nazarov, Yu.I., 1964. Printsipy i metodika sostavleniya krupno- i srednemashtabnykh prognozno-metallogenicheskikh kart. Zakon. Razm. Polez. Iskop. VII, 246-270.

Neale, J., 1975. Mount Goldsworthy iron ore deposit, W.A. In C.L. Knight, ed., Economic Geology of Australia and Papua New Guinea, v.1. Aust. Inst. Min. Metall., 932-935.

Nekrasoff, B., 1935. Copper-ore regions of the Union of Soviet Socialist Republics. 16th In. Geol. Congr., Washington D.C., Copper Resources of the World, v.2, 649-670.

Nelson, C.H. and Hopkins, D.M., 1972. Sedimentary processes and distribution of particulate gold in the northern Bering Sea. U.S. Geol. Surv. Prof. Paper 689, 27 pp.

Němec, D., 1965. Geologické a paragenetické poměry ložiska formace Pb-Zn-Ag u Bartoušova na Havlíčkobrodsku. Sbor. Geol. Věd, LG, Prague, 47-86.

Netterberg, F., 1969. The interpretation of some basic calcrete types. S. Afr. Arch. Bull. 24, 117-122.

Neumann, E.-R. and Ramberg, I.B., eds., 1978. Petrology and Geochemistry of Continental Rifts. Reidel, Dordrecht, 296 pp.

Neumann, H., 1944. Silver deposits of Kongsberg. Norg. Geol. Unders. No. 162, 142 pp.

Neumann-Redlin, C., Walther, H.W. and Zitzmann, A., 1976. The iron ore deposits of the Federal Republic of Germany. In A. Zitzmann, ed., Iron Ore Deposits of Europe. Bundesanst. f. Geowiss., Hannover, 165-186.

Newberry, R.J., 1982. Tungsten-bearing skarns of the Sierra Nevada, I, The Pine Creek Mine, California. Econ. Geol. 77, 823-844.

Newell, N.D. and Rigby, J.K., 1957. Geological studies on the Great Bahama Bank. In Soc. Econ. Paleont. Miner., Tulsa, Spec. Publ. 5, 15-72.

Newhouse, W.H., ed., 1942. Ore Deposits as Related to Structural Fetaures. Princeton Univ. Press, 280 pp.

Newnham, L.A., 1976. Renison Bell tinfield. In C.L. Knight, Economic Geology of Australia and Papua New Guinea, v.1. Austr. Inst. Min. Metall., 604-619.

Ney, C.S., 1954. Monarch and Kicking Horse mines, Field, British Columbia. Guidebook, 4th Ann. Field Conf., Alberta Soc. Petr. Geol., Edmonton, 119-135.

Ney, C.S. and Hollister, V.F., 1976. Geologic setting of porphyry

deposits of the Canadian Cordillera. Canad. Inst. Min. Metall., Spec. Vol. 15, Montreal, 21-29.
Nicolini, P., 1970. Gîtologie des Concentrations Minerales Stratiformes. Gauthier-Villars, Paris, 792 pp.
Nielsen, B.L., 1973. A survey of the economic geology of Greenland (exclusive fossil fuels). Grønl. Geol. Unders. Rapp. Nr. 56, 45 pp.
Nielsen, R.L., 1976. Recent developments in the study of porphyry copper geology-a review. Canad. Inst. Min. Metall. Spec. Vol. 15, Montreal, 487-500.
Nikiforov, N.A., 1968. Geotectonic types of mercury-antimony ore fields in southern Ferghana. Int. Geol. Rev., 10, 1371-1382.
Nikiforov, N.A., 1970. Osobennosti geologicheskovo stroyeniya i razmeshcheniya orodeneniya rtutno-sur'myannykh mestorozhdenii yuzhno-ferganskovo poyasa. In Ocherki po Geologii i Geokhimii Rud. Mestorozhd., Nauka, Moscow, 191-214.
Nikiforov, N.A., Pavlyukovich, Ye.A. and Ponomarev, F.I., 1962. Zakonomernosti razmeshcheniya bogatykh rtutnykh i surmyannykh rud na mestorozhdeniyakh yuzhnoi Fergany. Zakon. Razm. Polez. Iskop. V, 207-228.
Nikko Explor. Devel. Company, 1970. The geology and ore deposits of the Hanawa Mine. Unpublished, 55 pp.
Nilsen, O., 1978. Caledonian sulphide deposits and minor iron formations from the southern Trondheim region, Norway. Norges Geol. Unders. 340, 35-85.
Nippon Mining Co., 1966. Outline of Hitachi Mine. Unpublished, 13pp.
Nishiwaki, C., Iwafune, T., Shiobara, K., Sakuma, T. and Tono, A., 1970. Geology and ore deposits of the Kamioka and Hamayokokawa Mines. IMA-IAGOD Meeting, Japan 1970, Guidebook 7. 40 pp.
Noakes, L.C. and Jones, H.A., 1975. Mineral resources off-shore. In C.L. Knight, ed., Economic Geology of Australia and Papua New Guinea, v.1. Aust. Inst. Min. Metall., Monogr. 5, 1093-1106.
Noble, J.A., 1950. Manganese on Punta Concepción, Baja California, Mexico. Econ. Geol. 45, 771-785.
Noble, J.A., 1970. Metal provinces of the western United States. Geol. Soc. Amer. Bull. 81, 1607-1624.
Nokleberg, W.J. and Winkler, G.R., 1982. Stratiform zinc-lead deposits in the Drenchwater Creek area, Howard Pass Quadrangle, north-western Brooks Range, Alaska. U.S. Geol. Surv. Prof. Paper 1209, 23 pp.
Nolan, T.B., 1962. The Eureka mining district, Nevada. U.S. Geol. Surv. Prof. Paper 406, 78 pp.
Noranda Mines, Ltd., 1974. The Point Leamington sulphide deposit. In Strong, D.F., ed., Plate Tectonic Setting of Newfoundland Mineral Occurrences. A guidebook, NATO Adv. Study Inst., 60-61.
Northcote, K.E. and Muller, J.E., 1972. Volcanism, plutonism and mineralization, Vancouver Island. Canad. Inst. Min. Metall. Bull., October 1972, 49-58.
Northern Territory Geol. Survey, 1975. Amadeus, Ngalia, Georgina, Wiso, Daly River, Ord and Bonaparte Gulf basins-sundry mineralization. In C.L. Knight, ed., Economic Geology of Australia and

Papua New Guinea, v. 1, Aust. Inst. Min. Metall., 531-534.
Norton, D.L. and Cathles, L.M., 1973. Breccia pipes-products of exsolved vapor from magmas. Econ. Geol. 68, 540-546.
Notholt, A.J.G., 1979. The economic geology and development of igneous phosphate deposits in Europe and the U.S.S.R. Econ. Geol. 74, 339-350.
Novikova, A., 1964. The Russian Plate. In Tectonics of Europe, Nauka, Moscow, 54-69.
Novokhatskii, I.P., 1972. Zhelezisto-kremnistye formatsii Paleozoia Kazakhstana. In Geol., Genezis Dokembr. Zhele.-Kremnist. i Marg. Formatsii Mira, Nauk. Dumka, Kiev, 164-175.
Nunes, AdeB., Lima, R.FdaF., Filho, C.N.B., 1973. Levantamento de Recursos Naturaís, v.2, Folha SB 23 Teresina e parte da Folha SB.24 Jaguaribe. Dep. Nac. Prod. Min., Proj. RADAM, 33 pp.
Nyambok, I.O. and Gaciri, S.J., 1975. Geology of the fluorite deposit in Kerio Valley, Kenya. Econ. Geol. 70, 299-307.
Oberc, J. and Serkies, J., 1968. Evolution of the Fore-Sudetian copper deposits. Econ. Geol., 63, 372-379.
Obr, F., 1980. Akumulace kovů v severní části Sokolovské Pánve. Sbor. Geol. Věd, Prague, LG, 21, 83-100.
O.E.C.D., 1979. Uranium resources, production, demand. December 1979, Paris, 69 pp.
Oen, I.S., Fernandez, J.S. and Manteca, J.I., 1975. The lead-zinc and associated ores of La Unión, Sierra de Cartagena, Spain. Econ. Geol., 70, 1259-1279.
Oftedahl, C., 1958a. Oversikt over Grongfeltets skjerp og malmforekomster. Norge Geol. Unders., 202, 76 pp.
Oftedahl, C., 1958b, A theory of exhalative sedimentary ores. Geol. Foren., Stockholm, 80, 1-19.
Ohmoto, H. and Rye, R.O., 1970. The Bluebell Mine, British Columbia, I. Mineralogy, paragenesis, fluid inclusions, and the isotopes of hydrogen, oxygen and carbon. Econ. Geol., 65, 417-437.
Ohmoto, H. and Skinner, B.J., eds., 1983. The Kuroko and related volcanogenic massive sulphide deposits. Econ. Geol. Monograph 5, 604 pp.
Olade, M.A., 1980. Geochemical characteristics of tin-bearing and tin-barren granites, Northern Nigeria. Econ. Geol. 75, 71-82.
Oleinikov, B.V., 1979. Geokhimiya i rudogenez platformennykh baritov. Nauka, Sib. Otdel., Novosibirsk, 264 pp.
Ollier, C., 1975. Weathering. 2nd ed., Longman, London, 304 pp.
Olson, J.C., Shawe, D.R., Pray, L.C. and Sharp, W.N., 1954. Rare-earth mineral deposits of the Mountain Pass district, San Bernardino County, California. U.S. Geol. Surv. Prof. Paper, 261, 75 pp.
Olson, P.E. and Fyles, J.T., 1968. Mineral King Mine. In Brit. Columbia Minister of Mines Petr. Res. Rept. for 1967, 267-269.
Omel'yanenko, B.I., 1974. Zonal'nost' torii-niobievoi mineralizatsii v al'bititakh. In Zonal'nost' Gidrotermal'nykh Rudnykh Mestorozhdenii, v.1, Nauka, Moscow, 278-284.
Onuki, Y., 1967. Upper Paleozoic formations of coal fields in North China. In T. Ogura, ed., Geology and Mineral Resources of the Far

East. Univ. of Tokyo Press, 395-437.
O'Rourke, J.E., 1961. Paleozoic banded iron formations. Econ. Geol., 56, 331-361.
O'Rourke, P.J., 1975. Maureen uranium fluorine molybdenum prospect, Georgetown. In C.L. Knight, ed., Economic Geology of Australia and Papua New Guinea, v.1. Aust. Inst. Min. Metall., 764-768.
Osatenko, M.J. and Jones, M.B., 1976. Valley Copper. Canad. Inst. Min. Metall. Spec. Volume 15, Montreal, 130-143.
Oshima, T., 1964. Geology and ore deposits of the Yanahara Mine, western Japan. Japan. Journ. Geol. and Geoph., 35, 81-100.
Osika, R., 1969. Les richesses minerales de la Pologne. Annales des Mines, Nov. 1969, Paris, 35-54.
Osterwald, F.W. and Dean, B.G., 1961. Relation of uranium deposits to tectonic pattern of the central Cordillera Foreland. U.S. Geol. Surv. Bull. 1087-I, 337-390.
Ostwald, J., 1981. Evidence for a biogeochemical origin of the Groote Eylandt manganese ores. Econ. Geol., 76, 556-567.
Ottemann, J. and Augustithis, S.S., 1967. Geochemistry and origin of "platinum nuggets" in lateritic covers from ultrabasic rocks and birbirites of W. Ethiopia. Miner. Deposita, 1, 269-277.
Overstreet, W.C., 1967. The geologic occurrence of monazite. U.S. Geol. Surv. Prof. Paper 530, 327 pp.
Overstreet, W.C., White, A.M., Whitlow, J.W., Theobald, P.K., Jr., Caldwell, D.W. and Cuppels, N.P., 1968. Fluvial monazite deposits in the southeastern United States. U.S. Geol. Surv. Prof. Paper 568, 85 pp.
Ovtracht, A. and Tamain, G., 1972. La ceinture mineralisée varisque dans le sud de la Meseta Iberique. 24th Int. Geol. Congr., Montreal, Sect. 4, 101-109.
Owen, H.B., 1954. Bauxite in Australia. Bur. Min. Res., Geol. Geoph., Canberra, Bull. 24, 234 pp.
Oyarzún, M.J. and Frutos, J., 1980. Metallogenesis and porphyry deposits of the Andes (Southeastern Pacific region). Mém. du B.R.G.M., No.106, 50-62.
Ozerova, N.A., 1981. New mercury ore belt in Western Europe. Geol. Rud. Mestorozhd., 1981, No. 6, 49-56.
Page, B.M., 1958. Chamositic iron ore deposits near Tajmište, western Macedonia, Yugoslavia. Econ. Geol., 53, 1-21.
Page, N.J., Cassard, D. and Haffty, J., 1982. Palladium, platinum, rhodium, ruthenium and irridium in chromitites from the Massif du Sud and Tiébaghi Massif, New Caledonia. Econ. Geol., 77, 1571-1577.
Palabora Min. Comp. Ltd. Mine Geol. and Miner. Staff, 1976. The geology and economic deposits of copper, iron and vermiculite in the Palabora Igneous Complex: a brief review. Econ. Geol., 71, 177-192.
Palmason, G. and Saemundsson, K., 1974. Iceland in relation to the Mid-Atlantic Ridge. Ann. Rev. Earth and Planet. Sci., 2, 25-50.
Palmer, K.G., 1976. The Cleveland tin deposit. In M. Solomon and G.R. Green, eds., Ore Deposits of Western Tasmania. 25th Int.

Geol. Congr., Australia, Excursion Guide No. 31AC, 26-31.
Panayiotou, A., ed., 1980. Ophiolites. Proc. Intern. Ophiolite Symp., Cyprus, 1979. Cyprus Geol. Surv. Dept., Nicosia, 781 pp.
Panayiotou, A., 1980. Cu-Ni-Co-Fe sulphide mineralization, Limassol Forest, Cyprus. In A. Panayiotou, ed., Ophiolites. Cyprus Geol. Surv. Dept., Nicosia, 102-116.
Panek, S. and Szuwarzynski, M., 1975. Kopalne jamy krasowe z kruzccami w okolicach Chrzanova. Ann. Soc. Geol. Pologne, 45, 177-189.
Panteleyev, A. and Pearson, D.E., 1977. Kutcho Creek Map area. In Geology, Mining, Explor. in Brit. Columbia, 1975. B.C. Minist. of Min. Petr. Res., Victoria, G87-G93.
Paone, J., 1970. Germanium. Mineral Facts and problems, U.S. Bureau of Mines Bull. 650, 563-571.
Pardee, J.T. and Park, C.F., Jr., 1948. Gold deposits of the Southern Piedmont. U.S. Geol. Surv. Prof. Pap. 213, 156 pp.
Paris, J.P., 1981. Géologie de la Nouvelle Calédonie. Mém. du B.R.G.M., No. 113, 278 pp.
Park, C.F., 1946. The spilite and manganese problems of the Olympic Peninsula, Washington. Amer. J. Sci., 244, 305-323.
Park, C.F., Jr. and McDiarmid, R.A., 1975. Ore Deposits. 3rd ed., Freeman, San Francisco, 529 pp.
Parker, R.L., 1923. Alpine Minerallagerstätten. Schweiz. Miner. Petr. Mitteil., 3, 1923, 298 pp.
Parker, R.L. and Sharp, W.N., 1970. Mafic-ultramafic igneous rocks and associated carbonatites of the Gem Park Complex, Custer and Fremont Counties, Colorado. U.S. Geol. Surv. Prof. Pap. 649, 24 pp.
Parrish, I.S. and Tully, J.V., 1971. Molybdenum, tungsten and bismuth mineralization at Brunswick Tin Mines Ltd. Paper presented at 73rd Ann. Gener. Meet. of C.I.M., Quebec.
Parsons, G.E., 1961. Niobium-bearing complexes east of Lake Superior. Ont. Dept. Min., Geol. Rept. 3, 51-69.
Patterson, D.J., Ohmoto, H. and Solomon, M., 1981. Geologic setting and genesis of cassiterite-sulfide mineralization at Renison Bell, Western Tasmania. Econ. Geol. 76, 393-438.
Paterson, I.A., 1977. The geology and evolution of the Pinchi Fault Zone at Pinchi Lake, central British Columbia. Canad. J. Earth Sc., 14, 1324-1342.
Patterson, S.H., 1967. Bauxite reserves and potential aluminium resources of the world. U.S. Geol. Surv. Bull. 1228, 176 pp.
Patterson, S.H., 1971. Investigations of ferruginous bauxite and other mineral resources on Kauai and a reconnaisance of ferruginous bauxite deposits on Maui, Hawaii. U.S. Geol. Surv. Prof. Paper 656, 74 pp.
Pavillon, M.J., 1964. Paleogeographie devoniene et minéralisations ferrugineuses de Dielette (Manche) et plombozincifère de Surtainville (Manche). Soc. Geol. Franc. Bull.,6, 1, 121-126.
Pavillon, M.J., 1969. Les minéralisations plombo-zincifères de Cartagène (Cordillères bétiques, Espagne). Miner. Deposita, 4, 368-385.
Pavlenko, A.S., 1974. The Mongol-Tuva Province of alkaline rocks.

In H. Sørensen, ed., The Alkaline Rocks. Wiley, London, 271-293.
Pavlides, L., 1962. Geology and manganese deposits of the Maple and Hovey Mountains area, Aroostook County, Maine. U.S. Geol. Surv. Profess. Paper 362, 116 pp.
Pavlides, L., Gair, J.E. and Cranford, S.L., 1982. Central Virginia volcanic-plutonic belt as a host for massive sulfide deposits. Econ. Geol. 77, 233-272.
Pavlov, N.V., 1960. Strukturno-geologicheskie osobennosti magnetitovykh mestorozhdenii Tungusskoi Sineklizy. In Osnovnye Voprosy I Metody Izucheniya Struktur Rudnykh Polei i Mestorozhdenii. Gosgeoltekhizdat', Moscow, 456-468.
Pavlov, N.V. and Grigor'eva, I.I., 1974. Deposits of chromium. In V.I. Smirnov, ed., Ore Deposits of the U.S.S.R., Engl. Transl., Pitman, London, v.1, 179-236.
Pavlova, I.G., 1978. Mednoporfirovannye Mestorozhdeniya. Nedra, Moscow, 315 pp.
Pawlowski, S., Pawlowska, K. and Kubica, B., 1979. Geology and genesis of the Polish sulfur deposits. Econ. Geol. 74, 475-483.
Payne, J.G., Bratt, J.A. and Stone, B.G., 1980. Deformed Mesozoic volcanogenic Cu-Zn sulfide deposits in the Britannia District, British Columbia. Econ. Geol. 75, 700-721.
Pearre, N.C. and Heyl, A.V., Jr., 1960. Chromite and other mineral deposits in serpentine rocks of the Piedmont Upland, Maryland, Pennsylvania and Delaware. U.S. Geol. Surv. Bull. 1082-K, 707-833.
Pearson, F.E., 1974. Harrison, Lucky Jim. In Geol., Explor., Mining in Brit. Columbia in 1973. B.C. Minist. of Min., 125-129.
Pearson, W.N., 1978. Copper metallogeny, Lake Huron area, Ontario. Geol. Surv. Canada, Paper 78-1A, 263-268.
Pearson, W.N. and Clark, A.H., 1979. The Minto copper deposit, Yukon Territory: a metamorphosed orebody in the Yukon crystalline Terrane. Econ. Geol. 74, 1577-1599.
Peck, D.L. and Wones, D.R., 1980. Granite I: origin and evolution of granitic magmas. A Penrose Conf. Report. Geology, 8, 452-453.
Pek, A.V., Snezhko, Ye.A. and Kurbyukov, A.A., 1970. Nekotorye voprosy geologicheskovo stroyeniya Tyrnyauzskovo rudnovo polya. In Ocherki po Geologii i Geokhimii Rudnykh Mestorozhdenii. Nauka, Moscow, 242-257.
Pélissonnier, H., 1971. Le gisement de cuivre stratiforme de Cerro Negro (Aconcagua, Chili). Bull. du B.R.G.M., Sect.II, No.6, 1971, 43-50.
Pena, G.S. and Figueiredo, A. JdeA., 1972. Projeto Alcalinas, relatório final. Dep. Nac. Prod. Min., Brasil, Vol. 1, 143 pp.
Perfilieff, A. and Kheraskoff, N., 1964. The Urals. In A.A. Bogdanov et al., ed., Tectonics of Europe. Nauka, Moscow, 109-123.
Perhac, R.M. and Heinrich, E.W., 1964. Fluorite-bastnaesite deposits of the Gallinas Mountains, New Mexico, and bastnaesite paragenesis. Econ. Geol., 59, 276-239,
Perkins, J.M. and Lonsdale, J.T., 1955. Mineral resources of the Texas Coastal Plain. Texas Univ. Bur. Econ. Geol. Rept. for Bur. of Reclamation, 49 pp.

Perseil, E.A. and Grandin, G., 1978. Evolution minéralogique du manganèse dans trois gisements d'Afrique de l'Ouest: Mokta, Tambao, Nsuta. Miner. Deposita, 13, 295-311.
Peters, T. and Kramers, J.D., 1974. Chromite deposits in the ophiolite complex of Northern Oman. Miner. Depos. 9, 253-259.
Petersen, E.U. and Zantop, H., 1980. The Oxec deposit, Guatemala: an ophiolite copper occurrence. Econ. Geol. 75, 1053-1065.
Petersen, U., 1965. Regional geology and major ore deposits of Central Peru. Econ. Geol. 60, 407-476.
Peterson, N.P., 1962. Geology and ore deposits of the Globe-Miami district, Arizona. U.S. Geol. Surv. Prof. Pap. 342, 151 pp.
Petránek, J., 1974. Sedimentární železné rudy v ordoviku Krušné Hory. Sbor. Geol. Věd, Prague, LG, 16, 165-185.
Petrov, N.P., 1961. Molybdenum in a brown coal deposit of Uzbekistan. Int. Geol. Rev., 5, 335-339.
Petruk, W., 1977. Mineralogical characteristics of an oolithic iron deposit in the Peace River district, Alberta. Canad. Miner. 15, 1, 3-13.
Petrulian, N., 1973. Zăcăminte de Minerale Utile. Edit. Tehnică, Bucharest, 503 pp.
Pettijohn, F.J., 1975. Sedimentary Rocks. 3rd ed., Harper and Row, New York, 628 pp.
Peyve, A.V., et al., 1977. Development of continental crust in northern Eurasia. Geotectonics, 10, 5, 309-318.
Phillips, C.H., 1976. Geology and exotic copper mineralization in the vicinity of Copper Butte, Pinal County, Arizona. New Mexico Geol. Soc. Spec. Publ. No. 6, 174-179.
Phillips, C.H., Cornwall, H.R. and Rubin, M., 1971. A Holocene ore body of copper oxides and carbonates at Ray, Arizona. Econ. Geol. 66, 495-498.
Phillips, C.H., Cambell, N.A. and Fountain, D.S., 1974. Hydrothermal alteration, mineralization and zoning in the Ray deposit. Econ. Geol., 69, 1237-1250.
Phillips, K.A., 1959. Some interpretations arising from a remapping of the Katanga System southeast of Mumbwa, Northern Rhodesia. Assoc. Serv. Geol. Afric., 20th Int. Geol. Congr., Mexico, 213-223.
Picard, M.D. and High, L.R., Jr., 1972. Criteria for recognizing lacustrine rocks. Soc. Econ. Paleont. Miner. Spec. Publ. 16, 108-145.
Pidzhyan, G.O., 1975. Medno-molybdenovaya formatsiya rud Armyanskoi SSR, Yerevan. AN Arm. SSR, Yerevan, 312 pp.
Pierce, A.P., Gott, G.B. and Mytton, J.W., 1964. Uranium and helium in the Panhandle Gas Field, Texas, and adjacent areas. U.S. Geol. Surv. Prof. Paper 454-G, 57 pp.
Pinckney, D.M., 1976. Mineral resources of the Illinois-Kentucky mining district. U.S. Geol. Surv. Prof. Paper 970, 15 pp.
Pipino, G., 1980. Gold in Ligurian ophiolites. In A. Panayiotou, ed., Ophiolites. Cyprus Geol. Surv. Dept., Nicosia, 765-773.
Pipiringos, G.N., Chisholm, W.A. and Kepferle, R.C., 1965. Geology and uranium deposits in the Cave Hills area, Harding County, South Dakota. U.S. Geol. Surv. Prof. Paper 476-A, 64 pp.

Pirkle, E.C., Pirkle, W.A. and Yoho, W.H., 1974. The Green Cove Springs and Boulogne heavy-mineral sand deposits of Florida. Econ. Geol. 69, 1129-1137.

Plahuta, J.T., Lange, I.M. and Jansons, U., 1978. The nature of mineralization at the Red Dog prospect, western Brooks Range, Alaska (abs.). Geol. Soc. Am., Abstr. with Progr., 10, p. 142.

Platt, J.W., 1977. Volcanogenic mineralization at Avoca, Co. Wicklow, Ireland and its regional implication. In Volc. Processes in Ore Genesis, Inst. Min. Metall./Geol. Soc. London, 163-170.

Plimer, I., 1975. Wolfram Camp wolframite-molybdenite-bismuth-quartz pipes, North Queensland. In C.L. Knight, ed., Economic Geology of Australia and Papua New Guinea, v. 1. Austr. Inst. Min. Metall. Monogr. 5, 760-762.

Plimer, I.R., 1980. Exhalative Sn and W deposits associated with mafic volcanism as precursors to Sn and W deposits associated with granites. Miner. Deposita, 15, 275-289.

Plumridge, C.L., 1975. Mount Carbine tungsten reef swarm. In C.L. Knight, ed., Economic Geology of Australia and Papua New Guinea, v.1. Austr. Inst. Min. Metall., 610-613.

Pogrebitskii, Ye. O., et al., 1968. Poiski i Razvedka Mestorozhdenii Poleznykh Iskopaemykh. Moscow, 460 pp.

Pokalov, V.T., 1974. Deposits of molybdenum. In V.I. Smirnov, ed., Ore Deposits of the U.S.S.R., Engl. Transl., Pitman, London, vol. 3, 125-179.

Poldervaart, A. and Sukheswala, R.N., 1958. Deccan basalts of the Bombay area, India. Geol. Soc. Amer. Bull., 69, 1475-1494.

Popov, A., 1976. Les gisements de fer en l'Algerie. In A. Zitzmann, ed., Iron Ore Deposits of Europe. Bundesanst. f. Geowiss., Hannover, 83-89.

Popov, V.M., 1962. O blagopriyatnykh i ekraniyurushchikh gorizontakh v plastovykh mestorozhdeniyakh tsvetnykh metallov. Zakon. Razm. Polez. Iskop., 5, 353-384.

Popov, V.M., Asanaliev, U., Davydov, G.I., Dzhumaliev, T. and Khusainov, U., 1967. Problema osadochnovo rudoobrazovaniya na primere plastovykh mestorozhdenii svintsa, tsinka i drugikh elementov v osadochnykh formatsiyakh Kirgizii. In Problemy Geol. Sred. Azii i Kazakhstana, Nauka, Moscow, 97-115.

Popov, V.Ye., 1979. Vulkanogenno-osadochnye Mestorozhdeniya. Nedra, Leningrad, 295 pp.

Popović, A., 1954. Noble metals in the ash of some coals of the Timok Basin. J. Chem. Soc. Belgrade, 19, 305-307.

Porrenga, D.H., 1965. Chamosite in recent sediments of the Niger and Orinoco deltas. Geol. en Mijnbow, 44, 400-403.

Potter, R.R., Bingley, J.M. and Smith, J.C., 1972. Appalachian stratigraphy and structure of the Maritime provinces. 24th Int. Geol. Congr., Montreal, Guide to Excursion A57-C57, 48 pp.

Pouit, G., Bouquet, C. and Bois, J.-P., 1979. Les principaux niveaux mineralisés (Zn,Pb,Cu,Ba) du Paléozoïque des Pyrenees centrales: éléments de synthèse. Bull. du B.R.G.M., Sect. II, 1, 23-34.

Preston, J., 1981. Tertiary igneous activity. In C.H. Holland, ed.,

A Geology of Ireland. Scot. Acad. Press, Edinburgh, 213-223.
Preto, V.A., 1972. Lode copper deposits of the Racing River-Gataga River area. Geol., Explor., Mining in Brit. Columbia in 1971, Victoria, 75-107.
Proctor, P.D., 1953. Geology of the Silver Reef (Harrisburg, Washington County, Utah) mining district. Utah Geol. and Min. Surv. Bull. 44, 169 pp.
Prokin, V.A., Rudakov, V.M. and Solodkii, N.N., 1961. Gipogennaya zonal'nost' okolorudnykh izmenennykh porod kolchedannykh mestorozhdenii Sibai, Kul'-Yurt-Tau i Bakr-Tau. Vopr. Geokhr. i Geokh. Dokembr. i Paleoz. Yuzh. Urala, etc., Ufa, Bashk. AN SSSR, 33-45.
Pronin, A.A., 1962. Glavneishie metallogenicheskie epokhi i rudnye formatsii Urala. Zakon. Razm. Polez. Iskop. 5, 130-158.
Puchelt, H., Schock, H.H., Schroll, E. and Hanert, H., 1973. Rezente marine Eisenerze auf Santorin, Griechenland. Geol. Rundschau, 62, 786-812.
Puffer, J.H. and Peters, J.J., 1974. Magnetite veins in diabase of Laurel Hill, New Jersey. Econ. Geol. 69, 1294-1299.
Pumo, E., Melo, V. and Ostrosi, B., 1982. Albania. In F.W. Dunning, W. Mykura and D. Slater, eds., Mineral Deposits of Europe, 2, Southeast Europe. Miner. Soc./Inst. Min. Met., London, 203-214.
Purser, B.H., ed., 1973. The Persian Gulf. Springer Verlag, Berlin, 471 pp.
Purvis, J.G., 1975. Report on special prospecting licence C 115, Gold Ridge, Guadalcanal, Brit. Solomon Islands Protectorate. C.R.A. Explor. Pty. Ltd., Unpubl. Report, 25 pp.
Pustovalov, L.V., ed., 1965. Metally v Osadochnykh Tolshchakh. Nauka, Moscow, 390 pp.
Putzer, H., 1962. Geologie von Paraguay. Borntraeger, Berlin, 179 pp.
Putzer, H., 1976. Metallogenetische Provinzen in Südamerika. Schweizerbart, Stuttgart, 299 pp.
Quade, H., 1976. Genetic problems and environmental features of volcano-sedimentary iron ore deposits of the Lahn-Dill type. In K.H. Wolf, ed., Handbook of Stratiform and Strata-Bound Ore Deposits, v.7, Elsevier, 255-294.
Quinlan, J.F., 1972. Karst-related mineral deposits and possible criteria for the recognition of paleokarsts: a review of preservable characteristics of Holocene and older karst terranes. 24th Int. Geol. Congr., Montreal, Sect. 6, 156-168.
Quinn, A.W. and Glass, H.D., 1958. Rank of coal and metamorphic grade of rocks of the Naragansett Basin of Rhode Island. Econ. Geol. 53, 563-576.
Quiring, H., 1936. Ein Profil durch die Grube Goldberg bei Silberg. Zeitschr. f. prakt. Geol., 44, 59-65.
Rackley, R.I., 1976. Origin of Western-States type uranium mineralization. In K.H. Wolf, ed., Handbook of Stratiform and Strata-Bound Ore Deposits, v.7, Elsevier, Amsterdam, 89-156.
Radabaugh, R.E., Merchant, J.S. and Brown, J.M., 1968. Geology and ore deposits of the Gilman (Red Cliff, Battle Mountain) District, Eagle County, Colorado. In J.D. Ridge, ed., Ore Deposits of the

United States 1933-1967. A.I.M.E., New York, 641-664.
Rakchayev, A.D., 1962. Trends in the location of pyrite bodies in the Karabash group of the Urals. Intern. Geol. Rev., 6, 303-316.
Ramdohr, P., 1927. Die Eisenerzlager des Oberharzer (Osteroder) Diabaszuges un ihr Verhalten im Bereich des Brocken kontaktes. N. Jb. Miner., Beil. Bd. 55A, 33-392.
Ramos, J. R.de A., and Fraenkel, M.O., 1974. Uranium occurrences in Brazil. In Formation of Uranium Ore Deposits. Intern. Atom. En. Agcy, Vienna, 637-645.
Ramović, M. and Kulenović, E., 1964. Novi rezultati na rudarskim istragama i geološkoj studiji rudišta žive Draževici kod Čevljanovića i polimetalnog rudišta Borovica kod Varešа. Geološki Glasnik, Sarajevo, 10, 181-196.
Randall, J.A., 1972. Metallization sequence in the Tayoltita region, San Dimas, Durango, Mexico. 24th Int. Geol. Congr., Montreal, Sect. 4, 309-317.
Ransom, D.M. and Knight, J.A., 1975. Golden Plateau gold lodes. In C.L. Knight, ed., Economic Geology of Australia and Papua New Guinea, v.1. Austr. Inst. Min. Metall., 773-778.
Ransome, F.L., 1909. Geology and ore deposits at Goldfield, Nevada. U.S. Geol. Surv. Profess. Paper 66, 258 pp.
Raufuss, W., 1973. Struktur, Schwermineralführung, Genese und Bergbau der sedimentären Rutil-Lagerstätten in Sierra Leone (West-Afrika). Geol. Jb., Hannover, D5, 52 pp.
Raup, O.B., Guda, A.J., 3rd and Groves, H.L., 1967. Rare-earth mineral occurrence in marine evaporites, Paradox Basin, Utah. U.S. Geol. Surv. Prof. Paper 575-C, 38-41.
Rayment, B.D., Davis, G.R. and Willson, J.D., 1971. Controls to mineralization at Wheal Jane, Cornwall. Trans. Inst. Min. Metall., London, Sect. B, v. 80, B224-B237.
Razin, L.V., 1974. Deposits of platinum metals. In V.I. Smirnov, ed., Ore Deposits of the U.S.S.R., Engl. Transl., Pitman, London, vol. 3, 100-124.
Razmakhnin, Yu. N. and Razmakhnina, E.M., 1973. Sistematika, zonalnost' i metallogenicheskoe znacheniye metasomatitov olovonosnykh polei Sikhote-Alin'ya. Geol. Rud. Mestor., No. 1, 1973, 52-63.
Razmakhnin, Yu. N., Razmakhnina, E.M., Vasilenko, V.P., Lavrik, N.I., and Lakhnyuk, V.S., 1974. Exploration-estimation criteria for tin on the basis of regional and local metasomatic zonation (Sikhote-Alin). Int. Geol. Rev., 17, 1290-1298.
Rea, W.J., 1982. The Lesser Antilles. In R.S. Thorpe, ed., Andesites. Wiley, Chichester, 167-185.
Reading, H.G., 1972. Global tectonics and the genesis of flysch successions. 24th Int. Geol. Congr., Montreal, Sec. 6, 59-66.
Reading, H.G., ed., 1978. Sedimentary Environments and Facies. Elsevier, New York, 557 pp.
Rebek, R.J., 1975. Eddie Creek and Wau gold lodes. In C.L. Knight, ed., Economic Geology of Australia and Papua New Guinea, v. 1. Austr. Inst. Min. Metall., Monogr. 5, 867-871.
Reed, D.F., 1949. Investigation of Talladega gray iron ores, Talla-

dega County, Alabama. U.S. Bur. Mines Rept. Inv. RI 4426, 29 pp.
Reeve, A.F., 1977. The Goz Creek zinc deposit, Yukon Territory. North of 60, Miner. Indust. Rept. 1976, Yukon Terr., EGS 1977-1, Ottawa, 6-19.
Reid, J.A., 1907. The ore deposits of Copperopolis, Calaveras Co., California. Econ. Geol. 2, 380-420.
Reid, K.O., 1975. Mount Lyell copper deposits. In C.L. Knight, ed., Economic Geology of Australia and Papua New Guinea, v.1. Austr. Inst. Min. Metall., Monogr. 5, 604-618.
Rein, G., 1960. Ein Pechblendevorkommen bei Menzenschwand im Südschwarzwald. N. Jb. Miner., Abh., 94, 720-732.
Reineck, H.-E. and Singh, I.B., 1975. Depositional Sedimentary Environments. Springer Verlag, New York, 439 pp.
Reinsbakken, A., 1980. Geology of the Skorovass mine: a volcanogenic massive sulphide deposit in the central Norwegian Caledonides. Norges Geol. Unders., No. 360, 123-154.
Renfro, A.R., 1974. Genesis of evaporite-associated stratiform metalliferous deposits-a sabkha process. Econ. Geol., 33-45.
Reshit'ko, V.A., 1967. Platinovoe orudeneniye v brachyantiklinalyakh Kachanarskovo gabbro-peridotovovo massiva na Urale. Geologiya i Geofizika, 1967, No. 5, 33-42.
Reyes. F.C. and Salfity, J.A., 1973. Consideraciones sobre la estratigrafia del Cretacico (Subgrupo Pirqua) del noroeste argentino. Actas Quinto Congr. Geol. Arg., Buenos Aires, 3, 355-385.
Reynolds, D.G., 1965. Geology and mineralization of the Salsigne gold mine, France. Econ. Geol., 60, 772-791.
Reynolds, M.A., Best, J.G. and Johnson, R.W., 1980. 1953-57 eruption of Tuluman volcano: rhyolitic volcanic activity in the northern Bismarck Sea. Geol. Surv. Papua New Guinea Mem. 7, 44 pp.
Rich, R.A., Holland, H.D. and Petersen, U., 1977. Hydrothermal Uranium Deposits. Devel. in Econ. Geol. 6, Elsevier, 264 pp.
Richardson, J.M.G., Spooner, E.T.C. and McAuslan, D.A., 1982. The East Kemptville tin deposit, Nova Scotia. An example of a large tonnage, low grade greisen-hosted deposit in the endocontact zone of a granite batholith. Geol. Surv. Canada Paper 82-1B, 27-32.
Richey, J.E., Macgregor, A.G. and Anderson, F.W., 1961. British Regional Geology: The Tertiary Volcanic Districts (3rd ed.). 120 pp.
Rickard, D.T., Willden, M.Y. and Marinder, N.E., 1979. Studies on the genesis of the Laisvall sandstone lead-zinc deposit, Sweden. Econ. Geol., 74, 1255-1285.
Rigby, J.K. and Hamblin, W.K., eds., 1972. Recognition of Ancient Sedimentary Environments. Soc. Econ. Paleont. Miner., Tulsa, Spec. Publ. 16, 340 pp.
Riggs, S.R., 1979. Petrology of the Tertiary phosphorite system of Florida. Econ. Geol. 74, 195-220.
Riley, C., 1961. The River Jordan lead-zinc deposit, Revelstoke mining division, B.C. Canad. Inst. Min. Metall. Transact., v.64, 268-272.
Ringwood, A.E., 1975. Composition and Petrology of the Earth's Mantle. McGraw-Hill, 618 pp.

Ripley, E.M. and Ohmoto, H., 1977. Mineralogic, sulfur isotope, and fluid inclusion studies of the stratabound copper deposits at the Raul Mine, Peru. Econ. Geol., 72, 1017-1041.
Rivas, S.V., 1977. Geologia de los principales minas de estaño. Simpos. Intern. del Estaño, Nov. 1977, La Paz, Proc. No. 15, 35 pp.
Robbins, E.I., 1983. Accumulation of fossil fuels and metallic minerals in active and ancient rift lakes. Tectonophysics, 94, 633-658.
Roberts, A.E., 1973. The geological setting of copper orebodies at Inyati Mine, Headlands District, Rhodesia. Spec. Publ. Geol. Soc. S. Africa, 3, 189-196.
Roberts, R.J., Radtke, A.S., Coats, R.R., Silberman, M.L. and McKee, E.H., 1971. Gold-bearing deposits in North-Central Nevada and Southwestern Idaho. Econ. Geol., 66, 14-33.
Robertson, A.H.F., 1975. Cyprus umbers: basalt-sediment relationships on a Mesozoic ocean ridge. Jour. Geol. Soc. London, v. 131, 511-531.
Robertson, J.A., 1978. Uranium deposits in Ontario. In M.M. Kimberley, ed., Short Course in Uranium Deposits. Univ. of Toronto Press, 229-280.
Robertson, J.M., 1975. Geology and mineralogy of some copper sulphide deposits near Mount Bohemia, Keweenaw County, Michigan. Econ. Geol., 70, 1202-1224.
Robin, C., 1982. Mexico. In R.S. Thorpe, ed., Andesites. Wiley, Chichester, 137-147.
Robinson, B.W., 1974. The origin of mineralization at the Tui mine, Te Aroha, New Zealand, in the light of stable isotope studies. Econ. Geol., 69, 910-925.
Robinson, D.N., 1975. Magnetite-serpentine-calcite dykes at Premier Mine and aspects of their relationship to kimberlite and to carbonatite of alkalic carbonatite complexes. In Physics and Chemistry of the Earth, v.9, Pergamon Press, Oxford, 61-70.
Robinson, R.F. and Cook, A., 1966. The Safford copper deposit, Lone Star Mining District, Graham County, Arizona. In S.R. Titley and C.L. Hicks, eds., Geology of the Porphyry Copper Deposits, South-Western North America. Univ. of Arizona Press, Tucson, 251-266.
Roche, H. de la, Marchal, J. and Delbos, L., 1956. Prospection de monazite dans l'extreme sud-est de Madagascar. Madagascar Direct. Mines et Géol., Serv. Géol., 147-156.
Roethe, G., 1975. Silikatische Kupferlagerstätten in Nordchile. Geol. Rundschau, 64, 421-456.
Roger, G., 1972. Un type de minéralisations épigénétiques familières: les filons à antimoine du Massif Central français. Hypothèse de la sécretion latérale. Miner. Deposita, 7, 360-382.
Rogers, J.J.W., Ragland, P.C., Nishimori, R.K., Greenberg, J.K. and Hauck, S.A., 1978. Varieties of granitic uranium deposits and favourable exploration areas in the eastern United States. Econ. Geol., 73, 1539-1555.
Rogers, L.M. and Enders, M.S., 1982. Syenite-hosted gold deposits of the Little Rocky Mountains, Montana. Paper presented at the

88th Ann. N.W. Mining Assoc. Convent., Spokane, 22 pp.
Rona, P.A., 1978. Criteria for recognition of hydrothermal mineral deposits in oceanic crust. Econ. Geol., 73, 135-160.
Rona, P.A., Akers, L., Muse, L. and Hood, D., 1959. Determination of manganese in sea water. 1st Intern. Oceanogr. Congr., Washington, D.C., 1959.
Ronov, A.B. and Yaroshevsky, A.A., 1969. Chemical composition of the Earth's Crust. Amer. Geoph. Union Monogr. 13D, 37-57.
Roos, E., 1976. The cassiterite-silicate formation of tin deposits in the western part of the Krušné Hory Mts. (Erzgebirge), western Bohemia. Věstník Ústř. Úst. Geol., Prague, 51, 27-35.
Roots, E.F., 1954. Geology and mineral deposits of Aiken Lake map-area, British Columbia. Geol. Surv. Canada, Memoir 274, 246 pp.
Rose, E.R., 1973. Geology of vanadium and vanadiferous occurrences of Canada. Geol. Surv. Canada Econ. Geol. Rept. 27, 130 pp.
Rösler, H.J., Baumann, L. and Jung, W., 1968. Postmagmatic mineral deposits of the northern edge of the Bohemian Massif (Erzgebirge-Harz). 23th Int. Geol. Congr., Prague, Guide to Excursion 22 AC, 57 pp.
Rösler, H.J. and Lange, H., 1972. Geochemical Tables. Elsevier, Amsterdam, 468 pp.
Ross, C.P., 1953. The geology and ore deposits of the Reese River district, Lander County, Nevada. U.S. Geol. Surv. Bull. 997, 132 pp.
Ross, C.P., 1956. Quicksilver deposits near Weiser, Washington County, Idaho. U.S. Geol. Surv. Bull. 1042-D, 79-104.
Ross, C.P. and Yates, R.G., 1943. The Coso quicksilver district, Inyo County, California. U.S. Geol. Surv. Bull. 936-Q.
Ross, J.V. and Kellerhals, P., 1968. Evolution of the Slocan Syncline in south-central British Columbia. Canad. J. Earth Sc., 5, 851-872.
Rossovskiy, L.N. and Chmyrev, V.M., 1976. Distribution patterns of rare metal pegmatites in the Hindu Kush (Afghanistan). Intern. Geol. Rev., 19, 1977, 511-520.
Roth, Z., 1962. Vysvětlivky k přehledné geologické mapě ČSSR 1: 200,000 M-33-XXIV, Olomouc. Geofond, Prague, 226 pp.
Roubault, M., 1956. The uranium deposits of France and French Overseas Territories. Proc. Int. Conf. on Peaceful Uses of Atom. Energy, Geneva, 1955. Vol. 6, 152-161.
Routhier, P., 1963. Les Gîsements Métallifères. Géologie et Principes de Recherche. Mason, Paris, 1283 pp.
Routhier, P., 1980. Oú Sont les Métaux Pour L'avenir ? Mém. du B.R.G.M., No. 105, 410 pp.
Rowe, R.B., 1958. Niobium (columbium) deposits of Canada. Geol. Surv. Canada, Econ. Geol. Ser. 18, 108 pp.
Roy, S., 1976. Ancient manganese deposits. In K.H. Wolf, ed., Handbook of Stratiform and Strata-Bound Ore Deposits, v.7. Elsevier, Amsterdam, 395-476.
Rui, I.J., 1973. Structural control and wallrock alteration at Killingdal mine, central Norwegian Caledonides. Econ. Geol., 68, 859-883.

Rui, I.J. and Bakke, I., 1975. Stratabound sulphide mineralization in the Kjøli area, Røros district, Norwegian Caledonides. Norsk Geol. Tidsskrift, 55, 1, 51-75.

Ruitenberg, A.A. and Fyffe, L.R., 1982. Mineral deposits associated with granitoid intrusions and related subvolcanic stocks in New Brunswick and their relationship to Appalachian tectonic evolution. Canad. Inst. Min. Metall. Bull., vol. 75, 87-97.

Ruíz, C.F., 1965. Geología y Yacimientos Metalíferos de Chile. Inst. Invest. Geol., Santiago, 305 pp.

Ruíz, C., Aguilar, A., Egert, E., Espinosa, E., Peebles, F., Quezada, R. and Serrano, M., 1971. Strata-bound copper sulphide deposits of Chile. Soc. Min. Geol. Japan, Spec. Iss. 3, 252-260.

Rundkvist, D.V., ed., 1978. Kriterii Prognoznoi Otsenki Territorii na Tverdye Poleznye Iskopaemye. Nedra, Leningrad, 607 pp.

Runnels, D.D., 1969. The mineralogy and sulfur isotopes of the Ruby Creek copper prospect, Bornite, Alaska. Econ. Geol. 64, 75-90.

Rupke, N.A., 1978. Deep clastic seas. In H.G. Reading, ed., Sedimentary Environments and Facies. Elsevier, New York, 372-415.

Russell, M.J., 1968. Structural controls of base metal mineralization in Ireland in relation to continental drift. Trans. Inst. Min. Metall., Sect. B., 77, B117-B128.

Russell, M.J., 1975. The lithogeochemical environment of the Tynagh base metal deposit. Transact. Inst. Min. Metall., London, Sect. B, v. 84, 128-133.

Rusell, R.T. and Trueman, N.A., 1971. The geology of the Duchess phosphate deposit, northwestern Queensland, Australia. Econ. Geol., 66, 1186-1214.

Rutland, R.W.R., 1974. Andes: Antofagasta Segment. In A.M. Spencer, ed., Mesozoic-Cenozoic Orogenic Belts. Geol. Soc. London, Spec. Publ. 4, 733-743.

Rutter, E.H., Chaplow, R. and Matthews, J.E., 1967. The geology of the Løkken area, Sør-Trøndelag. Norges Geol. Unders. Nr. 255, 21-36.

Ruzhentsev, S.V., 1976. Kraevye ofiolitovye allokhtony. AN SSSR, Trudy, vyp. 283, Nauka, Moscow, 171 pp.

Ruzicka, V., 1971. Geological comparison between East European and Canadian uranium deposits. Geol. Surv. Canada, Paper 70-48, 196 pp.

Ryan, C.W., 1957. A guide to the known minerals of Turkey. Min. Res. and Expl. Inst. of Turkey, Ankara, 196 pp.

Rytuba, J.J. and Glanzman, R.K., 1979. Relation of mercury, uranium and lithium deposits to the McDermitt caldera complex, Nevada-Oregon. In J.D. Ridge, ed., Papers on Mineral Deposits of Western North America. Nevada Bur. Min. Geol., Rept. 33, 109-117.

Saegart, W.E., Sell, J.D. and Kilpatrick, B.E., 1974. Geology and mineralization of La Caridad porphyry copper deposit, Sonora, Mexico. Econ. Geol., 69, 1060-1077.

Sagunov, V.G., 1971. Geologiya agronomicheskikh rud Kazakhstana. AN Kazakh SSR, Alma Ata, 192 pp.

Sahama, T.G., 1974. Potassium-rich alkaline rocks. In H. Sørensen, ed.,

The Alkaline Rocks. Wiley, London, 96-109.
Sahasrabudhe, Y.S., 1978. Bauxite deposits of Gujarat, Maharashtra, and parts of Karnataka. Bull. Geol. Surv. India, Ser. A, No.39, New Delhi, 163 pp.
Said, R., 1962. The Geology of Egypt. Elsevier, Amsterdam, 377 pp.
Sainfeld, P., 1956. The lead-zinc bearing deposits of Tunisia. Econ. Geol., 51, 150-177.
Sainsbury, C.L. and MacKevett, E.M., Jr., 1965. Quicksilver deposits of southwestern Alaska. U.S. Geol. Surv. Bull. 1187, 89 pp.
Sainsbury, C.L., 1969. Geology and ore deposits of the Central York Mountains, western Seward Peninsula, Alaska. U.S. Geol. Survey Bulletin 1287, 101 pp.
Saito, M., ed., 1960. Geology and Mineral Resources of Japan, 2nd ed. Geol. Surv. of Japan, 304 pp.
Salop, L.I. and Scheinmann, Yu. M., 1969. Tectonic history and structures of platforms and shields. Tectonophysics, 7, 565-597.
Samama, J.C., 1968. Contrôle et modèle génétique de minéralisation en galène de type "Red-Beds". Miner. Deosita, 3, 261-271.
Samonov, I.Z. and Pozharisky, I.F., 1974. Deposits of copper. In V.I. Smirnov, ed., Ore Deposits of the U.S.S.R., v.2, English transl., Pitman, London, 106-181.
Samoylov, V.S. and Afanas'yev, B.V., 1978. New data on the carbonatites of the alkaline-ultramafic complex on Turiy Peninsula. Int. Geol. Rev., 22, 39-50.
Sangster, D.F., 1970. Metallogenesis of some Canadian lead-zinc deposits in carbonate rocks. Geol. Assoc. Canada, Proc., 22, 27-36.
Sangster, D.F., 1979. Evidence of an exhalative origin for deposits of the Cobar district, New South Wales. B.M.R. Journ. of Austral. Geol. and Geoph., 4, 15-24.
Sangster, D.F. anf Liberty, B.A., 1971. Sphalerite concretions from Bruce Peninsula, Southern Ontario, Canada. Econ. Geol., 66, 1145-1152.
Santos-Ygiño, L., 1952. Geologiya zhelezorudnykh mestorozhdenii Filipin. Zhelezorud. Mestorozhd. Mira, v.1, Izdat. Inostran. Liter., Moscow, 371-409.
Sass-Gustkiewicz, M., Dzułyński, S. and Ridge, J.D., 1982. The emplacement of zinc-lead sulfide ores in the Upper Silesian district-a contribution to the understanding of Mississippi Valley-type deposits. Econ. Geol., 77, 392-412.
Sato, T., 1974. Distribution and geological setting of the Kuroko deposits. In S. Ishihara, ed., Geology of Kuroko Deposits, 1-8.
Sattran, V., and others, 1966. Problémy metalogeneze Českého Masívu. Sborník Geol. Věd, Prague, LG, v.8, 7-12.
Sattran, V., Klomínský, J., Vejnar, Z. and Fišera, M., 1970. Petrometallogenic series as a source of metals of endogene ore deposits. In Z. Pouba and M. Štemprok, eds., Problems of Hydrothermal Ore Deposition. Schweizerbart'sche, Stuttgart, 78-81.
Saunders, A.D., Tarney, J., Stern, C.R. and Dalziel, J.W.D., 1979. Geochemistry of Mesozoic marginal basin floor igneous rocks from southern Chile. Geol. Soc. Amer. Bull. Pt.1, 90, 237-258.

Saupé, F., 1967. Note preliminaire concernant la genèse du gisement de mercure d'Almaden. Miner. Deposita, 2, 26-33.
Sauvé, P., Cloutier, J.P. and Genois, G., 1972. Base metal deposits of southeastern Quebec. 24th Int. Geol. Congr., Montreal, Excursion B-07, guidebook, 26 pp.
Sawkins, F.J., 1964. Lead-zinc ore deposition in the light of fluid inclusion studies, Providencia mine, Zacatecas, Mexico. Econ. Geol., 59, 883-919.
Sawkins, F.J., 1966. Ore genesis in the North Pennine Orefield, in the light of fluid inclusion studies. Econ. Geol., 61, 385-401.
Sawkins, F.J., 1976. Massive sulphide deposits in relation to geotectonics. Geol. Assoc. Canada Spec. Paper 14, 221-240.
Sawkins, F.J., 1977. Fluid inclusion studies of the Messina copper deposits, Transvaal, South Africa. Econ. Geol., 72, 619-631.
Sawkins, F.J., 1978. Some aspects of the metallogeny of continental rifting events. In E.-R. Neumann and I.B. Ramberg, eds., Petrology and Geochemistry of Continental Rifts. Reidel, Dordrecht, 51-54.
Sawkins, F.J., 1984. Metal Deposits in Relation to Plate Tectonics. Springer Verlag, Berlin, 325 pp.
Sawkins, F.J. and Rye, R.O., 1979. Additional geochemical data on the Messina copper deposits, Transvaal, South Africa. Econ. Geol., 74, 684-689.
Schau, M., 1970. Stratigraphy and structure of the type area of the upper Triassic Nicola Group in South-Central British Columbia. Geol. Assoc. Canada Spec. Pap. No.6, 123-135.
Schau, M., 1973. Alteration of an old island arc. In I. Ermanovics, ed., Volcanism and Volcanic Rocks. Geol. Surv. Canada Open File 164, 73-86.
Scheibner, E. and Markham, N.L., 1976. Tectonic setting of some strata-bound massive sulphide deposits in New South Wales, Australia. In K.H. Wolf, ed., Handbook of Stratiform and Strata-Bound Ore Deposits, v. 6. Elsevier, Amsterdam, 55-77.
Schermerhorn, L.J.G., 1966. Terminology of mixed coarse-fine sediments. J. Sedim. Petrol., 36, 831-835.
Schlanger, S.O. and others, 1963. Subsurface geology of Eniwetok atoll. U.S. Geol. Surv. Prof. Paper 260-BB, 991-1066.
Schlee, J.S., 1980. A comparison of two Atlantic-type continental margins. U.S. Geol. Surv. Prof. Paper 1167, 21 pp.
Schmidt, H.L., 1976. Rohstoffwirtschaftliche Länderberichte, X-Indonesien. Bundesanst. f. Geowiss., Hannover, 118 pp.
Schmidt, R.G., 1978. The potential for porphyry copper-molybdenum deposits in the Eastern United States. U.S. Geol. Surv. Profess. Paper 907-E, 31 pp.
Schmitt, J.-M. and Thiry, M., 1977. Minéralisation en plomb par évolutions pédogénétiques d'une serie arkosique du trias (Zeida, Haute Moulouya, Maroc). Bull. du B.R.G.M., Sect.II, 2, 113-133.
Schneider, H.-J., 1964. Facies differentiation and controlling factors for the depositional lead-zinc concentration in the Ladinian Geosyncline in the Eastern Alps. In G.C. Amstutz, ed., Sedi-

mentology and Ore Genesis. Elsevier, Amsterdam, 29-45.
Schneider, H.-J. and Lehmann, B., 1977. Contribution to a new genetic concept on the Bolivian Tin Province. In D.D. Klemm and H.-J. Schneider, Time and Strata-Bound Ore Deposits, Springer Verlag, Berlin, 153-168.
Schneiderhöhn, H., 1941. Lehrbuch der Lagerstättenkunde, 1. Fischer, Jena, 858 pp.
Schneiderhöhn, H., 1955. Erzlagerstätten. 3rd ed., Fischer, Jena, 375 pp.
Schneiderhöhn, H., 1961. Die Erzlagerstätten der Erde, v. II, Die Pegmatite. Fischer, Stuttgart, 720 pp.
Schnyukov, Ye.F., Orlovskii, G.N., Usenko, V.P., Grigor'ev, A.V. and Gordievich, V.A., 1974. Geologiya Azovskovo Mor'ya. Naukova Dumka, Kiev, 247 pp.
Schoell, M., Backer, N. and Baumann, Q., 1974. The Red Sea geothermal systems-new aspects on their brines and associated sediments. In Problems of Ore Deposition, 4th IAGOD Sympos., Varna, 1, 303-313.
Scholl, D.W., 1974. Sedimentary sequences in the North Pacific trenches. In C.A. Burk and C.L. Drake, eds., The Geology of Continental Margins. Springer Verlag, New York, 493-505.
Scholle, P.A., 1978. Carbonate Rock Constituents, Textures, Cements and Porosities. Amer. Assoc. Petr. Geol. Mem. 27, 241 pp.
Scholle, P.A., Bebout, D.G. and Moore, C.H., 1983. Carbonate Depositional Environments. Amer. Assoc. Petrol. Geol., Tulsa, 708 pp.
Scholtz, D.L., 1936. The magmatic nickeliferous deposits of East Griqualand and Pondoland. Trans. Geol. Soc. S. Africa 39, 81-210.
Schorscher, H.D., Santana, F.C., Polonia, J.C. and Moreira, J.M.P., 1982. Quadrilatero Ferrífero, Minas Gerais state: Rio dos Velhas greenstone belt and Proterozoic rocks. Intern. Symp. on Arch. and early Proteroz. Geol. Evol. and Metall., Excurs. Annex, 44 pp.
Schott, W., 1976. Mineral (inorganic) resources of the oceans and ocean floors: a general review. In K.H. Wolf, ed., Handbook of Stratiform and Strata-Bound Ore Deposits, v.3, Elsevier, Amsterdam, 245-296.
Schott, W. (compiler), 1980. Die Fahrten des Forschungschiffes "Valdivia" 1971-1978; Geowissenschaftliche Ergebnisse. Geol. Jahrb., Hannover, D 38, 201 pp.
Schrader, F.C., 1911. Gold-bearing ground moraine in northwestern Montana. U.S. Geol. Surv. Bulletin 470, 62-74.
Schroder, E., 1954. Zur Paleogeographie des Mittleren Bundsandsteins bei Mechernich, Eifel. Geol. Jahrbuch, 69, Hannover, 417-428.
Schultz, R.W., 1971. Mineral exploration practice in Ireland. Transact. Inst. Min. Metall., London, Sect. B, 77, B238-B258.
Schultz, L.G, Tourtelot, H.A., Gill, J.R. and Boerngen, J.G., 1980. Composition and properties of the Pierre Shale and equivalent rocks, northern Great Plains region. U.S. Geol. Surv. Prof. Paper 1064-B, 114 pp.
Schulz, O., 1972a. Unterdevonische Baryt-Fahlerz-Mineralisation und ihre steilachsige Verformung im Grosskogel bei Brixlegg (Tirol). Tscherm. Min. Petr. Mitt., 18, 114-128.

Schulz, O., 1972b. Horizontgebundene altpaleozoische Kupferkiesvererzung in der Nordtiroler Grauwackenzone, Österreich. Tscherm. Min. Petr. Mitt., 17, 1-18.

Schumacher, F., 1956. Die bergbauliche Entwicklung der Turkei in den letsten 20 Jahren. Z. Erzbergb. u. Metallhütt., 1956, No.10.

Schuylenborgh, J., 1971. Weathering and soil-forming processes in the tropics. In Soil and Tropical Weathering. UNESCO, Paris, 39-50.

Searle, D.L., 1972. Mode of occurrence of the cupriferous pyrite deposits of Cyprus. Transact. Inst. Min. Metall., London, Sect.B, v.8, B189-B197.

Searle, D.L and Panayiotou, A., 1980. Structural implications in the evolution of the Troodos massif, Cyprus. In A. Panayiotou, ed., Ophiolites. Cyprus Geol. Surv. Dept., Nicosia, 50-60.

Searls, F., Jr., 1952. Karst ore in Yunnan. Econ. Geol. 47, 339-346.

Segerstrom, K., 1960. Geologia del cuadrangulo Quebrada Paipote. Inst. Inv. Geol., Carta Geol. de Chile, v.2, No.1, Santiago, 35 pp.

Segerstrom, K. and Ryberg, G.E., 1972. Geology and placer-gold deposits of the Jicarilla Mountains, Lincoln County, New Mexico. U.S. Geol. Surv. Bull. 1308, 25 pp.

Selin, P.F., 1981. O pogrebennykh ostatochno-karstovykh mestorozhdeniyakh Gornovo Altaya. Geol. Rud. Mestor., 2, 128-129.

Semenov, Ye.I., 1974. Economic mineralogy of alkaline rocks. In H. Sørensen, ed., The Alkaline Rocks. Wiley, London, 543-552.

Seraphim, R.H., 1975. Denali-a nonmetamorphosed stratiform sulfide deposit. Econ. Geol., 70, 949-959.

Sestini, G., 1970. Flysch facies and turbidite sedimentology. Sedim. Geol., 4, 559-597.

Sestini, G., 1974. Northern Appenines. In A.M. Spencer, ed., Mesozoic-Cenozoic Orogenic Belts. Geol. Soc. London, 61-84.

Sevastyanov, V.F. and Volkov, I.I., 1966. Chemical composition of iron-manganese concretions of the Black Sea. Doklady A.N. U.S.S.R., 166, 701-704.

Shakhov, F.N., et al., 1972. Problemy Obrazovaniya Rudnykh Stol'bov. Nauka, Novosibirsk, 436 pp.

Sharfman, V.S., 1969. Priznaki zherlovykh i prizherlovykh zon na erodirovannykh paleovulkanakh. Sovet. Geol., 1969, No.4, 120-132.

Sharp, W.N., 1978. Geologic map of the Silver Cliff and Rosita volcanic centres, Custer County, Colorado. U.S. Geol. Surv. Misc. Invest. Ser., Map I-1081, 1:24,000.

Sharp, W.N. and Cavender, W.S., 1962. Geology and thorium-bearing deposits of the Lemhi Pass area, Lemhi County, Idaho and Beaverhead County, Montana. U.S. Geol. Surv. Bull. 1126, 76 pp.

Shawe, D.R., 1966. Arizona-New Mexico and Nevada-Utah beryllium belts. U.S. Geol. Surv. Prof. Paper 550-C, C206-C213.

Shawe, D.R., 1976. Sedimentary rock alteration in the Slick Rock District, San Miguel and Dolores Counties, Colorado. U.S. Geol. Surv. Prof. Paper 576-D, 51 pp.

Shcheglov, A.D., 1967. Osnovnye cherty metallogenii zon avtonomnoi aktivizatsii. Zakon. Razm. Polez. Iskop. VIII, 95-138.

Shcheglov, A.D. and Butkevich, T.V., 1974. Deposits of tungsten.

In V.I. Smirnov, ed., Ore Deposits of the U.S.S.R., v.3, Engl. transl., Pitman, London, 180-229.

Shcherba, G.N., 1968a. Deposits of the Atasu type in Kazakhstan. 23th Int. geol. Congr., v.7, Prague, 151-163.

Shcherba, G.N., 1968b. Greisens. Intern. Geol. Rev., 12, 114-150.

Shcherba, G.N. et al., 1972. Geotektonogeny Kazakhstana i Redkometal'noe Orudeneniye. Nauka Kazakh. SSR, Alma Ata, 218 pp.

Sheppard, F.P., 1973. Submarine Geology. 3rd ed., Harper and Row, New York, 551 pp.

Sheppard, R.A. and Gude, A.J., 1969. Authigenic fluorite in Pliocene lacustrine rocks near Rome, Malheur County, Oregon. U.S. Geol. Surv. Prof. Paper, 650D, 69-74.

Sherborne, J.E., Jr., Buckovic, W.A., Dewitt, D.B., Hellinger, T.S. and Pavlak, S.J., 1979. Major uranium discovery in volcaniclastic sediments, Basin and Range Province, Yavapai County, Arizona. Amer. Assoc. Petr. Geol. Bull., 63, 621-646.

Sheridan, D.M., Maxwell, C.H. and Collier, J.T., 1961. Geology of the Lost Creek schroeckingerite deposit, Sweetwater County, Wyoming. U.S. Geol. Surv. Bulletin 1087-J, 391-478.

Sheridan, D.M., Maxwell, C.H. and Albee, A.L., 1967. Geology and uranium deposits of the Ralston Buttes District, Jefferson County, Colorado. U.S. Geol. Surv. Prof. Pap. 520, 121 pp.

Sheridan, R.E., 1974. Atlantic continental margin of North America. In C.F. Burk and C.L. Drake, eds., The Geology of Continental Margins. Springer Verlag, New York, 391-407.

Sheridan, R.E., 1981. Recent research on passive continental margins. Soc. Econ. Paleont. and Miner., Tulsa, Spec. Publ. 32, 39-56.

Shikazono, N., 1975. Mineralization and chemical environment of the Toyoha lead-zinc vein-type deposits, Hokkaido, Japan. Econ. Geol., 70, 694-705.

Shimizu, M. and Iiyama, J.T., 1982. Zinc-lead skarn deposits of the Nakatatsu Mine, Central Japan. Econ. Geol., 77, 1000-1012.

Shkol'nik, E.L., 1972. Zhelezorudnaya effuzivno-yashmenaya formatsiya i zheleznye rudy Udskovo raiona Khabarovskovo Kraya. In Geol. i Genezis Dokembr. Zhelez.-Kremn. i Margants. Formatsii Mira. Nauk. Dumka, Kiev, 196-203.

Shoemaker, E.M., 1956. Occurrence of uranium in diatremes on the Navajo and Hopi reservations, Arizona, New Mexico and Utah. U.S. Geol. Surv. Prof. Paper 300, 179-185.

Siegl, W., 1974. Ein Beitrag zur Genese der Vererzung des Grazer Paläozoikums. Miner. Deposita, 9, 289-295.

Siemiatkowska, K.M. and Martin, R.F., 1975. Fenitization of Mississagi Quartzite, Sudbury area, Ontario. Geol. Soc. Am. Bull. 86, 1109-1122.

Siever, R., 1968. Sedimentological consequences of a steady-state ocean-atmosphere. Sedimentology, 11, 5-29.

Silant'ev, S.A. and Chernysheva, V.I., 1981. Metamorfizm giperbazit-gabbro-bazaltovovo kompleksa khrebta Karlsberg. Izvest. A.N. SSSR, Ser. Geol., No.12-1981, 47-55.

Sillitoe, R.H., 1973a. The tops and bottoms of porphyry copper de-

posits. Econ. Geol., 68, 799-815.
Sillitoe, R.H., 1973b. Geology of the Los Pelambres porphyry copper deposit, Chile. Econ. Geol., 68, 1-10.
Sillitoe, R.H., 1975. Lead-silver, manganese and native sulfur mineralization within a stratovolcano, El Queva, northwest Argentina. Econ. Geol., 70, 1190-1201.
Sillitoe, R.H., 1977. Metallic mineralization affiliated to subaerial volcanism: a review. In Volc. Processes in Ore Genesis. Inst. Min. Metall./Geol. Soc., London, 99-116.
Sillitoe, R.H., Halls, C. and Grant, J.N., 1975. Porphyry tin deposits in Bolivia. Econ. Geol., 70, 913-927.
Sillitoe, R.H. and Khan, S.N., 1977. Geology of Saindak porphyry copper deposit, Pakistan. Trans. Inst. Min. Metall., London, Ser. B, v.86, B27-B42.
Simons, F.S. and Straczek, J.A., 1958. Geology of the manganese deposits of Cuba. U.S. Geol. Surv. Bull. 1057, 289 pp.
Simpson, T.A. and Gray, T.R., 1968. The Birmingham red-ore district, Alabama. In J.D. Ridge, ed., Ore Deposits of the United States 1933-1967. A.I.M.E., New York, 187-206.
Sims, P.K., Drake, A.A., Jr. and Tooker, E.W., 1963. Economic geology of the Central City district, Gilpin County, Colorado. U.S. Geol. Surv. Prof. Paper 359, 231 pp.
Sims, P.K. and others, 1963. Geology of uranium and associated ore deposits, central part of the Front Range Mineral Belt, Colorado. U.S. Geol. Surv. Prof. Paper, 120 pp.
Sims, S.J., 1968. The Grace mine magnetite deposit, Berks County, Pennsylvania. In J.D. Ridge, ed., Ore Deposits of the United States 1933-1967. A.I.M.E., New York, 108-124.
Sinclair, W.D., Morin, J.A., Craig, D.B. and Marchand, M., 1976. North of 60. Mineral Industry Rept., 1975, Yukon Territ., 210 pp.
Singewald, Q.D., 1950. Mineral Resources of Colombia. U.S. Geol. Surv. Bull. 964-B, 53-204.
Singh, R.N. and Srivastava, S.N.P., 1980. Ophiolites and associated mineralization in the Naga Hills, north-eastern India. In A. Panayiotou, ed., Ophiolites. Cyprus Geol. Surv. Dept., 758-764.
Sivarajasingham, S., Alexander, L.T., Cady, J.G. and Cline, M.G., 1962. Laterite. Advances in Agronomy. Academic Press, N.Y., 1-60.
Sivoplyas, A.P., 1979. Formation of alluvial placers on the shelf of the Sea of Japan. Int. Geol. Rev., 22, 1088-1093.
Skall, H., 1975. The paleoenvironment of the Pine Point lead-zinc district. Econ. Geol., 70, 22-47.
Skinner, B.J. anf Peck, D.L., 1969. An immiscible sulfide melt from Hawaii. Econ. Geol. Monogr. 4, 310-322.
Skoček, V., Al-Qaraghuli, N. and Saadallah, A.A., 1971. Composition and sedimentary structures of iron ores from the Wadi Husainiya area, Iraq. Econ. Geol., 66, 995-1004.
Skripchenko, N.S., 1972. Gidrotermal'no-osadochnye sul'fidnye rudy bazal'toidnykh formatsii. Nedra, Moscow, 212 pp.
Škvor, V., 1979. The reactivation of the ancient structure and corresponding changes in the Earth's crust. An example of the Bohemian

Massif. Geol. Rundsch., Bd. &8, 793-804.
Slater, D. and Highley, D.E., 1976. The iron ore deposits of the United Kingdom of Great Britain and Northern Ireland. In A. Zitzmann, ed., 1976, The Iron Ore Deposits of Europe. Bundesanst. f. Geowiss., Hannover, 393-409.
Smirnov, S.S., 1951. Zony Oksidatsii Sulfidnykh Mestorozhdenii. A.N. SSSR, Moscow, 335 pp.
Smirnov, V.I., ed., 1968. Genezis Endogennykh Rudnykh Mestorozhdenii. Nedra, Moscow, 719 pp.
Smirnov, V.I., 1969. Geologiya Poleznykh Iskopaemykh. 2nd ed., Nedra, Moscow, 687 pp.
Smirnov, V.I., Borodayev, Yu.S. and Starostin, V.I., 1968. Pyritic ores and deposits of Japan. Int. Geol. Rev., 11, 8, 845-856.
Smirnov, V.I. and Gorzhevsky, D.I., 1974. Deposits of lead and zinc. In V.I. Smirnov, ed., Ore Deposits of the U.S.S.R., v.2, Engl. transl., Pitman, London, 182-256.
Smith, C.H., 1958. Bay of Islands igneous complex, western Newfoundland. Geol. Surv. Canada, Memoir 290, 132 pp.
Smith, D. and Silver, L.T., 1975. Potassic granophyre associated with Precambrian diabase, Sierra Ancha, central Arizona. Geol. Soc. Amer. Bull., 86, 503-513.
Smith, G.I., 1979. Subsurface stratigraphy and geochemistry of late Quaternary evaporites, Searles Lake, California. U.S. Geol. Surv. Prof. Paper 1043, 130 pp.
Smith, J.W. and Milton, C., 1966. Dawsonite in the Green River Formation of Colorado. Econ. Geol., 61, 1029-1042.
Smith, R.E., 1974. The production of spilitic lithologies by burial metamorphism of flood basalts from the Canadian Keweenawan, Lake Superior. In G.C. Amstutz, ed., Spilites and Spilitic Rosks, Springer Verlag, Berlin, 403-416.
Smith, R.M., 1970. Treasure Hill reinterpreted. Econ. Geol., 65, 538-540.
Smith, W.C., 1956. A review of some problems of African carbonatites. Quart. J. Geol. Soc. London, 112, 189-219.
Snelgrove, A.K. and Baird, D.M., 1953. Mines and Mineral Resources of Newfoundland. Newf. Geol. Surv. Inf. Circ. 4, 149 pp.
Snyder, G.L. and Fraser, G.D., 1963. Pillowed lavas, I: intrusive layered lava pods and pillowed lavas, Unalaska Island, Alaska. U.S. Geol. Surv. Prof. Paper 454-B,C, B1-B30.
Snyder, F.G. and Gerdemann, P.E., 1968. Geology of the Southeast Missouri Lead District. In J.D. Ridge, ed., Ore Deposits of the United States 1933-1967. A.I.M.E., New York, 326-359.
Sobolev, N.V., 1977. Deep-seated inclusions in kimberlites and the problem of the composition of the upper mantle. Amer. Geophys. Union, Washington, 279 pp.
Sobolev, V.S. et al., 1978. Poiskovye kriterii sul'fidnykh rud Noril'skovo tipa. A.N. SSSR, Sibir. Otd., Trudy 418, 320 pp.
Socolescu, M., 1972. Phénoménes métallogéniques dans la province de Baia Mare. Geologija, Ljubljana, 15, 287-297.
Soister, P.E., 1968. Stratigraphy of the Wind River Formation in

south-central Wind River Basin, Wyoming. U.S. Geol. Surv. Prof. Paper, 594-A, 50 pp.
Sokolov, G.A. and Grigor'ev, V.M., 1974. Deposits of iron. In V.I. Smirnov, ed., Ore Deposits of the U.S.S.R., v.1, Engl. transl., Pitman, London, 7-113.
Somm, A.F., 1975. Gove bauxite deposits, N.T. In C.L. Knight, ed., Economic Geology of Australia and Papua New Guinea, v.1. Aust. Inst. Min. Metall., Monogr. 5, 964-967.
Soregaroli, A.E. and Sutherland Brown, A., 1976. Characteristics of Canadian Cordilleran molybdenum deposits. Canad. Inst. Min. Metall. Spec. Volume 15, Montreal, 417-431.
Sørensen, H., ed., 1974. The Alkaline Rocks. Wiley, London, 622 pp.
Soulé de Lafont, D., 1967. Les gîtes de fluorine stratiforme de la bordure Nord du Morvan. Chron. Mines Rech. Min., mars 1967, no. 361, 85-107.
Sözen, A., 1977. Geological investigations on the genesis of the cinnabar deposits of Kaleçik/Karaburun (Turkey). In D.D. Klemm and Schneider, H.J., eds., Time and Strata-Bound Ore Deposits. Springer Verlag, Berlin, 205-219.
Spencer, A.C., 1906. The Juneau gold belt, Alaska. U.S. Geol. Surv. Bulletin 287, 161 pp.
Spencer, A.C., 1908. Magnetite deposits of the Cornwall type in Pennsylvania. U.S.Geol. Surv. Bulletin 359, 102 pp.
Spencer, A.M., ed., 1974. Mesozoic-Cenozoic Orogenic Belts. Geol. Soc. London Spec. Public. 4, 809 pp.
Spooner, E.T.C. and Fyfe, W.S., 1973. Sub-seafloor metamorphism, heat and mass transfer. Contr. Miner. Petrol 42, 287-304.
Spratt, R.N., 1975. Adau River sulphide nickel prospect. In C.L. Knight, ed., Economic Geology of Australia and Papua New Guinea, v.1. Austr. Inst. Min. Metall., Monogr. 5, 877-879.
Spurr, J., 1905. Geology of the Tonopah mining district, Nevada. U.S. Geol. Surv. Profess. Paper 42, 295 pp.
Staatz, M.H., 1972. Thorium-rich veins of Hall Mountain in northernmost Idaho. Econ. Geol., 67, 240-248.
Staatz, M.H., 1974. Thorium veins in the United States. Econ. Geol. 69, 495-507.
Staatz, M.H., 1978. I and L uranium and thorium vein system, Bokan Mountain, Southeastern Alaska. Econ. Geol. 73, 512-523.
Staatz, M.H., 1979. Geology and mineral resources of the Lemhi Pass thorium district, Idaho and Montana. U.S. Geol. Surv. Prof. Pap. 1049-A, 90 pp.
Staatz, M.H. and Carr, W.J., 1964. Geology and mineral deposits of the Thomas and Dugway Ranges, Juab and Tooele Counties, Utah. U.S. Geol. Surv. Prof. Paper 415, 188 pp.
Stach, E., Mackowsky, M.T., Teichmüller, M., Taylor, G.H., Chandra, D. and Teichmüller, R., 1975. Coal Petrology. Borntraeger, Berlin, 428 pp.
Stadnichenko, T. et al., 1953. Concentration of germanium in the ash of American coals-a progress report. U.S. Geol. Surv. Circular 272, 34 pp.

Stager, H.K., 1960. A new beryllium deposit at the Mount Wheeler Mine, White Pine County, Nevada. U.S. Geol. Surv. Prof. Pap. 400-B, B70-B71.

Stamper, J.W., 1970. Gallium. In Mineral Facts and Problems. U.S. Bureau of Mines, Bull. 650, 555-561.

Stanton, R.L., 1955. Lower Paleozoic mineralization near Bathurst, N.S.W. Econ. Geol., 50, 681-714.

Stanton, R.L., 1972. Ore Petrology. McGraw Hill, New York, 713 pp.

Stanton, R.L. and Ramsay, W.R.H., 1980. Exhalative ores, volcanic loss, and the problem of the island arc calc-alkaline series: a review and an hypothesis. Norges Geol. Unders. No. 360, 9-57.

Staritskii, Yu.G. and Tuganova, Ye.V., 1971. Metallogeniya Trappov Sibirskoy Platformy. In Trappy Sibirskoi Platformy i ikh Metallogeniya. A.N. SSSR, Sibir. Otd., Irkutsk, 48-52.

Stark, J.T. and Hay, R.L., 1963. Geology and petrography of volcanic rocks of the Truk Islands, East Caroline Islands. U.S. Geol. Surv. Profess. Paper 409, 41 pp.

Starostin, V.J., 1970. Bor and Majdanpek copper deposits in Yugoslavia. Intern. Geol. Rev., 12, 370-380.

Steinmann, G., 1926. Die ophiolitische Zonen in den mediterranean Kettengangebirgen. 14th Intern. Geol. Congr., F.2, 637-666.

Štemprok, M., Burnol, L. and Tischendorf, G., eds., 1978. Metallization Associated With Acid Magmatism. 3 volumes, Czech Geol. Surv., Prague.

Štemprok, M. and Šulcek, Z., 1969. Geochemical profile through ore-bearing lithium granite. Econ. Geol., 64, 392-404.

Stephens, M.B., 1980. Spilitization, element release and formation of massive sulphides in the Stekenjokk volcanites, central Swedish Caledonides. Norges Geol. Unders., No. 360, 159-193.

Steven, T.A. and Lipman, P.W., 1976. Calderas of the San Juan volcanic field, southwestern Colorado. U.S. Geol. Surv. Prof. Paper 958, 35 pp.

Steven, T.A. and Ratté, J.C., 1960. Relation of mineralization to caldera subsidence in the Creede district, San Juan Mountains, Colorado. U.S. Geol. Surv. Prof. Paper 400-B, B14-B18.

Steven, T.A. and Ratté, J.C., 1965. Geology and structural control of ore deposition in the Creede district, San Juan Mountains, Colorado. U.S. Geol. Surv. Prof. Paper 487, 90 pp.

Stevenson, J.S., 1948. Twin "J" mine. In Brownel, G.M., ed., Structural Geology of Canadian Ore Deposits, 88-93.

Stewart, F.H., 1963. Marine Evaporites. In Data of Geochemistry. U.S. Geol. Surv. Profess. Paper 440-Y, 52 pp.

Stewart, J.H., Poole, F.G. and Wilson, R.F., 1972. Stratigraphy and origin of the Chinle Formation and related upper Triassic strata in the Colorado Plateau region. U.S. Geol. Surv. Prof. Paper 690, 336 pp.

Stockford, H.R., 1972. The James Bay pyrochlore deposit. Canad. Inst. Min. Metall. Bulletin, June 1972, 61-69.

Stockley, G.M., 1945. The Liganga (titaniferous) magnetite deposits. Miner. Magaz. 73, 5, 265-274.

Stoiber, R.E. and Rose, W.I., 1970. The geochemistry of Central American volcanic gas condensates. Geol. Soc. Amer. Bulletin 81, 2891-2912.

Stoiber, R.E. and Rose, W.I., Jr., 1874. Fumarole incrustations at active Central American volcanoes. Geoch. Cosmoch. Acta, 38, 495-516.

Stoll, W.C., 1961. Tertiary channel gold deposits at Tipuani, Bolivia. Econ. Geol., 56, 1258-1264.

Stone, J.G., 1959. Ore genesis in the Naica district, Chihuahua, Mexico. Econ. Geol., 54, 1002-1034.

Stout, W., 1944. The iron ore-bearing formations of Ohio. Geol. Surv. of Ohio, 4th Ser., Bull. 45, Columbus, 230 pp.

Strahler, A.N., 1971. The Earth Sciences. 2nd ed., Harper and Row, New York, 824 pp.

Strakhov, N.M., 1962. Principles of Lithogenesis, v.2, Engl. transl. Oliver and Boyd, Edinburgh, 609 pp.

Strakhov, N.M., Shterenberg, L.Ye, Kalinenko, V.V. and Tikhomirova, Ye.S., 1968. Geokhimiya osadochnovo margantsevorudnovo protsessa. A.N. SSSR, Trudy, vyp. 185, 493 pp.

Strauss, G.K. and Madel, J., 1974. Geology of massive sulphide deposits in the Spanish-Portuguese pyrite belt. Geol. Rundsch., Bd. 63, 191-211.

Strauss, G.H., Madel, J. and Alonso, F.F., 1977. Exploration practice for strata-bound volcanogenic sulphide deposits in the Spanish-Portuguese pyrite belt: geology, geophysics and geochemistry. In D.D. Klemm and H.-J. Schneider, eds., Time and Strata-Bound Ore Deposits. Springer Verlag, Berlin, 55-93.

Strauss, G.K., Roger G., Lecolle, M. and Lopera, E., 1981. Geochemical and geologic study of the volcano-sedimentary sulfide orebody of La Zarza, Huelva Province, Spain. Econ. Geol., 76, 1975-2000.

Streckeisen, A., 1976. To each plutonic rock its proper name. Earth Sc. Review, 12, 1-33.

Strona, P.A.L., 1978. Glavnye Tipy Rudnykh Formatsii. Nedra, 199 pp.

Strong, D.F., 1977. Volcanic regimes of the Newfoundland Appalachians. Geol. Assoc. Canada Spec. Pap. 16, 61-90.

Strong, D.F., 1980. Granitoid rocks and associated mineral deposits of eastern Canada and western Europe. Geol. Assoc. Canada Spec. Paper 20, 741-770.

Štrucl, I., 1966. Some ideas on the genesis of the Karavanke lead-zinc ore deposits with special regard to the Mežica ore deposit. Min. Metall. Quart., 2, 25-34.

Sturt, B.A., Thon, A. and Furnes, H., 1980. The geology and preliminary geochemistry of the Karmøy ophiolite, S.W. Norway. In A. Panayiotou, ed., Ophiolites. Cyprus Geol. Surv. Dept., 538-554.

Subramanian, K.S., 1978. Bauxite in Tamil Nadu in the southern part of the Indian Peninsula. In Third Reg. Conf. on Geol. and Min. Res. of S.E. Asia, Bangkok, Nov. 1978, 367-359.

Suk, M., 1983. Petrology of Metamorphic Rocks. Devel. in Petrol., 9, Elsevier, Amsterdam, 322 pp.

Summons, T.G., Green, D.C. and Everard, J.L., 1981. The occurrence of chromite in the Andersons Creek area, Beaconsfield, Tasmania. Econ. Geol. 76, 505-518.
Sunarić, O. and Olujić, J., 1968. Osnovne geološke karakteristike ležišta hromita Duboštica. Geol. Glasnik, Sarajevo, 12, 261-270.
Sundblad, K., 1980. A tentative "volcanogenic" formation model for the sediment-hosted Ankarvattnet Zn-Cu-Pb massive sulphide deposit, central Swedish Caledonides. Norges Geol. Unders. 360, 211-227.
Superceanu, C.I., 1967. Die Geosynklinallagerstätten Provinzen Rumaniens. Geol. Rundsch., 56, 949-972.
Suppel, D.W., 1974. Girilambone anticlinorial zone. In N.L. Markham and H. Basden, eds., The Mineral Deposits of New South Wales. Geol. Surv. N.S.W., Sydney, 119-131.
Surdam, R.C., 1968. Origin of native copper and hematite in the Karmutsen Group, Vancouver Island, B.C. Econ. Geol., 63, 961-966.
Surgai, V.T., 1970. Osnovnye cherty regional'noi geokhimii i metallogenii Tyan-Shanya. Zakon. Razm. Polez. Iskop. IX, 118-144.
Surgai, V.T., 1980. O prirode sur'myanno-rtutnovo orudeneniya. Geol. Rud. Mestor., No.3, 1980, 3-14.
Sutherland Brown, A., 1955. Red Rose (Western Tungsten Copper Mines, Ltd.). In Brit. Columbia Minister of Mines Rept. for 1954, Victoria, A86-A95.
Sutherland Brown, A., 1957. Geology of the Antler Creek area, Cariboo district, British Columbia. Brit. Columb. Dept. Min. Bulletin 38, 105 pp.
Sutherland Brown, A., 1968. Geology of the Queen Charlotte Islands, British Columbia. Brit. Columb. Dept. Min. Bull. 54.
Sutherland Brown, A., 1974a. Aspects of metal abundances and mineral deposits in the Canadian Cordillera. Canad. Inst. Min. Metall. Trans., 77, 14-21.
Sutherland Brown, A., 1974b. Gibraltar mine. Geol., Expl., Min. in Brit. Columbia, Victoria, 299-318.
Sutherland Brown, A., 1976. Morphology and classification. Canad. Inst. Min. Metall. Spec. Volume 15, Montreal, 44-51.
Sutherland Brown, A., ed., 1976. Porphyry Deposits of the Canadian Cordillera. Can. Inst. Min. Metall. Spec. Vol. 15, 510 pp.
Sutherland Brown, A. and Merrett, J.E., 1965. Texada Island. Brit. Columbia Min. of Mines and Petr. Res. Rept. for 1964, 146-151.
Suzuki, J. and Ohmachi, H., 1956. Manganiferous iron ore deposits in the Tokoro district of North-Eastern Hokkaido, Japan. 20th Int. Geol. Congr., Mexico, Sympos. del Manganeso, v.IV, 199-204.
Svoboda, J., ed., 1966. Regional Geology of Czechoslovakia, Pt. I, The Bohemian Massif. Ústř. Úst. Geol., Prague, 668 pp.
Swaine, D.J., 1962. Boron in New South Wales Permian coals. Austral. J. Sci., 25, 265-266.
Swanson, E.R., Keizer, R.P., Lyons, J.I. and Clabaugh, S.E., 1978. Tertiary volcanism and caldera development near Durango City, Sierra Madre Occidental, Mexico. Geol. Soc. Amer. Bull. 89, 1000-1012.
Swanson, D.A. and Wright, T.L., 1981. The regional approach to stu-

dying the Columbia River basalt group. Geol. Soc. India Memoir 3, Bangalore, 58-80.

Swanson, E.A. and Brown, R.L., 1962. Geology of Buchans orebodies. Canad. Inst. Min. Metall. Transact., v. 85, 618-626.

Swanson, R.W., 1973. Geology and phosphate deposits of the Permian rocks in central western Montana. U.S. Geol. Surv. Prof. Paper 313-F, 779-833.

Swanson, V.E., 1961. Geology and geochemistry of uranium in marine black shales, a review. U.S. Geol. Surv. Prof. Paper 356-C, 67-112.

Swanson, V.E., Love, A.H. and Frost, I.C., 1972. Geochemistry and diagenesis of tidal marsh sediment, northeastern Gulf of Mexico. U.S. Geol. Surv. Bull. 1360, 83 pp.

Sweeting, M.M., 1973. Karst Landforms. Macmillan, London, 362 pp.

Syka, J., Čadek, J., Blažek, J., Herčík, F., Chabr, P., Studničná, B. and Veselý, T., 1978. Charakteristické rysy uranových a zirkonium-uranových akumulací ve svrchní křídě severních Čech. Sbor. Geol. Věd, LG, Prague, 19, 7-33.

Szalay, S., 1954. Enrichment of uranium in some brown coals in Hungary. Acta Geol. Acad. Sci. Hungariacae, 2, 299-311.

Szekely, T.S. and Grose, L.T., 1972. Stratigraphy of the carbonate, black shale and phosphate of the Pucará Group (upper Triassic-lower Jurassic), Central Andes, Peru. Geol. Soc. Amer. Bull. 83, 407-428.

Szubert, E.C., Grazia, C.A. and Shintaku, I., 1981. Importancía economica dos mineralizacões de cobre associadas aos basaltas da formacão Serra Geral. Mineracão, Metalurgía, 45, No. 429, 24-30.

Tabor, R.W. and Cady, W.M, 1978a. The structure of the Olympic Mountains, Washington-analysis of a subduction zone. U.S. Geol. Surv. Prof. Paper 1033, 1-75.

Tabor, R.W. and Cany, W.M., 1978b. Geologic map of the Olympic Peninsula, Washington. U.S. Geol. Surv. Map 1-994, 1:125,000.

Tailleur, I.L. and Brosgé, W.P., 1974. Zone 7, Northern Alaska (Brooks Range Orogen). In A.M. Spencer, ed., Mesozoic-Cenozoic Orogenic Belts. Geol. Soc. London Spec. Publ. 4, 582-585.

Takeda, H., 1970. Ore Deposits. In IMA-IAGOD Meeting Guidebook 9, Excursion B5, Tokyo, 14-25.

Takeuchi, T., Takahashi, I. and Abe, H., 1966. Wall-rock alteration and genesis of sulphur and iron sulphide deposits in nortern Japan. Science Rept. Tohoku Univ., Sendai, Ser. III, v.IX, No.3, 381-483.

Tankut, A., 1980. The Orhaneli massif, Turkey. In A. Panayiotou, ed., Ophiolites. Cyprus Geol. Surv. Dept., Nicosia, 702-713.

Tarasov, A.V., 1968. Structural control of copper and nickel mineralization in Noril'sk I deposit. Int. Geol. Rev., 12, 933-941.

Tarasov, A.V., 1974. Struktura yugo-zapadnoi chasti Talnakhskovo medno-nikelevovo mestorozhdeniya. Geol. Rud. Mestor., 2, 1974, 29-41.

Tarney, J., Dalziel, I.W.D. and deWit, M.J., 1976. Marginal basin "Rocas Verdes" complex from S. Chile: A model for Archean greenstone belt formation. In B.F. Windley, ed., The Early History

of the Earth. Wiley, New York, 131-146.
Taylor, G.R., 1976. Residual volcanic emanations from the British Solomon Islands. In R.W. Johnson, ed., Volcanism in Australasia. Elsevier, Amsterdam, 343-354.
Taylor, G.R., 1977. The ophiolite terrain and volcanogenic mineralisation of the Florida Islands, Solomon Islands. Ph.D. Thesis, Univ. of New England, Armidale, 173 pp.
Taylor, G.R. and Hughes, G.W., 1975. Biogenesis of the Rennell Bauxite. Econ. Geol., 70, 542-546.
Taulor, H.P., Jr., 1967. The zoned ultramafic complexes of southeastern Alaska. In P.J. Wyllie, ed., Ultramafic and Related Rocks. Wylie, New York, 97-121 pp.
Taylor, H.P., Jr. and Noble, J.A., 1960. Origin of the ultramafic complexes in southeastern Alaska. 21st Int. Geol. Congr., Norden, Pt. 13, 175-187.
Taylor, H.P. and Noble, J.A., 1969. Origin of magnetite in the zoned ultramafic complexes of southeastern Alaska. Econ. Geol. Spec. Monogr. IV, 209-231.
Taylor, P.S. and Stoiber, R.E., 1973. Soluble material on ash from active Central American volcanoes. Geol. Soc. Amer. Bull., 84, 1031-1042.
Taylor, R.G., 1979. Geology of Tin Deposits. Devel. in Econ. Geol. 11, Elsevier, Amsterdam, 543 pp.
Taylor, R.G. and Steveson, B.G., 1972. An analysis of metal distribution and zoning in the Herberton Tinfield, North Queensland, Australia. Econ. Geol., 67, 1234-1240.
Taylor, S.R., 1964. Abundance of chemical elements in the continental crust. Geoch. and Cosmoch. Acta, 28, 1280-1281.
Tazieff, H., Varet, J., Barberi, F. and Giglia, G., 1972. Tectonic significance of the Afar (or Danakil) Depression. Nature, 235, 144-147.
Tempelman-Kluit, D.J., 1972. Geology and origin of the Faro, Vangorda and Swim concordant zinc-lead deposits, central Yukon Territory. Geol. Surv. Canada Bulletin 208, 73 pp.
Tempelman-Kluit, D., 1981. Geology and mineral deposits of southern Yukon. In Yukon Geol. and Explor., 1979-80, Ottawa, 7-31.
Tempelman-Kluit, D.J., Gordey, S.P. and Read, B.C., 1976. Stratigraphic and structural studies in the Pelly Mountains, Yukon Territory. Geol. Surv. Canada Paper 76-1A, 97-106.
Temple, A.K. and Grogan, R.M., 1965. Carbonatite and related alkalic rocks at Powderhorn, Colorado. Econ. Geol., 60, 672-692.
Teuscher, E.O., 1982. Thorium concentrations in the Bodenmais, Bavaria, sulphide deposit. In G.C. Amstutz, ed., Ore Genesis, The State of the Art. Springer Verlag, Berlin, 460-468.
Thacker, J.L. and Anderson, K.H., 1977. The geologic setting of the Southeast Missouri lead district-regional geologic history, structure and stratigraphy. Econ. Geol., 72, 339-348.
Thayer, T.P., 1942. Chrome resources of Cuba. U.S. Geol. Surv. Bull. 935-A, 74 pp.
Thayer, T.P., 1964. Principal features and origin of podiform chro-

mite deposits, and some observations on the Guleman-Soridağ district, Turkey. Econ. Geol., 59, 1497-1524.

Thayer, T.P.,1968. Gravity differentiation and magmatic reemplacement of podiform chromite deposits. Econ. Geol. Spec. Monogr. IV, 132-147.

Thiry, H.B., Lenoble, J.-P. and Rogel, P., 1977. French exploration seeks to define mineable nodule tonnages on Pacific floor. Engin. and Min. Journ., July 1977, 86-87.

Thomassen, B., Clemmensen, L.B. and Schonwandt, H.K., 1982. Stratabound copper-lead-zinc mineralization in the Permo-Triassic of Central-East Greenland. Grønl. Geol. Unders. Bull. 143, 1-42.

Thompson, R.I. and Panteleyev, A., 1976. Stratabound mineral deposits of the Canadian Cordillera. In K.H. Wolf, ed., Stratiform and Strata-Bound Ore Deposits, v.5, Elsevier, Amsterdam, 37-108.

Thorarinsson, S., 1967. Hekla and Katla; the share of acid and intermediate lava and tephra in the volcanic products through the geological history of Iceland. In S. Bjørnsson, ed., Iceland and Mid-Ocean Ridges. Soc. Sci. Íslandica, 38, 190-199.

Thorpe, R., 1969. Metallogenesis and exploration in the Northwest Territories. Canad. Min. Journ., April 1969, 78-102.

Thorpe, R.I. and Thomas, G.M., 1976. Gold content of greywacke and slate of the Goldenville Formation, Nova Scotia, as determined by neutron activation analysis. Geol. Surv. Canada, Paper 76-1A, 319-326.

Thorpe, R.S., Francis, P.W., Hammill, M. and Baker, M.C.W., 1982. The Andes. In R.S. Thorpe, ed., Andesites. Wiley, Chichester, 187-206.

Thurlow, J.G., Swanson, E.A. and Strong, D.F., 1975. Geology and lithogeochemistry of the Buchans polymetallic sulfide deposits, Newfoundland. Econ. Geol., 70, 130-144.

Till, R., 1978. Arid shorelines and evaporites. In H.G. Reading, ed., Sedimentary Environments and Facies. Elsevier, New York, 178-206.

Tipper, H.W. and Richards, T.A., 1976. Jurassic stratigraphy and history of north-central British Columbia. Geol. Surv. Canada, Bull. 270, 73 pp.

Tischendorf, G., Schust, F. and Lange, H., 1978. Relation between granites and tin deposits in the Erzgebirge, GDR. In Metallization Associared with Acid Magmatism, v.3, 123-137.

Tissot, B.P. and Welte, D.H., 1978. Petroleum Formation and Occurrence. Springer-Verlag, Berlin, 538 pp.

Titley, S.R., 1975. Geological characteristics and environment of some porphyry copper occurrences in the southwestern Pacific. Econ. Geol., 70, 499-514.

Titley, S.R. and Beane, R.E., 1981. Porphyry copper deposits. Econ. Geol. 75th Anniv. Volume, 214-269.

TKE (Tomskaya Kompleksnaya Ekspeditsiya), 1964. Zapadno-Sibirskii zhelezorudnyi bassein. Sibir. A.N.-SSSR, Novosibirsk, 350 pp.

Tolbert, G.E., 1966. The uraniferous zirconium deposits of the Poços de Caldas Plateau, Brazil. U.S. Geol. Surv. Bull. 1188-C, 28 pp.

Tómasson, J. and Kristmannsdottir, H., 1972. High-temperature alteration minerals and thermal brines, Reykjanes, Iceland. Contrib. Miner. Petrol. 36, 123-134.
Tomkins, R.V., 1954. Natural sodium sulphate in Saskatchewan. Sask. Dept. Min. Res. Rept. No.6, 71 pp.
Tomson, I.N., Polyakova, O.P. and Konstantinov, R.M., 1970. Mesozoic stages of ore formation in the Transbaikalian metallogenic provinces and sources of mineralization within various ore-bearing belts. In Z. Pouba and M. Štemprok, eds., Peoblems of Hydrothermal Ore Deposition, 82-85.
Tooms, J.S., 1976. Ethiopian mineral deposits and their relationship to the opening of the Red Sea. Geol. Assoc. Canada Spec. Paper 14, 185-186.
Tourtelot, E.B., 1970. Selected annotated bibliography of minor element content of marine black shales and related sedimentary rocks, 1930-1965. U.S. Geol. Surv. Bull. 1293, 118 pp.
Trask, P.D. and Rodriguez, J.C., Jr., 1946. Manganese deposits of Mexico. U.S. Geol. Surv. Bull. 954-F, 209-315.
Trendall, A.F., 1975. Geology of Western Australian iron ore. In C.L. Knight, ed., Economic Geology of Australia and Papua New Guinea, v.1. Austr. Inst. Min. Metall., 883-891.
Trescases, J.J. and Guillon, J.H., 1977. Notice explicative sur la feuille Yaté. Territ. Nouv. Calédonie B.R.G.M., Carte Géol. a l'échelle du 1:50,000. Nouméa, 35 pp.
Trescases, J.J., Melfi, A.J. and de Oliveira, S.M.B., 1981. Nickeliferous laterites of Brazil. In Lateritization Processes. Balkena, Rotterdam, 170-184.
Trofimov, L.A., 1973. Behaviour of uranium and thorium during deposition of the Boshchekul' ores. Intern. Geol. Rev., 15, 183-186.
Trudinger, P.A., 1981. Origin of sulphide in sediments. Baas Becking Geobio. Res. Lab. Symposium, Canberra, May 1981 (abs.), p. 16.
Trümpy, R., 1960. Paleotectonic evolution of the central and western Alps. Bull. Geol. Soc. Amer., 71, 843-908.
Trümpy, R., 1982. Earth history as moral science. U.S. Geol. Surv. Prof. Paper 1193, 343-345.
Trushkov, Yu.N., 1971. Usloviya formirovaniya i zakonomernosti raspredeleniya rossypei v mesozoidakh Yakutii. Nauka, Moscow, 262 pp.
Tsusue, A. and Ishihara, S., 1975. "Residual" iron sand deposits of southwest Japan. Econ. Geol., 70, 706-716.
Tsypukov, Yu.P. and Vladykin, N.V., 1976. Samorodnoe zoloto v granitoidakh Mongolii. Yezhegodnik Inst. Geokh. Sibir. Otd. A.N. SSSR za 1975, Irkutsk, 42-45.
Tuach, J., 1974. The Rambler mines. In Strong, D.F., ed., Plate Tectonic Setting of Newfoundland Mineral Deposits. Guidebook, NATO Adv. Stud. Inst., 116-126.
Tuach, J. and Kennedy, M.J., 1978. The geologic setting of the Ming and other sulfide deposits, Consolidated Rambler Mines, Northeast Newfoundland. Econ. Geol., 73, 192-206.
Tufar, W., 1982. A new type of sulphosalt mineralization in the

Myrthengraben gypsum deposit, Semmering, Lower Austria. In G.C. Amstutz, ed., Ore Genesis, The State of the Art. Springer, Berlin, 131-140.
Tull, J.F. and Stow, S.H., 1982. Geologic setting of the Hillabee metavolcanic complex and associated strata-bound sulfide deposits in the Appalachian Piedmont of Alabama. Econ. Geol., 77, 312-321.
Turekian, K.K., 1977. The fate of metals in the oceans. Geoch. et Cosmoch. Acta, 41, 1139-1144.
Turneaure, F.S., 1960. A comparative study of major ore deposits of Central Bolivia. Econ. Geol., 55, 217-154 and 574-606.
Turneaure, F.S. and Welker, K.K., 1947. The ore deposits of the eastern Andes of Bolivia, the Cordillera Real. Econ. Geol., 42, 595-625.
Turner, P., 1980. Continental Red Beds. Devel. in Sedim. 29, Elsevier, Amsterdam, 562 pp.
Turner, F.J., 1980. Metamorphic Petrology, 2nd ed. Hemisphere Publ., Washington, D.C., 524 pp.
Turner, P., Vaughan, D.J. and Whitehouse, K.I., 1978. Dolomitization and the mineralization of the Marl Slate (N.E. England). Miner. Deposita, 13, 245-258.
Tuttle, O.F. and Gittins, J., eds., 1966. Carbonatites. Interscience, New York, 591 pp.
Tweto, O., 1968. Leadville District, Colorado. In J.D. Ridge, ed., Ore Deposits of the United States 1933-1967. A.I.M.E., New York, 681-705.
Uchida, E. and Toshimichi Iiyama, J., 1982. Physicochemical study of skarn formation in the Shinyama iron-copper ore deposit of the Kamaishi Mine, Northeastern Japan. Econ. Geol., 77, 809-822.
UNESCO, 1981. An International Symposium on Metallogeny of Mafic and Ultramafic Complexes. Proc., v.1-3, Athens, 432, 436 and 242 pp.
Upadhyay, H.D. and Smitheringale, W.G., 1972. Geology of the Gullbridge copper deposit, Newfoundland: volcanogenic sulfides in cordierite-anthophyllite rocks. Canad. J. Earth Sc., 9, 1061-1073.
Upadhyay, H.D. and Strong, D.F., 1973. Geological setting of the Betts Cove copper deposits, Newfoundland: an example of ophiolite sulfide mineralization. Econ. Geol., 68, 161-167.
Upton, B.G.J., 1967. Alkaline pyroxenites. In P.J. Wyllie, ed., Ultramafic and Related Rocks. Wiley, New York, 281-288.
Upton, B.G.J., 1974. The alkaline province of south-west Greenland. In H. Sørensen, ed., The Alkaline Rocks. Wiley, London, 221-238.
Urabe, T. and Sato, T., 1978. Kuroko deposits of the Kosaka mine, Northeast Honshu, Japan; products of submarine hot springs on Miocene Sea floor. Econ. Geol., 73, 161-179.
Urquhart, G., 1967. The Rex Hill Mine. Tasmania Dept. of Min., Geol. Surv. Rept. No. 9, 33 pp.
Urvantsev, N.N. et al., 1975. Medno-nikelevye rudy severozapada Sibirskoi Platformy. Nauch. Issl. Inst. Geol. Arktiki, Leningrad, 138 pp.
Vaasjoki, O., 1947. Microstructure of titaniferous iron ore of Otanmaki. Comm. Geol. Finl. Bull. 40, 107-114.

Vadász, E., 1960. Magyarország Földtana. Akademiai, Budapest, 646 pp.
Vakhromeev, I.S., 1969. Secondary volcanic formations in ores of pyrite mineralization in the southern Urals. Doklady A.N. SSSR, 1969, v. 187, 35-38.
Vakhromeev, S.A., 1961. Mestorozhdeniya Poleznykh Iskopaemykh, ikh Klassifikatsiya i Usloviya Obrazovaniya. Gosgeoltekhizdat', Moscow, 436 pp
Vakrushev, V.A. and Vladykin, N.V., 1979. Apatite-titanomagnetite ores of the carbonatite complex of Southern Mongolia. Intern. Geol. Rev., 22, 395-398.
Valée, M. and Dubuc, F., 1970. The St-Honoré carbonatite complex, Quebec. Canad. Inst. Min. Metall. Transact., 73, 346-356.
Valeton, I., 1972. Bauxites. Devel. in Soil Sci., 1, Elsevier, New York, 226 pp.
Valsardieu, C.A., Cocquio, D.S. and Bauchau, C., 1980. Uranium-molybdenum mineralization at Ben Lomond, Hervey Range, North Queensland, Australia. Proc. Australas. Inst. Min. Metall., No. 273, 27-35.
Van Alstine, R.E. and Schruben, P.G., 1980. Fluorspar resources of Africa. U.S. Geol. Survey Bull. 1487, 25 pp.
Van Bemmelen, R.W., 1949. The Geology of Indonesia. 2 vols., Govt. Printing Office, The Hague.
Van de Poll, H.W., 1978. Paleoclimatic control and stratigraphic limits of synsedimentary mineral occurrences in Mississippian-early Pennsylvanian strata of eastern Canada. Econ. Geol., 73, 1069-1081.
Van Eden, J.G., 1978. Stratiform copper and zinc mineralization in the Cretaceous of Angola. Econ. Geol., 73, 1154-1161.
Van Houten, F.B., 1973. Origin of red beds. Ann. Rev. Earth and Planet. Sc., v.1, 39-61.
Van Houten, F.B., 1974. Northern Alpine molasse and similar Cenozoic sequences of southern Europe. Soc. Econ. Paleont., Miner., Tulsa, Spec. Publ. 19, 260-273.
Van Houten, F.B. and Travis, R.B., 1968. Cenozoic deposits, Upper Magdaleña Valley, Colombia. Amer. Assoc. Petr. Geol. Bull., 52, 675-702.
Van Wambeke, L., 1977. The Karonge rare earth deposits, Republic of Burundi: new mineralogical-geochemical data and origin of the mineralization. Miner. Deposita, 12, 373-380.
Varentsov, I.M. and Rakhmanov, V.P., 1974. Deposits of manganese. In V.I. Smirnov, ed., Ore Deposits of the U.S.S.R., v.1. Engl. transl., Pitman, London, 114-178.
Varnes, D.J., 1963. Geology and ore deposits of the South Silverton mining area, San Juan County, Colorado. U.S. Geol. Surv. Prof. Paper 378-A, 56 pp.
Vartiainen, H. and Paarma, H., 1979. Geological characteristics of the Sokli carbonatite complex, Finland. Econ. Geol., 74, 1296-1306.
Veits, B.I., 1972. Mineralogiya glavneishikh mestorozhdenii i rudoproyavlenii Tekeliiskoi zony Dzhungarskovo Alatau. Nauka, Alma-Ata, 135 pp.

Velasco, J.R., 1966. Geology of the Cananea district. In S.R. Titley and C.L. Hicks, eds., Geology of the Porphyry Copper Deposits of Southwestern North America. Univ. of Arizona Press, 245-249.
Velikoslavinsky, D.A., Sokolov, Yu.M. and Glebovitsky, V.A., 1968. Zonation of progressive regional metamorphism and metallogenic metamorphic zones. 23th Int. Geol. Congr., Prague, v.4, 157-169.
Vennum, W.R., 1980. Evaporite encrustations and sulphide oxidation products from the southern Antarctic Peninsula. N. Zealand Journ. Geol. Geoph., 23, 499-505.
Verwoerd, W.J., 1967. The carbonatites of South Africa and South West Africa. S. Africa Geol. Surv. Handbook 6, 452 pp.
Verwoerd, P.J. and Cleghorn, J.H., 1975. Kanmantoo copper orebody. In C.L. Knight, ed., Economic Geology of Australia and Papua New Guinea, v.1. Austr. Inst. Min. Metall., 560-566.
Vickers, R.C., 1956. Airborne and ground reconnaisance of part of the syenite complex near Wausau, Wisconsin. U.S. Geol. Surv. Bull. 1042-B, 25-44.
Vikre, P.G., 1981. Silver mineralization in the Rochester District, Pershing County, Nevada. Econ. Geol., 76, 580-609.
Vilenskii, A.M., 1967. Petrologiya intruzivnykh trappov severa Sibirskoi Platformy. Nauka, Moscow, 272 pp.
Vine, J.D., 1962. Geology of uranium in coaly carbonaceous rocks. U.S. Geol. Surv. Prof. Paper 356-D, 113-167.
Vine, J.D. and Tourtelot, E.B., 1970. Geochemistry of black shale deposits-a summary report. Econ. Geol., 65, 253-272.
Vinogradov, A.P., 1977. Vvedeniye v Geokhimiyu Okeana. Nauka, Moscow.
Vinogradov, A.P., ed., 1968. Atlas litologicheskikh-paleogeograficheskikh kart SSSR, v. 1. G.U.G.K., Moscow, 52 maps.
Vlasov, K.A., Kuzmenko, M.Z. and Yeskova, Ye.M., 1959. The Lovozero Alkali Massif. Engl. Transl., Oliver and Boyd, Edinburgh, 627 pp.
Vokes, F.M., ed., 1960. Mines in south and central Norway. Norges Geol. Unders. no. 212 m, 73 pp.
Vokes, F.M., 1963. Geological studies on the Caledonian pyritic zinc-lead orebody at Bleikvassli, Nordland, Norway. Norges Geol. Unders., No. 222, 12-84.
Vokes, F.M., 1973. Metallogeny possibly related to continental break-up in Southwest Scandinavia. In D.H. Tarling and S.K. Runcorn, eds., Interpretation of Continental Drift to the Earth Sciences. Acad. Press., 573-579.
Vokes, F.M., 1976. Caledonian massive sulphide deposits in Scandinavia-a comparative review. In K.H. Wolf, ed., Handbook of Strataform and Strata-Bound Ore Deposits, v.6, Elsevier, 79-127.
Vokes, F.M., 1978. Introduction. In S.H.U. Bowie, A. Kvalheim and H.W. Haslam, eds., Mineral Deposits of Europe, v. 1. Inst. Min. Metall., London, 1-38.
von Backström, J.W., 1974. Other uranium deposits. In Formation of Uranium Ore Deposits. Int. Atom. Energy Agcy, Vienna, 605-624.
von Backström, J.W., 1976. Thorium. In Mineral Resources of the Republic of South Africa, 5th ed. S. Afr. Geol. Surv., Pretoria.
von Behrend, F., 1950. Die Blei- and Zinkerz führende imprägnations

Lagerstätten im Bundsandstein am Nordrand der Eifel und ihre Enstehung. 18th Int. Geol. Congr., London, Pt.7, 325-341.
von Bubnoff, S., 1956. Einführung in die Erdgeschichte. Akademie-Verlag, Berlin, 3rd ed., 808 pp.
von Eckermann, H., 1948. The alkaline district of Alnö Island. Sver. Geol. Unders., Ser. Ca, 36, 176 pp.
von Eckermann, H.V., 1966. Progress of research on the Alnő carbonatite. In O.F. Tuttle and J. Gittins, eds., Carbonatites. Interscience, New York, 3-32.
von Hiene, R., 1974. Modern trench sediments. In C.A. Burk and C.L. Drake, eds., The Geology of Continental Margins. Springer Verlag, New York, 207-211.
Wadsworth, W.B., 1968. The Cornelia Pluton, Ajo, Arizona. Econ. Geol., 63, 101-115.
Wagner, H., 1977. Rohstoffwirtschaftliche Länderberichte, XIII, Rumänien. Bundesanst. f. Geowiss. u. Rohst., Hannover, 176 pp.
Wagner, H. and Berthold, G., 1979. Rohstoffwirtschaftliche Länderberichte, XXIII- Comecon. Bundesanst. f. Geowiss., Hannover, 253 pp.
Wagner, P.A., 1914. The Diamond Fields of Southern Africa. Van Struik, Cape Town, 355 pp.
Walde, D.H.G., Gierth, E. and Leonardos O.H., 1981. Stratigraphy and mineralogy of the manganese ores of Urucúm, Mato Grosso, Brazil. Geol. Rundsch., 70, 3, 1077-1085.
Walker, F. and Poldervaart, A., 1949. Karoo dolerites of the Union of South Africa. Bull. Geol. Soc. Amer., 60, 591-706.
Walker, G.P.L., 1960. Zeolite zones and dike distribution in relation to the structure of the basalts of Eastern Iceland. Journ. Geol., 68, 515-528.
Walker, G.P.L., 1963. The Breiddalur central volcano, Eastern Iceland. Quart. Journ. Geol. Soc. London, v. 119, Pt.1, 29-63.
Walker, T.T., 1975. Red beds in the western interior of the United States. U.S. Geol. Surv. Prof. Paper 853, 49-56.
Wallace, S.R., Muncaster, N.K., Jonson, D.C., Mackenzie, W.B., Bookstrom, A.A. and Surface, V.E., 1968. Multiple intrusion and mineralization at Climax, Colorado. In J.D. Ridge, ed., Ore Deposits of the United States 1933-1967. A.I.M.E., New York, 605-640.
Wallace, S.R., MacKenzie, W.B., Blair, R.G. and Muncaster, N.K., 1978. Geology of the Urad and Henderson molybdenite deposits, Clear Creek County, Colorado, with a section on a comparison of these deposits with those at Climax, Colorado. Econ. Geol., 73, 325-368.
Walshe, J.L. and Solomon, M., 1981. An investigation into the environment of formation of the volcanic-hosted Mt Lyell copper deposits using geology, mineralogy, stable isotopes and a six-component chlorite solid solution model. Econ. Geol., 76, 246-284.
Waltham, A.C., 1968. Classification and genesis of some massive sulfide deposits in Norway. Transact. Inst. Min. Metall., London, Sect. B., v. 77, B 153-B 161.
Wanless, H.R., 1975. Distribution of Pennsylvanian coal in the United States. U.S. Geol. Survey Prof. Paper 853, 33-47.

Ward, P.L., 1971. New interpretation of the geology of Iceland. Geol. Soc. Amer. Bull., 82, 2991-3012.
Warden, A.J., 1970. Genesis of the Forari manganese deposit, New Hebrides. Transact., Inst. Min. Metall., London, v.79, B30-B41.
Warden, A.J. and Mitchell, A.H.G., 1974. New Hebrides. In A.M. Spencer, ed., Mesozoic-Cenozoic Orogenic Belts. Geol. Soc. London, 433-443.
Waring, G.A., Blankenship, R. and Bentall, R., 1965. Thermal springs of the United States and other countries of the world- a summary. U.S. Geol. Surv. Profes. Paper 492, 383 pp.
Warner, L.A., Holser, W.T., Wilmarth, V.R. and Cameron, E.N., 1959. Occurrence of nonpegmatite berrylium in the United States. U.S. Geol. Surv. Prof. Paper 318, 198 pp.
Watkinson, D.H., Mainwaring, P.R. and Pertold, Z., 1978. The Cu-Zn Obrázek ore deposit, Czechoslovakia: a volcanogenic deposit included in the Ransko Intrusive Complex. Miner. Deposita, 13, 151-163.
Watson, K.D., 1956. Mindamar Mine. In Structural Geology of Canadian Ore Deposits, v. 2, 495-502.
Watts, A.D., Weissel, J.K. and Larson, L.R., 1977. Sea-floor spreading in marginal basins of the Western Pacific. Tectonophysics, 37, 167-181.
Weaver, J.D., 1958. Utuado Pluton, Puerto Rico. Geol. Soc. Amer. Bull. 69, 1125-1142.
Weber, C.R., 1974. Woolomin-Texas Block plutonic rocks and intruded sediments. In N.L. Markham and H. Basden, eds., The Mineral Deposits of New South Wales. Geol. Surv. N.S.W., Sydney, 350-391.
Weber, C.R., 1979. The Nundle gold field, New South Wales. Quart. Notes Geol. Surv. N.S.W., 1st Oct. 1979, 2-14.
Webster, S.S. and Skey, E.H., 1979. Geophysical and geochemical case study of the Que River deposits, Tasmania, Australia. Canada Geol. Surv., Econ. Geol. Rept. 31, 697-720.
Wedepohl, K.H., 1964. Untersuchungen am Kupferschiefer in Nordwestdeutschland; ein Beitrag zur Deutung der genese bituminöser Sedimente. Geoch. et Cosmoch. Acta, 28, 305-364.
Wedepohl, K.H., 1971. "Kupferschiefer" as a prototype of syngenetic sedimentary ore deposits. Soc. Min. Geol. Japan, Spec. Issue 3, 268-273.
Wedow, H., Jr., 1967. The Morro do Ferro thorium and rare-earth ore deposit, Poços de Caldas district, Brazil. U.S. Geol. Surv. Bull. 1185-D, 34 pp.
Wedow, H., Jr., Kiilsgaard, T.H., Heyl, A.V. and Halle, R.B., 1973. Zinc. U.S. Geol. Surv. Profess. Paper 820, 697-711.
Wedow, H., Jr. and Hobbs, R.G., 1968. Zirconium. U.S. Geol. Surv. Prof. Pap. 580, 361-364.
Weeks, A.D. and Eargle, D.H., 1960. Uranium at Palangana Salt Dome, Duval County, Texas. U.S. Geol. Surv. Prof. Pap. 400-B, 48-52.
Wegmann, C.E., 1935. Zur Deutung der Migmatite. Geol. Rundschau, 26, 305-350.
Weissberg, B.G., 1969. Gold-silver ore-grade precipitates from New

Zealand thermal waters. Econ. Geol. 64, 95-108.
Welch, B.K., Sofoulis, J. and Fitzgerald, A.C.F., 1975. Mineral sand deposits of the Capel area, W.A. In C.L. Knight, ed., Economic Geology of Australia and Papua New Guinea, v.l. Austr. Inst. Min. Metall., Monogr. 5, 1070-1087.
Wells, J.D., 1960. Petrography of radioactive Tertiary igneous rocks, Front Ranges Mineral Belt, Colorado. U.S. Geol. Surv. Bulletin 1032-E, 223-272.
Wells, J.D., Stoiser, L.R. and Elliott, J.E., 1969. Geology and geochemistry of the Cortez gold deposit, Nevada. Econ. Geol. 64, 526-537.
Welsh, T.C., 1975. Cadia copper-gold deposits. In C.L. Knight, ed., Economic Geology of Australia and Papua New Guinea, v.l. Austr. Inst. Min. Metall., Monogr. 5, 711-715.
Wernecke, L., 1932. Alaska Juneau-geology of the ore zones. Engin. Min. Journ., v. 133, 493-499.
Westra, G. and Keith, S.B., 1981. Classification and genesis of stockwork molybdenum deposits. Econ. Geol. 76, 844-873.
Whalen, J.B. and Britten, R.M., 1982. Geochronology and geochemistry of the Frieda River Prospect area, Papua New Guinea. Econ. Geol., 77, 592-616.
Wheatley, C.J.V., 1971. Aspects of metallogenesis within the southern Caledonides of Great Britain and Ireland. Transact. Inst. Min. Metall., London, Sect. B., v. 80, 211-223.
White, D.E., 1967. Mercury and base-metal deposits with associatd thermal and mineral waters. In H.L. Barnes, ed., Geochemistry of Hydrothermal Ore Deposits. Holt, Rinehart and Winston, New York, 576-631.
White, D.E., 1968. Hydrology, activity and heat flow of the Steamboat Springs thermal system, Washoe County, Nevada. U.S. Geol. Surv. Prof. Paper 458-C, 109 pp.
White, D.E., 1981. Active geothermal systems and hydrothermal ore deposits. Econ. Geol. 75th Anniv. Volume, 392-423.
White, D.E. and J. Gonzaler R., 1946. San José antimony mines near Wadley, State of San Luis Potosi, Mexico. U.S. Geol. Surv. Bull. 946-E, 131-153.
White, D.E. and Roberson, C.E., 1962. Sulphur Bank, California, a major hot-spring quicksilver deposit. In Petrologic Studies (Buddington Volume), Geol. Soc. Amer., 397-428.
White, D.E. and Waring, G.A., 1963. Data of Geochemistry, 6th ed., Chapter K-Volcanic Emanations. U.S. Geol. Surv. Prof. Paper 440-K, 29 pp.
White, D. and Thiessen, R., 1913. The Origin of Coal. Bull. U.S. Bureau of Mines 38, 390 pp.
White, W.H., Bookstrom, A.A., Kamilli, R.J., Ganster, M.W., Smith, R.P., Ranta, D.E. and Steininger, R.C., 1981. Character and origin of Climax-type molybdenum deposits. Econ. Geol. 75th Anniv. Vol., 270-316.
White, W.C. and Warin O.N., 1964. A survey of phosphate deposits in the South-West Pacific ans Australian waters. Bur. Min. Res.,

Geol., Geophys., Canberra, Bulletin 69, 173 pp.
White, W.S., 1968. The native copper deposits of Northern Michigan. In J.D. Ridge, ed., Ore Deposits of the United States 1933-1967, A.I.M.E., New York, 303-325.
Whitebread, D.H., 1976. Alteration and Geochemistry of Tertiary volcanic rocks in parts of the Virginia City Quadrangle, Nevada. U.S. Geol. Surv. Profess. Paper 936, 43 pp.
Whitehead, W.L., 1919. The veins of Chañarcillo, Chile. Econ. Geol., 14, 1-45.
Whiteman, A.J., 1971. The Geology of the Sudan Republic. Clarendon Press, Oxford, 290 pp.
Wilkins, J.D., 1971. The Benson Lake Mine-Operating Practice. Canad. Inst. Min. Metall. Bull., April 1971, 71-77.
Wilkinson, J.F.G., 1974. The mineralogy and petrography of alkali basaltic rocks. In H. Sørensen, ed., The Alkaline Rocks. Wiley, London, 67-95.
Wilkinson, J.F.G., 1981. Continental and mid-ocean ridge tholeiites: some similarities and contrasts. In Geol. Soc. India Memoir 3, Bangalore, 340-361.
Wilkinson, W.J., Stevenson, R.W. and Garnett, J.A., 1976. Lorraine. Canad. Inst. Min. Metall. Spec. Vol. 15, Montreal, 397-401.
Williams, E., 1978. Tasman fold belt system in Tasmania. Tectonophysics, 48, 159-205.
Williams, F.A., et al., 1956. Economic geology of the decomposed columbium-bearing granites, Jos Plateau, Nigeria. Econ. Geol. 51, 303-332.
Williams, G.J., 1965. Economic Geology of New Zealand. 8th Commonw. Min. Metall. Congr., 4, 384 pp.
Williams, H. and McBirney, A.R., 1979. Volcanology. Freeman, San Fransicso, 397 pp.
Williams, L., 1964. Titanium deposits in North Carolina. N. Carol. Div. Min. Res., Inf. Circ. 19, 51 pp.
Williams, L.A.J., 1969. Volcanic associations in the Gregory Rift Valley, East Africa. Nature, v.224, 61-64.
Wilson, G.I., 1965. Gold deposits of Junction Reefs, Mandurama. In In J. McAndrew, ed., Geology of Australian Ore Deposits, 420-422.
Wilson, I.F., 1955. Geology and mineral deposits of the Boléo copper district, Baja California, Mexico. U.S. Geol. Surv. Prof. Paper 273, 134 pp.
Wilson, I.F. and Rocha, V.S., 1948. Manganese deposits of the Talamantes district near Parral, Chihuahua, Mexico. U.S. Geol. Surv. Bull. 954-E, 181-208.
Wilson, I.F. and Veytia, M., 1949. Geology and manganese deposits of the Lucifer district, northwest of Santa Rosalia, Baja California, Mexico. U.S. Geol. Surv. Bull. 960-F, 177-233.
Wilson, J.L., 1975. Carbonate Facies in Geologic History. Springer, New York, 471 pp.
Wimmenauer, W., 1966. The eruptive rocks and carbonatites of the Kaiserstuhl, Germany. In O.F. Tuttle and J. Gittins, eds., Carbonatites. Interscience, New York, 183-204.

Wimmenauer, W., 1974. The alkaline province of central Europe and France. In H. Sørensen, ed., The Alkaline Rocks. Wiley, London, 238-271.
Windley, B.F., 1977. The evolving continents. Wiley, London, 385 pp.
Winkler, H.G.F., 1974. Petrogenesis of Metamorphic Rocks. Springer-Verlag, New York, 320 pp.
Winterhalter, B., 1966. Iron-manganese concretions from the Gulf of Bothnia and the Gulf of Finland. Geotekn. Julk., 69, 1-77.
Winters, E., 1938. Ferromanganiferous concretions from some podzolic soils. Soil Sci., 46, 33-40.
Wisser, E., 1966. The epithermal provinces of Northwest Mexico. Nevada Bur. of Mines Rept. 13, Reno, 63-92.
Witkind, I.J., 1973. Igneous rocks and related mineral deposits of the Barker Quadrangle, Little Belt Mountains, Montana. U.S. Geol. Surv. Prof. Paper 752, 58 pp.
Wolf, K.H., 1973. Conceptual models, II. Sediment. Geol. 9, 235-260.
Wolf, K.H., 1976a. Introduction. In K.H. Wolf, ed., Handbook of Stratiform and Strata-Bound Ore Deposits, v.1, Elsevier, 1-9.
Wolf, K.H., 1976b. Ore Genesis Influenced by Compaction. In G.V. Chilingarian and K.H. Wolf, eds., Compaction of Coarse-Grained Sediments, II. Elsevier, Amsterdam, 475-676.
Wolf, K.H., ed., 1976, 1981, 1985. Handbook of Stratiform and Strata-Bound Ore Deposits, vols. 1-7 (1976), 8-10 (1981) and 11-14 (1985). Elsevier, Amsterdam.
Wolf, K.H., Chilingar, G.V. and Beales, F.W., 1967. Elemental composition of carbonate skeletons, minerals and sediments. In G.V. Chilingar, H.J. Bissell and R.W. Fairbridge, eds., Carbonate Rocks, vol.B, Elsevier, Amsterdam, 23-150.
Wolfe, J.A., 1973. Tectonic fingerprint in Philippine porphyry deposits. Soc. Min. Eng. A.I.M.E., Preprint No. 73-S-37, 31 pp.
Wolff, E.N. and Heiner, L.E., 1971. Mineral resources of Southeastern Alaska. Miner.Ind.Res.Lab.Rept. 28, College, 63 pp.
Wolff, F., 1978. Philippinen. Rohstoffwirtschaftliche Länderberichte, vol. XV, Bundesanst. f. Geowiss., Hannover, 190 pp.
Wood, H.B., 1968. Geology and exploitation of uranium deposits in the Lisbon Valley area, Utah. In J.D. Ridge, ed., Ore Deposits of the United States 1933-1967. A.I.M.E., New York, 770-789.
Woodcock, J.R. and Carter, N.C., 1976. Geology and geochemistry of the Alice Arm molybdenum deposits. Canad. Inst. Min. metall. Spec. Vol. 15, 462-475.
Woodland, A.W., 1956. The manganese deposits of Great Britain. 20th Int. geol. Congr. Mexico, Sympos. del Manganeso, 5, 197-218.
Woodring, W.P. and Daviess, S.N., 1944. Geology and manganese deposits of Guisa-Los Negros area, Oriente Province, Cuba. U.S. Geol. Surv. Bull. 935-G, 357-386.
Woodward, L.A., Kaufmann, W.H., Schumacher, O.L. and Talbott, L.W., 1974. Strata-bound copper deposits in Triassic sandstone of Sierra Nacimiento, New Mexico. Econ. Geol. 69, 108-120.
Woolley, A.R. and Garson, M.S., 1970. Petrochemical and tectonic relationship of the Malawi carbonatite-alkaline province and the

Lupata-Lebombo volcanics. In T.N. Clifford and I.G. Gass, eds., African Magmatism and Tectonics. Oliver and Boyd, Edinburgh, 237-262.

Worthington, J.E., 1964. An exploration program for nickel in the southeastern United States. Econ. Geol., 59, 97-109.

Worthington, J.E. and Kiff, I.T., 1970. A suggested volcanigenic origin for certain gold deposits in the Slate Belt of the North Carolina Piedmont. Econ. Geol., 65, 529-537.

Wright, J.B., 1970. Controls of mineralization in the Older and Younger tin fields of Nigeria. Econ. Geol., 65, 945-951.

Wright, W.B., Guild, P.W., Fish, G.E., Jr. and Sweeney, J.W., 1968. Iron and Steel. U.S. Geol. Surv. Profess. Paper 580, 396-416.

Wright, W.S., 1964. Types of lead and zinc deposits in Iran. In CENTO Sympos. on Min., Geol. and the Base Metals, Ankara, 89-100.

Wrucke, C.T. and Armbrustmacher, T.J., 1975. Geochemical and geologic relations of gold and other elements at the Gold Acres open pit mine, Lander County, Nevada. U.S. Geol. Surv. Prof. Paper 860, 27 pp.

Wu, I. and Petersen, U., 1977. Geochemistry of tetrahedrite and mineral zoning at Casapalca, Peru. Econ. Geol. 72, 993-1016.

Wyllie, P.J., ed., 1967. Ultramafic and Related Rocks. Wiley, New York, 464 pp.

Wyllie, P.J., 1981. Plate tectonics and magma genesis. Geol. Rundschau, 70, 1, 128-153.

Wyzykowski, J., 1971. Cechsztynska formacja miedzionosna w Polsce. Przegl. Geol., 19, 117-122.

Yakovlev, G.F., Zarayskii, G.P. and Starostin, V.I., 1965. Subvolcanic bodies of acid composition and copper pyrite mineralization of Blyava district (South Ural). Int. Geol. Rev. 8, 1017-1028.

Yakovlev, Yu.N., Goncharov, Yu.V., Neradovskiy, Yu.N. and Alekseyev, A.I., 1970. Nickel-sulfide ores in ultrabasites of Khabarnyi massif (South Urals). Int. Geol. Rev., 13, 1062-1068.

Yates, R.G., 1942. Quicksilver deposits of the Opalite district, Malheur County, Oregon and Humboldt County, Nevada. U.S. Geol. Surv. Bull. 931-N, 319-348.

Yates, R.G., Kent, D.F. and J. Fernandez C., 1951. Geology of the Huancavelica Quicksilver district, Peru. U.S. Geol. Surv. Bull. 975-A, 45 pp.

Yates, R.G. and Thompson, G.A., 1959. Geology and quicksilver deposits of the Terlingua district, Texas. U. S. Geol. Surv. Profess. Paper 312, 114 pp.

Yazeva, R.G., 1971. Sodic acidic volcanites of the Urals and plagio-rhyolites of recent island arcs. Int. Geol. Rev., 20, 1009-1020.

Yeend, W.E., 1974. Gold-bearing gravel of the ancestral Yuba River, Sierra Nevada, California. U.S. Geol. Surv. Prof. Paper 772, 44 pp.

Yefremov, A.V., 1934. Tsvetnye Metally. In Mineral'nye Resursy Urala. Sverdlovsk, 237-295.

Yegorov, Ye. V. and Timofeeva, M.V., 1972. Effuzivnye zhelezisto-kremnistye formatsii i zhelezorudnye mestorozhdeniya Malovo Khingana. In Geol. i Genezis Dokembr. Zhel.-Kremn. i Marg. Form.

Mira, Nauk. Dumka, Kiev, 188-195.
Yermakov, Yu. N., 1977. K voprosu o vozraste rudonosnoi kory vyvetrivaniya Voronezhskoi anteklizy. In Perspektivy Obnaruzh. Pogreb. Rudn. Mestor. v Tsentr. Rayonakh Russkoi Platformy. Moscow, 53-58.
Yesenov, Sh.E., ed., 1972. Geologiya i metallogeniya Severnovo Pribalkhash'ya. Nauka, Kazakh SSR, Alma Ata, 267 pp.
Yildiz, M. and Bailey, E.H., 1978. Mercury deposits in Turkey. U.S. Geol. Surv. Bulletin 1456, 80 pp.
Young, E.F., 1979. Genesis of the Schwartzwalder uranium deposit, Jefferson County, Colorado. Wyoming Univ. Contrib. Geol., 17, Laramie, 179-186.
Young, E.J. and Segerstrom, K., 1973. A disseminated silver-lead-zinc sulfide occurrences at Hahn's Peak, Routt County, Colorado. U.S. Geol. Surv. Bull., 1367, 33 pp.
Young, E.J. and Sims, P.K., 1961. Petrography and origin of xenotime and monazite concentrations, Central City District, Colorado. U.S. Geol. Surv. Bull. 1032-F, 273-297.
Ypma, P.J.M. and Simons, J.H., 1969. Genetical aspects of the tin mineralization in Durango, Mexico. In A Second Techn. Conf. on Tin, Bangkok, v.1, 179-191.
Yudin, B.A. and Zak, S.I., 1971. Titanium deposits of northwestern U.S.S.R. (eastern part of Baltic Shield). Int. Geol. Rev., 13, 864-872.
Yun, S. and Einaudi, M.T., 1982. Zinc-lead skarns of the Yeonhwa-Ulchin district, South Korea. Econ. Geol. 77, 1013-1032.
Zabolotnaya, N.P., 1974. Deposits of beryllium. In V.I. Smirnov, ed., Ore Deposits of the U.S.S.R., v.3, Engl. transl., Pitman, London, 320-371.
Zachrisson, E., 1971. The structural setting of the Stekenjokk ore bodies, central Swedish Caledonides. Econ. Geol., 66, 641-652.
Žák, L., 1951. Opál s germaniem z Března u Chomůtova. Sborník Ústř. Úst. Geol., Prague, 18, 641-644.
Zambonini, F., 1910. Mineralogia Vesuviana. Atti. R. Accad. Sci. Fis. Math., 14, 359 pp.
Zantop, H., 1978. Geologic setting and genesis of iron oxides and manganese oxides in the San Francisco manganese deposit, Jalisco, Mexico. Econ. Geol., 73, 1137-1149.
Zavaritsky, V.A., 1946. The spilite-keratophyre formation in the region of the Blyava deposit in the Ural Mountains. Intern. Geol. Rev., 2, 1960, 551-576 and 645-686.
Zeegers, H. and Leprun, J.-C., 1979. Evolution des concepts en altérologie tropicale et conséquences potentielles pour la prospection géochimique en Afrique occidentale Soudano-Sahélienne. Bull. du B.R.G.M., Sect. II, 2-3, 229-239.
Zeissinck, H.E., 1969. The mineralogy and geochemistry of a nickeliferous laterite profile (Greenvale, Queensland, Australia). Mineralium Deposita, 4, 132-152.
Zhabin, A.G., 1978. Syngenesis and metamorphism of carbonatites. Proc. of the First Intern. Sympos. on Carbonatites, Brazil, D.N.P.M., 191-195.

Zhabin, A.G., Sharfman, V.C. and Samsonova, N.S., 1974. Rekonstruktsiya obstanovki devonskovo vulkanogenno-osadochnovo sul'fidootlozheniya. Geol. Rud. Mestor., 2, 1974, 60-75.

Zharikov, M.G., 1974. Deposits of antimony. In V.I. Smirnov, ed., Ore Deposits of the U.S.S.R., v.2, Engl. transl., Pitman, London, 283-297.

Ziegler, P.A., 1978. North-western Europe: tectonics and basin development. Geologie en Mijnbouw, 57, 589-626.

Zies, E.G., 1929. The Valley of 10,000 Smokes, the fumarolic incrustations and their bearing on ore deposition. Nat. Geogr. Soc. Tech. Paper 1, 61-79.

Zimmerle, W., 1973. Fossil heavy mineral concentrations. Geol. Rundschau, Bd. 62, 536-548.

Zimmermann, R.A., 1969. Stratabound barite deposits in Nevada. Miner. Deposita, 4, 401-409.

Zitzmann, A., ed., 1976. The Iron Ore Deposits of Europe and adjacent areas. Bundesanst. f. Geowiss., Hannover, 418 pp.

Zitzmann, A. and Neumann-Redlin, C., 1976. The genetic types of iron ore deposits in Europe and adjacent areas. Ibid., 13-35.

Zolotukhin, V.V., 1964. Osnovnye zakonomernosti prototektoniki i voprosy formirovaniya rudonosnykh trappovykh intruzii. Nauka, Moscow, 176 pp.

Zolotukhin, V.V. and Vilenskii, A.M., 1978. Petrologiya i perspektivy rudonosnosti trappov severa Sibirskoi platformy. A.N. SSSR, Sibir. Otd., Novosibirsk, Vyp. 357, 217 pp.

Zonenshain, L.P., Kuz'min, M.I. and Moralev, V.M., 1976. Global'naya Tektonika, Magmatizm i Metallogeniya. Nedra, Moscow, 232 pp.

Zubovic, P., Stadnichenko, T. and Sheffey, N.B., 1964. Distribution of minor elements in coal beds of the eastern Interior region. U.S. Geol. Surv. Bull. 1117-B, 1-41.

Zuffardi, P., ed., 1969. Remobilization of ores and minerals. Convegno Sulla Rimobilizzazioni dei Minerali Metallici e Non Metallici. Univ. of Cagliari, 322 pp.

Zuffardi, P., 1976. Karsts and economic mineral deposits. In K.H. Wolf, ed., Handbook of Stratiform and Strata-Bound Ore Deposits, v.3, Elsevier, Amsterdam, 175-212.

Zuffardi, P., 1977. Ore-mineral deposits related to the Mesozoic ophiolites in Italy. In D.D. Klemm and H.-J. Schneider, eds., Time and Strata-Bound Ore Deposits. Springer, Berlin, 314-323.

"A ready reference book is only as good as is its index".

INDICES

GENERAL INDEX

C

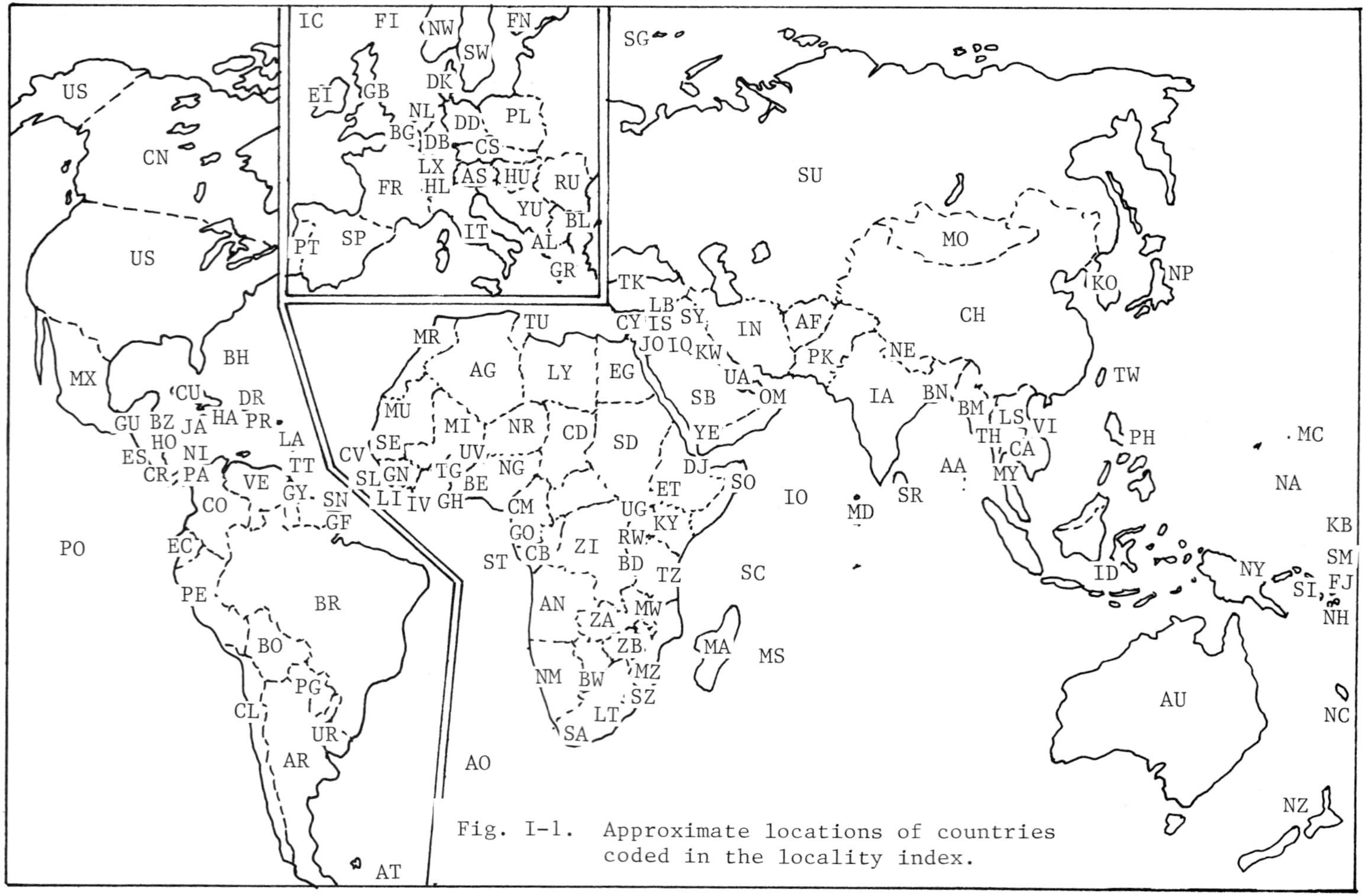

Fig. I-1. Approximate locations of countries coded in the locality index.

LOCALITY INDEX

NOTE: MOST LOCALITIES ARE FOLLOWED BY A COUNTRY:COMMODITY CODE (SEE FIG. I-1 AND TABLE I-1 FOR CODE LOCATIONS AND EXPLANATIONS).

Table I-1. List of country codes used in the locality index.

Code	Country	Code	Country	Code	Country
AA	Andaman-Nicobar Islands	FJ	Fiji	MX	Mexico
AF	Afghanistan	FN	Finland	MZ	Mozambique
AG	Algeria	FR	France	NA	Nauru
AL	Albania	GB	Great Britain	NC	New Caledonia
AN	Angola	GF	French Guyana	NE	Nepal
AO	Atlantic Ocean	GH	Ghana	NG	Nigeria
AR	Argentina	GL	Greenland	NH	New Hebrides
AS	Austria	GN	Guinea	NI	Nicaragua
AT	Antarctica	GO	Gabon	NL	Neatherlands
AU	Australia	GR	Greece	NM	Namibia
BD	Burundi	GU	Guatemala	NP	Japan
BE	Benin	GY	Guyana	NR	Niger
BG	Belgium	HA	Haiti	NW	Norway
BH	Bahamas	HL	Switzerland	NY	New Guinea
BL	Bulgaria	HO	Honduras	NZ	New Zealand
BM	Burma	HU	Hungary	OM	Oman
BN	Bangladesh	IA	India	PA	Panama
BO	Bolivia	IC	Iceland	PE	Peru
BR	Brazil	ID	Indonesia	PG	Paraguay
BW	Botswana	IN	Iran	PH	Philippines
BZ	Belize	IO	Indian Ocean	PK	Pakistan
CA	Cambodia	IQ	Iraq	PL	Poland
CB	Congo-Brazzaville	IS	Israel	PO	Pacific Ocean
CD	Chad	IT	Italy	PR	Puerto Rico
CF	Central African Republic	IV	Ivory Coast	PT	Portugal
		JA	Jamaica	RU	Rumania
		JO	Jordan	RW	Rwanda
CH	China	KB	Kiribati	SA	South Africa
CL	Chile	KO	Korea	SB	Saudi Arabia
CM	Cameroon	KY	Kenya	SC	Seychelles
CN	Canada	KW	Kuwait	SD	Soudan
CO	Colombia	LA	Lesser Antilles	SE	Senegambia
CR	Costa Rica	LB	Lebanon	SG	Spitzbergen
CS	Czechoslovakia	LI	Liberia	SI	Solomon Islands
CB	Cuba	LS	Laos	SL	Sierra Leone
CV	Cape Verde Islands	LT	Lesotho	SM	Samoa
CY	Cyprus	LY	Libya	SN	Suriname
DB	West Germany	LX	Luxembourg	SO	Somalia
DD	East Germany	MA	Madagascar	SP	Spain
DJ	Djibouti	MC	Micronesia	SR	Sri Lanka
DK	Denmark	MD	Maldives	ST	São Tome-Príncipe
DR	Dominican Rep.	MI	Mali	SU	U.S.S.R.
EC	Ecuador	MO	Mongolia	SY	Syria
EG	Egypt	MR	Morocco	SW	Sweden
EI	Ireland	MS	Mauritius	SZ	Swaziland
ES	El Salvador	MU	Mauritania	TG	Togo
ET	Ethiopia	MW	Malawi	TH	Thailand
FI	Farøe Islands	MY	Malaysia	TK	Turkey

TG	Togo	TZ	Tanzania	VE	Venezuela
TH	Thailand	UA	United Arab Emir.	VI	Vietnam
TK	Turkey	UG	Uganda	YE	Yemen
TT	Trinidad-Tobago	UR	Uruguay	YU	Yugoslavia
TU	Tunisia	US	United States	ZA	Zambia
TW	Taiwan	UV	Upper Volta	ZB	Zimbabwe
				ZI	Zaïre

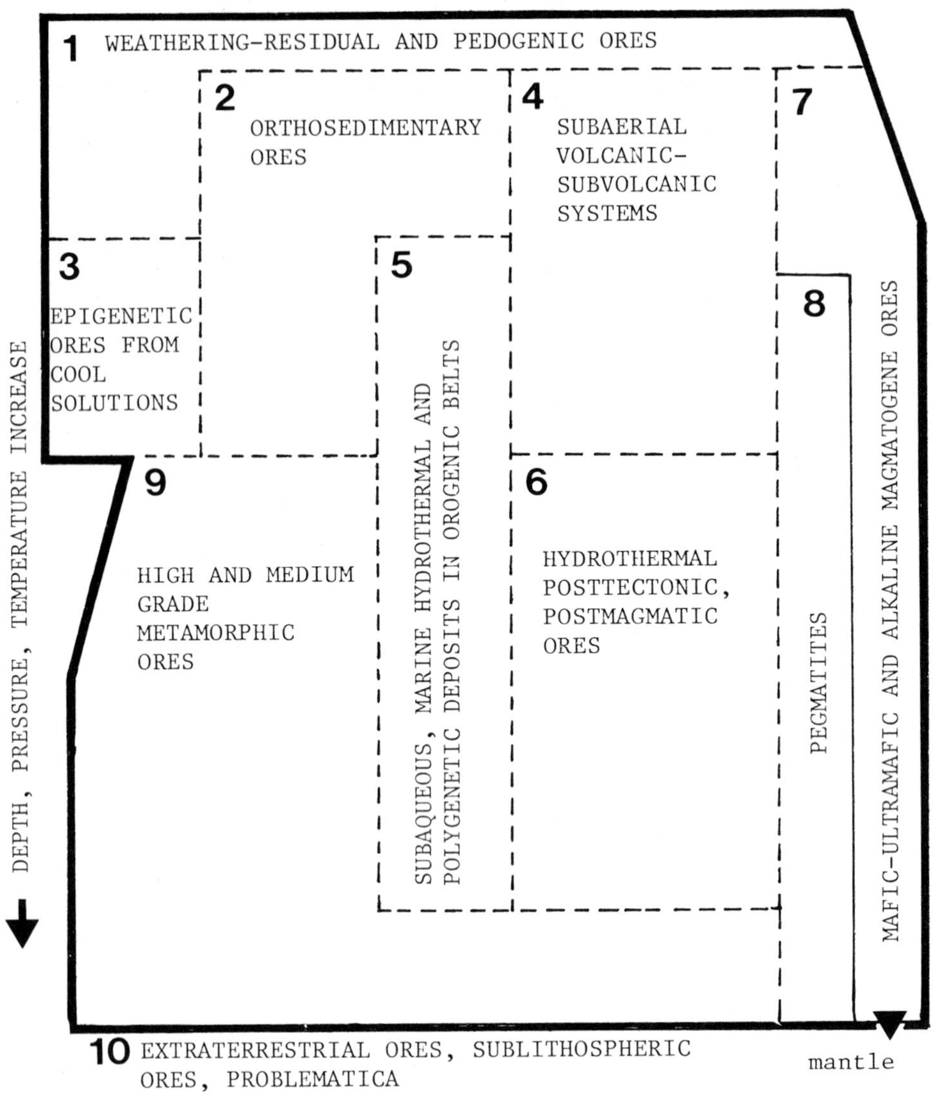

Fig. I-2. Utilitarian organization of metallic deposits (Laznicka, 1984). Most categories are transitional and overlapping.

GENETIC INDICES

A UTILITARIAN ORGANIZATION OF METALLIC DEPOSITS (Laznicka, 1984)
(NOTE: not a classification, because categories overlap and merge)

1. WEATHERING-RESIDUAL AND PEDOGENETIC ORES
(* denotes fossil ores, buried or recently exhumed)

BOREAL, TEMPERATE

-"limonite" in zoned soil profiles, 682
-gold nuggets, 682-3

HUMID TROPICS

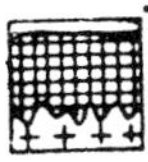
1A: tropical weathering profiles over "rocks"
-general, 199-205
-bauxite (Al-laterite), 39, 74-5, 377, 715-8*, 1488, 1432
-high Fe, Ti laterites, 38-9
-Sn,Au,Pt,U, 712-3, 1488

1B: enriched zones over metallic "protores"
-Fe oxides, 703-5
-Mn oxides, 706-10

1C: relic resistates (eluvial placers), Sn,Au, 692-718

KARSTED CARBONATE BASE

1D: ore lumps in residual clay, 729, 731-3
-phosphates, 591, 594-5

1E: residual metalliferous clays
-over carbonate, 596-603
-over carbonatite, 1576-7

SULPHIDE BASE

1F: gossans (Au), 338,343

ARID AND SEMI-ARID CLIMATES

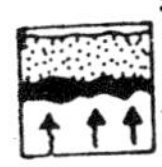
1G: calcrete, 566, 719-24, 914
silcrete, 725

2. (ORTHO)SEDIMENTARY ORES (syngenetic and diagenetic, mostly epiclastic derivation, unconsolidated to consolidated)

MOSTLY DETRITAL ORES

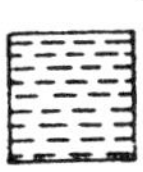
2A: clay to silt-size ore sediments
-Al,Fe,Ti,V, 87, 807-12, 834-35, 1489
-Mn oxides, 537-41
-superfine gold, 104, 765

sand-size ore minerals

2B: alluvial, colluvial placers, 91-100, 751-5, 759-67, 205, 209-11, 1532-3

2C: beach placers, 39, 46-7, 76-8, 91, 99, 209
offshore placers, 109-112
dune placers, 86, 795

2BC paloplacers (undifferent.) 209, 244-50, 806-66 525-29, 556-9

2D: granule to block-size ore component
-ore gravellite, 272, 806
-Fe beach granules, 100-2
-glaciers, glac. sediments 744-5, 1551
-talus, scree, debris flows, 746-7

MOSTLY CHEMICAL PRECIPITATES

2E: bedded floor sediments
-marine euxinic muds, 115
-limnic alunite, 792

3. EPIGENETIC ORES PRECIPITATED FROM COOL (SUB-HYDROTHERMAL) METEORIC, GROUND, FORMATIONAL, ETC. WATERS.

4. SUBAERIAL (PALEO)VOLCANIC-SUBVOLCANIC SYSTEMS

4I: impregn., repl. in volc.
-bertrandite, 995-7
-alunite, 908-9
-miner. miaroles, 902-3
-volc.-diagenetic ores, 889-91, 909-10

4J: "metalliferous volcanics"
-dispers. trace Au, 903
-ore phenocrysts, 58, 153, 157, 902

5. SUBAQUEOUS, MARINE, HYDROTHERMAL AND POLYGENETIC (TECTONOMAGMATICALLY MOBILIZED) DEPOSITS IN OROGENIC BELTS

5AB: complex ("lens and stockwork") hydr.-sedim. seafloor massive ore deposits

-Cu in metabasalts (Cyprus t.) 255, 186-90
-Cu-Zn in bimodal assoc., 321-347
-Cu in andesites, 305-6, 369-72
-Zn,Pb,Cu in seq. diff. and felsic volc., 295-7, 403-16
-mass.sulph. in sedim., 434-7

5B: "distal" massive ores (stratabound lens only)
-Fe (Lahn-Dill) oxides, 263-6
-siderite, 1292-6
-Mn, 295, 373-5

5C: "lens and stockw.", seiments-hosted
-Zn,Pb,Cu, 451, 459, 469, 1257

5D: stratabd. lens in sedim.
-PbZn, 313-7, 566-8 (remob.)

5E: stratabd. dissem. ores
-Sb, 287
-W,Hg,Sb, remob., 488-500
-Cu, 362-9, 609-10

5F: ores control. by deform. (mainly cleavage)
-py, Cu, 301-4
-py, Au, 319-21

5G: ores in shears
-Au, 281, 277-86, 442-6
-Ni,Cu, 181

5H: miscell. peneconc. mantos
-nat. Cu, 157, 253-9
-Hg, 552-4

6. HYDROTHERMAL POSTTECTONIC, POSTMAGMATIC ORES

a) close association with granitic intrusions

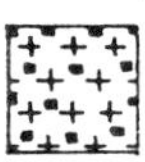
6A: "metalliferous granites" 1063-66
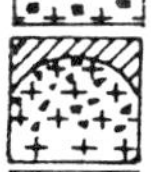
6B: feldspathized apogranites, 1496-9
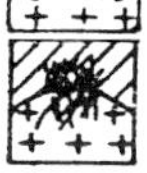
6C: dissem.,stockw.,breccia miner. in altered non-carb.
-porph. Cu-Mo, 325-49, 380-8, 968-78, 1018-9, 1088-1144
-stockw. Mo, 43, 1033, 1214-1226
-greisen Sn,W,Be, 1157-62, 1197
-Bolivian "porphyry Sn", 560-2, 978-86
-stockwork Au, 1033, 1275
-Pb-Zn, 1238

6D: fissure fill. and replac. veins, 1072-5
-Sn,W,Be, 560, 562, 985-6, 1157-85, 1197-1206, 1495-6
-Cu, 261, 439, 966-7, 1018-19, 1150-4
-Au, 261, 1035, 1039-43, 1046, 1275-80
-Sb, 1046, 1314-19
-Pb-Zn, 439, 1046, 1238-60
-Mo, 1227
-U, 1046

6E: skarns, 390-93, 1070-1, 1080-84
-Fe, 261-2, 1293-1303

6b) no obvious connection with granitic intrusions

7. MAGMATOGENE ORES, a) mafic-ultramafic association

7b) alkaline association

8. PEGMATITES

9. METAMORPHOSED AND METAMORPHOGENIC DEPOSITS IN HIGH-GRADE METAMORPHIC AND ULTRAMETAMORPHIC TERRAINS.

10. EXTRATERRESTRIAL ORES, SUBLITHOSPHERIC ORES, PROBLEMATICA

B SOME POPULAR "ORE TYPES" - A RAPID PARTIAL INDEX

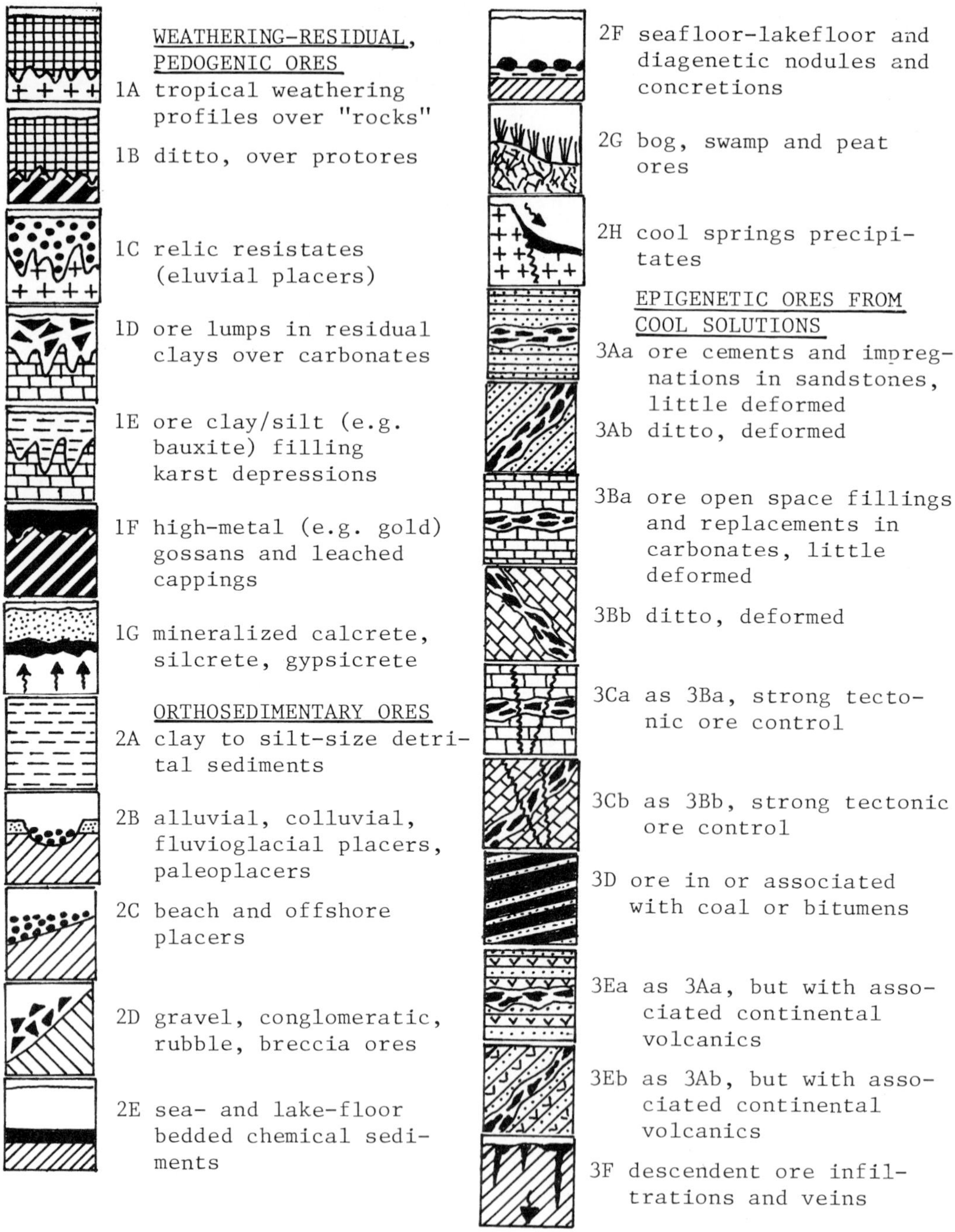

Fig. I-3. Legend to the ore style symbols from Laznicka (1984), used in the genetic and metal index.

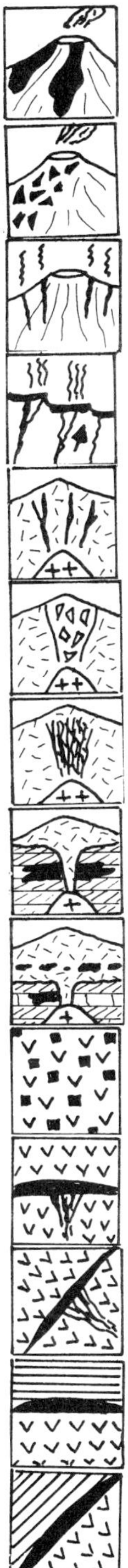

SUBAERIAL VOLCANIC-SUBVOLCANIC SYSTEMS

4A metalliferous lavas

4B metalliferous pyroclastics

4C fumarolic mineralizations

4D hot springs, epithermal brines, precipitates

4E epithermal veins

4F high-level hydrothermal ores in breccias

4G high-level hydrothermal stockworks

4H high-level hydrothermal carbonate replacements

4I high-level hydrothermal impregnations in volcanics, miner. basement

4J "metalliferous volcanics", phenocrysts, miaroles

SUBAQUEOUS, MARINE-HYDROTHERMAL AND POLYGENETIC DEPOSITS IN OROGENIC BELTS

5Aa complex (lens-and-stockwork) deposit in volcanics, little disturbed

5Ab ditto, deformed and metamorphosed

5Ba "distal" (lens only) orebody, little deform.

5Bb ditto, deformed and metamorphosed

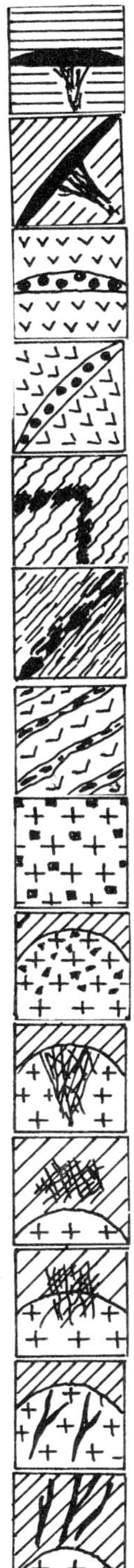

5Ca as 5Aa, but (meta)sediment hosted

5Cb as 5Ab, but (meta)sediment hosted

5Ea disseminated ore in mantos in (meta)volcanics, little deformed

5Eb ditto, deformed, metamorphosed

5F remobilized and hydrothermal-metamorphics ores

5G ores in shears and high strain zones subparallel with schistosity

5H peneconcordant or discordant ore sheets in (meta) volcanics and sediments.

HYDROTHERMAL-POSTTECTONIC POSTMAGMATIC ORES

6A metalliferous granites

6B disseminated ores in apogranites

6Ca stockworks and miner. breccias in endocontact

6Cb ditto, in exocontacts

6Cc ditto, along both sides of contacts or undifferentiated

6Da fissure and replacement veins in endocontacts

6Db ditto, in exocontacts

6Dc as 6Db, along both sides of the contact or undifferentiated

6E ores associated with skarns

6F ores in non-skarn carbonate replacements

ORES WITHOUT CLEAR ASSOCIATION WITH GRANITIC ROCKS

6G stockworks, disseminations

6H fissure, replacement veins

6Ia carbonate replacements

6Ib skarns at gabbro-diabase contacts

MAGMATOGENE ORES

a) MAFIC-ULTRAMAFIC

7A cumulus segregated ores

liquidomagmatic or wallrock interaction ores

7Ba within intrusion

7Bb along contacts

7Bc in exocontacts

7C podiform chromite, magnetite in ultramafic tectonites

b) ALKALINE ROCKS

7Da ore disseminations in layered intrusions

7Db ditto, in massive intrusions

7Ea veins, metasomatites, etc. in endocontacts

7Eb ditto, in exocontacts

7Ec ditto, along both sides of the contact or undifferentiated

7F mineralized diatremes

PEGMATITES

8Aa granitic pegmatites of shallow to intermediate depths

8Ab ditto, of great depths

8B alkaline pegmatites

HIGH TO MEDIUM-GRADE METAMORPHIC ORES

9A in high-pressure metamorphics

9B ores in largely statically metamorphosed terrains

9C ores in dynamometamorphic hosts

9D ores in granitized terrains or associated with granitization

Fig. I-3 (continued). Legend to the ore style symbols from Laznicka (1984), used in the genetic and metal index

METALS INDEX

Ag

Al

As

Au

2BC alluvial to beach placers
83, 95, 99, 110-13, 209, 282-3, 376, 444-5, 744-5, 753-7, 759-67, 874, 1046, 1039-41
-deep leads, 745, 767-70
-paleoplacers, 528-9, 805

2E in Kupferschiefer, 115*, 444 5*, 537 77

3D coal, 831*

4D hot springs, 63-67, 241

4E bonanza epithermal veins 1485-8

4F mineralized breccias, 1485-88

4I 376

4J metallif. volcanics, dike 903*, 913*, 1069*

5A 304, 318-21

5B Au mass. sulphide, 416

5D auriferous schists, 281-86

5F axial cleavage, saddle reefs, linear stockworks, 301-6, 435-7, 442-6, 479

5G shear lodes, 198, 277-86, 1039-41

6C stockw., brecc., 1018*, 1275, 1508-11*

6D veins, 184-5, 261, 286-7, 390-3, 444-5, 555, 1035-39 1275-80, 1377

6E skarns, 391

6F carb. replac., 1085, 1288-9, 1508*

6G stockworks, 376-88, 1071-2, 1319-21

6H veins, 1400, 1506

6I Carlin "type", 479-81, 566-71, 1289, 1319

7B 244-5

barite

1D resid. clay, lumps 628, 652, 731

2E black shale, 314-15

2F seafloor concretions, 35

3B anhydr. repl., 317, 665
carb. repl., 628, 631, 634

5B 299

5D 459, 466-7, 471

6H veins, 372, 567, 1401-2

Be

2B arid placers, 1163

3D coal*

4E epithermal veins, 995

4I tuffite repl. by bertrandite, 995-7

4J miaroles, 962

6A Be granite*, 1065

6B apogranites, 1163

6Ca greisen stocks, 1230-35, 1504-6

6D veins, 1231-4, 1506

6E skarn, 1187-8, 1230-4

7E alkal. complexes, 1525-31	8Ab deep-seated pegmatites, 1385-6
8Aa high-level pegm., rare m. pegm., 1230-5, 1387-94	

Bi

1D lumps in resid. clay, 718	4G stockworks, 985
4C fumaroles, 60	6E skarn, 1187-8
4E epithermal veins, 1322	6H veins, 1268-72

Cd

in sphalerites (various styles), 1324-5	4E epithermal veins, 945, 1325

Cr

1C eluvial placers, 204-9	2C paleoplacers, 209-11
2A ochers, 210-14	7A cumulus in ultramafics, 248
2BC placers, 205-9, 772-3	7C podiform chromite, 174-80, 234-6, 1041

Co

1A laterites (asbolite), 203-4, 236	3E 367
2A reworked laterite, 214, 223	6H veins, 391, 1268-72
2F oceanic nodules, 33-35	6Ib gabbro contact skarn, 1466-7
3Ba, MVT deposits, 617	7B gabbro contacts, 256

Cu

2A oceanic clays, 32*	3B 631-4, 652, 665
2D moraine, talus, ice 741, 746	3D coal, 826*, 831*
2E in banded iron formation 839	3E volc. red-beds, 875-80, 885-9, 906-7
2F oceanic Fe-Mn nodules, 33-35; Cu nodules, 875	4D hot springs, brines, 63
2G swamp, bog, peat 104, 782-3	4E enargite, luzonite veins 968-9
3A chrysocolla, 755-6, 796-7 sulph., 875-80, 887	4G stockw., 363-5, 380-4, 420-1, 969-72, 1441-2

F

Fe

Ga Ge

Hg

In

Li

Mn

Mo

Nb

Pt

Re

REE

Sb

Se

3Ba in roll U deposits, 856
3D in coal, 836
4C fumaroles, 39

4E epithermal veins, 954
6 Se in hydrothermal deposits 1324

Sn

1C eluvial placers, 712-13
1B glaciofl. plac., 744-5
colluvial placers, 729, 758-9
alluvial pl., 762-3, 770-3 1047-8, 1495-99
2C beach placer, 83
offshore, 108-111
? paleoplac., 556-60
2D moraine, talus, 741-2, 746-7
3A mantos, 885
3D in coal, 831*
4C fumaroles, rec., 60*
wood tin, 61
4G stockw., vein, 560-2 979-86

4J cassiterite in miaroles, 903
6Ca endocont. stockw., veins, 1157-74, 1495-9, 1503-4
6Db exocont. vein, stockworks 338, 556-62, 966-7, 1074, 1175-85, 1027-8, 1047-8
6E skarn, 1048, 1186-90
6F carbonate replacements, 1186-91
8A pegmatites, 1067-8, 1387-92
9C 1342-48

Ta

6B apogranites, 1162-64, 1382
6Ca stockworks, 1504-5

7Da layered intrus., 1517-21, 1547, 1549
8A pegmatites, 1381-3, 1387-92

Te

6 hydrothermal deposits, 199, 1324

6Ca stockworks, 1504-5

Th

1B resid. ochers, 1576-7
2C paleoplacers, 528
6A metallif. granites, 1063-4, 1066, 1202

7Bc 1469-70
7Da agpaitic intrusions, 1517, 1521, 1523
7Db carbonatite, 1568-9

Ti

Tl

U

7Db carbonatite, 1568-77

7E veins, 1526-7, 1530, 1540

7F diatreme infiltrations, 1489

9D U mobilizates, 1371

V

1B laterite, 713

1E karst, 734

2A claystone, 1489

2E black shale, phosphorite, 482-8, 544
marine ironstone, 531

3Aa SUV, 856-71

3D asphalt., coke, 675-6
coal, 831*

4C fumaroles, 60*

5Aa hydrothermal mounds, 25*

5H 259

7A Ti-magnetite cumulus, 184, 249

7Eb fenite zone, 1530

W

2B alluv. plac.,hubnerite, 759, 773; scheelite, 987

2E 490

2H playa brines, 789*
Mn spring prec. 798

3D coal, 830*

4D hot springs, 66*

4E veins, 986-7, 1205

4G stockworks, 985-6

5Db 288-9, 489, 1344

5G scheelite, 282-3

6Ca exocont. stockw., veins 1074, 1087

6Cc stockw., scheelite 1072, 1206-8
wolframite, 1197-1206

6D veins, 563, 966-7

6E scheel. skarn, 490-1, 1039-41, 1066, 1087-8, 1208-14, 1376

8A pegmatite, 1067-8

Zn

1B exotic, infiltr., 734-5

2E 482*, 544*, 566-68, remob. 605-7, 833*, 1413
Kupferschiefer, 1413

2H oil brines, 675

3Aa sandst., 582-4, 880-1, 884

3Ba MVT, 608-30

3Bb, APT, 608, 611, 638-51

3C Irish "type", 631-4

3D coal, 823-3*

4C fumaroles, 60*

4D hot springs, 63*

5Aa model, 158-62, Red Sea muds, 1414-7, black smokers, 25; Kuroko, 403-16

5Ab mass. sulphides, 273-76, 297-300, 321-47, 372

5Db 313-17, 458-71

5F cleavage oreb., 434-37

6C stockw., dissem., 1238

6Db, veins, 474-7, 563-6, 1018-9, 1238-44, 1252-4, 1256-60

6E skarns, 1244-6, 1255, 1263-4

6F repl. mantos, chimneys, 395, 654-62, 1244-52, 1254-55, 1261-2

6H veins, 1046

6Ia willemite mantos, 662

9C mass. sulphides, ZnCu, 1352, 1354-7 PbZn, 1353, 1357-9

Zr

1C caldasite, 1531

1G silcrete, 725*

2C beach plac., 83, 91-100 shelf plac., 111-2 paleoplacers, 527-8

2E phosphorite, 487*

7Da agpaitic intrus., 1516-21

7Db carbonatite, 1569-77

7E 1527-8

Abbreviations: *high trace metal content, low grade technological ore or potential "ores of the future"

m modified, mobilized orebodies